6th International School and Conference "Saint Petersburg OPEN" 2019

Optoelectronics, Photonics, Engineering and Nanostructures

Journal of Physics: Conference Series Volume 1410

St. Petersburg, Russia
22 - 25 April 2019

Part 1 of 2

ISBN: 978-1-7138-1317-0
ISSN: 1742-6588

Printed from e-media with permission by:

Curran Associates, Inc.
57 Morehouse Lane
Red Hook, NY 12571

CURRAN ASSOCIATES INC. proceedings.com

Some format issues inherent in the e-media version may also appear in this print version.

Institute of Physics
Dirac House, Temple Back
Bristol BS1 6BE UK

Phone: 44 1 17 929 7481
Fax: 44 1 17 920 0979

techtracking@iop.org

Additional copies of this publication are available from:

Curran Associates, Inc.
57 Morehouse Lane
Red Hook, NY 12571 USA
Phone: 845-758-0400
Fax: 845-758-2633
Email: curran@proceedings.com
Web: www.proceedings.com

TABLE OF CONTENTS

PART 1

INVITED

CRYSTAL GROWTH AND STRUCTURAL PROPERTIES OF SEMICONDUCTOR MATERIALS AND NANOSTRUCTURES

NANOBIOTECHNOLOGY, BIOPHYSICS AND BIOPHOTONICS

LASERS, SOLAR CELLS AND OTHER OPTOELECTRONIC DEVICES

NANOPHOTONICS, SPECTROSCOPY, MICROCAVITIES, OPTICS, PLASMONICS

PART 2

SPINTRONICS, ELECTRO- AND MAGNETO- OPTICS

ELECTRIC, MAGNETIC AND MICROWAVE DEVICES

OTHER ASPECTS OF NANOTECHNOLOGY

6th International School and Conference

Saint Petersburg OPEN 2019

Optoelectronics, Photonics, Engineering and Nanostructures

Saint Petersburg, April 22–25, 2019

Acknowledgements

RFBR
Russian Foundation for Basic Research

RSF
Russian Science Foundation

SPIE. STUDENT CHAPTER
SAINT PETERSBURG ACADEMIC UNIVERSITY RUSSIAN ACADEMY OF SCIENCES

IOP Institute of Physics

Editorial Preface

Dear Colleagues,

6th International School and Conference "Saint Petersburg OPEN 2019" on Optoelectronics, Photonics, Engineering and Nanostructures was held on April 22 - 25, 2019 at St. Petersburg Academic University. The School and Conference included a series of invited talks given by leading professors with the aim to introduce young scientists with actual problems and major advances in physics and technology. The keynote speakers were

Nikolay Maleev *(Institution for Microelectronics Research and Engineering RAS, Russia)*

Alexey Zhukov *(St Petersburg Academic University, Russia)*

Sergey Kukushkin *(Institute for Problems in Mechanical Engineering, Russia)*

Ivan Mitropolsky *(St. Petersburg Nuclear Physics Institute "Kurchatov Institute", Russia)*

Alexei Kuzmin *(University of Latvia, Latvia)*

Nikolay Ledentsov *(VI Systems GmbH, Germany)*

Andrei Lavrinenko *(Technical University of Denmark, Denmark)*

Joel Bellessa *(University of Lyon 1, France)*

Slaven Garaj *(National University of Singapore, Republic of Singapore)*

During the poster session all undergraduate and graduate students attending the conference presented their works. Sufficiently large number of participants with 287 student attendees from all over the world allowed the Conference to provide a fertile ground for the fruitful discussions between the young scientists as well as to become a perfect platform for the valuable discussions between student authors and highly experienced scientists. The best student papers, which were selected by the Program Committee and by the invited speakers basing on the theses and their poster presentation, were awarded with diplomas of the conference and prizes – see the photos.

This year "**Saint Petersburg OPEN 2019**" is organized by **St Petersburg Academic University**. School and Conference is supported by **Russian Foundation for Basic Research** (*Project # 19-02-20031*), **Peter the Great St. Petersburg Polytechnic University, Russian Science Foundation** and **SPIE**. It is a continuation of the annual schools and seminars for youth on topical problems of physics and technology that is organized by the Academic University since 2009. We invite all the students and young scientists to attend "**Saint Petersburg OPEN**" in 2020!

More details at http://spbopen.spbau.com

The head of Organizing Committee
Professor *Alexey Zhukov*

Participants

The invited speakers

Nikolay Maleev

Ivan Mitropolsky

Sergey Kukushkin

Alexey Zhukov

Alexei Kuzmin

Nikolay Ledentsov

Journal of Physics: Conference Series **1410** (2019) 011001 doi:10.1088/1742-6596/1410/1/011001

The invited speakers

Andrei Lavrinenko

Joel Bellessa

Slaven Garaj

Journal of Physics: Conference Series **1410** (2019) 011001 doi:10.1088/1742-6596/1410/1/011001

Elevator speech session

Journal of Physics: Conference Series **1410** (2019) 011001 doi:10.1088/1742-6596/1410/1/011001

Poster Session

Poster Session

Award ceremony
Prof. *Alexey Zhukov* and *Andrey Lipovskii* with participants

Peer review statement

All papers published in this volume of *Journal of Physics: Conference Series* have been peer reviewed through processes administered by the proceedings Editors. Reviews were conducted by expert referees to the professional and scientific standards expected of a proceedings journal published by IOP Publishing.

Microlasers based on GaAs and Si

A E Zhukov, M V Maximov, N V Kryzhanovskaya

St Petersburg Academic University, 194021, St Petersburg, Russia

Abstract. Results are presented on microlasers with quantum dot active region grown on either GaAs or Si substrates. The optical resonator of these lasers represents a microdisk supporting whispering-gallery modes with a Q-factor in excess of 30'000. When operating in continuous-wave mode without temperature stabilization, the microlasers have the threshold currents of mA-range with the minimal threshold current density of 250 A/cm^2. Under direct small-signal modulation, the -3dB modulation bandwidths were measured to be above 6 GHz with a little effect of self-heating on the performance. The consumed electric power was estimated to be about 2 pJ/bit. The achieved performance is a significant breakthrough in realization of low-threshold high-speed temperatrure-stable operation of microlasers compared with the previously reported parameters for the devices of comparable sizes.

1. Introduction

Quantum dot (QD) based lasers emerged in the mid of 1990s [1] owing to the search for semiconductor heterostructures with a quantum dimensionality lower than that of already existing quantum wells. Studies in the field of QDs were greatly motivated by the paper of Arakawa and Sakaki [2], who predicted the infinite temperature stability of a laser with zero-dimensional active region. Appearance of QDs, originally in the form of In(Ga)As self-organized islands on GaAs substrates, has been met with great enthusiasm as they do not require significant complication of the epitaxial growth technology. In semiconductor QD structures, high optical and structural quality is readily achieved, whereas etching or other polluting processes are not required for their fabrication. Quite soon the researchers have faced with the fact that it is extremely important to achieve deep localization of charge carriers and a high value of optical modal gain in order to realize low-threshold lasing at room and elevated temperatures. It was found, the former can be implemented by stacking several (typically 10) planes of QDs separated with sufficiently thick spacers of unstrained material [3], whereas the latter can be realized by covering QDs with a narrow-bandgap quantum well [4].

The resulting InAs/InGaAs/GaAs QD diode lasers made it possible to reach a record-low threshold current densities of about 10 A/cm^2, a high output power in spatially-singlemode regime of more than 1 W, an infinite characteristic temperature, and many other attractive features [5]. Among them, a capability of emission in the wavelength interval around 1.3 μm, which corresponds to the near-zero chromatic dispersion and one of the minimum attenuation windows in the standard single-mode silica fiber. This wavelength is also very promising for silicon photonics, as silicon is transparent there. The combination of these unique properties makes QD lasers very attractive for optical data transmission. For example, data transmission with a bit rate of 10 Gb/s was attained with a QD edge-emitting laser operating under fixed modulation and bias currents across a wide temperature range from 20 to 90°C [6]. Advantages of photonics over electronics (such as absence of Joule heating, higher speed, possibility of spectral multiplexing, and many others) motivate a gradual transition from wire to optical communication. This change, which began with the introduction of long-haul optical fiber communication systems and continued with the deployment of optical metropolitan-area and local-area networks, is characterized by a clear trend to reduce a connection distance. This solution, which is usually referred to as optical interconnect, in its ultimate form will involve a signal transmission from one integrated circuit or its part to another using light. Eventually, optical transmission will be used for signal exchange between boards, microchips and even inside the integrated circuit [7].

Edge-emitting laser diodes of traditional design, obviously, can not be used for this purpose because of inappropriately large footprint and high power consumption. Vertical cavity surface emitting lasers (VCSELs) [8], which were the first to make possible scaling the laser active region, seem too complicated for use in on-chip optical systems, especially those integrated with a silicon platform. A feasible alternative is a microlaser with a microdisk (MD) or microring optical resonator [9], where optical confinement and a high Q-factor are achieved due to the total internal reflection of light circulating at the resonator periphery (whispering gallery modes) rather than using very thick distributed Bragg reflectors and lateral aperture. Importantly, the epitaxial layers sequence intended for a microdisk laser is very similar to that of a separate confinement laser heterostructure and its total thickness is much smaller than that for VCSELs. It was discovered some time ago, but until recently it was not actually exploited that quantum dots suppress the migration of charge carriers along the active region plane []. This feature makes QD microlasers virtually insensitive to non-radiative recombination, which otherwise unavoidable for devices with small diameters. As a result, the QD microdisk laser can be simply fabricated by etching through the active region, and passivation of the side walls is not needed. Such microlasers have proved their ability to operate in continuous wave mode up to 100°C heatsink-free [10]. The minimal diameter at which lasing was achieved at room temperature (optical pump) is as small as 1 μm [11].

It turned out that the strong carrier localization within individual QDs is also favorable for their resistance to non-radiative recombination inevitable in III-V-on-Si structures. A significant progress has recently been demonstrated in QD structures monolithically grown on Si, including implementation of edge-emitting lasers, which offer threshold current densities very similar to their counterparts grown on GaAs [12]. Recently, microring [13] and micordisk [14] QD lasers on silicon have been demonstrated as well. Some representative images and an example of the microlaser characteristics are collected in Figure 1. The combination of small size, low threshold, and high-temperature operation makes QD-on-Si micolaser compatible to Si-based CMOS integrated circuits.

2. Experimental

An epitaxial structure suitable for microdisk laser fabrication can be grown by molecular beam epitaxy (MBE) or metal-organic vapor phase epitaxy (MOVPE). We use a separate confinement laser heterostructure with AlGaAs cladding layers and a GaAs waveguide comprising a QD-based active region. We use either MBE-grown InAs/InGaAs Stranski-Krastanow (SK) QDs emitting near 1.3 μm or non-SK InGaAs islands that formed by MOVPE on misoriented substrates and emit near 1.1 μm. The former sort of the active region is advantageous due to its deep localization energy of the charge carriers, which completely prevents carrier migration to the surface / defects. The lowest threshold current densities we achieved in microlasers when using this type of QDs. The disadvantage is the low optical gain caused by the low surface density of the SK QDs. The latter type of the active region, which we referred to as the quantum well-dots (QWD), is characterized by an order of magnitude higher density and, consequently, much higher optical gain as compared to that of SK QDs. However, a lower localization energy of charge carriers gives rise to an increased sensitivity to non-radiative surface recombination.

In case of Si substrate, a special III-V buffer layer comprising dislocation filters [15] is grown prior to the laser structure with the aim to reduce a dislocation density with simultaneous relaxation of the lattice mismatch strain. Different polarity of III-V materials and silicon can induce the formation of antiphase boundaries - crystallographic defects caused by the contact of miroregions with different order of the atomic layers (antiphase domains). Use of silicon (001) substrate with 4° offcut to the

[011] plane [16] is one of the methods for the formation of single-domain III-V-on-Si materials. Another significant challenge is the difference of the thermal expansion coefficients, which results in micro thermal cracks and can lead to the peeling of epitaxial layers from the substrate. Figure 1 shows an electron microscopic image of the surface of the laser heterostructure gown on a Si substrate with micro-cracks marked with arrows. A typical size of the defect-free regions is 50-100 μm, which emphasizes the need for an appropriate choice of the dimensions of optoelectronic devices made from such wafers.

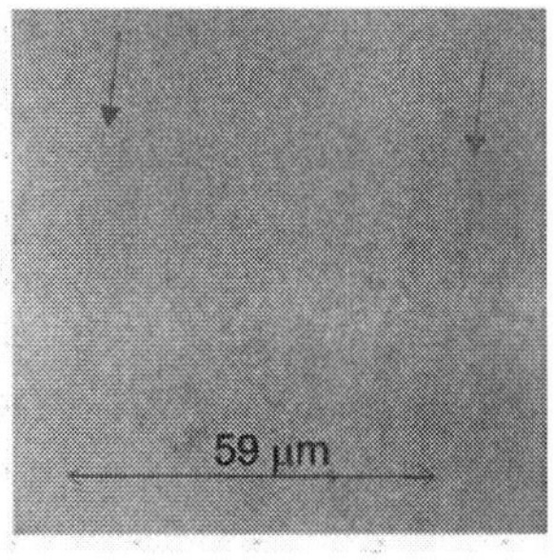

Figure 1. SEM image of III-V laser structure monolithically grown on Si substrate. Arrows indicate micro-cracks.

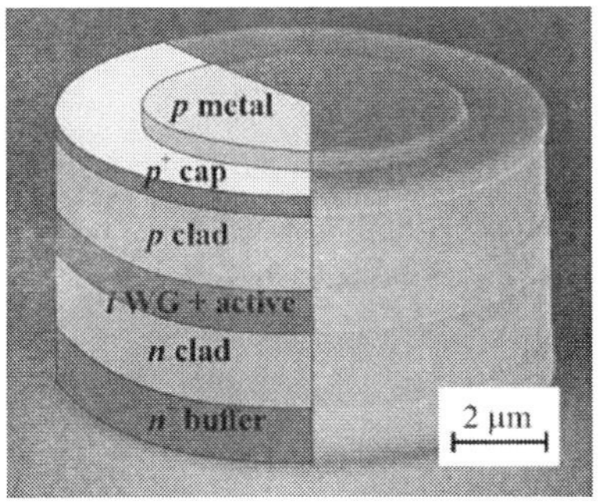

Figure 2. SEM image of microdisk laser with p-metal on the top.

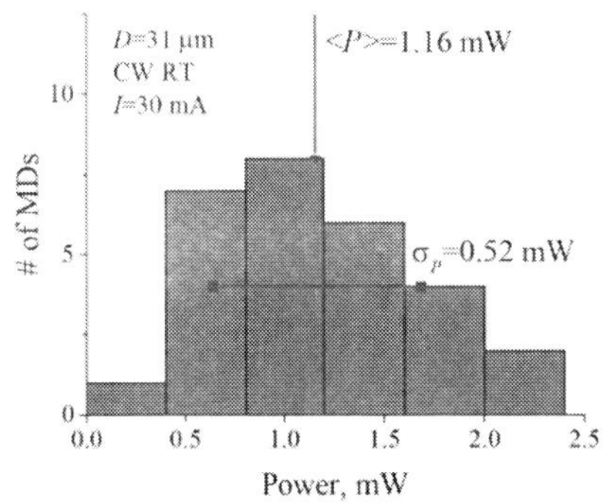

Figure 3. Histogram of distribution by output power of microdisk lasers manufactured in one process.

Microdisk resonators are formed using photolithography and subsequent deep plasma-assisted chemical etching. The mesas are several micrometers tall in order to ensure full prevention of the mode leakage to the substrate (Figure 2). No passivation, encapsulation, or any other sidewall treatment is usually applied. The upper circular-shaped p-ohmic contacts are placed on top of the MDs, and a common n-ohmic contact is deposited onto the etched surface between the MDs or onto the back side of the substrate. In some cases, planarization with a dielectric coating is used to facilitate the formation of the contact pads to small microlasers. The reproducibility of microlaser fabrication process still needs improvement, as can be seen from the histogram of the microlaser power distribution across a number of nominally identical devices (Figure 3). We suggest the reason is some damage of the microresonator sidewalls, which causes light scattering of varying degree. For device evaluation, microdisk lasers are soldered p-side up onto a copper holder and tested at room or elevated temperatures. Typically, no temperature stabilization or external cooling is used. The threshold, power, and spectral characteristics are measured in CW regime. Dynamic characteristics are measured under small-signal direct modulation with the RF-signal combined with the DC bias through a high frequency bias-tee and fed to microlaser by a high frequency ground-signal-ground probe head.

3. Results and discussion

Figure 4 illustrates an evolution of the intensity and linewidth of the optical mode emitted from the microdisk with QDs grown on a Si substrate. At low injection currents the mode intensity is low whereas the linewidth is relatively large indicating that the emission proceeds in the spontaneous regime. As soon as the current reaches its threshold value, the intensity starts to grow superlinearly and the linewidth drops down to 30 pm or so. A lasing spectrum (inset to Figure 4) reveals several side modes close to the dominant one with the side mode suppression ration over 20 dB (a quasi-single-mode spectrum). With increasing the current, the intensity tends to saturate, while the FWHM remains nearly constant or slowly increases. Further growth of the injection leads to suppression of the mode intensity, and the lasing jumps to a next longer-wavelength mode (extra inset to Figure 4). The mode hopping is caused by microdisk self-heating, which results the gain spectrum to be red-shifted with a rate of about 0.4 nm/K, whereas the micro-resonator mode is several times more stable against temperature (about 0.08 nm/K). Actual temperature of the microdisk laser operating in CW regime can be extracted from the current-induced modeline shift (solid line in the extra inset to Figure 4). Depending on the diameter and other parameters of the microdisks, the temperature increment ranges from a few to tens of degrees Celsius.

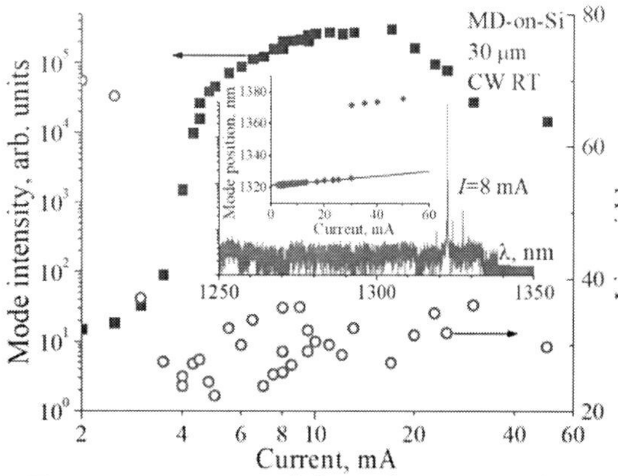

Figure 4. Integrated intensity (squares) and FWHM (circles) of the dominant mode line against injection current. Inset: Emission spectrum taken at 8 mA. Extra inset: mode position vs current.

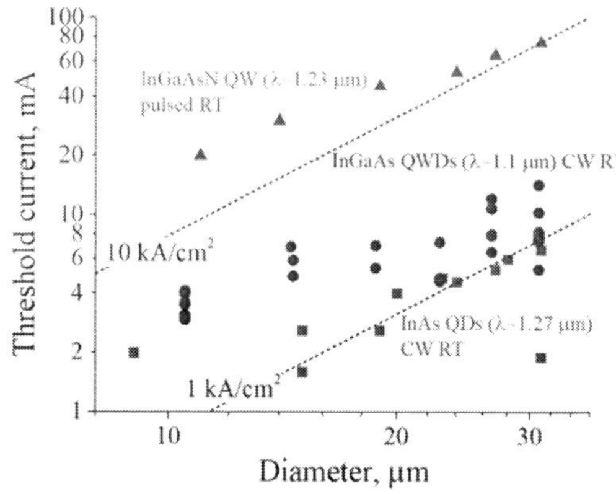

Figure 5. Threshold current of microdisk lasers with active region of various types vs microdisk diameter.

Figure 5 summarizes CW threshold current data for microdisk lasers of different diameters operating at room temperature with the active region based on either SK QDs (squares) or QWD (circles). Despite the apparent scatter of data, there is a clear tendency to decrease the threshold current with decreasing diameter of the microdisk. In case of SK QDs the threshold current varies approximately in proportion to the squared diameter, i.e. the threshold current density remains nearly constant. This means that the non-radiative surface recombination is negligible or, at least, significantly suppressed in microlasers with these deeply localizing quantum dots. For a given diameter, the data scatter can be attributed to irreproducibility of the quality of the microresonator surface, as discussed in relation to Figure 3. For example, the threshold current of the SK QD microdisks with a diameter of 31 μm ranges from 6.7 to 1.9 mA. The latter value corresponds to the threshold current density of 250 A/cm^2, which is, to the best of our knowledge, the lowest one among any QD-based microdisks/microrings operating at room temperature. This low threshold current density is only 6% higher than the transparency current density estimated in broad-area edge-emitting lasers made of the same epitaxial wafer.

As opposed to the SK QDs, in microlasers based on QWDs (circles in Figure 5) the surface component of the threshold current is more significant. The impact of the surface recombination is especially strong for the smaller diameters (< 20 μm or so), where I_{th} varies approximately in proportion to the first power of the diameter. Similar behavior is characteristic of microlasers based on 2D quantum wells (data for InGaAsN QWs are shown by triangles). We conclude that deep 3D localization of charge carriers is essential for achieving low thresholds in microdisk lasers.

Until recently, there have been only a few reports on the dynamic characteristics of QD-based microdisk/microring lasers. It was found that the modulation frequency f_{3dB}, at which the amplitude modulation response drops down to –3 dB, can exceed 5-6 GHz. An example of small-signal direct modulation characteristic is shown in Figure 6 for a QWD microdisk laser operating under temperature-stabilization condition at 18°C (open symbols). The modulation characteristic is strongly damped, which is typical for QD-based lasers. Although the modulation frequency is not a record-high compared to VCSELs, microdisk lasers, being a much simpler in fabrication, allow data transmission at speeds B of about 10 Gb/s ($B \approx f_{3dB}$). What is more intrigue is that microdisk lasers tolerate uncooled operation with a marginal effect on high-frequency performance. As seen in Figure 6, the –3 dB frequency decreases very insignificantly when temperature stabilization is no longer kept up.

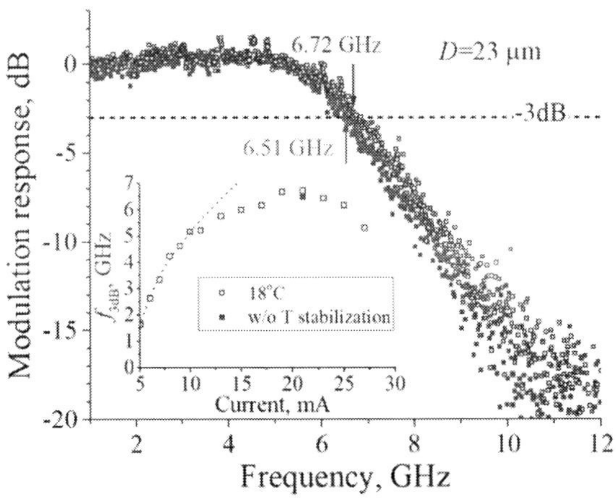

Figure 6. Small-signal modulation response of QWD microdisk laser operating at 18°C or without temperature stabilization. Inset: f_{3dB} frequency against bias current.

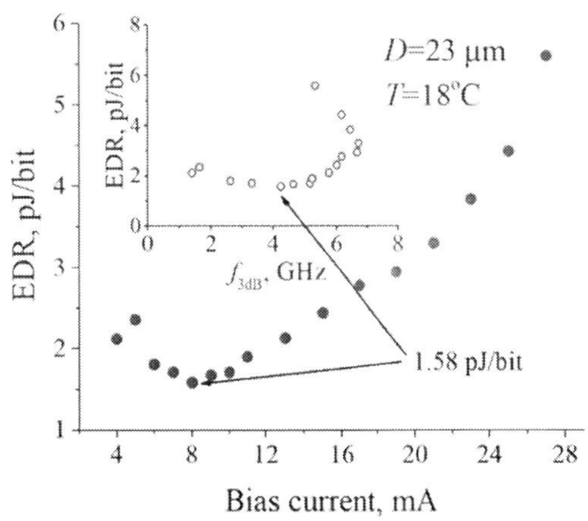

Figure 7. Energy-to-data ratio as a function of bias current (main panel) or f_{3dB} frequency (inset).

A laser source intended for optical data communication and related applications have to be energy-efficient. According to the definition, the electrical energy-to-data ratio (EDR) is the total electrical energy consumed per bit of information. EDR below 100 fJ/bit has been demonstrated for VCSELs [17]. The values of EDR are presented in Figure 7 for the MD laser of Figure 6. For an optimal bias current, the minimal EDR is achieved, which is 1.58 pJ/bit for the given microlaser. If the microlaser is biased to its maximum modulation speed, the EDR of about 3 pJ/bit is estimated. For comparison, energy consumption per bit at 10 Gbit/s modulation for AlGaInAs/InP microspiral disk lasers of comparable size ranges from 6.9 to 27.4 pJ/bit [18].

4. Conclusion

In conclusion, on-chip microlasers, especially those integrated with silicon, can provide a new opportunity for quantum dots, as they can hardly be implemented with an alternative active region. Deep localization of charge carriers in QDs is essential for the suppression of their migration along the plane. Insensitivity to surface defects, inevitable for deep etching, as well as to epitaxial defects inherent in III-V growth on Si, is achieved. These quantum-dot based microlasers are capable of operating in a CW regime or under high-frequency direct modulation without external cooling at room and elevated temperatures as it is required for integration with Si-based microelectronics platform. No modification of the layer/composition design is required, compared to conventional edge-emitting lasers. A simple post-growth process with a single-step lithography and a standard dry etching procedure are applied. No sidewall passivation or other protection can be used. Monolithic integration with Si substrate can be used with minimal impact on device performance.

Acknowledgments
This work was supported by the Russian Science Foundation, agreement 19-72-30010.

References

[1] Ledentsov N N, Ustinov V M, Egorov A Yu, *et al* 1994 *Semiconductors* **28** 832
[2] Arakawa Y, Sakaki H, 1982 *Appl. Phys. Lett.* **40** 939
[3] Egorov A Yu, Zhukov A E, Kop'ev P S, *et al* 1996 *Semiconductors* **30** 879
[4] Ustinov V M, Maleev N A, Zhukov A E, *et al* 1999 *Appl. Phys. Lett.* **74** 2815
[5] Zhukov A E, 2011 Quantum dot diode lasers, In: *Landolt-Börnstein: Numerical Data and Functional Relationships in Science and Technology - New Series*, vol. VIII/1B3, chap. 11.1, Springer-Verlag.
[6] Ishida M, Hatori N, Otsubo K, *et al* 2007 *Electron. Lett.* **43** 219
[7] Sun C, Wade M T, Lee Y, *et al* 2015 *Nature* **528** 534
[8] Towe E., Leheny R F, Yang A, 2000 *IEEE J. Selected Topics Quantum Electron.* **6** 1458
[9] McCall S L, Levi A F J, Slusher R E, *et al* 1992 *Appl. Phys Lett* **60** 289
[10] Kryzhanovskaya N V, Moiseev E I, Kudashova Yu V, *et al* 2015 *Electron. Lett.* **51** 1354
[11] Kryzhanovskaya N V, Zhukov A E, Maximov M V, *et al* 2015 *IEEE J. Selected Topics Quantum Electron.* **21** 1900905
[12] Chen S, Li W, Wu J, *et al* 2016 *Nature Photon.* **10** 307
[13] Wan Y, Jung D, Norman D, *et al* 2017 *Opt. Express* **25** 026853
[14] Kryzhanovskaya N, Moiseev E, Polubavkina Yu, *et al* 2017 *Opt. Lett.* **42**, 3319
[15] Tang M, Chen S, Wu J, et al 2016 *IEEE J. Selected Topics Quantum Electron.* **6** 1900207
[16] Fischer R, Neuman D, Zabel H, *et al* 1986 Appl. Phys. Lett **48** 1223
[17] Moser P, Lott J A, Wolf P, *et al* 2012 *Electron. Lett.* **48** 1292
[18] Yang Y-D, Zhang Y, Huang Y-Z, *et al* 2014 *Opt. Express* **22** 824

Anomalous lasing in VCSEL with double oxide current apertures

M A Bobrov[1], S A Blokhin[1], N A Maleev[1], A G Kuzmenkov[2], V M Ustinov[2]

[1] Physics of Semiconductor Heterostructure Laboratory, Ioffe Institute, Saint Petersburg 194021, Russia
[2] Submicron Heterostructures for Microelectronics, Research and Engineering Center of RAS, Saint Petersburg, 194021, Russia

Abstract. Detail measurements and full-vector three-dimensional FDTD technique were used to analyze anomalous mode behavior in double-aperture oxide-confined 850 nm VCSELs with large gain-to-cavity detuning.

1. Introduction

Modern high-speed (10–25 Gbit/s) short-distance (<100-200 m) optical interconnects in data centres and supercomputers based on multimode optical fibers and surface-emitting semiconductor lasers (VCSELs). As compared with conventional edge-emitting semiconductor lasers the important VCSEL advantages are symmetrical output beam, low power consumption, on-wafer testing, high reliability and high-speed operation under direct-current modulation [1]. Commercial 850 nm VCSEL technology based on design with doped AlGaAs distributed Bragg reflectors (DBRs), GaAs/AlGaAs multi-quantum well (MQW) active media and selectively-oxidized aperture for current confinement. VCSEL direct current modulation response limited both intrinsic properties (differential gain and damping rate) and parasitic parameters (capacitances and resistances). The effective approach to increase VCSEL data rate above 20 Gbit/s is a combination of strained InGaAs/AlGaAs MQW active media and low parasitic capacitance design based on double oxide aperture combined with thick low-k dielectric planarization layer under anode contact pad [2]. Since uncooled operation is required in most high-speed applications the temperature stability of VCSEL characteristics is a critical issue. A spectral misalignment between the cavity resonance wavelength and the gain spectrum maximum wavelength of the QW active region (the gain-to-cavity detuning) was proposed to improve a VCSEL's temperature stability [3]. However in case of large gain-to-cavity detuning double-aperture VCSELs can demonstrate anomalous mode behavior [4]. Particularly such behavior was observed for 850 nm-range top-emitting double-aperture VCSELs with current injection via doped DBRs [5]. Indeed, the devices with small-aperture (~ 3-5 μm per side) demonstrate classical lasing via low-order transverse optical modes over all operation currents, while the devices with large-aperture (> 7 μm per side) devices started multi-mode lasing via high-order modes at low bias current with the subsequent switching to multi-mode lasing via low-order modes at high bias current.

In this paper we present measurement results of lasing behavior for double-aperture oxide-confined 850 nm-range VCSELs with large gain-to-cavity detuning (~ 20 nm) and their analysis based on numerical modeling of laser optical modes in order to clarify the reasons of such anomalous lasing behavior. The schematic cross-section of the investigated oxide-confined 850 nm-range VCSEL is shown in Figure 1. More details about the structure of the lasers can be found in [5].

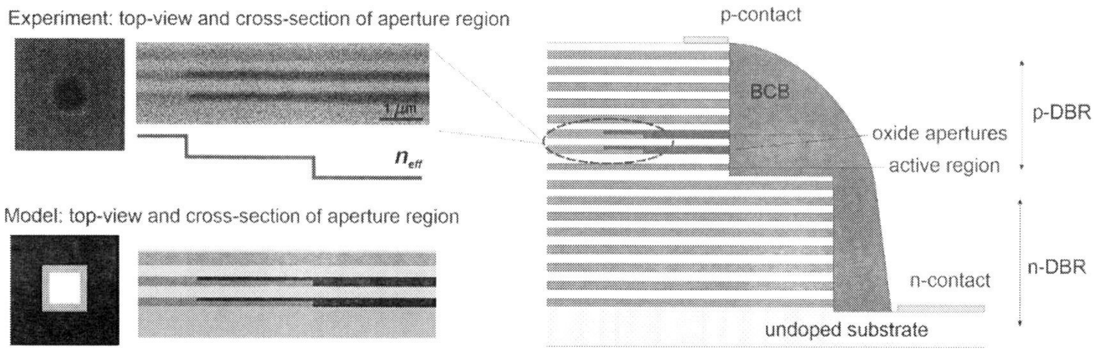

Figure 1. Schematic cross-section of the double-aperture oxide-confined 850 nm-range VCSEL. Insets: top-view IR-microscopy image and cross-section scanning electron microscopy image of oxide aperture region and corresponding model

2. Results and discussion

Figure 2a shows the lasing spectra for large-aperture device measured in pulsed regime (without current induced self-heating, pulse width 0.2 μs) and in quasi-CW regime (with strong current induced self-heating, pulse width 100 μs) at about two threshold current pumping level. At short pulse width multi-mode lasing via high-order modes were observed. The rise of device temperature caused by increasing of current pulse duration leads to multi-mode lasing via low-order modes. The estimated over-heating temperature for VCSEL active region is about +40°C (using ~0.06 nm/K fundamental mode wavelength shift with temperature). The spectral distance between low-order and high-order modes is about 7-8 nm. Figure 2b shows the lasing spectra measured at different bias current in CW regime. In this case, multi-mode lasing starts via high-order modes at low bias current and switch to lasing via low-order modes at high bias current.

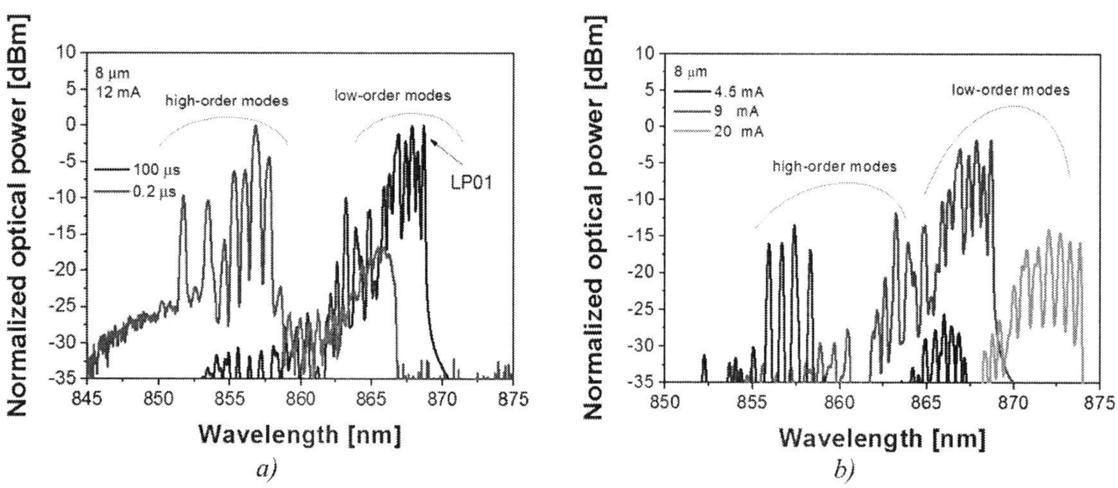

Figure 2. Lasing spectra for large-size (~ 8.0 μm per side) double-aperture 850nm-range VCSEL measured: (a) in pulsed (dashed line) and quasi-CW (solid line) regime at bias current 12 mA; (b) in CW regime at bias current of 4.5 mA, 9 mA and 20 mA

Near-field images were measured in CW-regime for large-aperture aperture VCSELs over their entire operational current range to investigate the laser mode structure in details. The images are coded in a grey scale colour scheme, where the highest intensity corresponding to white and the lowest intensity corresponding to black. At small currents, which correspond low pumping rate (I/I_{th} < 0.5), the nearly square spontaneous emission area is clearly seen (figure 3a). A further increase in current (pumping rate I/I_{th} <1) leads the appearance of the additional bright emission region along the oxide current aperture edges (figure 3b). As a result, the lasing starts via high-order optical modes spatially localized near the aperture edges in contrast to conventional large-aperture oxide-confined VCSEL (see figure 3c). With further increasing pumping rate, low-order optical modes appear, localized inside aperture area. At moderate pumping rate (I/I_{th} ~ 3.5), the near-field image seems to contain both high-order and low-order modes (see figure 3d). Finally, at high currents, which correspond high pumping rate (I/I_{th} > 5), the high-order modes disappears in the near-field image and the lasing switch to the low-order transverse modes (see figure 3e). The existence of the optical modes localized at the periphery of the oxide current aperture is possible due to the specific cross-section profile of the oxide current aperture, which have an additional extended region with an intermediate oxide thickness at its end (see insets in figure 1). The reason of such oxide profile can be associated with the use of aperture layers with a gradient Al-composition and a strong compositional dependence on the oxidation rate of AlGaAs layers.

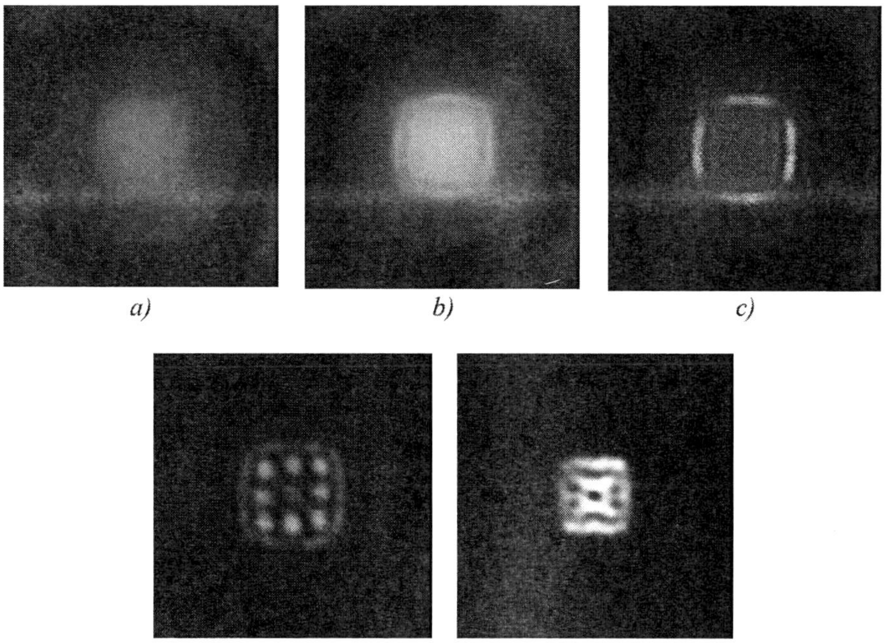

a) *b)* *c)*

d) *e)*

Figure 3. Near-field images for large-size (~ 8.0 μm) double-aperture 850nm VCSEL measured in CW-mode at bias current of 0.8 mA (a), 1.7 mA (b), 2.0 mA (c), 6.8 mA (d) and 10 mA (e).

As the first approximation, VCSEL transverse modes may by analyzed using effective index technique [7]. Figure 4 illustrates optical field intensity for investigated VCSEL structure calculated for non-oxidized area (cross-section A) and two parts of oxide aperture layers (cross-sections B and C for thin and thick oxide, respectively). The cross-section A corresponds to spatial localization of the low-order modes (particularly fundamental LP01 mode), whereas the cross-section B corresponds to the high-order modes spatially localized near the aperture edges (so called, oxide modes). The calculated spectral splitting between resonance wavelength (~ 7 nm) is corresponds well to experimental value obtained from lasing spectra (~ 7-8 nm between fundamental and high-order modes). Such large spectral splitting can be associated with the spatial location of oxide aperture layer location closed to the optical field maxima (which caused huge index contrast in the effective optical waveguide). The resonance wavelength for the cross-section C is shifted to short-wavelength by more than 18 nm as compared to that for the cross-section A and oxide aperture prevents the injection of carriers in this area.

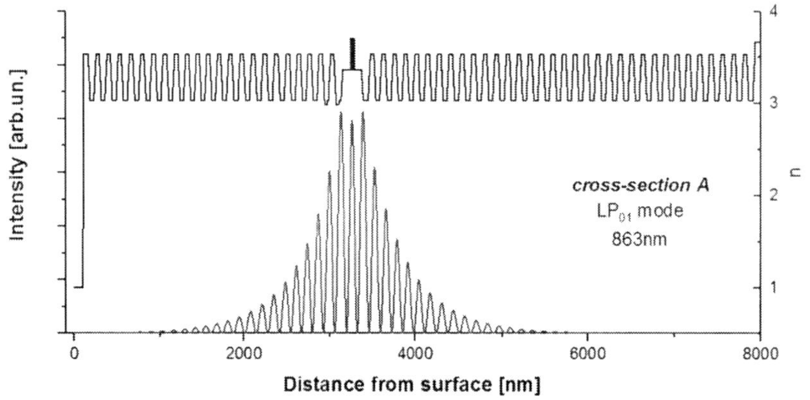

a)

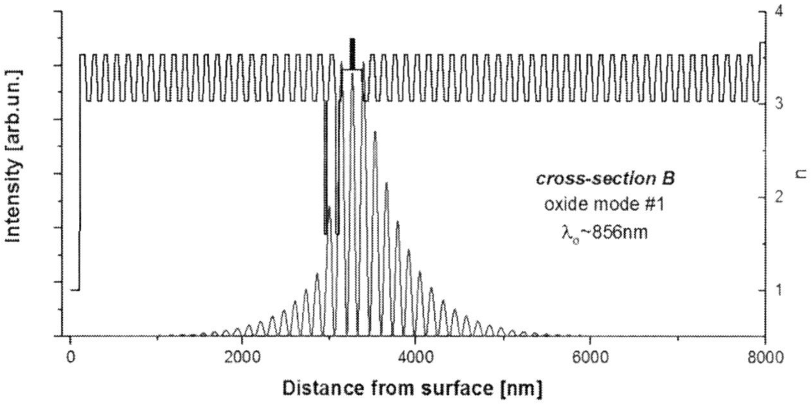

b)

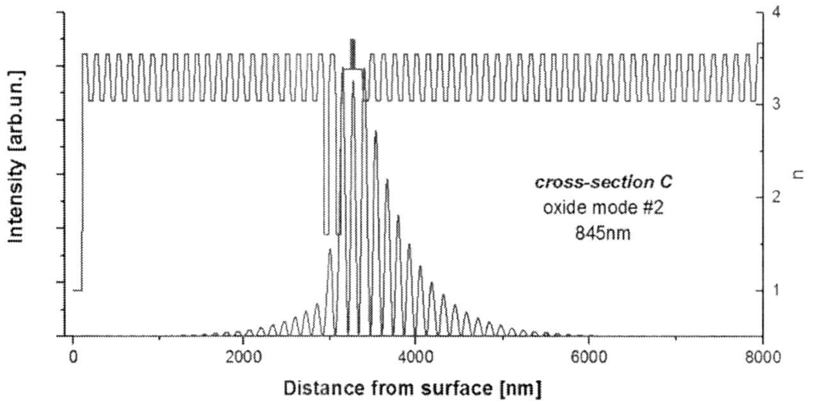

c)

Figure 4. Calculated optical field intensity in vertical direction for investigated VCSEL structure: non-oxidized area (a), thin (b) and thick (c) parts of oxide aperture layers

For more detail optical field analysis in VCSELs microcavity the full-vector three-dimensional (3-D) FDTD technique was used. Compared to other techniques, the FDTD method is a simple algorithm for solving time-dependent electromagnetic problems and it provides fairly reliable results. Figure 5 illustrates measured and calculated near-field profiles for spontaneous emission, fundamental and oxide-modes in microcavity. The optical modeling results are in good agreement with experimental data.

Calculation with FDTD method provides information about quality factor (Q-factor) for each of optical modes in microcavity. According to general form of VCSEL threshold condition [1] the lasing will occurs via optical mode with maximum value of gain multiplied by Q-factor, $g \cdot Q$. Moreover, the dependence of optical gain on temperature and current level should be taken into account [8]. In our case, at room temperature VCSEL structure has large gain-to-cavity detuning and large spectral distance between fundamental and oxide modes (see figure 6). As a result, at room temperature the $g \cdot Q$ value for oxide modes (~1560) is larger than that one for fundamental mode (~560). Therefore lasing via high-order modes is preferable at room temperature. At the device self-heating conditions (for VCSEL active region is about +40°C at CW regime) the corresponding $g \cdot Q$ values is ~1680 for fundamental mode and ~1350 for oxide modes. Therefore we obtain switching to multi-mode lasing via low-order modes.

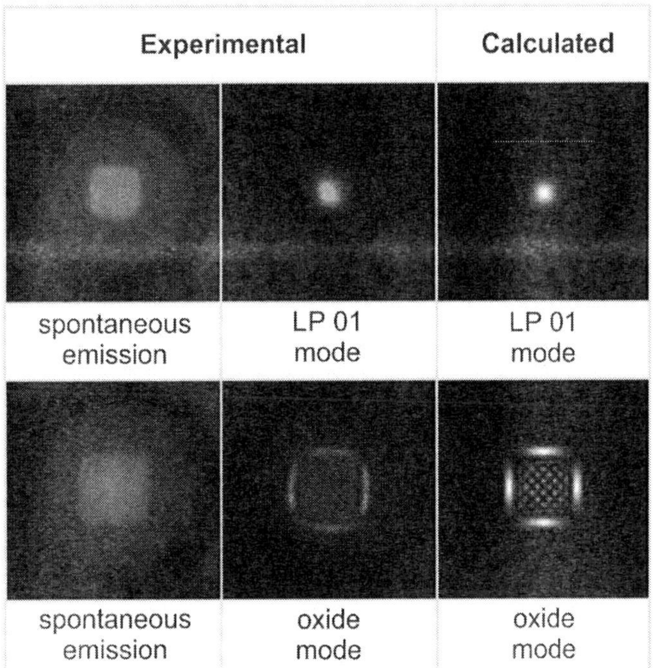

Figure 5. The measured near-field profiles for spontaneous and lasing emission in VCSEL microcavities and corresponding simulations.

3. Conclusion

In conclusion, the to clarify reasons of anomalous lasing behavior for double-aperture oxide-confined 850 nm-range VCSELs with large gain-to-cavity detuning is analyzed using detail measurements and numerical modeling of laser optical modes. The model predictions both for optical mode positions and lasing behavior are in good agreement with experimental data. According to the modeling results, initial lasing via high-order modes with following switching to lasing via low-order modes caused by thermal gain-to-cavity detuning in combination with large spectral distance between fundamental and oxide modes.

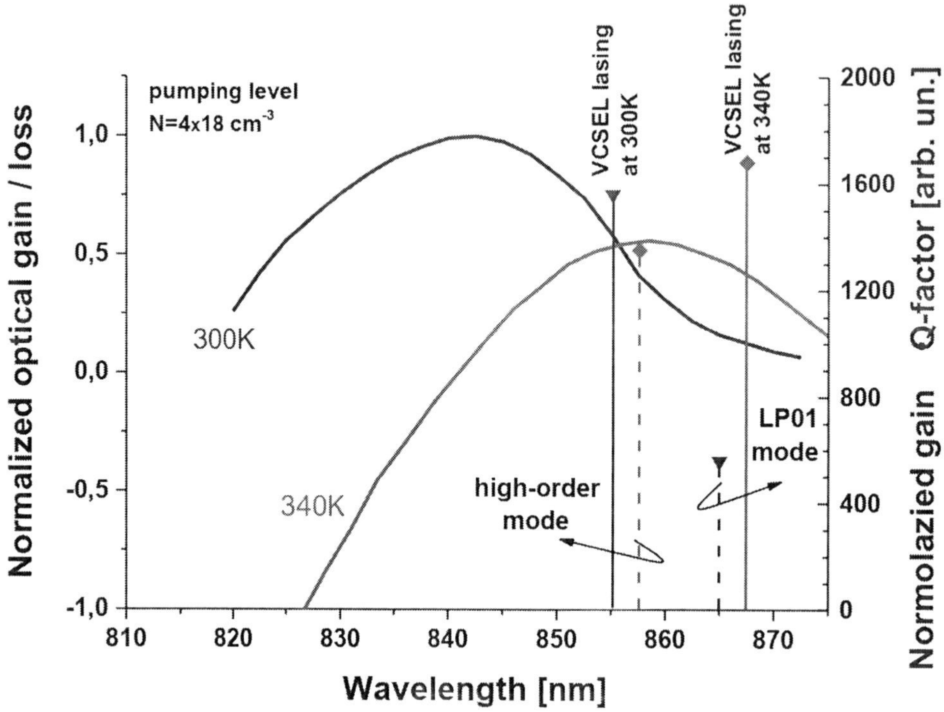

Figure 6. Calculated optical mode positions and corresponding $g \cdot Q$ values for double-aperture 850nm VCSEL with large gain-to-cavity detuning.

References

[1] Michalzik R., *VCSELs: Fundamentals, Technology and Applications of Vertical-Cavity Surface-Emitting Lasers* (Springer, Berlin, 2013)

[2] Blokhin S A, Maleev N A, Bobrov M A, Kuzmenkov A G, Sakharov A V, Ustinov V M, 2018, *Technical Physics Letters*, **44**, 1

[3] Young D B, Scott J W, Peters F H, Peters M G, Majewski M L, Thibeault B J, Corzine S W, Coldren L A, *IEEE J. Quantum Electron.*, 1993, **29**, 2013

[4] Lott J A, Maleev N A, Kuzmenkov A G, Kulagina M M, Zadiranov Yu M, Gladyshev A G, Nadtochiy A M, Nikitina E V, Ledentsov N N, Ustinov V M, 22nd IEEE International Semiconductor Laser Conference, Kyoto, Japan, 26-30 September 2010, paper WC6, pp. 184-185

[5] Blokhin S A, Bobrov M A, Maleev N A, Kuzmenkov A G, Sakharov A V, Blokhin A A, Moser P, Lott J A, Bimberg D, Ustinov V M, 2014, *Applied Physics Letters*, **104**, 163508

[6] Degen C, Fischer I, Elsaer W, 1999, *Optics Express*, **5**, 38

[7] Hadley G R, 1995, *Opt. Lett.* **20**,1483

[8] Debernardi P, Kroner A, Rinaldi F, Michalzik R, 2009, *IEEE J. Selected Topics Quantum Electron.*, **15**, 828

Strength and structural properties of AlN films grown on SiC/Si substrates synthesized by atomic substitution

A S Grashchenko[1], S A Kukushkin[2] and A V Osipov[1]
[1] Institute for Problems in Mechanical Engineering RAS, Saint Petersburg, 199178, Russia
[2] ITMO University, Saint Petersburg, 197101, Russia

Abstract. In the present work, we studied the strength and structural characteristics of the layers of the AlN/SiC/Si heterostructure. The surface morphology of the AlN film and the SiC/Si substrate was studied using atomic force microscopy. The thickness of the AlN and SiC layers was determined by analyzing the data of ellipsometry. The hardness of the films and the substrate was measured by the method of nanoscratch testing. It was experimentally shown that the films under study have a high crystalline quality.

1. Introduction

In the papers [1,2], a unique method was proposed to produce a new type of silicon Si substrates with a single crystal layer of silicon carbide SiC. This method allows nanoscale assembling of the SiC crystal inside the Si matrix simultaneously over the entire area of the substrate. This process occurs due to the chemical reaction of a Si crystal with carbon monoxide CO:

$$2Si(cr) + CO(gas) = SiC(cr) + SiO(gas) \uparrow$$

In this chemical reaction, a part of the Si atoms is substituted by carbon atoms C, and a nanoscale SiC layer is formed, as a result. Moreover, the SiC films grown in this way are practically unstressed, since the stress in the film relaxes because of a microporous structure that forms under the film during the synthesis [3]. With that, SiC films obtained by the method of substitution of atoms are low-defect, which makes them ideal substrates to grow wide-bandgap semiconductors, such as aluminium nitride AlN and gallium nitride GaN. In this work, we carried out a comprehensive study of structural and strength properties of the hybrid SiC/Si substrate and an AlN film grown on its basis.

2. Experimental methods

The substrate for the growth of the AlN film was synthesized by the method of substitution of atoms inside a silicon wafer of the KDB mark with the (111) crystallographic orientation without deviation from the base plane. The synthesis lasted for 20 minutes at the CO pressure of 0.6 Torr and a temperature of 1250 °C. Then the resulted substrate was cut into two parts. One of the parts was left in its original state, while on the other one, the AlN film was grown by the method of chloride-hydride vapor phase epitaxy (HVPE), according to the procedure described in [4]. Further, both samples were investigated identically. The structural properties of the surface of the SiC and AlN films were explored by atomic force microscopy using the Nanosurf EasyScan microscope. The surface morphology of the films was measured in the semicontact mode. The thickness and composition of the SiC and AlN films were analyzed by spectral ellipsometry using the Woollam M-2000D ellipsometer. The strength properties were studied by nanosclerometry on the MicroMaterials NanoTest 600 hardness tester. The nanosclerometry was carried out in a triple scan mode with a conical indenter with a radius of 25 μm. In this mode, the experiment follows three stages. At the first stage, the surface is scanned at the site of the future scratch, this is done to eliminate the influence of the natural inclination of the surface of a sample on the calculations. Then the scratching itself takes place, and, at the third stage, the residual deformation is scanned. The scanning of the surface topography and slope of the sample before and after scratching was carried out with a force of 0.1 mN. As a result of the analysis of the nanosclerometry data, the relative hardness was measured, which is determined as follows:

$$H_s = H_{ref}\frac{F_s}{F_{ref}}\left(\frac{W_{ref}}{W_s}\right)^2$$

where s and ref are indices related to the test and reference materials, respectively; H, F, and W are the hardness, applied force, and scratch width, respectively. The nanosclerometry method allows to estimate the strength properties of brittle materials, which is very important in the case of a microporous substrate.

3. Results

The Fig. 1a,b shows images of a surface of SiC and AlN films with an area of 10 μm^2, obtained by the atomic force microscope. It is clearly seen that the surface structure of both films is uniform. Morphological analysis of these images shows that the roughness is 22 and 25 nm for the SiC and AlN films, respectively.

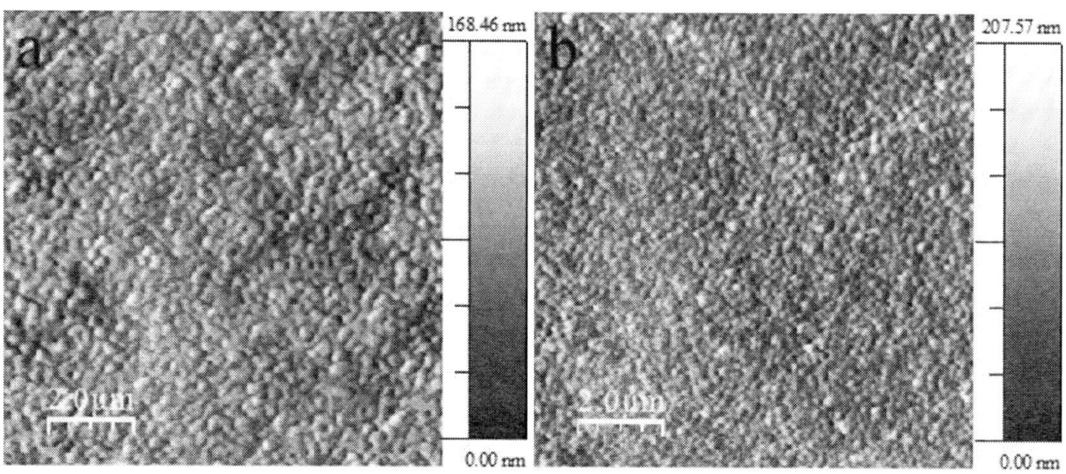

Figure 1(a, b). AFM images of surfaces. **(a)** SiC/Si substrate and **(b)** AlN film.

The Fig. 2a,b shows the ellipsometric spectra, i.e. the dependences of the real and imaginary parts of the dielectric constant on the photon energy. The solid line is the real part of the dielectric constant, the dotted line is the imaginary part of the dielectric constant. The spectrum of the SiC film on Si is well described by a two-layer theoretical model, while a single-layer ellipsometric model can't make the theoretical and experimental curves fit each other. In our case, a layer of pure SiC with a thickness of 135 nm is separated from the layer of Si with voids by a layer of a mixture of SiC with Si having a thickness of 45 nm. Considering the AlN film, the analysis of the ellipsometry data shows that the thickness of the AlN film is 610 nm, it is single-layer and contains 1.5% excess of aluminium.

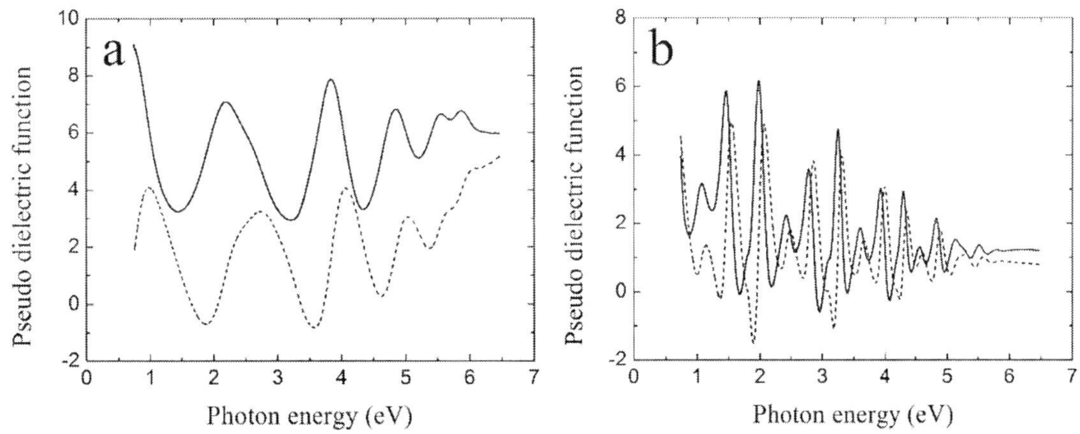

Figure 2(a, b). Ellipsometric spectra. **(a)** SiC/Si substrate and **(b)** AlN/SiC/Si heterostructure. The solid line is the real part, the dotted line is the imaginary part of dielectric permittivity.

The Fig. 3a shows the dependences of the applied normal load on the immersion depth during scratching, taking into account the natural inclination of the samples. The curves 1, 2, 3 correspond to the Si substrate, hybrid SiC/Si substrate, and AlN film.

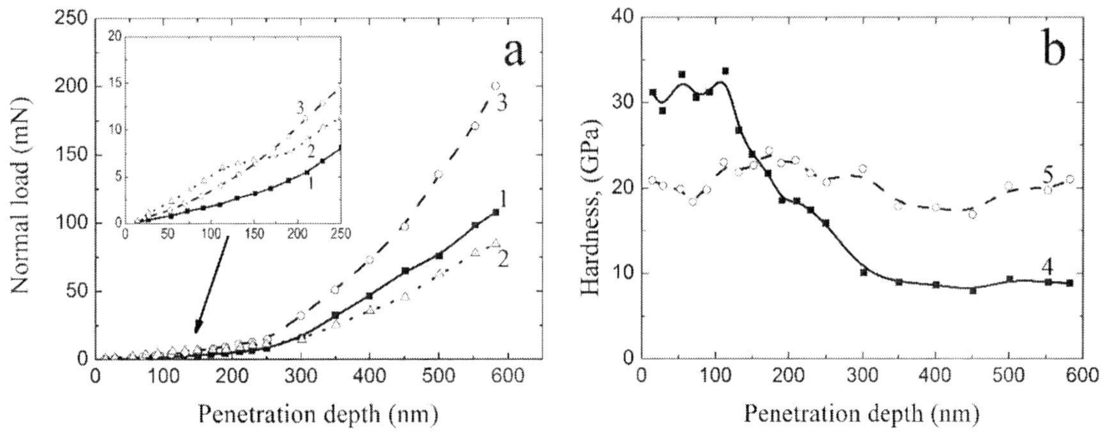

Figure 3(a, b). **(a)** Dependence of the applied normal load on the immersion depth during scratching (a) curves 1, 2, 3 correspond to the Si substrate, the SiC/Si substrate and the AlN film; **(b)** dependencies of hardness on immersion depth, curves 4, 5 correspond to the SiC/Si substrate and the AlN film.

The Fig. 3b shows the dependences of the relative hardness on the immersion depth for the hybrid SiC/Si substrate (curve 4) and AlN film (curve 5) calculated relative to the hardness of the Si substrate, which is 11.3 GPa according to the nanoindentation data. In the case of the SiC film grown on Si by the method of atomic substitution, it can be seen that at an immersion depth less than 130 nm, the hardness is almost constant and equal to 30.8 ± 2.3 GPa, which is only 7% lower than the hardness of bulk silicon carbide 33 GPa [5]. Then the hardness smoothly decreases to the value of 8.7 GPa, which is 22% less than the hardness of the calibration silicon substrate 11.3 GPa. In the case of the AlN film grown on the hybrid SiC/Si substrate, the hardness in the entire depth range is 20.5 ± 3.5 GPa. Note that, in contrast to nanoindentation [6, 7], the method of nanosclerometry made it possible to explore the strength characteristics of the film formed on the porous substrate over almost the entire thickness of the film.

4. Conclusions

The structural characteristics of an AlN film grown by the HVPE method on a hybrid SiC/Si substrate synthesized by the method of substitution of atoms have been studied. Atomic force microscopy data show that the surface structure of the AlN film and the SiC buffer layer is uniform, and their surface roughness is 22 and 25 nm, respectively. By the method of ellipsometry, it has been determined that the thickness of the AlN film is 610 nm, it is single-layer and contains 1.5% excess of aluminium. The relative hardness of the AlN film has been measured, its value is constant over the entire depth of the film and equal to 20.5 ± 3.5 GPa. Thus, it can be concluded that the AlN film grown on the SiC/Si hybrid substrate is homogeneous in its structural and strength properties, and stoichiometry of its chemical composition is close to ideal.

References

[1] Kukushkin S. A., Osipov A. V. Physics of the Solid State 2008, 50, 1238.
[2] Kukushkin S. A., Osipov A. V. J. Appl. Phys. 2013, **113,** 4909.
[3] Redkov A. V. et al. Physics of the Solid State 2019, 61, 299.
[4] Koukitu A. et.al. J.J.Appl.Phys. 1998, 37, 762.
[5] Quay R. Gallium Nitride Electronics. Berlin: Springer-Verlag, 2008, 6 p.
[6] Grashchenko A.S., Kukushkin S.A., Osipov A.V. Technical Physics Letters, 2014, 40, 1114.
[7] Grashchenko A.S., Kukushkin S.A., Osipov A. V., Redkov A.V. Journal of Physics and Chemistry of Solid, 2017, 102,151.

Reactor materials for high purity HVPE GaN growth: a thermodynamic analysis

Vladislav Voronenkov, Natalia Bochkareva, Andrey Zubrilov, Andrey Leonidov, Yuri Shreter
Ioffe Institute, Politehnicheskaya 26, St. Petersburg, 194021, Russia
E-mail: voronenkov@mail.ioffe.ru

Abstract. The chemical resistance of quartz, aluminum oxide, boron nitride, molybdenum, and platinum in the environment of the hydride vapor phase epitaxy of gallium nitride was estimated by chemical equilibrium calculation. The interaction of materials with hydrogen, hydrogen chloride, chlorine, ammonia, and gallium chlorides was analyzed. The conditions in which the use of each material is permissible were determined.

1. Introduction

Hydride vapor-phase epitaxy is today the primary method for producing GaN substrates. The low level of impurity doping is necessary to obtain non-compensated semi-insulating substrates, to achieve high thermal conductivity [1, 2] and to grow drift layers of high-voltage devices. To obtain highly pure epitaxial layers, it is necessary to minimize the level of unintentional doping, in particular, to minimize the emission of impurities from the materials of the reactor. Most of the HVPE GaN growth reactors described in the literature, starting from the pioneering study by Maruska and Tietjen [3, 4], contain quartz parts [5–13]. A significant disadvantage of quartz is its insufficient chemical resistance. During the early development of III-V crystal growth technology, quartz was identified as a source of silicon and oxygen contamination, [14–18], and decomposition of quartz reacting with liquid gallium, hydrogen, and hydrogen chloride was studied [14, 18]. Quartz–related contamination can be significantly reduced by substituting quartz with more chemically resistant materials [17, 19] or by excluding quartz from the hot zone of the reactor [20].

In this paper, the chemical resistance of quartz, aluminum oxide, boron nitride, and molybdenum is estimated by chemical equilibrium calculation.

2. Calculation procedure

Schematic diagram of a conventional HVPE GaN growth system is shown in figure 1. Nitrogen, argon, or hydrogen is used as a carrier gas. Gallium is transported to the growth chamber in the form of gallium monochloride [4] or gallium trichloride [21–23]. Gallium monochloride is obtained by chlorination of gallium [4] or trimethylgallium [24, 25] at a temperature of about 600–800 °C. Gallium trichloride is delivered as vapor from a bubbling evaporator [22] or is formed by the reaction of monochloride and an excess of chlorine [23]. Ammonia is fed through a separate channel and mixed with gallium chloride in the growth zone near the substrate.

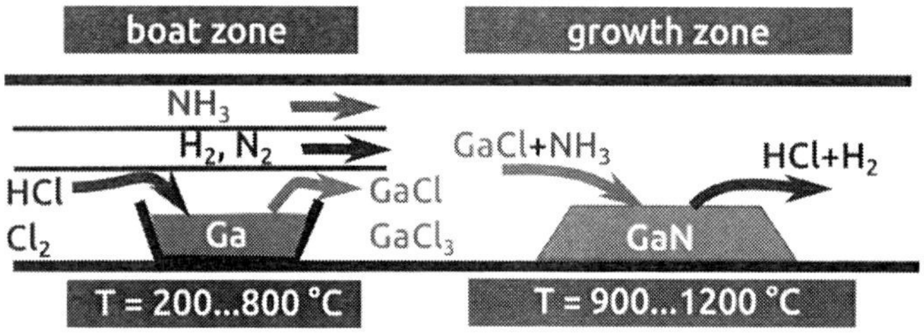

Figure 1. Schematic diagram of a typical HVPE reactor for GaN growth. The main chemically active components of the gas phase are H_2, NH_3, HCl, Cl_2, GaCl and $GaCl_3$.

Table 1. Substances considered in calculation. The aggregate state of substance is indicated by a superscript: s – solid, l – liquid, g – gas.

Element	Substances
H, N, O, Cl	H_2^g, H^g, N_2^g, N^g N_3^g, O_2^g, O^g, O_3^g, Cl_2^g, Cl^g, H_2O^g, OH^g, $H_2O_2^g$, HO_2^g, HCl^g, $HOCl^g$, NH_2^g, NH_3^g, NH^g, $N_2H_2^g$, $N_2H_4^g$, N_3H^g, $N_2O_3^g$, $N_2O_4^g$, $N_2O_5^g$, N_2O^g, NO_2^g NO_3^g, NO^g, HNO^g, HNO_2^g, HNO_3^g, $NH_2NO_2^g$, NH_2OH^g, NO_2Cl^g, $NOCl^g$, Cl_2O^g, ClO_2^g, ClO^g
Ga	Ga^l, Ga^g, $Ga_2Cl_2^g$, $Ga_2Cl_4^g$, $Ga_2Cl_6^g$, $GaCl_2^g$, $GaCl_3^g$, $GaCl^g$, GaH^g $GaOH^g$, GaO^g, $Ga_2O_3^s$, Ga_2O^g, GaN^s
Al	Al^s, Al^l, Al_2^g, Al^g, $Al_2Cl_6^g$, $AlCl_3^g$, $AlCl_2^g$, $AlCl^g$, AlH^g, AlH_2^g, AlH_3^g, AlO_2^g, Al_2O^g, AlO^g, $Al_2O_2^g$, $Al_2O_3^s$, $Al_2O_3^g$, $Al(OH)_2Cl^g$, $Al(OH)_2^g$, $Al(OH)_3^g$, AlH_2Cl^g, $AlHCl_2^g$, $AlHCl^g$, $AlOCl_2^g$, $AlOCl^g$, $AlOHCl_2^g$, $AlOHCl^g$, $AlOH^g$, $HAlO_2^g$, $HAlO^g$, AlN^g, AlN^s
B	B^s, B^l, B^g, B_2^g, $B_2Cl_4^g$, $B_2H_2^g$, $B_2H_3^g$, $B_2H_4^g$ $B_2H_5^g$, $B_2H_6^g$, B_2H^g, $B_3H_7^g$, $B_3H_9^g$, $B_3N_3H_6^g$, $B_4H_{10}^g$, $B_4H_{12}^g$, $B_4H_4^g$, $B_5H_9^g$, BCl_2^g, BCl_3^g, BCl^g, BH_2Cl^g, BH_2^g, $BH_3NH_3^g$, BH_3^g, BH_4^g, BH_5^g, $BHCl_2^g$, $BHCl^g$, BH^g, BN^g, BN^s
Mo	Mo^s, Mo^g, $MoCl_4^g$, $MoCl_5^g$, $MoCl_4^s$
Pt	Pt^s, $PtCl_3^g$, $PtCl_2^g$, $Pt_6Cl_{12}^g$, $PtCl_3^s$, $PtCl_2^s$, $Pt_6Cl_{12}^s$

Following gases and gas mixtures were chosen to perform comparison of material resistivity in HVPE reactor atmosphere: Hydrogen, which is used as a carrier gas, alone or mixed with nitrogen; HCl, HCl diluted with H_2 and Cl_2 used to obtain gallium chlorides in the boat, to control the supersaturation in the growth process and for dry cleaning of the growth chamber of the reactor; NH_3, the nitrogen precursor; $GaCl_3$, a gallium precursor in tri-halide HVPE; a mixture of GaCl:H_2:HCl in a ratio of 4:4:1 - a typical composition of the products of gallium chlorination carried out in the boat using hydrogen chloride.

To calculate the chemical equilibrium, a hierarchical [26–28], method was used with some modifications to increase the convergence rate [29]. In comparison with the nonstoichiometric methods [30, 31], the hierarchical method shows steady convergence in problems where one or several condensed species and gaseous species with low equilibrium pressure are taken into account in the calculation.

Thermochemical properties of substances from [32–38] were used. No preliminary assumptions were made about the composition of the system; the calculation accounted for all species for which thermochemical data were available in the literature. The total pressure in all the calculations performed was assumed to be 1 bar; ideal gas behavior was assumed.

The interaction of quartz and Al_2O_3 with ammonia was considered in a non-equilibrium condition: the decomposition of ammonia into nitrogen and hydrogen was assumed to be kinetically inhibited; all the ammonia decomposition reactions, except for reactions of ammonia with quartz and Al_2O_3, were excluded from the calculation.

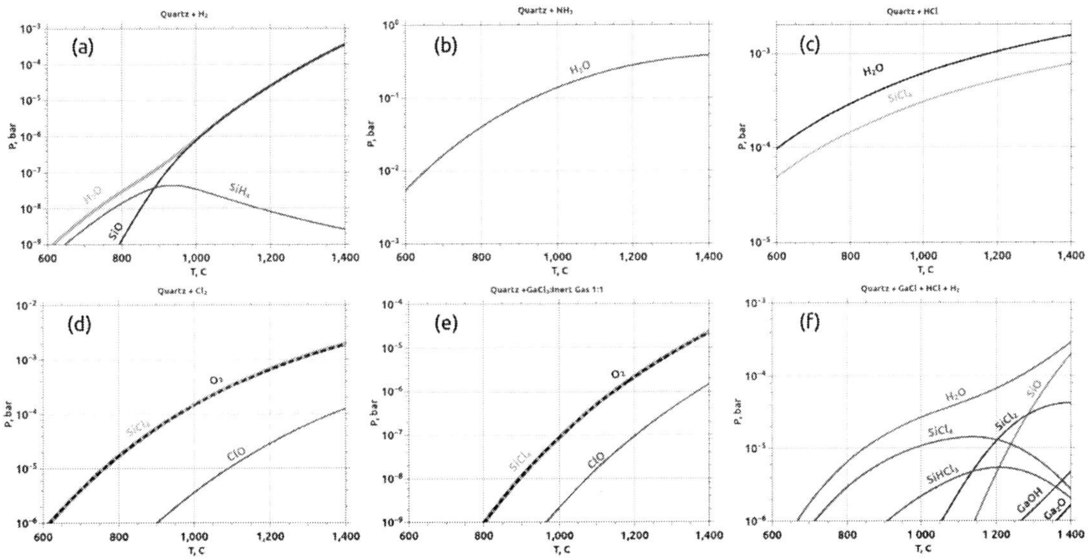

Figure 2. Equilibrium partial pressure of gaseous decomposition products of quartz in hydrogen (a), ammonia (b), hydrogen chloride (c), chlorine (d), gallium trichloride (e), and in a mixture of GaCl:H_2:HCl in a ratio of 4:4:1 (f).

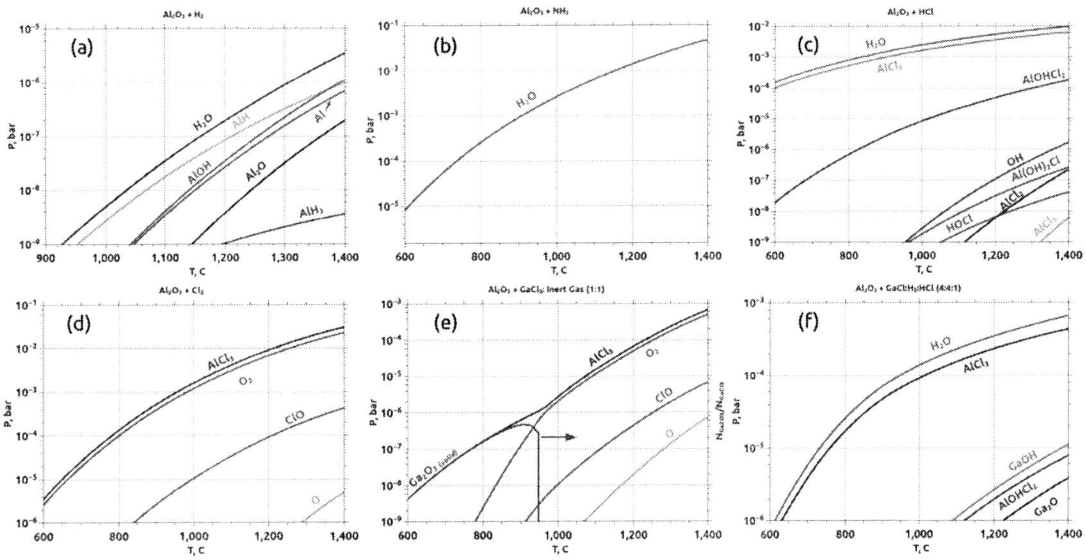

Figure 3. Equilibrium partial pressure of gaseous decomposition products of sapphire in hydrogen (a), ammonia (b), hydrogen chloride (c), chlorine (d), gallium trichloride (e), and in a mixture of GaCl:H_2:HCl in a ratio of 4:4:1 (f).

Chemical equilibrium analysis does not allow to determine the exact amount of impurities emitted by the materials of the reactor since the reaction kinetics is not taken into account. However, this analysis allows estimating the upper limit of contamination caused by the decomposition of the materials of the reactor.

The following was chosen as an arbitrary criterion for chemical resistance of the material: the

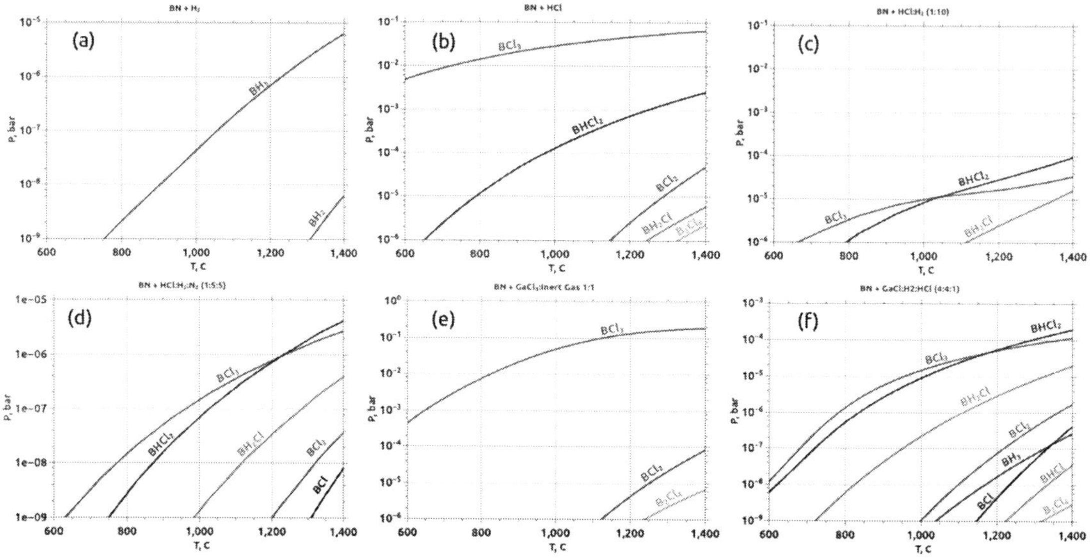

Figure 4. Equilibrium partial pressure of gaseous decomposition products of boron nitride in inert gas (a), hydrogen (b), hydrogen chloride (c), chlorine (d) gallium trichloride (e), and a mixture of GaCl:H_2:HCl in a ratio of 4:4:1 (f).

total pressure of the decomposition products should not exceed 10^{-6} bar, which corresponds to a contamination level of 1 ppm.

The substances that were considered in the calculations are listed in table 1.

3. Results and discussion

The results of the chemical equilibrium calculations performed are shown in the figures 2 – 6. For the sake of clarity, the graphs show the partial pressures of only those material decomposition products that introduce impurities to the growth process.

3.1. Quartz (SiO_2)

In a hydrogen atmosphere, quartz can be considered resistant at temperatures up to 1000 °C. The main decomposition products in the high-temperature region are H_2O and SiO, the partial pressure of which reaches 10^{-4} at a temperature of 1300 °C (figure 2a).

With ammonia, quartz reacts to a large extent to form solid silicon oxynitride and water vapor (figure 2b).

In hydrogen chloride, the main products of quartz decomposition are silicon tetrachloride and water vapor, the partial pressure of which exceeds 10^{-4} bar already at a temperature of 600 °C (figure 2c).

In chlorine, quartz meets the chosen criterion of chemical resistance at temperatures below 600 °C; the main decomposition products are silicon tetrachloride and oxygen (figure 2d).

The main products of the reaction of quartz with gallium trichloride are the same as in the reaction with chlorine, but the partial pressure of decomposition products is approximately three orders of magnitude lower (figure 2e).

In a mixture of hydrogen GaCl and chlorine hydrogen, the chemical resistance criterion is satisfied at T<650 °C. In the low-temperature region, the main decomposition products are H_2O and $SiCl_4$, in the high-temperature region – H_2O and SiO (figure 2f).

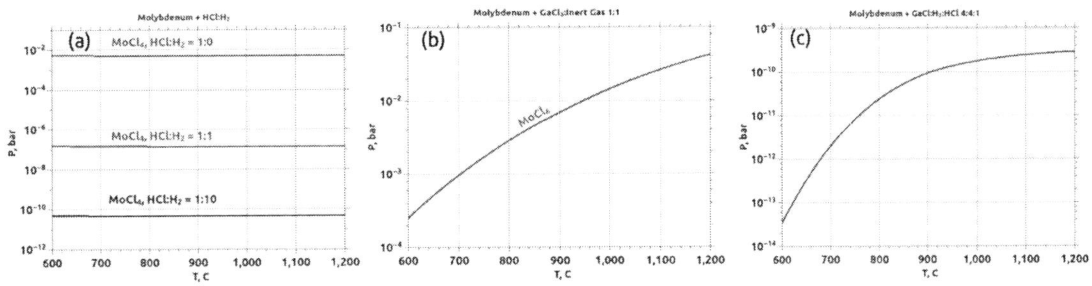

Figure 5. Equilibrium partial pressure of gaseous decomposition products of molybdenum in hydrogen chloride (a), gallium trichloride (b), and a mixture of GaCl:H_2:HCl in a ratio of 4:4:1 (c).

3.2. Aluminum Oxide (Al_2O_3)

Compared with quartz, aluminum oxide is significantly more resistant in hydrogen (figure 3a) and somewhat more resistant in ammonia, in the reaction with which aluminum nitride is formed (figure 3b). However, in the atmosphere of chlorine, hydrogen chloride, $GaCl_3$ and a mixture of GaCl:H_2:HCl, the equilibrium pressures of decomposition products are even higher than that above quartz under similar conditions (figure 3c,d,e,f). Experimental studies of Al_2O_3 decomposition [39–41] show a kinetically limited but observable etching rate.

3.3. Boron Nitride (BN)

In hydrogen, BN meets the criterion of resistance at temperatures below 1200 °C (figure 2a) In ammonia, BN is inherently resistant. With chlorine, BN reacts completely over the entire temperature range, BCl_3 is the main decomposition product (not shown). In pure hydrogen chloride, BN is unstable, decomposing into BCl_3, N_2 and H_2, but diluting the HCl with hydrogen (figure 4c) or hydrogen:nitrogen mixture (figure 4d), shifts the reaction equilibrium and significantly reduces the decomposition of BN.

BN is less resistant than Al_2O_3 and quartz in $GaCl_3$ ambient (figure 4e), but more resistant than Al_2O_3 and quartz in a mixture of GaCl:HCl:H_2 (figure 4f).

3.4. Molybdenum (Mo)

The main product of the interaction of molybdenum with hydrogen chloride is molybdenum tetrachloride $MoCl_4$, the concentration of which substantially depends on the hydrogen content in the gas phase [38]. When diluting the hydrogen chloride with hydrogen, in a 1:1 ratio, the equilibrium pressure of $MoCl_4$ decreases below 10^{-6} (figure 5a).

In gallium trichloride, molybdenum is unstable (figure 5b). On the contrary, in a mixture of GaCl:H_2:HCl, the equilibrium pressure of decomposition products is below 10^{-9} over the entire temperature range (figure 5c).

3.5. Platinum

Platinum reacts with chlorine to form volatile chlorides of $PtCl_2$ and $PtCl_3$, the equilibrium pressure of which exceeds 10^{-5} bar in the entire considered range (figure 6a). When diluting chlorine with an inert gas in the ratio of 1:10, the equilibrium pressure of platinum chlorides decreases, but remains above 10^{-6} bar at temperatures above 700 °C (figure 6b). In hydrogen chloride, platinum is stable over the entire temperature range considered (figure 6); in a mixture of hydrogen chloride and hydrogen, the equilibrium pressures of platinum chlorides decrease to negligibly small values (figure 6d). Platinum is also stable in $GaCl_3$ at temperatures up

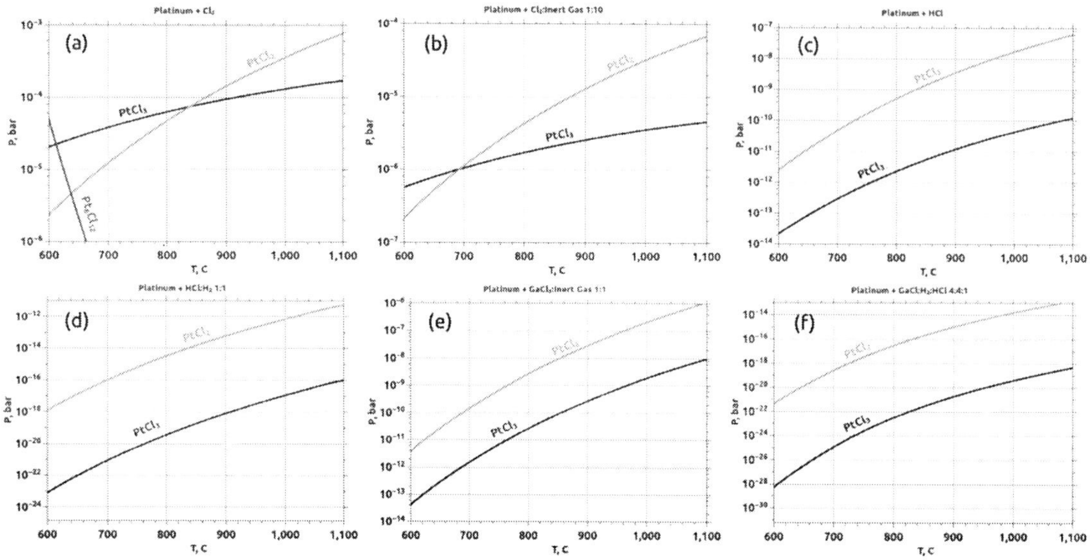

Figure 6. Equilibrium partial pressure of gaseous decomposition products of platinum in Cl_2 (a), Cl_2 diluted with an inert gas in a ratio of 1:10 (b), hydrogen chloride (c), a mixture of $HCl:H_2$ in a ratio of 1:1 (d), gallium trichloride (e), and a mixture of $GaCl:H_2:HCl$ in a ratio of 4:4:1 (f).

Table 2. Materials chemical resistance calculation summary: (+) total vapor pressure of decomposition products $< 10^{-6}$ bar, material is considered to be chemically resistant, (—) total vapor pressure of decomposition products $> 10^{-6}$ bar.

	SiO_2	Al_2O_3	BN	Mo	Pt
H_2	(+) T<1000 °C	(+) T<1300 °C	(+) T<1200 °C	(+)	(+)
NH_3	(—)	(—)	(+)	(+)	(+)
HCl	(—)	(—)	(—)	(—)	(+)
$HCl:H_2$	(—)	(—)	(+) T<1200C, $N_2:H_2$ / T<650C, H_2	(+)	(+)
Cl_2	(+) T<600 °C	(—)	(—)	(—)	(—)
$GaCl_3$	(+) T<1100 °C	(+) T<900 °C	(—)	(—)	(+) T<1100 °C
$GaCl:HCl:H_2$	(—)	(—)	(+), T<700 °C	(+)	(+)

to 1080 °C (figure 6e), and in a mixture of $GaCl:H_2:HCl$ over the entire temperature range (figure 6f).

4. Conclusion

The chemical resistance of quartz, aluminum oxide, boron nitride, molybdenum and platinum was estimated using chemical equilibrium analysis. Summary data on the durability of the materials studied are given in the table 2. None of the materials studied is chemically resistant to all gases involved in the HVPE process. An individual selection of material for each part of the reactor is necessary to minimize the decomposition of materials and to reduce the associated impurity contamination.

References

[1] Slomski M, Paskov P P, Leach J H, Muth J F and Paskova T 2017 *Phys. Status Solidi B* **254** 1600713

[2] Jezowski A, Churiukova O, Mucha J, Suski T, Obukhov I A and Danilchenko B A 2015 *Mater. Res. Express* **2** 085902

[3] Tietjen J J and Amick J A 1966 *J. Electrochem. Soc.* **113** 724–728

[4] Maruska H P and Tietjen J J 1969 *Appl. Phys. Lett.* **15** 327–329

[5] Safvi S, Perkins N, Horton M, Matyi R and Kuech T 1997 *J. Cryst. Growth* **182** 233 – 240

[6] Tavernier P R, Etzkorn E V, Wang Y and Clarke D R 2000 *Appl. Phys. Lett.* **77** 1804–1806

[7] Dam C, Grzegorczyk A, Hageman P, Dorsman R, Kleijn C and Larsen P 2004 *J. Cryst. Growth* **271** 192 – 199

[8] Segal A, Kondratyev A, Karpov S, Martin D, Wagner V and Ilegems M 2004 *J. Cryst. Growth* **270** 384 – 395

[9] Monemar B, Larsson H, Hemmingsson C, Ivanov I and Gogova D 2005 *J. Cryst. Growth* **281** 17 – 31

[10] Richter E, Hennig C, Weyers M, Habel F, Tsay J D, Liu W Y, Brckner P, Scholz F, Makarov Y, Segal A and Kaeppeler J 2005 *J. Cryst. Growth* **277** 6 – 12

[11] Williams A D and Moustakas T 2007 *J. Cryst. Growth* **300** 37 – 41

[12] Fujito K, Kubo S, Nagaoka H, Mochizuki T, Namita H and Nagao S 2009 *J. Cryst. Growth* **311** 3011 – 3014

[13] Amilusik M, Sochacki T, Lucznik B, Bockowski M, Sadovyi B, Presz A, Dziecielewski I and Grzegory I 2013 *J. Cryst. Growth* **380** 99 – 105

[14] Cochran C N and Foster L M 1962 *J. Electrochem. Soc.* **109** 149–154

[15] Stearns R I and McNeely J B 1966 *J. Appl. Phys.* **37** 933–934

[16] Greene P D 1973 *J. Phys. D: Appl. Phys.* **6** 1550–1554

[17] Callaghan M, Patterson E, Richards B and Wallace C 1974 *J. Cryst. Growth* **22** 85 – 98

[18] Weiner M E 1972 *J. Electrochem. Soc.* **119** 496–504

[19] Long G and Foster L M 1959 *J. Am. Ceram. Soc.* **42** 53–59

[20] Fujikura H, Konno T, Yoshida T and Horikiri F 2017 *Jpn. J. Appl. Phys.* **56** 085503

[21] Nickl J, Just W and Bertinger R 1974 *Mater. Res. Bull.* **9** 1413 – 1420

[22] Lee H and Harris J J S 1996 *J. Cryst. Growth* **169** 689–696

[23] Hirasaki T, Asano K, Banno M, Ishikawa M, Sakuma F, Murakami H, Kumagai Y and Koukitu A 2014 *Jpn. J. Appl. Phys.* **53** 05FL02

[24] Kumagai Y, Murakami H, Seki H and Koukitu A 2002 *J. Cryst. Growth* **246** 215 – 222

[25] Kryliouk O, Reed M, Dann T, Anderson T and Chai B 1999 *Mater. Sci. Eng., B* **66** 26–29

[26] Villars D S 1959 *J. Phys. Chem.* **63** 521–525

[27] Colonna G and D'angola A 2004 *Comput. Phys. Commun.* **163** 177–190

[28] Colonna G 2007 *Comput. Phys. Commun.* **177** 493–499

[29] Voronenkov V 2015 *Epitaxial growth of bulk gallium nitride layers: technological parameters optimization (in russian)*. Ph.D. thesis

[30] Gordon, Sanford and McBride, Bonnie J 1976 Computer Program for Calculation of Complex Chemical Equilibrium Compositions, Rocket Performance, Incident and Reflected Shocks, and Chapman-Jouguet Detonations. Interim Revision, March 1976 Tech. rep. NASA Lewis Research Center

[31] Smith W R 1980 *Industrial & Engineering Chemistry Fundamentals* **19** 1–10

[32] Gurvich L V and Veyts I 1990 *Thermodynamic Properties of Individual Substances: Elements and Compounds* vol 2 (CRC press)

[33] Chase M W (ed) 1998 *NIST-JANAF Themochemical Tables, Fourth Edition* (American Chemical Society)

[34] Przhevalskii I, Karpov S Y and Makarov Y N 1998 *MRS Internet J. Nitride Semicond. Res.* **3** 1–16

[35] Zinkevich M and Aldinger F 2004 *J. Am. Ceram. Soc.* **87** 683–691

[36] Schafer H 1980 *Z. Anorg. Allg. Chem.* **469** 123–127

[37] Schonherr E, Wojnowski M, Rabenau A and Lacher S 1988 *J. Less Common Met.* **137** 211–286

[38] Quill L (ed) 1950 *Chemistry and Metallurgy of miscellaneous materials: Thermodynamics*

[39] Reti F, Bertoti I, Mink G and Szekely T 1987 *React. Solid.* **3** 329 – 336

[40] Akiyama K, Araki T, Murakami H, Kumagai Y and Koukitu A 2007 *Phys. Status Solidi C* **4** 2297–2300

[41] Akiyama K, Ishii Y, Murakami H, Kumagai Y and Koukitu A 2009 *J. Cryst. Growth* **311** 3110 – 3113

Initial growth state MoS_2 thin film by magnetron sputtering

A I Belikov, A Dolomanzhi, V N Kalinin, Do Thi Nhan, Kyaw Zin Phyo, A I Syomochkin

Department of electronic technologies in mechanical engineering, Bauman Moscow State Technical University (BMSTU), Moscow 105005, Russia

Abstract. MoS_2 thin films were grown on sapphire substrates by the MoS_2 target magnetron sputtering, using the DC mode in Ar atmosphere at the substrate temperature of 200 °C for 10, 20 and 30 sec. The results of AFM studies have shown that the morphology of MoS_2 thin films changes significantly with the increase of the process time. At the initial stage of growth, after 10 sec of deposition, a continuous film about 3.5 nm thick was observed.

1. Introduction

Recently, molybdenum disulfide (MoS_2) has attracted considerable attention because of their potential for future applications in nanoelectronics, due to its outstanding electronic and optical properties of MoS_2 thin films. These unique properties makes MoS_2 suitable for various applications, such as field-effect transistors with low power consumption and high quality switching, touch and optical devices for the next generation of nano- and optoelectronics [1, 2]. In this regard, there is a problem of reproducible production of high-quality MoS_2 ultra-thin films using modern industrial-oriented methods. Among the deposition methods [3-5] magnetron sputtering is the most common method of forming a continuous high-quality ultra-thin films at a relatively low substrate temperature [6]. However, to use the magnetron method for MoS_2 ultrathin films, it is necessary to determine the relationship between the technological conditions of their formation and structural characteristics. These conditions determine the degree of homogeneity and phase composition, which, in turn, affect the morphology and electronic properties of the film. One of the main advantages of a magnetron sputtering is the ability to control the properties of the growing film and change the deposition rate by selecting operating parameters (power mode sputtering, the working pressure, the ion current density, etc.). The aim of the present work was to study the initial growth stage of MoS_2 ultrathin films.

2. Experimental

The experiments were carried out on technological equipment placed inside a vacuum chamber with dimensions of 46 cm in height and 40 cm in diameter. Previously, experiments were performed on the study of MoS_2 thin films, obtained by magnetron sputtering at low temperatures [7] and various technological parameters of the process [8]. In the present series of studies, films were deposited by sputtering a molybdenum disulfide target with a diameter of 75 mm, purity of 99.9% and high purity argon (99.9995%) was used as a plasma-forming gas. Sapphire plates about 1 cm^2 in size with a surface roughness of 0.877 nm were used as substrates. Before deposition, the substrate surface contamination was cleaned in an ultrasonic bath in an alkaline solution and ethanol for 10 min. The process chamber was pumped out to a residual pressure of 10^{-3} Pa. The main technological parameters of all MoS_2 deposition processes were constant, their values are given in Table 1. To estimate the initial stage of film growth, the deposition of a flow sputtered from the target surface for 10, 20, and 30 seconds was used. The substrate temperature was controlled by a thermostat with feedback of the heater and temperature sensor and maintained at 200 °C. The distance between the target and the substrate was 10 cm. The surface morphology and film thickness were measured by using atomic force microscopy.

Table 1. Deposition process parameters applied for the samples.

№	Deposition time, seconds	Substrate temperature, °C	Discharge power, W	Discharge current, A	Argon pressure, Pa
1	10	200	22	0.05	5×10^{-1}
2	20	200	22	0.05	5×10^{-1}
3	30	200	22	0.05	5×10^{-1}

3. Results and discussion

The results of AFM scanning of the MoS_2 films surface deposited on a sapphire substrate during different deposition times are shown in Figure 1. Ultrathin films with a thickness of 3.5 nm were obtained for samples No.1 (Figure 1a), which were deposited for 10 sec. The surface of the film for sample No.1 had a low roughness and high planarity, the measured value of the root-mean-square roughness (RMS) was only 0.2 nm. It is known that the value of surface energy for a MoS_2 crystal along the planes of the sulfur layer (γ_f) is less than the surface energy of the sapphire substrate (γ_s), therefore, at the initial stage, a film can be formed by a two-dimensional growth mechanism. In this case, molecules arriving on the film growth surface, if they have sufficient energy, can travel considerable distances in order to adsorb to areas with high potential energy, which are the vacancies or steps of the growing film. Judging by the smooth surface and high planarity of the films obtained at the minimum time of deposition and the parameters used in the initial stage, the mechanism of layer-by-layer growth of the MoS_2 film is realized.

With an increase in deposition time of up to 20 seconds, a film is formed with a slightly larger roughness. Deposition time further increase up to 30 seconds, results in multiple increase of roughness, which can be interpreted as a change in the growth mechanism from “layer-by-layer” to “layer-plus-island”. In addition, large islands formation (Figure 2c) was observed on the sample No.3 surface. The surface morphology of MoS_2 films for samples No.2 and No.3, which were obtained for 20 and 30 sec, respectively, shown in Figure 2b and 2c, the values of surface roughness were approximately 0.25 and 0.65 nm.

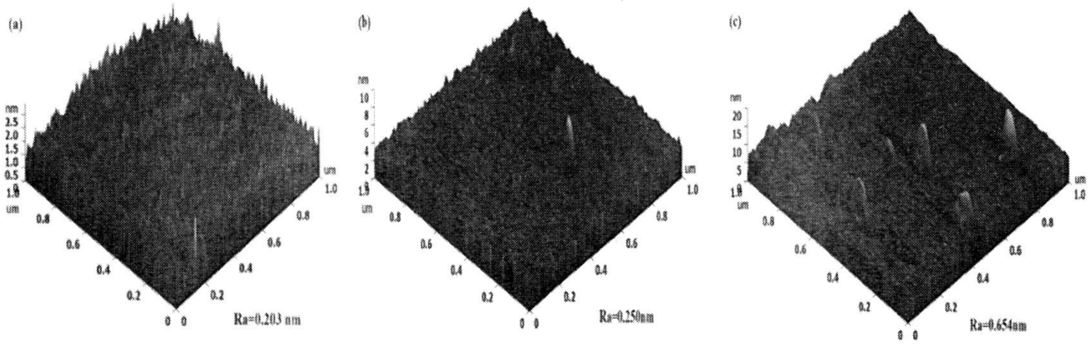

Figure 1(a, b, c). MoS_2 films deposited on sapphire substrates during different process times AFM 3D images ($1\times1 \mu m^2$): **(a)** sample No.1 (10 sec); **(b)** sample No.2 (20 sec); **(c)** sample No.3 (30 sec).

With an increasing of deposition time, a proportional increase in the film thickness was observed from 3.5 nm to 7 nm, and 12 nm (Table 2). Using the software for analyzing the grain structure on the scan surface, images of grain boundaries and their distribution over the surface were obtained (Figure 2), and the values of the average grain size for film samples were determined (Table 2). With an increasing of deposition time, an enlargement of the grains and an increase in their average size was observed. Summary results of all films parameters measurement are given in table 2.

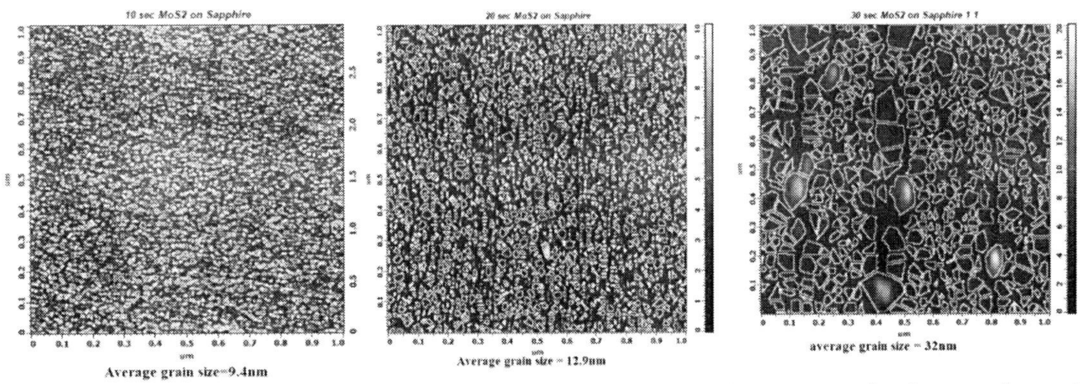

Figure 2(a, b, c). AFM 2D images (1μm×1μm) of MoS_2 films grains structure for the experimental samples: **(a)** sample 1 (10 sec); **(b)** sample 2 (20 sec); **(c)** sample 3 (30 sec).

From the analysis of the obtained images, it shows that, a structure is formed consisting, presumably, of planar two-dimensional grains, with observed slight differences in their surface level in height with low deposition time (10 seconds). As the deposition time increases to 20 seconds, there is a slight increase in the average grain size, which may be due to the appearance of some small number of larger island formations, probably caused by the formation of islands already on the surface of a continuous MoS_2 film. On the surface of samples obtained by deposition for 30 seconds, island formation on the surface of the film become larger and their number increases.

Table 2. MoS_2 thin films experimental values of thickness, roughness and grain size.

Sample	Thickness, nm	RMS, nm	Average grain size, nm
1. MoS_2 Depos.10 sec.	3.5	0.203	9.4
2. MoS_2 Depos.20 sec.	7.0	0.250	12.9
3. MoS_2 Depos.30 sec.	12.0	0.654	32.0

4. Conclusions

Thus, as a result of deposition and film growth, MoS_2 films with a thickness of 3.5 nm formed on the sapphire surface during the first 10 seconds, which corresponds to approximately five molecular layers of MoS_2, characterized by low roughness and high surface planarity, the "layer-by-layer" film growth is implemented at this stage. With an increase in the deposition time, the "layer-by-layer" growth mechanism is replaced by a "layer-plus-island" growth mechanism. With a film thickness of 12 nm, which was formed in 30 seconds, on the surface of the film, relatively large island formations are observed, which significantly increase the average roughness value recorded by the AFM method.

References

[1] X. Li, H. Zhu, 2015 J. Mater. **1** 33–44.

[2] R. Ganatra, Q. Zhang, 2014 ACS Nano **8** 4074–4099
[3] Yumeng Shi, Henan Li and Lain-Jong Li,2015 Chem. Soc. Rev **44**, 2744
[4] Martha I. Serna, et al.,2016 ACS Nano **10,** No 6, 6054–6061
[5] Jung Joon Pyeon, et al.,2016 Nanoscale **8** 10792
[6] C. Muratore, et al.,2014 Appl. Phys. Lett. **104** 261604
[7] Belikov A I, et al.,2018 Journal of Physics:IOP Conf. Series, **1124** 051002
[8] Belikov A I et al.,2019 Proc. SPIE, **11022** 110221F

Study of the influence of the electrophysical properties of ZnO nanorods grown on glass on their gas-sensitive properties

V V Petrov[1], Kh A Abdullin[2], A P Starnikova[1], Yu N Varzarev[1]
[1] Southern Federal University, Research and Education and Centre "Microsystem technics and multisensor monitoring systems", Taganrog, 347922, Russia
[2] Al Farabi Kazakh National University, National Nanotechnology Laboratory of Open Type, Almaty, 050040, Kazakhstan

Abstract. The article discusses the sensory properties of an array of ZnO-based nanorods, which were used as sensitive elements of carbon (II) oxide sensors. By hydrothermal synthesis, arrays of ZnO nanorods were grown on glass, which had an average transverse size of up to 40 nm and a height of up to 660 nm. Formed rods had a predominantly vertical orientation. Metal contacts were formed over the array of nanorods. Studies of the electrophysical properties showed that the temperature dependence of the conductivity of the formed structure has a hysteresis at temperatures above 160 ° C, and the current-voltage characteristic is close to linear. The gas sensitive properties of the ZnO array of nanorods were investigated to carbon monoxide (II) with a concentration of 20-100 ppm at operating temperatures from 20 to 270 ° C. It has been shown that the sensor response is strongly influenced by the rate of blowing the array with nanorod air or the test gas.

1. Introduction

The properties of nanostructured materials based on oxide semiconductors are currently being intensively studied, since these materials have great potential for practical application. ZnO-based nanostructures are used to manufacture sensitive elements of gas sensors, piezo transducers and energy harvesters [1-4]. The principle of operation of gas sensors is based on the property of ZnO to change its electrical parameters during the adsorption of gas molecules.

2. Experiment

Arrays of ZnO nanorods were synthesized on p-type silicon and quartz glass by the hydrothermal method. Previously, a seed layer was applied to the surface of thoroughly cleaned substrates by the sol-gel method. Sol was obtained by dissolving zinc acetate in ethanol. A uniform distribution of the sol on the substrate surface was achieved by centrifuging at a rotational speed of about 2000 rpm, further drying at 130 ° C and annealing at 350° C [1]. The hydrothermal synthesis of ZnO nanorods was carried out in an aqueous solution of zinc nitrate and hexamethylenetetramine ($C_6H_{12}N_4$) in a glass beaker, in which the substrate with the seed layer was placed in a vertical position on a fluoroplastic holder. Hydrothermal treatment was carried out in the temperature range 90-97 ° C for 3 hours with vigorous stirring. Samples were washed with deionized water and dried. The formed ZnO nanorods are predominantly vertical in orientation with a height of 590–660 nm and have an average transverse size of about 30–40 nm — figure 1 (a.) Then, contact metallization of V-Cu-Ni 0.2-0.3 μm thick was applied on them. The resistance of the obtained samples of gas sensors ranged from hundreds of kOhms to several tens of MOhm.

(a)

(b)

(c)

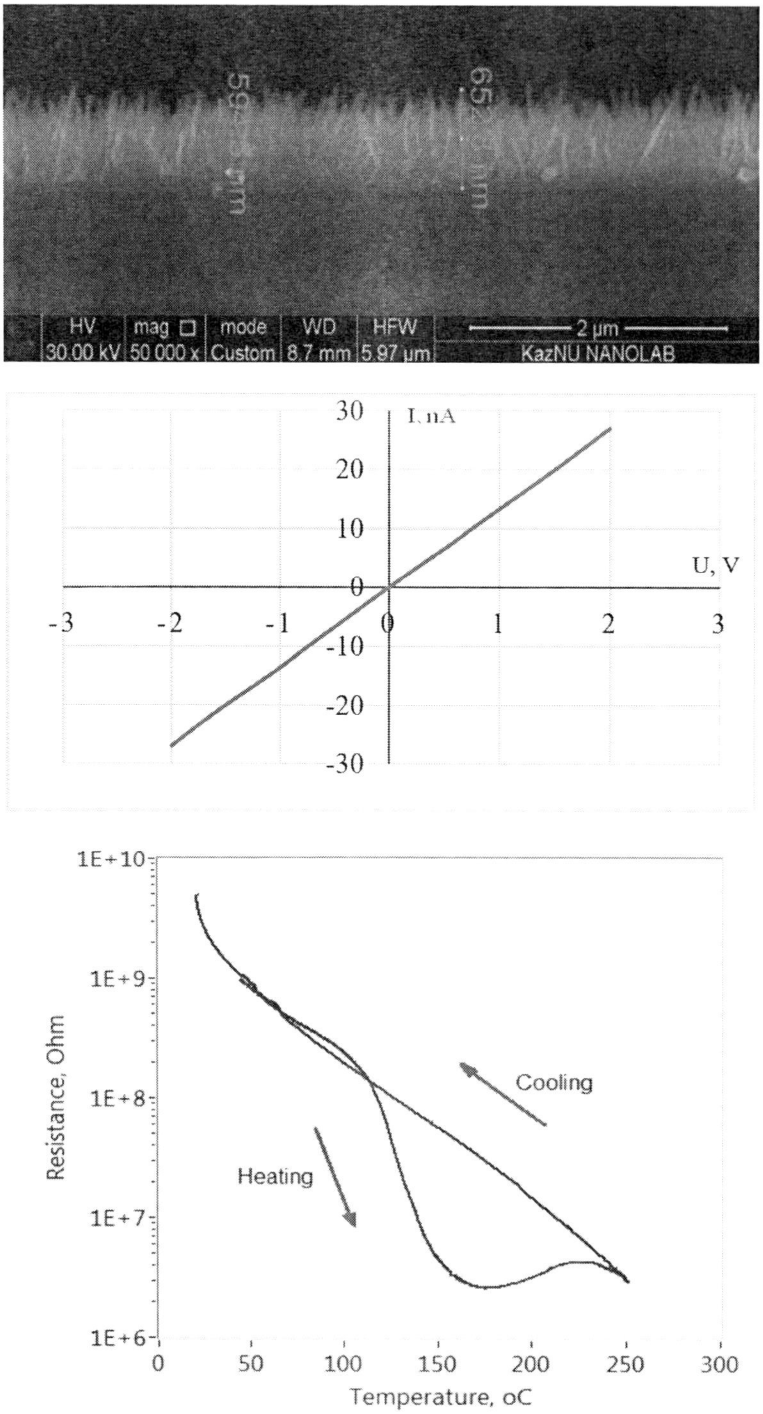

Figure 1 (a, b, c). (a) SEM images of the grown ZnO nanorod on glass; (b) CVC of the sensor structure at 200 ° C; (c) Temperature dependence of the resistance of the sensor structure.

Measurements of electrophysical and gas-sensitive properties were carried out on an automated bench for gas calibration of the collective use center "Microsystem Engineering and Integrated Sensors" — figure 2.

3. Results and discussion

Figure 2 (b) shows the current-voltage characteristic (VAC) of a formed structure heated to a temperature of 200 ° C. It can be seen that it indicates the presence of a weak potential barrier arising from the contact of ZnO nanorods with each other. A similar view of the IVC was shown in [3]. It is known that the electrical conductivity of zinc oxide is determined by oxygen vacancies, which are electron donors [4], and, accordingly, the activation energy of conductivity is determined by donor levels formed by vacancies in the forbidden zone ZnO.

On the temperature dependence of the resistance of the sensor structure presented in Figure 1, there is a temperature hysteresis with a portion of the positive temperature coefficient in the range of 170 – 225 °C. This can be explained by the interaction of the surface of zinc oxide with atmospheric oxygen. Oxygen molecules in the atmosphere are adsorbed on the surface of zinc oxide and become negatively charged due to the capture of electrons from near-surface ZnO layers, which leads to the formation of a lunch layer in the near-surface region and a decrease in conductivity [5]:

$$O_2(gas) \leftrightarrow O_2(ads) \tag{1}$$

$$O_2(ads) + e^- \leftrightarrow O_2^-(ads) \tag{2}$$

$$O_2^-(ads) + e^- \leftrightarrow 2O^-(ads) \tag{3}$$

Thus, the change in conductivity during heating and cooling is caused by two competing processes — thermal generation and recombination of electrons and their capture by oxygen adsorbed on the surface. Moreover, if in the first case, the equilibrium concentration is established almost instantly, then in the second it depends on the speed of the processes of adsorption and desorption. The presence of hysteresis in the temperature dependence of the conductivity indicates a low rate of these processes.

In addition, blowing air or gas ZnO nanorods can lead to a decrease in the temperature of their free ends, which can affect the kinetics of the response. Estimation of the temperature decrease of the free end of ZnO nanorod due to blowing with air or gas under isothermal heating / cooling conditions was carried out according to the formula [6]:

$$T - T_0 = (T_1 - T_0)\frac{\text{ch}\, a(l-x)}{\text{ch}\, al} \tag{4}$$

where $a = \sqrt{\mu p / k\sigma}$, T_0, T_1, T, are the ambient temperature, the temperature of the fixed end of the ZnO nanorod and the temperature at any point x ZnO of the nanorod, respectively; l, σ, p are the length, area and perimeter of the cross section of a ZnO nanorod, respectively; k is the coefficient of thermal conductivity of the rod; μ - coefficient of heat transfer from the rod to the environment, can take values of 5 ÷ 30 W / m · k.

Calculations according to equation (4) showed that the temperature of the free end of the nanorod can decrease by 20 - 40 ° C, depending on the experimental conditions. All this, in the end, determines the magnitude of the potential barrier when ZnO contacts the nanorods, and also affects the type of response of the sensor structure to the effect of gas. The study of gas-sensitive properties was carried out at a concentration of carbon monoxide (CO) of 20-1000 ppm at operating temperatures of 100-270 ° C. The air supply and the CO supply of the required concentration, obtained by dilution in air, was carried out several times according to the following cycle: gas supply for 2 minutes with a flow rate of 0.3 l / min - air blowing - 2 minutes with a flow rate of 0.03 l / min - Figure 2 (a) and at an air flow rate of 0.3 l / min - Figure 2 (b). As can be seen from Figure 2, when exposed to CO, the resistance does not decrease, as it should be when the n-semiconductor affects the reducing gas, but increases. A similar response when exposed to CO on a gas sensor based on ZnO nanorods was also observed in [7].

(a)

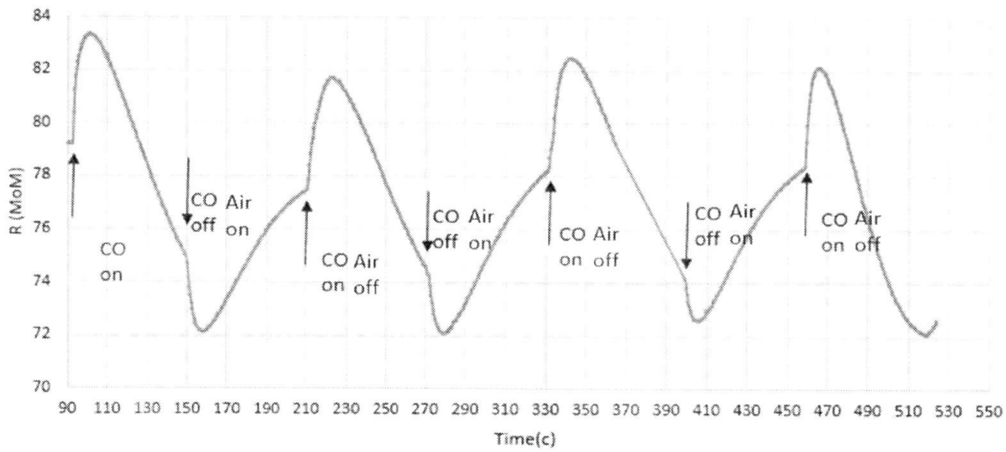

(b)

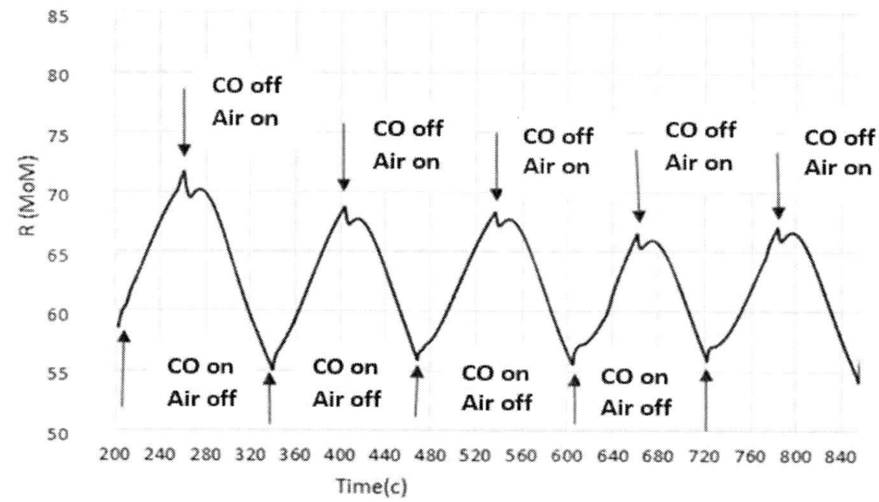

Figure 2 (a, b). Sensor response based on ZnO array of nanorods at 200 ° C to CO exposure with a concentration of: (a) 20 ppm and a gas flow rate of 0.3 l / min and air of 0.03 l / min; (b) 50 ppm and gas and air feed rates of 0.3 l / min.

Apparently, the main reason for this response is an increase in the height of the potential barrier during the adsorption of CO molecules on the surface of ZnO nanorods. The dynamics of the sensor response in Figure 2 (a) and 2 (b) shows the effect of the velocity difference of the blower. When the gas is supplied at a higher rate by an order of magnitude than the air supply after the end of the CO supply, the resistance of the sensor abruptly increases due to an increase in the temperature of the free ends of ZnO nanorods. Upon further air blowing, the resistance of the sensor was restored to its original values. During the second gas supply cycle, the sensor resistance initially fell due to a decrease in the temperature of the free ends of ZnO nanorods, and then increased due to the influence of CO molecules. To avoid the effect of this effect, the rate of blowing gas and air in subsequent experiments was the same (0.3 l / min). So in Figure 2 (b) it can be seen that there is practically no jump in resistance after at the time of the end of the gas supply and the start of air blowing.

In addition, it is necessary to avoid the influence of the hysteresis of the temperature dependence of the resistance presented in Figure 1 (c). For this, the temperature of the gas sensitivity experiment needs to be increased to 250 ° C or higher. In this case, the response of the sensory structure becomes 30–40%

higher than the response of a similar sensory structure based on an array of ZnO nanorods, but grown on silicon [8].

The maximum response of the sensory structure at 270 ° C was 1.36, when exposed to CO from concentrations of 100 ppm, 1.28 and 1.19 at 50 and 20 ppm, respectively.

Thus, the results of the study of the electrophysical properties of the sensor structure based on an array of ZnO nanorods grown on glass made it possible to adjust the method for conducting a study of their gas-sensitive properties. Studies have shown that temperature hysteresis is observed on the temperature dependence of the resistance of the sensor structure. When heated to 225 ° C, a temperature hysteresis is observed with a portion of the positive temperature coefficient in the range of 170 ÷ 225 ° C. In addition, the flow rate of gas or air affects the appearance of the sensor structure. In this regard, the methodology for conducting experiments to study the gas-sensitive properties of sensory elements was adjusted. The effect of carbon monoxide (II) depends on the rate of purging the sensitive element. and after exposure to CO and sample heating temperature. You can also see that using gas sensors based on ZnO nanorods grown on glass, you can record the concentration of carbon monoxide at the level of 20-50 ppm.

Acknowledgements
Thanks to design-engineer Kamentsev A.S. for assistance in measuring gas sensitive properties.

References

[1] Abdullin Kh A, Bakranov N B, Ismailov D V, Kalkozova J K, Kumekov S E, Podrezova LV Cicero G 2014 Composite Materials Based on Nanostructured Zinc Oxide// Semiconductors 48 No 4, p 471-475

[2] Kamentsev A S, Chetverikov S E, Petrov VV Abstracts & Schedule 2018 International Conference on "Physics and Mechanics of New Materials and Their Applications"(PHENMA 2017August 9 – 11th, 2018, Busan, Republic of Korea.p 183

[3] Park W J, Choi K J, Kim M H, Koo B H, Lee J L, and Baik J M, 2013 ACS Applied Materials & Interfaces 5, 6802

[4] Wang J, Chenb R, Xiangb L and Komarnenic S 2018 Ceram. Internt. 44 7357–77

[5] Rambu A P, Iftimie N and Rusu G I 2012 Mater. Sci. Eng. B-Adv 177 157– 163

[6] Yanpolsky AR, Hyperbolic Functions PHIZMATGIZ, 1960 p 195

[7] Nikolaev L, Kamencev A S, Lyanguzov N V, Petrov V V and Kaidashev E M, Proceedings of 2017 International Conference on "Physics, Mechanics of New Materials and Their Applications". Ed. by Ivan A. Parinov, Shun-Hsyung Chang, et al. (Nova Science Publ. Inc, 2018), p 441

[8] Petrov V V, Starnikova A P, Abdullin Kh A, Makarenko D P 2018 *Features of the mechanism of gas sensitivity of the zinc oxide nanorods arrays to carbon monoxide// IOP Conf. Series: Journal of Physics: Conf. Series* **1124** 022017

Effect of morphology features of patterned surface on the nucleation processes of In/GaAs nanostructures during droplet epitaxy

N E Chernenko[1,2], **S V Balakirev**[3], **M M Eremenko**[1], **M S Solodovnik**[1,3]
[1]Research and Education Center "Nanotechnologies", Southern Federal University, Taganrog 347922, Russia
[2]Department of Radio Engineering Electronics, Southern Federal University, Taganrog 347922, Russia
[3]Department of Nanotechnologies and Microsystems, Southern Federal University, Taganrog 347922, Russia

Abstract. We present the results of theoretical studies of the self-organization processes of nanoscale metal In droplets on GaAs(001) substrates with artificial structural heterogeneities of various types – with a rectangular, trapezoidal, and triangular shapes. The study showed that to improve the accuracy of nanostructure positioning and homogeneity, it is necessary to use patterns with triangular grooves. In order to ensure the full groove filling by the material and to suppress the undesirable structure formation outside modified areas, it is necessary to provide sufficient diffusion length of adatoms taking into account the peculiarities of the patterned surface morphology.

1. Introduction

Modern semiconductor technology is inexorably approaching the threshold of miniaturization and, consequently, an increase in its efficiency. This necessitates a transition to fundamentally new computing and telecommunication platform architectures. From this point of view, the creation of solid-state schemes for the implementation of such platforms based on the use of A3B5 nanoheterostructures with self-organizing structures, the creation of which requires the implementation of effective self-organization processes, looks most promising [1-17]. In turn, the droplet epitaxy technology has a high flexibility in the formation of self-organizing nanostructures [18-24]. Studying the physics and mechanisms of the processes underlying droplet epitaxy allows significant progress in the implementation of the technology for controlling the epitaxial synthesis and self-organization of A3B5 nanostructures, which ensures effective control of the properties of the structures obtained and their positioning. This opens up broad prospects in improving the efficiency of photodetectors and photoelectric converters based on A3B5 heterostructures due to surface plasmon effects [25-27], creating sources of single [28, 29] and entangled photons [30], functional blocks of cellular automata [31] and quantum computing [32], etc. One of the promising approaches to ensure the management of processes of self-organization, including with droplet epitaxy, is the use of structured substrates [33, 34].

The purpose of this work is theoretical studies of the self-organization processes of nanoscale In droplets on the GaAs growth surfaces with complex morphology, taking into account the main governing parameters of the droplet epitaxy technique. In this case, we used the previously developed hybrid mathematical model of droplet epitaxy processes, a feature of which is the use of a combination of calculations using the Monte Carlo kinetic method and analytical calculations according to the thermodynamic theory of nucleation [35, 36]. For the study of growth processes, we used GaAs(001) structured substrates with different shapes, sizes and densities of specified inhomogeneities.

2. Results and discussion

The calculation results showed that on substrates with complex morphology, the probability of island nucleation is distributed non-uniformly and depends on the surface area. On GaAs substrates with rectangular holes, nucleation can occur on flat areas, bottom or walls of the hole or on the edges of the faces. We have shown that on the As-stabilized GaAs substrate with rectangular holes, adatoms linger mainly on (001) faces, which are parallel to the planes substrate terraces, while the walls of the holes remain unfilled due to enhanced diffusion from faces (110). Due to the increase in the amount of material around the walls of the holes due to diffusion from the (110) faces, as well as an increase in the binding energy of the adatom with the substrate in the corner of the modified region, nucleation can also occur in the vicinity of the intersection of the hole bottom with its walls (**Figure 1**). Therefore, to achieve the best homogeneity of the geometric characteristics of nanostructures, the diameter of the rectangular holes should be reduced, adjusting the height so as to achieve the required effective volume and aspect ratio.

Figure 1. The morphology of In/GaAs(001) nanostructures array after deposition of 3 ML In on an As-stabilized substrate with rectangular holes at $T = 150$ °C and $v = 0.1$ ML/s. The length of the simulated area is 100 nm, the diameter and height of the hole are 20 and 10 nm, respectively, the distance between the hole centers is 50 nm.

The intermediate shape between the holes of a rectangular and triangular shape are trapezoidal ones. On substrates with this modification, the positions for the most likely nucleation will be determined based on the combination of technological regimes and diffusion parameters, which have a higher intensity on the (111) faces than on the (001) ones. The our study showed that with a large ratio of trapezoid bases, the In droplets forms approximately in the center of the hole, whereas with a small ratio and relatively large sizes of hole, the preferred position for nucleation shifts towards one of the corners as is the case with the straight shape (**Figure 2**).

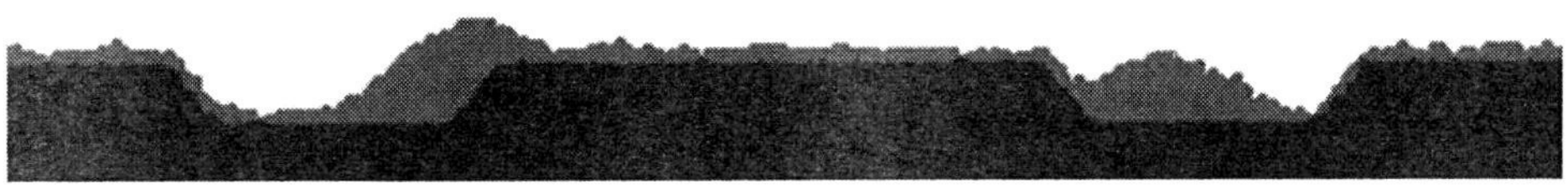

Figure 2. The morphology of In/GaAs(001) nanostructures array after deposition of 4 ML In on an As-stabilized substrate with trapezoid holes with bases of 20 and 13 nm at $T = 150$ °C and $v = 0.1$ ML/s. The length of the simulated area is 100 nm, the height of the recess is 5 nm, the distance between the hole centers is 50 nm.

An analysis of the results of theoretical studies has shown that the use of a shape close to a triangular one is distinguished by more pronounced positions for nucleation, which makes it possible to increase the uniformity in size and accuracy of positioning of nanostructures. On the such surface, it is possible to achieve full or partial stabilization of gallium or arsenic (**Figure 3**). As can be seen from the **Figure 3**, in all cases, the preferred position for the island nucleation is the hole bottom due to the increase in the number of substrate atoms surrounding the adatom under consideration, and, consequently, the increase in its binding energy with the current environment.

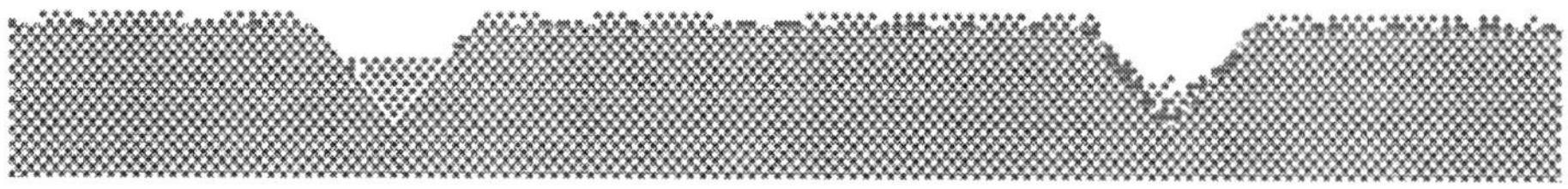

Figure 3. The morphology of In/GaAs(001) nanostructures array after deposition of 2 ML In on the substrate with triangular holes with Ga- (left) and As-stabilization (right) at T = 150 °C and v = 0.1 ML/s The length of the simulated area is 50 nm, the diameter and height of the recess are 5 and 5 nm, respectively, the distance between the centers of the recesses is 25 nm.

This phenomenon extends to a triangular holes with any angle between the walls. At the same time, as the angle between the walls increases, the probability of removing the adatom from the most preferable position increases, which can lead to nucleation on the wall or outside the structural inhomogeneity (**Figure 4**).

Figure 4. The morphology of In/GaAs(001) nanostructures array after deposition of 3 ML In on an As-stabilized substrate with triangular holes at T = 150 °C and v = 0.1 ML/s. The length of the simulated area is 100 nm, the diameter and height of the recess are 10 and 5 nm, respectively, the distance between the centers of the recesses is 50 nm.

In order to ensure 100% filling of the holes and suppress the formation of undesirable structures outside the modified areas, it is necessary to ensure sufficient diffusion length of adatoms taking into account morphological features. The probability of island nucleation outside the hole increases with decreasing temperature and increasing growth rate, which is associated with a decrease in the probability of adatoms to take the most energetically advantageous positions as a result of a decrease in the intensity of the diffusion surface and as a result of a decrease in the average diffusion length of adatoms, respectively. At temperature T = 150 °C and growth rate v = 0.1 ML/s, the distance between the triangular shape holes with a diameter and depth of 20 nm, at which 100% ordering of the nanostructure array is reached (in the calculation of the ratio of the number of complexes "1 nanostructure in 1 hole" to the number of holes) is 50 nm. With an increase in the distance between the holes, nucleation is observed outside the predetermined centers. But increasing the temperature to 250 °C and above allows to eliminate unwanted nucleation at distances between the holes exceeding 100 nm.

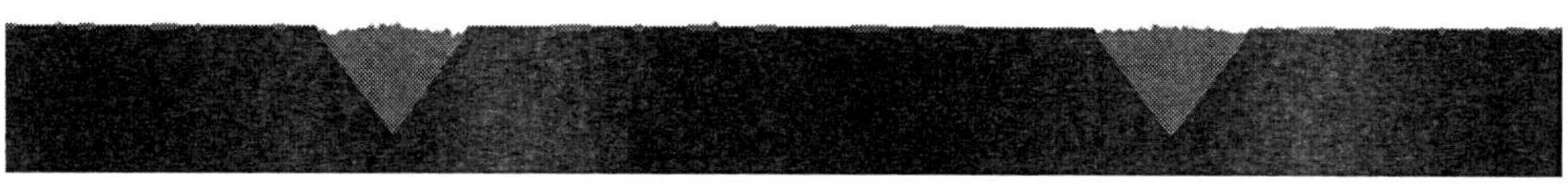

Figure 5. The morphology of In/GaAs(001) nanostructures array after deposition of 5 ML In on an As-stabilized substrate with triangular holes with a diameter and depth of 20 nm at T = 400 °C and v = 0.1 ML/s. The length of the simulated area is 200 nm, the distance between the hole centers is 100 nm.

Thus, the use of modified surfaces allows to adjust the density of nanostructures, providing the possibility of obtaining, including laterally coupled nanostructures. The transition to the incomplete condensation mode can further reduce the size of the structures while maintaining their density. So, an increase in temperature from 400 to 500 °C leads to a decrease in the size of nanostructures from 19 to 16 nm, all other things being equal (**Figure 5**).

Analysis and systematization of the obtained theoretical results showed the ability to control the size of nanostructures while maintaining the specified array configuration by reducing the thickness of the deposition material. Compared with a decrease in droplet size on a flat surface (limited due to the presence of a non-zero critical thickness of droplet formation), a significantly smaller nanostructure can be achieved on a modified substrate, which is associated with a decrease in the nucleation activation barrier and an increased probability of droplet nucleation in the region of structural inhomogeneities and, therefore, a decrease in the deposition thickness at which nucleation begins. Thereby, due to this effect, a reduction in the wetting layer thickness is achieved, which has a significant impact on the electronic and optical characteristics of the devices, as well as a reduction in the size of nanostructures.

Acknowledgments

This work was supported by Grant of the President of the Russian Federation No. MK-477.2019.8. The results were obtained using the equipment of the Research and Education Center and Center for Collective Use "Nanotechnologies" of Southern Federal University.

References

[1] Ageev O A, Solodovnik M S, Balakirev S V and Mikhaylin I A 2016 *J. Phys.: Conf. Ser.* **681** 012036
[2] Ageev O A, Solodovnik M S, Balakirev S V and Eremenko M M 2016 *Phys. Solid State* **58** 1045–1052
[3] Balakirev S V, Eremenko M M, Mikhaylin I A, Klimin V S and Solodovnik M S 2018 *J. Phys.: Conf. Ser.* **1124** 022018
[4] Vurgaftman I and Meyer J R 2001 *J. Appl. Phys.* **89** 5815–75
[5] Ozdemir S, Suyolcu Y E, Turan S and Aslan B 2017 *Appl. Surf. Sci.* **392** 817–25
[6] Dilger M, Haug R, Eberl K and Von Klitzing K 1996 *Semicond. Sci. Tech.* **11** 1493
[7] Klimin V S, Solodovnik M S, Lisitsyn S A, Rezvan A A and Balakirev S V 2018 *J. Phys.: Conf. Ser.* **1124** 041024
[8] Ageev O A, Solodovnik M S, Balakirev S V, Eremenko M M and Mikhaylin I A 2017 *J. Crys. Growth* **457** 46-51
[9] Luscombe J H, Randall J N and Bouchard A M 1991 *P. IEEE* **79** 1117–30
[10] Solodovnik M S, Karenkikh O G, Balakirev S V, Petrov S I, Ryzhuk R V, Alexeev A N and Ageev O A 2018 *J. Phys.: Conf. Ser.* **1124** 081024
[11] Balocco C, Song A and Missous M 2004 *Appl. Phys. Lett.* **85** 5911–3
[12] Balakirev S V, Solodovnik M S, Eremenko M M, Mikhaylin I A and Ageev O A 2017 *J. Phys.: Conf. Ser.* 2017 **917** 032034
[13] Ageev O A, Solodovnik M S, Balakirev S V, Eremenko M M and Mikhaylin I A 2016 *J. Phys.: Conf. Ser.* **741** 012012
[14] Geller M, Marent A, Nowozin T, Bimberg D, Akçay N and Öncan N 2008 *Appl. Phys. Lett.* **92** 092108
[15] Ageev O A, Solodovnik M S, Balakirev S V and Eremenko M M 2016 *J. Vac. Sci. Technol. B* **34** 041804
[16] Lisitsyn S A , Balakirev S V, Avilov V I, Kolomiytsev A S, Klimin V S, Solodovnik M S, Konoplev B G and Ageev O A 2018 *Nanotech. in Russia* **13** 26-33
[17] Ageev O A, Solodovnik M S, Balakirev S V and Mikhaylin I A 2016 *Technical Physics* **61** 971–7
[18] Balakirev S V, Eremenko M M, Mikhaylin I A and Solodovnik M S *J. Phys.: Conf. Ser.***1124** 022025

[19] Balakirev S V, Solodovnik M S, Eremenko M M, Mikhaylin I A and Ageev O A 2018 *J. Phys.: Conf. Ser.* **1124** 022001
[20] Somaschini C, Bietti S, Koguchi N and Sanguinetti S 2009 *Nano Lett.* **9** 3419–24
[21] Solodovnik M S, Balakirev S V, Eremenko M M, Mikhaylin I A, Avilov V I, Lisitsyn S A and Ageev O A 2017 *J. Phys.: Conf. Ser.* 2017 **917** 032037
[22] Kim J S and Koguchi N 2004 *Appl. Phys. Lett.* **85** 5893–5
[23] Balakirev S V, Solodovnik M S and Ageev O A 2017 *J. Phys.: Conf. Ser.* 2017 **917** 032033
[24] Mano T, Watanabe K, Tsukamoto S, Fujioka H, Oshima M and Koguchi N 1999 *Jpn. J. Appl. Phys.* **38** L1009–11
[25] Krishna S 2005 *J. Phys. D: Appl. Phys.* **38** 2142
[26] Perera A 2006 *Opto-Electron. Rev.* **14** 99–108
[27] Le Ru E, Howe P, Jones T and Murray R 2003 *Phys. Rev. B* **67** 165303
[28] Schneider C, Heindel T, Huggenberger A, Weinmann P, Kistner C, Kamp M, Reitzenstein S, Höflinga S and Forchel A 2009 *Appl. Phys. Lett.* **94** 111111
[29] Kuroda T, Abbarchi M, Mano T, Watanabe, Yamagiwa M, Kuroda K, Sakoda K, Kido G, Koguchi N, Mastrandrea C, Cavigli L, Gurioli M, Ogawa Y and Minami F 2008 *Appl. Phys. Express* **1** 042001
[30] Dousse A, Suffczyn'ski J, Beveratos A, Krebs O, Lemaıtre A, Sagnes I, Bloch J, Voisin P and Senellart P 2010 *Nature* **466** 217
[31] Khosroshahyab M B, Moaiyeri M H, Angizi S, Bagherzadeh N and Navi K 2017 *Microprocess. Microsy.* **50** 154–63
[32] Cavigli L, Abbarchi M, Bietti S, Somaschini C, Sanguinetti S, Koguchi N, Vinattieri A and Gurioli M 2011 *Appl. Phys. Lett.* **98** 103104
[33] Dias N L, Garg A, Reddy U, Young J D, Verma V B, Mirin R P and Coleman J J 2011 *Appl. Phys. Lett.* **98** 141112
[34] Kim J S, Kawabe M and Koguchi N 2006 *Appl. Phys. Lett.* **88** 072107
[35] Balakirev S V, Solodovnik M S and Ageev O A 2018 *Phys. Status Solidi B* **255** 1700360
[36] Balakirev S V, Solodovnik M S, Eremenko M M, Konoplev B G and Ageev O A 2019 *Nanotechnology* **30** 505601

Thermal-activated granulation of ultrathin Au films. Structural and phase transformations

O A Tomilina, V N Berzhansky, S V Tomilin, E T Milyukova
Institute of Physics and Technologies, V.I. Vernadsky Crimean Federal University, Crimea, Simferopol, 295007, Russian Federation

Abstract. This paper represents the results of time, temperature, and dimension factors influence on the formation dynamics and structural-phase transformations in ultrathin and islands metallic films during their synthesis by methods of thermal-activated granulation and gradient coatings deposition. It's shown that the efficiency of metallic coatings granulation depends on the temperature and time of the TAG and determined by the effective thickness of the granulated coating. At a certain effective thickness, the second-order phase transition occurs in the film during the granulation process (percolation transition).

1. Introduction

Ultrathin and nanoislands metallic films are often used in planar microelectronic technologies, in plasmon and magnetoplasmon sensing systems, to create the control elements of fiber-optic communication lines, in optical systems, and nanocatalysis [1–7]. It's known that a structural-phase transition of the second type (percolation transition) is observed at thermal-activated granulation of ultrathin films. This phase transition leads to the formation of ensembles of self-assembled nanoparticles (nanoisland structures).

Many factors influence on formation dynamics and parameters of self-assembled nanoparticles: such as method of islands formation, the materials of film and substrate, surface type and structure, system temperature, coating thickness, time of the process, etc. [8,9]. So, an actual problem is to study the dynamics of self-assembled nanoparticles formation and also structural and phase transformations in thin films during the percolation transition from continuous coating structure to nanoisland structure.

2. Experimental technique

The samples of Au ultrathin and nanoisland films were synthesized by thermal vacuum deposition on a GGG (gallium-gadolinium garnet) substrates. Before deposition of coatings, the substrates were cleaned with ethanol and treated with vacuum annealing (the pressure is not more than $6 \cdot 10^{-4}$ Pa, $T = 400°C$, $\tau = 30$ min) and with ion-beam cleaning in glow HF-discharge.

In this work two methods for the synthesis of self-assembled nanoparticles were used. There are vacuum annealing method and thin-screen method (Figure 1). The first method was based on coatings thermal-activated granulation (TAG) caused by surface self-diffusion. The second method was based on the formation of coatings with gradient of effective thickness [10]. In this case in the ultrathin area of the gradient the coating has a nanoisland structure, i.e. composed of self-assembled nanoparticles.

The using of both these methods together is the most interesting in practice. It makes possible to synthesize a gradient coating on one sample in a single technological process and this coating is granulated by annealing then. This technique is very effectiveness to make investigations of the dimensional factors influence on the physical properties of nanoisland coatings.

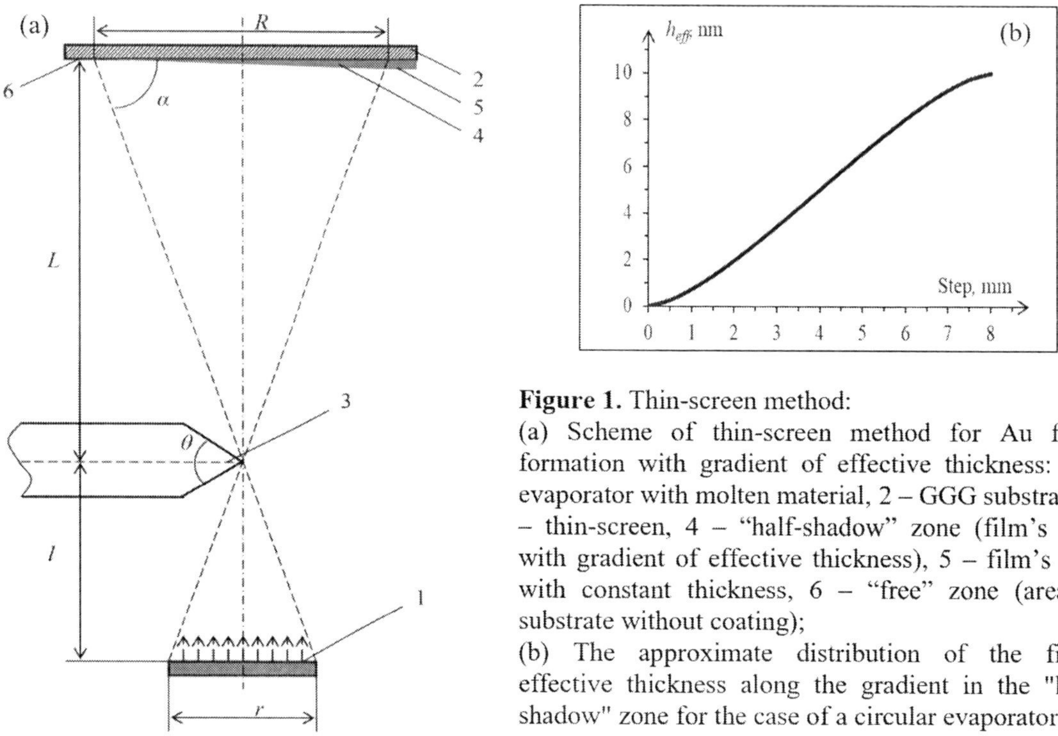

Figure 1. Thin-screen method:
(a) Scheme of thin-screen method for Au films formation with gradient of effective thickness: 1 – evaporator with molten material, 2 – GGG substrate, 3 – thin-screen, 4 – "half-shadow" zone (film's area with gradient of effective thickness), 5 – film's area with constant thickness, 6 – "free" zone (area of substrate without coating);
(b) The approximate distribution of the film's effective thickness along the gradient in the "half-shadow" zone for the case of a circular evaporator.

3. Results and discussion

At the firstly let's consider the temporal dynamic of film's thermal-activated granulation (TAG). In this experiment the Au film on GGG substrate has continuous thickness about 10 nm. We synthesized 6 identical samples in single deposition process and annealed them on air at T_{TAG} = 680°C during different time τ_{TAG}, such as: 1, 2, 5, 10, 30 and 60 min.

The results of surface morphology investigation by scanning electronic microscopy (SEM) are shown on figure 2. It's shown SEM-images of coating's nanoisland structure (self-assembled gold nanoparticles) and diagrams of nanoparticles size distribution. As can see, at low annealing time (1 and 2 min.) the complex chain-clusters consisting of three or more islands are presented in granulated coating. In this case the real nanoparticles size distribution (dotted line) is different from the symmetric Gaussian (solid line) and has asymmetric in the bigger clusters side. So, as nanoparticles size, we mean the parameter d associated with the planar projection area of a complex chain-cluster S through the equation:

$$d = 2\sqrt{S/\pi} \tag{1}$$

At annealing time increasing (5 and 10 min.) as complex only binary clusters are observed and the nanoparticles size distribution became symmetric (better approximated with Gaussian). At high τ_{TAG} (30 and 60 min.) only single nanoparticles are observed and as longer the annealing time, as smaller the average particles size.

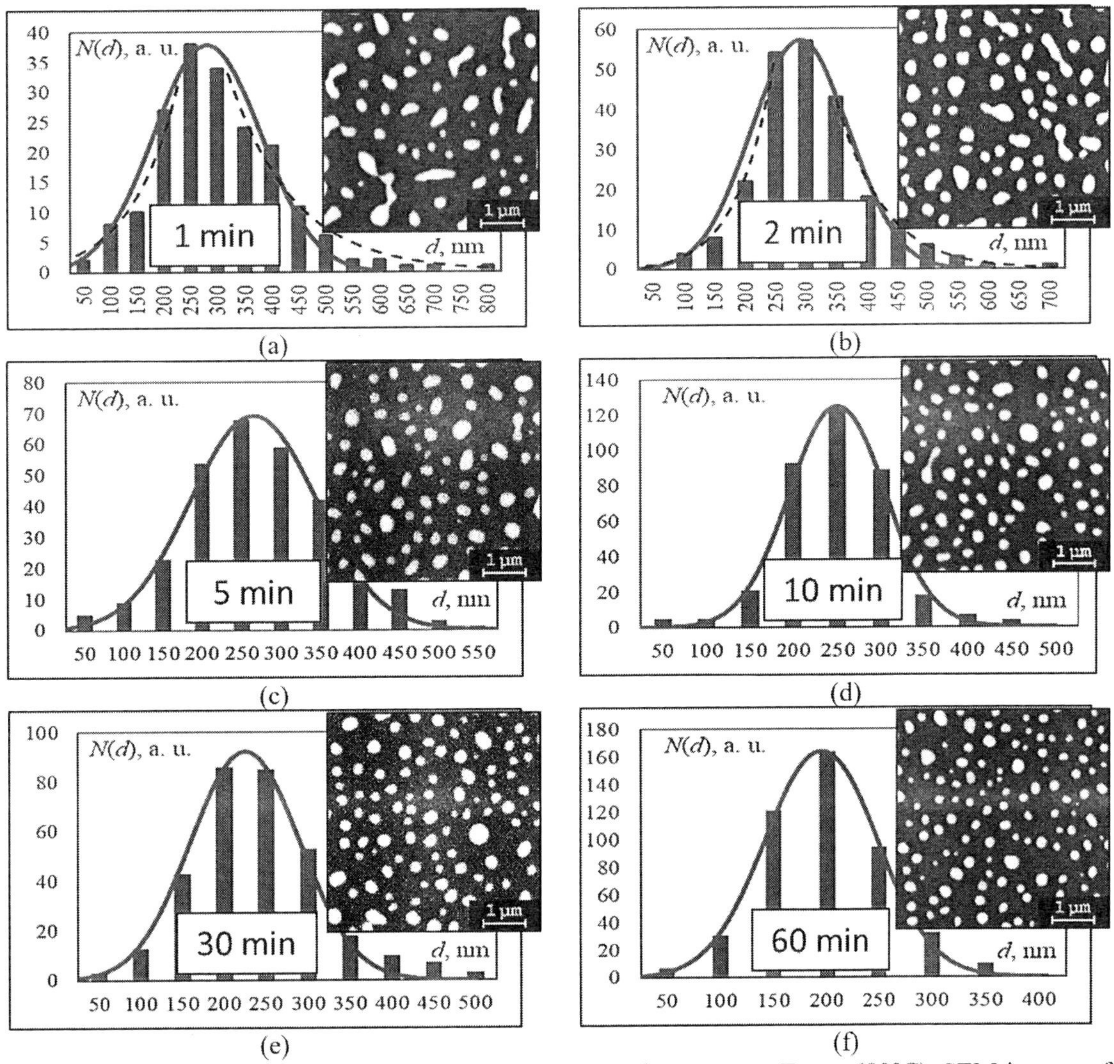

Figure 2. Temporal dynamics of TAG. Film $Au_{(NP)}$/GGG (h_{eff} = 10 nm, T_{TAG} = 680°C). SEM-images of the granular coatings surface and size distribution of self-assembled nanoparticles are presented. Insets shown the granulation time.

At figure 3 the dependences of distribution asymmetry index Δ and most probability particles size d_0 from annealing time τ_{TAG} are shown. The distribution asymmetry index Δ can be determined as:

$$\Delta = \frac{d_{\max} - d_0}{d_0 - d_{\min}}, \quad (2)$$

where $d_{\max}$ – the maximum size of observed nanoparticles, $d_{\min}$ – the minimum size, respectively.

Dependencies on figure 3 are demonstrated the classical percolation transition (a second-order phase transition) [11], during which became the destruction of complex chain clusters into individual nanoparticles.

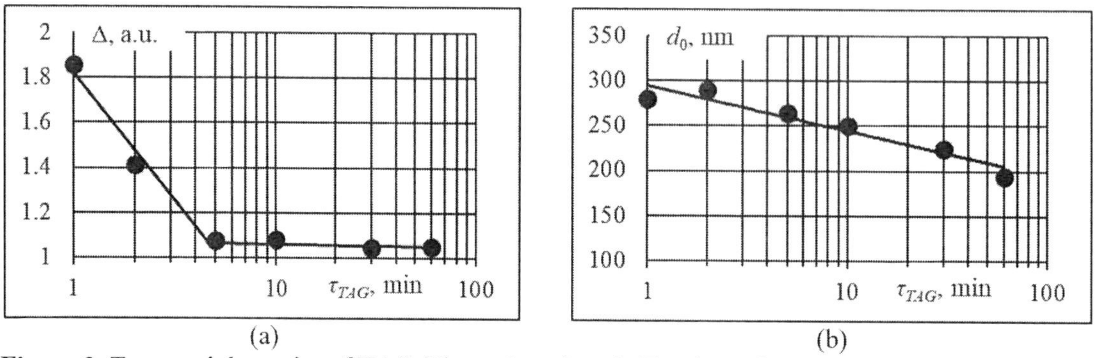

Figure 3. Temporal dynamics of TAG (Figure 2 analyses). The dependence of distribution asymmetry index Δ (a) and most probability particles size d_0 (b) from annealing time τ_{TAG}.

At the secondly let's consider the temperature dynamics of film's thermal-activated granulation (TAG). In this experiment the Au film on GGG substrate has continuous thickness near 10 nm. We synthesized 4 identical samples in the single deposition process and annealed them on air during 10 min at different temperatures T_{TAG} such as 680°C, 750°C, 850°C and 950°C. The results of surface morphology investigation shown on figure 4.

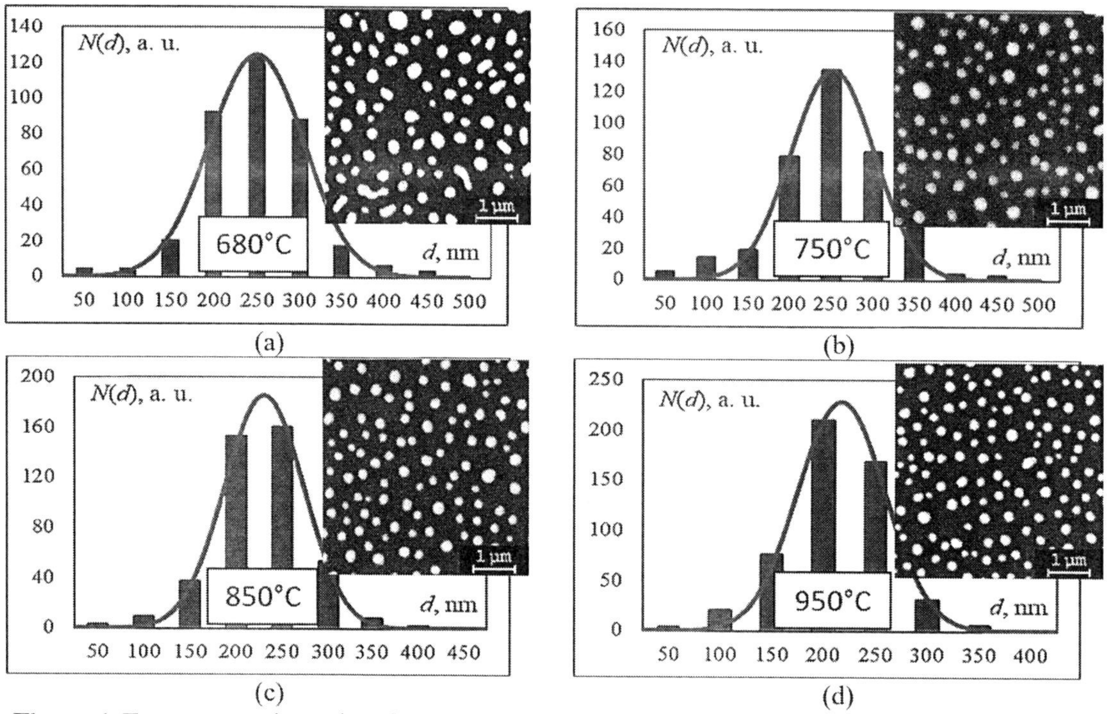

Figure 4. Temperature dynamics of TAG. The film $Au_{(NP)}$/GGG (h_{eff} = 10 nm, τ_{TAG} = 10 min.). SEM-images of the granular coatings surface and size distribution of self-assembled nanoparticles are presented. Insets shown the granulation temperature.

As can see at low annealing temperature binary clusters are observed. At T_{TAG} increasing only single nanoparticles are presents. Symmetrical size distribution of nanoparticles (well approximation with Gaussian) is observed in all cases. At figure 5 the dependences of standard deviation σ and most probability particles size d_0 from T_{TAG} are shown.

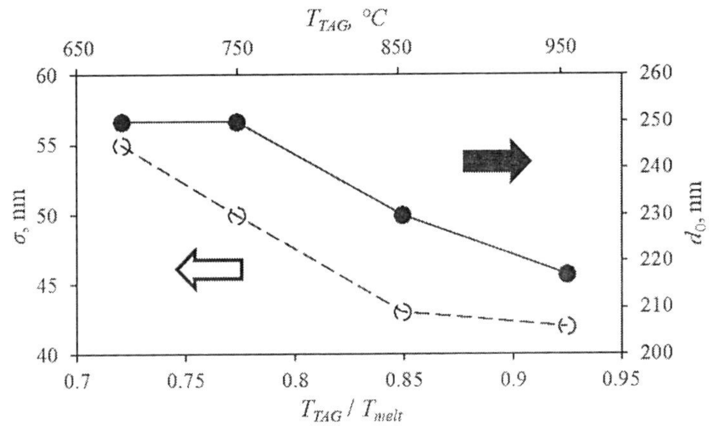

Figure 5. Temperature dynamics of TAG (Figure 4 analyses). The dependence of standard deviation σ and most probability particles size d_0 from annealing temperature T_{TAG}.

As can see the standard deviation and most probability particles size are decreases when the annealing temperature is increased. It's obvious that when approaching the melting point (T_{TAG} / T_{TAG} = 1), the parameters d_0 and σ will tend to some constant values which characteristic for gold in the liquid state (a first-order phase transition).

At the thirdly let's consider the dimensional percolation transition in film with gradient of effective thickness h_{eff} at TAG. In this experiment the effective thickness of Au coating on GGG-substrate was variable along the gradient from 5 nm (in the "thick" area) to zero thickness (clean substrate). After deposition the film was treated by vacuum annealing at 450°C during 120 min. The film was annealed immediately after deposition without removing the samples from the vacuum. This technique avoids additional contamination of the coating. It should be emphasized that the granulation temperature in this experiment was significantly lower than the melting point of gold.

As can see on presented SEM-images (Figure 6), at the "thickest" area of the film ($h_{eff} \approx$ 5 nm) after the TAG some separated defects began to forming in the coating continuity. These defects have type as "cracks" with wide up to 20 nm and long up to 150 nm (Figure 6a). But in this area the coating retains its integral continuity.

On figure 6b shows film area with a smaller effective thickness ($h_{eff} \approx$ 4 nm) than in the previous area. As can see the film in this area has many extended defects with type as "crack" which made a chaotic grid on the surface. In this area defects have a greater length up to 200–250 nm and a complex branched structure, with a crack width smaller than in the previous area (2–10 nm). It should focus that the metal coating areas still form continuous percolation channels.

On figure 6c showed the film surface morphology at area with $h_{eff} \approx$ 3 nm. As can see in this area defects with the "crack" type form a continuous connected grid. In this case the percolation channels are destroyed and discrete flat islands with irregular shape are formed. Images on figures 6b and 6c visually demonstrate the phenomenon of percolation transition in the coating structure (a second-order phase transition).

Figures 6d, 6e, and 6f demonstrate the successive stages of coating island structures evolution when its effective thickness decreases. On figure 6d ($h_{eff} \approx$ 2 nm) are shown isolated discrete islands with irregular form which separated by "wide" intervals up to 20 nm. On figure 6e ($h_{eff} \approx$ 1 nm) individual discrete islands are also visible, but these islands have smaller sizes (about 5 – 50 nm) and more regular form (it's nearly rounds or may be approximated as a system of two or more rounded elements). In the end of gradient (h_{eff} tends to 0 nm) we can observe on figure 6f separate islands with form become to round and with size 20 – 50 nm, which located from each other on considerable distance 0.3 – 1 μm.

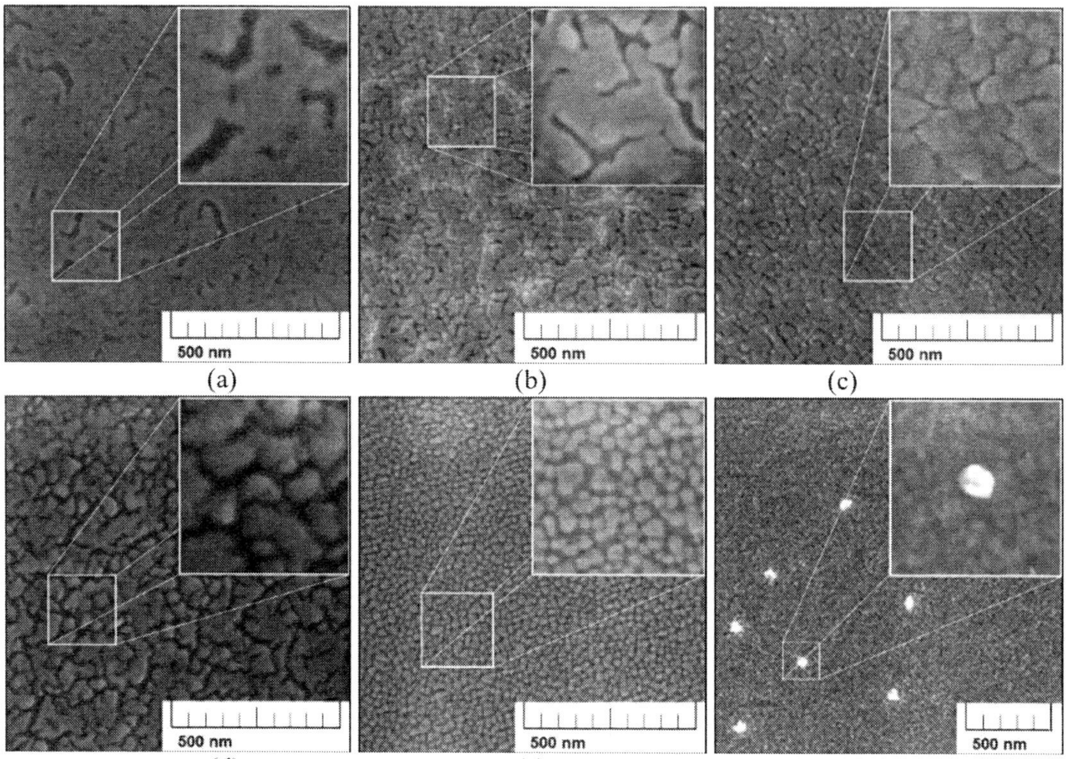

(a) (b) (c) (d) (e) (f)

Figure 6. Surface morphology of the Au/GGG film with a thickness gradient h_{eff} after annealing at 450°C during 120 min. (a–f) – different areas of the film along the gradient (insets – area 200 × 200 nm).

4. Conclusions

In the presented work it has been experimentally shown that the efficiency of metallic coatings granulation depends to the temperature and time of the TAG and determined by the effective thickness of the granulated coating.

When the TAG time changes, a second-order phase transition (percolation transition) is observed in the structure of a thin-film coating. At "low-temperature" granulation (the annealing temperature is significantly lower than the melting temperature), complex clusters of two or more islands are present in the structure of Au/GGG granulated coating (h_{eff} = 10 nm). It's leads to a significant dispersion and asymmetry of the nanoparticles size distribution.

It has been also experimentally established that the effective thickness parameter h_{eff} significantly affects to the processes of "low-temperature" granulation in ultrathin films. In experiment clearly demonstrated the processes of percolation transition in a granular film. It is shown that effective "low-temperature" granulation with formation of a discrete island structure occurs in films with $h_{eff} \leq 3$ nm. For such thicknesses at diffusion motion the kinetic energy of Au atoms in the TAG process becomes higher than sum of interatomic bonds energy and adhesion energy at the GGG/Au interface.

Acknowledgments

This work was supported by the RF Ministry of Science and Highest Education in the framework of the base part of the state task (Project No. 3.7126.2017/8.9) and partially supported by the V.I. Vernadsky Crimean Federal University Development Program (Projects No. ВГ22/2018).

References

[1] Gladskikh I A, Leonov N B, Przhibel'skii S G, Vartanyan T A. 2013. *Nanosystems: Phys., Chem., Math.* **4 4** 524.
[2] Agustsson J S, Arnalds U B, Ingason A S, at. all. 2008. *J. of Phys.: Conf. Ser.* **100** 082006.
[3] McPeak K M, Jayanti S V, Kress S J P, at. all. 2015. *ACS Photonic* **2 3** 326.
[4] Kake Zh, Donghai W, Jun L. 2009. *Nano Res.* **2** 29.
[5] Fujikawa R, Baryshev A V, Kim J, Uchida H, Inoue M. 2008. *J. of App. Phys.* **103** 07D301.
[6] Uchida H, Masuda Y, Fujikawa R, Baryshev A V, Inoue M. 2009. *J. of Magnetism and Magnetic Materials* **321** 843.
[7] Peng Sh, Lee Yo, Wang Ch, at all. 2008 *Nano Res* **1** 229.
[8] Mizutani Y, Uchida H, Masuda Y, Baryshev A V, Inoue M. 2009. *J. Magn. Soc. Jpn.* **33** 481.
[9] Tomilin S V, Berzhansky V N, Shaposhnikov A N, at al. 2016. *J. of Phys.: Conf. Ser.* **741** 012113.
[10] Tomilin S V, Yanovsky A S. 2013 *J. of Nano- and Electronic Physics* **5 3** 03014.
[11] Ruffinoa F, Grimaldi M G. 2010. *J. of App. Phys.* **107** 074301.

Genetic algorithm application for solving inverse problems of X-ray diffraction on systems with quantum dots

D V Sivkov[1,2], **V I Punegov**[1]
[1]IPM of Komi SC UrB RAS, Syktyvkar, 167982, Russia
[2]Immanuel Kant Baltic Federal University, Kaliningrad, 236041, Russia

Abstract. For solving inverse problem of X-ray diffraction on GaAs-AlAs superlattices with InAs quantum dots within the statistical theory of X-ray diffraction the Genetic Algorithm in the form of Differential Evolution method was used. The calculation results were compared with experimental data.

1. Introduction

Modern nanotechnologies allow to create artificial materials with new properties for different applications in physics, chemistry, biology, and medicine. Creating nanostructures using templates is recognized as a very promising way to achieve accurate positioning of nano-objects, in particular, arrays of quantum dots. Epitaxial self-organized quantum dots (QDs) are unique objects that a widely used in the electronic, optical, and structural properties of semiconductor systems [1]. High-resolution x-ray diffraction methods have been successfully used to characterize different types of epitaxial structures, including multilayer structures with QDs [2].

In addition to developing a theory describing the interaction of X-ray with matter, an effective algorithm for minimizing the residual functional is needed. The feature of this problem is in the large number of local minima in the parameter space and large number of parameters. Therefore, traditional iterative optimization methods are not well suited due to the considerable time required to find a solution. Evolutionary algorithms, including genetic ones [3], showed their high efficiency for searching solutions in the space of a large number of parameters in a wide range, particularly in problems of x-rays scattering [4].

Within the work to analyze the experimental diffraction data, both coherent and diffuse x-ray scattering components bounded together by a static Debye-Waller factor was taking into account.

2. Experiment

The studied structures were grown on the GaAs (001) substrate and GaAs buffer layer with a thick (~2 μm) bottom $Al_{0.3}Ga_{0.7}As$ barrier layer, a GaAs waveguide with the thickness of 240 nm and multilayers of InGaAs layers with InAs quantum dots 5 nm thickness separated by GaAs layers 15 nm thickness, repeated 20 times, finally capped with the same GaAs waveguide and a 450 nm thick $Al_{0.3}Ga_{0.7}As$ barrier layer. The main structures were grown on a 300 nm thick GaAs buffer layer.

The x-ray measurements were performed on a high-resolution x-ray diffractometer X'Pert MRD (PANalytical) with multilayer focusing mirror, Ge (011) Bartels monochromator and a three-bounce Ge (011) analyzer crystal. Double- and triple- crystal ω-2θ rocking curves (RC) and triple-crystal ω RCs at the maximum of the main diffraction peaks of the GaAs substrate and AlGaAs layer, and the superlattice (SL) peaks of the periodic QDs multilayer structure ("0SL", "± 1SL "and others) were measured. Reciprocal space maps (RSM) of the scattered radiation were recorded around symmetric (004) scattering spot.

3. Diffraction theory

In the X-ray scattering experiment, the total intensity obtained at the output of the detector is written as

$$I_{sim}(q_x, q_z) = K \cdot \overline{I_h}(q_x, q_z) + I_{bg}, \quad (1)$$

where I_{bg} – background intensity, K – intensity scaling parameter. The expression for the scattering intensity at the input of the detector, taking into account the influence of the instrumental function of the monochromator and the analyzer, has the form [1]

$$\overline{I_h}(q_x, q_z) = \frac{\int_{-\infty}^{+\infty}\int_{-\infty}^{+\infty} dq'_x dq'_z R^M(q'_x, q'_z) R^A(q'_x, q'_z) I_h(q_x - q'_x, q_z - q'_z)}{\int_{-\infty}^{+\infty}\int_{-\infty}^{+\infty} dq'_x dq'_z R^M(q'_x, q'_z) R^A(q'_x, q'_z)}, \quad (2)$$

where R^M, R^A – monochromator and analyzer reflection coefficients, I_h – beam scattering intensity. For multilayer heterostructure with buried QDs the intensity can be represented in two terms, reflecting the nature of the scattering, connected through the Debye-Waller factor $f = f_{QD} f_S$, where f_{QD} - a static factor, caused by the presence of QDs without accounting for the elastic deformation near them, and f_S is static factor caused by elastic strain in the matrix around the QDs.

$$I_h(q_x, q_z) = (1 - f^2) I_h^d(q_x, q_z) + f^2 I_h^c(q_x, q_z) F_L(q_x, q_y) F_V(q_z), \quad (3)$$

I_h^c – the coherent part, describing the ideal crystal scattering. It contains information on the thicknesses and periods of the layers in the superlattice structure; I_h^d – the diffuse part, containing information on the parameters of structural defects, such as quantum dots, pores, dislocations of different types (their size, shape, orientation, distribution in the sample, etc.). The diffuse scattering intensity was calculated using the ellipsoidal quantum dots model [5]. $F_L(q_x, q_y)$ – lateral structural factor, depending on the average QDs spatial distribution in superlattice period; $F_V(q_z)$– vertical interference structural factor, depending on the superlattis period.

The scattering intensity calculations were performed within the dynamical theory of X-ray diffraction expanded to the special case of transversely restricted wavefronts of the incident and reflected waves taking into account the effect of the diffractometer's instrumental function [4].

4. Residual functional

For the minimization procedure in the paper we considered the residual functional in the form

$$\rho_{sqr}(q_x^i, q_z^i) = \frac{1}{N_q}\sum_{i=1}^{N_q} \frac{[I_{sim}(q_x^i, q_z^i; \mathbf{x}) - I_{exp}(q_x^i, q_z^i)]^2}{I_{sim}(q_x^i, q_z^i; \mathbf{x})}, \quad (4)$$

where $\mathbf{x}$ – vector of parameters in the parameter space, I_{sim} – diffracted intensity theoretical data, I_{exp} – experimental data. For our calculations the number of angle points at which scattered radiation was collected was 50 for ω scans and 460 for ω-2θ scans. Minimization of the residual function was carried out for ω scans of all superlattice satellites. Total number of points for search was 450.

The root-mean-square deviation between theoretical and experimental data (4) is effective for analyzing highly noisy signals because it is slightly affected by the statistical noise, and represents the best compromise between speed and accuracy compared to comparison in absolute value, logarithms and the root-mean-square deviation of the logarithms comparisons [6].

5. Differential Evolution algorithm

Differential Evolution (DE) algorithm is described in detail in [7]. For the X-ray scattering intensity calculation DE-strategy in the form "rand/1/exp" was used. That means that population vector to be mutated was chosen randomly, one difference vector was used and trial vector was generated using exponential crossover.

For our task in obtaining mutant vectors, the absolute values of the corresponding expression were taken in order to exclude the search for a solution among the parameter vectors with negative parameters. DE algorithm parameters were chosen as the best for searching their values $F = 0.1$ and $C = 0.4$ were chosen according to the lowest goodness of fit (residual functional) ρ_{sqr} [3]. The target vector was chosen randomly from the population. Number of cycles was restricted by the value of 1000.

6. Results

In the work, the calculation of X-ray scattering on an 20-period GaAs(001)-AlGaAs-{InAs QDs-GaAs}SL superlattice was performed for (004) GaAs reflection σ-polarized Cu $K_{\alpha 1}$ radiation. Searching parameters were: layers thicknesses l_{InGaAs}, l_{GaAs} (characterize I_h^c); lattice mismatch $\Delta\varepsilon$, the average height and radius of the QDs H_{QD}, R_{QD} (characterize I_h^d); the average distance between the centers of the QDs in lateral plane d_{xy}, variances distance in lateral plane and vertical direction σ_{xy}, σ_z and the average number of QDs in stack n, determine vertical correlation length $l_V = n(l_{InGaAs} + l_{GaAs})$ – the distance between stacked QDs (characterize $F_L(q_x, q_y)$ and $F_V(q_z)$); static Debye-Waller factors of layers in periodic structure, substrate and barrier layers f_{QD}, f_S, f_{sub}, f_{AlGaAs} (characterize total intensity).

The statistical noise was rated and calculated according to the [3]. Calculations were performed using C++. Deviation of the calculated results from the experimental data was about 5%. Structure parameters using for modeling of the scattering experiment, their initial values search ranges, and obtained solutions are shown in the table 1.

Table 1. Structure parameters, their expected values, search ranges and obtained solutions.

Structure parameters	Expected values	Search ranges	Solutions
l_{InGaAs}	5 nm	(4-6) nm	5.2 nm
l_{GaAs}	15 nm	(14-16) μm	14.8 nm
$\Delta\varepsilon$	-	$(1\text{-}3)\times10^{-3}$	2.2×10^{-3}
H_{QD}	5 nm	(3-6) nm	5 nm
R_{QD}	-	(5-30) nm	12.5 nm
d_{xy}	-	(15-100) nm	65 nm
σ_{xy}	-		29 nm
σ_z	-		47 nm
n	-	1-20	7 nm
f_{QD}	-	0.7-1	0.927
f_S	-	0.7-1	0.917
f_{sub}	-	0.7-1	0.96
f_{AlGaAs}	-	0.7-1	0.95

The results of solving the inverse problem of X-ray diffraction within the framework of the proposed approach are presented in figure 1.

 1410 (2019) 012009 doi:10.1088/1742-6596/1410/1/012009

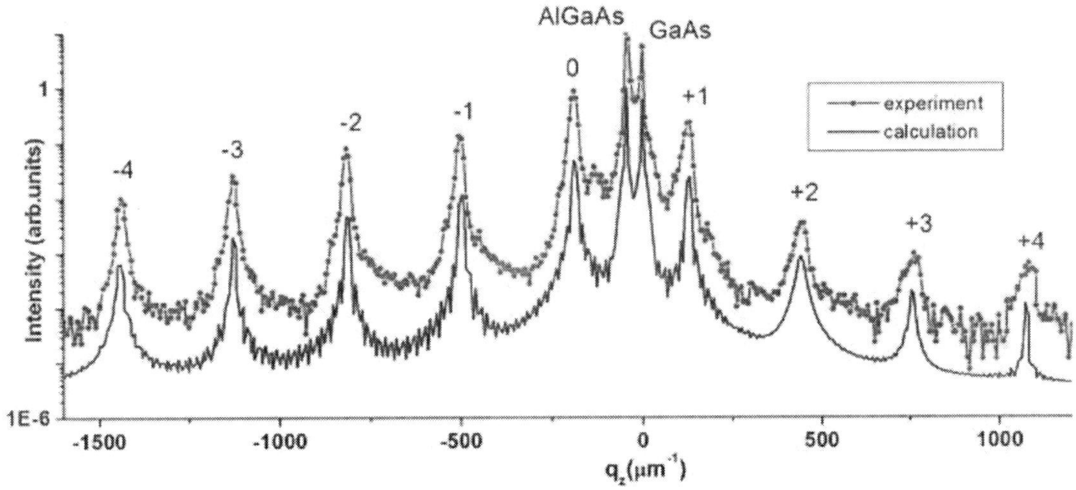

Figure 1. ω-2θ rocking curves. Experimental results (solid dots on the red line) shifted by an order in the intensity scale and residual functional minimization (solid blue line) ω-2θ rocking curves near the (004) GaAs reflection for [GaAs(001)-AlGaAs-{InAs QDs-GaAs}SL]x20 structure.

7. Conclusion

Periodic structures with quantum dots were quantitatively analyzed by simultaneous simulation of the experimentally observed coherent and diffuse scattered x-ray radiation. A priori technological information about the investigated structures, the parameters of the epitaxial layers and the QDs has been used as initial approximations for numerical simulations. The developed method may be successfully used for x-ray diffraction investigation of different nano-structured objects.

Acknowledgments

This research was supported by the Russian Academic Excellence Project at the Immanuel Kant Baltic Federal University, Program of UB RAS No 18-10-2-23, grant RFBR 17-02-00090.

References

[1] Mitchell M 1996 *An introduction to genetic algorithms* (London: MIT Press) p 158
[2] Wormington M, Pannaccione C, Matney K M, Bowen D K 1999 *Phil. Trans. R. Soc. Lond. A.* **357** 2827
[3] Hannon A F, Sunday D F, Windover D, Kline J 2016 *J. Micro/Nanolith. MEMS MOEMS* **15(3)** 034001
[4] Punegov V I , Pavlov K M, Karpov A V, Faleev N N 2017 *J. Appl. Cryst.* **50** 1256
[5] Punegov V I, Sivkov D V, Klad'ko V P 2011 *Tech. Phys. Lett. 37* 364
[6] Sivkov D V, Punegov V I 2018 *Journal of Physics: Conference Series* **1124** 002040.
[7] Storn R, Price K 1997 *Journal of Global Optimization* **11** 341

The study on agglomerates of WC, TiC, TaC nanopowder mixtures obtained from hard-alloy waste products

V.Gavrish[1], T.Chayka[1], G.Baranov[1]
[1]Scientific and educational center "Advanced technologies and materials", Sevastopol State University, 299053, Sevastopol, Russian Federation

Abstract. The paper presents findings of the nanopowder obtained from hard-alloy waste (a mixture of tungsten, titanium, and tantalum carbides) under a unique technology. Dispersion, particles' morphology have been studied. Chemical purity of this nanopowder has been confirmed by X-ray spectral microanalysis.

1. Introduction

Today introduction of nanopowders into various industries offers great opportunities to improve properties and characteristics of various materials. However their production on industrial scale is constrained by the need to use expensive equipment.

One of the promising and needed materials on the nanopowder market is high-melting compounds and/or their mixtures (tungsten W, tungsten carbide WC, titanium carbide TiC, etc.). It's known that carbides of these metals are characterized by high hardness, ruggedness at high temperatures, high-resistance to oxidation, abrasion resistance, etc.

In [1] there was proposed a method to produce nanopowders of high-melting metals and their carbides under a unique technology using hard-alloy metal scrap. This technology has no world analogues. This method opens up opportunities for production of nanopowders in production quantities, as it needs minimal technological equipment, environmental safety, relatively low operating costs and high purity of the powder.

This paper is a continuation of the papers [2-6]. Previous studies have shown availability to use powders of high-melting metals and their compounds as modifying additives in lubrication compositions [2], building materials [3], epoxy resins [4], plastics [5], radioprotective materials [6], and so on.

2. Experimental details

As a starting material to produce nanopowders we've used hard-alloy wastes (tungsten carbide WC – titanium carbide TiC – tantalum carbide TaC –cobalt Co).

Dispersion and particles' morphology were studied with the help of a PHENOM proX scanning electron microscope with an integrated system of energy-dispersion analysis. The elemental analysis of the hard-alloy wastes and studied powders was obtained using the Phenom Element Identification program. Particle Metric software was used to analyze size, shape, and morphology of the particles.

Figure 1 shows micrographs of alloy, used as a raw material for the production of powders.

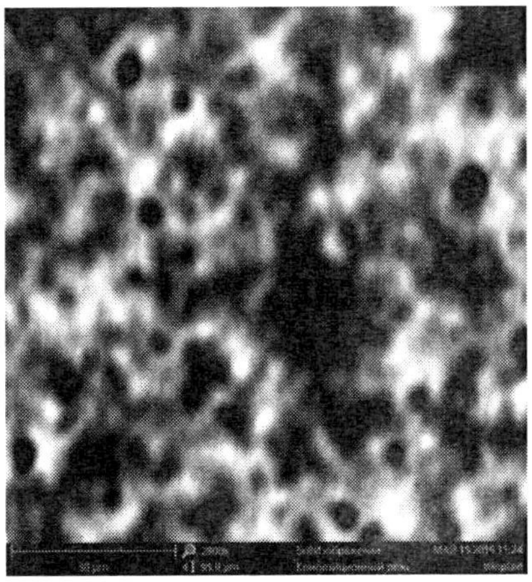

Figure 1. Micrograph of hard alloy.

Almost the entire volume of the structure is occupied by the carbide phase, which consists of grains of tungsten carbide, titanium carbide, tantalum carbide, which is confirmed by elemental analysis (table 1). Grain sizes are 1 - 7 microns.

Table 1. The chemical composition of alloy.

Element symbol	W	Ta	Ti	C	O	Co
Weight conc.	78.0	7.8	1.9	1.5	2.8	8.0

The technology to produce nanopowders of high-melting metals and their carbides from hard-alloy wastes (fig. 2) is described in [1, 2].

Figure 2. Alloys before/after processing.

This method is based on destruction of alloy matrix by specially prepared solutions using adapted microorganisms.

3. Results

The results of the study of sample elemental composition (table 2) showed that the powder obtained is a mixture of tungsten, titanium, and tantalum carbides (fig.3).

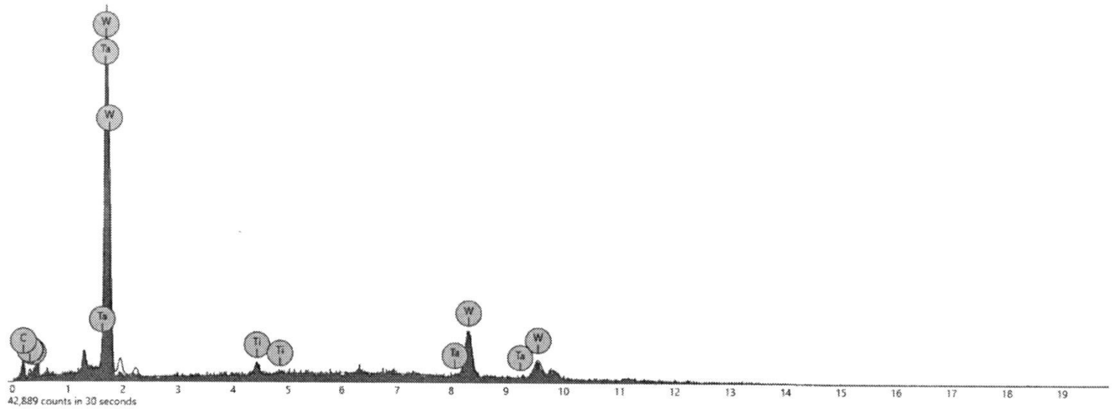

Figure 3. Spectra of the powder under study.

Table 2. Elemental composition of the powder under study.

Element symbol	W	Ta	Ti	C
Weight conc.	83.75	1.22	7.3	7.73

Figure 4 shows micrographs of particles (and their agglomerates) of the sample powder that is a mixture of tungsten, titanium, and tantalum carbides at different magnifications.

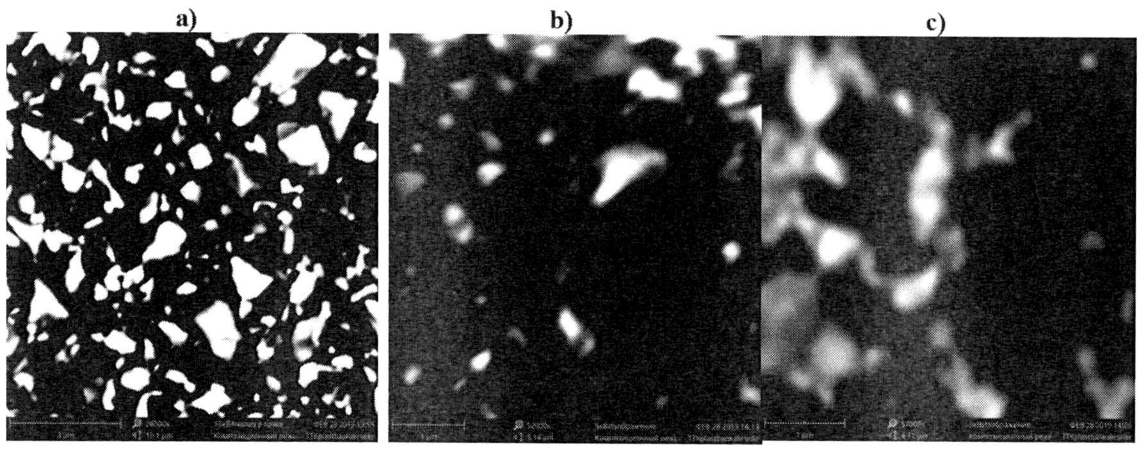

Figure 4. Micrographs of parcticles of the sample powder at different magnifications: **a)** 26500 times as much **b)** 52000 times as much , **c)** 57000 times as much.

From the presented photos (fig.4) it is clear that most particles have an irregular shape. It resulted in larger surface and therefore interaction reaction proceeds more actively. The powder is prone to agglomeration during storage (Fig.4, c). Multiple accretion of agglomerates among themselves is observed, boundaries of accreted points are uneven, tortuous.

The diagrams of particles (distributed in size) are shown in fig. 5. According to the diagram particles' (agglomerates') size of the powder under study 65-500 nm.

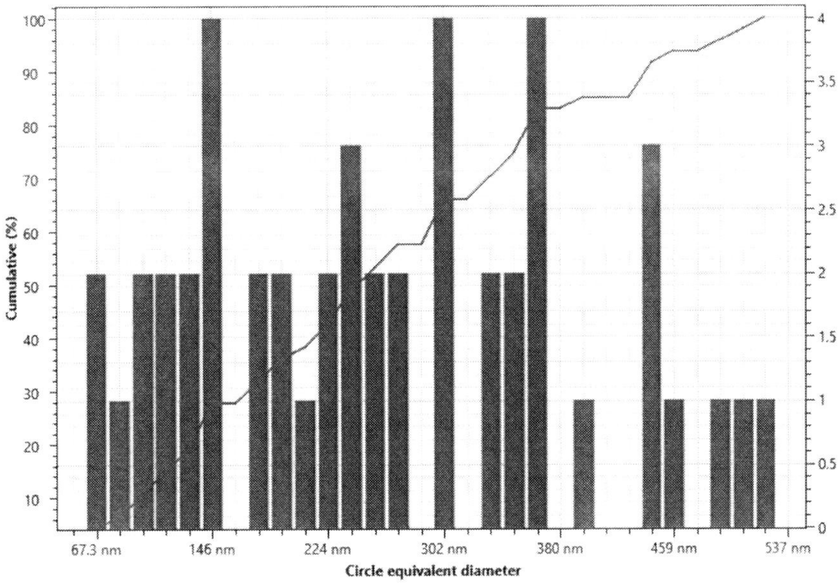

Figure 5. Sizing of the sample powder particles.

Thus, high purity of the nanopowder (agglomerates) that is a mixture of tungsten, titanium, and tantalum carbides powder was obtained as a result of processing of production hard-alloy wastes.

4.Conclusion

Based on the results of the experiment to produce nanopowders out of hard-alloy waste (a mixture of tungsten, titanium, and tantalum carbides) we can conclude the following:

-the powder obtained is a mixture of tungsten, titanium, tantalum carbides.

-powder particles are of irregular shape, some of them are agglomerates.

-the particles are 65-500 nm by size.

Further studies will involve ultrasonic destruction of agglomerates structure.

References

[1] EP 3138932 (A1) EC

[2] Gavrish V, Chayka T, Baranov G et. el. 2013 *J. Collection of studies, SNUNEI* 3 (47) 49 – 55

[3] Gavrish V, Baranov G, Chayka T 2017 *IOP Conference Series: Materials Science and Engineering*. 168. p. 1–6.

[4] Gavrish V, Baranov G, Chayka T et.el. 2016 *J. Power plants and technologies* 2(3) 64-69.

[5] Burya A, Gubskaya M,Baranov G 2011 *J. East Ukrainian National University named after Dahl.* 8 (162) 56 – 61

[6] Gavrish V.M., Baranov G.A., Chayka T.V. et.el. 2016 *IOP Conference Series: Materials Science and Engineering.* 110. p. 1–6.

Study of the mechanical characteristics of the Al_2O_3 film on the $LiNbO_3$ substrate

A Shumilin[1], **I Shumilina**[1], **A Fomin**[1]
[1] Yuri Gagarin State Technical University of Saratov, Saratov 410054, Russia

Abstract. The microhardness of a thin aluminum oxide film with a thickness of 80 to 240 nm on the surface of a lithium niobate single crystal was determined by the method of nanoindentation. The hardness and Young's modulus for a film deposited by dc-magnetron sputtering of an aluminum target in a gas ($Ar+O_2$) medium were more than 30 GPa and 300 GPa, respectively. The dependences of the film thickness on the mechanical properties of a layered system, in particular hardness and elastic modulus, were established.

1. Introduction

Aluminum oxide (Al_2O_3) is a widespread and inexpensive material that is used in various branches of science and technology. Aluminum oxide is applied in the form of a powder of various sizes as an abrasive material and reinforcing filler in metallurgy of alloys and the manufacture of ceramic products [1]. Synthetic and natural single crystals of aluminum oxide with different phase composition and impurities are used in the jewelry industry and high-energy optics (laser optics) [2].

Thin films of this material obtained by various methods have nanoscale elements of the structure and surface morphology. For example, aluminum oxide layers produced by the sol-gel method usually have a porous surface and a thickness of the order of hundreds of nanometers, which is necessary in the manufacture of filters and catalysts [3]. Modern technologies for the deposition of thin films allow the production of continuous, non-porous layers of dielectric materials with a thickness from a few nanometers to several micrometers. These methods include vacuum condensation techniques of the deposition of thin films, in particular PVD and CVD. One of the varieties of the PVD process is magnetron sputtering, which can be performed at direct (DC) or alternating (AC) current [4]. Modern PVD processes are characterized by reproducibility of parameters, high adhesion of deposited materials to the substrate, wide possibilities for the creation of composite materials, the possibility of alloying and relative simplicity of the process [5].

With all the variety of methods of production and applications of thin films of aluminum oxide, the question of studying their mechanical characteristics, in particular the hardness and modulus of elasticity of the "film-base" layered structure, remains relevant. According to a brief review, the microhardness of a $LiNbO_3$ single crystal reaches 5–10 GPa [6,7]. The value of microhardness of aluminum oxide depends on the method of its production and is in the range from 3 to 30 GPa [8,9].

The characteristics of the aluminum oxide film are studied in detail, which is required when creating high-quality dielectrics, protective and anti-reflective layers in solar cell elements [10]. However, the mechanical characteristics of the resulting layers are given insufficient attention. For example, in the layered system under consideration, hardness, elasticity, and wear resistance are important characteristics that determine their possible use in aggressive operating conditions, in particular in the open air or aerospace, as coatings for solar panels.

2. Methodology

The aluminum oxide film was deposited on the polished surface of lithium niobate single crystals ($LiNbO_3$), the shape of which was represented by parallelepipeds sized 10×11×6 mm. The normal to the deposition plane of the film was an angle of 36° with the positive direction of the crystallographic axis *Y*.

For PVD, an aluminum target (99.99 %) was used, which was installed in a DC magnetron. In this case, the gas mixture ($Ar+O_2$) with the 1:1 ratio of the partial pressure of gases was applied at a pressure inside the vacuum chamber P = 1.47 Pa. The deposition was conducted at a discharge current I = 50 mA, the discharge current density $j \approx 50$ A/m^2, and the substrate heating temperature t = 80 °C. The maximum value of magnetic induction on the target surface was about 25 mT. The films were deposited at a duration τ of 24, 48, and 72 min, which corresponded to a film thickness of 80, 160, and 240 nm.

Samples of the series "1", "2" and "3" prepared for the study of hardness and elasticity modulus were fixed on a rectangular profile using thermoplastic glue (Figure 1).

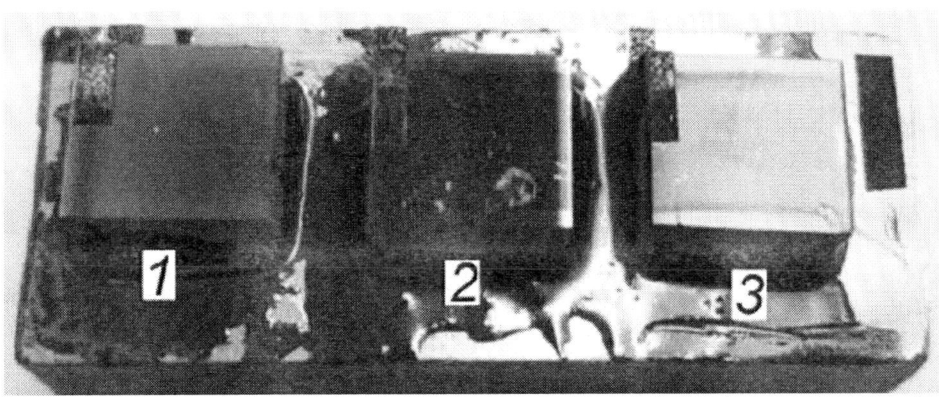

Figure 1. Samples of the layered "$LiNbO_3 - Al_2O_3$" system with different film thicknesses of 80 nm (*1*), 160 nm (*2*) and 240 nm (*3*) placed on an aluminum substrate to test the mechanical properties.

The hardness *H* and elasticity modulus *E* were determined according to the Oliver-Farr method (ISO 14577-1: 2015) [11,12]. The data of mechanical properties were presented in diagrams of applied force P and penetration depth of the indenter *h*, which were obtained on the mechanical tester "NANOVEA Ergonomic Workstation" with loads *P* on the Berkovich indenter of 5, 20 and 200 mN.

3. Results

Analysis of the loading diagrams showed that the hardness data referred to the "$LiNbO_3 - Al_2O_3$" layered system. The maximum penetration depth of the indenter *h* exceeded the oxide film thickness in samples "1", "2" and was comparable to the film thickness for sample "3" (Figure 2).

With a minimum load P = 5 mN and penetration of the indenter into the layered structure to a depth of not more than h = 100 nm, the hardness *H* reached maximum values above 32 GPa for all three samples (Figure 3). A maximum hardness of about 104±15 GPa was achieved for the film with a thickness of 240 nm. Thus, the resulting layered structure was characterized by the presence of a thin and hard oxide film on a softer substrate.

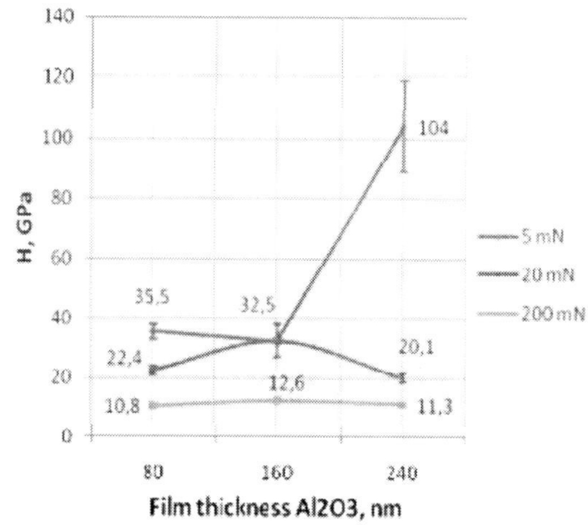

Figure 2. A graph showing dependencies of hardness of sample surfaces with different thicknesses of the oxide film at various loads.

The increase in hardness for sample "3" was associated with a lower influence of the substrate, since the depth of penetration of the indenter in this case was about 1/3 of the thickness of the oxide film. This was indirectly confirmed by the growth of the elasticity modulus E at a constant indentation depth for all samples. With a load on the indenter of 20 mN, the Young's modulus gradually fell with the growing film thickness and took values from 342±17 to 306±21 GPa. At a load of 5 mN, a smooth growth of the Young's modulus values with increasing film thickness was observed and for samples "1", "2" and "3" it was 141, 220 and 301 GPa, respectively.

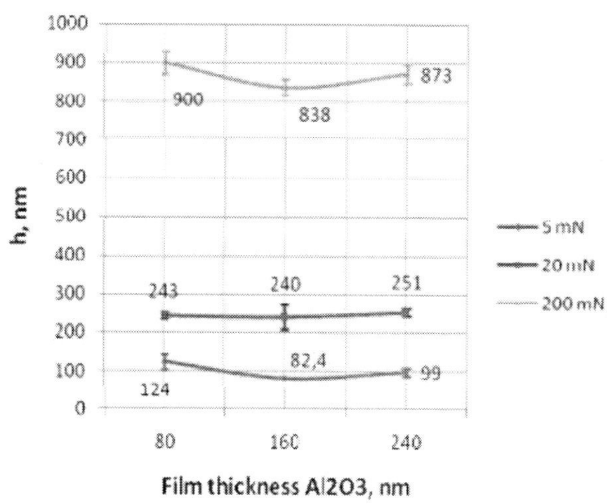

Figure 3. A graph showing the maximum depth of penetration of the indenter into the samples depending on the oxide film thickness under various loads.

Of practical interest was the structure of sample "2" with an oxide layer thickness of 160 nm. Such film is used as a quarter-wave anti-reflection single-layer coating on a $LiNbO_3$ crystal for a wavelength of light λ = 632.8 nm. The hardness of the resulting coating at a level of 30 GPa allows the prediction of high wear resistance and the strength of the coating to external mechanical effects and indicates the predominant content of aluminum oxide in the form of a stable solid α-phase without additional stabilization by impurities [13].

4. Conclusion

Thus, thin Al_2O_3 films were deposited on $LiNbO_3$ crystals by magnetron sputtering at a direct current. The hardness and elasticity modulus of the "$LiNbO_3$ – Al_2O_3" layered system were determined at different loads and film thicknesses. The results led to the conclusion about the possibility of their use in the form of thin anti-reflection Al_2O_3 films and protective layers for softer crystalline materials, which are the basis of modern solar energy devices.

Acknowledgments

The authors express their gratitude to Dr. Zakharevich A.M. and Dr. Skaptsov A.A. for the conducted nanoindentation tests and discussion of the results. The oxide films were deposited using the equipment of NPF "Piezon" of Yuri Gagarin State Technical University of Saratov, Russia.

References

[1] Ouyang Y, Bai L, Sun Z, Ding F and Yuan F 2019 *Ceram. Int.* **42** 2012
[2] Wiesner M and Ihlemann J 2011 *Physics Procedia* **12** 239
[3] Rachmadena D, Faizal M and Said M 2018 *Int. J. Adv. Sci. Eng. Inf. Technol.* **8** 694
[4] Pulker H 1999 *Surf. Coat. Techol.* **112** 250
[5] Liao B and Lee C 2011 *Opt. Express* **19** 7507
[6] Wang H, Zhang Y, Xiang D and Xu J 2018 *Optik* **164** 385
[7] Palatnikov M, Shcherbina O, Masloboeva S, Efremov V, Ivanenko D and Vladimirova S 2017 *Physical and chemical aspects of the study of clusters, nanostructures and nanomaterials: Interuniversity collection of proc.* **9** 356
[8] Szindler M, Dobrzański L, Szindler M, Pawlyta M and Jung, T 2017 *J. Achiev. Mater. Manuf. Eng.* **82** 49
[9] Cherneva S, Guergova D, Iankov R and Stoychev D 2019 *J. Eng. Mater. Technol.* **141** 011006
[10] Swatowska B 2018 *Microelectron. Int.* **35** 177
[11] Oliver W, Pharr G and Mater J. 1992 *J. Mater. Res.* **7** 1564
[12] Fomin A, Dorozhkin S, Fomina M, Koshuro V, Rodionov I, Zakharevich A, Petrova N and Skaptsov A 2016 *Ceram. Int.* **42** 10838
[13] Yamada-Takamura Y, Koch F, Maier H and Bolt H 2001 *Surf. Coat. Techol.* **142** 260

New polymorphic varieties of fluorographene forming during fluorine functionalization of 4-8 graphene layers

Maxim E Belenkov[1], Vladimir M Chernov[1] and Evgeny A Belenkov[1]

[1]Department of Physics, Chelyabinsk State University, Chelyabinsk 454001, Russia

E-mail: me.belenkov@gmail.com

Abstract. Using the methods of the density functional theory in the generalized gradient approximation, the structure and electronic properties of six new polymorphic varieties of fluorographene, formed during the chemical adsorption of fluorine by a 4-8 graphene layer, are calculated. The sublimation energy of 4-8 polymorphs of fluorographene varies from 13.36 to 14.09 eV $(CF)^{-1}$. The range of variation of the band gap of new CF-$L_{4\text{-}8}$ polymorphs of fluorographene is greater than the similar interval for CF-L_6 compounds.

1. Introduction

It is possible to modify the properties of graphene from metallic to semiconductor due to chemical adsorption of hydrogen, fluorine, chlorine, oxygen and other atoms on its surface [1-4]. The most promising is the functionalization of graphene by fluorine, since the fluorographene compounds are more thermally stable than others. It is possible to obtain various structural modifications of fluorinated graphene differing by the order of addition of fluorine atoms [5-7]. These polymorphic varieties may have different electronic properties, but their variation intervals are rather narrow. It is possible to obtain new structural varieties of fluorographene with a large range of changes in properties by chemically adsorbing fluorine on the surface of various polymorphs of graphene [8]. In this work, we studied compounds of fluorographene with a layered structure that can be obtained by fluorination of 4-8 graphene layers.

2. Methods

Theoretical analysis of possible ways of chemical adsorption on the surface of $L_{4\text{-}8}$ graphene layers showed that there are six main possible polymorphic varieties of fluorographene, differing by the order of attachment of fluorine atoms to the graphene layer. Calculations of the geometrically optimized structure of CF-$L_{4\text{-}8}$ layers and their electronic properties were performed by the density functional theory (DFT) method in the generalized gradient approximation (GGA). Calculations were carried out for bulk structures – stacks of fluorographene layers with a spacing of 12 Å between the layers. This distance ensured the absence of influence of adjacent layers on each other. The calculations used a set of k-points: 12 × 12 × 12. The dimension of the set of basis functions was limited by the value E_{cutoff} = 1 keV.

3. Results

During the geometrical optimization of the structure of six theoretically constructed varieties of CF-$L_{4\text{-}8}$ fluorographene, five layers (structural types T1, T2, T4-T6) turned out to be stable, and one layer (structural type T3) collapsed after breaking up into separate atoms. The geometrically optimized structure of the layers and their unit cells are shown in figure 1. For a layer of T3 type, the structure is shown after partial optimization, before the start of fracture. The crystal lattices of the fluorographene layers of the T1-T5 types are tetragonal, and the T6 layer has a rhombic type (table 1). The unit cells of the layers contain from 8 to 32 atoms. The lengths of the elementary translations vectors for crystal lattices of different types are given in table 1. The layer density of the polymorphs of the fluorographene varies from the minimum value of 1.558 mg m^{-2} (T1 layer type) to the maximum value 1.683 mg m^{-2} (T4 layer type).

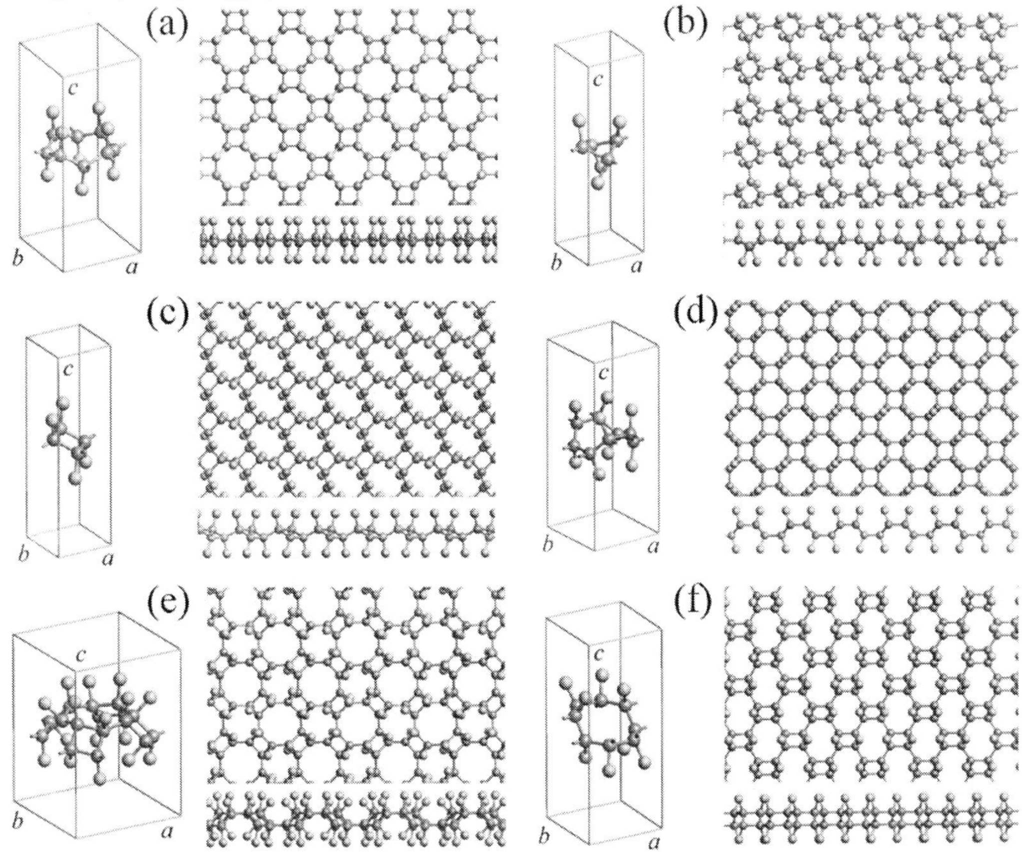

Figure 1. The unit cells and the geometrically optimized structure of polymorphic varieties of fluorographene layers based on the graphene layer L_{4-8}: (a) T1; (b) T2; (c) T3; (d) T4; (e) T5; (f) T6.

The results of measurements of the lengths of interatomic bonds are given in table 2. The first bond having the length L_1 is the bond between the fluorine and the carbon atoms. The length of this bond varies from 1.4196 to 1.4869 Å. In the T1-T4 layers and the T6 layer, all carbon atoms are in equivalent structural positions and, therefore, the lengths of the CF bonds are the same for each layer. In the T5 type layer, there are two different structural positions of atoms and, therefore, the length of CF bonds is different (table 2). The second carbon-carbon bond having a length L_2 is located between adjacent octagons of L_{4-8} layers. This bond is shorter than the carbon-carbon bond lengths of L_3 and L_4, which are sides of quadrangles. The lengths of the L_3 and L_4 bonds in each of the layers have close values, i.e. the shape of the quadrilateral is close to square. For the T3 type layer, which was not stable, the L_4 bond length is minimal compared to other layers and a significant difference from L_3. This may cause instability of the structure of the T3 type layer. For carbon-carbon bonds, table 2 also shows the mutual orientation of fluorine atoms at the ends of the corresponding bond. Only in the T1 type layer for each of the three carbon-carbon bonds is the orientation opposite.

Table 1. Structural parameters and properties of polymorphic varieties of the fluorographene layer L_{4-8} (a, b – translation vectors, N – number of atoms in the unit cell, ρ – layer density, E_{total} – total energy, E_{sub} – sublimation energy, E_F – Fermi energy, Δ – band gap).

Parameters	Type of CF-L_{4-8} layers					
	T1	**T2**	**T3**	**T4**	**T5**	**T6**
Crystal family	Tetragonal	Tetragonal	Tetragonal	Tetragonal	Tetragonal	Orthorhombic
a (Å) / b (Å)	5.0674	3.6607	3.5002	4.8759	7.0879	4.8955 / 4.8753
N (atom)	16	8	8	16	32	16
ρ (mg m^{-2})	1.558	1.493	1.633	1.683	1.593	1.676
E_{total} (eV $(CF)^{-1}$)	-800.30	-799.68	-800.41	800.37	799.88	800.33
E_{sub} (eV $(CF)^{-1}$)	13.98	13.36	14.09	14.05	13.56	14.01
E_F (eV)	-6.042	-6.019	-5.914	-6.305	-5.889	-5.943
Δ(eV)	3.2105	4.9582	3.0516	3.9464	4.6864	4.8774

Table 2. The lengths of interatomic bonds (Li) for fluorographene L_{4-8} layers of various types.

Type	L_1. Å	L_2. Å		L_3. Å		L_4. Å	
T1	1.4400	1.5064	↑↓	1.6081	↑↓	1.6081	↑↓
T2	1.4869	1.5388	↑↑	1.6032	↑↓	1.6032	↑↓
T3	1.4204	1.5246	↑↓	1.6089	↑↑	**1.5737**	↑↓
T4	1.4196	1.5343	↑↓	1.6083	↑↑	1.6082	↑↑
T5	1.4747	1.5888	↑↑	1.6126	↑↓	1.6028	↑↑
	1.4209	1.5888	↑↑	1.6268	↑↑	1.6028	↑↓
T6	1.4259	1.5606	↑↑	1.6091	↑↑	1.5768	↑↓

The values of the angles between the interatomic bonds in the structure of the L_{4-8} fluorographene layers are given in table 3. The values of the angles vary in the interval from 83.80 to 134.59 °. The atoms in the structure of the fluorographene layers are in sp^3 hybridized states and the equilibrium value of the angles between the bonds should be 109.47 °. This value is characteristic for interatomic bonds in the diamond structure. To estimate the degree of deformation of the structure of graphene layers compared to the equilibrium structure, the values of the deformation parameter Def were calculated. This parameter was found as the sum of the absolute values of the difference in the angles between the bonds in the layer and the angle between the bonds in the diamond. The values of the parameter Def are given in table 3. The minimum value of Def equal to 48.54 ° is observed for the T3 layer, and the maximum value is 113.03 ° for the layer of the T2 type.

Table 3. Angles (φ_{ij}) between interatomic bonds and deformation parameters (Def) for fluorographene L_{4-8} layers of various types.

Type	φ_{12}. °	φ_{13}. °	φ_{14}. °	φ_{23}. °	φ_{24}. °	φ_{34}. °	Def. °
T1	103.28	110.92	110.92	123.11	123.11	84.94	60.9
T2	98.39	89.59	121.47	142.35	120.99	83.80	113.03
T3	106.36	107.41	112.07	124.62	115.62	90.00	48.54
T4	107.28	105.97	105.97	122.73	122.73	90.00	55.18
T5	100.55	91.24	96.67	134.03	127.99	93.78	98.72
	101.69	101.84	124.60	134.59	110.29	86.23	79.72
T6	101.84	105.69	108.39	122.44	126.91	90.00	62.37

The calculated sublimation energies of the CF-L_{4-8} layers with a stable structure are in the range from 13.36 to 14.09 eV $(CF)^{-1}$. The dependence of the sublimation energy of the fluorographene layers on the values of the deformation parameters characterizing their structure is observed – the greater the value of the deformation parameter, the lower the sublimation energy (figure 2).

As a result of the DFT-GGA calculations, the band structures and densities of electronic states of L_{4-8} fluorographene polymorphs were found (figure 3, 4). All the investigated layers of fluorographene turned out to be semiconductors with a band gap from 3.21 eV for the T1 layer to 4.96 eV for the T2 layer. The range of variation of the band gap in polymorphic varieties of L_{4-8} layers of fluorographene turned out to be longer than the same interval for L_6 fluorographene layers [6]. New structural types of fluorographene, theoretically investigated in this paper, can find practical application in the design of nanoelectronic devices.

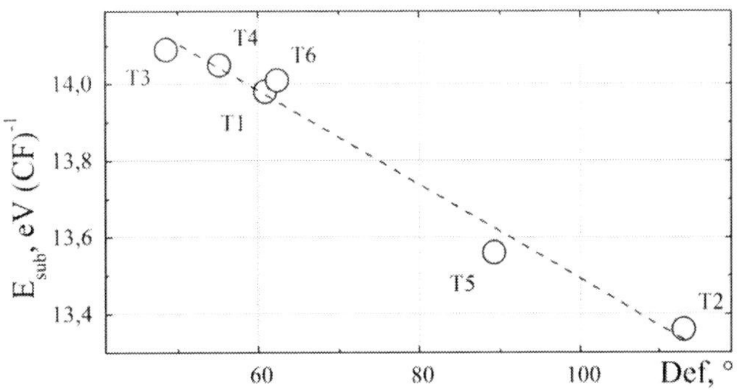

Figure 2. The dependence of the sublimation energy (E_{sub}) on the deformation parameter (Def) for polymorphic varieties of L_{4-8} layers of fluorographene.

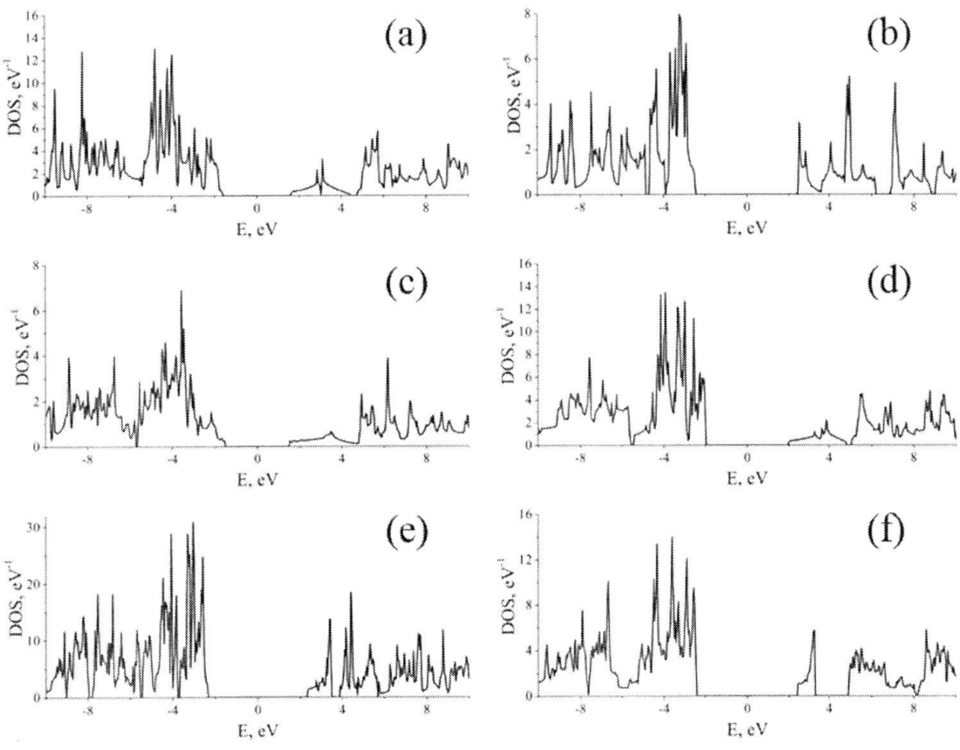

Figure 3. The density of electronic states (DOS) of polymorphic varieties of fluorographene layers based on the graphene layer L_{4-8}: (a) T1; (b) T2; (c) T3; (d) T4; (e) T5; (f) T6.

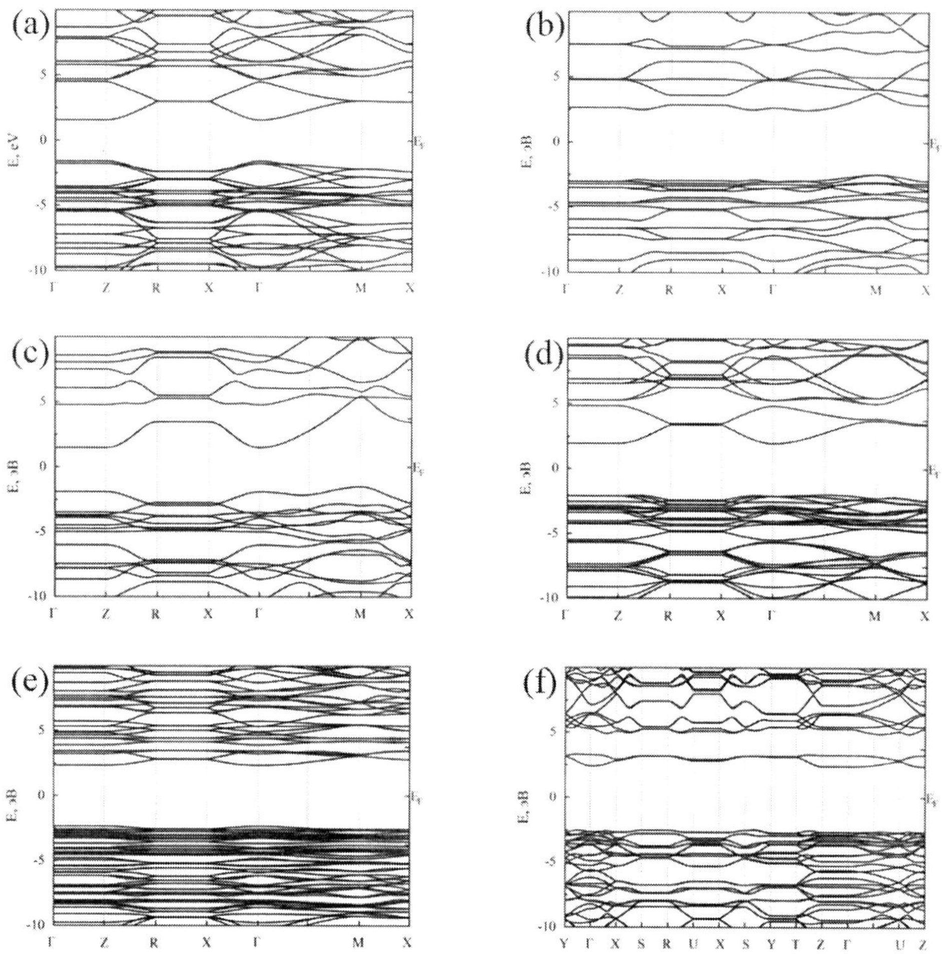

Figure 4. The band structure of polymorphic varieties of fluorographene layers based on the graphene layer L_{4-8}: (a) T1; (b) T2; (c) T3; (d) T4; (e) T5; (f) T6.

Acknowledgments

This research was supported by Foundation for Perspective Scientific Research of Chelyabinsk State University.

References

[1] Elias D C, Nair R R, Mohiuddin T M G, Morozov S V, Blake P, Halsall M P, Ferrari A C, Boukhvalov D W, Katsnelson M I, Geim A K and Novoselov K S 2009 *Science* **323** 610

[2] Robinson J T, Burgess J S, Junkermeier C E, Badescu S C, Reinecke T L, Perkins F K, Zalalutdniov M K, Baldwin J W, Culbertson J C, Sheehan P E and Snow E S 2010 *Nano Letters* **10** 3001

[3] Li B, Zhou L, Wu D, Peng H, Yan K, Zhou Y and Liu Z 2011 *ACS Nano* **5** 5957

[4] Chen D, Feng H and Li J 2012 *Chem. Rev.* **112** 6027

[5] Belenkov M E, Chernov V M and Belenkov E A 2018 *Materials research proceedings* **9** 148

[6] Belenkov M E, Chernov V M and Belenkov E A 2018 *IOP J. Phys. Conf. Ser.* **1124** 022010

[7] Grishakov K S, Katin K P, Maslov M M and Prudkovskiy V S 2019 *Appl. Surf. Sci.* **463** 1051

[8] Belenkov E A and Kochengin A E 2015 *Phys. Solid State* **57** 2126

Biosensors based on nanowire field effect transistors with Schottky contacts

T E Smolyarova[1,3], A V Lukyanenko[2,3], A S Tarasov[2,3], L V Shanidze [1,3], F A Baron[2], F V Zelenov[2], I A Yakovlev[2] and N V Volkov[1].

[1]Krasnoyarsk Science Center SB RAS, Krasnoyarsk 660036, Russia
[2]Kirensky Institute of Physics, Krasnoyarsk 660036, Russia
[3]Siberian Federal University, Krasnoyarsk 660041, Russia

*smol.nano@yandex.ru

Abstract. A top-down nanofabrication approach was used to obtain silicon nanowires from silicon-on-insulator wafers using direct-write electron beam lithography and plasma-reactive ion etching. Fabricated with designed pattern silicon nanowires are 0.4, 0.8, 2 μm in width and 100 nm in height. The devices can be applied in future medical diagnostic applications as novel biosensors with detection principle based on the changes in electrical characteristics of the silicon nanowires functionalized with thiol-containing molecules.

1. Introduction

Nanometer-scale structures, in particularly silicon nanowires (NW), attract considerable attention due to their possible application in nanotechnology and nanomedicine since their sizes are comparable to the sizes of chemical and biological species for detection. At that time in integrated nanoscale electronics the considerable attention is also paid to devices based on Si NWs with Schottky contacts [1,2]. In a point of fact, Schottky barrier FETs are of great interest on their own account as an alternative to traditional doped source and drain device structures, because sub-100 nm ranges scaling encounters fundamental problems including high leakage current and parasitic resistance. Many research groups have demonstrated different methods for the implementation of such devices in a nanoscale [3-8]. The sensing mechanism of such devices may be described in terms of the change in charge density, which induces a change in the electric field at the silicon nanowire surface. The most important applications of the SiNW FET biosensor are in disease diagnosis using specific probe molecules for nanowire functionalization these devices may operate as selective sensors for detection of nucleic acids [9], proteins [10], protein-DNA interactions [11], cells [12], and viruses [13]. The most popular method for fabrication of such devices in nanoscale is "top-down" synthesis approach [14-19], which allow to regulate device geometry and electrical characteristics due to precise control of lateral sizes of nanowire and enables direct contact with other structures on device. In comparison with "bottom-up" approach which usually use chemical methods to obtain nanostructures [20, 21] which are incapable to control geometrical characteristics. Taking into account the advantages of "top-down" approach, we report nanofabrication technology of biosensor based on Schottky barrier nanowire FET using electron beam lithography (EBL) process and chemical wet and plasma-reactive ion etching (RIE), the functionalization process of nanowire surface using the tip of atomic-force microscope and the results of thiol molecules detection.

2. Materials and methods

SiNW sensors are typical FET-based devices with Schottky contacts and have a number of advantages including good suppression of short channel effects, simple and low temperature processing and the elimination of doping and subsequent activation steps. These features are particularly desirable for SiNW devices because they can circumvent difficult fabrication issues such as an accurate control of the doping type and the formation of reliable ohmic contacts. SiNW FET sensors include source, drain and gate electrodes connected with the semiconductor nanowire, which serves as the sensing component of the device. Nanofabrication technology of silicon nanowires (nanowire FET) using "top-down" approach represents a multistep process, which will be briefly described in this paper.

2.1. Fabrication of biosensors based on SiNW FETs

Silicon nanowires were fabricated from SOI (100) wafers with 200 nm of the buried oxide (BOX) and 100-nm monocrystalline top boron doped silicon layer (with resistivity of 18 Ω·cm). Firstly, the SOI wafer was cleaned to remove contaminants and native oxide. After the cleaning procedure, thermal evaporation in ultra-high vacuum was used to obtain the thin film of Fe and then the EBL process was used to obtain Fe contact pads to allow the carrying out of the investigations of the electric properties. The metal contact pads for the source (S) and drain (D) were connected to each end of the fabricated nanowire. Further, the positive tone resist was spin coated on the sample and dehydrated on a hotplate. The next step was the EBL exposure process, after the sample was treated in nitric acid solution for etching of Fe film in areas not covered with the photoresist. Then reactive ion etching (RIE) was used with the help of NORDSON MARCH RIE-1701 plasma system to etch Si film. The result of described technological process is an array of silicon nanowire structures on a dielectric layer with metal pads for the source (S) and drain (D) connected to each end of the fabricate d nanowires and allow to carry out the measurements of the electrical properties. In addition, besides of the metal pads in the fabricated structure a back gate was obtained on the backside of the SOI substrate using doping with indium (In). A top view SEM image of the silicon nanowires after the dry etch process is shown in Fig 1(a). The results indicate that the silicon nanowires were formed with good uniformity, high resolution and good pattern placement. AFM images show the width of the silicon nanowires are correspondingly 0.4, 0.8, 2 μm Fig 1(b, c, d).

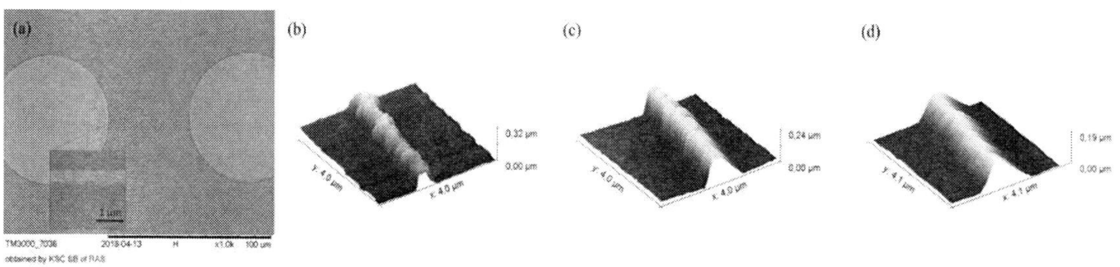

Figure 1. (a) SEM and (b, c, d) AFM images of fabricated devices with nanowire 0.4, 0.8, 2 μm in width correspondingly.

2.2. Nanowire functionalization

Functionalization of the silicon nanowire surface with a probe molecule is necessary step to evaluate the performance of manufactured devices. There are two approaches to attach the detectable molecule on the surface of silicon nanowire: electrostatic adsorption and covalent binding [22]. Electrostatic adsorption uses the attractive force responsible for adsorbing ionic solute on an oppositely charged adsorbent. In our case there is the coordinate covalent bond, which is based on the charge transfer between the probe molecule and SiNW surface molecules without the formation of a chemical bond. Since oxide can grow on the SiNW surface naturally, a number of methods rely on the functionalization of the oxide layer [23, 24]. For this purpose, the organic polymer 1-octadecanethiol (ODT) dissolved in acetonitrile was deposited on the nanowire surface [25]. This polymer is CH_3-$(CH_2)_{17}$-SH compound, where the first part (CH_3-$(CH_2)_{17}$) is organic substituent and the second part (SH) is the positively charged thiol group, which binds with the nanowire surface [26]. Using the probe of atomic-force microscope it is possible to deposit molecules on surface through the diffusion of particles through a water meniscus connecting the probe and the substrate (Fig. 2(a)). The environmental conditions during the experiment were supported with environment control chamber, so the temperature was stable at 24 - 26°C and relative humidity was about 44 - 49%. When the detectable molecules are deposited on the nanowire surface they start to form self-assembled layer due to Van der Waals interactions, [27] as a result reducing its own free energy (Fig. 2(b)) [28]. Changes occurring between the ODT molecules and the silicon surface were converted into an electrical signal. This signal was measured with the help of a probe station Lakeshore EMPX-HF 2 and Keithley 2634b SourceMeter at room temperature for pure nanowire and with deposited ODT-molecules.

3. Results and discussions

The quality of fabricated devices especially silicon nanowires is characterized by measuring of transfer characteristics with various source-drain bias V_{SD}. These characteristics demonstrates device current changing during gate voltage sweep for 0.4 μm nanowire in Fig. 2(c).

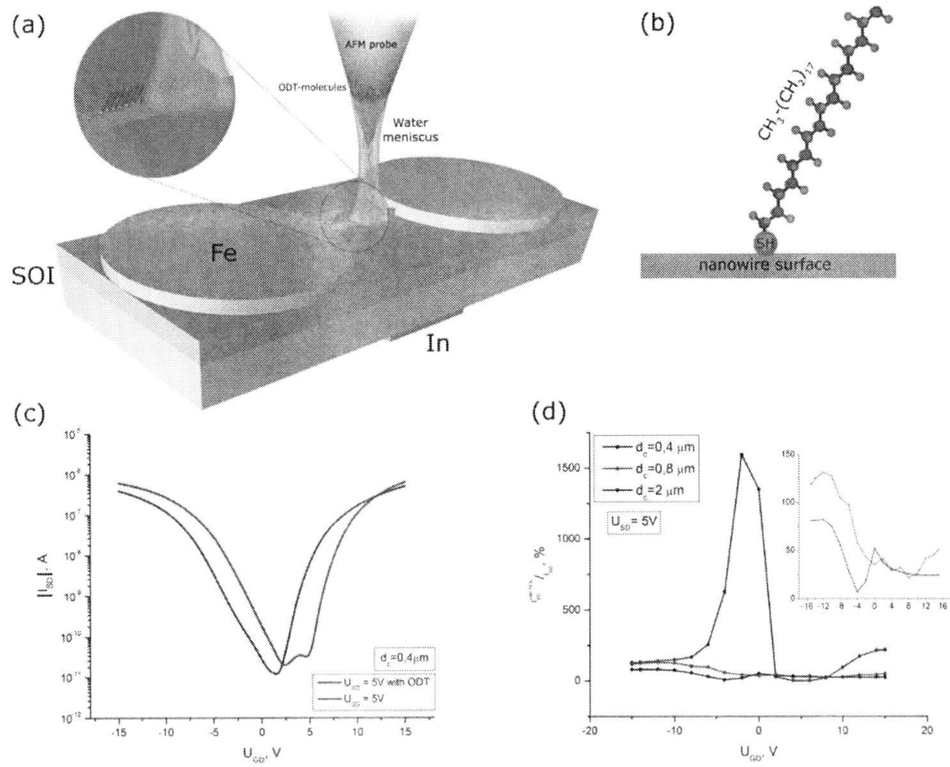

Figure 2. (a) Image of ODT-molecules deposition on the nanowire surface using AFM probe. (b) Scheme of ODT-molecular configuration. (c) Transfer characteristics of a SiNW FET with a nanowire width of 0.4 μm at different drain-source voltages for a pure nanowire and nanowire with deposited ODT. (d) Detection sensitivity for 400 nm nanowire at different biases at the back gate.

It is noted that the deposition of ODT-molecules does only the shift of transfer characteristic towards positive bias, herein the shape of the graph is almost unchanged. Thus, it is expected that the ODT-molecules don't influence on the operating modes in radical way that means main physical mechanisms of charge carrier transport in nanowires remain the same. The influence of ODT-molecules on the transport in nanowire may be considered as the enabling of an effective additional bias voltage at the gate or, that is more logically, as an appearing of virtual top gate. As shown in paper of Naumova et al [29] the shift of transfer characteristics towards higher positive bias may be caused by negative surface-charge density in silicon nanowire that is possible at the positive bias on the top gate. In our case ODT-molecule represents as a compound with positively charged thiol group able to bind with SiO_2 that covers nanowire surface. When the detectable ODT-molecule is deposited on the nanowire surface the coordinate covalent bonding appears and, as a result, the negative surface-charge density occurs in silicon nanowire. Thus, ODT-molecules operate as a top field gate of transistor. This result made possibility for electrical detection of organic molecules.

The efficiency represents the ability of fabricated devices to detect. The measurements of their characteristic show that the most sensitive is SiNW FET with 0.4 μm nanowire (Fig. 2(d)). Maximal value of sensitivity is defined as the relative nanowire current change after deposition of ODT-molecule ($Sens = \square 100\% \cdot I_{SD}^{TIOL} / I_{SD}^{Clean}$), that is more than 1000%, and is observed at relatively low voltages (V_{SD}=5V, V_{GD}=4V). For nanowires with larger widths the sensitivity roughly decreases. This fact may by caused by area of ODT-molecules deposition, which is tiny in comparison with the surface area of wide nanowires (0.8 and 2 μm). These investigations confirmed that devices fabricated using top-down nanofabrication process are of high quality and have a great potential for further development as biosensors.

4. Conclusions

Top-down approach was used to fabricate biosensors based on Schottky barrier SiNW FETs with different nanowire widths (0.4, 0.8, 2 μm) using multistep process including thermal evaporation in ultra-high vacuum, EBL and RIE processes. The results of SEM and AFM measurements of fabricated devices present good quality of technological process. Electrical investigations, in particular, transfer characteristics of the fabricated SiNW FETs demonstrates device current changing during gate voltage sweep after ODT- molecules deposition that is caused by covalent binding of positively charged thiol part of deposited molecule. Maximal value of sensitivity is defined for nanowire of 0.4 μm in width. According the carried out investigations it is possible to conclude that the fabricated nanowire biosensors have great potential for biomedical diagnostic applications.

Acknowledgements

This study was supported by the Russian Foundation for Basic Research, project no. 18-32-00035 and supported in part by the Ministry of Education and Science of the Russian Federation and the Siberian Branch of the Russian Academy of Sciences, project II.8.70, and the Presidium of the Russian Academy of Sciences, Fundamental Research Program no. 32 «Nanostructures: Physics, Chemistry, Biology, Basics of Technologies».

References

[1] Zhang, Guo-Jun, Yong Ning, *Analytica Chimica Acta* **749**, 1.
[2] Thomas Mikolajick, Walter M. Weber, 2015 Anisotropic Nanomaterials, 1-25.
[3] Farahidah Za'bah Nor et al. 2012, Journal of Applied Physics, 112.
[4] Vu XT, Ghosh Moulick R et al. 2010, Sensors and Actuators B, 144, 354–360.
[5] Rius G, Llobet J et al. 2011, AIP Conf. Proc, 1336.
[6] Gao A, Lu N et al. 2014, *Nanoscale*, **6**(21), 13036–13042.
[7] Adam Tijjani et al, 2013, *African Journal of Biotechnology*, **12**(36), 5486–5495.
[8] Shaurya Prakash, Junghoon Yeom, 2014, *Nanofluidics and Microfluidics*; In Micro and Nano Technologies: 171–239.
[9] J.-I. Hahm, C.M. Lieber, 2004, *Nano Lett.*, **4**, 51–54
[10] G. Zheng, F. Patolsky et al, 2005, *Nat. Biotechnol.*, **23**, 1294–1301.
[11] G.-J. Zhang, M. Huang et al., 2010, *Biosens. Bioelectron.*, **26**, 365–370.

[12] F. Patolsky, B.P. Timko et al., 2006, *Science*, **313**, 1100–1104
[13] Patolsky, G. Zheng et al., 2004, *Proc. Natl. Acad. Sci. U. S. A.*, **101**, 14017–14022.
[14] Chen KI, Li BR, Chen Y., 2011, *Nano Today*, **6**, 131–154.
[15] Park I, Li Z, Pisano AP et al., 2010, *Nanotechnology*, **21**, 1–9.
[16] M.Nuzaihan MN, Hashim U et al., 2015, *Current Nanoscience*, 11, 239–244.
[17] Teo Boon K, Sun XH, 2006, *Journal of Cluster Science*, **17**(4), 529–540.
[18] Tran DP, Wolfrum B et al., 2014, *J. Mater. Chem. C*, **2**(26): 5229–5234.
[19] Nor MNM, Hashim U et al., 2010, *Am. Inst. Phys. Conf. Proc.*, **1217**, 272–278.
[20] Cui Y, Wei Q, Park H et al., 2001, *Science*, **293**, 1289–1292.
[21] Hsiao CY, Lin CH et al., 2009, *Biosensors and Bioelectronics*, **24**, 1223–1229.
[22] Zhang, Guo-Jun, Yong Ning, 2012, *Analytica Chimica Acta*, **749**, 1-15.
[23] G.-J. Zhang, J.H. Chua et al., 2009, *Biosens. Bioelectron.*, **24**, 2504–2508.
[24] J. Chua, R.E. Chee et al, 2009, *Anal. Chem.*, **81**, 6266–6271.
[25] R. D. Piner, J. Zhu et al., 1999 *J. Science* **283**(5402), 661.
[26] Johannes G. et al., 2003, *Interfacial Supramolecular Assemblies. Wiley*, 88–94.
[27] Kaifer, Angel, 2001, *Supramolecular Electrochemistry. Coral Gables. Wiley VCH*, 191–193.
[28] Love, J. Christopher, et al., 2005, *Chemical reviews,* **105.4**, 1103-1170.
[29] Naumova et al., 2010, *Sci. Technol.* **25**, 055004

Selective area epitaxy of n^+-GaN layers on SiO_2 patterned GaN/c-Al_2O_3 templates by PA MBE

K Yu Shubina[1], A M Mizerov[1], S N Timoshnev[1], D V Mokhov[1], E V Nikitina[1], I Kim[1,2], A D Bouravleuv[1]
[1]Nanoelectronics Lab, St. Petersburg Academic University, St. Petersburg 194021, Russia
[2]Department of Electronics and Nanoengineering, Aalto University School of Electrical Engineering, P.O. Box 13500, FI-00076 Aalto, Finland

Abstract. The n^+-GaN epilayers were synthesised by PA MBE on the SiO_2 patterned GaN/c-Al_2O_3 templates, grown by MOCVD. Formation of the polycrystalline GaN atop of the SiO_2 mask during PA MBE was observed. It was found that macroscopic voids at the interface polycrystalline GaN/SiO_2/n-GaN template appeared during the PA MBE process. The polycrystalline GaN film was completely removed by etching in hot aqueous KOH solution. Hall measurements have shown that the value of electron concentration in n^+-GaN contact layer is about n_e~4.6×10^{19} cm^{-3}.

1. Introduction

Wide bandgap semiconductors, such as gallium nitride (GaN), play a special role in a progress of modern electronics. Because III-nitrides have a set of unique structural, electrical, optical and mechanical properties, devices, based on GaN are successfully implemented in optoelecronics and have a great perspective for the high-power and high frequency electronic applications [1].

At the same time, there are several serious obstacles for further development of the nitride electronics. The first is a well-known problem, which is connected with the lack of low-cost natural substrates for GaN. So, silicon carbide (SiC), sapphire (c-Al_2O_3) and silicon (Si (111)) substrates are typically used for epitaxial growth of GaN. Heteroepitaxial growth of the material on the substrates with lattice mismatch leads to the formation of structural defects such as threading dislocations, inversion domains, etc. Another well-known challenge for further development of modern nitride electronics is the formation of the low resistance ohmic contacts [2]. The contacts are usually formed by the rapid thermal annealing of the sputtered contact material, so the morphology of the contact pad may suffer from high temperature annealing [3]. Nowadays a special attention is paid to the study of non-alloyed ohmic contacts. They can be formed on the heavily doped n^+-GaN contact layers, synthesized by selective area growth (SAG) [4, 5].

Here we report on the results of the studies of n^+-GaN contact layer synthesis by plasma-assisted molecular beam epitaxy (PA MBE) on SiO_2 patterned GaN/c-Al_2O_3 templates.

2. Experimental details

In this work the GaN layers, grown on c-Al_2O_3 substrate by metalorganic chemical vapour deposition (MOCVD) were used as templates. After the MOCVD growth the template surface was patterned with SiO_2 mask about 160 nm thick, formed by plasma-enhanced chemical vapour deposition (PECVD) and standard photolithography process. Thereafter the preliminary selective MOCVD growth of GaN layer in the window openings of SiO_2 mask was carried out.

Prior to the PA MBE growth several preliminary actions were made. The backside of c-Al_2O_3 substrate was covered with Ti film by e-beam vacuum deposition. Then the template was cleaned with organic solvents CCl_4 and isopropyl alcohol and rinsed in deionized water. In order to remove excess water from the template surface, a preliminary annealing at T ≈ 200°C was carried out in the UHV load lock chamber of PA MBE setup. After that, *in-situ* pre-epitaxial cleaning of GaN/c-Al_2O_3 template surface of atoms adsorbed from the atmosphere (O, C etc.) was performed. The details of pre-epitaxial cleaning of the template surface were previously discussed [6]. Finally, the PA MBE selective regrowth of the upper n^+-GaN contact layer was carried out using Veeco Gen 200 setup. The schematic view of the obtained sample is shown in figure 1.

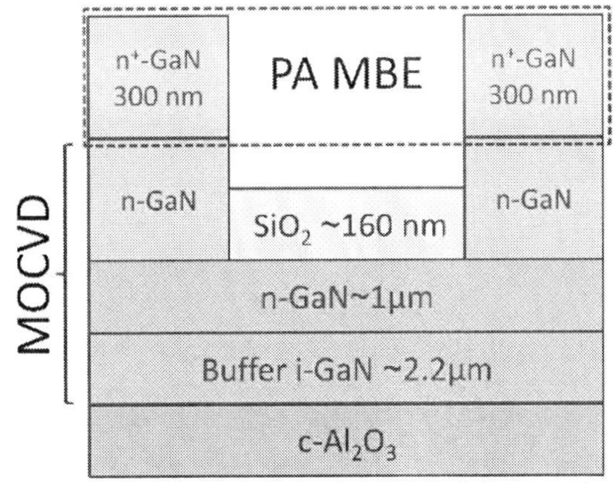

Figure 1. The schematic view of the sample.

The grown sample was studied using scanning electron microscopy (SEM) and Hall measurements. The polarity of the sample was identified by etching in hot aqueous KOH solution [7].

3. Results and discussion

The study of the synthesized sample by SEM has shown that the n^+-GaN contact layer, grown on the template GaN/c-Al_2O_3 patterned with SiO_2 mask, has inhomogeneous morphology and there are 3 different areas (figure 2a). The first consists of n^+-GaN layer, grown on the template in SiO_2 mask window openings. It demonstrates quite homogeneous surface morphology and smooth transition from MOCVD template to the n^+GaN contact layer, grown by PA MBE (figure 2b). The second area is the edge of the mask window. It was observed, that after the SiO_2 mask formation, the GaN layer formation occurred only at a distance of 15-17 µm from the mask (figure 2a). Finally, the third area consists of polycrystalline GaN (poly-GaN) layer, which was formed atop of the SiO_2 mask (figure 2c). However, some another works demonstrate, that the using of SiO_2 avoids GaN formation in the mask region [8, 9]. Moreover, it was found that the SiO_2 mask was destroyed during the PA MBE process, so multiple macroscopic voids were formed at the interface n-GaN/SiO_2/Poly-GaN and Poly-GaN layer was partially suspended (see figure 2c). It was suggested, that formation of the voids was caused by etching the mask either with Ga droplets due to high solubility of Si in Ga which takes place at a temperature T>400-650°C in ultra-high vacuum [10] or with high energy activated nitrogen species [11].

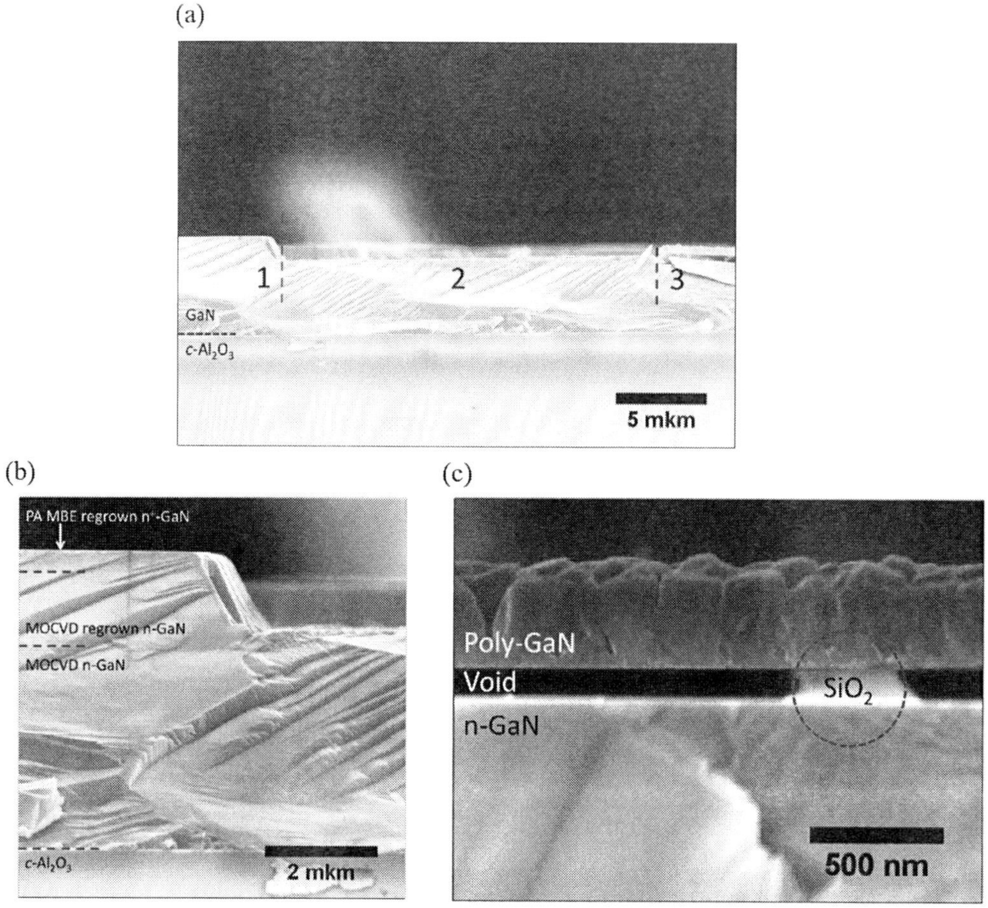

Figure 2(a, b, c). (a) cross-sectional SEM image of the grown sample. 1 - n^+-GaN layer grown directly on the template surface, 2 - the edge of the mask window, 3- poly-GaN atop of SiO_2 mask; **(b)** larger SEM image of the region 1; **(c)** larger SEM image of the region 2.

Then the sample was chemically etched in hot (70°C) aqueous KOH solution for 5 min in order to identify the crystallographic polarity of the GaN layer. During this process the polycrystalline GaN layer was slightly etched (figure 3b), thus it was decided to continue etching. After 10 min etching in hot KOH solution the poly-GaN was completely removed from the sample surface and and the remains of SiO_2 mask were found (figure 3c). At the same time, it was found that the n^+-GaN contact layer, formed by PA MBE, and the MOCVD template morphology remained smooth. Thus, they demonstrate Ga-polarity. Hexagonal etch pits, which were found on the GaN template surface (figure 3b, c), indicate the presence of dislocations. The studies of the grown sample by Hall measurements demonstrated that the value of electron concentration in n^+-GaN contact layer is about $n_e \approx 4.6 \times 10^{19}$ cm^{-3}.

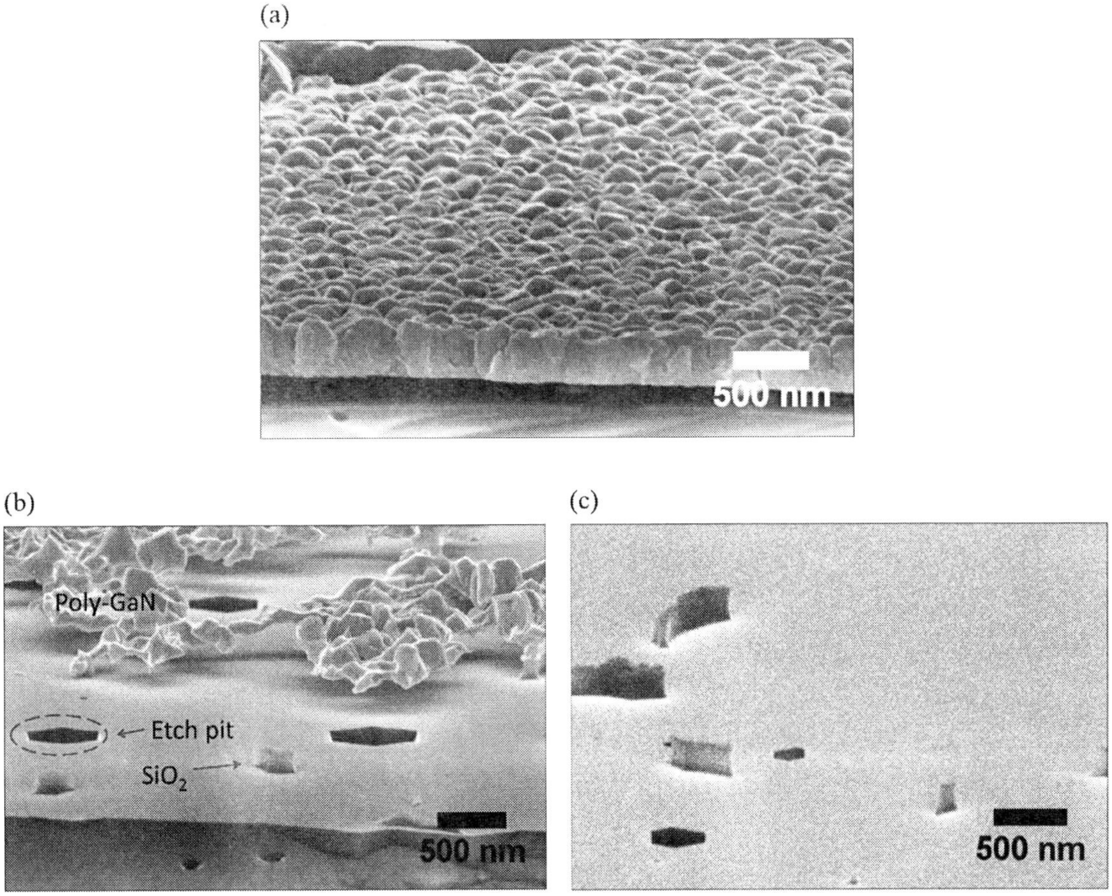

Figure 3(a, b, c). **(a)** Polycrystalline GaN before etching in KOH; **(b)** Etch time 5 min; **(c)** Etch time 10 min.

4. Conclusion

In this work the results of the studies of n+-GaN contact layer synthesis by SAG on SiO_2 patterned GaN/c-Al_2O_3 templates are presented. The n^+-GaN epilayers were synthesised by PA MBE on the templates, grown by MOCVD. The formation of the polycrystalline GaN atop of the SiO_2 mask during PA MBE was observed. It was found that SiO_2 mask can be partially destroyed during the PA MBE process with the consequent formation of the multiple macroscopic voids at the interface poly-GaN/SiO_2/MOCVD-GaN. The polycrystalline GaN film was completely removed by etching in hot aqueous KOH solution. Hall measurements have shown that the value of electron concentration in n^+-GaN contact layer is about $n_e \approx 4.6 \times 10^{19}$ cm^{-3}.

Acknowledgments

This work is supported by the grant of the Ministry of Education and Science of the Russian Federation № 16.9789.2017/BCh and Skoltech (agreement no. 3663-MRA, project №4).

References

[1] Amano H et al 2018 *J. Phys. D: Appl. Phys.* **51** 163001
[2] Wang W, Xie W, Deng Z, Yang H, Liao M, Li J, Luo X, Sun S, and Zhao D 2019 *Coatings* **9** 291

[3] Erofeev E V, Fedin I V, and Fedina V V 2018 *IOP Conf. Series: Journal of Physics: Conf. Series* **1115** 032068
[4] Pang L, Kim K 2014 *IOSR Journal of Engineering* **4** 8
[5] Zhelannov A V, Seleznev B I 2018 *IOP Conf. Series: Mat. Sci. and Engineering* **441** 012045
[6] Mizerov A M, Timoshnev S N, Sobolev M S, Nikitina E V, Shubina K Yu, and Bouravleuv A D 2019 *Proc. XXIII Int. Symp. "Nanophysics and Nanoelectronics" (Nizhny Novgorod)* vol 2 (Nizhny Novgorod: Publisher of Lobachevsky State University of Nizhni Novgorod) p 752
[7] Shubina K Yu, Berezovskaya T N, Mokhov D V, Mizerov A M, and Nikitina E V 2017 *Tech. Phys. Lett.* **43** 976
[8] Kruse J E *et al.* 2016 *J. Appl. Phys.* **119** 224305
[9] Calleja J *et al.* 2007 *Phys. Stat. Sol. (b)* **244** 2816–37
[10] Sugavaneshwar R P, Nagao T, and Nanda K K 2012 Materials Express 2 245
[11] Morkoç H 2009 *Handbook of Nitride Semiconductors and Devices* (Weinheim: Wiley – VCH)

Cylindrical thermal invisibility cloak based on transformation thermodynamics

V Y Soboleva[1], **A S Tukmakova**[2], **M V Sharaevsky**[1], **A V Vozianova**[1], **M K Khodzitsky**[1]

[1]Dept. of Photonics and Optical Information, ITMO University, Saint-Petersburg 197101, Russia

[2]Dept. of Cryogenic Engineering, ITMO University, Saint-Petersburg 197101, Russia

Abstract. The paper provides a theoretical proof that the methods of transformation optics can be used for thermodynamics. Thermal conductivity tensor is calculated for the development of the thermal invisibility cloak, which directs the heat flux around a certain area. The finite element method was used to simulate the cloak with parameters close to real. The simulation was performed by the finite element method for two cases: with gradient k and with copper/mica layered structure.

1. Introduction

Mathematical methods of transformational optics were proposed independently by two research groups: Ulf Leonhardt [1] and John Pendry [2, 3]. Using these methods, several exotic devices have been calculated, including the invisibility cloak, which directs the beam of a certain wavelength around the area, thereby making it invisible. Transformation optics concept was determined by Fermat's principle. The light propagation direction depends on the material refractive index. According to this, at present, transformation optics is a unique scientific tool that allows to combine the mathematical mapping of desired distortions of space with the actual special distribution of refractive index in physical space to control light propagation.

In 2012, Sebastien Guenneau proposed to use transformation optics for thermodynamics [4], where thermal conductivity acts as analogue of the optical properties (permittivity and permeability). In this work we describe the mathematical tools to calculate the thermal conductivity tensors for creation a cloak similar to the optical invisibility cloak, which directs the heat flow around the area, and provide a simulation of the device operation.

2. Mathematical methods and calculation

2.1. Transformation optics

Transformation optics uses physical space with arbitrary spatial coordinate system $\{x^i, i = 1,2,3\}$ and virtual one $\{x^{i'}, i' = 1,2,3\}$. Any optical medium distorts the geometry perceived by light (the same for medium with non-uniform thermal conductivity). This distortion is described by curved coordinate grid – physical space – in which the light follows along the coordinate lines (coordinate lines are isothermal contours for transformation thermodynamics). Physical space is transformed into a virtual space in which the effect of the medium has been fully accounted for by the shifting of the points. Virtual space demonstrates view, which is seen by observer (namely empty space). Physical and virtual spaces can be described by arbitrary coordinate system. Fig. 1 shows physical and virtual spaces of invisibility cloak in Cartesian and cylindrical coordinates.

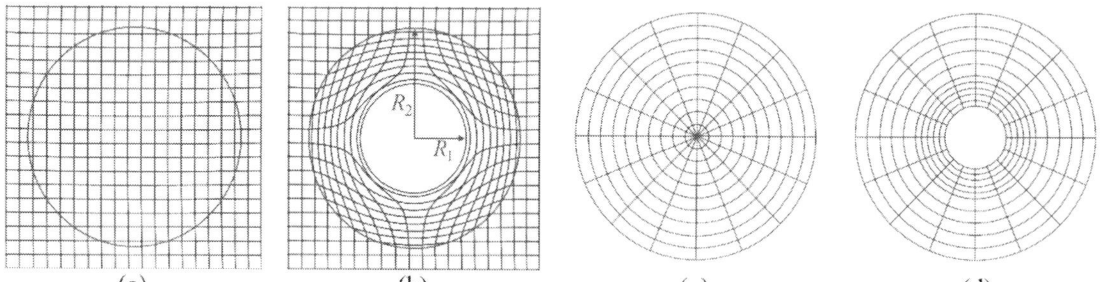

Figure 1(a, b, c, d). (a) The virtual space of the cloak in the Cartesian coordinates; **(b)** The physical space of the cloak in the Cartesian coordinates; **(c)** The virtual space of the cloak in the cylindrical coordinate system; **(d)** The physical space of the cloak in the cylindrical coordinate system.

2.2. Thermal conductivity tensor

To derive the thermal conductivity tensor $\boldsymbol{k}$, the two-dimensional heat equation for temperature distribution T with a source p is considered [4]. First, the equation is written for the virtual space in Cartesian coordinates $x^{i'}=\{x',y'\}$, where metric tensor is $\boldsymbol{g}'$=diag(1,1,1) (Fig. 1(a)):

$$\rho(\boldsymbol{x}')c(\boldsymbol{x}')\frac{\partial T}{\partial t} = \nabla\cdot(\boldsymbol{k}'(\boldsymbol{x}')\nabla T) + p(\boldsymbol{x}',t), \tag{1}$$

where $\rho(\boldsymbol{x}')$ – a density, $c(\boldsymbol{x}')$ – a heat capacity, $k'(\boldsymbol{x}')$ – a thermal conductivity, T – a temperature. To write the equation for the physical space in Cartesian coordinates $x^i=\{x, y\}$, where metric tensor g is a metric of curved space (fig. 1(b)), some differential geometry equations are used [5]:

$$\text{transformation matrix:}\quad \Lambda^i_{i'} = \frac{\partial \boldsymbol{x}}{\partial \boldsymbol{x}'} \tag{2}$$

$$\text{matrix equation:}\quad \boldsymbol{g}' = \boldsymbol{\Lambda}^T\boldsymbol{g}\boldsymbol{\Lambda} \tag{3}$$

$$\text{divergence of a vector field } \boldsymbol{V}\text{:}\quad \nabla\cdot\boldsymbol{V} = \frac{\mathbf{1}}{\sqrt{\det\boldsymbol{g}}}(\sqrt{\det\boldsymbol{g}}\,V^i)_{,i} \tag{4}$$

$$\text{components of the gradient vector of scalar field } \psi\text{:}\quad (\nabla\psi)^i = g^{ij}\psi_{,j} \tag{5}$$

where a comma means partial differentiation, with the following index giving the coordinate with respect to which the derivative is taken. It is possible now writing heat equation in curved physical space using formulas (2)-(5):

$$\sqrt{\det\boldsymbol{g}}\,\rho(\boldsymbol{x})c(\boldsymbol{x})\frac{\partial T_i}{\partial t} = (k_i(\boldsymbol{x})\sqrt{\det\boldsymbol{g}}\,g^{ij}T_{,j})_{,i} + \sqrt{\det\boldsymbol{g}}\,p(\boldsymbol{x},t), \tag{6}$$

Comparing this two equations, the equation for the thermal conductivity tensor for Cartesian coordinates can be obtained:

$$k^{ij} = \sqrt{\det\boldsymbol{g}}\,g^{ij}. \tag{7}$$

But in some cases it is convenient to use another coordinate systems, not only Cartesian. Then in equations (1) and (6) it is necessary to use also metric tensor of coordinate system $\boldsymbol{\gamma}$, which is calculated by the formula (3). For curvilinear system the thermal conductivity tensor is:

$$k^{ij} = \frac{\sqrt{\det \boldsymbol{g}}}{\sqrt{\det \boldsymbol{\gamma}}} g^{ij}. \quad (8)$$

From matrix equation (3) it can be obtained:

$$\sqrt{\det \boldsymbol{g}} = \frac{\sqrt{\det \boldsymbol{g}'}}{\det \boldsymbol{\Lambda}} \quad (9)$$

Using formulas (3) and (9) equation (8) is derived in matrix form:

$$\boldsymbol{k} = \frac{\sqrt{\det \boldsymbol{g}'}}{\sqrt{\det \boldsymbol{\gamma}}} \frac{\boldsymbol{\Lambda}(\boldsymbol{g}')^{-1}\boldsymbol{\Lambda}}{\det \boldsymbol{\Lambda}} k' \quad (10)$$

The similar equation are used for calculation permittivity and permeability in optics [5]. Here $\boldsymbol{g'}$ – a metric tensor of the virtual space, which coincides with the metric tensor of the coordinate system used, because virtual space isn't curved, $\boldsymbol{\gamma}$ – a metric tensor of the physical space coordinate system, $\boldsymbol{\Lambda}$ – a transformation matrix for transformation from physical to virtual space (Jacobian), $\boldsymbol{k'}$ – a thermal conductivity of the virtual space.

2.3. Thermal invisibility cloak

For the thermal cloak it is most convenient to use a cylindrical coordinates $\{r, \varphi, z\}$ (fig. 1(c, d)) with metric tensor γ of physical space and metric tensor g' of virtual space:

$$\boldsymbol{\gamma} = \mathrm{diag}(1, \frac{1}{r^2}, 1) \quad (11)$$

$$\boldsymbol{g}' = \mathrm{diag}(1, \frac{1}{r'^2}, 1) \quad (12)$$

Then the relationship between coordinates of the physical space and the virtual one is:

$$r(r') = \frac{(R_2 - R_1)}{R_2} r' + R_1,$$

$$\varphi(\varphi') = \varphi' \quad (13)$$

$$z(z') = z'$$

where R_2 – an exterior cloak radius, R_1 – an interior cloak radius (fig.1). Then the thermal conductivity tensor calculated by equation (10) is:

$$\boldsymbol{k} = \mathrm{diag}\left(\frac{r - R_1}{r}, \frac{r}{r - R_1}\right) \quad (14)$$

3. Simulation

The simulation was performed by the finite element method in the COMSOL Multiphysics software. The scheme of simulation is shown in fig. 2 (a). Firstly, temperature distribution for gradient anisotropic thermal conductivity (Eq. (14)) for $R_1 < r < R_2$ was simulated. A silicon with $k' = 130$ W·(m·K)$^{-1}$ was chosen for substrate ($r > R_2$). The object radius is $R_1 = 2$ μm and the cloak radius is $R_2 = 4$ μm.

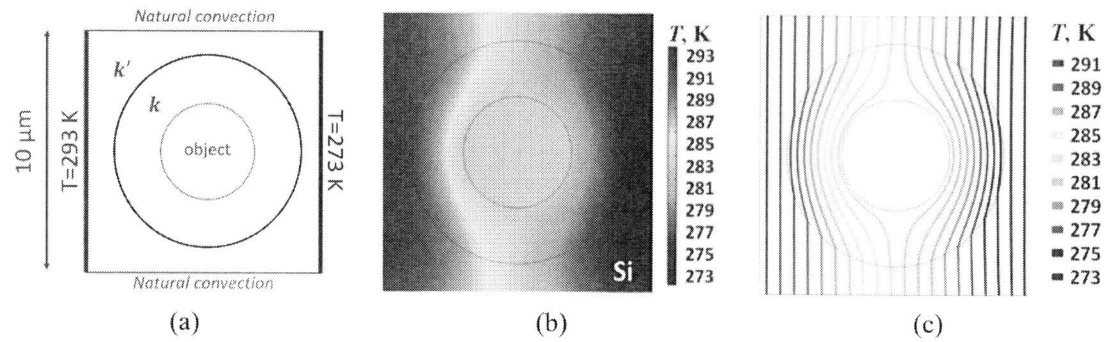

Figure 2(a, b, c). **(a)** The scheme of the simulation; Invisibility cloak for gradient thermal conductivity tensor *k*: **(b)** Temperature distribution; **(c)** Isothermal contours.

The object inside thermal invisibility cloak keeps constant temperature 283 K in device with gradient *k* calculated with transformation thermodynamics mathematical methods.
It is obvious that material with anisotropic gradient thermal conductivity is difficult to make so it was suggested in the paper [6] to use layered structure with real materials (fig. 3(a)). It is material with thermal conductivity lower then substrate on the inner radius $R_1 < r < R_2$ ($k_1 < k'$) and material with thermal conductivity higher than substrate on the outer radius $R_2 < r < R_3$ ($k_2 > k'$).

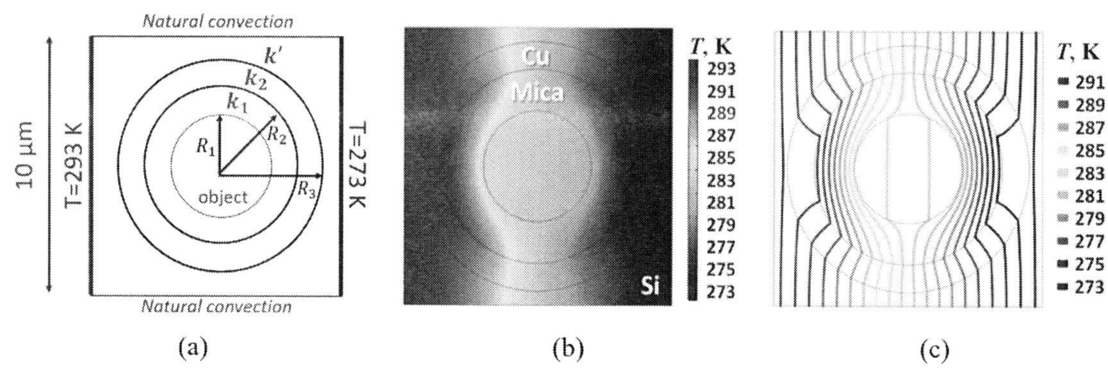

Figure 3(a, b, c). **(a)** Scheme of simulation; Invisibility cloak for layered structure with difference thermal conductivities k_1 and k_2: **(b)** Temperature distribution; **(c)** Isothermal contours.

In this work mica and copper are used in the inner and outer radii respectively and the silicon was used for the substrate. Mica thermal conductivity is $k_1 = 0.5$ W·(m·K)$^{-1}$ and copper $k_2 = 402$ W·(m·K)$^{-1}$. k_1 and k_2 agrees with conditions, that $k_1 < k'$ and $k_2 > k'$. The object radius is $R_1 = 2$ μm, the inner radius is $R_2 = 3.5$ μm and the outer radius is $R_3 = 4$ μm. Temperature distribution and isothermal contours demonstrated in fig. 3 (b, c). In device with layered structure (fig. 3) the object inside thermal invisibility cloak has temperature gradient of 3 K.

4. Conclusion

The mathematical methods of transformation optics were applied for thermodynamics. Thermal conductivity coefficient from thermodynamics is analogy of the optical refractive index. Thermal conductivity tensor was calculate for thermal invisibility cloak, which directs the heat flux around a certain area. The simulation was performed by the finite element method for two cases: with gradient *k* and with layered structure: material with $k < k'$ (mica) on the inner radius and another one with $k > k'$ on the outer radius (copper). The object in the thermal invisibility cloak keeps the constant temperature in the first case with gradient thermal conductivity *k* and it has temperature gradient of 3 K in the second case with the layered cloak.

References

[1] Leonhardt U 2006 *Science* **312** 1777-1780
[2] Pendry J B, Schurig D, Smith D R 2006 *Science* **312** 1780-1782
[3] Schurig D, Pendry J B, Smith D R 2006 *Opt. Express* **14(21)** 9794-9804
[4] Guenneau S, Amra C, Veynante D 2012 *Opt. Express* **20(7)** 8207-8218
[5] Leonhardt U and Philbin T 2010 *Geometry and Light: The Science of Invisibility* (Mineola, New York: Dover Publications, Inc.)
[6] Han T, Bai X, Gao D, et al. 2014 Physical Review Letters 112(5) 054302.

Structure and electronic properties of 4-8 and 4-6-12 layered varieties of boron nitride

Dmitry S Ryashentsev[a] and Evgeny A Belenkov[b]
[a,b]Department of Chemistry, Chelyabinsk State University, Chelyabinsk 454001, Russia

E-mail: ryashentsev_dmitry@mail.ru[a], belenkov@csu.ru[b]

Abstract. The structure of two new layered polymorphic varieties of boron nitride was calculated by the density functional theory method using the gradient approximation. The structure of layers BN-L_{4-8} and BN-L_{4-6-12} differs from the structure of the hexagonal layer of boron nitride in that their surface is formed not only from hexagons. As a result of the calculations, it was found that the sublimation energy of the new layers (17.36 and 17.14 eV $(BN)^{-1}$ for the layers BN-L_{4-8} and BN-L_{4-6-12}, respectively) is less than the sublimation energy of hexagonal boron nitride (18.14 eV $(BN)^{-1}$). The band gap of new boron nitride polymorphs is about 3.9 eV, which is less than the calculated width of the band gap in BN-L_6 – 4.69 eV.

1. Introduction

Boron nitride compounds have a structure similar to the structure of carbon compounds, since interatomic bonds in boron nitride compounds are predominantly covalent [1,2]. Therefore, there are BN-phases with a structure similar to a cubic and hexagonal diamond, as well as structural varieties of boron nitride with a layered structure similar to the structure of graphene [3-7]. To date, only a hexagonal layered polymorph of boron nitride, similar to hexagonal graphene, has been synthesized. However, the existence of various polymorphic graphene varieties is possible, such as: L_{4-8}, L_{3-12} and L_{4-6-12} [8-10]. In this work, the possibility of the existence of similar structural species of boron nitride was theoretically investigated.

2. Methods

Theoretical analysis of the structure of layered boron nitride compounds showed that, unlike graphene, only two new basic structural varieties are possible: BN-L_{4-8} and BN-L_{4-6-12}. The BN-L_{3-12} layer from boron and nitrogen atoms cannot be formed, since for the formation of a stable structure it is necessary that each atom of one type forms bonds with three adjacent atoms of the another type. In the BN-L_{3-12} layer, this is impossible due to the presence of triangular structures. Geometric optimization of new polymorphic varieties of boron nitride was performed using the density functional theory (DFT) method. The calculations used the generalized gradient approximation (GGA). Calculations of the structure and electronic properties were performed for three-dimensional crystals consisting of layers of boron nitride packed in stacks. The distance between the layers was chosen to be 10 Å. The calculations were performed for a set of k-points: $12\times 12\times 12$. The sublimation energy was calculated as the difference between the total energy per BN molecular group in the boron nitride layer and the energy of isolated boron and nitrogen atoms.

3. Results

As a result of the DFT-GGA calculations, the structural parameters of the boron nitride layers were found. The crystal lattice of the BN-$L_{4\text{-}8}$ layer belongs to the Bravais square lattices, and the BN-$L_{4\text{-}6\text{-}12}$ layer to the hexagonal lattices. The lengths of the translation vectors are 4.942 and 6.853 Å; the unit cells contain 8 and 12 atoms for the BN-$L_{4\text{-}8}$ and BN-$L_{4\text{-}6\text{-}12}$ layers, respectively (figure 1). The atomic coordinates in the unit cells are shown in table 1.

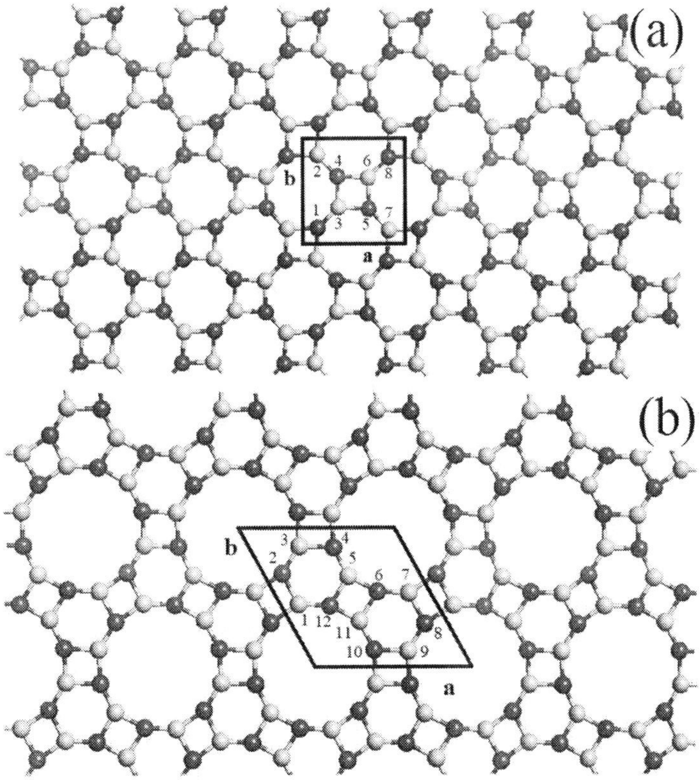

Figure 1. Structure of layers and unit cells of polymorphic varieties of boron nitride: (a) BN-$L_{4\text{-}8}$; (b) BN-$L_{4\text{-}6\text{-}12}$.

Table 1. Atomic coordinates in the unit cells of layered polymorphic varieties of boron nitride.

№	Atom	X, a	Y, b	№	Atom	X, a	Y, b
			BN- $L_{4\text{-}8}$				
1	N	0.1358	0.1357	5	N	0.6358	0.3235
2	B	0.1235	0.8358	6	B	0.6235	0.6234
3	B	0.3359	0.3358	7	B	0.8359	0.1234
4	N	0.3235	0.6357	8	N	0.8235	0.8235
			BN-$L_{4\text{-}6\text{-}12}$				
1	B	0.1183	0.4548	7	B	0.8787	0.5439
2	N	0.1256	0.6799	8	N	0.8714	0.3188
3	B	0.3341	0.8796	9	B	0.6629	0.1192
4	N	0.5519	0.8723	10	N	0.4451	0.1265
5	B	0.5431	0.6638	11	B	0.4539	0.3350
6	N	0.6790	0.5528	12	N	0.3180	0.4459

The layer density of the new layered structural varieties of boron nitride BN-L_{4-8} and BN-L_{4-6-12} is 0.675 and 0.608 mg m^{-2}, which is less than the analogous characteristic for the BN-L_6 layer (0.754 mg m^{-2}) (table 2). The specific energy of sublimation calculated for the L_{4-8} layer is 17.36 eV $(BN)^{-1}$, and for the L_{4-6-12} layer – 17.14 eV $(BN)^{-1}$. These values are less than the sublimation energy 18.14 eV $(BN)^{-1}$ calculated for the layer of hexagonal boron nitride. Apparently, this is due to the fact that in non-hexagonal BN layers, the angles between the bonds are significantly deformed compared with the values characteristic of the hexagonal layer. Therefore, hexagonal boron nitride should be the most stable structural variety. However, the sublimation energies of new BN polymorphs are high enough so that these layers can stably exist under normal conditions.

Table 2. Structural parameters and properties of polymorphic varieties of layered boron nitride (a – translation vectors, N – number of atoms in the unit cell, ρ – layer density, E_{total} – total energy, E_{sub} – sublimation energy, E_F – Fermi energy, Δ – band gap).

Parameters	BN-L_{4-8}	BN-L_{4-6-12}	BN-L_6
Crystal family	Tetragonal	Hexagonal	Hexagonal
a (Å)	4.942	6.853	2.512
N	8	12	2
ρ (mg m^{-2})	0.675	0.608	0.754
E_{total} (eV $(BN)^{-1}$)	-352.66	-352.43	-353.43
E_{sub} (eV $(BN)^{-1}$)	17.36	17.14	18.14
E_F (eV)	-4.163	-4.168	-3.764
Δ (eV)	3.894	3.895	4.686

In addition, the band structure and density of electronic states of BN-L_{4-8} and BN-L_{4-6-12} polymorphic varieties of boron nitride were calculated (figures 2, 3). The band gap at the Fermi energy level for hexagonal boron nitride is 4.69 eV, while for the BN-L_{4-8} and BN-L_{4-6-12} layers, the band gap is less – 3.89 eV and 3.90 eV, respectively. Thus, the electronic properties of polymorphic varieties of boron nitride can vary significantly depending on the characteristics of their structure.

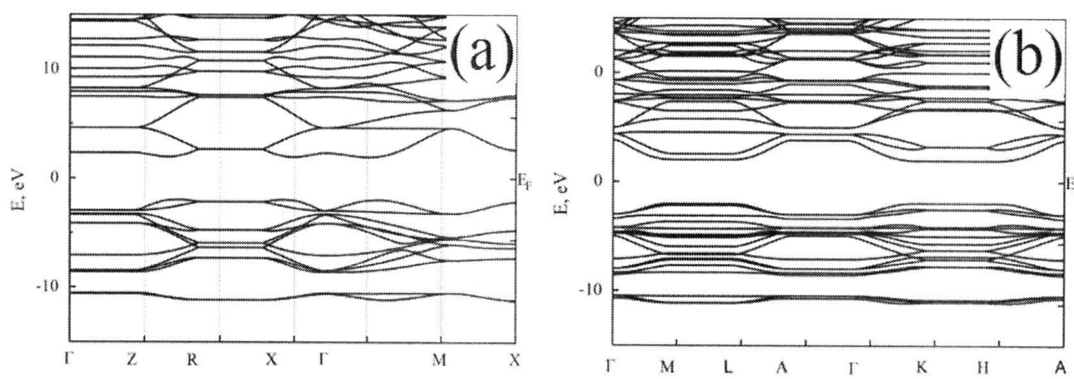

Figure 2. Band structure of polymorphic varieties of boron nitride: (a) BN-L_{4-8}; (b) BN-L_{4-6-12}.

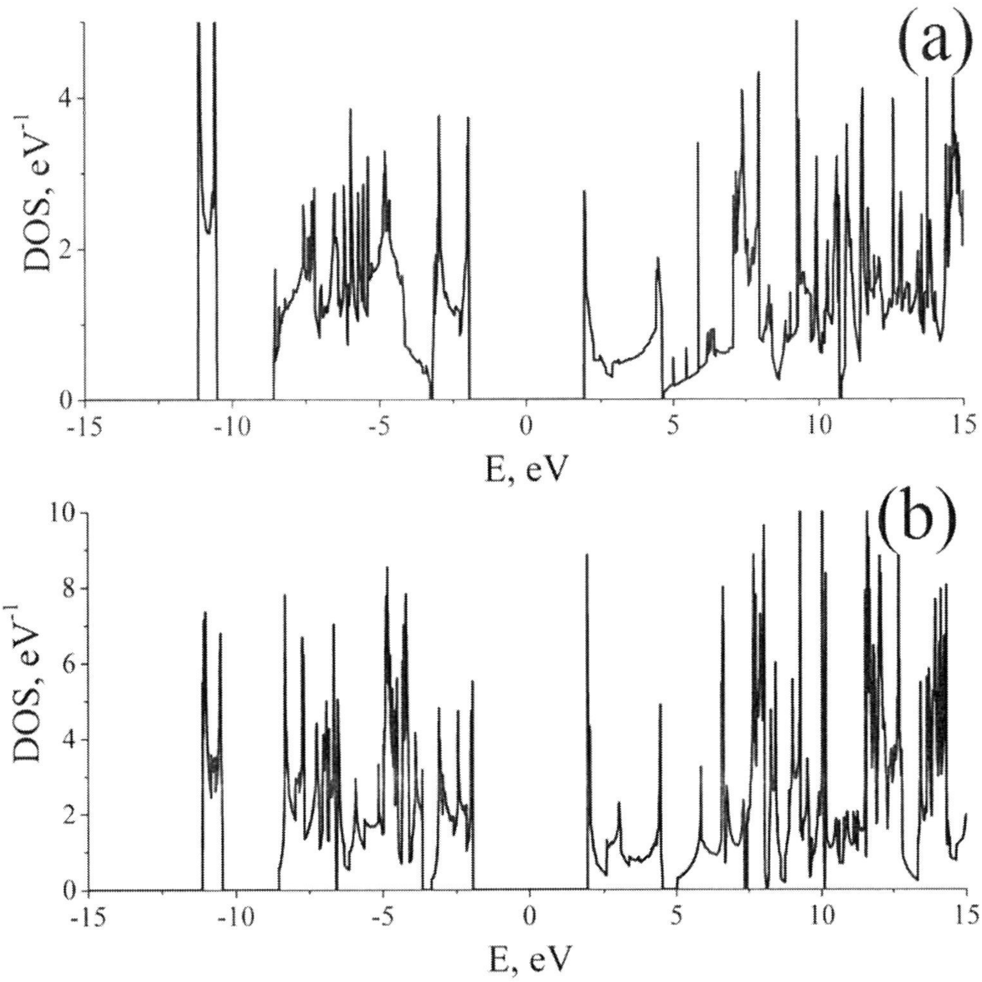

Figure 3. Density of electronic states (DOS) of polymorphic varieties of boron nitride: (a) BN-$L_{4\text{-}8}$; (b) BN-$L_{4\text{-}6\text{-}12}$.

Acknowledgments
This research was supported by Foundation for Perspective Scientific Research of Chelyabinsk State University.

References

[1] Neuberger M 1971 *III–V Semiconducting Compounds* (New York: Springer US)
[2] Weng Q, Wang X, Bando Y and Golberg D 2016 *Chem. Soc. Rev.* **45** 3989
[3] Chrenko R M 1974 *Solid State Commun.* **14** 511
[4] Huang M and Ching W Y 1985 *J. Phys. Chem. Solids* **46** 977
[5] Ferhat M, Zaoui A, Certier M and Aourag H 1998 *Physica B* **252** 229
[6] Christensen N E and Gorczyca I 1994 *Phys. Rev. B* **50** 4397
[7] Hoffmann D M, Doll G L and Ekiund P C 1984 *Phys. Rev. B* **30** 6051
[8] Belenkov E A and Greshnyakov V A 2013 *Physics of the Solid State* **55** 1754
[9] Belenkov E A and Kochengin A E 2015 *Physics of the Solid State* **57** 2126
[10] Belenkov M E, Chernov V M, Belenkov E A and Morilova V M 2018 *IOP Conference Series: Materials Science and Engineering* **447** 012005

Synthesis and transport properties of FET based on Heusler alloy thin films formed by rapid thermal annealing

A V Lukyanenko[1,2], A S Tarasov[1,2], L V Shanidze[1,2], I A Yakovlev[1], F V Zelenov[1], A N Masugin[3], A B Ivanov[1], F A Baron[1] and N V Volkov[1]
[1] Kirensky Institute of Physics, Federal Research Center KSC SB RAS, Krasnoyarsk 660036, Russia
[2] Institute of Engineering Physics and Radio Electronics, Siberian Federal University, Krasnoyarsk 660041, Russia
[3] Reshetnev Siberian State University of Science and Technology, Krasnoyarsk 660037, Russia

Abstract. In this work we show a preparation technique of Co_2FeSi full-Heusler alloy thin films on silicon-on-insulator (SOI) substrates, employing rapid thermal annealing (RTA). The films of the Co_2FeSi alloy were formed by a silicidation reaction, caused by RTA, between the ultrathin SOI (001) layer and the Fe/Co layers deposited on it. It is assumed that this technology is compatible with the process of formation of a half-metal source-drain in an advanced CMOS and SOI technology and will be applicable for the manufacture of a source-drain of a field-effect transistor. Schottky barrier field-effect transistors (FET) with a back-gate, based on silicon nanowires with source and drain of a Co_2FeSi film, synthesized on an SOI substrate, were manufactured. The transport properties of the device were investigated.

1. Introduction

Silicon nanowires (SiNW) and devices based on it continue to attract researchers for a long time. The properties of NW allow creating a variety of devices on its basis for a wide range of applications. The use of NW as components for future nanoelectronic devices seems very attractive from the point of view of the possibility of integration into the modern silicon technology. Silicon nanowires (FET NW) field-effect transistors are increasingly used not only in the basic research of physical processes in small sizes, but also as highly sensitive sensors for detecting a wide range of molecules. Today MOS-based sensors are widely distributed as commercial ones. But considered in this work, FET-type sensors have more advantages compared to traditional ones, due to their reduced shape, size and lower price.

For the formation of nanoscale wire current channel, there are two main approaches "bottom-up" and "top-down." The bottom-up approach implies the use of self-assembly processes. Among a number of experimental studies [1, 2], devices based on silicon nanowires were fabricated using a bottom-up approach, which often encounters difficulties in controlling the doping process and the formation of contacts. But the manufacture of NW on site, without the need for their transposition is possible with the use of modern infrastructure of silicon production. Devices based on SiNW, made using the top-down approach, show better control and reproducibility in these parameters. And within the framework of simple integration, as well as the possibility of subsequent implementation, the most promising is the use of a top-down approach, using existing silicon technologies.

At the moment, the focus is on traditional structures of the field-effect transistor with *n-p-n*-doping, which requires complex technological stages of doping [3, 4]. As part of this work, we will demonstrate that nanowires with Schottky contacts can be used as field effect transistors. They are of great interest as an alternative to traditional structures with doped source and drain. This approach avoids the need for doping by replacing source / drain [7] transitions with metal / semiconductor transitions, leaving the nanowire undopped [8]. Field-effect transistors with a Schottky barrier have a number of advantages, including simple and affordable manufacturing technology, as well as bypassing complex manufacturing problems, such as precise control of the type and level of doping, as well as the formation of reliable ohmic contacts.

The use of spin degrees of freedom for the transfer, storage and processing of information will allow to make another qualitative leap in electronics. The open question is the creation of the active element of spin electronics. From the point of view of technology and integration, it would be attractive to use FET geometry for the implementation of a spin transistor. Ferromagnetic metal should provide high spin polarization of conduction electrons. It is technologically most convenient to use 3d metals and their alloys. Iron silicides or Heusler alloys containing Si are good candidates to be used as a FM injector [5]. One of the advantages is the possibility of the formation of the drain and source of the field-effect transistor using a controlled silicidation process, induced by rapid thermal annealing (RTA). When using silicide, abrupt metal / nanowire transitions can be achieved. Consequently, local changes in the contact area as well as in geometry are avoided. For charge transfer this means that two energy barriers are introduced in the current path. The thickness of these barriers is effectively controlled by the gate due to the enhancement of the electric field in the area of the end metal electrode. The on state for a sufficiently small gate length is specified by tunneling through the Schottky barriers, while the off state is controlled by thermionic emission through a high and thick energy barrier. The possibility of the formation of silicon nanowires in the downstream process from top to bottom using bulk silicon substrates or silicon on an insulator gives this technology the possibility of full integration into integrated electronic systems.

SiNW FET for spintronic devices will be very sensitive to the quality of the interfaces between the drain / source and nanowires. A necessary condition for the presence of ferromagnetic elements leads to a number of restrictions. Using the top-down process of forming the current channel, in combination with the RTA process of forming the drain / source of FET, provide the necessary quality and ease of manufacture of devices. The method of forming drain / source contacts from Heusler Co_2FeSi alloys on silicon-on-insulator-type substrates using RTA will bypass the problem of introducing ferromagnetic elements into semiconductor technology and the quality problems of interfaces between ferromagnetic and semiconductor elements.

2. Experimental

Nanowire (NW) devices are formed on commercially available boron doped silicon on insulator (Simox SOI) substrates (with resistivity of 18 Ω cm) using Raith VOYAGER e-beam writer. The formation of the conductive channel and contact pads was carried out from the upper (device) silicon layer, using the common AZ 2035 nLof resist at different exposure doses for current channels and areas of the contact pads of Figure 1 (a). Then, to achieve the desired silicon topology, a dry chemical reactive-ion etching (RIE) stage followed. The etching of silicon over the entire depth (Si - 100 nm) to the dielectric layer (SiO_2 - 200 nm) occurred pulsed for 10 seconds in the plasma of a CF_4 gas. To thin the silicon layer, preliminary thermal annealing was performed at a temperature of 1000-1100 °C in an atmosphere of dry O_2 at atmospheric pressure, to form SiO_2. The stage of optical lithography is necessary for the formation of windows on the contact pads (drain / source) before the Co and Fe deposition process. Previously, an etching of a SiO_2 film in an HF solution in the area of contact pads was performed. Then, in one technological cycle, Co and Fe films (45 nm and 24 nm, respectively) were deposited by electron beam sputtering (EBS) according to the choice of the stoichiometric composition of the compound in Figure 1 (b). After the lift-off lithography, the substrate was thoroughly cleaned of residual resistive mask. Co_2FeSi alloy films are formed by a silicidation reaction induced by rapid thermal annealing (RTA) between the ultra-thin Si (001) layer and the Fe / Co layers deposited on it. The RTA process was carried out at 700-800 °C in N_2 atmosphere for 4 minutes to form Co_2FeSi Figure 1 (c). The layer of buried SiO_2 oxide SOI substrate prevents the diffusion of metals into the substrate during the silicidation process [9]. As a result, the devices represent itself silicon nanowires with half-metal drain and source (Figure 1 - (c)). A series of devices with a current channel width from 300 nm to 3000 nm was manufactured. A typical atomic force microscope (AFM) image of the NW current channel of 500 nm is shown on inset to Figure 1. AFM data exhibit good NW smoothness and the verticality of the side walls of the current channel. To create the back gate of the field-effect transistor, an ohmic contact of indium was formed on the back side of the SOI substrate.

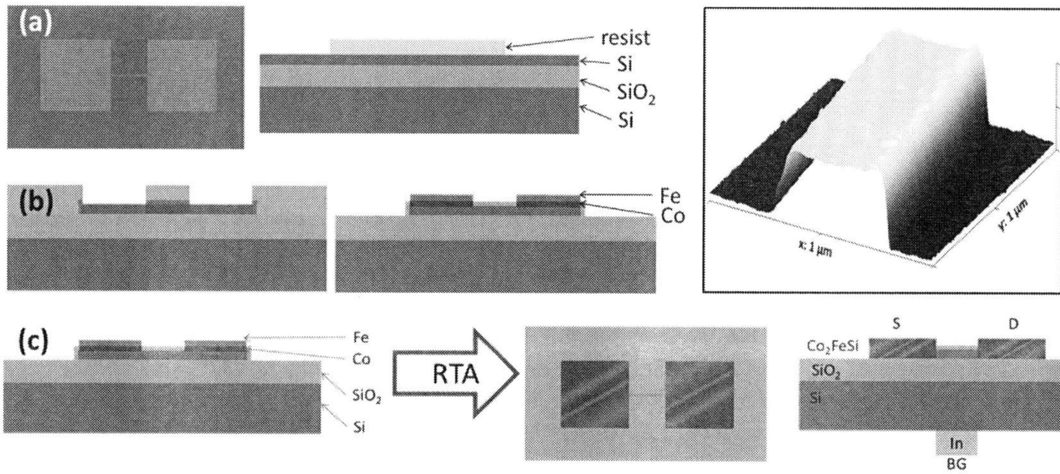

Figure 1(a, b, c) The main technological stages of the formation of devices (a) - (c). Inset: AFM image of a 500 nm wide device.

3. Results and discussion

The devices obtained measured the transport properties in EMPX-HF probe station (LakeShore cryotronics). For Co_2FeSi /SOI back-gate nanowire transistor transport properties were studied. Transfer characteristics demonstrate behavior like pseudo MOSFET or nanowire FET (NWFET) with Schottky source / drain contacts. The transport measurements of the device obtained, as can be seen from Figure 2, demonstrate the typical behavior of an NWFETs. In the accumulation mode, the Drain current between on-state and off-state differs by about 4 orders of magnitude for all devices.

The devices show ambipolar characteristics with both *p*- and *n*-type conductions depending on the gate bias. However, at a gate voltage of 13V, the drain current in the inversion mode is approximately 3 orders of magnitude less than the drainage current in the accumulation mode. This is clearly seen in Figure 2(b), where presented drain current versus drain voltage characteristics.

In addition, from Figure 2(a) it can be seen that the conductivity of the nanowire is not monotonously dependent on its width. The maximum drain current is observed for a device with a 700 nm wide nanowire. This result was unexpected for us and requires additional research beyond the scope of this article.

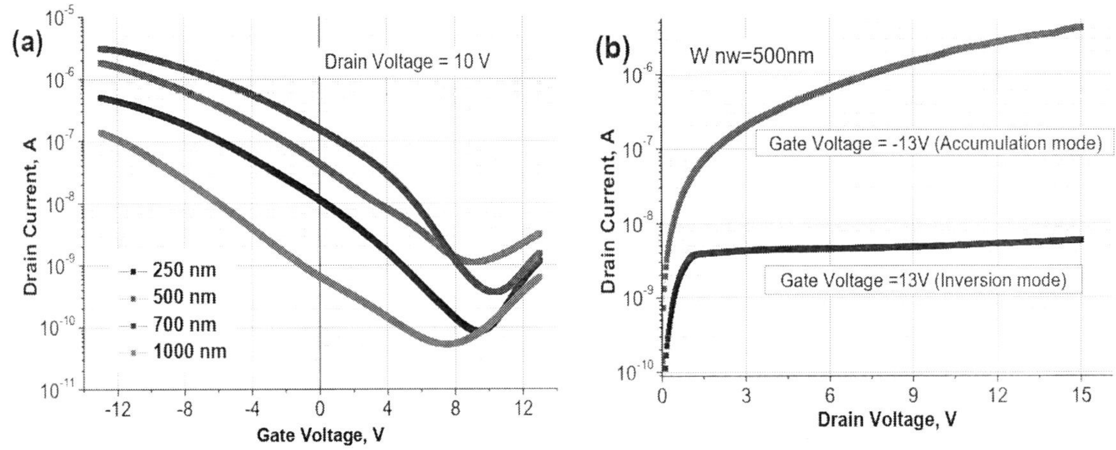

Figure 2(a, b). (a) The transfer characteristics Co_2FeSi/SOI back-gate nanowire transistor with a channel width of width of 250 nm to 1000 nm plotted on a semilog scale at different voltages on source-drain; **(b)** Output characteristics device in inversion and accumulation mode with a channel width of 500 nm.

The use of geometry with a back gate leaves the current channel open and allows the use of NWFET as a highly sensitive electrical sensor. The main sensitive mechanism of a simple sensor based on NWFET, includes the adsorption and isolation of molecules on the surface of the nanowire, which affects the electrical characteristics of the transistor [10].

The use of a top-down process of current channel formation combined with the RTA process of formation of a source/drain of a transistor provides the necessary quality and simplicity of device creation.

Acknowledgments

This study was supported by the Russian Foundation for Basic Research, project no. 17-02-00302 and supported in part by the Ministry of Education and Science of the Russian Federation and the Siberian Branch of the Russian Academy of Sciences, project II.8.70, and the Presidium of the Russian Academy of Sciences, Fundamental Research Program no. 32 «Nanostructures: Physics, Chemistry, Biology, Basics of Technologies».

References

[1] Cui Y et al 2003 Nano Lett. 3 149
[2] MaD D D et al 2003 Science 299 1874
[3] Allibert F et al 2001 Solid State Electron. 45 559
[4] Hu S F et al 2004 IEEE Trans. Nanotechnol. 3 93
[5] K. Hamaya, et al. 2008 Appl. Phys. Lett. **93** 132117
[6] K. Inomata, S. Okamura, A. Miyazaki, et al, 2006 J. Phys. D **39**, 816
[7] W. Weber et al., 2006 Nano Lett. **6**(12), 2660–2666
[8] W. Tang et al., 2012 Nano Lett. **12**(8), 3979–3985
[9] Y. Takamura, et al. 2008 Appl. Phys. **103** 07D719.
[10] Y. Huang, al., 2001, Science 294 1313–1317

Laser modification of high-pressure high-temperature nanodiamonds

K G Mikheev[1], T N Mogileva[1], A E Fateev[1], N Nunn[2], O A Shenderova[2], G M Mikheev[1]
[1]Institue of Mechanics, Udmurt Federal Research Center, UB RAS, Izhevsk 426067, Russia
[2]Adamas Nanotechnologies, Inc., Raleigh, North Carolina 27617, United States

Abstract. Micro-thin semitransparent films of purified high pressure high temperature nanodiamonds were produced on a glass substrate. A focused 632.8 nm He-Ne laser irradiation is shown to cause an emission of light in the green region of the optical spectrum accompanied with laser-induced blackening of the obtained films. Possible mechanisms of the observed phenomenon are discussed.

1. Introduction

Recently, investigations of laser radiation interactions with carbon nanostructures revealed interesting and useful phenomena [1–4]. One of most interesting carbon nanostructures is nanodiamonds (NDs), which are attractive for many applications due to their unique mechanical, chemical, thermal and optical properties [5]. For example, use of focused laser radiation can trigger physical and chemical processes that lead to a local laser-induced modification of the surface of the ND material [6]. Such modification may be of interest for obtaining various microstructured optical plates, which are diffractive optical elements, various scales and grids, limbs, rasters, etc [7–9].

It is well known that NDs synthesized using hydraulic presses at high pressure and high temperature (HPHT) [10] contain nitrogen-vacancy [N-V] defects (neutral $[N\text{-}V]^0$ and negatively charged $[N\text{-}V]^-$), which consist of one nitrogen atom N in a substitutional position and one carbon vacancy [V] [11]. These centers show particularly strong electronic transitions, allowing optical detection of individual centers [12]. The HPHT NDs can also contain a number of nickel related optical centers [13,14]. These centers can affect the laser modification of HPHT NDs, depending on the wavelength. In this paper, we show that low-power laser irradiation of HPHT ND films at 632.8 nm leads to transformation of the sp^3-carbon of NDs into sp^2-carbon. Such transformation is accompanied by luminescence in the green spectral region and a significant decrease in the luminescence level in the Stokes region. This phenomenon can be explained by photoionization of [Ni-N] and [N]-centers leading to the formation of electrons in the conduction band of the ND film and an increase of the local temperature of the sample transforming the NDs into sp^2-carbon.

2. Characterization of the films

HPHT NDs with an average size of 150 nm were used in the experiments. The particles obtained from the vendor (Diamond Innovations, USA) were first purified from sp^2-carbon by treatment in air at 500 °C for 2 hours followed by treatment in HCl at 70 °C for 2 hours. The particles were dispersed in deionized water at 1 wt% concentration and the suspensions were stored in plastic vessels. Films with a thickness of 20 μm (see figure 1a) were prepared by drying the ND suspension on a glass substrate at ambient conditions. Figure 1b shows the extinction spectrum of the 0.5 μm-thick HPHT ND film on a quartz substrate, recorded relative to an identical quartz substrate. The spectrum is determined by the absorption and scattering of light by the film nanoparticles and film inhomogeneities which arise due to the method of production not ensuring the uniformity of the film thickness. It is seen that the extinction spectrum has a maximum at a wavelength of 208 nm, due to single-photon absorption between the valence band and the conduction band of diamond. The photon energy corresponding to this maximum is 5.96 eV, which is higher than the energy of the forbidden band gap of a bulk diamond (5.6 eV). Figure 1b also shows that with increasing wavelength, the extinction of the film decreases monotonically, which is explained by the decrease in Rayleigh light scattering by ND particles and film inhomogeneities with the increasing light wavelength.

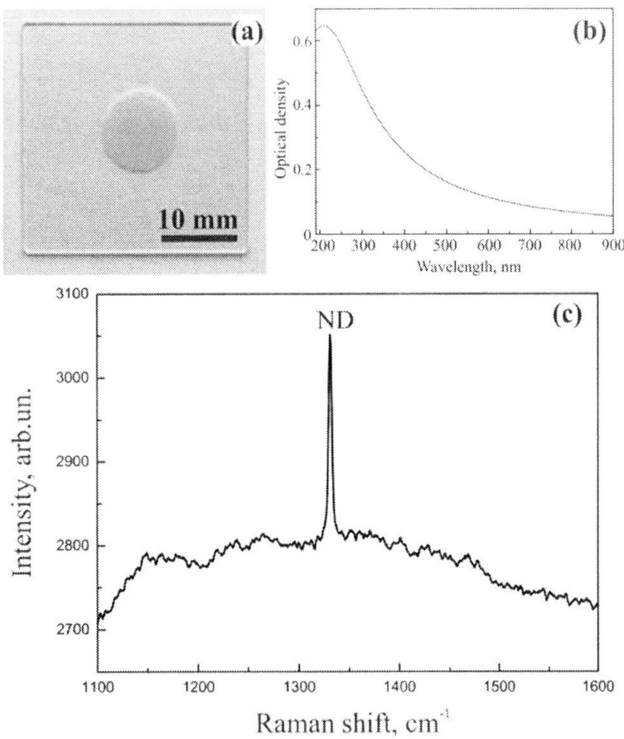

Figure 1 (a, b, c). (a) Photo of 20 μm thick HPHT ND film on a quartz substrate; **(b)** Extinction spectrum of the 0.5 μm-thick HPHT ND film on a quartz substrate, recorded relative to an identical quartz substrate; **(c)** Raman spectrum of ND film.

Figure 1c shows the Raman spectrum of an arbitrarily selected point on the ND film measured by a Horiba HR800 Raman spectrometer at $P = 5$ kW/cm^2, where P is excitation power density of 632.8 nm laser radiation. It can be seen that the presented spectrum consist of a broad luminescence band and one sharp peak with a frequency shift of about 1331 cm^{-1}, which coincides with the frequency shift of bulk crystalline diamond with sp^3-hybridization [15]. Note that Raman shifts of NDs depend on the exciting laser wavelength and the average size of nanoparticles [16,17], and varies in the range 1325-1331 cm^{-1}[18].

Figure 2a shows a typical image of the surface of the ND film, obtained with a scanning electron microscope (SEM) Inspect S50. One can see the granular structure of the film surface. The characteristic size of the particles in the suspension used to prepare the film measured by dynamic light scattering using a Malvern Zetasizer ZS device is about 150 nm (see figure 2b).

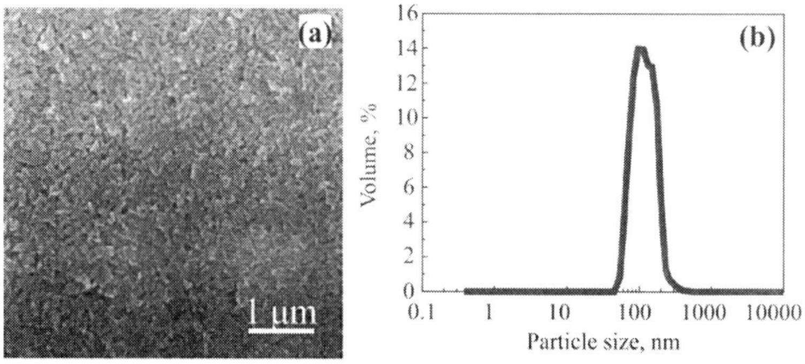

Figure 2 (a, b). (a) SEM image of the HPHT ND film surface; **(b)** the ND particles size distribution.

3. Experimental section

Experiments on laser modification of HPHT ND films were carried out using a He-Ne laser at the wavelength of 632.8 nm, embedded in a Horiba HR800 Raman spectrometer in a backscattering geometry. In these experiments the HPHT ND film was located on the coordinate stage (see figure 3a). Using an objective, laser radiation was focused on the surface of the film. Laser modification of the films was carried out while the film was moved relative to the focused laser beam using the coordinate stage. The maximum P of laser radiation on the surface of the films under study did not exceed 65 kW/cm^2.

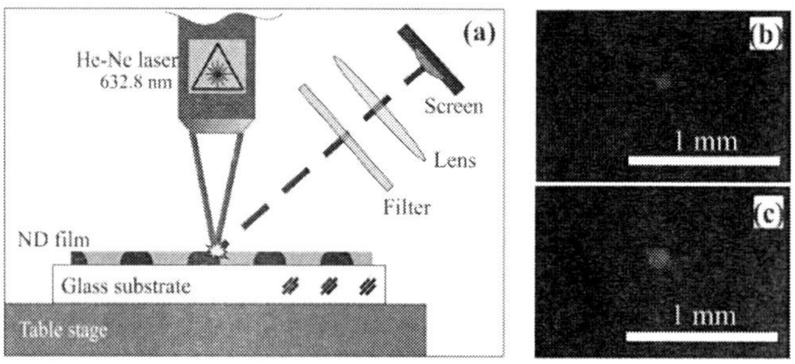

Figure 3 (a, b, c). (a) The sketch of the experiment on laser modification of HPHT ND film using focused laser radiation; images of laser exposure area at **(b)** $P = 14$ kW/cm^2 and **(c)** $P = 65$ kW/cm^2.

The emission from the surface of the film during laser modification was recorded using a Canon photocamera after passing through a color optical filter to attenuate the 632.8 nm laser radiation, reflected from the film surface (see figure 3).

4. Results and discussion

Experiments have shown that when the ND film is irradiated by laser radiation with $P > 33$ kW/cm^2, blackening can occur on the film surface in the area of the laser beam waist. We note that this blackening occurs only when the film is moving relative to the laser beam, i.e. when the film is motionless and irradiated by laser radiation with $P > 33$ kW/cm^2 no blackening is observed. Visual observation of the area irradiated by the laser beam through a glass filter, according to the scheme presented in figure 3a, showed that the blackening of the film is accompanied by blue-green emission (see figure 3c). Photographs of the waist region of a focused laser beam taken with the camera (see figure 3a) at $P = 14$ and 65 kW/cm^2 are shown in figure 3b,c. It was found that at $P < 33$ kW/cm^2, the camera registers only the red glow of the laser, and at $P > 33$ kW/cm^2, green emission is observed from the waist area, accompanied by blackening of the film. It should be particularly noted that repeated long-term irradiation of the blackened area of the film at the maximum $P = 65$ kW/cm^2 does not generate any emission.

Figure 4a illustrates the Raman spectra measured for three different points located on the x-axis perpendicular to the symmetry line of the laser-modified (blackened) region of the film (the symmetry line is parallel to the y-axis) (see figure 4b, inset). Here, the coordinate $x = 0$ is the intersection point of the x-axis and the line of symmetry. The Raman spectra were recorded at $P = 5$ kW/cm^2 using a 100$^\times$ objective. It can be seen that the spectra obtained at $| x | > 2$ μm, contain a pronounced diamond peak with a frequency shift of 1331 cm^{-1}. In the Raman spectrum of the point $x = 0$, the diamond peak is absent (see figure 4a, curve 3). Instead, the Raman spectrum shows two distinct peaks with frequency shifts of 1320 and 1580 cm^{-1}, characteristic for sp^2-carbon. Figure 4(b) shows the ND peak intensity as a function of the x coordinate. It can be seen that this dependence in the first approximation is symmetrical relative to the point $x = 0$. In addition, it is clearly seen that the decrease in the ND peak intensity occurs for points with coordinate$| x | < 10$ μm.

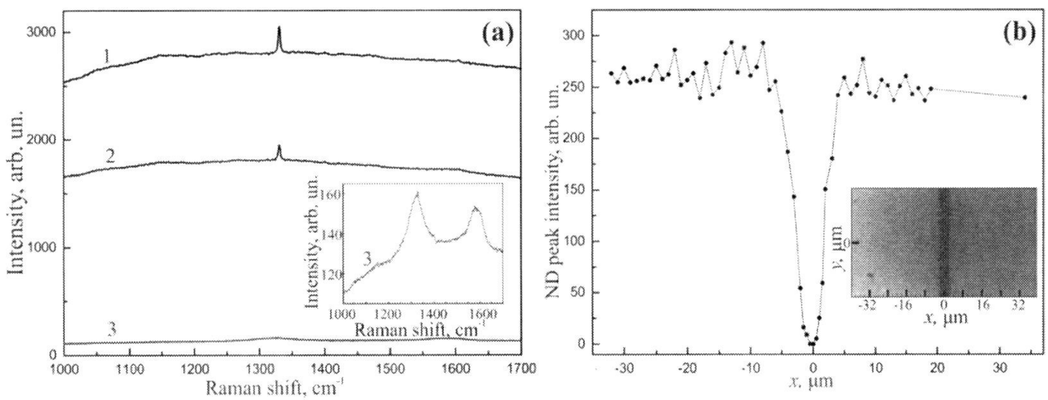

Figure 4 (a, b). **(a)** The Raman spectra of film points located on the x-axis perpendicular to the laser-modified (blackened) line on the surface of the film, the line being parallel to the y-axis, with coordinates $x = -25$ (curve 1), -3 (curve 2) and 0 (curve 3) μm; **(b)** the intensity of the nanodiamond Raman peak as a function of the x coordinate; inset – a photograph of a blackened line, with x and y axes being marked.

Thus, during the movement of a laser beam spot with $P > 33$ kW/cm^2 relative to the HPHT ND film the transformation of ND particles into sp^2-carbon can occur which is accompanied by a green emission. Partial transformation of ND particles into sp^2-carbon also occurs in the region adjacent to the waist, the size of which is several times larger than the waist diameter.

Though the structural transformation of sp^3- to sp^2-carbon was previously observed [19] while irradiating powders of detonation nanodiamonds [20,21] with continuous radiation from a 325 nm He-Cd laser ($P = 1300$ W/cm^2), such transformation was due to high-temperature heating with strong absorption of the incident radiation by the ND particles. In our case the result could not be explained by simple high-temperature heating. Firstly, in figure 5, the Raman spectra are presented of arbitrarily

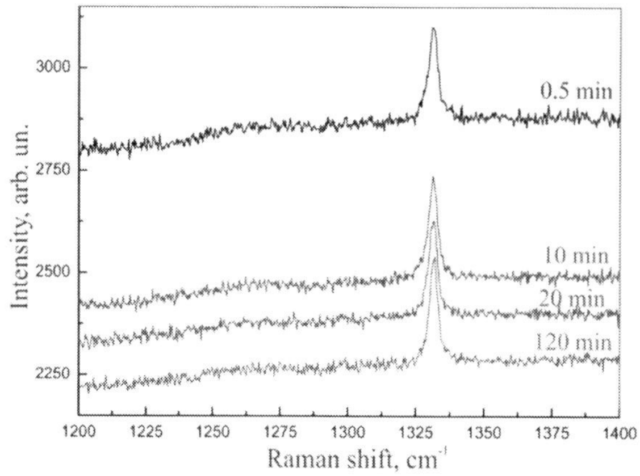

Figure 5. Raman spectra of an arbitrary selected point of the film irradiated by laser radiation with $P = 65$ kW/cm^2 for different periods of time (before measuring a Raman spectrum) measured at $P = 65$ kW/cm^2.

selected points on the motionless film, which were irradiated for different periods of time by laser radiation with $P = 65$ kW/cm^2 (before measuring the Raman spectrum). The spectra were all measured at the same $P = 65$ kW/cm^2. All of the spectra contain the ND peak, with approximately the same intensity. This means that irradiation of the motionless film with focused laser radiation at the maximum $P = 65$ kW/cm^2 for a long time does not lead to blackening and structural changes. Secondly, the process of laser modification of the film surface is accompanied by a blue-green emission (see figure 3c). Thirdly, we found that after blackening of the film has occurred repeated irradiation of the same segment of the film at the maximum P does not lead to any further visible changes. Furthermore, it is known that the HPHT NDs can contain Ni and N impurity complexes [13,22–25]. This means that in the studied HPHT NDs there is a set of energy levels that provide the possibility of the absorption of one or two photons at 633 nm. When absorption occurs, the upper-levels of the energy transitions of the Ni centers and the Ni – N complexes get populated, which can lead to the emission in the blue-green region due to various mechanisms of up-conversion luminescence. Simultaneously with this process, ionization of impurity centers, similar to the ionization of the [N-V]$^-$ - centers during laser irradiation may occur [12,26]. As a result, free electrons are generated in the zone of laser exposure, which begin to effectively absorb the incident laser radiation. These processes lead to the fact that the temperature of the material in the zone of laser exposure increases sharply to significant temperatures, leading to the transformation of the diamond material into sp^2-carbon, i.e. to blackening of the material under study. Subsequent laser irradiation of the blackened area devoid of impurity centers does not result in multi-photon absorption, and due to the low P of the laser radiation, the film cannot be heated up to a high temperature resulting in thermally induced light emission, as was reported by Osswald et al. [19]. Obviously, the Ni – N complexes scattered randomly on the film surface that explain the fact that blackening occurs only when the film is moving relative to the laser beam. In addition, recently, it was reported on intense broadband white emission with maximum at 620 nm obtained from the irradiation of HPHT diamonds by focused beam of 975 nm laser diode lead, suggesting that the white light emission may originate from the laser-induced $sp^2 \leftrightarrow sp^3$ hybridization change as a result of the mechanism analogous to the intervalence charge transfer in Sr_2CeO_4 nanocrystals [27].

5. Conclusions

Thus, we have shown the possibility of modifying films obtained from HPHT NDs using low-power He-Ne laser radiation at 632.8 nm. Laser modification occurs due to the transformation of the diamond material into sp^2-carbon and is accompanied by the emission of light in the blue-green region of the optical spectrum. The possibility of laser modification of HPHT NDs is explained by photon absorption at 633 nm, leading to the ionization of the impurity complexes of Ni and Ni – N contained in diamond nanoparticles.

Acknowledgments
This work was supported by the government assignment of IM UdmFRC, UB RAS (st. reg. number AAAA-A19-119021400126-5).

References

[1] Mikheev G M, Kuznetsov V L, Mikheev K G, Mogileva T N, Shuvaeva M A and Moseenkov S I 2013 *Tech. Phys. Lett.* **39** 337–40
[2] Mikheev G M, Mikheev K G, Anoshkin I V. and Nasibulin A G *Tech. Phys. Lett.* **41** 887–90
[3] Narayan J, Bhaumik A, Sachan R, Haque A, Gupta S and Pant P 2019 *Nanoscale* **11** 2238–48
[4] Iakovlev V Y, Sklyueva Y A, Fedorov F S, Rupasov D P, Kondrashov V A, Grebenko A K, Mikheev K G, Gilmutdinov F Z, Anisimov A S, Mikheev G M and Nasibulin A G 2018 *Diam. Relat. Mater.* **88** 144–50
[5] Mochalin V N, Shenderova O, Ho D and Gogotsi Y 2012 *Nat. Nanotechnol.* **7** 11–23
[6] Mikheev G M, Mikheev K G, Mogileva T N, Puzyr A P and Bondar V S 2014 *Quantum Electron.* **44** 1–3
[7] Rothschild M, Arnone C and Ehrlich D J 1986 *J. Vac. Sci. Technol. B* **4** 310–4
[8] Komlenok M S, Kononenko V V., Ralchenko V G, Pimenov S M and Konov V I 2011 *Phys. Procedia* **12** 37–45
[9] Konov V I 2012 *Laser Photonics Rev.* **6** 739–67
[10] Bundy F P, Hall H T, Strong H M and Wentorf R H 1955 *Nature* **176** 51–5
[11] Jelezko F, Tietz C, Gruber A, Popa I, Nizovtsev A, Kilin S and Wrachtrup J 2001 *Single Mol.* **2** 255–60
[12] Manson N B and Harrison J P 2005 *Diam. Relat. Mater.* **14** 1705–10
[13] Collinss A T and Spear P M 1983 *J. Phys. C.Solid State Phys.* **16** 963–73
[14] Kupriyanov I N, Gusev V A, Borzdov Y M, Kalinin A A and Pal'yanov Y N 1999 *Diam. Relat. Mater.* **8** 1301–9
[15] Solin S A and Ramdas A K 1970 *Phys. Rev. B* **1** 1687–98
[16] Mochalin V, Osswald S and Gogotsi Y 2009 *Chem. Mater.* **128** 273–9
[17] Chung P H, Perevedentseva E, Cheng C L 2007 *Surf. Sci.* **601** 3866–3870
[18] Zolotukhin A A, Ismagilov R R, Dolganov M A and Obraztsov A N 2012 Morphology and *J. Nanoelectron. Optoelectron.* **7** 22–8
[19] Osswald S, Behler K and Gogotsi Y 2008 *J. Appl. Phys.* **104** 074308 (1-7)
[20] Osswald S, Yushin G, Mochalin V, Kucheyev S . O and Gogotsi Y 2006 *J. Am. Chem. Soc.* **128** 11635–42
[21] Portet C, Yushin G and Gogotsi Y 2007 *Carbon* **45** 2511–8
[22] Lu H-C, Peng Y-C, Lin M-Y, Chou S-L, Lo J-I and Cheng B-M 2015 *Anal. Chem.* **87** 7340–4
[23] Collins A T 1992 T *Diam. Relat. Mater.* **1** 457–69
[24] Yelisseyev A, Babich Y, Nadolinny V, Fisher D and Feigelson B 2002 *Diam. Relat. Mater.* **11** 22–37
[25] Indblom J, Holsa J, Papunen H and Hakkanen H 2005 *Am. Mineral.* **90** 428–40
[26] Aslam N, Waldherr G, Neumann P, Jelezko F and Wrachtrup J 2013 *New J. Phys.* **15** 013064 (1-17)
[27] Strek W, Olejniczak A, Tomala R, Cichy B, Lukowiak A, Zhaludkevich A, Konovalova A, Ignatenko O 2018 *Proc. SPIE* **10683**

Engineering the synthesis procedure for PbS nanosheets with controlled optical properties in the near-infrared region

T K Kormilina[1], I D Skurlov[1], S A Cherevkov[1], A S Mudrak[1], A P Litvin[1], E V Ushakova[1], A V Baranov[1], A V Fedorov[1]
[1]Center of Information Optical Technologies, ITMO University, St. Petersburg 199034, Russia

Abstract. This work discloses routes of obtaining two-dimensional PbS nanostructures by chemical synthesis methods. Using the techniques of absorption and luminescence spectroscopy, their optical characteristics are revealed. Special attention is paid to the evolution of nanoplates in a colloidal solution. The processes of growth and degradation of the structure of nanoplatelets, as well as the corresponding changes in the optical properties of nanocrystals are considered.

1. Introduction

There is currently a trend in Materials Science to pay closer look to the two-dimensional materials. That is especially important for optoelectronics since it aims for better compatibility with the existing device designs. In colloidal photonics this addresses two-dimensional analogues of well-proven semiconductor nanocrystals. For example, in the visible range, nanoplatelets and nanoribbons of cadmium chalcogenides thanks to their unique features, such as thickness-dependent optical transitions [1], rapid charge carrier cooling [2], and anisotropic electron-phonon interaction [3], have now replaced quantum dots for many applications [4,5]. A promising task is to obtain two-dimensional analogues for semiconductor nanocrystals of the near-IR range as well. Lead and mercury compounds are excellent for this purpose. In the previous years, numerous nanostructures of lead and mercury chalcogenides were reported [6] as well as their alloys [7] and core/shell structures [8]. Their practical applications are abundant and breath-taking [9,10]. However, the growth mechanisms and synthesis procedures of the two-dimensional nanostructures for these materials are in a certain way more complex and significantly less studied than for cadmium compounds [11,12]. Development of the efficient and reproducible synthesis protocols for lead sulfide nanoplatelets and nanosheets of the desired optical parameters, is thus a field in the urgent need for improvement. Such structures are in demand to enrich the creation of LEDs, solar cells, as well as the study of nonlinear effects in the near IR range [13]. In this work we demonstrate the reliable synthesis conditions for obtaining PbS nanosheets with two different exciton transition energies, look into their stability issues and their influence on the optical responses of nanostructures. Among our goals was also expanding the wavelength range available for two-dimensional structures to shorter wavelengths. Thus, the thinner structures described here are new to the family of PbS nanosheets and were never reported before.

2. Synthesis

PbS nanosheets were obtained using a procedure described in [14] with significant modifications for the smaller thickness. In a 50 ml three-neck flask a mixture of 10 ml diphenyl ether, 3.5 ml oleic acid and 860 mg of lead acetate trihydrate was prepared. The flask was kept under vacuum at 85 °C for 1 h, and then cooled to 35 °C under argon flow. At this temperature the 1.5 ml of chloroalkane (dichloromethane or chloroform) was added to the reaction flask. Subsequently the solution was heated under Ar atmosphere until it reached the reaction (65 or 90 °C). At this temperature sulfur precursor was injected as a solution containing 12 mg of thioacetamide, 930 μl of trioctylphosphine and 100 μl of dimethylformamide. The reaction mixture was kept at this temperature for 30 minutes in the case of dichloromethane and 7 minutes for chloroform, after which it was slowly cooled to room temperature. The reaction mixture was redissolved twice in double volume of toluene and centrifuged at 6000 rpm for 10 minutes.

3. Optical properties

It was determined experimentally that at the reaction temperatures in the ranges of 60-70 and 71-90 °C, two different populations of nanocrystals with constant optical responses are formed. For an accurate assessment of the optical parameters of two-dimensional lead sulphide nanocrystals, their electronic energy structure was characterized by absorption and photoluminescence spectroscopy. Detailed parameter values for four syntheses are given in Table 1.

Table 1. Spectral parameters of PbS nanoplatelets in CCl4.

	T, °C	Abs. peak, nm	Abs. edge, nm	PL peak, nm	FWHM, nm	Stokes shift, nm
PbS1	90	1383	1510	1462	190	79
PbS2	75	1380	1520	1482	160	102
PbS3	69	1290	1380	1380	210	90
PbS4	65	1240	1380	1335	160	95

Typical spectra for two different populations of nanosheets represented by samples PbS1 and PbS4 are shown in Figure 1.

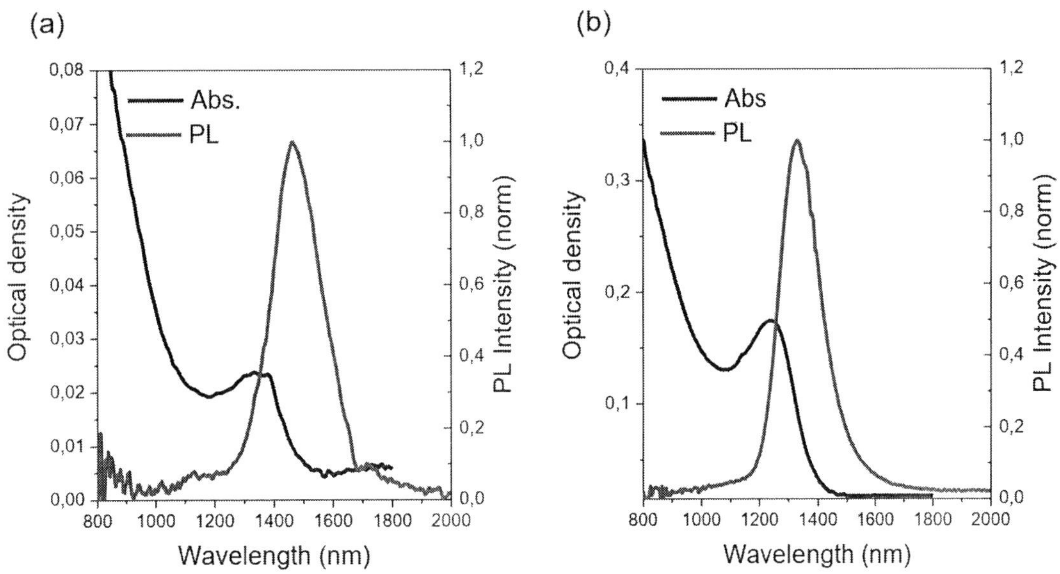

Figure 1(a, b). Photoluminescence (PL) and Absorption (Abs.) spectra of **(a)** PbS1 and **(b)** PbS4 nanosheets.

As can be seen from the graphs, the clearly defined edges of the absorption band at 1510 or 1380 nm with maxima at 1383 or 1240 nm overlap with the photoluminescence bands. The photoluminescence bands of nanosheets are characterized by maxima at 1462 nm and 1335 nm and full width at half maximum (FWHM) of 190 and 160 nm. Nanosheets PbS1 and PbS2 are in good accordance with published data for PbS nanosheets with thickness of 1.2 nm. Samples PbS3 and PbS4 correspond to thinner nanosheets that have not been reported before. Estimated size using the fitting curves is around 1.0 nm [14]. For each pair of the nanosheets the sample obtained at a lower temperature demonstrates narrower PL band and larger Stokes shift. That indicates the temperature conditions to be better suitable for nanocrystals of this particular thickness.
These results are fully consistent with the literature data for this class of nanoobjects. They demonstrate a number of fundamental differences from cadmium compound nanoplatelets, due to the different energy structures of nanocrystals having a zinc blende crystal structure in the case of cadmium compounds and a cubic salt one in the case of lead.

4. Morphology and degradation dynamics

Lead sulphide nanosheets present a certain difficulty in maintaining their stability. Without special measures, the stock solution degrades with the transition of nanosheets into quantum dots. The reason for this behaviour lies in the fundamental mechanisms of growth of two-dimensional nanocrystals. Cadmium chalcogenide nanoplatelets are formed during crystal growth due to the ongoing reaction of precursors [15]. In contrast, in the case of lead chalcogenides, nanosheet growth occurs by the directional attachment of grains or smaller islands formed by nanocrystal grains [11] with the subsequent rearrangement of surface atoms. The main difference of these structures from a well-studied self-assembled highly ordered superlattices lies in them being substrate-free, but more importantly in the formation of the homogeneous crystal structure of the nanosheets rather than a [16–18]. This process is easily reversible; therefore, for early samples, we observed the degradation of nanoplates into smaller clusters and individual nanocrystals. This process is illustrated on images from an scanning electron microscope (Merlin, Carl Zeiss).

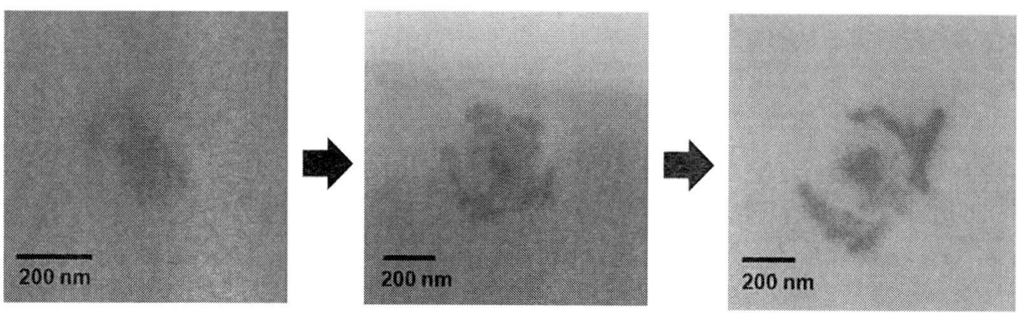

Figure 2. Images of nanosheets in PbS2 sample taken with an electron microscope, which reflect different stages of degradation of nanosheets.

Such a crucial change in the morphology of the nanocrystal leads to a change in the conditions of quantum confinement and, therefore, the optical responses. The graphs on the Figure 3 (a) show the spectra of PbS nanosheets on day 1 right after preparation and taken after 7 days. By this time, as represented on Figure 3 (b) there can be almost no nanosheets found in the sample, only small nanocrystals 8-9 nm in diameter and their 60 nm large aggregates. The solution to this problem lays in finding the right storage environment. Transferring nanosheets to carbon tetrachloride solution, preserves the position of the PL band, as can be seen from Figure 3 (c). The contribution to the PL from quantum dots, indicating their amount, has grown, but on a negligible scale.

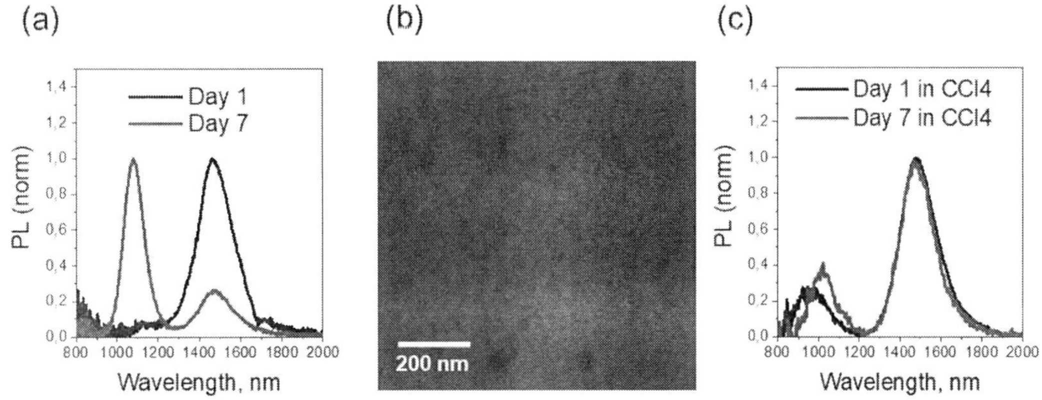

Figure 3. a) The photoluminescence spectra of the stock unwashed solution of PbS1 nanosheets on the first day after the synthesis and a week later; b) Image of nanostructures in the stock solution of the PbS1 sample after 7 days of storage; c) Spectra of a sample of PbS nanosheets, which were redissolved in carbon tetrachloride taken at the time of transfer and 7 days after it.

5. Conclusion

Samples of nanosheets in a colloidal solution were obtained by organic synthesis methods which allowed their optical properties to be tuned in accordance with study goals. Their optical properties were studied using absorption and luminescence spectroscopy techniques and suggest promising nanomaterial for IR range. We also showed dynamics of the morphology degradation of nanostructures in a sample that was carried out by means of electron microscopy and linked it with a change in their optical responses. Using the information obtained in this work, it is possible to create high-quality nanostructures with controlled optical properties in the near-IR range, which would be stable in a colloidal solution and present great interest for solving various tasks of modern optoelectronics and photovoltaics.

Acknowledgements
The work was carried out as a part of RFBR project no. 18-32-00027.

References

[1] Ithurria S, Tessier M D, Mahler B, Lobo R P S M, Dubertret B and Efros A L 2011 *Nat. Mater.* **10** 936–41
[2] Pelton M, Ithurria S, Schaller R D, Dolzhnikov D S and Talapin D V. 2012 *Nano Lett.* **12** 6158–63
[3] Cherevkov S A, Fedorov A V., Artemyev M V., Prudnikau A V. and Baranov A V. 2013 *Phys. Rev. B - Condens. Matter Mater. Phys.* **88** 041303
[4] Kormilina T K, Cherevkov S A, Fedorov A V. and Baranov A V. 2017 *Small* **13** 1702300
[5] Nasilowski M, Mahler B, Lhuillier E, Ithurria S and Dubertret B 2016 *Chem. Rev.* **116** 10934–82
[6] Rogach A L, Eychmüller A, Hickey S G and Kershaw S V. 2007 *Small* **3** 536–57
[7] Lifshitz E, Brumer M, Kigel A, Sashchiuk A, Bashouti M, Sirota M, Galun E, Burshtein Z, Le Quang A Q, Ledoux-Rak I and Zyss J 2006 *J. Phys. Chem. B* **110** 25356–65
[8] Shen G and Guyot-Sionnest P 2019 *Chem. Mater.* **31** 286–93
[9] Litvin A P, Martynenko I V., Purcell-Milton F, Baranov A V., Fedorov A V. and Gun'ko Y K

2017 *J. Mater. Chem. A* **5** 13252–75
[10] Zhao P, Xu Q, Tao J, Jin Z, Pan Y, Yu C and Yu Z 2018 *Wiley Interdiscip. Rev. Nanomedicine Nanobiotechnology* **10** e1483
[11] Schliehe C, Juarez B H, Pelletier M, Jander S, Greshnykh D, Nagel M, Meyer A, Foerster S, Kornowski A, Klinke C and Weller H 2010 *Science (80-.).* **329** 550–3
[12] Zhang H, Savitzky B H, Yang J, Newman J T, Perez K A, Hyun B R, Kourkoutis L F, Hanrath T and Wise F W 2016 *Chem. Mater.* **28** 127–34
[13] Lhuillier E and Guyot-Sionnest P 2017 *IEEE J. Sel. Top. Quantum Electron.* **23** 1–8
[14] Bhandari G B, Subedi K, He Y, Jiang Z, Leopold M, Reilly N, Lu H P, Zayak A T and Sun L 2014 *Chem. Mater.* **26** 5433–6
[15] Ithurria S, Bousquet G and Dubertret B 2011 *J. Am. Chem. Soc.* **133** 3070–7
[16] Baranov A V., Ushakova E V., Golubkov V V., Litvin A P, Parfenov P S, Fedorov A V. and Berwick K 2015 *Langmuir* **31** 506–13
[17] Wang Z, Schliehe C, Bian K, Dale D, Bassett W A, Hanrath T, Klinke C and Weller H 2013 *Nano Lett.* **13** 1303–11
[18] Williams K J, Tisdale W A, Leschkies K S, Haugstad G, Norris D J, Aydil E S and Zhu X-Y 2009 *ACS Nano* **3** 1532–8

Synthesis and energy structure of optical transitions of the nitrogen and sulfur co-doped carbon dots

P D Khavlyuk[1], I A Arefina[1], D P Bondarenko[1], K V Bogdanov[1], E A Stepanidenko[1], A V Baranov[1], A V Fedorov[1], E V Ushakova[1], A Dubavik[1], A L Rogach[1,2]

[1]Center of Information Optical Technologies, ITMO University, Saint Petersburg 197101, Russia

[2]Department of Materials Science and Engineering, City University of Hong Kong, 83 Tat Chee Avenue, Kowloon, Hong Kong SAR

Abstract. A green and efficient strategy was developed for one-step synthesis of sulfur and nitrogen co-doped carbon dots using urea, thiourea and thioacetamide as nitrogen and sulfur source and citric acid as carbon source. The influence of nitrogen and/or sulfur doping of the carbon dots on their optical transitions is investigated. The as synthesized carbon dots are demonstrated to possess redshifted photoluminescence with relatively high intensity. It is demonstrated that the presence of heteroatoms leads to the red shift of the photoluminescence band and increase of the crystalline phase inherent to amorphous carbon.

1.Introduction

Carbon dots (CDs) are novel fluorophores in the family of carbon nanomaterials – graphene, fullerenes, nanotubes, nanodiamonds, etc. [1] CDs possess unique optical properties, such as tunable photoluminescence, high quantum yield, relatively high photostability. Moreover, the ability of doping CD core and surface by different heteroatoms is a very essential characteristic and may expand the optical transitions range.

Single atom doping can effectively improve the emission properties of CDs. The content of heteroatom in CDs plays an important role in generating high PL quantum yield efficiency, the higher the heteroatom content on CDs, the more surface states get formed on CDs resulting enhancement of optical properties. Nitrogen atoms are well known as dopants of the CDs and introduce additional energy levels into the electronic structure resulting in the CDs emission band shift [2]. Nitrogen and carbon atoms have very close similarity, due to this N-doping is the most widely used way to improve the photoluminescence (PL) properties of CDs. The nitrogen atom changes the inner structure by injecting electrons into CDs, which substantially improve their optical properties. Recently, it was shown [3], that the emission of the N-doped CDs interacting with sulfur compounds is further shifted to the red spectral region. Comparing to the intensively researched N-doped CDs, CDs containing S atoms have been rarely studied. In recent years, S-doped CDs were synthesized from different precursors: sulfuric acid [4], sodium citrate [5], etc. The S-doped CDs possess a relatively high fluorescence QY of ~12% without any surface passivation at the 350 nm excitation. At the same time, efficient routes to enhance the fluorescent properties of CDs remained a challenge. Therefore, combination of heteroatoms has gained much attention in recent years, because it can create a unique electronic structure, due to the interdependent effect between the doped atoms in CDs [6].

Other approach of the heteroatom doping is the implementation of the co-doping multiplex heteroatoms within one precursor molecule. In particular, N and S co-doped CDs have been actively investigated recently and found to be the best combination of heteroatoms to shift PL to the red spectral region [1]. Considering the time-saving process, low toxicity and enhanced fluorescence properties enable the as-prepared CDs further to be used in various applications, such as sensing and photonics field [7]. However, the energy-level structure of doped CDs is still under discussion. In this work we investigate the influence of the heteroatoms on the optical transitions in CDs synthesized from various precursors, containing N and S atoms.

2.Experimental

Chemicals were purchased from Merck and used without additional purification. Poly(p-phenol)-lined stainless steel autoclaves for solvothermal reaction were purchased from Komviks. Raman spectra were obtained by InVia (Renishaw) with 514 nm excitation. Absorption (Abs) spectra were obtained by a spectrophotometer UV-3600 (Shimadzu). Photoluminescence (PL) and PL excitation (PLE) spectra were obtained by spectrofluorometer FP-8200 (Jasco).

3.Results

CDs were synthesized by a modified procedure based on [8]. Urea, thiourea and thioacetamide were used as a heteroatom source. These compounds were mixed with citric acid and dissolved in dimethylformamide (DMF) under solvothermal conditions. The solution was heated in a poly(p-phenol)-lined stainless steel autoclave at 160°C during 6 h, following by cooling down to room temperature. Thus, three samples of CDs, termed as CD-u, CD-tu, and CD-ta, were obtained from urea, thiourea and thioacetamide, respectively. As a reference sample, (termed CD-e), CDs synthesized from citric acid and ethylenediamine by procedure described in [9] was used. For further investigation the synthesized CD solution were diluted to obtain appropriate CD optical density.

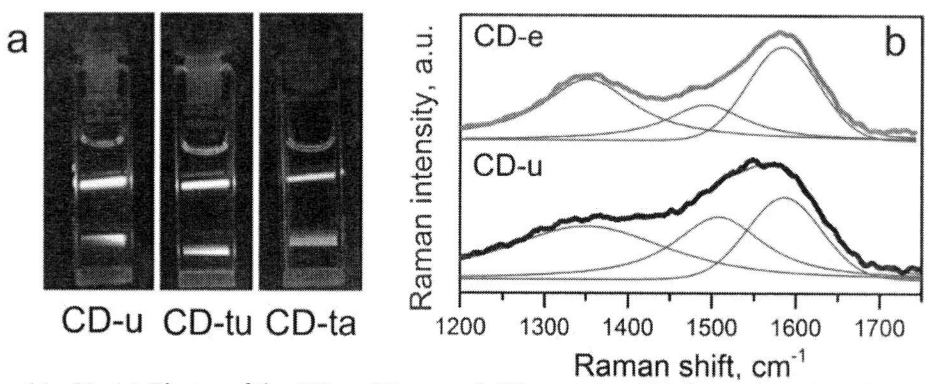

Figure 1(a, b). (a) Photos of the CD-u, CD-ta, and CD-tu under 405 (lower beam) and 532 (upper beam) nm excitation; **(b)** Typical Raman spectra of CD-e and CD-u.

All CDs synthesized in DMF possess an excitation-dependent emission, as can be seen in the Fig.1a, where the samples demonstrate emission at different wavelengths excited by 450 and 514 nm diode lasers. Typical Raman spectra (Fig.1b) of CD-e and doped CDs possess well known vibrational modes of disordered graphitic materials [10], which are assigned to D band (~1350 cm^{-1}), G band (~1585 cm^{-1}) and amorphous carbon (area of 1480-1530 cm^{-1}). Increased D/G ratio for CD-u, CD-tu, and CD-ta samples suggests that the heteroatom doping can introduce more defects into the graphitic structure of the CD core together with formation of additional surface groups. Therefore, we believe that the defects introduced by implementation of urea, thiourea and thioacetamide instead of ethylenediamine as heteroatoms source influence the CD energy structure and the optical transitions contributing to the PL signal.

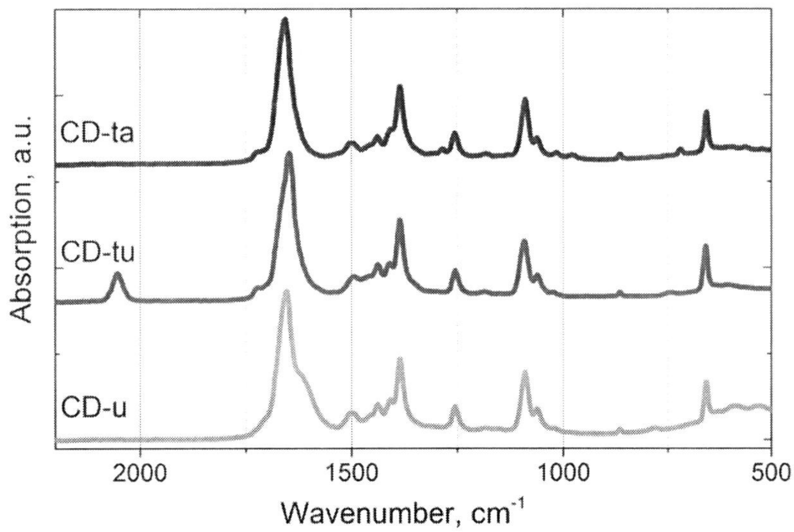

Figure 2. FTIR spectra CD-ta (blue), CD-tu (green), CD-u (orange)

FTIR spectra (Fig. 2) contain a set of peaks: 658, 1062, 1091, 1253, 1386, 1498, 1653, 2865, and 2930 cm^{-1}, which are common for all the CD samples. This set is typical for the cyclic organic compounds with amino and hydroxyl groups. In the FTIR spectra of CD-tu and CD-ta the peak at 1718 cm^{-1}, which can be attributed to the C=O vibration in carboxylic group, is more pronounced compared to CD-u sample. Peaks in the area of 1400-1500 cm-1 may be attributed to the stretching of N or S single bonds with C atoms. It also should mentioned, that the FTIR spectra of CD-u and CD-ta possess the peaks at 977 and 2775 cm^{-1} together with the absence of the peak at 2055 cm^{-1}, compared to CD-tu spectrum. Thus, we can assume that the optical parameters of the CD-u and CD-ta should be similar compared to the CD-tu sample.

The Abs spectra of all samples presented in Fig. 3a possess the band at 350 nm which is typical for the CDs and may be attributed to the $n - \pi$ transition in the CD core according to Ref. [2]. The position and optical density of low-energy bands attributed to the edge and/or surface states, e.g. originated from the dopants, differ for all samples. For the CD-ta, CD-u, and CD-tu the second Abs band with increased optical density is observed at 440, 455, and 500 nm, respectively. For the CD-u there is an additional wide band at 600 nm. Comparison of the PLE spectra (Fig. 3b) monitored around 540-550 nm shows that the PL signal is determined by the additional energy levels. The PLE spectrum of CD-tu is more structured and contains set of narrow peaks around 450-520 nm, which can be related to the molecular moieties within CD, unlike the CD-e and CD-ta spectra. The PL band (Fig. 3c) of the doped CDs is redshifted compared to the CD-e, which is in accord with Abs and PLE spectra. The largest PL shift is demonstrated by CD-u and CD-ta while excited at 405 and 500 nm, respectively.

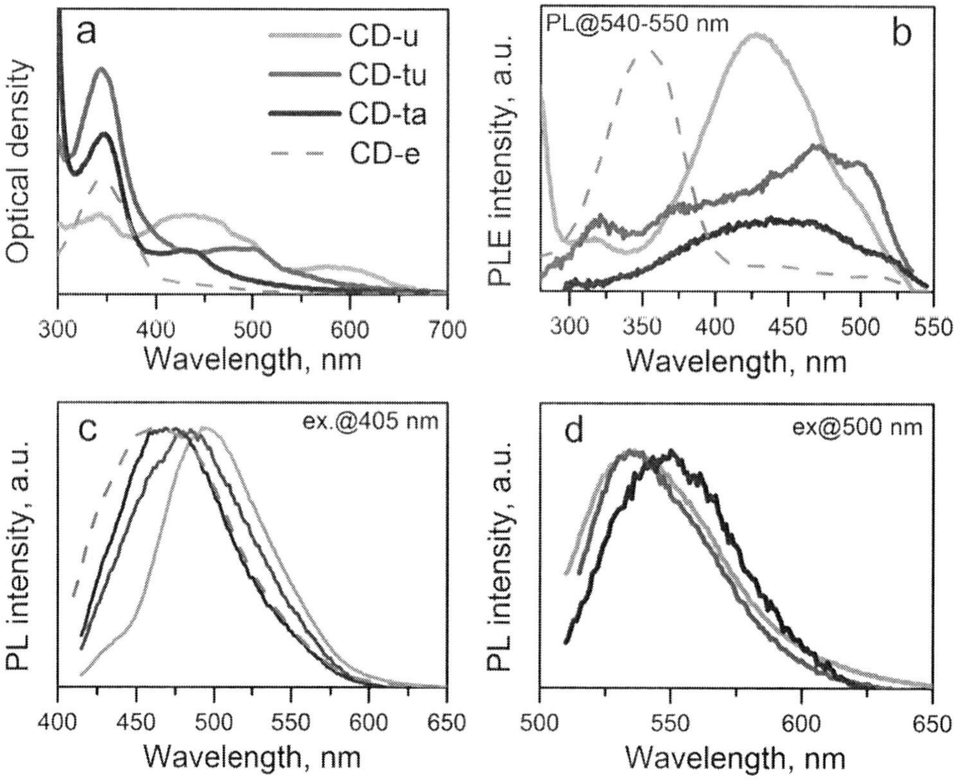

Figure 3(a, b, c, d). **(a)** Abs, **(b)** PLE, and **(c, d)** PL spectra of CD-ta (blue), CD-tu (green), CD-u (orange), and CD-e (dashed grey) excited at 405 **(c)** and 500 **(d)** nm.

In conclusion, we investigated the impact of N- and S-doping of CDs on their photophysical properties. In our study it is shown that the presence of nitrogen or/and sulfur heteroatoms in the CDs leads to the increased volume of amorphous carbon phase and to the appearance of additional energy levels, which in turn results in the red shift of the PL band.

Acknowledgements
This work has been supported by the Ministry of Science and Higher Education of the Russian Federation (Grant 14.Y26.31.0028).

References

[1] Ren X, Liang X, Wang P, Bunker C E, Coleman M, Teisl L R, Cao L, Sun Y 2018 *Carbon* **141** 553–560
[2] Fu M, Ehrat F, Wang Y, Milowska K Z, Reckmeier C, Rogach A L,Stolarczyk J K, Urban A S, Feldman J 2015 *Nanoletters* **9** 6030–6035
[3] Qu S, Zhou D, Li D, Ji W, Jing P, Han D, Liu L, Zeng H, Shen D 2016 *Adv. Mater.* **28** 3516–3521
[4] Zhu H, Wang X, Li Y, Wang Z, Yang F, Yang X, 2009 *Chem. Commun.***34** 5118
[5] Liu S, Tian J, Wang L, Zhang Y, Qin X, Luo Y, Asiri A M, Al-Youbi A O, Sun X 2012 *Adv. Mater.* **24** 2037
[6] Xu Q, Liu Y, Gao C, Wei J, Zhou H, Chen Y, Dong C, Sreeprasad T S, Li N, Xia Z, 2015 *J. Mater. Chem. C* **3** 9885
[7] Sun Y, Zhou B, Lin Y, Wang W, Fernando K A S, Pathak P, Meziani M J, Harruff B A, Wang X, Wang H, Luo P G, Yang H, Kose M E, Chen B, Veca L M, Xie S 2006 *J. Am. Chem. Soc.* **24** 7756-7757

[8] Li D, Jing P, Sun L, An Y, Shan X, Lu X, Zhou D, Han D, Shen D, Zhai Y, Qu S, Zboril R, Rogach A L 2018 *Adv. Mater.* **30** 1705913
[9] Zhu S, Meng Q, Wang L, Zhang J, Song Y, Jin H, Zhang K, Hongchen S, Wang H, Yang B 2013 *Angew. Chem.* **125** 4045
[10] Bogdanov K, Fedorov A, Osipov V, Enoki T, Takai K, Hayashi T, Ermakov V, Moshkalev S, Baranov A 2014 *Carbon* **73** 78–86

Low-dimensional heterostructures obtaining from an intelligent material and carbon material on a silicon

S. A. Gurin[1], E. A. Pecherskaya[2], A.B. Fimin[2], K.O. Nikolaev[2], Y.V. Shepeleva[2], D.V. Artamonov[2]
[1]CJSC Medtekhnika, Penza 440018, Russia
[2] Department of Information and measuring equipment and metrology, Penza State University, Penza 440026, Russia

Abstract. The results of the study of the technological regimes of the formation of a microcantilever beam of layered heterostructures based on the integration of smart material and silicon carbide are presented. Layers of silicon dioxide (SiO_2), silicon nitride, silicon carbide, and platinum (lower electrode) are successfully formed on a silicon substrate. The results of an experimental study of the Pt / PZT / Pt / SiC heterostructures parameters confirmed the technological compatibility of the layers and the promise of using these structures in sensors of dynamic deformation when creating new generation sensors, including those operating in heavy rocket conditions of space technology.

1. Introduction

The new generation MEMS allow miniaturizing sensor equipment, but due to physical limitations, increased reliability requirements, the problem of finding materials and synthesizing new structures with improved physical and mechanical parameters that can react to changes in external or internal conditions by changing dynamic behavior is relevant [1, 2].

This article proposes the use of intelligent materials as an element of transformation, which can significantly reduce the threshold sensitivity of dynamic deformation sensors based on piezoelectric films; increase the sensitivity of sensors based on MEMS technologies by two orders while maintaining long-term stability [3]

2. Technological conditions for the formation of a microcantilever beam of layered heterostructures based on the integration of smart material and silicon carbide

The film obtained by the CVD method is characterized by a coarse-grained polycrystalline structure, which, in turn, is the cause of the high surface roughness. Figure 1 shows the surface of a diamond film obtained by a Hirox 7700 optical microscope with a magnification of 1,400 times.

Silicon-carbide-PTS thin-film heterostructures are formed on a silicon substrate by the method of high-frequency magnetron sputtering of a $PbZr_{1-x}Ti_xO_3$ ceramic target in an oxygen atmosphere and a silicon carbide target in an argon atmosphere. This composition corresponds to a rhombohedral solid solution adjacent to the morphotropic phase boundary. In this case, the piezoelectric properties reach optimal values, which was confirmed in [4].

Figure 1. The surface of the diamond film at a magnification of 1400 ×, obtained by an optical microscope Hirox 7700.

On a silicon substrate films of silicon dioxide (SiO2), silicon nitride, silicon carbide, and platinum (lower electrode) are formed sequentially. A layer of silicon dioxide with a thickness from 0.2 to 0.5 μm is obtained by thermal oxidation. Silicon nitride was obtained by CVD method by deposition from the gas phase in the plasma of the ratio of the gases of ammonia, monosilane and argon. A layer of silicon carbide with a thickness of 1 μm was obtained by high-frequency magnetron sputtering with sequential ionic support to obtain the required surface roughness, with the result that the interface with the PZT film became monophasic. The lower platinum electrode with a thickness of 0.8 μm is applied by resistive evaporation. The thickness of the PZT ferroelectric film at a deposition temperature of 450 °C was 0.3 μm. Before use, the platinum surface was treated with an ion source. Then the process of deposition of the ferroelectric film alternates with the treatment of the deposited surface with an ion source. This made it possible to increase the film continuity. The top electrode with a thickness of 0.8 μm was formed by resistive evaporation.

3. Method for detecting defects in silicon structure

The well-known optical, electronic, X-ray versions of microscopy, topography and tomography, spectroscopic methods for identifying defects in silicon structures are widely applicable. The low density of structural defects in silicon indicates the feasibility of using the etching method to reliably identify individual defects. This method is based on a visual and microscopic analysis of the surface relief of a defective plate after it has been treated with etchant solutions (etchant Sirtle-Adler, Dash, Secco), specially designed to detect dislocations. After a similar surface treatment, a relief is formed on it, representing traces of micro-defects in the form of hills and craters. The method has a limitation due to a not high enough resolution for studying the crystal structure, as well as the difficulty of identifying the detailed structure of the defect and the impurities present. At the same time, the alternation of plate etching and microscopic studies makes it possible to obtain information on the distribution of defects over the sample depth [5].The method of identification of swirl defects is based on the difference in etching rates of perfect crystallographic regions of the plate and its areas with micro-defects. The etching rate varies in areas with micro-defects, which leads to the formation of a flat-bottomed pit, its geometry is due to the type of micro-defects and the orientation of the plane under study [6].

The image of a plate with swirl defects is shown in Figure 2 (a), Figure 4 (b) shows etching pits inherent in micro-defects forming a swirl pattern (magnification 800 times).

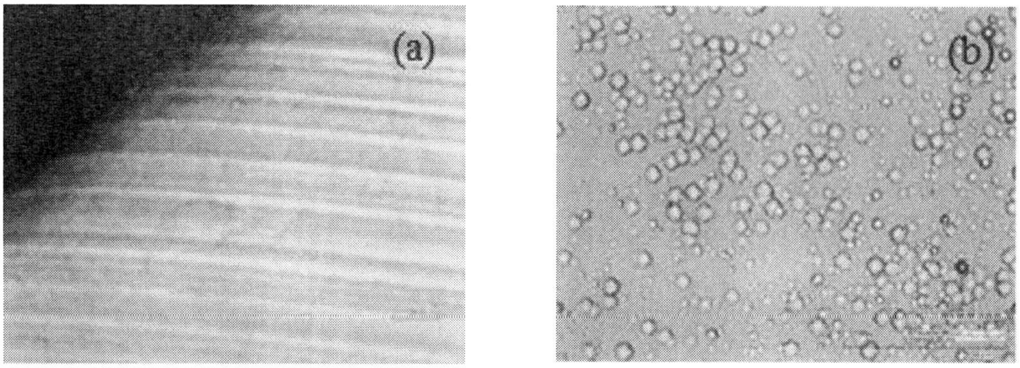

Figure 2(a, b). (a) Swirl - distribution of silicon wafer defects (multiplicity x200); **(b)** Etching pits corresponding to micro-defects forming a swirl pattern (multiplicity x200).

To identify swirl defects, a HF solution was used: an aqueous solution of CrO_3 (1200 g / l) = 1: 4. The duration of etching was 25 minutes.

4. Results of experimental studies of Pt / PZT / Pt / SiC heterostructure

A number of studies have been devoted to the study of the electro-physical properties of the hetero-structures [7, 8]. In [9, 10], the results of the study of the electro-physical parameters of thin PZT films with the help of the software and hardware measuring complex proposed by the authors are presented. As a result of the analysis of experimental data on the choice of the optimal technological regimes of hot polarization of a piezoelectric film integrated in a heterostructure with a carbon material, it was found that the use of such a polarization method is quite stable from the point of view of practical use of film samples as converters.

The temperature dependence of the residual piezomodule d_{33} was obtained (fig. 3), from which it follows that when heated to 250 °C d_{33} decreases by no more than 10% and, most importantly, the sample is not depolarized, which is consistent with [11].Note that as a general principle, for large tables font sizes can be reduced to make the table fit on a page or fit to the width of the text.

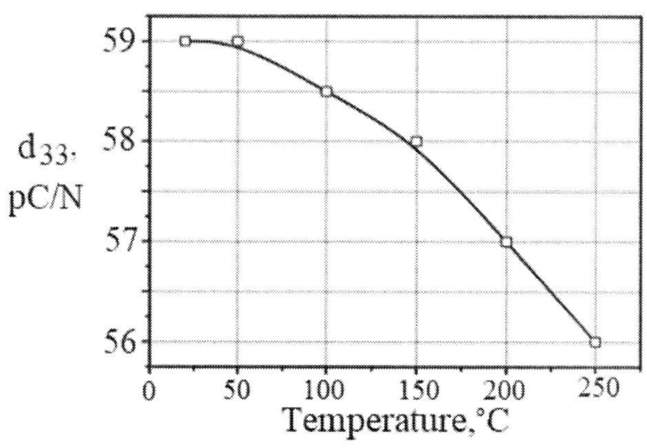

Figure 3. Temperature dependence of the residual piezomodule d_{33} of the hot-polarized PZT film.

 1410 (2019) 012021 doi:10.1088/1742-6596/1410/1/012021

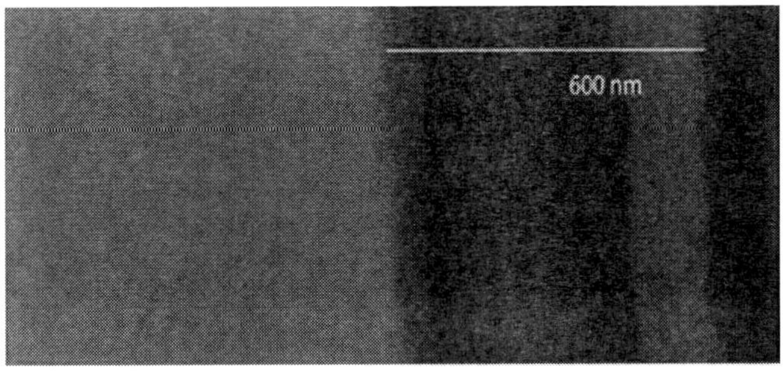

Figure 4. PEM image of the cross section of the Pt / PZT / Pt / SiC heterostructure.

The effect of increasing the residual piezomodule immediately after the creation of a stable polarized state in films due to the ongoing processes of its polarization in the charge region trapped in traps in the bulk of the film in accordance with [3] is established.

The study of the interface of the obtained heterostructure was carried out on a scanning electron microscope (Fig. 4). Analysis of this image confirms the absence of mutual diffusion between the integrated thin heterostructure layers.

The work uses the technology of manufacturing sensitive elements of the beam type, which is implemented through the selection of the optimal technological modes of applying thin layers with a thorough inter-operative control. There is a technological compatibility of hetero-structures made on the basis of high-strength carbon-containing materials and smart materials in combination with conductive layers of metals. MEMS, manufactured using the considered technology of creating micro-cantilever beams based on smart materials, can serve as a starting platform for highly intelligent sensors for detecting new-generation aerospace structures designed for operation in harsh environments.Tables should be centred unless they occupy the full width of the text.

5. Conclusion

Heterostructures made on the basis of high-strength carbon-containing materials and smart materials in combination with conductive layers of metals are technologically compatible. MEMS, made with the use of the considered technology based on smart materials, can serve as a launching platform for highly intelligent detection sensors of a new generation, working in hard operating conditions of rocket and space technology.

References

[1] Barinov I N 2010 *Components and technologies* **8** 64–71
[2] Pecherskaya E A, Zinchenko T O, Golubkov P E, Pechersky A V, Fimin A V, Nikolaev K O 2018 *Moscow Workshop on Electronic and Networking Technologies (MWENT)* 1– 4
[3] Mishina E D, Sigov A S, Mukhortov V M 2009 *Integrated Ferroelectrics* **106** 1–10
[4] Shibata T, Unno K, Makino E, and Shimada 2004 *A: Physical* **114** 398
[5] Fan T W et al. 2000 *Journal of Crystal Growth* **213** 276 – 282
[6] Föll H et al. 1975 *Applied physics* **8** (4) 319–331
[7] Fan T W et al. 2000 *Journal of Crystal Growth* **213** 276 – 282
[8] Senkevich S V, Kiselev D A, Osipov V V, Pronin V P, Kaptelov E Yu and Pronin I P 2015 *Ferroelectrics* **477** 84–92
[9] Pecherskaya E A, Artamonov D V, Kondrashin V I, Golubkov P E, Karpanin O V and Zinchenko T O 2017 *IOP Conference Series: Materials Science and Engineering* **225** 012254

[10] Pecherskaya E A, Golubkov P E, Fimin A V, Zinchenko T O, Pechersky A V, Shepeleva J V 2018 *Procedia Computer Science* **132** 1163–1170
[11] Jaffe B, Cook W, Jaffe G 1974 *Piezoelectric ceramics* (Moscow: Mir) p 288

Structure of V-defects in a-GaN films grown on r-sapphire substrate

A Savchuk[1,2], Yu Akhmerov[1], A Chelny[1], O Abdullaev[1], O Rabinovich[2], M Mezhenny[1], A Zharkova[1,2], A Aluyev[1], M Zakusov[1], N Kourova[1,2], V Murashev[1,2], S Didenko[2]

[1]Optron JSC, Moscow 105187, Russian Federation
[2]National university of science and technology "MISiS" Moscow, 119049, Russian Federation

Abstract. The a-GaN films were successfully grown on the r-sapphire substrate by MOCVD method. The structure of V-defects was investigated by AFM and SEM. The dependence of V-defects density on growth temperature of a-GaN film at a constant hydrogen flow through a TEG source was studied. The influence of V/III ratio on V-defects structure was investigated. Methods of V-defects density minimisation were purposed.

1. Introduction

InGaN/GaN films growth in nonpolar a-orientation is an important scientific and technological goal. It is known that the polarization effect leads to GaN bulk and surface changes in the band structure [1]. The films grown in a-nonpolar a orientation usage can reduce the effect of the built-in field on the injected charge carriers distribution, reduce the nonradiative recombination rate, and improve the emitting devices efficiency, including laser diode. However, there were few difficulties in a-GaN growing and one of them is high density of structural defects due to significant mismatch between a-GaN and r-sapphire, mechanical strains and other factors. One type of such defects are V-defects. V-defects are pits on the surface, usually with a crystalline faceting. V-defects are well studied for the c-plane [2,3], and also V-defects were studied for semipolar GaN on a SiC substrate [4]. One the goals for this work was to expand the understanding of the formation of V-defects for a-GaN grown on r-sapphire.

In this research was investigated the structure of V-defects by AFM and SEM. Also, the dependence of V-defects density on growth temperature of a-GaN film at a constant hydrogen flow through a TEG source and the influence of V/III ratio on V-defects structure was studied. Optimized conditions for V-defects density minimization were purposed.

2 Experement

a-GaN films were grown by MOCVD on r-sapphire substrates under reduced pressure (80 torr) with preliminary surface nitridisation. The source for III and V components were triethilgallium (TEG) and ammonia (NH_3), respectively. The layered gas flow in the quasi-laminar regime in the horizontal reactor was carried out through a porous input. The upper gas mixture flow considered of N_2, and NH_3 and the lower flow – H_2 and TEG. As source of donor and acceptor impurities, hydrogen sulphide(H_2S) (0.5% mixture in argon) and biscyclopentadienyl magnesium (Cp_2Mg), respectively, were used. Preliminary, the sapphire surface was nitride at temperature 950 °C for 10 minutes.

3 Results and discussion

Using the X-ray and Raman scattering spectrum, it was confirmed that the grown films are single-crystal a-GaN. The measured half-width of the rocking curve is 1200-1400 arcsec.

For defects identification SEM and AFM were used. The form, symmetry and depth of the most common defects allow to identify them as V-defects.

Unlike V-defects on c-plane, on the a-plane, the direction of such defects is perpendicular to the growth direction. On the figure 1 the SEM imagines of V-defects structure before and after wet etching is represented. It can be seen, that there were not facets observed. This can be explained by intense overgrowth of c-directed facets of defects.

Figure 1 represents appearance of the V-defects before and after wet etching. The structure of this defect can be explained by its formation mechanism. The V-defects formation mechaism is associated with two competing processes – growth and decomposition of the film. On a semi-polar orientation, the V-defects nucliation place is the dislocations emergence that formed the nanotube. In such place, there is a change in the equilibrium conditions of growth. According to the crystals etching theory [4] at the site dislocation emergence on the surface, the free energy of the etching pits formation ecreases, respectively, with the crystal growth. In this place the V-defects appearance probability is significant.

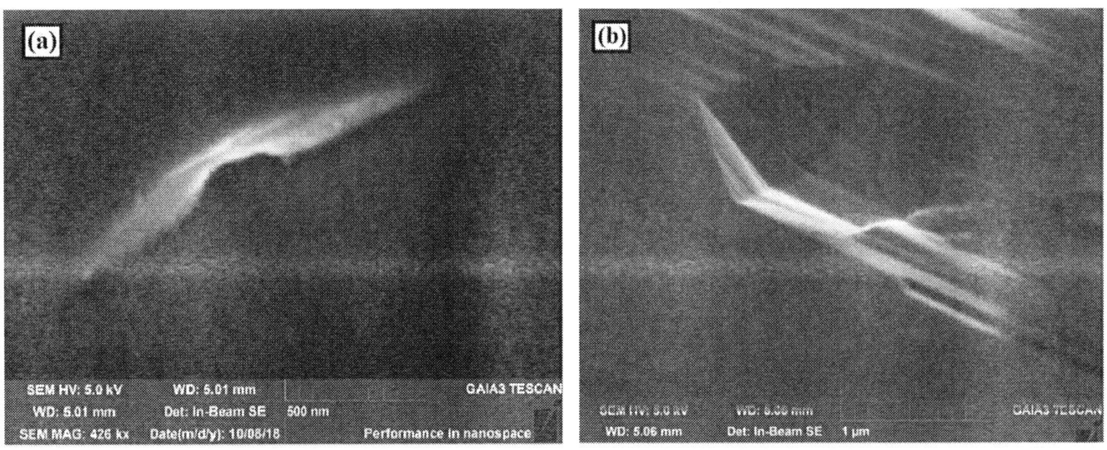

Figure 1. The SEM V-defect imagine: (a) before wet etching, (b) after wet etching in KOH

. Figure 2 shows the image of this part of the defect obtained with the help of AFM. After etching, a structure appears in the form of terraces similar to those observed in the semi-polar direction [5]. However, the length of these terraces is much higher. It can be assumed that the angle of nanotubes, formed V-defects on a-GaN, is greater. This explains the great length of the terraces and the tail of the defect that appeared after etching. Also a possible explanation is the formation of groups of oblique V-defects by analogy with what is observed for lateral V-defects on InGaN.

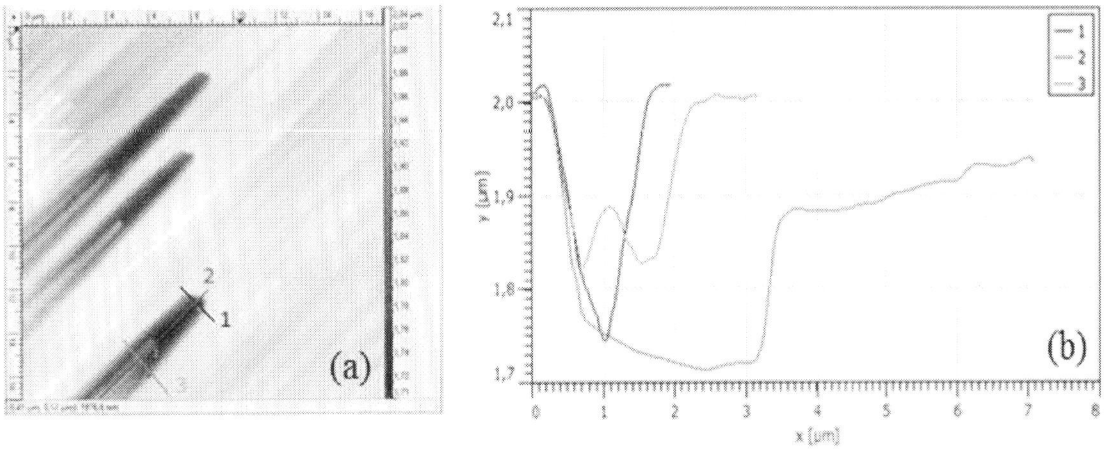

Figure 2. AFM for V-defects, (a) plan view, (b) cross-section

According to the theory and experimental investigation[6,7], the V-defects concentration depends on the V / III ratio and decreases with its increase. In this paper, the confirmation for it was detected for a-GaN. The lowest V-defects concentration was obtained, at all other parameters being equal, and at a maximum V / III ratio – 1000, it was $2\cdot10^{4}$ cm^{-2}.

4 Conclusion

a-GaN was successfully grown on r-sapphire by MOCVD. Using the X-ray and Raman scattering spectrum, it was confirmed that the grown films are a-GaN single-crystal films. The dominant type of defects on the film surface was investigated by SEM and AFM: they were the defects similar by nature to V-defects on the c-plane. Using the wet etching, the defects internal structure was investigated. It was shown that, by analogy with semi-polar films, these defects were formed at the emergence areas inclined dislocations formed into nanotubes, while the geometry and internal structure of V-defects differ from those for c- plane and semi-polar films. The V-defects concentration dependence versus the V / III ratio was investigated and it is similar to the results obtained for the c-plane, which confirms once more the general nature for these defects. Additionally, the investigations of V-defects formation and influence on the electrical and crystallographic characteristics of the a-GaN films improve the injection efficiency and reduce the proportion of non-radiative carrier recombination and

References

[1] Zh. Wu, P. Chen, G. Yang , Zh.Xu, F. Xu ,F. Jiang, T. Zhang, Y. Zheng, Superlattices and Microstructures 2015, 83, 22.
[2] D. Florescu, S. Ting, J. Ramer, D. Lee, V. Merai, A. Parkeh, D. Lu, E. A. Armour, L. Chernyak, Appl. Phys. Lett. 2003, 83(1), 128937.
[3] M. Shiojiri, C. Chuo, J. Hsu, J. Yang, H. Saijo, J. Appl. Phys. 2006, 99, 073505.
[4] K. Sangwal, Theory, Experiment and Application, 2009, 562, 1
[5] V. Bessolov, E. Konenkova, A. Zubkova, A. Osipov, T. Orlova, S. Rodin, S. Kukushkin Materials Physics and Mechanics 2014, 21, 266.
[6] A. Bag, S. Das, R. Kumarbc, D. Biswascd Cryst Eng Comm 2018, 2, 34.
[7] P. Venne, B. Diaby , H. Kim-Chauveau, L. Bodiou, H. Schenk, E. Frayssinet, R. Martin, I. Watson, J. Crys. Growth 2012, 353, 108.

Atomic layer etching of Silicon Oxide

V. Kuzmenko[1,2], **A. Miakonkikh**[1,2], **K. Rudenko**[1,2]
[1]Valiev Institute of Physics and Technology of Russian Academy of Sciences, Moscow, Russia
[2]Moscow Institute of Physics and Technology, Dolgoprudny, Russia

Abstract. The atomic layer etching process for Silicon oxide is studied. The process consists of deposition step and activation step separated by pumping and purging steps. The deposition step is performed in C_4F_8 plasma and produces a polymer fluorocarbon film with 0.2-3 nm thickness. Then activation step is performed in Argon or Oxygen plasma with DC bias applied to the waferholder. Accelerated ions enhance reaction of fluorine atoms of the polymer film with underlaying Silicon oxide which leads to controllable saturated etching.

1. Introduction

Modern semiconductor manufacturing requires advanced technologies to produce structures with atomic precision. The trend of scaling of logic and memory structures leads to the sub-10 nm nodes which means that some elements, like gate dielectric in MOSFET, are becoming of few nanometers thickness. In addition, atomic accuracy is required for various other applications, such as opening windows for contacts to two-dimensional semiconductors, the formation of spacers and others. This makes a new approach to etching — atomic layer etching (ALE) [1,2] — relevant for research, especially for dielectrics.

Depending on the goals of the process, an approach to atomic layer etching can be chosen. As a rule, processes requiring high anisotropy are realized in a plasma-stimulated or ion-stimulated process, where the direction of motion of accelerated ions determines the direction of preferential etching. If it is necessary to implement an isotropic etching process, a thermally activated process is usually chosen. The common feature of atomic layer etching is that (1) the chemical processes of delivering chemically active particles of the same type to the surface are temporarily separated and (2) the activation of their reaction with the underlying layer. In this case, provided there is no spontaneous chemical reaction with the substrate, its saturability, and no substrate etching in the second process in the absence of the reagent from the first, the so-called synergy process is provided. In this case, only the combined action of the two half-reactions leads to etching. This quality of the process should lead to obtaining the possibilities of developing etching processes with ultrahigh selectivity, and high homogeneity over the wafer.

Feeding gases are used which in the process of etching can give volatile compounds with the material to be etched, as a rule, usually compounds of chlorine and fluorine.

ALE technique has been reported for about 30 materials and most of them use non-commercial equipment [1]. Especially many works report on the implementation of atomic layer etching processes with high selectivity in the settings of ion and neutral beams. The disadvantage of such systems is the inability to scale them for commercial use for large scale wafers. On the other hand, the success of the implementation of atomic layer etching in such tools is due to the fact that there is a good capabilities for mass separation of ions and precise determination of the ion energy of the incident beam, which is not the case with the use of wide-aperture plasma etching tools.

We study plasma ALE process with two repeatable steps: deposition polymer film on the surface of target material, followed by activation of the reaction between deposited film and target material promoted by energetic particles from plasma [2].

In this work a following routine was realized: polymer film from C_4F_8 plasma was deposited as etchant film on the first step and then surface reaction was activated by accelerated ions from O_2 plasma. Both these steps are self-limiting, which makes possible to reach stable single-atomic etching process. This is not so well-studied approach to ALE because it is difficult to achieve appropriate condition in conventional plasma etching equipment [3], but it can allow whole wafer processing as opposed to the ion beam approach. As a result, this approach is perspective for industry.

2. Experimental

The experiment was carried out in PlasmaLab 100 Dual (OIPT) two-chamber RIE cluster. Etching was performed on 1x1 cm^2 samples of Silicon with deposited CVD films of SiO_2 with initial thicknesses of 350 nm. Thicknesses of oxide were measured before and after etching by spectral ellipsometer on J.A. Woollam Co M-2000X.

It is known that, in fluorocarbon plasma, depending on the parameters of the process, both etching the surface of silicon oxide and forming a polymer film on it are possible. For the process being implemented in our case it is especially important that at the first stage plasma-chemical etching does not occur, but a polymer film is formed. The equilibrium shifts towards the formation of the polymer in the general case when the temperature of the sample decreases, the relative share of carbon in the fluorocarbon feed gas increases, and the bias voltage decreases. We investigated several plasma ICP modes which, under the conditions typical for our installation, led to etching, so it was decided to implement this process in the mode with a lower density of fluorine radicals in the plasma. In our experiment, the power is not applied to the ICP coil, the RF power is applied only to the wafer holder, which creates a plasma similar in properties to the CCP. The disadvantage of this approach is that the possibility to separately regulate the plasma density in the reactor and the displacement is lost.

Deposition process was studied in details: for different sample temperature (0°C, +20°C, +50°C) (fig 1) and for different DC bias (50 V – 136 V) of C_4F_8 plasma (fig 2). CCP C_4F_8 plasma was excited for 2-20s.

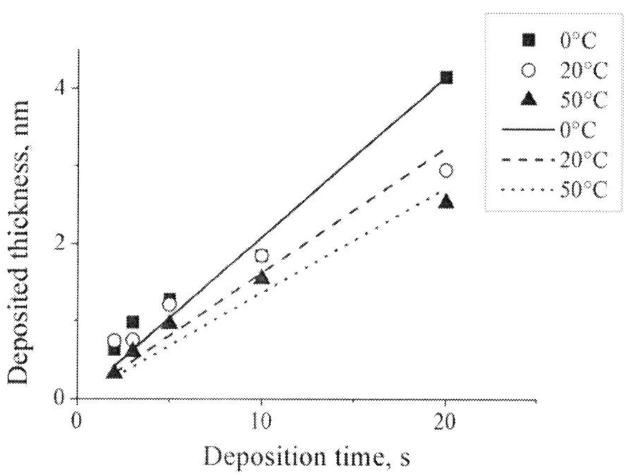

Figure 1. Deposited polymer thickness dependence on deposition step time for different temperatures.

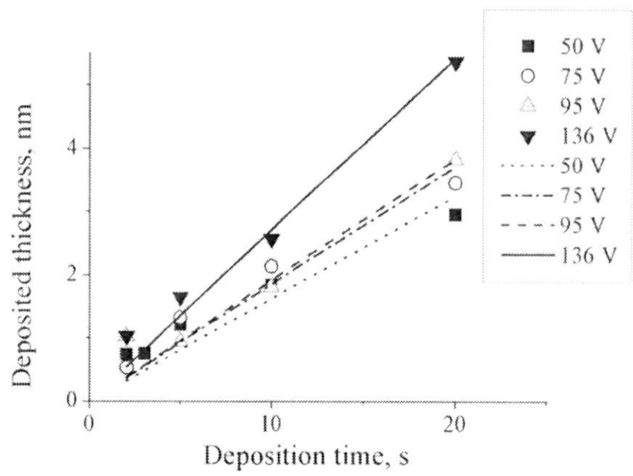

Figure 2. Deposited polymer thickness dependence on deposition step time for different DC bias of C_4F_8 CCP.

The growth rate of the film does not depend on the process time, which looks like a linear growth. It can be seen that as the temperature of the sample decreases, the growth rate of the film increases, as expected. As the bias voltage increases, the speed of the film growth increases. In our case, this is explained by the fact that with an increase in the input power in the CCP plasma, the concentration of fluorocarbon radicals will also increase.

Cyclic processes of etching were studied for sample temperature +20°C. A cycle consists of 2 steps: 1) deposition of a polymer film C_xF_y from C_4F_8 plasma at pressure 10 mTorr, which was excited by applying of RF bias (10W, DC bias of 47V) at 13.56 MHz to the wafer chuck with zero ICP power, 2) 10 second etching step by O_2 Inductively Coupled Plasma (ICP) with 15 W RF bias on the wafer chuck and 2500 W of ICP power. During the etch step DC bias was about 22V. Etching rate dependence on deposition time was obtained (fig 3).

Processes with an etching time of less than 3 seconds have not been investigated because of the nonstationarity during plasma ignition and the lack of the required installation speed. It can be seen that when a plasma wafer is exposed for 3 seconds or more, the polymer film reacts with the surface and there is a recession of silicon oxide. Further increase in the activation cycle leads to saturation of the process. This is explained by the fact that when the oxygen is exposed to the surface, an effective etching of the film occurs and it is removed, at the same time partially reacting with silicon oxide due to the energy of the incident ions, this leads to etching.

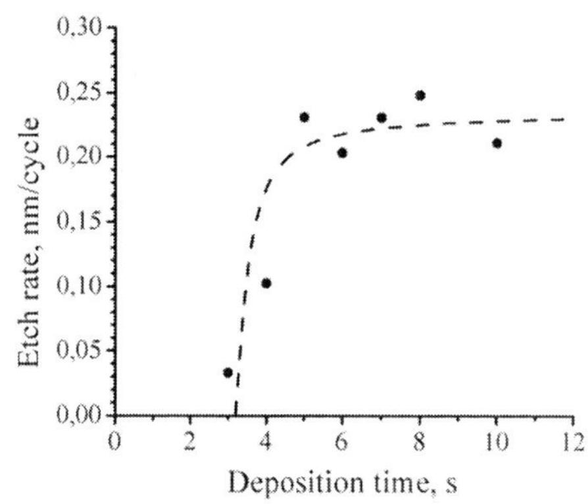

Figure 3. Etch rate dependence on deposition time for activation with O_2 plasma.

3. Conclusion

The study of polymer film deposition showed, that deposition rate decreases with the increase of sample temperature, which is associated with increasing of desorption. The growth of deposition rate with increase of DC bias can be attributed with decrease of ion flux in CCP regime, as in CCP there is no possibility to control DC bias and ion flux independently. Results of deposition rate for different sample temperature (tab 1) and different bias (tab 2)

Table 1. The table with deposition rate for different temperature.

Sample temperature, °C	Deposition rate, nm/s
0	0.21
20	0.16
50	0.14

Table 2. The table with deposition rate for different DC bias.

DC bias, V	Deposition rate, nm/s
50	0.16
75	0.18
95	0.19
136	0.27

The etching experiment showed that etch rate saturates with deposition time on 0.22 nm/cycle at approximately 5 s. That means that, if deposition step lasts less than 5 s, polymer film covers sample incompletely. And if deposition time more than 5 s activated reaction removes one monolayer of sample material. The process exhibits ALE properties of self-limiting and monolayer etch rate.

References

[1] Kanarik K J et al. 2018 *J. Phys. Chem. Lett.* **9** 4814
[2] Tsutsumi T et al. 2016 *J. Vac. Sci. Technol. A* **35(1)** 01A103
[3] Goodyear A and Cooke M 2016 *J. Vac. Sci. Technol. A* **35(1)** 01A105
[4] Rudenko K V and Miakonkikh A V 2018 VII International Symposium on Theoretical and Applied Plasma Chemistrys (Ivanovo) p 150

The effect of lead and nickel cations on $AgInS_2$/ZnS nanocrystals photoluminescence in aqueous solution

O A Korepanov, D S Mazing, O A Aleksandrova, V A Moshnikov
Micro- and nanoelectronics department Saint Petersburg Electrotechnical University "LETI" St. Petersburg, Russia

Abstract. A sensing fluorescent probe based on the photoluminescence (PL) quenching of the $AgInS_2$ colloidal nanocrystals (NCs) with a wide bandgap ZnS shell was studied for lead ions detection. The results were compared with the work where the influence of nickel ions presence on PL was studied. Addition of lead cations was accompanied by decreasing of PL intensity of the $AgInS_2$/ZnS NCs and redshift of maxima of PL spectra. The quenching effect was described by the Stern–Volmer relationship. The obtained results could be useful for the development of new express-methods for heavy metal detection in aqueous media.

1. Introduction

Environmental pollution with heavy metals is one of the leading problems of ecology. All heavy and transition metals are pollutants, the observation of which is mandatory in all environments since at high concentrations these elements have a negative impact on human health.

Traditional methods of the heavy metal ions concentration detection require a lot of time, expensive equipment and qualified personnel. In recent years, express-methods for analyzing the presence of heavy metal ions in environmental and biological samples have been actively developed. One of the promising directions is the development of fluorescent sensors based on the quenching effect of luminescence in the presence of heavy metal ions [1].

In this work, the quenching effect of PL of water-soluble $AgInS_2$/ZnS nanocrystals (NCs) and the shift of the PL spectra in the presence of Pb^{2+} and Ni^{2+} ions were studied. The main advantages of using $AgInS_2$/ZnS NCs are a wide range of emission spectra and photochemical stability. Their water solubility makes them potentially compatible with biological applications. This paper explores the possibility of using $AgInS_2$/ZnS NCs as active media for heavy metal ions sensing.

2. Experiments and results

$AgInS_2$/ZnS NCs were synthesized in aqueous solution [2,3]. $AgNO_3$ (0.2 mmol) and 3-mercaptopropionic acid (4 mmol) were dissolved in 5 ml of distilled water. The solution of 25% NH_4OH (0.45 ml) was added followed by an aqueous solution of $InCl_3$ (0.8 mmol). At room temperature 1 mmol of $Na_2S{\cdot}9H_2O$ were injected rapidly resulting in instantaneous nucleation. The mixture was heated up to 100°C for 30 minutes.

To coat thus obtained nanoparticles with ZnS shell complexes of Zn^{2+} with 3-mercaptopropionic acid (1 mmol) were added to the nanoparticle solution followed by dropwise addition of $Na_2S{\cdot}9H_2O$ (1 mmol). A certain amount of NaOH was used to raise pH and make the mixture translucent. The solution was then heated up to 100°C for 30 minutes. Nanoparticles were precipitated by addition of 2-

propanol followed by centrifugation and redispersed in distilled water or phosphate-buffered saline for further experiments and studies.

An analogous approach was used for the preparation of $AgInS_2$ nanoparticles using L-Glutathione as a surfactant with various ratios of cation precursors [Ag]:[In]. For the ratio [Ag]:[In] = 1:4 0.005 mmol of $AgNO_3$ and 0.02 mmol of $In(NO_3)_3 \cdot 4.5H_2O$ were dissolved in 5 ml of H_2O. 0.01 mmol of L-Glutathione and 0.08 mmol of sodium citrate were added to the solution. Nucleation was performed by injecting of 0.04 mmol of $Na_2S \cdot 9H_2O$. The solution was heated up to 95 ° C for 40 minutes. The same protocol was used for synthesis with [Ag]:[In] = 1:20 and [Ag]:[In] = 1:1

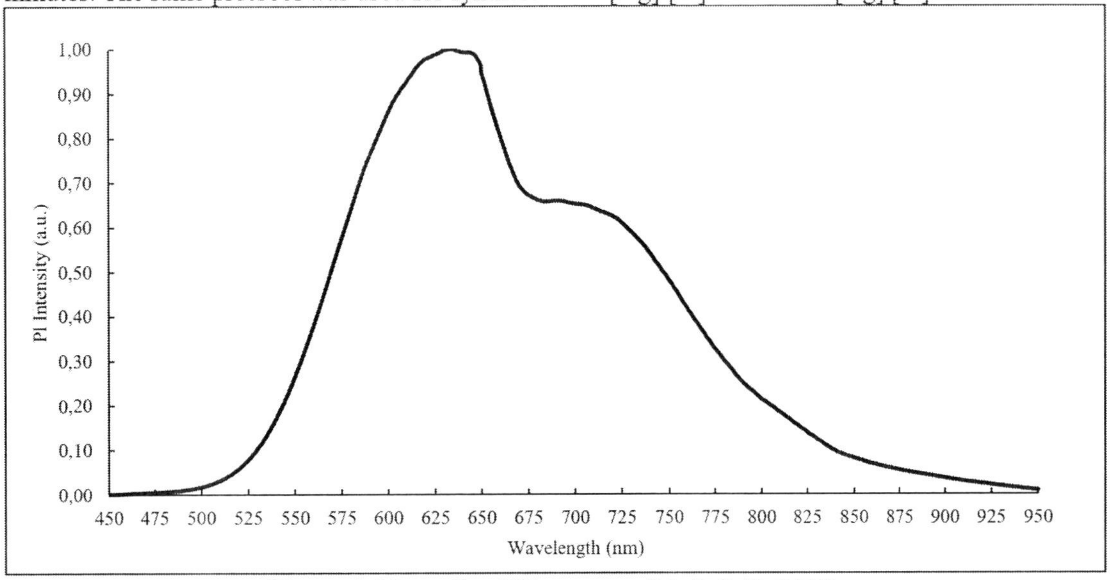

Figure 1. Normalized PL spectra of $AgInS_2/ZnS$ NCs

PL measurements were carried out on spectrofluorometer based on MDR-206 monochromator using laser excitation on $\lambda = 405$ nm. As can be seen from fig. 1, the PL spectra $AgInS_2/ZnS$ NCs are characterized by a wide PL spectrum (100–150 nm) in the wavelength range of 500–800 nm and have two characteristic maxima.

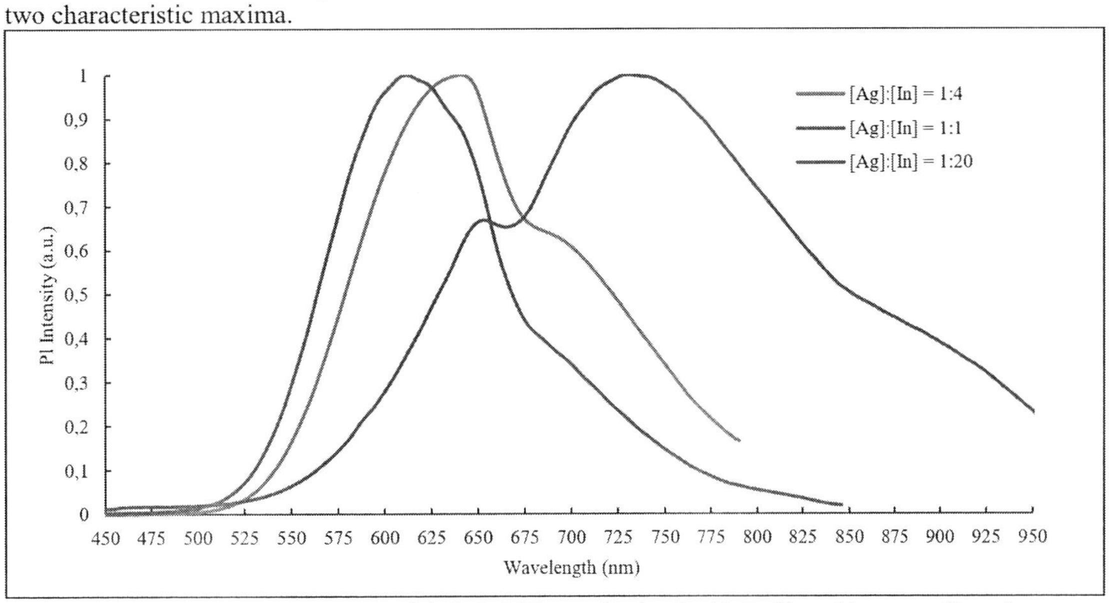

Figure 2. PL spectra of $AgInS_2$ NCs synthesized with L-Glutathione as a ligand.

By changing the ratio of the constituent components of the $AgInS_2$ NCs, in particular [Ag] and [In], it is possible to achieve the most desirable shape of the spectrum (the position of the PL maxima and their intensity) [4]. With increasing Ag concentration in the solution, the position of the low-energy luminescence peak shifts to the long-wavelength region, which indicates that NCs with a high Ag content have narrower band-gap, which is attractive from the point of view of the application in biomedical research. The nature of multiple PL bands is not clear. The short-wave peak is presumably associated with the surface-related states of NCs and the long-wave peak – with internal defects [5].

In sensorics, complex forms of the spectrum are preferable, since they allow a qualitative analysis of the metal cations present in the NCs solution with greater accuracy due to the redistribution of the intensities of the maxima of the PL spectra.

In these experiments to assess the effect of heavy metal ions on the PL quenching of $AgInS_2/ZnS$ NCs in aqueous solution, $AgInS_2/ZnS$ NCs with a molar ratio of [Ag]:[In] = 1:4 stabilized by 3-mercaptopropionic acid were used.

The source of the Pb^{2+} ions was lead (II) chloride, which was dissolved in distilled water to obtain different concentrations. The experiments were performed in the medium of phosphate-buffered saline, containing a constant amount of $AgInS_2/ZnS$ NCs. The final volume in each case was 1 ml; the pH of the medium in the investigated concentration range was equal to 7.4. The addition of Pb^{2+} ions to the solution resulted in a decrease of the PL intensity. Nickel ions were introduced in a similar way, where nickel (II) chloride was used as an ion source. Addition of nickel cations from nickel chloride in the solution results in changing of transparency and colour of the medium. The solution became yellowish and turbid [6]. As a result due to the low solubility of lead chloride in water, the experiment was performed in a narrow range of concentrations.

The quenching effect of Pb^{2+} and Ni^{2+} is described using the Stern-Volmer equation (1):

$$I/I_0 = 1 + K_{sv}/C_{Me}{}^{2+} \tag{1}$$

I_0 и I are PL intensities for the pure solution of $AgInS_2/ZnS$ NCs and in the presence of different concentrations $C_{Me}{}^{2+}$ of divalent metal cations, K_{sv} is the Stern–Volmer quenching constant.

Figure 3 (left) shows the Stern-Volmer relationship (1) for $AgInS_2/ZnS$ in the presence of nickel and lead cations. The approximating functions coincide at low concentrations of lead ions obtained using lead chloride. The Stern-Volmer constant is 7.87. The same results were obtained using lead cations from lead chloride with a concentration up to 0.48 mmol/L. Due to the low solubility of lead chloride in water at room temperature, it was not possible to investigate the effect of lead cations on the PL of $AgInS_2/ZnS$ NCs in a wide concentration range. Therefore, it wouldn't be possible to distinguish these metals in the solution. In addition to quenching of the PL, a blueshift of the PL spectrum is observed for nickel. This shift was 32 nm in the range of nickel ions concentration from 0 to 3.74 mmol/L. Using lead (II) nitrate as a source of lead cations, the PL spectrum was obtained at a concentration of 3.62 mmol/L. In this case, the shift was towards longer wavelength accompanied by the opposite PL spectrum rearrangement. Figure 3 (right) shows the PL spectrum of a solution without heavy metal ions, a solution containing lead cations with concentration 3,41 mmol/L, and a solution containing 3,62 mmol/L of nickel cations [6].

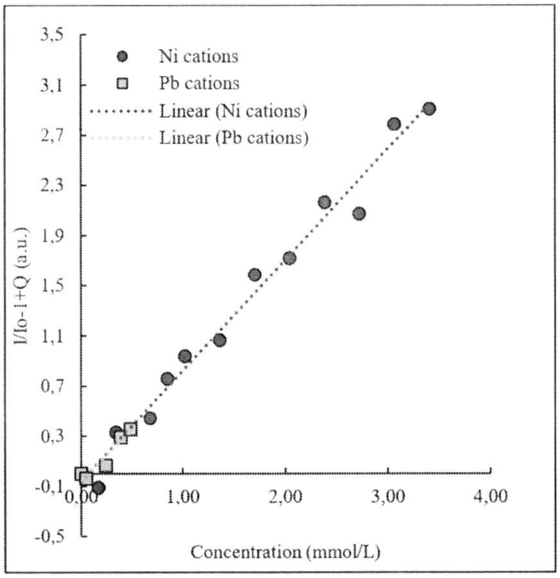

a

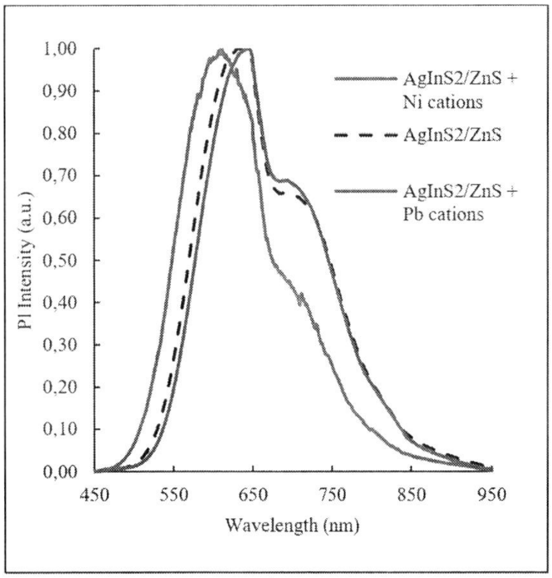

b

Figure 3. (a) The Stern–Volmer relationship for aqueous solutions of $AgInS_2/ZnS$ NCs with Ni^{2+} and Pb^{2+} ions; **(b)** Normalized PL spectra of $AgInS_2/ZnS$ NCs in presence of Ni^{2+} and Pb^{2+}.

The decomposition of complex shapes of the spectra could be done using at least four Gaussians.

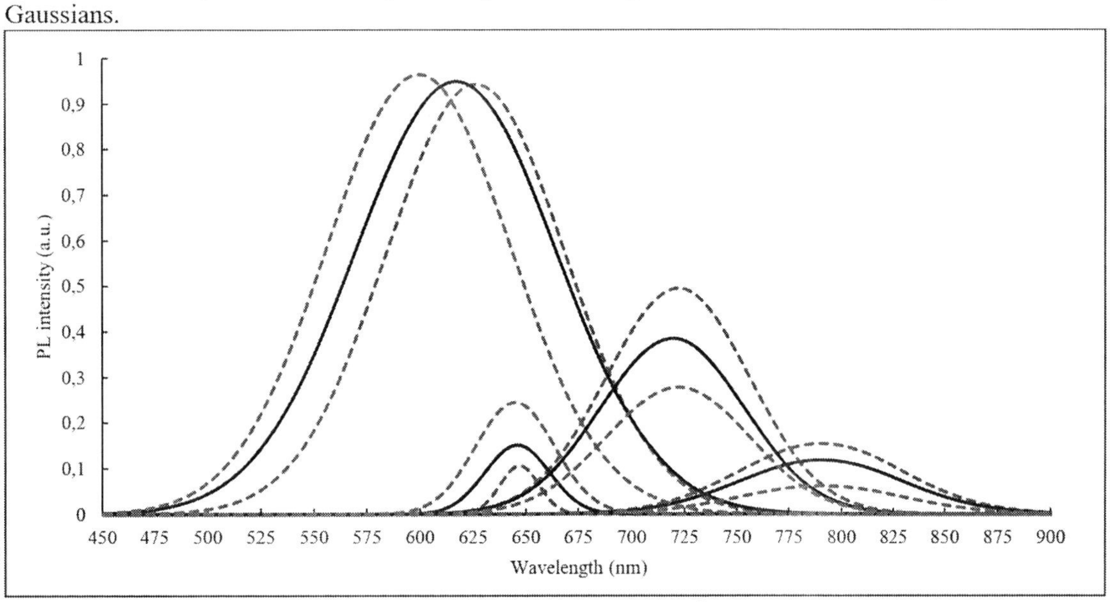

Figure 4. The decomposition results of normalized PL spectra of $AgInS_2/ZnS$ NCs.
Black solid lines correspond to the pure solution; blue dash lines correspond to the solution in the presence of Ni^{2+} cations; red dash lines - to the solution in the presence of Pb^{2+} cations.

From fig. 4, it can be seen that the most significant shift in the PL spectra occurs for a short-wave maximum. This can be related to the interaction of heavy metal cations with free surface states, having an impact on the process of luminescence quenching.

3. Conclusion

As a result, with the addition of various metal ions to the $AgInS_2/ZnS$ NCs solution, the quenching of the PL is observed. Furthermore, there is a change in the shape of the spectrum and its position. The addition of nickel cations leads to a blueshift of the PL spectrum with the decreasing of the low energy peak of the PL band. The addition of lead nitrate causes the opposite effect: there is a PL redshift with a relative increase of the low energy peak intensity. It can be assumed that the addition of nickel ions primarily changes the maximum associated with surface states, which results in a shift of high energy maximum towards shorter wavelengths. The addition of lead ions leads the opposite results, primarily due to the formation of deep-level centres in NCs or the occurrence of additional paths for nonradiative recombination.

These effects suggest possible selectivity of the method for heavy metal ions detection in aqueous solutions. However, further studies are needed to study the physicochemical processes of the interaction of heavy metal ions with $AgInS_2/ZnS$ NCs.

References

[1] Vázquez-González M, Carrillo-Carrion C, 2014 *J. Biomedical Optics* **19.10** 101503

[2] Mazing D S, Korepanov O A, Aleksandrova O A, Moshnikov V A, 2018 *Optics and Spectroscopy* **125** (5) 773-776

[3] Raevskaya A, Lesnyak V, Haubold D, Dzhagan V, Stroyuk O, Gaponik N, Dietrich RT Zahn, Eychmüller A 2017 *The Journal of Physical Chemistry C* **121** (16) 9032-9042.

[4] Mazing D S, Chernaguzov I S, Shulga A I, Korepanov O A, Aleksandrova O A, Moshnikov V A 2018 Journal of Physics: Conference Series 1038 (1) 012050

[5] Mao B, Chuang C H, Wang J, Burda C 2011 *The Journal of Physical Chemistry C* **115** (18) 8945-8954.

[6] Korepanov O A, Mazing D S, Aleksandrova O A, Moshnikov V A 2019 Proceedings of the 2019 IEEE Conference of Russian Young Researchers in Electrical and Electronic Engineering, ElConRus 2019 771-773.

Synthesis and Luminescent Properties of Perovskite Quantum Dots

A G Son[1,2], E V Krivogina[2], N V Romanov[1,2]
[1]Department of Molecular and Chemical Physics, Moscow Institute of Physics and Technology, Moscow 141701, Russia
[2]Kurnakov Institute of General and Inorganic Chemistry RAS, Moscow 119991, Russia

Abstract. In this paper, we obtained nanoparticles based on ABX_3 type of metal halide perovskites using the growth method of colloidal quantum dots in non-polar environment. The development of technological processes allowed us to synthesize the studied nanoparticles with room temperatures. The dependence of the photoluminescence intensity of the concentration of sols dots was determined using the photoluminescence method in the range from 400 to 700 nm. The effect of various stabilizers on the optical properties of nanoparticles was studied and the optimal composition of the stabilizer was defined. The method of dynamic light scattering was used to estimate the approximate dimensions of the synthesized ABX_3 type of metal halide nanoparticles.

The studies of solar cells focused on organic - inorganic lead trihalide perovskites. The most studied and common materials are lead methylammonium triiodide ($CH_3NH_3PbI_3$ or $MAPbI_3$) and triiodide formamidium lead ($NH_2CHNH_2PbI_3$ or $FAPbI_3$), which are three-dimensional hybrid organic-inorganic perovskite semiconductors with the general chemical formula ABX_3. These kind of materials demonstrate large distances of electron-hole transitions, connection energy of exition, high carrier mobilities, and suitable dad zones for practical use in photovoltaic cells. Inorganic perovskites are attractive by promising high heat resistance from a material with a band gap suitable for converting solar energy and exhibit significant hysteresis in current and voltage measurements. In connection with such good properties of inorganic perovskites, many were interested in how good their properties would be in nanocrystals [1].
In the modern world, special attention is paid to semiconductor nanocrystals associated with their physico-chemical data. Such nanoparticles are called quantum dots (QDs). QDs can be used in LEDs [2], lasers [3], photosensitizers, and as biolabels [4]. Semiconductor nanocrystals are promising materials for the development of photovoltaic cells (PE). Obtaining QDs often doesn't require expensive and sophisticated equipment, which makes it possible to use efficient and cheap EFs, and, moreover, solar energy may become more accessible and promising.
The most studied system is cadmium-based nanocrystals; however, there are difficulties with their use in large-scale production. The disadvantages of cadmium-based nanocrystals are high reaction temperature and surface passivation, which can't provide a high quantum yield of photoluminescence [5-6]. The first perovskite quantum dots were obtained by the Kovalenko group [7]. Scientists have used the hot injection method for the synthesis. The nanoparticles emit light in the entire range of the

visible spectrum, and the emission is controlled by the size and composition of the quantum dots. For the first time, bright LEDs based on perovskite quantum dots were made showing customizable color emission [8]. The hot injection method didn`t solve the problem of high temperatures in the synthesis and high dispersion of nanocrystals.

The method of synthesis at room temperature came to replace the hot injection method. This method doesn`t solve the problem of high dispersion of the obtained nanoparticles and requires an inert atmosphere. But the indices of the quantum yield of photoluminescence of nanoparticles are not inferior to the indices of QDs synthesized by the hot injection method [9-11]. Synthesis is based on oversaturated recrystallization, which works at room temperature by transferring ions from soluble solvents to insoluble ones [9]. This process takes a few seconds, and is easily reproducible and inexpensive. QDs synthesized by this method exhibit excellent optical properties, which makes them a promising material in the field of optoelectronic devices [9]. The synthesis mechanism is shown in Figure 1.

We obtained quantum dots of the ABX_3 type with the emission band in the region of the visible spectrum at 450 nm and 540 nm. The emission band at 450 nm matched to the blue emission in the visible region of the spectrum, and the emission at 540 nm conformed to the green emission (Fig 2 (a)). We also determined the size of the synthesized nanoparticles using the method of dynamic light scattering, they are 10, 20 and 200 nm. We assume that the luminescence emission band of nanoparticles depends on the size of the quantum dots. So quantum dots with a size of 20 nm predominantly have a blue glow of the visible spectrum, and nanocrystals with sizes of 200 nm have a green glow. These results are consistent with the literature [9-11].

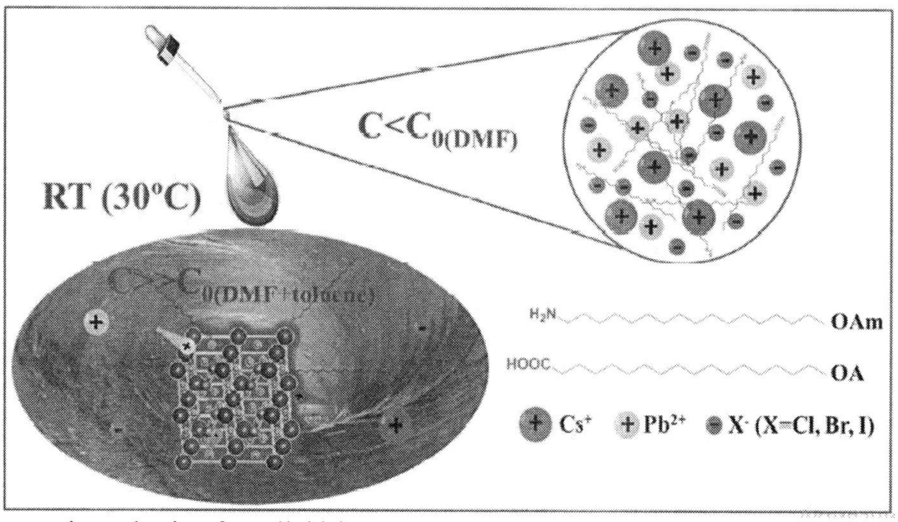

Figure 1, growth mechanism for colloidal quantum dots based on ABX_3 type of metal halide perovskites

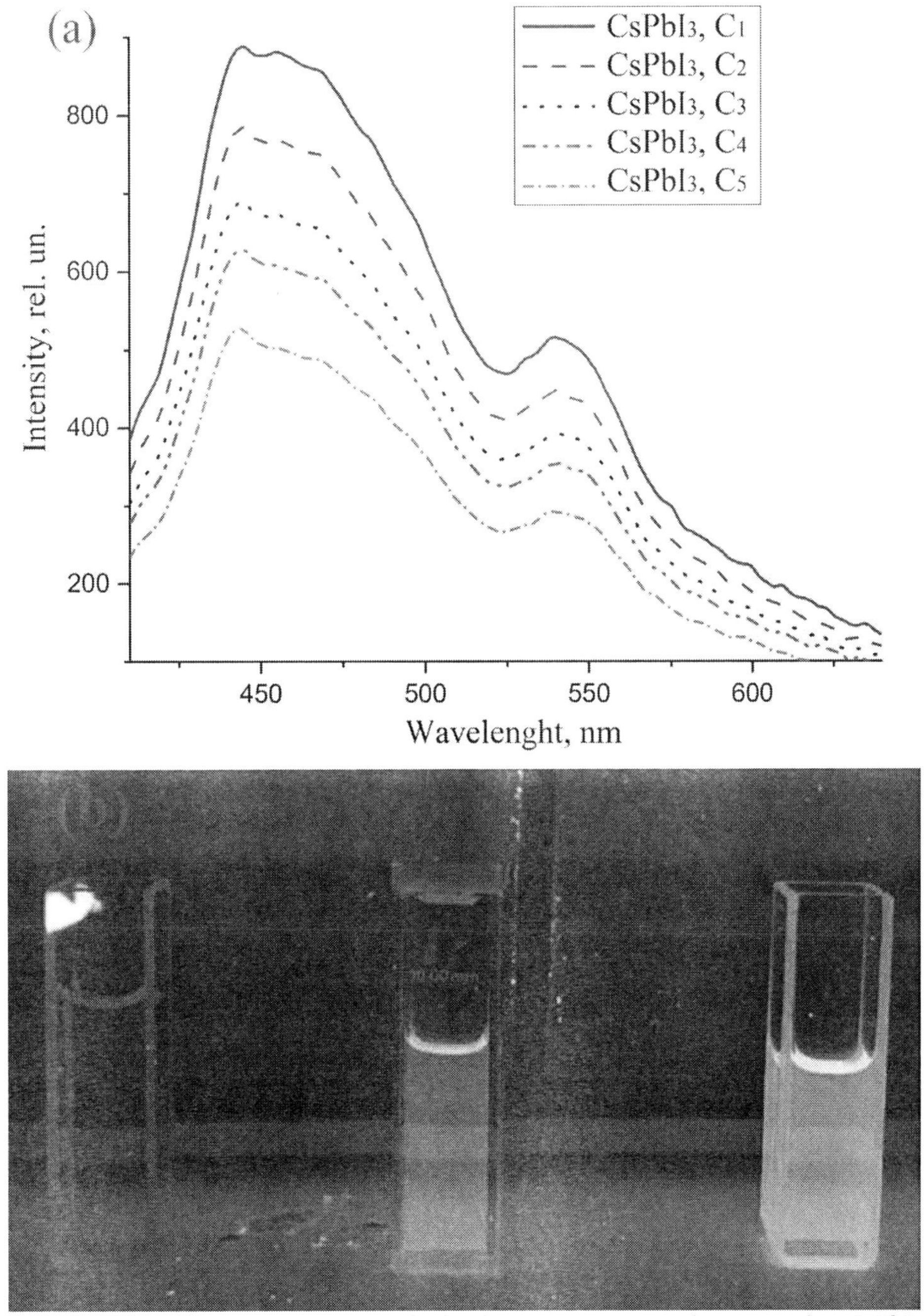

Figure 2(a, b). (a) The figure shows the luminescence spectra of quantum dots $CsPbI_3$ depending on the concentration ($C_1>C_5$), λ_{ex}=390 nm; **(b)** The figure shows the comparison of quantum dots $CsPbI_3$ of different sizes with water (from left to right is water, $CsPbI_3$ 20 nm, $CsPbI_3$ 200 nm)

In our work, we determined the relative quantum yield of luminescence. Such a quantum yield is determined with respect to a substance with a known quantum yield.

$$\varphi_i = \frac{(1 - 10^{-Ds}) \cdot S_i \cdot n_i^2}{(1 - 10^{-Di}) \cdot S_s \cdot n_s^2} \cdot \varphi_s \,,$$

where φ_i, φ_s – luminescence quantum yields of the test substance and standard; S_i, S_s - areas under the emission spectrum of the test substance and standard; D_i, D_s – absorptions of these substances at the excitation wavelength; n_i, n_s – refractive indices of solvents in which studies occur.

References

5.1. For papers:

[1] G E Eperon, G M Paterno, R J Sutton, A Zampetti, A A Haghighirad, F Caciallibc, H J Snaith, 2015, *J. Mater. Chem. A*, 3, 19688
[2] V. L. Colvin, M. C. Schlamp, A. P. Alivisatos, 1994, *Nature* **370**, p 354–357
[3] Klimov VI, Mikhailovsky AA, Xu S, Malko A, Hollingsworth JA, Leatherdale CA, Eisler H, Bawendi, 2000, *Science* 290, p 314-317
[4] W J Parak, D Gerion, T Pellegrino, D Zanchet, C Micheel, S C Williams, R Boudreau, M A Le Gros, C A Larabell, A P Alivisatos, 2003, *Nanotechnology,* 14, R15–R27
[5] H Han, Z Sheng, J Liang, 2007, *Analytica chimica acta.*, p 73-78
[6] G Wang, X Su, S Yang, Y Jia, D Li, 2012, *Journal of Luminescence*, p 2505-2511
[7] M V Kovalenko, L Manna, A Cabot, Z Hens, D V Talapin, C R Kagan, V I Klimov, A L Rogach, P Reiss, D J Milliron, P Guyot-Sionnnest, G Konstantatos, W J Parak, T Hyeon, B A Korgel, C B Murray, W Heiss, 2015, *ACS Nano, 9* , 1012
[8] J Song, J Li, X Li, L Xu, Y Dong, H Zeng, *Adv. Mater.*, *27*, 7162
[9] X Li, Y Wu, S Zhang, B Cai, Y Gu, J Song, H Zeng, *Adv. Mater.*, 2016, *26*, 2435–2445
[10] Y Tong, A Manzi, V A Hintermayr, A S Urban, L Polavarapu, J Feldmann, *Angew. Chem. Int. Ed.* 2016, 55, 1 – 7
[11] S Wei, Y Yang, X Kang, L Wang, L Huang, D Pan, *Chem. Commun.,* 2016, 52, 7265

On the dendrite growth simulation during multitrack selective laser melting process

A V Dubrov [1], **F Kh Mirzade** [1], **V D Dubrov** [1]
[1]ILIT RAS – Branch of the FSRC «Crystallography and Photonics» RAS, Svyatoozerskaya 1, 140700, Shatura, Moscow Region, Russia

Abstract. The evolution of the microstructure in the selective laser melting (SLM) process is investigated in the framework of a multiscale approach. The kinetic equation for the phase-field (order parameter) coupled with the energy transfer equation is used to describe the crystallization processes. To study the effect of technological parameters on the microstructure of the synthesized product, and to establish a connection between the models of macro- and micro-scale through the temperature gradient, numerical studies of macro processes in the formation of a series of parallel SLM tracks were carried out. It is shown that the adjustment of the laser power level changes the temperature dynamics in the test areas, which, in turn, through the changed temperature gradients affects the evolution of the microstructure in the micro-scale model. The calculations showed that the implemented numerical model qualitatively describes the kinetics of crystallization and formation of microstructures in the SLM process.

1. Introduction

Additive manufacturing (AM) has received increased attention in recent years because of its ability to produce parts directly from CAD data [1]. Selective laser melting (SLM) is a method of AM, in which the creation of a metal part is carried out through the layer-by-layer addition of powder materials. The SLM process uses a high-intensity laser source to create a melt pool in the powder layer. The process flow is determined by a large number of factors such as laser beam power, laser spot size, scanning speed and strategy, powder layer thickness, particle shape and distribution function, powder physical and chemical properties, etc. It should be noted the relationship of factors affecting the process and the presence of many interacting processes: the absorption and scattering of laser radiation by powder particles, heat conduction and convection, the evolution of the free surface of the melt due to capillary forces, evaporation, shrinkage, crystallization, the formation of the microstructure of the synthesized object [1].

From numerous experiments it is well known that a variety of microstructures (columnar, cellular, dendritic, cellular-dendritic), arising at the stage of crystallization in SLM, and the mechanisms of their formation and changes largely determine the physical and mechanical characteristics of the synthesized product. At the same time, the use of a single direct numerical simulation of dendritic crystallization to predict the properties of the material presents serious difficulties. The main problems here are related to the description of the complex interaction of nonlinear processes occurring at different scale levels. Recently, new directions of research concerning multilevel (two-level) models are emerging, considering jointly the processes at different scale levels: micro- and macro-levels [2]. In [3], a physical and mathematical model of the dynamics of the liquid-solid phase boundary movement and prediction of the microstructure in the SLM of the powder compact using a two-level (macroscopic–microscopic) approximation was developed.

In this paper, the growth dynamics and morphology of columnar dendritic structures are numerically investigated. For this purpose, the simultaneous growth of several crystallites under temperature gradient is considered. In addition, to establish the connection of macro- and micro-scale models through a local temperature gradient, numerical studies of macro processes in the formation of a series of parallel SLM tracks were carried out.

2. Numerical simulation of the evolution of the microstructure

The evolution of the microstructure (micro-level) is described by the phase-field method, which relies on the use of a differential equation for the order parameter coupled with the equations of thermal conductivity [3]. The phase-field model is very effective in the description of nonequilibrium dendritic evolution. In numerical studies based on them, all control equations are written for the entire region without separating the interface from the solid and liquid phases and no direct tracking of the interface position is required. Simulation of solidification in SLM by the phase-field method involves the growth of grains in the supercooled liquid phase. According to this hypothesis, it is necessary to take into account three regions: a crystalline nucleus, a liquid phase, and a solid / liquid interface. We introduce the characteristic of the phase state of matter - the phase-field: $\phi(\mathbf{r},t)$. A variable $\phi(\mathbf{r},t)$ can be interpreted as the degree of orderliness of a material in a microvolume with a radius vector $\mathbf{r}$ at time t. The phase-field variable $\phi=0$ for liquid, and $\phi=1$ for solid and varies smoothly between those values over the solid–liquid interface, so $\phi\subset(0,1)$ in the diffuse interface. The evolution of the microstructure is modeled by the kinetic equation (1) for the phase field (ϕ), which is coupled with the heat equation (2)

$$\tau\dot{\phi}=\varepsilon_0^2\nabla\cdot\left[\tilde{\varepsilon}(\theta)\left(\tilde{\varepsilon}(\theta)I+\tilde{\varepsilon}'(\theta)J\right)\cdot\nabla\phi\right]-\omega_g g'(\phi)+30g(\phi)\left[L_0\left(1/T-1/T_m\right)\right], \quad (1)$$

$$c\dot{T}=\chi\nabla^2 T-\left[p'(\phi)L+\omega_e g'(\phi)\right]\dot{\phi}, \quad (2)$$

$m_\phi=\tau^{-1}$ is the mobility of the phase boundary, L_0 and T_m are the heat of fusion and melting point, respectively. The constant ω_g characterizes the height of the energy barrier between phases: $\omega_g=3\sigma/\sqrt{2}T_m\delta$ (σ is the surface energy, δ is the width of the front of the phase transition), $\alpha^2=6\sqrt{2}\sigma\delta/T_m$ characterizes the width of the diffuse interface, $\delta=\varepsilon/\sqrt{4\omega}$, $\chi=M_eT^{-2}$ is the thermal conductivity, c is the heat capacity. ε is the coefficient taking into account the anisotropy of the surface energy: $\varepsilon=\varepsilon_0\tilde{\varepsilon}(\theta)$ ($\varepsilon_0>0$). The variable θ is the angle between the x-axis and the normal vector of the interface $\mathbf{n}=\nabla\phi$. For 2D geometry: $\tilde{\varepsilon}(\theta)=1+a\cos 4\theta$, $\theta=\arctan[\partial\phi/\partial y/\partial\phi/\partial x]$ [4]. The interpolation functions $g(\phi)=\phi^2(1-\phi)^2$ and $p(\phi)=\phi^3\left(10-15\phi+6\phi^2\right)$ are chosen in such a way as to provide a description of the interface of a finite width (the free energy potential has minima at $\phi=0$ and $\phi=1$).

The formation of the crystal structure of the material occurs as a result of the formation and growth of a set of crystalline nuclei. The rate of crystal formation is determined by the local degree of supercooling. Depending on this, the number of simultaneously growing crystallites in this area may vary. Materials obtained by laser AM characterized by high temperature gradients and high cooling rates. The most typical microstructure in AM of austenitic stainless steel is the cellular structure and columnar dendrites [5].

This section numerically studies the growth dynamics and morphology of columnar dendritic structures. For this purpose, the simultaneous growth of several crystallites under temperature gradient is considered. The development of several crystallites in the adjacent space leads to their competition [6]. If their crystallographic orientation is close, then due to the close location only the lateral branches are suppressed, and the primary branches grow in a parallel direction, forming a columnar structure. A

constant value of the temperature gradient was used, a fixed value of dimensionless supercooling $9 \cdot 10^{-4}$ was set at the left boundary and $7 \cdot 10^{-4}$ at the right boundary. These values are typical for the SLM process of metallic materials. Iron-specific data were used as material properties.
Calculation of dendritic microstructures was carried out in a two-dimensional area of 15*15 microns, which was divided into 2000 cells in each dimension. The total number of calculated cells in the mesh was 4 million. Before the start of the calculation, round "crystallites" with a radius of 8 cells were placed in the required places of the calculated area. A static dimensionless time step was used, equal to $0.5 \cdot 10^{-3}$. To carry out calculations in parallel, the spatial domain was decomposed into 16 subregions. The duration of the calculation was about 2 hours.

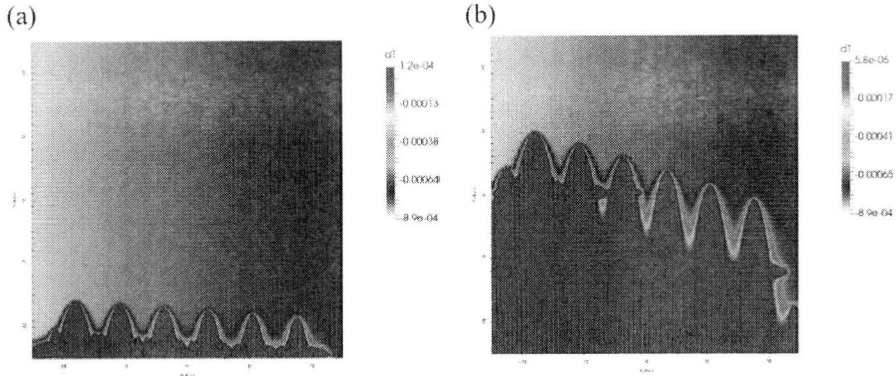

Figure 1. The distribution of temperature and phase fields during the evolution of six dendrites. Dimensionless time: **(a)** 0.15, **(b)** 0.6. Parameters: Gradient coefficient - 0.03, Thermal noise amplitude - 0.

Figure 1 shows the development of six crystallites uniformly placed at the initial time on the lower boundary of the computational domain. Evolution within 30 μs is demonstrated, which corresponds to the value of the dimensionless time 0.6. The color indicates the temperature distribution. Due to the dense packing of crystallites, almost complete suppression of the growth of secondary branches occurs, and a columnar microstructure develops. In the conditions of temperature gradient the difference in growth rates is observed. The development of dendrites, which are at a greater supercooling closer to the left border, is ahead of structures in less favorable conditions.
The morphology of columnar dendrites obtained by simulation qualitatively correspond to that observed in experiments, for example [7]. The calculations have shown that the numerical model qualitatively describes the physics of crystallization and formation of microstructures in the SLM process.

3. Influence of macro-conditions on microstructure development
To link the macro- and micro-scale models, the temperature gradient used in the calculation of the microstructure evolution can be taken from the corresponding sub-domain of the macro-scale. To study the effect of technological parameters on the final quality of the synthesized product, parametric studies of macro processes were carried out in the formation of a series of parallel SLM tracks. SS304L stainless steel data were used as substrate and powder material properties.
Figure 2 shows the temperature dependences in the test areas when scanning three parallel tracks for two laser power levels. The temperature measurement test areas are located at the substrate level in the middle of the respective tracks. The temperature peaks on the graphs correspond to the moments of the beam passing through the corresponding points. Each region experiences 3 heating-cooling cycles, which indicates that the material is subjected to cyclic processes of temperature change, which can lead to high residual stresses [8]. With the use of higher power, there is a slight increase in the maximum achievable temperature at the time of passage of the laser beam through the temperature measurement area (about 50-100 degrees). At the same time, the temperature drop occurs with a delay of about

1-1.5 ms. Preheating of the material during the processing of adjacent areas also increases by 50-100 degrees, but the dynamics of heating to melt point does not change. Also, the heating pattern of the first track and the subsequent ones, the material of which was preheated, does not differ.
It is shown that the adjustment of the laser power level changes the temperature dynamics in the test areas on the substrate surface, which, in turn, through the changed temperature gradients will affect the calculation of the microstructure evolution in the micro-scale model.

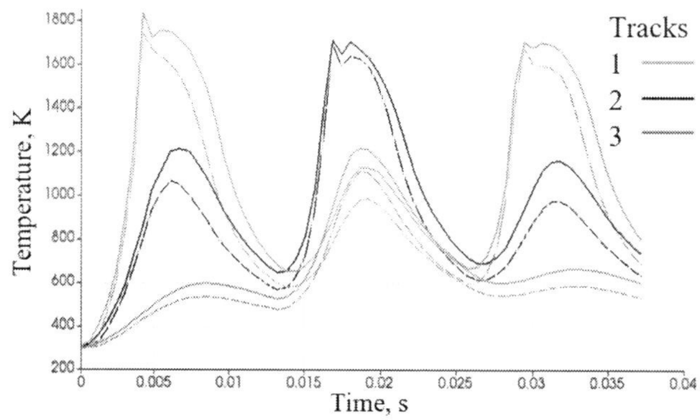

Figure 2. Evolution of the temperature in the test areas when scanning three parallel tracks with different laser power: dashed line – 200 W; solid line – 300 W. Process parameters: scanning speed 60 mm/s, the thickness of the powder layer 130 microns, the distance between the tracks 154 microns, particle size distribution 40-100 microns.

Acknowledgments
This work was supported by the Ministry of Science and Higher Education within the State assignment FSRC «Crystallography and Photonics» RAS in part of macroscale effects analysis, RFBR (Project No.16-29-11743 ofi_m) in part of microscale simulation.

References
[1] Brandt M 2016 *Laser Additive Manufacturing: Materials, Design, Technologies, and Applications* (Woodhead Publishing)
[2] Markl M, Korner C 2016 *Annu Rev Mater Res* **46** 1
[3] Mirzade F Kh, Dubrov A V 2017 *CEUR Workshop Proc* **1900** 108
[4] Warren J A, Boettinger W J 1995 *Acta Metall Mater* **43**(2) 689
[5] DebRoy T, Wei H L, Zuback J S et al. 2018 *Progr Mater* Sci **92** 112
[6] Clark D, Bache M R, Whittaker M T 2010 *Metal Mater Trans B* **41**(6) 1346
[7] Wang D, Song C, Yang Y et al. 2016 *Mater Des* **100** 291
[8] Dai K, Shaw L 2002 *Rap Prototyp J* **8**(5) 270

The two-stage electrochemical deposition of a manganese-doped silicon-carbon film onto the silicon (100) substrate

T S Mikhailova[1], **M N Grigoryev**[2], **T N Myasoedova**[1]
[1]Institute of Nanotechnologies, Microelectronics and Equipment Engineering, Southern Federal University, Taganrog 347900, Russia
[2] JSC "Taganrog Scientific-Research Institute of Communication", Taganrog 347913, Russia

Abstract. Manganese-doped silicon-carbon films with a thickness about 100 nm were deposited on a silicon substrate by using the electrochemical route from the methanol, hexamethyldisilazane and manganese sulfate (CH_3OH, $(CH_3)_3$-Si-NH-Si-$(CH_3)_3$ (HMDSN), $MnSO_4 \cdot 5H_2O$) solution. At the first stage, the current density decreased from 51 mA/cm^2 to less than 37 mA/cm^2 with increasing the deposition time and slightly increased at the second stage when manganese sulfate was added. The analysis derived from the X-ray diffraction pattern indicates that Mn atoms interact with Si atoms and form mainly Mn_5Si_2 and $Mn_{22,6}Si_{5,4}C_4$ phases.

1. Introduction

The synthesis of silicon-carbon films has been widely studied due to their attractive properties, such as wear resistance, chemically inertness, and wide band gap, which have already provided optical, electronic and other applications in aggressive ambient [1].

A serious difference between nanocomposites based on silicon-carbon films from other carbon coatings is that such films can be modified by high concentrations of metals (up to 50 at % of metal), with their silicon-carbon matrix remaining in an amorphous state [2]. This fact makes it possible to vary the metal content in a wide range. The chemical modification with metals provides new properties of the films that make it possible to use them in supercapacitors [3], Li-ion batteries [4], gas sensors [5], magnetic devices [6] and etc.

Silicon-carbon films derived from organosilicon substances have high adhesion to various substrates (silicon, sitall, glass, corundum etc.) and thermal stability (at least 300°C). The films can be obtained with a thickness of up to 100 microns [7, 8].

Traditional methods for silicon-carbon films deposition consist are vapor phase physical or chemical processes (typically PVD or CVD) and their variants. However, these techniques require high temperature, high vacuum and they are hard to be conducted in components of complex geometry. In this context, electrochemical deposition, because of its flexibility in operating parameters, maybe conducted at room temperature and it favors the industrial applications, becoming an alternative to traditional methods [9].

There is experimental evidence that most materials which can be deposited from the vapor phase can also be deposited in the liquid phase using electroplating techniques and vice versa [10, 11].

In this paper, the possibility of obtaining manganese-doped silicon-carbon films by a simple electrochemical route is investigated [12].

It is known, that Mn-doped SiC is a promising ferromagnetic material [13]. Also manganese oxides, due to their high specific capacity, availability, and environmental safety are of the greatest interest as active electrode materials for electrochemical supercapacitors [14].

2. Experimental

The electrolytic deposition system was used to obtain silicon-carbon films. It was substantiated and shown in [15]. Silicon-carbon film was deposited on a silicon substrate (100), with a size of 12x17x0.4 mm^3 and resistivity of 4.5 Om·cm. First, the substrate was dipped in the HF solution for a few minutes. Silicon substrate was mounted on the negative electrode and graphite was mounted on the positive electrode. The distance between the substrate and the positive electrode was set to 10 mm [15]. The deposition was carried out in two stages. At the first stage, a silicon-carbon film was precipitated from a solution of methanol with HMDSN (ratio 9:1) for about 30 minutes. Then manganese sulfate was added to the solution (0.05%), and the Mn-doped film was deposited for 15 minutes. The two-stage deposition process was due to the low adhesion of the manganese-doped film to the silicon substrate. The deposition time at the second stage was reduced due to the higher conductivity of the metal-containing solution.
The deposition potential was 300 and 100 V at the first and at the second stage, respectively.
Figure 1 shows the curves of current dependence on the deposition time for two types of solution.
During the deposition for the film, we found that the current density decreased from 51 mA/cm^2 to less than 37 mA/cm^2 with increasing the deposition time for a methanol solution of HMDSN and slightly increased for the solution with manganese sulfate. In both cases, a film was deposited on the cathode.

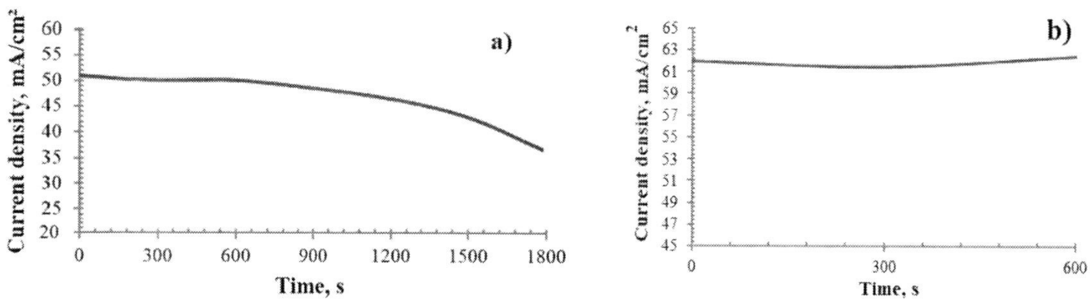

Figure 1 (a, b). (a) Current dependences on deposition time for a methanol-HMDSN solution; **(b)** Current dependences on deposition time fora methanol-HMDSN solution with manganese additive.

The thickness of the Mn-doped film after the two-stage deposition process was about 100 nm.
The silicon-carbon films investigated in this work are complex heterogeneous objects.
The Raman spectra (Figure 2) contain both the lines belonging to the silicon substrate and the lines in the range that is characteristic of the SiC polytypes. The samples are characterized by the presence of the hexagonal 6H SiC polytype with the impurities of rhombohedral 15R SiC phase. Also, the bands attributed to Si-C bond were observed at 580 cm^{-1}, 584 cm^{-1} and 612,5 cm^{-1}, 617,5 cm^{-1} for undoped samples and Mn-doped samples, respectively.
Figure 2 shows that D band was observed in both Raman spectra, while the G band was observed only at spectra for Mn-doped samples at ~1605-1607 cm^{-1}. A G band represents sp^2 bonds, corresponding to those observed in graphite. A D band was observed at ~1331-1372 cm^{-1}, which was at a lower wave number than the G band and represents sp^3 bonds, corresponding to those observed in diamond [16]. The peaks at 1196 cm^{-1}, 1190, 1118 and 1430 cm^{-1}, 1454 cm^{-1} have been inferred to be nanocrystalline diamond and sp^3 bonds [17].
The ratio of the intensity of the D peak to the G peak (I_D/I_G) is related to the cluster diameter of the free carbon domains [18]. The smaller ratio corresponds to smaller free carbon clusters. The I_D/I_G

ratios for the Mn-doped samples deposited from methanol and ethanol solution are 0.92 and 0.82, respectively.
Before surface modification of silicon-carbon films with Mn, the intensities of D and G bands were higher than those of modified films. The high intensity of the D band confirms the existence of unsaturated hydrocarbons on the surface of SiC nanoparticles [19].

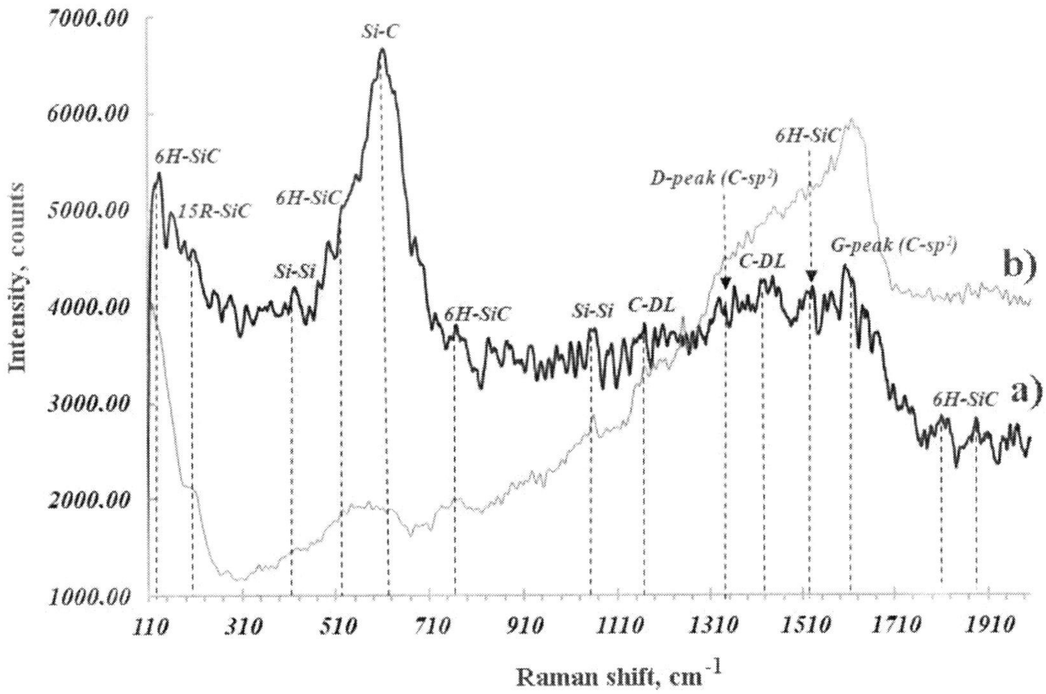

Figure 2(a, b). **(a)** Raman spectra of Mn-doped Si-C film; **(b)** Raman spectra of Non-alloy Si-C film.

The presence of manganese compounds and silicon-carbon bonds in the structure of the film was proved by X-ray diffraction (Figure 3). It is evident that only the peaks corresponding to 15R-SiC and 33R-SiC phases exist within the sensitivity of XRD diffractometer. Mn as an impurity element disturbs the phase equilibrium and induces the formation of Mn_5Si_2 and $Mn_{22,6}Si_{5,4}C_4$ phases. $Mn_{22,6}Si_{5,4}C_4$ is a good electronic conductor.

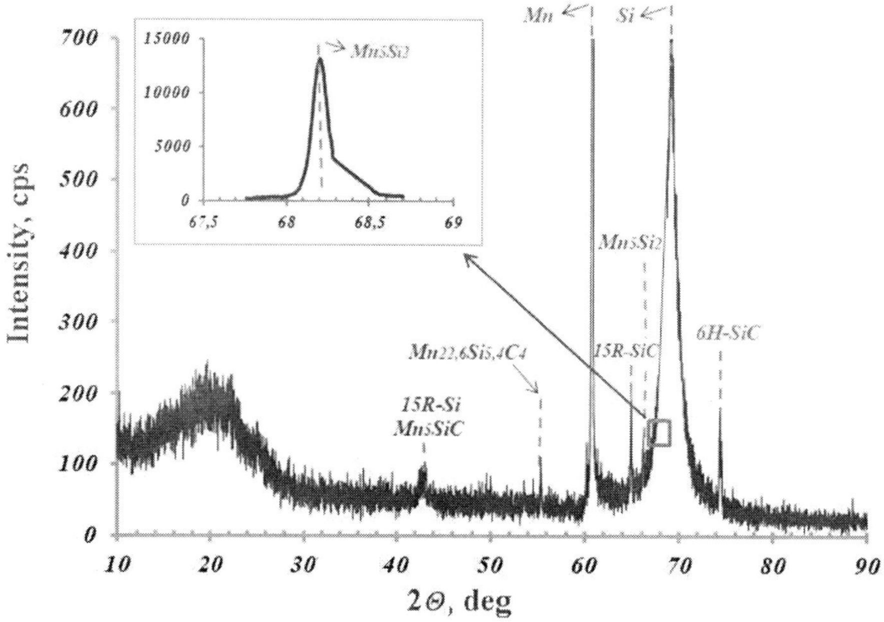

Figure 3. XRD-pattern of the manganese-doped silicon-carbon film.

3. Conclusions

It was shown that manganese-doped silicon-carbon films can be obtained from methanol-HMDSN-manganese sulfate solutions by the two-stage electrochemical deposition.The thickness of the Mn-doped film after the two-stage deposition process was about 100 nm.

The deposition potential was varied from 300 to 100 for deposition of pure and manganese doped solution, respectively. The Raman spectroscopy showed that silicon-carbon films investigated in this work are complex heterogeneous objects. The samples are characterized by the presence of the hexagonal 6H SiC polytype with the impurities of rhombohedral 15R SiC phase.The XRD analysis showed the presence of Mn, Mn_5Si_2,Mn_5SiC anda good electronic conductor $Mn_{22,6}Si_{5,4}C_4$ phase.

Acknowledgments

The work was supported by the Ministry of Science and Education Russia, under Contract № 14.575.21.0126 (the unique identifier of the contract is RFMEFI57517X0126).

References

[1] Yan X B 2006 *Electrochemistry Communications* **8** 740
[2] Barinova A D, Popova A I, Presnyakov M Yu 2017 *Inorganic Materials* **53(7)** 690
[3] Alper J, Kim M, Vincent M, Hsia B, Radmilovic V, Carraro C, Maboudian R 2013 *J. of Power Sources* **230** 302
[4] Jang K H, Hyun Y, Park Y H, Lee C S 2017 *J. of Nanoscience and Nanotechnology* **17(11)** 8500
[5] Singh N, Kumar A, Kaur D 2018 *Sensors and Actuators B Chemical* **262** 162
[6] Wang S, Li X, An Y, Wu Z 2014 *J. of Alloys and Compounds* **584** 339
[7] Sirk Aislinn H C, Sadoway R D 2008 *J. of The Electrochemical Society* **155** 49
[8] Randeniya L K, Bendavid A, Martin P J, Amin M S, Rohanizadeh R, Tang F, Cairney J M 2010 *Diam. Relat. Mater* **19** 329
[9] Falcade T, Gomes dos Reis O, Shmitzhaus T E, Vargas A L M 2012 *Applied Surface Science* **263** 18
[10] Maissel L L, Glang R 1970 *Handbook of Thin Film Technology* **5** 5

[11] Wang H, Shen M R, Ning Z Y, Ye C 1996 *Appl. Phys. Lett.* **69** 1074
[12] Yu G H 2013 *J. Nano Energy* **2** 213
[13] Kang C, Cai H, Zhang X, Tang J, Xu P, Li M, Zong H 2015 *Results in Physics* **5** 178
[14] Abdur R, Kim K, Kim J H, Lee J 2015 *Electrochimica Acta* **153** 184
[15] Grigoryev M N, Myasoedova T N, Mikhailova T S 2018 *J. of Physics: Conference Series* **1124** 1
[16] Ferrari A C, Robertson J 2004 *Phil. Trans. R. Soc. Lond. A* **362** 2477
[17] Chu P K, and Li L 2006 *Mater. Chem. & Phys* **96** 253
[18] Mehr M, Esquivel-Elizondo J R, Moore D T, Nino J C 2015 *J. of Materials Science* **50(21)** 7000
[19] Iijima M, Kamiya H 2008 *J. Phys. Chem.C* **112** 11786

Substrate-removed flip-chip photodiode array based on *InAsSbP/InAs* double heterostructure

S A Karandashev[2,3], A A Klimov[1,2,3], R E Kunkov[1,3], A A Lavrov[2,3], T S Lukhmyrina[2], B A Matveev[2,3], M A Remennyi[2,3], A A Usikova[2,3]
[1]Peter the Great Saint-Petersburg Polytechnic University, Russia
[2]IR Optoelectronics Laboratory, Ioffe Institute RAS, St.Petersburg 194021, Russia
[3]IoffeLED Ltd., St. Petersburg 194064, Russia

Abstract. N-*InAsSbP/InAs/*P-*InAsSbP* double heterostructures have been grown onto n^+-InAs substrate and further processed into 2×2 photodiode array containing no n^+-*InAs*. C-V, spectral response as well as mid-IR photoluminescence and electroluminescence in the 77-300 K temperature range have been measured and used for photodiode characterization including D*(λ) and BLIP temperature evaluation.

1. Introduction

There is an increased demand for robust and efficient mid-IR (3-5 μm) LED-photodiode pairs that are useful for a number of analytical instruments [1]. *InAs* based detectors sensitive to wavelengths around 3.4 μm are still in the top list of extensive research efforts due to high performance parameters recently reported both for heterojunction unipolar [2] and homo p-n junction [3, 4] devices. The *InAsSbP/InAs* p-n heterostrutures are also good candidates for fabrication of mid-IR photodiodes (PDs) and multielement back-side illuminated (BSI) PD arrays as BLIP temperature as high as 180 K (2π FOV) [5] and room temperature Johnson noise limited operation at frequencies above 100-200 Hz [6] have been already demonstrated for single heterostructure (SH) PDs. Some recently obtained data suggests that *InAsSbP/InAs* double heterostructure (DH) PDs are superior to the SH ones in terms of the zero-bias resistance-area product R_0A - figure of merit of any diode. However, there have been no attempts to fabricate linear or 2D PD arrays based on *InAsSbP/InAs* DHs.

The paper presents first results on 2×2 PD array based on *InAsSbP/InAs* DH with broad spectral response, enhanced responsivity, individual addressing and suppressed optical and electrical cross talk between elements. These performance improvements have been achieved due to elimination of semi-transparent n^+-*InAs* substrate and implementation of flip-chip bonding method for the PD array fabrication.

2. Experimental details

Our DH samples initially contained 4 μm P-*InAsSbP(Zn)* ($p{\sim}5\times10^{17}$ cm^{-3}) wide gap contact layer, 6 μm n-*InAs* undoped absorbing active layer with donor concentration $n= 5\times10^{16}$ cm^{-3}, 2 μm N-*InAsSbP* undoped wide gap cladding layer, and finally a 350 μm heavily doped n^+-*InAs* (100) substrate (n^+ $\sim10^{18}$ cm^{-3}) semitransparentin the spectral range of interest (see fig.1). It follows from fig.1 that PDs with radiation transmitted through n^+-*InAs* substrate (back-side illumination mode – BSI) suffer from sufficient absorption at important for gas analysis wavelength of 3.4 μm. According to transmission data presented in fig.1 and in [7] electron concentration n^+ in our substrates is in the $(2\text{-}3)\times10^{18}$ cm^{-3} range.

The photoluminescence (PL) spectrum measured at 77 K in a reflection mode showed strong longwave emission peak around 0.4 eV which is a common value for the undoped n°-*InAs*; PL spectrum contains also weak shortwave peak at 0.49 eV obviously related to recombination in P-*InAsSbP* (see fig. 2 (a)). Close to the above spectrum was recorded from reference N-*InAsSbP*/n^+-*InAs* sample also shown in fig.2 (a). Room temperature (RT) PL spectrum contain only single band corresponding to recombination in n-*InAs* active layer. Electroluminescence (EL) peaks matched the corresponding PL peaks at 77 K (hv = 0.4 eV) and 300 K (hv = 0.36 eV) (see fig 2 (b)). The RT EL peak value was typical for LEDs with n°-*InAs* active layer, while the spectral shape contains Fabry-Perot fringes typical for thin LEDs with reflective anode and mirror out coupling surface [8].

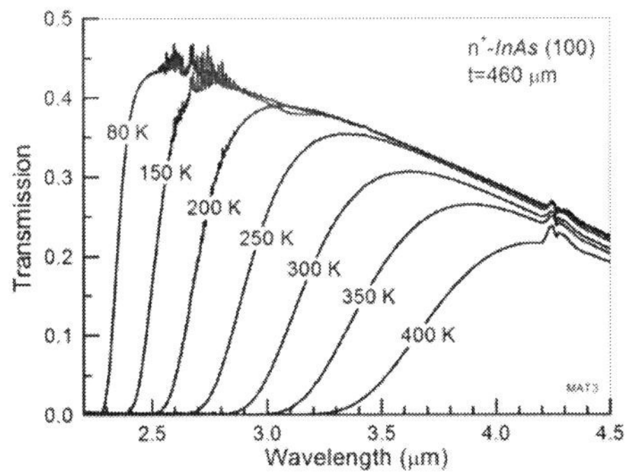

Figure 1. Transmission spectra of n^+-*InAs* substrate in the 80-400 K temperature range.

(a) (b)

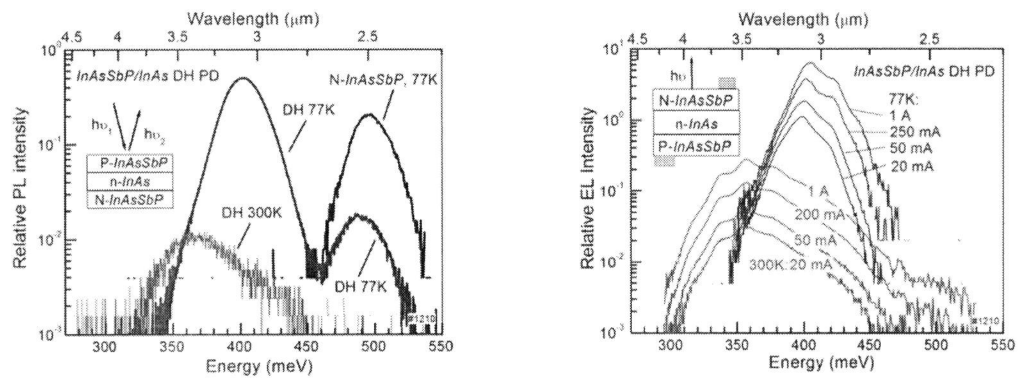

Figure 2. ''Reflection'' photo- (a) and electroluminescence (b) spectra in *InAsSbP/InAs* DH PD at 77 and 300 K.

Epitaxial wafers were subjected to standard multistage photolithography and wet chemical etching, which resulted in the formation of rectangular chip with a total size of ~1000×1000 μm (see fig. 3 (a)). The chip contained 4 photodiodes of 500×500 μm size each separated by ~20 μm deep grooves. Each photodiode had "cornertriangle" metal cathode denoted as "C" in fig.3(a) and broad metal anode denoted as "A"; the latter covered most of the 500×500 μm^2 element area. Both anode and cathode were formed during evaporation of *Cr, Ni* and *Au* at vacuum with further "enhancement" by electrochemical deposition of *Au* with final thickness of about 2μm. The chip was mounted onto ceramic read-out board as shown in fig.3 (b). Prior measurements the ceramic submount surface was protected by a photoresist and n^+-*InAs* substrate shown by dashed rectangular in fig.3 (b) was completely removed by mechanical polishing and further chemical etching. The groove depth exceeded the total *InAsSbP/InAs/InAsSbP* DH thickness and because of this the above process resulted in getting 4 individual PDs mounted close to each other on a single submount without any electrical and mechanical interconnection and thus with negligible cross-talk between elements (see fig.3(c)). Contact pads in fig.3 are denoted as Ai and Ci for each of the 4 PD elements (i=1, 2, 3, 4).

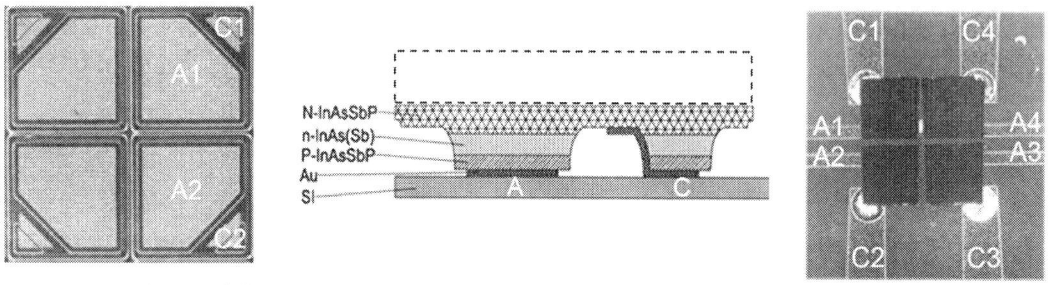

Figure 3 (a). Photo of the chip contact side.

Figure 3 (b). Schematic of the diagonal cross section of the individual PD element mounted onto submount. Dashed rectangle – location of the n^+-*InAs* substrate prior removal.

Figure 3 (c). Photo of the chip onto ceramic submount with anode and cathode bonding pads. View from the N-*InAsSbP* side.

Capacitance-voltage (C-V) characteristics exhibited linear dependence in the C^{-3}-V coordinates (see fig.4) indicating linear impurity distribution in the space charge region similar to the *InAsSbP/InAs* SH and DH described previously [5, 11]. However, the zero bias unit area values C_0/A are at least 2-fold smaller than the lowest values published for the *InAs* DH PDs in [11]. This makes good basis to creation in future high speed *InAs* matrix PDs.

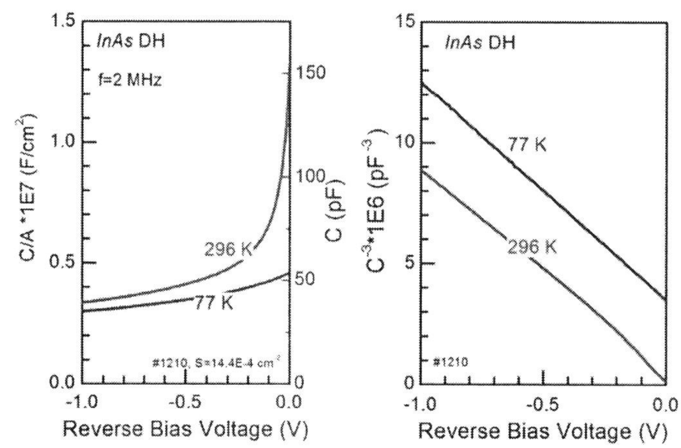

Figure 4. C-V characteristics at 77 and 296 K.

3. Results and discussion

At T>220 K the zero bias resistance R_o of the individual PDs exhibited temperature dependence that is typical for diodes with domination of the diffusion current as the activation energy of the exponential growth ($R_o \sim \exp(E_a/kT)$) is close to that of the energy gap value ($E_a = E_g = 0.41$ eV (see fig. 5)). At low temperatures (T<220 K) the activation energy is roughly a half of the E_g indicating domination of the generation-recombination or tunnel current.

Spectral response near 2.7 μm was partly distorted by presence of water vapour and carbon dioxide in an optical path while the peak wavelength region contained Fabry-Perot fringes corresponding to the total DH thickness of L=11 μm and refractive index n=3.42: $\Delta\lambda=\lambda^2/2nL$ (see fig.6 with $\Delta\lambda_{mean}=0.15$ μm and also fig.2 (b) with fringes in the EL spectra). No fringes associated with reflection from semiconductor/air interface and anode contact have been observed at the shortwave shoulder of the spectral response because of high absorption in P-*InAsSbP* and in n-*InAs* layers. The responsivity slope at the shortwave shoulder at $\lambda>2.6$ μm and at T<360 K matches that of the ideal photodetector revealing independence of collection efficiency on wavelength. Thus we can conclude that the diffusion length of holes generated by radiation in N-*InAsSbP* is much larger than 2 μm - the thickness of the N-*InAsSbP* layer. The reason for the spectral response degradation at $\lambda<2.6$ μm and T<200 K (see the step at λ=2.6 μm) is not clear at the moment. At the same time the longwave shoulder exhibits no uncommon features, e.g. the temperature variation of the red cut-off wavelengths ($\lambda_{0.1}$ and $\lambda_{0.5}$ respectively) matches well with the known variation of *InAs* energy gap (see fig. 7): $E_g = 415 - 0.28\times T^2/(T+83)$ [9, 10].

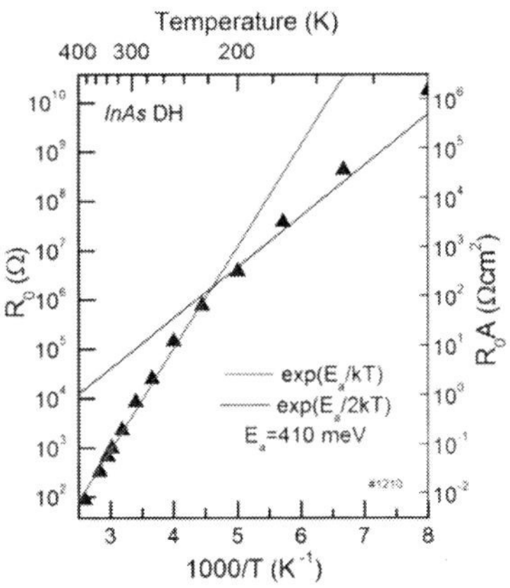

Figure 5. Zero bias resistance R_o vs temperature in single *InAsSbP/InAs* DH PD element.

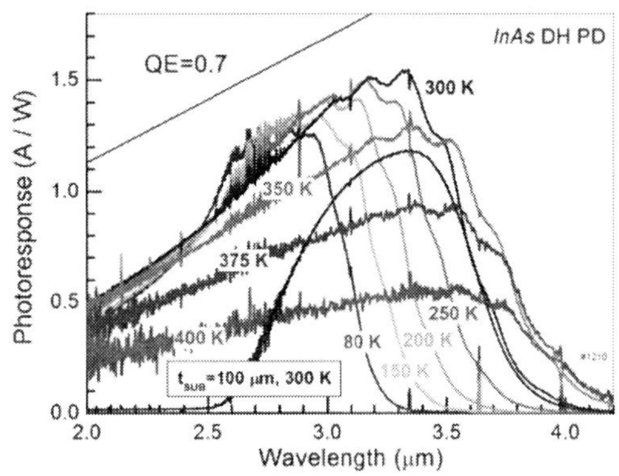

Figure 6. Spectral response of the individual PD element at several temperatures. Solid line presents responsivity of an ideal detector (QE=0.7). Narrow band response of the PD with t_{sub}=100 μm thick substrate is given for comparison.

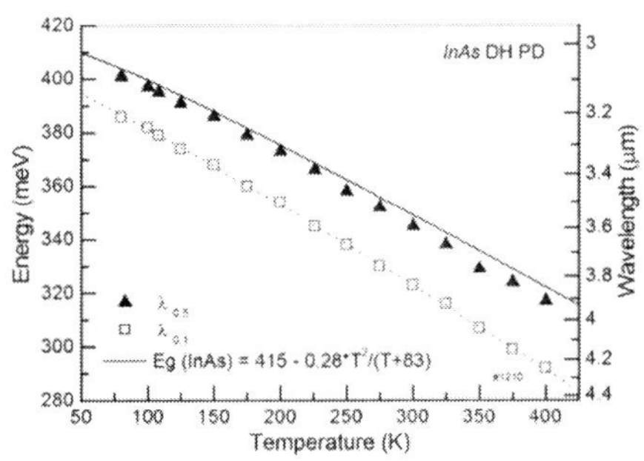

Figure 7. Red cut-off wavelength and energy vs temperature in *InAsSbP/InAs* DH PD.

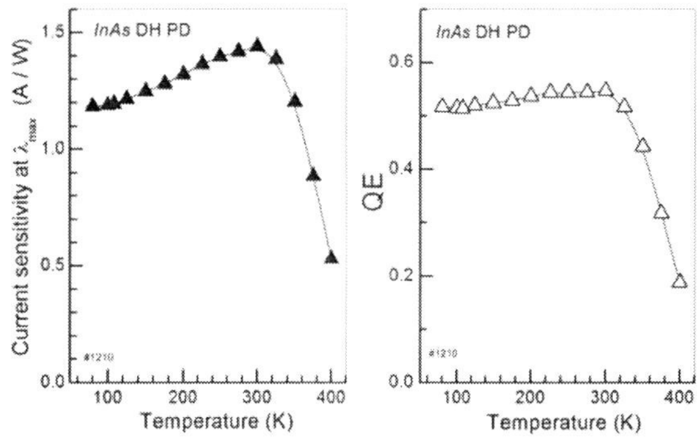

Figure 8. Current sensitivity and quantum efficiency at maximum vs temperature in *InAsSbP/InAs* DH PD.

Current sensitivity S_I and quantum efficiency at maximum in the DH PD free of substrate are shown in fig.8. Degradation of both S_I and QE at T>300 K are probably originate from N-*InAsSbP* layer transmission change.
Superposition of data presented in fig.5 and fig.6 enable us to simulate Johnson noise limited specific detectivity D* shown in fig.9. It follows from fig.9 that the D* value of the 2×2 array elements at room temperature is fairly close to the best values of the single element *InAsSbP/InAs* PDs known from literature and the BLIP operation mode is achieved at around 170 K for the 2π field of view which is close to data published for the *InAsSbP/InAs* SH PD [5]. However, spectral response of current array is much broader than that in [5] which may be essential in a curtain amount of applications.

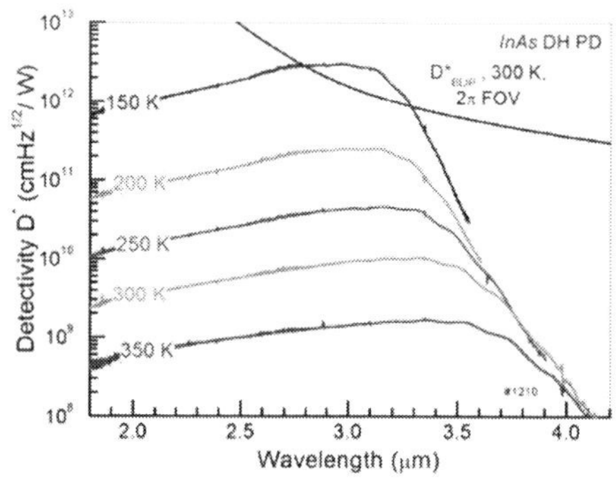

Figure 9. D* vs wavelength at several temperatures of the *InAsSbP/InAs* DH PD free of substrate.

Acknowledgments
We thank Il'inskaya N D for technical assistance in performing this research. The work performed at IoffeLED, Ltd. has been supported by the RF state program (ID: RFMEFI57618X0104).

References
[1] Matveev B A, Sotnikova G Yu, 2019 Optics and Spectroscopy, **127**, n.8 (in press)
[2] Savich G R, Pedrazzani J R, Sidor D E, Wicks G W 2013 Infrared Physics & Technology, **59**, 152 DOI: 10.1016/j.infrared.2012.12.031
[3] Sandall I C, ZhangSh, TanCh H 2013 OPTICS EXPRESS **21** 25780
[4] Maddox S J, Sun W, Lu Z, Nair H P, Campbell J C, Bank S R 2012 Appl. Phys.Lett., **101,** 151124
[5] Brunkov P N, Il'inskaya N D, Karandashev S A, Lavrov A A, Matveev B A, Remennyi M A,Stus' N M, Usikova A A, 2016 Infrared Physics & Technology, **78**, 249 http://dx.doi.org/10.1016/j.infrared.2016.08.013
[6] Dyakonova N, Karandashev S A, Levinshtein M E, Matveev B A, Remennyi M A, 2019 Semicond. Sci. Technol., **34,** 015013, DOI: 10.1088/1361-6641/aaf0c6
[7] Komkov O S, Firsov D D, Kovalishina E A, Petrov A S, 2014 PrikladnayaFizika, № 4 93 (in Russian).
[8] Zotova N V, Il'inskaya N D, Karandashev S A, Matveev B A,Remennyi M A, Stus' N M, 2008 Semiconductors, **42,** 625.
[9] Fang Z M, MaK Y, JawD H, Cohen, Stringfellow G B, 1990 J. Appl. Phys., **67,** 7034.
[10] http://www.matprop.ru/InAs_bandstr
[11] Brunkov P N, Il'inskaya N D, Karandashev S A, Lavrov A A, Matveev B A, Remennyi M A, Stus' N M, Usikova A A 2016 Infrared Physics & Technology. **76** 542. http://dx.doi.org/10.1016/j.infrared.2016.04.002

In situ study of structural transformation of the $(TeO)_x(Mo,V,Nb)_5O_{14}$ mixed oxide nanocatalyst with a tunnel structure

E. Derevyannikova[1], **T Kardash**[1], **A Marchuk**[1,2], **A. Simanenko**[1,2], **D. Svintsitsikiy**[1,2], **E. Lazareva**[1]
[1]Boreskov Institute of Catalysis SB RAS, Novosibirsk 630090, Russia
[2]Novosibirsk State University, Novosibirsk 630090, Russia

Abstract. The nano oxides $(TeO)_x(Mo,V,Nb)_5O_{14}$ with so-called M1 structure are highly attractive objects due to unsurpassed catalytic properties. In situ X-ray powder diffraction method with high time resolution (0.4 seconds per diffraction pattern) was applied to study the structure stability and the mechanism of phase transformation under heating in the reductive conditions. It was shown that M1 structure gradually transforms into the rutile-type $(Mo,V,Nb)O_2$ phase without intermediate phases and/or amorphisation at temperatures >480°C in a reductive media. The process is initiated by tellurium reduction and its sublimation to the gas phase.

1. Introduction
Nanosystems based on mixed Mo oxides attract high scientific attention due to unsurpassed catalytic properties in transformations of light alkanes into the valuable chemicals. $(TeO)_x(Mo,V,Nb)_5O_{14}$ mixed oxide with a tunnel structure (the so-called M1 phase) show the highest among other studied oxide systems activity and selectivity in the ethane oxidative dehydrogenation, selective propane oxidation to acrylic acid and propane ammoxidation to acrylonitrile processes [1-3]. Despite the extended knowledge on the M1 phase synthesis and structural features, the stability of this oxide is still not understood. Previous studies show that the catalysts deactivation under ethane oxidative dehydrogenation conditions (C_2H_6+O_2 mixture) occur due to reduction of Te^{4+} cations. Reduced Te^0 are formed on the catalyst's surface and are easily sublimated to the gaseous phase at temperatures above 450°C [4-6]. Therefore, reducing conditions cause the most severe deactivation of the M1 phase-based catalysts. However, the ability of the catalysts to preserve the active structure under severe operation conditions (e.g. a reductive mixture, temperatures >400°C) is one of the most important characteristics necessary for the commercialization of developed catalysts.
In this work the structural transformations of the $(TeO)_x(Mo,V,Nb)_5O_{14}$ mixed oxide during the heating in hydrogen flow using time-resolved in situ X-ray powder diffraction (XRPD).

2. Experimental
The mixed oxide with a bulk composition $(TeO)_{0.39}Mo_{3.5}V_{1.1}Nb_{0.4}O_{14}$ were obtained according the procedure described elsewhere [7].Synthesis steps included mixing of aqueous solutions of the starting components, drying of the resulting liquid precursor in B-290 (Buchi, Switzerland) laboratory spray drier at 220 °C, and a subsequent stepwise heat treatment of the dry precursor in flowing helium at 560 °C with intermediate heat treatment in flowing air at 310 °C.

In situ XRD experiments were performed at European Synchrotron Radiation Facility (ESRF), the ID22 Beamline (Grenoble, France). The oxide sample in powder form was placed in a 0.5 mm quartz mark tube (Hilgenberg mbH, Germany). The mark tube was opened from both sides and connected to a gas flow (1% H_2 in helium, 0.5 ml min^{-1}). A capillary was heated using a hot air stream up to 600 °C at the ramp of 10 K min^{-1}. The temperature was calibrated using silver powder as an external standard. The wavelength (λ = 0.2066 Å) and sample-to-detector distance were calibrated using LaB_6 as an external standard. Data were obtained every second using a Perkin Elmer flat detector. Parametric sequential refinements were performed using the TOPAS v 4.3 software. Profile parameters for the Lorentzian functions, cell parameters, and phase fractions were refined simultaneously for all phases.

3. Results

The mixed oxide $(TeO)_{0.39}Mo_{3.5}V_{1.1}Nb_{0.4}O_{14}$ is characterized as a single phase with M1-type structure. The diffraction data for the initial sample is refined using the structural data for this phase, by the orthorhombic cell, space group *Pba2*, and lattice parameters a=21.133 Å, b=26.644 Å and c=4.014 Å. Figure 1 shows the experimental and calculated pattern obtained on the final stage of the full profile Rietveld refinement. The good match of experimental and calculated pattern was achieved using a structural model described in [8] and the lattice parameters, the size of the coherent scattering domain, thermal parameters as refined parameters. The atomic coordinates and site occupancies were not refined.
The structure is built from the extended grids that are stacked on the top of each other in the [001] direction. Such grids in the (001) plane consist of the MoO_6 and VO_6 octahedra connected by the corners, as well as the pentagonal bipyramids NbO_7 connected with five octahedrons by edges.

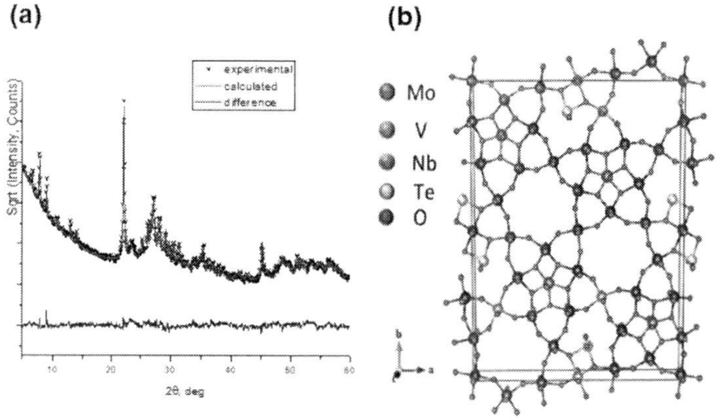

Figure 1. (a) Full-profile refinement by the Rietveld method for $(TeO)_{0.39}Mo_{3.5}V_{1.1}Nb_{0.4}O_{14}$ oxide within the framework of the M1 phase structure (b) the schematic presentation of the M1 structure in the *ab* plane

To obtain a good time resolution and monitor the structural transformation in detail, in situ Figure 1 shows time-resolved in situ XRPD experiments were performed using a synchrotron radiation with high energy (λ = 0.2066 Å) and two-dimensional detector. Figure 2 shows the typical experimental scan measured during the oxide heating in the gas flow (1%H_2 in helium). Thus, the total 2θ range of the diffraction pattern was measured simultaneously with a counting time of 1 s. There were 500 experimental points measured for the heating range from 400 to 600°C with a heating ramp 10°/min. For a data analysis, detector mask was applied, the experimental scans were integrated to get a 1D diffraction pattern and then a Rietveld refinement of each scan was performed. The refinement strategy was similar to those described above used for a characterization of the initial sample.

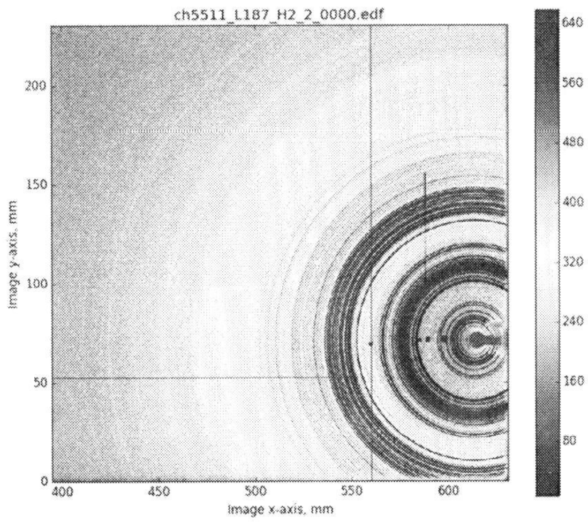

Figure 2. The example of the experimental scan measured for 1 second using 2D detector.

Figure 3(a) shows the top-view of the obtained temperature dependence of the diffraction patterns measured, when the sample was heated from 400 up to 600°C. The heating below 400°C did now reveal any structural changes of the sample. Figure 3(b) demonstrates the set of diffraction patterns obtained when the sample was heated from 476 to 506°C.

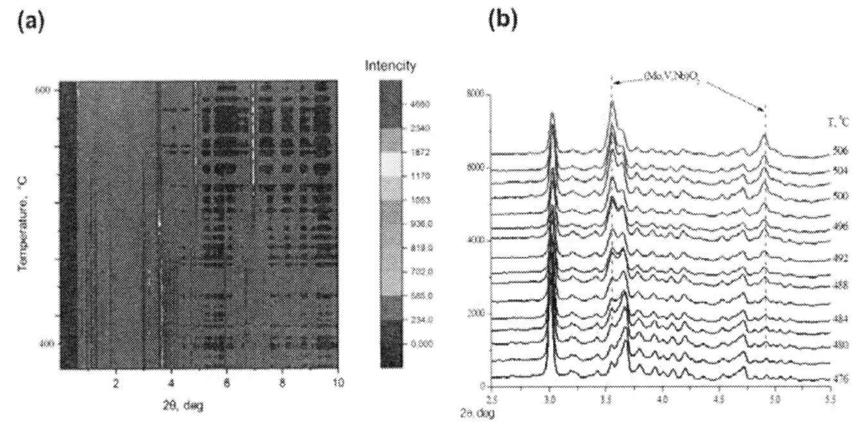

Figure 3. Temperature dependence of PXRD patterns during the reduction of the $(TeO)_{0.4}Mo_{3.5}V_{1.1}Nb_{0.4}O_{14}$ mixed oxides. (a) The top-view representation of the experimental data (b) the characteristic patterns obtained in a small temperature range.

Figure 4 shows how the phase composition of the sample changed: the mixed oxide with a monoclinic MoO_2-type of structure appears at temperatures above 480 ° C. Further heating leads to a gradual transformation of the initial M1 phase into the MoO_2-type of structure. According to the literature data, V and Nb cations can enter the cell of the MoO_2 phase forming a solid solutions [9-10]. However, Te cations do not enter the phase. This fact demonstrates that observed structural transformations start when all Te in the sample is reduced into a Te^0 state and sublimated into the gas phase.

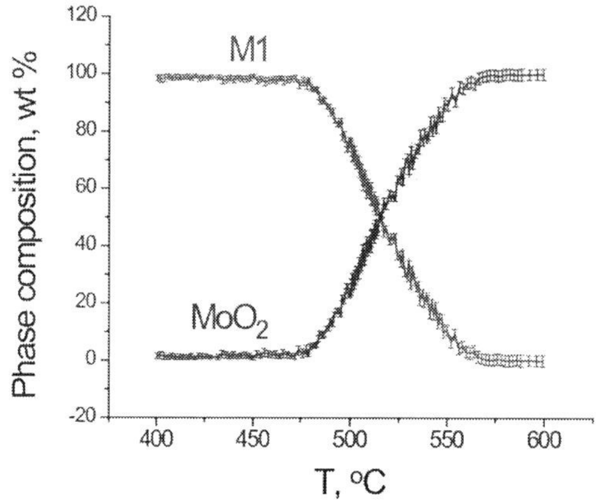

Figure 4. Changes in the phase composition of the sample when heated from 400 to 600 ° C in a mixture of 1% H_2 in helium

Monoclinic MoO_2 is a distorted rutile structure constructed from MoO_6 octahedra connected by vertices and edges. Analysis of the cell parameters of this phase after cooling to room temperature showed that the volume of its unit cell is 132.9 ± 2 Å^3, which is larger than the cell volume for pure molybdenum oxide MoO_2 [11]. The change in volume is associated with the introduction of vanadium and niobium cations into the MoO_2 structure, which confirms the formation of a mixed oxide (Mo,Nb,V)O_2 as a result of decomposition of the M1 phase.
The data obtained show that under the influence of a reducing atmosphere and the presence of local overheating, the active VMoNbTeO catalysts with the M1 phase structure can be destroyed. In the case of a catalytic process, the reducing agent can be ethane or propane, which are present in the gas feed.

Acknowledgments
Authors wishing to acknowledge Russian Science Foundation (RSF, grant 17-73-20073) for financial support.

References
[1] Lopez Nieto J.M., Solsona B., Concepcion P., Ivars, F., Dejoz, A., Vazquez, M. I. 2010 *Catal. Today*, **157**, 291
[2] Bondareva V.M., Kardash T.Y., Ishchenko E.V., Sobolev V.I. 2015 *Catal. Ind.* **7** 104
[3] Lintz H.-G., Müller S.P. 2009 *Appl. Catal. A Gen.* **357** 178
[4] Valente J., Armendáriz-Herrera H., Quintana-Solórzano R., del Ángel P., Nava N., Masso A., Lopez-Nieto J.M. 2014 *ACS Catal.*, **4,** 1292
[5] Mishanin I.I., Kalenchuk A.N., Maslakov K.I., Lunin V.V., Koklin A.E., Finashina E.D., Bogdan V.I. 2016 *Russ. J. Phys. Chem. A.,* **90,** 1132
[6] Ishchenko E.V., Gulyaev R.V., Kardash T.Y., Ishchenko A.V., Gerasimov E.Y., Sobolev V.I., Bondareva V.M. 2017 *Appl. Catal. A Gen.*, **534,** 58
[7] Ishchenko E.V., Andrushkevich T.V., Popova G.Y., Chesalov Y.A., Plyasova L.M., Ishchenko A.V., Kardash T.Y., Dovlitova L.S. 2011 *Catal. Ind.*, **2,** 291
[8] Li X., Buttrey D.J., Blom D.A., Vogt T. 2011 *Top. Catal.*, **54,** 614
[9] Marinder B.O. 1975 *Mater. Res. Bull.*, **10,** 909
[10] Marinder B.O., Dorm E., Seleborg M., Motzfeldt K., Theander O., Flood H. 1962 *Acta. Chem. Scand.*, **16,** 293
[11] Bolzan A.A., Kennedy B.J., Howard C.J. 1995 *Aust. J. Chem.*, **48,** 1473

Influence of two-stage growth during the formation of GaAs/Si heterostructures for the creation of optoelectronic devices

A A Geldash[2], V N Djuplin[2], V S Klimin[1,2], M S Solodovnik[1,2], O A Ageev[1,2]
[1]Department of Nanotechnology and Microsystems, Southern Federal University, Taganrog 347922, Russia
[2] Southern Federal University, Research and education center "Nanotechnologies", Taganrog, 347928, Russia

Abstract. The development of GaAs/Si heterostructures can be an important stage for use as solar cells, LEDs and lasers based on silicon substrates. At present, A3B5 compounds grown epitaxial on Si substrates are of great interest because of the monolithic integration of optoelectronic devices with Si-based microelectronics. High-quality epitaxial growth of heterostructures will not only ensure high mobility of the carrier materials, but also preserve the advantages of lightweight and inexpensive Si substrates with high mechanical strength and excellent thermal parameters. However, obtaining A3B5 compounds with high quality crystals of GaAs/Si heterostructures is a difficult task due to the formation of an antiphase domain boundary as a result of the growth of polar GaAs on a non-polar Si substrate. Because of this, a high density of penetrating dislocations arises due to the mismatch of the lattice parameters (4.1%), as well as the coefficient of thermal expansion (62%).

1. Introduction
The development of electronic technology is currently associated with the development of efficient and at the same time inexpensive technologies for producing advanced semiconductor materials and devices based on them [1-7]. The method of forming high-quality GaAs/Si structures is molecular beam epitaxy [8-13]. We select a two-stage growth scheme for high-quality GaAs/Si structures. The formation of a low-temperature GaAs germinal layer helps to remove the misfit stresses, thereby reducing the number of misfit dislocations at the adjacent GaAs/Si interface [14-21].

2. Materials and Methods
Crystal growth on another crystal (with a different structure) is called epitaxial if both crystals have a fixed ratio of structure orientations. Such epitaxy is observed only in the case of a flat boundary between the crystals and, moreover, if they have at least approximately the same two-dimensional cell in this plane [22-26]. The discrepancy between the sizes of the lattices of the upper and lower layers gives rise to mechanical stresses causing shifts in the crystal rows - dislocations. In our case, the pair that forms the epitaxial structure are silicon single crystal and semiconductor GaAs, which grows in high vacuum with layer-by-layer deposition [27-33]. Before the first stage of the two-stage growth, a thin SiO_2 layer with a thickness of 600 nm was created using plasma chemical deposition. The SiO_2 layer was deposited at T = 230^0 C, the deposition time is t = 9 minutes, with a flow of gases SiH_4 = 15 cm^3/min and N_2O = 60 cm^3/min. The power flux of the inductively coupled plasma source was W_{ISP} = 500 W, the power source flux capacitive plasma (reactive ion etching) W_{RIE} = 5 W. The

formation of round holes in the oxide layer of silicon was carried out by photolithographic etching. Earlier, a template for the photolithography process was developed and created. The main task at this stage of creating the structure was the formation of a "frame" of the future structure in the form of holes in the SiO_2 layer, which are guaranteed to have pure Si at the bottom as the basis for the growth of GaAs. Fig. 1 **(a)**, **(b)** shows the AFM images of the structure with the holes formed using photolithographic etching.

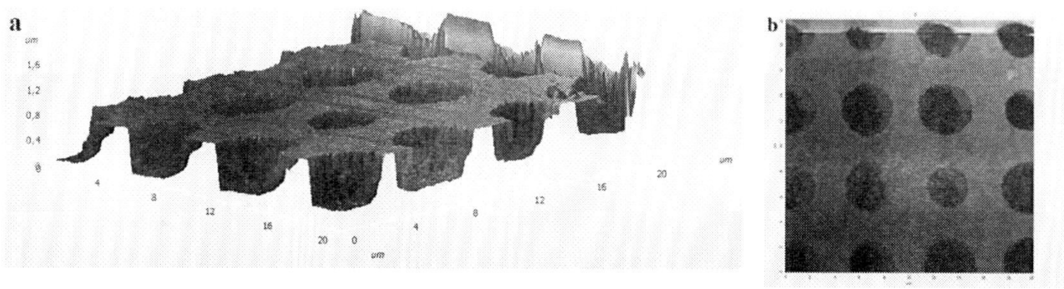

Figure 1(a, b). AFM image of the SiO_2 surface with an array of holes up to Si. **(a)** 3D AFM image of an array of holes in SiO_2, **(b)** AFM image of an array of holes in SiO_2 - top view.

At the first stage of a two-stage growth at low-temperature, a thin layer of GaAs grows, then high-temperature growth of GaAs on Si takes place, but already of a much greater thickness. The sequence of operations for MBE is that at the first stage, high-temperature annealing occurs at a temperature of $T = 810^0$ C and a duration of $t = 30$ min using a Si flow of $3 \cdot 10^{18}$ cm^{-3}. It is made to clean the surface of Si from oxide residue. It is important for us that the exposed Si surface is the basis for the growth of GaAs single crystals. At the second stage, the amorphous low-temperature layer grows at a temperature $T = 200^0$-250^0 C, thickness h = 5 nm, and a concentration of Si = $3 \cdot 10^{18}$ cm^{-3}. At the third stage, the growth of the crystalline low-temperature layer occurs at a temperature $T = 300^0$-350^0 C, thickness h = 25 nm. At the final stage, the high-temperature layer grows, at a temperature $T = 600^0$ C, thickness h = 650 nm. Thus, the total thickness of the layer caused by molecular beam epitaxy is about 700 nm. The second sample was created on the technological stages, different from those for which the first sample was created. At first, at a low-temperature, a thin layer of GaAs grows, then high-temperature growth of GaAs on Si occurs, but already of a much greater thickness. The sequence of operations for this sample at MBE is that at the first stage, high-temperature annealing occurs at a temperature $T = 830^0$ C, a duration t = 30 minutes and using a stream of Si with a concentration of $3 \cdot 10^{18}$ cm^{-3}. At the second stage, the amorphous low-temperature layer grows at a temperature of 200^0-250^0 C, thickness h = 50 nm. At the third stage, high-temperature annealing occurs at a temperature of $T = 650^0$ C, with a duration of t = 30 min and using an arsenic stream. At the final stage, the high-temperature layer grows at a temperature $T = 600^0$ C, thickness h = 650 nm.

3. Results

As a result of the experiment, structures with a layer of GaAs on Si and SiO_2 were obtained. Fig. 2 **(a)**, b shows an SEM image of the surface of the sample created by the first technology with the formation of a thin low-temperature crystalline layer of GaAs in the SiO_2 windows on Si.

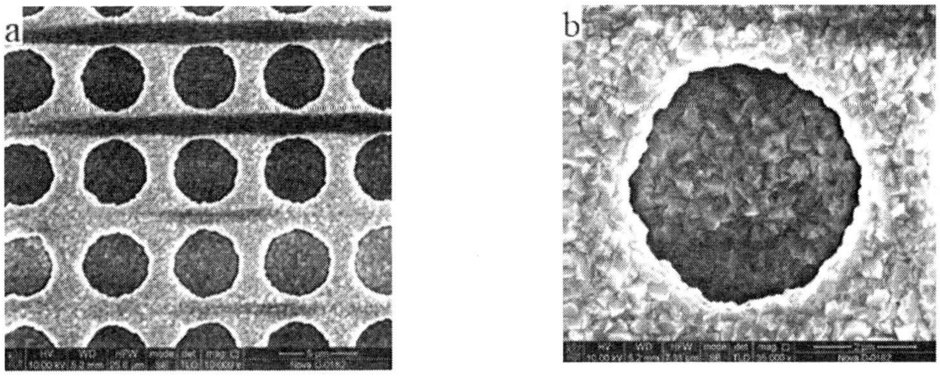

Figure 2(a, b). SEM image structure GaAs/Si (SiO_2), formed according to the first technology **(a)** an array of dimples with a deposited layer of GaAs on Si **(b)** a dimple in the SiO_2 layer from GaAs to Si

It can be seen from the photographs taken that the diameter of the holes in the SiO_2 layer with the GaAs and Si deposited on top is 4–4.5 μm. The edges of the circle do not have perfectly spherical geometry, since the entire surface outside the holes and in them is formed by separate single-crystal blocks that float on each other, thereby forming a continuous GaAs film on SiO_2 and Si. Fig. 3 shows the SEM images of a sample formed by the second technology.

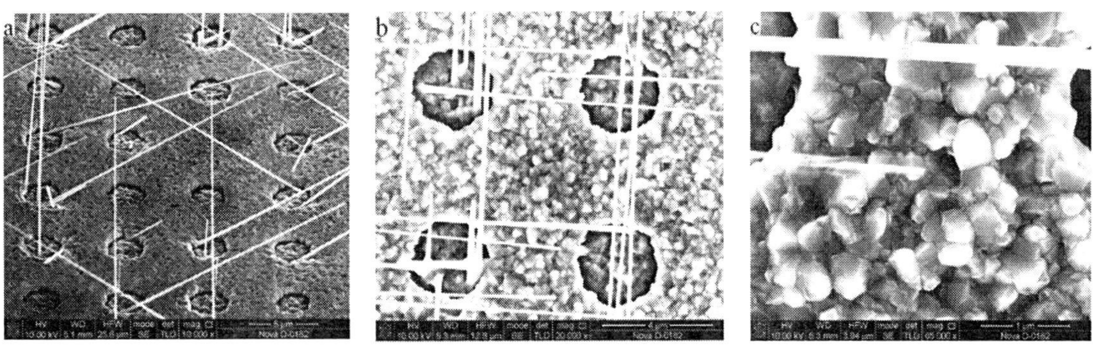

Figure 3(a, b, c). SEM image structure GaAs/Si (SiO_2), formed by the second technology **(a)** Array of nanowhiskers formed on the tops of GaAs blocks **(b)** Orientation of nanowiskers formed on the tops of GaAs blocks **(c)** Thicknesses of nanowhiskers formed on the tops of GaAs blocks.

On figure 3 **(a)**, it can be seen that nanowhiskers emanating from the tops of the crystals forming the GaAs film were formed in the cavities for the growth of GaAs on Si. These whiskers are about 17 μm long. It should be noted that nanowhiskers were not formed in all the grooves, an example of this is figure 3 **(a)**. In Fig. 3 **(b)**, it can be seen that the whiskers are perpendicular to each other. An important role in the orientation of nanowiserers is played by the crystallographic orientation of the silicon substrate. In Fig. 3 **(c)**, it can be seen that the thickness of nanowhiskers ranges from 140 to 200 nm. And also there is a "vibrating" of nanowhiskers on the search for tips. In fig. 4 **(a)** shows the SEM images of the structure formed by the second technology, without the introduction of a thin crystalline layer. There are an order of magnitude less such deepenings than those by deepenings with nanowhiskers.

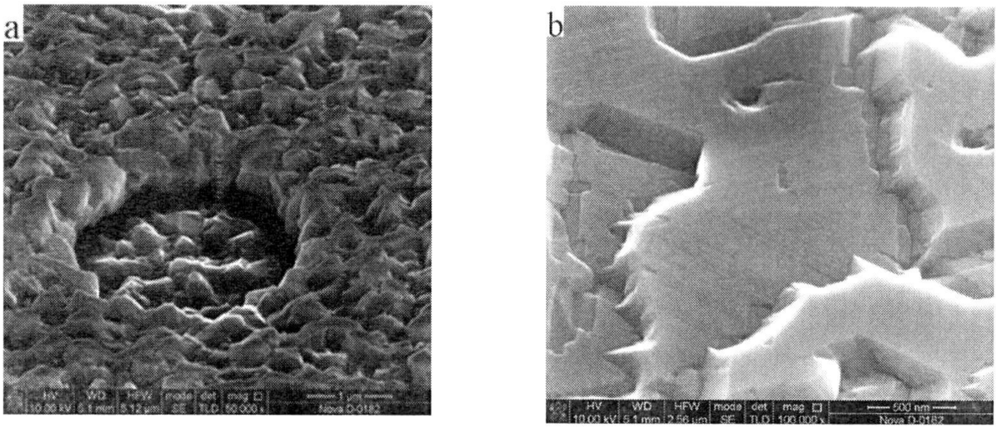

Figure 4(a, b). SEM image of the structure of GaAs/Si (SiO_2), created by the technology **(a)** The thickness of the GaAs layer in the recesses of the SiO_2 **(b)** Crystal structure of the GaAs film.

As we can see, the thickness of the GaAs layer is about 660 nm. GaAs on the surface of SiO_2 and at the bottom of the recesses with Cu at the base has the same structure – a continuous GaAs film consisting of mono-structural blocks and merged together. In Fig. 4 **(b)** shows the image of the GaAs/Si structure, which was created earlier, as a result of the refinement of the modes of deposition of GaAs on the Si substrate. Comparing the structure of the GaAs film in Fig. 4 **(a)** and Fig. 4 **(b)**, it can be concluded that the structures of the films are identical. The dimensions of the crystals in their height in diameters, their roughness, are identical for films grown on the surface of SiO_2 and at the bottom of the recesses, i.e. in both cases, the surface on which the films grew was SiO_2. Both structures with GaAs have a continuous film consisting of separate monocrystalline blocks, which merged during growth with each other.

4. Discussion and Conclusions.

On the Si substrate, GaAs originates in the form of islands, but depending on the growth mechanism, at the initial stage, both island growth, layer-by-layer film growth, and continuous film growth at the initial growth stage, which later changes to island growth, can occur. However, a thin continuous layer remains – a wetting layer. Passivation of the Si surface by elements of the third or fifth group is necessary for the growth of A3B5 semiconductor compounds on Si, which, as a result, leads to island growth of GaAs at the initial stage of epitaxial. All this is aimed at changing the conditions of deposition of GaAs on Si. The efficiency of high-temperature annealing is due to the fact that, with a wide range of temperature fluctuations, the GaAs film, due to a significant difference in thermal expansion coefficients with the substrate, is either compressed or in a stretched state, which causes the dislocation density to move. When encountered in different combinations, dislocation densities with oppositely directed Burgers vectors actively annihilate. The dimensions of the GaAs crystals on SiO_2 and Si are the same, they are 0,2 – 1 μm, and their surface is the basis for the growth of nanowhiskers. Nanowhiskers were formed due to the low-temperature amorphous growth of the GaAs layer. Nanowhiskers were not in the first technology due to the fact that both the low-temperature growth of the amorphous layer and the crystalline one were present. In the development of this technique, it is necessary to create structures only on the Si surface without being filled with GaAs molecules on the SiO_2 surface.

Acknowledgments

This work was supported by the Southern Federal University (grant VnGr-07/2017-02) and Grant of the President of the Russian Federation No. MK-3512.2019.8 The results were obtained using the equipment of the Research and Education Center "Nanotechnologies" of Southern Federal University.

References

[1] Gusev E Y, Jityaeva J Y, Geldash A A, Ageev O A 2017 *J. Phys.: Conf. Ser.* **917** 032029
[2] Schmidt B, Oswald S, Bischoff L, 2005 *J. Electrochem.* Soc. **152** 875-879
[3] Murdick D A, Wadley H N G and Zhou X W 2007 *Phys. Rev. B* **75** 125318
[4] Lisitsyn S A, Balakirev S V, Avilov V I, Kolomiytsev A S, Klimin V S, Solodovnik M S, Konoplev B G, Ageev O A 2018 *Nanotechnologies in Russia* **13(1-2)** p 26-33
[5] Kots I N, Klimin V S, Polyakova V V, Rezvan A A, Vakulov Z E, Ageev O A 2018 *J. Phys.: Conf. Ser.* **1124** 081035
[6] Shiraishi K and Ito T 1998 *Phys. Rev. B* **57** 6301
[7] Amrani A, Djafari Rouhani M and Mraoufel A 2011 *Appl. Nanosci.* **1** 59
[8] Tominov R V, Smirnov V A, Avilov V I, Fedotov A A, Klimin V S, Chernenko N E 2018 *IOP Conf. Ser.: Mater. Sci. Eng.* **443** 012036
[9]Murdick D A, Wadley H N G and Zhou X W 2007 *Phys. Rev.* B**75** 125318
[10] Vakulov Z E, Ivonin M, Zamburg E G, Klimin V S, Volik D P, Golosov D A, Zavadskiy S M, Dostanko A P, Miakonkikh A V, Clemente I E, Rudenko K V, Ageev O A 2018 J. Phys.: Conf. Ser. 1124 022032
[11]Shiraishi K and Ito T 1998 *Phys. Rev.* B**57** 6301
[12] Klimin V S, Rezvan A A, Ageev O A 2018 *J. Phys.: Conf. Ser.* **1124** 071020
[13]Amrani A, DjafariRouhani M and Mraoufel A 2011 *Appl. Nanosci.* **1** 59
[14]Kangawa Y, Ito T, Taguchi A, Shiraishi K, Irisawa T and Ohachi T 2002 *Appl. Surf. Sci.* **190** 517
[15]LiuY, LehmanV, WangL.2015 *Com. Net.* **83** 85-99
[16] Klimin V S, Rezvan A A, Kots I N, Naidenko N A 2018 *J. Phys.: Conf. Ser.* **1124** 071019
[17]Chekurov N, Grigoras K, Peltonen A, Franssila S, TittonenI 2009 *Nanotechnology.* **20** 5
[18] Nishinaga T and Shen X Q 1994 *Appl. Surf. Sci.* **82-83** 141
[19] Klimin V S, Rezvan A A, Ageev O A 2018 *J. Phys.: Conf. Ser.* **1124** 022035
[20] Higuchi Y, Uemura M, Masui Y, Kitada T, Shimomura S, Hiyamizu S 2003 *J. Cryst. Growth* **251** 80
[21] Schmidt M, Johari Z, Ismail R, Mizuta H, Chong H 2012 Microelectron. Eng. 98 313
[22] Balakirev S V, Eremenko M M, Mikhaylin I A, Klimin V S, Solodovnik M S 2018 *J. Phys.: Conf. Ser.* **1124** 022018
[23] Tseng A 2004 *J. Micromech. Microeng.* **14** 15
[24] Han J, Lee H, Min B, Lee S 2010 *Microelectron. Eng.* **87** 1
[25] Klimin V S, Solodovnik M S, Lisitsyn S A, Rezvan A A, Balakirev S V 2018 *J. Phys.: Conf. Ser.* **1124** 041024
[26] Klimin V S, Rezvan A A, Ageev O A 2019 *Proc. of SPIE* **11022** 1102220-1
[27] Nagase M, Nakamatsu K, Matsui S, Namatsu H 2005 *Japanese J. of Appl. Phys.* **44(7)** 5409
[28] Orloff J 2009 *CRC Press New York*
[29] Orloff J, Swanson L W, Utlaut M 2003 *Springer New York*
[30] Klimin V S, Kots I N, Polyakova V V, Rezvan A A, Ageev O A 2019 *Proc. of SPIE* **11022** 110221R-4
[31] Schaffer M, Schaffer B, Ramasse Q 2012 *Ultramicroscopy* **114** 62
[32] Kots I N, Kolomiitsev A S, Lisitsyn S G, Polyakova V V, Klimin V S, Ageev O A 2019 *Russian Microelectronics* **48(2)** p 72-79
[33]Salvati E, Brandt LR, Papadaki C, Zhang H, Mousavi SM, Wermeille D, Korsunsky AM, 2018 *Mater. Lett.* **213** 346–349

Investigation of a new C_{24} cluster for obtaining diamond-like phases: first-principle calculations

Vladimir A Greshnyakov[1] **and Evgeny A Belenkov**[1]

[1] Physics Department, Chelyabinsk State University, Chelyabinsk 454001, Russia

E-mail: greshnyakov@csu.ru

Abstract. In this work, first-principle calculations of the structure of a new fullerene-like C_{24} cluster are carried out. The surface of this cluster is formed of four- and six-membered atomic rings. It is established that these clusters should be stable under normal conditions. A simple cubic fullerite based on the C_{24} clusters can be used to synthesize the diamond-like CA4 phase under bulk compression up to 4.8 GPa. In addition, X-ray diffraction patterns of the «fullerite C_{24} → CA4 phase» structural transition are calculated.

1. Introduction

Carbon nanostructures are used as catalysts, adsorbents, structural elements of nanoelectronic devices and nanocomposites. They can also be used as precursors for new diamond-like and hybrid materials. Such precursors include graphene [1-3], carbon nanotubes [4-6], fullerenes [7-9], and a large number of hypothetical graphite-like nanostructures [10, 11]. One of the promising hypothetical nanostructures is a cubic cluster C_{24} similar to a truncated octahedron [11-13]. Various porous diamond-like phases with equivalent atomic positions can be obtained from these clusters [12-14]. The theoretical analysis, which was carried out in [11], showed that the most likely to obtain the porous diamond-like CA4 phase from a fullerite C_{24}. Therefore, first-principle calculations of the thermal stability of the C_{24} cluster and the process of obtaining the diamond-like CA4 phase from the fullerite C_{24} are performed in this work.

2. Methods

Geometric optimization of the structures and calculations of the total energy of carbon clusters and phases are carried out in the Quantum ESPRESSO program package [15] using the density functional theory (DFT) method [16] in the generalized gradient approximation (GGA) [17]. Modelling of the direct structural transitions of graphite-like C_{24} phase into diamond-like CA4 phase under uniform bulk compression of fullerite was performed using the technique described in [18, 19]. The norm-conserving pseudopotentials and the $10 \times 10 \times 10$ grids of k-points were used for the calculations. The cutoff energy for the basis of plane waves was 60 Rydberg.

The thermal stability of the C_{24} cluster was investigated using the molecular dynamics method with the time step of 0.5 fs and the $6 \times 6 \times 6$ k-point grid. The X-ray diffraction patterns of the «fullerite C_{24} → diamond-like CA4 phase » structural transition were calculated by the standard technique for the characteristic $K_{\alpha 1}$ copper radiation.

3. Results

As a result of DFT calculations, it is found that the optimized structure of the C_{24} cluster has the shape of a truncated octahedron (the point symmetry group is $m3m$) (figure 1a). The Cartesian coordinates of the atomic positions of this cluster are given in table 1. Each atomic position is characterized by three lengths of interatomic bonds (L_1 = 1.395 Å, L_2 = L_3 = 1.505 Å) and three angles between bonds (β_{12} = β_{13} = 120°, β_{23} = 90°).

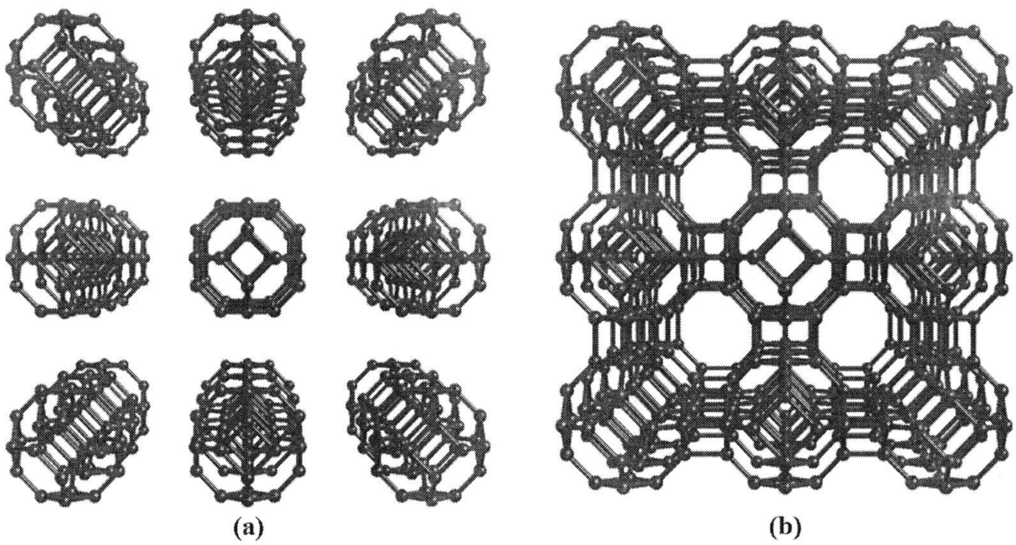

Figure 1. The structures of carbon compounds: **(a)** simple cubic fullerite C_{24}; **(b)** diamond-like CA4 phase.

Table 1. Absolute atomic coordinates in the C_{24} cluster.

Atomic number	X (Å)	Y (Å)	Z (Å)	Atomic number	X (Å)	Y (Å)	Z (Å)
1	0.987	2.051	0.000	13	0.000	2.051	0.987
2	3.116	0.000	2.051	14	2.051	0.000	3.116
3	0.987	2.051	4.102	15	4.103	2.051	0.986
4	3.116	4.102	2.051	16	2.051	4.102	3.115
5	2.051	0.987	0.000	17	0.987	0.000	2.051
6	0.000	3.116	2.051	18	3.116	2.051	0.000
7	2.051	0.987	4.102	19	0.987	4.103	2.051
8	4.102	3.116	2.051	20	3.116	2.051	4.102
9	0.000	0.987	2.051	21	2.051	0.000	0.987
10	2.051	3.116	0.000	22	0.000	2.051	3.116
11	4.103	0.987	2.051	23	2.051	4.103	0.986
12	2.051	3.116	4.102	24	4.102	2.051	3.115

The calculated cohesive energy of the C_{24} cluster is 6.70 eV/atom, which is 14.8 % less than the corresponding value for cubic diamond calculated in [11, 19]. To assess the thermal stability of the new fullerene-like cluster, its structure was annealed at 300 K using the molecular dynamics method. The graph of the total energy change (E_{total}) versus the annealing time is shown in figure 2. As a result of the annealing over 7 ps, neither destruction nor strong deformation of the structure is observed, which indicates the stability of the C_{24} cluster under normal conditions. Only the C_{24} cluster rotation relative to the equilibrium position in the simple cubic lattice is observed; therefore, the graph characterizing this process differs significantly from the standard relaxation graph.

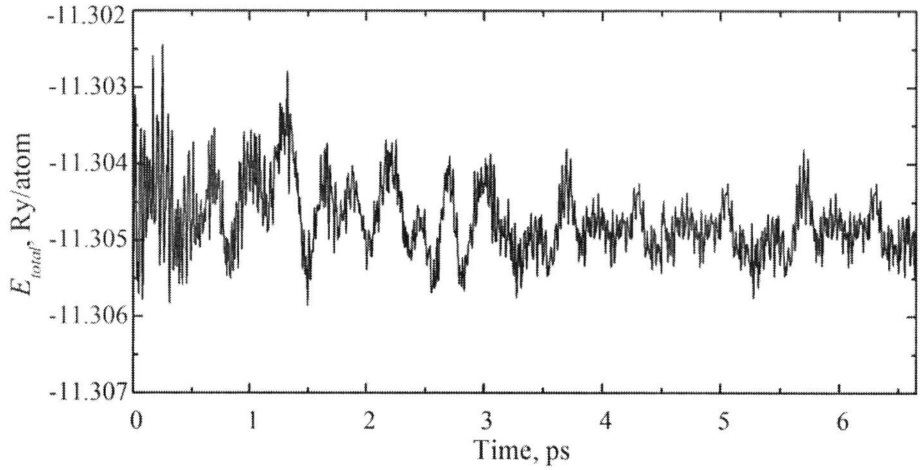

Figure 2. Dependence of the total energy on the annealing time for the cluster C_{24} at 300 K.

At the next stage of the study, the simulation of the direct phase transition of a simple cubic (sc) fullerite C_{24} (figure 1a) to the low-density diamond-like CA4 (figure 1b) phase is performed. The crystal lattices of the initial fullerite C_{24} and final CA4 phase belong to the space group *Pm*3*m*. In the process of bulk compression of the fullerite C_{24}, the formation of the CA4 phase occurs when the energy barrier of 0.36 eV/atom is overcome and the pressure reaches 4.8 GPa (figure 3). For this phase transition, X-ray diffraction patterns were calculated during compressing the fullerite and decompression of the formed CA4 phase. This first-order phase transition is accompanied by a significant change in the diffraction pattern, in which many of the original low-intensity maxima disappear, and many new maxima appear (figure 4). The theoretical X-ray diffraction pattern of the new diamond-like CA4 phase can be used for its experimental identification.

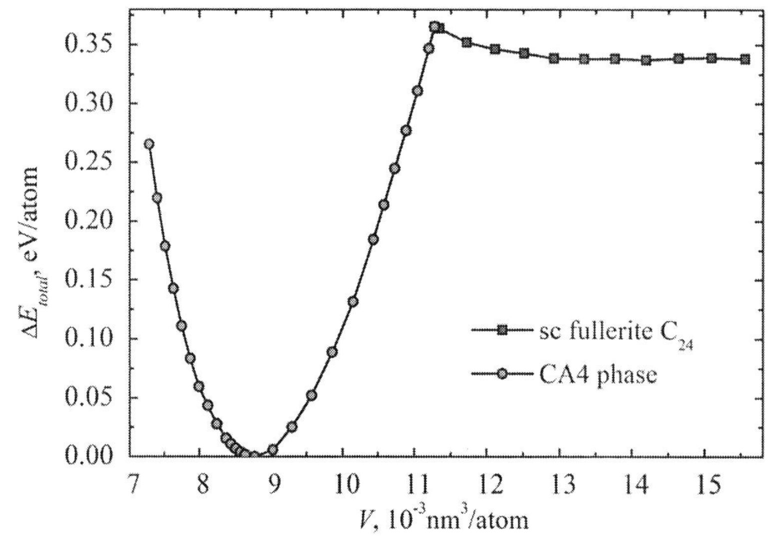

Figure 3. Dependence of the total energy difference on the atomic volume for the simple cubic fullerite C_{24} and diamond-like CA4 phase.

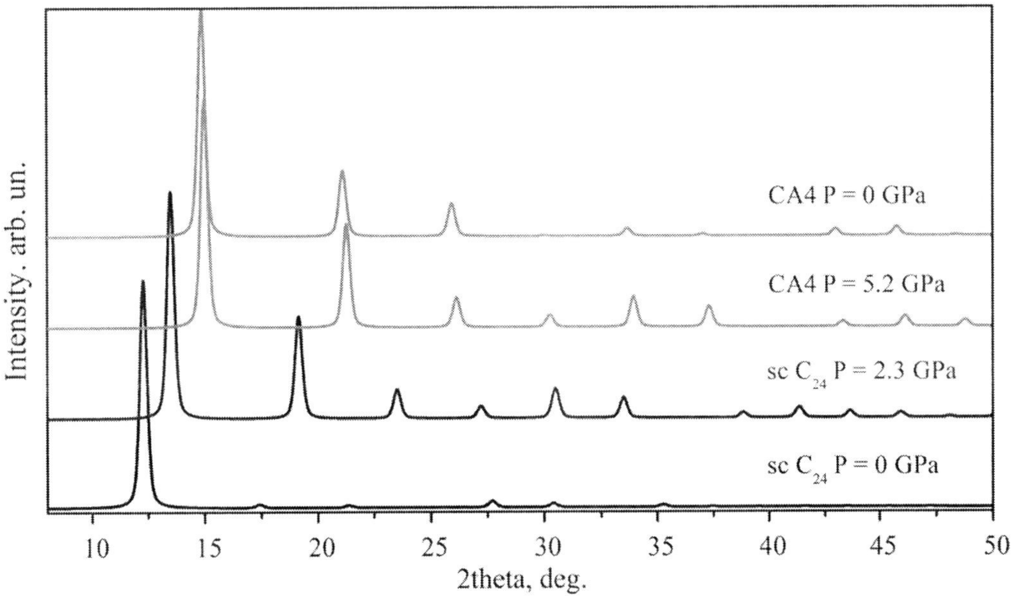

Figure 4. Theoretical powder X-ray diffraction patterns for the «simple cubic fullerite C_{24} → diamond-like CA4 phase» structural transition.

4. Conclusions

Thus, theoretical calculations of the stability and the possible method of applying the new fullerene-like C_{24} cluster, the surface of which is formed from four- and six-membered atomic rings, are carried out. The DFT calculations showed that these clusters should be stable at room temperature. It is also found that the simple cubic fullerite from C_{24} clusters can be used to synthesize the porous diamond-like CA4 phase during bulk compression up to 4.8 GPa. For experimental identification of the sc fullerite C_{24} and CA4 phase, the calculated X-ray diffraction patterns can be used.

Acknowledgments

The authors thank Foundation of Perspective Scientific Research of Chelyabinsk State University for the financial support.

References

[1] Novoselov K S, Geim A K, Morozov S V, Jiang D, Zhang Y, Dubonos S V, Grigorieva I V and Firsov A A 2004 *Science* **306** 666
[2] Galashev A E and Rakhmanova O R 2014 *Phys. Usp.* **57** 970
[3] Baimova J A and Rysaeva L Kh 2018 *J. Struct. Chem.* **59** 884
[4] Iijima S and Ichihashi T 1993 *Nature* **363** 603
[5] Dresselhaus M S, Dresselhaus G and Avouris Ph 2001 *Carbon nanotubes: Synthesis, structure, properties, and applications* (Berlin: Springer-Verlag) 453 p
[6] Prudkovskiy V S, Iacovella F, Katin K P, Maslov M M and Cherkashin N 2018 *Nanotechnology* **29** 365304
[7] Kroto H W, Heath J R, O'Brien S C, Curl R F and Smalley R E 1985 *Nature* **318** 162
[8] Sheka E 2011 *Fullerenes: Nanochemistry, Nanomagnetism, Nanomedicine, Nanophotonics* (Boca Raton: CRC Press) 312 p
[9] Rysaeva L Kh, Lobzenko I P, Baimova J A, Dmitriev S V and Zhou K 2018 *Rev. Adv. Mater. Sci.* **57** 143

[10] Belenkov E A and Greshnyakov V A 2013 *Phys. Solid State* **55** 1754
[11] Belenkov E A and Greshnyakov V A 2016 *Phys. Solid State* **58** 2145
[12] Belenkov E A and Greshnyakov V A 2014 *J Struct. Chem.* **55** 409
[13] Belenkov E A and Greshnyakov V A 2015 *Phys. Solid State* **57** 2331
[14] Greshnyakov V A and Belenkov E A 2018 *IOP Conf. Ser.: Mater. Sci. Eng.* **447** 012018
[15] Giannozzi P, Baroni S, Bonini N et al. 2009 *J. Phys.: Condens. Matter.* **21** 395502
[16] Hohenberg P and Kohn W 1964 *Phys. Rev. B* **136** 864
[17] Perdew J P, Burke K and Ernzerhof M 1996 *Phys. Rev. Lett.* **77** 3865
[18] Greshnyakov V A and Belenkov E A 2017 *Letters on Materials* **7** 318
[19] Belenkov E A and Greshnyakov V A 2018 *Phys. Solid State* **60** 1294

Suppression of Stark effect in ultra-thin stress-free GaN/AlN multiple quantum well structures grown by plasma-assisted molecular beam epitaxy

O A Koshelev, D V Nechaev, P N Brunkov, K N Orekhova, S I Troshkov, S V Ivanov and V N Jmerik
Ioffe Institute, 26 Politekhnicheskaya, St. Petersburg 194021, Russia

Abstract. We report on suppression of the Stark effect in (1.5-2)-monolayer(ML)-thick $(GaN/AlN)_{100}$ multiple quantum well (MQW) structures grown on AlN/c-Al_2O_3 templates by plasma-assisted molecular beam epitaxy. Different stress relaxation mechanisms are revealed in these structures by using a multi-beam optical stress sensor in comparison with the 5ML-MQW structure. The former (with well thicknesses ≤2MLs) demonstrate the nearly stress-free growth, whereas the latter structure with thicker wells exhibits the slow stress evolution from the high initial compressive stress to the nearly relaxed state with zero stress. Moreover, the former structures demonstrate a bright room-temperature cathodoluminescence (CL) with the single peak at the shortest wavelength 240 nm (1.5ML-QWs), while the latter shows much weaker multi-peak CL spectra in the spectral range of 270-360 nm.

1. Introduction

Semiconductor optoelectronics devices based on $Al_xGa_{1-x}N$ compounds with high x (>0.4) emitting in a middle ultraviolet (mid-UV) range (214-300nm) are of great interest for numerous applications in water/air purifications, medical diagnostics, security etc. [1-2]. These devices, having a small size, high efficiency, low operating voltage, and long lifetime, should replace the widely used Hg-based UV vacuum lamps. However, the efficiency and the output power of mid-UV LEDs are much lower as compared to InGaN-based LEDs operating in the visible and near-UV (>365nm) ranges. There exist several problems leading to a decrease in the efficiency of radiative recombination in AlGaN-based quantum well (QW) structures, including the strong Stark effect in the relatively thick QWs (1.5-2.5nm), the TE/TM polarization switching for the output emission from AlGaN alloy layers with high Al-content (x>0.25) [3,4]. In addition, non-uniform distribution of stress in multiple QW (MQW) structures with a large number of wells can lead to widening the output emission bands due to different Stark effect in the wells.

One of possible effective ways to overcome these problems is the development of ultra-thin GaN/Al(Ga)N MQW structures with a typical well thickness of a few monolayers (MLs) (1ML=0.25nm), which are intensively studied by several theoretical and technological groups [5-7]. Using this approach, both UV LEDs and e-beam pumped UV emitters operating in the sub-240nm spectral range have recently been fabricated. In these work, plasma-assisted molecular beam epitaxy (PA MBE) was successfully employed due to its unique ability of controllable growth of ultra-thin QWs with sharp interfaces owing to the fast (<0.5s) switching of growth fluxes from effusion cells and the low substrate temperatures (<730°C), which suppress any segregation and interdiffusion phenomena in the heterostructures.

Despite it is generally known that elastic stress in (Al,Ga)N MQW structures play an important role in their efficiency and luminescence spectra, the stress appearance and development have not been studied enough so far. Only Bellet-Amalric et al.[8], using reflection high-energy electron diffraction (RHEED) and X-ray diffraction (XRD) measurements, revealed the complex behavior of the *in plane* lattice parameter (stress) during PA MBE of AlN/GaN and GaN/AlN binary heterostructures with large layer thicknesses. However, these results cannot be used to analyze stress evolution and its effect in the ML-thick GaN/AlN MQW structures. In addition, the most advanced modern analytical method for observing the stress evolution in heterostructures is based on *in situ* measurements of substrate curvature, from which the average and incremental stresses can be calculated using the Stoney's equations [7]. Therefore modern PA-MBE and MOVPE setups are equipped with a multi-beam optical stress sensor (MOSS), however the use of this method to study thin GaN/AlN MQWs is still limited.

This paper reports on the different types of the stress evolution in binary AlN/GaN MQW structures with a well thickness varying from 1.5 to 5MLs, grown by PA MBE on AlN/*c*-Al_2O_3 templates. The possible influence of surface morphology on stress in these structures is discussed. A strong suppression of the Stark effect in the cathodoluminescence spectra of MQW structures with the thinnest (1.5-2ML) wells is demonstrated.

2. Experiment

GaN/AlN MQW heterostructures, schematically shown in Fig.1, were grown using PA MBE setup Compact21T on *c*-Al_2O_3 templates. First, AlN templates were grown at a growth temperature of 780°C using migration enhanced epitaxy (MEE) for initial growth of the 65nm-thick nucleation layer [10] and metal-modulated epitaxy (MME) for the next 1.45 μm-thick buffer layer [11]. Upper AlN BLs with a thickness of 0.3 μm were grown in a standard PA MBE mode at slightly Me-rich conditions with an Al/N flux ratio of about 1.1 and using a post-growth surface exposure under a plasma-activated nitrogen flux. One AlN template was grown without the MQW structure. Three studied GaN/AlN MQW structures were grown by a combination of MEE and standard MBE growth, using controlled Me-rich conditions for both GaN and AlN growth at a temperature of 700°C, as reported in detail in our previous paper [7]. They included 100-period MQWs with the same thickness of the AlN barrier layers of 4.5 nm (18MLs) and differed in the thickness of the GaN wells as 5, 2 and 1.5 ML, which were used to denote the structures.

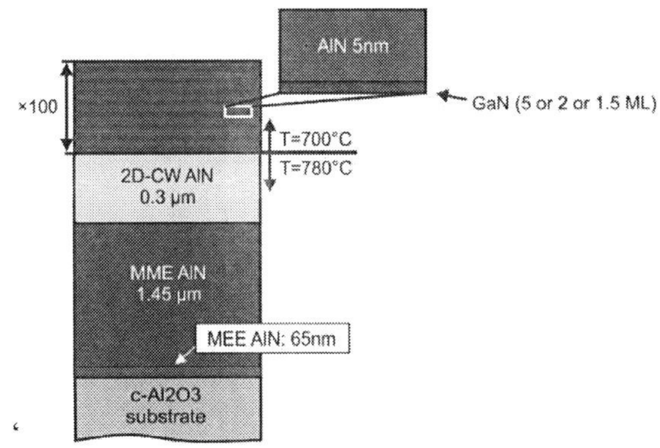

Figure 1. Schematic view of the samples design.

A homemade MOSS system based on a 10 mW solid-state laser (λ=532 nm) and a standard CCD-camera, was used to measure *in situ* the substrate curvature at a frequency of 15 Hz [9]. Thickness and growth rates of the AlN layers was controlled *in situ* by laser reflectometry (LR) (λ=532 nm). The surface morphology of the AlN layers and MQW structures was monitored using RHEED and studied *ex situ* by using atomic-force (AFM) and scanning electron (SEM) microscopies. The optical properties of MQW structures were studied by measuring cathodoluminescence (CL) spectra using an electron beam with a spot diameter of 2 μm and an energy(current) of 10 keV(30 nA).

3. Results and discussion

Figure 2 shows various evidences of the smooth surface morphology of the AlN templates, including the streaky RHEED patterns (Fig.2a), observed during almost the entire template growth, beginning from the upper parts of the NLs. The SEM and AFM images (Fig.2b,c) exhibit an atomically smooth and droplet-free surface morphology with a root-mean-square (RMS) roughness of about 0.7 nm over an area $1\times1\mu m^2$. Moreover, the AFM image shows that the template surface consists of grains with an average diameter of about 1 μm, which have an upward terrace-like morphology due to the step-flow spiral growth at the metal-enriched conditions used in MME.

AFM images showing the surface morphologies of the MQW structures are presented in Fig.3. Although the AFM images exhibit the conservation of the step-flow growth mode for all samples, the diameter of the terrace-like grains in the studied structures is different. If the 5ML-structure has a grain

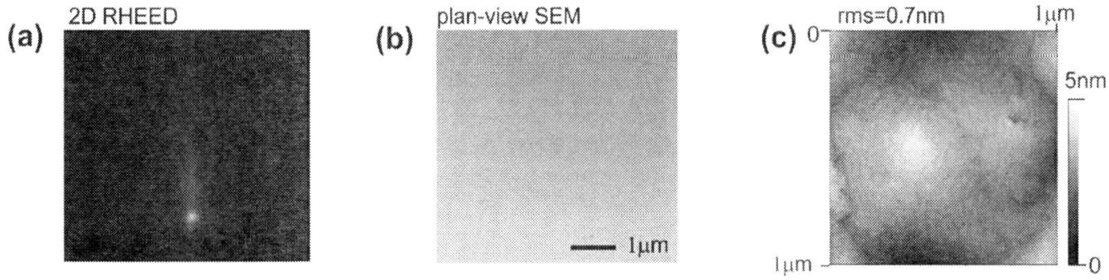

Figure 2. RHEED patterns during finishing the AlN template growth (a). Plan-view SEM (b) and AFM (c) images of AlN templates.

diameter of ~0.8μm, which is only slightly smaller than that of the AlN template, the 2ML- and 1.5ML--structures have more pronounced boundaries of the grains with average sizes of ~0.3 and ~0.2 μm, respectively. One should note that the RMS roughness in all the MQW structures exceeds the initial value in the AlN BL and varies between 1.6 and 2.7 nm. This dependence may be explained preliminary by a thicker Ga surfactant adlayer during AlN barrier growth in the 5ML-structure, which improves the low-temperature growth of AlN, though it needs additional experimental confirmation.

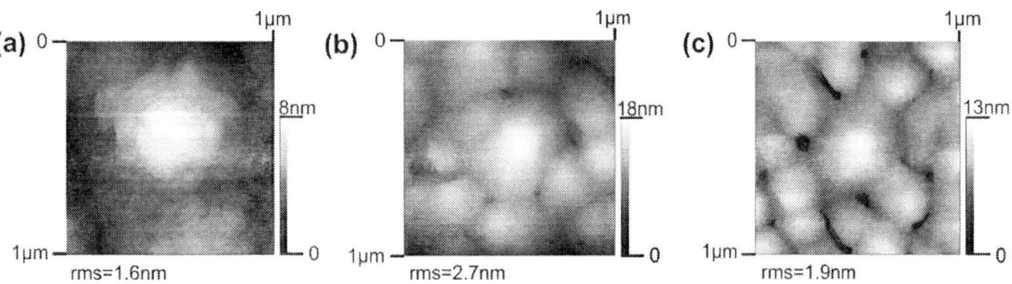

Figure 3. AFM images of MQW structures surfaces with QW thickness of 5 ML **(a)**, 2 ML **(b)** and 1.5 ML **(c)**.

Figure 4 shows the different character of stress evolution during growth of the MQW structures. First, the 5ML-structure demonstrates a gradual transition from a maximum compressive stress of -2.3GPa to an almost relaxed growth, when the structure thickness exceeds ~500 nm. It should be noted that the initial stress observed in this MQW structure with an average composition of $t_{AlN}/(t_{GaN}+t_{AlN})=0.8$ (where $t_{GaN(AlN)}$ are the thicknesses of GaN(AlN) layers in the MQW structure) corresponds to pseudomorphic growth of $Al_{0.8}Ga_{0.2}N$ layer on AlN template (the dashed red line in Fig.4a). Such a gradual relaxation of stress is typical for the growth of the (Al,Ga)N layers and heterostructures in low-temperature PA MBE [8]. In contrast, much weaker stresses with small or almost negligible changes in their magnitudes were observed in the 2ML- and 1.5ML-structures, respectively. Similar behavior of the stress in the GaN/AlN MQW structures with various well thicknesses was observed using MOSS in our previous work [7]. However the X-ray diffraction (XRD) analysis carried out earlier [7] and XRD reciprocal space mapping made recently (not shown here) revealed for the MQW structures with thin QWs (narrower than 2ML) practically unchanged vertical and lateral average lattice constants as compared to the AlN template. This implies that a noticeable compressive stresses should be accumulated in the MQW structures, which disagrees with the MOSS results showing nearly stress-free growth.

The contradiction between the MOSS-data demonstrating a stress-free growth of the GaN/AlN MQW structures and the results of XRD analysis indicating the same lattice constant of the AlN BL and MQW with the thin QWs was explained by the formation of ultra-thin spatially separated GaN platelets with a thicknesses of few MLs instead of a continuous GaN layer in growing MQWs. Indeed, the possibility of elastic relaxation of GaN(AlN) ultra-thin layers grown atop of the AlN(GaN) thick layers by PA MBE at

Me-rich conditions via development of discontinuities in the thin layers was proposed by Bourret et al. [12]. We applied this idea to explain the complete stress relaxation in the GaN/AlN MQW structures with the ultra-thin wells (≤2ML), in which the elastic lateral relaxation of the compressive stress induced in the GaN platelets at the lower interface causes the tensile stress relaxation in AlN barrier at the top interface of the platelets, which result in preserving both vertical and lateral average lattice constants [7].

Thus, it is expected that the same mechanism related to formation of nanoplatelets may be responsible for stress-free or weak-stress growth of the 1.5ML- and 2ML-structures, respectively, observed by MOSS in this study. One should mention that formation of the relatively small grains in 2- and 1.5ML structures, shown in Fig.3, could be considered as an alternative relaxation mechanism for the stress in the grained MQW structures. However, keeping in mind a large size of the grains (several hundred nanometers) as compared to the Ga-rich platelets (below 160 nm) [13], the possible contribution of this mechanism in the stress relaxation was preliminary estimated by us as negligible. In addition, usually the development of such grains, resulting in implementation of this mechanism of stress relaxation, occurs gradually during heteroepitaxial growth that contradicts the observed absence of any stress induced at the beginning of the 1.5ML-strcucture growth.

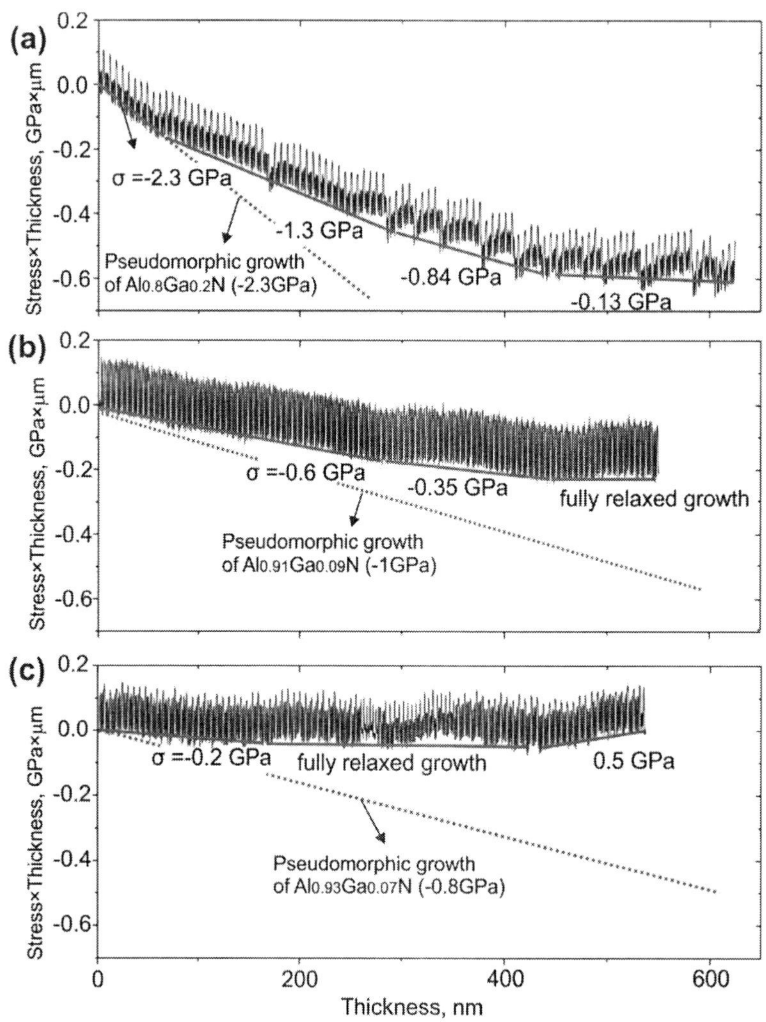

Figure 4. The measured by MOSS values of *(stress×thickness) vs thickness* in MQW structures with the QW thicknesses of 5ML (a), 2ML (b) and 1.5ML (c). The curves with oscillations demonstrate experimental data, while the red solid and dashed lines show the average stress values and the pseudomorphic growth lines for the effective alloy composition of MQWs, respectively.

To separate finally the contributions of the formation of platelets and grains in the MQW structures to relaxation of the compressive stress the detailed XRD analysis and CL studies should be performed. The XRD results will be reported elsewhere, whereas the analysis of the CL spectra of the MQW structures is presented below in Fig.5.

Figure 5a shows the CL spectrum of the 5ML-structure, consisting of several peaks distributed between the highest one at 360 nm towards the weakest ones at shorter wavelengths till 270nm. Thus, the full width at half maximum (FWHM) of the envelope curve for this fairly weak CL band is about 50 nm. In contrast, the 2ML- and 1.5ML structures exhibit much more intense and narrower single peaks at 270 and 240nm with the FWHM values of 21 and 13 nm, respectively. If position of the main CL peaks in the MQW structures is determined mainly by the quantum confinement effect, their spectral width depends on the elastic stress and thickness uniformity in the MQW structure.

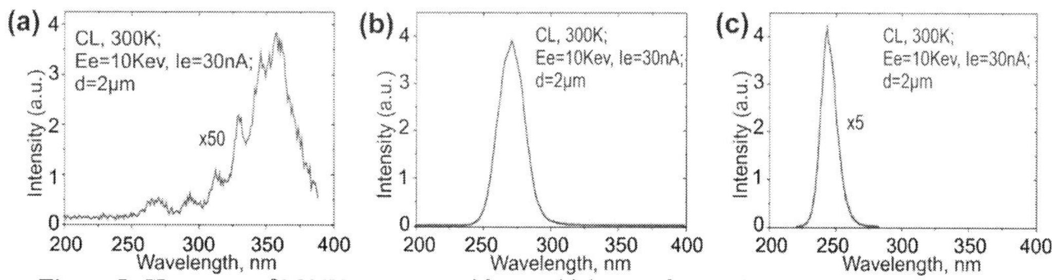

Figure 5. CL spectra of MQW structures with QW thickness of 5 ML (a), 2 ML (b) and 1.5 ML (c).

Consequently, the observed multi-peak CL spectrum apparently corresponds to rather high and strongly varying stress in the 5ML-thick structure, which changes the internal electric field and, hence, the magnitude of the Stark effect in the GaN QWs, causing the blue shift of the CL peaks from the initial MQW region under the compressive stress. The narrower and single CL peaks from the MQW structures with thin QWs provide the convincing evidence of uniform and low stress level in the GaN QWs across the structures. Thus, the observed formation of the terrace-like grains does not cause significant stress relaxation in the MQW structures with the thinnest QWs, where the grain density is higher, but the formation of the small nanoplatelets inside these grains is rather responsible for the stress-free growth of the structures.

4. Conclusion

We demonstrated that the $(GaN/AlN)_{100}$ MQW structures with extremely thin GaN QWs (1–2 MLs) grown by PA MBE on AlN have a constant and small average substrate curvature that means a stress-free growth. In contrast, the 5ML-MQW structure with the average Al content of x_{av}=0.8 exhibits the standard gradual stress relaxation from high level of the compressive stress (-2.3GPa), corresponding to the pseudomorphic growth, to the nearly fully relaxed growth at the surface. Although, the MQW structures with ultra-thin wells (<2ML) demonstrate the formation of the grains with the typically smaller sizes of about 200-300nm than in the 5ML-MQW structure, their contribution to stress relaxation seems to be negligible. The formation of small nanoplatelets is proposed as the main mechanism of elastic lateral relaxation of the local compressive and tensile stresses in the thinnest QWs, which provides a stress-free growth of the QWs and reduced Stark effect. This results in the strengthening and narrowing the single CL peaks from the structures with the shortest observed wavelength of 240nm. For the 5ML-MQW structure, the described mechanism of elastic stress relaxation through formation of separate platelets is apparently broken, that leads to a significant gradient of the compressive stress in the structure and, as a result, to widening the CL spectra consisting of several low-intensity peaks. Thus, development of the GaN/AlN MQW structures with ultra-thin (≤2MLs) wells is a promising way to develop high-power narrow-band sources of UV radiation in the sub-250 nm spectral range.

Acknowledgments
The work was supported in part (structure characterization) by Russian Science Foundation Project (Grant #19-72-30040).

Reference

[1] Hirayama H, Tsukada Y, Maeda T, Kamata N 2010 Appl. Phys. Express 3 031002.
[2] Shatalov M, Sun W, Lunev A, Hu X, Dobrinsky A, Bilenko Y, Yang J, Shur M, Gaska R, Moe C, Garrett G, Wraback M 2012 Appl. Phys. Express **5** 082101.
[3] Ng H M, Harel R, Chu S N G, Cho A Y 2001 J.Electron Mater. **30** 134.
[4] Nam K B, Li J, Nakarmi N L, Lin Y I, Jiang H X 2004 Appl. Phys. Lett. **84** 5264.
[5] Jmerik V N, Shubina T V, Mizerov A M, Belyaev K G, Sakharov A V, Zamoryanskaya M V, Sitnikova A A, Davydov V Yu, Kop'ev P S, Lutsenko E V, Rzheutskii N V, Danilchik A V, Yablonskii G P, Ivanov S V 2009 J. Cryst. Growth **311** 2080.
[6] Verma J, Islam S M, Protasenko V, Kandaswamy P K, Xing H, Jena D 2014 Appl. Phys. Lett. **104** 021105.
[7] Jmerik V N, Nechaev D V, Toropov A A, Evropeitsev E A, Kozlovsky V I, Martovitsky V P, Rouvimov S, Ivanov S V 2018 Appl. Phys. Express **11** 091003.
[8] Bellet-Amalric E, Adelmann C, Sarigiannidou E, Rouvière J L, Feuillet G, Monroy E, Daudin B 2004 J. Appl. Phys. **95** 1127.
[9] Hearne S, Chason E, Han J, Floro J A, Figiel J, Hunter J, Amano H, Tsong I S T 1999 Appl. Phys. Lett.**74** 356.
[10] Nechaev D V, Aseev P A, Jmerik V N, Brunkov P N, Kuznetsova Y V, Sitnikova A A, Ratnikov V V, and Ivanov S V 2013 J. Cryst. Growth **378** 319.
[11] Jmerik V N, Mizerov A M, Nechaev D V, Aseev P A, Sitnikova A A, Troshkov S I, Kop'ev P S, Ivanov S V 2012 J. Cryst. Growth **354** 188.
[12] Bourret A, Adelmann C, Daudin B, Rouvière J-L, Feuillet G and Mula G 2001 Phys. Rev. **63** 245307.
[13] Sviridov D E, Jmerik V N, Rouvimov S, Nechaev D V, Kozlovsky V I, and Ivanov S V 2019 Appl. Phys. Lett. **114** (6) 061601.

Crystal lattice tilt analysis in gradient composition layers by electron and X-ray diffraction

L A Snigirev[1,2], D A Kirilenko[2], E.V. Pirogov[3], M V Baidakova[2], N A Bert[2]
[1]ITMO University, Saint Petersburg 197101, Russia
[2]Ioffe Institute, Saint Petersburg 194021, Russia
[3]St. Petersburg Academic University RAS, Saint Petersburg 194021, Russia

Abstract. The combination of X-ray tilt-azimuth figure, reciprocal space map and transmission electron microscopy in selected area electron diffraction mode was applied to analyse the tilt of the MBE-grown 1.2-μm thick convex-graded $In_xAl_{1-x}As$ (x=0.05-0.80) metamorphic buffer layer on singular GaAs (001) substrate. The combination of the techniques is demonstrated to enable the determination of the tilt angle as well as the tilt direction together with monitoring of the tilt development along the growth direction. For the studied metamorphic buffer layer as an example, the tilt direction was deduced to be 20° off the (110) GaAs substrate crystallographic plane. The tilt angle of (002) buffer layer planes was observed to rapidly rise to 0.6° near the layer-substrate interface and then to slowly increase to 1.4° with the approach toward the surface.

1. Introduction

The epitaxial growth of perfect semiconductor solid solutions is possible only within the narrow composition range due to restricted set of lattice constants provided by available elemental or binary substrates. To grow device structures of which the lattice constant does not match that of the available substrates, metamorphic buffer layers (MBL) are used as a transition region between the substrate and an active structure e.g. [1-4]. Due to an essential difference in the lattice constant between the substrate and MBL, arising mismatch strain necessarily relaxes via generation of misfit dislocations. The MBL design and growth conditions have to be optimized as to achieve the complete mismatch strain relaxation without propagating the dislocations into the upper-lying active region.

The key issue in the epitaxial deposition of MBL is the control of the mismatch strain and its relaxation which detrimentally affects the electronic properties. For the general characterization of the strain relaxation of MBL, X-ray reciprocal space mapping (RSM) is widely used [5, 6]. The diffracted intensity distribution around a reflection plotted on the RSM is sensitive to the composition of MBL, to the accumulated strain and to the MBL tilt. Unfortunately, for the compositionally graded MBL with nonuniform strain relaxation along the growth direction (i.e. over the layer thickness), correlation of the RSM intensity distribution to the depth scale becomes questionable. Space resolved information on the strain state, defect types and distribution along the layer thickness can be provided by transmission electron microscopy (TEM) [7-9].

For the analysis of the strain relaxation in compositionally graded mismatched layers, combined RSM and TEM data representation in the form of RSM with superimposed lattice parameter profiles (m-profiles) measured by cross-section TEM was shown to be very informative [10]. The profiling procedure was applied to about 1-μm thick linearly-graded $In_xGa_{1-x}As$ (x=0.05-0.38) [11] as well as to

1-μm thick convex-graded $In_xAl_{1-x}As$ (x=0.05-0.79) [12] MBLs, both MBE-grown on the GaAs (001) substrate. The RSM with plotted m-profile was designated as the depth profiled RSM (DpRSM).
In conjunction with the mismatch strain relief through the dislocation generation and propagation, a tilt in mismatched layers occurs if dislocation densities differ in glide systems [13]. In this paper we apply DpRSM and additionally involve tilt-azimuth figure (TAF) recording to monitor the tilt behavior in 1.2-μm thick convex-graded $In_xAl_{1-x}As$ (x=0.05-0.80) MBL.

2. Experimental

The sample containing convex-graded $In_xAl_{1-x}As$ MBL was grown by molecular beam epitaxy on semi-insulating GaAs (001) substrate. The growth procedure is described in more detail elsewhere [12]. The complete layer sequence of the sample under study is presented in Fig. 1.

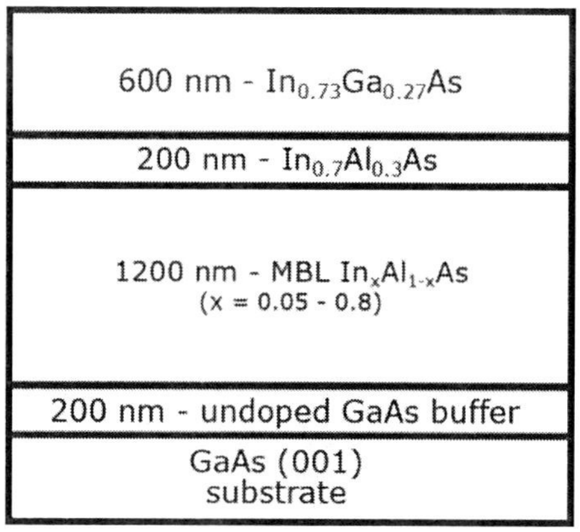

Figure 1. Design of the sample under study with $In_xAl_{1-x}As$ (x=0.05-0.8) metamorphic buffer layer on GaAs (001) substrate.

High-resolution XRD measurements were carried out using multifunctional triple-crystal diffractometer D8 Discover (Bruker, Germany) in a parallel X-ray beam geometry. The θ - 2θ scan as well as RSM around symmetric (004) GaAs reflection were used for the analysis of crystalline properties of the MBL. The TAF was recorded by collecting intensities during ω-scans at different azimuthal angles φ.
Cross-section specimens for TEM were prepared by conventional procedure of mechanical grinding-polishing and final Ar^+ ion milling.
For TEM study, JEM2100F electron microscope operating under an accelerating voltage of 200 kV was exploited in the conventional modes of diffraction contrast imaging and selected area electron diffraction (SAED).

3. Results and discussion

The TAF from the InAlAs MBL was collected at the angle $2\theta = 64.6°$, which corresponds to the diffraction conditions for $In_{0.7}Ga_{0.3}As$ composition. The correlation of the TAF collection region with the layer depth scale can be seen in Fig. 2. The geometry of the TAF recording is shown in Fig. 3a. During the experiment the sample was positioned in such a way that $[\bar{1}10]$ axis of GaAs substrate coincided with the azimuthal direction to 0° of the figure.
The collected TAF is represented in Fig. 3b. The diffuse intensity cloud about 0.5° wide clearly appears within the azimuthal angle range 90°-270°. At the azimuthal angles 0-90° and 270-360° the intensity attenuates, but the intensity behavior being still distinguishable. The intensity depletion results from a deviation of the (002) MBL atomic planes from the exact diffraction conditions due to

the presence of a tilt, of which the magnitude is greater than the rocking curve full width at half maximum $\Delta\omega$.

The measured tilt angle α is known to depend on the in-plane sample orientation azimuthal angle φ in respect to the incident X-ray beam, obeying to the sinusoidal dependence $\alpha = \alpha_0\cos(\varphi-\varphi_0)$, where α_0 and φ_0 are the true tilt and the tilt azimuthal angles respectively [14]. Consequently, a maximal tilt angle of 0.9° deduced from the figure was accepted as the true tilt angle α_0. It was observed at the azimuthal angles approximately 110° and 290°, i.e. 20° off the (220) crystallographic plane of the sample.

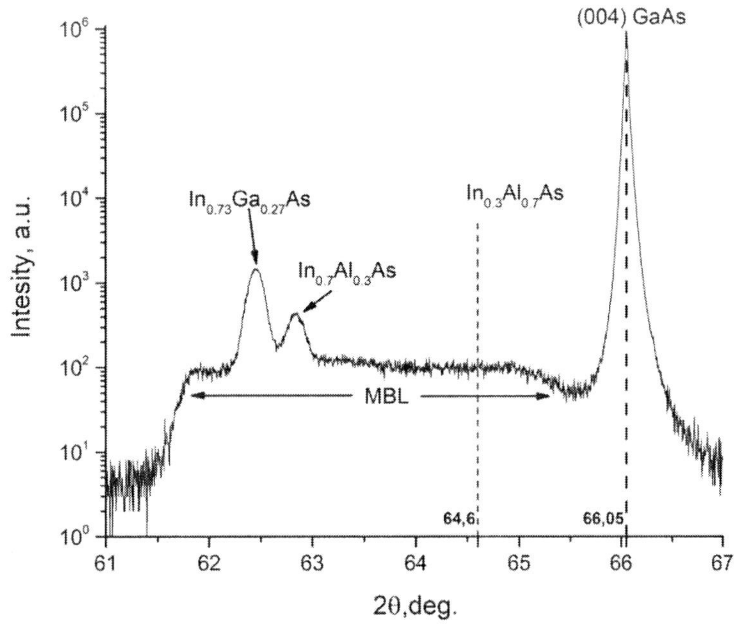

Figure 2. $\omega-2\theta$ - scan of the sample near (004) GaAs reflection. The section, which the TAF was recorded from, is marked by vertical dashed line at $2\theta = 64.6°$.

To monitor the tilt behavior along the layer thickness, a set of SAED patterns registered from (110) cross-section with the selected-area aperture size as small as 100 nm was obtained shifting the sample to a regular interval of approximately 30 nm along the growth direction. Two angles describing layer lattice tilt were measured with the reference to GaAs substrate: $\alpha^{\perp}$ for the tilt of (002) planes and $\alpha^{||}$ for the tilt of (220) planes. Simultaneously, the 002 and 220 diffraction vector length $q^{\perp}$ and $q^{||}$ were measured, and the in-plane $m^{\perp}$ and out-of-plane $m^{||}$ mismatch was derived from a simple equation $m=\Delta q/q_s=(a_l - a_s)/a_s$ where subscripts "l" and "s" stand for the MBL and substrate correspondingly. The data are represented in Fig. 4.

As it can be seen from Fig. 4, the layer tilt occurs quite close to the layer-substrate interface, its value rapidly reaches 0.6-0.8° already at a distance of 100 nm. Note that here the measured tilt angles $\alpha^{\perp}$ and $\alpha^{||}$ become perceptibly different. Then, two smooth steps are noticeable in the tilt behavior. The first step appears around 300 nm from the interface (point 1 on the graph) and results in both tilt angles increasing to the same value of approximately 1.0°. The second increase in the tilt starts at around 780 nm from the interface (point 2), and the layer tilt arrives to 1.1-1.4° at 1000 nm from the interface (point 3), both tilt angles becoming different again.

It should be noticed that the azimuthal figure (Fig. 3b) is recorded from the bottom part of MBL, where In concentration reaches x=0.3, i.e. at a distance of about 300 nm from the layer-substrate interface. One more remark is that the tilt angle measured by SAED is slightly less than the true tilt angle since the tilt vector deviates 20° off the (110) plane. With this in mind, the tilt angle measured by SAED at the point 1 coincides rather well with that deduced from TAF.

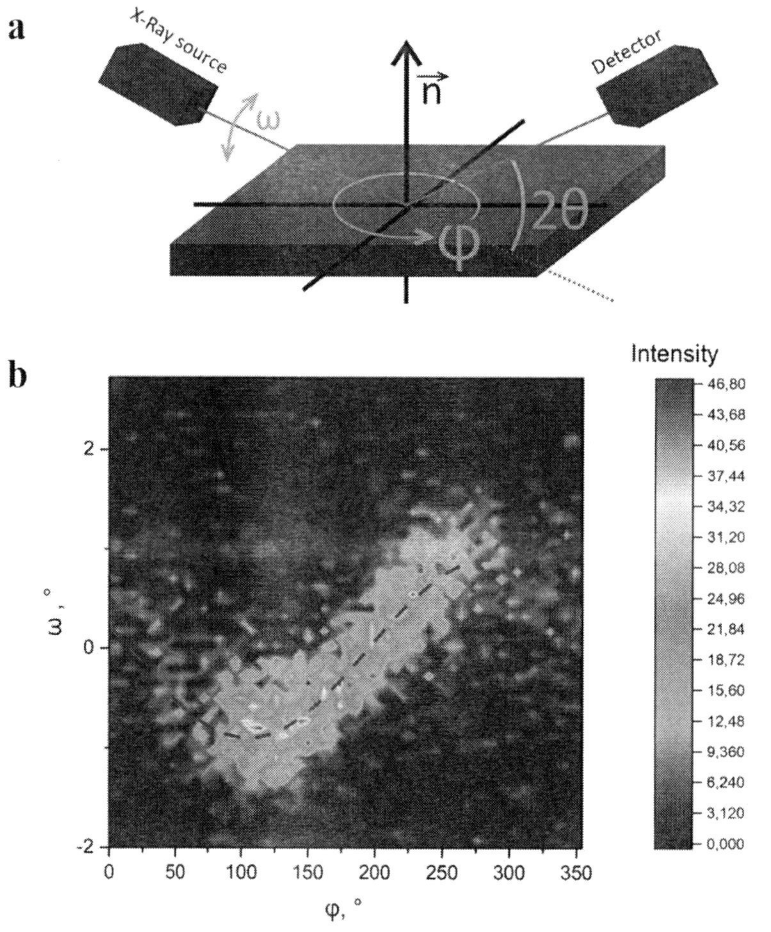

Figure 3. (a) Schematic of the experimental setup geometry and (b) tilt-azimuth figure recorded at $2\theta = 64.6°$ from a bottom part of the metamorphic buffer layer

The combined data representation in the form of RSM and the lattice parameter profiles (m-profiles) has been shown to be a useful visual representation of the quantitative data on the distribution of the strain and composition of MBL along the growth direction [10]. We collected the symmetrical (004) DpRSM which is known to reflect purely the tilt but not the strain. The DpRSM is replicated in Fig. 5 and shows the deviation of the layer node from the surface normal (q_z direction) due to presence of a tilt. The profile reflecting out-of-plane mismatch $m^{\perp}$ obtained by SAED is plotted on RSM with the reference to the GaAs substrate. The $m^{\perp}$ profile is represented by vertical red line in RSM and, as it can be seen, does not provide any valuable information on the mismatch behavior. The tilt correction can be introduced by comparison of the RSMs for symmetric and asymmetric reflections. However, the procedure for collecting few RSMs and performing large data treatment is rather complex and time consuming. Instead, to account for the MBL tilt we used the data obtained by SAED. The corrected profile is represented by blue line and allows one to trace both lattice mismatch and tilt angle behavior.

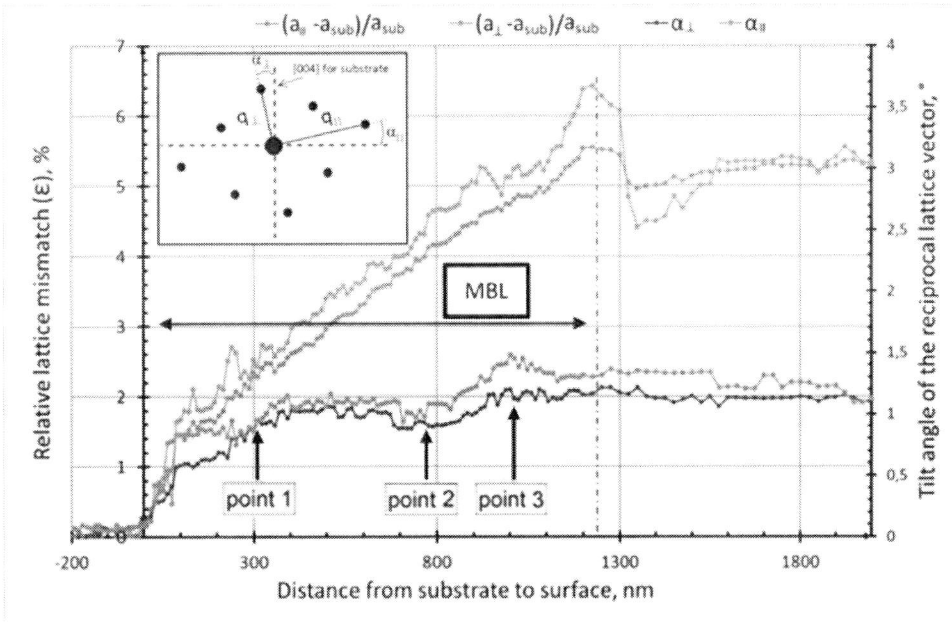

Figure 4. Behavior of out-of-plane $m^{\perp}$ and in-plane $m^{||}$ lattice mismatch and tilt angles $\alpha^{\perp}$ and $\alpha^{||}$ with increasing distance from GaAs substrate.

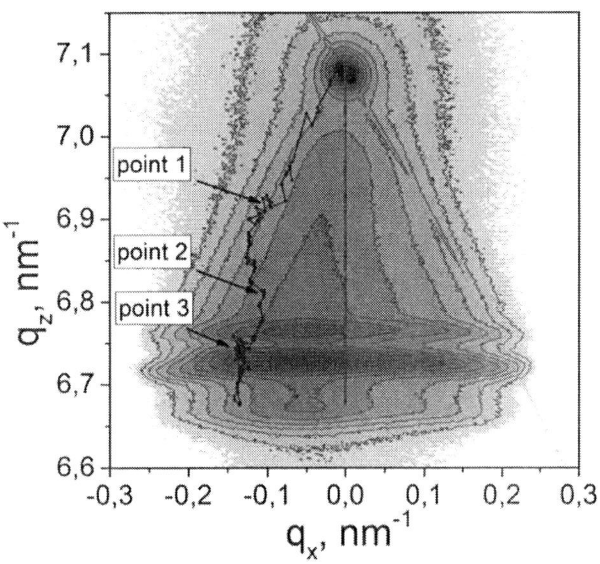

Figure 5. Reciprocal space map around symmetrical 004 node of the reciprocal lattice of InAlAs metamorphic buffer layer grown on GaAs (001) substrate

The characteristic points associated with the tilt behavior features observable in Fig. 4 are reproduced in the RSM.

4. Summary

We applied the procedure of X-ray TAF and RSM constructions combined with TEM SAED profiling to study the tilt of the MBE-grown $In_xAl_{1-x}As$ (x=0.05-0.80) MBL in respect to GaAs (001) substrate. The combination of the techniques is demonstrated to enable the determination of the MBL tilt angle as well as the tilt direction together with monitoring of the tilt development along the growth direction. As a result, the tilt direction of 1.2-μm thick convex-graded $In_xAl_{1-x}As$ (x=0.05-0.80) MBL was deduced to be 20° off the [110] GaAs substrate crystallographic axis. The tilt angle of (002) MBL planes was observed to rapidly rise to 0.6° near the interface and then to slowly stepwise increase to 1.4° with the approach toward the surface.

Acknowledgments

Authors are grateful to V.A. Solov'ev and M.Yu. Chernov (Laboratory of Quantum-size heterostructures, Ioffe Institute) for providing the sample. A.A. Sitnikova and M.A. Yagovkina are acknowledged for their help in conduction of this work.
The measurements have been made using equipment of the Federal Joint Research Center "Materials science and characterization in advanced technology".

References:

[1] Hoke W E, Leoni R E, Whelan C S, Kennedy T D, Torabi A, Marsh P F, Zhang Y, Xu C, Hsieh K C 2004 *J. Vac. Sci. Technol. B* **22** 1554
[2] Tangring L, Wang S, Sadeghi M, Larsson A 2006 *Electron. Lett.* **42** 691
[3] Cavus A, Monier C, Cox C, Pascua D, Gutierrez-Aitken A, Noori A, Hayashi S, Goorsky M 2006 *J. Vac. Sci. Technol. B* **24** 1492
[4] Belenky G, Wang D, Lin Y, Donetsky D, Kipshidze G, Shterengas L, et al *Appl. Phys. Lett.* 2013 **102** 111108
[5] Gu Y, Zhang Y, Chen X, Ma Y, Zheng Y, Du B, Zhang J 2017 *J. Phys. D: Appl. Phys.* **50** 385105.
[6] Lee D, Park M S, Tang Z, Luo H, Beresford R, Wie C R *J. Appl. Phys.* 2007 **101** 063523
[7] Chauveau J M, Cordier Y, Kim H J, Ferr D, Androussi Y, Di Persio J 2003 *J. Cryst. Growth* **251** 112
[8] Zhang M, Guo Z, Zhao L, Yang S, Zhao L 2018 *Materials* **11** 975
[9] Gangopadhyay A, Maros A, Faleev N, Smith D J 2019 *Acta Materialia* **162** 103
[10] Baidakova M B, Kirilenko D A, Sitnikova AA, Yagovkina M A, Klimko G V, Sorokin S V, Sedova I V, Ivanov S V, Romanov A E 2016 *Tech. Phys. Lett.* **42** 464
[11] Sorokin S V, Klimko G V, Sedova L V, Sitnikova A A, Kirilenko D A, Baidakova M V, Yagovkina M A, Komissarova T A, Belyaev K G, Ivanov S V 2016 *Journal of Crystal Growth* **455** 83
[12] Solov'ev V A, Chernov M Yu, Baidakova M V, Kirilenko D A, Yagovkina M A, Sitnikova A A, Komissarova T A, Kop'ev P S, Ivanov S V 2018 *Superlattices and Microstructures* **113** 777
[13] Ghanad-Tavakoli S, Hulko O, Thompson D A 2008 *J. Appl. Phys.* **103** 103527
[14] Olsen J A, Hu E L, Lee S R, Fritz I J, Howard A J, Hammons B E, Tsao J Y 1996 *J. Appl. Phys.* **79** 3578

Formation peculiarities of silver dendritic structures for photocatalysts of the visible radiation range

A P Sigaev[1], I A Averin[1], I A Pronin[1], A A Karmanov[1], N D Yakushova[1], V A Moshnikov[1, 2]
[1]Department of Nano- and Microelectronics, Penza State University, 440026, Penza, Russia
[2]Department of Micro- and Nanoelectronics, Saint-Petersburg Electrotechnical University, 197376, Saint-Petersburg, Russia

Abstract. The techniques for synthesis of silver structures in the form of a polyhedral array, as well as in the form of self-similar objects, spatially organized by the type of dendrites, are presented. The mechanisms that potentially determine the formation peculiarities of such structures are considered. Prospects of using this type of structures alongside with wide band gap semiconductor oxides, as highly efficient photocatalysts of the visible radiation range, are demonstrated.

1. Introduction
In recent years, much attention has been paid to the development of new materials, which, when transitioning to nanoscale characteristic sizes, acquire unique photocatalytic properties [1, 2]. In this case, nanomaterials that demonstrate high photocatalytic activity in light irradiation, both of ultraviolet and visible wavelength ranges, are of particular interest [3, 4]. To create photocatalysts of this type, various methods and approaches have been proposed, including those based on the use of surface plasmon polaritons (SPPs), excited at the interface boundary with different dielectric constant. The main idea here is to use a composite metal structure (platinum, palladium, gold, silver, etc.), namely, a wide band gap semiconductor (titanium dioxide, zinc oxide, etc.) to generate, collect and separate high-energy charge carriers causing high photocatalytic activity of the nanomaterial. According to this approach, with strong light and matter interaction in the nanostructured materials, binding of incident light in plasmon resonances can be observed. Under certain conditions, these photogenerated plasmon polaritons break up into a collective of non-equilibrium high-energy charge carriers, which later can be used to stimulate a chemical reaction [5]. For example, in the well-known work [6], this phenomenon was used to increase the efficiency of hydrogen generation process in light irradiation of nanostructured photocatalysts with near-zero dielectric constant.

However, as is well known, plasmon resonance for nanomaterials of a simple geometric shape, for example, for spherical metal nanoparticles, corresponds to a certain frequency, and as a consequence, to the fixed wavelength of incident radiation. In accordance with this, research related to the study of SPPs generation in nanomaterials with a complex spatial structure, including self-similar ones becomes relevant [7-9]. The purpose of this work was to study the formation peculiarities of silver dendritic structures for photocatalysts of the visible radiation range. In this case, the self-similar structure of the samples provides the fundamental possibility of generating surface plasmon polaritons not at a strictly defined frequency, but in the frequency range that covers both the ultraviolet and the visible area.

2. Experiment

In the framework of this work, the synthesis of structures based on Ag was carried out according to two main techniques, taking into account the known experimental works [10, 11]. According to the first technique, a solution of silver nitrate was prepared by dissolving 1 mmol salt in 100 mL of nitric acid with a concentration of 0.1 mol. Next, the solution was continuously stirred using a magnetic stirrer, and 3 mmol L-ascorbic acid dissolved in 10 mL of distilled water was added. As a result of the reaction carried out at a temperature of 25°C, silver particles were formed, being precipitated for 60 minutes. At the last synthesis stage, sediment separation by centrifugation, its double washing with water and acetone, and also drying at room temperature were carried out.

According to the second technique, solutions A and B were prepared at the first stage. At the same time, to prepare solution A, 0.2 mmol $AgNO_3$ was dissolved in 0.5 mL of distilled water, followed by the addition of 9.5 mL of acetone. Solution B was obtained by dissolving 0.4 mmol L-ascorbic acid in 0.5 mL of distilled water, and then 9.5 mL of acetone was added. At the second stage, solution B was dripped into solution A for 3 minutes at a temperature of 25°C. At the third stage, the resulting mixture was kept for 10 minutes at room temperature, after which the sediment was separated by centrifugation and washed with distilled water several times.

3. Results and Discussion

Figure 1 presents the data of scanning electron microscopy (SEM) of samples synthesized in accordance with the first technique. Figures 1a and 1b correspond to an increase of 10.000x and 50.000x, respectively.

Figure 1(a, b). SEM images of silver structures in the form of a polyhedral array with an increase of (a) 10.000x and (b) 50.000x

An analysis of the images presented in Figure 1 shows that in the solution of water and nitric acid, Ag structures are formed as a polyhedral array, the average size of which does not exceed 2 µm. Moreover, small size dispersion is characteristic for individual particles inside the array; silver particles have the highest concentration, the size of which is ≈ 1.2 µm along the longest axis. For such nanomaterials, even if they have a complex geometrical shape, plasmon resonance is observed in a narrow frequency range corresponding to the characteristic size of polyhedra with allowance for dispersion. This significantly limits the spectrum of the incident radiation, in which the composite metal-wide band gap semiconductor structures exhibit photocatalytic activity. In particular, for composite structures based on zinc oxide and an array of silver polyhedrons, photocatalytic decomposition of model pollutants is observed mainly under radiation of the ultraviolet wavelength range.

Figure 1 presents the data of scanning electron microscopy of samples synthesized in accordance with the second technique. Figures 1a and 1b correspond to an increase of 20.000x and 100.000x, respectively.

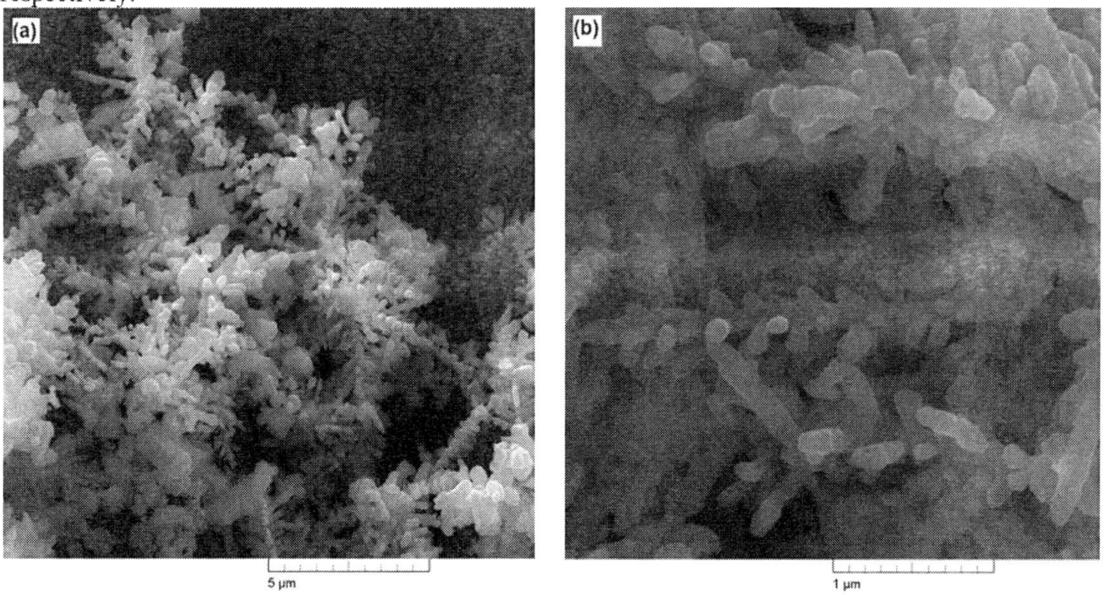

Figure 2(a, b). SEM images of silver dendritic structures with an increase of 20.000x **(a)** and 100.000x **(b)**

An analysis of the images presented in Figure 2 shows that in the solution of water and acetone, arrays of self-similar Ag structures are formed according to the type of dendrites, the trunk and main branches of which have a characteristic size of 1-5 µm, and the side ones are of 100-500 nm. This size of the side branches of the dendrites, as well as their spatial arrangement, ensures the equivalence of the studied structure of the metal surface with convexities and concavities, which realize the locking of electromagnetic energy in the volume, generating strong quasi-static localization [12]. This localization, in turn, ensures the efficient generation of high-energy charge carriers that stimulate catalytic reactions, for example, photocatalytic decomposition of organic pollutants.

4. Conclusions

Thus, the present work considers the formation peculiarities for two types of silver structures: 1) in the form of a polyhedral array; 2) in the form of self-similar objects, spatially organized by dendritic type. We assume that the growth mechanism for silver dendrites, synthesized in accordance with the second techique, is similar to the mechanism for the formation of nanotrees described in [13]. In our case, acetone performs a function similar to hydroxylamine, i.e. its molecules are adsorbed on the faces corresponding to the crystallographic planes (110) and (100), preventing the growth of Ag particles along the spatial directions corresponding to these faces. Being essentially self-similar, the growth mechanism under consideration determines the existence of dendritic branches of different generations. The results obtained may be of interest in the development of composite photocatalysts on the basis of such structures that are active in the visible radiation range.

Acknowledgments

The work has been financially supported by the Ministry of Education and Science of the Russian Federation within the framework of the project part of the state assignment for Penza State University No. 16.897.2017/4.6, as well as in the framework of the Presidential Grant and the Presidential

Scholarship (project No. MK-1882.2018.8, SP-84.2018.1 and SP-3800.2018.1). The authors also thank Kisin V.V. and Smirnov A.V. from Yuri Gagarin State Technical University of Saratov (SSTU) for assistance in the experimental study of silver structures using scanning electron microscopy.

References

[1] Sun L, Li R, Zhan W, Yuan Y, Wang X, Han X, Zhao Y 2019 *Nat. Commun.* **10** 2270
[2] Pronin I A, Averin I A, Kaneva N V, Bozhinova A S, Papazova K I, Dimitrov D Ts, Moshnikov V A 2014 *Kinetics and Catalysis* **55** 167
[3] Zhang Z, Ma Yg, Bu X, Wu Q, Hang Z, Dong Z, Wu X 2018 *Scientific Reports* **8** 10532
[4] Duan Z, Huang Y, Zhang De, Chen S 2019 *Scientific Reports* **9** 8008
[5] Linic S, Christopher P, Ingram D B 2011 *Nat. Mater.* **10** 911
[6] Tian Y, de Arquer F P G, Dinh C-T, Favraud G, Bonifazi M, Li J, Liu M, Zhang X, Zheng X, Kibria Md. G., Hoogland S, Sinton D, Sargent E H, Fratalocchi A 2017 *Adv. Mat.* 1701165
[7] Ma X-C, Dai Y, Yu L, Huang B-B 2016 *Light: Science & Applications* **5** e16017
[8] Moshnikov V A, Maksimov A I, Aleksandrova O A, Pronin I A, Karmanov A A, Terukov E I, Yakushova N D, Averin I A, Bobkov A A, Permyakov N V 2016 *Technical Physics Letters* **42** 967
[9] Pronin I A, Averin I A, Yakushova N D, Karmanov A A, Moshnikov V A, Ham M-H, Cho B K, Korotcenkov G 2017 *Arabian Journal for Science and Engineering* **42** 4299
[10] Martínez-Castañón G A, Niño-Martínez N, Loyola-Rodríguez J P, Patiño-Marín N, Martínez-Mendoza J R, Ruiz F 2009 *Mat. Let.* 63 1266
[11] Han Y, Liu S, Han M, Bao J, Dai Z 2009 *Crystal Growth & Design* **9** 3941
[12] Silveirinha M G, Engheta N 2007 *Phys. Rev. B* **76** 245109
[13] Yang C, Cui X, Zhang Z, Chiang S W, Lin W, Duan H, Li J, Kang F, Wong C-P 2015 *Nat. Commun.* **6** 8150

Critical pressure during the formation of carbon nanotubes by the method of plasma chemical vapor deposition

V S Klimin[12], A A Rezvan[1], J V Morozova[1]
[1]Department of Nanotechnology and Microsystems, Southern Federal University, Taganrog 347922, Russia
[2]Research and Education Center "Nanotechnologies", Southern Federal University, Taganrog 347922, Russia

Abstract. The critical pressure of acetylene required for the formation of nanotubes on nickel catalytic centers of a given size was calculated by plasma chemical vapor deposition. It has been established that for a stable growth of carbon nanotubes on a catalytic center with a diameter of 20 nm, the critical pressure of acetylene should be 1.9 Pa, while the growth time of carbon nanotubes in length of 5 μm is 131 sec. Nickel was used as a material for the formation of catalytic centers. In the calculations, the averaged parameters of carbon nanotubes were used. This analysis allowed to form a number of criteria for the materials of the underlayer and catalytic centers, to obtain an array of carbon nanotubes with a high degree of verticality.

1. Introduction

The increased interest in the study of carbon nanotubes is due, on the one hand, to their unique physicochemical properties, thanks to which they are an attractive object of basic research, and, on the other hand, to broad prospects for applied use [1-5]. The development of technologies for producing arrays oriented carbon nanotubes with controlled parameters opens up wide possibilities for their application in microelectronic sensors for creating sensitive elements of gas sensors. Such sensitive elements can reduce the operating voltage and improve the weight and size characteristics of the devices [6-12]. To implement practical developments using carbon nanotubes, selective synthesis of nanotubes with a controlled structure and properties is required [13-18]. One of the most promising methods of such synthesis is chemical vapor deposition initiated by plasma. The plasma chemical vapor deposition method allows to obtain oriented arrays on various substrates. A feature of this method is the use of catalytic centers. However, to obtain carbon nanotubes with the required geometric and electrical parameters requires a clear understanding of the processes and their precise control. In this connection, the study of the critical pressure of acetylene and its influence on the growth of carbon nanotubes on nickel catalytically layers is important [19-25].

2. Materials and Methods

This technique is consistent with the following points, namely, silicon substrates were used in the studies, the surface of which was cleaned by the method of liquid polishing. Nickel was used as a material for the formation of catalytic centers. The calculation method is based on the Knudsen theory and the basic equations of solid-state physics, which allows analyzing the diffusion of the carbon flow presented in Figure 1.

The calculations were performed using the averaged parameters of carbon nanotubes, which resulted in the calculation of the resulting single carbon nanotube (1).

$$V = \frac{\pi * l}{4}(d_2^2 - d_1^2), \quad (1)$$

There l is the length, d_1, d_2 are the inner and outer diameters, respectively. The inner diameter of the nanotube is assumed to be equal to the size of the catalytic center dk.

The number of atoms that make up a carbon nanotube is defined as (2):

$$N_H = \frac{M_N}{M_C} N_A \quad (2)$$

$M_N = V_M \cdot M_C$ – carbon nanotube molar mass, $V_M = (V/V_C)$ – carbon nanotube molar volume, V_C – molar volume of carbon, M_C – molar mass of carbon; N_A – Avogadro's number.

On fig. 1 (a, b) shows a scheme of a diffusive flow of carbon through a nickel catalytic center with the growth of carbon nanotubes.

Determine the molar concentration of carbon that is released during the dissociation of acetylene and dissolves on the surface of the nickel catalytic center:

$$\Delta C = 10^{-6} \frac{n_c}{V_{Ni}} \frac{M_{Ni}}{M_C} \quad (3)$$

V_{Ni} – molar volume of Ni; M_{Ni} - molar mass of Ni; n_C – carbon limit concentration, which, without entering into chemical interaction, dissolves in nickel at the temperature of growth of carbon nanotube (1023 K). Know the concentration of carbon dissolved on the surface of the catalytic center, one can estimate the molar flux of carbon to the base of the growing nanotube:

$$F \cong D * \frac{\Delta C}{L_D} \quad (4)$$

There D - carbon diffusion coefficient and L_D – diffusion length.

Using (3) and (4), the time of nanotube formation can be estimated as:

$$\tau = \frac{V_M}{F * A} \quad (5)$$

$$A \cong \pi d^2/4 \quad (6)$$

There A – the area of the catalytic center through which carbon diffuses to the base of the nanotube, assuming that the catalytic center has a spherical shape. Then the growth rate of carbon nanotube can be estimated as:

$$\omega = l/\tau \quad (7)$$

The number of carbon atoms arriving at the base of the nanotube per unit of time is determined by the formula:

$$\omega_A = D * \frac{\Delta C}{L_D} * N_A * A \quad (8)$$

There the area of the base of the nanotube is equal to the area of the catalytic center AH = A.

For a stable growth of carbon nanotubes, it is necessary that the acetylene molecules on the surface of the catalytic center dissociate at the same rate. Considering that as a result of the decomposition of acetylene $C_2H_2 \rightarrow 2C + H_2$, two carbon atoms are released, the steady growth of carbon nanotubes will be maintained if the dissociation rate balances the outflow of carbon to the base of carbon nanotube:

$$\omega_d = \omega_A/2 \quad (9)$$

From this it follows that per unit of surface area of the catalytic center per unit of time during the dissociation of acetylene appears

$$\omega^* = \omega_d/A \quad (10)$$

. Then it is possible to calculate the total number of dissociation centers N* on the surface of the catalytic center.

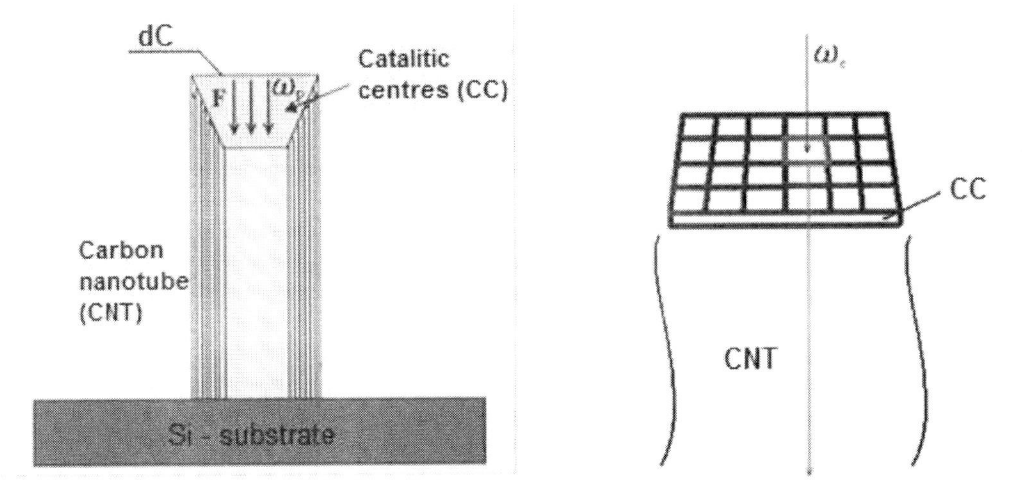

Figure 1 (a, b). Diagram of the diffusion flow of carbon through the nickel catalytic center with the growth of carbon nanotube

In this case, V_d dissociation will occur on the surface of the catalytic center per unit of time, and after each interval of time t_d two new carbon atoms will appear, that is C_2H_2 dissociation reaction is carried out. Thanks to a series of transformations, the following form was obtained for determining the magnitude of the critical pressure (11).

$$P = \frac{w}{2}(2\pi mkT)^{-1} \tag{11}$$

3. Results

The obtained values of the average parameters, as well as the constructed mode of formation of carbon nanotubes, were used to obtain an array on the catalytic nickel centers in the Ni/Ti/Si and Ni/V/Si system.

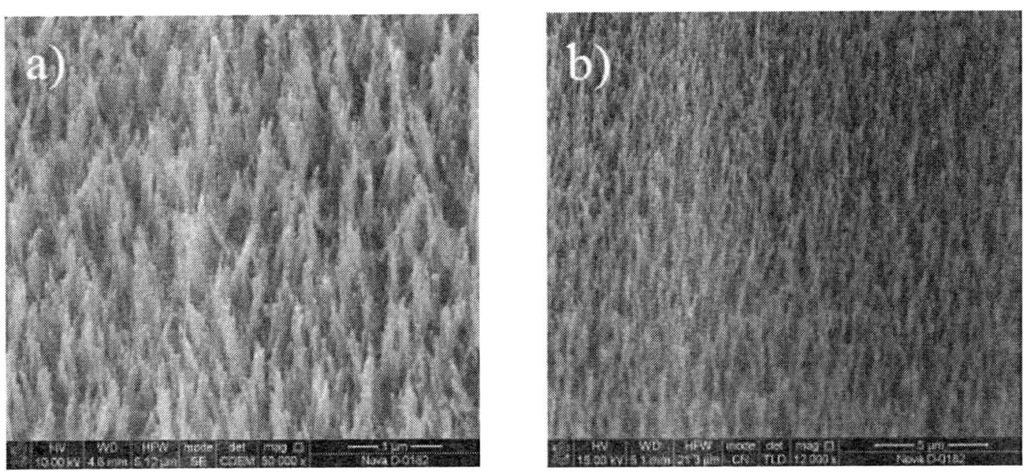

Figure 2 (a, b). SEM images of CNT arrays obtained on structures with different underlayer materials: **a)** Ni/Ti/Si; **b)** Ni/V/Si

4. Discussion and Conclusions

At the end of the analysis, taking into account the results obtained, the mode of the process of growing carbon nanotubes by the method of plasma chemical vapor deposition was formed and tested. Also, this analysis allowed us to form a number of criteria for the materials of the underlayer and catalytic centers, to obtain an array of carbon nanotubes with a high degree of verticality.

Calculations showed that for sustainable growth of carbon nanotubes on catalytic centers with a diameter of 20 nm, the critical pressure of acetylene should not be less than 1.9 Pa. The number of dissociation of acetylene molecules per unit of time per unit area of the catalytic center was $1.777*10^{22}$ $m^{-2}s^{-1}$, the formation time of carbon nanotubes was 131 seconds with an average speed of 2.3 μm/min.

Acknowledgments

This work was supported by Grant of the President of the Russian Federation No. MK-3512.2019.8 and the Russian Foundation for Basic Research Project № 16-29-14023 ofi_m. The results were obtained using the equipment of the Research and Education Center "Nanotechnologies" of Southern Federal University.

References

[1] Shiraishi K and Ito T 1998 Phys. Rev. B 57 6301
[2] Avilov V I, Polupanov N V, Tominov R V, Smirnov V A and Ageev O A 2017 *IOP Conf. Ser.: Mat. Sci.&Eng.* **256** 012001
[3] Jeong H E, Lee S H, Kim P and Suh K Y 2008 Colloid Surface A 313-314 359
[4] Il'Ina M V, Blinov Y F, Il'In O I, Klimin V S, Ageev O A 2016 *Proc. of SPIE* **10224** 102240U-1
[5] Qu L and Dai L 2007 *Adv. Mater.* **19** 3844
[6] Klimin V S, Rezvan A A, Kots I N, Naidenko N A 2018 *J. Phys.: Conf. Ser.* **1124** 071019
[7] Murdick D A, Wadley H N G and Zhou X W 2007 *Phys. Rev. B* **75** 125318
[8] Klimin V S, Solodovnik M S, Lisitsyn S A, Rezvan A A, Balakirev S V 2018 *J. Phys.: Conf. Ser.* **1124** 041024
[9] Daweritz L and Ploog K 1994 *Semicond. Sci. Tech.* **9** 123
[10] Kangawa Y, Ito T, Taguchi A, Shiraishi K, Irisawa T and Ohachi T 2002 *Appl. Surf. Sci.* 190 517
[11] Tominov R V, Smirnov V A, Avilov V I, Fedotov A A, Klimin V S, Chernenko N E 2018 *IOP Conf. Ser.: Mater. Sci. Eng.* **443** 012036
[12] Rouhrig M, Thiel M, Worgull M and Houlscher H 2012 *Small* **8** 3009
[13] Klimin V S, Rezvan A A, Ageev O A 2019 *Proc. of SPIE* **11022** 1102220-1
[14] Autumn K, Liang Y A, Hsieh S T, Zesch W, Chan W P, Kenny Th W, Fearing R and Full R J 2000 *Nature* **405** 681
[15] Morgan C G, Kratzer P and Scheffler M 1999 *Phys. Rev. Lett.* **82** 4886
[16] Klimin V S, Tominov R V, Avilov V I, Dukhan D D, Rezvan A A, Zamburg E G, Smirnov V A, Ageev O A 2019 *Proc. of SPIE* **11022** 110220E-1
[17] Modi, A., Koratkar, N., Lass, E., Wei, B., Ajayan, P.M. (2003) *Nature* **424 (6945)** pp. 171-174.
[18] Amrani A, Djafari Rouhani M and Mraoufel A 2011 *Appl. Nanosci.* **1** 59
[19] Klimin V S, Kots I N, Polyakova V V, Rezvan A A, Ageev O A 2019 *Proc. of SPIE* **11022** 110221R-4
[20] Kots I N, Kolomiitsev A S, Lisitsyn S G, Polyakova V V, Klimin V S, Ageev O A 2019 *Russian Microelectronics* **48(2)** p 72-79
[21] Klimin V S, Vakulov Z E, Tominov R V, Varzarev Y N, Clemente I E, Miakonkikh A V, Rudenko K V, Ageev O A 2019 *Proc. of SPIE* **11022** 110221E-2
[22] Tok E S, Neave J H, Zhang J, Joyce B A and Jones T S 1997 *Surf. Sci.* **374** 397
[23] Kley A, Ruggerone P and Scheffler M 1997 *Phys. Rev. Lett.* **79** 5278
[24] Klimin V S, Rezvan A A, Ageev O A 2018 *J. Phys.: Conf. Ser.* **1124** 071020
[25] Smirnov V A 2015 *Piezoelectrics and Nanomaterials: Fundamentals, Developments and Applications* 85-103

Effects of plasma on the formation of bulk nanoscale structures with lateral dimensions on the surface of GaAs

V S Klimin[1,2], A A Rezvan[1], M S Solodovnik[1,2], R V Tominov[1,2] and O A Ageev[1,2]

[1]Department of Nanotechnology and Microsystems, Southern Federal University, Taganrog 347922, Russia

[2]Research and Education Center "Nanotechnologies", Southern Federal University, Taganrog 347922, Russia

Abstract. In this work, the effect of plasma on the formation of bulk nanoscale structures with lateral dimensions on the GaAs surface was studied. The relative change in the area of the bulk structures in different crystallographic directions is demonstrated, in comparing the before and after the operations of modification and profiling with a successive change in the etching time from 0.5 to 2 minutes. The obtained data in the course of the study showed that when applying this technology for producing nanoscale structures, it is necessary to take into account a number of parameters, such as the crystallographic direction of the substrate, the etching time, and the formation voltage of oxide nanostructures.

1. Introduction

At present, in modern micro and nanoelectronics, the question remains open about the accuracy of modifying the surfaces of semiconductor materials for the subsequent formation of high-resolution structures. In connection with the trend of development of modern electronics, each increase in the density of the elements on the chip and the reduction of their technological size requires the development of new methods for the precise formation of the functional nanoelectronic element base [1-4]. One of the main problems in the technology of forming electronic devices based on GaAs is the need to search for new, effective and improving existing methods for processing, profiling and modifying the surface of nanoscale structures [5-9].

Those technologies should provide the required surface quality, high accuracy and resolution of the processes of formation of instrument elements with linear dimensions of less than 100 nm [10-14]. Analysis of the literature in the field of nanolithography showed that the existing traditional methods of submicron profiling, based on a combination of lithographic processes with liquid or plasma etching, have several disadvantages. Since lithographic processes are limited by diffraction, the properties of the resistors and masks used, as well as the low uniformity of the subsequent etching, significant deviations from the specified parameters of the required elements appear during the formation of structures less than 100 nm in size [15-19].

The combination of methods of local anodic oxidation and plasma chemical etching is one of the most promising areas for the modification and profiling of near-surface layers of solids. The advantage of this combination compared to the above is the ability to vary the modes of submicron profiling with a sufficient degree of accuracy [20-24]. In addition, at the same time, this technology does not require auxiliary operations to remove both the reaction products from the surface and external impurities that are contained in the initial reagents. This scientific work is devoted to the study of the use of local anodic oxidation followed by plasma chemical etching to modify the surface topology of A3B5 structures [25-28].

2. Materials and Methods

The essence of this study was to carry out combination techniques for modifying and profiling the surface of GaAs using local anodic oxidation and plasma chemical etching. The surface of the GaAs substrate was cleaned using standard liquid polishing operations. Oxide was formed by the method of local anodic oxidation at a probe displacement speed of 2.5 μm/s, air humidity of 90% and current of 1 nA. The voltage of formation of nanostructures ranged from 7 to 10 V (fig. 1 a, c).

The obtained bulk oxide structures were exposed to chloride plasma using plasma chemical etching with parameters: pressure in the reactor P_R = 2 Pa, gas flows N_{Ar} = 100 cm^3/min, N_{BCl3} = 10 cm^3/min, power of the capacitive plasma source was W_{RIE} = 35 V, power of the source inductively coupled plasma W_{ICP} = 400 V, etching time ranged from 0.5 to 2 minutes.

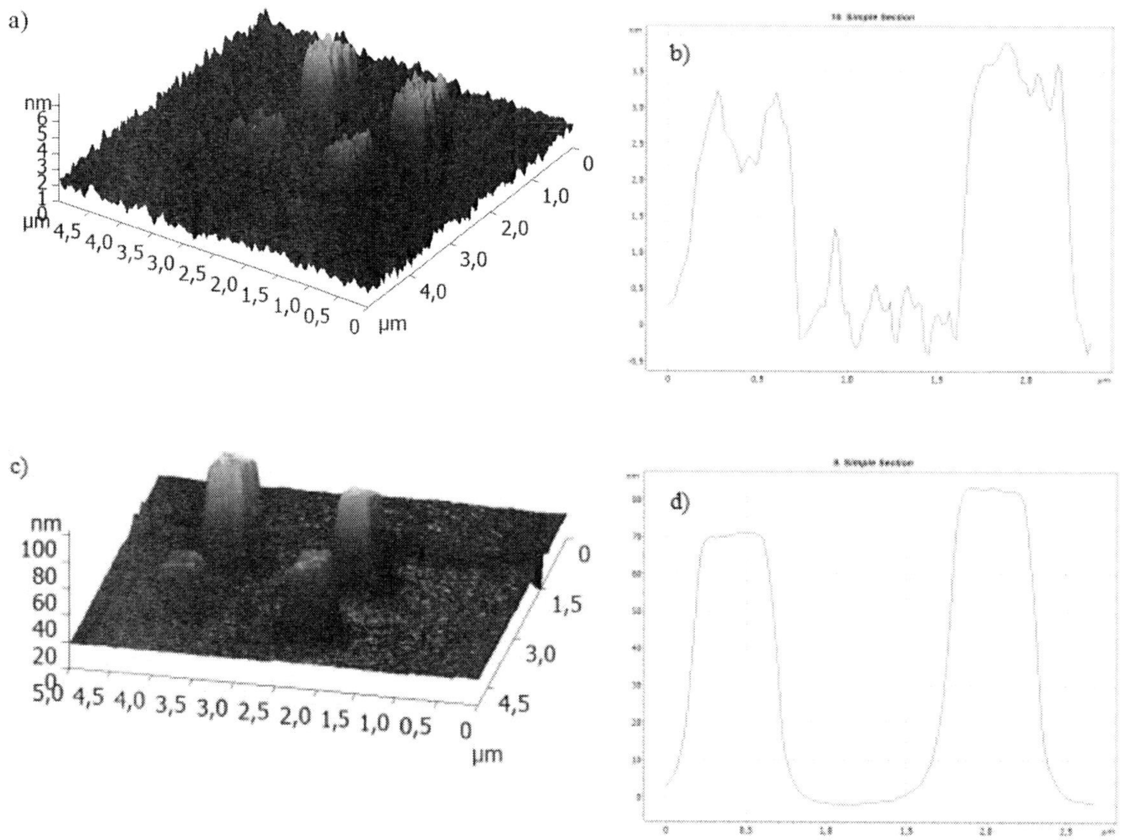

Figure 1 (a, b, c, d). AFM-scans of volumetric structures **(a, c)** and profiles of resulting throw those high structures **(b, d)**

To determine the effect of the parameters of both methods on the submicron formation and change in the topology of the surface layer of structures based on GaAs, the samples were oxidized at different oxide formation stresses and different etching times.

Fig. 1 b, d shows the profiles of the oxide structures and etched surface layer at 9 to 10 V and etching for 2 minutes.

3. Results

Series of experiments were carried out to study the effect of plasma on the process of formation, consisting in studying the etching rate, expanding the dimensions carried out and the angle of deviation from the vertical on two crystallographic directions [111] and [110] (fig. 2, a, b, c, d)

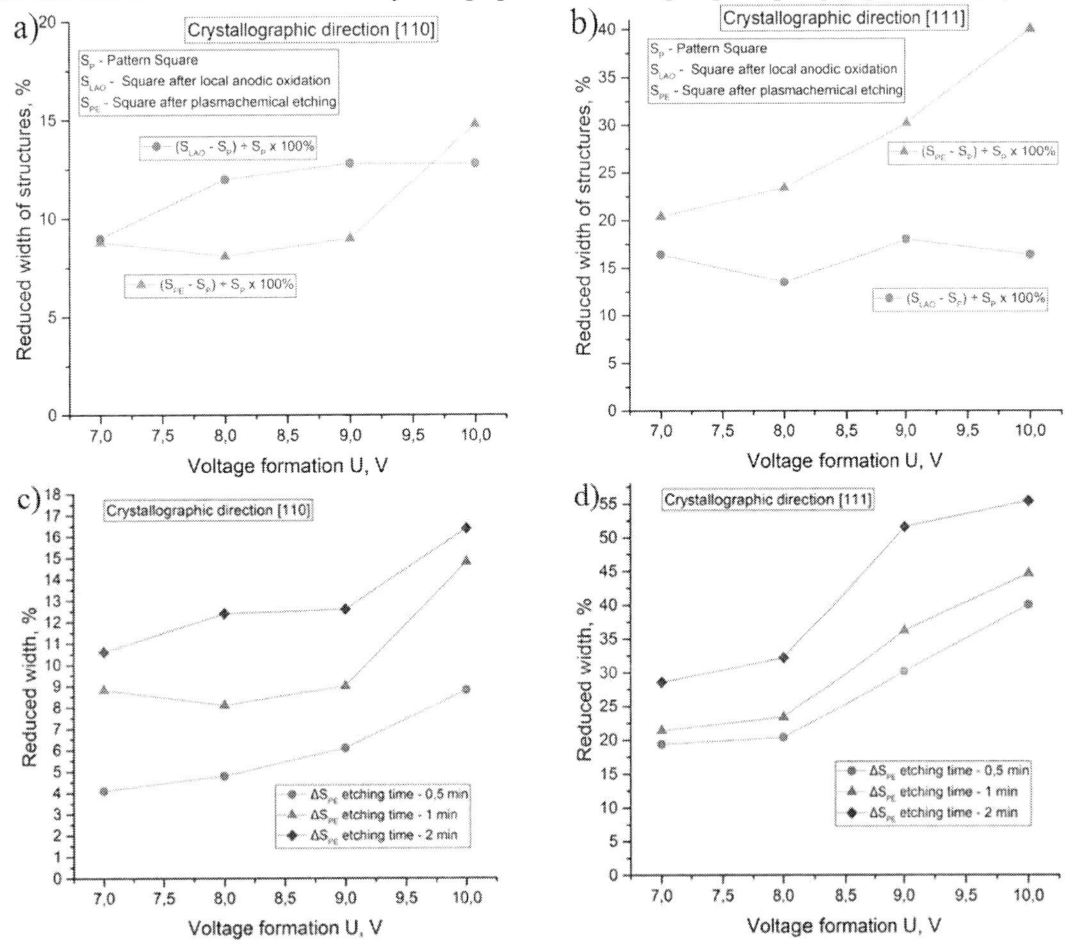

Figure 2(a, b, c, d). The relative change in the area of bulk structures in different crystallographic directions: **a)**, **c)** crystallographic direction [110]; **b)**, **d)** crystallographic direction [111]; **a)**, **b)** the area of structures in comparison before and after the modification; **c)**, **d)** the area of structures at different times of etching

From these dependencies, it can be concluded that bulk structures have walls with high degree of smoothness, as well as a low roughness of the entire surface. It demonstrates that this mode has a high degree of anisotropy of the process of plasma chemical etching. These dependences show that the use of this mode allows the formation of high dimensional volume structures, the verticality of which depends on the duration of the chemical effect of the reaction gases of the plasma chemical etching process. In addition, using this mode, it is possible to obtain etched three dimensional structures with different geometrical sizes and shapes, depending only on the oxide form.

4. Discussion and Conclusions

In this experimental work, we studied the effect of formation regimes on the geometric parameters of the resulting bulk structures on the GaAs surface. It is shown that the use of plasma chemical etching

allows to obtain bulk structures with relatively high smoothness of vertical walls. The effect of the parameters of local anodic oxidation and plasma-chemical etching is shown.
In addition, the data obtained during the study showed that when applying this technology to obtain nanoscale structures, it is necessary to take into account a number of parameters, such as the crystallographic direction of the substrate, the etching time, and the formation voltage of oxide nanostructures.

Acknowledgments
This work was supported by the Russian Science Foundation Grant No. 15-19-10006. The results were obtained using the equipment of the Research and Education Center a "Nanotechnologies" of Southern Federal University.

References
[1] Morgan C G, Kratzer P and Scheffler M 1999 *Phys. Rev. Lett.* **82** 4886
[2] Vakulov Z E, Ivonin M, Zamburg E G, Klimin V S, Volik D P, Golosov D A, Zavadskiy S M, Dostanko A P, Miakonkikh A V, Clemente I E, Rudenko K V, Ageev O A 2018 *J. Phys.: Conf. Ser.* **1124** 022032
[3] LiuY, LehmanV, WangL.2015 *Com. Net.* **83** 85-99
[4] Chekurov N, Grigoras K, Peltonen A, Franssila S, TittonenI 2009 *Nanotechnology*. **20** 5
[5] Il'In O I, Rudyk N N, Il'Ina M V, Klimin V S, Fedotov A A, Ageev O A 2018 *J. Phys.: Conf. Ser.* **1135** 012033
[6] Nishinaga T and Shen X Q 1994 *Appl. Surf. Sci.* **82-83** 141
[7] Higuchi Y, Uemura M, Masui Y, Kitada T, Shimomura S, Hiyamizu S 2003 *J. Cryst. Growth* **251** 80
[8] Klimin V S, Solodovnik M S, Lisitsyn S A, Rezvan A A, Balakirev S V 2018 *J. Phys.: Conf. Ser.* **1124** 041024
[9] Schmidt M, Johari Z, Ismail R, Mizuta H, Chong H 2012 Microelectron. Eng. 98 313
[10] Tseng A 2004 J. Micromech. Microeng. 14 15
[11] Balakirev S V, Eremenko M M, Mikhaylin I A, Klimin V S, Solodovnik M S 2018 *J. Phys.: Conf. Ser.* **1124** 022018
[12] Han J, Lee H, Min B, Lee S 2010 Microelectron. Eng. 87 1
[13] Orloff J, Swanson L W, Utlaut M 2003 Springer New York
[14] Nagase M, Nakamatsu K, Matsui S, Namatsu H 2005 Japanese J. of Appl. Phys. 44(7) 5409
[15] Klimin V S, Rezvan A A, Kots I N, Naidenko N A 2018 *J. Phys.: Conf. Ser.* **1124** 071019
[16] Schaffer M, Schaffer B, Ramasse Q 2012 Ultramicroscopy 114 62
[17]Salvati E, Brandt LR, Papadaki C, Zhang H, Mousavi SM, Wermeille D, Korsunsky AM, 2018 *Mater. Lett.* **213** 346–349
[18] Il'Ina M V, Blinov Y F, Il'In O I, Klimin V S, Ageev O A 2016 *Proc. of SPIE* **10224** 102240U-1
[19]Korsunsky AM, Salvati E, Lunt AGJ, Sui T, Mughal MZ, Daniel R, Keckes J, Bemporad E, Sebastiani M 2018 *J. Matdes***145** 55–64
[20]Sabouri A, Anthony C J,BowenJ, VishnyakovV, Prewett P D ,2014 *Microelectron. Eng.***121** 24-26
[21] Klimin V S, Rezvan A A, Ageev O A 2018 *J. Phys.: Conf. Ser.* **1124** 022035
[22] Taillon J A, Pellegrinelli C, Huang Y-L, Wachsman E D, Salamanca-Riba L G 2018 *Ultramicroscopy* **184** Part A 24-38
[23]Ostadi H, Jiang K, Prewett P D 2009 *Microelectron. Eng.***86** (4–6) 1021-1024
[24]Wang Y-c, Xie D-g, Ning X-h, Shan Z-w 2015 *Appl. Phys. Lett.***106** 081905
[25] Klimin V S, Rezvan A A, Ageev O A 2018 *J. Phys.: Conf. Ser.* **1124** 071020
[26] Murdick D A, Wadley H N G and Zhou X W 2007 *Phys. Rev.* B **75** 125318
[27] Lisitsyn S A, Balakirev S V, Avilov V I, Kolomiytsev A S, Klimin V S, Solodovnik M S, Konoplev B G, Ageev O A 2018 *Nanotechnologies in Russia* **13(1-2)** p 26-33
[28] Amrani A, Djafari Rouhani M and Mraoufel A 2011 *Appl. Nanosci.* **1** 59

Mechanism of formation of light-emitting silicon hexagonal phase 9R-Si

A A Nikolskaya, D S Korolev, A N Mikhaylov, A I Belov, A A Konakov, Yu I Chigirinsky, A A Sushkov, D A Pavlov and D I Tetelbaum

Lobachevsky University, 23/3 Gagarin prospect, Nizhny Novgorod, 603950, Russia

E-mail: alena.nikolskaya.1994@mail.ru

Abstract. A method of photoluminescence (PL) spectroscopy has been used to study the mechanism of formation of light-emitting hexagonal 9R-Si phase by krypton ion implantation into thermally grown oxide layer on silicon substrate with subsequent annealing. The PL band at ~ 1246 nm previously assigned to this phase appears at isochronous step-by-step annealing temperatures of 600 °C and higher as well as for one-step annealing. In addition, the PL bands at ~1324 and ~1408 nm previously observed in ion-implanted silicon and assigned to self-interstitial complexes are present in our case. The decrease in their intensities and simultaneous enhancement of the 9R-Si band are observed with increase in annealing temperature. It is concluded that the mechanical stresses arising in SiO_2/Si system during implantation are responsible for the formation of the 9R-Si phase.

1. Introduction

Until now, the increase of computing performance has been achieved by reducing the feature size of integrated circuits leading to the increase in transistor number per chip and clock speed. However, we cannot further rely on this trend because of fundamental limitations related to the insufficiently high signal rate transmission over metal interconnections. Active search for alternative approaches is underway now. An approach based on the replacement of electrical interconnections by optical ones, i.e. the creation of optoelectronic integrated circuits, is considered to be very promising. At the same time, the use of silicon as a main active material would be desired. Unfortunately, silicon is poor light emitter due to its indirect band structure. Therefore, it is of great importance to find the ways to improve the luminescent properties compatible with traditional CMOS technologies.

It has been established recently that some non-diamond-like silicon phases possess luminescence, the intensity of which exceeds the luminescence intensity of diamond-like silicon [1, 2]. This is one of the approaches to solve the problem, which consists in the creation on diamond-like silicon surface of a layer or regions of the corresponding allotropic modification of Si, for example hexagonal Si. However, most of the methods for synthesizing hexagonal silicon phases [3, 4, 5] are hardly compatible with modern silicon technology. Ion synthesis does not have this disadvantage. Another benefit of ion synthesis over other synthesis methods is also a high degree of controllability of the concentration of embedded atoms, the possibility of local phase synthesis, in particular by using focused ion beams.

There are few articles on the ion-beam formation of inclusions of hexagonal wurtzite-like (2H) Si phase under ion irradiation of diamond-like (3C) silicon [6, 7, 8], but the studies are not systematic. In some cases, there are special implantation conditions, for example, the use of sample heating by ion beam [7] or heating of a substrate holder [8], when it is rather difficult to maintain a constant temperature. The luminescent properties of such inclusions are not presented. Thereby, the systematic studies of the Si "hexagonalization" processes under ion implantation and the mechanisms of hexagonal phase formation, as well as the investigation of luminescent properties are very important tasks.

It has been previously shown by us [9, 10, 11] that, at double implantation of Ga^+ and N^+ ions as well as at Kr^+ implantation into silicon or dielectric films on silicon with subsequent annealing, the silicon regions form with another packing of the (111) planes compared to diamond-like silicon. Analysis of electron diffraction patterns shows that these regions are of the 9R-Si hexagonal type (hexagonalization degree is 2/3). Moreover, for the samples, in which this phase was obtained by implantation of krypton ions into SiO_2/Si substrate with subsequent annealing at 800 °C, the photoluminescence (PL) band peaked at 1242 ± 3 nm was found. According to the DFT band structure calculations, this PL emission was related to an indirect optical transition in 9R-Si. It was suggested (for the SiO_2/Si structure) that the formation of 9R phase is due to a relaxation of mechanical stresses at the interface with silicon subsurface layer. These stresses are associated with the action of ion-implanted SiO_2 film. In [11], the annealing was carried out by one step at the temperature of 800 °C. In the present work, for a more reliable judgment about the hexagonalization mechanism, the annealing was performed under different conditions – both one-step and step-by-step (isochronous) mode – combined with PL measurements after each annealing step.

2. Experimental

SiO_2 film with a thickness of 160 nm thermally grown on *n*-Si (100) wafer was implanted by Kr^+ ions at the energy of 80 keV and dose of $5\cdot10^{16}$ cm^{-2}. In accordance with the SRIM calculation [12], the mean projected ion range in SiO_2 was ~ 50 nm. The isochronous post-implantation annealing was carried out in vacuum chamber step-by-step at temperatures ranging from 600 to 950 °C and also at one-step at different temperatures. The time of each step and time of one-step annealing were 30 min. The PL spectra were measured at liquid nitrogen temperature with excitation by light-emitting diode at a wavelength of 530 nm and recorded in a standard lock-in scheme.

3. Results and discussion

PL spectra for the samples annealed in the step-by-step mode are shown in Figure 1. At the annealing temperature of 600 °C, three PL bands with maxima at 1324, 1408 and 1246 nm are observed. The position of the 1246 nm band coincides well with the PL band previously assigned to the emission of 9R-Si phase [11]. Other two bands were not observed in [11]. Their positions are very close to the positions of PL bands reported in [13] for silicon irradiated with Si^+ ions. These PL bands were identified as related to defect complexes, which include self-interstitial atoms [13]. In our case, the presence of such complexes is quite possible: according to the SRIM calculation [12] for the SiO_2 thickness of 160 nm, Kr^+ ion energy of 80 keV and ion dose of $5\cdot10^{16}$ cm^{-2}, the concentration of Frenkel pairs ~ 10^{21} cm^{-3} is produced in Si substrate near the interface with SiO_2 film, which exceeds significantly the one occurred in [13] (~ 10^{17}– 10^{19} cm^{-3}) under the irradiation of Si with Si^+ ions at the energy of 145 keV and dose of $1\cdot10^{12}$ - $5\cdot10^{13}$ cm^{-2}.

At the annealing temperature of 600 °C, the intensity of PL band related to the 9R-Si phase is significantly weaker than that for the other two temperatures. With increase in temperature of step-by-step annealing, the intensity of defect PL bands decreases with simultaneous enhancement of the 9R-Si PL band. The maximum intensity of the latter band is reached at the annealing temperature of 900 °C. After subsequent annealing at the temperature of 950 °C, this band disappears.

The behavior of PL spectra for the samples with one-step annealing is shown in Figure 2. As mentioned above, for the first annealing at 600 °C, three peaks are observed: two from defects and one from the 9R-Si phase. With increase in annealing temperature, the defect PL bands are weakened or even disappeared. However, the 9R-Si band also disappears at the temperatures higher than 700 °C in contrast to the step-by-step mode.

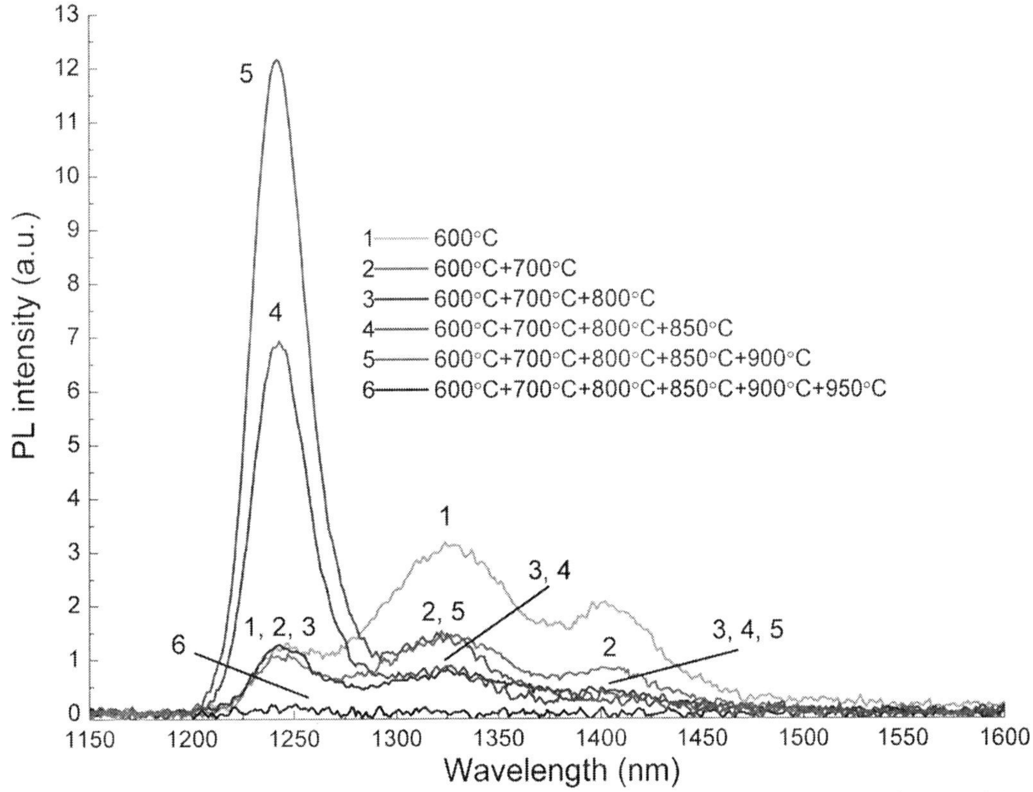

Figure 1. PL spectra of silicon samples with thermal oxide (160 nm) irradiated with Kr^+ ions with subsequent step-by-step annealing.

Let's consider the described regularities of PL behavior from the viewpoint of a hypothesis that the 9R phase forms under the influence of mechanical stresses. From this point of view, the hexagonalization occurs as a kind of relaxation of mechanical stresses localized at the interface layer of silicon. These stresses are created by the action of SiO_2 film implanted with krypton ions. This relaxation process in Si can occur only at sufficiently high temperatures, at which silicon becomes sufficiently plastic. On the other hand, with increase in annealing temperature, the stress relaxation also occurs in the SiO_2 film (viscous flow); therefore, the moving force of phase transition decreases. The resulting final degree of phase transformation is determined by the competition between these two processes, the result of which depends on annealing temperature. The absence or weakly pronounced hexagonalization at high temperatures for one-step annealing apparently indicates that in this case the stress relaxation process in SiO_2 is ahead of the hexagonalization process. (The difference from the result obtained in [11] for one-step annealing at 800 °C is apparently due to the difference in heating and / or cooling rates). On the other hand, in the case of step-by-step annealing, the process of hexagonalization at the second and next steps is promoted by the fact that this process no longer requires the passage of the nucleation stage of a new phase: its volume fraction increases by the growth of inclusions formed at previous steps; therefore, the 9R-Si PL is observed after final annealing at those temperatures, for which such PL is not observed at one-step annealing.

It is not surprising that PL from the 9R-Si phase vanishes at the highest used temperature of step-by-step annealing: the 9R-Si phase, unlike the diamond-like Si, is metastable, and at sufficiently high temperatures it should turn into the diamond-like phase. In general case, in addition to the "impurity" stresses associated with the implantation of krypton, thermal stresses also play a role. Since the atomic radius of Kr (0.198 nm) exceeds the atomic radius of Si (0.118 nm) and O (0.074 nm), and the temperature expansion coefficient of SiO_2 ($6.1 \cdot 10^{-6}$ K^{-1} at 400 K) is less than that of Si ($3.5 \cdot 10^{-6}$ K^{-1} at 400 K) [14], the action of implanted krypton is opposite to the action of thermal stresses.

The thermal desorption or aggregation of Kr atoms may be additional factors determining the final degree of hexagonalization. The heating and cooling rate can also play important role. The influence of these and other factors should be a subject of further investigation.

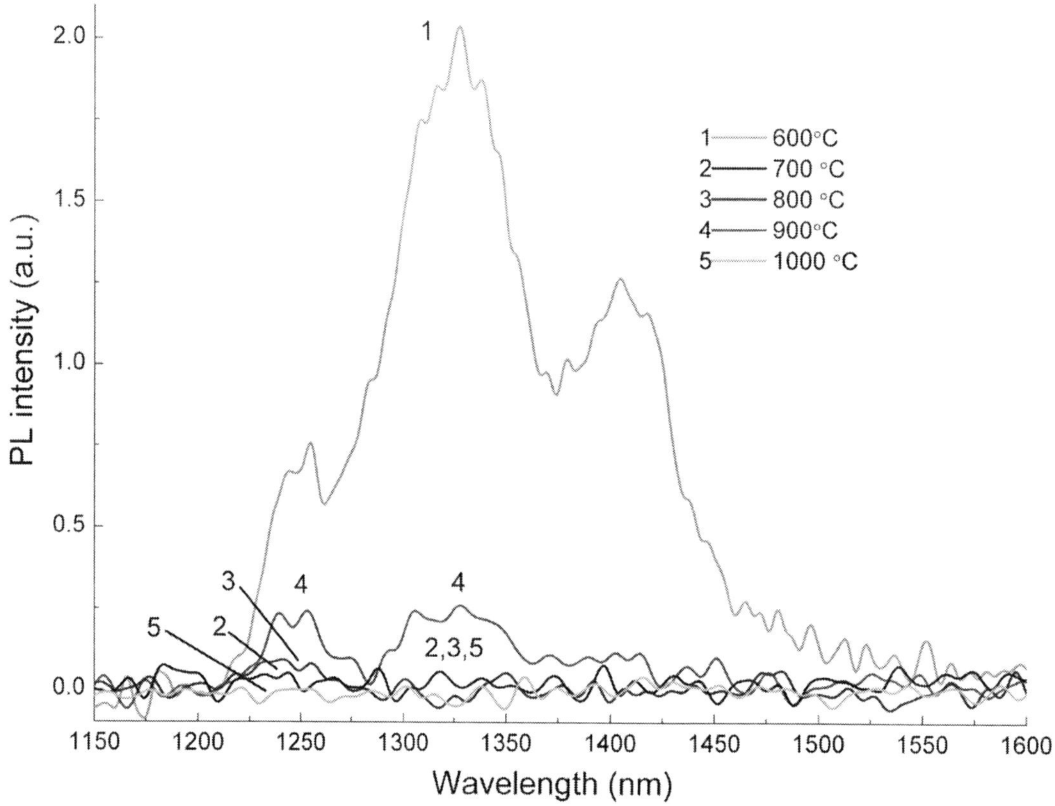

Figure 2. PL spectra of silicon samples with thermal oxide (160 nm) irradiated with Kr^+ ions with subsequent one-step annealing.

4. Conclusions

The obtained results confirm the proposed model, according to which mechanical stresses are the main moving force for the formation of 9R-Si phase when ions are implanted into SiO_2 film on Si. The formation of a thin strained layer in Si under the action of SiO_2 film is consistent with the results of molecular dynamics calculation for silicon covered with a layer of natural silicon dioxide [15]. Hopefully, the optimization of implantation and annealing conditions will improve the luminescence intensity during the ion-beam formation of the 9R-Si light-emitting phase. This will create the background for the application of ion-beam method in the manufacturing technology of silicon optoelectronic chips. The excellent compatibility of this method with traditional CMOS technology makes it particularly attractive.

Acknowledgments

The work was partially supported by Russian Ministry of Science and Higher Education (State Assignment No.16.2737.2017/4.6) and RFBR (Grant No. 18-32-20168-mol_a_ved). A.A. Nikolskaya also acknowledges the support in the framework of UMNIK program (№ 13192GU/2018).

References

[1] Fabbri F, Rotunno E, Lazzarini L, Cavalcoli D, Castaldini A, Fukata N, Sato K, Salviati G, Cavallini A 2013 *NanoLett.* **13** 5900
[2] Fabbri F, Rotunno E, Lazzarini L, Fukata N and Salviati G 2014 *Scientific Reports* **4** 3603
[3] Liu X, Wang D 2009 *Nano Res* **2** 575
[4] Dacorogna M M, Chang K J, Cohen M L 1984 *Phys. Rev. B.* **32** 1853
[5] Gu L, Yu Y, Sigle W, Usami N, Tsukimoto S, Maier J, Ikuhara Y, Aken P A 2010 *Appl. Phys. Lett.* **97** 213102
[6] Gerasimenko N N, Dvurechensky A V, Romanov S I, Smirnov L S 1973 *Semiconductors* **7(11)** 2195
[7] Tan T Y, Föll H 1981 *Phil. Mag. A* **44** 127
[8] Servidori M, Cannavo S, Ferla G, Ferla A, Campisano S U, Rimini E 1987 *Nuclear Instruments and Methods in Physics Research B* **19/20** 317
[9] Korolev D S, Nikolskaya A A, Krivulin N O, Belov A I, Mikhaylov A N, Pavlov D A, Tetelbaum D I, Sobolev N A and Kumar M 2017 *Technical Physics Letters* **43** 767
[10] Nikolskaya A A, Korolev D S, Mikhaylov A N, Belov A I, Sushkov A A, Pavlov D A and Tetelbaum D I 2018 . *Phys.: Conference Series* **1124** 022007
[11] Nikolskaya A A, Korolev D S, Mikhaylov A N, Belov A I, Sushkov A A, Krivulin N O, Muhamatchin K R, Elizarova A A, Marychev M O, Konakov A A, Tetelbaum D I and Pavlov D A 2018 *Appl. Phys. Lett.* **113** 182103
[12] Ziegler J F, Ziegler M D and Biersack J P 2010 *Nucl. Instrum. Methods Phys. Res. B* **268** 1818
[13] Giri P K 2005 *Semicond. Sci. Technol.* **20** 638
[14] Grigoriev I S, Meilikhov E Z 1991 *Physical Values, The Handbook* (Energoatomizdat, Moscow) p 1232 [in Russian]
[15] Stepanov A V, Tetelbaum D I 2017 *Journal of Surface Investigation: X-Ray, Synchrotron and Neutron Techniques* **11(4)** 756

Optically transparent conductive layer based on oriented metal networks

I R Nizameev[1,3], G R Nizameeva[2], M K Kadirov[1,2]
[1]Arbuzov Institute of Organic and Physical Chemistry, FRC Kazan Scientific Center, Russian Academy of Sciences, Kazan 420088, Russia
[2]Kazan National Research Technological University, Kazan 420015, Russia
[3]Kazan National Research Technical University named after A.N. Tupolev - KAI, Kazan 420111, Russia
E-mail: inizameyev@iopc.ru

Abstract. The work is aimed at solving the problem of finding a promising alternative to indium tin oxide as an optically transparent electrode used in many modern devices such as solar cells, touch screens, transparent triboelectric generators, etc. In this paper, we propose the use of an original method for synthesizing oriented metal nano networks to create an electrically conductive layer on a surface of a glass substrate. Due to the orientation of metal networks, it is possible to significantly reduce the use of metal and increase open areas of the coating, which leads to preservation of electrical conductivity of the coating with a high degree of transparency. The method of depositing the proposed coating is based on the method of chemical precipitation from a liquid using a self-organized micellar template.

1. Introduction

The use of conductive optically transparent coatings has greatly expanded due to the development of optoelectronic devices operating on transparent electrodes, such as touch screens and sensors, smart windows and solar panels. Coatings of this kind should combine such properties as transparency in the visible range of radiation and high conductivity. Basically, indium tin oxide (ITO) is used to create transparent conductors. This material has a low resistance and high permeability of radiation in the optical frequency range. However, besides the obvious advantages, there are a number of significant drawbacks. First of all, these are high costs in production, low mechanical strength and impossibility of use in flexible electronics, since in the folded state it loses its conductivity. In addition, ITO is precipitated at high temperatures. This causes damage to the substrate. In addition indium is a rare metal whose reserves are limited.

Perspective material that can become an alternative to the existing indium tin oxide can be developed on the basis of metallic nanonetworks, which are rather actively discussed in modern scientific papers [1-7].

The basic principle of conduction on transparent surfaces is due to the application of a special conductive transparent thin coating. This layer is either entirely or partially contains particles of a conductive or semiconducting material. For current to flow on a surface, there must be a closed circuit, that is, a conductive channel must be created for the current. If the conductive particles are not enough, as in Fig. 1a, conductivity does not occur. It is necessary to increase the quantities of these particles on the surface, as in Fig. 1b. Here a conduction channel appears (particles are highlighted in red). However, this closes most of the surface, i.e. deteriorating transparency of the material. Here there are two ways: either use transparent particles (for example, the ITO mentioned above or zinc oxide) or arrange the particles in such a way that a tiny amount is used to create the channel. So here metal nanowires are an excellent option. This direction is actively developing. Because of the problem of oxidation, noble metals are usually used for these purposes [8-10].

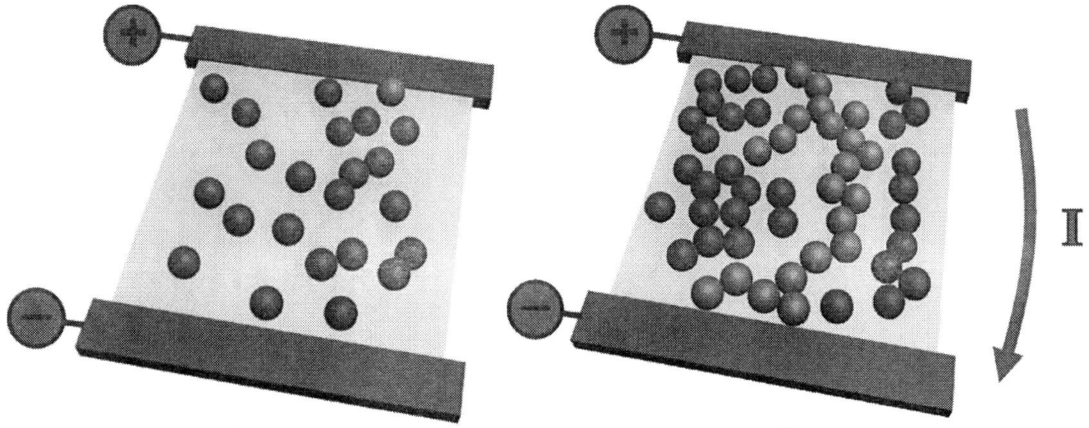

(a) (b)

Figure 1(a, b). (a) No conduction channel; **(b)** contacting particles form a conduction channel

Currently, one of the main directions in the field of creating optically transparent electrodes is the use of various metals nanonetworks or metal systems. Chaotically located conductors can form the necessary conduction channel when a specific value of their amount per unit of surface is reached (overcoming the percolation threshold). This paper proposes the use of oriented nanowires to reduce the percolation threshold (Fig. 2). Orientation and ordering of nanonetworks is not an easy task. The use of the unique technique for creating ordered and oriented structures of metal nanowires is an essential feature of this work.

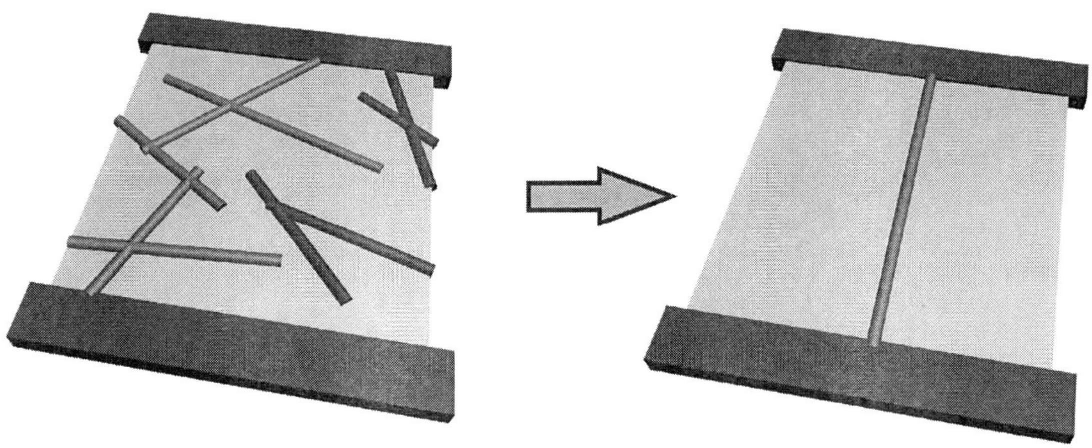

Figure 2. The transition to oriented metal networks and reducing the amount of material used

This paper discusses the use of oriented platinum nanonetworks deposited on a glass surface using a micellar template of the surfactant cetyltrimethylammonium bromide (CTAB). The use of such a template initiates the orientation of metal chains, which leads to the more rational use of the metal. The technique was described in more detail in previous works [11-13].

2. Results and discussion

To obtain oriented platinum networks, chemical precipitation from the liquid phase was used in the work. The shape of the deposited structure was controlled using the micellar template of surfactant cetyltrimethylammonium bromide (CTAB).

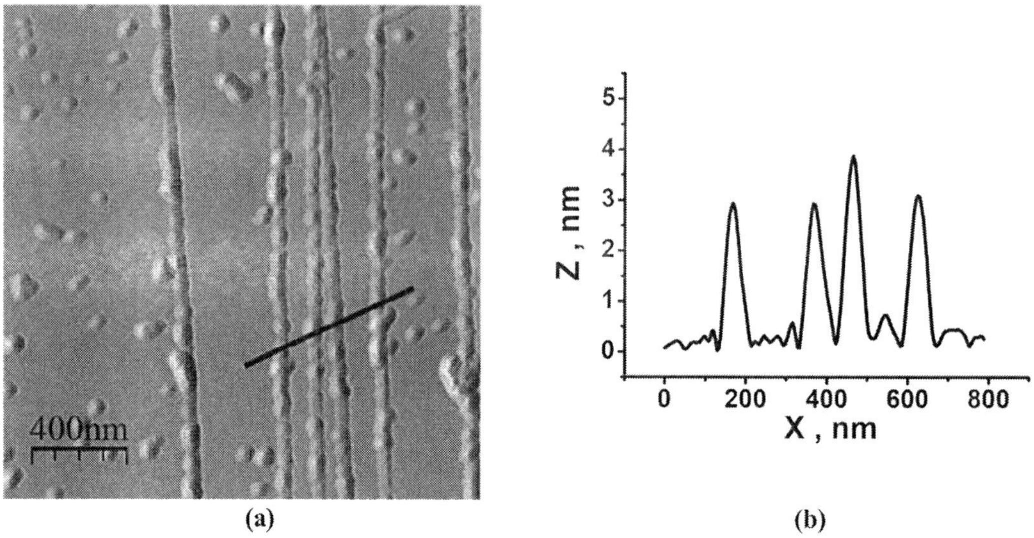

Figure 3(a, b). **(a)** AFM image of platinum strips on a glass surface; **(b)** section profile along the black line

Figure 3 shows a part of a surface of the glass onto which a system of platinum metal networks is applied (the image was taken with an atomic-force microscope MultiMode V in the tapping mode). Despite the heterogeneity in the form of inclusions of single nanoparticles, in general, the surface is covered with oriented metal nanowires. The network's topographic characteristics are as follows: the width of the bands varies in the range from 50 to 60 nm, and their height is 3-6 nm (Fig. 3b).

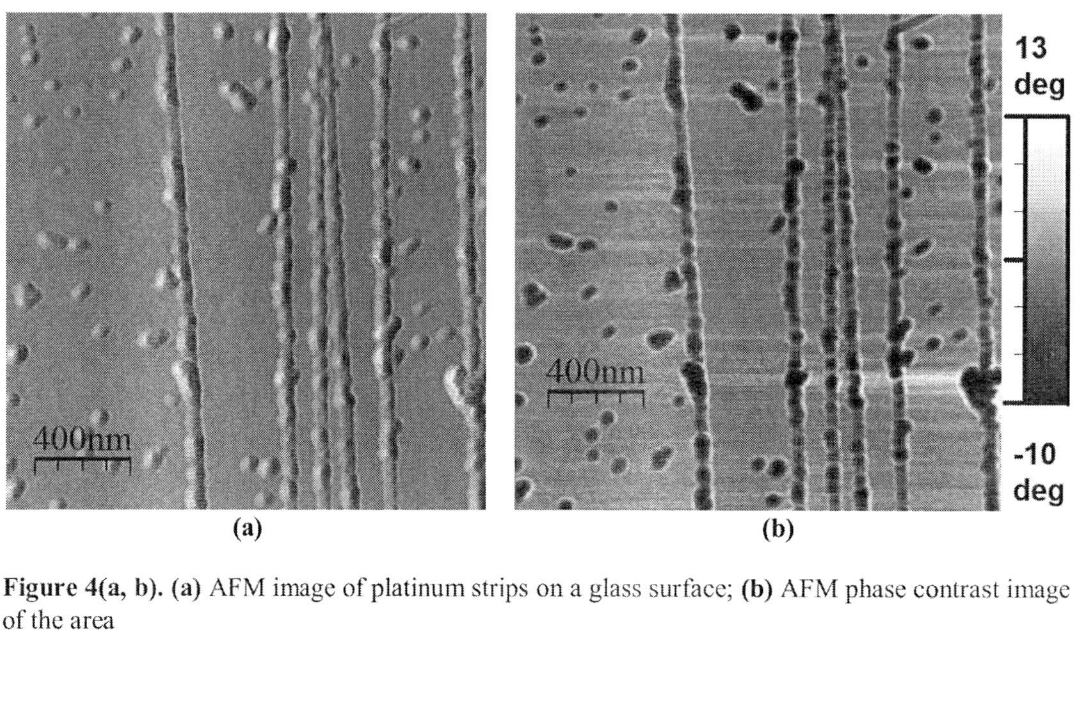

Figure 4(a, b). **(a)** AFM image of platinum strips on a glass surface; **(b)** AFM phase contrast image of the area

Atomic force microscopy (AFM) was used to visualize the surface. AFM has the advantage: in intermittent-contact mode, it is possible to get a picture of phase contrast. It is theoretically proved that the phase shift of the probe oscillation at the end of the cantilever from the oscillation phase of the master oscillator is directly proportional to the probe interaction strength with the surface under study. This phenomenon makes it possible to visualize the phase composition of the surface. In our case, the phase contrast map clearly shows areas covered with metal. Fig. 4b represents phase contrast. Shaded areas correspond to platinum in this case. The picture clearly traces both nanowires and nanoparticles (Fig. 4).
The optical transparency of the coating was studied on a Specord 50 optical spectrophotometer. It was found that the transmittance of the coating in the wavelength range of 400-1600 nm is close to the value of 90% and practically does not depend on the wavelength of the incident radiation. Four point probe method was utilized for the sheet resistance (R_s) measurement. The average value of R_s is 870 Ω/sq.

Acknowledgments
The reported study was funded by RFBR according to the research project № 19-03-00345 and the government assignment for FRC Kazan Scientific Center of RAS (the characterization of the transmittance was carried out by I.R. Nizameev and M.K. Kadirov).

References
[1] Ye S. et al. 2014 *Advanced Materials* **151**. 6670.
[2] Deng B. et al. 2015 *Nano letters* **15** 4206.
[3] An, Byeong Wan, et al. 2015 *Nano letters* **16.1** 471.
[4] Zhang, Shupeng, et al. 2016 *Organic Electronics* **32** 149.
[5] Kim, Jiye, et al. 2016 *The Journal of Physical Chemistry C* **120.2** 956.
[6] Chou, Tsu-Ruey, et al. *Organic Electronics* 2017 **48** 223.
[7] Xiong, Zhaoting and Changqing Liu. 2012 *Organic Electronics* **13.9** 1532.
[8] Milind Jagota and Nelson Tansu. 2015 *Scientific Reports* **5** 10219.
[9] Ji Hoon Yoo et al. 2015 *ACS Appl. Mater. Interfaces* **7** 15928.
[10] Qiaojing Xu et al. 2015 *ACS Appl. Mater. Interfaces* **7** 3272
[11] Kadirov M K, Nizameev I R, Zakharova L Ya 2012 *J. Phys. Chem. C.* **116** 11326.
[12] Kadirov M K, Litvinov A I, Nizameev I R, Zakharova L Ya 2014 *J. Phys. Chem. C.* **118** 19785
[13] Nizameev I.R. et al. 2019 *Nano-Structures & Nano-Objects* **17** 1.

Growth of a multicomponent crystal via Chernov's mechanism

A V Redkov[1], S A Kukushkin[1], A V Osipov[2]

[1]Institute for Problems in Mechanical Engineering RAS, St. Petersburg 199178, Russia
[2]ITMO University, St. Petersburg 197101, Russia
E-mail: avredkov@gmail.com

Abstract. The work is devoted to the growth of a multicomponent crystal via Chernov's mechanism, namely, due to diffusion of its own components from vapor or solution and their subsequent incorporation directly into steps. Equations describing the rate of advance of a group of equidistant steps are derived taking into account individual properties: diffusion coefficients, atomic volumes of all components.

1. Introduction

One of the mechanisms, playing key role during the growth of a crystal at low supersaturation is growth due to terrace-kink-step mechanism, at which atoms adsorbed to surface diffuse one-by-one to terraces and kinks and then incorporate into crystal lattice. This mechanism was described in the classical works of Burton, Cabrera, and Frank (BCF) and related generalizations in numerous reviews and books [1-3]. A theory has been proposed for describing the growth of a single-component crystals by this mechanism. The theory made it possible to predict the dependence of the crystal growth rate on various growth conditions, and to explain why the crystals grow even at very low supersaturations (<1%), when nucleation of nanoislands on the surface is impossible, and therefore growth by the mechanism of nucleation is suppressed. A little bit later A. Chernov proposed similar but slightly different mechanism of growth, which can take place in certain conditions (for example, when surface diffusion is not effective). The key idea was that components diffuse in gas (or solution) towards the steps and kinks and incorporate directly into them. This causes the crystal growth. In paper [4] Chernov developed a theory, which described exactly this mechanism and allowed one to find growth rate of a crystal. However, in general, all theory was developed mainly with respect to single-component crystals growing from its own vapors, atomic fluxes or solutions, and as far as we know, there is no generalization of the theory for multicomponent systems, which would allow one to fully describe the growth of an arbitrary single-component or multicomponent crystal from multicomponent media due to chemical reaction. This work is aimed at the extension of Chernov's theory onto multicomponent systems and it considers the growth of a multicomponent crystal due to diffusion of components from vapor or solution directly toward steps.

2. Growth of a multicomponent crystal via Chernov's mechanism

Let us consider a multicomponent crystal growing from vapors or solution of its own components. In general, the reaction can be written as:

$$\nu_1 A_{1(g,l)} + \nu_2 A_{2(g,l)} + \cdots + \nu_N A_{N(g,l)} \rightarrow S_{(s)}$$

where $A_{i(g,l)}$ are the crystal components, $S_{(s)}$ is a solid crystalline phase; ν_i are the stoichiometric coefficients. As a result of this reaction, $S_{(s)}$ adatoms are delivered to the steps on the crystal surface and cause their movement. We assume that the reaction of formation of the solid phase occurs directly on the surface, and there is no reaction in the gas phase, which is true for many growth processes. We also assume, that mass transfer is not effective to maintain a constant concentration directly at the surface and there is a diffusion layer of a thickness d, at one of the boundaries of which, due to mixing, constant concentrations of the components are maintained (see Fig. 1).

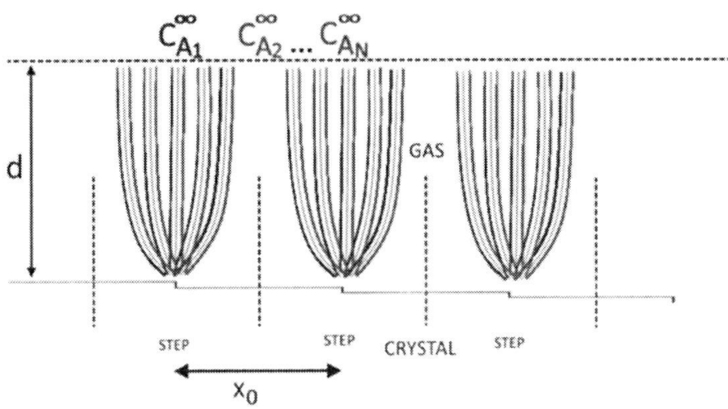

Figure 1. Schematic representation of the growing surface of the crystal with a set of equidistant steps. Diffusion of different components from bulk gas phase proceeds along color lines towards steps.

Following by Chernov to find distribution of each component C_i around the step we should solve the Laplace equation [4]:

$$\Delta C = 0$$

taking into account boundary conditions:

$$C_i|_{z=d} = C_{i\infty}$$

$$-D_i \frac{d\,C_i}{dr} = \beta_i V_i (C_i - C_{i0})\Big|_{z=a/\pi}$$

at x=0 and periodicity at x direction, where D_i is diffusion coefficient, β_i – incorporation rate, V_i – volume of the i-th component, x_0 is the distance between the steps [4], and d is the thickness of the boundary layer in which diffusion of components take place, C_{i0} – is equilibrium concentration of the i-th component, $C_{i\infty}$ - is concentration at the border of boundary layer, a is the height of the step.

It is easy to derive, that the solution for distribution of i-th component would be [4]:

$$C_i(x,y) = F_i + G_i \ln \sqrt{(\sin^2 \frac{\pi}{x_0} x + \mathrm{sh}^2 \frac{\pi}{x_0} y)}$$

where $F_i = C_{i\infty} + G_i \ln sh \frac{\pi d}{x_0}$, and $G_i = \frac{\beta_i V_i a (C_{i\infty} - C_{i0})}{D_i + \beta_i V_i a \ln(\frac{x_0}{a} sh \frac{\pi d}{x_0})}$

Knowing these distribution functions, we should find the fluxes towards steps taking into account that they are stoichiometric (it can be done using the method proposed in [5-7]). By omitting the intermediate calculations we write the resulting formula for the rate of advance of a group of equidistant steps ϑ_∞:

$$\vartheta_\infty = \frac{D^{os} \beta^{os}}{\beta^{os} a \ln\left(\frac{x_0}{a} sh \frac{\pi d}{x_0}\right) + D^{os}} \xi V$$

where

$$D^{os} = \left(\sum_{i=1}^{N+M} \frac{\nu_i^2}{\pi D_i C_{i0}}\right)^{-1}, \quad \beta^{os} = \left(\sum_{i=1}^{N+M} \frac{\nu_i^2}{\beta_i V_i C_{i0}}\right)^{-1}$$

$$\xi = \sum_{i=1}^{N+M} \nu_{si} \sigma_i$$

β_{os} is averaged incorporation rate, D_{os} - is averaged diffusion coefficient, ξ is the multicomponent supersaturation [5], $\sigma_i = C_{i\infty}/C_{i0} - 1$ – supersaturation of the i-th component. In the limit of one component the formula almost coincides with the one obtained in [4]. One can note, that in case when the growth is limited by incorporation of adatoms into steps ($\beta^{os} \ll D^{os}$), the rate of advance of a group of steps depends linearly on supersaturation:

$$\vartheta_\infty = \beta^{os} \xi V$$

In case, when growth is limited by diffusion of components ($\beta^{os} \gg D^{os}$), the rate of advance of a group of equidistant steps also depends strongly on interstep distance x_0 and on thickness of boundary layer d. The greater the thickness, the lower the rate of advance:

$$\vartheta_\infty = \frac{D^{os}}{a \ln\left(\frac{x_0}{a} sh \frac{\pi d}{x_0}\right)} \xi V$$

A typical dependence of the rate of advance on the interstep distance and the thickness of boundary layer is presented in Fig.2. One can note that at large interstep distances comparable with the thicknesses of boundary layer the rate of advance saturates, because the diffusion fields of the neighbor steps no longer affect each other.

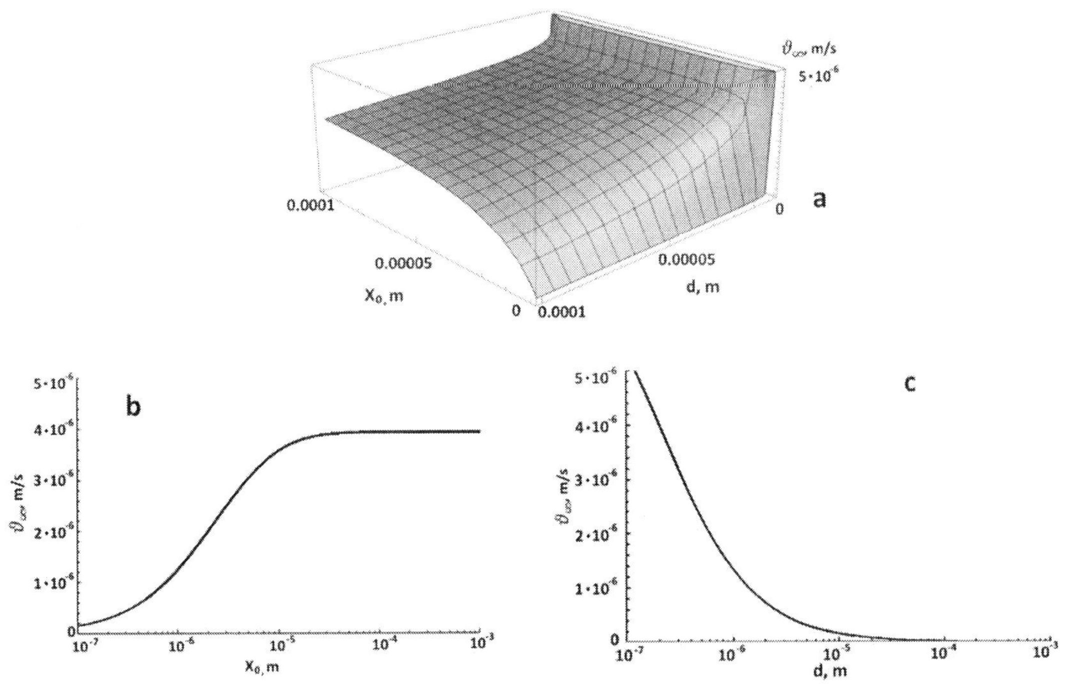

Figure 2. The typical dependences of the rate of advance of a group of equidistant steps on the interstep distance x_0 and on the thickness of the boundary layer d (a) and cross-sections of the 3D-graph in log-scale (b,c) at $D \sim 10^{-4}\ m^2/sec$, $\xi = 0.05$, $V = 10^{-23}\ m^3$, $x_0 = 10^{-7}\ m$, $d = 10^{-5}\ m$, $a = 10^{-10}\ m$.

One should also note, that during the real crystal growth, eg. by the mechanism of spiral growth, the interstep distance is dependent on the overall supersaturation. The greater the supersaturation the smaller the interstep distance. In this sense, the growth rate of the crystal will have more complicated dependence on supersaturation than in derived expressions.

3. Conclusions

The growth of a crystal from multicomponent media via Chernov's mechanism is considered. The model is very similar to classical one. It is shown, that the equation for the rate of advance of a group of equidistant steps is the same as for the single-component crystal, but instead of diffusion coefficient and incorporation rate in multicomponent "version" one should use averaged coefficients which are dependent on individual properties of all components.

Acknowledgements

A.V. Redkov thanks the Russian Science Foundation (grant RSF №19-72-00082) for the support of this work.

References

[1] Bennema P 1984 *Journal of Crystal Growth* **69**(1) 182-197

[2] Uwaha M 2016 *Progress in Crystal Growth and Characterization of Materials* **62**(2) 58-68

[3] Chernov A A 2012 *Modern crystallography III: crystal growth (Vol. 36)*. Springer Science & Business Media.

[4] Chernov A A 1961 *Sov. Phys. Usp* **4** 129

[5] Kukushkin S A and Slyozov V V 1996 Disperse Systems on the Surface of Solids (Evolution Approach): Mechanisms of Thin Films Formation. Nauka, St.-Petersburg

[6] Kukushkin S A, Osipov A V and Redkov A V 2014 *Physics of the Solid State* **56** 2530-36

[7] Redkov A V, Kukushkin S A and Osipov A V 2018 *Journal of Physics: Conference Series* **1124** p. 022006

Monolith GaAsP/Si dual-junction solar cells grown by MBE

M S Sobolev[1], I V Ilkiv[1], A A Lazarenko[1], A M Mizerov[1], E V Nikitina[1], E V Pirogov[1], S N Timoshnev[1], A I Baranov[1], A D Bouravleuv[1]

[1]St. Petersburg Academic University, St. Petersburg, Russia

Abstract. The growth of monolith GaAsP/Si dual-junction solar cells on Si substrates by molecular-beam epitaxy is demonstrated. The technological method for the formation of a highly doped tunnel p^+/n^+ junction was developed. An obvious increase in the value of the open-circuit voltage indicating the contribution of the top-junction in the total open-circuit voltage monolithic dual-junction solar cells was found.

1. Introduction

Currently, among renewable energy sources, photoelectric conversion of solar energy is recognized as the most promising. One of the main indicators of all solar cells is their conversion efficiency - a value that shows what percentage of the power of solar radiation incident on the Earth can be converted into electrical energy. There are various ways to increase the energy conversion efficiency of solar cells and the use of monolithic nanoheterostructures is one of the most successful areas of their development. An important advantage of monolithic nanoheterostructural solar cells based on A3B5 compounds is their ability to efficiently convert concentrated solar radiation, which reduces the consumption of semiconductor materials in proportion to the concentration ratio of solar radiation and significantly reduces the cost per watt of solar energy. [1]

Nevertheless, the modern approach to instrument making is based on reducing the cost of precursors (substrates, materials used for growth, etc.), proven technology of cheap and mass technological production, and integration of several devices on a single chip, which leads to the transition to a technological platform for manufacturing devices based on Si. In the last decade, significant progress has been made in the development of a technological platform for the manufacture of passive devices, such as filters, waveguides, splitters, adders and photoelectric converters, based on Si for integrated photonics. Integration of active devices remains a very difficult task.

Currently, active photonic devices on a silicon platform are produced by bonding, either a chip-on-chip, or on the basis of an A3B5 substrate. [2-5] However, in this case the part of III-V materials is lost during substrate removal.

Direct epitaxy on silicon overcomes these problems, but in most cases there are problems with the mismatch of the lattice parameter and differences in thermal expansion coefficients. [6-8] Part of the problem can be solved by thick relaxation layers (buffer layers), using active regions of the quantum dots to reduce threading dislocations [9] or using complex compositions of materials, such as Ga (N, As, P), which provide the lattice parameter matching. [10]

The density of threading dislocations in the direct growth of III-V on Si can be significantly reduced through the use of selective growth and geometric effects in nanostructured materials. [11, 12] In this case it is possible to obtain high quality crystalline material, but there are limitations in the accuracy and the size of the device. Though, for all the above-mentioned methods, separate manufacturing techniques for III-V and silicon materials are required, which cannot be performed on the same equipment due to cross-contamination.

Here we report on the growth of monolith GaAsP/Si dual-junction solar cells on Si substrates by molecular-beam epitaxy and the study of their properties.

2. Experiment

Epitaxial growth of the studied samples was carried out on MBE set up with cracker-type solid-state sources of arsenic and phosphorus and molecular sources of gallium, aluminum, indium, as well as effusion sources of silicon and beryllium alloying materials. To eliminate the problem associated with the formation of antiphase domains during the growth of III–V compounds on silicon, we used the vicinal surface of the silicon substrate with miscut angle of 4° towards [110] direction. Prior to the growth, the silicon substrates were subjected to a cycle of chemical treatment by the Shiraki method [13]. During the course of chemical treatment, the native oxide overlayer was removed from the substrate surface and a passivating layer of non-stoichiometric SiO_2 was formed. Subsequent annealing of the substrate in the MBE growth chamber at 850°C was performed to remove the passivation layer and the remaining traces of chemical compounds.

To form the n-GaP nucleating layer on a p-Si(100) substrate, we used the migration-enhanced epitaxy (MEE) technique [14]. After the formation of the n-GaP nucleating layer, the n-GaP buffer layer was formed by the standard MBE method. The state of the surface during epitaxy was controlled by the method of reflected high-energy electron diffraction (RHEED).

Figure 1 shows a schematic representation of a monolithic nanoheterostructure GaAsP/Si dual-junction solar cells grown on a silicon substrate. A series connection of photocells in a monolithic GaAsP/Si SC was carried out by means of a p^+/n^+ tunnel junction (Fig. 1). Usually, p^+/n^+ tunnel junction is present p–n junction with a high doping level corresponding to the semiconductor degeneration state, and is specially formed between stages to ensure low ohmic losses between the "upper" and "lower" elements of a dual-junction SC. The quality of the interfaces of the obtained samples was investigated using a Supra25 Zeiss electron scanning microscope.

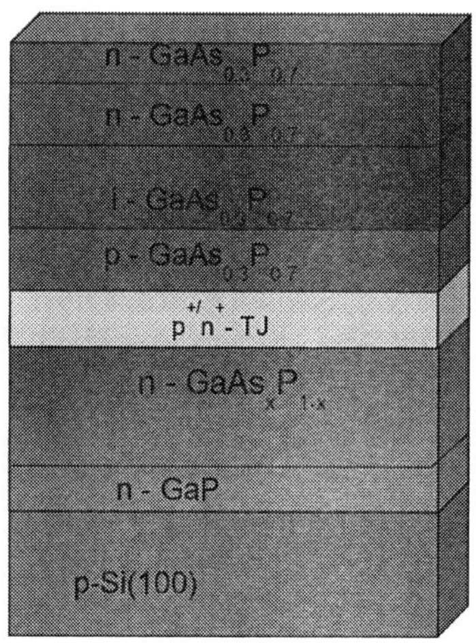

Figure 1. Schematic representation of a monolithic nanoheterostructure GaAsP/Si dual-junction solar cells.

3. Results and discussion

The formation of nucleating and buffer GaP layers with thicknesses of 10–20 and 150–200 nm, respectively, on a silicon substrate leads in the end to the appearance of a distinct streaky RHEED pattern and a comparatively smooth surface of the resulting crystal. The optimal epitaxial temperatures of the GaP nucleating layer are 350°C. In this case, a dashed-line RHEED pattern is formed and no transition to 3D growth occurs (Fig. 2a). During growth of the GaP buffer layer, the substrate temperature gradually increases to 620°C. As the GaP buffer layer grows, the GaP surface becomes smoother and the structure of the RHEED pattern changes from dashed-line to that of the streaky type (Fig. 2b).

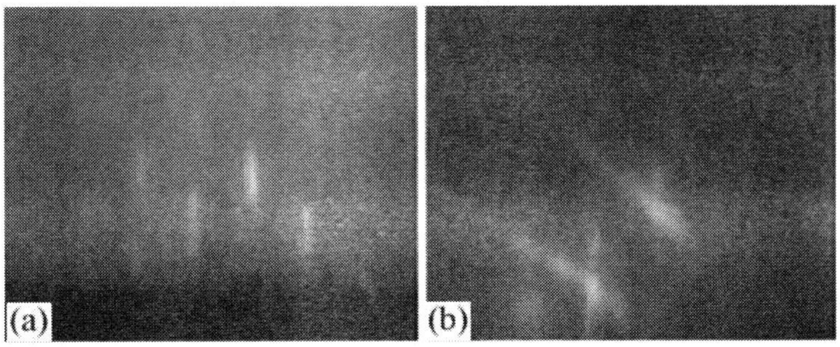

Figure 2(a, b). RHEED pattern of the GaP surface during the growth of a nucleating layer on a silicon substrate.

After the buffer layer a $GaAs_xP_{1-x}$:Si gradient layer was grown. The x changed from 0.1 to 0.3, is necessary for the transition from the silicon lattice parameter to the lattice parameter of the upper $GaAs_{0.3}P_{0.7}$ sub-element. For the formation of a highly doped p^+/n^+ tunnel junction, a technological method for doping of $GaAs_{0.3}P_{0.7}$ layers by silicon to a carrier concentration of more than $1 \cdot 10^{19}$ cm^{-3} was developed. This technological method consisted in reducing the temperature of the substrate and special δ-Si inserts into the $GaAs_{0.3}P_{0.7}$ layers. During the growth of the heavily doped $GaAs_{0.3}P_{0.7}$:Be layer in the tunnel junction, the substrate temperature was reduced to 550°C to create a carrier concentration above $1 \cdot 10^{19}$ cm^{-3} and prevent beryllium segregation on the layer surface.

Figure 3 shows a cross-sectional scanning electron microscope (SEM) image of a monolithic nanoheterostructure GaAsP/Si dual-junction solar cells. Despite the fact that the RHEED pattern of the layer has a streaky structure, an undulating microprofile is clearly seen on the surface. We found the formation of cones which can be associated with the annihilation of antiphase domains. They were formed at the Si/GaP heterointerface and were penetrated into the GaP layer to a depth of up to 100 nm (Fig. 3 circumscribed areas). Visible defects in the upper part of the nanoheterostructure GaAsP/Si dual-junction solar cells are not observed.

A theoretical calculation of the quantum efficiency spectrum of the SC obtained as the upper subelement of a dual-junction SC on a silicon substrate was performed. The simulated structure is a p-i-n junction with an active i-layer of GaPAs 400 nm thick and heavily doped n- and p-type plates. The calculation was carried out in the AFORS-HET 2.5 free software environment [http://www.hmi.de/bereiche/SE/SE1/projects/aSicSi/AFORS-HET], which automatically solves the Poisson and continuity equations in the one-dimensional case using drift and diffusion currents, which allows you to build a band diagram and extract the carrier concentration and other important parameters within the structure. Figure 4 shows the spectra of external quantum efficiency (QE) at a reflection of 30%, depending on the value of the lifetime in the i-layer, which was varied by introducing effective nonradiative recombination centers in it. The best agreement with experimental data is achieved at a lifetime of $5 \cdot 10^{-12}$ s. Obviously, this is a low value; therefore, the use of i-GaPAs layer for photoconverting applications is justified, since it allows to achieve a better collection of carriers compared with doped. However, according to calculations, even for lifetimes longer than 1 ns, the integral and maximum QE in the absence of reflection does not reach high values. This leads to calculated short-circuit current values of several mA/cm^2, which is insufficient for use in high-performance monolithic SCs.

Figure 3. SEM image of a monolithic nanoheterostructure GaAsP/Si dual-junction solar cells. Circumscribed areas show the annihilation of APDs. Scale bar corresponds to 200 nm.

The current-voltage characteristic of the monolith GaAsP/Si dual-junction solar cell without an antireflection coating is shown in Figure 5. This SC has the following characteristics measured under AM1.5 illumination: open circuit voltage - 0.9 V; short circuit current - 0.8 mA/cm^2; fill factor (FF) - 27.8%. Since monolithic dual-junction SC are a series connection of two current sources, the total current flowing in the external circuit will be equal to the minimum current generated by each p–n junction. The current value of the SC will be the highest when the magnitudes of the photocurrents of the upper and lower elements are consistent. Such coordination is achieved by selecting the width of the optical gap and the thickness of each of the i-layers. Therefore, a non-high short-circuit current value is determined by the upper element of the GaPAs/Si dual-junction solar cell and corresponds to the calculated data. In addition, a noticeable increase in the open circuit voltage value, compared with the value for a GaP/Si-based photocell (insert Fig. 5), indicates of the contribution of the upper element to the total open circuit voltage of a monolithic nanoheterostructure GaAsP/Si dual-junction solar cells. Thus, the technological method used to form a tunnel junction allows one to synthesize the monolith GaAsP/Si dual-junction solar cell on a silicon substrate by the MBE method.

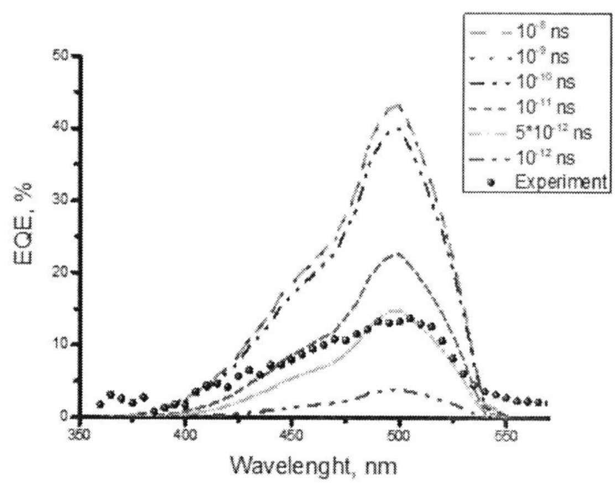

Figure 4. Experimental and calculated (for different values of the lifetime) spectra of the external quantum efficiency of the upper subcell with the active i-GaPAs layer.

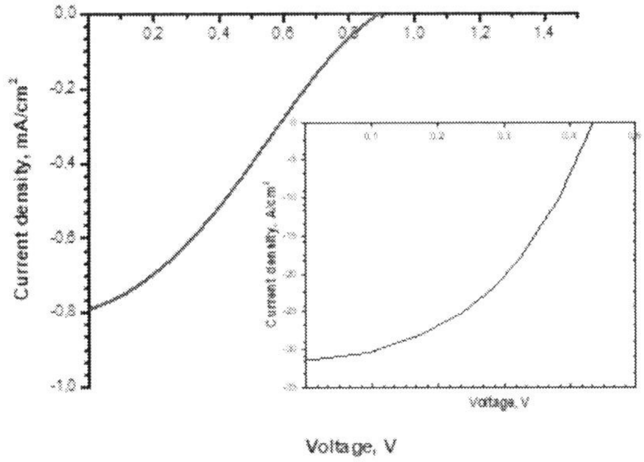

Figure 5. The current-voltage characteristic of monolith dual-junction GaPAs/Si solar cell. The inset shows the current-voltage characteristic of the GaP/Si photovoltaic cell.

4. Conclusion

The formation of initial GaP layers was performed by migration-enhanced epitaxy (MEE) method. The technological method for the formation of a highly doped tunnel p^+/n^+-junction have been developed and optimized to reach low ohmic losses between top and bottom sub-elements of dual-junction solar cells. The scanning electron microscopy studies of the samples obtained demonstrate that single crystalline GaPAs/Si layers were formed. The investigation of current-voltage characteristics exhibited the formation of dual p-n junction. Furthermore an evident increase in the value of the open-circuit voltage indicating the contribution of the top-junction in the total open-circuit voltage monolithic dual-junction solar cells was demonstrated.

Acknowledgments

The work was partially supported by the grant of Minobrnauki № 16.9789.2017/BCh and Skoltech (agreement № 3663-MRA).

References

[1] Michelle Vaisman, Shizhao Fan, Kevin Nay Yaung, Emmett Perl, Diego Martín-Martín, Zhengshan Jason Yu, Mehdi Leilaeioun, Zachary C Holman, and Minjoo Larry Lee 2017 ACS Energy Lett. 2(8) 1911–1918
[2] M. Lamponi, S. Keyvaninia, C. Jany, F. Poingt, F. Lelarge, G. de Valicourt, G. Roelkens, et al. 2012 IEEE Photon. Technol. Lett. 24 76-78
[3] G.-H. Duan, C. Jany, A. Le Liepvre, A. Accard, M. Lamponi, D. Make, P. Kaspar, et al 2014 Proc. SPIE 9002 90020X
[4] G.-H. Duan, S. Olivier, C. Jany, S. Malhouitre, A. Le Liepvre, A. Shen, X. Pommarede, G. Levaufre, N. Girard, D. Make, G. Glastre, J. Decobert, F. Lelarge, R. Brenot, and B. Charbonnier 2016 IEEE Journal of Selected Topics in Quantum Electronics 22(6) 379–389
[5] M.J.R. Heck, H.-W. Chen, A. W. Fang, B.R. Koch, DiLiang, H. Park, M.N. Sysak, J.E. Bowers 2011 IEEE J. Sel. Top. QE 17 333
[6] H. Liu, T. Wang, Q. Jiang, R. Hogg, F. Tutu, F. Pozzi and A. Seeds 2011 Nature Photonics 5 416
[7] A.Y. Liu, C. Zhang, J. Norman, A. Snyder, D. Lubyshev, J.M. Fastenau, A.W.K. Liu, A.C. Gossard, J.E. Bowers 2014 Appl. Phys. Lett. 104 041104
[8] A.Y. Liu, R.W. Herrick, O. Ueda, P.M. Petroff, A. C. Gossard, J.E. Bowers, 2015 IEEE J. Sel. Top. QE 21 1900708

[9] P. Bhattacharya, Z. Mi, J. Yang, D. Basu, D. Saha 209 J. Cryst. Growth 311 1625
[10] S. Liebich, M. Zimprich, A. Beyer, C. Lange, D. J. Franzbach, S. Chatterjee, N. Hossain, S. J. Sweeney, K. Volz, B. Kunert, and W. Stolz 2011 Appl. Phys. Lett. 99 071109
[11] B. Schwartz, M. Oehme, K. Kostecki, D. Widmann, M. Gollhofer, R. Koerner, S. Bechler, I.A. Fischer, T. Wendav, E. Kasper, J. Schulze, and M. Kittler 2015 Opt. Lett. 40 3209-3212
[12] S. Wirths, R. Geiger, N. von den Driesch, G. Mussler, T. Stoica, S. Mantl, Z. Ikonic, M. Luysberg, S. Chiussi, J.M. Hartmann, H. Sigg, J.Faist, D. Buca and D. Grützmacher 2015 Nature Photonics 9 88
[13] A.Ishizaka, Y.Shiraki 1986 J. Electrochem. Soc. 133 666
[14] Sobolev M.S Lazarenko A.A., Mizerov A.M., Nikitina E.V., Pirogov E.V., Timoshnev S.N., Bouravleuv A.D 2018 Semiconductors 52(16) 2128-2131

Application of the atomic layer etching technique for the formation of SiC-based field emitters

V S Klimin[1,2], I N Kots[2], R V Tominov[2], A A Rezvan [1], Y N Varzarev[1,2], O A Ageev[1,2]
[1]Department of Nanotechnology and Microsystems, Southern Federal University, Taganrog 347922, Russia
[2]Southern Federal University, Research and education center "Nanotechnologies", Taganrog, 347928, Russia

Abstract. This paper presents the technology of obtaining graphene films on the surface of SiC using a combination of methods of focused ion beams and plasma-chemical etching in a fluorine containing environment. For the formation of nanoscale structures with lateral dimensions of 280 nm, the method of focused ion beams was used. Nano sized structures were formed on the surface of silicon carbide by an ion beam, which were subsequently placed in the reactor. In the reactor, the structures on silicon carbide were processed in fluoride plasma. The depth of the structures was 900 nm. SF_6 was used for atomic layer etching operations. The study of the surface topology at each iteration was carried out using scanning electron microscopy. Electrical characteristics were measured by atomic force microscopy.

1. Introduction

Graphene is one of the most promising materials for micro and nanoelectronic. The promise of its use is taken up by a number of advantages, the main ones being high carrier mobility and resistance to ionizing effects [1–3]. The main method of obtaining graphene films on the SiC surface is thermal decomposition (thermal destruction) of the SiC surface [4–7]. The advantage of the method of thermal decomposition of silicon carbide is the low cost and the possibility of obtaining homogeneous nanocarbon films of a large area based on both conducting and semi-insulating substrates [8–13]. However, there is a significant disadvantage of this method-annealing at very high temperatures is required, which can lead to the formation of high mechanical stresses in the structure. In this connection, it is important to study the use of a combination of focused ionic and plasma chemical etching to obtain graphene films on SiC [14–19].

Currently, optical lithography, due to physical limitations, has reached its limit in the field of hard ultraviolet radiation. In this regard, the development of promising methods for nanoscale profiling of the surface of solids is an important task [20–23]. One such method is the method of focused ion beams. This method consists in the interaction of accelerated ions with the surface of solids [24–26]. With this interaction, a number of processes occur, such as sputtering the substrate material, the introduction of acting ions into the surface layer of the substrate material, etc. Using these qualities of the method of focused ion beams, one can obtain various structures on the surface of solids with a nanometer resolution [27-29].

2. Materials and Methods

This experimental work is aimed at considering the use of a combination of methods of focused ion beams and plasma-chemical etching to obtain field emission structures based on graphene films on a SiC surface.

At the initial stage, experimental studies on the formation of nanoscale structures were carried out by the method of focused ion beams. For this purpose, a raster electron microscope with an ionic column Nova NanoLab 600 (FEI Company, the Netherlands) was used. The built-in tools of the microscope control program have formed templates for the subsequent etching of structures with a focused ion beam. The templates were formed in the form of a torus with an outer diameter of 2 microns, and inner diameters from 600 to 800 nm. These dimensions allow measurement of the characteristics of the structure obtained. The ion beam current was 30 pA. The accelerating voltage was 30 keV. The exposure time at the point of 1 μs. Subsequently, the structures were etched on the silicon carbide substrate according to these patterns.

Experimental samples, which are plates of purified silicon carbide, were placed in the vacuum chamber of the module of focused ion beams and oriented so that the flow of accelerated ions would fall on the substrate in the direction of the normal. The working vacuum during the exposure of the beams was maintained at the level of $1 \div 2 \times 10^{-4}$ Pa. At this stage, an array in the form of pointed cathodes was formed on the SiC surface (fig. 1, b).

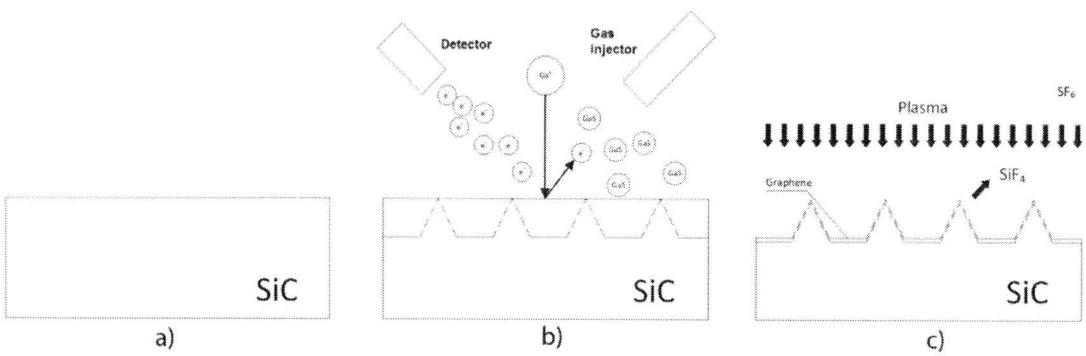

Figure 1(a, b, c). Schematic representation of the technology for producing a graphene layer on the surface of the SiC: **a)** SiC substrate; **b)** focus ion beams ; **c)** graphene on SiC

Carbon nanoscale layer was formed using the atomic layer etching technique in fluoride plasma. SF_6 was used as a fluorine-containing gas, thanks to which it was possible to etch the near-surface SiC layer. At the same time, their crystal lattice removed only Si and formed a thin carbon layer on the surface of the samples (fig. 1, c).

AFM images and current-voltage characteristics (CVC) of structures were obtained using Probe Nanolaboratory Ntegra (NT-MDT, Russia). SiC substrate was grounded, and a cantilever HA_HR with a W2C conducting coating was used. As result, 10 CVC at voltage from 5 to 20 V were obtained, and, according to the results, average CVC was built. Curves analysis was implemented using Origin 8.1 software. The AFM-image processing was performed using Image Analysis software.

3. Results

As a result of experimental studies, structures with a depth of 900 nm and a minimum structure size of 280 nm were formed. SEM images of these structures are presented in Fig. 2. The ion beam current was chosen to be 30 pA, based on the etching time, chosen in such a way that the structures were of minimal dimensions and there was no drift. The accelerating voltage was 30 keV, due to the higher energy of the ions, their greater speed and smaller diameter of the ion beam. The exposure time at the point of 1 μs is the optimal parameter for this system.

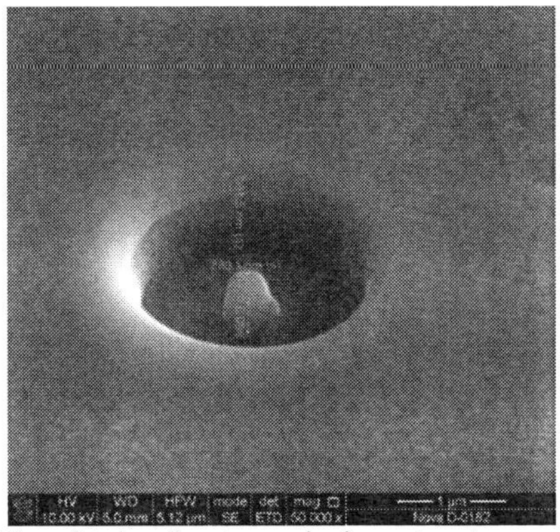

Figure 2. SEM - the image of the obtained auto emission structure on the surface of SiC

The formed structures were placed in a plasma reactor. The surface of the structures was processed in the plasma of combined capacitive and inductively coupled discharges. The processing modes of the structure P_R = 2 Pa, gas flows N_{Ar} = 70 cm^3/min, N_{SF6} = 15 cm^3/min, power of the capacitive plasma source was W_{RIE} = 20 V, power of the source inductively coupled plasma W_{ICP} = 300 V, etching time 1 minute. The analysis of AFM image (fig. 3) shows, that diameter of structures is 0.87±0.15 μm.

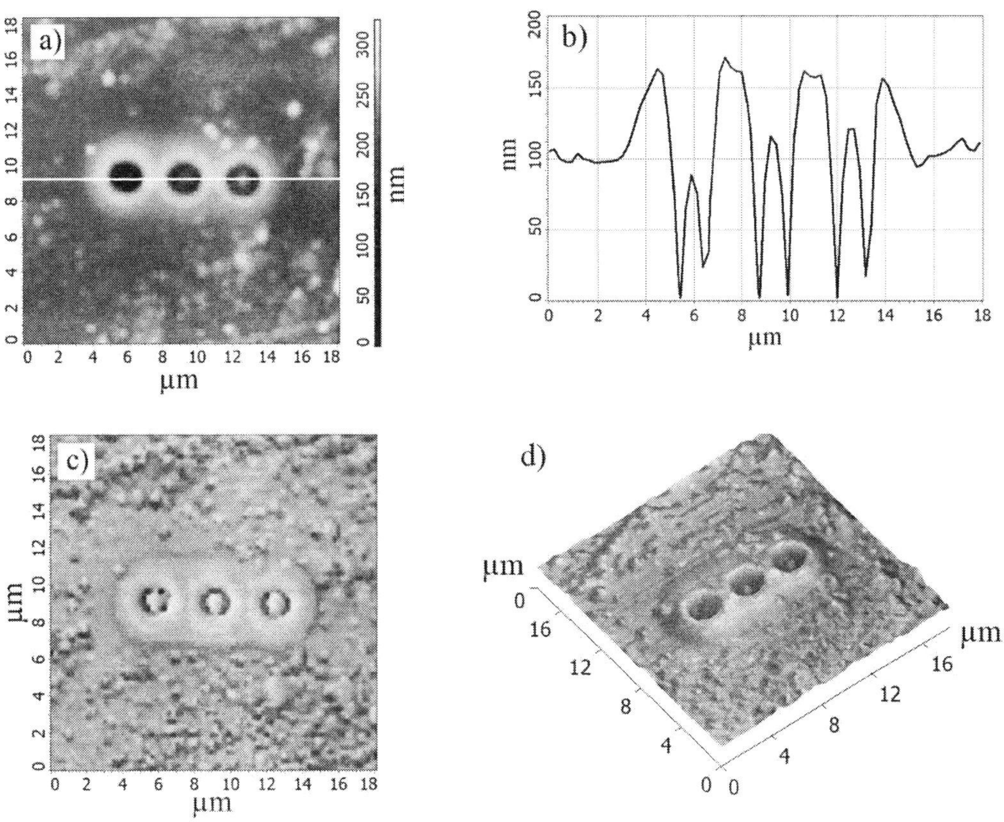

Figure 3 (a, b, c, d). Investigation structures, formed by FIB: **(a)** AFM-image; **(b)** cross-section profile along line on (a); **(c)** phase; **(d)** 3D AFM-image.

Structures obtained in the course of the experiments were examined using a probe microscope. It was shown, voltage increase from 5 to 20 V leads to current increase from 0.15±0.03 nA to 6.3±0.4 nA (fig. 4, a) and electric field increase from 60 to 180 MV/cm leads to current density increase from 96.32±8.81 A/cm^2 (fig. 4, b).

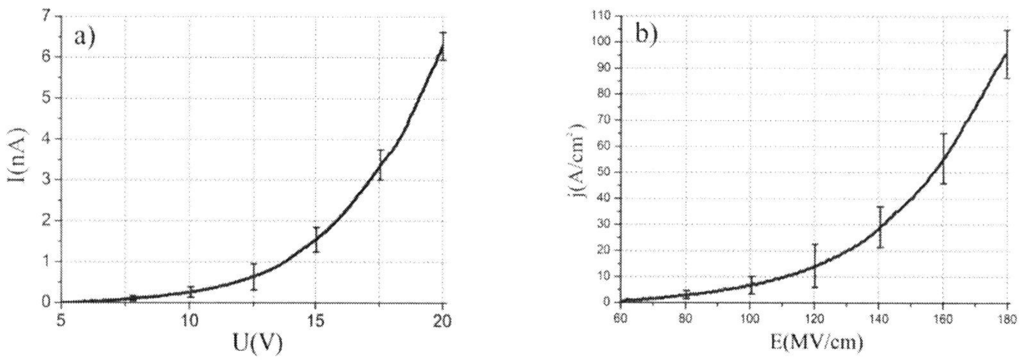

Figure 4 (a, b). Electric measurements of formed structure: **(a)** current voltage characteristic; **(b)** j-E plot.

Dependences obtained show that the structures formed by a combination of methods of focused ion beams and plasma chemical etching can have auto emission characteristics.

4. Discussion and Conclusions.

In this work, experimental studies have been carried out to obtain carbon nanostructures on the surface of silicon carbide. Structures with a depth of 900 nm and a minimum structure size of 280 nm were formed using the method of focused ion beams. Experiments on the formation of graphene on the surface of silicon carbide by plasma etching with fluoride plasma were performed. The characteristics of the emission cell have been investigated; it has been shown that the use of carbon nanostructures as an active element of the emission cell is promising. This technology can be applied to form modern vacuum microelectronic devices and to form pressure and gas sensors.

Acknowledgments

This work was supported by the Grant of the President of the Russian Federation No. MK-3512.2019.8 and Southern Federal University (grant VnGr-07/2017-02). The results were obtained using the equipment of the Research and Education Center "Nanotechnologies" of Southern Federal University.

References

[1] Kangawa Y, Ito T, Taguchi A, Shiraishi K, Irisawa T and Ohachi T 2002 *Appl. Surf. Sci.* **190** 517
[2] AgeV S, Solodovnik M S, Eskov A V, Krasnoborodko S Y 2016 *J. Phys.: Conf. Ser.* **741** 012178
[3] Daweritz L and Ploog K 1994 *Semicond. Sci. Tech.* **9** 123
[4] Kots I N, Kolomiitsev A S, Lisitsyn S G, Polyakova V V, Klimin V S, Ageev O A 2019 *Russian Microelectronics* **48(2)** p 72-79
[5] Tok E S, Neave J H, Zhang J, Joyce B A and Jones T S 1997 *Surf. Sci.* **374** 397
[6] Gusev E Yu, Avdeev S P 2018 *J. Phys.: Conf. Ser.* **1124** 022026
[7] Tominov R V, Zamburg E G, Khakhulin D, Klimin V S, Smirnov V A, ChuY H, Ageev, O.A. 2017 *J. Phys.: Conf. Ser.* **917** 032023
[8] Kley A, Ruggerone P and Scheffler M 1997 *Phys. Rev. Lett.* **79** 5278
[9] Shiraishi K and Ito T 1998 *Phys. Rev.* B **57** 6301
[10] Rudyk N N, Il'In O I, Il'Ina M V, FedotovA A, Klimin V S, Ageev O A 2017 *J. Phys.: Conf. Ser.* **917** 082008
[11] Morgan C G, Kratzer P and Scheffler M 1999 *Phys. Rev. Lett.* **82** 4886

[12] Murdick D A, Wadley H N G and Zhou X W 2007 *Phys. Rev.* B **75** 125318
[13] Amrani A, Djafari Rouhani M and Mraoufel A 2011 *Appl. Nanosci.* **1** 59
[14] Il'Ina M V, Blinov Y F, Il'In O I, Klimin V S, Ageev O A 2016 *Proc. of SPIE* **10224** 102240U-1
[15] Sun, X., Xu, D., Xiong, B., Wu, G., Wang, Y. (2013) *Sensors and Actuators, A: Physical*, 201, pp. 428-433.
[16] Foxon C T and Joyce B A 1977 *Surf. Sci.* **64** 293
[17] Gavignet, E., Lanzetta, F. (2014) Vacuum, 100, pp. 18-21.
[18] Lisitsyn S A, Balakirev S V, Avilov V I, Kolomiytsev A S, Klimin V S, Solodovnik M S, Konoplev B G, Ageev O A 2018 *Nanotechnologies in Russia* **13(1-2)** p 26-33
[19] Hou, Z., Xu, D., Cai, B. (2006) *Appl. Phys. Lett.*, 89, pp. 21-24.
[20] Modi, A., Koratkar, N., Lass, E., Wei, B., Ajayan, P.M. (2003) *Nature*, 424 (6945), pp. 171-174.
[21] Jitschin, W., Ludwig, S. (2004) *Vacuum*, 75 (2), pp. 169-176.
[22] Kots I N, Klimin V S, Polyakova V V, Rezvan A A, Vakulov Z E, Ageev O A 2018 *J. Phys.: Conf. Ser.* **1124** 081035
[23] Tseng A 2004 J. Micromech. Microeng. 14 15
[24] Han J, Lee H, Min B, Lee S 2010 Microelectron. Eng. 87 1
[25] Klimin V S, Rezvan A A, Ageev O A 2018 *J. Phys.: Conf. Ser.* **1124** 022035
[26] Avilov V I, Ageev O A, Blinov Y F, Konoplev B G, Polyakov V V, Smirnov V A and Tsukanova O G 2015 *Tech. Phys.* 60 717-23
[27] Orloff J, Swanson L W, Utlaut M 2003 Springer New York
[28] Nagase M, Nakamatsu K, Matsui S, Namatsu H 2005 Japanese J. of Appl. Phys. 44(7) 5409
[29] Ageev A O, Konoplev B G, Polyakov V V , Svetlichnyi A M, Smirnov V A 2007 *Russian Microelectronics* **36** 353-357

Formation of SiO_2 buffer layer for $LiNbO_3$ thin films growth

Z Vakulov[1], V S Klimin[1,2], A A Rezvan[2], R V Tominov[1], K Korzun[3], I N Kots[1], V V Polyakova[1], O A Ageev[1,2]
[1]Research and Education Center «Nanotechnology», Sothern Federal University, Taganrog 347922, Russia
[2]Institute of Nanotechnologies, Electronics and Equipment Engineering, Southern Federal University, Taganrog 347922, Russia
[3]Department of Applied Physics, Eindhoven University of Technology, 5600 MB, The Netherlands

Abstract. This paper shows the results of study of the effect of SiO_2 buffer layer thickness on the morphological parameters of nanocrystalline $LiNbO_3$ films formed by pulsed laser deposition. It has been established that with increasing in the thickness of SiO_2 buffer layer from 10 nm to 50 nm, the roughness of $LiNbO_3$ films decreases from 5.1 nm to 4.4 nm. The minimum value of the grain diameter (118 nm) corresponds to the thickness of the buffer layer equal to 50 nm. The results obtained can be used in the design and manufacture of integrated acousto-optic and piezoelectric devices, as well as sensitive elements of sensors using various effects of surface acoustic waves.

1. Introduction

Modern electronics faced with necessity of creation new principles of power elements formation, since the capabilities of electrochemical power sources cannot fully meet its growing demands [1-4]. Moreover, there are certain difficulties associated with the disposal of waste batteries. In this regard, active research on the modernization of existing structures and technologies of formation, as well as the development of fundamentally new energy sources based on nanomaterials that meet the requirements of efficiency, versatility and environmental safety are being conducted [5-7].

A promising direction in the field of battery development is the development of piezoelectric transducers using mechanical energy of the environment as a source of power supply for wearable electronics (smart watches, mobile phones, pacemakers) [8-13]. Using of such piezoelectric transducers can expand the capabilities of wearable electronics devices, significantly increasing the time of their autonomous operation and reducing the weight and dimensions of the power supply elements.

The creation of devices based on ferroelectric and piezoelectric films at the present stage of technology development is difficult due to the problems associated with the insufficient development of the technology for their production, as well as the need to use additional buffer layers. Most ferroelectric and piezoelectric materials ($BaTiO_3$, $PbTiO_3$, $LiNbO_3$) are multicomponent oxides, which significantly complicates the technological process of their production and their integration with silicon technology of micro- and nanoelectronics [14-18]. Pulsed laser deposition is one of the promising method for the formation of multicomponent oxides films, due to a large number of technological parameters, which allows control the composition of the formed films, as well as their morphological and electrical parameters [19].

The purpose of this work is determination of influence regularities of SiO_2 buffer layer thickness on the morphological parameters of $LiNbO_3$ nanocrystlline films fabricated by pulsed laser deposition.

2. Materials and Methods

For fabrication of naocrystalline $LiNbO_3$ films we used nanotechnological cluster complex NANOFAB NTK-9 (NT-MDT, Russia), comprising PLD module Pioneer 180 (Neocera Co., USA). $LiNbO_3$ target were ablated by excimer KrF laser (λ=248 nm). Energy density on target surface was maintained at 2.0 J/cm^2. The quantity and frequency of laser pulses were 50 000 and 10 Hz respectively. Target-substrate distance was 115 mm.

$LiNbO_3$ films were deposited on Si substrates (1 cm x 1 cm) with buffer layer of SiO_2. All substrates were subjected to preliminary purification in inorganic solvents to remove surface contamination. The morphology of the obtained film was researched by Ntegra probe nanolab (NT - MDT, Russia) and scanning electron microscop NOVA Nanolab 600 (FEI Co., Netherlands).

The formation of Si and SiO_2 films was carried out by plasma chemical deposition in a combined discharge plasma. The combined capacitive and inductively coupled plasma discharge allows controlling the granularity and roughness of the resulting dielectric surface. Film growth modes P_R = 2 Pa, gas flows N_{SiH4} = 5 cm^3/min, N_{N2O} = 60 cm^3/min, power of the capacitive plasma source was W_{RIE} = 10 V, power of the source inductively coupled plasma W_{ICP} = 400 V, deposition time 54 minutes.

3. Results

Figure 1 shows the results of AFM images of $LiNbO_3$ films obtained on SiO_2 buffer layer with different thickness.

(a)

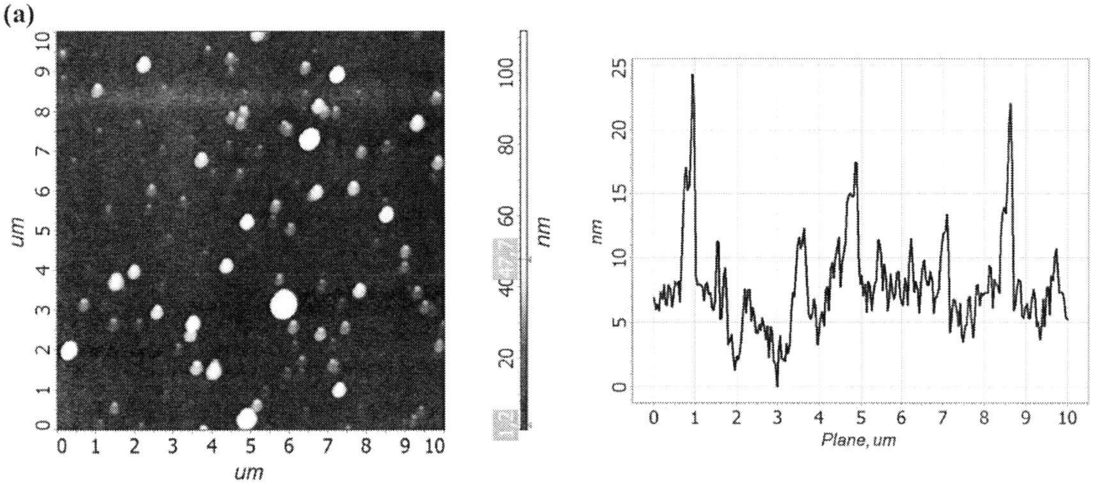

(b)

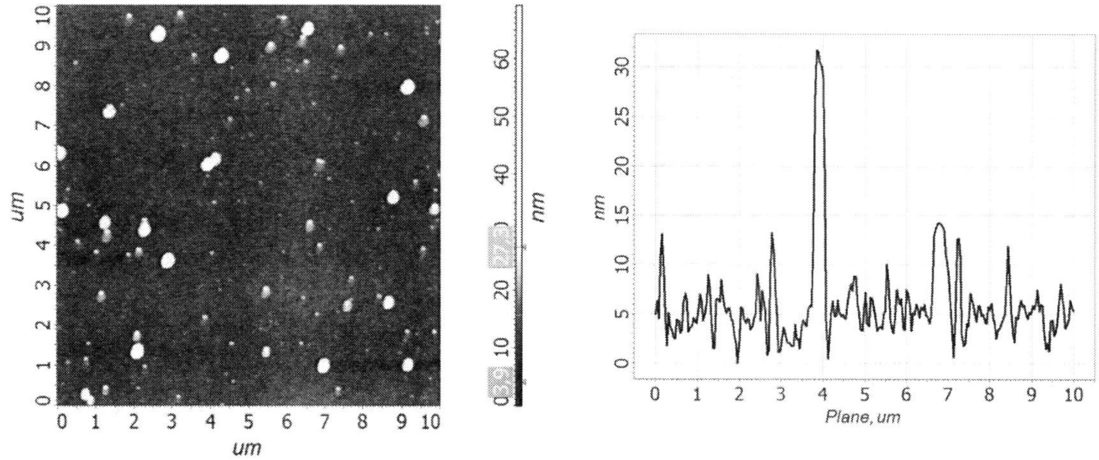

(c)

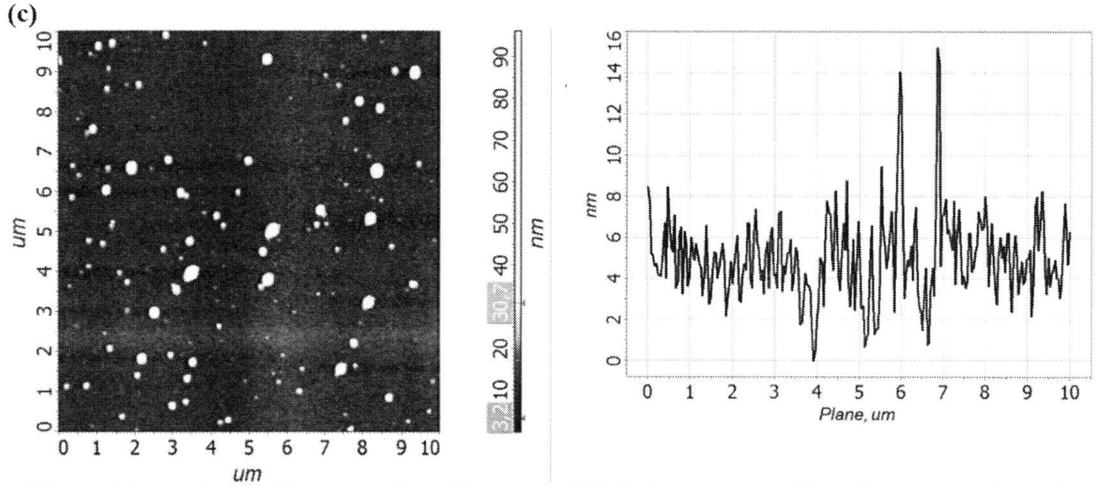

Figure 1 (a, b, c) AFM images and profilograms of $LiNbO_3$ nanocrystalline films obtained on SiO_2 buffer layer with different thickness: 10 nm **(a)**, 50 nm **(b)**, 100nm **(c)**

Figure 2 shows the values of surface roughness and grain diameter of nanocrystalline $LiNbO_3$ films obtained on SiO_2 buffer layer with different thickness.

(a)

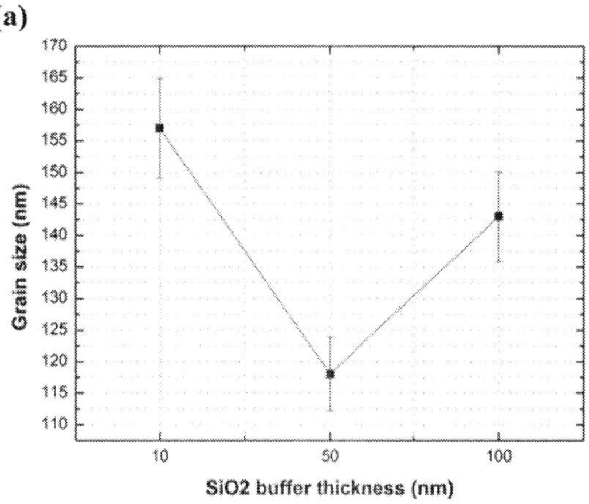

(b)

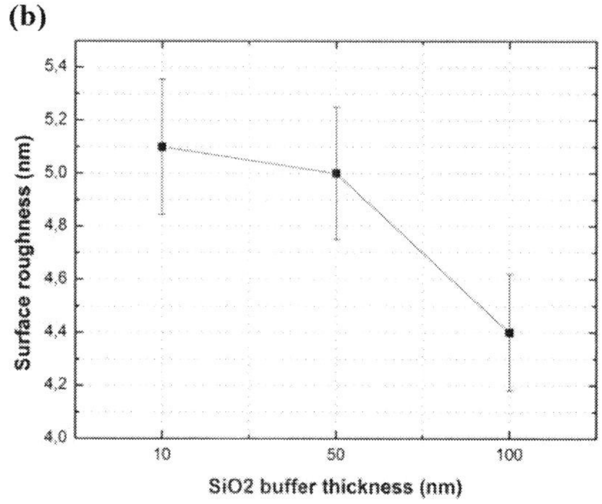

Figure 2 (a, b) Dependences of the morphological parameters of $LiNbO_3$ nanocrystalline films on the thickness of the SiO_2 buffer layer: grain size **(a)** and surface roughness **(b)**

As the thickness of the buffer layer increases from 10 to 100 nm, the roughness of the nanocrystalline $LiNbO_3$ films decreases due to the activation of the process of interatomic interaction of the materials of the formed film. Increasing in the grain size of $LiNbO_3$ films with an increase in the thickness of the buffer layer may be associated with an increase in the grain diameter of the SiO_2 layer and the effect of "inheriting" the morphology of the buffer layer. Figure 3 shows SEM images of $LiNbO_3$ nanocrystalline films obtained on SiO_2 buffer layer with different thickness.

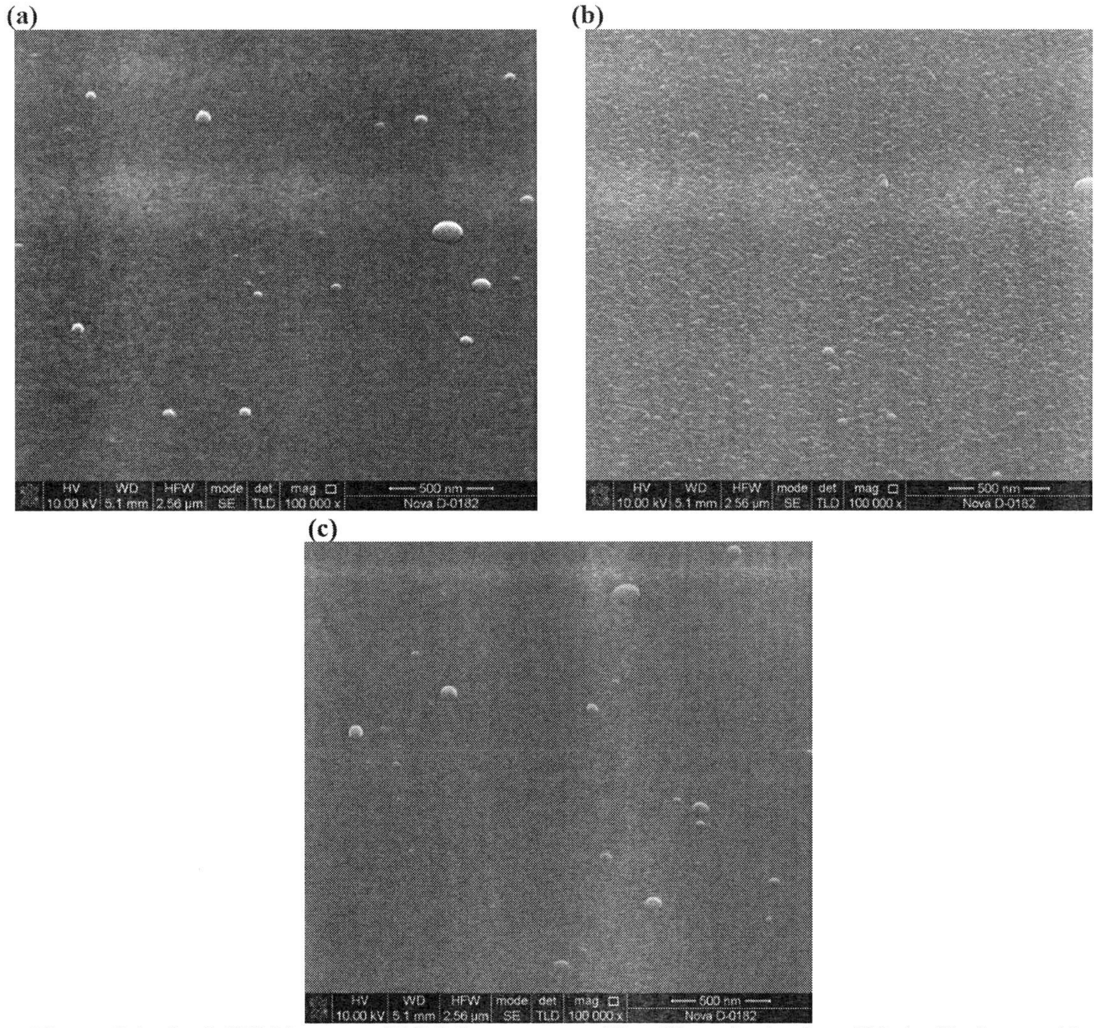

Figure 3 (a, b, c) SEM images of $LiNbO_3$ nanocrystalline films obtained on SiO_2 buffer layer with different thickness: 10 nm (a), 50 nm (b), and 100 nm (c)

4. Discussion and Conclusions

In this work, experimental studies were carried out on the deposition of a buffer layer of SiO_2 by plasma chemical deposition. $LiNbO_3$ films were obtained by pulsed laser deposition. It has been established that with an increase in the thickness of the buffer layer of SiO_2 from 10 nm to 50 nm, the roughness of $LiNbO_3$ films decreases from 5.1 nm to 4.4 nm. The effect of the thickness of the buffer layer of SiO_2 on the morphological parameters of $LiNbO_3$ nanocrystalline films formed by pulsed laser deposition was studied. The obtained films with these roughness indicators can be used in the formation of solar cells, as well as elements of nanophotonic devices.

Acknowledgments
This work was financially supported by the Russian Foundation for Basic Research (project № 18-29-11019 mk) and Grant of the President of the Russian Federation No. MK-3512.2019.8. The work was done on the equipment of the Research and Education Centre «Nanotechnology», Southern Federal University.

References
[1] Schlachter F 2013 *Proc. Natl. Acad. Sci. USA* **110** 5273
[2] Granqvist, C.G. 2007 Solar Energy Materials and Solar Cells **91 (17)** pp. 1529-1598
Jayaraj, M.K., Antony, A., Ramachandran, M. 002 Bulletin of Materials Science, **25 (3)** pp. 227-230
[3] Kots I N, Kolomiitsev A S, Lisitsyn S G, Polyakova V V, Klimin V S, Ageev O A 2019 Russian Microelectronics **48(2)** p 72-79
[4] Kon, M., Song, P.K., Mitsui, A., Shigesato, Y. 2002 *Japanese Journal of Applied Physics, Part 1: Regular Papers and Short Notes and Review Papers*, **41 (10)** pp. 6174-6179
[5] Liu, D.-S., Wu, C.-Y., Sheu, C.-S., Tsai, F.-C., Li, C.-H. 2006 *Japanese Journal of Applied Physics, Part 1: Regular Papers and Short Notes and Review Papers*, **45 (4 B),** pp. 3531-3536
[6] Krivchenko, V.A., Lopaev, D.V., Pashchenko, P.V., Pirogov, V.G., Rakhimov, A.T., Suetin, N.V., Trifonov, A.S. 2008 *Technical Physics*, **53 (8),** pp. 1065-1069
[7] Kim H S, Kim J H, Kim J 2011 Int. J. Pr. Eng. Man. **12** 1129
[8] Klimin V S, Vakulov Z E, Tominov R V, Varzarev Y N, Clemente I E, Miakonkikh A V, Rudenko K V, Ageev O A 2019 Proc. of SPIE **11022** 110221E-2
[9] le Hung, N., Ahn, E., Jung, H., Kim, H., Kim, D. 2010 Journal of the Korean Physical Society, **57 (61)**, pp. 1784-1788.
[10] Dang, W.L., Fu, Y.Q., Luo, J.K., Flewitt, A.J., Milne, W.I. 2007 *Superlattices and Microstructures*, **42 (1-6),** pp. 89-93
[11] Klimin V S, Tominov R V, Avilov V I, Dukhan D D, Rezvan A A, Zamburg E G, Smirnov V A, Ageev O A 2019 Proc. of SPIE **11022** 110220E-1
[12] Youssef, S., Combette, P., Podlecki, J., Al Asmar, R., Foucaran, A. 2009 *Crystal Growth and Design*, **9 (2),** pp. 1088-1094
[13] Minami, T., Miyata, T., Yamamoto, T., Toda, H. 2000 *Journal of Vacuum Science and Technology A: Vacuum, Surfaces and Films*, **18 (4 II),** pp. 1584-1589.
[14] Klimin V S, Rezvan A A, Ageev O A 2019 Proc. of SPIE **11022** 1102220-1
[15] Vakulov Z E, Varzarev Y N, Gusev E Y, Skrylev A V, Panich A E, Miakonkikh A V, Klemente I E, Rudenko K V, Konoplev B G, Ageev O A 2019 Russ. Microelectron. **48** 59
[16] Klimin V S, Kots I N, Polyakova V V, Rezvan A A, Ageev O A 2019 Proc. of SPIE **11022** 110221R-4
[17] Chang, Z., Gaoyuan, M., Dostanko, A.P., Golosov, D.A., Zavatskiy, S.M. 2010 *J. Appl. Opt.*, **31**, p. 855.
[18] Golosov, D.A., Zavadski, S.M., Svadkovski, I.V., Melnikov, S.N. 2010 *Vakuum. Tekh. Tekhnol.*, **20**, p. 227
[19] Subramanyam, T.K., Srinivasulu Naidu, B., Uthanna, S. 2000 *Crystal Research and Technology* **35 (10)** pp. 1193-1202.

Study of the ITO sublayer influence on the tubular carbon nanostructures formation

O I Il'in[1], N N Rudyk[1], M V Kuzhelev[2], M V Il'ina[1], A A Fedotov[1]

[1] Southern Federal University, Institute of Nanotechnologies, Electronics and Electronic Equipment Engineering, Taganrog, 347922, Russia
[2]«Rehm RUS» LLC, Moscow, 127018, Russia

oiilin@sfedu.ru

Abstract. The study of the formation of catalytic centers and the growth of CNT-based TCN was carried out on the Ni (10 nm)/ITO (100 nm)/Si (370 μm) structure. Shown that the temperature at the stage of heating and growth has a significant impact on the parameters of the formed TCN. Promising for instrumental use, the TCN were formed at a temperature of 750 °C, a heating time of 20 minutes, an activation time of 3 minutes. The effect of plasma ions at the stages of «activation» and growth was also demonstrated.

1. Introduction

The development of nanotechnology and the ability to create nanostructures with controlled properties have given rise to a wide interest in the synthesis of new materials. One of the most promising materials with unique electrical and mechanical properties is considered to be tubular carbon nanostructures (TCN), such as carbon nanotubes (CNT) and nanofibers (CNF), which allows them to be used as the basis for creating modern elemental base of nanoelectronics, nano- and microsystem technology [1-4]. However, the use of tubular carbon nanostructures is hampered by the lack of technologies for the mass production of these materials, the high cost of the finished product, as well as the complexity of the process for their production with given parameters [5, 6]. In addition, to date, the mechanisms and factors of their formation that affect the geometric parameters of the properties of tubular carbon nanostructures are not fully understood [7, 8]. At the same time, for instrument application, the need arises for growing TCN on a sublayer, providing electrical contact to nanostructures [9, 10].

The aim of this work is to study the effect of the conductive ITO sublayer on the parameters of the formed tubular carbon nanostructures based on carbon nanotubes.

2. Experiments and methods

The study of the formation of catalytic centers (CC) and the growth of CNT-based TCN was carried out on the Ni (10 nm) / ITO (100 nm) / Si (370 μm) structure. In this structure, an ITO film (indium tin oxide $(In_2O_3)_{0,9}$ - $(SnO_2)_{0,1}$) serves as a conductive sublayer, and a Ni film as a catalytic material.

The samples were heated to a temperature of 650 - 750 °C for 20 minutes in a plasma enhanced chemical vapor deposition module (PECVD), in the chamber of which argon and ammonia were simultaneously fed. Argon and ammonia fluxes were 40 sccm and 15 sccm, respectively. During heating, a pressure of 4.5 Torr was maintained in the chamber.

Then, CC were activated for 1-5 minutes in an ammonia atmosphere with and without plasma initiation. An array of obtained catalytic centers (CC) was studied by atomic force microscopy (AFM) using the Ntegra Nanolaboratory (NT-MDT, Russia) and scanning electron microscopy (SEM) using a Nova Nanolab 600 microscope (FEI, Netherlands).

The growth of the TCN was carried out at temperatures of 650–800 °C for 15 min in a flow of ammonia (210 sccm) and acetylene (70 sccm) with plasma initiation. Heating to the set temperatures was carried out in 20 and 45 minutes.

3. Results and discussion

The SEM studies of samples obtained after the heating stage confirm that the initial formation of Ni catalytic centers from a continuous film occurs exactly at the structure heating stage. At the same time, act such mechanisms as film corrugation due to thermal stresses in the structure during heating, as well as the Gibbs-Thompson effect.

Following the heating stage, the "activation" stage serves to reduction the CC from oxides, and also allows additional processing of the CC obtained during the heating phase, thereby expanding the possibilities for managing their geometrical parameters and chemical composition. SEM images of CC formed at the activation stage in the Ni/ITO/Si structure during 1, 3, 5 minutes with and without presence of plasma are shown in Figure 1.

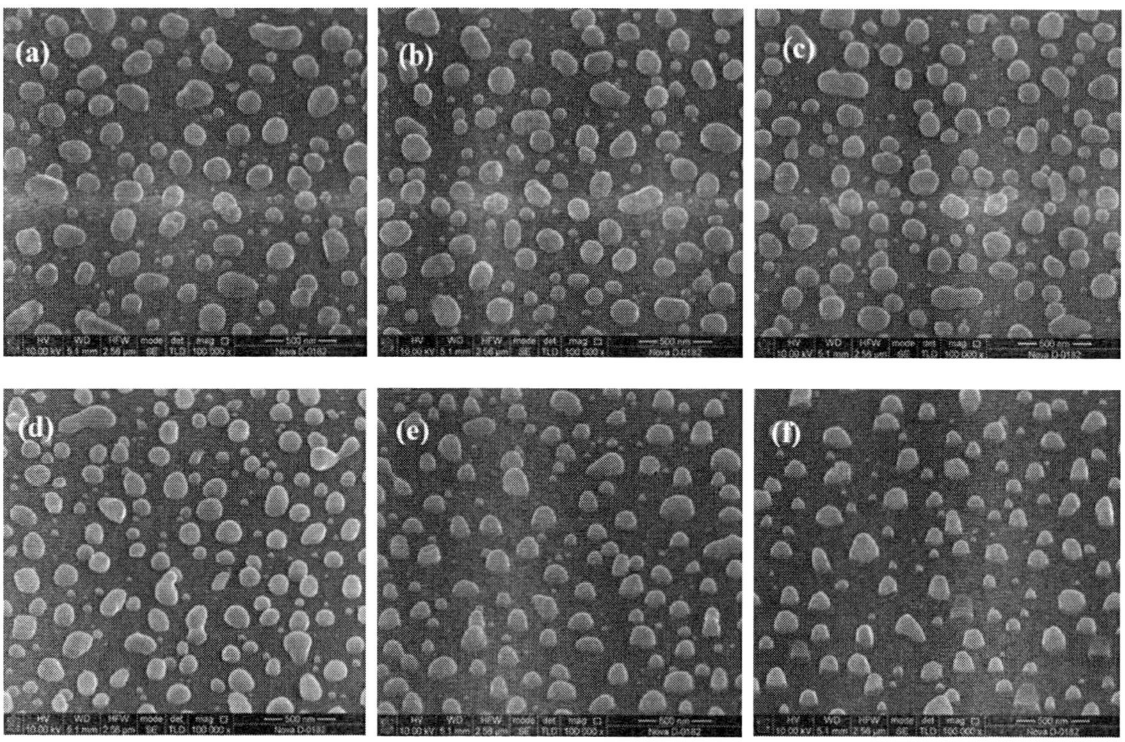

Figure 1(a,b,c,d,e,f). SEM- images of the surface of the Ni/ITO/Si structure after activation for various times: **(a)** 1 min, w/o plasma; **(b)** 3 min, w/o plasma; **(c)** 5 min, w/o plasma; **(d)** 1 min, w/ plasma; **(e)** 3 min, w/ plasma; **(f)** 5 min, w/ plasma.

According to the data obtained as a result of statistical processing, the dependences of the diameter, height, and density of the obtained TCN on time are presented in Figure 2.

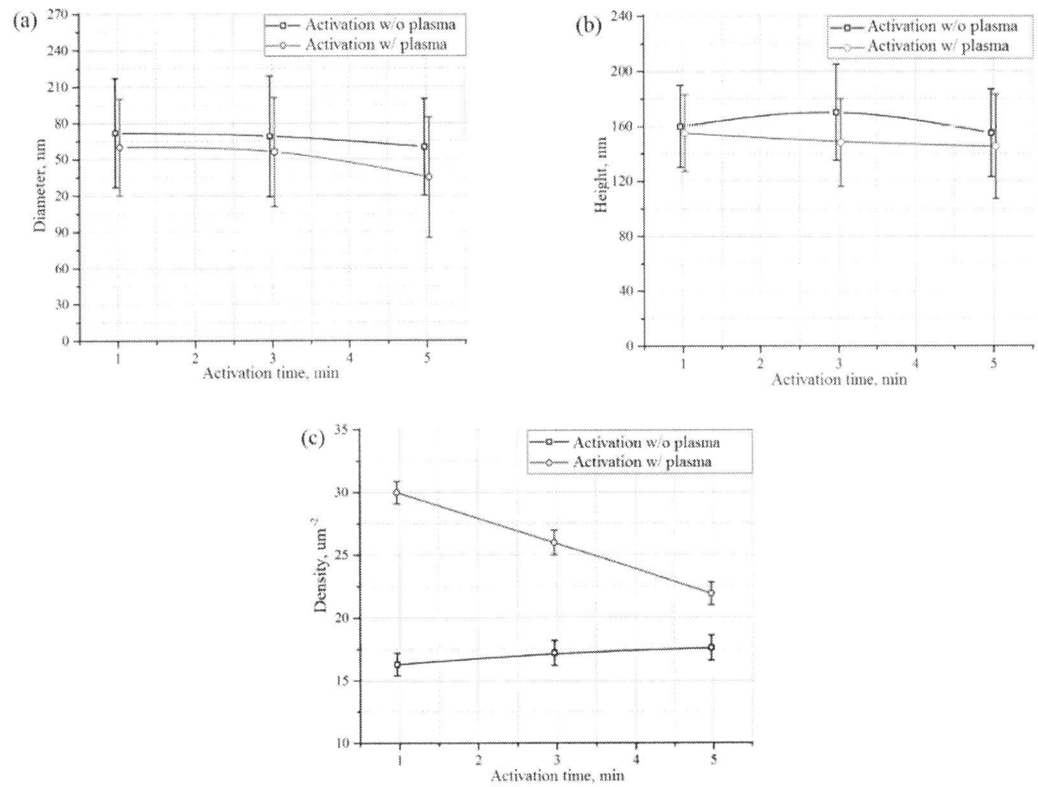

Figure 2(a,b,c). Geometric parameters of catalytic centers obtained at the stage of activation: (a) diameter; (b) height; (c) density.

From the obtained dependences, it can be seen that the diameter and height of the CC decrease with increasing activation time. This fact can be explained by the fact that in the process of activation, CC are etched with the reduction of metal oxides by hydrogen formed during the decomposition of ammonia. In the presence of plasma during activation, the CC is additionally etched with ions, so the etching rate of the CC increases. Formed CC in this case, smaller and their size decreases faster with increasing activation time. Also, in the case of activation in plasma, small CC are completely removed by etching with ammonia and plasma ions. In this case, the CC density decreases with increasing activation time.

The SEM images of the TCN formed at the growth stage at different temperatures with different heating rates are presented in Figure 3.

The analysis showed that during heating from 650 °C to 750 °C, the heating rate has practically no effect on the geometric characteristics of the formed TCN. From the obtained results, it can be seen that TCN grown at a temperature of 800 °C with different heating rates differ in height and diameter (Fig. 4).

At the tips of the obtained TCN there are CC, the diameter of which is much larger than the diameter of the TCN themselves. In these modes, at a temperature of 650 °C TCN, they are not vertical, even with growth in plasma.

In the case of growing at a temperature of 750 °C, TCN appear with a diameter of from 20 nm to 40 nm and a length of up to 1.5 μm. This mode allows to grow TCN with the smallest diameter in the system Ni/ITO/Si.

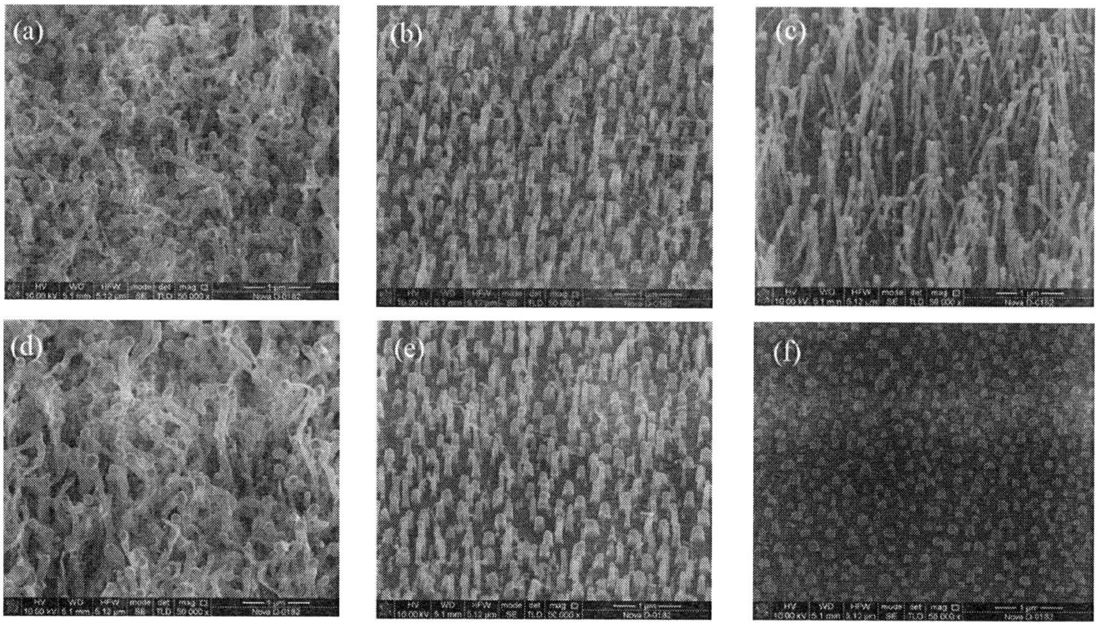

Figure 3(a, b, c, d, e, f). SEM images of the TCN grown by heating for: **(a)** 650 ^{O}C, 20 min; **(b)** 750 ^{O}C, 20 min; **(c)** 800 ^{O}C, 20 min; **(d)** 650 ^{O}C, 45 min; **(e)** 750 ^{O}C, 45 min; **(f)** 800 ^{O}C, 45 min.

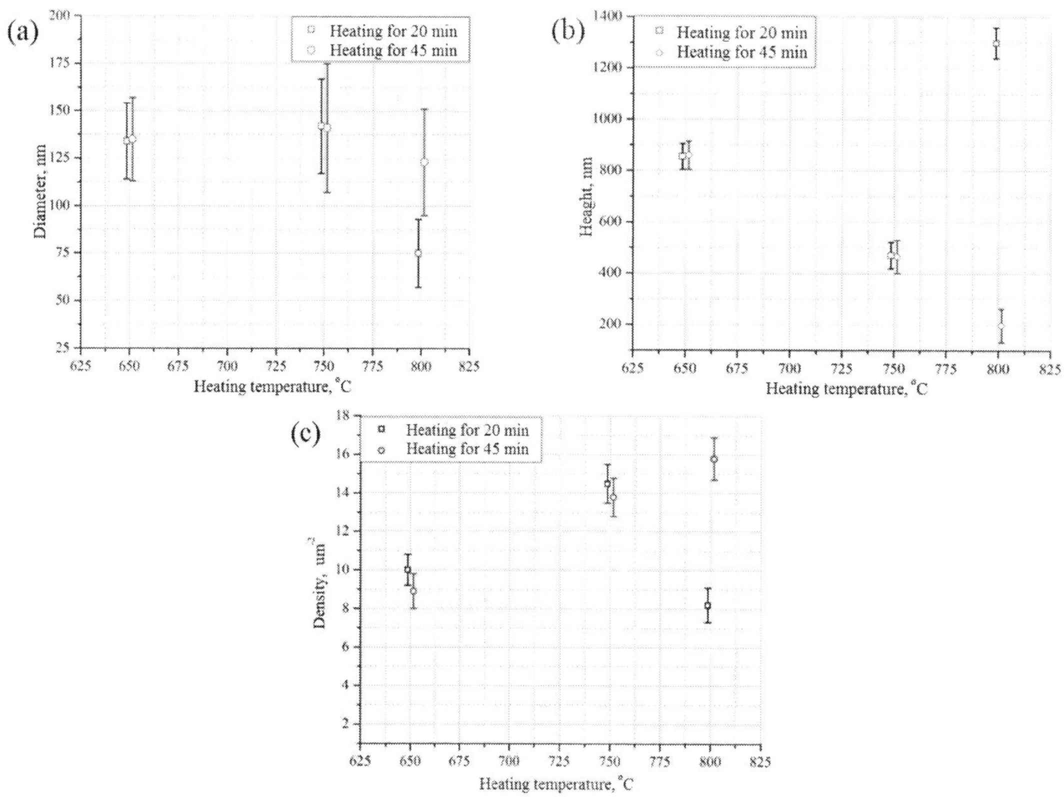

Figure 4(a, b, c). Geometric parameters of grown TCN: (a) diameter; (b) height; (c) density.

4. Conclusion

This paper presents the data of experimental studies of the formation of catalytic centers, as well as the growth of CNSs based on them in the Ni/ITO/Si structure. The results showed that the temperature at the stage of heating and growth has a significant impact on the parameters of the formed TCN. Promising for instrumental use, the TCN were formed at a temperature of 750 °C, a heating time of 20 minutes, an activation time of 3 minutes. The effect of plasma ions at the stages of «activation» and growth was also demonstrated, and it was shown that with its help one can somewhat reduce the dispersion of the geometric dimensions of the structures obtained. This results can be used to create elements and microelectronics sensors.

The results were obtained using the equipment of the Research and Educational Center and the Center for Collective Use "Nanotechnology" of the Southern Federal University.

Acknowledgments

The financial support was provided by Russian Foundation for Basic Research (project № 18-32-00652 mol_a) and Internal grant of the Southern Federal University (project VnGr-07/2017-26)

References

[1] Biswas C, Hee Lee Y 2011 *Adv. Funct. Mater.* **21** 3806–3826

[2] Il'ina M V, Il'in O I, Blinov Yu F, Smirnov V A, Ageev O A 2018 *Technical Physics* **63** pp. 1672–1677

[3] Brady G, Way A, Safron N, Evensen H, Gopalan P, Arnold M 2016 *Sci. Adv.* **2** e1601240

[4] Shulaker M, Hills G, Park R, Howe R, Saraswat K, Wong P, Mitra S *Nature* **547** 74-78

[5] Qiuchen Zhao, Ziwei Xu, Yue Hu, Feng Ding, Jin Zhang 2016 *Sci. Adv.* **2** (5) e1501729

[6] Clancy A, Bayazit M, Hodge S, Skipper N, Howard C, Shaffer M 2018 *Chem. Rev.* **118** 7363–7408

[7] Moore K , Tune D, and Flavel B 2015 *Adv. Mater.* **27** 3105–3137

[8] Il'ina M V., Il'in O I, Blinov Y F, Smirnov V A, Kolomiytsev A S, Fedotov A A, Konoplev B G, Ageev O A 2017 *Carbon* **123** 514–524

[9] Ageev O A, Blinov Y F, Ilina M V, Ilin O I, Smirnov V A 2016 *J. Phys. Conf. Ser.* **741** 012168

[10] Qing Cao, Shu-Jen Han, Tersoff J, Franklin A D, Zhu Yu, Zhang Z, Tulevski G S, Tang J, Haensch W 2015 *Science* **350** 6256

Chemical synthesis and optical properties of composite materials containing PbI_2 nanoparticles

A S Kulagina[1] **S K Evstropiev**[2,3] **A S Saratovskii**[2]
[1]Epitaxial Nanotechnology Lab, St. Petersburg Academic University, St Petersburg 194021, Russia
[2]Research Center for Optical Materials Science, ITMO University, St Petersburg 197101, Russia
[3]St Petersburg State Institute of Technology, St. Petersburg 190013, Russia

Abstract. The aqueous sols and composite organic-inorganic coatings containing PbI_2 nanoparticles (size ~ 4 nm) have been prepared. The absorption, photoluminescence and nonlinear transmittance (at 532 nm wavelength) have been studied. The electronic absorption spectra of PbI_2/PVP sols showed major peak of ground state near 406 nm, which slightly blue shift with decreasing size. Obtained materials are characterized high transparency in visible spectral range and demonstrate quantum confinement effect. PbI_2/PVP composite films manifested dependence of radiation resistance with PVP content at ns-laser exposure and size dependence of nonlinear transmittance.

1. Introduction

Lead iodide is a wide-band-gap semiconductor (E_g ~ 2.34 eV) which is considered to be used as room-temperature radiation detector with high environmental stability [1]. Spatial confinement of charge carriers in PbI_2 nanocrystals (NCs) whose dimensions are comparable to the carrier Bohr radius (~ 1.9 nm [2]) is known to produce substantial changes in their optical properties [2-4]. Surface states of small PbI_2 nanoparticles strongly affect on photoactivity of material and consequently on carriers transport in NPs [2-4]. It was shown in [5], that the absorption peaks are due to different orders of interband transitions of a nominally single size of particles rather than the first-order transitions of a number of discrete sizes of particles. Due to high photoconductivity the organic-inorganic composites based on PVP polymer matrix have been used in different optical and medical applications [6]. To date, the optical properties of PbI_2 in PVP have not been presented. The aim of this work is the study of the features of absorption, luminescence and non-linear optical properties of aqueous sols and composite coatings containing PVP and PbI_2.

2. Methods

The aqueous solutions of KI ($8.6\cdot10^{-2}$ M), $Pb(NO_3)_2$ ($4.3\cdot10^{-2}$ M) and high-molecular PVP ($4.6\cdot10^{-5}$ M) were used for the synthesis of PbI_2 NPs. Firstly the solutions of PVP and $Pb(NO_3)_2$ were mixed at room temperature. Then KI solution was slowly added into this mixture by dropping at the intense stirring. To prepare the composite layer on the glass surface we use the method consisting in applying a small amount of a mixed solution to one of the sides of a glass sample, followed by drying and obtaining a thin uniform composite coating. Chemical composition of mixed solutions and composite coatings are presented in the table 1. PbI_2 content in prepared materials was calculated basing on the chemical composition of mixed solutions.

Table 1. Chemical composition of mixed solutions and composite films.

Sample	Chemical composition of mixed solution, wt.%					Chemical composition of composite coatings, wt.%			
	PVP	$Pb(NO_3)_2$	KI	H_2O	PbI_2*	PVP	$Pb(NO_3)_2$	KI	PbI_2*
1	0.95	0.46	0.05	98.54	0.06 (1.4 mM)	65.53	31.37	3.10	4.37
2	1.89	0.45	0.04	97.62	0.06 (1.4 mM)	79.18	18.95	1.87	2.64
3	2.81	0.45	0.04	96.70	0.06 (1.4 mM)	85.29	13.58	1.13	1.89
4	3.71	0.44	0.04	95.81	0.06 (1.4 mM)	88.37	10.58	1.05	1.47
5	2.89	0.35	0.34	96.42	0.48 (11 mM)	80.77	9.67	9.56	13.46
6	3.46	0.28	0.27	95.99	0.38 (8.7 mM)	86.30	6.90	6.80	9.61

*Calculated data

In spectral investigations were used the next equipment: spectrophotometer Perkin Elmer LABDA 650, fluorescence spectrometer Perkin Elmer LS-50B. Laser measurements were prepared by using standard z-scan technique at 532 nm wavelength excitation and 5 ns pulse duration at 1 Hz frequency regime. Beam with Gaussian profile after 3x-beam expander was focused by the lens with F=15 cm, the minimum beam size at the focus of exposure was about 60 μm. There are two photometers (OPHIR) was used for energy detecting, the output energy was measured in two regimes "closed aperture" and "open aperture". All investigations were prepared at room temperature.

3. Results

Fig.1a presents absorption spectra of obtained PbI_2 sols stabilized by PVP. The main absorption band near 405 nm is observed. The intensity of this band was growing with PVP content increase. The detail consideration of the spectral positions of this peak shows that the increase of PVP content slightly shifts the maximum from 400 to 406 nm. Based on the relationship, given in [7], between the spectral position of this peak and PbI_2 particles sizes, we estimated the sizes of our NPs 3.9-4.4 nm. The results are summarized in the table 2 beolw.

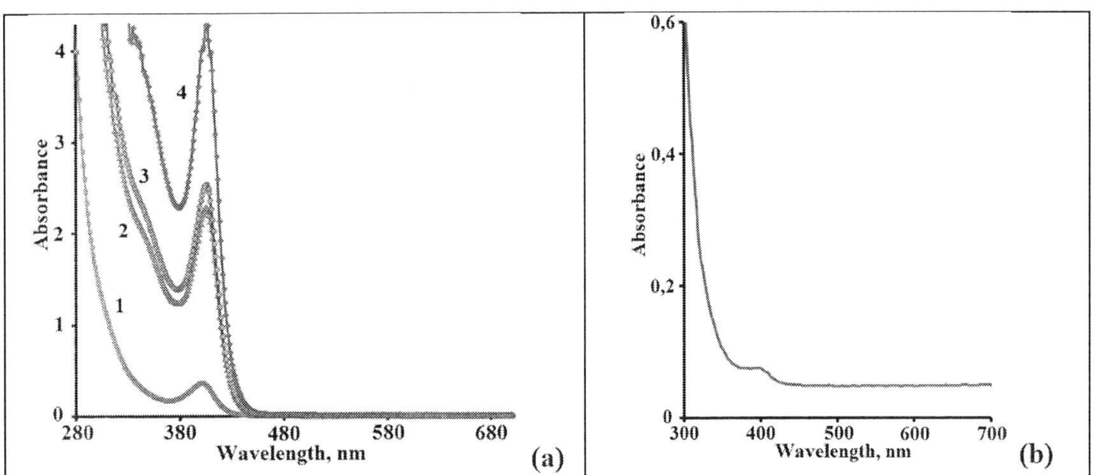

Figure 1(a,b). Absorption spectra of PbI_2-containing materials: **(a)** aqueous sols 1-4 (from 1 to 4 the increase of PVP contents); **(b)** composite coating, prepared from the sol 1.

Table 2. Absorption peak positions and PbI_2 crystal sizes in prepared sols.

Sample	PVP content	λ max, nm	λ max, eV	PbI_2 crystal size, nm
1	$2.3 \cdot 10^{-6}$ M	400	3.10	3.9
2	$4.6 \cdot 10^{-6}$ M	405	3.05	4.3
3	$6.9 \cdot 10^{-6}$ M	406	3.06	4.4
4	$9.2 \cdot 10^{-6}$ M	406	3.06	4.4

Some band is observed at λ ~ 345 nm in the absorption spectra of the sols 2-4. The light absorption in this spectral region could be attributed both to PbI_2 particles [7] and to iodide anions I^- and I^{3-}[8]. The absorption peak at 407 nm is seen in the spectra of composite PbI_2/PVP films prepared from the sols (Fig.1b). It's noticeable, that the spectral positions of this peak in the sol (Fig.1a) and in the film (Fig.1b) are the same. This fact suggests that the film deposition which is accompanied the water evaporation and significant increase of the volume concentration of PbI_2 nanoparticles doesn't change their sizes. The blue shift relative to the main absorption band of the bulk material indicates a quantum-size effect.

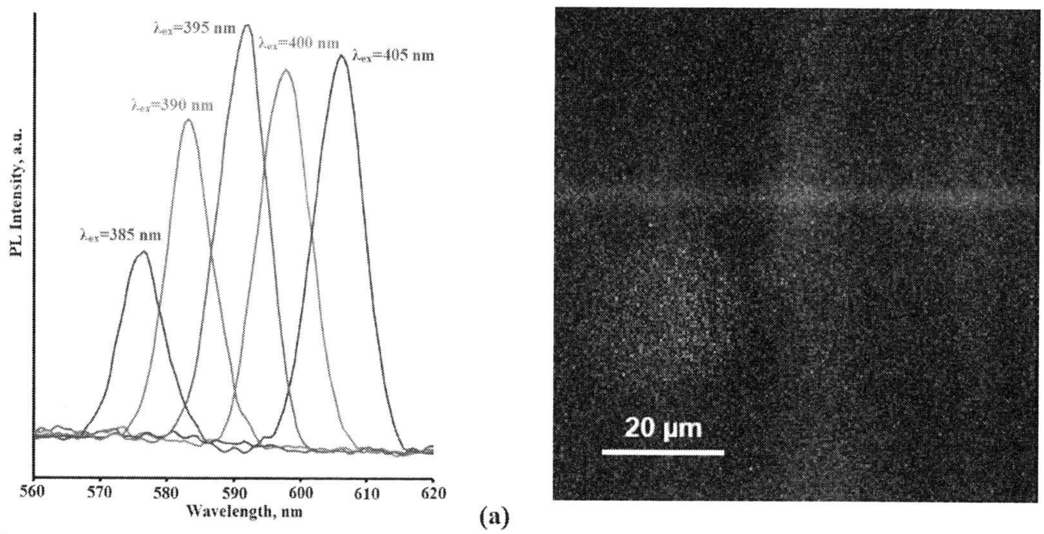

(a) (b)

Figure 2(a,b). (a) PL spectra of PbI_2/PVP coating "4" depending on the excitation wavelength 380-410 nm; **(b)** The image of coating "4" in focus (40x lens) of laser fluorescence microscope PicoQuant MicroTime 100 (λ_{ex} = 405 nm).

PL spectra of composite PbI_2/PVP coating (Fig.2a) show intensive broad peak near 580-610 nm. It position depended on excitation wavelength, the increase of the wavelength led to the red shift of the emission peak. It suggests that prepared material contains PbI_2 nanoparticles having different sizes. PL in the spectral range 500-600 nm was observed in [3,4] and was assigned to the excitation of excitonic states of PbI_2 nanoparticles. The inhomogeneous broadening of PbI_2 luminescence band observed in these works was related to the mixing of the pure excitonic states of PbI_2 nanoparticles with the surface ones [4]. However, this conclusion was made in [3,4] basing on the experiments with the PL excitation only by laser radiation (He-Cd laser, 325 nm) that complicates the study of the mechanism of observed broadening of the emission band. On fig. 2b cluster is shown containing luminescent PbI_2 nanoparticles, indicating that the coatings are micro-heterogeneous.

The second harmonic of the YAG:Nd laser (532 nm) is widely used in optical applications in the visible range. Because synthesized nanoparticles showed a wide absorption band in the VIS range (Fig. 1), we used 532 nm-laser irradiation in the transparency region to study multiphoton processes.

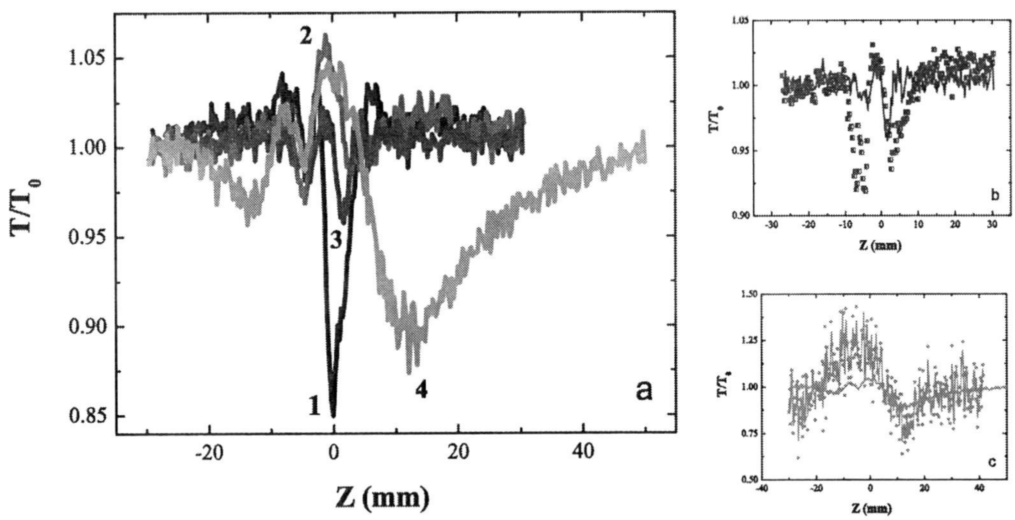

Figure 3.(a) Curves of laser transmittance of PbI_2/PVP coatings in z-scan measurements in open-aperture mode (I_{focus}≃0.14 GW/cm^2, λ = 532 nm, τ = 5 ns): 1 is coating "1", 2 is coating "2", 3 is coating "3", 4 is coating "4"; **(b)** transmittance of coating "3" in open-aperture mode at I_{focus}≃0.14 GW/cm^2 (lines) and at I_{focus}≃0.55 GW/cm^2 (dots) **(c)** transmittance of coating "2" in open-aperture mode (lines) and close-aperture mode (lines with dots), I_{focus}≃0.14 GW/cm^2.

Fig. 3 presents the results of laser transmittance studies of PbI_2/PVP coatings upon excitation into the transparency region. The measurements were carried out at two energies of 18.5 μJ and 55.5 μJ. With more energy, the position of the peaks and dips did not change and their amplitudes increased only (fig. 3b). In order to exclude the influence of high-order nonlinearities, we will consider the data only at lower energy. We expected to see either symmetric bleaching, which characterizes, e.g. the evaporation of the sample, or darkening, which characterizes scattering, absorption, or melting at the sample illumination in the transparency region in the open aperture mode. At low PVP concentration, the sample on curve 1 probably degraded with a gradual increase in intensity, which follows from the asymmetry of the dip relative to the focal plane. Due to the high initial transparency after passing through the focus, a kind of transmission recovery was observed. Curves forms of the remaining films did not allow us to specify the mechanism of nonlinear transmission in the open aperture mode. The data obtained resemble a diffraction pattern, as if there was a disk or a circular hole in the path of the laser beam. Considering the data obtained on a fluorescent microscope, it can be assumed that diffraction took place on PbI_2 NCs clusters with a size of about 10–20 μm (Fig. 2b). From the position of the curves in fig. 3A, it follows that the clusters had a different diameter and opened up a different number of Fresnel zones. This fact indirectly indicates the effect of PVP concentration on the size of nanoparticles formed during the synthesis. The data obtained in the closed-aperture mode were repeated in shape with the curves in Fig. 3A but had greater contrast (Fig.3C). In this mode, high noise was observed due to spatial instability of the laser beam and scattering by the coatings micro-inhomogeneities. Only "4" coating had a distinct amplitude from noise. In conclusion, we note that PVP itself did not manifest any nonlinear properties at the studied frequency [6]. With high PVP concentration prepared coatings can withstand about 5 seconds of 10 Hz-laser exposure at $I \simeq$ 0.14

GW/cm^2 and radiation parameters 532 nm, 5 ns without loss of optical quality. Therefore, PVP increases the resistance of nanoparticles to laser irradiation.

Acknowledgments

The synthesis of the materials and their spectral measurements were made with the support of the project 19-19-00596 of Russian Scientific Fund. The authors are grateful to I M Belousova and S I Vavilov State Optical Institute for technical support of the experiments on laser measurements. Part of the analysis of laser measurements was carried out with the support of the project 16.9791.2017/8.9 of Ministry Education and Science of Russia.

References

[1] Ahuja R, Arwin H, Ferreira da Silva A, Persson C, Osorio-Guillén J M, Souza de Almeida J, MoysesAraujo C, Veje N, Veissid E, An C Y, Pepe I, Johansson B 2002 J. Appl. Phys. 92 7219
[2] Lifshitz E, Yassen M, Bykov L, Dag I 1996 J. Lumin. 70 421
[3] Artemyev M V, Rakovich Yu P, Yablonski G P 1997 J. Crystal Growth 171 447
[4] Artemyev M V, Yablonski G P, Rakovich Yu P 1995 ActaPolonica A 87 523
[5] Mallik K, Dhami T S 1998 Phys. Rev. B 58 13055
[6] Kulagina A S, Evstropiev S K, Rosanov N N, Vlasov V V 2018 Semiconductors 52 997
[7] Gulia V, Vedeshwar A S 2007 Phys. Rew. B 75 045409
[8] Mu R, Tung Y S, Ueda A, Henderson D O 1996 J. Phys. Chem. 100 19927

XPS analysis of metallic wetting layer in In/GaAs system obtained at different growth temperatures

M M Eremenko[1], S V Balakirev[1,2], N E Chernenko[1,3], O A Ageev[1], M S Solodovnik[1,2]

[1]Research and Education Center "Nanotechnologies", Southern Federal University, Taganrog 347922, Russia

[2]Department of Nanotechnologies and Microsystems, Southern Federal University, Taganrog 347922, Russia

[3]Department of Radio Engineering Electronics, Southern Federal University, Taganrog 347922, Russia

Abstract. In this paper we investigate the processes of nucleation and growth of In/GaAs(001) nanostructures by droplet epitaxy. We determined the temperature dependence of the wetting layer thickness. Using the X-ray photoelectron spectroscopy technique to examine of samples with In/GaAs droplet nanostructures formed under different conditions we experimentally confirm an increase in the metallic wetting layer thickness with a decrease in the deposition temperature. Analysis of the data obtained shows that droplet nanostructures consist of In are without Ga impurity.

1. Introduction

The formation of nanoscale structures can be carried out by various methods, but one of the most promising is molecular beam epitaxy (MBE), the technology of which allows semiconductor structures, in particular, based on GaAs [1-10], to be obtained for the subsequent manufacture of nanophotonics devices, optoelectronics and nanoelectronics [11-16]. For a long time, the Stranski-Krastanov method was used to form structures with quantum dots (QDs), during which growth components of group III and V were deposited on the surface of the substrate simultaneously. However, this method limits the ability to control the parameters of nanostructures, which is associated with the impossibility of forming QDs in unstressed systems, the interdependence of size, shape and surface density of nanostructures, the technological complexity of controlling the thickness of the wetting layer, and the size and uniformity of QDs.

To further improve the functional parameters of devices based on semiconductor QDs, it is necessary to control the characteristics of nanostructures. Droplet epitaxy, due to the separate deposition of components, significantly expands the possibilities in the field of control and variation of the parameters of the obtained nanostructures. Previous complex studies of samples obtained by the method of droplet epitaxy show that changes in main control parameters can accurately vary the geometric characteristics of the droplets [17-28]. However, the role of the wetting layer formed in the process of droplet epitaxy still needs to be clarified as well as dynamics of its thickness during the growth. Because the wetting layer has a negative influence on the parameters of the nanostructures and devices based on them [29, 30], it is necessary to study the methods of accurate and reproducible control of the wetting layer thickness with the possibility of its complete removal.

2. Experimental procedure

For our experimental study, we grew a series of samples with In/GaAs(001) nanostructures using droplet epitaxy at various temperatures and deposition thicknesses. Growth temperature varied in the range of 100 – 300°C. The effective deposition thickness of In for each temperature corresponded to subcritical (without droplet formation) and supercritical (with droplet formation) values (**Figure 1(a, b)**). Then all obtained samples were investigated by the XPS technique.

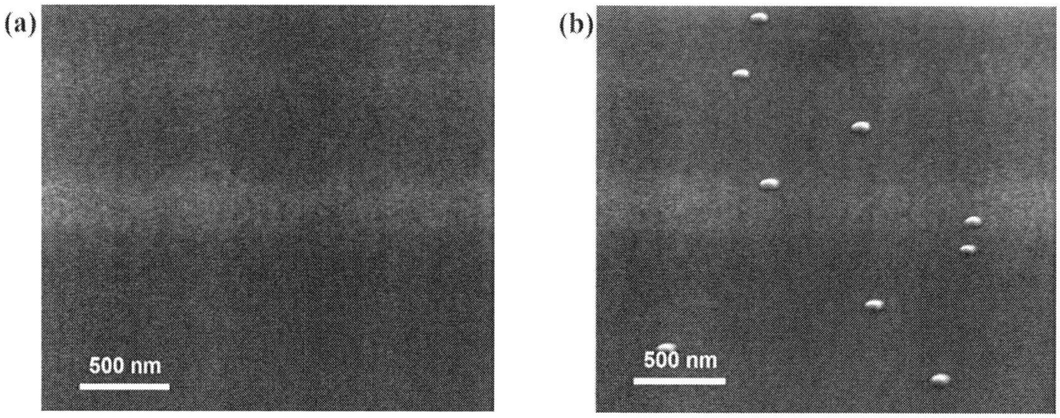

Figure 1(a, b). SEM images of droplet arrays after deposition of indium at various growth conditions: **(a)** $T = 100°C$, $H = 3$ ML, **(b)** $T = 300$°C, $H = 1.5$ ML.

3. Results and discussion

The results of experimental studies of samples with In/GaAs nanostructures by the XPS method, obtained by etching with a beam of Ar ions with an energy of 3 keV, are ambiguous (**Figures 2, 3**). Changes in the X-ray photoelectron spectra and the distribution of elements in depth, despite the high sensitivity of the method in surface studies, are insignificant from sample to sample and even for extreme points. Quantitative characteristics can be given only when comparing extreme points of the range and only for the initial stages of etching (0–60 s), and they correlate with the difference in the density of ensembles of In droplets, or rather, with the total effective area occupied by the droplets on the sample surface: there is a 4-fold increase in signal intensity when moving from a sample obtained at 300°C (**Figure 3**) to a sample obtained at 150°C (**Figure 2**).

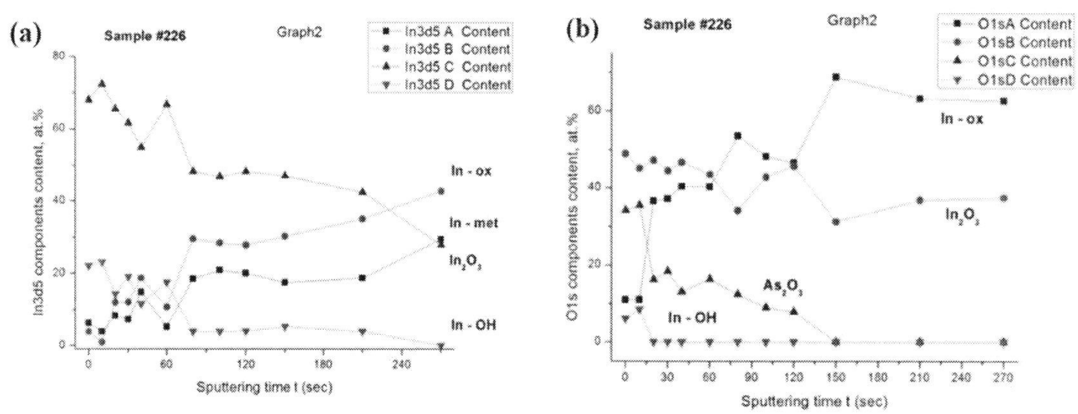

Figure 2 (a, b). The content profiles of the various components of the In spectra obtained on the In/GaAs structure formed at 150 °C based on **(a)** In-line and **(b)** O-line.

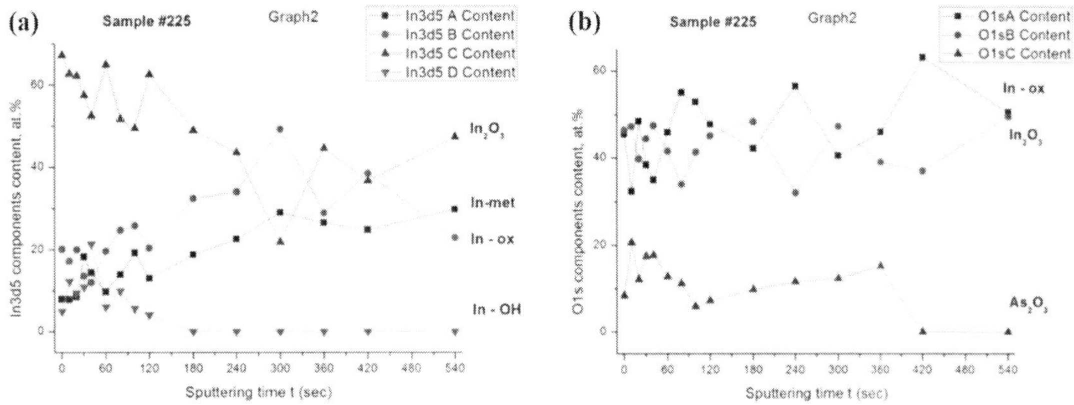

Figure 3 (a, b). The content profiles of the various components of the In spectra obtained on the In/GaAs structure formed at 300 °C based on **(a)** In-line and **(b)** O-line.

The thickness of the wetting layer using the selected method can only be estimated at a qualitative level. First, an analysis of the nature of the surface conductivity of the samples allows us to conclude that with a decrease in the growth temperature, the nature of conductivity changes from n-type impurity conductivity with localization of the Fermi level near the bottom of the conduction band (300 °C) to quasimetallic character (150 °C). This behavior is caused not only by an increase in the density of metal droplets at lower temperatures, but also by a 6-fold increase in the thickness of the wetting layer partially oxidized. The specificity of conductivity of high-temperature samples is due to the formation of an In wetting layer of submonolayer thickness on an As-stabilized surface (that is, in fact, the formation of an InAs monolayer terminated by In atoms) followed by disordering and the formation of a mixed oxide phase $(In(Ga),As)_xO_y$.

Secondly, a comparative analysis of the dynamics of changes in the intensity ratios of the In, In_2O_3, In_xO_y phases taken from the In3d5 (**Figures 2a, 3a**) and O1s (**Figures 2b, 3b**) spectral lines shows that, in low-temperature samples, the intensity of the transitional In_xO_y oxide phase begins to dominate relatively rapidly over the In_2O_3 phase. At the same time, in high-temperature samples, the ratios of these intensities remain unchanged. In addition, on samples formed at low temperature, the metal component begins to predominate over the oxide much earlier (270 versus 510 s). This behavior may be due to the fact that as the etching and removal of the oxide layer from the surface of the ensemble of nanostructures, a metal sublayer is exposed (in the presence of a relatively thick, oxidized wetting layer), which leads to a change in the intensity ratios. At the same time, as the three-dimensional metal droplets are "turned", the metal core and the oxide shell are exposed, which in the case of high-temperature samples (taking into account large droplet sizes) leads to a constant phase relationship for a very long time.

However, reducing the energy of the sputtering beam to 1 keV makes it possible to significantly increase the resolution of the method with respect to solving the problems posed within the framework of the work. **Figure 4** shows the comparative profiles of changes in the intensity of the In3d5A line upon etching samples with In/GaAs nanostructures formed at 100°C, 3 ML (curve 1) and 300°C, 1.5 ML (curve 2).

From the presented data (**Figure 4(a, b)**) it is clear that the intensity of the In3d5A line on curve 1 (low-temperature sample without droplets) drops to zero much earlier than the lines on curve 2 (high-temperature sample with droplets) – by 360 s and 840 s, respectively. This may indicate a different thickness of the deposited material on the samples (a change of 2 times). At the same time, the lines differ from the different samples qualitatively: curve 1 decreases conventionally monotonously, while curve 2 has two distinct parts: a sharp decline (0 – 120 s) followed by a relatively gentle decrease to zero (120 – 840 s). The conventionally monotonous nature of the decrease in the intensity of the

In3d5A spectral line in the first case indirectly indicates the relatively homogeneous structure of the In wetting layer. The complex nature of the intensity curve in the second case is due to the complex structure on the surface. It should be noted that the 3-fold difference in this case also correlates with the previously obtained experimental data.

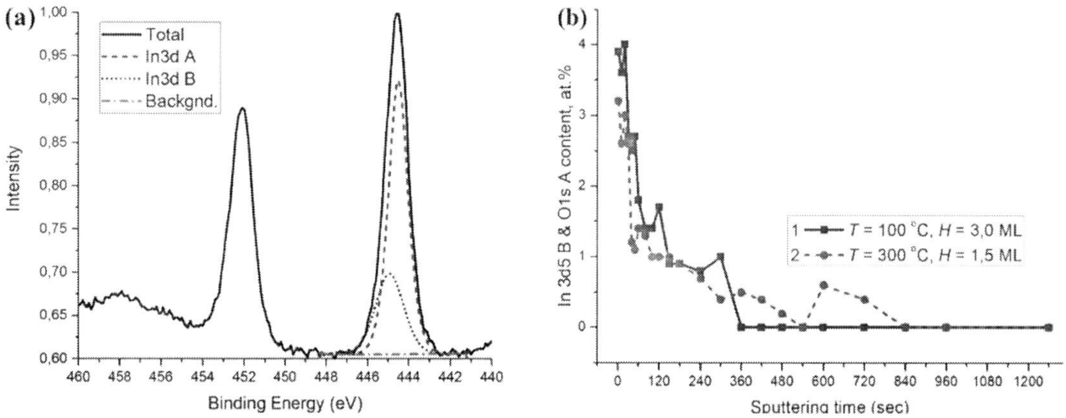

Figure 4(a, b). XPS spectra **(a)** from samples obtained at $T = 100°C$, $H = 3$ ML and $T = 300°C$, $H = 1.5$ML and change of In3d5A line intensity during etching of sample surfaces **(b)**.

4. Conclusions

Thus, the results of X-ray photoelectron spectroscopy of In/GaAs samples formed under different conditions confirm an increase in the wetting layer thickness with a decrease in the formation temperature of the systems. A quantitative analysis of the data obtained shows that droplet nanostructures consist of In, without Ga impurity, which suggests that in the range of temperatures and deposition modes under consideration the etching of the substrate with a metal droplet can be neglected.

Acknowledgments

This work was supported by the Russian Science Foundation Grant No. 15-19-10006. The results were obtained using the equipment of the Research and Education Center and Center for Collective Use "Nanotechnologies" of Southern Federal University.

References

[1] Ageev O A, Solodovnik M S, Balakirev S V and Mikhaylin I A 2016 *J. Phys.: Conf. Ser.* **681** 012036

[2] Ageev O A, Solodovnik M S, Balakirev S V and Eremenko M M 2016 *Phys. Solid State* **58** 1045–1052

[3] Vurgaftman I and Meyer J R 2001 *J. Appl. Phys.* **89** 5815

[4] Balakirev S V, Eremenko M M, Mikhaylin I A, Klimin V S and Solodovnik M S 2018 *J. Phys.: Conf. Ser.* **1124** 022018

[5] Riel B J, Hinzer K, Moisa S, Fraser J, Finnie P, Piercy P, Fafard S and Wasilewski Z R 2002 *J. Cryst. Growth* **236** 145

[6] Balakirev S V, Solodovnik M S, Eremenko M M, Mikhaylin I A and Ageev O A 2017 *J. Phys.: Conf. Ser.* 2017 **917** 032034

[7] Ageev O A, Solodovnik M S, Balakirev S V, Eremenko M M and Mikhaylin I A 2016 *J. Phys.: Conf. Ser.* **741** 012012

[8] Ledentsov N N *et al.* 1996 *Sol. St. Electron.* **40** 785

[9] Ageev O A, Solodovnik M S, Balakirev S V and Mikhaylin I A 2016 *Technical Physics* **61** 971–7

[10] LaBella V P, Bullock D W, Ding Z, Emery C, Harter W G and Thibado P M 2000 *J. Vac. Sci. Tech.* A **18** 1526
[11] Lin Y-M, Dimitrakopoulos C, Jenkins K A, Farmer D B, Chiu H-Y, Grill A and Avouris P 2010 *Science* **327** 662
[12] Duan X, Huang Y, Agarwal R and Lieber C M 2003 *Nature* **421** 241
[13] Semonin O E, Luther J M, Choi S, Chen H, Gao J, Nozik A J and Beard M C 2011 *Science* **334** 1530
[14] Lisitsyn S A , Balakirev S V, Avilov V I, Kolomiytsev A S, Klimin V S, Solodovnik M S, Konoplev B G and Ageev O A 2018 *Nanotech. in Russia* **13** 26-33
[15] Wu J, Shao D, Dorogan V G, Li A Z, Li S, Decuir E A, Manasreh M O, Wang Z M, Mazur Y I and Salamo G J 2010 *Nano Lett.* **10** 1512–6
[16] Solodovnik M S, Karenkikh O G, Balakirev S V, Petrov S I, Ryzhuk R V, Alexeev A N and Ageev O A 2018 *J. Phys.: Conf. Ser.* **1124** 081024
[17] Balakirev S V, Solodovnik M S and Ageev O A 2018 *Phys. Status Solidi B* **255** 1700360
[18] Noda T, Mano T and Sakaki H 2011 *Crystal Growth & Design* **11** 726–728
[19] Noda T and Mano T 2008 *Appl. Surf. Sci.* **254** 7777–7780
[20] Balakirev S V, Eremenko M M, Mikhaylin I A and Solodovnik M S *J. Phys.: Conf. Ser.***1124** 022025
[21] Balakirev S V, Solodovnik M S, Eremenko M M, Mikhaylin I A and Ageev O A 2018 *J. Phys.: Conf. Ser.* **1124** 022001
[22] Lee J H, Wang Z M and Salamo G J 2007 *J. Phys. Condens. Matt.* **19** 176223
[23] Somaschini C, Bietti S, Koguchi N and Sanguinetti S 2009 *Nano Lett.* **9** 3419–3424
[24] Solodovnik M S, Balakirev S V, Eremenko M M, Mikhaylin I A, Avilov V I, Lisitsyn S A and Ageev O A 2017 *J. Phys.: Conf. Ser.* 2017 **917** 032037
[25] Kim J S and Koguchi N 2004 *Appl. Phys. Lett.* **85** 5893–5895
[26] Balakirev S V, Solodovnik M S and Ageev O A 2017 *J. Phys.: Conf. Ser.* 2017 **917** 032033
[27] Somaschini C, Bietti S, Koguchi N and Sanguinetti S 2010 *Appl. Phys. Lett.* **97** 203109
[28] Mano T, Watanabe K, Tsukamoto S, Fujioka H, Oshima M and Koguchi N 1999 *Jpn. J. Appl. Phys.* **38** L1009–11
[29] Bhattacharya A and Bansal B 2015 *Handbook of Crystal Growth: Thin Films and Epitaxy*, ed. T F Kuech (Boston : North-Holland) pp 1057–99
[30] Deppe D and Huffaker D 2000 *Appl. Phys. Lett.* **77** 3325–7

On the stress relaxation mechanism of GaN thin films grown on Si(111) substrates

A A Koryakin[1,2], **A M Mizerov**[1], **A D Bouravleuv**[1]

[1]St. Petersburg Academic University, St. Petersburg 194021, Russia
[2]SPbD JSCC RAS - Branch of SRISA, Saint Petersburg 194021, Russia
E-mail: koryakinaa@spbau.ru

Abstract. Gallium nitride (GaN) thin films grown on Si(111) substrates via plasma-assisted molecular beam epitaxy were investigated. The morphology of GaN films was studied by scanning electron microscopy. It was found that GaN films at nanoscale contain a large amount of cavities. The Raman measurements of GaN/Si(111) system were performed and showed that the position of the GaN E_2 (TO) band is close to that of unstrained GaN. We proposed a stress relaxation mechanism to explain the low stress state of GaN films.

1. Introduction

Gallium nitride (GaN) thin films have been the object of intensive study for decades because of their potential applications in electronics and optoelectronics. At present, high electron mobility transistors, light-emitting diodes, detectors of ultraviolet radiation have been fabricated based of GaN thin films. The epitaxial growth of GaN thin films on silicon substrates is particularly of great interest due to the availability of high-quality silicon substrates of large size. Moreover, the use of nonpolar silicon substrates allows the growth of N- and Ga-polar GaN films. The main problem of GaN film growth on silicon is the thermal mismatch. As a result, the maximal thickness of crack-free GaN film on Si(111) is usually less than 1 μm [1].

Raman spectroscopy has been extensively used for studying stress in GaN thin films [2–4]. The mechanical stress in thin films is an important factor because it can influence the energy band structure and vibrational properties. The sensitivity of the E_2 (TO) phonon frequency to strain allows the finding of stress in thin films. The method used in paper [3] for finding the Raman stress factor of GaN thin film on Si(111) is as follows. Firstly, the micro-Raman measurements in the vicinity of a crack in the GaN film were performed. As a result, the dependence of the E_2 (TO) phonon frequency on the distance from the crack was found. The residual stress was evidenced by the red shift of the E_2 (TO) phonon frequency with respect to the value of ω_0. This value was obtained by measuring of the Raman spectra of bulk GaN. Secondly, the shear-lag model was used to calculate theoretically the biaxial stress field in the vicinity of a crack, namely, the dependence of the σ_{xx} component of stress tensor on the distance from the crack. Thirdly, the Raman stress factor K was found by comparison of these two dependencies. As a result, the following formula was obtained:

$$\Delta\omega = K\,\sigma_{xx}\ cm^{-1}\ GPa^{-1}, \quad (1)$$

where $\Delta\omega = \omega_0 - \omega$ and ω is the E_2 (TO) phonon frequency of GaN thin film on Si(111). The sign of $\Delta\omega$ indicates whether the stress in a film is tensile (the sign is positive) or compressive (the sign is negative).

In this paper, we continue the series of works [4–6] devoted to investigation of the growth mechanism and physical properties of GaN thin films grown on Si(111) substrates via plasma-assisted molecular beam epitaxy (PA-MBE). We study the morphology of GaN films that at nanoscale contain cavities by scanning electron microscopy (SEM). In order to characterize a stress state of these GaN films Raman microscopy is used.

2. Experiment

The GaN thin films were grown on semi-insulating Si(111) substrates (R > 10000 $\Omega\, cm$) by PA-MBE using Veeco Gen 200 setup. The high-frequency plasma source (13.56 MHz) RFN 50/63 (Riber) was used to activate nitrogen. The Si(111) substrates were chemically cleaned by the Shiraki method and annealed in the growth chamber at the temperature of $\sim 850\ °C$ for 30 min to remove the oxide layer and form the Si(111)7x7 surface reconstruction. Then, the Si(111) substrates were nitrided for 30 min at the temperature of $\sim 850\ °C$ with a fixed Ga flux corresponding to the GaN equivalent growth rate of 0.1 $\mu m / h$. The GaN thin films were grown at the temperature of $\sim 700\ °C$, the material fluxes were equal to $F_{Ga} = 0.5\, ML / s$ and $F_N = 0.1\, ML / s$, respectively, for gallium and nitrogen. The reflection high-energy electron diffraction (RHEED) was used to monitor the growth process.

The morphology of GaN films was investigated by SEM (Figure 1). It is seen that the GaN films

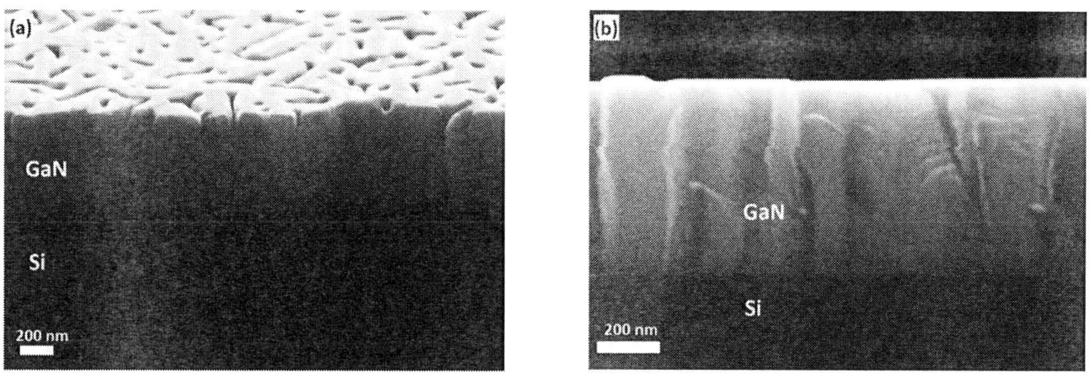

Figure 1(a,b). SEM images of GaN film on the Si(111) substrate: **(a)** the tilted view at the angle of 20° and **(b)** the cross-sectional view.

at nanoscale contain a large amount of cavities with the surface density of $\sim 20\ \mu m^{-2}$. These cavities widen at the distance of $\sim$ 60 nm from the GaN top surface. The depth of the cavities often equals to the thickness of GaN layer as seen on the cleavage surfaces. Such a structure of the GaN films is likely a result of the coalescence of nanocolumns that were observed at the beginning stages of GaN growth on Si(111) [5].

The vibrational properties of GaN were studied by Raman spectroscopy. Raman spectrum was obtained using Witec alpha 300R microscope with an excitation wavelength of 532 nm and a 100x objective lens. A typical Raman spectrum of the sample GaN/Si(111) is shown in Figure 2. The E_2 (TO) phonon frequency of our samples equals to $568.2 \pm 0.2\ cm^{-1}$.

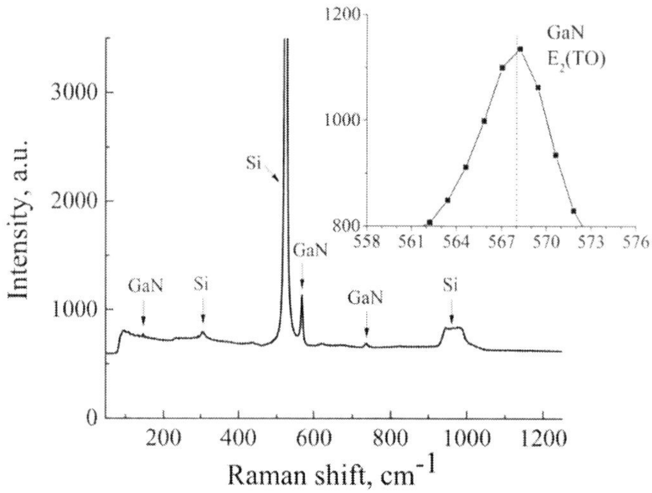

Figure 2. Raman spectrum of the GaN/Si(111) system. The position of the GaN E_2 (TO) band in the sample and its position in unstained GaN (568 cm^{-1}) are shown in the insert.

3. Results and discussion

The The E_2 (TO) phonon frequency obtained by Raman measurements is close to that of unstained GaN (568 cm^{-1}) found in [7]. However, the GaN E_2 (TO) band of the GaN/Si(111) system is usually shifted to longer wavelength [3,4]. The red shift corresponds to the presence of tensile stress in GaN films. In paper [4], for instance, the tensile stress in GaN film grown on Si(111) equals to ~0.15 *GPa*. This value was obtained using equation (1) with K = 4.2 $cm^{-1}GPa^{-1}$, $\omega_0 = 568\,cm^{-1}$ and $\omega \approx 567.37\,cm^{-1}$.

We suppose that the presence of the low stress state in our samples can be explained, at least partly, by the specific morphology of GaN film. The large amount of cavities allows the efficient stress relaxation on free surfaces. To validate this assumption, we find the ratio of the elastic energy stored in continuous GaN film and GaN film with cavities provided that the effective lattice mismatch between GaN and Si(111) is the same in both cases. For a simplified modeling, the model geometry of the GaN film shown in the insert in Figure 3 is considered instead of the real geometry with randomly distributed cavities. The size l corresponds to the average cavity size calculated by using SEM images (the area of 13 μm^2 was analyzed). The size s is calculated to satisfy the condition $1/s^2 \approx 20\ \mu m^{-2}$. We use the following values of parameters in our computation: l = 140 *nm*, L = 600 *nm*, s = 224 nm, δ = 0.5–10 nm.

We calculate the dependence of the elastic energy per GaN pair on the effective lattice mismatch between GaN film and Si substrate that includes the thermal mismatch and the difference in crystal lattice constants. The elastic constants of wurtzite GaN were taken from the paper [8]. The result of computation by the finite element method (FEM) is shown in Figure 3. This graph is plotted by using the value δ = 1 nm. The variation of δ leads to a negligible variation of the elastic energy. We find that the elastic energy of GaN film is about two times smaller than that of continuous GaN film. Consequently, the σ_{xx} component of stress tensor and the shift of the position of GaN E_2 (TO) band $\Delta\omega$ (equation(1)) is approximately two times smaller in the first case. If we consider the GaN film grown in [4] as an approximation for the continuous GaN film we find that the GaN E_2 (TO) band

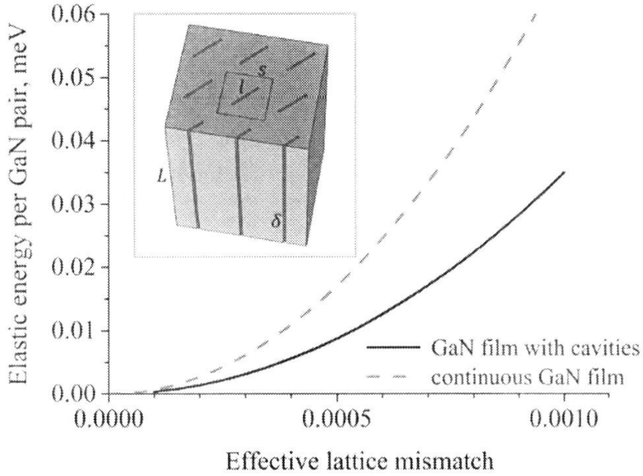

Figure 3. The elastic energy dependence on the effective lattice mismatch. The insert shows the model geometry of the GaN film used in the FEM computation.

shifts on ~0.3 cm^{-1}. Therefore, we can conclude that the low stress state of the GaN layer might be associated with the presence of cavities with a high surface densities. Although, the presence of other defects (misfit dislocations etc.) can also cause a decrease of stress in the film.

To summarize, the morphology and vibrational properties of GaN thin films grown on Si(111) substrates via PA-MBE were investigated. The position of the GaN E_2 (TO) band was determined (568.2 ±0.2 cm^{-1}). If we take into account the value of 568 cm^{-1} for the E_2 (TO) band in unstrained GaN the obtained value of the E_2 (TO) phonon frequency indicates that the stress in GaN film is relatively small. Finally, it was shown that the presence of a large amount of cavities in the GaN thin film observed by SEM caused effective stress relaxation. However, further studies are required to fully explain the stress relaxation mechanism in the GaN/Si(111) system.

Acknowledgments
The samples were fabricated with support from the Ministry of Education and Science of the Russian Federation (the grant № 16.9789.2017/BCh). Raman measurements were performed using the equipment of the Unique scientific facility (USF) «Physics, chemistry, and mechanics of crystals and thin films» of the Institute of Problems of Mechanical Engineering, Russian Academy of Sciences (St. Petersburg). The numerical calculations were performed in the framework of the research project №0065-2018-0016 (state registration number AAAA-A18-118041890053-2).

References
[1] Dadgar A, Bl.asing J, Diez A, Alam A, Heuken M, Krost A 2000 *Jpn. J. Appl. Phys.* **39** L1183
[2] Kisielowski C, Kruger J, Ruvimov S, Suski T, Ager III J W, Jones E, Liliental-Weber Z, Rubin M, Weber E R 1996 *Phys. Rev. B* **54** 17745
[3] Tripathy S, Chua S, Chen P, Miao Z 2002 *J. Appl. Phys.* **92** 3503
[4] Kukushkin S A, Mizerov A M, Osipov A V, Redkov A V, Timoshnev S N 2018 *Thin Solid Films* **646** 158
[5] Mizerov A M, Timoshnev S N, Sobolev M S, Nikitina E V , Shubina K Yu , Berezovskaia T N., Shtrom I V, Bouravleuv A D 2018 *Semiconductors* **52** 1529
[6] Timoshnev S N, Mizerov A M, Sobolev M S, Nikitina E V 2018 *Semiconductors* **52** 660

[7] Wang D, Jia S, Chen K J, Lau K M, Dikme Y, van Gemmern P, Lin Y C, Kalisch H, Jansen R H, Heuken M 2005 *J. Appl. Phys.* **97** 056103
[8] Polian A, Grimsditch M, Grzegory I 1996 *J. Appl. Phys.* **79** 3343

Combined scanning probe nanolithography and liquid etching techniques for profiled nanostructures formation

V V Polyakova, I N Kots, V I Avilov, N V Parshina, V A Smirnov

Institute of Nanotechnologies, Electronics and Equipment Engineering, Southern Federal University, Taganrog, 347928, Russia

Abstract. The results of studies of the effect of technological modes of nanoscale silicon substrate surface profiling by local anodic oxidation (LAO). The effect of relative humidity and the amplitude of the voltage applied to the tip-substrate system with LAO on the geometric parameters of the oxide nanostructures (ONS) and profiled nanostructures (PNS). It has been shown that increasing the voltage amplitude to the probe substrate 10 to the system 20 causes an increase in the height of the ONS 0.6 ± 0.2 to 2.0 ± 0.3 nm at 70% relative humidity. Investigations of the possibility of using ONS as a mask during the silicon etching liquid. It has been shown that increasing the silicon etching time from 10 to 80 seconds in the KOH + IPA causes increased silicon PNS height of 100 ± 10 nm and 350 ± 30 nm. The results can be used in the development of technological processes of manufacturing of nanoelectronics components based on silicon base using the probe nanotechnologies.

1.Introduction

The development of the element base of nanoelectronics imposes increased requirements on the resolution and accuracy of technological operations. In this case, the improvement of the structures of the elements of nanoelectronics leads to a decrease in their size[1]. One of the promising elements of nanoelectronics is memristor structures that can be used to create a new generation of computers based on neuromorphic systems[1-5]. One of the main problems in the manufacture of memristor structures is the need to improve the methods of lithography, which should ensure the accuracy and reproducibility of the manufacture of nanoscale structures[6-8].

Among the lithography methods, one of the most promising is local anodic oxidation (LAO) using an atomic force microscope (AFM), which provides high spatial resolution and reproducibility, as well as the ability to profile the substrate surface without additional operations, photolithography[9-15]. The LAO method using AFM makes it possible to form oxide nanostructures (ONS) on the surface of various materials that can be used in the development and creation of micro and nanoelectronics elements, resistive memory elements based on memristor structures, lithographic masks, and catalytic centers for growing threadlike nanostructures. In addition, the combination of the LAO method and liquid etching allows profiling the surface of the substrates to form nanostructures, which are used to create element of neuromorphic systems[15-24].

However, despite the fairly large number of scientific publications on this issue[9-24], the patterns of influence of the technological modes of local anodic oxidation and liquid etching on the geometric parameters of profiled nanostructures on the silicon surface remain insufficiently studied. The purpose of this work is to study the parameters affecting the process of nanoscale profiling of the silicon surface by methods of local anodic oxidation and liquid etching.

2. Experiment details

For experimental studies of the profiling of the silicon surface, two techniques have been developed, schematically presented on figure 1. Both methods are based on the preliminary formation of oxide nanoscale structures on the silicon substrate surface by the method of local anodic oxidation (figure 1, b). According to the first technique, the formation of profiled nanoscale structures on the silicon surface is performed by etching ONS in a solution of hydrofluoric acid ONS (figure 1 c.). For the formation of PNS silicon according to the second technique, ONS are used as a mask for liquid etching of the silicon surface in a KOH solution (figure 1 d).

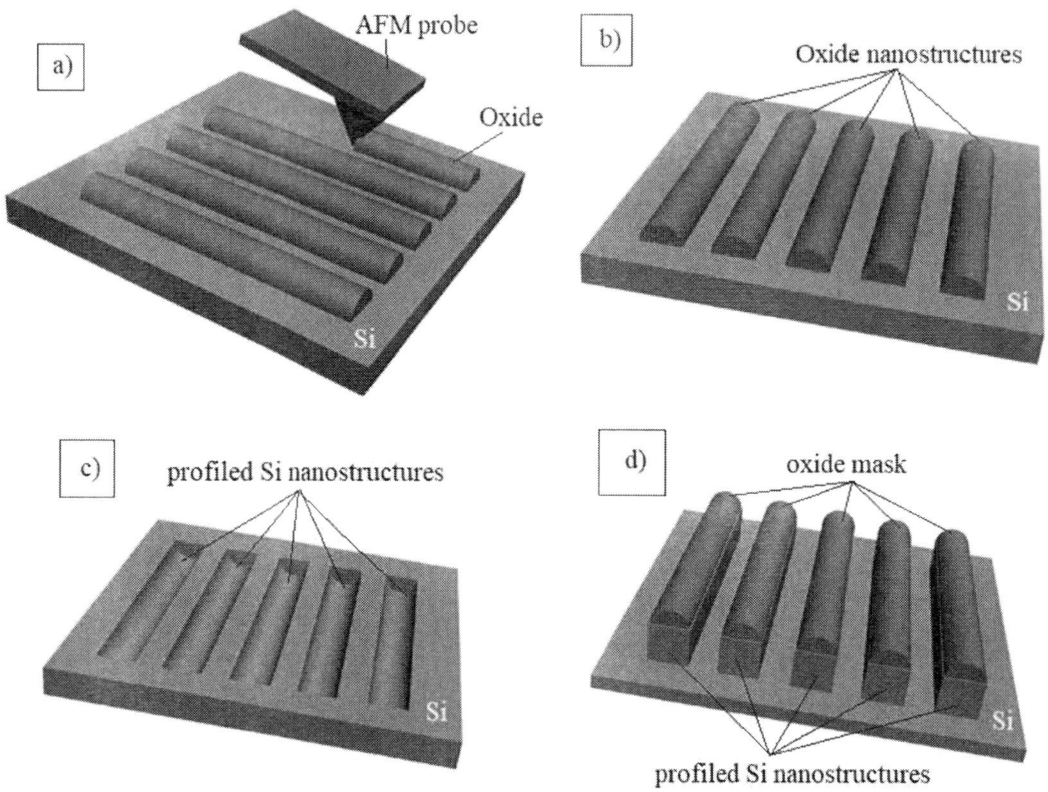

Figure 1. Schematic representation of the developed techniques for nanoscale profiling of the silicon surface: **(a)** LAO; **(b)** oxide nanostructures formed by LAO; **(c)** 1st technique; **(d)** 2nd technique

When conducting research, silicon substrates were used. Then, using the probe nanolaboratory (PNL) Ntegra (NT-MDT, Russia), a nanolithography was performed using the LAO method of silicon surface in the contact mode AFM with NSG 10 cantilevers with a conductive Pt coating, with the following parameters: voltage pulse amplitude changed from 5 up to 20 V, the voltage pulse duration was 100 ms, the oscillation frequency of the probe was 0.03 Hz, the current of the feedback circuit (in the PNL control program, the Set Point parameter) was 0.3 nA. The relative air humidity at LAO was monitored using an Oregon Scientific ETHG913R moisture meter and varied from 30 ± 1 to 90 ± 1%. As a result, matrices of 49 ONS (figure 2 a, c), as well as linear structures (figure 3 a, b), were formed on the surface of the silicon substrate.

To study the modes of profiling of silicon according to the first technique (figure 1, c), obtained by the LAO ONS method were removed by liquid etching in an aqueous HF solution (1:3) at room temperature. As a result, matrices of 49 profiled nanostructures, as well as linear structures with an AFM-image of which are shown on figures 2 and 3 c, d, respectively, were formed on the silicon surface.

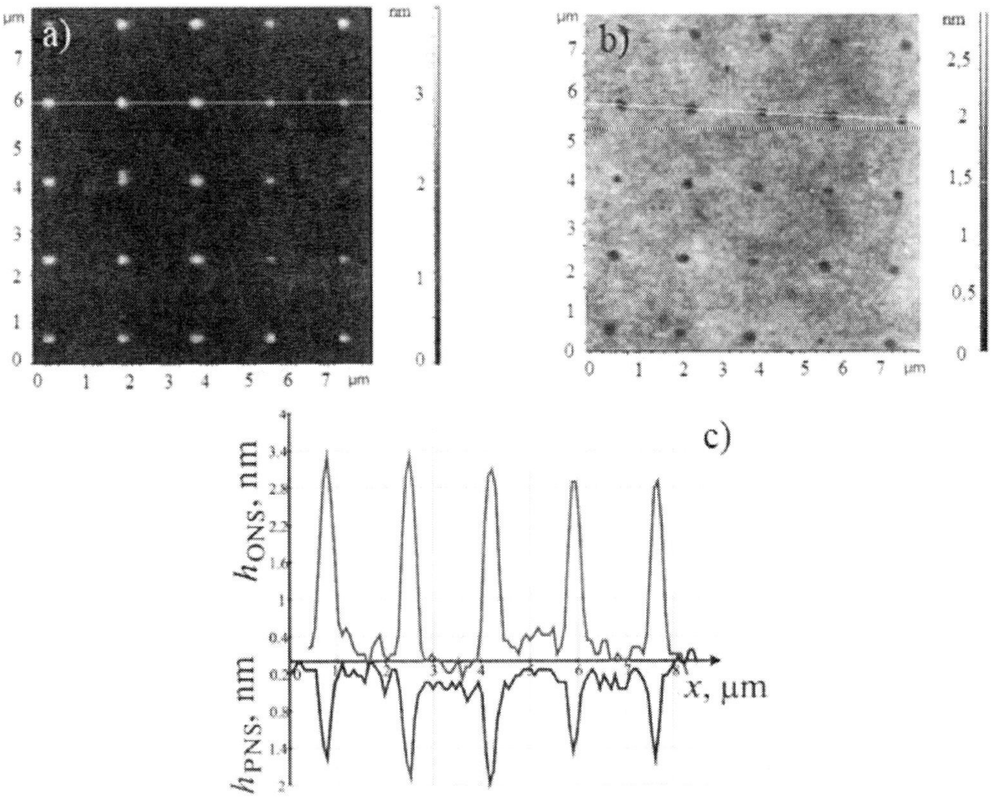

Figure 2. AFM-images of the silicon surface: **(a)** after the formation of ONS by the LAO; **(b)** after the formation of PNS; **(c)** profilograms along the lines on a and b

To study the silicon profiling regimes according to the second technique (figure 1 d), obtained by the LAO ONS silicon method were used as a mask for liquid etching of silicon in KOH solution (60%) + IPA (5: 1) at a temperature of 70 °C. As a result, profiled nanoscale structures were formed on the silicon surface, an AFM image and a profilogram of one of which are shown on figure 3 e, f.

3. Results and discussion

Statistical processing of the obtained AFM images was performed using the Image Analysis 3.5 software package using the developed technique. According to the data obtained, dependences on the amplitude of voltage pulses of the geometric parameters of ONS (height and diameter) and PNS (depth and diameter) of silicon obtained at different relative humidity with LAO, shown on figure 4.
Figure 5 shows the dependences of the height and diameter of oxide nanoscale structures obtained at a voltage of 15V on the air humidity. The analysis shows that an increase in the level of relative humidity leads to an increase in the geometric dimensions of the silicon ONS, which is explained by an increase in the concentration of the oxidizing agent.

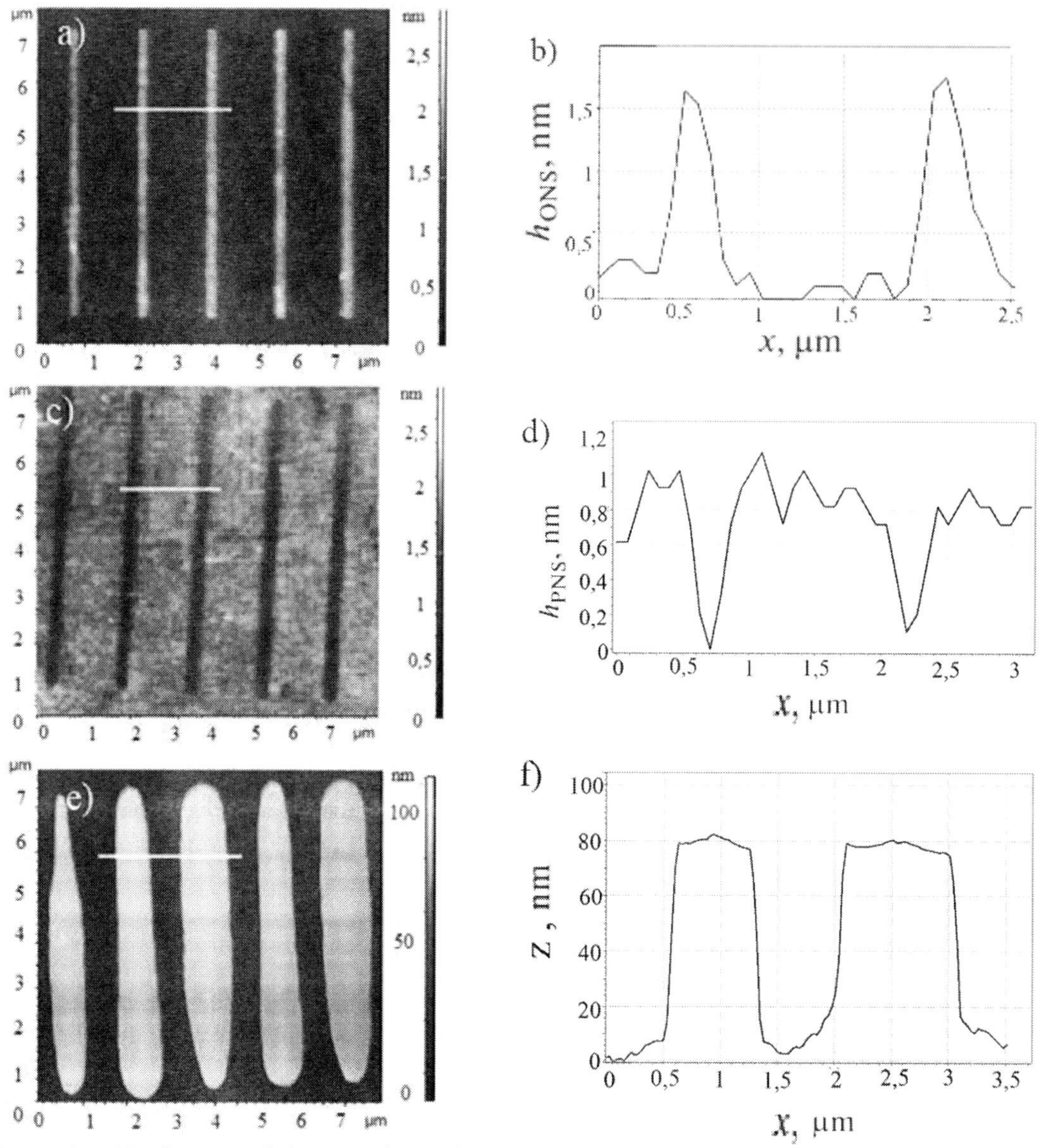

Figure 3. AFM-images of the experimental results of nanoscale profiling of the silicon surface according to the developed techniques: **(a)** the initial oxide mask; **(b)** profilogram along the line on (a); **(c)** PNS formed by the first technique; **(d)** profilogram along the line on (c); **(e)** PNS formed by the second technique; **(f)** profilogram along the line on (d)

To reduce the diameter of the obtained nanostructures using KOH + IPA solution, it is necessary to decrease the humidity value, increase the SetPoint value, increase the distance between the probe and the sample.

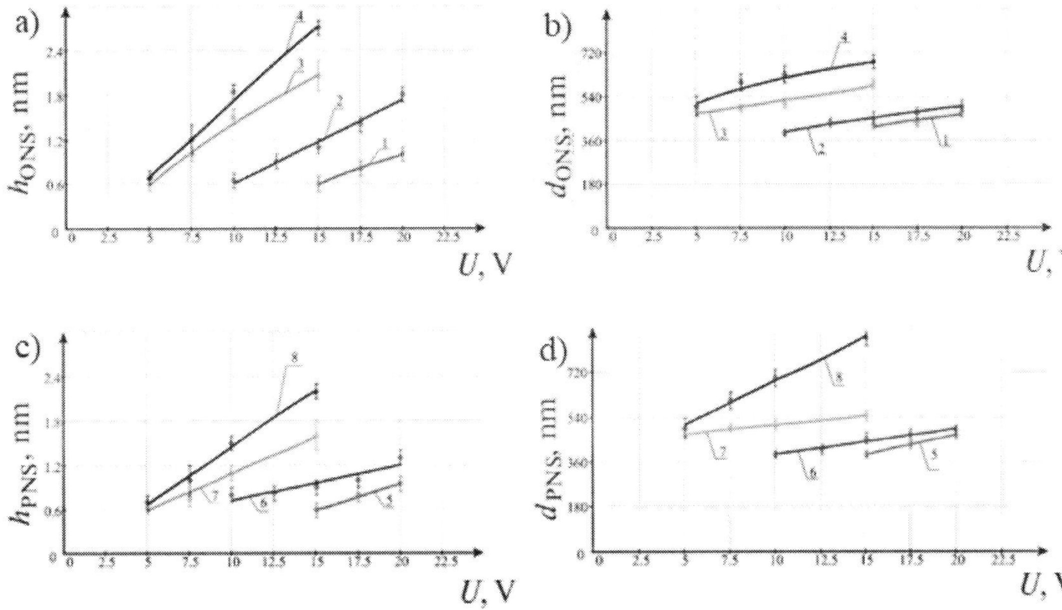

Figure 4. The dependences of the geometric parameters of formed nanostructures on the LAO voltage amplitude at different relative humidity (line 1 and 5 - 30 ± 1%; line 2 and 6 - 50 ± 1%; line 3 and 7 - 70 ± 1%; line 4 and 8 - 90 ± 1%): **(a)** heights of ONS (1-4); **(b)** diameters of ONS (1-4); **(c)** depths of PNS (5-8); (d) diameters of PNS (5-8)

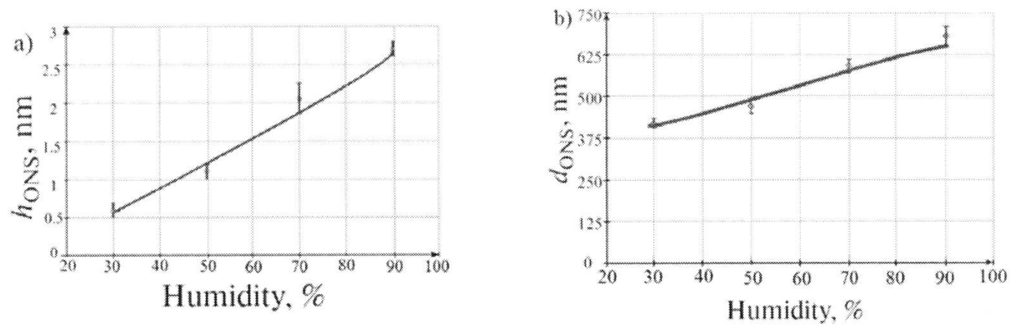

Figure 5. The dependencies of the geometric parameters of silicon oxide nanostructures with the application of a voltage of 15V at different relative humidity: **(a)** the height; **(b)** diameter

4. Conclusion

As a result of the research, a nanosized silicon surface profiling techniques based on the LAO method was developed, the dependences of the geometrical parameters of oxide and profiled nanostructures on the silicon surface on relative humidity and voltage pulses applied to the probe-substrate system during local anodic oxidation in the AFM contact mode were obtained. It is established that an increase in the impulses of the applied voltage with LAO leads to an increase in the height and diameter of the ONS, as well as an increase in the depth and diameter of the PNS on the silicon surface. Thus, it has been shown that local anodic oxidation is a promising method of nanolithography, the use of which allows surface profiling with nanometer resolution. The results can be used in the development and study of elements of nanoelectronics and nanosystem technology using probe nanolithography methods, which will allow the formation of memristor structures that can be used to create a crossbar architecture to create neuromorphic systems.

Acknowledgments
This work was supported by Grant of the President of the Russian Federation No.MK-2721.2018.8, by RFBR according to the research project № 18-37-00299 mol_a. The results were obtained using the equipment of the Research and Education Center and Center for Collective Use "Nanotechnologies" of Southern Federal University.

References
[1] Mohammed A Zidan, John Paul Strachan, Wei D Lu 2018 *Nature Electronics* **1** 22
[2] Haider Abbas, Yawar Abbas, Son Ngoc Truong, Kyeong-Sik Min, Mi Ra Park, Jongweon Cho, Tae-Sik Yoon and Chi Jung Kang 2017 *Semicond. Sci. Technol.* **32** 065014
[3] Avilov V I, Polupanov N V, Tominov R V, Smirnov V A and Ageev O A 2017 *IOP Conf. Ser.:* Mat. Sci.&Eng. **256** 012001
[4] Khakhulin D A, Vakulov Z E, Smirnov V A, Tominov R V, Yoon J G, Ageev O A 2017 *Journal of Physics: Conference Series* **917** 092008
[5] Ting-Chang Chang, Kuan-Chang Chang, Tsung-Ming Tsai, Tian-Jian Chu and Simon M. Sze 2016 *Materials Today* **19(5)** 254
[6] Avilov V I, Ageev O A, Kolomiitsev A S, Konoplev B G, Smirnov V A, Tsukanova O G 2013 *Semiconductors* **106** 50
[7] Tominov R V, Zamburg E G, Khakhulin D A, Klimin V S, Smirnov V A, Chu Y H, Ageev O A 2017 *Journal of Physics: Conference Series* **917** 032023
[8] Smirnov V A 2015 *Piezoelectrics and Nanomaterials: Fundamentals, Developments and Applications* 85-103
[9] Polyakova V.V., Smirnov V.A., Ageev O.A. 2018 *Nanotechnologies in Russia* **13(1-2)** 84-89
[10] Ageev O A, Alyab'eva N I, Konoplev B G, Polyakov V V, Smirnov V A 2010 *Semiconductors* **44** 1703
[11] Tominov R V, Bespoludin V V, Klimin V S, Smirnov V A, Ageev O A 2017 *IOP Conf. Ser.: Mater. Sci. Eng.* **256** 012023
[12] Avilov V I, Ageev O A, Smirnov V A, Solodovnik M S, Tsukanova O G 2015 *Nanotechnologies in Russia* **10** 214
[13] Ageev A O, Konoplev B G, Polyakov V V , Svetlichnyi A M, Smirnov V A 2007 *Russian Microelectronics* **36** 353
[14] Smirnov V A, Tominov R V, Chernenko N E, Ageev O A 2017 *Nanotechnologies in Russia* **12** 11
[15] Avilov V I, Ageev O A, Jityaev I L, Kolomiytsev A S, Smirnov V A 2016 *Proceedings of SPIE* **10224** 102240T
[16] Ageev O A, Konoplev B G, Rubashkina M V, Rukomoikin A V, Smirnov V A, Solodovnik M S 2013 *Nanotechnologies in Russia* **8(1-2)** 23
[17] Smirnov V A, Tominov R V, Alyab'eva N I, Il'ina M V, Polyakova V V, Bykov Al V, Ageev O A 2018 *Tech. Phys.* **63** 1236-1241
[18] Klimin V S, Rezvan A A, Ageev O A 2019 *Proc. of SPIE* **11022**1102220-1
[19] Kots I N, Klimin V S, Polyakova V V, Rezvan A A, Vakulov Z E, Ageev O A 2018 *J. Phys.: Conf. Ser.* **1124** 081035
[20] Avilov V I, Ageev O A, Blinov Y F, Konoplev B G, Polyakov V V, Smirnov V A, Tsukanova O G 2015 *Technical Physics* **60** 5
[21] Avilov V I, Ageev O A, Konoplev B G, Smirnov V A, Solodovnik M S, Tsukanova O G 2016 *Semiconductors* **50** 5
[22] Polyakova V V, Kots I N, Smirnov V A, Ageev O A 2019 *Russian Microelectronics* **48(2)** 66
[23] Ageev O A, Smirnov V A, Solodovnik M S, Rukomoikin A V, Avilov V I 2012 *Semiconductors* **46** 1616
[24] Karen'Kih, O.G., Avilov, V.I., Smirnov, V.A., Fedotov, A.A., Sharapov, N.A., Polupanov, N.A. 2018 *IOP Conf. Ser.: Mater. Sci. Eng.* **443(1)** 012013

[25] Tominov R V, Smirnov V A, Avilov V I, Fedotov A A, Klimin V S, Chernenko N E 2018 *IOP Conf. Ser.: Mater. Sci. Eng.* **443(1)** 012036

Investigation of indium antimonide nanoparticles, obtained by the method of liquid chemical etching

A I Mikhailov, V F Kabanov, M V Gavrikov
Department of Nano and Biomedical Technologies, Saratov State University, Saratov 410012, Russia

Abstract. The possibility of obtaining indium antimonide nanoparticles by method of liquid chemical etching was analyzed. The dependence of obtained nanoparticles size on the etching time and composition of the etcher has been investigated. The nanoparticles were etched in solutions prepared on the basis of the peroxide-ammonia mixture; cetyltrimethylammonium bromide, a cationic surfactant, was used to stabilize the surface of the nanoparticles. The obtained quantum dots of indium antimonide were studied by the methods of differential normalized tunnel current-voltage characteristics, electron microscopy, particle size analysis and spectral dependence of the absorption coefficient.

1. Introduction

Nowadays practical application of quantum dots (QD) includes nanoelectronics, optoelectronics (displays, photovoltaics, light sources), visualization of biological objects, sensors, and etc. [1-6].
Indium antimonide (*InSb*) QD causes particular interest due to the unique properties of indium antimonide: ultra-high electron mobility (up to 78000 $cm^2V^{-1}s^{-1}$), direct and narrow (0.18 eV) forbidden band, small effective electron mass ($0.013m_0$, where m_0 - mass of a free electron) [7] and large de Broglie wavelength of electrons – up to 55 nm.
The various obtaining technologies allows us to solve problems of physical modeling of electronic processes in QD, investigate the influence of composition and structure on their properties, their interaction during segregation. One of the ways of obtaining nanoparticles is using the approach, which is called «top-down» synthesis. In this case, monocrystalline plates of these materials are subjected to grinding, chemical (often electrochemical) etching and sedimentation, etc. In a result are nanoparticles - the raw material for quantum dots [8].
In this paper, *InSb* nanoparticles obtained by liquid chemical etching of *InSb* powders were investigated.

2. Model representations and evaluations

It is known that quantum-size effects in nanoparticles can be observed under the necessary conditions:
1. characteristic size of nanoparticle should be about the de Broglie wavelength (quantization of the energy spectrum of QD);
2. interval between discrete levels $\varepsilon_{i+1}-\varepsilon_i$ must be at least 3÷4 of kT value (for example, about 4 kT, which corresponds to 0.1 eV at room temperature).
Realisation of the necessary conditions for observing quantum-size effects is possible with characteristic sizes of *InSb* nanoparticles less than 30 nm (de Broglie wavelength is about 55 nm, interval between discrete levels $\varepsilon_2-\varepsilon_1$ is bigger than 0.1 eV). In this case, in the range of variable sizes of QD from 10 to 30 nm, the energy gap $\varepsilon_{c1}-\varepsilon_{v1}$ will vary from 0.3 to 1.0 eV considering the width of

the forbidden band of bulk material (and the position of maximum absorption coefficient will vary from 1.2 to 4.1 μm). This allows us to reasonably assume the possibility of a significant effect of the characteristic size of the *InSb* QD on their optical and electrical properties in a wide range.

3. Samples obtaining technology

The initial *InSb* powders were obtained by grinding monocrystalline plates in a planetary ball mill PULVERISETTE (Fritsch, Germany). After that, the obtained powders were etched in a mixture of etchants similarly to [9].

To work out the technique of etching *InSb* powders, the most known in the electronics industry oxidative peroxide-ammonia mixture (pH = 11), consisting of deionized water, an aqueous solution of ammonia and hydrogen peroxide. Etching solutions were prepared with the volume ratios: H_2O, NH_4OH, H_2O_2 – 30: 6: 9. To improve the quality of the surface, a surfactant, that forming micelles in a water (cetyltrimethylammonium bromide), was used for etching, which also plays the role of a stabilizer, which prevents the aggregation of the resulting nanoparticles.

During the etching of *InSb*, an exothermic reaction was observed in the first few seconds when the temperature of the reaction mixture rose to t = 50°C. Then for 10 min the reaction mass cooled to room temperature. For a detailed study of the etching process, sampling for further analysis was carried out after 1, 5, 15, 30, 150 minutes and 24 hours after the start of the reaction.

4. Research methods and results

The particle size analysis was performed by using a laser particle size analyzer Zetasizer Nano ZS (the results are shown in Fig. 1 (a)), which showed typical nanoparticle sizes of about 20 nm, which is less than the de Broglie wavelength for an electron in *InSb*.

An analysis of the energy absorption spectrum for the obtained nanoparticles was performed using a Shimadzu Irrafinity-1 IR spectrometer (the results of the dependence of the absorption coefficient α on the wavelength are shown in Fig. 1 (b); the absorption peak corresponds to the energy value (0.24 eV) of the first level of quantum dot, measured from the bottom of the conduction band of the bulk material). Estimation of the size of nanoparticles using the QD model of a cubic form [10] showed values of the characteristic size (cube edge) of 18–20 nm.

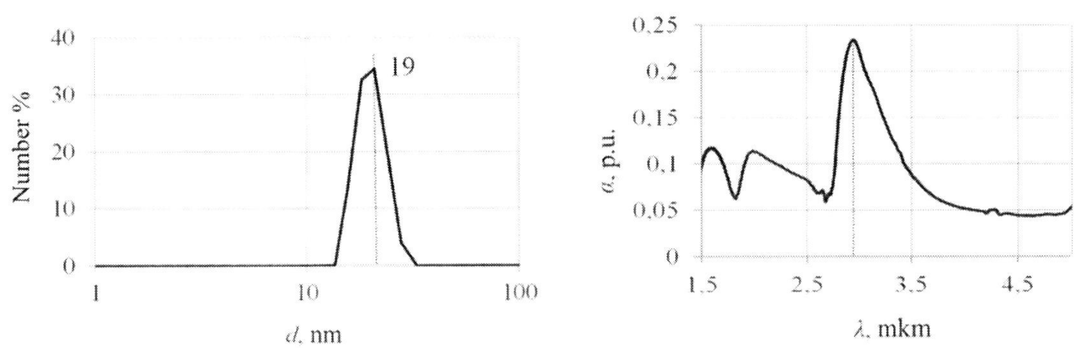

Figure 1(a, b). (a) Size estimation of obtained nanoparticles; **(b)** Dependence of absorption coefficient on the wavelength for obtained nanoparticles.

For more complete analysis of electrophysical properties of the obtained film samples with QD, in particular, the electronic spectrum, the scanning tunneling microscopy (STM) technique was used. The studies were carried out using an SPM SOLVER Nano scanning probe microscope. Before the investigation of the tunneling CVC of an individual nanoparticle, scanning of the film surface by STM

methods in the stabilized current regime was carried out. After analysis of the obtained STM image of the surface at least 10 points were selected to obtain the CVC. In automatic mode, at least 10 CVC per point were measured. CVC measurements were made within the current values from 10^{-11} to 10^{-9} A, voltage - from 0 to 1 V. By reproducibility of the measurement results, points with stable characteristics were selected, after which averaging of the CVC were performed.
For the analysis of experimental tunneling current-voltage characteristics, we used the method of normalized differential current-voltage characteristics $(dI/dV)/(I/V)$ as a function of voltage V. In addition, as shown in [6, 12], this method can be used for analysis of conductivity mechanisms of obtained structures, calculations of their parameters and other important electronic processes.
In this research, the normalized differential tunneling current-voltage characteristics with a negative bias potential on the substrate relative to the probe were examined and analyzed. In this case, tunneling of electrons occurs from the electrode through discrete levels of a quantum-size object into a probe microscope [10,11]. The discrete energy spectrum of conduction electrons of a quantum-size object determines the peaks on the normalized differential CVC (Fig. 2 (a)), they are indicated by arrows). The experimentally obtained values of the applied voltage on the peaks were set in accordance with the calculated values of the electron energy levels of the QD (Fig. 2 (b)). That way we determined the range of characteristic sizes of QD, which were compared with the available results. The analysis of experimental data on the position of the peaks on the differential normalized tunnel current-voltage characteristics also made it possible to estimate the linear size of a quantum object in the range of 18–21 nm.

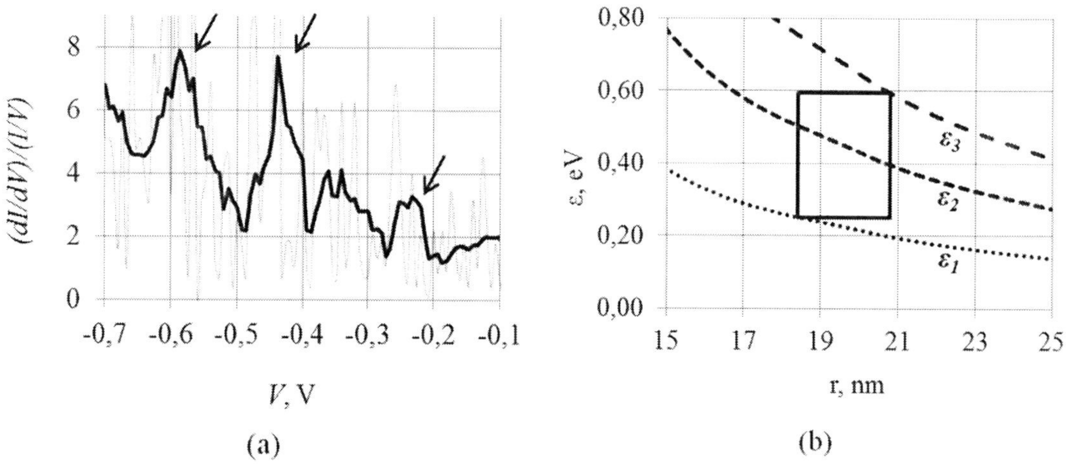

Figure 2(a, b). (a) Typical differential tunnel CVC of *InSb* QD; **(b)** Compliance with calculated energy spectrum levels.

To clarify the validity of these estimates, direct measurements were made by scanning electron microscopy (SEM) using a MIRA 2 LMU autoemission scanning electron microscope. The results presented in Fig. 3 show good agreement with the estimated QD sizes.

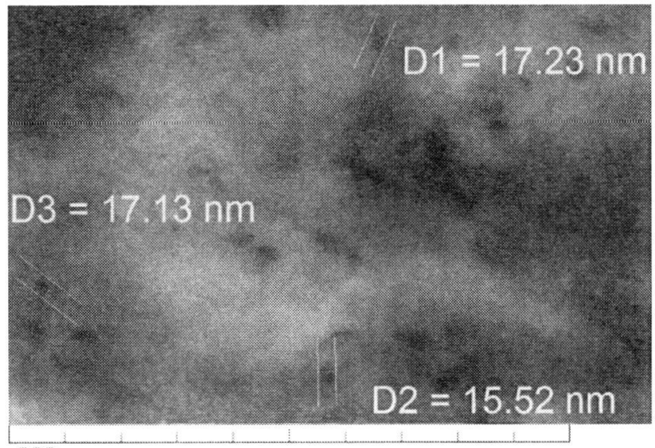

Figure 3. Typical SEM images of *InSb* QD.

5. Conclusions

Thus, it can be concluded that the method of colloid synthesis of *InSb* quantum dots allows to obtain nanoparticles in a wide range of sizes up to few nanometers with the required reproducibility and accuracy, suitable for the production of semiconductor quantum dots. Also, the phase and structural composition of the obtained nanoparticles almost does not differ from the initial bulk monocrystalline material.

Size estimates of obtained samples using different approaches (particle size analysis, analysis of the spectral dependence of the absorption coefficient, analysis by the method of differential normalized tunneling current-voltage characteristics and direct measurements using SEM) demonstrate qualitatively and quantitatively consistent results with an error less than 15%.

Acknowledgments

This work was supported by grants from the Russian Foundation for Basic Research № 19-07-00087 and № 19-07-00086

References

[1] Reiss P, Carriere M, Lincheneau C, Vaure L, Tamang S 2016 Chem. Rev. **116** 10731
[2] Yoffe A D 2010 Advances in Physics **50:1** 1
[3] Algar W R, Tavares A J, Krull U J 2010 Analytica Chimica Acta **673** 1
[4] Brichkin S B, Razumov V F 2016 Russ. Chem. Rev **85(12)** 1297
[5] Li N, Zhang X, Chen S, Hou X 2011 Journal of Physics and Chemistry of Solids **72(11)** 1195
[6] Mikhailov A I, Kabanov V F, Gorbachev I A, Glukhovsky E G 2018 Semiconductors **52** 750
[7] Abautret J, Perez J P, Evirgen A, Rothman J, Cordat A, Christol P 2015 J. Appl. Phys. **117** 244502
[8] Vasiliev R B, Dirin D N 2007 *Quantum dots: synthesis, properties, application* (Moscow: Moscow State University Press) p 34
[9] Tomashik Z F, Shelyuk I A, Tomashik V N, Okrepka G M, Moravec P, Stratiichuk I B 2012 Inorganic Materials **48(9)** 985-989
[10] Mikhailov A I, Kabanov V F, Zhukov N D, Glukhovskoy E G 2017 Nanosystems: Physics, Chemistry, Mathematics **8(5)** 596-599
[11] Mikhailov A I, Kabanov V F, Glukhovskoy E G, Shishkin M I, Gavrikov M V 2018 Nanosystems: Physics, Chemistry, Mathematics **9** 464–467
[12] Wang T 2015 ACS Nano **9(1)** 725-732

The model for in-plane and out-of-plane growth regimes of semiconductor nanowires

Y Berdnikov[1], **N V Sibirev**[1], **R R Reznik**[1], **A V Redkov**[1]
[1]ITMO university, St. Petersburg, 197101, Russia

Abstract. In this work we present the model capable of prediction whether the vapor-liquid-solid growth would preferably result in the formation of in-plane (horizontal) or out-of-plane (vertical or inclined) nanowires. Within the model, we analyze the particular case of gold-catalyzed germanium nanowire growth on Ge(111), Ge(110) and Si(100) substrates. We focus on two aspects of the growth process: detachment of the catalyst from the substrate and stabilization of horizontal growth by nucleation at the nanowire-substrate-liquid line.

1. Introduction

III-V semiconductor nanowires (NWs) and heterojunctions on their base are widely considered as the promising building blocks for novel photonic and optoelectronic applications [1–4]. However, their use in the device applications usually requires control of the direction of NW growth, which strongly depends on the growth mechanism. Nowadays, one of the most common approaches to the synthesis of III-V and group IV NWs is the epitaxial growth via the vapor-liquid-solid (VLS) mechanism[5,6]. VLS growth may result in different shapes and forms of NWs. Mainly, two types of possible NW growth: in-plane (horizontal) and out-of-plane (vertical or inclined) were observed for different III-V [7–11], II-VI[12] and group IV[13–17] materials. In planar growth, liquid catalyst remains in contact with the substrate, while in vertical or inclined growth the droplet detaches from the substrate and stabilizes on the top of NW. Although this feature of VLS epitaxial growth is not entirely understood.
In this work, we discuss two aspects of VLS process, which directly impacts the growth direction: the droplet detachment at the initial stage of NW formation and stabilization of in-plane growth due to nucleation at the nanowire-substrate-liquid line. We present the model which describes two aspects of the growth process and consider the specific cases of germanium NWs on Ge(111), Ge(110) and Si(100) substrates. We use our modeling results to explain the experimentally observed in-plane and out-of-plane Au-catalyzed Ge NW and discuss the factors which determine the growth direction.

2. Model and results

2.1 Droplet detachment

In the studies of droplet detachment or contacting the substrate, we rely on previous considerations of the droplet stability [18–20] to discuss the condition for droplet detachment from the substrate. The equilibrium state of the system, consisting of the droplet on top of the NW stem of length L and radius R is given by the minimum of the surface energy G_{drop} [18–20]:

$$G_{drop} = \frac{2\pi R^2}{1+\cos\beta}\gamma_{L-V} + \pi R^2\gamma_{Tip-L} + 2\pi Ry\gamma_{S-L} + 2\pi R(L-y)\gamma_{S-V}\,, \tag{1}$$

where y is the length of the sidewall wetted by the droplet, β – droplet contact angle, γ_{L-V}, γ_{Tip-L}, γ_{S-L}, γ_{S-V} are surface energies of liquid-vapor, NW tip – liquid, NW sidewalls – liquid and NW sidewalls-vapor interfaces. Previously it was demonstrated that under the constrain of constant volume the derivative of G_{drop} as the function of contact angle β can be written as:

$$\frac{dG_{drop}}{d\beta} = \frac{2\pi R^2}{(1+\cos\beta)^2}(\gamma_{L-V}\sin\beta + \gamma_{S-L} - \gamma_{S-V}). \tag{2}$$

One can notice that G_{drop} has the extremum, which corresponds to the Young's equation on the sidewalls

$$\gamma_{L-V}\cos\psi + \gamma_{S-L} = \gamma_{S-V}. \tag{3}$$

Here ψ is contact angle of droplet lying on NW sidewalls. Thus extremum condition $dG_{drop}/d\beta = 0$ is equivalent to $\sin\beta = \cos\psi$. The corresponding extremum is maximum when $\beta > \pi/2$ and minimum when $\beta < \pi/2$.

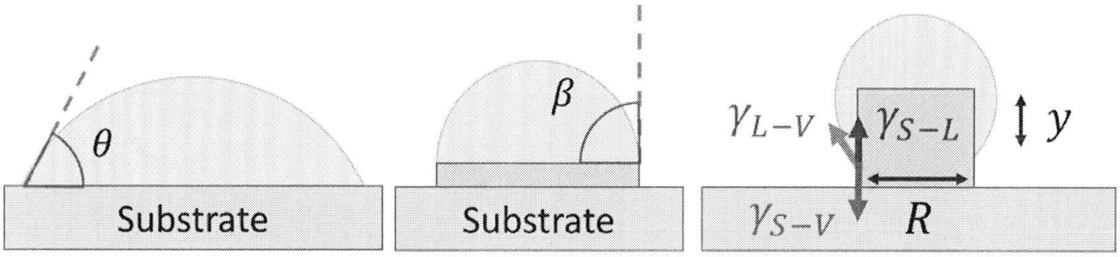

Figure 1. Schematics of the droplet on the substrate and after detachment from the substrate.

Meanwhile, it is widely considered in previous works [18,20] that the droplet does not wet the sidewalls when the condition $\gamma_{L-V}\sin\beta + \gamma_{S-L} > \gamma_{S-V}$ is satisfied. Which, in view of Eq. (7), is equivalent to $\sin\beta > \cos\psi$. In the opposite case, the droplets is unstable and slides down when $\gamma_{L-V}\sin\beta + \gamma_{S-L} < \gamma_{S-V}$ or $\sin\beta < \cos\psi$.
The initial value of β is given by the substrate contact angle, which we denote θ. If $\sin\theta > \cos\psi$ ($\theta < \psi + \pi/2$) the minimization of G_{drop} after the formation of the first monolayer will require the droplet detachment from the substrate and further increase of the contact angle. Otherwise, if $\sin\theta < \cos\psi$ ($\theta > \psi + \pi/2$) the minimization of G_{drop} corresponds to the decrease of the contact angle and droplet spreading.

2.2 Nucleus position in in-plane nanowire
VLS growth is generally considered to proceed in layer-by-layer regime [19–23]: the formation of each monolayer depletes the droplet and thus makes the nucleation of a new monolayer (ML) less improbable. Subsequent droplet refilling increases the concentration of the growth species and thus induces the nucleation of new ML. The 2D nucleus of size i is described by formation energy [21,22,24]:

$$G_{nucl} = -\Delta\mu\, i + A\, i + B\, i^{1/2} \tag{4}$$

where $\Delta\mu$ is the chemical potential difference per atom (or pair of atoms for binary materials). In the second term, A is proportional to the nucleus surface energy. The third term in Eq. (4) gives the energy of the side facets, which is proportional to the square root of the size i.
The nucleation barrier is given by the maximum of the formation energy:

$$G^*_{nucl} = B^2/4(\Delta\mu - A). \tag{5}$$

The nucleation probability $p \sim \exp(-G^*_{nucl}/T)$ decreases exponentially with G_* and thus the preferable growth regime supposed to has the lowest nucleation barrier G_*.
Liquid catalyst allows larger supersaturation in comparison to the vapor phase and thus increases $\Delta\mu$ and typically facilitate the nucleation [25]. The formation of 2D nucleus inside the droplet on top of substrate requires the formation of top and bottom interfaces as well as the elimination of the liquid-substrate interface, thus

$$A = \alpha\,(\gamma_{top} + \gamma_{bot} - \gamma_{sub}), \quad (6)$$

where γ_{top}, γ_{bot} and γ_{sub} are energies of the top, bottom and substrate-liquid interfaces respectively, as shown in Figure 1. The coefficient α here depends on the geometry of the nucleus. One can notice that $A = 0$ in the case of homoepitaxy or nucleation of new ML on top of the previous one. Thus, according to Eq. (2), the nucleation barrier in layer-by-layer regime is controlled by the energies of nucleus side interfaces which contribute to B.
The side facet of the nucleus may contact liquid, vapor or the substrate and form the interfaces with the energies γ_{S-L}, γ_{S-V} or γ_{S-Sub} respectively. Thus

$$B = \kappa_{S-L}\,\gamma_{S-L} + \kappa_{S-V}\,(\gamma_{S-V} - \gamma_{L-V}) + \kappa_{S-Sub}(\gamma_{S-Sub} - \gamma_{Sub-L}), \quad (7)$$

where different κ correspond to fractions of nucleus side facets contacted to liquid, solid or substrate.

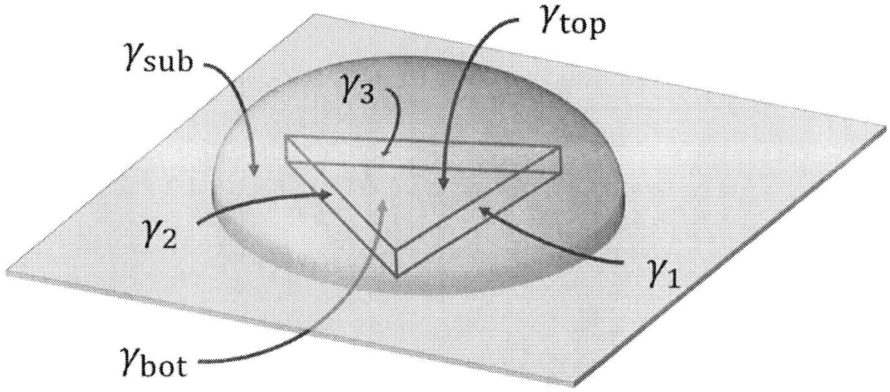

Figure 2. Schematic of the 2D nucleus and corresponding surface energies.

Next, we consider three possible positions of the nucleus on the previous ML: at the center of the droplet (C) or the triple line contacted either to vapor phase (VTPL) or the substrate (STPL). In general, (C) and (VTPL) positions are relevant both for planar and non-planar NWs, while the realization of (STPL) nucleation is possible in horizontal NWs only. We assume that the lowest formation energy defines the preferred growth mode. In calculations, for simplicity, we assume the triangle shape of the nucleus and, thus, $\kappa_{S-L} = \kappa_{S-V} = \kappa_{S-Sub} = 1/3$.
Table 1 summarizes the values of B in each regime of ML nucleation for germanium NWs on Ge(110), Ge(111) and Si(100) substrates grown with golden catalyst. In the growth on Ge(110) and Si(100) substrates NW are typically observed to grow in [110] direction with {111} sidewall facets. Thus, in this case γ_{S-L} = 0.542 J/m² [26], while γ_{S-V} = 1.149 J/m² [26] and γ_{L-V} = 1.15 J/m² [27]. γ_{S-Sub} = 0 in the case of homoepitaxy and approximated to γ_{S-Sub} = 0.64 J/m² in the growth of germanium on silicon substrate [28]. In the growth on Ge(111) we assume γ_{S-L} = 0.626 J/m² and γ_{S-V} = 1.1 J/m² [26].

Table 1. Expressions and calculated values (in J/m^2) of B for nucleation of a new ML at the central position (C), at the contact with vapor phase (VTPL) or at the contact with the substrate (STPL).

	C 	VTPL 	STPL 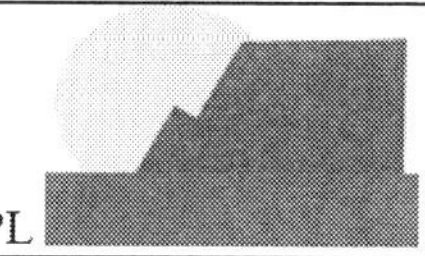
B	γ_{S-L}	$\gamma_{S-L}(1-\kappa_{S-V}) + \kappa_{S-V}(\gamma_{S-V}-\gamma_{L-V})$	$\gamma_{S-L}(1-\kappa_{S-Sub}) + \kappa_{S-Sub}(\gamma_{S-Sub}-\gamma_{Sub-L})$
Ge (110) substrate	0.542	0.411	0.153
Ge (111) substrate	0.626	0.575	0.209
Si (100) substrate	0.542	0.411	0.441

3. Conclusions

We consider two factors which impact the character of NW growth: (i) the detachment of the droplet from the substrate at the initial stage of NW growth and (ii) preferred nucleation position during the regular layer-by-layer growth.
On the one hand, the use of Young's equation (Eq. (7)) and surface energies in the previous section gives $\sin\theta > \cos\psi$ for both Ge (110) and Ge (111) substrates. Experimentally inclined and vertical Ge NWs were observed in the growth on deoxidized substrates[15,29].
On the other hand, in both considered examples of homoepitaxy the minimal B (and minimal G^*_{nucl}) corresponds to nucleation next to the substrate (STPL). Assuming that the preferred STPL nucleation insures the horizontal (in-plane) NW growth, one expect Ge NWs on Ge (110) and Ge (111) substrates to grow horizontally. Indeed, Ge NWs were systematically observed on clean and especially smooth substrates together with the decrease of droplet contact angle [30]. In contrast, the etching of the germanium substrate by gold droplet together with intensive surface passivation results in the increase of contact angle and droplet detachment from the substrate[31].
In the case of Si substrate, the minimal B and nucleation barrier corresponds to VTPL nucleation, which, in principle, does not guarantee that NWs will growth horizontally. Meanwhile, the Au droplet is usually observed to detach from the Si (001) [26,32] and Si(111) [33,34] substrates.
In general, large contact angles correspond to droplet detachment from the substrate and formation of vertical or inclined NW. At the same time, the detachment of the catalyst droplet at the initial stage of NW can be avoided in the case small contact angles of the droplet on the bare substrate. In the case of Ge NWs, the horizontal growth is stable due to the preferred nucleation next to the substrate. Meanwhile, it can be realized when the substrate preparation allows reduction of the contact angle of the catalyst droplet.

Acknowledgments

Authors thank the Ministry of Education and Science of the Russian Federation for financial support under grant 14.587.21.0040 (project ID RFMEFI58717X0040).

References

[1] Kishino K and Ishizawa S 2015 Selective-area growth of GaN nanocolumns on Si(111) substrates for application to nanocolumn emitters with systematic analysis of dislocation filtering effect of nanocolumns *Nanotechnology* **26** 225602
[2] Riedel M, Hölzel S, Hille P, Schörmann J, Eickhoff M and Lisdat F 2017 InGaN/GaN nanowires as a new platform for photoelectrochemical sensors – detection of NADH *Biosens. Bioelectron.* **94**

298–304
[3] Zhao S, Nguyen H P T, Kibria M G and Mi Z 2015 III-Nitride nanowire optoelectronics *Prog. Quantum Electron.* **44** 14–68
[4] Kuykendall T R, Schwartzberg A M and Aloni S 2015 Gallium Nitride Nanowires and Heterostructures: Toward Color-Tunable and White-Light Sources *Adv. Mater.* **27** 5805–12
[5] Glas F 2017 Comparison of Modeling Strategies for the Growth of Heterostructures in III-V Nanowires *Cryst. Growth Des.* **17** 4785–94
[6] Ambrosini S, Fanetti M, Grillo V, Franciosi A and Rubini S 2011 Vapor-liquid-solid and vapor-solid growth of self-catalyzed GaAs nanowires *AIP Adv.* **1**
[7] Fortuna S A, Wen J, Chun I S and Li X 2008 Planar GaAs Nanowires on GaAs (100) Substrates: Self-Aligned, Nearly Twin-Defect Free, and Transfer-Printable *Nano Lett.* **8** 4421–7
[8] Zhang X, Zou J, Paladugu M, Guo Y, Wang Y, Kim Y, Joyce H J, Gao Q, Tan H H and Jagadish C 2009 Evolution of Epitaxial InAs Nanowires on GaAs (111)B *Small* **5** 366–9
[9] Breuer S, Hilse M, Trampert A, Geelhaar L and Riechert H 2010 Vapor-liquid-solid nucleation of GaAs on Si(111): Growth evolution from traces to nanowires *Phys. Rev. B* **82** 075406
[10] Wallentin J, Kriegner D, Stangl J and Borgström M T 2014 Au-Seeded Growth of Vertical and in-Plane III–V Nanowires on Graphite Substrates *Nano Lett.* **14** 1707–13
[11] Cao Y Y and Yang G W 2012 Vertical or Horizontal: Understanding Nanowire Orientation and Growth from Substrates *J. Phys. Chem. C* **116** 6233–8
[12] Ben-Zvi R, Burrows H, Schvartzman M, Bitton O, Pinkas I, Kaplan-Ashiri I, Brontvein O and Joselevich E 2019 In-Plane Nanowires with Arbitrary Shapes on Amorphous Substrates by Artificial Epitaxy *ACS Nano* **13** 5572–82
[13] Mataev E, Rastogi S K, Madhusudan A, Bone J, Lamprinakos N, Picard Y and Cohen-Karni T 2016 Synthesis of Group IV Nanowires on Graphene: The Case of Ge Nanocrawlers *Nano Lett.* **16** 5267–72
[14] Zhang C, Miao X, Chabak K D and Li X 2017 A review of III–V planar nanowire arrays: selective lateral VLS epitaxy and 3D transistors *J. Phys. D. Appl. Phys.* **50** 393001
[15] Schmidtbauer J, Bansen R, Heimburger R, Teubner T, Boeck T and Fornari R 2012 Germanium nanowire growth controlled by surface diffusion effects *Appl. Phys. Lett.* **101** 043105
[16] Alonso M I, Ruiz A, Alonso M, Bailo E, Garriga M, AlejandroMolero, Vaccaro P O and Goñi A R 2015 Growth and Characterization of Epitaxial In-plane SiGe Alloy Nanowires *Mater. Today Proc.* **2** 548–56
[17] Zhang J J, Rastelli A, Schmidt O G, Scopece D, Miglio L and Montalenti F 2013 Self-organized evolution of Ge/Si(001) into intersecting bundles of horizontal nanowires during annealing *Appl. Phys. Lett.* **103** 083109
[18] Nebolsin V A and Shetinin A A 2003 Role of Surface Energy in the Vapor–Liquid–Solid Growth of Silicon *Inorg. Mater.* **39** 899–903
[19] Glas F, Harmand J C and Patriarche G 2007 Why does wurtzite form in nanowires of III-V zinc blende semiconductors? *Phys. Rev. Lett.* **99** 3–6
[20] Dubrovskii V G, Cirlin G E, Sibirev N V., Jabeen F, Harmand J C and Werner P 2011 New Mode of Vapor−Liquid−Solid Nanowire Growth *Nano Lett.* **11** 1247–53
[21] Sibirev N V and Dubrovskii V G 2004 A modified Kolmogorov model and the growth rate of a crystal face of arbitrary size *Tech. Phys. Lett.* **30**
[22] Dubrovskii V G and Sibirev N V. 2004 Growth rate of a crystal facet of arbitrary size and growth kinetics of vertical nanowires *Phys. Rev. E* **70** 031604
[23] Sakong S, Du Y A and Kratzer P 2013 Atomistic modeling of the Au droplet-GaAs interface for size-selective nanowire growth *Phys. Rev. B* **88** 1–9
[24] Kukushkin S A and Osipov A V. 1996 New phase formation on solid surfaces and thin film condensation *Prog. Surf. Sci.* **51** 1–107
[25] Wallentin J, Kriegner D, Stangl J and Borgström M T 2014 Au-seeded growth of vertical and in-plane III-V nanowires on graphite substrates *Nano Lett.* **14** 1707–13

[26] Li Y, Wang Y, Ryu S, Marshall A F, Cai W and McIntyre P C 2016 Spontaneous, Defect-Free Kinking via Capillary Instability during Vapor–Liquid–Solid Nanowire Growth *Nano Lett.* **16** 1713–8
[27] Sakong S, Du Y A and Kratzer P 2013 Atomistic modeling of the Au droplet–GaAs interface for size-selective nanowire growth *Phys. Rev. B* **88** 155309
[28] Koman B P and Yuzevich V M 2015 Energy Parameters of Interfacial Layers in Composite Systems: Graphene – (Si, Cu, Fe, Co, Au, Ag, Al, Ru, Hf, Pb) and Semiconductor (Si,Ge) – (Fe, Co, Cu, Al, Au, Cr, W, Pb) *J. Nano- Electron. Phys.* **7** 04059
[29] Zhu Z, Song Y, Zhang Z, Sun H, Han Y, Li Y, Zhang L, Xue Z, Di Z and Wang S *The Vapor-Solid-Solid Growth of Ge Nanowires on Ge (110) by Molecular Beam Epitaxy*
[30] Bansen R, Schmidtbauer J, Gurke R, Teubner T, Heimburger R and Boeck T 2013 Ge in-plane nanowires grown by MBE: influence of surface treatment *CrystEngComm* **15** 3478
[31] Kolíbal M, Pejchal T T, Tomáš Vystavě T, Tomáš T and Ikola S 2016 The Synergic Effect of Atomic Hydrogen Adsorption and Catalyst Spreading on Ge Nanowire Growth Orientation and Kinking *Nano Letters* **16** 4880-4886
[32] Boudaa F, Blanchard N P, Descamps-Mandine A, Benamrouche A, Gendry M and Penuelas J 2015 Structure and morphology of Ge nanowires on Si (001): Importance of the Ge islands on the growth direction and twin formation *J. Appl. Phys.* **117** 055302
[33] Koto M, Marshall A F, Goldthorpe I A and McIntyre P C 2010 Gold-Catalyzed Vapor-Liquid-Solid Germanium-Nanowire Nucleation on Porous Silicon *Small* **6** 1032–7
[34] Dailey J W, Taraci J, Clement T, Smith D J, Drucker J and Picraux S T 2004 Vapor-liquid-solid growth of germanium nanostructures on silicon *J. Appl. Phys.* **96** 7556–67

Dissociation and synthesis of target components during laser ablation of $LiNbO_3$

Z Vakulov[1], E G Zamburg[2], V S Klimin[1], A V Miakonkikh[3], I E Clemente[3], K V Rudenko[3], O A Ageev[1]
[1]Research and Education Centre «Nanotechnology», Southern Federal University, Taganrog 347922, Russia
[2]Department of Electrical & Computer Engineering, National University of Singapore, Singapore 119077, Singapore
[3]Valiev Institute of Physics and Technology RAS, Moscow 117218, Russia

Abstract. This paper shows the results of thermodynamic analysis of interfacial interaction in the lithium-niobium-oxygen system. The temperature dependences of the change in the Gibbs free energy calculated, taking into account the non-linear temperature dependences of the thermophysical properties of materials. Additionally, estimated the possibility of dissociation reactions of lithium and niobium oxides, as well as the reactions of lithium and niobium with oxygen. The results obtained can be used in the design and manufacture of integrated acousto-optic and piezoelectric devices, as well as for sensitive elements of sensors using various effects of surface acoustic waves.

1. Introduction

Over the past few years, the range of wireless wearable sensors and portable electronic devices has significantly expanded, in most cases, electrochemical batteries are used to power them [1-3]. Even though the performance of electronic devices is increasing every year [4-7]. However, significant progress in the creation of promising batteries has not yet been achieved. A promising approach in the field of energy harvesting is the development of piezoelectric transducers, which converts the mechanical energy of the environment into electrical energy [8]. These transducers can be used as the power sources for wearable electronics devices [9].

Due to the unique combination of $LiNbO_3$ piezoelectric, optical and ferroelectric properties, it can be considered that the backbone of promising piezoelectric energy converters, which are distinguished by high efficiency and environmental safety and fabricated by microelectronic technology. Nevertheless, at the current stage of technological development, it is challenging to build devices based on ferroelectric films due to the problems associated with insufficient development of the production technology. Most ferroelectric and piezoelectric materials ($BaTiO_3$, $PbTiO_3$, $LiNbO_3$) are multicomponent oxides, which significantly complicates the process of creation and their integration with the technology of micro- and nanoelectronics.

Pulsed laser deposition (PLD) is one of the most promising methods for the fabrication of multicomponent oxides films, due to a large number of technological parameters, which makes it possible to control the composition of the films formed, as well as their morphological and electrical properties. In the literature, the stoichiometry transfers during PLD process studied insufficiently. It is considered that incongruence ablation does not have a significant impact on the deposition. Nevertheless, laser ablation is a high-energy process and, therefore, when a laser pulse is applied to a multi-component target, the molecules of the target can dissociate, and as a result, the properties of grown films might be different from initial target properties (structure, composition) [10].

Therefore, the goal of this work is studying thermodynamics regularities of target material dissociation and synthesis processes during laser ablation of $LiNbO_3$ target.

2. Theoretical analysis

In order to study the phase formation processes during laser ablation of lithium niobate target, the possibility of chemical dissociation and synthesis of $LiNbO_3$ in the passage space at temperatures from 550 °C to 8500 °C is estimated. The Gibbs free energy change (ΔG) dependences on temperature were calculated using the chemical analysis software package FactSage 6.2 (GTT-Technologies, Germany), which has a regularly updated electronic database in the field of chemical thermodynamics.

In the study of thermodynamic regulareties, we took into account the possibility of interaction between the target components between themselves and the background atmosphere in the growth chamber. To simplify the perception of the results, stoichiometric coefficients in the equations of chemical reactions were taken into account but were omitted.

3. Results and discussion

In order to analyze possible dissociation reactions of $LiNbO_3$ during laser ablation, decomposition reactions of lithium niobate in a vacuum (10^{-5} Torr) and oxygen atmosphere (10^{-2} Torr) were identified.

Figure 1 shows the temperature dependences of change in the Gibbs free energy for $LiNbO_3$ dissociation reactions of target components during laser ablation of $LiNbO_3$ target.

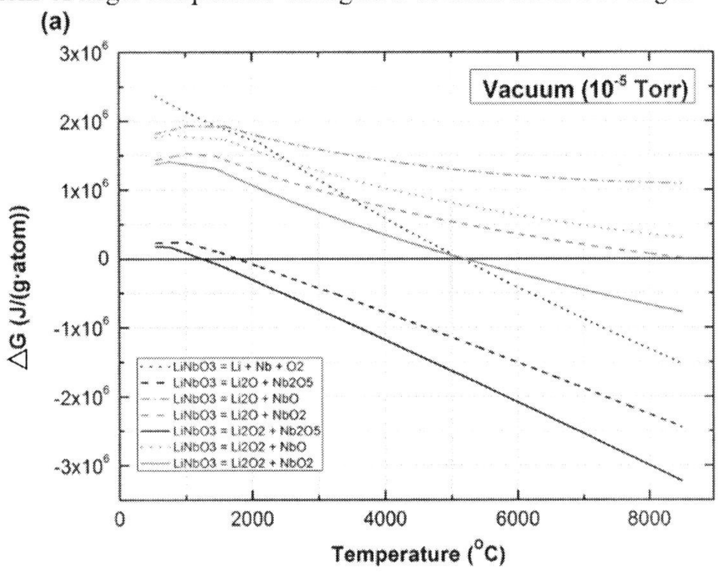

(b)

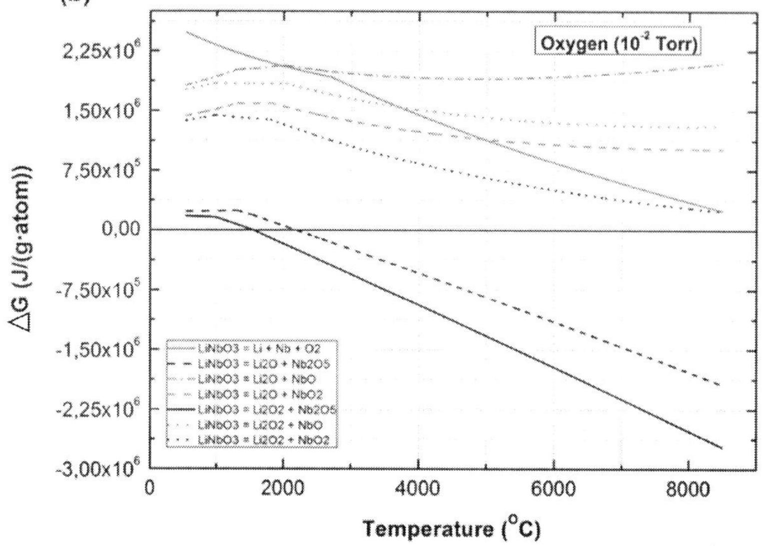

Figure 1 (a, b) Temperature dependences of ΔG for dissociation reactions of $LiNbO_3$ in **(a)** vacuum and **(b)** oxygen atmosphere

The analysis of the presented dependences showed that the most probable dissociation reaction of lithium niobate in vacuum and oxygen atmosphere at an oxygen pressure of 10^{-2} Torr is leading to the reaction $LiNbO_3 = Li_2O_2 + Nb_2O_5$, which begins at temperatures above 1230 °C and 1540 °C, respectively. If the temperature increases to 1760 °C (for vacuum) and 2189 °C (for oxygen atmosphere, pressure of 10^{-2} Torr), $LiNbO_3$ can dissociate on Li_2O and Nb_2O_5. With a further increase in temperature up to 5140 °C, $LiNbO_3$ dissociated into individual elements. At a temperature of 5190 °C in a vacuum, lithium niobate can decompose on Li_2O_2 and NbO. When the oxygen pressure is 10^{-2} Torr, the ΔG value of the other reactions is positive over the entire considered temperature range. Their flow in the forward direction in the temperature range from 550 °C to 8500 °C is impossible.

Obtained during the decomposition of $LiNbO_3$ oxides of niobium and lithium can also dissociate into individual elements. Figure 2 shows the temperature dependences of ΔG for dissociation reactions of lithium oxides in vacuum and oxygen atmosphere.

(a)

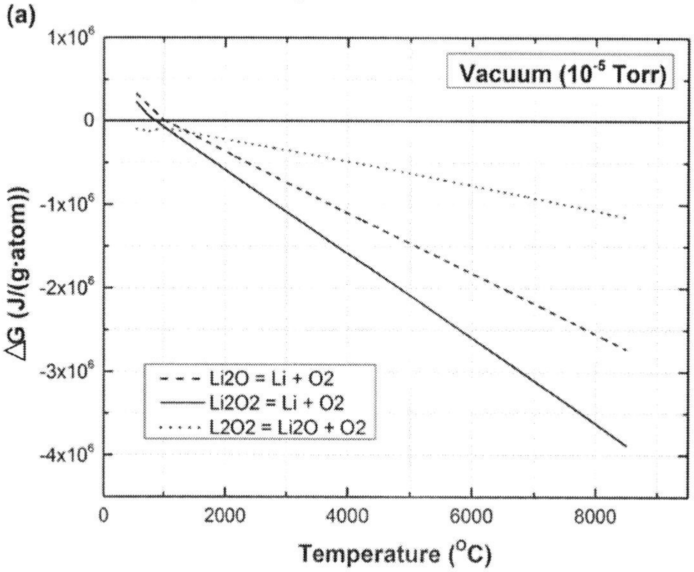

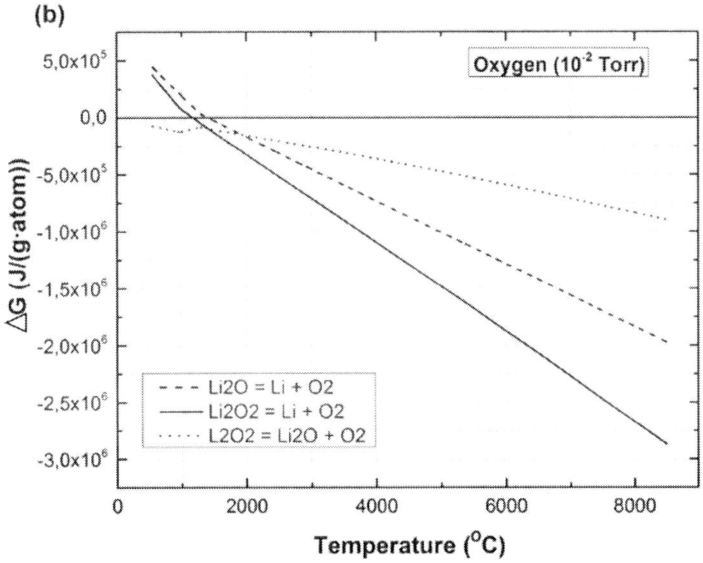

Figure 2 (a, b) Temperature dependences of ΔG for dissociation reactions of lithium oxides in **(a)** vacuum and **(b)** oxygen atmosphere

During the analysis of the obtained dependences, it was established that the Li_2O_2 decomposition into Li_2O and oxygen proceeds in the forward direction in the entire considered temperature range both in a vacuum and in a oxygen atmosphere at a pressure of 10^{-2} Torr. In the temperature range above 1500 °C, the competing reactions are the dissociation reactions of Li_2O_2 and Li_2O into individual elements. It should be noted that the presence of oxygen stabilizes lithium oxides, which prevents their thermal dissociation.

Figure 3 shows the temperature dependences of ΔG for dissociation reactions of niobium oxides in vacuum and oxygen atmosphere.

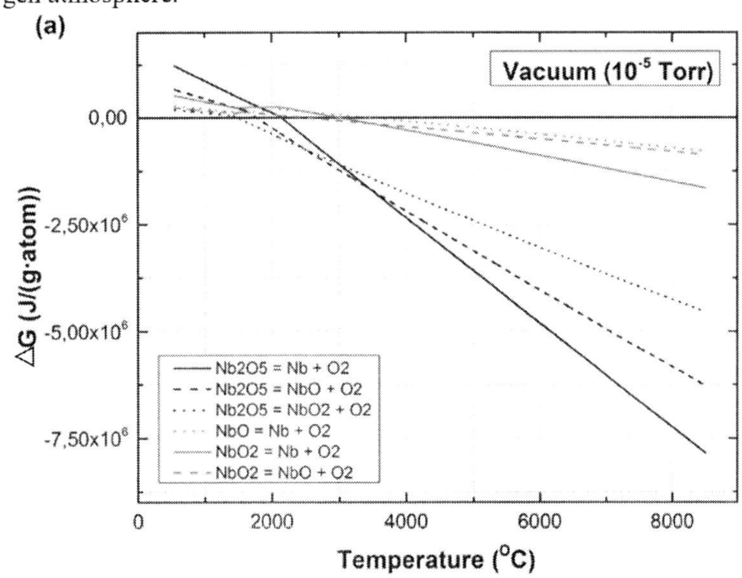

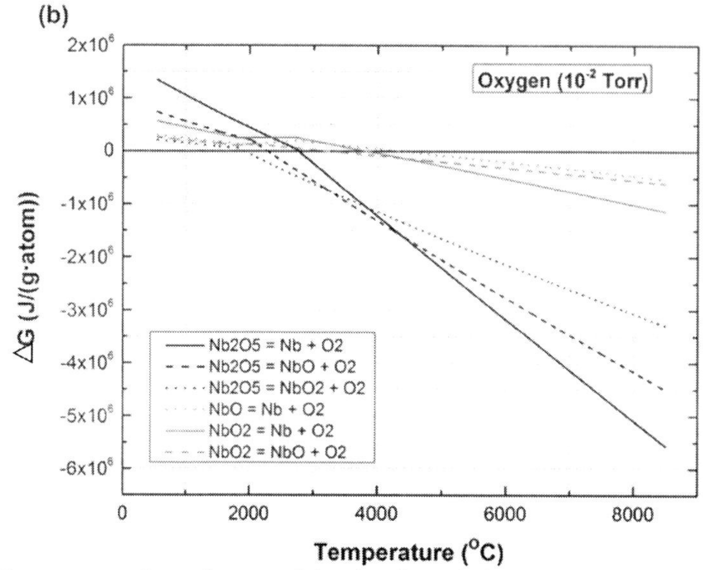

Figure 3 (a, b) Temperature dependences of ΔG for dissociation reactions of niobium oxides in **(a)** vacuum and **(b)** oxygen atmosphere

When analyzing the dependences of the reactions of decomposition of niobium oxides, it is possible to identify temperature ranges in which the occurrence of a particular reaction will be more likely. For a vacuum in the temperature range from 8500 °C to 3100 °C, the most suitable reaction will be the decomposition of Nb_2O_5 into individual elements. In the temperature range from 3000 °C to 1800 °C, competing for Nb_2O_5 decomposition reactions are possible.
In the process of the laser plume propagation from the target to the substrate, the temperature in it decreases, creating favorable conditions for the reactions of the interaction of components from the atmosphere of the growth chamber.

4. Conclusion

As a result of the thermodynamic analysis, the temperature ranges of formation of lithium and niobium compounds with oxygen were determined. Noticeably, the creation of oxides Li_2O and Nb_2O_5 is most feasible compared to the formation of other oxide phases. It was theoretically calculated that the dissociation of $LiNbO_3$ is a multistage process, which correlated to the results presented in [11]. In the first stage of the ablation process, lithium niobate dissociates into oxides with lower oxygen content. At temperatures above 1777 °C, the oxides decompose and complete dissociation of the target material into Li, Nb, and O_2. When the plasma plume expands, the processes of interaction of Li and Nb with O_2 occur with the formation of oxides, caused by a decrease in temperature. Obtained results can be used in the design and manufacture of piezoelectric energy harvesters based on $LiNbO_3$ films. The region of high temperatures corresponds to the base of the laser plume arising from the ablation of the target, the region of low temperatures is the tip of the plume.

The results obtained can be used in the design and manufacture of integrated acoustic-optic and piezoelectric devices, as well as sensitive elements of sensors using various effects of surface acoustic waves.

Acknowledgments

This work was financially supported by the Russian Foundation for Basic Research (project № 18-29-11019 mk). The work was done on the equipment of the Research and Education Centre «Nanotechnology» and Collective Use Centre «Nanotechnology», Southern Federal University.

References

[1] Zhang G, Li M, Li H, Wang Q, Jiang S 2018 *Energy Technol.* **6** 791
[2] Qian R C, Long Y T 2018 *Open Chem.* **7** 118
[3] Heo J S, Eom J, Kim Y H, Park S K 2018 *Small* 14 1703034.
[4] Klimin V S, Rezvan A A, Ageev O A 2018 *J. Phys.: Conf. Ser.* 1124 071020
[5] Klimin V S, Rezvan A A, Kots I N, Naidenko N A 2018 *J. Phys.: Conf. Ser.* **1124** 071019
[6] Klimin V S, Rezvan A A, Ageev O A 2018 *J. Phys.: Conf. Ser.* **1124** 022035
[7] Klimin V S, Rezvan A A, Ageev O A 2019 *Proc. of SPIE* **11022** 1102220-1
[8] Ferrari M, Ferrari V, Guizzetti M, Andò B, Baglio S, Trigona C 2010 *Sens Actuators A Phys.* **162** 425
[9] Lallart M, Guyomar D, Jayet Y, Petit L, Lefeuvre E, Monnier T, Guy P, Richard C 2008 *Sens Actuators A Phys.* **147** 263
[10] Vakulov Z, Zamburg E, Golosov D A, Zavadskiy S M, Dostanko A P, Miakonkikh A V, Klemente I E, Rudenko K V, Ageev O A 2017 . *Phys.: Conf. Ser.* **917** 032024
[11] Uda S, Tiller W A 1992 *J. Cryst. Growth* **121** 155

Arsenic background pressure effect on In droplet morphology

I A Mikhaylin, S V Balakirev[2], M M Eremenko[1], N E Chernenko[1,3], M S Solodovnik[1,2]
[1]Research and Education Center "Nanotechnologies", Southern Federal University, Taganrog 347922, Russia
[2]Department of Nanotechnologies and Microsystems, Southern Federal University, Taganrog 347922, Russia
[3]Department of Radio Engineering Electronics, Southern Federal University, Taganrog 347922, Russia

Abstract. The paper presents the results of an experimental study of the effect of arsenic background pressure on the characteristics of droplet nanostructures formed by droplet epitaxy. We have shown that with an increase in the initial size of the droplets, the sensitivity of the system to the background pressure decreases. It is shown that the residual pressure of arsenic can be used as an additional control parameter of the droplet epitaxy technique.

1. Introduction

Geometric characteristics of self-assembled A3B5 semiconductor nanostructures largely determine their functional properties [1-14]. The size, shape, relative position of the structures in the array directly affect their band structure. Together with the chemical composition of the nanostructures themselves, the composition of the surrounding matrix, and crystal perfection, this determines the totality of the functional properties of the systems based on them – the electrophysical and optical characteristics, etc. In this regard, the task of precision control of the parameters of self-organizing A3B5 nanostructures is relevant [15-16]. Since it largely predetermines the functional characteristics of devices based on them, the task of controlling the parameters of self-organizing nanostructures looks especially relevant [17-21].

In this regard, droplet epitaxy looks extremely promising technologies, allowing, in contrast to the Stranski-Krastanov method, not only to form a wide range of nanostructures (from metal nanoclusters to quantum dots and different types of complexes based on them) in a wide range of compositions and material systems (including lattice matched), but also significantly expand the range of control parameters of nanostructures due to the significant number of process control parameters [22-28]. Nevertheless, despite active research in this area, the influence of only the main parameters of the droplet epitaxy process (deposition rate and thickness, growth temperature, crystallization parameters) is relatively well studied [29-33]. At the same time, the influence of secondary parameters of this technique is still not well understood [34].

The purpose of this work is to experimentally study the influence of such a secondary parameter as the As residual (background) pressure of the working environment on the morphology of metallic In nanodroplets formed on GaAs(001) substrates by the droplet epitaxy.

2. Experiment details

Experimental studies were carried out on the SemiTEq STE 35 installation equipped with solid-state molecular sources of group III elements and a valve source of arsenic. For epitaxial growth we used

GaAs(001) epi-ready substrates. After thermal removal of the GaAs native oxide with the subsequent annealing of the substrate in the As flux, a GaAs buffer layer 250 nm thick was deposited at a temperature $T = 580°C$. Then growth was interrupted by cut-off growth component fluxes, and the substrate was cooled to the temperature of droplet nanostructure formation. It should be noted that the cooling parameters were selected in such a way that in the As flux absence on the surface the (2×4) reconstruction is maintained. Initially, a series of samples was formed at various temperatures $T = 150...300°C$ and deposition thicknesses $H = 1...3$ ML and a fixed As residual pressure of 1×10^{-7} Pa. In the second part of experiment the time and rate of substrate cooling was varied so that the deposition of droplets occurred at different values of the As residual pressure. We grew two series of samples at the same growth temperature $T = 300°C$, effective deposition rate of In (InGaAs ternary growth calibrated) $v = 0.25$ ML/s and different equivalent deposition thicknesses 1.5 and 3.0 ML, respectively. The exposure time of droplet structures in different series was also the same. In addition, in each series, the As background pressure was also changed, at which the deposition was carried out. Then the samples were investigated by the methods of AFM and SEM.

3. Results and discussion

Fig. 1 presents the results of experimental studies of the effect of growth temperature on the critical thickness of droplet formation and the minimum diameter of nanostructures observed in this case, which is achievable with the given parameters of the droplet epitaxy. As can be seen from the **Fig. 1**, an increase of deposition temperature T from 150 ° C to 300 ° C leads to an increase in the minimum diameter from 15 to 65 nm, respectively. At the same time, a decrease in the deposition thickness below the critical value to reduce the droplet size leads to the suppression of the formation of droplet nanostructures.

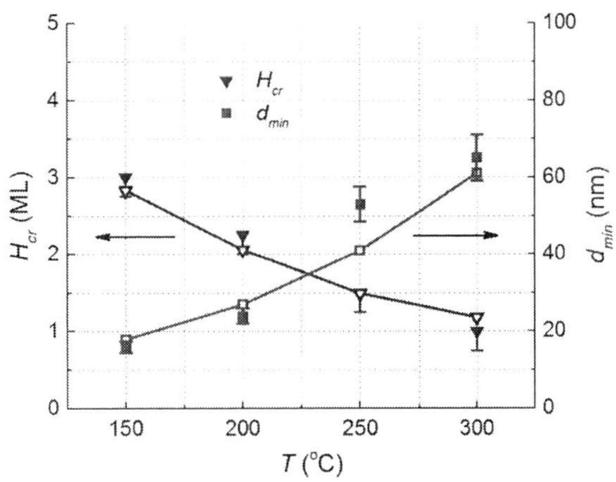

Figure 1. Temperature dependences of critical thickness and minimum achievable size during In/GaAs(001) droplet epitaxy at equivalent deposition rate $v = 0.25$ ML/s.

Analysis of the data obtained during the second part of the experiment showed that the As background pressure can have a significant impact on the final size of the droplet nanostructures. So, as we can see from **Fig. 2**, when the thickness of the deposition of 1.5 ML increase in background pressure from 8.7×10^{-8} Pa to 1.3×10^{-7} Pa results in decreasing In droplet size by more than 200% – from 65 nm to 30 nm. As follows from comparison with the results presented above, using only basic control parameters of droplet epitaxy, obtaining nanodroplets with such dimensions is not achievable in principle.

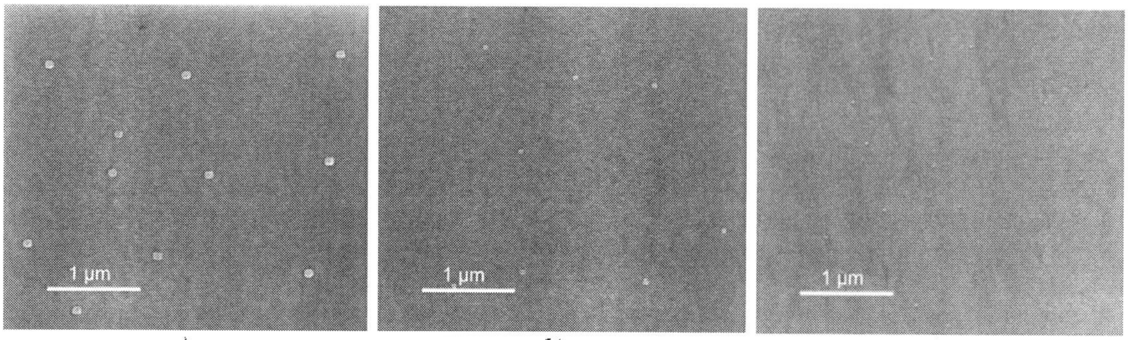

a) *b)* *c)*

Figure 2(a, b, c). SEM images of In nanodroplets obtained at growth temperature $T = 300$°C, deposition thickness $H = 1.5$ ML and different values of background pressure P: **(a)** 8.7×10^{-8} Pa; **(b)** 1.0×10^{-7} Pa; **(c)** 1.3×10^{-7} Pa.

At the same time, as follows from **Fig. 3** and **Fig. 4**, with an increase in the initial size of the nanostructures, the sensitivity of the system to the background pressure decreases. So, at deposition thickness of 3 ML In, the droplet size changes by only 50% – from 160 nm to 110 nm, while the background pressure of As varied over a significantly wider range of values – $(1.0 – 6.3)\times10^{-7}$ Pa.

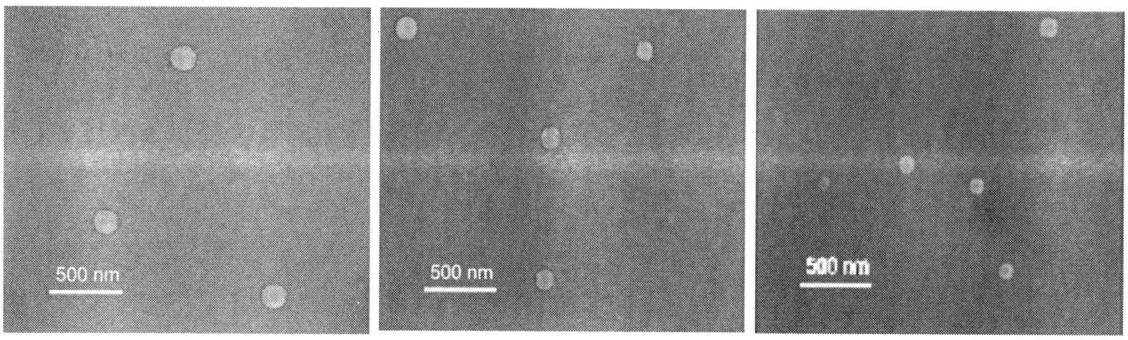

a) *b)* *c)*

Figure 3(a, b, c). SEM images of In nanodroplets obtained at growth temperature $T = 300$°C, deposition thickness $H = 3.0$ ML and different values of background pressure P: **(a)** 1.0×10^{-7} Pa; **(b)** 3.0×10^{-7} Pa; **(c)** 6.3×10^{-7} Pa.

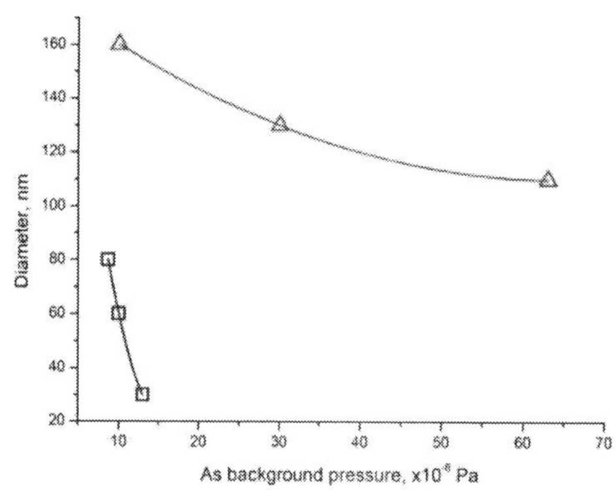

Figure 4. Dependences of In droplet size on As background pressure obtained at growth temperature of 300°C and different deposition thicknesses: 1.5 ML (blue) and 3 ML (red). Symbols – experimental data, lines – approximation curves.

This behavior of the system is due to a sharp increase in the volume of droplet nanostructures with an increase in the deposition thickness with a constant exposure time of the formed structures and a slight change in the As flux on the surface from the residual atmosphere. The decrease in the size of droplet structures with an increase in the As flux on the surface from the environment during exposure is due to In diffusion from the droplet in order to bind with arsenic adsorbed on the surface.

Thus, we have shown that in order to preserve the reproducibility of the results in the process of droplet epitaxy of In(As)/GaAs nanostructures, it is necessary to carefully control the residual pressure of arsenic. In addition, we have shown that the background pressure can potentially be used as an additional control parameter in order to modify the dimensions of the original structures.

Acknowledgments
This work was supported by Grant of the President of the Russian Federation No. MK-477.2019.8. The results were obtained using the equipment of the Research and Education Center and Center for Collective Use "Nanotechnologies" of Southern Federal University.

References
[1] Balakirev S V, Solodovnik M S, Eremenko M M, Mikhaylin I A and Ageev O A 2017 *J. Phys.: Conf. Ser.* 2017 **917** 032034
[2] Ageev O A, Solodovnik M S, Balakirev S V, Eremenko M M and Mikhaylin I A 2016 *J. Phys.: Conf. Ser.* **741** 012012
[3] Klimin V S, Solodovnik M S, Lisitsyn S A, Rezvan A A and Balakirev S V 2018 *J. Phys.: Conf. Ser.* **1124** 041024
[4] Abraham P, Garcia-Perez M A, Benyattou T, Guillot G, Sacilotti M and Letartre X 1995 *Semicond. Sci. Technol.* **10** 1585
[5] Kawamura Y, Kurisu H, Yoshimatsu K, Kamada A, Naito Y and Inoue N 1997 *Jpn. J. Appl. Phys.* **36** L757
[6] Ageev O A, Solodovnik M S, Balakirev S V and Eremenko M M 2016 *J. Vac. Sci. Technol. B* **34** 041804
[7] Ageev O A, Solodovnik M S, Balakirev S V and Mikhaylin I A 2016 *J. Phys.: Conf. Ser.* **681** 012036
[8] Kopf R F, Wei H P, Perley A P and Livescu G 1992 *Appl. Phys. Lett.* **60** 2386
[9] Ageev O A, Solodovnik M S, Balakirev S V, Eremenko M M and Mikhaylin I A 2017 *J. Crys. Growth* **457** 46-51
[10] Wakefield B, Halliwell M A G, Kerr T, Andrews D A, Davies G J and Wood D R 1984 *Appl. Phys. Lett.* **44** 341
[11] Ageev O A, Solodovnik M S, Balakirev S V and Eremenko M M 2016 *Phys. Solid State* **58** 1045–1052
[12] Anwar A F M and Webster R T 1998 *Solid-State Electron.* **42** 2101
[13] Ait Kaci H, Boukredimi D and Mebarki M 1997 *Phys. Status Solidi A* **163** 101
[14] Ageev O A, Solodovnik M S, Balakirev S V and Mikhaylin I A 2016 *Technical Physics* **61** 971–977
[15] Lisitsyn S A , Balakirev S V, Avilov V I, Kolomiytsev A S, Klimin V S, Solodovnik M S, Konoplev B G and Ageev O A 2018 *Nanotech. in Russia* **13** 26-33.
[16] Yamagiwa M, Mano T, Kuroda T, Tateno T, Sakoda K, Kido G, Koguchi N and Minami F 2006 *Appl. Phys. Lett.* **89** 113115
[17] Mano T, Kuroda T, Mitsuishi K, Nakayama Y, Noda T and Sakoda K 2008 *Appl. Phys. Lett.* **93** 203110
[18] Guti´errez M, Hopkinson M, Liu H Y, Ng J S, Herrera M, Gonz´alez D, Garcia R and Beanland R 2005 *Physica* E **26** 245
[19] Solodovnik M S, Karenkikh O G, Balakirev S V, Petrov S I, Ryzhuk R V, Alexeev A N and Ageev O A 2018 *J. Phys.: Conf. Ser.* **1124** 081024

[20] Scaccabarozzi A, Adorno S, Bietti S, Acciarri M and Sanguinetti S 2013 *Phys. Status Solidi RRL* **7** 173–6
[21] Wu J, Wang Z M, Dorogan V G, Li S, Zhou Z, Li H, Lee J, Kim E S, Mazur Y I and Salamo G J 2012 *Appl. Phys. Lett.* **101** 043904
[22] Balakirev S V, Eremenko M M, Mikhaylin I A and Solodovnik M S 2018 *J. Phys.: Conf. Ser.* **1124** 022025
[23] Koguchi N and Ishige K 1993 *Japan. J. Appl. Phys.* **32** 2052
[24] Balakirev S V, Solodovnik M S, Eremenko M M, Mikhaylin I A and Ageev O A 2018 *J. Phys.: Conf. Ser.* **1124** 022001
[25] Chikyow T and Koguchi N 1990 *Japan. J. Appl. Phys.* **29** L2093
[26] Solodovnik M S, Balakirev S V, Eremenko M M, Mikhaylin I A, Avilov V I, Lisitsyn S A and Ageev O A 2017 *J. Phys.: Conf. Ser.* 2017 **917** 032037
[27] Bietti S, Bocquel J, Adorno S, Mano T, Keizer J G, Koenraad P M and Sanguinetti S 2015 *Phys. Rev. B* **92** 075425
[28] Balakirev S V, Solodovnik M S and Ageev O A 2017 *J. Phys.: Conf. Ser.* 2017 **917** 032033
[29] Somaschini C, Bietti S, Trampert A, Jahn U, Hauswald C, Riechert H, Sanguinetti S and Geelhaar L 2013 *Nano Lett.* **13** 3607
[30] Balakirev S V, Solodovnik M S and Ageev O A 2018 *Phys. Status Solidi B* **255** 1700360
[31] Lee J, Wang Z M and Salamo G 2007 *J. Phys.: Condens. Matter* **19** 176223
[32] Balakirev S V, Eremenko M M, Mikhaylin I A, Klimin V S and Solodovnik M S 2018 *J. Phys.: Conf. Ser.* **1124** 022018
[33] Li C, Zeng Z Q, Fan D S, Hirono Y, Wu J, Morgan T A, Hu X, Yu S Q, Wang Z M and Salamo G J 2011 *Appl. Phys. Lett.* **99** 243113
[34] Wu J and Wang Z M 2014 *J. Phys. D: Appl. Phys.* **47** 173001

InGaN nanostructures of a branched morphology on silicon substrate: MBE synthesis and properties

R R Reznik[1-5], K P Kotlyar[1], A I Khrebtov[2], N V Kryzhanovskaya[1] and G E Cirlin[1-4]

[1]St-Petersburg Academic University – Nanotechnology Research and Education Centre RAS, Khlopina 8/3, 194021, St-Petersburg, Russia
[2]ITMO University, Kronverkskiy pr. 49, 197101, St-Petersburg, Russia
[3]Institute for Analytical Instrumentation RAS, Rizhsky 26, 190103, St-Petersburg, Russia
[4]Peter the Great St-Petersburg Polytechnic University, Polytechnicheskaya 29, 195251, St-Petersburg, Russia
[5]St Petersburg University, Universitetskaya Emb. 13B, 199034, St-Petersburg, Russia

E-mail: moment92@mail.ru

Abstract. A possibility of InGaN nanostructures of a branched morphology MBE growth on Si substrate has been demonstrated. The results of morphological studies have shown that InGaN synthesis occurs in several stages. InGaN nanostructures turned out to be optically active at room temperature and have a wide radiation visible range.

1. Introduction

Nowadays, semiconductor compounds of III-Nitrides are of great interest for modern optoelectronics due to its unique properties [1]. In particular, InGaN ternary compounds are promising for building devices for emitting in the visible range [2]. Interest in obtaining such structures is due, inter alia, to the prospects for integrating III-N semiconductors and silicon electronics, the possibility of using substrates of large (up to 150-200 mm) dimensions, their low cost, good electrical conductivity, as well as obtaining 'template' structures. However, the synthesis of high-quality direct-gap InGaN two-dimensional layers on silicon substrates leads to the appearance of a high density of defects in the InGaN layer due to the significant difference in the lattices constants and thermal expansion coefficients. Significant efforts are being made to successfully tackle these problems. In particular, to suppress the process of misfit dislocations occurrence and reduce the number of defects, barrier intermediate layers are formed in the epitaxial semiconductor nitride layer {SiC/Si(111), AlNGaN/SiC/Si(111) и GaN/AlN-GaN/SiC/Si(111)} [3,4]. This approach contributes to a decrease in the efficiency or in general impossibility of creating devices based on InGaN. At the same time, an alternative way to overcome the problem — the synthesis of nonplanar InGaN nanostructures of complex shape on the surface of a silicon substrate — is very promising. Due to the complex morphology, the mechanical stresses in semiconductor nanostructures relax and the number of defects decreases significantly [5]. However, it was possible to synthesize structures based on InGaN nanostructures of complex shape on silicon only using pre-growth surface preparation procedures [6].

This paper presents the results of experimental studies on the synthesis by molecular beam epitaxy (MBE) and studies of morphological and optical properties of InGaN nanostructures of branched morphology directly on the unprepared silicon surface.

2. Experimental procedures

Nanowires were grown by MBE using Riber Compact 12 setup equipped with Ga and In effusion sources and nitrogen plasma source First, Si(111) substrate was transferred into the growth chamber and the substrate temperature was set to 950°C for thermal treatment. Then the substrate temperature was lowered to the actual growth temperature of 700°C. The nitrogen plasma source is ignited and the N flux is set at 0.4 scam and 400 W. Finally, the gallium (Ga) and indium (In) sources were opened. Ga and In partial pressures were equal and set to 10^{-7} Pa as pre-specified by the Bayard-Alpert detector. The total growth time for a InGaN nanostructures was about 20 hours.

The surface morphology and optical properties were studied by scanning electron microscopy (SEM) and room-temperature photoluminescence (RT PL) techniques.

3. Results

Typical scanning electron microscopy image of InGaN nanostructures grown on Si(111) substrate are shown in Figure 1. It is seen that synthesized InGaN nanostructures have a complex morphology. The average height of the grown structures was 5 μm. It should be noted that, based on the figure, the synthesis of InGaN nanostructures occurred in several stages: first, nanocolumns were formed, then a transition layer, and, finally, nanostructures with more complex morphology. Probably, different stages of growth correspond effectively to different (despite the constancy of its maintenance by thermocouple measurements) temperature on the surface, which decreases with increasing height of the nanostructure. This may be due to a change in the thermal absorption coefficient of the growing structure due to the formation of comlex surface [7, 8].

Figure 1. SEM image of InGaN nanostructures grown on Si(111) substrate

Figure 2 shows the photoluminescence spectrum at room temperature of the grown sample. The measurement results showed that the InGaN nanostructures are optically active up to room temperature. This fact indicates the high optical quality of the grown structures. It can be seen from the figure that InGaN nanostructures emit in a wide range of wavelengths - from 400 to 900 nm. We assume that this is due to a change in the percentage ratio of In and Ga in a solid solution, which may be caused, inter alia, by an effective change in surface temperature during growth. Such nanostructures are promising for creating white LEDs based on a single material.

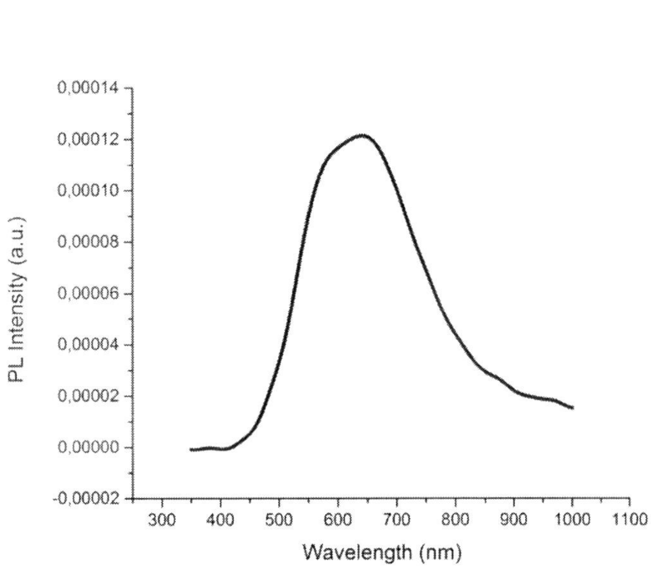

Figure 2. RT PL spectra K of InGaN nanostructures grown on Si(111) substrate.

4. Conclusions

In summary, we have demonstrated the possibility of synthesizing InGaN nanostructures by the method of molecular beam epitaxy on the unprepared surface of the silicon substrate. The developed growth technology makes it possible to synthesize InGaN nanostructures of branched morphology without the use of pre-growth surface preparation procedures, which contributes to a significant reduction in the price and time to create such nanostructures. The results of morphological studies have shown that growth occurs in several stages. The grown structures turned out to be optically active in a wide range of wavelengths at room temperature. All these facts indicate the promise of such structures for optical applications.

Acknowledgments

The study was supported by the Ministry of Education and Science of the Russian Federation under State assignment no. 16.24834.2017/4.6. The samples were grown under support of Russian Science Foundation (project No 19-72-30010).

References

[1] Ambacher O Growth and applications of group III-nitrides *Journal of Physics D: Applied Physics* **31** 2653 (1998)
[2] Moustakas T D, Iliopoulos E, Sampath A V, Ng H M, Doppalapudi D, Misra M, Korakakis D, Singh R Growth and device applications of III-nitrides by MBE *Journal of Crystal Growth* **13** 227 (2001)
[3] Bessolov V N, Botnaryuk, V M, Zhilyaev Y V, Konenkova E V, Poletaev N K, Raevskiĭ S D, Rodin S N, Smirnov S L, Sharofidinov Sh, Shcheglov M P, Park H S, Koike M Chloride vapor-phase epitaxy of gallium nitride on silicon: Structural and luminescent characteristics of epilayers *Technical physics letters* **32** 674 (2006)
[4] Reznik R R, Soshnikov I P, Kukushkin S A, Osipov A V, Talalaev V G and Cirlin G E The use of SiC/Si hybrid substrate for MBE growth of thick GaN layers *AIP Conference Proceedings* **2064** 040004 (2019)
[5] Dubrovskii V G, Cirlin G E and Ustinov V M Semiconductor nanowhiskers: Synthesis,

properties and applications *Semiconductors* **43** no 12 1539-1584 (2009)
[6] Hu F R, Ochi K, Zhao Y, Hanea K High-efficiency light-emitting column-crystallized InGaN/GaN quantum-well flower structure on micropillared Si substrate *Applied Physics Letters* **89** 171903 (2006).
[7] Glas F, Harmand J-C Calculation of the temperature profile in nanowhiskers growing on a hot substrate *Phys. Rev. B* **73** 155320 (2006)
[8] Sibirev N V, Soshnikov I P, Dubrovskii V G, Arshansky E Temperature profile along a nanowhisker growing in high vacuum *Technical Physics Letters* **32** 292 (2006).

Fabrication and luminescent properties of(In,Fe)Sb/GaAs/InGaAs diodes

M. V. Ved, M. V. Dorokhin, V. P. Lesnikov, A. V. Zdoroveyshchev, Yu. A. Danilov, P. B. Demina and A. V. Kudrin
Research Institute for Physics and Technology of Lobachevsky State University of Nizhny Novgorod, Nizhny Novgorod, 603950, Russia

Abstract. The electrical and luminescent characteristics of InGaAs/GaAs spin light-emitting diodes with an injector based on (In,Fe)Sb were investigated and the circular polarization of the electroluminescence was obtained for such structure. It has been established that the deposition of (In,Fe)Sb layer does not introduce any additional defects into the region of near-surface quantum wells, but directly affects the (In,Fe)Sb/GaAs interface. It was found that the application of a thin protective layer of MgO between the ferromagnetic (In,Fe)Sb injector and the light-emitting structure minimizes this effect.

1. Introduction

Diluted magnetic semiconductors (DMS) are materials doped with transition element atoms. It possesses both ferromagnetic and semiconducting properties. Such materials can be used as the elements of spintronic devices, for example, as an injector in spin light-emitting diodes (SLED) [1]. Today one of the most promising magnetic semiconducting materials are (A3,Fe)B5 DMS and, in particular, (In,Fe)Sb, since its Curie temperature exceeds 300K [2]. At the current stage of (A3,Fe)B5 technology level, the most important task is the effective integration of (A3,Fe)B5 into devices based on the most common optoelectronic material such as GaAs. In this work the fabrication of (In,Fe)Sb layer by the method of pulsed laser deposition (PLD) and its effect on the properties of an (In,Fe)Sb/InGaAs/GaAs spin light-emitting diodes was investigated.

2. Experimental technique

The samples were fabricated in four stages. At the first stage, two semiconductor heterostructures were grown by metal-organic vapor phase epitaxy at temperature of 600°C. The first structure was grown on the n-GaAs substrate and consisted of three $In_xGa_{1-x}As$/GaAs quantum wells (QW) with different indium content (x_1=0.12; x_2=0.15; x_3=0.2) which located at different distances from the surface of the structure (30 nm, 70 nm and 110 nm respectively). The second heterostructure was grown on the p-GaAs substrate and consisted of single InGaAs/GaAs QW, which was located at 15 nm from the surface of the structure. At the next stage, the structures were coated with a 1 nm protective layer of MgO to prevent exposure by the laser plasma during the following deposition of the (In,Fe)Sb layer using the PLD method, and also to prevent the diffusion of Fe atoms into the light-emitting structure. At the third stage, an (In,Fe)Sb layer with a thickness of 40 nm was deposited on samples with a protective MgO layer and on the initial sample of the first structure at a temperature of 200°C by PLD technique. At the last stage of diode fabrication, Au contacts were deposited onto samples with an (In,Fe)Sb injector by electron beam evaporation in vacuum, then 500 μm mesa structures were made by photolithography and chemical etching, and a basic Ohmic contact was formed. Thus, three samples were formed on the basis of the first structure with 3 QWs: 1 - SLED with an (In,Fe)Sb injector and a protective layer of MgO; 2 - SLED with an (In,Fe)Sb injector **without** a protective layer; 3 - the original light-emitting structure without contacts. In addition one sample of the second structure with one QW was formed: 4 – SLED with an (In,Fe)Sb injector and a protective layer of MgO. The schematic representation of the sample structures are shown at figure 1.

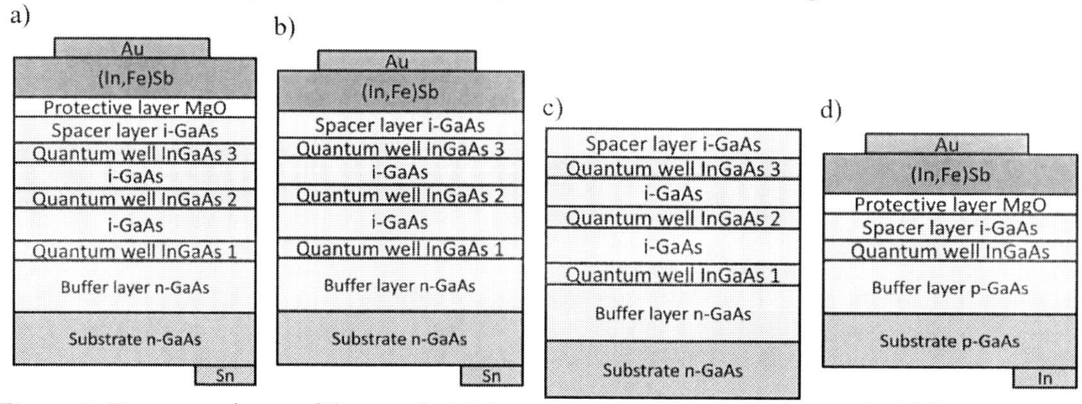

Figure 1. Structure scheme of the sample 1 (**a**); sample 2 (**b**); sample 3 (**c**); sample 4 (**d**).

The samples of the first structure (samples 1, 2 and 3) were investigated to determine the effect of the (In,Fe)Sb layer on the radiative characteristics of LEDs. The photoluminescence (PL) spectra were measured for samples 1, 2 and 3. For the excitation of photoluminescence, a He-Ne laser with a wavelength of 632 nm was used. For samples 1 and 2 the Au and (In,Fe)Sb layers were etched before measuring photoluminescence. The electroluminescence (EL) spectra were measured for diodes with an injector of (In,Fe)Sb (samples 1 and 2). For electroluminescence studies, a forward bias was applied to the samples (a positive potential with respect to the substrate was applied to Au contact). Excited EL radiation was detected from the substrate side. Measurement temperature was 77K.

The sample 4 was used to investigate the magnetic field dependence of the degree of circular polarization (P_{EL}) caused by a spin injection from a magnetized (In,Fe)Sb into GaAs. When structure was introduced into perpendicular magnetic field, the EL emission becomes partially circularly polarized. The degree of circular polarization P_{EL} is calculated by the formula:

$$P_{\text{EL}} = \frac{I(\sigma^+) - I(\sigma^-)}{I(\sigma^+) + I(\sigma^-)}, \quad (1)$$

where $I(\sigma^+)$, $I(\sigma^-)$ are intensities of EL components polarized along the left and right circles, respectively.

The P_{EL} magnetic field dependence of sample 4 was measured at a temperature 10 K in a closed-loop cycle cryostat.

3. Results and discussion

The photoluminescence spectra of the samples 1, 2 and 3 are presented at Figure 2a.

The PL spectra show 3 peaks corresponding to radiative transitions in quantum wells with the corresponding In content. It can be seen that the PL intensities are the same in samples with (In,Fe)Sb and in initial structure. This indicates that the concentration of non-radiative recombination centres in the quantum well region did not change significantly after the deposition of (In,Fe)Sb layers. Consequently, the deposition of (In,Fe)Sb layer did not introduce any additional defects in the region of near-surface quantum wells.

The EL spectra of samples 1 and 2 are shown in Figure 2b. As can be seen from the presented spectra, in the (In,Fe)Sb/GaAs diode without the protective MgO layer, the EL peak from the closest-to-the-surface QW (with the highest In content) is completely quenched.

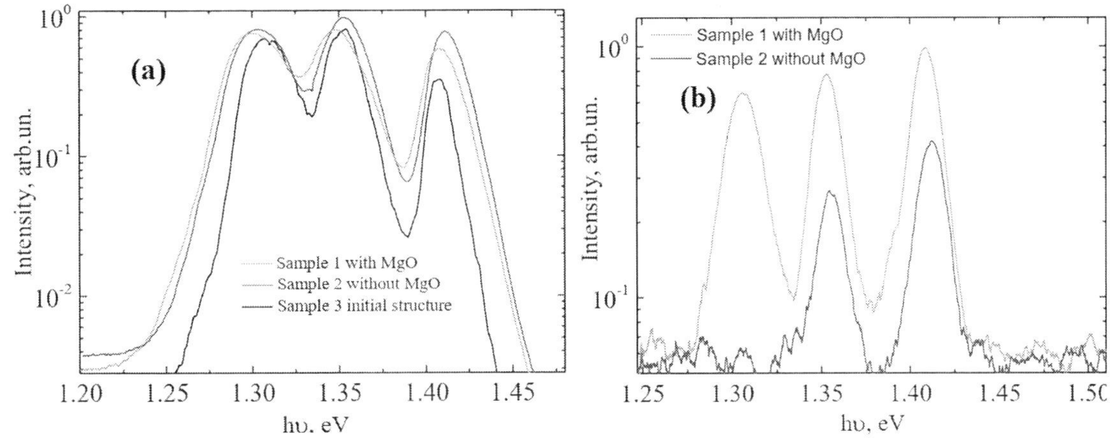

Figure 2. **(a)** PL spectra of samples 1, 2 and 3; **(b)** EL spectra of studied samples 1 and 2, current through the structures = 100 mA. Measurement temperature in both cases = 77K.

When applying a protective layer of MgO between the ferromagnetic injector and the light-emitting part of the SLED, the mentioned above peak is maintained, and the integral intensity of EL is also increased by several times. This leads us to a conclusion that (In,Fe)Sb deposition on the top of GaAs influences the properties of the surface. The MgO deposition prior to (In,Fe)Sb allows decreasing this influence.

Presumably, the quenching of EL from the nearest to the surface QW is associated with the formation of charged states in the surface region of GaAs, which create a strong electric field in the region of the first QW, even in the forward bias mode. In structures with MgO layer, the number of charged centers and the magnitude of inner electric field decrease, that significantly reduces the escape of carriers from the first QW and makes it possible to detect EL radiation.

Since the MgO deposition has shown the improvement of (In,Fe)Sb/GaAs interface quality, we have fabricated a spin-light emitting diode structure with (In,Fe)Sb/MgO injector (structure 4 at Fig.1).

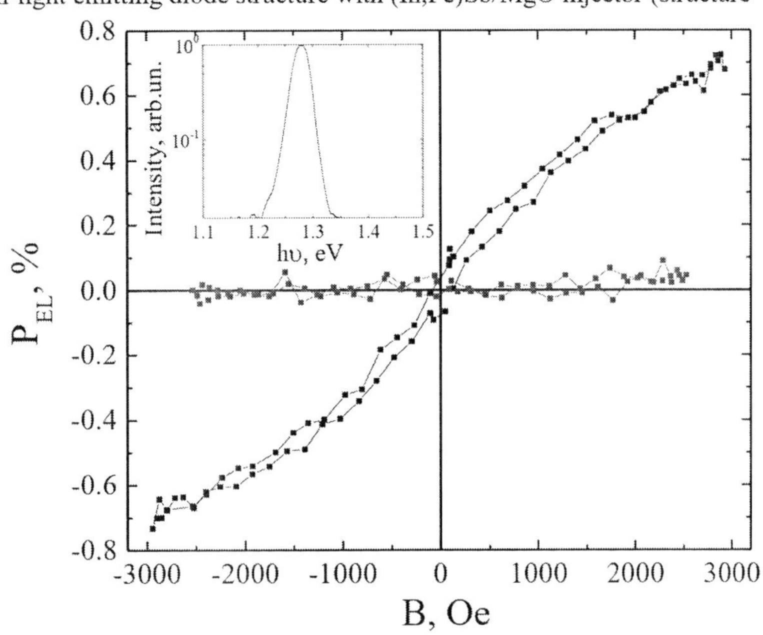

Figure 3. Magnetic field dependence of P_{EL} for investigated sample 4 and control structure without (In,Fe)Sb injector. The inset shows the electroluminescence spectrum for sample 4.

Figure 3 shows the magnetic field dependence of EL circular polarization degree (P_{EL}(B)) for sample 4 measured at temperature 10K.
As can be seen from Fig.3, the P_{EL}(B) is a non-linear function of the applied magnetic field and the circular polarization value at the magnetization saturation is rather high (~ 0.7%). Such type of dependence is associated with the injection of spin-polarized electrons from ferromagnetic layer. It should be noted, that in similar sample but without MgO the EL was not detected. This confirms a conclusion, that MgO is a protective layer, which prevents the negative impact of (In,Fe)Sb on the luminescence characteristic of the structure. Also it should be noted, that the circular polarization of EL of a similar structure without (In,Fe)Sb layer was investigated (Au contact was directly deposited on the GaAs spacer layer). Magnetic field dependence of P_{EL} for such a structure is shown on the figure 3. The value of the degree of circular polarization for control structure did not exceed 0.1%.

4. Conclusions

Thus, (In,Fe)Sb/GaAs/InGaAs SLED were formed and studied, and methods for increasing the intensity of EL by modifying the ferromagnet/semiconductor interface were considered. Also in this work, the spin injection of electrons from the DMS (In,Fe)Sb into the semiconductor structure based on GaAs with further recombination and emission of EL radiation from the QW region was obtained. The magnetic field dependences of the degree of circular polarization P_{EL} were measured at temperature 10K.

Acknowledgments

Authors would like to acknowledge Dr. B.N. Zvonkov for the fabrication of investigated samples.
This work was supported by the Russian Science Foundation - project No. 18-79-10088 (fabrication of samples) and the Ministry of Education and Science of Russian Federation - projects No. 8.1751.2017/PCh and SP-2450.2018.5 (luminescence measurements).

References

1. M. Holub, P. Bhattacharya // J. Phys. D: Appl. Phys., **40**, P.R179–R203 (2007).
2. A. V. Kudrin, *et al.*// Journal of Appl. Phys., **122**, 183901 (2017).
3. W.Hsu, R.Raj// Appl. Phys. Lett., **60**, 3105 (1992).

ZnO thin films synthesis by RF magnetron sputtering deposition

V.O.Gridchin[1], K.P.Kotlyar[1], A.V.Vershinin[1], N.V.Kryzhanovskaya[1], E.V.Pirogov[1], A.A.Semenov[2], P.Y.Belyavskiy[2], A.V.Nashchekin[3], G.E.Cirlin[4], I.P.Soshnikov[1,3]

[1]St. Petersburg Academic University, St. Petersburg 194021, Russia
[2]St. Petersburg Electrotechnical University, St. Petersburg 197376, Russia
[3]Ioffe Physical Technical Institute RAS, St. Petersburg 194021, Russia
[4]ITMO University, St. Petersburg 197101, Russia

Abstract. Investigation of ZnO thin films synthesis using RF magnetron sputtering deposition in the Ar/O_2 plasma gas mixture at different O_2 composition and at different growth temperatures is presented. The effect growth process on structural (morphology and orientation films, grain sizes, lattice parameters) and optical (transmittance, absorption, refractive index, photoluminescence) properties are examined. It is shown that synthesized ZnO thin films at a temperature from 200 to 300 °C are polycrystal and textured. Notice that ZnO thin films, which are synthesized using pure Ar plasma gas, are porosity. The optical transmittance and absorption spectra have an absorption edge near 380 nm. The PL spectra of synthesized ZnO films show exciton peaks at 370-400 nm and the wide emission at the yellow and red spectral regions.

1. Introduction

ZnO thin films attract the attention of researchers due to it promising properties such as direct bandgap with E_g = 3,3 eV at room temperature, transmittance for the visible light and radiation stability [1,2]. In addition, undoped ZnO films show n-type conductivity and the largest piezoelectric response d_{33} among the tetrahedrally bonded semiconductors [3]. Accordingly, ZnO thin films can be applied like transparent electrodes for solar cells [2,4], optoelectronics [5] and flexible electronics [4,6].
Radio frequency magnetron sputtering deposition (RF MSD) is perspective methods for the synthesis of ZnO structures (films, nanostructures) due to precisely control of the processes synthesis and material properties [7,8]. Depending on the objective, the method allows to synthesize ZnO thin films at both low and high temperatures. However, literary data about of implementation of the technological process allowing synthesis of high quality planar ZnO thin films is not enough.
Therefore, the work is devoted investigation of the effect of technological conditions (plasma gas composition) on structural, optical and electrical properties of ZnO thin films grown using RF MSD at growth temperature.

2. Experiments

ZnO thin films are synthesized using VUP7 equipment with RF MSD system (SELMI, Ukraine). Polycrystal ZnO wafer (diameter 3") manufactured by Pi-KEM Ltd is used as a sputtering target. The growth of ZnO films are performed on Si (111) and quartz glass substrates. The distance of sample - target is fixed about 85 mm. The residual and work pressures are constituted $5 \cdot 10^{-3}$ Pa and 1 Pa, respectively. The Ar/O_2 plasma gas mixture with the different O_2 composition (x = 0, 5, 10, 30 %) is exploited. The magnetron power is about 140 W. The thin films are grown at substrate temperatures and exposition from 200 to 300 °C and 30 min, correspondently.

The structural properties of the ZnO films are examined using scanning electron microscopy SUPRA 25 (C. Zeiss SMT Ltd), JSM 7001F (JEOL, Japan) and X-ray diffractometer DRON-8 with Cu Kα radiation. The optical transmission spectra of ZnO films are measured using spectrometer Solar Laser Systems S100. The refractive index is determined using ellipsometer Horiba Jobin Yvon with He-Ne laser (632.8 nm). The PL spectra are examined using 15 mW He-Cd laser (325 nm) at room temperature.

3. Results and discussion

3.1 Structural properties

Fig. 1 demonstrates morphology ZnO films which are fabricated at different oxygen composition x of the plasma gas mixture. It is shown that the ZnO films grown using pure argon plasma gas have a porosity and high roughness. The size of the grains is about 50 – 250 nm. The oxygen addition in the plasma gas results to growth of uniform polycrystal films with crystallites oriented along the normal to surface. ZnO films which are grown at $x = 5$ % have basalt texture nanocrystals. Size of nanocrystals is about from 50 to 100 nm in the lateral direction. The increasing of the oxygen composition up to 30% gives rise to the formation of planar continuous films.

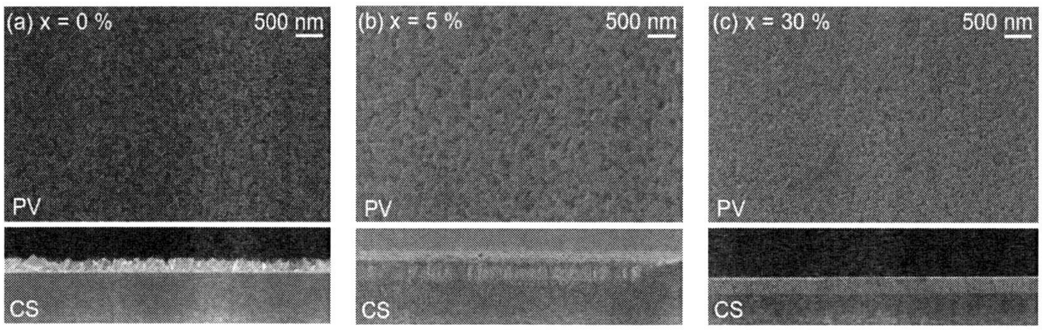

Figure 1. SEM images of the ZnO films grown at oxygen content $x = 0$ % **(a)**; 5 % **(b)**; 30 % **(c)**.

The diffraction angle 2θ and the value of the full-width at half-maximum (FWHM), the average crystallite sizes D, the lattice parameter *c*, the average uniform strain ε and the stoichiometric composition O/Zn are given in table 1.

Table 1. Structural parameters of the ZnO thin films

Sample	2θ (°)	FWHM (°)	D (nm)	*c* (nm)	ε (10^{-3})	Composition (X ray microanalysis) O/Zn
ZnO, x = 0 %	-	-	-	-	-	0.75
ZnO, x = 5 %	34.41	0.572	27	0.5208	-0.17	0.95
ZnO, x = 10 %	-	-	-	-	-	1.0
ZnO, x = 30 %	34.32	0.740	20	0.5221	2.369	1.0

The results show that peak at a diffraction angle of 2θ ~ 34.4° is exhibited for samples which are synthesized with $x = 5$% and $x = 30$ %. The diffraction peak position of 2θ correspond to the reflection from the (0002) plane ZnO (JCPDS card No. 36-1451). Accordingly, the ZnO films are polycrystalline and textured along the normal to the surface. At the same time, samples which are synthesized at the pure argon and at the oxygen composition of 10 %, shows low intensity (0002) peaks. This is probably due to the porous structure and crystal twinning in the observed samples, relatively.
The diffraction peak position of 2θ shifts towards lower angle compared to bulk ZnO value for samples which are grown under $x = 30$ %. The shift of the diffraction angle indicates the presence of residual strain in the films. The average uniform strain ε and the lattice parameter *c* show that the sample has tensile deformation along the c-axis direction. Whereas ZnO thin film deposited at oxygen composition

of 5 % shows properties close to a bulk crystal (c_{bulk} = 5.2075 Å [1]). The FWHM of the films deposited under oxygen content of 5 % and 30 % are 0.572° and 0.740°, respectively. The smaller FWHM value indicates that the ZnO films are more textured along the c-axis direction.

The stoichiometric composition O/Zn thin films synthesized at pure argon is deviated in the direction of increasing zinc. Thin films synthesized at oxygen composition of about 10 and 30 % has stoichiometric composition 1 to 1.

3.2 Optical properties

Fig. 2 (a) shows optical transmittance spectra of ZnO films deposited at different oxygen composition in the plasma gas. Fig. 2 (b) shows the plots of $(\alpha h\nu)^2 = h\nu$ of the ZnO thin films according to Tauc equation [9].

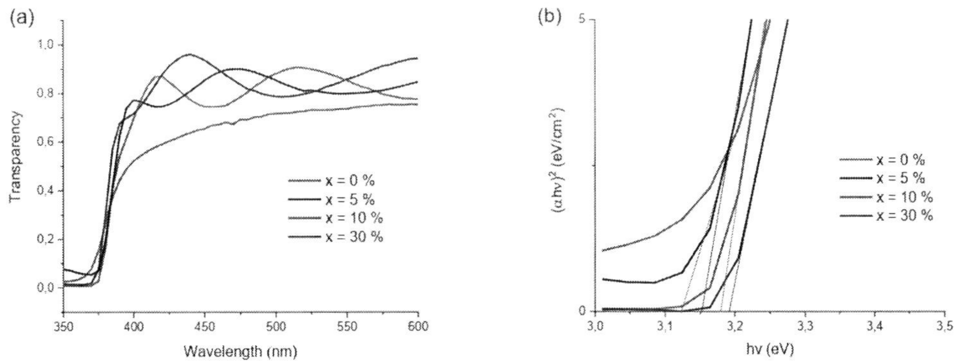

Figure 2. (a) Optical transmittance spectra of the ZnO films; **(b)** Plots of $(\alpha h\nu)^2 = h\nu$ for the ZnO films.

As illustrated in Fig. 2 (a), the optical transmittance spectra have an absorption edge in the area of 380 nm. Samples synthesized using pure argon plasma demonstrate a decreasing the optical transmittance in the visible spectral range. Apparently, the result is explained by the deviation of the stoichiometric composition in the direction of increasing zinc. The ZnO films synthesized at oxygen composition in the plasma gas shows high transparency.

The band gap is estimated by Tauc plot as illustrated in Fig. 2 (b). The optical band gap energy increases from 3.12 to 3.19 eV at increasing oxygen composition in the plasma gas. The increasing optical band gap is associated with reducing Zn interstitial atoms and O vacancy.

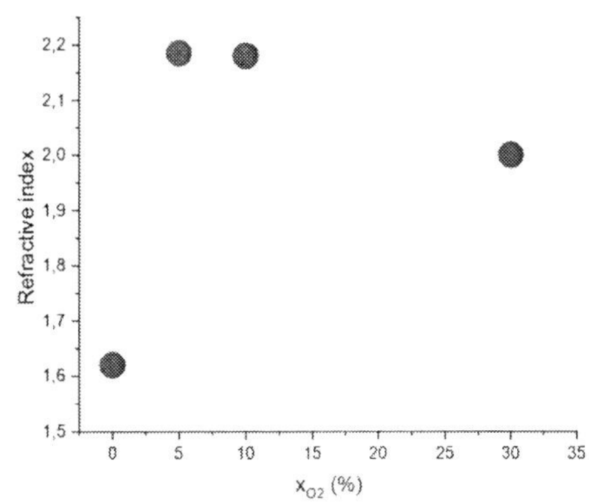

Figure 3. The refractive index vs the oxygen composition in the plasma gas mixture.

Fig. 3 shows the dependence of the refractive index on the oxygen composition in the plasma gas. The results of the ellipsometry study show that the samples synthesized using Ar/O_2 gas mixture at the oxygen composition of 30% have a refractive index is about 2.01 that near for ZnO bulk crystal. The samples synthesized at the pure argon gas show the refractive index about 1.74 that can be explained by the porosity of the structure. The refractive index has a maximum value of 2.19 on the samples synthesized in oxygen content of 5 and 10 %.

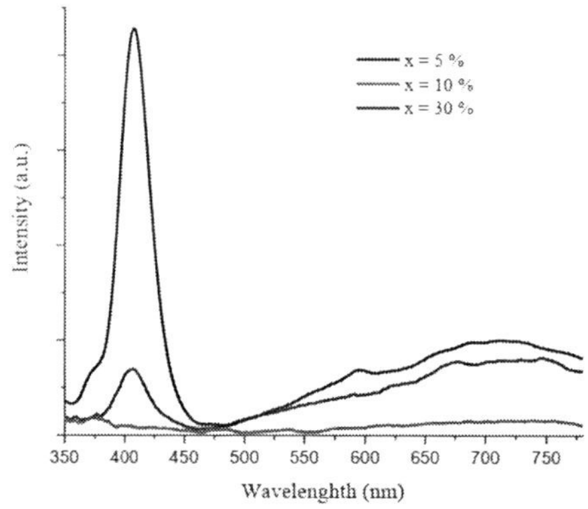

Figure 4. Photoluminescence spectra of the ZnO thin films at 77 K.

Fig.4 shows photoluminescence spectra (PL) of the ZnO thin films that are deposited at the oxygen composition of 5, 10, 30 % in the plasma gas. The PL spectra of the synthesized thin films contain the main emission peak in the region about 380 nm. The peak is associated with near bandgap emission. In addition, emission bands are found in the green and red regions. The luminescence lines are associated at different defects in the materials [1].

4. Conclusion

The effect of the growth temperature in diapason from 200 to 300 °C on structural and optical properties of ZnO films synthesized by RF MSD in O_2/Ar gas mixture is weak. It is shown that process with pure argon plasma gives non stoichiometric porosity ZnO films at reflectivity index about 1.65. The ZnO films synthesized by RF MSD in oxygen composition of the gas mixture are stoichiometric solid polycrystal at reflectivity index and energy band gap about 2.0-2.1 and 3.12-3.2 eV. The results can be applied for the synthesis ZnO films for different applications.

Acknowledgements

The study of the synthesis of ZnO thin films by RF magnetron sputtering deposition was financially supported by the Russian Foundation for Basic Research (project 18-07-01364 A). The electron microscopy investigations were carried out at the Federal Center for Collective Use "Materials Science and Diagnostics in Advanced Technologies" at the Ioffe Institute.

References

[1] Özgür Ü et al 2005 *J. Appl. Phys.* **98** 11
[2] Ellmer K, Klein A and Rech B vol 104 2007 *Transparent Conductive Zinc Oxide: Basics and Applications in Thin Film Solar Cells* (Springer Science & Business Media)
[3] Dal Corso A et al 1994 *Phys. Rev. B* **50** 10715
[4] Wang Z L, Wu W and Falconi C 2018 *MRS Bulletin* **43** 922–7
[5] Tang K et al 2018 *Appl. Surf. Sci.* **433** 177–80
[6] Zhang Y, Wu Z, Li P, Ono L K, Qi Y, Zhou J, Shen H, Surya C and Zheng Z 2018 *Adv. Energy Mater.* **8** 170156
[7] Suo X et al 2019 *Surf. Coat. Tech.* **357** 978–83

[8] Kumar G A, Reddy M V R and Reddy K N 2015 *IOP Conf. Ser. Mater. SCI. ENG* **73** 012133
[9] Tauc J 1968 *Mater. Res. Bull.* **3** 37–46

Simulation of AlGaAs-heterostructures CVC kinetic due to degradation

V D Shasurin[1], N A Vetrova[1], S A Meshkov[1], Y A Ivanov[1], E V Kuimov[1]
[1]Department of Radioelectronics and Laser Technologies, Bauman Moscow State Technical University (National Technological University), Moscow 105005, Russia Federation

Abstract. A model of degradation of AlGaAs-heterostructures has been developed. Parasitic resistance growth and diffusion of aluminum and donor impurities are considered as the main reasons of degradation of the structure. Current transfer models are generalized to the case of structures with continuous impurity distribution profiles. Algorithms based on numerical methods with reduced time complexity have been developed. RTS CVC was calculated before and after the annealing operation in order to verify the models. According to the results of the calculation, it was concluded that the developed algorithms are advisable for reliability prediction applications.

1. Introduction

Nanoelectronic devices based on semiconductor A_3B_5 nanoscale heterostructures are of interest as the basis of the new generation of superperfomance devices [1,2]. Research results [3,4] show that such devices have a potential for high reliability applications. Gallium-aluminum arsenide (AlGaAs) is the most common material for heterostructures of the A_3B_5 compounds. Silicon doping is usually used to control the conductivity of the contact layers in such structures.

Reliability prediction of AlGaAs-heterostructures task requires refusal to classical statistical reliability theory in favor of developing a gerontological physical and mathematical model of devices [5]. The main problems of creating such model is a simulation of degradation of AlGaAs-heterostructures due to aluminum and silicon diffusion in the heterostructure layers, simulation of ohmic contacts blurring [6,7] and generalization of the models of transfer of electric current to the structures suscseptible to degradation.

2. Simulation

The gerontological model of AlGaAs heterostructures is based on models of aluminum and silicon diffusion processes in gallium arsenide (which causes shifting and broadening of metastable electron energy levels) and metal of ohmic contacts penetration into the contact layers, as a result their resistance increases. In the framework of the method of equivalent circuits serial resistance took into account the conductivity of the contact layers.

Fick equation with zero flow boundary conditions was used to simulate diffusion. Such conditions are chosen as the mathematical implementation of the prohibition of diffusion of atoms out of the modeled region. Piece-constant distribution of impurities was used as initial condition.

The models for accounting for dissipative processes described in [8, 9] has been developed to study current transfer in structures with an aluminum profile with a jump discontinuity at heterointerfaces. Obviously, these models require generalization for structures subject to diffusion profile changes, when the heterointerface cannot be unambiguously determined. In the framework of the two-valley model of current transfer, the Schrödinger equations are used for describing the electron motion in the active region for each valley with a communication operator, which is determined using the coordinates of heterointerfaces. To generalize the two-band model to the case of a continuous aluminum profile, the following definition of a communication operator was used [10]

$$A = \alpha \left| \frac{dx}{dz} \right| \quad (1)$$

where α – intervalley interaction constant, x – aluminum molar fraction, z – coordinate along the axis perpendicular to the layers of the heterostructure.

The inelastic electrons scattering by phonons were simulated by including a complex optical potential into the Hamiltonian. This potential is zero everywhere, except for the layers that form the quantum well. Thus, inelastic scattering is included in the model as an additional mechanism that allows electrons to pass through potential barriers [11]. If there are no explicit heterointerfaces, it is difficult to determine the regions where the optical potential takes nonzero values. Diffusion blurring of the optical potential profile was simulated to solve this problem. From the assumption that the optical potential is related to the fraction of aluminum, it follows that the diffusion coefficient of the optical potential should be used to calculate the diffusion of the optical potential.

The finite difference method without reducing accuracy order while taking into account the boundary conditions was used for the numerical solution of differential equations. Finite difference schemes were optimized for five-diagonal Thomas algorithm to reduce the time complexity. Elements of finite difference equations are obtained by averaging the members of the Fick and Schrödinger equations on one grid, which allowed avoiding loss of accuracy when interpolating the aluminum molar fraction and the concentration of silicon from the nodes of one grid to another.

3. Results

In order to verify the developed algorithms, diffusion changes of resonant tunnel structures (RTS) were calculated at the technological stage of annealing and their current-voltage characteristics (CVC). Annealing for 30 s at a temperature of 800°C was simulated [12–15]. The simulation was carried out self-consistently taking into account dissipative processes (the number of self-consistency iterations is 20, the inter-valley interaction constant is 0.7 eV Å, the optical potential in the quantum well is 0.02 eV). The calculation results in comparison with the experimental data are shown in Figure 1.

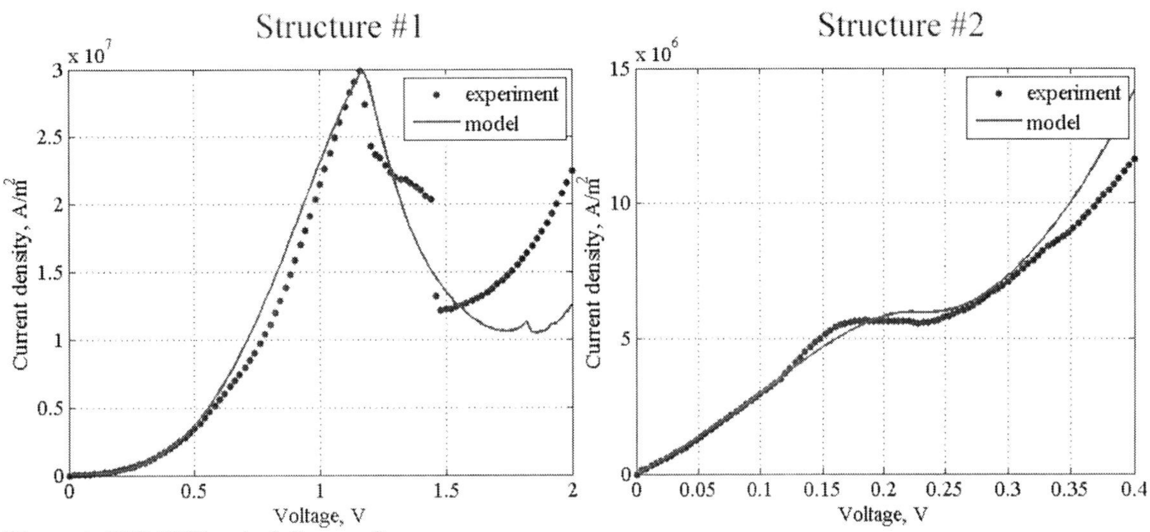

Figure 1. RTS CVCs calculation results.

Since the calculated structures were developed for use in radio signal mixers, which use the initial section of CVC, the relative deviation in the form of the initial section was used as an indicator of the discrepancy between the calculation results and the experiment. It's calculated by formula (2)

$$\Delta C = \int_{x_0}^{x_1} \left(\frac{1}{J_{pc}} J_c(xV_{pc}) - \frac{1}{J_{pe}} J_e(xV_{pe}) \right)^2 dx \cdot \left[\int_{x_1}^{x_2} \left(\frac{1}{J_{pe}} J_e(xV_{pe}) \right)^2 dx \right]^{-1} \quad (2)$$

where $J_c(V)$ and $J_e(V)$ – calculated and experimental CVC's, V_{pc} and V_{pe} – peak bias on calculated and experimental CVC's, J_{pc} and J_{pe} – peak current density on calculated and experimental CVC's.

Such evaluation of the relative deviation in the form of the initial section allows obtaining information on the discrepancy between the calculation and the experiment both in the entire initial section of the IVC and in its individual parts. The evaluation results of RTS CVCs in Figure 1 are presented in Figure 2.

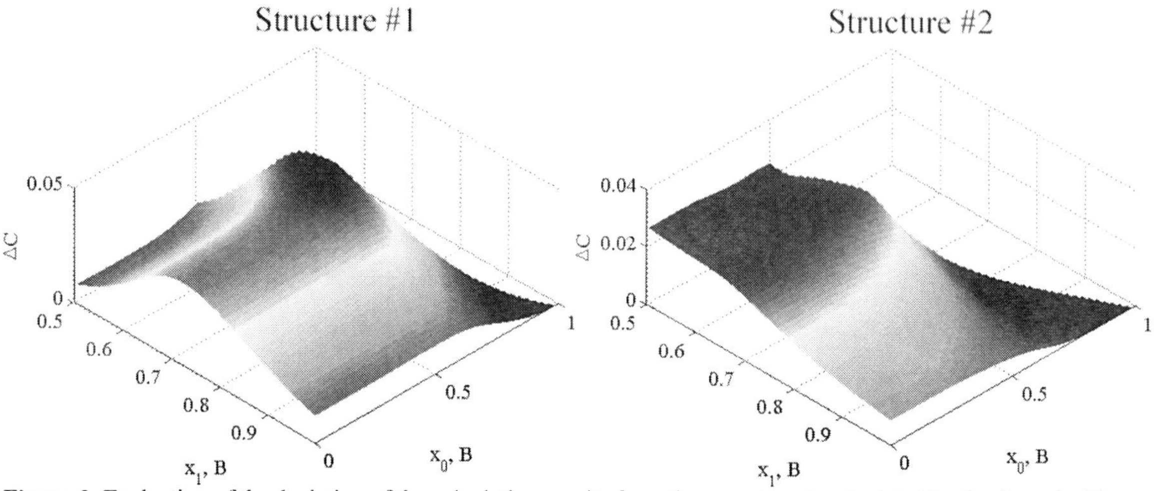

Figure 2. Evaluation of the deviation of the calculation results from the experiment, calculated by the formula (2)

As can be seen from Figure 2, the greatest deviation of the calculation results from the experimental data on the curvature of the initial section is less than 5% in different segments of the initial section. The relative deviation of the calculation results from the experiment on the curvature of the whole ($x_0 = 0$, $x_1 = 1$) of the initial segment does not exceed 3%.

4. Conclusion

Algorithms for calculating the kinetics of the CVC of AlGaAs-heterostructures under hard temperature stress have been developed. Current transfer models are adapted to the profile of aluminum profile to diffusion changes. By comparing the calculation results and experimental data, it was found out that the presented algorithms make it possible to calculate CVCs of the RTS based on AlGaAs-heterostructures with an error of less than 3% in the form of the initial section. Based on the results of verification, it was concluded that it is advisable to use the developed software to predict the reliability indicators of devices whose performance characteristics are determined by the initial part of heterostructures CVC.

Acknowledgments

This work was supported by the Ministry of Education and Science of the Russian Federation (project No. 16.1663.2017/4.6).

References

[1] Nadar S, Zaknoune M, Wallart X 2017 IEEE Trans. Terahertz Sci. Technol. **7** 780
[2] Magruder C K, Levi A F J 2012 Physica E **44** 1503
[3] Sin Y, Presser N, Lingley Z, Brodie M, Foran B, Moss S 2016 Proc. of SPIE **9733** (973304)
[4] Bonsendorf D, Schneider S, Meinschien J, Tomm J W 2016 Proc. of SPIE **9733** (973302)
[5] Lu Y, Nie Z, Zhang P, Wang Z, Xiong L, Wang S, Wu D, Liu X 2016 17th International Conference on Electronic Packaging Technology (Wuhan, China) pp 76-83
[6] Makeev M O, Ivanov Y A, Meshkov S A, Litvak Y N, Vetrova N A 2013 Engineering Journal: Science and Innovation **6** 48.
[7] Ushanov V I, Chaldyshev V, Preobrazhenskii V V, Putyato M A 2018 Semiconductors **52** 1704
[8] Abramov I I 2016 Osnovi modelirovania elementov micro- I nanoelectroniki (Saarbrücken, Germany)
[9] Bennabi Y, Hammou A B, Nouredine Z 2011 Mod. Phys. Lett. B **25** 1691
[10] Vetrova N A, Ivanov Y A, Kuimov E V, Makeev M O, Meshkov S A, Pchelintsev K P, Shashurin V D 2018 Radioelectronic. Nanosystems. Information Technologies **10**(1) pp 71-76
[11] Shashurin V D, Vetrova N A, Nazarov V V, Rudenko N R, Kuimov E V 2018 Journal Nanotechnology: development and applications - XXI Century **10**(1) pp 40-46

[12] Macherzynski W, Gryglewicz J, Stafiniak A, Prazmowska J, Paszkiewicz R 2016 Advances in Electrical and Electronic Engineering **14**(1) pp 83-88
[13] Piotrowska A 1993 Acta Physica Polonica Series A **84**(3) pp 491-504
[14] Nishi H, Inada T, Saito J, Ishikawa T, Hiyamizu S 1983 Jpn. J. Appl. Phys. **22** pp 401-404
[15] Yanxu Z, Weiwei C, Yuyu F, Ye D, Chen X 2014 Journal of Semiconductors **35**(2) p 6004

Optical activity and SEM of chiral chitosan ascorbates

O N Malinkina, Yu Yu Zhuravleva, I V Zudina, A B Shipovskaya
Department of High Molecular Compounds, Saratov State University, Saratov 410012, Russia

Abstract. The effect of the chiosan ascorbate chirality on its structural features in solution and in the solid state was studied using optical rotation dispersion and scanning electron microscopy. It has been established that the isomeric form of ascorbic acid has a significant impact on the stereomeric ordering of chiral chitosan ascorbate macromolecules in solution and solid-phase supramolecular structures isolated therefrom.

1. Introduction

Recently, scientific interest in chiral materials [1, 2], in particular, chitosan-containing ones, grows. Despite the large variety of bio-applications, chirality is the least studied property of chitosan (CS). The optical activity of this aminopolysaccharide is known to be due to the presence of asymmetrically substituted carbon atoms in glucopyranose rings, the conformation of the unit and the molecule as a whole, which predetermine the helicoidal self-ordering of macrochains in solution [2-7]. The influence of the molecular weight of chitoligomers and chitosan [3-5], the degree of deacetylation and the overlapping concentration of macromolecular coils in standard solvent mixtures (aqueous solutions of acetic and hydrochloric acids, and sodium acetate buffer) [6, 7], as well as the physical state of substance (solution or film) on the value of the specific optical rotation [α] was assessed [8-10]. The method of optical rotation dispersion (ORD) has shown that as the number of monomer units in glucosamine oligomers increases, the value of [α] decreases [3], undergoing a sign inversion at $n = 9$ [5]. Aqueous acidic solutions of high-molecular-weight samples of chitosan are characterized by negative values of [α] which are almost independent of molecular weight [9].

The use of aqueous solutions of biologically active organic acids, for example, ascorbic acid (AscA), as a chitosan solvent medium is promising for practical applications in medicine and pharmacology [11]. However, of the two commercially available diastereomeric forms of AscA (L and D), almost only the L-form (vitamin C) has been used to obtain chitosan ascorbates [11-17]. E.g., studies of chitosan L-ascorbate have shown its specific adsorption to fatty acids [12, 13], selective cytotoxicity against atypical cells [11], anti-inflammatory and other biological activities [14, 15]. Since obtaining chitosan ascorbate by the Schiff reaction in the presence of a reducing agent lowers the antibacterial activity of the polysalt by reducing the total charge of the macromolecule [16], ionic salt formation is most promising for biomedical applications through donor-acceptor interactions of protonated amino groups of chitosan [18] responsible for the biological activity of the polymer system, and the ascorbic acid anion. The use of just the L-isomer of ascorbic acid in these studies is explained by the classic idea of the pharmacological activity of pure vitamin C. However, biological activity has been found for D-isoascorbic (erythorbic) acid (D-AscA) in recent years. It is a more powerful enhancer of iron bioavailability [19] and differs from the L-form by its melting point and reactivity.

In previous studies, chiral salts of chitosan with L- and D-AscA as solutions, powders, glycerohydrogels and glycerohydrogel plates were obtained, significant differences in the total volume charge and size of macromolecules, chiro-optic and physicomechanical properties, gelation kinetics and biological activity were shown as depends on the enantiomeric form of ascorbic acid [20-22]. Chitosan D-ascorbate was found to slow down the sol-gel process, and the resulting thin-film plates based on chitosan D-ascorbate show higher mechanical strength and elasticity as compared to plates based on chitosan L-ascorbate [21, 22]. In addition, systems based on chitosan D-ascorbate significantly increased viability and accelerated the processes of proliferation and adhesion of dermal fibroblast cells; hydrogels based on chitosan D-ascorbate showed the highest in vitro antibacterial activity against gram-positive and gram-negative bacteria and higher wound healing ability in vivo [23]. In the whole, all this indicates a significant effect of the isomeric form of AscA upon the conformational state of chitosan macrochains.

The conformation of macromolecules in solution is known to affect the supramolecular ordering of substance when its solid phase is isolated by removing the solvent. Therefore, it would be interesting to estimate the effect of the chirality of chitosan ascorbates on their morphostructure in the solid state. Freeze drying was chosen as the fastest way to remove the solvent from an aqueous solution, which allows the conformation of macromolecules to be fixed relatively quickly during the solid phase formation [24-26]. Structural differences were visualized by scanning electron microscopy (SEM), which allows obtaining information on the surface morphology and relief of the material under study with sufficient accuracy [24-28].

The aim of the work was to assess the influence of the chirality of chitosan L-, D- and (L+D) ascorbates on their structural features in the solid-state state using ORD and SEM.

2. Methodology

Samples of chitosan (Bioprogress Ltd., RF) with a viscosity-average molecular weight of 50 (CS-50) and 700 kDa (CS-700) were used, with a close deacetylation degree DD = 80 ± 2 mol %; L-AscA (Meligen Ltd., RF); D-AscA (Base of Chemicals No. 1 Ltd., RF); and distilled water were used. Stock CS solutions were prepared by dispersing a polymer powder in a solvent mixture (L-, D- or (L+D)-AscA aqueous solution) on a magnetic stirrer for 30 min, followed by keeping for 1 day at 20 ± 2°C until complete dissolution of the polymer, then they were filtered through a porous glass filter with a pore diameter of 160 μm. We used freshly prepared solutions with a polymer concentration 0.1–3.0 wt% and with an acid concentration 0.1–3.0%, obtained at a 1:1 polymer : acid molar ratio.

Air-dry powders of chitosan and AscA salts (L-CS, D-CS and (L+D)-CS) were isolated from the corresponding freshly prepared solutions on a VirTis 2KBTES-55 laboratory lyophilizer (SP Scientific, US) at –55°C and 495-390 mtorr for 1 day.

ORD curves were recorded on an PolAAr 3001 (Optical Activity, GB) automatic spectropolarimeter in the wavelength range of 365–589 nm at 25°C. A 20W tungsten halogen lamp was the light source. The specific optical rotation of solutions $[\alpha]_{\lambda}^{25°C}$ [deg·mL/dm·g] was calculaed by the formula:

$$[\alpha]_{\lambda,\ nm}^{25°C} = \frac{\alpha \cdot 100}{C_P \cdot l},$$

where α is the measured angle of the optical rotation of the solution at the wavelength λ of the incident light, deg; CP the concentration of the polymer (the optically active substance) in the solution, g/dl; l the optical path length, dm. The error in measuring optical rotation angles in the range of λ = 405 – 589 nm was ± 0.001 deg, at λ = 365 nm it was ± 0.01 deg.

Experimental ORD curves were processed using the one-term Drude equation (in the case of registration of the ORD curves far from the absorbing electronic transition, i.e. when a single band contributes to the rotation):

$$[\alpha]_{\lambda,\ nm}^{25°C} = \frac{K}{\lambda^2 - \lambda_0^2},$$

where K and λ_0 are the rotational and dispersion constant, respectively. To calculate these constants, Lowry's (the $1/[\alpha] - \lambda^2$ dependence), Yang–Doty's ($[\alpha]\cdot\lambda^2 - [\alpha]$), and Heller's ($1/[\alpha]\cdot\lambda^2 - 1/\lambda^2$) graphical transformations were used. The least squares method was used to plot straight lines.

SEM images were taken on a MIRA\\LMU scanning electron microscope (Tescan, Czech Republic) at a voltage of 7 kV and a conducting current of 400 pA. A 5 nm thick gold layer was deposited on each sample using a K450X Carbon Coater (Germany).

3. Results and discussion

The specific optical rotation and the type of dispersion curves of aqueous solutions of AscA, CS in L-, D- and (L+D)-AscA were evaluated. Our analysis of the optical activity of individual AscA isomers in the visible range of the spectrum showed that aqueous solutions of L-AscA and D-AscA were characterized by opposite signs of specific optical rotation [α] and different types of their ORD curves (Fig. 1, curves *1* and *2*). The L-AscA solution showed dextrarotation $[\alpha]^{25°C}_{589\ nm} = +23$ deg·mL/dm·g, an anomalous ORD (the $[\alpha] = f(\lambda)$ dependence had a point of inflection at ~400 nm). The D-AscA solution showed laevorotation $[\alpha]^{25°C}_{589\ nm} = -17$ deg·mL/dm·g and normal dispersion. The aqueous solution of a mixture of diastereomers (L+D)-AscA had an intermediate position of the normal-type ORD curve, $[\alpha]^{25°C}_{589\ nm} \approx 0$ deg·mL/dm·g (curve *3*).

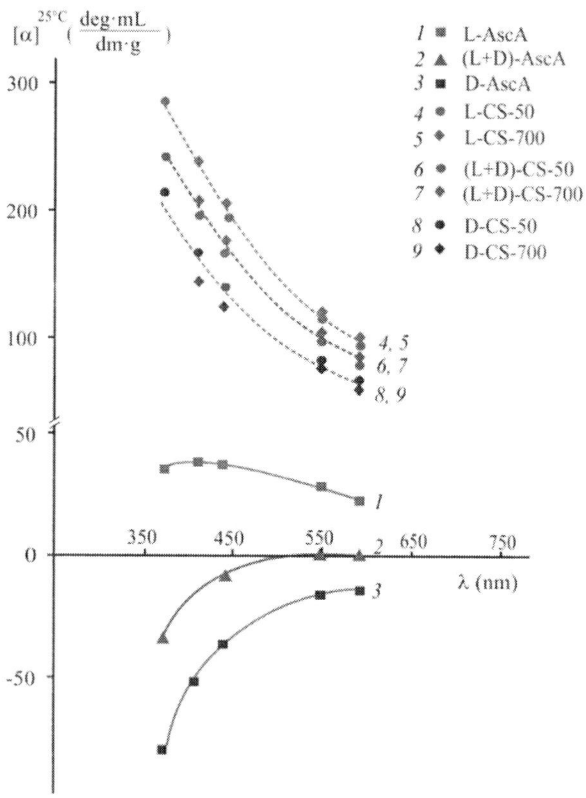

Figure 1. ORD-curves of aqueous solutions of L- (*1*), (L+D)- (*2*), D-AscA (*3*), the ascorbate salt of L-CS-50 (*4*), L-CS-700 (*5*), (L+D)-CS-50 (*6*), (L+D)-CS-700 (*7*), D-CS-50 (*8*), and D-CS-700 (*9*).

Aqueous solutions of L-CS, D-CS and (L+D)-CS are characterized by a positive direction of rotation of the polarization plane and normal-type ORD curves (Fig. 1, curves *4-9*). This fundamentally distinguishes chiral chitosan ascorbates from chitosan salts with non-chiral acids, for example, chitosan hydrochloride and acetate, which are characterized by levorotatory [3, 6, 7]. For L-CS, D-CS and (L+D)-CS, the molecular weight of chitosan has little effect on the observed values of specific optical rotation. A similar unchanged [α] with increasing molecular weight was previously obtained

for chitosan acetates [9]. There was also a difference in the values of [α] of solutions of chitosan ascorbates of different chirality: the ORD curves of L-CS solutions were always located higher than those of D-CS, and the ORD curves of aqueous solutions of (L+D)-CS had an intermediate position.
The use of the Drude equation and the corresponding Lowry, Yang-Doty and Geller transformations for processing ORD curves allowed the direct method to determine the wavelength of the optically active transition λ_0, which may be even in the UV region. Table 1 summarizes the calculated values of the rotational K and the dispersion constants λ_0. It can be noted that plotting the ORD curve in the Yang-Doty and Geller coordinates gives stronger convergence in the values of the calculated constants (Table 1). Our processing of the ORD curves of chitosan solutions allows us to state that they are satisfactorily described by the one-term Drude equation.

Table 1. The rotational and dispersion constants in the Drude equation for CS solutions in L-, (L+D)- and D-AscA.

Sample	Rotational constant, $K \cdot 10^{-7}$				Dispersion constant λ_0, nm			
	Lowry	Yang-Doty	Heller	$\overline{K}$	Lowry	Yang-Doty	Heller	$\overline{\lambda_0}$
L-AscA	1.1	1.1	1.2	1.1	236	249	270	250
L-CS-50	3.0	3.1	3.1	3.0	195	180	180	180
L-CS-700	3.0	3.1	3.1	3.0	205	196	196	200
(L+D)-AscA	−340.0	−450.0	−368.0	−390.0	357	347	354	350
(L+D)-CS-50	2.6	2.7	2.7	2.7	170	180	180	180
(L+D)-CS-700	2.6	2.7	2.7	2.7	208	201	201	200
D-AscA	−30.0	−31.0	−30.0	−30.0	240	240	240	240
D-CS-50	1.9	2.0	2.0	1.9	190	180	180	180
D-CS-700	2.1	2.1	2.1	2.1	205	198	199	200

Comparing the obtained values of λ_0 for the starting components and salts shows that as a result of salt formation of chitosan and ascorbic acid, a new chromophore is formed, which mainly contributes to the rotation and is located in the ultraviolet region of the spectrum (180-200 nm), which is consistent with CD [21]. The value of λ_0 for L-CS and D-CS differs by 20 nm. The effect of the enantiomeric form of ascorbic acid used to obtain the polysalt on the rotational K and dispersion constants λ_0 is also traced. The value of the rotational constant characterizing the rotational force of the electron transition of an optically active chromophore decreases in the row of chiral chitosan salts L-CS → (L+D)-CS → D-CS. This may be due to the difference in the total charge of macromolecular coils of these polysalts (like protein and DNA molecules), found in Ref. [20].
Differences in the stereomeric structure of chiral polysalts are also manifested themselves in the morphology of their air-dry samples, lyophilized from the corresponding solutions. Fig. 2 shows SEM images.
Samples of L-CS and D-CS are characterized by a complex surface relief and layered laying of the supramolecular elements of the polymer matrix in the sample bulk (Fig. 2 a and b, d and e), typical of lyophilized chitosan-containing materials [24-28]. However, differences in the supramolecular ordering of L-CS and D-CS are observed. E.g. the solid phase of the L-CS polysalt is represented by a more densely packed layered microstructure, which is most pronounced for the higher-molecular|-weight sample (Fig. 2 d). Looser packing of supramolecular formations is characteristic of D-CS salts. The surface morphology of the (L+D)-CS sample is significantly different from the solid phase of chitosan salts obtained using individual Asc diastereomers. The (L+D)-CS-50 sample has a fibrous structure made of anisodiametric formations (Fig. 2 c). For the (L+D)-CS-700 sample, a highly porous structure with a system of interconnected pores is observed (Fig. 2 f).

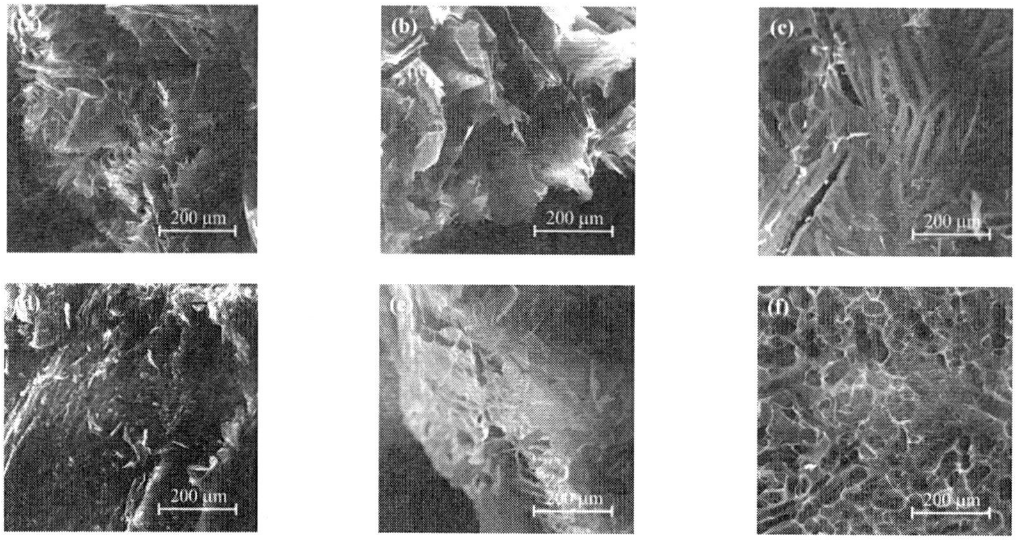

Figure 2(a-f). SEM-images of the particle morphostructures of air-dry powders based on CS-50 **(a-c)**, CS-700 **(d-f)** and L- **(a, d)**, D- **(b, e)**, (L+D)-AscA **(c, f)**.

Thus, the results obtained allow us to state that the isomeric form of AscA has a significant impact on the stereomeric ordering of the chitosan ascorbate molecules and the chiral supramolecular structures built therefrom.

References

[1] Iftime M M, Marin L 2018 *Ultrason. Sonochem.* **45** 238
[2] Lermontov S A, Sipyagina N A, Malkova A N, Baranchikov A E 2017 *Micropor. Mesopor. Mat.* **237** 127
[3] Shipovskaya A B, Fomina V I, Kazmicheva O F, Timofeeva G N 2007 *Polym. Sci. Ser. B* **49** 288
[4] Kumar S, Koh J 2012. *J. Appl. Polym. Sci.* **124** 4897
[5] Domard A, Cartier N 1989 *Int. J. Biol. Macromol.* **11** 297
[6] Koralewski M, Bodek K H, Marczewska K 2006 *Polish Chitin Soc.* **11** 29
[7] Koralewski M, Bodek K H, Wachowski T 2007 *Polish Chitin Soc.* **12** 79
[8] Shipovskaya A B, Malinkina O N, Fomina V I, Rudenko D A 2015 *Rus. Chem. Bul.* **64** 1172
[9] Shipovskaya A B, Fomina V I, Kazmicheva O F, Rudenko D A 2017 *Polym. Sci. Ser. A* **59** 330
[10] Shipovskaya A B, Shmakov S L, Gegel N O 2019 *Carbohyd. Polym.* **206** 476
[11] Sekar V, Rajendran K, Vallinayagam S, Deepak V, Mahadevan S 2018 *Ind. Eng. Chem. Res.* **62** 239
[12] Osman M, Fayed S A, Ghada I M, Romeilah R M 2010 *AJBAS* **4** 89
[13] Tsujikawa T, Kanauchi O, Andoh A, Saotome T, Sasaki M, Fujiyama Y, Bamba T 2003 *Nutrition* **19** 137
[14] Yanagiguchi K 2000 *Korean Chitin Chitosan Journal* **5** 182
[15] Lu B, Wang T, Li Z, Dai F, Lv L; Tang F, Yu K, Liu J, Lan G 2016 *Int. J. Biol. Macromol* **82** 884
[16] Hafsa J, Charfeddine B, Smach M A, Limem K, Majdoub H, Sonia R 2014 *IJPBCS* **4**
[17] Ogawa K, Nakata K, Yamamoto A, Nitta Y, Yui T 1996 *Chem. mater.* **8** 2349
[18] Berger J M, Reist J M, Mayer O F, Peppas N A, Gurny R 2004 *Eur. J. Pharm. Biopharm.* **57** 19
[19] Fidler M C, Davidsson L, Zeder C, Hurrell R F 2004 *Am. J. Clin. Nutr.* **79** 99
[20] Malinkina O N, Gegel N O, Shipovskaya A B 2019 *J Mol. Liq.* **284** 75
[21] Gegel N O, Zhuravleva Yu Yu, Shipovskaya A B, Malinkina O N, Zudina I V 2018 *Polymers* **10** 259
[22] Shipovskaya A B, Zhuravleva Yu Yu, Khonina T G, Malinkina O N 2019 *J. Sol-Gel Sci. Technol.* 1
[23] Zudina I V, Malinkina O N, Shipovskaya A B 2018 *Microbiology* **87** 732
[24] Ayensu I, Mitchell J C, Boateng J S 2012 *Int. J. Biol. Macromol.* **50** 905
[25] Silva A R P D, Macedo T L, Coletta D J, Feldman S, Pereira M D M 2016 *Matéria* **21** 964
[26] Pezeshki-Modaress M, Rajabi-Zeleti S, Zandi M, Mirzadeh H 2014 *J. Biomed. Mater. Res. A* **102** 3908
[27] Tang Y, Wang X, Li Y, Lei M, Du Y, Kennedy J F, Knill C J 2010 *Carbohydr. Polym.* **82** 833
[28] Shimojo A A M, Perez A G M, Galdames S E M, Brissac I C S 2016 *Mater. Sci. Eng. C* **60** 538

Estimation of particle size using the Debye equation and the Scherrer formula for polyphasic TiO_2 powder

A O Bokuniaeva[1,2]*, **A S Vorokh**[1]

[1]Institute of Solid State Chemistry of the Ural Branch of the Russian Academy of Sciences, Ekaterinburg, 620990, Russia
[2] Institute of Natural Sciences and Mathematics, Ural Federal University, Ekaterinburg, 620002, Russia

a.o.bokuniaeva@urfu.ru

Abstract. There are two methods to estimate the particle size from X-ray diffraction data: the Debye equation and the Scherrer formula. The main goal of this study is to describe the methodology of particle size estimation on the base of two these methods and to apply it to TiO_2 powder to determine the diameters and the mass content of anatase and brookite components. The studied nano-dispersed TiO_2 powder was synthesized by the sol-gel method. The proposed method of particle size estimation consists of several steps: 1. Approximation of diffraction peaks by Gaussians and calculation of initial values of particle size with the use of the Scherrer formula; 2. Iterations with the use of the Debye equation to obtain more accurate particle size values; 3. Calculation of the mass content of different components corresponding to the minimum R-factor.

1. Introduction

Titanium dioxide TiO_2 is an extensively applied compound due to high photosensitivity, non-toxicity, easy availability, pH and long-thermal and chemical stabilities, low cost [1]. In recent decade, advanced TiO_2-based photocatalysts have been promisingly suggested as the materials for removal of organic contaminants and disinfection of freshwater from algae [2], hygiene coatings to control and reduce populations of pathogenic bacteria [3], CO_2 photoreduction in the presence of water vapor to produce solar fuels (CO and CH_4) [4]. In the field of alternative energetics, TiO_2 is used as a matrix of photoactive layer in dye-sensitized solar cells [5]. There are three phases of titanium dioxide: rutile, anatase, and brookite. Photocatalytic activity can be enhanced through mixing of several phases of TiO_2: for an example, anatase/rutile, brookite/rutile [6], anatase/brookite [7]. In this work, we applied three different methods for particle size estimation of TiO_2 polyphasic powder: the direct crystallographic measurement, the use of the Scherrer formula, the use of combination of the Scherrer formula and the Debye equation.

Particle size remains among the most important parameters of substance. The majority of researchers use the Scherrer formula [8] to determine this parameter for the sample nanostructured powder. The result of particle size estimation derived from the Scherrer formula is approximate and based on peak broadening. Although there are numerous refinements of the formula, the issue on the limits of its applicability is still unsolved. The theoretical upper limit of applicability of the Scherrer formula was analyzed in detail and was found for up to 600 nm for LaB_6, Si and CeO_2 [9]. In general case, this value is 11.9 % of extinction length [10]. However, the use of the Scherrer formula provides a more accurate result in case of measurement of a single reflection [11]. The use of the Debye equation [12] for particle size estimation provides a result of higher accuracy. The Debye equation is used for diffraction pattern modelling on the basis of the intensity of X-ray scattering over interatomic distances. The result derived from this equation is rather precise as it deals with diffuse scattering at each diffraction pattern point, not only at Bragg angles. The Debye equation is applicable to amorphous clusters as well. However, using the Debye equation is rather time-consuming.

The main goal of the current study is to describe the method of estimation of particle size and to apply it to calculate TiO_2 particle size and mass content of two phases of TiO_2. In our case, TiO_2 powder represents a mixture of anatase and brookite. The proposed method lies in the combination of the Debye equation and the Scherrer formula and allows to estimate the particle size and phase ratio more precisely than using these relations separately.

2. Method

TiO_2 powder was synthesized by the sol-gel method on a *one-pot* principle and represented a mixture of brookite and anatase at a certain ratio. X-ray diffraction (XRD) patterns of all composites were obtained using a Shimadzu MAXima-X XRD-7000 automatic diffractometer (λ=1.5406 Å for CuKα radiation) in the angle interval 2θ of 5-80° with step 0.03° and an exposure time of 10 sec. Calculations at each stage of the method were performed with the use of different software facilities: Powder Cell 2.3, Igor Pro, Crystalmaker, and DIANNA [13].

3. Discussion

X-ray diffraction measurements of TiO_2 powder demonstrates the pure titania without any additives. The experimental diffraction pattern is shown on Fig. 1 with peak positions of anatase and brookite. The majority of peaks contain contributions of more than one reflection. The most of peaks can be referred to anatase, but these broad peaks overlap with peaks of brookite. The only one peak of low-intensity at 30° directly indicates the presence of brookite phase, reflection (211).

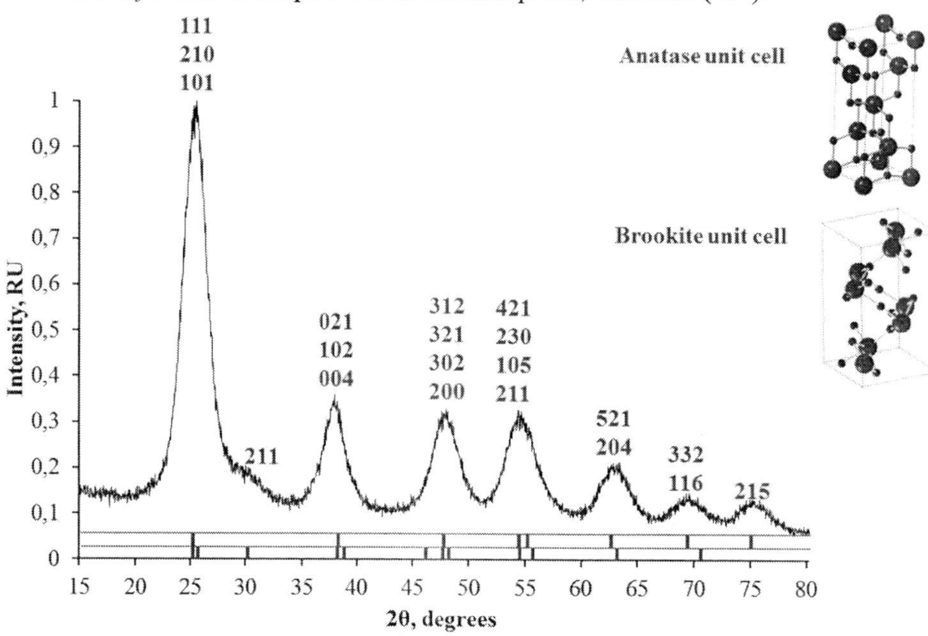

Figure 1. The experimental diffraction pattern of TiO_2 powder. The most intensive reflections of anatase and brookite are shown in the bottom by sticks. The anatase and brookite unit cells are shown in the right part of figure.

Direct calculations of particle size were carried out using software "PowderCell 2.3" and led to ambiguous results. On the Fig. 2a, calculated diffraction peaks (green curve) were obviously narrower than the experimental one (black curve). Moreover, they coincide poorly with the experimental ones in the region $2\theta \approx 30°$. The diameter of anatase particles was determined as 9.1 nm, and the diameter of brookite particles – 8.9 nm. The crystallographic approach could not allow to simulate the very broadening of peaks, and to determine size of very small particles. Thus, the method of particle size estimation based on direct application the Scherrer formula and XRD pattern simulation by the Debye formula is proposed.

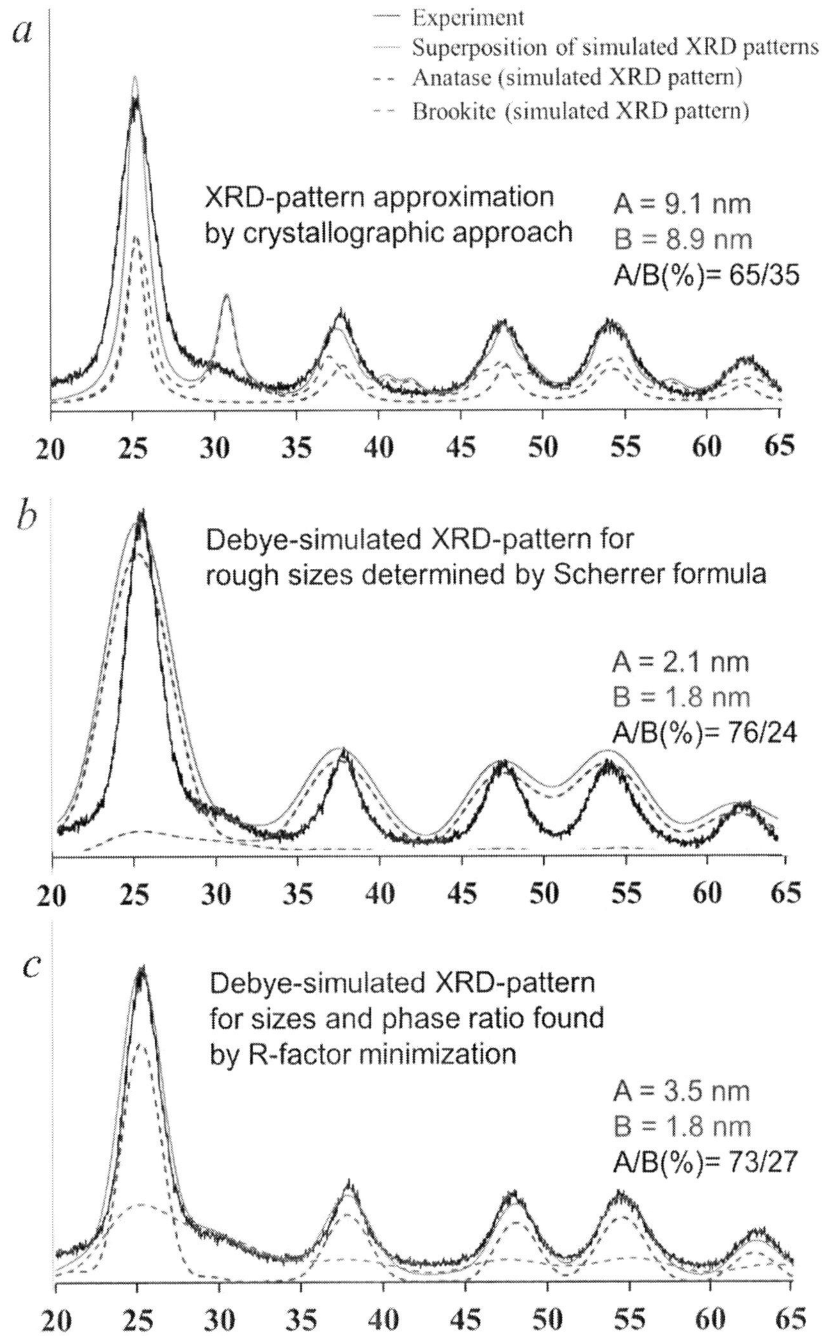

Figure 2. The experimental diffraction pattern (black line) and the theoretical one (green line). The contributions of anatase and brookite are designated by red and blue dash lines correspondingly. The particle size of anatase as A, brookite as B and phase ratio as A/B(%) is shown for each method. The theoretical diffraction pattern was obtained using: (*a*) by PowderCell 2.3, and by DIANNA for particle size and phase ratio: (*b*) determined by the Scherrer formula, (*c*) and determined by R-factor minimization.

The proposed method consists of three steps, which are shown schematically on the Fig. 3.

(*i*) Calculation of initial value of particle size. The most intensive peak was decomposed into three peaks: anatase reflection (101) and two brookite reflections (210) and (111). Anatase peak (101) and brookite peak (211) were described by Gaussians and calculated the half-width of both reflections. The initial values of the size of anatase and brookite particles were calculated according to the Scherrer formula:

$$D_{hkl} = \frac{K\lambda}{\beta_{hkl} \cdot \cos\theta} \tag{1}$$

where K – the particle shape factor, 0.9, λ – X-ray wavelength, β_{hkl} – the half-width of (hkl) reflection, $\theta = 2\theta/2$ – Bragg angle corresponding to (hkl) reflection. The anatase and brookite particle sizes were determined as 2.1 nm and 1.8 nm correspondingly.

Then, we tested the obtained values of particle sizes by simulating of diffraction pattern for these particle diameters using by the Debye equation. We used the software DIANNA [13]. On the Fig. 2b we can see not matching Debye-simulating XRD-pattern and experimental curve. The reason of divergency is the next: the Scherrer formula takes into account broadening of peaks only because of the particle size while in reality, a lot of factors contribute to broadening of peaks. The half-width in the experimental diffraction pattern depends not only on the particle size. It means that the share of the half-width derived from the experimental diffraction pattern, that responsible for broadening of peaks due to the particle size, is actually less than the overall value obtained from direct measurement.

(*ii*) Refinement of the obtained values. On this step the particle diameter was determined by XRD-pattern simulation using the Debye equation. This approach allowed the direct calculation of XRS-pattern from particle as cluster or, strictly, set of atomic positions. It is possible to calculate the diffraction pattern of any size particles. However, calculation time increases with the particle size. Simulation of the diffraction pattern was carried out with the use DIANNA software [13]. The Debye equation looks as follows:

$$I(s) = P(s) \sum_i \sum_j f_i(s)\, f_j(s) \frac{\sin(sr_{ij})}{sr_{ij}}, \tag{2}$$

where $s = \dfrac{4\pi\sin(\theta)}{\lambda}$ is the current reciprocal space coordinate, $P(s)$ – instrumental factor, $f_i(s)$ and $f_j(s)$ are the scattering factors of atoms i and j, respectively.

(*iii*) Determination of mass content. The sum of simulated pattern from anatase and brookite was compared with experimental pattern by the R-factor. The R-factor reflects the difference between theoretical and experimental diffraction data. The actual ratio corresponds to the minimum R-factor.

Several iterations according to the Debye equation were carried out in order to define more precisely the anatase particles diameter, 3.5 nm. We varied the particle size both anatase and brookite by increment 0.1 nm and compared Debye-simulated XRD-pattern with experimental curve by R-factor. Schematically this procedure is demonstrated on the Fig.3. The result of R-factor minimum searching is demonstrated on the Fig. 2c.The brookite particle diameter remains the same, 1.8 nm. The minimum R-factor corresponds to 27% brookite mass content.

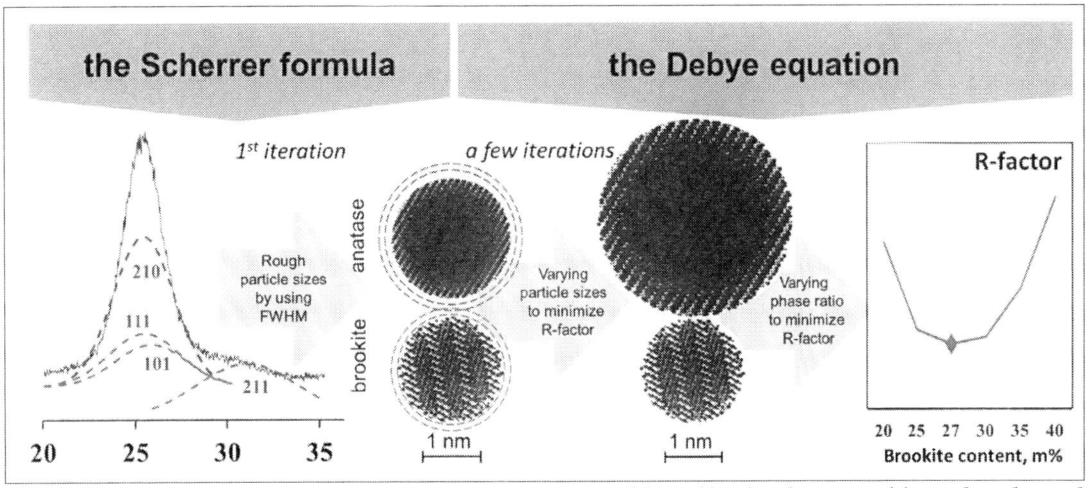

Figure 3. The main steps comprising the proposed algorithm: (*i*) The decomposition of peaks and calculation of rough values of particle size by using the Scherrer formula and then simulation XRD-pattern for obtained sizes; (*ii*) Several iterations of the size values variation using by the Debye equation; (*iii*) the searching of the R-factor minimum for three parameters: two particle sizes and phase ratio. The rough and final results of anatase and brookite particle sizes are shown in scaling. The dependence R-factor from brookite content is shown.

4. Conclusion

In this study, we proposed method which allows to obtain three parameters: diameters of brookite and anatase particles and mass content for TiO_2 polyphasic nanopowder. In the first, we found that the crystallographic approach led to ambiguous results. The anatase and brookite particle sizes are determined 9.1 nm and 8.9 nm correspondingly. Afterwards we calculated directly values of particle size from peak broadening using by the Scherrer formula and found them 2.1 nm for anatase and 1.8 nm for brookite. The nominal obtained values differs more than four times but the values ratio between anatase and brookite both the crystallographic measurement and the Scherrer formula are close to 1. This result pointed the necessity of another approach based on direct simulation of Debye formula. The application of the Scherrer equation and the Debye formula together provided more accurate data, Fig. 2c.

The result obtained by the proposed method was the following: 1.8 nm diameter of brookite particles, 3.5 nm diameter of anatase particles, mass content was 73% of anatase and 27% of brookite phase. The proposed method based on the combination of the Scherrer equation and the Debye formula can be applied to any XRD measurements of nanostructured materials as well.

References

[1] Ahmad A, Shah J A, Buzby S and Shah S I 2008 *Eur. J. of Inorg. Chem.* **6** 948
[2] Kim Y, Hwang H. M., Wang L, Kim I, Yoon Y and Lee H 2016 *Sci. Rep.* **6** 25212
[3] Leyland N S, Podporska-Carroll J, Browne J, Hinder S J, Quilty B and Pillai S C 2016 *Sci. Rep.* **6** 24770
[4] Zhao H, Liu L, Andino J M and Li Y 2013 *J. Mater. Chem. A.* **1** 8209
[5] Park N G, Schlichthörl G, van de Lagemaat J, Cheong H M, Mascarenhas A and Frank A J 1999 *J. Phys. Chem. B.* **103** 3308
[6] Cao Y, Li X, Bian Z, Fuhr A, Zhang D and Zhu J 2016 *Appl. Catalysis B:* **180** 551
[7] Kozhevnikova N S, Ulyanova E S, Shalaeva E V, Gorbunova T I, Bokunyaeva A O, Yushkov A S, Buldakova L Yu, Yanchenko M Yu, Kuznetsov M V, Pasechnik L A, Enyashin A N and Vorokh A S 2019 *J Mol. Phys.* **284** 29

[8] Scherrer P 1918 *Bestimmung der Grosse und der inneren Struktur von Kolloidteilchen mittels Rontgenstrahlen* (Gottingen: Nachrichten Gesellschaft Wissenschaft Gottingen) pp 98–100
[9] Muniz F T L, Miranda M A.R., dos Santos C M and Sasaki 2016 *J M Acta Crystallogr., Sect. A: Found. Crystallogr.* **72** 385
[10]Miranda M A R and Sasaki J M 2018 *J M Acta Crystallogr., Sect. A: Found. Crystallogr.* **74** 54
[11]Vorokh A 2018 *Nanosyst.: Phys., Chem., Math.* **9** 364
[12]Debye P 1915 *Ann. Phys. B* **46** 809
[13]Yatsenko D and Tsybulya S 2018 *Krist.-Cryst. Mater.* **233** 61

Acknowledgments
The research was supported by the Russian Science Foundation (Project 17-79-20165) and carried out in ISSC UB RAS.

Modeling the formation of InP/$Ga_xIn_{1-x}P$ axial nanowire heterostructures

A A Koryakin[1,2], E D Leshchenko[3,2], V G Dubrovskii[2]
[1]St. Petersburg National Research Academic University of the Russian Academy of Sciences, St. Petersburg 194021, Russia
[2]ITMO University, St. Petersburg 197101, Russia
[3]Solid State Physics and NanoLund, Lund University, Box 118, S-22100 Lund, Sweden
E-mail: koryakinaa@spbau.ru

Abstract. A model is developed to depict the formation of InP/$Ga_xIn_{1-x}P$ axial heterostructures in self-catalyzed $Ga_xIn_{1-x}P$ nanowires. The composition profiles of the InP/$Ga_xIn_{1-x}P$ axial heterostructure are calculated taking into account elastic stresses. It is shown that the InP/$Ga_xIn_{1-x}P$ axial heterojunction width at the growth temperature of 450°C is larger than 12 monolayers for nanowires with the radius larger than 10 nm. Also, the comparison with $Ga_xIn_{1-x}As$ system is performed and reveals that the InP/$Ga_xIn_{1-x}P$ axial heterojunction width is approximately two times smaller than the InAs/$Ga_xIn_{1-x}As$ axial heterojunction width.

1. Introduction

Semiconductor nanowires (NW) are often used as building blocks for nanodevices due to their unique geometry and physical properties [1]. The formation of axial and radial heterostructures in ternary III-V NW particularly allows the combination of materials with large lattice mismatch without generating defects. The modeling of heterostructure formation in NW synthesized via catalytic growth methods is usually performed within the framework of the following approaches [2,3]. The first approach is based on the binary nucleation theory and allows the finding of composition of the critical nucleus forming on the NW-catalyst interface. This composition is then used as an approximation of the monolayer composition. The second approach is based on the assumption that the dependence of the monolayer composition on the catalyst droplet composition during the NW growth is the same as in case of thermodynamic equilibrium between NW and droplet. Within the third approach (the irreversible growth model), the axial heterostructure formation is described in terms of large supersaturations when the size of the critical nucleus is close to unity. The second approach was developed in the paper [4] to describe the axial NW heterostructure formation in case of materials with large lattice mismatch. The model [4] accounts for elastic stresses produced by the lattice mismatch between consecutive monolayers in NW. In this paper, we use the model [4] to calculate the width of InP/$Ga_xIn_{1-x}P$ axial heterojunction in self-catalyzed $Ga_xIn_{1-x}P$ NW. Note, the advantage of the self-catalyzed NW growth is that there is no incorporation of foreign catalyst atoms into NW crystal lattice.

2. Theoretical model

We consider a cylindrical $Ga_xIn_{1-x}P$ NW with the radius R which grows along the [111] direction. The InP/$Ga_xIn_{1-x}P$ axial heterostructure is formed by switching the group III fluxes. In order to take into account elastic stresses, the NW is considered as a set of adjacent monolayers. The dependence of monolayer composition on the droplet composition can be found by using the equality of chemical potentials for species in the droplet and monolayer. In such a case, the monolayer composition depends on the composition of underlying monolayers and can be written in the following form [3,4]:

$$x_i = x_i(y_i, c_{tot,i}, x_{i-1}, x_{i-2}, ...), \tag{1}$$

where x_i is the fraction of GaP pair in the i monolayer, $y_i = c_{Ga,i} / c_{tot,i}$, $c_{tot,i} = c_{Ga,i} + c_{In,i}$, $c_{Ga,i}$, $c_{In,i}$ and $c_{Ga,i}$ are the concentrations of Ga, In and P atoms in the droplet such that $c_{Ga,i} + c_{In,i} + c_{P,i} = 1$. Following the paper [2], the catalyst droplet and monolayer materials are considered as regular liquid and solid solutions, respectively. The elastic stress terms to the chemical potentials of GaP and InP pairs are calculated by the following equations [4]: $\mu^{elas}_{GaP,i} = W_i + (1 - x_i)\partial W_i / \partial x_i$ and $\mu^{elas}_{InP,i} = W_i - x_i \partial W_i / \partial x_i$, where W_i is the average elastic energy per one III-V pair in the i monolayer. The elastic energy W_i as a function of x_i is calculated by the finite element method [5] and then fitted by the quadratic function. The details of the calculation of the function $x_i = x_i(y_i, c_{tot,i}, x_{i-1}, x_{i-2}, ...)$ can be found in the paper [4].

To obtain a simplified material balance equation for the species in droplet, we introduce the following assumption [2–4]. Firstly, the concentrations of Ga and P atoms in droplet are much less than the concentration of In atoms. Secondly, the group V flux is constant and determines the NW growth rate. Thirdly, we neglect the change in the NW radius during heterostructure formation. Within the assumptions made above, the material balance equation for Ga species can be only considered to calculate the composition profile of the InP/$Ga_xIn_{1-x}P$ axial heterostructure:

$$y_i(c_{tot,i}, x_i, x_{i-1}, ...) - y_{i-1}(c_{tot,i-1}, x_{i-1}, x_{i-2}, ...) = p(q - x_i), \tag{2}$$

here $p = N_{ML} / N$, $q = J / N_{ML} V$, N_{ML} is the number of III-V pairs in monolayer, N is the total number of atoms in the droplet, J is the Ga flux entering the droplet (in s^{-1}), V is the NW growth rate (in monolayers per second) and $y_i(c_{tot,i}, x_i, x_{i-1}, ...)$ is the inverse function of $x_i(y_i, c_{tot,i}, x_{i-1}, x_{i-2}, ...)$.

3. Results and discussion

The modeling of the InP/$Ga_xIn_{1-x}P$ axial heterostructure formation is performed by using the reference data [2,6-8]. The interaction coefficients in the droplet equal [6]: $\omega_{GaIn} = 4450.0 + 1.19185\ T\ J/mol$, $\omega_{GaP} = -9862\ J/mol$, $\omega_{InP} = 14124.08 - 10.17931T\ J/mol$. The interaction coefficient of GaP and InP pairs in solid is estimated in the paper [7]: $w = 15187.9\ J/mol$. The chemical potentials of pure components can be found as functions of temperature in [6]. The values of the model parameters p and q are taken from the paper [2] where the $Ga_xIn_{1-x}As$ system was considered. The elastic constants and the lattice parameter of $Ga_xIn_{1-x}P$ are calculated by using the Vegard's law and the data for bulk GaP and InP [8]: $C^{GaP}_{11} = 140.5\ GPa$, $C^{GaP}_{12} = 62.0\ GPa$, $C^{GaP}_{44} = 70.3\ GPa$, $C^{InP}_{11} = 101.1\ GPa$, $C^{InP}_{12} = 56.1\ GPa$, $C^{InP}_{44} = 45.6\ GPa$, $a_{GaP} = 0.54505\ nm$ and $a_{InP} = 0.58687\ nm$. Figure 1 shows the composition diagrams calculated for the case of R = 10 nm and the growth temperature of 450°C. The monolayers of the $Ga_xIn_{1-x}P$ section of NW are numbered from $i = 1$. It is seen from Figure 1 that there is no miscibility gap as opposed to the case without elastic stresses (the dashed region on the gray line, $x_i = 0.138 - 0.862$).

The composition profiles of InP/$Ga_xIn_{1-x}P$ axial heterostructure are shown in Figure 2. As for the $Ga_xIn_{1-x}As$ system [4] , the formation of atomically sharp heterojunction is hardly achieved due to elastic stresses. Even if the NW radius is very small (R = 5 nm) the heterojunction width is about of 8 monolayers. It is seen that the heterojunction width in the case of $Ga_xIn_{1-x}P$ system is shorter compared to the $Ga_xIn_{1-x}As$ system. This is because the ratio of the reaction rate constants for the self-catalyzed growth of $Ga_xIn_{1-x}P$ NW is higher than for the self-catalyzed growth of $Ga_xIn_{1-x}As$ NW:

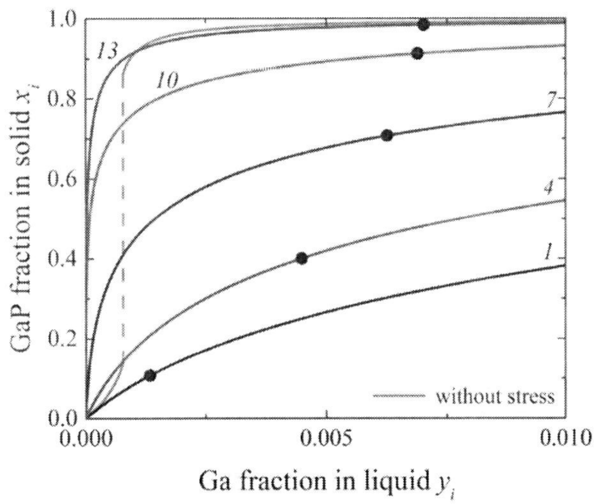

Figure 1. Composition diagrams for several monolayers of $Ga_xIn_{1-x}P$ NW (i = 1, 4, 7, 10, 13). The monolayer compositions obtained as a result of the self-consistent modeling are marked by dots. The composition diagram calculated without taking into account elastic stresses is shown by the gray line.

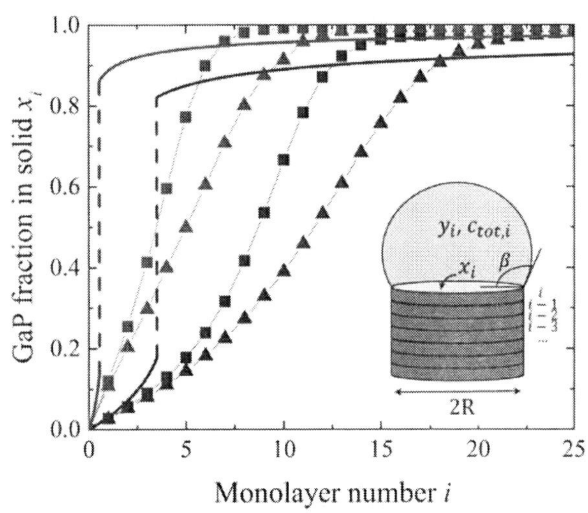

Figure 2. Composition profiles of $InP/Ga_xIn_{1-x}P$ axial heterostructure (the red lines and symbols) and $InAs/Ga_xIn_{1-x}As$ axial heterostructure [4] (the blue lines and symbols) at the growth temperature of 450°C. The rectangular and quadratic symbols correspond to the NW radius of 10 nm and 5 nm, respectively. The composition profiles are also plotted by using the formula for the dependence $y(x)$ derived in the paper [2], where elastic stresses were neglected (the lines without symbols). The insert shows the model geometry of the system.

$\ln K_{InP} / K_{GaP} \approx 6.3$ and $\ln K_{InAs} / K_{GaAs} \approx 4.4$, where K_{AB} is the reaction rate constant for the reaction $A + B \rightarrow AB$. Although, the values of elastic constants of $Ga_xIn_{1-x}P$ are larger than those of $Ga_xIn_{1-x}As$ [8]. It is important to note that the elastic constants used in our computation correspond to bulk materials. Elastic properties of monolayers can differ significantly from elastic properties of bulk materials. Thus, the monolayer elastic constants can be smaller or larger than those of bulk materials [9]. Therefore, the theory presented in our paper should be considered only as a first attempt to understand the role of elastic stresses in the formation of $Ga_xIn_{1-x}P$ axial nanowire heterostructures.

To summarize, the estimates of the InP/$Ga_xIn_{1-x}P$ axial heterojunction width in the self-catalyzed $Ga_xIn_{1-x}P$ NW were presented. The dependence of the monolayer composition on the droplet composition was obtained within the framework of the thermodynamic equilibrium model. The elastic energy terms in the chemical potentials of species in monolayers were calculated by using the finite element method. It was found that the InP/$Ga_xIn_{1-x}P$ axial heterojunction width at the growth temperature of 450°C is larger than 12 monolayers if the NW radius is larger than 10 nm. Also, it was shown that the axial heterojunction width in case of $Ga_xIn_{1-x}P$ NW is approximately two times smaller than that in case of $Ga_xIn_{1-x}As$ system.

Acknowledgments
The reported study was funded by RFBR according to the research project № 18-32-00559.

References

[1] Zhang A, Zheng G, Lieber C M 2016 *Nanowires: Building Blocks for Nanoscience and Nanotechnology* (Cham: Springer International Publishing)
[2] Dubrovskii V G 2017 *J. Phys. D: Appl. Phys.* **50** 453001
[3] Glas F 2017 *Cryst. Growth Des.* **17** 4785
[4] Koryakin A A, Leshchenko E D, Dubrovskii V G 2019 *Technical Physics Letters* (submitted)
[5] Chen C R, Li S X, Zhang Q 1999 *Mater. Sci. Eng. A* **272** 398
[6] Ansara I, Chatillon C, Lukas H L, Nishizawa T, Ohtani H, Ishida K, Hillert M, Sundman B, Argent B B, Watson A, Chart T G, Anderson T 1994 *CALPHAD* **18** 177
[7] Stringfellow G B 1974 *J. Cryst. Growth* **27** 21
[8] Levinshtein M E, Rumyantsev S L 1996 *Handbook Series on Semiconductor Parameters* (London: World Scientific)
[9] Cammarata R C 1994 *Progress in Surface Science* **46** 1

Role of the wetting layer in the crystallization stage during droplet epitaxy of InAs/GaAs nanostructures

S V Balakirev[1], M M Eremenko[2], N E Chernenko[2,3], O A Ageev[1], M S Solodovnik[1,2]
[1]Department of Nanotechnologies and Microsystems, Southern Federal University, Taganrog 347922, Russia
[2]Research and Education Center "Nanotechnologies", Southern Federal University, Taganrog 347922, Russia
[3]Department of Radio Engineering Electronics, Southern Federal University, Taganrog 347922, Russia

Abstract. In this paper, we present the results of experimental study of the role of the wetting layer in the crystallization stage of droplet nanostructures in the lattice-mismatched InAs/GaAs material system. We demonstrate that the thickness of the wetting layer can significantly affect the characteristics of nanostructures formed during crystallization in the arsenic flux. While the surface density of InAs nanostructures does not increase relatively to the initial droplet density obtained at 300 °C, a significant increase of the nanostructure density is observed after crystallization of droplets obtained at 200 °C. This behaviour is attributed to the additional stress-driven Stranski-Krastanov quantum dot formation and confirmed by the 2D-3D transformation of indium layer obtained at 100 °C.

1. Introduction

Droplet epitaxy is a promising technology that has many advantages over other molecular beam epitaxy (MBE) techniques, including those based on the Stranski-Krastanov growth mode [1-9]. The technique allows obtaining a wide variety of nanostructures: quantum dots [7,10-11], single and multiple quantum rings [12,13], nanoscale disks [14], hybrid nanostructures with metal clusters and complexes based on them [15-17]. At the same time, unlike the Stranski-Krastanov growth mode, droplet epitaxy does not require lattice mismatch to produce quantum dots, which makes it possible to use it for the quantum dot formation in almost any material system of $A^{III}B^{V}$ group which is remarkable for its unique electronic, optical and technological properties [18-25]. However, the influence of processes occurring in the wetting layer of lattice-mismatched systems during the crystallization stage has not yet been studied. This is especially important due to the presence of a pronounced temperature dependence of the wetting layer thickness [26].

The purpose of this work is to study the effect of the wetting layer thickness on the characteristics of InAs/GaAs quantum-size structure arrays during the stage of crystallization in the arsenic flux.

2. Experimental procedure

The samples were grown on GaAs(001) epi-ready substrates in a SemiTEq STE 35 MBE system with solid-state sources of arsenic and group III elements. The growth procedure was monitored by an *in situ* reflection high-energy electron diffraction (RHEED) system. After removal of a native surface oxide and annealing of the substrate under an excessive As flux, a flat GaAs surface was formed by growing a 250 nm thick buffer layer at a substrate temperature T of 580 °C and a nominal growth rate v of 1 ML/s which is typical for MBE growth of a GaAs buffer layer [27-30]. Then, the substrate temperature was gradually decreased to 520 °C and the valved As cell was closed to discontinue the supply of As_4.

After the background pressure was reduced to $\sim 1\cdot 10^{-7}$ Pa, the substrate temperature was decreased to a temperature of In deposition measured by thermocouple. The RHEED pattern showed a characteristic (2×4) surface reconstruction over the entire cooling period. Then, In atoms were deposited at a nominal growth rate of 0.25 ML/s (preliminarily calibrated by the growth of InGaAs ternary compounds) during 6, 10 and 12 s at deposition temperatures of 300, 200 and 100 °C, respectively, which yielded an amount of indium material corresponding to 1.5, 2.5 and 3.0 monolayers (ML) of InAs. Then, the samples were quenched and transferred out of the chamber for the scanning electron microscopy (SEM) and atomic force microscopy (AFM) characterization.

In order to study the influence of the wetting layer on the crystallization processes, we exposed the metallic layers under As flux equivalent to that used for the growth of GaAs buffer layer and annealed them during 10 min at $T = 400$ °C. Then, the samples were re-examined by SEM and AFM methods.

3. Results and discussion

Deposition of 1.5 ML of In at $T = 300$°C leads to the formation of droplets with an average diameter of 80 nm and surface density of $9.1\cdot 10^{7}$ cm^{-2} (Figure 1a). As demonstrated in our previous study [26], critical thickness of droplet formation at an increment of the deposition thickness equal to 0.25 ML is 0.75 ML, i.e. droplets do not form until at least 1 ML of indium is deposited. Therefore, we can conclude that approximately 0.75 ML of material is consumed to form droplets while 0.75 ML remains on the surface as a wetting layer.

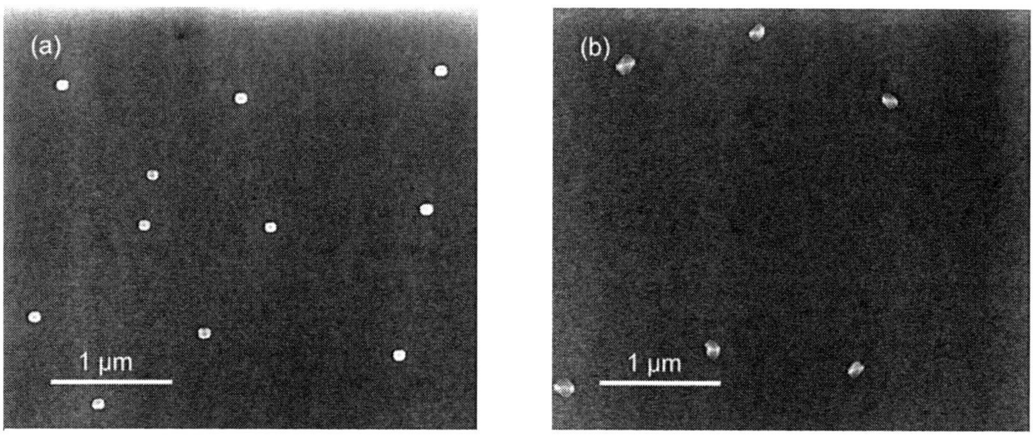

Figure 1(a, b). Morphology of the InAs/GaAs(001) droplet epitaxial system after **(a)** deposition of 1.5 ML of In at $T = 300$°C; **(b)** subsequent crystallization of metallic droplets in the As flux.

After crystallization of this system in the arsenic flux, a slight increase in the nanostructure size up to 90 nm is observed (Figure 1b). This increase is not unexpected since droplets absorb As atoms in an approximate ratio 1:1 to a number of In atoms in a droplet. However, droplets do not increase twice in size because an average interatomic distance in In crystal is larger than in InAs crystal. A similar size increase of InAs nanostructures in comparison with In droplets was observed elsewhere [11].

Arsenic crystallization of In droplets obtained at $T = 300$°C does not increase the surface density of nanostructures. Moreover, it leads to a decrease of their surface density from $9.1\cdot 10^{7}$ to $4.5\cdot 10^{7}$ cm^{-2} (Figure 1b). Although this phenomenon can be a result of an accidental error during droplet array formation, the difference in values can also be due to the diffusion decay of droplets under the As flux.

It was previously shown in many studies that In atoms migrate outwards from droplet when exposed by a low-value arsenic flux or at a high crystallization temperature [31-33].

At a lower temperature (200°C) a deposition thickness of 1.5 ML is not sufficient to form droplets [26], therefore 2.5 ML of indium was deposited with subsequent formation of metallic droplets (Figure 2a). Crystallization of this system in the arsenic flux led to a sudden change in the density of InAs/GaAs nanostructures by an order of magnitude – from $2{,}2{\cdot}10^8$ to $1{,}1{\cdot}10^9$ $см^{-2}$ (Figure 2). Due to the introduction of an additional component into the composition of droplet nanostructures, their average size also increases from 60 to 80 nm.

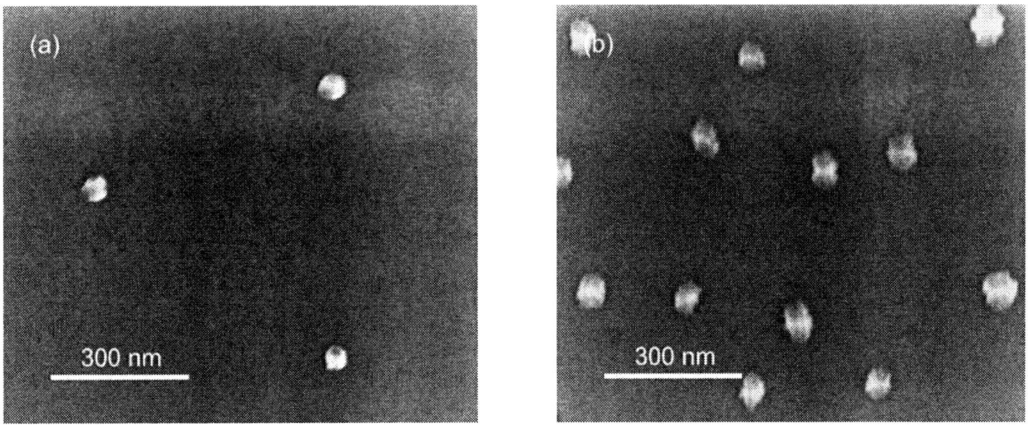

Figure 2(a, b). Morphology of the InAs/GaAs(001) droplet epitaxial system after **(a)** deposition of 2.5 ML of In at $T = 200$°C; **(b)** subsequent crystallization of metallic droplets in the As flux.

We suppose that the increase in the surface density is due to the fact that under these conditions the thickness of the wetting layer is about 2.0 ML. This value significantly exceeds the critical thickness for the Stranski-Krastanov growth mode. Therefore, under high temperature conditions, an additional contribution to the InAs nanostructure formation is observed, due to the transformation of an overstressed two-dimensional wetting layer into three-dimensional crystalline islands.

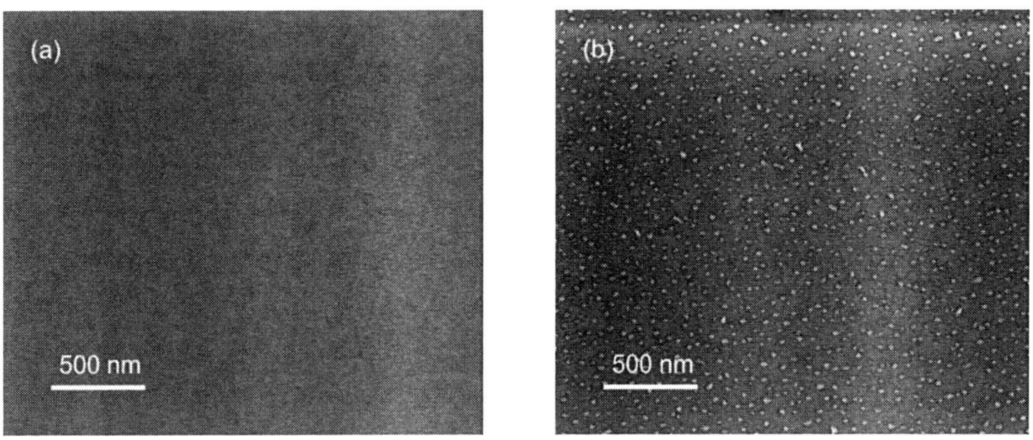

Figure 3(a, b). Morphology of the InAs/GaAs(001) droplet epitaxial system after **(a)** deposition of 3.0 ML of In at $T = 100$°C; **(b)** subsequent crystallization of metallic layer in the As flux.

In order to confirm the phenomenon of Straski-Krastanov nanostructure formation, we decreased the deposition temperature down to 100°C and deposit larger amount of material so that droplet did not form. A deposition thickness of 3 ML was sufficient to observe a continuous two-dimensional layer on the surface with no droplets formed (Figure 3a). Then, we crystallized the overheated indium layer in the arsenic flux at a high temperature of 500°C. As a result of this process, the formation of InAs/GaAs nanostructures with a high surface density of $2 \cdot 10^{10}$ cm^{-2} was observed on the surface (Figure 3b).

The most probable reason of this phemonomen is the strain-induced transformation of the wetting layer by the Stranski-Krastanov growth mechanism [34]. Formation of InAs nanostructures when no droplets were formed after deposition of indium was previously shown elsewhere [35]. Whereas indium two-dimensional layer is defective and presumably has an amorphous structure, InAs layer formed as a result of arsenic incorporation has a relatively similar lattice constant as compared to GaAs crystal. This leads to the emergence of stresses which are a driving force for the 2D-3D-transition of the wetting layer [36]. As a result, an array of InAs quantum dots is formed on the surface but it should be born in mind that the wetting layer may still exist on the surface.

Thus, the arsenic crystallization of In/GaAs systems obtained by droplet epitaxy may lead to both a decrease and an increase in the surface density of nanostructures depending on the wetting layer thickness. Furthermore, the formation of InAs nanostructures after crystallization of the two-dimensional layer is possible. Besides these phenomena must be taken into account when the crystallization of In nanostructures with specified parameters is carried out, they may be used intentionally to control the geometrical parameters of nanostructures.

4. Conclusions

In order to achieve maximum correspondence between the parameters of In droplet arrays in the first and of InAs crystalline nanostructure in second stages during droplet epitaxy, crystallization should be carried out under conditions under which the wetting layer thickness does not exceed 1.5 ML. Crystallization of an indium layer obtained at a low temperature and a large deposition thickness may lead to a formation of additional nanostructures which are not crystallized metallic droplets but nanostructures emerged from the stressed wetting layer.

Acknowledgments

This work was supported by the Russian Science Foundation Grant No. 15-19-10006. The results were obtained using the equipment of the Research and Education Center and Center for Collective Use "Nanotechnologies" of Southern Federal University.

References

[1] Somaschini C, Bietti S, Koguchi N and Sanguinetti S 2010 *Appl. Phys. Lett.* **97** 203109
[2] Mano T, Watanabe K, Tsukamoto S, Fujioka H, Oshima M and Koguchi N 1999 *Jpn. J. Appl. Phys.* **38** L1009–11
[3] Balakirev S V, Eremenko M M, Mikhaylin I A, Klimin V S and Solodovnik M S 2018 *J. Phys.: Conf. Ser.* **1124** 022018
[4] Solodovnik M S, Balakirev S V, Eremenko M M, Mikhaylin I A, Avilov V I, Lisitsyn S A and Ageev O A 2017 *J. Phys.: Conf. Ser.* **917** 032037
[5] Lee J H, Wang Z M and Salamo G J 2007 *J. Phys. Condens. Matt.* **19** 176223
[6] Balakirev S V, Solodovnik M S and Ageev O A 2017 *J. Phys.: Conf. Ser.* 2017 **917** 032033
[7] Somaschini C, Bietti S, Koguchi N and Sanguinetti S 2009 *Nano Lett.* **9** 3419–3424
[8] Lee J H, Wang Z M, Kim E S, Kim N Y, Park S H and Salamo G J 2009 *Nanoscale Res. Lett.* **5** 308–314
[9] Balakirev S V, Eremenko M M, Mikhaylin I A and Solodovnik M S 2018 *J. Phys.: Conf. Ser.***1124** 022025
[10] Mano T, Watanabe K, Tsukamoto S, Fujioka H, Oshima M and Koguchi N 2000 *J. Cryst. Growth*

209 504–508
[11] Kim J S and Koguchi N 2004 *Appl. Phys. Lett.* **85** 5893–5895
[12] Pankaow N, Thainoi S, Panyakeow S and Ratanathammaphan S 2011 *J. Cryst. Growth* **323** 282–285
[13] Noda T, Jo M, Mano T, Kawazu T and Sakaki H 2013 *J. Cryst. Growth* **378** 529–531
[14] Mano T, Watanabe K, Tsukamoto S, Koguchi N Fujioka H, Oshima M, Lee C D, Leem J Y, Lee H J and Noh S K 2000 *Appl. Phys. Lett.* **76** 3543–3545
[15] Nogues G, Merotto Q, Bachelier G, Lee E H and Song J D 2013 *Appl. Phys. Lett.* **102** 231112
[16] Lyamkina A A and Moshchenko S P 2013 *J. Phys. Chem.* **117** 16564–16570
[17] Balakirev S V, Solodovnik M S, Eremenko M M, Mikhaylin I A and Ageev O A 2018 *J. Phys.: Conf. Ser.* **1124** 022001
[18] Vurgaftman I and Meyer J R 2001 *J. Appl. Phys.* **89** 5815–5875
[19] Ageev O A, Solodovnik M S, Balakirev S V, Eremenko M M and Mikhaylin I A 2017 *J. Crys. Growth* **457** 46-51
[20] Lisitsyn S A , Balakirev S V, Avilov V I, Kolomiytsev A S, Klimin V S, Solodovnik M S, Konoplev B G and Ageev O A 2018 *Nanotech. in Russia* **13** 26-33.
[21] Ageev O A, Solodovnik M S, Balakirev S V and Mikhaylin I A 2016 *J. Phys.: Conf. Ser.* **681** 012036
[22] Ageev O A, Solodovnik M S, Balakirev S V and Eremenko M M 2016 *J. Vac. Sci. Technol. B* **34** 041804
[23] Balakirev S V, Solodovnik M S, Eremenko M M, Mikhaylin I A and Ageev O A 2017 *J. Phys.: Conf. Ser.* 2017 **917** 032034
[24] Klimin V S, Solodovnik M S, Lisitsyn S A, Rezvan A A and Balakirev S V 2018 *J. Phys.: Conf. Ser.* **1124** 041024
[25] Ageev O A, Solodovnik M S, Balakirev S V and Mikhaylin I A 2016 *Technical Physics* **61** 971–977
[26] Balakirev S V, Solodovnik M S and Ageev O A 2018 *Phys. Status Solidi B* **255** 1700360
[27] Chang L L, Esaki L, Howard W E, Ludeke R and Schul G 1973 *J. Vac. Sci. Technol.* **10** 655–662
[28] Ageev O A, Solodovnik M S, Balakirev S V, Eremenko M M and Mikhaylin I A 2016 *J. Phys.: Conf. Ser.* **741** 012012
[29] Solodovnik M S, Karenkikh O G, Balakirev S V, Petrov S I, Ryzhuk R V, Alexeev A N and Ageev O A 2018 *J. Phys.: Conf. Ser.* **1124** 081024
[30] Ageev O A, Solodovnik M S, Balakirev S V and Eremenko M M 2016 *Phys. Solid State* **58** 1045–1052
[31] Noda T, Mano T and Sakaki H 2011 *Crystal Growth & Design* **11** 726–728
[32] Noda T and Mano T 2008 *Appl. Surf. Sci.* **254** 7777–7780
[33] Noda T, Mano T, Kuroda T, Sakoda K and Sakaki H 2009 *J. Cryst. Growth* **311** 1836–1838
[34] Baskaran A and Smereka P 2012 *J. Appl. Phys.* **111** 044321
[35] Zhao C, Chen Y, Xu B, Jin P and Wang Z 2007 *Appl. Phys. Lett.* **91** 033112
[36] Prieto J E and Markov I 2002 *Phys. Rev. B* **66** 073408

Photovoltaic properties of InP NWs/p-Si heterostructure

K P Kotlyar[1,5], A V Vershinin[1], R R Reznik[6], S I Pavlov[3], D A Kudryashov[2], K S Zelentsov[2], A M Mozharov[2,6], A A Karaborchev[7], I S Mukhin[2,6], I P Soshnikov[1,3,4], G E Cirlin[1,4,6]

[1]Laboratory of epitaxial nanotechnologies, St. Petersburg Academic University, Khlopina 8/3, St. Petersburg 194021, Russia
[2]Laboratory of renewable energy sources, St. Petersburg Academic University, Khlopina 8/3, St. Petersburg 194021, Russia
[3]Ioffe Physical-Technical Institute of the RAS, Politekhnicheskaya 26, St. Petersburg 194021, Russia
[4]Institute for Analytical Instrumentation of the RAS, Rizhsky pr. 26, St. Petersburg 190103, Russia
[5]St. Petersburg State University, Universitetskaya nab. 7-9, St. Petersburg 199034, Russia
[6]ITMO University, Kronverkskiy pr. 49, St. Petersburg 197101, Russia
[7]St. Petersburg Electrotechnical University, ul. Professora Popova 5, St. Petersburg 197376, Russia

Abstract. The experimental and modelling results of the investigation of photovoltaic properties of the InP nanowires grown on p-Si substrate are presented. It is shown that InP 2D-layer between nanowires on Si is II-type heterojanction. The resulting structure exhibits an open-circuit voltage (Voc) of 0,37 V, a short-circuit current density (Jsc) of 10.6 mA/cm2, a fill factor (FF) of 0.61 and efficiency of 2.4% at 1 sun. The InP/p-Si structure can be applied as prototype for solar cells or the photosensitive component in an integrated circuit based on silicon technology.

1. Introduction

Nanowires (NWs) of III-V materials can be applied for fabrication of various optoelectronic devices (photonics, piezotronics, etc. [1-3]). For example, the AlGaAs/GaAs QD/AlGaAs nanowire array can be used to create a single photon source [4]. Previous experimental work [5] has shown the possibility of growth of coherent InP/InAsP QD/InP NWs on a Si substrate. An intense emission from NW array near 1.3 μm makes it possible to consider the given system promising for further integration of optical elements on a silicon platform with fiber-optic systems. The influence of growth parameters on the optical, structural, and morphological properties of arrays of InP/InAsP QD/ InP NWs grown on Si [5] and the transport properties of single InP NW [6] were investigated. However, the transport properties of the InP/Si structure must be studied for the development of the technology for the successful fabrication of various devices. In this paper, we study the photovoltaic and transport properties of the InP NWs/Si structure, in particular, the effect of the InP 2D-layer between NWs on the structure properties.

2. Experiment

The InP NW arrays were grown by molecular-beam epitaxy (MBE) on boron-doped Si(111) substrates (carrier concentration is $4{,}4 \cdot 10^{16}\ cm^{-3}$) by method described in [5]. The growth experiments were performed in Riber Compact 21 MBE machine. The surface morphology of the samples was studied using JSM 7001F (JEOL, Japan) and Supra 25 (Carl Zeiss, Germany) scanning electron microscopes (SEM). In the preliminary stages of contacts fabrication (figure 1a) for photovoltaic properties measurements, the samples were cleaned in $HF:H_2O$ (1:3) for silicon and HCl: H_2O (1:3) for InP. On the first step, Al layer was deposited on the back side of the sample using vacuum thermal deposition. On the next step, transparent ITO contact layer (100 nm) in the shape of 2 mm circle was deposited through metallic mask on the top surface of the sample use RF magnetron sputtering (BOC Edwards Auto 500 RF). Dependence of the external quantum efficiency on the light spectrum was measured in the range of 350–1200 nm with the use of M266 (Laser Solar) monochromator. Current-voltage characteristics (I-V) were measured using of Keithley 2400 SourceMeter multimeter and SunLight Solar Simulator (ABET Technologies) providing AM1.5 G, 100 mW/cm^2 spectrum. The photovoltaic properties of samples were simulated and analysed with the Silvaco software.

3. Results and discussion

Figure 1b shows typical SEM images of the NW arrays before and after the ITO layer deposition. The InP NWs average height is 5 μm with a surface density of $3 \cdot 10^5\ cm^{-2}$. The average diameter of the NWs is 100 nm near the bottom and about 20–40 nm on the top. Between NWs is 2D InP-layer (average thickness: 250 nm). This layer is formed in NWs growth process.

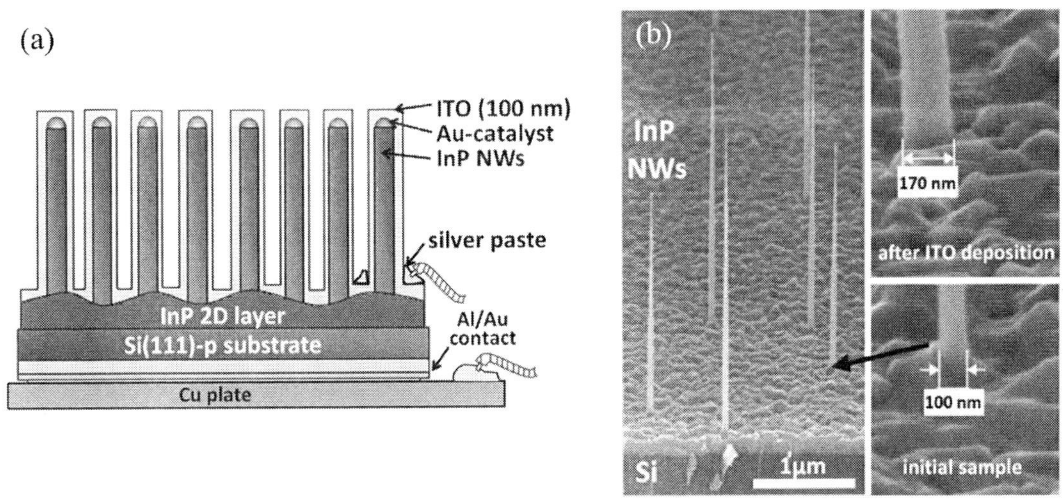

Figure 1(a, b). **(a)** Scheme of the InP NWs/Si device. **(b)** SEM images of InP NWs on Si(111) substrate. The inserts shown of the samples surface morphology before and after ITO layer deposition.

Figure 2 shows the I–V characteristics of InP NWs/Si-p structure under light. An experimental sample demonstrates a diode current-voltage characteristic. The photovoltaic parameters are an open-circuit voltage (V_{oc}) of 0.37 V, a short-circuit current density (J_{sc}) of 10.6 mA/cm^2, and a fill factor (FF) of 0.61 and efficiency of 2.4%. Simulation of the current-voltage characteristics and band structure of a thin-film InP/Si solar cell was carried out. In the process of modeling, the resistance of the structure is correlated to the experimental data by changing the resistance of the upper contact so that each point can be compared with the experimental result. Analysis of the simulation results shows that the band structure of the prototype solar cell corresponds to the type-II heterojunction (figure 3a). The light absorption occurs predominantly in a 2D layer InP in consequence low density of NWs.

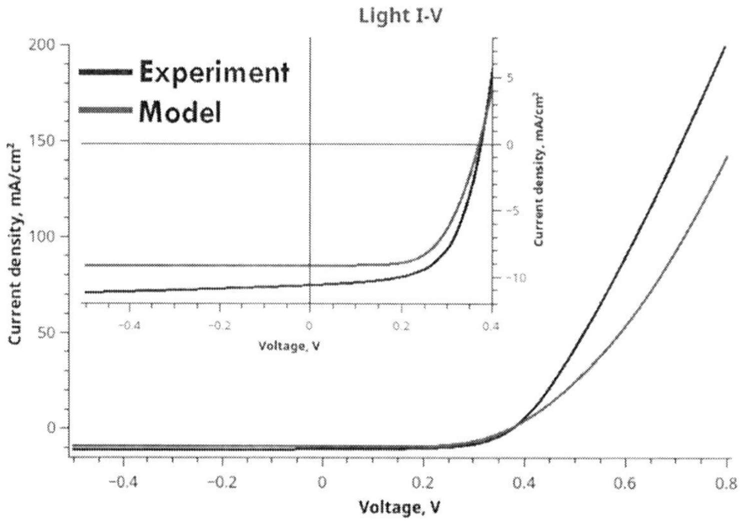

Figure 2. Light I-V characteristics of the InP NWs/Si device: experiment and model results.

The decrease of the parameters of the InP/Si structure and a slight discrepancy between the modeling and experimental results is explained by the high rate of light absorption in the InP layer (high defective layer) and a high density of interface states (InP-Si). The map (figure 3b) of the open-circuit voltage and the fill factor versus the doping level of the InP layer and the density of InP/Si interface states allows to determine the parameters of structure. According to the simulation data, the doping level of the InP layer is 10^{18} $см^{-3}$ and density of InP/Si interface states is 10^{12} cm^{-2}. In conclusion, the InP/p-Si structure can be applied as prototype for solar cells or the photosensitive component in an integrated circuit based on silicon technology.

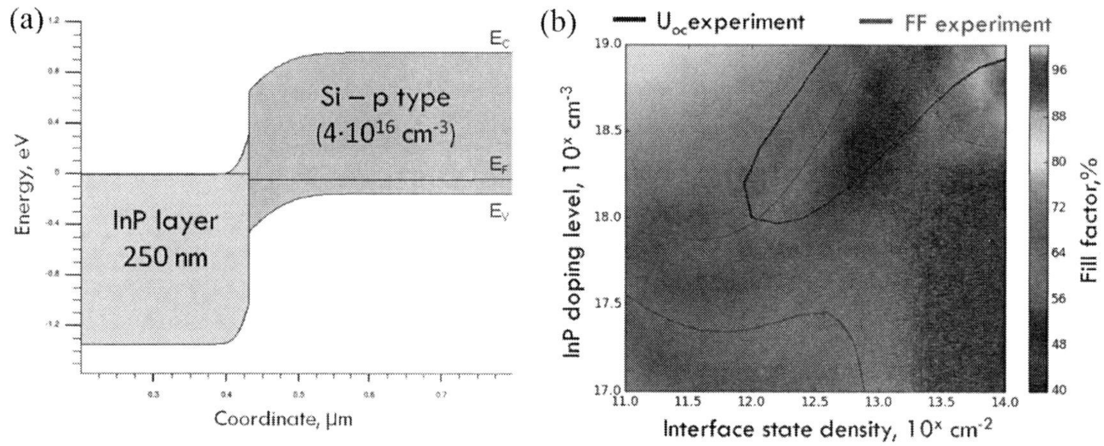

Figure 3(a, b). (a) Energy Band diagram of the InP/Si structure. **(b)** Map of the open-circuit voltage and the fill factor (experimental data) versus the doping level of the InP layer and the density of InP/Si interface states (model data).

Acknowledgments

The nanowire samples were grown under the support of Russian Science Foundation (Project № 19-72-30004). SEM characterization were performed using equipment owned by the Federal Joint Research Centre "Material science and characterization in advanced technology" with financial support by Ministry of Education and Science of the Russian Federation (id RFMEFI62117X0018). Reznik R.R. grateful the support of RFBR (grant № 18-32-00768 mol_a). The study of transport properties were financially supported by the RFBR (grant № 18-07-01364 A).

References

[1] Wallentin J et al., 2013 *Science* **339**(6123) 1057-1060
[2] Otnes G et al. 2018 *Nano Lett.* 2018
[3] Cavalli A et al. 2016 *Nanotechnology* **27** 454003
[4] Cirlin G E et al. 2017 *J. Phys. D: Appl. Phys.* **50**(48) 4003
[5] Reznik R R et al., 2018 *Technical Physics Letters* **44**(2) 112–114
[6] Sharov V A et al., 2019 *JPCS* **1199** 012021

Raman analysis of graphene on SiC

K Yu Shubina[1], V L Sharov[1], D V Denisov[1], I S Mukhin[1,2] and A V Babichev[1]
[1] Saint Petersburg National Research Academic University RAS, St. Petersburg 194021, Russia
[2] ITMO University, St. Petersburg 197101, Russia

Abstract. The paper presents the results of a study of the structural quality of grapheme produced by the thermal decomposition of 4H-SiC. Maps obtained by atomic force microscopy (AFM mapping) have demonstrated the existence of characteristic terraces on the surface. The Raman spectra have demonstrated the existence of G and 2D lines. A narrow Lorentzian 2D peak and low intensity of the G peak in addition to the reflectance mapping data confirms the conclusion that there is primarily monolayer graphene on the sample surface.

1. Introduction

Graphene is a monolayer material in electronics with a high mobility of charge carriers, which makes it a promising basis for the development of high frequency transistors. By the present time, results have been demonstrated of the formation of high frequency (HF) transistors based on exfoliated graphene (scotch-tape graphene) [1]. However, the features of exfoliated graphene make this method absolutely unsuitable for industrial applications (the dimensions of the samples do not exceed 1×1 mm^2, and the number of layers during the exfoliation process is uncontrollable). In addition to mechanical exfoliation of graphene (scotch-tape graphene), methods of chemical-vapor deposition on substrates of transition metals (Cu, Ni, etc.) and thermal decomposition of SiC have been developing lately. The recent developments in epitaxy have made it possible to grow high structural quality graphene with a controlled number of layers on a big area.

The output characteristics of graphene transistors differ from those of traditional silicon field-effect transistors because there is no band gap in graphene. No current saturation has been observed at drain voltages of under 2 V. At the same time, a low "on-off" signal ratio is not required for a number of HF applications including amplifiers and mixers, which supports the interest of researchers and the industry to developments in this research area.

IBM Research Center has presented results of HF transistor formation on SiC surface. The cut-off frequency was 100 GHz [2]. The dielectric applied was a electron resist of the deep-UV spectral range (157-248 nm) based on polyhydroxystyrene followed by HfO_2 deposition. Due to the research of the graphene/substrate and graphene/top dielectric interfaces and the optimization of the wetting layers and dielectric deposition conditions, the cut-off frequency could be increased up to 300 GHz. Graphene produced by thermal decomposition of silicon carbide has demonstrated n-type conductivity with a typical doping level of 10^{12} cm^{-2}, which is determined by the properties of the material itself and the buffer layer. As a result, it is necessary to develop techniques of doping (in-situ doping compensation) in order to approximate graphene to the Dirac point [3]. Moreover, reducing the channel doping is an effective method for increasing the slope of the device's output characteristics. At the same time, it should be noted that the research of the formation of transistor structures is at an

early stage yet. There are a number of works on the problem of the formation of high quality graphene/dielectric interfaces, dielectric deposition techniques, and changes of the doping level.
This paper presents the results of a study of the Raman spectra of commercially available graphene produced by thermal decomposition of Si-face 6H-SiC substrate.

2. Experiment

Maps obtained by atomic force microscopy (AFM mapping) were recorded on a characteristic scale of 10×10 μm^2 in a semi-contact scanning mode. The existence of terraces has been demonstrated (see Fig. 1a). The typical size of terrace regions is approximately 2 μm. The terrace regions are focused on the image in a vertical plane, and there are step edges. The scanning profile shows a height difference of approximately 6 nm on step edges. The number of layers was determined by reflectance mapping [4]. The results are shown in Fig. 1b. Graphene is primarily monolayer (on 98.4% of the scanning area).

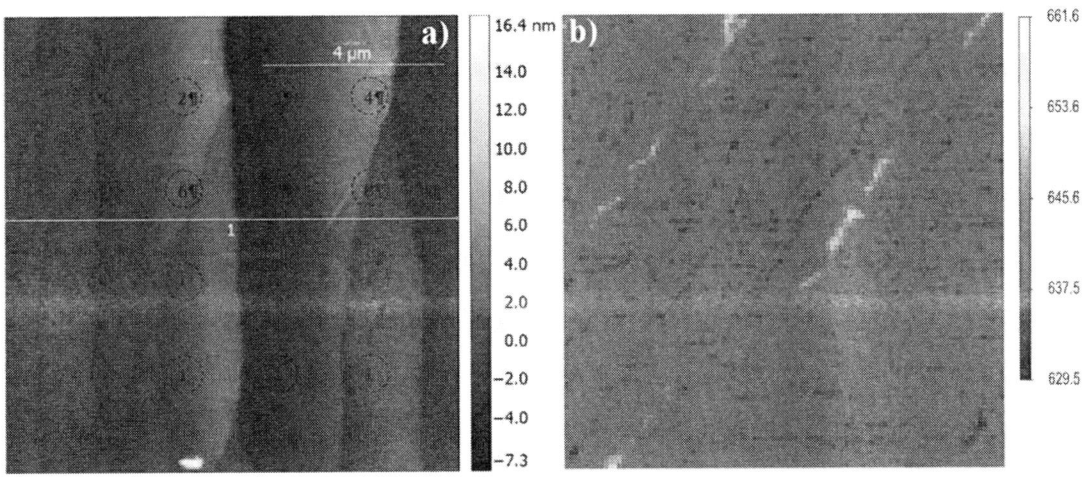

Figure 1(a,b). a) AFM mapping of graphene on SiC. The circles indicate the Raman spectra measurement points; **b)** 30×30 nm reflectance map with a 300 nm interval.

The Raman spectroscopy study was performed in the range of 1200÷3000 cm^{-1} with correlation of the spectra to the position on the AFM maps (measurements were made at 16 points). The pump laser wavelength is 532 nm, the lens is ×50, the confocal hole and the slit are 500 μm each, the attenuator is D0.6, the exposure time is 30 s, and the laser power is 2 mW. For evaluating the contribution of the substrate to the Raman spectra of graphene, the 4H-SiC Raman spectrum was measured on the SiC surface. The result is shown in Fig. 2a. Peaks in the spectrum have been demonstrated that are typical for 4H-SiC [5, 6]. The peak near 203.5 cm^{-1} is an E_2 planar or transverse acoustic (PA) mode, 196 cm^{-1} is an E_2 doublet (in brackets), 266.0 cm^{-1} is E_1 planar acoustic (PA), 610.5 cm^{-1} is A_1 axial or longitudinal acoustic (LA), 777.0 cm^{-1} is E_2 planar optical (PO), 797.5 cm^{-1} is an E_1 transverse optical (TO) mode, 967.0 cm^{-1} is A_1 longitudinal optical (LO) mode.

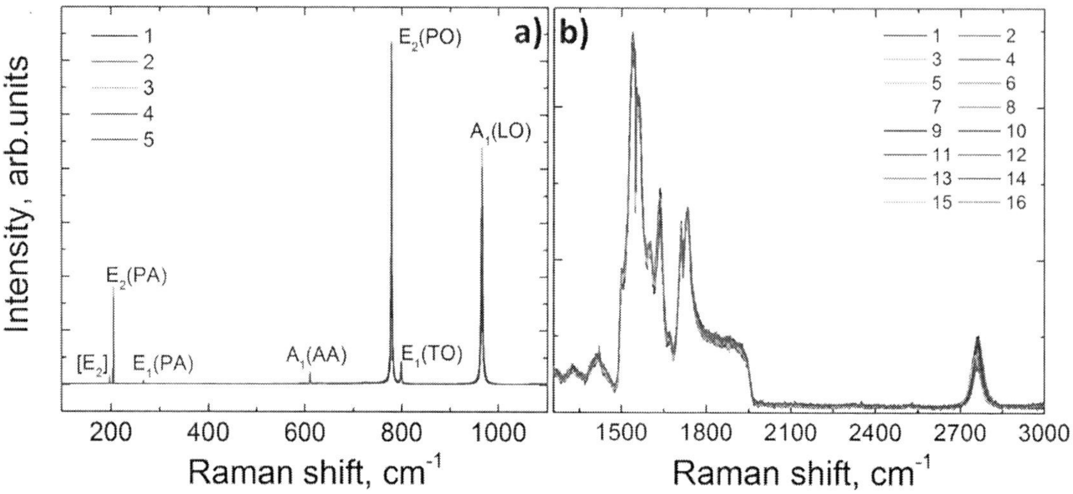

Figure 2(a,b). a) The Raman spectrum of 4H–SiC wafer taken at room temperature; **b)** The Raman spectrum of graphene at room temperature.

For estimating the contribution of possible overheating, the Raman spectra of graphene were measured in dependence on the optical pumping rate in the power range of a 0.1-10 W laser. For further experiments, the laser power was fixed at the rate of 2 mW. The Raman spectra of graphene with correlation to points on the AFM map (see Fig. 1a) are shown in Fig. 2b. The spectra have demonstrated the presence of G, D and 2D lines, which are typical for graphene. There are also lines typical for silicon carbide (2nd order SiC lines). The intensity of D line is not big, which evidences a low density of defects. The 2D peak of graphene is well described by single Lorentzian. The average full width at half maximum (FWHM) value is 38 cm^{-1}, which corresponds to the typical value of 2D peak FWHM of monolayer graphene [7]. The medium position of the 2D peak corresponds to 2747 cm^{-1}. It has been shown that the intensity of the 2D peak when measured in the step edges area is greater as compared to the terrace regions. A narrow Lorentzian 2D peak and low intensity of the G peak in addition to the reflectance mapping data confirms the conclusion that there is primarily monolayer graphene on the sample surface.

Acknowledgments
This work was carried out with support from a Grant of the President of the Russian Federation, no MK-5418.2018.8. D.V. Denisov is also grateful for partial support of the research by the Ministry of Education and Science of the Russian Federation, government order no.16.9789.2017. We are grateful to I.G. Ivanov for the mapping of optical reflectance.

References
[1] Liao L, Lin Y-C, Bao M, Cheng R, Bai J, Liu Y, Qu Y, Wang K L, Huang Y and Duan X 2010 *Nature* **467** 305–8
[2] Lin Y-M, Dimitrakopoulos C, Jenkins K A, Farmer D B, Chiu H-Y, Grill A and Avouris P 2010 *Science* **327** 662
[3] Wu Y, Lin Y, Bol A A, Jenkins K A, Xia F, Farmer D B, Zhu Y and Avouris P 2011 *Nature* **472** 74–8
[4] Ivanov I G, Hassan J U, Iakimov T, Zakharov A A, Yakimova R and Janzén E 2014 *Carbon*, **77**, 492–500
[5] Feldman D W, Parker J H, Choyke W J and Patrick L. 1968 *Phys. Rev.* **173** 787–93

[6] Burton J C, Sun L, Pophristic M, Lukacs S J, Long F H, Feng Z C and Ferguson I T. 1998 *J. Appl. Phys.* **84** 6268–73
[7] Grodecki K, Jozwik I, Baranowski J M, Teklinska D and Strupinski W 2016 *Micron* **80** 20–23

STRUCTAL AND OPTICAL PROPERTIES AlGaAs NANOWIRES GROWN BY MBE ON Si(111) SUBSTRATE

A V Vershinin[1,3], I V Shtrom[2,3], K P Kotlyar[1,3], N G Filosofov[3], A Yu Serov[3], Yu B Samsonenko[2], I V Ilkiev[1], R R Reznik[1,2,3] and G E Cirlin[1,2]
[1]St.-Petersburg Academic University, Russian Academy of Sciences, 194021 St. Petersburg, Russia
[2]Institute for Analytical Instrumentation RAS, Rizhsky pr. 26, 190103 St. Petersburg, Russia
[3]St. Petersburg State University, Universitetskaya nab. 79, St. Petersburg, 199034 Russia

Abstract. In this work, we present the results of photoluminescence measurements of wurtzite AlGaAs nanowires, coupled with the transmission electron microscopy structural analysis. AlGaAs NWs were grown by molecular beam epitaxy under the nominal aluminum content of 0.3 - 0.8.The results of TEM measurements demonstrated wurtzite crystal structure of AlGaAs nanowires.

1. Introduction

While bulk layers and thin films of GaAs and AlGaAs have the structure of zinc blende (ZB), regions with both ZB and wurtzite (WZ) crystal structures can be observed in GaAs and AlGaAs nanowires (NWs). Experimental and theoretical results show that the band gap values of A3B5 NWs in the WZ and ZB phases are different. The optical properties of NWs in the WZ phase can substantially differ from NWs in the ZB phase. Thus, WZ NWs are a new semiconductor material, and the study of its optical properties has fundamental interest in semiconductor physic. For AlGaAs NWs, the transmission electron microscopy data confirm the presence of the WZ phase in these NWs, but the results of experimental studies of WZ AlGaAs optical properties, in the current scientific literature, to the best of our knowledge, is not represented. It should also be noted that the optical properties of structures based on $Al_xGa_{1-x}As$, depending on the composition (x), have been studied in detail for bulk layers and thin films, while for $Al_xGa_{1-x}As$ NWs, similar studies have not been carried out. There are only a few papers describing the optical properties of the $Al_xGa_{1-x}As$ NWs [1,2,3]. In these literature data, the possible presence of a WZ phase in the formation of AlGaAs NWs is no considered. The possible influence of core-shell structure formed by the growth of the AlGaAs on the optical properties of the NWs is not discussed in the literature. Although it is known that with the growth of the AlGaAs NWs by molecular beam epitaxy, spontaneously formed NWs with one Al content in the middle – core and with a large Al content along the lateral faces - shell [4].

2. Experiment

In this work, studies of the optical properties of the NWs array $Al_xGa_{1-x}As$ are presented. The PL measurements are compared with the structural properties of $Al_xGa_{1-x}As$ NWs obtained using scanning electron microscopy (SEM) and transmission electron microscopy (TEM). Nominal content of Al in the solid solution was varied within the range x = 0.3-0.8 [4]. The $Al_xGa_{1-x}As$ NWs array was synthesized using the molecular beam epitaxy (MBE) method on a Si (111) substrate. The synthesis of the investigated structures was described in detail in our works published earlier [1,4].

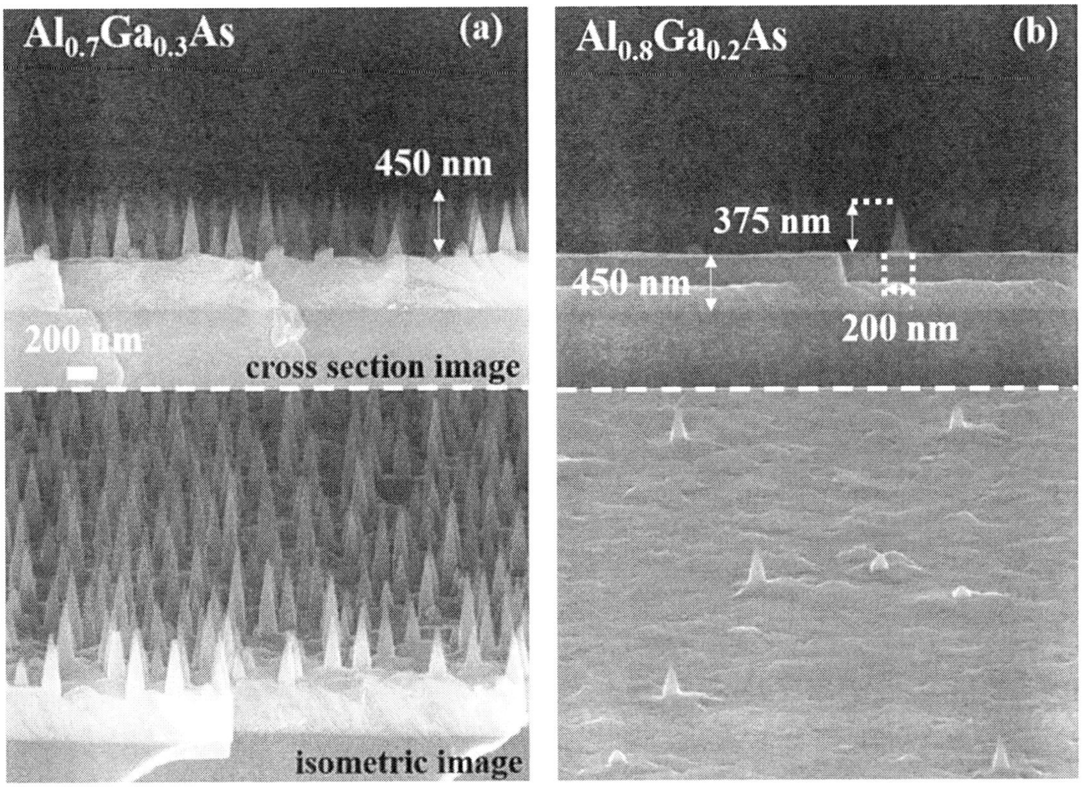

Figure 1(a,b). SEM images of AlGaAs NWs with nominal content of Al in the solid solution x =0.7**(a)** and x =0.8**(b)**.

SEM measurements showed that during the growth "pencil-like" Al_xGa_{1-x} As NWs are formed (Figure 1). The Al_xGa_{1-x} As NWs average height is 450 nm at a surface density of 10^8 cm^{-2} for x =0.7 (Figure 1(a)). The $Al_xGa_{1-x}As$ NWs average height is 375 nm at a more low surface density of 10^5 cm^{-2} for x =0.8 (Figure 1(b)). The average diameter of the NWs is 200 nm near the bottom. Between NWs is Al_xGa_{1-x} As layer (average thickness: 450 nm) growth in process of synthesis of Al_xGa_{1-x} As NWs. The results of TEM measurements demonstrated wurzite crystal structure of all $Al_xGa_{1-x}As$ nanowires with nominal content of Al in the range x = 0.3-0.7 [4]. As it was outlined in our previous paper [4] AlGaAs NWs synthesized using the MBE method formed a "core-shell" structure with a strong edge between «core» and «shell». The results obtained using high-angle annular dark-field imaging and energy dispersive X-ray spectroscopy demonstrate the Al content in the NWs «shell» is about y= 0.55, that is practically corresponded to the nominal Al content x = 0.6, the Al content in the «core» about y= 0.35 [4]. The numerical calculations of Al content in «core» and «shell» of $Al_xGa_{1-x}As$ NWs are presented in this work for the different nominal Al content x [4]. Known, that $Al_xGa_{1-x}As$ is a direct band gap semiconductor at x <0.4 and indirect band gap semiconductor at x ≥ 0.4 for AlGaAs in the ZB phase [5]. What happens to AlGaAs in the WZ phase is not known at the moment.

3. Results and discussion

Thus, we can investigate the optical properties of a new semiconductor material AlGaAs in the wurtzite phase with different compositions of Al. The Figure 2(a) shows the normalized PL spectra of AlGaAs NWs with a nominal content of Al x from 0.3 to 0.7. PL excitation was carried out at a wavelength of 532 nm at a temperature of 10 K. PL spectra significantly differ from AlGaAs layers of the same composition in Al [5]. The Figure 2 (b) shows the normalized PL spectra of AlGaAs NWs with a nominal content of Al x from 0.8. The PL spectrum is shifted to lower energy, in contrast to the previous values of x (Figure 2(a)). However, the density and height of nanowires are very small for x = 0.8. Thus, the PL signal detected from the AlGaAs layer.

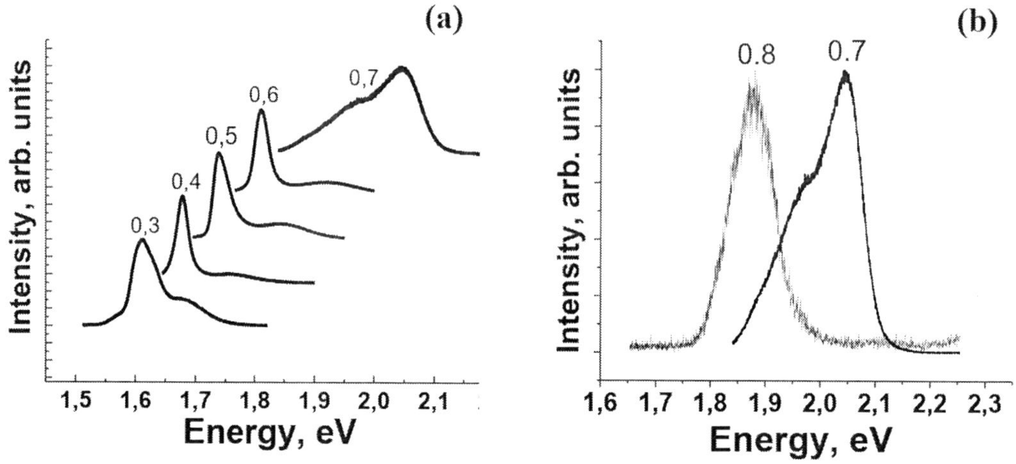

Figure 2(a,b). (a) PL spectra of AlGaAs NWs with nominal content of Al in the solid solution from x = 0.3 to 0.7. **(b)** PL spectra of AlGaAs NWs with nominal content of Al in the solid solution x = 0.7 and 0.8 obtained at 10 mW excitation power. Laser spot diameter is 0,3 mm.

In addition to difference between the band gap values of AlGaAs in the WZ and ZB phases, in the case of NWs, we have a heterointerface between the "core" and the "shell". Indirect optical transitions can exist on this interface, which can explain the presence of two PL maxima for all Al compositions. The numerical simulation predicted for AlGaAs NWs with nominal composition of x = 0.7 and x = 0.8 [4] that the composition of Al in the "core" is more than 0.4. In this case, semiconductor material AlGaAs in the ZB phase has indirect band gap [5]. In the case of NWs, we observe a significant change in the shape of the PL spectra Figure 2. It is possible that in the wurtzite phase AlGaAs is an indirect band gap semiconductor with an Al composition of more than 0.4, and does not always remain direct band gap, as theoretical predicted in [6]. Further optical investigations of AlGaAs NWs allow us to determine the difference of the WZ AlGaAs band gap vs ZB AlGaAs band gap, and to determine the band structure of WZ AlGaAs with different composition in Al.

Acknowledgments

We grateful the support the support of the Russian Science Foundation (Project No. 18-72-10047)

References

[1] George E Cirlin, Rodion Reznik, Igor Shtrom, et al., *Journal of Physics D: Applied Physics*.**50, 48** (2017)
[2] Z. H. Wu, M. Sun, X. Y. Mei, and H. E. Rudaa, *Appl. Phys. Lett.* **85**, 657 (2004)
[3] Chanchal G. , et al., *Appl. Phys. Lett.* **94**, 123105 (2009)
[4] V.G.Dubrovskii, I.V.Shtrom, et al., *Cryst. Growth Des.*, **v. 16**, p.7251 (2016)
[5] Lorenzo Pavesi1 and Mario Guzzi J. *Appl. Phys.* **75**, 4779 (1994)
[6] A. De and Craig E. Pryor, *PHYSICAL REVIEW B*, **v. 81**, p. 155210, (2010)

The method of intracellular fluorescent species *in situ* production within living cells by femtosecond laser pulses

A A Osychenko[1], A D Zalessky[1], A M Shakhov[1], A A Astafiev[1], M S Syrchina[1], U A Tochilo[1], A V Karmenyan[2] and V A Nadtochenko[1]
[1]N.N. Semenov Federal Research Center for Chemical Physics, Russian Academy of Sciences, RAS, Moscow, 119991, Russia
[2]National Dong Hwa University, No. 1, Sec. 2, Da Hsueh Rd., Shoufeng, Hualien 97401,Taiwan, R.O.C.

Abstract. Here we report on a novel non-invasive technique of *in situ* fluorescence species creation with the use of tightly-focused femtosecond laser beam with 800 nm wavelength. Fluorescent species can be created in a strictly defined area inside a living cell, without damaging the plasma membrane and surrounding structures. Species are characterized by bright fluorescence with a maximum at 520 nm, high stability and low toxicity. No significant effect of fluorescent species on cell viability and development is observed. The technique of *in situ* fluorescence species creation is an attractive tool for *in vivo* cell visualization and intracellular tracking because of its high accuracy and low damage effect.

1. Introduction

Studying of intracellular organization, organelle interaction, and dynamics of intracellular processes are fundamental problems in cell biology. The topical direction in cell imaging is developing non-invasive high-specific labels for the tracking of cellular components. Nowadays multiple methods of *in vivo* and *in vitro* imaging are available, and most of them work on the principle of selective chemical binding with specific molecules. Moreover, delivery of fluorescent species into the cell is required in the most methods. Another principle of fluorescent labeling is production of fluorescent species within biological material, and the most promising instruments for this purpose are tightly focused ultrafast lasers [1]. They allow us to create fluorescent particles within tissue from its material in a strictly defined area [2, 3] by the non-invasive manner, in a very localized volume with minimal chemical, mechanical and thermal stress [4, 5].

In this research we describe a novel technique for creating fluorescent species using femtosecond lasers inside a living cell. Fluorescent species retained for more than 24 hours and did not affect viability and development of mouse GV oocytes. These features make the method of *in situ* fluorescence species creation an attractive tool for intracellular visualization and tracking.

2. Methods

2.1. Experimental setup

In the experiments, we used the inverted optical Olympus IX71 microscope with the objective 60× and NA = 0.7. Fluorescent species creation was performed using a femtosecond laser Tsunami (Spectra Physics) generating the radiation at the wavelength 800 nm. The diameter of the beam waist was 1.39 μm. The laser pulse duration was 30 fs, pulse energy 100 nJ, and the pulse repetition rate was

1 KHz. The estimated power density for the pulse in the beam waist was $2.2\cdot10^{14}$ W·cm^{-2}. The visual control was implemented using the CMOS 1.3Mpix Thorlabs camera.

2.2. Fluorescence registration

We used two methods to record fluorescence signal from the sample. In the first method fluorescence was excited by single-photon absorption of laser diode radiation with λ = 462 nm (Nichia NDB7675), which was coupled to the Olympus microscope and focused by the same objective lens as femtosecond radiation. The laser diode beam was focused on the objective lens to give an about 500 µm wide and uniform field of illumination. The fluorescent image was also recorded by the same cameras as described above with exposure time up to 1 second.

In the second method, fluorescence was excited by two-photon absorption of femtosecond laser pulses focused by the objective lens. In order to avoid damage to the biological material or chemical reactions induced by laser radiation its average power was reduced below 10 mW. Fluorescence signal was collected by the same objective lens and directed to the side port of the microscope via a beamsplitter cube and then was coupled to an Acton SP300i monochromator and then to a PI-MAX 2 CCD camera (Princeton Instruments) used to record the fluorescence spectra.

2.3. Sample prerаration

Mouse germinal vesicle (GV) oocytes were obtained from 4-week-old C57Bl/CBA female mice. Mice were superovulated with 10 IU of pregnant mare`s serum gonadotropin intraperitoneally 48 h before preparation. The ovaries were removed from the mice, transferred to M2 medium («Sigma»), and were prepared with thin tweezers. Cumulus-oocyte complexes were collected with a glass capillary and placed into 0,1% hyaluronidase solution to remove cumulus. For the experiments, GV oocytes were placed in a 50 µl drop of M2 medium onto Petri dish with a cover glass and central hole («SPL Lifesciences»). Drop was covered with 2 ml of paraffin oil («Origio»).

2.4. In vitro maturation

Oocytes were cultivated in DMEM medium («PanEco») supplemented with 15% (*v/v*) fetal bovine serum (Gibco), 1 IU/ml gentamicin («Sigma») and 1 *IU*/ml of PMSG («Intervet») at 37°C, 90% humidity, 5% CO_2 for 24 h. After IVM, oocytes were examined for the presence of polar bodies and metaphase plates.

3. Results

Fluorescent species were created in the cytoplasm, nucleus and nucleolus of GV oocytes using tightly focused femtosecond laser beam (Fig. 1, A-C). Area of fluorescent species creation was set by focusing to a place of interest. Minimal spot size was estimated as 500 nm. Fluorescent particles did not diffuse through the cell, but retained their original shape for the long time and even after oocyte maturing.

The luminescence of fluorescent species were able to be excited both one-photon (at the wavelength 460 nm) and two-photon (at the wavelength 800 nm). The maximum of the luminescence corresponds to 520 nm and seems to be the same for cytoplasm, nucleus and nucleolus (Fig. 2).

Fluorescent particles persisted in a living cell for 24 hours and even more (Fig. 1, D-F). Presence of fluorescent species in the cell did not affect the ability of oocytes to develop up to the metaphase II (MII) stage during *in vitro* maturation. Oocytes of both experimental and control groups developed to the MII stage with 80-90% rate. This indicates that the method of *in situ* fluorescent species production has low toxicity for living cells.

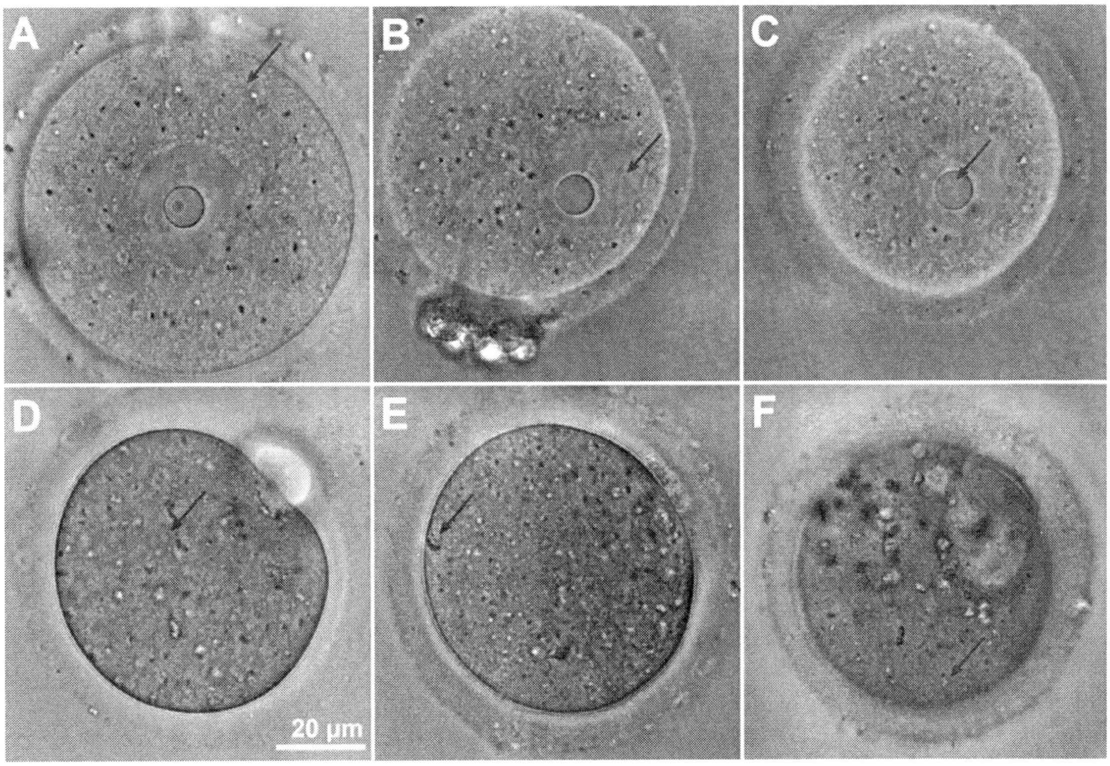

Fig. 1. Fluorescent species in the cytoplasm (A), nucleus (B) and nucleolus (C) of GV oocyte. Arrows indicate the fluorescent spots.

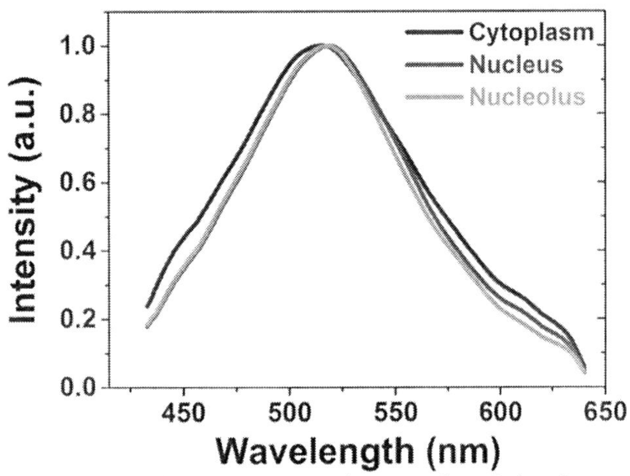

Fig. 2. Luminescence spectra of fluorescent species, produced in the cytoplasm, nucleus and nucleolus.

Intracellular tracking was performed on GV oocytes. Fluorescent spot created within a nucleus seemed to be anchored to the chromatin filament. Spot movement revealed by the time-lapse imaging shows chromatin restructuring (Fig. 3) during the experiment (~6 minutes). Bleaching of fluorescent species was not observed during this time.

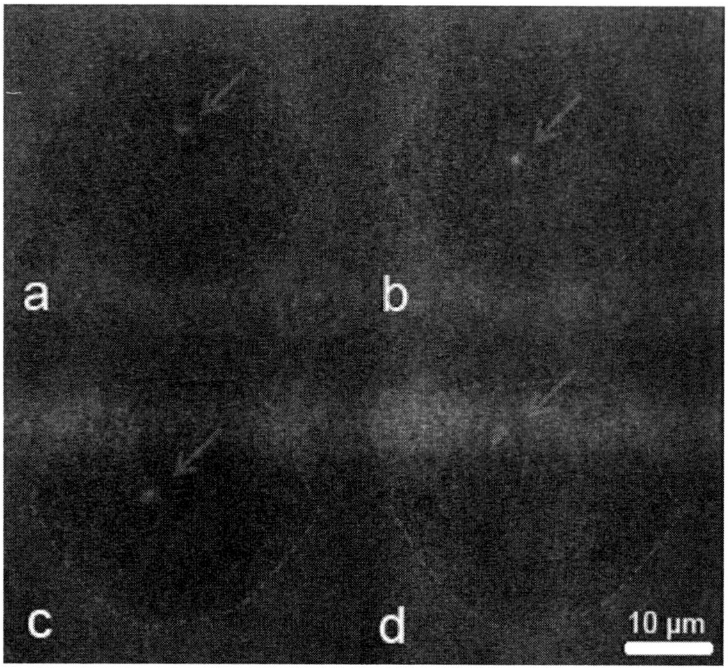

Fig. 3. Fluorescent spot migration within the nucleus of GV oocyte. Arrows indicate fluorescent particles, the dotted line indicates nucleus region. A – 0 s, B – 56 s, C – 231 s, D – 350 s.

4. Conclusion.

We have developed a novel method of non-invasive intracellular labeling, which allows to perform tracking of structures of the interest and does not require delivery to the cell. Tightly focused femtosecond laser beam creates fluorescent spot right in the focus area and does not disturb surrounding intracellular contents. It seems that fluorescent species are formed due to the carbonization of intracellular organics. Presumably, the composition of fluorescent particles is similar for species, produced from cytoplasm, nucleus and nucleolus, because its luminescence spectra are the same.

Acknowledgments
The work was supported into frameworks of the state task for FRCCP RAS, state registration number AAAA-A19-119012990175-9, in the part of fluorescence registration. Experiments were carried out in the FRCCP RAS Center of the Collective Equipment (no. 506694). The work has financial support from the Russian Foundation for basic research: grant number 19-53-52007 MNT_a in the part of intracellular tracking, grant number 18-03-01121 in the part of laser synthesis of fluorescent species.

References
[1] Damalakiene L., Karabanovas V., Bagdonas S., Valius M., Rotomskis R. 2013. International journal of nanomedicine, **8**, 555-568.
[2] Ding P, Wang H., Song B., Ji X., Su Yu, He Y. 2017. Anal Chem. **89 (15)**, 7861-7868
[3] Sun Q., Qin Z., Wu W., Lin Y., Chen C., He S., Li X., Wu Z., Luo Y., Qu J.Y. 2018. Biomed Opt Exp. **9 (2)**, 581-590
[4] Gattass R.R., Mazur E. 2008. Nature Photonics. **2 (4)**, 219-225
[5] Hovhannisyan V., Lo W., Hu C., Chen S.-J., Dong C. Y. 2008. Opt Exp. **16 (11),** 7958-7968

Reduced graphene oxide based nanobiocomposite as basis for flexible biosensors

I A Komarov[1], N S Struchkov[2], O M Antipova[3], A N Kalinnikov[1] and S N Shcherbin[1]

1 Interbranch Engineering Centre "Composites of Russia", Bauman Moscow State Technical University, Moscow, 2-nd Bauman str. 5, Russian Federation
2 Research and Educational Center for Probe Microscopy and Nanotechnology, National Research University of Electronic Technology, Moscow, Zelenograd, Shokin sq. 1., Russian Federation
3 Lomonosov Moscow State University, Leninskie Gory, 1-40, Moscow, 119991, Russian Federation
ikomarov@emtc.ru

Abstract. Flexible selective electrochemical biosensor based on reduced graphene oxide film and short oligonucleotides (aptamers) was developed. Laser scribing was applied for graphene reduction due to controllable reduction rate and simple devises pattering. Optimal parameters of film reduction were determined for effective aptamers coupling, by varying laser output power. Mild reduced graphene oxide, as was expected, revealed better reactivity for aptamers coupling. Resistive response to biosensors exposure to thrombin and albumin proteins was measured. As a result we developed bionanocomposite that can be used in a new generation of available low-cost biosensors.

1. Introduction

Personal healthcare is one of the powerful and prospective trends in device development. There are lots of different fitness trackers, smartwatches and even sensors in smartphones that gives a possibility to monitor such parameters as pulse, step number, stress level etc. However, up to nowadays one can't monitor personal health on different illnesses excluding cardiovascular disease or diabetes due to the lack of the appropriate sensors. One of the ways to solve problem of health monitoring is to create new sensor with the sensitivity to different illnesses. Main goal of this work is to show possibility of coupling biosensitive agent – aptamers to the reduced graphene oxide to form one of the main part of biosensor – transducer. Carbon nanomaterials, in particular, graphene [1] and its derivatives [2], and carbon nanotubes [3], meet well with requirements of flexibility, acceptable conductivity and presence of reactive functional groups [4] with which different biomaterials can be coupled.

Modified or reduced graphene oxide [1] is one the most promising material for transducer formation due to the presence of different functional groups including carboxyl, epoxy, amino etc. [5]. On the other hand, reduced graphene oxide have large surface area, flexibility, relatively good conductivity and possibility of reduction ratio management.

Two main mechanisms of graphene oxide reduction: chemical [6] and thermal [7] can be implemented by different techniques, with laser reduction as the most perspective for nanodevice fabrication. During the last decades, there appeared some other reduction methods including UV treatment in vacuum [8] or laser treatment [9], that based on photochemical [10] or thermal [11] reduction mechanism. In case of femtosecond laser pulses it can be realized both thermal and photochemical reduction mechanisms, the last one allows to attach or remove various functional groups [12]. Laser induced reduction contrary to thermal or chemical reduction have such advantages as possibility of local and easy patterning of graphene oxide and possibility of providing different photochemical reactions in case of femtosecond lasers.

In our work we used pulsed microsecond laser to locally pattern graphene oxide film with manageable reduction ratio and further aptamer coupling to form nanobiocomposite that works as transducer. Aptamers are ideal sensitive substrates for immobilization. In addition to high sensitivity and affinity, aptamers have such advantages over their nearest competitors - antibodies: higher stability, much easier and cheaper synthesis [13 - 15].

2. Materials and methods

Electronic part of a biosensor consists of flexible substrate (175 μm thick poly(ethylene terephthalate)) and GO/rGO film with about 1 μm thickness. Substrates with 20 x 20 mm dimensions were precleaned with 2-propanol and water. GO suspension (4,7 mg/ml, LLC "MIP Graphene",Yakutsk, Russia) [16] was diluted to 1,5 mg/ml and deposited by drop-casting method. Films were dried for 24 h after deposition under normal conditions.

Local laser treatment was carried out by the own-designed system with 445 nm solid-state laser facility with motorized table [17]. Output laser power was varied in 10-400 mW range by pulse width modulation (PWM) at 30 MHz frequency. Pulses width ranged from 1 to 20 μs. Treatment time in a spot (~1,5·103 μm2) was constant 30 ms.

For biocomposite formation, we used the amine-modified derivative of the thrombin binding aptamer AmTBA (5'-GGTTGGTGTGGTTGG-3 ') with amino group at the 5'-end. The amino-derivative aptamer AmTBA of 50 mkM coupled with carboxylic groups on the surface of the rGO by standard conjugation method using EDC (1-Ethyl-3-(3-dimethylaminopropyl)-carbodiimide, Sigma-Aldrich, USA) in 100mM MES buffer solution, pH = 6.0 (2-(N-morpholino)ethanesulfonic acid, Sigma-Aldrich, USA) with 25% ethanol addition, at room temperature in an inert gas atmosphere within 1 hour, after that the solution was collected, centrifuged and analysed by HPLC.

We performed Raman spectroscopy for characterization of reduction depth by estimating relative intensities of specific GO and rGO Raman peaks. We got Raman spectra with Ntegra Spectra (NT-MDT Ltd, Moscow, Russia).

Structures resistance before aptamer coupling and sensor response were measured by IPS-16 (Pracktic NC Ltd., Russian Federation) sensor parameter measurement system that allows performing long-term measurements up to 48 hours long. We also used the self-made platform with the OMRON XF2J connector that allows high stability of electric contact and thus resistance and response measurement.

3. Results and discussion

3.1. Graphene oxide reduction and biosensor design

In our work we processed graphene oxide film by laser irradiation with different fluence parameters to find out optimal reduction ratio. We also performed aptamer coupling to structures with low, medium and high fluence (25, 35 and 42 J/cm2) and investigated approximate amount of functional groups (table 1). We also found a reduction limit at about ~ 50 J/cm2 when the film resistance starts to increases, which indicates the onset of rGO ablation and oxidation by atmosphere oxygen. When fluence excess about 60 J/cm2 substrate degradation begins thus such fluence value considered as unacceptable for sensors fabrication.

Investigation of aptamer coupling showed that low fluence (25 J/cm2) is better for aptamer coupling due to presence of a larger number of functional groups that is discussed further.

Based on the obtained data we choose low fluence value to perform GO film reduction and made a set of biosensor structures with 13 × 5 mm dimensions and П-shaped pattern of reduced graphene oxide area (figure 1a). After formation of set of biosensor structures we conducted a reaction of aptamer coupling to reduced graphene oxide.

We performed measurement of sensor structures resistance stability for 120 minutes. Statistics of resistance values of different structures is of figure 1b. We can notice that 19 of 30 (i.e. 63%) of sensor structures have resistance not more than 200 kOhm. On the other hand, resistance of some structures can achieve more than 5 MOhm that we attribute to superposition of such factors as local film roughness, nonuniformity of laser movement and local non-flatness of the substrate. Thus, we can predict about 60% yield of sensor structures ready for aptamer coupling.

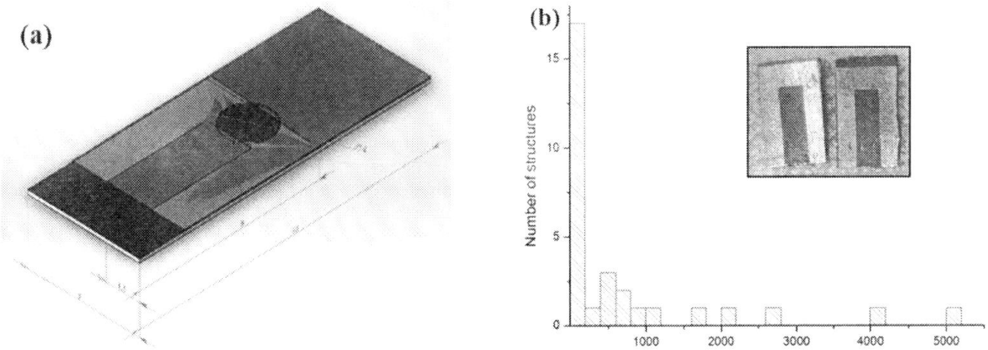

Figure 1. (a) 3D model of a biosensor based on reduced graphene oxide, (b) histogram of the resistances of the obtained sensors (photo in the insert image).

3.2. Biocomposite formation

To confirm the presence of functional groups and the possibility of aptamer coupling we used well-studied thrombin binding aptamer with amino-modification (Am-TBA) [18]. TBA recognizes the thrombin with a very high affinity using its TT-loops, and the core structure of the G-quadruplex stabilizes exact loops arrangement [19]. The functional amino-group of Am-TBA reacted with carboxylic groups of GO film preactivated by carbodiimide (EDC) giving conjugation product.

This approach is an extremely widespread practice for bioconjugation: such reaction proceeds in quantitative yield. However, when we applied this method for the amino-modified aptamer derivative conjugation to the film of reduced graphene oxide, we observed the hydrophobic effect of reduced graphene oxide surface appear to play an important role. We showed that the addition of 25% v/v ethanol to the reaction mixture improves the coupling reaction. The method allows visualizing reactive carboxylic groups only. However, the number of active carboxyl groups determines the number of conjugated sensing molecules on the surface of the reduced graphene oxide and therefore it determines the effectiveness of the biosensor.

In practice, a biosensor is often produced as a construct, one of the elements of which is the conductive layer containing sensing molecules for binding the target. Realization the reaction of covalent attaching an aptamer to the whole construct differs from attaching an aptamer to a substrate that is not packaged.

In particular, for whole constructs, it is difficult to carry out the conjugation reaction in solution, since some parts of the biosensor constructs may be damaged during prolonged contact with aqueous solutions of buffers. The conjugation reaction in the case of whole biosensors should be carried out locally. Therefore, it is the reason to increase the concentration of the EDC activator to increase the efficiency of the conjugation reaction. The reaction of the aptamer conjugation with reduced graphene oxide within constructs demands more mixing to conduct the reaction in an inert gas atmosphere. An important factor that ensures high conjugation efficiency is the mixing regime of the reaction.

Based on amount of reacted amino-modified aptamer AmTBA we calculated that the specific amount of carboxyl groups reaches 29 pmole per 1 mm2, of maximum fluence 42 J/cm2, 32 pmole per 1 mm2, of fluence 35 J/cm2 and 43 pmole per 1 mm2 for the film of minimum fluence 25 J/cm2 (Table 1).

Table 1. Results of aptamer conjugation.

Laser fluence, J/cm^2	The percentage change in the concentration of AmTBA,%	Quantitative change in the quantity of AmTBA, nmol	The specific amount of reactive carboxyl groups per unit area, 10^{12}/nm^2
25	32	3.2	43
35	29	2.9	32
42	26	2.6	29

Thus, we observe a trend on increasing of functional groups present with decreasing of laser fluence and identified that laser treatment with fluence near 25 J/cm^2 is suitable for aptamer coupling and, so, for the whole sensor.

3.3. Structure characterization

We made series of sensor structures with coupled aptamers. Typical Raman spectra from these samples before and after aptamer coupling presented on figure 2. In these spectra, we can observe D band (~1350 cm^{-1}), G band (~1600 cm–1) and 2D band (~2700 cm-1). It is known that the D band arises due to A1g symmetry and originates from the zone boundary phonons. The D band is usually attributed to various defects in the graphite lattice. The G band on the other hand is first-order scattering, related to E2g symmetry and corresponds to the Brillouin zone of crystalline sp2 lattices in graphite [20].

Higher ID/IG ratio in the reduced graphene oxide indicates larger defect density [21, 22]. It has been reported that the 2D peak profile is sharp for pristine graphene, whereas low and wide 2D peak intensity as compared to D and G peaks indicates higher disorder rate in the GO structure [23].

From figure 2 and abovementioned information on the nature of Raman bands we can make next observations and deduction. One can see that there are 2 main features in spectra after coupling: change in ID/IG ratio (0,98-0,99 before to 0,86 - 1,13 after aptamer coupling) and decrease of 2D peak intensity.

As we assume these effects could be corresponded to increase of defect number due to presence of aptamer-rGO conjugates with the trend and values ID/IG ratio before and after aptamer coupling very similar to the effects obtained in [24, 25]. Authors of [25] corresponded this effect to the coupling of polyethylenimine (PEI) to the graphene oxide. In our case, such comparison is legal due to low reduction rate of graphene oxide that thus contains different functional groups.

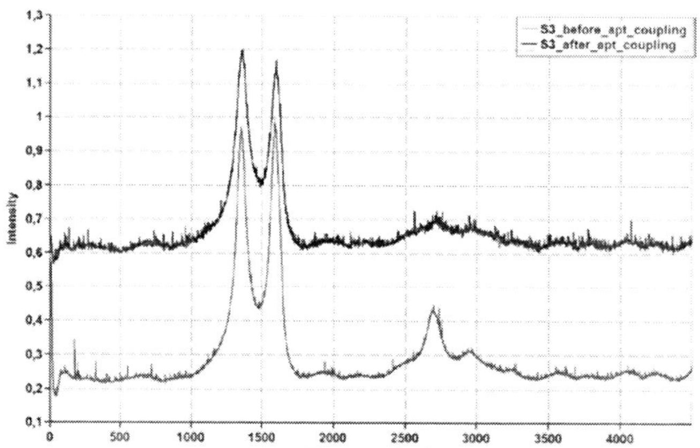

Figure 2. Raman spectra of reduced graphene oxide surface before and after aptamer coupling (S7 sensor structure).

3.4. Response measurements

We performed exposition of a sensitive area with thrombin (target) and albumin (reference) proteins to get response from sensor. Sensors with coupled thrombin binding aptamer were exposed to thrombin (figure 3a) and albumin (figure 3b) proteins. We observed the significant difference between sensor signals exposed to different proteins. Sensors also showed different signs of change in resistance, the response to albumin was about 0.9%, whereas to thrombin it was ten times higher, 8.5%.

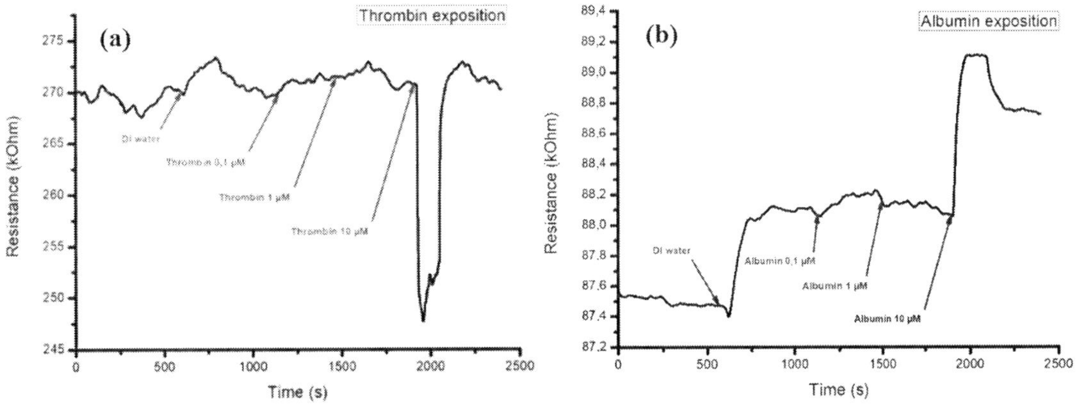

Figure 3. Sensor response on (a) thrombin, (b) albumin exposition.

We attributed such a difference in the signals to the specificity of the aptamer to the target protein. The decrease in resistance of the reduced graphene oxide film modified by an aptamer can be explained by the p-type conductivity of this material in normal conditions due to the presence of carboxyl groups on its surface [11]. The low change in resistance of the sensor exposed to albumin can be explained by a nonspecific reaction of the protein with the surface of the reduced graphene oxide. We also showed that the lower detectable concentration is 10 µM for this sensor design.

4. Conclusion

In this work we showed possibility of local manageable reduction of graphene oxide and pattern graphene oxide film to prepare biosensor drafts. We made series of sensor structure with about 60% yield of structures with relatively low resistance less than 200 kOhms. We also showed possibility of covalent coupling of aptamers to the reduced graphene oxide that is one of main challenges in development of a new generation of electrochemical biosensors.

Acknowledgments

This work is financially supported by Ministry of Education and Science of Russian Federation grant number 14.574.21.0182 (grant unique I.D. RFMEFI57417X0182).

References

[1] Celik N, Balachandran W, Manivannan N 2015 *IET Circuits Devices Syst.* **9** 434

[2] Young J O, Jung J Y, Yong I K 2014 *Electrochimica Acta* **116** 118

[3] Kumar D, Chaturvedi P, Saho P, Jha P, Chouksey A, Lal M, Rawat J S B S, Tandon R P and Chaudhury P K 2017 *Sensors and Actuators B: Chemical* **240** 1134

[4] Cui Y, Lee Y H, Yang J W 2017 *Scientific Reports* **7** 3146

[5] Chen J, Zhang Y, Zhang M, Yao B, Li Y, Huang L, Li C and Shi G 2016 *Chemical science* **7** 1874

[6] Stankovich S, Dikin D D, Piner R D, Kohlhaas K A, Kleinhammes A, Jiac Y, Wu Y, Nguyen SB T, Ruoff R S, 2007 *Carbon* **45** 1558

[7] McAllister M J, Li J-L, Adamson D H, Schniepp H C, Abdala A A, Liu J, Herrera-Alonso M, Milius D L, Car R, Prud'hommeIlhan R K, Aksay A 2007 *Chem. Matter* **19** 4396

[8] Rabchinskii M K, Shnitov V V, Dideikin A T, Aleksenskii A E, Vul' S P, Baidakova M V, Pronin I I, Kirilenko D A, Brunkov P N, Weise J, Molodtsov S L, Baidakova 2016 *The Journal of Physical Chemistry C* **120** 28261
[9] Zhang Y-L, Guo L, Xia H., Chen Q-D, Feng J, Sun H-B 2014 *Adv. Optical Mater.* **2** 10.
[10] Gengler R Y N, Badali D S, Zhang D 2013 *Nature Communications* **4** 2560
[11] Sokolov D A, Shepperd K R, Orlando T M 2010 *J. Phys. Chem. Lett.* **1** 2633
[12] Bobrinetskiy I I, Emelianov A V, Smagulova S A, Komarov I A, Otero N, Romero P M 2017 *Materials Letters* **187** 20
[13] Röthlisberger P, Gasse C, Hollenstein M 2017 *Int J Mol Sci.* **18** E2430
[14] Park K S 2018 *Biosens Bioelectron.* **102** 179
[15] Pu F, Ren J, Qu X 2018 *Chem Soc Rev.* **47** 1285
[16] Alexandrov G N, Smagulova SA, Kapitonov A N, Vasil'eva F D, Kurkina I I, Vinokurov P V, Timofeev V B and Antonova I V 2014 *Nanotechnologies in Russia* **9** 363368
[17] Struchkov N S, Kondrashov V A, Rozanov R Y and Nevolin V K 2017 *Journal of Physics: Conference Series* **816** 012014
[18] Bock L C, Griffin L C, Latham J A, Vermaas E H, Toole J J, 1992 *Nature* **355** 564
[19] Pica A, Krauss R I, Merlino A, Nagatoishi S, Sugimoto N, Sica F 2013 *FEBS J.* **280** 6581
[20] Yang D, Velamakanni A, Bozoklu G, Park S, Stoller M, Piner R D, Stankovichc S, Jung I, Field D A, Ventrice Jr. C A, Ruoff R S 2009 *Carbon* **47** 145
[21] Ferrari A C, Meyer J C, Scardaci V, Casiraghi C, Lazzeri M, Mauri F, Piscanec S, Jiang D, Novoselov K S, Roth S and Geim A K 2006 *Phys. Rev. Lett.* **97** 187401
[22] Kotakoski J, Krasheninnikov A V, Kaiser U, Meyer J C 2011 *Phys. Rev. Lett.* **106** 105505
[23] Malard L M, Pimenta M A, Dresselhaus G and Dresselhaus M S 2009 *Phys. Rep.* **473** 51
[24] Kurapati R, Bonachera F, Russier J, Sureshbabu A R, Ménard-Moyon C, Kostarelos K and Bianco A 2018 *2D Mater.* **5** 015020.
[25] Rodriguez-Silva A A, Movil-Cabrera O, Oliveira dos Anjos C T, Staser J A 2016 *Journal of The Electrochemical Society* **163** B256

Dynamic characteristics of totally glycosylated human glucocerebrosidase carrying N370S substitution

A A Cheblokov[1], **G N Rychkov**[1,2]
[1]Department of Molecular and Radiation Biophysics, Petersburg Nuclear Physics Institute named by B.P. Konstantinov of National Research Centre "Kurchatov Institute", Gatchina, 188300, Russia
[2]Institute of Physics, Nanotechnologies and Telecommunications, Peter the Great Saint-Petersburg Polytechnic University, Saint-Petersburg, 195251, Russia

Abstract. Full-atom models of a totally glycosylated wild type human glucocerebrosidase (GBA) and its mutant form N370S were constructed. Molecular dynamics simulations in explicit water box have revealed high flexibility and mobility of glycans bonded to the enzyme; glycans was able to spread a rather large volume determined by their length. Amino acid substitution N370S decreased the flexibility of GBA polypeptide chain region forming Loop 1 (residues 311-319), while the mobility of catalytic residues remained at the similar level both in wild type and N370S GBA mutant form.

1. Introduction

Gaucher disease (GD) is a hereditary lysosomal storage disease associated with glucocerebrosidase enzyme deficiency, most often caused by mutations in the *GBA1* gene, leading to amino acid substitutions L444P and N370S [1]. As a result, the enzyme's undegraded substrate, glucocerebroside, accumulates in lysosomes. Macrophages are the most affected cells in GD patients. Swelled macrophages can infiltrate the spleen, liver, and bone marrow, and cause inflammation and organomegaly [1]. Along with enzyme-replacement and substrate-reducing therapies, the use of small chemical compounds capable of restoring the activity of the mutant form of the enzyme is one of the promising methods for reducing the symptoms of GD. Virtual screening of such compounds requires construction of full-atom model most closely reconstituting the native structure of the enzyme [2].

Mature GBA enzyme is a glycoprotein containing 497 amino acid residues and having molecular weight varying from 59 to 69 kDa, considering to differential glycosidic remodeling [3]. There are four glycosylation sites on the GBA surface (amino acid residues Asn19, Asn59, Asn146 and Asn270). Glycosylation seems to be critically needed for proper spatial organization of amino acid residues in the GBA active center, and consequently for the enzyme activity [3]. The previous study on effect of wild-type enzyme (WT) glycosylation level on its polypeptide chain flexibility showed the importance of glycosylation at the first site (Asn19) [3].

2. Methods

Enzyme models were constructed based on spatial structure of human glucocerebrosidase deposited in Protein Data Bank (entry 2NT1). Asparagine at position 370 was substituted with serine to produce the mutant form of the enzyme. Glycan chains with the following sequence DNeup5Aca2-3DGalpb1-4DGlcpNAcb1-6[DNeup5Aca2-3DGalpb1-4DGlcpNAcb1-2]DManpa1-6[DNeup5Aca2-3DGalpb1-4DGlcpNAcb1-2DManpa1-3]DManpb1-4DGlcpNAcb1-4DGlcpNAcb1-OH that was previously described by a combined Smith degradation and chromatography study [4], were added using Glycoprotein builder at Glycam web server (www.glycam.org). The influence of amino acid substitution N370S on mobility of the enzyme polypeptide chain and catalytic amino acid residues was studied by the method of molecular dynamics (MD) simulations in explicit solvent with periodic boundary conditions (AMBER16 software package, ff14sb force fields for amino acids and GLYCAM for carbohydrate part, TIP3P water molecule model, 300K, 1 atm, PME periodic boundary conditions, 10 Å cutoff distance for nonbonded interactions, LINCS, 2 fs integration step). Two independent 100 ns simulations were run for each of the models.

3. Results

In this study, we investigated the effect of glycosylation on the dynamic characteristics of the polypeptide chain of GBA mutant form N370S. We have constructed full-atom models of WT GBA and its mutant form N370S bounded with triantennary, trisialylated, complex-type oligosaccharide at all four sites mentioned above.

Analysis of MD trajectories has shown that, due to their high flexibility and mobility, glycans do not occupy fixed positions on the surface of the enzyme, but spans a certain volume restricted by the length of the carbohydrate chain. The glycan chains occupy different conformations and interact with distinct parts of the GBA surface. The glycan attached to the residue N146 is located in proximity to both the active site and the predicted membrane associated region of the GBA. None of the glycans were found to block the entrance to neither the active site nor the membrane binding site.

On the modelled time frame (100 ns), glycosylation does not cause any sufficient conformational changes of the GBA tertiary structure. However, it should be noted that the N370S amino acid substitution provokes positional stabilization of amino acid residues 316 and 317 associated with the Loop 1 (composed of residues 311-319 according to conventional classification [2]), and Loop 2 (composed of residues 345-349) located next to the active site (Figure 1). This result is in good agreement with the X-ray crystallography data [5] and validates constructed model. Apart from this stabilization, the rest of enzyme's secondary structures do not show noticeable changes.

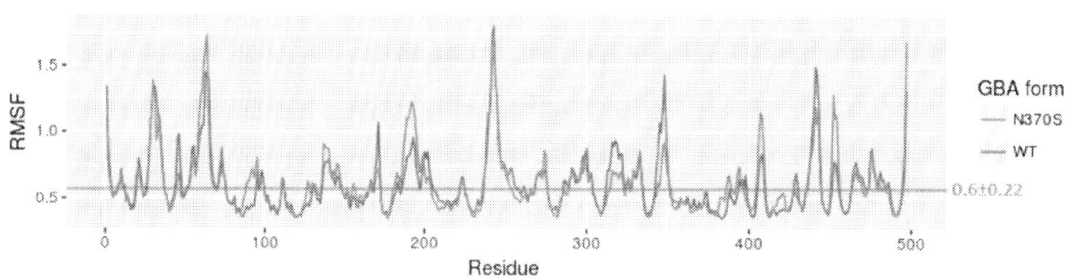

Figure 1. RMSF curves of WT (blue) and N370S mutant (red) forms of GBA. Grey horizontal line displays RMSF average value.

Both WT GBA and N370S mutant form have distinguishable root-mean-square fluctuations (RMSF) of a number of backbone regions that are higher than the sum of average RMSF value and its deviation 0.6+0.22 Å (Figure 1). For the WT enzyme, these are amino acid residues: 1, 29-34, 60-64, *171*, 240-245, *345*, *347-349*, 439-443, 497. For the mutant form N370S, these are residues: 1, 29-34, 58-64, *189-194*, 239-246, *407-408*, 441-443, *452-454*, 497. Regions with high rate of fluctuations specific to WT or N370 mutant form are marked by italic. Indicated residues compose not only unstructured loops connecting an elements of secondary structures, but termini of alpha-helices of the domain II and beta-sheets of the domains I and III.

Introduction of substitution N370S can lead to significant changes in mobility of several backbone regions (Figure 2) displayed by molecular dynamics simulations. Main chain atoms of residues 58-63, 88-94, 127-128, 137, 187-196, 239, 274, 383, 386-387, 393, 406-409 and 451-497 increased, while that of residues 1, 2, 32, 45-47, 64-65, 171, 201-202, 242-243, 254, 300, 313-319, *343-349*, 398, 429 and 439-442 decreased their mobility in comparison with WT enzyme. Residues 313-319 and 343-349 compose Loop 1 and Loop 2 respectively forming the entrance to the active site of the enzyme.

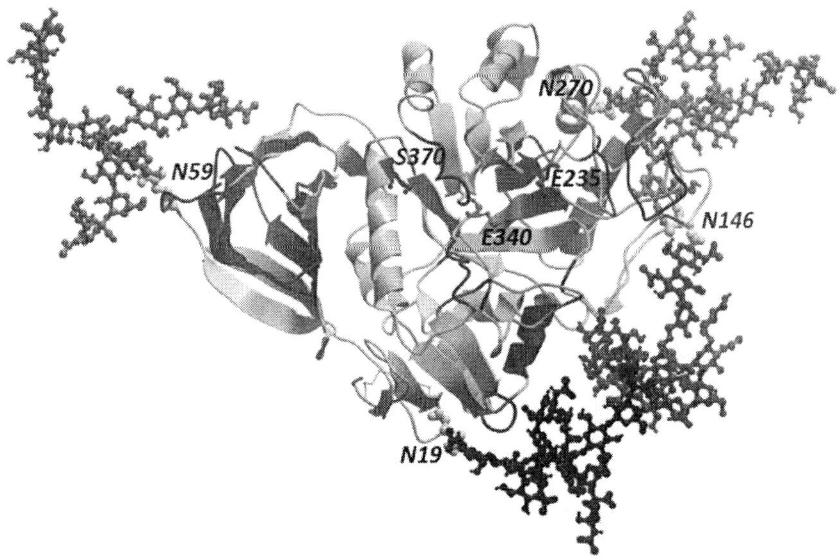

Figure 2. The median structure of a totally glycosylated N370S mutant form of GBA obtained by cluster analysis of the MD trajectory. Depending on changes of RMSF value backbone regions were considered more (red) or less (blue) flexible in comparison with WT enzyme. Glycan chains and essential amino acid residues are shown in balls and sticks representation. Catalytic residues (E235, E340) are colored in green, residues bonded with glycans (N19, N59, N146, N270) are in yellow and serine at position 370 is in magenta.

The distances between carbon atoms of carboxylic groups of the catalytic residues (E235 and E340) are 7.93±0.96 Å and 7.54±1.42 Å for WT and mutant form of GBA, respectively. RMSFs of E235 are 28.0 Å and 32.9 Å and RMSFs of E340 are 16.2 Å and 10.3 Å for WT and mutant form of GBA, respectively.

4. Conclusion

Since the catalytic amino acid residues have the similar mobility both in the WT human glucocerebrosidase and its mutant form N370S, it can be assumed that decay of enzyme activity resulting from amino acid substitution is primarily caused by decrease of Loop 1 flexibility, located at the entrance to the active site of the enzyme.

Acknowledgements

This work was supported by the Russian Science Foundation under grant 17-75-20159. The authors acknowledge the use of resources of the of Peter the Great Saint-Petersburg Polytechnic University Supercomputing Center (www.scc.spbstu.ru).

References

[1] Rosenbloom B.E. and Weinreb N.J. 2013 Critical reviews in oncogenesis **18**(3) 163
[2] Lieberman R.L. 2011 Enzyme Res. **2011** ID 973231
[3] Pol-Fachin L., et al. 2016 Glycoconjugate J. **33**(3) 309
[4] Takasaki S., et al. 1984 J Biol Chem. **259** 10112 – 10117
[5] Lieberman R., et al. 2007 Nature Chem Bio. **3(2)** 101–107

Nondestructive 3D-evaluation of human dentin by microtomography using synchrotron radiation

Z V Gudkina[1,2], **T S Argunova**[2], **M Yu Gutkin**[1,3,4]
[1]Peter the Great St. Petersburg Polytechnic University, St. Petersburg, Russia
[2]Ioffe Institute, RAS, St. Petersburg, Russia
[3]Institute of Problems of Mechanical Engineering, RAS, St. Petersburg, Russia
[4]ITMO University, St. Petersburg, Russia

E-mail: gudkinazhanna@mail.ru

Abstract. Results of Synchrotron Radiation (SR) microtomography of human dentin specimens subjected to compression tests are presented. A special attention is paid to cracks formed in the samples under compression. In all the samples, the dominant crack paths were observed to deviate and deflect around certain areas at or near to dentinal tubules. It is confirmed that secondary (satellite) cracks play an important role in dominant crack blunting. It is shown that microtomography is an effective method for investigation of the formation of satellite cracks and crack-tubule interactions over surface areas and across the whole depth of a dentin specimen. These processes can be visualized and evaluated using reconstructed 2D slices. A qualitative model for the formation of secondary cracks opening at stress concentrators that are dentinal tubules is suggested. It is shown that on the surface of the tubules, which are the closest to the tip of a mode II crack, the tensile circumferential stress can reach extremely high values that are more than enough for generation of secondary cracks.

1. Introduction

Dentin is a natural porous material with a hierarchical structure which includes at least seven scale levels. For example, the dentin tubules are microscopic quasi-cylindrical channels of 3–5 μm in diameter, which pass through dentin matrix and are spaced apart a distance of 10 μm, whereas the characteristic diameter of nanoscopic collagen fibrilles, which compose the dentin matrix, is about 100 nm. For many decades, specific strength and toughness properties of dentin have been in the focus of extensive studies, however their origin is not clear so far. It is explained in part by the fact that no imaging technique can display at least several structural levels of dentin at the same time.

Among other important topics related to human dentin, there has been much interest in the processes of crack propagation in it. Research in this area has shown that the salient toughening mechanism is crack bridging by uncracked ligaments [1–5]. The so called 'bridges' of intact material were observed directly by optical and scanning electron (SEM) microscopies [1, 2, 4] and X-ray tomography [1, 5]. On the surface of a dentin specimen (treated with conditioning chemical agents) SEM, optical and atomic force microscopies show the openings of tubules. Improved image resolution achieved in SEM and micro-tomography techniques allows to observe microcracks in the vicinity of the main crack, which form deflections and branching in the crack wake. The evidence for tubule interactions with a growing microcrack were provided [4].

However, no mechanism has been proposed to explain how secondary (satellite) microcracks are formed and how a microcrack is attracted to a tubule.

In the present work, synchrotron phase sensitive tomography has been used to study the development of microcracks in dentin specimens tested under uniaxial compression. Based on the experimental results, we propose a model for the generation of secondary microcracks in the vicinity of the initial crack tip. The tensile circumferential stress over the surfaces of the tubules in the vicinity of the crack tip is evaluated and shown to be high enough to promote the generation of satellite cracks.

2. Experimental procedure

The experiments were made with dentin samples cut from the roots of caries free human teeth. The samples with 0.5 mm thickness (h) had length (l) and width (w) dimensions equal to: $l = 5$ mm, $w = 3$ mm, so the ratio $d/h = 12$, where $d = \sqrt{l^2 + w^2}$. Strength properties of dentin cuboids tested under uniaxial compression were shown to depend on the d/h ratio [3]. Specifically, in the compressed samples with the ratio of d/h equal to 10–12, cracks were detected, although the samples were not separated under the test until the compressive stress reached $\sigma = P/(lw) = 840$ MPa. To the samples tested in this study, a compressive stress of $\sigma = 370$ MPa was applied. The plastic deformation was 15%.

Post-test observations of crack paths inside a specimen were obtained using X-ray tomography. Nowadays two ways to implement X-ray tomography are available: conventional absorption and phase contrast modes. In the absorption mode, tomography has a serious limitation in the visualization of micrometre-size features present in the sample like holes, pores, inclusions, *etc.* Absorption decreases with decreasing a feature size. At the same time, the refraction of X-rays increases with respect to the feature size. In the light provided by a coherent SR source the increase comes to the point where the feature image is determined only by the phase. X-ray phase contrast allows visualization of tiny objects. However, in order to obtain real-space parameters of the object, it is necessary to solve the inverse problem. Up to now, no such solution has been attempted for dentin material.

Our objective was to investigate dentin specimens whose surfaces were treated with no chemical agents. The damaged layer that appeared during cutting was removed with fine-grain (10 μm) abrasive paper. The incoming beam, which was not limited to a selected coherent fraction, illuminated a large sample. Experiments were performed on the third generation SR source: Pohang Light Source, Pohang city, Korea. On 6c biomedical-imaging beamline a multipole wiggler port provides high intensity radiation with the lateral coherence length of 31 μm in the vertical direction. X-ray images were obtained with 25 keV monochromatic

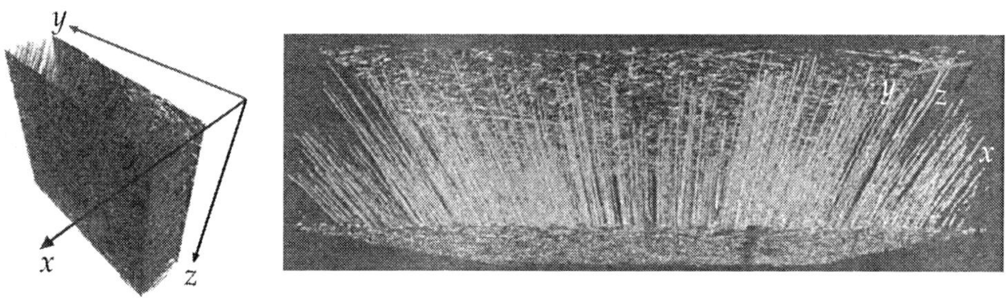

Figure 1. Left panel: rendered 3D-image of the dentin specimen. Beam direction is antiparallel to the x axis. Rotation axis is parallel to the z axis. The sample-to-detector distance is 5 cm. Detector is parallel to the yz plane. Right panel: reconstruction region size along the z axis is reduced from 702 to 181 μm.

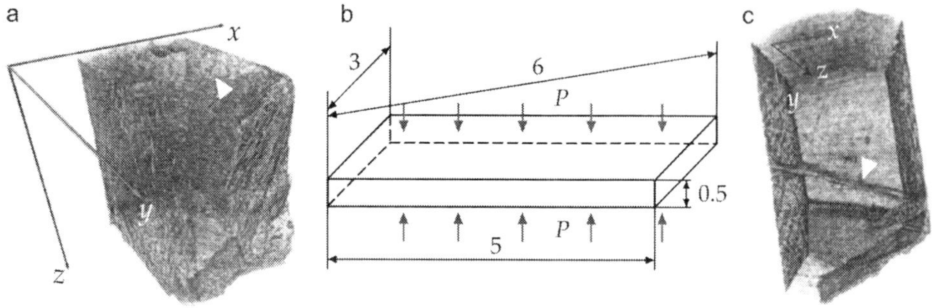

Figure 2. 3D images and the loading geometry used for the uniaxial compression tests of dentin specimens. (*a, c*) The direction of compressive loading is parallel to the x axis. A white marker shows the crack plane. The size of the reconstruction region is $832 \times 832 \times 702\,\mu m^3$. (*b*) The specimen size is $3 \times 5 \times 0.5$ mm^3; $d/h = 12$. The direction of the load P is indicated by red arrows.

radiation and converted into visible light images using a YAG:Ce scintillator crystal of 30 μm thickness. A magnified light image had an effective pixel size of 0.32 μm.

When the specimen is rather thick, there are many tubules in the beam path. The wave function of radiation is variable and a phase-contrast approach (see, e.g. [6, 7]) is not applicable. In this study, microtomography is used to visualize the directions of tubules, rather than to evaluate the tubules themselves. Separated images from several tubules can be recorded until the image size of a tubule exceeds the distance between them. Conditions for observing the arrangement of tubules are met in the near field. When the specimen-to-detector distance is $z = 5$ cm, the diameter of the first Fresnel zone is $D_1 = 3.2$ μm. It is smaller than the diameter of a tubule $D = 4$ μm. With increasing the distance z, the image size may exceed the intertubular spacing and tubule images overlap.

Samples were mounted on a rotation stage so that $z = 5$ cm and projections were recorded every 0.2°. Normalization and reconstruction of the projection images were obtained using Octopus V8.7. Visualization and 3D-rendering were performed with the aid of Amira V5.4.

3. Results

Rendered three-dimensional (3D) image of original dentin specimen before testing is shown in Fig. 1. In the left panel, a full field of view has the sizes 832, 832 and 702 μm along the axes x, y and z, respectively. In the right panel, the amount of contrasts in the reconstruction region is reduced by decreasing the size of the region along the z axis from 702 μm to 181 μm. The linear contrasts are observed to extend between the surfaces of the sample. They substantially deviate from the straight lines by which tubule directions are usually approximated (see, e.g., [1, 6, 7]). We believe that the linear contrasts show the directions of tubules. However, we note that these contrasts do not contain information about the tubules themselves.

After a compression test, cracks appear at the surface of the material and penetrate the entire volume of the sample. Scheme of uniaxial compression of the specimen (that has the d/h ratio of 12) is shown in Fig. 2. The compressive load P is parallel to the x axis. The 3D images display the main plane of the crack, whose detailed behavior in the interior of the specimen is examined using 2D reconstructed slices (tomograms).

On tomograms, tubules appear as dark spots and cracks appear as black and white lines (Fig. 3). Since in our experiments the tubules are inclined with respect to the axis of rotation in a random manner, they seem to be randomly distributed on a tomogram. In addition, their cross-sections have random shapes which are not at all like circular voids [6, 7]. Moreover, in

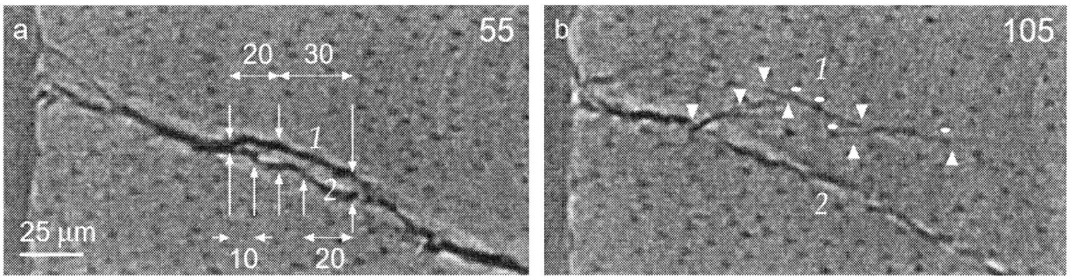

Figure 3. X-ray tomograms computed for a dentin specimen containing cracks. Tubule cross-sections are represented with dark spots. Some tubules are highlighted with rounded shapes. (*a*) Dominant crack starts from the sample edge and splits into two components: 1 and 2. The crack paths deviate around certain areas with separation distances of the order of 10 μm or a multiple of 10 μm. (*b*) This tomogram locates 50 μm 'below' the one shown in (*a*). The crack 1 splits into several satellite cracks, as highlighted by white arrowheads for the locations of crack tips.

the vicinity of crack images the visibility of the tubules is poor because the cracks show always a brighter contrast. However, despite these obstacles and complications, the examination of the crack shapes suggests that the tubules are involved in the crack paths. On the one hand, the paths deviate around certain areas. In Fig. 3*a* both the upper 1 and lower 2 cracks deflect around the areas separated by a distance of 10 μm or a multiple of 10 μm. On the other hand, the value of 10 μm does not contradict the conventional tubule spacing in dentin, and does not disagree with SEM data [4]. We believe that the deflections of dominant cracks (Fig. 3*a*) occur around the tubules when the latter are intersected by propagating cracks.

Figure 3*b* represents the differences which occur with depth. The tomogram in Fig. 3*b* locates 50 μm below the one shown in (*a*). The depth values (from the edge of the view field of the detector) are given in micrometres in tomograms. One can see the initiation of secondary (satellite) cracks ahead of the crack 1. In other words, the formation of bridges of intact material impedes the opening of the crack 1. SEM observations of similar phenomenon performed on the dentin surface treated with chemical agents revealed, in addition, the difference between the tubules. Some tubules can cause the crack deflection, while others do not [4]. Microtomography suffers from some uncertainty caused by overlapping images. In the meantime, it is a method capable of non-destructive depth analysis of shapes and sizes of microcracks.

4. Model

Consider a qualitative theoretical model which explains, in the first approximation, the reasons for the formation of secondary (satellite) cracks in dentin under compression. Figure 4 illustrates the opening of a mode II crack in a dentin sample which is compressed across the average orientation of dentin tubules. The crack starts from the 'top' surface of the sample under a compressive stress σ and propagates (until blocking) to a distance L under the shear stress τ (Fig. 4*a,b*). The crack plane is parallel to dentin tubules and inclined to the sample surface. As a result of the crack opening, a step of height B is formed on the surface.

To analyze the stressed state at the tip of the crack, it is convenient to use its dislocation model (Fig. 4*c*) which is a pile-up of N edge dislocations with the Burgers vector $\mathbf{b}$ [8]. The number N of the dislocations in such a pile-up is determined by the values of the shear stress τ and the crack length L [8]: $N = \pi(1-\nu)L\tau/(Gb)$, where ν is the Poisson ratio, G is the shear modulus, and $b = |\mathbf{b}|$. In the case under consideration, the elastic moduli ν and G are taken in the sense of overall effective elastic constants of a porous material with unidirectional packing

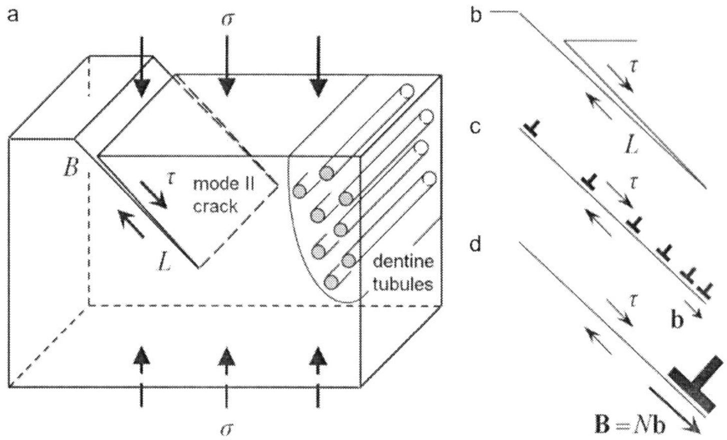

Figure 4. Sketch of mode II crack opening in a dentin sample under compression. (*a*, *b*) The crack starts from the 'top' surface of the sample under a compressive stress σ and propagates to a distance L under the shear stress τ in a plane parallel to the average orientation of dentin tubules and inclined to the sample surface; as a result, a step of height B is formed on the surface. (*c*) The crack is modeled by the corresponding pile-up of N edge dislocations with the Burgers vector **b**. (*d*) The pile-up of N dislocations is modeled by a superdislocation with the Burgers vector $\mathbf{B} = N\mathbf{b}$.

of long cylindrical pores [9].

In the first approximation, the pile-up of N dislocations can be modeled as an edge superdislocation (Fig. 4*d*) with the Burgers vector $\mathbf{B} = N\mathbf{b}$ with magnitude $B = \pi(1-\nu)L\tau/G$ (equal to the surface step of height B shown in Fig. 4*a*). In this case, one can qualitatively

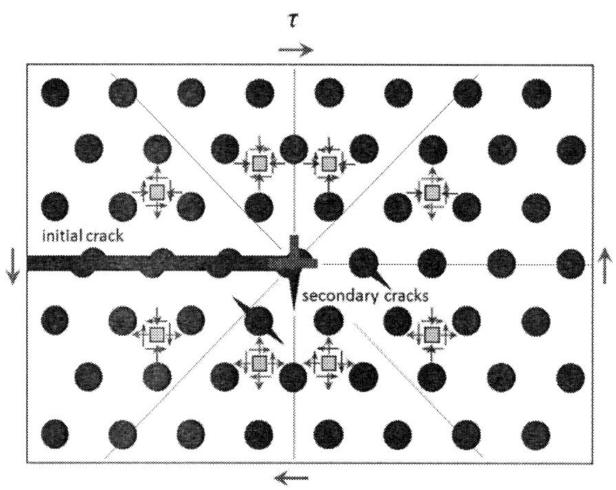

Figure 5. Model of a blocked initial mode II crack in the plane parallel to dentin tubules under remote shear stress τ. The stress field in the vicinity of the crack tip is modeled by that of an edge superdislocation. Near the crack tip, the signs of acting normal and shear stresses are sketched. The secondary cracks are opened at dentin tubules, where the tensile stresses reach their maximum values, and propagate in directions different from the direction of the initial crack growth.

image the stressed state around the crack tip with using the well-known stress field of an edge dislocation in an elastically isotropic infinite solid [8]. In doing so, we sketch the signs of acting normal and shear stresses near the crack tip in Fig. 5. Based on this sketch, we suppose that the secondary cracks are opened at stress concentrators that are dentin tubules, where the tensile stresses reach their maximum values. Therefore, the generation of secondary cracks is supposed at the dentin tubules just near the initial crack tip, in the areas 'ahead' and 'under' the superdislocation (Fig. 5).

To provide a deeper consideration of secondary crack generation, we have used the solution of the boundary-value problem in the theory of elasticity for a straight edge dislocation placed near a cylindrical void [10]. In this approach, the edge dislocation models the crack tip, while the void is a model of an individual dentin tubule. It is worth noting that we have not taken into account the effect of surrounding dentin tubules on the stress distribution around the probe dentin tubule, which is correctly enough when the average distance between the dentin tubules is much larger than their diameters (Figs. 3 show that this is true for our samples of dentin). Thus, using the solution [10], we have calculated the distribution of tensile circumferential stress over the surfaces of the tubules in the vicinity of the crack tip. For some exemplary tubules, which are the closest to the crack tip and are distant by 7 μm from it, the tensile stress has been shown to reach extremely high values about of 1.5 GPa (for the remote compressive stress $\sigma = 370$ MPa) that seems to be more than enough for generation of secondary cracks in dentin sample with characteristic Young modulus about of 2 GPa (in the case of compression test for $d/h = 12$, see [11] for details).

5. Summary

We used the results SR microtomography of human dentin specimens subjected to compression tests to shed some light onto the crack evolution in dentin under compression. We observed that the dominant crack paths deviate and deflect around certain areas at or near to dentinal tubules and showed that SR microtomography is an effective method for investigation of the formation of satellite cracks and crack-tubule interactions. We also developed a qualitative model for the formation of secondary cracks opening at stress concentrators that are dentinal tubules, and showed that on the surface of the tubules closest to the dominant mode II crack tip, the tensile circumferential stress can reach extremely high values, thus providing generation of secondary cracks.

Acknowledgements

The authors are thankful to Dr. D. V. Zaytsev for providing the samples of human dentin used in this work. M. Yu. G. thanks the Russian Foundation of Basic Research for the support (grant No. 18-38-20097).

References

[1] Nalla R K, Kinney J H, Ritchie R O. 2003 *Biomaterials* **24** 3955.
[2] Kruzic J J, Nalla R K, Kinney J H *et al.* 2005 *Biomaterials* **26** 1195.
[3] Zaytsev D, Ivashov A S, Mandra J V, Panfilov P 2014 *Mater. Sci. Eng.* C **41** 83.
[4] Koester K J, Ager J W III, Ritchie R O. 2008 *Biomaterials* **29** 1318.
[5] Kinney J H, Nalla R K, Pople J A, *et al.* 2005 *Biomaterials* **26** 3363.
[6] Zabler S, Riesemeier H, Fratzl P, Zaslansky P. 2006 *Opt. Express* **14** 8584.
[7] Kohn V G, Argunova T S. 2018 *Phys. Status Solidi* B **255** 1800209 (7).
[8] Hirth J P, Lothe J 1982 Theory of Dislocations (New York, USA: Wiley).
[9] Kachanov M, Sevostianov I 2018 Micromechanics of Materials, with Applications (Berlin, Germany: Springer).
[10] Dundurs J, Mura T 1964 *J. Mech. Phys. Solids* **12** 177.
[11] Zaytsev D V, Grigoriev S S, Panfilov P E 2017 The Origin of Strength in Dentin and Enamel of Human Teeth (Novosibirsk, Russia: Izd-vo SO RAN, in Russian).

A new method for monitoring the health condition based on nondestructive signals of laser radiation absorption and scattering

R V Davydov[1], **M S Mazing**[1], **V V Yushkova**[2], **V V Davydov**[1, 3], **A P Glinushkin**[3], **A V Stirmanov**[3] **and V Yu Rud**[1,3]
[1]Peter the Great St. Petersburg Polytechnic University, Saint Petersburg 195251, Russia
[2]Saint Petersburg University of Management Technologies and Economics, 190109, Russia
[3]All Russian Research Institute of Phytopathology, Moscow Region, 143050, Russia

Abstract. The article describes a new method for processing a pulse wave, which is formed from the recorded absorption and scattering signals of laser radiation on a blood vessel or human tissues. A new method of tuning the optical part of the pulse oximeter has been developed to increase the reliability of the results in the rapid diagnosis of the human condition. Experimental data on studies of various people are presented.

1. Introduction

One of the most popular methods for the rapid diagnosis of human health are contactless measurements that do not make irreversible (even temporary) changes in the physical structure and chemical composition of organs [1-7]. Several such methods are currently being developed for the study of various human organs. Among them it is necessary to distinguish magnetic resonance imaging (MRI). The use of MRI allows obtaining a large amount of information both on the state of the whole human body and on its individual organs [7-12].

Most of these methods are implemented in instrumentation, which can work only in stationary laboratories and require special maintenance [9-16]. The possibilities of diagnosing health condition on these devices are limited. Therefore, new ones are being developed and contactless methods are being improved, which an ordinary person can apply for the operational control of his condition [16-22]. One of such methods is pulse oximetry [23-26].

The results of research by various scientists have shown that a huge amount of information is in human blood. If you examine this volume, you can get data on the status of many organs. Therefore, the control of a person's health state using a pulse wave, which is formed using the signals of absorption and scattering of laser radiation on blood vessels or a vein, is extremely important [4, 5]. Its capabilities are currently not fully used in determining the state of a person. In addition, when using pulse oximeters currently in use to monitor human health, a number of difficulties arise. When solving them, it is required to take into account many features of pulse measurement and to search for new methods to increase the reliability of the results of rapid diagnostics of the human condition. One of such solutions is proposed in our work.

2. Features of the registration of the pulse wave

Despite the large number of developed models of sensors for the pulse oximeter and methods for processing the received information from the pulse wave, difficulties arise in the diagnosis of the human condition [23-26]. In most cases, they are associated with the appearance of artifacts in the recorded pulse wave image. Our studies have shown that some patterns remain that need to be clarified, and in some cases offer an additional method for diagnosing a person's health using an image of a pulse wave to increase the reliability of the results. In Fig. 1 shows the scheme of filling the blood vessel or vein with the release of blood from the left ventricle of the heart.

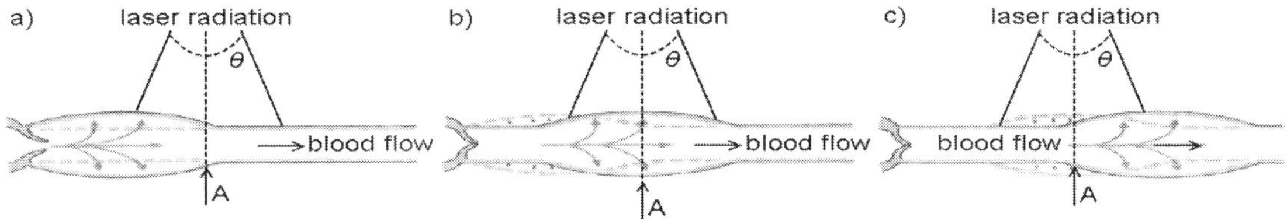

Figure 1(a, b, c). (a) Corresponds to the initial expansion of the vessel in the zone of exposure to laser radiation; **(b)** Corresponds to the greatest expansion of the vessel in the area of exposure to laser radiation; **(c)** Corresponds to the process of attenuation of the expansion in the zone of exposure to laser radiation to the complete filling of the zone of exposure to laser radiation.

Spreading from the aorta to the capillaries, the pulse wave fades. Its front is spreading. The expansion of the vessel at the time of arrival of the pulse wave will depend on its rigidity and density. For this reason, and also because people have a large variation of parameters in the blood composition, especially in concentrations of hemoglobin, leukocytes, erythrocytes, it is necessary for each person to choose laser radiation with a specific wavelength λ for research. In addition, people often suffer from various non-dangerous blood diseases, for example, due to poor ecology, etc.

Figure 2 shows the experimental dependencies of the pulse wave amplitude on different wavelengths of laser radiation in the red region of the spectrum obtained in the diagnosis of the state of various people. For the experiments, a standard pulse oximeter sensor was used, in which semiconductor laser diodes with different wavelengths were placed with radiation power P = 0.2 mW with a flat angle of the radiation pattern from 10 to 12 degrees [27-29]. All laser diodes were made based on heterostructures $In_xGa_{1-x}P$.

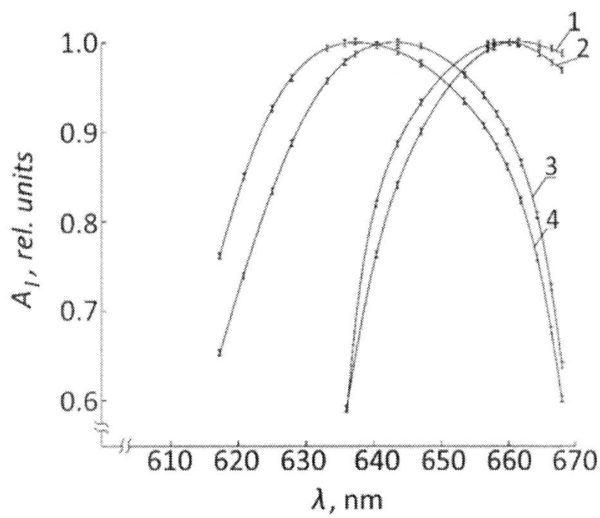

Figure 2. The dependencies of the pulse wave amplitude A_l on different wavelengths λ of laser radiation. Curves 1, 2, 3, 4 correspond to patients of different sex and age: a 56 year old man, a 21 year old woman, a 47 year old woman, a 54 year old woman.

The analysis of the dependences presented in Fig. 2, allows us to draw the following conclusion: for each person, the most appropriate solution to prevent the occurrence of artifacts will be to use a laser source with a particular λ when conducting research on their state. We found that the choice of λ can be the signal-to-noise ratio in the recorded pulse wave. This, with the results we obtained earlier and the results of other scientists [26, 30 - 35], made it possible to make insignificant the influence of a number of artifacts on the shape of the recorded pulse wave and made it possible to develop a new method for monitoring human health using the pulse wave.

3. The new technique, experimental results and discussion

In Fig. 3 shows the results of the registration of the pulse wave with and without our developments.

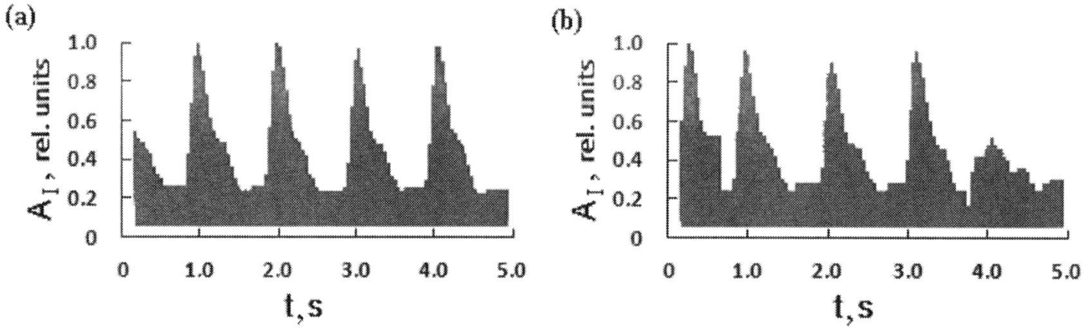

Figure 3(a, b). (a) The shape of the pulse wave of women aged 23 – the effect of artifacts on signal registration is not significant; **(b)** There are significant errors in measurements.

The results show that the use of our developments in registering a pulse wave made it possible to eliminate errors that contribute to the appearance of artifacts in a pulse wave. The presented results in Fig. 3 confirm this. The influence of artifacts on measurement results became insignificant.

Our experiments have shown that for each person, because of his individual body structure, the structure of the rise and fall front of a pulse wave is different. By changing the structure of the pulse wave (the appearance of distortions on the fronts), it is possible to establish deviations in the work of the heart. These abnormalities can be associated with both heart disease and the disease of other organs of the body that affect its work. In Fig. 4 shows pulse waveforms as an example.

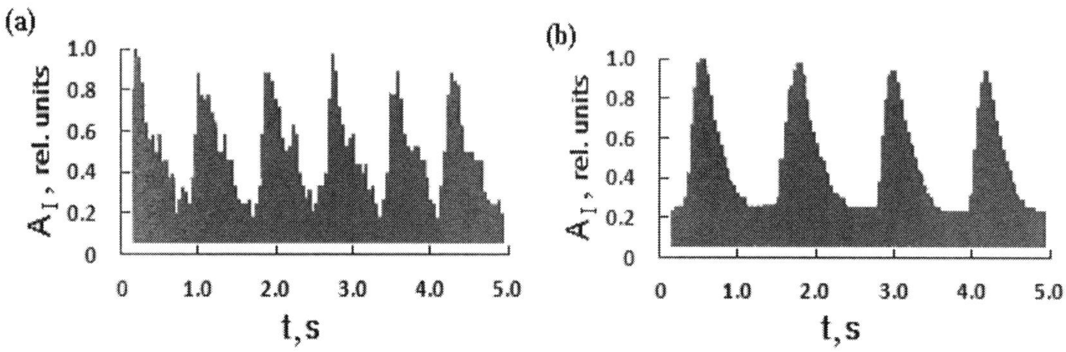

Figure 4(a, b). (a) The pulse waveform of 22-year-old man; **(b)** The pulse waveform of 24-year-old man.

Image analysis of the pulse wave in Fig. 4 allows to conclude that the rise and fall fronts of the pulse wave signal are stable. Registered pulse wave peaks periodically repeat. The value of the human pulse is stable. The oxygen content in the blood is determined with an error of less than 2%. A person's condition can be qualified as good and stable. With a more thorough study of the pulse waves of these people, differences in the rates of rise and fall of the various peaks of which the pulse wave is composed were noted. To consider in the studies of these processes, we propose a new method for processing the image of a pulse wave. To implement it, it is necessary to perform the approximation of the rise front of the pulse wave peak by the following function $-\ exp\ (Ft)$, and the decay front $-\ exp\ (-Ct)$, since the appearance of these dependencies is close to exponential. In Fig. 5 presents an example of the approximation of one of the pulse wave peaks.

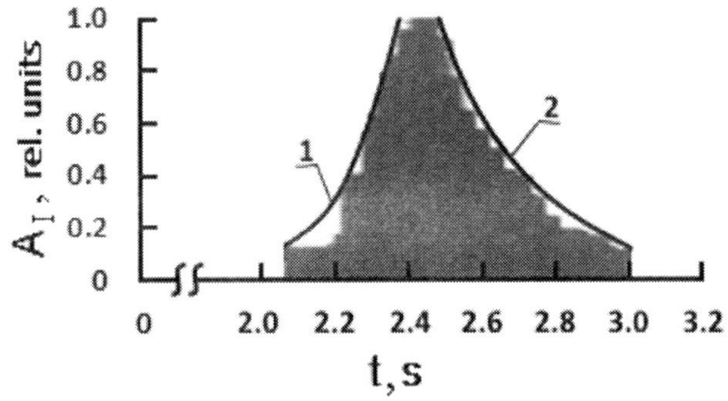

Figure 5. The peak of the pulse wave of men aged 24 years. Curve 1 –approximation of the rising front, curve 2 - recession.

Studies of the health status of a man at the age of 24 years during a month (the person felt well, other devices and clinical examinations did not record deviations in his health) allowed to establish that the values of the F and C coefficients can vary within certain limits ΔF and ΔC. Therefore, we propose to monitor the change in the state of human health according to the new coefficient KF, which is related by the shape of the pulse wave peaks as:

$$K_F = F_0/C_0 \quad (1)$$

where F_0 and C_0 are the coefficients measured during the approximation of the pulse wave signal at a given time.

The diagnostics of the health state of people according to the dynamics of changes in the K_F coefficient, measured pulse values and oxygen content made it possible in some cases to establish the early stages of heart disease with very good health. The results of clinical examinations of these people confirmed that they have these diseases.

4. Conclusion

The proposed method allows to make the effect of artifacts on the pulse wave image that we register insignificant. The obtained results showed the validity of using the dynamics of changes in the coefficient K_F to determine the state of a human. Additional information obtained by the analysis of changes in K_F values, as well as data on the oxygen content in the blood and pulse, allow to reveal at an early stage several diseases associated with the work of the heart and other organs.

Acknowledgments

This work partly supported by Grant of RFBR № 18-29-25071.

References

[1] Daev Zh A 2016 *Measurement Techniques* **59** 243-246

[2] Grebenikova N M, Smirnov K J, Artemiev V V, Davydov V V and Kruzhalov S V 2018 *Journal of Physics: Conference Series* **1038 (1)** 012089

[3] Davydov V V, Kruzhalov S V, Grebenikova N M and Smirnov K J 2018 *Measurement Techniques* **61** 365-372

[4] Karabegov M A 2004 *Measurement Techniques* **47** 1106-1112
[5] Nepomnyashchaya E K, Savchenko E A, Velichko E N and Bogomaz T A 2018 *Journal of Physics: Conference Series* **956(1)** 012009
[6] Baranov M A, Velichko E N and Aksenov E T 2017 *Journal of Physics: Conference Series* **917(1)** 062059
[7] Nepomniashchaia E K, Cheremiskina A V, Velichko E N, Aksenov E T and Bogomaz T A 2016 *Journal of Optical Technology* **83(5)** 305-308
[8] Zhernovoi A I and Sharshina L M 2003 *Biofizika* **48** 68-72
[9] Davydov V V 2016 *Optics and Spectroscopy (English translation of Optika i Spektroskopiya)* **121(1)** 18-24
[10] Dalitz F, Maiwald M and Guthausen G 2012 *Chemical Engineering Science* **75** 318
[11] Osan T M, Olle J M, Carpinella M, Cerioni L M C, Pusiol D J, Appel M, Freeman J and Espejo I 2011 *Journal of Magnetic Resonance* **209** 116
[12] Davydov V V and Myazin N S 2017 *Measurement Techniques* **60** 183–189
[13] Davydov V V, Dudkin V I and Karseev A Yu 2015 *Measurement Techniques* **58** 317-322
[14] Vitkovskii O S and Marusina M Y 2016 *Measurement Techniques* **59(3)** 247-251
[15] Davydov V V, Dudkin V I and Myazin N S 2016 *Journal of Communications Technology and Electronics* **61(10)** 1159–1165
[16] Davydov V V 2017 *Measurement Techniques* **59(11)** 1202–1209
[17] Haner R L, Llanos W and Mueller L 2000 *Journal of Magnetic Resonance* **143** 69
[18] Davydov V V and Myazin N S 2017 *Measurement Techniques* **60** 491–496
[19] Zhernovoi A I and Latyshev G D 1965 *Nuclear magnetic resonance in a flowing liquid* (New York: Consultants Bureau)
[20] Nepomnyashchaya E K, Cheremiskina A V, Velichko E N, Aksenov E T and Bogomaz T A 2015 *Journal of Physics: Conference Series* **643(1)** 012018
[21] Davydov V V, Dudkin V I, Karseev A Yu and Vologdin V A 2016 *Journal of Applied Spectroscopy* **82** 1013-1019
[22] Myazin N S, Logunov S E, Davydov V V, Rud' V Yu, Grebenikova N M and Yushkova V V 2017 *Journal of Physics: Conference Series* **929 (1)** 012064
[23] Yu H and Liu B 2007 *Anaesthesia* **62** 734
[24] Gople R 1999 *Sleep* **22** 932
[25] Hizbyllin R N 2017 *Fotonika* **1** 144
[26] Grevtseva A S, Smirnov K J, Davydov V V and Rud' V Yu 2018 *Journal Physics: Conference Series* **1135(1)** 012056
[27] Davydov R V and Antonov V I 2016 *Journal of Physics: Conference* Series **769 (1)** 012060.
[28] Davydov R V and Antonov V I 2017 *Journal of Physics: Conference Series* **929 (1)** 012040
[29] Myazin N S, Smirnov K J, Davydov V V and Logunov S E 2017 *Journal of Physics: Conference Series* **929 (1)** 012080
[30] Neronov Yu I and Seregin A N 2011 *Measurement Techniques* **54(1)** 103-107
[31] Rykin E V, Moroz A V, Smirnov K J, Davydov V V and Yushkova V V 2018 *MATEC Web of Conference* **245** 12002
[32] Davydov R V, Antonov V I and Moroz A V 2018 Proceedings of the 2018 *IEEE International Conference on Electrical Engineering and Photonics, EExPolytech 2018 (Saint-Petersburg)* **8564378** p. 236-239
[33] Davydov V V, Dudkin V I, Petrov A A and Myazin N S 2016 *Technical Physics Letters* **42(7)** 692-696
[34] Marusina M Ya, Volgareva A P and Sizikov V S 2015 *Journal of Optical Technology (A Translation of Opticheckii Zhurnal)* **82(10)** 673-687
[35] Davydov V V, Myazin N S and Davydova T I 2017 *Russian Journal of Nondestructive Testing* **53(7)** 520-529

Improvement of properties of cellulose dielectrics by their structure modification with nanocellulose produced of wastes of agricultural crops

N Zhuravleva[1], A Reznik[1], D Kiesewetter[1], A Stolpner[1], E Smirnova[2], V Budaeva[3]

[1]Institute of Power Engineering and Transportation, Peter the Great Saint Petersburg Polytechnic University, Saint Petersburg 195251, Russia

[2]State University of Industrial Technologies and Design, Saint Petersburg, Russia

[3]Institute for Problems of Chemical and Energetic Technologies of the Siberian Branch of the Russian Academy of Sciences, Biysk, 659322, Russia

Abstract. A comparative study of laboratory castings of insulating paper was fulfilled: the traditional composition made of insulating cellulose of pine and cellulose, modified with the nanocellulose of the agricultural waste (composite). The positive effect of the introduction of this structure-forming modifier on the electrical and mechanical strength of the cellulose dielectric, as well as its heat resistance was revealed. Confirmation of expediency of development of works of this direction for expansion of raw material base for producing of the bio component promoting increase of working capacity of the cellulose dielectrics used in high-voltage insulation of power devices is received.

1. Introduction

Increasing electrical and mechanical strength, as well as resistance to prolonged exposure to elevated temperatures at the conditions of destructive processes in the components of paper-impregnated insulation (PII) is one of the priorities for the production of power transformers [1]. It is the aging of insulation, the resource of which is determined by the state of the cellulose dielectric, in most cases causes the failure of power transformers (PT), often with serious consequences. High quality electrical insulation paper (EIP) in its original state, in particular, good parameters of electrical and mechanical strength that strive to achieve in the production of paper, do not guarantee good heat resistance of the material. The gradual degradation of cellulose during operation of PT leads to a decrease in the degree of polymerisation of its macromolecules and mechanical destruction of paper, which is aggravated by the dynamic effects on the transformer windings at the event of short-circuit modes. Among the many ways to reduce the intensity of EIP degradation, the structural modification of the paper web, which provides strengthening of the cellulose base by introducing components that have an affinity for cellulose [1-3], is of particular practical interest.

One of the ways of such modification is the use of so-called bacterial cellulose (BC). However, the synthesis of bacterial cellulose is not only complex, but also expensive process, in particular, due to the use of food raw materials as a nutrient medium for the used strains of bacteria. Therefore, the creation of technologies that reduce the cost of a valuable product is an urgent and practically significant task. A possible direction for solving this problem is the use of oat hulls [4] for the synthesis of biopolymer from non-food renewable raw materials. The present research is devoted to the study of electrophysical and mechanical properties, as well as the heat resistance of prototypes of a composite made of EIC and bacterial cellulose derived from oat waste (BC MG) in accordance with the technology described in the patent [2]. If successful, we can expect to create technologies that provide cost-effective conditions for the production of bacterial cellulose in industrial volumes.

2. The objects of study and technique of the measurements

In the course of comparative tests, laboratory castings of electrical insulating paper of traditional manufacturing from industrial EIP of pine sulphate processing (sample No. 1) and experimental composite material (sample No.2) at the manufacture of which 7.5% of biopolymer was introduced into the paper mass and was disintegrated with the Valley roll, were studied. This modifier was nanocellulose (NC) derived from the waste of a certain type of agricultural products, in particular, oat hulls. The electrical (E_b, kV/mm) and mechanical (σ_b, MPa) strength of these materials in the initial state were studied; as well as their heat resistance (by assessing the intensity of the reduction of the mechanical tensile strength in the process of thermal aging in an air-dry environment at the temperature of 140^0 C). Fragments of the samples (15x60 mm) were subjected to thermal aging during which 10 strips of each material has been tested. Preparation of test samples for measurements was carried out in accordance with the existing standard [5].

Morphological features of cellulose materials were investigated by scanning electron microscope (SEM) "Supra 55VP-25-78". Ash content (A_{sc}) of the studied paper samples and pH of their aqueous extract (AE) in accordance with the existing standards [6] were also measured. The sorption capacity of the standard EIP and EIP with the modified structure was determined by comparing the relative light transmission coefficients at the wavelength of 458 nm ($K_{rt\,458}$, %) of transformer oil samples after its thermal aging with the test paper. Measurements were carried out using the MKMF-1 micro colorimeter in standard measuring cell (10x10 mm). More detailed information about the measurement methods used is given in [7-8].

3. Obtained results

The images of the electrical insulating paper structure obtained using SEM presented in the Fig. 1 and Fig. 2. Figure 2 is illustrating the structure-forming function of the NC in the composite where are shows the finest fibers filling the voids of the web of standard cellulose insulating paper. This modification of the cellulose dielectric web, in our opinion, led to a significant increase in its characteristics compared with the EIP made of 100% EIC. Information about the structure of the standard insulating paper and its parameters are also presented in the works of the authors [3,9-10].

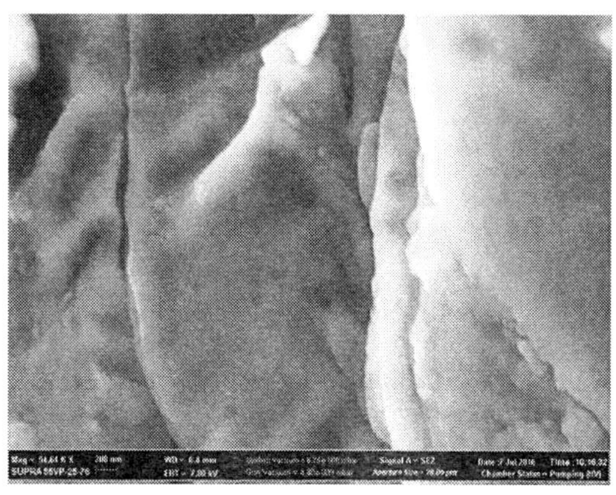

Figure 1. The obtained image of the structure of the standard electrical insulating paper. Magnification – 54.64 Kx.

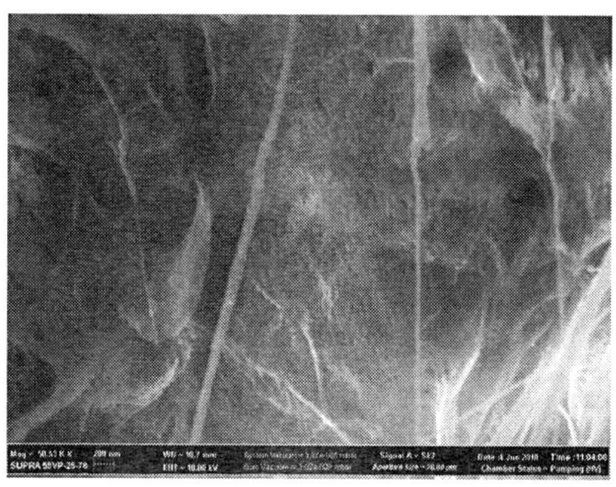

Figure 2. The obtained image of the structure of the EIP modified with BC MG. (sample No. 2). Magnification – 50.53 Kx.

It was found that the E_b of the composite (sample №2) in the initial state is 1.5 times higher than the same parameter of traditional EIP. It was also found that at the process of thermal aging (studies were conducted up to 861 hours), the breakdown electric field strength decreased slightly (less than 5% for sample No.1 and less than 1% for sample No. 2).

The measured electrical and mechanical characteristics of the two insulating paper samples at initial state are summarized in Table 1. From Table 1 it follows that the mechanical characteristics of the composite (sample No. 2) exceed the characteristics of EIP of traditional manufacture (sample No. 1) in the value of the mechanical tensile strength 2.5 times, and the resistance to extrusion (P_p) – 2 times. It should also be noted that the pH of the aqueous extract of the composite is lower than the maximum permissible value – 9.0 (units pH), but the ash content is slightly higher than the acceptable limit by standard [6] (not more than 0.6 %).

Table 1. Electrical and mechanical characteristics of the studied EIP.

The parameter	Sample No. 1	Sample No. 2
E_b (kV/mm)	6.4 ± 0.3	9.6 ± 0.6
σ_p (MPa)	15.5	41.5
P_p (KPa)	2.4	4.7
A_{sc} (%)	0.55	0.8
pH AE (units pH)	-	6.5

The results of measuring changes in the mechanical strength of the EIP samples during thermal aging are shown in the Fig. 3. As can be seen from the Fig. 3 at the initial stage of thermal aging, the limit of the mechanical strength of the paper samples increased. This known effect can be explained by the gradual removal of water from the EIP samples and the formation of additional hydrogen bonds between the cellulose macromolecules. This effect is most pronounced for standard EIP samples.

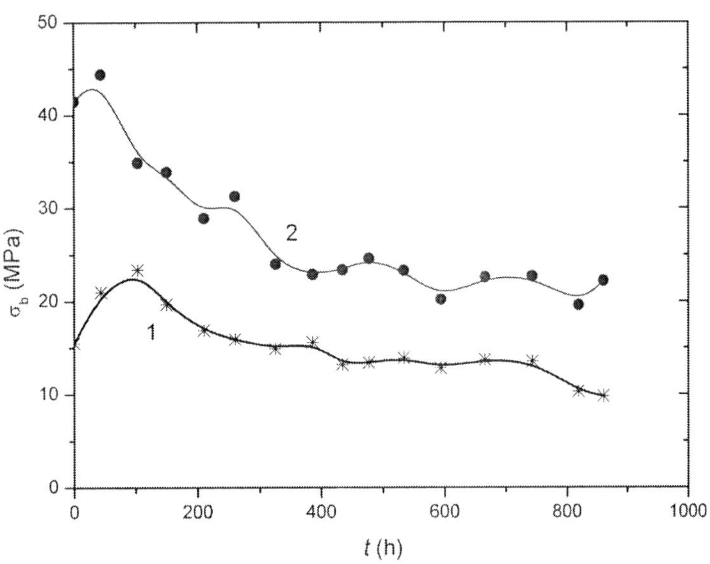

Figure 3. The dependence of mechanical strength limit (σ_b) on thermal aging time (t): 1 – for the standard EIP, 2 – for EIP with modified structure (experimental points and spline approximation).

Throughout the thermal aging process, the mechanical tensile strength limit of the sample No. 2 was more than 1.5 times higher than that of the sample No. 1. At the end of aging, the value of σ_b of the composite (s. No. 2) was at least 2 times higher than that for standard (not modified) insulating paper (s. No. 1). It is possible to assume, as at aging time more than 1000 hours the specified regularity to remain.

The results of the coefficient $K_{rt\ 458}$ measurement are summarized in the Table 2. Tests have shown that at the initial stage of aging, the light transmission coefficient of the oil aging of which was carried out with a copper catalyst and composite EIP (sample No. 2) more than in the case of aging oil with copper catalyst and unmodified EIP (sample No. 1). The data in Table 2 describe the aging of petroleum transformer oil "GK" when heated it in the presence of a copper catalyst. Row 1 of Table 2 corresponds to the presence of standard EIP (s. No. 1) in the oil, line 2 – to EIP with modified structure (s. No. 2) and line 3 ("+Cu") – for aging oil without EIP. Thus, the measurements confirm the high sorption capacity of the insulating paper with the modified structure, especially in the early stages of aging of transformer oil.

Table 2. Light transmission coefficient measured the aged oil.

t (h)	0	80	130	162	188	208	230	244	283	306	332	354	400
No. 1	88.5	76.5	68.0	29.5	20.5	16.5	14.0	12.5	10.0	8.0	6.0	5.0	4.5
No.2	88.5	78.0	71.0	57.0	28.5	20.5	14.0	12.0	9.0	6.5	5.0	4.0	2.5
+ Cu	88.5	23.5	13.0	11.5	9.5	6.5	5.0	4.8	1.0	1.0	0.5	0.2	0

4. Conclusion

The presented results indicate the feasibility of research aimed at the use of renewable non-food raw materials (in particular, oat hulls) in the development of technologies for the synthesis of bacterial cellulose for its use, in particular, as a modifier of cellulose dielectrics, characterized by increased heat resistance and improved electrophysical properties.

References

[1] Zhuravleva N, Kiesewetter D, Reznik A, Smirnova E, Khripunov A 2018 *St. Petersburg polytechnic university journal of engineering science and technology* **24** 1 75 (in Russian)
[2] Zhuravleva N, Sazhin B, Smirnova E, Khripunov A, Tkachenko A 2011 *Pat. RU 2415221*
[3] Zhuravleva N, Reznik A, Kiesewetter D, Stolpner A, Khripunov A 2018 *IOP Journal of Physics: Conf. Series* **1124** 031008 1
[4] Sakovich G, Skiba E, Budaeva V, Gladysheva E, Aleshina L 2017 *Doklady Biochemistry and Biophysics* **477** 1 357
[5] Standard 13523-78 *Fiber semi-finished products, paper and board. Method for conditioning of samples* 1978, rev. 1989 USSR Moscow 3 (in Russian)
[6] Standard 30180.2-99 *Paper of electrical insulating cellulose. Specifications. Part 2. Test method* 2000 Minsk 20 (in Russian)
[7] Zhuravleva N, Reznik A, Kiesewetter D, Tukacheva A, Smirnova E 2016 *Proc. 57th International Scientific Conference on Power and Electrical Engineering of Riga technical University* 1
[8] Zhuravleva N, Reznik A, Kiesewetter D, Tukacheva A, Smirnova E 2016 *Proc. ELEKTRO-2016, High Tatras* 649
[9] Kiesewetter D, Malyugin V, Reznik A, Yudin A, Zhuravleva N 2017 *IOP Journal of Physics: Conf. Series* **917** 042020 1
[10]. Zhuravleva N, Reznik A, Kiesewetter D, Tukacheva A 2017 *Proc. 2017 IEEE North West Russia Young Researchers in Electrical and Electronic Engineering Conference* 1220

Synthesis of glycidoxy spacer on the surface of magnetic nanoparticles and immobilization of albumin

D V Korolev[1], K G Gareev[2], V N Zorin[2], N V Evreinova[1,4], V N Postnov[1,3] T N Romanova[2]
[1]Almazov National Medical Research Centre, Saint-Petersburg, 194156, Russia
[2]Saint Petersburg Electrotechnical University , Saint Petersburg, 197376, Russia
[3]Saint Petersburg State University, Saint Petersburg, 199034, Russia
[4] Saint Petersburg State Institute of Technology, Saint Petersburg, 190013, Russia

Abstract. The glycidine spacer was synthesized from three different solvents. It is shown that the samples synthesized from benzene have the highest efficiency. Human albumin was immobilized onto a glycidine spacer. A mechanism for the interaction of albumin with a glycine spacer is proposed. An analysis of the determination of albunin adsorbed by magnetic nanoparticles was performed in this work.

1. Introduction

The reaction of the glycidine group with amines has been known [1] for a long time and is used to crosslink various, mainly polymeric materials [2, 3]. It passes with the disclosure of the cycle (Fig. 1). There are various methods [3] for the synthesis of an amino spacer (Fig. 2). At the same time, any preparation having an amino group can be immobilized on the glycidine spacer, or the corresponding conjugate can be synthesized. Thus, in [4], nanoparticles of a gelatin conjugate and (3-glycidyloxypropyl) trimethoxysilane (GPMS) with a diameter of 200 nm were synthesized. The nanoparticles were modified with S-nitrosothiol. As a result of this, a preparation of efficient delivery of NO for the regulation of vascular cells was obtained, releasing 0.12 mmol/mg of nitric oxide for 7 days. Enzyme-functionalized silicon dioxide nanoparticles were obtained by immobilizing HRP antibodies and alpha-fetoproteins on the surface of silica nanoparticles with the help of linking GPMS [5].

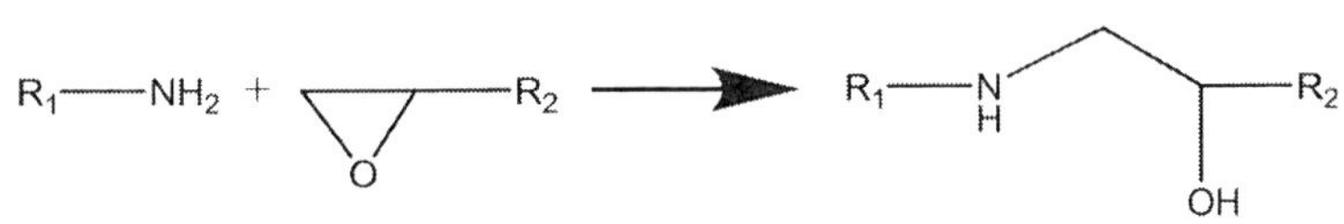

Figure 1. Reaction of glycidine group with amines.

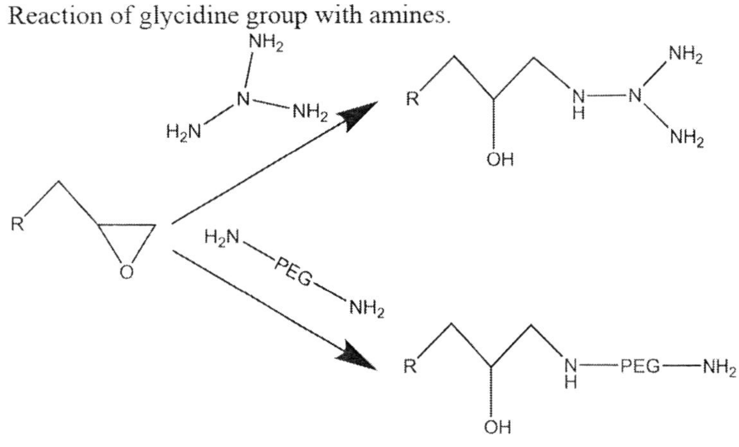

Figure 2. Synthesis of amino-spacer.

Since the blood protein albumin consists mainly of amino acids, a completely logical idea arose to sew them with the help of a glycidine spacer on the surface of a nanoparticle. Also, due to the fact that albumin is a transport protein, it is also, crosslinked on the surface of the MNP, may be able to chemisorb drugs.

2. Materials and methods

2.1. Synthesis of MNPs
The synthesis of the magnetic nanoparticles (MNPs) was carried out as follows [6,7].

2.2. Synthesis of the glycidine spacer
For the synthesis of the glycidine spacer, a (3-glycidyloxypropyl) trimethoxysilane (GPMS) reagent (Sigma Aldrich, 440167) was used. The reaction was carried out in a 50 ml pear-shaped glass flask. 2 grams of lyophilized magnetic nanoparticles (MNPs) were filled with 23.75 ml of solvent and 1.25 ml of GPMS was added. The reaction was carried out in a thermostatic cell at a temperature of 80 ° C for 2 hours with stirring on a magnetic stirrer. The immobilization scheme is shown in Figure 3. Benzene, toluene and cyclohexane were used as the solvent.
The resulting product was washed five times with cyclohexane, and then freeze-dried at –50 ° C and an absolute pressure of 3 Pa in a freeze-dried VaCo 2 (ZirBus, Germany).

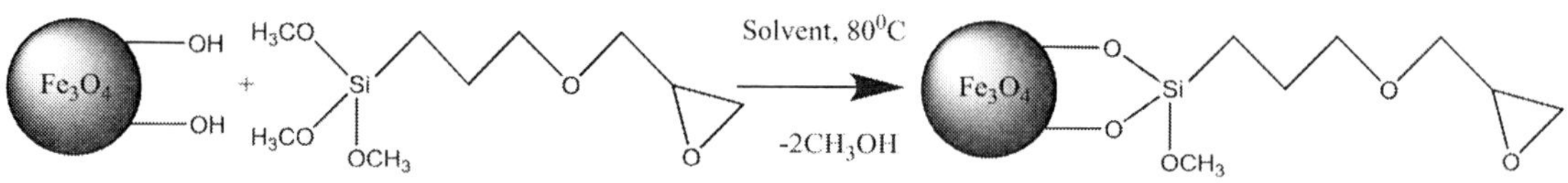

Figure 3. Synthesis scheme of glycidine spacer.

2.3. Determination of the number of spacer by the silicon content
The number of the spacer can be calculated from the silicon content in the synthesis using GPMS, as a chemical substance containing Si in its composition. The analysis was carried out as follows. 50 mg of nanoparticles modified by GPMS were taken and their mineralization was carried out by wet ashing, for which they were boiled in 20 ml of concentrated nitric acid diluted with distilled water in a 1: 1 ratio until completely dissolved in a 250 ml flat-bottomed conical flask. The cooled mineralization was adjusted to 100 ml with distilled water. The mineralizate thus obtained was analyzed for the content of silicate ion by a spectrophotometric method using molybdenum blue [8]. For this, a 2 ml aliquot sample was placed in a 100 ml volumetric flask, 10 ml of iron chloride II solution at a concentration of 2 g/l and 10 ml of ammonium heptamolybdate solution with a concentration of 0.05 g/ml were added. The pH was set at 1.3 ± 0.1 and left for 15 minutes. Then 5 ml of sulfuric acid was added to neutralize the phosphate complex, 5 ml of reducing agents each - a solution of oxalic acid with a concentration of 0.08 g/ml and a mixture of 2.5 g of citric acid and 0.5 g of ascorbic acid per 100 ml of distilled water. The solution was stirred and left for 15 minutes for color development, adjusted to the mark with distilled water, and the relative optical density was measured on a Unico 2502S spectrophotometer (USA) at a wavelength of 830 nm in cuvettes with a layer thickness of 1 cm.

2.4. Formation of albumin coating
Albumin was immobilized on an LS-220 stirrer at a frequency of 300 min^{-1} for 2 hours. To this end, 50 mg of MNP with the glycidine spacer were placed in a 15 ml polypropylene tube and 2 ml of a 20% aqueous solution of albumin was added. Human albumin was used (Baxter AG, Vienna, Austria). After sorption, the resulting preparation was washed five times with magnetic separation. Presumably, the reaction proceeds with ring opening and the interaction of the resulting radical with the primary amine of the terminal amino group. The meaning of the interaction can be illustrated by the scheme of interaction of the spacer with a single alpha amino acid tyrosine (Figure 4).

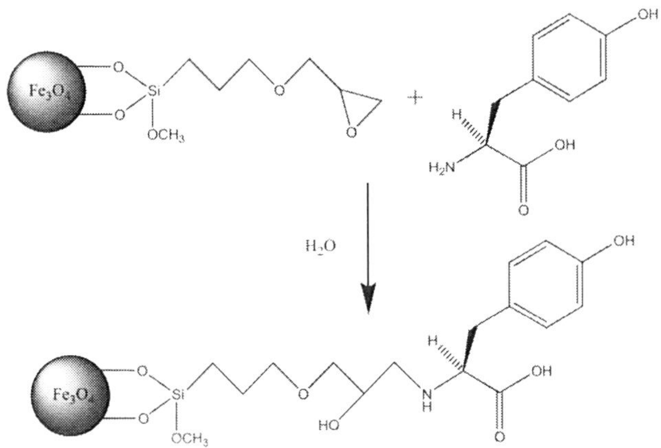

Figure 4. Diagram of the interaction of tyrosine with a glycidine group.

2.5. Determination of the amount of albumin in the coating by thermogravimetry

The amount of spacer and chemisorbed albumin was determined by thermogravimetric method when heated to a temperature of 900 °C in air at a rate of 10 °C/min on a Setaram Setsys Evolution instrument. It was assumed that the whole organic shell consists of a spacer or albumin and burns, interacting with oxygen in the air.

3. Experimental results

3.1. The results of determining the amount of spacer by silicon

The result of determining the amount of spacer by silicon is shown in Table 1. The largest amount of spacer is observed during the synthesis from cyclohexane. The value of the specific surface area of this sample is the lowest. Apparently, this suggests the formation of a thick polymer shell, including agglomerates of nanoparticles. Despite the high formal density of vaccination, one should expect low effectiveness in the immobilization of drugs on this sample. During the synthesis from benzene, a high specific surface area is observed and, relative to the sample from toluene, a high content of the spacer. A high efficacy should be expected in the immobilization of drugs on this sample.

Table 1. The results of determining the amount of spacer by silicon.

	Specific surface area, m^2/g	Silicon content, μg	Concentration of the spacer, mmol/g	Concentration of the spacer per unit surface, μmol/m^2
Fe_3O_4 initial	63	—	—	—
Fe_3O_4-GPMS from benzene	70	9.714	0.35	4.9
Fe_3O_4-GPMS from toluene	77	4.000	0.14	1.8
Fe_3O_4-GPMS from cyclohexane	42	35.429	1.26	30.0

3.2. Determination of the amount of albumin in the coating by thermogravimetry

The results of thermogravimetric analysis of samples obtained from benzene are shown in Figure 5 (a). As the thermogram shows, there are two sections. One of them is low-temperature, corresponds to the burning out of the organic shell containing a glycine spacer, the second, at a higher temperature, corresponds to the additional oxidation of magnetite to a higher oxide. The results of the analysis of samples with immobilized albumin are shown in Figure 5 (b). As can be seen from the thermogram, with the immobilization of albumin in the low-temperature region of the destruction of the polymer shell, another stage is added. Apparently, it corresponds to protein denaturation - albumin. This corresponds to the literature data [9], where it is stated that all protein denatures at temperatures below 100 °C. Then, there is a two-stage burnout of denaturation products, which can continue up to 450 °C [10]. The average albumin content ranges from 0.5 to 3%.

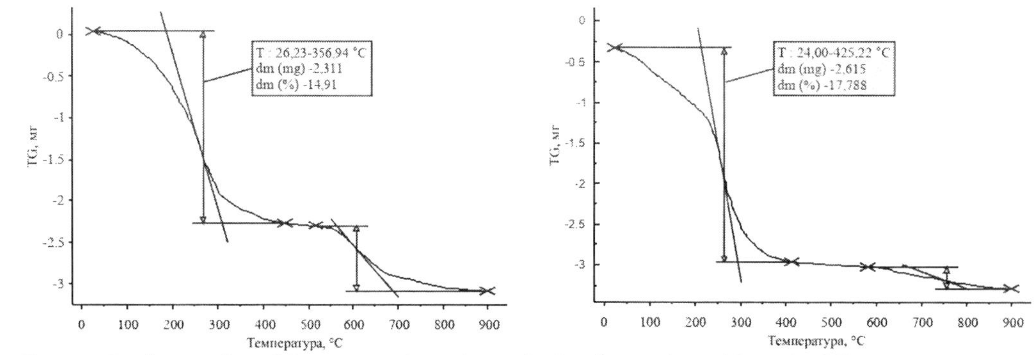

Figure 5. The results of thermogravimetric analysis of samples with a glycidin spacer: **(a)** synthesis from benzene; **(b)** synthesis from benzene followed by immobilization of albumin.

Acknowledgments
The work is based on the equipment of the Resource Center of St. Petersburg State University "Innovative Technologies of Composite Nanomaterials".

References

[1] Shechter L, Wynstra J, P. Kurkjy R 1956 J. *Ind. Eng. Chem.* **48 (1)** 94
[2] Fryauf K, Strehmel V, Fedtke M 1993 *J. Polymer* vol 34 № 2 p 323
[3] Kimmins S D, Wyman P, Cameron N R 2014 *J. Polymer* vol 55 p 416
[4] Zhang Q-Y, Wang Z-Y, Wen F, Ren L, Li J, Teoh S H, Thian E S 2015 *J. Biomed. Mater. Res. Part A* vol 103 p 929
[5] Wu Y, Chen C, Liu S 2009 *J. Anal Chem.* **81(4)** 1600
[6] Toropova Y G, Golovkin A S, Gareev K G, Malashicheva A B, Gorshkov A N, Afonin M V, Korolev D V, Galagudza M M 2017 *Int. J. Nanomedicine* **12** 593
[7] V N Zorin et al 2018 *J. Phys.: Conf. Ser.* **1124** 031009
[8] Williams W J *Handbook of Anion Determination* 1982 (Moscow: Khimiya) p 199
[9] Kulikova G A, Ryabinina I V, Guseynov S S, Parfenyuk E V 2010 Thermochim. Acta **503-504** 65
[10] Parfenyuk E V, Kulikova G A, Ryabinina I V 2010 *J. Therm. Anal. Calorim.* vol 100 № 4 p 987

Study of refractive index of human stomach cancer tissue in THz frequency range

A V Kuzikova[1], R O Grigorev[1], A P Kurasova[1], P S Demchenko[1], A V Senyuk[2], A A Zakharenko[2], J R Belolipetskaya[2], A H Khamid[2], M K Khodzitsky[1]
[1]International Laboratory Terahertz Biomedicine, ITMO University, Saint Petersburg, 199034, Russia
[2]Pavlov First St. Petersburg State Medical University, Saint Petersburg, 197022, Russia

Abstract. In this paper, the samples of healthy and cancer tissues of the internal and external sides of a human stomach were studied by time-domain terahertz reflectometry in the frequency range from 0.2 to 1 THz. Since healthy and cancer tissues have different morphological structure, and also contain different amounts of water, their optical properties and spectral characteristics differ. It was found that the refractive index of internal side of stomach cancer tissue is greater than for the healthy one. The refractive index of the external side of stomach cancer tissue is less than for the healthy one. This effect should be taken into account for development of stomach cancer diagnosis.

1. Introduction

Terahertz (THz) spectroscopy is used in many biomedical applications [1-2], such as therapy [3-5], glucose level control [6-7], water concentration determination [8] as well cancer diagnosis [9-11]. THz radiation has low energy and less scattering compared to visible and IR radiation. It is non-ionized and harmless to biological tissues. The THz frequency range is remarkable in problems of biomedical diagnostics, since it has characteristic absorption peaks for various biomolecules. One of the application areas of THz radiation in medical research has become the diagnosis of a cancer as the most dangerous and common disease. In the early stages, the cancer can be surgically removed. It is necessary to remove the entire tumor for successful treatment. For this reason it is needed to clearly differentiate the borders of the tumor during surgery [9,12]. At the moment, there is the problem to find a safe, affordable and effective method that will allow to perform interoperative diagnosis. The purpose of this work was to study the possibility of using THz time-domain spectroscopy in reflection mode to identify tumor tissues of the gastrointestinal tract. Cancer and healthy tissue have different morphologies. Therefore, their optical properties are different. The border of the tumor can be determined by the difference in optical properties of healthy and tumor tissues. The diagnosis of gastric cancer can be carried out in two ways: endoscopic and interoperative. For this purpose, the optical properties of healthy and tumor tissues of the human stomach were determined for the external side (serous tissue) and the inner side (mucous tissue).

2. Materials and methods

The samples were investigated by terahertz time-domain spectrometer in the reflection mode. The schema of the spectrometer is shown in Fig. 1. The setup uses a femtosecond laser Yb: KYW with the wavelength of 1040 nm, the pulse duration of 200 fs, the pulse repetition rate of 70 MHz. The beam-splitter divides the laser pulses into pump and probe beams. The pump beam passes through the optical delay line and the mechanical chopper and then it hits the InAs crystal, where THz radiation is generated using the Dember effect. THz pulse is collimated and focused on the sample using parabolic mirror. The beam reflects of the sample and focuses on the detector (crystal CdTe). The probe beam passes through the λ/2 plate and Glan prism and hits the detector. The CdTe crystal becomes birefringent and alters the polarization state of the probe beam into elliptical polarization. After that, the quarter-wave plate makes the probe beam circularly polarized and the Wollaston prism divides the probe beam into two orthogonally polarized beams. The balance detector calculates a difference between two orthogonal polarizations of the probe beam. The THz broadband pulsed radiation has the following parameters: the spectral range of 0.2–1 THz, the pulse duration of 2.7 ps, the average power of 30 μW, and the power density of 60 $\mu W/cm^2$.

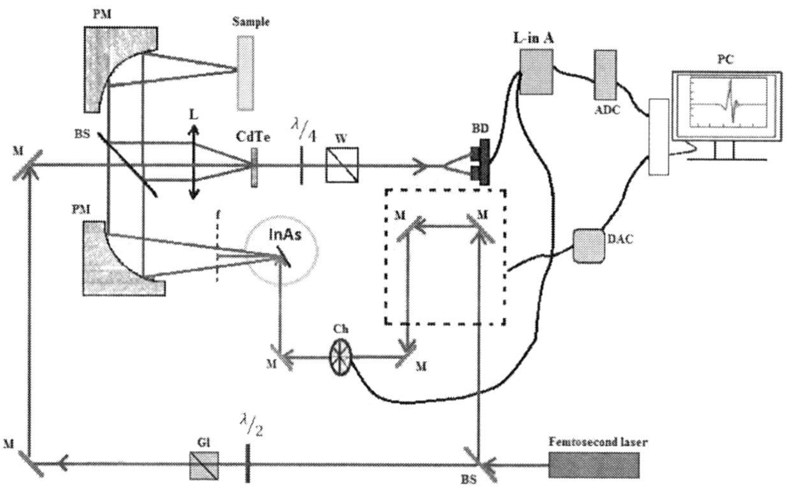

Figure 1. Principal scheme of THz time-domain spectrometer in reflection mode. BS – beam splitter, M – mirror, Ch – chopper, PM – parabolic mirror, ADC – analog-to-digital converter, DAC – digital-to-analog converter, BD – balanced detector, L-in A – Lock-in Amplifier, W – Wollaston prism, f – filter, L – lens, λ/2 - half-wave plate, λ/4 - quarter-wave plate, CdTe- crystal cadmium telluride, InAs - crystal indium arsenide, Gl- Glan prism, PC- personal computer .

Fresh stomach tissues with tumor were surgically extracted from 45- 67 years old patient with a gastric malignant neoplasm. The tissues were placed in physiological saline and were delivered from the Pavlov First St. Petersburg State Medical University to the THz Biomedicine Laboratory within an hour (Fig.2(a)). The samples were placed in a cell, where they were attached to the silicon plate for order to ensure the normal incidence. A special cell was designed and manufactured for tight adherence of the sample and the silicon window. The scheme of the cell is shown in the Figure 2(b). The temperature of the laboratory was kept constant during all experiments (18°C).

a) **b)**

Figure 2(a, b). (a) Photo of the samples; **(b)** Box for samples fixation during experiment.

The configuration of double reflection is used to obtain the dispersion of optical properties [4]. For this, the sample is pressed tightly against the reference dielectric (silicon). In this configuration, the recorded signal consists of two peaks: the first is the reflection from the air / silicon interface (reference signal), and the second is the reflection from the silicon / sample interface (object signal). The signal is processed using Fast Fourier Transform with a Hemming window function. As a result, we have two complex amplitudes: the reference spectrum and the sample spectrum. The complex amplitudes can be expressed from the phase shift in the reference dielectric and the reflection and transmission coefficients at the boundary of two media, which depend on the complex refractive indices. The ratio of complex amplitudes is a transfer function, which can be expressed as:

$$\hat{H}(\omega) = \frac{\hat{E}_{sam}}{\hat{E}_{ref}} = \frac{\tau_{aw}\tau_{wa}}{\rho_{aw}} \exp\left[-2jn_w \frac{\omega l}{c}\right] \frac{n_w - \hat{n}_s}{n_w + \hat{n}_s}, \tag{1}$$

where $\hat{E}_{sam}(\omega)$ is the sample spectrum, $\hat{E}_{ref}(\omega)$ is the reference spectrum, l is the window thickness, τ_{aw} and τ_{wa} are the transmission coefficients for the air-window and window-air interfaces, respectively, ρ_{aw} is the reflection coefficient of the air-window interface, $\hat{\rho}_{ws}$ is the complex reflection coefficient of the window-sample interface, n_w is the refractive index of the window, $\hat{n}_s = n_s + jk_s$ is the complex refractive index of the sample, c is the speed of light.

The transmission and the reflection coefficients are given by:

$$\tau_{aw} = \frac{2}{1+n_w}, \tag{2}$$

$$\tau_{wa} = \frac{2n_w}{1+n_w}, \tag{3}$$

$$\hat{\rho}_{ws} = \frac{n_w - \hat{n}_s}{n_w + \hat{n}_s}. \tag{4}$$

The formula (1) can be converted to the form:

$$\frac{n_w - \hat{n}_s}{n_w + \hat{n}_s} = \frac{\rho_{aw}}{\tau_{aw}\tau_{wa}} \hat{H}(\omega) \exp\left[2jn_w \frac{\omega l}{c}\right] = A\exp[j\varphi], \tag{5}$$

The refractive index can be deduced from (5) as

$$\hat{n}_s(\omega) = \frac{n_w(1-A^2)}{1+A^2+2A\cos\varphi}, \tag{6}$$

where A and φ are the amplitude and phase of the reflection coefficient for window-sample interface.

3. Results and conclusion

The results of the study demonstrate the pronounced differences in the optical properties of cancer and normal tissues of the human stomach both sides (external and internal) in the frequency range of 0.2 - 1 THz (Fig. 3). For the stomach external side, the cancer tissue refractive index is less than the healthy tissue refractive index. For the stomach internal side, the cancer tissue refractive index is higher than the healthy tissue refractive index. This study confirms the potential of THz time-domain spectroscopy for the gastric cancer diagnosis. However, it should be taken into account the differences in the dispersion of the refractive index for the external and internal sides of the stomach tissue.

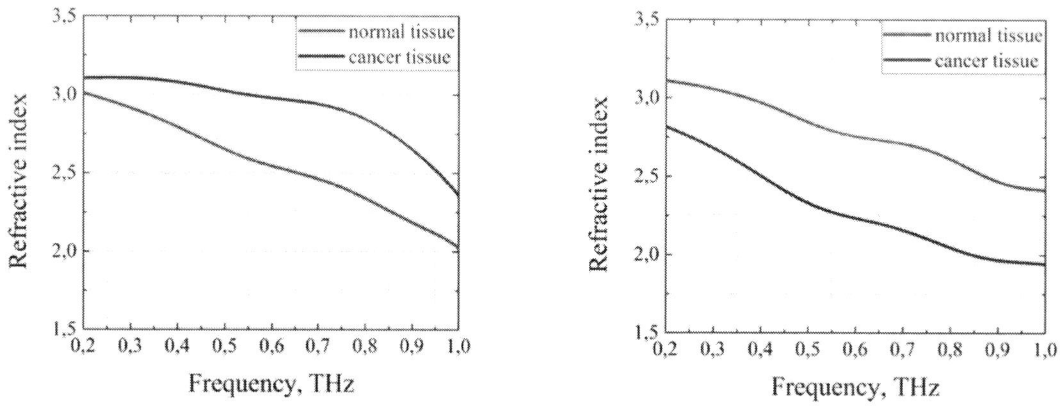

a) b)

Figure 3(a, b). **(a)** The refractive index of cancer and normal tissue of the internal side of the stomach tissue; **(b)** the refractive index of cancer and normal tissue of the external side of the stomach tissue.

Acknowledgements

This work was financially supported by Government of Russian Federation (Grant 08–08).

References

[1] Bombardieri 2007 *Advances in nuclear oncology: diagnosis and therapy* (CRC Press).
[2] Sun Q. He Y., Liu K., Fan S., Parrott E.P. and Pickwell-MacPherson E. 2017 *Quantitative imaging in medicine and surgery* **345** 7(3).
[3] Duka M V, Dvoretskaya L N, Babelkin N S, Khodzitskii M K, Chivilikhin S A, Smolyanskaya O A 2014 *Quantum Electronics* **44**(8) 707.
[4] Geyko I A, Smolyanskaya O A, Sulatsky M I, Parakhuda S E, Sedykh E A, Odlyanitskiy E L, Khodzitsky M K, Zabolotniy A G 2015 Proceedings of SPIE - The International Society for Optical Engineering **9542** 95420E.
[5] Borovkova M, Serebriakova M, Fedorov V, Sedykh E, Vaks V, Lichutin A, Salnikova A, Khodzitsky M 2017 *Biomedical Optics Express* **8**(1) 273-280.
[6] Gusev S I, Demchenko P S, Fedorov V I, Khodzitsky M K, Cherkasova O P 2018 *Chinese Optics* **11**(2) 182-189.
[7] Gusev S I, Borovkova M A, Strepitov M A, Khodzitsky M K 2015 *Proceedings of SPIE - The International Society for Optical Engineering* **9537** 95372A.
[8] Borovkova M, Khodzitsky M, Demchenko P, Cherkasova O, Popov A, Meglinski I 2018 *Biomedical optics express* **9**(5) 2266-2276.
[9] Goryachuk A, Simonova A, Khodzitsky M, Borovkova M, Khamid A 2017 *In IEEE International Symposium on Medical Measurements and Applications (MeMeA)* **7985863** 134-137.
[10] Smolyanskaya O A, Kravtsenyuk O V, Panchenko A V, Odlyanitskiy E L, Guillet J P, Cherkasova O P, Khodzitsky M K 2017 *Quantum Electronics* **47**(11), 1031.

[11] Strepitov E A, Prozheev I V, Balbekin N S, Sulatsky M I, Khodzitsky M K, Smolyanskaya O A, Trulioff A S, Serebryakova M K 2014 *Progress In Electromagnetics Research Symposium Proceedings* 1707-1710.
[12] Park C. H., Kim B., Chung H., Lee, H., Park J. C., Shin S. K., Lee S. K. and Lee Y. C 2015 *Digestive diseases and sciences* **38** 60(1)
[13] Naftaly M. 2015 *Terahertz metrology* (London: Artech House).

Highly hydrophilic and stable polylactic acid surface achieved by argon plasma treatment

A V Romashkin[1], Yu A Polikarpov[1], N S Struchkov[1], G O Silakov[2], V A Petukhov[1], D D Levin[1] and V K Nevolin[1]
[1] Scientific and Educational Center of Probe Microscopy and Nanotechnology, National Research University of Electronic Technology, Moscow 124498, Russia
[2] Institute of Advanced Materials and Technologies, National Research University of Electronic Technology, Moscow 124498, Russia

Abstract. The effect of modifying additives such as low molecular weight polylactide (PLA) and benzyl alcohol (BA) on the degree of hydrophilicity of high molecular weight PLA before and after plasma treatment with low density and specific power per area was investigated. The lowest water contact angle (WCA) after modification was about 17°. A comparative evaluation of the formed functional groups amount, the hydrophilicity and stability of the plasma modification was carried out. WCA measurements of ageing of modified samples over time up to one year was carried out. It was found that plasma treated PLA films, obtained from tetrahydrofuran solution of high-molecular weight PLA and BA, demonstrated one of the highest ageing stability at high hydrophilicity. At the same time, the increase of the oxygen-containing functional groups amount on the modified PLA surface (about 40%, estimated by XPS) is close to the critical value in point of view of ensuring both high hydrophilicity and stability of the PLA surface after the modification.

1. Introduction

Poly(lactic acid) (PLA) based polymers have a high potential for orthopaedics and dentistry application as biocompatible coatings that improves osteointegration [1] and as the base material of bioresorbable implants [2]. Low hydrophilicity is one of the PLA disadvantages that reduces cell adhesion, growth and therefore osteointegration [3]. PLA surface can be hydrophilized by addition of oxygen-containing polar functional groups by plasma treatment. High dose of plasma exposure allows to obtain a large density of functional groups, however, leads to a significant degradation of the surface with a molecular weight decrease. Exceeding a maximum dose specific for each polymer significantly changes the molecular structure that leads to surface instability to rinsing with a polymer nonsolvent removing of measurable amounts of surface layer [4, 5]. Such processing is unacceptable from the point of view of aging stability and resistance to biological mediums.

Moreover it leads to unacceptable modification for cells cultivation especially in the case when the hydrophilicity that is acceptable for cell adhesion and spreading significantly and rapidly degrades in several hours or first days of cultivation [3]. An optimal ratio of hydrophilicity, stability and morphology has to be achieved for better cells fixation and grow on the PLA surface. These properties can be adjusted by optimization the film composition and plasma treatment parameters both at once.

This work is devoted to the study of influence of the polymer composition and plasma treatment parameters, such as power density and treatment time, on the hydrophilicity of the PLA surface that can improve cells fixation and growth [3].

2. Materials and methods

High molecular weight PLA (high-M_w PLA, M_w ~ 80 kDa) was dissolved in tetrahydrofuran or dichloromethane (THF or DM). Concentration of PLA was 45 mg/ml. Benzyl alcohol as a low evaporation rate modifying additive, as well as low molecular weight PLA (low-M_w PLA, M_w ~ 5 kDa) were added to the solution in selected experiments in the amount of 1.5-3 volume % relative to solvent volume and 15 weight % relative to the high-M_w PLA weight respectively.

The PLA films were dip-coated (figure 1) at low drawing rate in nitrogen atmosphere on the cover glasses covered with 20 nm titanium layer.

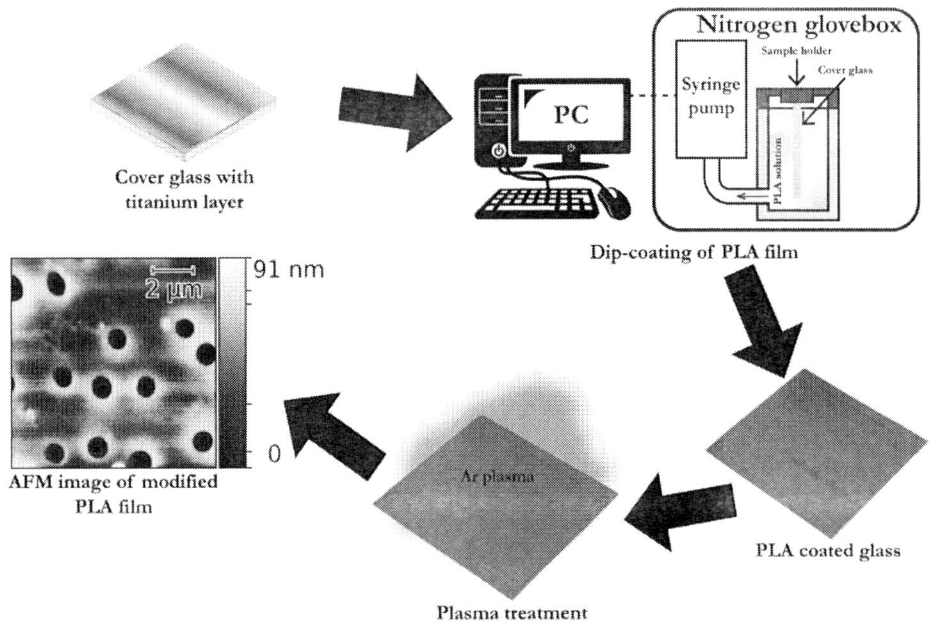

Figure 1. Illustration of the process of PLA film obtaining and modifying.

Obtained samples were annealed at 60 °C in atmosphere after deposition. The amount of residual solvent as well as benzyl alcohol was reduced to the value that is not detectable by Raman spectroscopy. Relatively little change of the water contact angle (WCA) of the obtained PLA surface was the only evidence of residual solvents and modifying additive. The different solvents used are also slightly affect the WCA of PLA surface as was previously reported [6].

PLA films were modified by RF argon plasma treatment for 100 s at low plasma and power density (less than 50 mW/cm^2). Samples were analysed by x-ray photoelectron spectroscopy (XPS) and atomic force microscopy (AFM) before and after modification.

The obtained XPS spectra were fitted with Gauss-Lorentz curves by specialised software. To estimate the amount of functional groups the area under the peaks corresponding to R-C-O and R-C=O bonds (about 287 eV and 289 eV, respectively) was calculated and normalized to the C-C bonds peak (285 eV). The ratio of the values obtained before and after plasma treatment was used as estimation of the functional groups amount increase. To investigate the stability of the modification the WCAs of initial, plasma treated and aged (for several days) surfaces were measured. AFM and WCA measurements were carried out as described previously [6].

3. Results and discussion

According to the results of WCA measurement the most hydrophilic films were obtained by plasma treatment of dip-coated PLA samples containing a low-M_w PLA (26°) or BA (17°) (figure 2a). The best of obtained values of WCA are significantly less than those described in the literature [3].

The high hydrophilicity obtained is due not only to the composition of the material, but also to the used plasma treatment parameters. Plasma treatment was carried out with low power density (2-5 times less compared to literature sources [7, 8]), that significantly increased modification stability over a long period (figure 2a, figure 3). However, the presence of both low-M_w PLA and BA reduced hydrophilicity measured immediately after modification and demonstrated lowest stability of modification over time (figure 2a, figure 3). The observed decrease of the modification stability when both components were used can be explained by a decrease in the average molecular weight of the main component of the film material, which is similar to the effect of plasma with high power density and treatment time reported in the literature [5, 8].

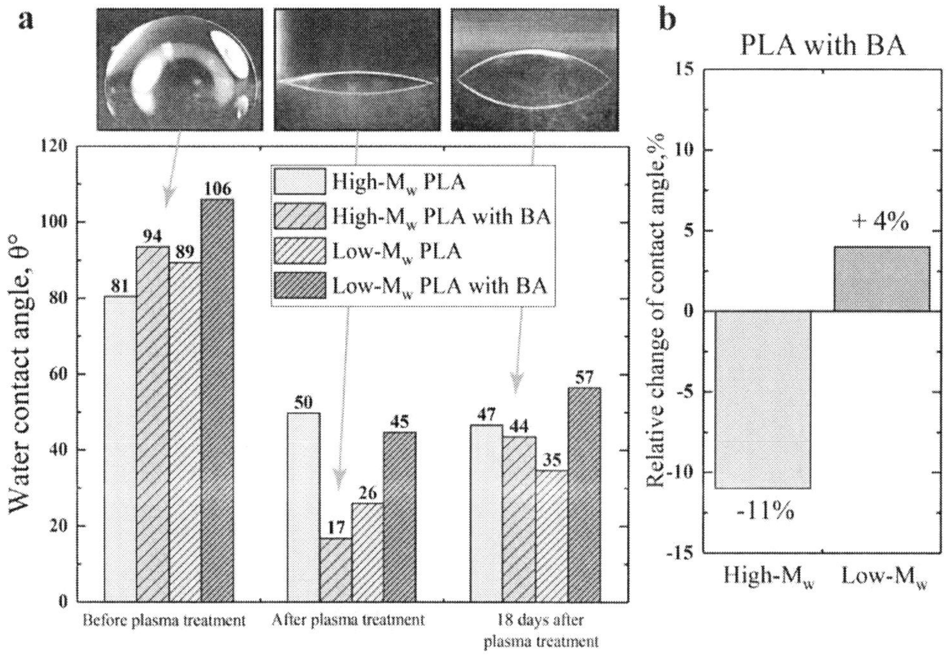

Figure 2 (a, b). (a) Evolution of WCA before and after treatment and with ageing of PLA samples obtained from different solutions with DM. **(b)** Effect of ethyl alcohol rinsing on WCA of PLA films obtained with using different molecular weight PLA.

Additionally, we demonstrate that when we use a high-M_w PLA with a BA additive, WCA decreases after sequential rinsing with ethyl alcohol and water and film wettability degradation slows down (figure 2b). At the same time, in the case of low-M_w PLA and BA, the effect of additional degradation with rinsing is observed, which is consistent with the literature data [8]. It indicates that the low molecular weight component accelerates wettability degradation over time, therefore the formation of too short PLA molecules is unacceptable as a result of plasma treatment, which imposes restrictions on the plasma treatment parameters.

A similar effect of a low-M_w PLA or BA addition was observed for THF samples: WCA reduced to 37° and 38°, respectively, while samples that hadn't containing additives demonstrated significantly higher WCA - about of 45-50° after similar plasma treatment (figure 3). However, the presence of both modifying components at the same time leads to the relatively high WCAs similar to theirs absence. It indicates that the presence of both additives doesn't additionally increase functional groups amount on the surface (figure 3). It should be noted that addition of a low-M_w PLA doesn't change the chemical structure of polymer and only reduces the average molecular weight of the polymer molecules that cannot change the mechanisms of functional groups formation on the surface during plasma treatment. Functionalization by plasma treatment is always accompanied by etching.

Therefore, the effect observed in the experiment apparently due to increased contribution of the etching process in plasma modification. Taking into account that fact, etching process apparently start to play a significant role in case of the simultaneous presence of both modifying additives due to increase of etching rate of polymer with lower molecular weight together BA content. So the effect of plasma treatment was reduced (figure 2).

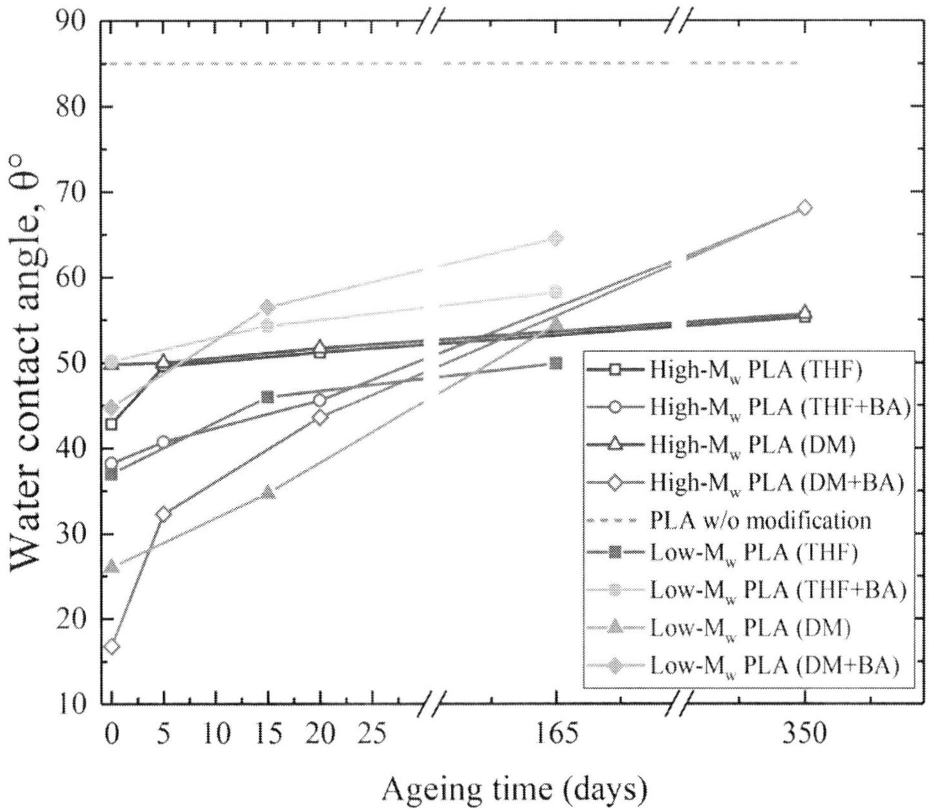

Figure 3. The degradation of WCA of the PLA films with different molecular weights and from various solvents over time after plasma treatment. The filled markers correspond to the experimental points of high-M_w PLA, empty markers - low-M_w PLA. The dashed line corresponds to WCA of the PLA without plasma treatment.

A similar effect of reducing the hydrophilicity of the surface is observed in another work [3]. Slepicka et al. used a different power plasma (3 W and 8 W), while the etching rate remained rather high. When plasma power decreased from 8 W to 3 W (approximately 2.67 times), the thickness of the etched material changed from 140 nm to 80 nm with a process time of 240 s, which corresponds to a change in the etching rate from 0.58 nm/s to 0.33 nm/s (decreasing only 1.75 times) [3]. A smaller change in the etching rate of the material compared to a change in plasma power indicates an increase the contribution of the etching process in plasma treatment process with a decrease in plasma power to 3 W. This leads to a greater hydrophilicity of the material with increasing plasma power due to the formation of a larger amount of functional groups per unit time. Therefore, probable reason of that effect in the case of using low-M_w PLA and BA simultaneously is also relative increase of etching rate with equal or lower efficiency of functional groups forming at surface by plasma treatment. According to the results of the measurements of the film thickness obtained by AFM before and after plasma treatment with increased treatment time to 420 seconds, the etching rate was about 0.07 nm/s. No more than 7 nm of PLA film was etched at 100 seconds of plasma treatment. Etching effect was investigated also by AFM. All pores, initially closed with a thin layer of PLA and formed during film coating process with subsequent heat treatment, lost this layer and became open (figure 4).

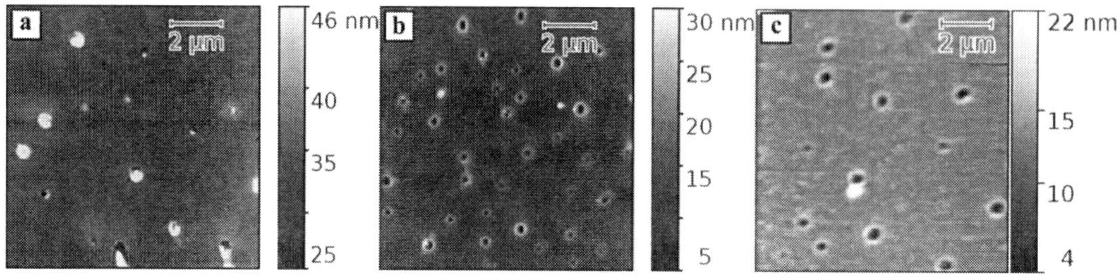

Figure 4 (a, b, c). AFM-images of PLA film surfaces obtained with THF and BA solution with low-M_w PLA additive before **(a)** and after plasma treatment **(b)** and also with THF and BA only in solution **(c).**

At the same time, the number of pores per surface area unit significantly increases and their size significantly decreases when a low molecular weight polylactide in the composition of the solution was used. It is similar to BA usage as an additive to the solution [6], that can be seen by comparison of AFM images of high-M_w PLA in THF without BA (figure 1), containing both BA and low-M_w PLA, and high-M_w PLA containing BA only (figure 4). In addition, anomalously high hydrophobicity of the as deposited films was observed when both at once low-M_w PLA and BA were added (figure 2a). That in conjunction with the changed pore size and their amount per area on the surface, indicates the difference in mechanisms of solvents evaporation from film layer and their residual amount, which inevitably affects the achieved value of surface hydrophilicity and its stability over time.

To assess the effectiveness of the formation of functional groups in the surface layer by plasma treatment XPS was carried out. XPS data (figure 5) demonstrated an appreciable increase in the intensity of the peaks corresponded to the formation of functional groups, although several times lower than achieved in the literature [7].

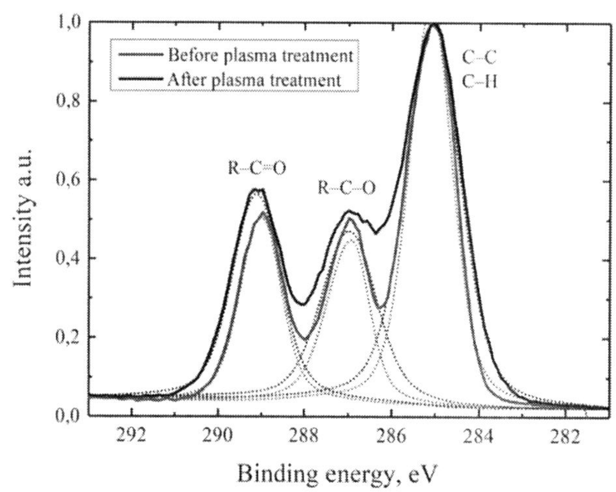

Figure 5. XPS spectra of C1s peaks of PLA film obtained with solution with low-M_w PLA before and after plasma treatment.

The achieved increase of functional groups percent in the surface layer was up to 47 % (compared to 225% in literature [7]). At the same time, the achieved WCA stability over time under normal conditions is 0.34° per day for the high-M_w PLA with BA sample (figure 3), while it was at least 1.5°

per day in the analysed literature [7] (determined by the ratio of WCA change to the time when the WCA degradation rate reduce to less than 0.2° per day). Taking into account the obtained low WCA and its stability, concentration of functional groups on the surface that was achieved is close to a limit, at which the modification is still stable for a long time.

Obtained results are extremely significant, since culturing cells on the surface or in vivo implantation are accompanied by the polymer surface exposure to elevated temperatures, grow medium or intercellular fluids, causing degradation of the surface and its hydrophilicity. High rate of degradation of modified surface may result in decrease of cells grow rate on the surface after a long period (about 5-7 days) [3], that is unsuitable for implant material. Demonstrated high stability of modification is a good prerequisite for suppression mentioned effect.

4. Conclusion

Addition of low-molecular weight PLA as well as BA in the composition of PLA films can significantly improve the hydrophilicity of the surface modified by plasma treatment. Additionally, the low molecular weight polylactide and BA reduce the diameter of pores that are forming in the polylactide film during deposition. Plasma-modified high-molecular PLA containing BA as an additive demonstrates sufficiently high hydrophilicity in conjunction with great stability over time and resistance to rinsing in ethyl alcohol and water. Achieved ageing and rinsing stability is extremely important for implants fabrication application, since the cell detachment due to significant degradation of the PLA surface and its hydrophilicity reduction during the first few days is undesirable and can nullify the benefits of surface modification. Comparison of the WCA stability over time with XPS data, as well as literature data, allows to conclude that the treatment with low plasma density and low power density per area that was used, provides sufficiently high hydrophilicity even without high density of formed functional groups in opposition to other works. The density of functional groups obtained is close to optimal to ensure both high hydrophilicity and stability over time and can be implemented to implants modification.

Acknowledgments

This work was supported by the Ministry of Science and Higher Education of the Russian Federation, agreement № 14.575.21.0125 (unique ID RFMEFI57517X0125).

References

[1] Pawar R P, Tekale S U, Shisodia S U, Totre J T and Domb A J 2014 *Recent patents on regenerative medicine* **4** 40
[2] Schaller B, Burkhard J P M, Chagnon M, Beck S, Imwinkelried T and Assad M 2018 *J. Oral Maxillofac. Surg.* **76** 2138
[3] Slepička P, Trostová S, Slepičková Kasálková N, Kolská Z, Sajdl P and Švorčík V 2012*Plasma Process. Polym.* **9** 197
[4] Richard M and Robert D 1998 *Langmuir* **14** 4827
[5] Izdebska-Podsiadłya J and Dörsamb E 2017 *Vacuum* **145** 278
[6] Romashkin A V, Struchkov N S, Polikarpov Y A, Petukhov V A, Levin D D, and Nevolin V K 2018 *Journal of Physics: Conference Series* **1124** 031027
[7] Jacobs T, Declercq H, De Geyter N, Cornelissen R, Dubruel P, Leys C, Beaurain A, Payen E and Morent R 2013 *J. Mater. Sci. Mater. Med.* **24** 469
[8] Inagaki N, Narushima K, Tsutsui Y and Ohyama Y 2002 *J. Adhes. Sci. Technol.* **16** 1041

Optical spectral sensor for the assessment of uric acid kinetics during hemodialysis treatment

G A Konoplev[1], O S Stepanova[1], G A Zemchenkov[2], A Frorip[3]
[1]Department of Photonics, St. Petersburg Electrotechnical University, St. Petersburg, 197376, Russia
[2]Nevsky Nephrology Center, St. Petersburg, 193318, Russia
[3]LDIAMON AS, Tartu, 51014, Estonia

Abstract. The dual-wavelength direct spectrophotometric method for uric acid determination in effluent dialysate and deep UV LED based optical spectral sensor for non-invasive online hemodialysis monitoring were developed. The time profiles of uric acid concentration in effluent dialysate were recorded for a group of 12 patients and 88 hemodialysis sessions by means of the sensor. The parameters of the double-pool kinetic model for each session were assessed and analyzed. The feasibility of online optical monitoring in the UV region as a useful and convenient tool for the experimental investigation of uric acid kinetics during hemodialysis was demonstrated.

1. Introduction

Hemodialysis (HD) is an extracorporeal method of blood purification for treating patients with chronic renal failure in the terminal stage. During this life-supporting medical procedure low molecular weight metabolic wastes and excessive water are eliminated from the human body. In dialysis machines (artificial kidneys) blood continuously circulates through a dialyzer containing the semi-permeable membrane functioning as a molecular sieve. Small molecules of metabolic wastes easily pass through the pores of the membrane to dialysis fluid (dialysate) and with effluent dialysate go to the outlet of a dialysis machine while proteins and other big molecules remain intact. HD treatment is carried out 3 times a week; every session lasts 3-5 hours.

Various mathematical models of uremic markers kinetics are used to assess the efficiency of waste products elimination during HD treatment [1]. The most widely used approach is the single-pool urea kinetics, which with some limitations can be applied to other uremic markers such as creatinine, uric acid (UA), β-2-microgloubuline etc. According to this model the time dependence of urea concentration during HD is as follows:

$$C(t) = C_0 e^{-t/\tau_1}, \tag{1}$$

where τ_1 – the time constant that characterizes the transfer of uremic markers via a dialysis membrane. More accurate double-pool kinetic model additionally takes into account the intercompartment transfer of eliminating substances from extracellular fluid to blood [1]:

$$C(t) = C_1 e^{-t/\tau_1} + C_2 e^{-t/\tau_2} + C_3, \tag{2}$$

where τ_1 – the time constant that characterizes the transfer of uremic markers via a dialysis membrane; τ_2 – the time constant that characterizes the intercompartment transfer.

The areas under the curves for the first S1 and the second S2 terms in the formula (2) reflect the role of each process (the transfer of uremic markers via a dialysis membrane and the intercompartment transfer) in the elimination of a uremic marker. They can be calculated by the following integrals:

$$S_1 = \int_0^T C_1 e^{-t/\tau_1} dt \quad S_2 = \int_0^T C_2 e^{-t/\tau_2} dt , \tag{3}$$

where T – the duration of the treatment.

2. Experiment and results

To experimentally estimate the parameters of the double-pool model (C_1, C_2, C_3, τ_1, τ_2) for a particular uremic marker it is necessary to take numerous (at least 5…7 per session) blood or effluent dialysate samples during the treatment, take them to a clinical laboratory, determine the concentration in every sample by standard biochemical assays, and fit the model curve (fig. 1) to the experimental points. The procedure is very expensive, cumbersome and time-consuming, so it is more preferable to use special sensors for on-line monitoring of uremic markers concentration in effluent dialysate [2, 3], but existing methods of HD monitoring are intended for the control of urea concentration only and require disposable materials or high-cost equipment.

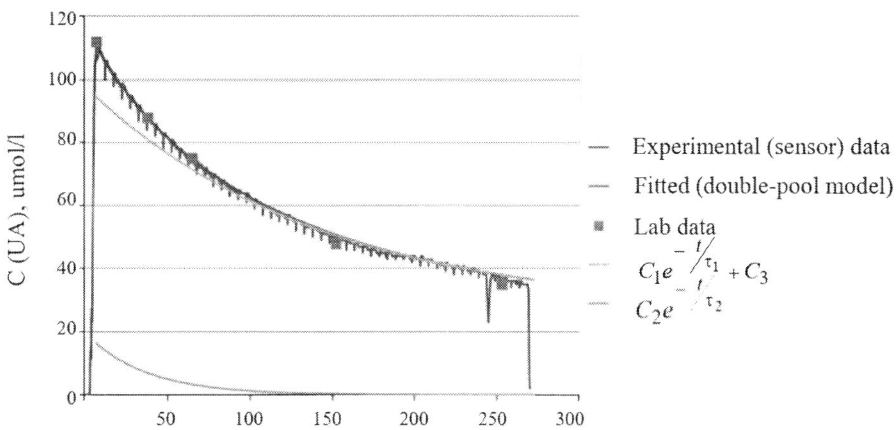

Figure 1. The time profile of UA concentration in effluent dialysate measured using the optical sensor, results of the lab tests for effluent dialysate samples taken during HD treatment and curve fitting according to the double-pool model for one of the patients.

The authors have developed the dual-wavelength direct spectrophotometric method for uric acid determination in effluent dialysate and optical spectral sensor for non-invasive online hemodialysis monitoring. The sensor incorporates UV LEDs with emission wavelength 260 nm and 285 nm, solar-blind UV photodetectors and a quartz flow-cell which is connected to the outlet of a dialysis machine. It is fully automatic, reagentless and capable of measuring the time profiles of UA concentrations with extremely high temporal resolution (down to 5 sec). More detailed description of the sensor can be found elsewhere [4].

The time profiles of uric acid concentration in effluent dialysate were recorded for a group of 12 patients by means of the sensor. For each patient measurement were carried out at least 5 times over a 3 month period; in total 88 HD sessions has been monitored. For validation purposes during some (not all) sessions the samples of effluent dialysate were taken at 0, 30, 60, 150 and 240 minutes of a session and lab tests for UA were made by clinically approved biochemical assay with the automated analyzer "Beckman Coulter AU680". Statistical analysis by Bland-Altman method [5] showed a good agreement between the results obtained with the sensor and the standard assay; the relative error of determination for UA concentration was less than 10% (fig. 2).

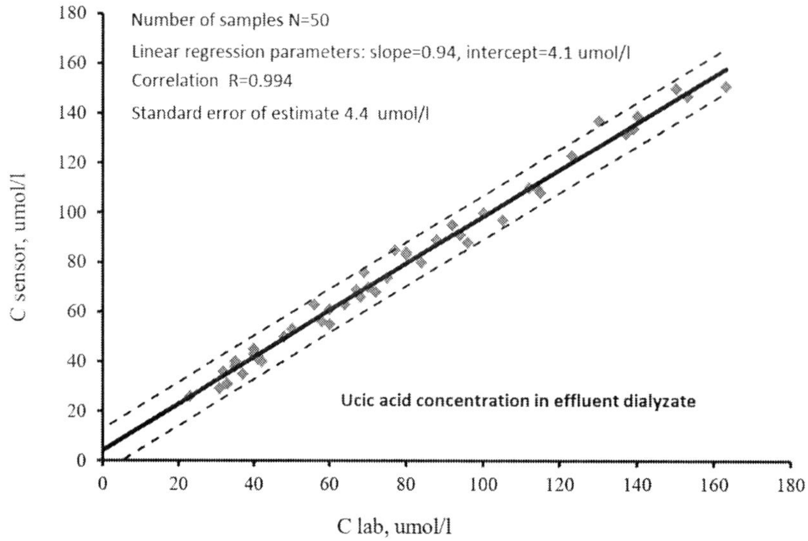

Figure 2. The scatter diagram for uric acid concentration in the effluent dialysate samples measured with the optical sensor and with the automated biochemical analyzer "Beckman Coulter AU680".

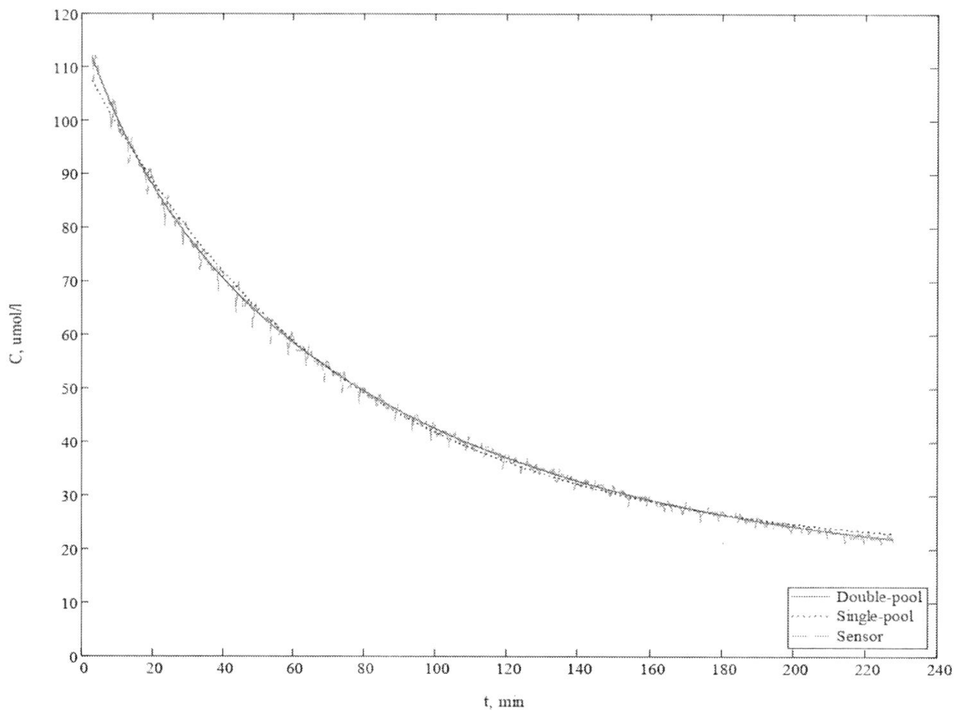

Figure 3. Approximation the UA concentration time profile according to the double-pool and single-pool models.

The parameters of the single- and double-pool models of UA kinetics according to the expressions (1) and (2) correspondingly were assessed and analyzed for each HD session using Mathcad software. It was confirmed that the results of double-pool modelling slightly better fit experimental data compared to the single-pool model, especially at the beginning and at the end of a HD session (fig. 3).

Table 1. Parameters of the double-pool model for patient #4 for several HD treatments.

Date of treatment	τ_1, min	τ_2, min	C_1, umol/l	C_2, umol/l	C_0, umol/l
17.09.2018	**94**	**13**	71	18	12
19.09.2018	**100**	**19**	66	16	10
24.09.2018	**91**	**14**	63	20	12
26.09.2018	**97**	**11**	67	17	11
29.09.2018	**94**	**16**	78	16	14
01.10.2018	**94**	**13**	86	43	13
06.10.2018	**88**	**17**	84	18	16
10.10.2018	**98**	**15**	69	14	14
13.10.2018	**101**	**18**	84	16	13
27.10.2018	**102**	**21**	90	21	12

Table 2. Average parameters of the double-pool model for 12 patients.

Patient #	τ_1, min	τ_2, min	C_1, umol/l	C_2, umol/l	C_0, umol/l
1	**90**	**22**	97	36	15
2	**131**	**17**	110	11	19
3	**82**	**19**	127	53	20
4	**96**	**16**	76	20	13
5	**149**	**23**	93	48	7
6	**103**	**4**	205	34	25
7	**108**	**13**	131	28	16
8	**126**	**31**	81	107	12
9	**188**	**18**	95	21	13
10	**117**	**18**	109	37	15
11	**115**	**40**	86	70	14
12	**149**	**25**	76	35	10

To give an example of the obtained results a set of parameters of the double-pool model for patient #4 assessed for 10 HD treatments conducted in 45 days period presented in table 1. Model parameters for the whole group of 12 patients averaged for several HD treatments presented in table 2. It can be seen from the data in table 1 and table 2 that τ_1, τ_2 time constants are subject to significant variations both for particular patient between different HD treatments and between different patients. To reveal the clinical significance of these variations and find the way to individualize dialysis treatment regime based on the individual character of elimination kinetics more in-depth analysis on larger group of patients is needed.

3. Conclusion

It was shown that online optical monitoring in the UV region by means of deep UV LED based spectral sensors is a useful and convenient tool for the experimental investigation of uric acid kinetics during HD which is necessary for improving methods for assessing the adequacy of HD and adjusting the working regimes of dialysis machines.

Acknowledgments

Authors wishing to acknowledge A.Yu. Zemchenkov and R.P. Gerasimchuk for invaluable help in conducting pre-clinical trials of the optical spectral sensor.

References

[1] Schneditz D, Daugirdas J T 2001 *Sem. Dial.* **14** 271.
[2] Burgelman M, Vanholder R, Fostierx H, Ringoir S 1997 *Med. Eng. Phys.* **19** 69.
[3] Daugirdas J T, Tattersall J E 2010 *Kidney Int.* **78** 833.
[4] Vasilevsky A M, Gerasimchuk R P, Zemchenkov A Y, Konoplev G A, Rubinsky A V, Stepanova O S, Frorip A 2015 *Biomed. Eng.* **49** 125.
[5] Bland J M, Altman D G 1986 *The Lancet* **327** 307.

Age and gender recognition from speech signals

Assim Ara Abdulsatar[1,3], V V Davydov[1, 3], V V Yushkova[4], A P Glinushkin[3] and V Yu Rud[1, 3]

[1]Peter the Great St. Petersburg Polytechnic University, St. Petersburg 195251, Russia
[2]University of Salahaddin, Erbil 44001, Iraq
[3]All Russian Research Institute of Phytopathology, Moscow Region 143050, Russia
[4]Saint Petersburg University of Management Technologies and Economics, 190109, Russia

E-mail: araasimyahya@gmail.com

Abstract. The aim of this research is to identify gender and age from speech, the system consists of two parts. The first part is called pre-processing and future extraction. The second part is called classification. This research investigates an automatic gender and age recognizer from speech. First four formant frequencies and twelve MFCCs are used to extract relevant features to recognize the gender. K-NN has been used as a classifier for the age recognizer model, stimulated using MATLAB. A special selectin of solid feature is used in this work to improve the accuracy of the gender and age classifiers based on the frequency range that the features represent.

1. Introduction

The difficulties that rise during the process of automatic speech estimation based on age are: Firstly, speech is affected by speaker's weight, height and mood, these characteristics may interact with age [1, -3]. Secondly, a huge database is needed for different speakers of different ages. Lastly, most of speeches in our daily life are from noisy environments, filters and high-quality microphones are required. Using different recording tools for recording the same sentence from the same speaker leads to different results.

Age and gender recognition have many practical applications, among them are: Commercial advertisements, the ads can become more relevant and target a specific group of age and gender, leading to an increase in sales. Another application is in forensic science, number of suspects can be reduced if there is an evidence such as a telephone call. This system can also be used for user authentication based on their speech.

There are different approaches to perform automatic gender and age recognition. For example, cepstral features, like Mel Frequency Cepstral Coefficients (MFCC). For age recognition. MFCC is known for producing poor results for gender and age classification with recorded signals [4]. To avoid this problem, the MFCC features are enhanced by analyzing the parameters that affect the process of extracting the features. MFCC has been used in many speech applications from speech recognition to language identification, and there is another acoustic feature that can be extracted, it is format frequency.

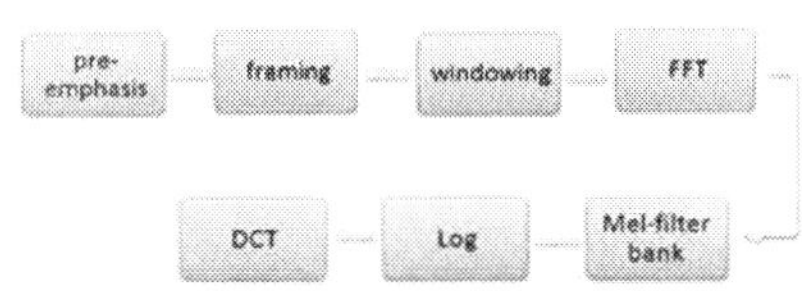

Figure 1. MFCC steps in speech analysis

The first stage of the process is pre-emphasis, this process is needed to increase the energy of the signal at high frequencies because the signal at higher frequencies has less energy compared to low frequency signals, this step is needed because the spectrum for voiced segments has more energy at lower frequencies than higher frequencies, this is called spectral tilt, spectral tilt is caused by the nature of the glottal pulse, therefore boosting high-frequency energy gives more info to acoustic model and consequently improves recognition performance, output response of pre-emphasis filter in n-domain is given below:

$$y(n) = x(n) - 0.95\, x(n-1) \quad (1)$$

Following pre-emphasis (1), the signal is divided into frames, this is known as framing. It is advantageous to frame signals to guarantee stationarity. The width of the frames is generally about 30ms with an overlap of about 20 ms (10 ms shift). Each frame carries N sample points of the speech signal. Overlap rate of frames is between 30 % and 75 % of the length of the frames, M=100 and N=256 that is why we frame the signal into 20-40 ms frames. If the frame is much shorter then we won't have enough samples to get a reliable spectral estimate, if it is longer then the signal will change too much throughout the frame. It is assumed that although the speech signal is non-stationary, but it is stationary for a short duration of time.

After framing, the window function is used to smooth the signal for the computation of DFT (discrete Fourier transform). There are different windowing functions. Commonly used ones are: Hamming, Blackman, Gauss, rectangular, and triangular. The hamming window is usually used in speech analysis because its spectrum falls off rather quickly, so the resulting frequency resolution is better, which is suitable for detecting formants. The equation is given below:

$$w(n) = 0.54 - 0.46 \cos\left(\frac{2\pi n}{M}\right) \quad (2)$$

Where M is the order of the filter, which is equal to the filter length – 1 and n is the index, the result of this process (2) is shown in the figure below:

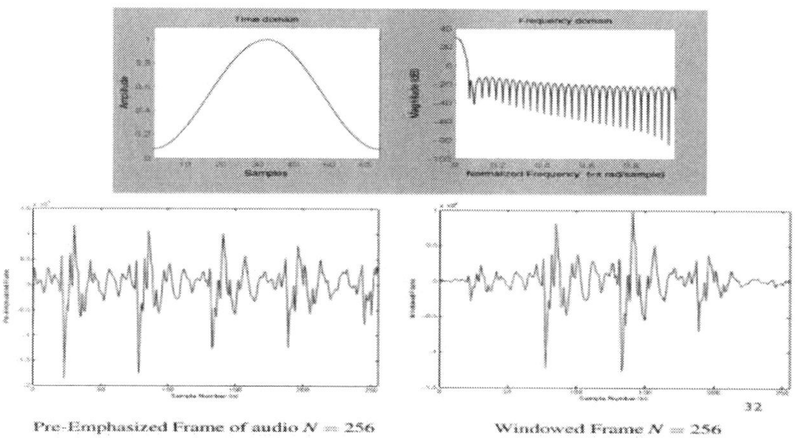

Figure 2. Original pre-emphasized frame (on left) vs. windowed frame (on right)

Fast Fourier Transform is used to convert each frame of N samples from time domain into frequency domain. This statement is supported the equation below:

$$F(x) = \sum_{n=0}^{N-1} f(n).\, e^{-j2\pi(x\frac{n}{N})} \quad (3)$$

Where N is the length of the discrete signal, $F(x)$ is the signal in frequency domain and $f(n)$ is the signal in time domain (3), FFT size can be 512, 1024 or 2048.

Human hearing is not sensitive to all frequency bands. Humans are less sensitive at higher frequencies, roughly > 1000 Hz. In other words, our hearing perception is non-linear. Human ear is like a filter that only concentrates on certain frequency components thus Mel-filter bank is implemented, this statement is shown in the figure below:

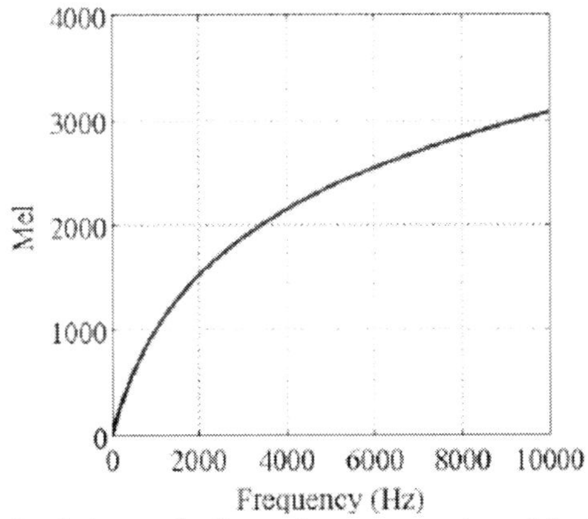

Figure 3. Mapping between the linear frequency scale and the Mel scale

Next step, the logarithm of the square magnitude of the output of Mel-filter bank is computed, dynamic range of values is compressed through logarithm, human response to signals is logarithmic. Human hearing is less sensitive to small differences in amplitude at high amplitudes than at lower amplitudes, this makes frequency estimates less sensitive to slight variations in the input, the picture below shows the signal before and after taking the logarithm:

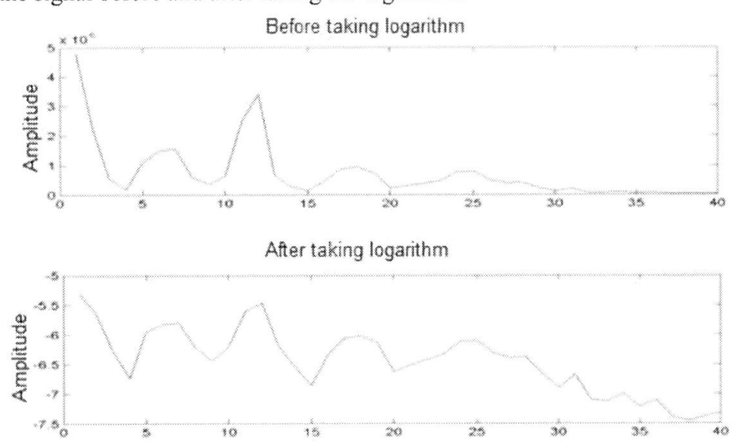

Figure 4. Amplitude of the signal before and after taking its logarithm

The last stage of MFCC is DCT, it's the process of converting the log Mel spectrum into time domain using DCT [2,6]. The result of this conversion is called Mel Frequency Sepstrum coefficients, this set of coefficients is called acoustic vectors. In fact, DFT should be applied because we are transforming the signal from frequency domain back to time domain using the following equation:

$$yt(k) = \sum_{m=1}^{M} \log(|Yt(m)|) \cos\left(k(m-0.5)\right)\frac{\pi}{M} \;, k = 0, \ldots .J \quad (4)$$

Once the main features are extracted from the speech signal (4), classifiers are used. In this research, KNN (K-Nearest Neighbor) is used. KNN algorithm is one of the simplest machine learning algorithms. Data processing is carried out in MATLAB, which is actively used in various scientific fields [5-7]. In addition, a developed signals processing method can be used to obtain information in direction finding systems, radio monitoring and different communication systems [8-14].

2. Experimental results

The experimental results obtained for the prediction of age and gender using 40 clean speech samples in English. The model classes (seniors, adults and children from both genders), the gender recognition system reached 66 % recognition accuracy. While age recognition reached 55 % recognition accuracy.

Table 1. Rates of age recognition and rates of gender recognition using MFCCs only

MFCCs	Gender recognition rate	Age recognition rate
MFCC 1 2 3	55.55%	44.44%
MFCC 4 5 6 7 8 9	66.66%	44.44%
MFCC 10 11 12	22.22%	33.33%
All MFCCs	66.66%	44.44%

Table 2. Rates of recognition (age and gender merged) using MFCCs only

MFCCs	Recognition rate (age and gender together)
MFCC 1 2 3	33.33%
MFCC 4 5 6 7 8 9	44.44%
MFCC 10 11 12	11.11%
All MFCCs	44.44%

When using MFCC feature the best result was obtained using MFCC 4 to 9. The following tables are the results obtained using formants:

Table 3. Rates of age recognition and rates of gender recognition using formant frequencies only

Formants	Age recognition rate	Gender recognition rate
f3,f4	22.22%	22.22%
f3	38.88%	27.77%
f4	27.77%	44.44%
f1,f4	38.88%	55.55%
f1,f2,f3	50%	55.55%
f1,f2,f4	44.44%	50%
f1,f2	55.55%	61.11%
f1,f3,f4	33.33%	44.44%
f1,f3	50%	55.55%
f1	38.88%	38.88%
f2,f3	55.55%	50%

f2,f3,f4	38.88%	50%
f2,f4	50%	38.88%
f2	38.88%	38.88%
f1,f2,f3,f4	50%	66.66%

Table 4. Rates of recognition (age and gender merged) using formant frequencies only

Formants	Recognition rates (for age and gender together)
f1,f2,f3,f4	50%
f1,f4	33%
f1,f3	39%
f1,f2	55%
f1	28%
f1,f2,f3	39%
f1,f2,f4	39%
f1,f3,f4	28%
f2,f3,f4	33%
f2,f4	33%
f2	33%
f4	17%
f3	22%
f2,f3	39%
f3,f4	22%

Table 5. Rates of recognition using formant frequencies and MFCCs only

Formant frequencies	MFCCs	Age & gender	Gender	Age
All	All	50%	66.66%	50%
f1 f2	All	50%	55.55%	50%
f1 f2	MFCC 1 2 3	50%	55.55%	50%
f1 f2	MFCC 10 11 12	50%	55.55%	50%
f1 f2	MFCC 4 5 6 7 8 9	50%	55.55%	50%

When using formant frequencies feature the best result was obtained by using formant 1 and 2, best result for gender recognition obtained by using all formants together.
Gender recognition rate for different formants is shown in the table, excluding children [15, 16], we achieved a much higher rate (91%).

Table 6. Gender recognition rates using formant frequencies (without considering children)

Gender recognition (without children)	
Formants	**Gender**
f1 f2 f3 f4	83.33%

f1 f2	83.33%
f3 f4	91.66%
f3	66.66%
f4	66.66%
f1 f4	66.66%

3. Conclusion
In conclusion, gender and age can be extracted from speech signals. Using formant frequencies, the best result was obtained using formant frequency 1 & 2 because the main information of speech signal is concentrated at the low frequency part, the same is true for MFCC but the results were not improved when using MFCC because MFCC is not active with data that has been recorded at different places, with different computers and microphones, in order to improve the results, MFCC should be combined with another acoustic feature especially the fundamental frequency. In future the results can be improved further by using a smart classifier instead of the KNN, for example SVM or ANN.

Acknowledgments
I would like to express my deep gratitude to my lecturer (Fatima K. Faek) in the University of Salahaddin as well as my current professor (V. V. Krasnoshchekov) in Peter the Great Saint Petersburg Polytechnic University who helped me and motivated me in writing this paper which also helped me in doing a lot of research. I would also like to express my deep gratitude to my parents for their encouragement and support. This work partly supported by Grant of RFBR No. 18-29-25071.

References
[1] Bahari M H and Van hamme H 2011 *Proceedings of the 2018 IEEE Workshop on Biometric Measurements and Systems for Security and Medical Applications* (Italy) 123-127
[2] Faek F K and Al-Talabani A K 2013 *International Journal of Computer Applications* **70(20)** 11-14
[3] G Dobry, M Hetch, M Avegal and Y Zigel 2011 *IEEE Trans. Audio, Speech and Language Processing* **19(7)** 1975–1985
[4] M Li, K J Han and S Narayanan 2012 *Computer Speech and Language* **27** 151-167
[5] R Davydov, V Antonov and N Kalinin 2015 *Journal of Physics: Conference Series* **643(1)** 012107
[6] J Stenis, W Hogland, M Sokolov, V Rud' and R Davydov R. 2019 *IOP Conference Series: Materials Science and Engineering* **497(1)** 012061
[7] R V Davydov, V I Antonov and D V Molodtsov 2018 *Journal of Physics: Conference Series* **1135(1)** 012088bnh
[8] Podstrigaev A S, Davydov R V, Rud' V Yu and Davydov V V 2018 *Lecture Notes in Computer Science (including subseries Lecture Notes in Artificial Intelligence and Lecture Notes in Bioinformatics)* **11118 LNCS** 624-630
[9] Moroz A V, Davydov V V, Rud V Yu, Rud Yu V, Shpunt V Ch and Glinushkin A P 2018 *Journal Physics: Conference Series* **1135(1)** 012060
[10] Davydov R V, Saveliev I V, Lenets V A, Tarasenko M Yu, Yalunina T R, Davydov V V and Rud' V Yu 2017 *Lecture Notes in Computer Science (including subseries Lecture Notes in Artificial Intelligence and Lecture Notes in Bioinformatics)* **10531 LNCS** 177-183
[11] Petrov A A, Davydov V V and Grebenikova N M 2018 *Lecture Notes in Computer Science (including subseries Lecture Notes in Artificial Intelligence and Lecture Notes in Bioinformatics)* **11118 LNCS** 641-648
[12] Logunov S E, Davydov V V, Vysoczky M G, Koshkin A Y and Rud' V Yu 2017 *Journal of Physics: Conference Series* **917(5)** 052058
[13] Davydov V V, Sharova N V, Fedorova E N, Gilshteyn E P, Malanin K Y, Fedotov I V, Vologdin V A and Karseev A Yu 2015 *Lecture Notes in Computer Science (including subseries Lecture Notes in Artificial Intelligence and Lecture Notes in Bioinformatics)* **9247** 712-721

[14] Myazin N S, Logunov S E, Davydov V V, Rud' V Yu, Grebenikova N M and Yushkova V V 2017 *Journal of Physics: Conference Series* **929 (1)** 012064
[15] Hugo M and Isabel T 2011 Project ACM: Transactions on Speech and Language Processing **7(4)** p. 16
[16] V Tiwari, G Ganga, J Singhai and M Azad 2011 *An International Journal (SPIJ)* **5(2)** 52

Shellac hologram

A Toropova[1], M Fokina[1]
[1]Faculty of applied optics, ITMO University, Saint Petersburg 197101, Russia

Abstract. The results of experimental research of rainbow holograms made of edible shellac, analysis of diffraction efficiency and analysis of the hologram lifetime in various environmental conditions are reported in this work. The properties of shellac as a food biopolymer allow to use it for creation edible holographic coatings on food and pharmaceutical products. Edible shellac holograms can be use for creation images not using food dyes, original appearance, labels, sensors of shelf life and observance of storage regulations of products.

1. Introduction

The idea of creation edible rainbow holograms appeared at the end of the 20 century and remains relevant today [1]. The holograms can be applied to food products in order to provide them an original appearance. The advantage of rainbow holograms is the ability to create bright color images on the surface of edible products without food dyes. The holograms can be applied to pharmaceutical products as a label. They are formed directly on the pill surface and protect the products from falsification [2]. Some organic polymers such as agar, gum arabic, pectin, gelatin, silk fibroin, albumin can be a good material for edible rainbow holograms creation [1-4].

In this research shellac as the material for creation edible rainbow holograms is presented. Shellac is used in the food and pharmaceutical industries as coating and glazing agent.

2. Materials and methods

Shellac is compounds of aleuritic acid, shellolic acid, jalaric acid and other aliphatic acids. Shellac has the form of thin flakes, the color of shellac depends on the degree of purification and can has color from yellow to dark brown. Shellac is water-insoluble, highly soluble in alcohol, has a melting temperature 75-80°C [5].

Some physical properties of shellac were measured. Viscosity of alcohol shellac solution was measured by viscometer «Visco Basic Plus» at temperature 22,5°C. Optical characteristics were obtained using refractometer «Refractometr PTR 46» and «Shimadzu UV-1800», microscope "Olympus STM-6" with a magnification of 100× was used to observe the samples. To estimate diffraction efficiency of the samples the intensities of the diffracted and transmitted rays of a diode laser with a wavelength of 625 nm were measured by luxmeter "TKA-PCM-06". Diffraction efficiency is estimated as the ratio (in %) of light intensity in the first diffraction order to the sum of the intensities in the first and zero diffraction orders [6]. Moisture capacity of shellac films was estimated as percentage of water absorbed by the film in relation to film mass, films were in an environment with a humidity 90% and temperature 35°C during the day. Measurement results are presented in table 1.

Table 1. Shellac properties

Viscosity of shellac solution, mPa/s	9.5
Moisture capacity of shellac films, %	0.6
Refractive index of shellac films	1.44
Average thickness of shellac films, µm	8
Transmittance coefficient of shellac films in 350–900 nm	70–83%

Hologram samples were prepared by the nanoimprinting method [7]: a drop of shellac solution was placed on a substrate, covered with a silicone mold, dried at room temperature, than the mold was removed. The diffraction grating of silicone mold has a period of 600 lines per mm. The thickness of the film with a diffraction relief is near 8 microns, the film is breakable, so it is difficult to use it as an isolated holographic element. A holographic coating was formed directly on the surface of different substrates like as glass, caramel, chocolate and marzipan.

3. Experimental results

The samples of diffraction gratings made of shellac on the glass surface were exposed to temperature and humidity, the diffraction efficiency was controlled. A graph of the diffraction grating degradation is presented in figure 1 (a,b).

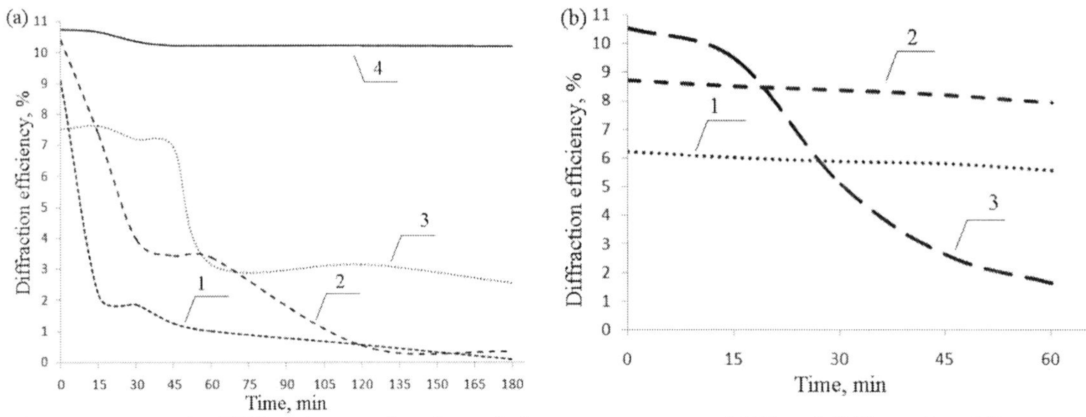

Figure 1(a, b). (a) Diffraction grating degradation at temperature 35°C and different humidity: 1 – 90%, 2 – 70%, 3 – 50%, 4 – 30%; **(b)** Diffraction grating degradation in humidity 10% and different temperature: 1 – 35°C, 2 – 40°C , 3 – 50°C.

The diffraction grating degradation was not observed at temperature of 35°C and a humidity of 30% (line 4) and significant degradation occurs at a temperature of 35°C and a humidity of 90% (line 1) during one hour. Under the influence of temperature 50°C degradation occurs within an hour. Lifetime of the shellac diffraction gratings can vary from an hour to a year depending on the storage conditions. Lifetime of grating stored in conditions with temperature 35°C and humidity 10% is 75 days. At the same temperature and increasing humidity the lifetime is reduced: in humidity 50% lifetime is 3 days, in humidity 70% – 240 minutes, in humidity 90% – 180 minutes.

Infrared spectrums of shellac films before exposure of humidity and after exposure of humidity are shown in figure 2. When shellac absorbs water molecules, the main changes are observed in the region of valence oscillations of the C=H groups 1735cм$^{-1}$ and 1712cм$^{-1}$. Changes in the region of deformational oscillations of C–O bonds in –C–C–O and –C–O–C groups (1250cм$^{-1}$ and 1035cм$^{-1}$) indicate the formation of associates of water molecules with shellac.

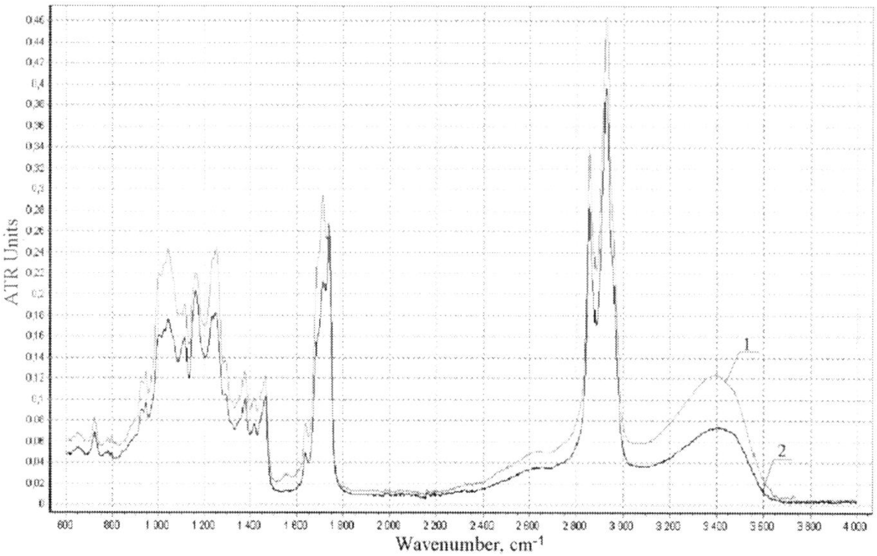

Figure 2. Infrared spectrums of shellac films: 1) before exposure of humidity; 2) after exposure of humidity.

Despite the low moisture capacity of shellac films, the absorbed water affects on the preservation of the diffraction grating, what is confirmed by decrease of the diffraction efficiency. The surface changes of the shellac diffraction gratings due to exposure of 90% humidity for an hour are shown in figure 3 (a, b).

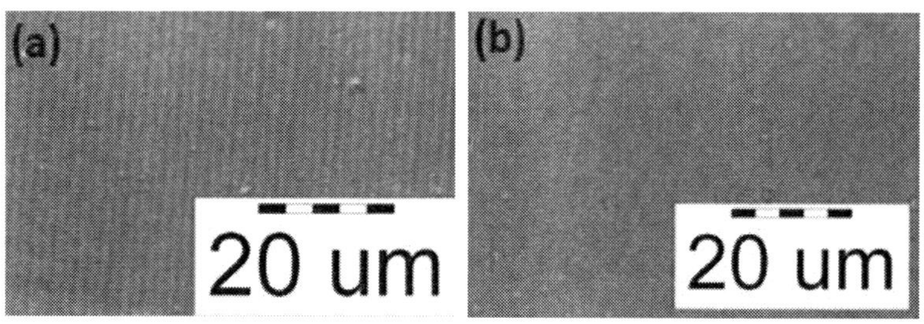

Figure 3 (a, b). (a) The surface of the shellac diffraction gratings before exposure of humidity, the diffraction efficiency – 9%; **(b)** The surface of the shellac diffraction gratings after exposure of humidity, the diffraction efficiency – 1%.

When shellac gratings stored at room temperature about 18-25 and humidity not higher 50% lifetime is about one year. This feature of shellac diffraction grating can be used to creation holographic sensors – indicators of compliance with the storage conditions of food and pharmaceutical products.

Shellac holograms can be use for decoration food products. In previous work rainbow holograms were made of chocolate [8] and caramel [9]. The lifetime of caramel holograms was about 40 hours, and the lifetime of chocolate holograms was about 3 months. In this research shellac holograms were applied on surface of caramel, chocolate and marzipan (figure 4). The diffraction efficiency of the grating on the caramel surface in transmitted light was 4,5%, in the reflected light – 3%. The diffraction efficiency of the grating on chocolate surface and marzipan surface was 2.1% in reflected light. The iridescent color of the samples is clearly distinguishable by an eye despite the small values of diffraction efficiency. Using of shellac holograms allowed to increase the lifetime to a year and to create hologram on porous surface of marzipan.

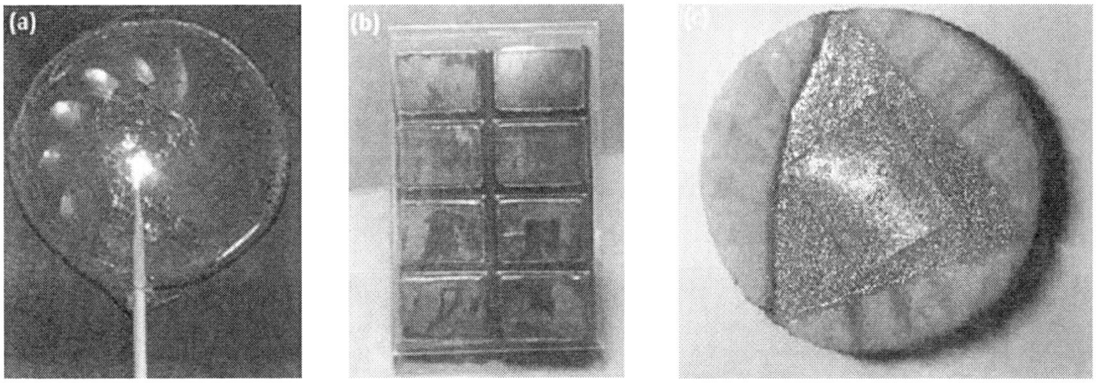

Figure 4 (a, b, c). (a) Shellac hologram on the surface of caramel; **(b)** Shellac hologram on the surface of chocolate; **(c)** Shellac hologram on the surface of marzipan.

4. Conclusion

Shellac can be used as an optical biopolymer for creation edible rainbow holograms. These holograms have sufficient diffraction efficiency to observe a bright visual effect. The peculiarity of the diffraction gratings made of shellac to destruction under the influence of temperature and humidity of the environment creates conditions for the creation of holographic sensors – indicators of storage regulations for food and pharmaceutical products.

Acknowledgments

The research was performed with the financial support of the Innovation Promotion Fund in science and technology

References

[1] Begleiter E. Holographic Products. Patent USA № 4,668,523 from 26.05.1987

[2] Schnieper M, Walter H., Pharmaceutical tablet with a diffractive microstructure comprises a mass of single powder particles. Patent Germany №DE102007026957A1 from 20.12.2007

[3] Pérez-Salinas P, Mejias-Brizuela N.Y., Olivares-Perez A., Grande-Grande A., Páez-Trujillo G., Hernández-Garay M.P, Fuentes-Tapia I. 2008 Proc. of SPIE Vol. 6912, 691213

[4] Omenetto F.G., David L. Kaplan D.L. Edible holographic silk products. International patent №WO2009155397A2 from 23.12.2009

[5] Sharma S.K., Shukla S.K., Vaid D.N. Shellac-structure, characteristics and modification. Def Sci J, Vol 33, No.3, 1983, p 261

[6] Vorzobova N., Bulgakova V., Moskalenko A., Pavlovets I., Denisyuk I., Burunkova Y., Development of periodic and three-dimensional structures in acrylic-monomer photopolymer materials by holographic methods., Radiophysics and Quantum Electronics, T.57, №8-9, 2015, pp. 595-602

[7] Denisyuk I., Aref'eva N. Using the imprint method for single replication of polymeric Fresnel optics and micro-optics, Journal of Optical Technology, T.75, №7, 2008, pp. 462-464

[8] Toropova A.P., Fokina M.I., Pozdnyakova S.A. Peculiarities of diffraction gratings application on chocolate surface. Scientific and Technical Journal of Information Technologies, Mechanics and Optics, 2018, vol. 18, no. 1, pp. 166–168

[9] Toropova A.P., Santalina I.Y., Fokina M.I. Peculiarities of rainbow holograms application on the surface of the caramel. IOP Conf. Series: Journal of Physics: Conf. Series 1062 (2018) 012016, Russia, 2018, 4p.

Automated pulmonary nodule detection system in computed tomography images based on Active-contour and SVM classification algorithm

F Shariaty[1, 2], V V Davydov[1, 3], V V Yushkova[4], A P Glinushkin[3] and V Yu Rud[1, 3]
[1]Department of Electrical Engineering, University of Qom, Qom, Iran
[2]Peter the Great St. Petersburg Polytechnic University, St. Petersburg 195251, Russia
[3]All-Russian Research Institute of Phytopathology, Moscow Region 143050, Russia
[4]Saint Petersburg University of Management Technologies and Economics, 190109, Russia

Abstract. Lung cancer is a common type of cancer that requires early diagnosis. Computer systems by particular different image processing techniques can use for increase the speed and accuracy of lung nodule detection. CT images used in this work in order to process medical images. In this paper proposed an automatic lung nodule detection algorithm using active contour method and SVM classification method. In proposed method, at first in order to achieve better results, lung CT image pre-processing is performed. Then the lung area is segmented by thresholding method followed by some reconstruction techniques to transfer non-isolated nodules into isolated ones. In the next step the nodule candidates are determined using active contour method. Then, nodules are detected by the support vector machine (SVM) classifier using efficient 2D stochastic and 3D anatomical features. In the result, nodules are detected with an overall detection rate of 87%; the number of false positive is 7.5/scan and the location of all detected nodules are recognized correctly.

Shariaty3@gmail.com

1. Introduction

Lung cancer is the most common reason for cancer death, which leads to 26% of cancer death in the United States [1]. Small cell lung cancer (SCLC) and non-SCLC (NSCLC) are two types of lung cancer but 85% of lung cancer is related to NSCLC. Unfortunately, just 17% of the patients will survive 5 years after diagnosing [2].

Early detection of lung nodule considerably increases the survival rate, If the tumor detects while it is smaller than 3cm, the 5-year survival rate may be increased 60%-80% after surgical resection. A Computed-Tomography (CT) image of the lung consists of several slices (2-D images) and the manual examination of these slices by physicians is a time-consuming process. On the other hand, computing accurate values of parameters has the inherent error and is prone to failing to discover all data and imaging information [3-6].

With the advancement of computer systems, they can acquire the experience of radiologists to extract data from medical images. Human analysis is typically subjective and qualitative that involve carelessness. In the meantime, a comparative analysis is needed between an image with nodule and another nodule pattern, and the human observer typically provides a qualitative response. The usage and extraction of quantitative or numerical features of images definitely need exploitation computers. on condition that most analyses that conducted by humans are based on qualitative judgment; they depend on time for a specific observer or from an observer to a different observer [7]. Firstly, because of the lack of diligence or due to inadequate information and, second, due to the range of education and also the level of understanding or ability. Computers can apply a given procedure frequently in brief time and with high exactness. additionally, data of the many specialists might be completed in a machine method, therefore computers are trained within the specific field by a team of many human specialists [8-11].

Computer Aided Detection (CADe) and Computer Aided Diagnosis (CADx) systems are two types of CAD systems. CADe is a system for locating lesions in medical pictures. While the CADx system performs a diagnosis of lesions, for example, the distinction between benign and malignant tumors. The CADe and CADx systems for the diagnosis of lung cancer in recent decades have been important areas of research. The CADe system does not show the radiological profile of the tumors, and the CADx system does not detect nodules and does not have a good level of automation, as a result, these systems are not widely used in clinical settings [12]. Some CAD systems can be both CADx and CADe; in total, this paper examines CADe systems. The diagnosis of a nodule by the CAD system consists of five important steps [7]: database access, preprocessing, segmentation, analysis, and classification. The block diagram of these steps shown in Fig.1.

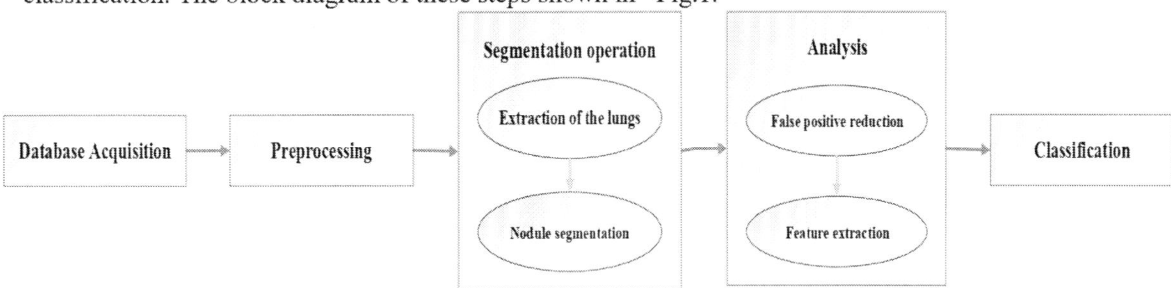

Fig. 1. Five important steps of CAD systems

2. Materials and Methods

In this paper proposed an automatic lung nodule detection algorithm using classification methods. The steps of the proposed algorithm: 1. Preprocessing of CT images, 2. Separate the lung from the surrounding area, 3. Nodule candidates detection, 4. Feature extraction from nodule candidates. 5. Use an SVM classification algorithm in order to detect nodules.

2.1. Preprocessing of CT images

In order to achieve better results in the diagnosis of the lung nodule, lung CT image pre-processing is performed. In pre-processing algorithm, the median filter used with a smoothing algorithm. These two algorithms used to remove the salt and pepper noise and pulse interference and also effectively avoid the fuzzy edges in the image noise removing process.

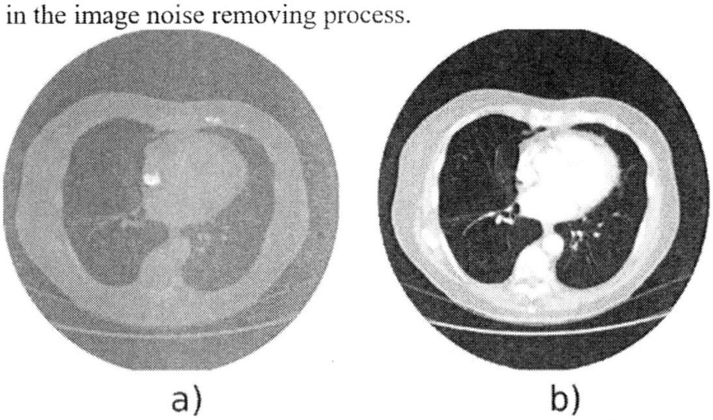

a) b)

Figure 2 (a. b). Preprocessing CT Lung Image Processing. **a)** Original image
b) Image with changes in opacity, color and gradient and applying a gradient filter to enhance image quality [13]

2.2. Separate the lung from the surrounding area

Lung extraction is one of the important steps in CAD systems because the lung is a complex organ that includes various structures such as vessels and gaps. Lung extraction also leads to speeds up operations in the system, as well as increasing the accuracy of nodule diagnosis and segmentation. A completely automatic algorithm for recognition lungs in 3D pulmonary X-ray CT images used in order to lung segmentation [14]. This method includes the following major processing steps: First the lung lesion segmented from surrounding area using optimal thresholding method and make a 3D mask (Fig. 3). Then each mask slice compared with the previous slice in order to find nodules attached to the lung wall. In presence of juxta-pleural nodules a reconstruction algorithm consists of several morphological operations has been applied to rectify the mask image (Fig. 4).

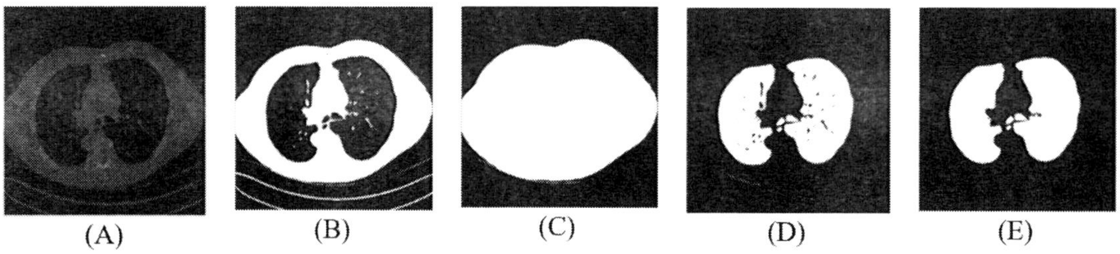

Figure 3 (a, b, c, d, e). The steps of thresholding algorithm to reach the lung mask: **a)** Input Image; **b)** Segmentation with optimal threshold; **c)** Fill the holes and select largest area; **d)** Logical "AND" between C and inverse B; **e)** Fill the holes

The advantage of this method is separation of attached nodules to the lung wall which are removed in ordinary lung segmentation methods.

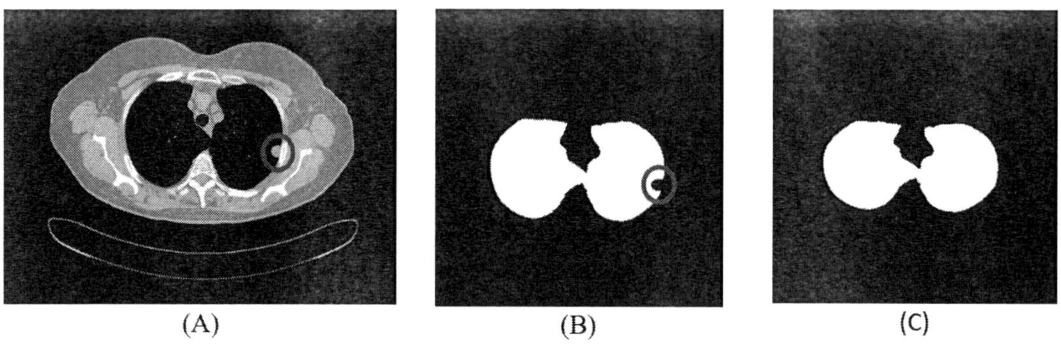

Figure 4 (a, b, c). Segmentation of the lung with juxta-pleural nodule: **a)** input image; **b)** segmented image with error; **c)** segmented image after reconstruction operation.

2.3. Nodule candidates detection

The nodule candidates segmentation is performed by active contour algorithm. In the active contour algorithm, active lines or snakes generate curves that move inside the image to find out the boundaries of the object under the influence of internal and external energies [15]. Active contour algorithm needs an input mask to start segmentation operation. In this work, we want to segment automatically the similar structures to lung nodule. For this propose, the active contour algorithm starts with circular masks which have very small radiuses. After applying these steps, the similar shapes to the lung nodule (nodule candidates) specified (Fig. 5).

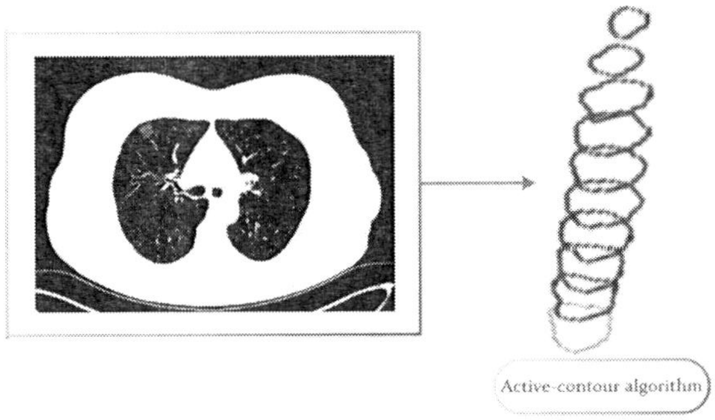

Figure 5. The segmented lung nodule in CT image with active contour method. This nodule appeared in 10 slices and Shown in three-dimensional form.

2.4. Feature extraction from nodule candidates.

The features of nodule candidates should extract in the next step, in order to classify them into two categories of nodules and non-nodules. Features which are extracted in this work divided to three sets: two-dimensional, three-dimensional and grayscale. Two-dimensional features: Area, Diameter, Perimeter, circularity. Three-dimensional features: volume, compactness, Mean breadth, x-y plane projection compactness. Grayscale features: minimum intensity value inside, mean contrast, variance intensity inside.

2.5. Use an SVM classification algorithm in order to detect nodules.

Many articles have been presented in the field of automatic nodule extraction by computer systems, which mainly use classification methods to divide nodule candidates into two different types of nodule and non-nodule. Classification algorithms typically use features, which described in the previous step, to detect nodules. The SVM classification algorithm used to classify nodule candidates [16]. SVM finds the maximum-margin hyper-plane that divides the data points. The SVM learned with train data, in which the nodules are determined by doctor. The proposed method was verified and evaluated with 20 datasets provided by LIDC (Lung Image Database Consortium).

3. Results and Conclusions

The experimental results indicate high performance of our method with high accuracy of nodule segmentation and an overall detection rate of 87% and FP value of 7.5/scan.

Using computer systems and image processing for medical CT images analysing has progressed a lot, and in general, many published literatures have the potential to use in medical practice. This method also can be used in the clinical practice to help radiologists more accurately and consistently detect and quantitatively assess lung cancers at an early state. It is desirable to use this method with the nuclear magnetic resonance method and optical methods, which are more safety for health compared to CT scans [17-20]. Since there are restrictions for a person on the number of CT scans per month.

Acknowledgment:

This work partly supported by Grant of RFBR No 18-29-25071

References

[1] Tsao A S, Scagliotti G V, Bunn Jr P A, Carbone D P, Warren G W, Bai C, ... & Adusumilli P S 2016 *Journal of Thoracic Oncology* **11(5)** 613-638

[2] Molina J R, Yang P, Cassivi S D, Schild S E and Adjei A A 2008 In Mayo Clinic Proceedings **83(5)** 584-594 Elsevier

[3] Nesbitt J C, Putnam Jr J B, Walsh G L, Roth J A and Mountain C F 1995 *Annals of thoracic surgery* **60(2)** 466-472

[4] Humphrey L L, Teutsch S and Johnson M 2004 *Annals of Internal Medicine* **140(9)** 740-753

[5] Flehinger B J, Kimmel M and Melamed M R 1992 *Chest* **101(4)** 1013-1018

[6] Zhang Y, Oikonomou A, Wong A, Haider M A and Khalvati F 2017 *Scientific reports* **7** 46349

[7] Shariaty F and Mousavi M 2019 *International journal of Informatics in Medicine Unlocked* **15(C)** 100173

[8] Firmino M, Angelo G, Morais H, Dantas M R and Valentim R 2016 *Biomedical engineering online* **15(1)** 2
[9] Sprindzuk M V, Kovalev V A, Snezhko E V and Kharuzhyk S A 2010 *Polish journal of radiology* **75(1)** 67
[10] Doi K 2005 *The British journal of radiology* **78(1)** 3-19
[11] Van Ginneken B, Schaefer-Prokop C M and Prokop M 2011 *Radiology* **261(3)** 719-732
[12] Firmino M, Angelo G., Morais H, Dantas M R and Valentim R 2016 *Biomedical engineering online* **15(1)** 2
[13] Kim H, Nakashima T, Itai Y, Maeda S, kooi Tan J and Ishikawa S 2007 *Proceedings of the 2007 International Conference on Control, Automation and Systems* p. 1274-1277
[14] F. Shariaty, S. Hosseinlou1, V. V. Davydov, V. Yu. Rud (2019). Automatic Lung Segmentation Method in Computed Tomography Scans. International Conference "Emerging Trends in Applied and Computational Physics 2019".
[15] Vese L A and Chan T F 2002 *International journal of computer vision* **50(3)** 271-293
[16] Vapnik V 1998 *Statistical Learning Theory* (John Wiley&Sons. Inc., New York)
[17] Davydov V V, Dudkin V I, Velichko E N and Karseev A Yu 2015 *Journal of Optical Technology (A Translation of Opticheckii Zhurnal)* **82(3)** 132-135
[18] Davydov R V, Atonov V I and Moroz A V 2018 *Proceedings of the 2018 IEEE International Conference on Electrical Engineering and Photonics, EExPolytech 2018* (Saint-Petersburg) **8564378** p. 236-239
[19] Myazin N S, Davydov V V, Yushkova V V and Rud V Yu 2018 *Journal of Physics: Conference Series* **1038 (1)** 012088
[20] Davydov V V, Velichko E N, Myazin N S and Rud' V Yu 2018 *Instruments and Experimental Techniques* **61** 116–122

Study of human skin radiation in the terahertz frequency range

K. Baksheeva[1], A. Vdovydchenko[2], K. Gorshkov[1], R. Ozhegov[1,3], N. Kinev[4], V. Koshelets[4], G. Goltsman[1,3]
[1]Moscow State Pedagogical University, Moscow, 119992, Russia
[2]Moscow Institute of Physics and Technology, Dolgoprudny, 115184, Russia
[3]National Research University Higher School of Economics, Moscow, 101000 ,Russia
[4]Kotel'nikov Institute of Radio Engineering and Electronics, Moscow 125009, Russia

Abstract. The radiation of human skin in the terahertz frequency range under the influence of mental stresses has been studied in the current work. An experimental setup for observation of changes in human skin radiation, which occur under the influence of psychological stresses, by means of a superconducting integrated receiver has been developed. More than 30 volunteers participate in these studies, which allows us to verify presence of correlation between the signals from the superconducting integrated terahertz receiver and other sensors that monitor human mental stress.

1. Introduction

In 2007, the image of the sweat gland was obtained for the first time using optical coherence tomography [1]. In Figure 1 it can be seen that the upper part of the sweat duct has a spiral structure. Later, in 2008, based on the geometric dimensions and the fact that the ducts are filled with a conducting liquid, it was suggested that the sweat ducts can work as low-Q terahertz helical antennas [2].

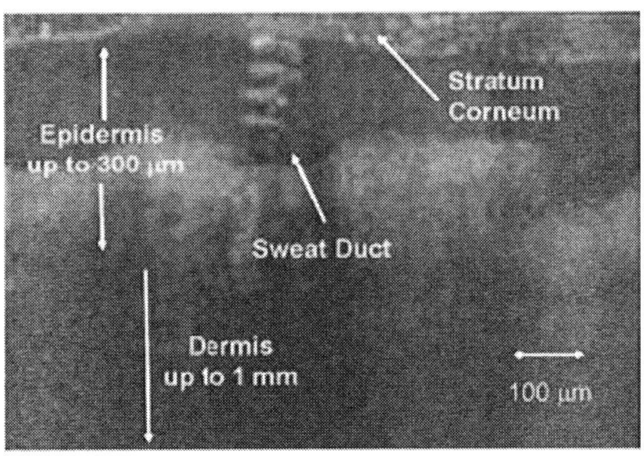

Figure 1. An image of the top layer of the skin obtained using optical coherence tomography.

This hypothesis is developed by a group of scientists led by J. Feldman, who works at the University of Jerusalem [3,4]. In their studies, they measure the characteristics of the duct antenna by removing the terahertz signal reflected from the skin of the palm. However, if one assumes that the sweat duct acts like an antenna and the human skin is an absolutely black body, then such antennas can affect the power radiated from the skin as a function of wavelength. In this case, these antennas can be studied by measuring the self-radiation of human skin, which is the main goal of our work. The sweat glands are directly controlled by the sympathetic nervous system (SNS). The change in the work of the SNS occurs under the influence of both intense physical exertion, and under the influence of mental activity, as well as from stress and the emotional state of a person. In this case, stress and mental activity can serve as a modulating factor for the antenna, which is the sweat duct. We can change the level of human stress and study the radiation of such antenna.

2. Experiment

To solve this problem, our group has developed an experimental setup, which enables us to probe the radiation of human skin under the condition of mental stresses. A superconducting integrated receiver (SIR), which is under the development at IRE RAS [5, 6], has been chosen as the main receiver element of our setup. SIR is a unique receiver that combines a sensitive terahertz frequency mixer and a highly stable terahertz oscillator equipped with a phase-locked loop and allows recording brightness temperature changes of 10 mK. With the help of this receiver, the brightness temperature of human skin was recorded at a frequency of 550 GHz.

A galvanic skin response sensor (GSR - Galvanic Skin Response) was used to detect stress of human beings. This sensor records the conductivity of human skin. It allows us to track changes in the sympathetic nervous system of the person under stress. A thermometer and a photoplethysmograph were also used to monitor the pulse and the blood oxygen level. These sensors were attached to the person's arm, from which the radiation was detected. The hand itself during the experiment was fixed with the holder.

A test report was developed and prepared. This protocol can be divided into five parts. The first part - a period of relaxation, lasts for five minutes. At this time, views of nature are shown on the monitor in front of the subject, and relaxing melodies are fed into the headphones that are worn on him. The second part associates with to the period of mental stress at the subject. During this period, different types of mental stress alternate: the Stroop test and the figure test. In the Stroop test, a person is asked to correlate the color of a word, which means some other color, with one of the suggested options. So for example, if the word "red" is written in green, a person must find among the listed options "green" and select it. In the test with the figures, the subject is shown various shapes that follow each other, and the person must choose whether this image corresponds to what was previously. All this, as well as in the Stroop test, should be done in a limited time, accompanied by intense music. This stage also lasts for five minutes, one minute is allotted for each task. Then a five-minute period of relaxation takes place. The fourth stage is physical activity, during which the subject must compress the ball. This stage lasts for two minutes. And the test report also ends with a relaxation period that lasts for one minute. The entire protocol of measurements on one person takes eighteen minutes. Since the procedure was automated, the human factor during the tests was minimized.

3. Discussion

Currently, studies have been taken on 32 volunteers aged 18-35 years of both genders. According to the data obtained, correlations of the signals from the superconducting integrated receiver and the GSR sensor were calculated. Figure 2 (a, b) presents an example of the dependence of the normalized readings of the SIR and GSR signals on time and a graph of the obtained correlation of these signals. As can be seen from Figure 2 (a), the graph of GSR versus time (the blue line on the graph) decreases during the stages of relaxation, while under mental and physical loads, the graph increases dramatically, indicating a change in the state of the subject. Also, the graph of the dependence of SIR on time has its' peaks. These data differ from the readings taken with a thermometer, which indicates the presence of some additional signal component.

The primary results of the correlations were inconsistent. The results obtained during data processing showed that the correlation of signals is inhomogeneous both in time and from person to person. For all subjects, the correlation of the signals had a different shape and meaning, for some people the correlation was positive, for the others it was negative, while for third group it was equal to zero.

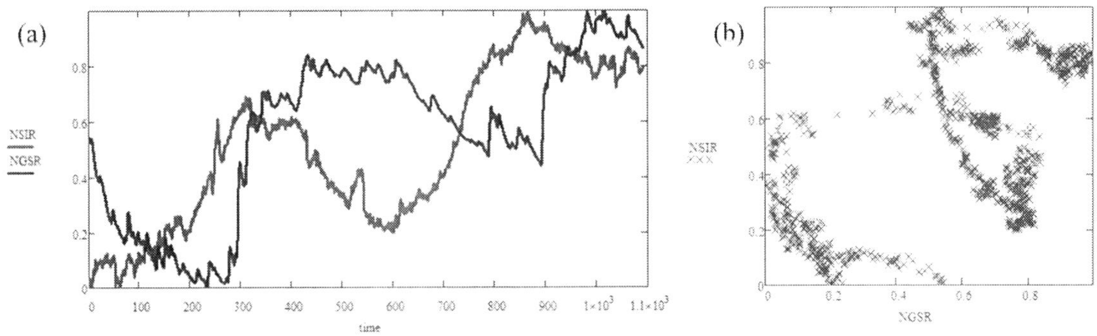

Figure 2(a, b). (a) Readings from the GSR and the integral receiver as the function of time; **(b)** Correlation between the readings from the integrated receiver and the GSR sensor.

From Figure 2 (a, b) it is clear that the correlation of signals is low and corresponds to 0.461. This may be caused by the difference in the relaxation processes of the galvanic response of the skin, which were described in [7], and its terahertz radiation. This fact leads to the necessity in further data processing. A matrix of correlations of signals GSR and SIR was constructed and considered, with additional processing of SIR data. Namely, it was considered how the correlation of the signals of these two receivers will change, after compression-defragmentation and a time shift analysis of the SIR data. The resulting correlation matrix is presented in Figure 3. The vertical axis represents the compression-tension of the raw SIR data over time. In this case, 0 corresponds to data stretching by approximately 30%, and 30 corresponds to data compression by approximately 30%. The horizontal axis represents the time delay of the received SIR data relative to the GSR data. In the resulting picture, the red areas correspond to the positive correlations, the blue ones to the negative, and the green ones to the absence of correlation.

The correlations obtained at very large time shifts were discarded from the consideration, since in such cases the data sets is too short for the correct correlation analysis. Thus, we considered the correlations of the signals which delay didn't exceed half of the total test time (400 seconds). In the figure 3, the region of maximum obtained correlation of signals with additional data processing is highlighted. When plotting the dependencies of the GSR and SIP data, taking into account the additional processing, which corresponds to the maximum correlation (Figure 4 (a)), we can see that the shapes of the data sets are very similar.

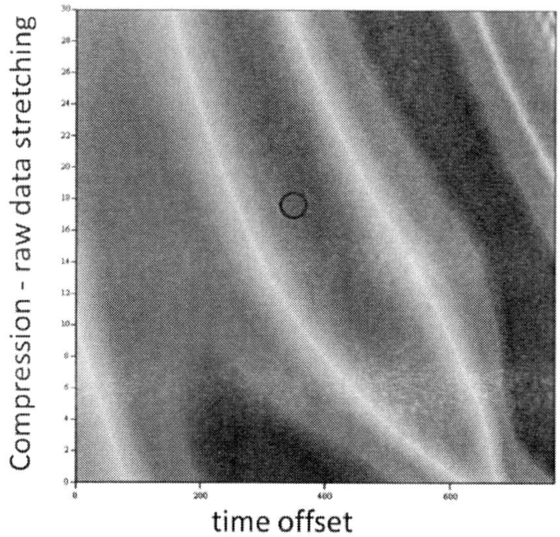

Figure 3. The correlation matrix of GSR and CIP signals, with the additional data processing

3. Results

The correlation achieved with additional data processing was 0.968. (Figure 4 (a, b)) Such correlation level was achieved when the CIP graph was shifted by 375 s. Such long signal delay may be caused by the difference in the mechanisms of the occurrence of signals recorded by the sensors. The difference in the response mechanisms leads to the difference in the response and relaxation time as the reaction to stress. These mechanisms are not well understood and require additional research.

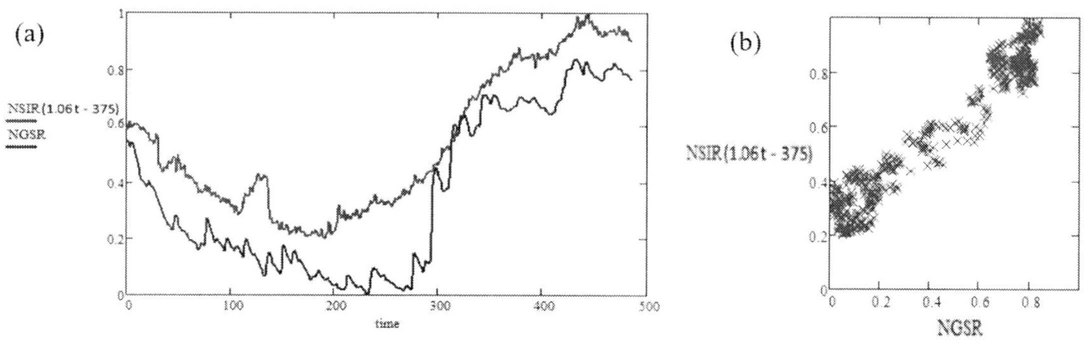

Figure 4(a, b). **(a)** Readings from the GSR and the integrated receiver as the function of time after additional data processing; **(b)** Correlation between the readings from the integrated receiver and the GSR after additional data processing.

Acknowledgments

These studies are carried out with the financial support of the Russian Foundation for Basic Research (grant # 17-29-02354), K. Baksheeva thanks the support of the grant of the Russian Science Foundation (grant # 17-72-30036).

References

[1] Lademann J., Otberg N., Richter H., Meyer L., Audring H., Teichmann A., Thomas S., Knüttel A., Sterry W. 2007 *Skin Res. And Tech.* **13**.
[2] Feldman Yu., Puzenko A., Ben Ishai P., Caduff A., A. J. Agranat A. J. 2008 *Phys. Rev. Lett.* **100**
[3] Safrai E., Ishai P.B., Caduff A., Puzenko A., Polsman A., Agranat A.J., Feldman Y. 2012 *Bioelectromagnetics* **33** 5
[4] Betzalel N., Feldman Y., Ishai P.B. 2017 *IEEE Transactions on Terahertz Sci. and Technol.* **7** 5
[5] Koshelets V.P., Shitov S.V., Filippenko L.V., Baryshev A.M., Golstein H., De Graauw T., Luinge W., Schaeffer H., Van De Stadt, H. 1996 *Applied Phys. Lett* **68**
[6] Koshelets V.P., Shitov S.V. 2000 *Superconductor Sci. and Technol* **13**.
[7].J. Bakker, M.Pechenizkiy, N. Sidorova 2011 *11th IEEE International Conference on Data Mining Workshops*

Semiconductor resonant all-optical temperature sensor and thermal release trigger of encapsulated anti-cancer drugs for *in vitro* studies

G P Zograf[1], M I Petrova[1], M I Petrov[1], P A Belov[1], F E Komissarenko[1], E K Makarova[1], A P Pushkarev[1], Y Sun[1], P Ghosh[2], Q Li[2], M Qiu[2], S V Makarov[1], M V Zyuzin[1]

[1]Department of Physics and Engineering, ITMO University, Saint-Petersburg, 197101, Russia

[2]State Key Laboratory of Modern Optical Instrumentation, College of Optical Science and Engineering, Zhejiang University, Hangzhou, 310027, China

Abstract. Nanostructures made of high-index semiconductors ($n > 2.5$) with the several orders of magnitude lower losses than noble metals possess (Au, Ag, Pt etc.) are gaining high interest in recent years for various advanced optical applications, including nanothermometry and optical heating. Owing to the high-Q of Mie-optical modes that are being supported by the nanostructures, such nanoresonators can be considered as a powerful tool for light absorption at the nanoscale. However, high potential of resonant semiconductor nanoparticles in drug release and delivery applications has not been studied yet. In this work, we prove both experimentally and theoretically that optically resonant iron-oxide (III) (Fe_2O_3) can be employed as a remote optical trigger for thermal rupture of polymer microcontainers used as a drug delivery platform. Such approach allows one to measure real-time temperature via thermally sensitive Raman signal of iron-oxide nanoparticles, that were embedded into the walls of microcontainers. We believe that, suggested system consisting of resonant semiconductor nanoparticles and polymer carrier capsules can serve as a highly efficient drug delivery platform with controllable remote release with temperature feedback.

1. Introduction

Recent studies demonstrated considerable interest in plasmonic nanoparticles as nanoscale heat sources for different applications, including drug delivery. However, there are limitations with their applicability governed by either limited resonant performance or lack of direct thermometry at the nanoscale, suitable for real-time applications. Recently, it was discovered that non-plasmonic (dielectric and semiconductor) resonant nanoparticles are highly efficient light-to-heat energy converters in the visible and near-IR [1], making them attractive for numbers of bio-applications, including real-time tracing of protein unfolding process [2].

In this study, we reveal theoretically and prove experimentally novel approach concept for non-plasmonic effective optical heating and thermometry with resonant nanoparticles embedded into polymer drug carrier microcapsules walls for remote thermal rupture and subsequent release of the antitumor drug [3]. High-Q optical Mie-resonances, which occur upon laser illumination of the nanoresonators, are crucial not only for the enhanced absorption, but as well as for the enhanced thermally sensitive Raman signal [4], thus reducing the temperature measurement time.

2. Results and discussion.

We start our consideration from steady-state heat diffusion equation of a single spherical nanoparticle in homogeneous media illuminated by electromagnetic plane-wave. We believe that such approach is reasonable for optical heating of resonant nanoparticles in aqueous media (i.e. water, liquids for *in vitro* studies).

The power absorbed by the nanoparticle of radius a can be expressed in terms of the absorption cross section (C_{abs}) [5] and the intensity of the incident radiation (I) [6]:

$$Q = C_{abs} I \quad (1)$$

To determine the temperature distribution inside the particle $T_1(\mathbf{r})$ and outside $T_2(\mathbf{r})$, it is necessary to solve the heat diffusion equations:

$$\begin{gathered} \nabla\big(\kappa_1(\mathbf{r})\nabla T_1(\mathbf{r})\big) = -q(\mathbf{r}) \\ \nabla\big(\kappa_2(\mathbf{r})\nabla T_2(\mathbf{r})\big) = 0\,, \end{gathered} \quad (2)$$

where κ_1 and κ_2 are the thermal conductivities of the particle and the environment, respectively, q is the power density (in this case it is equal to:$q = \frac{Q}{V} = \frac{3Q}{4\pi a^3}$). Assuming the thermal conductivities κ_1 and κ_2 are homogeneous in space, and taking into account the boundary conditions, the solutions of the equations are written in the form:

$$\begin{gathered} T_1 = \frac{Q}{8\pi\kappa_1 a}\left(1 - \frac{r^2}{a^2}\right) + \frac{Q}{4\kappa_2 a} + T_0 \\ T_2 = \frac{Q}{4\pi\kappa_2 a} + T_0, \end{gathered} \quad (3)$$

where T_0 is the medium temperature before excitation of the nanoparticle. Assuming that the thermal conductivity of the material of the particle is much larger than the thermal conductivity of the environment ($\kappa_1 \gg \kappa_2$), the temperature change of the nanoparticle will be:

$$\Delta T = \frac{C_{abs} I}{4\pi\kappa_2 a}. \quad (4)$$

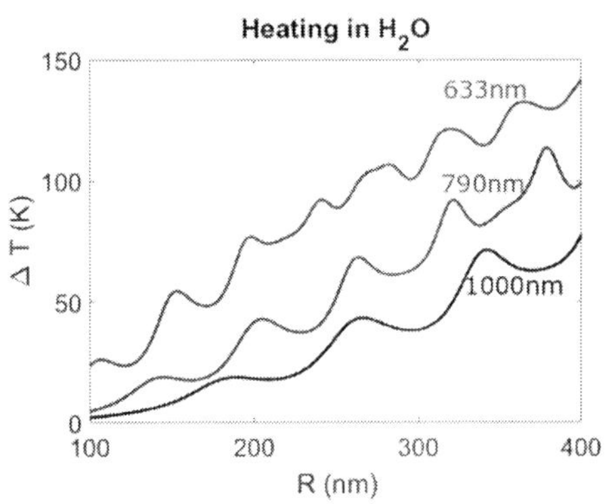

Figure 1. Optical heating of spherical iron-oxide nanoparticle in aqueous homogeneous media for different nanoparticles radii and incident wavelengths in visible and near-IR region. Incident wavelength was of 633 nm, 790 nm and 1000 nm. Considered intensity is 1 mW/μm^2. Calculation is done using eq.4.

Figure 1 depicts analytical calculation for optical heating of spherical semiconductor iron oxide (III) nanoparticles in homogeneous aqueous media using eq.4. The results demonstrate worsening of optical heating efficiency upon red-shifting of incident wavelength to the near-IR. However, optical heating remains sufficient even in the region of 1000 nm, thus making these nanoparticles promising for bio-medical application, since it corresponds to biological tissue transparency window. Moreover, spherical semiconductor nanoparticles with thermally sensitive Raman scattering already demonstrated the ability not only in light-to-heat conversion at nanoscale but also serving as nanothermometer [1].

On the other hand, polymer microcarriers attracted a lot of interest among scientist due to ability for remote targeted release of medical drugs [7]. In this work, we propose combining the ability for optical heating and thermometry of semiconductor nanoparticle and the carrier systems based on polyelectrolyte microcapsules. Thus, making drug delivery systems with remote thermally induced all-optical release trigger.

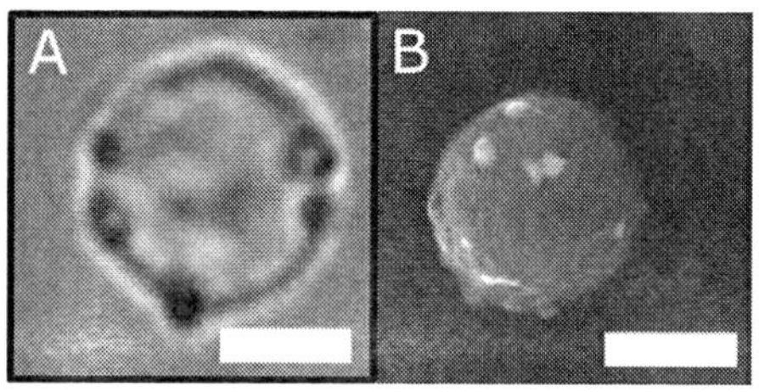

Figure 2. (A) Bright-field microscopy image of the fabricated microcapsule with iron oxide nanoparticles embedded into walls. Scale bar is 2μm. (B) Scanning electron microscopy (SEM) image of the fabricated polymer microcapsule with embedded into walls iron oxide nanoparticles. Scale bar is 2μm.

Figure 2 demonstrates images of fabricated microcapsules with iron-oxide nanoparticles embedded into walls carried out by means of optical elastic bright-field microscopy (Figure 2(A)) and scanning electron microscopy (Figure 2(B)). The nanoparticles under laser illumination serve both as temperature sensor and nanoheater for thermal rupture of the capsules wall.

Therefore, we believe that suggested approach of combining semiconductor resonant nanoparticles with active thermally sensitive Raman response and polyelectrolyte microcarriers of medical drugs can serve as an effective platform for number of bio-medical applications.

Acknowledgments
This work was financially supported by the Ministry of Science and Higher Education of the Russian Federation (Project №14.584.21.0024 with unique identificator RFMEFI58417X0024).

References

[1] Zograf GP et al. 2017 *Nano Letters* **17**(5) 2945-2952
[2] Milichko VA, Zuev DA, Baranov DG, Zograf GP et al. 2018 *Laser & Photonics Reviews* **12**(1) 1700227
[3] Zograf GP et al. 2019 *Laser & Photonics Reviews* (under review)
[4] Dmitriev PA et al. 2016 *Nanoscale* **8**(9) 5043-5048

[5] Bohren CF and Huffman DR 2008 *Absorption and scattering of light by small particles* (New York - John Wiley & Sons)
[6] Baffou G and Quidant R 2013 *Laser & Photonics Reviews* **7**(2) 171-187
[7] Zyuzin MV et al. 2017 *Bioconjugate chemistry 28*(2) 556-564

Development of a measuring device for the study of thermal processes during the polymerase chain reaction

V S Reznik[1], V A Kruglov[1], A I Petrov[2], V V Davydov[1], A P Glinuchkin[3] and V Yu Rud'[3]
[1]Higher School of applied physics and space technologies, Peter the Great St. Petersburg Polytechnic University, St. Petersburg 195251, Russia
[2]Institute for Analytical Instrumentation of RAS, St. Petersburg 190103, Russia
[3]All Russian Research Institute of Phytopathology, Moscow Region 143050, Russia

Abstract. In this paper, we consider the problem of studying transient temperature processes in thermal blocks of analyzers of nucleic acids based on real-time polymerase chain reaction, in particular, the ANK-32 device. Characteristics of transient temperature processes were obtained using sensors built into the ANK-32. The result of the work is an original measuring device that is able to clarify the characteristics is obtained.

1. Introduction

Currently, devices based on the polymerase chain reaction (PCR) have been actively distributed in medicine, science, health care and forensic science. This is due to the fact that studies carried out with their help, in contrast to NMR spectrometers and other various optical meters and sensors, provide the necessary information in a shorter time [1-10]. In addition, the use of PCR allows to obtain a series of data that cannot be obtained by other methods [8-14].

Polymerase chain reaction is a method based on multiple selective copying of the necessary for the study of the DNA segment under the influence of various enzymes in artificial conditions. In figure 1 shows a diagram of the formation of the necessary DNA for research. On each subsequent amplification cycle, previously synthesized DNA fragments are again copied using DNA polymerases. As a result, the number of specific DNA fragments increases many times, which in turn greatly simplifies the subsequent analysis.

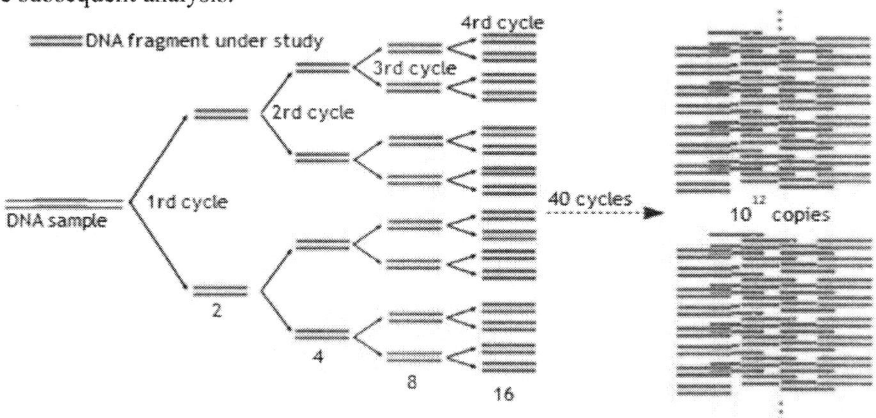

Figure 1. Polymerase chain reaction scheme.

The formation process consists in the fact that on each subsequent amplification cycle, the previously synthesized DNA fragments are again copied using DNA polymerases. As a result, the number of specific DNA fragments increases many times, which in turn greatly simplifies the subsequent analysis. And also increases its accuracy and reliability compared with other optical methods [5, 7, 9-16].

For carrying out the polymerase chain reaction, the reaction mixture is used, consisting of the following components: primers, Taq - to decipher DNA polymerases, mixtures of deoxynucleotide triphosphates (dNTPs), buffer, and also the sample under study. The reaction mixture passes several phases of heating and cooling sequentially during PCR. As a result, when carrying out PCR, it becomes necessary to have an accurate idea of the change in temperature of the reaction mixture placed in a test tube [17]. The data on the value of T is necessary for the subsequent identification of DNA chains (comparing them with test samples). In our work is considered one of the solution to this problem.

2. The design of measuring device and its principles of operation

The temperature sensors developed at the present time either do not satisfy the accuracy of measuring T, or cannot be placed in a test tube to the sample under study. In devices, which are in operation for implementation of PCR, it is possible to obtain data on the temperature of the amplifier (precision thermostat) [18, 19], using the built-in temperature sensor. However, this temperature characterizes the state of the outer wall of the tube, which, in turn, does not allow to draw accurate conclusions about the temperature of the reaction mixture inside the tube. In figure 2 shows the dependence of temperature change measured by these sensors.

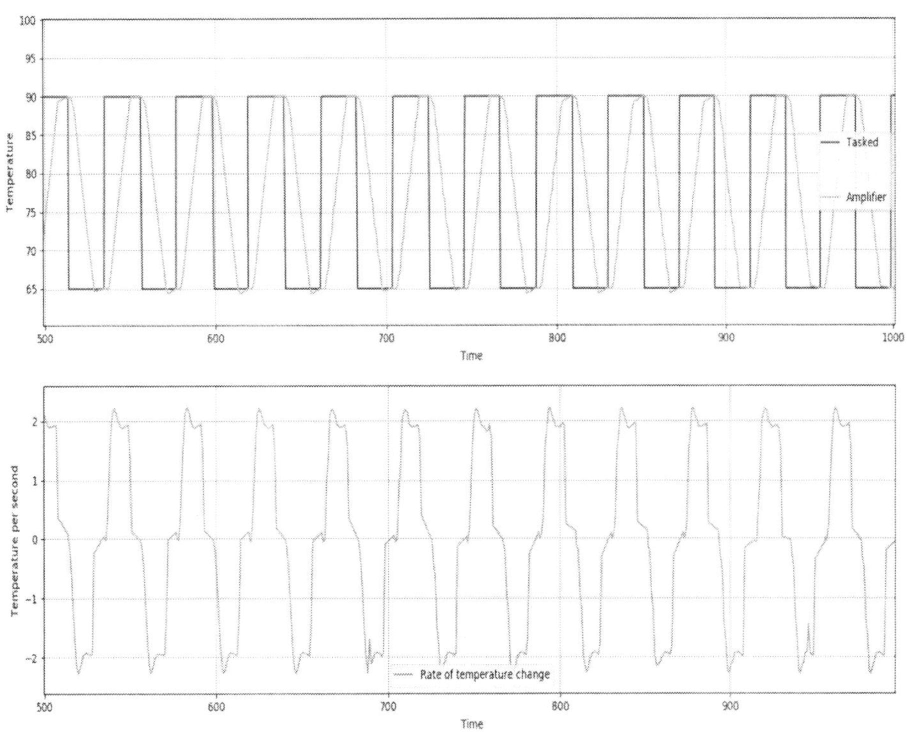

Figure 2. Data received from the device ANK-32.

The analysis presented in figure 2 dependences show that it is necessary in some cases to specify the results using this measuring device in order to reproduce the temperature measurements of the reaction mixture itself. This creates additional difficulties. Therefore, the development of a measuring device for measuring the temperature of the reaction mixture itself is very relevant.

As a result of the research, the parameters of the developed measuring device were determined. The size of the temperature sensor should not exceed the dimensions of the test tube with the reaction mixture, which are 21 mm in height and 6 mm in diameter, temperature measurement should be implemented with an accuracy of 0.01 K, the time resolution should be at least 0.01 s.

Most modern measuring devices are implemented as ready-made printed circuit boards using various microchips. This fact dictates the need to implement the developed measuring device in this form.

In this case, it is more expedient to use a thermistor as a temperature sensor, due to its small size, a significantly higher signal than that of a thermocouple, as well as a higher accuracy of measurement.

The choice of a thermistor as a temperature sensor determines the principle of construction of the concept of the measuring device, namely the presence of analog and digital parts. The main problem in measuring the temperature in the sensors on the thermoelectric effect is the compensation of the parasitic resistance of the connecting wires. Especially when measuring T with high accuracy. In figure 3 shows the connection diagram of the thermistor.

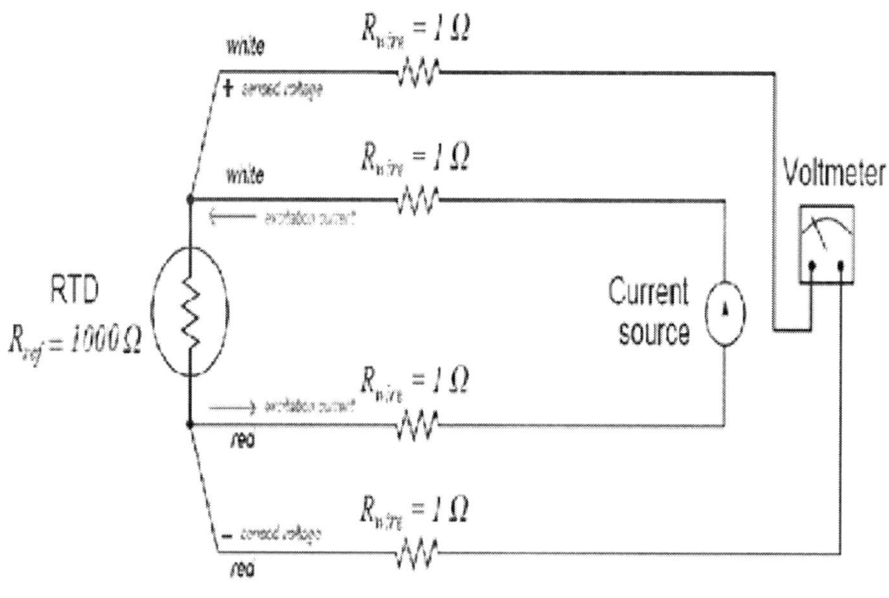

Figure 3. Thermistor connection diagram using the Kelvin method.

In this scheme (fig. 3), we proposed a new solution to the problem of measuring the current flowing through a thermistor using a resistive voltage divider using a resistor with a very low value of the temperature coefficient of resistance (<50 ppm). The voltage applied to the divider is distributed between the resistor with a low temperature coefficient of resistance and the thermistor is proportional to the values of their resistances. Since the voltage divider is used, the current on both resistors will be the same, which will allow determining the resistance of the thermistor at the time of measurement with high accuracy. This made it possible to keep the dependence of resistance RTD on the ambient temperature of the thermistor close to linear. In figure 4 this dependence for used sensor is presented.

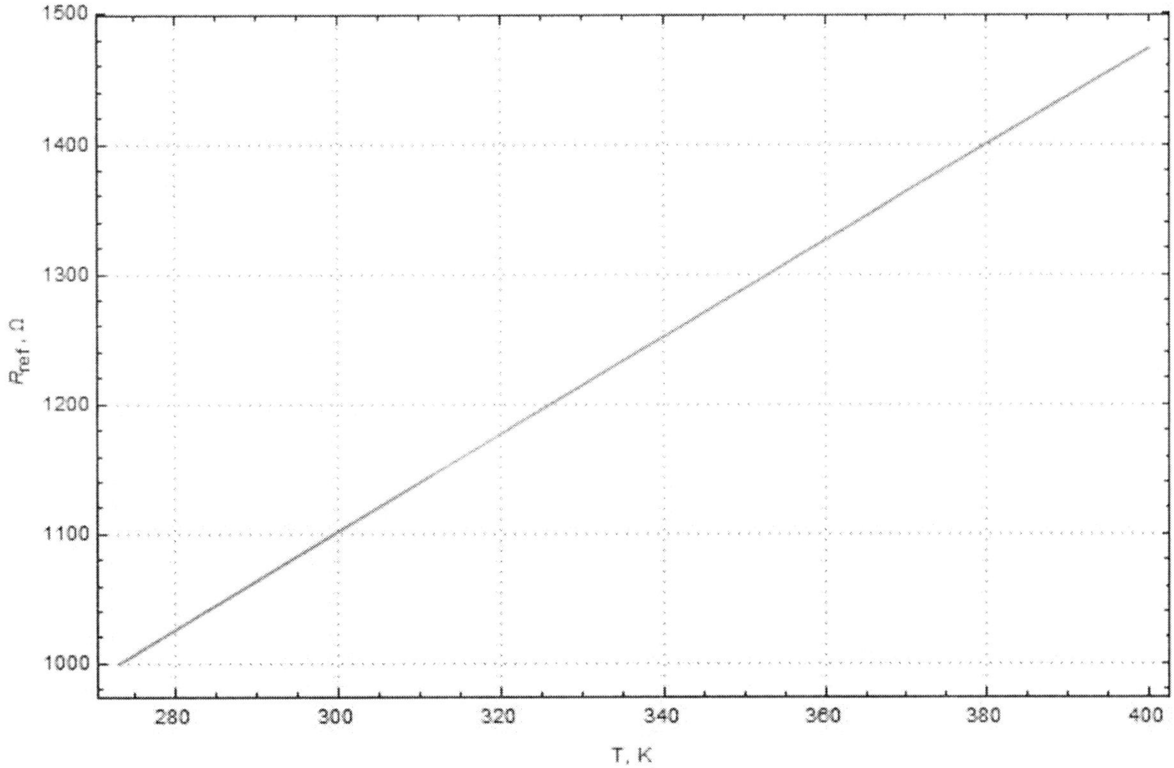

Figure 4. The dependency R_{ref} from temperature T of thermal sensor.

This allows you to use the following dependence for measurements:

$$R_{ref} = R_0(1 + A(T - 273{,}15) + B(T - 273{,}15)^2) \tag{1}$$

Where, R_{ref} is the resistance value of the thermistor in ohms at a certain value of temperature, measured in degrees Kelvin; R_0 is the resistance value of the thermistor at ambient temperature components of 0 K; A and B constants of value:

$$A = 3{,}81 \times 10^{-3}\ \mathrm{K}^{-1} \tag{2}$$

$$B = -6{,}02 \times 10^{-7}\ \mathrm{K}^{-2} \tag{3}$$

The use of an analog signal implies its further conversion to a digital code. As a result, the need arises to use an analog-to-digital converter (ADC).

The need for accurate temperature measurements implies the use of low-noise and highly stable elements. As a consequence, the sigma-delta ADC was chosen - AD7714. The presence of two measuring channels in it allows to realize the input circuit of the measuring device in the form of a resistive voltage divider.

Temperature measurements are carried out as follows: the first channel is used to measure the voltage across a resistor with a constant resistance value, which in turn allows determining the current flowing in the circuit, and the second channel is necessary to measure the voltage drop across the thermistor. As a result, this allows to obtain the exact value of the resistance of the thermistor at the time of the measurement. Next, the temperature of the reactive mixture is determined by converting the obtained resistance value in accordance with the technical documentation of the thermistor.

To visualize the received data, it is necessary to implement on a printed circuit board a device that would be able to communicate with a computer. A microcontroller of the ARM STM32F303VCT6 family was chosen as such a device. The choice of this particular microcontroller is explained by its high speed, 32-bit resolution, as well as the presence of the necessary interfaces for working with peripheral devices.

Data transfer from the microcontroller is implemented through the UART, through the CP2102 chip, which was chosen due to its high stability, as well as low current consumption.

The principle of operation of the selected microcontroller is based on interruptions of the system timer called every 10 milliseconds. During each of the interruptions, the values of the voltages on the resistive voltage divider are digitized, their further mathematical processing and sending the obtained temperature value of the reaction mixture in text form.

Next, the information goes to a data processor written in Python 3 versions, which visualizes temperature characteristics using the serial, matplotlib, and drawnow libraries. In figure 5 shows the appearance of the measuring device developed by us.

Figure 5. Prototype of measuring device.

3. Conclusion

The developed measuring device will allow to obtain the temperature characteristics of the transients of the reaction mixture in a test tube with a given accuracy, which will optimize the performance of devices based on PCR. The temperature measurement error is no more than 0.01 K. The measurement time for one temperature measurement is no more than 0.01 s.

Acknowledgments

The work was performed in the framework of the State Task No. 075-00780-19-00 on the topic No. 0074-2019-0013 of the Ministry of Science and Higher Education of the Russian Federation.

References

[1] Myazin N S, Davydov V V, Yushkova V V, Davydova T I and Rud' V Yu 2017 *Journal of Physics: Conference Series* **917(4)** 042017

[2] Davydov V V, Dudkin V I, Petrov A A and Myazin N S 2016 *Technical Physics Letters* **42** 692–696

[3] Davydov V V 1999 *Russian Physics Journal* **42(9)** 822-825

[4] Myazin N S, Davydov V V, Yushkova V V and Rud V Yu 2018 *Journal of Physics: Conference Series* **1038 (1)** 012088
[5] Rykin E V, Moroz A V, Smirnov K J, Davydov V V and Yushkova V V 2018 *MATEC Web of Conference* **245** 12002
[6] Davydov V V, Myazin N S and Davydova T I 2017 *Russian Journal of Nondestructive Testing* **53(7)** 520-529
[7] Davydov R V, Antonov V I and Moroz A V 2018 *Proceedings of the 2018 IEEE International Conference on Electrical Engineering and Photonics, EExPolytech 2018 (Saint-Petersburg)* **8564378** p. 236-239
[8] Davydov V V, Myazin N S and Velichko E N 2017 *Technical Physics Letters* **43** 607-610
[9] Nepomnyashchaya E K, Velichko E N and Aksenov E T 2016 *Journal of Physics: Conference Series* **769(4)** 012025
[10] Nepomnyashchaya E K, Cheremiskina A V, Velichko E N, Aksenov E T and Bogomaz T A 2015 *Journal of Physics: Conference Series* **643(1)** 012018
[11] Grebenikova N M, Smirnov K J, Artemiev V V, Davydov V V and Kruzhalov S V 2018 *Journal of Physics: Conference Series* **1038 (1)** 012089
[12] Grebenikova N M, Smirnov K J, Davydov V V , Rud V Yu and Artemiev V V 2018 *Journal of Physics: Conference Series* **1135 (1)** 012055
[13] Grebenikova N M, Smirnov K J, Artemiev V V, Davydov V V and Kruzhalov S V 2018 *Journal of Physics: Conference Series* **1038 (1)** 012089
[14] Grebenikova N M, Myazin N S, Rud V Y and Davydov R V 2018 *Proceedings of the 2018 IEEE International Conference on Electrical Engineering and Photonics, EExPolytech 2018* (Saint-Petersburg) **8564409** p. 295-297
[15] Davydov V V, Kruzhalov S V, Grebenikova N M and Smirnov K J 2018 *Measurement Techniques* **61** 365-372
[16] Myazin N S, Smirnov K J, Davydov V V and Logunov S E 2017 *Journal of Physics: Conference Series* **929(1)** 012080
[17] Belov Yu V, Petrov A I, Lavrov V V and Kurochkin V E 2011 *Scientific instrument engineering* **21(1)** 44-49
[18] Alekseev J I, Belov Yu V, Varlamov D A, Konovalov C V, Kurochkin V E, Marakushin N F, Petrov A I, Petryakov A O, Rumyanzev D A, Skoloblilov S Yu, Sokolov V N, Fesenko V N and Chernushov A V 2006 *Scientific instrument engineering* **16(3)** 132-136
[19] Antonova O S, Rudnizkay G E, Tupik A N, Bylianiza A L, Evstrapov A A and Kurochkin V E 2011 *Scientific instrument engineering* **21(4)** 5-21

Fluorescent detection upon droplet amplification of nucleic acids

A Tupik[1], A Bulyanitsa[1], G Rudnitskaya[1], Y Posmitnaya[1], D Varlamov[2], A Evstrapov[1]
[1]Laboratory of Bio-& Chemosensors microsystems, Institute for Analytical instrumentation RAS, Saint Petersburg 190103, Russia
[2]All-Russia Research Institute of Agricultural Biotechnology, Moscow, 127500, Russia

Abstract. Droplet microfluidics with amplification of nucleic acids allows to achieve sensitivity at the level of single molecules and is in demand in medical diagnostics, forensic science, and genetic research. The results of detection of nucleic acid amplification in liquid and water-in-oil emulsion are compared on the example of model solutions. Since the results of quantitative analysis are based on the calculation of the proportion of drops with a positive reaction result, one of the tasks was to study the features of automatic counting of drops in different zones of the reaction chamber.

1. Introduction

Droplet microfluidics allows manipulation of analyte droplets in an inert oil medium [1-2]. Special microfluidic devices with a droplet generator are used to "break" the sample into hundreds (thousands) of drops. At the same time, each drop can be effectively used for carrying out reactions, both at the molecular level, and in the study of cells or microorganisms. The advantages of the "droplet microfluidics" are: low consumption of reagents and samples, high droplet formation rates (with generation frequencies up to kHz), the ability to conduct high-performance studies in an array of identical droplets. In addition, it is possible to use both real-time polymerase chain reaction (PCR) techniques for high analyte concentrations and principles of digital PCR at low one. However, reducing the reaction volume to a droplet reduces the signal-to-noise ratio. The drop detection rate is limited when scanning with high resolution over a large area (array) of droplets.

The aim of the work was to compare the results of the analysis with amplification of nucleic acids in a liquid and in an emulsion using the example of model solutions. For this purpose, the oligonucleotide probe with fluorescent dye (without quencher) was used. Also of interest is the assessment of the legitimacy of selective analysis of digital PCR in emulsion from fragments of the image of the reaction chamber.

2. Experimental technique

Microfluidic chips for water-in-oil emulsion generation were made of polydimethylsiloxane by soft lithography [3]. Emulsions were generated by flow focusing method. Mineral oil with the addition of ABIL EM 180 (Evonik, Germany) surfactant was used as the continuous phase (of transport flow) for droplet generation. The flow rate of the continuous phase was 2.4 μl/min, and the discrete one was 0.6 μl/min. A solution of oligonucleotide probe (27 bp long) was used as amplified DNA fragments

obtained as a result of PCR. These probes were coupled to the FAM fluorescent dye and did not have a quencher.
The final concentration of amplicons after successful PCR is determined by the initial concentrations of the primers, which are the main components of the amplification reaction. The selected range of oligonucleotide probe concentrations (200 – 800 nM) corresponded to the real concentration of primers for amplification in droplets [4].
Emulsions were injected into a special reaction chamber, in which the droplets were distributed in one layer. For comparison, similar reaction chambers were filled with appropriate probe solutions and measured under the same conditions. To simulate the results of digital amplification, droplets with a fluorescent probe were mixed with non-fluorescent droplets from a buffer solution in a ratio (1 : 9).
A prototype fluorescent scanner with an excitation wavelength of 473 nm, was used for detection. The scanner was produced by Institute for Analytical instrumentation RAS. The result of the scan was a set of individual images of (460 × 350) μm in size that could be analyzed together or separately. For image processing was used special software.

3. Results and discussion

Figure 1 shows images of emulsions obtained from solutions with different concentrations of a fluorescent probe, as well as from a buffer solution. The average diameter of the resulting droplets for all images was 16 μm, which corresponds to a volume of ~ 2 picoliters. Observed large-diameter droplets make up less than 10% of the total number of droplets in the reaction chamber. Apparently, when the emulsion was injected into the reaction chamber by laboratory dispenser (pipette), some drops were combined under the action of external pressure, forming larger droplets.

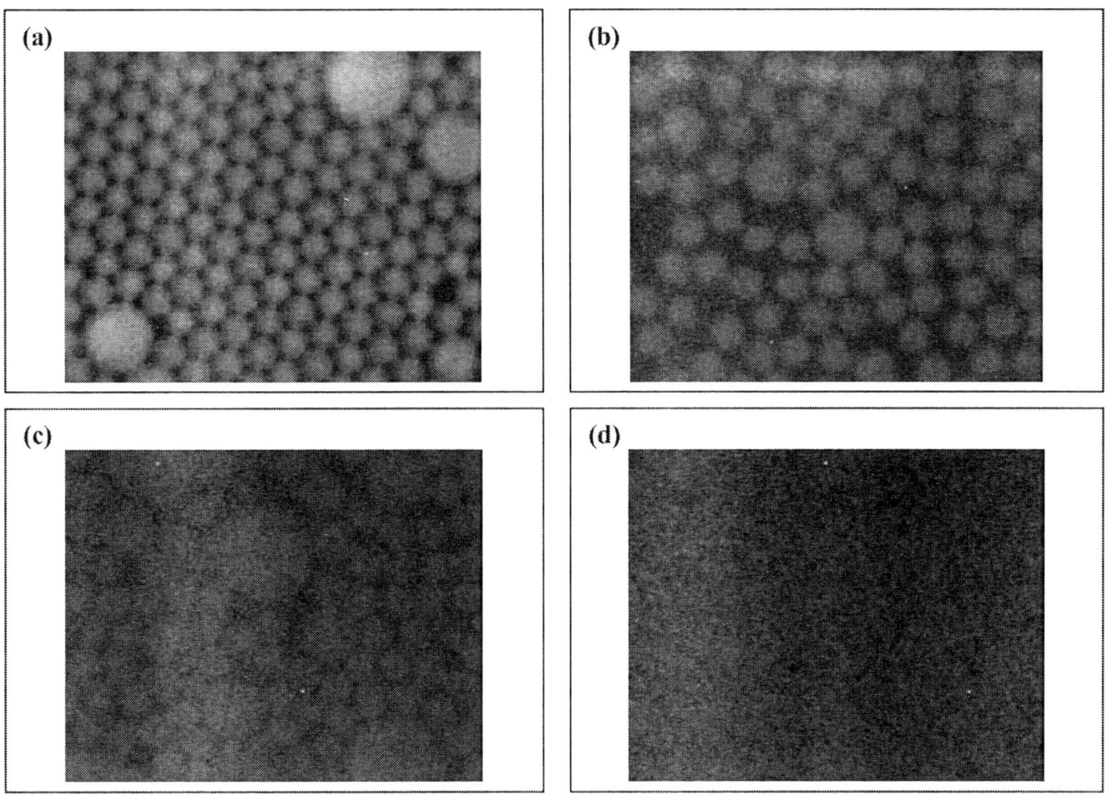

Figure 1(a, b, c, d). Emulsions from solutions with different probe concentrations: **(a)** 800 nM; **(b)** 400 nM; **(c)** 200 nM; **(d)** buffer solution (without dye).

The value of the fluorescence signal was measured at the center of random 30 drops from three images. For the analysis of fluid data, a line was selected on the image frame (460 μm in length). Then, for each experiment, the average value of the fluorescence signal was calculated from the results of processing data from the lines from five images. The results obtained are presented in Table 1.

Table 1. Measured fluorescence signal at different probe concentrations (mean and standard deviation).

	Oligonucleotide probe concentration, nM					
Measurement conditions	0 (buffer solution)	50	100	200	400	800
Liquid	214 ± 8	228 ± 9	247 ± 11	277 ± 11	392 ± 17	539 ± 30
Emulsion	233 ± 7	-	-	265 ± 6	296 ± 9	476 ± 27

The threshold values of the fluorescence signal and the corresponding concentration value, interpreted as the detection limit, were determined. Based on the data in the table, the fluorescence signals from three low concentrations were modeled by a normal distribution with the integral characteristics presented in the table: mean and standard deviation. For intermediate concentrations, the characteristic values were obtained by linear interpolation.

For liquids, the average value of the fluorescence signal (228 relative units) at the lowest measured concentration of 50 nM corresponded to the upper limit of the 96.3% confidence interval for the buffer solution and at the same time the lower limit of the 96.3% confidence interval for concentration 100 nM. If a concentration of 100 nM was taken as the detection limit and 228 relative units was taken as the detection threshold, then the total probability of a false-negative and false-positive analysis was obtained around 7.5%. If we set the total probability of a false positive and false negative analysis result not exceeding 1%, then the detection limit is 160 nM (with a minimum signal value with a positive result of 235 relative units).

If the same value of 160 nM was set as the limiting concentration for detecting emulsions, the total probability of a false-positive and false-negative result of the analysis increased to 4.5%. By reducing it to 1%, the concentration limit became the measured concentration of 200 nM (the minimum value of the signal with a positive result would be 251 relative units).

On the other hand, for detecting amplification in an emulsion, it is also important not only to identify the fluorescent droplets, but also to calculate their ratio to the total number of droplets. It was of interest to evaluate the legitimacy of the selective analysis of several individual images of reaction chamber.

A mixture of fluorescent and non-fluorescent drops was analyzed. The total number of drops was more than 20 000. The results of the automatic counting of drops were rectangular matrices of (14 × 18) cells. Each cell of this matrix contained information on one field of view measuring (460 × 350) μm. Two matrices were necessary for the analysis: one with the number of only fluorescent drops, the other with the total number of both fluorescent and non- fluorescent drops.

To assess the homogeneity of the samplings both matrices were divided into two parts (89 and 79 cells) by a vertical section. For each of the parts, the proportion of fluorescent droplets was calculated relative to their total number. In the first sample, the mathematical expectations was 0.165 and the standard deviation was 0.077; in the second sample - 0.161 and 0.062, respectively. According to the generalized Student's criterion, these samples was homogeneous with high probability ($P > 90\%$). This allowed us to conclude that the total number of droplets and the number of fluorescent droplets recorded in different areas of the reaction chamber are random variables with a sufficiently large coefficient of variation. Therefore, the mathematical expectations of division does not coincide with the division of mathematical expectations: if we consider the ratio of mathematical expectations (average number of fluorescent drops to the average number of drops) then it gives a value of 0.155. This value is not too different from previous estimates of mathematical expectations.

Evaluation of the law of the distribution of the number of drops in different regions of the reaction chamber showed that it approximately satisfies the normal (Gaussian) distribution with the mathematical expectation and variance estimated from the sampling. 92 cells were distributed over 7 intervals. Pearson's Chi-square criterion of agreement showed the confidence probability of accepting the hypothesis of a normal distribution of about 14-15 %, which is sufficient reason to accept the hypothesis. Combining the two extreme intervals with a small number of drops will increase this probability.

Combinations of 9 cells with automatic counting of drops according to the (3 × 3) square scheme (essentially, smoothing of the selection elements) were considered. An estimate of the ratio of the number of fluorescent drops to the total number has a mathematical expectation of 0.146 and a standard deviation of 0.044. The generalized Student's criterion in assessing the homogeneity of the results gives a value less than 0.8, which indicates the homogeneity of the sampling. Consequently, combining areas with a corresponding change in the scanner settings will not lead to a significant change in the analysis result.

The ratio of fluorescent droplets to all droplets calculated over the entire reaction chamber is slightly higher than the expected ratio of 0.10. The reason may be the combination of droplets, which leads to an underestimation of the total number of droplets and an overestimation of the proportion of fluorescent droplets.

It is interesting how the calculated ratio of the proportion of fluorescent droplets depends on the location of the analyzed cell. 10 enlarged locations with different distances from the center of the reaction chamber were selected. This ratio ranges from 0.10 to 0.22. There is a positive correlation (the sample correlation coefficient is +0.51) between the distance from the center of the chamber and the proportion of fluorescent droplets — it increases in the regions corresponding to the edges of the reaction chamber. Fewer drops are recorded at the edges of the reaction chamber, and the counting error has a large effect. The decrease in the number of droplets recorded at the edges can also be explained by the characteristics of hydrodynamics. Additional experimental confirmation and theoretical substantiation of the observed effect is the subject of further research.

When analyzing another reaction chamber, it was proposed to combine the cells into groups and analyze the mean values, medians and standard deviation calculated by groups and by all cells. It turned out 9 groups: in the first group the cells contained 40 or less drops, in the latter - more than 180 drops. It is noted that if we exclude the measurements of the first group with the minimum number of drops (about 20% of measurements), then for the other groups the average estimates practically coincide with the median. It also significantly increased the average value for the sample (from 112 to 135) and reduced the standard deviation (the variance decreased more than twice). The measurement exclusion threshold can be selected as 100, 80 or 60. At the same time, the mean (and median) estimates changed by less than 10% (from 149 to 138), and the standard deviation — from 27.5 to 36.4 (less than 33%).

4. Conclusion

The possibility of fluorescence detection and automatic counting of a large number of drops using the developed devices and special software was shown. The same analysis algorithms for fluorescent droplets on the background of a large number of non-fluorescent ones can be used in analyzing the results of a digital polymerase chain reaction of a droplet at the end point.

According to the experimental data, the threshold value of the fluorescence signal and the corresponding value of the limiting concentration were determined. For the detection of emulsions under given conditions, the limiting concentration was 160 nM, and when detecting a liquid, it was about 100 nM.

The data obtained indicate that the main reliable information on the number of drops was concentrated in the central part of the reaction chamber. From this point of view, edge images having the interface between media (walls of the reaction chamber) can be neglected. This reduces the measurement area and reduces the scan time.

Acknowledgments
The work was carried out and financially supported within the framework of the State contract (no. 075-00780-19-00).

References
[1] Guo M, Rotem A, Heymanab J, Weitz D 2012 *Lab Chip* **12** 2146.
[2] Marre S, Jensen K Synthesis 2010 *Chemical Society Reviews* **39** 1183.
[3] Tupik A, Posmitnaya Y, Rudnitskaya G, Varlamov D, Evstrapov A 2018 *JPCS* **1135** 012110.
[4] Schuler F, Siber C, Hin S, Wadle S, Paust N, Zengerleab R, Stetten F 2016 *Anal. Methods* **8** 2750.

Infection of aerosol concentration on the radon decay products fractions

Mostafa Y A Mostafa[1,3], Hyam N B Khalaf [1,3], M Zhukovsky[1,2]
[1]Ural Federal University, Yekaterinburg 620002, Russia
[2]Institute of Industrial Ecolog, Ekaterinburg, Russia
[3]Minia University, Faculty of Science, Department of Physics, El-Minia, Egypt

Abstract. In this work, an experimental setup has been constructed to study the effect of aerosol concentration on the size distribution and unattached fraction of radon decay products. The experiments are conducted with 2 m^3 standard radon chamber under controlled conditions (aerosol concentration, temperature and humidity). Number concentration and size distribution of aerosols inside radon chamber are measured with a Diffusion Aerosol Spectrometer (DAS 2702M, range from 5 nm to 10 μm). The activity size distribution of radioactive aerosols is estimated with diffusion battery. Number distribution diameter at low aerosol concentration has interval 50-100 nm. When additional aerosols injected to the radon chamber, distribution peak shifted to large diameters 70-150 nm at high aerosol particle concentration. At high aerosol concentration the unattached fraction fell sharply to 10-15% of the total activity compared to 50-70% when the aerosol concentration is low. The ratio between the fractions of the unattached and attached is in the range from 0.9 to 1.2 at low aerosol concentration. At high aerosol concentration, this ratio becomes from 0.09 to 0.14.

1. Introduction

The size distribution of radon-222 decay products consists of ultrafine clusters with median diameters below 4nm (unattached activity) and decay products associated with ambient aerosol particles in size range 100 and 1000 nm (attached activity) [1]. The ultrafine clusters or unattached activity fraction generated with the birth of Po-218 from radon-222. During the short half-lives time of radon decay products, they react and attach to surrounding aerosol. In this case the aerosol concentration and size distribution have a significant effect on radon decay products.

The unattached fraction is deposited nearly completely in the respiratory tract during inhalation, where as 80% of the attached are exhaled without deposition [2,3]. In ambient air, the amount of unattached activities up to about 10% of the total activity. This value present about 50% of the total radiation dose[1–3]. Therefore, it is necessary to determine the size of the unattached fraction of radon decay products and the effect of surrounding aerosols.

In this work, the effect of aerosol concentration on the size distribution and unattached fraction of radon decay products is studied. Aerosols are generated in standard radon chamber. The size distribution and unattached fraction of radon decay products are measured at low and high aerosol concentration with known size parameters.

2. Method

The experimental work was carried out with standard radon box 2 m^3[4–6]. Before start the experiments, there were no additional sources of aerosols in the box. Therefore, the dispersion composition of aerosols inside the box was primarily determined by the natural processes of aerosol formation in the air inside the box with radon gas and its decay products.
The diffusion aerosol spectrometer (DAS 2702M) is used to measure the aerosol concentrations and the particle size spectrum, covering a range of sizes from 5 nm to 10 μm. Also, the total concentration, air temperature, humidity and atmospheric pressure are given in the output file[5,7]. The aerosols parameters are measured continuously with flow rate 0.5 l/min. Additional aerosol particles are injected with especial aerosol source. Aerosol was released from e-cigarette with using smoking machine with flow rate 1 l/min. More details about the used e-cigarette can be found in these papers [4–8].
The activity size distributions of ultrafine fraction (≤ 200 nm) are measured with screen diffusion battery (SDB). The SDB has been used with 10 brass elements for measuring radioactive aerosol size distribution in the range from 1 to 100 nm. This SDB followed by three analytical filters (AFA-RSP-20) to catch all other sizes.

Alpha activities on the elements of diffusion battery and the analytical filters are consequently measured with scintillation alpha radiometer during 30 min after the finishing of the sampling. Then radon equivalent equilibrium concentration (EEC) of radon at the moment of finishing of the sampling was calculated by the modified Kuznets methods [9]. The activity size distributions are estimated from the alpha energy spectrum analysis of SDB modes obtained by using mathematical method, expectation maximization method of Maher and Laird (EMM)[10,11].

3. Results

Number Size distribution of the investigated aerosol (0-200 nm) at low and high aerosol concentrations presented in figure 1. These results are measured with DAS. At low aerosol concentrations, the number median diameter, NMD, has range from 50-70 nm with broad size. This value shift to 70-150 nm in the case at high aerosol concentration. The difference of size distribution is clear with the effect of high aerosol concentration.

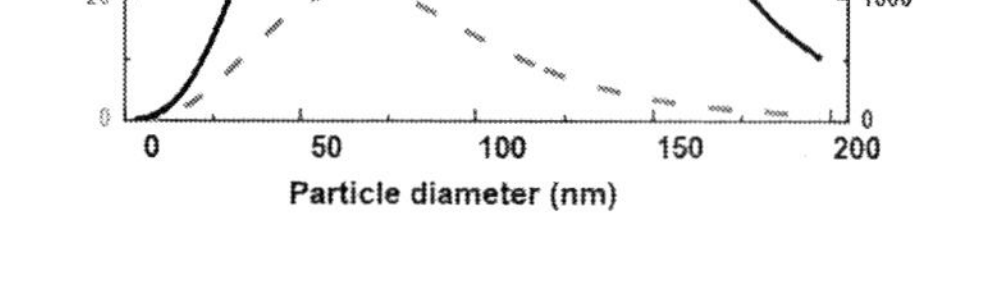

Figure 1. Number Size distribution of the ultrafine investigated aerosol. The left scale for low aerosol concentrations and the right one for high concentrations

Activity Size distribution at low and high aerosol concentrations measured with SDB is presented in figure 2. These results demonstrated that at the low aerosol concentration 50-70 % of radon decay product activity connected with aerosol particles with AMTD ~ 1 nm (so-called unattached fraction). Additionally, a mode with a diameter of ~ 10 nm was registered with small fraction. After aerosols injection in the experimental radon box, the activity of unattached fraction with AMTD ~ 1 nm is sharply decreased to zero due to attachment process of radon decay products with injected aerosols. At high aerosol concentration, activity size distribution mode with AMTD ~ 10-20 nm is observed with half relative activity of the unattached fraction of 1 nm in the case of low aerosol concentration. The residual activity is collected with backup analytical filters with size more than 200 nm which out of the SDB range.

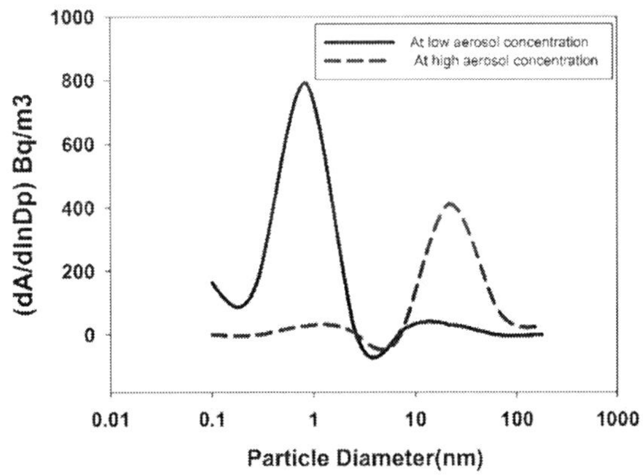

Figure 2. Activity Size distribution with SDB, at low and high aerosol concentrations.

Radon Equilibrium Equivalent Concentration, EEC, at low and high aerosol concentration is calculated and compared. Figure 3 shows EEC at low and high aerosol concentration using diffusion battery and 3 AFA-RSP filters for attached (at filters) and unattached (SDB) radon decay products. These results are the average of 10 series of experiments. The total activity is divided equally at low aerosol concentration the ratio between attached fraction nearly 1. While the aerosols add into the radon chamber the, high aerosol concentration, fraction of unattached not more 10 % compared to 50-70 % at low aerosol concentration. Ninety presents of the total activity are collected in the analytical filters. Its mean that at high aerosol concentration the chance of attachment of radon decay products is very high. Although, there is no additional aerosols source inside radon chamber, but the activity size distribution with median diameters more than 200 nm have ~ 50 % of the total activity. This mean that the gross of particles inside the box has different factors not only aerosol concentration.

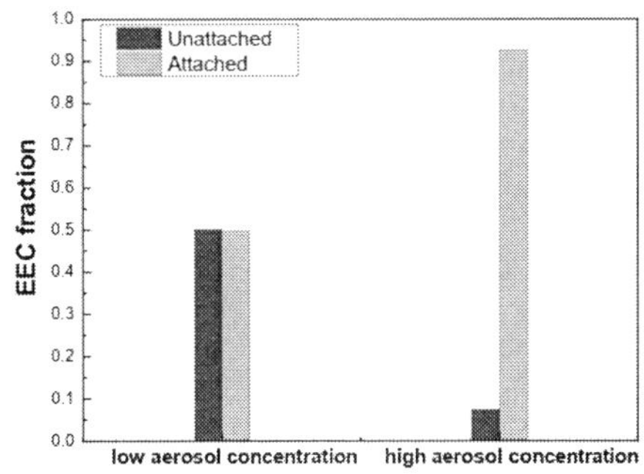

Figure 3. The radon Equilibrium Equivalent Concentration, EEC, at low and high aerosol concentration using diffusion battery and 3 AFA-RSP filters.

4. Conclusions

- At high aerosol concentration, fraction of unattached radon decay products not more 10 % compared to 50-70 % at low aerosol concentration
- 1 nm size mode of radon decay products completely disappear due to the attachment process at high aerosol concentration
- At high aerosol concentration the chance of attachment of radon decay products is very high.
- Although, there is no additional aerosols source inside radon chamber, but the activity size distribution with median diameters more than 200 nm have ~ 50 % of the total activity.
- The ratio between the fractions of the unattached and attached is in the range from 0.9 to 1.2 at low aerosol concentration and becomes from 0.09 to 0.14 at high aerosol concentration.
- To understand the movement and distribution of radon decay products, whole activity size distribution measurement system should be established with size range from 1nm to 10 μm.

References

[1] Yuness M, Mohamed A, AbdEl-hady M, Moustafa M and Nazmy H 2015 Effect of indoor activity size distribution of222Rn progeny in-depth dose estimation *Appl. Radiat. Isot.* **97** 34–9

[2] Yuness M, Mohamed A, Abd El-Hady M, Moustafa M and Nazmy H 2015 Indoor Activity of Short-Lived Radon Progeny as Critical Parameter in Dose Assessment *Solid State Phenom.* **238** 151–60

[3] Yuness M, Mohamed A, Nazmy H, Moustafa M and Abd El-hady M 2016 Indoor activity size distribution of the short-lived radon progeny *Stoch. Environ. Res. Risk Assess.* **30** 167–74

[4] Khalaf H N, Mostafa M Y A and Zhukovsky M 2019 Effect of electronic cigarette (EC) aerosols on particle size distribution in indoor air and in a radon chamber *Nukleonika* **64** 31–8

[5] Khalaf H N, Mostafa M Y A, Vasyanovich M and Zhukovsky M 2019 Comparison of radioactive aerosol size distributions (Activity, number, mass, and surface area) *Appl. Radiat. Isot.* **145** 95–100

[6] Khalaf H N B, Mostafa M Y A and Zhukovsky M 2018 Radiometric efficiency of analytical filters at different physical conditions *J. Radioanal. Nucl. Chem.*

[7] Nazmy H, Mostafa M Y A and Zhukovsky M 2018 Particle Size Distribution of E-Cigarette Aeroso ls in Indoor Air *J. Radiat. Nucl. Appl. An Int. J.* **3** 111

[8] Khalaf H N B, Mostafa M Y A and Zhukovsky M 2019 Radioactive aerosol permeability through Russian radiometric analytical (PF) filters *J. Radioanal. Nucl. Chem.* **319** 1283–9

[9] Nazaroff WW. 1980 An improved technique for measuring working levels of radon daughters in residences. *Health Phys.* **39** 683–8

[10] Zhukovsky M, Rogozina M and Suponkina A 2014 Size distribution of radon decay products in the range 0.1-10 NM *Radiat. Prot. Dosimetry* **160** 192–5

[11] Rogozina M, Zhukovsky M, Ekidin A and Vasyanovich M 2014 Thoron progeny size distribution in monazite storage facility *Radiat. Prot. Dosimetry* **162** 10–3

Test object as a subject of biometric studies of eye reaction

N Bikberdina[1], D Bobrova[1], M Boronenko[1]
[1]Yugra State University, Khanty-Mansiysk 628012, Russia

Abstract. As you know, a person does not pay attention to any object, but on the basis of the importance of the object or circumstance at the current time. Attention is concentrated on the object the more often, the higher is subjective significance. The significance is due to the motivational force and information saturation of the object. For preventive measures to identify people at risk, you can use test objects that are carriers of well-known information (representing images, audio and video files). Test objects revoke an emotional response from everyone, but the level of emotions experienced will be different. Unambiguous reaction is the presence of emotional components in the waves of attention.

1. Introduction

In the modern world, high technologies are rapidly developing, and this applies to forecasting technologies. Interest in detecting the psychophysical state of a person by various biometric data is becoming more widespread every year. The methods of forecasting include pupillography. Previously it was thought that the appointment of pupil - it is the adjustment of the luminous flux incident on the surface of the eye. Modern studies have shown the complexity of the reactions of the visual nervous system. However, because of the relative ease of implementation and high reliability of the results, the application of pupillography quite common. For example, the pupillograph records changes in pupil size during speech [1]. If a person deceives, he experiences stress, the size of the pupils changes, which registers the pupillograph. With the help of pupillography learned to identify dangerous people who are in a state of drug and alcohol intoxication. Ahmetvaliev A. M., Karasev A. S. introduced the concept of contactless identification of persons in the state of alcoholic and narcotic intoxication presenting a threat to public safety [2]. However, those who are the bearers of evil intentions are in the shadows. The standard method of determining emotions, focused on mimic changes is good, but if a person is good at controlling himself, it is difficult to recognize evil intentions, and sometimes it is too late. Therefore, the purpose of the work is to develop a test object that will allow you to transfer certain emotions to the subject, to register an uncontrolled reaction.

2. Experimental method

We considered the reaction of the pupil to the test objects with a certain thematic focus, which have an emotional impact in proportion to the degree of the individual's internal attitude to the topic. At the first stage of the experiments the images shown on the laptop monitor were used. The subject of information can be determined by the actual problems of society, urgent psychological problems. The average value of the level of intensity of emotions was considered a normal reaction, because the hypothesis was accepted that most people are mentally balanced and tolerant. In General, the intensity of the evoked emotions on a scale from 1 to 10 did not exceed 4, where 10 is the strongest emotion that can be. The following hypothesis is used. For each person there are topics that cause ultra-high emotional responses. The information that concerns them is significant-important for a person, and causes involuntary attention in case of appearance in the field of perception of the individual. If we assume that such information may contain a test object, the reaction of the pupils to such a test object is proportional to the intensity of the emotions experienced at the same time Ψ. In the selection of test objects it is necessary to consider the following factors:

- to identify which category of persons the efforts are directed;
- if the illumination of the surface of the eye during the recording of pupillage different, it is necessary to introduce a correction factor;
- the depth of the emotions experienced should not push a person to the beginning of action. As a result, the intensity of the impacting stimulus should belong to the interval $\Psi < \Psi_{activation}$.

For the study, 70 images of various subjects and emotional coloring were selected: education, family values, joy, happiness, cultural traditions of different peoples, belonging to different faiths, loss, violence, etc. the Images were selected so that it was possible to identify existing problems in society. In addition to active test objects, neutral test objects are required. They do not carry any information, or contain insignificant information for people of the target category. Such test objects allow you to "adjust" the system to the selected category of persons, allow you to take into account the spontaneous fluctuations of the pupil and random noise.

The study involved groups of two age groups of 16-25 years and 45-50 years of 10 and 5 people, which were attended by representatives of different sexes. Eye diseases in people were absent, vision was normal or corrected. Entry pupillograph carried out on the installation shown in figure 1. The process of video shooting was carried out while minimizing the effect of illumination from the test object (taking into account its color) on the size of the pupils. Additional control of illumination is carried out by the skin tone of the subject (in the gray scale, normalized by the average value).

Figure 2(a, b). (a) Experimental setup; **(b)** Test object and track of the center of mass (focus) of the monitor reflection in the pupil.

For this test, a participant with sensitivity to scandals in the family was selected. The hardware part includes UPS: Galvanic Skin Response Sensor (GSR - Sweating), which measures resistance between two points; Camcorder T7 Camera Astro Astronomical Astronomy Planetary High Speed Electronic Digital Eyepiece Telescope Lens for Guiding Astrophotograp, movie mode 30 fps, lens with optical zoom, 1X-100X; Specially designed helmet, creating a rigid coordinate connection between the camera and the head; Arduino UNO R3; Images were analyzed in the freely distributed program FiJi it is an improved and expanded distribution ImageJ, combining many plugins that allow you to conduct a complete scientific analysis of the image. Pupillography and oculogram received in the program OriginPro 2019. Figure 2a shows a typical pupillogram. Synchronization of the track of the center of attention and record pupillography during reaction to a specific test object allows us to specify the element in the test object, which happened emotional reaction..

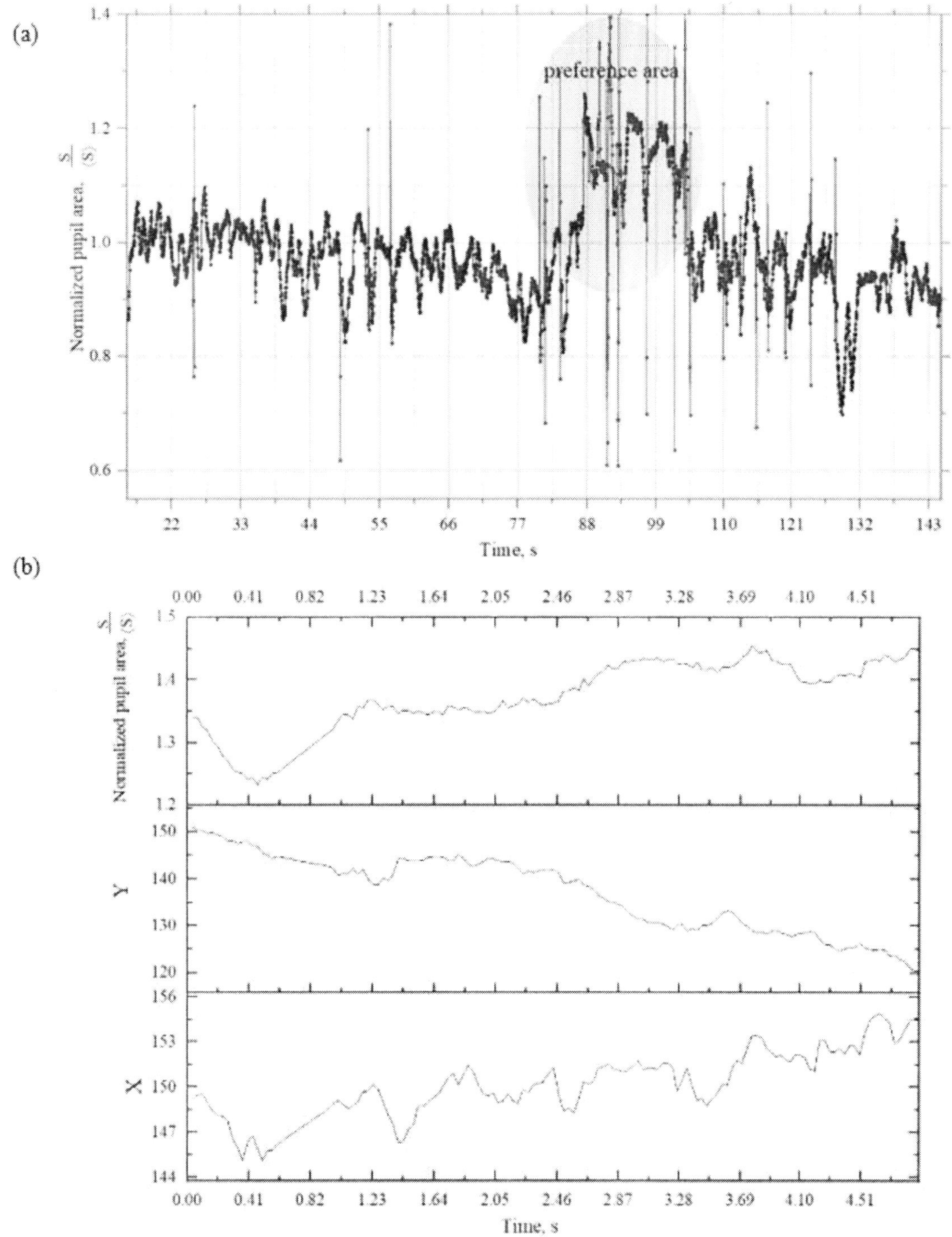

Figure 2(a, b). **(a)** Change in pupil size over the entire observation period; **(b)** Synchronized pupillogram and coordinates of the center of attention in the area with emotion.

For this purpose, the image of the monitor reflected in the pupil was isolated on the obtained video frames. Then the centre of mass of the selection was monitored in each frame and its coordinates were determined. Plots with S\<S>. more than 1.2 correspond to the manifestation of emotions or attachments. In our case, as an irritant, stimulating the response of the pupil, complex images were used, the details of which can also be considered under certain conditions as independent stimuli.

3. Results and discussion

In all experiments pupillography participants were allowed to monitor the focusing of the eyes on the desired test object. The structure of the wave of attention in pupillograph includes a saccade when you move the view, micro-saccade focusing on the individual elements of these objects and the emotional component, provided that the test-object is significant for the individual. Evaluation of the level of intensity of emotions caused by information was carried out by statistical method. Participants had to evaluate the emotions that arise when viewing each image on a 10-point scale, in which 1 – the image is very unpleasant, and 10 – the image is very nice. After the survey and data were obtained, the normal distribution was tested using the Kolmogorov-Smirnov test to determine the discriminativeness of the test objects. As a result of the inspection, 26 images were selected that meet the requirements of the law of normal distribution. Then factor analysis was carried out to group the images. As a result of the factorization of the data set by the principal components method, 3 factors were identified, covering 62.58% of the share of the explained variance (table. 1).

Table 1. Factor matrix.

№ images	Factors			№ images	Factors		
	1	2	3		1	2	3
и8			0,870	и34		0,864	
и10	0,809			и35		0,851	
и14		0,657		и36	0,684		
и16	0,795			и38	0,757		
и21	0,668			и39	0,665		
и22			0,792	и41	0,786		
и23		0,575		и43	0,630		
и25			0,589	и48		0,613	
и29	0,694			и49	0,713		
и31	0,629			и56		0,592	
и34		0,864		и58	0,693		
и66	0,750			и60			0,701

The first factor describes 44.25% of the sample. The images of external belonging to the Eastern culture (clothes, rituals) have the greatest load. Taking into account the magnitude of the factor loads, we denote this factor "Belonging to the Muslim culture". The second factor describes 9.65% of the sample. The biggest load is the image of a woman at the head of a team of employees. It also includes images of success and career achievements. Taking into account the values of factor loads, we denote this factor as "career Success". The third factor describes 8.68% of the sample. The greatest burden was a variable with the image of the Christian ritual. Given the values of factor loadings, this factor is denoted "Belonging to the Christian culture". The reliability of the scales was checked by the internal consistency of the items included in them using the Cronbach alpha coefficient: "Belonging to Muslim culture" -0,949; "career Success" -0,877; "Belonging to Christian culture"-0,821. The observed size changes of the pupil resulting from the effect of all the factors (like emotional or any other). All changes are controlled by the potentials of the brain.
The pupillogram in figure 2 belongs to the person whose gaze is directed at the photo of Jim Kirk's face (star Trek movie). From pupillography and oculogram deleted the moment of blinking. In this case, the focus moved vertically from bottom to top (in FiJi, the origin (0;0) is in the upper left corner of the working field). The coordinates had changed little. The amplitude of this pupillogram is weakly dependent on the change in the position of the centre of attention, and consistently exceeds the average value. At the same time, bright emotions are not observed and the wave of attention is steadily positive, without special emotional outbursts. Pupillography well approximated by a Gaussian function having the form:

$$y = y_0 + \frac{A}{w}\sqrt{\frac{4\ln(2)}{\pi}}\,exp\left(\frac{-4\ln(2)(x-xc)^2}{w^2}\right). \qquad (1)$$

To find regularities, we found the inter-frame difference between video files containing the pupillary response to the test objects. Then we found the power spectrum (Fig.3, 4). The periodogram allows us to estimate the power from the amplitude of the Fourier transformed data. Blustein's algorithm correctly calculates complex 2D\3D-FFT images of arbitrary size.

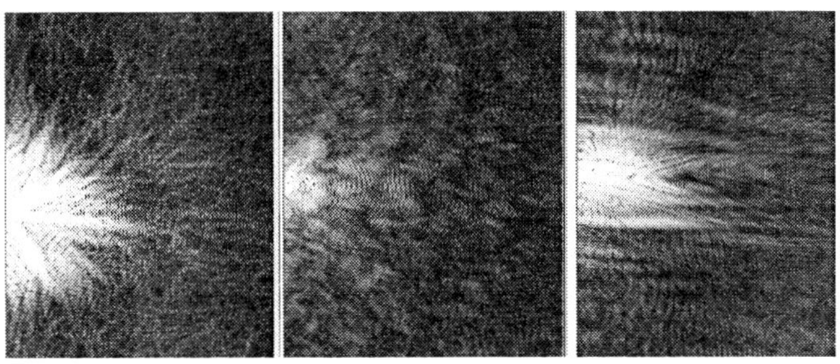

Figure 3. FFT inter-frame pupil image difference when viewing (from left to right) neutral, emotionally-negative, emotionally-positive test objects.

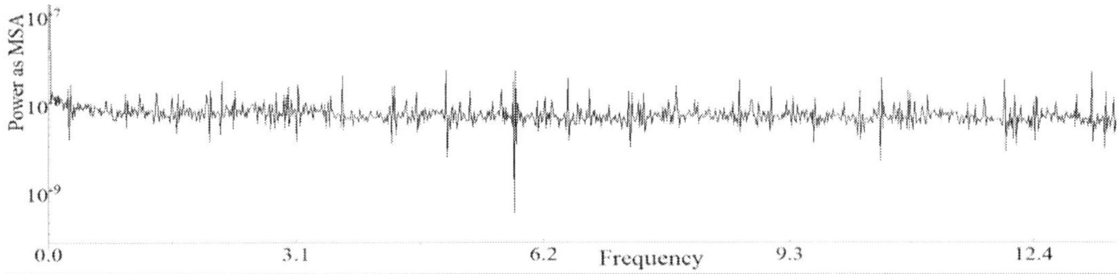

Figure 4. Power spectrum of differential chronogram [3], pupillary response to emotionally negative test objects

Power spectra indicate a different frequency and amplitude of the pupillary response at different emotional stress test objects. This suggests the possibility of identifying common patterns of pupillary reaction, with the similarity of the polar properties of the emotions experienced.
In the process of development are test objects, which are images, accompanied by short text messages Text information will more accurately convey the meaning and specify the "pain point". Their purpose is to cause short-term, but bright emotional outbursts. One of the intended applications is to identify people with suicidal intentions. Before the study of the reaction of the pupil to the depressive state of a person, you need to create a room with a home environment, it is desirable that he was in a normal state (not too cheerful, not too sleepy), the subject is desirable to distract from heavy thoughts, should feel comfortable in his clothes, not hungry, as the internal mood affects the passage of the test. To register the reaction of the pupil, it is desirable to use an inconspicuous camera that will not create unnecessary anxiety.

4. Conclusion

Created a set of test objects that cause involuntary human reaction, uniquely interpreted:
- Thematic test objects were developed, the degree of emotional colouring of test objects was estimated by the method of statistical survey. The reliability of the scales was checked by the internal consistency of their items with the application of the Cronbach alpha coefficient. The results suggest a high degree of reliability of each of the scales;
- It was found that the complete replacement of video files with sound test objects is impractical, because in the field it becomes impossible to track the pupillary response and its specification. Therefore, research is aimed at the development of images, video images.
- Developed non-specific test objects that are carriers of well-known information (representing images, sound and video files). Test objects evoke an emotional response in everyone, but the level of emotions experienced depends on the magnitude of the emotion experienced.

Unambiguous reaction is the presence of emotional components in the waves of attention. The level of the registered reaction at which it can be considered exceeded the threshold value must be set.

Acknowledgments

The reported research was funded by Russian Foundation for Basic Research and the government of the region of the Russian Federation, grant № 18-47-860018 p_a.

References

[1] Kolbeneva M G, Changes in the electrical resistance of the skin when updating the experience associated with various senses //Vopr. the course of studies. - 2014. - no. 2. S. 93-106.
[2] Pashkov A A, Bakhtin I S, Kharisova N C, Electroencephalographic biomarkers of experimentally induced stress / / Bulletin of South Ural state University. Series: Psychology. - 2017. - Vol. 10. - 4.
[3] Stern R. M. et al. Psychophysiological recording. – Oxford University Press, USA, 2001.

The empowerment of biometric systems of psychophysical research

D Bobrova[1], M Boronenko[1], N Bikberdina[1]
[1]Yugra State University, Khanty-Mansiysk 628012, Russia

Abstract. Interest in detecting the psychophysical state of a person according to various biometric data (KGR, pupillography and profiling, etc.) is becoming more widespread every year. The most statistically reliable and resistant to counterfeiting systems are the identification of a person by the iris and face. Biometric identification systems can be combined. This paper presents proposals for improving the existing systems of recognition of the psychophysical state of a person.

1. Introduction

In today's world, special procedures are increasingly used to measure the unique behavioral and physical characteristics of a person. Biometric technologies are considered most often as automated or automatic methods by means of which recognition of the person's face according to his individual characteristics is carried out. The most common method of biometrics is facial and fingerprint recognition. [1] the Standard method of determining emotions, focused on mimic changes is good, but if a person is good at controlling himself, then it is difficult to recognize evil intentions, and sometimes it is too late. The advantage of using pupillography is an unambiguous biochemical uncontrolled reaction of the human body to the heard (seen) information, including changes in the size of the pupil [2]. Terrorists, sadists have a set of preferences, values, pleasures that differ (opposite) from ordinary people. The spatial and temporal resolution of modern digital video cameras allows video shooting with the desired resolution [3]. Been successfully applied to recognition system, the diameter of the pupils of potentially dangerous people, which represents an open threat. In these cases, the pupil identifies people who are in a state of drug or alcohol intoxication. The aim of the work is to expand the capabilities of such systems through the use of recognition of people who pose a hidden threat. In this case, the pupillary response of an individual who has to solve the tasks requires a comprehensive approach that combines several methods.

2. Experimental method

Basic provisions, which should correspond to the means of display and reception of information:
- It is known that for reliable reproduction of a compact object in a digital image, its size must be at least four pixels. In our case, the requirements are tightened. The resolution of the recording camera should not be worse than 5 MPCs. This allows you to reliably monitor changes in pupil size is not worse than 0.1 S.
- The image on the screen should be clear and full color when changing the viewing angle.
- If the test objects are shown from the screens of collective use (receiving information simultaneously by a large number of people), the observation distance can be significant. The brightness of the radiation from the image of useful information should be at least 3000 CD [4].

- The design observation distance should depend on the screen size and the distance between the emitting cluster elements and the gap between the composite screen modules.

- The design observation distance is determined by tests on the results of the change in the greatest distance between the emitting elements of the cluster (screen) and the theoretical angular resolution of the eye, with the pupil diameter and wavelength corresponding to the maximum value of the relative spectral efficiency of the eye.

$$\iota = \psi \cdot r_{pr} \quad (1)$$

$$\psi = (1{,}22 \cdot \lambda) \,/\, d \quad (2)$$

$$r_{pr} = \iota \cdot d \,/\, (1{,}22 \cdot \lambda) \quad (3)$$

where r_{np} is the smallest distance from the screen to the observer, at which the emitted elements are indistinguishable separately.

- When changing the viewing angle by ±60°, the brightness of any color should not change more than twice.

Thus, there is an acute problem of taking into account the dependence of pupil size on illumination.

To study the reaction of the pupil in idealized conditions, a helmet was developed, creating a rigid coordinate connection between the video camera and the head (Fig.1.). Rigid coordinate communication facilitates the processing of images received by the camera. At this stage, it is necessary to avoid ambiguous interpretation of the results. For video filming used video camera T7 Astro Camera Astronomical Astronomy Planetary High Speed Electronic Astrophotograph, video mode 30 fps, microscope lens with optical magnification 1X-100X. Images were analyzed in the freely distributed program FiJi-it is an improved and expanded distribution ImageJ, combining many plugins that allow you to conduct a complete scientific analysis of the image.

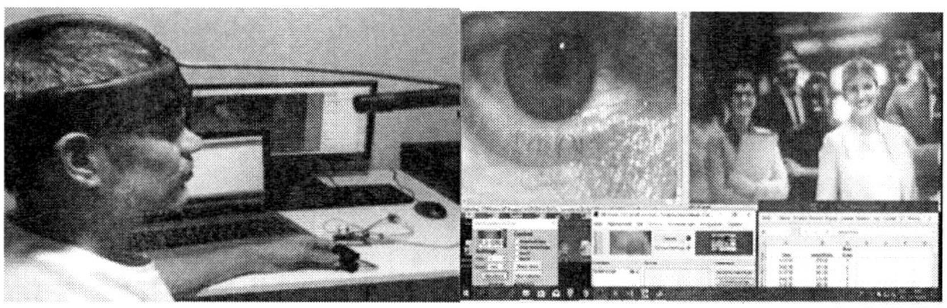

Figure 1. Installation for receptions pupillogram.

The helmet is a frame welded from a copper tube. The frame is wrapped with a layer of foam. In the occipital part of the frame there is an adjusting screw that allows you to loosen or tighten the helmet on the head. A person wearing a helmet is located at a distance at which the difference in the illumination of the eye is insignificant. Preliminary selection of the distance was carried out on the basis of measurements of illumination created by monochrome frames, each of which had a different color. The footage was shown from a monitor located at a distance varying from 0.75 to 4 m from the photometric head of the luxmeter. Figure 2 shows a graph of the light produced by the monitor at different distances.

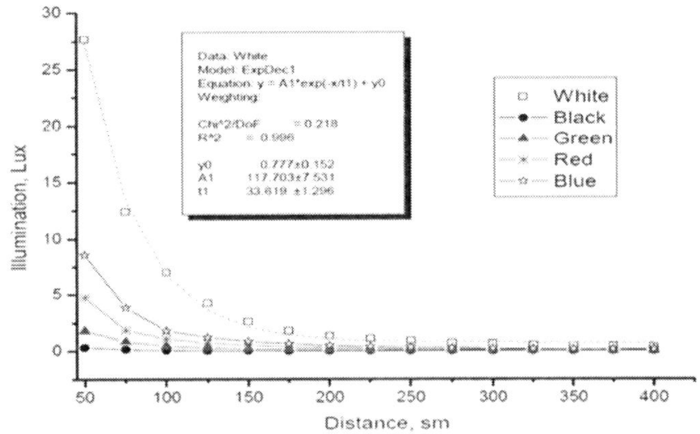

Figure 2. The illumination created by the test object on the photosensitive element of the meter.

The experiments determined the distance at which the stimuli do not contribute to the change in pupil size. In human physiology, retinal illumination is important. The size of the pupil, as well as the level of illumination, is very important to ensure sufficient illumination of the retina and, therefore, to feel the brightness. For this reason, the intensity of illumination of the pupil L was introduced as a measure of illumination of the retina in the daytime. The value L is the product of illumination E on the surface area of the pupil S:

$$L = S \cdot E \tag{4}$$

It is known that in field conditions the person of rigid coordinate communication of a video camera and the head won't be. In this case, the illumination of the retina will depend on the angle. Therefore, light control (Fig. 2) will be carried out in brightness of skin area near the eye. Let the pupil area S1, the average brightness of the pixels included in the selected area of the skin near the pupil (in grayscale) I1 be at the illumination of the pupil E1. Then, with an unplanned change in the illumination E2, the pupil area will become S2, the brightness of the skin I2. It is known that the transformation of light by the photomatrix is linear. Therefore, the brightness of the pixels is proportional to the illumination of the photomatrix. Then the change in illumination in E2\E1=I2\I1 is taken into account by introducing a correction factor:

$$k = I_1 \setminus I_2 \tag{5}$$

Another problem occurs when processing images. The problem of finding a circle is one of the most well-known in image processing. To solve it created a large number of different methods, some of which are applicable in this particular case. At the first stage of experiments it is enough to apply the contouring "manually" what is happening in semi-automatic mode. For images with very poor quality, we used the following algorithm:

1.Trim file. You must select the area of the image that contains the pupil. The selection must be large enough so that the pupil does not leave the selected area on any of the frames. Cropping is necessary, because the files have a high resolution, and the PC will not be able to process it entirely.
2. Run the subtract background command.
3. We carry out masking on the threshold, the most suitable for contouring.
4. Remove outliers (Remove Outlier). The radius defines the area (uncalibrated, that is, in pixels) used to calculate the median. The threshold determines how far the pixel must deviate from the median to be replaced, in raw (uncalibrated) units.
5. If you are satisfied with the result, then in each of the above items the action is automatically applied to the entire stack by pressing "OK".

Getting oculogram occurs in several stages:

1.Uploading an image to ImageJ;

2."Masking" is performed, i.e. setting the boundaries of the pupil;
3.Next, you need to build a graph of the coordinates of the center of mass of the time in the program Origin.

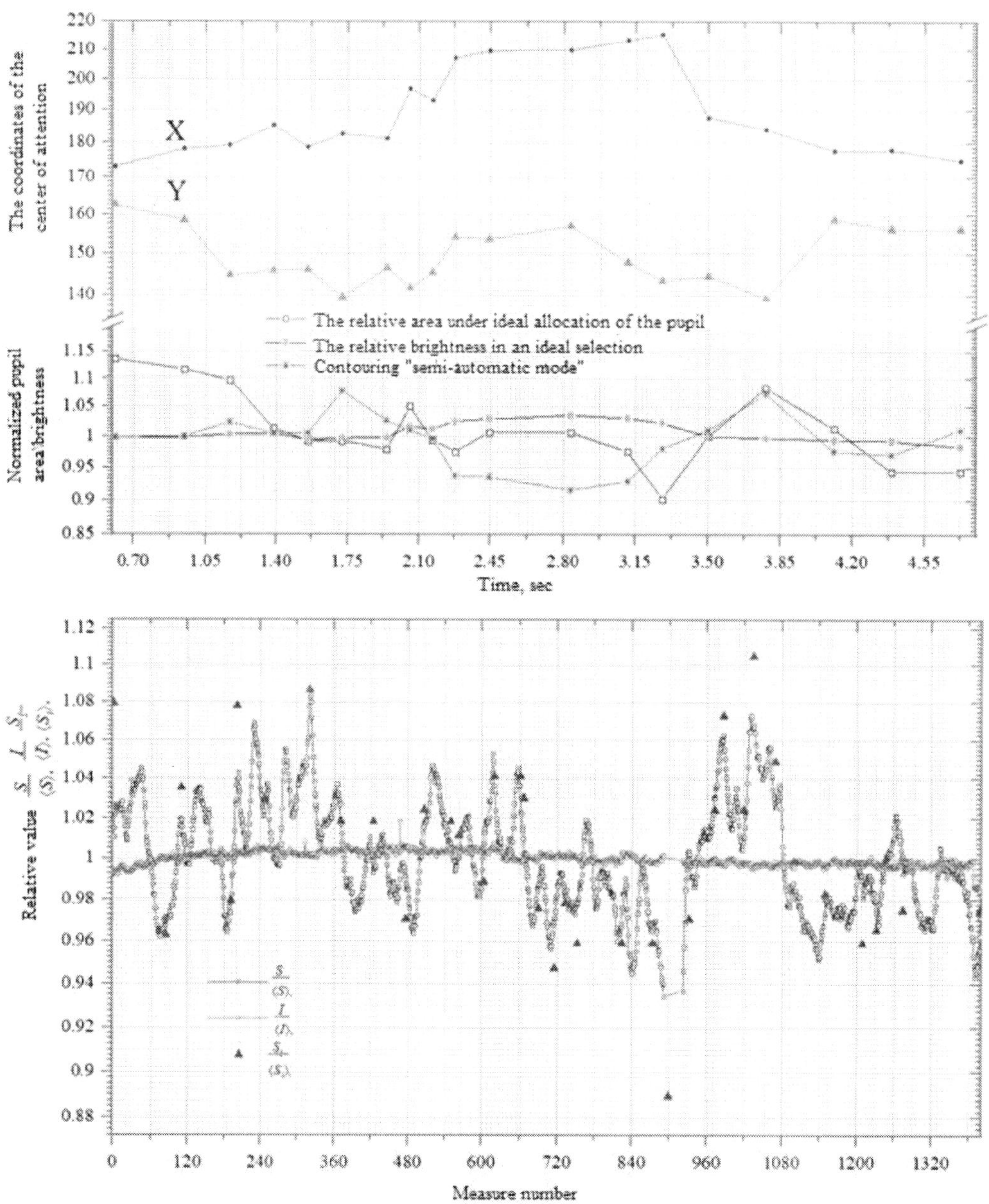

Figure 3. The measurement results of the relative changes of pupil diameter using different methods (two of the ejection - outline on the slides in a moment of blink)

Comparing the result of contouring with different methods, you can choose the most accurate result. Based on statistical analysis and as can be seen from the graphs (Fig.4) the closest results are obtained by a semi-automatic method based on the developed algorithm. The standard deviation of the normalized values Si\⟨Si⟩=1.0049±0.0461; S\⟨S⟩=1.0125±0.0677; I\⟨I⟩1.037±0.1999. It can be seen from the graphs that at constant illumination the method of measuring the relative change in the pupil area by the level of gray is the least sensitive to changes in the pupil size. Selection of contours manually and in semi-automatic mode gives almost the same result. In this case, the error of contouring is an order of magnitude less than the change in the area of the pupil.

3. Approbation of the method

Approbation of the method was carried out when registering the pupillary response in response to viewing test objects. As test objects we used images exposed by Internet users to the public access. The distance between the monitor and the eye is 2.5 m. the intensity Level of emotionally colored video files, on a scale from 1 to 10, does not exceed 4, where 10 is the strongest emotion that can be. The control of illumination was carried out by the skin tone of the test subject (in the gray scale, normalized by the average value), Fig.4. For confidence in interpreting pupillages in each frame followed the line of sight. For this purpose, the image of the monitor reflected in the pupil was isolated on the obtained video frames. Then the center of mass of the selection was monitored in each frame and its coordinates were determined. The coordinates of the center of mass of the reflection of the monitor in the pupil and measured in each frame, the pupil diameter used to build pupillage and oculogram.

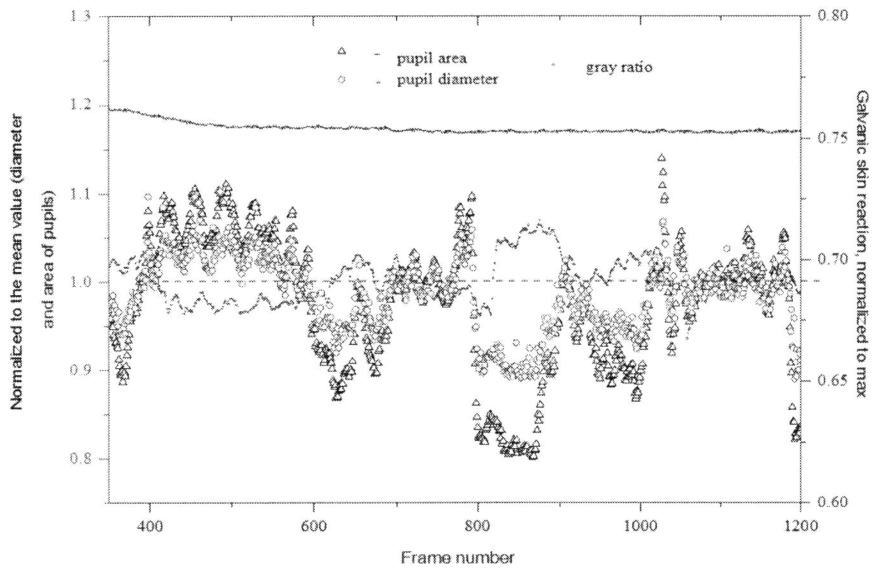

Figure 4. Control of the stability of the illumination of the eye surface

4. Conclusion

In the course of the work, the main results can be considered: the introduction of a correction factor for unplanned changes in illumination; the development of a method to track the coordinates of the center of attention. Method of registering the waves of attention are based on synchronous recording of pupillage and oculogram (not swinging saccades). The method can be used as a selective indicator of the presence of low-intensity reactions to specific test objects, provided that the stimulus is significant for the individual. The possibility of specifying the element of the test object to which the emotional reaction occurred.

Acknowledgments

The reported research was funded by Russian Foundation for Basic Research and the government of the region of the Russian Federation, grant № 18-47-860018.

References

- [1] https://habr.com/ru/post/126144/ Modern biometric identification methods
- [2] Lanata, A., Armato, A., Valenza, G., and Scilingo, E. P. (2011). "Eye tracking and pupil size variation as response to affective stimuli: a preliminary study," in 2011 5th International Conference on Pervasive Computing Technologies for Healthcare (PervasiveHealth) (Dublin: IEEE), 78–84.
- [3] Andreassi George. L. Psychophysiology: human behavior and physiological response. - Psychology of the press, 2010.
- [4] GOST R 52870-2007 Means of display of information of collective use. Requirements for visual display of information and methods of measurement
- [5] de Gee J. W., Knapen T., Donner T. H. Decision-related pupil dilation reflects upcoming choice and individual bias //Proceedings of the National Academy of Sciences. – 2014. – T. 111. – №. 5. – C. E618-E625.
- [6] Gulyaev P.Yu., Gulyaev I.P., Milyukova I.V., Cui H.Z. In-situ selfpropagating-hightemperature-synthesis controlled by plasma // Bulletin of Ugra state University. 2012. № 2 (25). C. 28-33.
- [7] Gulyaev P.Yu., Gulyaev I.P., Milyukova I.V., Cui H.-Z. Temperature measurements for Ni-Al and Ti-Al phase control in SHS Synthesis and plasma spray processes // High Temperatures – High Pressures. 2015. T. 44. № 2. C. 83-92.
- [8] Garkol' D.A., Gulyaev P.Y., Evstigneev V.V., Mukhachev A.B. A new high-speed brightness pyrometry method to investigate self-propagating high-temperature synthesis // Combustion, Explosion, and Shock Waves. 1994. Volume 30. Issue 1. pp 72-76. DOI:10.1007/BF00787888
- [9] Development prospects of SHS technologies in Altai state technical university/ V.V. Evstigneev, P.J. Guljaev, I.V. Miljukova, at al // International Journal of Self-Propagating High-Temperature Synthesis. 2006. T. 15. № 1. C. 99-104.
- [10] Gulyaev I.P., Ermakov K.A., Gulyaev P.Yu. New High-Speed Combination of Spectroscopic And Brightness Pyrometry For Studying Particles Temperature Distribution In Plasma Jets // European researcher. 2014. № 3-2 (71). C. 564-570.
- [11] Dolmatov A.V., Gulyaev I. P., Imamov R. R. Spectral pyrometer for temperature control in thermosynthesis // Bulletin of Yugra state University. 2014. № 2 (33). P. 32-42.
- [12] Gulyaev I. P., Gulyaev P. Yu., Jordan VI resolution virtual temperature control devices of particles in the plasma flow at total spectrum // polzunovskii almanac. 2008. №. 2. C. 13-14.

Formation of Structurally Ordered Nanoscale Complexes of DNA with Nuclear Proteins HMGB1 and H1

E Chikhirzhina[1], N Romanov[2], A Polyanichko[1,2]
[1] Institute of Cytology RAS, Saint-Petersburg 194064, Russia
[2] Faculty of physics, Saint Petersburg State University, Saint Petersburg 199034, Russia

Abstract. The structural organization of DNA in complex with linker histone H1 and non-histone chromosomal protein HMGB1 have been studied using atomic force microscopy and vibrational spectroscopy. The possibility of formation ordered nanoscale DNA-protein complexes is reported.

1. Introduction
hromatin of the eukaryotic cells comprises of DNA and variety of nuclear proteins. The most abundant nuclear proteins are linker histone H1 and non-histone chromosomal proteins HMGB1/2 **[1, 2]**. These proteins interact with linker DNA the different grooves of the helix **[2-6]**. Both of these proteins interact with DNA and play critical role in the structural organization of the chromatin. Linker histone H1 is one of the best studied chromatin proteins. It binds to linker DNA at the entrance or exit of the nucleosome and plays an essential role in formation of the 30-nm fibril **[2-5,7]**. HMGB1/2 are the members of HMG superfamily (*H*igh *M*obility *G*roup), which are characterized by high electrophoretic mobility **[1,6]**. The HMGB proteins have structural functional motif – HMGB-domain, which is responsible for the interactions with DNA. HMGB-domains were identified in number of eukaryotic regulatory factors **[1,8]**. HMGB proteins take an active part in the transcription, DNA reparation, replication and etc. **[1,6,9]**. The most numerous members of the family, the HMGB1 and HMGB2 proteins, have two DNA-binding HMGB domains and an unordered negatively charged C-terminal region **[6]**. We have previously shown that the C-terminal fragment modulates HMGB1 binding to DNA affecting the strength of protein binding to DNA, as well as the structure and stability of the resulting complexes **[10-12]**. Although the particular role of HMGB proteins remains unclear, they were assigned mostly structural functions **[6]**. In some cases they act together with other proteins, which function as a part of rather large DNA-protein complexes. Histone H1 and HMGB1 protein interact with linker DNA in chromatin and there were several attempts to study their mutual influence on DNA binding **[1-4,6]** resulted in somewhat controversial conclusions **[4,5,13,14]**. Yet structural information about their complexes with DNA is still not available. Here we demonstrate a possibility to form artificial structurally ordered nanoscale complexes based on DNA and nuclear proteins H1 and HMGB1. The complexes were characterized using atomic force microscopy (AFM), electronic and vibrational spectroscopy.

2. Results and Figures
Using AFM and molecular absorption spectroscopy in mid-IR regions we have studied structural organization of the complexes between DNA and linker proteins HMGB1 and H1. For FTIR measurements DNA samples were placed in a demountable cell composed of two BaF_2 windows

separated by a 50 mkm Teflon spacer. The IR spectra were registered using Tensor 27 (Bruker) spectrometer equipped with MCT-detector. The spectra obtained are the average of 264 accumulations at the 2 cm-1 resolution. Data preprocessing was performed using OPUS (Bruker) software provided with the spectrometer. Calf thymus DNA lyophilized sodium salt (Sigma) was used without further purification. The concentration and nativity of DNA were determined spectrophotometrically after hydrolysis by 6% perchloric acid **[15]**. For FTIR measurements isotopic substitution of H_2O by D_2O was performed **[16-20]**. HMGB1 and H1 proteins were isolated from calf thymus as described earlier **[10]**. DNA-proteins complexes were prepared by direct mixing of appropriate volumes of component stock solutions.

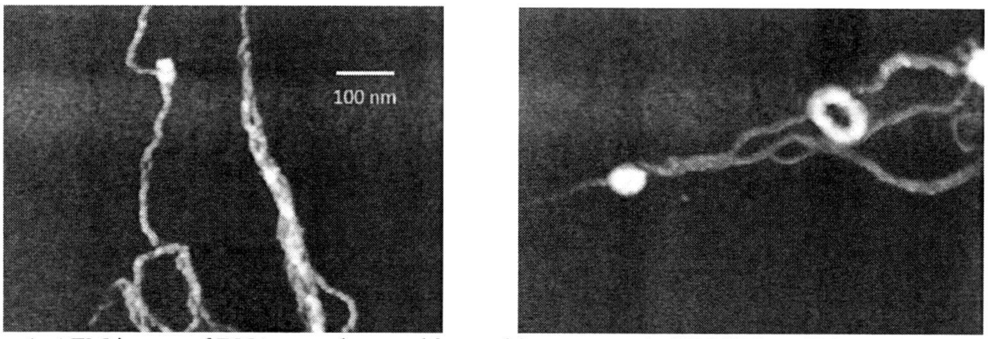

Figure 1. AFM image of DNA complexes with non-histone protein HMGB1 and histone H1 at protein to DNA ratio r = 1 and [HMGB1]:[H1] molar ratio is 1 : 1.

We have demonstrated that depending on the protein to DNA ratio the structural organization of the complexes between DNA and nuclear proteins HMGB1 and H1 can be rather different. It was found that the nanoscale ordered complexes can be obtained at equimolar H1 to HMGB1 ratios (Figure 1). Fibril-like structures are formed in the HMGB1 / H1 complex with DNA, each of which is formed by several DNA molecules. At the same time, protein molecules bind with each other individual DNA molecules in the fiber. Multiple protein – protein and DNA – protein interactions make such structures very stable.

Our earlier studies of the interaction of DNA with histone H1 and HMGB1 protein using circular dichroism and absorption spectroscopy showed that the properties of DNA–protein complexes depend on the protein content and cannot be considered as a mere summing up of the effects of individual protein components **[21]**. The combined action of HMGB1 and H1 proteins stimulates DNA condensation with the formation of anisotropic DNA–protein complexes with typical ψ-type CD spectra. Based on the analysis of vibrational spectra (Figure 2) we conclude, that the structural transition resulted in formation of highly ordered fibrillar structures is accompanied by considerable changes in DNA-protein and protein-protein interaction pattern. The assignment of the major bands in FTIR spectra of the DNA–HMGB1-H1 complex is presented in the Table. Changes in the spectra of the DNA bases at 1693 and 1643 cm^{-1} show dominant interactions between DNA and HMGB1, which occur in the minor groove. Weaker interactions of the histone H1 in the major groove can also be detected as slight changes in intensities at 1678 cm^{-1} as well as those at 1590, 1572, and 1560 cm^{-1} (guanine C=N and C-ND_2). Protein binding is also accompanied by considerable DNA unwinding and bending, resulting in distortions in the mutual orientation of the bases. Changes in the sugar-phosphate backbone vibrations indicate intensive binding of the histone H1 with phosphate groups and changes in geometry of the sugar-phosphate backbone.

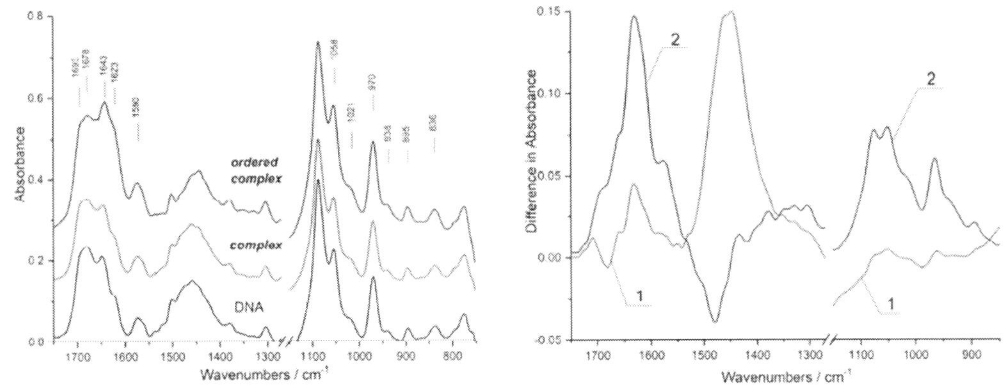

Figure 2. FTIR spectra of DNA complexes with non-histone protein HMGB1 and histone H1 at different protein to DNA ratios: (1) regular unordered complex (red line); (2) ordered complexes corresponding to the AFM image (Figure 1) (blue line).

Structural reorganizations in the system leading to the formation of large complexes of DNA-HMGB1-H1 might be mainly due to the action of the positively charged protein, i.e. histone H1. Those bands in spectra, which were attributed to the interactions of HMGB1's groups with DNA stay mostly stable in position and sign. Therefore, the main role at this stage belongs to the unbound to the bases protein molecule. Moreover, these 'HMGB1'-bands and 'H1'-bands increase in intensity with different sign, correlating with the difference in the charges of the proteins. AFM images illustrate the structure of supramolecular complexes (H1 to HMGB1 ratio is 1).

Table. The assignment of the major bands in FTIR spectra of the DNA–proteins complex. Assignments are based on Refs. **[16-18]**.

Band position in complex, cm^{-1} absorption	Attributions
1730	C=O stretching vibration in Asp side chain
1705	C=O stretching vibration in Glu side chain
1693	Stretching vibrations of C2=O in thymine
1690 1678	Superposition of 1688/1678 cm-1 peaks (unbound) and 1693/1678 cm-1 peaks (complex)
1678	Stretching vibrations of C=O groups in guanine and cytosine
1646, 1643, 1639	In-plane ring vibrations of cytosine, C2=O stretching of cytosine, and/or C4=O4 of thymine
1641	Amide I. In the complex, VCD peak is overlapped with DNA
1623	Adenine, thymine, and (?) cytosine ring vibrations
~1575	Asymmetric stretching of O=C-O- group in Asp/Glu side chains
1502	C=N stretching in cytosine
1452	Amide II
1086	Symmetrical vibrations of O=P=O Bonds
1058	C-OD in plane bending vibration of H-bonded Asp and Glu
1053	Stretching vibrations of C-O bonds in sugar ring
1021	Deoxyribose vibration
970	Sugar ring vibrations
938, 895,836	Sugar ring vibrations; B-form DNA markers

Thus our experimental data indicate that the binding of one of the proteins to DNA stimulates binding to it and the second. When they are simultaneously present in a complex with DNA, the positively charged histone H1 interacts mainly with the phosphate groups of DNA and the C – terminal domain of HMGB1, screening their negative charges and stabilizing the resulting supramolecular DNA – protein complexes. In the absence of the C-terminal domain of HMGB1 both DNA-binding domains of protein interact with DNA **[10-12]**. Studies of the structure of the complex formed by DNA with histone H1 in the presence of HMGB1 have shown that binding of individual proteins is observed both with the sugar-phosphate backbone and with the bases of DNA.

3. Acknowledgments
The authors acknowledge financial support from the Russian Foundation for Basic Research (RFBR) (grant 18-08-01500). The authors express their special gratitude to the Microscopy and Imaging Facility at the University of Calgary for providing instrumentation for AFM imaging. Part of the work was performed using equipment of the Research Park of Saint-Petersburg State University (Center for Optical and Laser Materials Research and Centre for Diagnostics of Functional Materials for Medicine, Pharmacology and Nanoelectronics, Cryogenic Dapartment).

References
[1] Reeves R 2015 *DNA Repair* **36** 122
[2] Crane-Robinson C 2016 *Biochim Biophys Acta* **1859** 431
[3] Fyodorov D V, Zhou B R, Skoultchi A I, Bai Y 2018 *Nat Rev Mol Cell Biol* **19** 192
[4] Chikhirzhina E, Chikhirzhina G, Polyanichko A 2014 *Biomed Spectr Imaging* **3** 345
[5] Chikhirzhina E, Starkova T, Polyanichko A 2018 *Biophysics* **63** 858
[6] Stros M 2010 *Biochim Biophys Acta* **1799** 101
[7] Chikhirzhina E, Starkova T, Polyanichko A 2019 *Biophysics* In press
[8] Raucci A, Maggio S D, Scavello F, D'Ambrosio A, Bianchi M E., Capogrossi M C 2019 *Cell Mol Life Sci* **76** 211
[9] Liu Y, Prasad R, Wilson S H 2010 *Biochim Biophys Acta* **1799** 119
[10] Chikhirzhina E V, Polianichko A M, Skvortsov A N, Kostyleva E I, Houssier C, Vorob'ev V I 2002 *Mol Biol (Mosk)* **36** 525
[11] Polyanichko A M, Chikhirzhina E V, Skvortsov A N, Kostyleva E I, Colson P, Houssier C, Vorob'ev V I 2002 *J Biomol Struct Dyn* **19** 1053
[12] Chikhirzhina E, Polyanichko A, Leonenko Z, Wieser H, Vorobyev V 2010 *Spectroscopy: An International J* **24** 361
[13] Postnikov Y V, Bustin M 2016 *Biochim Biophys Acta* **1859** 462
[14] Cato L, Stott K, Watson M, Thomas J O 2008 *Mol Biol* **384** 1262
[15] Spirin A.S. 1958 *Biokhimiya* **23** 657
[16] Andrushchenko V, Tsankov D, Wieser H 2003 *J Mol Struct* **661–662** 541
[17] Andrushchenko V, van de Sande J, Wieser H 2003 *Biopolymers* **69** 529
[18] Polyanichko A M, Andrushchenko V V, Chikhirzhina E V, Vorob'ev V I, Wieser H 2004 *Nucl Acids Res* **32** 989
[19] Polyanichko A M, Andrushchenko V V, Bouř P, Wieser H 2011 *In: Circular Dichroism: Theory and Spectroscopy*, pp 67-126
[20] Polyanichko A M, Romanov N M, Starkova T Y, Kostyleva E I, Chikhirzhina E V 2014 *Cell Tiss Biol* **8** 352
[21] Chikhirzhina E V, Polyanichko A M, Vorob'ev V I 2011 *Mol Biol (Mosk)* **45** 318

Indications on self mode-locking in a broad area single-section quantum dot laser

D Auth[1], V V Korenev[2], A V Savelyev[2], M V Maximov[2], A E Zhukov[2], and S Breuer[1]

[1] Institute of Applied Physics, Technische Universität Darmstadt, Schlossgartenstraße 7, 64289 Darmstadt, Germany

[2] St. Petersburg Academic University RAS, ul. Khlopina 8/3, 194021 St. Petersburg, Russia

E-mail: stefan.breuer@physik.tu-darmstadt.de

Abstract. Broad-area edge-emitting monolithic mode-locked semiconductor quantum dot lasers emitting at 1.26 μm could potentially serve as ideal sources for the generation of high power broad optical frequency combs for short-reach inter and intra data-center links. In this contribution, the inter-mode beat frequency of a 2 mm long InAs/InGaAs quantum dot laser with a broad-ridge waveguide are studied experimentally. Laser output power, radio-frequency and spectral domain analysis is performed. -3 dB spectral widths ranging from 2 nm to 5.3 nm and the existence of an inter-mode beat frequency at 20.4 GHz with a signal-to-noise ratio from 2 dB up to 11 dB are experimentally confirmed for injection currents from 0.225 A to 1 A. Our results indicate a potential way towards high output power optical frequency comb generation by electrically injected monolithic semiconductor lasers.

1. Introduction

Self mode-locked or frequency-modulated operation of near-infrared and mid-infrared edge-emitting semiconductor lasers allow for the generation of optical frequency combs [1, 2, 3]. In contrast to amplitude-modulated comb formation by passive mode-locking of two- or multi-section semiconductor lasers, the longitudinal modes generated by single-section Fabry-Pérot semiconductor frequency comb lasers are phase-locked without the need for a saturable absorber, or any active optical or electrical modulation and a continuous wave laser emission is generated. As an initial indicator for the coherence among the lasing modes, the appearance of a inter-mode beat signal and a threshold behavior of the inter-mode beat frequency line width suggests self-locking of the longitudinal modes due to internal non-linear effects as experimentally and numerically studied recently for a single-section narrow-ridge InAs/InGaAs quantum dot laser [4]. Above lasing threshold at a certain gain current, multi-modal emission switched from uncorrelated modes to self locked modes, corresponding to a reduction in inter-mode beat frequency line width and integrated intensity noise. The peak signal-to-noise ratio of the inter-mode beat frequency or pulse repetition rate is considered as one of the indicators of mode locking in two or multi-section semiconductor lasers [5]. The underlying physical effects causing self mode-locking or frequency modulated comb generation are yet to be fully understood although a fundamental role is played by four-wave mixing [1, 2, 6, 7, 8, 9, 10, 11]. Yet, the existence of an inter-mode beat frequency signal of a broad-area semiconductor laser and its evolution with biasing conditions, indicating self mode-locked operation and thus optical frequency comb formation, has yet not been reported.

2. Quantum dot broad area laser and experimental set-up

The investigated single-section broad-area quantum dot laser consists of 10 layers of InAs/InGaAs quantum dots separated by GaAs barriers. A light microscope picture is depicted in the inset of Figure 1(a). The waveguide width is 50 μm and both facets are left as-cleaved leading to a reflectivity of around 32 %. The total laser length is 2 mm corresponding to a free spectral range of 20.4 GHz. The laser is mounted on a copper cooling block and temperature stabilized at 21°C. Following light collimation, the laser emission passes an optical isolator (>60 dB isolation ratio) to prevent unwanted optical feedback. Then, the light is fiber-coupled and optical spectra are acquired with an optical spectrum analyzer (spectral resolution: 10 pm). Radio-frequency domain measurements are carried out by a direct detection technique consisting of a fast photo-diode (electrical bandwidth: 45 GHz) and an electrical spectrum analyzer (electrical bandwidth: 50 GHz). The emitted average optical output power is measured by a power meter.

3. Experimental results and discussion

The laser output power characteristics is depicted in Figure 1 for increasing (blue) and decreasing (green) gain injection currents indicating the absence of an optical output power hysteresis. The lasing threshold amounts to 0.13 A (0.13 kA/cm^2) and a slope efficiency of 0.19 W/A is indicated. An exemplary optical spectrum centered at around 1260 nm is depicted in Figure 1(b) for a gain injection current of 0.875 A (0.875 kA/cm^2). At this operation point, the -3 dB optical spectra width corresponds to 5.33 nm covering 50 optical modes.

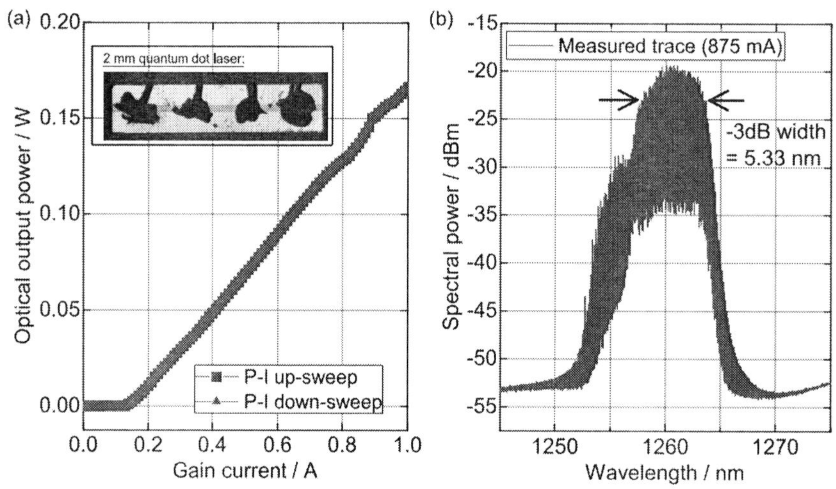

Figure 1. (a) Light output power characteristics of the broad-area quantum dot laser. The inset shows a light microscope picture of the laser. (b) Measured optical multimodal spectrum at a gain current of 0.875 A depicting a -3 dB width of 5.33 nm.

An exemplary full-span radio-frequency spectrum is depicted in Figure 2(a) for a biasing current of 0.875 A. A strong radio-frequency beat note signature denoting the inter-mode beat frequency is apparent at 20.4 GHz with a signal-to-noise floor ratio of 11 dB. The evolution of the inter-mode beat frequency peak signal-to-noise floor ratio is depicted in Figure 2(b) in dependence on the gain injection current. Up to an injected gain current of 0.2 A, the

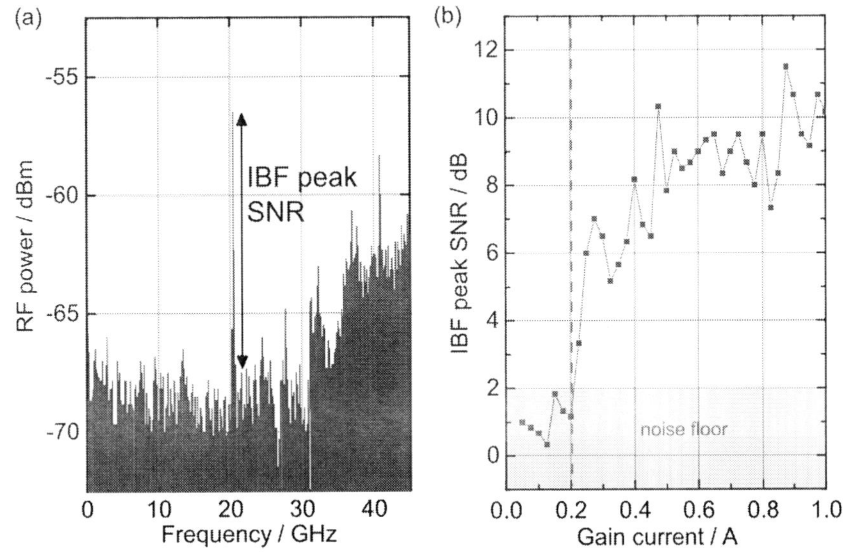

Figure 2. (a) Radio-frequency spectrum of the broad-area quantum dot laser exhibiting an inter-mode beat frequency (IBF) at 20.4 GHz with an inter-mode beat frequency peak signal-to-noise floor ratio (SNR) of 11 dB at a gain injection current of 0.875 A. (b) Inter-mode beat frequency signal-to-noise ratio in dependence on the gain injection current indicating the presence of strong beat frequency beat notes above 0.2 A.

inter-mode beat frequency is below the noise floor of the measurement instrument. For gain injection currents above 0.2 A however, strong inter-mode beat frequencies up to 11 dB of inter-mode beat frequency peak signal-to-noise floor ratio are recovered. The existence of such a threshold behavior in inter-mode beat frequency peak power has also been found experimentally in narrow-ridge quantum dot lasers and corresponding line width and noise characteristics have been linked to self mode-locking operation of a 1 mm long narrow-ridge single-section quantum dot laser [4, 2]. Figure 3(a) depicts the colour-coded optical spectra evolution in dependence on

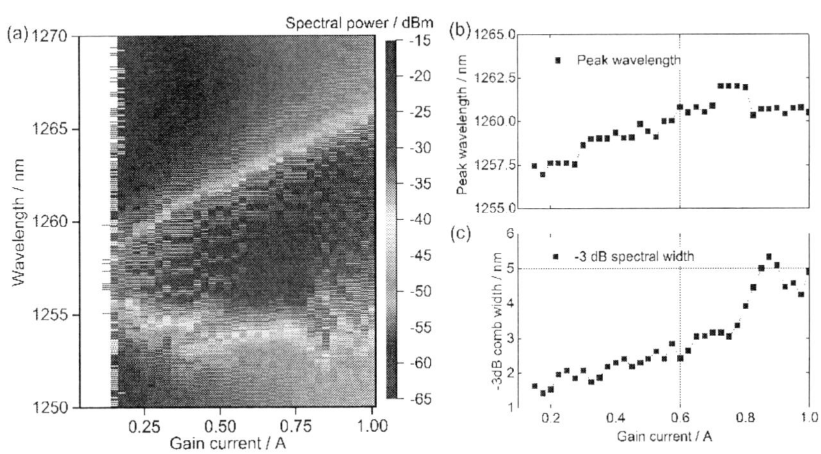

Figure 3. (a) Colour-coded optical spectra evolution in dependence on the gain injection current. (b) Peak center wavelength and (c) -3 dB spectral width.

the gain injection current in biasing current steps of 25 mA. The emitted multimodal spectra are centered at around 1260 nm. Figure 3(b) depicts the recorded emission wavelengths of the optical mode in the optical spectrum with highest spectral power as a function of applied gain current. A red-shift of the peak wavelengths with increasing gain injection currents by 4.7 nm/A are obtained. The -3 dB spectral widths of the optical spectra emitted by the broad-area quantum dot laser are presented in Figure 3(c). With increasing gain injection current, the optical spectral width increases up to a maximum value of 5.33 nm covering 50 optical lines.

4. Conclusion

We have experimentally studied the radio-frequency and spectral emission characteristics of a 2 mm long edge-emitting Fabry-Pérot monolithic InAs/InGaAs quantum dot laser with a 50 μm broad-area waveguide. For injection currents up to 1 A, the laser emits multi-modal emission centered at 1260 nm and with -3 dB spectral widths ranging from 2 nm to 5 nm, covering up to 50 modes. At injection currents below 0.225 A, the inter-mode beat signal is below the noise floor of the measurement equipment. Starting at an injection current of 0.225 A and up to 1 A, a sharp radio-frequency signature at 20.4 GHz, the inter-mode beat frequency, has been observed. A corresponding threshold behaviour has recently been found experimentally in narrow-ridge quantum dot lasers and related inter-mode beat frequency line width and intensity noise threshold characteristics had been associated to a transition from uncorrelated modes to self locked mode. Within the injection current range investigated in this contribution, the inter-mode beat frequency peak signal-to-noise ratio increased from 6 dB up to 10 dB. This threshold behaviour with increasing injection current suggests an initial indication for self mode-locking or frequency modulated comb generation in broad-area quantum dot lasers. Future studies will focus on the injection current dependence of the inter-mode beat line widths.

Acknowledgements

This work is supported by the Russian Foundation for Basic Research (project #18-502-12081), Deutsche Forschungsgemeinschaft (project #389193326) and by the Ministry of Higher Education and Science of the Russian Federation (project #3.9787.2017/8.9).

References

[1] Piccardo M, Kazakov D, Schwarz B, Chevalier P, Amirzhan A, Hillbrand J, AlMutairi S Z, Wang Y, Xie F, Lascola K, Becker S, Hildebrandt L, Weih R, Belyanin A and Capasso F 2019 *IEEE Journal of Selected Topics in Quantum Electronics* **25** 1–12 ISSN 1077-260X

[2] Bardella P, Columbo L L and Gioannini M 2017 *Opt. Express* **25** 26234–26252

[3] Hillbrand J, Auth D, Piccardo M, Opacak N, Strasser G, Capasso F, Breuer S and Schwarz B 2019 In-phase and anti-phase synchronization in a laser frequency comb (*Preprint* 1908.08504)

[4] Weber C, Columbo L L, Gioannini M, Breuer S and Bardella P 2019 *Opt. Lett.* **44** 3478–3481

[5] Moskalenko V, Latkowski S, Tahvili S, de Vries T, Smit M and Bente E 2014 *Opt. Express* **22** 28865–28874

[6] Rosales R, Merghem K, Martinez A, Akrout A, Tourrenc J P, Accard A, Lelarge F and Ramdane A 2011 *IEEE Journal of Selected Topics in Quantum Electronics* **17** 1292–1301

[7] Lelarge F, Dagens B, Renaudier J, Brenot R, Accard A, van Dijk F, Make D, Gouezigou O L, Provost J G, Poingt F, Landreau J, Drisse O, Derouin E, Rousseau B, Pommereau F and Duan G H 2007 *IEEE Journal of Selected Topics in Quantum Electronics* **13** 111–124

[8] Faugeron M, Lelarge F, Tran M, Robert Y, Vinet E, Enard A, Jacquet J and Dijk F V 2013 *IEEE Journal of Selected Topics in Quantum Electronics* **19** 1101008–1101008

[9] Akrout A, Shen A, Brenot R, Dijk F V, Legouezigou O, Pommereau F, Lelarge F, Ramdane A and Duan G H 2009 *IEEE Photonics Technology Letters* **21** 1746–1748

[10] Gosset C, Merghem K, Martinez A, Moreau G, Patriarche G, Aubin G, Ramdane A, Landreau J and Lelarge F 2006 *Applied Physics Letters* **88** 241105

[11] Hugi A, Villares G, Blaser S, Liu H C and Faist J 2012 *Nature* **492** 229–233

Simulation of a THz radiation source on AlGaAs/GaAs superlattices

A S Dashkov[1], L I Goray[1]
[1]St. Petersburg Academic University, Khlopin 8/3, Let. 'A', Saint-Petersburg 194021, Russian Federation

Abstract. We report on the results of simulations of the carrier transport processes in THz radiation sources. The current-voltage characteristic, dipole matrix elements and LO-phonon relaxation time were obtained using the Ensemble Monte Carlo method, the shooting and effective mass methods. The electron-longitudinal phonon interaction assumed as a dominant mechanism of the carrier scattering for the current-voltage characteristic. The weakly coupled AlGaAs/GaAs superlattices considered for simulations.

1. Introduction

Light emitting devices based on semiconductor superlattices nowadays have a broad range of applications: from medical diagnostics up to optical communication systems. First ideas of a light emission phenomenon in the semiconductor superlattices based on the intersubband transitions were proposed by R. Kazarinov and R. Suris in 1971 [1] and implemented in the first quantum cascade laser (QCL) by a group of researchers led by F. Capasso in 1994[2].

Since then the QCL production has expanded and enhanced, but challenges such as the development of THz radiation sources remain to be accomplished. In order to improve device in the early stages of the development, computer simulations might be applied to predict operational characteristics. One of the most valuable applications of simulation is computing current-voltage characteristics and dipole matrix elements of the structure of a device.

In this research, the Monte Carlo, shooting and effective mass methods are used. The weakly coupled semiconductor superlattices were considered as the initial design for THz sources due to the fact that they experimentally demonstrate negative differential conductivity regions on their current-voltage characteristics. Two types of GaAs/AlGaAs superlattices (Table 1) with doping $5 \cdot 10^{14} cm^{-3}$ were investigated at temperature 65 K. The current-voltage characteristic, dipole matrix elements and LO-phonon relaxation time were obtained.

Table 1. The GaAs/AlGaAs investigated structures description.

Structure #	Barrier thickness, nm	Well thickness, nm	Al fraction in AlGaAs, %
1	1.5	18.1	10
2	2	10	30

2. Calculations methods

The carrier transport is assumed as a sequence of transitions of electrons between energy states. Therefore, an initial stage of the simulation technique is to obtain eigenvalues and eigenvectors, i.e., a numerical solution of Schrodinger-Poisson equation system for a given conduction band profile. For this matter, the shooting and effective mass methods applied. At the second stage, the scattering rates are calculated. In a particular case, only two dominant scattering mechanisms are considered for THz radiation sources:

- electron-longitudinal phonon (LO) interaction;
- electron-electron scattering.

The theory of scattering on LO phonons is introduced with the Frohlich Hamiltonian and using the Fermi's golden rule the phonon lifetime obtained (Figure 1b) as the following [3]:

$$\tau_{ij}^{-1} = \frac{m^* e^2 E_{LO}}{2\hbar^3 \varepsilon_p} \int_0^{2\pi} \frac{\int dz dz' \psi_i(z)\psi_j(z) e^{-Q|z-z'|} \psi_i(z')\psi_j(z')}{Q} d\theta, \qquad (1)$$

where E_{LO} – is the LO phonon energy, m^*– electron effective mass, Q – phonon wavevector, ψ_i – a wavefunction for the energy state with index i, ε_p – relative permittivity. The numerator of (1) also referred to as a form-factor $F(Q)$.

Electron-electron scattering can be considered in two ways:

- screened Coulomb potential (short-range interaction);
- electron-plasmon coupling (long-range interaction).

The second one will be neglected, due to the fact that it becomes significant only at higher doping levels. Considering electron-electron scattering as a two-electron process, the scattering rate can be written in the form [4]:

$$W_{ik,j} = \frac{m^* e^2}{32\pi\hbar^3 S\varepsilon} \sum_{i_0, j_0, k_0} f_{i_0}(k_0) \int_0^{2\pi} \frac{d\theta}{(q+q_s)^2} |F_{i i_0 j j_0}(q)|^2, \qquad (2)$$

where $q = k - k'$ and k, k' – initial and final wavevectors of an electron, k_0 – init. wavevector of the second electron, q_s – screening wavenumber in the approximation of single band [4], $f_{i_0}(k_0)$ – carrier distribution function, i, j – initial and final subband numbers of the first electron, i_0, j_0 – of second electron, θ – angle between $k_0 - k$ and $k_0' - k'$.

The form-factor defined in a slightly different way:

$$F(q) = \left| \int_{-\infty}^{+\infty} dz \int_{-\infty}^{+\infty} dz' \psi_i(z)\psi_{i_0}(z') e^{-q|z-z'|} \psi_j^*(z)\psi_{j_0}^*(z') \right|^2. \qquad (3)$$

The second electron is chosen randomly from the ensemble during the scattering simulation process. In (2) the 1/2 factor was introduced to consider only the interaction of electrons with antiparallel spin since it prevails over the interaction between electrons with parallel spin.

The derived values of scattering rates considered in the third simulation stage that exploits the ensemble Monte Carlo (EMC) method, i.e., statistical sampling of the scattering events for a large ensemble ($N_e = 10^5$) of carriers [4]. The EMC method considers periodic boundary conditions, which implies simulation of the scattering between states of the central and two nearest periods. The scattering event from one state to the other was simulated using a pseudo-random generator that exploits normalized probabilities of the scattering.

3. Results and the proposed structure

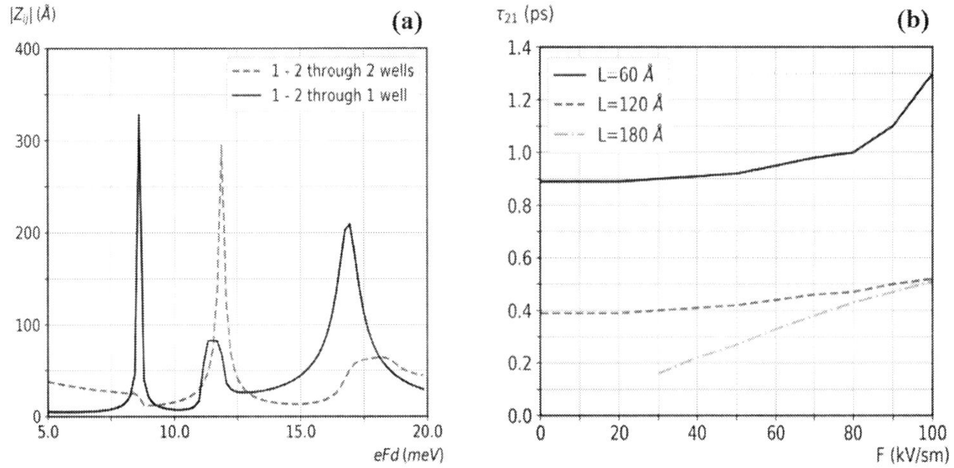

Figure 1(a, b). **(a)** Dipole matrix elements of structure 1; **(b)** The LO-phonon lifetime for $k_{init} = 0$. L – the width of the well, F – field applied to the structure, d – the thickness of the structure.

The first structure - the superlattice proposed in the literature [5] used as the benchmark for the simulation methods. Its dipole matrix elements (Figure 1a) exhibit the resonance behavior for transitions with tunneling through one and two quantum wells near 10 kV/cm. The dipole matrix elements remain high even beyond resonance regions, which is in the good agreement with the different theoretical approach [5].

The LO-phonon lifetimes (Figure 1b) were calculated for different width of the well. The relaxation time increases with the field due to the increase of transition and LO-phonon energy difference, that weakens scattering rate through Q in (1). The demonstrated behavior was also suggested in previous theoretical consideration [3]. The obtained current-voltage characteristic (Figure 2a) exhibits satisfactory agreement with experimental data. The obtained results demonstrate plateau regions similar to the experimentally observed.

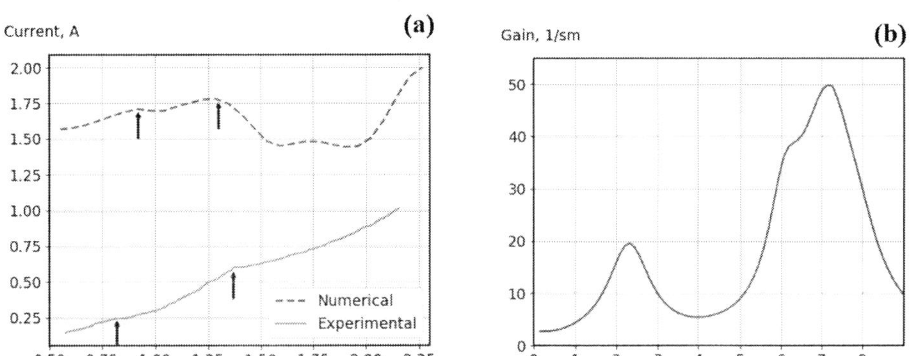

Figure 2(a, b). **(a)** Simulation results and experimental data of current-voltage characteristic of structure 2. The arrows emphasize plateau areas; **(b)** Simulation results for spectral gain.

The second structure is the proposed structure of a THz radiation source. For the described structure the transition occurs between first energy level and second energy level through three quantum wells. For this transition to happen the barriers of the structure shall be weak. The weakness of barriers implies reasonable barrier thicknesses and low amount of Aluminum fraction. The latter is also important for obtaining less strained structures. The new design demonstrates sufficient gain for the wavelengths near expected transition (Figure 2b).

4. Conclusion

We described simulation techniques for GaAs/AlGaAs superlattice investigation. Obtained dipole-matrix elements, LO-phonon lifetimes show good agreement with theoretical estimations proposed earlier and current-voltage exhibit satisfactory agreement with experimental data. For the proposed structure the gain coefficient was calculated showing the possibility of light emission.

References

[1] Kazarinov R F and Suris R A 1972 Electric and electromagnetic properties of semiconductors with a superlattice Sov. Phys. Semicond. 6 120
[2] Faist J, Capasso F, Sivco D L, Sirtori C, Hutchinson A L and Cho A Y 1994 Quantum cascade laser Science 264 553–6
[3] Ferreira R and Bastard G 1989 Evaluation of some scattering times for electrons in unbiased and biased single-and multiple-quantum-well structures *Phys. Rev. B* 40 1074
[4] Jacoboni C and Lugli P 2012 *The Monte Carlo method for semiconductor device simulation* (Springer Science & Business Media)
[5] Andronov A A, Dodin E P, Zinchenko D I, Nozdrin Y N, Ladugin M A, Marmalyuk A A, Padalitsa A A, Belyakov V A, Ladenkov I V and Fefelov A G 2015 Stimulated emission at transitions between Wannier–Stark ladders in semiconductor superlattices *JETP Lett.* **102** 207–11

Upconversion luminescence of CaF_2-SrF_2-ErF_3 single crystals upon 1.5 μm laser excitation

A A Lyapin[1*], A S Ermakov[1], S V Kuznetsov[2], S V Gushchin[1], P A Ryabochkina[1], V A Konyushkin[2], A N Nakladov[2], P P Fedorov[2]

[1] Institute of Physics and Chemistry, National Research Ogarev Mordovia State University, 430005, Saransk, Russia
[2] Prokhorov General Physics Institute of the Russian Academy of Sciences, 119991, Moscow, Russia

*Corresponding author: andrei_lyapin@mail.ru

Abstract. In present work, the spectral-luminescent properties of CaF_2-SrF_2-ErF_3 (CaF_2-SrF_2:Er) single crystals were investigated. The upconversion luminescence (UCL) of CaF_2-SrF_2: Er in the visible and infrared spectral regions upon excitation of the $^4I_{13/2}$ level Er^{3+} ions were studied for the first time. The absolute energy yields of the UCL upon laser excitation at 1.5 μm are determined for CaF_2-SrF_2:Er. The possible application of fluoride crystals for increasing the efficiency of solar cells was discussed.

1. Introduction

Today, photon upconversion is a rapidly developing area of photonic with a large direction of potential applications, for example, solid-state laser, laser beam visualizers, bioimaging, biometrics for laser scanning microscopy, temperature sensors, solar cell and etc [1-6]. In present paper we will focus on the investigation of UCL of fluoride crystals for application to enhance the efficiency of solar cells. For example, absorption threshold of the solar cell cannot allow to crystalline silicon solar cells utilized 20% of the energy incident from the sun [1, 7]. Upconversion materials can decrease these losses by generating one high-energy photon from two or three lower-energy photons.
Previously UCL of fluorides MF_2 (M = Ca, Sr) doped with Er^{3+} ions with fluorite-type structure upon 1.5 mcm laser excitation was investigated in different papers [8-15]. Fluoride materials have a high value of transmittance in a wide spectral range (from 0.16 to 11 microns), low phonon energy, clustering effect et al. Here we presented results of investigation UCL of mixed fluoride crystals CaF_2-SrF_2:Er upon excitation of $^4I_{13/2}$ level.

2. Experimental

CaF_2-SrF_2 single crystals with 0.5 mol%, 1 mol%, 1.5mol% and 2 mol% Er^{3+} ions doping were grown by the vertical directed crystallization method (Bridgman method) (Table 1). Diameter and thickness of samples were 0,8 cm and 0,17 cm, respectively. Absorption spectra of single crystals were recorded with a Perkin–Elmer spectrophotometer Lambda 950. The luminescence of the Er^{3+} ions excited by fiber laser at 1531.8 nm was recorded using a Horiba FHR1000 spectrometer. The focused excitation beam diameter on the samples was 212 μm. The integrating sphere method was used to measure the absolute photoluminescence energy yield. The system consists of the OL IS-670-LED integrating sphere, an OL-770 UV/VIS (Gooch & Housego) spectroradiometer, and a monochromator-spectrograph M833 (Solar LS).

Table 1. UCL energy yields, chromaticity coordinates and color temperatures of the luminescence of CaF_2-SrF_2:Er phosphors.

Compositions	Contraction	B_{en}, %	X	Y	T (K)
$0{,}6CaF_2$-$0{,}395SrF_2$-$0{,}005ErF_3$	CaF_2-SrF_2:Er(0,5%)	5,1	0,5681	0,4195	1737
$0{,}6CaF_2$-$0{,}39SrF_2$-$0{,}01ErF_3$	CaF_2-SrF_2:Er(1%)	9,1	0,5073	0,4770	2585
$0{,}6CaF_2$-$0{,}385SrF_2$-$0{,}015ErF_3$	CaF_2-SrF_2:Er(1,5%)	9,9	0,4756	0,5067	3136
$0{,}6CaF_2$-$0{,}38SrF_2$-$0{,}02ErF_3$	CaF_2-SrF_2:Er(2%)	11,4	0,4331	0,5464	3904

3. Results

Upconversion materials can be used as materials for efficiency enhancement of solar cell. To develop UC phosphors for the above application field, the absorption properties, luminescence properties, photoluminescence energy yield and color temperature were investigated in present paper.

Absorption spectra allow to estimate the spectral region of the potential upconversion of solar irradiance. Figure 1 presents the 300-K absorption spectra of the Er^{3+} ion in the CaF_2-SrF_2:Er single crystal. The absorptions are due to the transitions from the $^4I_{15/2}$ ground level to the $^4G_{11/2}$, $^2H_{9/2}$, $^4F_{3/2}$+$^4F_{5/2}$, $^4F_{7/2}$, $^2H_{11/2}$, $^4S_{3/2}$, $^4F_{9/2}$, $^4I_{9/2}$, $^4I_{11/2}$, $^4I_{13/2}$ excited levels.

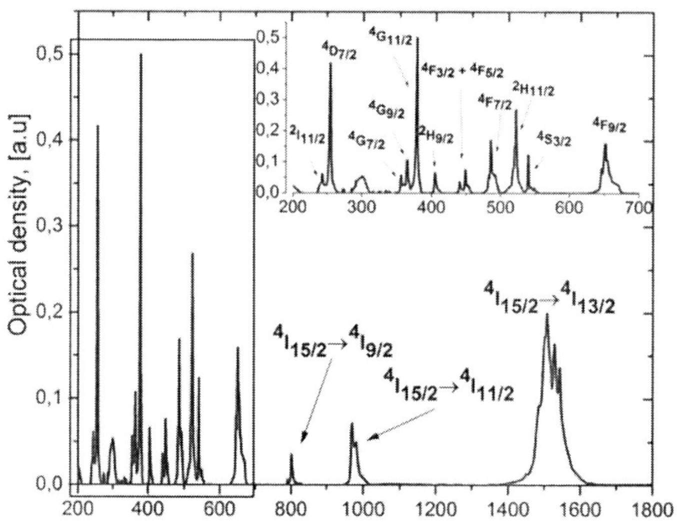

Figure 1. Absorption spectra of CaF_2-SrF_2:Er(2%) in the spectral range 200-1800 nm.

The spectral dependences of the absorption cross-sections for the transition $^4I_{15/2}\rightarrow{}^4I_{13/2}$ of Er^{3+} ions were showed in Fig.2. The peaks absorption cross-sections of CaF_2-SrF_2 crystals doped with 0.5 mol.%, 1 mol.%, 1.5 mol.% and 2 mol.% of Er^{3+} ions at wavelength 1508 nm are $7{,}1\times10^{-21}$, $6{,}0\times10^{-21}$, $6{,}5\times10^{-21}$, and $6{,}6\times10^{-21}$ cm^2, respectively. Small difference in the absorption peaks can be explained by the accuracy of determining the concentration of Er^{3+} ions. From analyses absorption of CaF_2-SrF_2:Er crystals and solar irradiance spectrum above atmosphere and at surface [16-19], we conclude that CaF_2-SrF_2:Er crystals potentially can convert low energy photons in wide spectral range 1450-1600 nm.

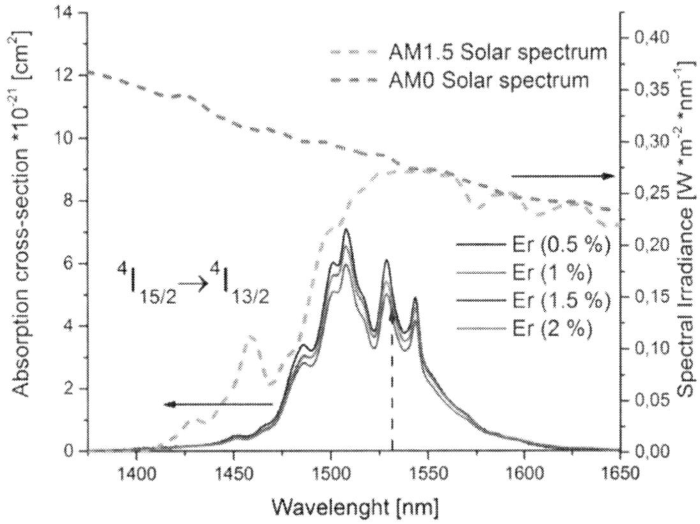

Figure 2. The spectral dependence of the absorption cross-sections for the transition $^4I_{15/2}\rightarrow^4I_{13/2}$ of Er^{3+} ions for CaF_2-SrF_2:Er crystals and solar irradiance spectrum above atmosphere (AM0 Solar spectrum) and at surface (AM1.5 Solar spectrum).

Upon 1531.8 nm laser excitation UCL of CaF_2-SrF_2:Er was observed in visible ($^2H_{9/2}\rightarrow^4I_{15/2}$, $^2H_{11/2}\rightarrow^4I_{15/2}$, $^4S_{3/2}\rightarrow^4I_{15/2}$, $^5F_{9/2}\rightarrow^4I_{15/2}$) and infrared ($^4I_{9/2}\rightarrow^4I_{15/2}$, $^4I_{11/2}\rightarrow^4I_{15/2}$) spectral ranges. The most intensive luminescence was observed at around 980 nm which corresponding to the $^4I_{11/2}\rightarrow^4I_{15/2}$ transition of Er^{3+} ions. In the visible spectral range red luminescence dominates on the green and blue band for all samples. The intensity ratio of the UCL bands of CaF_2-SrF_2:Er has good agreement with the spectral sensitivity of silicon solar cells [20-21]. From Fig. 3a we can conclude that the spectral power of the UCL in range 360-1100 nm increases with the increase of the concentration of the Er^{3+} ions until to 2%. The investigation of chromaticity coordinates showed that the UCL of CaF_2-SrF_2:Er crystals is characterized by color temperatures of 1737– 3904 K at an incident power density of 283 W/cm^2 (Fig. 3b).

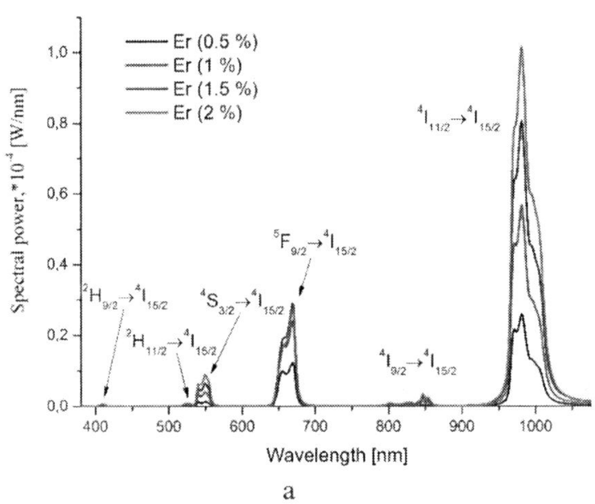

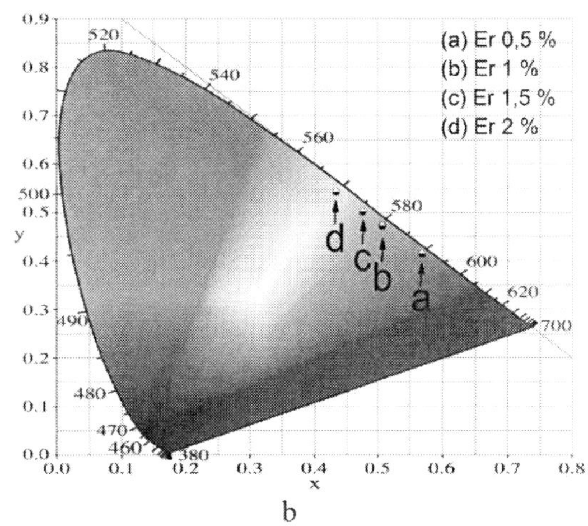

Figure 3. (A) Spectral power of UCL of CaF_2-SrF_2:Er crystals upon excitation at 1532 nm. (B) The CIE chromaticity diagram of CaF_2-SrF_2: Er.

Next we estimated efficiency of UCL of CaF_2-SrF_2:Er crystals (Table 1). The energy yield is defined as the UCL power per that of absorbed power by luminescence materials. The maximum energy yield of the UCL in the spectral range of 380-1100 nm for the CaF_2-SrF_2:Er (2%) crystal was 11.4%. Interesting absorption and luminescence properties and high values of energy yields allow to us in the future investigate UCL of Er^{3+} heavily doped CaF_2-SrF_2:Er single crystals.

4. Conclusion

In present paper spectral-luminescence properties of CaF_2-SrF_2:Er crystals were investigated. Wide absorption spectrum of CaF_2-SrF_2:Er crystals allows to upconvert radiation with a wavelength range 1450-1600 nm. Intensive UCL at around 980 nm and 650 nm good appropriate for spectral sensitivity of crystalline silicon cells. The investigated samples showed sufficiently high values of the UCL energy yields 11.4 %. The present results indicate that CaF_2-SrF_2:Er single crystals is a promising upconversion material for increasing the efficiency of crystalline silicon solar cells.

Acknowledgments

This work was supported by the Russian Foundation for Basic Research and Government of Republic of Mordovia, grant no. 18-42-130011.

References

[1] Goldschmidt J, Fischer S 2015 *Adv. Optical Materials* 3 510-537
[2] Burbano D, Naccache R, Capobianco J 2015 *Handbook on the Physics and Chemistry of Rare Earths* 47 273-347
[3] Dong N, Pedroni M, Piccinelli F, Conti G, Sbarbati A, Enrique Ramı´rez-Herna´ndez J, Martı´nez Maestro L, Carmen Iglesias-de la Cruz M, Sanz-Rodriguez F, Juarranz A, Chen F, Vetrone F, Capobianco J, Garcı´a Sole J, Bettinelli M, Jaque D and Speghini A 2011 *ACS Nano* 5 8665-8671
[4] Wang F, Banerjee D, Liu Y, Chen X, Liu X 2010 *Analyst* 135 1839-1854
[5] Ende B , Aarts L, Meijerink A 2009 *Phys. Chem. Chem. Phys.* 11 11081-11095
[6] Trupke T, Shalav A, Richards B, Würfel P, Green M 2006 *Solar Energy Mater. Solar Cells* 90 3327–3338
[7] Fischer S, Ivaturi A, Jakob P, Krämer K, Martin-Rodriguez R, Meijerink A, Richards B, Goldschmidt J 2018 *Optical Materials* 84 389-395
[8] Lyapin A , Gushchin S , Kuznetsov S , Ryabochkina P , Ermakov A , Proydakova V , Voronov V, Fedorov P, Artemov S, Yapryntsev A and Ivanov V 2018 *Opt. Mater. Express* 8 1863-1869
[9] Lyapin A, Ryabochkina P, Gushchin S, Kuznetsov S, Chernov M, Proydakova V , Voronov V, Fedorov P 2018 *Optics and Spectroscopy* 125 537-542
[10] Ivanova S, Pelle F, Tkachuk A, Joubertd M, Guyotd Y, Gapontzeve V 2008 *Journal of Luminescence* 128 914-917
[11] Maciel G, Araujo C 1997 *Physical Review* 55 6335-6342
[12] Pelle F, Ivanova S, Guillemoles J 2011 *EPJ Photovoltaics* 2 20601
[13] Pollack S, Chang D, Moise N 1986 *Journal of Applied Physics* 60 4077-4086
[14] Pollack S, Chang D 1988 *Journal of Applied Physics* 64 2885-2893
[15] Pollack S, Chang D 1990 *Optical and Quantum Electronics* 22 S75-S93
[16] Gueymard C 2001 *Solar Energy* 71 325–346
[17] Gueymard C 2004 *Solar Energy* 76 423-453
[18] Wehrli C 1985 *World Radiation Center (WRC)* Publication No. 615
[19] Neckel H, Labs D 1981 *Solar Physics* 74 231-249
[20] Green M 1987 Seventh E.C. *Photovoltaic Solar Energy Conference* 681-687
[21] Field H 1997 *26rd IEEE Photovoltaic Specialists Conference* 471-474

Memristive properties of solar cells based on perovskite and polymer composite

M A Masharin[1], D S Gets[1], G S Verkhoglyadov[2], S V Makarov[1], A A Zahidov[3]
[1] Department of Physics, ITMO University, St. Petersburg 197101, Russian Federation
[2] Institute of Physics, Nanotechnology and Telecommunications, Peter the Great St. Petersburg Polytechnic University, St. Petersburg 195251, Russian Federation
[3] Alan G. MacDiarmid NanoTech Institute, University of Texas at Dallas, Richardson, TX 75083, USA

Abstract. Organic-inorganic halide perovskites recently have emerged as a perspective material for using in solar cells. This material has many perspective properties, one of them is the ion migration. Modern ionic memristors, electric circuit elements that can change its resistance depending on the current flowing through it, is based on this phenomenon. In this paper, we present the experimentally obtained memristic properties in the perovskite solar cell, which contains composite of perovskite and polymer as absorbing layer instead of pure perovskite.

1. Introduction

Nowadays, everyone knows three main elements of an electrical circuit: a resistor, a capacitor, and an inductor that bind four physical quantities: electrical current, charge, voltage, and charge flux. The resistor connects the voltage and current; capacitor - charge and voltage; inductor - current and charge flux. In 1971, Leon Chua [1] suggested that there is a fourth element - the "memristor". Memristor is a nonlinear, bipolar, passive circuit element capable to change its resistance depending on the charge flowing through it. After power off, the memristor remembers its last state, that can be used as a non-volatile memory cell [3]. In the future, it is possible to implement computing systems using memristors, in which information will be stored and processed by the same device [1].

There are two types of memristors on the principle of operation: ionic thin-film memristors (Titanium dioxide memristors [1], Polymeric (ionic) memristors [2], layered memristor [3] and others) and memristors based on spin-magnetic interaction (Spintronic Memristors [4], Spin Torque Transfer (STT) [5]). Ionic memristors are based on the phenomenon of ion migration through the film. The latter rely on the spin separation of electrons due to the change of the external magnetic field. All this leads to the appearance of the hysteresis phenomena of I – V characteristics. Memristors have two distinct states: low- and high-resistance, one of which is called on-state, the second is called off-state [1]. Each of these states is characterized by set voltage and reset voltage, respectively. The current voltage (I-V) characteristic of the memristor has clearly pronounced surges at these voltages, forming a hysteresis loop, which is an evidence of the memristive properties.

Usually, ionic memristors (for example TiO_2-based memristors) consists of two layers: with high and low oxygen concentration. The high-resistance state is obtained by moving oxygen ions from its oxygen rich part to the deplete part of the device (Figure 1a) under the electric field. The material doped with oxygen ions has very high-resistance and ensures the "off" state of the memristor. When the applied voltage is inverse to the previous one, oxygen ions move backward, reducing the sample resistance

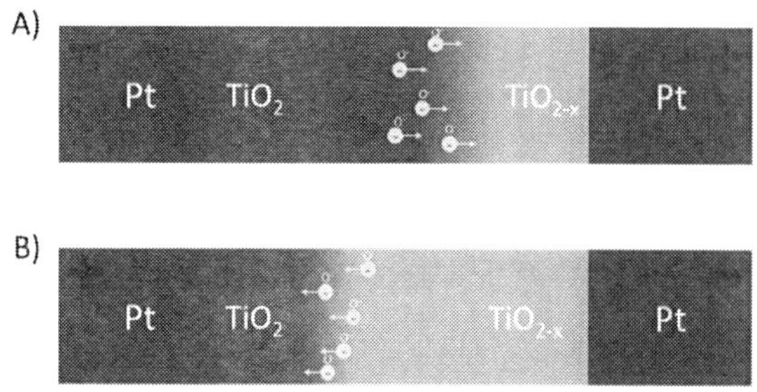

Figure 1. Illustration of the oxidation / reduction process of an ionic memristor. **a** Aplication of external voltage makes oxygen ions move towards cathode and dope TiO_{2-x} layer and therefore lower the device resistance. **b** Aplication of reverse bias makes oxygen ions move back and deplete the TiO_{2-x} layer with consequent increase of device resistance.

(Figure 1b) and switching the memristor to the on state.

Organometal halide perovskites (OHP) have emerged 10 years ago as a family of materials prospective for optoelectronics, photovoltaic and photonic application [6]. From this point, OHP attracted tremendous attention from scientific community since new material demonstrate many interesting properties like high absorption, high exciton energy at room temperature, lasing and many more. But the most important feature of OHP is solution processing which enables smooth band gap tuning and production of low cost and highly efficient solar cells and light emitting diodes [2].

The record efficiency of perovskite solar cells (PSC) reached more than 23% [7] surpassing its senior competitors: organic solar cells (record PCE 17.3%) [8] and dye-sensitized solar cells (record PCE 14.3%) [9]. And its characteristics, has already surpassed the record indicators of silicon solar cells (a record PCE of 24.7%) [10]. Along with the development of PSC, scientists revealed hysteresis phenomena of the I-V characteristics, which are associated with the phenomenon of ion migration in perovskite. Under applied external electric voltage ions of cations (methylammonium, MA; formamidinuim, FA; cesium, Cs) and anions (Cl^-, Br^-, I^-) migrate towards perovskite interfaces [4,5] (Figure 2). Thus, by maintaining the PSC under voltage, anions and cations dope the interfaces of the perovskite with transport layers, which leads to formation of a p-i-n structure. However, if we turn off the voltage, the ions will migrate back, and the p-i-n structure will disappear.

The presence of ion migration opens the way for the implementation of perovskite-based memristors [11]. Perovskite based device in the states with the formed p-i-n structure and without it have different resistances and, thus, IV characteristics will have a hysteresis loop. However, to implement such a

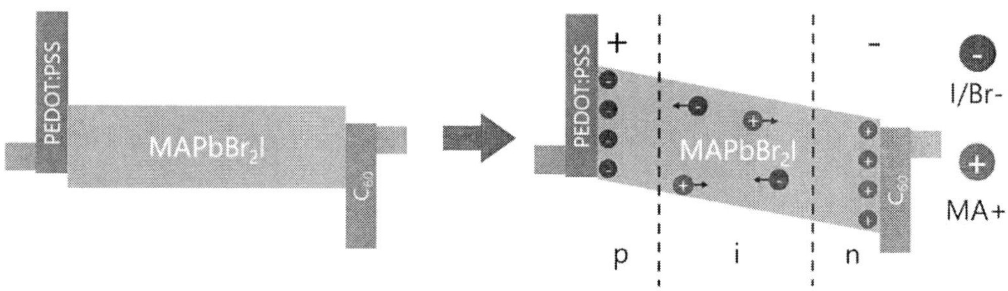

Figure 2. Ion migration in $MAPbBr_2I$-based perovskite solar cell. **a** Non-biased device. **b** Device under voltage biasing. Ions of perovskite (MA^+, Br^-, I^-) move towards perovskite interfaces which form a p-i-n structure inside the provskite layer.

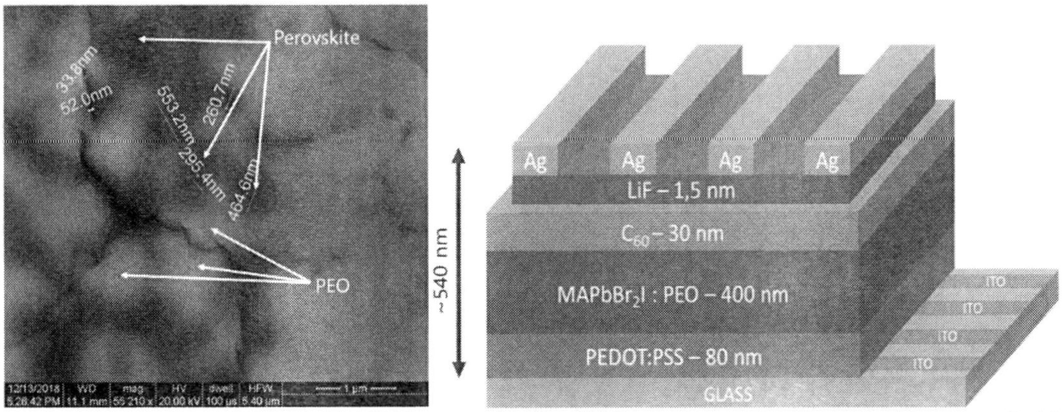

Figure 3. **a** SEM image of perovskite:PEO composite film. White areas correspond to perovskite covered with PEO and grey areas correspond to the pristine perovskite. **b** Design of perovskite solar cell.

device, it is necessary to resolve a number of problems, one of which is the backward ion migration in the absence of an external electric voltage. This leads to absence of fully developed both memristor loops since p-i-n structure which was formed under high voltage will disappear. To solve this problem we added a small amount of polyethylene oxide (PEO) into the perovskite layer. PEO is an ion conducting polymer and it is usually used as host for development of polymer batteries. In our case, PEO fills the space between the perovskite grains. Figure 3a demonstrates a SEM image of the $MAPbBr_2I$:PEO film light areas correspond to perovskite covered with PEO since it has high electric resistance and dark areas correspond to pristine perovskite. PEO allows to restrain ions as they move under applied external voltage and slow down their migration. The presence of the PEO can help maintain the p-i-n structure after removal of external voltage.

2. Results

The investigated solar cells had the structure: ITO / PEDOT:PSS / $MAPbBr_2I$:PEO / C60 / LiF / Ag. (Figure 3b). The absorbing layer is a composite of $MAPbBr_2I$ and PEO. The mass of the polymer was 0.004% of the perovskite mass. Perovskite precursors (MAI and $PbBr_2$) were dissolved in a DMF: DMSO solvent mixture (1:1) in a molar concentration of 0.8M. PEDOT: PSS was used as a hole transport layer and it was deposited on a glass substrate covered with ITO by spincoating and annealed at 150 C during 10 minutes. On the PEDOT:PSS, a $MAPbBr_2I$:PEO composite was deposited by two-step spincotaing technique with pouring of diethyl ether at the tenth second of the second step. Then in a vacuum deposition chamber a 30 nm of fullerene (C_{60}), 1.5 nm layer of LiF and 60 nm of Ag were deposited on the perovskite film. The I-V characteristics of the devices were obtained under the AM1.5G irradiation (100mW/cm^2). Before measuring the I-V characteristics, the devices were biased at small forward voltage ($V_{bias} \approx 1 - 1.5$V). Measurement of I-V characteristics were conducted from $V_{bias} \rightarrow 0 \rightarrow V_{bias}$.

Figure 4a shows the I-V characteristics of devices with PEO and without it, measured after 1.5 V biasing for 1 minute. In the I-V characteristic of the perovskite-polymer composite device, a hysteresis loop is observed, which is almost absent in the I-V characteristics of the device without the polymer. During voltage biasing the ions in perovskite film gradually migrate to the boundaries, form p-i-n structure and penetrate into the PEO. With the start of measurement and voltage decrease, the perovskite ions try to migrate back to the bulk of the perovskite layer, but the polymer between perovskite grains restrain them, and the ions detach from the boundaries later then in case of with pure perovskite, where nothing can slow down the backward ion migration.

With decreasing the voltage, we observe state switching to high-resistance state (off-state) at 0.7V that is probably associated with backward ion migration in film and p-i-n structure disappearance. At low voltage device is in off state with high resistance. With increasing voltage, at 1.3 V perovskite ions

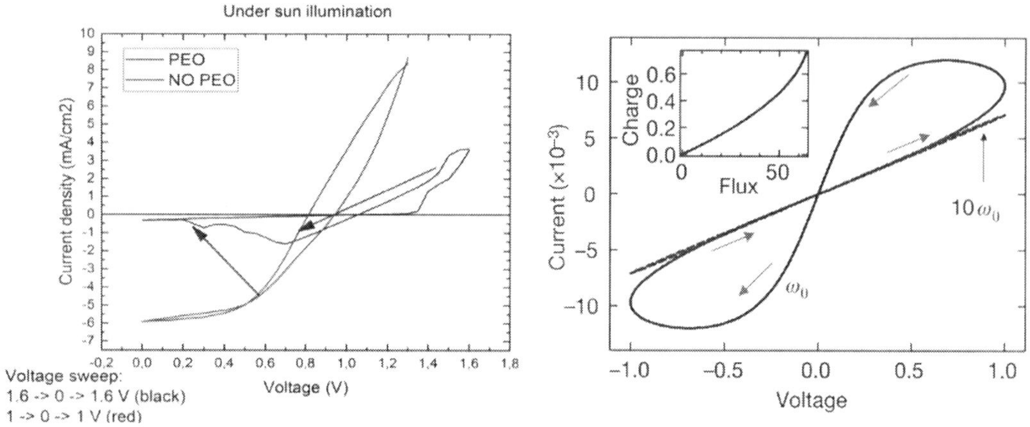

Figure 4. **a** PV characteristics of PCE with polymer in the perovskite and without it;
b Theoretical I-V characteristic of the memristor [1]

become able to migrate through PEO to boundaries and form p-i-n structure inside the perovskite layer and switch device to the low-resistance state (on-state). This I-V dependence forms a two hysteresis loop (Figure 4a) that is very similar to I-V of a memristor (Figure 4b).

We demonstrated memristive properties of perovskite devices by introduction of small amount of ion conducting polymer into the perovskite layer and utilization of device biasing for development of p-i-n structure. Finding a ways to control ionic migration and utilization it for development new types of devices are more preferable than just cupping it by introduction of multiple cations, since it can provide a cheap realization of 4th main element of electrical circuit.

Acknowledgments
The work was supported by the Russian Science Foundation (Project 19-73-30023)

References

[1] Strukov D, Snider G, Stewart D, Williams R 2008 *Nature* **453** 7191

[2] Krieger, J. H.; Spitzer, S. M. 2004 *Proceedings of the 2004 Non-Volatile Memory Technology Symposium*, p. 121

[3] Bessonov A, Kirikova M, Petukhov D , Allen M , Ryhänen T, Bailey M 2014 *Nature Materials* **14** 2

[4] Wang X, Chen Y, Xi H, Dimitrov D, 2009, *IEEE Electron Device Letters* **30** 3

[5] Huai, Y. 2008 *AAPPS Bulletin*, **18** 6

[6] Lee M, Teuscher J, Miyasaka T, Murakami T, Snaith H 2012, *Science* **338** 6107

[7] Jeon N, Na H, Jung E, Yang T, Lee Y, Kim G, Shin H, Seok S, Lee J, Seo1 J 2018, *Nature Energy* **3** 682

[8] Meng L, Zhang Y, Wan X, Li C, Zhang X, Wang Y, Ke1 X, Xiao Z, Ding L, Xia R, Yip H, Cao Y, Chen Y 2018, Science **361** 6407

[9] Kakiage K, Aoyama Y, Yano T, Oya K, Fujisawa J, Hanaya M, *2015 Chemical Communications* **51** 88

[10] Taguchi M, Yano A, Tohoda S, Matsuyama K, Nakamura Y, Nishiwaki T, Fujita K, Maruyama E, *2013 IEEE Journal of Photovoltaics* **4** 96

[11] Eames C., Frost J.M., Barnes P.R., O'regan B.C., Walsh A., Islam M.S., 2015 *Nature communications* 6, 7497.

[12] Gets D, Saranin D, Ishteev A, Haroldson R, Danilovskiy E, Makarov S, Zakhidov A, 2019 *Applied Surface Science* **476** 486

[13] Yan K, Dong B, Xiao X, Chen S, Chen B, Gao X, Hu H, Wen W, Zhou J and Zou D 2017 *Scientific reports* **7** 6025

Polarization sensitivity of ZnSe single crystals based structures

I A Zharikov[1], V Yu Rud'[1, 2], Yu V Rud'[2, 3], V V Davydov[1, 2] and E I Terukov[2,3]

[1]Peter the Great Saint Petersburg Polytechnic University, Saint Petersburg 195251, Russia
[2]Department of Ecology, Russian Research Institute for Phytopathology, 143050, Moscow Region, Odintsovo district, B. Vyazyomy, Russia
[3]Departament Solid State Physics, A.F. Ioffe Physico-Technical Institute

Abstract. The work is devoted to study of polarization photosensitivity of structure on the basis of ZnSe single crystals. To produce such photosensitive structure we used heat treatment of ZnSe substrates on dry air. The polarization photosensitivity was detected and studied in relation to their manufacturing conditions and photoregistration configuration. We conclude that such photosensitive structure can be used as polarimetric short wavelength photodetector.

1. Introduction
The compound of ZnSe is among promising II-VI compounds and is widely used in developing short wavelength semiconductor electronic devices and also as data displaying systems [1, 2, 3]. Moreover, ZnSe is one of the typical materials of phases of variable composition; therefore, its properties "keep track" of thermodynamic equilibrium parameters, which requires precise control of the material existence conditions during growth. In the present work we report the initial results of an experimental study of the photoelectric structures based on ZnSe single crystals in linearly polarized light, which have allowed us to draw conclusions regarding their potential for application in polarization photonics.

2. The method
To date, growth methods from vapor have been applied to grow high-quality crystals. However, growth method from melt has attracted attention because it is more suitable to obtain large bulk crystals for relatively short growth duration. $A_{II}B_{VI}$ single crystals have a number of unique properties which significantly depend on nano-scale defect type and concentration. Native point defects generating as a result of deviation from stoichiometric composition influence on luminescence, conductivity, and other structure sensitive properties similar to dopants. Electrical and optical properties of ZnSe crystals are affected by deviations from stoichiometry [4] which can varied within the homogeneity limits of ZnSe. ZnSe is II-VI Semiconductors with zinc-blende structure straight-band semiconductors forming tetrahedrally directed covalent bonds from hybridized s and p atomic orbitals. In an elementary unit cell containing one anion and a cation, the diatomic basis contains a cation at and the anion at relative to the lattice point (Fig. 1). Zinc selenide (ZnSe) is an intrinsic semiconductor yellow color (Fig. 2) with a band gap of about 2.72 eV at 300 K.

Photosensitive structures were derived from n-ZnSe single crystals grown from a melt with the composition close to the stoichiometry of the compound. Single crystals has a sphalerite structure and a lattice parameter the corresponding value [4].

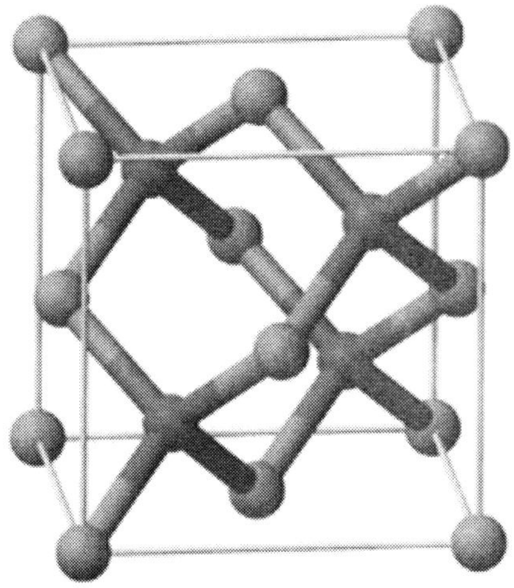

Figure 1. Crystal structure of ZnSe lattice.

To produce the photosensitive structures of the n-type single cristals substrates via crystallographic plane of splitting (100) with average dimensions of 5 × 4 × 2 mm^3 were subjected to for heat treatment at a temperature of 723 K in dry air. As the result reproduces the colored p-type layer. It was observed that the heat treatment of ZnSe single cristals on dry air lead to the change of color in ZnSe with a sub-surface region. So such changes during heat treatment in the air seems to be related to interference of the formed oxide layers. We used the well-known method of polarization photoabsorption spectroscopy to evaluate the quality of the frontal plane of such photosensitive structures and their possible application in polarization photonics.

3. Results and discussion

The photovoltaic effect was detected in all the studied structures. Linearly polarized light along the normal to the photodetector surface the photosensitivity of Schottky barriers as well as type structures A and B are practically independent of the spatial orientation of the electric field vector of the light wave E relative to crystallographic axes in a crystal substrates. This fact corresponds to the isotropic nature of the photoactive absorption in the substrates ZnSe having a sphalerite structure. Therefore, natural photopleochroism in structures based on ZnSe not observed [10].

In all structures derived from n-ZnSe, only "induced photopleochroism" is detected when angle of incidence of linearly polarized radiation at the surface becomes different from 0 degree [11].

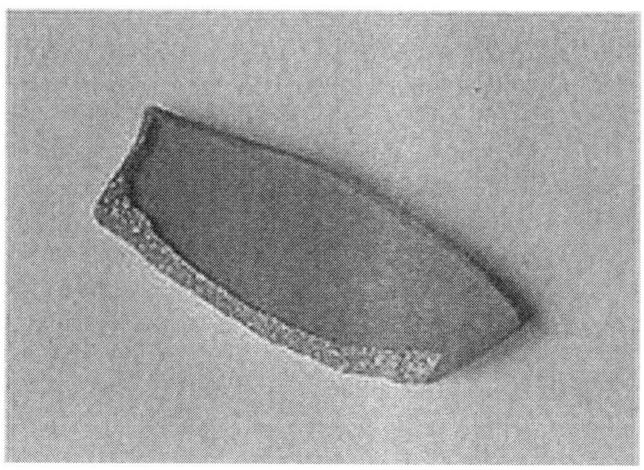

Figure 2. The external form of ZnSe crystal.

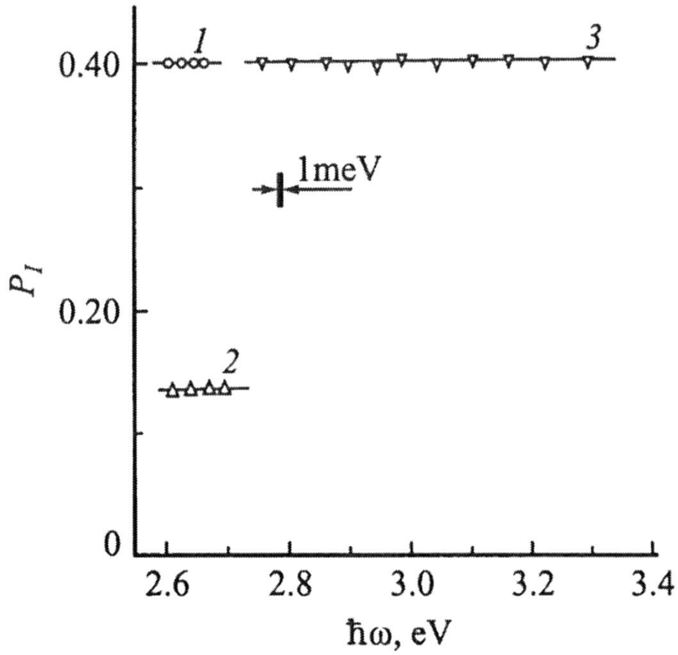

Figure 3. Induced photopleochroism coefficient spectras for different structure basen on ZnSe with different time of treatment (curve 1- 7 min, curve 2- 10 min, curve 3 -20 min).

Only induced photopleochroism was detected in all photosensitive structure on the basis of ZnSe single crystals. This type of photopleochroism observed when the angle of incidence of linearly polarized light different from 0.

When structures illuminated along (001), the photocurrent is polarization –independent, such n-p transitions corresponding functionally to GaAs diodes. Zn Se structures prepared on (100) plates exhibit polarization indicatrices of photocurrent in agreement with the generalized Malius law [5, 6, 7, 8, 9]. The main laws for the polarization photosensitivity investigations of such ZnSe based structures consists in following. All structures demonstrates the angular dependencies of short circuit

Changing the geometry of photodetection also affects markedly the polarization photosensitivity of such ZnSe based conversion devices. Spectral dependences of the induced polarization photosensitivity for such ZnSe structures plotted on Fig. 3. Coefficient of induced photopleochroism reach 40 % as well as the light enter to the active region of structures. This correspond to the analysis carried out in [9-21].

It was concluded ZnSe photosensitive structure based on ZnSe single crystals can be used as polarimetric photodetector for short wavelength range in polarization photonic.

References

[1] Itoh S, Nakano K and Ishibashi A 2000 *J. Cryst. Growth* **214–215** 1029
[2] Bairamov B Ch, Rud' V Yu and Rud' Yu V 1998 *MRS Bull* **23** 91
[3] Nesheva D, Aneva Z, Scepanovic M J, Bineva I, Levi Z, Popovic Z V and Pejova B 2010 *J. Phys.: Conf. Ser.* **253** 12035
[4] *Simashkevich A V 1980 Heterojunctions Based on II-VI Semiconductor Compounds (Shtiintsa, Kishinev)*
[5] Kesamanly F P, Rud' V Yu and Rud' Yu V 1996 *Semiconductors* **33** 5 483
[6] Kesamanly F P, Rud' V Yu, Rud' Yu V and Shock G V 1999 *Semiconductors* **33**(3) 483

[7] Rud V Y, Rud Y V and Schock H.W. 1999 *Sol. St. Phenomena* **67–68** 421
[8] Wal'ter T., Rud V Y, Rud Y V and Schock H. W. 1997 *Semiconductors* **31** 681
[9] Rud V Y, Rud YV and Terukov E I 2016 *J. Opt. Technol.* **83(5)** 275
[10] Rud` V Y, Rud`Y V, Shpunt V C and Terukov E I 2016 *Optical Memory & Neural Networks (Information Optics)* 25(1) 40
[11] Myazin N S, Smirnov K. J, Davydov V V, Logunov S E 2017 *Journal of Physics: Conference Series* **929**(1) 012064
[12] Davydov R V, Antonov V I, Moroz A V 2018 *Proceedings of the 2018 IEEE International Conference on Electrical Engineering and Photonics* **8564378** p 236-239
[13] Davydov R V, Antonov V I 2017 *Journal of Physics: Conference Series* **929 (1)** 012040
[14] Rykin E V, Moroz A V, Smirnov K J, Davydov V V, Yushkova V V 2018 *MATEC Web of Conference* **245** 12002
[15] Moiseev E I, Kryzhanovskaya N V, Polubavkina Yu S, Maximov M V, Kulagina M M, Troshkov S I, Zadiranov Yu M, Lipovskii A A, Mukhin I S, Guina M, Niemi T, Zhukov A 2017 *Journal of Physics: Conference Series* **769** 012056
[17] Grebenikova N M, Smirnov K J, Artemiev V V, Davydov V V, Kruzhalov S V 2018 *Journal of Physics: Conference Series* **1038(1)** 012089
[18] Myazin N S, Logunov S E, Davydov V V, Rud' V Yu, Grebenikova N M and Yushkova V V 2017 *Journal of Physics: Conference Series* **929 (1)** 012064
[19] Grebenikova N M, Smirnov K J, Davydov V V , Rud V Yu and Artemiev V V 2018 *Journal of Physics: Conference Series* **1135 (1)** 012055
[20] Moroz A V, Davydov V V, Rud V Yu, Rud Yu V, Shpunt C Ch and Glinushkin A P 2018 *Journal Physics: Conference Series* **1135(1)** 012060
[21] Podstrigaev A S, Davydov R V, Rud' V Yu and Davydov V V 2018 *Lecture Notes in Computer Science (including subseries Lecture Notes in Artificial Intelligence and Lecture Notes in Bioinformatics)* **11118 LNCS** 624-630

Ultraviolet cathodoluminescence of pure zinc aluminate $ZnAl_2O_4$

S V Lisovskii[1], A V Meganov[1], V R Khrustov[1,2], V V Ivanov[1]
[1]Phystech School of Electronics, Photonics and Molecular Physics, Moscow Institute of Physics and Technology, Dolgoprudny 141701, Russia
[2]Institute of Electrophysics, Ural Branch of the RAS, Ekaterinburg 620016, Russia
E-mail: lisovskii@phystech.edu

Abstract. The features of the intrinsic cathodoluminescence of pure zinc aluminate $ZnAl_2O_4$ having a radiation maximum in the range 250-270 nm are studied. The kinetics of the afterglow of a sample excited by short (~ 2 ns) pulses is analyzed. Analysis and possible explanation of the influence of the type and composition of the atmosphere, in which zinc aluminate was annealed, is given. Possibilities for a multiple increase in the cathodoluminescence intensity of zinc aluminate for use in bactericidal lamps are demonstrated.

1. Introduction

Zinc aluminate ($ZnAl_2O_4$) is well known for use as a matrix for a phosphor with an activator emitting in the visible range. However, it has recently been discovered that pure zinc aluminate has also a rather intense intrinsic cathodoluminescence (CL) in the ultraviolet region (250-270 nm) [1]. It is shown that the device based on this phosphor and cold cathode has bactericidal properties comparable to the UV mercury lamp [2]. Zinc aluminate is chemically stable, sufficiently resistant to overheating and charging, and can be used to create radiating devices with a long service life. To create competitive devices based on zinc aluminate, it is necessary to achieve high CL intensity of zinc aluminate and the position of the spectrum most effective for sterilization tasks.

According to the assumption of scientists who discovered the phenomenon of intrinsic CL of zinc aluminate, this phenomenon is associated with the appearance of oxygen vacancies in zinc aluminate crystals [1, 2]. This assumption is based on the enhancement of the luminescence of a phosphor synthesized in a reducing atmosphere, i.e. in conditions of oxygen deficiency. There is other method for obtaining oxygen vacancies in spinel [3]. At the same time a certain shift in the spectrum is observed, depending on the synthesis conditions, on the composition of the atmosphere and on the artificially created excess of zinc or aluminium. It is known that zinc aluminate contains three types of own defects on which the corresponding luminescence is possible: oxygen vacancies and zinc substitutions with aluminum or aluminum with zinc [4], also associated with the formation of oxygen vacancies.

Another group of researchers found the dependence of the intensity of useful cathodoluminescence on the purity of zinc aluminate as well as on the particle size of the phosphor powder [5]. In addition, there are reports of the enormous influence of the composition of the auxiliary substances used during the synthesis on the intensity of cathodoluminescence [6].

This work continues exploring the features of the cathodoluminescence of zinc aluminate: the influence of the quasistationary and pulsed excitation regime on the intensity of cathodoluminescence, the effect of the annealing atmosphere on the cathodoluminescence spectrum, the afterglow kinetics, and the phenomenon of sample burning.

2. Materials and Methods

Zinc aluminate powders were synthesized by solid-phase synthesis from chemically pure zinc oxide and aluminum oxide powders with a 2% excess of zinc oxide (air, 5 hours, 1200 °C). The XRD analysis confirmed that the sample was zinc aluminate. Then the samples were further annealed in a reducing atmosphere (H_2 and CO, 5 and 10 hours, 1200 °C). Samples were studied in a scanning electron microscope (SEM) Quanta 200MK2 (FEI). The CL spectra were collected with the help of a special Chroma CL mirror (Gatan) and HR4000CG-UV-NIR (Ocean Optics). The electron beam standing time at a single point varied from 50 ns to 30 μs to estimate the effect of the quasistationary excitation regime of zinc aluminate on the intensity of CL. The applied accelerating voltage of the electron beam varied from 10 to 30 kV. In addition, the samples were studied with a pulsed cathodoluminescent analyzer of materials (pulse duration ~ 2 ns, accelerating voltage 140-160 kV) [7], and the kinetics of the afterglow of the samples was studied with the help of a streak camera (Princeton Instruments).

3. Results

Although for all samples cathodoluminescence was observed in the ultraviolet range, the radiation spectrum of different samples was somewhat different (Figure 1). It can be seen that samples that have undergone additional annealing in CO show a shorter wavelength spectrum, while annealing in H_2, on the contrary, shifts the spectrum to the right. Also worth noting is the steepness of the left edge of the spectrum in CO and its some bimodality.

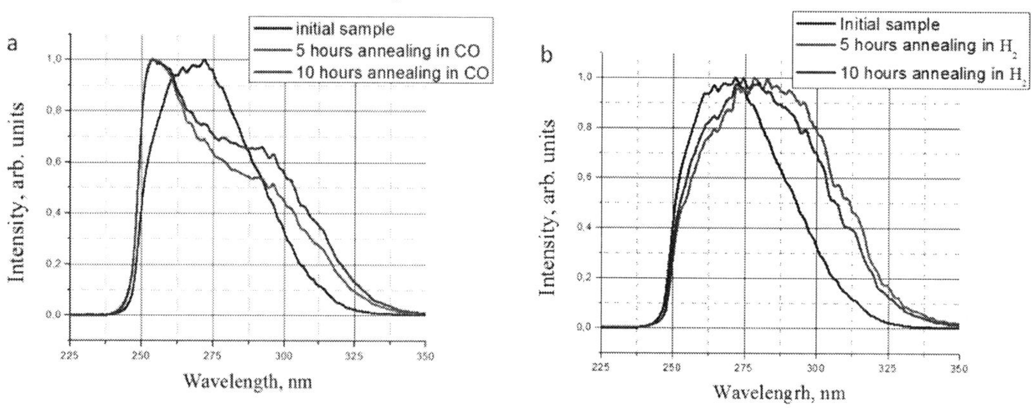

Figure 1 (a, b). Cathodoluminescence spectra of zinc aluminate samples synthesized in an air atmosphere and subjected to additional annealing for 5 and 10 hours (a) in CO; (b) in H_2.

From the point of view of the change in the spectrum, the result demonstrated by burning the sample (sample temperature reached 1000 °C or more) in a vacuum of the SEM under the intense electron beam current of 1 μA is also of interest. The sample displayed a gradually attenuating cathodoluminescence within a few minutes after burning out; the spectrum of cathodoluminescence was shifted to a large extent in the long-wave region (Figure 2). Visually, the place of burning was blackened.

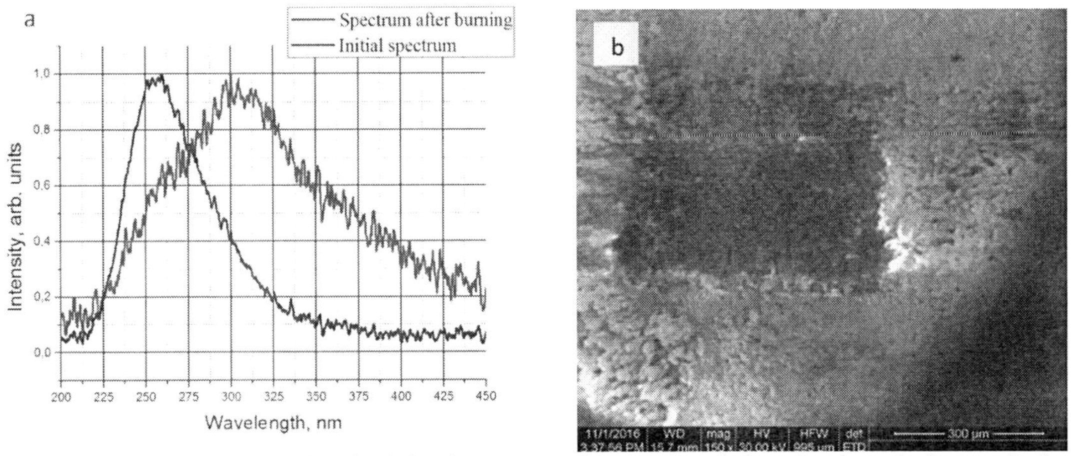

Figure 2 (a, b). (a) Spectra of cathodoluminescence of a sample of zinc aluminate before and after burning; (b) an image of burning in the SEM.

Remaining measurements of cathodoluminescence in the SEM were carried out at lower electron beam currents (10 100 nA), which do not lead to a sample burning (sample temperature did not reach 300 °C). The averaged cathodoluminescence intensity as a function of the standing time at each point (from 50 ns to 30 μs) from a certain fixed scanning site yielded the result shown in Figure 3. It can be seen that the increase in the intensity of cathodoluminescence occurs inversely proportional to the standing time at the point (excitation time), i.e. the dependence is close to hyperbolic.

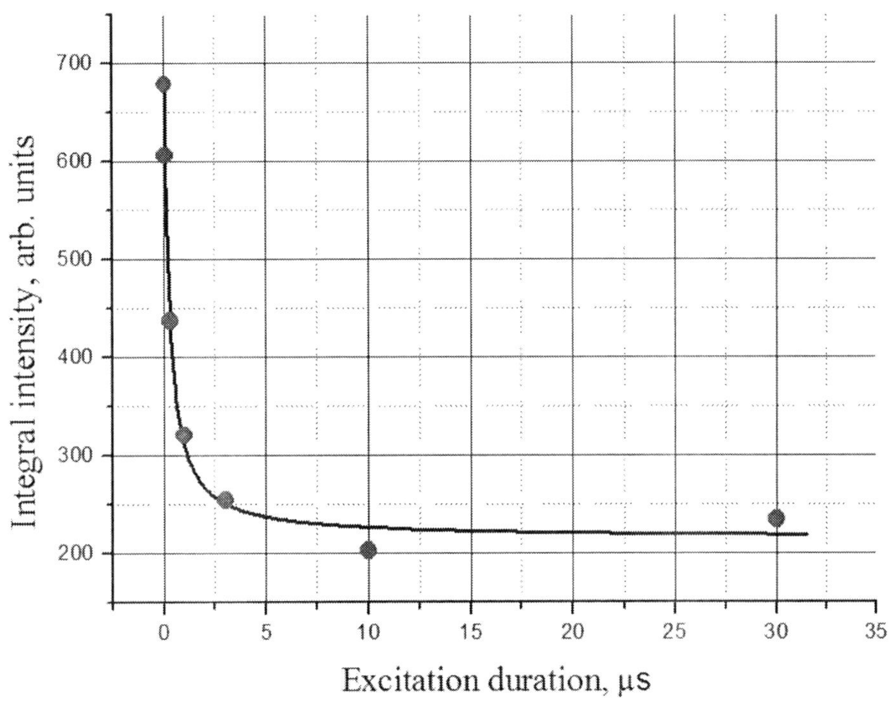

Figure 3. Integral intensity of cathodoluminescence from the sample area of zinc aluminate as a function of the time of electron beam standing at a point.

Measurements with the help of a streak camera and a pulsed cathodoluminescent analyzer of materials made it possible to evaluate the nature of the afterglow kinetics and compare the cathodoluminescence intensities for various samples. It was found that the sample with the highest intensity is prepared in an atmosphere of CO – 5 hours of additional annealing increased intensity by 1.5 times, 10 hours – by 3.5 times. At the same time, additional annealing in the H_2 atmosphere led to the opposite results – after 5 hours the intensity decreased by 8 times, after 10 hours by 7 times. As for the afterglow kinetics, the sample annealed in the air has the largest characteristic afterglow time – ~30 μs (the time after which the cathodoluminescence intensity falls by a factor of e). For CO samples, characteristic afterglow time was 13 to 21 μs, for H_2 it was 5-10 μs. After a short increase in the intensity of the luminescence, it begins to subside. At the time of decline, the character of the dependence of the intensity of the afterglow on time is close to hyperbolic (Fig. 4, for example); although in the early stages (less than 0.1 ms) the data may be fitted with the exponent as well.

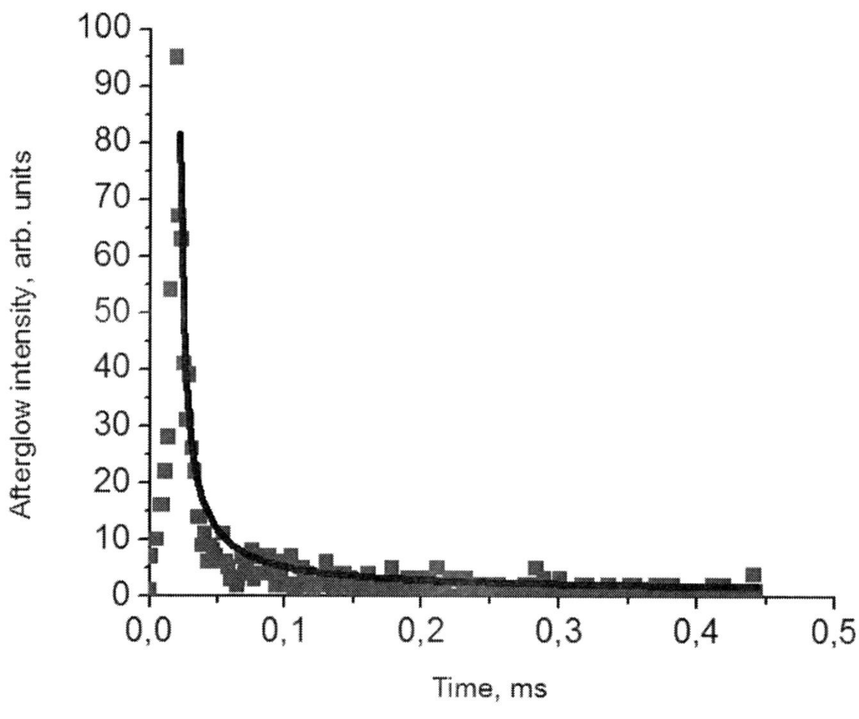

Figure 4. Intensity of afterglow of zinc aluminate as a function of time.

4. Discussion

The hyperbolic character of the dependence of the afterglow intensity on time will indicate that the luminescence does not occur inside a single luminescence center and is not recombination but occurs between charge carriers localized on electron and/or hole traps and free charge carriers in the valence band/conduction band [7].

The time scale of the afterglow can also be estimated from the dependence of the integrated cathodoluminescence intensity on the sample area of the zinc aluminate on the time of the electron beam standing at the point. Dependence is hyperbolic because the number of afterglow excitations for individual points (pixels) depends on the scanning speed of the section. Respectively, the integral intensity of the afterglow from the site is inversely proportional to the time at the point. If we assume that the integral value of the current that excites the section does not depend on the standing time at the point, then the total intensity of the cathodoluminescence at the moment of excitation and after will be a hyperbola, shifted upward by a constant $I = b / t + a$, where t is equal to the time at a point, a characterizes the intensity of cathodoluminescence at the instant excitation, and b characterizes the afterglow. The value $b / a = \tau$ corresponds to the characteristic afterglow time. For Figure 3 $\tau = 0.4$ μs, which is two orders of magnitude lower than the characteristic time measured with the help of a streak camera. Such discrepancies may be due to incomplete illumination of the afterglow of cathodoluminescence in the quasi-stationary excitation regime (pulse time greater than 50 ns) in comparison with the pulsed excitation regime (pulse time ~ 2 ns).

A different shift in the spectrum and a different change in the intensity of the cathodoluminescence for various reducing annealing atmospheres indicates that not only the reducing character of the atmosphere matters, but also its composition does. Taking into account the fact that the intensity grew and the spectrum shifted to the left during annealing in the CO atmosphere, we can assume an increased value of carbon proper in these phenomena. First, this is consistent with a decrease in intensity and a shift in the right-hand side of the cathodoluminescence spectrum after annealing in an H_2 atmosphere in which a part of the carbon obtained by annealing in air could be lost. Secondly, it conforms to a decrease in intensity and a shift to the right side when the sample is burned, which judging from the visual blackening is accompanied by the release of carbon. Thirdly, this is in agreement with the data of the researchers who found a 25-fold increase in the intensity of zinc aluminate in the case when the carbon containing Li_2CO_3 was used as a flux in the course of chemical synthesis [6].

The mechanism of carbon influence can be the following: a carbon atom due to high temperatures is able to dope zinc aluminate, penetrating some depth of the crystal lattice and localizing near oxygen vacancies as, for example, here [8, 9]. Then the luminescence shift is explained by the change in the energy level of the trap in the band structure due to the carbon localized nearby; the bimodal character of the spectrum can be associated with the presence of both types of oxygen vacancies – with and without a localized carbon series, and the ratio of these types of oxygen vacancies should vary with depth; an increase in the intensity of the cathodoluminescence may be due to the immobilization by the carbon of the oxygen vacancy and, as a consequence, the shift of the thermodynamic balance of the number of oxygen vacancies in the sample in the direction of increase (the carbon-immobilized oxygen vacancy becomes incapable of vanishing, but carbon does not affect the creation of new oxygen vacancies). The ability of carbon to immobilize the oxygen vacancy in some other oxides is discussed, for example, here [10].

5. Conclusions

Ultraviolet cathodoluminescence of zinc aluminate occurs from own defects, which are traps for free charge carriers. Probably, these are oxygen vacancies.

The use of a quasistationary cathodoluminescence excitation regime makes it possible to increase the intensity of cathodoluminescence by a factor of at least three, but this increase is several orders of magnitude lower than that which can be achieved in the pulsed excitation regime according to the kinetics of the afterglow of cathodoluminescence of zinc aluminate.

The annealing atmosphere affects the spectrum and intensity of cathodoluminescence, and the effect is not only on the reducing nature of the atmosphere, but also on its composition. Probably the cathodoluminescence of zinc aluminate is affected by carbon, which interacts with oxygen vacancies.

Acknowledgements

This work was supported by the Ministry of Education and Science of the Russian Federation (project № RFMEFI57517X0160).

References

[1] Kominami H *Journal of Advanced Research in Physics* 2011 **2** 2
[2] Ishinaga T, Iguchi T, Kominami H, Hara K, Kitaura M, Ohnishi A 2015 *Physica Status Solidi C*, **12** 797
[3] Seeman V, Feldbach E, Kärner T, Maaroos A, Mironova-Ulmane N, Popov A I, Shablonin E, Vasil'chenko E, Lushchik A 2019 *Optical Materials* **91** 42
[4] Cornu L, Gaudon M, Jubera V *Journal of Materials Chemistry C* 2013 **1** 5419
[5] Ehmenina I V, Sheshin E P, Chadaev N N *Nano- and Microsystems Technology* 2010 **2** 39
[6] Bakhmet'ev V V, Malygin V V, Podsypanina N S, Lebedev L A, Keskinova M V, Sychev M M 2017 All-Russian Scientific and Technical Conference *Innovative Materials and Technologies in Design* p 53
[7] Solomonov V I, Michailov S G, Lipchak A I, Osipov V V, Shpak V G, Shunailov S A, Yalandin M I, Ulmaskulov M R 2006 *Laser physics* **16** 126
[8] Benemanskaya, G V, Dement'ev P A, Kukushkin, S A, Osipov A V, Timoshnev S N 2019 **45** 201
[9] Kukushkin S A, Osipov A V, Feoktistov N A 2019 *Physics of the Solid State* **61** 456
[10] Vykhodets V B, Kurennykh T E, Kesarev A G, Kuznetsov M V, Kondratiev V V, Davletshin A E, Kester J, Hulsen K *Physico-chemical kinetics in gas dynamics* 2010 **10** 28

The study of the optical properties of transparent conductive oxides SnO_2:Sb, obtained by spray pyrolysis

T O Zinchenko, E A Pecherskaya, K O Nikolaev, P E Golubkov, Y V Shepeleva, D V Artamonov
Department of Information and measuring technology and metrology, Penza State University, Penza 440026, Russia

Abstract. The study results of the dependences of the refractive index and samples transmittance of transparent conductive films of tin dioxide depending on wavelength are presented. Films with different doping levels, with different solutions volumes were synthesized on an automated installation by spray pyrolysis. It is shown that the transmittance practically does not depend on the amount of substance sprayed onto the substrate; the transmittance practically does not depend on the solution volume, but depends on the impurity concentration. The chemical composition and temperature of film deposition influence on its refractive index is established. The research results can be used in the development of technologies for the manufacture of optoelectronic structures containing transparent conductive coatings.

1. Introduction

Transparent conductive oxides (TCO) of metals have high transparency and conductivity. Most widely used today are TCS based on metal oxides (MeO). Most of them are binary compounds (In_2O_3, ZnO, SnO_2 and CdO) containing one metallic element. In the stoichiometric composition, these compounds are dielectrics, however, due to the large number of internal defects, which are expressed in the presence of oxygen vacancies or in the presence of interstitial metal atoms, they can become semiconductors with a wide forbidden band (Eg> 3 eV). The formation energy of vacancies and atoms in the interstitial is very low; therefore, these defects are easily formed, which explains the relatively low resistance of non-stoichiometric metal oxides [1]. In addition to high conductivity, TCS also have a good optical transparency (transmittance T> 80%) in the visible and near infrared regions of the electromagnetic spectrum. Therefore, the TCS transmission window is in the range λ from 400 nm to 1500 nm. This is explained by the fact that in the region of long waves (λ> 1500 nm) light is reflected as a result of the appearance of a plasma edge when the light frequency ω coincides with the frequency of collective oscillations of charge carriers in a material (plasma frequency ωp). While the light transmission in the near ultraviolet (λ <350 nm) is limited by the forbidden band, since photons with energy ω> Eg are absorbed. Also, tin dioxide has important physical properties for use in TCO [2]. For the promising TCO application in the field of transparent electronics, for example in thin-film transparent transistors [3] or hetero-structures to create light-emitting diodes, it is necessary to measure and study the dependences of the complex refractive index on the wavelength, which allow to obtain data on the optical width of the forbidden zone, defect levels, phonon and plasma frequencies. To control the optical characteristics, it is necessary to investigate the laws governing the optical TCO properties of the formation obtained by the spray pyrolysis method depending on the deposition parameters, and also to take into account the interrelation of the optical and electrical TCO properties [4].

2. Experiment

The deposition of tin dioxide films with different doping levels was carried out using the developed spray pyrolysis installation [5]. The conditions for applying all the films are the same and are summarized in the Table 1.

Table 1. Technological parameters of the TCS obtaining process.

Parameter	Value
Precursor №1	SnCl4 · 5H2O
Precursor №2	SbCl3
Precursor №2, concentration Npec2	0, 0.1, 0.25 mol.%
Vrast (volume of solutions)	5-20 ml

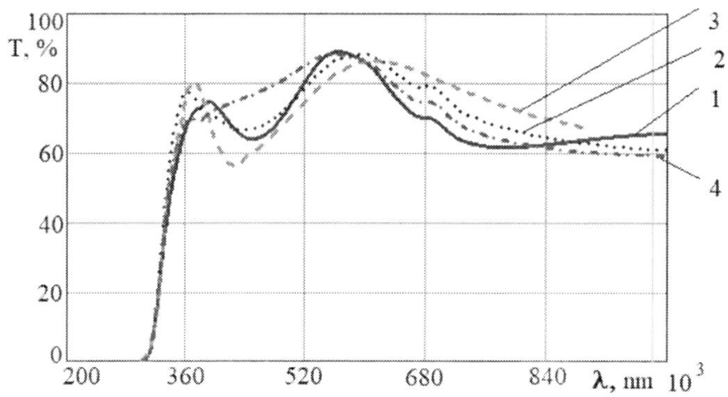

Figure 1. Dependence of the samples transmittance obtained with different volume of solutions: 1) solution volume 20 ml; 2) solution volume 15 ml; 3) solution volume 10 ml; 4) solution volume 5 ml on the wavelength.

Figure 1 shows the dependence of the films transmittance on the wavelength for pure tin dioxide with a different volume of the sprayed solution.
When an impurity is introduced into the semiconductor, the charge carrier concentration in the TCS will lead to a change of the plasma frequency. As the charge carrier concentration in the TCS increases, the plasma frequency will shift towards the visible range, and the corresponding transparency window will decrease significantly. The plasma frequency and charge carrier concentration in the material is related as follows [6]:

$$\omega_p = \sqrt{\left(\frac{ne^2}{m^* \varepsilon_0 \varepsilon_\infty}\right)} \tag{1}$$

where n is the concentration of free charge carriers, e is the electron electric charge ($1.6 \cdot 10^{-19}$Cl); m* is the effective electron mass in the material, ε_0 is the electric constant ($8.85 \cdot 10^{-12}$F/m), ε_∞ is the dielectric constant of the material at high frequencies.
Electromagnetic waves of a lower frequency are absorbed by the material, and for waves of a higher frequency, it becomes much more transparent. In practice, it is more convenient to use the plasma length, which is easier to count as:

$$\lambda = \sqrt{\frac{\pi}{n * re}} \tag{2}$$

where *re* is the classical electron radius, equal to $2.818 \cdot 10^{-15}$. If we take the working wavelength more than the critical one, then there will be no reflection, and the layer for this wavelength will be transparent. The results are described in detail in [7].

Reducing the transparency window is an undesirable effect in solar photovoltaic applications [8, 9]. This change in transmittance is due to the Burstein-Moss effect, which is associated with the filling of the conduction band with electrons. The absorption and emission of light by any substance is determined by three factors — the energy structure of the substance, the population of the energy levels, and the probabilities of optical transitions. Any effect on a substance, leading to a change in its spectra or other absorption and emission characteristics, primarily affects the value of these factors. When impurities are introduced into a semiconductor, especially in large quantities, all three characteristics of a substance can change significantly. First, new levels and even zones of permissible energy values appear in the forbidden zone. Secondly, the distribution of electrons and holes in levels changes dramatically; the Fermi level is shifted to the forbidden zone in the transition from one type of semiconductor to another. Third, because of the perturbation of the wave functions describing the electrons motion in the zones, the probabilities of optical and non-optical transitions change, new channels of light absorption and recombination radiation are opened [10].

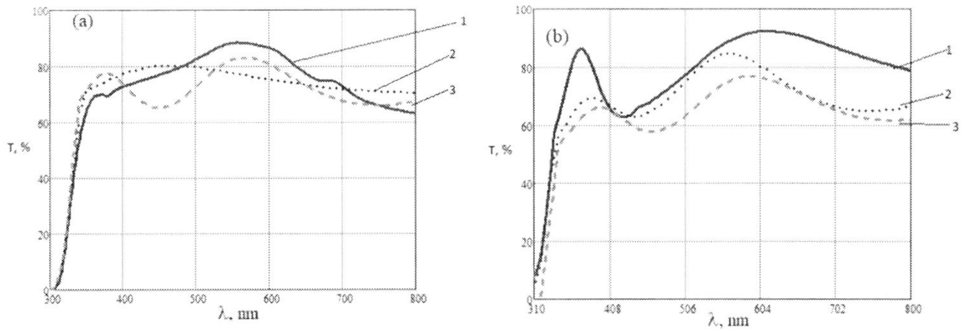

Figure 2(a, b). (a) Dependences of the samples transmittance with different doping levels (volume 5 ml on the wavelength); **(b)** Transmittance of samples with different doping levels (volume 10 ml): 1) impurity concentration 0%; 2) impurity concentration 0.1%; 3) impurity concentration 0.25 %.

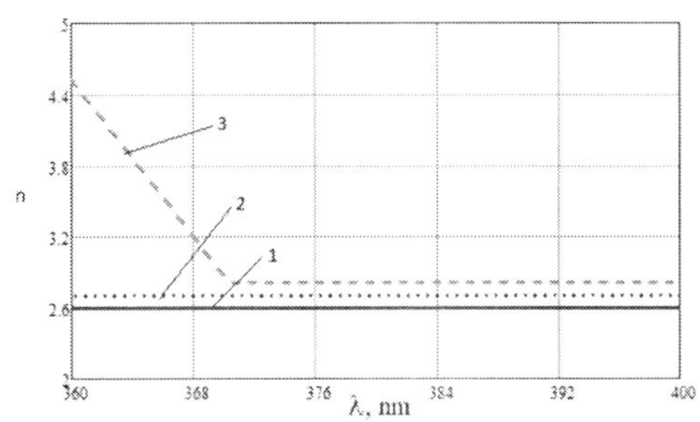

Figure 3. Dependences the refractive index of samples at different solution volumes on the wavelength: 1) solution volume 10 ml; 2) solution volume 15 ml; 3) solution volume 20 ml.

The refractive index of experimental samples differs only in the region of ultraviolet radiation, and in the visible region the refractive index tends to one value, being in the range of 2.4-2.7. This is due to the fact that the refractive index depends on the material being sprayed, as well as on the deposition temperature. As the deposition temperature increases, the refractive index decreases. This is explained by a decrease in the sizes of sub-grains and the formation of a polycrystalline, rather than an amorphous film.

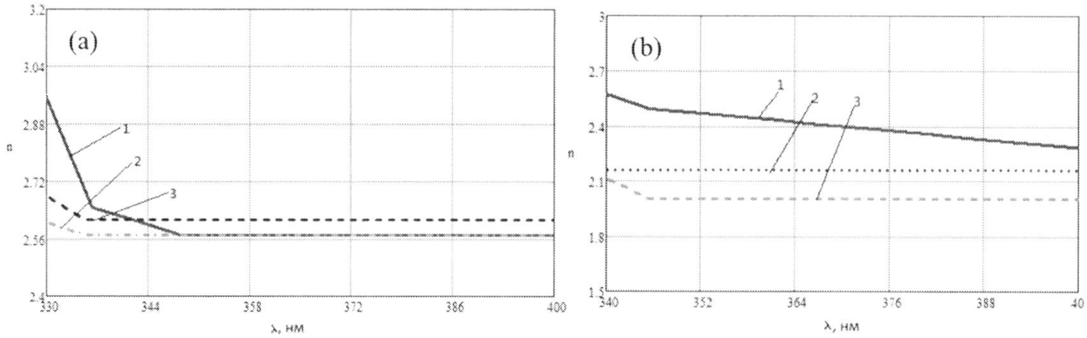

Figure 4 (a, b) - Dependences of the samples refractive index at different doping levels (volume. 5 ml) on the wavelength b) Dependencies of the samples refractive index at different doping levels (volume. 10 ml) on the wavelength 1) impurity concentration 0% 2) concentration of impurities 0.1% 3) concentration of impurities 0.25%.

Figure 4 shows that the refractive index does not change linearly. The reason for the decrease in the refractive index is the decrease in the number of defects and the compaction of films as a result of doping with antimony [11]. When spraying 10 ml of the solution, a thin defect-free film with a crystal structure is obtained (as evidenced by the correspondence of the refractive index to the theoretical data). With an increase of the solution volume, the film acquires an inhomogeneous structure: due to the long application of a large volume of solution, the substrate surface cools to a temperature below 450 °C, which accordingly leads to the formation of an amorphous structure [12].

3. Conclusion

The transmittance is almost independent on the amount of substance sprayed onto the substrate. However, the transmittance is strongly influenced by the chemical films composition. The transmittance also weakly depends on the sprayed solution volume. In this case, a clear dependence of the transmittance on the impurity concentration in the films is observed. This phenomenon is associated with a change in the frequency of the plasma, which depends on the charge carriers concentration. The higher the concentration of charge carriers in the semiconductor, the more the plasma frequency will shift towards the visible range, which accordingly reduces the range in which the TCS is transparent. The refractive index of the samples also depends on the impurity concentration in the initial solution. With increasing impurity concentration, the refractive index decreases slightly and is in the range from 2 to 2.7. The minimum refractive index is observed for samples obtained by spraying 10 ml of the precursor solution.

References

[1] Pecherskaya E A, Zinchenko T O, Kondrashin V I, Kozlyakov A S, Nikolaev K O and Shepeleva J V 2017 *IOP Conference Series: Materials Science and Engineering* **225** 012255
[2] Pecherskaya E A, Kondrashin V I, Raksha S V, Nikolaev K O 2015 *Journal of nano- and electronic physics* **7** 04062
[3] Fortunato E, Barquinha P, Martins R *2012 Advanced Materials* **24** 2945
[4] Pecherskaya Y A, Zinchenko T O, Kondrashin V I, Kozlyakov A S, Shepeleva Y V 2017 *18th International Conference of Young Specialists on Micro/Nanotechnologies and Electron Devices* p. 320
[5] Katerinchuk VN, Kovalyuk M Z 2010 *Semiconductors* **44** 1176
[6] A. R. Babar, S. S. Shinde, A. V. Moholkar, C. H. Bhosale, and K. Y. Rajpure 2011 *Journal of Semiconductors* **10** 102001
[7] Lia Q L, Zhang X H, Lin T, Gao K H 2018 *Journal of Alloys and Compounds* **76**
[8] Meng L.J., Santos M.P. 1997 *Thin solid films* **303** 151

[9] Hass G., Francomba M., Hoffman R. 1978 *Physics of thin films* 320.
[10] Klein A., Korber C., Wachau A., Sauberlich F., Gassenbauer Y., Harvey S. P., Proffit 2010 *Materials* **3** 4892
[11] Mirko Weidner Junjun Jia Yuzo Shigesato Andreas Klein 2016 *Physica status solidi* **253** 923
[12] Timur Zinchenko, Ekaterina Pecherskaya, Dmitriy Artamonov. 2019 *AIMS Materials Science* **6** 276

Design and investigation of GaPAsN/Si light-emitting diode

A Lazarenko, A Gudovskikh, A Baranov, M Sobolev, E Pirogov, E Nikitina
St.Petersburg Academic University, 8/3 Khlopina str., St.Petersburg, 194021, Russia

E-mail: alexashpigun@yandex.ru

Abstract. The heterostructure of light-emitting diode on a silicon substrate with an active region based on A3B5N was synthesized and investigated. Light-emitting diode demonstrates effective electroluminescence at 645 nm up to the 360 K. This indicates the high crystalline and optical quality of the light-emitting structure, as well as the prospects of this approach for the development of silicon integrated photonics

1. Introduction

Epitaxy of A3B5 materials on a silicon wafers is one of the most promising approaches to creating efficient light-emitting devices for integrated silicon photonics.

GaP is a A3B5 binary semiconductor that has the smallest lattice mismatch with Si (0.37% for GaP/Si). However, with a non-zero mismatch of the crystal lattices, the layer begins to plastically relax when the thickness exceeds the critical thickness. For GaP/Si, this value was estimated in the range from 40 nm to 100 nm [1-3]. The thickness of the laser structure is usually from 1 to 3 microns. Thus, compounds with an even smaller mismatch of the crystal lattices with Si are necessary.

The ideal lattice matching can be achieved by adding a small amount of nitrogen (2.2% in GaP at room temperature); moreover, the addition of nitrogen of only 0.5% leads to the formation of a direct structure of the electron zones of $GaP_{1-x}N_x$ solid solutions. Besides the problems associated with the lattice mismatch of the GaP and Si, there are problems caused by different thermal expansion coefficients of A3B5 materials and Si, as well as different numbers of valence electrons. To solve these problems, during the growth of the heterostructure of the light-emitting diode based on A3B5N on Si substrate, the technological approaches described earlier were used [4,5].

2. Experiment

The heterostructure of light-emitting diode (LED) on a silicon substrate with an active region based on A3B5N was synthesized and investigated. The heterostructure was grown by molecular beam epitaxy (MBE) system with solid-state sources of III and V groups of materials and a source of nitrogen with plasma activation. To overcome the problem of formation of anti-phase regions during the epitaxy of III-V compounds on silicon [4], the vicinal silicon substrates, misoriented by 4^0 relative to the (001) surface in the [110] direction was used.

Before the epitaxial growth, silicon substrates underwent a chemical treatment cycle [6], and then they were heated in a pre-growth chamber to a temperature of 350^0 C, and next to a temperature of 850^0 C in an epitaxial reactor of the MBE.

For the formation of a high-quality GaP initial layer on a Si substrate, the technique of "epitaxy with enhanced migration" (MEE) was used [7]. The MEE technique represents the successive interaction of the substrate with the stream of gallium atoms and phosphorus molecules. The temporary absence of phosphorus molecules on the epitaxial surface allows gallium atoms adsorbed on the surface to migrate over the surface for a longer time without forming a chemical bond. In practice, the use of MEE technique leads to the suppression of three-dimensional island growth and the formation of a crystal with a smooth surface [8].

Thereafter, on the obtained MEE GaP initial layer, the LED heterostructure was grown by the standard MBE mode, its design and band diagram are presented on figure 1.

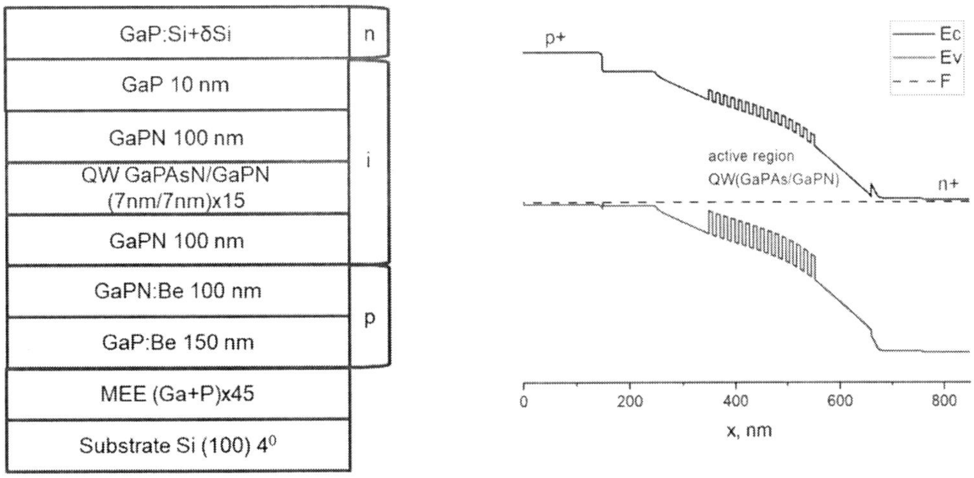

a **b**

Figure 1. **a** Design and **b** band diagram of LED GaPAsN/Si heterostructure (black line - conduction band, red line – valence band, blue dotted line – Fermi level)

LED GaPAsN/Si heterostructure consist of active region and contact layers. Bottom contact layer contains GaP and GaPN layers doped by Be.

A quantum-well (QW) heterostructure consisting of 15 periods of $GaP_{0.982}N_{0.018}$/$GaP_{0.832}As_{0.150}N_{0.018}$ layers is used as the active region. The $GaP_{0.982}N_{0.018}$ layers play the role of barriers, and the $GaP_{0.832}As_{0.150}N_{0.018}$ layers - quantum wells. The upper contact layer is doped with silicon to obtain n-type conductivity. Bulk doping of GaP with silicon does not allow obtaining a high concentration of carriers in the GaP:Si layer. Therefore, to obtain a highly doped GaP:Si layer, a periodic structure (20 periods) was used, consisting of a Si (δSi) monolayer and GaP:Si 10 nm thick [9].

In our early paper it was shown that the the dislocation yield density on the surface of the heterostructure consisting of the initial MEE GaP layer and the GaP buffer layer 200 nm thick grown on a Si (100) substrate with a misorientation by 4^0 relative to the (001) surface in the [110] direction is $\sim 2 \cdot 10^8$ cm^{-2} [10]. Good morphology of the surface of the grown GaP layer is achieved due to the fact that the sprouting dislocations arising at the Si/GaP heterojunction are bent during the growth of GaP buffer layer and do not reach the surface. However, a large number of dislocations at the GaP / Si heterointerface does not allow making contact through the substrate. Another obstacle for creating a contact through the substrate is the diffusion of Ga and P atoms into the silicon substrate during the epitaxial growth [8]. As a result of this diffusion, an undesirable p – n junction occurs, which can lead to losses during electrical pumping of the LED structure. This problem can be solved by using intra-cavity contacts, which are created by etching and passivating heterostructure layers, electrodes are brought through the active layers to the doped p- and n-type layers [9].

3. Results and discussion

The current-voltage characteristic of investigated GaPAsN/Si-based LED heterostructure is shown in figure 2. The cut-off voltage at room temperature is ~3.2 V.

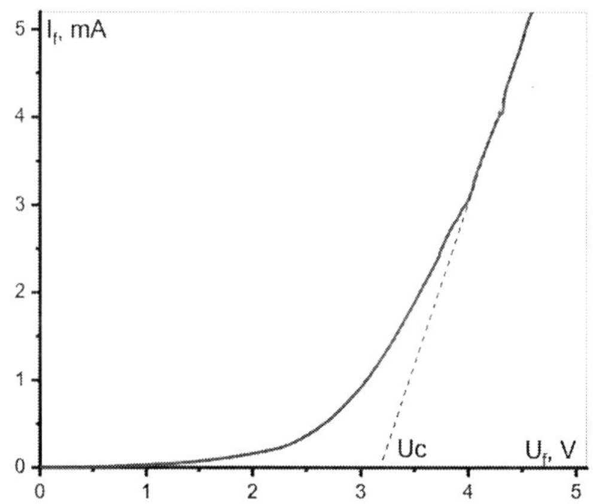

Figure 2. Current-voltage characteristic of investigated GaPAsN/Si-based LED heterostructure

Figure 3 shows the electroluminescence (EL) spectra in a wide temperature range, obtained at forward bias U = 5.4 V (at temperatures of 80–260 K (a) and 300–360 K (b)).

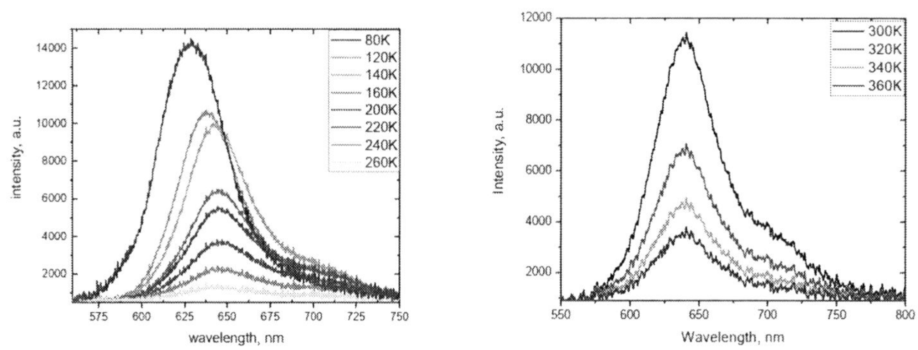

a **b**

Figure 3. EL spectra observed at temperature 80-260 K (a) and 300-360 K (b) at U = 5.4 V

A visible intensive EL in the region of 645 nm up to a temperature of 360 K is observed. As the temperature increases, the intensity of EL decreases and the wavelength of the EL maximum shifts to the long wavelength region.

As the temperature increases from 80 K to 160 K, a smooth shift of the position of the EL peak to the long wavelength region is observed. With further increase of the temperature of the LED heterostructure, the position of the maximum does not change and remains at the value of 645 nm. This behavior is associated with internal local heating of the heterostructure, which exceeds the external temperature.

4. Conclusion

The article describes the stages of creating LED on silicon, and also describes the design of the device. Current-voltage characteristic at room temperature, as well as electroluminescence spectra in a wide temperature range (80-360 K) for grown LED heterustructure were obtained.

Our LED demonstrates effective electroluminescence at 645 nm up to the 360 K. This indicates the high crystalline and optical quality of the light-emitting structure, as well as the prospects of this approach for the development of silicon integrated photonics.

Acknowledgments

This work is supported by RFBR Grant № 17-08-00840

References

[1] Takagi Y, Furukawa Y, Wakahara A and Kan H 2010 *J. Appl. Phys.* **107** 063506
[2] Skibitzki O *et al* 2012 *J. Appl. Phys.* **111** 073515
[3] Almosni S, Robert C, Thanh T Nguyen *et al* 2013 *J. Appl. Phys.* **113** 123509
[4] Ohlsson B J, Malm J-O, Gustafsson A and Samuelson L 2002 *Applied physics letters* **80** 4546-4548
[5] Yonezu H, Furukawa Y and Wakahara A 2008 *Journal of Crystal Growth* **310** 4757–4762
[6] Ishizaka A and Shiraki Y 1986 *J. Electrochem. Soc.* **133** 666
[7] Horikoshi Y, Yamaguchi H, Rao T S, Ramesh S and Kobayashi N 1991 *AIP Conference Proceedings* **227** 29
[8] Sobolev M S, Lazarenko A A, Nikitina E V, Pirogov E V, Gudovskikh A S and Egorov A Yu 2015 *Semiconductors* **49** 559–562
[9] Lazarenko A A, Sobolev M S, Pirogov E V and Nikitina E V 2018 *J.Phys.: Conf. Ser.* **1124** 022022
[10] Kryzhanovskaya N V, Polubavkina Y S, Nevedomskiy V N, Nikitina E V, Lazarenko A A, Egorov A Yu, Maximov M V, Moiseev E I and Zhukov A E 2017 *Semiconductors* **51** 267–271

Measurement and analysis of the recombination coefficients distribution on the area of light-emitting InGaN/GaN heterostructures

I V Frolov, O A Radaev, V A Sergeev
Ulyanovsk Branch of Kotel'nikov Institute of Radio-Engineering and Electronics of Russian Academy of Sciences, Ulyanovsk 432071, Russia

Abstract. A method and an installation for evaluation the profile of distribution of recombination coefficients and current density over the area of light-emitting heterostructures at low injection levels based on local values of the 3 dB frequency are described. Using the example of measuring the parameters of commercial blue InGaN/GaN-based LEDs, it is shown that the current density is higher in those areas of the chip where the 3 dB frequency is lower.

1. Introduction
A fundamental characteristic of light-emitting InGaN/GaN heterostructures is the spatial inhomogeneity of the electrical and optical properties due to fluctuations in the composition of the InGaN solid solution and the presence of a system of extended defects penetrating the material. The compositional fluctuations of the band gap in the active region of the InGaN/GaN light-emitting heterostructure lead to the localization of charge carriers in regions with a high content of indium (with a minimum potential energy) [1]. As shown in [2], indium clusters can reach sizes of the order of 5 ... 10 μm, and the spatial dispersion of the central luminescence wavelength for green light-emitting heterostructures is several times greater than for blue light-emitting heterostructures. The non-uniform distribution of the centers of radiative and nonradiative recombination in the heterostructure causes a non-uniform distribution of the current density, temperature, and electroluminescence intensity. A local increase in the current density during operation leads to a local heating of the structure, which in turn causes an increase in the rate of the heterostructure degradation [3]. To measure the inhomogeneity parameters of the composition of light-emitting heterostructures, the methods of near-field mapping of photoluminescence [4], near-field scanning optical microscopy [5], and dynamic photoelectric spectroscopy [6, 7] are used. The aim of the work is to develop and experimentally test an automated complex for measuring spatial inhomogeneity of the distribution of the Shockley-Read-Hall (SRH) and radiative recombination coefficients and current density over the area of the LED chip using dynamic electroluminescence parameters.

2. Measuring installation
To measure the distribution profile of the dynamic electroluminescence parameters over the area of light-emitting heterostructures, an automated installation has been developed, consisting of an AG 1022F functional generator, a Levenhuk D320L microscope with magnification up to 1600x, incorporating a digital camera C310T NG with a maximum resolution of 2048x1536, and a personal computer with LabView software.

The principle of operation of the measuring setup is as follows. From the output of the generator to the LED under investigation, a sequence of current pulses is supplied with a duty cycle of 2. The frequency of the current pulses is tuned in the range of 1 kHz to 5 MHz with a logarithmic step. For each frequency set, the digital camera registers a photo of the chip, and the brightness of the image of the chip is directly proportional to half the amplitude of the variable optical signal. The exposure time of the camera is set in such a way as to minimize the effect of non-linearity of the brightness characteristic of a digital camera. The resulting array of images is stored in the computer's memory, after which the program calculates the amplitude-frequency characteristic of each image pixel and calculates the 3dB modulation frequency, at which the brightness of the pixel drops 50% relative to the value measured at 1 kHz.

3. Experimental results

The operation of the automated measuring installation was tested on the example of measuring the distribution of the f_{3dB} frequency over the chip area of commercial blue InGaN LEDs with a chip size of 240 × 240 μm. Measurements are performed at currents of 10 μA and 20 μA. Fig. 1 shows the distribution profile of the 3dB frequency values over the chip area at a current of 10 μA.

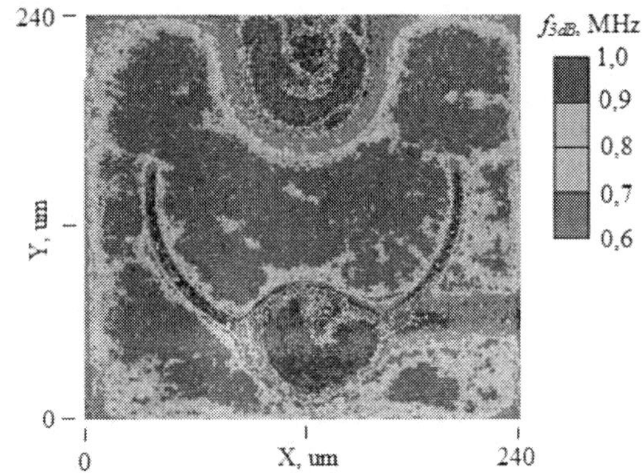

Figure 1. The distribution profile of the f_{3dB} frequency over the area of the LED chip at the 10 μA current

If the capacitance of the LED is small, the 3dB frequency f_{3dB} is determined by the charge carriers lifetime in heterostructure τ [8, 9]:

$$f_{3dB} = \frac{\sqrt{3}}{2\pi\tau}. \tag{1}$$

Using the values of the lifetime of charge carriers, one can estimate the non-radiative recombination coefficients of SRH (A), radiative (B) and Auger recombination (C) [10]:

$$\tau = \left(A + 2Bn + 3Cn^2\right)^{-1}, \tag{2}$$

where n is the concentration of charge carriers in the active region.

In the absence of carrier leakage from the active region of the structure, the current density in structure J is associated with the recombination coefficients by the expression:

$$J = qd(An + Bn^2 + Cn^3), \tag{3}$$

where q is the electron charge, d is the thickness of the active region.

It is known that at low currents the effect of Auger recombination ($Cn^3 \approx 0$) can be neglected. To determine the coefficients A and B, it is sufficient to measure the boundary frequencies $f_{3dB\_1,2}$ of the modulation of electroluminescence at two currents J_1 and J_2, respectively:

$$A = \frac{2\pi}{\sqrt{3}}\left(\frac{J_1 f_{3dB\_2}^2 - J_2 f_{3dB\_1}^2}{J_1 - J_2}\right)^{1/2}, \quad (4)$$

$$B = \frac{\pi^2 qd}{3}\frac{f_{3dB\_2}^2 - f_{3dB\_1}^2}{J_2 - J_1}. \quad (5)$$

Based on the results of measurements of the 3dB frequency at currents of 10 μA and 20 μA, the distributions of A and B recombination coefficients over the chip area at a current of 10 μA are calculated (Fig. 2).

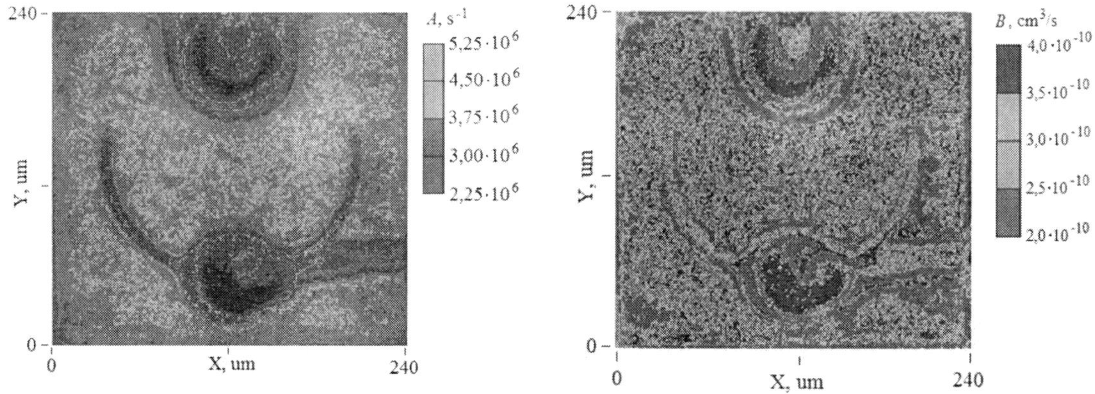

(a) (b)

Figure 2. The distribution profile of the values of the coefficients A (a) and B (b) over the LED chip area

According to the results of measurements of the 3 dB frequency, the carrier concentration in the active region was calculated [11]:

$$n(I) = \frac{1}{qV}\int_0^I \tau(I')dI' \quad (6)$$

where q is electron charge; V is active area volume. For the calculation, the value of V was taken to be 10^{-10} cm^3.

By substituting the values of the coefficients A and B and the carrier concentration n in the expression (3) under the condition $Cn^3 \approx 0$, we calculate the current density distribution over the LED chip (Fig. 3).

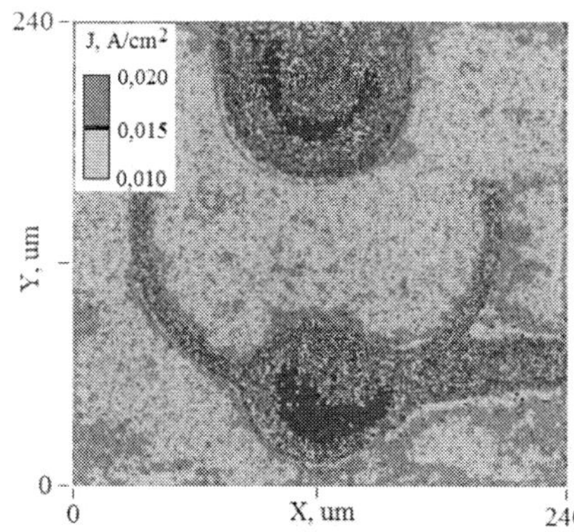

Figure 3. The distribution profile of the current density on the area of the LED chip

The current density is on average higher in those regions of the chip in which the 3dB frequency is lower.

4. Conclusions.

A method and installation have been developed for measuring the profile of the distribution of the 3dB frequencies of the electroluminescence modulation, the radiative and non-radiative recombination coefficients of Shockley-Reed-Hall and the current density over the area of light-emitting heterostructures at low injection levels. The installation was tested on the example of measuring the distribution of the 3dB frequencies of electroluminescence modulation over the chip area of commercial blue InGaN-based LEDs. The presence of local chip regions with an increased current density is revealed. It is determined that the current density is higher in those regions of the chip in which the 3 dB frequency is less. The results obtained will be used to develop methods for assessing the degree of imperfection and diagnosing the quality of light-emitting heterostructures using the parameters of the spatial inhomogeneity of the static and dynamic characteristics of electroluminescence.

Acknowledgments
The work was carried out within the framework of the state task and partially was supported by the Russian Foundation for Basic Research, the project No. 19-07-00562 A.

References

[1] Marutsuki G, Narukawa Y, Mitani T, Mukai T, Shinomiya G, Kaneta A, Kawakami Y, Fujita Sg 2002 *Physica Status Solidi (a)* **192** 110–116
[2] K. Gelžinytė, R. Ivanov, S. Marcinkevičius, Zhao Y, Becerra D L, Nakamura S, DenBaars S P, and Speck J S 2015 *J. Appl. Phys.* **117**, 023111
[3] Peng Z, Lu Y, Gao Y, Chen G, Zheng J, Guo Z, Lin Y, Chen Z 2018 *IEEE Photonics Journal* **10** 8201908
[4] Tateishi K, Wang P, Ryuzaki S, Funato M, Kawakami Y, Okamoto K, Tamada K 2017 *Applied Physics Letters* **111** 172105
[5] Kim M, Choi S, Lee J-H, Park C, Chung T-H, Baek J H, Cho Y-H 2017 *Scientific Reports* **7**
[6] Sergeev V A, Vasin S V, Frolov I V, Radaev O A 2018 *Measurement Techniques* **61**, 914
[7] Frolov I V, Radaev O A, Sergeev V A, Vasin S V 2018 *Journal of Physics: Conference Series* **1124** 041019
[8] Schubert E F 2006 *Light Emitting Diodes* (Cambridge University Press)
[9] Zhao L X, Zhu S C, Wu C H, Yang C, Yu Z G, Yang H, Liu L 2016 *Sci.China-Phys. Mech. Astron.* **59**, 107301
[10] Reklaitis I, Nippert F, Kudzma R et al. 2017 *J. Appl. Phys.* **121**, 035701
[11] Meneghini M, Trivellin N, Meneghesso G, Zanoni E, Zehnder U, Hahn B 2009 *J. Appl. Phys.* **106** 114508

Nanosphere lithography on silicon oxide for high aspect ratio cryo etching of vertically-align silicon structures.

I.A. Morozov [1] A.S. Gudovskikh [1, 2], D. A. Kudryashov[1]

[1]Saint-Petersburg Academic University RAS, 194021, St.-Petersburg, Russia
[2]Saint-Petersburg Electrotechnical University "LETI", 197376, St.-Petersburg, Russia

Abstract. Nowadays, classical silicon solar cell achieved efficiencies close to theoretical limit and new design of silicon solar cell is required. Vertically aligned silicon solar cells seems to be a promising design of future silicon multijunction solar cells. These structures can be produce with dry etching technics. However, dry etching requires high quality mask. Mask can be archived with classical photolithography, but photolithography is multistep process and cannot be used in mass production of solar cells, because of the price. Nanosphere lithography is a powerful one-step method to produce periodic mask on silicon, but polystyrene spheres does not have enough selectivity in silicon etching process. Etching depth in silicon that can be archived in cryo process with 900 nm polystyrene sphere is about 3 μm. For deep silicon etching, we have to use hard mask. In this paper, we investigated nanosphere lithography on 500 nm thick PECVD deposed silicon oxide on silicon. Influence of isopropyl content in nanosphere solution were investigated. The effect of process conditions is discussed in the paper.

1. Introduction

Vertically aligned silicon structures is a promising way to increase a solar cell performance [1,2]. Etching of vertically oriented silicon structures archived either using wet chemical etching or by dry plasma etching. One of the methods of wet etching is metal assisted chemical etching (MACE), which consists in significantly increasing the rate of anisotropic etching of silicon under a metallic catalyst (Ag) in an HF / H_2O_2 solution. MACE was successfully used to create silicon nanofibers (SiNW) [3–5]. The main disadvantage of this MACE method is the need for additional technological steps to remove traces of the catalyst, which can act as additional sources of recombination on the surface vertically aligned silicon nanostructures. ICP is an another powerful method of dry etching of silicon [6]. ICP allows one to independently control the concentration of ions and their energy, and has a high etching rate and anisotropy. The most suitable method for the formation of vertically oriented silicon structures is cryo etching.[7,8]. The feature of this method is that the silicon substrate is at a very low temperature from -80°C to -150°C. At this temperature, the side surface of Si is passivating due to the formation of a non-volatile SiOF compound. This is continuous process that allows one toto avoid the micro roughness obtained by Bosch process and to adjust the angle of etching walls with high accuracy, it also allows one to increase the resistance of the etching mask due to low temperature. Traditional photolithography, interference lithography and nanosphere lithography can be used as a mask for vertically oriented silicon structures etching [9]. Nanosphere lithography has advantages in fewer steps, simple manipulations, and it does not require sophisticated technical equipment. Unfortunately, nanosphere lithography does not possess sufficient mask durability. Therefore, it was proposed to use the intermediate SiO_2 layer as a hard mask for etching. For this purpose we have developed a simple one-step method for deposition of nanospheres on SiO_2 surface.

2. Experimental details.

At the beginning, 500 nm of SiO_2 were deposed on 4 inch (100) silicon substrate using Oxford PlasmaLab 100 PECVD setup. Deposition was made after HF-dip pretreatment. Initial polystyrene solution was 10 wt. % 0.89 μm sphere in water. Before spin coating, substrates were cleaned in oxygen plasma. Spin coating was used for nanosphere lithography. Following spin coating parameters were used: deposition speed of 300 rpm; drying speed of 900 rpm; low acceleration. Polystyrene spheres were dried in centrifuge and diluted in isopropyl/water solution before deposition. In this work, we fix the spin coating parameters and vary isopropyl/water ratio. Properties of nanosphere mask were investigated by optical microscopy and SEM (Carl Zeiss Supra).

3. Results

Isopropyl/water volume ratio was varied from 1/4 to 4/1. After deposition, samples were inspected by optical microscope. Images of different water/isopropyl ratios are shown in figure 1 and figure 2.

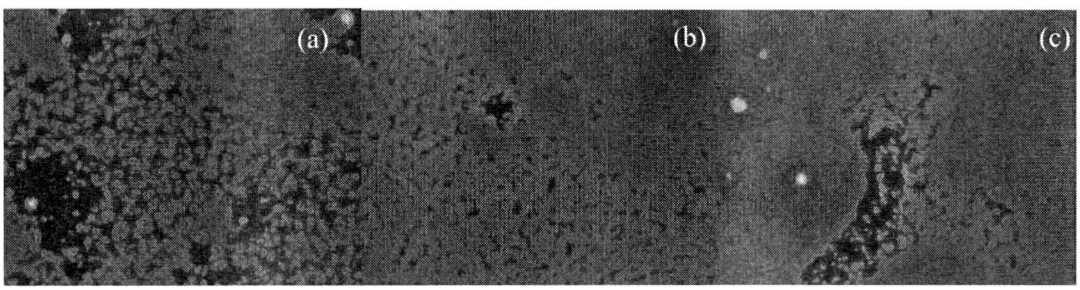

Figure 1. Polystyrene mask on silicon oxide for different water/isopropyl volume ratio: (a) 3/1, (b) 2/1, (c) 1/1.

As it seen on figure 1a, isopropyl/water ratio of 1/4 leads to loosening of mask. In this case mask covered area is about 30% of total area. This is due to the lower viscosity of the solution. In the next step, we increased isopropyl concentration to 2/1 water/isopropyl ratio. It leads to increase of surface latex sphere concentration (figure 1b). Less distances between latex spheres is observed water/isopropyl ratio of 1/1 as can be seen in figure 1c. The further increasing isopropyl concentration in solution leads to better coverage as presented in figure 2.

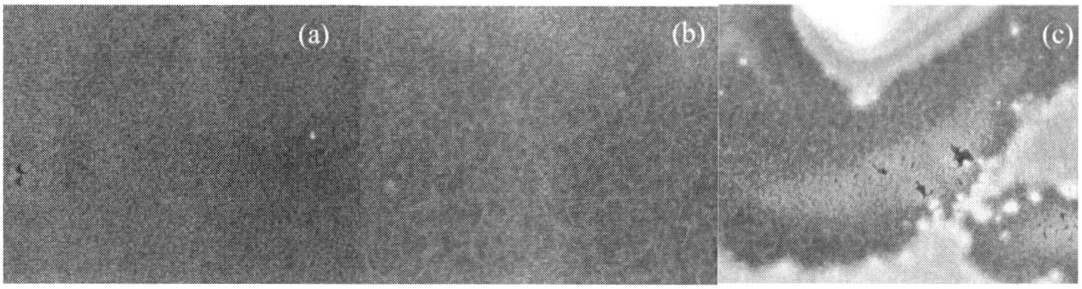

Figure 2. Polystyrene mask on silicon oxide for different water/isopropyl volume ratio: (a) 9/11, (b) 7/13, (c) 1/3.

For water/isopropyl ratio of 9/11 the latex spheres have uniform distribution but still not closed packed (figure 2a). Best result may be archived with 7/13 water isopropyl ratio as demonstrated in figure 2b. One close pack monolayer is observed. Four inch Si substrate fully covered by one monolayer of close pack polystyrene spheres is shown in figure 3. Further increasing of isopropyl concentration leads to two or more monolayers formation (figure 2c). Summarized results are presented in table 1.

Table 1. Surface coverage vs water/isopropyl ratio.

	Water/isopropyl ratio, volume	Coverage, percentage of monolayer
Sample 1	3/1	30
Sample 2	2/1	40
Sample 3	1/1	60
Sample 4	9/11	80
Sample 5	7/13	95
Sample 6	1/3	150

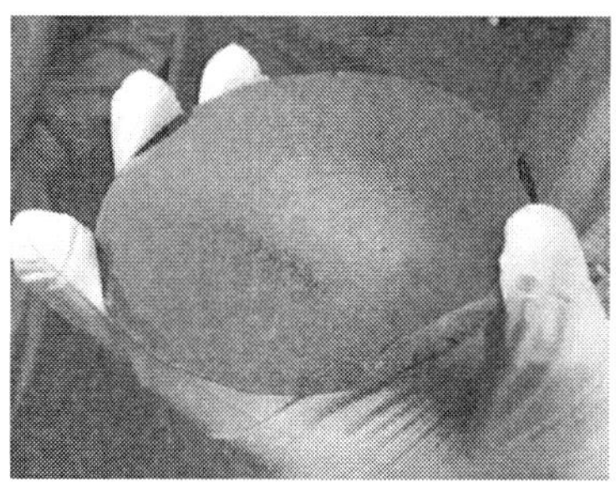

Figure 3. 4 inch Si substrate fully covered by one monolayer of close pack polystyrene spheres

In conclusion, the influence of isopropyl content in nanosphere solution were investigated. The effects of process conditions were discussed in this paper. An increase in the concentration of isopropyl leads to an increase in the surface concentration of the latex spheres. The optimal water / isopropyl ratio was obtained being equal to 7/13 for our experimental conditions. Using this optimal water/isopropyl solution allows us to get close pack one monolayer for total substrate coverage. Also, this method allows one to adjust the surface concentration of the spheres by changing the concentration of isopropyl in solution.

Acknowledgements
This work was supported by the Russian Scientific Foundation under Grant No. 18-79-10059.

References

[1] R.R. Bahabry, A.T. Kutbee, S.M. Khan, A.C. Sepulveda, I. Wicaksono, M. Nour, N. Wehbe, A.S. Almislem, M.T. Ghoneim, G.A. Torres Sevilla, A. Syed, S.F. Shaikh, M.M. Hussain, Adv. Energy Mater. 8 (2018). doi:10.1002/aenm.201702221.
[2] I. Hwang, H.D. Um, B.S. Kim, M. Wober, K. Seo, Energy Environ. Sci. 11 (2018) 641–647. doi:10.1039/c7ee03340k.
[3] V. Sivakov, G. Andrä, A. Gawlik, A. Berger, J. Plentz, F. Falk, S.H. Christiansen, Nano Lett. 9 (2009) 1549–1554. doi:10.1021/nl803641f.
[4] Z. Huang, N. Geyer, P. Werner, J. De Boor, U. Gösele, Adv. Mater. 23 (2011) 285–308. doi:10.1002/adma.201001784.
[5] W.K. Choi, T.H. Liew, M.K. Dawood, H.I. Smith, C. V. Thompson, M.H. Hong, Nano Lett. 8 (2008) 3799–3802. doi:10.1021/nl802129f.
[6] J.H. Keller, J.C. Forster, M.S. Barnes, J. Vac. Sci. Technol. A Vacuum, Surfaces, Film. 11 (1993) 2487–2491. doi:10.1116/1.578597.
[7] S. Tachi, K. Tsujimoto, S. Okudaira, Appl. Phys. Lett. 52 (1988) 616–618. doi:10.1063/1.99382.
[8] R. Dussart, T. Tillocher, P. Lefaucheux, M. Boufnichel, J. Phys. D. Appl. Phys. 47 (2014) 123001. doi:10.1088/0022-3727/47/12/123001.
[9] P. Colson, C. Henrist, R. Cloots, J. Nanomater. 2013 (2013). doi:10.1155/2013/948510.

Performance of InGaAs metamorphic laser power converters at different conditions.

N A Kalyuzhnyy[1], V V Evstropov[1], SAMintairov[1], MAMintairov[1], R A Salii[1], M Z Shvarts[1]
[1]Ioffe Institute, St.-Petersburg 194021, Russia

Abstract. The topic of the work is the metamorphic InGaAs heterostructures for laser power converters (LPC) and their performance at different measurement conditions. Three different illumination conditions were studied: monochromatic Xe-lamp uniform illumination, laser with quasi-uniform illumination (with laser beam unfocusing) and non-uniform focused laser illumination. For simulating the experimental *IV*-curves as well as FF-J_g, V_{oc}-J_g and η-J_g dependences, a tube model developed earlier and describing the charge carrier transport along current lines was used for all three illumination regimes. In the absence of losses (Xe-lamp and unfocused laser), the perfect model simulation model exhibiting an experimental efficiency greater than 50% (for λ=1064nm) has been demonstrated. As a result of simulation of LPC characteristics obtained under focused laser illumination, the losses associated with uniform light distribution as well as with an additional resistance, which decrease the LPC efficiency, have been found.

1. Introduction

There are two main practical cases for laser power converters (LPCs) using. First, it is LPCs for fiber optics, when laser radiation wavelength should correspond to one of the fiber optics line transparency windows. Second, it is LPCs for the atmospheric transmission window and can be practically used for wireless energy transmission systems, for example, for recharge household unmanned aerial vehicles. The most studied GaAs photodetectors demonstrating efficiency of up to 60% [1, 2] are suitable only for local networks due to large attenuation of near-IR radiation (λ up to 850nm) in the optical fiber. LPCs tuned to the wavelength range beyond the GaAs absorption edge (λ=1-1.1 μkm) can be used for both fiber optics and wireless energy transmission systems.

In the work, laser power converters (LPCs) for high power monochromatic light based on metamorphic $In_xGa_{1-x}As$ heterostructures have been obtained and investigated. The metamorphic growth technology is successfully used for creating high-efficient multijunction heterostructure solar cells [3]. Depending on the composition of the active region, such single p-n junction heterostructures can be tuned to a specific laser radiation wavelength and cover a wide range of the tasks for transmitting energy or information via a direct air channel or fiber optics.

The main purpose of the paper is to study the influence of the laser irradiation conditions on LPC performance. Two different cases were studied: a focused laser (when the laser spot is smaller than the LPC photoactive area), what corresponds to signal transmission with a fiber-optics, and an unfocused one corresponding to transmitting radiation through the atmosphere with a laser beam dispersion. Comparison with the results obtained with uniform illumination with a Xe-lamp was carried out.

Previously [4] we determined that, in using a continuous high-power laser, the main losses are associated with sample overheating. And the thermal coefficient $\alpha \approx 1.2$ mV/K of overheating was calculated for continuous laser using V_{oc}-J_g dependence. In this work, all LPCs were mounted on an improved heat sink and were only irradiated with a pulsed laser (focused or defocused), what allowed leveling thermal losses. Thus, the main factor affecting the characteristics of the LPCs was only the uniformity of laser light distribution over their surface.

2. Experimental details

LPC structures were grown by the MOVPE and based on $In_{0.23}Ga_{0.77}As$ solid alloys grown over MB on GaAs-n substrates. The MBs consisted of a sequence of layers slowly increasing in lattice-mismatch with respect to a GaAs wafer to form step-compositional-graded profile of In composition. The developed MB technology allows suppressing threading dislocations in the photoactive region at least at the level of $<1\times10^6$ cm^{-2}[5]. The LPCs were manufactured 3×3.4nm^2 (photoactive area is 0.0784cm^2). The LPC *IV* characteristics were measured for λ=1064nm in the following regimes: pulsed fiber-guided laser (with 1.5msec pulse duration), the unfocused pulsed laser (this situation simulates realistic wireless energy transmission through an open optical channel) and a Xe flash lamp (uniform illumination over sample surface and absence of overheating at *IV*-curve recording). Effective heat sinks were used for all LPCs.

3. Results and discussion

In the work, LPCs with improved heat sinks were used, what allowed completely removing the effect of sample overheating. This is confirmed by coincidence of the photovoltaic characteristics, in particular the V_{oc}-J_g (J_g – photogenerated current), for a LPC uniformly illuminated with Xe-lamp and a LPC illuminated with an unfocused laser (Figure 1). Herewith the photovoltaic characteristics of the LPC illuminated with a focused laser do not coincide (Figures 1, 2 and 3).

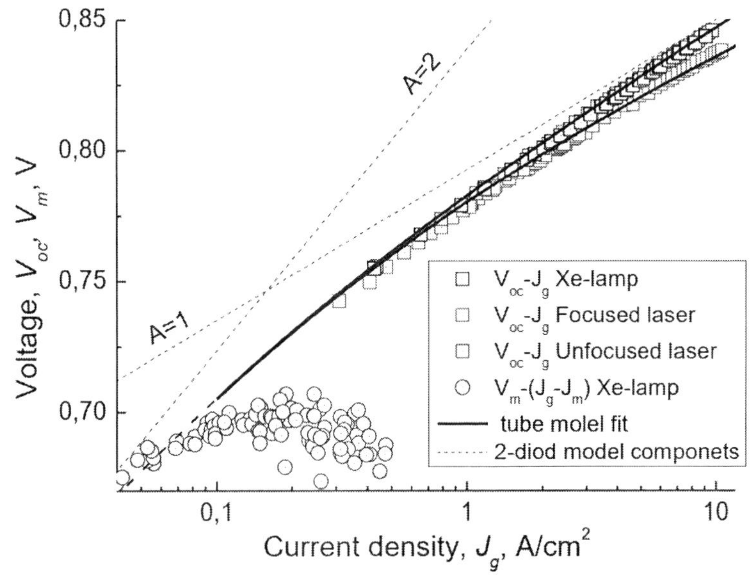

Figure 1. Experimental V_{oc}–J_g (squares) and V_m-(J_g-J_m)(circles) photovoltaic dependences for a $In_{0.23}Ga_{0.77}As$ LPC measured at different regimes: focused 1064nm high-power pulsed laser (red), unfocused 1064nm pulsed laser (blue) and uniform Xe flash lamp (black). The black thick solid lines are the tube model simulation; the black dashed lines show the diffusion (ideality factor A=1) and recombination (A=2) components of the current flow mechanism.

Contribution non-uniformity effect of laser radiation on the LPC efficiency was determined at absence of overheating. For this purpose, the dependences V_{oc}–J_g and V_m-(J_g-J_m) (V_m and J_m are voltage and current at the maximum load point) have been obtained from experimental dependencies and analysed (Figure 1). Note that the V_{oc}–J_g dependence coincides with the non-resistive dark IV-curve, and the V_m-(J_g-J_m) dependence deviates from it due to the series resistance [6, 7]. The analysis of the non-resistive part of the V_m-(J_g-J_m) dependence and the non-resistive V_{oc}–J_g dependence allow determining the parameters of the dark *IV*-curve.

The non-resistive dark *IV*-curve was well described by a two-exponential model and absolutely coincided with the non-resistive V_{oc}–J_g dependence recorded at Xe flash lamp. The saturation currents for both diffusion and recombination current flow mechanisms were $J_{01} = 1.7\times10^{-14}$A/cm^2 and $J_{02} = 5.2\times10^{-8}$A/cm^2, correspondently.

The obtained values of the saturation currents were used to calculate IV-curves in taking into account the process of current spreading. The calculation was based on the previously developed tube model of current spreading [8]. In the model, current lines through the photovoltaic structure were presented as small thickness tubes, and the resistance parameters of each tube were calculated. In our case of the uniform illumination, the photogenerated current values were set for each tube individually (but not the same for all tubes, as in the case of uniform illumination considered in [8]).

Figure 2 shows the experimental *I–V* curves (symbols) for all three cases of illumination for the same photogenerated currents. It can be seen that the I–V curve obtained in the case of a focused laser illumination deviates from other curves by voltages in the region near the optimal load point. As a fact, such a deviation can be explained by an increase in resistive losses. However, calculation of the *IV*-curves by the tube model taking into account the nonuniform illumination has shown that such a deviation can be described by changing the processes of spreading charge carriers between contact bars.

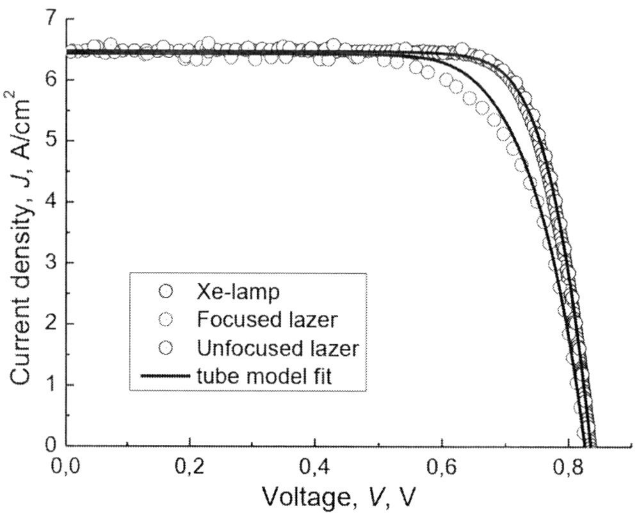

Figure 2. Experimental (circles) *I–V* curves for a $In_{0.23}Ga_{0.77}As$ LPC measured at different regimes: focused 1064nm high-power pulsed laser (red), unfocused 1064nm pulsed laser (blue) and uniform Xe flash lamp (black). The black thick solid lines are the tube model simulation.

All calculated results are presented in Figures 1, 2 and 3 by the black thick solid lines. In all calculations, the number of tubes was 30, the lateral resistance parameter was R_L= 1.3·10^{-2}Ohm · cm^2 and the vertical resistance parameter was R_V= 1.1·10^{-3}Ohm·cm^2. The photogenerated current in the tubes were specified differently for different illumination conditions. For the case of uniform illumination, it was specified by uniform distribution, for the case of focused laser it was specified by a normal distribution. Parameters of the standard distribution were chosen for the best fit of the experimental *IV*-curve and normalized in such a way that the total current in the tubes was equal to the experimental photogenerated current of PLC.

For both Xe-lamp uniform illumination and unfocused laser, the calculation agrees well with the experiments. For a focused laser, the calculation is slightly deviated from the experiential data, but in general it describes the observed deviation. Not perfect agreement between the calculation and the experiment can be explained by additional resistive losses for the unfocused laser. The geometry of the current lines seems to change slightly under nonuniform illumination, and the current lines became not optimal. And, consequently, the resistance parameters of each tube can change.

The *IV*-curves were calculated over the entire illumination range. Based on the calculation, the V_{oc}-J_g dependences (Figure 1 - the black thick solid line), as well as the FF-J_g (Figure 3 a - the black thick solid line) were obtained. Instead of the more fundamental V_m–J_g dependence, a more practical one FF-J_g was used, because efficiency was calculated being based on the *FF* values.

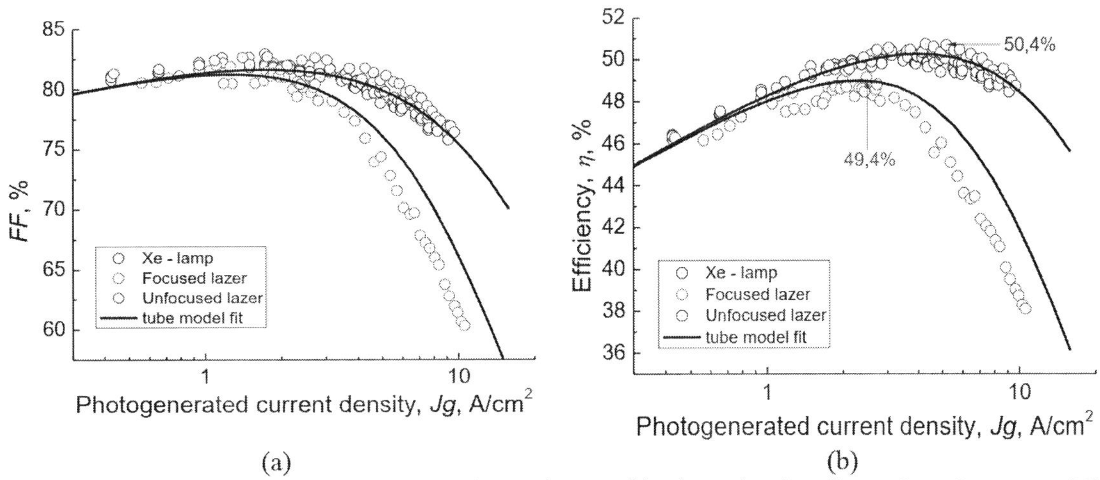

Figure 3. Experimental (circles) FF-J_g(a) and η-J_g (b) dependencies for a $In_{0.23}Ga_{0.77}As$ LPC measured at different regimes: focused 1064nm high-power pulsed laser (red), unfocused 1064nm pulsed laser (blue) and uniform Xe flash lamp (black). The black thick solid linesare the tubemodel simulation.

It can be seen (Figure 3a) that the calculation describes well the FF-J_g dependency for unfocused laser and slightly deviates for focused laser, especially after the maximum. At the same time, the V_{oc}-J_g dependence ideally coincides with the experimental one (Figure 1). All this indicates that, in addition to the influence of the illumination nonuniformity on the characteristics, there are also the influence of the current tube geometry distortion and, consequently, the changes in their resistance parameters. Indeed, resistive losses maximally influence on the position of the optimal load point, and, therefore on resistive characteristics, such as FF-J_g and V_m-J_g. At the same time, resistive losses practically have no effect on the resistiveless V_{oc}-J_g dependence.

The calculated η-J_g dependences are similar to the FF-J_g dependences for all three measurement conditions (Figure 3b). In the case of an unfocused pulsed laser, there are no additional losses in a LPC with a heat sink. In this case, the efficiency is the highest possible and equal to the monochromatic efficiency under Xe-lamp illumination. The efficiency for both these cases is more than 48% in a wide range of photogenerated currents (1-10 A/cm^2) and a laser power densities (1.3–13W/cm^2) with a maximum value of 50.4% at 6 W/cm^2 (λ = 1064 nm). The losses discussed above associated with both illumination nonuniformity and additional resistive losses lead to the efficiency decrease in the case of a focused laser down to 49.4% for the maximum point and a significant decrease in the efficiency at more high powers.

4. Conclusions

The influence of the laser irradiation conditions on the InGaAs metamorphic laser power converter performance was studied. Different measurement conditions were compared: a focused pulsed laser with nonunoform illumination distribution, an unfocused pulsed laser and uniform illumination with a Xe-lamp.

It is shown that with a pulsed laser and a good heat sink levelling thermal losses, the main factor affecting the characteristics of solar cells, including its efficiency, is the uniformity laser illumination distribution, in the case, when the laser spot is smaller than the LPC photoactive area.

For simulating the experimental *IV*-curves, FF-J_g, V_{oc}-J_g and η-J_g dependences, the tube model describing the charge carrier transport along current lines was used for all three illumination modes. A perfect fit of simulation results to the experiment in cases of uniform radiation (Xe-lamp and unfocused laser) was demonstrated. In case of the focused laser, it was shown that for the characteristic fit, it is required to set the illumination distribution over the device surface by a normal distribution function assuming that the resistance parameters of the current tubes did not change. However, even in this case, the simulation of the *IV*-curves under the focused laser will not be accurate near the optimal load point. This indicates that we must take into account an additional resistance increase of the tubes, for example, due to unlighted areas of the photoactive surface leading to distortion of the current lines. As a result, resistive characteristics (FF-J_g, η-J_g) are not accurately simulated at high laser powers.

In the absence of losses, in the case of an unfocused laser the efficiency comparable to the conversion of uniform monochromatic radiation was demonstrated (the maximum efficiency was 50.4% at $6 W/cm^2$). This is the record result of laser conversion for λ = 1064nm according to the literature data. The unfocused laser mode simulates the practical case of energy transfer through an open optical channel.

For another practical case of energy transfer through a fiber optics (focused laser), the losses caused by radiation nonuniformity and resistivity rise reduce the efficiency by almost 1%. For such an energy transfer method, it is necessary to decrease the PLC photoactive area to the laser beam spot square.

Acknowledgements
The work has been supported by the Russian Foundation for Basic Research (grant No. 19-08-00881).

References

[1] Khvostikov V P, Kalyuzhnyy N A, Mintairov S A, Potapovich N S, Shvarts M Z, Sorokina S V, Luque A, Andreev V M 2014*AIP Conf. Proc.* (7-9 April 2014, Albuquerque, NM, USA), vol 1616 pp 21-24
[2] Khvostikov V P, Kalyuzhnyy N A, Sorokina S V, Potapovich N S,Emelyanov V M, TimoshinaN Kh, Andreev V M 2016 *Semiconductors* **50** 1220
[3] France R M et al. 2015 *IEEE Journal of Photovoltaics* **54** 32
[4] Kalyuzhnyy N A, Emelyanov V M, Evstropov V V, Mintairov S A, Mintairov M A, Nahimovich M V, Salii R A, Shvarts M Z 2019 *AIP Conf. Proc.* (25-27 March 2019, Fez, Morroco) *to be published*
[5] Kalyuzhnyy N A, Mintairov S A, Nadtochiy A M, Nevedomskiy V M, Rybalchenko D V, Shvarts M Z 2017 *Electron. Lett.* **53** 173
[6] Mintairov M A, Evstropov V V, Kalyuzhnyy N A, Mintairov S A, Timoshina N Kh, Shvarts M Z, Lantratov V M 2012 *Semiconductors* **46** 1051
[7] Mintairov M A, Kalyuzhnyy N A, Evstropov V V, Lantratov V M, Mintairov S A, Shvarts M Z, Andreev V M, Luque A 2015 *IEEE Journal of Photovoltaics* **5** 1229
[8] Mintairov M A, Evstropov V V, Mintairov S A, Timoshina N Kh, Shvarts M Z and Kalyuzhnyy N A 2016 *Semiconductors* **50** 970

Spectral method for determining quantum wells thickness and the composition in strained heterostructures GaAs/InGaAs

N V Pavlov[1], V V Evstropov[2], M A Mintairov[1], S A Mintairov[2], and N A Kalyuzhnyy[2]
[1]Submicron Heterostructures for Microelectronics, Research & Engineering Center RAS, St Petersburg, Russia
[2]Ioffe Institute RAS, Saint Petersburg 194021, Russia

Abstract. A spectral method for determining the composition and width of quantum wells in strained GaAs/InGaAs heterostructures, based on the effect of separating of radiation absorption peaks of light and heavy holes, is proposed. Within the framework of the Van de Walle and Krijn model-solid theory for the band edges calculating based on the Bir and Pikus theory of deformation potentials, graphical and numerical methods were developed for determining the composition and width of the GaAs/InGaAs quantum wells with an indium content of around 0.4.

1. Introduction

Low dimensional heterostructures (HS) based on GaAs/$In_xGa_{1-x}As$ solid solutions are one of the most important materials of modern semiconductor optoelectronics. One of the features of this system is a significant mismatch of the layers in the lattice parameter, which leads to large mechanical stresses in the HS in pseudomorphic growth mode and to a significant shift of the energy band edges. In the valence band, there is a significant difference in the energy values of light and heavy hole subband edges, which leads to the separation of radiation absorption peaks involving heavy and light holes. An important example of such HSs is a quantum well (QW) with x=0.4, which has great perspectives for application in multi-junction solar cells.

An important feature of this system is a significant shift of the band edges due to elastic stresses, and light and heavy holes band edges are shifted in opposite directions. Figure 1a shows the band diagram of GaAs/$In_{0.4}Ga_{0.6}As$ QW, and also schematically shows the shift of the band edges due to hydrostatic compression and shear deformation.

A significant shift of the band edges of light and heavy holes in opposite directions leads to the fact that the absorption peaks involving heavy and light holes become significantly separated in energy, and this feature of the absorption spectrum can be used to determine the parameters of the HS. The purpose of this work is to develop a method that allows determining the composition and thickness of a GaAs / InGaAs QW quantum cell from known energy values of absorption peaks involving heavy and light holes.

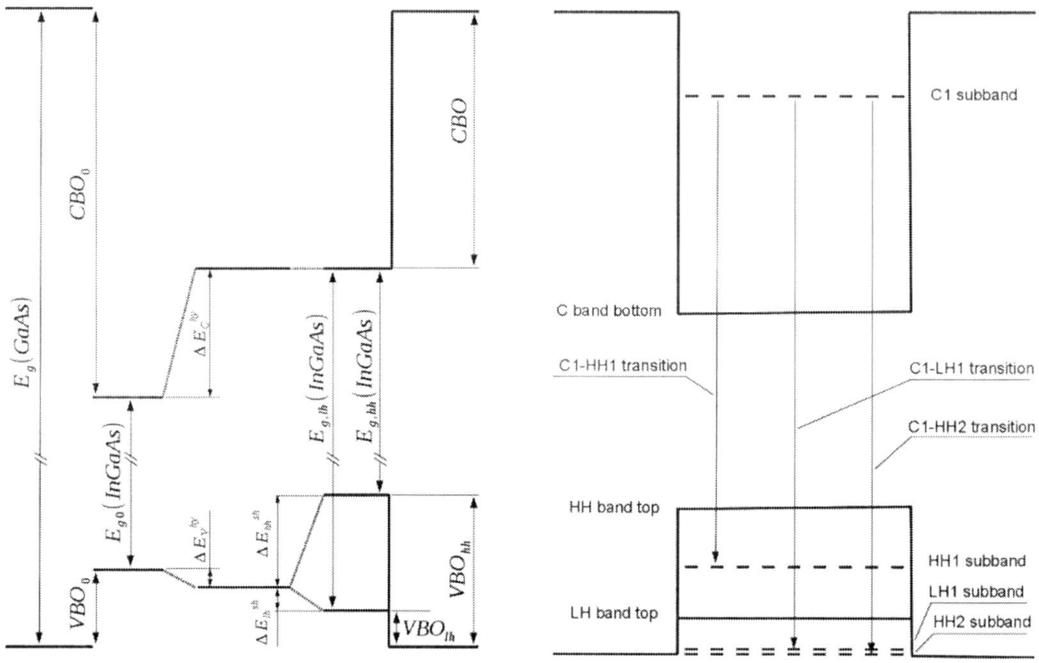

Figure 1(a, b). (a) Band diagram of GaAs/$In_{0.4}Ga_{0.6}As$ QW; **(b)** Optical transitions in GaAs/$In_{0.4}Ga_{0.6}As$ QW.

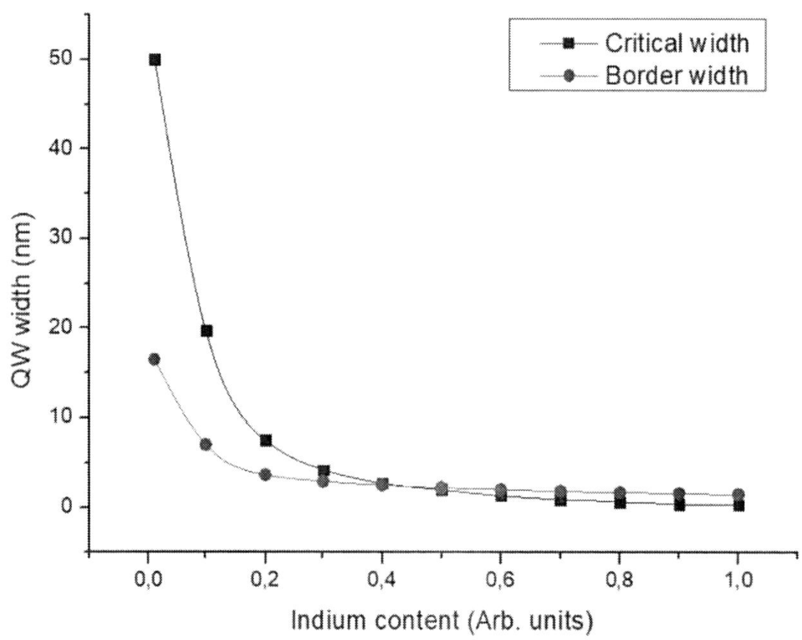

Figure 2. Dependence of the QW thickness corresponding to excited levels arising and of critical width on the indium content.

It should be noted that as the QW thickness increases, additional absorption peaks will appear associated with optical transitions from excited charge carrier levels. Figure 1b shows the scheme of optical transitions in the GaAs/$In_{0.4}Ga_{0.6}As$ HSs with a QW width a=2.5 nm. It can be seen that the absorption peaks associated with the C1-LH1 and C1-HH2 transitions will overlap each other, which will make it difficult to analyse the characteristics of the HS. Therefore, the object of the research is, first of all, to take the range of parameters of the HS at which only the main levels of size quantization of charge carriers will exist. Figure 2 shows the dependence of the maximum QW thickness at which this condition will be satisfied (border width), on the indium content. The figure also shows the dependence of the critical sample thickness on the indium content, calculated using the Matthews formula [1]. It can be seen from the figure that, at x>0.4, the object of research will be limited by the critical sample thickness, and at x<0.4, by the thickness corresponding to the appearance of excited size quantization levels.

2. Calculation of quantum well width and composition for absorption peaks by heavy and light holes

In Fig. 3a the photocurrent spectra (internal quantum efficiency) of a 15 GaAs/$In_xGa_{1-x}As$ QWs grown by MOCVD method with different nominal thickness represented in monolayers (ML) are shown. In the spectra, there are two well distinguishable peaks associated with transitions with the participation of heavy and light holes; therefore, it becomes possible to determine the QW width (a) and the x value from the energy values of these peaks. The calculation of the band edge shifts due to the influence of elastic stresses was made according to the Van de Walle [2,3] and Krijn [4] model-solid theory based on deformation potentials theory of Bir and Pikus [5]. All constants for the InGaAs solid solution are taken from [4].

For x=0.4, the mismatch of the lattice parameters is at room temperature in a plane parallel to the interface $\varepsilon_{par} = -2.8\%$ and along the crystal growth axis $\varepsilon_{per} = 2.7\%$. Hydrostatic shifts of the valence and the conduction band edges are, respectively: $\Delta E_v^{hy} = -31.5meV$, $\Delta E_c^{hy} = 181.9meV$, where positive energy values correspond to a shift up the energy scale. The band edges shifts due to shear deformation are: $\Delta E_{lh}^{sh} = -32.1meV$, $\Delta E_{hh}^{sh} = 95.4meV$, $\Delta E_{so}^{sh} = -63.3meV$. The total band edge shifts are: $\Delta E_{hh} = 63.9meV$, $\Delta E_{lh} = -63.6meV$, $\Delta E_{so} = -94.8$.

Barrier heights are: $U_c = 290.4meV$, $U_{lh} = 37.9meV$, $U_{hh} = 166.1meV$, $U_{so} = 46.8meV$ for the conduction band, for the heavy, light and so-hole bands, respectively.

The size quantization levels of charge carriers can be obtained using Bastard's boundary conditions [6]: $\Psi(a/2+0) = \Psi(a/2-0)$ and $\frac{1}{m_B}\frac{\partial \Psi}{\partial z}(a/2+0) = \frac{1}{m_W}\frac{\partial \Psi}{\partial z}(a/2-0)$.

The dispersion equations for size quantization levels obtaining are:

$$\left(\tan\frac{ka}{2} - \frac{m_W}{m_B}\frac{\kappa}{k}\right)\left(\cot\frac{ka}{2} + \frac{m_W}{m_B}\frac{\kappa}{k}\right) = 0$$

Here $k = \sqrt{\frac{2m_W E}{\hbar^2}}$ and $\kappa = \sqrt{\frac{2m_B(U-E)}{\hbar^2}}$ are the wave vectors of charge carriers in the well and barrier regions, respectively. m_W and m_B are effective masses of charge carriers in the well and barrier regions, respectively.

The first method for determining a and x is graphical. In the axes $E_{c\text{-}lh}$ (electron-light hole transition energy) and $E_{c\text{-}hh}$ (electron-heavy hole transition energy), a grid of lines corresponding to the same composition and the same QW thickness should be plotted. The location of the point corresponding to the experimentally obtained optical transition energy values determines a and x. The grid for GaAs/$In_xGa_{1-x}As$ QWs is shown in Fig. 3b.

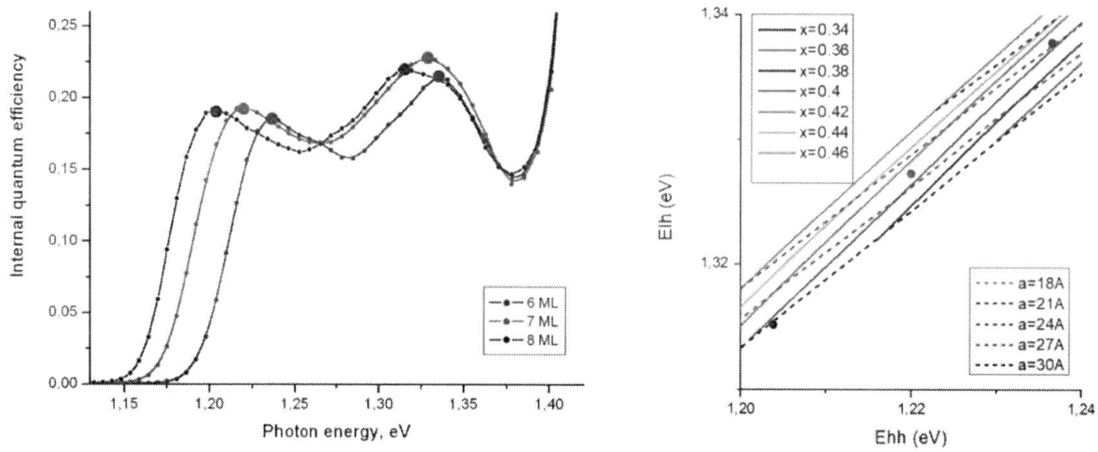

Figure 3(a, b). (a) Internal quantum efficiency InGaAs/GaAs QW HS with different nominal thickness; **(b)** Calculated data for photon energy of light and heavy hole transitions depending on QW thickness and indium content.

The second method is a numerical calculation algorithm. First the root of the equation $E_{c-hh}(x,a)=E_{c-hh}^{\exp}$ should be found. Considering one of the variables as a parameter (for example - a), the dependence $x(a)$ can be obtained. This dependence should be substituted into equation $E_{c-lh}(x(a),a)=E_{c-lh}^{\exp}$, here $E_{c-hh}^{\exp}$ and $E_{c-lh}^{\exp}$ are experimentally measured energy values. Thus the value of a can be found. Substituting the obtained value into the equation, we can get x. Results of the calculation are presented in Table 1:

Table 1. Results of the GaAs/InGaAs QWs width and composition calculating.

Number of MLs	x_{sup}, %	a_{sup}, nm	$E_{c-hh}^{\exp}$, meV	$E_{c-lh}^{\exp}$, meV	x_{calc},%	a_{calc}, nm
6	40	2.46	1236.5	1335.5	39.6	2.38
7	40	2.87	1219.9	1328.8	40.7	2.65
8	40	3.28	1203.8	1315.2	40.3	3.03

Here x_{sup} and a_{sup} are supposed values of x and a, x_{calc} and a_{calc} are calculated values of x and a, respectively. It can be noted that the method used gives good results in determining the composition of the quantum well, but it gives slightly undervalued values for the width. This is explained by the fact that, firstly, the boundaries of the quantum wells are not sharp, and, secondly, by partial compensation of elastic stresses by GaAs layers.

3. Conclusions

The graphical and numerical methods for determining the composition and thickness of strained GaAs/InGaAs QWs from experimentally measured energy values of interband optical transitions involving heavy and light holes are presented. These methods have good accuracy and can be used for express diagnostics of HSs.

It should be noted that these methods are applicable only for $x>0.18$, since for smaller values of the indium content there will be no potential well for light holes, however, the barrier height for light holes will be only a few meV, therefore, this HS can be attributed to type b of Bastard's classification [6].

Acknowledgments
The work has been supported by the Russian Science Foundation (Grant № 17-72-20146).

References
[1] Matthews J W, Blakeslee A E 1974 *J. of Crystal Growth* **27** 118
[2] Van de Walle C G, Martin R M 1987 *Phys. Rev. B.* **35** 8154
[3] Van de Walle C G, 1987 *Phys. Rev. B.* **39** 1871
[4] Krijn M P C M 1991 *Semicond. Sci. Technol.* **6** 27
[5] Bir G L, Pikus G E 1974 *Symmetry and Strain Induced Effects in Semiconductors* (Wiley, New York)
[6] Bastard G 1990 *Wave Mechanics Applied To Semiconductor Heterostructures* (Wiley, New York)

Simulation of current-voltage characteristics and power-voltage characteristics of «space» solar cells based porous silicon in the Comsol Multiphysics package.

I A Shishkin[1], **D A Lizunkova**[1], **N V Latukhina**[1]
[1]Samara National Research University, Samara 443086, Russia

Abstract. To analyze the characteristics of solar modules based on porous silicon involved in the flight experiment aboard a small spacecraft AIST 2D, the method of computer simulation of solar cell characteristics was used. The paper presents the procedure for developing a simulation model of photovoltaic modules in Comsol Multyphysics. The developed model allows us to investigate the current-voltage and voltage-watt characteristics of solar modules and determine the main parameters. The discrepancy between the experimental data obtained from the telemetry data from the AIST-2D with the simulation results does not exceed 5%.

1. Introduction

Each process under study can be described by different models, using the universal simplified model, reflecting the main characteristics of the photocell, allows you to see the relationship of cause and effect, quickly draw the necessary conclusions, approximate the experimental dependencies to an inaccessible measurement area [1]. In this paper, a mathematical model was presented for analyzing the characteristics of solar cells placed on board a small AIST-2D spacecraft in COMSOL Multiphysics with the Semiconductor module. Photosensitive structures with a working layer of porous silicon are promising for use in space-based solar cells [2], and were included in the flight experiment program to study the degradation of their parameters in open space. Due to the limited telemetry data, it became necessary to create a model solar cell based on such structures that adequately describe their characteristics so that it's possible to determine the missing parameters and evaluate their change during operation in open space.

2. Computation methodology

Designing this structure with porous silicon properties involves determining the electrostatic potential by solving the Poisson and continuity equations for a given electron concentration n and hole concentration p and depth of the p-n junction.

$$\nabla(\varepsilon_s \nabla\varphi) = -(n - p + N_a - N_d),$$
$$\frac{\partial n}{\partial t} = \frac{1}{q}\nabla J_n - U_n + G_n,$$
$$\frac{\partial p}{\partial t} = \frac{1}{q}\nabla J_p - U_p + G_p,$$

where ε_s – semiconductor permittivity, U_n и U_p – recombination rates, G_n and G_p – the rate of generation of electrons and holes, respectively, N_a and N_d – concentrations of acceptor and donor impurities. As a rule, mainly boron and phosphorus alloyed. The current densities, J_n и J_p, have both drift and diffusion components, defined as:

$$J_n = -q\mu_n n\nabla\varphi + qD_n\nabla n,$$
$$J_p = -q\mu_p p\nabla\varphi + qD_p\nabla p,$$

where μ – charge carrier mobility, $D = \mu\frac{kT}{q}$ – diffusion coefficient with Boltzmann constant k and temperature T, which in this model is $T = 300\ K$. The total current density: $J = J_n + J_p$. The boundary conditions for currents depend on the type of metallic contacts in question. In our case, the Schottky contacts, where the dependence is on the carrier concentrations and recombination rates, U_n or U_p on the metal / semiconductor border:

$$J_n \cdot \hat{n} = -qU_n(n - n_0),$$
$$J_p \cdot \hat{n} = -qU_p(p - p_0),$$

where $\hat{n}$ – unit normal at the border, n_0 и p_0 – equilibrium concentrations of electrons and holes. Electric potential at the border $\varphi = -\Phi_m + V_a$, where Φ_m – function of work function in metal and V_a – applied voltage. The above set of equations are often called "semiconductor equations", which are used in the numerical method. In a semiconductor, charge carriers can be generated or lost due to recombination through various physical properties that depend on the material parameters, in this case, on the parameters of porous silicon, external conditions and energy states (defects) available in the band gap. The Comsol Multiphysics semiconductor module includes several built-in options for recombination mechanisms. The appropriate mechanism for our conviction is the Shockley-Reed-Hall method, which is used in the Semiconductor module. This recombination mechanism is the most likely process in the case of semiconductors with an "indirect" band gap, such as Si and Ge. Then the rate of change in the concentrations of electrons and holes in the corresponding allowed energy states in the process of recombination with the participation of energy levels in the band gap of a semiconductor is written in the form:

$$-\frac{\partial n}{\partial t} = c_n n N_t(1 - f_t) - e_n N_t f_t,$$
$$-\frac{\partial p}{\partial t} = c_p p N_t f_t - e_p N_t(1 - f_t),$$

where c_n is the electron capture coefficient in the conduction band to the local level, c_p is the hole capture coefficient in the valence band to the local level, N_t is the concentration of charge capture centers, e_n and e_p are the thermal emission coefficients of electrons and holes from the local level to the conduction band and the valence conduction band, respectively, f_t is the probability of finding an electron at a local level (non-equilibrium Fermi function). The solar cell itself looks like a straight line, where the points at the ends of the straight line are metal contacts. The resulting characteristics after modeling in the Comsol Multiphysics package are presented in Fig. 1 and Fig. 2.

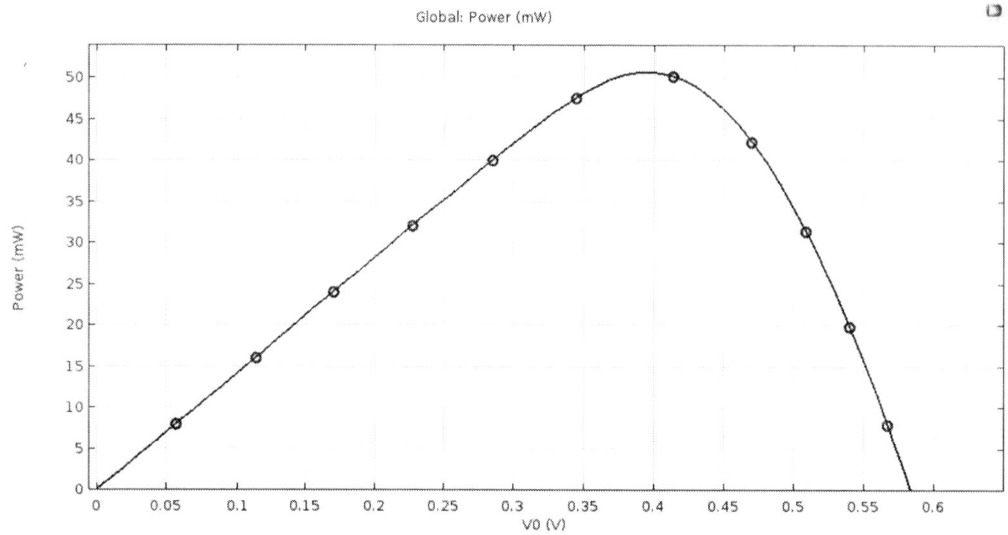

Figure 1. Watt-voltage characteristics of solar cell model in the Comsol Multiphysics.

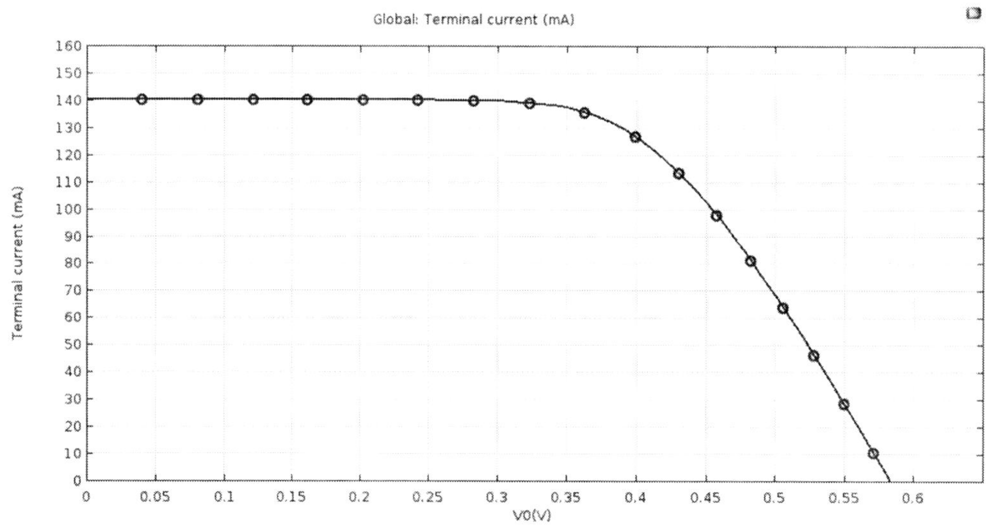

Figure 2. Current-voltage characteristics of solar cell model in the Comsol Multiphysics.

3. Results and discussion

This model was compared with experimental data obtained from photovoltaic cell №1 located on board a small spacecraft AIST-2D. Solar cell №1 is a photosensitive structure with a polished surface, containing layers of porous silicon and zinc sulfide. By changing the concentration of charge carriers and the depth of the p - n junction, the current–voltage and power–voltage characteristics were obtained, which satisfy the data obtained (Fig. 3). For a given solar cell, the concentration of electrons and holes: $N_d = 1.9 \times 10^{16}\ 1/cm^3$, $N_a = 2.2 \times 10^{16}\ 1/cm^3$. Depth of the p - n junction – 0.58 μm. This solar cell has a short circuit current of order 140.558 mA, and the open circuit voltage – 0.5836 V. This can be seen from the I-V characteristic given in Fig.3. The maximum power received from this structure is 50.729 mW. Also for the solar cell found fill factor of the I-V characteristic and energy conversion efficiency.

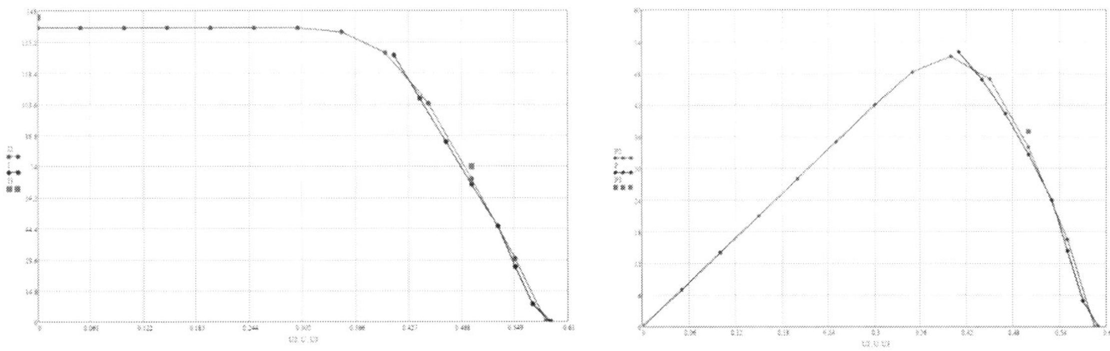

Figure 3 (a, b). Current-voltage **(a)** and watt-voltage **(b)** characteristics of solar cell №1 from board the AIST-2D: blue line - experimental data, red line - simulation results in Comsol Multiphysics, pink dots - experimental data obtained from the solar module, on Earth.

4. Conclusion

An imitation model was obtained that allows displaying current–voltage and power–voltage characteristics solar modules depending on a given impurity concentration and depth of the p - n junction. This model has a universal character and allows us to describe more complex processes of current transport in structures. A comparison of the experimental characteristics of the solar module on board the AIST-2D and the characteristics obtained by modelling in Comsol Multiphysics indicated that the discrepancy between the simulation results and the experimental characteristics does not exceed 5%, which is acceptable for engineering calculations. This confirms the adequacy of the proposed simulation model.

Acknowledgments
This work was supported by the Foundation for Assistance to Small Innovative Enterprises, grant № 12980GU/2018.

References

[1] L.S. Lunin, A.S. Pashchenko. Simulation and study of the characteristics of photoelectric converters based on GaAs and GaSb // Journal of Technical Physics, 2011, vol. 81, no. 9. - P.71-76.
[2] Latukhina, N. Efficient Silicon Solar Cells for Space and Ground-Based Aircraft / N.Latukhina, A. Rogozin, G. Puzyrnaya, D. Lizunkova, A. Gurtov, S. Ivkov // Procedia Engineering 12/2015; V.104, 31. P. 157-161.

Relation between energy gap and saturation currents in GaInAs homo p-n junctions

M.A. Mintairov[1], V.V. Evstropov[1], S.A. Mintairov[1], M.Z. Shvarts[1] & N.A. Kalyuzhnyy[1]
[1] Ioffe Institute, St.-Petersburg, Russia, Polytechnicheskaya str. 26
E-mail: mamint@mail.ioffe.ru

Abstract. The paper is devoted to the relationship between the energy gap and the saturation currents (diffusion and recombination ones) of GaInAs homo p–n junctions. Such a relationship is required for developing multi-junction solar cells and photodetectors for a given wavelength. The saturation currents and the energy gap have been determined experimentally, and an empirical exponential formula was obtained for the required relationship. The results of the work allow predicting the saturation current for a GaInAs p-n junction with a given energy gap.

1. Introduction

Research and development of photovoltaic devices with homo *Ga(1-x)In(x)As* p-n junctions is currently relevant. This is due to the development of multi-junction solar cells [1] and to the creation of high efficiency photodetectors with a given energy gap [2,3]. Generally for creating such photovoltaic devices, the technology for producing a *GaInAs* metamorphic buffer is used, which allows obtaining *Ga(1-x)In(x)As* layers in a wide range of *x*. A set of Ga(1-x)In (x)As photo-detectors with different compositions x has been investigated. For all samples, the energy gap - E_g and saturation currents - J_0 (pre-exponentials) for the diffusion (ideality factor $A=1$) and the recombination ($A=2$) current flow mechanisms have been determined. The relationship between the E_g and J_0 values have been determined. It has been shown that such a relationship is exponential one.

2. Obtaining and analyzing energy gap –saturation current dependences

Investigated structures were grown by metal-organic vapor-phase epitaxy and based on *GaInAs* solid alloys grown over metamorphic buffer on *GaAs-n* (100) substrates misoriented by 6° towards to the [111] direction. The In concentration in the photoactive region have been varied in the range of 22-29%. For all structures, electroluminescence and photo-response spectra have been used to determine their energy gap value. Also light IV-curves have been obtained using pulsed simulator of solar radiation. It allows obtaining the open circuit voltage (V_{oc}) vs photogenerated current (J_g) dependences. Figure 1 shows the V_{oc}-J_g dependencies (open circles) combined with the *IV*-curve (solid circles). As is well known, the V_{oc}-J_g characteristic coincides with the resistanceless dark *IV*-curve. Using of the V_{oc}-J_g allows determining the dark *IV*-curve values at high currents, where the influence of the series resistance is significant. Experimental dependencies (Fig. 1) have been approximated by the standard two-diode model:

$$J = J_{01}\exp\left(\frac{V}{kT}\right) + J_{02}\exp\left(\frac{V}{2kT}\right) \qquad (1)$$

where k – Boltzmann constant, T – absolute temperature, J_{01} and J_{02} – diffusion (unary) and recombination (binary) current pre-exponents. It is seen (Fig. 1, solid lines) that the model adequately describes the experimental data. Note that, for the *GaAs* sample (E_g = 1.4224 *eV*), a tunnel-trap current was observed, which is absent in other samples. This is probably due to the positive effect of the presence of a metamorphic buffer in other *GaInAs* based photoconverters.

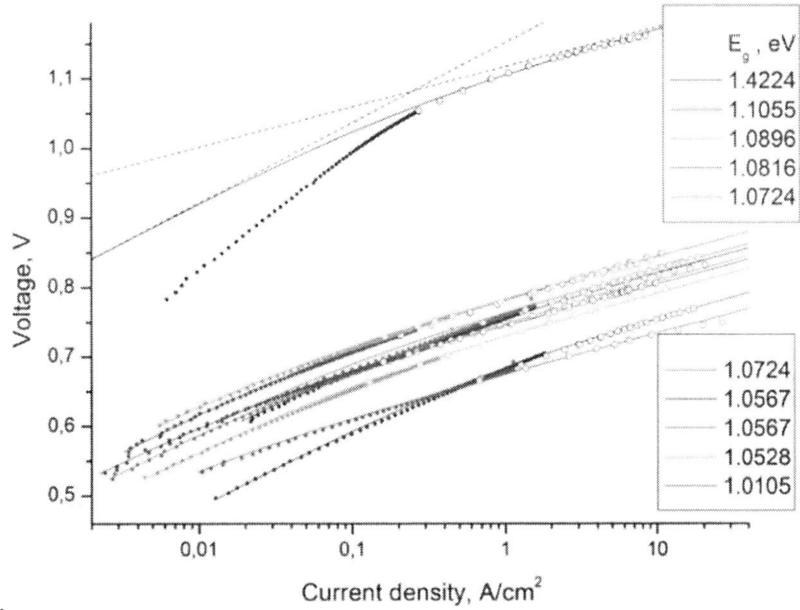

Figure 1. Determination of pre-exponents J_{01} and J_{02} by approximation with two-diode model (solid lines) of the experimental data: *IV*-curves (solid circles) and V_{oc}-J_g dependencies (open circles). Two-diode model components are shown by dashed lines

Figure 2 illustrates the two methods used in the work for energy gap determination. In the first one, – by the electroluminescence peak position, in the second – by analyzing the absorption edge in the external quantum efficiency (EQE) spectrum. The absorption edge was determined by the intersection of two extrapolated lines. One – the horizontal line (dotted line in Figure 2) – characterizes the plato in the absorption spectrum with the energy greater than E_g, the other line (linear on the logarithmic scale) characterizes the band edge blurring due to the Urbach tail. The spectral position of the lines intersection was interpreted as the energy gap. It can be seen (Figure 2) that the values of E_g determined by the two methods almost coincide.

Table 1. The GaInAs p-n junction parameters: E_g – energy gap, J_{01}, J_{02} – saturation currents (pre-exponents)

E_g, *eV*	J_{01}, *A/cm²*	J_{02}, *A/cm²*
1.4224	$4.0\cdot10^{-20}$	$1.0\cdot10^{-10}$
1.1055	$2.0\cdot10^{-14}$	$4.0\cdot10^{-8}$
1.0896	$2.0\cdot10^{-14}$	$3.0\cdot10^{-8}$
1.0816	$5.0\cdot10^{-14}$	$9.0\cdot10^{-8}$
1.0724	$4.0\cdot10^{-14}$	$6.5\cdot10^{-8}$
1.0567	$6.0\cdot10^{-13}$	$6.3\cdot10^{-7}$
1.0567	$1.0\cdot10^{-13}$	$5.0\cdot10^{-8}$
1.0528	$1.6\cdot10^{-13}$	$1.2\cdot10^{-7}$
1.0105	$1.7\cdot10^{-12}$	$1.5\cdot10^{-7}$

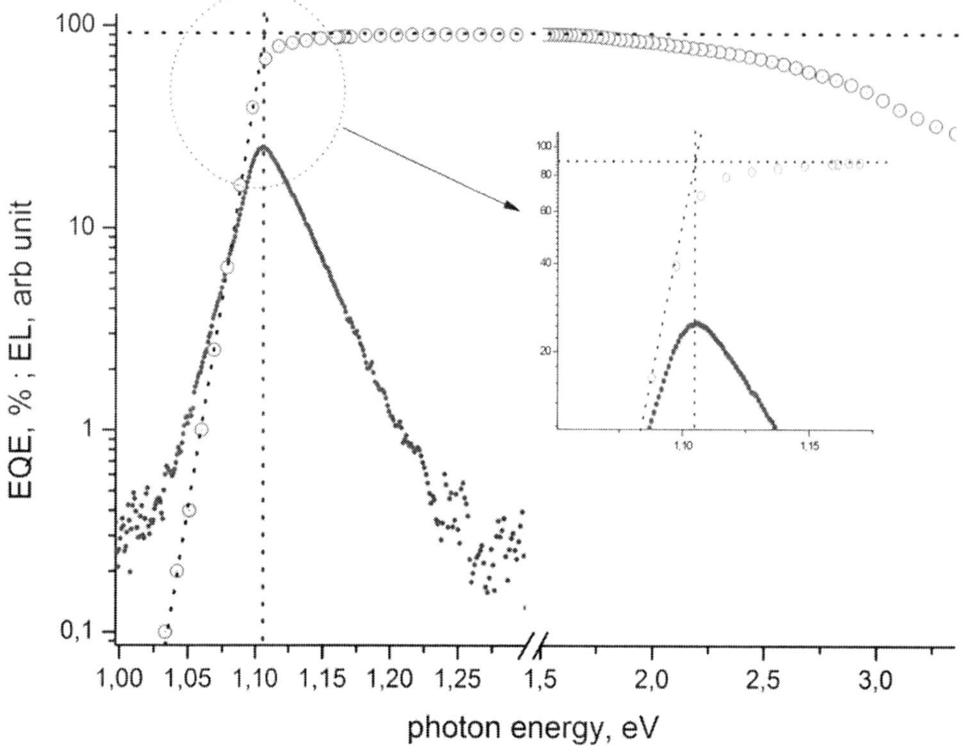

Figure 2. Comparison of electroluminescence spectrum (green dots) and EQE spectrum (red open circles). The result of the E_g determination by EQE analysis (the intersection of black dashed lines) coincides with the EL peak position (blue dashed line).

The obtained J_{01} and J_{02} vs E_g are shown in Table 1 and on Figure 3 (solid squares). The data for InAs (open circles) is also shown. These data are based on the experiment presented in [4]. It can be seen that the obtained dependences are well described by straight lines on a semi-log scale. In general, the saturation current J_0 can be described by the formula:

$$J_{0A} = J_{zA} \exp\left(\frac{-E_g}{AkT}\right), \qquad A = 1 \text{ or } 2 \qquad (2)$$

where J_{zA} characterizes the dependence of the saturation current on the lifetime, diffusion length, doping level, etc. It is obvious that the observed linearity (Figure 3) means that J_{z1} and J_{z2} are constants. Approximation of the experimental dependences allowed us to obtain their values $J_{z1} = 2.5 \cdot 10^5\ A/cm^2$, $J_{z1} = 1.4 \cdot 10^2\ A/cm^2$. Note that the straight lines (Figure 3) intersect at a current $J_{21} \approx 0.1\ A/cm^2$. The physical meaning of this current is the conditional current boundary between the unary (first term in eq. 1) and binary (second term in eq. 1) components. This means that, on average, the current boundary is practically independent of the Ga(x)In(1-x)As energy gap.

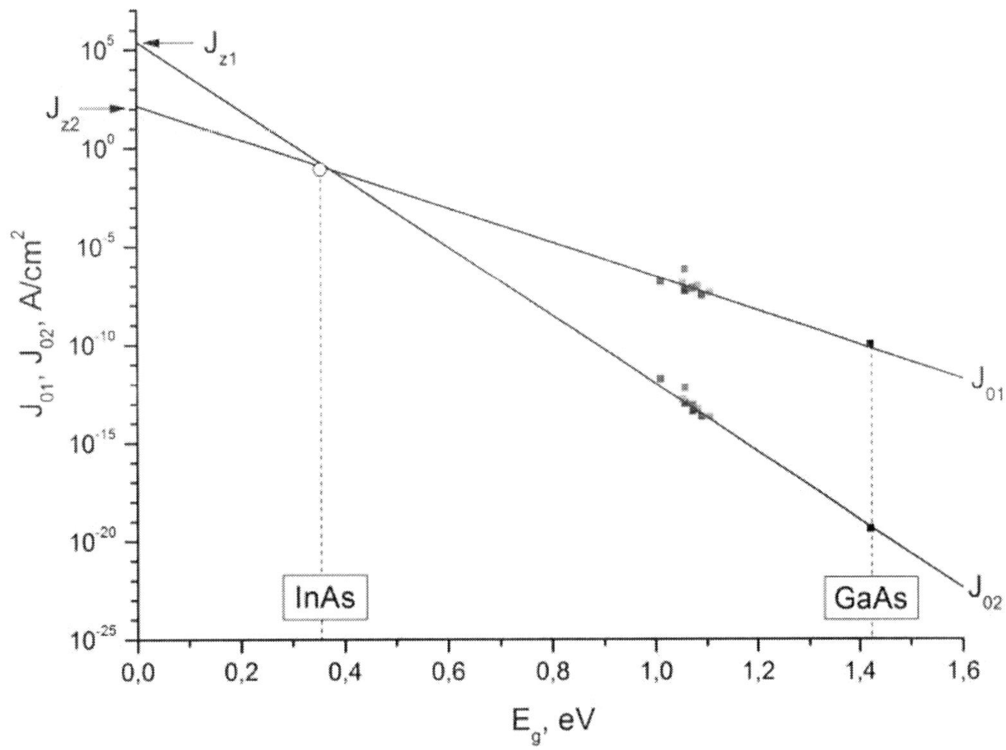

Figure 3. The relationship between the GaInAs p – n junction energy gap and the saturation currents (diffusion and recombination). The points are the experimental data, the lines are the approximations using equation (2).

3. Conclusion

The dependences of the saturation currents (pre-exponents) on the energy gap of *GaInAs* solid alloys ware determined experimentally. It is established that the desired dependence is linear on a semi-log scale and is described by the formula (2). From this linearity, it follows that the conditional current boundary between the unary (diffusion) and binary (recombination) component J_{21}, as well as the values J_{z1} and J_{z2} are constants. The obtained result allows predicting the saturation currents (pre-exponents) and, consequently, the efficiency of *GaInAs* photoconverters with different energy gaps

References

[1] Patel, P., et al 38th IEEE Photovoltaic Specialists Conference, Austin, TX, 2012.
[2] D.V. Rybalchenko, S.A. Mintairov, R.A. Salii, et al., Semiconductors 51, 93-99 (2017)
[3] N.A. Kalyuzhnyy, V.M. Emelyanov, S.A. Mintairov, et al., AIP Conf. Proc. 2012, 110002 (2018)
[4] G Lucovsky, Br. J. Appl. Phys,12, 311 (1961)

Signal processing scheme for error control in visible light communication data transmission system

A I Borodkin[1], E Z Gareev[1], I A Krasavtsev[1], O A Kozyreva[1], I S Polukhin[1,2], M A Odnoblyudov[2], V E Bougrov[1]
[1]ITMO University, Kronverksky Pr. 49, Saint Petersburg 197101, Russia
[2]Peter the Great St. Petersburg Polytechnic University, Polytechnicheskaya 29, Saint Petersburg 195251, Russia

Abstract. We developed wireless visible light simplex data transmission channel and error correction algorithm for the reciever. The system was design specially for using modern smartphone device as a receiver and a conventional domestic LED illuminator as transmitter. We calculated highest probable data transfer speed in our communication channel, fabricated the transmitter modulator and developed the client application for an android smartphone equipped with stock camera module, perform a series of data speed measurements and conducted the analysis of both analog and digital signals. The system allowed to transmit data with rates up to 8 kbps on distance up to 1.5 meters.

1. Introduction

In recent years, technologies that allow to combine advanced functionality in conventional things are emerging [1]. This trend has not bypassed the illumination systems, an important task of which became the wireless communication through modulation luminous flux. Thus, arose VLC technology, using LED as an optical signal transmitter and a photodetector as a receiver [1]. The main signal transmission principle by such a system is based on the light emission modulation with a frequency of the order of several kHz and higher. The human eye is not able to perceive the frequency of flicker above 100 Hz, which allows you to make data transfer using VLC technology imperceptible to a user. A conventional modulation of LED illumination On-Off Keying is based on changes of luminous flux intensity. On-Off Keying is the modulation principle that the developing modulator designed to integrate into existing systems of urban, industrial and domestic lighting without the need to create special systems for matching and integrating additional photodetectors into existing lamps, and thereby expanding the functionality of such systems, providing the opportunity for creating wireless data transfer over the luminous flux. However, a very important functionality degree indicator of VLC technology-based systems is their bandwidth, increasing which, apart from everything else, allows to raise the transmitted data protection level, maintaining channel data transfer speed [1]. Using a mobile device as a receiver in such data transmission systems allows you to make information accessible to a large number of users, due to the absence of the need to have a special separate client module acting as a receiver, however, a mobile device camera module shooting frame rate is often limited to 120-240 Hz, due to the lack of a camera matrix possession ability, which makes it necessary to apply special processing methods of shots. The target of this work was the development of the algorithm that allows to use the maximum of channel capacity by special methods of working with camera frames.

2. System description

A prototype of an office-standard luminaire was used as a signal source. The emitting module consisted of white phosphors LEDs with a modulator was developed by the authors. Luminaire luminous flux was measured amounted to 1680 lm. In order to maintain the intensity of the luminous flux PWM modulation was used. LED luminaire was modulated with 50% duty ratio squarewave signal, the maximum received signal value was set for «1» and the minimum for «0», respectively. The Modulator had a maximum modulation frequency of 1 MHz and was designed to work with a up to 43 Watts LED luminaires. The CMOS censor of the mobile device camera module was used for signal receiving and the mobile device itself for the processing and visualizing transferred data. Mobile device equipped 1/2.6-inch size 12 megapixel resolution sensor and 1/1.8 aperture size camera module was used. The highest frame rate supported by the mobile device was 240 frames/sec. A software package was developed that includes part decoding signal shot by camera and an application designed for a mobile device having a user-friendly interface for reading received data. In order to obtain oscillograph of received signal photorecieving module was used. The module was developed by the authors and used 4 THORLABS FDS100 Si photodiodes with the 350-1100nm wavelength and rise/fall time of 10 ns. The figure 1 is illustrating the experimental setup, which was used to perform series of algorithm tests.

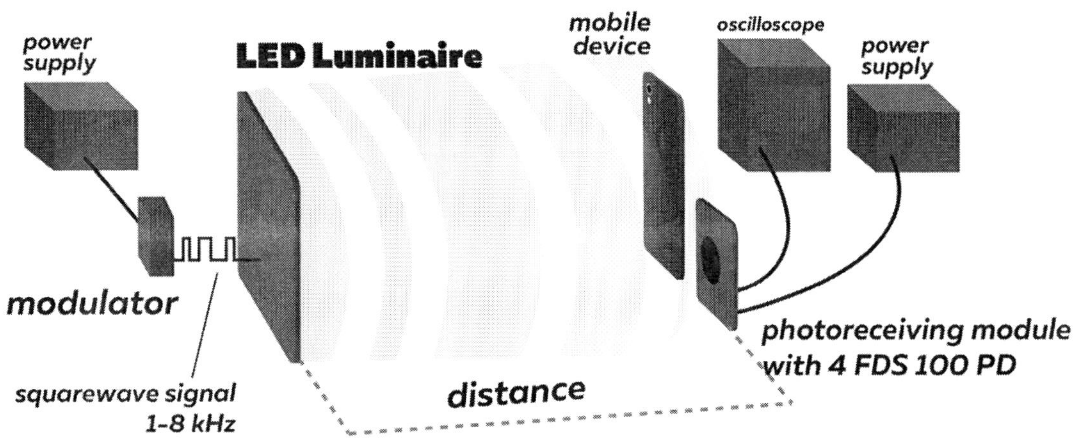

Figure 1. Experimental setup of the developing system.

At the first stage, the authors created an algorithm that uses the "rolling shutter" effect. This effect is described in detail by N. Rajagopal et al. in [2]. The developed algorithm allowed to process data at speeds up to 2.5 kbps when shooting at 15 frames / sec. We used modulation frequencies from 1 to 7 kHz. However, a significant disadvantage of this algorithm was the loss of at least of 2/3 transmitted data. Only 1/3 of data transmitted was processed, as depicted in figure 2 (a), that can be explained by the fact that the process of exposing the CMOS matrix and the processing of the received signal are performed sequentially. At the second stage. We developed algorithm, so that the processes described above took place in parallel, thus it made possible to minimize the loss of transmitted information, as shown in Figure 2 (b,c). The figure 2 (c) illustrated the main principal of rolling shutter effect and the implementation of this effect into developed parallel algorithm.

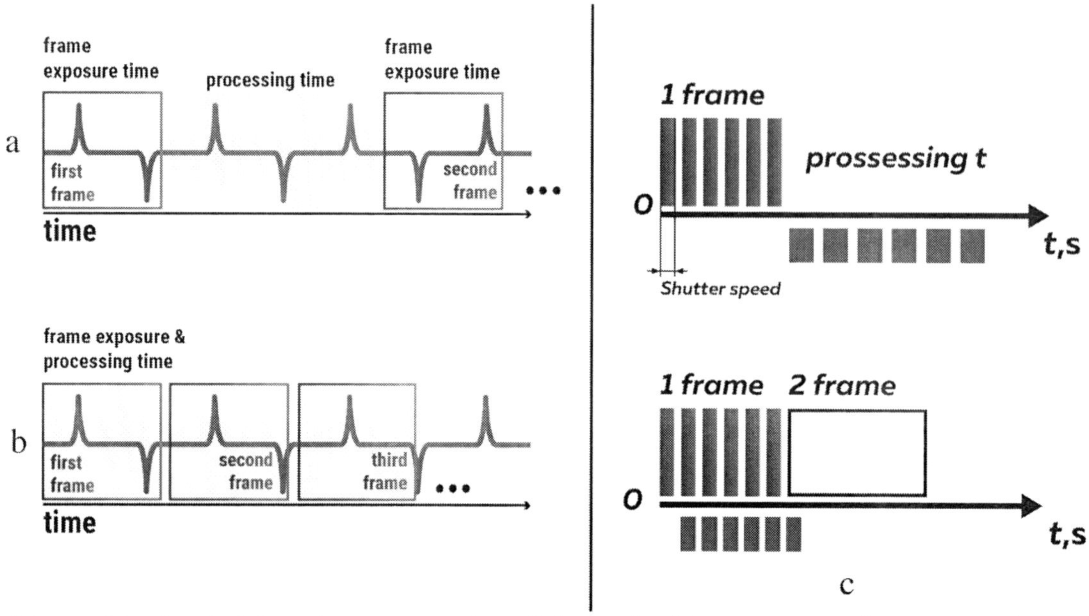

Figure 2 (a, b, c). Schematic representation of the work of the processes of exposure and image processing. **(a)** sequentially, **(b)** in parallel, **(c)** sequential (top diagram) and parallel (bottom diagram) algorithm with the rolling shutter effect implementation, where blue lines are the group of camera module matrix rows that were exposed sequentially, according to the rolling shutter effect, and pink lines depict the time periods needed to process each of the matrix rows group.

3. Results&discussion

Throughout the experiment, all the devices were in a dark room. First, we estimated the channel capacity and measured the data transmission speed in the system.
Bandwidth achieved with the use of developed algorithm can be calculated using equation (1). and maximum possible data transmission can be calculated by the equation (2).

$$B = mNt^{-1}, \tag{1}$$

where B is bandwidth, m is the number of separate exposed censor lines, N is the number of frames processed, t is the processing time.

And the maximum possible data transmission can be calculated as follows (2):

$$B_{max} = s^{-1}, \tag{2}$$

where B_{max} is the maximum available bandwidth and s is the camera module shutter speed of a mobile device.

According to the (2) the highest theoretical bandwidth with our setup using the developed algorithm was 24 kbps, as the shutter speed of a mobile device used during the experiments was 1/24000 sec, while the highest frame rate supported by smartphone was 240 frames / sec.
Then we modulated the luminous flux from LED luminaire with 0&1 sequence by squarewave signal. The modulation frequency was from 1 kHz to up to 7.5 kHz. We perform the experiment for 3 different distances to the mobile device. We also use the oscilloscope and photorecieving module in order to

analyze the optical analog signal we receive by the camera module of the mobile device. The results are shown in the figure 3.

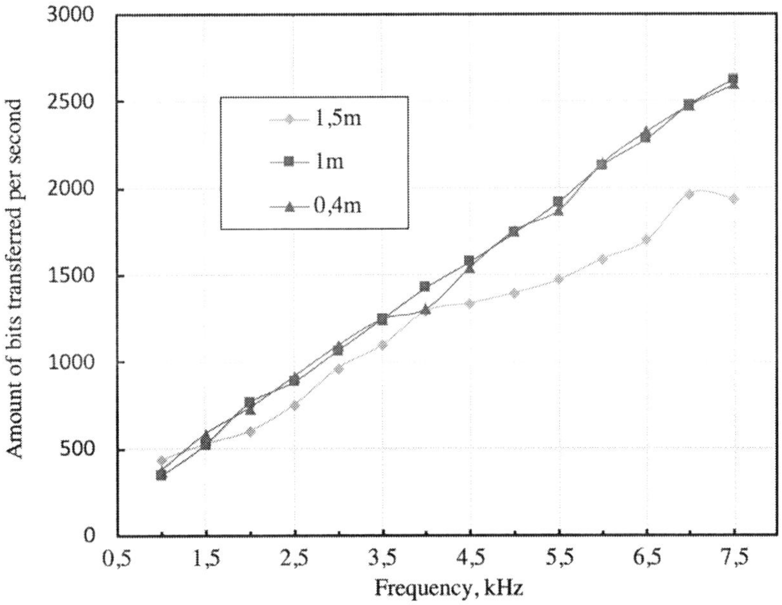

Figure 3. Amount of processed data per second to the modulation frequency at different distances with the first version of the algorithm. The mobile device had 15 frames per second during the experiment.

From the figure 3 it can be seen that the blue and red line, illustrating the dependences at 0.4m and 1.006m, respectively, increase linearly in the modulation section from 1 to 7.5 kHz. The loss of 2/3 of data transferred can be observed at all distances. However, as the distance increases (the green curve), the graph becomes non-linear, which can be explained by a decrease in the received optical power by the matrix of the mobile device, as shown in Figure 4 (a), which causes individual light lines to "merge" into one wider line, as shown in the Figure 4 (b), and as a result, part of the data is lost. The authors also observed a similar problem with a modulation frequency of more than 10 kHz at a distance of more than a meter.

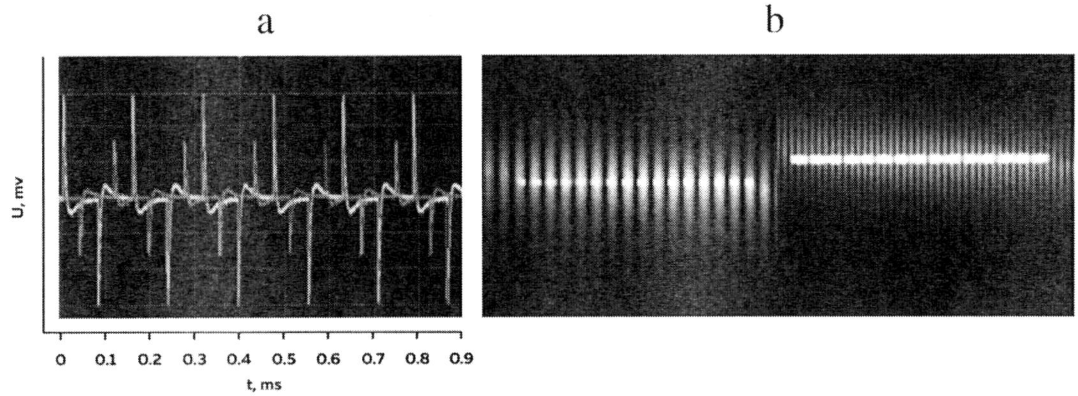

Figure 4 (a, b). (**a**) Waveform of an optical signal received by a photorecieving module on a distance of 1.006m and 1.566m displayed by the yellow and blue curve, respectively, the waveform is not a squarewave signal, due to the design of the module (**b**) screenshots of the "white" and "black" lines received by a mobile device (left — 4 kHz, right — 11 kHz).

During the experiment we ascertain three main speed determining parameters are: sensor exposure time — a function that determines the duration of each pixel exposure, ISO sensitivity — a measure of the camera's ability to capture light and fps. The first two parameters define the contrast of the white and black lines. The results of series of experiments performed by the authors show that the minimal exposure time and maximum ISO value should be used in order to obtain the highest contrast. ISO — 3200 and exposure time — 56311 ns for the mobile device we used during the experiment.

In the second stage we conducted a series of experiments with the optimized algorithm so that the processes of exposure and image processing were parallel. As can be seen from the figure 5, the number of processed bits of information with the optimized algorithm is significantly higher than in the previous experiment, illustrated by the green line in the Figure 5. Moreover, the use of the parallel algorithm allowed us to increase the FPS, which also led to the increase of speed in channel. In addition, we also carried out real dataset transmitting measurements. We used Manchester coding for real dataset experiment. This type of coding allowed us to minimize strobe effect and determine the headline of a code definitely. The results are shown in Figure 5. The drop in the number of processed bits of information on 4.5 kHz can be explained by the fact that using real dataset, causes the sequences of bits contains combinations of several light (or dark) lines in a row, and with the modulation frequency increase, these line rows "merge" into one line, in the same way as shown in Figure 4 (b). The dataset was a pseudo random binary sequence (PRBS).

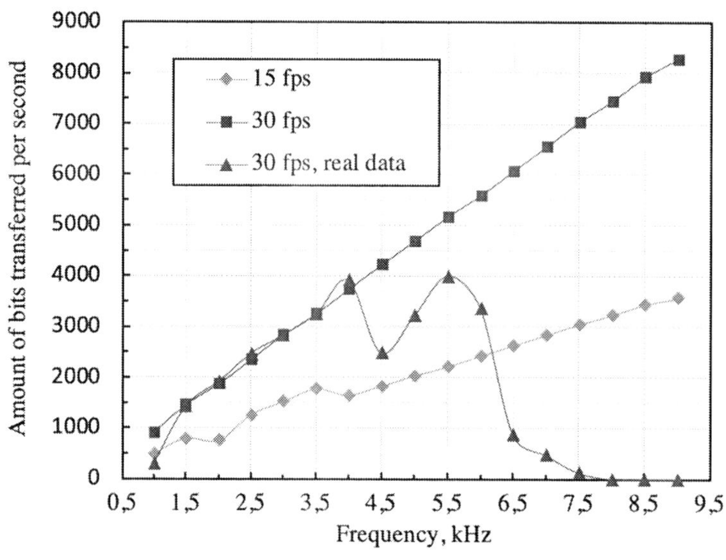

Figure 5. Dependence of amount of processed data per second on the modulation; green curve —non-rebuilt algorithm; red curve — rebuild algorithm, blue curve — rebuilt algorithm, using Manchester coded OOK PRBS

4. Conclusion

In this work we demonstrated the data transmission via LEDs and mobile device camera and developed the processing algorithm for camera capable to decode signals with modulation frequencies up to 9 kHz with the 0&1 PRBS and up to 4 kHz with the real Manchester coded OOK PRBS data. The future plan is the increasing of data transmission speed due to the introduction of special correction factors in the data processing algorithm, which makes camera better solve sequences with the same repeated symbols. Changing the design of the luminaire can also increase the speed of data transmission. For instance, an additional special-form diffuser will increase the uniformity of the light spot of the luminaire, which, in turn, will increase the uniformity of the light lines, thereby preventing them from merging. We are also planning to add the ability to detect the brightest zones during the direct exposure to our algorithm. That will allow us to increase the possible speed in the channel by adding the correction index. In addition, we assume that the signal receiving in such a system will often occur from the reflected illumination, which also allows to increase the uniformity of light spot, and, as a result, reduce losses.

Acknowledgments
This work was supported by the Ministry of Science and Higher Education of the Russian Federation in the framework of the Federal Target Program "Research and development on priority directions of scientific-technological complex of Russia for 2014-2020", the code 2017-14-582-0001, agreement № 14.581.21.0029 of October 23, 2017, unique ID RFMEFI58117X0029.

References
[1] Haas H, Yin L, Wang Y, and Chen C 2016 What is Li-Fi? *J. of Lightw. Tech.* **34**, 1544
[2] Rajagopal N, Lazik P and Rowe A 2014 *Visual Light Landmarks for Mobile Devices* (IPSN-14 Proceedings of the 13th International Symposium on Information Processing in Sensor Networks)

Spectral analysis of the electroluminescence and photoresponse of heterostructures with InGaAs quantum objects

R A Salii[1], V V Evstropov[2], S A Mintairov[1], M A Mintairov[2], M Z Shvarts[2] and N A Kalyuzhnyy[2]
[1]Submicron Heterostructures for Microelectronics, Research & Engineering Center RAS, St Petersburg, Russia
[2]Ioffe Institute RAS, Saint Petersburg 194021, Russia
E-mail: r.saliy@mail.ioffe.ru

Abstract. In this work, physical and optical properties of $In_{0.8}Ga_{0.2}As$ quantum dots (QDs) embedded in the structure of a single-junction GaAs solar cell (SC) grown by MOVPE technique were investigated using spectral characteristics of external quantum yield (EQE) and electroluminescence (EL). It has been found that, in characterizing QD physical parameters, simplified model of a thin stressed quantum well can be applied to a wetting layer (WL). It has been demonstrated that the EL spectra allows determining the absorption energy of photons in WL and QDs more accurately compared to the EQE spectra. Energies of "E_e-E_{hh}" and "E_e-E_{lh}" transitions in WL have been determined and were 1.325 eV and 1.388 eV respectively. The calculated values of the WL thickness (5.63 Å) and In composition of QDs (xIn = 80%) coincide with the technological parameters used in the epitaxial growth of the device.

1. Introduction

Today, possibilities of using various quantum-dimensional objects (QOs), such as quantum dots (QDs) [1-3], quantum wells (QWs) [4], their combination and other related objects [5] are being actively investigated to improve characteristics of semiconductor optoelectronic devices. The study of the possibilities of their use in solar cells (SC) began relatively recently, when it was found that the spectral photosensitivity of heterostructures with QDs is significantly extended to the long-wave region of the spectrum. The spectral dependence of the quantum efficiency (QE) of such SCs, as a rule, has several spectral peaks beyond the GaAs absorption edge, which represents the contribution of the QO array to the photogenerated current. Identification of these peaks allows obtaining optical properties and physical parameters of QDs as localization centers of charge carriers, and, therefore, is one of the main tasks in describing the processes of light absorption in SCs with QDs in the i-region.

The authors of [1, 2], by comparing the photoluminescence spectra and QE spectral dependences of GaAs SCs with embedded InAs and InGaAs QDs have shown that the main contribution to the photogenerated current is made by the wetting layer (WL). However, there is an issue in determining the depth of energy levels and peak positions of the optical spectra depending on the composition and amount of the material in such QOs. It is necessary to take into account the shifts of the band edges in stressed (compressed) material and also the solution of the Schrödinger equation in the effective mass approximation. In [6], for such SCs with QDs, the spectral dependence of QE was modeled, and QD parameters (according to the representation of a parallelepiped shape) were calculated using the empirical k·p hamiltonian method [7, 8]. However, for InGaAs/GaAs QDs, the issue of peak

determination is more complicated due to the lack of information on real In composition in QDs and WL after QO formation in the Stranski-Krastanov mode.
Nevertheless in [9], we have already demonstrated that $In_{0.8}Ga_{0.2}As$ QDs have several advantages in comparison to classical InAs QDs, when they are embedded in the structure of a single junction GaAs SC. In particular, better relaxation of a QD array in the GaAs matrix is achieved, which makes it possible to increase the volume of the absorption region without loss the quality of a SC structure. Also in [9], an approach based on the representation of WL through the model of a thin, mechanically compressed QW was proposed for estimating the contribution of QDs and WL to the device photocurrent. Using this model the theoretical calculation of band transitions in structures with QDs allows characterizing the peaks of the QE spectral dependence and determining the thickness and composition of WL. However, the calculation accuracy within this model depends on the accuracy of determining the maxima peak position of the QE dependence in the QD absorption region, which characterizes, in particular, the electron-heavy hole (E_e-E_{hh}) and electron-light hole transitions (E_e-E_{lh}) in a WL. The QE spectral characteristic of a SC, in this case, has several disadvantages, which do not allow determining accurately the absorption energy in the QDs and the position of the spectral peak for "E_e-E_{hh}" energy transition. In addition, the blurred edge of the absorption of GaAs [10] also does not allow estimating the position of the peak for "E_e-E_{lh}" energy transition. In this work, to determine peak positions for WL and QDs, electroluminescence (EL) spectra of SC structures with embedded $In_{0.8}Ga_{0.2}As$ QD arrays were obtained and compared with the QE spectral characteristics. The obtained data were used for theoretical calculation of WL thickness and composition according to the model-solid theory [11].

2. Experimental procedure and modeling

The experimental SC structure (figure 1) was grown by the MOVPE technique on an R&D installation with a horizontal type reactor at low pressure (100 mbar).

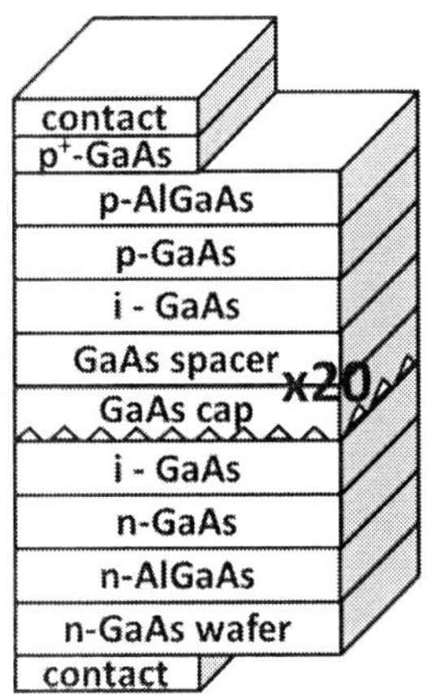

Figure 1. GaAs SC with 20 layers of InGaAs QDs experimental structure.

The growth was carried out according to the following technological process: after growing the $Al_{0.3}Ga_{0.7}As$ rear potential barrier, the n-GaAs base and a part of the GaAs i-region were grown at 700 °C, and then the reactor was cooled down to QD material deposition temperature – 520 °C. Then, $In_{0.8}Ga_{0.2}As$ material was deposited for nucleation of coherent islands in the Stranski-Krastanov mode with forming a WL. After that, the QDs were covered with GaAs layer at the same temperature in order to protect them from degradation during the subsequent reactor heating to the GaAs spacer layer growth temperature (600 °C). The GaAs spacer thickness was 35 nm. The process was repeated according to the number of embedded QD layers (20 in this work), and then the rest of the i-GaAs, the p-GaAs emitter, $Al_{0.8}Ga_{0.2}As$ "window" and n+-GaAs contact layer were grown at 700 °C. The $In_{0.8}Ga_{0.2}As$ material amount for the QDs formation was chosen according to the previous study, which was devoted to determination of the maximum photoluminescence (PL) intensity of $In_{0.8}Ga_{0.2}As/GaAs$ heterostructures and was 2 ML (~6 Å) [12].

A multifunctional installation has been used for measuring the external quantum efficiency (EQE) of the SCs. The measuring hardware included an unblocked ultra-violet halogen light source, a grating monochromator with 2nm/mm dispersion within the 300–1200 nm wavelength scanning range, and an optical chopper of 90 Hz and high sensitivity lock-in electronics. The EQE measurement was controlled by a laptop. The lock-in technique allows damping output signals at the QD sensitivity range to be detected confidently.

Electroluminescence spectra have been measured from the photo-receiving surface of a SC sample with a fiber-optic spectrophotometer based on a 2048 elemental CCD detector matrix with high photometric sensitivity in the spectral range of 200–1100 nm and an optical resolution of 0.04–20 nm. The luminescence signal occurred, when direct current in the range of 0.001–1A from a stabilized current source passed through a SC. To prevent uncontrolled heating, the sample was mounted on a thermostatic surface with a digital temperature controller.

The investigated QOs are mechanically stressed due to InGaAs - GaAs lattice mismatch. Therefore, the model-solid theory [13] was applied to WL to determine shifts of the band edges and, consequently, depths of the electron-, heavy hole- and light hole potential wells. To determine the energy levels and the phononless energy transitions in QOs, the Schrödinger equation was solved in the effective mass approximation [14-16]. The equation has an exact solution, if the well is rectangular and symmetrical.

3. Results and discussion

The EQE dependence of GaAs SC with InGaAs QDs has a long-wavelength shoulder with energy of ~ 1.24 eV, which obviously represents the contribution of QDs to the photosensitivity of the device (figure 2). In addition, the EQE curve has one distinct peak with energy of 1.33 eV, which represents the energy transition "E_e-E_{hh}" in WL according to the model of a thin stressed QW. Following this model the EQE dependence should also have a peak close to the GaAs absorption edge, which represents the "E_e-E_{lh}"energy transition in the WL. In this case, it is observed on the EQE curve as a shoulder in the energy range from ~1.37 to 1.4 eV.

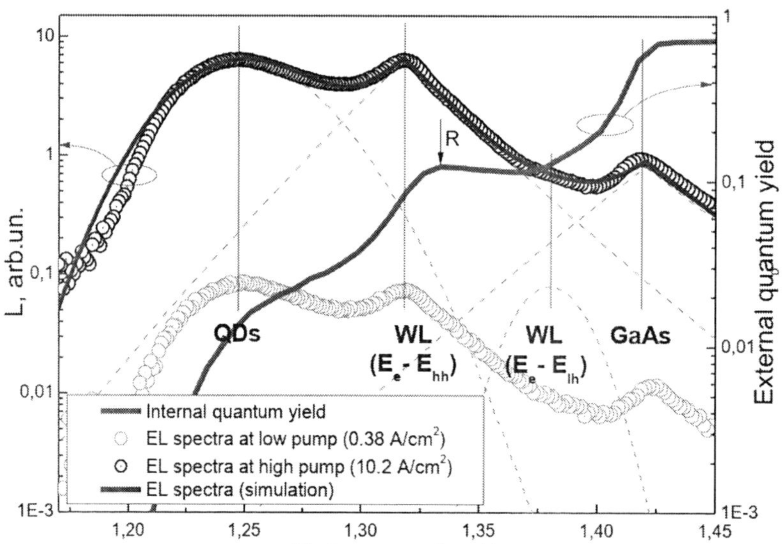

Figure 2. EL and EQE spectra comparisons for a GaAs SC with 20 layers of $In_{0.8}Ga_{0.2}As$ QDs.

This shoulder on the EQE spectrum allows estimating the "E_e-E_{lh}" transition energy in WL during photon absorption very roughly. The experimental EL spectrum, in contrast, has three distinct peaks and represents the energy transitions in QDs, "E_e-E_{hh}" and the GaAs matrix, and also has a straight section on the curve in the energy range from ~1.37 to 1.4 eV. A mathematical decomposition of the EL spectrum into four spectral bands has been carried out. Wherein the Gaussian functions were used to describe the shape of the spectral bands related to QDs and the "E_e-E_{lh}" transition in WL. For the transition "E_e-E_{hh}" in the WL and the GaAs matrix the Laplacian functions were used due to their exponential behavior. The decomposition results are showed in figure 2 (blue line). The total resulting curve coincides quite well with the experimental curve (figure 2, black circles). As a result of the decomposition, the peak position, which represents the "E_e-E_{lh}" energy transition in the WL was determined (table 1, WL (E_e-E_{lh})).

Table 1. EL peak positions in the QD adsorption area at high current density (10.2 A/cm^2).

Layer	QDs	WL (E_e-E_{hh})	WL (E_e-E_{lh})	GaAs matrix
Peak position, eV	1.245	1.317	**1.380**	1.417

It is important to note that, to calculate the thickness and composition of WL according to the model-solid theory, it is necessary to know the energies of the phononless "E_e-E_{hh}" and "E_e-E_{lh}" transitions. As a result of comparative analysis of the EL and photoresponse spectra, the Stokes shift for the energy transition "E_e-E_{hh}" is observed. In this work, the assumption that the spectral peaks of EL correspond to phononless transitions is used. The basis for this assumption is the experimental fact that the EL peak of the GaAs matrix at low current density (figure 2, grey circles) is exactly equal to the phonon-free band gap (1.425 eV). Therefore, for further theoretical calculations we used the energy values of the peak positions determined by decomposition of the EL spectrum which were increased by 0.008 eV (half of the Stokes shift value) and amounted to 1.388 and 1.325 eV for the "E_e-E_{hh}" and "E_e-E_{lh}" transitions, respectively.

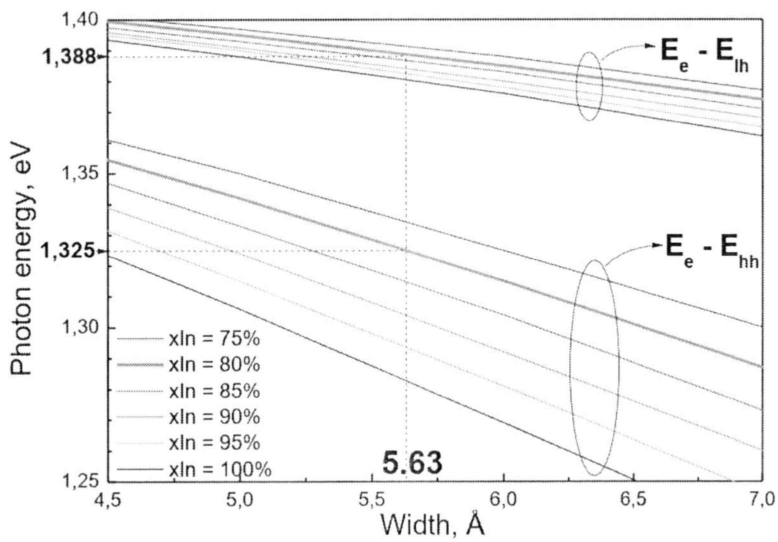

Figure 3. Theoretical calculation of energy transitions in a InGaAs/GaAs QW.

Based on the model-solid theory, a theoretical calculation of phononless energy transitions in $In_xGa_{1-x}As/GaAs$ QWs for a wide range of In composition (from 0.75 to 1) and QW thicknesses (from 4.5 to 7.5 Å) was carried out. The results are presented in figure 3 as two sets of curves for the energy

transitions "E_e-E_{hh}" and "E_e-E_{lh}", respectively. Using the values of peak positions obtained by the decomposition of the EL spectrum, one can determine the composition and thickness of WL. For this, it is needed to choose the In composition value in QW, at which the transition energies "E_e-E_{hh}" and "E_e-E_{lh}" are equal to 1.325 eV and 1.388 eV, respectively, at the same QW thickness. In this case, these conditions are satisfied for xIn = 80% and QW thickness of 5.63 Å (dotted lines in figure 3).
The performed calculation shows that the model of a thin stressed QW applied to WL allows determining the values of its physical parameters quite accurately. The calculated parameters of WL (In composition (x) and thickness) correspond to the growth technology: x = 80% and the amount of deposited $In_{0.8}Ga_{0.2}As$ material does not exceed 2 ML (6 Å), in accordance with the optimal thickness used during epitaxial growth for the formation of QD coherent islands.

Conclusions
The spectral characteristics of the single-junction GaAs SC with embedded arrays of $In_{0.8}Ga_{0.2}As$ QDs have been studied. It has been shown that photon absorption in WL dominates in photoresponse spectrum for such SC. It has been found that when physical determining the spectral dependences of the photon absorption in a SC with QDs the simplified model of a thin QW can be applied to WL. Wherein, using the EL spectra one can accurately estimate the energy transitions in WL and QDs. It has been demonstrated that, with a separate estimation the QDs and WL contributions to the photocurrent of the device, the EL spectra have a significant advantage over the EQE spectra due to the more distinct spectral peaks beyond the absorption edge of the GaAs SC matrix. The energies of the "E_e-E_{hh}" and "E_e-E_{lh}" transitions in WL have been determined to be 1.325 eV and 1.388 eV respectively. The calculated values of WL thickness (5.63 Å) and In composition of QDs (xIn = 80%) coincide with the technological parameters used in the epitaxial growth of the device.

Acknowledgements
The work has been supported by Russian Science Foundation, agreement no. 17-72-20146.

References

[1] Kalyuzhnyy N A et al. 2016 *Progress in Photovoltaics* **24** 9 pp 1261–1271

[2] Lantratov V M et al. 2010 *Advances in Science and Technology* **74** pp 231–236

[3] Bailey C G et al. 2012 *Journal of Photovoltaics* **2** pp 269–275

[4] Browne B et al. 2013 *AIP Conference Proceedings* **1556** pp 3–5

[5] Mintairov S A et al. 2017 *Semiconductor Science and Technology* **32** 015006

[6] Panchak A et al. 2016 *Solar Energy Materials & Solar Cells* **145** pp 180–184

[7] Luque A et al. 2011 *Solar Energy Materials & Solar Cells* **95** pp 2095–2101

[8] Voon L C L Y and Willatzen M 2009 *The k p Method* (Springer, Berlin)

[9] Salii R et al. 2018 *Journal of Physics: Conference Series* **1135** pp 012078

[10] Moss T S, Burrell G J and Ellis B 1973 *Semiconductor Opto-Electronics* (Butterworth & Co.)

[11] Van de Walle C G 1989 *Physical Review* **39** pp 1871–1883

[12] Salii R et al. 2017 *Semiconductors* **51**(1) pp 870–876

[13] Krijn M P C M 1991 *Semiconductor Science and Technology* **6** pp 27–31

[14] Hiroshima T and Lang R 1986 *Applied Physics Letters* **49** pp 639–641

[15] Nakamura K et al. 1989 *Journal of Quantum Electronics* **25**(5) pp 889–895

[16] Fritz I J 1987 *Journal of Applied Physics* **61** pp 2273–2276

Experimental study of power-limiting factors of 1.1 μm range edge-emitting lasers based on InGaAs/GaAs quantum well-dot nanostructures

A A Serin[1], A S Payusov[1], G O Kornyshov[2], Yu M Shernyakov[1], S A Mintairov[1], N A Kalyuzhnyy[1], M M Kulagina[1], A E Zhukov[2], N Yu Gordeev[1], M V Maximov[2]

[1]Ioffe Institute, 26 Polytechnicheskaya, St Petersburg 194021, Russia
[2]St. Petersburg Academic University, 8/3 Khlopina, St Petersburg 194021, Russia
e-mail: *spbgate21@gmail.com*

Abstract. We present a study of high-power characteristics of edge-emitting lasers based on quantum well-dots (QWD) in a pulsed regime. QWD-based lasers with 2 and 5 active layers emitting at ~1.1 μm provided maximal optical power of 39 W limited by the pulse current source available. We have investigated the lasing spectra and shown that under the injection current above 20 kA/cm^2 the active region overheats approximately by 0.5°C per 1 kA/cm^2 during 100 ns current pulse. The active region overheating correlates well with the reducing of the differential efficiency of our devices. We believe that maximal pulse optical power of the QWD-based lasers is limited mainly by our laser wafer design rather than by the QWD active media properties.

1. Introduction

GaAs-based high-power high-brightness diode lasers emitting in the wavelength range of 1.1-1.2 μm have a great potential for a number of applications such as direct material processing, pumping of the rare earth element doped fiber lasers [1], nonlinear frequency conversion to green-yellow and yellow-orange spectral range. Currently *InGaAs* quantum wells (QW) are the typical choice for the active region for such devices. However, in order to shift the lasing wavelength beyond 1.1 μm one needs to increase the indium (*In*) content in the QWs, which results in extra strain in the active region. This strain limits the lifetime of the QW-based lasers to several thousand hours [2].

Recently a new type of *InGaAs/GaAs* nanostructures – quantum well-dots (QWD) have been proposed [3]. These *InGaAs* nanostructures allow obtaining the *In* content of 30-50% without strain-induced dislocations. QWD-based edge-emitting lasers have been successfully demonstrated [4]. We believe that QWD could be considered as an alternative to *InGaAs* QWs for high-power 1.1 μm range lasers. In this paper, we demonstrate that maximal pulse optical power of the QWD-based lasers is limited mainly by the overheat under high currents rather than by QWD active media properties.

2. Laser wafer design and sample processing

The laser wafers were grown by metal-organic vapor phase epitaxy (MOVPE) on n-(100) *GaAs* substrates misoriented by 6° toward [111] direction. Laser structures consist of an undoped 0.78 μm *GaAs* waveguide sandwiched between p- and n-type doped 1.5 μm $Al_{0.4}Ga_{0.6}As$ cladding layers. In order to minimize the internal loss in the claddings their doping level of $2\cdot10^{18}$ cm^{-3} was reduced down to $7 \cdot 10^{17}$ cm^{-3} in the regions adjoining the waveguide. Zinc and silicon were used as p- and n-dopants respectively. The active regions consisted of 2 or 5 layers of $In_{0.4}Ga_{0.6}As$ QWDs (2xQWD and 5xQWD respectively) separated with 40 nm-thick *GaAs* spacers. The effective thickness of each QWD layer was 2 nm.

The wafers were processed into 100 μm broad-area shallow mesa lasers using standard photo-lithography. Devices with various cavity length were cleaved and mounted p-side down onto copper heatsinks with an indium solder in order to reduce overheating and improve current spread in long (>2mm) devices. No facet coatings were used.

3. Experiment details

First, the devices with different cavity length were characterized at room temperature in a pulsed regime with 300 ns pulse width and the duty cycle of 0.1%. Then the values of the internal efficiency (η_i) and the internal loss (α_i) were extracted from the dependence of the reciprocal differential efficiency on the cavity length. The pulsed voltage supply (Avtech A4-B) capable to deliver 100V was used to drive the devices. The driving current was measured as a voltage drop over 1Ω resistor connected in series with the device under test. For the light-current and spectral measurements, we used 100 ns pulses and the duty cycle of 10^{-3}%. An integrating sphere with a fast Ge photodiode connected to a 100 MHz preamp was used to measure the pulse light-current characteristics. Amplitudes of the driving current and the photocurrent were registered with a 300 MHz sampling oscilloscope.

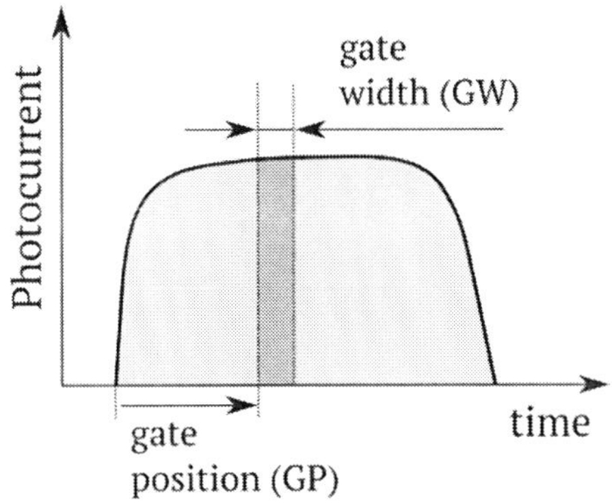

Figure 1. Schematic representation of the photocurrent pulse.

The output emission from the rear facet was used to measure the lasing spectra. It was collected with 200 μm fiber and then focused onto the input slit of a 0.6 m monochromator. A fast InGaAs receiver providing 2.5 ns rise time connected to the gated integrator SR250 from Stanford Research was used to detect the signal from the output of a monochromator. The gate width (GW) and the gate position (GP) against the photocurrent pulse (figure 1) were controlled using the oscilloscope. The spectral measurements were performed with averaging over 100 pulses. The diode laser temperature was stabilized during measurements using a Peltier-based temperature controller. A thermistor was placed on the laser heatsink in the near proximity of the laser chip.

4. Results and discussion

Both laser structures demonstrated the internal efficiency over 80%, the internal losses were estimated as 1.0 cm^{-1} for 2xQWD and 1.5 cm^{-1} and 5xQWD lasers. Obtained parameters are similar to ones reported in [5] for QW-based devices with the same lasing wavelength. Figure 2 a,b shows the pulse light-current (LI) characteristics for 2 and 5xQWD devices respectively. Dashed lines show slopes for the initial parts of the curves up to 15 A. Due to the higher internal loss the slope for the 5xQDW lasers is less than that for 2xQWD devices. All tested lasers demonstrated a decrease of the differential efficiency under currents over 20 A. However, the LI characteristics of the 5xQWD lasers saturate slower than 2xQWD lasers. The 2xQWD and 5xQWD laser demonstrated practically the same maximal output power of 38 W and 39 W respectively (figure 2 a,b, red dots) which was limited by our current supply. The effect of improvement in linearity of light-current characteristic with an increasing number of active layers has been previously analyzed and reported in [5].

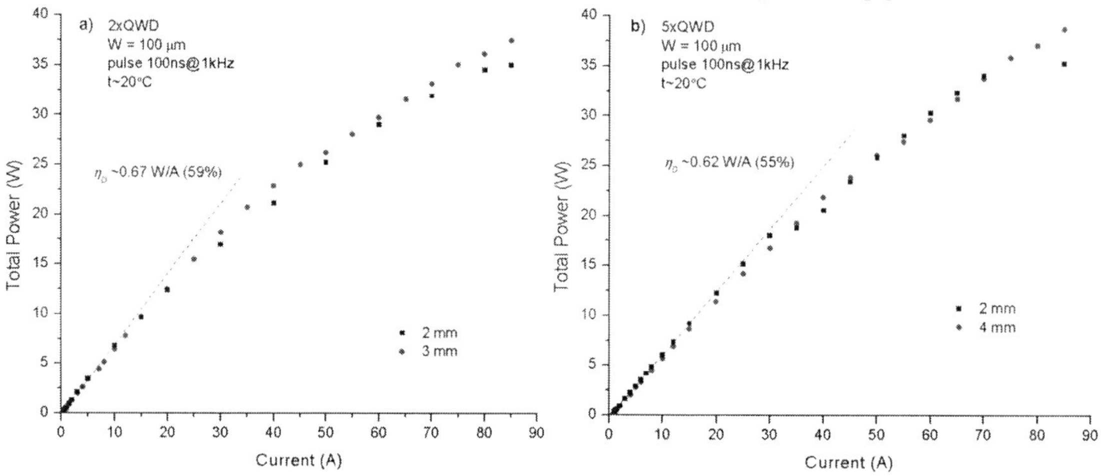

Figure 2. Light-current characteristics of 2xQWD (a) and 5xQWD (b) lasers with various cavity length measured in pulsed regime at 20°C.

In order to investigate the cause of the output power saturation, we measured lasing spectra at different currents in the pulsed regime from 2 mm long devices. At first, measurements were carried out with GW set to 120 ns in order to integrate a signal over the whole photocurrent pulse. The results are shown in figure 3 a,b. There is no evident saturation of the ground state emission under the high currents. Instead, one can see the long wavelength tale emerging with the increasing current. A redshift is usually associated with the active region overheat. To confirm this we measured spectra under 70 A pump current with 10 ns GW and different GPs over the photocurrent pulse. Figure 4 shows the results for 5xQWD laser with 2 mm long cavity. There is an evident red shift during the pulse without significant change in the line shape. The linewidth of the spectrum taken with 10ns GP is smaller due to the 30 ns rise time of the pump current pulse. However, this fact is unimportant for using this spectrum as a reference one for the redshift detection.

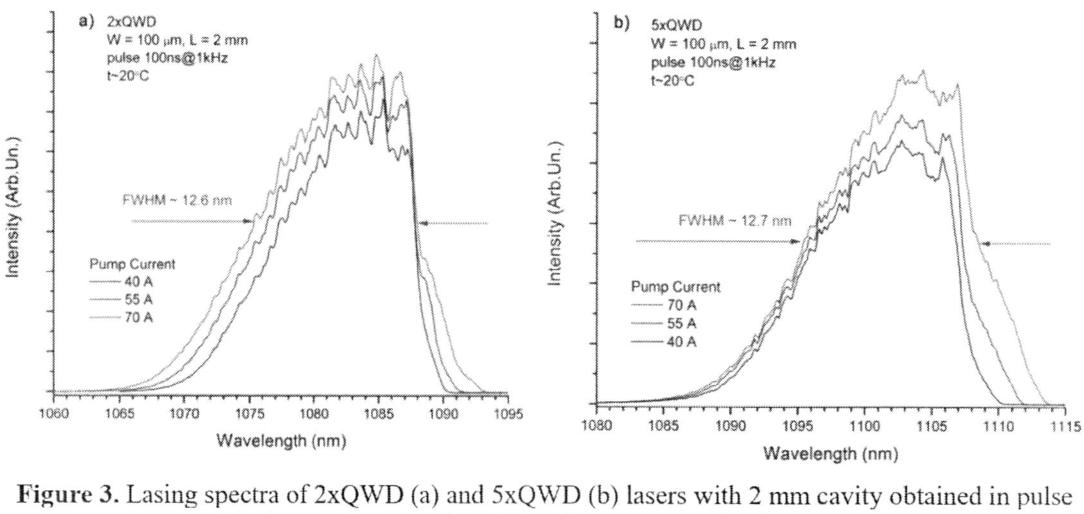

Figure 3. Lasing spectra of 2xQWD (a) and 5xQWD (b) lasers with 2 mm cavity obtained in pulse regime integrating over the whole photocurrent pulse (GW 120 ns).

In order to quantify the discovered overheat, we measured the lasing spectra under 40 A pump current with 10 ns GW and 50 ns GP at different temperatures. Temperature-induced wavelength shift was found to be 0.34 nm/°C and 0.33 nm/°C for 2xQWD and 5xQWD lasers respectively. Using these data along with the redshift we calculated the overheat under different currents (Table 1). One can conclude that overheat builds up at an approximate rate of 0.5°C per kA/cm^2. The overheat build-up correlates well with output power saturation.

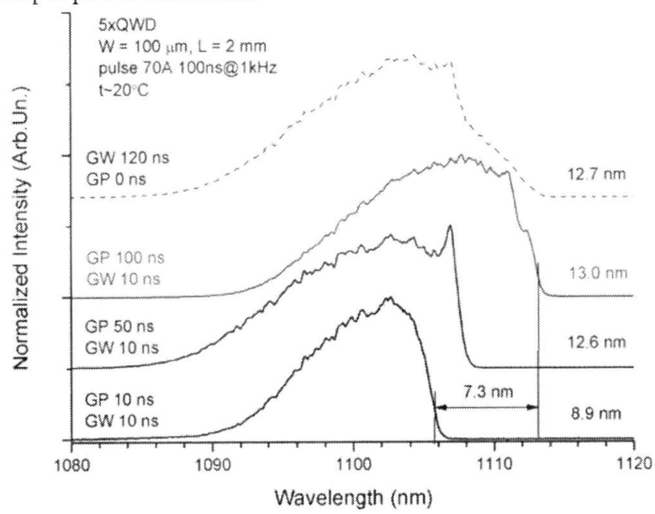

Figure 4. Lasing spectra of the 5xQWD device obtained at 70A driving current with GW of 10 ns at different GP with respect to photocurrent pulse. On the right side full width oh half maximum of the lasing line is shown.

Table 1. The overheat over 100 ns at different pump current densities for 2 mm long lasers calculated from the spectral red shifts.

Current, A	Current density, kA/cm^2	Δt_{2xQWD}, °C	Δt_{5xQWD}, °C
40	20	9	13
55	27.5	11	17
70	35	17	21

The power saturation mechanisms at high current densities in the pulsed regime have been previously studied both theoretically and experimentally. The carrier accumulation in waveguide layers is considered to be the main reason for increasing the optical loss and subsequent power saturation [6-8]. Meanwhile, most authors presume the absence of overheat in the pulsed regime and omit it as a factor affecting carrier escape to waveguide layers. However, the obtained results show that this assumption is not accurate at least for the devices with the designs similar to ours. We would like to point out that lasing intensity from the QWD ground state does not saturate with the increasing current, meaning that the QWD active media properties are irrelevant to the observed output power saturation. Thus, we assume that the wafer design determines the active region overheating. The main factors affecting the heat spread are the thickness and thermal conductivity of the layers in between the active region and the surface on which the device is mounted. The design of our laser waveguide is rather conventional. Therefore, we assume that the results obtained can be used to improve pulse high power properties of a great number of laser designs.

5. Conclusions

We have investigated high-power characteristics of the edge-emitting lasers based on quantum well-dots (QWD) in pulsed regime. QWD-based lasers with 2 and 5 active layers emitting at ~1.1 μm provided maximal optical power of 39 W limited by the pulse current source available. We have investigated the lasing spectra and shown that under the injection current above 20 kA/cm^2 the active region overheats approximately by 0.5°C per 1 kA/cm^2 during 100 ns current pulse. The active region overheating correlates well with the reducing of the differential efficiency of our devices. We believe that maximal pulse optical power of the QWD-based lasers is limited mainly by our laser wafer design rather than by the QWD gain properties.

Acknowledgments

Authors acknowledge the support from the Russian Foundation for Basic Research (project no. 18-02-01123).

References

[1] Jackson S D *et al* 2007 *Opt. Lett.* **32** 2496
[2] Bugge F *et al* 2018 *J. Cryst. Growth* **491** 31–5
[3] Mintairov S A *et al* 2015 *Nanotechnology* **26** 385202
[3] Payusov A S *et al* 2018 *J. Phys. Conf. Ser.* **1135** 012071
[4] Pietrzak A *et al* 2009 *Semicond. Sci. Technol.* **24** 035020
[5] Ryvkin B and Avrutin E 2006 *Electron. Lett.* **42** 1283
[6] Wenzel H *et al* 2009 *Opt. Quantum Electron.* **41** 645–52
[7] Vinokurov D A *et al* 2007 *Semiconductors* **41** 984–90

Development of antireflective coatings based on SiO_2/Si_3N_4 on textured silicon for heterojunction solar cells

Y Nevenchannyy[1], K Khoruzhiy[1], D Kudryashov[2]

[1]Lyceum "Physical-Technical High School" (PTHS), St. Petersburg 194021, Russia
[2]St. Petersburg Academic University, St. Petersburg 194021, Russia

Abstract. The technology of antireflective coatings formation on textured silicon by magnetron sputtering was developed. The growth rates of individuals thin films were determined as well as their refractive indexes. According to the results obtained an antireflective coating based on SiO_2/Si_3N_4 on textured silicon with a thickness of SiO_2 68 nm and Si_3N_4 of 74 nm showed the lowest reflection.

1. Introduction

Silicon-based heterojunction solar cell (SHSC) is one of the most attractive solar cell architectures due to its numerous advantages such as low fabrication temperature, source materials availability, scalable for the mass production, low surface recombination. To realize the whole potential of the SHSC limiting factors are needed to be eliminated such as an optical reflection from the front surface of solar cell. To reduce the reflection, such approaches as texturing the surface of solar cells [1] and the use of antireflective (AR) coatings [2] are known. Due to texturing, the incident light is repeatedly reflected from the faces of the micro-pyramids, thus more light penetrates into the material and the reflection from the solar cell is reduced. AR coatings are based on effect optical interference. Under certain conditions the amplitude of the reflected electromagnetic waves decreases, causing the reflection coefficient to decrease accordingly. The aim of this work is to combine these two approaches and to develop a technology of manufacturing of AR coatings based on SiO_2/Si_3N_4 layers on textured silicon for heterojunction solar cells.

2. Experiment details

SiO_2/Si_3N_4 layers were grown by RF-magnetron sputtering on BOC EDWARDS Auto 500RF Sputter Coater. The deposition was conducted with a magnetron power of 200 W and a chamber pressure of 10^{-4} - 10^{-3} mbar. Pure argon (99.999 %) was used as a working gas for plasma maintenance. To grow the oxide and nitride layers, oxygen or nitrogen were additionally supplied to the growth chamber. A silicon target with a diameter of 3" was used as a Si source. The layers were deposited on a flat and a textured monocrystalline Si(100) substrates and also on glass slides to check the transmittance of the applied layers. The control of the thickness of the obtained layers was carried out using a profilometer AMBiOS XP-1. The refractive index at a wavelength of 633 nm was determined using ellipsometer HORIBA JOBIN YVON. Total reflection spectra of the polished samples were recorded using a S100 spectrometer. Total reflection spectra of the textured samples were recorded using integrating sphere.

In the first phase of the research one-layer films were grown with different magnetron power to find out the growth rate and the conditions when the refractive indexes of the layers were closest to the monocrystal structures of SiO_2 and Si_3N_4 (Fig 1). Rise of magnetron power leads to an increasing of film growth rate. Refractive index of Si_3N_4 thin film non-linearly depends on sputtering power and after the magnetron power value of 200 W remains nearly constant.

Then using the obtained data, theoretical calculation of the SiO_2/Si_3N_4 pair of thicknesses, which would give the smallest average reflection for wavelengths from 400 to 1000 nm, for two-layer AR coating were carried out.

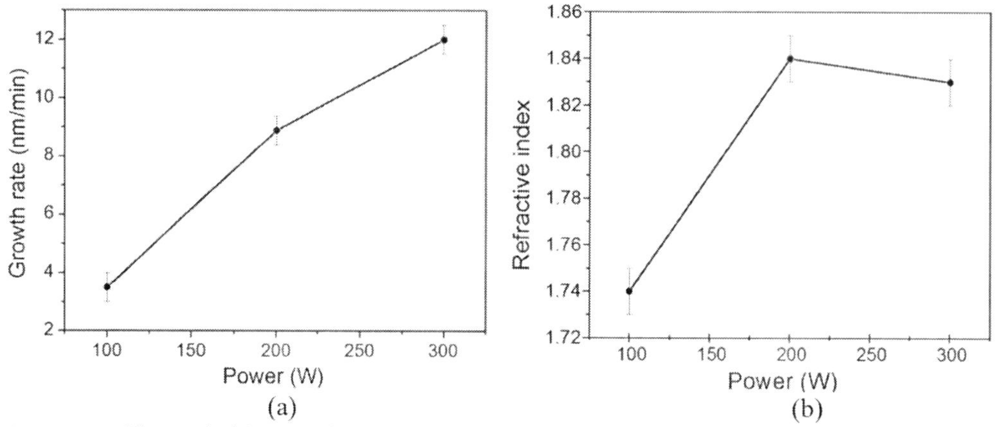

Figure 1. Silicon nitride growth rate (a) and its refractive index (b) as a function of magnetron power.

A series of films depositions by magnetron sputtering were carried out with different pairs of thicknesses of silicon nitride and oxide, based on theoretical calculations, to find the optimal combination which would give the minimal average reflectance.

Finally, the AR deposition was carried out on textured silicon surface. The following method was used to texture silicon. The polished silicon wafer (100) was divided into several samples (4x4 cm^2) and were set in a teflon holder. The samples were placed in an aqueous solution of isopropyl alcohol and KOH. The temperature of the solution was maintained at 85 °C. The samples were kept under such conditions for 60 minutes, after which they were rinsed with distilled water to remove alkali residues. SEM image of textured silicon wafer is shown on Fig 2.

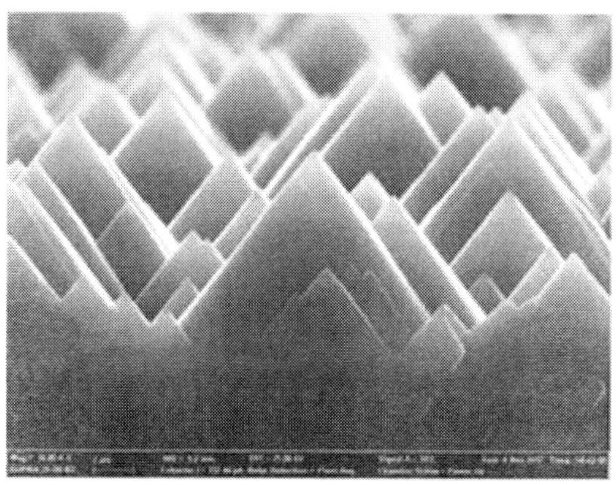

Figure 2. SEM image of silicon surface after texturing process.

The textured wafer along with a polished one were used as substrates for AR coating deposition. The samples were mounted on a substrate holder. The deposition was conducted without additional substrate heating.

3. Results and discussion

Figure 3 demonstrates calculated integral reflection coefficients for SiO_2/Si_3N_4 AR coating formed on silicon as a function of the layers thickness.

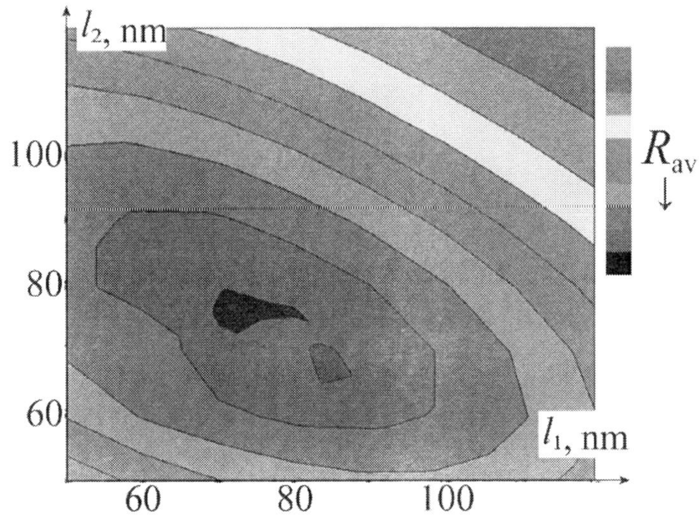

Figure 3. Calculated integral reflection coefficients for SiO_2/Si_3N_4 AR coating formed on silicon as a function of the layers thickness

The lowest reflection is observed at thicknesses of silicon oxide and nitride of 80 nm. Increasing of both layers thicknesses over 80 nm leads to a rising of integral reflection. Experimental data for measured total reflection spectra of polished samples with AR coatings are presented on figure 4, while the average reflectance coefficients are presented in table 1.

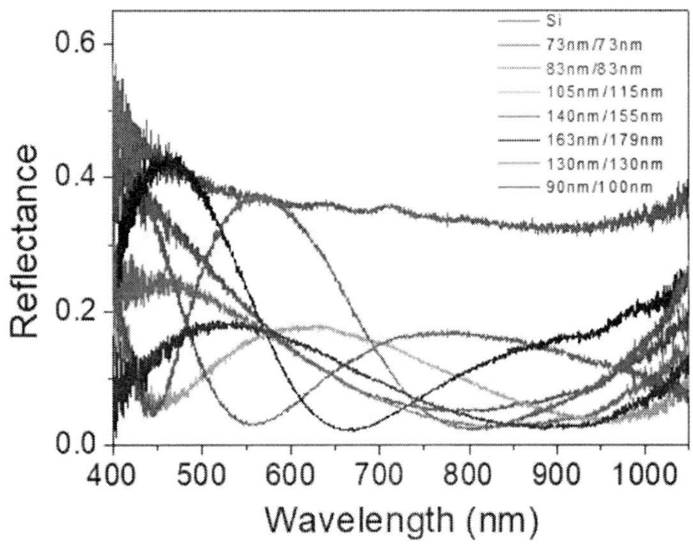

Figure 4. Measured reflection spectra of polished silicon samples with and without AR coatings.

Flat silicon sample without any deposited thin films show an average reflectance of a 37% in a wavelength range of 300-1100 nm. This is a very large value for the top surface of solar cell. For this reason an AR coating is needed to reduce optical reflection. Deposition of silicon oxide and nitride thin films leads to a changing of reflectance spectra. In one wavelength region the reflectance is decreasing but in other – on the contrary – it tends to get close to initial, corresponding to the bare silicon.

Experimental data for measured total reflection spectra measured on textured samples with AR coatings are presented on figure 5, and the average reflectance coefficients in table 2.

Table 1. Characteristics of grown AR coatings formed on polished silicon surface.

Thickness of Si_3N_4 layer (nm)	Thickness of SiO_2 layer (nm)	R_{av} (%)
0	0	36
73	73	15.3
83	83	11.7
90	100	10.3
105	115	10.9
130	130	17.7
140	155	13.5
163	179	18.3

The lowest reflection from textured samples is observed at the thickness of SiO_2 68 nm and Si_3N_4 of 74 nm. Small deviation from the simulated earlier values could be connected with the effect of thinning of film thickness while depositing on inclined surface like textured pyramid.

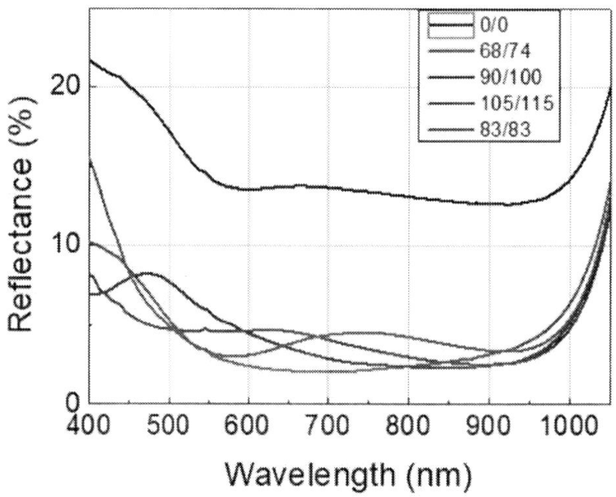

Figure 5. Measured total reflection spectra of textured samples with AR coatings.

The reflectance spectra for the textured silicon show more smooth form without very pronounced peaks. Its form is more suitable for solar cells fabrication.

The lowest reflection from polished samples was observed at the thickness of SiO_2 90 nm and Si_3N_4 of 100 nm with the average reflectance of 10.3%.

Table 2. Characteristics of an AR coating formed on textured silicon surface.

Thickness of Si_3N_4 layer (nm)	Thickness of SiO_2 layer (nm)	R_{av} (%)
0	0	14.7
68	74	3.64
83	83	4.3
90	100	3.95
105	115	3.76

The technology of AR coatings formation on textured silicon by magnetron sputtering was developed. According to the results obtained an antireflective coating based on SiO_2/Si_3N_4 on textured silicon with a thickness of SiO_2 68 nm and Si_3N_4 of 74 nm showed the lowest reflection.

Acknowledgements
This work was carried out with the support of the grant of government of the Russian Federation (16.2593.2017/8.9)

References

[1] Kudryashov D, Gudovskikh A, Rodin A, Pinaev N. 2018 J. Phys: Conf. Series 1124 041010
[2] Kotlikov E N 2009 Design, Manufacturing and Investigation of Interference Coatings: Tutorial (St. Petersburg: GUA Press) p 188

Experimental study of self-induced transparency mode- locking in Ti:sapphire laser

M V Arkhipov[1,2], R M Arkhipov[1,2,3], A A Shimko[1], I Babushkin[4], N N Rosanov[2,3,5]

[1] St. Petersburg State University, 199034 St. Petersburg, Russia
[2]ITMO University, St. Petersburg 197101 Russia
[3]Ioffe Institute, St. Petersburg 194021, Russia
[4]Institute of Quantum Optics and PhoenixD, Leibniz University Hannover, Hannover 30167 Germany
[5]Vavilov State Optical Institute, St. Petersburg 199053 Russia

Abstract. In this paper, passive mode-locking in Ti:sapphire laser with a coherent absorber cells with rubidium vapor placed in the cavity is demonstrated experimentally. For the best of our knowledge these experiments are the first time experimental demonstration of passive mode-locking based on self-induced transparency regime in the coherent absorber. Up to now, such a regime has not been observed experimentally and was predicted only theoretically.

1. Introduction

Passive mode-locking (PML) in lasers occurs when an absorbing medium having a nonlinear dependence of the transmission on the intensity is placed in the laser cavity [1]. In standard PML lasers it arises due to gain/absorption saturation in amplifying/absorbing media. In such lasers pulse duration cannot be shorter than the polarization relaxation time T_2 of the active medium. If the absorber operates in the regime of coherent light-matter interactions (pulse duration is shorter than T_2) the relaxation times of the medium will not impose restrictions on the duration of the generation pulse, which can reach the value in one cycle of light oscillations [2]. In this case, the so-called 2π pulse of the self-induced transparency (SIT) can occur in the laser cavity when pulse propagates in the absorber without losses as 2π pulse [3]. Mode-locking based on SIT phenomenon is called coherent mode-locking or SIT mode-locking [4-6]. To date there is no experimental demonstration of mode-locking in lasers arising due to SIT. CML was studied only theoretically in the situation when both absorber and gain operates in the coherent regime [2, 4-6]. In the situation when mode-locking arises in the laser with a coherent absorber (gain medium remains in the incoherent regime) due to SIT phenomenon was studied theoretically in [7].

In this paper, we report the peculiarities of experimental demonstration of SIT based mode-locking in Ti:sapphire laser with rubidium absorbing cell placed in the cavity, which arises only due to SIT in rubidium vapor. A detailed study of this regime is described in Refs. [8-9] and review [10].

2. Experimental setup

The setup is shown in Fig. 1. Ti:Sa laser with several cavity designs was assembled on an optical table using the standard components of a laser of the company «Technoscan». To these components belong a sapphire crystal (1) and spherical and plane mirrors (M1-M6). For the wavelength tuning a Lio filter and Fabry-Perot etalon were used. The output of the pump laser VERDI V10 (Coherent) was focused, using the lens (L), through the mirror M1 into the Ti:Sa crystal (1). The Rb vapor cell with the natural mixture of isotopes and high quality windows oriented at the Brewster angle to the beam path to minimize the losses. The cell was placed before the output mirror of the cavity M5, and could be warmed and cooled.

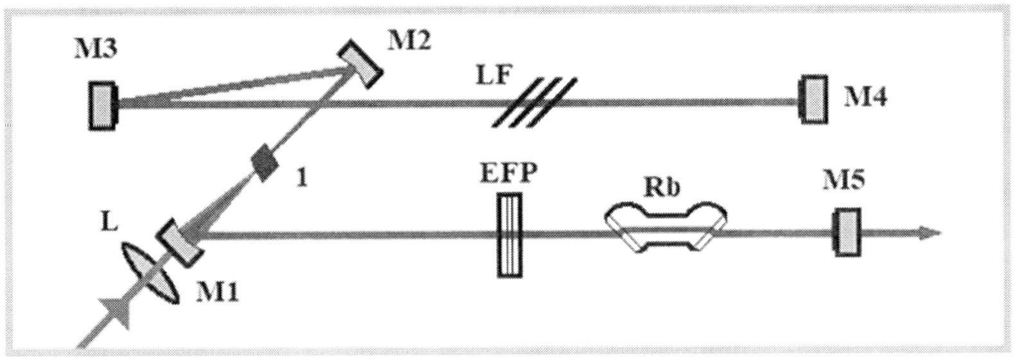

Figure 1. Ti:Sa laser with a Rb vapor cell in a cavity 1 Ti:Sa crystal, M1,2 are the confocal mirrors of the cavity, M3 is an auxiliary plane mirror, M4,5 are the fully reflecting plane mirror and the output mirror (transparency around 8%), L is a focusing lens. LF is the Lio filter, and EFP is the Fabry-Perot etalon for tuning the lasing wavelength.

Laser intensity in dependence on time was registered using a high speed photodiode and an oscilloscope DSO 9104A (Agilent Technologies) with the temporal resolution near 300 ps. The wavelength was determined by a spectrometer, the power was measured by the power meter «Maestro» (Standa). To estimate the pulse durations we used an interferometric autocorrelator based on the Michelson interferometer, as well as the Fabry Perot interferometer with variable base.

3. Results

In the experiment several regimes were observed. In the first case, there was a regular mode-locking. The amplitude of the pulses was stable during long time. An example of such a mode0locking regime is shown in Fig. 2.

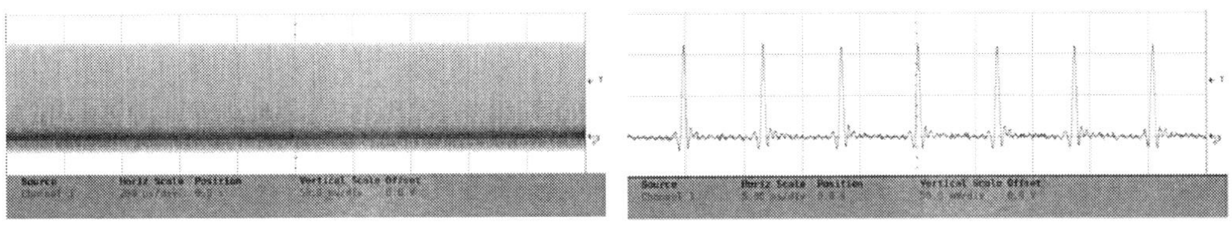

Figure 2. Examples of mode-locking oscillogram due to the effect of self-induced transparency on the DII line of rubidium. On the left oscillogram the time scale is 200 μs/div, on the right 5 ns/div.

In such regime [8-10] the generation pulse duration was obtained from 2 ns to 60 ps. It is shown that the pulse duration is determined only by the generation power. Generated pulse duration decreases with increase of generation power and the pulse area determined according to McCall and Hanh as integral on the pulse envelope over the time [3] remains constant and equals 2π. Our analysis revealed that pulses in the cavity are pulses of self-induced transparency [8-10]. All observed pulse durations were shorter than the polarization relaxation time T_2 of rubidium vapors. The minimum duration achieved was less than T_2 by two orders of magnitude.

The second regime was with pronounced irregular pulses. An example of arising such regime from regular mode-locking below in Fig. 3. The fragment with a duration of 0.325 ms is shown in Fig. 3 (a). Waveforms of pulses in places marked with numbers 1, 2, 3, 4, 5 in fig. 3 (b) - (f).

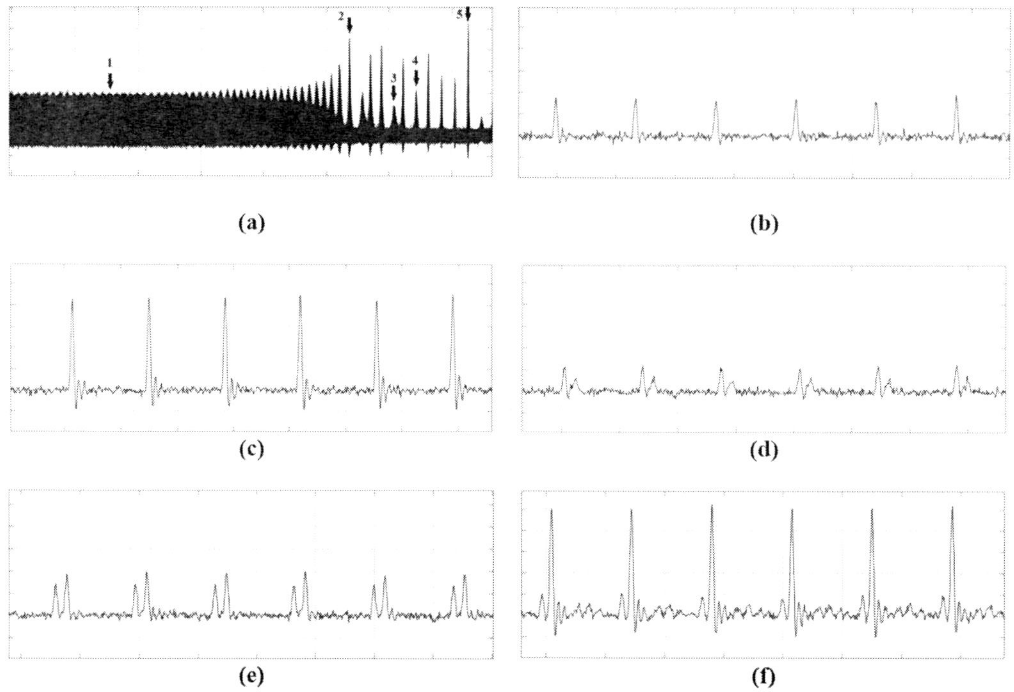

Figure 3. Examples of arising irregular regime. (a) – time scale 50 μs/div. (b)-(f) time scale 5ns/div. Oscillations at the trailing edge of the pulses occurred due to transients in the input electrical circuit photodetector - oscilloscope.

In the third regime, there were switching between longtime quasi-stable fragments with different pulse parameters. The example is given in Fig. 4.

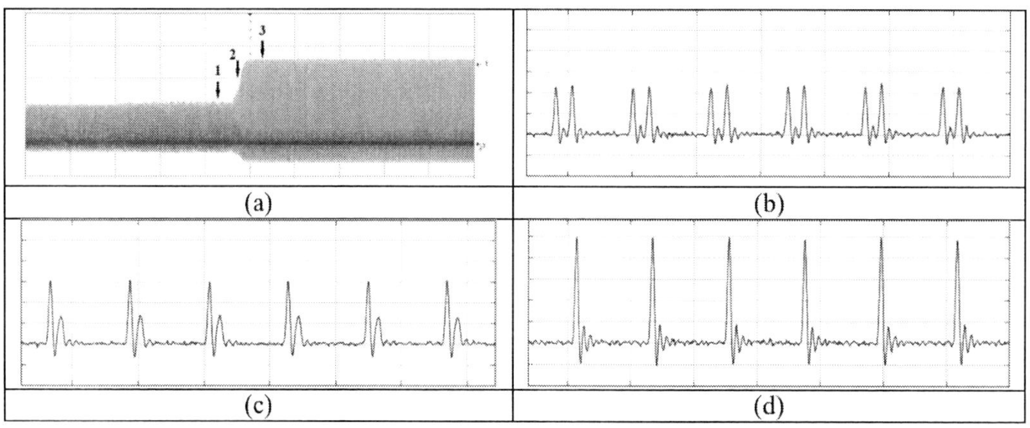

Figure 4. Examples of mode-locking with switching between quasi-stable regimes. Waveforms of pulses in places marked on (a) with numbers 1, 2, 3 are in (b) - (d). (a) – time scale 200 μs/div. (b)-(d) time scale 5 ns/div. Oscillations at the trailing edge of the pulses occurred due to transients in the input electrical circuit photodetector - oscilloscope.

The existence of such irregular and switching regimes raises the question of their nature. Here it is necessary to take into account one important circumstance. To implement SIT mode locking, it is necessary that the pulse area is 2π. If the beam cross-section into absorber changes, this condition cannot be satisfied. Coherent absorber creates losses. Laser generation at the transition frequency in rubidium becomes impossible. The reasons of these regimes may be any change in the beam cross section due to the configuration of the resonator and the self-action of the beam in the absorber, which also creates a change in the beam cross section. In this case, the area of the pulses will vary. Then it can be assumed that the only option in which there are minimum losses in a coherently absorbing medium is the case of propagation of a zero-area pulse. During the experiments, we noted that the transitions between the regular and irregular regimes could be obtained by changing the tuning of the cavity mirrors. This partially confirms the assumption made above. When changing the installation angles of the cavity mirrors, the beam shape in the absorber changes, which can lead to the specified effects. There are either fast or slow switching between generation modes.

Note that in our experiments 0π pulses with a zero area were observed. Below in Fig.5 is an example of such pulses. They have two maxima, which are characteristic of such pulses. The spectrum has a dip in the center. These dips were recorded on interferograms, which were obtained using a Fabry-Perot interferometer.

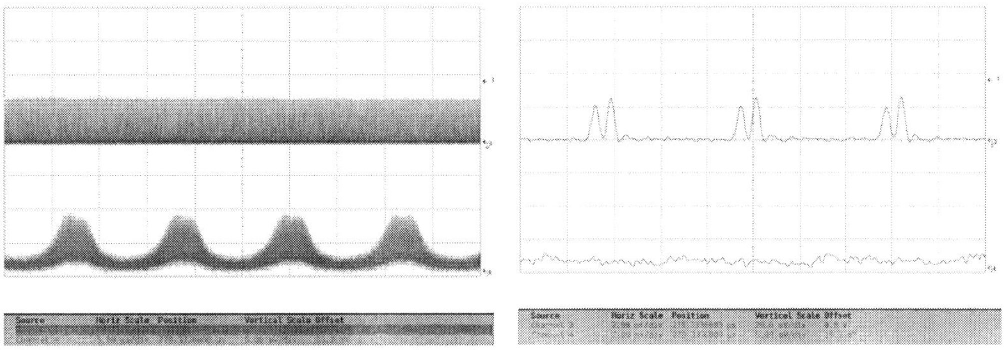

(a) (b)

Figure 5. Examples of mode-locking with 0π pulses. (a) – generation waveform on upper trace, interferogram with deep in center on low trace. (b) – part of (a) data, upper line 0π pulses with two maxima

In the case of stable SIT pulses generation it was possible to obtain the harmonic mode locking. In the cavity there was not one pulse, but several pulses were observed. The maximum number of pulses that we observed was seven. We received this when we placed the rubidium cell in the resonator of the commercial laser of the Mira Optima 900 D (Coherent). In this laser, the output power generation reached 1.4 watts with cell in cavity. This is 10 times more than in the experiments described in our publications [8, 9]. The high lasing output power and long cavity length made possible to exist several SIT pulses in the cavity simultaneously. Examples of waveforms are given in Fig. 6

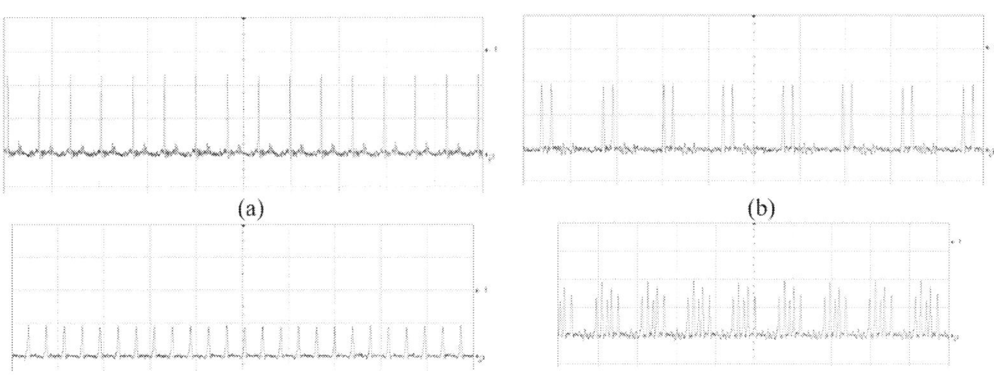

(a) (b) (c) (d)

Figure 6. Examples of SIT harmonic mode-locking in Ti:Sa laser Mira Optima (Coherent). (a) fundamental mode-locking, cavity round trip time 17 ns, time scale 20 ns/div. (b) – two pulses in cavity, cavity round trip time 17 ns, time scale 10 ns/div. (c) – three pulses in cavity, cavity round trip time 12 ns, time scale 10 ns/div (d) – five pulses in cavity, cavity round trip time 12 ns. Time scale 10 ns /div.

4. Conclusions

In the presence of a coherent absorber in a Ti: Sa laser and tuning of the generation wavelength to the absorption line, different regimes can occur: regular SIT coherent mode locking, regimes with irregular pulsation, switching between quasi-stationary mode-locking regimes. Regime of zero-area pulses was also observed. In the case of SIT, the existence of harmonic mode locking is shown. Simultaneously, from one to seven pulses were observed in the cavity. Note that the listed regimes (harmonic mode-locking, chaotic and switching regimes) with the exception of the zero-area pulses case are also observed in lasers with saturable absorbers.

The Investigations were performed at the Center for Optical and Laser Materials Research St. Petersburg State University.

Acknowledgments
R.A. thanks Russian Science Foundation (project 19-72-00012) for the financial support.

References

[1] Keller U 2010 *Appl. Phys. B* **100** 15.
[2] Kozlov V V, Rosanov N N, Wabnitz S 2011 *Phys. Rev. A* **84** 053810
[3] McCall S L, Hahn E L 1969 *Phys. Rev.* **83** 457.
[4] Arkhipov R M, Arkhipov MV, Babushkin I V 2015 *JETP Letters* **101** 149
[5] Arkhipov R M, Arkhipov M V, Babushkin I 2016 *Opt. Comm.* **361** 73
[6] Arkhipov R M, Arkhipov M V, Babushkin I, Rosanov N N *Opt. Lett.* 2016 **41** 737
[7] Kalosha V P, Müller M, Herrmann J 1998 *Opt. Lett.* **23** 117
[8] Arkhipov R M, Arkhipov MV, Shimko A A, Babushkin I, Rosanov N N 2019 *JETP Lett.* **109** 634
[9] Arkhipov M V, Arkhipov R M, Shimko A A, Babushkin I, Rosanov N N , submitted (arXiv preprint arXiv:1906.04587)

[10] Arkhipov R M, Arkhipov MV, Shimko A A, Pakhomov A V, Rosanov N N 2019 *JETP Lett.* **110** 15

The influence of the method of applying a smearing layer on the surface characteristics of steel X12CrNiTi18-10 after laser alloying

V Proskuryakov[1], I Rodionov[1], V Koshuro, A Fomin, S Borodina[1]
[1]Yuri Gagarin State Technical University of Saratov, Saratov 410054, Russia

Abstract. A comparative analysis of the influence of the method of applying a smearing layer of graphite paste and anatase powder on the morphology and hardness of stainless chromium-nickel steel X12CrNiTi18-10 was carried out. It has been established that for hardening the surface of stainless steel it is advisable to use a smearing consisting of two layers: 1 - graphite paste and 2 - anatase nanoparticles powder. Laser pulsed doping of the surface with the specified coating structure allows forming a hardened modified layer with a thickness of 60 μm on steel, with a hardness value of up to 6±0.1 GPa with a total porosity of 44%.

1 Introduction

Stainless chromium-nickel steel grades X12CrNiTi18-9, X12CrNiTi18-10 due to the increased corrosion resistance and availability are widely used in chemical, vacuum, engineering, aircraft manufacturing and other industries [1,2]. Products made of these alloys, operating under conditions of contact friction or exposure to aggressive media, may be subject to premature wear. In this regard, there is a need to increase their service life and functional reliability during operation. To improve the performance characteristics of metal products used methods of surface hardening. The most common methods are based on diffusion saturation with elements from a solid, gas or liquid medium. Methods of chemical heat treatment, for example, nitriding, cementation, boration, chrome plating and cyanation, are characterized by a number of disadvantages:

- duration of the process;
- high energy intensity;
- the possibility of deformation and warping of products;
- need for additional machining [3].

Today, the technology of laser modification of materials is developing most intensively [4,5]. Laser processing is characterized by the possibility of forming a highly hard, therefore wear-resistant surface. Due to its relative simplicity, the method of laser doping with a previously prepared surface has become widespread [6]. Preparation involves applying to the material of the coating a certain component composition, on which the complex of the obtained properties of the treated surface depends. To date, a list of chemical elements and compounds used for alloying steel X12CrNiTi18-10 is known. For example, to increase the hardness and wear resistance of nickel-chromium alloys, doping is carried out with titanium carbide particles. However, there are no industrially manufactured

coatings [7–9]. For each technological process of processing, the composition of the coatings is selected separately in order to impart the required surface properties.
The aim of the work was to study the influence of the method of applying a coating layer consisting of graphite paste and anatase nanoparticles on the morphology and hardness of stainless chromium-nickel steel X12CrNiTi18-10.

2 Methodology

Studies were performed on samples of stainless chromium-nickel steel X12CrNiTi18-10 in the form of plates with dimensions of 10 × 10 × 3 mm. The samples were preliminarily polished with moisture-resistant sandpaper with grit P230 and subsequent purification from technological contamination using an ultrasonic bath "Crystal-2.5" in an aqueous solution of ethyl alcohol for 20 min.
The alloy coating was made of graphite paste (GOST 8295-73) and titanium dioxide powder (anatase) with a dispersity of 20-25 nm. In the studies, three different methods were used to form the component composition of the coating on the steel surface:
1. A layer of graphite paste with a thickness of 100–200 μm was applied on the metal surface, followed by a layer of titanium dioxide powder with a thickness of 100 ± 50 μm.
2. A layer of anatase powder with a thickness of 100 ± 50 μm was formed on the steel surface by pressing at a pressure of 245 MPa. A paste with a thickness of 100-200 μm was applied to the layer of powder ceramics.
3. Graphite paste and titanium dioxide powder were mixed at a powder-paste ratio (1:2) until a homogeneous mass was formed. Then it was uniformly applied to the metal surface with a layer thickness of 300 ± 50 μm.
Modification of the surface layer was carried out on an automated installation for thermophysical coherent modification of the surface of the LRS-50A at a pulse voltage of 300 V and a duration of 1 ms. The beam was focused on a spot with a diameter of 0.5 mm on the metal surface. Processing was carried out at a pulse repetition rate of 20 Hz and an overlap factor of 0.1.
Microhardness was measured using a PMT-3M hardness tester with a Vickers indenter under a load of 1.961 N (ISO 6507-1: 2005). The surface structure was studied using the microhardness meter optical system. The analysis of the structure of the surface layer was carried out according to a known technique using the Metallograf software package [10].

3 Results

The surface morphology of the X12CrNiTi18-10 steel after laser doping, depending on the method of doping formation, has significant differences (Figure 1). It was possible to obtain a uniformly distributed structure of the surface layer on the sample, where the alloying component was deposited according to the first method (Figure 1 a). On samples of steel, the coating of which was prepared in two other ways, the melting and randomness of crystallized surface particles was visualized (Figure 1 b and 1 c).

Figure 1. The morphology of X12CrNiTi18-10 steel after laser doping of the surface with a coating applied in various ways: a - graphite paste layer + anatase powder layer; b - anatase powder layer + graphite paste layer; c - a mixture of graphite paste and anatase powder (see method)

According to the results of the program evaluation of the total open porosity, no significant changes were found in the values of this indicator. For all three laser doping methods, the average value of porosity was about 45%. According to the data of particle size analysis, it was found that as a result of laser doping, the method of applying a dopant coating has a significant impact on the size and number of pores and grains (Table 1).

Table 1. The results of software processing of the morphological characteristics of the modified surface of the steel X12CrNiTi18-10 with various methods of applying a coating layer (the area of the investigated surface S = 5 mm^2)

The method of applying the smearing	Surface characteristics of steel						
	Particles			Pores			
	Quantity, pcs	Average value, μm	Dispersion, μm^2	Quantity, pcs	Average value, μm	Dispersion, μm^2	Total porosity, %
1	1856	6,68	12,32	2887	6,74	11,63	44
2	495	12,37	25,22	1926	10,04	17.41	46
3	1085	13,43	24,42	591	13.17	29.06	43

The surface hardness of stainless chromium-nickel steel X12CrNiTi18-10 increased as a result of laser alloying. The formation of the coating layer according to the first method made it possible to increase the hardness from 2.1 GPa to 5.1 GPA. The application of the coating according to the second method led to an increase in hardness up to 4.9 Gpa. The surface hardness after doping, where the coating was applied, as a mixture of components, was 3.43 Gpa.

Studies of the structure of the modified steel surface using optical microscopy showed that in addition to the formation of the modified surface layer 1, a diffusion layer 2 and a heat-affected zone 3 are formed, the hardness of which is higher than the base metal 4 (Figure 2).

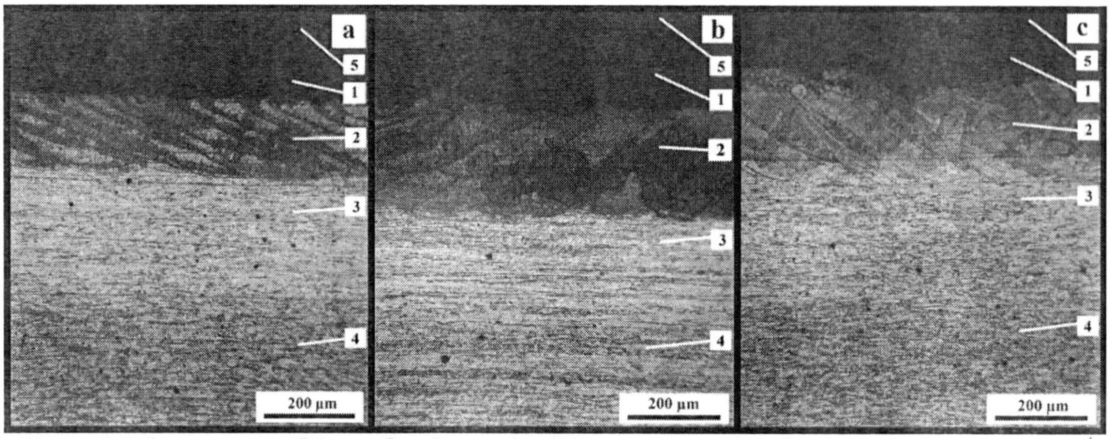

Figure 2. The structure of the surface layer of stainless steel samples after laser doping using the 1st method of applying the coating (a), the 2nd method (b), the 3rd method (c), where: 1 – the resulting modified layer; 2 – diffusion layer; 3 – heat-affected zone; 4 – metal base; 5 – epoxy resin (650 μm optical field of view)

In addition, measurements were made of the hardness and thickness of the zones of the surface layer of transverse sections of stainless steel. According to the obtained results, a table was constructed of the average values of hardness and thickness of the three zones formed for laser treatment with different methods of applying a coating material on the samples (Table 2). The highest hardness value (4-6 Gpa) is characteristic of modified surface layer 1 (Figure 2). The average hardness value at the border of the second and third zones for each method is higher than the hardness of the metal base of the samples.

Table 2. The results of the average values of microhardness and thickness of the three zones for steel X12CrNiTi18-10, formed as a result of laser doping with different methods of applying alloying components

The method of applying the smearing	Surface area	Average thickness, μm	Hardness H, GPa
1	1	60	6.08
	2	120	2.74
	3	210	2.1
2	1	165	5.3
	2	170	6.08
	3	200	2.84
3	1	80	4.04
	2	150	1.79
	3	210	1.76

4 Conclusions

It is established that as a result of laser doping of X12CrNiTi18-10 steel, a layered high-hard surface structure is formed. The application of a layer of graphite-containing paste and the subsequent layer of anatase powder nanoparticles makes it possible to form areas on the steel surface that are characterized by hardness up to 6 ± 0.1 GPa and total porosity of 44%. The depth of the modified layer reaches 60 μm. In the case when the first layer of coating is anatase, and the second is graphite paste, after laser treatment, an area of hardness up to 5.3 ± 0.1 GPa, porosity 46% and thickness 165 μm is formed. The use of a uniform coating layer consisting of graphite paste and anatase powder does not lead to a significant increase in surface hardness (H = 4 ± 0.1 GPa). The value of porosity with this method of coating is 43%, thickness - 80 μm. Therefore, for hardening the surface of stainless steel X12CrNiTi18-10, it is advisable to use a coating prepared according to the first method, described in the method, and consisting of two layers: 1 - graphite paste and 2 - anatase nanoparticle powder.

Acknowledgments
The research was supported by the Ministry of Education and Science of the Russian Federation in the framework of the Program of Scientific Research in Universities (project No. 11.1943.2017/4.6).

References

[1] Rodionov I 2013 *Fundamental problems of modern materials science* **10(1)** 98.
[2] Romakhin A, Rodionov I, Fomin A and Posivalova E 2014 *Sat. proceedings II Intern. Scientific-practical conf. Engineering Engineering. Saratov: Right-Expo Publishing House* **2** 49.
[3] Grigoryants A, Shiganov I and Misyurov A 2006 *Studies. manual for universities // M .: MSTU. NE Bauman* 664.
[4] Bernatsky A 2013 *Automatic welding* **12** 3.
[5] Kim V, Bashkov O, and Sataeva I 2016 *Scientific notes of Komsomolsk-on-Amur State Technical University* **1 (2)** 64.
[6] Ignatov M, Kalashnikova M and Belova S 2002 *Vestn. Perm State Technical University: Mechanics and technology of materials and structures* **5** 154.
[7] Burakov V, Brover G and Burakova N 1985 *Metal science and heat treatment of metals* **11** 2.
[8] Tarasova T, Belashova I, Kuzmin S and Kazakov A 2013 *Materials of the XIX International Symposium "Dynamic and Technological Problems of Mechanics of Structures and Continuum" named after A.G. Gorshkova* 38.
[9] Chung D D L 2010 *Composite Materials: Science and Applications* 1.
[10] Fomina M, Koshuro V, Papshev V, Rodionov I and Fomin A 2018 *Data in Brief* **20** 1409

Temperature performance of InGaAs/InGaAlAs laser diodes with δ-doping active region

S S Rochas[1], I I Novikov[1], A G Gladyshev[1], E S Kolodeznyi[1], M V Maximov[2,3], F I Zubov[2], Yu M Shernyakov[2,3], L Ya Karachinsky[3,4], A E Zhukov[2,5], D V Denisov[4,5], A Yu Egorov[1]
[1]ITMO University, 197101 St Petersburg, Russia
[2]St Petersburg National Research Academic University of RAS, 194021 St Petersburg, Russia
[3]Ioffe Institute of RAS, 194021 St Petersburg, Russia
[4]Connector Optics LLC, 194292 St Petersburg, Russia
[5]St Petersburg State Electrotechnical University "LETI", 197376 St Petersburg, Russia

Abstract. The optical gain performance of 1530-1565 nm laser diodes with active regions containing p-doped barrier layers has been investigated. We have studied the threshold current density and differential quantum efficiency in wide temperature range and compared modal gain behaviour of laser diodes made of heterostructure with delta-doped barrier layers by carbon at level of 10^{12} cm^{-2} and heterostructure with undoped barrier layers.

1. Introduction

One of the main approaches to create the 1300-1550 nm range vertical cavity surface emitting laser (VCSEL) is a wafer fusion. The wafer fusion belongs to the hybrid method of VCSELs production. It allows integrating the active region made of InGaAlAs/InP heterostructure and distributed Bragg reflectors (DBR) based on AlGaAs/GaAs layers in one heterostructure [1]. An essential characteristic affecting on semiconductor lasers modulation speed is a differential optical gain [2]. High differential gain can be achieved using elastically strained InGa(Al)As quantum wells (QW) in active region. Besides elastically strained QWs, the heavy holes' dispersion effects on the differential gain and can be controlled by changing the lattice mismatch parameter between QWs and InP substrate.

Although the rise of strain in QWs leads to the growth of differential gain in an active region [3], which increases VCSEL modulation frequency, the use of this approach is limited. It is impossible to fabricate heterostructures with multiple QWs and the lattice mismatch parameter between InGa(Al)As QWs and InP substrate more than 2-2.5% [4]. Thus, for further increasing of modulation frequency it is necessary to combine this approach with others. Selective doping of the active region near QWs is one of promising approaches to increase the VCSEL modulation speed [5].

In our case, we assume δ-doping of a thin layer in the InGa(Al)As barrier between QWs by carbon, i.e. acceptor impurity. As a result, the charge state of QWs, the charge carriers distribution between p-i-n emitters of the VCSEL's structure, electron and hole fluxes change. The raise of resonance frequency f_r determined by:

$$f_r \propto \left(\Gamma G_{diff} S\right)^{1/2} \quad (1)$$

where Γ – is the rate confinement factor, S – is the number of photons, G_{diff} – is the differential gain. The increase of differential gain leads to the limiting resonance frequency of the laser due to charge neutrality in QWs, taking into account the different density in energy of the charge carriers distribution [6].

2. Heterostructures

Two types of strained InGaAlAs/InGaAs heterostructures have been grown on InP substrate by molecular beam epitaxy (MBE). Active region of both types of heterostructures consisted of 7 $In_{0.74}Ga_{0.26}As$ QWs sandwiched between $In_{0.53}Al_{0.20}Ga_{0.27}As$ barrier layers. First type of heterostructure (S1) contained active region with barrier layers doped by carbon with doping level of $1 \cdot 10^{12}$ cm^{-2} in the centre of InAlGaAs barrier layer (Fig.1). Second type of heterostructure (S2) contained active region with undoped InAlGaAs barrier layers.

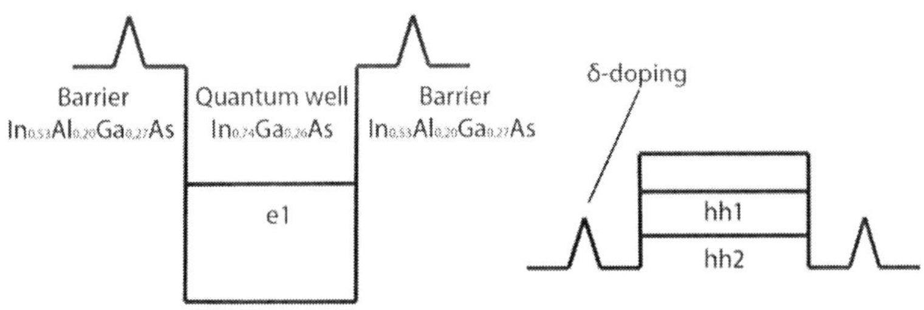

Figure 1. Schematic band diagram of the δ-doping barriers for S1 heterostructure's $In_{0,74}Ga_{0,26}As/In_{0,53}Al_{0,20}Ga_{0,27}As$ heterojunction.

Dependences of the integrated emission intensity on the excitation power for both heterostructures are shown in the figure 2a. The intensity is higher for doped heterostructure, comparing with undoped heterostructure, at excitation powers below 9 mW. At the higher excitation level, the integrated emission intensities for both heterostructures become similar and practically indistinguishable. The analysis of the full width at half maximum (FWHM) of PL spectra as a function of the excitation level is shown in the figure 2b. The values of FWHW lies approximately in the range of 60-100 meV and increases with a pumping level up for the S2 structure. The S1 heterostructure's change of FWHM is very weak and FWHM lies in the range of 78-90 meV. This relates to a fact that in doped heterostructure holes states are always occupied due to holes from δ-doped layers. Captured by QWs electrons always can find holes for effective radiative recombination [7].

3. Experiment

Temperature stability of laser characteristics is one of the main issues for InP based laser diodes. InGaAs/InGaAlAs heterostructures can enhance temperature performance of 1300-1550 nm range lasers due to higher conduction band offset in comparison with InGaAsP heterostructures [8]. The active regions similar to S1 and S2 heterostructures were grown by MBE for laser heterostructures. The active region was placed between two $In_{0.52}Al_{0.48}As$ cladding layers, top cladding layer was 1.5 um thick and bottom cladding layer was 1 um thick. P-doped InGaAs layers used as a p-contact. The 100-um wide stripe laser diodes with various cavity lengths were fabricated using the standard post-growth process. No coatings were deposited on the laser diodes facets. Measurements of threshold current density and light-current characteristics of the laser diodes were performed in pulse laser operation. Temperature was controlled by thermal electric cooler.

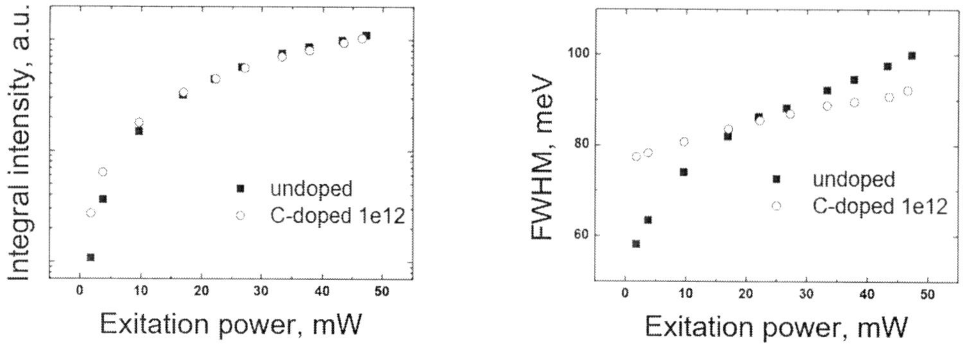

(a) (b)

Figure 2 (a,b). (a) Dependences of integrated emission intensity on excitation power for S1 and S2 heterostructure; **(b)** Dependences of FWHW on excitation power for S1 and S2 heterostructures.

Threshold current density (J_{th}) and differential quantum efficiency (η_{diff}) for the laser diodes with cavity lengths of 0.5, 1.0 and 2.0 mm were measured in the temperature range from 16 to 70 °C. Using equation (2) characteristic temperatures T_0 and T_1 were estimated as:

$$J_{th}(\Delta T) = J_{th}(0)\exp\left(\frac{\Delta T}{T_0}\right), \eta_{diff}(\Delta T) = \eta_{diff}(0)\exp\left(-\frac{\Delta T}{T_1}\right) \quad (2)$$

where ΔT – temperature range, T_0 – characteristic temperature of threshold current density and T_1 – characteristic temperature of differential quantum efficiency. Obtained results are shown in figure 3(a) and in figure 3(b).

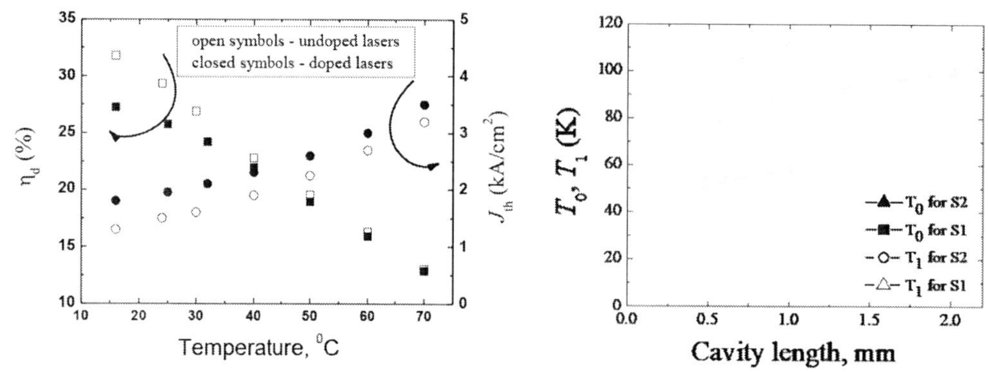

(a) (b)

Figure 3(a,b). (a) Threshold current density and deferential quantum efficiency dependences on temperature for S1 and S2 heterostructures; **(b)** Characteristic temperatures T_0 (closed symbols) and T_1 (open symbols) for S1 and S2 heterostructures depending on cavity length.

Figure 3(b) shows the values of T_0 and T_1 extracted from the analysis of the temperature dependences of the threshold current and the differential quantum efficiency. With temperature higher than 50 °C threshold current density J_{th} and differential quantum efficiency η_{diff} in lasers with strained QWs and doped barriers are close to the values in lasers with undoped barriers. Laser diodes with δ-doped active region are characterised by T_0 in a 77-99 K range and T_1 in a 102-110 K range. The S2 heterostructure demonstrates 51-71 K and 53-73 K for T_0 and T_1 respectively.

The modal gain dependence (g_{mod}) on the injection current can be obtained using the fact the modal gain is equal to the total laser losses at the threshold:

$$\frac{g_{mod}}{\eta_{int}} = \frac{\alpha_{out} + \alpha_{in}}{\eta_{int}} = \left(\frac{\alpha_{out} + \alpha_{in}}{\eta_{int} \cdot \alpha_{out}}\right) \cdot \alpha_{out} = \frac{\alpha_{out}}{\eta_{diff}} \quad (3)$$

where α_{out} - output optical losses, α_{out} – internal optical losses, η_{diff} – differential quantum efficiency. Figure 4 shows the threshold modal gain dependence on temperature for both structures and different cavity lengths, plotted using the equation (3), and temperature dependence of η_{diff}. The maximum obtained value of the effective modal gain (i.e. g_{mod}/η_{int}) exceeded the 140 cm^{-1}.

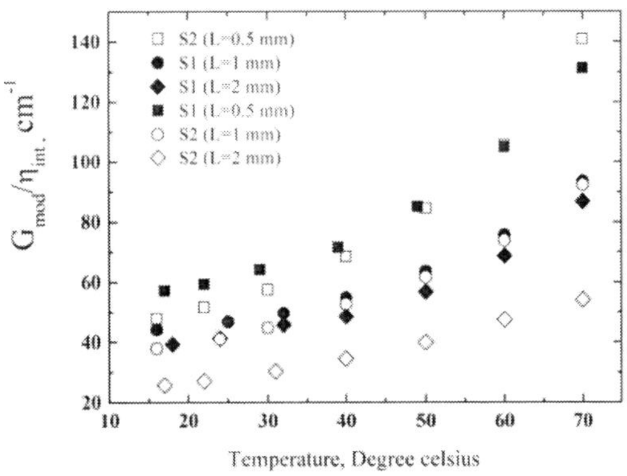

Figure 4. Modal gain depending on temperature for different cavity lengths: S1(closed symbols) and S2(open symbols).

4. Conclusion

The laser diodes of the spectral range about 1.55 um made of InGaAs/InGaAlAs heterostructures grown by MBE were investigated. Structures with an undoped active region are characterized by a strong temperature sensitivity of both the threshold current and the differential quantum efficiency ($T_0 \sim T_1 \sim$ 50-70 K). Doping of barriers with carbon at the level of $1 \cdot 10^{12}$ cm^{-2} per quantum well increases T_0 to 77-99 K and T_1 to 102-110 K while simultaneously increasing the threshold current and reducing the differential efficiency near room temperature. With temperatures >50 °C the effect of modulated doping disappears. The obtained results can be used for design and optimization of semiconductor laser active regions.

Acknowledgments

This work was supported by the Ministry of Science and Higher Education of the Russian Federation in the framework of the Federal Target Program "Research and development on priority directions of scientific-technological complex of Russia for 2014-2020", agreement № 14.578.21.0253, unique ID RFMEFI57817X0253.

References

[1] Liau Z L and Mull D E 1990 *Appl. Phys. Lett.* **56** 737
[2] Babichev A V, Karachinsky L Ya, Novikov I I, *et al* 2017 *IEEE J. Quant. Electr.* **53** 1
[3] Spiga S, Andrejew A, and Amann M C 2016 *18th ICTON* 1
[4] Thiis P J A, Tiemeijerb L F, Binsma J J M and Van Dongen T 1994 *IEEE J. Quant. Electr.* **30** 477
[5] Vahala K J and Zah C E 1988 *Appl. Phys. Lett.* **52** 1945
[6] Uomi K. 1990 *Japan. J. Appl. Phys.* **29** 81
[7] Kolodeznyi E S, Kurochkin A S, Rochas S S, *et al* 2018 *Semiconductors.* **52** 1156
[8] Yu E T, McCaldin J O and McGill T C 1992 *Solid State Phys.* **46** 1

The effect of pretreatment of X12CrNiTi18-10 steel on surface characteristics as a result of laser hardening

N Makhov[1], V Proskuryakov[1], I Rodionov[1], V Koshuro[1], A Fomin[1], S Borodina[1]
Yuri Gagarin State Technical University of Saratov, Saratov 410054, Russia

Abstract. The results of studies of the effect of pretreatment of stainless chromium-nickel steel X12CrNiTi18-10 on the morphological characteristics and microhardness of the surface as a result of laser hardening are presented. It was established that with laser modification, a uniformly distributed surface microrelief is formed with a microhardness value of up to 12.58 ± 0.1 GPa.

1 Introduction

Increased values of corrosion resistance, resistance to mechanical stress make stainless chromium-nickel steel X12CrNiTi18-10 (Russian National Standard: GOST 5632-72) the most rational material for use in various industries. However, products made from this alloy, operating under conditions of contact friction or under the influence of various corrosive media, may be subject to premature wear. To extend the service life of such products, methods of surface hardening are used [1].

Today, the most versatile method of surface modification is laser processing. Its main advantages are: the possibility of local impact, the formation of a hardened surface layer with the specified parameters of the structure. It should be noted that in some cases, this method requires the creation of a preliminary surface roughness to provide an increased ability to absorb the energy of laser radiation [2-4].

The purpose of this work was to determine the effect of preliminary surface preparation of steel X12CrNiTi18-10 on the morphology and hardness as a result of laser hardening.

2 Methodology

The studies used samples of stainless steel X12CrNiTi18-10 with dimensions 10 × 10 × 3 mm. The surface of the samples was pretreated in two ways: 1 - by abrasive blasting (ASO) at an air pressure of 0.5-0.6 MPa with particles of electrocorundum powder with a dispersion of 200-250 μm, 2 - by grinding (G) with moisture resistant sandpaper P230 grit.

Laser processing was performed using the installation for thermophysical coherent modification of the surface of the LRS-50A at a voltage of U 250, 300, 350, 400 and 450 V. The duration τ of the impulse effect was 1 ms. The beam was focused on a spot with a diameter of 1 mm. The surface was scanned at a pulse repetition rate of 10 Hz and an overlap factor of 0.25.

The magnitude of the open porosity, the linear dimensions of the grains and pores, and the thickness of the coating were determined using the metallographic micrographic image processing computer program according to the well-known method [5]. At the same time, areas of a coating with an area of 5 mm^2 were analyzed. Microhardness was measured using a PMT-3M hardness tester with a Vickers indenter with a load on the indenter of 1.961 N (ISO 6507-1: 2005).

3 Results

The morphology of the surface of steel modified with laser radiation is structurally heterogeneous (Figure 1).

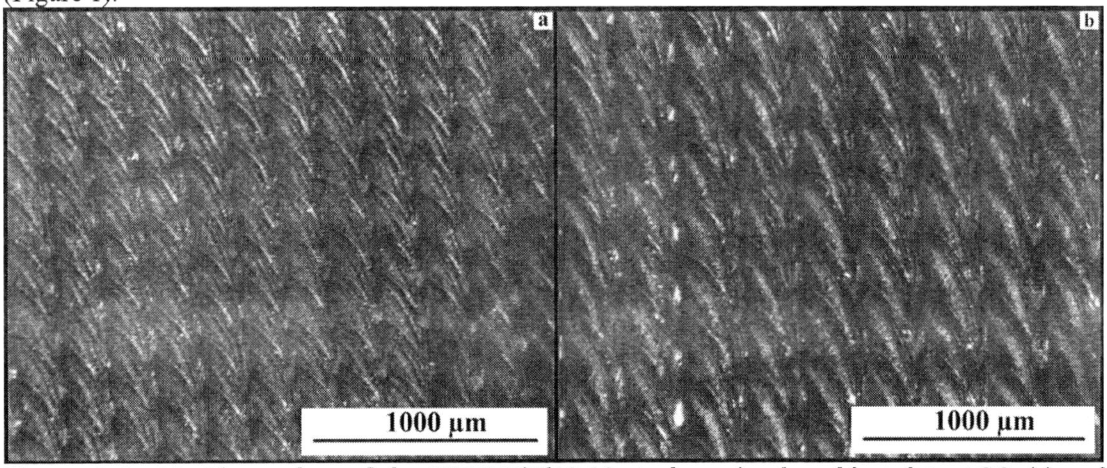

Figure 1 (a, b). The surface of the X12CrNiTi18-10 steel previously subjected to ASO (a) and grinding (b) after laser modification at U = 400 V, τ = 1 ms

The results of the study of the microhardness of transverse thin sections showed that in addition to the formation of the surface layer 1, a heat-hardening zone 2 with a microhardness of up to 6.8 ± 0.1 GPa is formed. It should be noted that the average value of the microhardness of the metal base 3 is 2.4 ± 0.1 GPa (Figure 2).

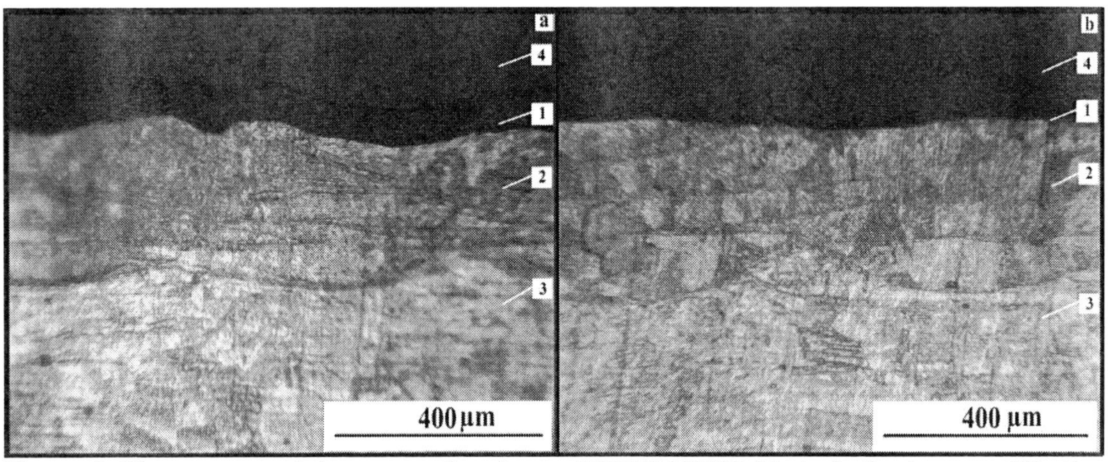

Figure 2. The structure of the near-surface layer of X12CrNiTi18-10 steel after laser modification with a surface that has previously undergone abrasive blasting is a, with pre-ground surface – b, where: 1 – is the resulting modified layer; 2 – heat-hardening zone; 3 – metal base; 4 – epoxy resin (850 μm optical field of view)

In addition, microhardness measurements were carried out of the most treated steel surface. The maximum microhardness value H=12.58±0.1 Gpa was obtained at a voltage of 450 V and a pulse duration of 1 ms on the surface that had previously undergone abrasive blasting.

Table 1. The results of the average values of the microhardness of the surface of samples of steel X12CrNiTi18-10, obtained as a result of laser hardening in various modes

τ, ms	Sample pretreatment method	U, V	H, GPa
1	G	250	3,34±0.1
	G	300	5,38±0.1
	G	350	6,81±0.1
	G	400	6,68±0.1
	G	450	11,21±0.1
	ASO	250	7,42±0.1
	ASO	300	7,51±0.1
	ASO	350	8,25±0.1
	ASO	400	9,09±0.1
	ASO	450	12,58±0.1

Analysis of the particle size distribution of the steel surface after laser treatment showed a significant increase in the number of pores and grains with increasing pulse voltage. At the same time, the size of pores and grains remains practically unchanged (Table 2).
When evaluating the total open porosity P after laser modification, no significant differences were found depending on the method of preliminary surface treatment. According to the obtained data of particle size analysis, it was established that as a result of laser modification, the pretreatment method has a significant impact on the size and number of pores and grains (Table 2).

Table 2. The results of the software processing of the morphological characteristics of the modified surface of the X12CrNiTi18-10 steel under various laser modification modes (the surface area under study S = 5 mm^2)

Surface pretreatment method	Laser hardening modes		Surface characteristics of steel						
			Particles			Pores			
	τ, ms	U, V	Quantity, pcs	Average value, μm	Dispersion, μm^2	Quantity, pcs	Average value, μm	Dispersion, μm^2	P, %
ASO	1	250	364	4,64	23,84	426	3,45	1,13	25
		300	608	4,46	18,45	686	4,51	17,38	24
		350	662	4,12	17,69	667	3,44	1,03	24
		400	1787	4,05	10,86	2588	4,42	9,17	24
		450	1064	3,88	1,67	781	3,79	1,51	22
G		250	225	6	30,29	240	4,02	1,52	14
		300	789	4,65	16,24	856	4,64	15,61	26
		350	1214	5,02	2,52	1434	5,35	2,97	28
		400	2553	4,92	9,22	1823	4,28	1,70	30
		450	1621	3,83	1,64	1407	3,75	1,55	25

4 Conclusions
It has been established that laser modification of X12CrNiTi18-10 steel with preliminary ASO allows to increase the surface hardness to 12.58±0.1 GPa, with preliminary grinding - to 11.21±0.1 GPa. The formation of a heat-hardening zone with a hardness of up to 6.87±0.1 GPa for an ASO surface and up to 5.82±0.1 GPa for a surface treated by grinding was also revealed.

Acknowledgments
The research was supported by the Ministry of Education and Science of the Russian Federation in the framework of the Program of Scientific Research in Universities (project No. 11.1943.2017/4.6).

References
[1] Rodionov I 2012 *Prospective materials* **4** 36
[2] Akimov O and Alaa F 2015 *Newsletter of the National Technological University Kharkiv Polytechnical Institute. Seriya: New Solutions in Modern Technologies* **39** 3
[3] Fedorov V, Shlykova A and Yakovlev A 2017 *Vestnik Tambovskogo universiteta* **22(5-2)** 1100
[4] Zhu Y, Fu J, Zheng C and Ji Z 2016 *Optics & Laser Technology* **83** 21
[5] Fomina M, Koshuro V, Papshev V, Rodionov I and Fomin A 2018 *Data in Brief* **20** 1409

NdYAG surface modification of ZnO for solar cells

D Redka[12], N Mukhin[2], A Grishkanich[12], E Terukov[23], S Hirsch[4], L Ignateva[2]
[1]ITMO University, Saint-Petersburg, Russia; [2]St.Petersburg State Electrotechnical University, Saint-Petersburg, Russia; [3]Ioffe Physico-Technical Institute of the Russian Academy of Sciences, Saint-Petersburg, Russia; [4]Technische Hochschule Brandenburg, Brandenburg, Germany

Abstract. Zinc oxide films are used as a photovoltaic transparent electrodes for current collection. They are a good substitute for expensive transparent electrodes on the basis of Indium tin oxide. The ZnO films fabricated with a certain morphology of the structure are able to act as a light diffuser.

1. Introduction

Large size (~1,4 m2) silicon-based thin-film solar module (SM) consists of elements connected between each other in a way to get the maximum efficient solar energy conversion in the output. In solar modules production the technological process of separating elements is realized using laser scribing. The research includes analysis of possibility to decrease the «dead zone» and to increase the parameters of electrical energy output.

2. Experiment

The glass film with deposited thereon zinc oxide was used as an experimental sample. Processing of the sample was performed by laser radiation with a wavelength of 355 nm, the laser pulse energy ranged from 14 mJ to 21 mJ (± 0.05 mJ). There were seven groups for different cuts of focus distance that vary in increments in 0.1 mm. Each group included five consecutive scribes obtained with the power of laser pulses of 14 mJ, 16mJ, 18 mJ, 20 mJ, 21mJ, respectively. Pulse repetition rate was 60 kHz and a pulse duration was 30 ns. There was detected scribe, which has the following characteristics: a minimum width on the surface of 27 ± 0,5μm scribe, scribe at the bottom (on the side of the glass substrate), the width is 25 ± 0,5 μm; width of cut on the surface and at the bottom is constant in different coordinates of the cutting length; minimum number of chips and no shunt elements.

The glass film with deposited thereon zinc oxide was used as an experimental sample. Processing of the sample was performed by laser radiation with a wavelength of 355 nm, the laser pulse energy ranged from 14 mJ to 21 mJ (± 0.05 mJ). There were seven groups for different cuts of focus distance that vary in increments in 0.1 mm. Each group included five consecutive scribes obtained with the power of laser pulses of 14 mJ, 16mJ, 18 mJ, 20 mJ, 21mJ, respectively. Pulse repetition rate was 60 kHz and a pulse duration was 30 ns.

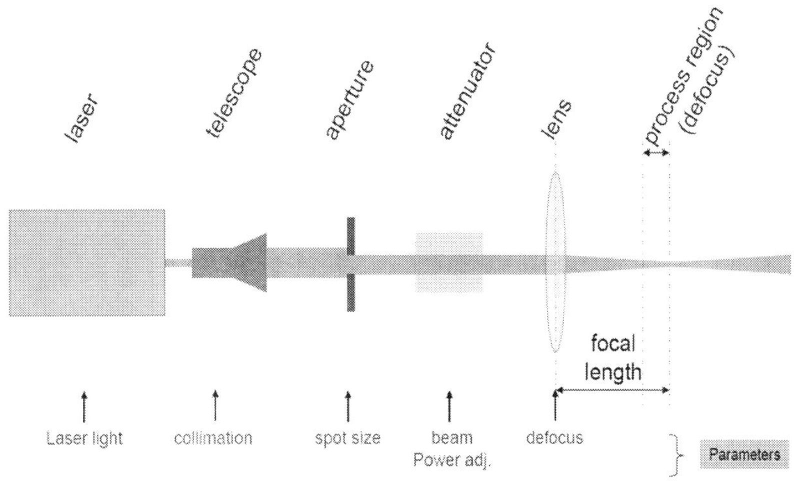

Figure 1. The experimental scheme.

The intensity of the laser pulse has a shape close to a Gaussian distribution with a maximum at the center of focus of the beam. Data matrix was obtained during the experiment. Scribe control was performed using optical microscopy. Photographing of groups of scribes was conducted in different coordinates of the sample. There was detected scribe, which has the following characteristics: a minimum width on the surface of 27 ± 0,5μm scribe, scribe at the bottom (on the side of the glass substrate), the width is 25 ± 0,5 μm; width of cut on the surface and at the bottom is constant in different coordinates of the cutting length; minimum number of chips and no shunt elements. This scribe was obtained by defocusing value (the distance from the focus to the treated surface) of the laser beam by 1.5 mm and a laser pulse energy of 20 mJ. When passing by through the glass beam was focused to the treated layer ZnO.

Table 1. ZnO main characteristics.

	Bandgap (eV)	Conductivity (S cm^{-1})	Electron concentration (cm^{-3})	Mobility ($cm^2V^{-1}c^{-1}$)
ZnO	3.35	8000	> 1021	20

3. Results

During the experiments it was found that at low energies of the laser pulse (less than 3 mJ) it is impossible to obtain uninterrupted scribe, as a result there are shunts because is not enough energy to remove material by laser ablation. The value of the width of the resulting "track" is an important factor. Increasing the width of cut reduces the active area of the solar module, as a consequence, the decrease of the photocurrent values of the solar module.

The power of the laser radiation used in the processing has a direct influence on the cutting depth (scribe). Scribe depth determined by the equation:

$$h \approx Pt / (\pi r_0^2 L_u) \quad (1),$$

where P-power of laser beam, t- duration of the laser pulse, r_0 - the radius of the laser spot on the surface to be treated, L_u - hide heat of vaporization.

Laser spot size will also affect the width scribe. The dependence of the width of the scribe on the laser spot size is described by equation:

$$l \approx 2\left(r_0 + Ptg\gamma/\pi r_0 L_u\right) \qquad (2),$$

If using equation (1) we determine r_0 and put the value obtained in the equation 2, we obtain the following relation:

$$l \approx 2(\sqrt{Pt/h\pi L_u} + Ptg\gamma/\pi r_0 L_u) \qquad (3),$$

where l - scribe width - half angle of the focused beam impinges on the sample, h- thickness of ZnO layer in place of the laser beam.

The best results for laser treatment have been achieved by annealing the films of zinc oxide ZnO with a laser beam at power density ranging 0.163 ÷ 0.260 W / cm^2. Laser radiation with the power density less than 0.163 W / cm^2 causes almost no change, and when the power density was over 0.260 W / cm^2, sections with significant areas of melting and cracking of the film were obtained. More over it should be noted here that due to the energy distribution in the laser beam Gaussian surface annealed films unevenly and a large enough portion of the film was hardly processed.
Study of the samples micro relief was conducted using atomic force microscope Ntegra Therma (NT-MDT), the results are shown in Figure 1, 2. It is revealed that with an increase in power incident on the sample the surface roughness increases and this leads to an increase in Haze-factor.

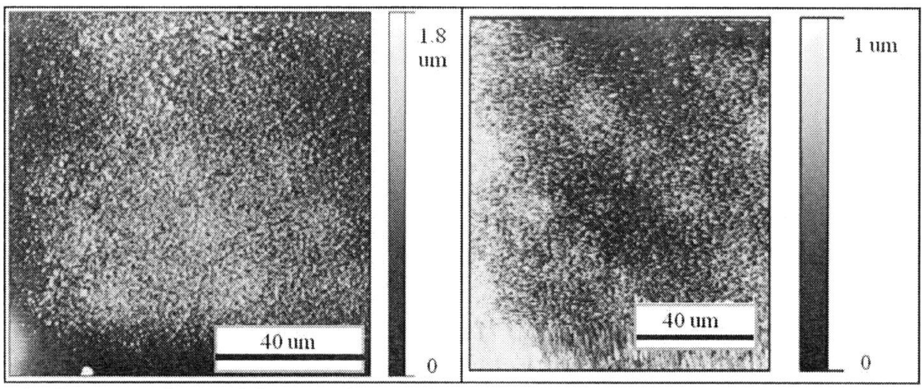

Figure 2. The image surfaces of the samples number 11 (left) and 12 (right) obtained by an atomic force microscope.

The films were treated by radiation with a wavelength of 355 nm. The structure of the films was examined by Raman spectroscopy (Fig. 3). To study the vibrational properties of ZnO crystal lattice the Raman spectroscopy was used. Raman spectra were recorded in the reverse scatter geometry at room temperature with a spectrometer LabRam HR800, combined with a confocal microscope (manufactured by Jobin-Yvon Horiba). As the excitation source was used second harmonic of Nd: YAG-laser (wavelength 532 nm). The laser beam is focused to a spot with a diameter of ~ 1-2 μm on the sample surface.

In the hexagonal symmetrical structures C46v, such as ZnO, there are six branches of optical phonons at Γ (the center of the Brillouin zone) Γ = A1 + E1 + 2B1 + 2E2 [12-15]. Phonon modes of symmetry A1 and E1 are polar at the point Γ and split into longitudinal and transverse optical vibrations (LO and TO, respectively). Non-polar phonon modes symmetry E2 shares two frequencies: E2(high) related to the vibrations of the atoms of oxygen and E2(low) related to the vibrations of the atoms of zinc. [16] Mode B1 is «silent mode» and does not appear in the optical spectra of [16,17].

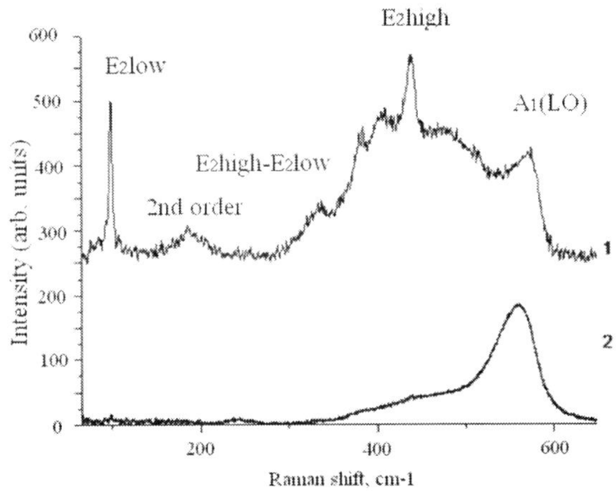

Figure 3. The Raman spectra of the film ZnO: 1 - before modification; 2 - after modification.

4. Conclusion

Processing with of optimum focus settings allows to obtain correct geometrical form of cuts, the width of which is almost independent on the ZnO film thickness unevenness (in the case cutting width variations differ in less than 0.5 microns), with a minimum number of chips and absence of the shunt elements. Implementation of the results achieved in the TFSM production technology will increase the output current about 2-5 % by reducing the area of laser micromachining (compared with typical TFSM produced by this technology).
Summing up the results of the analysis, it can be concluded that the polycrystalline ZnO structure is destroyed by modifying laser radiation, while in the center of the spot amorphous zinc oxide is formed. The results can be used to create thin-film solar modules of a large size.

Acknowledgments

The work is also financially supported by RF Ministry of Education and Science, Grant RFMEFI57518X01800 05.575.21.0180.

References

[1] Iakovlev, A., Redka, D., Afanasjev, V., Mukhin, N., Grishkanich, A., Terukov, E., & Hirsch, S. (2017, February). Laser surface modification of ZnO for solar converters. In *Laser Applications in Microelectronic and Optoelectronic Manufacturing (LAMOM) XXII*(Vol. 10091, p. 100911B). International Society for Optics and Photonics.

[2] Shah, Arvind, ed. Thin-film silicon solar cells. EPFL press, 2010.

[3] Kuwano, Y. "Amorphous Silicon Integrated Cell Modules (I), 1st Photovoltaic Sci. & Engg. Conf." (1979).

[4]Buzas, A., and A. Nemeth. "Selective cutting of ZnO: Al contact layers." Proc. of 24th European Photovoltaic Solar Energy Conference. Hamburg. 2009.

[5] Redka, D. N. "Research of the laser treatment of zinc oxide." Smart Nanocomposites 7.1 (2016): 41.

[6] Haas, S., et al. "Patterning of thin-film silicon modules using lasers with tailored beam shapes and different wavelengths." Proc. 23rd European Photovoltaic Solar Energy Conference Valencia, Spain. 2008.
[7] Canteli, D., et al. "Nanosecond laser ablation processes in aluminum-doped zinc-oxide for photovoltaic devices." Applied Surface Science 258.23 (2012): 9447-9451.
[8] Booth, Heather. "Laser processing in industrial solar module manufacturing." J. Laser Micro/Nanoeng 5.3 (2010): 183-191.
[9] Redka, D. N. "Research of the laser treatment of zinc oxide." Smart Nanocomposites 7.1 (2016): 41.
[10] Yang, Fan. "Large Area a-Si/μc-Si Thin Film Solar Cells." Solar Cells-Thin-Film Technologies (2011): 338-339.
[11] Shi, Shaobo, et al. "Effect of annealing on the structural and luminescent properties of ZnO nanorod arrays grown at low temperature." Journal of Applied Physics 109.10 (2011): 103508.
[12] Singh, Sukhvinder, R. S. Srinivasa, and S. S. Major. "Effect of substrate temperature on the structure and optical properties of ZnO thin films deposited by reactive rf magnetron sputtering." Thin Solid Films 515.24 (2007): 8718-8722.
[13] Song, Dengyuan. "Effects of rf power on surface-morphological, structural and electrical properties of aluminium-doped zinc oxide films by magnetron sputtering." applied surface science 254.13 (2008): 4171-4178.
[14] Fan, Hong Jin, Yang Yang, and Margit Zacharias. "ZnO-based ternary compound nanotubes and nanowires." Journal of Materials Chemistry 19.7 (2009): 885-900.
[15] Charpentier, C., et al. "X-Ray diffraction and Raman spectroscopy for a better understanding of ZnO: Al growth process." EPJ Photovoltaics 2 (2011): 25002.
[16] Afanasjev, V. P., et al. "The formation of zinc and copper oxides thin films for heterostructure solar cells." SPIE Photonics Europe. International Society for Optics and Photonics, 2016.
[17] Mukhin, N. V. "Investigation of the formation kinetics of grain boundary inclusions of lead oxide in lead zirconate–titanate films." Glass Physics and Chemistry 42.1 (2016): 64-69.

AlGaAs/GaAs solar cell with CNT transport layer: numerical simulation

S A Raudik[1], A M Mozharov[1,4], D M Mitin[1,2], A D Bolshakov[1,4], V V Fedorov[1,4], V V Neplokh[1], P M Rajanna[3], A G Nasibulin[3] and I S Mukhin[1,4]
[1]St.Petersburg Academic University, Khlopina 8/3, 194021 St. Petersburg, Russia
[2]Ioffe Institute, 194021 St. Petersburg, Russia
[3]Skolkovo Institute of Science and Technology, 121205 Moscow, Russia
[4]ITMO University, 197101 St. Petersburg, Russia

Abstract. AlGaAs/GaAs solar cells (SC) was modeled. Conventional SC was compared with structure with the thinner emitter. SC with additional CNT transport layer was simulated and compared with SC without it. The simulation was carried out with different geometry of the contacts and light flux. CNT transport layer has allowed achieving an increase in SCs efficiency up to 1.7% compared with metal contact grid.

1. Introduction

The advantages of GaAs-based solar cells (SCs) is higher efficiency in comparison with Si-based [1] but they traditionally are used in aerospace and military applications because of their high cost. Moreover, new technology can be tested on single junction SC to further implementation in multi-junction SCs

Today the efficiency of GaAs single-junction SCs is close to the theoretical limit due to last developments in optimization of the thin-film structure and epitaxial growth conditions. Therefore, the role of the contact layer in solar cells has increased. Contact layer determines the efficiency of charge carriers collecting and affects the absorption of the light. These factors have a direct impact on the SC efficiency which stimulates intensive research in this area. To reduce ohmic losses, it is necessary to increase the contact area, but it leads to an increase in shading area and a decrease in short-circuit current. Another way to reduce lateral resistance is by using a transparent layer with high conductivity, such as ITO, ZnO, CNT, etc.[2] In our previous work, it was shown that SCs with CNT layer can have better efficiency compared with other materials [3]. In this paper, we concentrate our attention on SC simulation with CNT layer.

The GaAs-based SCs often were used with light concentrator in order to increases open-circuit voltage and reduce square of the photoactive element due to its high cost [4]. Furthermore, CNT transport layer application can be more advantageous in the area of light concentrator SCs which use high currents thus conductivity of the upper layer becomes more important.

2. Calculation

A heterostructural AlGaAs/GaAs solar cell with CNT transport layer and metal contact grid have been modeled. The SC structure is presented on Fig 1. Since the electrical field component which is perpendicular to the growth direction, predominantly exist in the emitter and window, as well as transport equations in the first approximation, can be considered linear in the electric field, the model was separated into two parts: one-dimensional model of the semiconductor structure and two-dimensional model of the surface current spreading. Continuity equation and Maxwell equations were solved by the finite element method for one-dimensional simulation along the growth direction. The Shockley-Reed-Hole, optical and Auger recombinations were taken into account in our model [5] [6] [7] to computation current and voltage losses. Mobility model with velocity saturation was used to correct calculation in the high internal electrical fields such as in the p-n junction.

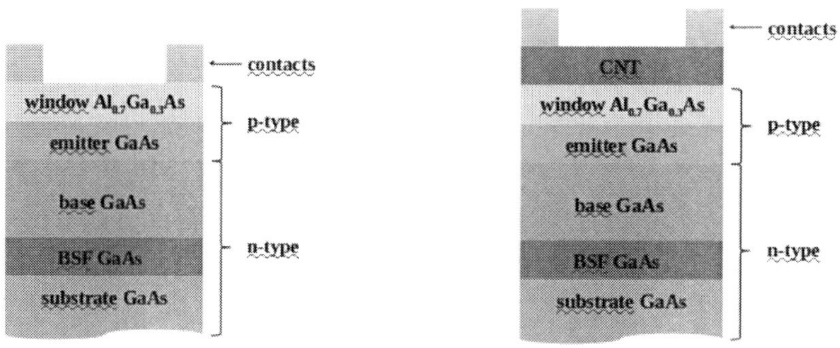

Figure 1(a, b). SC structure **(a)** without transport layer; **(b)** with transport layer

Anti-reflection coatings with near-zero refractive have been made by nowadays [8]. Therefore in our work only processes of light absorption and currents generating were consider in our research, the ideal anti-reflection was used.

In conventional CSs with only metal grid contacts, emitter thickness is about 0.5 um [9] [10]. Our simulation showed that in solar cells with CNT layer, using 30 nm thickness emitter is preferable. In table 1 are presented comparing of the recombination rate for structures with 30 nm and 400 nm emitter. It can be seen, that the total recombination rate is less in the structure with thinner emitter . For this reason, current in the structures with thinner emitter is greater and the efficiency consequently is greater too. In the structures without transport layer, the thicker emitter is necessary to provide sufficient lateral conductivity. In our structure lateral conductivity is provided by the transport layer. It can be seen on Fig 2.

Table 1. Recombination characteristics of the modeled SCs

	SC with 30 nm emitter	SC with 400 nm emitter
Recombination rate in emitter	$1.68 \times 10^{14} s^{-1} cm^{-2}$	$2.10 \times 10^{16} s^{-1} cm^{-2}$
Recombination rate in base	$2.12 \times 10^{16} s^{-1} cm^{-2}$	$1.15 \times 10^{16} s^{-1} cm^{-2}$
Total recombination rate	$2.13 \times 10^{16} s^{-1} cm^{-2}$	$3.25 \times 10^{16} s^{-1} cm^{-2}$
Short-circuit current	27.35mA cm^{-2}	25.66 mA cm^{-2}

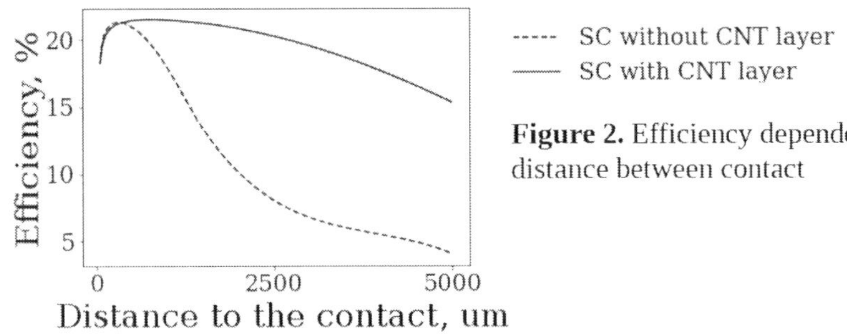

Figure 2. Efficiency dependency on distance between contact

Series of simulation with different contacts, CNT layers and light flux was performed. Efficiency dependency on light flux and contacts is presented on Fig 3. Figures 3(a) and 3(b) show the dependences for the cases of SC with and without CNT, respectively. The efficiency gain after adding CNT transport layer was 0.2% in case AM 1.5 spectrum and 1.7% in case of using 200 times light concentration. With decreasing the distance to contacts efficiency decline due to reflexion from the contacts. In the area of the graph with a bigger distance to the contacts, the efficiency decrease because of growing recombination rate inside the SC and decrease the effective working area of SC. The reason for this effect is a non-zero voltage drop due to lateral currents and consequently, increase the voltage to the values close to open-circuit voltage in the area which is far from contacts. The increased voltage leads to the growth of the recombination rate and rapid drop of the generated current density as it can be seen on Fig 4. The voltage and generated current density distribution is presented on Fig 5. Efficiency growth with increasing light flux explain by increase open-circuit voltage [9], moreover, there is opposite effect: currents magnification leads to increasing ohmic losses and for this reason optimal distance between contacts decrease. After 200 times light concentration for this reason efficiency declines.

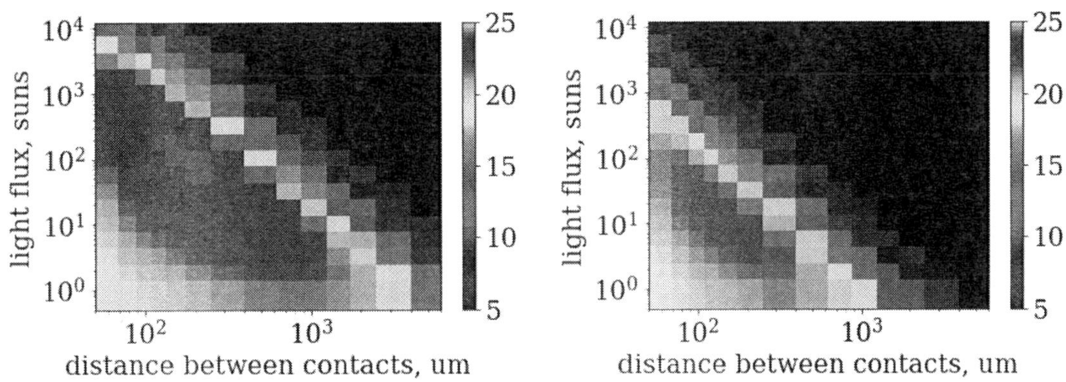

Figure 3(a, b). Efficiency dependency on distance between contact and light flux.

(a) SC with 98% transparency CNT **(b)** SC without CNT

 1410 (2019) 012107 doi:10.1088/1742-6596/1410/1/012107

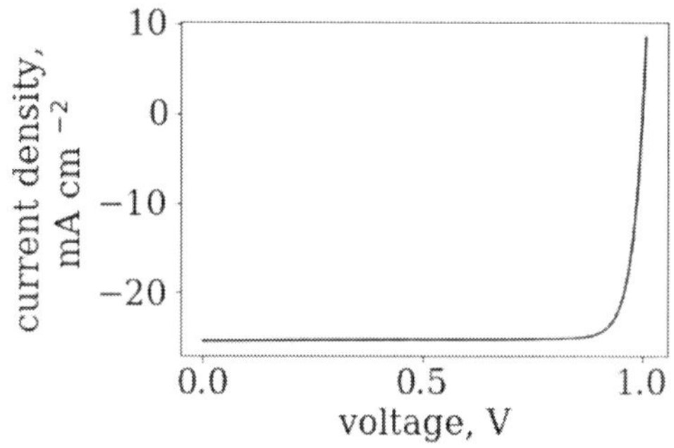

Figure 4. The current–voltage characteristic of one-dimensional structure.

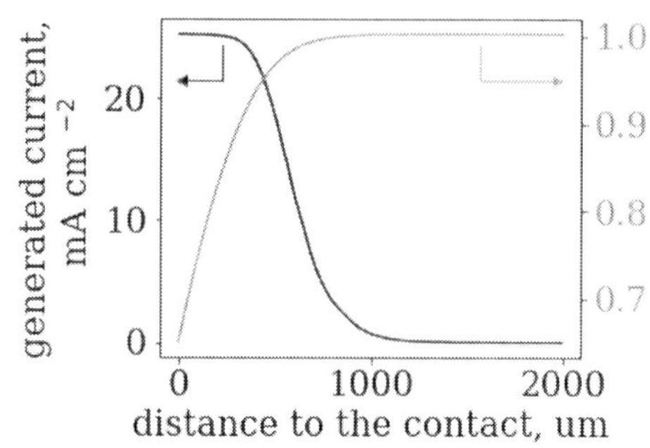

Figure 5. Voltage and generated current distribution vs distance to the contact, SC without CNT layer, AM1.5 spectrum, distance between contacts 4000 um

3. Summary

The GaAs/AlGaAs SC was modeled. Optimization of the one-dimensional model showed that emitter in SCs with CNT layer should be thinner than in conventional SCs. Although SC with thin emitter has insufficient conductivity, adding CNT layer solves this problem. Moreover, there is efficiency gain 0.2% in the case AM 1.5 spectrum and 1.7% in case of using 200 times light concentration despite absorption increases.

Acknowledgments

This work was carried out with the support of Skolkovo Institute of Science and Technology (Agreement No. 3663-MRA dated December 25, 2017), The Russian Federation President Council for grants (grant MK- 1458.2019.2, SP-2324.2018.1), The Ministry of Science and Higher Education of the Russian Federation (grant 3.9796.2017/8.9, grant 16.2593.2017/4.6).

References

[1] Green M A et al 2018 *Progress in Photovoltaics: Research and Applications*, ***26*(7)**, 427-436.
[2] Marinho B et al 2012 *Powder Technology* **221** 351–58
[3] Raudik S et el 2018 *J. Phys.: Conf. Ser.* 1124 041040
[4] Sze S M 1981 *Physics of Semiconductor Devices* (New York: Wiley–Interscience)
[5] Shockley W and Read W T 1952 *Phys. Rev.* 87 835–42
[6] Hall R N 1952 *Phys. Rev.* 87 387.
[7] Selberherr S 1984 *Analysis and Simulation of Semiconductor Devices* (New York: Springer-Verlag Wien)
[8] Yang T, Wang X, Liu W, Shi Y, and Yang F 2013 *Optics express* **21(15)** 18207-15.
[9] Saletes A et al 1986 *Solar cells*, ***17*(2-3)**, 373-381.
[10] Tobin S P et al 1990 *IEEE Transactions on Electron Devices* **37(2)** 469-477.

Φ-OTDR based on tunable Yb-Er:phosphate-glass laser

T V Choban[1], A A Zhirnov[1], A O Chernutsky[1], K V Stepanov[1], A B Pniov[1], G Galzerano[2], V E Karasik[1], C Svelto[3]
[1]Bauman Moscow State Technical University, Moscow 105005, Russia
[2]IFN-CNR, Milano 20133, Italy
[3]DEIB, Politecnico di Milano, Milano 20133, Italy

Abstract. In recent times, phase-sensitive optical time domain reflectometers became widely common for monitoring of extended objects and providing necessary measurements in optical telecommunications. In this paper we present our investigations and results in the development and characterization of a new type of laser for phase-sensitive optical time domain reflectometry.

1. Introduction

Phase-sensitive optical time domain reflectometers (ϕ-OTDR) are widely used in remote monitoring systems, providing a great amount of measurements for plants installation and security (pipelines, railway tracks, perimeter security, etc.), in proximity of an optical fiber [1-5]. The most important component of a ϕ-OTDR is its laser source, which defines the quality of the system. Different requirements arise for a laser providing efficient and well-precise operation of the ϕ-OTDR. The Yb-Er:glass solid-state laser [6] presents a great interest as it has very good short-term wavelength stability and possibility of wavelength tuning for compensation of ϕ-OTDR sensitivity 'dead zones'. One significant question is about finding the most adequate structure of the laser system, achieving the accuracy requirements for ϕ-OTDR with a simple and practical minilaser setup.

2. Laser system for ϕ-OTDR

In previous works we have studied a microchip Yb-Er:glass laser, measuring amplitude and phase noise of this laser setup [7]. Its main problems were high RIN peak level and frequency, due to relaxation oscillations in the short and lossless laser cavity, and the complex wavelength/cavity tuning. Based on this experience, we modified the structure of the laser to solve these problems. Now the resonator is formed by a 1 mm-thick Yb-Er:glass gain medium and an external concave mirror. Pumped by a fiber-coupled semiconductor laser diode at 976 nm, this plane-spherical laser operates in the range 1532..1570 nm. The structure of the resonator is shown in Fig. 1: first surface of the gain medium reflects laser radiation and is transparent for the pump ($R_{1@1550nm}$>99.9 %, $R_{1@976nm}$<2 %); second surface of gain medium is transparent for laser radiation and reflects the pump ($R_{2@1550nm}$<0.02 %, $R_{2@976nm}$>80 %); concave mirror reflects laser radiation providing 0.5 % output coupling ($R_{3@1550nm}$=99.5 %). Maximum cavity length is limited by the radius of curvature of the mirror (now *ROC*=50 mm). A GaAs plate is an additional component that can be placed placed inside the resonator in order to provide better wavelength stability and RIN reduction.

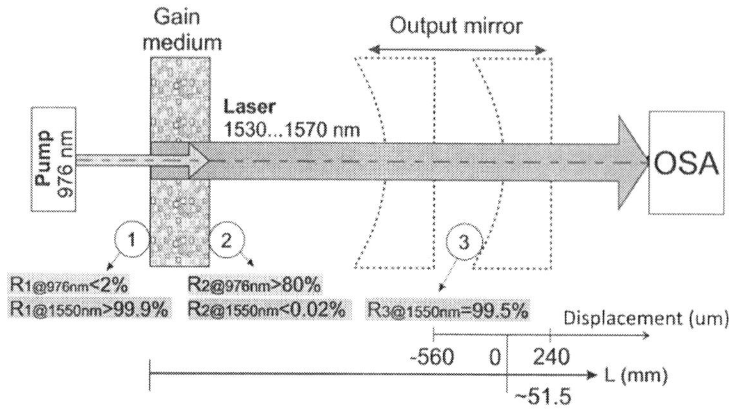

Figure 1. Structure of the laser resonator used in experiments and output mirror displacements.

Example of laser multi-longitudinal-mode output spectrum, depending on cavity length, is shown in Fig. 2. As the position of output mirror is changeable, we have experimentally studied how laser output spectrum depends on resonator construction in order to find the most adequate lengths of the resonator. One can see that the laser with maximum cavity length generates the widest output spectrum, which is important for wavelength tuning. Due to the spherical mirror, different cavity lengths provide laser beams with different waist diameters and thus working in different net gain conditions (gain in pumped region minus losses in unpumped regions minus other losses). Shorter cavity provides for larger waist diameter, which is shown in Fig. 3 (a), and in this case low population inversion zones appear within the laser mode volume. Fig. 3 (b) shows the gain spectrum of Er^{3+}. One can see that large waist diameter beams create reabsorption losses at 1535 nm in unpumped regions of active medium and thus generation in region 1540..1565 nm becomes more probable.

One of the emitted wavelengths is chosen by an optical band-pass filter. By coarsely and finely adjusting laser cavity length, it is possible to provide laser wavelength tuning which is important for matching with optical filter in ϕ-OTDR scheme.

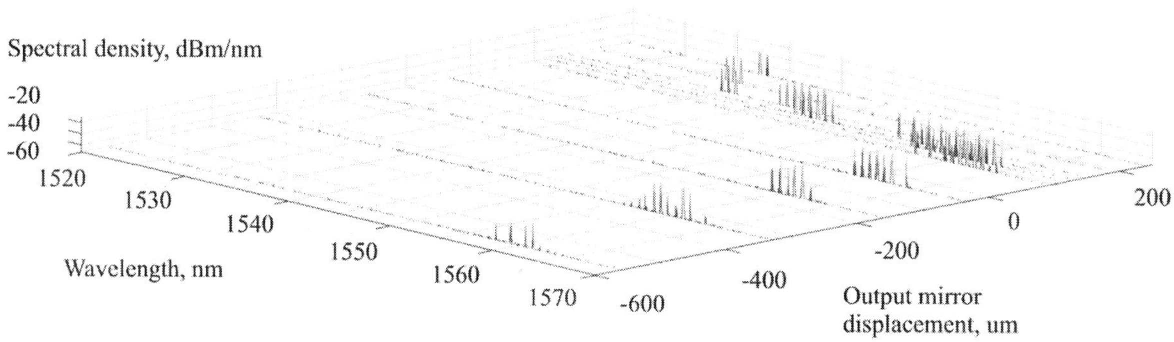

Figure 2. Multi-longitudinal-mode output spectrum of the laser depending on cavity length.

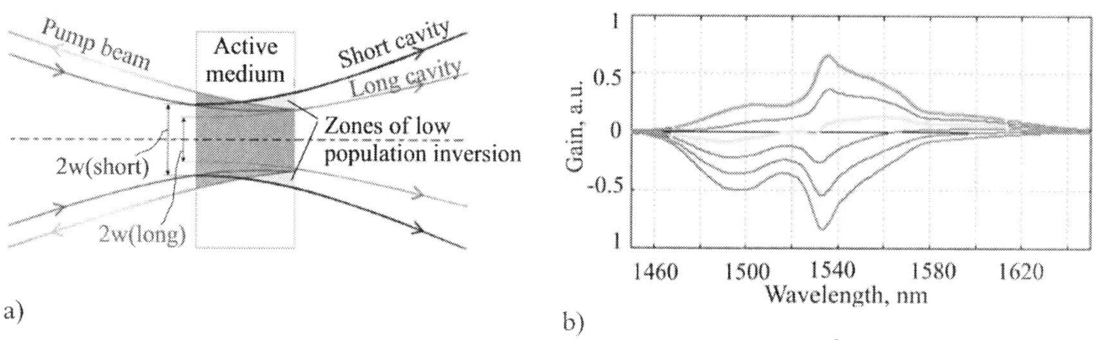

Figure 3 (a, b). a) Population inversion in Er-Yb:glass active medium; **b)** Er^{3+} gain spectrum. Yellow refers to low population inversion and orange refers to high population inversion.

While developing a laser for ϕ-OTDR, the most important requirements are wavelength and amplitude stability. In order to decrease the amplitude of relaxation oscillations, we used a two-photon absorber GaAs plate inside of the resonator, as GaAs energy levels configuration allows for laser output power stabilization. RIN peak frequency in this system (Fig. 4 (a)) decreased compared to the microchip laser [7] due to the now much longer cavity. Fig. 4 (b, c) show frequency deviations of microchip laser and plane-spherical laser respectively. Reducing amplitude of relaxation oscillations leads to significant increase of Signal-to-Noise Ratio (SNR), providing for better ϕ-OTDR operation.

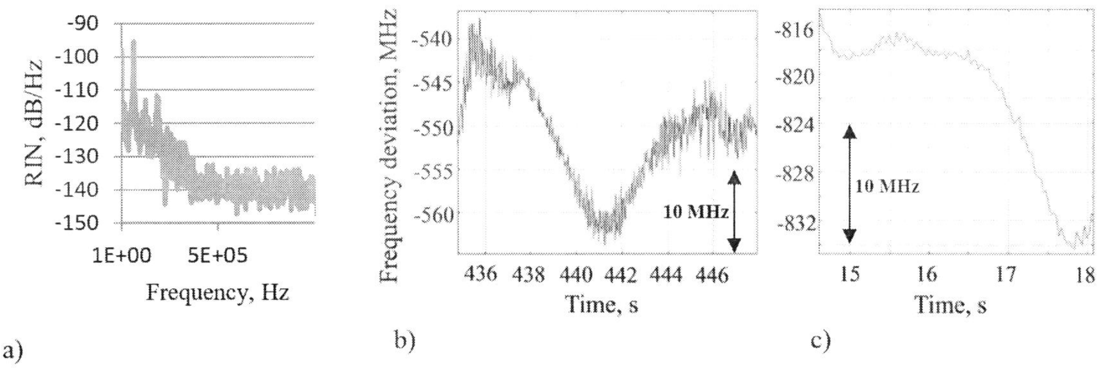

Figure 4 (a, b, c). a) Plane-spherical laser RIN spectrum; **b)** Microchip laser frequency deviations; **c)** Plane-spherical laser frequency deviations.

3. Φ-OTDR experimental scheme

The newly developed plane-spherical Yb-Er:glass laser was embedded in ϕ-OTDR experimental system, which is shown in Fig. 5 (a), quite different from common ϕ-OTDR scheme [8]. Light beam, emitted by the laser system is coupled into optical fiber (1). Optical circulator (2) delivers the light to a tunable filter (3), which we used for choosing just one longitudinal mode in (2)→(4) light propagation, as well as for suppressing ASE noise of the optical preamplifier (10) in (4)→(2) light propagation. Due to optical circulator (4) light is amplified by EDFA-booster (5) and is modulated by AOM (6). Passing through the third optical circulator (7), a light pulse goes to a sensing fiber (8). Propagating through sensing fiber, the light pulse is scattered on refractive index inhomogeneities. Back-scattered light returns to the system, so optical circulator (7) delivers it to EDFA-preamplifier (10). As mentioned earlier, light passes again through filter (3) and is detected by photoreceiver (11), being then processed by ADC (12) and personal computer (13). To observe output signal changes in time, we influenced the sensing fiber (8) with a PZT actuator (9). The PZT is a cylinder with 20 m of fiber coiled on it. We performed experiments with sine-wave electrical signal driving the PZT, with different frequencies: 10, 30, 50 and 70 Hz – most typical for real signals – and amplitude 3 V. In Fig. 5 (b) one can see how ϕ-OTDR output intensity changes with and without operating PZT (at 70 Hz), which is emulating a disturbance to the monitored structure along which the sensing fiber is buried. In plot of standard deviation (orange line) the region of PZT work can be easily distinguished.

Fig. 6 (a) shows an obtained reflectogram sequence ("waterfall"), having good visibility and sharpness, as obtained by novel plane-spherical laser. For comparison, in Fig. 6 (b) is presented waterfall obtained by previous laser with not enough wavelength stability.

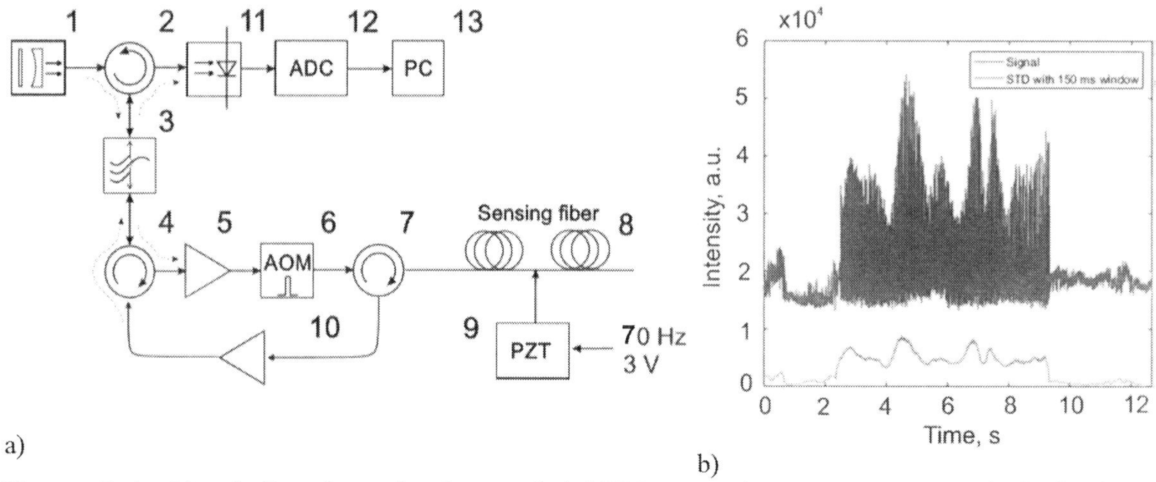

Figure 5 (a, b). a) Experimental scheme of ϕ-OTDR. 1 – laser; 2, 4, 7 – optical circulator; 3 – tunable filter; 5 – EDFA-booster; 6 – Acousto-Optic Modulator; 8 – sensing fiber; 9 – PZT actuator; 10 – EDFA-preamplifier; 11 – photoreceiver; 12 – ADC; 13 – personal computer; **b)** Intensity of an output signal in time with operating PZT from 2.51 s to 9.34 s.

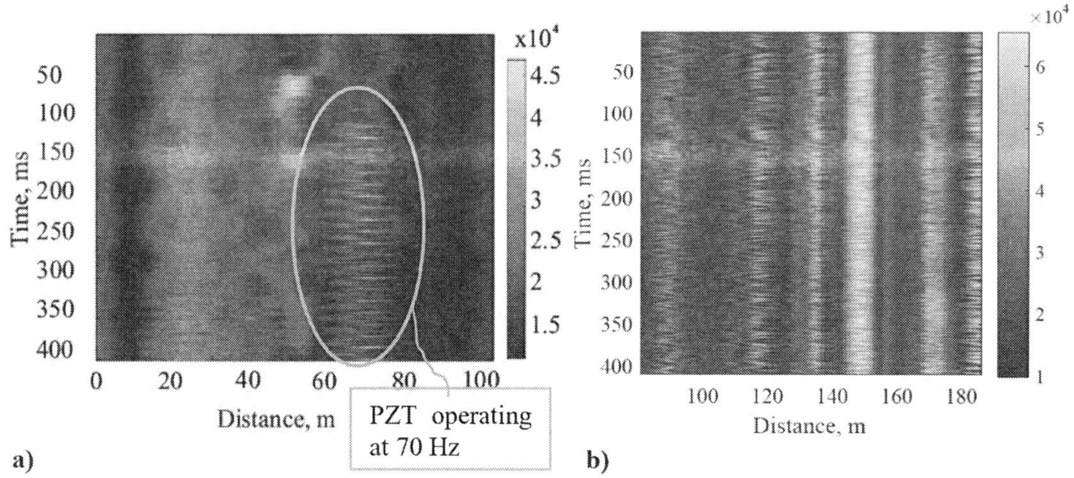

Figure 6 (a, b). Obtained ϕ-OTDR "waterfall": start point in time (t=0 ms) is equivalent to the stimulus start (t=2.51 s) in Fig. 5 (b). Output intensity is shown with colours. **(a)** "Waterfall" from laser with enough wavelength stability. **(b)** Waterfall from laser with not enough wavelength stability.

To assess ϕ-OTDR performance, we estimated SNR as the ratio of signal standard deviation to noise standard deviation:

$$\boldsymbol{SNR = 20\,log\left({}^{STD_S}/_{STD_N}\right)}$$

We have observed output signal changing in time under 10, 30, 50 and 70 Hz influence, emulating a disturbance to the monitored structure. For each frequency we choose regions with duration 2 s under PZT operating (*e.g.* from 4 to 6 s in Fig. 5 (b)) and calculated standard deviation $\boldsymbol{STD_S}$ during this time interval, which refers to signal (with actually added noise) standard deviation. For $\boldsymbol{STD_N}$ calculations we choose a 2 s time interval from the same sensor point without PZT operating, so standard deviation calculated in this interval referred to just-noise standard deviation in each of the experiments for the four frequencies of the influence signals. Results for different frequencies are given in Table 1.

Table 1. SNR for different external influences (corresponding to different PZT frequencies).

Frequency, Hz	*SNR*, dB
10	12.6
30	12.2
50	15.6
70	17.8

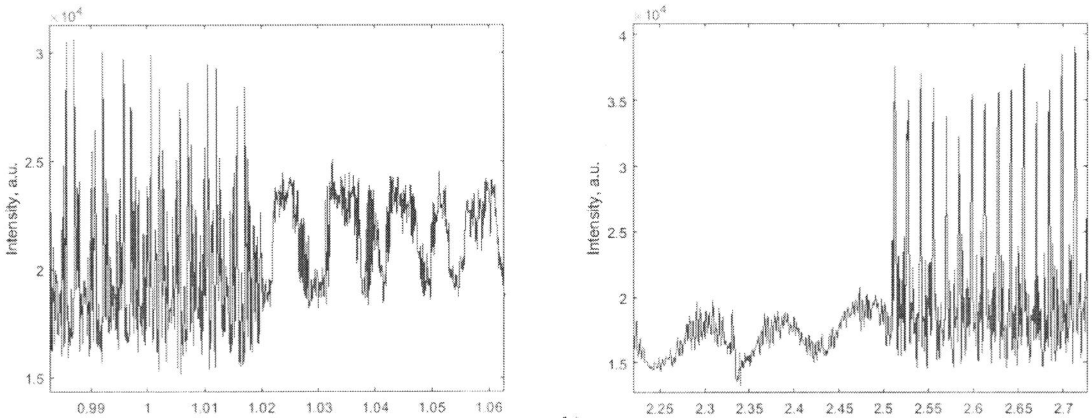

a) b)

Figure 7 (a, b). (a) Signal with oscillations from impact at 10 Hz (before 1.02 s) and just from laser frequency instability (after 1.02 s); **(b)** Signal with oscillations from impact at 70 Hz (after 2.5 s) and just from laser frequency instability (before 2.5 s).

Fig. 7 (a) shows two regions of signal instability: due to PZT disturbance in left part and due to laser frequency fluctuations in right part. Besides, Fig. 7(b) clearly shows that laser frequency fluctuations have a shape corresponding to some sine-wave disturbance. It is worth noting that oscillations (due to mechanical vibrations) in the laser wavelength deviations lead to output signal fluctuations, causing SNR decrease. The way to solve this problem is to rigidly connect the resonator components, reducing such oscillations. Our further investigations are aimed to develop a monolithic cavity and first step will be creation of glued connection between active medium and GaAs plate. This will allow to avoid additional vibrations from plate and decrease RIN level due to its two photon absorption [9].

4. Conclusion

In this paper we proposed a structure of an Yb-Er:glass laser system for ϕ-OTDR, meeting the requirements of high enough stability operation. An adjustable cavity length allows for wavelength tuning, so we reported about the most adequate structure of the resonator by choosing proper cavity length. An experimental setup of ϕ-OTDR was assembled with the plane-spherical laser embedded. The proposed scheme provided for reflectogram sequences with good visibility and sharpness, which confirms suitable operation of this ϕ-OTDR system using the erbium minilaser.

Acknowledgments

The reported study was funded by RFBR according to the research project № 18-32-00688.

References

[1] Honglan L, Zhenhai Z, Fei J 2018 Int. Conf. on Optical Instruments and Technology: Advanced Optical Sensors and Applications (Beijing) vol 10618 (SPIE-Intl Soc Optical Eng) 1061804
[2] Ölçer İ, Öncü A 2018 Fiber Optic Sensors and Applications XV (Orlando) vol 10654 (SPIE) 106540C
[3] Alekseev A E, Gorshkov B G, Potapov V T 2019 *Laser Phys.* 29(5) 055106
[4] Nesterov E T, Treshchikov V N, Ozerov A Z, Sleptsov M A, Kamynin V A, Nanii O E and Sus'yan A A 2011 *Technical Physics Letters* 37(5) 417-420
[5] Hartog A, Liokumovich L B, Ushakov N A, Kotov O I, Dean T, Cuny T and Constantinou A 2018 *Geophysical Prospecting* 66(S1) 192-202
[6] Taccheo S, Laporta P, Longhi S, Svelto O, Svelto C 1996 *Applied Phys. B Laser and Optics* 63 (5) 425–436
[7] Pniov A, Zhirnov A, Shelestov D, Stepanov K, Nesterov E, Karasik V, Laporta P, Galzerano G, Taccheo S, Piroddi L, Norgia M, Pesatori A, Svelto C 2016 *ACTA IMEKO* 5 (4) 24–28
[8] Zhirnov A A, Pnev A B, Svelto C, Norgia M, Pesatori A, Galzerano G, Laporta P, Shelestov D A, Karasik V E 2017 *J. Phys.: Conf. Ser.* 917 052032
[9] Van Leeuwen R, Xu B, Watkins L S, Wang Q, Ghosh C 2008 Enabling Photonics Technologies for Defense, Security, and Aerospace Applications IV (Orlando) vol 6975 (SPIE) 69750K

Stabilizing effect of injection in broad-area lasers

E A Yarunova [1,2], A A Krents[1,2], N E Molevich[1,2], D. A. Anchikov[1]
[1]Department of Physics, Samara National Research University, Samara 443086, Russia
[2]Department of Theoretical Physics, Lebedev Physical Institute, Samara 443011, Russia

Abstract. In this paper, spatiotemporal dynamics of broad-area lasers with external optical injection is investigated. We demonstrate theoretically that relatively small optical injection suppresses unstable transverse modes and stabilizes laser output. In this paper, the general case of a mismatch in the frequency of the radiation injected and generated by a laser was numerically investigated. It is shown that the frequency mismatch does not destroy the stabilizing effect.

1. Introduction

One of the most actual problem of the modern optics is to improve the radiation quality of miniature solid-state microchip lasers and semiconductor vertical cavity surface emitting lasers (VCSELs), which are of great technological importance in data processing systems and optical communication channels. However, a short cavity length leads to a rather small output power. The most natural way to increase output power without abandoning all the advantages of a short-cavity configuration is to use lasers with a large transverse size. An increased transverse size leads to the excitation of a large number of transverse modes, the nonlinear interaction between them leads to the development of spatio-temporal chaos [1-6]. This leads to multiple decrease in laser radiation brightness and the loss of spatial and temporal coherence of a laser beam having a fundamental value for most applications. The appearance of filamentation is a consequence development of unstable transverse spatial components of the optical field. Definition of these components allows us to explore the threshold of filamentation and methods of their suppression in order to improve the parameters of the laser. We have previously shown that external optical injection suppresses short-wave transverse spatio-temporal instability in broad-area lasers of dynamic class B and makes it possible to achieve spatially uniform generation [7]. In the present work, it is shown that a slight frequency mismatch does not destroy the stabilizing effect.

2. Mathematical model and numerical results

2.1. Mathematical model

Mathematical modeling of the spatio-temporal dynamics of a broad-area laser with injection of external optical radiation was carried out using the Maxwell-Bloch system of equations (schematic representation is provided by figure 1 [8]):

$$\begin{cases} \dfrac{\partial E}{\partial t} = ia\Delta_{\perp}E + \sigma\left(P - (1 - i\delta)E + E_{inj}e^{i\theta t}\right), \\ \dfrac{\partial P}{\partial t} = -(1 + i\delta)P + DE, \\ \dfrac{\partial D}{\partial t} = -\gamma\left[D - r + \dfrac{1}{2}\left(E^{*}P + P^{*}E\right)\right], \end{cases} \quad (1)$$

where E, P, D are the dimensionless envelopes of the electric field, active media polarization, and population inversion, respectively. $\gamma = \gamma_{\parallel}/\gamma_{\perp}$, $\sigma = k/\gamma_{\perp}$, where $\gamma_{\perp}$, $\gamma_{\parallel}$, and k are the coherence relaxation rate, population inversion relaxation rate, and electric field decay rate, respectively. $\delta = (\omega_{21} - \omega)/(\gamma_{\perp} + k)$ is the rescaled detuning between the optical transition frequency ω_{21} and the cavity resonant frequency ω_{cav}, $a = c^2/(2\omega\gamma_{\perp}d^2)$ is the diffraction parameter, where d is the transverse spatial scale of the laser; r is the pumping value normalized to its threshold value. External injection is characterized by two parameters: E_{inj} denotes the injection strength and θ corresponds to the frequency mismatch for the injected emission.

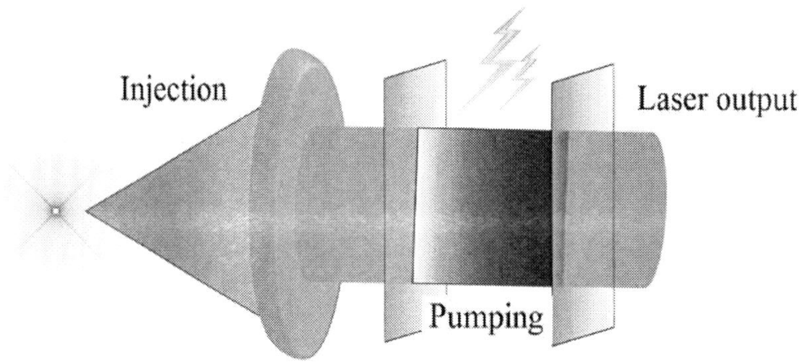

Figure 1 Schematic representation of an optically-injected broad-area laser

2.2. Numerical results

In a free-running broad-area laser in the absence of injection ($E_{inj} = 0$), the mode of spatially uniform paraxial generation is unstable with respect to small space-time perturbations. In semiconductor and solid-state lasers, wave instability is realized, which leads to spatial-temporal modulation in the beam cross section. The characteristic spatial dimensions of the modulation and the frequency of the temporal modulation are determined by the expressions:

$$q_{unst} = \sqrt[4]{\frac{2\gamma\sigma I_{st}}{(1+\delta^2)a^2}}, \quad \Omega_{unst} = \sqrt{\frac{4\gamma\sigma I_{st}}{1+\delta^2}} \quad (2)$$

where $I_{st} = |E_{st}|^2 = r - 1 - \delta^2$ is the stationary intensity value. It is clearly seen that the characteristic spatial and time scales (2) are related to the frequency of relaxation oscillations (3)

$$\omega_{rel} = \sqrt{\frac{2\gamma\sigma I_{st}}{1+\delta^2}}. \quad (3)$$

The injection of external optical radiation changes the expression for a stationary intensity value:

$$E_{inj} = \frac{rE_{st}(1-i(\theta+\delta))}{|E_{st}|^2+1+(\theta+\delta)^2} - E_{st}\left(1-i\left(\delta-\frac{\theta}{\sigma}\right)\right). \tag{4}$$

In (4), the stationary intensity-is given by an implicit function. Figure 2 shows typical curves of the stationary intensity dependence on the magnitude of the injected radiation.

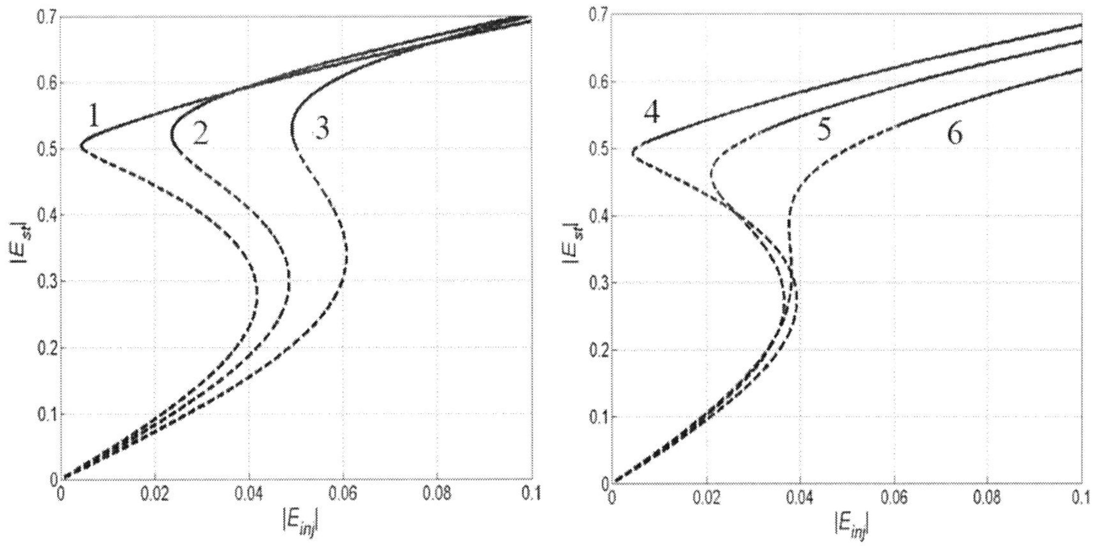

Figure 2. Dependence of the stationary intensity on the injected radiation for $r = 1.2$, $\delta = -0.15$ *u 1)* $\theta = -0.0001$ *2)* $\theta = -0.0005$ *3)* $\theta = -0.001$ *4)* $\theta = 0.0001$ *5)* $\theta = 0.0005$ *6)* $\theta = 0.001$

It has been demonstrated that coherent injection can completely eliminate transverse instability. The suppression of filamentation has a threshold character, that is, it occurs when the amplitude of the injected radiation exceeds a certain threshold value, what is shown in figure 3.The threshold value of the amplitude of the injected radiation is small in comparison with the amplitude of the optical field of the output radiation. The latter condition corresponds to the optical resonator of the laser, in which the input radiation does not play a decisive role in its operation. The possibility of radiation stabilization even at such a relatively weak injection allows us to consider the proposed approach as a potentially effective way to improve the output characteristics of high-power wide-aperture lasers.

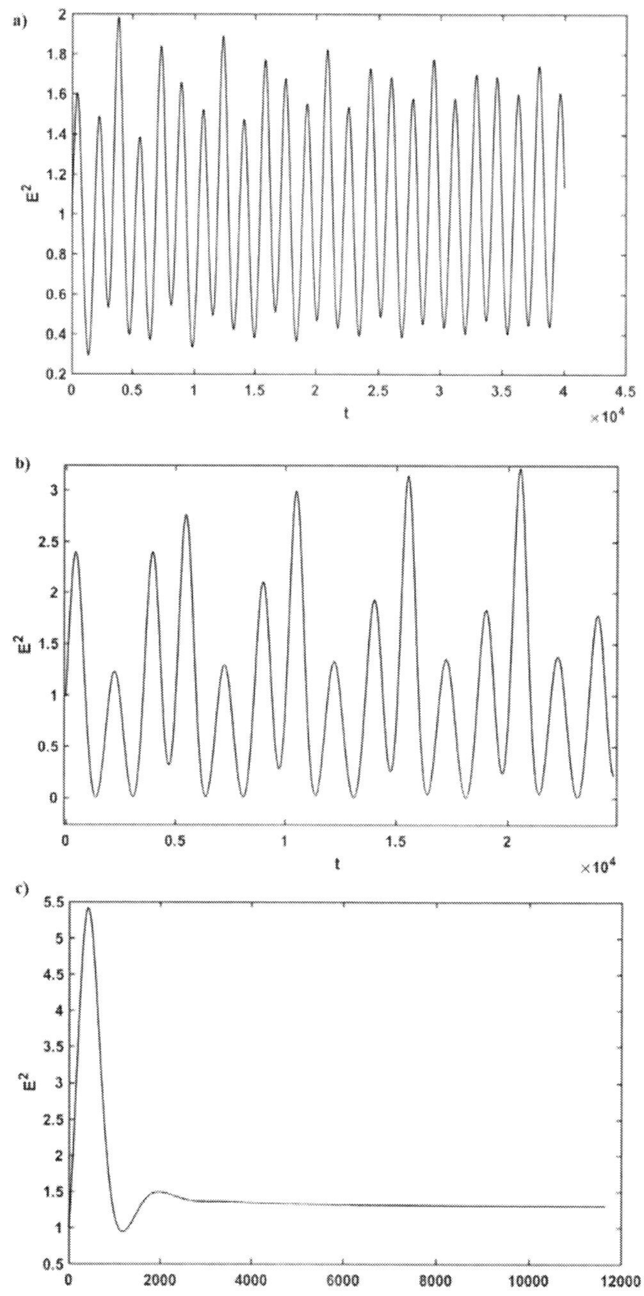

Figure 3. Temporary dependence of intensity for parameters $\delta = -0.15,\ \theta = 0, r = 2, \gamma = 5\cdot10^{-5}, a = 6\cdot10^{-4}, \sigma = 0.025$ and for a) $Einj = 0.001$; b) $Einj = 0.002$; c) $Einj = 0.005$

In a free-running broad-area laser in the absence of injection ($E_{inj} = 0$), the mode of spatially uniform paraxial generation is unstable with respect to small spatio-temporal perturbations. The dynamics is simulated with parameters corresponding to the broad-area semiconductor lasers: $\delta = -0.15, r = 1.1, \gamma = 5\cdot10^{-5}, a = 6\cdot10^{-4}, \sigma = 0.025$.

Figure 4(a) shows an example of the dispersion curves (black line) containing unstable wavenumbers with the positive increment Λ for the laser without injection. The presence of the injection makes the laser generation stable (red line) even if there is the frequency mismatch of the injected and generated radiation. In figure 4(b), the bifurcation range of values of the strength and the frequency mismatch of the injected radiation, at which stabilization occurs, is built (grey domain).

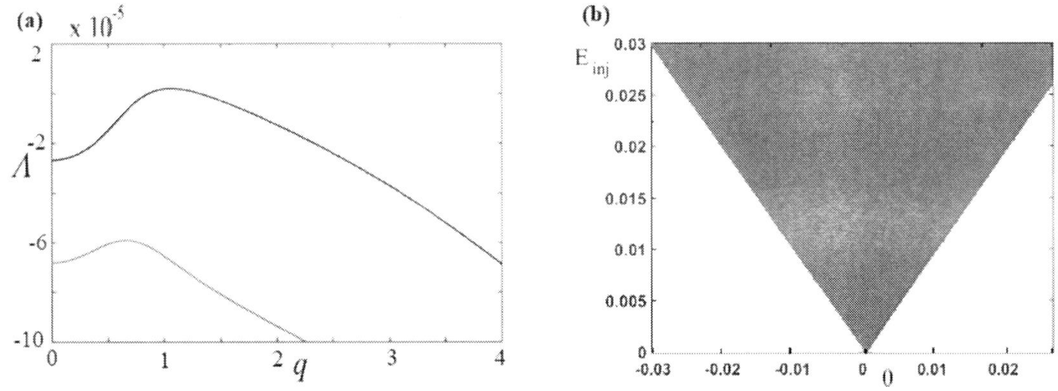

Figure 4(a, b). **(a)**Dispersion curves for the parameters $E_{inj} = 0$ (black line), $E_{inj} = 0.001$ (red line); **(b)** a bifurcation diagram representing the dependence of the dynamics on θ and E_{inj}.

3. Conclusion

In the present paper, we performed a study of the spatio-temporal dynamics of a broad-area laser with injection of external optical radiation and taking into account the frequency detuning between the frequency of the generated and injected radiation. It is shown that the frequency detuning between the frequency of the generated and injected radiation does not destroy the stabilizing effect. There is the synchronization region (grey domain in Figure 2b) in which non-coherent injection effectively suppresses unstable transverse modes and stabilizes the radiation of a broad-area laser.

Acknowledgments

The work was supported by the Russian Foundation for Basic Research under grant 18-32-00704 mol_a.

References

[1] Fischer, I., Hess, O., Elsäßer, W. and Gobel, E. 1996 Europhysics Letters **35** 579
[2] Otsuka K., Chu S. 2009 Optics Letters **34** 10
[3] Pakhomov A. V., Molevich N. E., Krents A. A., Anchikov D. A. 2016 Optics Communications **372** 14
[4] Babushkin, I. V., Loiko, N.A, Ackermann, T. 2004 Physical Review E **69** 066205
[5] Gensty, T., Becker, K., Fischer, I., Elsäßer, W., Degen, C., Debernardi, P. and Bava, G. P. 2005 Physical Review Letters **94** 233901
[6] Krents A. A., Molevich N. E., and Anchikov D. A., 2017 J. Opt. Soc. Am. **B 34** 1733
[7] Anchikov D. A., Krents A. A, Molevich N. E., Pakhomov A. V. 2016 Computer Optics **40** 31
[8] Jacobsen P., Moloney J., Newell A., Indik R. 1992 Phys. Rev. A. **45** 8129.

Investigation of ITO magnetron deposition effect on carrier lifetime degradation in silicon wafer

D Kudryashov[1], **A Baranov**[1] , **A Gudovskikh**[1,2]
[1]St. Petersburg Academic University, St. Petersburg 194021, Russia
[2]St. Petersburg Electrotechnical University "LETI", St. Petersburg 197376, Russia

Abstract. The effect of ITO magnetron deposition on carrier lifetime in silicon was demonstrated. The method of 2D photoluminescence decay time measurements revealed the lifetime degradation in silicon after magnetron deposition of thin ITO film which is used for top contact in modern solar cells.

1. Introduction

Modern solar cells are mainly produced from silicon due to its high availability and the developed Si technology. Silicon-based solar cell efficiency is determined principally by the quality of silicon wafer which acts as a volume where the carriers photogeneration and separation occures. The record efficiency of 26.7% [1] was reached for the silicon solar cell produced from float-zone silicon. It's a high-purity silicon wafer with extremely low concentration of defects. Photogenerated charge carriers in such wafer have high value of lifetime (more than several ms). High lifetime values promote the carriers to reach the silicon surface without significant recombination in the bulk. However the surface by itself consists of numerous abrupt bonds that act as a recombination centres. Nanosized passivation films (for example a-Si:H) on the both side of silicon wafer eliminate surface recombination resulting in more efficient current output. Hydrogen in such films saturates abrupt silicon bonds however the Si-H bond strength is quite low. Thermal heating, UV irradiation or plasma treatment can break off the Si-H bond and result in passivation degradation. Transparent conductive oxide (TCO) deposition is one of the stages for solar cells fabrication. Principally magnetron sputtering is used for the process. During the sputtering the sample (silicon covered by a-Si:H) is exposed to strong UV irradiation and to high kinetic particles collision. This could potentially destroy the passivation and moreover - create defects in the silicon surface layer. As a result the efficiency of the solar cell can be decreased. The aim of this work is to reveal the effect of indium tin oxide (ITO) magnetron deposition on the silicon passivation efficiency.

2. Experiment details

Cz as-cut n-Si(100) 1-7 Ohm·cm wafer with a thickness of 210 μm and a size of 40×40 mm^2 was used. Texturing and cleaning procedure was described earlier [2]. After an HF dip a 100 nm of a-Si:H layer was deposited on each silicon side by PECVD with Oxford Plasmalab equipment at a temperature of 250 °C. Then a two-dimensional (2D) photoluminescence (PL) decay measurements were carried out. Two-dimensions PL decay time measurements were performed on as-passivated by a-Si:H textured silicon wafer with the experimental setup shown in figure 1. The 850 nm IR laser was used to excite the sample during PL decay measurements. PL response from silicon was detected by low capacitance InGaAs photodiode (GAP500) with ultrafast amplifier from Analog Devices coupled with analog to digital converter managed by microcontroller and connected to PC. PL decay curves were measured with a fixed laser intensity of 70 mW/cm^2. Controllable moving of the measured sample allows to provide 2D measurements, i.e. to measure PL decay time locally on the whole wafer with a resolution of the order of 0.5 mm limited by laser spot and the photogenerated carriers diffusion length.

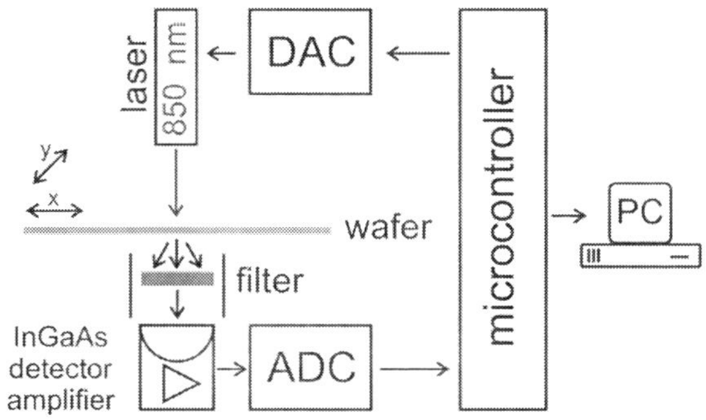

Figure 1. Schematic of experimental setup used to perform PL decay measurements.

To investigate the effect of ITO sputtering on the front surface of the passivated silicon sample a 100 nm thick ITO layer was deposited by RF magnetron sputtering with BOC EDWARDS AUTO 500 RF Sputter. To separate the influence of different potential sources of degradation during the sputtering three different type of mask were used. Stainless steel mask with open areas of 4x4 cm^2 was used to form the separated ITO areas on the top sample surface. Mask from boron glass and fused silica were used to investigate the effect of strong UV irradiation during the first stages of TCO sputtering. The ITO deposition was carried out with a fixed magnetron power of 50 W and a chamber pressure of 10^{-3} mbar. Pure argon (99.999 %) was used as a working gas. The distance between magnetron source and the substrate was 8 cm. The deposition process duration was 10 min.
As an ITO layer absorbs a part of incident light thus prior to the PL measurements the deposited film was selective etched in $HCl:H_2O$=1:3 solution.

3. Results and discussion

Figure 2 shows the result of 2D PL decay measurements for the 4x4 cm^2 textured silicon wafer passivated by a-Si:H.

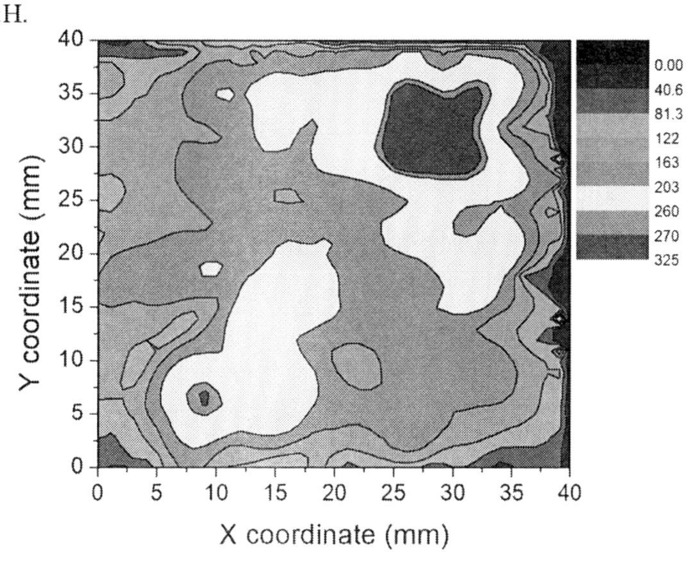

Figure 2. PL decay mapping of silicon wafer passivated by a-Si:H.

In general an average PL decay value of about 200 μs is observed except the one region where the value is much bigger and reaches of 300 μs. Probably it is related to the inhomogeneities in silicon wafer attributed to the different impurities distributions.
Figure 3a shows a mask placement on the passivated silicon substrate and the result of ITO magnetron deposition (Fig. 3b). Dark areas are the ITO thin layers deposited on silicon and due to refractive index of 2.0 the transparent conductive layer acts as a single layer antireflective coating leading to optical reflection decreasing.

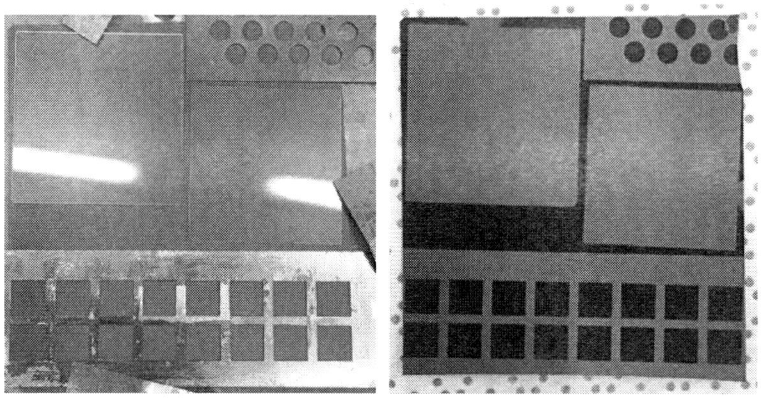

(a) (b)

Figure 3. Photo of a mask placement on a-Si:H passivated silicon wafer (a) and the result of the ITO deposition (b).

Next, the deposited ITO layer was selective etched in $HCl:H_2O=1:3$ solution to eliminate its influence on PL decay measurements. After the etching visible patterns from the metallic mask in the top part appeared (Fig 4a). This points out on some degradation of a-Si:H thin layer. The corresponded 2D PL decay measurements are shown in figure 4b.

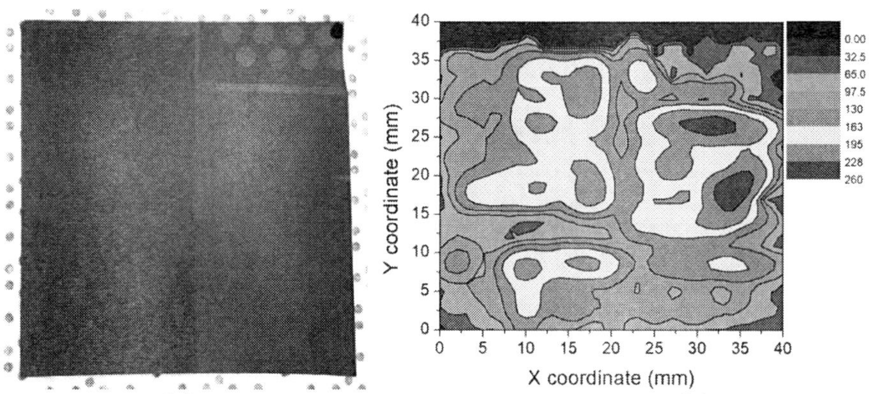

(a) (c)

Figure 4. Photo of a top a-Si:H passivated silicon surface after ITO etching (a) and the corresponding PL decay mapping (b) of the sample.

It can be seeing that after ITO deposition and dissolving the maximum value of the PL decay time was decreased from 325 to 260 μs. As the initial PL decay distribution was not homogeneous there was a precise calculation realized to show the difference between the initial PL map and the final (Fig. 5). It is clearly visible that at the areas where the ITO film was deposited and then etched a strong PL decay time decreasing is observing. The highest degradation is observed at the top areas where were seen the visible patterns form metallic mask (Fig. 4a).

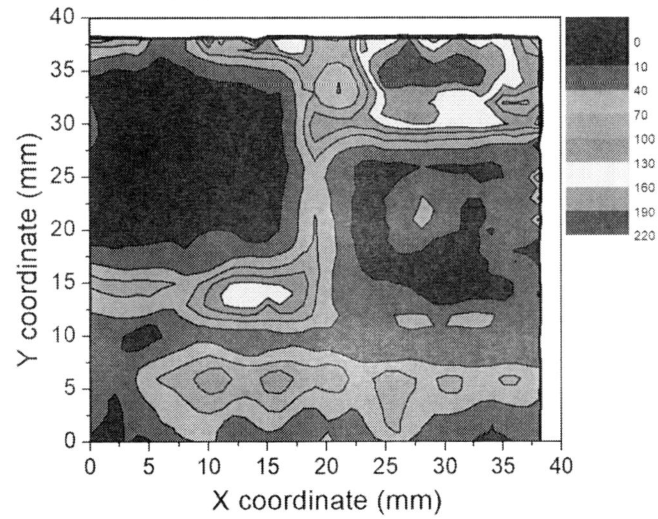

Figure 5. PL decay mapping difference between the measured on the a-Si:H passivated silicon and after the ITO deposition/etching

Metallic mask was placed at top and the bottom parts of the sample but the degree of passivation degradation is different. Taking into account the visible patterns on the top part of the sample after the ITO etching it can be assumed that the ITO deposition conditions during magnetron sputtering are vary by the Y coordinate. The distance between magnetron source and the substrate was 8 cm. The sample size was 4×4 cm^2. For such low distance and considering the size of the sample there is a probability that at outer areas the sputtered particles (In, O and Sn) will have different local concentration and kinetic energy. Figure 6 shows measured values of sheet resistance (R_s) for 100 nm ITO film deposited on glass substrate in regimes similar to this work.

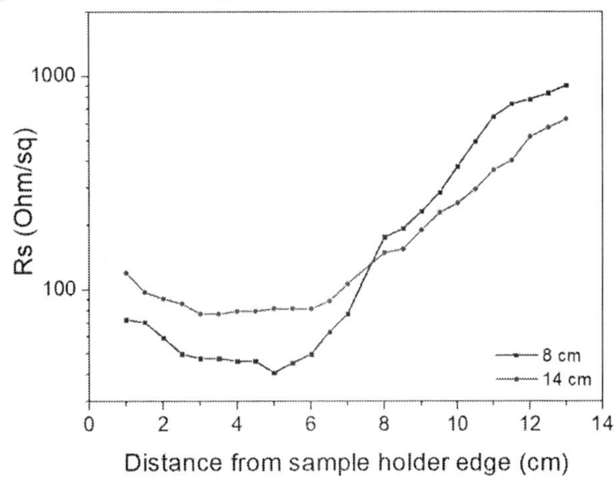

Figure 6. ITO sheet resistance as a function of the sample placement on the sample holder for two different magnetron source to sample distances.

Sheet resistance in ITO film is defined by the concentration of the three elements and the defects formed during the deposition. The investigated sample was located inside the region between 2 and 6 cm where the R_s value is minimal. The most degraded area on a-Si:H passivated silicon sample was located at a distance of 6 cm from the sample holder edge. According to figure 6 at this area R_s is starting to increase but is still low. So the reason for such high degradation of PL decay in the top region is still unclear. As an assumption it could be an excess of oxygen particles which reacts with hydrogen from a-Si:H layer leading to the passivation degradation that visually appears due to refractive index change [3].

As it shown in figure 5 areas masked by boron glass or fused silica didn't show any distinct degradation of PL decay time after ITO magnetron deposition. Thus can be concluded that strong UV radiation during ITO deposition didn't affect on passivating properties of 100 nm a-Si:H layer. And the described earlier effect of passivation degradation is related only to chemical or kinetical interactions between sputtered ITO and a-Si:H layer.

Acknowledgments

This work was supported by the Russian Scientific Foundation under grant number 18-79-10059.

References

[1] Green M A et al. 2019 *Prog Photovolt Res Appl.* **27** 3
[2] Kudryashov D, Gudovskikh A, Rodin A, Pinaev N 2018 *J. Phys.: Conf. Series* **1124** 041010
[3] Silvio P et al. 2015 *International Journal of Photoenergy* **1** 7

Investigation of cryogenic dry etching of silicon on I-V curves of Schottky diodes

A Maximova[1,2], **A Gudovskikh**[1,2], **D Kudryashov**[2]
[1]Department of Fotonics, Saint Petersburg Electrotechnical University "LETI", St. Petersburg, 197376 Russia
[2] St. Petersburg Academic University, St. Petersburg 194021, Russia

Abstract. I-V characteristics of silicon wafer with Schottky contacts were investigated to find out influence of dry plasma etching on defects emerging in the structure. Wafer was divided into two pieces, one of them has affected to cryogenic etching using SF_6/O_2 (in ratio 5:1) and a pressure 5 mTor for two minutes. The second part of the wafer was used as a reference. It was founded, that for the same contact diameters, etched samples demonstrate higher slope of the I-V curve at the direct voltage. A similar effect was observed in the case of the existence of a defective near-surface region, where carrier recombination is enhanced, which results a significant increase in the direct current.

1. Introduction

Recently, there has been an increased interest in renewable energy sources due to global warming from emissions of hydrocarbon fuel combustion products. According to calculations, the Earth receives about 800 W/m^2 from the Sun. To convert solar radiation into electricity, photovoltaic cells or solar cells are used as a special case. At the moment there are many types of solar cells, however, silicon solar cells are most prevalent. The achieved efficiency of a silicon-based solar cells with a heterojunction is 26.6% and has not increased since 2016. To further improve the efficiency of solar cells, new approaches are required. One such approach is to form another transition on top of the silicon cell, which more effectively transforms the shortwave part of the spectrum of solar radiation. For its formation can be used amorphous hydrogenated silicon having a large band gap. However, due to the low conductivity of this material, it is impossible to increase its thickness over a wide range without an increase in resistance. Therefore, the approach involves the use of vertically oriented silicon nanostructures as a template for further deposition of the transition from amorphous hydrogenated silicon on it. In this way, the length of transmission of optical radiation in the material increases while maintaining its thickness. For the formation of vertically oriented silicon nanostructures, the method of low-temperature plasma chemical etching is used. However, plasma is a source of high-energy particles and hard ultraviolet radiation, which can potentially cause defects on the surface of a silicon substrate.

The purpose of this research is to study the effect of low-temperature plasma-chemical etching on the formation of point defects in silicon in the manufacture of a solar cell with vertically oriented nanostructures. This requires studying the characteristics of silicon before and after plasma exposure using the method of volt-amperometry.

2. Experiment details

The volt-amper characteristics of the n-Si structure with a Schottky barrier were investigated in order to reveal the effect of etching conditions on the formation of defects in the structure.

Samples were made on the silicon n-Si (100) wafer of the company TopSil double-sided polishing. The plates were processed using the Shiraki method, then a layer of amorphous hydrogenated silicon (n-a-Si:H) was deposited on one side with the method of plasma-chemical deposition. Then processed substrate in hydrofluoric acid (HF). Silicon dioxide SiO_2 was deposited on the layer of amorphous silicon to protect it from subsequent etching. Next, the substrate was divided into two parts, one of which was subjected to low-temperature (cryo) etching in SF_6 / O_2 in a ratio of 5:1 at a plasma pressure of 5 mTor for two minutes. Then both plates were again treated with hydrofluoric acid. After that, round gold contacts are applied on the front side, and a layer of silver is applied on the back side, and then gold is applied to protect silver from oxidation. Figure 1 presents a schematic representation of the final structure

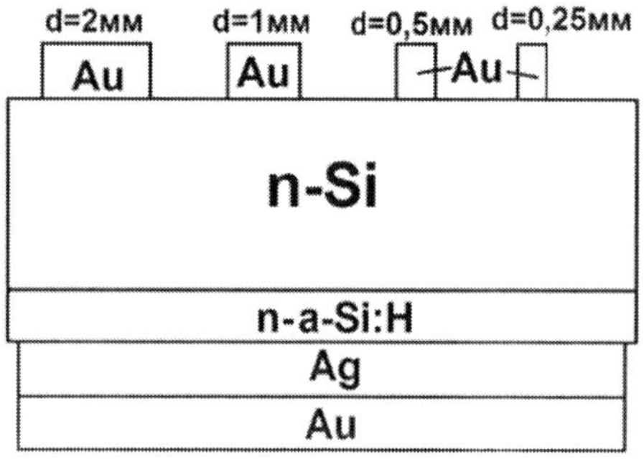

Figure 1. Schematic representation of the structure.

Figure 2 provides a photo of the front side of the control sample (a) - not subjected to cryogenic etching (b) - subjected one, taken with an optical microscope camera.

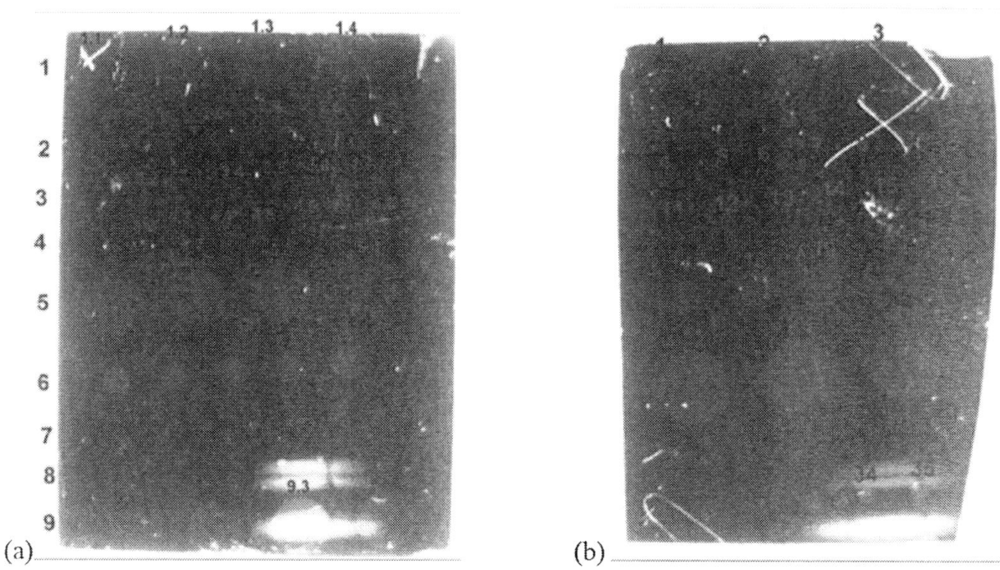

Figure 2(a,b). (a) Photo of the sample not subjected to cryogenic etching; (b) subjected one.

3. Results and discussion

When measuring current-voltage characteristics, an array of contacts with a number of about 35-50 is investigated, since a large sample is required for the statistical reliability of the experiment. Contacts are numbered for more convenient display of results.

Measurements of the I-V characteristics were carried out both before and after annealing at a temperature of 170 ° C for 20 minutes.

The measured I-V characteristics of the sample for contacts of diameter 1mm before and after annealing are presented in Figure 3.

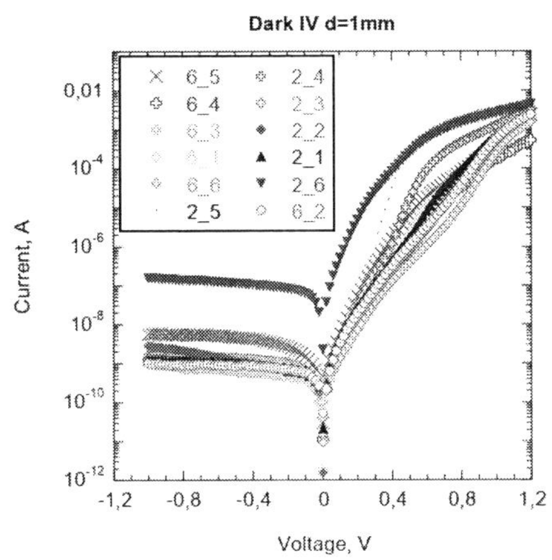

(a)

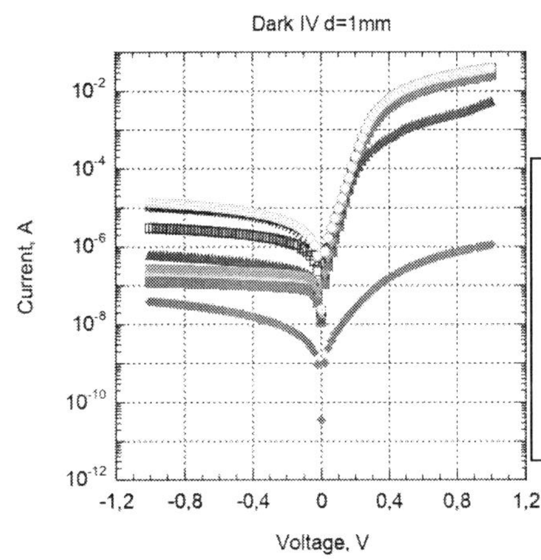

(b)

Figure 3(a,b). IV characteristics of the reference sample for d = 1 mm before(a) and after annealing(b)

From the general analysis of the graphs it can be seen that there is a certain discrepancy between the characteristics of the structures under study within one sample, which manifests itself in different amounts of reverse currents. This effect may be due to the presence on the original surface of traces of dirt or incomplete removal of the oxide layer (SiO_2).

The greatest divergence of reverse currents is observed in structures with the lowest and highest diameters of contact pads.

Annealing of structures at a temperature of 170 ° C contributes to some averaging of the values of reverse currents. However, in some particular cases it leads to a sharp decrease in the currents on the straight branch, which indicates the appearance of a potential barrier.

The measured current-voltage characteristics of the control sample for contacts of diameter 1mm before and after annealing are presented in Figure 4.

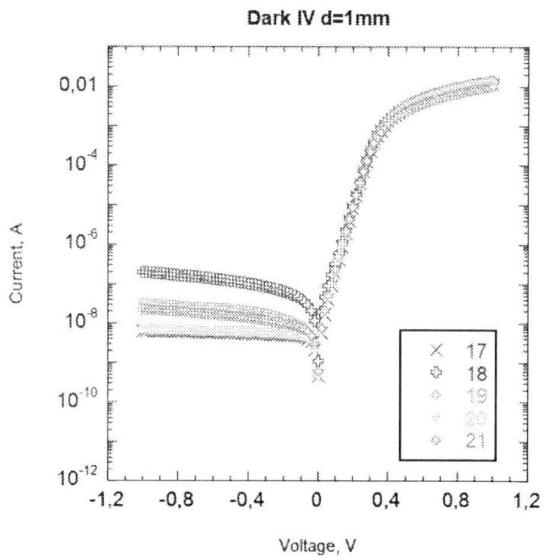

(a)

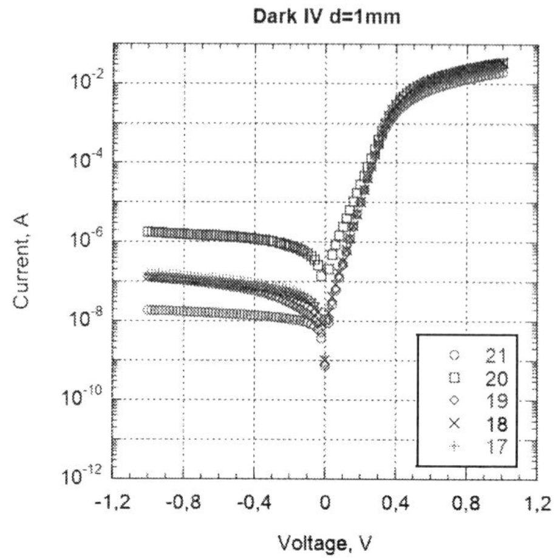

(b)

Figure 4(a,b). IVC of the etched sample for d = 1 mm before(a) and after annealing(b)

From the general analysis of the graphs, it can be seen that, after low-temperature plasma-chemical etching, the direct branches of the I-V characteristics practically do not differ within the limits of one sample. For reverse currents, the spread of values also became smaller. Annealing of these samples practically did not lead to a change in the form of the I-V characteristic.

At the same contact sizes, the samples after etching show a sharper slope of the curve on the straight branch of the IVC. A similar effect is observed in the case of the existence of a defective near-surface region, where carrier recombination is enhanced, which results in a significant increase in the forward current. With a further increase in voltage, a slowdown in the current growth rate is observed.

Acknowledgments

This work was supported by the Russian Scientific Foundation under grant number 18-79-10059.

References

[1] Cohen J. D., Lang D. V. Calculation of the dynamic response of Schottky barriers with a continuous distribution of gap states // Phys. Rev. B. 1982. Vol. 25. P. 5321–5350.
[2] W Shockley and W T Read, "Statistics of the Recombinations of Holes and Electrons .," Phys. Rev., vol. 87, no. 5, pp. 835-842, 1952.
[3] Sze S M 1969 *Physics of Semiconductor Devices* (New York: Wiley–Interscience)

Shapes of polarization sensitive photocurrent pulses in Ag/Pd nanocomposite films: influence of firing temperature

A S Saushin[1,2], V M Styapshin[1], E V Aleksandrovich[1], Yu P Svirko[2] and G M Mikheev[1]

[1]Institute of Mechanics, Udmurt Federal Research Center of the UB RAS, Izhevsk 426067, Russia

[2]Institute of Photonics, University of Eastern Finland, Joensuu 80101, Finland

Abstract. The results of investigation of the shapes of polarization-sensitive photocurrent pulses in nanocomposite Ag/Pd films with different temperatures of firing are presented. Ag/Pd nanocomposite films were produced in accordance with thick film technology at temperatures T = 773 and 1013 K. Photocurrent excitation was carried out by femtosecond laser pulses. The obtained polarization photocurrent dependencies are characteristic for photocurrent due to photon drag and surface photogalvanic effects. It was shown that longitudinal photocurrent pulse duration in 773 K films (26 ns) is significantly longer than in 1013 K films (3 ns).

1. Introduction

It is known that silver-palladium (Ag/Pd) thick films possess temporally stable electrical characteristics and have long been used in electronics [1]. The electronic properties of the Ag/Pd films vary in a wide range depending on thermodynamics and kinetics of Pd oxidation, Ag diffusion and migration, properties of inorganic and organic additives, and other factors. These films are widely used in production of hybrid microcircuits, multicrystal modules, integrated circuit assemblies and passive electronic components [2]. Recently, it has been shown that nanosecond pulses of polarization-sensitive photocurrent depending on the direction of wave-vector can be generated in these films upon nanosecond and femtosecond laser irradiation [3,4]. According to [5,6], photocurrent generation in these films occurs in a wide wavelength range (266-3500 nm) of incident radiation due to the photon drag (PDE) [7–9] and the surface photogalvanic effects (SPGE) [10,11]. This allows one to employ the Ag/Pd films to visualize polarization of powerful laser beams by the methods described in [12,13]. Usually, the temporal shape of photocurrent generated due to the PDE and SPGE repeats the shape of a nanosecond excitation laser pulse (e.g. [14,15]), because the momentum relaxation time of charge carriers lies in the interval from $\sim 10^{-12}$ to 10^{-13} s [16]. However, the photocurrent pulse duration in the Ag/Pd nanocomposite films is substantially longer than the duration of exciting nanosecond laser pulses [5,6]. One of the reasons for the elongation of polarization-sensitive photocurrent pulses may be the peculiarities of the electrical conductivity in these films. It is obvious that the peculiarities of electrical conductivity of the material depend on its phase composition and structure. According to [17,18], the structure, phase composition, and electrical properties of the Ag/Pd nanocomposite films essentially depend on the firing temperature. However, experiments for the investigation of the effect of the firing temperature on the pulse duration of polarization-sensitive photocurrent in these films have not yet been carried out. In this regard the purpose of this work is investigation of influence of the Ag/Pd film firing temperature on time parameters of polarization-sensitive photocurrent pulses.

2. Experimental

2.1. The Ag/Pd films

The Ag/Pd films are produced using technologies [17,19] based on firing of a special paste, which contains silver oxide (Ag_2O) and Pd, at high temperatures on the insulating substrate surface. The films studied in our experiments were fabricated at temperatures T = 773 and 1013 K from the resistive paste consisting of the following components, wt. %: Ag_2O – 24.2, Pd – 30.8, glass frit – 45 and organic binder – 25. The size of the obtained films was 5 × 5 mm. The thickness of the films was about 10 μm. To measure the photocurrent, the films were provided with two parallel electrodes which were arranged along the opposite sides of the Ag/Pd film between the aluminum oxide (Al_2O_3) ceramic substrate and the film. To estimate the composition of the Ag/Pd films the X-ray diffraction analysis was performed using a D2 PHASER diffractometer with monochromatized CuK_α radiation at the wavelength of 0.1541 nm.

2.2. Photocurrent excitation

The sketches of longitudinal and transverse photocurrent measurement setups are shown in figures 1a and 1b, respectively. Longitudinal photocurrent flows in a direction parallel to the plane of incidence σ of laser radiation and transverse photocurrent is perpendicular to the σ. The photocurrent was excited using femtosecond pulsed laser radiation at λ = 795 nm with pulse duration of τ = 120 fs. The pulse energy was in range of 150-300 μJ. In the experiments the linearly polarized laser radiation was directed through the quarter-wave or half-wave phase plate. The laser beam with diameter of 4 mm was passed through the phase plate and directed onto the Ag/Pd film surface at the angle of α = 45°. The polarization state of the beam was determined by the angle φ between slow axis of the phase plate (n_e) and the polarization azimuth of the initial beam (x'). When quarter-wave plate was used, the beam was left- and right-circularly polarized at φ = 45° and 135°, respectively, and had elliptical or linear polarization at other φ. We controlled the polarization azimuth Φ = 2φ of linearly polarized incident beam by rotating the half-wave plate by the angle φ. A digital oscilloscope with bandwidth of

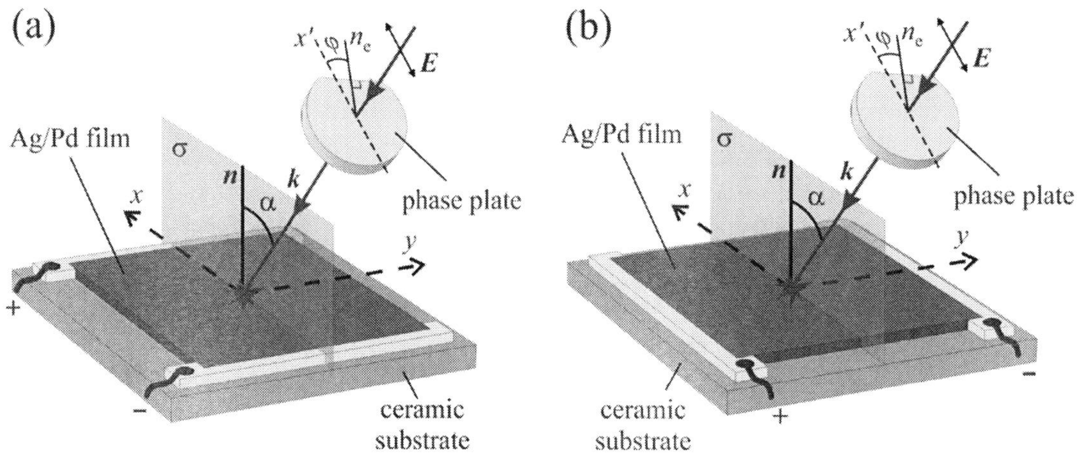

Figure 1(a, b). Sketch of the experimental setup for **(a)** longitudinal and **(b)** transverse photocurrent registration in Ag/Pd nanocomposite film: (σ) plane of incidence, (α) angle of incidence, (***n***) film normal, (***k***) wavevector, (x') polarization azimuth of the initial beam, (n_e) slow axis of the phase plate, (***E***) electric field vector, (***E***, x' ∥ σ). Electrodes are deposited on the sample's edges **(a)** perpendicular and **(b)** parallel to the σ, which coincides with the (xz) plane of the laboratory Cartesian frame.

300 MHz and an input impedance of $r = 50\ \Omega$ was used for the photocurrent registration. The energy of laser pulses (E_{in}) incident on the film was measured using a pyroelectric energy meter. A fast response SIR-5 *p-i-n* photodiode (ThorLabs) with a rise time of about 60 ps connected to the same digital oscilloscope was used as a reference photoelectric converter. The temporal evolution of the photocurrent pulses was characterized by measuring the rise time and fall time, which were defined with respect to the 0.1 and 0.9 of the pulse amplitude. The duration of the photocurrent pulse was defined with respect to the 0.5 of the amplitude.

3. Experimental results and discussion

A femtosecond laser pulses produced an electric current directed parallel or perpendicular to the plane of incidence σ (figure 1) in the film. The longitudinal and transverse photocurrents manifested themselves as an unipolar nanosecond electric pulses of voltage with extreme values U_x or U_y, respectively.

Figure 2 shows the temporal shape of the SIR-5 photoresponse and the corresponding temporal shapes of the longitudinal (figure 2a, 2c) and transverse (figure 2b, 2d) photocurrent pulses excited in the Ag/Pd films produced at $T = 773$ K (upper oscillograms) and $T = 1013$ K (lower oscillograms). The longitudinal photocurrent was registered at *p*-polarized excitation, and the transverse one was excited by left-circularly polarized beam.

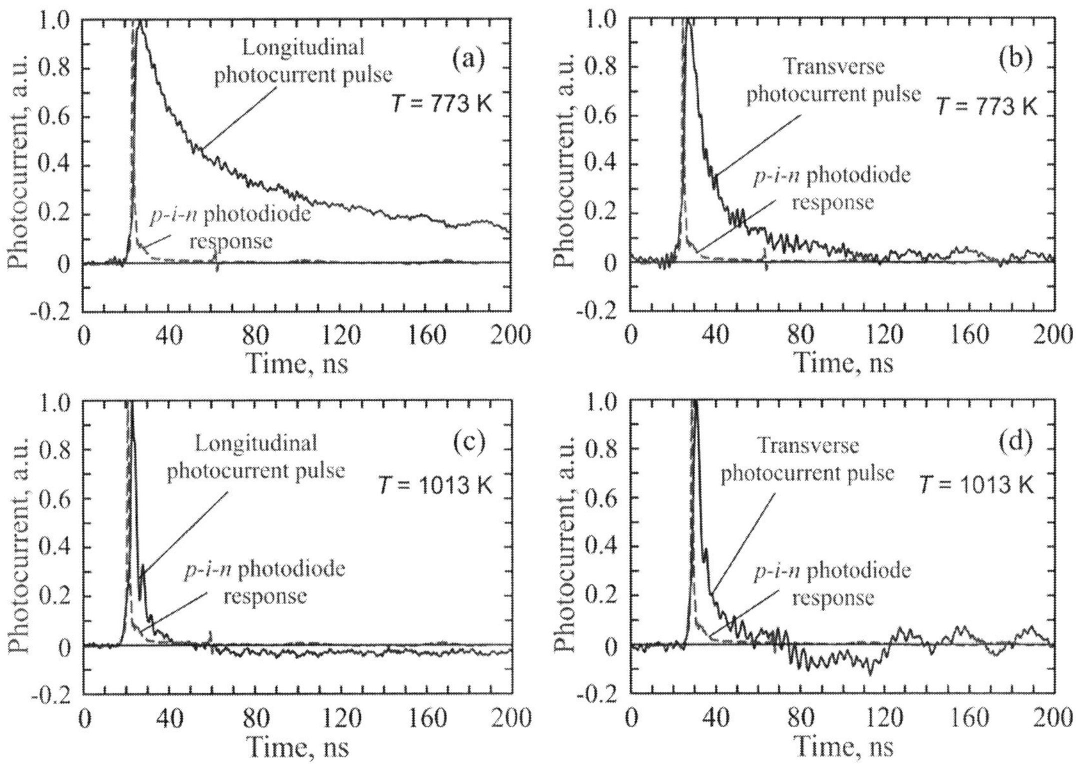

Figure 2(a-d). Oscillograms of **(a, c)** longitudinal and **(b, d)** transverse photocurrent pulses induced by the femtosecond laser pulses in the Ag/Pd nanocomposite films produced at **(a, b)** $T = 773$ and **(c, d)** 1013 K, including *p-i-n* photodiode response oscillograms.

It can be seen from table 1 that under 120 fs pulse excitation time parameters (pulse duration and fall time) of the longitudinal and transverse photocurrent pulses are significantly different for the Ag/Pd films produced at $T = 773$ and 1013 K. Thus, for *p*-polarized radiation, the pulse duration of the longitudinal photocurrent for the films obtained at 1013 K is only 3 ns, which is almost nine times less than in samples obtained at 773 K. It is noteworthy that the fall times of the photocurrent pulses presented in figure 2 differ from each other by a higher factor.

Table 1. Time parameters of the longitudinal and transverse photocurrent pulses in Ag/Pd films synthesized at $T = 773$ and 1013 K.

	Rise time (ns)		Fall time (ns)		Pulse duration (ns)	
	773 K	1013 K	773 K	1013 K	773 K	1013 K
Longitudinal photocurrent	3.6	3.6	295.8	8.8	26.4	3
Transverse photocurrent	3.6	3.2	44.8	14.4	8.8	3.6

According to our previous studies [4], the dependence of the photocurrent in Ag/Pd films on the laser energy is linear, so it is convenient to represent the photocurrent as conversion efficiency: $\eta_x = i_x/E_{in}$ and $\eta_y = i_y/E_{in}$ for the longitudinal and transverse photocurrents, respectively, where $i_x = U_x/r$ and $i_y = U_y/r$. The photocurrent polarization dependencies for the Ag/Pd film synthesized at $T = 1013$ K are presented in figure 3. One can see from figure 3a that the dependence $\eta_x(\Phi)$ is an even function, positive at all polarization azimuth values. It reaches its maximum and minimum values at *s*- and *p*-polarized laser beams, respectively, which is in good agreement with the theory of simultaneous generation of the photocurrent due to the PDE and SPGE described in [6]. The dependence $\eta_y(\Phi)$ (see figure 3b) is an odd function, and has its zero values at *p*- and *s*-polarizations of the incident radiation and the maximums at $\Phi = \pm 45°$. By rotating the quarter-wave plate, we registered a superposition of linear and circular photocurrents [5] generated simultaneously (figure 3c). All presented polarization dependences correspond to those obtained earlier for the films produced at a lower temperature. Therefore, the photocurrent in the films produced at $T = 1013$ K can be considered to be generated due to PDE and SPGE mechanisms. The difference is that the conversion efficiency for those films decreased by factors of 1.4 and 1.8 compared to the films obtained at $T = 773$ K for *s*-polarized and circular polarized incident radiation, respectively.

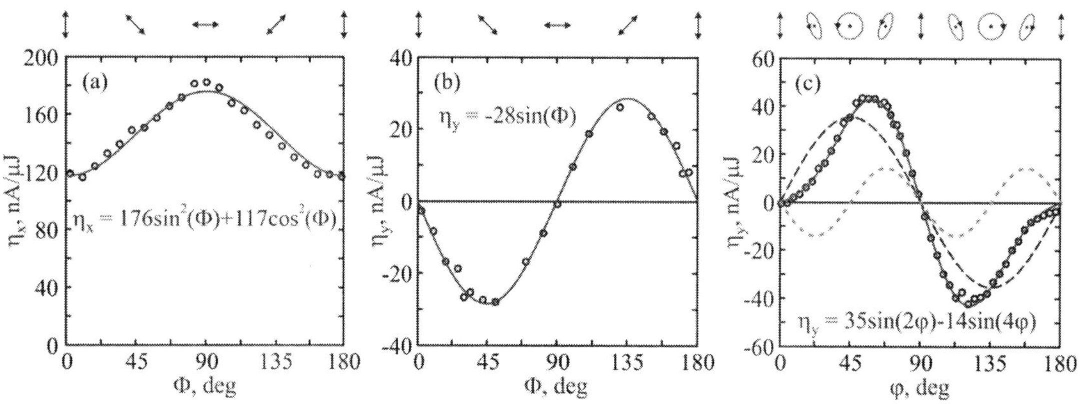

Figure 3(a-c). **(a)** – longitudinal and **(b)** – transverse photocurrent dependences on polarization azimuth Φ, **(c)** – transverse photocurrent dependence on the quarter-wave plate orientation angle φ. The red solid curves are fittings of the experimental data, the blue dashed and green dotted curves denote circular and linear contributions, respectively. Orientations of the electric field vector are shown at the top.

Figure 4 shows the X-ray diffractogram fragments of the Ag/Pd film samples obtained at different firing temperatures. Analysis of the diffractograms shows that the main components of the films in both cases are nanocrystallites of the Ag-Pd solid solution and palladium oxide (PdO). Herewith, the Ag-Pd solid solution is represented by two phases both for low and high firing temperatures. All solid solution phases have different proportions of Ag and Pd and different parameters of the crystal lattice. Films nanocrystallite sizes are in range of 28-98 nm. With an increase in the firing temperature, the Ag-Pd peaks shift to the right, which may indicate an increase in the Pd content in the solid solution. At the same time the ratio of the heights of the peaks of PdO and Ag-Pd decreases with the increase in temperature, that is, the relative content of PdO is also reduced. Thus, it can be assumed that the additional Pd released at a higher firing temperature from the PdO passes into the solid solution. This is in agreement with previously obtained results [17].

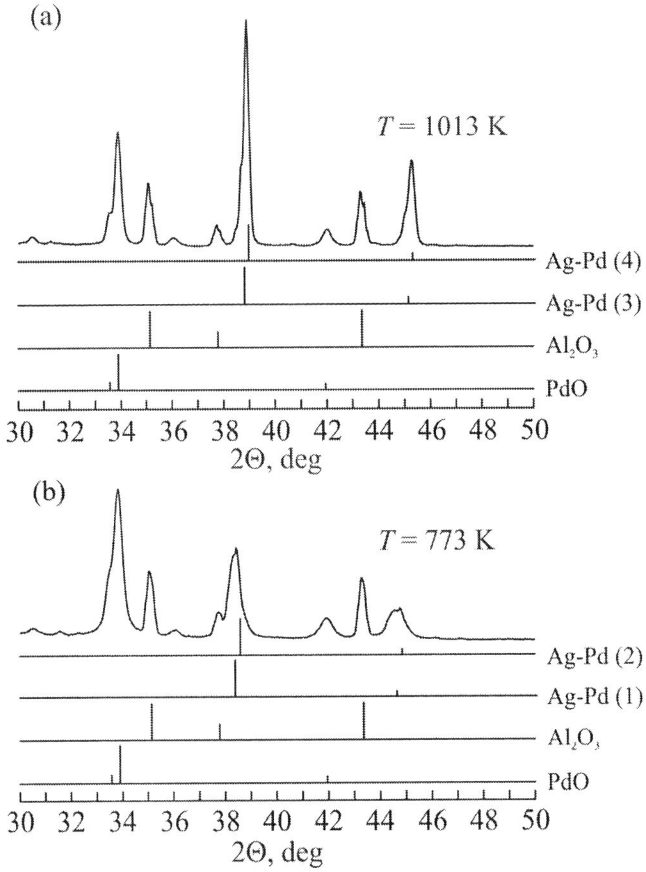

Figure 4(a, b). Fragments of diffractograms of the Ag/Pd nanocomposite films produced at **(a)** $T = 1013$ and **(b)** 773 K, and main X-ray diffraction patterns of detected phases (CuK_{α} radiation)

It follows from the presented experimental results that the change in the phase composition of the films leads to shortening of the photocurrent pulses for the films with higher temperature of the firing. According to our previous research, Ag/Pd films have porous nanocomposite structure [17] (size of pores depends on the firing temperature [18]), formed by a mix of Ag-Pd metal and PdO semiconductor nanocrystallites in an amorphous dielectric glass matrix. Nanocrystallites of Ag-Pd solid solution and PdO act as resistors, and dielectric barriers made of amorphous glass act as capacitors, accumulating charge. Furthermore, PdO inclusions in the matrix of the film form Schottky barriers with the Ag-Pd solid solution. Thus, Ag/Pd nanocomposite can be represented as a complex system containing a set of randomly distributed RC circuits. When photocurrent is generated on the surface [4], part of it flows in short-circuits in the bulk of the sample. Charge is accumulated on the "capacitors" in the course of photocurrent generation. When the photoexcitation ends, this "capacitors" will discharge, with part of the current passing through the measuring oscilloscope, and the other part will charge adjacent RC circuits of the nanocomposite, leading to the subsequent charge and discharge. Therefore, charge and discharge processes, which can persist even after the end of photoexcitation, will occur in the bulk of the film. As the result, the fall time of the photocurrent pulse, registered by the oscilloscope, will be significantly higher than the relaxation time of the photoexcited electrons. Charge and discharge times depend on capacitance and resistance of the elements of RC circuits, therefore structure and composition of the Ag/Pd films influences the photocurrent pulse duration. This mechanism, presumably, is the main reason behind elongation of the PDE and SPGE photocurrent pulses in the Ag/Pd films.

4. Conclusion

The nanocomposite Ag/Pd films were produced at firing temperatures of 773 and 1013 K. The main phase components of the films are the solid solution Ag-Pd and PdO. The films produced at higher temperatures have a lower PdO content relative to the Ag-Pd. The content of Pd in the Ag-Pd solid solution increases with the increase in the firing temperature. The amplitude of the photocurrent excited by femtosecond laser pulses decreases at the same time. The nanosecond polarization-sensitive longitudinal and transverse photocurrent pulses are significantly shorter in the Ag/Pd films produced at 1013 K than in the films produced at 773 K. The research results can be used to develop and create fast response polarization-sensitive sensors based on the Ag/Pd films.

Acknowledgments

This study was supported by the RFBR (project no. 18-32-00224) and by the Academy of Finland (grant nos. 298298 and 323053).

References

[1] Melan E H 1967 *Microelectron. Reliab.* **6** 53–65
[2] Wang S F, Dougherty J P, Huebner W and Pepin J G 1994 *J. Am. Ceram. Soc.* **77** 3051–72
[3] Mikheev G M, Zonov R G and Aleksandrov V A 2010 *Tech. Phys. Lett.* **36** 675–8
[4] Mikheev G M, Saushin A S, Vanyukov V V, Mikheev K G and Svirko Y P 2017 *Nanoscale Res. Lett.* **12** 39
[5] Mikheev G M, Saushin A S and Vanyukov V V 2015 *Quantum Electron.* **45** 635–9
[6] Mikheev G M, Saushin A S, Styapshin V M and Svirko Y P 2018 *Sci. Rep.* **8** 8644
[7] Gibson A F, Kimmitt M F and Walker A C 1970 *Appl. Phys. Lett.* **17** 75–7
[8] Kovalev V M, Miroshnichenko A E and Savenko I G 2018 *Phys. Rev.* B. **98** 165405
[9] Shalygin V A, Moldavskaya M D, Danilov S N, Farbshtein I I and Golub L E 2017 *J. Phys. Conf. Ser.* **864** 012072
[10] Al'perovich V L, Belinicher V I, Novikov V N and Terekhov A S 1981 *Sov. Phys. JETP* **53** 1201–08
[11] Gurevich V L and Laiho R 2000 *Phys. Solid State* **42** 1807–12
[12] Mikheev G M and Styapshin V M 2012 *Instruments Exp. Tech.* **55** 85–89
[13] Akbari M and Ishihara T 2017 *Opt. Express* **25** 2143–52
[14] Mikheev G M, Nasibulin A G, Zonov R G, Kaskela A and Kauppinen E I 2012 *Nano Lett.* **12** 77–83
[15] Alaruri S D 2014 *Optik* **125** 4964–67
[16] Valov P M, Goncharenko K V, Markov Yu V, Pershin V V, Ryvkin S M and Yaroshetskii I D 1977 *Sov. J. Quantum Electron.* **7** 50–4
[17] Mikheev G M, Saushin A S, Goncharov O Yu, Dorofeev G A, Gil'mutdinov F Z and Zonov R G 2014 *Phys. Solid State* **56** 2286–93
[18] Mikheev K G, Saushin A S, Zonov R G, Nasibulin A G and Mikheev G M 2016 *J. Nanophotonics* **10** 012505
[19] Larry J R, Rosenberg R M and Uhler R O 1980 *IEEE Trans. Compon., Hybrids, Manuf. Technol.* **3** 211–25

New optical method for studying a magnetic track from a moving object

S E Logunov[1], V V Davydov[1, 2], A P Glinuchkin[2], I Yu Podkovyrov[2, 3] and V Yu Rud'[2]

[1]Higher School of applied physics and space technologies, Peter the Great Saint-Petersburg Polytechnic University, Saint Petersburg 195251, Russia
[2]All-Russian Research Institute of Phytopathology, Moscow Region 143050, Russia
[3]Volgograd State University of Architecture and Civil Engineering, Volgograd 400074, Russia

Abstract. A new method for detecting a magnetic track from a moving magnetic object is presented. A technique has been developed to study the nature of changes in the magnetic field in a magnetic track using ferromagnetic fluid. The results of experimental studies are presented.

1. Introduction

The study of the structure of magnetic fields and physical phenomena associated with them is one of the urgent problems of fundamental and applied physics [1–8]. Using a magnetic field, carry out various studies and measurements [6-14]. The magnetic field is associated with the formation of a number of physical phenomena [15-19]. The obtained data on the value of the magnetic field induction and its structure help to understand many physical processes and related phenomena [1–4, 7, 20–23]. One of the most difficult in this area for the study of phenomena is the magnetic track, which occurs when a magnetic object moves in different environments (a comet or meteor in space, a ship at sea, etc.). The appearance of a magnetic track is associated with a change in the structure of the magnetic field lines in the zone of movement of the magnetic object.

Of particular interest are studies of the magnetic track at a depth in the marine area. The influence of external factors (water flow, wind, etc.) on the path destruction processes is insignificant. The currently relevant methods of monitoring the situation in various areas of the water area do not allow detecting a moving object at depths of more than 500-550 m [1–4, 20, 21–23]. The only effective method for detecting objects (magnetic) at such depths is the use of magnetometers, especially quantum ones, which are highly sensitive to variations in the magnetic field [1–4, 7, 24–27]. In some cases, the position of a moving object at such depths can be determined using fiber-optic antennas [28, 29]. When using a magnetometer (by immersing it in depth), it is possible to establish the presence of a magnetic track from a moving object in a marine environment. Also, on the basis of the readings of this device, it is possible to investigate the variation of the magnetic field variations in different parts of the magnetic track. Determine the rate of its destruction.

Conducting such experiments in real conditions is a rather complicated process. Therefore, we have developed a new optical method for the study of the magnetic track in the laboratory. In addition, experiments have shown that finding a magnetic track in sea space with a single magnetometer is an extremely difficult task. A reliable detection system is required that is resistant to diving to great depths. Its development requires additional research.

2. The method of studying the magnetic track and the method of its detection

With the movement of a moving object on the surface of the water, a peculiar wave (crest) is formed. The crest is formed due to the turbulent motion of water molecules during the movement of a moving object. The formation of this ridge is described by the Bernuli equation:

$$p + \frac{pv^2}{2} + pgh = const, \quad (1)$$

where ρ - the density of the fluid, υ - the flow velocity, h - the height, p - the pressure, g - the acceleration of gravity.

This wave crest on the surface of a water body is determined by radar methods [23, 30]. If a moving object is at a depth of more than 550 m, the waves caused by the turbulent motion of the fluid, attenuate at the boundary of two water layers. This boundary is located at a depth of 500 - 530 m. A wave crest with a height h when moving a moving object at a depth of 550 m and more does not form on the surface of the sea water area. In addition, it is necessary to take into account that the turbulent movement of water at the indicated depths occurs at a pressure of more than 55 atm. Therefore, the process of forming a magnetic track behind a moving magnetic object that has left a certain zone is very fast. Empty space is filled with water at high speed.

The magnetic track that we study is formed in the process of moving molecules, which in turn affect the change in the structure of the magnetic field lines. The magnetic field lines are closed on the magnetic object and open when it moves. Based on previous studies of the structure of the magnetic field lines in the interpolar space of the magnetic system [2, 4, 7, 20, 21], it was decided to create a magnetic track in an aqueous solution of single-domain magnetite nanoparticles (ferromagnetic liquid) with an average particle size of 14 nm. The concentration of the liquid was 0.025 particles per unit volume. In the role of the surface substance (surfactant) acted oleic kilot. Magnetic fluid is placed in a quartz cell [2, 4, 7, 20, 21]. The nature of the motion of a magnetic object in a ferromagnetic fluid was reproduced by a point magnetic field of the "magnetic pen" with an induction at the end of the order of 1 T. The volume of the disturbing magnetic field is 1 mm3. Laser radiation is applied to the transparent face of the cuvette. The other side of the cuvette is covered with an opaque, poorly reflective material. "Magnetic pen" was moving along this face. The ferromagnetic liquid cuvette was in a laboratory magnetic field. A speckle pattern is recorded in reflected light by a specialized video camera. The operation of the experimental setup for registration of speckle structures was considered in more detail in [2, 4, 20, 27, 31]. Laser radiation with λ = 632.8 nm, various photodetectors and optical systems were used to register speckle pictures [2, 4, 20, 27, 31-36].

In Figure 1 shows the data of the pattern without the influence of the magnetic pen on the ferromagnetic fluid and when it moves along the opaque face of the cuvette. The movement of a moving magnetic object in a magnetic fluid is simulated by moving along an opaque cell wall of a one-sided magnetic needle system with an induction on the tip of the order of 1 T. In the area of a magnetic needle, nanoparticles are placed on the magnetic field lines of its magnetic field (fig. 1.b). Formed agglomerates (speckle structure). It should be noted that at a high speed of movement of the magnetic "needle" along the wall, a wave arises on the surface of the magnetic fluid. This is due to the fact that under the action of a strong alternating magnetic field, particles quickly tend to occupy a certain position in space.

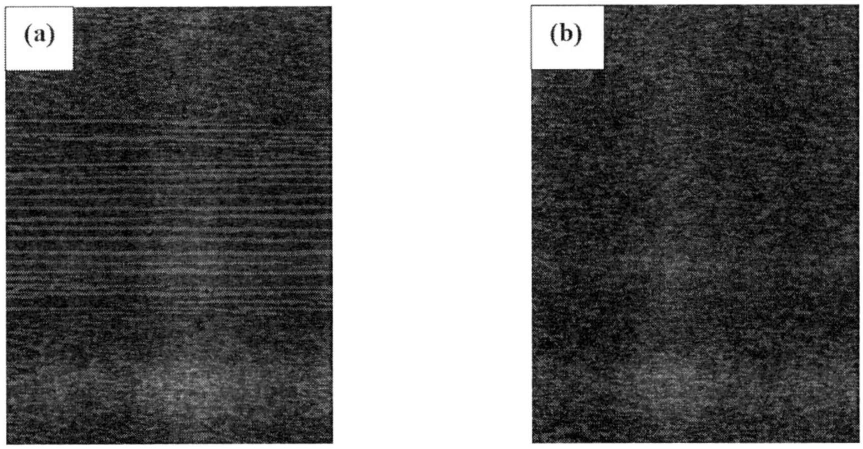

Figure 1 (a, b). The diffraction pattern of the laser radiation in the case of the magnetic liquid placing: (a) in magnetic field Земли; (b) in field movable magnetic needle.

The experimental results obtained, as well as the data in previous studies [2, 4, 20, 27, 31], allowed us to develop a new method for determining the presence of a magnetic track in the marine water area. For the practical implementation of our proposed method, it is proposed to place optical sensors based on ferrofluidic cells [2, 4, 20, 27, 31] developed earlier in a construction in the form of squares with a side of 8-10 m. To confirm the possibilities of determining the presence of a magnetic track in a water area with the use of an optical sensor, we have developed and assembled an experimental setup. Its structural diagram is presented in figure 2.

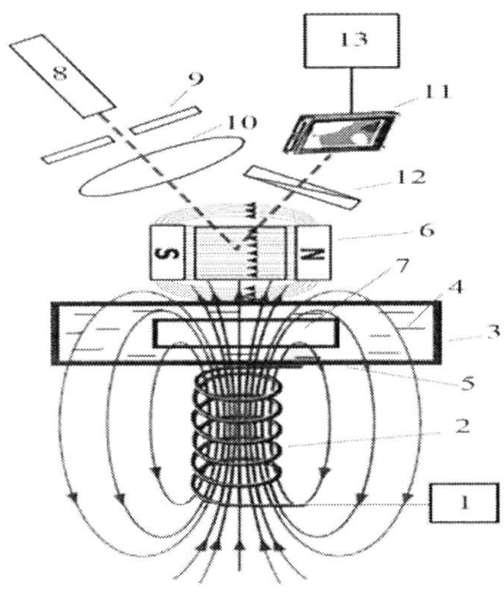

Figure 2. Structural scheme of experimental setup: 1 - power supply; 2 – solenoid; 3 - rectangular vessel; 4 – water; 5 - non-magnetic material; 6 - ferrofluid cell; 7 – the body; 8 – the laser; 9 - diaphragm; 10 – the lens; 11 - photosensitive element; 12 - polarizer; 13 - processing device;

Magnetic fluid (an aqueous solution of magnetite nanoparticles with a concentration of 0.03 - surfactant - oleic acid) is placed on a stand made of non-magnetic material 5 in a ferrofluid cell 6. The cell is placed in a magnetic field of a solenoid 2. The induction of a magnetic field B1 in the cell area of 6.8 mT. Between the solenoid and the ferrofluid cell is a rectangular vessel 3 with a liquid medium 4 (water). The length of the vessel 3 moves the body 7, made of permalloy. To register the reflected laser radiation from speckle structures, the same registration system is used as in figure 1. In figure 3.a shows the recorded image of the camera 11 of the reflected laser radiation from speckle structures formed by nanoparticles on the magnetic field lines of the solenoid.

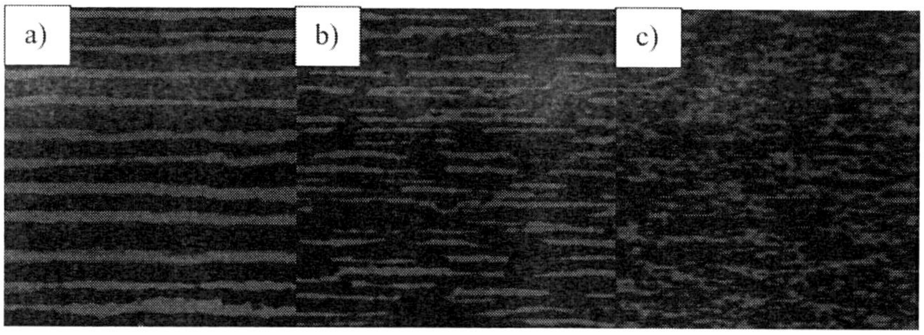

Figure 3 (a, b, c). The diffraction pattern of the laser radiation in the case of the magnetic liquid placing: (a) in a uniform magnetic field; (b) a magnetic field force lines are closed on the body; (c) 2 minutes passed after the closure of all the magnetic field lines on the body.

When the body 7 moves through the vessel, the magnetic field lines of the solenoid close to 7. The agglomerates that were formed in the magnetic field begin to collapse under the action of thermal motion, and the recorded speckle pattern changes. Fig. 3.b shows the image of the reflected radiation from the speckle pattern of nanoparticles 120 seconds after the magnetic field lines of the solenoid close to the body. This is the average transit time of a moving magnetic object through a point in space at a depth in the zone of formation of a magnetic track.

Analysis of the results shows the possibility of registering changes in field variations in a magnetic track with an optical sensor. By processing images from optical sensors spaced at equal distances, it is possible to construct diagrams of variations in the magnetic field at a depth of the water area. Experiments have shown that according to these diagrams, it is possible to establish the presence of a magnetic track at the depth of the water area.

3. Conclusion

The results obtained confirm the possibility of studying the magnetic track with optical sensors. The data obtained using the developed system using optical sensors allows us to establish the presence of a magnetic track in the marine area. Experimental data using the developed system in a specialized pool coincided with measurements performed by quantum magnetometers.

The experiments have shown that the data on the amplitude of magnetic variations, taking into account their nature of changes in the track volume, allow determining the time of its formation and the direction of movement of the magnetic object with an error of no more than 30%. The accuracy of the definition of this information meets the requirements for carrying out activities to control the water area.

Acknowledgments

This work partly supported by Grant of RFBR No 18-29-25071.

References

[1] Vershovskii A K, Dmitriev S P and Pazgalev A S 2013 Technical Physics Journal **83** 879
[2] Aleksandrov E B and Vershovskii A K 2009 Physics-Uspekhi 52 573-601
[3] Logunov S E, Davydov V V, Vysoczky M G and Titova O A 2018 Journal of Physics: Conference Series **1135(1)** 012069
[4] Logunov S E, Davydov V V, Vysoczky M G and Mazing M S 2018 Journal of Physics: Conference Series 1038(1) 012093
[5] Davydov V V and Myazin N S 2017 *Measurement Techniques* **60** 491–496
[6] Petrov A A, Davydov V V and Grebenikova N M 2018 *Lecture Notes in Computer Science (including subseries Lecture Notes in Artificial Intelligence and Lecture Notes in Bioinformatics)* **11118 LNCS** 641-648
[7] Myazin N S, Logunov S E, Davydov V V, Rud' V Yu, Grebenikova N M and Yushkova V V 2017 *Journal of Physics: Conference Series* **929 (1)** 012064
[8] Prokof'ev A V, Pleshakov I V, Bibik E E and Kuz'min Y I 2017 *Technical Physics Letters* **43** 194-196
[9] Davydov R V, Antonov V I and Moroz A V 2018 *Proceedings of the 2018 IEEE International Conference on Electrical Engineering and Photonics, EExPolytech 2018 (Saint-Petersburg)* **8564378** p. 236-239
[10] Myazin N S, Smirnov K J, Davydov V V and Logunov S E 2017 *Journal of Physics: Conference Series* **929(1)** 012080
[11] Davydov V V, Myazin N S, Fadeenko V B and Logunov S E 2018 *Technical Physics Letters* **44(2)** 153-156
[12] Rykin E V, Moroz A V, Smirnov K J, Davydov V V and Yushkova V V 2018 *MATEC Web of Conference* **245** 12002
[13] Davydov V V, Myazin N S and Davydova T I 2017 *Russian Journal of Nondestructive Testing* **53(7)** 520-529
[14] Davydov R V, Antonov V I and Moroz A V 2018 Proceedings of the 2018 *IEEE International Conference on Electrical Engineering and Photonics, EExPolytech 2018 (Saint-Petersburg)* **8564378** p. 236-239
[15] Marusina M Ya, Volgareva A P and Sizikov V S 2015 *Journal of Optical Technology (A Translation of Opticheckii Zhurnal)* **82(10)** 673-687
[16] Vitkovskii O S and Marusina M Y 2016 *Measurement Techniques* **59(3)** 247-251
[17] D'yachenko S V, Kondrashkova I S and Zhernovoi A I 2017 *Technical Physics* **62** 1602–1604
[18] Kashaev R S and Gazizov E G 2010 *Journal of Applied Spectroscopy* **77** 321–328
[19] Davydov V V 1999 *Russian Physics Journal* **42(9)** 822-825
[20] Logunov S. E., Davydov V.V., Vysoczky M.G., Titova O A 2018 Proceedings of the 2018 IEEE International Conference on Electrical Engineering and Photonics, EExPolytech 2018 (Saint-Petersburg) 8564402 p. 282-284
[21] Logunov S E, Koshkin A Yu and V V Davydov V V 2018 *Journal Physics: Conference Series* **1124(1)** 041025
[22] Myazin N S, Neronov Yu I, Dudkin V I, Davydov V V and Yushkova V V 2018 *MATEC Web of Conferences* **245** 11013
[23] Lenets V A, Tarasenko M Yu, Davydov V V, Rodygina N S and Moroz A V 2018 *Journal of Physics: Conference Series* **1038 (1)** 012037
[24] Vershovskii A K and Dmitriev A K 2015 Technical Physics Letters **41** 393-396
[25] Davydov V V, Dudkin V I and Karseev A Yu 2015 *Technical Physics* **60** 456-460
[26] Davydov V V, Myazin N S and Velichko E N 2017 *Technical Physics Letters* **43** 607-610
[27] Logunov S E, Koshkin A Yu, Davydov V V and Petrov A A 2016 *Journal of Physics: Conference Series* **741(1)** 012092

[28] Ushakov N A and Liokumovich L B 2014 *Optical Engineering* **53** 114103
[29] Bisyarin M A, Kotov O I, Yartog A H, Liokumovich L B and Ushakov N A 2017 *Applied Optics* **56** 354-364
[30] Podstrigaev A S, Smolyakov A V, Davydov V V, Myazin N S and Slobodyan M G 2018 *Lecture Notes in Computer Science (including subseries Lecture Notes in Artificial Intelligence and Lecture Notes in Bioinformatics)* **11118 LNCS** 509-515
[31] Logunov S E, Vysocky M G, Davydov V V, Koshkin A Yu and Rud' V Yu 2017 *Journal of Physics: Conference Series* **917(5)** 052058
[32] Davydov V V, Kruzhalov S V, Grebenikova N M and Smirnov K J 2018 *Measurement Techniques* **61** 365-372
[33] Grebenikova N M, Smirnov K J, Artemiev V V, Davydov V V and Kruzhalov S V 2018 *Journal of Physics: Conference Series* **1038 (1)** 012089
[34] Grebenikova N M, Smirnov K J, Davydov V V , Rud V Yu and Artemiev V V 2018 *Journal of Physics: Conference Series* **1135 (1)** 012055
[35] Grebenikova N M, Smirnov K J, Artemiev V V, Davydov V V and Kruzhalov S V 2018 *Journal of Physics: Conference Series* **1038 (1)** 012089
[36] Grebenikova N M, Myazin N S, Rud V Y and Davydov R V 2018 *Proceedings of the 2018 IEEE International Conference on Electrical Engineering and Photonics, EExPolytech 2018* (Saint-Petersburg) **8564409** p. 295-297

Direct GaInP-to-Ge optical interaction in the triple-junction solar cells with thinned intermediate GaAs subcell

S A Levina, E D Filimonov, M A Mintairov and M Z Shvarts

Ioffe Institute, St. Petersburg, 194021, Russia
e-mail: levina@mail.ioffe.ru

Abstract. In the present work processes of direct light coupling in the triple-junction solar cells between top GaInP (wide band gap) and bottom Ge (narrow band gap) subcells are considered, when parameters of intermediate optical medium (comprised of middle GaAs p-n junction layers) are changed. Conditions of luminescent coupling gain and blocking are studied in multijunction structures with selectively tuned built-in photonic structures such as Bragg reflectors. An ability to manage the efficiency of optical interaction between subcells is investigated.

1. Introduction

Effect of luminescent (optical, light) coupling appears inside multijunction GaInP-GaAs-Ge solar cells when narrow band gap subcell absorbs a part of recombination irradiance that comes from adjacent p-n junction with wider band gap. Such a coupling can influence on solar cell photovoltaic characteristics (such as spectral dependencies of external quantum efficiency), and, in some extent, on accuracy of their determination [1-3]. Efficiency of optical coupling, or in other words, the measure of impact of luminescent light of one subcell on the photocurrent of generated by another one, is usually characterized by the dimensionless quantity γ via equation [4]:

$$\gamma = \frac{J_N}{J_W - J_N}, \qquad (1)$$

where J_W, J_N –photocurrents in wide and narrow bad gap subcells, respectively.

In practice the description of optical coupling phenomenon in triple-junction GaInP-GaAs-Ge solar cells are constrained only to adjacent pairs of subcells (GaInP→GaAs and GaAs→Ge) and interaction between top→bottom p-n junctions is neglected: assumed that recombination photons are completely absorbed in optically thick GaAs subcell layers [4, 5]. The present research shows that such a simplification inapplicable for multijunction structures with built-in Bragg reflectors. In that type of solar cells GaAs layers are considerable thinner rather than in the same structures without photonic structures, and could be transparent for luminescent photons forthcoming from top GaInP subcell.

2. Experimental results

In the paper, the optical coupling effect between top GaInP (wide band gap) and bottom Ge (narrow band gap) subcells is investigated as a function of characteristics of the optical medium, is the middle

GaAs subcell layers and Bragg reflector (BR). As objects of research the following triple-junction solar cells were chosen:

1) SC#1 with BR, $d = 1.7$ μm;
2) SC#2 with BR, $d = 2.2$ μm;
3) SC#3 without BR, $d = 4$ μm.

Parameter d corresponds to the thickness of the middle GaAs p-n junction. All samples have a high radiative recombination yield. BR of the SC#1 and SC#2 have the same composition and number of photonic layers, which were built-in between the middle and bottom subcells. BR has photonic band gap, that coincides with the wavelength of the GaAs luminescent radiation. Thus, BR can reflect recombination photons and block optical coupling between middle and bottom junctions in the wide range of temperatures 80-300K [7]. In the solar cell without BR SC#3 the leakage of luminescent light from GaAs to Ge subcell is sufficient, whereas optical coupling in GaInP-Ge pair is extremely weak due to thick GaAs middle layers.

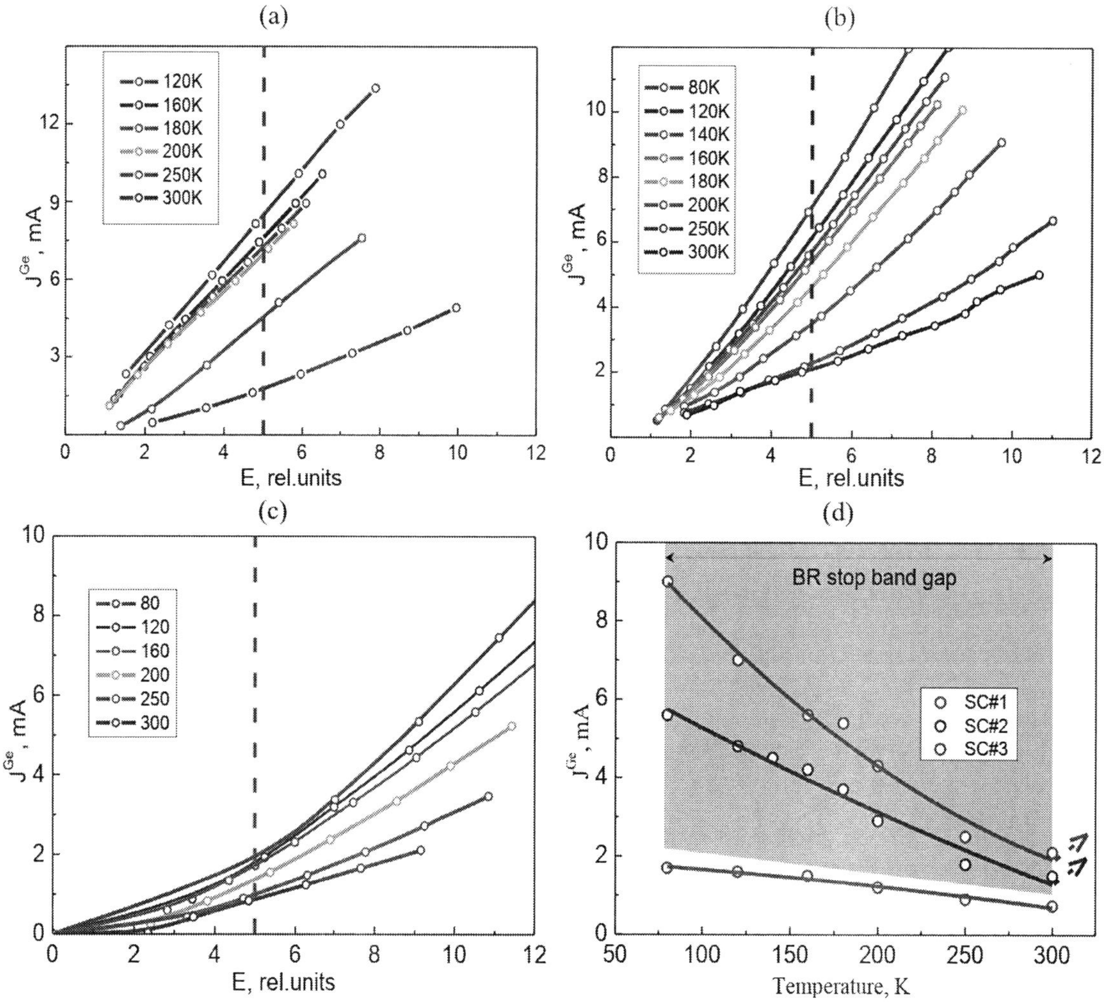

Figure 1. Temperature dependencies of induced luminescent photocurrents of the Ge subcell: SC#1 (a), SC#2 (b), SC#3 (c), under external light illumination of blue laser (450nm) with the intensity E. Vertical dashed lines (red (a), blue (b), green (c)) correspond to the equal conditions of illumination, for which the comparison of photocurrents was made on (d). Grey region illustrates the photonic band gap of Bragg reflectors with optical coupling blocking conditions for the samples SC#1 and SC#2. Red and blue arrows after 300K show further direction of the dependencies.

A special feature of these studies is the use of external illumination from a single source of narrow-band radiation (blue laser 450 nm), which is in the range of the spectral sensitivity of the top wide-band gap subcell. Therefore, inside the semiconductor structure the cascade of luminescent coupling mechanism was able to initiate [8]. As a result, tree coupled pairs of subcells (GaInP→GaAs, GaAs→Ge, and GaInP→Ge) were formed.

Experimental studies at different temperature and illumination regimes for tested samples have demonstrated that in blocking mode for optical interaction between GaAs→Ge (SC#1 Fig.1a и SC#2 Fig.1b), the Ge subcell photocurrent is characterized by linear dependency from external light bias. The thinner the thickness d is, the stronger $J^{Ge}(E)$ inclines from linearity. For the sample SC#3 (Fig.1c) Ge subcell photocurrent is described by sub-linear law and is mostly determined by luminescent light coming from adjacent GaAs due to the much higher efficiency of optical coupling γ between middle-bottom junctions, rather than in top-bottom pair.

In this case, the magnitude of the induced current J^{Ge}, under equal conditions of the external illumination, is minimal for the sample SC#3 with coupled GaAs-Ge subcells (green line Fig. 1d). This behavior can be explained by the energy that transferred to the bottom junction by luminescent photons from GaAs (1.4 eV) and GaInP (1.8 eV). Obviously, the contribution to the photocurrent from the latter will be much higher, even with a small number of them. With an increase in the thickness of the middle subcell, the intensity of such radiation is decreased exponentially, and already at a depth of 4 μm it turns out to be so low that it cannot have a significant effect on the contribution to the photocurrent of the bottom subcell (Fig. 2a).

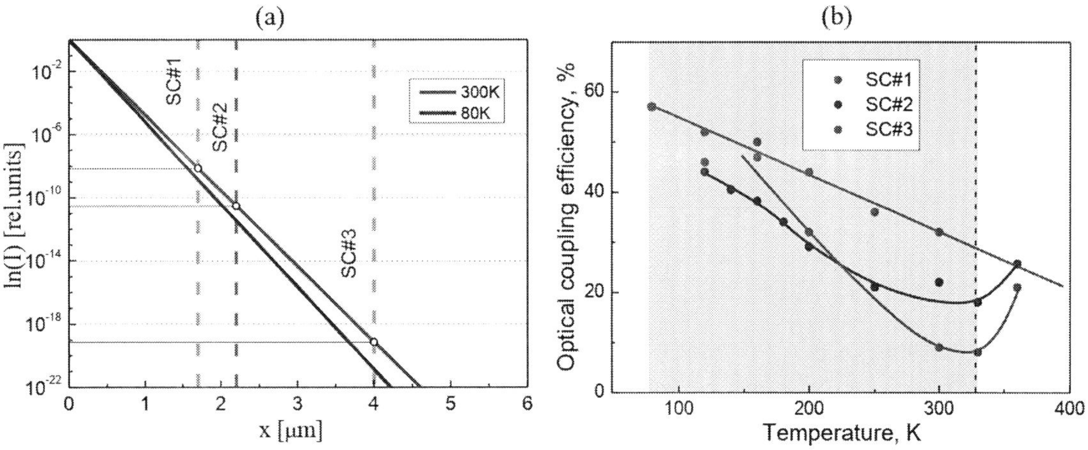

Figure 2. (a) – GaInP luminescent intensity decay depending on penetration depth in semi-logarithmical scale. Vertical lines correspond to the thicknesses of GaAs subcell layers: green SC#1, magenta SC#2, orange SC#3. Blue and red dependencies are determined for temperatures 80K and 300K. (b) - Temperature dependency of light coupling efficiency for multijunction solar cells with (red curve for SC#1 and blue one for SC#2) and without (green line for SC#3) built-in Bragg reflector. Grey region corresponds to photonic band gap of Bragg reflector for SC#1 and SC#2.

An increase in temperature leads to the GaInP luminescence wavelength shift to a longer wavelength region of the spectrum (from 640 to 660 nm at the heating by 220K), and to the change of the absorption coefficient of the GaAs p–n junction [9]. The absorption coefficient during heating increases by approximately an order of magnitude, which reflects in the increase of the photocurrent of the Ge subcell. However, instead of the expected growth, the drop of the J^{Ge} is observed for all studied samples (Fig. 1d). The heating of the structure provokes a replacement of the dominant recombination mechanism from radiative to nonradiative, which is assumed causes a decrease in the photocurrent of the narrow-gap Ge subcell.

During the investigation it was found that for solar cells with BR for each new temperature mode exists some superposition of optical interaction between three junctions (determined by material band gap, optical thickness of semiconductor layers, radiative and absorption properties) and photonic structure characteristics. Adjusting sample temperature, it is possible to set different conditions of luminescent interaction including the ones when optical coupling in GaAs-Ge pair is fully blocked. On the temperature dependency of the light coupling efficiency the blocking state is characterized by the minimum of the function (Fig. 2b).

3. Conclusion

In the present work a processes of direct light coupling between top GaInP (wide band gap) and bottom Ge (narrow band gap) subcells are considered when parameters of intermediate optical medium (comprised of middle GaAs p-n junction layers) are changed. The values of light coupling efficiency for pairs of subcells incorporated in multijunction structures of different composition and architecture are analyzed and compared. To tune the reflectivity of BR, wavelength of luminescent light and coupling efficiency level the temperature of the devise has been varied gradually.

References

[1] H Yoon, R King, G Kinsey, S Kurtz, D Krut 2003 Radiative coupling effects in GaInP/GaAs/Ge multijunction solar cells Proc. of 3rd World Conference on Photovoltaic Energy Conversion v 1 pp 745–748

[2] D Friedman, J Geisz, M Steiner 2014 Effect of Luminescent Coupling on the Optimal Design of Multijunction Solar Cells IEEE Journal of photovoltaics v 4 n 3 pp 986-990

[3] S Lim, J Li, E Steenbergen, Y Zhang 2011 Luminescence coupling effects on multi-junction solar cell external quantum efficiency measurements Progress in Photovoltaics: Research and Applications v 10 p 102

[4] M Shvarts, V Emelyanov, V Evstropov, M Mintairov, E Filimonov, S Kozhukhovskaia 2016 Overcoming of Luminescent Coupling Effect in Experimental Search for True Quantum Efficiency Value in MJ SCs AIP Conf. Proc. v 1766 p 060005

[5] M Shvarts, V Emelyanov, V Evstropov, M Mintairov, N Timoshina 2015 Temperature influence on luminescent coupling efficiency in concentrator MJ SCs AIP Conference Proceedings v 1679 p 120003

[6] V Emelyanov, N Kalyuzhniy, S Mintairov, M Shvarts, V Lantratov 2010 Multijunction GaInP/GaInAs/Ge solar cells with Bragg reflectors Semiconductors v 44(12) pp 1600-1605

[7] S A Levina Emelyanov, A V Malevskaya, E D Filimonov, M Z Shvarts 2018 J. Phys.: Conf. Ser. 1135 012075

[8] Shvarts M, Emelyanov V, Evstropov V, Mintairov M, Filimonov E and Kozhukhovskaia S 2016 Overcoming of Luminescent Coupling Effect in Experimental Search for True Quantum Efficiency Value in MJ SCs AIP Conf. Proc. 1766 060005

[9] http://www.ioffe.ru/SVA/NSM/rintroduction.html

Organic light-emitting diodes with colloidal quantum dots in the active layer

A E Degterev, I I Mikhailov, I A Lamkin, S A Tarasov
Department of Photonics, Electrotechnical University "LETI", Saint Petersburg 197376, Russia

Abstract. In this research we have created light-emitting structures that contain colloidal quantum dots of cadmium selenide in the active layer. To do so, we have used methods of vacuum thermal deposition and spin coating for the formation of organic layers. We have compared photo- and electroluminescence spectra of the structures, and have revealed necessary conditions for the creation of high quality layers.

1. Introduction
Organic Light Emitting Diodes (OLED) are semiconductor devices created from a number of thin organic films. The structure of OLED consists of a transparent substrate, a transparent anode, transport layers with hole and electronic conductivity, an active layer, and a cathode [1].
Due to the high quantum yield of fluorescence, it is possible to increase the efficiency of the structure by adding colloidal quantum dots (CQD) to the composition of the active layer, as well as to control the emitting range.

2. Research
We have created organic light-emitting structures FTO/PEDOT:PSS/TPD/Alq3/Al as it is shown on figure 1(a). To do so, we used glass substrate with an FTO layer on it, deposited transport layers of PEDOT:PSS, TPD and Alq3 on top of it, and finished with aluminium contacts.

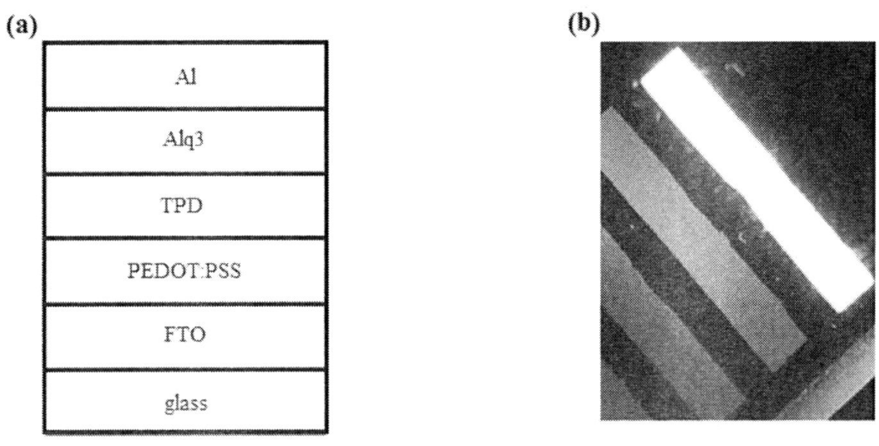

Figure 1(a, b). **(a)** Structure of created OLED; **(b)** Photo of working OLED.

Figure 1(b) demonstrates obtained organic light emitting diode when we apply voltage of 7 V and the current is 3 mA.

2.1. Change of substrate material

To increase efficiency we have created two types of structures with same composition but with different substrates. In the first structure we have deposited PEDOT:PSS, TPD, CQD, Alq3 and aluminium contacts on a glass substrate with a layer of indium tin oxide doped with fluorine (FTO). In the second structure we have made the films of these materials on a flexible substrate with a layer of indium tin oxide (ITO).

Relation between emitting efficiency and applied current is shown on figure 2. It has demonstrated properties of created structures. Taking into account that the work function of ITO is more than that of FTO [2], difference in emitting efficiency could be explained by disruption of uniformity of the layers created on flexible substrate.

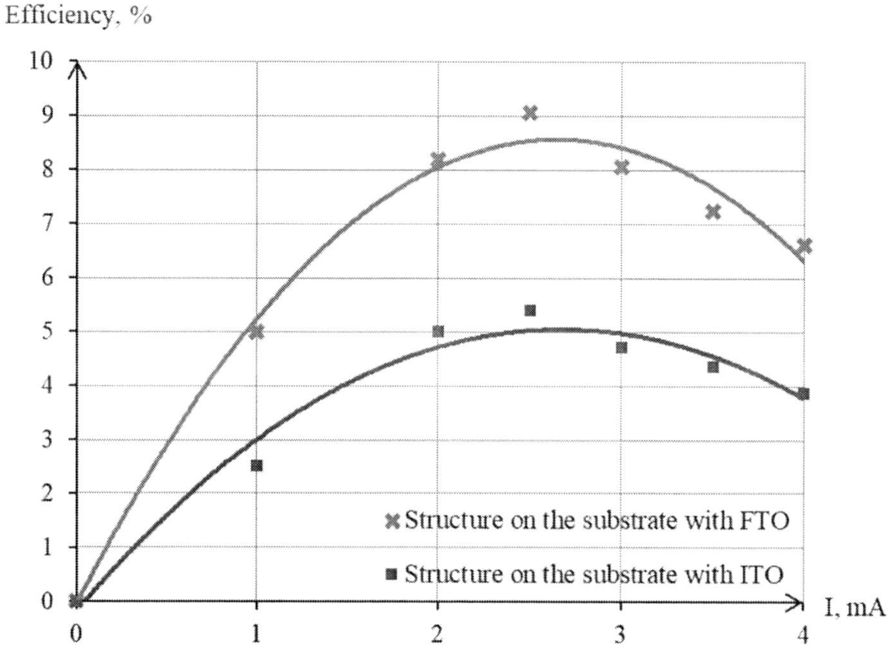

Figure 2. Relation between structure efficiency and applied current

2.2. The influence of CQD on the spectral characteristic

The most important goals of our research was to demonstrate the change in OLED electroluminescence spectrum after the introduction of colloidal quantum dots (CQD) into the active layer. To do this we have created a series of light-emitting structures of FTO/PEDOT:PSS/TPD/Alq3/Al and FTO/PEDOT:PSS/TPD/TPD+CQD/Alq3/Al that contain colloidal quantum dots of CdSe in ZnS shell.

Figure 3 shows spectral characteristics of obtained structures at a voltage of 25 V. The OLED with the CQD has a second peak due to electroluminescence of the CQD. We have obtained peaks at a wavelength of 650 nm and 680 nm by using relevant colloidal quantum dots. The intensity of this maximums depends on applied voltage and on concentration of CdSe/ZnS nanoparticles. In Alq3 electroluminescence peak shifts, which is associated with effects of internal absorption in the array of CQD [3].

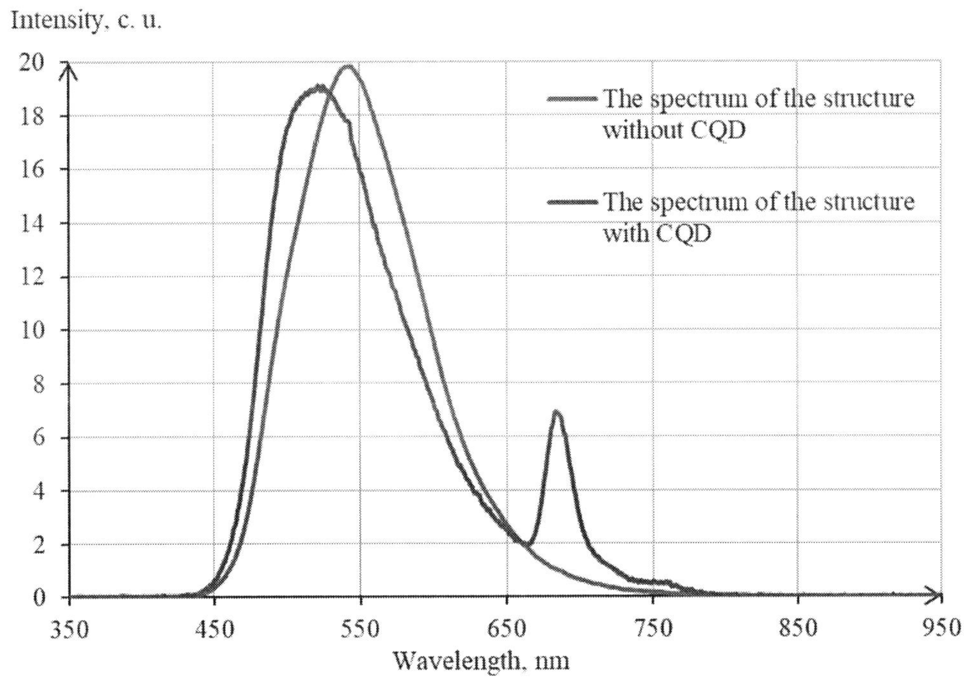

Figure 3. Electroluminescence spectra of structures

3. Conclusion

In this work we have created organic light-emitting structures FTO/PEDOT:PSS/TPD/Alq3/Al and FTO/PEDOT:PSS/TPD/TPD+CQD/Alq3/Al. We have demonstrated that the inclusion of CQDs in the composition of the active OLED layer leads to a significant change in the spectral characteristics.

We have also showed that a structure created on a substrate with the FTO as anodic contact is more efficient than a similar one on a substrate with the ITO. Emitting efficiency of organic LEDs with former substrate is almost double of that in the latter.

References

[1] Stepanov E M, Mikhailov I I, Tarasov S A, Solomonov A V 2016 Proceedings of the 2016 IEEE North West Russia Section Young Researchers in Electrical and Electronic Engineering Conference ElConRusNW p 82

[2] Pavlova M D, Lamkin I A, Tarasov S A and Solomonov A V 2018 Organic photodetective device based on metal phthalocyanine Journal of Physics: Conference Series 1038 012104

[3] Mikhailov I I, Tarasov S A, Lamkin I A, Tadtaev P O, Kozlovich L I, Solomonov A V, Stepanov E M 2016 Optical parameters of ITO/TPD/Alq3/Al luminescent structures, containing arrays of CdSe/ZnS colloidal quantum dots Journal of Physics: Conference Series 741 012103

Capacitance characterization of GaP/Si superlattice grown by time-modulated PECVD

A I Baranov[1], I A Morozov[1], A V Uvarov[1], A S Gudovskikh[1,2]
[1]St Petersburg Academic University of RAS, 194021 St Petersburg, Russia.
[2]St Petersburg Electrotechnical University "LETI", 197376 St Petersburg, Russia.

Abstract. The work is devoted to study of GaP/Si superlattice grown on n-type GaP wafer by time-modulated plasma-enhanced chemical vapor deposition. Deep-level transient spectroscopy showed appearance of defects in μc-GaP layer. Addition of silicon layers 0.5 nm thick leads to decreasing of deep defect concentration. However, further increasing of silicon content in superlattice leads to appearance of new deep defects in structure.

1. Introduction

According to theoretical estimation the optimal values of bandgap energy are 1.1 eV for bottom subcell and 1.7 eV for top subcell in double-junction solar cell [1]. However, even epitaxial methods have not allowed to grow lattice-matched to silicon semiconductor alloys with Eg=1.7 eV. For example, dilute nitrides of GaPAsN have very poor photoelectrical properties [2,3]. Recently, the time-modulated plasma-enhanced chemical vapor deposition (or atomic-layer deposition) was successfully applied to the growth of thin layers of GaP on silicon wafers [4]. It is based on alternation of P and Ga sources at low temperature in PECVD chamber, and an additional flow of silane allows to obtain n-type doping in GaP to form anisotype heterojunction n-GaP/p-Si as perspective bottom subcell in multi-junction solar cell. Furthermore, capacitance properties of GaP/n-Si and GaP/p-Si heterojunctions were studied by different techniques (capacitance-voltage characteristic, admittance spectroscopy and deep-level transient spectroscopy) [5,6], where authors concluded that increase of RF hydrogen plasma lead to defect formation near to interface in both types of heterojunction, but bulk properties of silicon wafers are not deteriorated.

Here, the PE-ALD technology will be applied for growth of periodic nanoheterostructures GaP/Si as superlattice on GaP wafers. According to theoretical prediction, such superlattice can have direct bandgap structure with E_g=1.7 eV [7]. Therefore, it can be used in top subcell for future application in suggested double-junction solar cells. In paper, electrical and capacitance properties of such structures with Schottky barriers will be presented.

2. Experiments and methods

Three different samples were grown at on n-GaP ($n=5\times10^{17}$ cm^{-3}) wafers using an Oxford PlasmaLab System 100 PECVD (13.56 MHz) setup. The sample #1 is a layer of microcrystalline GaP (μc-GaP) 60 nm thick grown by PE-ALD mode. It is used as reference sample to study influence of silicon incorporation on properties of structures. The sample #2 is a 20 period superlattice of nanoheterostructure GaP(3nm)/Si(0.5nm), where GaP layer was grown by PE-ALD mode, and silicon layers were grown by classical continuous PECVD mode. Then, thicknesses of silicon layers were increased from 0.5 nm to 1 nm in the sample #3. More details of growth processes can be found elsewhere [8]. Gold contact dots with diameter of 0.5 mm were evaporated on the top of the structures to form Schottky barrier diodes to μc-GaP and GaP/Si superlattices in BOC Edwards Auto500 setup. An ohmic contact was formed by evaporation of Au/Ge metallic alloys on the bottom side of GaP wafer. Standard I–V curves were measured using a Keithley 2400 source-meter. Capacitance measurements were performed using a precision LCR-meter Agilent E4980A. Measurements of classical capacitance deep-level transient spectroscopy (DLTS) [9] were performed using an automated installation based on a Boonton-7200B capacitance bridge in the temperature range of 80-360 K.

3. Results

Current-voltage characteristics for samples are presented in Figure 1a. According to curves sample #1 and #2 have classical rectifying behavior at forward bias voltage and low current at reverse bias voltage. Therefore, it means good quality of Schottky barrier formed by gold to GaP and superlattice GaP(3nm)/Si(0.5nm). On the other hand, behavior of sample #3 is different: its I-V curve is almost symmetric. It can be explained by poor quality of interface gold/semiconductor, and it should be more defective. Nevertheless, value of current at reverse voltage is low so these structure can be studied by capacitance methods.

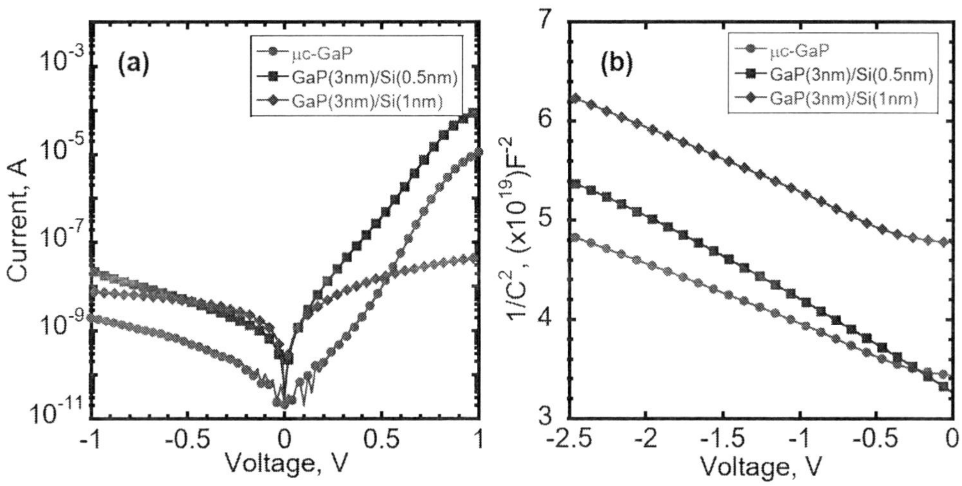

Figure 1 (a, b). Current-voltage characteristics **(a)** and Mott-Schottky plots **(b)** for samples.

According to obtained capacitance-voltage characteristics all grown layers should be fully depleted even at zero applied voltage. Increase of reverse bias voltage leads to capacitance decrease so space charge region exists in structures, but it expands only in n-GaP wafer. It is confirmed by Mott-Schottky plots (Figure 1b) where $1/C^2(V)$ is straight line, and estimated concentration of free charge carriers corresponds to donor concentration in GaP wafer. In this case, DLTS measurements were performed at following conditions: V_{init}=0 V, V_{pulse}=+2 V, t_{pulse}=50 ms. So forward bias voltage is applied during the filling pulse: it allows to fill possible defect levels which are located closer to interfaces metal/semiconductor and in grown layers. DLTS spectra for different rate windows are presented in Figure 2. According to them there are series of peaks related to responses from defects. In sample #1 with μc-GaP two defects were detected (Figure 2a), and first one with activation energy E_a=0.055±0.005 eV and capture cross-section σ_T=(1–10)×10^{-21}cm^{-2} should be impurity level. On the other hand, the second defect is deep-level with following parameters: E_a=0.60 eV, σ_T=8.5×10^{-15}cm^{-2} and concentration N_T=3.8×10^{16}cm^{-3}. In this case, it can be considered as effective non-radiative recombination centre responsible for low lifetime of charge carriers in microcrystalline GaP. Then, response from low-energy defect with E_a=0.035±0.005 eV, σ_T=(4–40)×10^{-23}cm^{-2} and N_T=(5–10)×10^{16}cm^{-3} was detected in superlattice GaP(3nm)/Si(0.5nm), but such small values are not allow consider it as an effective non-radiative recombination centre. Also, decreasing of amplitude peaks at 260–360 K means partially inhibition deep defect formation. Further increasing of silicon layer from 0.5 nm to 1 nm in sample #3 leads to the absence of responses from defect with 0.03–0.05 eV, but another response appears from defect with following parameters: E_a=0.44 eV, σ_T=1.2×10^{-15}cm^{-2} and N_T=6.3×10^{16}cm^{-3} (Figure 2c) so it can reduce lifetime of charge carriers in superlattice GaP(3nm)/Si(1nm). Therefore, addition content of silicon in microcrystalline GaP allows to reduce concentration of deep levels in superlattice GaP(3nm)/Si(0.5nm). However further increase is not effective due to appearance of new defects in superlattice GaP(3nm)/Si(1nm) as well as the deterioration of the structural properties observed before [7].

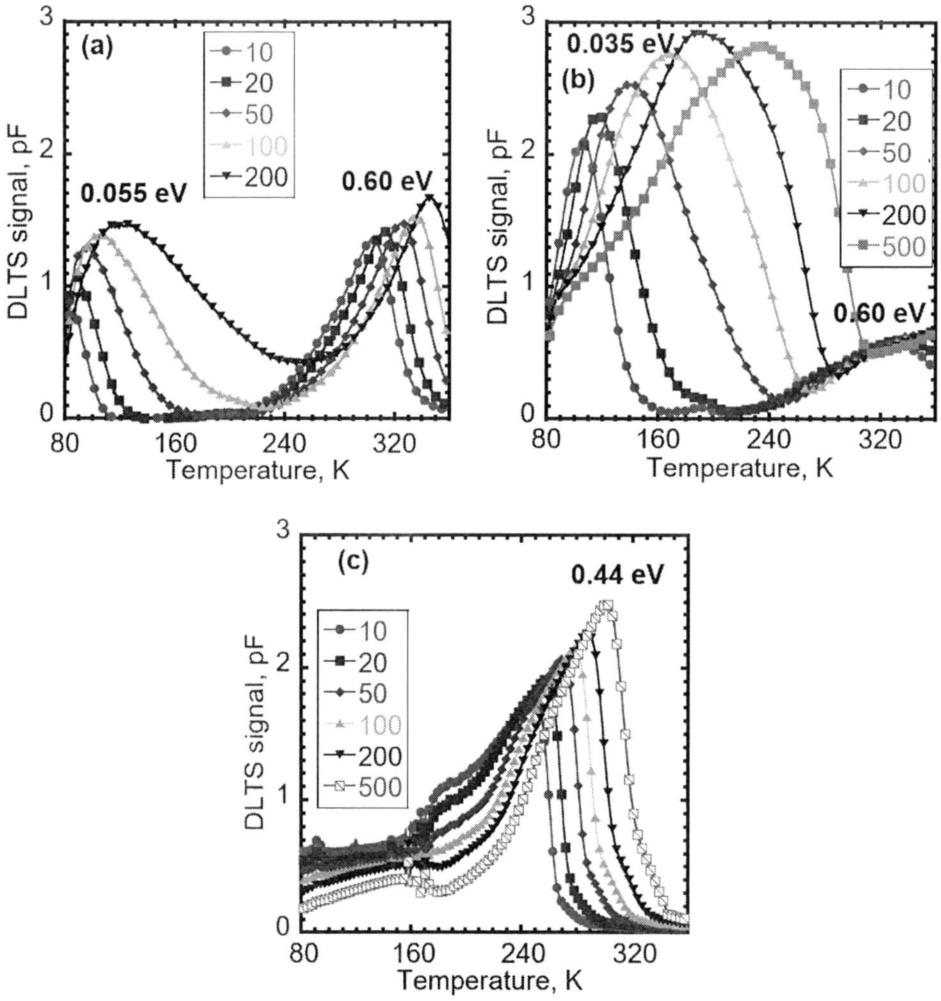

Figure 2 (a, b, c). DLTS spectra for different rate windows (in s^{-1}) for structures with μc-GaP **(a)**, GaP(3nm)/Si(0.5nm) **(b)** and GaP(3nm)/Si(1nm) **(c)**.

Acknowledgments
This work was supported by grant from the President of Russian Federation № СП-1207.2019.1 (measurements of electrical and capacitance properties) and by the Russian Scientific Foundation under Grant No. 17-19-01482 (defect characterization by deep-level transient spectroscopy).

References

[1] Kurtz S R, Faine P and Olson J M 1990 *J. Appl. Phys.* **68** 1890–5
[2] Baranov A I, Gudovskikh A S, Nikitina E V and Egorov A Y 2013 *Tech. Phys. Lett.* **39** 1117–20
[3] Geisz J F, Friedman D J and Kurtz S 2002 *Conf. Rec. 29th IEEE Photovolt. Spec. Conf. 2002.* 864–7
[4] Gudovskikh A S, Morozov I A, Uvarov A V, Kudryashov D A, Nikitina E V, Bukatin A S, Nevedomskiy V N and Kleider J 2018 *J. Vac. Sci. Technol. A Vacuum, Surfaces, Film.* **36** 021302
[5] Gudovskikh A S, Uvarov A V., Morozov I A, Baranov A I, Kudryashov D A, Zelentsov K S, Jaffré A, Le Gall S, Darga A, Brezard- Oudot A and Kleider J-P 2019 *Phys. status solidi* **216** 1800617
[6] Baranov A I, Gudovskikh A S, Darga A, Le Gall S and Kleider J-P 2017 *Journal of Physics: Conference Series* **917** 052027
[7] Gudovskikh A S, Uvarov A V, Morozov I A, Baranov A I, Kudryashov D A and Zelentsov K S 2019 *Mater. Today Proc.*
[8] Gudovskikh A S, Uvarov A V., Morozov I A, Baranov A I, Kudryashov D A, Zelentsov K S, Bukatin A S and Kotlyar K P 2018 *J. Vac. Sci. Technol. A Vacuum, Surfaces, Film.* **36** 02D408
[9] Lang D V. 1974 *J. Appl. Phys.* **45** 3023–32

Current induced mode competition in microdisk lasers

I K Boikov[1], A V Savelyev[2], S Breuer[3], V V Korenev[2], K E Kudryavcev[4], N V Kryzhanovskaya[2], M V Maximov[2] and A E Zhukov[2]

[1] Department of Physics and Semiconductor Techonology, St. Petersburg Academic University of the Russian Academy of Sciences, St. Petersburg 194021, Russia
[2] Nanophotonics Lab, St. Petersburg Academic University of the Russian Academy of Sciences, St. Petersburg 194021, Russia
[3] Institute of Applied Physics, Technical University of Darmstadt, Darmstadt, 64289, Germany
[4] Institute for Physics of Microstructures of RAS, Nizhny Novgorod, 603950, Russia

E-mail: ik.boikov@gmail.com

Abstract. A phenomenon of mode competition in microdisk lasers under various injection currents was studied. The increase of the injection current causes rise of the internal temperature. The modeling has shown that this increase leads to the shift of the gain spectrum. That may be enough to cause drastic redistribution of emitted power between microdisk modes. This effect has been studied in the experiment and by the numerical simulation.

Keywords: microdisk lasers, thermal mode shift, mode competition.

1. Introduction

The light has several advantages over electricity in data transmission thus today long-range communications are using light exclusively. Short-range (such as chip-to-chip) transmission is still being worked on, as it requires compact and energy efficient photonic devices with high response rate and in-plane signal extraction. Semiconductor lasers with microdisk resonators are a promising solution to this problem and its high Q-factor together with small effective mode volume allow for low threshold current [1].

It is well known that as temperature increased, energies of modes decreased nearly linearly. For example, lasing mode spectral shift under ambient temperature variation was observed [2]. However, our experiment has demonstrated similar behavior with constant external temperature, but variable injection current. Moreover, the emitting power transfer between neighboring laser modes was observed.

Many practical applications such as wavelength-division multiplexing (WDM) [3] require power stability of radiated light into each mode. Thus, observed mode switching is undesirable and so we aim to investigate the underlying physics of that phenomenon.

The general idea of the paper is schematically illustrated in the Figure 1. We discuss two modes that are close to each other in spectral domain. Due to thermal shift of the material gain spectrum, the maximum modal gain that can be achieved for each mode is also varying. In fact, at different temperatures of the active area of the laser the material gain spectrum can be favourable for lasing at different modes. Since the modes 'compete' for the same carriers in the active area the favorable conditions for one mode may results that another mode would cease completely. When we consider laser heating with the increase of the injection current the close picture of mode competition can be obtained. Below we provide theoretical modeling relevant for this scheme and compare numerical results with the experiment.

2. Theory

2.1. General rate equations

In our experiment laser's active area contains multiple QWDs layers (see [4] and references wherein). Due to inhomogeneous broadening QDs have different sizes (and thus, optical transition energies). Transition energy distribution can be approximated by a Gauss function [5]:

$$\rho(E) = \frac{1}{\sigma\sqrt{2\pi}} \exp\left(-\frac{(E-E_0)^2}{2\sigma^2}\right). \tag{1}$$

Here E_0 is an energy of the transition with the highest probability. Each QD with transition energy E creates a Lorentzian gain profile around E [5]:

$$F(\delta E) = \frac{1}{1+(\delta E/\Gamma)^2}, \tag{2}$$

where Γ is a homogeneous broadening parameter.

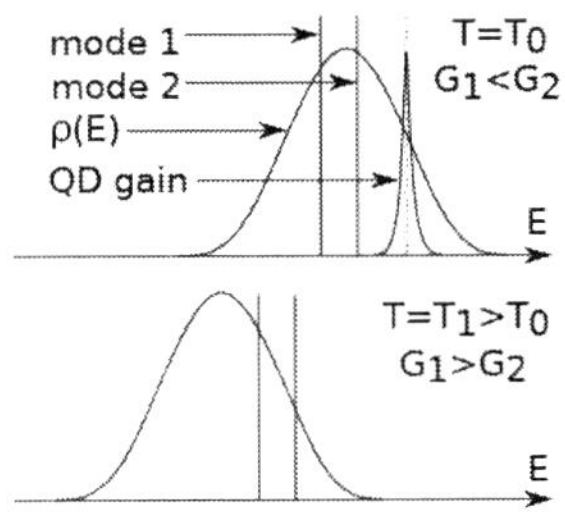

Figure 1. Gain spectra and laser modes shift under temperature variation. $G_{1,2}$ are modal gains of modes 1 and 2.

The theory of charge carrier statistics in QWD structures are not well developed up to now and detailed discussion of QWD filling factor $f(E)$ could be complicated. Hence, we will assume that all QDs are filled equally and that concentrations of electrons and holes are equal. This way $f(E) = \langle f \rangle$ and the carrier dynamics boils down to the one parameter - captured electron concentration n:

$$n = n_{QD}\langle f \rangle / d, \tag{3}$$

where n_{QD} is surface density of QDs over all layers and d is effective active area thickness. In this case phenomenological approach [6] to laser dynamics can be used:

$$\frac{dn}{dt} = \frac{\eta J}{ed} - \frac{n}{\tau} - V_A^{-1}(G_1 S_1 + G_2 S_2). \tag{4}$$

where η — injection efficiency, that also considers effective surface of the microdisk mode, J — injection current density, d — active area effective thickness, V_A — the volume of active area, τ — electron lifetime, G_k — gain of mode k (measured in s^{-1}) and S_k is the number of photons in mode k.

If we denote rate of spontaneous emission into the mode k as $R_{sp,k}$ then rate equations for modes photon number can be written as [5]:

$$\begin{cases} \frac{dS_k}{dt} = \Big(G_k - \alpha(E_k)\Big)S_k + R_{sp,k} \\ G_k = G_{max}\frac{\sigma}{\Gamma}\sqrt{\frac{2}{\pi}}\int dE\rho(E)F(E-E_k)\Big(2f(E)-1\Big) \\ R_{sp,k} = G_{max}\frac{\sigma}{\Gamma}\sqrt{\frac{2}{\pi}}\int dE\rho(E)F(E-E_k)f(E). \end{cases} \tag{5}$$

where α — internal and extraction losses, and G_{max} — gain spectra peak, which is usually known from experimental data. We have no means of evaluating α since numerous technological factors have an impact on it. Let us assume that losses α_k are equal for all modes and its values will be taken from similar experimental setups [4].

Typically, $\sigma \approx 20$ meV and $\Gamma \approx 5$ meV at $T = 300$ K in the structures studied. Using that values, and following $f(E) = \langle f \rangle$ approximation one can simplify integrals in (5) via Taylor expansion by Γ/σ and discarding terms of higher order of magnitude than linear.

$$\int dE\rho(E)F(E-E_k) \approx \pi\Gamma\rho(E_k) = \sqrt{\frac{\pi}{2}\frac{\Gamma}{\sigma}}\exp\left(-\frac{1}{2}\left[\frac{E_k - E_0}{\sigma}\right]^2\right). \tag{6}$$

2.2. Shift of laser modes and gain spectrum

The experiment has shown that for both modes wavenumber reduces linearly as the injection current increases (see Figure 2). Thus mode energy reduces linearly as well — $E_k(J) = \chi(J - J_k) + E_k(J_k)$.

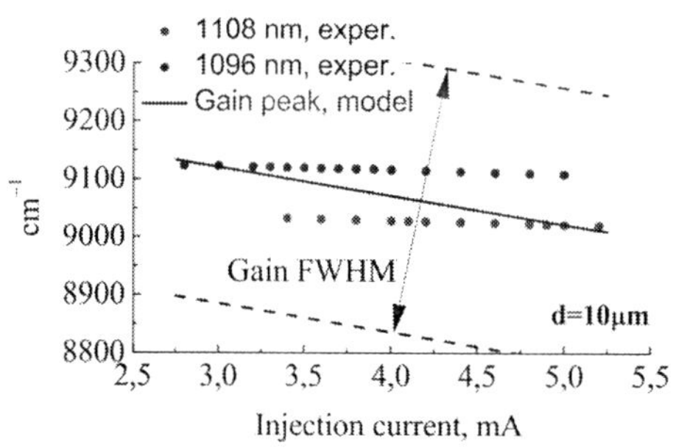

Figure 2. The wavenumber ($1/\lambda$, cm^{-1}) of laser modes at different injection current (dots). Position of gain maximum and FWHM limits of gain spectra are also shown by solid and dashed lines respectively.

Similar shift was observed in [2], but ambient temperature was varied instead. Also it was concluded that the gain spectra shifts with temperature as well due to band gap variation. This way, we introduce $E_0(T)$ and $E_k(T)$ functions and together with (5) and (6) we can write down for G_k and $R_{sp,k}$ following expressions:

$$G_k(T) = G_{max}\left(2\frac{n}{n_{QD}/d} - 1\right)\exp\left(-\frac{1}{2}\left[\frac{E_k(T) - E_0(T)}{\sigma}\right]^2\right) \tag{7}$$

$$R_{sp,k}(T) = G_{max}\frac{n}{n_{QD}/d}\exp\left(-\frac{1}{2}\left[\frac{E_k(T) - E_0(T)}{\sigma}\right]^2\right). \tag{8}$$

The experiment in [2] has shown that the temperature dependency of wavelength λ_k can be approximated by a linear function $\lambda_k(T) = \gamma(T - T_0) + \lambda_k(T_0)$. Here $(T_0; \lambda_k(T_0))$ is a point in the experimental data and $\gamma \approx 0.07$ nm/K. Also it was estimated that the wavelength of a gain peak λ_0 changes linearly $\lambda_0(T) = \gamma_0(T - T_0) + \lambda_0(T_0)$ and $\gamma_0 \approx 0.4$ nm/K. All these changes are small relative to a common wavelength $\lambda \approx 1080$ nm, so corresponding energies E_0 and E_k change linearly with coefficients $-(hc/\lambda^2)\gamma_0$ and $-(hc/\lambda^2)\gamma$ respectively. This way, we can write estimates for $E_0(T)$ and $E_k(T)$

$$E_0(T) = -\frac{hc}{\lambda^2}\gamma_0(T - T_0) + E_0(T_0) \tag{9}$$

$$E_k(T) = -\frac{hc}{\lambda^2}\gamma(T - T_0) + E_k(T_0) \tag{10}$$

Since E_k reduces linearly as T and J increase, one can conclude that T increases linearly as J increases.

$$T(J) = -\frac{\chi\lambda^2}{hc\gamma}(J - J_0) + T_0 \tag{11}$$

Most likely, it happens via Joule heating in the laser that is linear with the current that is accompanied with non-radiative recombination in the active area. The value T_0 is found via (10) and measured $E_k(J_0)$ as J_0 corresponds to T_0.

2.3. Calculation results

Overall, in order to find the laser state under given injection current, the following system of nonlinear equations should be solved:

$$\begin{cases} \dfrac{\eta J}{ed} - \dfrac{n}{\tau} - V_A^{-1}(G_1 S_1 + G_2 S_2) = 0 \\ (G_k - \alpha) S_k + R_{sp,k} = 0 \\ G_k = G_{max}\left(2\dfrac{n}{n_{QD}/d} - 1\right)\exp\left(-\dfrac{1}{2}\left[\dfrac{E_k(T) - E_0(T)}{\sigma}\right]^2\right) \\ R_{sp,k} = G_{max}\dfrac{n}{n_{QD}/d}\exp\left(-\dfrac{1}{2}\left[\dfrac{E_k(T) - E_0(T)}{\sigma}\right]^2\right) \\ E_0(T) = -\dfrac{hc}{\lambda^2}\gamma_0(T - T_0) + E_0(T_0) \\ E_k(T) = -\dfrac{hc}{\lambda^2}\gamma(T - T_0) + E_k(T_0) \\ T(J) = -\dfrac{\chi\lambda^2}{hc\gamma}(J - J_0) + T_0 \end{cases} \tag{12}$$

At Fig.3 experimental data are shown for 10 μm diameter microdisk with electrical injection [4]. The active region is of conventional type with 5 layers of QWD emitting near 1.1 μm. At the figure the mode's power are shown as a dependence on an injection current density. Spectral analyses (not shown here) let us conclude that these are two adjacent modes with spectral spacing of 12 nm. The experiment was carried out at the room temperature.

Since we are interested in the relation between S_1 and S_2 but not their absoulte values, the accuracy of d and G_{max} does not matter much. On the other hand, that is not the case for γ_0, γ, τ, α and especially $E_0(T_0)$ as they are directly related to the currents for onset and cease of the lasing. The accuracy of $E_0(T_0)$ should be of order of 1 meV.

Table 1. Parameter values in the numerical solution of the system

Parameter	Value	Source
η	$5{\cdot}10^{-2}$	threshold current and microdisk geometry
τ	0.25 ns	threshold current
d	50 nm	typical thickness of 5 QWD layers
n_{QD}	$5 \cdot 10^{11}$ cm^{-2}	QWD surface density 10^{11} cm^{-2}
γ_0/γ	5.3	[2]
J_0	4000 A/cm^2	experiment
$E_0(J_0)$	1129.3 meV	model adjustment
σ	25 meV	[5]
G_{max}	510 ns^{-1}	[7]
α	45 ns^{-1}	FWHM of linewidth in [4]

Despite the simplicity, the model, described above, managed to closely capture the experimentally observed effect. The switching between modes are well seen at the current density around 5300 A/cm^2. That corresponds to injection current value 4.15 mA where according to Fig. 2 peak of the gain spectrum becoming closer to the mode with higher wavelength. This shows that an injection current may indeed cause a microdisk temperature to increase and change the lasing wavelength through the modes competition.

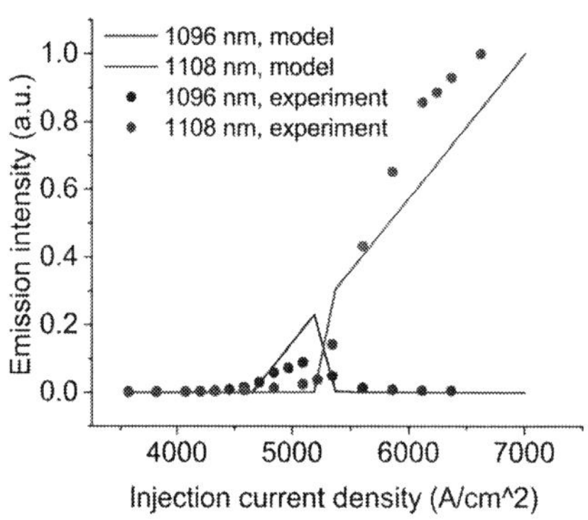

Figure 3. Dependence of emitting power via two laser modes on the injection current density: theory (dots) and experiment (lines).

3. Conclusion

We have proposed the cause of the observed current induced mode competition. Injection current increase leads the microdisk temperature to rise, that, in turn, causes laser modes and gain spectra to shift. As a result, we observe the behavior similar to one in [2]. However, in this work we have relied on the experimental data and provide rough estimates of microdisk internal parameters, which are fairly specific and are tied to a particular design.

As a conclusion, we can state that temperature shift alone, without other supporting effects, may cause switching between modes in microdisks and should be taken into account in relevant multimode models of the laser with electrical injection.

Acknowledgement

The work has been supported by RFBR-DFG grant 18-502-12081 and Russian Ministry of Science and Higher Education (3.9787.2017/8.9).

References

[1] McCall S, Levi A, Slusher R, Pearton S and Logan R 1992 *Applied Physics Letters* **60** 289–291
[2] Kryzhanovskaya N, Blokhin S and Gladyshev A 2006 *Semiconductors* **40** 1101–1104
[3] Agrawal G P 2005 *Lightwave technology: telecommunication systems* (John Wiley & Sons)
[4] Moiseev E, Kryzhanovskaya N and Maximov M 2018 *Opt. Letters* **43** 4554–4557
[5] Savelyev A, Korenev V, Maximov M and Zhukov A 2015 *Semiconductors* **49** 1499–1505
[6] Coldren L A, Corzine S W and Mashanovitch M L 2012 *Diode lasers and photonic integrated circuits* vol 218 (John Wiley & Sons)
[7] Mintairov S, NA K and Lantratov V 2015 *Nanotechnology* **26** 385202

Transverse mode switching in quantum well-dot lasers triggered by gain saturation

A S Payusov[1], A A Kharchenko[2], A A Serin[1], G O Kornyshov[2], Yu M Shernyakov[1], S A Mintairov[1], N A Kalyuzhnyy[1], M V Maximov[2], A E Zhukov[2], N Yu Gordeev[1]

[1]Ioffe Institute, St.Petersburg, 194021, Russia
[2]St.Petersburg Academic University, St.Petersburg, 194021, Russia

Abstract Having investigated narrow-ridge lasers based on a single layer of InGaAs quantum well-dots emitting at the wavelength of 1.06 μm we have shown anomalous transverse multimode lasing not observed in the broad-area lasers. The effect is shown to be caused by the wavelength shift triggered by the ground state gain saturation.

1. Introduction
Nowadays, the vast majority of diode lasers is based on either a quantum well (QW) active region or a quantum dot (QD) one. The former allows obtaining high optical gain and is technologically simpler, the latter enables stacking dozens of active layers without lattice stress accumulation. The QDs also extend the laser emission range [1]. Recently we have proposed and experimentally shown a new type of active region morphologically combining certain properties of both QWs and QDs [2]. Properties of these so-called quantum-well-dot (QWD) nanostructures are still under intensive study. In this paper, we investigate the impact of InGaAs/GaAs QWD gain saturation on transverse modes in narrow-ridge lasers.

2. Experimental results and discussion
The lasers under study contain a single InGaAs QWD active layer with a nominal thickness of 2 nm emitting at the wavelength of ~1060 nm, an undoped GaAs waveguide core with the thickness of 0.78 μm not exceeding the second mode cut-off. The core is sandwiched between 1.5 μm-thick $Al_{0.39}Ga_{0.61}As$ claddings. The wafer was grown by metal-organic chemical vapor deposition (MOCVD) on an n-GaAs (100) substrate and processed into 50 μm wide shallow-mesa lasers. The as-cleaved laser samples were mounted p-side down on indium-covered copper heatsinks. The laser parameters were measured at room temperature in pulsed (400 ns) and CW (continuous-wave) modes.
The broad-area lasers show the threshold current density of ~105 A/cm^2 in 2 mm cavities, the internal quantum efficiency of 80% and intrinsic loss of 0.7 cm^{-1}. At all excitation levels, the lasers show stable lasing on the fundamental transverse mode with the divergence of ~47° FWHM. The lasers with the cavity length of 3 mm have shown CW optical power as high as 6 W limited by catastrophic optical damage (COD) [3]. Since the broad-area lasers demonstrated very competitive parameters, the laser wafer seemed to be very promising for narrow-ridge lasers. We have processed 4 μm wide ridge lasers with a residual p-cladding thickness of 200 nm. These waveguide parameters usually allow obtaining single-mode lasing. Like the broad-area lasers, the narrow ridge ones were mounted p-side down on

heatsinks and measured under pulse pumping current. In contrast to the 50 μm stripe lasers, the narrow-ridge lasers demonstrate completely different transverse far-field patterns. The lasers with longer cavities start to operate on the fundamental mode and switch to the second mode when the pumping current increases (figure 1). The lasers with the cavities shorter than 1.5 mm operate on the second transverse mode merely. Existence of the second transverse mode contradicts with our pre-growth calculations showed that 0.78 μm $GaAs/Al_{0.39}Ga_{0.61}As$ waveguide does not support the second-order mode. To explain this discrepancy, we have measured and analyzed a current dependence of the lasing spectra.

Shallow-mesa narrow-ridge lasers typically have a much higher threshold current density than broad-area lasers processed from the same wafer. One of the main reasons for that is a lateral current spreading. The current spreading length can be as large as dozens of microns and worsens internal differential efficiency, internal optical loss and threshold current therefore in narrow-ridge lasers [4]. On this account, our 4 μm lasers have much higher threshold current densities than the broad-area ones (figure 2). The experiment revealed that the increase of the pumping current in the longer-cavity lasers triggers wavelength shift from the QWD ground state (GS) ~1060 nm to the excited state (ES) ~980 nm (figure 3) while the shorter-cavity lasers operate purely via the ES (figure 2). We associate this shift with the GS gain saturation. The effect is well known for QD active media [4]. Quantum dots possess much lower GS gain than QWs. To provide the modal gain required for lasing stacking of several QD layers is usually needed even for broad-area lasers. The laser may switch to operate via the ES state having higher optical gain when the available GS gain is insufficient for overcoming the losses and cannot be compensated by increasing the pumping current. In this sense, the active media under study demonstrates a similar behavior as a single InGaAs QWD layer produces enough GS gain for the broad-area lasers but not enough for the narrow ones.

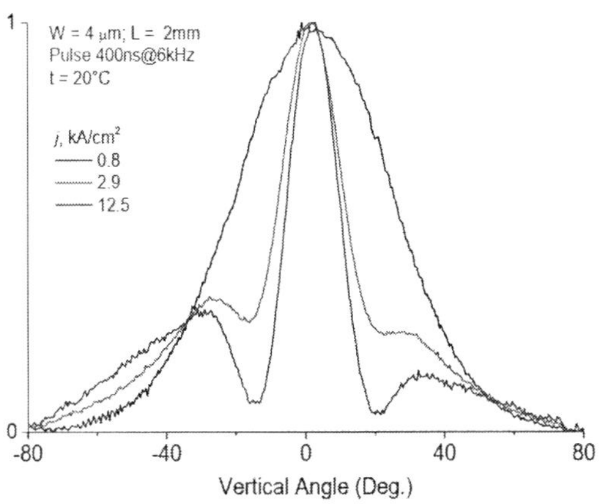

Figure 1. Far-field patterns at different pumping currents.

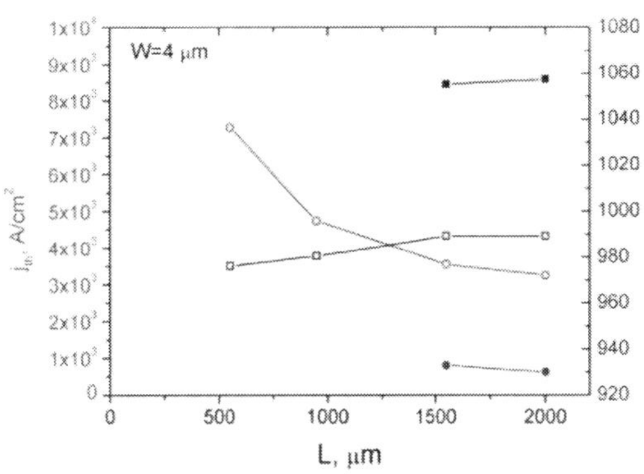

Figure 2. Cavity length dependence of the lasing wavelength and threshold current density for 4 μm ridge QWD lasers.

To clear up the appearance of the second transverse mode we have carried out 2D numerical simulations of the laser waveguide for two wavelengths 1060 nm and 980 nm using the FIMMWAVE mode solver from Photon Design. We have previously shown that in narrow ridge lasers fundamental modes tend to have lower lateral confinement which in some cases can promote high-order mode lasing [6]. The optical mode modelling shows (figure 4) that the given waveguide at the wavelength of 1060 nm support only fundamental and first modes. We have calculated 2D optical confinement factors (Γ-factors) for the active region underneath the ridge. We note here that the first transverse mode would not lase as it has a node at the active region located in the waveguide center. The fundamental mode TE00 has the Γ-factor of 0.18%. The second mode exists only as a leaking mode having the Γ-factor of 0.05%. At the wavelength of 980 nm due to the refractive index dispersion, the waveguide has a different set of the optical modes. The second mode is a guided one with the Γ-factor of 0.14% that is almost equal to the fundamental mode Γ-factor of 0.13%. As a result, at the shorter wavelength, the laser can operate on the high-order mode, which was observed in the experiment.

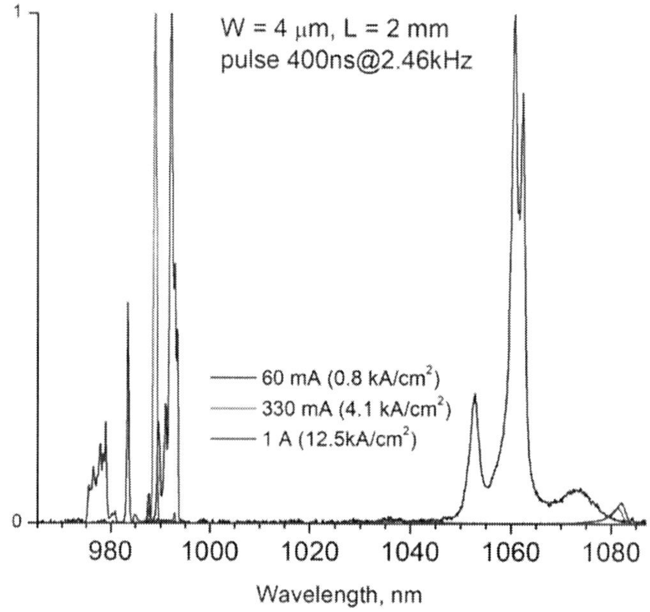

Figure 3. Dependence of the lasing spectra on the pumping current.

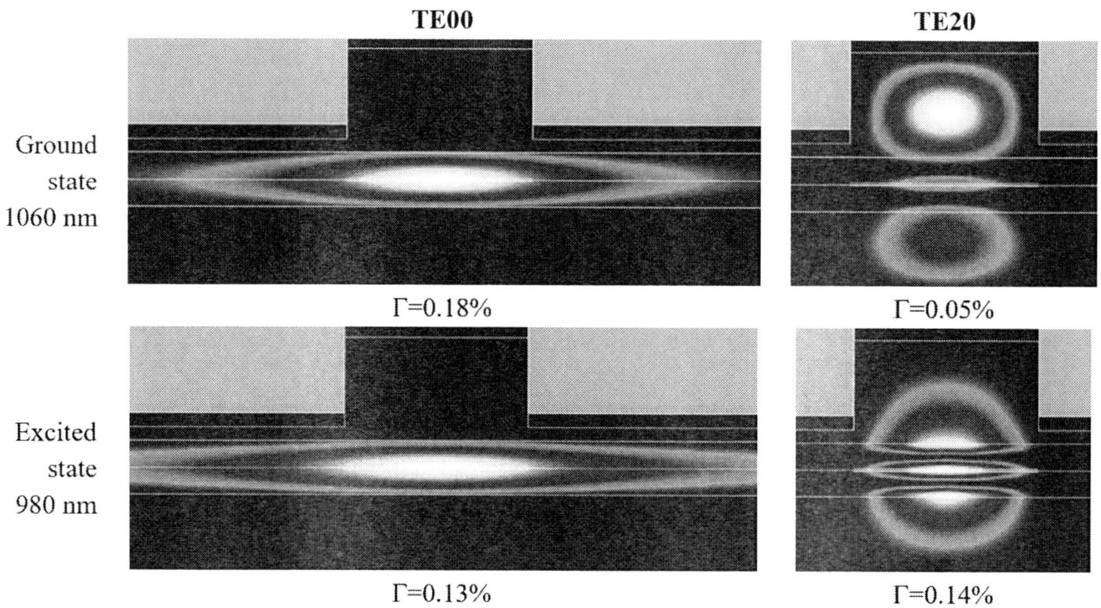

Figure 4. Calculated intensities and Γ-factors for TE_{00} and TE_{20} modes for ground and excited states. Lateral and transverse sizes are in different scales.

3. Conclusions

In conclusion, having investigated narrow-ridge lasers based on a single layer of InGaAs QWD we have revealed its GS gain saturation, the effect typical for quantum dot active media. Gain-saturation-induced switch from GS to ES lasing affects the transverse mode behavior. The corresponding wavelength shift promotes lasing on the second transverse mode. The effect should be carefully accounted in designing optical waveguides and active regions of QWD lasers. For narrow-ridge lasers using of several QWD layers would be preferable as they would provide sufficient modal gain and allow avoiding GS gain saturation.

Acknowledgment

The authors A.S. Payusov and A.A. Serin acknowledge the support from the Russian Science Foundation (project No 17-72-10060).

References

[1] Ledentsov N N, Bimberg D and Alferov Z I 2008 *J. Light. Technol.* **26** 1540–55
[2] Mintairov S A, Kalyuzhnyy N A, Lantratov V M, Maximov M V, Nadtochiy A M, Rouvimov S and Zhukov A E 2015 *Nanotechnology* **26** 385202
[3] Payusov A S *et al* 2018 *J. Phys.: Conf. Ser.* **1135** 012071
[4] Wenzel H, Erbert G, Knauer A, Oster A, Vogel K and Tränkle G 2000 *Semicond. Sci. Technol.* **15** 557–60
[5] Asryan L V, Grundmann M, Ledentsov N N, Stier O, Suris R A and Bimberg D 2001 *J. Appl. Phys.* **90** 1666–8
[6] Gordeev N Y *et al* 2019 *Laser Phys.* **29** 025003

A fair comparison of spectral properties of Slot and Hybrid plasmonic micro-ring resonators

MA Butt[1], **SA Fomchenkov**[1,2], **NL Kazanskiy**[1,2]
[1]Department of Technical Cybernetics, Samara National Research University, Samara, Russia
[2]Institute of RAS-Branch of the FSRC "Crystallography and Photonics" RAS, Samara, Russia

Abstract. In this work, we presented micro-ring resonator structures based on slot and hybrid plasmonic waveguides. The slot waveguide is formed by keeping a narrow air gap between two silicon strip waveguides. Whereas, the hybrid waveguide is composed of Metal-Air-Silicon structure, where the E-field is considerably enhanced in the narrow air gap. The transmission spectrum and E-field distribution of these structures are simulated using the finite element method (FEM). Micro-ring resonator based on hybrid plasmonic waveguide shows the best sensitivity of 333.3 nm/RIU. Whereas, slot waveguide based micro-ring resonator showed the best performance in Q-factor and the figure of merits.

1. Introduction

Recently, silicon photonics has been exploited in several bio-sensing applications by taking advantage of a remarkably sensitive evanescent field generated near silicon-on-insulator (SOI) photonic waveguides due to a high refractive index contrast between core (Si) and cladding (SiO_2)[1]. Si photonic devices such as Si micro-ring resonators [2,3], slot waveguides [4,5], hybrid plasmonic waveguides [6,7], Mach-Zehnder interferometers [8] and photonic crystal waveguides [9] show potential for label-free detection of target molecules and for real-time monitoring of binding events that arise near the device surface. Micro-ring resonators are flexible structures that have been used in optical communication circuits as wavelength filters switches [10, 11, 12, 13] and frequency converters [14]. Numerous unique and promising designs of micro-ring resonators have been proposed up-to-date [15-17].

The optical waveguides are modelled and fabricated in such a way that portion of the light energy known as an evanescent field, extends beyond the surface of the waveguide that interacts with the analytes in the close vicinity of the waveguide [18, 19]. When the refractive index (RI) of the ambient media changes, the resonant wavelength (λ_{res}) changes accordingly. In recent times, a new type of plasmonic waveguides known as hybrid plasmonic waveguides has developed a significant interest due to its potential to provide both sub-wavelength confinement and long propagation lengths [20, 21]. In this work, we investigated the performance of micro-ring resonators based on two different kinds of waveguides such as slot waveguide and hybrid plasmonic waveguide. The ring radius (R=1.5 μm) is kept constant in both the devices in order to compare their spectral characteristics. Moreover, the sensitivity, Q-factor and figure of merits are also evaluated. This can provide a set of information for selecting a suitable waveguide structure while designing a ring resonator sensors.

2. Design parameters of ring resonators

We presented two micro-ring resonator designs based on Si slot waveguide and hybrid plasmonic waveguide. The dimensional parameters of both the designs are kept similar throughout the paper in order to have a fair evaluation for their spectral behaviour while comparing. The length, width and height of bus waveguide is denoted by L (5 μm), W_{Wg} (0.4 μm) and H_{Wg} (0.22 μm), respectively. The gap between the bus waveguide and the ring is fixed and is labelled as G (0.1 μm). The Si ring radius is R (from the outer edge) which is identical for both the designs. For slot and hybrid plasmonic waveguides, the slot width is represented by S (0.04 μm) which is also fixed at a certain value. The Si ring waveguide width is labelled as W_{ring} in case of the hybrid plasmonic waveguide ring resonator which is varied at 0.3, 0.31, 0.32, 0.33, 0.34 and 0.35 μm. Whereas, in the case of slot waveguide, the outer and inner WG width is denoted as O-W_{ring} (0.3, 0.31, 0.32, 0.33, 0.34, 0.35 μm) and I-W_{ring} (0.3, 0.29, 0.28, 0.27, 0.26, 0.25 μm), respectively as shown in figure 1.

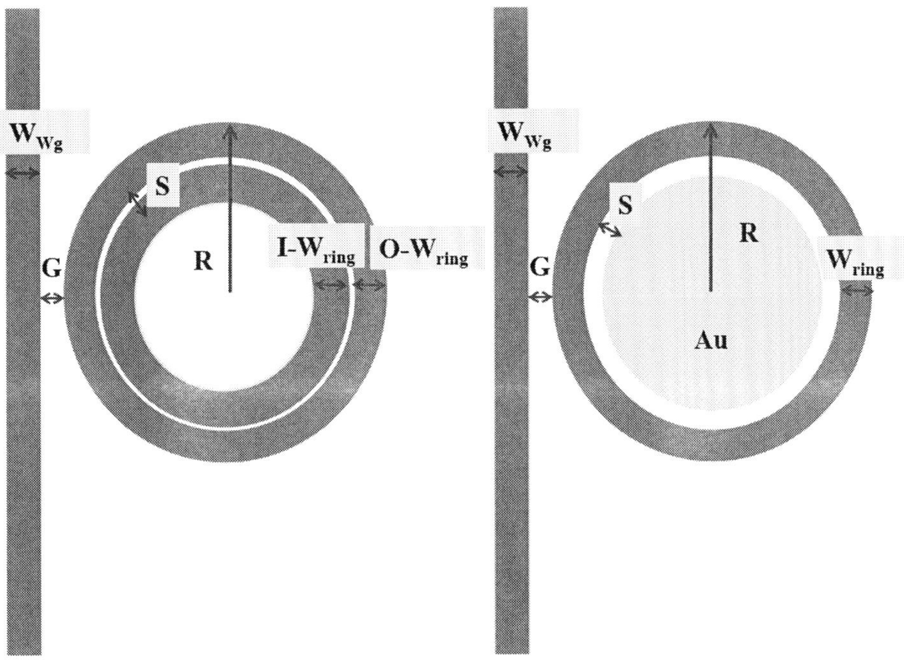

Figure 1. Schematic of a ring resonator design, a) Slot ring resonator, b) Hybrid plasmonic ring resonator.

For hybrid plasmonic ring resonator, the Si circular strip encircling a metal disk in the centre. Gold (Au) is chosen as the metal layer because of its biocompatibility and resistant to oxidation as compared to silver (Ag). Figure 2 (a and b) shows the E-field confinement and transmission spectrum of a slot and hybrid plasmonic waveguide ring resonator at their respective λ_{res}. The total width of the Si slot waveguide ring is equal to I-W_{ring}+O-W_{ring}+S=640 nm. To improve the confinement of EM-waves in a ring resonator based on slot waveguide, an asymmetric slot waveguide structure is proposed. S is shifted towards the inner periphery of the ring by keeping the total width of the slot waveguide steady. In the case of a hybrid plasmonic waveguide, it consolidates the characteristics of the dielectric and plasmonic waveguides [22]. Theoretical studies on slot waveguide show that it can support a low loss compact mode whose propagation length mainly depends on slot width (S) [23]. The excitation of surface electron oscillation in the hybrid waveguide concentrates the photons to stay close to the metal surface as a result, the radiation loss is reduced. Ring resonator based on slot-waveguide has a smallest full width at half maximum (FWHM) value which is ~2.66 nm at λ_{res}.

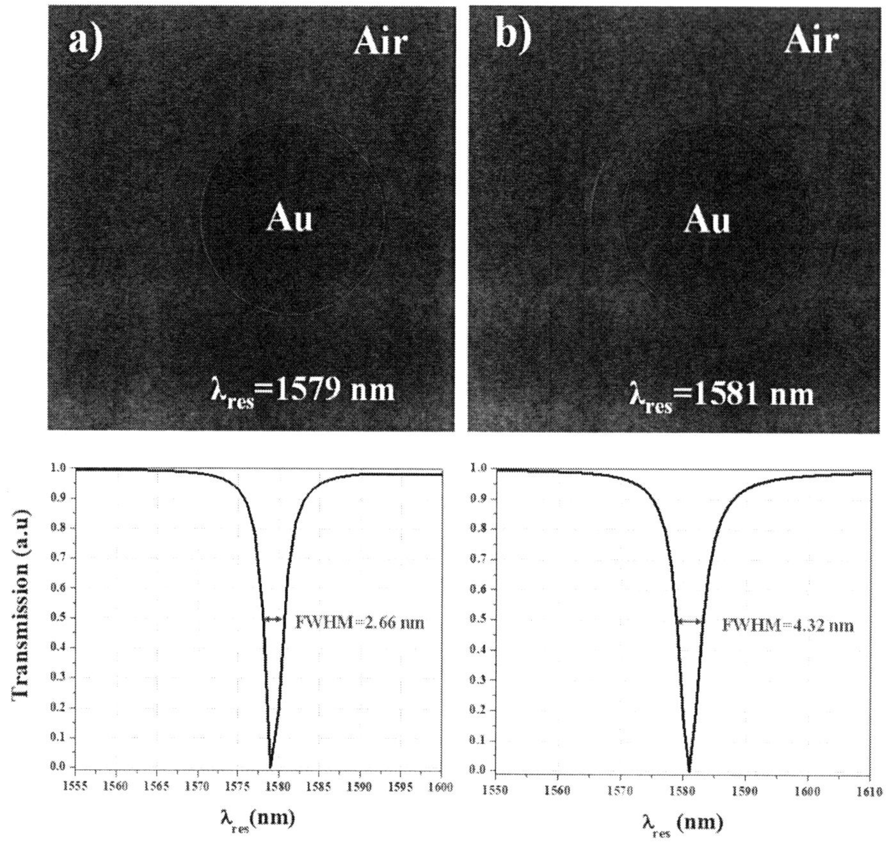

Figure 2. E-field confinement in the ring, a) Si slot ring resonator, where O-W_{ring}=350 nm and I-W_{ring}= 250 nm, b) Hybrid plasmonic ring resonator, where W_{ring}=340 nm. The picture below shows the transmission spectrum of the respective design.

3. Results analysis

In this section, we examined the sensing capabilities of both designs along with the fabrication tolerance and provide a fair comparison. In order to obtain the optical resonances and RI sensitivity, the medium was filled with a dielectric having n=1.03 for both designs and for all values of W_{ring}. The transmission spectrum and E-field distribution are simulated using Finite Element Method (FEM). Sensitivity is defined as the ratio between the change in λ_{res} and change in ambient refractive index. The Q-factor is defined as λ_{res}/FWHM. The figure of merit (FoM) is another parameter which should also be considered while designing the ring resonator sensor. FoM is expressed as Sensitivity/FWHM. Therefore it can be seen that FWHM has direct control on the Q-factor and FoM.

In figure 3a, the sensitivity of ring resonators is plotted with respect to W_{ring} for Δn=0.03. It can be seen that hybrid plasmonic ring resonator has higher S as compared to slot ring resonator at W_{ring}=300 nm which deteriorates pretty fast as the W_{ring} increases. The S_{max} drops from 333.33 nm/RIU to 170 nm/RIU when W_{ring} changes from 300-350 nm. This fact signifies that hybrid plasmonic ring resonator has less fabrication tolerance as compared to other Si slot waveguide design.

In terms of FWHM, Si slot ring resonator has the lowest value=2.66 nm at W_{ring}= 350 nm. FWHM has a direct relationship with the ring waveguide size. When the W_{ring} increases the FWHM decreases which signifies the better confinement of light in the ring and a decrease in an evanescent field as shown in figure 3b.

Despite the fact that the hybrid plasmonic ring resonator has the highest S but its broad FWHM makes its FoM and Q- factor lower than Si slot ring resonator design. As we have mentioned earlier that FoM is the ratio between sensitivity and FWHM, therefore, design with narrow FWHM can have a larger FoM as shown in figure 3c. The maximum calculated FoM is approx. 75.2 and 58.5 for Si slot and hybrid plasmonic ring resonator, respectively. The Q-factor of Si slot resonator is higher than the hybrid plasmonic ring resonator as shown in figure 3d. The maximum calculated Q-factor for Si slot and hybrid plasmonic ring resonator is 593.6 and 396.5, respectively.

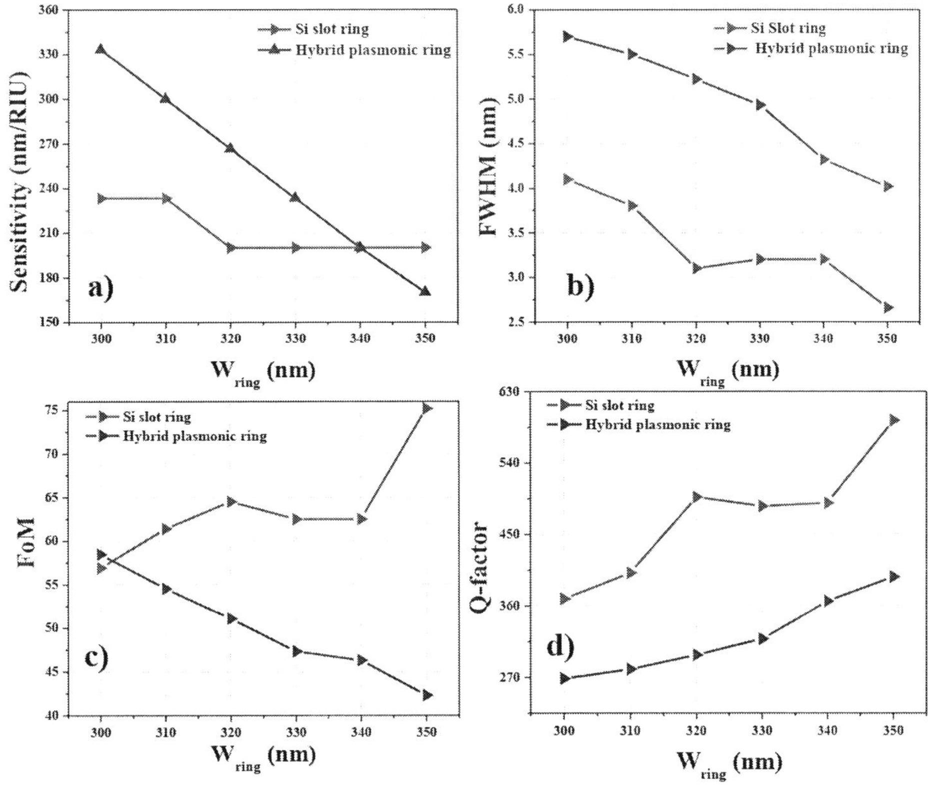

Figure 3. Spectral analysis of slot and hybrid plasmonic ring resonator: a) Sensitivity at $\Delta n=0.03$, b) FWHM, c) FoM, d) Q-factor. Note: For Si slot ring resonator, the x-axis represents the O-W_{ring}.

4. Conclusion

We presented micro-ring resonators based on slot waveguide and hybrid plasmonic waveguides. The variation in geometric parameters of the ring resonator is taken into consideration for their spectral analysis. Each design shows its exclusive property and fabrication tolerance. Hybrid plasmonic ring resonator has the highest sensitivity of 333.3 nm/RIU at W_{ring}=300 nm. But it reduces when the W_{ring} increases which make it highly vulnerable to fabrication tolerance. In contrast, silicon slot ring resonator shows promising results when it comes to FWHM, Q-factor and FoM. Therefore, a particular sensor design should be selected depending on the application and available fabrication facilities. Ring resonator design based on hybrid plasmonic waveguide is advantageous when high sensitivity is required.

Acknowledgements

This work was financially supported by the Russian Foundation for the Basic Research (grant No. 16-29-09528_ofi_m, 16-29-11744_ofi_m) for numerical calculations and by the Ministry of Science and

Higher Education within the State assignment FSRC <<Crystallography and Photonics>> RAS (No. 007-GZ/Ch3363/26) for theoretical results.

References

[1] Guider R, Gandolfi D, Chalyan T, Pasquardini L, Samusenko A, Pederzolli C, Pucker G, Pavesi L 2015 *Sensing and Bio-Sensing Research* **6** 99.
[2] Li A, Huang Q, Bogaerts W. 2016 *Photon. Res.* **4** 84.
[3] Butt MA, Khonina SN, Kazanskiy NL. 2019 *Laser Phys.* **29** 046208.
[4] Vivien L, Marris-Morini D, Griol A, Gylfason KB, Hill D, Alvarez J, Sohlstrom H, Hurtado J, Bouville D, Cassan E 2008 *Opt. Exp.* **16** 17237.
[5] Degtyarev SA, Butt MA, Khonina SN, Skidanov RV 2016 *International conference on Computing, Electronics and Electrical Engineering, ICE CUBE* 7495222. DOI. 10.1109/ICECUBE.2016.7495222.
[6] Butt MA, Khonina SN, Kazanskiy NL 2018 *Waves in Random and Complex Media*. DOI. 10.1080/17455030.2018.1506191.
[7] Butt MA, Khonina SN, Kazanskiy NL. 2018 *Journal of Modern Optics* **65** 1135.
[8] Guha B, Godarenko A, Lipson M 2010 *Opt. Exp.* **18** 1879.
[9] White TP, O'Faolain L, Li J, Andreani LC, Krauss TF 2008 *Opt. Exp.* **16** 17076.
[10] Takahashi K, Kanamori Y, Kokubun Y, Hane K 2008 *Opt. Exp.* **16** 14421.
[11] Kazanskiy NL, Serafimovich PG, Khonina SN 2010 *Optical Memory and Neutral Networks* (Allerton Press) **19** 318.
[12] Kazanskiy NL, Serafimovich PG, Popov SB, Khonina SN 2010 *Computer Optics* **34** 162.
[13] Butt MA, Khonina SN, Kazanskiy NL 2019 *Journal of Modern Optics* **66** 1172.
[14] Turner Ac, Foster MA, Gaeta AL, Lipson M 2008 *Opt. Exp.* **16** 4881.
[15] Sun X, Dai D, Thylen L, Wosinski L 2015 *Photonics* **2** 1116.
[16] Butt MA, Degtyarev SA, Khonina SN 2018 Au-SiO_2-Si hybrid plasmonic waveguide micro-ring resonator sensor. *J. Phys.: Conf. Ser.* 1124, 051001.
[17] Pernice WHP, Li M, Tang H. 2009 *Opt. Exp.* **17** 12424.
[18] Butt MA, Khonina SN, Kazanskiy NL 2018 *Journal of Modern Optics.* **65** 174.
[19] Butt MA, Degtyarev SA, Khonina SN, Kazanskiy NL 2017 *Journal of Modern Optics* **64** 1892.
[20] Oulton RF, Sorger VJ, Genov D et al 2008 *Nat. Photon.* **2** 496.
[21] Kim JT, JU JJ, Park S, et al.2010 *Opt. Exp.* **18** 2808.
[22] Wu M, Han Z, Van V.2010 *Opt. Exp.* **18** 11728.
[23] Dai D, He S. 2009 *Opt. Exp.* **17** 16646.

Polarization properties of single-walled carbon nanotube thin film on silicon substrate in terahertz frequency range

A Kvitsinskiy[1, 2], P Demchenko[1], E Litvinov[1], M Masyukov[1, 3], I Anoshkin[1], A Vozianova[1] and M Khodzitsky[1, 3]

[1] Terahertz Biomedicine Laboratory, ITMO University, 3b Kadetskaya Liniya, Saint Petersburg 199034, Russian Federation

[2] Center for Bioengineering, ITMO University, 49 Kronverksky Prospekt, Saint Petersburg 197101, Russian Federation

[3] Radiation Medicine Laboratory, ITMO University, 3b Kadetskaya Liniya, Saint Petersburg 199034, Russian Federation

E-mail: anatolykvitsinskiy@gmail.com

E-mail: khodzitskiy@yandex.ru

Abstract. Terahertz time-domain spectroscopic polarimetry (THz-TDSP) method was used to study of polarization properties of a randomly oriented single-walled carbon nanotube (SWCNT) thin film on a silicon (Si) substrate in terahertz (THz) frequency range under an external optical pumping (OP) and an external static magnetic field (MF). Frequency dependencies of azimuth and ellipticity angles of a polarization ellipse and the polarization ellipse at various frequencies of the Si substrate and the SWCNT thin film on the Si substrate were obtained experimentally. The results confirm the fact that, based on carbon nanotubes, it is possible to devise efficient tunable THz polarization modulators for use in the latest security and telecommunication systems.

1. Introduction

Terahertz (THz) radiation [1] is widely used in physics and astronomy [2–4], chemistry and medicine [5–7], security and telecommunication systems [8–10], and other fields of science and technology [11–13].

Currently, an important area of THz photonics is study of carbon nanotubes under an external optical pumping (OP) [14,15] and an external static magnetic field (MF) [16,17] to devise efficient tunable THz polarization modulators based on them [18–23].

THz frequencies have been proposed for security and telecommunication systems, for example, for communication between geostationary satellites. And amplitude, frequency, and polarization modulators are the basic components in this systems. Also, in THz spectroscopy, fast polarization modulators have a fundamental importance for the investigation of properties of novel materials.

One of the main methods for studying polarization and magneto-optical properties of carbon nanotubes is the THz time-domain spectroscopic polarimetry (THz-TDSP) [24–28].

2. Material and Methods

The randomly oriented single-walled carbon nanotube (SWCNT) thin film on a silicon (Si) substrate and a sample of the Si substrate were studied. SWCNT thin film was synthesized on a nitrocellulose micropore filter by the catalytic ethanol-chemical vapor deposition [29] and then transferred to a Si substrate. SWCNTs diameters were calculated using the Kataura plot [30] and were 1.3–2.0 nm. SWCNTs length was visualized by the transmission electron microscopy [31] and was ~15 µm. SWCNT film thickness was calculated from the optical absorbance at 550 nm [32] and was ~110 nm. Si substrate thickness was ~1 mm.

The samples were studied using the THz-TDSP system based on the THz time-domain spectrometer [33], two polarizers, 980 nm laser for creating an external OP of $\sim 1.0\ \mathrm{W \cdot cm^{-2}}$ and NdFeB axially magnetized magnet for an external static MF of ~1.3 T. Scheme of the experimental THz-TDSP setup and pictures of the experimental samples are shown in Figure 1.

Infrared femtosecond laser beam of Yb:KYW 1040 nm laser is divided into a pump beam and a probe beam. The pump beam passes through the delay line, optomechanical modulator and falls onto the semiconductor indium arsenide crystal surface. Due to mobility of electrons in semiconductor is more than mobility of holes, thereby forming electric dipoles, which oscillate in terahertz frequency range. Constant magnetic field used for rotate dipoles and amplify terahertz radiation. Next, the THz beam is focused into a sample and two polarizers. First polarizer passes vertical component or superposition of vertical and horizontal components, depending on azimuth rotation. Second polarizer is fixed and passes vertical component only. Beams are detected by a balanced detector, which gives an analog signal of the difference in beam intensities. The received signal passes through the lock-in-amplifier and analog-to-digital converter and finally goes to the control device.

By recording THz radiation spectra for two different rotation angles of the polarizer, it is possible to calculate the azimuth and ellipticity angles of polarization ellipse of the samples for various frequencies of interest.

3. Results

For each sample, temporal waveforms of the transmitted THz signals under various external influences were recorded using LabVIEW software at parallel and crossed by 45° positions to the transmission direction of the polarizers. Experimental data processing was done using MATLAB software, 4th order coiflet-based denoising technique and rectangular signal windowing [34], which allowed to obtain noise-free spectra. The windowing was done to exclude the influence of the water vapor absorption.

Frequency dependencies of the azimuth ψ and the ellipticity χ angles of the polarization ellipse of the samples were calculated from the Stokes parameters [35]:

$$S_0 = E_1^2 + E_2^2, \tag{1}$$
$$S_1 = E_1^2 - E_2^2, \tag{2}$$
$$S_2 = 2E_1E_2\cos\delta, \tag{3}$$
$$S_3 = 2E_1E_2\sin\delta, \tag{4}$$
$$\psi = \frac{1}{2}\sin^{-1}\left(\frac{S_3}{S_0}\right), \tag{5}$$
$$\chi = \frac{1}{2}\tan^{-1}\left(\frac{S_2}{S_1}\right), \tag{6}$$

where E_1 and E_2 are amplitudes of parallel and perpendicular components of electric field vector $\boldsymbol{E}$, and δ is the phase difference between them.

The results are shown in Figures 2, 3 and 4.

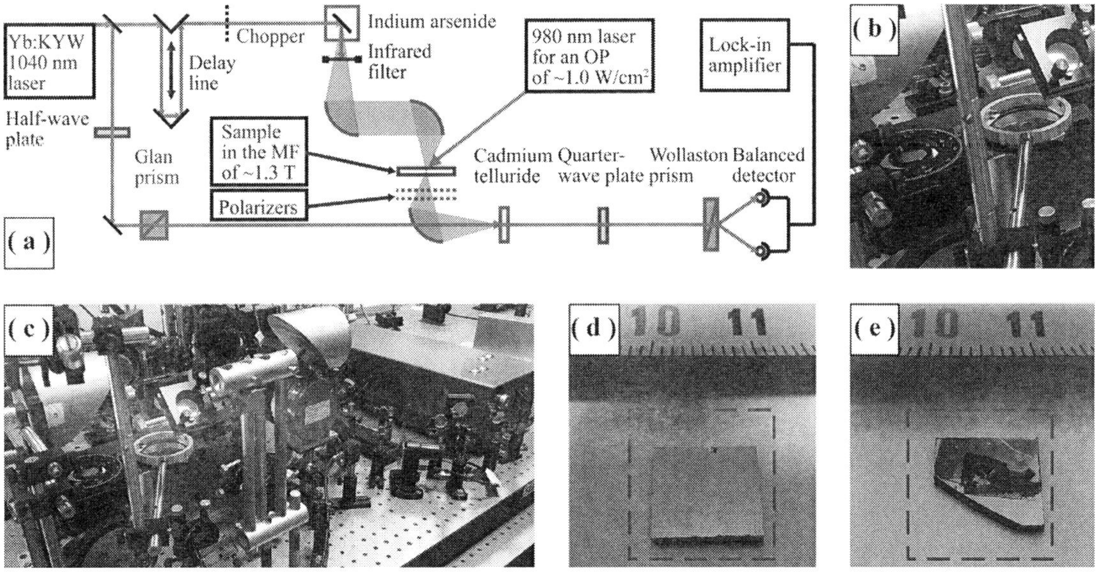

Figure 1. (**a**) Scheme of the experimental THz-TDSP setup, (**b**) picture of a cuvette for the experimental samples, (**c**) picture of the assembled experimental THz-TDSP setup, (**d**) picture of the Si substrate (highlighted in red), and (**e**) picture of the SWCNT thin film on the Si substrate (highlighted in red).

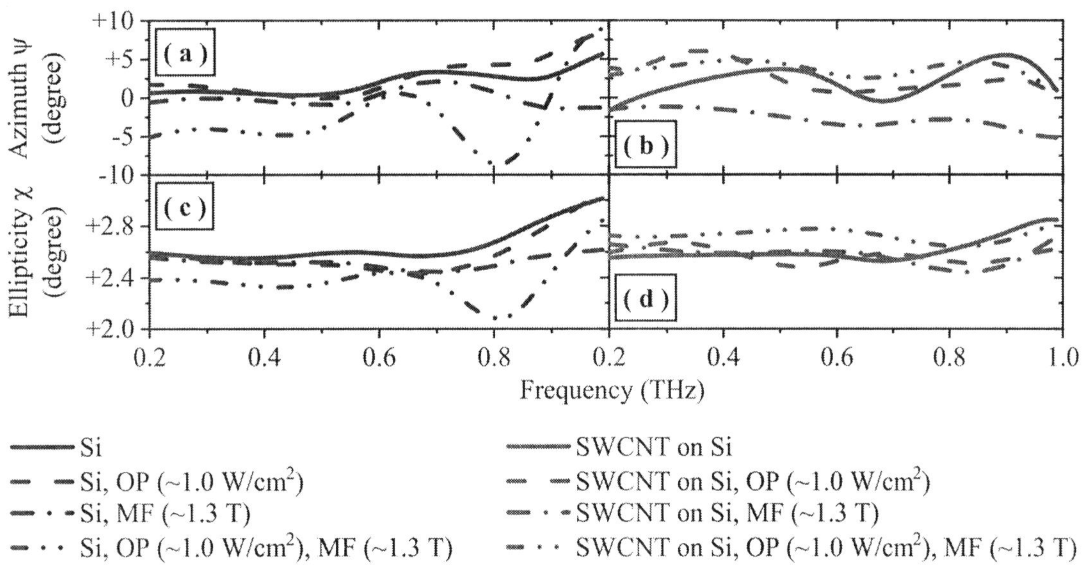

Figure 2. Frequency dependences of (**a**) azimuth angle ψ of polarization ellipse of the Si substrate, (**b**) azimuth angle ψ of polarization ellipse of the SWCNT thin film on the Si substrate, (**c**) ellipticity angle χ of polarization ellipse the Si substrate, and (**d**) ellipticity angle χ of polarization ellipse the SWCNT thin film on the Si substrate under external influences.

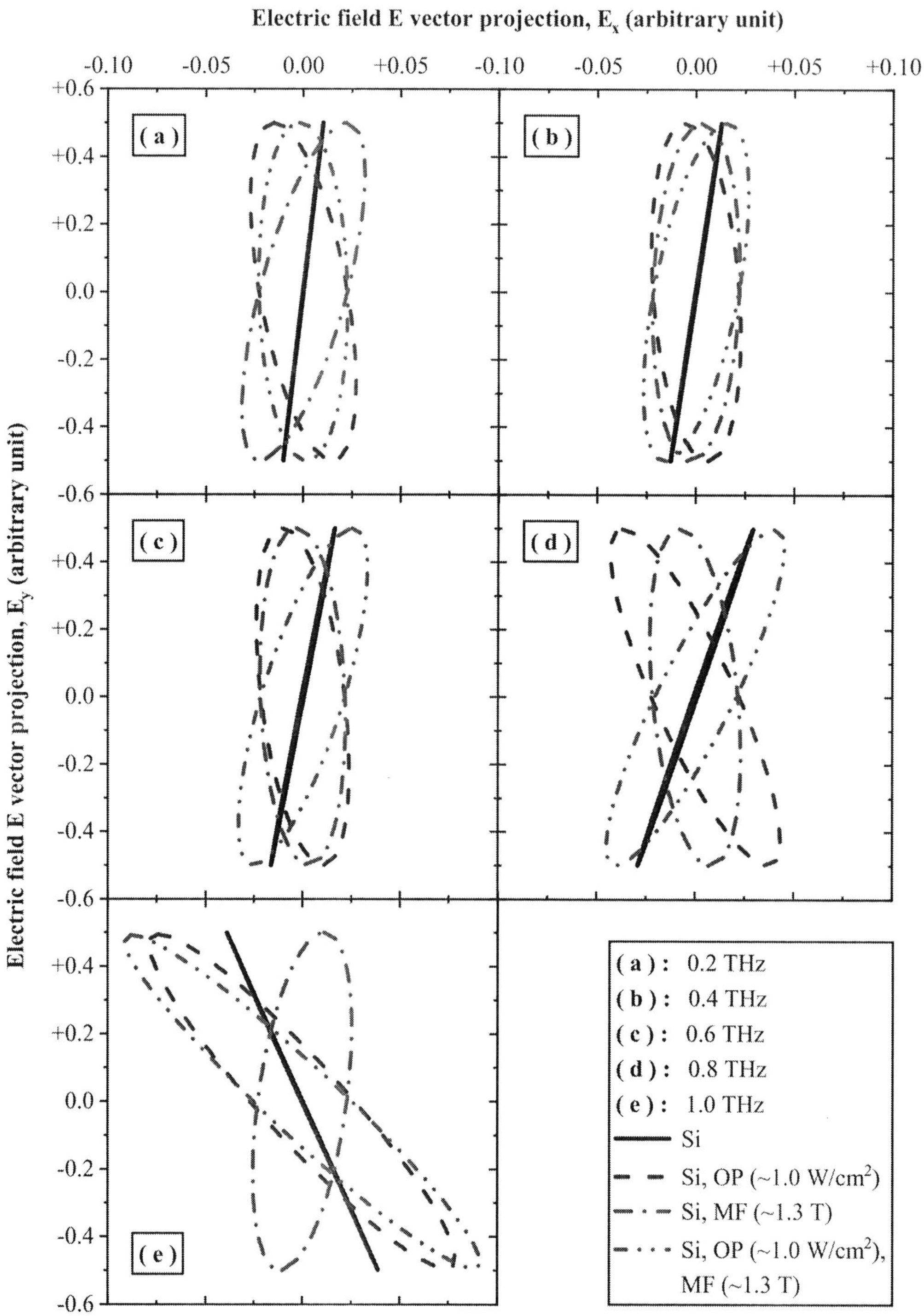

Figure 3. Polarization ellipse of the Si substrate at frequencies of (a) 0.2 THz, (b) 0.4 THz, (c) 0.6 THz, (d) 0.8 THz, and (e) 1.0 THz under external influences.

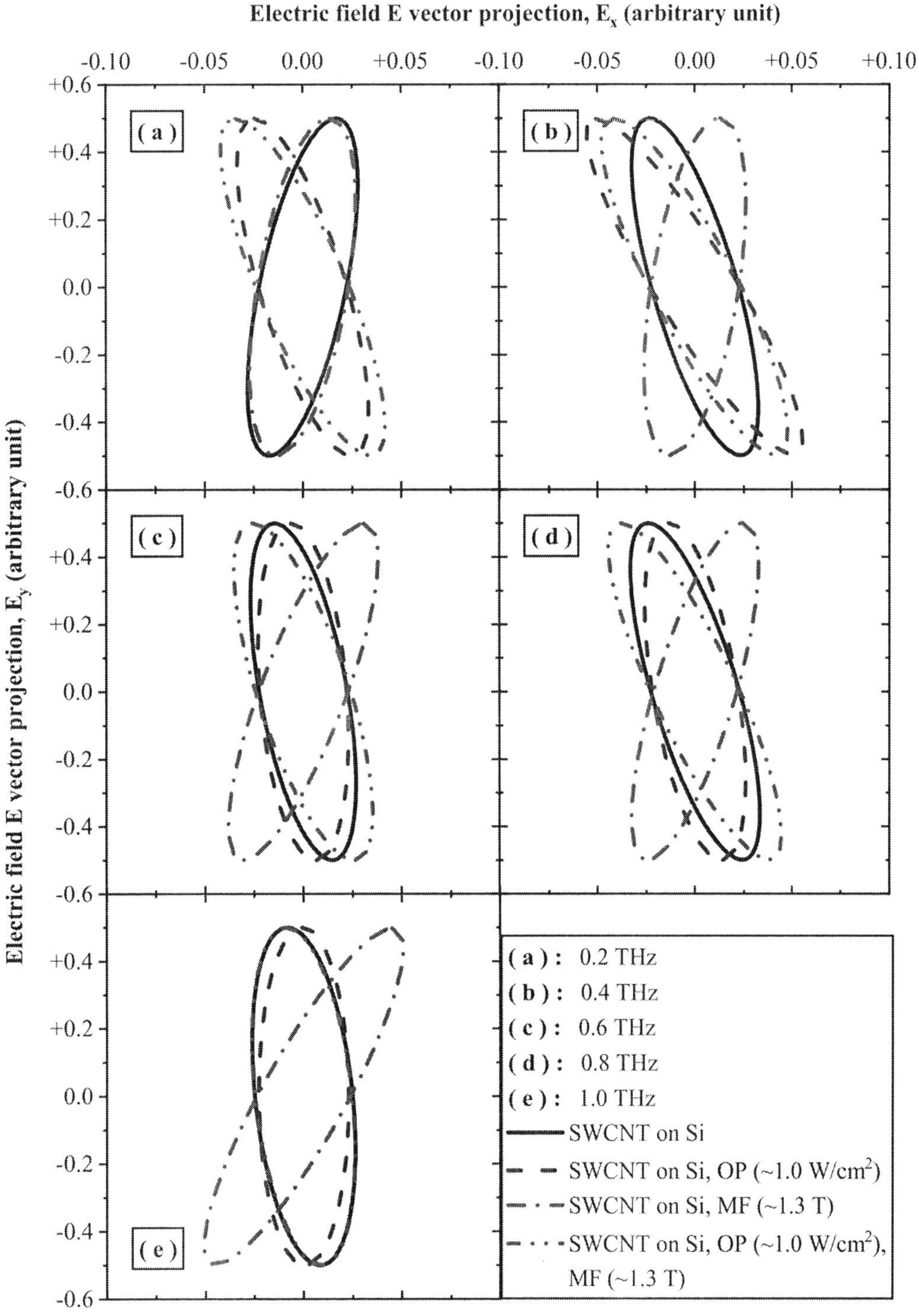

Figure 4. Polarization ellipse of the SWCNT thin film on the Si substrate at frequencies of (**a**) 0.2 THz, (**b**) 0.4 THz, (**c**) 0.6 THz, (**d**) 0.8 THz, and (**e**) 1.0 THz under external influences.

4. Discussion and Conclusion

The changes in the angles are the result of the magneto-optical Faraday effect. Applying MF to isotropic material cause anisotropy, and as a result a circular birefringence. Taking this into account, the vector of the transmitted linearly polarized wave is not the same as it was before the sample. The azimuth polarization angle rotation caused by Faraday effect depends on free carrier concentration, which can be increased by OP, and the value of the applied MF.

The results confirm the fact that, based on SWCNTs, it is possible to devise efficient tunable THz polarization modulators for use in the latest security and telecommunication systems.

References

[1] Dhillon S S *et al.* 2017 *J. Phys.* D. Appl. Phys. **50**(4) 043001
[2] Hafez H A, Chai X, Ibrahim A, Mondal S, Férachou D, Ropagnol X and Ozaki T 2016 *J. Opt.* **18**(9) 093004
[3] Consolino L, Bartalini S and De Natale P 2017 *J. Infrared Millim. Terahertz Waves* **38**(11) 1289–1315
[4] Grebenchukov A N, Azbite S E, Zaitsev A D and Khodzitsky M K 2019 *J. Magn. Magn. Mater.* **472** 25–8
[5] Yin M, Tang S and Tong M 2016 *Appl. Spectrosc. Rev.* **51**(5) 379–96
[6] Yang X, Zhao X, Yang K, Liu Y, Liu Y, Fu W and Luo Y 2016 *Trends Biotechnol.* **34**(10) 810–24
[7] Soboleva V Y, Gusev S I and Khodzitsky M K 2018 *Sci. Tech. J. Inf. Technol. Mech. Opt.* **18**(3) 377–83
[8] Guerboukha H, Nallappan K and Skorobogatiy M 2018 *Adv. Opt. Photonics* **10**(4) 843–938
[9] Litvinov E A, Vozianova A V and Khodzitsky M K 2018 *J. Phys.* Conf. Ser. **1062** 012010
[10] Masyukov M S, Vozianova A V, Grebenchukov A N and Khodzitsky M K 2018 *J. Phys.* Conf. Ser. **1124** 051053
[11] Mittleman D M 2017 *J. Appl. Phys.* **122**(23) 230901
[12] Ren A, Zahid A, Fan D, Yang X, Imran M A, Alomainy A and Abbasi Q H 2019 *Trends Food Sci. Technol.* **85** 241–51
[13] Cosentino A 2019 *Technologies* **4**(1)
[14] Smirnov S, Anoshkin I V, Demchenko P, Gomon D, Lioubtchenko D V, Khodzitsky M and Oberhammer J 2018 *Nanoscale* **10**(26) 12291–6
[15] Kar S and Sood A K 2019 *Carbon* **144** 731–6
[16] Makarova T L *et al.* 2016 *Compos.* B. Eng. **91** 505–12
[17] Wang L, Hong Y, Dai H and Ni Q 2016 *Acta Mech. Solida Sin.* **29**(6) 567–76
[18] Hashemi M R, Cakmakyapan S and Jarrahi M 2017 *Rep. Prog. Phys.* **80**(9) 094501
[19] Wang R, Xie L, Hameed S, Wang C and Ying Y 2018 *Carbon* **132** 42–58
[20] Ren L, Pint C L, Arikawa T, Takeya K, Kawayama I, Tonouchi M, Hauge R H and Kono J 2012 *Nano Lett.* **12**(2) 787–90
[21] Docherty C J, Stranks S D, Habisreutinger S N, Joyce H J, Herz L M, Nicholas R J and Johnston M B 2014 *J. Appl. Phys.* **115**(20) 203108
[22] Zubair A, Tsentalovich D E, Young C C, Heimbeck M S, Everitt H O, Pasquali M and Kono J 2016 *Appl. Phys. Lett.* **108**(14) 141107
[23] Xu S T, Mou L L, Fan F, Chen S, Zhao Z, Xiang D, de Andrade M J, Liu Z and Chang S J 2018 *Opt. Express* **26**(22) 28738–50
[24] Paul M J *et al.* 2012 *Appl. Phys. Lett.* **101**(11) 111107
[25] Aschaffenburg D J, Williams M R C and Schmuttenmaer C A 2016 *J. Chem. Phys.* **144**(17) 174705
[26] Mosley C D W, Failla M, Prabhakaran D and Lloyd-Hughes J 2016 *Sci. Rep.* **7**(1)
[27] Sanjuan F, Gaborit G and Coutaz J L 2018 *Appl. Opt.* **57**(21) 6055–60
[28] Zhao G, Savini G, Yu Y, Li S, Zhang J and Ade P 2019 *Sci. Rep.* **9**(1)
[29] Tonkikh A A, Pozharov A S, Arutyunyan N R, Anoshkin I V, Kauppinen E I and Obraztsova E D 2012 *J. Nanoelectron. Optoelectron.* **7**(1) 99–101
[30] Kataura H, Kumazawa Y, Maniwa Y, Umezu I, Suzuki S, Ohtsuka Y and Achiba Y 1999 *Synth. Met.* **103**(1) 2555–8
[31] Krause B, Boldt R and Potschke P 2011 *Carbon* **49**(4) 1243–7
[32] Nasibulin A G *et al.* 2011 *ACS Nano* **5**(4) 3214–21
[33] Bespalov V G, Gorodetskii A A, Denisyuk I Y, Kozlov S A, Krylov V N, Lukomskii G V, Petrov N V and Putilin S E 2008 *J. Opt. Technol.* **75**(10) 636–42
[34] Ferguson B and Abbott D 2001 *Microelectronics J.* **32**(12) 943–53
[35] Born M and Wolf E 2005 *Principles of Optics: Electromagnetic Theory of Propagation, Interference and Diffraction of Light* 7th ed (Cambridge: Cambridge University Press) ISBN 0-521-64221 hardback

Tunable upconversion luminescence of SrF_2:Er,Tm phosphors

A A Lyapin[1*], S V Gushchin[1], S V Kuznetsov[2], A S Ermakov[1], P A Ryabochkina[1], V Yu Proydakova[2], V V Voronov[2], A D Yapryntsev[3], V K Ivanov[3], P P Fedorov[2]

[1]Institute of Physics and Chemistry, National Research Ogarev Mordovia State University, Saransk 430005, Russia
[2]Prokhorov General Physics Institute of the Russian Academy of Sciences, Moscow 119991, Russia
[3]Kurnakov Institute of General and Inorganic Chemistry of the Russian Academy of Sciences, Moscow 119991, Russia

* Corresponding author: andrei_lyapin@mail.ru

Abstract. In preset paper, for the first time upconversion luminescence (UCL) of SrF_2:Er,Tm phosphors upon excitation of $^4I_{13/2}$ level of Er^{3+} were investigated. Fluorite-type SrF_2:Er,Tm phosphors were synthesized by co-precipitation from aqueous nitrate solution technique. Tuning the UCL by SrF_2:Er through Tm^{3+} ions co-doping was demonstrated. Also we estimated the UCL energy yield in the visible and near-infrared spectral range.

1. Introduction

Upconversion phosphors play important role in the modern photonics because of different applications in biophotonics, medicine, laser physics, anti-counterfeiting etc [1-5].

Alkaline earth fluorides MF_2 (M = Ca, Sr) doped with rare-earth ions with fluorite-type structure show highly efficient UCL due to the low phonon energy and the tendency to form multiple cluster configurations of dopant ions [6-8]. Also wide absorption spectral bandwidths of fluorides doped rare-earth (RE) ions allow to visualized laser radiation from 760 to 2100 nm [9-17].

Currently, there are a large number of articles devoted to the study of UCL of RE-doped MF_2 materials upon excitation of different energy levels of RE ions [9-26]. Previously we investigated UCL of SrF_2:Er solid solutions upon excitation of $^4I_{13/2}$ level of Er^{3+} ions [14]. In present paper we studied effect of Tm^{3+} co-doping on UCL of SrF_2:Er,Tm solid solutions upon 1531.8 nm laser excitation.

2. Experimental

The SrF_2:Er,Tm phosphors were synthesized by co-precipitation from aqueous nitrate solution technique. Precipitates were washed a several time in bidistilled water and annealed for 1 hour at 600 °C. The crystalline phases of the synthesized samples were examined by X-ray powder diffraction (XRD) patterns recorded on a Bruker D8 Advanced diffractometer (Cu Kα radiation). Scanning electron microscopy (SEM) images and chemical composition be energy dispersion analysis (EDX) were determined using a Carl Zeiss NVision 40 microscope. The diffuse reflection spectra for Er^{3+} and

Tm^{3+} ions from 300 to 1800 nm spectral range were recorded using Lambda 950 Perkin–Elmer spectrophotometer. The UCL in spectral range from 380 to 1100 nm was recorded using a Horiba FHR1000 spectrometer. UCL energy yields were measured using an OL IS–670–LED integrating sphere, OL–770 VIS/NIR (Gooch & Housego) spectroradiometer and monochromator–spectrograph M833 Solar LS [27, 28].

3. Results

XRD pattern of the SrF_2:Er(2.5%),Tm(0.7%) powder after annealing at 600 °C presents on Figure 1. The XRD pattern of the synthesized powders corresponds to SrF_2 structure from JCPDS database 06–0262 without any impurity phases. The peaks position was shifted from JCPDS data because of solid solution formation.

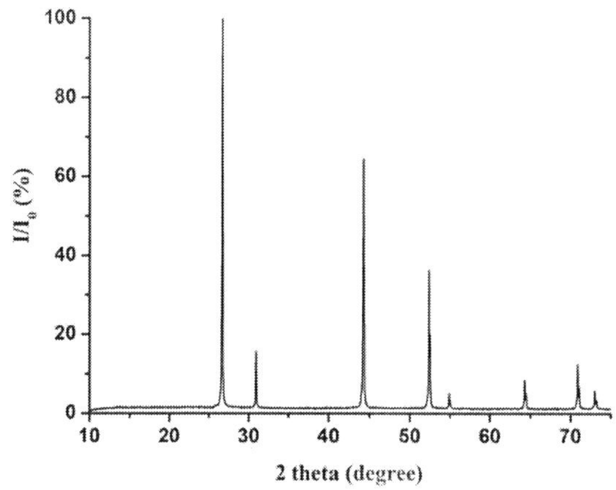

Figure 1. XRD pattern of the SrF_2:Er(2.5%),Tm(0.7%) powder after annealing at 600 °C.

Scanning electron microscopy shows that average diameter of SrF_2:Er,Tm particles with spherical shape is about 100 nm. The results of energy-dispersive analysis of SrF_2:Er,Tm solid solution presented in Table 1. The Er^{3+} and Tm^{3+} ions contents in solid solutions are similar to the initial aqueous solutions and higher for Er single-doped samples. It means the congruent and incongruent crystallization manner for SrF_2:Er,Tm and SrF_2:Er solid solutions, respectively.

Table 1. The chemical and luminescence characteristics of SrF_2:Er,Tm powders.

Composition in initial aqueous solution*	Composition based on EDX*	B_{en}, % 380-780 nm	B_{en}, % 380-1100 nm	X	Y	T, K	GRR
SrF_2:Er(0.5%), Tm(2.0%)	SrF_2:Er(0.5%), Tm(2.5%)	-	-	0.4717	0.4526	2846	0.18
SrF_2:Er(1.0%), Tm(2.0%)	SrF_2:Er(1.1%), Tm(2.7%)	-	0.2	0.5352	0.4417	2102	0.12
SrF_2:Er(2.0%), Tm(1.0%)	SrF_2:Er(2.3%), Tm(1.3%)	0.3	1.2	0.6303	0.3576	2173	0.04
SrF_2:Er(2.0%), Tm(0.5%)	SrF_2:Er(2.5%), Tm(0.7%)	1.0	4.4	0.6414	0.3471	2522	0.04
SrF_2:Er(7.4%)	SrF_2:Er(8.8%)	0.2	2.1	0.4172	0.5512	4151	0.32
SrF_2:Er(11.5%)	SrF_2:Er(14.2%)	0.2	2.1	0.4146	0.5480	4177	0.3

* - Er^{3+} and Tm^{3+} concentrations are presented in mol. %

The room-temperature diffuse reflection spectra of Er^{3+} and Tm^{3+} ions in the range 300–1800 nm for the SrF_2:Er(2.5%),Tm(0.7%) phosphor presents on Figure 2a. Absorption bands correspond to the transitions $^4I_{15/2} \rightarrow {}^4G_{11/2}$, $^4I_{15/2} \rightarrow {}^2H_{9/2}$, $^4I_{15/2} \rightarrow {}^4F_{5/2}$, $^4I_{15/2} \rightarrow {}^4F_{7/2}$, $^4I_{15/2} \rightarrow {}^2H_{11/2}$, $^4I_{15/2} \rightarrow {}^4S_{3/2}$, $^4I_{15/2}$

$\rightarrow$ $^4F_{9/2}$, $^4I_{15/2} \rightarrow {}^4I_{9/2}$, $^4I_{15/2} \rightarrow {}^4I_{11/2}$, $^4I_{15/2} \rightarrow {}^4I_{13/2}$ of Er^{3+} ions and $^3H_6 \rightarrow {}^1G_4$, $^3H_6 \rightarrow {}^3F_2$, $^3H_6 \rightarrow {}^3F_3$, $^3H_6 \rightarrow {}^3H_4$, $^3H_6 \rightarrow {}^3H_5$, $^3H_6 \rightarrow {}^3F_4$ of Tm^{3+} ions. The dashed arrows indicate the excitation of the $^4I_{13/2}$ level of Er^{3+} ions. Wide absorption infrared bands of Er^{3+} and Tm^{3+} ions potentially allow SrF_2:Er,Tm powders to visualize laser radiation in the 960 – 1000 nm and 1450 – 1800 nm areas.

UCL spectra of SrF_2:Er,Tm powders at 300 K in the visible and infrared spectral ranges upon laser excitation at 1531.8 nm were recorded (Figure 2b). These luminescence bands in spectra correspond to the transitions $^2H_{9/2} \rightarrow {}^4I_{15/2}$, $^4F_{5/2} \rightarrow {}^4I_{15/2}$, $^4F_{7/2} \rightarrow {}^4I_{15/2}$, $^2H_{11/2} \rightarrow {}^4I_{15/2}$, $^4S_{3/2} \rightarrow {}^4I_{15/2}$, $^4F_{9/2} \rightarrow {}^4I_{15/2}$, $^4I_{9/2} \rightarrow {}^4I_{15/2}$ and $^4I_{11/2} \rightarrow {}^4I_{15/2}$ of Er^{3+} ions. The relative intensity of the red luminescence (the $^4F_{9/2} \rightarrow {}^4I_{15/2}$ transition of Er^{3+}) is higher than the intensity of the spectral lines in the green and blue spectral regions for the $^4S_{3/2} \rightarrow {}^4I_{15/2}$ and $^2H_{9/2} \rightarrow {}^4I_{15/2}$ transitions, respectively (Figure 2b).

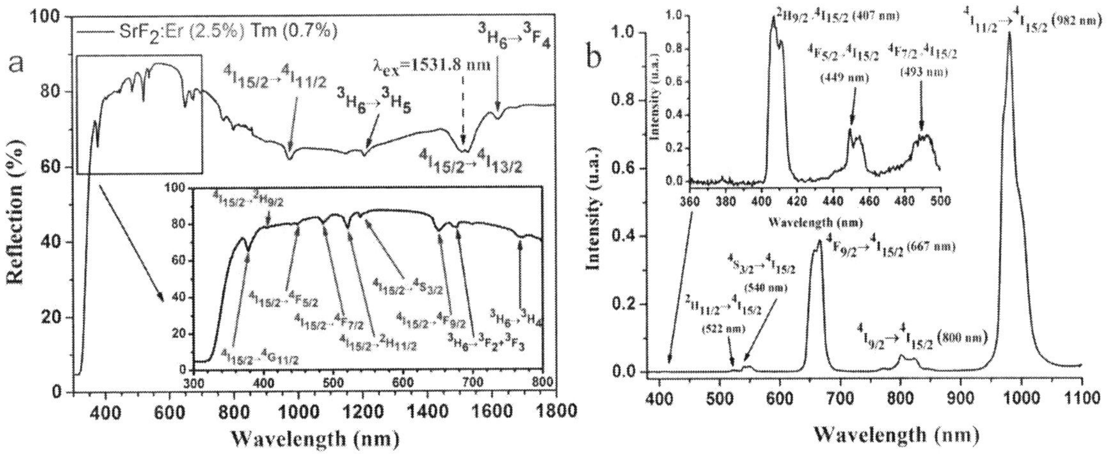

Figure 2(a, b). (a) The diffuse reflection spectra of Er^{3+} and Tm^{3+} ions in the 300 – 1800 nm spectral range for SrF_2:Er(2.5%),Tm(0.7%) phosphors. **(b)** UCL spectra of the SrF_2:Er(2.5%),Tm(0.7%) phosphors in visible and near-infrared spectral range.

The spectral power of the UCL of Er^{3+} ions in the visible and infrared spectral range upon laser excitation at 1531.8 nm for SrF_2:Er,Tm powders were recorded (Figure 3a). The most intensive luminescence in visible and infrared spectral range was observed for SrF_2:Er(2.5%),Tm(0.7%) powder. We also observed that the intensity ratio of green to red (GRR) UCL of the Er^{3+} ions can be tuned by changing concentration of Er^{3+} and Tm^{3+} ions (Figure 4). The intensity ratio of green to red was calculated by dividing the integrated UCL from $^4S_{3/2}$ and $^2H_{11/2}$ levels to the $^4I_{15/2}$ level of Er^{3+} ions by the integrated UCL from $^4F_{9/2}$ to $^4I_{15/2}$ level. This parameter shows the change in the spectrum of luminescence depending on various conditions. The intensity ratios of SrF_2:Er,Tm powders can be changed from 0.04 to 0.18 (Table 1). Previously [14], we show that the intensity ratio of green to red luminescence of Er^{3+} ions is the similar for all Er-doped strontium fluoride samples and is equal to 0.3 (Table 1). Intensity ratio of red to green UCL of SrF_2:Er,Tm are strongly depend on the concentration of Er^{3+} and Tm^{3+} ions. This result can be explained by upconversion mechanisms responsible for the UCL of SrF_2:Er,Tm powders. The energy level diagram of SrF_2:Er,Tm with absorption, luminescence and upconversion processes is shown in Figure 3b [24, 29]. The ETU6 process begins to populate the $^4F_{9/2}$ level of Er^{3+} ions and significantly change the intensity ratio of green to red UCL at Tm^{3+} co-doping.

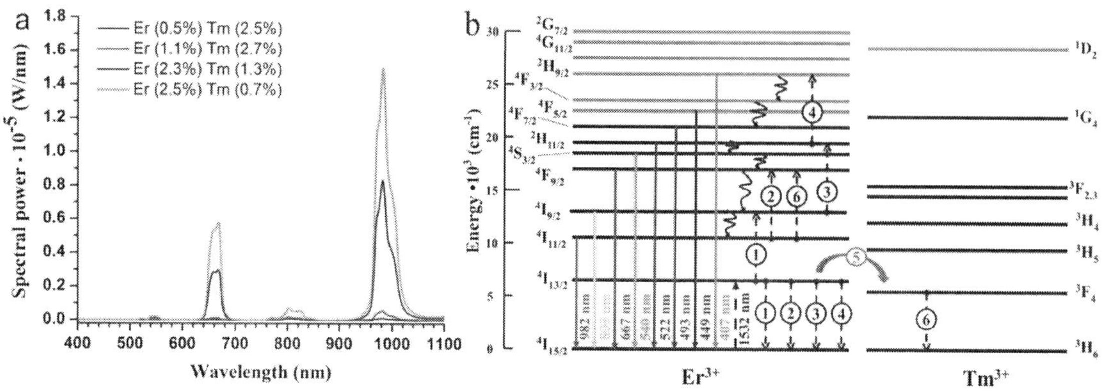

Figure 3(a, b). (a) The spectral power of the UC luminescence of SrF_2:Er,Tm phosphors. **(b)** Partial energy-level diagram of Er^{3+} and Tm^{3+} in SrF_2.

SrF_2:Er,Tm phosphors can be used as infrared laser beam visualizers. For these applications UCL energy yield, chromaticity coordinates and color temperature of UCL of SrF_2:Er,Tm powders were estimated (Table 1). Tuning ratio of green to red UCL was also showed on the CIE chromaticity diagram (Figure 4). The color temperatures of UCL of SrF_2:Er,Tm were 2102 – 2846 K.

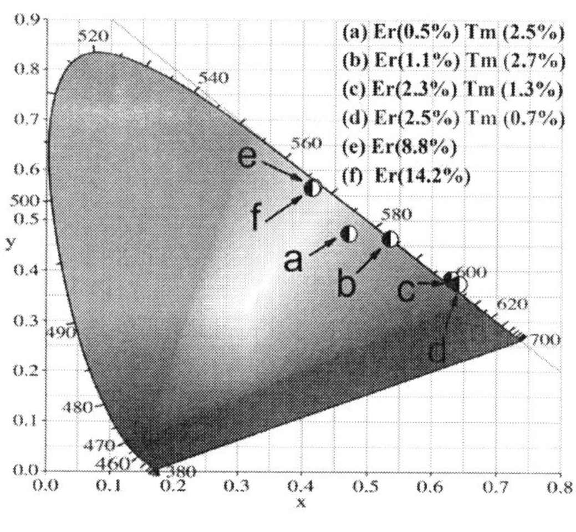

Figure 4. The CIE chromaticity diagram of SrF_2:Er,Tm powders. Excitation power density is 709 W/cm^2.

The photoluminescence energy yield B_{eu} is defined as the ratio of the UCL power to the absorbed power by materials. Maximum energy yields of UCL of 1.0% and 4.4% were achieved for the SrF_2:Er(2.5%),Tm(0.7%) in the spectral regions 380 – 780 nm and 380 – 1100 nm, respectively.

SrF_2:Er,Tm powders are promising upconversion materials due to the intensive visible UCL, high energy yield and wide absorption bands.

4. Conclusion

In summary, SrF_2:Er,Tm powders with different concentration of Er^{3+} and Tm^{3+} ions have been successfully prepared by using a co-precipitation from aqueous solution technique. For the first time UCL of SrF_2:Er,Tm powders upon excitation of $^4I_{13/2}$ level of Er^{3+} ions were investigated. Co-doping Tm^{3+} ions allows significantly change the intensity ratio of green to red UCL from 0.04 to 0.18.

Maximum energy yields of UCL of 1.0% and 4.4% were achieved for the SrF_2:Er(2.5%),Tm(0.7%) in the spectral regions 380 – 780 nm and 380 – 1100 nm, respectively. The color temperatures of UCL for the SrF_2:Er,Tm were 2102 – 2846 K. The results show that SrF_2:Er,Tm phosphors can be used as infrared laser beam visualizers.

Acknowledgments
This work was supported by the Russian Science Foundation under grant 17-72-10163.

References

[1] Vetrone F, Naccache R, Zamarro´n A, Juarranz de la Fuente A, Sanz-Rodrı´guez F, Martinez Maestro L, Martı´n Rodriguez E, Jaque D, Garcı´a Sole´ J and Capobianco J A 2010 ACS Nano. 4(6) 3254

[2] Dong N N, Pedroni M, Piccinelli F, Conti G, Sbarbati A, Enrique Ramı´rez-Herna´ndez J, Martı´nez Maestro L, Carmen Iglesias-de la Cruz M, Sanz-Rodriguez F, Juarranz A, Chen F, Vetrone F, Capobianco J A, Garcı´a Sole J, Bettinelli M, Jaque D and Speghini A 2011 *ACS Nano.* **5(11)** 8665

[3] Gnach A, Lipinski T, Bednarkiewicz A, Rybka J and Capobianco J A 2015 *Chem. Soc. Rev.* **44(6)** 1561

[4] Silversmith A J, Lenth W and Macfarlane R M 2015 *Appl. Phys. Lett.* **51(24)** 1977

[5] Gao G, Busko D, Joseph R, Turshatov A, Howard I A and Richards B S 2019 *Part. Part. Syst. Charact.* **36(3)** 1800462

[6] Bevan D J M, Greis O and Strahle J 1980 Acta Crystallogr. A 36 889

[7] Kazanskii S A, Ryskin A I, Nikiforov A E, Zaharov A Y, Ougrumov M Y and Shakurov G S 2005 Phys. Rev. B 72 014127

[8] Seelbinder M B and Wright J C 1979 *Phys. Rev. B* **20** 4308

[9] Rozhnova Yu A, Kuznetsov S V, Luginina A A, Voronov V V, Ryabova A V, Pominova D V, Ermakov R P, Usachev V A, Kononenko N E, Baranchikov A E, Ivanov V K and Fedorov P P 2016 *Mater. Chem. Phys.* **172** 150

[10] Lyapin A A, Ryabochkina P A, Chabushkin A N, Ushakov S N and Fedorov P P 2015 *J. Lumin.* **167** 120

[11] Fedorov P P, Luginina A A, Kuznetsov S V, Voronov V V, Lyapin A A., Ryabochkina P A, Chernov M V, Mayakova M N, Pominova D V, Uvarov O V, Baranchikov A E, Ivanov V K, Pynenkov A A and Nishchev K N 2017 *J. Fluorine Chem.* **202** 9

[12] Lyapin A A, Ryabochkina P A, Ushakov S N and Fedorov P P 2014 *Quantum Electron.* **44(6)** 602

[13] Lyapin A A, Kuznetsov S V, Ryabochkina P A, Merculov A P, Chernov M V, Ermakova Yu A, Luginina A A and Fedorov P P 2017 *Laser Phys. Lett.* **14(7)** 076003

[14] Lyapin A A, Gushchin S V, Kuznetsov S V, Ryabochkina P A, Ermakov A S, Proydakova V Yu, Voronov V V, Fedorov P P, Artemov S A, Yapryntsev A D and Ivanov V K 2018 *Opt. Mater. Express* **8(7)** 1863

[15] Lyapin A A, Gushchin S V, Ermakov A S, Kuznetsov S V, Ryabochkina P A, Proydakova V Yu, Voronov V V, Fedorov P P and Chernov M V 2018 *Chin. Opt. Lett.* **16(9)** 091901

[16] Lyapin A A, Ryabochkina P A, Gushchin S V, Kuznetsov S V, Chernov M V, Proydakova V Yu, Voronov V V and Fedorov P P 2018 *Opt. Spectrosc.* **125(4)** 537

[17] Fedorov P P, Luginina A A, Kuznetsov S V, Voronov V V, Lyapin A A, Ermakov A S, Pominova D V, Yapryntsev A D, Ivanov V K, Pynenkov A A and Nishchev K N 2019 *Cellulose* **26(4)** 2403

[18] Kuznetsov S V, Ermakova Yu A, Voronov V V, Fedorov P P, Busko D, Howard I A, Richards B S and Turshatov A 2018 *J. Mater. Chem. C* **6(3)** 598

[19] Pak A M, Ermakova J A, Kuznetsov S V, Ryabova A V, Pominova D V and Voronov V V 2017 *J. Fluorine Chem.* **194** 16

[20] Wang Z, Han W and Liu H 2019 *CrystEngComm* **21(11)** 1691
[21] Bullock S R, Reddy B R, Venkateswarlu P and Nash-Stevenson S K 1997 *J. Opt. Soc. Am. B* **14(3)** 553
[22] Seelbinder M B and Wright J C 1979 *Phys. Rev. B* **20(10)** 4308
[23] Ivanova S, Pellé F, Tkachuk A, Joubert M F, Guyot Y and Gapontzev V P 2008 *J. Lumin.* **128(5)** 914
[24] Pollack S A, Chang D B and Moise N L 1986 *J. Appl. Phys.* **60(12)** 4077
[25] Pollack S A and Chang D B 1988 *J. Appl. Phys.* **64(6)** 2885
[26] Balabhadra S, Debasu M, Brites C, Ferreira R and Carlos L 2017 *J. Lumin.* **189** 64
[27] Ryabova A V, Pominova D V, Krut'ko A V, Komova M G and Loschenov V B 2013 *Photon. Lasers Med.* **2(2)** 117
[28] Kobayashi A 2010 Absolute Measurements of Photoluminescence Quantum Yields of Organic Compounds Using an Integrating Sphere (Japan: Gunma University) p. 109
[29] Chan E M, Gargas D J, Schuck P J and Milliron D J 2012 *J. Phys. Chem. B* **116(35)** 10561

Spectral Properties of Magneto-plasmonic Nanocomposite. Vertical Shift of Magneto-Optical Hysteresis Loop

S V Tomilin, V N Berzhansky, A N Shaposhnikov, S D Lyashko, T V Mikhailova, O A Tomilina
Institute of Physics and Technologies, V.I. Vernadsky Crimean Federal University, Crimea, Simferopol, 295007, Russian Federation

Abstract. In present work the results about study of spectral features of the magneto-optical hysteresis loop (Faraday rotation) vertical shifting effect in the magneto-plasmonic nanocomposite Bi:YIG/$Au_{(NP)}$/GGG are shown. It is shown that in the spectral range near localized plasmon-polariton resonance (LPPR) the magneto-optical hysteresis loop has a vertical shift (Faraday "delta"). The loop shifting at the left and right side of the LPPR have a different sign and in the resonance the loop shifting become null. It is shown that the spectral locations of Faraday "delta" maxima are due to the width of the LPPR resonant line.

1. Introduction

Magneto-optical sensing systems and structures are widely used in control devices of fiber-optic communication lines, in manufacturing of sensitive magnetic field sensors for magnetocardiography and low-field magnetic tomography, in visualizing of magnetic fields for defects detection by eddy current, etc. [1]. One way to enhance the sensitivity of such systems and structures is enhancing the specific magnitude of magneto-optical effects (in particular the Faraday effect). There is a method to enhancing the Faraday effect by injecting the plasmon subsystem into the magneto-optical dielectric medium, which becoming to significantly enhancement of the Faraday rotation on the range of plasmon resonance [2]. A similar effect was also observed by the authors of this article but at sample remagnetization not only increasing of the magneto-optical hysteresis loop (MOHL) height (enhancement of the effect) was detected, but also an effect of the vertical shift of the loop (Figure 1), which named Faraday "delta".

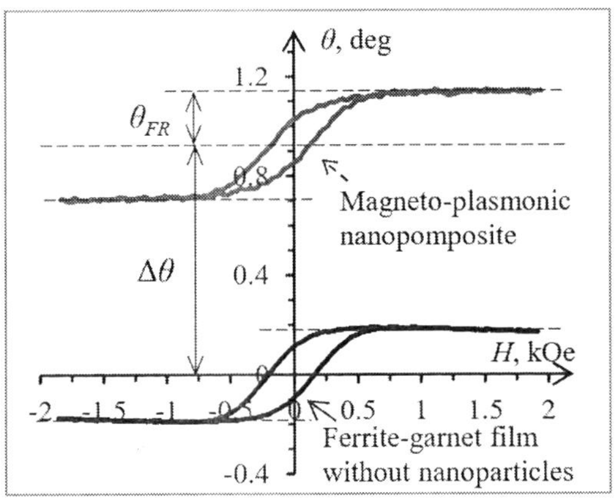

Figure 1. The magneto-optical hysteresis loop (MOHL) in pure bismuth-substituted ferrite-garnet film Bi:YIG/GGG (without plasmonic Au nanoparticles) and magneto-plasmonic nanocomposite Bi:YIG/$Au_{(NP)}$/GGG (with plasmonic Au nanoparticles):
θ_{FR} – angle of Faraday rotation at film magnetic saturation (half-height of the magneto-optical hysteresis loop);
$\Delta\theta$ – "Faraday delta", additive vertical shifting of the MOHL.

2. Experimental technique

The magneto-plasmonic nanocomposite Bi:YIG/$Au_{(NP)}$/GGG (Figure 2a) was synthesized by method of vacuum deposition with following heat treatment. It's a layer of self-assembled gold nanoparticles $Au_{(NP)}$ which formed on a gallium-gadolinium garnet (GGG) substrate and covered with a continuous layer of bismuth-substituted ferrite-garnet Bi:YIG. During materials deposition an original technique for obtaining films with a thickness gradient along the sample surface was used [3]. This technique allows to carry out studies of films with different thickness at one sample which was obtained in one technological cycle.

In this experiment we synthesized a sample with gradient of Au film effective thickness h_{eff}(Au). This sample was investigated in three areas where h_{eff}(Au) is 1.6 nm, 1.3 nm and 1 nm, respectively. Layer of bismuth-substituted ferrite-garnet Bi:YIG has constant thickness 100 nm for all sample area. Bi:YIG has $Bi_{1.5}Gd_{1.5}Fe_{4.5}Al_{0.5}O_{12}$ stoichiometric composition.

After Au film deposition, it has been treated by annealing on air at 950°C during 10 min for formation of self-assembled gold nanoparticles $Au_{(NP)}$/GGG. SEM-image on figure 2b demonstrate the surface morphology of granulated Au film. As can see, the form of self-assembled gold nanoparticles is ellipsoidal.

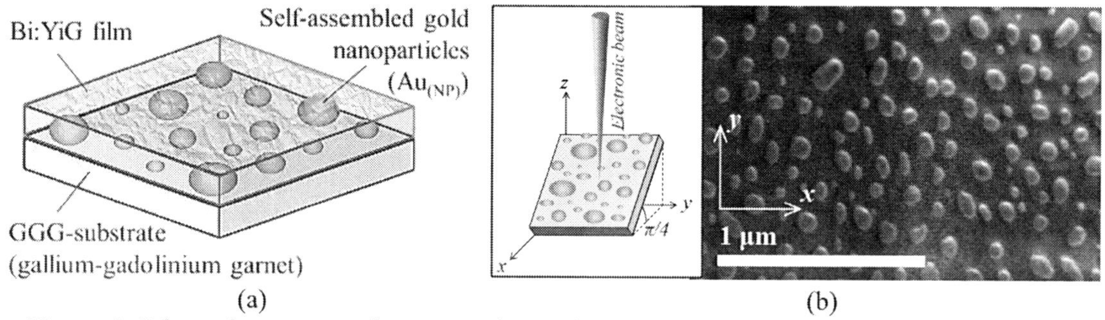

Figure 2. Schematic structure of magneto-plasmonic nanocomposite Bi:YIG/$Au_{(NP)}$/GGG (a); SEM-image of self-assembled gold nanoparticles $Au_{(NP)}$/GGG (b), image was making at 45° degree (π/4) along the direction of OX-axis (as shown inset).

The investigation of optical, plasmonic and magneto-plasmonic properties was carried out using the spectral photometer-magneto-polarimeter. Scheme of this device is shown on figure 3. The sample of magneto-plasmonic nanocomposite locate between two electromagnets on the optic way of light beam. The wave-vector of light beam ***k*** and the vector of magnetic field ***H*** are collinear. The optic scheme of magneto-polarimeter consist of the light source (halogen lamp), the monochromator with diffraction grating, some focusing lenses for formation of beam geometry, the unmoving polarizer and the rotatable analyzer for investigation of Faraday rotation, the semiconductor photo-detector. The optical range of monochromator is 400 – 1000 nm, the optical range of polarizer-analyzer is 470 – 750 nm.

The magneto-optical rotation spectra of the polarization angle of transmitted radiation (total rotation) were obtained by magnetization the sample to saturation in the ***H***+ field, directed along the radiation wave vector ***k***, and in the ***H***- field, directed against the wave vector. In this case, the Faraday rotation angle was determined as half of distance between the spectra in the ***H***+ and ***H***- field:

$$\theta_{FR} = [\theta_{TR}(\boldsymbol{H}+) - \theta_{TR}(\boldsymbol{H}-)]/2, \quad (1)$$

Accordingly, the shifting of the magneto-optical hysteresis loop (Faraday "delta") was determined as the value of the midline between the spectra in the ***H***+ and ***H***- field:

$$\Delta\theta = [\theta_{TR}(\boldsymbol{H}+) + \theta_{TR}(\boldsymbol{H}-)]/2 \quad (2)$$

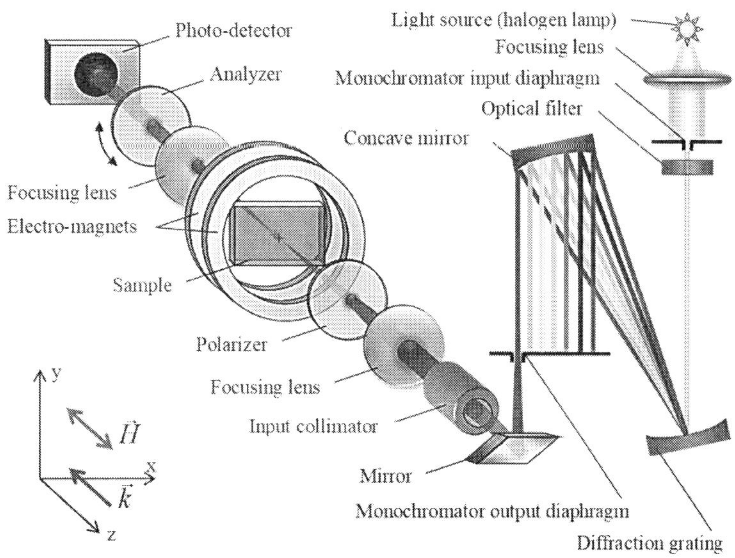

Figure 3. Scheme of spectral photometer-magneto-polarimeter for investigation of magneto-optical Faraday effect in magneto-plasmonic nanocomposite Bi:YIG/$Au_{(NP)}$/GGG.

3. Results and discussion

The investigation results of optical properties of magneto-plasmonic nanocomposite and his components are shown on figure 4. So, on figure 4a the transmittance spectra of Au/GGG film after deposition (before treatment) are presents. It's spectra for sample's areas with different effective thickness (1.6 nm – green line, 1.3 nm – red line, 1 nm – blue line).

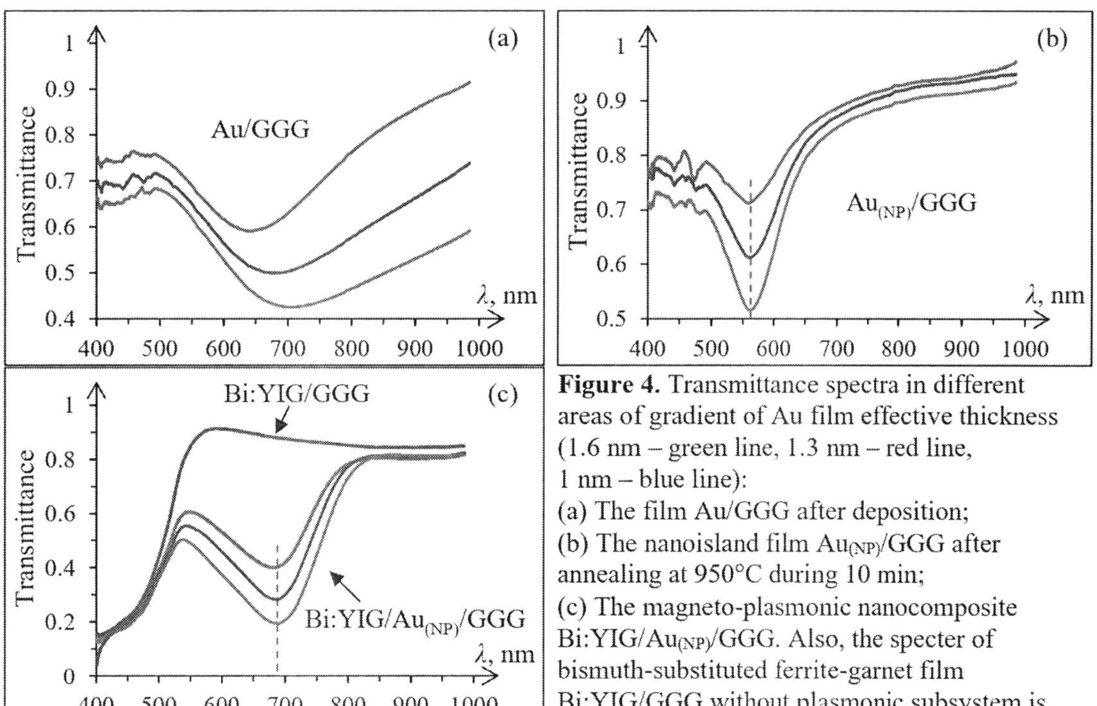

Figure 4. Transmittance spectra in different areas of gradient of Au film effective thickness (1.6 nm – green line, 1.3 nm – red line, 1 nm – blue line):
(a) The film Au/GGG after deposition;
(b) The nanoisland film $Au_{(NP)}$/GGG after annealing at 950°C during 10 min;
(c) The magneto-plasmonic nanocomposite Bi:YIG/$Au_{(NP)}$/GGG. Also, the specter of bismuth-substituted ferrite-garnet film Bi:YIG/GGG without plasmonic subsystem is shown (violet line).

The transmittance spectra of granulated nanoisland film $Au_{(NP)}/GGG$ after annealing on air at 950°C during 10 min are shown on figure 4b. As can see, the local minima are presents in spectra. These minima are associated with energy absorption for resonant excitation of localized plasmon-polaritons in self-assembled gold nanoparticles. The resonant wavelength λ_{LPPR} of localized plasmon-polariton resonance (LPPR) is 565 nm for all areas of Au effective thickness gradient.

On figure 4c the transmittance spectra of magneto-plasmonic nanocomposite $Bi{:}YIG/Au_{(NP)}/GGG$ are shown. Also, the specter of bismuth-substituted ferrite-garnet film Bi:YIG/GGG without plasmonic subsystem is shown for comparation (violet line). As can see, after deposition of 100 nm bismuth-substituted ferrite-garnet film on surface of self-assembled Au nanoparticles, the resonant wavelength λ_{LPPR} is shifted to the long-wave spectral range. The new resonant wavelength λ_{LPPR} = 685 nm. The spectral dip in short-wave range (less than 550 nm) is due to absorption of ferrite-garnet film.

The spectral features of magneto-optical properties of magneto-plasmonic nanocomposite $Bi{:}YIG/Au_{(NP)}/GGG$ are shown on figure 5.

For comparation, on figure 5a the magneto-optical properties of pure ferrite-garnet film Bi:YIG/GGG without plasmonic subsystem are presents. The spectra of magneto-optical total rotation of the transmitted radiation polarization angle $\theta(\lambda)$ are shown there. They were obtained during sample magnetization in the ***H***+ field, directed along the radiation wave vector ***k***, and in the ***H***- field, directed against the wave vector. The Faraday rotation specter, determined by equation (1), coinside with specter of total rotation in saturation field ***H***+, because the spectra in fields ***H***+ and ***H***- are symmetrical relatively the λ-axis. In this case, the Faraday "delta" $\Delta\theta$ equal to null in all spectral range.

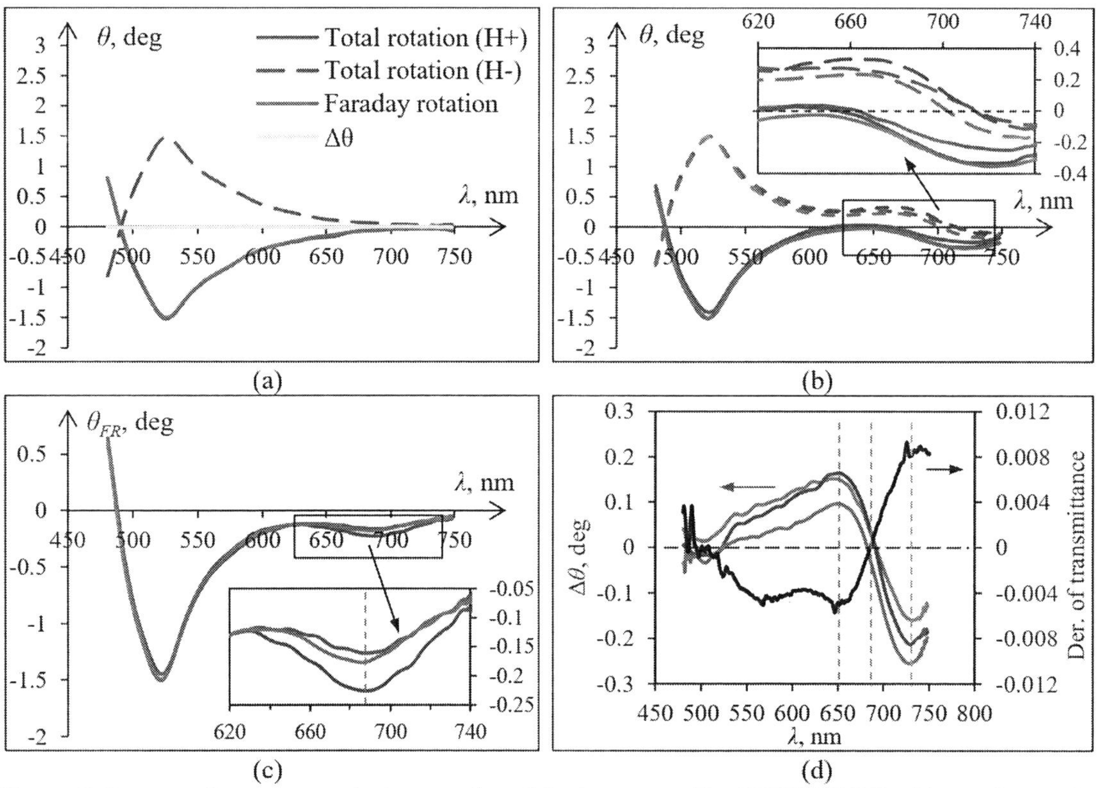

Figure 5. Spectra of magneto-optical properties of ferrite-garnet film Bi:YIG/GGG without plasmonic subsystem (a) and nanocomposite $Bi{:}YIG/Au_{(NP)}/GGG$ in different area of gradient of Au film effective thickness (1.6 nm – green line, 1.3 nm – red line, 1 nm – blue line): (b) The angle of total rotation of polarization plane (solid lines – in field ***H***+, dotted lines – in field ***H***-); (c) The angle of "true" Faraday rotation; (d) The "Faraday delta" and the derivative of transmittance (black line).

On figure 5b the spectra of magneto-optical total rotation in fields ***H***+ (solid lines) and ***H***- (dotted lines) are shown for magneto-plasmonic nanocomposite Bi:YIG/$Au_{(NP)}$/GGG. In the range of LPPR the asymmetric of spectra is observed. This asymmetric is due to vertical shifting $\Delta\theta$ of magneto-optical hysteresis loop.

The Faraday rotation spectra $\theta_{FR}(\lambda)$ of nanocomposite Bi:YIG/$Au_{(NP)}$/GGG are shown on figure 5c. Those spectra are determined by equation (1). As can see, the enhancement of Faraday rotation effect is observed on λ_{LPPR} (magenta dotted line). For example, at the area of the sample with Au effective thickness 1.3 nm (red line) the angle of Faraday rotation enhanced almost 3.5 times on λ_{LPPR} = 685 nm. (θ_{FR} = -0.225 deg for Bi:YIG/$Au_{(NP)}$/GGG and θ_{FR} = -0.065 deg for Bi:YIG/GGG).

On figure 5d the spectra of Faraday "delta" $\Delta\theta(\lambda)$ are presents. They were obtained by equation (2) as the average line between spectra of total rotation at fields ***H***+ and ***H***-. As can see, the specter of Faraday "delta" (vertical shift of magneto-optical hysteresis loop) has two extremums: the positive maximum at the "left" side of λ_{LPPR} (650 nm, brown dotted line) and the negative minimum at the "right" side of λ_{LPPR} (730 nm, orange dotted line). At the wavelength of plasmonic resonance the Faraday "delta" equal to null $\Delta\theta(\lambda_{LPPR})$ = 0 (685 nm, magenta dotted line).

For comparation the derivation of transmittance specter dT/dλ also shown on figure 5d (black solid line). As can see, both maximum and minimum of Faraday "delta" are located symmetrically to extremums of dT/dλ specter relatively the λ-axis. So, the spectral location of maximum and minimum of Faraday "delta" is determined by LPPR resonant line width.

This effect of MOHL shifting (Faraday "delta") is associated in [4, 5] with localized plasmonic resonance (SPR-induced rotation) for s- and p-polarized waves in anisotropic system of gold nanoparticles. But authors noted that the anisotropic don't clearly observing on SEM-images.

We think that this effect of MOHL vertical shifting is due to the constant additive to the angle of the polarization plane rotation. This additive doesn't depend on the magnitude and sign of the external magnetic field, and its magnitude is determined by the spectral position relatively to plasmonic resonance in self-assembled gold nanoparticles. The sign changes of the Faraday "delta" and the location of its maxima, which due to the width of the LPPR resonance line, can be explained as follows.

As it was shown in [3], the size distribution of self-assembled gold nanoparticles $Au_{(NP)}$/GGG has a quasisymmetric form and described well by the Gaussian (Figure 6a). At resonant frequency ν_{LPPR} (absorption maximum) the resonance conditions are satisfied for nanoparticles having the most probable size (distribution maximum). For particles that are smaller than the "resonant" (red region), the self-frequency exceed the frequency of the exciting field ν_{LPPR} ("advancing" frequency), at this case the particle polarizability is always directed against the field. For particles that are larger than the "resonant" (green region), the self-frequency will be lower than the frequency of the exciting field ν_{LPPR} ("lagging" frequency), so the particles will be polarized along the field [6].

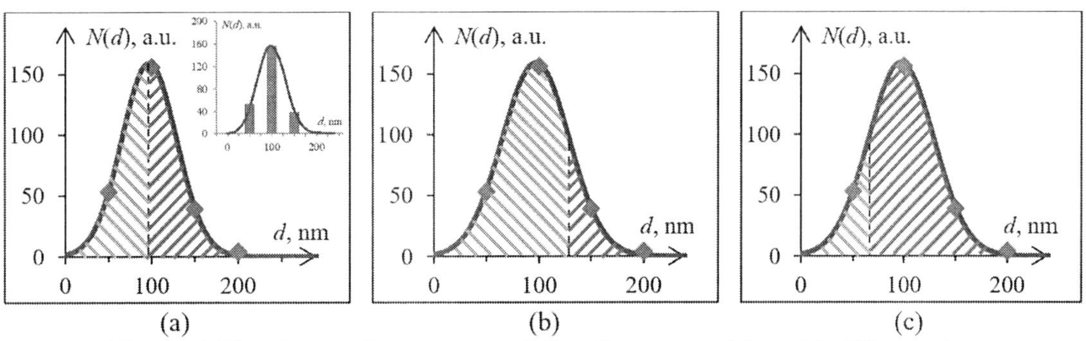

Figure 6. The change of resonant conditions for nanoparticles with different size (for explanation of Faraday "delta" features).

It is known, that near the LPPR condition, a resonating dipole (in our case it's a nanoparticle) induce a strong electromagnetic field around itself [7, 8]. This field is localized around the resonator and penetrated into the surrounding dielectric medium to distance of about 100 nm. It's obvious that the particles whose having "advancing" and "lagging" self-frequency (relative to the frequency of the exciting field) will induce resonant fields with the opposite direction.

Let's consider the state of the resonating system when the frequency of the exciting field is ν_{LPPR} (Figure 6a). At this case the number of particles with an "advancing" self-frequency is equal to the number of particles with a "lagging" self-frequency (the red and green shaded areas are equal). Thus, the resonant fields of the resonating nanoparticles will be mutually compensated and their affecting on the ferrite-garnet layer will be minimal. It explains the null value of the Faraday "delta" in the LPPR condition.

When the frequency of the exciting field is less then ν_{LPPR}, the strict resonant conditions are satisfied for larger particles (Figure 6b). In this case, the number of particles with an "advancing" self-frequency (red shaded area) significantly exceeds the number of particles with a "lagging" self-frequency (green shaded area). Thus, the uncompensated resonant field appear near the particles which polarized against the exciting field, and a negative MOHL shifting is observed.

When the frequency of the exciting field is more then ν_{LPPR}, the strict resonant conditions are satisfied for smaller particles (Figure 6c). Wherein the number of particles with an "advancing" self-frequency (red shaded area) significantly less than the number of particles with a "lagging" self-frequency (green shaded area). In this case, the uncompensated resonant field appear near the particles which polarized along the exciting field, and a positive MOHL shifting is observed.

Based on the above, the relations between the position of the Faraday "delta" maxima and the width of the LPPR resonant line becomes obvious. The maximum MOHL shifting is observed at the condition that the maximum number of nanoparticles is involved in the resonance and the number of particles with an "advancing" self-frequency are maximality exceeds the number of particles with a "lagging" self-frequency (or vice versa).

4. Conclusions

The existence of the plasmon subsystem in the magneto-plasmonic nanocomposite Bi:YIG/$Au_{(NP)}$/GGG leads to enhancement of the Faraday effect at the LPPR frequency in 3.5 times collation with a pure ferrite-garnet film Bi:YIG/GGG. The effect of vertical shifting of the magneto-optical hysteresis loop $\Delta\theta$ in magneto-plasmonic nanocomposite Bi:YIG/$Au_{(NP)}$/GGG was experimentally detected. This effect was observed near the condition of plasmon resonance in the $Au_{(NP)}$ plasmonic subsystem, at that the loop shifting change the sign on the left and right side from resonance, and in the resonance $\Delta\theta = 0$. The spectral locations of Faraday "delta" maxima (positive and negative) are due to the width of the LPPR resonance line.

Acknowledgments

This work was supported by the RF Ministry of Science and Highest Education in the framework of the base part of the state task (Project No. 3.7126.2017/8.9) and partially supported by the V.I. Vernadsky Crimean Federal University Development Program (Project No. ВГ22/2018).

References

[1] Bremer J, Vaicikauskas V, Hansteen F, Hunderi O. 2001. *J. of App. Phys.* **11** 6177.
[2] Tkachuk S, Lang G, Krafft C, Rabin O, Mayergoyz I. 2011. *J. of App. Phys.* **109** 07B717.
[3] Tomilin S V, Berzhansky V N, Shaposhnikov A N, at al. 2016. *J. of Phys.: Conf. Ser.* **741** 012113.
[4] Fujikawa R, Baryshev A V, Kim J, Uchida H, Inoue M. 2008. *J. of App. Phys.* **103** 07D301 ().
[5] Baryshev A V, Uchida H, Inoue M. 2013. *J. of Optic. Soc. of America* **9** 2371.
[6] Alu A, Engheta N. 2006. *Journal of the Optical Society of America B* (Special issue on "Metamaterials") **23** **3** 571
[7] Axelevitch A, Apter B, Golan G. 2013. *Optics Express* **21** **4** 4126.
[8] Kelly K L, Coronado E, Zhao L L, Schatz G C. 2003. *J. Phys. Chem.* B **107** 668

Synthesis, structure and spectroscopy of Fe^{2+}:$MgAl_2O_4$ transparent ceramics

L Basyrova[1], S Balabanov[2], A Belyaev[2], V Drobotenko[2], A Volokitina[1], V Vitkin[1], O Dymshits[3], and P Loiko[1]
[1]ITMO University, 49 Kronverkskiy Pr., 197101 St. Petersburg, Russia
[2]G.G. Devyatykh Institute of Chemistry of High-Purity Substances (IChHPS) of the Russian Academy of Sciences, 49 Tropinin St., 603950 Nizhny Novgorod, Russia
[3]S.I. Vavilov State Optical Institute, 36 Babushkina St., 192171 St. Petersburg, Russia

Abstract. Transparent ceramics based on magnesium aluminium spinel, $MgAl_2O_4$, doped with iron ions (Fe^{2+}) are synthesized by hot pressing (at 1500 °C / 50 MPa) of powders obtained by the sol-gel method. The ceramics are characterized by X-ray diffraction, SEM, Raman and optical spectroscopy. The ceramics exhibit a broad (1.3-3.5 μm) absorption due to the $^5E \rightarrow {}^5T_2$ (^5D) transition of Fe^{2+} ions in T_d sites and they are promising for applications as saturable absorbers of mid-infrared lasers.

1. Introduction

Divalent iron ions (Fe^{2+}) embedded in II-IV group materials such as ZnS or ZnSe are known for their intense and broadband absorption and emission bands in the mid-infrared (MIR) originating from the $^5E \leftrightarrow {}^5T_2$ (^5D) transitions for ions in tetrahedral (T_d) crystal field [1]. For Fe^{2+}:ZnS, the maximum absorption is at 2.8 μm and the emission band is centered at 3.94 μm. This makes Fe^{2+}-doped II-IV chalcogenides suitable for laser gain media [2,3] operating in this spectral range with multiple applications in gas sensing, molecular spectroscopy, range finding, medicine. Due to the high ground-state absorption cross-sections for Fe^{2+} ions in T_d sites (σ_{GSA} = 0.92×10^{-18} cm^2 for Fe^{2+}:ZnS), Fe^{2+}-doped materials are also interesting for saturable absorbers (SA) of mid-infrared lasers [4].

Due to difficulties in synthesis of high-optical-quality II-IV chalcogenides, it is promising to find other hosts for Fe^{2+} doping. One of the possible candidates for this is cubic spinel, $MgAl_2O_4$, featuring excellent thermo-mechanical properties. The $MgAl_2O_4$ spinel demonstrates widely variable occupations of two (tetrahedral and octahedral) sites by various cations and disordering of the cations between these sites. It is known for doping with transition-metal ions such as Co^{2+} and Ni^{2+} [5,6]. When doped with Co^{2+} ions, it is recognized as a SA for eye-safe lasers operating at ~1.5 μm [7].

Nowadays, transparent ceramic technology is becoming competitive to the traditional single crystal growth methods because is less technologically demanding and allows for size-scalable production and good optical quality. Recently, multiple studies were dedicated to the development of transparent spinel ceramics, such as $MgAl_2O_4$ and $ZnAl_2O_4$, both undoped or Co^{2+}-doped [8,9].

In this work, we aimed to fabricate and to study the microstructure, vibronic and optical properties of novel transparent ceramics based on Fe^{2+}:$MgAl_2O_4$ spinel.

2. Fabrication of ceramics

The magnesium aluminium spinel powders doped with Fe^{2+} ions were synthesized by the sol-gel method. Metallic magnesium and aluminum (>99% purity) and isopropyl alcohol (99.9% purity) were used as starting materials. Mg and Al metals (32 and 68 wt%, respectively) were melted at 600 °C under nitrogen. The alloy was crushed, mixed with 0.1 wt% of $SnCl_2$ and 0.2 wt% of NH_4Cl and heat treated in an enclosed stainless steel vessel. $MgAl_2(OPr^i)_8$ was synthesized in a tube reactor at 80–150 °C by an addition of small portions of alcohol to the granulated magnesium–aluminum alloy. After that, the final product was purified by vacuum distillation. The hydrolysis of 50 vol% solution of $MgAl_2(OPr^i)_8$ in isopropyl alcohol was conducted by using 1200 ml of isopropyl alcohol-water azeotropic mixture per 1 mol of alkoxide. The $MgAl_2(OPr^i)_8$:H_2O ratio was 1:8, which is an equimolar ratio required for complete hydrolysis. The azeotropic mixture contained iron (II) chloride as a source of iron ions. The addition of LiF was carried out as follows. 72 ml of suspension prepared by interaction of $LiNO_3 \cdot 3H_2O$ and NH_4F aqueous solutions was added to 1200 ml of Pr^iOH–H_2O azeotrope and further added into 1 mol of $MgAl_2(OPr^i)_8$ under stirring. After hydrolysis, the suspensions were cooled down to room temperature (RT) in air. Obtained suspensions were dried at 150 °C under ~10 torr pressure followed by calcination in air at 750 °C for 3 h. Resulted spinel powders contained 0.1–0.2 mol% Fe^{2+} ions and about 1.5 wt% of LiF.

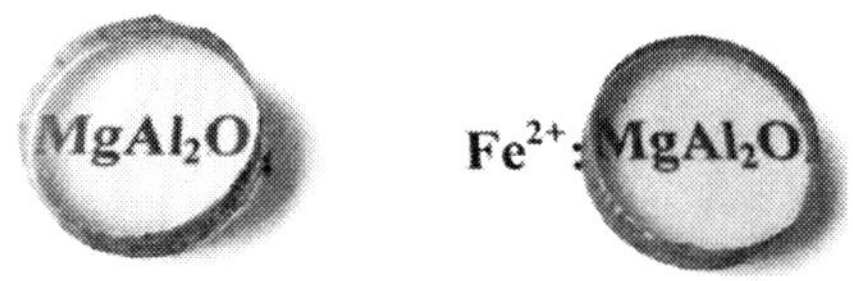

Figure 1. Photographs of the laser-grade-polished undoped $MgAl_2O_4$ (*left*) and 0.1 mol% Fe^{2+}:$MgAl_2O_4$ (*right*) ceramic disks.

To obtain Fe^{2+}:$MgAl_2O_4$ optical ceramics, the synthesized powders were pre-shaped, loaded into a graphite mold and heated from the RT up to 1600 °C with a heating rate of ~7.5°C/min and held for 6 h before cooling down. The load was applied at a temperature of 800 °C and then increased to ~50 MPa in 30 min. The pressure was released before free cooling of the furnace. The Fe^{2+}:$MgAl_2O_4$ transparent ceramic was laser-grade polished on both surfaces using a diamond suspension. The sample had dimensions of Ø13 × 2.8 mm and a weak grey coloration, Figure 1.

For comparison, undoped $MgAl_2O_4$ ceramics were prepared. They were transparent and colorless.

3. Results and discussion

The phase purity and the structure of ceramic were confirmed by X-ray powder diffraction (XRD), see Figure 2(a). The XRD patterns were measured with a Shimadzu XRD-6000 diffractometer, CuKα radiation with a Ni filter. The XRD pattern is typical for cubic spinel (sp. gr. O^7_h – $Fd\bar{3}m$) and does not contain traces of any impurity phase. The determined lattice constant a = 8.10 Å.

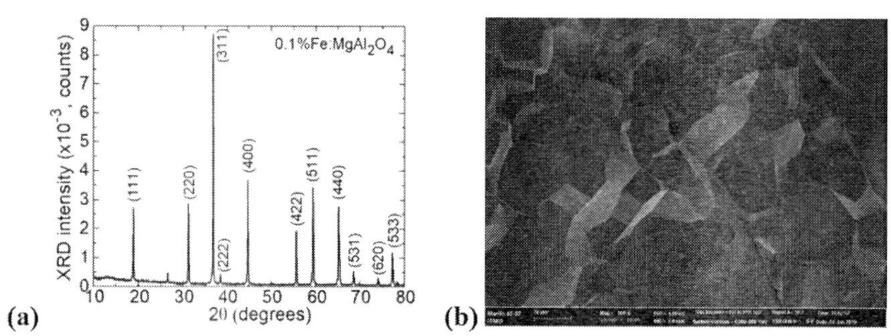

Figure 2(a, b). (a) X-ray powder diffraction (XRD) pattern of 0.1 mol% Fe^{2+}:$MgAl_2O_4$ ceramic; *numbers* denote the Miller's indices (*hkl*); **(b)** SEM image of the fracture surface of this ceramic. The scale bar is 20 μm.

The morphology of the fracture surface was characterized by Scanning Electron Microscopy (SEM) using a MERLIN SEM microscope (Carl Zeiss), Figure 2(b). The ceramic is almost pore-free. The grain boundaries are clean and there is no secondary phase at the boundaries. The average grain size is about 50 μm, see Figure 3.

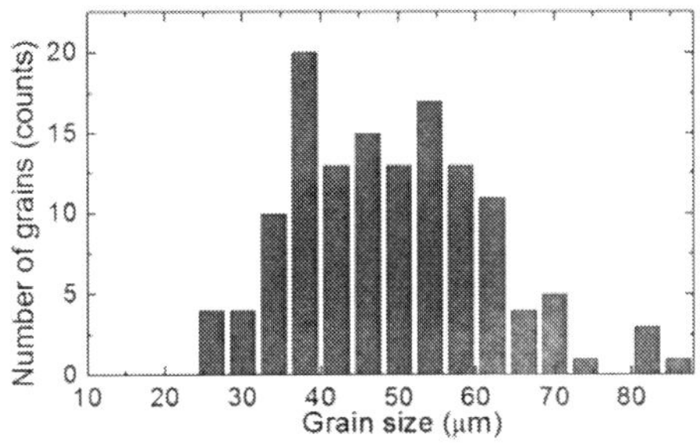

Figure 3. Grain size distribution of the 0.1 mol% Fe^{2+}:$MgAl_2O_4$ ceramic.

The vibronic properties of ceramics were studied by Raman spectroscopy at RT, Figure 4. A Renishaw inVia confocal Raman microscope with an excitation wavelength of 514 nm (Ar^+ laser line) was used. The characteristic spinel vibrations are observed at 303, 405, 496, 667, 721 and 764 cm^{-1}. The broad asymmetric band at 405 and the band at 721 cm^{-1} are typical features of synthetic spinel [10]. The band at 721 cm^{-1} is assigned to symmetric Al-O stretching vibration of AlO_4 groups created by redistribution of some Al^{3+} ions from octahedral to tetrahedral sites and evidences a certain disorder of the spinel structure.

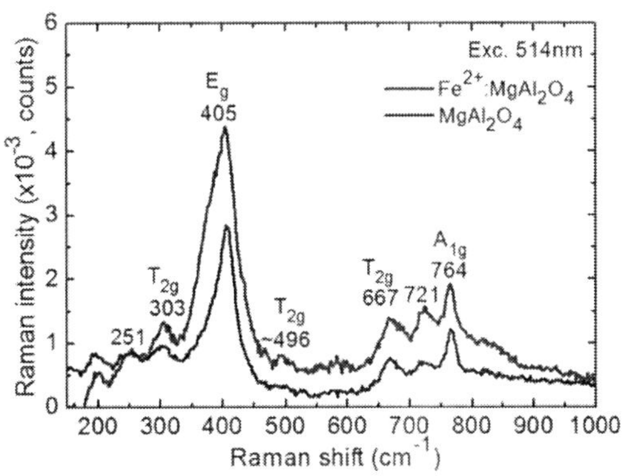

Figure 4. Unpolarized Raman spectra of Fe^{2+}:$MgAl_2O_4$ and undoped $MgAl_2O_4$ ceramics, λ_{exc} = 514 nm, *numbers* denote the positions of the Raman peaks in cm^{-1}.

The RT transmission spectra of the polished Fe^{2+}:$MgAl_2O_4$ ceramic samples are presented in Figure 5 and compared with that for the undoped $MgAl_2O_4$ ceramic. The spectra were measured using a FTIR spectrometer Bruker Tensor 27. In the spectra of Fe^{2+}-doped ceramics, a broad band spanning from 1.3 to 3.5 μm is observed due to the $^5E \rightarrow {}^5T_2$ (^5D) transition of Fe^{2+} ions in tetrahedral sites [11]. The intensity of this band increases with the Fe concentration. The shape of the absorption band is similar to that in Fe^{2+}:ZnSe but it is blue-shifted by about ~1.2 μm.

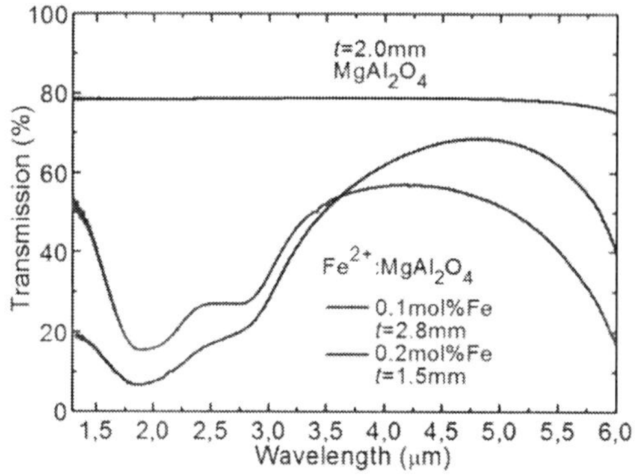

Figure 5. Infrared transmission spectra of laser-grade polished Fe^{2+}:$MgAl_2O_4$ and undoped $MgAl_2O_4$ ceramic disks (*t* - thickness).

4. Conclusions

We report on the synthesis of novel transparent Fe^{2+}-doped spinel $MgAl_2O_4$ ceramic promising for saturable absorbers of mid-infrared lasers (2-3 μm). Further studies will focus on the measurements of absorption saturation and laser damage threshold of this ceramic, as well as its possible applications for fabrication of saturable absorbers. It is also promising to study transparent glass-ceramics containing nanosized Fe^{2+}:$MgAl_2O_4$ spinel nanocrystals which may allow better optical quality of the host matrix.

Acknowledgments

This work was supported by Russian Science Foundation (Agreement No. 18-13-00200). The work at IChHPS was performed within the state assignment No. 0095-2018-0016. O.D. acknowledges financial support of the RFBR (Grant 19-03-00855). P.L. acknowledges financial support from the Government of the Russian Federation (Grant 074-U01) through ITMO Post-Doctoral Fellowship scheme.

References

[1] Mirov S, Fedorov V, Moskalev I, Martyshkin D, Kim C 2010 *Laser Photon. Rev.* **4** 21
[2] Voronov A A, Kozlovskii V I, Korostelin Y V, Landman A I, Podmar'kov Y P, Skasyrskii Y K, Frolov M P 2008 *Quantum Electron.* **38** 1113
[3] Doroshenko M E, Jelínková H, Koranda P, Šulc J, Basiev T T, Osiko V V, Komar V K, Gerasimenko A S, Puzikov V M, Badikov V V, Badikov D V 2009 *Laser Phys. Lett.* **7** 38
[4] Voronov A A, Kozlovskii V I, Korostelin Y V, Landman A I, Podmar'kov Y P, Polushkin V G, Frolov M P 2006 *Quantum Electron.* **36** 1
[5] Kuleshov N V, Mikhailov V P, Scherbitsky V G, Prokoshin P V, Yumashev K V 1993 *J. Lumin.* **55** 265
[6] Kuleshov N V, Shcherbitsky V G, Mikhailov V P, Kück S, Koetke J, Petermann K, Huber G 1997 *J. Lumin.* **71** 265
[7] Volk Y V, Malyarevich A M, Yumashev K V, Matrosov V N, Matrosova T A, Kupchenko M I 2007 *Appl. Phys. B* **88** 443
[8] Goldstein A, Loiko P, Burshtein Z, Skoptsov N, Glazunov I, Galun E, Kuleshov N, Yumashev K 2016 *J. Amer. Ceram. Soc.* **99** 1324
[9] Loiko P, Belyaev A, Dymshits O, Evdokimov I, Vitkin V, Volkova K, Tsenter M, Volokitina A, Baranov M, Vilejshikova E, Baranov A 2017 *J. Alloys Compd.* **725** 998
[10] Cynn H, Sharma S K, Cooney T F, Nicol M 1992 *Phys. Re. B* **45** 500
[11] Gaffney E S 1973 *Phys. Rev. B* **8** 3484

AUTHOR INDEX

AUTHOR INDEX

AUTHOR INDEX

AUTHOR INDEX

AUTHOR INDEX

AUTHOR INDEX

AUTHOR INDEX

AUTHOR INDEX

AUTHOR INDEX

6th International School and Conference "Saint Petersburg OPEN" 2019

Optoelectronics, Photonics, Engineering and Nanostructures

Journal of Physics: Conference Series Volume 1410

St. Petersburg, Russia
22 - 25 April 2019

Part 2 of 2

ISBN: 978-1-7138-1317-0
ISSN: 1742-6588

Printed from e-media with permission by:

Curran Associates, Inc.
57 Morehouse Lane
Red Hook, NY 12571

proceedings.com

Some format issues inherent in the e-media version may also appear in this print version.

Institute of Physics
Dirac House, Temple Back
Bristol BS1 6BE UK

Phone: 44 1 17 929 7481
Fax: 44 1 17 920 0979

techtracking@iop.org

Additional copies of this publication are available from:

Curran Associates, Inc.
57 Morehouse Lane
Red Hook, NY 12571 USA
Phone: 845-758-0400
Fax: 845-758-2633
Email: curran@proceedings.com
Web: www.proceedings.com

TABLE OF CONTENTS

PART 1

INVITED

CRYSTAL GROWTH AND STRUCTURAL PROPERTIES OF SEMICONDUCTOR MATERIALS AND NANOSTRUCTURES

NANOBIOTECHNOLOGY, BIOPHYSICS AND BIOPHOTONICS

LASERS, SOLAR CELLS AND OTHER OPTOELECTRONIC DEVICES

NANOPHOTONICS, SPECTROSCOPY, MICROCAVITIES, OPTICS, PLASMONICS

PART 2

SPINTRONICS, ELECTRO- AND MAGNETO- OPTICS

ELECTRIC, MAGNETIC AND MICROWAVE DEVICES

OTHER ASPECTS OF NANOTECHNOLOGY

Interface Layers of Niobium Nitride Thin Films

A V Lubenchenko[1], V A Iachuk[1], S Krause[2], A B Pavolotsky[2], D A Ivanov[1], O I Lubenchenko[1] and O N Pavlov[1]
[1]National Research University "Moscow Power Engineering Institute", Moscow, 111250, Russia
[2]Chalmers University of Technology, Göteborg, 41296, Sweden

Abstract. Intermediate layers formed by thin NbN films are studied. A surface phase of NbN different from the bulk one under the oxide layer and a layer consisting of NbN_x-SiO_y between the film and the substrate are found.

1. Introduction

Thin films of niobium nitride are widely used in hot electron bolometer mixers (HEB) which are the most sensitive detectors for spectroscopy in the terahertz frequency range above ca. 1 THz [1]. Performance of HEB's depends on uniformity of the 3-6 nm thin NbN layer [2]. It was found [3] that by air oxidation of NbN films a transition layer between niobium nitride and oxide is generated, about few monolayers thick. However, its phase compound was not clear. One of the purposes of this work is a study of this intermediate layer for films of different thicknesses. Operation properties of films depend on this interface significantly. Another purpose is to investigate the lower interface, between the NbN film and the substrate.

2. Experimental Details

NbN films were deposited on a silicon substrate by the method of magnetron sputtering in the experimental setting AJA Orion-5-U-D. The film thickness during sputtering was controlled by the known sputtering rate (the sputtering rate was approved by investigation of the films by TEM).

XPS spectra were recorded with the Nanofab 25 (NT-MDT) electron-ion spectroscopy platform. The pressure in the analysis chamber was of 10^{-7} Pa provided by the titanium sublimation pump and the ion pump. The residual pressure and composition of residual gas were monitored by a Bayard-Alpert vacuum gauge and a secondary ion mass-spectrometer. For XPS analysis, the X-ray source SPECS XR 50 with a dual Al/Mg anode was used providing 1486.6 eV and 1253.6 eV photons. The X-ray source was located at the angle of 54.7° relative to the analyzer axis. The spectra were recorded with the electrostatic hemispherical energy analyzer SPECS Phoibos 225. The energy resolution estimated as the full width at half maximum (FWHM) of the spectrometer at the Ag3d5/2 line (peak) was of 0.78 eV for non-monochromatic X-radiation Mg Kα. The energy positions of the spectra were calibrated with reference to the Cu2p3/2 (binding energy 932.62 eV), Ag3d5/2 (368.21 eV) and Au4f7/2 (83.95 eV) peaks. All sur-vey spectra scans were taken at a pass energy of 80 eV. The detailed scans of high intensity lines were in most cases recorded as wide as needed just to encompass the peak(s) of interest and were obtained with the pass energy of 20 eV. The energy analyzer was operated in Fixed Analyzer Transmission (FAT) mode.

The ion source SPECSIQE 12/38 was used for sputtering the samples. The ion source had differential pumping and was fed with 99.9995% pure Ar. The ion beam scanned the area of 2.8 mm × 4.0 mm at the incidence angle of 70° to the surface normal, the ion beam energy was of 1 keV. Ion profiling included total 5 cycles of sputtering with recording of XPS spectra before sputtering and after each cycle. For sputtering cycles #1–#5, each cycle was 20 min long.

3. Depth profiling

One of none-destructive methods of thin and ultra-thin films analysis is X-ray photoelectron spectroscopy (XPS). With the help of XPS chemical and phase depth profiling of surface is performed. By the standard XPS method, relative atomic concentrations are calculated assuming a homogeneous target by all the depth of profiling. However, real surfaces are always inhomogeneous and multicomponent by depth. Disregarding this leads to significant errors and depreciates information about relative atomic concentrations.

As a rule, a surficial region is not only multilayer but multicomponent and multiphase. Depth profiling based on interpretation of photoelectron spectra from such targets is a complex inverse problem with many a priori unknown parameters. For correct solution of this problem paper [4] offers: 1) a method of background subtraction considering the difference of energy losses on surface and in bulk; 2) using constant parameters in all the range of the photoelectron spectrum for background and line profile calculation; 3) using of line profile parameters from the spectra of Handbook of X-Ray Photoelectron Spectroscopy [5] for pure homogeneous targets; 4) simultaneous interpretation of different lines of the same element using the same model.

The reconstruction method of layer-by-layer profile is based on a target model that results from the history of its production and life. The principle statements of the target model used in this paper:

1) The target consists of few plane layers on a substrate. Each layer is homogeneous and may be multicomponent. Such statement is reasonable because an XPS signal is recorded from a surface area which dimensions are by orders greater than the probing depth. This area depends on focusing of the energy analyzed and geometry of the experiment and it is about 0.1 mm^2. The probing depth is about 10 nm. So the signal will be measurement area-averaged.

2) Inhomogeneities (islets, interlayer asperities, inclusions, etc.) are layer-averaged. Inhomogeneity rate will define relative concentrations of element phases.

3) After unload from the chamber, the surface is oxidized and oxide and suboxide layers are generated. As far as oxidation progresses from the surface, the oxidation level decreases with depth. Besides that, at the top a hydrocarbon layer will be precipitated.

To decrease the uncertainty of number of layers, we propose to introduce the minimum homogeneous layer thickness equal to the mean distance between monolayers in solid (0.5 nm). Let start the calculation from the maximum possible number of layers assuming each layer be of the constant oxidation rate. If calculation results that thickness of a layer be less than the minimum one, this layer is added to the next one with a higher oxidation rate of the main element. The newly formed layer will be multicomponent. This layer can be regarded as an inhomogeneous one averaged to a homogeneous layer of some effective thickness. The inhomogeneity degree may be estimated with the help of relative atomic concentration of the element in this layer. After that the layer thicknesses are recalculated. Such approach enables to decrease the number of alternate calculations significantly and to regard the alternates as profiles calculated with different depth detailing.

Layer thicknesses will be calculated by the formula [4]:

$$d_i = \lambda_i \cos\theta \ln\left(\frac{I_i/(n_i\omega_i(\gamma)\lambda_i)}{\sum_{j=0}^{i-1} I_j/(n_j\omega_j(\gamma)\lambda_j)} + 1\right) \quad (1)$$

where d_i is the thickness of the i-th layer, n is the atomic concentration, $\omega_i(\gamma)$ is the differential cross-section of photoelectron production [6], γ is the angle between the directions of incident radiation and to the energy analyzer, λ is the IMFP (IMFP is calculated by the TPP2M formula [7]), θ is the angle between the direction to the energy analyzer and the surface normal, I_i is the i-th peak intensity. The layers will be numerated upwards from the substrate. Number 0 means the substrate.

4. Results and Discussion

In this work, 5 nm film of NbN on a silicon substrate were studied. Standard XPS analysis of all samples identified presence of chemical elements like C, O, Nb, Si. Figure 1 shows survey spectra for a NbN 5 nm target after various stages of sputtering. Estimated relative atomic concentrations by XPS are presented in table 1.

Table 1. Relative atomic concentration.

	c, %					
	0	1	2	3	4	5
C	35.6	8.4	1.8			
O	33.0	30.3	18.6	40.4	14.4	16.0
N	6.1	15.6	24.2	12.8	26.0	21.0
Nb	23.1	42.8	48.4	41.3	45.3	29.5
Si	2.2	2.9	7.0	5.5	14.3	33.5

The XPS spectra can be used to analyze chemical bounds of elements and the chemical composition of the matter. For that an XPS line of a certain element is decomposed into corresponding peaks. Structure of these peaks may be quite complex due to chemical shift, satellite peaks, shift caused by sample charging etc. Additionally, shape and width of the peaks depend on various factors, and the peaks can over-lap. For deconvolution of XPS peaks, we used the method described in [8].

The shape of a spectral line of photoelectrons will be determined by a convolution of functions describing the natural line shape and instrumental broadening. The natural line shape will be described the Doniach-Šunjić expression and the instrumental broadening function by the Gauss function. We suggest to take the binding energy and spin-orbital interaction energy for chemically pure elements from the experimental data of the Handbook of X-ray Photoelectron Spectroscopy. The chemical shift energy is almost linearly proportional to the oxidation level, so it is enough to find the chemical shift energy for the most oxidized element. For example, we used a value of 5.31 eV for niobium oxide Nb_2O_5.

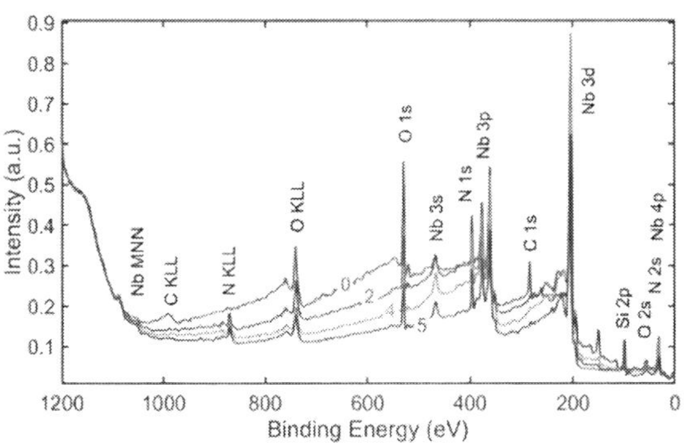

Figure 1. Survey XPS spectra. Target: NbN, 5 nm. Numbers: number of sputtering cycles.

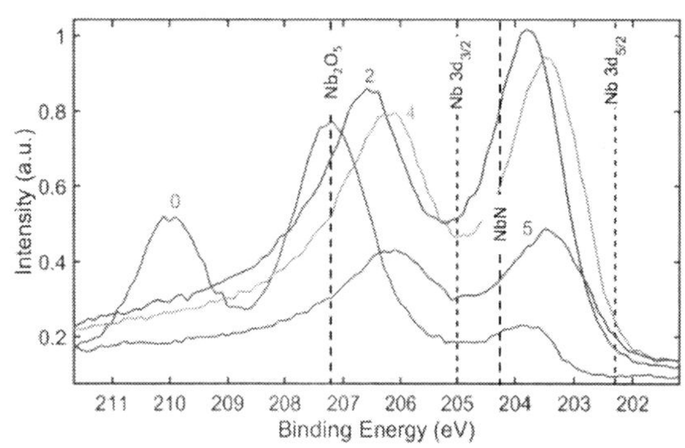

Figure 2. XPS spectra of line Nb 3d. Target: NbN, 5 nm. Numbers: number of sputtering cycles.

The chemical shift energy of nitrogen-niobium compound depends significantly on its stoichiometry. Paper [8] shows a dependence of chemical shift for lines Nb 3d and N 1s on the stoichiometric coefficient x in compound NbN_x. After sputtering positions of NbN peaks in line Nb 3d changes to lesser binding energies (see 'figure 2'). This indicates that the phase compound of a NbN film changes to lesser x in NbN_x.

The obtained XPS spectra are shown in figure 3-5. Circles show recorded detailed spectra, the solid lines: calculated using method [4], the dotted line: separate calculated peaks. Information on chemical and phase composition on the layers was obtained by the analysis of peaks produced by elastically scattered electrons.

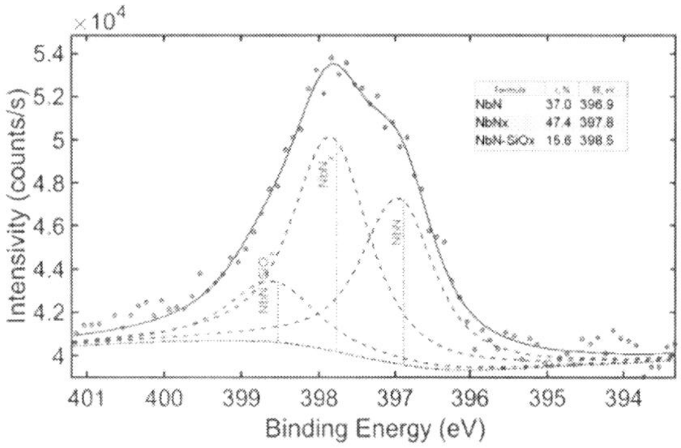

Figure 3. XPS spectra of line N 1s. Target: NbN, 5 nm. Solid lines: calculation. Circles: experimental data.

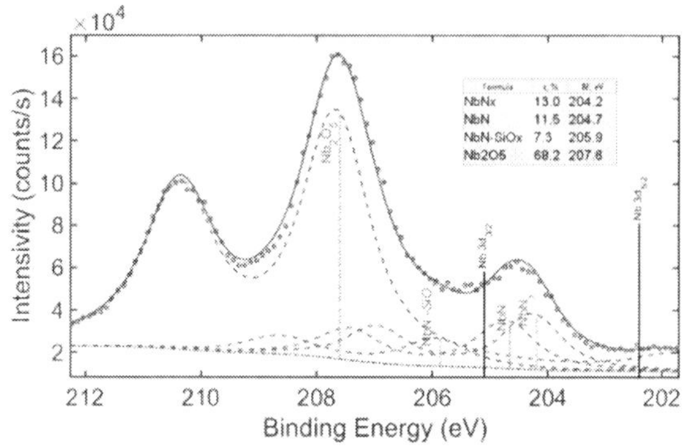

Figure 4. XPS spectra of line Nb 3d. Target: NbN, 5 nm. Solid lines: calculation. Circles: experimental data.

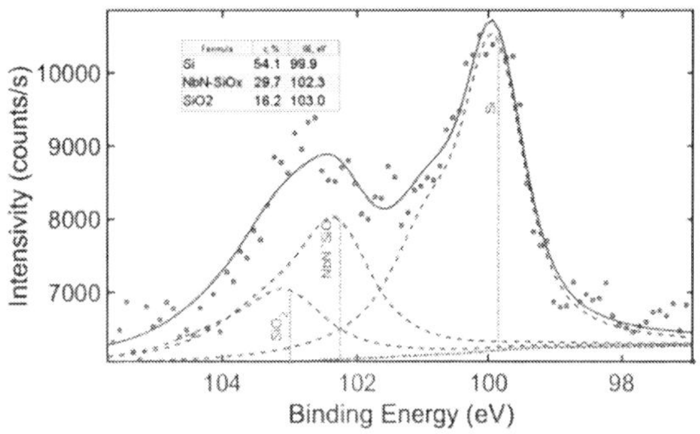

Figure 5. XPS spectra of line Si 2p. Target: NbN, 5 nm. Solid lines: calculation. Circles: experimental data.

As a result of this study, it was found that line N1s was decomposed into three peaks corresponding to two different phases of niobium nitride and complex compound of NbN_x and SiO_y 'figure 3'. The peaks of these phases were observed in line Nb 3d 'figure 4' and line Si 2p 'figure 5'. Following the data from [8] the phases of niobium nitride were determined: NbN и NbN_x, $x \approx 0.79$ that corresponded to Nb_5N_4. So before sputtering another phase was of lower oxidation rate of N. We assume these results show that by air oxidation under the oxide layer an intermediate layer is formed with a surface phase different from the phase of NbN in the bulk. The thickness of that layer is about one or two monolayers (1 nm). By sputtering of niobium nitride due to prevailing sputtering of nitrogen by argon ions (the sputtering yield of nitrogen is more than that of niobium) the phase of niobium nitride shifted to lesser x. After the 5th sputtering stage $x \approx 0.73$ that corresponded to Nb_4N_3.

Table 2. Chemical and Phase Depth Profiling of a $NbN/SiO_2/Si$ Target.

	d, nm		
	0	5	
6	0.8	-	hydrocarbons
5	2.9	-	Nb_2O_5
4	0.8	0.9	NbN_x
3	0.9	0.4	NbN
2	2.3	1.5	NbN_x-SiO_y
1	1.3	0.6	SiO_2
Substrate			Si

Interpretation of XPS spectra before and after sputtering enabled to determine thickness and phase compound of the transition layer between the NbN film and the substrate formed by film growth. Our

measurements showed the thickness of this layer about 0.5 nm. This layer consists of a complex compound of NbN_x and SiO_y. This conclusion results from decomposition of the line Si2p: it is decomposed into three doublets corresponding to Si, NbN_x-SiO_y and SiO_2 (see 'figures 3-5').
Table 2 displays computing results for thicknesses following the formula (1) from work which accounts for partial intensities obtained by decomposition of the lines N1s, Si2p and Nb3d.
Our investigation shows that the structure of NbN films is always complex, depending on conditions of growth and oxidation. Influence of interface onto HEB's performance requires further research.

References

[1] Meledin D, Pavolotsky A, Desmaris V et. all. 2009 *IEEE Transactions on Microwave Theory and Techniques* **57** 89–98.
[2] Sergeev A V and Reizer M Y 1996 *International Journal of Modern Physics B* **10** 635–67.
[3] Lubenchenko A V, Batrakov A A, Pavolotsky A B et. all. *EPJ Web of Conferences* vol 132, p 03053
[4] Lubenchenko A V, Batrakov A A, Pavolotsky A B, Lubenchenko O I and Ivanov D A 2018 *Applied Surface Science* **427** 711–21.
[5] Moulder J F, Stickle W F, Sobol P E and Bomben K D 1979 *Handbook of X Ray Photoelectron Spectroscopy: A Reference Book of Standard Spectra for Identification and Interpretation of XPS Data* (Physical Electronics).
[6] Yeh J J and Lindau I 1985 Atomic subshell photoionization cross sections and asymmetry parameters: $1 \leqslant Z \leqslant 103$ *Atomic Data and Nuclear Data Tables* **32** 1–155.
[7] Tanuma S, Powell C J and Penn D R 2003 *Surface and Interface Analysis* **35** 268–75.
[8] Prieto P, Galán L and Sanz J M 1991 *Surface Science* **251–252** 701–5.

Study of the effect of carbon nanotube lengths on their conductivity in the terahertz frequency range during optical pumping

R Baimagambetova, D Gomon, P Demchenko, I Anoshkin, M Khodzitsky
THz Biomedicine Laboratory, Department of Photonics and Optical Information Technologies, ITMO University, St. Petersburg, 197101, Russian Federation

Abstract. In this work, the conductivity of carbon nanotubes with the different lengths with / without optical pumping in the frequency range of 0.2 - 1 THz was experimentally studied. The results show that the restructuring of the geometric properties of carbon nanotubes will allow to create THz devices with tunable spectral characteristics.

1. Introduction

Terahertz (THz) waves offer unprecedented applications in imaging, spectroscopy, and high speed communications. Critical components, such as s metalenses [1-2], magneto-tunable devices [3-6], tunable reflectors [7], switchers [8], filters [9-11], phase shifters [12], perfect absorbers [13], left-handed media [14-15], polarization converters[16-18] are still open for improvement and integration, which will have a direct enhancement on the performance on THz systems. At the moment there are no commercially implemented controlled devices (polarizers, filters, etc.) of THz frequency range, therefore the need to create THz tunable devices is an important issue. To solve this problem, such materials as graphene [19], vanadium oxide (VO_2) [20], liquid crystals [21], and carbon nanotubes (CNTs) [22] can be used. CNTs are effective materials with a wide range of properties for a number of applications (large-scale structures in automobiles and nanometer-scale electronics). On the basis of carbon nanotubes, it is possible to develop devices for THz frequency range, including tunable ones under various influences (electrical, optical, etc.).

2. Experimental

Carbon nanotubes were synthesized by aerosol chemical vapor deposition with precise control of their morphology, formation of arrays and layer thickness [23]. This method allows obtaining CNT with very high quality. The length of the CNT was visualized by transmission electron microscopy, and the length distributions were measured using image analysis [24]. The diameters of the CNT were calculated from the position of the S_{11} peak from the Kataura plot [25]. The conductivities of CNT samples were studied using THz time–domain spectrometer with/without optical pumping at wavelength of 980 nm. The radiation was generated by an InAs crystal using Dember photoeffect. The detector is an electrooptic crystal which has a linear electrooptic effect (Pockels' effect). The calculation of complex conductivity was carried out using the thin film method according to the Drude model:

$$\hat{\sigma}(\omega) = \frac{\left[\frac{(\hat{n}_{sub}(\omega)+1)\hat{E}_0(\omega)}{\hat{E}(\omega)} - \hat{n}_{sub}(\omega) - 1\right]}{Z_0}, \quad (1)$$

where ω is the circular frequency, $\hat{n}_{sub}(\omega)$ is the complex refractive index of the substrate, $\hat{E}_0$ and $\hat{E}$ are the complex amplitudes of the electric field of THz wave transmitted through the substrate and the substrate with CNT, respectively, $Z_0 = 377$ Ohm is the impedance of free space.

3. Results and discussion

Using the formula (1) the dispersions of the real and imaginary parts of the conductivity of CNTs with different lengths (20-40 μm; 20μm; 10-15 μm) were obtained (Figure 1 (a, b)). It is seen that the real part of conductivity increases, and the imaginary part of the conductivity decreases while increasing the CNT length.

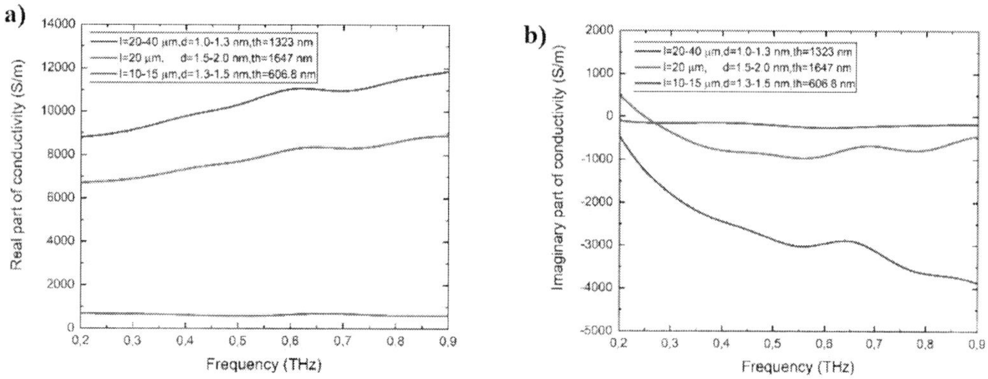

Figure 1 (a, b). Dispersion of the real **(a)** and imaginary **(b)** part of the conductivity at the different lengths of CNTs.

The real and imaginary parts of the CNTs permittivity increase while increasing the CNT length. It is shown in Figure 2 (a, b).

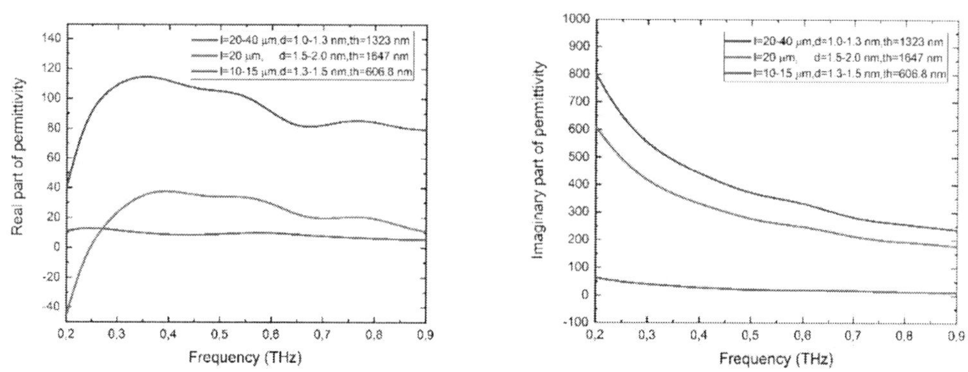

Figure 2 (a, b). Dispersion of the real **(a)** and imaginary **(b)** part of the permittivity at different lengths of nanotubes.

Then the influence of optical pumping on the complex conductivity of CNTs was studied. The dispersions of the real (a) and imaginary (b) parts of the conductivity of CNTs with length of 20-40 μm, diameter of 1.0-1.3 nm, the thickness of 1323 nm at the different optical pumping intensities were obtained (Figure 3 (a, b)). The real part of the CNTs conductivity decreases and the imaginary part of the CNTs conductivity decreases as the optical pumping radiation intensity at wavelength of 980 nm increases.

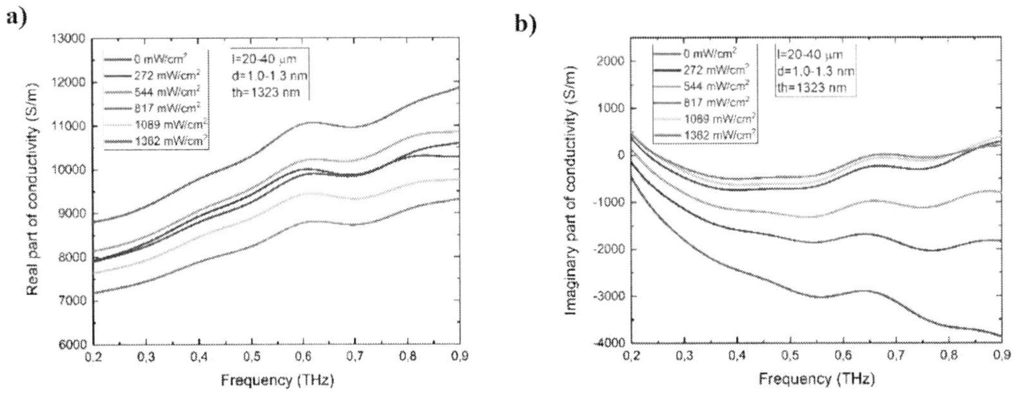

Figure 3 (a, b). Dispersion of the real **(a)** and imaginary **(b)** part of the conductivity at the different intensities of optical pumping.

Thus, we experimentally demonstrated that the changing of geometric parameters of CNTs and optical pumping can control the CNTs conductivity in the THz frequency range. This gives us the opportunity to create tunable terahertz devices based on CNTs.

Acknowledgements

This work was financially supported by Government of Russian Federation (Grant 08–08).

References

[1] Khorasaninejad M, Chen W T, Zhu A Y, Oh J, Devlin R C, Rousso D, Capasso F 2016 *Nano letters* **16**(7), 4595-4600.
[2] Khodzitsky M K, Vozianova A V, Gill V V, Chernyadiev A V, Grebenchukov A N, Minin I V and Minin O V 2016 *Proceedings of SPIE - The International Society for Optical Engineering* **9918** 99182X.
[3] Tarapov S I, Khodzitskiy M, Chernovtsev S V, Belosorov D, Merzlikin A M, Vinogradov A P and Granovsky A B 2009 *Solid State Phenomena* **152-153** 394-396.
[4] Tarapov S I, Khodzitsky M, Chernovtsev S V, Belozorov D, Merzlikin A M, Dorofeenko A V, Vinogradov A P, Inoue M and Granovsky A B 2010 *Physics of the Solid State* **52** 1427-1431.
[5] Girich A, Khodzitsky M, Nedukh S, Tarapov S 2011 *Terahertz and Mid Infrared Radiation. NATO Science for Peace and Security Series B: Physics and Biophysics* 159-164.
[6] Grebenchukov A N, Azbite S E, Zaitsev A D, Khodzitsky M K 2019 *Journal of Magnetism and Magnetic Materials* **472** 25-28.
[7] Plum E 2016 *Applied Physics Letters* **108** 241905.
[8] Grebenchukov A, Zaitsev A, Khodzitsky M 2018 *Chinese Optics* **11**(2) 166-173.
[9] Denisultanov A Kh and Khodzitsky M K 2013 *Proceedings of SPIE - The International Society for Optical Engineering* **8806** 88062I.

[10] Soboleva V Y, Gomon D A, Sedykh E A, Balya V K, Khodzitskii M K 2017 *Journal of Optical Technology* **84**(8) 521.
[11]Terekhov Y E, Khodzitsky M K, Grachev Y V, Sedykh E A, Belokopytov G V and Zhang X-C 2013 *Proceedings of SPIE - The International Society for Optical Engineering* **8806** 88062Q.
[12] Smirnov S, Anoshkin I V, Demchenko P, Gomon D, Lioubtchenko D V, Khodzitsky M, Oberhammer J 2018 *Nanoscale* **10**(26) 12291-12296.
[13] Gomon D A, Sedykh E A, Rodriguez S, Monroy I T, Zaitsev K I, Vozianova A V, Khodzitsky M K 2018 *Chinese Optics* **11**(1) 47-59.
[14] Khodzitsky M K, Kharchenko A A, Strashevskyi A V and Tarapov S I 2009 *Telecommunications and Radio Engineering* **68** 561-566.
[15] Litvinov E A, Demchenko P s, Sheklanova E B, Vozianova A v, Khodzitsky M K 2018 *Proceedings of SPIE - The International Society for Optical Engineering* **10719** 1071939.
[16] Korolenko S Yu, Grebenchukov A N, Masyukov M S, Vozianova A V, Khodzitsky M K *2016 J. Phys.: Conf. Ser.* **735** 012089.
[17] Masyukov M S, Vozianova A V, Grebenchukov A N, Khodzitsky M K 2018 *J. Phys.: Conf. Ser.* **1062** 012012.
[18] Chebotarev V S, Solov'ev A N, Trofimov A D, Khodzitskiĭ M K 2017 *Journal of Optical Technology* **84**(8) 525-527.
[19] Sensale-Rodriguez B et al. 2012 *Nature communications* **3** 780.
[20] Liu M et al. 2012 *Nature* **487**(7407) 345.
[21] Yu Jian-Ping et al. 2018 *Optics express* **26**(2) 663-673.
[22] Nefedova I et al. 2016 *IEEE Transactions on Terahertz Science and Technology* **6**(6) 840-845.
[23] Zhang Q, Wei N, Laiho P and Kauppinen E I 2017 *Topics in Current Chemistry* **375**(6) 90.
[24] Krause B et al. 2011 *Carbon* **49**(4) 1243-1247.
[25] Kataura H et al. 1999 *Synthetic metals*, **103**(1-3), 2555-2558.

Flat-top laser beams over an extended range

Andra Naresh Kumar Reddy[1], Vishwa Pal[2], Simon Mahler[1], Asher A. Friesem[1], and Nir Davidson[1]
[1]Department of Physicsof Complex Systems, Weizmann Institute of Science, Rehovot 7610001, Israel
[2]Department of Physics, Indian Institute of Technology Ropar, Rupnagar 140001, Punjab, India

Abstract. Designs based on single diffractive-optical-elements for obtaining flat-top laser intensity distributions that remain constant over a long range during free-space propagation are presented. Flat-top beams with different orders *n* exhibit a different range of propagation. For various working distances z, the resulting flat-top beam yields a different depth of focus. By controlling spectral properties of laser distributions, it is possible to maintain invariant flat-top intensity distributions for relatively long propagation distances.

1. Introduction

Flat-top laser distributions have been incorporated into several applications, such as annealing of photovoltaics, micromachining, laser scanning, laser radar antenna arrays, and lithography [1-3]. The formation of flat-top beams (uniform intensity profile) involves intensity and phase modulations with either transmissive or refractive or diffractive optical elements [4-5]. Such elements redistribute the intensity distribution of a propagating beam.

Perfect flat-top beams have sharp edges, resulting in strong diffraction and distortions during propagation. Such strong diffraction and distortions can be alleviated by reducing the flatness requirement of the propagating beams. For applications where long working distances are required, it is essential to extend the depth of focus. For achieving the long depth of focus, consider the flat-top beam with round edges, which have a continuous conversion from uniform intensity to zero intensity [5-6]. Usually, n stands for the order of super-Gaussian flat-top beam, and the order $n > 2$ yields nearly flat-top laser distribution. For instance, the flat-top laser distribution with $n = 6$, whose intensity profile remains constant over a long propagation distance whereas the flat-top distribution with $n = 12$ distorts rapidly during free-space propagation.

In this study, we convert Gaussian distributions into flat-top distributions that maintain relatively uniform phase distribution. The much needed phase-only diffractive optical elements (DOEs) are designed through a modified Grechberg-Saxton (GS) algorithm, to form nearly ideal flat-top laser distributions that have relatively uniform phase distributions. The range of propagation of flat-top beams depends on different parameters such as the order n, working distance z, beam waist W_{out}, operating wavelength λ [5-6]. All these parameters considered into account while calculating intensity distributions of propagated flat-top beams.

2. Basic procedure

In the modified GS algorithm that we used, we begin with a well-known Gaussian field distribution having random phases and then propagate it over a working distance z. After propagation, a new amplitude and phase distributions are obtained, onto which we superimpose non-Gaussian flat-top distribution. The combined new distribution is relayed back to the initial position after spatial filtering, and the procedure continues iteratively until finding the needed phase profile for the DOE. Some representative results are shown in Figure 1. Figures 1(a) shows the intensity distribution of a typical Gaussian input beam (n = 2) with beam waist W_{in} = 1.5 mm and λ = 808 nm, Fig.1(b) shows the phase profile of a DOE, and Fig. 1 (c) the corresponding nearly flat-top beam, which is a super-Gaussian beam (n = 6). The Gaussian input beam, incident on DOE at z = 0, is transformed into a super-Gaussian flat-top output beam after propagating a working distance z = 30 cm.

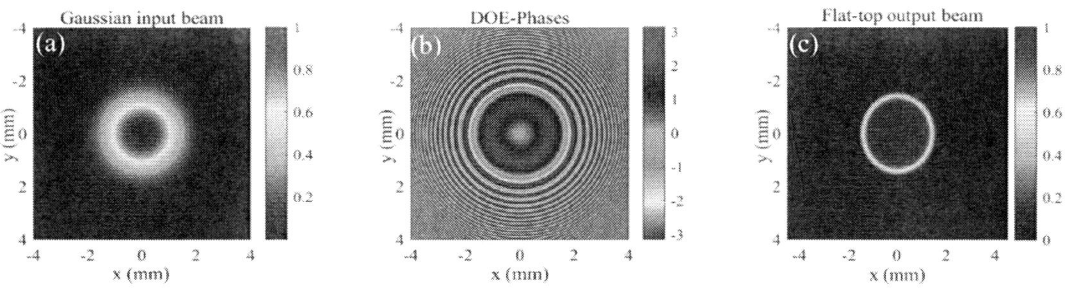

Figure 1(a,b,c). Transforming a Gaussian input beam to a flat-top output beam over a working distance of z = 30 cm. **(a)** Gaussian input intensity distribution; **(b)** DOE phase profile at z = 0; **(c)** Flat-top output intensity distribution at z = 30 cm.

3. Results and Discussion

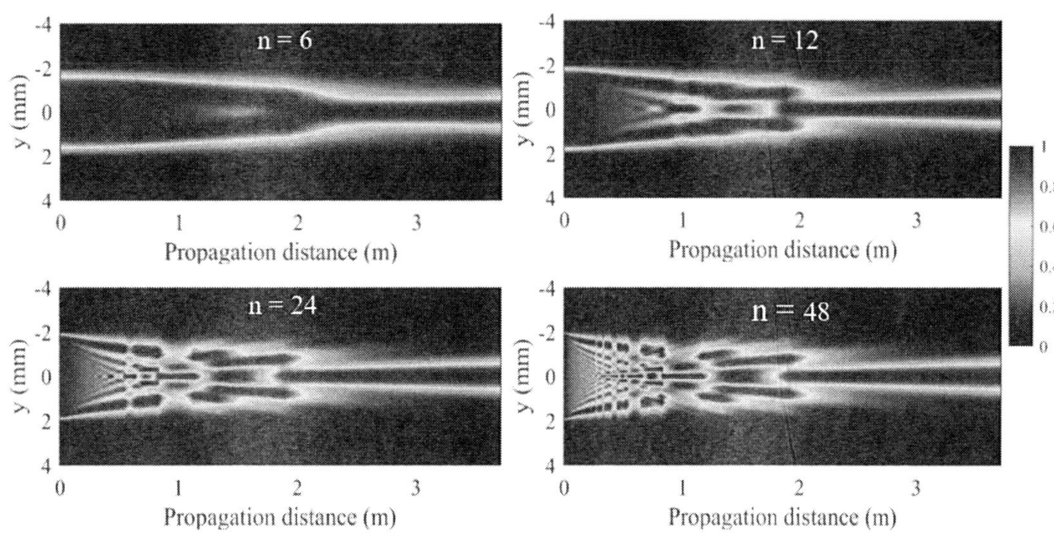

Figure 2. Calculated intensity cross-section of flat-top beams with different orders n (= 6, 12, 24 & 48) as a function of propagation distance. The size of the beam waist is W_{out} = 2 mm, wavelength λ = 1350 nm and the working distance z = 30 cm.

Figure 2 shows the propagation of flat-top beams with different super-Gaussian powers n. It has shown that flat-top beams with large n value distorted very fast upon propagation as compared to flat-top beams with small n value. Accordingly, flat-top beam with $n = 6$ has a longer depth of focus than that of flat-top beams with larger n (> 6) value. However, the flatness of a flat-top beam increases with increasing the order n.

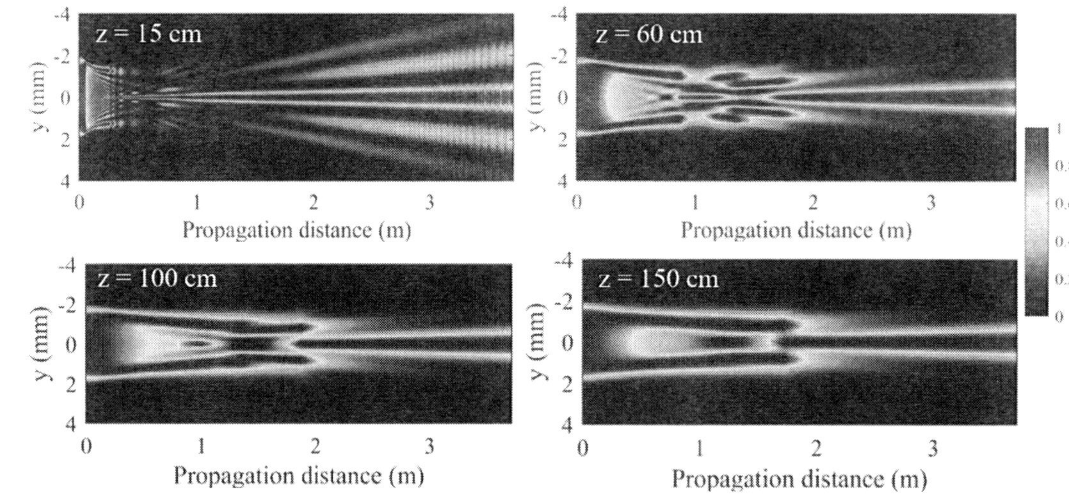

Figure 3. Calculated intensity cross-section of flat-top beams with $n = 8$ as a function of propagation distance for various working distances z. The size of the beam waist is $W_{out} = 2$ mm, wavelength $\lambda = 1350$ nm.

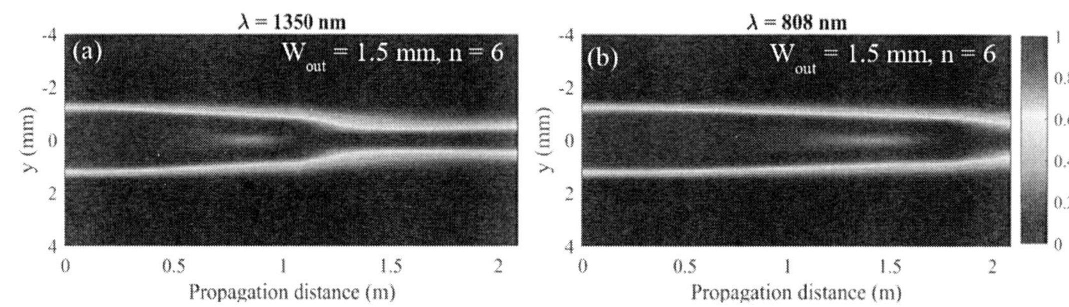

Figure 4(a,b). Cross-section intensity distributions of flat-top beams of different wavelengths (λ) as a function of propagation distance.The order $n = 6$ and the working distance $z = 30$ cm.

The range of propagation and the quality of flat-top laser distributions generated by diffractive phase elements depend on the working distance z, as seen in Fig. 3. Figure 3 shows the propagation of a flat-top beam with n = 8 for various working distances z. The calculated intensity cross-section in figure 3 shows that a flat-top beam propagates a short distance for small working distances z. As the working distance z increases, the propagation distance of a flat-top beam also increases. For longer working distances z, the resulting small phase range DOE forms a flat-top beam with improved uniform phase distribution, giving rise to propagation distance. Figure 4 shows the intensity cross-section of flat-top beams of two different near-infrared wavelengths as a function of propagation distance. The results of figures 4(a) and 4(b) indicate that the propagation distance over which the laser intensity distributions that remain constant and robustly depends on wavelength λ, and increases as wavelength decreases. Specifically, flat-top beams of short wavelength λ maintain invariant cross-section shape for relatively longer propagation distance than that of flat-top beams with longer wavelength λ, as shown in Fig. 4(b).

4. Conclusions

The present study concluded that single-DOE designs and results for generating super-Gaussian flat-top laser beams that can propagate without distortions over a long distance.The designs take into account the order n, working distance z, operating wavelengths, and beam waist. Note that, the propagation distance of invariant flat-top beams varies with the wavelength λ. As evident, a flat-top beam with a shorter wavelength λ yields a longer depth-of-focus. By increasing the working distance z, the propagation distance of a flat-top beam increased significantly. It concludes that as the order n increases the flatness of a flat-top beam also increases, but, it decreases the depth-of-focus of a propagating flat-top beam.

Acknowledgments

The work supported by the Israel Science Foundation (ISF); Magnet project on Advanced Laser Technology for Industrial Applications (ALTIA).

References

[1] Dickey F M, Holswade S C 2000 *Laser beam shaping – Theory and Techniques* (Marcel Dekker)
[2] Romero L A, Dickey F M 1996 J. Opt. Soc. Am. A ***13*** 751.
[3] Shelay D, Holswade S C, Dickey F M 2005 *History of beam shaping in Laser Beam Shaping Applications* (CRC Press) p 307.
[4] Li S, Ding L, Du P, Lu Z, Wang Y, Zhou L, and Yan X 2018 *Appl. Opt.***57** 7060.
[5] Pal V, Tradonsky C, Chiriki R, Kaplan N, Brodsky A, Attia M, Davidson N, and Friesem A A 2018 *Appl. Opt.***57** 4583.
[6] Naresh Kumar Reddy A, Pal V, Friesem A A, Davidson N 2019 *Unpublished.*

Total reflection of near infrared range wave from subwavelength silicon 1D photonic crystal with small packing density

R T Minnullin[1,2], **M Yu Barabanenkov**[2,3], **A G Italyantsev**[2]
[1]Moscow Institute of Physics and Technology, Dolgoprudny, 141701, Russia
[2]JSC Molecular Electronics Research Institute, Moscow, Zelenograd, 124460, Russia
[3]Institute of Microelectronics Technology RAS, Chernogolovka, 142432, Russia

Abstract. One-dimensional (1D) photonic crystals have been widely used in silicon photonics due to their simple structure and multiple working regimes. In this paper we present the computation of reflection and transmission spectra and spatial electric field patterns for 1D photonic crystal with packing density less than 10 percent with use of matrix Riccati equation technique in the theory of multiple electromagnetic wave scattering in inhomogeneous media. Narrow ($\Delta\lambda \lesssim 30$ nm with maximum at $\lambda = 1550$ nm) reflectivity peaks reaching 100 percent maximum were observed for photonic crystals represented as periodically located silicon threads with rectangular or round cross section. The shape and dimensions of the threads were altered to investigate their influence on the peaks width.

1. Introduction

Progress in nanoelectronics through the scaling of the traditional planar technology faces certain problems [1], the Moore's law in its traditional form seems to come to an end, and at this moment other and new approaches are coming to the research forefront. One of the prominent examples is silicon photonics. It has been of large interest in past decades, which resulted in an extensive research in design, fabrication and integration of various photonic devices for different applications such as communication networks, optical computing, medical diagnostics and sensing [2, 3]. Among the most notable appliances, for instance, in computing is the analog-to-digital converter allowing for high-precision signal processing [4].

Component base of silicon photonics includes a number of basic units starting from conventional strip-like waveguides and fibers and ending with more complex ring structures or periodical ones like diffraction gratings and different types of photonic crystals. The subject of our research is one dimensional photonic crystal (PhC) represented as an array of periodically located silicon threads with round or rectangular cross section (Fig. 1). The exact and fast approach [5] to the theory of electromagnetic wave multiple scattering in inhomogeneous dielectric media, which uses the Riccati type equation for the matrix wave reflection coefficient and associated differential equation for the matrix wave transmission coefficient, is applied to calculate the reflection and transmission spectra and corresponding spatial electric field distributions. This method allows taking into account the effects of strong energy transformations between homogeneous and evanescent waves in the near wave zone of the PhC, appearing due to the subwavelength or comparable to the wavelength dimensions of the considered structures.

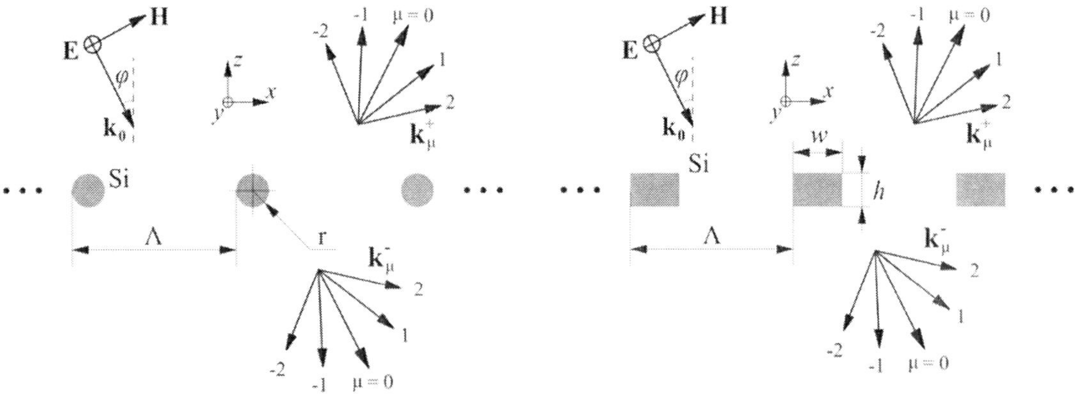

Figure 1(a, b). Silicon threads elongated along the y-axis with round (a) and rectangular (b) cross section and illustration of incident, reflected and transmitted waves.

2. Matrix Riccati equation approach

Let us assume that a plane TE polarized wave of wavelength λ is incident on the periodic structure with period Λ, height h and dielectric permittivity $\varepsilon(\mathbf{r}, \omega)$ at an angle φ (see Fig. 1). The method that we use allows representing the electric field of reflected and transmitted radiation as a superposition of plane waves with certain coefficients and wave vectors:

$$E_y(r) = \begin{cases} \exp\left(i\mathbf{k}_0\left(\mathbf{r} - h\mathbf{n}_z\right)\right) + \sum\limits_{\mu=-\infty}^{\infty} R_{\mu 0}(h)\exp\left(i\mathbf{k}_\mu^+\left(\mathbf{r} - h\mathbf{n}_z\right)\right), & z > h \\ \sum\limits_{\mu=-\infty}^{\infty} T_{\mu 0}(h)\exp\left(i\mathbf{k}_\mu^-\mathbf{r}\right), & z < 0 \end{cases}, \tag{1}$$

$$\mathbf{k}_\mu^\pm = \begin{pmatrix} \beta_\mu \\ 0 \\ \pm\sigma_\mu \end{pmatrix} = \begin{pmatrix} k_0 \sin\varphi + 2\pi\mu/\Lambda \\ 0 \\ \pm\sqrt{k_0^2 - \beta_\mu^2} \end{pmatrix}. \tag{2}$$

Reflection and transmission coefficients are derived from the solution of the following matrix differential equations:

$$\mathrm{R}_z' = \mathrm{R}\mathrm{A}(z)\mathrm{R} + \left(\mathrm{A}(z) + \mathrm{C}\right)\mathrm{R} + \mathrm{R}\left(\mathrm{A}(z) + \mathrm{C}\right) + \mathrm{A}(z), \tag{3}$$

$$\mathrm{T}_z' = \mathrm{T}\left(\mathrm{A}(z) + \mathrm{C} + \mathrm{A}(z)\mathrm{R}(z)\right), \tag{4}$$

$$\mathrm{R}, \mathrm{T}\big|_{z\to 0} = \mathrm{R}_0, \mathrm{T}_0, \tag{5}$$

where R, T, A, C are $(2n+1)\times(2n+1)$-dimensional matrices with complex values and n is selected in a special way. Components of the matrices A and C are defined as follows:

$$A_{\mu\nu}(z) = i\frac{2\pi}{\lambda^2\sigma_\mu} f_{\mu-\nu}(z), \tag{6}$$

$$C_{\mu\nu} = i\sigma_\mu\delta_{\mu\nu}, \tag{7}$$

and $f_\alpha(z) = \frac{\pi}{\Lambda}\int_0^\Lambda \varepsilon(x,z)\exp\left(-i\frac{2\pi\alpha}{\Lambda}x\right)\mathrm{d}x$ is a transformation function, it describes the interaction between propagating and evanescent waves at multiple scattering in periodical inhomogeneous media.

3. Results and discussion

In this report reflection and transmission spectra of 1D PhC and electric field patterns are obtained (Fig. 2) for near infrared range waves at different parameters (structure period, shape and dimensions of threads cross section). Typical values for the wavelength, period and thread sizes are 1550 nm, 1500 nm and 100 nm respectively (see Fig. 1). Thus PhC duty cycle (or filling factor) defined as a ratio of thread's width to the period is usually less than 10 percent. Similarly the packing density (*PD*) is introduced as a ratio of thread's volume to the one-period cell volume. For threads with rectangular cross section duty cycle and packing density coincide, for round threads packing density is less than duty cycle.

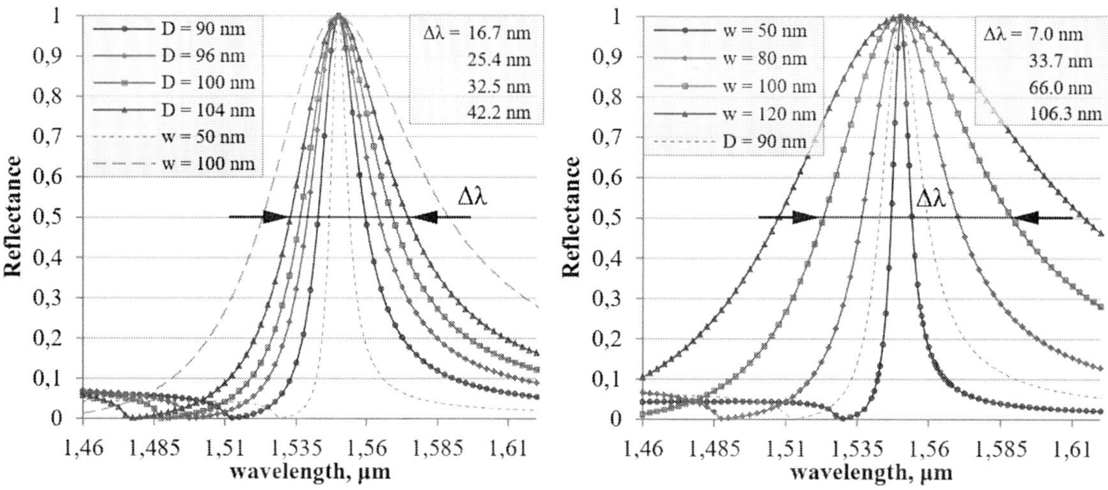

Figure 2(a, b). Reflection spectra for threads with round (a) and rectangular (b) cross section at normal incidence. *D* and *w* stand for the diameter and the width of the threads respectively. Height *h* is maintained 100 nm for threads with rectangular cross section. Dashed lines on the left represent dependencies for rectangular threads ($w = 50$, 100 nm), on the right – dependencies for round shaped threads ($D = 90$ nm).

Distinct peaks in reflectance up to 100 percent occur not far from the moments of the so called modes opening. They are also known as Wood–Rayleigh diffraction grating anomalies and defined by the relation between the wavelength and the period of the structure. At normal incidence when these two parameters become equal, modes of order $\mu = \pm 1$ turn from evanescent to homogeneous if we move from longer wavelengths or shorter periods. These moments are clearly visible on graphs of the Fig. 2 – the points where dependencies have no derivative and change their type of convexity. The period Λ is selected for each structure in such a way that maximum is achieved at a wavelength of 1.55 μm.

Electric field distributions are calculated at points of maxima (Fig. 3). They look very similar for different cross sections as well as reflection spectra, but field pattern for PhC comprised of round threads with diameter of 112 nm is more like that of PhC with square threads of 100 nm in size. Considering the full widths of the peaks at half maximum (FWHM) round threads show better results than rectangular ones at the same size parameters ($D = w = h = 100$ nm) with $\Delta\lambda = 32.5$ nm versus 66.0 nm (green line and wide dashed line on Fig. 2(a)), and, furthermore, they have smaller packing density. However varying separately width and height of rectangular threads gives more degrees of freedom, which helps to soften precision requirements for these and other geometrical parameters.

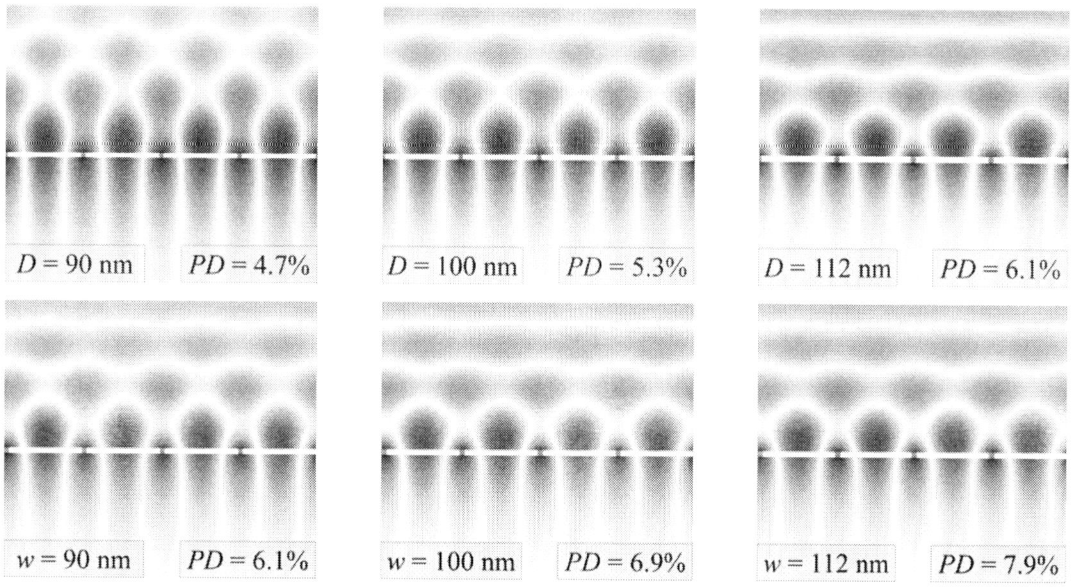

Fig. 3. Spatial electric field distributions for photonic crystals consisting of round threads with diameter D (first row) and rectangular ones with width w and height $h = 100$ nm (second row) at 1.55 μm. The more intense color means the greater absolute value of electric field. PD – packing density.

It is possible to further decrease FWHM in order to reach more sharp resonances by reducing the sizes of threads. Examples of such cases are demonstrated on Fig. 4. The quality factor can reach 10^3–10^4 for the peaks with $\Delta\lambda \approx 0.1$–$0.4$ nm. However, such structures require extremely high precision on geometrical parameters, especially on the period of the photonic crystal, which deviation could not be greater than the peak width itself. The use of rectangular threads slightly softens this requirement, but it is still not enough to allow for simple manufacturing process.

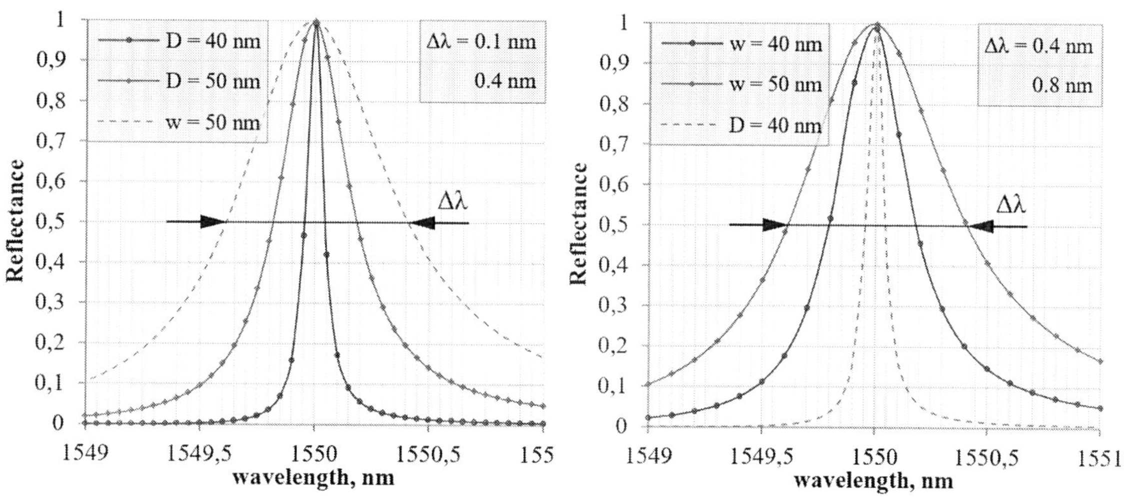

Fig. 4(a, b). Reflection spectra for round (a) and rectangular (b) threads with sharp resonances at a wavelength of 1550 nm.

It is worth mentioning that considering their small packing density such structures can be regarded as mirrors from almost nothing. Varying different parameters of the photonic crystal it is possible to achieve certain pattern of electric field, for example in application of sensing, or to obtain resonant dependence in reflection or transmission spectrum at a given wavelength to construct different devices like optical filters and resonators.

References
[1] Krasnikov G Ya, Zaitcev N A, 2009 *Nano and Microsystem Technique*, **1** (102), pp. 2–5
[2] Fang Zh, Zhao C Zh, 2012 *ISRN Optics*, vol. 2012, article ID 428690
[3] Scarcella C et al., 2016 *Photonics*, **3**, 1
[4] Svidzinsky K K, 2016 *Electronic Engineering. Series 3: Microelectronics,* **3**, pp. 41–52
[5] Barabanenkov Yu N, Barabanenkov M Yu, 2006 *PIERS Proceedings*, **2**, pp. 10–12

Evolution of photoemission properties on Cs/GaAs and GaAs(Cs,O) surfaces under thermal cycling

V S Khoroshilov[1,2], **D M Kazantsev**[1,2], **A G Zhuravlev**[1,2]

[1]Rzhanov Institute of Semiconductor Physics, 630090 Novosibirsk, Russia
[2]Novosibirsk State University, 630090 Novosibirsk, Russia

E-mail: zhuravl@isp.nsc.ru

Abstract. The photoemission properties on Cs/GaAs and GaAs(Cs,O) surfaces prepared at room temperature were studied under thermal cycling. The evolution of electron affinity and escape probability to vacuum was measured using photoemission quantum yield spectroscopy for the surfaces with various Cs-O overlayer compositions. It was found that an increase in the oxygen exposure led to the improvement in the thermal stability of electron affinity.

1. Introduction

The non-equilibrium processes, which are associated with the adatom adsorption (sticking, diffusion and accommodation) on semiconductor surfaces, are of scientific and practical interest. The observation of non-monotonic cesium dose dependence and relaxational kinetics of photoemission current on the Cs/GaAs surface [1, 2] proves that the non-equilibrium processes play a significant role at room temperature (RT), at which *p*-GaAs(Cs,O) photocathodes with the state of negative electron affinity are prepared [3]. These photocathodes are widely used in photomultipliers and sources of ultra-cold and spin-polarized electrons [4]. So, the temporal and thermal instabilities of their photoemission properties are a problem that hinders these practical applications. Surfaces with relatively small positive electron affinities also attract attention in relation with the opportunity to increase the efficiency of solar energy converters by using the photon-enhanced thermionic emission (PETE) [5, 6, 7]. The direct photoemission and PETE are illustrated by the pathway 1 and pathway 2 in figure 1, respectively. According to theoretical estimations [5, 6], the working temperature of PETE converters should be relatively high, exceeding 200°C. Thus, for these converters, the issue of the thermal stability of emission properties is highly relevant.

The temperature influence on the Cs/GaAs surface photoemission properties was experimentally studied earlier [8, 9]. In Reference [8], two types of experiments were carried out. First, it was found that the Cs deposition on the *p*-GaAs surface heated to a moderate temperature ($T \sim 100$°C) leads to substantial qualitative changes in the shape of Cs coverage dependences of photoemission current and surface band bending, along with the changes of relaxational kinetics after the Cs deposition. Second, the photoemission thermal stability was studied by preparing Cs/GaAs surfaces at room temperature followed by measuring photoemission quantum yield spectra under the heating from RT up to $T \sim$ 100°C and a subsequent cooling down to RT. Substantial irreversible variations in the effective affinity and escape probabilities were observed under such thermal cycling. A possible way to increase the thermal stability of Cs/GaAs photoemission properties consists in adding some oxygen to the adsorption layer and, thus, forming a GaAs(Cs,O) surface. It is known from the results of the thermal desorption experiments that an oxygen exposure leads to an increase of the cesium binding energy on the GaAs surface [10]. It was proved earlier that the codeposition of Cs and O_2 on the *p*-GaAs surface at elevated temperatures qualitatively changed the shape of the photoemission current relaxational kinetics and substantially reduced its amplitude [9]. However, the question about the oxygen influence on the photoemission stability of the GaAs(Cs,O) surface prepared at RT under subsequent thermal cycling remained open. The present paper is aimed at answering this question experimentally. It is shown that the increase of the oxygen dose leads to the overall improvement of the *p*-GaAs(Cs,O) photoemission stability under thermal cycling, as compared to the Cs/GaAs surface without oxygen.

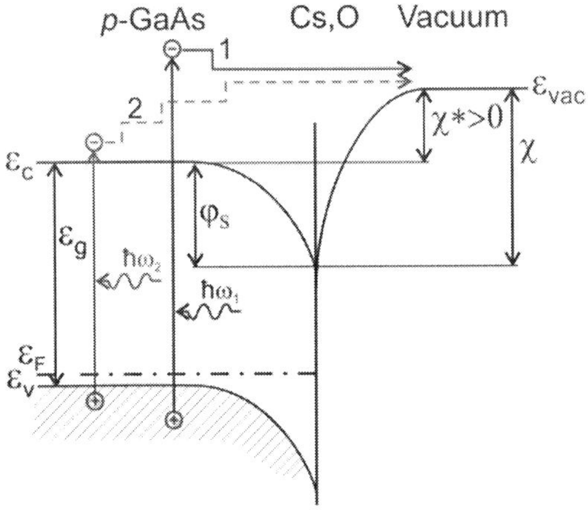

Figure 1. Band diagram of the GaAs(Cs,O) surface with positive effective electron affinity $\chi^* > 0$. Thermalization, diffusion, and emission pathways are shown schematically by the stepwise arrows for the direct photoemission (1) and PETE (2).

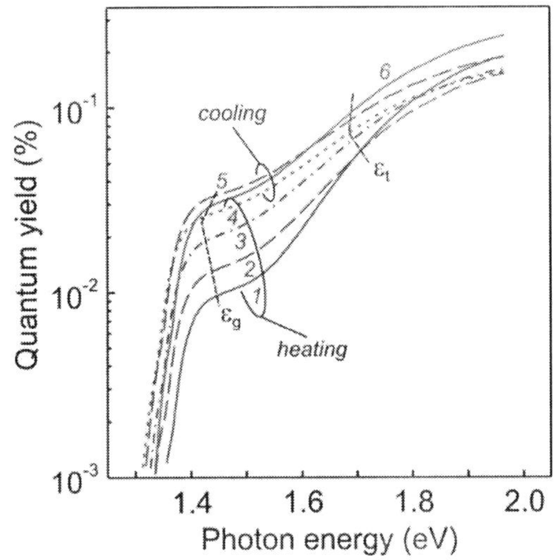

Figure 2. Photoemission quantum yield spectra measured at the GaAs(Cs,O) surface under heating (curves 1–4) and cooling (curves 4–6). Curve (1) T = 20°C; (2) 40°C; (3) 60°C; (4) 80°C; (5) 40°C; (6) 20°C. Dashed lines indicate the evolution of the thresholds at ε_g and ε_t.

2. Experimental

The measurements were carried out on the epitaxial *p*-GaAs(001) layers with hole concentration $p \approx 7\times10^{18}$ cm^{-3}. Atomically clean GaAs(001) surfaces were prepared by removing oxides in the solution of HCl in isopropyl alcohol under dry nitrogen atmosphere, transferring to an ultrahigh vacuum set-up without air contact, and a subsequent heating in vacuum at $T \approx 540$°C [11, 12, 13]. In the experiments, the temperature was controlled by diffusive reflectance spectroscopy with an accuracy better than ± 5°C [11, 14, 15]. Cesium and oxygen were deposited from carefully outgassed dispensers filled with cesium chromate and barium peroxide, respectively. The cesium coverage was determined by the calibration based on the results of X-ray photoelectron spectroscopy [16]. The oxygen exposure was estimated from the time dependence of the oxygen pressure [17, 18].

The evolution of photoemission properties was studied using photoemission quantum yield spectroscopy. To record the photoemission quantum yield spectra, the surface was illuminated by a monochromatic light beam, and the photocurrent was measured. The quantum yield values were obtained after the photocurrent normalization by the incident light intensity. Effective electron affinity χ^* and the probability of electron escape to vacuum were determined from the comparison of experimental and calculated quantum yield spectra, as described in References [19, 20].

3. Results and discussion

In order to study the thermal stability of electron affinity and electron escape probability on the GaAs surface with cesium and oxygen overlayers, we measured the quantum yield spectra under thermal cycling. The initial GaAs(Cs,O) surface was prepared at room temperature by the deposition of 0.7 monolayer (ML) of cesium and 0.02 Langmuir (L) of oxygen on the atomically clean GaAs surface. After cesium and oxygen depositions, the surface was kept at room temperature for one hour to complete the major relaxation processes [1, 2, 9]. Then the photoemission quantum yield spectra were measured at elevated temperatures. The average rate of heating and cooling was 10°C every 10 minutes. Figure 2 shows the photoemission quantum yield spectra measured at various temperatures increasing stepwise from T = 20°C up to 80°C (lines 1-4) and then decreasing from 80°C down to 20°C (lines 4-6). In this experiment the electron affinity on the GaAs(Cs,O) surface remains positive ($\chi^* > 0$). Therefore, two thresholds are observed in the quantum yield spectra [19, 20]. The high energy threshold ε_t corresponds to the onset of the emission of "hot" photoelectrons light-excited with the initial energy above the vacuum level (pathway 1 in figure 1). The low energy threshold at the GaAs band gap ε_g corresponds to the emission from the "Maxwellian tail" of the energy distribution of photoelectrons that are excited by light below the vacuum level near the bottom of the conduction band and are thermalized upward in energy owing to the interaction with phonons (pathway 2 in figure 1). Thermal cycling leads to the reversible variations in threshold ε_g caused by the temperature dependence of the GaAs band gap, and to the irreversible variations in threshold ε_t caused by the change in the work function on the GaAs(Cs,O) surface due to the thermally stimulated structural relaxation or partial desorption of the Cs-O layer.

Figure 3 shows the evolution of the electron affinity (a) and the escape probability (b) on the GaAs(Cs,O) surfaces under the thermal cycling, which were obtained from the measured spectra. For comparison, in figure 3, the temperature dependences of electron affinity and escape probability on the Cs/GaAs surface from Reference [8] are also shown. As seen in figure 3(a), on the Cs/GaAs surface, effective affinity χ^* increased during both heating and cooling. So, the final value of χ^* was ~ 100 meV higher than the initial value. The electron escape probability decreases by an order of magnitude under heating and subsequent cooling (figure 3(b), squares). It is seen that the oxygen deposition on the Cs/GaAs surface leads to the increased stability of electron affinity and escape probability. The electron affinity χ^* on the GaAs(Cs,O) surface decreased by only ~ 20 meV and ~ 10 meV under heating and cooling, respectively. The observed affinity variations can be caused by the irreversible restructuring of Cs-O overlayer (diffusion, accommodation or, maybe, partial desorption). At the same time, the electron escape probability does not change significantly. Thus, the photoemission properties of the GaAs(Cs,O) surface are more stable under thermal cycling, in comparison with the Cs/GaAs surface.

The thermal cycling experiments presented in figure 3 were performed up to 80°C, while the proposed PETE devices should operate at higher temperatures $T \geq 200$°C [5, 6]. The increase of the oxygen dose in the Cs-O overlayer enabled us to increase the maximal temperature in our further studying the thermal stability of emission properties of the GaAs(Cs,O) surface. In these experiments, prior to each heating, 0.75 ML of cesium was deposited on the clean GaAs surface, and then the Cs/GaAs surfaces were exposed to various oxygen doses at room temperature. Figure 4 shows the evolutions of electron affinity and electron escape probability of the GaAs(Cs,O) surfaces with various oxygen concentrations under heating up to $T \sim 160$°C. The average heating rates were 10°C and 20°C every 10 minutes for opened and closed symbols, respectively. As shown earlier in Reference [19], under the oxygen adsorption on the Cs/GaAs surface, the electron affinity decreases, becomes negative, passes through the minimum at an oxygen dose of about 5 mL, and then increases back to the state of positive effective electron affinity. Therefore, as seen in figure 4(a), the initial value of affinity at room temperature increases with the increasing oxygen doses from 0.01 to 0.04 L.

Although the average heating rates of the GaAs(Cs,O) surfaces with various oxygen exposure values were different, it is seen that, for the low-temperature region (20°C < T < 100°C), the increase in the oxygen dose leads to a higher thermal stability of the effective affinity. At high temperatures T >

100°C we observed the increase in the electron affinity even for the highest oxygen dose. This increase can be caused by the onset of a significant desorption of the Cs-O overlayer [10]. It should be noted that, at elevated temperatures, the photoemission quantum yield dramatically decreases, and the interpretation of affinity evolution under cooling these surfaces was complicated by the high noise level in the measured photoemission spectra. Therefore, the cooling stage of the thermal cycle is not shown in figure 4.

As seen in figure 4(b), at elevated temperatures, large variations of the electron escape probability for all oxygen doses were observed. Apparently, the amplitude of these variations is determined by the universal affinity dependence of the escape probability, which consists in a sharp decrease of the escape probability with the increasing affinity above $\chi^* > 0.2$ eV [19, 20].

It should be noted that the decrease in the escape probability and, hence, photoemission quantum yield Y at high χ^* did not allow us to investigate the thermal stability of the photoemission properties for oxygen doses over 0.04 L. Earlier, in Reference [10], the increase in the bond energy of cesium adatoms was observed for the oxygen doses from 0.2 L up to 10 L. In this work it was found that the

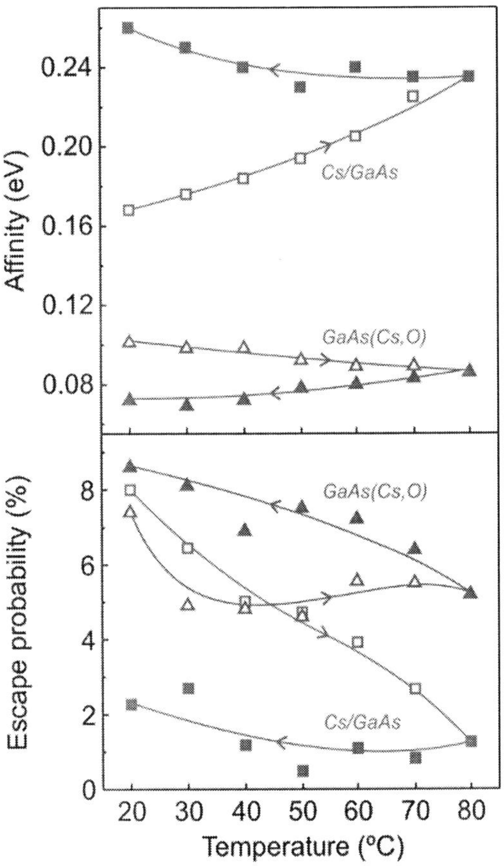

Figure 3. Temperature dependences of effective electron affinity χ^* (a) and electron escape probability (b) under the heating (empty symbols) and subsequent cooling (filled symbols) of the Cs/GaAs (squares, [8]) and GaAs(Cs,O) (triangles) surfaces. The lines are visual guides. The arrows on the lines show the heating and cooling direction.

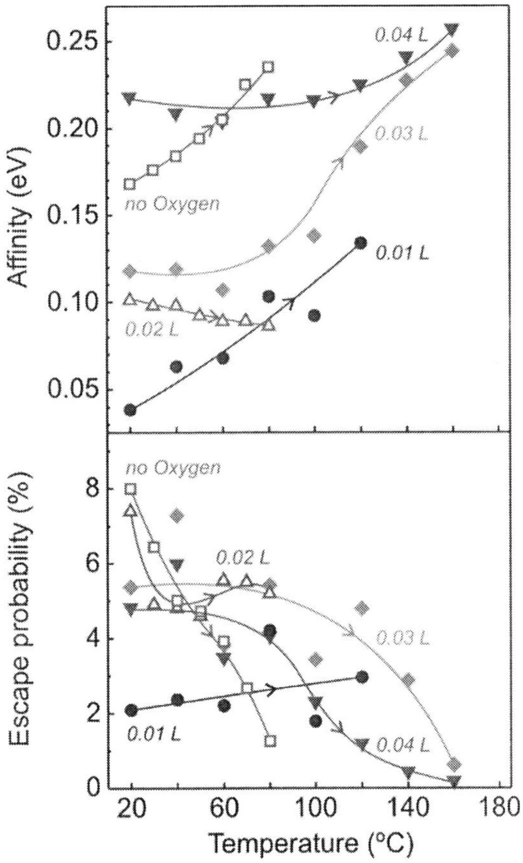

Figure 4. Temperature dependences of effective electron affinity χ^* (a) and escape probability (b) under the heating of the GaAs(Cs,O) surfaces with various oxygen doses. The lines with empty symbols, being obtained at a different heating rate, were copied from figure 3.

increase in the thermal stability of photoemission properties occurs at much smaller doses. However, increasing oxygen doses causes an increase in the electron affinity and a decrease in the escape probability [19]. For possible PETE applications, it is still necessary to find a compromise between the electron affinity, escape probability and thermal stability of photoemission properties.

4. Summary

In summary, in this paper, we experimentally studied the evolutions of electron affinity and electron escape probabilities on the GaAs(Cs,O) surface at elevated temperatures. It was shown that, due to oxygen exposure, the photoemission properties of the GaAs(Cs,O) surface are more stable under thermal cycling, in comparison with the Cs/GaAs surface. Furthermore, the affinity evolution was studied under heating GaAs(Cs,O) surfaces with various doses of deposited oxygen, and it was found that the increase in the oxygen dose leads to a further improvement of the thermal stability. However, the obtained GaAs(Cs,O) surfaces are still unstable at high temperatures $T > 100°C$. Therefore, for the possible practical PETE applications of these surfaces, their thermal stability should be substantially increased.

Acknowledgments

In the part related to measuring the evolution of the photoemission properties under thermal cycling, this work was supported by the Russian Foundation for Basic Research and the Novosibirsk Regional Government (project no. 18-42-543005).

References

[1] Zhuravlev A G, Savchenko M L, Paulish A G and Alperovich V L 2013 *JETP Lett.* **98** 455
[2] Zhuravlev A G and Alperovich V L 2018 *Appl. Surf. Sci.* **461** 10
[3] Bell R L 1973 *Negative Electron Affinity Devices* (Oxford: Clarendon)
[4] Karkare S, Dimitrov D, Schaff W, Cultrera L, Bartnik A, Liu X, Sawyer E, Esposito T and Bazarov I 2013 *J. Appl. Phys.* **113** 104904
[5] Schwede J W, et al 2010 *Nat. Mater.* **9** 762
[6] Schwede J W, et al 2013 *Nat. Commun.* **4** 157
[7] Kribus A and Segev G 2016 *J. Opt.* **18** 073001
[8] Zhuravlev A G and Alperovich V L 2017 *Appl. Surf. Sci.* **395** 3
[9] Khoroshilov V S, Kazantsev D M and Zhuravlev A G 2019 *J. Phys. Conf. Ser.* **1199** 012011
[10] Kamaratos M 2001 *Appl. Surf. Sci.* **185** 66
[11] Tereshchenko O E, Chikichev S I and Terekhov A S 1999 *J. Vac. Sci. Technol.* A **17** 2655
[12] Alperovich V L, Tereshchenko O E, Rudaya N S, Sheglov D V, Latyshev A V and Terekhov A S 2004 *Appl. Surf. Sci.* **235** 249
[13] Zhuravlev A G, Kazantsev D M, Khoroshilov V S, Kozhukhov A S and Alperovich V L 2018 *J. Phys. Conf. Ser.* **993** 012007
[14] Wang Z, Kwan S L, Pearsall T P, Booth J L, Beard B T and Johnson S R 1997 *J. Vac. Sci. Technol.* B **15** 116
[15] Thompson P, Li Y, Sato D L, Flanders L and Lee H P 1997 *Appl.Phys.Lett.* **70** 1605
[16] Tereshchenko O E, Alperovich V L and Terekhov A S 2004 *JETP Lett.* **79** 131
[17] Bakin V V, Pakhnevich A A, Zhuravlev A G, Shornikov A N, Akhundov I O, Tereshechenko O E, Alperovich V L, Scheibler H E and Terekhov A S 2007 *e-J. Surf. Sci. Nanotech.* **5** 80
[18] Bakin V V, Toropetsky K V, Scheibler H E and Terekhov A S 2015 JETP Lett. 101 380
[19] Zhuravlev A G, Khoroshilov V S and Alperovich V L 2017 *JETP Lett.* **105** 686
[20] Zhuravlev A G, Khoroshilov V S and Alperovich V L 2019 *Appl. Surf. Sci.* **483** 895

Theoretical optimization of the photolithography through array of 1.2 μm silica microspheres

L N Dvoretckaia[1], A M Mozharov[2], A D Bolshakov[1,2], V V Fedorov[1], A A Vasiliev[1] and I S Mukhin[1,2]
[1]Renewable Source Laboratory, St Petersburg Academic University 194021, Russia
[2]ITMO University, St. Petersburg 197101, Russia

Abstract. Interest in heteroepitaxy of III-V compounds on Si has been growing rapidly in recent years due to the potential of the optoelectronic components integration on silicon. However, most of the semiconductor compounds conventional in optoelectronics cannot be easily integrated on Si substrates due to the formation of the lattice defects. In this paper, we consider the fabrication of the mask consisting of the ordered nanoscale holes with the use of microsphere photolithography for selective epitaxial growth – promising approach for nanostructures fabrication on mismatched substrates. We have carried out the calculation of electromagnetic wave absorption in the photoresist layer through 1.2 μm microspherical silica lenses. The theoretical optimization of the photoresist thickness parameter allowed to obtain a value at which the minimum holes diameter in the photoresist is achieved. These data are necessary for carrying out the process of lithography through microspherical lenses to create a patterned growth mask.

1. Introduction

There are three main problems in the technology of A3B5 semiconductor compounds synthesis on Si: high dislocations density due to the mismatch of the growing material lattice with the substrate, formation of antiphase domains due to the presence of polar/non-polar interface and lattice cracking due to the difference in thermal expansion coefficients between two materials [1]. Due to the large surface area/volume ratio, low dimensional structures can be used to suppress the mentioned effects. Self-assembled nanostructures are easy to synthesize, can have different morphology providing versatility of their properties and can be grown with variety of materials providing control over their electronic properties. In recent years there has been an increasing demand on mass production of ordered and periodic nanostructures since such nanostructures have found wide application in many high-performance devices, such as LEDs [2], photodetectors, data storage elements, nanoantennas, photonic crystals [3].

Arrays of highly ordered nanostructures can be achieved by several methods [4], such as immersion lithography [5], electron beam lithography [6], focused ion beam lithography [7], deep ultraviolet lithography (DUV) [4], nanoimprint lithography [8] and other less developed methods. The main flaw of these methods is their limited application, mostly governed by the low manufacturing capability on large scale due to the high cost or low production capacity. Thus, today, there is a need of new cheaper and high-performance methods development that will allow manufacture of periodic and highly

ordered arrays of nanostructures over large areas [9]. One of the promising approached is the microsphere photolithography [10, 11].
This work is aimed at development of the fabrication technique of the mask with nanometer-sized openings, acting as windows in the oxide layer obtained by the photolithography on a Si substrate surface for the selective area epitaxy. It is assumed that the use of nano-sized openings in the oxide layer allows to obtain planar structure or ordered array of nanostructures, without dislocations and the formation of antiphase domains at the heterojunction between Si and the semiconductor material [12].
We performed the numerical calculation of the electromagnetic wave propagation and absorption in the photoresist of different thickness covered with SiO_2 1.2 μm microlenses to study correlation between the size of the openings and the photoresist thickness.

2. Modeling

We carried out 3D simulation of the plane incident electromagnetic wave propagation through SiO_2 microspherical lenses into the positive photoresist layer. The modeled structure consists of Si substrate with a 50 nm layer of silicon oxide, covered with a photoresist layer and array of hexagonally ordered microspheres with a diameter of 1.2 μm. The choice of the sphere diameter was governed by their commercial availability. The simulation was performed in the CST Microwave Studio. The package uses differential forms of the Maxwell equations for the calculation. Using the calculation in the time domain, the absorption of the electromagnetic field in the photoresist layer with different thickness was studied.
The results of the numerical calculations showed that the depth of the radiation penetration into the resist and the quality of focusing (diameter of the exposed area) depend on the photoresist thickness and on the wavelength used. For the simulation, the wavelength of the plane incident wave of 405 nm and 365 nm was considered, corresponding to the radiation used for the photoresist exposure.
It was established experimentally that when using a positive photoresist thickness of less than 200 nm, there is instability of its thickness. It leads to a variation in the mask pattern size. Thereby, the calculation was carried out with the photoresist thickness in the range from 200 to 500 nm with a 50 nm step. Figure 1 (a - e) shows the electromagnetic wave absorption efficiency in the photoresist layer.

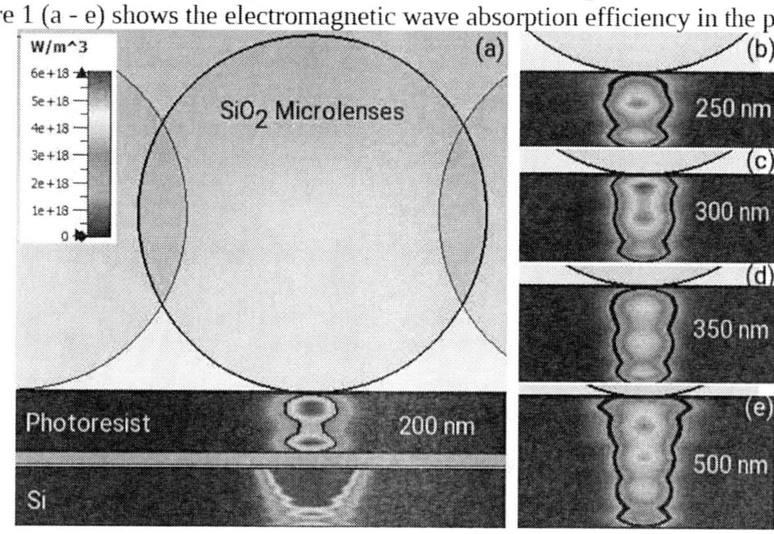

Figure 1(a, b, c, d, e). Absorption of the incident wave having the wavelength of $\lambda = 405$ nm in the positive photoresist covered with ordered array of 1.2 μm SiO_2 microlenses on a Si substrate coated with a 50nm layer of SiO_2: **(a)** 200 nm, **(b)** 250 nm, **(c)** 300 nm, **(d)** 350 nm, **(e)** 500 nm.

Journal of Physics: Conference Series **1410** (2019) 012129 doi:10.1088/1742-6596/1410/1/012129

The black contours corresponds to the minimum light intensity required for exposure of the photoresist. As can be seen, the profiles of these contours have maxima due to the interference of the incident and reflected from the silicon substrate waves. Analyzing the contours we obtain values of the top and bottom (on the boundary with the oxide layer) exposed area diameter and evaluate their dependencies on the resist thickness shown in Figure 2. The black curve in Figure 2 corresponds to the projected area diameter in the top region. The red curve shows the area diameter in the bottom. From Figure 2 it can be seen that for a wavelength of 405 nm the optimum thickness of the photoresist is 200 nm, and the corresponding size of the holes in the photoresist should be about 180 nm.

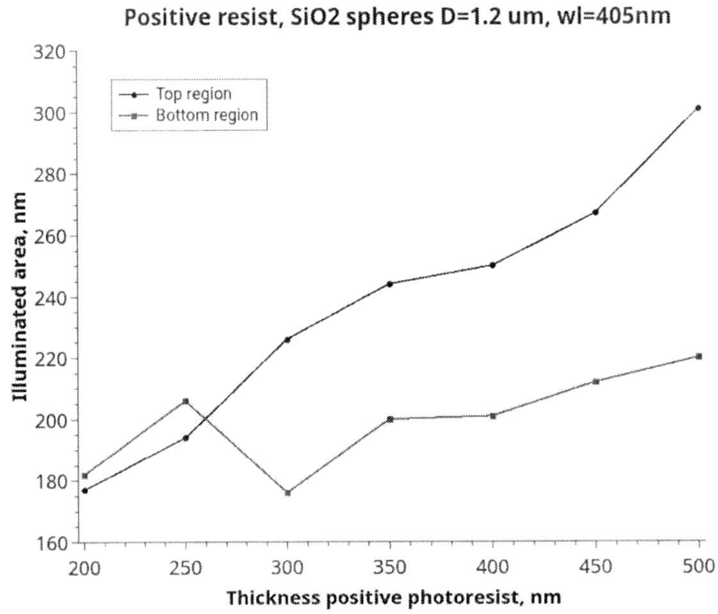

Figure 2. Plot of the exposed area diameter of the positive photoresist layer on its thickness, for the exposure wavelength of 405 nm, “top region” is the diameter of the exposed area on the surface of the photoresist, "Bottom region" is the diameter of the exposed region of the resist near the surface of the SiO_2 layer.

The modeling was also carried out for the incident wave length of 365 nm. The thickness of the photoresist was varied from 200 to 500 nm with a step of 50 nm. Figure 3 shows the light absorption for several photoresist thicknesses. It can be seen that at small thickness of the photoresist (200 nm), the exposed area increases. At large thicknesses (from 300 nm), the diameter of the exposed area also increases. At the same time, the power required to expose the photoresist increases 2 times when the thickness increases from 200 nm to 400nm. The radiation with a power used for 200nm resist exposure is completely absorbed in the 400nm photoresist and cannot pass through the photoresist layer.

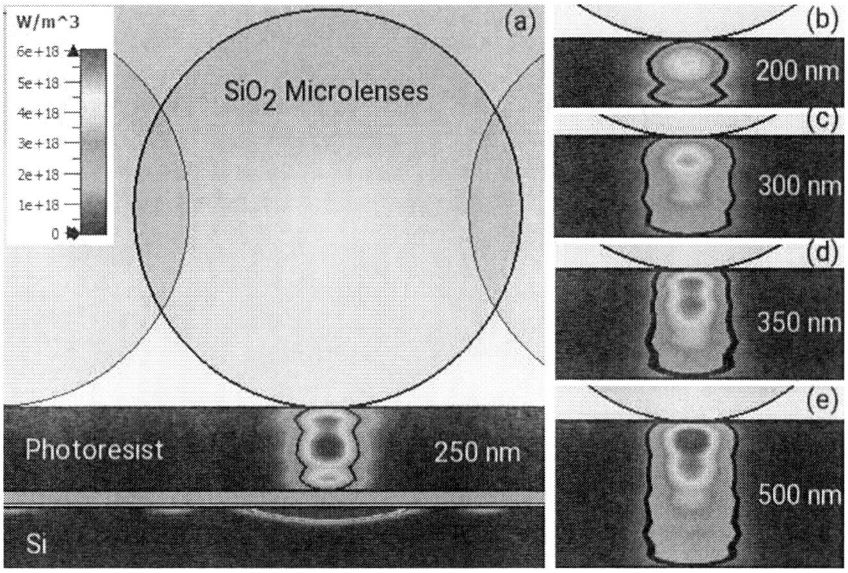

Figure 3(a, b, c, d, e). Absorption of the incident wave having the wavelength of λ = 365 nm in the positive photoresist covered with ordered array of 1.2 μm SiO_2 microlenses on a Si substrate coated with a 50nm layer of SiO_2: **(a)** 250 nm, **(b)** 200 nm, **(c)** 300 nm, **(d)** 350 nm, **(e)** 500 nm.

The dependence of the positive photoresist exposed area diameter as a function of its thickness is presented in Figure 4. Analyzing the results presented in the graph, it can be seen that for a wavelength of 365 nm, the photoresist thickness of 250 nm corresponds to the minimum focusing region of the incident radiation. With these parameters, the size of the holes should be as small as 170 nm.

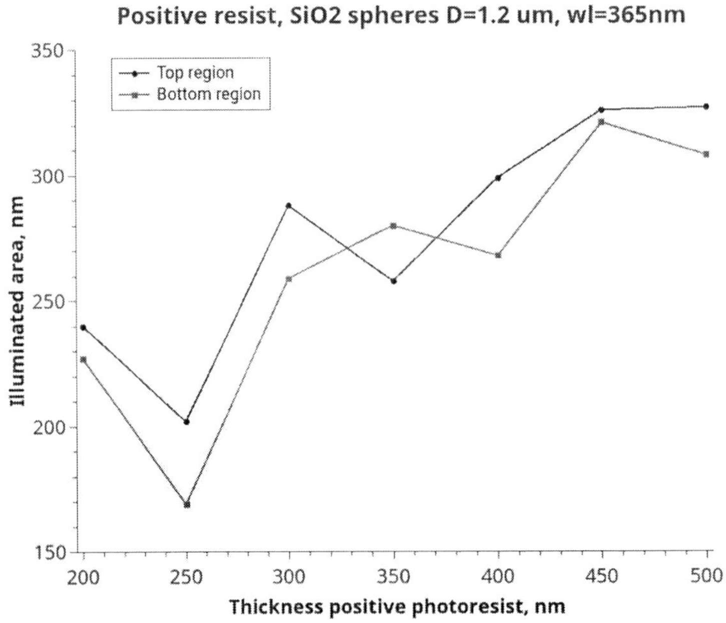

Figure 4. Plot of the exposed area diameter of the positive photoresist layer on its thickness, for the exposure wavelength of 365 nm, "top region" is the diameter of the exposed area on the surface of the photoresist, "Bottom region" is the diameter of the exposed region of the resist near the surface of the SiO_2 layer.

3. Conclusion

In this work, we have studied numerically propagation of the electromagnetic wave with a wavelength of 405 and 365 nm and its focusing under array of microspherical lenses in the volume of varied thickness positive photoresist. The 50nm silica sublayer that can be used as a mask material for selective epitaxy is taken into account in the model.

According to the calculated data, it was obtained that at a wavelength of 405 nm a layer of positive photoresist with a thickness of 200 nm is suitable for the most efficient focusing of SiO_2 by microspherical lenses with a diameter of 1.2 μm. In this case, the diameter of the openings in the developed photoresist can be as small as 180 nm. For the wavelength of 365 nm, the optimum value of the photoresist thickness was 250 nm, while the size of the holes in the photoresist should be about 180 nm.

The obtained calculated data allows to determine the parameters of the photolithography process for fabrication of the phtotresist pattern having subwavelength features. The results are aimed at creation of a large scale mask with ordered nanoholes in the oxide layer for epitaxial synthesis of A3B5 semiconductor nanostructures on Si.

Acknowledgments

This work was carried out with the support of the Russian Foundation for Basic Research 18-32-00899 mol_a, the research was partially supported by FASIE. V.V.F. thanks for support of the modeling results analysis the Russian Science Foundation (Grant 18-72-00219).

References

[1] Fang S F, Adomi K, Iyer S, Morkoc H, Zabel H, Choi C, Otsuka N 1990 J. Appl Phys. **68**(7):R31-58.
[2] Zhang Q, Li K H, Choi H W 2012 J. Appl Phys. **100**(6):061120.
[3] Huang G, Yin Y, Pan Z, Chen M, Zhang L, Liu Y, Zhang Y, Gao J 2014 Biomacromolecules **15**(12):4396-402.
[4] Zhang Z, Geng C, Hao Z, Wei T, Yan Q 2016 Advances in colloid and interface science **228**:105-22.
[5] Gil D, Brunner T A, Fonseca C, Seong N, Streefkerk B, Wagner C, Stavenga M 2004 Science & Technology B: Microelectronics and Nanometer Structures Processing, Measurement, and Phenomena **22**(6):3431-8.
[6] Corbierre M K, Beerens J, Lennox R B 2005 Chemistry of materials **17**(23):5774-9.
[7] Phillips G N, Siekman M, Abelmann L, Lodder J C 2002 Applied physics letters. **81**(5):865-7.
[8] Guo L J 2007 Advanced materials. **19**(4):495-513.
[9] Dvoretckaia L N, Mozharov A M, Fedorov V V, Bolshakov A D, Mukhin I S 2018 Journal of Physics: Conference Series Vol. 1124, No. **2**, p. 022042
[10] Wu M H, Whitesides G M 2001 Applied physics letters. **78**(16):2273-5.
[11] Wu M H, Park C, Whitesides G M 2003 Journal of colloid and interface science. **265**(2):304-9.
[12] Renard C, Cherkasin N, Jaffré A, Vincent L, Michel A, Molière T, Hamouche R, Yam V, Alvarez J, Fossard F, Mencaraglia D 2013 Applied Physics Letters. **102**(19):191915.

On the need to control the state of the flowing media by the values of relaxation constants

N S Myazin[1], **V V Davydov**[1, 2], **V V Yushkova**[3], **A P Glinushkin**[2], **N I Taranda**[4], **V Yu Rud'**[2]
[1]Peter the Great St. Petersburg Polytechnic University, St. Petersburg 195251, Russia
[2]All-Russian Research Institute of Phytopathology, Moscow Region 143050, Russia
[3]Saint Petersburg University of Management Technologies and Economics, 190109, Russia
[4]Grodno State Agrarian University, Grodno, 230008, Belarus

Abstract. The article substantiates the necessity of measuring the longitudinal T_1 and transverse T_2 relaxation times of the flowing media to control its state. For such measurements, a new method for measuring T_1 in a flowing fluid is proposed. Therefore, the results of experimental studies are presented. In addition, measured values of relaxation constants were compared to the data obtained on industrial devices was performed. The results of this comparison are also presented.

1. Introduction

Currently, the phenomenon of nuclear magnetic resonance (NMR) has many different applications. This includes scientific researches (determination of the composition or molecular structure of the media, etc.) and for monitoring the condition of condensed media in production, services, etc. [1–7]. Industrial devices, the principle of which is based on the phenomenon of nuclear magnetic resonance, are among the most accurate and reliable measuring devices [6–10]. They have several undeniable advantages over other devices. Their measurements do not cause irreversible changes in the physical structure and chemical composition of the medium under study [8–13], unlike optical, ultrasonic and other devices [14–20]. This makes it possible to use devices based on the NMR method in biology, medicine, etc.

The most difficult studies are studies of condensed media flow, especially turbulent flow [21–23]. When conducting various experiments with the use of such flow, in most cases it is necessary to control not only the flow of the medium, but also its state. The state of the flowing liquid medium also needs to be known in order to ensure the efficient functioning of various physical processes. This may be, for example, the heat exchange between the media in the different circuits of the nuclear reactor in power plants [24–26] or change of technological process of fuel combustion for higher heat output. One of the possible solutions to the problem of effective and reliable control of the flowing media is considered in this paper.

2. Peculiarities of flowing media state control by NMR

In practical applications of nuclear magnetic resonance (NMR), many methods have been developed for measuring the times of longitudinal T_1 and transverse T_2 relaxation of condensed media. It is assumed that these media are in a stationary state. The results of the measurements mentioned earlier are successfully used for express control of the state of media in solving various problems [3, 11, 12, 27-29]. This is because relaxation times T_1 and T_2 may change due to various reasons. These reasons include any changes in the molecular structure of the medium (ingress of impurities into it, a concentration change of the components, etc.), as well as a change in medium temperature. Knowing the values of T_1 and T_2, corresponding to the standard state of the medium, you can monitor its state at the place of sampling in real time [3, 11, 12, 27–32]. In most cases, when conducting research using a flowing liquid medium, it is necessary not only to monitor its state, but also to measure its flow rate q. The same applies to the control of technological processes in the production required to monitor its state. Many different NMR flowmeter models have now been developed [1, 2, 6, 21–26, 30–33]. A number of these devices are designed to measure the transversal relaxation time T_2. The results of the studies have shown that there are many cases, especially for liquid media mixtures, when the measurement of T_2 time alone is not enough to determine the state of the medium [2, 11, 12, 23, 27, 28]. Both relaxation constants must be measured for reliable and efficient monitoring of the flowing media. The same problem is relevant for express control of condensed medium by NMR [3, 4, 11–13].

There are many different impulse methods used to measure the values of T_1 and T_2 of a condensed medium in a stationary state. However, the results of our experiments have shown that it is extremely difficult to use these methods to measure the relaxation constants of the flowing medium. This is because the residence time of the liquid in the registration coil of the NMR flowmeter-relaxometer must be greater than the total time of a single measurement. This time consists of the durations of the pulses acting on the magnetized current fluid, as well as the duration of the intervals between them and the time of the registration of the free precession signal (its processing allows us to determine the relaxation constants). Considering that the possible range of changes in the values of q and relaxation constants can be at least two orders of magnitude, the need to develop new methods for measuring T_1 and T_2 for NMR flowmeter-relaxometer is evident.

Figure 1 shows the scheme of the experimental setup developed by us for conducting research of flowing media by NMR method.

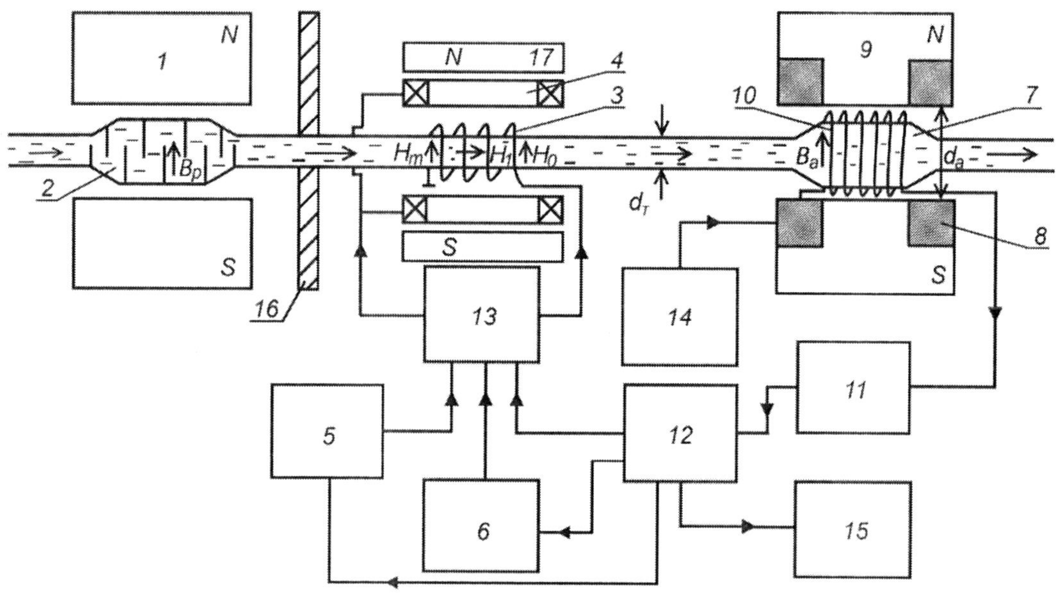

Figure 1. Structural scheme of experimental setup. 1 — magnet-polarizer; 2 — vessel-polarizer; 3 — nutation coil; 4 — modulation coils; 5, 6 — nutation and modulation coils generators; 7 — vessel-analyzer; 8 — magnet-analyzer field modulation coils; 9 — magnet-analyzer; 10 — NMR signal registration coil; 11 — NMR signal registration circuit; 12 — control and processing unit; 13 — electronic keys; 14 — RF generator; 15 — indication circuit; 16 — magnetic shield.

In contrast to the setup for the study of the stationery liquid medium, in this experimental setup the spatial separation of the zone of flowing liquid magnetization (magnet-polarizer 1) and the zone of NMR signal registration (magnet-analyzer 9) is performed. To measure the flow rate q of a liquid medium, the effect of rotation of the magnetization vector [22, 23, 30, 34, 35] in the nutation coil 3 is used. Full inversion of magnetization M (i.e. rotation of the magnetization vector by the angle $\varphi_n = 180°$) occurs at the resonant frequency f_n of the H_1 field. To measure q, signals with and without magnetization inversion are used. Figure 2 shows, as an example, the registered NMR signals from feed water at $T = 299.3$ K, $f_m = 50$ Hz.

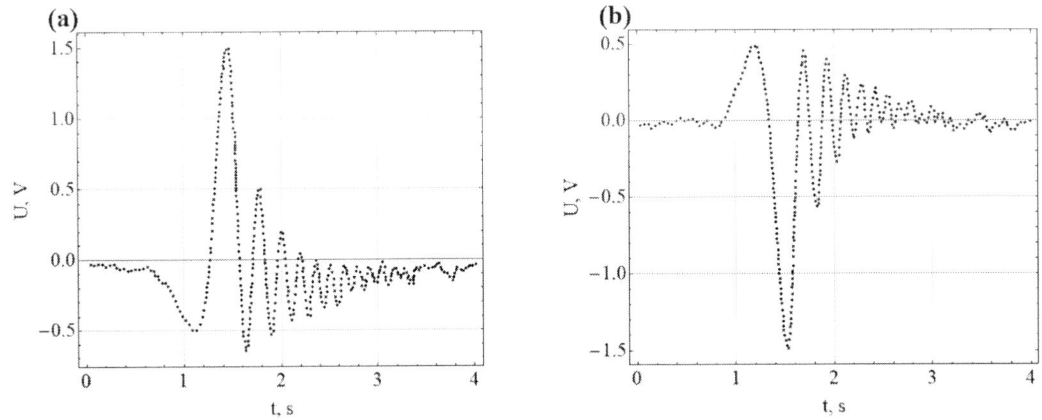

Figure 2(a, b). NMR signal line shape for feed water **(a)** without magnetization inversion; **(b)** with magnetization inversion at $f_n = 1788204$ Hz, $H_1 = 15.2$ A/m.

Figure 2.a corresponds to $\varphi_n = 0$, while figure 2.b corresponds to $\varphi_n = 180°$ (the NMR signal with the magnetization inversion). In case $\varphi_n = 90°; 270°$the magnetization vector components are equal to $M_z = M_x = M_y = 0$, so the NMR signal amplitude in the registration device is equal to zero. The obtained NMR signals have a high signal-to-noise ratio, which allows them to be used to measure relaxation constants (both the signal with and without magnetization inversion). The use of two signals to measure relaxation constants makes it possible to increase the accuracy of measurements compared to methods that use only one NMR signal (without inversion of magnetization).

It should be noted that due to the peculiarities of using NMR flowmeters-relaxometers, the registration of NMR signals in a wide range of liquid flow rates is possible only with the use of modulation technique. When using it, it is necessary to take into account the time t_r when the flowing liquid is in the registration coil 10, as well as the modulation period (modulation frequency f_m) of the magnetic field B_a. The results of our studies allowed us to establish that in order to provide the maximum signal-to-noise ratio (S/N) and to obtain the recorded NMR signal in the form of "wiggles" (figure 2) the following ratio should be performed:

$$t_r > {}^{2}/_{f_m}, \tag{1}$$

If ratio (1) is not met, the S/N ratio is reduced, and the shape of the recorded NMR signal is distorted. In several cases it creates insurmountable problems when measuring relaxation constants of the flowing medium.

3. Results of experimental studies and their discussion

The results of our research showed that the developed design of the experimental setup allows to register NMR signals from different nuclei of the flowing medium. This is a prerequisite for the use of NMR methods in various experiments with flowing media. Figure 3 shows, as an example, the registered NMR signals from the flowing aqueous solution of sodium hydroxide NaOH.

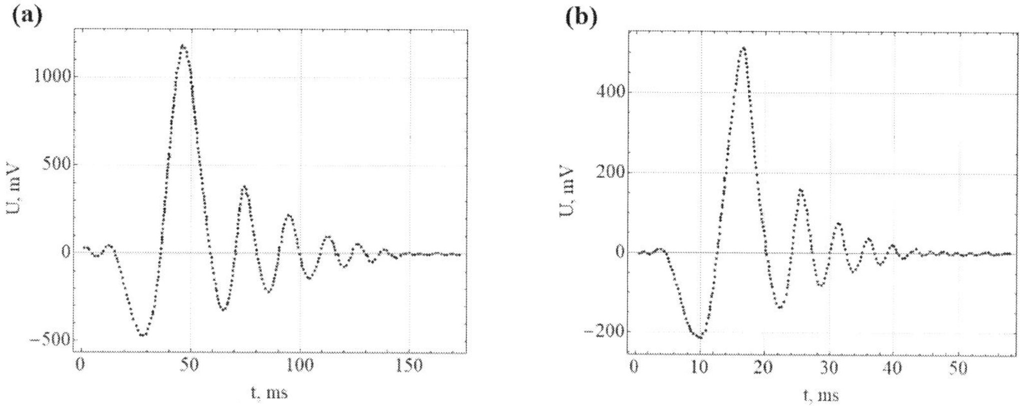

Figure 3(a, b). NMR signal line shape for aqueous solution of sodium hydroxide. **(a)** signal at protons resonance frequency f_p; **(b)** signal at sodium resonance frequency f_{Na}.

The registration was performed at resonance frequencies of protons f_p = 19287097 Hz (figure 3.a) and sodium nuclei f_{Na} = 5108578 Hz (figure 3.b) at optimal flow rate q = 0.01 m^3/s, temperature T = 292.4 K, and modulation frequency f_m = 50 Hz. Fluid velocity in the connecting section of the pipeline is equal to υ = 3.538 m/s.

In the previously developed design of the NMR flowmeter-relaxometer [21-26, 31] the measurement of sodium hydroxide flow rate q was possible only with the use of the NMR signal recorded at the resonant frequency of protons f_p. The analysis of the NMR signals presented in figure 3 shows that when registering the NMR signal at the resonant frequency of sodium nuclei f_{Na}, the ratio S/N more than 100 without using the NMR signal accumulation scheme.

When using measuring devices, which have a structural scheme and the principle of NMR signal registration like the one we are considering, there are several difficulties in developing new methods for determining T_1 and T_2. This is especially true for measuring T_1. This is because NMR flowmeters-relaxometers use a modulation technique to register the NMR signal. The NMR waveform represents damped oscillations (figure 4), also known as “wiggles”. Figure 4.b shows the envelope of the NMR signal waveform, drawn by peaks of “wiggles”.

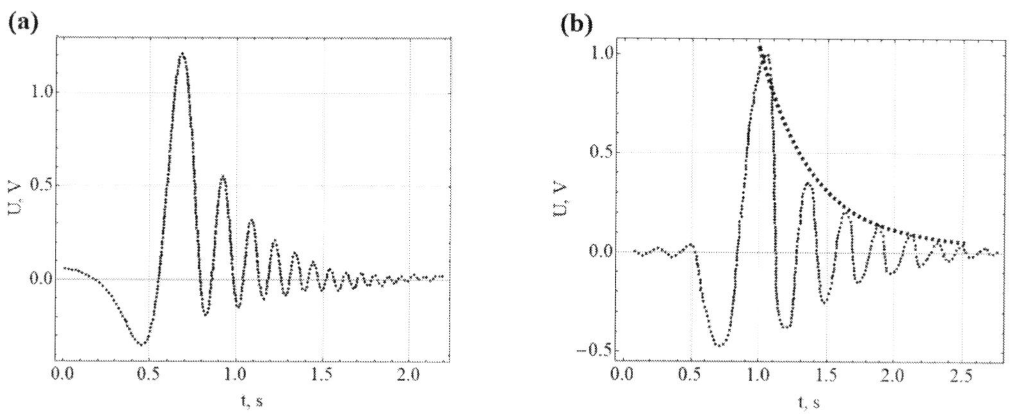

Figure 4(a, b). NMR signal waveform. **(a)** corresponds to tap water at temperature T = 276.3 K; **(b)** corresponds to tap water at temperature T = 279.9 K.

The shape of the recorded NMR signal at a single pass through the resonance can be approximated by the following dependence:

$$u(t) = U_0 \exp\left(-\frac{t}{T_2^*}\right) F(t), \tag{2}$$

where $F(t)$ is the function describing the harmonic nature of the recorded signal, T_2^* is the effective time of transverse relaxation, U_0 is the maximum value of the amplitude of the recorded NMR signal.

Using the ratio (2), the effective transverse relaxation time T_2^* is determined by the envelope decline. Relaxation time T_2 can be found from the T_2^* time, using the value of the magnetic field heterogeneity in the placement zone of the registration coil.

Using Bloch's phenomenological equations and theoretical postulates of the Julotto method for stationary condensed media, we studied the processes occurring in the experimental setup (figure 1). The results of these studies allowed to obtain the ratio for the change of magnetization M in the flowing liquid in the registration coil 10 (figure 1). The recorded amplitude value of the NMR signal U_0 is proportional to the value of magnetization M, the change of which is determined by the following relationship:

$$M = M_0\left(1 - \frac{1 - \exp\left(-\frac{\tau}{T_1}\right)}{1 + \exp\left(-\frac{\tau}{T_1}\right)}\right) \tag{3}$$

where M_0 is the magnetization of the liquid medium in the absence of modulation of the magnetic field.

It follows from (3) that to measure T_1, it is necessary to measure the amplitude of the NMR signal at two different modulation frequencies (i.e., for two different τ), and then calculate the value T_1. Modulation frequencies must correspond to the ratio (1). It should be added that in order to obtain a reliable result and estimate the error in accordance with standard methods, it is desirable to carry out measurements at least 10 times.

For tap water at T = 276.3 K the measured values of relaxation times were $T_1 = 1.436 \pm 0.013$ s and $T_2 = 0.896 \pm 0.007$ s (measurements were made 10 times). A sample of this water was investigated at the same temperature on a stationary NMR relaxometer Minispec mq 20, the measured values were $T_1 = 1.4338 \pm 0.0028$ s and $T_2 = 0.8942 \pm 0.0028$ s. The obtained results of T_1 and T_2 values coincided within the limits of measurement error.

4. Conclusion

The experimental results obtained for the T_1 and T_2 values confirmed the possibility of monitoring the state of the flowing medium in real time by the proposed method in a wide range of flow rate q changes. This significantly expands the possibilities of using NMR flowmeters-relaxometers both for scientific research and for industrial applications in power engineering, pharmacology, etc.

This work partly supported by Grant of RFBR No 18-29-25071.

References

[1] D'yachenko S V, Kondrashkova I S and Zhernovoi A I 2017 *Technical Physics* **62** 1602–1604

[2] Kashaev R S and Gazizov E G 2010 *Journal of Applied Spectroscopy* **77** 321–328

[3] Myazin N S, Neronov Yu I, Dudkin V I, Davydov V V and Yushkova V V 2018 *MATEC Web of Conferences* **245** 11013

[4] Rykin E V, Moroz A V, Smirnov K J, Davydov V V and Yushkova V V 2018 *MATEC Web of Conference* **245** 12002

[5] Davydov R V, Atonov V I and Moroz A V 2018 *Proceedings of the 2018 IEEE International Conference on Electrical Engineering and Photonics, EExPolytech 2018* (Saint-Petersburg) 8564378 p. 236-239

[6] Marusina M Ya, Bazarov B A, Galaidin P A, Silaev A A, Marusin M P, Zakemoskya E Yu, Gilev A G and Alekseev A V 2014 *Measurement Techniques* **57** 461

[7] Alexandrov A S, Archipov R V, Ivanov A A, Gnezdilov O I, Gafurov M R and V.D. Skirda V D 2014 *Applied Magnetic Resonance* **45** 1275-1287

[8] Filippov A V, Artamonova M R, Rudakova M F, Gimatdinov R G and Skirda V D 2012 *Magnetic Resonance in Chemistry* **50** 114 – 119
[9] Chizhik V I and Tagirov M S 2017 *Applied Magnetic Resonance* **48** 621-623
[10] Alakshin E M, Gazizulin R R, Klochkov A V, Kuzmin V V, Sabitova A M, Safin T R and Tagirov M S 2013 *Magnetic resonance in solid* **15(1)** 13104
[11] Myazin N S, Davydov V V, Yushkova V V, Davydova T I and Rud' V Yu 2017 *Journal of Physics: Conference Series* **917(4)** 042017
[12] Myazin N S, Logunov S E, Davydov V V, Rud' V Yu, Grebenikova N M and Yushkova V V 2017 *Journal of Physics: Conference Series* **929(1)** 012064
[13] Davydov V V, Dudkin V I, Petrov A A and Myazin N S 2016 *Technical Physics Letters* **42** 692–696
[14] Nepomnyashchaya E K, Velichko E N and Aksenov E T 2016 *Journal of Physics: Conference Series* **769(4)** 012025
[15] Nepomnyashchaya E K, Akenov E T, Bogomaz T A and Velichko E N 2015 *Journal of Optical Technology (A Translation of Opticheskii Zhurnal)* **82** 162-165
[16] Nepomnyashchaya E K, Cheremiskina A V, Velichko E N, Aksenov E T and Bogomaz T A 2015 *Journal of Physics: Conference Series* **643(1)** 012018
[17] Grebenikova N M, Smirnov K J, Artemiev V V, Davydov V V and Kruzhalov S V 2018 *Journal of Physics: Conference Series* **1038 (1)** 012089
[18] Grebenikova N M, Smirnov K J, Davydov V V , Rud V Yu and Artemiev V V 2018 *Journal of Physics: Conference Series* **1135 (1)** 012055
[19] Grebenikova N M, Myazin N S, Rud V Y and Davydov R V 2018 *Proceedings of the 2018 IEEE International Conference on Electrical Engineering and Photonics, EExPolytech 2018* (Saint-Petersburg) **8564409** p. 295-297
[20] Grebenikova N M, Smirnov K J, Davydov V V and Rud V Y 2018 *Journal of Physics: Conference Series* **1124 (4)** 041011
[21] Davydov V V, Dudkin V I, Velichko E N and Karseev A Yu 2015 *Journal of Optical Technology (A Translation of Opticheckii Zhurnal)* **82(3)** 132-135
[22] Davydov V V, Dudkin V I and Karseev A Yu 2015 Russian Physics Journal **58** 146 – 152
[23] Davydov V V 1999 *Russian Physics Journal* **42(9)** 822-825
[24] Davydov V V, Dudkin V I and Karseev A Yu 2014 *Optical Memory and Neural Networks (Information Optics)* **23** 170-176
[25] Davydov V V, Dudkin V I and Karseev A Yu 2014 *Optical Memory and Neural Networks (Information Optics)* **23** 259-264
[26] Davydov V V, Dudkin V I and Karseev A U 2013 *Optical Memory & Neural Networks (Information Optics)* **22** 112 – 117
[27] Myazin N S, Davydov V V, Yushkova V V and Rud V Yu 2018 *Journal of Physics: Conference Series* **1038 (1)** 012088
[28] Davydov V V, Myazin N S and Velichko E N 2017 *Technical Physics Letters* **43** 607-610
[29] Davydov V V, Myazin N S, Logunov S E and Fadeenko V B 2018 *Russian Journal of Nondestructive Testing*. **54(3)** 213-221
[30] Davydov V V, Dudkin V I and Myazin N S 2016 *Journal of Communications Technology and Electronics* **61** 1159–1165
[31] Davydov V V, Dudkin V I and Karseev A Yu 2014 *Measurement Techniques* **57(8)** 912-918
[32] Marusina M Y, Bazarov B A, Galaidin P A, Marusin M P, Silaev A A, Zakemovskya E Y and Mustaev Y N 2014 *Measurement Techniques* **57(5)** 580-586
[33] Davydov V V 2016 *Optics and Spectroscopy (English translation of Optika i Spektroskopiya)* **121(1)** 18-24
[34] Davydov V V, Dudkin V I and Karseev A Yu 2015 *Thechnical Physics* **63** 456-460
[35] Davydov V V, Dudkin V I, Velichko E N and Karseev A Yu 2015 *Measurement Techniques* **58(5)** 556-561

Spectroscopic study of defect states in CdSe nanoplatelets

M A Miropoltsev[1], Y A Gromova[2], V N Smelov[1], S A Cherevkov[1], T K Kormilina[1], A P Tkach[1], V G Maslov[1]
[1]Center of Optical Information Technologies, ITMO University, St. Petersburg 197101, Russia
[2]School of Chemistry, Trinity College Dublin, Dublin 2, Ireland

Abstract. Here we report investigation of red-shifted defect photoluminescence observed in two-dimensional colloidal cadmium selenide nanoplatelets. We used absorption, photoluminescence and photoluminescence excitation measurements to show that red-shifted photoluminescence bands are related directly to nanoplatelets and may originate from surface imperfections as well as be a result of interaction between different nanoplatelets in the solution. We also found that the defect photoluminescence may be easily altered by simple mechanical treatment, including sonication and stirring, thus disclosing the ways to control nanoplatelets emission over time.

1. Introduction

Colloidal two-dimensional (2D) nanoplatelets (NPLs) emerged more than 10 year ago [1] and have been extensively studied for the past 10 years. Strong two-dimensional confinement results in special physical and optical properties of these structures, including narrow-line emission with very short lifetime [2]. Convenient 2D geometry, along with the ability to control nanoplatelets thickness on the atomic scale, made them a promising material for various photonic applications [3].

A standard colloidal high-temperature synthesis of CdSe NPLs is quite good established [4]. Nevertheless, chemical liquid-phase methods could never be as precise as, for instance, epitaxial methods, which were used to produce quantum wells yet before colloidal synthesis was developed. Impurities and small changes in synthesis conditions could lead to nanocrystals with different imperfections in arrangement of inner and/or surface atoms, which reflects on its electronic structure and thus optical properties. For instance, it is well known that quantum dots surface is far from ideal crystal lattice, and it has already been reported, that an opening of the subband gap state may occur due to surface relaxations [5]. Concerning the nanoplatelets, not only structure, but also morphology and architecture may affect their behavior. Due to pronounced 2D geometry NPLs tend to form stacks in the solution [6], which changes their emission properties. For example, stacked CdSe NPLs were found to expose red-shifted LO phonon line emission at low temperatures [7]. Another example is that stacking of NPLs of the same thickness may lead to energy transfer between them, resulting in the decrease of photoluminescence (PL) quantum yield [8]. Expectedly, when NPLs with different thicknesses are involved, nonradiative fluorescence resonance energy transfer (FRET) within standard donor-acceptor scheme could be observed [9].

In this paper we report spectroscopic study of additional photoluminescence bands with the lower energies compared to exciton emission line in CdSe nanoplatelets of various morphology. We describe several observed band morphologies: slide-shaped PL, multiple peaks and broad background. We also confirm FRET between two NPLs populations within one solution. In some cases, simple mechanical treatment reduces the intensity of defect PL and increases the intensity of main excitonic peak, thereby showing its interaction-based nature.

2. Materials and Methods

We investigated photoluminescence of three different CdSe nanoplatelets samples (thicknesses are expressed here in monolayers, ML):

- 6 ML nanoplatelets, emitting at 513 nm;
- 5 ML nanosheets with giant lateral dimensions, emitting at 464 nm;
- mixture of 5 and 6 ML nanoplatelets.

NPLs with thicknesses of 5 and 6 monolayers and sizes which are not exceeding 100 nm were prepared by the standard synthetic procedure which could be found elsewhere [10]. The approach to obtain 5 ML nanosheets (i.e. nanoplatelets with lateral dimensions up to 700 nm) is referred to [11]. NPLs in all cases were dissolved in pure chloroform and had oleic acid as a stabilizing agent.

We used simple mechanical treatment to change NPLs conditions in every sample and find out how it could affect its optical properties. The treatment included precipitation/redissolution, as well as stirring and sonication. All these actions are widely used for colloidal nanocrystals to manipulate interaction of dissolved particles. Precipitation/redissolution was performed by adding isopropanol to nanoplatelets and subsequent centrifuging at 13000 rpm for 5 minutes until NPLs precipitate. After that, the supernatant was removed and NPLs were dissolved again. This technique is analogous to those used for ligand exchange or nanocrystal separation as described here [12,13]. Where mentioned, stirring was performed at the rate of 800 rpm for approximately 10 minutes.

Absorbance, photoluminescence and photoluminescence excitation measurements were held for every sample before and after treatment. Absorbance spectra were registered on Shimadzu UV-3600 spectrophotometer with integrating sphere, while PL and PLE spectra were registered on Cary Eclipse (Agilent) spectrofluorometer. For all presented absorbance spectra baseline was subtracted, and all PL spectra were divided by the absorbance at the excitation wavelength (400 nm for all measurements).

3. Results and Discussion

3.1. 6 ML NPLs with slide-shaped defect PL

On the Figure 1(a) one can see absorbance, emission and PLE spectra of 6 ML thick NPLs. The first pronounced absorption peaks at 479 and 507 nm correspond to the excitonic transitions from light and heavy hole subbands, respectively [14]. A long featureless tail is observed at lower energies, which has already been reported in [15]. Authors note that it could be attributed to the light scattering by NPLs aggregates. Interestingly, the slide-shaped tail can also be seen on PL spectrum. PLE spectrum registered at 555 nm does not expose any transition-like features, indicating that the slide-shaped emission could not be attributed to individual nanoplatelets.

After 60 seconds of sonication NPLs emission increased by 39% without any change of slide-shaped tail intensity, as follows from Figure 1(b), while absorbance spectrum stayed absolutely the same. To explain the effect of PL enhancement, let us consider the stacked form of NPLs in the solution. According to [8], when NPLs are stacked, energy transfer decreases the excitonic emission. Sonication leads to destruction of stacks, which results in higher quantum yield. Since nothing happened to the low energy tail in absorbance, one may conclude that it is not scattering which is responsible for.

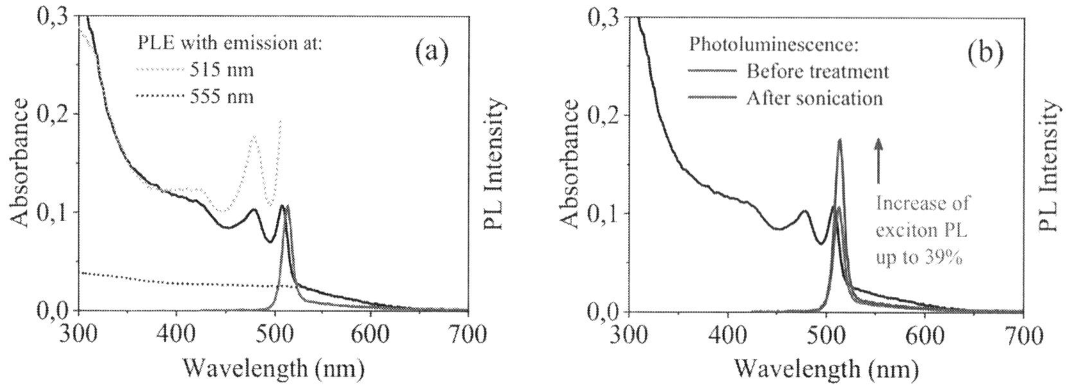

Figure 1(a, b). (a) Absorbance (solid black line), PL (solid red line) and PLE (dotted lines, shifted for clarity) spectra of 6 ML thick CdSe NPLs; **(b)** PL spectra of the same sample before and after mechanical treatment. See legend for details.

3.2. 5 ML nanosheets with defect PL originating from thicker domains

Nanosheets are nanoplatelets with lateral dimensions which could be two orders of magnitude greater than their thickness. Because of increased area, behavior of nanosheets in solution may be different from that of small NPLs. For example, it is known that nanosheets tend to form scrolls [11,16]. Our study shows yet another one effect related to large surface area of nanosheets. Absorbance, PL and PLE spectra of 5 ML nanosheets sample are presented on Figure 2(a). Absorbance spectrum contains small peaks at lower wavelengths, which could be interpreted as originating from different populations of NPLs. The same is true for PL spectrum: it shows several "impurity" peaks, corresponding to different NPLs thicknesses [14], and a broad background band at 500-700 nm. However, PLE measurements reveal that "impurity" emission always gains energy from the main nanosheets in the ensemble, emitting at the shorter wavelengths.

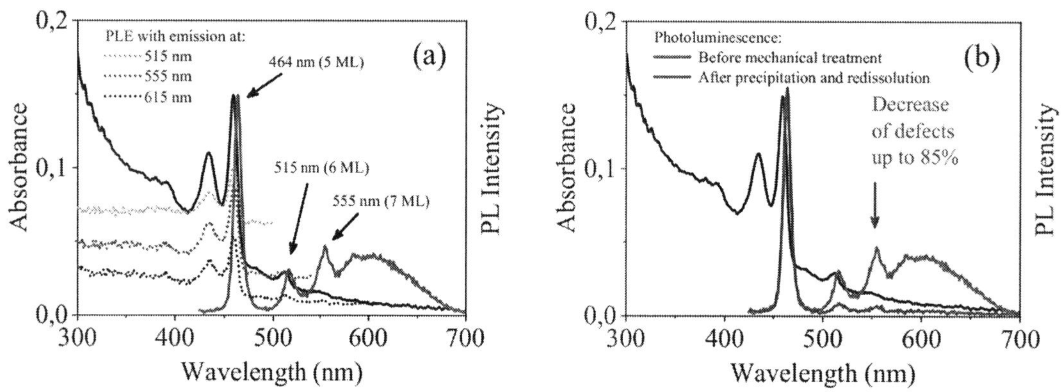

Figure 2(a, b). (a) Absorbance (solid black line), PL (solid red line) and PLE (dotted lines, shifted for clarity) spectra of 5 ML thick CdSe nanosheets; **(b)** PL spectra of the same sample before and after mechanical treatment. See legend for details.

To find out whether "impurities" are individual particles, we used a standard precipitation/redissolution technique, which was described earlier. After the procedure, absorbance spectrum of the sample stayed absolutely the same, but PL spectrum changed drastically, as can be seen on Figure 2(b).

Defect PL bands intensity decreased up to 85%, while exciton PL corresponding to the main population of nanosheets insignificantly increased. Demonstrated results lead us to hypothesis that that the "impurity" bands could originate from small island-like CdSe particles bound to the large surface of nanosheets. When mechanically destabilized (e.g. due to precipitation/redissolution), these particles unstuck from the nanosheets surface, and do not emit light, although continue to absorb it. When bound to nanosheets, they form thicker domains, thus resulting in appearance of PL bands corresponding to thicker NPLs. Described idea, however, could not be treated more that just an assumption and requires additional research to approve or reject it.

3.3. Mixture of 5 and 6 ML NPLs with energy transfer

Finally, we studied a mixture of NPLs with thicknesses of 5 and 6 ML, where non-radiative energy transfer is observed [9]. Absorbance, PL and PLE spectra of the sample, presented on Figure 3(a), clearly show existence of two populations: 5 ML thick NPLs emitting at 466 nm and 6 ML thick NPLs emitting at 513 nm. PL and PLE spectra, registered at different wavelengths, prove the presence of energy transfer. Broad PL band at 500-650 nm probably originates from surface defects and is related to thicker particles. To demonstrate it, we used precipitation/redissolution method. As follows from the previous section, this method leads to nanoplatelets separation, and thus reduces FRET efficiency.

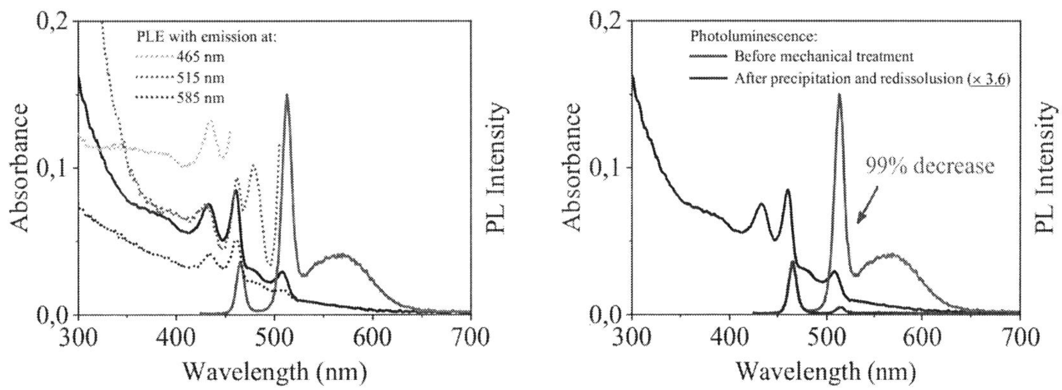

Figure 3(a, b). (a) Absorbance (solid black line), PL (solid red line) and PLE (dotted lines, shifted for clarity) spectra of sample, containing 5 and 6 ML thick CdSe NPLs; **(b)** PL spectra of the same sample before and after mechanical treatment. See legend for details.

Absorbance spectrum of the sample after the treatment did not change, and PL spectra are presented on Figure 3(b). It is seen that the emission of 6 ML thick NPLs has almost disappeared, as well as broad defect PL band. This confirms that 513 nm band and broad defect band are originating from the same particles in the solution. The fact that these particles can emit much more efficiently when FRET requirements are met denotes that interaction between NPLs is a complicated process which may affect quantum yield of both donor and acceptor.

4. Conclusions

In summary, colloidal CdSe nanoplatelets of various morphology exhibit additional PL bands, which may originate from surface defects or be a result of NPLs interaction. We have confirmed that, in general, NPLs interaction affects greatly their luminescent properties, especially when there are several NPLs populations in the solution. Moreover, we have demonstrated how simple mechanical treatment may alter primary and defect PL bands, thus disclosing the ways to control NPLs emission. To obtain more information on the nature of CdSe NPLs defect states, additional research is required.

Acknowledgments

This study was supported by the Federal Target Program for Research and Development of the Ministry of Education and Science of the Russian Federation, grant no. 14.587.21.0047 (ID RFMEFI58718X0047). Y G thanks Irish Research Council for financial support (IRC Project 208153). Authors also wish to acknowledge Professor A V Baranov, the head of Laboratory of Optics of Quantum Nanostructures at Center of Optical Information Technologies, ITMO University, for providing all experimental equipment.

References

[1] Ithurria S and Dubertret B 2008 Quasi 2D colloidal CdSe platelets with thicknesses controlled at the atomic level *Journal of the American Chemical Society* **130** 16504–16505
[2] Nasilowski M, Mahler B, Lhuillier E, Ithurria S and Dubertret B 2016 Two-dimensional colloidal nanocrystals *Chemical reviews* **116** 10934–10982
[3] Kormilina T K, Cherevkov S A, Fedorov A V and Baranov A V 2017 Cadmium Chalcogenide Nano-Heteroplatelets: Creating Advanced Nanostructured Materials by Shell Growth, Substitution, and Attachment *Small* **13** 1702300
[4] Ithurria S, Bousquet G and Dubertret B 2011 Continuous transition from 3D to 1D confinement observed during the formation of CdSe nanoplatelets *Journal of the American Chemical Society* **133** 3070–3077
[5] Puzder A, Williamson A, Gygi F and Galli G 2004 Self-healing of CdSe nanocrystals: first-principles calculations *Physical review letters* **92** 217401
[6] Jana S, Phan T N, Bouet C, Tessier M D, Davidson P, Dubertret B and Abécassis B 2015 Stacking and colloidal stability of CdSe nanoplatelets *Langmuir* **31** 10532–10539
[7] Tessier M D, Biadala L, Bouet C, Ithurria S, Abecassis B and Dubertret B 2013 Phonon line emission revealed by self-assembly of colloidal nanoplatelets *ACS nano* **7** 3332–3340
[8] Guzelturk B, Erdem O, Olutas M, Kelestemur Y and Demir H V 2014 Stacking in colloidal nanoplatelets: Tuning excitonic properties *ACS nano* **8** 12524–12533
[9] Guzelturk B, Olutas M, Delikanli S, Kelestemur Y, Erdem O and Demir H V 2015 Nonradiative energy transfer in colloidal CdSe nanoplatelet films *Nanoscale* **7** 2545–2551
[10] Prudnikau A, Chuvilin A and Artemyev M 2013 CdSe–CdS Nanoheteroplatelets with Efficient Photoexcitation of Central CdSe Region through Epitaxially Grown CdS Wings *Journal of the American Chemical Society* **135** 14476–14479
[11] Bouet C, Mahler B, Nadal B, Abecassis B, Tessier M D, Ithurria S, Xu X and Dubertret B 2013 Two-dimensional growth of CdSe nanocrystals, from nanoplatelets to nanosheets *Chemistry of Materials* **25** 639–645
[12] Dong A, Ye X, Chen J, Kang Y, Gordon T, Kikkawa J M and Murray C B 2010 A generalized ligand-exchange strategy enabling sequential surface functionalization of colloidal nanocrystals *Journal of the American Chemical Society* **133** 998–1006
[13] Chemseddine A and Weller H 1993 Highly monodisperse quantum sized CdS particles by size selective precipitation *Berichte der Bunsengesellschaft für physikalische Chemie* **97** 636–638
[14] Ithurria S, Tessier M, Mahler B, Lobo R, Dubertret B and Efros A L 2011 Colloidal nanoplatelets with two-dimensional electronic structure *Nature materials* **10** 936
[15] Antanovich A, Prudnikau A, Matsukovich A, Achtstein A and Artemyev M 2016 Self-assembly of CdSe nanoplatelets into stacks of controlled size induced by ligand exchange *The Journal of Physical Chemistry C* **120** 5764–5775
[16] Vasiliev R B, Sokolikova M, Vitukhnovskii A G, Ambrozevich S, Selyukov A and Lebedev V S 2015 Optics of colloidal quantum-confined CdSe nanoscrolls *Quantum Electronics* **45** 853

Optical properties of Fe- and Nd-doped ZnSe crystals measured by pump-probe spectroscopy

M O Zhukova[1], A O Nabilkova[1], S E Putilin[1], V P Chegnov[2], O I Chegnova[2], A N Tcypkin[1]
[1]Laboratory of Femtosecond Optics and Femtotechnology, ITMO University, Saint Petersburg, 197101, Russia
[2]Research Institute of Materials Science and Technology, Moscow, 124460, Russia

Abstract. We present the excited-state dynamics and optical properties of Fe- and Nd-diffusion doped zinc selenide polycrystals with different concentration measured by pump-probe spectroscopy in the visible and NIR range (400-1000 nm). The introduction of dopants allows one to control the optical properties of samples, which then can be used to create effective optoelectronic components for wide spectral ranges.

1. Introduction

The wide bandgap zinc-blende semiconductor crystals are attractive materials for various optoelectronic applications. ZnSe is transparent in a wide spectral range, it is a chemically inert, non-hygroscopic and high-purity product that is effective in many optical applications due to its extremely low volume loss, high resistance to thermal shock and stability in almost all environments, and is also easily processed. Zinc selenide with different dopants is a promising material for highly efficient, tunable [1] and pulsed [2] mid-infrared lasers. It is also used for spintronics applications [3] and pulsed terahertz radiation generation [4-5] and electro-optic sampling [6]. Doping with the transition and rare-earth elements provides the appearance of new crystal properties caused by changes in the energy levels. The better understanding of recombination processes and optical properties of newly synthesized materials is required. For characterization of excited-state dynamics, the pump-probe spectroscopy can be used. This powerful time-resolved technique allows measuring the sample absorbance variation after the excitation given by the pump pulse.

In this work, we observe the excited-state dynamics and optical properties of Fe- and Nd-diffusion doped zinc selenide polycrystals with different concentration measured by femtosecond pump – supercontinuum probe spectroscopy in the visible and NIR spectral range (400-1000 nm). The introduction of dopants (Fe^{2+} ions with a concentration 0.05-0.45 wt.%. and Nd^{2+} ions varying from 0.08 to 0.48 wt.%.) allows one to control the optical properties of samples which then can be used to create effective optoelectronic components for wide spectral ranges.

2. Experimental methods

2.1. The crystal growth and doping

The experiments were carried out on ZnSe plates alloyed with iron by thermal diffusion. The plates were cut from boules grown from the melt by the Bridgman method in graphite crucibles at a pressure of high-purity argon of 100 atm. The loading material was CVD grown ZnSe with 99.9996 wt. %

purity which, as shown by ICP-MC analysis, provided an ingot of 99.9992 wt.% purity [7]. Impurities of Fe and Nd were introduced into ZnSe plates by thermal diffusion in a quasi-closed quartz reactor at a temperature of 1000 °C. The samples thickness varies around 1 mm.

The first group of samples was ZnSe crystals doped with an iron over the entire thickness of the plate. The concentration of iron, determined by x-ray fluorescence method was 0.23 wt.% and 0.15 wt.%. The measurements were carried out using a Bruker Mistral M-1 spectrometer.

The second group of samples consisted of ZnSe: Fe plates from the first group was additionally doped with Nd from the one side by thermal diffusion. The surface concentration of Nd, determined by the X-ray fluorescence method was 0.1 wt.%

The third group of samples consisted of undoped ZnSe plates was additionally simultaneously doped with Fe and Nd by thermal diffusion. The surface concentration of Fe and Nd, determined by the X-ray fluorescence method, was, respectively, 0.05 and 0.48 wt.%, 0.7 and 0.2 wt.%, 0.45 and 0.08 wt.%. The overview of all samples properties is shown in Table 1.

Table 1. The ZnSe samples doping concentrations.

	Fe concentration, wt.%	Diffusion depth	Nd concentration, wt.%	Diffusion depth
Group 1	0.23 0.15	through	-	-
Group 2	0.23 0.15	through	0.10	on surface
Group 3	0.05 0.70 0.45	on surface	0.48 0.20 0.08	on surface

2.2. The pump-probe spectroscopy setup

The experiments were carried out on the pump-probe supercontinuum spectrometer shown on Fig. 1. It is based on a Ti-sapphire regenerative amplifier (*Ti:Sa RA*) Regulas 35F1K (Avesta), which produces 30 fs pulses with energy up to 2.3 mJ and repetition rate of 1 kHz. The beam is split into two parts by *BS*1. One is used as a pump beam and its intensity is varied by the polarization attenuator (*A*) in the reflection geometry. The second part of the split beam is used for supercontinuum generation (*SCG*). To produce broad (350-1400 nm) SC in the visible – NIR spectral region, the water jet is applied [8-9]. The transmitted SC spectra are registered by the compact spectrometer ASP-100M (200-1000 nm).

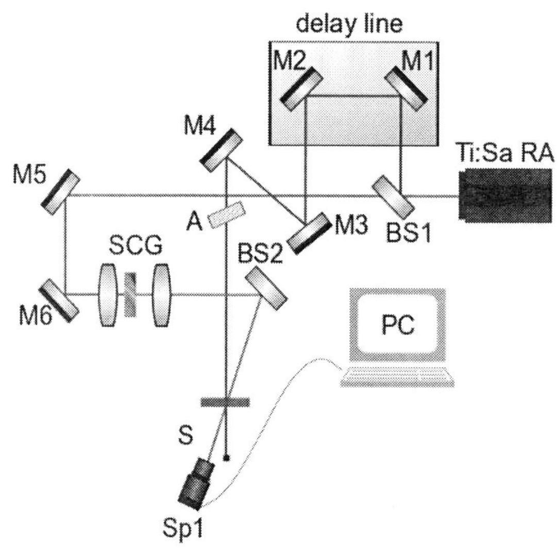

Figure 1. The femtosecond pump supercontinuum probe experimental set-up.

3. Results and Discussions

To explain the optical absorption and changes in the band structure of undoped, Fe- and Nd-doped ZnSe samples pump-probe scans were recorded probing at different frequencies. The pump power at 800 nm was varied from 100 to 300 mW that corresponds to 3.6 and 10.9 GW/cm^2, respectively. The wavelength evolutions of induced absorption peaks extracted from various power dependences help to understand the energy levels' positions in the band structure and the carrier redistribution in zinc selenide polycrystals with different dopant concentration. The examples of typical pump-probe dynamics are presented in Fig. 2. The wavelength evolution of absorption peak position is used to extract the refractive index dispersion, as was shown in [10]. Relaxation of induced excitation wavelength evolution can be analyzed by the bandwidth of peak at half-maximum.

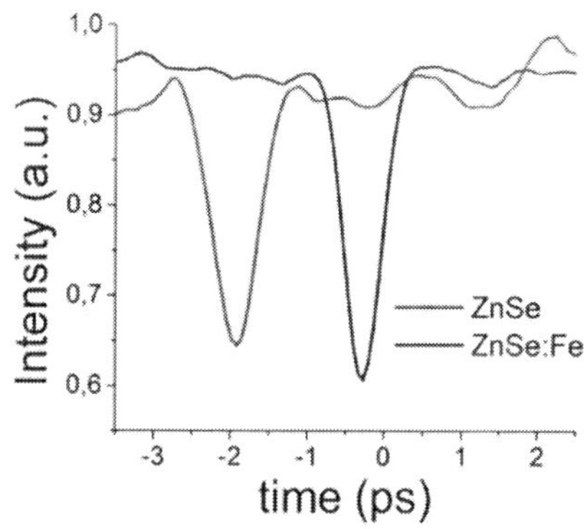

Figure 2. Transmission dynamics of 700 nm probe wavelength for 10 GW/cm^2 pump pulse.

Fig.3 illustrates transmission of supercontinuum spectra by ZnSe and ZnSe:Fe crystals from Group1 after the pump beam excitation with different power levels. The effect of the short wavelength absorption edge shift for the doped sample confirms the known fact that this shift depends on the impurity concentrations and also on the annealing temperature during the diffusion of iron ions. Also, new absorption lines associated with the impurity level appear. The doping with iron change the transmitted supercontinuum spectra shifted to the red region.

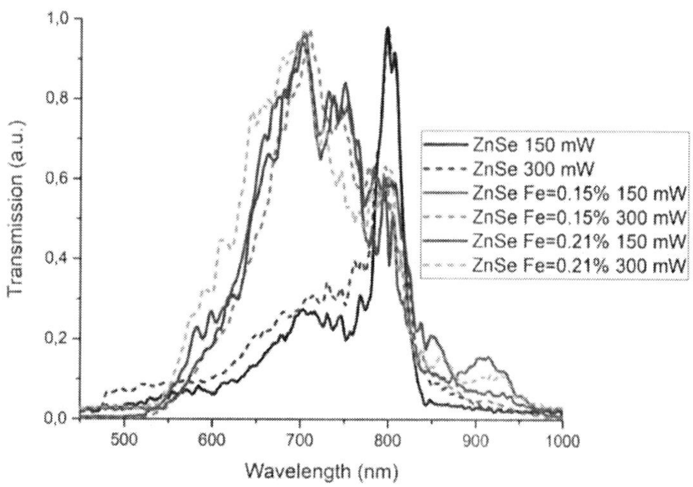

Figure 3. Transmission of supercontinuum spectra (normalized) after 150 and 300 mW pump beam excitation for Fe-doped ZnSe samples.

Figure 4 shows the results for (a) 150 mW and (b) 300 mW pump power level excitation for samples from all three groups under study. It can be noticed that doping of Nd ions into the ZnSe:Fe samples sift the transmission spectra in the middle position between pure and doped samples shown in Fig.3. The increase of pump power also has an influence on the transmitted spectrum curve shape but not as strong as in the case of doping.

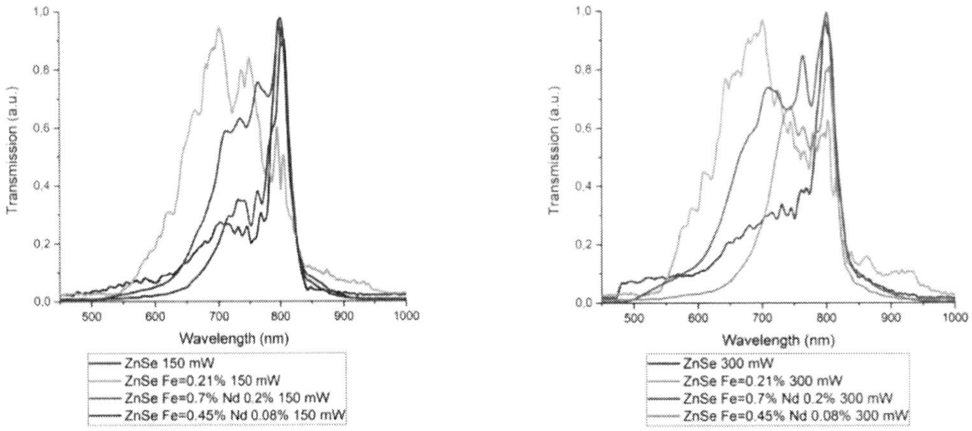

Figure 4(a, b). Transmission of supercontinuum spectra (normalized) after **(a)** 150 mW and **(b)** 300 mW pump beam excitation for Fe- and Nd- doped ZnSe samples.

The next study will help us to understand the way how we can vary the optical properties by doping ZnSe crystals with the transition and rare-earth elements. This can be used to create effective optoelectronic components for wide spectral ranges, for example, for improvement of terahertz electro-optical detection systems.

References

[1] Dormidonov A E, Firsov K N, Gavrishchuk E M, Ikonnikov V B, Kazantsev S Y, Kononov I G, Kotereva T V, Savin D V, Timofeeva N A 2016 *Appl. Phys. B* **122** 211

[2] Wang Y, Fernandez T T, Coluccelli N, Gambetta A, Laporta P, Galzerano G 2017 *Opt. Express* **25** 25193

[3] Behloul M, Salmani E, Ez-Zahraouy H, Benyoussef A 2016 *J. Magn. Magn. Mater.* **419** 233–9

[4] Tcypkin A N, Putilin S E, Kulya M C, Melnik M V, Drozdov A A, Bespalov V G, Kozlov S A 2018 *Bull. Russ. Acad. Sci.: Phys.* **82** 1547

[5] Wu X, Chen X, Zhao F, Jia T, Wang G. 2007 *Jpn. J. Appl. Phys.* **46** 1497

[6] Zhukova M, Makarov E, Putilin S, Tsypkin A, Chegnov V, Chegnova O, Bespalov V 2017 *J. Phys. Conf. Ser.* **917** 062021

[7] Khanh T, Mozhevitina E N, Khomyakov A V, Avetisov R I, Davydov A, Chegnov V P, Antonov V, Kobeleva S P, Zhavoronkov N V, Avetissov I Ch 2014 *J. Cryst. Growth* **401** 686

[8] Tcypkin A N, Putilin S E, Melnik M V., Makarov E A, Bespalov V G, Kozlov S A 2016 *Appl. Opt.* **55** 8390

[9] Melnik M V, Tcypkin A N, Kozlov S A 2018 *Romanian J. Phys.* **63** 203

[10] Zhukova M O, Makarov E A, Putilin S E, Tsypkin A N, Chegnov V P, Chegnova O I, Bespalov V G 2018 *J. Phys. Conf. Ser* **1062** 012009

Excitation of unidirectionally propagating plasmon in the periodic graphene structure

K V Mashinsky [1], V V Popov [1], D V Fateev [1,2]
[1] Kotelnikov Institute of Radio Engineering and Electronics of the Russian Academy of Sciences (Saratov Branch), 410019 Saratov, Russia
[2] National Research Saratov State University, 410012 Saratov, Russia

Abstract. The transformation of normally incident electromagnetic wave into propagating plasmon in periodical graphene structure with double grating gate is theoretically investigated. The most effective transformation appears in case of excitation plasmon modes with the wavevectors quantized with structure period. Blocking the reflection and the transmission of electromagnetic wave through the graphene structure by metal back gate and Fabry-Perot resonator are used. We found the 72% transformation of the incident wave radiation power into the power of unidirectionally-travelling plasmon for 1 ps momentum relaxation time in graphene.

1. Introduction
Graphene is a perspective media for excitation and investigation of the plasmons in terahertz (THz) frequency range [1]. Charge carriers mobility in graphene reaches around 100000 $cm^2/(Vs)$ [2] and its momentum relaxation time reaches the values of 2 ps [3] at room temperatures.

Periodical graphene may be used for amplify [4], detect [5], and transform [6]. The periodical 2D gratings are used for coupling incident THz electromagnetic wave with plasma waves in graphene [7,8]. Plasmon excitation in such graphene structures can be identified via absorption of the incident electromagnetic wave energy. The perfect absorption was found in periodic graphene structure with back gate due to utilizing the Fabry-Perot conditions for reflected wave [9].

In this work, we proposed the periodic graphene structure with dual-grating gate (DGG) and asymmetric unit cell as an effective transformer of normally incident THz electromagnetic wave into propagating unidirectionally-travelling plasmon.

2. Results and discussion
Let us consider a periodic graphene structure shown in Figure 1. The graphene monolayer screened by interdigitated metal DGG and homogeneous metal back gate. Graphene separated from DGG by Al_2O_3 with thickness d and from back gate by SiO_2 with thickness h. The DGG consists of two periodical interdigitated sub-gratings, which fingers alternate. Two sub-gratings are laterally shifted in respect to each other to introduce an asymmetry into the structure unit cell of the periodic DGG-graphene structure. The THz wave with the electric field polarized across the DGG fingers is normally incident on the DGG. The incident wave can induce noncentrosymmetric electric fields in graphene [5], which produces the plasmonic effect of electron drag [10]. Structure thickness $d+h$ picked for using Fabry-Perot resonance at frequency of 5 THz. Fabry-Perot resonance produce destructive interference which suppress reflected THz waves.

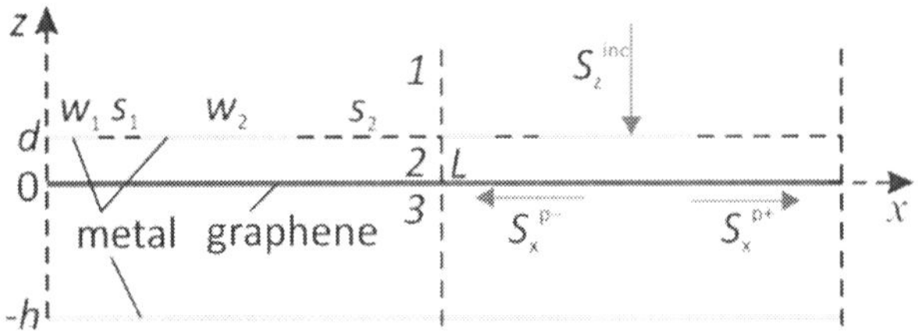

Figure 1. Schematic view of the DGG graphene structure.

We solve the problem of the excitation of plasmon modes in s DGG-graphene structure within a self-consistent electromagnetic approach [11]. The THz response of graphene is described by the dynamic conductivity [12] with the Fermi energy of 150 meV. All results shown here were calculated for the DGG openings $s_1 = 0.27\ \mu$m, $s_2 = 0.08\ \mu$m, the finger width $w_2 = 0.5\ \mu$m, dielectric slab thicknesses $d = 0.105\ \mu$m, $h = 5.895\ \mu$m and momentum relaxation time was 1 ps. The dielectric slab thickness comparable with gate finger widths allows exciting plasmon modes with wave numbers quantized as $k(n) = 2\pi n / L$, with n is integer. In case of DGG graphene structure under investigation the plasmon modes with wave numbers $k(n) = 2\pi n / L$ can be effectively excited due to Fabry-Perot blocking of electromagnetic energy inside the DGG structure.

We consider the transformation coefficient of incident THz wave into propagating plasmon as

$$T_{\mathrm{P}} = \frac{S_x^{\mathrm{P}}}{\left|S_z^{\mathrm{inc}}\right| L},$$

where

$$S_x^{\mathrm{P}} = \frac{1}{L} \int_{-h}^{+\infty} \int_{0}^{L} S_x dx dz$$

is spatially averaged over the period length and integrated over the z-direction x-component of Poynting plasmon flux. The incident flux of normally incident THz wave E_x^{inc} averaged over the spatial period of the structure is described by

$$S_z^{\mathrm{inc}} = -0.5 \sqrt{\frac{\varepsilon_1 \varepsilon_0}{\mu_0}} \left|E_x^{\mathrm{inc}}\right|^2 .$$

The plasmon spatially averaged flux S_x^{P} consists of two opposite fluxes $S_x^{\mathrm{P}} = S_x^{\mathrm{P+}} + S_x^{\mathrm{P-}}$, where $S_x^{\mathrm{P+}}$ is plasmon Poynting flux in positive direction of x-axe and $S_x^{\mathrm{P-}}$ is in the negative one. Then we introduce the positive transformation coefficient of incident wave into the plasmon propagating in x-direction as

$$T_{\mathrm{P}}^{+} = \frac{S_x^{\mathrm{P+}}}{\left|S_z^{\mathrm{inc}}\right| L}$$

and negative transformation coefficient of incident wave into the plasmon propagating in negative direction of x-axis as

$$T_{\mathrm{P}}^{-} = \frac{S_x^{\mathrm{P-}}}{\left|S_z^{\mathrm{inc}}\right| L}.$$

Figure 2 demonstrates calculated negative $T_x^{\mathrm{P-}}$ (a) and positive $T_x^{\mathrm{P+}}$ (b) transformation coefficients depending on frequency and DGG finger width w_1. We consider these graphs in order to find maximal values of positive transformation coefficient and minimal of negative transformation coefficient. Difference by an order of magnitude between positive and negative transformation coefficients leads to unidirectional plasmon energy flux, which means excitation of unidirectionally travelling plasmon.

It is achieved in case of interaction of two different plasmon modes, excited in spatially asymmetric DGG-graphene structure. The most successful excitation of unidirectionally travelling plasmon is achieved in case of DGG finger width $w_1 = 0.672\ \mu$m.

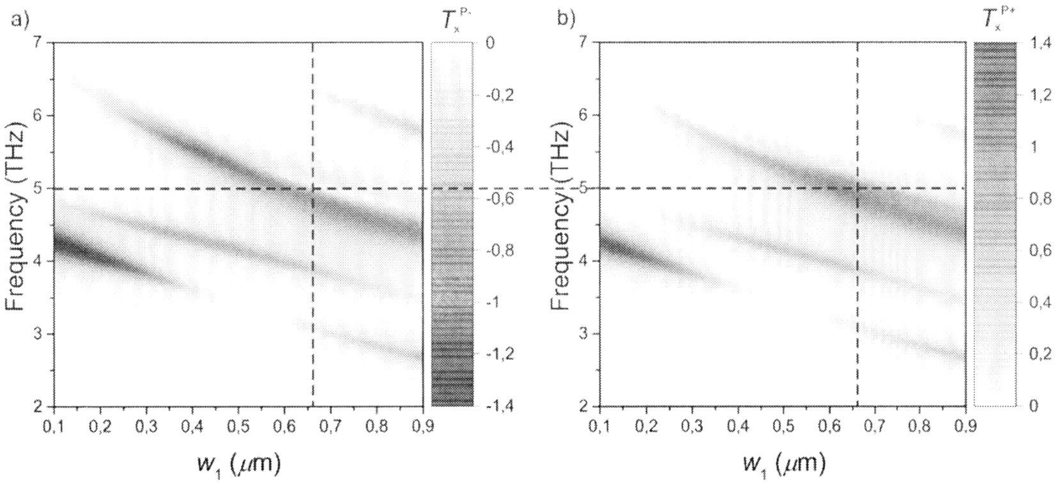

Figure 2. Negative T_x^{P-} (a) and positive T_x^{P+} (b) transformation coefficients depending on frequency and DGG finger width w_1.

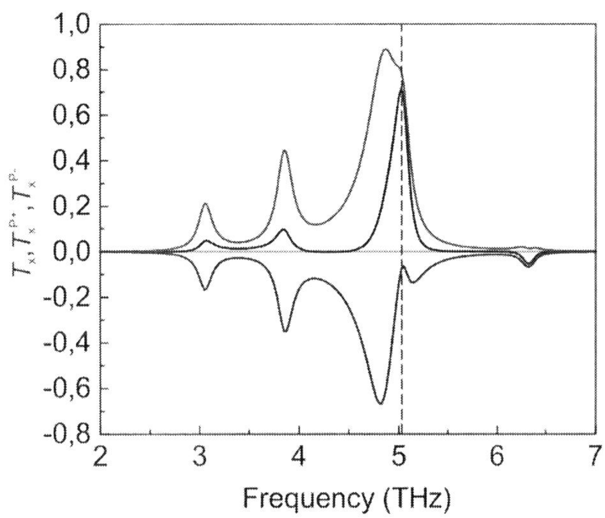

Figure 3. Total T_x, positive T_x^{P+} and negative T_x^{P-} transformation coefficients depending on frequency at DGG finger width $w_1 = 0.672\ \mu$m.

Figure 3 demonstrates calculated total T_x, positive T_x^{P+} and negative T_x^{P-} transformation coefficients depending on frequency at DGG finger width $w_1 = 0.672\ \mu$m. Maximal difference between positive and negative transformation coefficients reaches 72% at frequency 5 THz. The difference between the plasmon fluxes oppositely propagating in DGG structure exceed order of magnitude, which is actually excitation of the unidirectionally travelling plasmon in graphene structure.

In this work we show that it is possible to transform normally incident electromagnetic wave into propagating unidirectionally travelling plasmon using DGG structure with asymmetric unit cell. Fabry-Perot blocking of reflectance from the DGG structure allows for increasing the efficiency of transformation incident wave into propagating plasmon.

Acknowledgement
This work was supported by the Russian Science Foundation through the Grant No. 18-79-10041.

References

[1] F. H. L. Koppens et al., Nat. Nanotech. 9, 780 (2014).
[2] A. S. Mayorov et al., Nano Lett. 11, 2396 (2011).
[3] D. Svintsov et al., J. Appl. Phys. 111, 083715 (2012).
[4] O. V. Polischuk et al., Appl. Phys. Lett. 111, 081110 (2017).
[5] D. V. Fateev et al., Appl. Phys. Lett. 110, 061106 (2017).
[6] O.V. Polischuk et al., Appl. Phys. Lett. 109, 131101 (2016).
[7] V. Ryzhii, Japan Journ. Appl. Phys. 45, L923 (2006).
[8] H. Yan et al., Nature Nanotech. 7, 330334 (2012).
[9] R. Alaee et al., Optics Express 20, 28017 (2012).
[10] V. V. Popov et al., Phys. Rev. B 91, 235436 (2015).
[11] O. V. Polischik et al., Semiconductors 51, 1460 (2017).
[12] H. A. Atwater et al., Phys. Rev. B 90, 165409 (2014).

Numerical simulation of a nitride single quantum well laser

E E Medvedeva[1], S S Konoplev[2], A V Savelyev[2]
[1]Department of Physics of Semiconductors & Nanoelectronics, Peter the Great St.Petersburg Polytechnic University, St. Petersburg 195251, Russia
[2]Nanophotonic Lab, St. Petersburg Academic University, St. Petersburg 194021, Russia

Abstract. Numerical simulation of a quantum well on nitrides , optical interband transitions in low dimensional semiconductor structures (semiconductor quantum wells), conduction-heavy hole and transitions in quantum wells, refinement of the gain for a quantum well on nitride materials with allowance for the overlap integral and approximate this.

1. Introduction

The optical gain is a key parameter in the design of quantum well lasers. The design optimization necessitates a large degree of numerical computation because there are a large number of laser parameters involved, such as the quantum well barrier composition, the number of QW's, the cavity length, and the facet reflectivity. In the dynamic performance of QW lasers, the differential gain and the line width enhancement factor are key parameters in addition to the gain. The efficiency of numerical calculations becomes a serious factor when analyzing the structures of 2D or 3D quantum-well lasers, such as nitride lasers, in which medium deformations due to polarization effects play a significant role. Furthermore, in the design optimization, some of the important physics becomes obscured because of the large number of laser parameters. Therefore, simple analytical formulas are desired that can clearly delineate the relations between design parameters and predict device performance characteristics.

2. Analytical approach

Alloys of GaN, which are discussed in our modeling, are mainly used to achieve increased device performance or to achieve a certain radiation wavelength, but many important parameters, such as energy of the forbidden zone, depend on the composition of the alloy. In our work, this parameter was also one of the most important. For the calculation, Vegard's law was used, with which most of the experimental results for the width of the forbidden zone of solid solutions recently agree. Vegard's law for $In_xGa_{1-x}N$:

$$Eg(In_xGa_{1-x}N) = xEg(InGa) + (1-x)Eg(InN) - bx(1-x),$$

where b is the bending parameter determined by the structure.

Nitride materials, as already noted above, are of particular interest because of their internal spontaneous P_s and external P_{pz} piezoelectric effect. For example, the piezoelectric field (E_{pz}), induced by voltage, along the axis [0001], in the absence of free charges, is then defined as:

$$E_{pz}^{z} = -\frac{P_z}{r\varepsilon_0} = -\frac{2d_{31}}{r\varepsilon_0}\left[C_{11} + C_{12} - 2\frac{C_{13}^2}{C_{33}}\right]\varepsilon_{xx},$$

where ε_r and ε_0 are the dielectric constant of the material and the dielectric constant of the vacuum, ε_{xx} is the deformation component along the x axis, which can be defined in terms of lattice constants as $\varepsilon_{xx} = (a_b - a_e) / a_e$, where a_e and a_e are the lattice constants of the epitaxial layer and bulk material, respectively Thus, the direction of the piezoelectric field depends on whether the epitaxial layer is in a state of compression or tension, which determine the variable ε_{xx}.

To check the results obtained by other scientific groups when trying to verify the truth of the study, we attach a table with structural parameters used in our research (Table 1). The parameters of the crystal structure of these compounds are given in the Table. 1, where *a,c* - constants of the hexagonal lattice, and P_s - spontaneous polarization.

Table 1. Structural parameters and calculated spontaneous polarization of nitrides of elements of the nitride group.

	GaN	InN	AlN
a, °A	3,189	3,54	3,112
c, °A	5,185	5,705	4,982
a / c	1,627	1,612	1,601
Ps, C/m2	−0,029	−0,032	−0,081

A numerical model was developed for solving the one-dimensional stationary Schrödinger equation and the search for allowable energy values for a quantum well on nitrides. During the calculations, spontaneous and piezoelectric polarizations of nitrides were taken into account, leading to a distortion of the profile of the quantum well on nitrides (in particular, GaN/$In_{0.7}Ga_{0.2}N$/GaN). The overlap integral of the wave functions was calculated, which makes it possible to more accurately calculate the characteristics of nitride lasers (including amplification, the probability of spontaneous emission and absorption, spontaneous emission).

The rate of stimulated absorption $R\uparrow$ and emission $R\downarrow$ can be calculated by the formulas below:

$$R_{\uparrow,\downarrow} = \left(\frac{q}{m}\right)\left(\frac{\pi m_p}{n n_g^M{}_0}\right)\left|\int dx[\psi_e]^* \psi_{hh}\right|^2 \frac{mE_p}{4}(n_x{}^2 + n_y{}^2) f_{v,c}(1 - f_{c,v}) \frac{m_{r2}}{\pi h^2}\theta(h\omega - Ec - Ep);$$

Where $\left|\int dx[\psi_e]^* \psi_{hh}\right|^2$ is integral of overlap *I*, n_x, n_y- components of the electric polarization vector, f_c, f_v - Fermi function for electrons and holes, Ec, Ep - quantization energy level for electrons and holes, n_p- photon density, m- electron mass, n, n_g^M, E_p- determined by the material parameters,m_{r2} - reduced mass, q- electron charge, ε_0- dielectric constant, h- ''h-bar'' Planck's constant.

We take the traditional model, the optical gain $g(\omega)$ model of Asada et al. [1]. It is assumed that all levels are parabolic and that the optical transitions obey the *k*-selection rules. In order to have a general idea of the nature of gain, we write the empirical expression for optical gain without taking into account spontaneous radiation for single QW, which is expressed as:

$$g(\omega) = \frac{R_\downarrow(\omega) - R_\uparrow(\omega)}{n_p v_g^M}$$

The absorption coefficient of a pure photon is the difference between the speed of stimulated absorption and the speed of stimulated emission. The semiconductor absorption coefficient can then be associated with the photon group velocity v_g^M and with the photon density n_p. Through the values of the stimulated emission and absorption rates in this paper, we have calculated the attenuation and other laser characteristics. The work echoes the themes that affect the articles in the list of references [2]-[4]. The numerical simulation results are presented in Figure 1-3.

a) **b)**

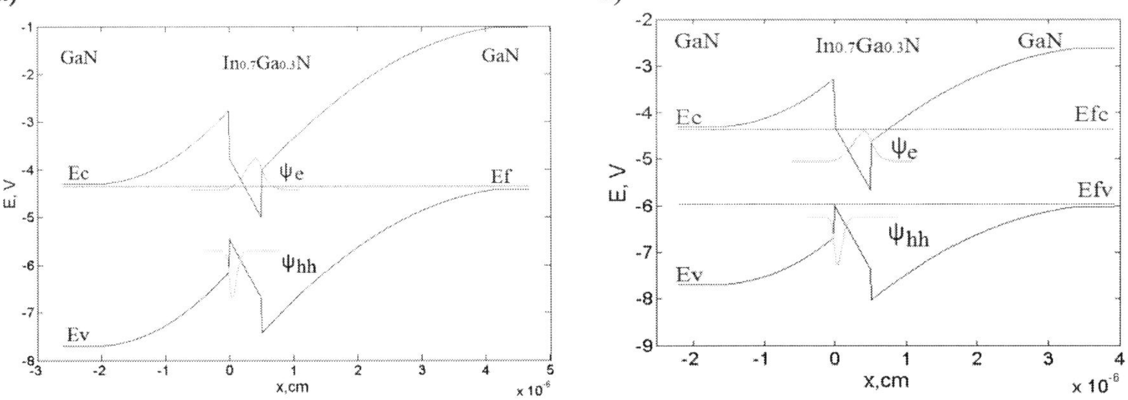

Figure 1. (a,b) The potential well profile (dependence of the quantum well energy on the coordinate) is distorted by the polarization in the GaN/$In_{0.7}Ga_{0.2}N$/GaN structure. By the shooting method, the wave functions for electrons Ψ_e and heavy holes Ψ_{hh} are calculated. (a) - is in equilibrium (*Ef* - the equilibrium Fermi level), Integral of overlap I=0.124, (b) - at an applied voltage of 1.6V (*Efc*, *Efv* - the quasi-Fermi level), I=0.103.

For a better understanding of the configuration of the drawings Figure 1 (a, b), Figure 2 is also attached. with the image of a simulated zone diagram without calculating the Schrödinger equation, which is calculated taking into account information on the internal and external piezoelectric effect of the nitride. The figure shows the regions of the quantum well (radiating region), and the p, n-doped regions (mirrors) doped out along the bottom of the conduction band and the valence band top.

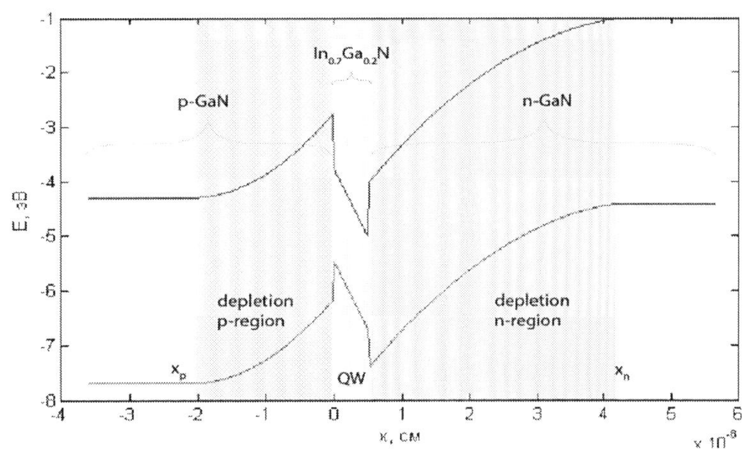

Figure 2. The energy diagram of a nitride laser with a quantum well, taking into account the distortion of the diagram under the action of polarization of nitrides.

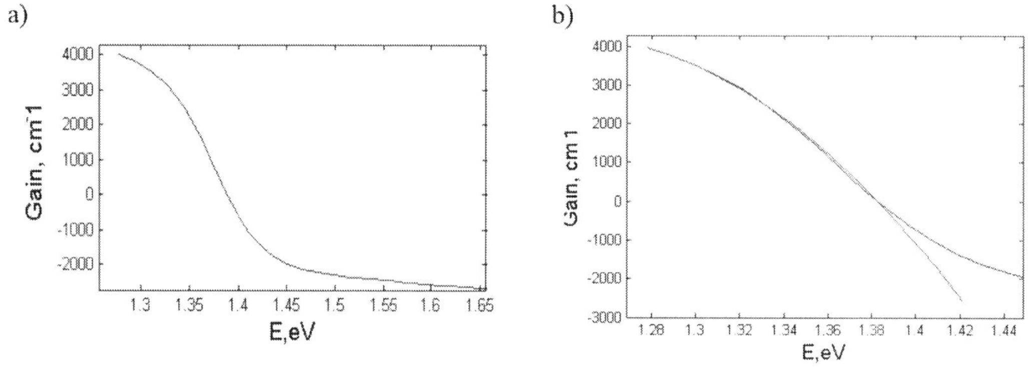

Figure 3. Calculation results **a** - is a numerical calculation of the optical gain in a nitride quantum laser with a heterostructure shown above, **b** - analytical approximation by a parabola of optical gain. The carrier densities are 3.10^{16} cm^{-2}.

3. Result

An analytical expression was obtained for optical amplification in nitride quantum-well lasers using explicit quadratic approximations of approximations of the parabolic optical gain part. It was shown that approximate formulas provide not only an efficient way to calculate the parameters associated with the gain, but also a convenient way to obtain physical information for the general interaction of the SQW parameters. It is expected that the presented analytical approach will demonstrate its significant advantage, especially in terms of optimizing the design of practical SQW nitride lasers, which require consideration of material deformation parameters.

Acknowledgments

This work was supported by Russian Ministry of Science and Higher Education (grant No. 3.9787.2017/8.9).

References

[1] M. Asada, A. Kameyama, and Y. Suematsu, "Gain and intervalence band absorption in quantum-well lasers" ZEEE J. Quantum Electron., vol. QE-20, pp. 745-753, July 1984.
[2] O. Ambacher et al., Pyroelectric properties of Al(In)GaN/GaN hetero-and quantum well structures, J. Phys.: Condens. Matter 14, 3399–3434 (2002).
[3] Pelá, R. R. at al., Accurate band gaps of AlGaN, InGaN, and AlInN alloys calculations based on LDA-1/2 approach, Appl. Phys. Lett., 2011, 98, 151907.
[4] A. T. Meney, E. P. O'Reilly and A. R. Adams, Optical gain in wide bandgap GaN quantum well lasers, IOP Science, Semicond. Sci. Technol. 11 (1996) 897–903.

Energy transfer in heteronuclear rare gas clusters under multiphoton excitation and ionization

I Balmaev [1], P Serdobintsev [1,2], A Melnikov [2], A Pastor [1], M Khodorkovskiy [2]
[1] Saint-Petersburg State University, Saint-Petersburg, Russia
[2] Peter the Great St. Petersburg Polytechnic University, Saint-Petersburg, Russia

Abstract. The ionization processes of the Ar_M, Xe_N and Xe_NAr_M clusters in a supersonic beam with multiphoton excitation were studied. The third 263 nm harmonics of a femtosecond Ti : sapphire laser were used for excitation and ionization. Kinetic energy spectra of photoelectrons has been recorded by a magnetic bottle time-of-flight electron spectrometer. Analysis of the electronic spectra of multiphoton ionization of clusters shows that, in heteronuclear clusters, the transfer of excitation energy between xenon and argon atoms plays a significant role in ionization.

1. Introduction

The study of non-radiative relaxation processes of electronically excited clusters of rare gases is of considerable interest from both fundamental and applied points of view. Indeed, clusters are a transitional stage from atoms and large molecules to a solid, and the study of relaxation processes in such nanoparticles sheds light on the ratio of bulk and surface relaxation processes. The practical significance of such research consists primarily in elucidating the role of the processes of relaxation of excitation and ionization, as the initial stage of the so-called "Coulomb explosion" of ionized clusters, which serves as a source of multiply charged ions to produce intense VUV radiation. Earlier [1,2], in our group, nonradiative relaxation processes of xenon clusters in an electronically excited states were investigated. Clusters were created in a supersonic beam during expansion into a vacuum and ionized by absorbing several photons of a femtosecond laser. The rates of the following processes were measured: desorption of excited atoms from a cluster, relaxation during electron-phonon interaction, and interatomic Coulomb decay (ICD). The rates of the above nonradiative relaxation processes of xenon clusters are in the range 10^{11} - 10^{12} s^{-1}. In this work, we measured the mass composition of xenon, argon, and xenon-argon mixed clusters. Using the time-of-flight spectrometer, photoelectron spectra of clusters were obtained. It was found that transfer of excitation energy from an excited xenon center to argon is observed in heteronuclear Xe_N Ar_M clusters.

2. Experimental set up

The pulsed supersonic cluster beam was produced by the adiabatic expansion of the gas mixture through a 200 μm diameter sonic nozzle at room temperature. The beam was skimmed at 20 mm distance from the nozzle by a 1.5 mm diameter skimmer. The cluster beam crossed the focus of a laser radiation. The composition of cluster beam was measured by time-of-flight mass spectrometer equipped with electron ionization source with energy of 30 eV, since at higher energies the cluster fragmentation increases significantly [3].

The third 263 nm (UV, hν=4.71 eV) harmonic of a femtosecond Ti : sapphire laser with a repetition rate of 10 Hz was used for excitation. The laser delivered pulses with duration of 50 fs. To avoid above-threshold ionization, the power density in the interaction region was governed by the position of the lens and did not exceed 10^{12} W/cm^2.

Kinetic energy spectra of photoelectrons emitted from the interaction zone has been recorded by a magnetic bottle time-of-flight electron spectrometer. The resolution of the spectrometer did not exceed 30 mV. A photoelectron spectrum of multiphoton ionized xenon atoms was used for energy calibration.

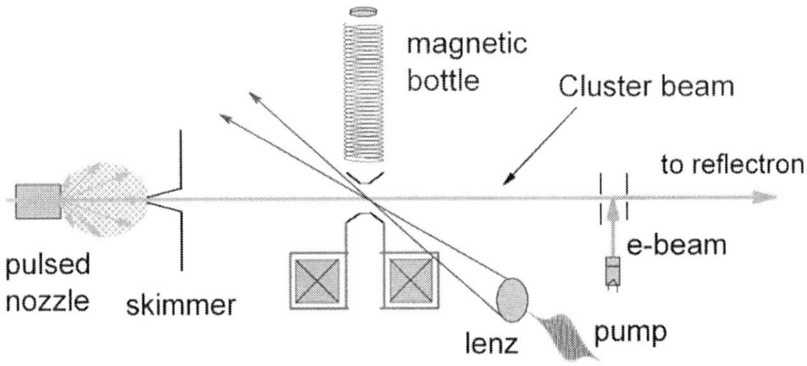

Figure 1. The scheme of the experimental setup.

3. Results and discussion

To create Ar_M, Xe_N and Xe_NAr_M clusters, pure argon, helium-xenon mixture (8% xenon in helium) and argon-xenon mixture were used in a ratio of 98:2, respectively. Figure 2 shows the mass spectra of the beam in argon, xenon, and argon-xenon mixtures at a pressure of ~ 4 bar. It can be seen that the peaks of xenon clusters are much wider than the peaks of argon clusters. In argon, 99.6% of the isotopic composition refers to the ^{40}Ar isotope. The mass spectrum of the Ar_M clusters actually consists of a single peak. Seven xenon isotopes from ^{128}Xe to ^{136}Xe have abundance above 1%. The mass spectrum of the Xe_N cluster is a set of peaks corresponding to clusters of different isotopic composition, which overlap each other in fig.2.

Part of the spectrum 358-442 Da for Xe_N and Xe_NAr_M cluster beams on a large scale is shown in Fig. 3. Only the Xe_3 cluster is contained in the xenon cluster beam in this region of the spectrum. It is seen that the mass spectrum of this cluster is a set of more than 20 peaks corresponding to different combinations of isotopes in the cluster. The spectrum of the xenon-argon cluster beam contains lines corresponding to the Ar_9, Ar_{10}, Ar_{11}, Xe_3 clusters, as well as to all possible heteronuclear clusters: Ar_6Xe, Ar_7Xe, Ar_3Xe_2, Ar_4Xe_2, $ArXe_3$. The mass range of the isotopic distribution for each cluster is indicated in Fig. 3 by a black line. It is seen that the isotropic distributions for different ArXe clusters overlap even for relatively small clusters. The observed mass set for Xe_NAr_M is much wider than Xe_N and Ar_M. As a result, it is possible to distinguish individual clusters only for masses not exceeding 1000 Da.

Despite the fact that the initial mixture contains only 2% xenon, the number of Ar_M clusters in the beam is small compared with the number of heteronuclear Xe_NAr_M clusters. A similar effect of a significant enrichment of the cluster beam of argon-xenon clusters containing xenon atoms as compared to the ratio of components in the initial mixture was observed in [4]. Note that an increase in the fraction of xenon in the initial mixture by 1-2% leads to the complete disappearance of argon clusters in the mass spectrum.

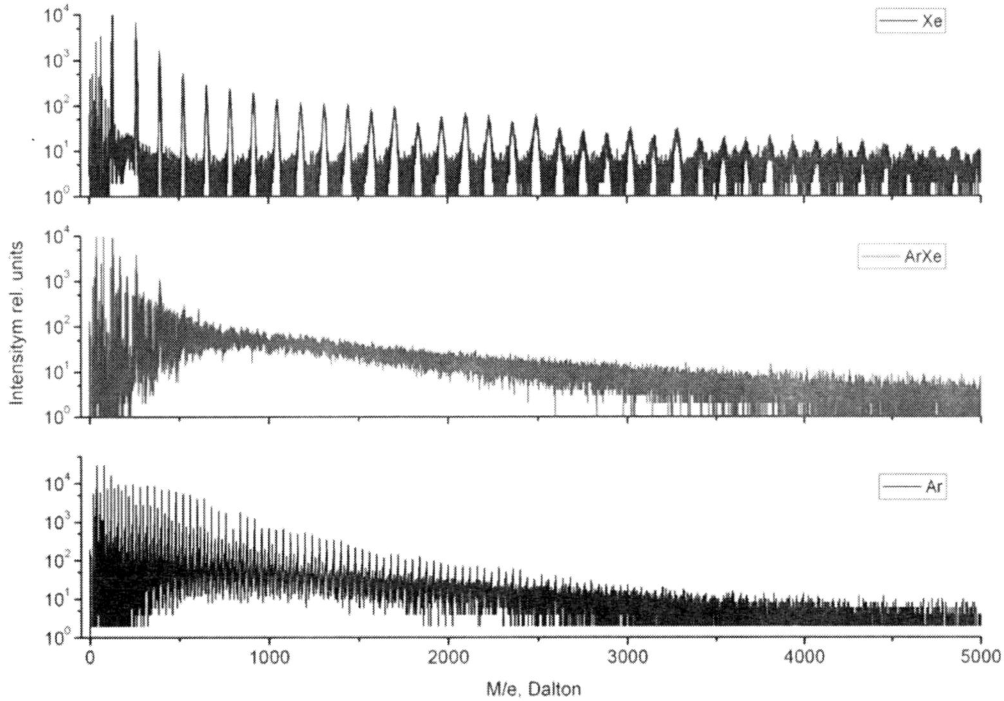

Figure 2. Mass spectra of a cluster beam for xenon, argon and xenon-argon mixture at stagnation pressure ~ 4 bar.

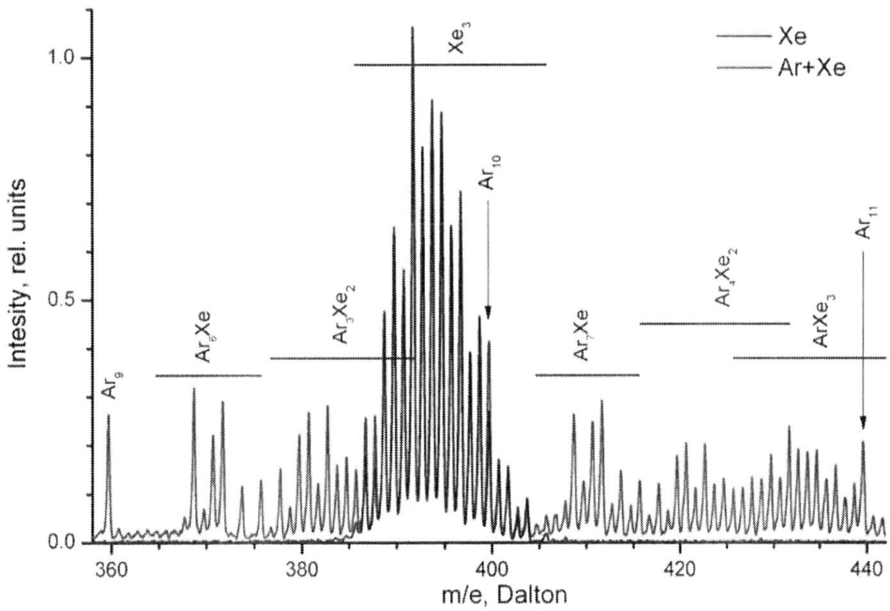

Figure 3. Isotope distribution for Xe and Xe+Ar beam in mass region 358-442 Dalton. Black lines indicate the mass range of the isotopic distribution with the intensity of individual lines of more than 1%.

Note that the comparison of the electronic ionization spectra for beams of different compositions is not trivial. The mass distribution in the beam depends on the stagnation pressure and composition of the mixture. At the same stagnation pressure, the mass distribution in the xenon and xenon-argon beams is different. By varying the pressure, it is possible to achieve approximately the same mass distribution in beams of pure xenon and xenon-argon mixtures. Obviously, in this case, the size distribution of the clusters in the beams will be different.

In [4,5], it was shown that the Xe_NAr_M cluster consists of a nucleus formed by xenon atoms and an outer shell of argon atoms. Taking into account that the ionization potential of argon atoms is higher than that of xenon atoms, it can be expected that the ionization processes of the Xe_NAr_M cluster will depend only on the xenon core of the cluster. Thus, the ionization spectrum of Xe_NAr_M clusters should correspond to the ionization spectrum of Xe_N clusters. However, in the experiment, significant variations are observed in the photoionization spectra of Xe_NAr_M and Xe_N clusters. Figure 4 shows the spectra of multiphoton ionization of a supersonic cluster beam for argon, xenon and an argon-xenon mixture. The stagnation pressures were chosen so that the mass distribution of clusters in the xenon and xenon-argon mixtures beams was approximately the same.

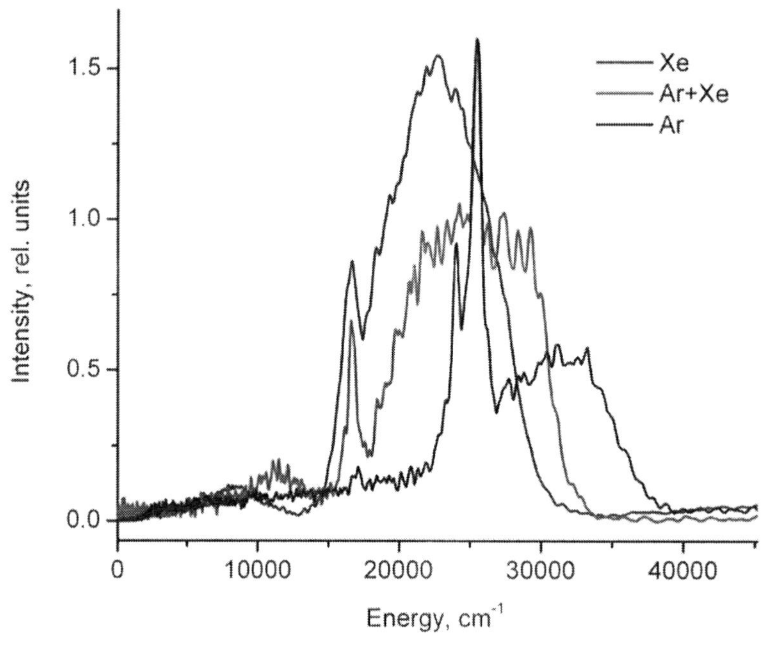

Figure 4. Photoelectron spectra of a cluster beam irradiated with the third-harmonic (263 nm) radiation of a femtosecond laser for xenon, argon and xenon-argon mixture. Stagnation pressure is about 8 bar for Xe, 5 bar for Ar+Xe mixture and 6 bar for Ar.

In the spectrum of the xenon beam, the intense line (electron energy 14530 cm^{-1}) correspond to three-photon ionization of xenon atoms to the $Xe^+(^2P_{3/2})$ states. The energy of two quanta (9.4 eV) is in the region of the effective two-photon absorption of xenon clusters. The ionization of these excited states of clusters by the third UV quantum leads to the formation of a broad, weakly structured ionization band in the energy range of 16000–25000 cm^{-1} [1].

Narrow lines of the multiphoton ionization of argon atoms (24000 and 25400 cm^{-1}) are also observed in the spectrum of the argon beam. The signal from argon clusters is in the region of 26000-35000 cm^{-1}. The intensity of photoionization in an argon beam is noticeably lower than in a xenon beam, since in this case 4 photons are required for ionization.

Narrow lines corresponding to the ionization of xenon and argon atoms are also observed in the spectrum of an argon-xenon mixture. The wide photoionization band in the energy range of 18500–30500 cm^{-1} corresponds to the ionization of Xe_NAr_M clusters. Electrons with energies up to 25 000 cm^{-1} can be produced as a result of three-photon ionization of Xe_NAr_M clusters with the formation of xenon ions in a cluster in states $^2P_{1/2}$ and $^2P_{3/2}$. Electrons with energies greater than 25000 cm^{-1} cannot be produced as a result of three-photon ionization. Such electrons are born as a result of four- photon ionization with the formation of argon ions in a cluster. However, in Ar_M clusters, this process has very low efficiency. In Xe_NAr_M heteroclusters, this process effectively proceeds through the intermediate two-photon excitation of xenon centers. Thus, in this process, excitation transfer from the xenon center to argon inside the cluster with absorption of a UV quantum is observed.

Acknowledgments
This work was supported in part by Russian Federation Grant No. 3.8742.2017/8.9. The equipment of the Resource Center "Physical Methods of Surface Investigation" of the St. Petersburg State University was used in experiments.

References
[1] Serdobintsev P, Melnikov A, Pastor A, Timofeev N, Khodorkovskiy M 2018 *J. Chem. Phys.* **148**, 194301.
[2] Serdobintsev P, Rakcheeva L, Murashov S, Melnikov A, Lyubchik S, Timofeev N, Pastor A, Khodorkovskii M 2015 *J. Chem. Phys.* **143** 114302.
[3] Arseniev A, Serdobintsev P, Melnikov A, Rakcheeva L, Pastor A, Khodorkovskii M 2018 *J. of Physics: Conference Series* **1038**, 012073.
[4] Danylchenko O, Doronin Y, Kovalenko S, Samovarov V 2006 *JETP letters*, **84**, P.324.
[5] Tchaplyguine M, Lundwall M, Gisselbrecht M, Öhrwall G, Feifel R, Sorensen S, Svensson S, Martensson N, Björneholm O 2004 *Phys. Rev. A*, **69**, 031201.

Numerical modelling of an error of manufacturing of ion-exchange waveguide for the tasks of quantum computations

V S Gerasimenko[1], **N D Gerasimenko**[1], **F D Kiselev**[1]
[1]Saint Petersburg National Research University of Information Technologies, Mechanics, and Optics, Kronverkskiy 49, St. Petersburg, 197101, Russia

Abstract. This paper describes modeling of single-mode waveguide and 3dB directional coupler for integrated optical quantum circuits manufactured by ion-exchange process. We present results of diffusion ion exchange simulation for $Na^+ \leftrightarrow K^+$ in the R_2O-Sn_2O-SiO_2 glass and optical elements modeling performed by the beam propagation method. There was demonstrated that the refractive index profile in the overlapping area of directional coupler can be calculated by summing two separate profiles, and error would not exceed 3%. We were able to minimize light propagation looses in the device with about 20 mm length. It was done by reducing interaction length to zero and adjusting width of the bend. After that fabrication tolerance of the device was looked. The overall transmittance of the proposed directional coupler was evaluated as 0.96

1. Introduction

It is well known that quantum information processing can be implemented on linear optics. Distribution of a photon between two modes, that can be spatial, polarization or temporal, is suitable to encode a qubit. Photonic integrated circuits are primary physical platform for quantum computations due to dimensional stability over time and compact form factor in comparison with bulk optics. Modern devices of integrated quantum optics are manufactured commonly by growing on a crystal surface or using lithography [1-5]. Both these methods are expensive and complicated. In addition, grown waveguides are often unsuitable for thermal or electric control because of the requirement of optical insulation. In this research we consider ion-exchanged waveguides in glass. This process provides a relatively easy way to get quantum chip with buried waveguide [6]. In addition light effective mode area in such device is similar to one in a single-mode optical fiber, so it is suitable for fiber-to-chip coupling with no additional device.

2. Method

In the process of ion exchange, the longest stages are the diffusion of substituting ions into glass and the substituted ones from it [7].

The first task was to simulate these processes at the stages of forming (first stage) and burying (second one) of a waveguide. It was done by solving diffusion equation for concentration (C) of substituting ions for only in-glass area near the surface (Fig. 1a) with constant diffusion coefficient (D) due to high ions similarity:

$$\frac{\partial C}{\partial t} = D\nabla^2 C$$
.

Sources was simulated by boundary conditions. For first stage they were:

$$\begin{cases} C = 1, & \text{for non-masked area} \\ \dfrac{\partial C}{\partial t} = 0, & \text{for masked area} \\ \dfrac{\partial C}{\partial t} = DC, & \text{for inner boundaries} \end{cases}$$
,

and for second one:

$$\begin{cases} C = 0, & \text{for melt-glass boundary} \\ \dfrac{\partial C}{\partial t} = DC, & \text{for inner boundaries} \end{cases}$$
.

After it concentration distribution was converted to refractive index profile (n) [8,9]:

$$n = n_0 + \Delta n_{\max} C$$
,

where Δn_{max} is the change in the refractive index at $C = 1$ (and $\Delta n = \Delta n_{max} C$), n_0 is refractive index of the glass before ion exchange (at $C = 0$).
We achieved desired refractive index profile with 2 μm wide mask (Fig. 1b).

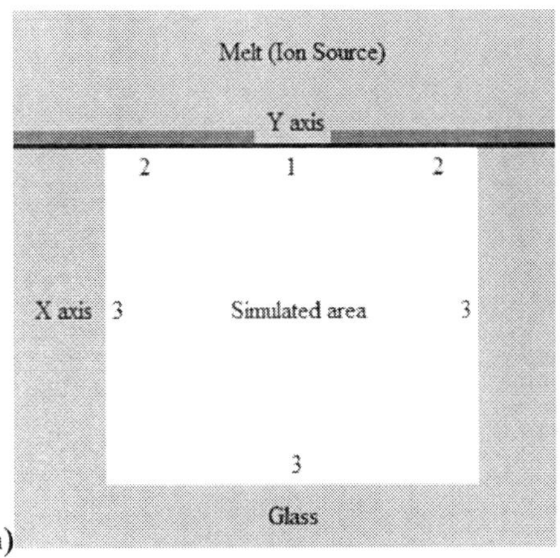

(a)

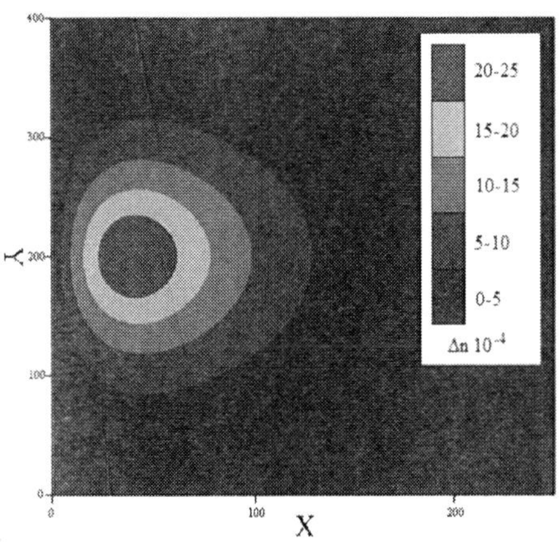

(b)

Figure 1(a, b). (a) Simulated area. Space only inside white rectangle was modelled. 1 is a mark for non-masked area, 2 for masked one when waveguide is manufacturing. When it is burying both of the marks denote melt-glass boundary. 3 is a mark for inner boundaries (regardless of manufacturing stage); **(b)** The result of modeling the profile of refraction. The step of counts on the axes is 0.1 μm.

We also showed that modelling of the overlapped region can be performed by directly summing two separate refractive index profiles (Fig. 2 a,b).

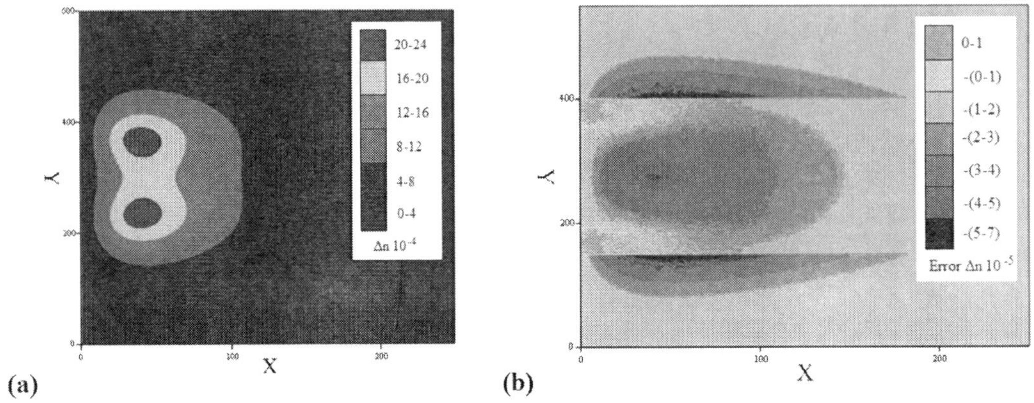

Figure 2(a, b). **(a)** The result of diffusion simulation for interaction area; **(b)** The comparison of diffusion simulation and single profiles summing. The steps of counts on the axes is 0.1 μm.

Since refractive index difference between the core of the waveguide and the substrate is quite small we used beam propagation method [10] to simulate the performance of our device.

At the beginning a linear waveguide was modelled. Single-mode regime was observed for the whole infrared telecommunication range. Important result about fundamental mode in an ion-exchanged waveguide is that this mode has a profile duplicates the refraction index gradient (Fig. 3).

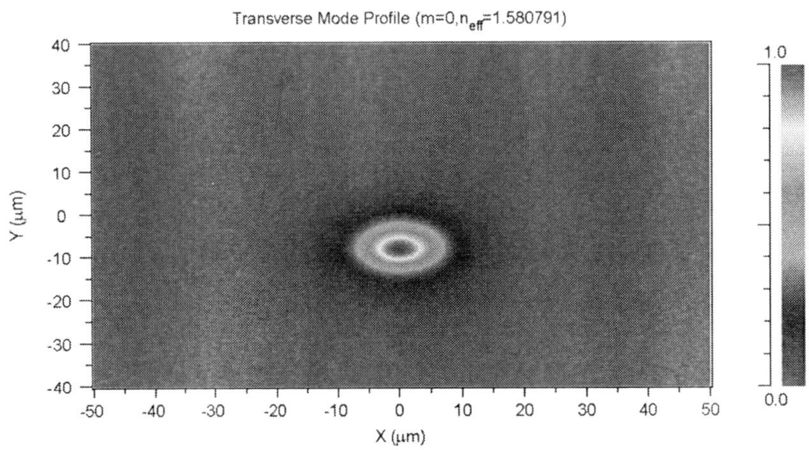

Figure 3. Calculated fundamental mode of the ion-exchange glass waveguide with refractive index profile shown on Fig. 1b.

After waveguide, the directional coupler was simulated. Geometry of the mask in general represents geometry of the device (Fig. 4). To reach correct coupling coefficient we need to overlap two waveguides, but refractive index profile complexity in interaction area might cause high order modes excitation. This process occur with light leaking from the device. We optimized our structure by adjusting width of the s-bends and reducing interaction length between the waveguides to zero for looses prevention (Fig. 5).

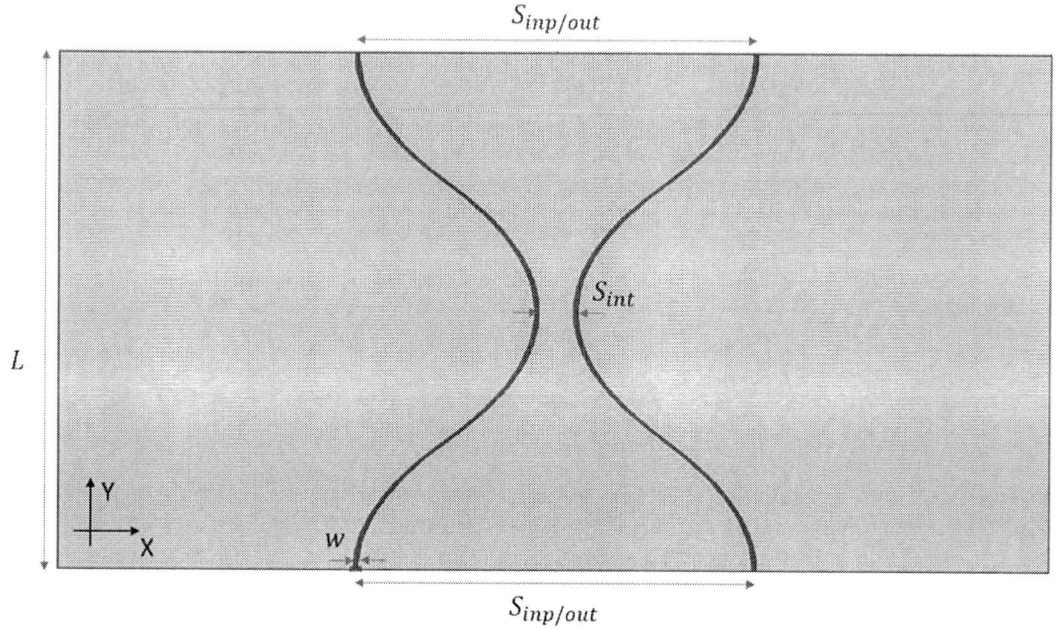

Figure 4. Geometry of the mask for the 3dB directional coupler. $S_{inp/out}$ is initial and S_{int} is the interaction region separations, *w* is a width of slits in the mask which is 2 μm. Each of the two slits correspond to left and right optical channels of the device.

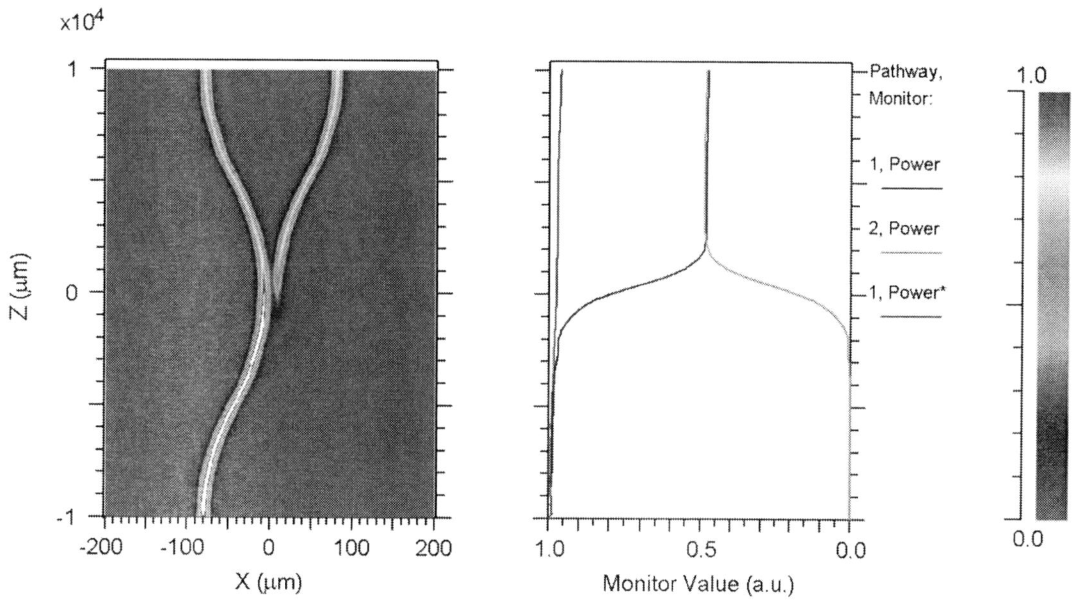

Figure 5. Simulation of the optimized 3dB directional coupler. Initial separation - 160 μm, separation in the interaction region – 15.82 μm, interaction length – 0 μm. Left is the field contour map along the device and right picture shows power in the left channel (blue), power in the right channel (green) and overall power in the region (red).

As a final stage we performed simulations of the directional coupler with some errors in manufacturing process to determine scheme fabrication tolerance (Table 1).

Table 1. Effect of manufacturing errors on splitting coefficient.

Type of error	Value	Splitting coefficient difference
Separation	+5%	-0.140
Separation	-5%	+0.125
Mask thickness	+5%	-0.057
Mask thickness	-5%	-0.057
Exposure time (first stage)	+5%	-0.042
Exposure time (first stage)	-5%	-0.025
Exposure time (second stage)	+5%	-0.018
Exposure time (second stage)	-5%	-0.060

3. Results and conclusion

We performed diffusion, optical and fabrication tolerance modeling of the 3dB directional coupler based on ion-exchanged waveguides in glass. It was shown that the refractive index profile in the regions where two waveguides overlap can be simulated by summing two separately calculated profiles which gave us a significant speed up in the optimization process. One of important ideas was to reduce coupler interaction length to zero to prevent light leaking from device. The only drawback is high sensitivity of the coupler to separation width error. Proposed optimized device has length of 20mm with maximum and minimum separation distances of 160 μm and 15.82 μm respectively. Its overall transmittance was evaluated as 0.96.

References

[1] Matthews J C F, Politi A, Stefanov A, O'Brien J L 2009 *Nat. Photonics* **3(6)** 346-350

[2] Crespi A, Ramponi R, Osellame R, Sansoni L, Bongioanni I, Sciarrino F, Vallone G, Mataloni P 2011 *Nat. Commun* **2** 566

[3] Politi A, Matthews J C F, Thompson M G, O'Brien J L 2009 *IEEE J. Sel. Top. Quantum Electron.* **15(6)** 1673-1684

[4] Politi A, Cryan M J, Rarity J G, Yu S, O'Brien J L 2008 *Science* **320(5876)** 646-649

[5] Zhang Y, McKnight L, Engin E, Watson I M, Cryan M J, Gu E, Thompson M G, Calvez S, O'Brien J L, Dawson M D 2011 *Appl. Phys. Lett.* **99(16)** 161119

[6] Hallett D, Foster A P, Hurst D L, Royall B, Kok P, Clarke E, Itskevich I E, Fox A M, Skolnick M S, Wilson L R 2018 Optica **5** 644-650

[7] Nikonorov N V, Aseev V A, Zhukov S N, Ignatiev A I, Kiselev S S, Rokhmin A S 2008 *WAVEGUIDE PHOTONICS* (St. Petersburg: SPSU ITMO)

[8] West B R, Madasamy P, Peyghambarian N, Honkanen S 2004 J. *Non-Cryst. Solids* **Nov 1** 347(1-3): 18-26.

[9] Zhabrev V A 1987 *Obtaining and applying protective coatings: proceedings of the 12th All-Union Conference on Heat-Resistant Coatings, Leningrad April 16-18, 1985* (Leningrad: Science) p 14-18

[10] Yevick D, Hermansson B 1990 *J. Quantum Electron.* **26** 109

Influence of neutron irradiation on optoelectronic properties of structures with the InAs/GaAs quantum dots

N S Volkova[1], A P Gorshkov[2], A N Trufanov[3], L A Istomin[1], S Levichev[1]
[1] Research Institute for Chemistry, Lobachevsky State University of Nizhny Novgorod, Nizhny Novgorod, 603950 Russia
[2] Department of Physics, Lobachevsky State University of Nizhny Novgorod, Nizhny Novgorod, 603950 Russia
[3] Sedakov All-Russian Scientific Research Institute of Measuring Systems, GSP-486 Nizhny Novgorod, 603950 Russia

Abstract. The effect of neutron irradiation on the photosensitivity of the InAs/GaAs quantum dots has been investigated. It was shown that after neutron irradiation with a fluence of 1.5×10^{15} cm^{-2} the photosensitivity at the room temperature has been decreased at 3 times, whereas the shape of the photosensitivity's temperature dependence didn't reveal any visible changes, despite an appearance of defects in quantum dot layer. The effect was explained by the difficulty of a motion of photoexcited carriers in quantum dot layer to the recombination centres arised after irradiation.

1. Introduction

The possibilities of using semiconductor devices based on quantum confined structures are determined, among other factors, by radiation durability in conditions of increased radiation exposure [1]. Thus, it is necessary to study the influence of various types of ionizing radiation on the physical properties of such structures. In this paper photoelectric spectroscopy [2] was used in order to investigate the influence of defect formation induced by the gamma-neutron irradiation on photo-electric properties of the InAs/GaAs quantum dot (QD) structure.

2. Experiment details

The structure was grown by Metal Organic Vapour Phase Epitaxy at the atmospheric pressure of hydrogen [3]. Initially, a 300 nm thick *n*-GaAs buffer layer with the electron density of ~ 10^{16} cm^{-3} on (100) *n*-GaAs substrate was formed at 650°C. Subsequently, an InAs QDs layer and a 30 nm thick GaAs cap layer were deposited at 520°C.

The heterostructure was irradiated with neutrons with an average energy of 1.89 MeV and a fluence of 1.5×10^{15} cm^{-2}. The exposure dose of the accompanying γ-ray irradiation was 3×10^{5} R. For a comparison, another InAs QDs sample with defects created by anodic oxidation was investigated. The oxidation was carried out at 10 VDC in the electrolyte consisted from 0.5 M solution of ammonium pentaborate and ethylene glycol in the ratio 1:1. Ohmic contact to the buffer layer and the substrate was formed by burning of the tin foil using electric sparks.

In order to characterise the defect influence's degree on QDs optoelectronic properties the spectral dependences of the photovoltage at surface barrier was studied in the wide temperature range of 77 - 350 K. The measurements were carried out by the "soft-contact" technique [4]. The Acton SpectraPro-500i monochromator equipped with a 250 W tungsten halogen lamp and a grating of

600g mm^{-1} was used as a monochromatic light source. The measurements were performed on an alternating signal, using a mechanical optical chopper (Stanford Research Systems SR540) operating at the 130 Hz frequency. Detection of the alternating electrical signal was carried out according to the standard lock-in technique using the Stanford Research Systems SR810 DSP lock-in amplifier. The spectral dependence of the relative photosensitivity $S_{ph}(hv)=V_{ph}(hv)/L(hv)$ was plotted, where $V_{ph}(hv)$ stands for a photovoltage, and $L(hv)$ stands for the light intensity (in arbitrary units).

A photoelectric signal in the spectral region of the interband optical absorption of QDs appears as a result of the escape of photoexcited electrons and holes from quantum confined levels into the semiconductor matrix and is determined by the ratio of the escape and recombination effective lifetimes. The method based on the analysis of the temperature dependences of the photosensitivity was found to be more informative in the study of defect formation. The photoelectric signal was found to be more sensitive to a defect formation, when the recombination lifetime was approximately equal to the escape lifetime of photoexcited charge carriers in QDs. In this case, the escape efficiency η, was roughly estimated as 50 %. The generation of defects into the structure substantially affects the recombination lifetime of charge carriers, whereas the temperature influences the escape lifetime. Therefore, the equality of these lifetimes at different defect concentrations is bound to be attained at different temperatures; consequently, it is expected that the corresponding temperature dependences will be shifted with respect to each other. Previously, this method was successfully applied in the study of defects originating from the cobalt deposition on the surface of the QD structure [5].

3. Results and discussion

The photosensitivity spectra of the QDs structure at 300 K before and after neutron irradiation are shown on the Fig. 1a. The peaks at 0.94 eV and 1.01 eV are related to optical transitions between holes' and electrons' discrete ground states and the first excited states in QDs, respectively. After the irradiation the photosensitivity S_{ph} at 300 K related to the interband QDs optical absorption was uniformly decreased at 3 times over the whole spectrum. This result is explained by the fact that defects, formed in the QD layer increased the effective carrier recombination rate.

The Fig. 1b shows the temperature dependence of photosensitivity of the InAs QDs ground state transition of the sample before and after irradiation. The shape of the photosensitivity temperature dependence is determined by the ratio between the recombination and escape lifetimes of carriers in the QD. The rapid drop of photosensitivity with temperature occurring under 220 K is associated with decreasing of the electron's escape rate from the QD ground state to the semiconductor matrix. It is worth to mention the fact that the shape of the temperature dependence of the photosensitivity in the region of the ground state transition in QDs after irradiation kept the same.

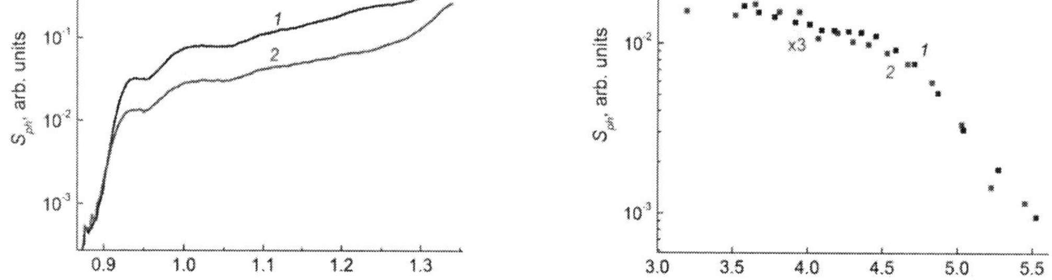

Figure 1. The influence of neutron irradiation on the photosensitivity spectrum from the QDs (a) and on the temperature dependence of photosensitivity in the region of the ground state transition in QDs ($hv \approx 0.94$ eV at 300 K) (b). *1* – before irradiation, *2* - after irradiation.

The effect of defect formation at anodic oxidation on the QD optoelectronic characteristics was founf to be substantially different. The influence of anodic oxidation on photosensitivity spectra (300 K) for the similar nanostructure with the QDs layer is displayed on the Fig. 2a. After an oxidation the photosensitivity of QDs ground state transition was decreased, which also associated with defects formation as well as in the case of neutron irradiation. The red shift of QD photosensitivity spectrum (60 meV) is related with elastic strain relaxation due to the decreasing of the GaAs cap layer thickness. A strong shift in the temperature dependence of the photosensitivity to the region of high temperatures took place (Fig. 2b). In this case, shift in the temperature dependence of photosensitivity and therefore the escape efficiency can be due both to a decrease in the recombination carrier lifetime in QD due to a defect formation, and to an increase in the height of the escape barrier in QD due to a change in the QD energy spectrum due to an elastic strain relaxation. Calculations using the theory [6] show that if the escape barriers for electrons and holes increase by 30 meV with an unchanged recombination lifetime, the temperature dependence of QD photosensitivity shifts by no more than 25 K, which is significantly less than the 60 K shift observed in this work. We note that after application of other methods of surface defect formation in such structures (ion beam implantation [7], deposition of a chemically active metal [5]) the same shift of the temperature dependence of the photosensitivity was observed without any changes in QD ground state transition energy.

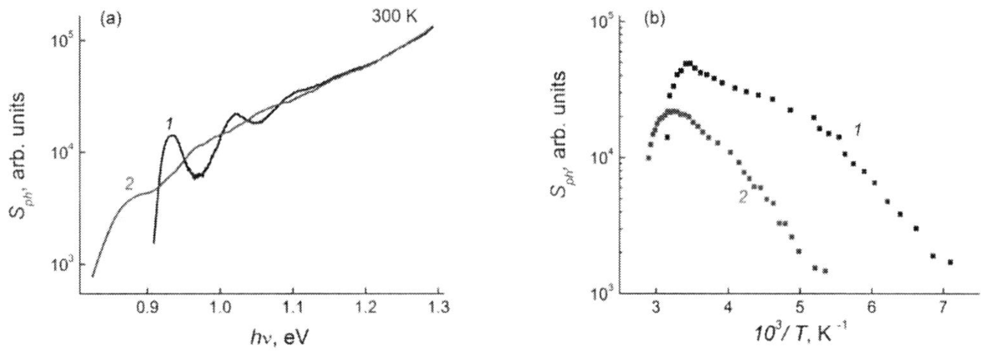

Figure 2. The influence of anodic oxidation on the photosensitivity spectrum from the QDs (a) and on the temperature dependence of photosensitivity in the region of the ground state transition in QDs (b). *1* – before oxidation, *2* - after oxidation.

This difference in features of defect influence is caused by a different nature of the defect formation. At low energy treatment, such as anodic oxidation, ion beam implantation [7], chemically active metal deposition [5], defects distribute uniformly over the area of the structure. In the case of neutrons, highly defective regions with a defect concentration of ~ 10^{20} cm^{-3} [8] are formed in the structure, but they are compact (the radius of the region is about 20 nm [8]) and are located at a considerable distance from each other. These disordered defective regions can't suppress the photosensitivity from the entire array of QDs. The photosignal completely disappears (even at 300 K) only from those QDs that directly located into these highly defective regions. In the remaining QDs the conditions for recombination stay the same as before irradiation, since there is spatial limitation of the lateral motion of carriers in the QD layer.

4. Conclusion

It was shown that after neutron irradiation the behavior of the temperature dependence of InAs/GaAs QDs photosensitivity does not change, despite the fact that defect formation is present in the QD layer.

The effect is explained by the difficulty in moving the photoexcited carriers in the quantum dot layer to the recombination centres that arise during irradiation.

Acknowledgments
We are grateful to B.N. Zvonkov (Research Institute for Physics and Technology, Lobachevsky State University of Nizhny Novgorod) for growing the structures. This study was supported by the State contract of the Ministry of Education and Science of the Russian Federation (4.8337.2017/BCh).

References

[1] Schramm A, Tukiainen A, Pessa M, Konetzni C, Heyn Ch, and Hansen W 2009 *J. Appl. Phys.* **105** 104308.
[2] Karpovich I A, Levichev S B, Morozov S V, Zvonkov B N, Filatov D O, Gorshkov A P, Ermakov A Yu. 2002 *Nanotechnology* **13** 445.
[3] Zvonkov B N, Karpovich I A, Baidus N V, Filatov D O, Morozov S V, Gushina Yu Yu 2000 *Nanotechnology* **11** 221.
[4] Datta Sh, Ghosh S, and Arora B M 2001 *Rev. Sci. Instrum.* **72** 177.
[5] Volkova N S, Gorshkov A P, Zdoroveyshchev A V, Istomin L A, Levichev S B 2015 *Semiconductors* **49** 1592.
[6] Volkova N S, Gorshkov A P, Filatov D O and Abramkin D S 2014 *JETP Lett.* **100** 156.
[7] Gorshkov A P, Karpovich I A, Pavlova E D, Volkova N S 2012 *Semiconductors* **46** 1506.
[8] Bobyl' A V, Kanakova R V, Kononov V K, Malinin V G, Malyshev M M, Prokopenko I V, Slutskii M I, and Tkhorik Yu A 1992 *Elektron. Tekh., Ser. Upravl. Kachestvom* **4 – 5** 31.

The size reducing of the focal spot with focusing short pulses using high-performance computer systems

D A Savelyev[1,2], S A Fomchenkov[1,2]
[1]Samara National Research University, 34, Moskovskoye Shosse, Samara, Russia, 443086
[2]Laboratory of Laser Measurements, Image Processing Systems Institute - Branch of the Federal Scientific Research Centre "Crystallography and Photonics" of Russian Academy of Sciences, Samara 443001, Russia

Abstract. We investigated the change in the type of focal spot when pulsed radiation passed through the diffraction axicon. The Gauss beam and the Gauss-Laguerre mode (1,0) were considered as the input laser radiation with linear polarization. Modelling of diffraction is numerically investigated by the finite difference time domain (FDTD) method. The formation of a shadow focal spot in the immediate vicinity of an optical element and amplification of the longitudinal component of the electric field were shown.

1. Introduction

The many papers recently have appeared that consider the possibility of reducing the size of the focal spot during sharp focusing. Most often, amplitude or phase apodization of the pupil of the lens [1-4] and special types of polarization of laser radiation [5-7] are proposed, as well as a combination of these characteristics [8-11]. D. Savelyev and S. Khonina [12] shown that polarization has the greatest effect on the longitudinal component of the light field. However, another factor affecting the field distribution in the focal area is the duration of the laser pulse [13-16]. Note that this factor is particularly important in applications such as laser printing, micro- and nano-textures [17, 18].

The analysis of the effect of the pulse duration on the field distribution in the focal area is carried out as a rule on the basis of the field expansion in plane waves and integral expressions [13-16, 19, 20]. This is due to the possibility of using analytical representations and reducing the time of the computational process. However, in some cases it is required to use a more accurate calculation model, in particular, a direct solution of Maxwell's equations. The FDTD method [21] is one of the most accurate approaches for modeling temporal processes in micro-optics and allows one to obtain correct results of the propagation and diffraction of short pulses.

The spatial structure of the laser beam is important in applications based on the interaction of electromagnetic radiation with matter. The classical elements offer little possibilities in controlling the spatial structure of the laser beam [22]. The widest range of spatial transformations of laser radiation is provided by means of diffractive optics. The most commonly used diffraction gratings and spatial light modulators (SLM) [23-25]. With all the dynamic advantages, spatial modulators lose to diffraction optical elements in efficiency and resolution. It was shown that focusing laser radiation by a diffraction axicon with a high numerical aperture makes it possible to achieve sharper focusing than with an aplanatic objective with the same numerical aperture value of the system [26].

The presence of a phase singularity in the beam, which can change the polarization properties of the radiation even for beams with a uniform polarization, is played an important role [27]. Various types of polarizations and their influence on the focusing of laser radiation were considered in [28, 29]. It is known that the phase singularity can be used to amplify the longitudinal component of the electric vector of laser radiation [30–33] in the case of uniform polarization, which many of modern lasers produce. The presence of a powerful longitudinal component in the field of focus improves optical resolution and is used for optical manipulation, material processing, microscopy, and other applications [34, 35].

In this paper, we explore the possibility of reducing the size of the focal spot when focusing short pulses by a diffraction axicon. To numerically simulate the diffraction of the laser radiation, the FDTD method using high-performance computations is used [36], implemented in the Meep software package [37]. Calculations were made on the computational cluster with power of 850 GFlops.

2. Investigation of focusing short pulses

Simulation parameters: the wavelength $\lambda = 0.532$ microns, the size of the computational domain x, y, z $\in [-3.8\lambda; 3.8\lambda]$. The thickness of the absorbing layer PML $\sim 1.3\lambda$, the sampling step of space – $\lambda/21$, the sampling step of time – $\lambda/(42c)$, where c is the velocity of light. The Gauss beam and the Gauss-Laguerre mode (1,0) were considered as the input laser radiation with linear polarization. The refractive index of the axicon and the substrate is $n_{ax} = 1.5$. The numerical aperture (NA) of the focusing binary axicon was 0.95. The input beams and the considered optical scheme is shown in figure 1.

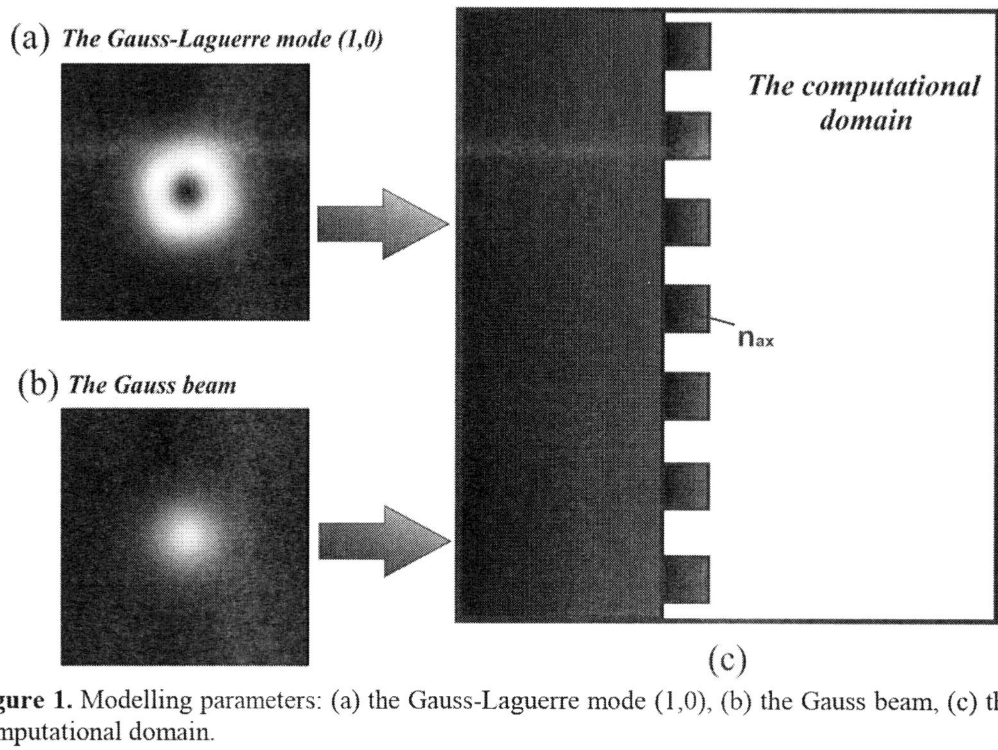

Figure 1. Modelling parameters: (a) the Gauss-Laguerre mode (1,0), (b) the Gauss beam, (c) the computational domain.

The results of numerical simulation in the xz plane at different times t are shown in figure 2. It should be noted that the focusing of laser radiation occurs in the immediate vicinity of an optical element. The formation of a shadow focal spot near the optical element is observed for the case of a Gaussian beam. The amplification of the electric field z-component on the optical axis should be noted for the case of the Gauss-Laguerre mode (1,0).

The intensity distributions of the longitudinal component of the electric field for the cases t = 8 and t = 9 for the Gaussian beam and Gauss Laguerre modes are shown in figure 3.

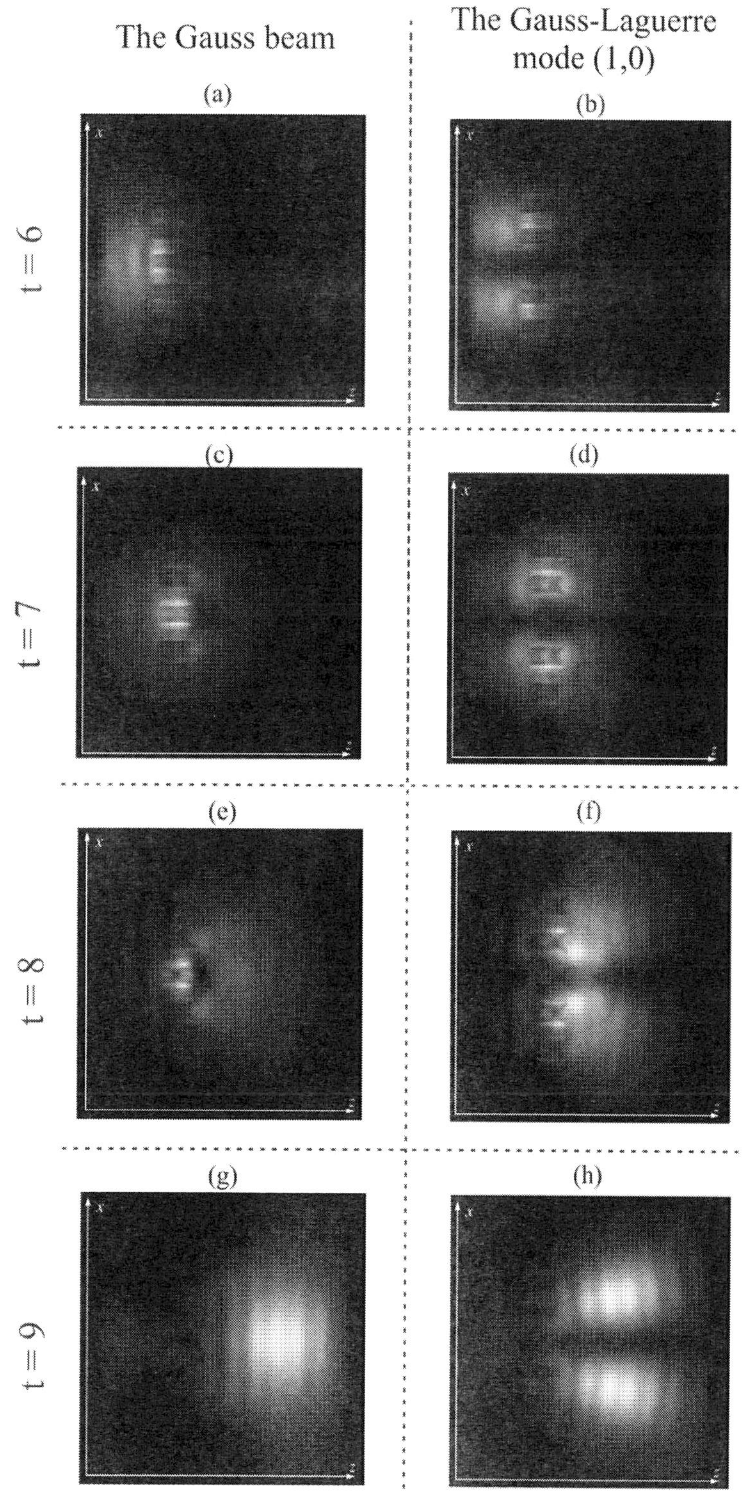

Figure 2. The longitudinal cross section (xz) of propagation short pulses at different times: the Gauss beam, (a) t = 6, (c) t = 7, (e) t = 8, (g) t = 9; and the Gauss-Laguerre mode (1,0), (b) t = 6, (d) t = 7, (f) t = 8, (h) t = 9.

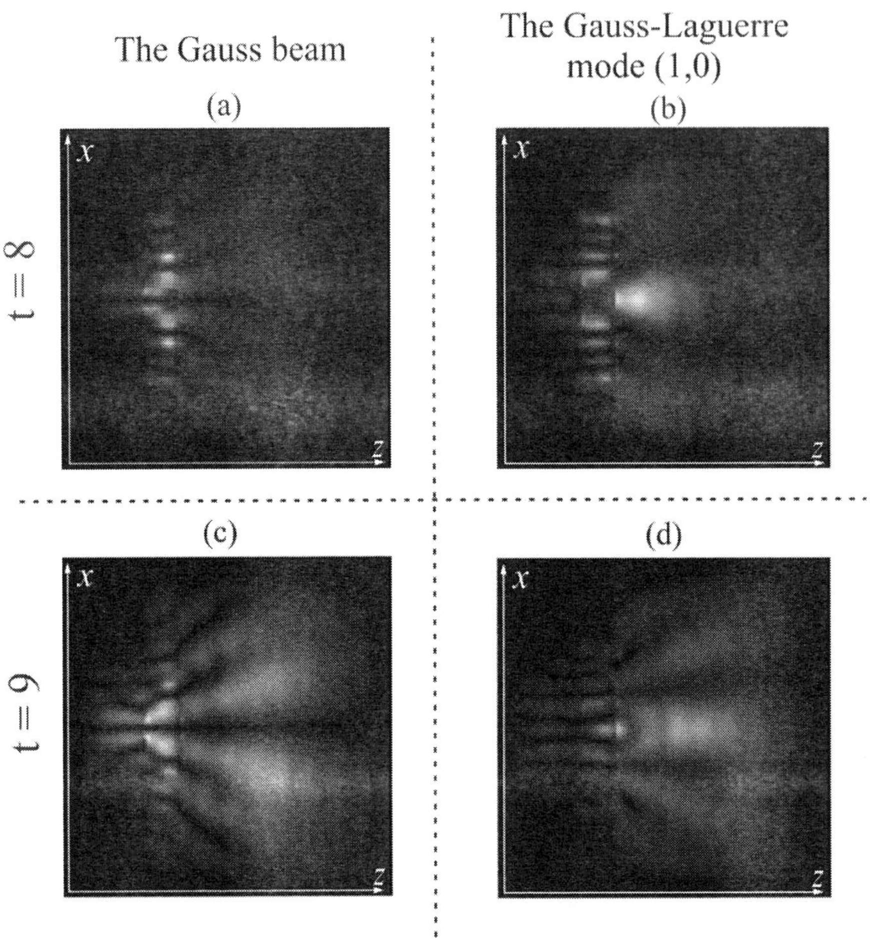

Figure 3. The longitudinal cross section (xz) of propagation short pulses at different times, z-component: the Gauss beam, (a) t = 8, (c) t = 9; and the Gauss-Laguerre mode (1,0), (b) t = 8, (d) t = 9.

It is worth noting that for the Gaussian beam, focusing occurs inside the element. The formation of a light needle can be observed for the Gauss Laguerre mode (1,0). The size of the light needle (full-width at half-maximum - FWHM) intensity for the case t = 8 is 0,68λ, for the case t = 9 is 0,55λ.

3. Conclusion

The formation of pulsed radiation (the Gauss beam and the Gauss-Laguerre mode (1,0)) with linear polarization passed through the diffraction axicon with numerical aperture 0.95 was investigated. To numerically simulate the diffraction of the laser radiation under consideration, the finite difference in the time domain method using high-performance computing is used. It should be noted the formation of a shadow focal spot in the immediate vicinity of an optical element. Amplification of the longitudinal component of the electric field also occurs, and the formation of a light needle near the optical element is observed (for the Gauss Laguerre mode (1,0)). The size of the light needle intensity for the case t = 8 is FWHM = 0,68λ, for the case t = 9 is FWHM = 0,55λ.

Acknowledgments

This work was financially supported by the Russian Foundation for Basic Research (grants 16-29-11698-ofi_m), grant of the President NSh-6307.2018.8 and by the Ministry of Science and Higher

Education of the Russian Federation (agreement No. 3.3025.2017/4.6) within the State assignment FSRC "Crystallography and Photonics" RAS (No. 007-GZ/Ch3363/26).

References

[1] Quabis S, Dorn R, Eberler M, Glockl O and Leuchs G 2000 *Opt. Commun.* **179** 1
[2] Helseth L E 2004 *Opt. Commun.* **229** 85
[3] Gao X, Wang J, Gu H and Xu W 2007 *Optik* **118** 257
[4] Khonina S N 2013 *Optical Engineering* **52** 091711
[5] Dorn R, Quabis S and Leuchs G 2003 *Phys. Rev. Lett.* **91** 233901
[6] Iglesias I and Vohnsen B 2007 *Opt. Commun.* **271** 40
[7] Khonina S N and Golub I 2012 *J. Opt. Soc. Am. A* **29** 1470
[8] Sun C-C and Liu C-K 2003 *Opt. Lett.* **28** 99
[9] Pereira S F, and van de Nes A S 2004 *Opt. Commun.* **234**, 119
[10] Sheppard C J R and Choudhury A 2004 *Appl. Opt.* **43** 4322
[11] Khonina S N and Golub I 2011 *Optics Letters* **36** 352
[12] Savelyev D A, Khonina S N 2015 *Computer Optics* **39**, 654
[13] Wong L J, Kärtner F X and Johnson S G 2014 *Opt Lett* **39** 1258
[14] Li X, Salamin Y I, Hatsagortsyan K Z and Keitel C H 2016 *J. Opt. Soc. Am. B* **33** 405
[15] Khonina S N and Golub I 2016 *J. Opt. Soc. Am. A* **33** 1948
[16] Khonina S N and Golub I 2018 *Journal of the Optical Society of America A* **35** 985
[17] Kuchmizhak A, Pustovalov E, Syubaev S, Vitrik O, Kulchin Y, Porfirev A, Khonina S, Kudryashov S I, Danilov P and Ionin A 2016 *ACS Appl. Mater. Interfaces* **8** 24946
[18] Syubaev S, Zhizhchenko A, Kuchmizhak A, Porfirev A, Pustovalov E, Vitrik O, Kulchin Yu, Khonina S and Kudryashov S 2017 *Optics Express* **25** 10214
[19] April A 2009 *Proc. of SPIE* **7386** 73862X
[20] Khonina S N, Ustinov A V and Volotovsky S G 2018 *Computer Optics* **42** 432
[21] Taflove A and Hagness S 2005 *Computational Electrodynamics: The Finite-Difference Time-Domain Method* (Artech House, Boston, MA)
[22] Salter P S 2011 *Optics Letters* **36** 2302
[23] Hayasaki Y 2005*Applied Physics Letters* **87** 031101
[24] Kuang Z 2009 *Applied Surface Science* **255** 6582
[25] Obata K 2010 *Optics Express* **18** 17193
[26] Khonina S N, Savelyev D A and Kazanskiy N L 2015 *Optics Express* **23** 17845
[27] Khonina S N, Kazanskiy N L and Volotovsky S G 2011 *J. Mod. Opt.* **58** 748
[28] Khonina S N, Savelyev D A and Kazanskiy N L 2016 *Optik-International Journal for Light and Electron Optics* **127** 3372
[29] Savelyev D A, Khonina S N and Golub I 2016 *AIP Conference Proceedings* **1724** 020021
[30] Ganic D, Gan X and Gu M 2003 *Optics Express* **11** 2747
[31] Chen Z, Pu J and Zhao D 2011 *Physics Letters A* **375** 29584
[32] Savelyev D A and Khonina S N 2012 *Computer Optics* **36** 511
[33] Khonina S N, Karpeev S V, Alferov S V and Savelyev D A 2013 *Computer Optics* **37** 76
[34] Zhan Q 2009 *Advances in Optics and Photonics* **1** 1
[35] Khonina S N and Golub I 2012 *Journal of the Optical Society of America A* **29** 2242
[36] Savelyev D A and Khonina S N 2014 *Journal of Physics: Conference Series* **490** 012213
[37] Oskooi A F, Roundy D, Ibanescu M, Bermel P, Joannopoulos J D and Johnson S G 2010 *Computer Physics Communications* **181** 687

Large active area superconducting single photon detector

M Moshkova[1,2], P Morozov[2], A Divochiy[2], Yu Vakhtomin[2,3], K Smirnov[1,2,3]
[1]National Research University Higher School of Economics, Moscow, 101000, Russia
[2]LLC "Superconducting Nanotechnology" (SCONTEL), Moscow, 119021, Russia
[3]Moscow State Pedagogical University, 119991, Russia

Abstract. We present development of large active area superconducting single-photon detectors well coupled with standard 50 μm-core multi-mode fiber. The sensitive area of the SSPD is patterned using the photon-number-resolving design and occupies an area of 40×40 μm^2. Using this approach, we have obtained excellent specifications: system detection efficiency of 47% measured using a 900 nm laser and low dark count rate of 100 cps. The main advantages of the approach presented are a very short dead time of the detector of 22 ns and FWHM jitter value of about 130 ps.

1. Introduction

Superconducting single photon detectors (SSPD) [1] are the best solution for many applications [2] as they positively outperform other types of single-photon detectors. At the same time, up to now all the SSPD-based receivers demonstrated were equipped with an SSPD the sensitive area of which was rather small and well suited to be coupled to a single-mode (SM) optical fiber through which the radiation was delivered to the receiver. Large active area detectors with ultimate specifications are required for many applications, such as aerospace and defense technologies, LIDAR and quantum communications, bio-medical sciences, etc. Unfortunately, attempts to create SSPDs with an enlarged detection area did not prove to demonstrate all the advantages of the SSPDs [2]. According to the SSPD detection mechanism [1], an SSPD with a large active area would require creation of a very long (several tens of millimeters) and narrow (about 100 nm) superconducting strip with an extremely small width deviation (less than 5 nm). This is a very complicated technological task. Also, there are additional drawbacks of this approach. First of all, such increase in the length of the strip leads to a significant increase in its kinetic inductance L_k [3,4] which determines the recovery time (dead time) growth and consequently limits the maximal counting rate of such detectors drastically [5,6]. A significant increase in L_k of the detector would inevitably lead to a degradation of its jitter, too. Several groups have developed large active area SSPDs with high system detection efficiency (SDE), but all of them have demonstrated a very large dead time and therefore low response counting rate [2,7]. Another common problem of the large-area SSPDs is a significant increase of the dark counting rate (DCR) which is determined by the background radiation [4] and is roughly proportional to the detection area. Compared to 9/125 SM fiber, 50/125 multimode (MM) fiber passes approximately 50 times more background-generated photons at room temperature onto the detector. Thereby, during development of a large-area SSPD, all the mentioned drawbacks should be addressed in order to preserve all the advantages of the SSPDs.

In this work, we present results of development and research of the SSPD detectors coupled to a MM fiber with the core diameter of 50 μm. The topology of the photon number resolving (PNR) detectors [8] was implemented for creation of the detector with a large sensitive area. PNR SSPD contains several parallel meander-shaped sections connected to a single contact pad via additional in-series resistors for preventing the cascade switching of these strips when a resistive domain (or "hot spot") is formed across any of the strips due to the photon detection event. Implementing this topology is beneficial for a number of reasons. First of all, the detection efficiency of each section does not depend on that of the other sections. Secondly, usage of such a topology leads to a decrease in the L_k of the structure by a factor of N^2, where N is the number of sections, and to a corresponding reduction of the detector's dead time. Also, as the jitter of the SSPD is partially determined by the superconducting strip length, significant decreasing in length of the strip due to the use of the PNR topology allows for reduction in the jitter, too.

2. Device fabrication and characterization

2.1. Device fabrication

Large area detectors were made using 7 nm thick NbN film by implementing a standard technology routine which is described in details elsewhere [9]. An optical cavity formed by 12 bilayers of Bragg reflector was optimized in order to maximize photon absorption by NbN structure at wavelength range of about 900±50 nm. The large area detector consists of four independent sections, where each of the sections is a meandering superconducting NbN strip of ~120 nm width (filling factor is 0.6) and is connected to a 4 Ohm series resistor made from a thin film of gold. The sensitive area of the detector is 40×40 μm² and each section is evenly distributed over this area, as shown in Fig. 1a. Detector was coupled to a standard MM fiber with 50 μm core diameter and installed onto the cold plate of a GM closed-cycle refrigerator (Sumitomo RDK-101D) with minimally reached temperature of 2.1 K. Equivalent circuit of the four-section PNR SSPD detector is presented in Fig. 1b. The operation principle of such a detector resolving the number of photons during a short optical pulse is given in [8,10,11]. Briefly, the absorption of a photon by one of the sections leads to generation of a voltage signal along the strip. At the same time, the current redistribution between the sections is hampered by the L_k of the section and the series resistance R_{is}. Therefore, when a photon is absorbed by any of the sections, the others still remain in the superconducting state and are capable to detect another photon. When two photons are simultaneously absorbed by two different sections, the resulting voltage pulse becomes roughly two times larger, and so on. In our case, it is not the ability to resolve the number of photons that is important but a possibility to decrease the large area detector's kinetic induction to a value comparable to that of a "standard" small area detector (i.e., detector with the sensitive element area coupled to SM fiber).

a)

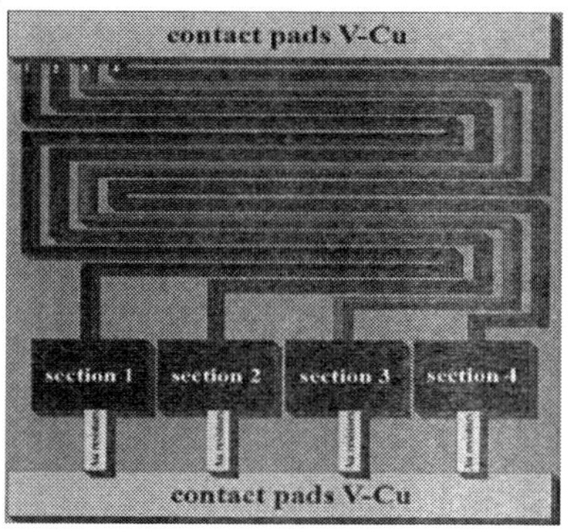

b)

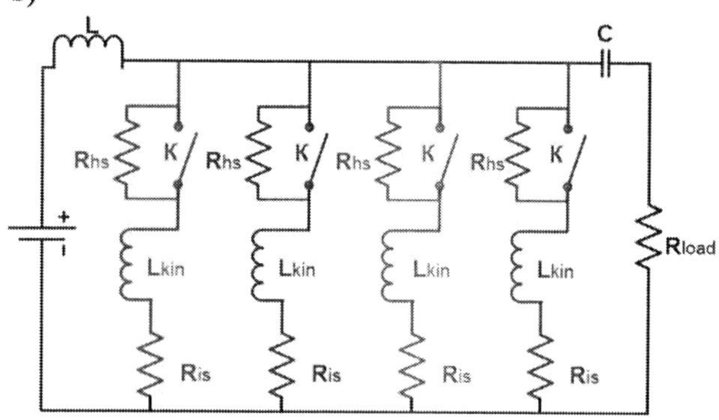

Figure 1(a, b). (a) Schematic view of the 4-section SSPD topology; **(b)** Equivalent circuit of the 4-section topology SSPD detector [Ref. 8, Fig. 2a.].

2.2. System detection efficiency and dark count rate

System detection efficiency (SDE, i.e. efficiency referred to the input of the MM fiber) and DCR were measured using a standard technique described in Ref. [9]. As the light source, an IR-spectrophotometer was used where a wideband light from a halogen lamp was transformed into a narrowband (~1nm) radiation with help of a diffraction grating with 600 lines/mm and a set of cut-off filters. The output power was collected by a coupler (collimating lens) and fed to a multi-mode 50/125 fiber. We used SPAD (Excelitas SPCM-NIR) and a universal frequency counter (Agilent, A53131) to measure the in-fiber power. The power was adjusted by the voltage applied to the halogen lamp and by changing the spectrometer output gap. The SDE was calculated by measuring the number of electrical pulses generated by the SSPD. The results of the experiment are presented in Fig. 2a. The maximal value of the SDE is about 50% at DCR of 300 cps and about 47 % at DCR of 100 cps. It can be seen that dependence of SDE (I_b) demonstrates a tendency to saturating at a current which is close to the critical current of the SSPD. It means that internal quantum efficiency, i.e. probability of the voltage pulse generation due to the photon absorption event by the superconducting structure is close to unity. Difference between the measured system detection efficiency and calculated absorption of ~0.65 is associated primarily with the insufficiently effective optical coupling between the SSPD with the active area of 40×40 μm^2 and MM fiber with core diameter of 50 μm.

a) **b)**

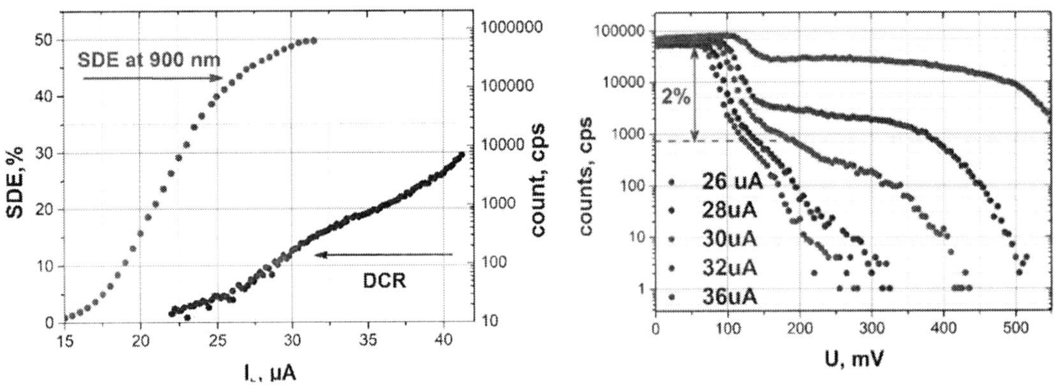

Figure 2(a, b). (a) Measured dependencies of SDE and DCR vs. bias current; **(b)** Measured dependence of photon counts vs. pulse counter trigger level (measured using CW-laser at 900 nm).

For measurement of the dependence *DCR (I_b)* the optical input to the SSPD was covered by a black cap. It has to be noticed that in order to reach a low DCR level, it is required to filter out the background radiation, which is irradiated by the room temperature (300 K) load and inevitably arrives to the detector through the fiber. Compared to the single-mode fiber, the number of the background photons incident onto the detector through the multimode fiber, is extremely increased. Unfortunately, an effective method of filtering out the background radiation, which was applied to receivers based on the single-mode fiber [12], cannot be used for receivers based on the multi-mode fiber. In such situation, we decided to decrease the quantum efficiency of the detector beyond the target wavelength range around 900 nm by two methods simultaneously. The first one diminishes absorption efficiency at wavelength above 1500 nm. For this, we used a substrate with the Bragg-reflector which has good reflection in the range 800-1000 nm (>95%) and is well transparent at wavelengths above 1500 nm. In addition, we intentionally reduced internal quantum efficiency of the detector at wavelengths greater than 1500 nm. As known, the internal quantum efficiency of the SSPD depends on the superconducting strip cross-section and for detectors, which operate well at longer wavelengths a smaller cross-section strip is required [13]. At the same time, the detectors with wider strip and the unchanged thicknesscan operate at short wavelengths sufficiently well. So, we fabricated detectors with different width of the superconducting strip using 7 nm thick NbN film and found out that detectors with width of 120 nm demonstrated an $SDE(I_b)$ dependence close to saturation at 900 nm and at the same time a very low internal detection efficiency (less than 5%) at wavelengths longer than 1500 nm.

Thanks to these two techniques, we managed to achieve a level of dark counting rate of less than 100 cps. The dependence of DCR vs. bias current is shown in Fig. 2a. Dark counts are completely due to the room temperature background but the DCR is still within a few hundred of counts per second, which is a reasonable value for a practical receiver. The region of the exponential growth of DCR vs. bias current of the detector's intrinsic dark counts corresponds to a bias current of more than 42 μA and is not represented in the figure.

Since the detector has a PNR structure, we investigated the cascade switching effect in order to analyze the crosstalk between the meander sections. For that, we used a CW laser source with power corresponding to a level at which the theoretical probability of simultaneous response of any two sections of the detector is four orders less than the probability of a single section response. We have measured the experimental probability of cascade switching by measuring the number of counts depending on the counter trigger level. These results are depicted in Fig. 2b where 5 dependencies corresponding to different bias currents are shown. The effect of cascade switching at bias current of 36 μA is big enough and is about 37 %. However, the cascade switching effect is decreasing with decrease of the bias current and at operation current of 30 μA where SDE ≈ 47 % and DCR is ~100 cps, probability of the cascade switching is less than 2 %. It should be noted that effect of the cascade switching plays an important role when the SSPD is operated as the PNR detector because this effect leads to a distortion of the measured photon statistics. In our case, we did not use the detectors in the PNR mode and the cascade switching could only influence the timing characteristics of the emerging voltage pulses, namely, jitter and dead time of the SSPD.

2.3. Dead time and jitter

The dead time of the detector was deduced from the oscilloscope traces of the electrical pulses generated by the detector. Fig. 3a depicts an oscilloscope trace of the electrical pulse when only one section of the detector was triggered. According to this trace, the dead time, determined as the time corresponding to the falling edge voltage decrease by a factor of *e*, was 22 ns. Similarly, dead time in case when two sections of the detector are triggered is 28 ns, and in case when three sections of the detector are triggered is 37 ns. As it was shown above, the probability of cascade switching of a second section at 30 μA bias current is about 2%; the cascade switching of a third section is less than 0.01%. Therefore, this neither should make a significant contribution to the jitter and dead time, nor have a significant impact on the measurement results when using such a detector.

The jitter was determined by a standard method [9] using TCSPC SPC-150NX from Becker&Hickl GmbH and a pulsed laser (pulse duration ~1 ps). The pulsed laser has a standard single-mode fiber (SMF-28e) as the output. So, pulsed radiation from the laser's SMF-28e fiber was fed to the multi-mode fiber coupled to the large area SSPD. Such approach does not allow to illuminate all the active areas of the detector uniformly, but we that this did not have a significant impact on the data obtained. We set the laser power to a level corresponding to ~0.1 photons per pulse and therefore the detector was operated at conditions where the probability of simultaneous triggering of several detector sections at simultaneous absorption of several photons was negligible. The jitter was measured at the counter trigger level of 50 mV, which is approximately a half of the voltage pulse amplitude when a single section of the detector is triggered. Of course, cascade switching pulses were also counted. The results of the jitter measurements are presented in Fig. 3b. The measured FWHM (full width at half maximum) jitter value was ~130 ps. This value is larger than the jitter of the SSPD coupled to the single-mode fiber [14,15], due to the fact that the length of the superconducting strip of each section of the large area detector is approximately two times longer than that of a standard "small area detector" which leads to a proportional decrease in the steepness of the rising edge of the pulse, and the bias current flowing through each section is just ~7.5 μA which inevitably ensures a significant contribution of the noise component to the measured jitter value [16]. Nevertheless, the measured jitter of the large-area SSPD was demonstrated to be smaller than that of an APD based single photon counters [17] and can be further improved by optimizing the length of the superconducting strips constituting the SSPD sections and the bias current applied.

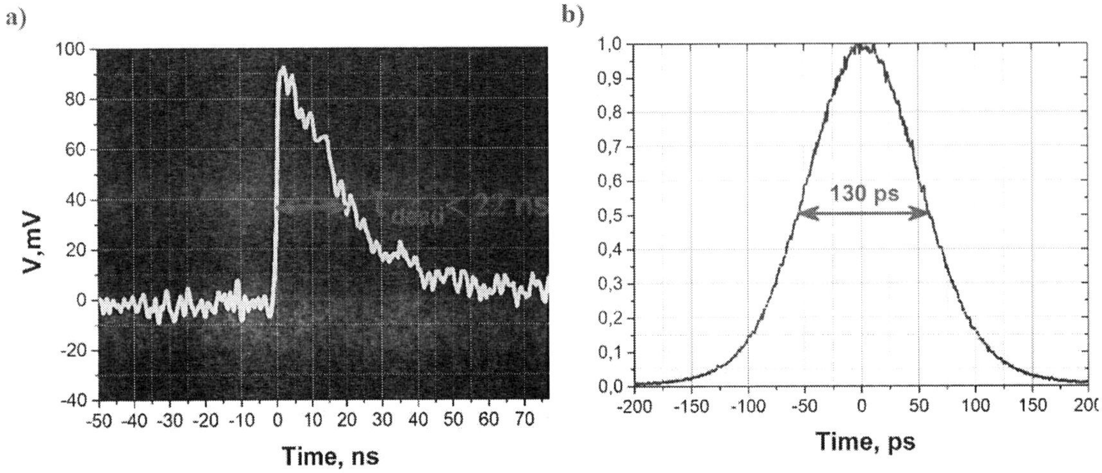

Figure 3(a, b). **(a)** Waveform of the pulse generated by the large active area SSPD; **(b)** Jitter of the detector with the large active area, measured at bias current $I_b = 32$ μA.

3. Summary

We have developed the SSPD detector with the large active area of 40×40 μm^2, which can be well-coupled to the standard 50-μm core multimode fiber. The overall performance of the devices (SDE, DCR, dead time, jitter, cascade switching) is only marginally inferior to that of conventional detectors coupled to SM fiber but explicitly exceeds previously obtained results for detectors with a large active area. In addition, the developed SSPD can be advantageously used in any application whenever the photon number resolving is required.

Acknowledgments

This work was supported by the Russian Science Foundation under grant № 18-12-00364.

References

[1] Gol'tsman G, Okunev O, Chulkova G, Lipatov A, Semenov A, Smirnov K, Williams C and Sobolewski R J Appl. Phys. Lett. 79 705–707

[2] Robert H. Hadfield, Göran Johansson 2016 Superconducting Devices in Quantum Optics (New York: Springer International Publishing Switzerland)

[3] Kerman A, Dauler E, Keicher W, Yang J, Berggren K, Gol'tsman G and Voronov B 2006 J Appl. Phys. Lett. 88, 111116

[4] Hadfield R, Miller A, Nam S, Kautz R, and Schwall R. 2005 J Appl. Phys. Lett. 87 203505

[5] Li H, Zhang L, You L, Yang X, Zhang W, Liu X, Chen S, Wang Z, and Xie X. 2015 Opt. Express 23 17301-17308

[6] Liu D, Miki S, Yamashita T, You L, Wang Z and Terai H 2014 Opt. Express 22, 21167-21174

[7] Miki S, Fujiwara M, Sasaki M, Baek B, Miller A, Hadfield R, Nam S, and Wang Z. 2008 J Appl. Phys. Lett. 92 061116

[8] Moshkova M, Divochiy A, Morozov P, Vakhtomin Yu, Antipov A, Zolotov Ph, Seleznev V, Ahmetov M and Smirnov K 2019 JOSA B, 36 Is3 B20-B25

[9] Smirnov K, Divochiy A, Vakhtomin Yu, Morozov P, Zolotov Ph, Antipov A, Seleznev V. 2018 Superconductor Science and Technology 31 N3

[10] Mattioli F, Zhou Z, Gaggero A, Gaudio R, Jahanmirinejad S, Sahin D, Marsili F, Leoni R and Fiore A. 2015 Superconductor Science and Technology 28 Is10 104001
[11] Divochiy A, Marsili F, Bitauld D, Gaggero A, Leoni R, Mattioli F et al. 2008 Nature Photonics 2 302-306
[12] Smirnov K, Vachtomin Y, Divochiy A, Antipov A, Goltsman G. 2015 Applied Physics Express 8, 022501
[13] Marsili F, Najafi F, Dauler E, Bellei F, Hu X, Csete M, Molnar R and Berggren K. 2011 Nano Lett. 11 2048–53
[14] Shcheslavskiy V, Morozov P, Divochiy A, Vakhtomin Y, Smirnov K, Becker W. 2016 Review of Scientific Instruments 87, 053117
[15] Sidorova M, Divochiy A, Vakhtomin Yu, and Smirnov K. 2015 Nanophotonics **9**, 093051
[16] Di Zhu , Marco Colangelo , Boris A. Korzh, Qing-Yuan Zhao , Simone Frasca, Andrew E. Dane, Angel E. Velasco, Andrew D. Beyer, Jason P. Allmaras, Edward Ramirez, William J. Strickland, Daniel F. Santavicca , Matthew D. Shaw, and Karl K. Berggren. Appl. Phys. Lett. 114, 042601 (2019);
[17] Excelitas Technologies' SPCM-NIR https://www.excelitas.com/product/spcm-nir

Fabrication and optical properties of Au:SiO_2 nanostructured thin films

I V Ilkiv[1,2], K M Morozov[1], K P Kotlyar[1], D. Amelchuk[3], A D Bouravleuv[1,3]
[1] St. Petersburg Academic University, St. Petersburg 194021, Russia
[2] St. Petersburg State University, St. Petersburg 199034, Russia
[3] Ioffe Physicotechnical Institute, St. Petersburg 194021, Russia

Abstract. We demonstrate synthesis and theoretical study of the optical properties of Au:SiO2 nanocomposite thin films. Localized surface plasmon resonance peak induced by Au NPs was clearly observed in optical spectra at wavelength range between 500 and 600 nm. Intensity and position of PL peak are demonstrated to depend on the Au NPs embedding depth.

1. Introduction

Noble metal nanoparticles (NPs) have been the subject of intense research in recent years due to their original structural, optical and electrical properties compared with their bulk counterparts [1]. Among them, Au NPs have drawn particular attention since they are capable of biological conjunction, as well as they demonstrate the nonlinear and broad absorption band in the visible range, which arise from localized surface plasmon resonance [2]. These properties make them highly suitable for the use in nanophotonics [2, 3]. For instance, there have been several demonstrations of chemical and biological sensing based on the plasmon absorption and scattering of Au NP [4] and their assemblies [5].

Typically, the realization of such devices relies on the deposition of NPs on a support, such as Al_2O_3, SiO_x substrates, polymers, etc. [6-9]. There are several preparation techniques that can be employed for NPs deposition onto the substrates. Doing so with nano-lithographic techniques allows to fabricate well-defined NPs [10, 11], but at the same time it is difficult, time-consuming and inefficient to create sparse patterns of small particles. In turn, thermal annealing of thin Au films is the simplest and low-cost method, but there are the dispersion in the size of NPs forming [12]. Nowadays colloidal gold have become an important alternative because of stability, ease of synthesis [3] and deposition of NPs onto substrates. Furthermore, colloidal solution containing NPs with desired shapes are commercially available in a wide range of sizes from a few to tens of nanometers. Unfortunately, a common problem in the preparation of reliable sensors is poor adhesion between the Au NPs and oxide substrates, which may result in detachment of the NPs from the substrate and, subsequently, deterioration in detection efficiency [13].

Meanwhile, recent studies [14, 15] demonstrate that Au NPs could be thermally embedded into subsurface layer by the thermal annealing technique. Therefore, it can facilitate the better stability of Au NPs on oxide surfaces. However, it has been pointed out that the optical response of Au NPs during the annealing step can be changed due to the shape as well as dielectric environment variation. Therefore, establishing the correlation between the NPs morphology and optical properties is a prerequisite for both understanding metal supported systems and for creating future photonic devices.

In this work, we report on the fabrication and investigation of optical properties of SiO_2 thin films with embedded Au NPs.

2. Experiment

The investigations were carried out using Au colloidal solutions having 40 nm nominal diameter NPs. The deposition of Au colloidal solutions was performed onto the Si(100) substrates with preliminary created by thermal oxidation 190 nm thick SiO_2 layer. Prior to the deposition, the surface of the samples was modified by a short time Ar plasma treatment. Plasma treatment, as it was demonstrated previously [16], leads to a global improvement of the NP adhesion and the hydrophilic properties of the Si substrates. NP deposition time was equal to 150 s, afterwards the solution remnants were blown away with a dry N_2 gas.

The heat treatments of the samples with deposited Au nanoparticles were carried out in the horizontal quartz furnace at 1000°C under ambient air. After the annealing the samples were cooled down passively to room temperature inside the oven.

The morphological properties of samples were investigated using a scanning electron microscope (SEM) Zeiss Supra 25 operating at 20 kV.

The morphology of the nanowires was investigated using Zeiss Supra 25 field-emission scanning electron microscope (SEM).

Numerical simulations of PL spectra were carried out in COMSOL Multiphysics modelling environment.

3. Results

The SEM image of the sample surface with the deposited colloidal Au NPs is shown in Figure 1. Spherical 40 nm in diameters Au NPs are distributed randomly, but rather homogeneously across the whole sample surface. The average surface density of NPs is about 8×10^8 cm^{-2}.

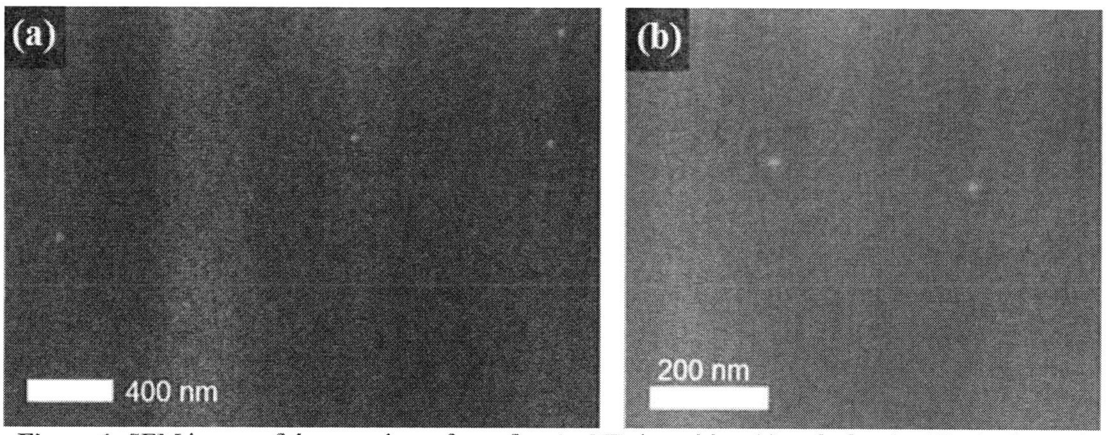

Figure 1. SEM image of the sample surface after Au NP deposition (a) and after the 25 min thermal annealing.

After the thermal annealing for 25 min at 1000°C in a horizontal furnace under ambient air, the smaller diameter NPs surrounded by rims were observed (see Fig. 1b). The results obtained are attributed to the thermally-induced Au NPs embedding into silicon dioxide and the similar results were observed in other studies [14, 17-19]. In should be emphasized that despite the simplicity of the fabrication method, the mechanism of Au NPs embedding remains an open question.

Figure 2 illustrates the proposed kinetic of the Au NP embedding. The heating of the sample to the high temperatures leads to the formation of liquid gold NPs, which wets the sample surface (see Fig. 2b). At the same time Au NPs begins to penetrate into SiO_2 layer. This process is followed by the material transport from under the Au NP to the triple point (metal-vapor-oxide), leading to the ridge formation. In addition, it should be pointed here, that gold atoms can be evaporate or be diffused into SiO_2, which can result in the NP size decreasing during the embedding process [14]. Thus, the thermal annealing can provide the controllable Au NPs embedding into SiO_2.

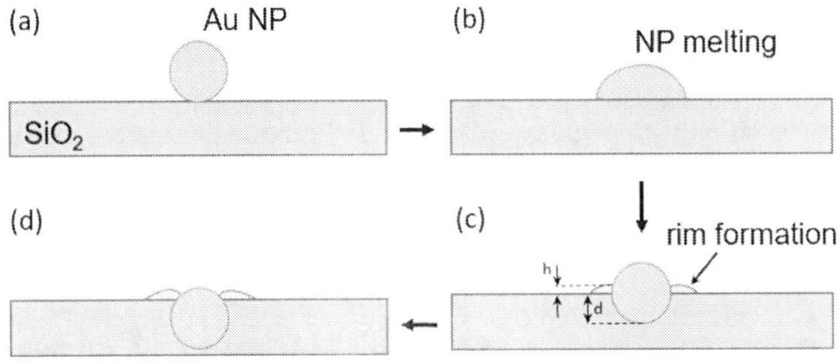

Figure 2. Schematical representation of the proposed Au NP embedding in SiO_2 surface.

The morphologies of the Au NPs demonstrated in Figure 2 were used for the simulations of optical properties of Au:SiO_2 nanostructured thin films obtained. Taking into account the size and the shapes of Au NPs, as well embedding depth and the ridge formation, numerical modelling of the spectra nanostructured films were performed. Numerical modelling studies were carried for the solid-state Au NPs having 40 nm diameter and surface density of 1×10^9 cm^{-2}. The model calculations assumed that the NPs preserve their volume during the annealing.

Figure 3 presents the calculated reflectivity spectra of obtained Au:SiO_2 nanostructured films. It is seen that the resonance peak of Au NPs deposited over SiO_2 surface is located near 500 nm. As the embedding depth increases, the peak shifts towards the red wavelengths. Simultaneously, since the NPs effective dielectric constant enhances, the reflectance intensity increases. It should be noted, this calculations did not take into account the formation of ridges. In turn, the formation of ridges, which were modeled by the addition of 5 nm height rims around the NP, leads to the increased reflection also.

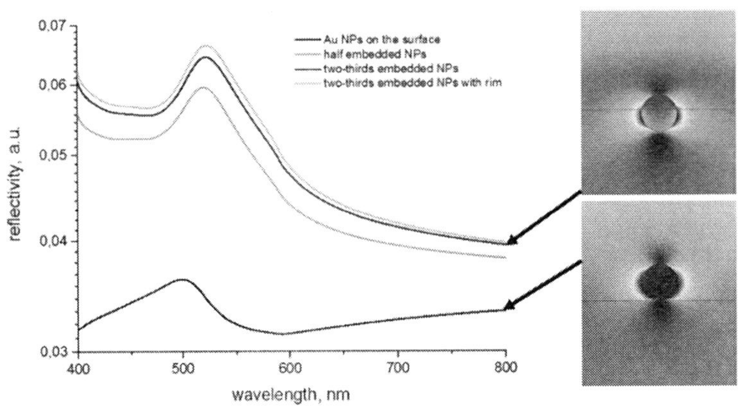

Figure 3. The calculated reflectance data of Au:SiO_2 nanostructured films. The right–hand panel shows the electric field distribution.

The electric field patterns of Au NPs at the resonance wavelengths (Fig. 3) show that the Au NPs behave roughly as a dipole with accumulated electric field on the sides of the NPs. This confirms that the SiO_2 surrounding of Au NPs strongly influence the resonance.

4. Conclusion

In summary, the embedding of Au NPs, which can provide its better stability on the SiO_2 surface, was performed with applied by thermal annealing technique. The ability to control the embedding depth and, as a consequence, to tune the optical properties may provide new opportunities for the fabrication novel plasmonic materials.

Acknowledgments

The work was partially supported by the grant of Minobrnauki № 16.9789.2017/BCh and Skoltech (agreement № 3663-MRA).

References

[1] Shang L, Dong S, Nienhaus G U 2011 *Nano today* **6** 401
[2] Liu J, He H, Xiao D, Yin S, Ji W, Jiang S, Luo D, Wang B, Liu Y 2018 *Materials* **11** 1833
[3] Kretschmer F, Muehlig S, Hoeppener S, Winter A, Hager M D, Rockstuhl C, Schubert U S 2014 *Particle & Particle Systems Characterization* **31** 721
[4] Raschke G, Kowarik S, Franzl T, Sönnichsen C, Klar T A, Feldmann J, Kürzinger K 2003 *Nano letters* **3** 935
[5] Pazos-Pérez N, Ni W, Schweikart A, Alvarez-Puebla R A, Fery A, Liz-Marzán L M 2010 *Chemical Science* **1** 174
[6] Prakash J, Kumar V, Kroon R E, Asokan K, Rigato V, Chae K H, Swart H C 2016 *Physical Chemistry Chemical Physics* **18** 2468
[7] Hiep H M, Yoshikawa H, Saito M, Tamiya E 2009 *ACS nano* **3** 446
[8] Rodríguez-González V, Zanella R, del Angel G, Gómez R 2008 *Journal of Molecular Catalysis A: Chemical* **281** 93
[9] Tan K H, Lee H W, Chen J W, Dee C F, Majlis B Y, Soh A K, Chang W S 2017 *The Journal of Physical Chemistry C* **121** 13487
[10] Bechelany M, Maeder X, Riesterer J, Hankache J, Lerose D, Christiansen S, Philippe L 2010 *Crystal Growth & Design* **10** 587
[11] David C, Guillot N, Shen H, Toury T, de La Chapelle M L 2010 *Nanotechnology* **21** 475501
[12] Farzinpour P, Sundar A, Gilroy K D, Eskin Z E, Hughes R A, Neretina S 2012 *Nanotechnology* **23** 495604
[13] Chen H, Zhao L, Chen D, Hu W 2015 *Journal of colloid and interface science* **460** 258
[14] Gosavi A A, Hedrick J L, Chen P C, Notestein J M, Mirkin C A 2019 *Nano Research* 1
[15] Ilkiv I, Kotlyar K, Amel'chuk D, Lebedev S, Cirlin G, Bouravleuv A 2017 *Acta Physica Polonica A* **132** 366
[16] Bouravleuv A, et al. 2017 *Nanotechnology* **29** 045602
[17] Sui M, et al. 2017 *Applied Surface Science* **393** 23
[18] de Vreede, Lennart J, Albert van den Berg, Eijkel J 2015 *Nano letters* **15** 727
[19] Karakouz T, et al. 2013 *Physical Chemistry Chemical Physics* **15** 4656

Gold nanoantennas for enhancement of photon emission from STM contact

V A Shkoldin[1,2], D V Permyakov[1], A O Golubok[1,3], A V Uskov[1,4], A K Samusev[1], I S Mukhin[1,2]
[1] ITMO University, St. Petersburg 197101, Russia
[2] St. Petersburg Academic University, St. Petersburg 194021, Russia
[3] Institute for Analytical Instrumentation RAS, St. Petersburg 198095, Russia
[4] P. N. Lebedev Physical Institute, Russian Academy of Sciences, 119991 Moscow, Russia

Abstract. Inelastic electron tunneling in a tunnel junction may be used as an electrical nanosource of surface plasmon polaritons and photons. In this work, we investigate emission from tunnel contact between the Platinum/Iridium tip and a thin Au film on glass. The experiment has shown that the intensity of this emission can be enhancement by use a gold nanoantenna located in the vicinity of tunnel contact.

1. Introduction

One of the way increasing of systems on chip (SoC) performance is transition to optical signal links between parts of SoC, such as RAM and CPU, for example. This step of technological evolution may be possible after creation of full range of electro-optical components. One of the necessary components is local light source compatible with modern semiconductor manufacturing processes. The most popular light sourses, such as semiconductor lasers with Fabry-Perot or microring resonators are not an appropriate option to be implemented on a chip, due to its large size equal or greater than the emitted wavelength.

An alternative way of light generation is based on tunnel junction use. Effect of light emission from the tunnel junction has been known for decades [1]. Development of devices based on this effect may accelerate the transition from electronic to photonic digital data processing. Future photon devices are expected with greater performance then traditional computing circuits based on electrical signals only. However, the quantum efficiency of the light emitting of tunnel junction is very low. One of the possible approaches to increase photon emission yield is to use of nanoantennas. As theoretically shown in [2], a resonant optical nanoantenna can improve light emission from junction.

Light emission may be observed in a tunnel junction with applied bias voltage (V_b). A bias voltage provide electrons in one electrode with excess energy eV_b with e being the elementary charge. The excess energy may be loosed for heating of another one electrode. This is elastic processes (Fig. 1 a). In addition, some of excess energy may be loosed to the excitation of an optical mode with energy $\hbar\omega$, where $\hbar$ is the reduced Planck constant and ω is the angular frequency of the optical mode (Fig. 1. b)

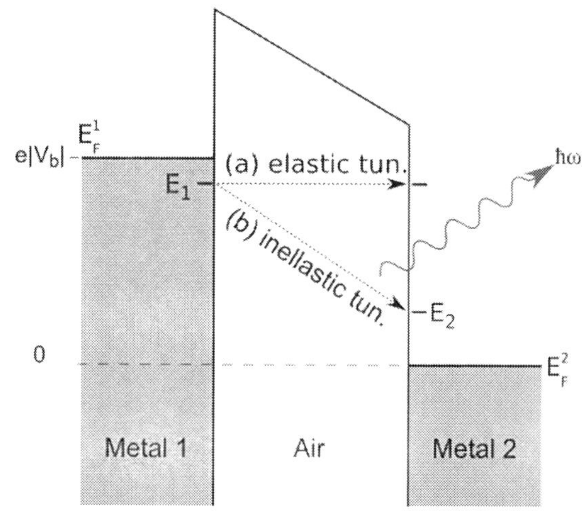

Figure 1. Band diagram of a tunnel junction between two metals with air as potential barrier region. EF1, EF2 – Fermi levels in metals 1 and 2, Vb – applied bias voltage. E1, E2 – electron energies before and after the tunneling. Process (a) corresponds to elastic tunneling and (b) inelastic tunneling with partial electron energy loss as photon with ħω energy.

2. Sample and experimental setup

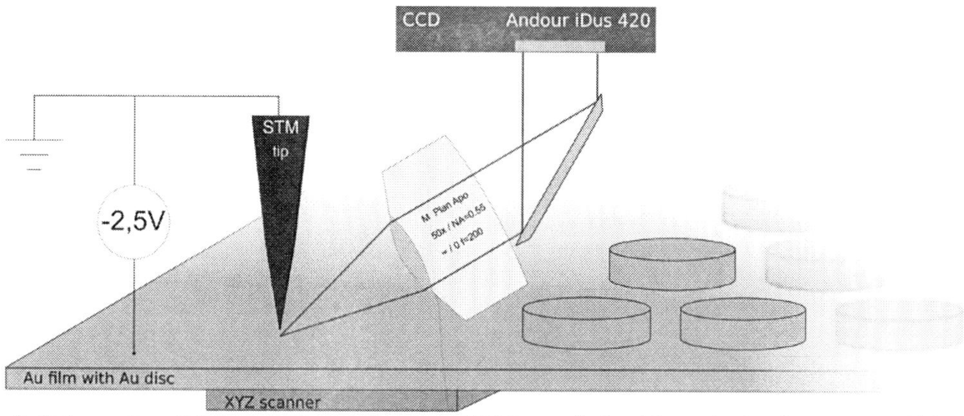

Figure 2. Schematic of the experimental setup. STM coupled with optical microscope. Objective focused onto tip-sample gap.

A scanning tunneling microscope (STM) is a powerful tool for the tunnel junction effects study. The experimental setup comprises a commercially available STM head operating in air (AIST-NT Combiscope) coupled with an optical microscope. The emitted light was collected with Mitutoyo 50x NA=0.55 objective in side channel (as shown in figure 2) and focused onto spectrometer (HORIBA LabRAM HR) with cooled detector (Andor iDus 420).

The tunnel contact was established between thick gold film on glass substrate and the STM tip. The tip was used commercial available Platinum/Iridium probes by Bruker. The thick gold films were deposited on 140-μm thick glass substrate with a thin chromium adhesion underlayer via the thermal evaporation. Gold nanodiscs with diameter of ~200 nm were produced by means of electron beam lithography (Fig. 3).

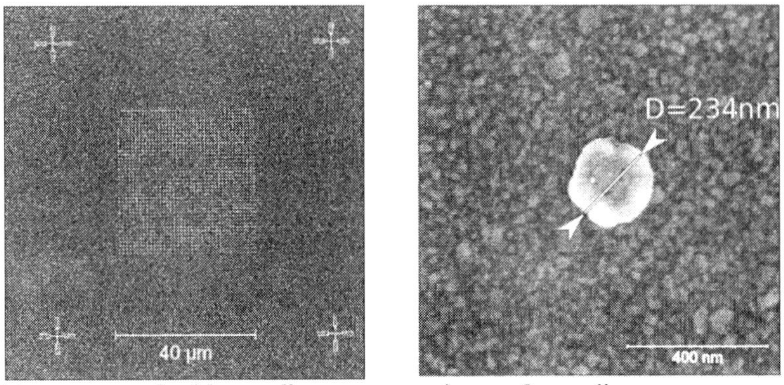

Figure 3. SEM image one of gold nanodiscs arrays and one of nanodiscs.

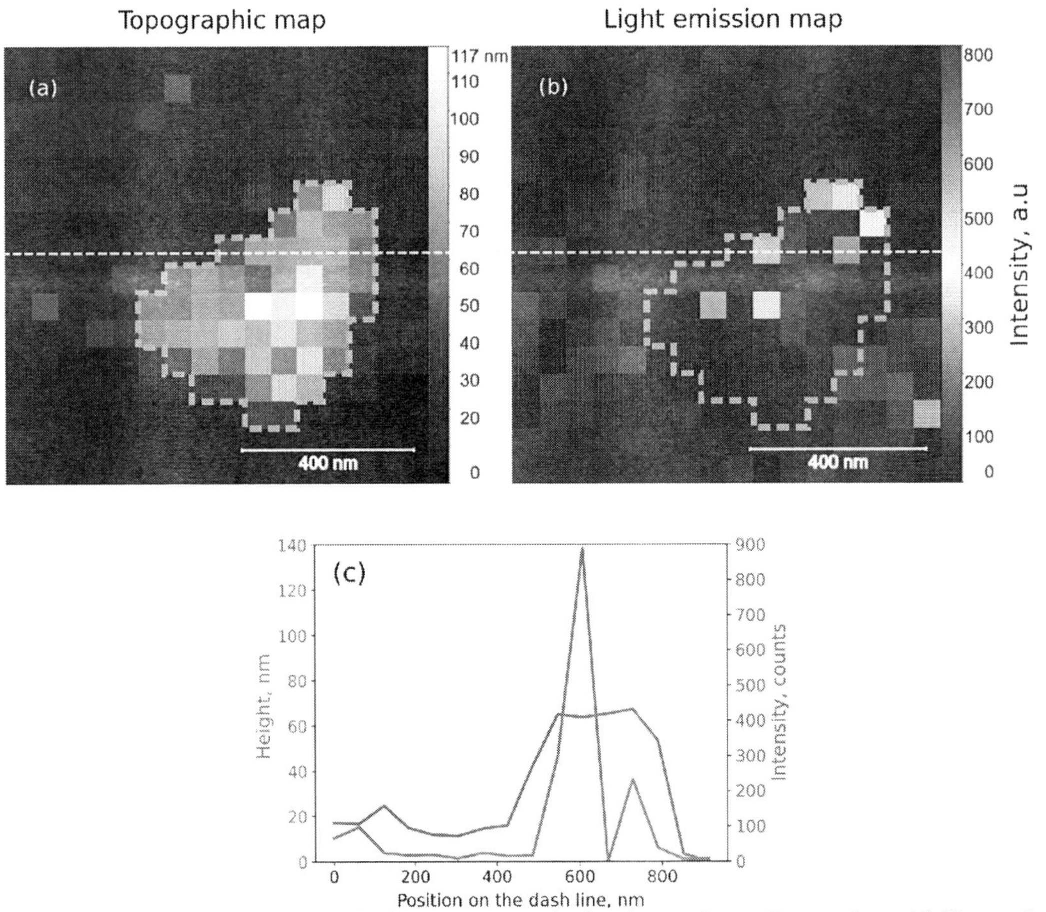

Figure 4(a,b,c). **(a)** STM image of single nanoantenna in the shape of nanodisc on the gold film surface; **(b)** mapping of light intensity from STM scanning area; **(c)** The section of STM topography and intensity map taken along dash line.

To obtain STM image and simultaneously record optical signal, STM work in constant current mode. We use high value of feedback loop gain in STM in order to implement the "saturation" regime, as recommend in [3] for high rate of photon emission. In that case the STM tip was has some deformation and doesn't provide true image of topography. The fig. 4a shows STM image of single nanoantenna in the shape of nanodisc on the gold film surface. We did synchro scan of the nanoantenna and recorded light emission (fig. 4b). The signal is collected during 500 ms for each pixel of the map. The figure 4c shows the section of STM topography and intensity map taken along dash line.

3. Conclusion

Table 1. The table with intensity comparison.

Probe location	Average intensity counts	Maximum intensity counts
Film	50	288
Disk	85	890

As we can see on figure 4c and table 1, the intensity signal shows the increase of light emission intensity, when the STM tip located on the nanoantenna. This enhancement of light emission rate is associated with the increase of local density of optical states of the nanoantenna in comparison with metal film [2]. Further studying of nanoantenna shape influence on light emission rate may push forward the development of an effective optical nanosource based on tunnel junction contact.

Acknowledgements
This work was carried out with the support of the Russian Science Foundation (Grant 17-19-01532).

References

[1] John Lambe, S. L. McCarthy 1976 *J. Physical Review Letters* **37**, 923
[2] F Bigourdan, J P Hugonin, F Marquier, C Sauvan, J J Greffet 2016 *J. Physical review letters*, **116**, 106803
[3] B Rogez, S Cao, G Dujardin, G Comtet, E Le Moal, A Mayne, E Boer-Duchemin 2016 *J. Nanotechnology* **27**, 465201

Propagation of short current pulses in Josephson transition line and ultrafast qubit control

M V Bastrakova[1,*], N V Klenov [2,3], V I Ruzhickiy[2,3] and A M Satanin[3,4]

[1] Lobachevsky State University of Nizhny Novgorod, Nizhny Novgorod 603950, Russia
[2] Faculty of Physics, Lomonosov Moscow State University, Moscow 119991, Russia
[3] Dukhov All-Russia Research Institute of Automatics (VNIIA), Moscow 101000, Russia
[4] Russia National Research University Higher School of Economics, 101000 Moscow, Russia
*Corresponding author: mar.denisenko@gmail.com

Abstract. The results of the study of the impact of single-quantum magnetic flux pulses moving along the Josephson transmission lines (JTL) on the dynamics of states of qubits magnetically coupled with such lines are presented. The JTL dynamics was calculated in the frame of resistively shunted junction model with different damping coefficients. This allowed us to find the form of control current pulses (fluxon) acting on a superconducting qubit. Numerical simulations of the simplest logic operations with superconducting qubit due to calculated fluxon impact are presented.

1. Introduction

In recent years, schemes based on Rapid Single Flux Quantum (RSFQ) logic have been actively studied for their applicability in reading weak magnetic and current signals from nanoscale structures: bolometers, single photon detectors, and superconducting quantum bits [1-3]. One of these schemes uses an extremely "rapid" ballistic method. In this method, single-quantum pulses (fluxon) corresponding to magnetic flux quanta (voltage pulses $V(t)$ is as follows: $\int V(t)dt = \Phi_0$, $\Phi_0 = h/2e$, h is the Planck's constant, e is the elementary charge) are propagating along the Josephson transmission line (JTL), which is magnetically coupled to the measured qubit [4, 5]. The sensitivity of the ballistic detector is limited by "natural" changes in the speed of movement of single-quantum pulses, which can be caused by thermal fluctuations of currents and technological variations in parameters of JTL elements. In this case, the magnetic energy associated with the field that penetrates from the JTL into the qubit, strongly depends on the specific topology on the chip.

In this paper, we investigate the influence of the voltage and current pulses ("real" fluxons) formed in the transmission lines on the evolution of the state of a superconducting qubit [6-10]. The calculation of the JTL with different damping coefficients allowed us to find the shape of fluxon. Based on the solution of the master equation of the qubit, we was shown that current pulses in the JTL with low damping can control the populations of the qubit by changing only the magnetic coupling energy with high fidelity. The influence of quantum noise in a qubit on the state switching processes is studied.

2. Single-quantum pulse in JTL

The main equations for JTL dynamics in the frame of resistively shunted junction model have the

IOP Publishing
1410 (2019) 012142 doi:10.1088/1742-6596/1410/1/012142

fallowing form (we use here normalization by the critical current I_c of Josephson junctions, connected in parallel with inductances L_{JTL}):

$$i_{pulse} + i_b - (\varphi_1 - \varphi_2)/l_{JTL} = \sin(\varphi_1) + \dot{\varphi}_1 + \beta_L \ddot{\varphi}_1, \quad (1)$$

$$i_b + (\varphi_{k-1} - \varphi_k)/l_{JTL} - (\varphi_k - \varphi_{k+1})/l_{JTL} = \sin(\varphi_k) + \dot{\varphi}_k + \beta_L \ddot{\varphi}_k, \quad (2)$$

$$i_b + (\varphi_{N-1} - \varphi_N)/l_{JTL} = \sin(\varphi_N) + \dot{\varphi}_N + \beta_L \ddot{\varphi}_N, \quad (3)$$

where i_{pulse} is the input normalized current pulse, i_b is is the normalized bias current, l_{JTL} is the normalized inductance, $\beta_C = (\omega_c/\omega_p)^2$ is the McCumber parameter which determines the effective normalized capacitance Josephson junction, ω_c and ω_p are the characteristic and plasma frequencies respectively, $\Delta\varphi/l_{JTL}$ is the current through the inductance, N is the length of JTL in the number of Josephson junctions. It is important for us to emphasize that the parameter $(\beta_C)^{-1/2}$ (damping coefficient) also determines the decay rate of plasma oscillations for Josephson contacts, determines the magnitude of the energy dissipation in the system.

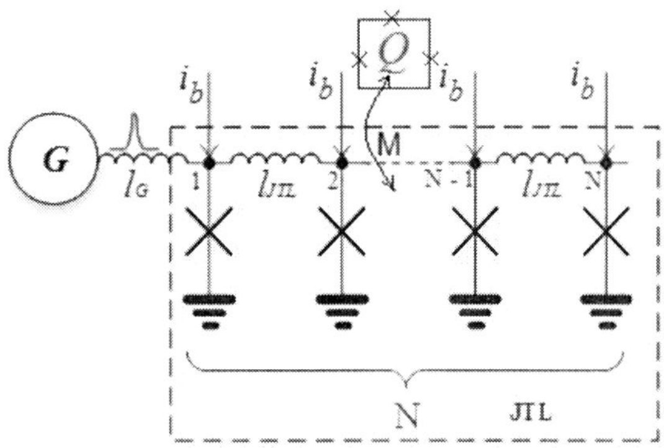

Figure 1. Josephson transmission line (JTL) with the phase generator, G, on the input. Josephson contacts are marked by crosses; qubit to measure/control denoted by Q (for instance, it may be realized as a superconducting loop interrupted by three Josephson junctions) and which is coupled with the JTL by means of mutual inductance M.

In this paper we consider the propagation of fluxons in the JTL (see figure 1) for the case, when bias current is less than the critical one, $i_b < 1$. Otherwise, the system switches to a resistive state. At the initial moment, the phase at the first Josephson junction was equal to 0. The current pulse converted the phase to the value 2π. Due to the second junction is inductively connected to the first junction by means of the inductance l_{JTL}, there is a "leaking" current equal to $2\pi / l_{JTL}$. If this current is less than the critical value, then it will be "frozen" between the two junctions, otherwise it will transfer the second junction to the state with phase 2π. In this case, the phase difference between the first and second junctions disappears (the current through the inductance is zero), but a similar situation occurs between the second and third junctions. Thus the fluxon propagates in the JTL.

The JTL is magnetically coupled with a superconducting qubit (this is shown schematically in figure 1). So the current impulse in the inductance located next to the qubit creates a magnetic impulse acting on the qubit. This current value I_L as a function of time was found numerically for various parameters of Josephson junctions and the relations between them.

Numerical simulations were performed with the following JTL parameters: N = 150, $l_{JTL} = 3$, $i_b = 0.5$, $\beta_C = 1; 10; 100$. The system of equations (1) - (3) was solved using the 4-th order Runge – Kutta – Fellberg method for each value damping β_C. The current through the inductance in the middle of the JTL was calculated to eliminate the influence of boundary effects. The currents through the inductance normalized to the magnitude of the critical current for different parameters β_C (for different values of energy dissipated near qubit) are shown in figure 2.

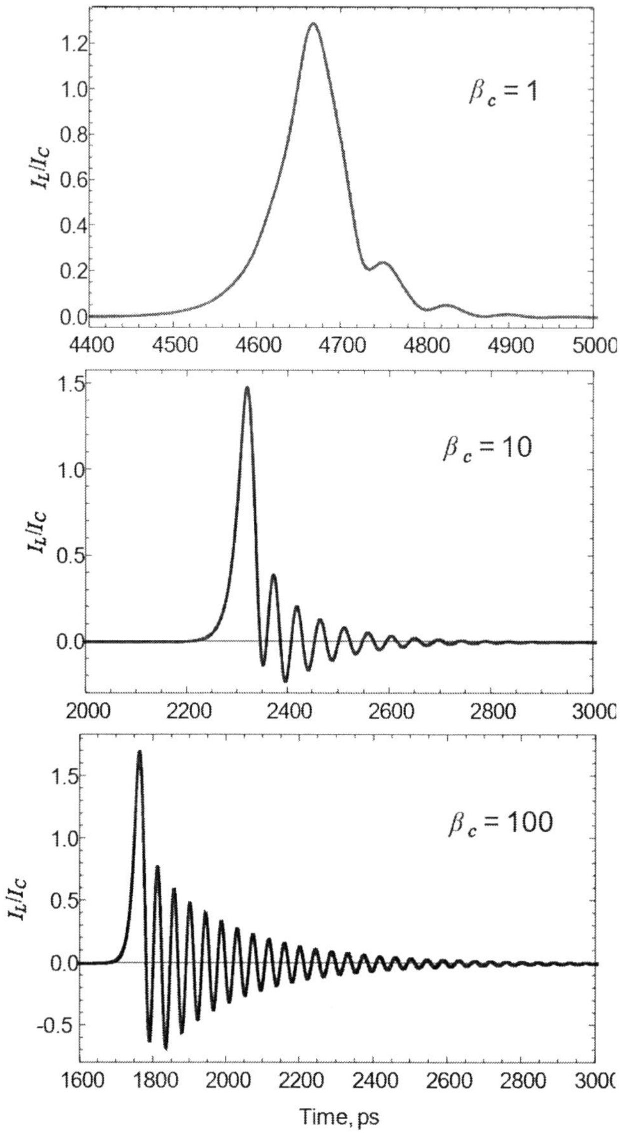

Figure 2. The currents induced in the JTL, which through a magnetic coupling affect the state of the qubit. Here are the JTL parameters: $N = 150$, $l_{JTL} = 3$, $i_b = 0.5$, $\beta_C = 1$; 10; 100.

3. Switching of qubit states by current pulses

The basic dynamical behavior of a superconducting artificial atom driven by magnetic field can be described by the Hamiltonian [10]:

$$H = \frac{h}{2}\begin{pmatrix} \omega_q & \varepsilon(t) \\ \varepsilon(t) & -\omega_q \end{pmatrix} \tag{4}$$

where $\varepsilon(t)$ describes the impact of the current pulses in JTL and ω_q is the frequency of the transition between the basis states of a qubit ($|0\rangle$ and $|1\rangle$).

The superconducting qubit is inductively coupled with the JTL, therefore, the shape of the driving pulse arriving at the qubit is determined by the inductance currents I_L in the JTL (see figure 1), and the amplitude of the external influence (energy) A depends on the coupling force of the qubit to the JTL line (which is defined by inductance M), that is:

$$\varepsilon(t) = A\left(\frac{I_L}{I_c}\right)^2. \tag{5}$$

In experimental conditions the interaction of the qubit with reservoir shows considerable effect on the qubit dynamics. These processes are described here by considering the interaction of a qubit with a bosonic reservoir. We calculated the equation for the density operator of the qubit ρ in the Markov approximation [11]:

$$\frac{\partial\hat{\rho}}{\partial t} = -\frac{i}{\hbar}\left[\hat{\rho},\hat{H}\right] + \frac{\Gamma_\phi}{2}(\hat{\sigma}_z\hat{\rho}\hat{\sigma}_z - \hat{\rho}) + \frac{\Gamma_e}{2}\left(2\hat{\sigma}_-\hat{\rho}\hat{\sigma}_+ - \hat{\sigma}_+\hat{\sigma}_-\hat{\rho} - \hat{\rho}\hat{\sigma}_+\hat{\sigma}_-\right) \tag{6}$$

where the rate Γ_ϕ characterizes the process of dephasing, and the parameter Γ_e is responsible for the rate of energy loss [12]. We assume that at the initial time the qubit is initialized in the ground state. Numerically solving equation (6) with (4) for each current control pulse characterizing by different damping β_C (see figure 2) we have found the amplitude coefficient A of pulses, which implements the process of switching the qubit states from the ground to the excited qubit state. Time dependences of switching probability $W(t)$ are shown in figure 3. Qubit with characteristic frequency ω_q = 5 GHz was chosen to simulate switching states under the influence of fluxon. Note that the moment of switching for the qubit states is mainly determined by the main Lorentz peak of the current pulse (see figure 3). For small values of damping $\beta_C = 1; 10$ of the pulse current fluctuations has a negligible contribution to the behaviour of populations of the qubit levels. However, it can be seen that for large values of the McCumber parameter ($\beta_C = 100$) plasma oscillations of current I_L is a parasitic phenomenon, which leads to significant errors when switching the qubit. Numerical analysis has shown that the best switching of qubit states for control current pulse with relatively small dissipation ($\beta_C = 100$, see figure 2) can be achieved at the amplitude of A = 0.02 GHz. However, we note that the switching error is about 20%, which is unacceptable accuracy for single qubit quantum operations. We conducted a further study and found that this type of pulse (figure 3 (c), black curve) can be used however for qubits with smaller frequencies $\omega_q \sim$ 100 MHz.

It is well known that the phase relaxation (loss of coherence) is due to the energy transfer from the ground state to the excited one. Dephasing leads to changes in the off-diagonal elements of the density matrix from Eq. (6) only. Since the phase fluctuations are random, then such kind of energy transfer is accompanied by a loss of coherence over time, $T_\phi = \Gamma_\phi^{-1}$, and leads to equilibrium population of levels (see figure 3, red and blue curves). In turn, the energy relaxation affects both the diagonal and off-diagonal elements and it is a stronger criterion for the destruction of the quantum state. Note that we cannot ensure the damping of the population of the upper level without changing coherence, because this would lead to the loss of positive definiteness of the density matrix. Experimental results show the characteristic relaxation times for superconducting qubit of a few dozens of microseconds [12]. However, since the duration of the control pulse is about of several ps it is sufficient to perform thousands of single-qubit operations [8-9]. This fact gives hope for long-term development of quantum computing through the use of the capabilities of superconducting digital logic circuits [1-3, 10].

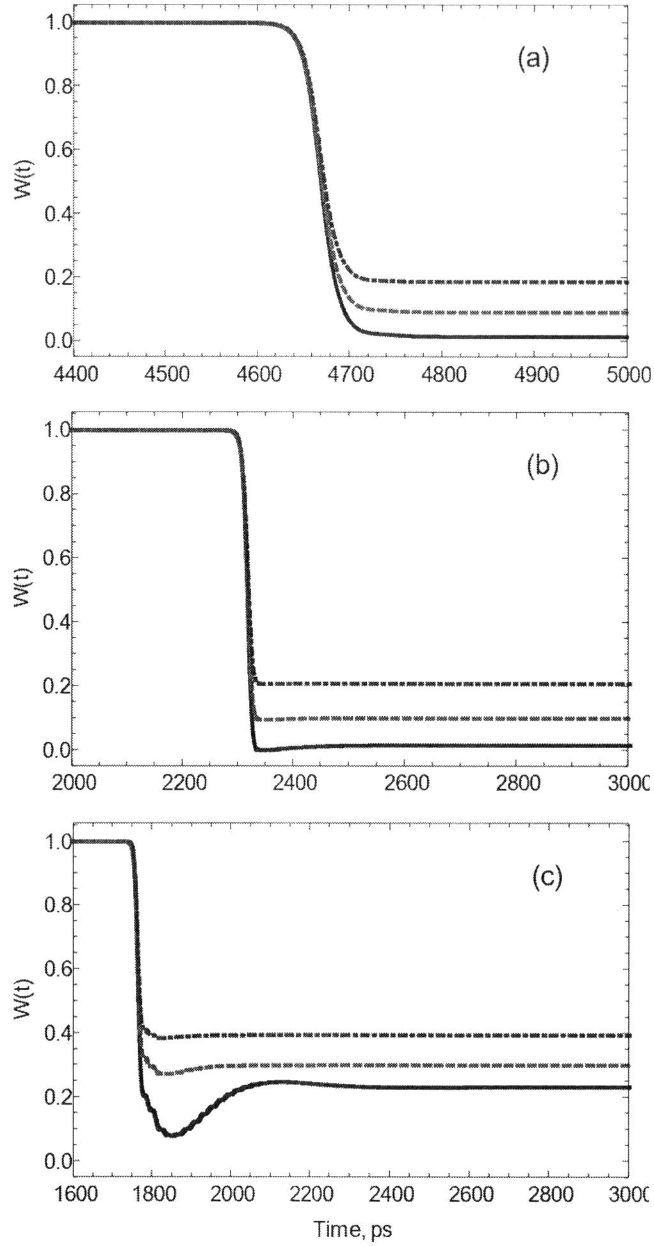

Figure 3. The time dependencies of the populations for qubit states switching for three different forms of fluxon control pulses, depicted in figure 1. The qubit frequency is $\omega_q = 5$ GHz, fluxon amplitudes are equal **(a)** $A = 0.15$ GHz, **(b)** $A = 0.083$ GHz, **(c)** $A = 0.02$ GHz. Here are the noise parameters in the system $T_e = 100\mu s$: black curve corresponds to $T_\phi = 100\mu s$, red – $T_\phi = 10\mu s$, blue – $T_\phi = 1\mu s$.

4. Conclusion

The propagation of the magnetic flux (fluxon) in superconducting Josephson circuits was studied depending on the McCumber parameter. When the fluxon passes through the Josephson array in the vicinity of the inductively coupled qubit, the local field pulse causes transitions between the levels of the quantum system. The switching processes for qubit states in the field of the fluxon were numerically simulated. The amplitudes of the current pulses that necessary to optimal switching of the qubit states are found. Thus we have shown the need to use control pulses, specially created in the circuits of energy-efficient upgrades of rapid single-flux-quantum logic [13].

Acknowledgments
This work was supported by the RSF No. 18-72-00158 and the RFBR grant No. 18-07-01206. V.I. Ruzhickiy acknowledges the Basis Foundation scholarship.

References

[1] Averin D V, Rabenstein K, and Semenov V K 2006 *Phys. Rev. B* **73** 094504
[2] Herr A, Fedorov A, Shnirman A, Ilichev E and Schon G 2007 *Supercond. Sci. Technol.* **20** S450
[3] Pankratov A L, Gordeeva A V and. Kuzmin L S 2012 *Phys. Rev. Lett.* **109** 087003
[4] Soloviev I I, Klenov N V, Pankratov A L, Il'ichev E and Kuzmin L S 2013 *Phys. Rev. E* **87** 060901(R)
[5] Soloviev I I, Klenov N V, Bakurskiy S V, Pankratov A L, Kuzmin L S 2014 *Appl. Phys. Lett.* **105** 202602
[6] McDermott R, Vavilov M G 2014 *Phys. Rev. Applied* **2** 014007
[7] Soloviev I, Klenov N V, Pankratov A L, Revin L S, Il'ichev E, Kuzmin L S 2015 *Phys. Rev. B* **92** 014516
[8] Klenov N V, Kuznetsov A V, Soloviev I I, Bakurskiy S V and Tikhonova O V 2015 *Beilstein Journ. of Nanotech.* **6** 1946
[9] Klenov N V, Kuznetsov A V, Soloviev I I, Bakurskiy S V, Denisenko M V and Satanin A M 2017 *Low Temperature Physics* **43** 789
[10] McDermott R, Vavilov M G, Plourde B L T, Wilhelm F K, Liebermann P J, Mukhanov O A and Ohki T A 2018 *Quantum science and technology* **3**(2) 024004
[11] Scully M O, Zubairy M S 1997 *Quantum optics* (Cambridge: Cambridge University Press)
[12] Yan F, Gustavsson S, Kamal A, Birenbaum J, Sears A P, Hover D, Gudmundsen T J, Rosenberg D, Samach G, Weber S,. Yoder J L, Orlando T P, Clarke J, Kerman A J and Oliver W D 2016 *Nature communications* **7** 12964
[13] Takeuchi N, Ozawa D, Yamanashi Y, Yoshikawa N 2013 *Supercond. Sci. Technol.* **26** 035010

Analysis of the charge state of nitrogen-vacancy centers in diamond driven by an infrared laser field

M A Smirnov[1], **I V Fedotov**[1,2,3,4], **S A Moiseev**[1,5], **A M Zheltikov**[1,2,3,4,6]
[1] Kazan quantum center, A N Tupolev Kazan National Research Technical University, Kazan 420011, Russia
[2] M V Lomonosov Moscow State University, Moscow 119991, Russia
[3] Texas A&M University, College Station, Texas, USA
[4] Russian Quantum Center, Skolkovo, Moscow Region 143025, Russia
[5] Kazan E K Zavoisky Physical-Technical Institute, Russian Academy of Sciences, Kazan 420029, Russia
[6] Moscow Institute of Physics and Technology, Dolgoprudny, Moscow Region 141701, Russia

Abstract. The methods of controlling the charge state of nitrogen–vacancy centers using pulsed infrared laser field are studied. The effect of radiation intensity at a wavelength of about 1 μm on the ionization of these color centers was shown. The obtained results can be useful for multiphoton microscopy of biological objects, high-resolution microscopy and sensory applications.

1. Introduction

The nitrogen-vacancy center (NV center) is one of the many defects in the crystal structure of diamond. This defect consists of a nitrogen atom and a vacancy resulting from the removal of a carbon atom from a diamond lattice site. Vacancies can be obtained by irradiating a diamond with high-energy particles, usually electrons. Subsequent annealing at high temperature causes the diffusion of vacancies and the formation of color centers with nitrogen, which is often present in the diamond as an impurity. Unique spin properties, high brightness and excellent photostability of NV centers causes their use as various nanoscale temperature sensors [1], sensors of magnetic [2] and electric [3] fields, as single-photon sources [4], in high resolution microscopy (STED) [5,6], as well as in other applications of quantum technologies [7].

This color centers can be in two charge states: in the negative charge state (NV^-) with a zero-phonon line at 637 nm which corresponds to an optical transition of 1.95 eV and in the neutral charge state (NV^0) with a zero-phonon line at 575 nm which corresponds to an optical transition of 2.16 eV. Most often a NV center means exactly a color center in a negative charge state. Detailed information on the charge states of these color centers, their optical characteristics and dynamics is necessary for all this applications described above. It is known that the charge state of nitrogen-vacancy color center in diamond is not photostable, but may transfer into a "dark" state in which it ceases to radiate. In a number of works [8, 9], it was shown that this state is most likely a transition of the center to a state with a neutral charge ($NV^- \rightarrow NV^0$). Ionization of the center and transition to the neutral state can occur in various ways: one-photon (1P), two-photon (2P) [9], or three-photon (3P) excitation [10]. At the same time, there is a reverse process with the center recombining into a state with a negative charge ($NV^0 \rightarrow NV^-$). It is important that this process is also photo-induced, an essential property of which is the need to initiate the transfer of NV^0 to the excited state [9]. The detailed studying of the full cycle of ionization-recombination of nitrogen-vacancy color centers arising in front of multiphoton absorption can be useful for applying these centers for two-photon/multiphoton microscopy of biological objects [11] and can improve control charge state of this centers. Images obtained by multiphoton excitation are a powerful tool for imaging tissues in living organisms with longer depths of penetration. In this case the anti-Stokes lines of the luminophore are excited at the focus of the spot that reduces the autoluminescence of the cell. This microscopy can potentially provide better contrast and low cell photodamage, especially using femtosecond infrared laser radiation. In some works [10, 12], the possibilities of 2P and 3P excitations in the range of 700–1250 nm have already been demonstrated. This paper presents a more detailed study of the behavior of centers at different intensities of excitation using femtosecond infrared laser radiation at different wavelengths.

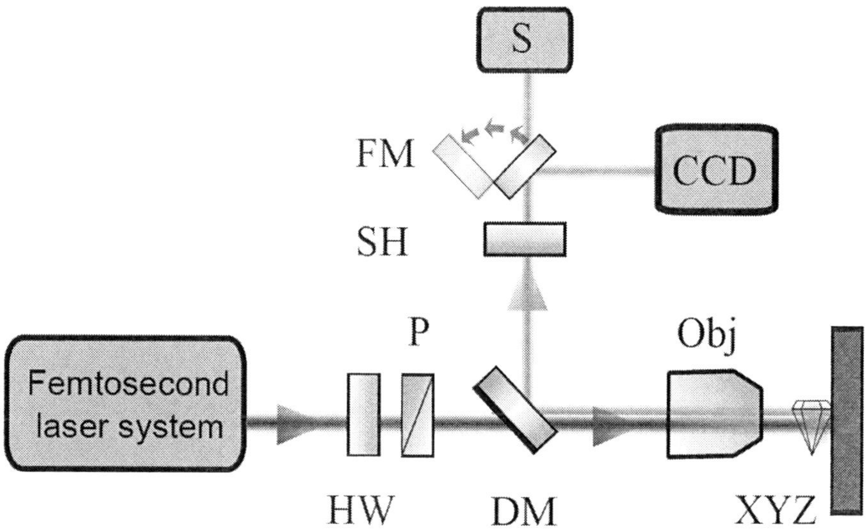

Figure 1. Schematic representation of the experimental setup: HW – half-wave plate, P – glan polarizer, DM – dichroic mirror, Obj – objective (40X), XYZ – XYZ-axis platform, SH – short-pass filter, FM – flip mirror, S – spectrometer, CCD – camera.

2. Description of the experiment

The schematic diagram of the experimental setup is shown on Fig. 1. For optical multiphoton excitation of color centers, we used femtosecond optical pulses with a wavelength of 810 nm and the pulse duration of about 50 fs from the Ti:Sa pulsed laser (Mira-HP, Coherent) and optical parametric oscillator (Mira-OPO, Coherent) with a wavelength of 1030 nm (1150 nm) and the pulse duration of about 50 fs. The parametric oscillator, in turn, was pumped by a Ti:Sa pulsed. The frequency of the laser pulses was about 76 MHz. Both lasers allow to obtain tunable laser radiation in the range from 750 to 1500 nm. Since multiphoton absorption is a less probable process, which depends nonlinearly on the intensity of optical radiation, the use of this laser system and focusing optics made it possible to achieve the necessary peak values of the intensity of laser radiation. A 40X objective lens (Olympus Plan Achromat Objective, 0.65 NA) was used to focus the pump radiation and simultaneously collect the photoluminescence radiation. A diamond crystal with a high content of NV centers (~ tens ppm) was used as the sample under study; its size is about 150 μm. To cut off the pump radiation from the luminescence radiation, a dichroic mirror (long pass 950 nm or 800 nm) and a cut-off filter (short pass 800 nm or 750 nm) were used. To control the location of the pump radiation on the diamond surface, a folding mirror was used, which directed the image to the CCD camera. The emission spectrum was recorded using a spectrometer. To regulate the power of the exciting radiation, a half-wave plate and a glane polarizer were used (see Fig.1). Measurement of the average power of laser radiation is carried out using a thermal power sensor (S302C, Thorlabs) and an optical power meter (PM100D, Thorlabs).

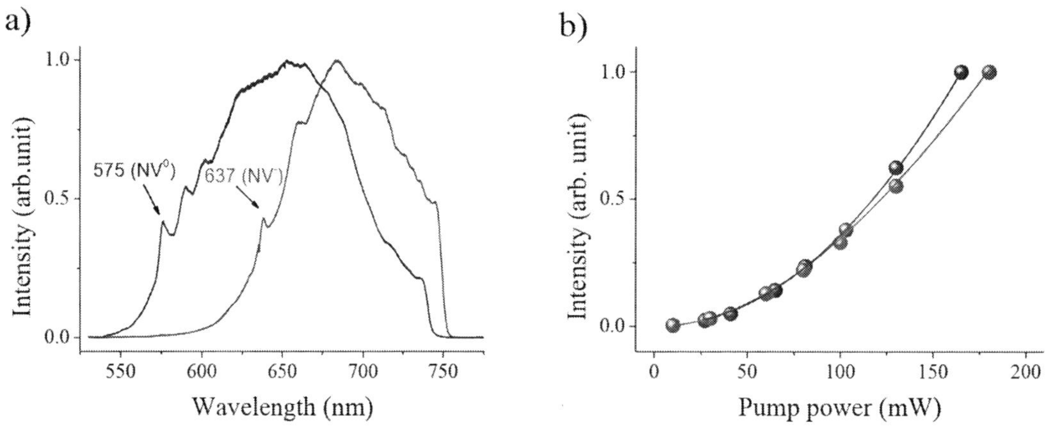

Figure 2. Experimental results. **(a)** Example of photoluminescence spectra obtained using excitation at 810 nm (blue line) and 1150 nm (red line); **(b)** the dependence of the luminescence intensity at 810 nm (blue circle) and 1150 nm (red circle) excitation wavelength on excitation power; this dependences were approximated by a polynomial function (blue and red line, respectively).

3. Experimental data and discussion

Fig. 2a shows the photoluminescence spectra obtained on our sample that was excited using laser radiation at 810 nm (blue line) and 1150 nm (red line). In the first case, we see the emission spectrum only from NV^0 centers with a peak at 575 nm from the zero-phonon line. This spectrum is due to luminescence after two-photon absorption at 810 nm, which is equivalent to an energy transition of 3.06 eV or optical excitation at a wavelength of 405 nm. Excitation with this energy can lead to two-photon absorption of NV^0 centers, as well as two-photon ionization of NV^- centers and a transition to a neutral state, because in this case the excitation energy is higher than the ionization energy, which is 2.6 eV [9]. In the second case, we see the opposite picture and observe the emission spectrum of only a NV^- centers with a small peak at 637 nm from the zero-phonon line. This spectrum is due to luminescence after two-photon absorption at 1150 nm, which is equivalent to an energy transition of 2.16 eV or optical excitation at a wavelength of 575 nm.

Such excitation mainly results in two-photon absorption of NV^- centers, since the energy obtained in this case is close to the absorption maximum of the NV^- center (about 520 nm) and is absorbed little by the NV^0 center, since it is far from the absorption maximum (about 460 nm).
Fig. 2b shows the dependence of the intensity for each of the types of excitation (at 810 and 1150 nm) on the power of the exciting radiation. The figure shows that these dependencies are very close at low power, but later they begin to diverge in their values. The experimental values obtained were approximated using the polynomial function of the form $I^{PL}=a(P)^2+b(P)^3$, where I^{PL} are the photoluminescence intensity values, a,b is the fitting parameters, and P is the power of the exciting field. The analysis of the parameters obtained from the approximation showed that with an increase in the pump power, the processes with a cubic dependence begin to affect the luminescence (see Fig. 2b). In the case of excitation of neutrally charged centers, the fitting parameter b has a positive value (for NV^0, a=3.5·10^{-5} and b=1.4·10^{-8}). In our case, this contribution has no significant effect on the behavior of the experimental curve and can be neglected. However, in the case of excitation of negative centers, the fitting parameter b has a negative value and cannot be neglected (for NV^- a=3.9·10^{-5} and b=-4.3·10^{-8}). Using the fitting parameters, it is easy to find that, at maximum pumping, the cubic contribution underestimates the photoluminescence values of the NV centers due to two-photon absorption by 20%. Most likely, this behavior is due to partial ionization of the NV centers and their transfer to a neutral state, in which the center is not excited at two-photon with the given pump parameters. It is worth noting that in the future with increasing pump intensity, a three-photon excitation of a neutral defect is possible, as was demonstrated in [10].

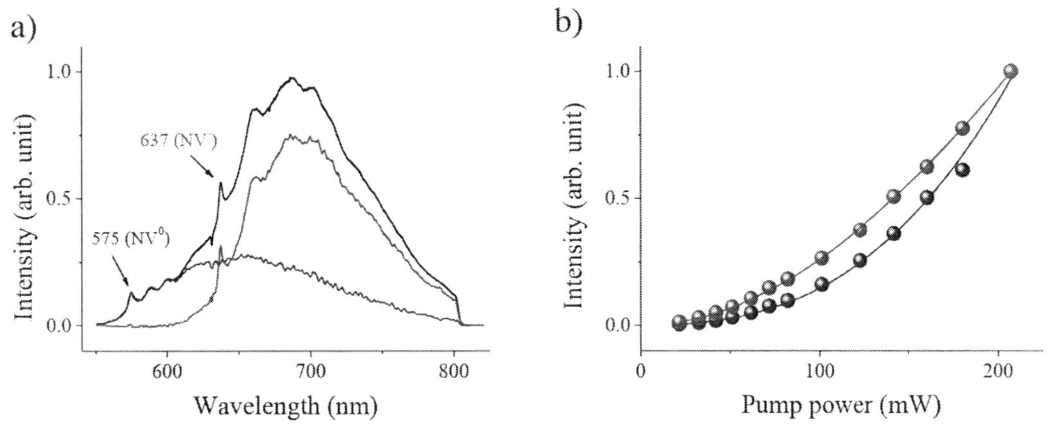

Figure 3. Experimental results. **(a)** The example of photoluminescence spectrum obtained using excitation at 1030 nm (black line) ; on this graphic is depicted two part of the spectrum – from NV^0 (blue line) and NV^-centers (red line); **(b)** the dependence of the luminescence intensity from the NV^0 part (blue circle) and the NV^- part (red circle) on excitation power; this dependences were approximated by a polynomial function (blue and red line, respectively).

Fig. 3a shows the photoluminescence spectra obtained on the sample that was excited using laser radiation at 1030 nm (black line). In this case, the photoluminescence spectrum is due to radiation from the centers in both charge states, which we can determine using the presence of zero-phonon lines on the graph, characteristic of each defect state. For clarity, both contributions were divided and shown on the graph separately for NV^0 (blue line) and NV^- (red line). This spectrum is due to luminescence after two-photon absorption at 1030 nm, which is equivalent to an energy transition of 2.41 eV or optical excitation at a wavelength of 515 nm. Excitation with such energy is efficiently absorbed by the center in both charge states and can lead to two-photon absorption of NV^0 and NV^- centers. Fig. 3b shows the dependence of the photoluminescence intensity on the power of the exciting radiation for each spectral component (NV^0 and NV^-) which were picked out from the experimental spectra. The experimental values obtained were approximated using the polynomial function of the form $I^{PL}=a(P)^2+b(P)^3$, where I^{PL} are the photoluminescence intensity values, a,b is the fitting parameters, and P is the power of the exciting field. Analysis of the parameters obtained from the approximation showed that with increasing pump power, the luminescence processes are significantly affected by processes with a cubic dependence (see Fig. 3b). The obtained fitting curves have the following parameters: for NV^0 $a=7.24\cdot10^{-6}$ and $b=7.44\cdot10^{-8}$; for NV^- $a=2.84\cdot10^{-5}$ and $b=-2.51\cdot10^{-8}$. Analysis of the fitting parameters shows that the dominant contribution for NV^0 centers with increasing power is the cubic contribution, which in our case reaches 70% for NV^0. Most likely, this cubic contribution is due to the ionization process NV^- centers and their transfer to a neutral state. This confirms the fact that for NV^- centers centers the cubic contribution turned out to be negative, which indicates a decrease in the number of the centers. The latter circumstance is extremely important, since it shows the ways to control the charge state of a defect with the help of laser radiation in the IR band. Fig. 4 shows the ratio between the intensities of zero phonon lines for NV^0 and NV^- color centers, which was found on the basis of the fitting curves for the normalized experimental data presented in Fig. 3b and the measured maximum line values for each of the contributions at maximum pump power. The curve clearly shows the almost linear growth of NV^0 centers with increasing power (intensity) of the exciting radiation.

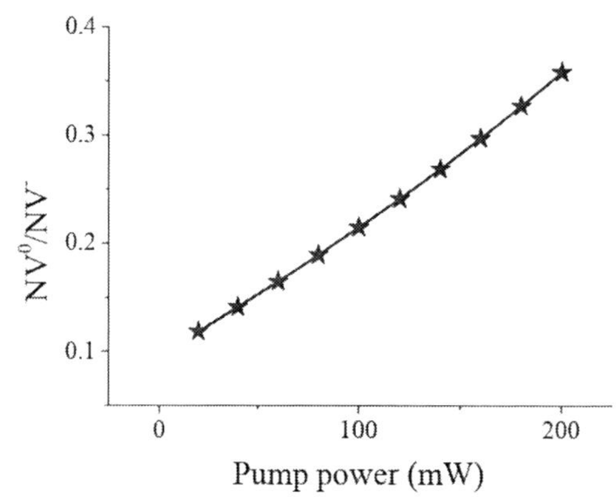

Figure 4. The ratio between the intensities of zero phonon lines for NV^0 and NV^- color centers.

4. Conclusion

In this work multiphoton excitation of NV centers was performed at different wavelengths of the exciting infrared femtosecond laser radiation. It was shown that at the wavelengths of about 810 and 1150 nm, the photoluminescence emission spectrum is mainly due to the two-photon absorption of the NV^0 and NV^- centers, respectively. However, at a wavelength of the exciting field of about 1030 nm, three-photon processes begin to dominate in the luminescence spectrum (especially in the case of NV^0), which is associated with the ionization of the center and its transition from a negative to a neutral charge state. The obtained data can be useful for the application of these centers for multiphoton microscopy of biological objects or high resolution microscopy.

Acknowledgments
The reported study was funded by RFBR according to the research project No. 19-32-80029. The study was also partially supported by the Russian Science Foundation project No. 18-72-00252 (data analysis).

References

[1] Aderhold J, [1] Kucsko G., Maurer P. C., Yao N. Y., Kubo M., Noh H. J., Lo P. K., Park H., and Lukin M. D., 2013 *Nature* **500** (7460) 54
[2] J. R. Maze et al 2008 *Nature* **455** (7213) 644
[3] Dolde F et al 2011 *Nature Phys.* **7**(5) 459
[4] Mizuochi N. et al., *Nat. Photonics*, 2012 **6** 299
[5] Han K Y, Kim S K, Eggeling C and Hell S W 2010 *Nano Lett.* **10** 3199
[6] Rittweger E., Han K. Y., Irvine S. E., Eggeling C., and Hell S. W. *Nat. Photonics* 2009 **3** 144
[7] Togan E et al 2010 *Nature* **466** 730
[8] Waldherr G, Beck J, Steiner M, Neumann P, Gali A, Frauenheim Th, Jelezko F and Wrachtrup J 2011 *Rev. Lett.* **106** 157601
[9] Aslam N., Waldherr G., Neumann P., Jelezko F., and Wrachtrup J. 2013 *New Journal of Physics* **15**(1) 013064
[10] Ji P., Balili R., Beaumariage J., Mukherjee S., Snoke D., and Dutt M. G. 2018 *Phys. Rev. B* **97**(13) 134112
[11] Alkahtani M. H., Alghannam F., Jiang, L., Almethen A., Rampersaud A. A., Brick R., ... and Hemmer P. R. 2018 *Nanophotonics* **7**(8) 1423
[12] Ivanov I. P., Li X., Dolan P. R., & Gu M. 2013 *Opt. Lett.* **38**(8) 1358

Spectral and luminescent properties of codoped Er-Yb fluorophosphates glasses with small content of phosphate

O Bogdanov[1], E Kolobkova[1,2]
[1]Department of chemical technology of silicate materials, Saint-Petersburg State Institute of Technology, Saint-Petersburg, 190013, Russia
[2] The Faculty of Photonics and Optical Information Technology, ITMO University, Saint Petersburg 197101, Russia

Abstract. Novel fluorophosphate glasses with small content of phosphate doped by various concentration of ErF_3 and 5 mol. % of YbF_3 were obtained. Absorption spectrum and experimental τ_{exp} lifetimes for all concentration of ErF_3 were measured. Radiation lifetime of the excited state τ_{exp} and emission cross-section σ_{emi} were calculated. It was shown that the highest quantum yield *q* (100%), τ_{exp} (9 ms) and σ_{emi} ($0.64x10^{-20}$ cm^{-2}) corresponds to 0.5 mol. % ErF_3.

1. Introduction

Among the numerous rare-earth doped fiber materials, the Erbium doped (Er^{3+}) fiber is of a special interest due to the strong emission band centered at 1535 nm, which coincides with the third fiber-optics communication window. An improvement frequently used in Er^{3+}-doped materials consists in the co-doping with an Yb^{3+} ions which act as a sensitizer of the active erbium ions. Since these materials behave as a three-level gain system at the wavelength λ = 1530 nm, the co-doping with Yb^{3+} enhances the pumping efficiency in a row of short-length active erbium materials employed in compact laser devices and high-power optical amplifiers [1]. Fluorophosphate and phosphate glasses are attractive host materials for RE lasers and amplifiers because of their transmittance from the ultraviolet to the infrared spectral region, their fluorescence properties, and their low non-linear refractive indices [2].

2. Experimental

Glasses with nominal composition $5Ba(PO_3)_2$-$10BaF_2$-(33-x)AlF_3-18.5CaF_3-10MgF_2-18.5SrF_2-5YbF_3-xErF_3 , где x=0.5, 1, 2, 3 mol.% were prepared by conventional melt quenching techniques in vitreous carbon crucibles. After melting without stirring at 1000°C in argon atmosphere for 1 hour, the glass liquid was cast on a vitreous carbon plate and annealed at T= for 2 hour. The quenched samples were annealed at transition temperature Tg=440°C to remove an internal stresses. The residual stress was examined by the polariscopic measuremens. Differential scanning calorimetric (DSC) measurements were performed using a Netzsch DSC apparatus on 25 mg glass pieces into Pt pans at a heating rate of

10°C* min^{-1} up to temperature of 700°C. The refractive index n(λ) was measure by Abbe refractometer and fitted using Sellmeier's equations.

Density was measured by Archimedes' principle using distilled water as an immersion liquid. Refractive index was measured on an Abbe refractometer using monobromonaphthalene as adhesive coating. UV-VIS-NIR absorption spectra of optically polished samples were measured in Perkin Elmer Lambda 900 UV/Vis/NIR spectrophotometer. The excitation source for infrared luminescence was a Nd:YAG frequency tripled laser pulse source (Solar System). Excitation wavelength at 945 nm was obtained with the passage of radiation through a parametric generator. The lifetime of the excited state was determined with Q-switched Nd:YAG laser LQ 529B by «Solar Laser» with emission wavelength 532 nm. The luminescence radiation was detected using a monochromator "Spectra Pro 300i" by «Acton Research Corporation» with InGaAs- ID-441 photodetector. The duration of the excitation pulses was about 15 ns. The luminescence decay curves were recorded by an Infinium HP54830 digital storage oscilloscope (Agilent Technologies).

3. Data analysis

The nine major bands were identified in absorption spectra as transitions from the ground energy level $^4I_{15/2}$ of Er^{3+} ions to upper ones for all glasses (Table. 1). For calculation Judd-Ofelt parameters, we need to know the experimental line strengths from initial state J to final states J', so-called S_{exp} (J-J'). S_{exp} (J-J') were calculated for each band with Eq.1[4-5]:

$$S_{ed}^{meas}(J \to J') = \frac{27hc(2J+1)n}{8\pi^3\bar{\lambda}e^2\left(n^2+2\right)^2}\int \sigma(\lambda)d\lambda - \frac{9n^2}{\left(n^2+2\right)^2}S_{md}, \quad (1)$$

where, c is the speed of light in vacuum, h is Planck's constant, *e* is the electron charge, J and J' are the quantum numbers of total angular momentum of initial and final states respectively. N corresponds to Nd^{+3} concentration (ions per cm^3), $\bar{\lambda}$ is the mean wavelength of absorption band in nanometer, n refractive index, α is absorption coefficient equals to $2.303D\times d^{-1}$, where *D* is the optical density, *d* is the absorption layer thickness.

For refractive index determination we used fitting by means of Sellmeier's equation (Eq. 2):,

$$n^2(\lambda) = 1 + B_1\lambda^2/\left(\lambda^2 - C_1\right) \quad (2)$$

where B_1 and C_1 are empirical coefficients equal to 1,035 and 6,359×10^{-3} [3].

For the transitions that comply with the selection rules $\Delta S = \Delta L = 0, \Delta J = 0, \pm 1$ there is a contribution of the magnetic-dipole transition (Eq. 1) [4, 5]. The wave functions in the intermediate-coupling approximation were used for the calculation of the oscillator strength associated with the magnetic-dipole mechanism.(Eq.4) [4]: $\boldsymbol{S_{md}}$ is the magnetic force of the line having a value other than zero, only for transitions that comply with the selection rules $\Delta S = \Delta L = 0, \Delta J = 0, \pm 1$. According to the theory of J-O calculated electric and magnetic force lines are taking as (Eq. 3, 4):

$$S_{ed}^{calc}(\psi J, \psi' J') = \sum_{k=2,4,6} \Omega_k \left|\left\langle \psi J \left\| U^{(k)} \right\| \psi' J' \right\rangle\right| \quad (3)$$

$$S_{md}(\psi J, \psi' J') = \left[\frac{h}{4\pi mc}\right]^2 \left|\left\langle \psi J \left\| L + 2S \right\| \psi J' \right\rangle\right|^2 \quad (4)$$

The values of three parameters Ωt (t = 2, 4, 6) can be obtained from absorption cross-sections by using the method of least squares fitting (Table1) According to Judd-Ofelt theory, values of three Ω_t and S_{cal} can be determined by assuming an equivalence of experimental and calculated line strengths $S_{exp} = S_{cal}$
The accuracy of the Judd-Ofelt fitting can be expressed in the term of the root mean square deviation (Δ_{rms}) with the usual Eq.5

$$\Delta rms = \left[\frac{\sum\left(S_{\exp} - S_{cal}\right)^2}{p-q}\right]^{\frac{1}{2}} \tag{5}$$

Where p is the number of absorption bands included in the Judd -Ofelt fit and q is the number of spectroscopic intensity parameters Ω_t. In this work, p and q takes values of 9 and 3 respectively. The reduced matrix elements U^k are found to be almost invariant to the environment and we have used the values given [5].
For the $^4I_{13/2}$–$^4I_{15/2}$ transition the difference in the total angular momentum is ΔJ = 1, therefore, for this transition it is necessary to take into account the magnetic-dipole contribution. The spontaneous emission probability $A(\psi J,\psi' J')$ of this transition is given by Eq.5:

$$A(\psi J,\psi' J') = \frac{64\pi^2 e^2}{3h(2J+1)\lambda^3}\frac{n(n^2+2)^2}{9}S_{ed}^{cal} + \frac{64\pi^2 e^2 n^3}{3h(2J+1)\lambda^3}S_{md} \tag{5},$$

The radiative lifetime (τ_r) of the excited metastable state was then calculated by using Eq.6:

$$\tau_{rad} = \left[\sum_{\psi' J'} A(\psi J,\psi' J')\right]^{-1} \tag{6}.$$

The equation for calculation of the stimulated emission cross-section spectrum by the Fuchtbauer-Landenburg equation (FL method) for isotropic media is:

$$\sigma_{emi} = \frac{\lambda^4 I(\lambda)}{8\pi n^2 c\tau_{\exp}\int I(\lambda)d\lambda} \tag{7}$$

Where I(λ) –luminescent spectrum, τ_{exp} - experimental life time of the excited ɔstate.

4. Results and Discussion

Experimental oscillator strength for transitions from ground state 4I15/2 to $^4I_{13/2}$, $^4I_{11/2}$, $^4I_{9/2}$, $^4F_{9/2}$, $^4S_{3/2}$, $^2H_{11/2}$ и $^4F_{7/2}$ were obtained by Eq.1. Obtained values of the Ω_t – parameters for all glasses studied are shown in Table1. According to the J-O theory, the line strength of the electric dipole components of the 1.55-μm transition is given by [6]

$$S_{ed}^{calc}(^4I_{13/2}, {^4I_{15/2}}) = \sum_{k=2,4,6}\Omega_k\left|\left\langle {^4I_{13/2}}\left\|U^{(k)}\right\| {^4I_{15/2}}\right\rangle\right| = 0.019\Omega_2 + 0.118\Omega_4 + 1.462\Omega_6 \tag{8}$$

It should be noted that among the *f–f* transitions of Er^{3+}, the radiative transition probabilities of $^4I_{13/2}\rightarrow{^4I_{15/2}}$ (1.5 μm) depend strongly on Ω_6, since the reduced matrix elements

$\langle \| U^{(2)} \| \rangle$ and $\langle \| U^{(4)} \| \rangle$ of these transitions are equal to zero or negligibly small. Parameter Ω_6 is the most sensitive to variations of the electron density.

Table 1. Comparison of Judd-Ofelt parameters of fluorophosphates glasses with various concentration of ErF_3 with other samples

Glasses	N_0, 10^{20} ion/cm^3	Ω_2, 10^{-20}	Ω_4, 10^{-20}	Ω_6, 10^{-20}	Ω_4/Ω_6	Ref.
0.5 ErF3, mol.%	0,98	3.28	1.17	1.71	0.68	This work
1 ErF3, mol.%	1,96	3.20	1.23	1.70	0.72	
2 ErF3, mol.%	3,9	2.33	0.77	1.26	0.61	
3 ErF3, mol.%	5,9	1.96	0.97	0.94	1.03	
YF3–BaF2–Ba(PO3)2	-	4.65	1.56	1.78	0.87	[8]
BBBM	-	4.03	1.34	0.53	2.52	[9]
FP 4	4	2.94	1.37	1.50	0.91	[2]

The absorption and stimulated emission cross sections of glasses with different ErF_3 concentrations are shown in Fig 1.

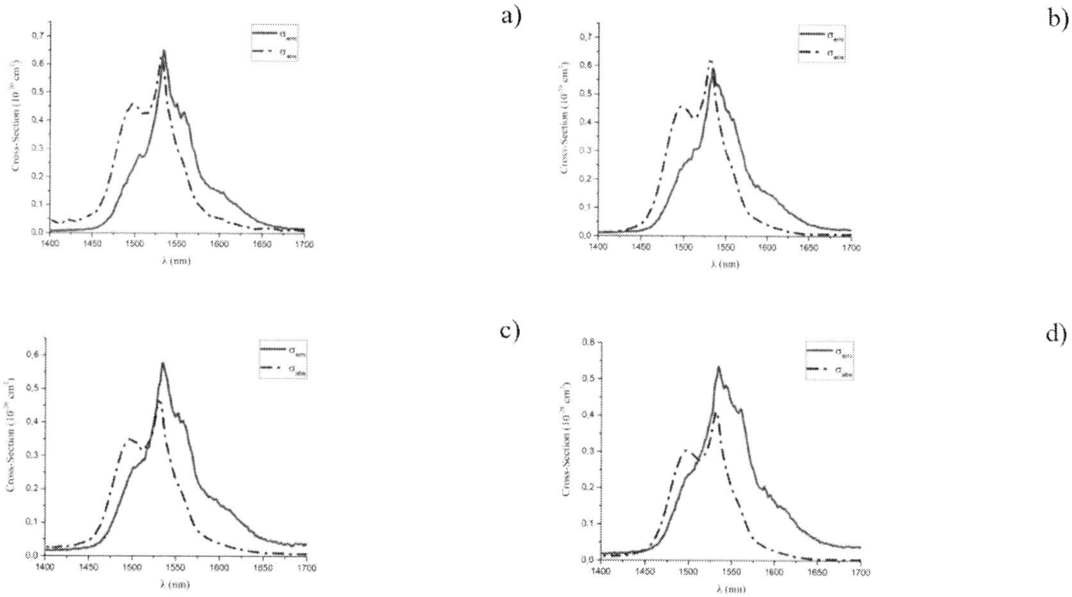

Figure 1. Cross-Section of emission and absorption of fluorophosphates glasses with various concentration of ErF_3: a) ErF_3 = 0.5 mol.%; b) ErF_3 = 1 mol.%; c) ErF_3 = 2 mol.%; d) ErF_3 = 3 mol.%.

Analysis of the results shows that the parameter Ω_6 is maximum and almost unchanged at concentrations up to 1 mol. % of erbium fluoride. It should be noted that the increase in the concentration of the activator is due to the decrease in the concentration of aluminum - which is a network former and determines the rigidity of the glass network. Thus, the change of J-O parameters indicates the influence of the structure rigidity on the spectroscopic characteristics of erbium ions

Table 2. Er^{3+} ion concentration (ion/cm³), refractive index n_D, spontaneous probabilities A (ψJ,ψ'J'), cross-section of emission σ_{emi}, radiative τ_{rad} and experimental τ_{exp} lifetimes, quantum yield q.

ErF3, mol.%	N_0, 10^{20} ion/cm^3	n_D	$^4I_{13/2}$-$^4I_{15/2}$ $\bar{\lambda}$, nm	A (ψJ,ψ'J'), s^{-1}	σ_{emi}, 10^{-20} cm^2	τ_{rad}, ms	τ_{exp}, ms	*q*, %
0.5	0.98	1.4526	1518.1	116	0.64	8.6	9	100
1	1.96	1.4530	1518.9	115.6	0.59	8.6	8.3	100
2	3.9	1.4550	1518.4	92.4	0.58	10.8	7.5	68
3	5.9	1.4660	1518.8	77.6	0.53	12.9	6	40

5. Conclusion

The longest lifetime of the excited state and the stimulated emission cross section were observed for glass with an ErF_3 concentration of $(1.0\text{-}4.0)*10^{20}cm^{-1}$. The decrease of the lifetime of the excited state and the cross section of emission were associated with the phenomenon of concentration quenching of luminescence. The quantum efficiency of the most important laser transition $^4I_{13/2}\rightarrow{}^4I_{15/2}$ ~1.54 nm of Er^{3+} in the fluorophosphates glass was found to be approximately 100%. These results indicate that the Er–Yb co-doped fluorophosphates glasses are good candidates for the fabrication of optical amplifiers and lasers.

Acknowledgement

This work was financially supported by Russian Scientific Foundation (Agreement # **19-13-00343)**

References

[1] Cruz Vicente S.G, Martınez Gamez M.A, Kir'yanov A.V, Kumar G.A 2005 *Optical Materials. 27 1563;*
[2] Ebendor-Heidepriem H, Ehrt D, Bettinelli M, Speghini A 1998 J. Non-Crystalline Solids **240** 66;
[3] E. Kolobkova, A. Alkhlef, B.M. Dinh, A.S. Yasukevich, O.P. Dernovich, N.V. Kuleshov, N. Nikonorov 2019, J. Luminescence. **206**, 523-529;
[4] A.A. Assadi, A. Herrmann, R. Lachheb, K. Damak, C. Rüssel, R. Maâlej 2016 J. of Luminescence, **176**, 212-219;
[5] W. T. Carnall, P. R. Fields, K. Rajnak 1968 J. of Chemical Physics **49**, 4424.
[6] M.J. Weber, 1967 Phys. Rev. **157** 262–272;
[7] E.W.J.L. Oomen, A.M.A. van Dongen, 1989 J. Non. Cryst. Solids. **111** 205–213;.
[8] Tassia S. Gonçalves, Raphaell J. Moreira Silva, Marcos de Oliveira Junior, Cynthia R. Ferrari, Gael Y. Poirier, Hellmut Eckert, Andrea S.S. de Camargo, 2015, Materials chemistry and physics, 1-15;
[9] Ju H. Choi, Frank G. Shi, Alfred Margaryan, Ashot Margaryan, Wytze van der Veer, 2008, J. of Alloys and Compounds, **450**, 540-545;

Simulation of infrared photonic crystal fibers based on crystals system AgBr – $TlBr_{0.46}I_{0.54}$

D. D. Salimgareev, A. A. Lashova, A. S. Shmygalev, A. E. Lvov, L. V. Zhukova
Ural Federal University named after the first President of Russia B.N. Yeltsin. Yekaterinburg, 620002, Russia.

Abstract. In this paper, a computer simulation of photonic crystal fibers (PCF) based on the AgBr – $TlBr_{0.46}I_{0.54}$ system was performed. The simulation was carried out in the SMTP program integrated into Matlab. As a simulation result, PCF with the hexagonal structure of PBG-inserts with an increased mode field diameter of up to 200 μm was obtained.

1. Introduction

Among the infrared materials suitable for the manufacture of optical fibers, monocrystals of silver halides and monovalent thallium are distinguished. One of such monocrystalline system is AgBr - $TlBr_{0.46}I_{0.54}$ crystals, on the basis of which using extrusion infrared fibers can be made that have unique optical properties: a wide transmission range from 2.0 to 25.0 μm, photo- and radiation resistance, small optical loss up to 0.5 dB/m. These properties of IR fibers based on AgBr - $TlBr_{0.46}I_{0.54}$ system crystals make them promising for use in various optical systems [1]. However, various applications require single-mode operation from optical fibers, as well as the transmission of high-power radiation with an increased mode field diameter. To solve these problems, photonic crystal fibers (PCF) of various types are used, in particular PCF with an active core (AC) and surrounding inclusions - photonic bandgaps (PBG) of optically less dense material than the main fiber material - matrix, and the central insert - AC from a more dense substance [2]. PCF AC allow transmitting high-power radiation in a single-mode with a Gaussian distribution of the mode field over the fiber section.

The manufacture of such PCFs is a long and resource-intensive process, which is carried out using a combination of “stack and drow” and “row in tube” technologies, which make it possible to obtain a clear interface and a strict structure of inserts in the fiber. In order to optimize the PCF production process, a computer simulation stage was introduced, necessary for determining the geometric parameters of the fiber and predicting its optical properties.

So computer modeling is used to preliminary determination the mode of operation of the PCF and determine the optimal geometrical parameters, the chemical composition of the matrix and inserts, and the nonlinear optical properties.

2. Materials and methods

Earlier, the authors of [3] developed photo- and radiation resistant, non-hygroscopic, plastic crystals of the AgBr-$TlBr_{0.46}I_{0.54}$ system, which don't have the effect of cleavage, from which optical products (windows, lenses) can be made by hot pressing and optical fibers for the mid-IR range are produced by extrusion. The phase diagram of the AgBr-$TlBr_{0.46}I_{0.54}$ system [3] was studied, based on which single-crystals of various compositions were grown (Fig. 1). It is established that, depending on the composition, the obtained crystals are transparent from 0.46 to 50.0 μm without absorption windows [3, 4], their refractive index varies from 2.165 to 2.389 at wavelengths from 3.0 to 14.0 μm [3 , 4], and also they possess high photo- [5] and radiation resistance [3]. Optical wares (IR fibers, windows, lenses, etc.) made of AgBr-$TlBr_{0.46}I_{0.54}$ crystals system are necessary to create fiber lasers and amplifiers, fiber-optic systems for IR spectroscopy, including in conditions of increased background radiation, low-temperature IR pyrometry, for the transfer of CO (5.3 - 6.2 μm) and CO_2 lasers (9.2 - 10.6 μm) and other applications.

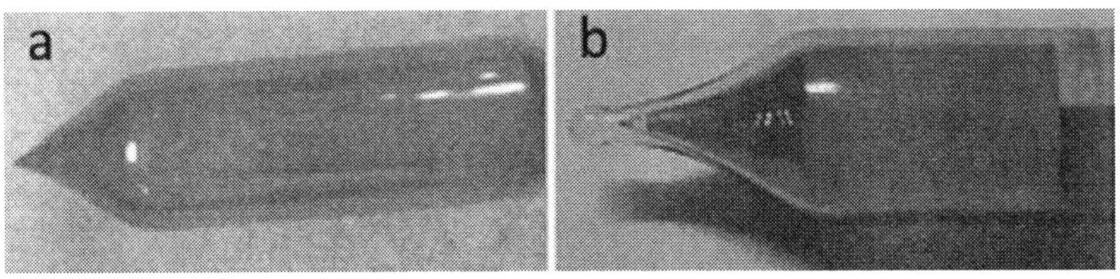

Figure 1. AgBr-$TlBr_{0.46}I_{0.54}$ system single-crystals grown by the Bridgman method: a - 3 mol. % $TlBr_{0.46}I_{0.54}$ in AgBr; b - 10 mol. % $TlBr_{0.46}I_{0.54}$ in AgBr

The fibers were modeled using the SMTP (Source Model Technique Package) software package developed by the authors [6]. This program allows to determine the number of modes passing through the fiber with high convergence of simulation results with experimental data at low computing power costs. The input data for the simulation are the geometry of the PCF, the dielectric and magnetic permeabilities of the media, the refractive indices and the wavelengths of the transmitted radiation. As a result of the simulation, the modes of transmitted radiation, their power, effective refractive indices and the distribution of electromagnetic radiation in the fiber cross section were obtained. When converting two-dimensional pictures into three-dimensional ones, the profile of the radiation distribution over the cross-section of the PCF was evaluated.

3. Experimental data

When simulating PCF based on AgBr-$TlBr_{0.46}I_{0.54}$ (KRS-5) system crystals, KRS-5 contents in silver bromide were taken from 3 to 29 wt.% and the compositions were selected so, that the matrix refractive index was greater than the PBG index, but less than the central insert index. The simulation was performed for a wavelength of λ = 10.6 μm, corresponding to the CO_2-laser emission, with a hexagonal arrangement of the photon inserts and with a central inclusion in the structure core. The fiber core was assumed to be 200 μm, the insert diameters d = 16 μm, the inter-insertion distance Λ = 100 μm. The outer diameter of the fiber was taken equal to 240 μm. In the simulation, the calculation error was set to no more than 10^{-3}, and also the calculation of both the real and imaginary parts of the effective refractive index was specified. The magnitude of the imaginary component of the refractive index was at a value of not less than 10^{-5}, since smaller values do not lead to substantial losses on mode hold.

As a result of the simulation, two-dimensional PCF models were obtained, as well as their effective refractive indices. For all the models obtained, only the valid parts of the index were calculated, which indicates insignificant loss on hold. In addition, photonic fibers with an increased mode field diameter were obtained for each composition. Thus, Figure 1 shows the obtained PCF models with a mode field diameter of up to 200 μm, operating in the single-mode, with the KRS-5 content in AgBr, presented in Table 1.

Table 1. Simulated infrared fiber compositions: KRS-5 content in AgBr in wt.%

№ PCF	PBG inserts	Matrix	Central insert	Effective refractive indices
1	5	10	21	2,192
2	10	21	29	2,203

For each of the selected compositions, single-mode modes of operation were obtained, but to estimate the intensity distribution over the cross section, these models were required to be converted into three-dimensional images (Fig. 2, 3). The obtained three-dimensional structures have a Gaussian radiation distribution in the cross section (Fig. 4) with a high transmittance of power in a continuous mode up to 30 kV / m^2.

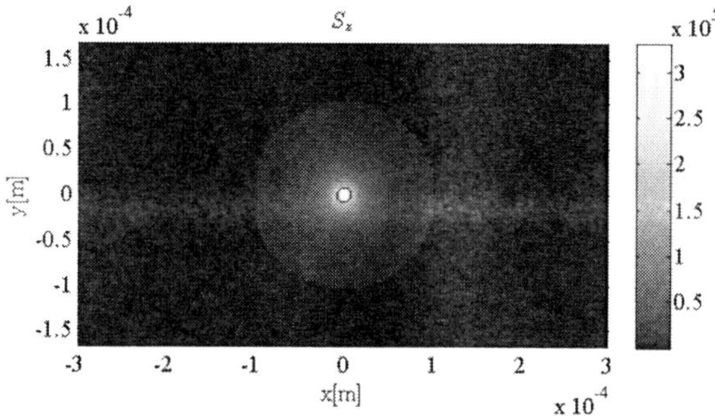

Figure 2. Simulation results of the radiation distribution in the PCF cross section for composition №1 with effective refractive index n_{eff} = 2.1919

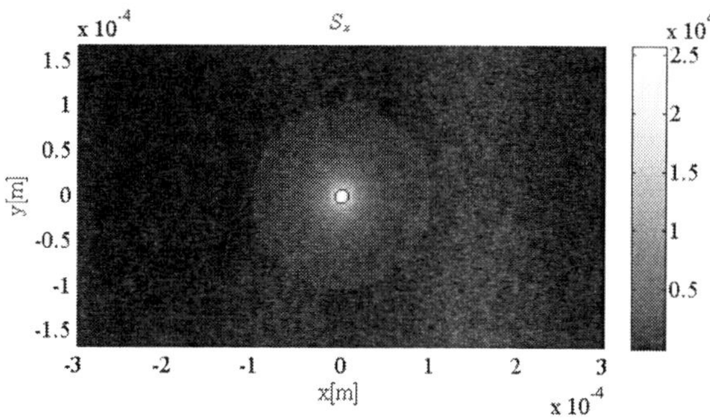

Figure 3. Simulation results of the radiation distribution in the PCF cross section for composition №2 with effective refractive index n_{eff} = 2.2025

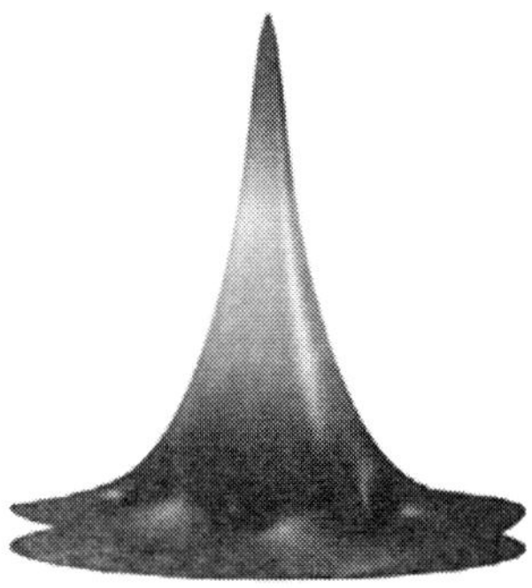

Figure 4. 3-D visualization of the radiation distribution in the PCF cross section with the composition (KRS-5 in AgBr): PBG inserts - 5 wt.%, Matrix - 10 wt.%, Central insert - 21 wt.%

Thus, on the basis of AgBr-$TlBr_{0.46}I_{0.54}$ monocrystals, photonic crystal fibers can be developed for the mid-infrared range of the spectrum, operating in single-mode with an increased mode field diameter. Modeling in the software package SMTP, integrated into the Matlab, confirmed the performance of these fibers.

4. Conclusion

Based on AgBr-$TlBr_{0.46}I_{0.54}$ system crystals, unique photonic-crystal fibers can be developed, transmitting infrared radiation from 2 to 25 μm without absorption windows with low optical losses of no more than 0.5 dB/m and a mode field diameter increased to 200 μm. The manufacture of such fibers is carried out according to the technology of stack-and-drow, which is complex and resource-intensive, and therefore computer simulation is introduced into the PCF production process, which makes it possible to carry out a preliminary assessment of the optical properties and geometrical characteristics of the future fiber. The simulation was carried out in the SMTP software package integrated into Matlab, which allows determining the transmitted radiation modes number, their intensity and effective refractive indices for various fiber configurations. Thus, the simulation allows you to effectively design a PCF structure with the required optical and geometric properties without additional material and time costs. As a result of the simulation, photonic-crystal fibers of two compositions were obtained, each of which provided a single-mode mode of operation at a wavelength of 10.6 μm and a Gaussian radiation distribution over the entire fiber cross section at a maximum radiation power of 30 kW / m^2.

Acknowledgments
This work was supported by the Russian Science Foundation under grant No. 18-73-10063

References

[1] Zhukova L, Korsakov A and Salimgareev D 2015 *Infrared crystals: theory and practice* (Ekaterinburg: Ural federal university) p 215
[2] Broeng D 1999 *Opt. Fiber Technol.* **5** 305
[3] Zhukova L, Lvov A, Korsakov A, Salimgareev D, Korsakov V 2018 *Optics and Spectroscopy* **125** 933
[4] Korsakov A, Salimgareev D, Lvov A, Zhukova L 2017 *Optics and Laser Technology* **93** 18
[5] Korsakov A, Salimgareev D, Lvov A, Zhukova L 2016 *Optical Materials* **62** 534
[6] Hochman A, Leviatan Y 2004 *J. Opt. Soc. Am. A.* **21-6.** 1073

Dynamics of cluster structures in the near-surface layer of distilled water under laser excitation

V I Shostka[1], N V Shostka[2]

[1]Physics and Technology Institute, V.I. Vernadsky Crimean Federal University, Simferopol 295007, Russia
[2]Department of the organization of research work of students and competitions, V.I. Vernadsky Crimean Federal University, Simferopol 295007, Russia

Abstract. We explore changes in the distilled water structure by optical methods in this work. The lack of rigorous ideas about the structure of water and its organization at the molecular level hinders the development of methods not only for quantitative analysis, but also for qualitative assessment of the structure of water. Optical methods of studying changes in the structure of water are used in present work, since it is a disordered fluid whose hydrogen bonds are short-lived. The mechanism for the formation of cluster structures near the surface of a water drop is proposed. It is shown that water has unique properties associated with the presence of order in its various states due to the cooperative effect of structural and dynamic changes in a heterogeneous hydrogen bond network.

1. Introduction

In recent years the interest paid on the structure of water not only from biologists, chemists, or physicians, but also from physicists has been growing. Water as a widespread solvent has unusual properties. Due to intermolecular hydrogen bonds it has different properties from other hydrides and is structurally a very sensitive system with a huge number of metastable states. The question of observation water and aqueous solutions structure has already moved to the plane of studying the effect of changes in the structure of water and aqueous solutions on various technological processes, as well as on the level of human health. The lack of rigorous ideas about the structure of water and its organization at the molecular level hinders the development of methods not only for quantitative analysis, but also for qualitative assessment of water structure.

Unique properties of water structure can be explained by the presence of order in its various states, which is a consequence of the cooperative effect of the structurally changing inhomogeneous network of hydrogen bonds. As a result, the structure of water can be associated with the presence of hydrogen bonds and the formation of hydrated ions that form a complex three-dimensional network. The number of possible ways to connect tetrahedral water molecules with each other and stable configurations based on them is enormous, and therefore non-stationary processes lead to the formation of clusters. Individual molecules connected by hydrogen bonds are combined into associates or clusters, the size of which depends on different conditions, primarily on temperature [1].

One of the important properties of the surface layer of water is a change in the transmission and refraction coefficients of laser radiation due to which it is possible to "see" clusters, determine their size and number [2]. It is believed that the lifetime of stable nanoclusters in the bulk phase of water does not exceed 0.01 nanoseconds [3]. Getting from the bulk phase to the near-surface one short-lived clusters stabilize in it, and their lifetime increases significantly. Stabilization of clusters in the near-surface layer of water is associated with a high structural ordering of the layer. As the distance from the interface into the liquid increases, the degree of order decreases. And vice versa, at some distance from the surface the order corresponds to the state of the liquid in the bulk phase.

As we mentioned above, water and aqueous solutions are highly sensitive to external influences, especially to the influence of temperature. Therefore, nondestructive research methods are necessary for their adequate study.

The purpose of this work is to explore changes in the structure of water as an unordered liquid by optical methods. Water despite the fact that its hydrogen bonds are short-lived is capable of self-organization due to the formation of new hydrogen bonds and cooperative effects.

2. Experimental part

A rather simple realizable laser interferometry method based on the use of amplitude-phase modulation of an extended laser beam passed through a layer of water droplets, which allows us to study the formation and changes of a non-uniform structure in the surface water layer, is considered in proposed work [4-6].

The scheme of the experimental setup for visualizing the fractal-cluster structure of inhomogeneities of double-distilled water in laser radiation is shown in Fig. 1.

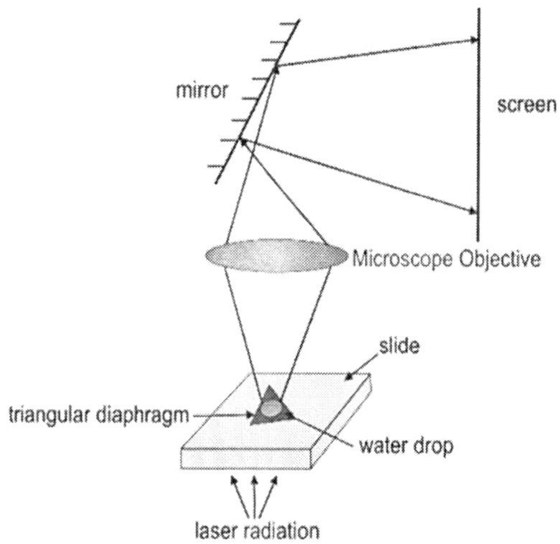

Figure 1. Scheme of experimental setup for visualization of fractal-cluster structure of double distilled water in laser radiation [4].

As a result, for the first time by the amplitude-phase laser interferometric method the water structures were fixed and models of cluster formation in the surface layer of double distilled water were proposed, which are given below:

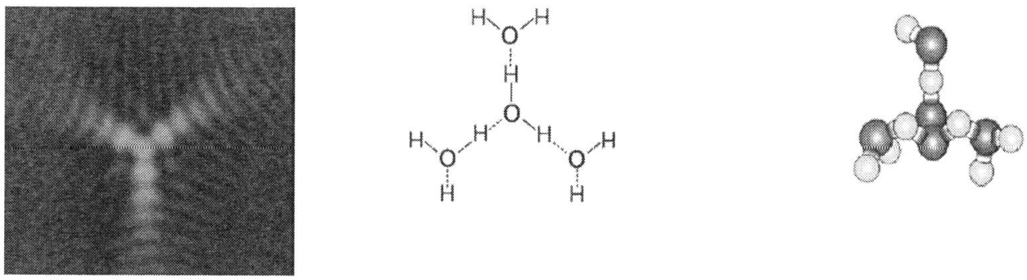

Figure 2. Structure and models of water clusters (associates) $H_{11}O_5$.

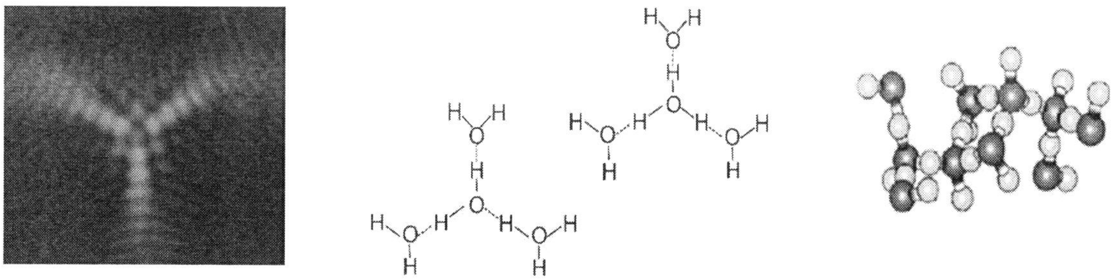

Figure 3. Structure and models of water clusters (associates) $H_{21}O_{10}$.

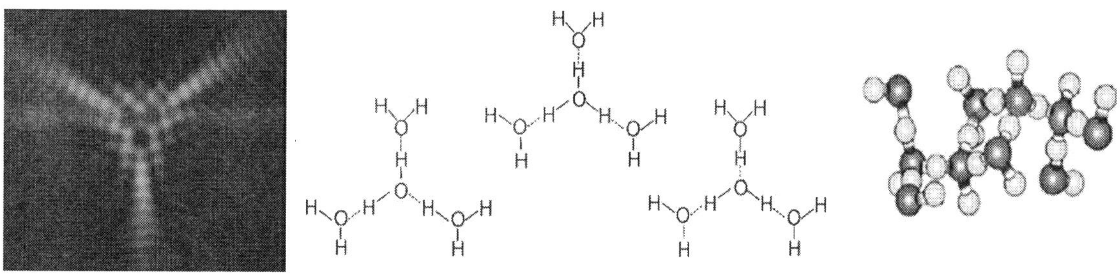

Figure 4. Structure and models of water clusters (associates) $H_{29}O_{14}$.

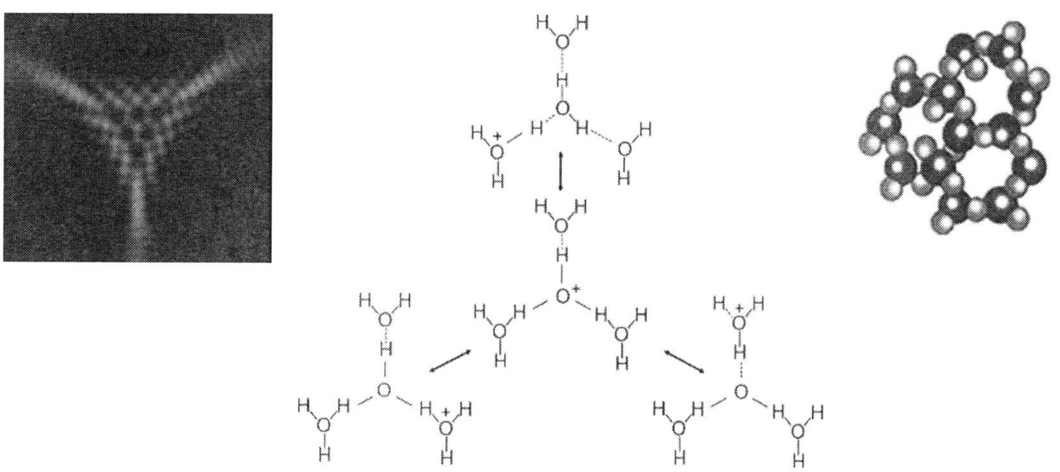

Figure 5. Structure and croquet mechanism of cluster formation in the surface layer of double distilled water.

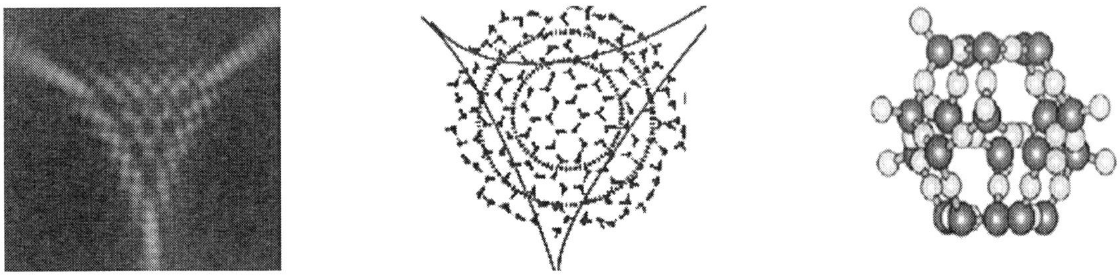

Figure 6. Structure and models of $H_{41}O_{20}$-endo-H-fullerene clusters in the near-surface layer of double distilled water.

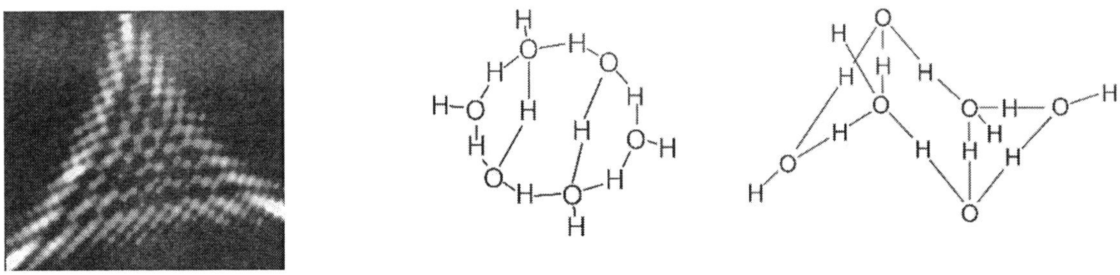

Figure 7. Cyclic forms of the structure and hexamer model of double distilled water with two intracyclic hydrogen bonds.

3. Conclusion

The obtained structures of the clusters of double distilled water rather convincingly testify in favor of the fact that stable polymeric states are present in liquid water.

It was confirmed, that water has unique properties that are associated with the presence of order in its various cluster states due to the cooperative effect of structural and dynamic changes in a heterogeneous network of hydrogen bonds. Clusters can have a different number of molecules that form associates of various shapes.

The results of this study can serve as the basis for new, simple to implement laser methods for quickly determining the structure of water.

Acknowledgments

The reported research was funded by RFBR and Council of Ministers of the Republic of Crimea according to the research project № 19-42-910010.

References

[1] Roldugin V I 2003 Fractal structures in disperse systems *Advances in chemistry* **72** (10) 931 – 959

[2] Polish A K, Pershina E D at al 2013 Modern representations on the structure and conductivity of water *Scientific and technical news* [in Russian – Naukovo - tekhníchní vístí] **13** (3) 24 – 40

[3] Zenin S V 1994 Hydrophobic model of the structure of water molecule associates *Journal of physical chemistry* **68** 634 – 641

[4] Lapaeva S N, Vershitskiy V I, Shostka V I 2004 Vortical structure of wave caustics *Proceedings of SPIE - The International Society for Optical Engineering* **5582** (26) 187-195

[5] Shostka V I, Shostka N V, Donenko I L 2017 Formation of cluster structures in water *young people's View on the problems of the regional economy – 2017.* (Materials of the all-Russian open competition of University students and young researchers, Tambov) 41 – 44

[6] Shostka V I, Shostka N V, Donenko I L 2017 Visualization of the cluster water structure with

laser radiation *Journal of Physics and Technology Institute of V.I. Vernadsky Crimean Federal University* [in Russian – Vestnik fiziko-tekhnicheskogo instituta] **1** (3) 44-52.

Superconducting photon counter for nanophotonics applications

N N Manova[1], Eu O Smirnov[1], Yu P Korneeva[1] A A Korneev[1,2], G N Goltsman[1,2]
[1] Institute of Physics, Technology, and Information Systems, Moscow State Pedagogical University, Moscow 119991, Russia
[2] National Research University Higher School of Economics, Moscow 101000, Russia

Abstract. We develop large area superconducting single-photon detector SSPD with a micron-wide strip suitable for free-space coupling or packaging with multi-mode optical fibres. The detector sensitive area is 20 μm in diameter. In near infrared (1330 nm wavelength) our SSPD exhibits above 30% detection efficiency with low dark counts and 45 ps timing jitter.

1. Introduction

One of the prospective detectors for nanophotonics and quantum optics is superconducting single-photon detector SSPD which is typically made of highly-disordered superconducting films of such materials as NbN, NbTiN, WSi, MoSi. These detectors became a device of choice in many optical applications ranging from non-invasive semiconductor integrated circuits debug to the study of single-photon sources, as well as quantum computing and quantum cryptography [1].

SSPDs are made from 3-to-5-nm-thick superconducting film in a shape of a strip, and are operated at a temperature which is well below the critical temperature but with the current close to the critical current of the strip I_c. For a long time the operation of the detector was explained by a 'geometrical' models [2] which predicted that for improved photon detection efficiency the strip should be comparable with the normal hot-spot produced by the absorbed photon, and this width is always smaller than 100 nm. For practical applications SSPDs are made as a meandering strip covering an area of about 100 μm^2 (typically as square of about 10x10 μm^2, or a circle with the diameter of 10 μm) which is sufficient for a good geometrical coupling to single-mode telecom optical fibres, such as Corning SM-28.

Meanwhile, practical quantum optics and photonics stimulate further development of advanced fast and efficient photon counters which are suitable for free-scape coupling or packaging with multi-mode fibres. The traditional approach consists in increasing the sensitive area of SSPD covered by narrow meandering strip [3,4]. Such treatment contains two disadvantages: (1) the strip length increases up to several millimeters that leads to a higher probability of defects, constrictions and weak links, which restricts critical current and reduces the yield of good devices, and (2) the long strip has a large kinetic inductance restricting the dead time to hundreds of nanoseconds time scale [3,4]. There are also a more complicated approaches such as connection of several meandering strips in parallel, or using lenses integrated with the multy-mode fibre to focus the light on smaller area SSPD. The first approach is hampered by the lack of scalability: strips connected in parallel are operated in either avalanche regime near I_c, or arm-trigger regime at lower current as described by Marsili et al [5]. The practical applications

a)

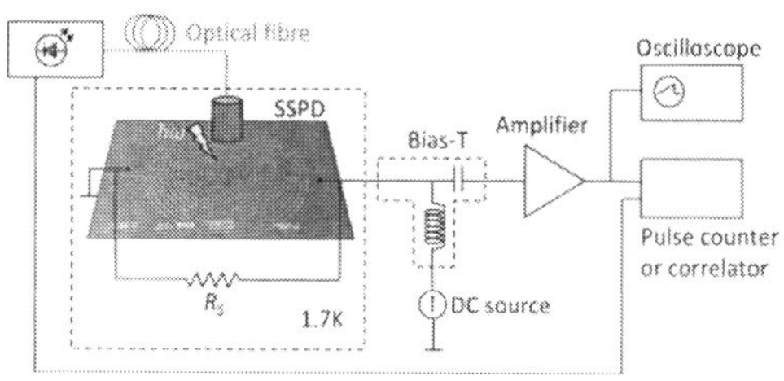

b)

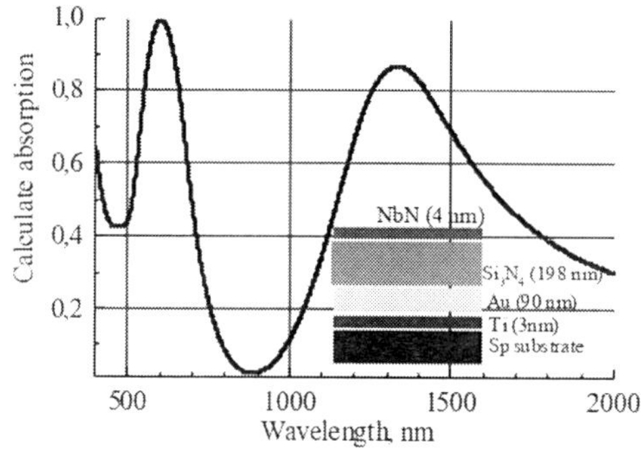

Figure 1. (a) Experimental setup, and design of the detector as a 20-μm-diameter double Archimedean spiral with the strip width of 1 μm. (b) The inset presents the layers of the $\lambda/4$ optical cavity which the detector is integrated in, and calculated absorption spectrum of such a cavity.

require operation in avalanche regime only, thus limiting the number of parallel strips to 3-4. The latter approach with fibre-integrated microlenses is much more vulnerable and is seldom used in practice.

Recently Vodolazov et al [6] theoretically predicted single-photon detection mechanism which is independent of the strip width. According to it, if the current in the strip is sufficiently close to the depairing current, the superconducting vortex-antivortex pairs, which appear in the hot-spot after photon absorption, start moving. Then Joule heating produced by the moving vortices drive a portion of the strip to the normal state. Ref. [6] predicts that if bias current is higher than 0.7 to 0.8 of the depairing current, and is uniform over the cross-section of the strip, the absorption of a photon leads to the destruction of the superconducting state independently of the strip width. Typically, superconductivity in a strip is destroyed at bias currents much smaller than the depairing current due to defects of the film (weak links with reduced superconducting gap Δ) or due to movement of vortices. Earlier, we have demonstrated single-photon response on 'bottle-neck' type bridges with gradually changing width [7], and in straight strips (will be published elsewhere) with the width 1 μm and wider.

Here we present our further progress towards large-area practical detector. As a first step we focused on 1-μm-wide strips as sufficiently wide from one hand, and on the other hand, demonstrating good sensitivity in visible and infrared. In addition, we abandoned meandering structures, as the bends of the strip lead to a decrease of the critical current due to current-crowding effect [8], and potentially prevent achievement of the minimum detection current as described by Vodolazov et al [6]. Therefore, for large area detectors, it is reasonable to choose the topology of the sensing element in the form of a spiral. Such spirals with about 100-nm-wide strip were proved to exhibit a significantly reduced current-crowding effect [9,10]. To achieve high optical coupling and high detection efficiency we make the detector as a 20-μm-diameter double Archimedean spiral, as shown in Fig 1a, with the strip width being equal to 1 μm. Here we present optical and timing characterization of such a detector.

2. Experimental methods and results

Our samples are made from NbN film on sapphire substrate with $\lambda/4$ optical cavity layers consisting of 80 nm thick-Au and 198 nm-thick Si_3N_4 layers as shown in the inset of Fig1b. This cavity is optimized for increased absorption in wavelength range 1300-1600 nm. Figure 1b presents the calculated absorption spectrum of the cavity. The films are deposited by DC reactive magnetron sputtering of Nb target in nitrogen atmosphere. We varied substrate temperature, deposition rate and partial pressure of N_2 to achieve maximum critical current in 1-μm-wide strips.

The film thickness d is determined from the film deposition rate and deposition time. The critical temperature T_c is determined from the midpoint of the resistivity transition curve (the temperature at which the resistance is the half of the resistance in normal state at 20K temperature). The resistivity ρ of the films was derived from the sheet resistance and film thickness. The diffusion constant D is experimentally determined from the temperature dependence of the second critical magnetic field Hc_2. The density of states at Fermi level N_0 as well as coherence length ξ(0) at zero temperature, are calculated from resistivity ρ and diffusivity D, which was extracted from dependence on critical temperature vs magnetic field measurements. The parameters of the devices are summarized in Tables I and II.

Table I. Superconducting parameters of studied samples. d- thickness film calculated from speed and deposition time, Tc is the critical temperature determined from the midpoint of the resistive transition, Rs(300K) and Rs(20K) are resistivity at T=300 K and T=20K, respectively. D-the diffusion constant is extract from the temperature dependence of the second critical magnetic field, ξ(0) – zero-temperature coherence length.

Sample ID	d nm	Tc K	Rs (300K) Ω/sq	Rs (20K) Ω/sq	k	D m^2/s	ξ(0) nm
2037#6	7.5	7.5	695	1177	0.59	0.46	5.16
2039#4	7.5	9.1	514	775	0.68	0.41	4.4
1762#1	5.4	8.58	453	559	0.81	0.2	3.18
1975#3	4.8	8.5	579	772	0.75	0.2	3.2
2115#3	5.7	8.4	625	961	0.65	0.46	4.88

Table II. Different currents of studied samples. Ic(4.2 K) and Ic(1.7K) are the critical current measured without shunt resistor at the T=4.2 K and T=1.7 K. Idep(4.2K) and Idep(1.7K) are the calculated depairing currents at the T=4.2 K and T=1.7 K using Eq.2 from [7].

Sample ID	Ic (4.2 K) μA	Ic(1.7 K) μA	Idep (0 K) μA	Idep (4.2 K) μA	Ic/Idep (4.2 K)	Idep (1.7 K) μA	Ic/Idep (1.7 K)
1975#3	101.2	140	244	161	0.63	230	0.61
2039#4	167	228	305	214	0.78	293	0.78
2115#3	95	146	230	150	0.63	216	0.65

Then from the optimized films we patterned spiral-shaped detectors and studied their performance: detection efficiency, dark count rate, and timing jitter. Our experimental setup is presented in Fig. 1(a). Detector under study is cooled down to 1.7 K temperature in a special insert for storage Dewar. The light from the pulsed picosecond laser is delivered by the optical fibre. Detector is DC biased from a home-made DC source through the bias-T. The shunt resistor R_S enables the recovery of the superconductivity after photon absorption, the value of the shunt resistor is adjusted experimentally and usually is in range from 1 to several Ohms. Amplified photon-count pulses are fed either to pulse counter (for dark count rate and detection efficiency measurements) or to time correlator board for jitter measurement.

Figure 2a shows the voltage transient of the photoresponse electrical pulses recorded with the digital oscilloscope. The falling edge exhibits clear exponential decay due to kinetic inductance L_k of the sample. From the geometry of the spiral and sheet resistance of the film we estimate the total kinetic inductance to be about 23 nH which is much smaller compared to the traditional meandering SSPD with 100-nm-wide strip. Nevertheless, the decay time is not as short as could be expected due to the shunt resistor connected in parallel giving the characteristic fall time (a time during which the voltage drops

a) b)

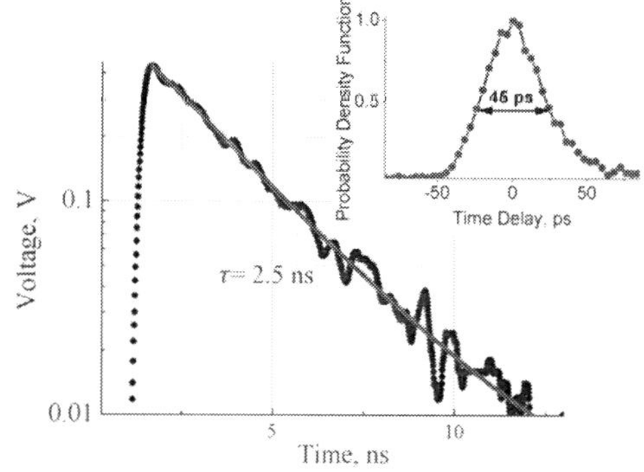

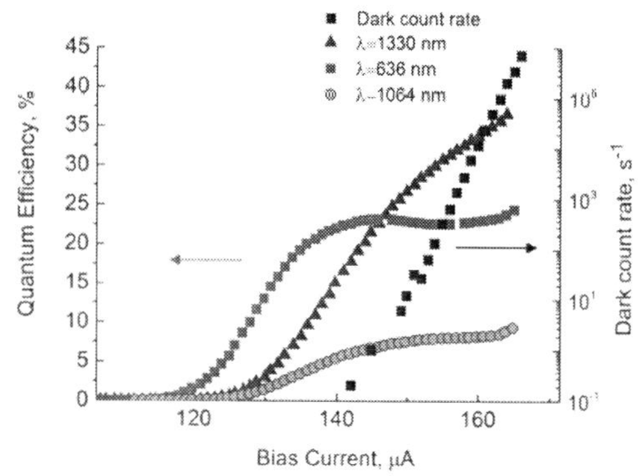

Figure 2. (a) The voltage transient of the detector, the clear exponential decay of the falling edge is due kinetic inductance. The characteristic decay time corresponds to the *e*-times (*e* is the base of the natural logarithm) decay of the falling edge. The inset shows the jitter, which is 45 ps. (b) Quantum efficiency measured at wavelengths of 636, 1064, and 1330 nm, and dark counts rate. The maximum quantum efficiencies correspond to the absorption curve of the optical cavity shown in Fig. 1b.

by a factor of *e*, where *e* is the base of the natural logarithm) $\tau = L_k / R_{shunt} = 2.5$ ns. The inset of Fig. 2a shows the probability density function of photon detection events measured with the correlator board and the laser producing 5-ps-long optical pulses at 1064 nm wavelength. The width of this function represents timing jitter of the detector, which is 45 ps and is limited by the electronics readout.

Figure 2(b) presents quantum efficiency (which is the probability of photon detection, also known as detection efficiency) of the detector at three wavelengths: 636, 1064, and 1330 nm. The quantum efficiency at 636 and 1330 nm clearly demonstrates saturation at high bias currents which hints that the maximum internal detection efficiency is achieved (i.e. each absorbed photon produces a 'click'). The non-monotonous behavior of the maximum quantum efficiency is due to the integration of the detector with the optical cavity. Although the ratios of quantum efficiencies at different wavelength correspond reasonably to the ratios of absorption according to the simulation in Fig 1b, the origin of low absolute value of quantum efficiency, which is 35%, is not clear.

3. Conclusion

We have demonstrated a novel approach to the development of large-area superconducting single-photon detector SSPD. We have made the detector as a double spiral of 1-μm-long superconducting strip covering a 20-μm-diameter circle. These detectors exhibit response time in nanoseconds scale with 45 ps jitter and 35% quantum efficiency.

Acknowledgments
This work is supported by the Russian Foundation for Basic Research grant No. 18-29-20100.

References
[1] C.M Natarjan, M.G. Tanner, R.H. Hadfield 2012 *Supercond. Sci. Technol.* **25** 063001
[2] A. Engel, J.J. Renema, K.Ilin, A.Semenov 2015 *Supercond. Sci. Technol.* **28** 114003
[3] C.L. Lv, H. Zhou, H. Li, L.X. You, Y.Wang, W.J. Zhang, S.J. Chen, Z. Wang, X.M. Xie 2017 *Supercond. Sci. Technol.* **30** 115018

[4] G. Min, Z. La-Bao, K. Lin, Z. Qing-Yuan, J. Tao, W. Chao, X. Rui-Ying, Y. Xiao-Zhong, W. Pei-Heng, Z. Yong, X. Jin-Song 2015 *Chin. Phys.B* . **24** 068501
[5] F. Marsil, F. Najafi, E. Dauler, F. Bellei, X. Hu, M. Csete, R. J. Molnar, K. Berggren 2011 *Nano Lett. Technol.* **11, 5** 2048
[6] D. Yu. Vodolazov 2017 *Phys. Rev. Appl.* **7** 034014
[7] Yu. Korneeva, D. Yu. Vodolazov, A. V. Semenov, I. Florya, N. Simonov, E. Baeva, A. A. Korneev, G. N. Goltsman, T. M. Klapwijk 2018 *Phys. Rev. Applied.* **9** 064037
[8] J. Clem, K. Berggren, 2011 *Phys. Rev. B.* **84** 174510
[9] A. Semenov, I. Charaev, R. Lusche, K. Ilin, M. Siegel, H.-W. Hubers, N. Bralovic, K. Dopf, D. Vodolazov 2015 *Phys. Rev. B.* **92** 1745118
[10] D. Henrich, L. Rehm, S. Dörner, M. Hofherr, K. Il'in, A. Semenov, M. Siegel 2013 *IEEE Trans. on Appl. Supercond*, **23(3)**, 220405

Second harmonic generation and charge relaxation of poled glasses

I V Reshetov [2], **V P Kaasik** [1], **A A Lipovskii** [1,2], **D K Tagantsev** [2]
[1]St. Petersburg Academic University RAS, St. Petersburg 194021, Russia
[2]Peter the Great St. Petersburg Polytechnic University, St. Petersburg 195251, Russia

Abstract. We measured thermal relaxation of the second order nonlinearity (SON) observed in a thermally-poled soda-lime silicate glass. It is shown that 50-min annealing of the glass at 400°C results in a complete SON degradation. Performed studies of thermo-stimulated depolarization current (TSDC) revealed that the poling-induced spatial electric charge relaxes above the glass transition temperature. This indicates that SON is not induced by this spatial charge.

1. Introduction

Studies of thermally-poled multicomponent silicate glasses are of interest due to the appearance of second order nonlinear optical phenomena, such as the second harmonic generation (SHG) after poling [1], contrary to virgin glasses. These effects could find applications in micro- and nanophotonics, integrated optics and others. It is a common point of view that the formation of a spatial electric charge in the course of the thermal poling and "frozen" electric filed generated by this charge are responsible for the SHG [2]. In our previous studies, it has been found that the relaxation of the most of spatial electric charge, which was formed in the glass by poling, takes place above the glass transition temperature [3]. Here we present the data on the SHG relaxation and the spatial charge relaxation based on thermally stimulated depolarization current (TSDC) spectra [4]. These provide the information about the relation of the processes occurring in the course of the thermal poling and the origin of the optical nonlinearity observed in thermally-poled glasses. In this paper, we show that the spatial charge relaxation and the observed SHG relaxation are independent processes.

2. Experimental

A soda-lime glass slides "Menzel" purchased from Agar Scientific were used in this research. The glass composition is presented in Table 1.

Table 1. Composition of Menzel glass in wt % of oxides [5].

SiO_2	Al_2O_3	Na_2O	K_2O	MgO	CaO	others
72.2	1.2	14.3	1.2	4.3	6.4	0.33

The polarization of 1 cm thick glass slides was performed at 300 °C for 50 min under DC electric voltage of 1 kV. After the poling, the samples were cooled down to room temperature under the voltage applied and voltage was off. Experimental layout for thermal poling is shown in fig. 1. We used stainless steel electrodes with 10x20 mm^2 area pressed to the glass surface.

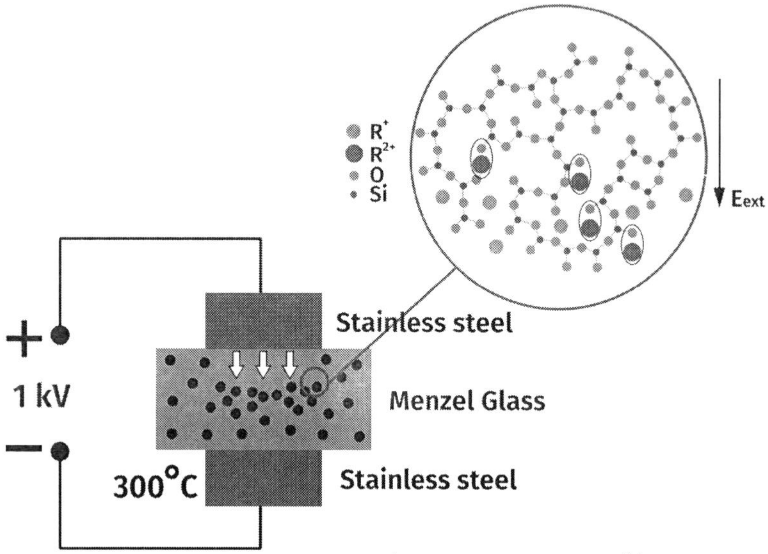

Figure 1. Experimental layout for thermal poling: R^+ - univalent cation, R^{2+} - bivalent cation, O – oxygen, Si – silicon, E_{ext} – external electric field.

Under the voltage applied, thermally activated alkali and alkali-earth cations move toward cathode from the anodic side of the glass. This results in the formation of a spatial charge, which electric field finally compensates the external field generated by the electrodes, and, as the result, the current through the glass drops. This process is characterized by the glass polarization curve (fig.2). After cooling under the voltage applied the spatial charge became "frozen" and does not relax because of the kinetic restrictions.

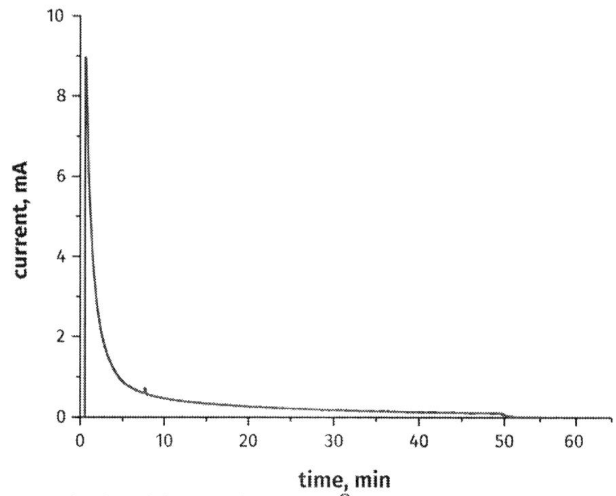

Figure 2. Polarization curve obtained from 1 kV, 300 ^{O}C poling.

The magnitude of SHG in the poled glass sample was evaluated using Maker fringes technique [6]. The experimental setup used to characterize the SHG is shown in fig.3. After measuring the Maker curve using 1064 nm wavelength laser generating 10 nm pulses, we fixed the angle of the laser beam incidence at the sample equal to 63 ^{O}, which provided maximal SHG signal magnitude. Then we heated this sample to 400 ^{O}C and measured time dependence of the SHG signal, that is, its relaxation (see fig. 4(a)). As seen, the SHG signal has almost completely relaxed in ~50 minutes and dropped down to the SHG signal generated by the surface of non-poled glasses.

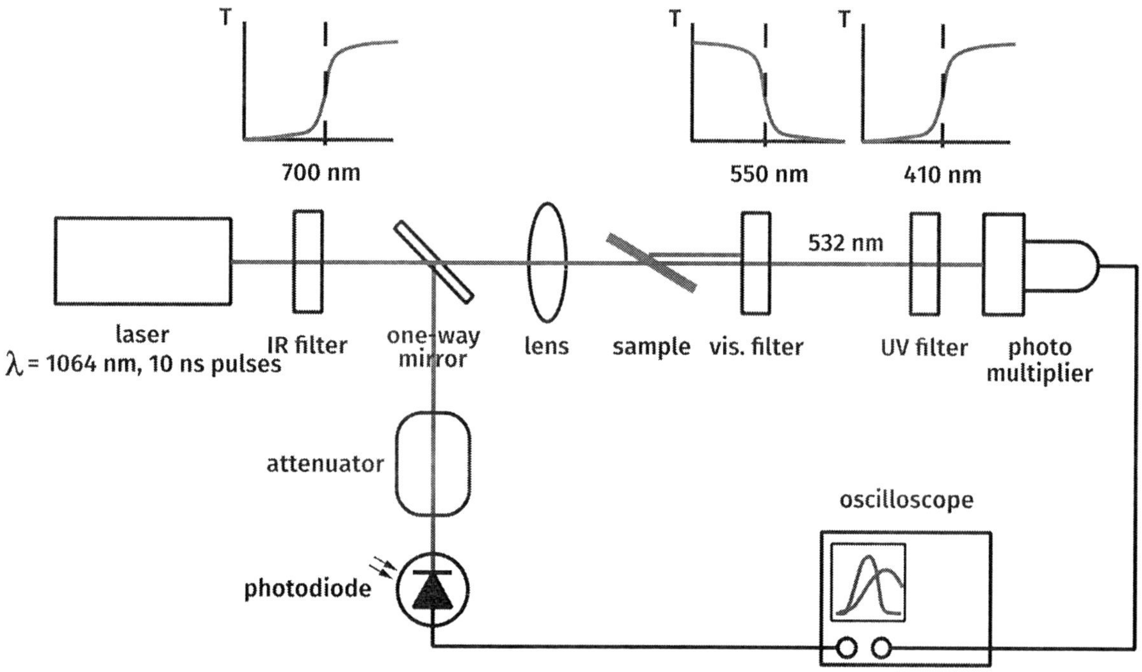

Figure 3. Experimental set-up used for the SHG measurements. The spectra above the filters represent their transmittance.

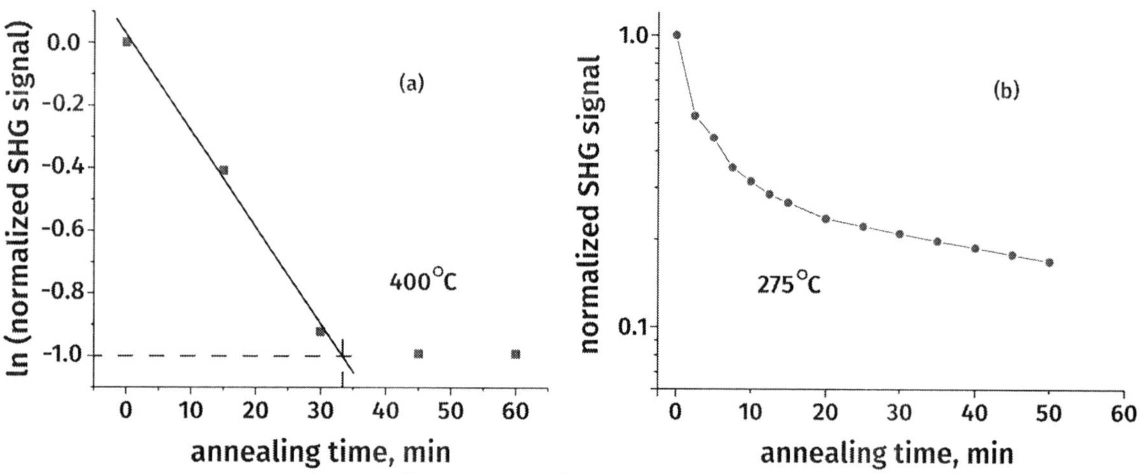

Figure 4. SHG relaxation: (a) at 400 °C, (b) at 275 °C.

The study of the SHG relaxation at temperature of 400 °C have shown that the decay of the SHG signal obeys the first-order kinetics (fig 4 a), with relaxation time being equal to about 34 minutes.

Similar measurements performed with the sample under annealing at lower temperature, 275 °C, have shown that the process of the SHG relaxation cannot be described by one relaxer only (see fig. 4(b)). In the relaxation at 275 °C, contrary to the anneal at 400 °C, one can observe that there is a process, which cannot be registered at 400 °C, for at this temperature it is much faster.

To characterize the spatial charge relaxation (glass depolarization when the shifted positive charge carriers move back towards the anodic side of the glass and compensate negative charges formed in the poling process) we have performed TSDC characterization of the poled and then annealed at 400 °C glass sample. The schematic of the measurements and the measured TSDC are shown in fig. 5. The TSDC was recorded at heating rate of 10 °/min. We used graphite electrodes pressed to the glass surface.

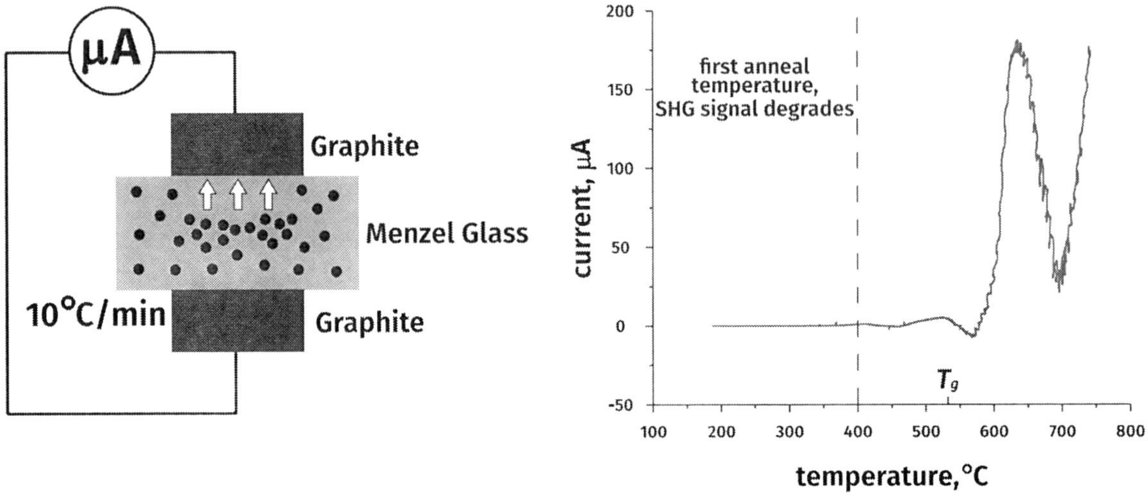

Figure 5. Schematic of the TSDC measurements (a) and the measured TSDC (b).

3. Discussion

The TSDC of the annealed at 400 ^{o}C glass sample, where the SHG has completely degraded, reveals the spatial charge relaxation peak at 640 ^{o}C and then an increase in the relaxation current, which starts at ~700 ^{o}C. This indicates that the SHG in the poled glass is not directly related to the spatial charge, which relaxes under the secondary heating of the glass above the glass transition temperature (530 ^{o}C [5]).

4. Conclusion

Our experiments have shown that 50-min annealing at 400 ^{o}C results in the disappearance of SHG in the poled soda-lime glass sample with the SHG relaxation time of ~34 min. The TSDC through this sample shows the independence of the poling-induced spatial charge relaxation and the second harmonic generation relaxation. These relaxation processes occur at essentially different temperatures due to a difference in their relaxation times.

Acknowledgments

This work was supported by Russian Science Foundation, grant #16-12-10044-P.

References

[1] Okada A, Ishii K, Mito K, Sasaki K 1992 *Appl. Phys. Lett.* **60** 2853
[2] Nasu H, Suzuki Y, Ohta H, Yamamoto Y, Ito Y, Hashimoto T, Kamiya K, Narasaki A, Tanaka K, Hirao K 2000 *Japan. J. Appl. Phys.* **39** 6530
[3] Lipovskii A, Morozova A, Tagantsev D 2016 *J. Phys. Chem.* **120** 23129
[4] Menczel J, Prime R 2014 *Thermal Analysis of Polymers: Fundamentals and Applications* (New Jersey: John Wiley & Sons) p 696
[5] http://www.agarscientific.com/microscope-slides.html
[6] Maker P, Terhune R, Nesenoff M, Savage C 1962 *Phys. Review Letters* **8** 21

Growth and spectroscopy of orthorombic Yb:KY(MoO_4)$_2$ laser crystal with a layered structure

A Volokitina[1], P Loiko[1], J M Serres[2], X Mateos[2], N Kuleshov[3], V Trifonov[4], and A Pavlyuk[4]
[1]ITMO University, 49 Kronverkskiy Pr., 197101 St. Petersburg, Russia
[2]Física i Cristal·lografia de Materials i Nanomaterials (FiCMA-FiCNA)-EMaS, Dept. Química Física i Inòrganica, Universitat Rovira i Virgili (URV), Campus Sescelades, E-43007 Tarragona, Spain
[3]Center for Optical Materials and Technologies (COMT), Belarusian National Technical University, 65/17 Nezavisimosti Ave., 220013 Minsk, Belarus
[4]A.V. Nikolaev Institute of Inorganic Chemistry, Siberian Branch of Russian Academy of Sciences, 3 Lavrentyev Ave., Novosibirsk 630090, Russia

Abstract. We report on the growth, structure, vibronic and spectroscopic properties of novel Yb^{3+}-doped orthorhombic potassium yttrium double molybdate crystal, Yb:KY(MoO_4)$_2$. The layered structure of this material determines a strong polarization-anisotropy of absorption and emission bands of Yb^{3+} ions. The maximum stimulated-emission cross-section is 3.7×10^{-20} cm^2 at ~1.01 μm for light polarization $\boldsymbol{E} \parallel \boldsymbol{b}$ and the lifetime of the $^2F_{5/2}$ state is 0.46 ms. Naturally cleaved Yb:KY(MoO_4)$_2$ plates are suitable for microchip lasers at ~1 μm.

1. Introduction

Among the oxide host crystals, potassium (rare-earth) double tungstates (DTs) and double molybdates (DMos) with a general chemical formula of KLn(XO_4)$_2$ (where X = W or Mo, respectively, and Ln = Y, Gd, Lu, etc.) are attractive for doping with laser-active trivalent rare-earth ions (RE^{3+}) [1,2] because they provide ordered structure, high available RE^{3+} doping concentrations, intense and strongly polarized absorption and emission bands, weak non-radiative relaxation and strong Raman response.

Ytterbium (Yb^{3+}) ions are well-known for their emission at ~1 μm according to the $^2F_{5/2} \rightarrow {}^2F_{7/2}$ electronic transition. They exhibit a simple energy-level scheme eliminating the unwanted processes such as energy-transfer upconversion or excited-state absorption leading to low heat loading and high laser efficiencies. Yb^{3+}-doped materials can be pumped by high-power InGaAs laser diodes emitting at ~0.98 μm. To date, Yb^{3+}-doped monoclinic DTs have been implemented for efficient continuous-wave (CW) [3], passively Q-switched [4] and especially mode-locked oscillators at ~1 μm [5,6].

Contrary to DT crystals, their molybdate counterparts, KLn(MoO_4)$_2$, are less widespread [2,7]. However, these crystals do not exhibit a polymorphic phase transition below the melting point (like in DTs) and thus they can be grown by the conventional Czochralski (Cz) method [8,9]. KLn(MoO_4)$_2$ crystals belong to the orthorhombic class and they feature a layered structure leading to a perfect cleavage feature. This allows one to fabricate thin crystalline films [10] suitable for microchip or thin-disk lasers.

In the present work, we report on the growth, structure, vibronic and spectroscopic characterization of a novel crystal in the orthorhombic DMo family – Yb^{3+}-doped potassium yttrium double molybdate, Yb:KY(MoO_4)$_2$.

2. Crystal growth and structure

$KY(MoO_4)_2$ crystals are chemically stable, they have a relatively low melting temperature (about 970 °C) and do not show any polychromic transformation below the melting point. In this work, a 3 at.% Yb:$KY(MoO_4)_2$ crystal was grown by the Cz method using a [100]-oriented undoped seed. 5–7 mol% of potassium trimolybdate ($K_2Mo_3O_{10}$) were added to the melt to prevent its partial dissociation. The seed rotation speed was 20 rpm, the pulling rate was 1-2 mm/h and the cooling rate was 2 °C/day. The temperature gradient in the melt was below 3 °C/mm. The growth rate was 5–10 g/day.

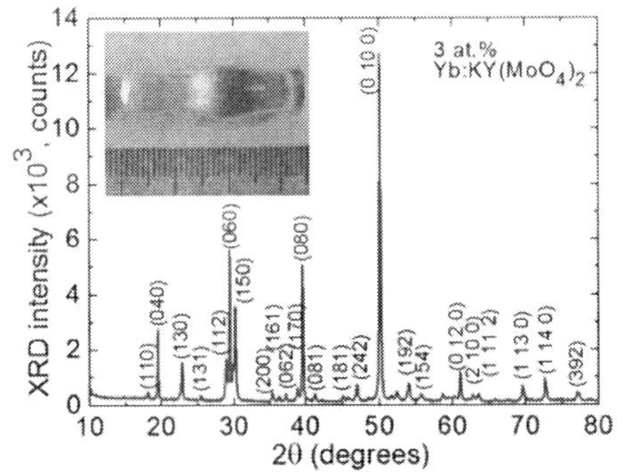

Figure 1. X-ray powder diffraction (XRD) pattern of 3 at.% Yb:$KY(MoO_4)_2$, *numbers* indicate the Miller's indices, (*hkl*), *inset* – photograph of the as-grown crystal; the growth direction is [100].

The as-grown crystal of good optical quality was transparent and free of cracks and inclusions, see inset in Fig. 1. No post-growth annealing was applied. The crystal was colorless. The phase purity and the structure of the as-grown crystal were confirmed by X-ray powder diffraction (XRD), see Fig. 1. Yb:$KY(MoO_4)_2$ is orthorhombic (sp. gr. *Pbna* – D^{14}_{2h}, No. 60). The Yb^{3+} ions in $KY(MoO_4)_2$ replaces the Y^{3+} ones in a single type of crystallographic sites (symmetry: C_2, VIII-fold O^{2-} coordination). Briefly, the structure of Yb:$KY(MoO_4)_2$ is as following [8]: continuous belts of edge-sharing distorted [Y|YbO_8] octahedrons are parallel to the ***b***-axis. In the ***a-b*** plane, they share corners with the [MoO_4] tetrahedra. The [Y|Yb$(MoO_4)_2$]$^-$ radicals form porous layers in the ***b-c*** plane. Because of this, the crystal exhibits a perfect cleavage along the (100) plane.

3. Raman spectra

All members of the DT and DMo crystal families are Raman-active [1]. This feature may be used for self-Raman conversion of the laser output. To characterize the vibronic properties of Yb:$KY(MoO_4)_2$, the room temperature (RT) Raman spectra were measured with polarized light using a Raman microscope (Renishaw inVia). The excitation wavelength was 488 nm (an Ar^+ laser line). A cleaved crystal plate (i.e., an ***a***-cut) was used and the studied excitation / collection geometries were ***a***(***bb***)***a***, ***a***(***bc***)***a***, ***a***(***cb***)***a*** and ***a***(***cc***)***a*** (according to Porto's notations).

The Raman spectra are shown in Fig. 2. They are strongly polarized with the most intense Raman response in the ***a***(***bb***)***a*** geometry. The observed bands are classified into three groups of vibrations [8]. The low-frequency range, 80-272 cm^{-1}, contains translational (T') and rotational (R') modes of the K, Y and Mo cations. Internal bending vibrations (δ) of the oxygen bridged $[MoO_4]^{2-}$ tetrahedral are observed in the intermediate range, 300–435 cm^{-1}. The high-frequency range, 726-944 cm^{-1}, contains intense stretching vibrations (ν) of these tetrahedra.

The most intense Raman mode is at 865 cm^{-1}. The full width at half maximum (FWHM) of this mode is 18.8 cm^{-1}. The maximum phonon frequency, $h\nu_{max}$ = 944 cm^{-1}.

Journal of Physics: Conference Series **1410** (2019) 012149 doi:10.1088/1742-6596/1410/1/012149

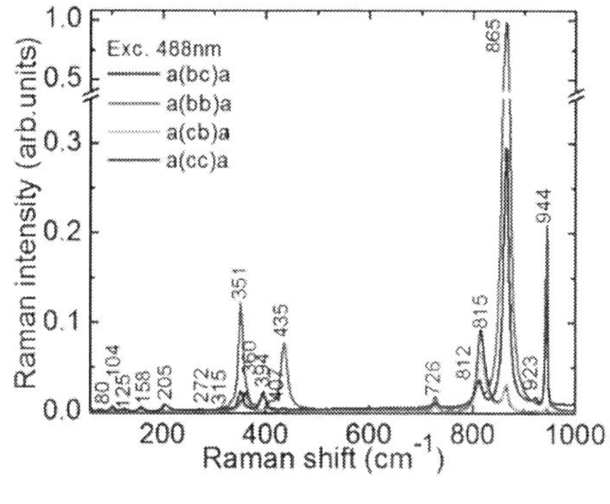

Figure 2. Polarized Raman spectra of an ***a***-cut Yb:KY$(MoO_4)_2$, λ_{exc} = 488 nm, *numbers* indicate the frequencies of the Raman peaks in cm^{-1}.

4. Optical spectroscopy

Orthorombic Yb:KY$(MoO_4)_2$ crystal is optically biaxial. The optical indicatrix axes coincide with the crystallographic ones while the assignment of the refractive indices is still unknown. We characterized the spectroscopic properties for the principal light polarizations indicated as ***E*** || ***a***, ***E*** || ***b*** and ***E*** || ***c***. All the studies were performed at RT (20 °C).

The absorption spectra were measured using a Varian CARY 5000 spectrophotometer and a Glan-Taylor polarizer. The absorption cross-sections, σ_{abs}, were then calculated using the Yb^{3+} ion density of 2.29×10^{20} cm^{-3}, as shown in Fig. 3(a). The maximum $\sigma_{abs} = 1.77\times10^{-20}$ cm^2 at 977.1 nm for ***E*** || ***b*** and the corresponding FWHM of the absorption peak is 19 nm which is advantageous for diode-pumping. Note that the absorption cross-sections for the light polarizations ***E*** || ***b*** and ***E*** || ***c*** are much higher than that for ***E*** || ***a*** (orthogonal to the cleavage plane) which is due to the layered crystal structure. The UV absorption edge λ_{UV} is at 334 nm and in the IR, the transparency window is until 5.3 μm.

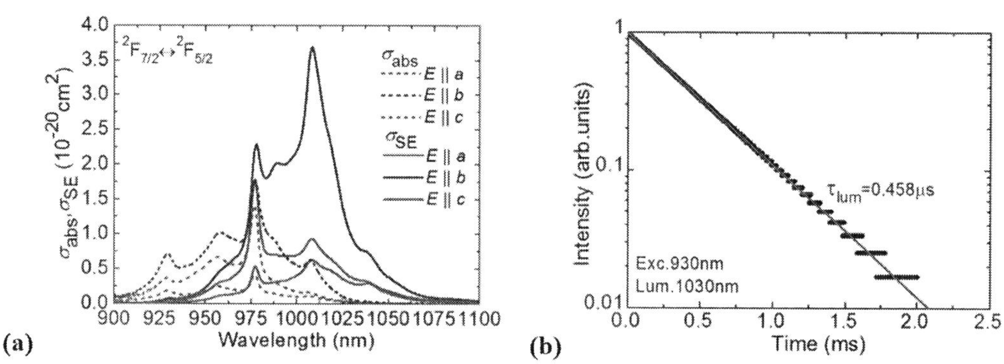

Figure 3(a, b). Spectroscopy of Yb^{3+} ions in orthorhombic KY$(MoO_4)_2$ crystal: **(a)** absorption, σ_{abs}, and stimulated-emission (SE) cross-sections, σ_{SE}, for the $^2F_{5/2} \leftrightarrow {}^2F_{7/2}$ transition, light polarizations are ***E*** || ***a***, ***E*** || ***b*** and ***E*** || ***c***; **(b)** the luminescence decay curve.

The luminescence spectra were measured using an optical spectrum analyser (OSA, Hamamatsu, model AQ6373) and a Glan-Taylor polarizer; a Ti:Sapphire laser tuned to 928 and 977 nm was used as an excitation source. The stimulated-emission (SE) cross-section were calculated by the Füchtbauer–Ladenburg (F-L) formula [11]:

$$\sigma_{SE}^{i}(\lambda)=\frac{\lambda^5}{8\pi\langle n\rangle^2 c\tau_{rad}}\frac{W_i(\lambda)}{1/3\sum_{i=a,b,c}\int\lambda W_i(\lambda)d\lambda}, \quad (1)$$

where, $i = a, b, c$ is the light polarization, λ is the light wavelength, $\langle n\rangle \approx 1.9$ is the mean refractive index of the crystal, c is the speed of light, τ_{rad} is the radiative lifetime of the emitting state ($^2F_{5/2}$, see below), $W_i(\lambda)$ is the measured luminescence spectrum for i-th polarization. The SE cross-section spectra are shown in Fig. 3(a). The maximum σ_{SE} for the $^2F_{5/2} \rightarrow {}^2F_{7/2}$ transition of Yb^{3+} ions is 3.70×10^{-20} cm^2 at 1008.0 nm for the light polarization $\boldsymbol{E} \parallel \boldsymbol{b}$. The FWHM of the emission band is >35 nm. The strong polarization anisotropy of the SE cross-sections is a prerequisite for linearly polarized laser emission from the Yb:$KY(MoO_4)_2$ crystal.

The luminescence decay curve for Yb:$KY(MoO_4)_2$ was measured under ns pulse excitation from an optical parametric oscillator using a fast InGaAs photodetector and an 8 GHz digital oscilloscope. A thin (~100 μm) cleaved film was used to avoid the effect of reabsorption. The excitation wavelength λ_{exc} was 930 nm, the luminescence from the $^2F_{5/2}$ Yb^{3+} multiplet was monitored at 1030 nm. The decay curve plotted in a semi-log scale is shown in Fig. 3(b). It is clearly single-exponential in agreement with a single type of sites for Yb^{3+} ions. The luminescence decay time (τ_{lum} = 458 μs) was determined according to a single-exponential law, $I_{lum}(t) = I_0 exp(-t/\tau_{lum})$.

5. Conclusion

Orthorhombic Yb:$KY(MoO_4)_2$ crystals featuring simple growth by the standard Cz method, easy Yb^{3+} doping, strongly polarized absorption and stimulated-emission cross-section spectra due to the layered structure and a perfect cleavage feature are very promising for microchip lasers based on crystalline films with a thickness of few hundreds of μm and emitting at ~1 μm. Further work will focus on laser characterization of this material.

Acknowledgments

This work was partly supported by Russian Science Foundation (Agreement # 18-13-00200). P.L. acknowledges financial support from the Government of the Russian Federation (Grant 074-U01) through ITMO Post-Doctoral Fellowship scheme. P.L. would like to thank Dr. Olga Dymshits for the XRD measurement.

References

[1] Petrov V, Pujol M C, Mateos X, Silvestre O, Rivier S, Aguiló M, Solé R M, Liu J, Griebner U, Díaz F 2007 *Laser & Photon. Rev.* **1** 179
[2]. Kaminskii A A, Sarkisov S E, Li L 1973 *Phys. Stat. Sol. (a)* **15** K141
[3] Liu J, Petrov V, Mateos X, Zhang H, Wang J 2007 *Opt. Lett.* **32** 2016
[4] Loiko P, Serres J M, Mateos X, Yumashev K, Yasukevich A, Petrov V, Griebner U, Aguiló M, Díaz F 2016 *Opt. Lett.* **41** 2620
[5] Griebner U, Rivier S, Petrov V, Zorn M, Erbert G, Weyers M, Mateos X, Aguiló M, Massons J, Díaz F 2005 *Opt. Express* **13** 3465
[6] Pekarek S, Fiebig C, Stumpf M C, Oehler A E H, Paschke K, Erbert G, Südmeyer T, Keller U 2010 *Opt. Express* **18** 16320
[7] Kaminskii A A, Klevtsov P V, Pavlyuk A A 1970 *Phys. Stat. Sol. (a)* **1** K91
[8] Volokitina A A, Loiko P A, Vilejshikova E V, Mateos X, Dunina E B, Kornienko A A, Kuleshov N V, Pavlyuk A A 2018 *J. Alloys Compd.* **762** 786
[9] Loiko P A, Kifle E, Serres J M, Mateos X, Aguiló M, Díaz F, Vilejshikova E V, Kuleshov N V, Pavlyuk A 2018 *Laser Phys. Lett.* **15** 065002
[10] Kaminskii A A, Bagayev S N 1944 *Quantum Electron.* **24** 1029
[11] Aull B F, Jenssen H P 1982 *IEEE J. Quantum Electron.* **18** 925

Multifocal Autofocused Airy Beam Metasurfaces

M. Mahdi Shanei[1], Mahdieh Hashemi[2], A. Naresh Kumar Reddy[3], Carlos. J. Zapata Rodríguez[4]
[1]Department of Electromagnetic Engineering, KTH Royal Institute of Technology, SE-100 44 Stockholm, Sweden
[2]Department of Physics, College of Science, Fasa University, Fasa 74617-81189, Iran
[3]Department of Technical Cybernetics, Samara National Research University, 34 Moskovskoye Shosse, Samara 443086, Russia
[4]Department of Optics and Optometry and Vision Science, University of Valencia, Dr. Moliner 50, Burjassot 46100, Spain

Abstract. To focus the incident light, both convex metalenses with hyperbolic phase profile and autofocused Airy (AFA) beams are widely used. To obtain AFA-based bifocal metalenses, we propose two methods in this article. One is to bring the two conjugate focal points of an AFA beam into the real space by applying a proper convex lens phase profile. The other is to use the inefficient central space between the two launched Airy beams with zero amplitude and constant phase profile as an independent metalens. We also show that combining the two introduced methods, leads the design to upgrade to a multifocal structure.

1. Introduction

The bent nature of the time-invariant Airy beams causes formation of two conjugate focal spots [1,2]. The one that is formed in front of the structure is the real focus. The virtual one, which is formed behind the structure, remains not accessible. Utilizing a convex lens in the way of propagation of the AFA beams will bring the virtual image into the real space [3]. To design a bifocal AFA-based metasurface which transforms the virtual focal spot to the real space we use a thin layer of metasurfaces for hybrid lens combination to the AFA beams [4]. With generating the two lunched mirror symmetric Airy beams in the Cartesian coordinate for light focusing, the space between the two Airy beams remains empty. We reach a bifocal design by using this ineffective space as an independent metalens [5].

We show that using the inutile space between the two Airy beam as an independent lens together with applying a converging lens profile to the AFA beam profile leads to the multifocal AFA generation [4]. We apply the required phase and amplitude profiles of the desired structure at the launching plane to an arrangement of metaatoms. With advantageous like reduced size, lower weight and ease of fabrication, metasurfaces are commonly used in the field of fabricating the photonic components [6]. In this paper we utilize C-shape silicon Metaatoms at the wavelength of 700 nm that enable us to have full control over the transmitted light phase and amplitude [7].

2. Bifocal AFA-Based Metasurfaces

To provide the required amplitude and phase control, we use C-shaped high-index silicon meta-atoms to support both electric and magnetic dipole resonances. Highly efficient designed silicon-based metasurfaces at the wavelength of 700 nm are totally compatible with CMOS technology. Suitable amplitude and phase coverage are achieved for the co-polarized transmitted electric field component with sweeping the inner radius and the opening angle of the c-shapes over the range of 0 to 130 nm and 0 to 160°, respectively. While keeping the outer radius, R_o, thickness, t, of each metaatom and period of the substrate, P_s, fixed as 140 nm, 200 nm and 360 nm, respectively.

2.1. Transformation of the virtual focal spot to the real space: bifocal AFA design

Multiplying the phase profile of the two mirror symmetric launched airy beams with the focal point around $z_f=70\lambda$ of the AFA to that of a convex metalens with $f=z_f/2=35\lambda$ leads to the transportation of the virtual focal spot to $z=23\lambda$, which is shown in Fig. 1(a) and is consistent with the lens-maker formula. With the phase distribution of the Airy beams that are oscillating between two values of 0 and 180°, multiplication of the AFA beam profile to that of the lens causes the lens phase to be mirrored at positions with 180° phases of the AFA beam. Phase profiles of the AFA beam, lens, and their multiplication are indicated in Fig. 1(b) by purple dotted, dashed green, and solid blue lines, respectively.

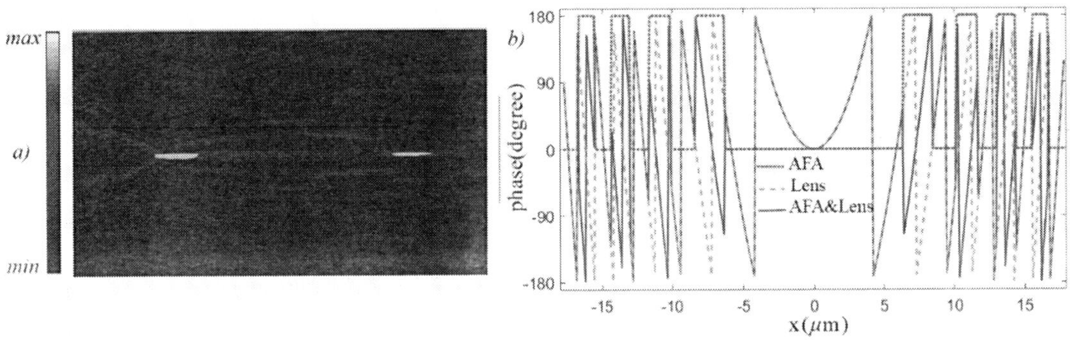

Figure 1(a, b). (a) Intensity distribution of the focused AFA beam (f =35λ); **(b)** Phase profile of the AFA beam, lens, and their multiplication.

2.2. Subjoining a metalens to the space between the two mirrored Airy beams of the AFA beam

The space between the two launched mirrored converging Airy beams which are used for light focusing is left ineffective with zero amplitude transmission and constant molding phase profile.

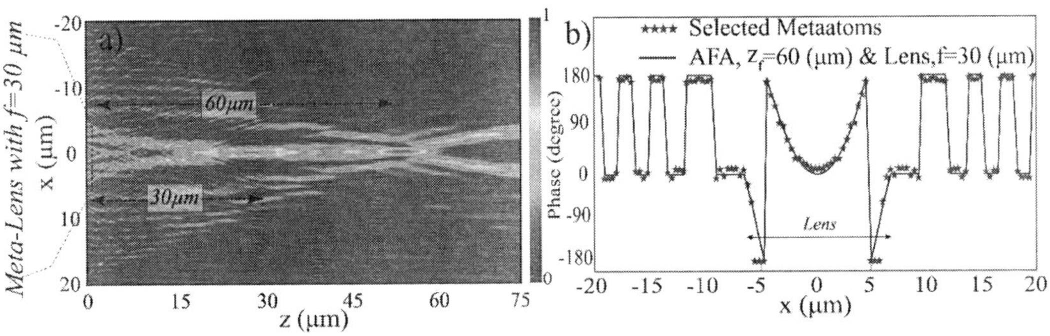

Figure 2(a, b). (a) Co-polarized electric field intensity distribution; **(b)** Spatial distribution of the required phase profile (solid black line) and selected metaatoms (red stars).

We show that exploiting this unused space as an independent metalens not only interfere in the focusing operation of the AFA beam, but it will add the flexibility in light focusing at the second focal spot. For example, we set the focal length of the included metalens to be $f = 30\ \mu m$ in Fig. 2(a) while AFA beams are focused at $z_f=60\ \mu m$. Parabolic phase profile of the central metalens which is added to the 0 and 180° phase distribution of the AFA beams is shown in Fig. 2(b). The red stars demonstrate selected metaatoms with the closest phase for fulfilling this phase distribution in the z=0 plane in Fig. 2(b).

3. Bifocal AFA-Based Metasurfaces

To increase the number of focal points to three by adding the focusing spot of the metalens to the two images of the focused AFA beam, amplitude modulation technique is necessary. To include focal spot of the metalens, amplitude distribution of the AFA beam should be modified toward the required unit amplitude of the metalens, at least, at the center of the metaatoms arrangement. It is important to note that the value of the transmitted light should be selected cautiously not to deteriorate the AFA image formation. Amplitude distribution of such design for the modified focused AFA (MFAFA) of Fig. 1 is illustrated in Fig. 3b as the dashed green line. Intensity distribution of this arrangement of metaatoms is included in Fig. 3a. Three distinct images at $Q_1= 70\lambda$, $Q_2 =23\lambda$ and the focal point of the metalens $f = 35\lambda$ can be distinguished clearly.

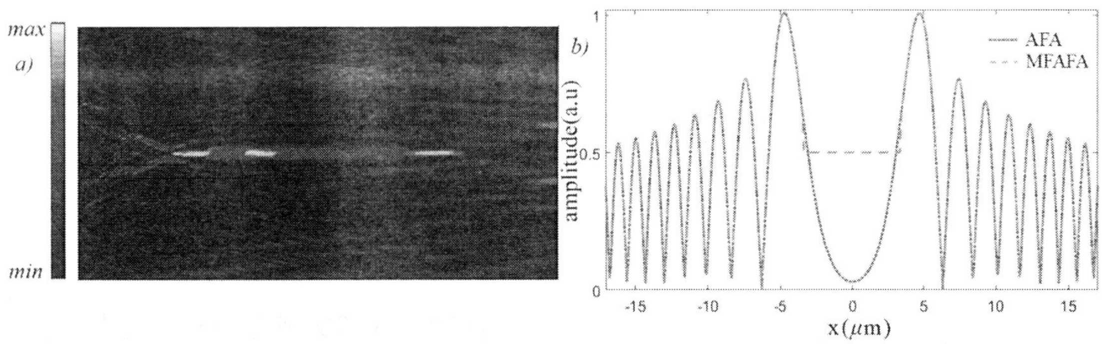

Figure 3(a, b). (a) Co-polarized electric field intensity distribution; **(b)** Spatial distribution of the required phase profile (solid black line) and selected metaatoms (red stars).

4. Conclusion

The two presented methods in this article illustrate bifocal AFA-based focusing with using single layer of thin metasurfaces. The first method is based on the AFA focusing with a convex lens which leads to the imaginary focal spot transformation to the real space. Applying necessary phase profile of an AFA and a convex lens in a single metasurface arrangement leads to the bifocal AFA-based design. Within the other method, central metaatoms in the unused space between the two launched mirrored Airy are set to play the role of an independent metalens.

Flexibility in designing a bifocal metalens with enhanced focusing efficiency is achievable by the introduced methods. To add the focal spot of the metalens to the two images of the focused AFA beam, amplitude profile of the AFA beam should be modulated to get close to the amplitude distribution of the lens. Such an amplitude modulation of the focused AFA beam will lead to the formation of three focal spots in front of the metaatoms arrangement. The introduced method in this paper is a promising way to replace bulky photonic imaging and holographic tools with subwavelength arrangement of metaatoms.

Acknowledgments
This work was financially supported by the Ministry of Education and Science of the Russian Federation in the framework of the government contract 3.5319.2017/8.9 and by the Federal Agency of Scientific Organizations (Agreement No 007-GZ/C3363/26), Ministry of Science, Research and Technology, Iran and the Russian Foundation for Basic Research (RFBR) (Grant Nos: 16- 29-11698, 16-29-11744).

References
[1] Siviloglou G A, Christodoulides D N 2007 *Opt. lett.* **32** 979
[2] Papazoglou D G, Efremidis N K, Christodoulides D N, Tzortzakis S 2011 *Opt. lett.* **36** 1842
[3] Papazoglou D G, Fedorov V Y, Tzortzakis S 2016 *Opt. Lett.* **41** 4656
[4] Hashemi M, Shanei M M, Zapata-Rodríguez C J 2018 *J. Opt. Soc. Am. B* **35** 2864
[5] Shanei M M, Hashemi M, Zapata-Rodríguez C J 2019 *J. Opt. Soc.* **21** 055101
[6] Naserpour M, Hashemi M, Zapata-Rodríguez C J 2017 *Recent progress in far-field optical metalenses. Metamaterials-Devices and Applications.*
[7] Forouzmand A, Mosallaei H 2017 *Adv. Opt. Mater* **5** 1700147

The interaction of an exciton and several optical modes in layer of GaAs

A V Belonovskii[1], **K M Morozov**[1,2], **M V Fetisova**[1]
[1]St-Petersburg Academic University, St. Petersburg 194021, Russia
[2]ITMO University, St. Petersburg 197101, Russia

Abstract. Theoretically, we calculated exciton fraction in polariton modes, which correlates with the intensity of the exciton radiation associated with these modes for several microcavities. Thus, we obtained the form of the dependence of the radiation probability on the eigenfrequencies of the structure.

1. Introduction

In 1946 Purcell discovered, that the intensity of the interaction of an emitter with light can be enhanced or suppressed by engineering its electromagnetic environment [1]. From this decisive observation sprung a whole field of research, today called cavity quantum electrodynamics. In 1992 Weisbuch et al the first observations of the vacuum-field Rabi splitting in semiconductor quantum microcavities [2]. These works have led to **extensive** researches in the field of strong coupling in the interaction between light and matter in semiconductor microcavities during the last three decades [3-6].

In the semiconductor microcavities, a strong field-matter interaction occurs between the optical modes and excitons, whose modes are relatively sharp, so the exciton-photon modes are often called cavity polaritons. These states emerge from coherent energy exchange between quantum well excitons and cavity mode. The reason for consideration to these quasi-particles is their low in-plane effective mass around the center of the first Brillouin zone in quantum well nanostructures. These considerations are important for understanding such effects as laser action and developing optoelectronic and photonic devices.

The strength of the interaction between exciton and optical modes is proportional to the probability of the transmission from the crystal ground state to the exciton state and is proportional to the Purcell factor. The factor Purcell, in turn, is proportional to the Q factor of the resonator. And the square root of the strength of the interaction between exciton and optical modes is proportional to the Rabi-splitting. Thus, optical modes with a high Q factor provide a strong coupling between an exciton and photons and a larger Rabi splitting than low Q factor modes.

We considered the interaction of an exciton with several optical modes with different Q-factors arising in the GaAs layer. This interaction occurs immediately in the strong and weak coupling regime, which is accompanied by multiple Rabi splitting.

2. Results and discussion

At first, we considered the layer of GaAs, more precisely the ideal square resonator. The size of these cavities corresponds to several wavelengths. And calculated the distribution of the intensity of the electromagnetic field of the modes inside of this cavity (Figure 1).

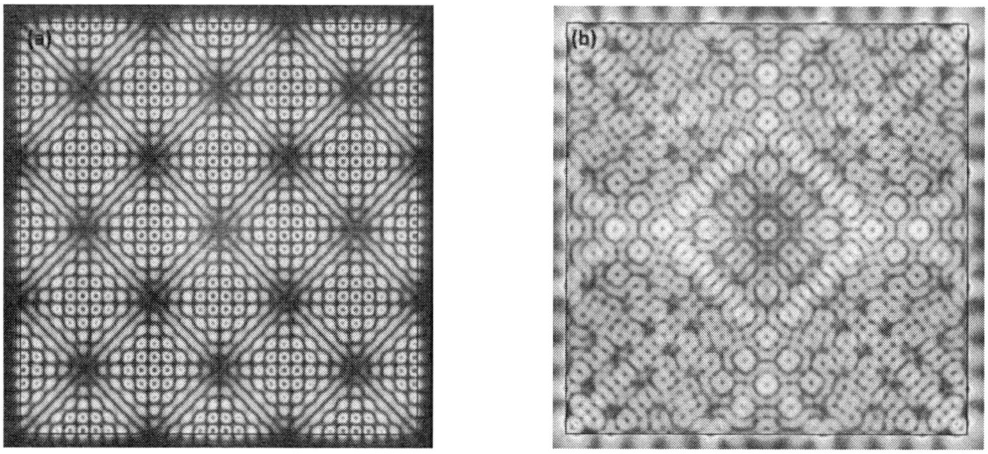

Figure 1 (a, b). (a) The distribution of the electromagnetic field intensity of the modes inside the ideal square resonator with a side 5 μm and the refractive index n = 3.458 (λ_0 = 800 nm), calculated using COMSOL. (a) Example of strong localized mode (1.55 eV); **(b)** Example of a weakly localized mode (1.42 eV). Surface: Electric field (v/m).

From Figure 1, we see that a strong localized mode (a) experiences a total internal reflection on the edges of the square. The intensity of weak mode (b) more uniformly distributed within the square.

Since the exciton energy in GaAs is equal to: $E_X = 1.5554$ eV [6], we have found all the eigenmodes of this structure around this energy, as well as the Q-factor of these eigenmodes. From the known Q-factors of eigenmodes and structure parameters, we can find the Purcell coefficient for each eigenmode [1]:

$$F = \frac{Q\lambda^2}{n^2 S} \quad (1)$$

where λ is the wavelength within the cavity material, n is refractive index and S is the square of the mode.

Figure 2 (a) shows the energies and quantities F for the eigenmodes localized in the ideal square structures. Clearly, while most of the modes are characterized by moderate values of F, there are still few modes with a rather high F. Figure 2 (b) shows the distribution of the mode probability depending on F for ideal square structures. There is a small probability of the appearance of modes with a high value of F for ideal square structures.

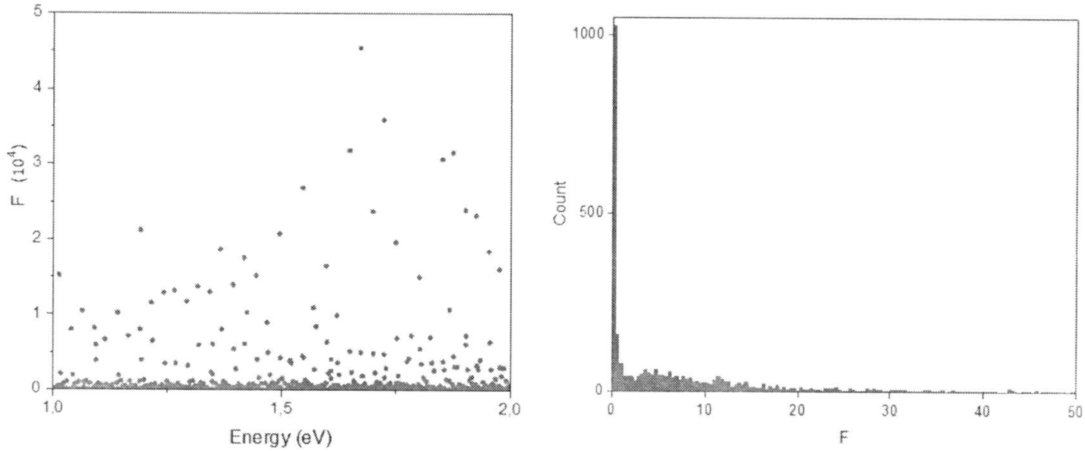

Figure 2 (a, b). (a) Energies and quantities F for the modes localized in the square GaAs; **(b)** Histograms illustrating distribution of cavity modes depending on the quantity F.

As the simplified model, that could explain the interaction between exciton and several modes of ideal square GaAs structure, we have carried out an analysis of the polaritonic mode structure using the following assumption. The Hamiltonian of interaction consists of exciton and several photon modes. This Hamiltonian can be simplified by using rotating wave approximation. In the case of exciton Hamiltonian, we will exploit second quantized form and field [7]:

$$\mathcal{H} = \hbar w_0 \hat{x}^+ \hat{x} + \sum_k \hbar w_k \hat{c}_k^+ \hat{c}_k + \sum_k \hbar (g_k \hat{c}_k \hat{x}^+ + g_k^* \hat{c}_k^+ \hat{x}) \quad (2)$$

where $\hat{x}^+, \hat{c}_k^+$ are the operators of the creation of exciton and photons, respectively, $\hat{x}, \hat{c}_k$ are the operators of the annihilation of exciton and photons, respectively, w_0 is the exciton frequency, w_k are the frequencies of cavity modes, g_k are the constants, describing the strength of the interaction between exciton and optical modes. The Hamiltonian (2) of the interaction in the matrix form reads as:

$$\mathcal{H} = \hbar[\hat{x}^+ \ \hat{c}_1^+ \ \hat{c}_2^+ \cdots \hat{c}_N^+] \cdot \begin{bmatrix} w_0 & g_1 & g_2 & \cdots & g_N \\ g_1^* & w_1 & 0 & \cdots & 0 \\ g_2^* & 0 & w_2 & \cdots & 0 \\ \vdots & \vdots & \vdots & \ddots & \vdots \\ g_N^* & 0 & 0 & \cdots & w_N \end{bmatrix} \cdot \begin{bmatrix} \hat{x} \\ \hat{c}_1 \\ \hat{c}_2 \\ \vdots \\ \hat{c}_N \end{bmatrix} = \hbar \hat{C}^+ W \hat{C} \quad (3)$$

where C^+ is a row, made up of operators of creation, C is a column, made up of operators of annihilation, W is a square matrix, which consists of energies of the system and constant g_k. We can diagonalize this Hamiltonian using a unitary transformation [8]. Figure 3 shows the diagonalization scheme. After the diagonalization of the matrix, we obtain new creation and annihilation operators. These are polariton state operators. The squares of the coefficients standing next to the exciton birth operator in the polariton state birth operator are the excitonic contribution to the polaritonic state. The eigenstates of the bound system superposition's of exciton and cavity modes are exciton-polariton modes $|i\rangle$. After diagonalization of the Hamiltonian we obtain the probability of the exciton radiation modes for each frequency.

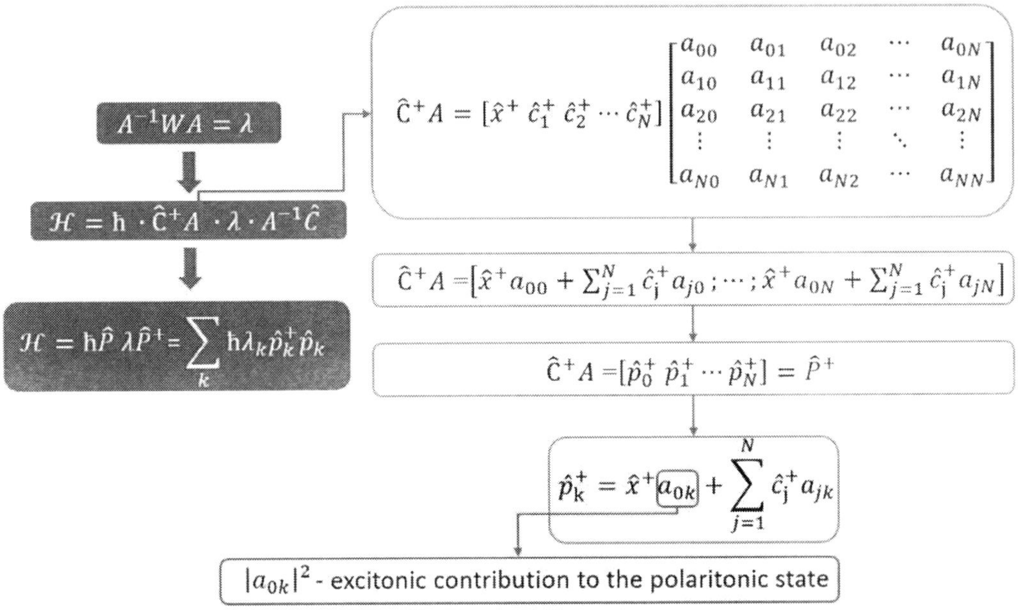

Figure 3. Diagonalization scheme.

The strength of interaction of exciton with the specific mode g_k is proportional to the Pursell factor and the probability of excitonic emission in the uniform media. The intensity of the excitonic emission associated with specific polariton mode $|i\rangle$ will correlate with its excitonic fraction described by $|a_{0i}|^2$. Figure 4 shows the energy dependence of the excitonic contribution to the polaritonic state.

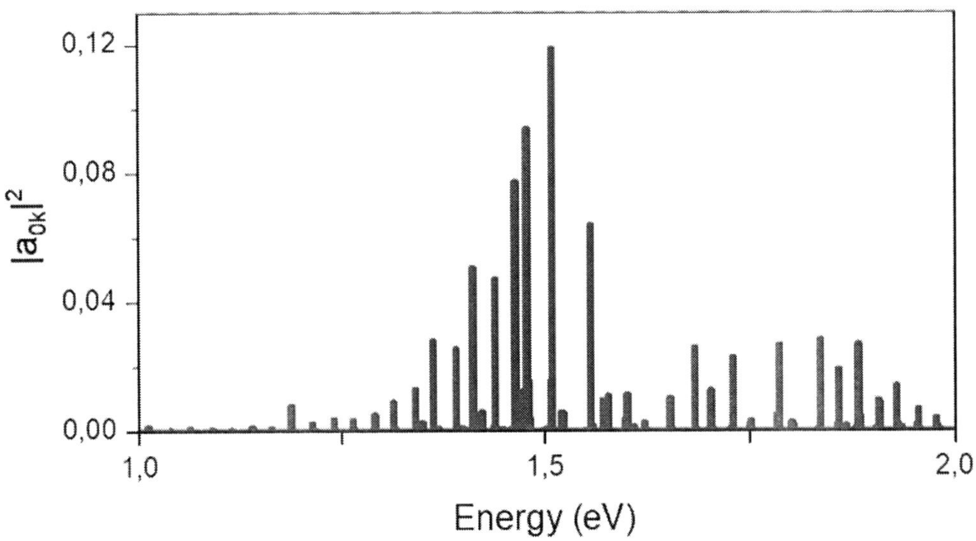

Figure 4. Energy dependence of the excitonic contribution to the polaritonic state.

3. Conclusions

We have found the eigenmodes of ideal square GaAs structure around of the exciton energy and the Q factor of these eigenmodes. There is a small probability of the appearance of modes with a high value of F for ideal square structures. And we have calculated energy dependence of the excitonic contribution to the polaritonic state. We have received multiple Rabi splitting due to the interaction occurs immediately in the strong and weak coupling regime. Thus, we have obtained the form of the dependence of the radiation probability on the eigenfrequencies of the ideal square GaAs structure.

4. Acknowledgments
This work has been supported by Russian Ministry of Science and Higher Education (3.9787.2017/8.9).

References

[1] Purcell E M, 1946 *Phys. Rev.* **69** 681
[2] Weisbuch C, Nishioka M, Ishikawa A and Arakawa Y 1992 *Phys. Rev. Lett.* **69** 3314
[3] Pau S, Björk G, Jacobson J, Cao H, Yamamoto Y 1995 *Phys. Rev. B* **51** 14437
[4] Fisher T A, Afshar A M, Whittaker D M, Skolnick M S, Roberts J S, Hill G and Pate M A 1995 *Phys. Rev. B* **51** 2600
[5] Vladimirova M R, Kavokin A V and Kaliteevski M A 1996 *Phys. Rev. B* **54** 14566
[6] Cotta E A and Roma P M S 2014 *Condens. Matter Phys.* **17** 23702
[7] Shan G, Zhao X and Huang W 2011 *J. Nanoelectron. Optoelectron.* **6** 138
[8] Loudon R 1973 *The Quantum Theory of Light* (Oxford: Clarendon Press) p 184

Towards an efficient light-emitting source based on self-implanted silicon with dislocation-related luminescence

D S Korolev, A A Nikolskaya, A N Tereshchenko*, A N Mikhaylov, A I Belov and D I Tetelbaum
Lobachevsky University, 23/3 Gagarin prospect, Nizhny Novgorod, 603950, Russia
*Institute of Solid State Physics, Russian Academy of Sciences, Chernogolovka, Moscow region, 142432, Russia

Abstract. The regularities of ion synthesis of dislocation-related luminescence centers in silicon have been investigated. By varying the conditions of additional irradiation with boron ions, as well as the conditions of subsequent annealing, we obtain an increase in the luminescence intensity, as well as a shift of maximum of the temperature dependence towards higher temperatures. It was found for the first time that, for the highest used dose of boron ions ($3 \cdot 10^{17}$ cm^{-2}) and additional heat treatment at 830 °C, it is possible to get the measurable luminescence at room temperature.

1. Introduction

Creating an efficient silicon-based light-emitting source is an important, but still not fully resolved problem. Due to the indirect band gap structure, the band-to-band luminescence intensity of silicon is extremely low. One of the ways to solve this problem can be the use of dislocation-related luminescence (DL), one of the spectral lines of which *D1* (1.5 μm) is in the region of maximum transparency of silicon optical fibers and has the highest temperature stability with respect to other DL lines. However, the traditional method of producing silicon structures with DL by the method of plastic deformation [1,2] is not suitable for practical use. An important result of recent years has been the discovery of possibility of synthesis of silicon layers with an intense DL by high-temperature annealing of silicon implanted with Si^+ ions (self-implanted silicon) [3]. This approach has such advantages as full compatibility with silicon technology, high reproducibility and the ability to control the properties of the formed structures by varying the parameters of ion irradiation and subsequent annealing. However, such structures are characterized by a significant temperature quenching of the DL intensity, and, as a result, very low intensity at room temperature.

One of the appropriate models of the *D1* line emission is based on the fact that centers responsible for luminescence are the centers that contain intrinsic interstitial atoms, as well as impurities and defects located in dislocation atmospheres [2]. By varying the composition of these atmospheres, for example, by using additional ion irradiation, it is possible to modify the light-emitting properties of the *D1* centers. We have previously found that, in the samples subjected to additional implantation of boron ions, the intensity of the *D1* line is higher than that in the samples not subjected to additional implantation but annealed in the same conditions [4,5]. In this paper, we present the results of additional studies aimed at the control of temperature dependence of the DL intensity by additional ion-beam and heat treatment.

2. Experimental

The initial samples were *n*-Si (100) substrates with a resistance of 4.5 Ω·cm. The formation of a dislocation structure was carried out by implantation of Si^+ ions with an energy of 100 keV and a dose of $1\cdot10^{15}$ cm^{-2} with annealing in an oxidizing atmosphere for 1 hour. Additional irradiation with boron ions was carried out with an energy of 50 keV and doses of $1\cdot10^{16}$ and $3\cdot10^{17}$ cm^{-2}. Subsequent annealing was carried out sequentially at temperatures of 600, 700 and 800 °C in a dried nitrogen atmosphere (30 min each).

The synthesized samples were subjected to additional heat treatment in a mixed atmosphere (Ar + H_2) at 830 °C (60 min), followed by uniform cooling to 300 °C for 3 h. The temperature dependencies of the photoluminescence (PL) were measured in the temperature range of 4.2–300 K by using the standard phase-sensitive technique with a cooled germanium photoresistance as a detector. PL was excited by photons with a wavelength of 920 nm at an optical excitation level of 10 mW/mm^2, corresponding to the linear part of the dependence of the PL intensity on the pump power.

3. Results and discussion

We start our consideration with the results of studying the effect of additional boron implantation with subsequent annealing into the samples, in which dislocations were formed by Si^+ ion implantation into silicon substrates with subsequent annealing in an oxidizing atmosphere. The study of temperature dependence of the *D1* line intensity reveals a change in the nature of temperature quenching in the samples implanted with boron with respect to the sample that was not subjected to additional boron irradiation (Figure 1). In particular, along with the presence of a maximum in the temperature dependence of the intensity of the *D1* line located at a temperature of ~ 40 K, characteristic of all samples studied, the implantation of boron ions leads to the appearance of an additional maximum at temperatures of ~ 60-70 K.

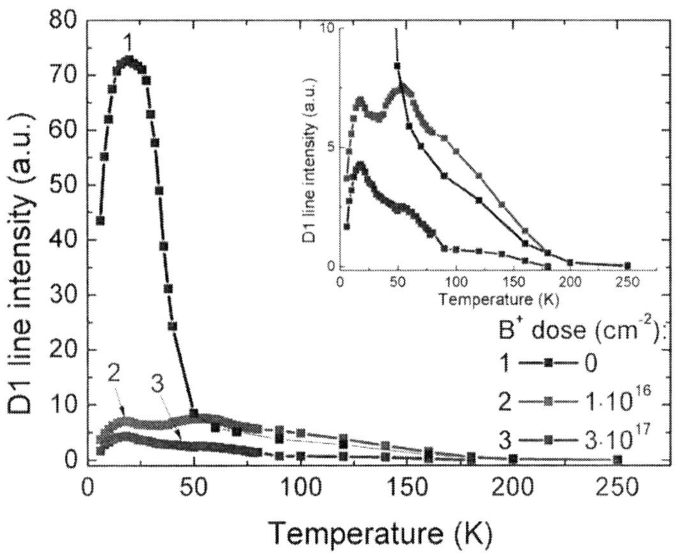

Figure 1. Dependence of the *D1* line intensity on temperature for the Si-DL samples irradiated with B^+ ions before additional annealing at 830 °C (reference samples). Enlarged image of the low-intensity region is shown in the inset.

The most pronounced effect is observed at a boron dose of $1\cdot10^{16}$ cm^{-2}; with increase in the dose of boron to $3\cdot10^{17}$ cm^{-2}, a significant weakening of the second maximum is observed. We also note the decrease in the absolute intensity of the low-temperature DL in samples irradiated with boron, which is associated with the formation of radiation defects, which are the centers of nonradiative recombination. The temperature dependence of the intensity of *D1* DL line in the samples irradiated with boron ions is discussed in detail in [6].

Further, the samples were subjected to additional heat treatment at 830 °C under the conditions of slow cooling of the samples, in contrast to preliminary annealing, in which the cooling of the samples occurred more quickly – when the container with samples was unloaded from the furnace immediately after the annealing was completed. Temperature dependences of the PL intensity of the samples before and after additional heat treatment are shown in Figure 2.

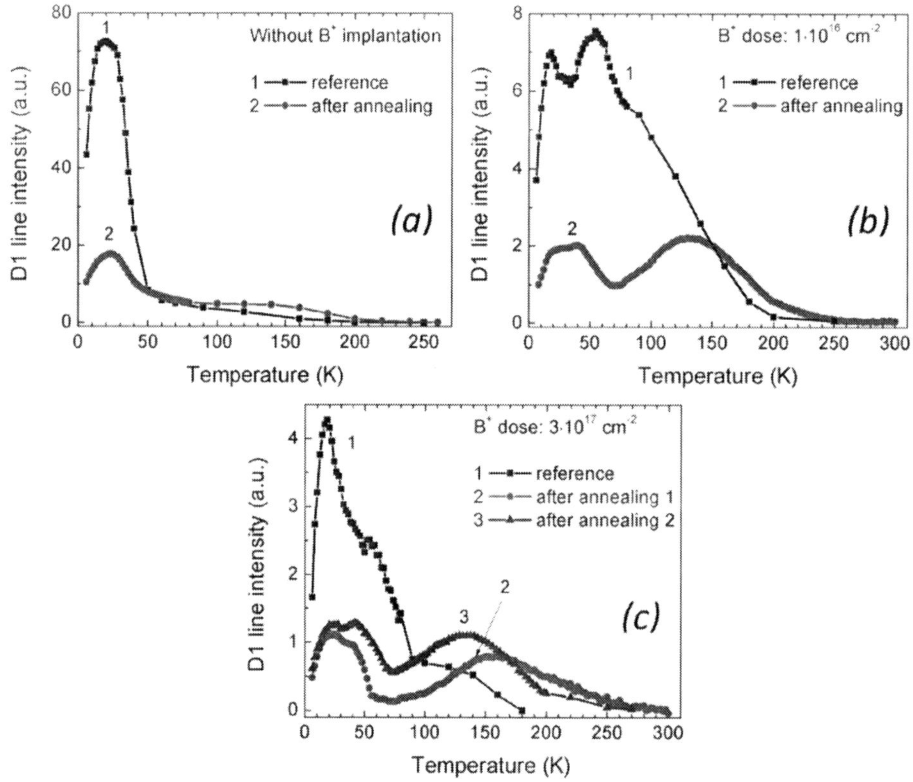

Figure 2. Dependence of the D1 line intensity on temperature for the Si-DL samples without B^+ irradiation (a), irradiated with B^+ ions with the dose of $1\cdot10^{16}$ cm^{-2} (b) and $3\cdot10^{17}$ cm^{-2} (c), before (reference samples) and after additional annealing at 830 °C (1 h).

As can be seen from the figures, additional heat treatment at 830 °C leads to quenching the DL intensity at low temperatures in both boron-implanted samples and samples without boron implantation (reference). This may be due to the contamination during annealing with random impurities that form nonradiative recombination centers [4]. However, in the high-temperature region, the effect of additional annealing at 830 °C for the boron-implanted samples differs significantly from the samples without implanted boron. In the sample without boron implantation, as a result of annealing, there is a slight improvement in the temperature dependence of the DL intensity in the temperature range of 80–200 K. In a sample irradiated with boron to a dose of $1\cdot10^{16}$ cm^{-2}, additional annealing leads to a significant shift of the high-temperature maximum from 60 to 140 K. It is important that the temperature quenching of DL is less pronounced as compared with the sample that was not subjected to additional heat treatment. The most interesting result was obtained by annealing the sample irradiated with B^+ to a dose of $3\cdot10^{17}$ cm^{-2}. As a result of additional heat treatment, a high-temperature peak with a maximum at ~ 160 K is observed in the temperature dependence of the *D1* line intensity; however, unlike the sample irradiated with boron to a dose of $1\cdot10^{16}$ cm^{-2}, the measurable PL is maintained up to room temperature. Next, this sample was subjected to additional annealing at 830 °C (annealing 2) under conditions completely identical to the first annealing. As a result, the luminescence intensity increased in both the low and high temperatures and the second annealing shifts the high-temperature maximum to the lower temperatures.

The observed behavior of the temperature dependence of *D1* line intensity can be explained as follows. During the heat treatment, the interaction of implanted boron atoms with DL centers occurs, which, as shown earlier [6], contributes to the appearance of an additional maximum on the temperature dependence at 70 K. At the same time, the effect at the maximum used dose of B^+ ions ($3\cdot10^{17}$ cm^{-2}) was much weaker than at lower doses, which was explained by the formation of a significant amount of boron precipitates, leading to the PL quenching. Additional heat treatment at higher temperatures (830 °C) and longer annealing time (1 h) can lead to a decomposition of such precipitates. This leads to a sharp increase in the concentration of free boron atoms and intrinsic interstitial atoms, and causes the formation of additional emission centers, as well as defects, from which, according to the model outlined in [6], thermal "pumping" of electrons to these centers occurs. We also note the fact that, at the used annealing temperature, silicon becomes plastic, which can be accompanied by a transformation of the system of dislocations responsible for DL. The role of boron atoms is confirmed by the data demonstrating the best result for a boron dose of $3\cdot10^{17}$ cm^{-2}. The behavior of the temperature dependence of DL intensity after the second annealing can be related to a competition between the formation of nonradiative centers introduced by uncontrolled impurities in the annealing process and an increase in the concentration of boron-containing centers, which increase the luminescence intensity.

An important feature of these experiments is that the cooling after additional annealing was carried out slowly, and this helps to eliminate the effect of introducing additional (thermal) defects associated with rapid cooling.

4. Conclusions

The effect of additional heat treatment on the temperature dependence of the photoluminescence intensity of the *D1* line of dislocation-related luminescence in self-ion-implanted silicon has been studied. The effectiveness of such heat treatment for the samples subjected to additional irradiation with boron ions is shown. A detailed study of the observed patterns requires a separate careful investigation. The discovered effect opens up new possibilities for obtaining an acceptable external quantum efficiency of dislocation-related emission for the use in silicon optoelectronics.

Acknowledgments

This work was supported by the Ministry of Science and Higher Education of the Russian Federation (State assignment 16.2737.2017/4.6), ISSP RAS - Russian Government contracts and also partially with the support of the Russian Foundation for Basic Research (No. 17-02-01070) and a fellowship from the President of the Russian Federation (SP-1147.2018.3).

References

[1] Kveder V V, Steinman E A, Shevchenko S A and Grimmeiss H G 1995 *Phys. Rev. B* **51** 10520
[2] Sauer R, Weber J, Stolz J, Weber E R, Küsters K, Alexander H 1985 *Appl. Phys. A* **36** 1
[3] Sobolev N A 2010 *Semiconductors* **44** 1
[4] Mikhaylov A N, Belov A I, Korolev D S, Timofeeva A O, Vasiliev V K, Shushunov A N, Bobrov A I, Pavlov D A, Tetelbaum D I, Shek E I 2014 *Semiconductors* **48** 199
[5] Tetelbaum D I, Mikhaylov A N, Belov A I, Korolev D S, Shushunov A N, Bobrov A I, Pavlov D A, Shek E I, Sobolev N A 2015 *Phys. Stat. Sol. C* **12** 84
[6] Tereshchenko A N, Korolev D S, Mikhaylov A N, Belov A I, Nikolskaya A A, Pavlov D A, Tetelbaum D I, Steinman E A 2018 *Semiconductors* **52** 843

Two-photon photoluminescence of a thin-film hybrid material based on CdSe(core)/ZnS/CdS/ZnS(multishell) semiconductor quantum dots

D V Dyagileva[1], V A Krivenkov[1], P S Samokhvalov[1], I Nabiev[1,2], Yu P Rakovich[1,3,4]
[1] National Research Nuclear University MEPhI (Moscow Engineering Physics Institute), 115409 Moscow, Russia
[2] Laboratoire de Recherche en Nanosciences, LRN-EA4682, Université de Reims Champagne-Ardenne, 51100 Reims, France
[3] Centro de Física de Materiales (MPC, CSIC-UPV/EHU) and Donostia International Physics Center, 20018 San Sebastian, Spain
[4] IKERBASQUE, Basque Foundation for Science, 48013 Bilbao, Spain

Abstract. Semiconductor quantum dots (QDs) are widely used as components of hybrid materials for development of efficient light emitters and convertors. Their unique nonlinear optical properties, such as two-photon absorption and two-photon photoluminescence from biexcitons, make them promising materials for photovoltaic and optoelectronic applications. In this study, thin-film hybrid materials based on the CdSe(core)/ZnS/CdS/ZnS(multishell) QDs have been fabricated, and the two-photon photoluminescence (PL) from the generated biexcitons have been studied. The results show that fabricated thin-film hybrid materials based on the QDs are efficient fluorophores in the one- and two-photon PL regimes for applications in optoelectronics and biosensing.

1. Introduction

Semiconductor quantum dots (QDs) are widely used as the basis of hybrid materials for novel optoelectronic devices [1,2]. Their high photostability [3] and quantum yield (QY) of photoluminescence (PL) reaching 100% [4], allow one to consider these nanocrystals as one of the most attractive material in the fields of nanophotonics, nanomedicine, and nano-biotechnology [5]. An interesting feature is that, unlike traditional molecular fluorophores, single QDs can have two and/or more excited states [6]. Thus, QD can generate two excitons simultaneously [7]. Biexcitons (BX) is a second exciton in the same QD; they are known for the ability to produce two photons as a result of exciton/biexciton cascade recombination. This unique optical process makes QDs a promising material for optoelectronic applications in the field of entangled photons generation [8], which can be used for quantum cryptography [9]. However, the QY of BX in CdSe-core QDs, with traditionally used inorganic ZnS or CdSe shells, is only about 4-5% [10,11]. This is due to the strong Auger interaction between the two excitons in the same QD, what leads to a high rate of non-radiative relaxation [12]. Moreover, even in the CdSe/CdS core/shell QDs with ultra-thick shells, the QY of BX is only about 10% [13]. CdSe-core QDs covered with an improved multicomponent ZnS/CdS/ZnS shell-structure are known for their high quantum yields reaching the values up to 100% [4]. These QDs are characterized by the small diameters of about 5 nm, what distinguishes them from the giant shell QDs with much larger volumes. The improved energy structure of CdSe(core)/ZnS/CdS/ZnS(multishell) QDs can improve the QY of BX, what will expand the QDs application for entangled photons generation for the field of quantum cryptography. Thereunder, the main goal of this study was to investigate the optical properties of excitons and biexcitons in thin-film nanohybrid material based on CdSe(core)/ZnS/CdS/ZnS(multishell) QDs embedded in a polymer matrix.

2. Materials and methods

2.1. Quantum dots

The QDs with a CdSe core and ZnS/CdS/ZnS multishells were synthesized according to the two-stages procedure and dissolved in a hexane solution. The first stage corresponded to the synthesis of QD-cores. The QDs with the wavelength of the maximum of PL band at 560 nm should have a core size of 2.3 nm. The synthesis of QDs cores was carried out according to the method of high-temperature reaction between cadmium hexadecylphosphonic and selenides of trioctylphosphine. The choice of these reagents allows to achieve a high QDs homogeneity in size and shape. The synthesis was carried out in a high boiling non-polar organic solvent 1-octadecene at the injection temperature of 240 °C [14].

The second stage corresponded to the growing of the epitaxial shell with additional intermediate purification, which provided an additional increase in the monodispersity of the QDs ensemble. Further, the purified CdSe-cores with the sizes of 2.3 nm were coated with the multi-component epitaxial shell of ZnS/CdS/ZnS structure by the method of layer-by-layer ionic adsorption and reaction (SILAR) [15]. This method allowed us to control the thickness of the shell by accurately calculating the required amount of reagents. After completion of the reaction, QDs were purified by coagulation and dispersion. After growing the shell, the average size has been estimated from TEM images and was found to be about 5 nm. Figure 1(a) shows the PL and absorption spectra of the synthesized QDs. The PL decay curve recorded from a solution of QDs (Figure 1(b)) showed the average PL lifetime of 30.1 ns. The relative QY of the QDs in hexane was determined by comparison of the QDs PL with a fluorescence of Rhodamine 6G in methanol and was found to be 90%.

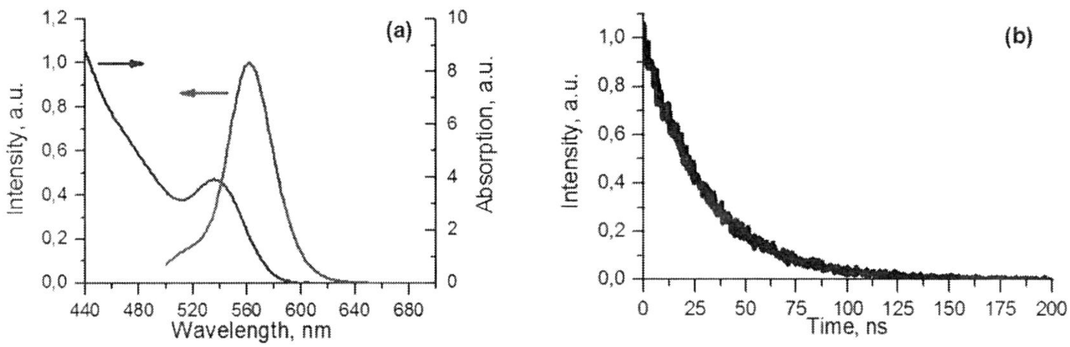

Figure 1. The optical properties of CdSe(core)/ZnS/CdS/ZnS(multishell) QDs. **(a)** The PL and absorption spectra of QDs; **(b)** The PL decay curve recorded from a hexane solution of QDs.

2.2. Fabrication of thin-film hybrid materials

To fabricate thin-film hybrid structures, the method of layer-by-layer spin-coating with the KW-4A Spin Coater has been used. To prevent interaction of QDs with the glass surface, the first layer of PMMA was deposited on a cover glass by spin-coating at a rotation speed of 1000 rpm, from a solution of polymethyl methacrylate (PMMA) in toluene (PMMA molar weight 120 000 Da, weight concentration 0.2 wt %). Further, the next layer of QDs was deposited by spin-coating from a solution in hexane with a molar concentration of 0.1 nmol/l. Then a second layer of PMMA was deposited on the QD-layer from a toluene solution with a weight concentration of 0.4 wt %, to fix the QDs in the PMMA matrix. The thickness of the obtained structure was estimated using the atomic force microscopy (AFM) height profiling method after cutting off a part of the deposited film. The measured thickness of the matrix structure was 6-8 nm, what corresponds to a monolayer of fixed QDs in PMMA matrix.

2.3. Setup and methods of measurement

To study the emission properties of single QDs, a scanning confocal time-resolved fluorescent microscope setup (Micro Time 200, PicoQuant) was used. The optical scheme of the setup is shown in Figure 2. The structure formed was excited by a diode laser with a wavelength λ=485 nm (P$\leq$0.3 μW) in a pulsed mode with a repetition frequency of 5 MHz and a pulse duration of 50 ps. The excitation beam was reflected from dichroic mirror and focused on the sample using UPLSAPO 60XW Plan Apochromat water-immersion objective lens with numerical aperture NA = 1.2. The PL signal was collected by the same lens, passed through a dichroic mirror and 50-μm pinhole and recorded by the avalanche photodiode connected to the photon counting system (HydraHarp 400) with a time resolution of 16 ps.

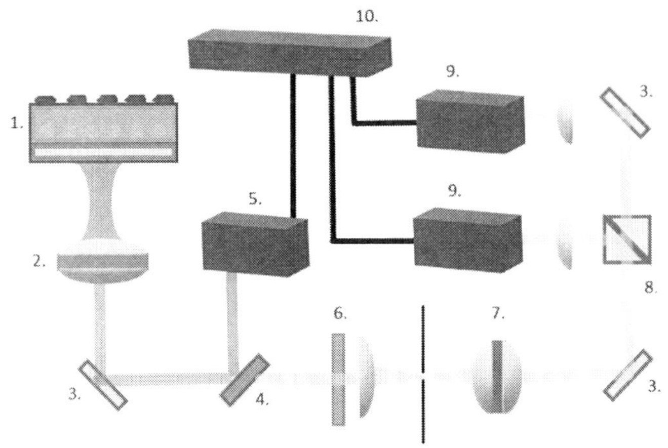

Figure 2. Optical scheme of the confocal time-resolved fluorescent microscope Micro Time 200 used for measurement of PL properties of single QD. 1- sample; 2 – water immersion objective with 1.2 NA; 3 – mirrors; 4 – dichroic mirror; 5 – 485 nm diode laser with 100 ps pulse duration; 6 - optical filter cutting off the laser line; 7 - 150 μm pinhole; 8 – beam splitter; 9 – two avalanche photodiodes; 10 – Hydra Harp 400 multichannel time-correlated single photon counting module.

The PL properties of biexcitons in QDs were studied using photon correlation spectroscopy methods. The second-order cross correlation function $g^{(2)}$ of single QDs was measured by splitting the PL signal into two detectors. This function is calculated by the following equation:

$$g^{(2)}(\tau) = \frac{<I_1(t)I_2(t+\tau)>}{<I_1(t)><I_2(t+\tau)>}, \quad (1)$$

where $I_1(t)$ is the temporal signal from the first detector, $I_2(t)$ is the signal from the second detector, and τ is the time delay between them. The equation (1) describes the probability of photon registration by one of two detectors. For a single photon emitter, a typical value of $g^{(2)}(0) = 0$, that means the absence simultaneous events on the detectors. In contrast, in the case of two-photon PL (biexciton PL), the amplitude of the central peak of the cross-correlation function is not zero. As is known from the literature, at low excitation intensities the area ratio of the central peak to the average side peak is equal to the ratio of the QY of exciton to the QY biexciton [16]:

$$\Phi_{XX} \approx \Phi_X \frac{\int_{-\Delta t}^{\Delta t} g^{(2)}(\tau)d\tau}{\int_{T-\Delta t}^{T+\Delta t} g^{(2)}(\tau)d\tau} = \Phi_X \frac{central\ peak}{average\ side\ peak}, \quad (2)$$

where Φ_{XX} and Φ_X are the QYs of the BX and single exciton, respectively. The equation (2) was used to estimate QY of the BX in single QD.

PL decay curves recorded from a single QD were approximated by the following equation:

$$I(t) = \sum_i A_i \cdot e^{-t/\tau_i}, \quad (3)$$

where I is the intensity, t is the time, and τ_i are the PL decay times. To compare the temporal characteristics of the PL in different samples, the parameter τ_{awl} was used, which is the amplitude-weighted average lifetime and can be expressed as:

$$\tau_{awl} = \frac{\sum_i A_i \cdot \tau_i}{\sum_i A_i}. \quad (4)$$

This parameter is proportional to the average value of QY at a condition of a constant rate of radiative relaxation [17].

3. Results and discussion

In order to analyze the PL properties of excitons (EX) in the single QDs, the area of the sample in which there was only one single QD has been found (see inset images in Figure 3). To determine QY of the EX states in the found single QDs, we have used the method based on the comparison of experimentally obtained PL average lifetimes in solution and in PMMA matrix. Using the measured value of QY of QDs in hexane solution (nearly 90%) and the average PL lifetime of 30.1 ns, one can calculate the radiative lifetime of QDs using the following equation:

$$QY = \frac{k_r}{k_r+k_{nr}} = \frac{\tau_{nr}}{\tau_r+\tau_{nr}}, \tag{5}$$

where k_{nr} is the non-radiative decay rate; k_r is the radiative decay rate; τ_{nr} and τ_r are the corresponding lifetimes. The obtained value of the radiative lifetime is 33.3 ns. Then, the average lifetimes of single QDs in the thin-film PMMA matrix were measured. The PL decay curves recorded from six single QDs are presented in Figure 3. The PL decay kinetics of QDs were approximated by equation (3) and the average lifetimes were calculated by equation (4). Using these data, the QYs of EX states in the PMMA matrix were calculated using the following equation, which can be derived from equation (5), assuming the constant radiative decay rate k_r:

$$\frac{QY_1}{QY_2} = \frac{k_{r2}+k_{nr2}}{k_{r1}+k_{nr1}} = \frac{\tau_{awl1}}{\tau_{awl2}}. \tag{6}$$

The obtained results are summarized in the Table 1.

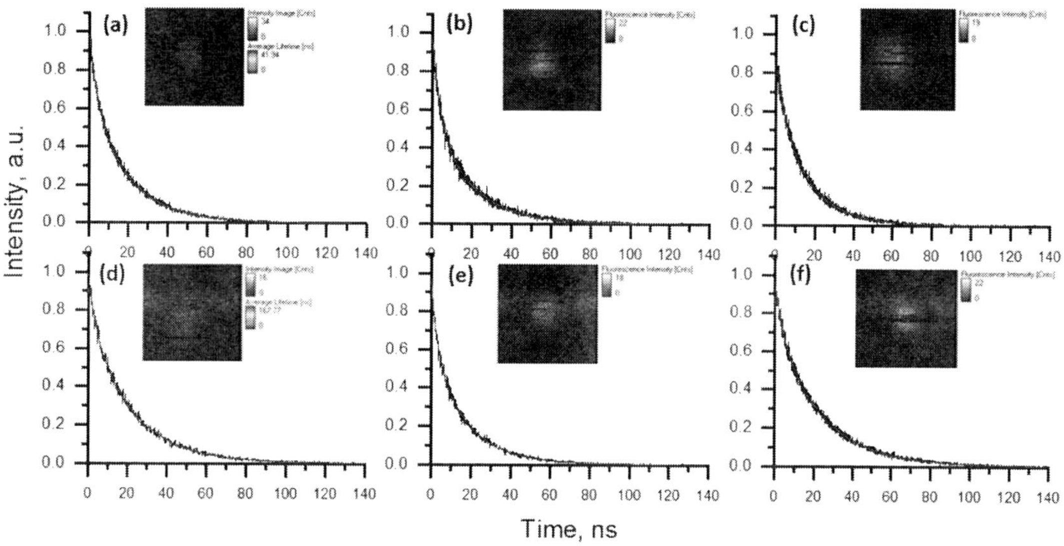

Figure 3. The PL decay curves recorded from six single QDs. Inset figures show the PL images of these QDs

To determine the QY of BX PL in these six single QDs, their second order cross-correlation functions were measured (Figure 4). In all cases, the amplitude of the central peak of $g^{(2)}$ was found to be not zero, but lower than 0.5, what confirms that the observation was carried out from a single QD and not from the aggregates of QDs [18]. It's well known, that the ratio of EX QY to BX QY is equal to the ratio of the area of the central peak from $g^{(2)}$ to the averaged area of side peaks (Equation 2) [16]. Using the obtained values of EX QY and ratio of $g^{(2)}$ peaks, the QYs of BX PL in six selected single QDs have been calculated and were found to be in the range from 8% to 16.3%. The results are presented in the Table 1. The obtained values are much higher than the BX QY of QDs with a conventional ZnS or CdS shell structures [9,10] and as high or even higher than the BX QY of QDs with the ultra-thick shells [12]. It is worth mentioning, that the size of our CdSe(core)/ZnS/CdS/ZnS(shell) QDs is much smaller in size than that for QDs with an ultra-thick shell. Our results show that CdSe QDs with ZnS/CdS/ZnS multishell are promising nanomaterials for fabrication of hybrid systems with two-photon (exciton/biexciton cascade) PL properties.

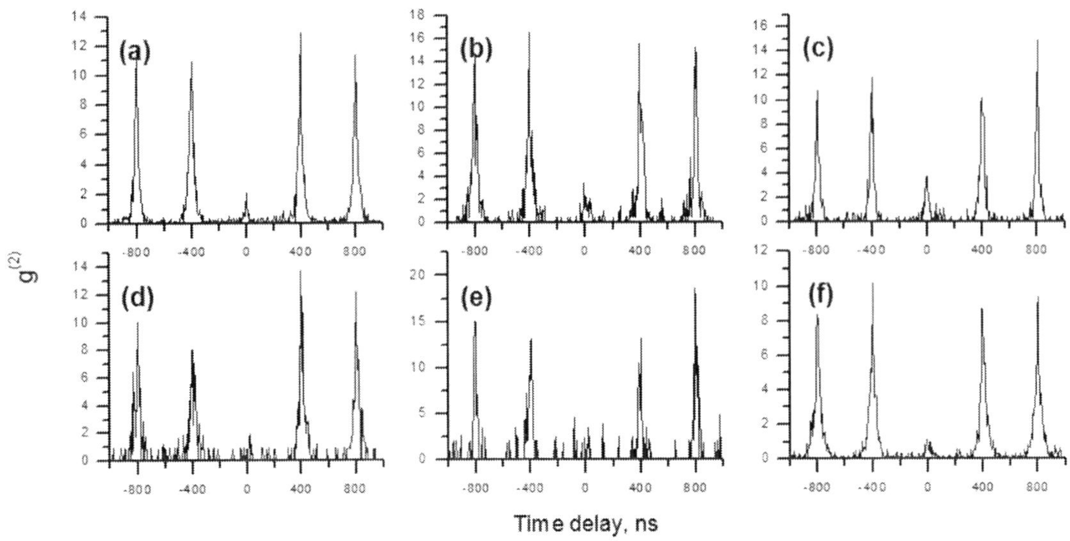

Figure 4. Cross-correlation functions of the single QDs. Panels **(a)** – **(f)** show the cross-correlation functions of six different single QDs.

Table 1. Calculated parameters for samples of single QD.

Sample	τ, ns	QY of EX, %	QY of BX, %
1 (a)	14.6	43.6	8.3
2 (b)	12.7	37.9	8.3
3 (c)	13.0	38.8	16.3
4 (d)	17.4	52.0	8.8
5 (e)	11.9	35.6	10.3
6 (f)	18.9	56.5	9.0

4. Conclusions

The purpose of this study was to investigate the two-photon PL properties of QDs with CdSe cores and ZnS/CdS/ZnS multishells, embedded into thin-film PMMA matrix. We have fabricated a thin-film structures to study the EX and BX PL of single QDs. The average lifetimes and QY values of EX PL were measured. To calculate the BX PL QY of the single QDs, we have measured the second order cross-correlation functions $g^{(2)}$ of six single QDs. The results confirm that the measurements were carried out from single QDs. The values of the BX QYs were calculated using the values of EX QY and ratios of the area of $g^{(2)}$ central peak to the averaged area of $g^{(2)}$ side peaks. The obtained biexciton QYs for all fabricated samples were found to be up to 16.3%. These values are much higher than that for the BX QYs of QDs with the traditional ZnS or CdSe shell structures and are as high or higher than the BX QYs of QDs with an ultra-thick shell. The obtained data show that QDs with improved ZnS/CdS/ZnS multishell structures are promising materials for the fabrication of the two-photon PL sources for use in biosensing, quantum cryptography and optical computing.

Acknowledgments

This study was supported by the Russian Science Foundation, grant no. 18-72-10143.

References

[1] Samokhvalov P, Artemyev M, and Nabiev I 2013 *Chem. Eur. J.* **19** 1534

[2] Vokhmintcev K, and Nabiev I 2016 *Nano Today* **11** 189
[3] Sark W, Frederix P, Bol A, Gerritsen H, Meijerink A 2002 *ChemPhysChem* **3** 871
[4] Linkov P, Samokhvalov P, Vokhmintsev V, Zvaigzne M, Krivenkov V, Nabiev I 2019 *JETP Letters* **109** 112
[5] Litvin A, Martynenko I, Purcell-Milton F, Baranov A, Fedorov A, Gunko Y 2017 *J. Mater. Chem.* **5** 13252
[6] Schaller R and Klimov V 2004 *Phys. Rev. Lett.* **92** 186601
[7] Krivenkov V, Goncharov S, Samokhvalov P, Sánchez-Iglesias A, Grzelczak M, Nabiev I, and Rakovich Yu 2019 *J. Phys. Chem. Lett.* **3** 481
[8] Akopian N, Lindner N, Poem E, Berlatzky Y, Avron J, Gershoni D, Gerardot B, and Petroff P 2006 *Phys. Rev. Lett.* **96** 7
[9] Jennewein T, Simon Ch, Weihs G, Weinfurter H, and Zeilinger A 2000 *Phys. Rev. Lett.* **84** 4729
[10] Yuan C, Yu P, Ko H, Huang J, Tang J 2009*ACS Nano* **3** 3051
[11] Zhao J, Chen O, Strasfeld D, Bawendi M 2012 *Nano Lett.* **12** 4477
[12] Park Y, Bae W, Pietryga J, Klimov V 2014 *ACS Nano* **8** 7288
[13] Galland Ch, Ghosh Y, Steinbrück A, Hollingsworth J, Htoon H, Klimov V 2012 *Nature Communications* **3** 908
[14] Jasieniak J, SmithJoel L, Embden J, Mulvaney P, Califano M 2009 *J. Phys. Chem.* **113** 19468
[15] Li J, Wang A, Guo W, Keay J, Mishima T, Johnson M, Peng X 2003 *J. Am. Chem. Soc.* **125** 12567
[16] Akopian N, Lindner N, Poem E, Berlatzky Y, Avron J, Gershoni D, Gerardot B, and Petroff P 2006 *Phys. Rev. Lett.* **96** 7
[17] Lakowicz, J. R. 2006 *Kluwer Academic: New York* **3** 954
[18] Mangum B, Ghosh Y, Hollingsworth J, and Han Htoon H 2013 *Optics Express* **21** 7419

Development of efficient polymer films with upconversion particles

A A Lyapin[1]*, M V Chernov[1], A M Kuzmin[1], S V Gushchin[1], A S Ermakov[1], P A Ryabochkina[1]

[1]National Research Mordovia State University, Saransk 430005, Russia

Abstract. In this work, the polymer film based on the upconversion particles doped Er^{3+} and Yb^{3+} ions and polyethylene was obtained byusing extrusion and hot pressing methods. Upconversion luminescence spectra of polymer films were recorded in the visible spectral range upon excitation by laser radiation at wavelengths of 972 nm and 1532 nm. The applications of a polymer film in various areas of photonics are discussed.

1. Introduction

Currently, there are a large number of works devoted to the study of upconversion phosphors doped with rare-earth ions since they have numerous applications in bioimaging, biometrics for laser scanningmicroscopy, temperature sensors, infrared quantum counters, solar cell, multimodal diagnostics andtherapy [1–6]. The aim of present paper is obtaining of polymer films dopedupconversion particles and investigation of spectral-luminescence properties of them. Upconversion polymer films can find application for visualization of laser radiation and increasing efficiency solar cells.

2. Experimental

The polymer films were synthesized by using the extrusion [7-9] and hot pressing methods.The initial reagents for the synthesis of films were butene linear low density polyethylene SABIC LLDPE 118NJ and upconversion powder yttrium oxysulfide doped Er^{3+} and Yb^{3+} ions (F(a)SD-546-1 RPF, Luminofor Corp.)Absorption spectra of films were recorded with a Perkin–Elmer spectrophotometer Lambda 950. The spectra of the upconversion luminescence (UCL) of the visible spectral range were recorded by using the Horiba FHR1000 spectrometer. Diodeand fiber lasers operating at 972 nm and 1532 nm, respectively, were used for excitation of upconversion luminescence of Er^{3+} and Yb^{3+} ions.

3. Results

Polyethylene and powder were mixed in the HAAKE PolyLabRheomix 600 OS laboratory mixer with Roller rotors at 150 ° C.At the first stage, the polyethylene (m_{pol}=20 g) was placed in the extruder and melted for 5 minutes at a rotor speed of 30 rpm. Next, the upconversion powder (m_{pow}=30 g) is added to the extruder. Stirring occurs within 10 minutes and at a speed of 50 rpm.At the third stage, the compounding process is accompanied by an increase of the melt temperature and the rotor torque.Upon reaching constant values of these parameters, the mixture was unloaded into a tray and cooled to room temperature. After that, polymer upconversion film was obtained by hot pressing on a Gibitre press at a temperature of 150 °C and a press force of 100 kN. As a result of extrusion and hot pressing processes, the luminescence films with dimensions of 20×20 cm and a thickness of 0.5 mm were obtained. The appearance of the composite material is shown in figure 1.

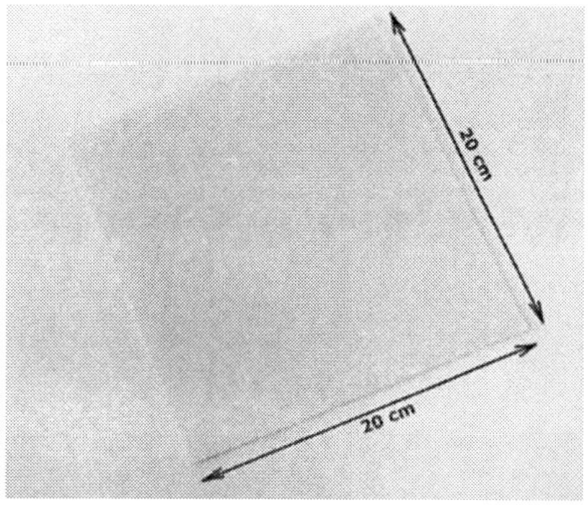

Figure 1. Appearance of polymer material with upconversion particles.

Next we investigated spectral-luminescence properties of upconversion film. Absorption spectra of upconversion films from 300 nm to 1800 nm was presented in figure 2. Observing bands correspond to the transitions $^4I_{15/2}\rightarrow{}^2H_{9/2}$, $^4I_{15/2}\rightarrow{}^4F_{3/2}$, $^4I_{15/2}\rightarrow{}^4F_{5/2}$, $^4I_{15/2}\rightarrow{}^4F_{7/2}$, $^4I_{15/2}\rightarrow{}^2H_{11/2}$, $^4I_{15/2}\rightarrow{}^4S_{3/2}$, $^4I_{15/2}\rightarrow{}^4F_{9/2}$, $^4I_{15/2}\rightarrow{}^4I_{9/2}$, $^4I_{15/2}\rightarrow{}^4I_{11/2}$, $^4I_{15/2}\rightarrow{}^4I_{13/2}$ of Er^{3+} ions and $^4F_{7/2}\rightarrow{}^4F_{5/2}$ transition of Yb^{3+} ions.

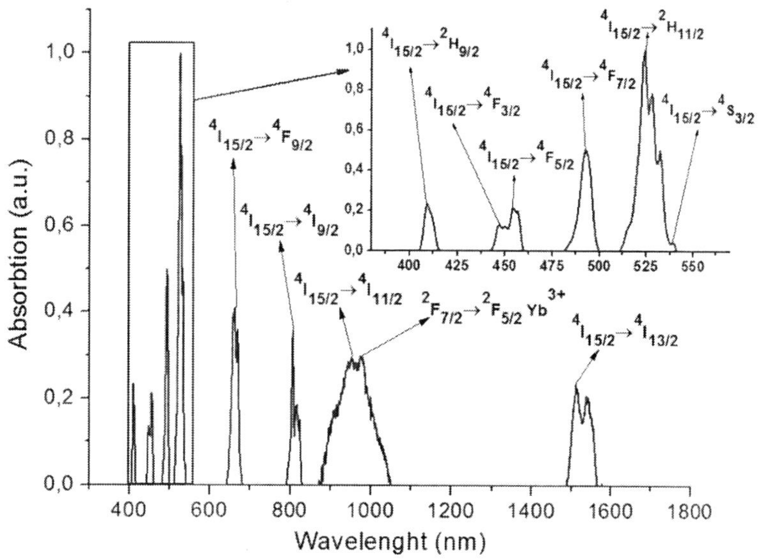

Figure 2. Absorption spectra of upconversion film in spectral range 300-1800 nm.

Upon excitation of the $^4I_{13/2}$and $^4I_{11/2}$levels of Er^{3+} ions, intensive UCL in visible spectral range corresponding to the was observed (fig.3). Luminescence bands correspond to the electronic transitions$^2H_{9/2}\rightarrow{}^4I_{15/2}$, $^2H_{11/2}\rightarrow{}^4I_{15/2}$, $^4S_{3/2}\rightarrow{}^4I_{15/2}$ and$^4F_{9/2}\rightarrow{}^4I_{15/2}$ of Er^{3+} ions. The most intensive luminescence was observed in the red spectral range. Also, for different excitation wavelengths, chromaticity coordinates and temperatures were calculated (fig.3c).The color temperatures of UCL of film upon excitation of 972 nm and 1532 nm were 4966 K and 3825 K, respectively.

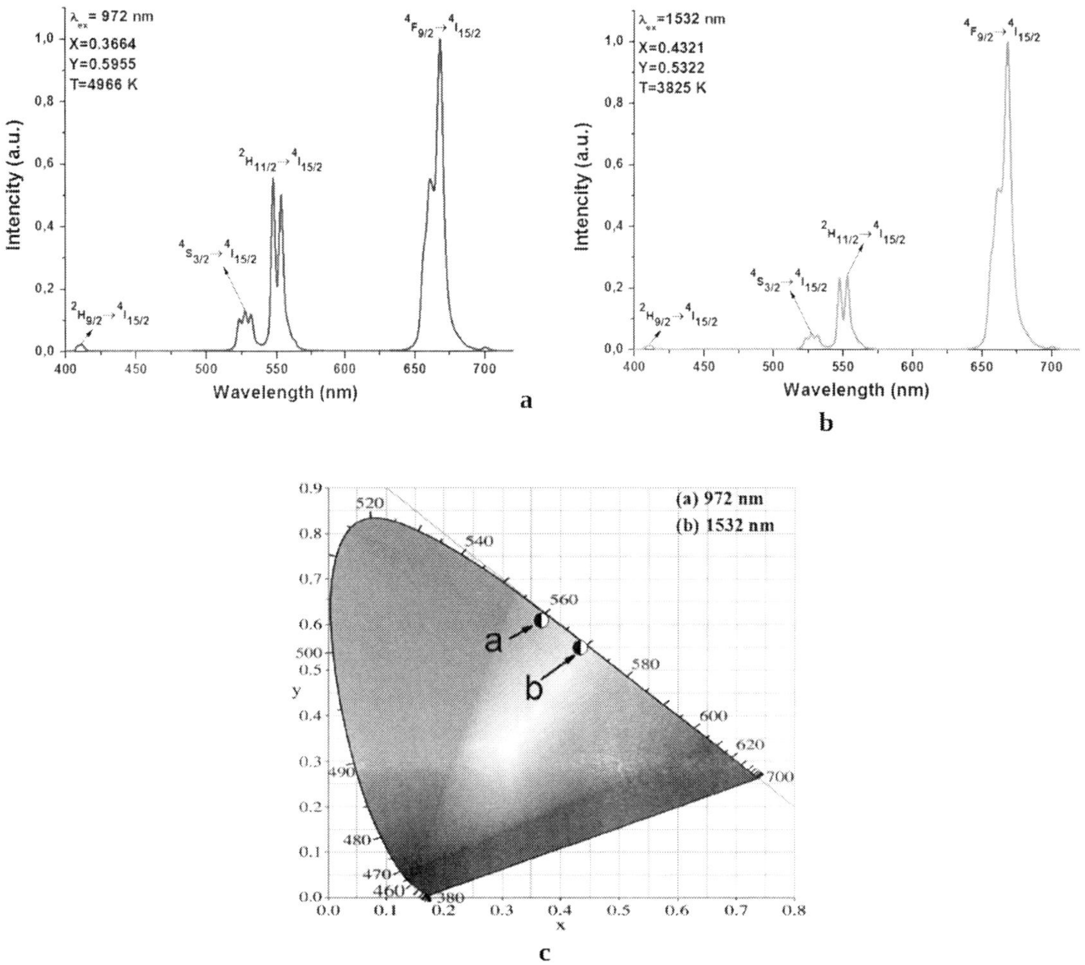

Figure 3.UCL of polymer film uponexcitation by laser radiation at wavelengths of 972 nm **(a)** and 1532 nm **(b)**. The CIE chromaticity diagram of upconversion film **(c)**.

Intensive UCL upon infrared laser excitation and wide absorption bans allow to use upconversion polymer films for visualization of laser radiation.Visualizers based on the obtaining polymer films have some advantages: high value of laser damage threshold, local damage of the polymer does not destroy the all visualizer and small thickness. Also upconversion polymer films can be used for increasing efficiency solar cells.

4. Conclusion

Thus, in present paper upconversion polymer film was successfully obtained by theextrusion and hot pressing methods.Results of investigation of absorption and luminescence properties show that obtaining polymer upconversion film can be used for visualization of laser radiation and increasing efficiency solar cells.

Acknowledgments

This work was supported by the Russian Foundation for Basic Researchand Government of Republic of Mordovia, grant no. 18-42-130011.

References

[1] Auzel F Chem. Rev. 2004 104139
[2] Rodriguez Burdano D C, Naccache R and Capobianco J A 2015 Handbook on the Phys. and Chem. of Rare Earths 47273
[3] Goldschmidt J C and Fischer S 2015 Adv.Opt.Mat. 34510
[4] Fischer S, Ivaturi A, Jakob P, Kramer K W, Martin-Rodriguez R, Mejerink A, Richards B S and Goldschmidt J C 2018 Opt. Mat. 84389
[5] Lyapin A A, Kuznetsov S V, Ryabochkina P A, Merculov A P, Chernov M V, Ermakova Yu A, Luginina A A and Fedorov P P 2017 Laser Phys. Lett. 147076003
[6] Lyapin A A, Gushchin S V, Kuznetsov S V, Ryabochkina P A, Ermakov A S, Proydakova V Yu, Voronov V V, Fedorov P P, Artemov S A, Yapryntsev A D and Ivanov V K 2018 Opt. Mat. Exp. 871863
[7] Rauwendaal C 2017 Polymer Extrusion (Amsterdam: North-Holland/American Elsevier)
[8] Giles Jr H F, Wagner Jr J R and Mount E M 2013 Extrusion: The Definitive Processing Guide and Handbook: Second Edition (New York: William Andrew Publishing)
[9] Steves M J and Covas J 2012 Extruder Principles and Operation: Second Edition (New York: Springer Science & Business Media)

The research of temperature instability influence of fiber optic communication line in phase direction finder channels on peleng accuracy

A S Podstrigaev[1, 2], A S Lukiyanov[1], A V Smolyakov[1, 2], A N Shishkov[1], V V Davydov[3, 4], A P Glinuchkin[4] and E A Sinicyna[3]

[1]Scientific-Research Institute Vector OJSC, st. Academician Pavlov 14 A, St. Petersburg, Russia, 197376
[2]Saint Petersburg Electrotechnical University "LETI", st. Professor Popova 5, St. Petersburg, Russia, 197376
[3]Higher School of applied physics and space technologies, Peter the Great Saint Petersburg Polytechnic University, Polytechnicheskya street 29, Saint Petersburg, Russia, 195251
[4] All Russian Research Institute of Phytopathology, Moscow Region 143050, Russia

Abstract. In the article the influence of optical fiber temperature change to phase shift between optical signals in transmission channels is considered. The method for influence assessment of temperature phase shift on bearing accuracy is proposed. The results of experimental researchers and calculations are presented.

1. Introduction

The phase direction finder is widely used to search for radio sources of microwave range [1-9]. As you know, phase and amplitude-phase direction finding methods are the most accurate. Therefore, they are widely used in practice in radio monitoring systems [1, 2, 6-11]. Currently, a large number of radio direction finder designs have developed. They are located on various objects, including space [12-17].

When placing the antenna elements on a large-sized carrier often requires significant removal of antenna elements relative to the receiving and processing equipment. This is necessary to create a large base of direction finding, etc. The use of microwave cables in this case leads to unacceptably large losses of signal power. Fiber-optic communication line (FOCL) has a low specific attenuation of the signal [5, 10, 11, 18-20]. Therefore, the following is expedient, it is advisable to transfer the received signal to the optical range, transmitting it via fiber optic and reverse the signal to the microwave range. For comparison: the specific attenuation of a microwave cable in the frequency range up to 18 GHz is 0.7 ... 1.7 dB/m, and up to 40 GHz - 1.7 ... 3.0 dB/m (fig. 1). The attenuation of the optical signal in the FOCL is about 0.1 dB/km.

At the same time, a number of problems arise with the use of FOCL [5, 10, 11, 18-22]. One of them is associated with the appearance of an additional phase shift between the signals transmitted by FOCL. The phase shift occurs due to changes in the temperature of the fiber optic line and equipment. This may lead to an increase in direction finding error. Especially with long-term operation of the direction finder in conditions where frequent changes in ambient temperature are possible.

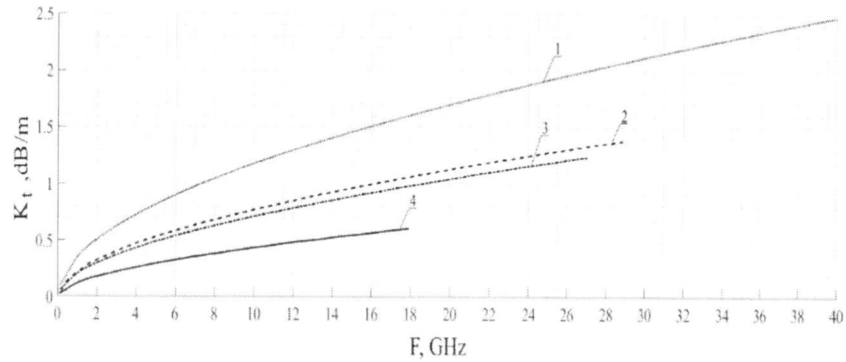

Figure 1. The dependence of the gain (attenuation) on the frequency of the microwave signal of the microwave cable of various brands at an ambient temperature of T = 293 K. Graph 1 corresponds to the brand of cable SF 240, graph 2 - SF 229, graph 3 - SF 404, graph 4 - SF 406.

Therefore, the aim of the work is to study the effect of changes in ambient temperature on the accuracy of direction finding when using FOCL to transmit a microwave signal.

2. The experimental setup and measurement methods

In the radio direction finder, thermal insulation can be performed both for the whole apparatus and separately for FOCL. It depends on the design and permissible weight and size characteristics of the radio direction finder equipment. Therefore, the temperature stability of the phase difference in the channels was investigated for two cases. The first case changes the temperature of the FOCL only. The second case changes the temperature of the entire optical equipment with FOCL. In the figure 2 the block diagram of the experimental setup is presented.

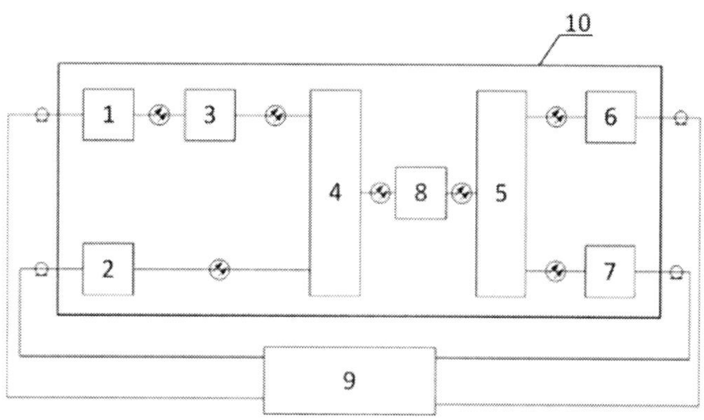

Figure 2. Structural diagram of the experimental setup: 1- laser module Optilab DFB-C-PM-M-1590-20; 2 – transmitter OTS-2T-0518; 3 – electro-optic modulator Optilab CMP-40-LD-V-PM-TQ; 4 and 5 - optical multiplexer (demultiplexer) CWDM-8-M-47-90-10-FC/APC-DK; 6 and 7 – photodetector SIRU3040-00; 8 – fiber-optic communication line; 9 - vector analyzer; 10 – optic electronic tract.

To optical converters 1, 3 and 2 from the output of the vector network analyzer ZVA40 9 are served the sine wave microwave signals having the same frequency. To transfer these signals to the optical frequency range in the first channel, a laser module 1 and an electro-optical modulator 2 are used. The second channel uses the transmitter 2 to solve this problem. For spectral multiplexing of optical signals, a multiplexer 4 (demultiplexer 5) is used. Using this device allows you to transmit signals from two channels over a single fiber optic cable 8. In the experiment, the length of this cable was 90 m. Signal transmission through a single cable minimizes the change in the phase difference in the channels due to temperature change. Optical signals are recorded by photodetectors 6 and 7. Analog signals from the photodetector outputs are fed to the inputs of a vector network analyzer 9.

This experimental setup allows you to measure the difference in phase incursions in FOCL channels. It should be noted that this phase difference is not reduced to the range of 0 - 360 degrees. This makes it possible to measure the exact value of the jumps in the phase difference between the channels, even if it exceeds 360 degrees. For the analysis, the selected frequency band is 17.9 - 18.0 GHz. This is due to the fact that the phase shift in the channels in this frequency band is maximum.

The measurements were carried out with a change in temperature for two cases. Changed the temperature of all elements of the optical tract or only FOCL. During the experiment, the ambient temperature increased from 284.5 K to 296.5 K. В ходе эксперимента температура окружающей среды повышалась от 284.5 K до 296.5 K. At the beginning of the measurements, the steady-state temperature of the optical cable and elements of the path was 284.5 K. В начале измерений установившаяся температура оптического кабеля и элементов тракта составляла 284.5 K. At the same time, the value of the difference of phase ramps in the fiber optic channels $\Delta\varphi_{1-2}$ was periodically recorded. The working wavelength is $\lambda = 1590$ nm. The working frequency band of the microwave path equipment is 2–18 GHz.

3. Experimental results and their discussion

The analysis of the experimental data showed that the value of the phase shift in the channels with increasing temperature varies significantly. It was found that the phase difference between the channels varied in small limits. With an increase in the temperature of the optical cable, the difference of phase raids in the channels $\Delta\varphi_{1-2}$ changed from minus 3 degrees to plus 8 degrees. For the case of increasing the temperature of all equipment, the difference in phase raids in the channels $\Delta\varphi_{1-2}$ varied from minus 6 degrees to plus 6 degrees. As a result of the experiments, it was found that the maximum change in $\Delta\varphi_{1-2}$ was 8 degrees for the case of fiber optic heating and 6 degrees and in the case of heating all the equipment. The transmission coefficient of the optical path in the process of heating changed slightly. This shows a good internal thermal stabilization in the devices used.

Experiments have shown that one of the main factors that affect the accuracy of direction finding is the value of the phase difference. Therefore, it is necessary to investigate the effect of the resulting phase difference between the $\Delta\varphi_{1-2}$ channels on the direction finding accuracy of the phase method.

To do this, consider the problem of determining the direction of arrival of a signal using two antennas. In figure 3 is shown a diagram which illustrating this task.

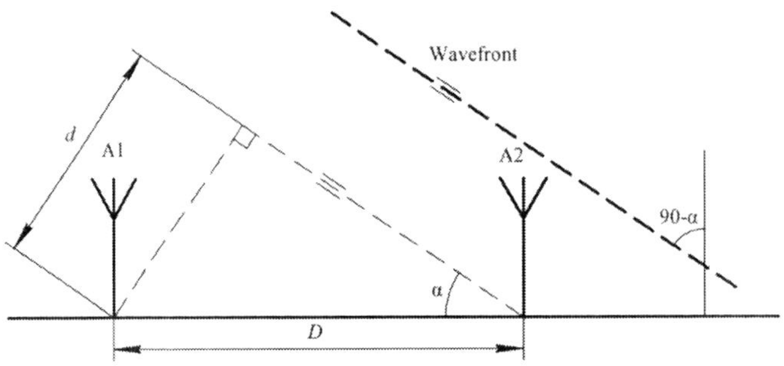

Figure 3. The direction determining of signal arrival by the phase method.

In accordance with figure 3, it is necessary to determine the angle α, indicating the direction of arrival of the signal. The angle α is determined from the measured value of the phase difference of the signals received by the antennas A1 and A2. This difference will be determined by the difference d of the distances that the wave front of the signal traveled in the process of propagation to the antenna points:

$$\Delta\varphi = 2\pi\frac{d}{\lambda} = 2\pi\frac{D\sin\alpha}{\lambda} \quad (1)$$

In the formula (1) the distance between the antennas (direction finding base), related to the wavelength of the signal $\tilde{D} = D/\lambda$ is, in fact, a scale factor. This scale factor relates the angle of arrival of the signal with the phase difference of the signals received by the antennas. This ratio is determined by the range of possible angles of arrival of the signal, within which it is necessary to provide an unambiguous measurement of the bearing. Unambiguity is ensured when, over the entire range of measured angles α, the phase difference Δφ does not exceed π in absolute value, that is:

$$\left|\frac{D\sin\alpha}{\lambda}\right| \leq \frac{1}{2}.$$

From this:

$$\tilde{D} = \frac{1}{2\cdot\left|\sin\alpha_{\max}\right|}.$$

The resulting ratio shows that it can be seen that the direction finding system is narrowband. The required direction finding base is uniquely determined by the operating frequency of the system. Therefore, in further calculations it is advisable to consider the ratio of the direction finding base to the wavelength. Absolute values of these quantities are impractical to consider.

For a radio direction finder, a characteristic can be obtained which makes it possible to determine the sine of the angle corresponding to the direction of arrival of the signal from the measured phase difference Δφ, can be obtained from (1):

$$\sin\alpha = \frac{\Delta\varphi}{2\pi\tilde{D}}$$

Its steepness

$$K_\alpha = \frac{d\sin\alpha}{d\varphi} = \frac{1}{2\pi\tilde{D}}$$

The error in determining the direction of arrival of the signal α_e for a given measurement error of the phase difference of the signals in the antennas φ_e can be determined by the value of K_c:

$$\alpha_e = \varphi_e K_\alpha. \quad (2)$$

The relation (2) shows that the error α_e, is due to the error φ_e, completely determined by the slope of direction finding characteristics. The error α_e does not depend on the direction of arrival of the signal. The resulting error in direction finding is the sum of the error α_e and errors caused by other factors. Consideration of these factors lies outside the scope of this work.

In table 1 presents the calculations α_e for different ranges of unambiguous measurement of the direction of signal arrival. Calculations are made for maximum measurement error of the phase difference (equal to 8) which was fixed during the experiments.

Table 1. . Calculations of error α_e

$\pm\alpha_{max}/2$, degree	$\tilde{D}$	K_α	φ_e, degree	α_e, degree
±15	1,93	0,08		0,64
±30	1	0,16	8	1,28
±45	0,71	0,22		1,76
±60	0,58	0,27		2,16

Based on the results presented in table 1 we can draw the following conclusions:

1. For radio systems review (providing direction finding in wide angles ranges from ± 45 to ± 60) additionally introduced error does not exceed 2.5 degrees. It is caused by phase distortions in the optoelectronic path.

2. In accurate direction finding systems (less than ± 15 degrees), the introduced error does not exceed 0.7 degrees.

3. In systems of radio technical review and accurate direction finding the additional phase incursions in the direction finder channels arising from the use of fiber optic lines slightly degrade the accuracy of direction finding

4. Conclusion

The obtained results show that a change in ambient temperature does not lead to a significant decrease in the accuracy of determining the bearing in the frequency band up to 18 GHz. when spectral seal channels and transfer them over a single fiber optic cable.

References

[1] Likhachev V P and Pasmurov A Ya 1999 *Journal of Communication Technology and Electronics* **44(3)** 275-280

[2] Bykov V V, Dushkin A V, Kupryashkin I F and Likhachev V P 2005 *Radiotekhnika* **7** 85-87

[3] Mel'nikov Yu P 2000 *Radiotechnika* **9** 27-32

[4] Podstrigaev A S, Smolyakov A V, Davydov V V, Myazin N S and Slobodyan M G 2018 *Lecture Notes in Computer Science (including subseries Lecture Notes in Artificial Intelligence and Lecture Notes in Bioinformatics)* **11118 LNCS** 509-515

[5] Lenets, V A, Tarasenko M Yu, Davydov V V, Rodygina N S and Moroz A V 2018 *Journal of Physics: Conference Series* **1038 (1)** 012037

[6] Mashkov G, Borisov E and Fokin G A 2016 *Proceedings - 2016 International Conference on Radar, Antenna, Microwave, Electronics, and Telecommunications, ICRAMET 2016 (Jakarta)* **7849572** p. 7-12

[7] Mashkov G, Borisov E and Fokin G A 2017 *Proceedings - 19th International Conference on Advanced Communications Technology, ICACT 2017 (Phoenix Park Pyeongchang)* **7890254** p. 979-984

[8] Podstrigaev A S, Lukashev V P and Ryazantsev L B 2016 *Measurement Techniques* **59(5)** 547-550

[9] Podstrigaev A S 2014 *Proceedings - 24th International Crimean Conference Microwave and Telecommunication Technology, CriMiCo 2014 (Sevastopol)* **6959682** p. 896-897

[10] Podstrigaev A S, Davydov R V, Rud' V Yu and Davydov V V 2018 *Lecture Notes in Computer Science (including subseries Lecture Notes in Artificial Intelligence and Lecture Notes in Bioinformatics)* **11118 LNCS** 624-630

[11] Davydov R V, Saveliev I V, Lenets V A, Tarasenko M Yu, Yalunina T R, Davydov V V and Rud' V Yu 2017 *Lecture Notes in Computer Science (including subseries Lecture Notes in Artificial Intelligence and Lecture Notes in Bioinformatics)* **10531 LNCS** 177-183

[12] Petrov A A, Davydov V V and Grebenikova N M 2018 *Lecture Notes in Computer Science (including subseries Lecture Notes in Artificial Intelligence and Lecture Notes in Bioinformatics)* **11118 LNCS** 641-648

[13] Petrov A A, Davydov V V and Myazin N S 2017 *Lecture Notes in Computer Science (including subseries Lecture Notes in Artificial Intelligence and Lecture Notes in Bioinformatics)* **10531 LNCS** 561-568

[14] Myazin N S, Smirnov K J, Davydov V V and Logunov S E 2017 *Journal of Physics: Conference Series* **929 (1)** 012080

[15] Rykin E V, Moroz A V, Smirnov K J, Davydov V V and Yushkova V V 2018 *MATEC Web of Conference* **245** 12002

[16] Davydov V V, Kruzhalov S V, Grebenikova N M and Smirnov K J 2018 *Measurement Techniques* **61** 365-372

[17] Davydov R V, Antonov V I and Moroz A V 2018 *Proceedings of the 2018 IEEE International Conference on Electrical Engineering and Photonics, EExPolytech 2018* (Saint-Petersburg) **8564378** p. 236-239

[18] Tarasenko M Yu, Davydov V V, Sharova N V, Lenets V A and Yalunina T R 2017 *Lecture Notes in Computer Science (including subseries Lecture Notes in Artificial Intelligence and Lecture Notes in Bioinformatics)* **10531 LNCS** 227-232
[19] Fadeenko V V, Kuts V A, Vasiliev D A and Davydov V V 2018 *Journal Physics: Conference Series* **1135(1)** 012053
[20] Davydov V V, Karseev A Yu, Nepomnyashchay H K, Petrov A A and Velichko E N 2014 *Lecture Notes in Computer Science (including subseries Lecture Notes in Artificial Intelligence and Lecture Notes in Bioinformatics)* **8638** 694-702
[21] Ivanov S I, Lavrov A P and Saenko I I 2017 *Lecture Notes in Computer Science (including subseries Lecture Notes in Artificial Intelligence and Lecture Notes in Bioinformatics)* **10531** 482-489
[22] Ivanov S I, Lavrov A P and Saenko I I 2016 *Lecture Notes in Computer Science (including subseries Lecture Notes in Artificial Intelligence and Lecture Notes in Bioinformatics)* **9870** 670-679

Properties of GaN optomechanical nanobeam cavity

K M Morozov[1,2], K A Ivanov[1,2], E I Girshova[1], A S Sokolovskii[2] and A P Mikitchuk[3]

[1] St.Petersburg Academic University, St.Petersburg, 194021 Russia
[2] ITMO University, St.Petersburg, 197101 Russia
[3] Belarusian State University, Minsk, 220030 Belarus

E-mail: morzconst@gmail.com

Abstract. Gallium nitride based nanobeam cavity system was theoretically analysed by using finite element method-based model. Optical and mechanical resonant localized modes structure of the nanobeam cavity was calculated. Depend on the geometrical properties there were found modes with high quality factor for the both cases. The maximum optical quality factor reaches value $5{\cdot}10^4$ in cavity with value $d = 90nm$.

1. Introduction

Optomechanics is a growing scientific area that explores the interaction between the mechanical motion and the localized electromagnetic fields [1]. A number of phenomena such as cooling the mechanical mode to the ground motion state open new possibilities and applications concepts [2]. The aim of the many types of research is to find the optimal condition for high-quality eigenmodes localized in the same cavity area. One of the classes of such systems is a freestanding semiconductor system with etched holes of peculiar geometry. It may be noted a "zipper" cavity [3] and nanobeam cavity [4, 5] due to their high optical quality factor and low effective mass. One of the most promising semiconductors in many application areas is a gallium nitride (GaN). It's optical (emission bandgap and refractive index) and mechanical (Young's modulus) properties are close to the widely considered materials for optomechanics such as diamond. However, GaN material is easier to work with in structures fabrication process (like etching of holes in nanobeam cavity). Also, recent studies demonstrate the possibility to fabricate complex GaN structure with peculiar geometry and high crystallography quality even without etching [6, 7]. Recently a design of nanobeam cavity optimized to GaN material and operating in UV spectral area [8] was demonstrated.

In the present study, we considered a quasi-1D GaN nanobeam cavity and theoretically analyzed the localization of both mechanical and optical eigenmodes in such system with a variety of geometrical parameters.

2. Results and discussion

Nanobeam cavity represents the strip of 170 nm width, fixed at both ends. The strip usually placed above the slit in the substrate. Elliptic holes in strip define the properties of the system to localize both optical and mechanical waves. Scheme of the nanobeam cavity shown in figure 1(a). Optimization of the geometry and size of the hole in the nanobeam cavity is a topic for multiple kinds of research in this area [4, 5].

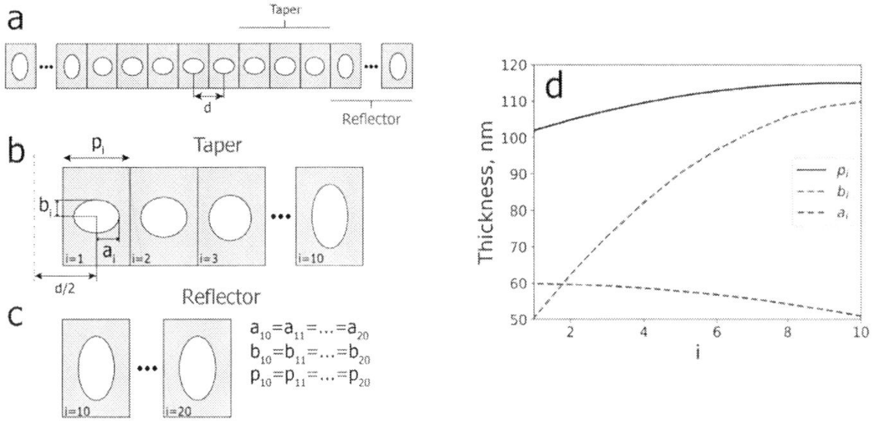

Figure 1. (**a**) Scheme (top view) of the GaN nanobeam cavity and detailed geometrical properties of cavity parts: taper (**b**) and reflector (**c**). The distribution (squared) of the nanobeam cavity parameters in the taper region (**d**).

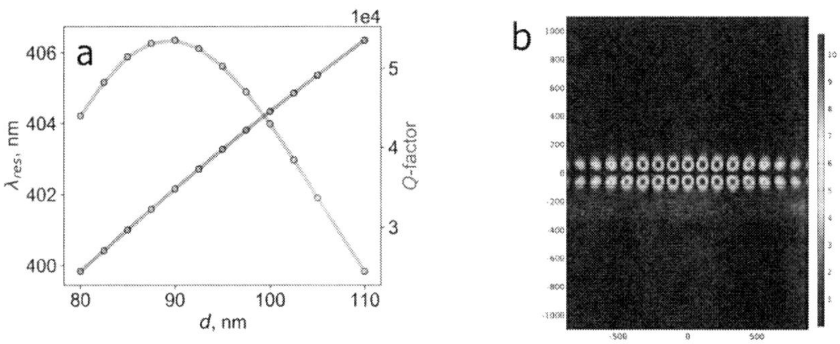

Figure 2. (**a**) Optical mode resonance wavelength dependence on the distance between two central holes parameter (left axis) and the quality factor dependence (right axis). (**b**) Electric field distribution (V/m) of optical eigenmode (λ = 402 nm, Q = 5.3·10^4) for the cavity with d=90 nm.

There are two main parts of the cavity: a taper (figure 1(b)) and a reflector (figure 1(c)). The taper region (10 holes) defined by the three parameters: a_i, b_i, p_i. There p_i - the period of the element, a_i and b_i - elliptic hole parameters and d - the distance between two central holes. Those values depend on position by the quadratic function (figure 1 (d)) as in [5] for the optimal operation of the system in the near UV ($\approx 400nm$) region. Size parameters for a refactor region is the same for all holes and equal to a_{10}, b_{10}, p_{10}. The distance between two central holes d allow tuning the resonant wavelengths in both optical and mechanical cases.

The model was based on a finite element method (FEM), that allow calculating eigenfrequencies and both optical mode and mechanical displacement distributions in structure. Perfect electric conductor boundary conditions around the nanobeam cavity area were used to calculate optical eigenstates (GaN refractive index was chosen as 2.55). Mechanical eigenmode problem was solved by the two-dimensional model with fixed ends of the strip (GaN Youngs modulus is 330 GPa).

Results are demonstrating high optical quality factor states supports in the considered system. Figure 2 shows the results of the optical case simulation. A number of modes were found, but only one has maximally localized in the cavity central region. Dependence of the modes quality factor on the distance between central holes (d) shown in figure 2 (a). There are maximum at $d = 90nm$ corresponding to the $Q = 5.3{\cdot}10^4$ at $\lambda = 402$ nm. Electric field distribution is shown in figure 2 (b). Figure 3 shows the spectral position dependence on the value of the d for the high localized mechanical modes. Insets show the displacement field for the symmetric (blue curve) and the antisymmetric (green curve) localized modes. It can be seen, that operation frequency can be widely tuned by variation of the parameter d ($\sim$ 0.4 GHz for the 10 nm) .

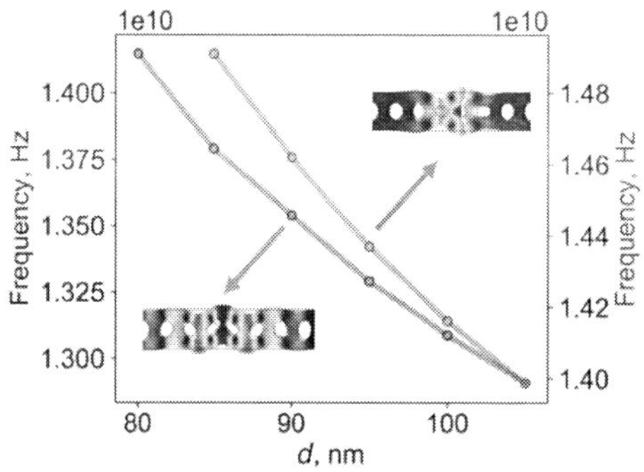

Figure 3. Mechanical mode resonance wavelength dependence on the distance between two central holes parameter. Mechanical displacement fields of the symmetric (12.8 GHz) and antisymmetric (14.9 GHz) modes are shown in the insets.

3. Conclusions

Results demonstrates significant localization of optical and mechanical modes inside the considered system. Dependence of mechanical and optical eigenstates spectral position on the geometrical parameter was demonstrated. The maximum optical quality factor reaches value around $5{\cdot}10^4$ in cavity with $d = 90$ nm.

Acknowledgments

The reported study was funded by RFBR and BRFBR according to the research project 19-51-04001.

References

[1] Aspelmeyer M, Kippenberg T J and Marquardt F 2014 *Rev. Mod. Phys.* **86** 1391–1452
[2] Metcalfe M 2014 *Appl. Phys. Rev.* **1** 031105
[3] Eichenfield M, Camacho R, Chan J, Vahala K J and Painter O 2009 *Nature* **459** 550–555
[4] Chan J, Safavi-Naeini A H, Hill J T, Meenehan S and Painter O 2012 *Appl. Phys. Lett.* **101** 081115
[5] Huang Z, Cui K, Li Y, Feng X, Liu F, Zhang W and Huang Y 2015 *Sci. Rep.* **5** 15964
[6] Pozina G, Gubaydullin A R, Mitrofanov M I, Kaliteevski M A, Levitskii I V, Voznyuk G V, Tatarinov E E, Evtikhiev V P, Rodin S N, Kaliteevskiy V N and Chechurin L S 2018 *Sci. Rep.* **8** 7218
[7] Pozina G, Ivanov K A, Mitrofanov M I, Kaliteevski M A, Morozov K M, Levitskii I V, Voznyuk G V, Evtikhiev V P and Rodin S N 2019 *Phys. Status Solidi Basic Res.* **256** 1800631
[8] Zhou W, Yu Z, Ma J, Zhu B, Tsang H K and Sun X 2016 *Sci. Rep.* **6** 37134

Investigation of optical properties of In(Ga)As/GaAs mesa structures with active region based on quantum wells, quantum dots, and quantum well-dots

A S Dragunova[1], N V Kryzhanovskaya[1,2], M V Maximov[1,3], S A Mintairov[3,1], N A Kalyuzhnyy[3,1], E I Moiseev[1], I S Mukhin[1,4], V G Tikhomirov[5], A E Zhukov[1,2]

[1]St. Petersburg Academic University, 8/3 Khlopina, St Petersburg 194021, Russia
[2]Peter the Great St. Petersburg Polytechnic University, 195251 St. Petersburg, Russia
[3]Ioffe Institute, 26 Polytechnicheskaya, St Petersburg 194021, Russia
[4]ITMO University, 197101, St. Petersburg, Russia
[5]Saint-Petersburg State Electrotechnical University "LETI", St. Petersburg, Russia

Abstract. In this work photoluminescence (PL) of mesa-structures that contain three different types of active regions based on InGaAs/GaAs quantum wells (QWs), InAs/InGaAs/GaAs quantum dots (QDs) and InGaAs/GaAs quantum well-dots(QWD) is studied. Comparative analysis of the PL intensity obtained at different temperatures and optical excitation powers on mesa diameter is done.

1. Introduction

In recent few years, the development of new semiconductor optoelectronic devices based on microcavities, such as single photon sources, miniature radiation sources, detectors, compact sensors, etc., has attracted increasing interest. During the processing of microcavities, etching of epitaxial layers is usually required through the active light-emitting region. This can lead to a significant effect of surface nonradiative recombination on the etched side walls on the optical characteristics. The effect of the nonradiative recombination of carriers in mesa-structures with InAs/InGaAs/GaAs QDs was considered in detail in [1]. It was shown that, due to the strong localization of carriers in InAs/InGaAs/GaAs quantum dots, the effect of nonradiative recombination on the mesa surface is insignificant. The use of QDs arrays as the active region of microlasers helps to achieve very small diameters [2] and high operation temperatures [3]. Despite the fact that lasers based on InAs/InGaAs/GaAs quantum dots showed record characteristics (high operating temperature, low threshold current, etc.), gain saturation in quantum dots stimulates studies of microlasers based on two-dimensional quantum wells or InGaAs quantum well-dots (QWD)[4].

2.Experimental details

All the structures studied were grown on GaAs substrates. Samples containing QDs was synthesized by molecular beam epitaxy (MBE). Three sheets of QDs were inserted into the middle of 0.2-µm thick GaAs cavity confined from the bottom and top sides with $Al_{0.35}Ga_{0.65}As$ barriers. Each QD plane was formed by InAs seeding islands covered with InGaAs capping layer. The structures with QWD and QWs were grown by metalorganic chemical vapour deposition (MOCVD). The active region was placed into a 0.6-μm thick GaAs optical cavity. The cavity was confined with $Al_{0.39}Ga_{0.61}As$ barriers. The QW structure contains two InGaAs QWs. Each QW has a thickness of about 8 nm, indium mole fraction is about 15%. The QWD structure comprises a single plane of QWD formed by deposition of InGaAs thin layer on slightly misoriented GaAs surface. The 50-nm thick AlGaAs barriers were used to prevent the diffusion of the optically excited non-equilibrium carriers into the surface or the substrate were they could recombined non-radiatively.

The epitaxial structures were then processed into arrays of deep mesas. Electron-beam lithography and dry etching were used. The mesa diameter varied from ~ 0.3 to ~ 10 μm. Etching was performed through the whole thickness of the cavity. The mesa height was about 1 μm. Photoluminescence (PL) was excited with a neodymium-doped yttrium lithium fluoride (YLF:Nd^{3+}) laser. Excitation power was about 1 mW. We investigated the dependences of the PL intensity on temperature and pump power for each array of mesa. The temperature interval was 77–290 K.

3. Results and discussion

Figure 1 shows a couple of representative scanning electron microscopy (SEM) images taken for the mesa arrays with the smallest diameters. Good reproducibility of etching and high verticality of the sidewalls are observed.

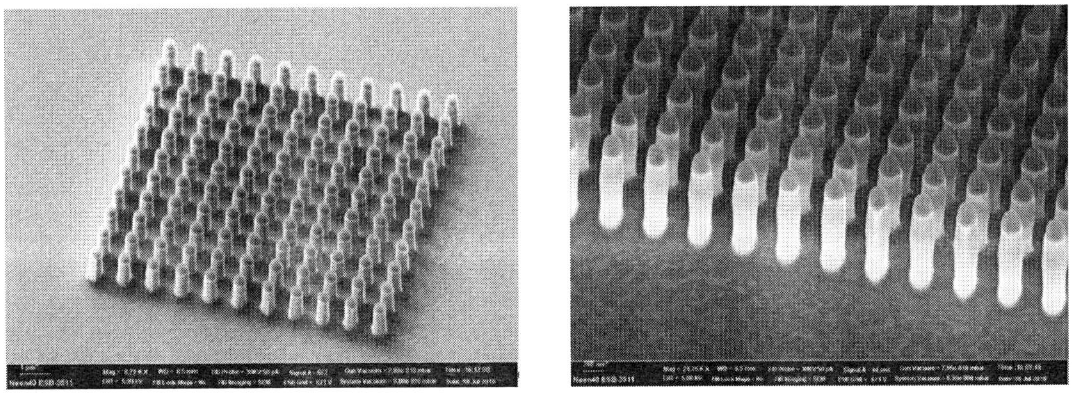

(a) (b)

Figure 1. SEM images of mesa-structures array with mesa diameters of 0.5 μm (a) or 0.3 μm (b)

First, we studied PL of the mesa-structures of the largest diameter (10 μm), where the effect of sidewalls is minimal compared to mesas of smaller diameters. PL signal was measured at various temperatures with the excitation power of 1 mW. The results are presented in Figure 2(a-c) for the mesas containing QWs (a), QWDs (b), and QDs (c). All the samples demonstrate comparable intensity of PL signal at liquid nitrogen temperature. The room-temperature emission (dominant peak) is centred around 0.97, 1.04, and 1.26 μm in QW-, QWD-, and QD-structures, respectively. Wavelength of luminescence is directly related to the localization energy of charge carriers in the active region: longer wavelength corresponds to deeper potential well with respect to surrounding matrix material (GaAs). Among the structures studied, QDs provide the deepest and QWs provide the shallowest localization with QWDs having an intermediate characteristic. In the spectra of QD-structure, several less intense peaks are clearly seen as a result of carrier localization within excited electronic states.

Another important feature, which can be extracted from PL spectra, is the linewidth. The QW structure has the narrowest peak with full width at half maximum (FWHM) gradually changes from 8 to 25 nm as temperature rises from 77 to 290 K. This indicates that the linewidth in this case of two-dimensional QW is predominantly determined by temperature-induced broadening of the Fermi function. In contrast, in the case of QDs the spectral width is practically independent of the temperature: FWHM of the ground-state peak changes from 40 to 49 nm within the temperature interval studied. We conclude that the linewidth of the QD structure reflects non-uniformity of quantum dots (size, chemical composition, etc) within the array. Meanwhile the full spectral interval of QD emission is much broader as it covers the wavelengths from ~1.05 to 1.3 μm as the result of the overlay of several quantum sub-levels. The QWD structure again demonstrates an intermediate behavior with FWHM that varies from 35 to 72 nm from liquid-nitrogen temperature to room temperature.

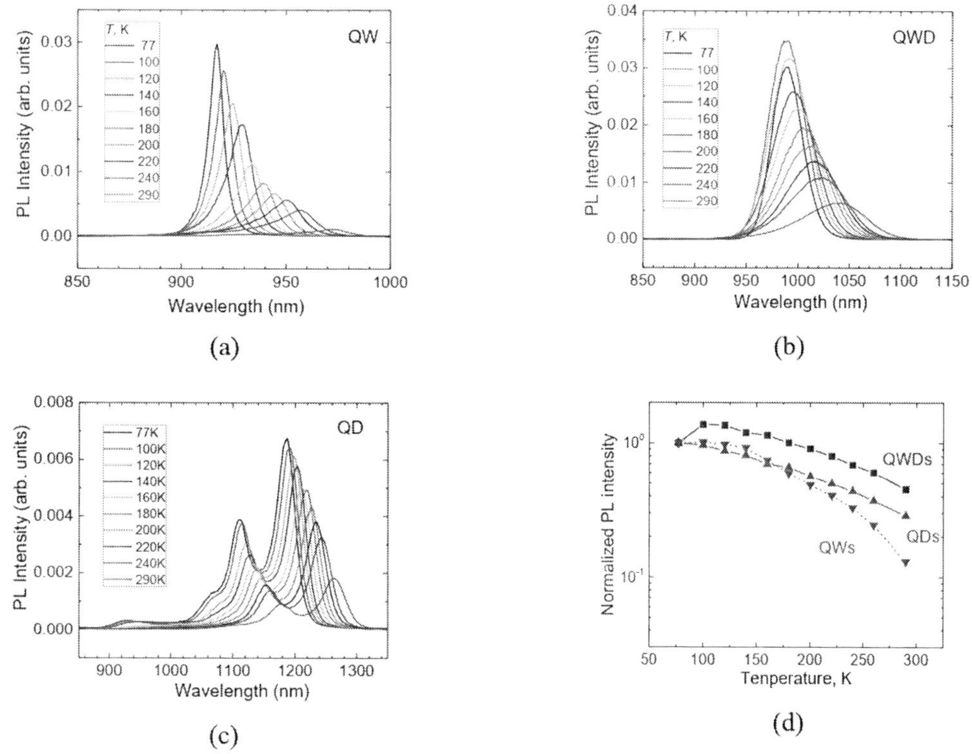

Figure 2. PL spectra of QW (a), QWD (b), and QD (c) mesa-structures with diameter of 10 μm taken at different temperatures; (d) normalized PL intensity as function of temperature

Figure 2(d) reveals temperature-induced variation of the integrated PL intensity of mesa-structures with diameter of 10 μm and different type of the active region. The intensities for each type of the structures are normalized to their values at the lowest measured temperature, i.e. the 77-K data are taken as 1. As the temperature rises to 290 K, the integrated intensity of luminescence decreases by 7.6, 2.2, or 3.5 times in QW-, QWD-, and QD-structures, respectively. We would like to emphasize that the QW- and QWD-structures have identical layer design (excepting the active region itself), were grown by the same method and in the same MOVPE apparatus. Therefore, it is hardly possible that these two structures are characterized by different concentrations of the growth defects. Taking into account that QWs are grown far from the border of the pseudomorphic growth, a noticeable quenching of PL signal at room temperature in the QW-based mesa-structure compared to the QWD-based counterpart can only be explained by stronger effect of non-radiative recombination at mesa sidewalls in the case of the QW mesas.

In order to further investigate this effect, we compare the luminescence properties of the mesa-structures having smaller diameters, where the contribution of the sidewalls should be more significant. The results are shown in Figure 3 where the integrated PL intensity measured at 77 K (a) or 290 K (b) is shown as a function of the mesa diameter for all three sorts of the active region. Again, the intensities for each type of structures are normalized to their 77-K values measured in the largest mesas.

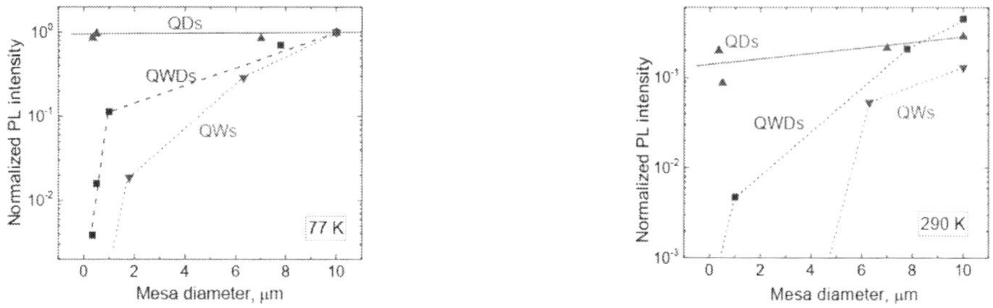

(a) (b)

Figure 3. Normalized luminescence intensity as function of mesa diameter as measured at 77 K (a) and 290 K (b).

We observed that in case of the QDs active region, the 77-K intensity is independent of the mesa diameter (up to the mesas as small as 0.3 μm). At room temperature, the intensity decreases with decreases the diameter by about 2 times. In the case of QWD active region, the PL intensity (both at 77 and 290 K) shows fast decrease with reducing the mesa diameter. The smallest mesas, where the PL signal was detected, was 1 μm (at room temperature) and 0.3 μm (at liquid nitrogen). In the case of QW active region, PL was observed only in 6-10 μm mesas, whereas in the smaller mesas the luminescence was practically absent.

4. Conclusions

According to the experimental results, the nonradiative recombination at the etched sidewalls can be crucial for characteristics of GaAs-based device with InGaAs QW active region. Free lateral diffusion of charge carriers along the QW plane gives them the possibility to reach the sidewalls. The negative effect of non-radiative recombination becomes more significant with increasing the temperature and decreasing the mesa size. Though InGaAs QWDs provide three-dimensional localization of the carriers, the shallow energy levels are thermalized at room temperature and nonradiative recombination starts play significant role. For these reasons the realization of the effective InGaAs QW or QWDs microdevices requires suppression of sidewall nonradiative recombination by surface passivation especially when the typical dimension become less than 10 μm. In case of InAs/InGaAs QDs, the etched sidewalls have negligible effect owing to the deep 3D localization potential. In its turn, this leads to a significant reduction of the diffusion length compared to quantum wells [5]. This makes QDs suitable for application in microdevices without passivation and/or in light-emitting devices grown on non-parent substrates, such as III-V on silicon.

Acknowledgments

This work was supported by the Russian Foundation for Basic Research (grants 16-29-03037, 16-29-03111), Russian Ministry of Science and Higher Education (3.9787.2017/8.9).

References

[1] M.V. Maximov, et al, Proc. 26th ICPS, IOP Publishing Conf. Ser. 171, p. 10, (2003).
[2] M.V. Maximov, et al, Nanoscale Research Letters 9 657 (2014).
[3] N.V. Kryzhanovskaya, et al, Electron. Lett. 51(17), 1354 (2015).
[4] E. Moiseev, et al, Optics Letters Vol. 43, No. 19, 4554-4557, (2018).
[5] A. Fiore, et al, Phys. Rev. B 70, 205311 (2004).

Topological plasmon-polariton on a Dirac magnet helical state

I V Iorsh[1], G R Rakhmanova[1] and M Titov[2, 1]
[1]ITMO University, Saint Petersburg 197101, Russia
[2]Radboud University, Institute for Molecules and Materials, NL-6525 AJ Nijmegen, The Netherlands

Abstract. A one-dimensional plasmon-polariton which is characterized by non-linear dispersion is supported by electromagnetic field interacting with topologically protected helical state. The electronic helical state arises at the surface of topological insulator in the close proximity of the ferromagnet. In a two-dimensional Dirac magnet these electro-optical excitations are limited by domain walls. Topological electronic imagining of domain walls can provide generally a novel tool to couple magnetic dynamics with both transport and optical properties of helical electronic states An exact dispersion relation for the topological plasmon-polariton were obtained.

1. Introduction

In 1987 edge magnetoplasmons excitations were investigated of an inhomogeneous 2D system in the quantum Hall effect regime [1]. We concidered a system, where similar quasiparticles arise. It is a surface of 3D topological insulator or topological semimetal in a close proximity to a ferromagnet thin film. A one-dimensional domain wall in the ferromagnet is imaged in the Dirac electron system as a zero mass line that supports a helical electronic state. The helical state at the domain wall results from the anomalous Hall effect in a Dirac magnet and the properties of such a state are similar to those of a quantum Hall edge state. Along the helical state one-dimensional plasmon-polariton excitations can be created by light polarized along the domain wall. We studied the properties of this hybrid light-matter quasiparticles. Moreover, these states can be used for the all-optical imaging of the domain wall structure of the ferromagnets using scattering SNOM technique (see Fig. 1). The physics can be realized in ZrSiS thin crystal that is weakly coupled to a ferromagnet film.

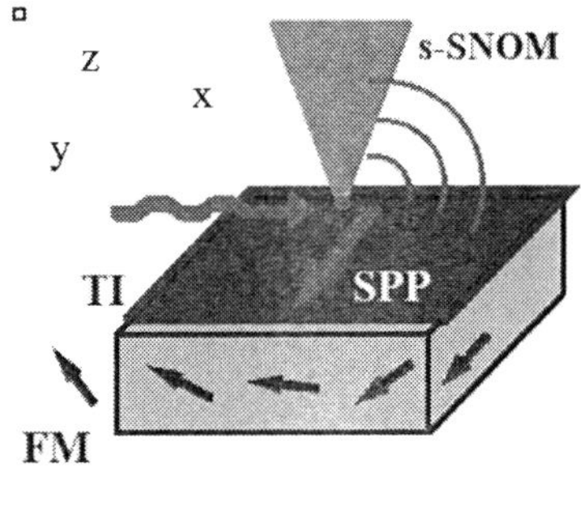

Figure 1. Optical imaging of the ferromagnetic domain wall via the excitation of the plasmon-polariton in the topological semimetal.

2. Dispersion of the topological plasmon polariton

The Hamiltonian of this system

$$H = m\sigma_z + v_F[\sigma \times p]_z \tag{1}$$

where the direction z is chosen perpendicular to the 2D surface, p is the momentum operator, σ is the three-dimensional vector of Pauli matrices, m is a magnetization of a ferromagnet, in the semimetal it is responsible for the energy gap, $v_F \approx 5 \times 10^5$ m/s is the effective velocity of Dirac quasiparticles.

Due to the interaction of the magnetization in the ferromagnet with the spins in the semimetal an energy band gap opens. Where ferromagnet has domain wall energy gap is closed.

Lets consider a domain wall that is elongated in y direction and has a characteristic width a_0 in x direction.

The edge states wavefunctions are

$$\psi_A = \psi_B = (4\pi a_0^2)^{-1/4}\left(\frac{\Gamma(1/2+\tilde{a})}{\Gamma(\tilde{a})}\right)^{1/2} \cdot \cosh^{-\tilde{a}}\left(\frac{x}{a_0}\right) = f(x)\ , \tag{2}$$

where $\tilde{a} = \frac{a_0 m}{\hbar v_F}$.

The state ψ is evidently an eigenstate of the current operator:

$$\hat{j} = \frac{dH}{dA} \tag{3}$$

$$\hat{j}_y = -ev_F\hat{\sigma}_x \tag{4}$$

$$\hat{j}_x = -ev_F\hat{\sigma}_y \tag{5}$$

All current components except along the direction are reset:

$$\langle\psi|\hat{j}_y|\psi\rangle = f^2(x)v_F \tag{6}$$

Conductivity can be calculated using the Kubo formula [2]. Assuming zero chemical potential, low temperature and light frequency smaller than the gap width, we can omit the contribution of the bulk states to the conductivity and take into account only the contribution from the localized state. The expression for the conductivity is given by

$$\sigma_{yy}(z, z', x, x', \omega, q_y) = \frac{ie^2}{\hbar} v_F \delta(z)\delta(z')f^2(x)f^2(x')) \int_{-m/\hbar v_F}^{m/\hbar v_F} dk_y \frac{n_f(\epsilon_{k_y-q/2}) - n_f(\epsilon_{k_y+q/2})}{q_y(w - v_F q_y)} \tag{7}$$

where n_f is the Fermi-Dirac distribution function, q_y is the wave-vector in y direction. The integral is almost a constant and equal to 1. The expression for the surface conductivity then simplifies. It can be seen that conductivity tensor has the only non-zero component, corresponding to both current and electric field aligned along the domain wall. This conductivity has a pole, where it goes to infinity, this pole corresponds to the excitation of a directed plasmon. Plasmon has linear dispersion ($w = v_F q_y$).

THz radiation applied to the system is capable of creating one-dimensional plasmon-polariton excitation along helical state. To find its dispersion it is necessary to solve the Maxwell equation:

$$\mathrm{rot\,rot}\,\vec{E}(\vec{r}, \omega) = k_0^2\vec{E} - \frac{4\pi}{c^2}\partial_t\vec{j}(\vec{r}, \omega) \tag{8}$$

where the linear response of the current density j to the electric field E is defined in the frequency domain by $\vec{j} = \int d^3\vec{r}'\hat{\sigma}(\vec{r}, \vec{r}', w)\vec{E}(\vec{r}', \omega)$

This equation has formal solution:

$$\vec{E}(\vec{r}, \omega) = i\frac{4\pi k_0}{c}\int d^3(\vec{r}')G(\vec{r}, \vec{r}', \omega)\int d\vec{r}''\sigma_{\alpha\beta}(\vec{r}', \vec{r}'', w)E_\beta(\vec{r}', \omega) \tag{9}$$

where G - is the dyadic Green's function in vacuum.
That allows us to get the linear algebraic equation system for $\Lambda(\omega, q_y) = \int dxdz\, f^2(x)\delta(z)E_y$:

$$\left[\hat{\mathrm{I}} - \mathrm{i}\frac{4\pi k_0}{c}\tilde{\sigma}(\omega, q_y) \int \mathrm{d}x\, \mathrm{d}x' f^2(x) f^2(x') G_{yy}(x, x', 0,0, q_y, \omega)\right] \Lambda(\omega, q_y) = 0 \quad (10)$$

The determinant of this matrix gives the dispersion equation for the topological plasmon-polariton. After some algebra, the dispersion equation can be written as

$$1 - \tilde{v}\tilde{q} = \alpha\tilde{v}(\tilde{q}^2 - 1) S(\omega, q_y, a_0, \tilde{a}) \quad (11)$$

Where $\tilde{v} = \frac{v_F}{c}, \tilde{q} = \frac{q_y}{c}$, $\alpha \approx 1/137$ - is the fine structure constant. The quantity S is given by

$$S = \int_0^\infty \mathrm{d}x \frac{|\Gamma(\tilde{a}(1+ix/2))|^4}{\Gamma^4(\tilde{a})\sqrt{\kappa^2+x^2}}, \quad (12)$$

where $\kappa = \frac{\hbar\tilde{v}\omega}{m}\sqrt{\tilde{q}^2 - 1}$

As $\tilde{a} \to 0$ the dispersion equation is simplified to

$$1 - \tilde{v}\tilde{q} = \frac{\alpha\tilde{v}(\tilde{q}^2-1)}{2}\left[\frac{\tanh^{-1}\kappa'}{\kappa'} - \frac{1}{\kappa'^2} + \frac{\tanh^{-1}\kappa'}{\kappa'^3}\right] \quad (13)$$

For the case of and some generic ferrromagnetic with saturation magnetization of 0.5 T and domain wall width $a_0 \approx$ 10 nm, we find that $m \approx$ 1.5meV, $v_F \approx$ 4.3×105m/s, $\tilde{a} \approx 0.06$ and thus approximate dispersion equation Eq. (13) is valid. In Fig. 2 we plot the dispersion of the plasmon polariton for different values of a_0 and thus $\tilde{a}$. It can be seen that the dispersion for the case of $\tilde{a} = 0.06$ almost fully coincides with the analytical expression for $\tilde{a} = 0$ given by Eq. (13). The discrepancies in the dispersion become visible at large $\tilde{a} \approx 1$. The dispersion has asymmetry. It depends on the ratio of the Fermi velocity and the speed of light. The faster the electrons move along the structure, the more asymmetric the picture.

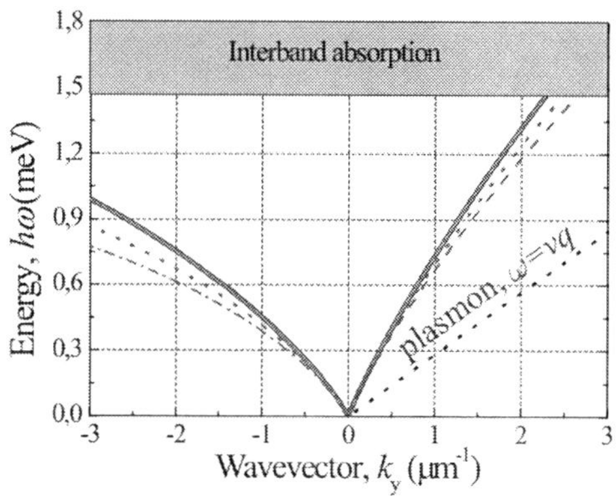

Figure 2. Dispersion of the topological plasmon polariton at different values of $\tilde{a}$. Black line corresponds to $\tilde{a} = 0$, green to $\tilde{a} = 0.06$, dotted red - to $\tilde{a} = 1$ and dashed blue to $\tilde{a} = 2$.

3. Experiment simulation

We considered excitation of a helical plasmon-polariton by a point source of terahertz radiation and looked as the scattered field of such structure looks (see Fig. 3). The dipole oriented along axis and placed at the position $r_0 = (x_0,0,z_0)$ in the vicinity of the domain wall.
The y component of the electric field is then given by

$$E_y(q_y, r) = G_{yy}(x - x_0, z - z_0, q_y) + \alpha\tilde{v} \int dx' G_{yy}(x - x', z - 0) f^2(x') \Lambda(q_y, \omega, x_0, z_0) \quad (14)$$

and Λ is given by

$$\Lambda(q_y, \omega, x_0, z_0) = \frac{\int dx'' dz'' G_{yy}(x''-x_0, z''-z_0) f^2(x'') \delta(z'')}{1-\tilde{v}\tilde{q}-\frac{\alpha\tilde{v}(\tilde{q}^2-1)}{2}\left[\frac{\tanh^{-1}\kappa'}{\kappa'}-\frac{1}{\kappa'^2}+\frac{\tanh^{-1}\kappa'}{\kappa'^3}\right]} \quad (15)$$

In Fig. 3 the dipole field that quickly fades out is shown and it excites a plasmon-polariton runs along a domain wall. As can be seen in the inset the interference between the incident field and the excited plasmon-polariton leads to occurring fast oscillations. The field from a source has its own wave vector, the phase changes with own speed, but the plasmon-polariton field has wave vector is much larger, the phase changes quickly. The spatial period of the oscillations can be estimated as $\frac{\pi}{k_0-q_p}$, where q_p is the plasmon-polariton wavevector. so it can be find out what the plasmon wave vector is equal to.

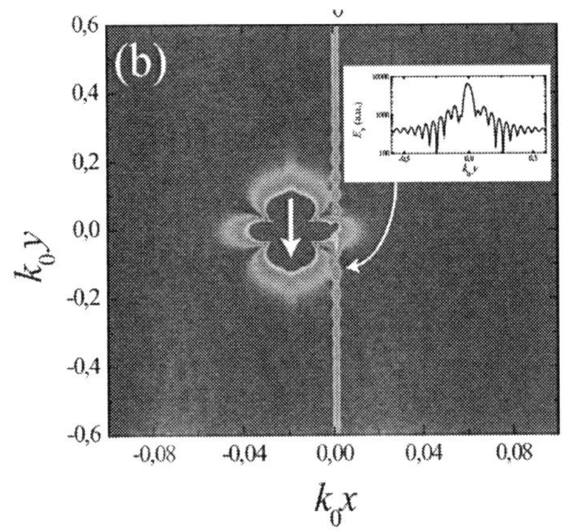

Figure 3. Real space map of y-component of the electric field (plasmon polariton is excited by a dipole and located in 10 microns from the domain wall).

To sum up we considered a helical electron state which supports edge plasmon-polariton, it can be formed along at a surface of topological insulator or topological semimetal that is in a close proximity to a ferromagnet thin film. Such a state is analogous to the quantum Hall edge state but arise due to the anomalous Hall effect. The exact dispersion equation for topological plasmon-polariton is derived. It is possible to get imaging of domain wall structure of the ferromagnetic via the expatiation of the edge plasmon polariton in the semimetal.

References

[1] Volkov V. A, Mikhailov S. A 1988 *Zh. Eksp. Teor. Fiz.* **94** 217-241
[2] H. Haug and S. W. Koch 2009 *Quantum Theory of the Optical and Electronic Properties of Semiconductors: Fivth Edition* (World Scientific Publishing Company)

The influence of iron concentration on the cathodoluminescence kinetics in the mid-IR range in ZnSe:Fe crystals

M Chukichev[1], R Rezvanov[2], V Chegnov[3], O Chegnova[3], S Mironov[4], A Gladilin[4], N Il'ichev[4], V Kalinushkin[4]
[1] Faculty of Physics, Lomonosov Moscow State University, Moscow 119991, Russia
[2] National Research Nuclear University 'MEPhI', Moscow 115409, Russia
[3] Research Institute of Material Science and Technology, Moscow 124460, Russia
[4] Prokhorov General Physics Institute of the Russian Academy of Sciences, Moscow 119991, Russia

Abstract. Integral intensity and kinetics dependences of cathodoluminescence of ZnSe:Fe crystals in mid-IR spectral range was studied. Concentration of iron varied from 0.01 to 14 wt.%. The experiments were carried out in 77 - 300 K temperature range. It is established that the luminescence pulse rise time is longer than excitation pulse at low iron concentration at 77 K temperature. The increasing ones lead to shortening of the luminescence pulse rise time, decreasing of the luminescence lifetime and increasing of the luminescence intensity in mid-IR. The annealed ZnSe:Fe concentration series in vapour Zn are reproduced the same trends. The nature of observed experimental effects is discussed.

1. Introduction

Zinc selenide crystals doped with iron are of interest as active laser media for the mid-IR range 3-5 μm. [1]. The ZnSe:Fe crystals demonstrate high lasing characteristics with optical pumping by HF or Er:YAG lasers [2-4]. However, the practical implementation of laser systems with such method is a difficult due to dimension and weight of excitation systems. The alternative pumping method of activator atoms is impact excitation with hot (dozen of keV) electrons. In [5] was reported the first study of cathodoluminescence (CL) spectral characteristics of ZnSe:Fe samples in the range of 3-5 μm. The CL spectrum coincides with luminescence spectrum obtained with optical pumping and is associated with $^5T_2$ и ^5E intracenter transition of Fe^{2+} ions [6]. In present investigation we demonstrated the influences of Fe concentration and annealing process in Zn atmosphere on the mid-IR CL kinetics of ZnSe:Fe.

2. Samples and experimental technique

The sample investigated were ZnSe plates doped with iron by thermal diffusion. The initial undoped ZnSe plates were cut of ingots produced using Bridgman method in graphite crucibles at the pressure of high-purity argon of 100 atm. The purity of ingot was 99.9992 wt.% observed by ICP-MC analysis.

The doping based on thermal diffusion of Fe from the surface. The thickness of the iron layer on the surface was about 150 μm. The doping carried out in quasi-closed quartz reactor at a temperature of 1000 °C for 20 hours.

In experiments we used the 2.5 mm thick ZnSe plate with the Fe concentration of 14 wt. % on the surface after diffusion. The concentration of iron was determined by the XRF using the Bruker Mistral M-1 spectrometer.

The samples with Fe concentration in the range of $5 \cdot 10^{18}$ to $3 \cdot 10^{21}$ cm^{-3} were produced by mechanical polishing of the doping surface controlled by 5-point concentration measurements within standard deviation of 0.05.

Thus, the studied range of iron concentrations was significantly expanded, since, the investigation with optical excitation of ZnSe:Fe are carried out at iron concentrations lower than $8 \cdot 10^{19} cm^{-3}$ [6, 8]. It should be noted that iron clusters and inclusions of other phases could be formed at concentrations higher than 0.5 wt. %. After CL experiments the samples were annealed in a quasi-closed volume in Zn atmosphere at a temperature of 1000 °C for 20 hours

The CL of crystals was excited by a pulsed electron gun at accelerated electron energy of 36 keV, the average electron penetration depth in the sample was ≈ 4 μm. The current of the electron beam in a pulse varied in the range from 0.01 to 0.2 A, the beam diameter was 8 mm, the average generation rate of electron-hole pairs in the excited layer in this case was 10^{24} - 10^{26} pairs $cm^{-3} s^{-1}$. The duration of the excitation pulse was 250 ns (non-annealed samples) and 800 ns (annealed samples. Electron irradiation and CL were observed at an angle of 45 degrees to the sample surface. The CL radiation was coupled out of the chamber through a CaF_2 window and focused on the input slit of an MS2004 (SOL Instruments Ltd) monochromator. As a detector, we used a PD42NB (IoffeLED Ltd) diode with a spectral sensitivity within the range 3.15 –4.75 mm. The time resolution of detection system (photodiode and amplifier) was about 50-100 ns. A signal from the detection system was sent to the input of a digital oscilloscope directly.

The CL kinetics were approximated by the exponent function:

$$y = A_1 e^{\left(\frac{-x}{t_1}\right)} + A_2 e^{\left(\frac{-x}{t_2}\right)},$$

where A_1 and A_2 are amplitudes, t_1 and t_2 are lifetimes. The intensity of the CL was determined by the maximum signal.

3. Experimental results and discussion

Figures 1(a, b) and 2(a, b) show the kinetics of the pulses of non-annealed and annealed samples at 77 and 300 K temperatures. The kinetics of CL at 300 K temperature was almost same of the non-annealed and annealed samples, was quite independent of the Fe concentration and was approximated by a single exponent with the lifetime in the range of 220 - 380 ns. The rise time varied in the range of 100 -150 ns and was close to the time constant of the detection system.

The CL kinetics of non-annealed sample with Fe concentration of 0.01wt.% observed at 77 K (fig. 1(a)) was approximated by two exponentials with lifetimes of 3.5 μs and 20 μs; the 0.05 wt.% sample – 750 ns and 3.8 μs; the CL of 0.1 wt.% sample approximate by single exponent with lifetime of 630 ns. The rise time for 0.01wt.% sample was about 450 ns, and 300-350 nm for 0.05 and 0.1 wt.% samples. The CL kinetics of the samples with concentration higher than 0.5 wt.% observed at 77 K temperature were similar to CL kinetics observed at 300 K (fig. 1(b)).

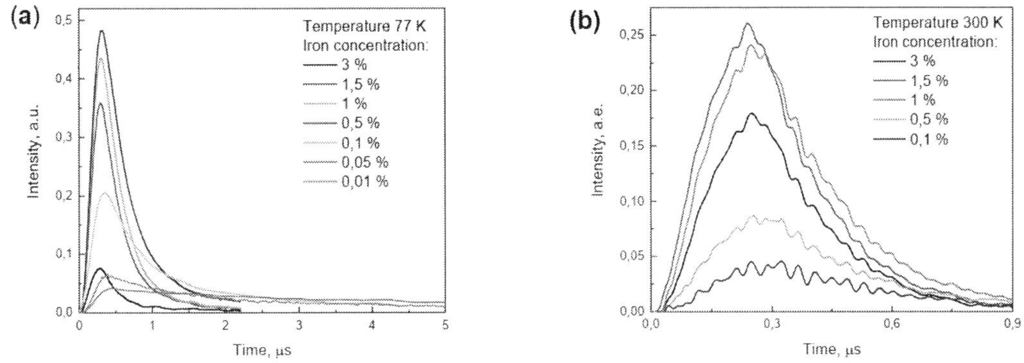

Figure 1(a,b). The cathodoluminescence kinetics of non-annealed samples at 77 K (a) and 300 K (b) temperatures.

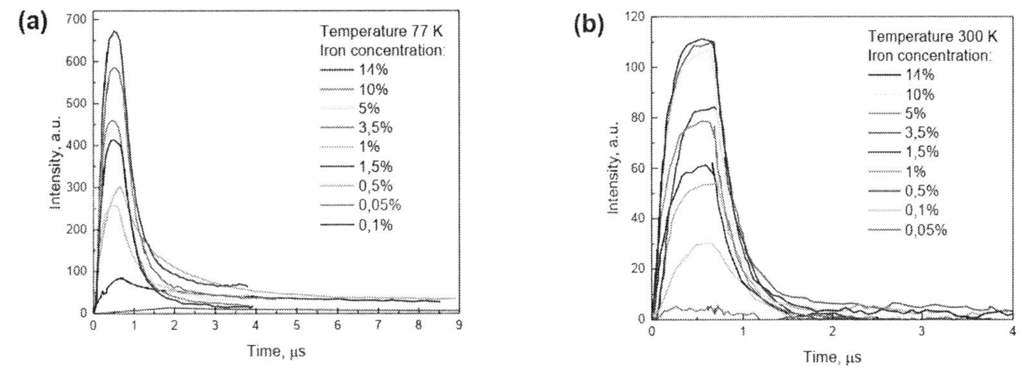

Figure 2(a, b). The cathodoluminescence kinetics of annealed samples at 77 K (a) and 300 K (b) temperatures.

Slightly differ results of CL kinetics of annealed samples were observed at 77 K temperature (fig. 2(a)). The CL lifetimes of samples with 1 wt.% and higher were about 350-380 ns.

The rise time of the CL was 100 – 150 ns and was less than the duration of the exciting pulse. Kinetics of the crystal with an iron concentration of 0.05 wt. % was close to that of the non-annealed samples with an iron concentration of 0.01 wt. %. The CL kinetics was approximated by two exponents with lifetimes of 2.8 μs and 19.5 μs. The rise time was about 3 μs. The CL kinetics of samples with an iron concentration in the range of the 0.05 wt. % and 1.5 wt. % were approximated by two exponents with lifetimes of 380 - 400 ns and 3.5-1.3 μs. The rise time of the luminescence intensity decreased from 600 ns to regular 100–150 ns with increasing iron concentration.

Therefore the rate of transition of carriers from the $^5T_2$ level to the ^5E level of Fe^{2+} ions at 77 K temperature strongly depends on the concentration of iron. Apparently, the interaction between iron ions significantly increases the transfer probability of carriers from those levels. The mechanism of this effect is unclear and requires further research. This effect in ZnSe:Fe crystals was observed for the first time in present paper. Consequently, the samples with non-homogenous distribution of iron ions in crystal volume (e.g. thermal diffusion based doping) had areas with different transition times from $^5T_2$ to ^5E at 77 K temperature. The observed decrease in the rise time of the luminescence intensity with increasing iron concentration at a temperature of 77 K can be associated with a change in the mechanism of excitation of iron ions. At low iron concentrations, the observed luminescence is the result of the recombination of non-equilibrium carriers generated by the electron flow through Fe^{2+} ions. The known mechanisms of such a process consist of several stages including the initial capture of an electron by a shallow donor and the subsequent transfer of excitation to the Fe^{2+} ion [9-11]. These processes can lead to extension of the rise time of the CL kinetics and that becomes longer than the duration of the excitation pulse. The iron concentration increasing leads to increase of the impact ionization probability of Fe^{2+} ions by "hot" electrons (including "primary" electrons). These processes are faster and stop after the end of the exciting pulse and after the "thermolization" of hot electrons. We suggest that at concentrations higher than 0.5 wt. % the mechanism of Fe^{2+} ions excitation by hot electrons impact ionization was dominated.

It should be noted that the results of CL kinetics excited with electron beam at 77 K were significantly differ from luminescence obtained with optical excitation into the absorption band of Fe^{2+} ions [6, 8]. The results obtained at 300 K (fig. 2(b)) temperature are similar. The kinetics of the luminescence with optical excitation of ZnSe:Fe samples with the maximum Fe concentration of $3 \cdot 10^{19}$ and $8 \cdot 10^{19}$ cm^{-3} at 77 K temperature was non-approximated by the exponent, and the lifetime decreasing was observed with iron concentration increasing.

The difference is that lifetime with optical pumping, obtained by the exponential part of the luminescence pulse, is much longer (about 40 μs) than the lifetime with electron pumping is 20 μs at samples with iron concentrations of $5 \cdot 10^{18} cm^{-3}$. Perhaps this trend associated with different experimental conditions. In case of electron excitation a lot of non-equilibrium carriers are generated. In the case of optical excitation of ions in their absorption band, the non-equilibrium carriers are not generated. The different rise time of CL in cases of non-annealed and annealed samples can be explained by changing of defective-impurity composition of the samples.

4. Conclusion

The results obtained in this paper allow us to draw the following conclusions:

The relaxation time of current carriers from level $^5T_2$ to level ^5E of iron ions at liquid nitrogen temperature strongly depends on the concentration of iron

The processes of excitation and relaxation of iron ions may be influenced by the impurity-defect composition and concentration of current carriers in ZnSe: Fe crystals.

At concentrations greater than 0.5 wt. % may be the change of the mechanism of iron ion excitation from recombination of non-equilibrium current carriers through them to impact ionization by hot electrons.

Acknowledgments
The reported study was funded by RFBR according to the research project № 18-29-20048.

References

[1] Fedorov V V, Mirov S B, Gallian A, Badikov D V, Frolov M P, Korostelin Y V, Kozlovsky V I, Landman A I, Podmar'kov Y P, Akimov V A, Voronov A A 2006 IEEE J. of Quantum Electronics **42** 907

[2] Martyshkin D V, Fedorov D V, Mirov M, Moskalev I, Vasilyev S, Mirov S B 2015 CLEO: OSA Technical Digest SF1F.2.

[3] Gavrishchuk E M, Ikonnikov V B, Kazantsev S Yu, Kononov I G, Rodin S A, Savin D V, Timofeeva N A, Firsov K N 2015 Quantum Electronics **45** (9) 823 –827

[4] Kozlovsky V I, Korostelin Y V, Podmar'kov Y P, Skasyrsky Y K, Frolov M P 2016 J. Phys.: Conf. Ser. 7**40** 012006

[5] Gladilin A A, Gulyamova E S, Danilov V P, Il'ichev N N, Kalinushkin V P, Odin I N, Pashinin P P, Rezvanov R R , Sidorin A V, Studenikin M I, Chapnin V A, Chukichev M V, 2016 Quantum Electronics **46** (6) 545 –547

[6] Myoung N S, Fedorov V V, Mirov S V, 2010 Proc. of SPIE **7578** 75781H-1

[7] Khanh T, Mozhevitina E N, Khomyakov A V, Avetisov R I, Davydov A, Chegnov V P, Antonov V, Kobeleva S P, Zhavoronkov N V, Avetissov I Ch 2017 Journal of Crystal Growth.**T. 457** 331 - 336

[8] Kernal J, Fedorov V V, Gallian A, Mirov S B, Badikov V V, 2005 Optics Express **13** 10608

[9] Surma M, Godlewski M, Surkova T R, 1994 Phys.Rev. B **50** 8319

[10] Tabei M, Shionoya S, Ohmatsu H 1975 Jpn. J. Appl. Phys. **14** 240

[11] Donnell K P O, Lee K M, Watkins G D 1983 J. Phys. C, Solid State Phys. **16** 723-728

Numerical modeling of the spectral and spatial distribution of the electromagnetic modes in a tunable microcavity for investigation of the light-matter interaction.

M A Lednev[1], **D S Dovzhenko**[1], **Yu P Rakovich**[1,2,3], and **I Nabiev**[1,4]
[1] National Research Nuclear University MEPhI (Moscow Engineering Physics Institute), 115409 Moscow, Russia
[2] Centro de Física de Materiales, (MPC, CSIC-UPV/EHU) and Donostia International Physics Center, 20018 San Sebastian, Spain
[3] IKERBASQUE, Basque Foundation for Science, 48013 Bilbao, Spain
[4] Laboratoire de Recherche en Nanosciences, LRN-EA4682, Université de Reims Champagne-Ardenne, 51100 Reims, France

Abstract. The light-matter interaction between a molecule and confined electromagnetic field is of great interest because it allows tuning the energy states and the spectral properties of the coupled matter. This effect offers a great number of applications in many areas, such as nonlinear physics, biosensing and lasing. The most widely used approach to achieve light-matter coupling is to place an ensemble of molecules inside an optical cavity. In order to maximize the effects of interaction, it is necessary to model the spectral properties of the cavity in order to find the optimal parameters for the experiments. In this study, the model for the numerical calculation of the spectral and spatial properties of electromagnetic modes of a tunable microcavity was developed and a mode analysis has been performed. The cavity transmission spectra and the electromagnetic field distribution were investigated. The results showed a good agreement with the experimental data obtained earlier.

1. Introduction

The resonant interaction between a confined electromagnetic field and molecules is of fundamental importance for a wide range of scientific fields and technology. One of the possible ways to confine the electromagnetic field is to employ an optical microcavity. When the rate of coherent energy exchange (coupling strength) exceeds the total losses of the hybrid system, the strong coupling regime is being reached [1]. This regime is characterized by the formation of the new states with the energy difference being proportional to the coupling strength [2]. In order to maximize the coupling strength, it is necessary to minimize the electromagnetic mode volume of the cavity. In practice, in order to find the optimal parameters of the experiment and to properly analyze its results, it is important to model the optical properties of a cavity.

Recently, in our group, tunable microcavity has been developed for the investigation of the light-matter coupling. Details on the design and implementation of the microcavity could be found elsewhere [3,4]. Briefly, the microcavity consists of unstable Fabry-Perot microresonator formed using the upper convex mirror and the lower plane metallic mirror placed on a piezoelectric actuator. To calculate the key optical parameters (Q-factor, mode volume, and field distribution) of the microcavity it is necessary to develop an appropriate mathematical model. This cavity with plane-convex mirrors have several advantages, such as high precision tunability of the resonance wavelength for coupling to the molecular transition which is provided by precise control of cavity length using the piezoelectric actuator and effortless sample replacement. The convex shape of the mirror ensures the confinement in transverse direction. Here, the numerical method of modeling should be used, as an exact analytical solution for this geometry seems to be impossible due to the convexity of one mirror and subsequent inapplicability of analytical methods developed for resolution of one-dimensional problems.

In the present study we propose a numerical approach for modeling the optical properties of the microcavity. We show that by accurately selecting the conditions and by taking into account the exact geometry of the microcavity, good agreement between the stimulated and experimental data can be reached. The modeled data allow very correctly estimate the key optical parameters of the microcavity modes.

2. Method

Our model relies on the finite elements method [5], which is based on the fragmentation of the calculating area, numerically solving the Maxwell's equations for each cell separately and connecting solutions in adjacent fragments. An important advantage of this method is simplicity of taking into account the material properties such as the refractive index and extinction coefficient.

The wavelength dependences of the refractive index and the extinction coefficient for silver films which were used as mirrors of our microcavity were taken into account using the data given in [6]. The space between the mirrors was filled with immersion oil. In our model we considered that the medium has no losses and its refractive index is constant and equal to 1.51.

3. Results and discussion

Firstly, the dependence of the electromagnetic field energy inside the microcavity on the wavelength of the electromagnetic wave was investigated. For the cavity length of 2.12 μm, the calculated spectrum was compared with the experimental transmission spectrum, that had been measured previously (Figure 1).

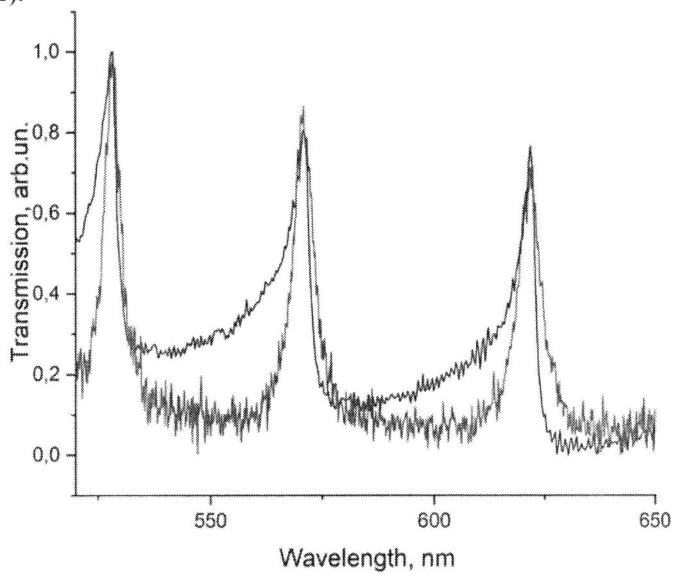

Figure 1. The calculated spectral distribution of electromagnetic field energy inside the microcavity (black), and the experimental microcavity transmission spectrum (red).

It can be seen that modeled spectrum adequately describes the positions of the transmission maxima which correspond to the longitudinal modes of the microcavity. However, in other spectral regions the modeled spectrum does not coincide with the experimental one. In particular, we can see that calculated peaks are wider and have an asymmetric shape. This fact is caused by the superposition of the accessory peaks corresponding to the higher transverse modes. Small difference in eigenenergies of these transverse modes leads to the merging into continuous high energy wing near each longitudinal mode.

For better agreement between the modeled and experimental spectra we took into account the square size of the measurement area of the experimental system in a confocal regime. The comparison of the corresponding calculated and experimental spectra is presented on figure 2.

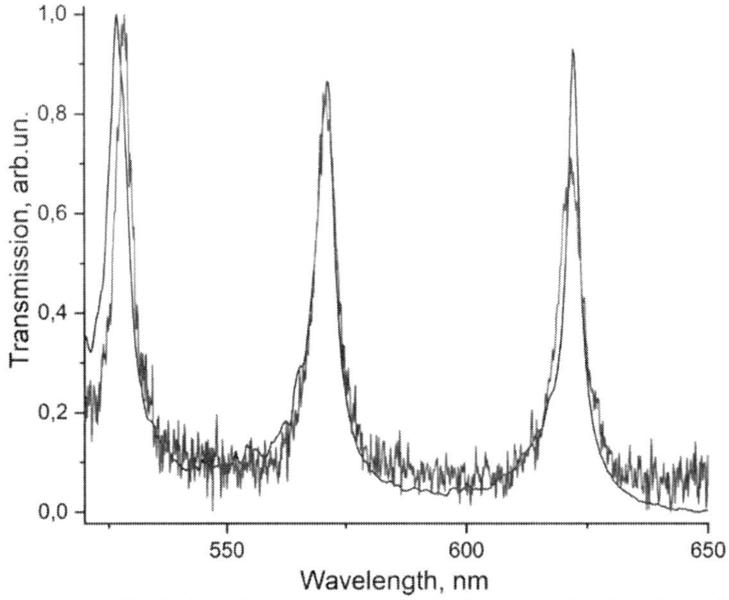

Figure 2. The calculated spectral distribution of the electromagnetic field energy inside the microcavity taking into account the confocal regime (black), and the experimental microcavity transmission spectrum (red).

As one can see reducing the area in transverse direction down to 400 nm for calculation the transmission spectrum allows to eliminate the accessory peaks in the modeled spectrum. Thus, decrease of calculating area made possible the mode selection by ignoring the higher transverse modes. After this correction (or procedure) the calculated spectrum agrees well with the experimental one. In particular, the positions and the widths of the modeled peaks coincide with the data that had been measured previously. This fact proves that the developed model correctly calculates the parameters of tunable microcavity such as the Q-factor which was estimated to be 110 for the central mode on the graph.

Also, for the central peak of the spectra, the spatial distribution of electromagnetic energy inside the cavity was computed (Figure 3).

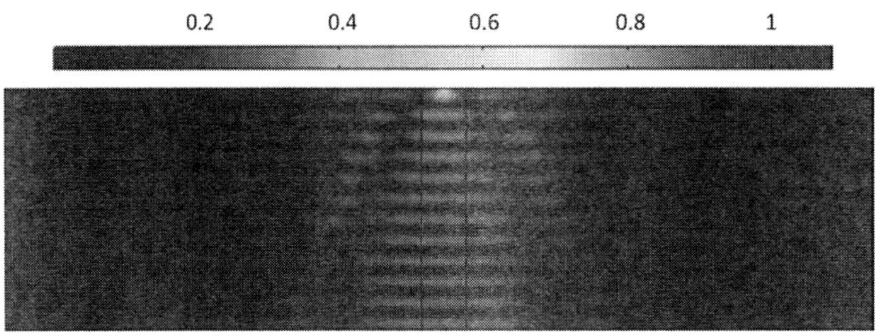

Figure 3. The spatial distribution of the electric field inside the cavity for the eigenmode with the resonant wavelength of 570nm. The top and bottom edges of the figure correspond to the convex and plane mirrors, respectively The vertical black solid lines limit the area used for the calculation of the transmission spectrum (the distance between them is equal to 400 nm). The horizontal image size characterizes the effective area of field localization, which is equal to 7 μm.

The figure shows that for the eigenmode with resonant wavelength of 570 nm the electromagnetic field inside the cavity is relatively highly confined. For the numerical evaluation of this confinement area the mode volume was calculated using the following formula:

$$V_m = \frac{\int_V |E|^2 dV}{|E_{max}|^2}, \tag{1}$$

where E is the electric field strength and E_{max} is the maximum of the electric field strength inside the cavity. The mode volume was estimated to be 70 in units of $(\lambda/n)^3$. For comparison, the typical values of the effective modal volume in optical microcavities are $\sim 10^5$ [2]. Due to the inverse square root dependence of the coupling strength on the mode volume, such a decrease in the value of V_m can lead to an increase in the coupling strength by more than an order of magnitude.

4. Conclusion

In this study, the model for the numerical calculations of the spectral properties of a tunable microcavity was developed and the mode analysis was performed. The cavity transmission spectra and the electromagnetic field distribution were investigated. The results showed a good agreement with the experimental data obtained earlier. Importantly, the calculated value of the mode volume of the eigenmode with a resonant wavelength of 570 nm has been found to be at least three orders of magnitude smaller than that of the typical planar Fabry-Perot microcavities, which are widely used for the investigation of the light-matter interactions.

Acknowledgments

The reported study was funded by RFBR according to the research project №18-29-20121.

References

[1] Skolnick M, Fisher T and Whittaker D 1998 *Semicond. Sci. Technol.* **13** 645
[2] Dovzhenko D, Ryabchuk S, Rakovich Yu and Nabiev I 2018 *Nanoscale* **10** 3589
[3] Mochalov K, Vaskan I, Dovzhenko D, Rakovich Yu and Nabiev I 2018 *Rev. Sci. Instr.* **89** 053105
[4] Dovzhenko D, Mochalov K, Vascan I, Kryukova I, Rakovich Yu and Nabiev I 2019 *Opt. Express* **27** 4077
[5] Jin J 1993 *The finite element method in electromagnetics* (Hoboken, New Jersey: John Wiley & Sons)
[6] Ciesielski A, Skowronski L, Trzcinski M, Szoplik T 2017 *Appl. Surf. Sci.* **421B** 349

Ultra high-speed all-optical coherent memory cell

S N Bagayev[1], I B Mekhov[2,4], V G Nikolaev[3], I A Chekhonin[2], M A Chekhonin[2]

[1]Institute of Laser Physics, Novosibirsk, Russia
[2]St.Petersburg State University, St.Petersburg, Russia
[3]ITMO University, St. Petersburg, Russia
[4]University of Oxford, UK

E-mail: chekhonin@mail.ru

Abstract. We study the interaction of two counterpropagating unipolar video pulses of electromagnetic radiation in a dense resonant two-level medium. The pulse durations are less than one oscillation period of an atomic transition. We show that a polariton cluster (i.e. the compact long-living strongly coupled state of electromagnetic field and matter polarization) is created, when two unipolar pulses collide in a resonant medium of the frequency ω_o (the pulses correspond to self-induced transparency solitons of the same amplitudes and opposite polarities).
We have studied for the first time the conditions of maximal excitation of a resonant polarization in the polariton cluster volume. We show that there exist optimal durations and amplitudes of the recording pulses. We studied the processes of double and multiple recording and erasing of an optical memory cell in a thin layer of a resonant medium (quantum dots). We demonstrated that the pulse rate of the recording-erasing pulses can reach 60 000 GHz and higher, which happens under the coherent regime of the interaction between the electromagnetic field and polarization of the medium.

1. Introduction

Recently, the generation, coherent propagation, and interaction in dense resonant media of ultrashort (few-cycle, one-cycle, sub-cycle) pulses of the electromagnetic field has attracted strong interest. Modern methods of generating unipolar video pulses are reviewed in [1,2], the effects of coherent interaction between counterpropagating pulses are considered in [3-6] as well.

Here we theoretically study the processes of formation and dynamics of *polariton clusters*, which are created during collisions between unipolar pulses of the amplitudes of opposite signs (solitons of the self-induced transparency).

The correct description of propagation and interaction of ultrashort (one-cycle, sub-cycle) solitons is possible only by eliminating the approximations imposed by the slowly varying envelope (SVEA) and rotating-wave approximations (RWA) both in time and spatial coordinate. The theory of propagation of such solitons in dense resonant media is well established, cf. for example [7-13].

Dynamics of a quantum two-level particle with the dipole moment d and transition frequency ω_o is described by the Bloch equations [7,8] for the pseudospin vector projections $\boldsymbol{s} = (s_1, s_2, s_3)$:

$$\dot{s}_1 = -\omega_0 s_2 - \frac{1}{T_2} s_1 \quad (1)$$

$$\dot{s}_2 = \omega_0 s_1 + 2\frac{d}{\hbar} E(t,z) s_3 - \frac{1}{T_2} s_2 \quad (2)$$

$$\dot{s}_3 = -2\frac{d}{\hbar}E(t,z)s_2 - \frac{1}{T_1}(s_3+1) \tag{3}$$

$$P(t,z) = N_o \cdot d \cdot s_1 . \tag{4}$$

Here $E(z,t)$ is the electric filed amplitude of a pulse propagating along the coordinate z, T_1 и T_2 are the longitudinal and transversal relaxation times, $P(z,t)$ is the medium polarization, N_o is the resonant particle density. The physical meaning of the pseudospin projection s_3 is the normalized population difference of a two-level medium: $s_3(z,t) = N(z,t)/N_o$.

Dynamics of the electromagnetic field propagation is described by Maxwell's equations:

$$\frac{1}{c}\frac{\partial H_y}{\partial t} = -\frac{\partial E_x}{\partial z} \tag{5}$$

$$\frac{1}{c}\frac{\partial E_x}{\partial t} = -\frac{\partial H_y}{\partial z} - \frac{4\pi}{c}\dot{P} \tag{6}$$

or by the wave equation

$$\frac{\partial^2 E}{\partial t^2} - c^2\frac{\partial^2 E}{\partial z^2} = -4\pi\ddot{P} . \tag{7}$$

Here E_x and H_y are the projections of electric and magnetic fields of the pulse on the transverse axes x and y, c is the speed of light in vacuum.

In the absence of relaxation ($T_1 = T_2 = \infty$), the Maxwell-Bloch equations have analytical single-soliton solutions of the self-induced transparency theory [7-13]:

$$E = \pm\frac{\hbar}{dt_p} sech\left(\frac{t - zV^{-1} - t_0}{t_p}\right) \tag{8}$$

$$V = \frac{c}{\sqrt{1 + \frac{2\alpha\tau_p^{\,2}}{1+\tau_p^{\,2}}}} \tag{9}$$

$$\alpha = \frac{4\pi}{\hbar}\cdot\frac{N_0 d^2}{\omega_0} = \frac{1}{t_c\omega_0} . \tag{10}$$

Here V is the soliton propagation velocity along the coordinate z, t_p is the soliton duration, $\tau_p = \omega_o t_p$.

2. Recording and erasing polariton clusters. All-optical coherent memory cell.

In this work we have studied the process of inelastic collision between two self-induced transparency solitons (8) in a resonant medium. We theoretically considered a problem of two identical counterpropagating unipolar solitons of either the same or different polarities. We solved numerically the Maxwell-Bloch equations (1)-(3) and (5)-(6) without using the slow amplitude approximations (SVEA, RWA) both in time and spatial coordinate. We took into account the finite relaxation times of a medium T_1 и T_2. The details of the simulation methods are described in [4].

We studied theoretically for the first time the optimal conditions for creation of polariton clusters, which appear in the process of collision between unipolar pulses of opposite signs. In particular, we considered a case, where in a thick layer (≈ 5λ, 3 mkm) of resonant medium, two pulses of the duration τ_p, equal to 1/3 of the atomic transition period (τ_p = 0.67 fsec), collide (cf. Figures 1 and 2). The light-induced stationary polariton cluster appears in the pulse collision point and it has a size of ≈ λ/5 (120 nm). Within the cluster, the medium polarization $P(z,t)$ oscillates with the amplitude close to the maximal one ($s_1 \approx 1$) at the atomic transition frequency ω_o, but its emission is forbidden in the far field (cf. Figure 1). The maximal lifetime of polariton cluster (optical memory time) is determined by the time T_2. The repeated action of two pulses of opposite polarities on the polariton cluster leads to its destruction (cf. Figures 1 and 2).

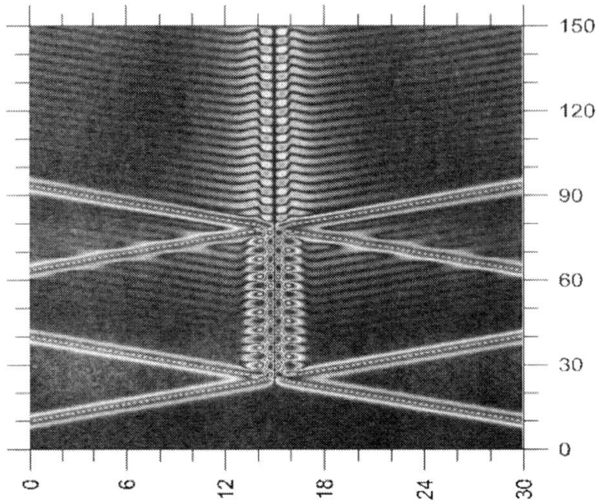

Figure 1. Dynamics of the polarization $P(z,t)$ in the polariton cluster during its recording and erasing. The horizontal axis represents the unitless propagation coordinate Kz, the vertical axis represents the unitless time $\omega_0 t$. $K = 2\pi/\lambda$ is the wave vector, $\lambda = 600$ nm. The pulse rate is 60 000 GHz (60 THz). The pulse duration is $\tau_p = 0.67$ fs ($\omega_0\tau_p = 2$), $T_1 = T_2 = 1$ ps.

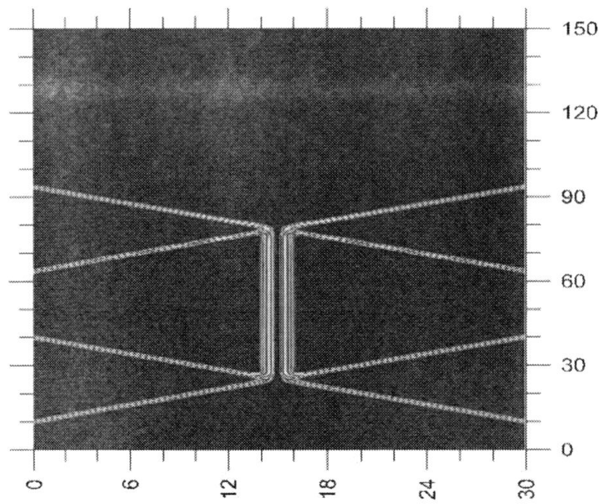

Figure 2. Dynamics of the population difference $N(z,t)$ in the polariton cluster during its recording and erasing.

Therefore, the polariton cluster has all necessary properties of the RAM memory cell: properties of the recording and erasing, enough information storage time. From a practical point of view, a convenient method to obtain two counterpropagating unipolar pulses of opposite polarities is based on the pulse reflection from a mirror [6].

The problem under consideration has two aspects. The first aspect is the study of physical processes leading to the suppression of the radiation of a polariton cluster in the far field in a dense medium. This question was discussed in details in [4] for the case of few-cycle pulses and in [6] for the case of the one-cycle pulses.

The second aspect is the study of such conditions of the cluster recording that simultaneously realize the radiationless regime of the cluster existence and the oscillation amplitude of the resonant polarization reaches the limiting value of $s_1 = 1$.

In Figure 2, we show dynamics of the normalized population difference $s_3 = N(z,t)/N_o$ during the collision of two unipolar solitons. One can see that the long-living polariton cluster is formed during the inelastic collision between the solitons of the opposite polarity near the collision point z_o (Kz_o = 15). The cluster has a double spatial structure. By varying the soliton duration t_p, we have shown that the quantity $<N(z)>/N_o$ averaged over the cluster existence area Kz = [14...16] and the temporal range of $\omega_o t$ = [37…40] has a well-defined maximum at the soliton duration $\omega_o t_p = 0.755$ (cf. Figure 3).

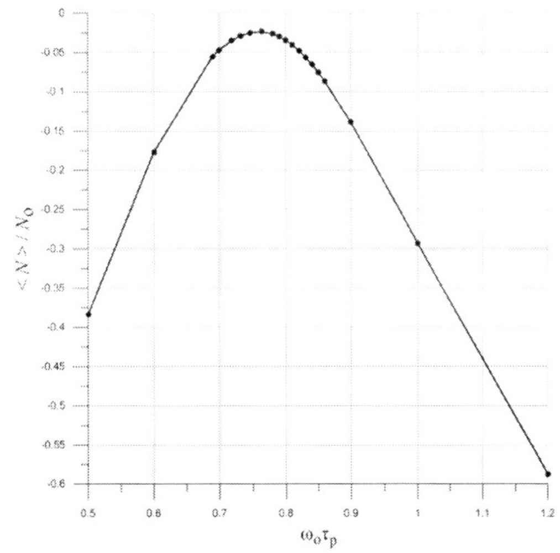

Figure 3. The dependence of $< s_3 > = <N(z,t)>/N_o$ on the duration of colliding solitons t_p. T_1 = 1 psec, T_2 = 1 psec, $d = 5D$, $\omega_o = 1\cdot 10^{15}$ rad/sec.

At the optimal soliton duration ($\omega_o t_p = 0.755$), the oscillation amplitude of the resonant polarization in polariton cluster reaches its maximal value of $s_1 = 1$.

As we see in Figure 2, the cluster has a double spatial structure with maxima of $N(z,t)$ at the points z' and z''. Figure 4 shows the dynamics of resonant polarization of a medium at the points z' and z''. One can see that the polariton cluster represents a very compact coupled structure of the resonant medium polarization and electromagnetic field: $z'' - z' \approx \lambda/20$.

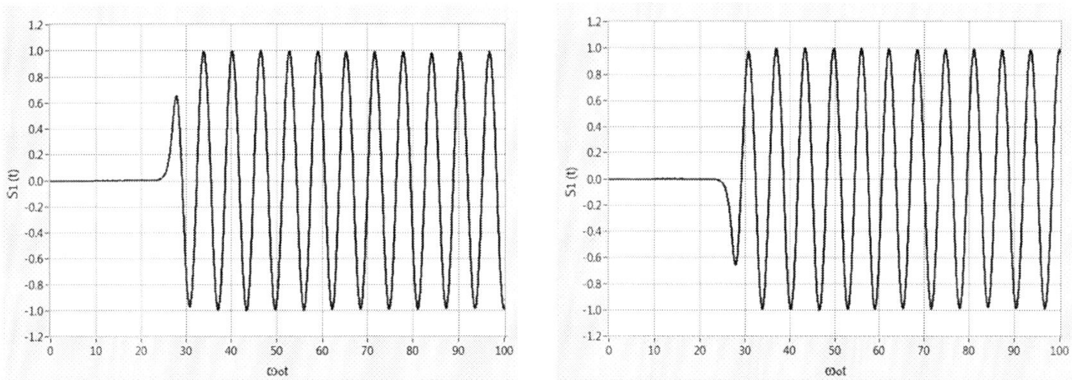

Figure 4. Temporal behaviour of the resonant polarization in two points of the polariton cluster z' и z'' near the point of the pulse collision z_o (Kz_o = 15). Left panel: $Kz' = Kz_o - 0.16$, right panel: $Kz'' = Kz_o + 0.16$.

At the change of the soliton duration t_p, its amplitude E_o changes as well self-consistently (cf. (8)). In this part of the work, we studied the dependence of the time of life of the polariton cluster and the polarization oscillation amplitude on the amplitude of the colliding pulses E_{op}, when the soliton duration is optimal and is fixed ($\omega_o t_p = 0.755$).

Figure 5 shows the amplitude dependence of the medium excitation in the polariton cluster on the ratio between the amplitude of the colliding pulses E_{op} and the amplitude of the solitons of self-induced transparency E_o (8).

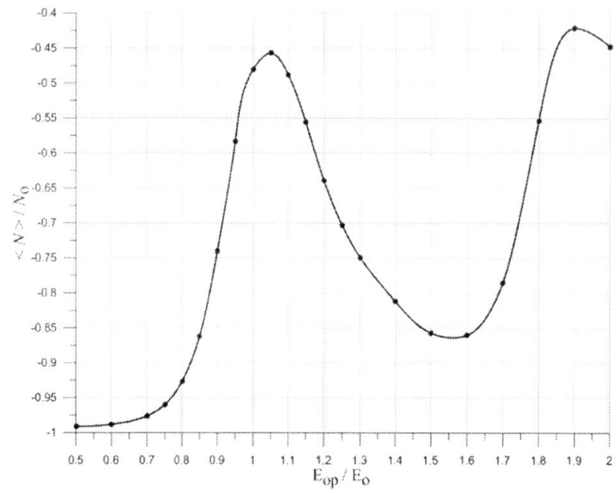

Figure 5. Dependence of $<N(z,t)>/N_o$ on the pulse duration E_{op}.
$T_1 = 1$ psec, $T_2 = 1$ psec, $d = 5D$, $\omega_o = 1 \cdot 10^{15}$ rad/sec, $\omega_o t_p = 0.755$.

The study of the radiative decay time of the polariton cluster also showed that it is maximal at $E_{op} / E_o \approx 1$ and $E_{op} / E_o \approx 2$.

Therefore, we demonstrated that the use of self-induced transparency solitons (2π-pulses) with the optimal duration ($\omega_o t_p = 0.755$) for recording information in the optical memory cell is optimal for three reasons. The colliding solitons are stable during the propagation in a resonant medium, they iduce in the polariton cluster resonant polarization with the maximal amplitude, and they create the polariton cluster with the maximal lifetime.

3. Conclusion

In conclusion, we have systematically studied an effect of creation of *polariton cluster* (i.e. a compact long-living strongly coupled state of the electromagnetic field and medium polarization). We performed numerical solution of the problem of inelastic collision between two counterpropagating solitons of the self-induced transparency effect.

We show that there exist three conditions for the efficient creation of polariton clusters (memory cells).

First, the colliding pulses should have the opposite signs. In this case, near the collision point of the unipolar pulses, the field $E(t)$ has a form of oscillating one-cycle pulses.

Second, the duration of colliding pulses τ_p should satisfy the condition of the resonance between the one-cycle pulse and atom frequency ω_o.

Third, the amplitude of unipolar pulses should be equal to the amplitude of solitons in the self-induced transparency theory (2π-pulses of the duration τ_p).

We demonstrated that the self-induced transparency solitons (2π-pulses) with the optimal duration ($\omega_o t_p = 0.755$) are the optimal pulses for recording information in a optical memory cell.

Acknowledgements
This work was supported by Russian Science Foundation (project № 17-19-01097).
I.B.M. acknowledges additional support by the Russian Foundation for Basic Research (project № 18-02-01095).

References

[1] Frantzeskakis D J, Leblond H, Mihalache D 2014 Rom. J. Phys. **59** (7-8) 767-784
[2] Arkhipov R M, Pakhomov A V, Arkhipov M V, Babushkin I, Tolmachev Yu A, Rosanov N N 2017 JETP Lett. **105** (6) 408-418
[3] Rosanov N N, Semenov V E, Vyssotina N V 2007 Laser Physics **17** (11) 1311–1316
[4] Bagaev S N, Egorov V S, Nikolaev V G, Chekhonin I A, Chekhonin M A 2015 Russian Journal of Physical Chemistry B **9** (4) 582-586
[5] Novitsky D V 2012 Phys. Rev. A **85** (4) 043813-1 - 043813-7
[6] Bagaev S N, Egorov V S, Nikolaev V G, Mekhov I B, Chekhonin I A, Chekhonin M A 2018 J. Phys.: Conf. Ser. **1124** 051018
[7] Allen L, Eberly J H, *Optical resonance and two-level atoms* (Wiley, New York, 1975)
[8] Bullough R K, Caudrey P J, Eilbek J C, Gibbon J D 1974 Opto-electronics **6** 121-140
[9] Eilbeck J C, Gibbon J D, Caudrey P J, Bullough R K 1973 J. Phys. A **6** 1337-1347
[10] Bullough R K, Jack P M, Kitchenside P W, Saunders R 1979 Phys. Scripta **20** 364-381
[11] Maimistov A I, Caputo J G Optics and Spectroscopy 2003 **94** (2) 245–250
[12] Kalosha V P, Herrmann J Phys. Rev. Lett. 1999 **83** 544-547
[13] Bullough R K, Ahmad F 1971 Phys. Rev. Lett. **27** (6) 330-333

Optical trapping and arrangement with reconfigurable "bottle" beam for digital holographic microscopy

N V Shostka[1], **B N Sokolenko**[1,2], **O S Karakcheva**[1,2], **D A Poletaev**[2], **A O Titova**[2], **A V Prisyazhniuk**[2], **I A Ismailov**[2]
[1]Department of Science and Research, V.I. Vernadsky Crimean Federal University, Simferopol 295007, Russia
[2]Institute of Physics and Technology, V.I. Vernadsky Crimean Federal University, Simferopol 295007, Russia

Abstract. The design and construction of optical tweezers based on uniaxial crystal anisotropy for generation of adjustable "bottle" beam trap carrying optical vortex with orbital angular momentum is considered. In coupling with digital holographic microscopy, optical trapping becomes a high precision instrument for visualization, shape definition and refractivity measurements of isolated micro structures and biological objects in-situ. The non-destructive and sterile non-contact tweezing of specimens or their parts in localized intensity minima of coherent vortex beam was performed with using of 200 mW semiconductor 532 nm trapping laser and $LiNbO_3$ crystal. Visualization and position control of trapped marine centric diatoms was performed by a lens-free axial digital holographic microscopy in liquid medium.

1. Introduction

One of the fundamental objects in the field of singular optics is an optical vortex, which has unique properties of maintaining continuous zero intensity on the phase uncertainty line and has ability to carry out its own orbital angular momentum, which is applicable in practice. These properties of singular beams are widely used in the interaction of the electromagnetic field with matter, in particular, high-resolution microscopy and in devices for trapping and manipulation of microparticles [1-3]. Optical traps have been used in biological and medical research, for example, in studying the mechanism of RNA polymerase movement during the synthesis of new DNA molecules [4]; the interaction of neutrophils with pathogens has been studied [5]. In [6] infrared optical tweezers were used to capture and manipulate red blood cells for cleaning of the blocked microvessel inside the subcutaneous capillaries of live mice, in such a way, this research significantly expanded the use of optical tweezers for studying the dynamics of living cells in animals. Special attention is paid to the laser tweezers and femtosecond scalpel manipulators for noncontact microsurgery laser-mediated polar body and biopsy of early mammalian embryos [7]. All mentioned above researches implement two-dimensional trapping in fluid media, thus the development of methods for three-dimensional manipulation in any surroundings become an actual task of modern research in optics and its application to microbiology, biophotonics and medicine.

Recently, optical tweezers with feedback have been used to apply arbitrary potentials to a colloidal particle. The feedback trap detects particle position, calculates the force based on the imposed "virtual potential" and shifts the centre of the trap to generate the desired force [8]. In 2018, it was demonstrated that with using of spatial light modulators it is possible to simultaneously form a large number of optical tweezers, and using them in a real time makes possible to perform various kinds of manipulations with position of trapped particles or living cells [9] and to capture absorbing particles for position stabilization and orientation, also to control over several specimens simultaneously [10].

The searching for ways of forming optical structures carrying a stable three-dimensional tweezers is in focus of attention of many research groups. The using of singular beams as optical traps expands the range of their application to the gaseous medium. In such environments, the trapping becomes much more complicated in comparison to capturing in liquids, especially due to the uncompensated gravitational force and significant impact of forces associated with the thermal interaction of the captured particle with the gaseous medium. The advantages of optical trapping is in non-invasive, sterile interaction, with no special specimen preparation requirements, but using of conventional microscopy for visualization of trapped objects imposes restrictions and eliminates some of them.

In optical microscopy, the question of creation of compact imaging devices with noncontact trapping of studying object become a research challenge. Especially, in recent years, a growing number of publications indicates an interest to that question in biophotonics and medical applications due to the limitations of widely used conventional light microscopies with lens systems. Since most of the specimens are transparent, from the point of view of optics – a phase objects, it makes difficulties in applying of bright-field light microscopy or requires more sophisticated technological solutions based on confocal, dark-field, polarization or interference microscopy [11-13]. Such equipment has complex optical systems with lenses and polarization elements, prisms etc., which brings to the image, in addition to useful, also negative qualities: aberrations and wavefront distortions. Moreover, it takes up a lot of space and has special requirements for operation, which limits its using in research field, where the specimens has linear dimensions which doesn't exceed a 100 um or limited position in space. In particular, more effective and versatile analytical devices are needed when studying the behaviour of specimens in their natural environment, such as atmospheric aerosols [14], the dynamics of the movement of marine plankton [15], cellular processes [16], where it is not always possible to transfer them into the microscope stage [17].

In combination with optical trapping by "bottle" beams, which, in contrast to trapping with gradient forces, allow moving the specimen within a few centimetres, the using of conventional microscopy would be difficult for observation of objects due to the shallow depth of field and relatively small numerical aperture. In this case, the optimal solution is the choice of digital holography as a method of imaging [18]. Digital holographic microscopy is used as a non-destructive and contactless method and allow performing visualization with only one holographic image with numerically focus adjustment in a range much larger than a mechanical tuning in optical microscope.

Thereby, optical trapping in liquid and gaseous medium allows contact-free observation of particles under relevant conditions and provide access to manipulation over long period of observation time, while the digital in-line holographic microscopy provides height resolution maps of spatial particle positions. In this paper, we show the flexible ways to control "bottle" beam trap parameters by tilting of birefringent crystal which cause the dependence of the loss of a symmetric field pattern in the focusing region on the efficiency of the optical trap. By adjustment of crystal angular position, the shape of the trapping zone can be controlled with high accuracy. Visual observation of captured specimens was performed with in-line digital holographic lens-less microscopy.

2. The "bottle" beam generation for optical trapping

Three-dimensional optical trapping can be performed using a "bottle" beams, a unique property of which is the maximum of the light intensity surrounds the central region which is "empty" from the light energy [19]. Once in such a beam, the absorbing particle is retained in its axial region, and under certain conditions cannot penetrate through the walls of bright light. Any shift of a particle leads to the heating of its part that is farther from the centre of the beam. The resulting pressure difference on the cold part of the particle closest to the centre of the beam and the hot peripheral returns the particle to a position of stable equilibrium inside the light beam with the "bottle" shape in sagittal section. Currently, great interest to singular optics is caused by studies of propagation and transformation of the laser beams structure after anisotropic crystals [19-22]. It arose due to the fact that anisotropic medium makes it possible to form not only the arrays of phase and polarization singularities in the beam, but also to control their shape and mutual arrangement. It is known [23] that the structure of a circularly polarized Gaussian beam propagating along the optical axis of a crystal changes in such a way that an optical vortex with a double topological charge appears on its axis in its orthogonal polarized component.

One of the highly efficient methods for generating a bottle beam is the method, based on the use of a birefringent crystal. Gaussian beam passing along of crystal optical axis represents a superposition of ordinary and extraordinary beams having different radii of the wave front curvature [19]. When outgoing beam is focused by a lens located at a certain distance from the exit face of the crystal, three focus areas will be formed: the first focus corresponds to the ordinary beam, the second one to the extraordinary beam, hence, a zone between them corresponds to the field on the output face of the crystal. Such a structure has the form of a classic "bottle" beam with a minimum on the axis, evenly surrounded in all coordinates by a high intensity zone.

Let us consider the effect of focusing a Gaussian beam propagating at a small angle to the optical axis of a uniaxial crystal and the formation of a "bottle" beam. The distance between two foci will depend in this case on the orientation of the optical axis of the crystal and the direction of beam propagation $\delta(\varphi) = L\sin^2\varphi(n_o^2 - n_e^2)/(n_e^2 n_o)$. Positions of foci are given by z_o and z_e that corresponds to ordinary "*o*" and extraordinary "*e*" beams respectively. When the inclination angle of the Gaussian beam is changed with respect to the optical axis of the crystal, the patterns of the intensity and polarization distributions begin noticeably change [24]. The distance δ between the two foci is not permanent and varies with the inclination. The calculated dependence of the distance between two foci on inclination angle of the crystal is illustrated in Figure 1.

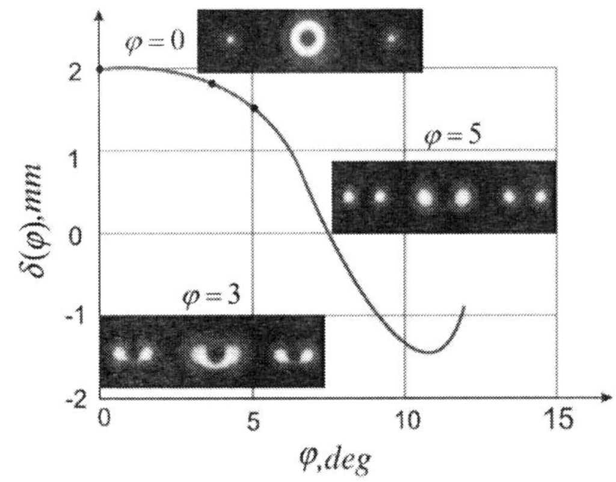

Figure 1. Computer modulated dependence of distance between foci δ on angle φ of tilt of optical axis of uniaxial crystal with respect to the beam axis.

From analysis of the plotted curve, we may notice the dependence of the distance between two beam waists on the inclination angle of the crystal optical axis relatively to the z-axis. It can be seen that while angle increases, the distance δ gradually decreased. When the angle is 5 degrees, ordinary and extraordinary beams are focusing in one plane. With further increasing of inclination angle, the beam waists change places with a further increase in the distance between each other. Moreover, when the inclination angle is more than three degrees, the axisymmetric picture is completely destroyed, therefore, such angles are not acceptable for rapping. Overall beam shape in sagittal section is shown in figure 2 (b) where two foci is separated by the distance $\delta(\varphi)$.

In our experiment we used diode laser with wavelength of λ = 532 nm, illuminating the uniaxial crystal of $LiNbO_3$ with n_o = 2.2863, n_e = 2.2030. The crystal was mounted on *z0x* rotational plane (with step of rotation *0.35 deg*) and initially aligned along the beam axis (φ = *0 deg*). Laser beam was focused by the lens L ($f = 25\ mm$) as illustrated in Figure 2 (a). The state of the intensity distribution in the beam waists can be changed by the second polarizer and a quarter-wave plate, cutting out orthogonally circularly polarized field components at the exit of the crystal. Microscopic objective MO_2 focuses the beam in a cuvette filled with a suspension of specimens in water.

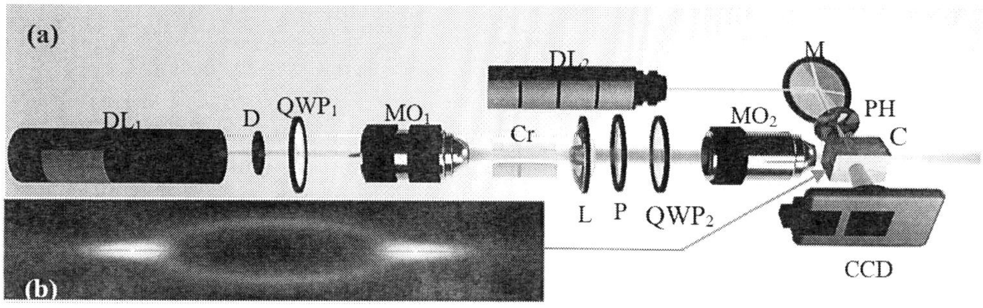

Figure 2 (a, b). (a) The scheme of experimental set-up: DL_1 – trapping 650 nm diode laser, D – diaphragm, $QWP_{1,2}$ – quarter-wave plate, MO_1 – 20^x microscope objective, Cr – $NbLiO_3$ crystal, DL_2 – imaging 532 nm diode laser, L – f25 mm lens, P – polarizer, MO_2 – 4^x microscope objective, M – mirror, PH – pinhole, CCD – viewing camera, C – cuvette; **(b)** Sagittal section of intensity profile in MO_2 focus area.

3. Digital holographic microscopy and image processing of trapped specimens

There are lot of modifications and optical schemes of digital holographic microscopes [25, 26], including combination with optical traps, where the trapping beam also served as an object beam for recording a hologram [27]. The advantage of using a single emitting source, in turn, limits the ability to control holographic recording, in particular, to change the wavelength of a light source, the optical scheme, etc. In addition, the using of such a scheme is impractical when large volume of media with specimens is recording on a digital hologram.

In this section, we demonstrate the possibility of using digital holographic microscopy with optical traps with discrete coherent light sources, which enables avoidance of mutual interference. For our task, we restrict ourselves to the axial recording scheme proposed by Gabor [28], built on a lens-less principle using a coherent radiation source and a pinhole [29-31].

The intensity distribution in the plane of the hologram can be described mathematically, using complex algebra. In this case, we calculate an intensity as a result of the interference of two waves is represented as follows [32]:

$$I = |o_1 + o_2|^2 = |o_1|^2 + |o_2|^2 + o_1 o_2^* + o_1^* o_2 = I_1 + I_2 + 2\sqrt{I_1 I_2}\cos(\varphi_1 - \varphi_2) \quad (1)$$

where o_1 and o_2 are the complex amplitudes of interfering waves, $o(x,y) = |o(x,y)|\exp[-i\phi(x,y)]$. o^* – complex conjugate, represent the initial phase of the wave. φ_1 and φ_2 represent the initial phase of the wave. Therefore, the optical intensity depends only on the phase difference.

In the process of recording a digital hologram, a specimen with a light scattering surface is located at a distance d from the CCD camera as shown in Figure 2 (a). To obtain the object wave of the test sample in digital holography, a numerical calculation of the optical field propagation in the form of a product from the hologram plane $o_2(x,y)I(x,y)$ to the object plane (ξ,η) is used.

Restored Diffracted Field $Q(\xi,\eta)$ in the image plane (ξ,η) at a distance d from the hologram plane can be represented in the paraxial approximation as follows [33-35]:

$$Q(\xi,\eta) = \frac{1}{i\lambda d}\int_{-\infty}^{\infty}\int_{-\infty}^{\infty} o_2(x,y)I(x,y)dxdy \exp\left[i\frac{\pi}{\lambda d}\left[(\xi - x)^2 + (\eta - y)^2\right]\right]\exp\left(i\frac{2\pi}{\lambda}d\right)dxdy \quad (2)$$

Intensity $I(x,y;d)$ and phase $\varphi(x,y;d)$ of reconstructed images can be obtained from the complex field $Q(\xi,\eta)$ calculated at a distance d using the following relations:

$$I(x,y;d) = |Q(x,y)|^2, \quad \varphi(x,y;d) = \arg\left(Q(x,y)\right) \quad (3)$$

In digital holography the reconstruction of the object wave field is performed numerically by simulating diffraction process with using scalar diffraction theory. We used free open source software ImageJ with implemented angular spectrum approach for numeric reconstruction of digital holograms.

The principle of the in-line configuration is that the laser light, passing through sample, splits into the two optical fields: diffracted by the sample and reference beam, without phase perturbations. As depicted in Figure 3 (a), a modified digital holographic recording set-up carrying 532 nm diode laser Ls emits coherent beam into the input of pinhole of 25 um in diameter. The chamber C with spacemen is located on the translation table. The set-up base BS enables rotation of a whole scheme around the chamber for adjustment of imaging plane. The hologram of rapped objects was recorded by CCD and after applied reconstruction, the image of specimen has been received as shown in Figure 3 (b, c).

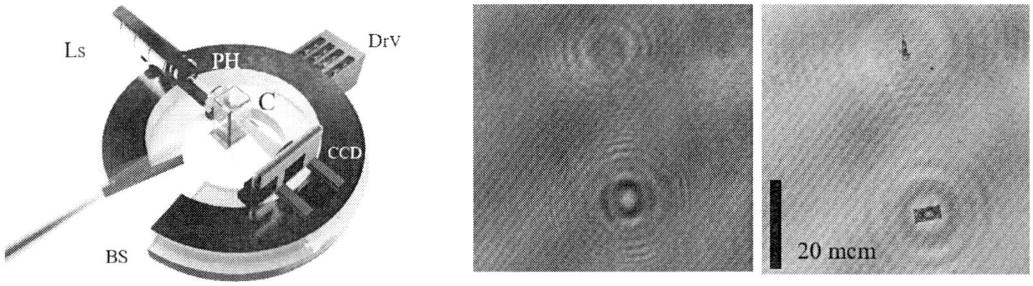

(a) (b) (c)

Figure 3 (a, b, c). (a) The sketch of rotational scanning holographic microscope: Ls is the 532 nm diode laser, PH – 25 mcm pinhole, C is the cuvette with the specimen, Dr is the motorised drive of rotational base stage BS and CCD – camera with 3 MPix resolution. Trapping laser beam is shown by red limes. The image of optically trapped objects represented as: digital hologram of marine centric diatoms in-situ **(b)** and their reconstructed image **(c)**.

The advantage of lens-less digital holography is an object can be placed close towards the CCD matrix, which increases the numerical aperture and improves resolution. For this purposes CCD is mounted on the movable along z-axis optical cage, in such a way we can change the distance between cuvette and imaging plane.

The experimental results of "bottle" beam trapping applied to marine centric diatoms is illustrated in Figure 4. Two frames sequentially depict the mutual positions of specimens before (Figure 4, a) and after trapping and spatial displacement (Figure 4, b). Red arrows show the direction of movement.

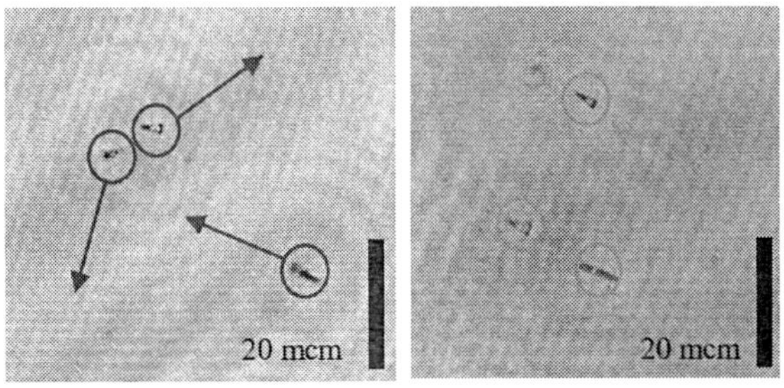

(a) (b)

Figure 4 (a, b). (a) The numerically reconstructed digital hologram of three diatoms in water, trapped by "bottle" laser beam; the cropped frame to 50×50 mcm was extracted from original hologram with dimensions of 4.30×3.50 mm. The configuration of the traps can be changed using crystal adjustment and chosen particles had been displaced **(b)** by movement of the cuvette.

4. Conclusion

In our research, we present an analysis of the capture options of micron-sized objects by an optical trap with real-time tunable properties. Moreover, in addition to the implementation of optical trapping, optical manipulations in three coordinates together with digital holographic microscopy make it possible to study various isolated particles with great flexibility. Also, the loss of beam field symmetry in the tweezing region depends on the efficiency of an optical trap which may be used for selective trapping microparticles of deferent mass and shape. Digital holographic microscopy allow to scan the cuvette in various depth of space from only one shot and precisely define specimen localisation. Experiments are able to record the evolution of the trapped particles in their natural environment, their size and shape in real time.

5. Acknowledgments

The reported research was partially supported by the V. I. Vernadsky Crimean Federal University Development Program for 2015 – 2024 and was funded by RFBR and Council of Ministers of the Republic of Crimea according to the research project 19-42-910010. The authors acknowledge to Khurchak Aleksandr from Marine Hydrophysical Institute of RAS for proposed specimens.

References

[1] Sokolenko B, Poletaev D 2017 *Proc. SPIE* **10350**
[2] Popiołek-Masajada A, Masajada J, Szatkowski M 2018 *Proc. SPIE* **10612**
[3] Ivanov M, Shostka N 2014 *Proc. SPIE* **9194**
[4] Abbondanzieri E et al. 2005 *Nature* **438** 7067
[5] Suzuki S et al. 2006 *Cell Res.* **16** 887
[6] Zhong M-C at al. 2013 *Nature Communications* **4** 1768
[7] Il'ina A V, Sitnikov D S at al. 2011 *Biophotonics: Proc. of SPIE* **8427**
[8] Kumar A, Bechhoefer J 2018 *Applied Physics Letters* **113** 18
[9] Sinjab F et all. 2018 *Optics Express* **26** 19
[10] Butaite U, Gibson G, Ho Y-D, Taverne M, Taylor J, Phillips D 2019 Nature Comm. **10** 1
[11] Peter de Groot 2015 *Adv. Opt. Photon.* **7**
[12] Reed J, Troke J, Schmit J, Han S, Teitell M, Gimzewski J 2018 *ACS Nano* **2**
[13] Groot P, Colonna de Lega X 2007 *Opt. Lett.* **32**
[14] David G, Esat K, Thanopulos I, Signorell R 2018 *Communications Chemistry* **1**
[15] Talapatra S, Sullivan J, Katz J at al. 2012 *Proc. SPIE* **8372**
[16] Miccio L, Memmolo P, Merola F at al. 2012 *Proc. SPIE* **8430**
[17] Belashov A V, Petrov N V, Semenova I V, Vasyutinskii O S2015 *Proc. SPIE* **9529**
[18] Schnars U, Jüptner W 2002 *Meas. Sci. Technol.* **13** 9
[19] Shvedov V, Hnatovsky C, Shostka N, Krolikowski W. 2013 *JOSA B: Optical Physics* **30**
[20] Sokolenko B, et al. 2011 *Proc. SPIE* **8338**
[21] Sokolenko B, Poletaev D, Rubass A, Volyar A 2016 *J. Phys.: Conf. Ser.* **735**
[22] Shvedov V, Krolikowski W 2018 *New J. Phys.* **20**
[23] Pogrebnaya A O, Halilov S I, Rubass A F, 2016 *J. Phys.: Conf. Ser.* **735**
[24] Ivanov M O, Shostka N V 2016 *J. Opt.* **18**
[25] Moradi A, Anand A, Kutub M, Javidi B, 2011 *Proc. SPIE* **8043**
[26] Miccio L, Finizio A, Puglisi R, Balduzzi D, Galli A, Ferraro P 2011 *Biomed. Opt. Express* **2**
[27] Memmolo P, Finizio A, Paturzo M, Miccio L, Ferraro P 2011 *Opt. Express* **19** 25
[28] Gabor D. A 1948 *Nature* **161** 4098
[29] Pedrini G, Schedin S, Tiziani H 1999 *Opt. Commun.* **171**
[30] Schnars U, Juptner W 1994 *Appl. Opt.* **33**
[31] Stahl R, Vanmeerbeeck G, Lafruit G et al. 2014 *Proc. SPIE* **8947**
[32] Schnars U, Werner P O Juptner 2002 *Meas. Sci. Technol.* **13**
[33] Hossain M, Singh Mehta D, Shakher C 2006 *Proc. SPIE* **6293**
[34] Wahba H, Kreis T 2009 *Applied Optics* **48** 8
[35] Mensah P et al. 2018 *J. Phys. Commun.* **2**

Nanostructures with magnetooptical and plasmonic response for optical sensors and nanophotonic devices

T V Mikhailova, A N Shaposhnikov, S V Tomilin, D V Alentiev
V.I. Vernadsky Crimean Federal University, Simferopol 295007, Russia

Abstract. It was proposed photonic nanostructures with magnetooptical and plasmonic properties that can be effectively used for optical sensors and nanophotonic devices. Optical, magnetooptical and functional properties of photonic crystals with magnetooptical structural elements in the form of films of bismuth-substituted iron garnets and gold plasmonic layer were considered. Simultaneous excitation of optical Tamm states (OTS) and Fabry-Perot resonance or OTS, Fabry-Perot resonance and surface plasmon-polaritons in proposed nanostructures was considered.

1. Introduction

Due to the achievements of functionality limits of semiconductor electronics devices an increasing number of developments are associated with the fundamentally new alternative areas, such as spintronics, nanophotonics and optomagnonics. Generally, the ideology of nanostructuring lays in the basis of many modern devices [1–3]. The idea of structuring optical media and term "photonic crystal" appeared in 1987 [2] and are still relevant and fashionable in many of the world's leading laboratories. Synthesis, investigation and application of photonic crystals form a huge number of fundamental and applied tasks.

In the work photonic crystal media with magneto-optical and plasmonic properties that can be effectively used in nanophotonic devices (circulators, insulators, switches, displays, etc.) and sensors (optical sensors) were proposed for the design. We considered Tamm structures, that allow to amplify the amplitude of electromagnetic wave in the active layer located on the border of photonic crystal and metal and excite the plasmon-polaritons propagating over a long distance [1, 4], and optimized the structures for a number of parameters.

2. Models and simulation methods

Three nanostructures with hybrid mode of OTS and Fabry-Perot resonance and OTS, Fabry-Perot resonance and surface plasmon-polaritons were proposed and designed by conventional 4×4 transfer matrix method [5]. Figure 1 shows schematically considered nanostructures:

No. 1 – $[TiO_2 / SiO_2]^m$ / M1 / M2 / SiO_2 / Au / $[SiO_2 / TiO_2]^m$;

No. 2 – $[SiO_2 / TiO_2]^m$ / SiO_2 / $[TiO_2 / SiO_2]^m$ / M1 / M2 / SiO_2 / Au;

No. 3 – $[SiO_2 / TiO_2]^{2m}$ / SiO_2 / M1 / M2 / SiO_2 / Au;

where m is repetition number of layers pairs, M1 and M2 are bismuth-substituted iron garnets of compositions of $Bi_{1.0}Y_{0.5}Gd_{1.5}Fe_{4.2}Al_{0.8}O_{12}$ and $Bi_{2.8}Y_{0.2}Fe_5O_{12}$, respectively. Substrate of gadolinium gallium garnet (GGG) with crystallographic orientation of (111) and the spectral dependence of components of permittivity tensors of layers were taken into account [6]. Components of permittivity tensors of layers were defined using the experimental data of single films and synthesized one-dimensional magnetophotonic crystals [7]. We considered that TM polarized light falls to the structure. The angle of incidence of ligth β on the structure was taken into account in the calculations. The case of total internal reflection in the Kretschmann geometry was considered in order to excite the surface plasmon-polaritons. The refractive index of the prism was considered equal to refractive index of GGG substrate.

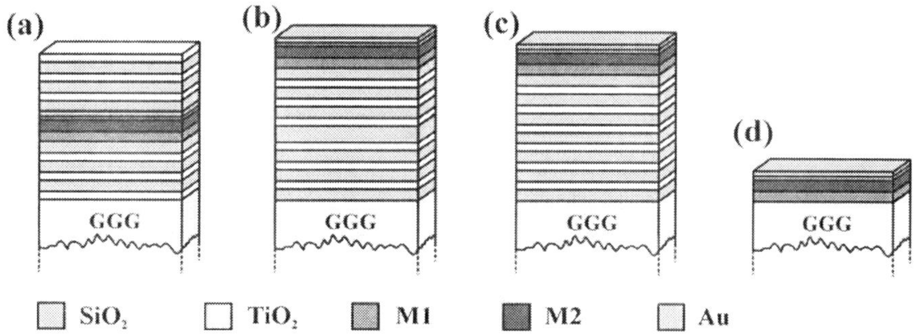

Figure 1 (a, b, c, d). Schematic diagrams of nanostructures No. 1 **(a)**, No. 2 **(b)**, No. 3 **(c)** and multilayer No. 4 **(d)**. For illustrated structures $m = 3$.

The configurations of structures were chosen in such a way that the resonant wavelength λ_{res} was about 700 nm at a normal incidence of light. The thicknesses of layers of Bragg mirrors TiO_2 h_{TiO2} and SiO_2 h_{SiO2} for all structures were 77 and 120 nm, respectively. The thickness of magnetic sublayer M1 was $h_{M1} = 70$ nm. The thickness of main magnetoactive layer M2 h_{M2} varied from 44 to 144 nm to change the excitation length of OTS inside a photonic band gap of Bragg mirrors. The layer of SiO_2 with thickness of $h_{SiO2T} = 30$ nm was taken into account as its application allows to use the technology of closed crystallization of garnet. The technic allows to reduce the roughness of interfaces during synthesis of structure [8]. The thickness of gold layer h_{Au} and number m varied in the ranges from 0 to 50 nm and from 1 to 10, respectively, to determine the optimal configuration of structures.

The calculations of magnetooptical quality factor Q and figure of merit F were carried out to assess the perspectivity of structures for practical implementation and subsequent applications [9, 10]. The values of Q and F for transmissive structures are defined as

$$Q = -2 \cdot |\Theta_F| / \ln T \quad [\text{deg}], \tag{1}$$

$$F = T \cdot \sin(2 \cdot \Theta_F) \quad [\%]. \tag{2}$$

There Θ_F is angle of Faraday rotation, T is transmittance in [%].

Optimization of these parameters (or similar them in the case of reflection structure) is the main problem of photonic design with magnetooptical layers for various applications, since expressions (1) and (2) determine the optimal ratio of the angle of rotation of polarization plane and the intensity of propagating light.

The main aim of structure optimizing for optical sensor is to find the configuration with the highest sensitivity, which is defined as $\Delta n/\Delta R$ (where Δn is the change in refractive index of external environment, ΔR is the change in reflectance or other measured quantity). We compared this parameter for proposed structures No. 2 and 3 and a multilayer with a top gold layer of composition M1 / M2 / SiO_2 / Au (No. 4, figure 1 *d*).

3. Results

3.1. Spectra and magnetooptical quality factor of structures

Figure 2 shows the transmittance and Faraday rotation spectra of three modelled structures with $m = 4$, $h_{Au} = 40$ nm and $h_{M2} = 84$ nm. Structures No. 1 and 3 have one resonance at the design wavelength. Structure No. 2 is distinguished by the presence of two resonances related to Fabry-Perot and OTS modes [11]. Although a gold layer and a central magnetic defect layer in configuration of structure No. 1 are present, we didn't observe two resonances in the spectra. However, the authors of [12] presented similar structure, at which two OTS and Fabry-Perot resonances are observed simultaneously. In this regard, we do not have a clear interpretation of observed resonance and the possibility of hybridization existence. Structure No. 3 has a single OTS resonance and typical high values of Faraday rotation angle at resonance wavelength. It should be noted that both resonances in structure No. 2 have almost identical non-zero values of Faraday rotation angle, although the central defect layer in configuration is nonmagnetic. This indicates the presence of strong hybridization between Fabry-Perot mode and OTS.

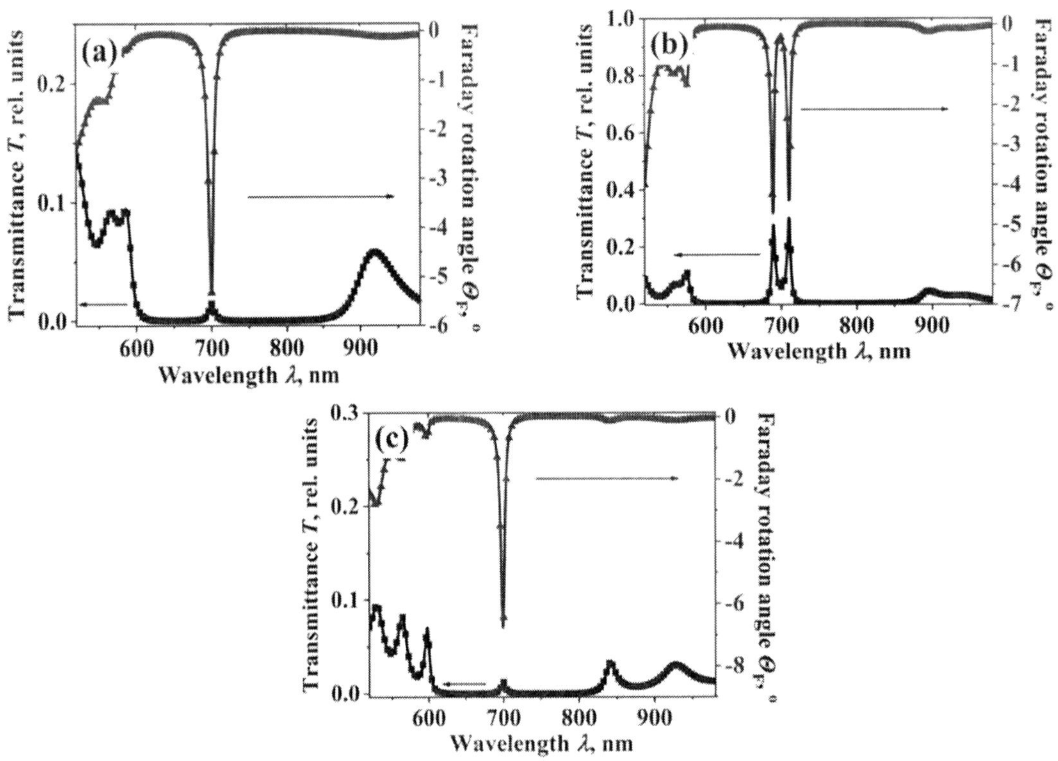

Figure 2 (a, b, c). Optical and magnetooptical spectra of nanostructures No. 1 **(a)**, No. 2 **(b)** and No. 3 **(c)** at normal incidence $\beta = 0°$. The parameters of structures are given in the text.

Figure 3 shows the comparison of magnetooptical quality factor Q at resonance wavelengths for different configurations of structures No. 1 (λ_{res} = 700 nm, a), No. 2 (λ_{resI} = 690 nm and λ_{resII} = 710 nm c and d, respectively) and No. 3 (λ_{res} = 700 nm, b) at fixed h_{M2} = 84 nm. Simulated dependences of magnetooptical figure of merit F are completely similar to given dependences of Q. Dependencies of structures No. 1 and 3 are bell-shaped or asymptotically approaching to zero. Dependencies of structure No. 2 are more complex due to the presence of strong hybridization of optical modes. Comparing the configurations of structures No. 2 and 3 and the calculated results to them (Figure 4 c, d and b), it is evident that the hybridization of modes and introduction to the structure *nonmagnetic* defect layer leads to an increase in quality factor Q and figure of merit F of structures with m in the range from 3 to 6. The functional properties of structure No. 1 with gold layer are much worse than the properties of the same structure without gold layer or the properties of structures No. 2 and 3 with identical layer parameters (for $m < 7$ and $h_{Au} > 20$ nm).

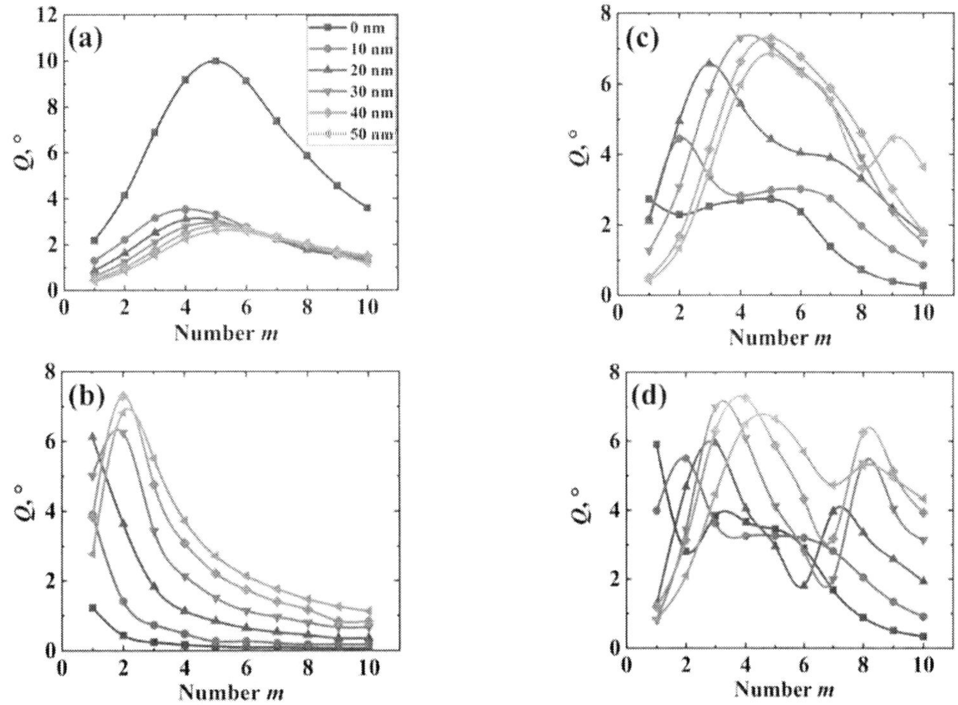

Figure 3 (a, b, c, d). Magnetooptical quality factor Q for resonant wavelengths as a function of the repetition number m and gold thickness h_{Au} of nanostructures No. 1 **(a)**, No. 3 **(b)** and No. 2 **(c, d)** with h_{M2} = 84 nm. Dependences were simulated using structure spectra at normal incidence β = 0°. The designations of thicknesses in all figures are the same.

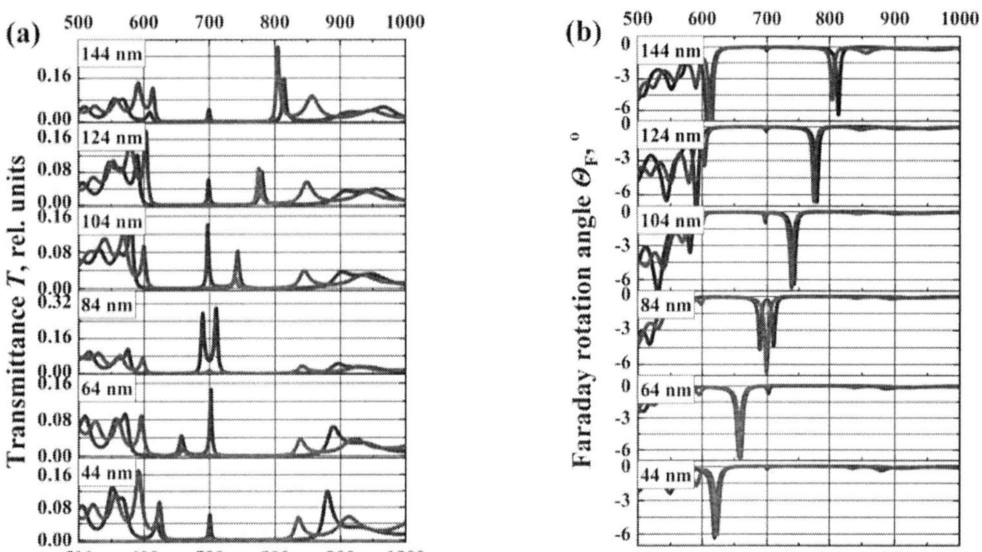

Figure 4 (a, b). Optical **(a)** and magnetooptical **(b)** spectra of nanostructures No. 2 (blue lines) and No. 3 (red lines) with m = 4 and h_{Au} = 40 nm at normal incidence β = 0° as a function of the thickness of magnetic layer M2 h_{M2}.

The following optimal configurations should be highlighted among all structures:

- No. 1 with $Q = 10.0\,^{\circ}$, $F = 6.2$ %, $m = 5$ and $h_{Au} = 0$ nm;
- No. 2 with average values of $Q = 6.9\,^{\circ}$ and $F = 4.3$ % for two resonances $m = 4$ and $h_{Au} = 40$ nm;
- No. 3 with $Q = 7.3\,^{\circ}$, $F = 4.3$ %, $m = 2$ and $h_{Au} = 40$ nm.

Magnetooptical parameters of proposed structures significantly exceed the magnetooptical parameters of known structure $[SiO_2 / Bi{:}\ IG]^5$ / Au ($\Theta_F = -0.4\,^{\circ}$, $T = 25$ %, $Q = 7.3\,^{\circ}$, $F = 4.3$ %) [1].
Comparison of optical and magnetooptical spectra of structures No. 2 and 3 with identical parameters of layers $m = 4$, $h_{Au} = 40$ nm and $h_{M2} = 84$ nm are shown in Figure 4. It can be seen that the functional properties of structure with hybrid state are better due to an increase in the transmittance by 15 times.

3.2. Sensitivity of structures in the case of excitation of surface plasmon-polaritons

The results of calculations of properties of structures No. 2 and 3 when surface plasmon-polaritons is excited are shown in Figure 5.

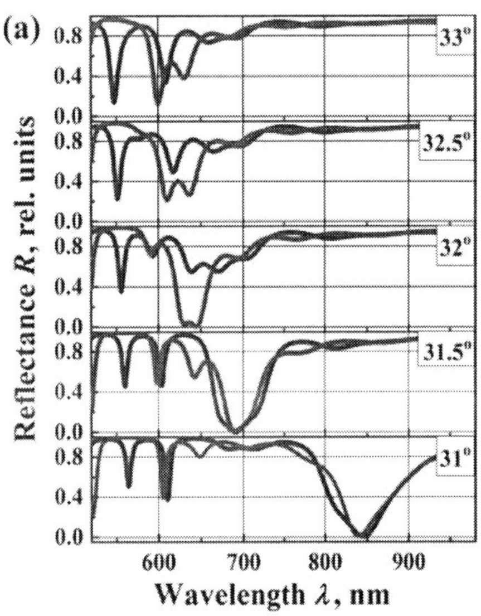

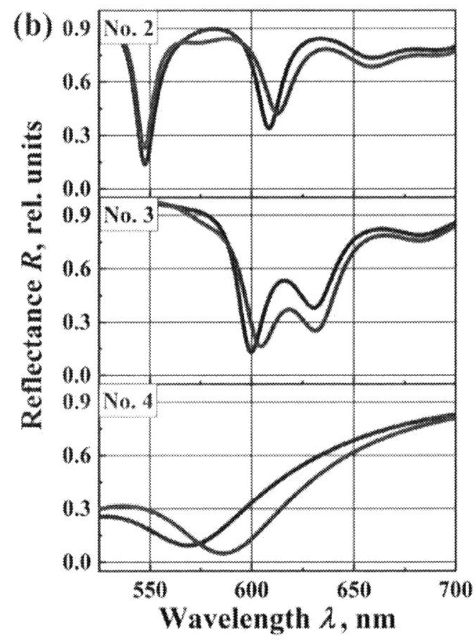

Figure 5 (a, b, c). (a) Spectra of reflectance of nanostructures No. 2 (blue lines) and No. 3 (red lines) as a function of the angle of light incidence β; Spectra of reflectance **(b)** and Kerr rotation angle **(c)** of nanostructures No. 2 and 3 and multilayer No. 4 at $\beta = 33\,^{\circ}$ and refractive index of the medium adjacent to the gold layer in Kretschmann geometry of 1.00 (blue line) and 1.01 (red line). The parameters of structures are $m = 5$, $h_{M2} = 84$ nm and $h_{Au} = 50$ nm. The thicknesses of layers in multilayer are the same.

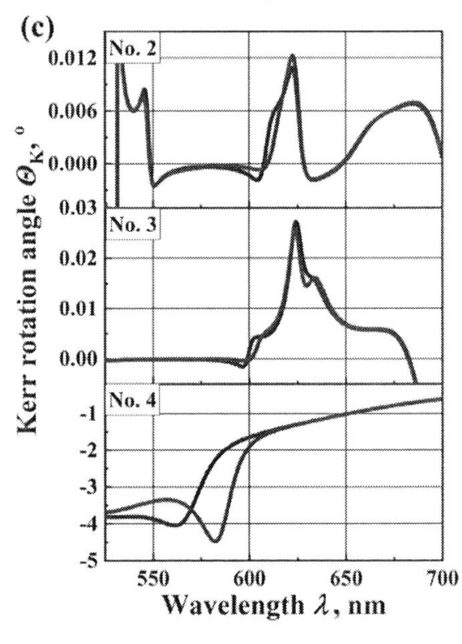

The blue shift of resonance of surface plasmon-polaritons occurs in the range of angles of light incidence β from 31 ° to 33 ° (figure 5, *a*). The resonance value *R* of OTS imposed on the surface plasmon-polaritons resonance decreases. Sensitivity comparison of spectra of structures No. 2, 3 and 4 to a change in the refractive index of medium ($\Delta n = 0.01$) is presented in figure 5 (*b*, *c*). Changes in reflectance in the vicinity of OTS wavelength for structures No. 2 and 3 are 6.6 % and 5.6 %, respectively. Changes in reflectance in the vicinity of resonance of surface plasmon-polaritons for multilayer are 5.8 %. Magnetooptical spectra indicate that Kerr effect measurements are more practical for multilayer system No. 4.

4. Conclusions

The obtained results can be used to design of magnetooptical devices promised to further application in optical information process, photonics, spintronics and sensors.

Acknowledgments
The authors acknowledge support by the RF Ministry of Science and Higher Education in the framework of the state task (project no. 3.7126.2017/8.9) and V. I. Vernadsky Crimean Federal University (Grant no. ВГ13/2018).

References

[1] *Magnetophotonics* 2013 (M. Inoue et al. eds.; Springer-Verlag Berlin Heidelberg) p 163
[2] Yablonovitch E 1987 *Phys. Rev. Lett.* **58** No 20 2059
[3] Chung K H, Kato T, Mito S et al. 2010 *J. Appl. Phys.* **107** 09A930-2
[4] Baryshev A V et al. 2013 *J. Phys. D: Appl. Phys.* **46** 125107
[5] Yin C P, Wang T B and Wang H Z 2012 *Eur. Phys. J.* B **85** 104
[6] Mikhailova T V, Berzhansky V N, Karavainikov A V, Shaposhnikov A N, Prokopov A R, Lyashko S D 2016 *Journal of Physics: Conference Series* **741** 012121
[7] Mikhailova T V, Berzhansky V N, Shaposhnikov A N, Karavainikov A V, Prokopov A R, Kharchenko Yu M, Lukienko I M, Miloslavskaya O V, Kharchenko M F 2018 *Opt. Mater* **78** 521
[8] Krikun A S, Shaposhnikov A N, Karavainikov A V, Prokopov A R, Mikhailova T V, Berzhansky V N Abstract Book of International conference "Scanning Probe Microscopy", Ekaterinburg, August 28-30, 2017 ; Youth conference "Application of probe microscopy in research and development", Ekaterinburg, August 27-30, 2017 (Ekaterinburg, Ural Federal University, 2017) p 124
[9] Takahashi K, Takagi H, Shin K H et al. 2007 *Phys. Stat. Sol. (c)* **12** 4
[10] Khartsev S I, Grishin A M 2007 *J. Appl. Phys.* **101** 053906
[11] Mikhailova T, Tomilin S, Lyashko S, Shaposhnikov A, Prokopov A, Karavainikov A, Bokova A, Berzhansky V 2018 *Journal of Physics: Conference Series* **1124** 051062
[12] Bruckner R, Sudzius M, Hintschich S I, Frob H, Lyssenko V G, Leo K 2011 *Physical Review* **B83** 033405

Formation of catalytic centers for the growth of carbon nanostructures for nanophotonics devices

V S Klimin[1,2], A A Rezvan[1], J V Morozova[1]
[1]Department of Nanotechnology and Microsystems, Southern Federal University, Taganrog 347922, Russia
[2]Research and Education Center "Nanotechnologies", Southern Federal University, Taganrog 347922, Russia

Abstract. In this paper, we studied the effect of heating temperature on the geometric parameters of catalytic centers. The temperature range was from 600 to 900 °C with an interval of 50 °C, heating was carried out for 20 minutes. At the end of the experimental studies, the AFM and SEM images of the catalytic centers formed at various heating temperatures on the Ni/V/Si structure are presented. Based on the data obtained, the corresponding dependences of the packing density, height range and areas of catalytic centers on temperature were constructed. It was determined that the catalytic centers have the smallest volume of 0.5 nm^3 and a diameter of 80 nm at a heating temperature in the range from 750 °C to 850 °C, which is caused by their uniformity and geometric shape.

1. Introduction

In recent years, there is an intensive development of science and technology; each year new discoveries are made in the field of materials and their formation [1-6]. More recently, allotropic modifications of carbon have been discovered, but which have already found their application in various branches of micro- and nanoelectronic [7-12]. The use of modern nanomaterials allows you to develop and form low-cost and high-effective nanophotonics devices [13-18]. Get a regular array of carbon nanotube needed method to do this. Such a method is chemical vapor deposition. A feature of this method is the use of the prepared surface for the growth of carbon nanostructures. On the substrate must be present catalytic centers [19-22].

2. Materials and Methods

Based on the analysis of literature and a series of experimental studies, materials were chosen for films of catalytic material and an underlayer on silicon to form catalytic centers that ensure the best growth of the carbon nanotube's array and the formation of a conductive layer to ensure electrical contact. In the process of experimental studies, the structures Ni/V/Si, Ni/Ti/Si, Ni/Cr/Si, Ni/Al/Si were formed. The experiments used samples formed on the substrate, the thickness of the materials of the sublayer, which varied within the specified limits; the thickness of the layer of catalytic material in all cases was 10 nm.

The focus is on conducting research on the effect of heating temperature on the geometric parameters of catalytic centers. The temperature range was from 600 to 900 °C with an interval of 50 °C, heating was carried out for 20 minutes [23-26]. The pressure of the process gases in the chamber was maintained with the help of an oil spray and was 4.5 Torr. The ammonia feed rate is Q_{NH3} = 15 cm^3/min, argon Q_{Ar} = 40 cm^3/min.

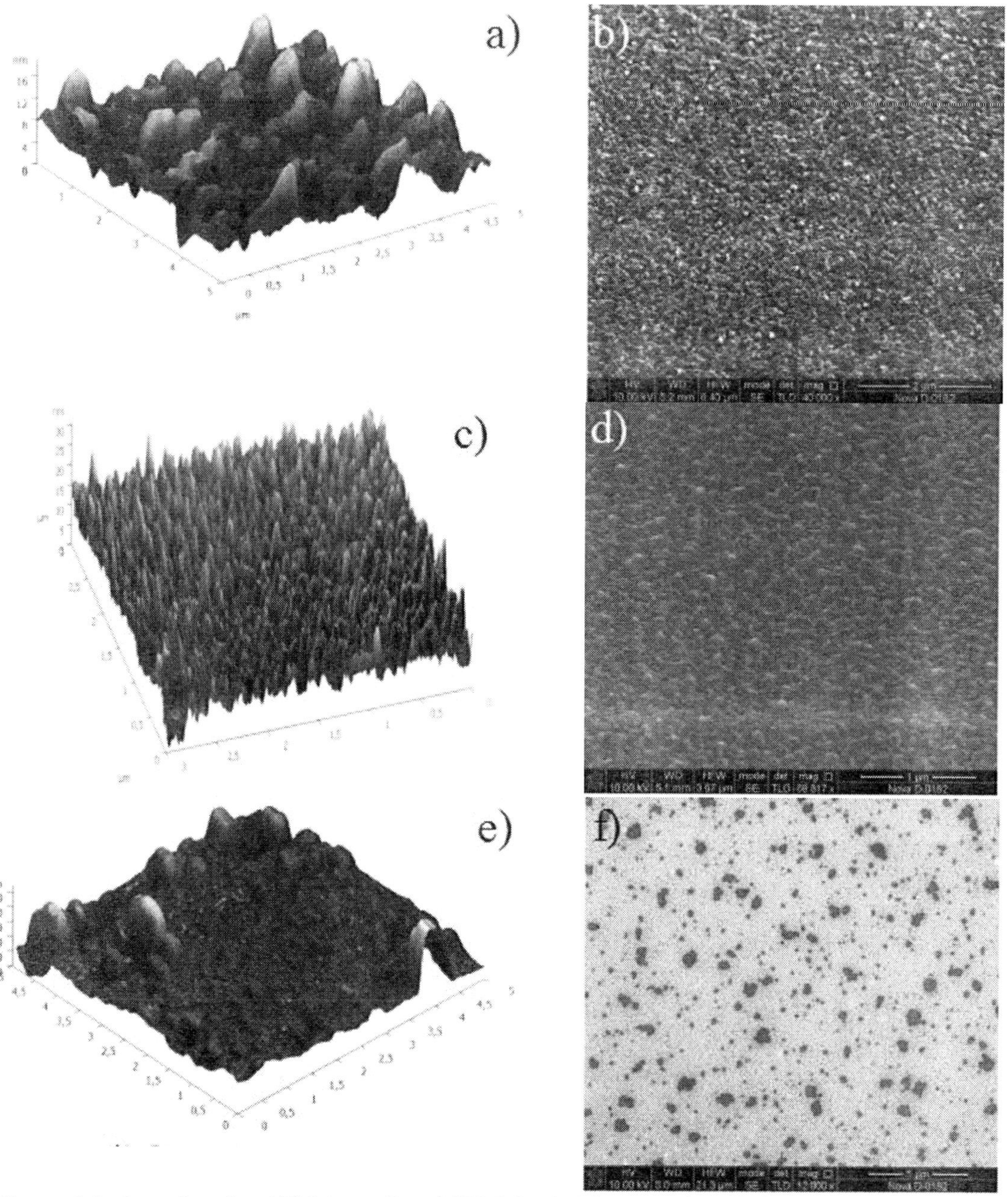

Figure 1 (a, b, c, d, e, f). AFM (**a, c, d**) and SEM (**b, d, f**) images of catalytic centers formed on the Ni/V/Si structure at various temperatures: **a, b)** 600 °C; **c, d)** 800 °C; **e, f)** 900 °C

Figure 1 (a, b, c, d, e, f) shows the AFM and SEM images of catalytic centers formed at different heating temperatures on the Ni/V/Si structure.

3. Results

Using the specialized software package Image Analysis 3.5.0, the statistical processing of the obtained AFM images was performed, based on the data of which the corresponding dependences of the geometric parameters of the catalytic centers on the heating temperature were constructed (Figure 2).

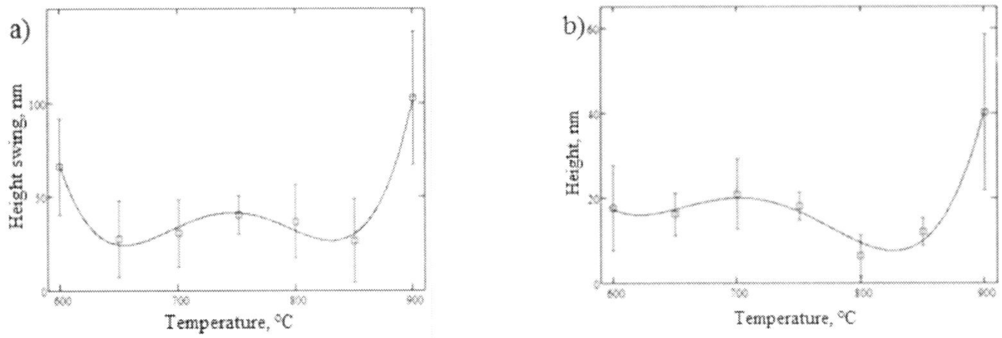

Figure 2 (a, b). The dependence of the geometric parameters of the surface of the structure with catalytic centers on the heating temperature: **a)** height range; **b)** height of catalytic centers

From the presented dependences, it is seen that at heating temperatures of 600 °C and 900 °C the greatest height range and roughness is observed, which means the particles are not uniform in size and when grown on them, carbon nanotubes will have a high dispersion along the diameter of nanotubes in the array. The optimum temperature range for heating to ensure minimum values and dispersion of the catalytic centers along the height range and roughness is the temperature range (700-800) °C. At temperatures from 750 °C and up to 850 °C, catalytic centers have the smallest height, the surface roughness is the smallest, and accordingly the smallest dispersion is observed. The smallest roughness is observed in the temperature range from 700 °C to 800 °C.

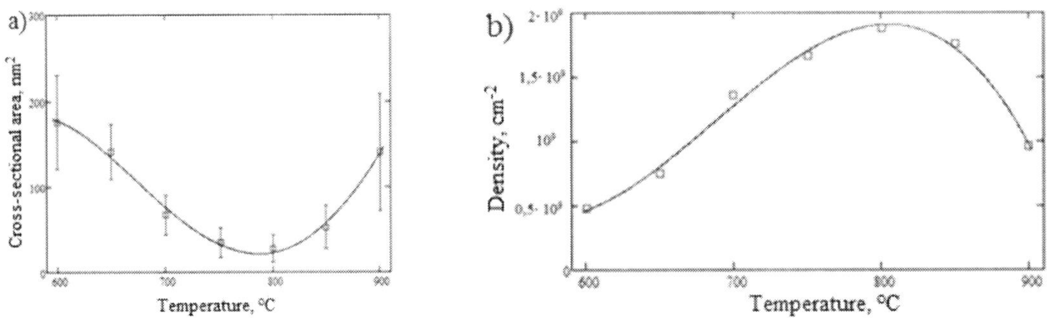

Figure 3 (a, b). The dependence of the geometric parameters of the catalytic centers on the heating temperature: **a)** sectional area; **b)** density

Figure 3 (a, b) shows the dependences of the geometric parameters of the formed catalytic centers on the heating temperature. From the dependences obtained, it can be seen that the cross-sectional area at temperatures of 600-900 °C is relatively large.

4. Discussion and Conclusions

According to the experiments performed at the heating stage, it can be concluded that the temperature range of 700-800 °C will provide the best parameters of the catalytic centers. By varying the technological regimes at this stage, it is possible to control the geometric parameters and density of the catalytic centers on the substrate, for a film of a catalytic layer of a fixed thickness.

It was determined that the optimal heating temperatures are in the range of 750-850 °C, in which the values of the cross-sectional area of 20 nm^2 with the greatest homogeneity. Catalytic centers have the smallest volume of 0.5 nm^3 and a diameter of 80 nm at a heating temperature in the range from 750 °C to 850 °C, which is caused by their uniformity and geometric shape. With the smallest diameter, with the same amount of matter, the catalytic centers will be distributed with the greatest uniformity over the surface, with a maximum density, in particular at a heating temperature of 800 °C, the density of the catalytic centers with a diameter of 80 nm is $1.9 \cdot 10^9$ cm^{-2}.

Acknowledgments
This work was supported by Grant of the President of the Russian Federation No. MK-3512.2019.8 and the Russian Foundation for Basic Research Project № 16-29-14023 ofi_m. The results were obtained using the equipment of the Research and Education Center "Nanotechnologies" of Southern Federal University.

References
[1] Amrani A, DjafariRouhani M and Mraoufel A 2011 *Appl. Nanosci.* **1** 59
[2] Kangawa Y, Ito T, Taguchi A, Shiraishi K, Irisawa T and Ohachi T 2002 *Appl. Surf. Sci.* **190** 517
[3] Tominov R V, Zamburg E G, Khakhulin D, Klimin V S, Smirnov V A, ChuY H, Ageev, O.A. 2017 *J. Phys.: Conf. Ser.* **917** 032023
[4] Daweritz L and Ploog K 1994 *Semicond. Sci. Tech.***9** 123
[5] Klimin V S, Rezvan A A, Kots I N, Naidenko N A 2018 *J. Phys.: Conf. Ser.* **1124** 071019
[6] Il'Ina M V, Blinov Y F, Il'In O I, Klimin V S, Ageev O A 2016 *Proc. of SPIE* **10224** 102240U-1
[7] Foxon C T and Joyce B A 1977 *Surf. Sci.***64** 293
[8] Klimin V S, Rezvan A A, Ageev O A 2018 *J. Phys.: Conf. Ser.* **1124** 071020
[9] Tok E S, Neave J H, Zhang J, Joyce B A and Jones T S 1997 *Surf. Sci.* **374**397
[10] Kley A, Ruggerone P and Scheffler M 1997 *Phys. Rev. Lett.***79** 5278
[11] Autumn K, Liang Y A, Hsieh S T, Zesch W, Chan W P, Kenny Th W, Fearing R and Full R J 2000 *Nature* **405** 681
[12] Rudyk N N, Il'In O I, Il'Ina M V, FedotovA A, Klimin V S, Ageev O A 2017 *J. Phys.: Conf. Ser.* **917** 082008
[13] Popov, V.N. 2004 Materials Science and Engineering R: Reports **43 (3)** pp. 61-102.
[14] Yang, Y., Liu, L., Jiang, K., Fan, S. 2012 Vacuum **86 (7)** pp. 885-888.
[15] Hao, X., Maenaka, K., Takao, H., Higuchi, K. 2014 Sensors and Actuators, A: Physical **205** pp. 92-102.
[16] Tominov R V, Bespoludin V V, Klimin V S, Smirnov V A, Ageev O A 2017 *IOP Conf. Ser.: Mater. Sci. Eng.***256** 012023
[17] Wilfert, S., Edelmann, C. 2012 Vacuum, **86 (5)**, pp. 556-571.
[18] Gavignet, E., Lanzetta, F. 2014 Vacuum **100** pp. 18-21.
[19] Górecka-Drzazga, A. 2009 *Vacuum* **83 (12)** pp. 1419-1426.
[20] Modi, A., Koratkar, N., Lass, E., Wei, B., Ajayan, P.M. 2003 *Nature* **424 (6945)** pp. 171-174.
[21] Hou, Z., Xu, D., Cai, B. (2006) *Appl. Phys. Lett.* **89** pp. 21-24.
[22] Jitschin, W., Ludwig, S. 2004 *Vacuum* **75 (2)** pp. 169-176.
[23] Kots I N, Kolomiitsev A S, Lisitsyn S G, Polyakova V V, Klimin V S, Ageev O A 2019 *Russian Microelectronics* **48 (2)** p 72-79
[24] Klimin V S, Rezvan A A, Ageev O A 2018 *J. Phys.: Conf. Ser.* **1124** 022035
[25] Lisitsyn S A, Balakirev S V, Avilov V I, Kolomiytsev A S, Klimin V S, Solodovnik M S, Konoplev B G, Ageev O A 2018 *Nanotechnologies in Russia* **13 (1-2)** p 26-33
[26] Ageev O A, Smirnov V A, Solodovnik M S, Rukomoikin A V and Avilov V I 2012 Semiconductors 46 1616–21

Synthesis of SiO_2@Ag core-shell photonic crystals for SERS application

M S Ashurov[1], **E A Eremina**[1,2], **B A Abdusatorov**[3], **S O Klimonsky**[1,2]
[1]Faculty of Materials Science, Lomonosov Moscow State University, Moscow, 119991 Russia
[2]Department of Chemistry, Lomonosov Moscow State University, Moscow, 119991 Russia
[3]Skolkovo Institute of Science and Technology, 143026, Russia

Abstract. Core-shell SiO_2@Ag composite spheres with dense, complete and nanoscaled silver shell were prepared by using facile chemical reduction method. Self-assembly of the composite microspheres into 3D photonic crystals (PhCs) was carried out using the vertical deposition method. The core-shell PhCs composites were characterized by X-ray powder diffraction (XRD), scanning electron microscopy (SEM), UV-Vis spectroscopy and energy dispersive X-ray spectroscopy (EDX). The results of amplification of Raman scattering by these composite substrates are presented.

1. Introduction

In recent years, three-dimensional (3D) metall-dielectric photonic crystals (PhCs) have attracted much attention due to the special features of PhCs (large surface area and ability to modulate light) and the plasmonic enhancement effect of noble metals [1]. They have promising potential applications in many fields, such as plasmonic enhanced emission [2] and surface-enhanced Raman scattering (SERS) [3]. 3D PhCs are periodic structures which can be generated by a controllable self-assembly method of the spherical colloidal particles [4, 5]. They have photonic band gaps (PBG) which forbid propagation of light in certain frequency ranges [6]. This work is aimed to provide a simple and effective way to significantly enhance Raman signals by the synergetic interaction of the surface plasmon resonance of metallic nanostructures and the photonic band gap (PBG) effect of the photonic crystal. For this purpose, we synthesized a new type PhC based on close-packed SiO_2@Ag core-shell microspheres. The obtained PhCs composites can be promising for SERS application.

2. Experimental

2.1. Materials

Tetraethyl orthosilicate (TEOS) for synthesis of SiO_2 microspheres, $AgNO_3$ for the synthesis of Ag nanoparticles and polyvinylpyrrolidone (PVP-can prevent the aggregation of the Ag nanoparticles) were acquired from Sigma-Aldrich enterprise. Methylene blue (MB) and glucose were obtained from Reachem. Menzel-Gläser glass coverslips of 24×24×0.4 mm, thoroughly washed in ethanol and acetone, were used as substrates for the deposition of SiO_2@Ag core-shell microspheres.

2.2. Preparation SiO_2@Ag core-shell microspheres and photonic crystals based on them

SiO_2 colloidal microspheres of 230-300 nm in diameter were synthesized by the method presented in [7]. Then, 20 ml of a 0.001 M solution of ammoniacal silver nitrate and 30 ml of a 7 μM solution of

PVP (can prevent the aggregation of the Ag nanoparticles) were added to the colloidal suspension with 0,1 g of SiO_2 microspheres. $[Ag(NH_3)_2]^+$ ions were adsorbed on the surface of the silica microspheres. At last, the SiO_2@Ag composite particles were formed by a simple reduction process in the presence of 10 ml 2 mM glucose at 50 ^{0}C for 3h. The obtained composite was collected by centrifugation and washed three times with deionized water. Self-assembly of core-shell nanocomposites into 3D PhC was implemented by the vertical deposition method [8]. The procedure for preparing SiO_2@Ag PhC composites is illustrated in Fig. 1.

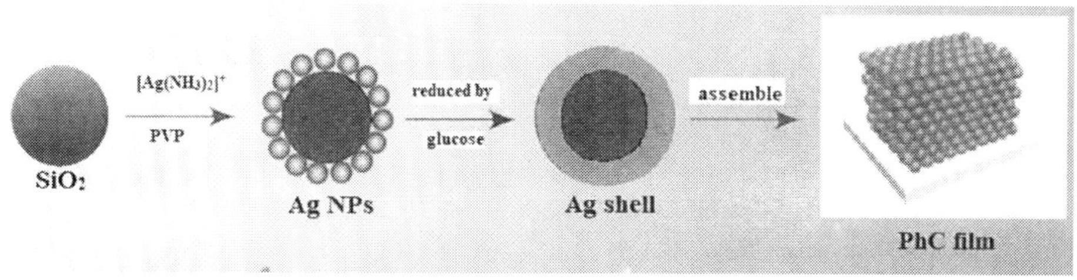

Figure 1. Illustration of the synthesis route for SiO_2@Ag PhC nanocomposites.

2.3. Characterization
The absorption spectrum of solutions and the reflection spectrum of photonic crystal films were recorded with a Lambda 950 spectrophotometer (Perkin Elmer). Images of the surface microstructure and EDX analysis of composite films were obtained using a scanning electron microscope (SUPRA 50 VP, Carl Zeiss, Germany) with the accelerating voltage of 7–21 kV (SE2 and VPSE detectors). Synthesized samples were characterized by X-ray diffraction (XRD) on a Rigaku D/MAX 2500 diffractometer (rotating anode, Cu Kα radiation, step scan mode with a step of 0.02° over 2θ = 10 – 70°, Rietveld refinement). Raman spectra were collected with DXRxi Raman Imaging Microscope (Thermo Fisher Scientific) equipped with a ×50 magnifying lens and a green solid-state laser (λ = 532 nm). Hundred accumulations were made for each spectrum at 2 mW laser power and 60 Hz operating frequency. For the study SERS of MB, PhC films were vertically immersed in a 50 μM ethanol-based MB solution and kept there for 1 minute, followed by air drying.

3. Results

The X-ray diffraction patterns of the obtained SiO_2 microspheres and SiO_2@Ag core-shell nanocomposites are shown in Fig. 2a. For pure silica particles, there is only a broad scattering maximum centered at 22.5°, corresponding to amorphous silica [9]. Yet, for core-shell composites, except amorphous silica characteristic diffraction peak, it also exhibits three well-resolved diffraction peaks at 2θ angles of 37.9°, 44.1° and 64.3° in the range of approximately 10° to 70°, which can be indexed to the (111), (200) and (220) reflections of face-centered cubic silver (JCPDS Card No. 87-0720), indicating that the Ag nanoparticles with high crystallinity could be obtained on the surface of SiO_2. In addition, we also can find that the peaks are a little broader than that of bulk silver because the silver grain size is relatively small. The UV-Vis spectroscopy is one of the most widely used techniques for structural characterization of silver nanostructures [10]. The absorption spectra of the prepared bare silica and core-shell structure are shown in Fig. 2b. The bare silica colloids do not show any UV-Vis absorption in the range of 300 nm to 800 nm, but the core-shell composites show an obvious absorption peak at around 438 nm due to the excitation of the Mie plasmon resonance associated with the silver nanoparticles (NPs) on the surface of SiO_2. Compared to pure Ag NPs, the absorption peak shifts from ~400 nm to 438 nm [11, 12]. The possible reason for the red shift can be attributed to much larger size of silver nanoparticles and higher coverage on the silica surface as indicated in Fig. 3 (SEM). The results are consistent with the previous reports [10, 13]. The strong dipole-dipole interactions between neighboring nanoparticles and Mie scattering of silver shell would promote red shift and broadening of the plasmon bands for silver clusters attached on silica spheres.

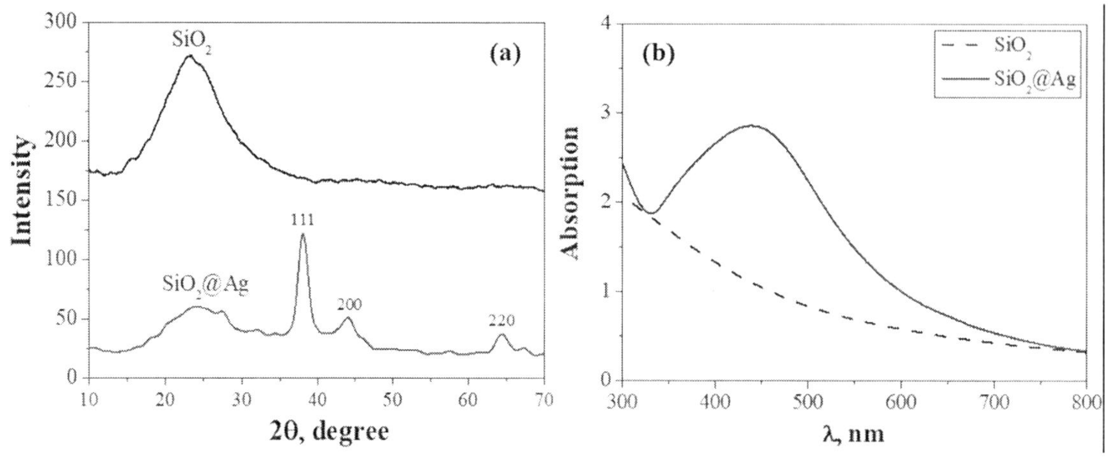

Figure 2(a, b). **(a)** X-ray diffraction patterns of SiO_2 and SiO_2@Ag composite microspheres; **(b)** UV-Vis absorption spectra of SiO_2 and SiO_2@Ag microspheres.

SEM image of synthesized nanocomposites SiO_2@Ag core-shell microspheres are shown in Fig. 3a. All of the particles are spherical in shape and the average diameter is about 235 nm. The chemical composition of resulting SiO_2@Ag composite has been analyzed by EDX elemental analysis as shown in Fig. 3a (inset). In this pattern, only Si, O, and Ag peaks are clearly shown and no other peaks are detected. The atomic ratio of Si and O is about 1:2, and the total content of Ag element is about 15 wt. %. This means that such a concentration of silver shell is enough for a successful coating of silica microspheres in the present study.

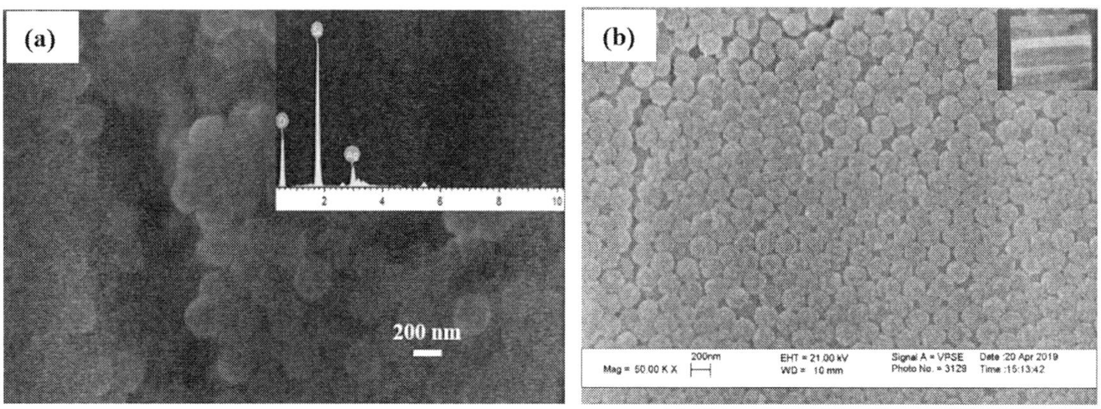

Figure 3(a, b). **(a)** SEM image of SiO_2@Ag nanocomposites (EDX spectrum is shown in the inset); **(b)** SEM image of self-assembled SiO_2@Ag into 3D PhC (photograph of PhC film shown at inset).

The SEM images of the surface of the PhC films (Figure 3b) confirmed that the SiO_2@Ag microspheres can self-assemble into an ordered PhC structure with the help of vertical deposition method [8].
Next, the optical properties of the obtained PhC films were investigated. The reflectance spectra (Figure 4) of PhC films fabricated from SiO_2@Ag microspheres with different average diameters were measured closer to normal incidence to detect the PBG arising from the periodicity of the SiO_2@Ag PhC films.

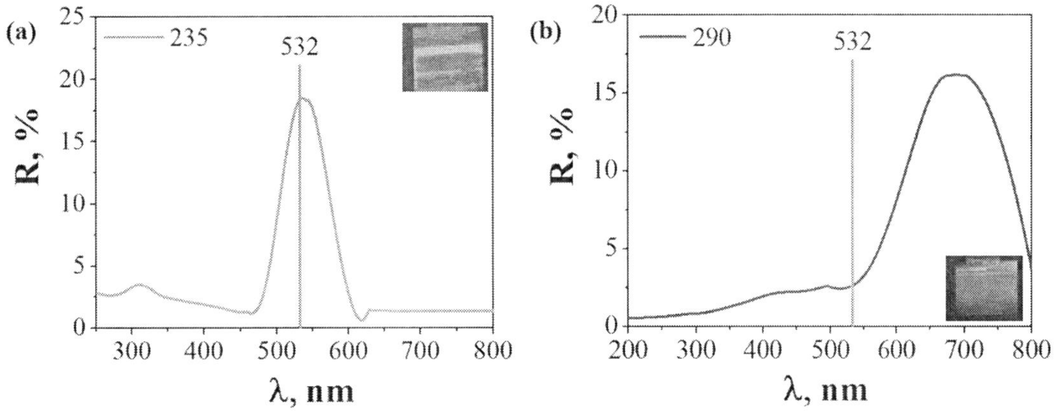

Figure 4(a, b). **(a)** Reflection spectra of the SiO_2@Ag PhCs films prepared from self-assembled microspheres with 235 nm average diameters; **(b)** With 290 nm average diameters (Insets: Digital photographs of the films).

The peak positions of SiO_2@Ag PhC films with different diameters of SiO_2@Ag microspheres (235 and 290 nm) were λ = 536 and 688 nm, respectively, and these films displayed the corresponding structure colors, as shown in the insets of Figure 4a-b. The vertical line in the reflection spectra indicates the position of the Raman laser wavelength (λ = 532 nm). Owing to the existence of a PBG, the PhCs can be used to modulate localized electric fields effectively, and it is expected that through a combination of the effects of plasmonic resonance and PBG, the Raman scattering signals can be enhanced more significantly. The Raman spectra of the two prepared PhC films (PBGs at λ = 536 and 688 nm) after absorption of MB (C = 50 μM) are illustrated by Figure 5.

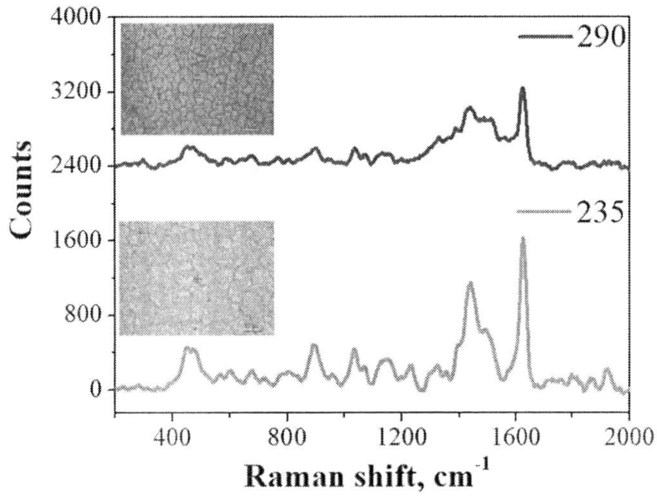

Figure 5. SERS spectra of MB adsorbed in porous PhC substrates with different PBGs. Numbers in the figure correspond to the diameters of the microspheres in nm. Insets: Optical microphotographs of surface areas where the Raman spectra were measured.

It is obvious that the SiO_2@Ag PC films with a PBG at λ = 536 nm showed the highest sensitivity, compared to the sample where PBG is located at λ = 688 nm. In the SERS measurement, the wavelength of laser is λ = 532 nm, which is located close to the center of the PBG at λ = 536 nm. This means that the higher density of electromagnetic modes at the PBG band center leads to a larger light–matter interaction and results in greater enhancement of Raman scattering.

4. Conclusions

In summary, monodisperse SiO_2@Ag core–shell microspheres, consisting of Ag nanoparticles coated by SiO_2, were obtained. The SiO_2@Ag PhC films were prepared by self-assembly of the SiO_2@Ag core–shell microspheres with the help of vertical-deposition method. The strong SERS effect was established for the SiO_2@Ag PhC films used as SERS substrates. The position of the PBG is one of the main factors determining the enhancement of Raman scattering.

Acknowledgments

This work was financially supported by the Russian Foundation for Basic Research (grant number 19-03-00556)

References

[1] Jiajie B, Genrui F, Suli W, Xin S, Hongbo X, Shu-Fen Zh 2017 *ChemistryOpen* **6** 637
[2] Niu W, Su L T, Chen R, Chen H, Wang Y, Palaniappan A, Sun H, Tok A I 2014 *Nanoscale* **6** 817
[3] Wu C Y, Huang C C, Jhang J S, Liu A C et al. 2009 *Opt. Express* **17** 21522
[4] Yablonovitch E 1987 *Phys. Rev. Lett.* **58** 2059
[5] Galisteo-Lopez J F, Ibisate M, Sapienza R et al. 2011 *Adv. Mater.* **23** 30
[6] Ge J, Yin Y 2011 *Angew. Chem. Int. Ed.* **50** 1492
[7] Klimonsky S O, Bakhia T, Knotko A V, Lukashin A V 2014 *Dokl. Chem.* **457** 115
[8] Jiang P, Bertone J F, Hwang K S and Colvin V 1999 *Chem. Mater.* **11** 2132
[9] Zhang Q S, Ye J W, Tian P, Lu X Y, Lin Y, Zhao Q, Ning Q L 2013 *RSC Adv.* **3** 9739
[10] Tang S C, Tang Y F, Zhu S P, Lu H M, Meng X K 2007 *J. Solid State Chem.* **180** 2871
[11] Khan Z, Al-Thabaiti S A, Obaid A Y, Alyoubi A O 2011 *Colloid. Surface. B* **82** 513
[12] Steinigeweg D, Schlucker S 2012 *Chem. Commun.* **48** 8682
[13] Xu C, Li W J, Wei Y M, Cui X Y 2015 *Mater. Design* **83** 745

Extracting hot-spot correlation length from SNSPD tomography data

M I Polyakova[1], I N Florya[2], A V Semenov[2,3] A A Korneev[1,2], G N Goltsman[1,2]
[1] National Research University Higher School of Economics, Moscow 101000 Russia
[2] Moscow State Pedagogical University, Moscow 119991
[3] Moscow Institute of Physics and Technology, Moscow region, Dolgoprudny, 141700, Russia

Abstract. We present data of quantum detector tomography for the samples specifically optimized for this problem. Using this method, we take results of hot-spot correlation length of 17 ± 2 nm.

1. Introduction

Since the first realization of the superconducting nanowire single-photon detector (SNSPD) in 2001, they have substantially evolved and now they demonstrate high detection efficiency, high count rate low timing jitter. During the time passed several theoretical models of SSPD operation have been proposed. There is a model according to which a local suppression of superconductivity is limited to a small area of the strip, called 'hot spot', and the transition of the film to the normal state occurs due to current crowding around it. The absorption of photons reduces the entrance barrier for vortices and contributes to the nucleation of a vortex-antivortex pair inside the film in the presence of a weak superconducting region. If the bias current is high enough the Lorenz force starts driving a vortex. The resistive area is formed and grows due to Joule heating, thus switching the superconducting strip in a resistive state [1]. However, this model does not answer the question about the size of the hot spot. For better use of such devices it is necessary to understand where and when the absorption event occurred, but the hot spot is too small and has a too short lifetime to be measured in a direct way. The closest to the direct measurement will be the quantum detector tomography (QDT). Here, we apply the QDT method to samples that were specially made in order to conveniently implement the protocol.

2. The method

The QDT method, which we use in our work, is described in detail in our previous publication [2]. In brief, if a photon is absorbed in the detector's strip, a hot spot is formed. The appearance of the hot spot can result in a voltage pulse, which we will call a "click", but can also result in nothing observable, depending on whether the bias current is large enough. The production of the click by the hot spot is a probabilistic process, with probability dependent on the bias current I. the probability to have a click when exactly one photon is absorbed is $\eta_1(I)$, we shall call it "single-spot efficiency". We consider the situation when the bias current is low and one hot spot is not enough to produce a click **(Fig.1a).** But two spots formed close to each other, say, in the same cross-section of the strip, are able to do it. We call click of this kind "double-spot count" and denote the double-spot efficiency by η_2.

To find $\eta_2(I)$, we proceeded as follows. First, we calculated mean number of absorbed photons M, just multiplying known mean number of incident photons by on-chip detection efficiency (OCDE) at saturation, assuming, that each absorbed photon produces a click. Then, we fitted the dependencies $P(M)$ by the 2-nd order polynomial

$$P(M) \approx \eta_0 + \eta_1 M + \frac{1}{2}(a_2 M^2) \tag{1}$$

Here, the term η_0 represents the probability of a dark count.

Repeating this fit for data at different bias currents, one can obtain $\eta_1(I)$, which is of course trivial, and $\eta_2(I)$ – which is the quantity of our interest. Our ultimate goal is to find *s-hot-spot correlation length* .

To do it, we have to observe saturation of $\eta_2(I)$ with the increase of I, at some level $\eta_2{}^{max}$, which can then be interpreted as s/L.

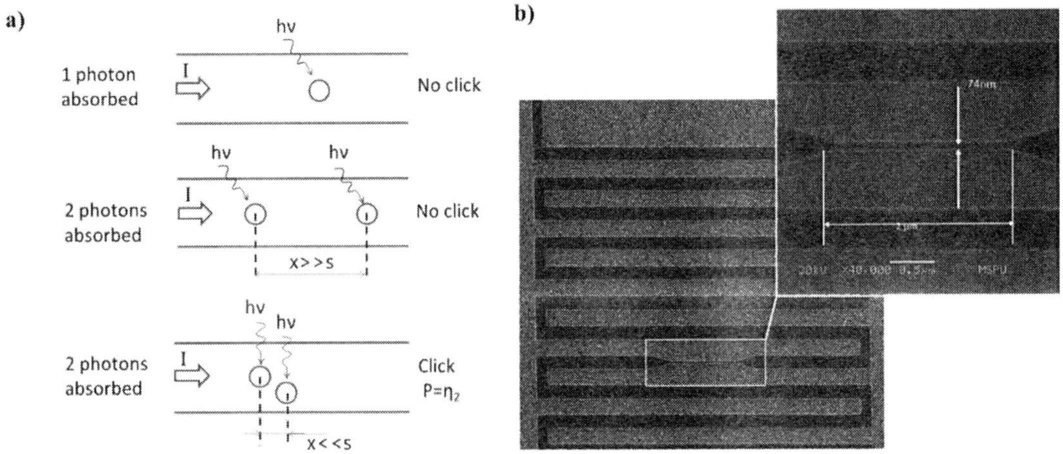

Figure 1. **a)** small current- two photons produce the click if they can be absorbed at the distance less then value called hot-spot correlation length. **b)** SEM-image of superconducting strip that is enclosed in a meander.

In our models we expect that $a_2 = \eta_2 - (\eta_1)^2 \equiv \eta_2 + a_2{}^{stat}$. With the increase of the bias current, single-spot efficiency increases too, the linear contribution becomes larger than quadratic and make it difficult to extract the coefficient a_2. There is also a parasitic effect of AC-biasing [2]. In order to check the contribution of this parasitic effect, we can write: $a_2{}^{bias} = 2I\eta_1\, d\eta_1 / dI\, \varphi$ hence

$$a_2{}^{stat} + a_2{}^{bias} = -(\eta_1)^2 + 2I\eta_1 \frac{d\eta_1}{dI}\varphi \tag{2}$$

Here $\varphi = f\tau$, where f is pulse repetition rate, and τ is the SSPD pulse duration. The φ parameter depends on length of device: the shorter the sample, the shorter the time $\tau = \varphi / f$.

The expected value of $\eta_2{}^{max}$ is inversely proportional to the length of the sample, so for the most accurate measurement it is preferable to have a short sample. At the same time, the length should not be too short, so that the detection event at the transition points from the wide NbN film to the detector strip can be neglected, detection in the contact areas and non-uniform current distribution in the sample strip can be excluded.

Based on these considerations, we fabricated samples 74 nm wide and 2 μm long. The contribution from areas of non-uniform current distribution was minimized and amounted to no more than 10%.

3. Sample (or experiment)

Superconducting strip was made out of 6 nm- thick niobium nitride (NbN) film by means of electron-beam lithography using negative resist MAN 24-01. Figure 1(a) shows superconducting strip having width of ~70 nm and length 2 μm. For better alignment with single-mode fiber (diameter of the light-bearing core is 9 μm) and preventing latching, the superconducting strip is enclosed in a meander (area of 10x10μm^2, width of 350 nm). For our protocol, we used a pulsed lasers, one at wavelength 1064 nm with 5ps pulse width, and the second at 1550 nm wavelength with ~50ps pulse width. The photons should be absorbed simultaneously. The measurements were carried out at a temperature of 1.7 K and critical current of the sample *Ic* was 16.4 uA.

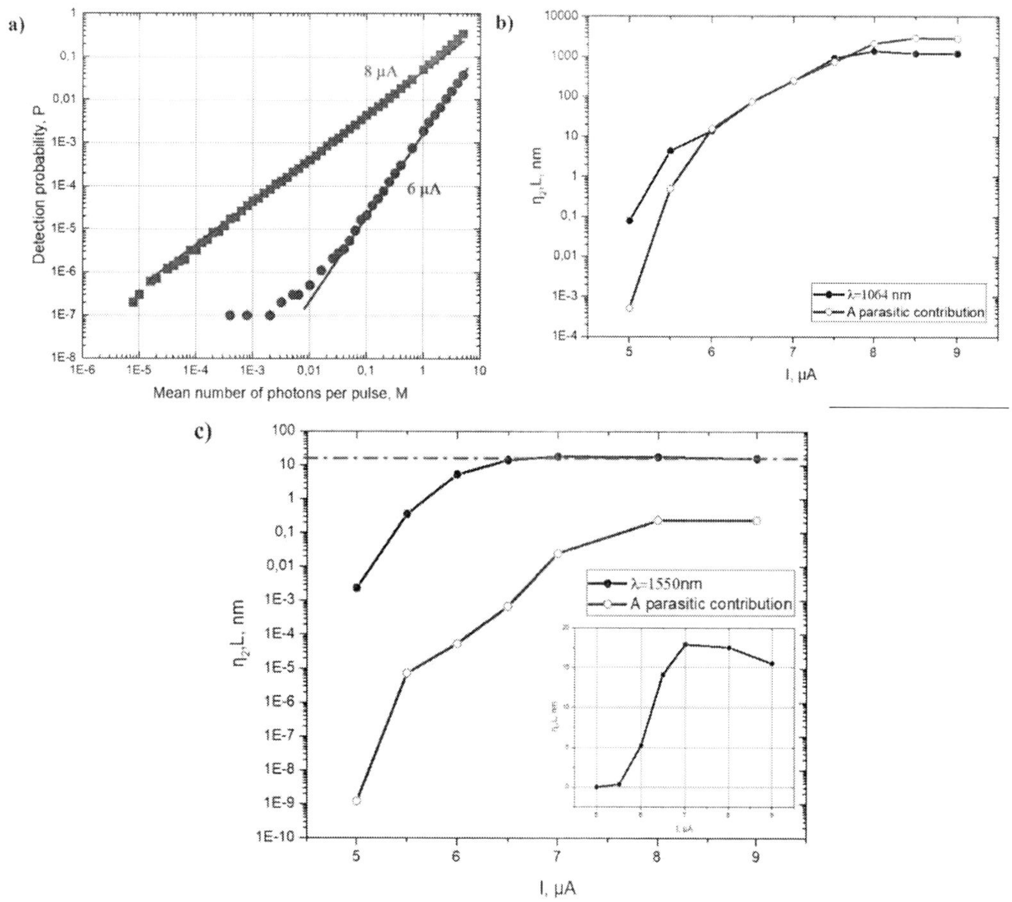

Figure 2 a) single-spot and double spot counts and how determine it. (b) hot-spot correlation length for wavelength of 1550 nm (black curve) and parasitic contribution (blue curve) of 1550 nm. The red line shows saturation which can be interpreted as the hot-spot correlation length **(c).** hot-spot correlation length for wavelength of 1550 nm (black curve) and parasitic contribution (blue curve) of 1550 nm

In order to determine the absorption coefficient, it is necessary to have one hundred percent internal quantum efficiency. In our experiment, a plateau was observed, which we interpret as one hundred percent internal quantum efficiency. We found an absorption coefficient of 4x10^{-7}. Further, we measured the dependence of the count per second vs power for different currents. With the same bias current, but different powers, SSPD can operate in different detection modes, while two-photon events are usually observed with a small current and high power [3]. Fig.2a shows the count rate vs mean number of photons per pulse for the current, at which the single-photon mode prevails (squares) — the linear slope of the curve, and the second curve (circles) measured at a higher current, at which the two-photon mode is observed — the quadratic slope of the curve prevails. The coefficients η_1 and η_2 were found by applying a fit of the experimental data taken at various currents: from 5 to 9 μA. By applying our protocol to this data, the single- and double-spot efficiencies were extracted as a function of current. The results are presented in Fig. 2 b and 2c.

4. Results

Figures 2b) and 2c) show the dependences obtained at the wavelength of 1064 nm with a pulse repetition rate of 200 MHz and 1550 nm with a pulse repetition rate of 10 MHz. As we assumed, starting at some current value of η_2 start to saturate. To make sure that we see exactly the saturation of the coefficient η_2 (black solid circles in the fig. 2b) and 2c) we analyzed parasitic contribution (blue open circles in the fig. 2b and 2c). It is seen that for the wavelength of 1550 nm, this contribution is small everywhere. We find hot-spot correlation length of 17 ± 2 nm (fig 2c). For a wavelength of 1064 nm, the value of the two-spot efficiency, multiplied by L reaches saturation at about 100 nm, but unfortunately, due to the laser repetition rate being too high, most of the data obtained is a parasitic contribution (Fig. 2b). results are presented in Fig. 2 b and 2c.

5. Conclusions

Using our QDT protocol and applying it to specially designed sample, we were able to accurately measure the correlation length at the wavelength of absorbed photon of 1550 nm and find that it equals 17 ± 2 nm for the sample with the width of 74 nm.. Made of 6nm-thick NbN strip.

Let us discuss what is the relation of the obtained correlation length to the size of the hot spot. There are two alternative interpretations of the hot-spot correlation length. The first one, is that the distance between two spots is of the order of the width of the strip. We can divide the strip into squares [4], then a double-spot count is occurring when two spots fall into one square of the strip. In this case, the correlation length should be proportional to the width of the sample. The second scenario – the two-spot count appears when the two hot-spots touched each other. Then the hot spot correlation length can be interpreted as the hot-spot size.

Although our resulting value of the correlation length is less than the width of the strip, it is of the order of the width, and therefore we cannot consider it to be contradiction to the first interpretation. As a speculation, we note, however, that our result supports rather the second interpretation. Besides the small absolute value of the extracted correlation length, this value is close to what was obtained in the work of Renema *et.al.*[5], 23 ± 2 nm, with the use of wider samples and another protocol. This indicates that the correlation length does not grow with the strip width, which agrees with the second scenario. Of course, to come to a more valid conclusion, a systematic study of the dependence of the correlation length on the width of the strip is needed.

Acknowledgments
The work was supported by RSF grant # 17-72-30036.

References

[1] Zotova A.N. *et al.* 2014 *Supercond. Sci. Technol.* **27** 125001
[2] Polyakova M. *et al.* 2019 *IEEE Transactions on Applied Superconductivity* 29 10518223
[3] Elezov M.S. *et al.* 2013 *Journal of Optical Technology* **80**
[4] Akhlaghi *et al.* 2009 *IEEE Transactions on Applied Superconductivity* 19 361-366
[5] Renema J.J. *et al.* 2012 *Optics express* **3** 2806-2813

Interpretation of photoluminescence spectra of metamorphic InAlAs/GaAs heterostructures

G A Denisova, D D Firsov and O S Komkov

Department of micro- and nanoelectronics, St. Petersburg Electrotechnical University "LETI", Prof. Popov str. 5, St. Petersburg 197376, Russia

Abstract. The results of analyzing the photoluminescence spectra of metamorphic $In_{0.75}Al_{0.25}As/In_{0.05-0.8}Al_{0.95-0.2}As/GaAs$ heterostructures grown by molecular-beam epitaxy method are presented. In the spectra of low-temperature photoluminescence, three emission peaks have been detected, which correspond to carrier recombination in different areas of the structure. These include a part of the $In_{0.75}Al_{0.25}As$ "virtual substrate" grown at high temperature, and the area of a triangular potential well formed by the metamorphic buffer layer and the "virtual substrate". Photoluminescence related to defects in the high-temperature part of the "virtual substrate" was also measured.

1. Introduction and samples

InAs and its aloys are some of the most demanded materials for creating microwave transistors, as well as devices for mid-wavelength infrared optoelectronics. The growth of the epitaxial layers of these materials on a GaAs substrate is carried out using a metamorphic technology [1]. GaAs substrates, compared with another substrates of the III-V group, have a number of advantages such as higher reliability, low cost and large available plate sizes (100-150 mm) [2].

The purpose of this study is to analyze the obtained photoluminescence spectra of the $In_{0.75}Al_{0.25}As/In_{0.05-0.8}Al_{0.95-0.2}As/GaAs$ metamorphic heterostructures and evaluate the effect of low-dimensional effects on the samples under study.

The samples were grown by molecular beam epitaxy (MBE) method in the Ioffe Institute (Russian Academy of Sciences). In table. 1, the sequence of layers of the structures under study can be seen. Due to the large mismatch of the lattice constants of the GaAs substrate and the subsequent $In_{0.75}Al_{0.25}As$ layer, called the "virtual substrate", it is impossible to grow the epitaxial layer directly, since the relaxation of strain will lead to the large number of defects in the active layer of the future device. Thereforea metamorphic buffer layer (MBL) $In_{0.05-0.8}Al_{0.95-0.2}As$ was used to grow such structures. This layer matches the substrate with the subsequent layer by gradually changing the lattice parameter. Metamorphic technology allows to obtain a "virtual substrate" with the required lattice parameter, which serves as a basis on which the active layers are grown [3]. In fig. 1 the band structure of one of the samples is shown, as provided by a calculation taking the strain into account [4].

Table 1. Structure of the studied samples

Sequence of layers	**Sample #1**	**Sample #2**	**Sample #3**
	Layer thickness, nm		
High-temperature part of the "virtual substrate" $In_{0.75}Al_{0.25}As$	140	-	1030
InAs	1	-	1
Low-temperature part of the "virtual substrate" $In_{0.75}Al_{0.25}As$	20	100	20
Metamorphic buffer layer $In_{0.05-0.8}Al_{0.95-0.2}As$	1200		
Buffer layer GaAs	200		
Semi-insulating GaAs (001) substrate	$3\cdot10^5$		

2. Experiment and results

The samples were examined by the low-temperature (from 10K to 80K) photoluminescence method using an experimental setup based on the VERTEX 80 Fourier transform infrared spectrometer (FTIR). This method is chosen because it is highly sensitive and non-destructive. It does not require special preparation of the samples for the measurements. In fig. 2 a typical PL spectrum of the sample #1 is shown.

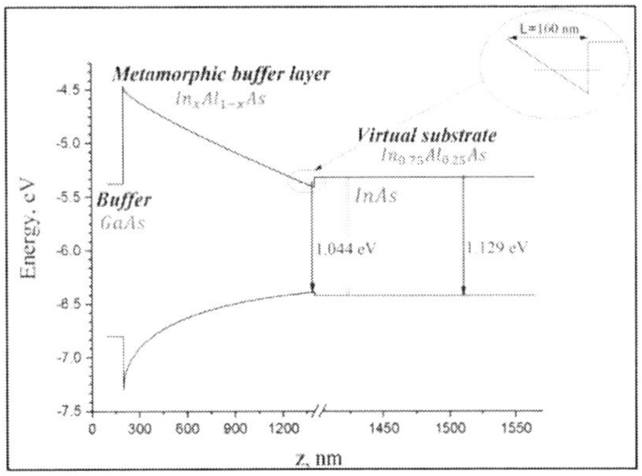

Figure 1. Band diagram of the sample #1 indicating the transition energies of the $In_{0.75}Al_{0.25}As/In_{0.05-0.8}Al_{0.95-0.2}As/GaAs$ heterostructure.

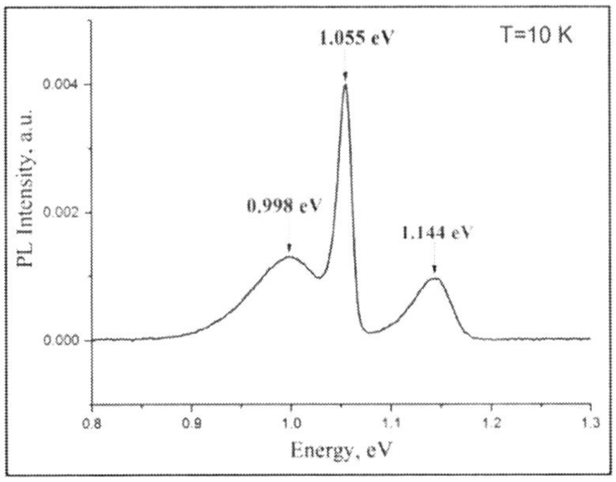

Figure 2. Low-temperature PL spectrum of the sample #1 indicating the transitions of the $In_{0.75}Al_{0.25}As/In_{0.05-0.8}Al_{0.95-0.2}As/GaAs$ heterostructure.

Three peaks of radiation characteristic for most of the samples, corresponding to charge carrier recombination in different areas of the structure, were observed.

2.1. Interpretation of the peak with the energy of 1.055 eV

It was assumed that this peak corresponds to the charge carrier recombination in the area of a triangular potential well formed by the upper part of the metamorphic buffer layer and the "virtual substrate"(see Fig.1). The transition energy is close to the calculated value of 1.044 eV. The shift towards higher energies as compared to the calculated value can be associated with the quantization of the charge carriers in the triangular potential well. The well width (L=160 nm) is comparable to the de Broglie wavelength in the corresponding material at T = 10 K. On an inset of Fig. 1, the first quantization level is shown. One upper level also exists, but it is situated close to the continuum. An unusual feature of this heterostructure is that, unlike the conventional single heterojunction and metal-oxide-semiconductor (MOS) structures, quantum wells are present for both electrons and holes. Therefore, the observed increase in the energy of the PL peak above the calculated value might be explained by low-dimensional effects. In addition, under the influence of unaccounted residual compressive strain [5], the band gap of the upper part of the MBL also increases.

In fig. 3 typical PL spectra of the structure under study depending on the pumping laser power is shown. This behavior of the peaks' intensity is typical for all the samples studied. It can be seen that the width at half-maximum of the 1.055 eV peak under consideration weakly depends on power: 17.4 meV at P=17 mW and 22 meV at P=380 mW, and the position of the peak does not change.

The peak intensity increases almost linearly in log scale with increasing pump power (Fig. 4), which is typical for direct interband transitions in quantum wells. Similar behavior was also observed in a number of other papers [6], [7].

In fig. 5, it can be seen that as the temperature increases, the spectrum is expected to shift to lower energies. The peak with an energy of 1.055 eV is absent at a temperature above 30 K. This is due to the fact that the charge carriers accumulated in the quantum well receive sufficient kinetic energy to overcome the potential barrier and leave it.

A comparison was made for the samples with different maximum composition x_{max} of the metamorphic buffer layer [3]: 0.77 for a sample #4, 0.79 for #5, 0.81 for #6 and 0.83 for #7. In fig. 6 the PL spectra of these samples are shown. In this series of samples, an active region was grown on the "virtual substrate" [8].

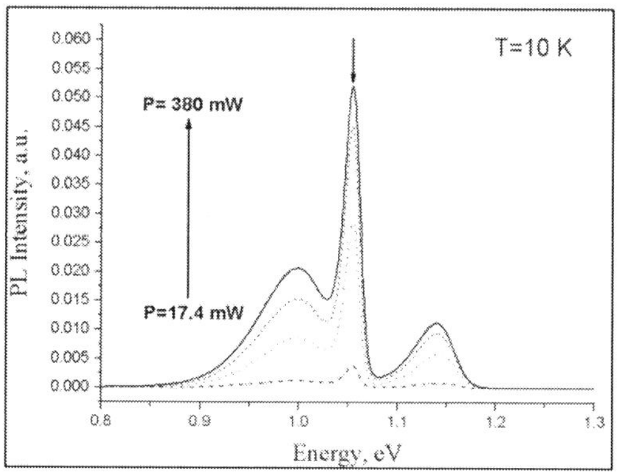

Figure 3. Dependence of the PL intensity on the laser power.

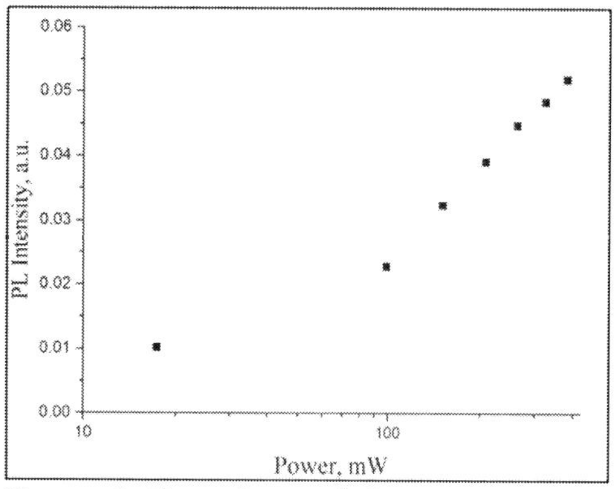

Figure 4. Dependence of the 1.055 eV peak intensity on the laser power.

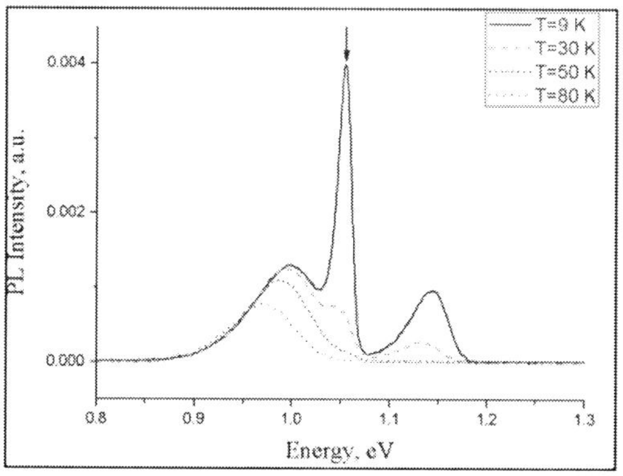

Figure 5. Temperature dependence of the typical PL spectrum of investigated metamorphic heterostructures $In_{0.75}Al_{0.25}As/In_{0.05\text{-}0.8}Al_{0.95\text{-}0.2}As/GaAs$.

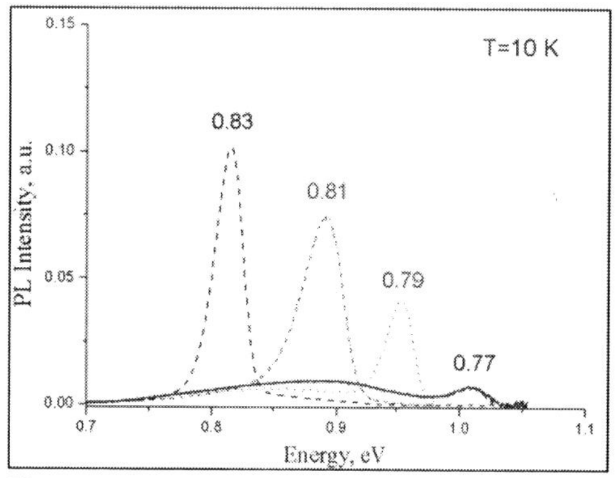

Figure 6. PL spectra of samples with different final composition of the metamorphic buffer layer.

As it can be seen in Fig. 6, the peak under consideration is the most intense in each of the PL spectra of these samples. The shift of the peak with the decrease in the inverse step towards higher energies is explained by an increase in the band gap of the solid solution of the final MBL composition. The decrease in the intensity of this peak with that in the maximum composition x_{max} can be explained by reduction of the quantum well depth.

2.2. Interpretation of the 1.144 eV peak

The samples shown in Fig. 7 differ from each other by the absence (sample #2) of the high-temperature part of the "virtual substrate" and its presence (sample #1).

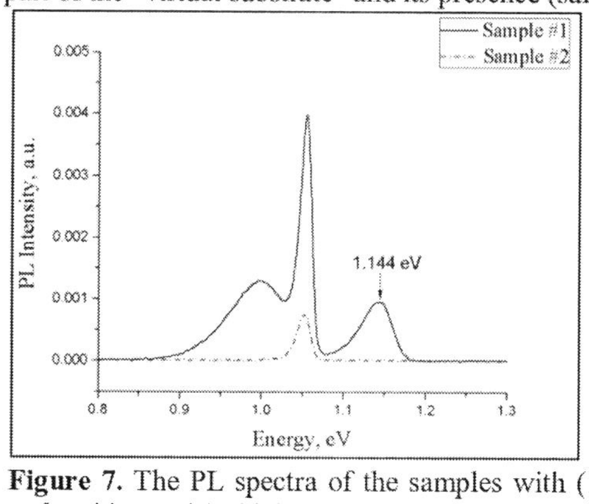

Figure 7. The PL spectra of the samples with (1) and without (2) high-temperature part of the "virtual substrate".

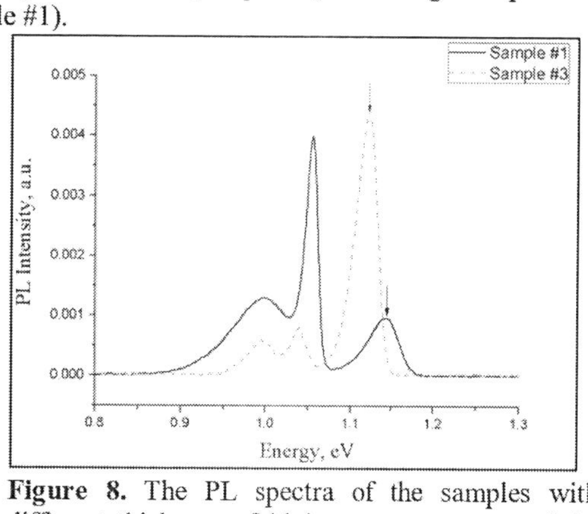

Figure 8. The PL spectra of the samples with different thickness of high-temperature part of the "virtual substrate".

The PL spectrum shows that for the sample #2 the peak is not observed. Thus, it can be assumed that the peak with an energy of 1.044 eV corresponds to carrier recombination in the high-temperature part of the "virtual substrate". In fig. 8 the PL spectra of the samples with different thickness of this layer are shown: 1030 nm for the sample #3 and 140 nm for the sample #1. As it can be seen, the peak of the sample with a thicker layer was 4.5 times more intense in amplitude than that of the sample with a thinner layer

The assumption that this peak corresponds to the high-temperature part of the "virtual substrate" is also confirmed by the calculation of transition energy. This value was 1.129 eV. The shift of the peak position of the PL spectrum towards higher energy is explained by the effect of residual compressive strain in the "virtual substrate". In the PL spectrum, a shift of the peak position to lower energies can be seen. It can be explained by a smaller effect of residual compressive strain on the thicker layer and, accordingly, a smaller increase in the band gap of the "virtual substrate" than in the sample #1.

This conclusion corresponds to previous studies by other authors. For example, in [9], it is seen in the PL spectra that for the sample in which an additional layer of the "virtual substrate" at a higher temperature was grown, another photoluminescence peak appears. In [10], [11], where identical InAlAs/GaAs structures were studied, the presence of a similar peak of the "virtual substrate" is mentioned.

2.3. Interpretation of the 0.998 eV peak

The power dependence of the 0.998 eV peak is shown in Fig. 9. It can be seen that this peak behaves sublinearly and at a sufficiently high power goes into saturation. The temperature dependence of the peak width at half-maximum (Fig. 10) has also been considered.

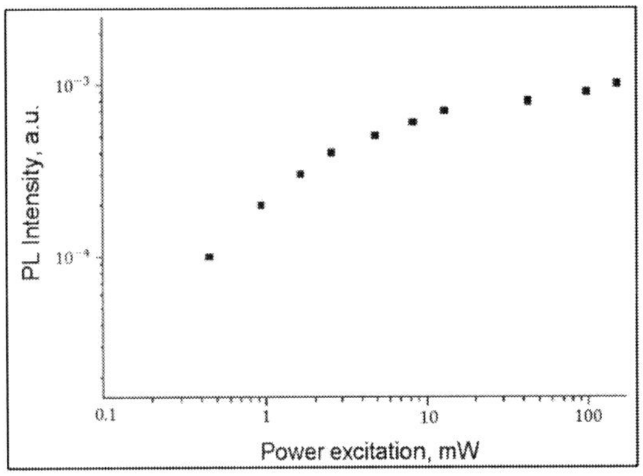

Figure 9. Dependence of the 0.998 eV peak intensity on the pump power for the sample #1.

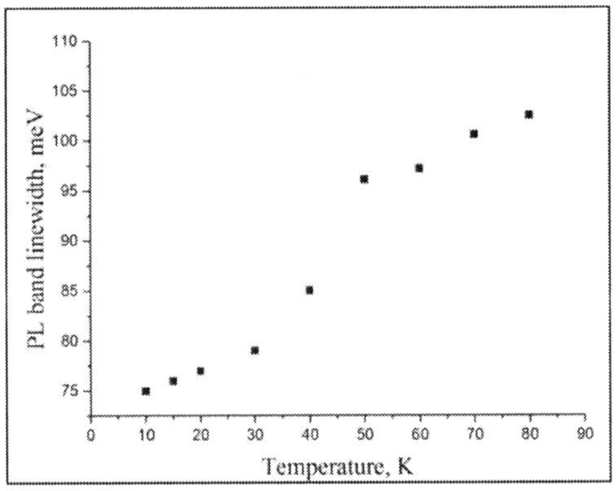

Figure 10. Dependence of the 0.998 eV peak width at half-maximum on the temperature for the sample #1

This dependence is typical for optical transitions through defect levels. In [7], it is concluded that the estimated peak corresponds to the PL caused by the presence of defects in the area of the metamorphic buffer layer. However, Fig. 7 shows that there is no peak with an energy of 0.998 eV for the sample #2 (for which the high-temperature part of the "virtual substrate" has not been grown). This allows us to suggest that this PL is related to the presence of a defect in the high-temperature part of the "virtual substrate".

3. Conclusion

Thus, we analyzed the photoluminescence spectra of the structures under study, which showed three typical PL peaks for all samples. The possible regions of the heterostructure in which carrier recombination occurs were determined. The observed shift in the experimental energy values in comparison to the calculated ones was explained by the effect of mechanical compressive strain, as well as low-dimensional effects in the structures.

Acknowledgments

The study was carried out with the partial financial support of the Russian Science Foundation (project No. 18-79-10161).

The authors are deeply grateful to S. V. Ivanov, V. A. Solov'ev and M. Yu. Chernov for the samples provided.

References

[1] Chernov M Yu, Komkov O S , Firsov D D, Meltser B Ya, Semenov A N, Terent'ev Ya V, Brunkov P N , Sitnikova A A , Kop'ev PS, Ivanov S V, Solov'ev V A 2017 *J. Cryst. Growth* **477** 97
[2] Tsai J-H , Lee C-S , Chiang C-C, Chao Y-T 2014 *Semiconduct. Sci. Technol.* **48** 809
[3] Komkov O S, Firsov D D, Chernov M Yu, Solov'ev V A, Sitnikova A A, Kop'ev P S, Ivanov S V 2018 *J. Appl. Phys.* **51** 055106
[4] Krijn M P C M 1991 *Semicond. Sci. Technol.* **6** 27
[5] Chernov M Yu, Solov'ev V A, Komkov O S, Firsov D D, Meltser B Ya, Yagovkina M A, Baidakova M V, Kop'ev P S, Ivanov S V 2017 *J. Appl. Phys.* **10** 121201
[6] Galiev G B, Pushkarev S S, Vasil'evskii I S, Klimov E A, Klochkov A N, Mal'tsev P P 2015 *Semiconduct. Sci. Technol.* **49** 1238
[7] Yahyaoui N, Aloulou S, Chtourou R, Sfaxi A, Oueslati M 2008 *Thin Solid Films* **516** 1694
[8] Ivanov S V, Chernov M Yu, Solov'ev V A, Brunkov P N, Firsov D D, Komkov O S 2019 *Prog. Cryst. Growth Charact. of Mater.* **65** 20
[9] Shang X Z, Wu J, Wang W C, Wang W X, Huang Q, Zhou J M 2007 *Solid-State Electron.* **51**, 85
[10] Galiev G B, Klimov E A, Klochkov A N, Lavruhin D V, Pushkarev S S, Maltsev P P 2014 *Semiconduct. Sci. Technol.* **48** 640
[11] Chen X Y, Gu Y , Zhang Y G, Ma Y J, Du B, Zhang J, Ji W Y, Shi Y H,Y Zhu Y 2017 *J. Cryst. Growth* **488** 51

Static light scattering for determination of physical parameters of macro- and nanoparticles

I Kolokolnikov, E Nepomnyashchaya, E Velichko
Institute of Physics, Nanotechnology and Telecommunications, Peter the Great St. Petersburg Polytechnic University, Saint Petersburg 195251, Russia

Abstract. The work is dedicated to determination of physical parameters of micro- and nanoparticles by static light scattering method. We present static light scattering device and theory to determine sizes of particles in suspensions. We analyze calculated distributions of scattering intensity for particles of different sizes and compare it with experimental data, received for spherical particles of different sizes. We also suggest theory to calculate form factor of scattering intensity for micro- and nanoparticles of different shapes. We also demonstrated that Rayleigh scattering theory is better to be applied for small particles investigation.

1. Introduction

Micro- and nanoparticles are widely used in biomedical science [1], metalworking industry, researching of environment and a great variety of different nanotechnologies [2]. Methods of static light scattering (SLS) are being developed for more than a century and nowadays contain an amount of analytical solutions for different particle geometries. Low cost laser radiation sources, such as laser diodes, made light scattering technologies available for basic researches and education [2].

Nevertheless nowadays, this method is still not common if we need to analyze shapes of non-spherical particles with unknown sizes. We suggest the theory, which describes static light scattering by the particles of different shapes and sizes, and compare the theoretical considerations with experimental data received by designed SLS-device.

2. Theory

Particles interact with electromagnetic field of light, and scatter light in different directions. The relation between scattered and incident light can be presented by the following matrix equation [3].

$$S_{scattered}(I_v, I_h, U, V)_{4\times1} = F_{4\times4} S_{incident}(I_{v0}, I_{h0}, U_0, V_0)_{4\times1}/{k_0}^2 r^2 \quad (1)$$

All 16 elements of transform matrix F are combinations of four complex amplitude functions S_1, S_2, S_3, S_4, which can be computed using different scattering theories.

2.1 Rayleigh theory

If particle size is much less than light wavelength, all elementary dipoles of particle emit coherently. This case of scattering is named after Lord Rayleigh who determined the dependence of the intensity of scattered light on the wavelength in 1871. Indicatrix can be described by the following expression [4]

$$I(\theta) = \frac{24\pi^3 N V^2 \left(n_p^2 - n_0^2\right)^2 I_0}{L^2 \left(n_p^2 + 2n_0^2\right)^2 {\lambda_0}^4} (1 + cos^2\theta) \quad (2)$$

where n_0 – solvent refractive index, n_p – particle refractive index, I_0 – incident light intensity, λ_0 – incident light wavelength in vacuum, θ – scattering angle, L – distance from the scattering volume to detection point, N – is concentration of particles, V – volume of the spherical particle.
With increasing particle size, dipole moments in different parts of it begin to emit in different phases, and interference will cause decrease of the total amplitude of electromagnetic radiation.

2.2 Scattering factors for particles of different shapes
When the dimension of particle is larger than that which can be treated as a single dipole, the approach is to treat the particle as an assembly of many tiny and structureless scattering elements. Each of these scattering elements gives rise to Rayleigh scattering independent of other scattering elements in the particle. The phase shift corresponding to any element in the particle is negligible and the phase difference between different elements in the particle is determined only by their positions and is independent of the material properties of the particle. Scattering in a given direction from all these elements results in interference because of the different locations of these elements in the particle. To satisfy the above assumptions, the particle has to have a refractive index close to 1 in addition to being small compared to the wavelength of light.

In the Rayleigh-Debye-Gans theory, the amplitude functions can be presented as [4]

$$\begin{pmatrix} S_1 & S_4 \\ S_3 & S_2 \end{pmatrix} = \frac{ik_0{}^3(m-1)}{2\pi} P^{1/2}(x) \begin{pmatrix} 1 & 0 \\ 0 & cos\theta \end{pmatrix} \tag{3}$$

Therefore scattering pattern can be described by the scattering factor P(x), which depends on a shape of the particle [4]

Thin disk:

$$P(x) = \frac{2}{x^2}(1 - \frac{1}{x}J_1(2x)) \tag{4}$$

where $x = K \cdot d/2$, d – diameter, radius of gyration $R_g = d^2/8$.

Thin rod:

$$P(x) = \int_0^{2x} \frac{sint}{t} dt - \frac{sin^2 x}{x^2} \tag{5}$$

where $x = K \cdot l/2$, l – length, $R_g = l^2/12$.

Ellipsoid of revolution:

$$P(x) = \int_0^{\pi/2} [\frac{3(sinx - xcosx)}{x^3}]^2 \, cos\beta \, d\beta \tag{6}$$

where $x = Ka[cos^2\beta + (\frac{b}{a}sin\beta)^2]^{1/2}$, a, b – semiaxes, β is the angle between figure axis and bisectrix. More formulas for different shapes can be found in [5].

2.3 Mie theory
For bigger particles, we should also take into account primary wave reemission by elementary dipoles. When constructing the Mie theory, the Maxwell equations are solved inside and outside the scattering particle; therefore, it is necessary to introduce boundary conditions on the particle surface. For spherical particles, Gustav Mie in 1908 obtained an analytical solution. The original Mie theory is restricted to plane wave scattering by a homogeneous, isotropic sphere in a nonabsorbing embedding medium [6].

If we assume the field at a point inside the sphere is what it would be if the sphere were not present (Rayleigh-Debye approximation) [7]. Mie scattering of unpolarized light can be approximated by the following expression [7]

$$I(\theta) = I_0\left[\frac{3|J_1(x)|}{x} + \gamma\right]^2 (1 + cos^2\theta)/2 \quad (7)$$

where $x = ka(1 + m^2 - 2m\cos\theta)^{1/2}$, $J_1 = (\sin x - x \cos x)/x^2$ – J_1 is often referred to as the first-order spherical Bessel function of the first kind, m – real relative index of diffraction, a – radius of a single sphere, $k = 2\pi/\lambda$ – wave vector, $I_0 = c\varepsilon_0 E_0^2(m^2-1)^2a^6k^4 / 18r^2$, c – velocity of light, ε_0 – permittivity of a free space; γ – parameter of approximation, which is necessary, when ka>1. For our experimental situation γ can be chosen as a function $x^{-3/2}$.

Mie scattering theory has no size limitations and converges to the limit of geometric optics for large particles. In small particles limit it matches to Rayleigh theory.

2.4 Fraunhofer diffraction

When particle size is more than light wavelength, the main process is diffraction. When the particle dimension is much larger than the wavelength of light or the material is highly absorptive, the edge effect of particle contributes much more to the total scattered intensity. The interference effect now arises mainly from particle's contour; i.e., diffraction produced by the bending of light at the particle boundary. In these instances, scattering from the inner part of the particle is less important and neglected as an approximation. The particle now behaves like a two-dimensional object instead of a three-dimensional particle. Only the projected area perpendicular to the light matters, not the volume of particle [4]. In this case scattering can be described by Fraunhofer diffraction theory.

3. SLS scattering measurements

The scheme of the SLS device is presented if Fig. 1. Light beam from semiconductor laser (wavelength 653 nm) is cleared by the diaphragm and focused to the center of cuvette. Measurements are consisted of two quantities: angle on the goniometer, which shows direction of dissipated radiation, and voltage from photomultiplier on the oscilloscope, which is proportional to the intensity of radiation scattered in this direction.

As a result, we have series of measurements, forming an indicatrix of light scattering, showing relative angular distribution of radiation energy. Main advantages of the actual method are simplicity, removability of the components of setup and non-invasiveness.

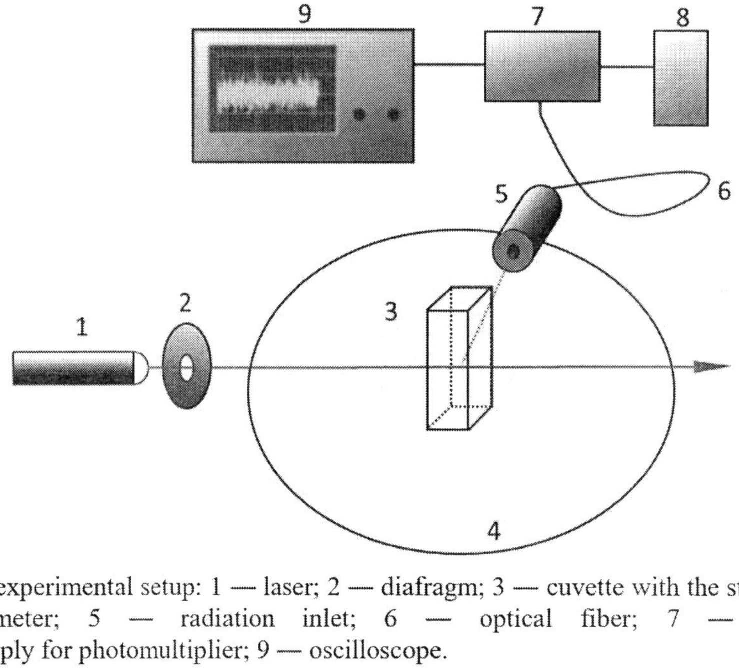

Figure 1. The experimental setup: 1 — laser; 2 — diafragm; 3 — cuvette with the studied suspension; 4 — goniometer; 5 — radiation inlet; 6 — optical fiber; 7 — photomultiplier; 8 — power supply for photomultiplier; 9 — oscilloscope.

Journal of Physics: Conference Series **1410** (2019) 012168 doi:10.1088/1742-6596/1410/1/012168

4. Results and conclusions

In the experiment, we measured scattering curves for solutions of monodisperse spheres of polystyrene latex (refractive index n = 1.586, relative to water refractive index m = 1.586/1.331 = 1.192) of the following diameters: 0.54 μm (solution concentration C = 0.2 mg/ml), 2.1 μm (C = 0.1 mg/ml), 11 μm (C = 0.012 mg/ml).

Following figures present theoretical and experimental scattering patterns for considered latex particles in logarithmic scale. For particles, which size is smaller, than wavelength we considered both, Rayleigh and Mie theories.

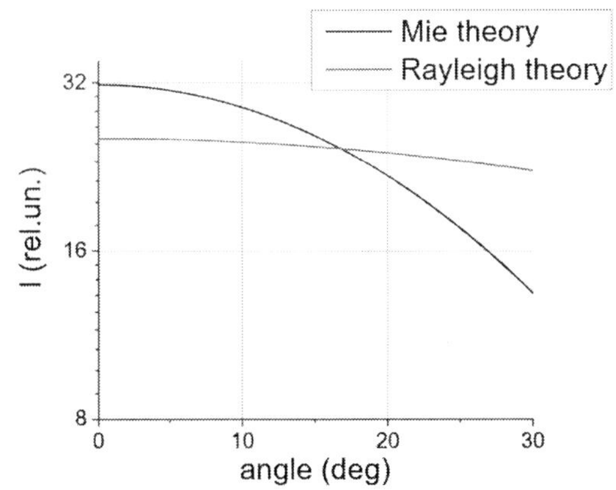

Figure 2. Theoretical scattering curves, calculated for monodisperce microparticles with diameter of 0.54 μm

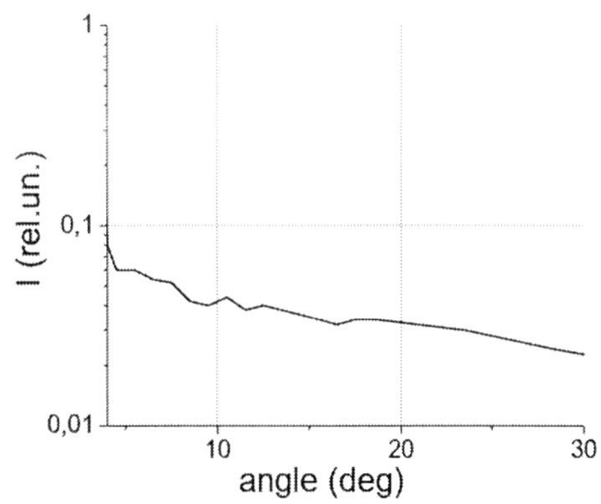

Figure 3. Experimental scattering curve, measured for monodisperce microparticles with diameter of 0.54 μm

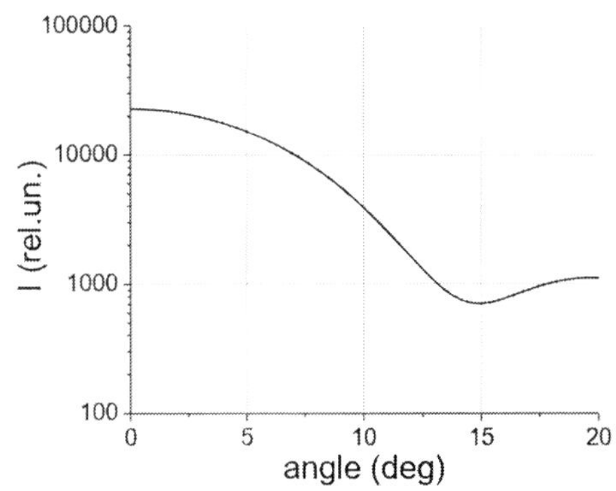

Figure 4. Theoretical scattering curve, calculated for monodisperce microparticles with diameter of 2.1 μm

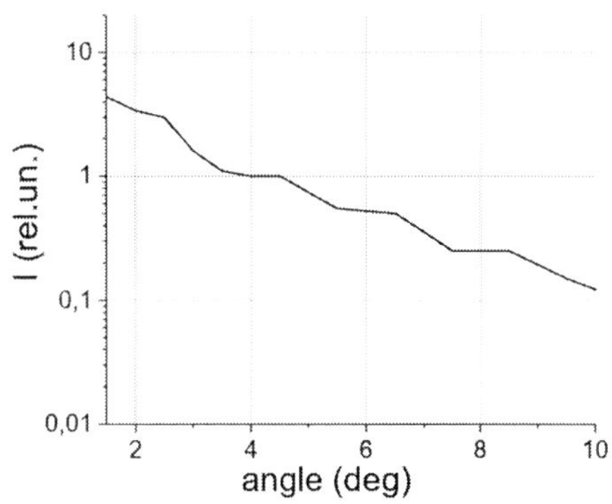

Figure 5. Experimental scattering curve, measured for monodisperce microparticles with diameter of 2.1 μm

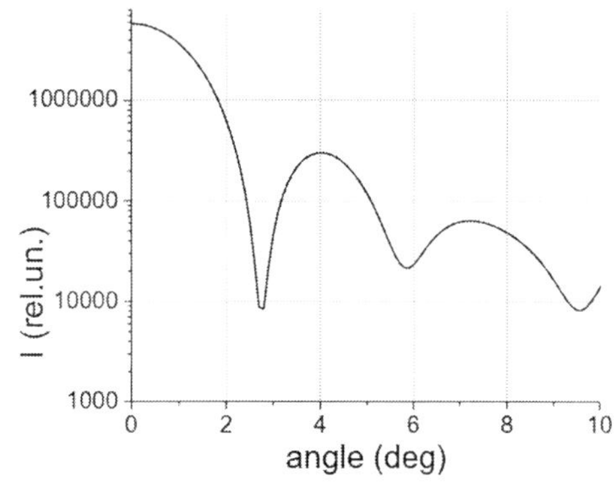

Figure 6. Theoretical scattering curve, calculated for monodisperce microparticles with diameter of 11 μm

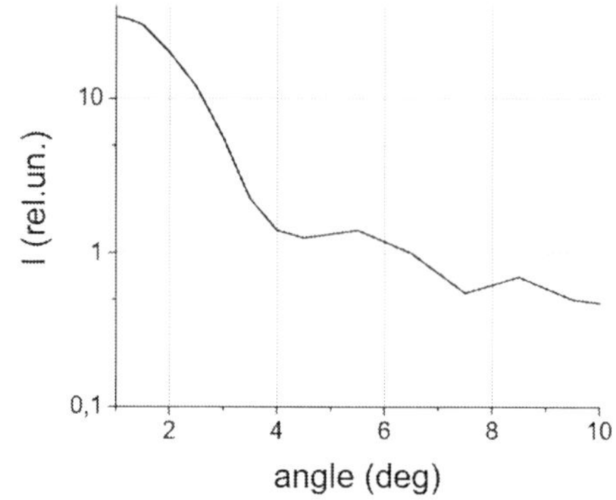

Figure 7. Experimental scattering curve, measured for monodisperce microparticles with diameter of 11 μm

It can be noticed that particle sizes can be distinguished using the level of scattered intensity in small angles. Concentration of solutions influence scattering, thus, method accuracy depends on accuracy of concentration measurements.

For small particles, with size lower than the light wavelength, we can conclude, that it is better to use Rayleigh theory for scattering curve approximation. Experimental scattering curve for solution of particles with size of 2.1 μm are not well approximated by Mie theory either. Therefore, we can conclude, that particles with sizes, which are close to the incident light wavelength are not well studied by SLS. In these cases, it is better to apply dynamic light scattering methods [8].

Scattering for 11 μm spheres fits Mie theory, because of the higher diameter/wavelength ratio. Mismatch of location of theoretical and experimental scattering extremums for it can be described, for example, by some polydispersity of solution.

Provided comparisons of theoretical and experimental data showed, that SLS could be successfully applied for solutions with microparticles with high diameter/wavelength ratio. In the next work, we will compare scattering curves for nonspherical particles with high diameter/wavelength.

References

[1] Dolgushin S A, Yudin I K, Deshabo V K, Shalaev P V, and Tereshchenko S A 2016 *Biomedical Engineering* **49** (6) 394–397

[2] Xu R 2015 *Particuology* **18** 11–21

[3] David W H 2009 *Light scattering theory*, p 13

[4] Xu R 2002 *Particle Characterization: Light Scattering Methods* Kluwer (USA: Academic Publishers) p 397

[5] Pederson J S 1997 *Advances in Colloid and Interface Science* (Denmark: Elsevier) **70**, 171-210

[6] Quinten M 2011 *Optical properties of nanoparticle systems* (Weinheim: Wiley-VCH)

[7] Drake R M, Gordon J E 1985 *American Journal of Physics* **53,** 955

[8] Nenomnyashchaya E, Antonova E, *2018 IEEE International Conference on Electrical Engineering and Photonics* 136–14

SERS induced by Au/ZnO plasmonic nanorods coupled to Au nanoparticles

V. Kaydashev[1,2], P. Zolotukhin[3], A.Belanova[3], A. S. Anokhin[4], T. Minasyan[1], A. Chernishev[5], and E. M. Kaidashev[1]

[1]Laboratory of Nanomaterials, Southern Federal University, Rostov-on-Don 344090, Russia
[2] Laboratory of 2D materials` Optoelectronics, Moscow Institute of Physics and Technology, Dolgoprudny 141701, Russia
[3]Academy of Biology, Southern Federal University, Rostov-on-Don 344090, Russia
[4]Southern Scientific Center RAS, Rostov-on-Don 344006, Russia
[5]Institute of Physical and Organic Chemistry, Southern Federal University, Rostov-on-Don 344090, Russia

Abstract. We study the surface enhanced Raman scattering of Methylene Blue (MB) dye molecules induced by 3D array of "hot spots" made of vertical Au/ZnO core-shell nanorods coupled to self-assembled plasmonic Au nanoparticles. SERS substrates based on array of nanorods coupled to nanoparticles show much better performance compared with bare Au/ZnO nanorods. The hybrid 3D SERS substrate perfectly resolve Raman spectra of MB molecules chemisorbed from solutions with analyte concentrations of ~10^{-7} M even upon washing of a sample i.e. when only chemisorbed molecules were remained. Raman signal enhancement results from the superposition of two effects, namely, the ability of a plasmonic coupled system to enhance the Raman scattering via local field enhancement and from the large 3D surface which provides more adsorption sites compared with traditional 2D surfaces.

1. Introduction

Urgent request for detection of ultralow concentrations of chemical and biological molecular species using surface enhanced Raman scattering (SERS) induced by plasmonic nanoobjects has stimulated rapid development of new approaches to design efficient concentrators of the electric field at nanoscale. Suspensions of nanoparticles including core-shells [1] with strong plasmon coupling provide unique possibilities to in vivo biological and medical techniques. Meanwhile in vitro methods need efficient substrates with large amount of "hot spots" on the surface. Variety of 2D surfaces of plasmonic coupled nanobjects such as bowtie antennas [2] were prepared by expensive e-beam lithography method. A number of more cost-efficient self-assembling methods were also demonstrated. Namely, nanoparticle chains [3], particle multilayer stacks of monolayers of isolated [4] and plasmonic coupled nanoparticles [5] were designed for these purposes. Nanoengineering of substrates based on 3D architecture concept is capable to boost the sensitivity of SERS technique even further. A superior efficiency of some 3D SERS substrates was demonstrated recently [6, 7]. Indeed, 3D nanoobjects such as arrays of vertical nanorods of several microns length with large surface area do provide several orders of magnitude more adsorption sites compared with 2D surfaces.

Plasmoic nanorods in contrast to nanoparticles have the preferential direction which results in additional resonances corresponding to dipole, quadrupole, octopole and other modes of higher multipolarity [8,9]. Particle-like or "absorbing" nanorods with size much smaller than the wavelength of incident light show maximal local field enhancement at the wavelength of a dipole resonance and minor enhancement at multipolar modes. With increasing the rods length-to-diameter aspect ratio the resonances show red shift. For instance, the major dipole resonance of a rod of 40 nm diameter is shifted from ~700 to 1800 nm when its length is increased from 120 to 480 nm [10]. Corresponding near field enhancement coefficient averaged over the surface area is increased from ~300 to 870 for the dipole mode. For larger nanorods the input of scattering to extinction becomes considerable and it should be taken into account when 3D SERS substrates are designed.

We report a robust CVD and PLD based method to prepare SERS substrates with 3D array of "hot spots" made of oriented Au/ZnO core-shell nanorod arrays coupled with self-assembled plasmonic Au nanoparticles, in which Raman signal enhancement results from a superposition of two effects, namely, the ability of a plasmonic coupled system to enhance the Raman scattering and the larger 3D surface with more adsorption sites compared to 2D surfaces.

2. Experimental

2.1. Text in figures

Vertically oriented arrays of Au/ZnO nanorods with plasmonic shells are prepared by using thr following procedure. First, ~40 nm ZnO buffer layer is deposited on the SiO_2 substrate by PLD method at 500° C in oxygen ambient at pressure of $2*10^{-2}$ mbar. Then, a thin Au film of a thickness which is slightly exceeding the percolation threshold is deposited by PLD method for 800-1000 laser shots using procedure described elsewhere [10]. Au film is annealed at 500 C for 1 hour to prepare an array of well separated Au nanoparticle cataysts. An array of highly oriented ZnO nanorods shown in Figure 1a is prepared by catbothermal CVD method at the temperature of 1000° C as described elsewhere [11]. Next, ZnO nanorods are covered by thin Au continuous ~30 nm film using PLD for 1500 laser shots. The prepared Au/ZnO nanorods are covered by ~5-10 nm chitosan based polymer layer. Finally, Au nanolarticles are deposited for 500 laser shots in argon [10, 12]. In all PLD procedures the KrF laser was focused on rotating target to give a fluence of ~2 J/cm^2.

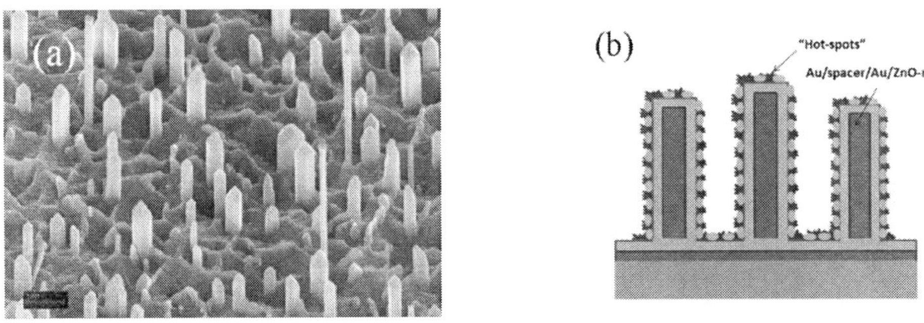

Figure 1(a, b). (a) SEM image; **(b)** and schematics of Au-particles/spacer/Au/ZnO-rods structure.

2.2. Optical transmission characterization and SERS measurements

Optical transmittance was studied by using UV/Vis Spectrophotometer Agilent Technologies HP-8453. Surface enhanced Raman scattering (SERS) of dye molecules induced by Au/ZnO nanorod arrays was studied for Methylene Blue (MB) molecules dissolved in water at concentrations of 10^5 - 10^7 M/l. The molecules were excited by 632.8 nm HeNe laser light and spectra were recorded using

Renishaw inVia Reflex Raman spectrometer with spectral resolution better than 1 cm^{-1}. The samples were excited by light of HeNe laser with power density less than ~5.4×10^5 W/cm^2 using ×50 long focal distance objective (NA=0.5). Raman scattering was collected in the backward direction.

3. Results and discussion

Optical transmission spectra of bare Au/ZnO nanorods and rods covered by ultrathin dielectric spacer and Au nanoparticles are shown in Figure 2. The ZnO buffer layer used for deposition of Au/ZnO nanorods shows ~80% transmission at photon energies larger than ~400 nm and a characteristic ZnO absorption edge near 365 nm. Upon the deposition of the Au/ZnO nanorods the optical transmission is gradually reduced and does not exceed ~4% in the range of 400-1000 nm. The extinction spectra of bare Au/ZnO nanorods reveal broad plasmon bands in visible/near IR range. As the plasmonic nanorods have size comparable with the wavelength of incident light the input of scattering to extinction spectra is significant. Upon the deposition of polymer spacer and Au nanoparticles the absorption is further decreased to be smaller than 1%.

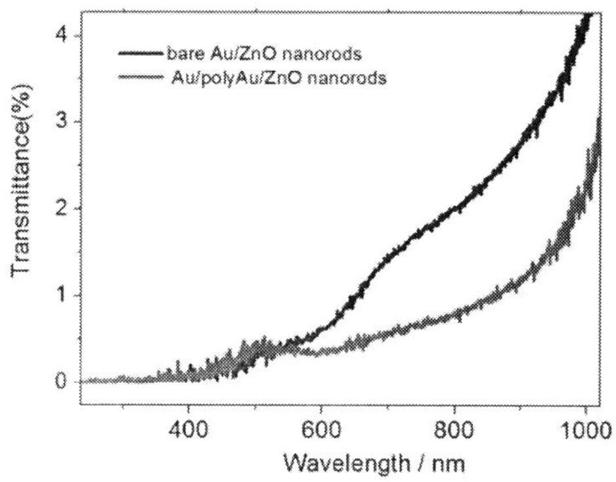

Figure 2. Optical transmission spectra of bare Au/ZnO core-shell nanorods (black) and Au particles/polymer spacer/Au/ZnO structures (red).

The following measurement protocol was applied for the SERS measurements. Solutions of Methylene Blue in bi-distilled water were used to characterize the performance of the SERS substrate. Upon the 5 min incubation in tested solution the sample was well washed in water and dried. Thus, only chemisorbed molecules could remain on the surface. At used analyte concentrations of 10^{-6} -10^{-8} M no Raman signal of MB molecules was detected on reference sample free of nanostructures after the washing, whereas the samples with nanorods reveal well resolved spectra of MB.

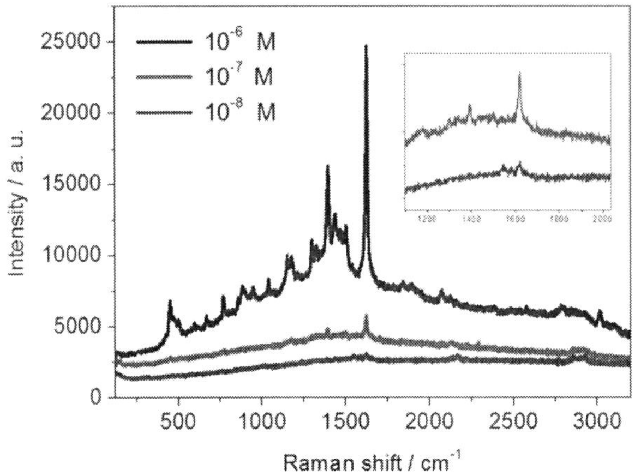

Figure 3. SERS response of Methylene Blue dye molecules adsorbed on the surface of bare Au/ZnO core-shell nanorods recorded at MB concentrations of 10^{-6} and 10^{-7} M. Smaller features of spectra are shown in the inset.

The fluorescence from molecules adsorbed on nanorods is gradually quenched whereas the Raman spectra are enhanced. Using the substrates with bare Au/ZnO plasmonic nanorods we could detect well resolved spectra at MB concentration of $\sim10^{-6}$ M and poorly resolved spectra at $\sim10^{-7}$ M as shown in Figure 3. Upon the deposition of Au nanoparticles onto polymer/Au/ZnO nanorods the SERS sensitivity is gradually enhanced compared with as prepared nanorods. At MB concentrations of $\sim10^{-7}$ M perfectly resolved spectra were observed as shown in Figure 4 even after SERS substrates thorough washing in bi-distilled water.

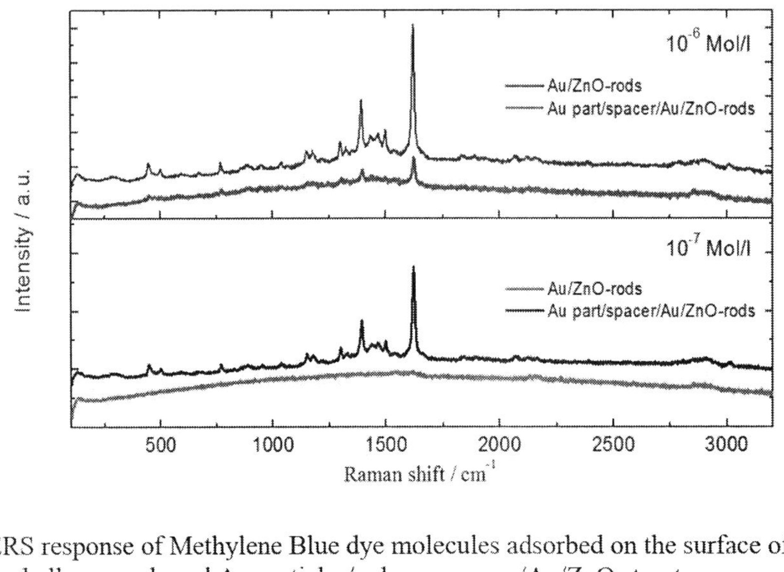

Figure 4. SERS response of Methylene Blue dye molecules adsorbed on the surface of bare Au/ZnO core-shell nanorods and Au particles/polymer spacer/Au/ZnO structures recorded at concentrations of 10^{-6} and 10^{-7} M.

4. Conclusions

SERS substrates based on array of Au/ZnO plasmonic nanorods coupled to Au nanoparticles show much better performance compared with bare Au/ZnO nanorods. The hybrid 3D SERS substrate perfectly resolve Raman spectra of MB molecules chemisorbed from solutions with analyte concentrations of ~10^{-7} M even upon washing of a substrate. The SERS substrate performance is expected to be one or two orders of magnitude higher if one would use more common measurement protocol in which the droplet of analyte is dried on the substrates surface, i.e. when all the deposited molecules make input to SERS signal.

5. Acknowledgments

The study was supported by RFBR according to the research project №18-02-00151A "Study of photothermal effect in plasmonic coupled nanosystems". The CVD growth of ZnO nanorods was supported by Russian Education and Science Ministry project No.16.5405.2017/8.9

References

[1] Jin X, Khlebtsov B N, Khanadeev V A, Khlebtsov N G, Ye J, 2017 ACS Appl Mater Interfaces. 9, 30387
[2] Zhang J, Irannejad M, Cui B. 2015 *Plasmonics* **10** 831
[3] Hanske C, Tebbe M, Kuttner C, Bieber V, Tsukruk V V, Chanana M, König T A F, and Fery A 2014 *Nano Lett.* **14** 6863
[4] Mühlig S, Cialla D, Cunningham A, März A, Weber K, Bürgi T, Lederer F, Rockstuhl C 2014 *J. Phys. Chem. C* **118** 10230
[5] Kaydashev V, Zolotukhin P, Belanova A, Anokhin A, Volodin A, Chernyshev A, Lyanguzov N, Kaidashev E, 2018 *J. Phys.: Conf. Series* **1124** 051024
[6] Liu L, Zhang Q, Lu Y, Du W, Li B, Cui Y, Yuan C, Zhan P, Ge H, Wang Z, Chen Y 2017 *AIP Advances* **7**, 065205
[7] Wu L, Wang W, Zhang W, Su H, Liu Q, Gu J, Deng T, Zhang D, 2018 *NPG Asia Materials*, 10, e462
[8] Encina E R, Coronado E A 2008 *J. Phys. Chem. C* **112** 9586
[9] Encina E R, Perassi E M, Coronado E A 2009 *J. Phys. Chem. A.* **113** 4489
[10] Kaydashev V E, Lyanguzov N, Zhilin D, Tsaturyan A, Raspopova E A, Kaidashev E M 2016 *J. Phys.: Conf. Series* **741** 012145
[11] Kaidashev E M, Lerer A M, Golovacheva E V, Kaydashev V E, Lyanguzov N V, Tsvetyanskii E A, 2017 *J Comm. Technol. and Electron.* **62** 1328
[12] Kaydashev V, Zolotukhin P, Belanova A, Anokhin A, Volodin A, Chernyshev A, Lyanguzov N, Kaidashev E M 2018 *J. Phys.: Conf. Series* **1124** 051024

Electron-microscopy study of ordered silver nanoparticles synthesized in a ZnO:Al polycrystalline film

L A Sokura[1,2], E V Shirshneva-Vaschenko[1], D A Kirilenko[1,2], Zh G Snezhnaia[1], P S Shirshnev[1] and A E Romanov[1]

[1]Saint Petersburg National Research University of Information Technologies, Mechanics and Optics (ITMO University), Saint Petersburg 197101, Russia
[2]Ioffe Institute RAS, Saint Petersburg 194021, Russia

Abstract. The optical absorption spectra and structural properties of samples series consisting of AZO oxide films with an intermediate layer of silver nanoparticles with different sol compounds were investigated. TEM analysis of the obtained coatings showed that the nanoparticles in the AZO film form an ordered structure, and the distance between them depends on the sol composition and the film annealing temperature. Optical measurements showed that a position of the nanoparticle plasmon peak in AZO also depends on them.

1. Introduction

Developing the approaches for obtaining ordered nanostructured composites with controlled physical properties remains a major challenge for photonics, biosensors, nanofluidics [1-4]. New physical properties were observed by use of ZnO, Al_2O_3, TiO_2 and SiO_2 as functional matrix materials in the preparation of metal–oxide nanocomposites [5-8]. For dispersed phases in nanocomposites silver and gold are among the most interesting metals because of their remarkable chemical reactivity in solution [9].

ZnO:Al oxide films (AZO) with metal nanoparticles (NP) exhibit unique optical and electrical properties induced by excitation of plasmon resonances [10-13]. The ability to manage these properties via varying film structure gives rise to many important features and applications.

In this work, composite coatings consisting of AZO films with an intermediate layer of silver NP obtained by sol-gel technology were investigated. The advantages of the sol-gel method are a low temperatures synthesis, a high degree of homogeneity in a multicomponent system, and the possibility of obtaining noncrystalline systems, which synthesis by traditional methods leads usual either to phase separation or crystallization [14]. The main aim was to reveal correlation between optical and structural properties of films produced from different sol compounds and for different annealing temperatures.

The correlation of the optical and structural properties of composite coatings consisting of ZnO:Al oxide films (AZO) with an intermediate layer of silver nanoparticles obtained by the sol-gel technology were investigated. The introduction of silver nanoparticles with dimensions of several tens of nanometers leads to the appearance of an additional peak in the optical absorption spectrum of AZO, corresponding to the plasmon peak of nanoparticle absorption [10]. A shift of the nanoparticle plasmon peak in AZO was observed depending on the AZO sol composition and the film annealing temperature with the same concentration of silver introduced into the sol.

For a series of samples consisting of AZO oxide films with an intermediate layer of silver nanoparticles with different sol compounds, optical absorption spectra were measured, and their structural properties were determined by transmission electron microscopy (TEM). The comparison of optical and structural properties measurements data of the samples allowed to derive some regularities of the silver nanoparticles formation in the AZO matrix that were consistent with the results of the optical properties of model structures (Figure 1, part 1-3) obtained from the simulation [15].

2. Fabrication of nanocomposites

For this study, series of AZO films with involved layers of silver nanoparticles samples were prepared. To fabricate nanocomposite structure five monalayers of AZO were deposited on a fused quartz substrate, than four layers of silver nanoparticles dissolved in the solvent were deposited. Finally, the obtained structure was covered by five monolayers of AZO. The series of samples were distinguished by temperature treatment conditions. All layers were deposited by spin-coating method. Each layer was dried at 300 °C.

The sol solution for AZO consisted of zinc acetate dehydrate [$Zn(O_2CCH_3)_2$, ZnAc] and aluminum nitrate [$Al(NO_3)_3$] was dissolved in 2-Methoexyethanol with concentration of 0.2 M. The solution of silver nanoparticles consisted of silver nitrate [$AgNO_3$] in 2-Methoexyethanol with concentration of 0.03 M.

In this work, two series of samples have been investigated: the first one, where layers of silver nanoparticles were deposited on non-annealed layers of AZO; the second one, where layers of silver nanoparticles were deposited on annealed at layers. In each series the non-annealed samples of nanocomposite, annealed at 570 °C, 650 °C and 700 °C have been studied.

3. Characterization

The morphology of AZO film with Ag nanoparticles (NP) was determined through transmission electron microscope (TEM) technique using a JEOL JEM-2100F with the accelerating voltage of 200 keV. The sample was prepared in a cross section in accordance with the standard procedure, including preliminary mechanical thinning and further polishing with an ion beam (argon ion beam energy - 4 keV).

The structure of samples also was investigated by scanning electron microscope (SEM) LYRA3 TESCAN with the accelerating voltage of 10-20 keV.

The optical density spectra measurements were performed with the fiber spectrometer AvaSpec-2048.

4. Results

The optical density spectra (Figure 1.1a-1.11a), the cross-section TEM images (Figure 1.1b-1.11b) and images of model composite structures corresponding to the investigated samples (Figure 1.1c-1.11c) are presented. Comparing the optical spectroscopy and TEM results for each sample, one can derive the following regularities of composite structures formation.

After several monolayers of silver nanoparticles solution deposition and their heat treatment at 300 °C the nanoparticles coalescence is observed. The size and shape of the formed nanoparticles depends on the heat treatment conditions of the AZO layers, on which they are deposited, and on the annealing conditions of the entire composite. At temperature drying of 300 °C, the nucleation of polycrystalline grains in AZO films occurs, the size of the grains is about 3-5 nm (Figure 1.3b, 1.3c, 1.8b, 1.8c). An increase in the annealing temperature of the AZO films leads to an increase in the grain size (Figure 1.2b, 1.2c, 1.5b, 1.5c, 1.6b, 1.6c). The diffraction pattern for AZO films annealed up to temperature of 650 °C, corresponds to a polycrystalline structure (Figure 2a). Upon annealing at 650 °C, a single-crystal thin-film structure or a structure with equal oriented AZO grains is formed (Figure 2b). In the case, when silver nanoparticles are deposited on the non-annealed AZO layers, silver nanoparticles coalesce into spherical particles of about 20–35 nm in size on the granular boundary of the layers (Figure 1.8b, 1.8c). An increase of the annealing temperature leads to the AZO grains and nanoparticle

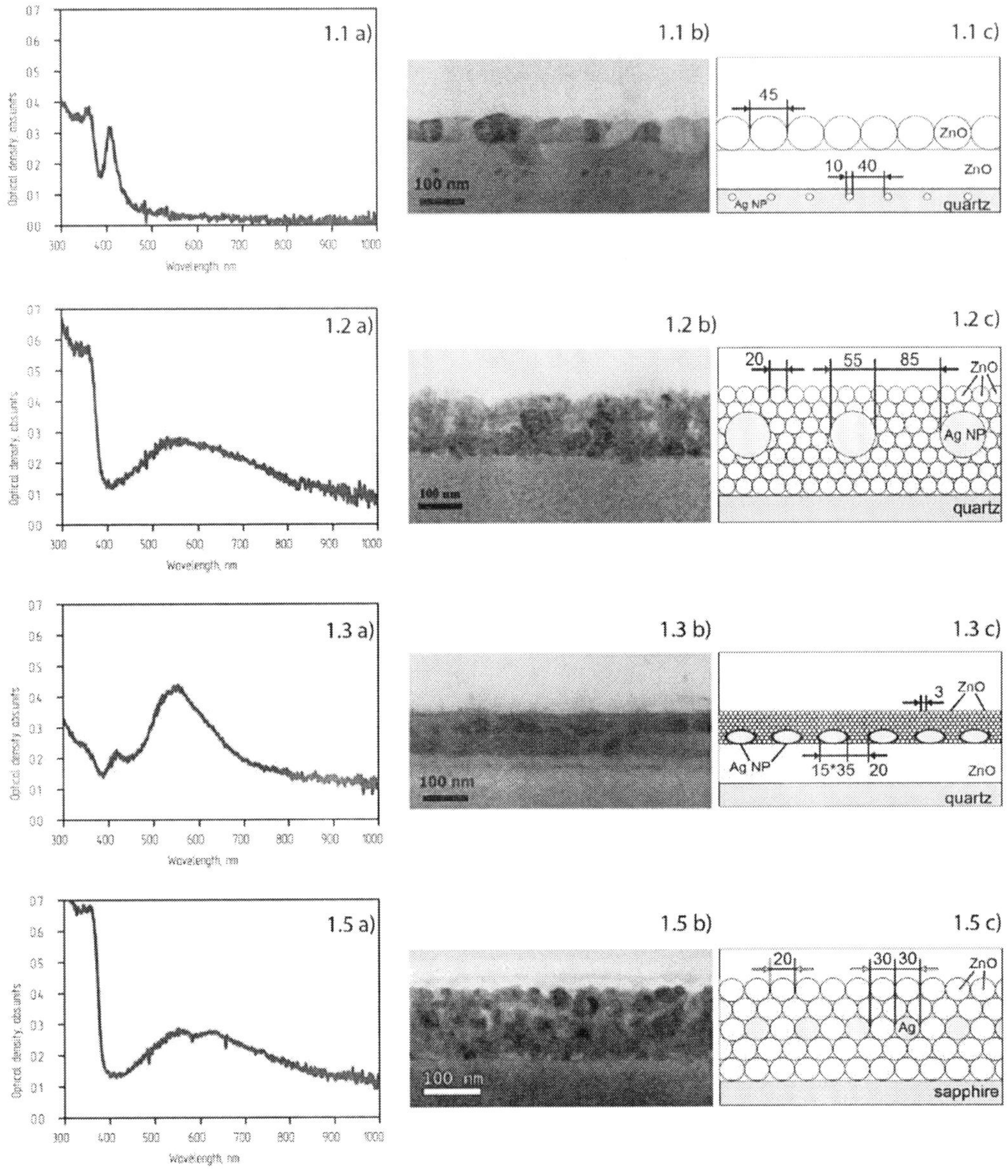

Figure 1. (part 1) Optical density spectra (a), light-field cross-section TEM image (002) (b) and model structures (c) of samples № 1, 2, 3, 5 polycrystalline AZO films with silver nanoparticles:
1.1, a-c correspond to the structure: fused quartz substrate - ZnO:Al (3%), 5 layers, annealing 580 °C- NP Ag, 5 layers without annealing- ZnO:Al (3%), 5 layers, annealing 580 °C;
1.2, a-c correspond to the structure: fused quartz substrate - ZnO:Al (3%), 5 layers, without annealing- Ag, 4 layers without annealing- ZnO:Al (3%), 5 layers, annealing 700 °C;
1.3, a-c correspond to the structure: fused quartz substrate -ZnO:Al (3%), 5 layers, annealing 650 °C - NP Ag, 4 layers, without annealing - ZnO:Al (3%), 3 layers, without annealing;
1.5, a-c correspond to the structure: sapphire substrate - ZnO:Al (3%), 5 layers, without annealing - NP Ag, 4 layers, without annealing - ZnO:Al (3%), 5 layers, annealing 570 °C;

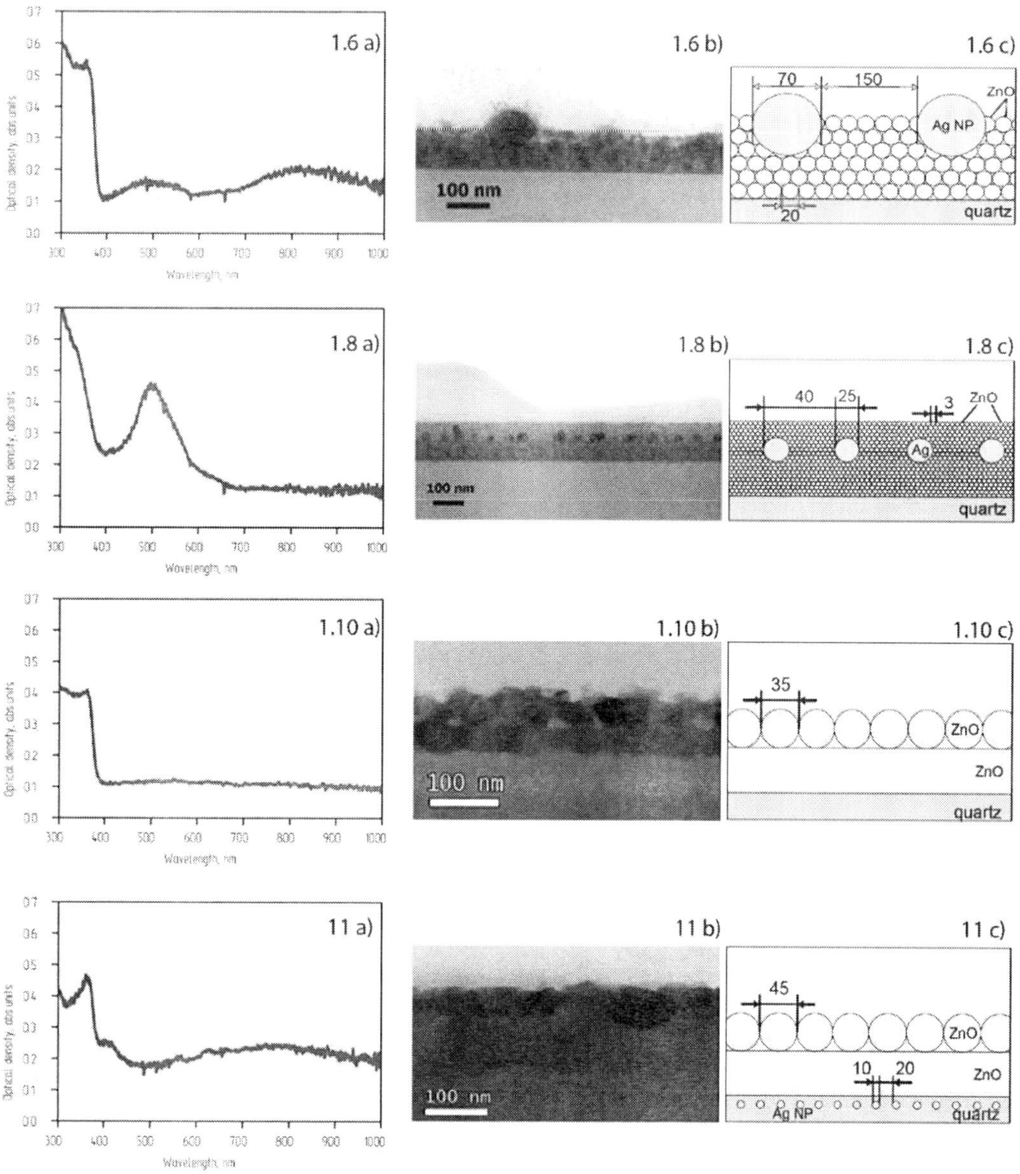

Figure 1. (part 2) Optical density spectra (a), light-field cross-section TEM image (002) (b) and model structures (c) of samples № 6, 8, 10, 11 polycrystalline AZO films with silver nanoparticles:
1.6, a-c correspond to the structure: fused quartz substrate- ZnO:Al (3%), 5 layers, annealing 650 °C - NP Ag, 4 layers, without annealing-ZnO:Al (3%), 5 layers, annealing 570 °C;
1.8, a-c correspond to the structure: fused quartz substrate- ZnO:Al (3%), 5 layers, without annealing- NP Ag, 5 layers, without annealing- ZnO:Al (3%), 5 layers, without annealing;
1.10, a-c correspond to the structure: fused quartz substrate-. ZnO:Al (3%), 5 layers, without annealing-ZnO:Al (3%), 5 layers, annealing 700 °C;
1.11, a-c correspond to the structure: fused quartz substrate - ZnO:Al (3%), 5 layers, without annealing - NP Ag, 8 layers, without annealing - ZnO:Al (3%), 5 layers, annealing 700 °C;

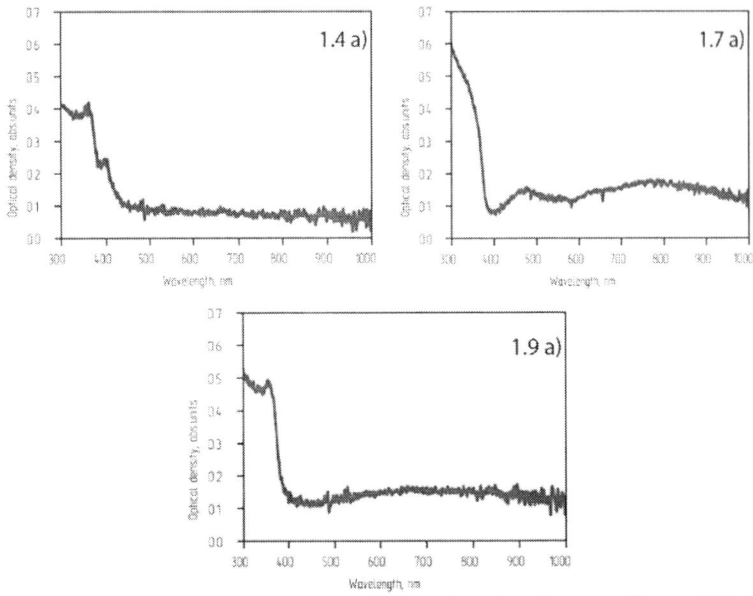

Figure 1. (part 3) Optical density spectra of samples № 4, 7, 9 polycrystalline AZO films with silver nanoparticles:
1.4, a correspond to the structure: fused quartz substrate - ZnO:Al (3%), 5 layers, annealing 650 °C - NP Ag, 4 layers, without annealing - ZnO:Al (3%), 4 layers, annealing 700 °C;
1.7, a correspond to the structure: fused quartz substrate- ZnO:Al (3%), 5 layers, annealing 650 °C- NP Ag, 4 layers, without annealing- ZnO:Al (3%), 5 layers, without annealing;
1.9, a correspond to the structure: fused quartz substrate- ZnO:Al (3%), 5 layers, without annealing.

sizes increase. In the case, when silver nanoparticles are deposited on the annealed AZO layers, they have the form of oblate ellipsoids (Figure 1.3b, 1.3c). The ellipsoidal observed in two plasmon peaks in the spectrum of optical density (Figure 1.3a). Annealing of the composite leads to the formation of a more spherical shape factor nanoparticles (Figure 1.2b, 1.2c). Upon annealing at 700 °C, silver drifts through composite with the release of spherical nanoparticles in a fused quartz substrate (Figure 1.1b, 1.1c, 1.11b, 1.11c). The drift of nanoparticles into a substrate at a temperature of 700 °C can be associated with the diffusion of silver along the grain boundaries due to the melting of the particles [16]. The peak in the optical absorption spectrum at 420 nm (Figure 1.1a, 1.11a) corresponds to the plasmon resonance wavelength for spherical silver nanoparticles surrounded by the quartz. For several samples, the effect of silver nanoparticles formation on the surface of AZO was observed (Figure 1.6b, 1.6c, 3).

Figure 3 presents the SEM image corresponding to the sample of Figure 1.6b. In this case, the particles of a spherical shape, with a size of 90 nm and a distance between them of 150-200 nm, are formed. A similar effect was observed in [17] during heat treatment in water vapor of glass with silver ions in the near-surface layer. Conceivable in our case, the AZO films were not sufficiently annealed, oxygen species at grain boundaries of AZO remained [18], and similar conditions were formed for the formation of nanoparticles on the surface of the composite. Although the mechanism for obtaining such structures is not yet completely clear, the composites with partly project out metal nanoparticles from the surface are nevertheless of interest as a sensor due to the surface enhanced Raman scattering activity [19].

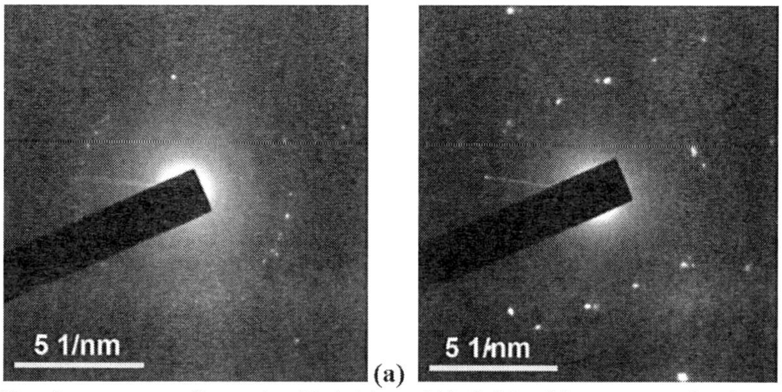

Figure 2. Diffraction pattern of samples № 1 **(a)** and № 5 **(b)** with mono- and polycrystalline AZO films, respectively.

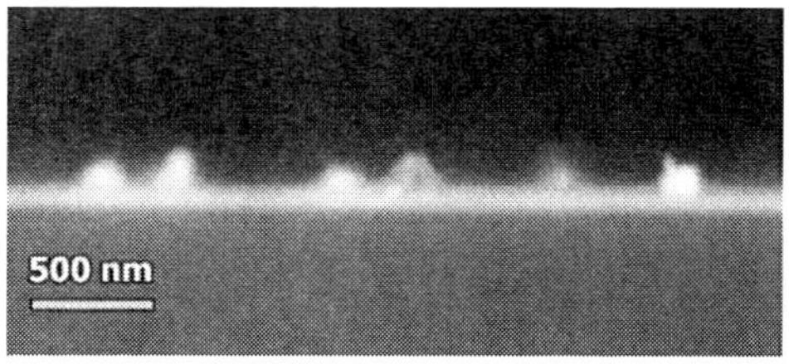

Figure 3. SEM image (002) corresponds TEM image 1.6b of polycrystalline AZO films with silver nanoparticles.

5. Conclusion

TEM analysis of the obtained coatings showed that the nanoparticles in the AZO film form an ordered structure, and the distance between the particles depends on the sol composition. The relationship determined between the sizes of AZO grains and silver particles in the AZO and the distance between them allowed to obtain controlled the ordered plasmon structures by varying the composition and synthesis mode of the AZO films.

An annealing process results in a gradual decrease and broadening of Ag NP absorption peak. This phenomenon corresponds to the reorganization of NP in the AZO film in a more disordered way and even silver escape into the substrate.

Acknowledgments

The work was supported by the Ministry of Education and Science of the Russian Federation within the framework of the Federal target program "Research and development in priority directions of the scientific-technological complex of Russia for 2014–2020", code 2017-14-576-0003, agreement № 14.575.21.0127 from September 26, 2017, unique ID RFMEFI57517X0127. The TEM studies were performed using equipment owned by the Joint Research Center "Material science and characterization in advanced technology" (Ioffe Institute).

References

[1] H A Atwater et al. 2010.*Nature materials* **9** (3) 205.
[2] M M Sk et al. 2016. *Journal of Power Sources* **308** 121140.
[3] M Ge et al. 2016. *Journal of Materials Chemistry A* **4** (18) 6772-6801.
[4] P S Brown et al. 2016.*Scientific reports* **6** 21048.
[5] B Houng et al. 2006. *Surface and Coatings Technology* **201** (6) 3188-3192.
[6] U Pal et al. 2004. *Solar energy materials and solar cells* **81** (3) 339-348.

[7] K Kawahara et al. 2005. *Physical chemistry chemical physics* **7** (22) 3851-3855.
[8] N Viart et al. 2003.*Thin Solid Films* **437** (1-2) 1-9.
[9] B Houng. 2006. *Surface and Coatings Technology* **201** (6) 3188-3192.
[10] O Szabo et al. 2015. *Facta Universitatis, Series: Electronics and Energetics* **29** (1) 7788.
[11] C S Hong et al. 2006. *Thin Solid Films, Proceedings of the 33rd International Conference on Metallurgical Coatings and Thin Films*, **515** (3) 95760.
[12] T Toyoda et al. 2009. *Journal of Applied Physics* **105** (3) 034314.
[13] F Sarto et al. 2003. *Rev. Adv. Mater. Sci* **5** (4) 329-336.
[14] S S Bochkareva. 2016. *Proceedings of Universities.Applied Chemistry and Biotechnology.*
[15] E V Shirshneva-Vaschenko et al. 2018. *Journal of Physics: Conference Series. – IOP Publishing* **993** (1) 012032.
[16] K S Moon et al. 2005. *Journal of Electronic Materials* **34** (2) 168-175.
[17] P A Obraztsov et al. 2013. *Physics of the solid state* **55** (6) 1272-1278.
[18] Q Nian et al. 2015. *Scientific reports* **5** 15517.
[19] P N Brunkov et al. 2015.*Technical Physics* **60** (2) 270-274.

Investigation of functional properties of metal surfaces after laser treatment

J Ruzankina[1,2], **O Vasiliev**[3], **S Tarasov**[2]

[1]Department of low-temperature energy, ITMO University, St. Petersburg 197101, Russia

[2]Department of Photonics, Electrotechnical University, St. Petersburg, 197376, Russia
[3]LLC 'Laser Center', St. Petersburg 195176, Russia

Abstract. This article presents the results of the study of the processes of formation of the metal surface, which has different hydrophysical properties and high chemical resistance to acidic media. The dependence of the contact angle of the surface (θ) on various parameters of laser processing material, the density of arrangement of the scanning vectors, the depth of the heated layer, the presence of oxide layer, etc. Determined optimum modes of laser processing, with which it is possible to obtain the greatest hydrophobicity, and maximum adhesion to the metal surface.

1. Introduction

The interaction of water and water vapor with the surface of various materials has a great influence on the functional and operational properties of objects. In particular, water and water-forming media contribute to the active corrosion of metals. For example, the mechanism of repulsion of water drops has long been used in nature. More than 200 species of plants (the well-known "Lotus effect") and various living organisms have surfaces whose microrelief is able to prevent interaction with water. A good illustration of the natural use of structured micro roughness is the leaf of the Tarot plant (Colocasia esculenta). The property of hydrophobicity of these species of plants and animals occurs due to the special geometric structure of their surface (figure 1).

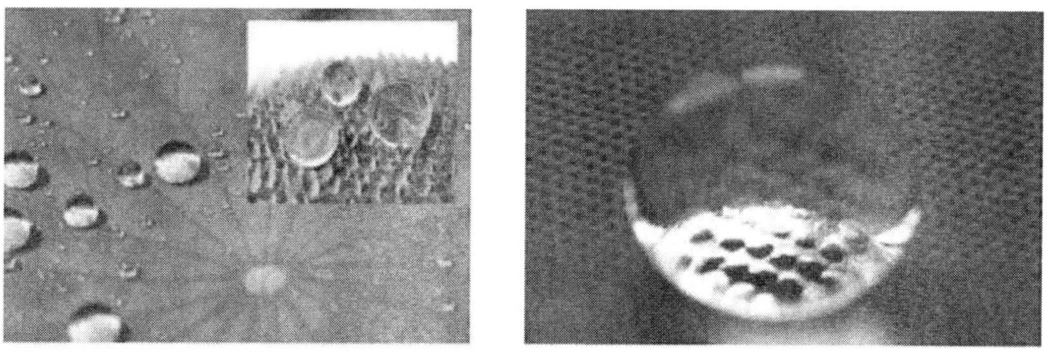

a b

Figure 1(a, b). Examples of natural and artificial hydrophobic surfaces **(a)** Lotus leaf surface; **(b)** artificially created water repellent surface [1].

In this connection, no less effective way to protect the metal from corrosion is the creation of hydrophobic surfaces with using laser radiation. Laser micromachining is a relatively new method of manufacturing microrelief of the surface but is already actively used in modern and innovative manufacturing enterprises. The main research on the creation of hydrophobic surfaces is carried out on pico - and femtosecond lasers (λ=1030 nm and λ=800 nm). Despite the significant amount of work on the creation of hydrophobic surfaces, today there are a number of unresolved problems that limit the widespread use of these technologies: insufficient wear resistance of microgeometry, the need for the use of consumables, weak chemical resistance of hydrophobic layers. In some cases, the created surface microstructures and coatings are characterized by insufficient mechanical strength. Therefore, reducing the rate of degradation of the hydrophobic state as a result of interaction with water molecules during underwater and underground operation of materials remains an urgent task.
The aim of the study is the possibility of forming a surface with high adhesion and hydrophobicity properties by a fiber nanosecond laser. This work is a continuation of the authors ' research [2-7] on the creation of anticorrosive coatings using fiber laser radiation.

2. Experiment and results

In our work we used a fiber nanosecond laser (operating in pulse-periodic mode) with the wavelength of 1070 nm and power up to 18, 4 W (figure 2). The power density of laser radiation was varied in the range from q_0=0,1*10^8 W/cm^2 to q_0=2*10^8 W/cm^2. The density of the vectors was from 1 to 20 line/mm. As a test sample, steel grade was used 10X18H10T. Surface treatment of metal was held at a room temperature of 23 ℃ and a relative humidity of air of 55%

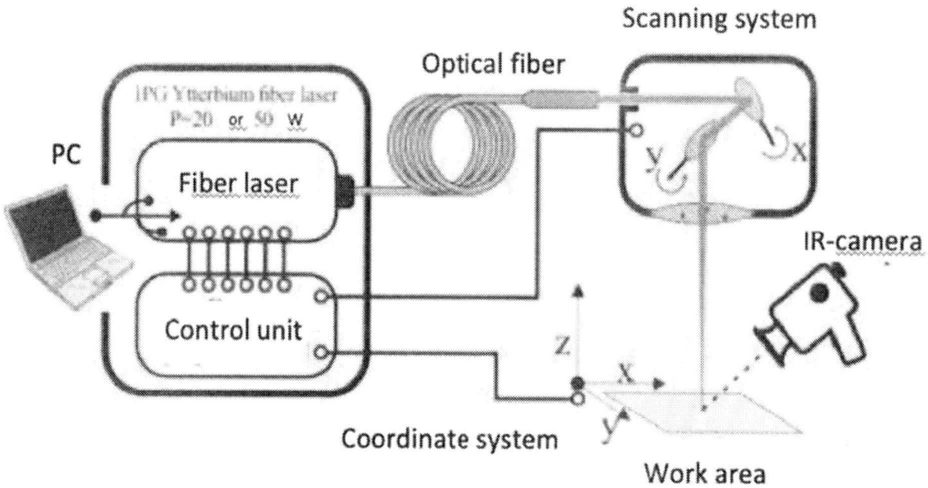

Figure 2. Schematic diagram of the installation for microstructuring the metal surface on the basis of a ytterbium fiber laser with a wavelength λ=1070 nm and a power density up to q=2·10^8 W/cm^2.

The analysis of the obtained results allows us to conclude that laser irradiation of the surface of steel with a fiber nanosecond laser can serve as an alternative to expensive lasers with ultrashort pulse durations in the creation of microstructured surfaces. Figure 3 and 4 show microphotography of the steel surface with the best wetting angle and the best adhesion properties.

Journal of Physics: Conference Series 1410 (2019) 012171 doi:10.1088/1742-6596/1410/1/012171

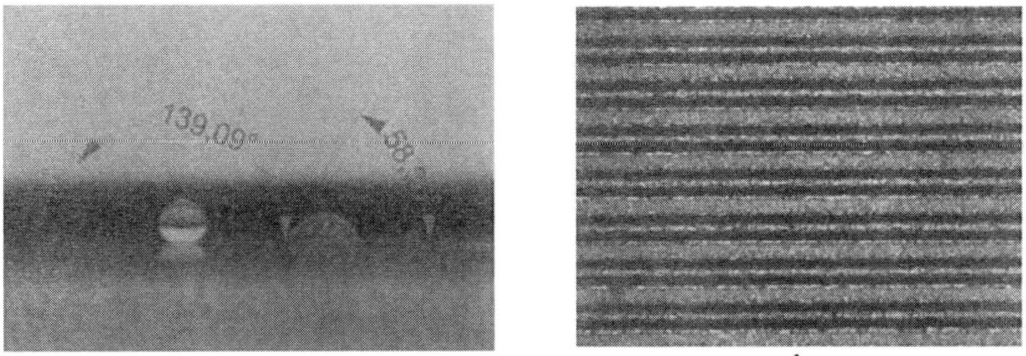

a b

Figure 3(a, b). Microphotography **(a)** maximum wetting angle of the sample surface; **(b)** linear metal surface fill vectors.

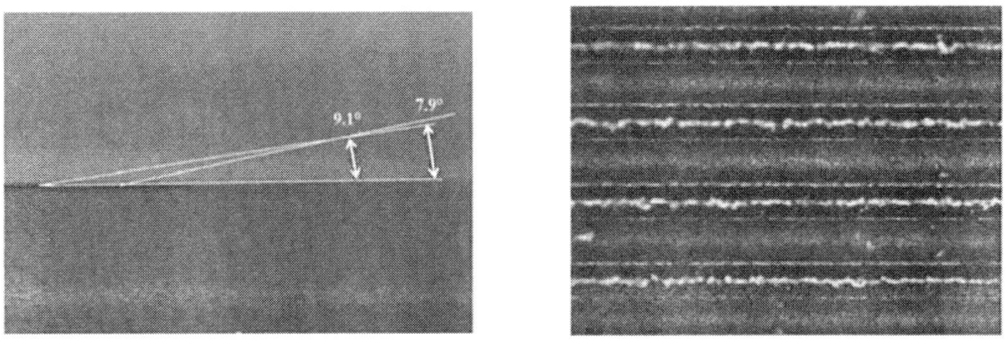

a b

Figure 4(a, b). Microphotography **(a)** sample with the best adhesive properties; **(b)** linear metal surface fill vectors.

Studies have shown that the greatest hydrophobicity of the surface corresponding to the highest wetting angles θ, measured immediately after the surface treatment with laser radiation, is obtained at a pulse duration τ = 4 ns, pulse repetition frequency f =20 kHz and scanning speed V to 100 mm / s. Increasing the repetition frequency of laser pulses f from 20 kHz to 100 kHz, increasing their duration τ to 100...200 ns, and a decrease in the scanning speed V to 150 mm / s leads to a sharp decrease in the wetting angle θ, which indicates a decrease in hydrophobicity – an increase in the hydrophilicity of the metal surface.

Figure 5 shows how the degree of wetting of the surface changed (the wetting angle of the untreated surface is indicated by a solid line). Some modes, such as increasing the fill density (Lin./mm) and increased scanning speed (V), give a positive hydrophobic result. At the same time, there is a set of parameters at which the value of the wetting angle could not be measured practically, since the angle $\theta<10°$. Such a value of the wetting angle indicates the complete absorption of the surface of the water drop is achieving a high adhesive ability.

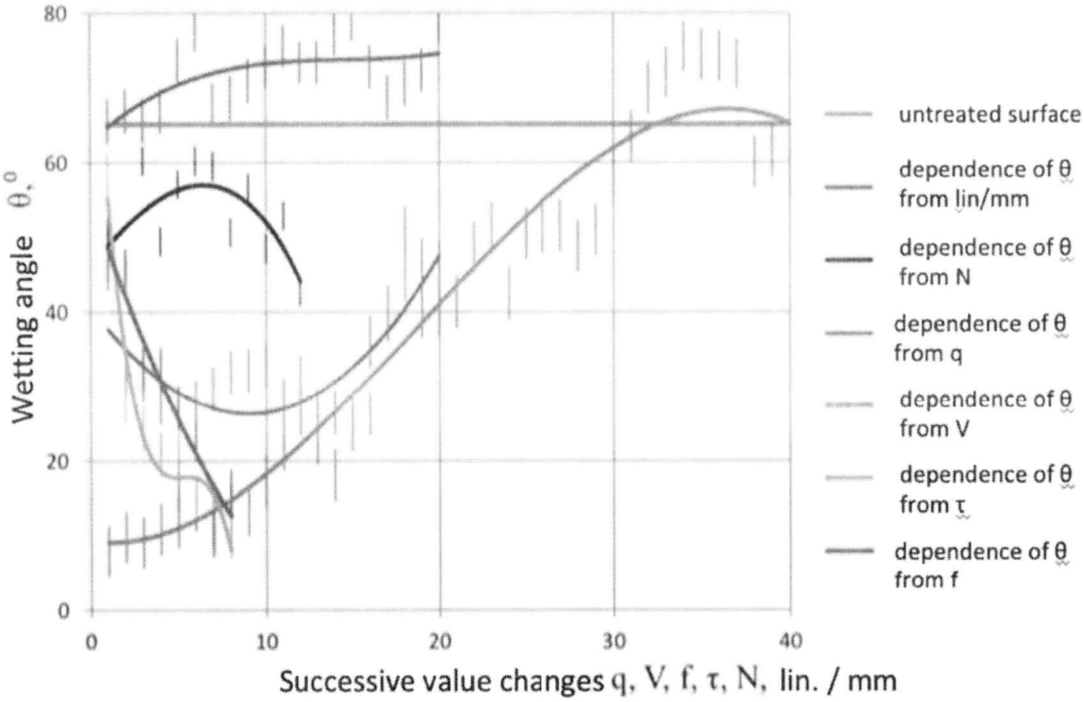

Figure 5. Graph of the dependence of the wetting angle θ of the treated surface on the successive changes in the various values of the laser system parameters q, V, f, τ, the number of passes N and the density of the fill immediately after the formation of the surface microtopology.

The samples were tested a month after the experiment. As can be seen from the results (figure 6), the picture changes over time. Some of the experiments carried out with high scanning speed and a low pulse repetition rate gave a positive hydrophobic effect. The wetting angle in them exceeds 120°, which allows them to be classified as hydrophobic surfaces (over 150° superhydrophobic surfaces). It is suggested that this change over time is due to the slow oxidation of the metal surface under normal storage conditions.

As can be seen from figures 5 and 6, a month later the total degree of hydrophobicity increased, which is confirmed by both experiments and literature [8]. Several modes of laser treatment allowed to approach the values of the wetting angle θ≈140°, which characterizes the metal surface as hydrophobic.

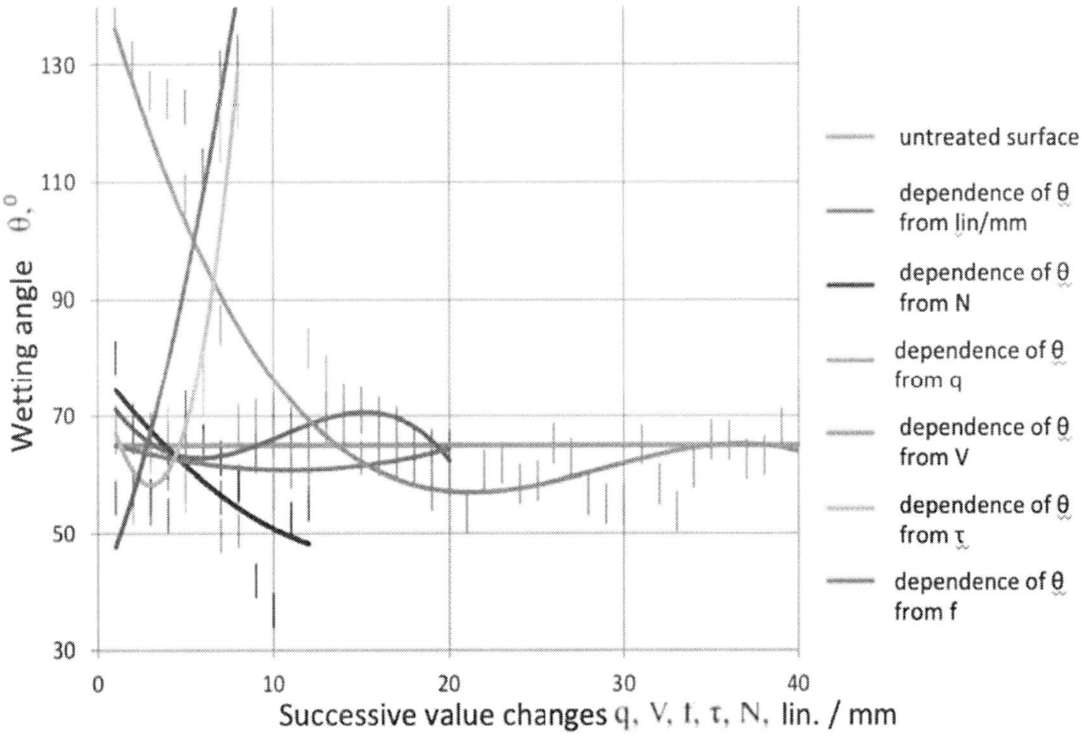

Figure 6. Graph of the dependence of the wetting angle θ of the treated surface on the successive changes in the various values of the laser system parameters q, V, f, τ, the number of passes N and the density of the fill in a month after the formation of the surface microtopology

Table 1 presents the laser processing parameters for best results.

Table 1. The parameters of the laser system and the values of the contact angle.

	θ	τ, ns	V, mm/s	q, W/cm^2	f, HGz	N, line/mm	Repeat	Type	θ'	$θ_0$
1	9.1	4	25	2*10^8	20	20	1	lin	139	64
2	7.9	4	25	2*10^8	20	10	6	lin	25	64

3. Conclusion

The article presents an experimental possibility of obtaining surfaces with a high degree of hydrophobicity immediately after laser treatment and over time.

By adjusting the parameters of the laser system, it is possible to produce a metal surface, which in its properties will simulate the hydrophobic surface of the Lotus leaf. Despite the fact that the experimental setup used can not provide the accuracy of the surface roughness control of the material necessary to obtain superhydrophobic surfaces, it allows to produce surfaces whose total stochastic roughness leads to an increase in the hydrophobicity of the surface, which was demonstrated in the article. In this connection, the widespread fiber nanosecond laser can serve as an alternative to expensive lasers with ultrashort pulse durations in the production of microstructured surfaces.

Studies have shown that the greatest hydrophobicity of the surface corresponds to the greatest wetting angles θ. It is also worth noting that the properties of manufactured surfaces with high hydrophobicity and a high degree of adhesion, change over time under the influence of changing environmental conditions. This phenomenon requires additional observations taking into account the time parameter. In addition, the obvious dependence of the wetting angle θ on the laser radiation parameters has not yet been detected. It is only possible to determine the nature of the change in the desired angle from the increase or decrease of a particular parameter of the laser system.The investigated technology of creation of a microrelief is not capable to replace all available methods of processing of a surface of metals but will help to solve a number of problems connected with a change of properties of a surface. In the future, it is planned to conduct research on mechanical stability and service life.

4. Acknowledgments

The work was carried out with the financial support of the Ministry of education and science of the Russian Federation in the framework of research of the project part of the GZ, project №
16.1750.2017/4.6

References

[1] Montemor M F 2014 *Surface and Coatings Technology* **258**
[2] Ruzankina J S, Parfenov V A, Vasiliev O S 2019 *Laser Physics* **29** 1
[3] Vasil'Ev O S, Veiko V P, Gornyi S G, Ruzankina Y S 2015 *Journal of Optical Technology* **82** 12
[4] Ruzankina J S and Vasiliev O S 2016 *Journal of Physics: Conference Series* **735** 1
[5] Iakovlev A, Ruzankina J, Kasheev S, Vasilyev O, Parfenov V, Grishkanich A 2017 *Proceedings of SPIE* **10097**
[6] Ruzankina J S, Vasiliev O S, Parfenov V A 2017 *Journal of Physics: Conference Series* **917** 6
[7] Vasiliev O S, Ruzankina J S 2016 *Journal of Physics: Conference Series* **735** 1
[8] Tang M, Shim V, Pan Z Y [et al.] 2011 *JLMN-Journal of Laser Micro/Nanoengineering* **6** 1

The discovered painting decoration of the Vorontsov` Palace in St. Petersburg: Spectroscopic characterization in support of conservation

M Becucci[1], C Lofrumento[1], V Parfenov[2], M Ricci[1], M Rogozny[3], A Pakhunov[4]

[1] Department of Chemistry 'Ugo Schiff', University of Florence, via della Lastruccia 3, Sesto Fiorentino (FI), 50019, Italy
[2] Departemt of Photonics, St.Petersburg Electrotechnical University, Prof. Popov str., 197376, Russia
[3] The St.Petersburg Stieglitz State Academy of Art and Design, Solyanoy lane 13, St Petersburg, 191028, Russia
[4] Centre for Paleoart Studies, Institute of Archaeology of Russian Academy of Sciences, Dm. Ulyanov str. 19, Moscow, 117036, Russia

Abstract. The paper reports about the unique discovery of fragments of wall painting of the middle of the XVIII century from the Vorontsov` Palace in St. Petersburg and its study by means of laser and IR spectroscopic techniques. The history of creation and reconstruction of the palace as well as technical and technological features of paintings are described.

1. Introduction

During the repair and restoration works, taken place in 2015-2017 in the complex of buildings of the St. Petersburg's Suvorov Military School, located at crossing of Moskovsky avenue and the Fontanka River, fragments of wall painting belonged to the R. I. Vorontsov's palace were found out. The palace was originally built by the member of one of the prominent families in XVIII Century Saint Petersburg. Roman Illarionovich Vorontsov (1717-1783) was the count of the Roman Empire, General-in-Chief, Senator, Governor-General of Vladimir, Penza and Tambov, and one of the first figures of Russian Freemasonry and the most famous personalities of the Vorontsov family. In 1757 R. I. Vorontsov acquired the estate on the banks of the Fontanka River and in 1759–1760 he built a country mansion there. The architect in charge of the project has not been positively identified, but a search on the documents on Vorontsov's house property suggested contributions from Francesco Bartolomeo Rastrelli and Giacomo Quarenghi[1].

Figure 1. A view of the Vorontsov Palace before 1863 reconstruction

After the death of R. I. Vorontsov, the country mansion was bought by the Czar Pavel I and was used for hosting of one of the established Military Orphan Homes. However, the house required repairs and alterations for the new destination. The adjustment of buildings was carried out according to the project of architect G. Paulson, and already in 1798 the new use was beginning. The exterior of the Vorontsov's former country mansion was almost unchanged but many internal alterations were made.[2]

In view of the constant lack of space for children, in 1806 the military engineer A.E. Shtaubert attached to the former palace some new buildings together with the Orthodox Church dedicated to the Resurrection of Christ, which were located along Tsarskoye Selo Avenue (now Moscow Avenue). The new buildings connected the new front housing with the former Vorontsov's Palace.[2] The house itself was not rebuilt at that time.

The next reconstruction was carried out in 1863, it affected more in both the external appearance of the Vorontsov's Palace (side wings were built up to three floors, and the triangular pediment was replaced by an attic) and the architectural look of the interior. Some halls were divided by partitions, additional staircases were built in, and interior decoration was partially destroyed. Also, the biggest hall of the palace was rebuilt, and its walls were decorated with illusory paintings imitating architecture. During this reconstruction, part of the hall was used to create a corridor and a separation wall erected. The ceiling was lowered both in the corridor and in the hall itself (Figure 2). As a result, many parts of the original decoration were preserved in the interfloor space. Large areas of the picturesque stucco, painted using the Grisaille technique, were found. Double attached columns of the Ionic order with cannelures and pilasters were depicted. In the walls between the pillars there were decorative mirrors framed with plats with rectangular enlarged ears. The alternating wreaths of laurel leaves and flower festoons were depicted in the mirrors. Samples from these paintings were selected to learn about the original building and decoration techniques. The sample used in this work (shown in Figure 3) was taken from that part of the mural where the characteristic color range was determined. The sample was selected from an area without later inclusions and minimum amount of dirt. For the safe extraction, it was collected from a region were the was already partially detached and along the cracks. The fragment was extracted carefully with the help of scalpel. At a preliminary check, painting was carried out on a two-layer lime-sand plaster. A thin layer of a fine-grained lime coating was applied on the first layer with the coarse silica filler. These samples were further investigated by optical and spectroscopic methods, including Raman and infrared Fourier spectroscopy, in order to identify the pigments on the paint layer and verify the possible presence of organic fixative, extraneous to the artwork. These were often placed on the surface some years after the completion of the paintings and were used in restoration until the beginning of the twentieth century.

These samples were also analyzed in order to understand the technical composition of the paintings, to examine the mortar and to identify the possible presence of salts.

Figure 2. The area upon which the pilaster sided entablement was leaned

2. Materials and Methods

Raman spectra were obtained on a Renishaw RM2000 spectrometer equipped with a Leica DMLM microscope and a semiconductor laser with emission wavelength at 785 nm. The objective lens used for all measurements was a 50x/0.80N.A. Fine grains obtained from the original samples have been analyzed directly, without any preliminary treatment.

Attenuated total reflectance-FTIR spectra were recorded by a Perkin Elmer Spectrum 100 spectrometer. The reflection system (the universal attenuated total reflectance) consists of a diamond crystal, a flat focusing lens made of KRS-5, and a movable device applying on the sample a pressure aimed at eliminating air. The detector is FR-DTGS (fast recovery deuterated L-alanine triglycine sulfate) and the spectral range is 4000–380 cm^{-1} with a spectral resolution of 4 cm^{-1}.

Polished cross sections were prepared embedding the sample in a bi-component epoxy resin (Epofix, Struers DK). The cross sections were then observed with optical microscope, a ZEISS Axio Scope.A1 polarized microscope equipped with a 5 Mpixel camera and a dedicated Axio Vision image analysis software was used.

3. Results and Discussion

The different spectra obtained from the samples revealed a wealth of information about the decoration process and the building materials. The identification of the different constituents was possible by a simple comparison with spectra of reference materials we have collected over time.

Raman spectroscopy data were mostly useful for identification of the materials used for the wall decoration [3] Figure 3 shows the image of the plaster sample taken from the decoration and Figure 4 shows Raman spectra obtained in the different colored area. The list of materials that we have found on the samples is presented in Table 1.

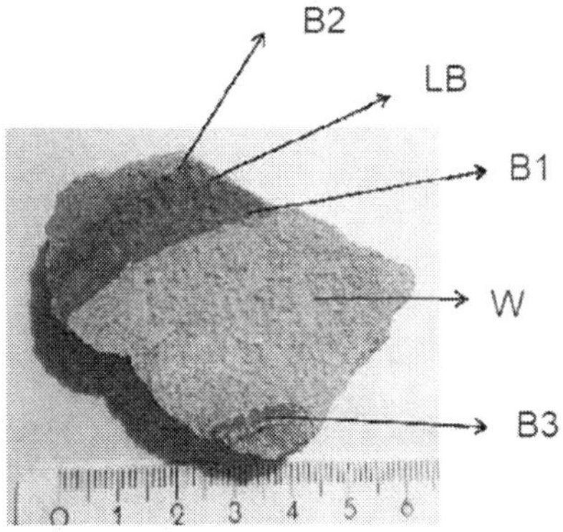

Figure 3. The plaster sample representative of the mural decoration used in this work

Table 1. List of materials identified by Raman spectroscopy

Description	**Mineral pahases**
Brown 1 area	
B1_Y1 (yellow grain)	goethite + amorphous carbon
B1_W1 (white grain)	hematite + calcite + amorphous carbon
B1_R1 (red grain)	hematite
Brown 3 area	
B3_B1 (black grain)	graphite
B3_R1 (red grain)	hematite + goethite + amorphous carbon + manganese oxides
B3_W1 (white agglomerate)	calcite + amorphous carbon
B3_W2 (defined white grain)	calcite
Light brown area	
LB_R1 (red grain)	hematite
White grain	calcite + amorphous carbon
Red grain	hematite + amorphous carbon
White area	
W_B1 (black grain)	calcite + amorphous carbon
W_W1 (white grain)	calcite

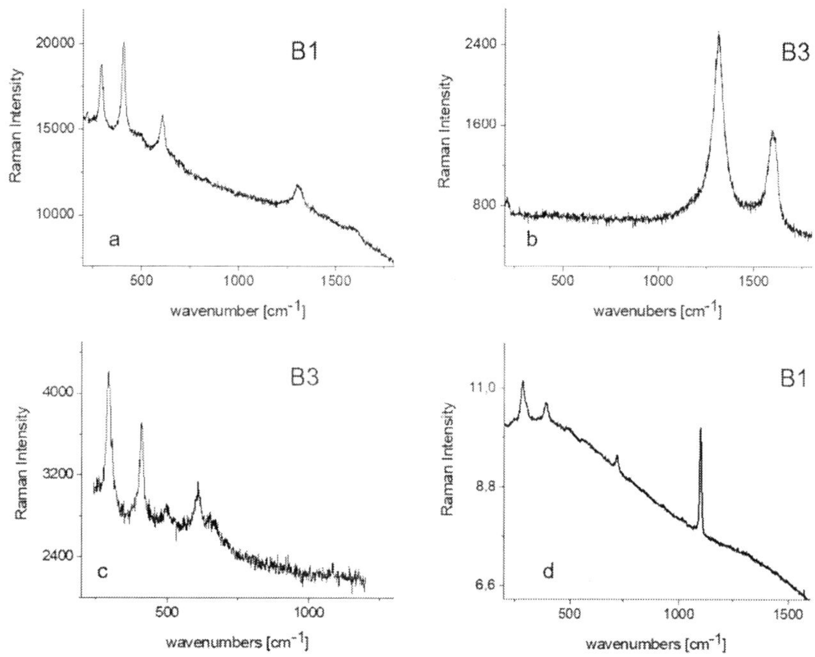

Figure 4. Some representative Raman spectra obtained from different points in the sample: (a) hematite, (b) amorphous carbon, (c) hematite and manganese oxides and (d) calcite

All brown paints revealed the presence of hematite (α-Fe_2O_3), which gives a well-recognizable Raman signal and the yellowish hydrated form goethite, α-FeOOH.[4] Some representative spectra are shown in Figure 4.

Sporadic black grains were observed in some brown drawings and were identified by micro-Raman as amorphous carbon. However, they were also observed on light brown and white areas. Thus, they appeared more likely to be impurities, rather than a component intentionally mixed to red and yellow ochre to give it a deeper shade.

Instead, in the spectra obtained in the brown bent strip decoration a broad band was observed in a spectral range which is typical of Mn-O and Mn-OH stretching vibration frequencies (450-750cm^{-1}).[5] The presence of manganese oxides–hydroxides were thus assumed. The identification of such manganese compounds by means of Raman microscopy is known to be quite problematic. No clear Raman signature is often obtained probably because of the low Raman activity of the manganese-based pigments, low crystallinity and high degree of chemical heterogeneity (different manganese compounds can be present simultaneously at the microscopic level). Firstly, manganese oxides/hydroxides tend to undergo thermal transformations under high laser power values, in a manner similar to analogous iron compounds, but in a stronger way. Secondly, manganese exists in several oxidation states and exhibits a lot of hydro-oxides modifications, many of them being non-stoichiometric and disordered compounds. Consequently, only a tentative assignment can be done. The stronger band at 585–610 cm^{-1} was assigned to bixbyite (Mn_2O_3), while the presence of the Raman signal at 643–652 cm^{-1} was attributed to a variable contribute coming from a spinel structure (Mn_3O_4, haussmanite). The spectrum in Figure 4c shows an additional band at 553 cm^{-1}, which may be ascribed to groutite sites (MnOOH). Moreover, black drawings appeared as homogeneous phases in which no distinctive grains could be observed, so that a fine mixture of different Mn(II), Mn(III) and Mn(IV) oxides and hydroxides has to be assumed.

Instead, infrared spectroscopy was used mostly to characterize the mortar which was used to create the plaster layer of mural paintings. The analysis performed on this preparatory layer of the painting has positively identified calcite in the binder structure (calcite is the result of reaction while drying, calcium hydroxide is what supposed to be actually used as binder), and the absence of harmful crystalline salts, such as nitrates and sulfates (see Figure 5).

The optical microscopy analysis of the sample stratigraphy was very helpful to gain insights on the preparation method of the plaster and its decoration. A simple analysis of these images reveal that the painting was made in layers (see Figure 6). These layers are partially mixed, which suggests that the top paint layer was applied on the incompletely dried out background color.

Thanks to the combination of optical images and spectroscopic analysis it was possible to understand the technical execution of the paintings: lime binding material was found in the pigment part, and the paint layer was applied directly on the lime plaster without additional layers of soil, we can conclude that the murals were created using the "*secco*" or "*semi-secco*" mural technique.

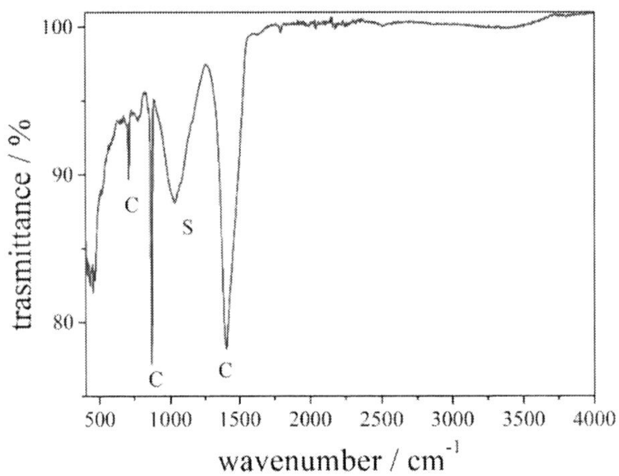

Figure 5. ATR-FTIR spectrum of mortar; the symbols C and S correspond to the principal bands of calcite and silicate, respectively

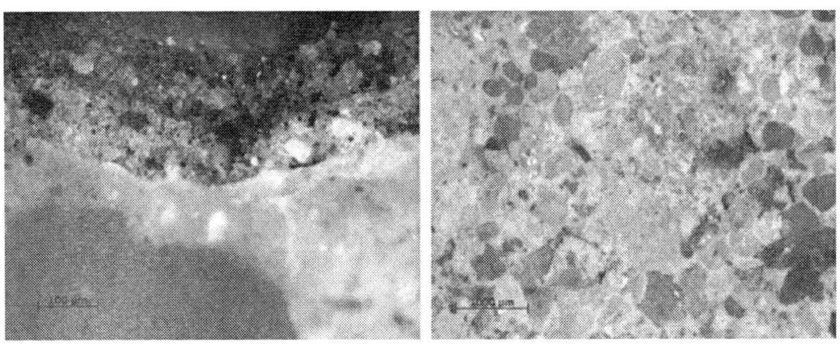

Figure 6. Optical microscope images of a samples's cross section at different magnification

Figure 7. A detail of the wall decoration before (left) and after (right) the cleaning and conservation

The results of the research were helpful to select the necessary materials for the conservation and the following cleaning of the painting surface from the persistent dust pollution. According to the investigations, the consolidation of the plaster structure was carried out with the materials found in the original lime binder. Given the mineral nature of the constituents of the decoration, volcanized latex sponges Akapad (hard and soft) were used for the dry cleaning of the mural from the contamination of dust and soot. Spots and smudges of later color layers were taken away mechanically, with scalpel. The mortar based on hydrate of white lime KalkInjektionsmörtel was used for the grouting of detached plaster layer. The solution was injected in the hollows of the stucco lags and that parts were then pressed to the basement. The edging repair of the exposed plaster along the lacunae was made with the lime mortar prepared with Kalk-Kontor lime filler and quartz sand 1:2. In Figure 7 we show a portion of the wall decoration before and after the cleaning and conservation process.

4. Conclusions

The spectroscopic characterization of the sample reveled the use of materials typically found in "*secco*" or "*semi-secco*" mural technique. We have found no evidence for possible presence of organic fixative, extraneous to the artwork.

These results were useful for the restorers of the Stiglitz Academy in order to carry out successfully their conservation works of fresco paintings, because most appropriate ways of cleaning and preservation were chosen. Some of such methods were already used earlier used for restoration of wall paintings in St. Petersburg, for instance, in the interiors of the General staff, the State Hermitage and the State Russian Museum. But a comprehensive work on the preservation of fresco painting the restorers of the Academy of Stieglitz conducted for the first time.

It should be noted that the definition of painting technique helped to attract public attention to the wall paintings of the Vorontsov Palace. It made possible to introduce them into scientific circulation as unique, as in the palaces of St. Petersburg fresco painting was rarely used. Moreover, due to multiple repairs and reconstructions most of wall paintings were lost and nowadays there are only few ones of XVIII century.

Acknowledgments

One of the authors (V.P.) gratefully acknowledges support from the University of Florence for research periods in Italy.

References

[1] Historical documentary of the Pavlovsk Military School, Pavlovsky Cadet Corps and the imperial military orphanage. 1798-1898. Under the main editorship of the Member-Scientific Committee, the General Staff of Major-General A.N. Petrov. Department of Military Orphan

Home. Drawn up on the basis of archival documents of the General Staff, General of Infantry D.G. Anuchin S- Pb., 1898. P. 124. {in Russian}
[2] Historic reference. Historical and cultural expertise. "LenStroyUpravleniye", St. Petersburg, 2014 P. 6. {in Russian}
[3] Zoppi A, Lofrumento C, Ricci M, Cantisani E, Fratini T and Castellucci E M 2012 A novel piece of Minoan art in Italy: the first spectroscopic study of the wall paintings from Phaistos. *J. Raman Spectroscopy*, 43: 1663-1670
[4] de Faria D L A, Venâncio Silva S, de Oliveira M T 1997 Raman microspectroscopy of some iron oxides and oxyhydroxides, *J. Raman Spectroscopy*, 28:873-878
[5] Bernard C, Hugot-Le Goff A, Vu Thi B, Cordoba de Torresi S 1993 Electrochromic reactions in manganese oxides. I. Raman Analysis. *J. Electrochem. Soc.* 140:3065–3070

Effective formation of diffraction-free laser beams based on the diaphragm of the annular spatial spectrum

PA Khorin[1]
[1]Samara National Research University, 34, Moskovskoye Shosse, Samara, Russia, 443086

Abstract. In this article, we simulate the formation of diffraction-free beams, which combine some of the properties required for technologies such as planar microscopy, information coding and optical manipulation. Analytical functions in combination with a diaphragmed annular spatial spectrum are proposed as an optical beam. The analysis is carried out at distribution in the free environment and lens systems.

1. Introduction

Despite the many works devoted to the formation of diffraction-free beams, this topic remains relevant in view of the large number and variety of practical applications of such beams. The best known among the diffraction-free beams are the Bessel modes. This is the solution of the Helmholtz equation in cylindrical coordinates [1, 2], the Mathieu beams for the elliptical coordinate system [3] and the parabolic beams in the parabolic coordinate system [4, 5]. The diffraction-free beams are also considered beams Airy [6], with additional properties of acceleration [7, 8] and autofocus [9].
Also, various generalizations of diffraction beams were considered and studied [10, 11]. Separately, we can mention the generalization considered in [12], where the angular spectrum is not an analytical function, but was obtained as a result of an iterative solution of the inverse problem. The Increasing interest in the development and formation of new types of diffraction-free beams linked to their huge success in various applications, including optical manipulation, the coding information, microscopy [13-15]. One of the methods for forming diffraction-free beams is the use of a narrow annular diaphragm superimposed on the lens [2, 16]. In [17, 18], a simple method for the energy-efficient formation of various diffraction-free laser beams using partial diaphragm of the annular spatial spectrum was considered.
This paper presents the results of a numerical study and experimental approbation of the proposed optical scheme [19] in order to form diffraction-free beams. Beam characteristics such as its length, thickness, and lack of pronounced side lobes are important. It is obvious that these requirements are at the same time difficult to achieve, so it is proposed to calculate some analytical functions in conjunction with a diaphragm ring spatial spectrum and analyze them when distributed in a free medium and lens systems.

2. Numerical simulation

It was proposed to calculate various types of optical beams based on analytical functions. For this purpose, it was proposed to submit a plane wave on the diffractive optical element with a predetermined topography and skip distribution obtained through an annular diaphragm. In a number of experiments, it was revealed that the beam either has diffraction-free properties, but retains pronounced additional petals, or quickly falls apart. Figure 1 schematically shows the proposed algorithm for the propagation of a given field through an annular diaphragm in free space.

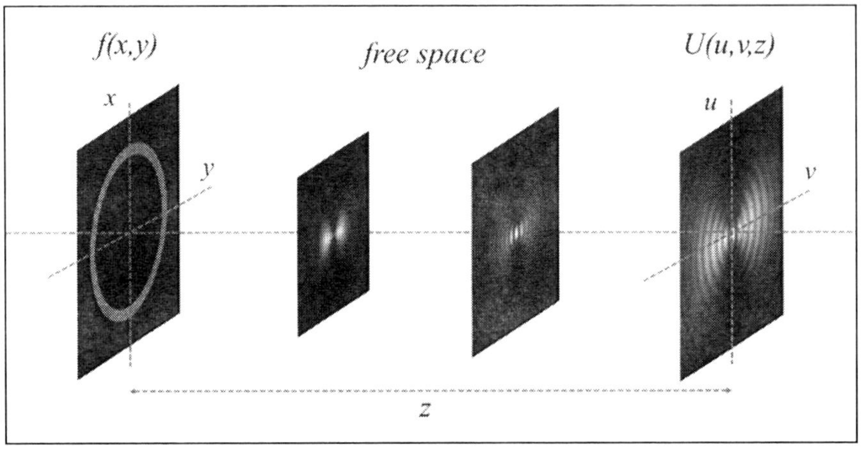

Figure 1. Schematic representation of field propagation in free space at a distance of z

The analysis of the beams propagating in a free environment and lens systems using a Fresnel transform has been carried out. The finite Fresnel transform is a decomposition in parabolic waves: $U(u,v,z) = {-ik}/{2\pi z}\exp[ikz]\int_{-a}^{a}\int_{-b}^{b} f(x,y)\exp\left[{ik}/{2z}((x-u)^2+(y-v)^2)\right]dxdy$, where $k=2\pi/\lambda$, λ – wavelength, z – wavefront propagation distance.

It was found that the annular diaphragm in combination with the function of the form cos (ax) allows the formation of a highly concentrated energy beam in the central region, but with distinct side lobes. The results of the propagation of a given field in free space (figure 1) and through a spherical lens (figure 2) – $f(x)=f(x)*\exp[ik/2f\,(x^2+y^2)]$ are presented in Table 1.

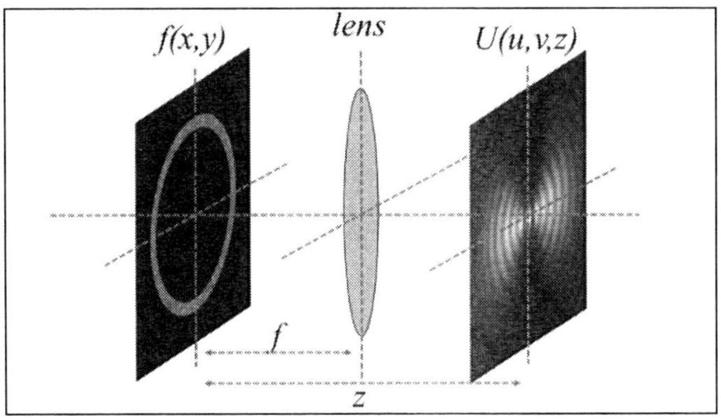

Figure 2. Schematic representation of the propagation of a field through lens with a focus *f* at a distance of z.

Table 1. Results of numerical modeling (amplitude) of the propagation of a type field – $\cos(0.1\pi x)$ at a distance of 250mm, f=150 mm

Input field	Propagation	Propagation through spherical lens
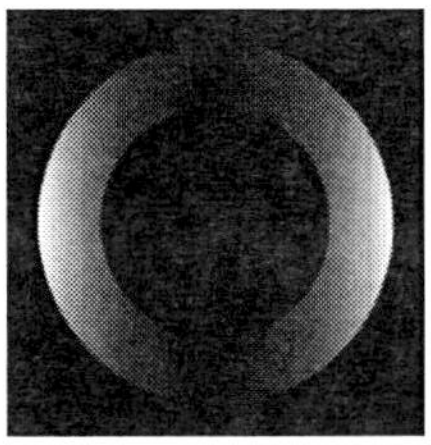	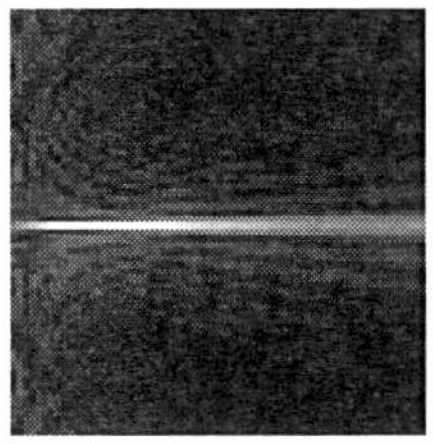	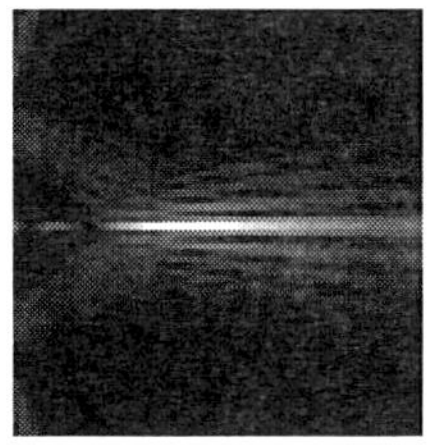

In a number of numerical experiments, it was found that the limitation of the Gaussian beam to an annular diaphragm in combination with a spherical lens gives the best result. than other analytic functions. Table 2 presents the results of numerical modeling of the propagate input field through a cylindrical and spherical lens.

Table 2. Results of numerical modeling (amplitude and phase) of the propagation of a type field Gauss beam with ring mask at a distance of 250mm

Input field	cilindrical lens OX	cilindrical lens OY	spherical lens
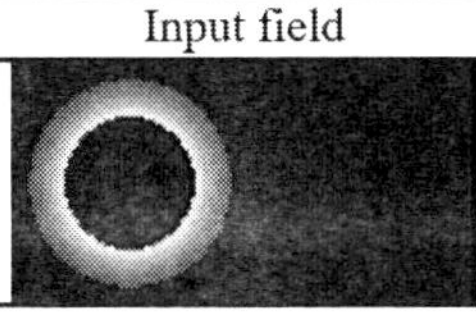	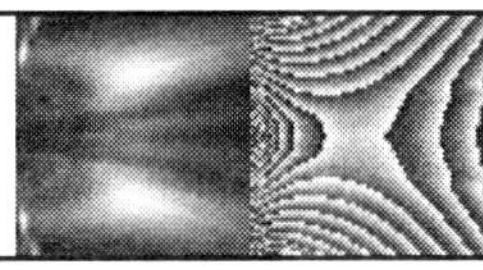	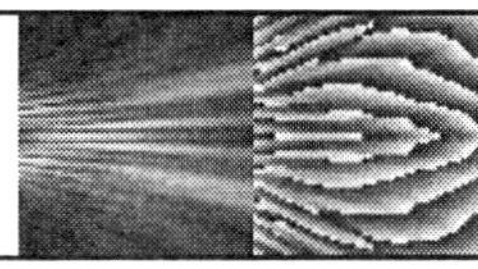	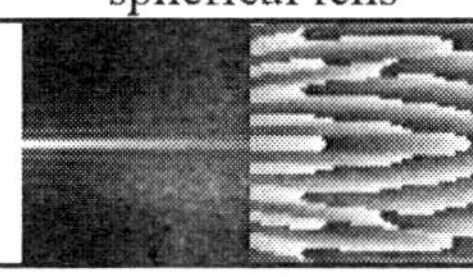

The best result was obtained using a spherical lens. Table 3 presents the cross-sections when the specified field is distributed..

Table 3. Results of numerical modeling (amplitude and phase) of the propagation of a type field Gauss beam with ring mask at a distance of 100, 150, 200 and 250 mm (cross-section)

100	150	200	250
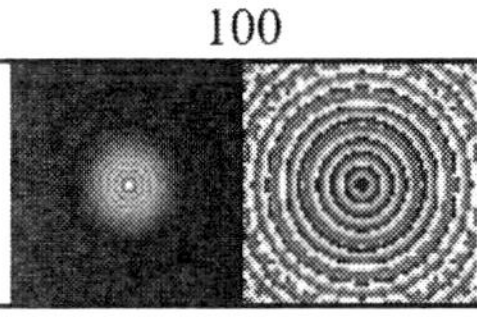	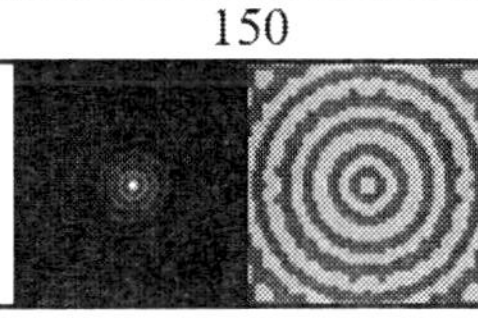	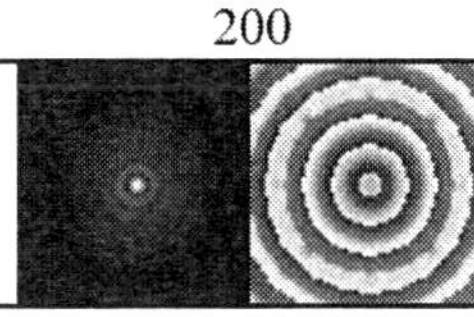	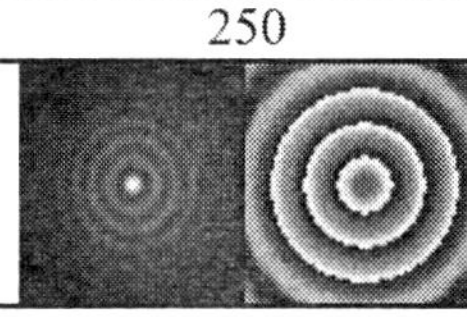

3. Conclusion

In this work, the effective formation of some diffraction-free laser beams was shown using the example of trigonometric functions in conjunction with a diaphragm ring spectrum. As a result, a diffraction relief was calculated and the diffraction-free properties of the calculated distributions were shown. After passing the field through a lens of the form exp $[ik / 2f (x^2 + y^2)]$, it becomes possible to focus the beam in a certain area, while the properties without diffraction are preserved. It should be noted that the resulting distribution in the longitudinal plane resembles Mathieu beams. In a number of experiments, it was revealed that a similar effect is achieved with the help of an annular diaphragm together with trigonometric functions or Gaussian beams. You may find that it has the properties of a light sheet. Further research is planned to be associated with the search for diffraction-free beams based on other types of aperture.

Acknowledgments
This work was partially supported by the Russian Foundation for Basic Research (grant No. 18-37-00056 (theoretical foundations), 20-07-00444 (numerical simulation)).

References
[1] Bereznyi A E, Prokhorov A M, Sisakyan I N, Soifer V A 1984 *DAN USSR* **274** 802-805
[2] Durnin J Phys. 1987 *Rev. Lett.* **58** 1499-1501
[3] Gutierrez-Vega J C 2000 *Opt. Lett.* **25** 1493– 1495.
[4] Bandres M A 2004 *Opt. Lett.* **29** 44–46.
[5] Khonina S N, Ustinov A V, Chávez-Cerda S 2018 *J. Opt. Soc. Am. A* **35** 1511-1517
[6] Grossman J G 1984 *Appl. Opt.* **23** 48–52
[7] Siviloglou G A, Christodoulides D N 2007 *Opt. Lett.* **32** 979–981
[8] Khonina S N 2011 *Optics Communications* **284** 4263–4271
[9] Efremidis N K, Christodoulides D N 2010 *Opt. Lett.* **35** 4045–4047
[10] Gutierrez-Vega J C, Bandres M A 2005 *J. Opt. Soc. Am. A* **22** 289–298
[11] Khonina S N, Ustinov A V 2017 *J. Opt. Soc. Am. A* **34** 1991-1999
[12] Lopez-Aguayo S 2010 *Phys. Rev. Lett.* **105** 013902
[13] Garces-Chavez V 2002 *Nature*, **419** 145-147
[14] Leitgeb R A 2006 *Opt. Lett.* **31** 2450-2452
[15] Khonina S N, Golub I 2015 *Opt. Lett.* **40** 4070-4073
[16] Ziolkowski R W, Besieris I M, Shaarawi A M 1993 *J. Opt. Soc. Am. A.* **10** 75-87
[17] Anguiano-Morales M 2008 *Optics Communications* **281** 401–407
[18] Khonina S N 2009 *Computer Optics* **33** 70-78
[19] Khorin P A 2019 *Journal of Physics: Conf. Series* **1096** 012104 10.1088/1742-6596/1096/1/012104

Journal of Physics: Conference Series **1410** (2019) 012174 doi:10.1088/1742-6596/1410/1/012174

The relationship of the geometric parameters of the surface with its properties (optical, emission, tribological)

O Vasiliev[1,2], J Ruzankina[2,3], S Tarasov[3]

[1]LLC 'Laser Center', St. Petersburg 195176, Russia

[2]Department of Low-Temperature Energy, ITMO University, St. Petersburg 197101, Russia

[3]Department of Photonics, Electrotechnical University, St. Petersburg, 197376, Russia

Abstract. This article presents the results of the possibility of manufacturing different types of microstructures. Special surface microreliefs have been created to solve a number of important problems related to the functional properties of the object surface and its relationship with the environment.

1. Introduction

In nature, there are often microstructures in the form of a correct convex four - or hexahedron, having minimum free energy, which provides maximum stability under any external influence – mechanical, chemical, thermal, etc. (figure 1). Such microstructures have important surface properties [1-4], such as increased shape stability (impact resistance) and corrosion resistance, increased contact stiffness, fatigue limit, slip resistance, density and hydraulic density of compounds, electrical strength, etc. The presence of such microstructures increase the adhesion of rubber coatings during prolonged contact "rubber– metal", improve the quality of galvanic coatings, speed up the cleaning of the surface in vacuum conditions, reduces the electron emission to the environment, etc. in the manufacture of a negative image of convex hemispheres by means of forging or rolled, the resulting surface improved adhesion properties. These and many other advantages of convex spherical microstructures make such reliefs in demand in modern industry and research activities.

The authors of the article attempted to recreate such a microstructure using fiber laser radiation and study the relationship of geometric parameters of the surface topography with its various properties.

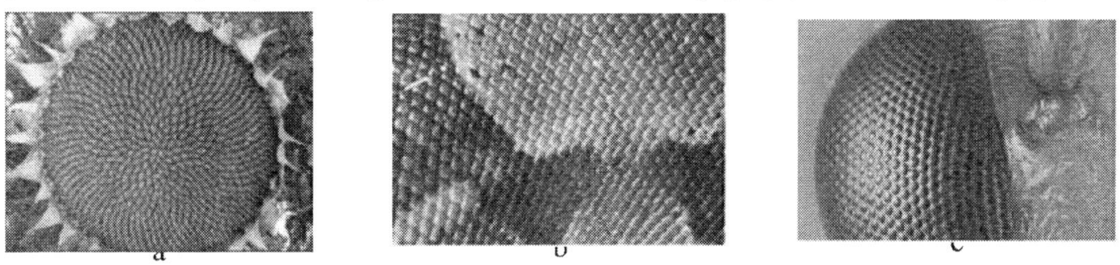

Figure 1 (a, b, c). Natural analogs of the regular microrelief: **(a)** sunflower; **(b)** snakeskin; **(c)** in the faceted eye of an insect

2. Experiment and results

In our work we used a fiber laser (operating in pulse-periodic mode) with the wavelength of 1070 nm and power up to 18, 4 W. As a test sample, steel grade was used 10, 9X and 10X18H10T. Surface treatment of metal was held at a room temperature of 23 °C and a relative humidity of air of 55%.

2.1. The relationship of the geometric parameters of the relief surface with its optical properties

The presence of microrelief (roughness) of the surface leads to a number of qualitatively new effects in the reflection of light. First, in addition to the specularly reflected wave, diffusely scattered waves appear. Secondly, when reflected from metal surfaces, the presence of roughness leads to the possibility of converting part of the incident wave into a surface electromagnetic wave (surface plasmon-polariton).

Optical properties of the surface are a priority in a number of applications of functional materials. Of course, in the first place here is photoelectricity, where the problem of weakly reflecting (black) surfaces determines the efficiency and General prospects of the entire direction. Optical properties of the surface play an important role in many products of space and military equipment (anti-reflective surfaces), in the jewelry industry, in the processes of laser marking of products, etc.

It becomes possible to control the spectral reflection coefficient of the incident light wave. Thus, the shape of the microstructure and the slope of the side faces of the pyramids can be adjusted at the stage of ablation, and the thickness of the oxide and its color can be adjusted at the stage of "polishing". For example, in the course of the experiments surface were created microstructures with an oxide layer which having a shade of purple (figure 2). Visually, the color of the microstructure varies from blue to red-purple. This is characterized by the difference in the course of the light wave in the thickness of the oxide layer, depending on the angle of illumination. The microstructure of the surface oxide films of different colors can be used in optical instruments, spectrum analyzers, and reflectors. The problem of controlling the optical properties of the surface can be solved both by creating specific microstructures – periodic reliefs and by chemical processes of obtaining (oxide) coatings on the surface of materials. Both of these methods can be effectively implemented by laser irradiation of the surface of most metals.

Figure 2. An example of oxidization microstructured surface steel grade 9XC

2.2. Connection of geometric parameters of the surface relief with its ability to electron emission

A very important property of the surface for many applications is its ability to enhance (emission reliefs) or suppress (anti-emission – electro-protective reliefs) electronic emission. Emission reliefs are necessary for the manufacture of a large range of specific elements of vacuum technology – cathodes, both metal (tungsten) and oxide.

The ability to control the value of the emission current for the cathode from a given material is associated with the effect of local amplification of the e field near the projections on the surface having the shape of a tip (blade). Field gain:

$$\beta = \frac{E}{E_0} \qquad (1)$$

(E_0 = V/d – the field strength in the absence of the tip at the same potential difference, V – the potential difference between the electrodes, d – the distance between them) in the simplest case, the cylindrical tip height H and the radius of R-vertex R is estimated from the ratio (when d >> H >> r):

$$\beta = \frac{H}{r} \qquad (2)$$

It follows from the above formula that the smaller the tip radius, the greater the local field gain an emission current. The reduction of r down to the nanometer size allows detecting currents of field emission by several orders of magnitude larger than for micron-tip-shaped emitter. It should also be noted that the resistive heating of emitters significantly decreases with a decrease in the radius of curvature of the tip, which allows recording record emission currents. Of course, in the transition region of the nanoscale is, in principle, violated the basic assumptions of the theory of Fowler - Nordem (the one-dimensionality of the potential barrier, the uniformity of the field distribution on the surface of the emitter, etc.). Obtaining simple analytical expressions, in this case, is difficult, however, at the qualitative level of the description, the noted regularities remain valid. In this connection, the emission properties of cathode surfaces will be significantly determined by their geometry. Thus for antialiasing terrain is typical that the radius of the protrusions of the relief (R1) is much greater than the radius of the depressions of the relief (R_2), and for monetary relief on the contrary. Figure 3 and 4 show the reliefs with antiemission and emission properties, which were obtained by surface treatment by laser radiation. For monetary relief best treatment mode corresponds to a power density q=1*10^8 W/cm^2, and antiemission elevation q=0,8*10^8 W/cm^2.

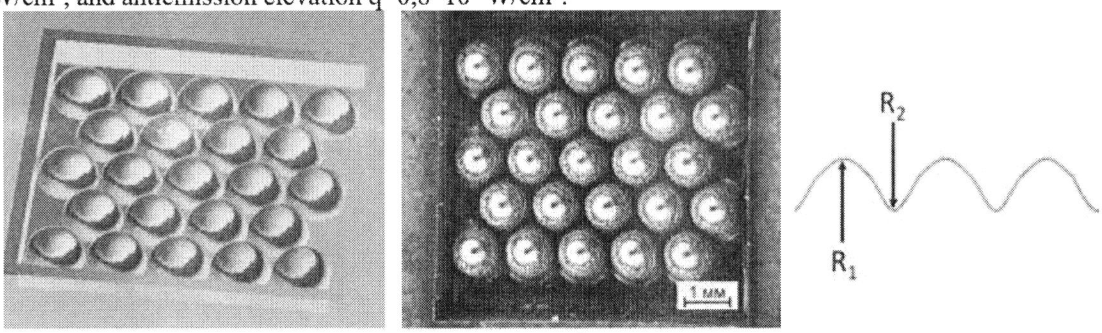

a b c

Figure 3(a, b, c). The characteristic shape of the experimentally obtained relief of the surface of the steel 10 to give it antiemission properties: **(a)** 3D model; **(b)** microstructure of steel; **(c)** a characteristic type of terrain $R_1 > R_2$.

a b c

Figure 4(a, b, c). Characteristic types of experimentally obtained surface relief of steel grade 10 to give it emission properties (a) 3D model; (b) microstructure of steel; (c) a characteristic type of terrain $R_1 < R_2$.

2.3. The relationship of the geometric parameters of the surface topography with its tribological properties

The rear surface of the object has a fairly rich set of topographic characteristics that can be adjusted depending on the requirements of production. They can be divided into two groups, among which:

- Macrogeometry – macro-relief, shape deviation, altitude, step-by-step, lateral, periodic and other surface characteristics.
- Microgeometry – surface roughness, porosity, microdefects, etc.

Each of these characteristics determines the size of the nominal contact area and, as a result, has a significant impact on the tribological behavior, performance and wear resistance of objects.
The proposed technology of microstructuring the metal surface allows you to adjust each of the described parameters: the first stage of metal ablation sets the macrogeometric, and the third stage of "polishing" microgeometric surface parameters. Table 1 presents the main characteristics of the microstructures presented (figure 5 and 6).

Table 1. The main characteristics of microstructures.

	The height of the structure h, mkm	Size L, mm	Density N, pcs /cm^2	Roughness parameter R_a, mkm	Roughness parameter R_z, mkm	The thickness of the remote H, mkm	Time of manufacturing t, min
Cylinders (figure 4)	950	0.90	16	0.26	1.10	1000	~30 (3 step)
Gutters (figure 5)	290	0.03	5000	1.10	6.94	300	~20 (2 step)
Untreated surface	-	-	-	0.6	2.77	-	-

The roughness of the surface "polished" by laser radiation was 2.5 times less than the roughness of the untreated surface. This result once again proves the effectiveness of the three-stage laser technology for the formation of microstructures to change the tribological properties of the surface, which is extremely necessary for the production of components of mechanisms in which there is active friction.

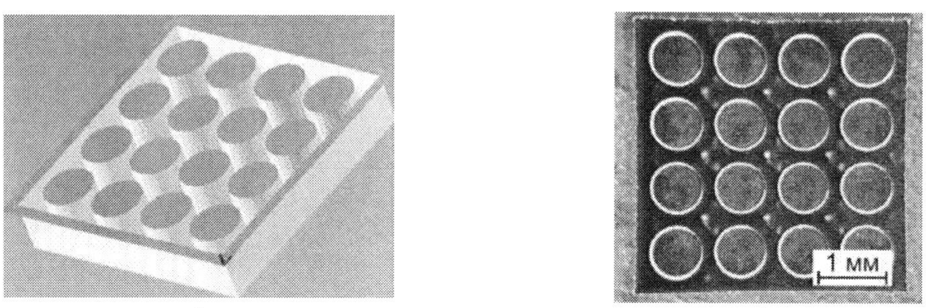

a b

Figure 5(a, b). Images of microstructure on the surface of steel grade 10: **(a)** graphics model of microstructure in the form of a matrix of cylinders (number of objects 16 PCs/cm^2); **(b)** the corresponding photo

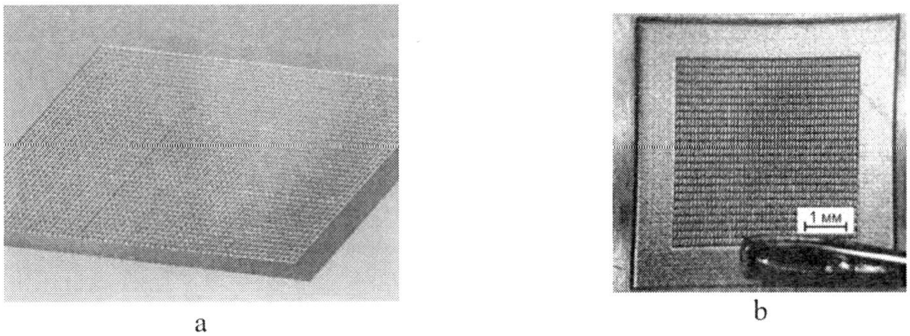

Figure 6(a, b). Images of the microstructure on the surface of steel grade 10: **(a)** graphical model of the microstructure in the form of a friction plane with gutters; **(b)** corresponding photo

The analysis of the obtained results allows us to include that laser irradiation of the surface of steel with a fiber laser can change the shape of the surface topology of the metal by adjusting the surface properties. Figure 7 shows the dependence of the roughness parameters of the manufactured microstructure on the gradual reduction of the pulse duration in the process of polishing the surface of steel 10 by laser emission.

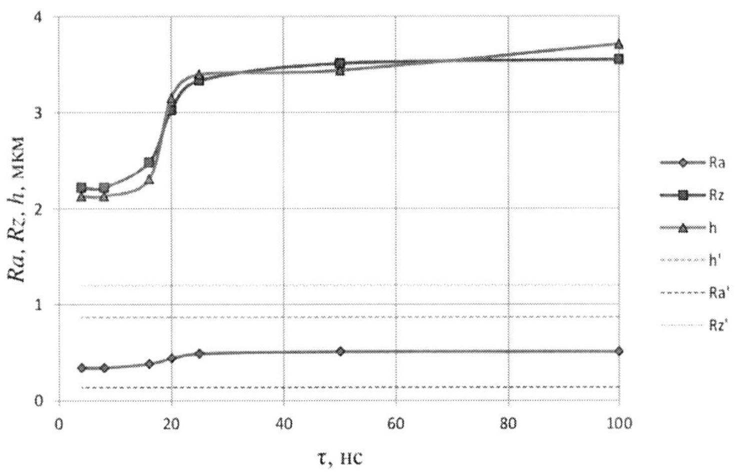

Figure 7. Graph of the dependence of the roughness parameters of the manufactured microstructure on the gradual reduction of the pulse duration in the process of polishing the surface of steel 10 laser radiation (Ra', Rz', h' – roughness parameters of the untreated surface)

3. Conclusion

The article proposes a method for the creation of microstructures on the surface of steel grades 10 and 9HS. The surface reliefs of various types and shapes are obtained: pyramids, cones, hemispheres, cylinders, images, etc., with characteristic dimensions of 30...1000 μm.

It can be concluded that by changing the shape of the surface topology of the metal, you can adjust the properties of the surface: mechanical, optical, electrical, etc. However, to obtain the required accuracy and roughness relief for each material requires the development of individual modes of laser processing.

The main factor affecting the quality of manufactured microstructures is the duration of laser pulses. It is experimentally confirmed that shorter pulse duration is effectively used for surface finishing, which

leads to a decrease in the degree of roughness of the produced microrelief. In this case, the minimum τ=4 ns, which is due to the technical capabilities of the fiber laser used in the work.
The proposed method of formation of the microgeometry of the surface using the volume-controlled ablation in the setting with scanning system restores the beam can be used for practicing the basics of laser surface engineering in a variety of industries for research activities.

4. Acknowledgments
The work was carried out with the financial support of the Ministry of education and science of the Russian Federation in the framework of research of the project part of the GZ, project № 16.1750.2017/4.6

References
[1] Ruzankina J S, Parfenov V A, Vasiliev O S 2019 *Laser Physics* **29** 1
[2] Vasil'Ev O S, Veiko V P, Gornyi S G, Ruzankina Y S 2015 *Journal of Optical Technology* **82** 12
[3] Ruzankina J S and Vasiliev O S 2016 *Journal of Physics: Conference Series* **735** 1
[4] Vasiliev O S, Ruzankina J S 2016 *Journal of Physics: Conference Series* **735** 1

X-Ray linear polarization degree for RGBL beamline BESSY II

O V Petrova[1], A E Mingaleva[1,2], D V Sivkov[1,3], S V Nekipelov[1,4], R N Skandakov[1], V N Sivkov[1,4].
[1]Institute of Physics and Mathematics FRC Komi SC UB RAS, Syktyvkar, Russia
[2]V. A. Fock Institute of Physics St Petersburg State University, St Petersburg, Russia
[3]Immanuel Kant Baltic Federal University, 236041, Kaliningrad, Russia
[4]Pitirim Sorokin Syktyvkar State University, 167001, Syktyvkar, Russia

Abstract. The results of the calculation of the X-Ray synchrotron radiation linear polarization degree of Russian-German dipole beamline at synchrotron BESSY II are discussed. It is shown that the integral synchrotron radiation linear polarization degree (λ, ψ_h) for incident photon energy ε=300eV reaches 80%

1. Introduction

One of the unique and useful property of synchrotron radiation for soft X-Ray spectral analysis with incident photon energy from 10 to 1200 eV is high polarization degree P [1]. High degree of this parameter is crucial for soft X-ray absorption coefficient determination of graphene-like structures or linear and planar molecules adsorbed on the surface. Such structures are characterized by stable position of free molecules orbitals with pi- and sigma-symmetry lying perpendicular to molecular plane and along the molecular axis, correspondingly.

The selection of high polarized beam or calculation of its linear polarization degree allow us to determine the adsorbed molecule orientation or to provide the experimental measurements of absorption coefficient in soft X-ray range for electron transition from core level to free molecular orbital of pi- and sigma- symmetry separately and independently of each other. The linear polarization degree of synchrotron source depends not only on synchrotron parameters but beamline equipment and may vary for given synchrotron depending on beamline.

In current work the linear polarization degree of synchrotron radiation from Russian-German dipole beamline (RGBL) [2] at synchrotron BESSY II is calculated. The RGBL is aimed to high resolution photoemission and photoabsorption research in the soft X-ray photon-energy region (30–1500 eV). The calculation was based on theoretical approach of Kunz [3] and performed for different photon energies (ε=40, 200, 300, 500, 800 eV).

2. The polarization of synchrotron radiation

It is well known, that synchrotron radiation, in general, is characterized by elliptical polarization included two electron field vector components E_{II} and $E_{\perp}$, which are parallel and perpendicular to electrons rotation orbit plane. The direction of elliptical polarization rotation is changed when the radiation passes through the rotation plane of radiating electrons and exactly in the electrons rotation plane the radiation is characterized by the highest degree of linear polarization. The theoretical and experimental researches show the parallel and perpendicular components distributed with maximum and minimum lying in orbit plane, respectively (Figure 1) [4], that leads to extremely high linear polarization degree with parallel component E_{II} exactly in the electron orbit plane. However, it is well seen from the Figure 1 that on practice perpendicular component doesn't drop to zero in orbit plane direction which means that even in the orbit plane the radiation is not strictly linearly polarized and include small amount of perpendicular component $E_{\perp}$.

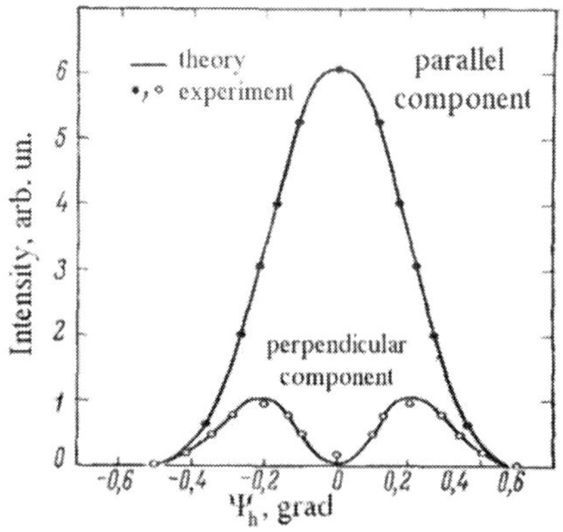

Figure 1 Parallel and perpendicular components of Synchrotron radiation depending on vertical angle ψ_h between radiation direction and orbit plane estimated theoretically (lines) and measured experimentally at synchrotron NBS with energy E=180 eV for radiation with wavelength λ=500 nm (dots) [4].

Figure 1 shows that in electron rotation orbit plane ($\psi=0$) the radiation include only parallel component and the angle increasing leads to the parallel component arises and then increases. At the same time with the photon wavelength increasing (or photon energy decreasing) the $I_{\perp}(\lambda, \psi)$ and $I_{II}(\lambda, \psi)$ dependencies become wider. In the other hand with the photon wavelength decreasing (or photon energy increasing) the radiation with booth polarization components become concentrating more close to electron rotation orbit plane ($\psi=0$), that mean the percentage of hard X-ray component in synchrotron beam increasing with decreasing of the entrance slit vertical angle ψ.

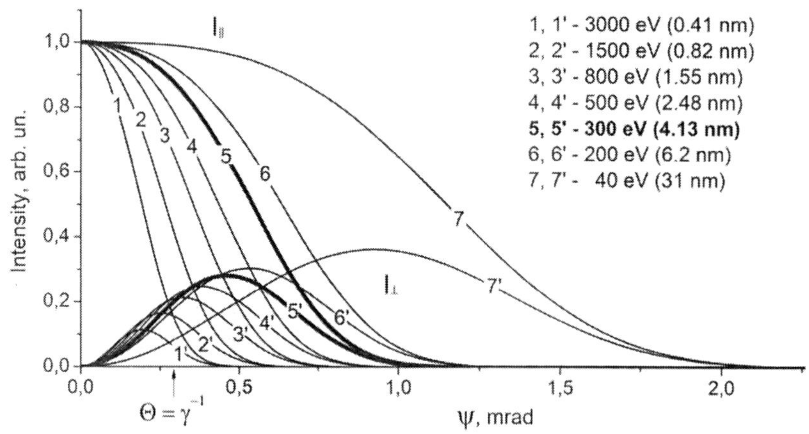

Figure 2. The radiation intensity with parallel I_{II} and perpendicular $I_{\perp}$ linear polarization component depending on vertical angle ψ for different photon energies.

The high linear polarization degree of about the 98% can be reached by the cutting out of radiation in a small vertical angle near the electron orbit plane by the entrance slit reduction [4, 5]. On the other hand in the cause of soft X-ray spectroscopy with a diffraction grating as the dispersing element such slit size reduction leads to decreasing of soft X-ray component intensity in incident radiation, increasing of hard X-Ray intensity contribution and as a result increasing of second and higher order diffraction background radiation. Such soft (low energy) and hard (high energy, more than 1200 eV) X-ray component distribution depending on the exit slit vertical angle is clearly observed in the Figure 2 where the radiation intensity with parallel and perpendicular polarization component depending on the vertical angle from electron rotation orbit plane for different photon energy (ε =40, 200, 300, 500, 800, 1500, 3000 eV) are shown.

3. Calculations details

The calculation of synchrotron radiation intensity $I(\lambda, \psi)$ and its parallel $I_{II}(\lambda, \psi)$ and perpendicular $I_{\perp}(\lambda, \psi)$ components dependent on the photon wavelength λ (or energy ε) and vertical angle ψ between radiation direction and orbit plane was performed by the formulas (1-3) [3]:

$$I(\lambda,\psi)=\frac{27}{32\pi^3}\cdot\frac{e^2c}{R^3}\left(\frac{\lambda_c}{\lambda}\right)^4\gamma^3\left[1+(\gamma\psi)\right]^2\left[K^2_{2/3}(\xi)+\frac{(\gamma\psi)^2}{\left[1+(\gamma\psi)^2\right]}K^2_{1/3}(\xi)\right],\quad \xi=\frac{\lambda_c}{2\lambda}\left[1+(\gamma\psi)^2\right]^{\frac{3}{2}},\tag{1}$$

$$I_{\perp}(\lambda,\psi)=\frac{27}{32\pi^3}\cdot\frac{e^2c}{R^3}\left(\frac{\lambda_c}{\lambda}\right)^4\gamma^3\left[1+(\gamma\psi)\right]^2\frac{(\gamma\psi)^2}{\left[1+(\gamma\psi)^2\right]}K^2_{1/3}(\xi),\tag{2}$$

$$I_{II}(\lambda,\psi)=\frac{27}{32\pi^3}\cdot\frac{e^2c}{R^3}\left(\frac{\lambda_c}{\lambda}\right)^4\gamma^3\left[1+(\gamma\psi)\right]^2K^2_{2/3}(\xi),\tag{3}$$

where $\gamma=\frac{E}{mc^2}$ – relativistic factor, $\lambda_c=\frac{0.4\pi R}{3\gamma^3}=\frac{0.559R}{E^3}$ is critical wavelength, R – electron orbit radius and $K_{2/3}(\xi)$ and $K_{1/3}(\xi)$ are Bessel functions of the second kind.

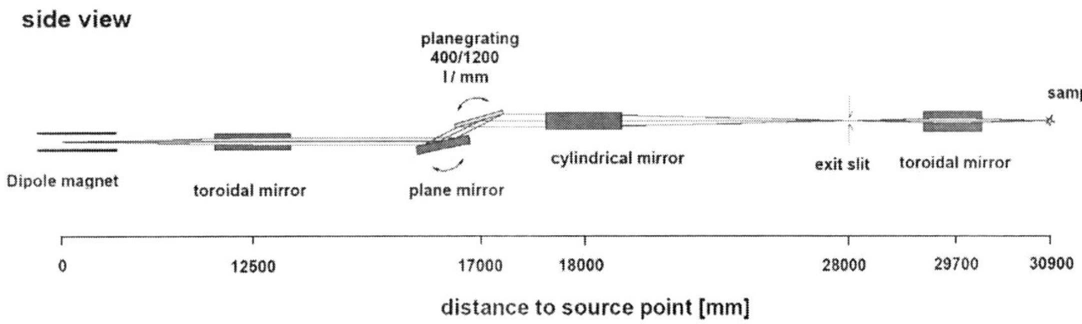

Figure 3. Optical layout of the Russian–German beamline at BESSY [4]

In the cause of RGBL beamline it's optical layout doesn't include entrance slit (see figure 3) and thus the equipment of this beamline doesn't allow reducing the linear polarization degree by entrance slit size reduction. The solution of this problem is to calculate theoretically the linear polarization degree by the formula (4) and then remove the effect of $E_{\perp}$-component from experimental absorption spectra.

The linear polarization $P(\lambda, \psi)$ of synchrotron radiation dipole source depends on emitted photon wavelength λ (or energy ε) and vertical angle ψ between radiation direction and orbit plane and may be calculated as [3]:

$$P = \frac{E_{\|}^2}{E_{\|}^2 + E_{\perp}^2} = \frac{I_{\|}}{I_{\|} + I_{\perp}}, \tag{4}$$

$$P = \frac{K_{2/3}^2(\xi)}{K_{2/3}^2(\xi) + \frac{(\gamma\psi)^2}{1+(\gamma\psi)^2} K_{1/3}^2(\xi)}, \quad \xi = \frac{\lambda_c}{2\lambda}\left[1+(\gamma\psi)^2\right]^{\frac{3}{2}}, \tag{5}$$

For synchrotron BESSY II the following values are used: E=1.7 GeV and λ_c=0.495 nm (ε_c=2.5 KeV).

It is clearly seen from figure 3 that at RGBL the focusing toroidal mirror of width w=750 mm and height h=60 mm positioned at an angle of 85° to the incident radiation beam is used as the entrance slit. The mirror cuts out the synchrotron radiation in the vertical and horizontal angles $\psi_h \approx 2.5$ and $\psi_w \approx 3$ mrad and focuses it to plane-grating monochromator.

4. Result and discussion

The results of the linear polarization $P(\lambda, \psi)$ of synchrotron radiation calculation for different photon energies (ε=40, 200, 300, 500, 800 eV) performed according to the formula (5) are shown in the figure 4. The integral linear polarization $P(\lambda, \psi_h)$ for vertical angle range 0–ψ_h and for different photon energies (ε=40, 200, 300, 500, 800 eV) are shown in Figure 5.

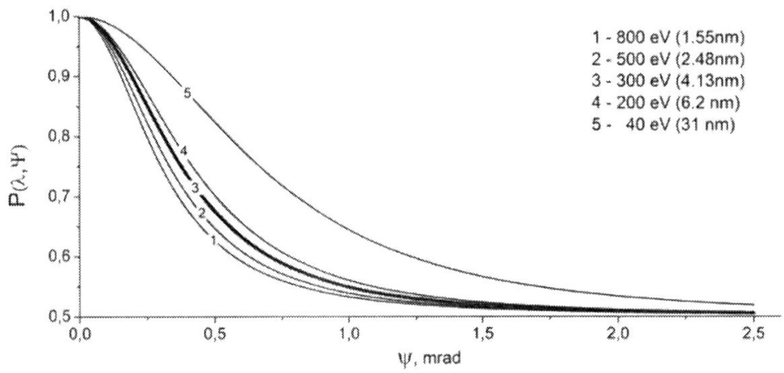

Figure 4. The linear polarization degree P (λ, ψ) depending on vertical angle ψ for different photon energies (ε=40, 200, 300, 500, 800 eV) for RGBL BESSY II.

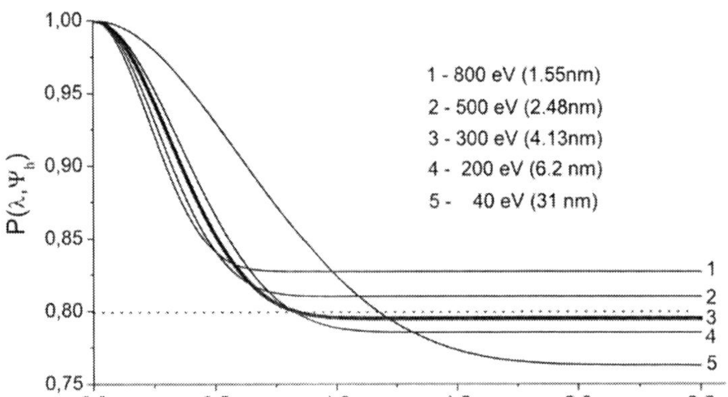

Figure 5. The integral linear polarization P(λ, ψh) depending on vertical angle range 0–ψh and for different photon energies (ε=40, 200, 300, 500, 800 eV) for RGBL BESSY II.

The calculated results (see figure 4, 5) demonstrate that for the energy range 200–800 eV the integral synchrotron radiation linear polarization of RGBL BESSY II lies in interval 0.75–0.85 (75-85%) and for photon energies ε=40, 200, 300, 500, 800 eV it is equal to *P(40)=0.76, P(200)=0.78, P(300)=0.79, P(500)=0.81, P(800)=0.83,* respectively. For the energy range of carbon atom 1s-absorbtion edge (280-320eV) the integral synchrotron radiation linear polarization of RGBL BESSY II reaches *80%*.

Acknowledgments
This work was supported by Program of UB №18-10-2-23, the Russian Academic Excellence Project at the Immanuel Kant Baltic Federal University, Grant RFBR № 19-32-5062/19 mol_nr and the Bilateral Program of the Russian-German Laboratory at BESSY II

References
[1] Stöhr J 1992 *NEXAFS Spectroscopy* (Berlin: Springer Verlag) p403
[2] Fedoseenko S I, Vyalikh D V, Iossifov I E et al. 2003 Nucl. Instrum. Methods Phys. Res. A **505** 718
[3] Kunz C.1979 *Synchrotron Radiation Techniques and Applications* (Berlin: Springer) p421
[4] Codling K, Madden R P 1964 J. Appl. Phys. **36** 380–387
[5] Sokolov A. A., Eggenstein F., Erko A. et al. 2014 Advances in Metrology for X-Ray and EUV Optics V, V.92060J

Monitoring of oil paintings by means of optical 3D scanning

A A Zhuravlev, V A Parfenov
St.Petersburg Electrotechnical University, Saint Petersburg 197376, Russia

Abstract. A novel technique of monitoring of painted surfaces is proposed. Experimental results concerned with use of optical 3D scanning for monitoring of oil paintings are presented. One of oil paintings with collection of The Military-Historical Museum of Artillery, Engineer and Signal Corps (St. Petersburg, Russia) was scanned by means of optical 3D scanner, which operates using structured light. The scanning was carried out 2 times with time interval of 4.5 years. In both cases we created computing 3D models of the same painting. These models were compared using software GeomagicStudio that allowed to get a precise quantified information about deformations of canvas and frame of painting.

1. Introduction

Artifacts of history and culture, regardless of whether they are exhibited (in the open air or stored in museum collections), are gradually destroyed over time. This happens under the influence of adverse environmental factors, whose impact has increased significantly in recent decades due to a dramatic environmental deterioration. In this situation, regular monitoring of the state of conservation of artworks is becoming an increasingly urgent task in the field of preserving cultural heritage. In case of paintings, the need for monitoring is associated with the fact that defects in the paint layer usually occur at the "micro level", and it is very difficult to detect them in a timely manner using photography (including high resolution one) which is traditionally used in museums to monitor the state of paintings.

One of the possible approaches to solving the problem of paintings monitoring is the use of laser and optical 3D scanning technologies. Currently, this technology is widely used to document the paintings. There are known cases where it was used to monitor stone monuments [1], [2], as well as mural paintings. In particular, one of the authors of this article showed the principal possibility of using 3D scanning to monitor the state of wall paintings to control the "swelling" and the width of the cracks of the plaster layer [3]. However, there is no information on the use of 3D scanning for monitoring oil paintings in the scientific literature. This suggests a scientific novelty of research conducted in this work.

We carried out experiments on monitoring the painting "Equestrian artillery company No. 13 barrages French columns during the retreat of the "great army" from Russi" (created by Russian artist P. Karyagin, 1912) from the collection of the Military-Historical Museum of Artillery, Engineer and Signal Corps in St. Petersburg. Its general view is shown in Fig 1. The choice of this painting as an object of study was connected with that fact its surface has a complex relief. This artist is characterized by the "volumetric" style of painting technique, which is manifested in the application of the paint layer by separate textured strokes of different thickness and width. It allows one to easily capture the 3D relief of the painting's surface and track the changes of the paint layer's state as the relief changes during the reference time interval.

Figure 1. Painting "Equestrian artillery company No. 13 barrages French columns during the retreat of the "great army".

2. Measurements

In the process of experiments, the surface of the painting within a separate control area of 149.3 cm cm^2 was scanned using an optical 3D scanner Cronos 3D (manufacturer – Open Technologies Ltd., Italy) in 2014. The surface geometry capture mode with texture (the surface geometry capture mode with texture was chosen) was used. This scanning mode was used to simplify the process of assembling the scans of the painting into a single computing 3D-model at the stage of processing the data obtained in the scanning process. Though the surface of the painting has an individual relief, but its geometry is complicated for aligning the scans. However, combining the same texture maps may sufficiently simplify the process of assembling 3D model and increase the accuracy of aligning the scans.

In 2019, we re-scanned the same section of the painting. In the course of this work optical scanner RangeVision Pro5M (manufacturer – RangeVision, Russia) was used

3. Experimental results

To verify the possible changes in the geometry of the painting over the past 4.5 years, a specialized computing software Geomagic Studio 2012 was used to proceed the data obtained in the process of 3D scanning. With the help of this program, a comparison of 3D models of the picture obtained in 2014 and 2019 was made. This comparison was made by combining the obtained 3D models, which allowed to visually and quantitatively estimate the deviation of the surfaces of the paint layer of the picture fixed in one and the other model. The program allows you to create a color scheme, which displays the value of different colors deviation 2 compared surfaces relative to each other. In this case, one of the surfaces is selected as a reference. In this case, the areas of the said color scheme, represented by shades of blue, are below the reference surface, and areas of red – above the reference surface (see Fig 2).

To obtain the most correct data, firstly the surface of the paint layer from the 3D model of 2014 and then 3D model of 2019 was chosen as the reference surface. In both cases, similar results were obtained.

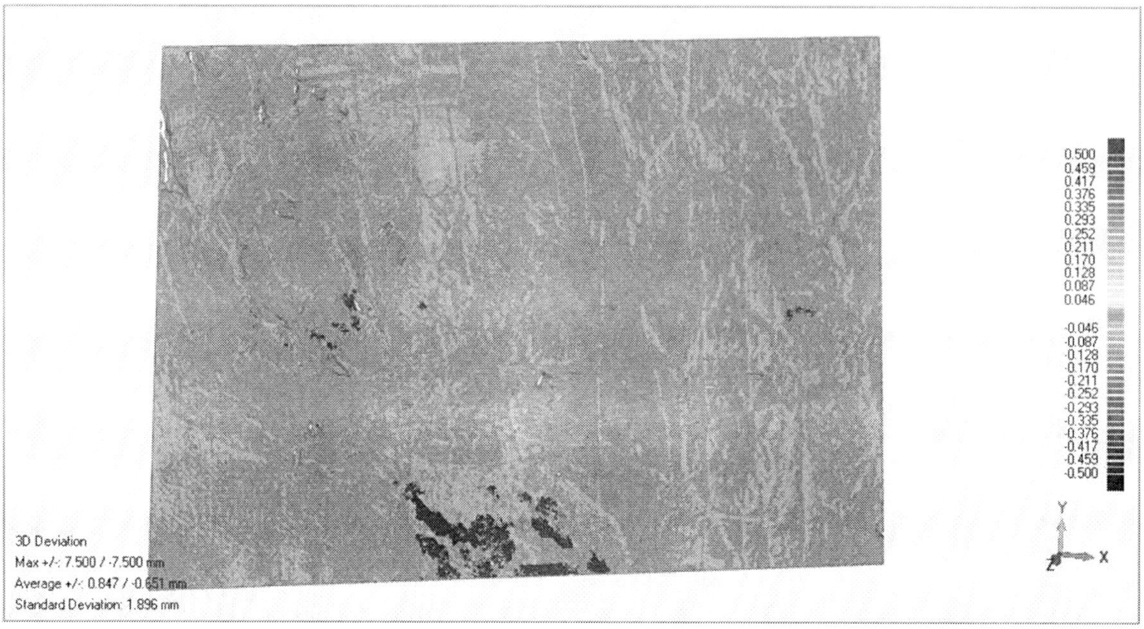

Figure 2. Visualisation of difference between two 3D-models of 2014 and 2019.

As a result of the experimental studies it was shown that the control section of the canvas has 7 zones of yellow and blue colour shades (each of them is an area of about 13.5 cm^2, one of them is shown in Fig 3 by red color), which indicates the presence of deviations of the order of 200 micrometres in greater or lesser sides (i. e., there are "swollen" and concave areas. Since these irregularities are distributed throughout the study surface, it can be concluded that the nature of the deformations of the canvas is wavy.

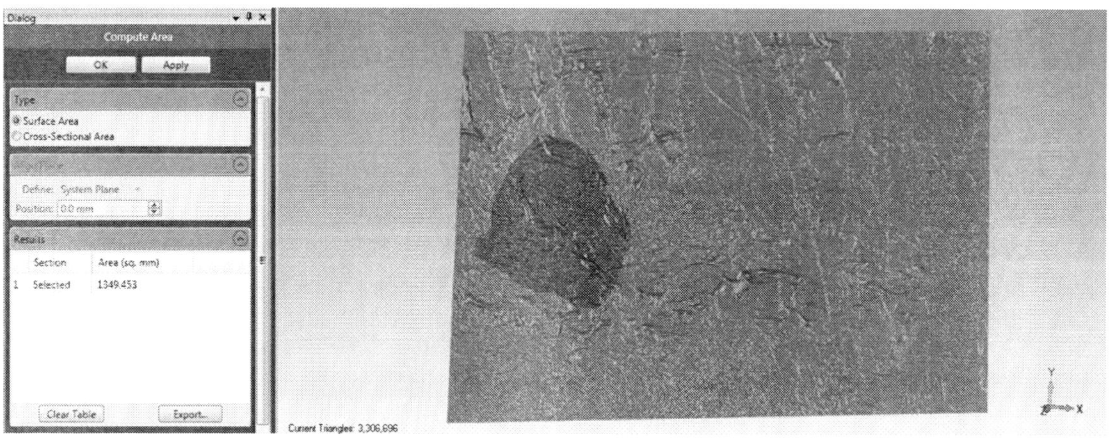

Figure 3. 3D surface of painted layer.

A similar method was used to monitor the wooden frame of the painting. As a control zone, the lower right corner of the picture was chosen, where there is a crack at the junction of the vertical and horizontal sections of the frame. During the control period of 4.5 years, the gap has increased significantly – when matching 3D models in the Geomagic program along the entire length of the slit, a blue area with a width of about 0.4 mm is clearly visible (see Fig 4). As for the frame itself, it should be pointed out that there is a "redness" of the surface of the polygonal grid closer to the edge of the frame as well as presence of a blue area near the joint of the frame and the canvas. Such a distribution of color may be evidence that the edge of the frame began to bend inward - toward the painted layer.

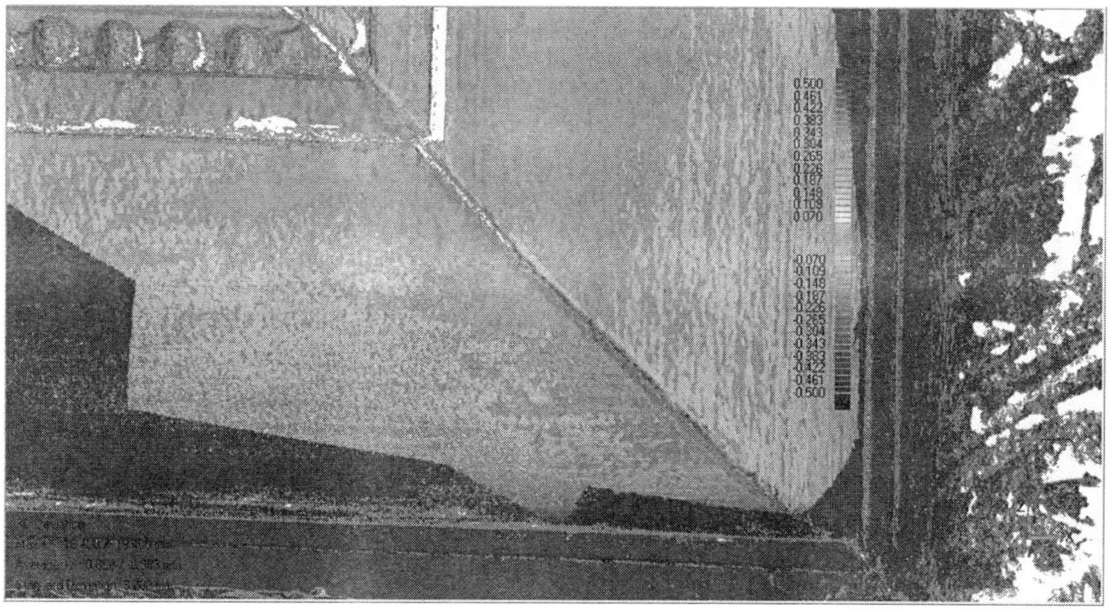

Figure 4. Visualisation of difference between two 3D-models of wooden frame

4. Conclusion

Thus, as a result of the experiments, the principal possibility of using 3D scanning for monitoring oil paintings was demonstrated. It was shown that this method allows one to control the state of the surface of the painted layer as well as the wooden frame with high accuracy (fractions of a millimeter). Such information can be useful for the conservators of paintings to decide on the need to change the conditions of storage and exposure of paintings.

Acknowledgments
Authors would like to thank Dr. V I. Kobyakova with the Military-Historical Museum of Artillery, Engineer and Signal Corps for availability of painting and for useful discussions.

References

[1]. Boochs F., Huxhagen U., Kraus K. 2008 Potential of high-precision measuring techniques for the monitoring of surfaces from heritage objects (Sesto Fiorentino) (Italy: Proceedings of the International Workshop SMW08) p 87–96.

[2]. Parfenov V. A., Frank-Kamenskaya O. V., Leonova I. A. Moshkina S. L., Mosnikov E. E. 2018 *Izvestiya SPbGETU «LETI»* p 73-79 {in Russian}

[3]. Maslov K. I, Parfenov V. A., Guzanov F. V., 2012 Wall paintings monitoring by laser 3D-scanning Preliminary results (Moscow) Vol 3 (Russia: Studies in cultural heritage conservation. International scientific-methodical conference) p 166-168 {in Russian}

Numerical Analysis Spectral Huygens Filter for pulsed broadband Terahertz Radiation

X Liu, N V Petrov, M S Kulya, A N Tsypkin, Ya V Grachev, S A Kozlov
ITMO University, Kadetskaya line 3b, St. Petersburg, Russia

Abstract. In this paper, we consider a Huygens filter for a specified frequency lying within a broad spectrum of a pulsed THz radiation which represents simplest two-zone plate. The condition of the infinite plane wave and the Gaussian beam are separately used as the initial source has been discussed. A property method of terahertz spectral filter has been explored.

1. Introduction

The THz band is an area of convergence between the electronics field (0.1 THz, 30m) and the photonics field (10 THz, 3mm). Nowadays, terahertz technology has become one of the most promising technology for wireless communication [1-4]. To modulate, to control, and to encode is always an interesting topic in THz area [5]. In order to encode a THz carrier with data, clearly some sort of modulation scheme is required [6, 7]. Fresnel zone plates (FZP) becomes a new solution for the development of compact passive components for THz devices [8, 9]. In this paper, a possible method of THz spectral modulation has been explored. Based on the analysis by Augustin-Jean Fresnel, we consider a Huygens filter for a specified frequency lying within a broad spectrum of a pulsed THz radiation which represents simplest two-zone plate

2. Working principle and initial conditions

The working principle diagram is shown in Fig.1, where an infinite plane wave (Case A) and a Gaussian beam (Case B) are separately used as initial sources. A mask with a circular aperture is used as a filter to modulate the THz spectrum. The radius of the circular aperture is R. The initial temporal form in the plane z = 0 of a single-period THz beam can be described by Eq. (1) [10, 11]:

$$E(t) = E_0 \frac{t}{t_0} e^{-\frac{t^2}{t_0^2}} \tag{1}$$

where E_0 is the amplitude of the electrical field, and t_0 is the pulse duration. This equation is an approximation of THz pulse which is generated by an optical rectification of femtosecond pulses through an optical crystal that is pulsed by the femtosecond lasers [12]. The corresponding spectral representation of this temporal form can be obtained by the Fourier integral with the analytic solution:

$$G(\omega) = \int_{-\infty}^{+\infty} E(t) e^{-i2\pi\omega t}\, dt = -i\pi^{\frac{3}{2}} E_0 \tau^2 \omega e^{-\pi^2\tau^2\omega^2} \tag{2}$$

The electric field amplitude Eq.1 and the amplitude of its spectrum Eq.2 are shown in Fig.2. The spatial distribution for this initial THz pulse will be considered in the axially symmetric Gaussian form with transverse spatial size a:

$$E = E_0 \exp(-\frac{x^2+y^2}{a^2}) \quad (3)$$

The details of the dynamic theory can be found in papers of A. Ezerskaya [13-16].

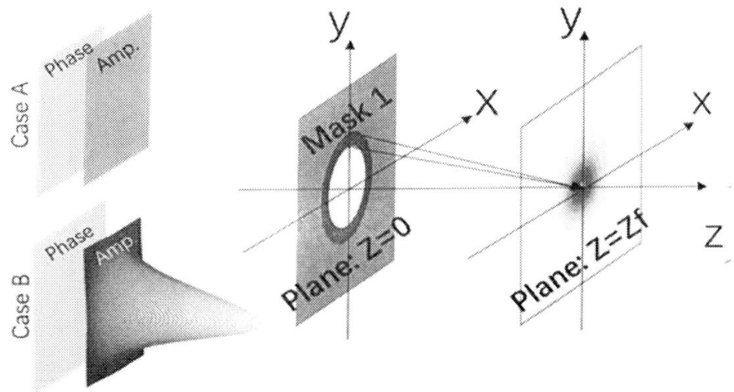

Figure 1. The diagram of working principle diagram.

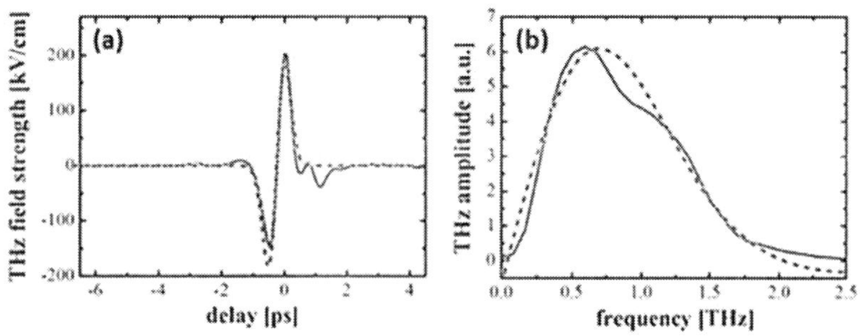

Figure 2(a,b). The initial THz pulse: **(a)** Electric field amplitude E(t). **(b)** The amplitude of its spectrum G(ω)

Journal of Physics: Conference Series **1410** (2019) 012177 doi:10.1088/1742-6596/1410/1/012177

3. Numerical results and analysis

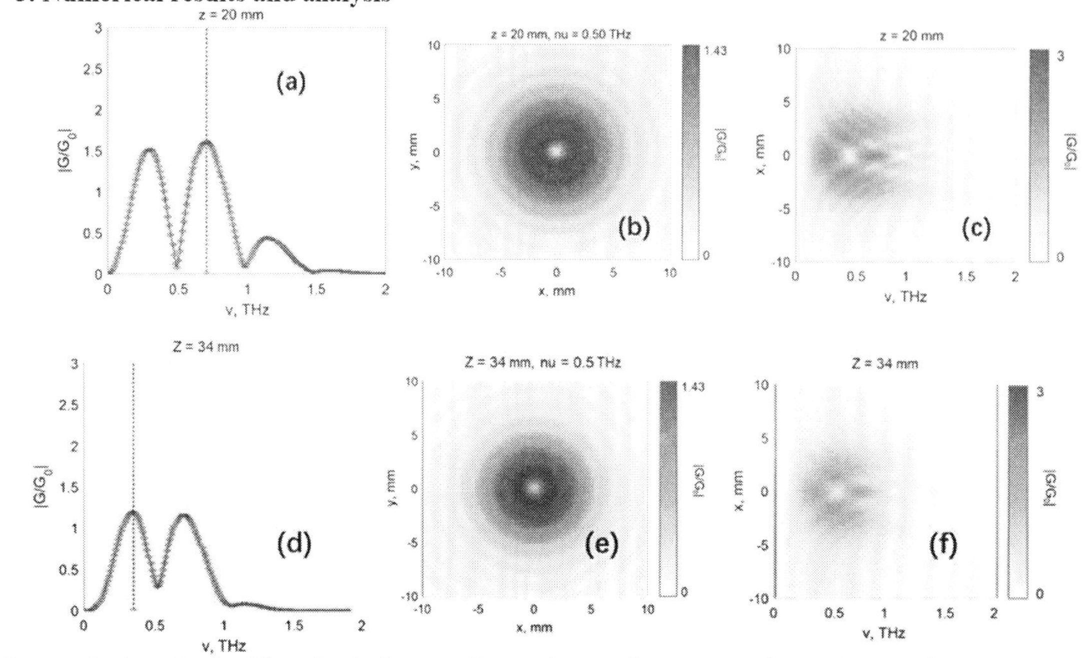

Figure 3(a,b,c,d,e,f). The simulation results under a plane wave input (Case A): **(a)** Spectral amplitude at x=0 **(b)** Spatial distribution of spectrum| **(c)** Spatio-spectral distribution. Under a Gaussian beam input (Case B): **(d)** Spectral amplitude at x=0 **(e)** Spatial distribution of spectrum . **(f)** Spatio-spectral distribution

We considered two case: a plane wave as the initial THz wave or a Gaussian beam as it is defined by Eq. 1-3. Due to the diffraction phenomenon, a specific frequency disappears near the z-axis at $z = z_f$ plane. To comply with the real experimental conditions, 0.5THz is chosen as the target frequency. The simulation results under a plane wave input (Case A) are shown in Fig. 3(a-c). It can be seen from Fig.3 (a), a spectral gap can be achieved for the specific frequency f = 0.5 THz, when the aperture radius R = 5 mm, at z_f = 20 mm position. Also, spatiospectral distribution (Fig. 3 (b)) and spectral distribution (Fig.3 (c)) confirm that 0.5 THz is indeed filtered out. The calculated distribution for the spectral amplitude at x=0, spatial distribution of spectrum and the spatio-spectral distributionof Gaussian beam is shown in Fig. 3(d-f) separatly. It is clearly that the frequency gap of the plane wave is tend to be zero deeper than the gap of Gaussian beam. To qualify the depth of the gap for both cases, we introduce a new parameter:

$$\boldsymbol{V} = \frac{|G|_{max} - |G|_{min}}{|G|_{max} + |G|_{min}} \quad (4)$$

Visibility, which define by Eq. (4) has been invesgated in both spectral and spatial view. Figure 4. is the spectral analysis of Gaussian beam. As shown in Fig 4. (c), the spectral visibility increases with "a". When "a" tends to infinity, the Gaussian beam becomes a plane wave, and the corresponding Spectral Visibility approaches to 1. Fig. 5 shows the spatial view of the Gaussian beam diffraction. Fig. 5 (a) shows the amplitude distribution at 0.5 THz in the plane y = 0 of Gaussian beam diffraction. Fig. 5(c) is the spatial visibility as a function of "R". The depth of the frequency gap first increases then decreases with R. It is can be seen that at R=6.4mm the best spatial visibility can be got.

Journal of Physics: Conference Series **1410** (2019) 012177 doi:10.1088/1742-6596/1410/1/012177

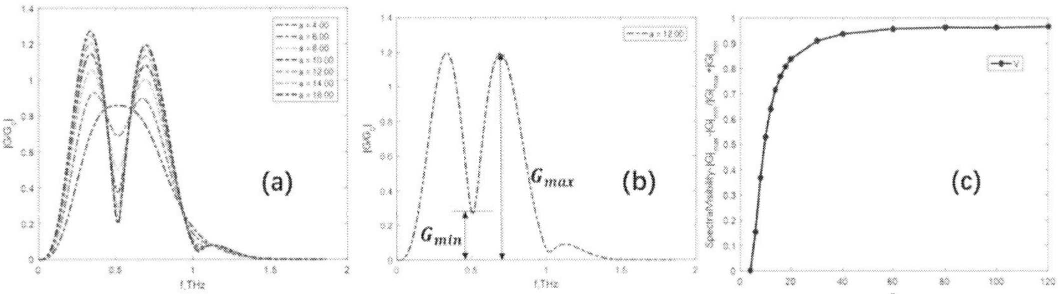

Figure 4(a,b,c). Spectral analysis of Gaussian beam **(a)** Spectral distribution at 0.5 THz at plane y=0 of one cycle Gaussian beam diffraction. **(b)** Spectral distribution of 0.5 THz of aperture R=6.4mm and Gaussian RMS width a=12. **(c)** Spectral visibility as the function Gaussian beam Gaussian RMS width.

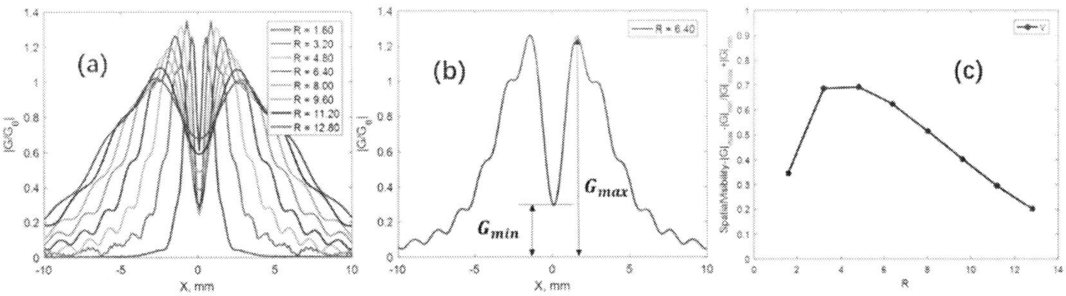

Figure 5(a,b,c). **(a)** Amplitude distribution at 0.5 THz at plane y=0 of Gaussian beam diffraction. **(b)** Amplitude distribution of 0.5 THz of aperture R=6.4mm. **(c)** Normalized amplitude gap as a function of the aperture

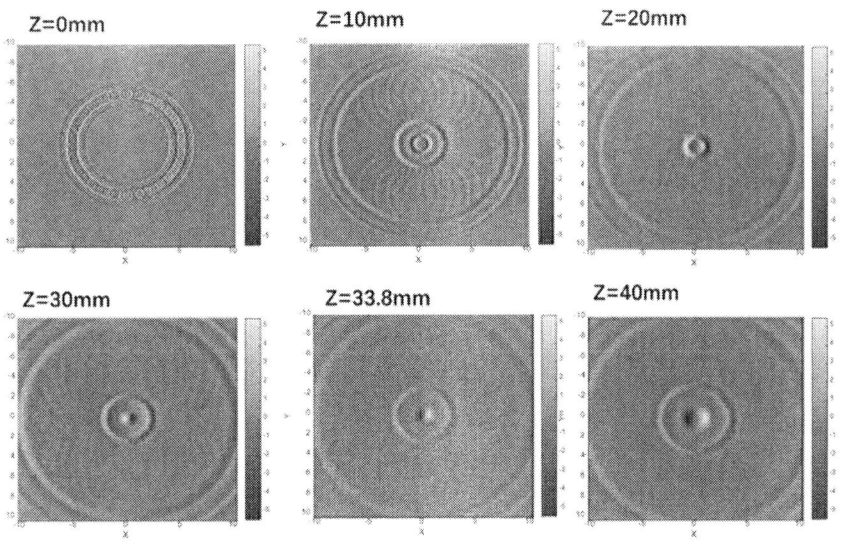

Figure 6. Longitudinal component of the electric field.

The numerical result of longitudinal component of the electric field of different distance utilizing Gaussian beam is shown in Fig.6. Diffraction phenomenon becomes weaker as the aperture radius increases, whereas the central beam energy increases.

4. Conclusions

In this paper, we have proposed a Huygens filter for a specified frequency lying within a broad spectrum of a pulsed THz radiation which represents two- or another even number of Fresnel zones plate. The condition of the Gaussian beam and plane wave are used as the initial source has been discussed. Spectral and spatial visibility have been investigated. The numerical result of longitudinal component of the electric field has been analysed as well. We have explored a feasible method of THz spectral modulation. This method may put forward a new point of view for THz modulation.

Acknowledgement

This work was supported by Russian Foundation for Basic Research № 17-00-00275 (17-00-00272).

References

[1]Thomas Kürner, Sebastian Priebe. "Towards THz Communications - Status in Research, Standardization and Regulation". - J Infrared MilliTerahz Waves, 2013.
[2]Hirata, A., T. Nagatsuma, T. Kosugi, et al., \10-Gbit/s wireless communications technology using sub-terahertz waves," Proc. SPIE, Terahertz Physics, Devices, and Systems II, Vol. 6772, 67720B,(2007).
[3]T. Nagatsuma, G. Ducournau, and C. C. Renaud, "Advances in terahertz communications accelerated by photonics," Nat. Photonics 10,371–379 (2016).
[4]V. Semenova and V. Bespalov, "Terahertz technologies," Photonics 51, 126–141 (2015).
[5] Y. V. Grachev, X. Liu, S.E. Putilin, A.N. Tsypkin, V.G. Bespalov, S.A. Kozlov, X.C. Zhang, Wireless data transmission method using pulsed THz sliced spectral supercontinuum IEEE Pho-tonic. Tech. L., 30, 103-106 (2018).
[6] M.Kulya, V. Semenova, A. Gorodetsky, V. G. Bespalov, N. V. Petrov, Spatio-temporal and spatiospectral metrology of terahertz broadband uniformly topologically charged vortex beams, Appl. Opt., 58, A90-A100 (2019)
[7] N.V. Petrov, M.S. Kulya, A.N. Tsypkin, V.G. Bespalov, A. Gorodetsky, Application of Terahertz Pulse Time-Domain Holography for Phase Imaging, IEEE Trans. Terahertz Sci. Technol., 6,464-472 (2016)
[8] M. I. B. Shams, Z. Jiang, S. M. Rahman, L.-J. Cheng, J. L. Hesler, P. Fay, and L. Liu, "A 740-ghz dynamictwo-dimensional beam-steering and forming antenna based on photo-induced reconfigurable fresnel zone plates,"IEEE Transactions on Terahertz Sci. Technol.7, 310–319 (2017).27.M. I. B.
[9] Shams, Z. Jiang, J. Qayyum, S. Rahman, P. Fay, and L. Liu, "A terahertz reconfigurable photo-induced fresnel-zone-plate antenna for dynamic two-dimensional beam steering and forming," in 2015 IEEE MTT-S InternationalMicrowave Symposium,(IEEE, 2015), pp. 1–4
[10] A. Koulouklidis, V. Y. Fedorov, and S. Tzortzakis, "Spectral bandwidth scaling laws and reconstruction of THz wave packets generated from two-color laser plasma filaments," Phys. Rev. A 93, 033844 (2016).
[11] Y. A. Kapoyko, A. A. Drozdov, S. A. Kozlov, and X.-C. Zhang, "Evolution of few-cycle pulses in nonlinear dispersive media: velocity of the center of mass and root-mean-square duration," Phys. Rev. A 94, 033803 (2016).
[12] V. G. Bespalov, A. A. Gorodetskii, I. Y. Denisyuk, S. A. Kozlov, V. N. Krylov, G. V. Lukomskii, N. V. Petrov, and S. E. Putilin, "Methods of generating superbroadband terahertz pulses with femtosecond lasers," J. Opt. Technol. 75, 636–642 (2008).
[13] A. Ezerskaya, D. Ivanov, V. Bespalov, and S. Kozlov, "Diffraction of single-period terahertz electromagnetic waves," J. Opt. Technol. 78,551–557 (2011).
[14] A. A. Ezerskaya, D. V. Ivanov, S. A. Kozlov, and Y. S. Kivshar, "Spectral approach in the analysis of pulsed terahertz radiation," J. Infrared Millim. Terahertz Waves 33, 926–942 (2012).
[15] A. A. Ezerskaya, D. V. Ivanov, V. G. Bespalov, S. A. Kozlov, "Diffraction of one-period terahertz electromagnetic waves with Gaussian transversal distribution,"
[16] Y. A. Shpolyanskiy, "Envelope, phase, and frequency of ultra-broadband signal in a transparent medium," J. Exp. Theor. Phys+. 111, 557–566 (2010)

Microdisk resonators as high-sensitive devices for biodetection

M V Fetisova[1], **A A Kornev**[1], **A S Bukatin**[1], **N A Filatov**[1], **N V Kryzhanovskaya**[1,2], **I V Reduto**[3,1,2], **E I Moiseev**[1], **M V Maximov**[1], **A E Zhukov**[1]

[1]St. Petersburg Academic University, St. Petersburg 194021, Russia
[2]Peter the Great St. Petersburg Polytechnic University, St. Petersburg 195251, Russia
[3]University of Eastern Finland, Joensuu 80101, Finland

Abstract. The single mode lasing of microdisks (MDs) with attached antibodies in aqueous medium is demonstrated. MDs comprise quantum dot active region emitting in the vicinity of 1.3 μm. The presence of antibodies on a MD surface leads to a number of effects, including the increase of threshold power of laser generation, likewise, changes in the linewidth of the resonance peak.

1. Introduction

The whispering gallery modes (WGMs) in microresonators are called by analogies with the acoustic waves propagating along the inner surface of the gallery in St. Paul's Cathedral's dome. WGM resonators support an ultra-small mode volume and extremely high *Q*-factor [1]. WGM sensors have demonstrated enviable sensitivity levels, thanks to the enhancing the interaction between light and matter effect [2], and thus have found a multitude of applications including low-threshold lasers [3] and biological sensing at a single molecule level [1], integrated optical devices [4,5], optical sensors [6–8] , as well as optomechanical [9,10] or nonlinear optics devices [11–13].

The major mechanism of detecting of single molecules by WGM biosensors is independent to a particular type of a biomolecule and is based on changes in wavelength and/or the linewidth of the resonance peak caused by a perturbation induced to a microresonator cavity by a particle. This effect is easy to detect and can allow one to simplify the laser construction in perspective. In passive WGM devices based on SiO_2 microspheres the possibility of biodetection was demonstrated [14]. The microsphere was located in the vicinity of an optical fiber, which is used for propagation of the external tunable laser radiation. The distance between the microsphere and the fiber should be extremely small (~ 100 nm) to ensure the interaction of the evanescent electromagnetic field of the WGM with the fiber mode. The interaction between the emission of a tunable laser propagating through the optical fiber is observed at frequencies corresponding to the WGMs of the microresonator. In the presence of biomolecules or gases on the microresonator's surface one can observe the frequency shift of the WGM and/or the *Q*-factor decrease. However, the relative position between the microsphere and the fiber must be precisely adjusted using a nanopositioning device, that increases the cost of the structure, its size and makes the system more complicated. The system can by simplified by using active resonator sensor configuration, where the fluorescence or a laser spectrum is collected by free-space optics making the sensing system more practical [2]. Semiconductor WGM resonators support strongly confined WGMs due to the large refractive index contrast between the resonator material and its surrounding. Lasers based on semiconductor WGM resonators represent a perfect platform for active small footprint high-sensitive devices for biodetection. Recently, we achieved room temperature lasing in optically pumped semiconductor MDs immersed in the aqueous medium [15].

In this work, we for the first time study the possibility to detect antibodies attached to MD surface via characterisation of the lasing parameters such as threshold power, resonator quality factor, etc. Specific biodetection is based on the analyte recognition by molecular receptors, which are protein molecules. For this purpose the MD surface is covered by receptor molecules that do not have significant effect on the resonator properties. Biomolecular detection is achieved by monitoring the binding of the analyte molecules to receptors.

2. Experiment details

An epitaxial structure was grown by molecular beam epitaxy on GaAs substrate. The active region comprises 5 layers of $InAs/In_{0.15}Ga_{0.85}As$ QDs separated by 30-nm-thick GaAs spacers. Microdisks were fabricated using photolithography and reactive ion etching. The outer diameter of MD laser was chosen to be 10.4 μm. Further, the $Al_{0.98}Ga_{0.02}As$ layer was transformed into an $(AlGa)_xO_y$ layer by the selective oxidation process to ensure the optical confinement from the substrate side. The MD surface was covered by 5 nm-thick SiO_2 layer using atomic layer deposition (ALD). This layer helps to adhere the proteins to a MD surface and does not influence the quality of resonator. MDs were tested after the SiO_2 layer deposition and no sign of lasing parameters degradation was observed. The sample with microdisk lasers was put into a microfluidic chamber made of PDMS by standard soft lithography technique [16].

The preparation of antibody-conjugated microlasers includes the following materials: N-Cetyltrimethylammonium bromide (CTAB), Tetraethylorthosilicate (TEOS), , and 3-(Trimethoxysilyl)propylmethacrylate (MPS), N-Hydroxysulfosuccinimide sodium salt (sulfo-NHS), N-(3-Dimethylaminopropyl)-N′-ethylcarbodiimide hydrochloride (EDC), 2-(N-Morpholino)ethanesulfonic acid (MES), phosphate-buffered saline (PBS) and TWEEN 20 were purchased from Sigma-Aldrich. We used commercially available chimeric monoclonal antibodies against the protein CD20 (rituximab) as a detecting layer on a microlaser's surface. To conjugate microdisk lasers with antibodies their surface was functionalized with COOH groups. After the ALD process microlasers were covered by functional groups using a sol-gel method with TEOS as a silica precursor and CTAB as a surfactant [17]. In brief, 0.12 g of CTAB was dissolved in 10 ml of 40% ethanol solution. Wafer with microlasers was put into the CTAB solution, 80 μl of 25% water ammonia solution and 160 μl of TEOS were added in a dropwise manner and stirred for 24 hours at room temperature. After that the wafer was washed 3 times in water and put into 10 ml of 40% ethanol solution with 0.5 ml of MPS and stirred for 12 hours at room temperature. The wafer was washed 3 times in ethanol. The anti-CD20 antibodies were conjugated on the surface of microdisk lasers via an EDC/NHS reaction [18]. Briefly, the wafer with functionalized microlasers was put into 0.5 ml of MES buffer (0.1M MES, 0.5M NaCl, pH 5,0) and 0.4 mg EDC was added. Anti-CD20 antibodies were dissolved in 0.5 ml PBS with 1.1 mg of sulfo-NHS. Final antibody concentration was 80 μg/ml. After 15 minutes antibody solution was added to microlasers solution and stirred for 2 hours at room temperature. Finally this solution was stirred for 12-24 hours at +4°C. Before the measurements the wafer was 3 times washed in PBS with 0.001% of TWEEN 20.

Next the chamber was filled with phosphate buffered saline (PBS) buffer with refractive index of 1.34 [19]. First the microlasers with primary antibodies were characterized at room temperature. For optical studies cw-operating YLF:Nd laser (λ=527 nm) was used to pump the MD by means of a confocal optical microscope (Ntegra Spectra, NT-MDT). An x100 Mitutoyo objective was used to collect the photoluminescence signal from a microlaser. The emission was detected by a monochromator MS5204i (Sol Instruments) and an InGaAs CCD-camera iDUS (Andor). To investigate microdisk lasing parameters before and after attachment of biomolecules on their surface in-house developed HRP conjugated mouse monoclonal antibodies to human IgG1 kappa (U5) and heavy (U4) chains were used as secondary antibodies (Fig. 1, a). Secondary antibodies at concentration 1 μg/ml in PBS with 0.001% TWEEN were introduced into the chamber with microlasers and kept for 30 minutes at room temperature. After that the chamber was washed by PBS with 0.001% TWEEN. Then the emission spectra and lasing parameters of microlasers were measured again for comparison. To visualize secondary antibodies Clarity Western ECL Blotting kit (BioRad) and ChemiDoc MP Imaging System (BioRad) were used.

3. Results

Fig. 1(b) represents the lasing spectrum obtained for a microresonator of 12 μm in diameter with secondary antibodies. The inset on Fig. 1(b) shows the dependence of the 1299 nm lasing line intensity on the optical pump power obtained for MD with primary antibodies and with secondary ones. Both dependencies demonstrate pronounced threshold behaviour. After coating the structure with secondary antibodies an increase in the threshold power from 1.36 mW to 1.8 mW was observed. The full width at half maximum of lasing line increased from 53 pm to 60 pm that can be explained by the additional losses introduced by secondary antibodies.

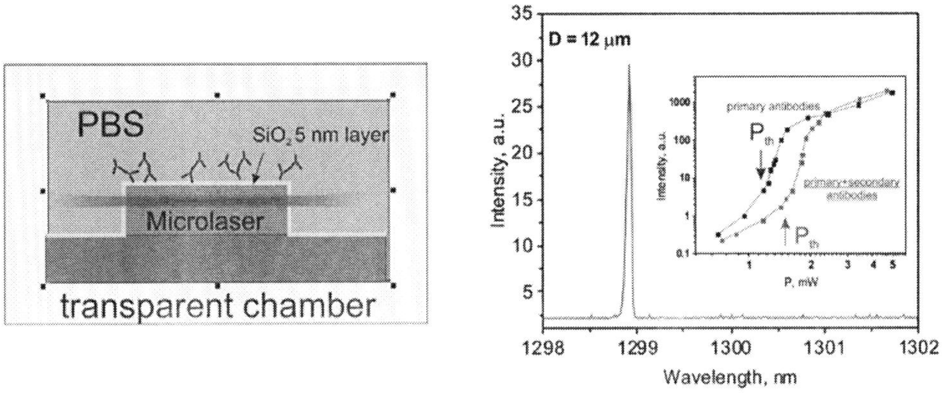

(a) (b)

Figure 1 (a,b). (a) A schematic image of the measured structure, (b) The room temperature lasing spectrum of the MD with secondary antibodies attached. Inset: laser emission intensity versus power. The lasing thresholds are marked with arrows.

To conclude, we demonstrated for the first time the possibility to detect the proteins with typical size of ~ 235×40 Å attached to the quantum-dot microdisk lasers immersed in phosphate buffered saline.

Acknowledgments
The work is supported by RFBR 18-02-00895, Program of fundamental studies of the Presidium of RAS and Russian Ministry of Education and Science (3.9787.2017/8.9 and 16.9790.2017/BCh).

References

[1] He L, Özdemir Ş K and Yang L 2013 Whispering gallery microcavity lasers *Laser Photonics Rev.* **7** 60–82
[2] Lu Q, Chen X, Fu L, Xie S and Wu X 2019 On-Chip Real-Time Chemical Sensors Based on Water-Immersion-Objective Pumped Whispering-Gallery-Mode Microdisk Laser *Nanomaterials* **9** 479
[3] Jiang X F, Zou C L, Wang L, Gong Q and Xiao Y F 2016 Whispering-gallery microcavities with unidirectional laser emission *Laser Photonics Rev.* **10** 40–61
[4] Yang K Y, Oh D Y, Lee S H, Yang Q F, Yi X, Shen B, Wang H and Vahala K 2018 Bridging ultrahigh-Q devices and photonic circuits *Nat. Photonics* **12** 297–302
[5] Levy J S, Gondarenko A, Foster M A, Turner-Foster A C, Gaeta A L and Lipson M 2010 CMOS-compatible multiple-wavelength oscillator for on-chip optical interconnects *Nat. Photonics* **4** 37–40

[6] Foreman M, Swaim J D and Vollmer F 2015 Whispering gallery mode sensors *Adv. Opt. Photonics* **7** 168–240
[7] Fan X, White I M, Shopova S I, Zhu H, Suter J D and Sun Y 2008 Sensitive optical biosensors for unlabeled targets: A review *Anal. Chim. Acta* **620** 8–26
[8] Yang J and Guo L J 2006 Optical sensors based on active microcavities *IEEE J. Sel. Top. Quantum Electron.* **12** 143–7
[9] Kippenberg T J . and Vahala K J . 2007 Cavity Opto-Mechanics *Opt. Lett.* **15** 17172–205
[10] Wang H, Dong C, Fiore V and Kuzyk M 2012 Optomechanical dark mode *Science (80-.).* **338** 1609–14
[11] Armani; D K, Kippenberg; T J, Spillane; S M and Vahala K J 2003 Ultra-high-Q toroid microcavity on a chip *Nature* **421** 925–8
[12] Jiang X, Shao L, Zhang S-X, Yi X, Wiersig J, Wang L, Gong Q, Lončar M, Yang L and Xiao Y-F 2017 Chaos-assisted broadband momentum transformation in optical microresonators *Science (80-.).* **358** 344–7
[13] Lu Q, Liu S, Wu X, Liu L and Xu L 2016 Stimulated Brillouin laser and frequency comb generation in high-Q microbubble resonators *Opt. Lett.* **41** 1736
[14] Ward J and Benson O 2011 WGM microresonators: sensing, lasing and fundamental optics with microspheres *Laser Photonics Rev.* **5** 553–70
[15] Fetisova M V., Kryzhanovskaya N V., Reduto I V., Moiseev E I, Blokhin S A, Kotlyar K P, Scherbak S A, Lipovskii A A, Kornev A A, Bukatin A S, Maximov M V. and Zhukov A E 2018 Room temperature lasing from microdisk laser in aqueous medium *J. Phys. Conf. Ser.* **1124** 051007
[16] Bukatin A S, Mukhin I S, Malyshev E I, Kukhtevich I V., Evstrapov A A and Dubina M V. 2016 Fabrication of high-aspect-ratio microstructures in polymer microfluid chips for in vitro single-cell analysis *Tech. Phys.* **61** 1566–71
[17] Yu Q Y, Hui J F, Wang P P and Wang X 2012 Anion-Exchange-Driven Disassembly of a SiO2/CTAB Composite Mesophase: the Formation of Hollow Mesoporous Silica Spheres *Inorg. Chem.* **51** 9539–43
[18] Qu W, Meng B, Yu Y and Wang S 2017 EpCAM antibody-conjugated mesoporous silica nanoparticles to enhance the anticancer efficacy of carboplatin in retinoblastoma *Mater. Sci. Eng. C* **76** 646–51
[19] Diéguez L, Darwish N, Mir M, Martínez E, Moreno M and Samitier J 2009 Effect of the Refractive Index of Buffer Solutions in Evanescent Optical Biosensors *Sens. Lett.* **7** 851–5

CWDM demultiplexer using anti-reflection, contra-directional couplers based on silicon nitride rib waveguide

E Zubkova[1,2], **A Golikov**[2], **P An**[2], **V Kovalyuk**[2,3], **A Korneev**[1,2,4], **S Ferrari**[5,6], **W Pernice**[5,6], **G Goltsman**[1,2]
[1]National Research University Higher School of Economics, Moscow 101000, Russia
[2]Department of Physics, Moscow Pedagogical State University, 119992, Russia;
[3]Zavoisky Physical-Technical Institute of the Russian Academy of Sciences, 420029, Russia;
[4]Moscow Institute of Physics and Technology (State University), 141700, Russia;
[5]Institute of Physics, University of Münster, 48149, Germany;
[6]CeNTech - Center for Nanotechnology, University of Münster, 48149, Germany;

Abstract. We report on the development and fabrication of a 9-channel coarse wavelength-division multiplexing for telecommunication wavelengths (1550 nm) using anti-reflection contra-directional couplers, based on silicon nitride (Si_3N_4) rib waveguide. The transmitted and reflected spectrum in each channel of the demultiplexer were measured. The average full width at half maximum of the transmitted (reflected) spectra is about 3 nm.

1 Introduction

Today, coarse wavelength-division multiplexing (CWDM) and dense wavelength-division multiplexing (DWDM) data transmission technologies are the basis for most fiber-optic communication lines [1]. Due to the active development of quantum photonic integrated circuits (QPICs), implementation of CWDM systems on a single chip represent a promising solution to obtain low-loss and compact building blocks which can be replicated on large scale. In this work we report on the development and fabrication of a 9-channel CWDM demultiplexer for telecommunication wavelengths using anti-reflection (AR), contra-directional couplers (CDCs) based on corrugated rib waveguide structure. The rib waveguide has been realized on silicon nitride platform, which combines low absorption at visible and IR wavelengths, a high refractive index ($n \approx 2$), good mechanical properties as well as CMOS compatible fabrication technology.

2 Device design and fabrication

The developed structure consists of a long curved comb waveguide with a periodic anti-phase Bragg grating (Fig. 1(a)). This waveguide is the main transmission line or the main channel in which the light propagates. The grating is not located along the entire length of the waveguide, but only in chosen areas. Along each of the sections, another comb waveguide with an anti-phase Bragg grating is located very closely. These waveguides do not intersect with the main channel and are the demultiplexer “escape” channels to which filtered (based on its wavelength) light is diverted. Together, both waveguides constitute a single CDC [2-3]. The grating periods (Δ) of both waveguides coincide in one channel, but vary for different channels. The designed and fabricated structure has 9 channels (sections). The period

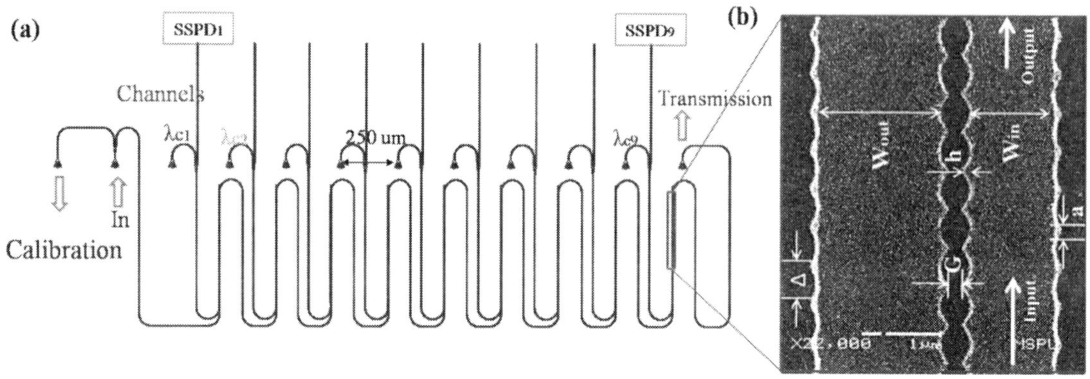

Figure 1 (a, b). (a) Schematic image of a fabricated 9-channel CWDM demultiplexer. **(b)** SEM image of CDC section based on the Bragg waveguide (Δ is a period of the Bragg waveguide, W_{in} is a width of the main waveguide, W_{out} is a width of the branch with reflected signal, h is a height of the teeth, a is a width of a tooth, G is a spacing between the teeth of the two waveguides).

increases by 0.625 nm. The length of each section is 1000 periods (Fig. 1(b)). Also on each outgoing channel there is a filtered output to the optical fiber with the help of a focusing grating coupler (FGC) [4] and a long section of the waveguide for the subsequent fabrication of superconducting single-photon detectors on the waveguide (WSSPDs) [5].

The range to be filtered out is determined by the parameters of the Bragg grating as

$$\lambda_c = \Delta \times 2n_{eff\_av}, \quad (1)$$

where λ_c is the central wavelength of the reflected spectrum, Δ is the Bragg grating period, n_{eff_av} is the average refractive index, respectively [2].

The structures were fabricated by means of electron-beam lithography using a positive resist ZEP 520A, providing good contrast during the lithography. NanoMaker software [6] was used for proximity effect correction and precise adjustment of the e-beam exposure time. Removing of the exposed material was done by reactive-ion-etching (RIE) in CHF_3-Ar mixture. The resist residuals were cleaned out with help of the oxygen plasma.

3 Experimental setup and obtained data

In order to characterize the fabricated structure, the total transmittance of the main channel of the demultiplexer and the reflection in each output channel were measured. Light from a tunable laser source through a polarization controller was brought into the structure using an optical fiber array, one of the outputs of which was located in front of the input focusing grating coupler (FGC) [4] structure, the other output of the array was coupled to the output FGC, which made it possible to measure the total transmittance of the structure. The other outputs of the optical fiber array were located above the output channel of the demultiplexer output channels; with the help of them the reflected light was recorded for each channel. In the measured transmission spectrum there is a signal corresponding to all of the 9 channels (Fig.2 (a-b)). The average FWHM of the output channel spectra is about 3 nm.

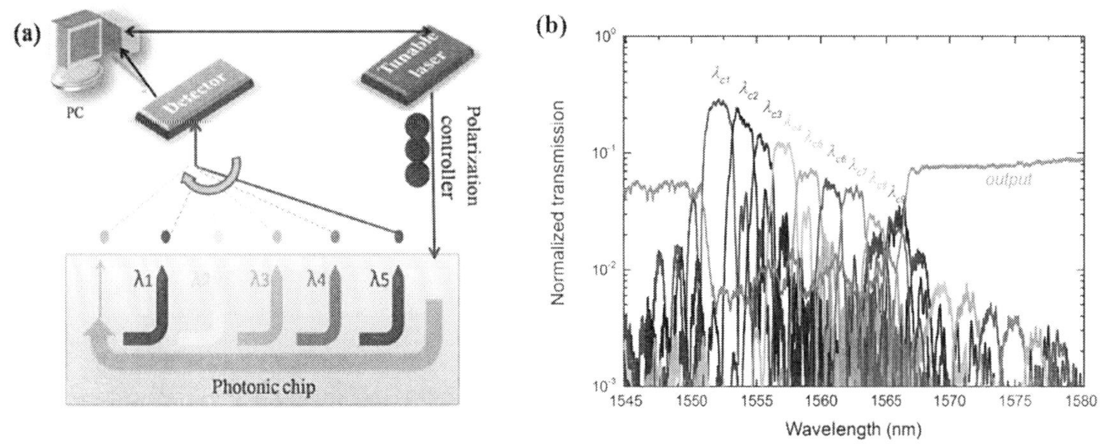

Figure 2 (a, b). (a) Schematic view of the experimental setup including: tunable laser source, multichannel photonic chip, low noise photodetector and PC. **(b)** Normalized transmitted and reflected spectra of the fabricated 9-channel CWDM demultiplexer.

4 Discussion

We observe that, in addition to the peak at the central wavelength (λ_c) of the spectrum of each channel, additional side-peaks at wavelengths ($\lambda_{r1,}$ and λ_{r2}) are present (Fig.3(a)). The presence of pronounced additional peaks results from a different behavior of the CDC demultiplexer on respect to the one observed from previously fabricated single CDC structures [3].

To explain this effect, we assume that the additional peaks at side wavelengths are associated with the non-uniformity of Bragg grating fabrication on the inner and outer faces CDC, that was due to the peculiarities of the technological processes. To confirm this assumption with the help of SEM image, we measured in detail the dimension of the Bragg grating teeth and carried out a numerical simulation of the effective refractive index n_{eff}. Figure 3b shows the result of simulation the effective refractive index for two different cases when the optical radiation power passes through a wider waveguide (TE_1, n_{eff_1}) and a less wide waveguide (TE_2, n_{eff_2}) [3].

In the general case, when the teeth outside the CDC are equal to the teeth inside, but shifted by half a period, a completely destructive interference of light reflected from different parts of the waveguide teeth occurs and no additional peaks are observed [2]. In our case the shift difference remains, but the teeth sizes on both sides of the waveguide do not match, only partial destructive interference of the reflected light occurs and additional peaks are seen corresponding to each effective refractive index in each waveguide (n_{eff_1} and n_{eff_2}) of the CDC. It was determined that additional peaks λ_{r1} and λ_{r2} at side wavelengths match n_{eff_1} and n_{eff_2} calculated for TE_1 and TE_2 modes, and λ_c match n_{eff_av} average effective refractive index.

Experimental and simulations results are in good agreement with the literature [2]), and further improvement of the demultiplexer is associated with both the improvement of the fabrication technology and the integration of superconducting single-photon detectors on a chip [5]).

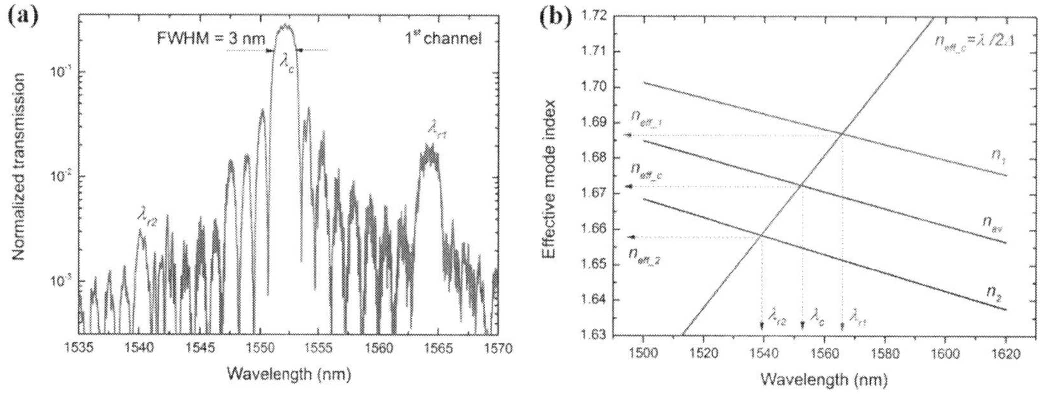

Figure 3 (a, b). (a) Normalized transmitted of a first channel CWDM demultiplexer. The main peak of the reflection in the channel (λ_c), an additional peaks in the channel (λ_{r1} and λ_{r2}). **(b)** Simulated effective refractive index of a first channel CWDM demultiplexer (the main peak of the reflection in the channel (λ_c) corresponds to n_{av}).

5 Conclusion

We fabricated a 9-channel CWDM demultiplexer for telecommunication wavelengths with a bandwidth of about 3 nm. We also report on the investigation of an additional peak in the transmission spectrum of each channel. To explain additional peaks at side wavelengths, a numerical simulation of the effective refractive index of the fabricated CDC was carried out. Further work will be directed to fabrication based on the fabricated waveguides of superconducting single-photon detectors fully integrated with the demultiplexer (WSSPD) [5].

Acknowledgments

We acknowledge support by Ministry of Education and Science of the Russian Federation No. 14.583.21.0065 (RFMEFI58317X0065) and thankful to Yu. Lobanov (MSPU) for helpful discussion and proof reading of the English version of the text.

References

[1] VlasovY 2012 *IEEE Communications Magazine* **50** 67-72

[2] Shi W, Yun H, Lin C, Greenberg M, Wang X, Wang Y, Fard S T, Flueckiger J, Jaeger N. A. F., Chrostowski L 2013 *Optics Express* **21** 6733–6738

[3] Zubkova E, An P, Kovalyuk V, Korneev A, Ferrari S, Pernice W, Goltsman G 2018 *IOP Conf. Series: Journal of Physics: Conf. Series* **1124** 051048

[4] Van Laere F, Claes T, Schrauwen J, Scheerlinck S, Bogaerts W, Taillaert D, O'Faolain L, Van Thourhout D, Baets R 2007 *Optics Express* **19** 21–24

[5] Kahl O, Ferrari S, Kovalyuk V, Goltsman G, Korneev A, Pernice W 2015 *Scientific Reports* **5** 10941

[6] http://www.nanomaker.com/

CMOS compatible nanoantenna-nanodiamond integration

S Komrakova[1], J Javadzade[4,7], V Vorobyov[3], S Bolshedvorskii[3,4], V Soshenko[3], A Akimov[3,7,6], V Kovalyuk[1,2], A Korneev[1,2,4], G Goltsman[1,2,5]
[1]Department of Physics, Moscow State Pedagogical University, 119992, Russia
[2]Zavoisky Physical-Technical Institute of the Russian Academy of Sciences, 420029, Russia
[3]P.N. Lebedev Physical Institute, 53 Leninskij Prospekt, Moscow, 119991 Russia
[4]Moscow Institute of Physics and Technology (State University), 141700, Russia
[5]National Research University Higher School of Economics, Moscow 101000, Russia
[6]Texas A & M University, 4242 TAMU, College Station, TX 77843, USA
[7]Russian Quantum Center, 100 Novaya St., Skolkovo, Moscow, 143025, Russia

Abstract. Here we demonstrate CMOS compatible method to deterministically produce nanoantenna with nanodiamonds systems on example of bull-eye antenna on top of on hyperbolic metamaterials. We study the statistics of the placement of nanodiamonds and measure the fluorescence lifetime and the second-order correlation function of NV-centers inside nanodiamonds.

1 Introduction

Quantum technology is a relatively young field that manipulates at the atomic or microscopic level, based on the laws of quantum mechanics. For the practical implementation of quantum technology, photon sources are of interest, since photons are ideal information carriers. The ideal source of photons is a device that emits one photon on demand. A promising candidate for single-photon source is a nanodiamond with a nitrogen-vacancy (NV) color center. Nitrogen-vacancy-center nanodiamonds will emit single photon after excitation. Previously nitrogen vacancies (NVs) shown to be one of the most promising quantum system, but their optical characteristics (the generation rate and directivity) can be significantly improved. One of the way to enhance optical properties of NV is its positioning on hyperbolic metamaterials (HMM), but in this case HMM mostly increase radiation into high *k*-modes which are absorbed inside the material. One of the way to change it is to use nanoantennas [2]. We were pursuing this way in order to outcouple high *k*-modes into the free-space.

In this paper we continue our previous study of an controlled on-chip NV placements [1] and demonstrate CMOS compatible method to deterministically produce nanodiamonds and a bull-eye nanoantenna.

2 Device design and fabrication

At the beginning of the integration process nanoantennas and nanodiamonds a TiN layer of 60 nm thick on a HMM using reactive magnetron sputtering was deposited. We used a spin-coated negative e-beam resist (ma-N 2403) spin coated for 30 seconds with a speed of 3000 rpm. The resulting layer was dried on a hot plate at a temperature of 90 °C for 1 minute, providing resist thickness approximately 270 nm. After e-beam lithography drawing bull-eye antennas, we developed the resist in ma-D 525 and stopped reaction in water during two minutes. The etching of TiN was carried out with reactive ion etching (RIE) procedure in SF_6 atmosphere with a plasma power of 50 Watts. Images of four nanoantennas obtained with an atomic force microscope (AFM) can be seen in Fig. 1(a). To determine the location for the subsequent positioning of nanodiamonds, we used spin-coated 100 nm thick polymethyl methacrylate (PMMA) positive resist and baked at 150 °C for 2 minutes on a hot plate. Using e-beam lithography, we created an array of 33 by 33 holes with a diameter of 170 nm. Each hole we combined with the center of the corresponding nano-antennas. The accuracy of the resulting alignment ranges from 5 to 50 nm in different arrays. Then we developed the chip in a solution of isopropanol and water (8:1) for 20 seconds (Fig. 1 (b)). The next step, that is illustrated in

Fig. 1 (c), we put a drop of nanodiamonds solution and left it to incubate for 30 minutes. To avoid unnecessary particles, we first removed a drop of the solution

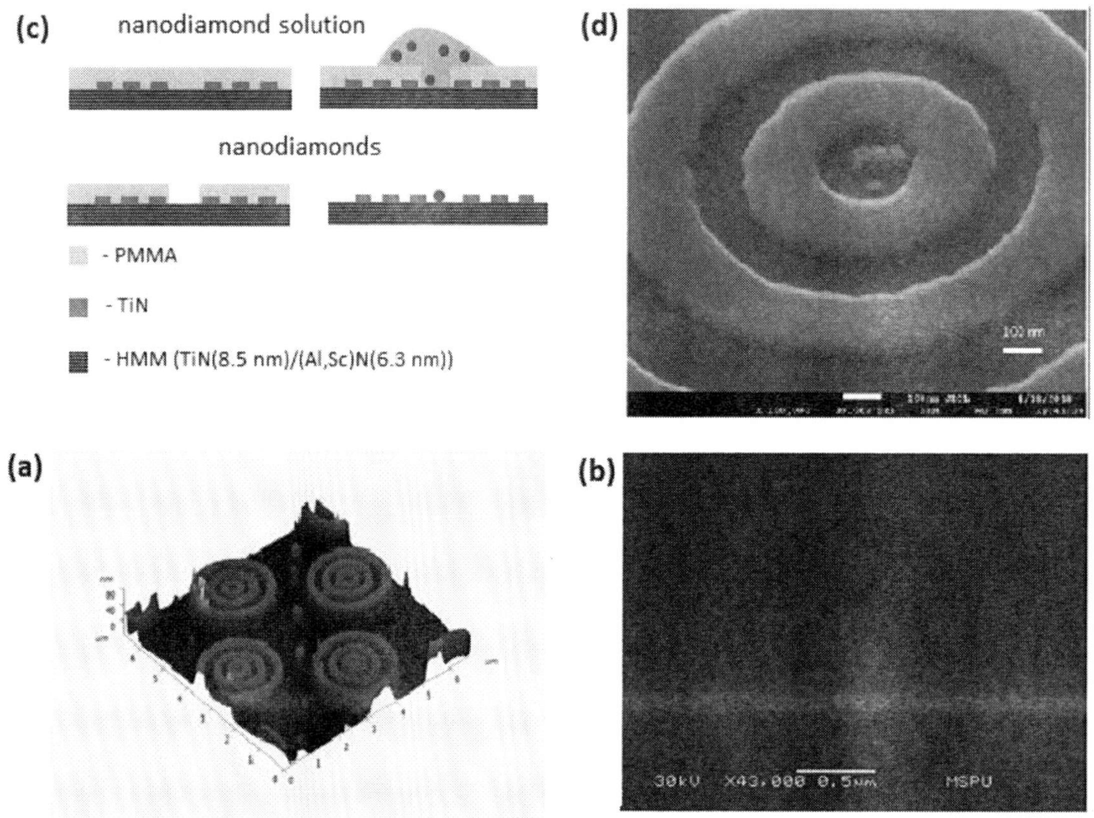

Figure 1 (a-b). (a) AFM image of four nanoantennas. **(b)** Scanning electron microscope (SEM) image of device after development of e-beam PMMA resist. **(c)** Schematic view of the process of controlled placement of nanodiamonds inside nanoantennas. **(d)** Scanning electron microscope image of nanodiamonds inside single nanoantenna.

with a pipette and sent a stream of isopropanol on the sample. Finally, for PMMA removing, we washed chip by an acetone stream heated to 55 °C, followed by washing in isopropanol and drying in stream N_2.

One of the fabricated devices is shown in Fig. 1 (d). Concentric circles represent a nanoantenna, in the center of which there is a nanodiamond with an NV center. The average size of nanodiamons is 50 nm. Also, only every eighth nanodiamond in solution has a NV-center, which helps to create a single-photon source, even if there are a lot of nanodiamonds left after removing PMMA.

3 Experimental study and results

The first stage of research is devoted to the study of the effectiveness of the placement of nanodiamonds inside the nanoantennas. To count placement efficiency we used a Scanning Electron Microscope (SEM) images (one of them is shown in Fig.2 (a)). Totally, we investigated 700 on-chip nanoantennas, which were classified into 3 categories:

1) single diamond in the center of the nanoantenna (42.8%);
2) single diamond in the center of the nanoantenna is absent (33.5%);

3) diamonds are not located in the center of the antenna or located as agglomerate of nanodiamonds (23.7%).

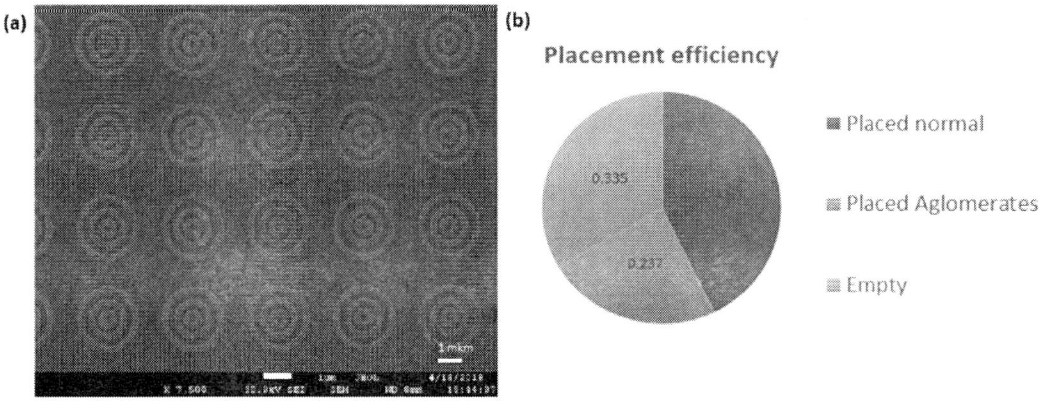

Figure 2 (a-b). (a) AFM image four nanoantennas; **(b)** Scanning electron microscope (SEM) image of device after development of e-beam PMMA resist.

Diagram of placement efficiency is shown in Fig.2 (b). It should be noted that we cannot reliably distinguish nanodiamond from dust particles of a similar size without confocal technique.

The second phase of the study is devoted to the study of nanodiamonds using confocal microscopy.

To analyze optical properties of NV-centers on the sample, we used home built confocal microscope employing high numerical aperture immersion oil objective (NIKON Apo TIRF 100X NA 1.49) with working distance of 120 μm (Fig. 3(a)). A continuous wave laser (Coherent Compass 300, wavelength 532 nm) served as the source of NV excitation and galvo mirrors (Cambridge Technologies) were used as a scanning element, enabling a 100 × 100 μm field of view for our microscope design (Fig. ° 1(b)). The collected fluorescence was coupled into a Hunbury-Brown-Twiss (HBT) interferometer that consisted of two avalanche photodiodes (PerkinElmer SPCM-AQRH-14-FC) and a 45:55 beam splitter. We used the combination of an optical notch filter with a stop-band centered at 532 ° nm and longpass optical filter with cut-off at 600 nm to remove the residual green excitation light and Raman signal from the collected emission. A time-correlated single photon counting module (Picoquant Picoharp 300) was used to obtain second-order photon correlation histograms from NV-centers (Fig. 4(b)).

For lifetime measurements, a picosecond diode laser (Picoquant LDH-P-FA-530 XL) was used with the same time-correlation single photon counting module. Due to the fact that lifetime of NV-center excited state is much longer than the length of the laser pulse, the fluorescence signal recorded after the end of the applied laser pulse should exhibit exponential decay (Fig. 4(a)). In Fig. 4(a), the actual signal observed consist of two exponents – fast and slow one. The fast exponential component corresponds to fluorescence of the sample and other similar background fluorescence signatures that do not have a slowly decaying exponential component. This was verified by focusing the microscope on a region of the samlpe that contained no diamonds and measuring the fluorescence decay time collected from this region. We therefore associate long-lived exponential tail in Fig. 4(a) with the NV-center luminescence.

Confocal microscope allow distinguish nanoantenna with NV-centers nanodimonds in the antennas array (Fig. 3(b)) and investigate the NV-centers optical properties. The orange circle shows the single NV-center nanodiamond (Fig 4(b)) inside the bull eye antenna. Figure 4 (b) demonstrates second-order autocorrelation function measurements deep $g^2(0) \approx 0.3$ which associated with single NV-center. Figure 4(a) demonstrates lifetime measurements of 8.3 ns for single NV-center inside bull eye antenna.

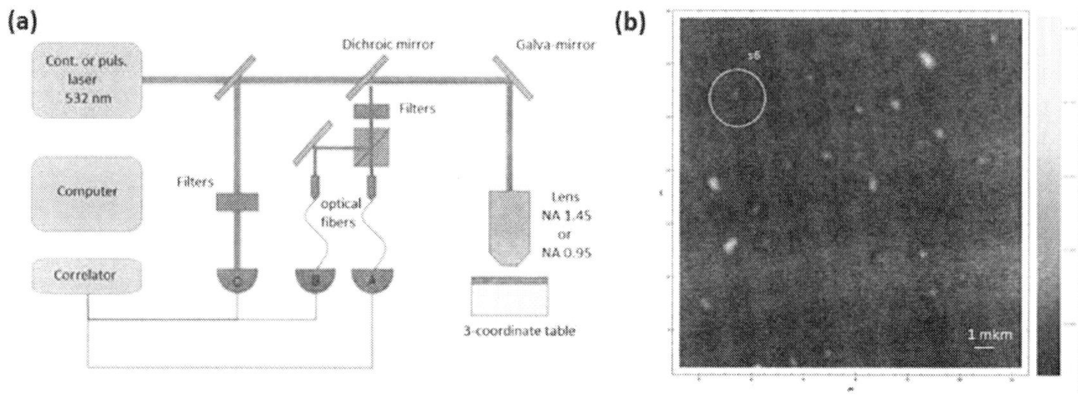

Figure 3 (a-b). **(a)** Scheme of the confocal microscope; **(b)** CLSM image of array of nanodiamonds inside nanoantennas.

Totally, we investigate 11 single NV-centers on glass and 5 single nanodiamonds on HMM nanostructures. The average lifetime for NV on nanostructures was 8.5 ns, while on glass it is 22.5 ns.

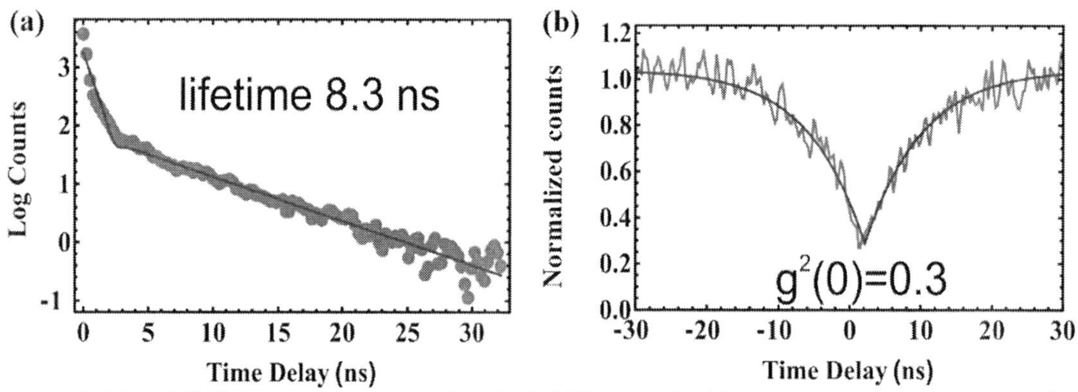

Figure 4 (a-b). **(a)** Lifetime measurements for single NV-center inside nanoantenna; **(b)** second-order autocorrelation function $g^2(\tau)$ for single NV-center.

Thus, the placement technology gives satisfactory results. At the same time, the enhancement of the radiation of nano-amps with antennas in the metamaterial is ambiguous, since only 5 cases were investigated there. The investigation of the antennas efficiency requires further research

4 Conclusion

We demonstarted nanodiamond with color centers integrated into the CMOS process to create integrated optical circuits. The placement technology gives satisfactory results when 42.8% of nanodiamonds were placed as desired. Further research will be directed to study optical and spin properties of NV [2-4] inside nanoantennas. Moreover, nanodiamonds with single NV-centers were found and their optical were characterized.

Acknowledgments
S. Komrakova, A. Korneev and V. Kovalyuk acknowledge support by Ministry of Education and Science of the Russian Federation № 14.586.21.0063 (RFMEFI58618X0063).

References

[1] S. Komrakova S, Javadzade J, Vorobyov V, Bolshedvorskii S, Soshenko V, Akimov A, Kovalyuk V, Korneev A, Goltsman G 2018 *J. Phys. Conf. Ser.* 1124, 051046, 1–4
[2] Galfsky T, Krishnamoorthy H N S, Newman W, Narimanov E E, Jacob Z, Menon V M 2015 *Optica* **2** 1
[3] Bermu'dez-Uren E, Gonzales-Ballestero C, Geiselmann, Marty R, Radko I P, Holmgaard T, Alavedyan Y, Moreno E, Garci'a-Vidal F J, Bozhevolnyi S I, Quidant R 2015 *Nature communications* **6** 1
[4] Shalaginov M Y, Vorobyov V V, Liu J, Ferrera M, Akimov A V, Alexei Lagutchev A, Smolyaninov A N, Klimov V V, Irudayaraj J, Kildishev A V, Boltasseva A, Shalaev V M 2015 *Laser Photonics Review* **9** 1

Efficiency of focusing grating couplers versus taper length and angle

A Kuzin[1,2], V Kovalyuk[1,3], A Golikov[1], A Prokhodtsov[1,2], A Marakhin[4,5], S Ferrari[6,7], W Pernice[6,7], N Gippius[4], G Goltsman[1,2,3]
[1]Department of Physics, Moscow State Pedagogical University, 119992, Russia
[2]National Research University Higher School of Economics, Moscow 101000, Russia
[3]Zavoisky Physical-Technical Institute of the Russian Academy of Sciences, 420029, Russia
[4]Skolkovo Institute of Science and Technology, 121205, Russia
[5]Moscow Institute of Physics and Technology, 141701, Russia
[6]Institute of Physics, University of Münster, 48149, Germany
[7]CeNTech - Center for Nanotechnology, University of Münster, 48149, Germany

Abstract. Here we experimentally studied dependence of a focusing grating coupler efficiency versus taper length and angle on silicon nitride platform. As a result, we obtained a dependence for the efficiency of a focusing grating coupler on the parameters of the taper length and angle.

1. Introduction

Today one of the most pressing problems of integrated optics is the search of ways for improving the coupling between optical fibers and waveguide structures [1]. In modern photonics, the most widely adopted methods for input and output optical radiation from a chip are based on end-face (or butt-coupling), focusing grating couplers (FGCs) [2] as well as near-adiabatic-mode-converters [3] approaches. Due to capabilities to study of hundreds of on-chip devices, ease of alignment and fabrication, as well as flip-chip technology [4] the last approach is the most popular and a large amount of research has been done to improve the efficiency of FGC, including geometry analysis [5], apodization [6], directionality [2] and so on. Additional solutions for efficiency improvement are the use of different materials [7] and individual layers [8, 9]. In addition, there are several works in literature report on studies of the FGC coupling efficiency with different geometrical solutions [10, 11, 12].

In this work, we focus on the study FGC coupling efficiency dependence on two geometrical parameters: the taper length (L) and taper angle (α).

2. Device designing and fabrication

We designed a 2D array of devices with two sweep parameters: L = 10÷30 μm and α = 30 ÷ 60 degrees (Figure 1(a)). The total number of devices in the array was 225. The typical element of the array consisted of two FGCs for input and output light, connected by rib waveguide bus with 1μm width for all the devices (Figure 1(b)). The FGC filling factor [13, 14] was fixed and equal to 0.33±.

Commercially available Si wafers with 450 nm Si_3N_4 surface as an optical layer and SiO_2 as an optical buffer between Si and Si_3N_4 were used. The fabrication process consisted of one e-beam lithography step. First, ZEP 520A resist was deposited by spin coating procedure. Then, the image of the structure was drawn in the resist using an electron beam lithography by CABL-9000C and resist was developed in O-Xylene, protecting covered silicon nitride under reactive ion etching (RIE) in Ar and CHF_3 gas mixture. Finally, using oxygen plasma the residual resist was removed.

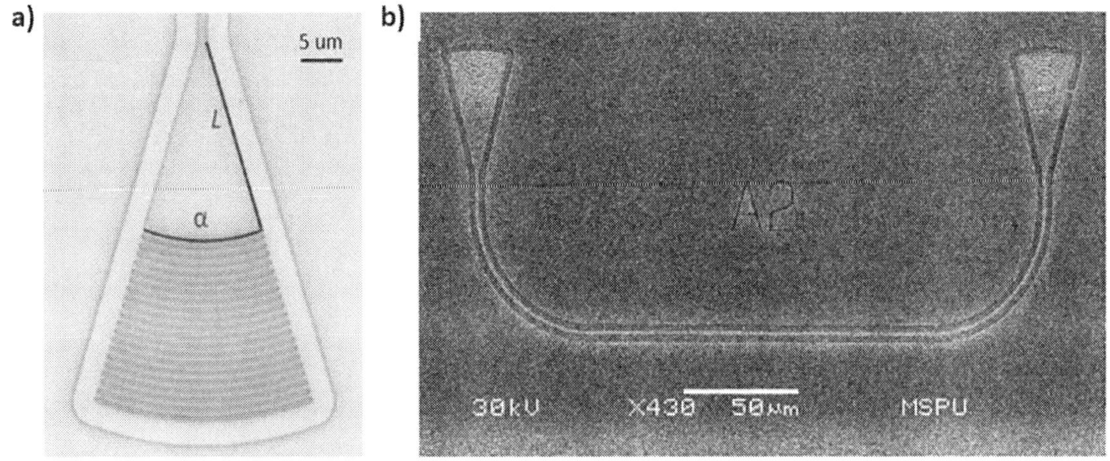

Figure 1 (a, b). (a) Optical micrograph of one of the fabricated focusing grating couplers (FGCs), where the taper length (L) and the taper angle (α) are highlighted by color; **(b)** Scanning electron microscope image of a nanophotonic device with two FGCs and the waveguide.

3. Experimental setup and results

For testing of nanophotonic devices in a spectral range of 1510 ÷ 1620 nm we used a tunable laser source (New Focus TLB-6600) and a fast photodetector (Hamamatsu G9801), controlled by a NI DAQ system.
We used a precision (X, Y, Z, angle) stage with piezoelectric picomotors (NewFocus 8303) and an optical microscope to align the nanophotonic structures with the two-channels optical fiber array.

Typical transmission spectra of two nanophotonic devices with different taper angles ($\alpha = 32°$ for black solid line, and $\alpha = 60°$ for red solid line) at a fixed taper length of $L = 22.86$ μm are shown in Figure 2a. We used a Gaussian fit (blue and green lines) to extract the output power at the central wavelength.We measured all 225 devices on the chip and found that the maximum of the spectra was distributed at the central wavelength $\lambda_c = 1543 \pm 1.55$ nm, demonstrating a good reproducibility of the period and the fill factor of FGCs. Optical output power (P_{out}) for fabricated nanophotonic devices can be found as:

$$P_{out} = P_{in} \times C^2 \times T_{wg}, \tag{1}$$

where P_{in} is input power, C is coupling efficiency and T_{wg} is the waveguide optical transmission. Considering the optical losses, previously found for the same nanophotonic architecture from the data obtained by O-ring resonators (< 1 dB/cm) [14], losses can be neglected ($T_{wg} \approx 1$) and coupling efficiency can be found as:

$$C = \sqrt{\frac{P_{out}}{P_{in}}}. \tag{2}$$

In Figure 3(b) we plot the experimentally obtained color contour map of the maximum FGC efficiency at a central wavelength λ_c at different L and α. The highest FGC efficiency corresponds to areas with a taper length from 21 to 23 μm and a taper angle from 33 to 36 degrees with the highest efficiency value equals to $C = 0.18$.

4. Discussion

In general, the coupling efficiency depends on phase and amplitude matching. The phase matching can be maximized due to the period of the diffraction grating as well as amplitude matching is determined by the compliance of the optical spot from the fiber (D) and the working area of the FGC (Figure 3(a)). This can be achieved with different configurations of taper length L and taper angle α:

$$S = \frac{D}{2} \cdot \cot\frac{\alpha}{2}, \quad (3)$$

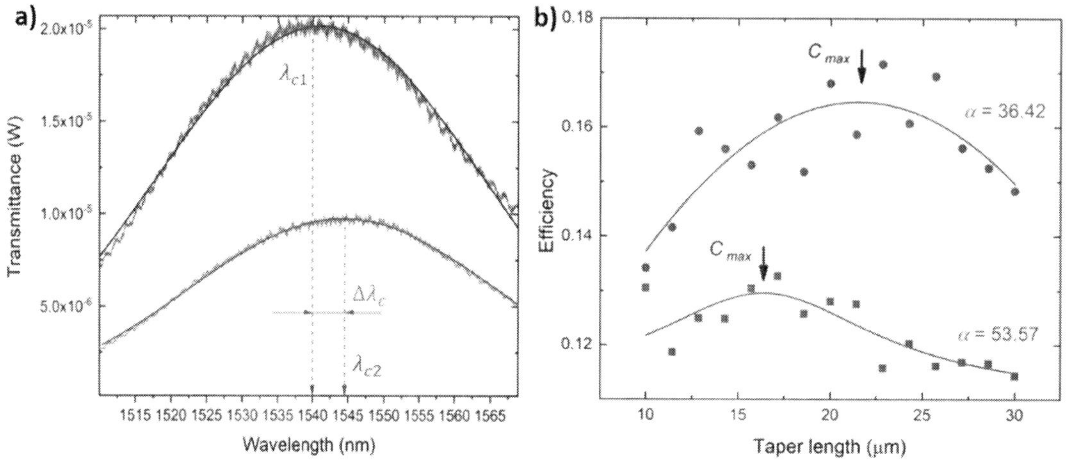

Figure 2 (a, b). (a) Measured transmission spectra of two focusing grating couplers (FGCs) with different geometrical parameters: L = 22.86 μm and α = 32° (black solid line), L = 22.86 μm and α = 60° (red solid line). Blue and green lines corresponds to Gaussian fit. **(b)** Dependence of the coupling efficiency on the taper length at different taper angle (α = 36.42°, 53.57°). Dots and squares are experimental data as well as solid lines correspond to Lorentz fit The arrows show the maximum efficiency (C_{max}) for each curve.

where $S = L + \Delta$ is the distance from the end of taper to the point where the output power has the maximum value and where optical fiber should be placed.

To analyze the experimental results, we plotted the dependence of the coupling efficiency on the taper length and fitted it by a Lorentz function (Figure 2 (b)). If the angle is fixed, then the efficiency reaches its maximum value (C_{max}) at a certain taper length. When the angle increases, the efficiency maximum shifts towards to shorter taper lengths.

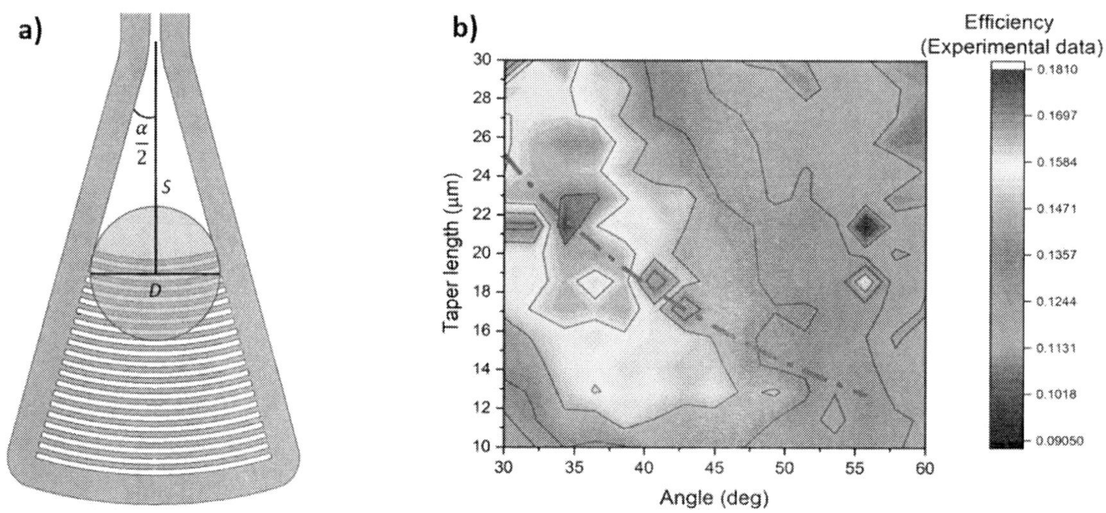

Figure 3 (a, b). (a) Focusing grating coupler design: D is mode field diameter of an optical fiber SMF-28, α is the taper angle and S is the distance from the end of taper to the point where the output power has the maximum value. **(b)** Color contour map of the efficiency on taper length and taper angle with fitted dependences of C_{max} (L) (purple dash-dot line).

Since the phase matching is the same for all the devices in the array (the period and the filling factor do not change), this behavior can be explained by the fact that for a given angle and length of the taper, amplitude matching is performed in the best way. In this case, the waveguide mode should fit completely into the FGC geometric region, and the maximum of the light scattering by the grating should coincide with the center of the optical fiber mode.

To test our guess, we extracted the dependences of C_{max} (L) and fitted it using equation 3. The result of the fit (purple dash-dot line) is shown together with the experimental data (Fig. 3b). The extracted optical spot dimension (D) data from the experimentally obtained data correspond to 13 μm, which is close to the modal diameter of the telecommunication fiber SMF-28 (D = 10.4±0.5 μm). We associate the increase in diameter due to the final height of the fiber above the FGC. Considering the numerical aperture NA = 0.14, the height of the optical fiber above the coupling can be reasonably estimated as 10 μm.

5. Conclusions

In conclusion, we studied dependence of grating couplers efficiency vs the taper length (L) and taper angle (α) for central wavelength λ_c = 1543 ± 1.55 nm. It was found that the maximum efficiency of FGCs for fixed λ_c efficiency is achieved at low taper angles and long lengths. Further work related to 3D modeling of FGCs in COMSOL Multiphysics, apodization in the FGC structure [15, 16] integrating the FGCs into resonator and antireflection layers can allow to significantly improve the coupling efficiency.

Acknowledgments
G. Goltsman and V. Kovalyuk acknowledge support by Ministry of Education and Science of the Russian Federation №14.583.21.0065 (RFMEFI58317X0065).

References

[1] Kopp C, Augendre E, Orobtchouk R, Lemonnier O, Fedeli J-M 2011 *Journal of lightwave technology* **29** 12
[2] Vermeulen D, Selvaraja S, Verheyen P, Lepage G, Bogaerts W, Absil P, Van Thourhout D, and Roelkens G 2010 *Opt. Express* Vol. **18** 17
[3] Gehring H, Blaicher M, Hartmann W, Varytis P, Busch K, Wegener M, Pernice W H P 2019 *APL Photonics* **4** 010801
[4] Soganci I, Porta A, Offrein 2013 *Optics Express* **21** 13
[5] Brazas J, Li L 1995 *Applied Optics* **34** 19
[6] Chen X, Li C, Fung C K Y, Lo S M G, Tsang H K 2010 *IEEE Photonics technology letters* **22** 15
[7] Scheerlinck S, Schrauwen J, Van Laere F, Taillaert D, Van Thourhout D and Baets R 2007 *Opt. Express* **15** 15
[8] Roelkens G, Van Thourhout D and Baets R 2006 *Opt. Express* **14** 24
[9] Taillaert D, Van L F, Ayre M, Bogaerts W, Van Thourhout D, Bienstman P and Beats R 2006 *Japanese Journal of Applied Physics* **45** 8A
[10] Chen X 2009 *IEEE Photonics Journal* **1** 3
[11] Chen X and Tsang H.K 2011 *Opt. Letters* **36** 6
[12] Mekis A, Gloeckner S, Masini G, Narasimha A, Pinguet T, Sahni S, De Dobbelaere P 2011 *IEEE journal of selected topics in quantum electronics* **17** 3
[13] Sure A, Dillon T, Murakowski J, Lin C, Pustai D and Prather D W 2002. *Opt. Express* **11** 26
[14] An P, Kovalyuk V, Golikov A, Zubkova E, Ferrari S, Pernice W, Korneev A, Goltsman G 2018 *J. Phys. Conf. Ser* **1124** 051047
[15] Soref R 2006 *IEEE J. Sel. Top. Quantum Electron* **12** 6
[16] Halir R, Cheben P, Janz S, Xu D-X, Molino-Fernandez I and Wanguemert-Perez J. G 2009. *Opt. Lett.* **34** 9

The impact of terahertz chiral metasurface resonator rotation on polarization properties of transmitted wave

M. S. Masyukov[1], A V Vozianova[1], A N Grebenchukov[1], M K Khodzitsky[1]
[1]Terahertz Biomedicine Laboratory, Department of Photonics and Optical Information Technologies, ITMO University, St. Petersburg 199034, Russia

Email:msmasyukov@itmo.ru

Abstract. In this paper the influence of one and two chiral metasurfaces on polarization of the transmitted wave has been studied. The numerical simulation in CST Microwave Studio confirms the influence of the resonator rotation in the unit cell on polarizing properties of the metasurface. The ability of changing the polarization of the terahertz wave by two layers of such metasurfaces was shown. Thus, such metasurface can be used in development of mechanically-tunable polarization converter.

1. Introduction

Metamaterials is a class of composite structured materials with specific properties which cannot be found in nature. Usually these materials consist of unit cells, or meta-atoms, consequently the properties of metamaterial depend on the array of such unit cells. The ability of transformation of the unit cell allows to design and scale metamaterials for the necessary frequency range. Applications of metamaterials in terahertz frequency range attract more and more attention due to an ability to cover a lack of simple and chip elements and systems for terahertz frequencies, such as multispectral imaging systems[1-2], magneto-tunable devices[3-6], tunable reflectors[7], switchers[8], filters[9-11], phase shifters[12], perfect absorbers[13], left-handed media [14-15], polarization converters[16-18], mirrors, etc. Such problem is being solved by using chiral metamaterials. In this work we studied the possibility of the transmitted wave polarization changing by using one and two layers of chiral metasurfaces with different angle position of the chiral resonators. The results of the research can be used for THz polarimetry biomedical diagnostics [19-20].

2. The structure under the investigation

The sketch of the unit cells of metasurfaces under study is shown in Figure 1-a). The resonant rosette consists of four planar aluminium half-rings rotated around the normal to the unit cell plane at its central point. The outer radius Rmax of the rosette petals is 150 μm, the inner one is 125 μm. The rosette is patterned on a dielectric (ε=3.5) substrate, which thickness is h=65 μm and the side size is a=600 μm. The angle α corresponds to the angle of rotation between the y-axis and the edge of the rosette.

In this paper, four different cases have been studied: α_1=0 degrees, α_2=15 degrees, α_3=30 degrees, α_4=45 degrees (Fig.1-b). The metasurfaces with angle of chiral element rotation α_1=0 degrees is considered to be a basic one. Also four types of metasurface stacks created by two metasurfaces under each other were studied (Fig. 1-c). In this case, the top metasurfaces chiral element was rotated by the same angles α_1=0 degrees, α_2=15 degrees, α_3=30 degrees, α_4=45 degrees, while the bottom metasurface was the basic one (α_1=0 degrees). The air gap between two metasurfaces was d=5μm.

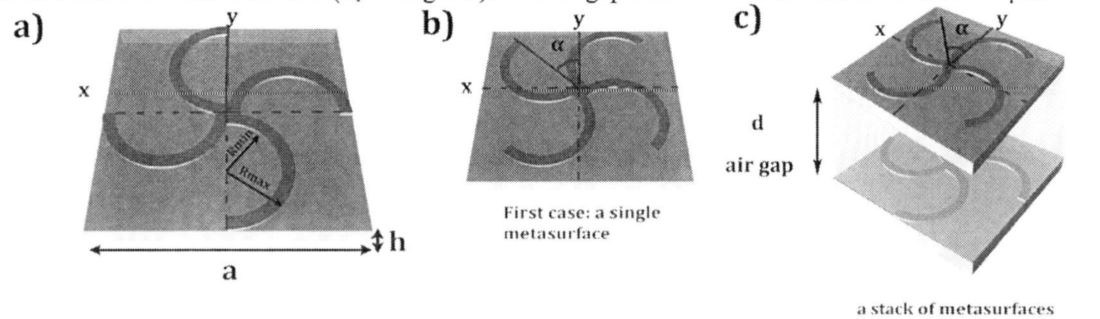

Figure 1. The unit cell of the metasurface: a) the original unit cell, b) rotated resonator in single metasurface; c) stack of two metasurface.

3. Numerical simulation and calculation approach

The numerical simulations were performed in CST Microwave Studio based on Finite Difference Time Domain (FDTD) method. The linearly polarized waves from Port 1 become elliptically polarized ones after the propagation through the metasurface (Fig.2), and then they are received by Port 2. (Fig. 2). To characterize the polarization state of the transmitted wave the co- and cross-polarization spectra T_{xx} and T_{xy} were found.

Transmission spectra (matrix $\mathbf{T}_{cir}$) for right-handed circularly waves (T_{++}) and left-handed circularly polarized waves (T--) can be found using the next formula [21]:

$$\boldsymbol{T}_{cir} = \begin{pmatrix} T_{++} & T_{+-} \\ T_{-+} & T_{--} \end{pmatrix} = \begin{pmatrix} T_{xx} + iT_{xy} & 0 \\ 0 & T_{xx} - iT_{xy} \end{pmatrix}, \quad (1)$$

where T-+ = T+-=0 due to four-fold symmetry of the unit cell.

After that, the ellipticity angle η is used to evaluate the transmitted wave polarization changes:

$$\eta = \frac{1}{2} sin^{-1}(\frac{|T_{++}|^2 - |T_{--}|^2}{|T_{++}|^2 + |T_{--}|^2}), \quad (2)$$

The ellipticity angle shows a type of polarization: thus, the value of |η|=45 degrees corresponds to the circular polarization of the electromagnetic waves, and sign "+" or "-" before the value refers to the right-handed or left handed type of polarization state respectively; values between η=±45 degrees are related to elliptical polarization state, except the case when η=0 for linear polarization.

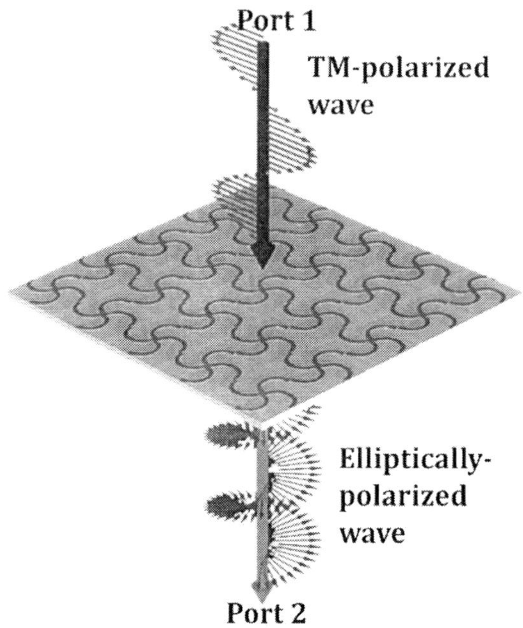

Figure 2. Scheme of numerical calculation.

4. Results

The ellipticity angle spectra for the single metasurface are shown in Fig. 3.It can be noticed that there are few resonance frequencies, i.e. frequencies where the ellipticity angle is not equal to zero and has extreme value (maximal or minimal). For each case of metasurfaces design the resonant frequencies are almost the same, meanwhile the ellipticity angle changes its values.

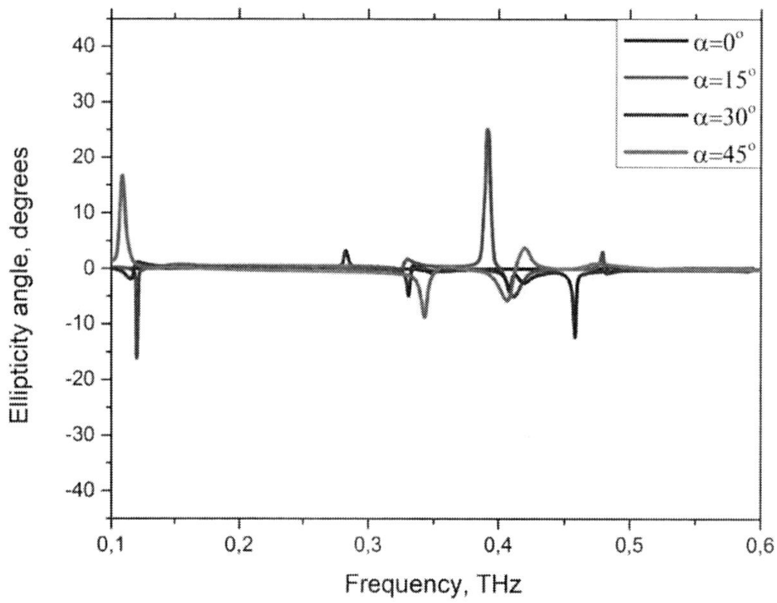

Figure 3. Ellipticity angle spectra for single chiral metasurface.

Then the influence of the combinations of metasurfaces (Fig. 1-c) with differently rotated chiral rosettes of the top metasurface on polarization state of the transmitted wave was studied. For this case the ellipticity angle spectra are shown in Fig. 4.

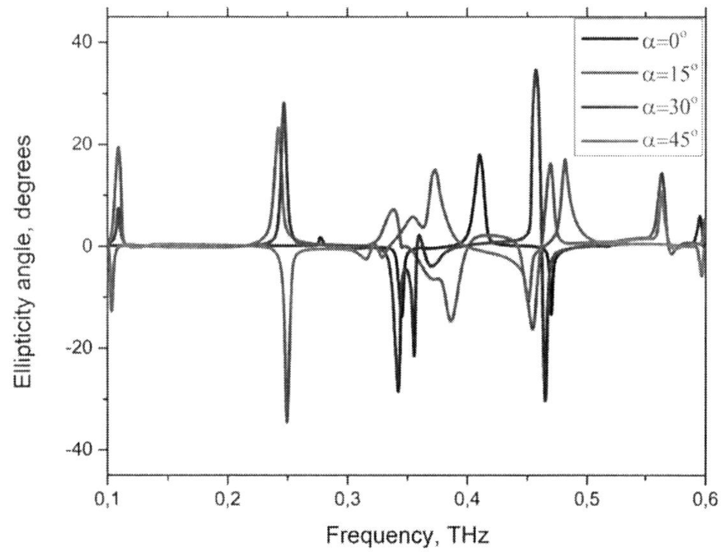

Figure 4. Ellipticity angle spectra for the stack of metasurfaces.

As can be seen from Fig. 4, the stack of two metasurfaces changes the polarizing properties of the transmitted wave more efficiently than the single metasurface. A few similarities in behavior of ellipticity angle spectra may be noticed: absolute values of ellipticity in four different cases are about the same at the frequencies of f_1=0.1 THz, f_2=0.25 THz, f_1=0.35 THz, while their signs are different. The nature of this phenomenon is connected with interaction between gammadion petals of the neighbor unit cells.

5. Conclusions

The single chiral metasurfaces and their stacks with different angles of rotation of the chiral unit cell were studied. It was shown that ellipticity angle spectra of single metasurface have several resonant frequencies, where linear polarization transforms into elliptical one. Ellipticity angle spectrum depends on rosette position at the unit cell. Hence, the rotation of the chiral element leads to the changing of the ellipticity angle at these frequencies. It was shown that the ellipticity angle η=25 degrees can be reached at frequency f=0.35 THz by rotation of the chiral resonators in the single metasurface. Significant increment of ellipticity angle at the resonant frequencies is achieved by adding the second metasurface. The ability of changing the handedness of polarization by combination of metasurfaces with differently rotated resonators was demonstrated. Thus, it is possible to make multiband polarization converter using several different chiral metasurfaces.

Acknowledgements

This work was financially supported by Government of Russian Federation (Grant 08–08). Student authors acknowledge SPIE and MTT-s societies for financial help in 2018 year.

References

[1] Khorasaninejad M, Chen W T, Zhu A Y, Oh J, Devlin R C, Rousso D, Capasso F 2016 *Nano letters* **16**(7), 4595-4600.
[2] Khodzitsky M K, Vozianova A V, Gill V V, Chernyadiev A V, Grebenchukov A N, Minin I V and Minin O V 2016 *Proceedings of SPIE - The International Society for Optical Engineering* **9918** 99182X.
[3] Tarapov S I, Khodzitskiy M, Chernovtsev S V, Belosorov D, Merzlikin A M, Vinogradov A P and Granovsky A B 2009 *Solid State Phenomena* 152-153 394-396.
[4] Tarapov S I, Khodzitsky M, Chernovtsev S V, Belozorov D, Merzlikin A M, Dorofeenko A V, Vinogradov A P, Inoue M and Granovsky A B 2010 *Physics of the Solid State* **52** 1427-1431.
[5] Girich A, Khodzitsky M, Nedukh S, Tarapov S 2011 *Terahertz and Mid Infrared Radiation. NATO Science for Peace and Security Series B: Physics and Biophysics* 159-164.
[6] Grebenchukov A N, Azbite S E, Zaitsev A D, Khodzitsky M K 2019 *Journal of Magnetism and Magnetic Materials* **472** 25-28.
[7] Plum E 2016 *Applied Physics Letters* **108**, 241905.
[8] Grebenchukov A, Zaitsev A, Khodzitsky M 2018 *Chinese Optics* **11**(2) 166-173.
[9] Denisultanov A Kh and Khodzitsky M K 2013 *Proceedings of SPIE - The International Society for Optical Engineering* **8806** 88062I.
[10] Soboleva V Y, Gomon D A, Sedykh E A, Balya V K, Khodzitskii M K 2017 *Journal of Optical Technology* **84**(8) 521.
[11]Terekhov Y E, Khodzitsky M K, Grachev Y V, Sedykh E A, Belokopytov G V and Zhang X-C 2013 *Proceedings of SPIE - The International Society for Optical Engineering* **8806** 88062Q.
[12] Smirnov S, Anoshkin I V, Demchenko P, Gomon D, Lioubtchenko D V, Khodzitsky M, Oberhammer J 2018 *Nanoscale* **10**(26) 12291-12296.
[13] Gomon D A, Sedykh E A, Rodriguez S, Monroy I T, Zaitsev K I, Vozianova A V, Khodzitsky M K 2018 *Chinese Optics* **11**(1) 47-59.
[14] Khodzitsky M K, Kharchenko A A, Strashevskyi A V and Tarapov S I 2009 *Telecommunications and Radio Engineering* **68** 561-566.
[15] Litvinov E A, Demchenko P S, Sheklanova E B, Vozianova A V, Khodzitsky M K 2018 *Proceedings of SPIE - The International Society for Optical Engineering* **10719** 1071939.
[16] Korolenko S Yu, Grebenchukov A N, Masyukov M S, Vozianova A V, Khodzitsky M K *2016 J. Phys.: Conf. Ser.* **735** 012089.
[17] Masyukov M S, Vozianova A V, Grebenchukov A N, Khodzitsky M K 2018 *J. Phys.: Conf. Ser.* **1062** 012012.
[18] Chebotarev V S, Solov'ev A N, Trofimov A D, Khodzitskiĭ M K 2017 *Journal of Optical Technology* **84**(8) 525-527
[19] Borovkova M, Khodzitsky M, Demchenko P, Cherkasova O, Popov A, Meglinski I 2018 *Biomedical Optics Express* **9**(5) 2266
[20] Gusev S I, Demchenko P S, Cherkasova O P, Fedorov V I, Khodzitsky M K 2018 *Chinese Optics* **11**(2) 182-189.
[21] Menzel C, Rockstuhl C, Lederer F 2010 *Physical Review A*, **82**, 053811.

Ultraspectral UV polychromator for SRS-isotopic sensor

A Zhevlakov[1], E Miharev[2], A Grishkanich[12], Y Chubchenko[13], L Ignateva[2], N Paklinov[4]
[1]ITMO University, Saint-Petersburg, Russia; [2]St.Petersburg State Electrotechnical University ETU, Saint-Petersburg, Russia; [3]D.I. Mendeleyev Institute for Metrology (VNIIM), Saint-Petersburg, Russia; [4] Tyumen Industrial University, Tyumen, Russia

Abstract. We have developed a spectral module for the selection of narrow characteristic spectral lines of isotopes in the ultraviolet Raman spectral channel. To allow resolving closely spaced isotope lines in the continuous input radiation spectrum, extremely high resolution λ / Δλ = 1000 is implemented. The total inverse linear dispersion at the output of the double polychromator (dλ / dl = 0.048 nm / mm) allows us to confidently separate the methane lines 283.31; 284.42 nm

1. Introduction

For spectral devices in general, and also for Raman sensors, which measure weak signals in particular, effective control with background highlights in the device itself is necessary while simultaneously suppressing Rayleigh scattering of airborne laser radiation by air molecules. The most radical means of dealing with diffused light is the use of double and even triple monochromators. Among the variety of spectrographs with a concave lattice, only a few have received the most widespread use, in particular, the Paschen – Runge scheme. In devices built according to this scheme, the entrance slit, the concave grating and the exit slit or photoreceiver are fixed at the Rowland circle, which is the dispersion line, and mechanically unrelated to each other. With the transition to the UV region in prisms and diffraction gratings, as is well known, the angular and linear dispersion increases substantially.

The complexity of developing a spectral module, designed to isolate narrow characteristic spectral IV lines from the continuous spectrum of input radiation, is due to the extremely high (ultra-spectral) resolution λ / Δλ = 1000. To resolve closely spaced IV isotopes, the following options are considered [1]:
1. Scheme of Cherni-Turner with many (2-3) diffractive elements in the mode of addition of wave fronts and dispersions
2. Cherne-Turner scheme with a diffraction grating in 3-4 diffraction order
3. Ebert's scheme based on a combined collimating-focusing mirror
4. High-performance multichannel spectrometer with a Paschen-Runge monochromator based on the Rowland circle and concave hologram gratings
5. High-performance spectrometer on Echelle type gratings.

2. Experiment

To obtain the minimum level of scattered radiation, a double polychromator can be constructed according to various schemes: with a horizontal plane of symmetry, or in a horizontal-vertical scheme.

As was said, the scattered light depends on the number of optical parts in the device: the smaller, the better. It is impossible to quantify the level of scattered radiation at the design stage, since scattered radiation depends on the quality of the optics and the culture of its production.
We will conduct a preliminary assessment of the polychromator scheme, as a key element of the spectral module (fig.1). We will carry out the calculation of the polychromator in the range of the UV spectrum, where the absorption cross sections in the Raman are maximum, 250 - 300 nm [1].
We believe that the polychromator is composed of two wide-angle stigmatic hologram gratings of the 2nd type and three slots: input, intermediate and output. Each grating is made with $N = 3600\ mm^{-1}$ lines, radius R = 500 mm and has a resolving power of 450,000 for wavelengths of about 280 nm and a spectral reflectance of 60%.

3. Results

The use of hologram optical elements in the spectral module will not only reduce the noise characteristics in comparison with rifled gratings with noticeable light scattering, but also allow you to work with a wide aperture, because unlike rifled ones they are free from spherical aberration and astigmatism. The angles of incidence ψ and diffraction φ are related by the known relation $\sin \psi + \sin \varphi = k \cdot \lambda \cdot N$, (43) where k = 1, 2, 3, ... - spectrum order; N is the number of grid lines per 1 mm. Angular dispersion $d\varphi / d\lambda = k / (b \cdot \cos \varphi)$ Linear dispersion $dl / d\lambda = f \cdot d\varphi / d\lambda$, where f is the focal length. The limiting resolution $\lambda / \Delta\lambda = Nb / \lambda = h / \lambda$, here b is the lattice period, h is the width of the shaded part of the lattice.
Thus, the resolving power is proportional to the number of lines on the entire area of the diffraction grating. As a result, the total inverse linear dispersion of a concave diffraction grating for a double polychromator is obtained equal to $d\lambda / dl = 0.224$ nm / mm. Such an ultra-spectral resolution makes it possible to confidently separate the lines C12H4 (283.31; 284.42 nm), N2 (278.69 nm).
In particular, when the difference in wavelengths of 4.62 nm for CH4 and N2 molecules, the linear distance in the polychromator was 20.6 mm.

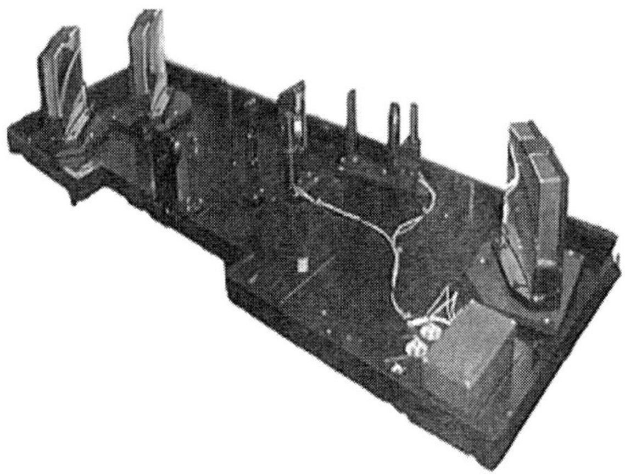

Figure 1.a. High resolution polychromator.

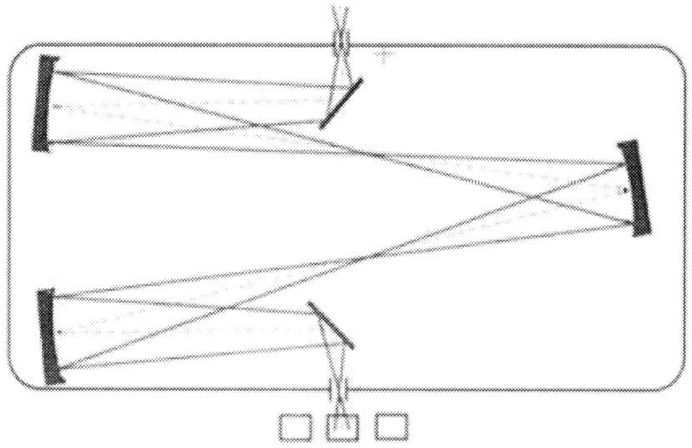

Figure 1. b. The scheme of high-resolution polychromator on two concave diffraction gratings in the first diffraction order with N = 3600 mm^{-1}strokes, radius R = 500 mm.

Characteristics of mentioned polychromator are given below in Table 1.

Table 1. Characteristics of high resolution polychromator.

High resolution polychromator	
Dispersive elements, type	Holographic grating, concave
Number of dispersion elements	3 pcs
Number of strokes	3600 mm^{-1}
Radius of curvature	500 mm
Diffraction order	1
Spectral range	264-294 nm
Reflectivity (280 nm)	0.6
Spectral resolution	>>1000
Dimensions	820x370x185 mm

4. Conclusion

Studies have shown the possibilities of application of Scheme of Cherni-Turner as a sufficient accuracy method for detection lines of isotopes in the ultraviolet Raman spectral channel. Further, we have used the hologram optical elements, which are expand the aperture. In addition, the resolving power was received enough to separate the lines C12H4 (283.31; 284.42 nm), N2 (278.69 nm).
The authors propose to continue their research to develop the technology for the selection closely spaced isotope lines in the continuous input radiation spectrum.

Acknowledgments
The study was supported by the grant RF Ministry of Education and Science, Grant RFMEFI57518X01800 05.575.21.0180.

References
[1] Borengasser M., Hungate W. S., Watkins R. Hyperspectral remote sensing: principles and

applications. – CRC press, 2007.

[2] Zhevlakov A. P. et al. Oil and gas deposits determination by ultraspectral lidar //Advanced Environmental, Chemical, and Biological Sensing Technologies XII. – International Society for Optics and Photonics, 2015. – T. 9486. – C. 94860V.

[3] Close D. H. Holographic optical elements //Optical Engineering. – 1975. – T. 14. – №. 5. – C. 145408.

[4] Kascheev S, Elizarov V, Grishkanich A, Bespalov V, Vasil'ev S, Zhevlakov A 2014 Proc. SPIE 92741K

Tamm plasmon structure with an optical Fibonacci lattice coating

A R Gubaydullin[1,2] **and K M Morozov**[1,2]

[1] St.Petersburg Academic University, St.Petersburg, 194021 Russia
[2] ITMO Univesity, St.Petersburg, 197101 Russia

E-mail: `gubaydullin.azat@gmail.com`

Abstract. We have theoretically investigated a Tamm plasmon structure covered by subwavelength grating with quasicrystal symmetry using a finite element method. The possibility to support a number of high-Q hybrid optical eigenstates in such systems was demonstrated. The model structure consists of a $GaAs/Al_{0.95}Ga_{0.05}As$ Bragg reflector covered with a F_6 subwavelength Fibonacci silver/PMMA coverage. The considered system with a quasicrystalline grating has interesting properties and can be used as light filtering and emission amplifying components of optoelectronic systems.

1. Introduction

Tamm plasmon (TP) is an electromagnetic state localized at the interface between the metal and the specially designed Bragg mirror. In last years, a huge amount of researchers are interested in such systems supporting TP state [1, 2, 3]. Tamm plasmon demonstrates lower losses than conventional surface plasmon modes [4] and provides new possibilities for utilizing metallic parts in modern photonics. To reduce losses and increase emission control of TP based systems a formation of the subwavelength elements out of the metallic layer [5, 6] can be done. Various types of metallic photonic crystals as a covering pattern has a high potential for applications [7], as well as photonic quasicrystals [8, 9]. Properties of these type systems such as negative refractive index [7], high-order symmetry and fractal density of photonic states [8] can significantly improve control over the light localizationin such hybrid TP systems. Recently, it was theoretically and experimentally demonstrated [6], that variation of the grating filling factor parameter allows to control the spectral position and increase the quality factor of the TP state.

This paper is aimed at the theoretical investigation of the TP mode emission control by using metallic photonic quasicrystal coverage.

2. Results and discussion

A Tamm plasmon structure with a quasicrystalline subwavelength structuration of the covering silver layer (figure 1) was studied. The model structure consists of 30 pair $GaAs/Al_{0.95}Ga_{0.05}As$ distributed Bragg reflector (DBR) with a quasicrystalline subwavelength grating formed by the silver (45 nm) and the PMMA resist material (90 nm). The scheme of the structure with a periodic metal/resist grating considered in [6] is demonstrated in figure 1(a). We will compare the properties of the TP cavity with periodic structuration with a quasicrystal (Fibonacci sequence based) covering.

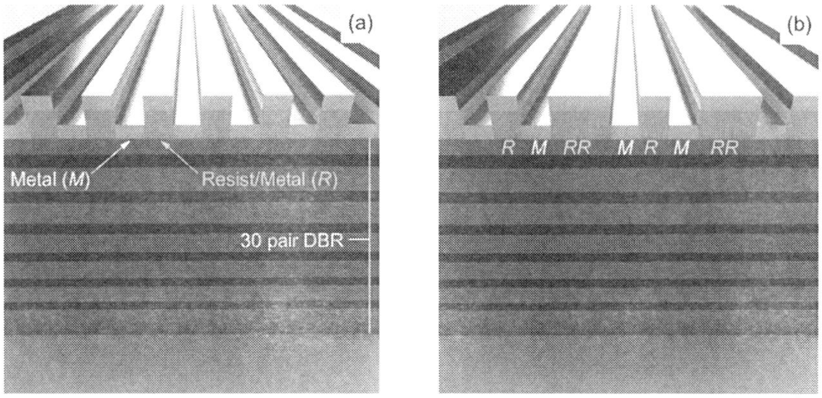

Figure 1. **(a)** Scheme of the TP structure with the periodic sub-wavelength grating. **(b)** Scheme of the TP structure with the sub-wavelength grating formed by the Fibonacci optical lattice sequence.

The simplest 1D photonic quasicrystal is an optical Fibonacci lattice. The Fibonacci sequence of order F_{i+1} is constructed using the recurrent rule $F_{i+1} = \{F_{i-1}, F_i\}$ with initial values $F_0 = M$ (metal layer) and $F_1 = R$ (resist and metal layer). The scheme of the TP structure with the Fibonacci sequence based metal/resist grating is demonstrated in figure 1(b). The structure under study is possible to fabricate by using electronic lithography as in [6]. Due to the self-similarity of the optical Fibonacci lattices, properties of complex structure with Fibonacci-based structuration can lead to more explicit features in eigenmode spectrum. Therefore, we are analyzing how the quasicrystalline sequence of order F_6 (with configuration: $MRRMRRMRMRRMR$) with different filling factors (ff) influences on the optical properties of the structure.

Two-dimensional electromagnetic simulations were carried out using a finite element method (FEM). Periodic boundary conditions in the lateral dimension were used. Real and imaginary parts of a silver refractive index were obtained by fitting the experimental data. Reflectivity spectra were calculated for the structures with different filling factors in case of TE polarized wave and normal incidence. From the obtained spectra (by fitting the spectrum dip using the Lorentz function), the resonant wavelength and the Q factor of the system eigenmodes were estimated.

Figure 2 demonstrates the calculation results. Figure 2(a) shows the examples of reflectivity spectra calculated for the F_6 type covered structure. When the ff parameter is close to 100% (metal covering almost all area) there is a single notable feature in the spectrum, corresponding to the TP state. On the other hand, with the low ff value, there are multiple dips, which correspond to the localized modes. Figure 2(b) shows the generalized results for the F_6 structure obtained from the reflectivity spectrum. Colour circles show eigenstates spectral position, colour and size correspond to the value of Q factor. The blue line figure demonstrates the spectral position of the TP mode depending on the filling factor for the periodical structuration case. In the quasicrystal covered structure, the high-Q eigenstates with a lower wavelength take place. The highest Q factor values appear in structure, defined by the ff parameter less than 40%. Figures 2(c) and 2(d) show the electric field distribution for the localized modes for the ff = 98.4% ff = 23.8%, correspondingly. It can be seen, that in high ff value case (ff = 98.4%) the mode field distribution looks like usual Tamm plasmon state (localization near the edge between metal

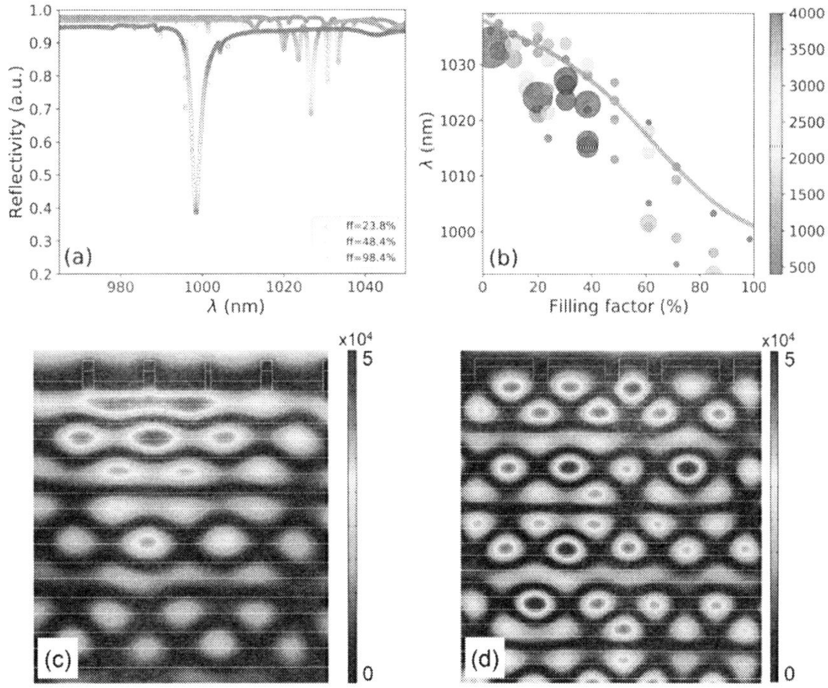

Figure 2. **(a)** Reflectivity spectra of the F_6 type covered system for the different ff parameter. **(b)** Results of reflectivity spectra fitting: eigenstates spectral position and Q factor. The blue line shows the same spectral position dependence on the ff parameter for a periodical grating coverage. Calculated electric field $|E|$ [V/m] distribution of the structures with ff = 98.4% and at $\lambda = 1000$ nm **(c)** and with ff = 23.8% and at $\lambda = 1033$ nm **(d)** for the F_6 case.

and DBR) and in low ff value case (ff = 23.8%) looks more peculiar. Thus, as well as a high number of modes in the spectrum, systems with ff<40% are characterized by more sophisticated localized fields.

3. Conclusions

Theoretical analysis of the Tamm plasmon structure optical properties formed by GaAs/$Al_{0.95}Ga_{0.05}As$ DBR with the quasicrystal subwavelength structuration was done. FEM based electromagnetic simulations were carried out with varied filling factor parameter for the F_6 Fibonacci based coverage structure. In the TP structure with subwavelength structuration with quasicrystal based elements with the F_6 sequence, a number of high-Q modes support was shown. The considered system with a quasicrystalline grating has interesting properties and can be used as light filtering and emission amplifying components of optoelectronic systems.

Acknowledgments

The reported study was funded by RFBR according to the research project No 18-32-00800.

References

[1] Afinogenov B I, Bessonov V O, Soboleva I V and Fedyanin A A 2019 *ACS Photonics* **6** 844–850
[2] Bikbaev R, Vetrov S and Timofeev I 2019 *Photonics* **6** 28
[3] Cheng H C, Kuo C Y, Hung Y J, Chen K P and Jeng S C 2018 *Phys. Rev. Appl.* **9** 064034

[4] Gubaydullin A R, Symonds C, Bellessa J, Ivanov K A, Kolykhalova E D, Sasin M E, Lemaitre A, Senellart P, Pozina G and Kaliteevski M A 2017 *Sci. Rep.* **7** 9014
[5] Azzini S, Lheureux G, Symonds C, Benoit J M, Senellart P, Lemaitre A, Greffet J J, Blanchard C, Sauvan C and Bellessa J 2016 *ACS Photonics* **3** 1776–1781
[6] Gubaydullin A R, Symonds C, Benoit J M, Ferrier L, Benyattou T, Jamois C, Lemaître A, Senellart P, Kaliteevski M A and Bellessa J 2017 *Appl. Phys. Lett.* **111** 261103
[7] Swift G P, Gallant A J, Kaliteevskaya N, Kaliteevski M A, Brand S, Dai D, Baragwanath A J, Iorsh I, Abram R A and Chamberlain J M 2011 *Opt. Lett.* **36** 1641
[8] Morozov K, Ivanov K, Gubaydullin A and Kaliteevski M 2017 *Opt. Spectrosc.* **122** 235–242
[9] Kaliteevski M A, Brand S, Abram R A, Krauss T F, De La Rue R M and Millar P 2000 *J. Mod. Opt.* **47** 1771–1778

Coupling of plasmon and grating resonances for SERS enhancement

S A Scherbak[1,2], **E S Babich**[2], **I V Reduto**[1,2,3], **A A Lipovskii**[1,2]
[1] Department of Physics and Technology of Nanostructures, St. Petersburg Academic University RAS, St. Petersburg 194021, Russia
[2] Institute of Physics, Nanotechnology and Telecommunications, Peter the Great St. Petersburg Polytechnic University, St. Petersburg 195251, Russia
[3] Institute of Photonics, University of Eastern Finland, Joensuu 80101, Finland

Abstract. We used thermal poling structuring of glass samples followed by chemical etching and subsequent out-diffusion growth of silver nanoparticles to fabricate periodic arrays of grooves etched in the glass with silver nanoparticles at their bottom. A numerical modeling of optical properties of these periodic structures was performed. We demonstrated that they support both plasmon and grating resonances, which interaction provides additional enhancement of an incident electric field. This can increase surface-enhanced Raman scattering by such biresonant structure.

1. Introduction

Metal nanoparticles, being able to greatly enhance electromagnetic field of an incident light wave because of surface plasmon resonance (SPR) phenomenon [1], are under extensive studies during last decades. Particularly, they have become almost irreplaceable for single-molecule sensing based on surface-enhanced Raman scattering (SERS) [2]. Nanostructures, which supports two or more different optical resonances (so-called biresonant structures), can provide additional enhancement of local electric field and, therefore, sensing efficiency. E.g., these are metal core - dielectric shell nanoparticles [3], bimetal nanostructures [4, 5], periodically arrayed plasmonic [6, 7] or dielectric [8] nanoparticles. Recently, nanostructures combining a plasmon mode of a metal nanoisland film with a grating resonance (GR) of an array of dielectric pillars were proposed for Raman enhancement [9]. However, the authors of [9] admit that the nature of this enhancement is unclear: it can be due to either the interaction of the SPR and the GR or the SPR and the spatial dielectric resonance of each pillar separately.

In the present study, we consider a similar nanostructure: a periodic array of grooves etched in glass, few hundred nanometers in depth, with silver nanoparticles at their bottom. By the results of a numerical modeling, we univocally identify plasmon resonances and grating resonances. We estimate the enhancement of the local electric field provided by the nanostructures and discuss possibilities of their applications in sensing. Undeniable advantage of these structures is that they can be fabricated within a unified technological process based on thermal poling structuring of the glass and out-diffusion growth of the silver nanoparticles [10, 11].

2. Fabrication technology

The technology that is used for fabrication samples under discussion consists of four subsequent steps. First is silver-to-sodium ion exchange of soda-lime glass in $AgNO_3$-$NaNO_3$ melt. This way, glass samples enriched with Ag^+ ions are obtained. Second, the samples are subjected to thermal poling using a periodically profiled anodic electrode. The applied electric field arranges the Ag^+ ions distribution in the glass. Third, we etch the samples in a NH_4F:$8H_2O$ solution. The poling also results in the decrease of the etching rate of the poled regions of the glass, therefore, we obtain periodical grooves in the glass. Fourth, the samples are annealed in H_2 atmosphere. Hydrogen reduces Ag^+ ions, and this leads to the self-assembled formation of Ag nanoparticles at the grooves bottom via out-diffusion process. The detailed and graphical description of the fabrication steps can be found in our previous works [10-12].

Using the described technology we fabricated a series of test samples and estimated their morphological properties with scanning electron microscopy (SEM). In Figure 1 we demonstrate examples of the SEM images obtained.

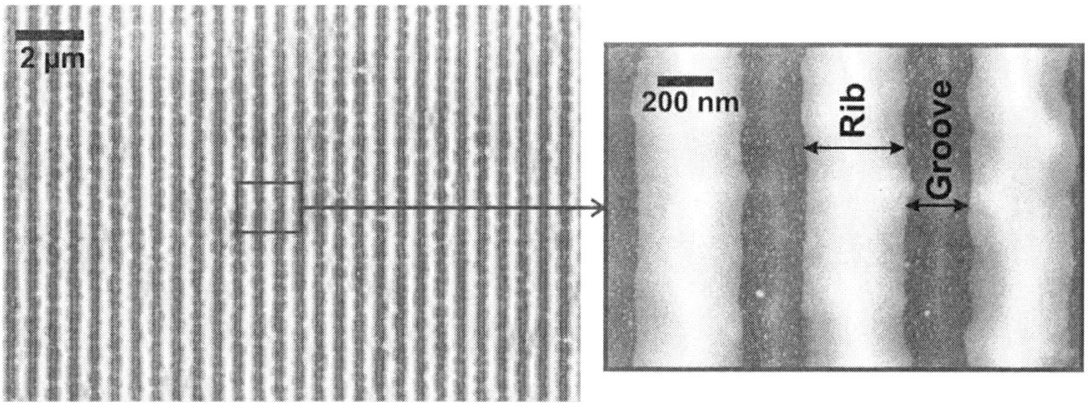

Figure 1. SEM images of a periodic array of grooves in glass with silver nanoparticles at the grooves bottom; the period of the structure ~600 nm.

The images in Figure 1 demonstrate decent quality of the samples fabricated. The grooves are reasonably even and periodic. Estimated depth of the grooves is about hundreds of nanometers and depends on the etching time. The width of the ribs and grooves depends on the design of anodic electrode used in the poling. In the zoomed image (the right one in Figure 1) one can see the ensemble of silver nanoparticles at the grooves bottom, the average diameter of which is less than 10 nm. However, the size of the particles can be controlled by the time and the temperature of the hydrogen annealing step of the technology described above. The favorable size of nanoparticle is 40-60 nm, since its SPR wavelength coincides with the wavelength of the laser commonly used in SERS (532 nm) [13]. Optimal fabrication regimes and SERS measurements are the subject of further studies.

3. Modelling and discussion

To simulate optical properties of the periodic structures under consideration we used finite elements method in COMSOL Multiphysics environment. We considered a single particle in each groove and neglected the interaction of the particles inside the same groove for simplicity since the aim of this model is to demonstrate the SPR and GRs interaction.

A spherical metal nanoparticle, 50 nm in radius, was placed at the centre of the bottom of a groove in glass with sidewalls 250 nm in height (see Figure 2). We used Drude dielectric function (λ_p=300 nm, $\gamma=10^{14}$ sec^{-1}) for the nanoparticle, dispersionless dielectric permittivity ε_{sub}=2.25 for the glass and ε_{out}=1 for the upper media (air). The incident light wave was directed normal to the groove bottom and polarized normal to the grating layers. Using periodic boundary conditions allowed us to model only a single cell of the array (it is marked with a dotted line in Figure 2).

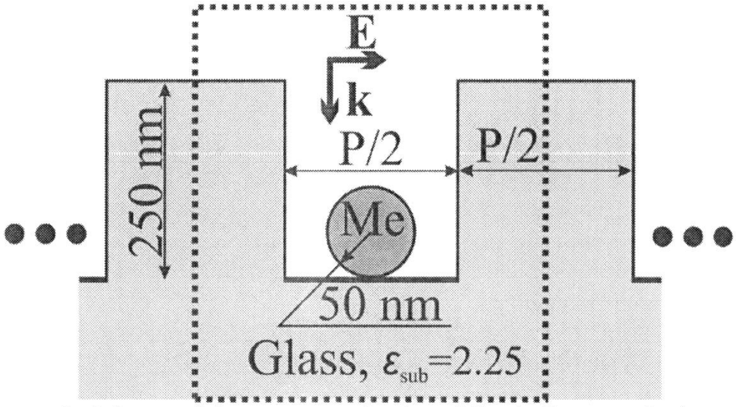

Figure 2. Scheme of the structure; the dotted line indicates a single period.

We varied the excitation wavelength, and calculated the electric field enhancement by a single isolated nanoparticle, see Figure 3a. Here and further by the enhancement of the electric field we mean resulting electric field averaged over the particle surface and normalized by the magnitude of the incident wave. Then we calculated the field enhancement by the periodic structures of different period with nanoparticles of the same size and the same dielectric function, and normalized these by the maximal enhancement calculated for the isolated nanoparticle. Thus, the results obtained are easily comparable with the enhancement by the single isolated nanoparticle. The field enhancement vs. the grating period (P) and the excitation wavelength (λ) is presented in Figure 3b.

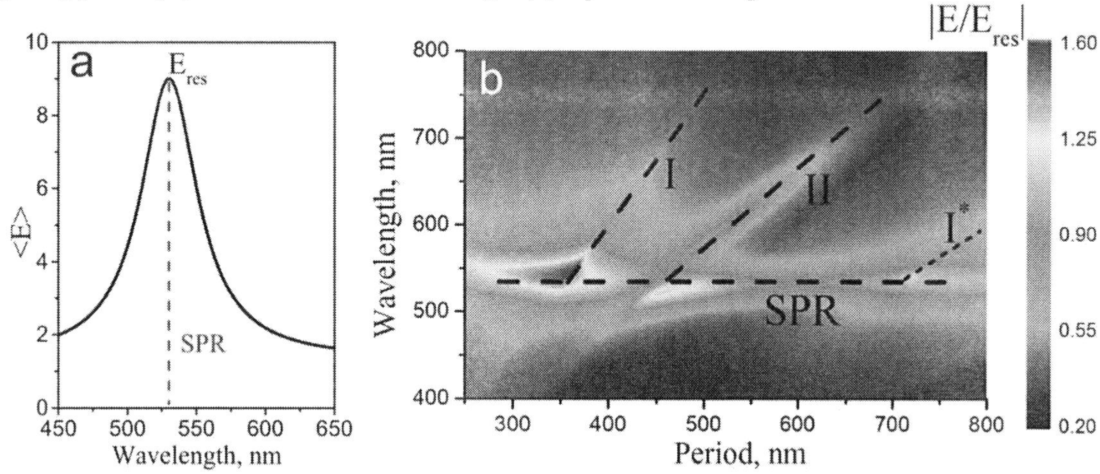

Figure 3. (a) Spectrum of the electric field enhancement by the single isolated nanoparticle (out of the grating). (b) The electric field enhancement in the periodic structure vs. the period (P) and the excitation wavelength (λ); The dependence is normalized by the maximal resonant enhancement of the electric field by the same nanoparticle.

In Figure 3b one can find regions of the field enhancement, which correspond to different optical resonances. The almost horizontal branch relates to the ordinary SPR (the same one is marked in Figure 3a), which wavelength is around ~530 nm for the particle of the considered size and having the abovementioned dielectric function. The slant branches labelled as I, II and I* correspond to different GRs. Branch I is the GR corresponding to the wave interaction through the glass substrate. This branch precisely coincides with the line $\lambda/n_{sub}=P$, where $n_{sub}=\sqrt{\varepsilon_{sub}}$. Branch I* is the second order of this resonance and it coincides with the line $2\lambda/n_{sub}=P$. This GR is faint since it occurs at high periods when the particles in the lattice interact weakly, and its impact will not be considered further. Branch II matches with the GR corresponding to the interaction through the upper environment: $\lambda/n_{up}=P$, where n_{up} is an effective refractive index of the upper medium that consists of glass layers and air spacers, therefore n_{up} is somewhere in between of n_{sub} and 1.0. Note that n_{up} depends on P, and for periods much larger than the height of the glass layers (700-800 nm and more) n_{up} tends to 1.0. However, at such periods nanoparticles barely feel each other in the lattice, and for the considered periods n_{up} can be roughly estimated as n_{up}~(1.0+1.5)/2=1.25.

The combination of the resonances described above can be used to strengthen a SERS signal. When the SPR coincides with a GR (intersection of the SPR branch with branches I and II), the noticeable field enhancement occurs: about 1.6-fold for branch I and 1.3-fold for branch II. Note that this was calculated relatively to the maximum enhancement of the isolated particle, e.g. if the isolated particle at the SPR 10-fold enhances the electric field, then in the present cases the enhancement will be by 16 and 13 times, respectively. This can result in 3-5 times increase of the SERS efficiency compared to isolated metal nanoparticles.

The other way is, contrary, to spectrally separate these resonances to enhance the electric field at different wavelengths. This is relevant to nonlinear processes, in which light waves of different frequencies participate. Particularly, power of a SERS signal is defined by the enhancement of the electric field at the excitation wavelength and at the wavelength shifted by the Raman offset $\Delta\lambda$:

$$I_{SERS} \sim |E(\lambda)|^2 |E(\lambda + \Delta\lambda)|^2 . \quad (1)$$

Commonly, only $E(\lambda)$ is plasmon-enhanced, whereas $E(\lambda+\Delta\lambda)$ is non-resonant. However, using such biresonant structures one can resonantly enhance both $E(\lambda)$ and $E(\lambda+\Delta\lambda)$, adjusting λ to λ_{SPR} and $\lambda+\Delta\lambda$ to λ_{GR}. The GRs only depend on the structure period and can be relatively easy tuned to the desired wavelength. Also, their magnitude is comparable to the one of the SPR (see Figure 3b). Therefore, this would greatly increase the selectivity of sensing and is applicable for SERS spectroscopy aimed at sensing of specific substances. The spectrum of the electric field enhancement by the structure with period of 510 nm is presented in Figure 4. The period P=510 nm was chosen for demonstration purposes to adjust the GR (branch II in Figure 3b) to the Raman-shifted wavelength of dye BPE ($\Delta\lambda$~1640 cm^{-1}) – a well-known substance that is typical for gauge of SERS setups.

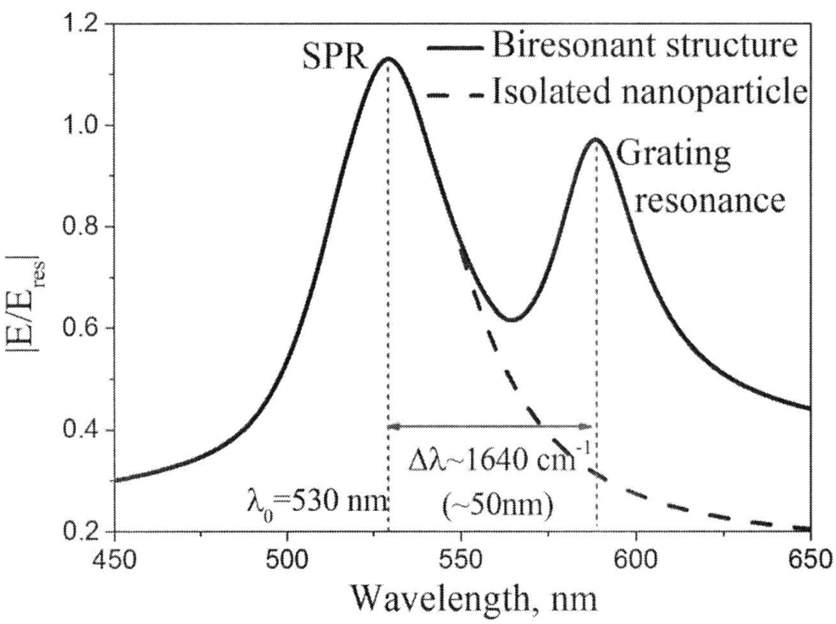

Figure 4. Spectrum of the electric field enhancement by a periodic structure with period P=510 nm (solid line); dashed line indicates single-resonant enhancement by the isolated particle.

According to Figure 4 $E_{\mathrm{bires}}(\lambda+\Delta\lambda)\sim 1$, whereas $E_{\mathrm{single}}(\lambda+\Delta\lambda)\sim 0.3$. Therefore, according to Eq. (1) and assuming that the isolated nanoparticle and the biresonant structure at ordinary SPR enhance the field approximately equally, the biresonant nanostructure provides about 10-fold increase of SERS signal ($|E_{\mathrm{bires}}(\lambda+\Delta\lambda) / E_{\mathrm{single}}(\lambda+\Delta\lambda)|^2\sim 10$).

4. Conclusion

We proposed biresonant nanostructures that comprise arrays of metal nanoparticles in periodic grooves in glass. Test samples were successfully fabricated using the unified technological process based on thermal poling structuring of the glass and out-diffusion growth of silver nanoparticles. The morphological analysis with SEM demonstrated their descent quality.

Performed numerical modelling revealed different kinds of optical resonances the structures under discussion possess: the ordinary surface plasmon resonance that only depends on nanoparticles' properties and the grating resonances that defined by the period of the nanostructures. The combination of SPR and GRs results in 1.3-1.6-fold increase of electric field enhancement compared to isolated nanoparticles. Therefore, such structures can be up to 3-5 times more efficient for SERS applications. Also, we demonstrated that in this biresonant structure the electric field at both fundamental and Raman-shifted wavelengths can be resonantly enhanced at the SPR and at the GR, respectively. This should provide at least more than an order of magnitude increase in SERS sensing of specific substances.

Acknoledgement

This study was funded by RFBR according to the research project № 18-32-00097

References

[1] Shahbazyan T V and Stockman M I 2013 *Plasmonics : Theory and Applications* (Dordrecht: Springer Netherlands)
[2] Prabowo B, Purwidyantri A and Liu K-C 2018 Surface Plasmon Resonance Optical Sensor: A Review on Light Source Technology *Biosensors* **8** 80
[3] Scherbak S A and Lipovskii A A 2018 Understanding the Second-Harmonic Generation Enhancement and Behavior in Metal Core–Dielectric Shell Nanoparticles *J. Phys. Chem. C* **122** 15635–45
[4] Babich E, Redkov A, Reduto I and Lipovskii A 2018 Self-Assembled Silver-Gold Nanoisland Films on Glass for SERS Applications *Phys. status solidi - Rapid Res. Lett.* **12** 1700226
[5] Okuno Y, Nishioka K, Kiya A, Nakashima N, Ishibashi A and Niidome Y 2010 Uniform and controllable preparation of Au–Ag core–shell nanorods using anisotropic silver shell formation on gold nanorods *Nanoscale* **2** 1489
[6] Zou S and Schatz G C 2006 Theoretical studies of plasmon resonances in one-dimensional nanoparticle chains: narrow lineshapes with tunable widths *Nanotechnology* **17** 2813–20
[7] Khlopin D, Laux F, Wardley W P, Martin J, Wurtz G A, Plain J, Bonod N, Zayats A V, Dickson W and Gérard D 2017 Lattice modes and plasmonic linewidth engineering in gold and aluminum nanoparticle arrays *J. Opt. Soc. Am. B* **34** 691-700
[8] Rendell R W, Shirey L, Long J P, Guo J, Glembocki O J, Caldwell J D, Owrutsky J C, Bezares F J and Kasica R 2013 Mie resonance-enhanced light absorption in periodic silicon nanopillar arrays *Opt. Express* **21** 27587
[9] Kukushkin V I, Grishina Y V, Egorov S V, Solov'ev V V and Kukushkin I V 2016 Combined dielectric and plasmon resonance for giant enhancement of Raman scattering *JETP Lett.* **103** 508–12
[10] Redkov A V, Zhurikhina V V and Lipovskii A A 2013 Formation and self-arrangement of silver nanoparticles in glass via annealing in hydrogen: The model *J. Non. Cryst. Solids* **376** 152–7

[11] Zhurikhina V V, Brunkov P N, Melehin V G, Kaplas T, Svirko Y, Rutckaia V V and Lipovskii A A 2012 Self-assembled silver nanoislands formed on glass surface via out-diffusion for multiple usages in SERS applications *Nanoscale Res. Lett.* **7** 1–5
[12] Redkov A, Chervinskii S, Baklanov A, Reduto I, Zhurikhina V and Lipovskii A 2014 Plasmonic molecules via glass annealing in hydrogen *Nanoscale Res. Lett.* **9** 606
[13] Babich E S, Redkov A V, Reduto I V, Scherbak S A, Kamenskii A N and Lipovskii A A 2017 Raman enhancement by individual silver hemispheroids *Appl. Surf. Sci.* **397** 119–24

The influence of optical density change on measurement error of refractive index in flowing liquid

N M Grebenikova[1], V V Davydov[1, 2] and V Yu Rud'[2]

[1]Higher School of applied physics and space technologies, Peter the Great Saint Petersburg Polytechnic University, Saint Petersburg, 195251, Russia
[2]All Russian Research Institute of Phytopathology, Moscow Region 143050, Russia

Abstract. The article discusses the physical processes in a flowing fluid that affect the change in its optical density. Changes in the refractive index and its errors in single-component and multicomponent media for different fluid flow regimes are investigated. A new model has been developed for calculating error of refractive index, considering the non-uniformity of optical density in a stream liquid. A technique is proposed for selecting a refractometer model for reliable monitoring the state of liquid environment.

1. Introduction

The deterioration of the ecological state of water bodies, the introduction of automatic processes in the production of different products and other factors require a person to constantly improve means of monitoring the state of condensed matter [1–8]. The greatest difficulties arise in monitoring the state of liquids flowing in pipelines of various diameters. Especially if the measurements to determine the liquid environment state should not make irreversible changes in its physical structure and chemical composition [8-16]. Therefore, the greatest preference when conducting such measurements is given to contactless methods [1-6, 9-12, 14-21]. Among them, the greatest advantages compared with others, has a method based on the use of the phenomenon of refraction [22-31]. The control of the medium state in the methods using the refraction phenomenon is carried out by the measured value of the refractive index n or the position of the light-shadow border [22, 26-28]. Usually for scientific research or to control the state of the technological process, for example, in the pharmaceutical industry, the values of n with an error of no higher than 10^{-4} should be measured [1, 17, 18, 23, 24].

There is factor that has been little studied, this factor has a significant effect on the measurement error n. This factor is the change in the optical density of the liquid. Especially, this factor is pronounced in turbulent flows. This greatly limits the use of refractometers for process control. Research is needed to develop a way to compensate for the effect of a change in optical density on the measurement error n of a flowing fluid. The research results and the method of calculating the change in the optical density of the flowing fluid to the value of n will increase the knowledge in the description of refraction.

2. The effect of optical density on error measurement of the refractive index

To measure the refractive index of a flowing fluid, various types of refractometers are used [17, 20, 23–31]. Generally, preference is given to differential type refractometers, in which compensation schemes are used to reduce the effect of various negative phenomena on the measurement error n [1, 23, 29-32]. The use of compensation schemes in the case of laminar flow of a liquid makes it possible to obtain measurement errors of n not worse than 10^{-4}. When measuring n in a turbulent fluid flow, the measurement error n increases. Therefore, the reasons that cause this process should be considered in more detail.

In the study of the liquid flow, the use of compensation devices in a refractometer is possible only if the device is built according to a differential scheme [29-32]. The developed designs of differential type automatic flow refractometers for measuring n use the Anderson difference prism method with a differential cell. The basic design of the differential (difference) cell consists of two tanks [29-31]. According to one container (flow cell) the investigated liquid moves. Its refractive index n_f must be controlled. In the second tank (closed cell) is a liquid medium with a known refractive index n_c.

In fig. 1 shows a block diagram of a differential cuvette (technical name - cuvette transducer (CT)) and the course of laser radiation in it.

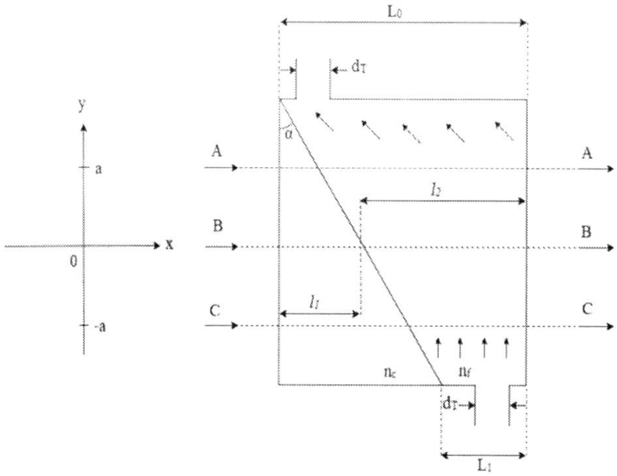

Figure 1. Propagation of laser beams in a differential cuvette.

The flowing fluid enters the flow cell 2. With the fluid flowing rapidly in the cell, the fluid layers are mixed. To make this process less intense, the diameter d_T is made much smaller than L_1. In addition, the entrance to the flow cell and the liquid outlet from it are located opposite each other. Our studies have shown that this leads to the formation of a stagnant zone (in Fig. 1 this is part of the flow cell next to the designation of the angle α). With long-term measurements, especially medical suspensions, this can lead to the formation of foreign structures hazardous to human health (for example, when controlling the process in pharmaceutical companies).

Therefore, we proposed a different principle for locating the position of pipelines, which is shown in Fig. 1. In this case, the mixing of the liquid layers is more significant than in the previously used CT schemes, but the probability of the formation of a stagnant zone in the CT is much less.

Consider the mechanism of the occurrence of the error, which is caused by fluctuations in the optical density of the measured liquid. According to the Lambert-Beer law, the light value at the output of a CT, if it is filled only with a comparative liquid, is determined by the following formula:

$$E_0 = \frac{\Phi_0}{2ab} 10^{-D_0} = \frac{\Phi_0}{2ab} 10^{-K_0 L_0}, \quad (1)$$

where Φ_0 – input light output CT; K_0 - specific optical density ratio of the reference fluid; L_0 – optical base CT; D_0 - optical density of a comparative fluid.

Consider the central beam B. For a simpler consideration, we choose the ideal case when the beam is almost not refracted on the partition between the flow cell and the closed cell (angle α is small). Then the central beam (B) passes the path l_1, in a cuvette filled with a comparative fluid, and the path l_2 in the flowing fluid. The specific coefficient of optical density, which can vary in the flowing fluid, is denoted as K_x. Extreme rays A and C pass through various paths in the measuring and comparative fluids and are attenuated differently, which causes an imbalance of the measuring circuit (MC) not caused by the displacement Δx and the occurrence of an error. The distances l_{1y} and l_{2y} traveled by an arbitrary beam lagging behind the middle beam B by the value of y ($-a \leq y \leq a$), in comparative and controlled fluids, are determined by the expressions:

$$\begin{cases} l_{1y} = l_1 - y \cdot tg\alpha \\ l_{2y} = l_2 + y \cdot tg\alpha' \end{cases} \quad (2)$$

where $l_1 + l_2 = L_0$.

Different distances traveled by the rays in liquids with different specific coefficients of optical density, lead to ambiguous absorption of the light beam in the CT along its width 2a.

Consequently, the illuminance E depends on the y coordinate (the laser beam propagates along the y axis), taking into account the variations in optical density $\Delta K = K_x - K_0$:

$$E(y) = \frac{\Phi_0}{2ab} 10^{-(K_0 l_{1y} + K_x l_{2y})} = \frac{\Phi_0}{2ab} 10^{-(K_0 l_1 + K_x l_2) - \Delta K y tg\alpha} \quad (3)$$

In accordance with the previously discussed theories to determine the error of the signal at the output of the MC expression (3) can be represented as follows:

$$(\Delta z_1)_0 = Lb\left(\int_{-a}^{0} E(y)\,dy - \int_{0}^{a} E(y)\,dy\right) == L\frac{\Phi_0}{2a} 10^{-(K_0 l_1 + K_x l_2)} \times \left(\int_{-a}^{0} 10^{-\Delta K y tg\alpha} dy - \int_{0}^{a} 10^{-\Delta K y tg\alpha} dy\right) \quad (4)$$

Substituting relation (3) into (4) and taking into account (1) and (2), we can obtain the following expression for the error signal provided that the refractive index of the liquid fluid does not change:

$$(\Delta z_1)_0 = L\frac{\Phi_0}{2a} 10^{-(K_0 l_1 + K_x l_2)} \times -\frac{ch[2{,}3025 \cdot \Delta K \cdot a tg\alpha] - 1}{2{,}3025 \Delta K \cdot tg\alpha}. \quad (5)$$

Relation (5) determines the error at the output of the MC of a differential automatic refractometer with a differential measurement circuit at zero. It should be noted that when $\Delta K \rightarrow 0$ in the formula (5) uncertainty arises, when revealed by the rule of L'Hôpital we get $(\Delta z_1)_0 = 0$, what corresponds $E(y) \rightarrow E_0$. This is confirmed by experiment..

The complex expression (5) for absolute error makes it difficult to use for practical purposes. The transition to relative values allows us to simplify the error formula, which can be represented by the following expression:

$$\delta(\Delta n) = \delta_1(0) + \delta_2(\Delta n), \quad (6)$$

where $\delta_1(0)$ – characterizes the offset "zero" when the deviation of optical density from the normalized value; $\delta_2(\Delta n)$ – functional term associated with the measured error value Δn.

For addend $\delta_1(0)$ в (6), which characterizes the offset, the following formula was obtained:

$$\delta_1(0) = 0{,}993 \frac{\Delta L^2 \Delta d}{l \Delta n_{max}}, \quad (7)$$

where ΔL – half of width light beam; l – distance from the differential cell to the photodetector; $\Delta d = d_x - d_0$ – change in the optical density of the measured liquid d_x relative to the initial (unreinforced) value d_0.

For another term in (6) the following formula is obtained:

$$\delta_2(\Delta n) = -\frac{1{,}15ltg^2\alpha\cdot(\Delta n)^2}{\Delta n_{max}}\Delta d, \qquad (8)$$

where α - refracting angle of the cuvette; Δn= n_f - n_c – the difference in refractive index measured (n_f) and comparative (n_c) liquid flows.

Our studies have shown that for refractometers with a compensation measurement scheme, formula (6) takes the following form

$$\delta(\Delta n) = \delta_1(0), \qquad (9)$$

In this case, the measurement error n in these refractometers is determined only by the offset of the “zero” caused by the change in optical density (Δd≠0).

3. The results of calculations of the effect of changes in optical density on the error in measuring the refractive index

For different values of the difference between the refractive indices Δn, changes in optical density, width of the light flux and other typical parameters of refractometers, the values of the components of the absolute error δ(Δn) were calculated. The results of the calculations are presented in tables 1 and 2.

Table 1.

Δn_{max}	Δd, см $^{-1}$	Δ L, см	$\delta_1(0)$, %		
			l=30 см	*l*=40 см	*l*=50 см
0,001	0,001 0,01 0,1	0,1	0,031	0,023	0,019
		0,2	0,124	0,092	0,076
		0,1	0,311	0,233	0,187
		0,2	1,244	0,932	0,748
		0,1	3,110	2,333	1,866
		0,2	12,440	9,992	7,464
0,01	0,001 0,01 0,1	0,1	0,003	0,002	0,002
		0,2	0,012	0,004	0,004
		0,1	0,031	0,23	0,019
		0,2	0,124	0,920	0,076
		0,1	0,311	0,233	0,187
		0,2	1,244	0,932	0,748
0,1	0,001 0,01 0,1	0,1	0,003	0,003	0,002
		0,2	0,012	0,008	0,008
		0,1	0,031	0,023	0,019
		0,2	0,124	0,092	0,076

Table 2.

Δn_{max}	Δd, см $^{-1}$	α_1, °	$-\delta_2(\Delta n_{max})$, %		
			l=30 см	*l*=40 см	*l*=50 см
0,001	0,001	45	0,003	0,004	0,006
	0,01	60	0,009	0,012	0,018
	0,1	70	0,023	0,030	0,045

		45	0,035	0,046	0,058
		60	0,105	0,138	0,174
		70	0,263	0,345	0,435
		45	0,345	0,460	0,575
		60	1,035	1,380	1,725
		70	2,588	3,450	4,313
		45	0,035	0,046	0,058
		60	0,105	0,138	0,174
		70	0,263	0,345	0,435
	0,001	45	0,345	0,460	0,575
0,01	0,01	60	1,035	1,380	1,725
	0,1	70	2,588	3,450	4,313
		45	3,45	4,60	5,75
		60	10,35	13,80	17,25
		70	25,88	34,50	43,19
		45	0,345	0,460	0,575
		60	1,035	1,380	1,725
		70	2,588	3,450	4,313
	0,001	45	3,45	4,60	4,75
0,1	0,01	60	10,35	13,80	17,25
	0,1	70	25,88	34,50	43,13
		45	34,5	46,0	57,5
		60	103,5	138,0	172,5
		70	258,8	345,0	431,3

For series-connected links of the differential refractometer, the values of $\sum(\Delta n)$ were calculated. For the calculations used the following formula:

$$\sum(\Delta n) = \frac{\delta_2(\Delta n)}{\delta_1(0)} = 1,2\left(\frac{l\, tg\alpha\Delta n_{max}}{\alpha}\right)k^2, \quad (10)$$

where $k = \frac{\Delta n}{\Delta n_{max}} \leq 1$. Δn_{max} – the maximum possible difference between the refractive indices of the liquid and the reference fluid.

The results of calculations $\sum(\Delta n)$ are presented in table 3.

Table 3.

Δn_{max}	k	α_1, °	$\sum(\Delta n)$, % *l*=30 см	*l*=40 см	*l*=50 см
		45	0,001	0,002	0,003
	0,1	60	0,003	0,006	0,009
		70	0,008	0,015	0,023
		45	0,027	0,048	0,075
0,001	0,5	60	0,081	0,144	0,225
		70	0,203	0,360	0,563
		45	0,108	0,192	0,300
	1	60	0,324	0,576	0,900
		70	0,810	1,440	2,250

		45	0,108	0,0192	0,300
	0,1	60	0,324	0,576	0,900
		70	0,810	1,440	2,250
		45	2,7	4,8	7,5
0,01	0,5	60	8,1	14,4	22,5
		70	20,3	36,0	56,3
		45	10,8	19,2	30,0
	1	60	32,4	57,6	90,0
		70	81,0	144,0	225,0
		45	10,8	19,2	30,0
	0,1	60	32,4	57,6	90,0
		70	81,0	144,0	225,0
		45	270	480	750
0,1	0,5	60	810	1440	2250
		70	2030	3600	5630
		45	1080	1920	3000
	1	60	3240	5760	9000
		70	8100	14400	22500

Analysis of the calculation results (Tables 1-3) allows us to draw the following conclusion. With increasing Δn, the sensitivity of the measurement error n_f to a change in optical density decreases. This fact is confirmed by the experiment.

4. Conclusion

The experimental results obtained by us confirmed the model proposed by us for calculating the error in measuring the value of n, which is related to the nonuniformity of optical density in a flowing fluid, and showed its universality. The model can be used for both stationary and flowing fluid.

Using this model allows us to estimate the magnitude of the measurement errors n for various liquids and determine the feasibility of using a differential type refractometer to monitor their status. Or use other types of measuring devices to monitor the state of the liquid environment, since the measurement error n does not satisfy, for example, the requirements of the experiment.

References

[1] Daev Zh A 2016 *Measurement Techniques* **59** 243-246

[2] Baranov M A, Velichko E N and Aksenov E T 2017 *Journal of Physics: Conference Series* **917(1)** 062059

[3] Nepomniashchaia E K, Cheremiskina A V, Velichko E N, Aksenov E T and Bogomaz T A 2016 *Journal of Optical Technology* **83(5)** 305-308

[4] Davydov V V 2016 *Optics and Spectroscopy (English translation of Optika i Spektroskopiya)* **121(1)** 18-24

[5] Davydov V V, Myazin N S and Davydova T I 2017 *Russian Journal of Nondestructive Testing* **53(7)** 520-529

[6] Myazin N S, Logunov S E, Davydov V V, Rud' V Yu, Grebenikova N M and Yushkova V V 2017 *Journal of Physics: Conference Series* **929 (1)** 012064

[7] Davydov V V, Dudkin V I, Petrov A A and Myazin N S 2016 *Technical Physics Letters* **42(7)** 692-696

[8] Rykin E V, Moroz A V, Smirnov K J, Davydov V V and Yushkova V V 2018 *MATEC Web of Conference* **245** 12002

[9] Davydov R V, Antonov V I and Moroz A V 2018 Proceedings of the 2018 *IEEE International Conference on Electrical Engineering and Photonics, EExPolytech 2018 (Saint-Petersburg)* **8564378** p. 236-239

[10] Myazin N S, Logunov S E, Davydov V V, Rud' V Yu, Grebenikova N M and Yushkova V V 2017 *Journal of Physics: Conference Series* **929 (1)** 012064

[11] Davydov V V, Dudkin V I, Karseev A Yu and Vologdin V A 2016 *Journal of Applied Spectroscopy* **82** 1013-1019

[12] Vitkovskii O S and Marusina M Y 2016 *Measurement Techniques* **59(3)** 247-251

[13] Zhernovoi A I and Sharshina L M 2003 *Biofizika* **48** 68-72

[14] Davydov V V 1999 *Russian Physics Journal* **42(9)** 822-825

[15] D'yachenko S V, Kondrashkova I S and Zhernovoi A I 2017 *Technical Physics* **62** 1602–1604

[16] Kashaev R S and Gazizov E G 2010 *Journal of Applied Spectroscopy* **77** 321–328

[17] Grebenikova N M, Myazin N S, Rud V Y and Davydov R V 2018 *Proceedings of the 2018 IEEE International Conference on Electrical Engineering and Photonics, EExPolytech 2018* (Saint-Petersburg) 8564409 p. 295-297

[18] Davydov V V, Dudkin V I, Velichko E N and Karseev A Yu 2015 *Journal of Optical Technology (A Translation of Opticheckii Zhurnal)* **82(3)** 132-135

[19] Davydov R V and Antonov V I 2016 *Journal of Physics: Conference* Series **769 (1)** 012060

[20] Myazin N S, Smirnov K J, Davydov V V and Logunov S E 2017 *Journal of Physics: Conference Series* **929 (1)** 012080

[21] Davydov R V and Antonov V I 2017 *Journal of Physics: Conference Series* **929 (1)** 012040

[22] Davydov V V, Kruzhalov S V, Grebenikova N M and Smirnov K J 2018 *Measurement Techniques* **61** 365-372

[23] Mishchenko Yu V 2007 *Measurement Techniques* **50** 1274–1281

[24] Leibengardt G I, Naidenov A S and Shur V L 2014 *Measurement Techniques* **47** 1211–1216

[25] Shur V I, Naidenov A S and Lukin A 2006 *Measurement Techniques* **49** 815–819.

[26] Grebenikova N M, Smirnov K J, Davydov V V , Rud V Yu and Artemiev V V 2018 *Journal of Physics: Conference Series* **1135 (1)** 012055

[27] Grebenikova N M, Smirnov K J, Artemiev V V, Davydov V V and Kruzhalov S V 2018 *Journal of Physics: Conference Series* **1038 (1)** 012089

[28] Grebenikova N M, Smirnov K J, Davydov V V and Rud V Y 2018 *Journal of Physics: Conference Series* **1124 (4)** 041011

[29] Karabegov M A 2007 *Measurement Techniques* **50** 619-628

[30] Karabegov M A 2012 *Measurement Techniques* **55** 478-484

[31] Karabegov M A 2012 *Measurement Techniques* **54** 1203-1212

[32] Mishchenko Yu V 1998 *Measurement Techniques* **41** 1120–1124

Advanced aperture cantilevers made by focused ion beam for both magnetic domain structure and morphology investigation in high resolution magneto optical microscopy

Yu E Vysokikh[1], S Yu Krasnoborodko[1], A S Kolomiytsev[2], A A Fedotov[2], A N Shaposhnikov[3], M F Bulatov[1], D V Churikov[1], V I Shevyakov[4]
[1]Scientific and Technological Center of Unique Instrumentation of the RAS, Moscow, 117342, Russia
[2]Southern Federal University, Rostov-on-Don, 344006, Russia
[3]V.I. Vernadsky Crimean Federal University, Simferopol, 295007, Russia
[4]National Research University of Electronic Technology, Zelenograd, Moscow, 124498, Russia

Abstract. Magneto-optical microscopy is a well-known and widely used method of magnetic domain structure investigation. Spatial resolution is limited by diffraction limit of the light and practically not exceed 0.5μm. Possibility to overcome diffraction limit by combination of magneto-optical microscopy and magnetic-force microscopy was showed earlier [1]. Aperture cantilever was used to light up the sample in near-field mode. In this paper, authors review most popular technology for aperture cantilever production with its advantages and disadvantages. Authors propose new technology based on focused ion beam treatment for creation aperture cantilevers with advanced parameters such as small aperture size and light throughput. Bi-substituted yttrium iron garnet (Bi: YIG) thin films with high magneto-optical activity and different period of domain structure were used as test objects. High resolution images of both morphology and domain structure obtained with new advanced aperture cantilevers are shown.

1. Introduction

Magneto-optical method has been used for a long time for magnetic domain structure investigation [1-4]. But since technology is growing up investigated domain structure became smaller than 0.5 μm. Magneto-optical method is limited by diffraction limit and in common case could not be used for domain structure investigation with the size less than 0.5 μm. Possibility to overcome diffraction limit by combination of magneto-optical microscopy and magnetic-force microscopy [5-8] was showed earlier [9], [10]. Nonmagnetic cantilever with aperture [11] is used for illuminating the sample in near-field magneto-optical mode. It allows to improve spatial resolution up to the size of aperture diameter.

Aperture cantilever should meet specific requirements to ensure reliability and quality of measurements. It is important to create as small aperture of the probe as possible to have a better spatial resolution. There are number technologies for creating aperture probes. The most common one is fiber aperture probes obtained by sharpening glass fibers by thermal pulling [12]. Tip diameter of the fiber probe is several hundreds of nanometres. This fact limited spatial resolution of optical methods. In addition, fiber optic aperture probes have low light throughput and poor reproducibility [13]. Silicon cantilevers theoretically might have very small aperture with size of several dozens of nanometres. In this case focused ion beam used to etch the silicon pyramid of the cantilever [14]. These probes have much better light throughput and reliability than fiber optic probes. But there are also some limitations. It is not possible to etch the probe from the tip side because of the high disturbance effect of ion beam on a tip so the hole size become much more than it should be. But it is also not easy to get the hole directly on the tip of pyramid in case of etching from the top side of cantilever. So poor reproducibility of this technology limits the usability of aperture cantilever for high resolution magneto-optical microscopy.

Development of new type of aperture cantilevers with possibility of creation desired small size hole, high light throughput and good reproducibility is important for high resolution magneto-optical microscopy implementation.

2. Experimental

Authors propose new technology of aperture cantilever creation. Standard silicon cantilevers NSG-11 (TipsNano) were used for creation new type of probes. Ion beam etching and ion beam assistance were implement using Nova NanoLab 600 tool. Gallium liquid metal was used as ion beam source. In the beginning tip is removed by ion beam etching in vacuum 5x10-7 Torr. High current mode of 20nA was used for higher speed process since high resolution was not necessary during the tip removing. Then round shape hole with diameter of 5 µm was created by ion beam etching. This hole is input for light illumination of the sample. Input diameter should be big enough for possibility of light illumination but from the other hand it should not exceed reasonable values in order to ensure a sufficient distance from the hole to the edge of the cantilever to prevent parasitic illumination from the laser to sample and save strength and rigidity of the cantilever (figure 1).

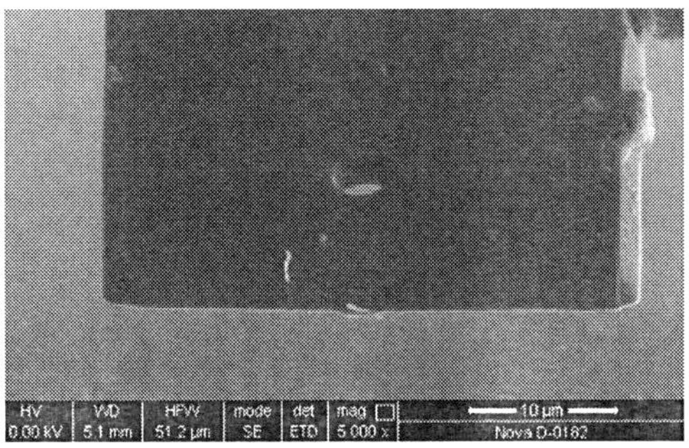

Figure 1. Round shape hole in silicon cantilever created by ion beam etching.

After the hole creation carbon pyramid formation has been started. Gas C6H14 was applied to the area of the hole thru the gas injection system. C6H14 interacting with a 30 keV gallium ion beam, decomposed into volatile components (which were immediately removed by the vacuum system of the microscope) and solid carbon, which was deposited around the hole. At an energy of 30 keV, the diameter of the ion beam is minimal (relatively lower energy values), moreover at this energy value the decomposition gas precursor efficiency is maximum, which ensures a higher deposition rate and minimizes the amount of organic impurities in the deposited material. The trajectory of the ion beam was determined to deposit carbon in an orderly manner from the base of the cone (5 µm in diameter) to its tip. As a result, a hollow tip with a height of about 5-6 µm was formed. After that, at the end of the cone an exit hole was formed for optical radiation with a diameter of 50–200 nm (figure 2).

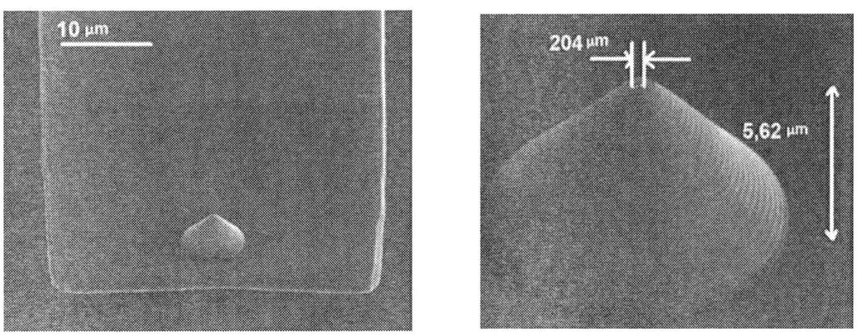

Figure 2. Carbon hollow pyramid deposited to the silicon cantilever.

Aperture cantilevers with different aperture size were created and tested in high resolution magneto optical mode. Bi: YIG films with uniaxial anisotropy (i.e. magnetization vector is normal to the film plane) and domain structure period of 4 - 9 μm were obtained by the method of liquid phase epitaxy [15], [16]. These films have a high magneto-optical activity in the visible and infrared wavelengths of light. The principle of studying the magnetic domain structure is based on recording the transmitted or reflected light through the sample in crossed polarization mode. Sample is illuminated through a nonmagnetic aperture probe at a distance significantly shorter than the wavelength – typically 10 nm or less. Figure 3 shows images of magnetic domain structure with the periods of 4 μm and 9 μm.. Opposite oriented domains rotate polarization in opposite direction which is registered by analyser (bright and dark stripes correspond to opposite oriented domains).

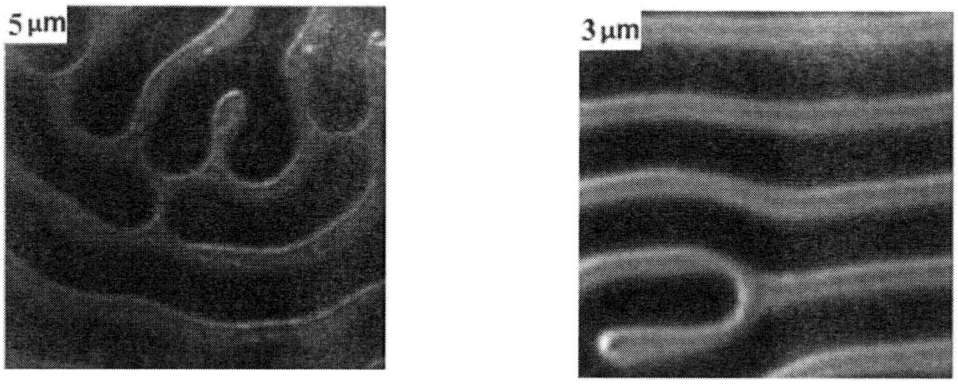

Figure 3. High resolution magneto-optical image of 9 μm (left) and 4 μm (right) of YIG thin film.

Bi: YIG thin films with small period (300 nm) of domain structure were obtained by reactive ion beam sputtering in argon-oxygen mixture [17]. High resolution images of both morphology and domain structure obtained with new aperture cantilevers are shown in figure 4.

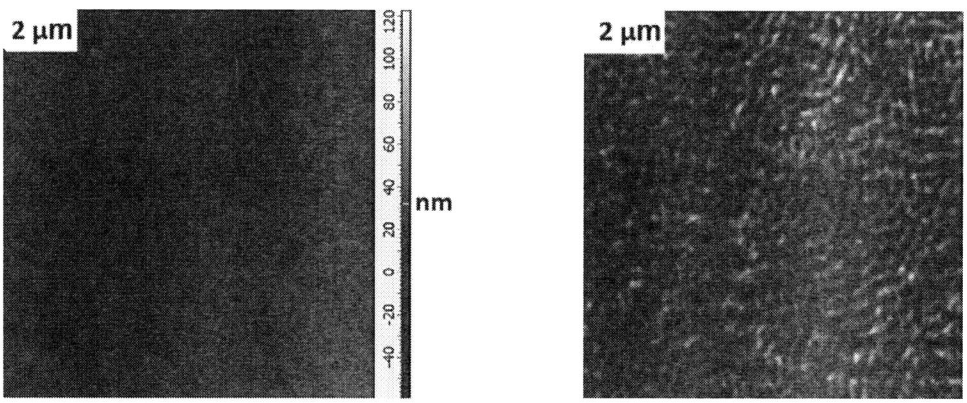

Figure 4. Topography (left) and high resolution magneto-optical image (right) of Bi: YIG thin film. Scan size is 20 x20 μm.

Light throughput rate was estimated for common used silicon aperture cantilever and proposed aperture cantilever with carbon tip. For same aperture size (~ 200 nm) the values are 2,1 x 10^{-3} and 1,5 x 10^{-1} correspondently. Light throughput rate for aperture of 50 nm (cantilevers with carbon tip) is 2 x 10^{-4}. It shows significant improvement of light throughput rate due to avoiding pyramid disturbance during cantilever etching which is possible by proposed technology of carbon tip creation with ion beam assistance method. Possibility to save light throughput rate for small aperture size allowing obtain high-resolution magneto-optical image with resolution down to 50 nm.

3. Conclusion

New technology for aperture cantilever creation was offered. Aperture cantilevers with carbon hollow pyramid tip were created by ion beam assistance method. Comparison of common used aperture cantilevers and new one was performed. New cantilevers showed better light throughput and reliability. Aperture size of 50 nm was achieved by proposed technology. Light throughput for 50nm aperture is 2×10^{-4} which allows to achieve spatial resolution down to 50 nm in magneto-optical mode. The images of magnetic domain structure with period of 300 nm were demonstrated by cantilevers with carbon hollow pyramid tip in high resolution magneto-optical mode.

References

[1] Fowler C A, Fryer E M 1956 *Physical Review* 104 2 552–553.
[2] Randoshkin V V, Chervonenkis A Y 1992 *Applied magnetooptics*.
[3] Williams H J, Foster F G, Wood E A 1951 *Physical Review* 82 1 119-120.
[4] Karpasyuk, V K, Bulatov, M F 1994 *IEEE Transactions on Magnetics* 30.
[5] Binnig G, Quate C, Gerber C 1986 *Physical Review Letters* 56 9. 930–933.
[6] Mironov V L 2004 *Fundamentals of Scanning Probe Microscopy*.
[7] Israelachvili J N *Intermolecular and Surface Forces* 2011 107–132.
[8] Hartmann U, Göddenhenrich T, Heiden C 1991 *Journal of Magnetism and Magnetic Materials* 101 1–3 263–270.
[9] Vysokikh Yu E, Shelaev A V, Shevyakov V I, Prokopov A R, Belov A N, Krasnoborod'ko S Y 2016 *J. Nanotechnologies in Russia* 815.
[10] Vysokikh Yu E, Shevyakov V I, Shaposhnikov A N, Prokopov A R, Shelaev A V, 2017 *IEEE Conference of Russian Young Researchers in Electrical and Electronic Engineering (EIConRus)*, 1445–1448.
[11] Betzig E, Lewis A, Harootunian A, Isaacson M, Kratschmer E 1986 *Biophysical journal* 269.
[12] Veerman J A, Otter A M., Kuipers L, Van Hulst N F 1998 *Applied Physics Letters* 3115.
[13] Dunn R C 1999 *Chemical reviews* 2891.
[14] Kolomiytsev A S, Shandyba N A, Panchenko I V, Lisitsyn S A 2018 *IOP Conf. Series: Materials Science and Engineering* 443 012015.
[15] Prokopov A R, Vetoshko P M, Shumilov A G, Shaposhnikov A N, Kuz'michev A N, Koshlyakova N N, Berzhansky V N, Zvezdin A K, Belotelov V I 2016 *Journal of Alloys and Compounds* 671 403–407.
[16] Leitenmeier S, Körner T, Griesbauer J, Herbort M, Heinrich A, Stritzker B 2008 *Journal of Crystal Growth* 310 24 5392–5401.
[17] Vysokikh Yu E, Shaposhnikov A N, Prokopov A R, Berzhansky V N, Karavainikov A V, Gerasimenko N N, Smirnov D I 2017 *Materials Research Bulletin* 95 115–122.

Magneto-optical properties of terbium-doped metaphosphate and borate glasses

E V Kulpina, A N Babkina, Y K Fedorov

Research Institute of Nano-Photonics and Optical Information Technology, ITMO University, Saint Petersburg 199034, Russia

e-mail: katrinakulpina@yandex.ru

Abstract. The paper shows the results of Faraday Effect study of borate and metaphosphate glasses doped with various concentrations of terbium under uniform longitudinal magnetic field. The Verdet constant of glass under study is obtained based on the experimental data for several wavelengths. The optical and spectral properties of glasses are also studied.

1. Introduction

Faraday Effect is a rotation of a polarization plane of light due to circular birefringence in a material caused by influence of an external magnetic field. In a non-absorbing or weakly absorbing medium, a linearly polarized monochromatic light beam passing through a material along direction of applied magnetic field experiences circular birefringence, resulting in rotation of polarization plane of incident light [1].

The Faraday Effect has a number of significant practical applications. Magneto-optical materials (crystals, glasses, films) are promising for use in modern high-tech devices, such as current sensors [2] and optical isolators [3].

In the most common magneto-optical materials, terbium ion (Tb^{3+}) is used as an ion possessing magnetoactive properties, which creates a large Verdet constant in the visible and near-IR spectral ranges. This was shown in terms of works devoted to both classical magneto-optical crystals TGG [4,5], TSAG [6,7], TAG [8] and terbium-containing magneto-optical glass. In most papers, the authors ultimately conclude that the higher the content of terbium, the better the magneto-optical characteristics in materials transparency window. However, with increasing terbium concentration, the intensity of its absorption bands increases. The main disadvantage of glasses doped with terbium is low transmittance in the wavelength range of < 400 nm and > 1400 nm.

Glass with high concentration of rare-earth elements has a large magneto-optical rotation with high transparency in the visible region [9]. However, the Verdet constant becomes insignificant in the long-wavelength area.

The present paper demonstrates the results of the investigation of optical, spectral and magneto-optical properties of heavily terbium-doped metaphosphate and borate glasses, including absorption spectra registration, refractive index studies, the rotation angle of the light polarization plane measurement and the Verdet constant calculation.

2. Methods and measurements

In the present work three types of glass were synthesized: 75 P_2O_5 – x Tb_2O_3 – (25-x) La_2O_3 glass system (where x varies from 5 to 25 mol %), 50 PbO – 50 B_2O_3 glass system doped with x Tb_2O_3 (x=25, 35 mol %) and 50 PbO – 50 B_2O_3 glass system doped with x TbF_3 (x=25, 35, 45 mol %).

The Verdet constant, which characterizes the magneto-optical properties of a material, cannot be measured directly. Therefore, it was necessary to measure the angle of rotation of the polarization plane of the radiation passing through the glass samples under study. For this purpose, a standard experimental setup was used, the scheme of which is shown in Figure 1.

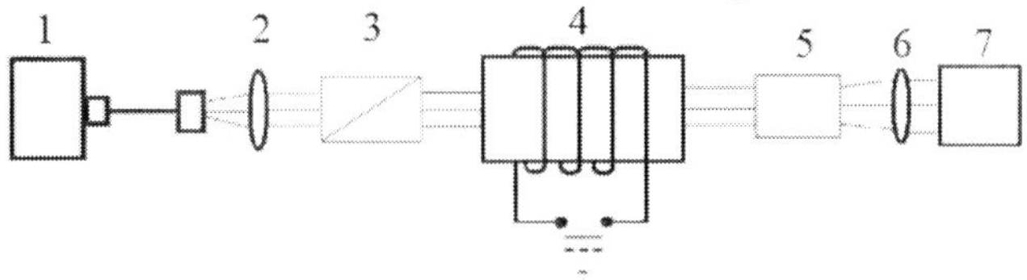

Figure 1. Schematic diagram of experimental setup: 1 - laser, 2 - lens, 3 - polarizer, 4 –system of magnetic coils, 5 - analyzer, 6- collimator, 7 - photodetector

Several coherent sources of radiation were used: He-Ne laser (633 nm), second harmonic of Nd:SSO laser (457 nm), and several semiconductor lasers with 532, 808, and 980 nm wavelengths. The polarizer and analyzer were crossed at an angle of 45 degrees. A magnetic field with an induction of 177 mT was created by a system of 4 magnetic coils, each of them consisted of 2050 turns of copper wire with a diameter of 1.16 mm. Measuring the angles of rotation of the polarization plane, as well as knowing the dimensions of the glass samples, allowed us to calculate the Verdet constant for each sample using the formula obtained from formula (1):

$$V = \frac{\theta}{H \cdot l}, \tag{1}$$

where l is the sample length (m), H is the magnetic field strength (T), θ is the angle of rotation of the polarization plane (deg).

Optical density spectra were recorded using a spectrophotometer Lambda 900. The measurements were carried out in the wavelength range of 250 - 3000 nm (Figures 4, 5). The obtained spectra of optical density were converted into the absorption coefficient by the formula:

$$\alpha = \frac{D}{l}, \tag{2}$$

where α is the absorption coefficient, D is the optical density, and l is the sample thickness.

The refractive index was measured on an Abbe IRF-454 refractometer for a wavelength of 589.3 nm with an accuracy of ± 0.0005. The results of the measurements are shown in Tables 1 and 2.

3. Results

Figures 2 and 3 present the results of the Verdet constant measurements at various wavelengths for metaphosphate and borate glasses doped with various concentrations of terbium. It can be seen from the graphs obtained, the Verdet constant increases with increasing concentration of terbium ions and decreasing incident radiation wavelength. Its values become insignificant in the long-wavelength region.

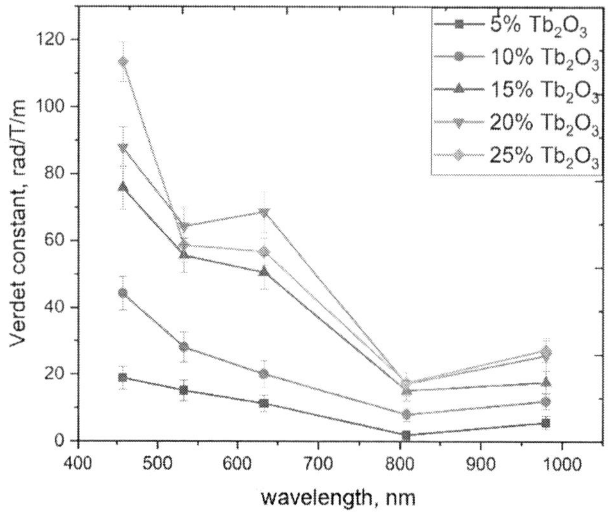

Figure 2. Verdet constant spectral dependence of metaphosphate glasses with different concentration of terbium-comprising dopants (mol%).

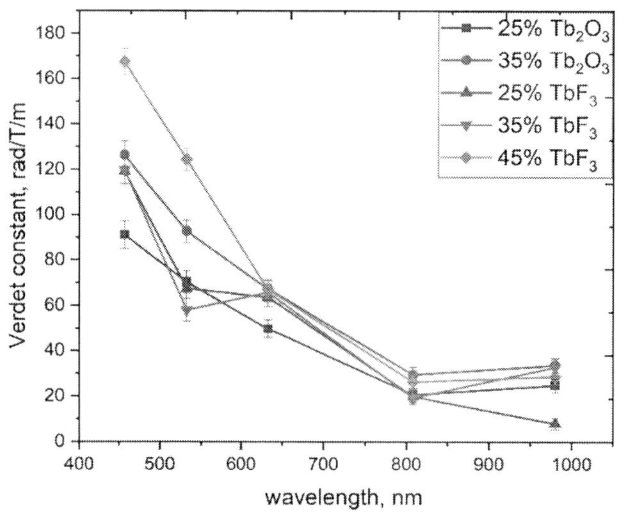

Figure 3. Verdet constant spectral dependence of borate glasses with different concentration of terbium-comprising dopants (mol%).

The results of the absorption spectra measurements are shown in Figures 4 and 5.

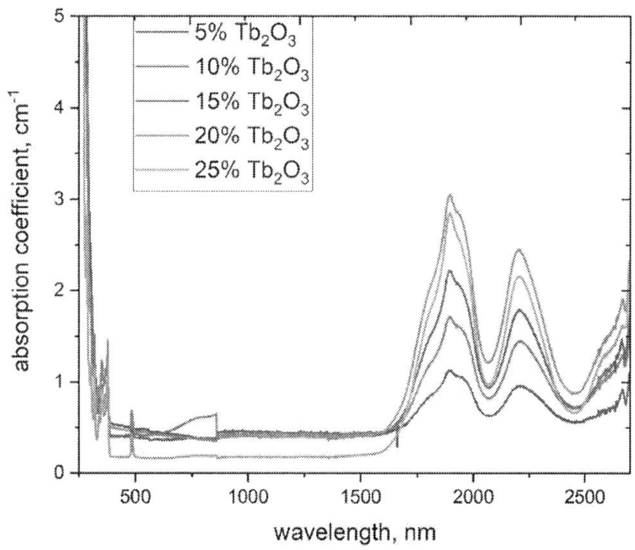

Figure 4. Absorption spectra of metaphosphate glasses with different Tb_2O_3 concentration (mol %).

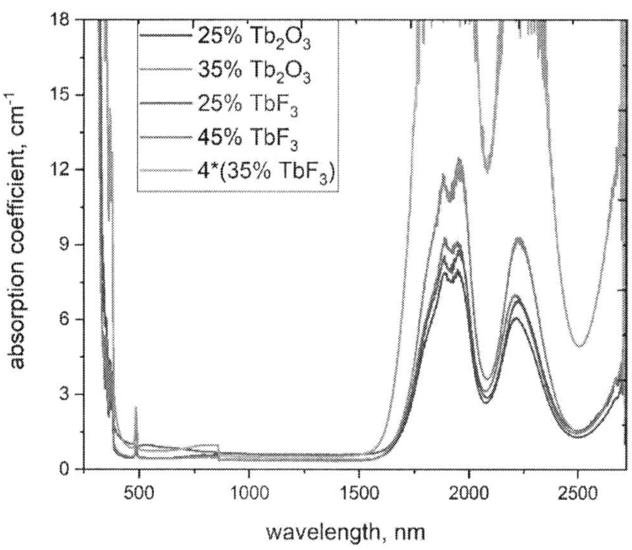

Figure 5. Absorption spectra of of borate glasses with different concentration of terbium-comprisisng dopants (mol %).

According to the literature [10], the set of the absorption bands of terbium ions is located in the wavelength region of 350-500 nm. For this reason, materials doped with terbium acquire a yellow tint. However, in our case these bands had a low intensity; therefore, the materials under study were colorless. The absorption spectra show the presence of two broad absorption bands in the region of 1500–2500 nm, which are associated with Tb^{3+} ions. With terbium concentration increase in the glass composition, the intensity of these bands increased.

Table 1. The refractive index of the metaphosphate glasses with different Tb_2O_3 concentration (mol %).

Samples	5% Tb_2O_3	10% Tb_2O_3	15% Tb_2O_3	20% Tb_2O_3	25% Tb_2O_3
Refractive index	1.5891	1.5831	1.5756	1.5723	1.5599

Table 2. The refractive index of the borate glasses under study.

Samples	25% Tb_2O_3	35% Tb_2O_3	25% TbF_3	35% TbF_3	45% TbF_3
Refractive index	1.6572	1.6574	1.6570	1.6571	1.6572

The refractive index of phosphate glasses decreases with an increase in the concentration of terbium, while in borate glasses we observe an inverse dependence.

4. Conclusion

In this work, phosphate and borate glasses with different concentrations of terbium ions were synthesized. The absorption spectra of glasses in the visible and near-IR spectral regions were obtained. All glasses have high transparency in the wavelength range from 300 to 1500 nm. The refractive index of phosphate glasses decreases, while in borate glasses it increases insignificantly with increasing terbium concentration. Measurements of the rotation angle of the light polarization plane and the subsequent calculation of Verdet constant showed that, with increasing registration wavelength, the value of Verdet constant decreases for all the glasses under study. Concentration studies of the magneto-optical characteristics showed a direct relationship between the content of the paramagnetic ions and the value of Verdet constant for all glasses doped with terbium ions.

References

[1] Jain A, Kumar J, Zhou F, Li L and Tripathy S 2002 *Am. J. Phys.* **67** 714–7
[2] Chen Q, Ma Q, Wang H, Wang Q, Hao Y and Chen Q 2017 *Bol. la Soc. Esp. Ceram. y Vidr.* **56** 1–12
[3] Starobor A V., Zheleznov D S, Palashov O V., Savinkov V I and Sigaev V N 2016 *Opt. Commun.* **358** 176–9
[4] Jin W, Ding J, Guo L, Gu Q, Li C, Su L, Wu A and Zeng F 2018 *J. Cryst. Growth* **484** 17–20
[5] Dentz D J, Puttbach R C and Belt R F 1974 *Magn. Magn. Mater. — 1973 Ninet. Annu. Conf.* **954** 954–8
[6] Yasuhara R, Snetkov I, Starobor A, Mironov E and Palashov O 2016 *Opt. Express* **24** 15486
[7] Kagamitani Y, Pawlak D A, Sato H, Yoshikawa A, Martinek J, Machida H and Fukuda T 2004 *J. Mater. Res.* **19** 579–83
[8] Ganschow S, Klimm D, Reiche P and Uecker R 1999 *Cryst. Res. Technol.* **34** 615–9
[9] Ballato J and Snitzer E 1995 *Appl. Opt.* **34** 6848–54
[10] Ebendorff-Heidepriem H and Ehrt D 2000 *Opt. Mater. (Amst).* **15** 7–25

Excitation of spin waves edge modes in chains of ferromagnetic pillars

S Osokin[1,2], A Sharaevskaya[2,3], A Safin[2], D Kalyabin[1,2]
[1]Moscow Institute of Physics and Technology, Dolgoprudny 141701, Russia
[2]Kotelnikov Institute of Radioengineering and Electronics of RAS, Moscow 125009, Russia
[3]Laboratory "Metamaterials", Saratov State University, Saratov 410012, Russia

Abstract. We report on studies of spin waves properties in finite waveguides, formed by discrete arrays of ferromagnetic pillars. Dispersion characteristics of spin waves in such structures can be altered by changing the magnetization configurations of chain of pillars, thus creating the conditions for the bulk and edge modes to exist. Using theoretical and numerical methods we study the frequencies and spatial distributions of different spin waves modes formed by the presence of the boundaries and shape anisotropies of the waveguide.

1. Introduction

The studies of magnetization dynamics in magnetic and ferromagnetic systems [1,2] are still relevant due to the perspective for using spin waves and spin currents for the data processing [3,4] and data storage [5]. For this purpose, a variety of magnetic nanostructures were developed and studied [6,7,8]. The properties of spin waves in such structures can be shaped by the geometry of such structures. In the spatially confined geometries such as thin films, wires and dots the shape effects and the interactions with external forces leads to the existence of new topological states of the spin waves. Devices designed on such principles use the properties of spin waves associated with the topology of magnetic structures, for example, the creation of dedicated directions for the propagation of spin waves and the exchange of energy between the spin waves in individual elements of the devices. Artificial structuring of magnetic films and material parameters of magnetic films allows one to create structures with configurable allowed or forbidden directions of propagation, or, by changing the boundary conditions, allows one to control the dispersion properties of propagating nanoscale spin waves [9]. The magnetic configuration can be further controlled by the application of the external bias (magnetic field, temperature gradient, electric field or strain in magnetic systems) [10]. The interaction between discrete ferromagnetic nanopillars and nanowires were studied in the number of recent papers [11,12]. The theoretical formalism was developed to describe the properties of spin waves propagating in the infinite periodical and semi-infinite clusters of pillars [13,14]. However, the topological effects of confined structures on spin wave properties are an area of interest for future research. In this work, we present the theoretical and numerical studies of spin wave properties in finite chain of ferromagnetic pillars, where the influence of edge effects and shape anisotropies on dispersion characteristics is taken into account.

2. Theoretical model and numerical calculations

The properties of finite-size magnetic structures [15,16] in spintronics and magnonics, such as waveguides and resonators, differ from infinite and semi-infinite structures, which are often encountered in the construction of theoretical models. We present a theoretical study of a finite waveguide for spin waves, formed by a discrete linear chain of ferromagnetic pillars [17]. For such structures, bulk and edge modes of spin waves modes exist due to the presence of a boundary in structures with finite spatial dimensions or in the presence of shape anisotropies in the discrete elements of structure.

In this way, resonant properties of such spin waves in the chain of ferromagnetic pillars in an external magnetic field are considered with the presence of the dipole-dipole interaction between them. A theoretical study of the properties of spin waves in such structures by solving the Landau-Lifshits-Gilbert's (LLG) equations in macrospin approximation is presented. Such an approach makes it possible to distinguish the edge modes of spin waves, whose frequencies depend on the magnetization configuration at the chain boundaries and defects. In addition to the resonance frequency of the bulk spin wave mode, the edge mode is localized near the edge of the chain with distinct frequency. Pillars with different magnetization can act as a defect in the chain of identical pillars, and the defect mode with its own frequency will be localized near it.

To take into account full spectrum of the interactions between pillars we numerically solve the system of LLG equations. The time evolution of the magnetization is obtained from the micromagnetic numerical calculations using the MuMax software. We chose to use permalloy (Py) as a ferromagnetic material (Py) for pillars (Fig.1). The equilibrium magnetization for cylindrical pillars with height $h=100$ nm and radius $R=25$ nm is non-uniformly distributed inside each pillar due to shape effects.

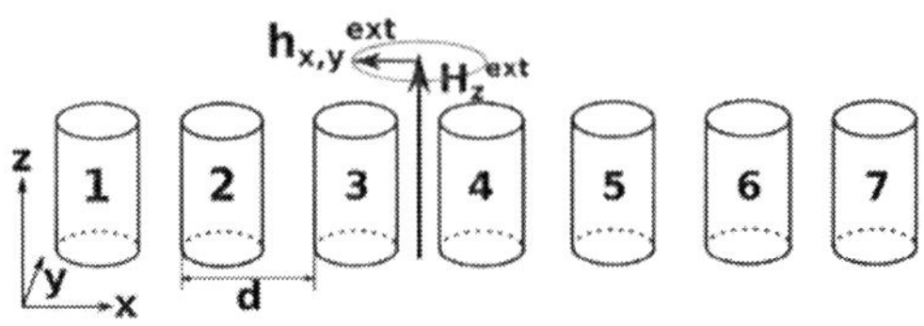

Figure 1. The chain of ferromagnetic pillars with numbers $i=1,..,7$ and constant and time dependent components of external magnetic field.

The effective magnetic field distribution is depicted in the Fig. 2 for 3 pillars in the middle of the chain and the area of free space near them. The static dipole-dipole interaction between pillars leads to the different effective magnetic field distribution inside pillars in the middle and at the edge of the chain. This distribution along the axis of the symmetry for each pillar is shown in the Fig. 3, where we one can see that x- and z-component of the effective magnetic field B^{eff} can be separated into 3 regions: top part ($h<z<2h/3$), middle part ($2h/3<z<h/3$), and bottom part ($h/3<z<0$). The shape anisotropy effects from edges of the pillar are localized in the top and bottom regions.

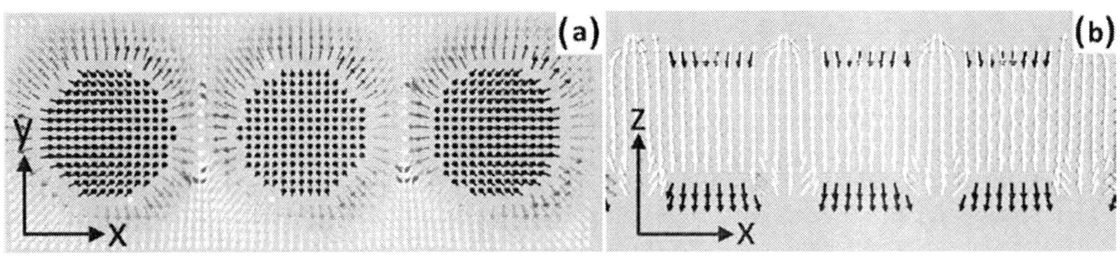

Figure 2. The distribution of the effective magnetic field in the top (x,y)-plane (a) and central (x,z)-plane (b). The arrows shows only direction of the field. Colors denotes the direction of the field. White and black colors are for +z and -z directions, other colors are for directions in (x,y)-plane).

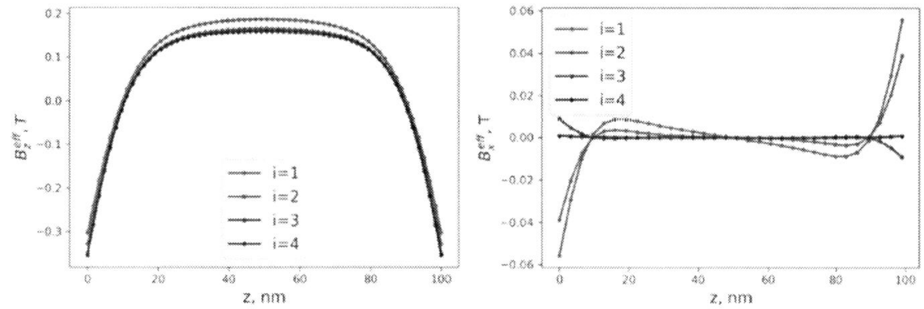

Figure 3. The distribution of the effective magnetic field inside the pillars with numbers $i=1,2,3,4$ along z-axis with the external magnetic field $B^{ext}_z=0.3\ T$.

From the magnetostatic approach the eigenfrequency of spin waves f can be formulated by analytical formula:

$$f=\gamma((H_{ext}+(N_x-N_z)M_{sat})(H_{ext}+(N_y-N_z)M_{sat}))^{1/2}$$

Where $N=(H_x,N_y,N_z)$ is the demagnetization tensor for region. Each region can be approximated by its own demagnetizing tensor. Further we can average the magnetization for each pillar with index i over each region and perform the Fourier transformation for the time evolution $m^i_x(t)$ in order to analyze frequency characteristics of the eigenmodes.

In our model we use the periodic external magnetic fields an external force of the spin wave excitation. The short pulse of this field is circularly polarized in (x,y)-plane and has carrying frequency $f=14\ GHz$. In Fig. 4 the time and frequency dependencies ($m_x(t)$ and $m_x(f)$) are calculated for single isolated cylindrical pillar. From the frequency spectrum we established that there are two resonance frequencies corresponding to different regions of the pillar. These frequencies are $f^{edge}=5\ GHz$ and $f^{bulk}=10\ Ghz$, and their amplitudes localized near the edges of the pillar and in the middle of the pillar correspondingly.

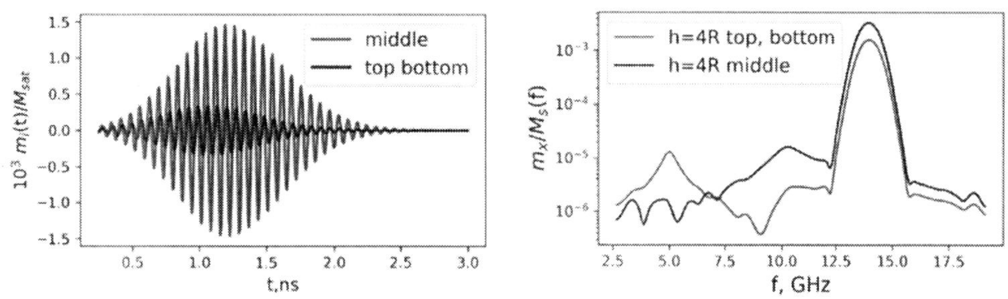

Figure 4. Time evolution of the magnetization $m_x(t)$ and frequency spectrum $m_x(f)$ for single isolated pillar averaged over middle, top and bottom regions.

3. Excitation of the spin waves in the chain of pillars

If the external magnetic field is applied to one pillar with index $i=1$ at the edge of the chain, then the excitation is transferred to neighboring pillars by the dipole-dipole interaction (Fig. 5).

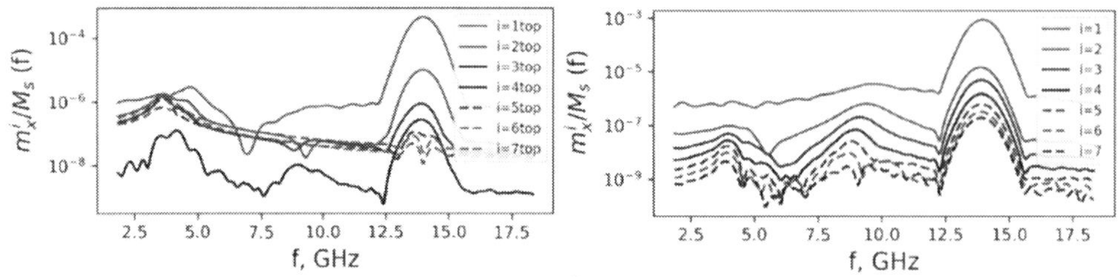

Figure 5. The frequency dependency of the amplitudes m^i_x for the pillars with numbers i, averaged over top (left picture) and middle (right picture) regions of the pillar.

In Fig. 5, 3 main resonance frequencies correspond to the frequency of the excitations f^{xt} of the external magnetic field and two modes of the spin waves. These modes are the bulk mode with frequency $f^{bulk}=9$ *GHz* and edge mode $f^{edge}=4$ *GHz*. Due to the coupling in the chain of pillars the resonance frequencies has lesser resonance values than single isolated pillar. The additional demagnetization at the edge of the chain leads to the frequency shift of the magnetic oscillation for the pillars $i=1,2$.

4. Summary

In conclusion, we present the results of theoretical investigations of frequency characteristics of spin waves in finite chains of ferromagnetic pillars. We considered the excitation of spin wave modes with the pulse of external magnetic field. For cylindrical pillars with finite height it was shown that the demagnetization and shape effects leads to the appearance of two resonance frequencies. The dipole-dipole interaction in the finite chain of such pillars causes the negative frequency shift. The additional resonance frequencies for the edge modes of the spin waves exists for the boundary elements due to the demagnetization at the edges of the chain.

Acknowledgments

Authors acknowledge Russian Foundation for Basic Research (Projects RFBR #18-37-20048, #18-29-27018, #18-07-00509) and Grant of the President of the RF (No. MK-3607.2019.9) for financial support.

References

[1] Sander D, Valenzuela S, Makarov D, Marrows C, Fullerton E, Fischer P, McCord J, Vavassori P, Mangin S, Pirro P,Hillebrands B, Kent A, Jungwirth T, Gutfleisch O, Kim C and Berger A. 2017 *Journal of Physics D: Applied Physics* **50** 363001.

[2] Chumak A, Serga A and Hillebrands B 2017 *Journal of Physics D: Applied Physics* ***50*** 244001.

[3] Nikitin A, Ustinov A., Alexander A., Chumak A.,Serga A.,Vasyuchka V., Lahderanta E, Kalinikos B, Hillebrands B 2015 *Applied Physics Letters* **106** 102405

[4] Sinova J and Žutić I 2012 *Nature materials* **11** 368

[5] Gertz F, Kozhevnikov A Filimonov Y, Nikonov D and Khitun A 2015 *IEEE Journal on Exploratory Solid-State Computational Devices and Circuits* **1** 67-75

[6] Wang X, Zhang H, Wang X 2018 *Phys. Rev. Applied* **9**024029

[7] Volker S, Tobias S, Rodolfo A et. al. 2019 Nature Nanotechnology **14** 02

[8] Ivanov O Nikitov S, Gulyaev Y, 2006.. Physics-Uspekhi. **49** 167

[9] Galkina E, Ivanov B 2016 *JETP Letters* **104**

[10] Krawczyk M, Grundler D, 2014 *Journal of physics. Condensed matter* **26** 123202.

[11] Lisenkov I, Tyberkevych V, Nikitov S, Slavin A 2016 *Physical Review B* **93** 214441.

[12] Meitzler T, Verba R Tiberkevich Vasil, Bankowski E, Melkov G, Slavin A 2013 *Applied Physics Letters*. 103.
[13] Centala G,. Sokolovskyy M, Davies C, Mruczkiewicz M, Mamica S, Rchły J, Klos W, Kruglyak, V, Krawczyk, M, 2019 arxiv
[14] Rychły J, Tkachenko V, Klos J, Kuchko A, Krawczyk M, *2018 Journal of Physics D: Applied Physics* **52**
[15] Barabanenkov Y, Osokin S, Kalyabin D, Nikitov S 2015 *Phys. Rev. B* **91** 214419
[16] Barabanenkov Y, Osokin S, Kalyabin D, Nikitov S 2016 *Phys. Rev. B* **94** 184409
[17] Osokin S, Safin A, Barabanenkov Y, Nikitov S 2018 *JMMM* **465** 519-523

Magneto-optical properties of easy-plane and easy-axis garnet ferrite films for eddy current testing

N.V. Lugovskoy[1], **V.N. Berzhansky**[1,2], **E.Yu. Semuk**[1,2], **S.D. Lyashko**[1], **A.N. Shaposhnikov**[2], **E.A.Evtukhova**[1], **T.V. Mikhailova**[1,2]
[1]Experimental Physics Department, Physics and Technology Institute, V.I. Vernadsky Crimean Federal University, Crimea, Simferopol 295007, Russian Federation.
[2]Research Centre for Functional Materials and Nanotechnologies, Physics and Technology Institute, V.I. Vernadsky Crimean Federal University, Crimea, Simferopol 295007, Russian Federation

Abstract. The paper presents the magnetic, optical and magneto-optical properties of easy-plane and easy-axis films on the base of Bi-garnet. The possibilities and features of the operation of these films as sensors in eddy current magneto-optical introscopy are shown on the example of visualization of linear defects in products made from aluminum alloys.

1. Introduction

Magneto-optical (MO) eddy-current (EC) introscopy is a relatively young type of flaw detection, but it has several advantages over other methods by which it can find its place among other types of non-destructive testing of products made from various conductive magnetic and non-magnetic materials [1 - 4]. The main sensory element in the MOEC introscopy is the MO of epitaxial film of garnet ferrite (FGEF) grown by the method of liquid phase epitaxy. These films should have the greatest possible effect values Faraday and specially selected magnetic properties, such as the type of magnetic anisotropy. In the literature already dealt with particular application as sensors with magnetic anisotropy of films like "easy axis" (LO) [1.2].The main disadvantage of these sensors is the stochastization of the domain structure (DS) at high excitation fields [2, 5]. In easy-plane (EP) of the films, in which there is practically no domain structure there is no such lack. When the EC magnetic field acts on the EA or EP sensor, different remagnetization mechanisms are working. In EO films it is the processes of displacement of the domain boundaries and in EP films - the magnetic moment rotation processes. This difference significantly affects the operation of magnetic films as sensors for dynamic magnetic fields. The aim of this work was the study of magnetic and magneto-optical properties of LP and LO films and evaluation of their effectiveness as sensors in EMCO defectoscopy.

2. Experimental results and discussion

Bismuth garnet ferrite films $(BiRe)_3(FeMe)_5O_{12}$, where Re - Y, Gd, Lu, Tm, Nd, Ca; Me - Al, Ga, Si, Ge, Be, Sc, synthesized by liquid phase epitaxy on single crystalline substrates $Gd_3Ga_5O_{12}$ (GGG) of orientation (111). Depending on the requirements for the magnetic properties of the samples, the necessary chemical elements were introduced into the composition, including in order to minimize the mismatch of the lattice parameters $\delta = (a_f - a_s)/a_s$, where a_f and a_s are the lattice parameters of the film and the substrate. Four films with anisotropy "easy plane" $(BiPrLuTmGd)_3(FeAlGa)_5O_{12}$ (EP1), $(BiTm)_3(FeAlGa)_5O_{12}$ (EP2), $(BiY)_3(FeAlSc)_5O_{12}$ (EP4) and two EA films $(BiSmLu)_3(FeGa)_5O_{12}$ (EA2) and $(BiLuTmNdCa)_3(FeSiGa)_5O_{12}$ (EA9) were selected for the study.

The magnetization and anisotropy in uniaxial films was adjusted by replacing Fe ions with Al and Ga ions, as well as with Be – Si, Ca – Ge pairs. The thickness of the films was measured interferometrically, and ranged from 1.9 to 12 μm.

The optical and magneto-optical properties of the films were investigated in the wavelength range of 500–950 nm. The maximum values of the Faraday specific rotation at a wavelength of 520 nm for EA samples is 1.6 deg / μm, and for EP it is 0.74 deg / μm. Figure 1 shows the spectra of the specific Faraday rotation (a) and optical transmission (b) for several EP and EA films.

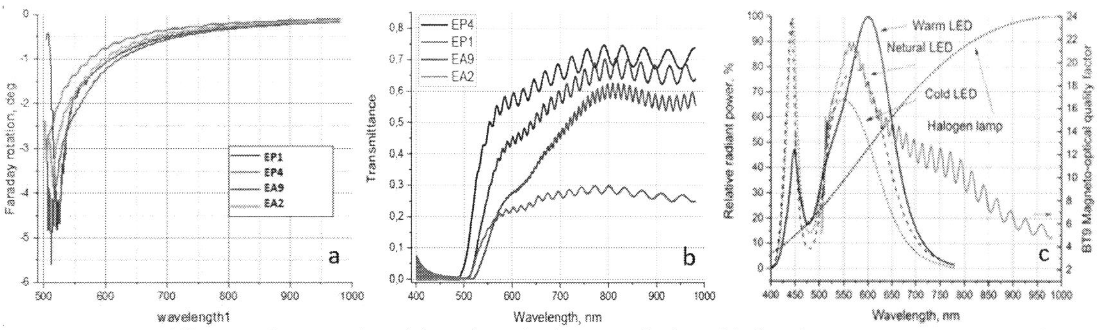

Figure 1. Specific Faraday rotation (a) and optical transmission (b) for the EP1, EP4, EA9 and EA2 samples. The spectrum of the EP1 film Q-factor and the emission spectra of various light sources (c).

The maximum of the Faraday Effect in the short-wave part of the spectrum is due to electronic transitions in tetra and octahedral Fe ions of garnet [6, 7]. The maximum of the optical absorption is also observed in the same region; therefore, an important parameter of the films is the magneto-optical quality factor, which can be estimated by the formula

$$Q = 2\theta_f/\ln T, \qquad (1)$$

where θ_f is the value of the Faraday rotation, T is the optical transmission.

Figure 1c presents the spectrum of MO of quality factor, on the basis of which recommendations can be made on the choice of the type of light source required for operation as part of an MOEC introscope. It is seen that the most optimal sources are Warm LED, the emission spectrum of which have the greatest overlap with the spectrum of MO of quality factor.

An important characteristic of films for operation as sensors in the MOEC of flaw detection are the field dependences of magnetization, which largely determine their sensitivity. The large values of the Faraday Effect, which the films of Bi-garnets possess, make it possible to measure these dependences by the MO method. Figure 2 presents similar dependences, measured at a wavelength of 640 nm, which for EA (Fig. 2a) and EP (Fig. 2b) films have a hysteresis and hysteresis-free character, respectively. The main parameters that determine the sensitivity of the films are the normalization field H_n for the EP films, the saturation field H_s, and the coercive force H_c for EA sensors.

The coercive force H_c and the type of domain structure are an important factor in the operation of the MO introscope because they determine the remagnetization process in alternating magnetic fields. The process of visualization of the distribution of magnetic fields of eddy currents depends on the dynamics of the domain structure in EA sensors. In all EA films, a labyrinth domain structure is observed with a period from 5 to 32 μm.

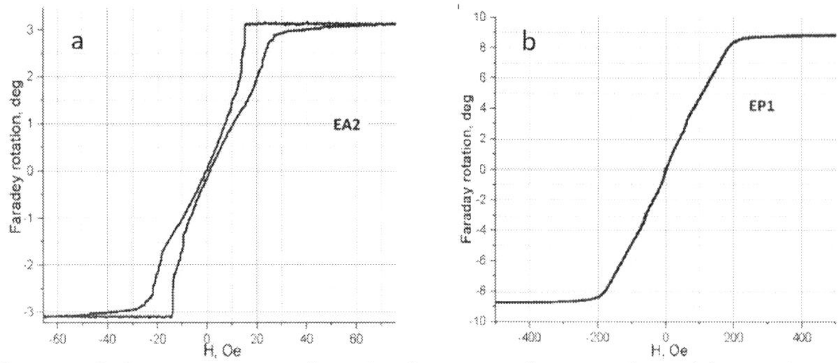

Figure 2. MO hysteresis loops at a wavelength of 640 nm for samples with EA (EA2) and EP (EP3) anisotropy.

The normalization and saturation fields are determined by the values of the effective magnetic anisotropy field

$$H^{a}_{eff}= H_{dem}+H_{u}+H_{k}, \quad (2)$$

where $H_{dem} = 4\pi M_s$, demagnetizing field of the sample form, $H_u = 2K_u/M_s$ is the field of uniaxial anisotropy, $H_k = -4/3K_1/M_s$ is the field of cubic magnetic anisotropy. The uniaxial anisotropy constant is determined to the greatest extent by the magnetoelastic contribution due to the magnitude and sign of δ. The cubic anisotropy constant in Bi-garnets is usually small, so that $H_k<H_{dem}$, H_u. The field of effective anisotropy H^{a}_{eff} can be measured by FMR.
The saturation fields for EA films are in the range of 20 – 70 Oe, and the normalization fields for EP films of 230 – 330 Oe.
When using FGEF as sensors in the MOEC introscopy, certain conditions are imposed on the surface quality of the samples - surface roughness of no more than 1-5 nm and the absence of larger defects. This is necessary for the rapid remagnetization of the film for EC magnetic field visualization, since it is known that domain walls are very sensitive to defects in the film structure and can linger on them.
The value of H_c is largely determined by the defectiveness of the films and in the studied EA films will vary from 0.5 to 2.7 Oe.
Figure 3a shows the labyrinth domain structure of the EA3 film obtained by standard magneto-optical microscopy. The surface defect of the films and a detailed study of the DS was carried out on the multifunctional complex NTEGRA.

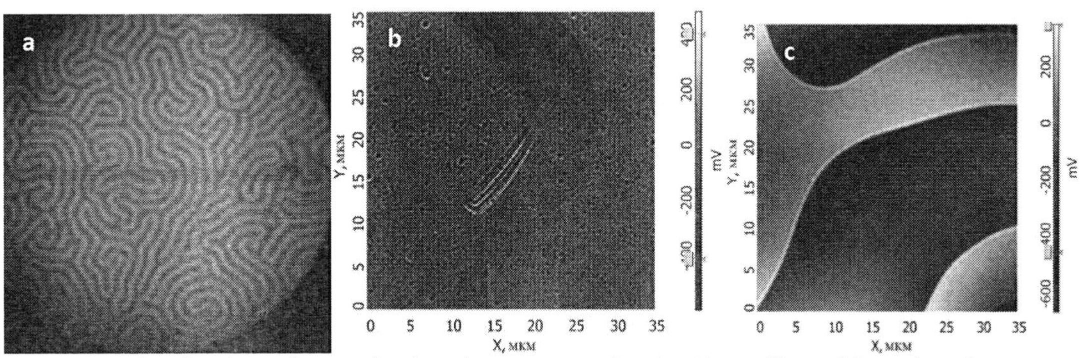

Figure 3. Visualization of a static domain structure for the EA3 film with a domain structure period 2w = 32 μm and thickness h = 4.3 μm using the methods: a - magneto-optical microscopy, b - scanning near-field magneto-optical microscopy, c - scanning laser magneto-optical microscopy.

So, using the atomic force microscopy method on a NTEGRA scanning probe microscope (SPM), it was shown that the average roughness of the films studied did not exceed 5 nm. Figure 3b demonstrates the effect of a relatively large defect on the position of the domain wall in the film. The image was obtained using the method of scanning near-field magneto-optical microscopy on SPM NTEGRA. A more contrast image of both domains and domain boundaries in EA films can be obtained using the SPM NTEGRA method of scanning laser magneto-optical microscopy (Fig. 3c).
To determine the effective magnetic anisotropy fields H_{eff}, as well as to assess the degree of film homogeneity, the FMR method was used. It was found that almost all the films had additional layers located, as a rule, at the interface of the film-substrate and associated with the features of the growth of the films. As an example, figure 4 shows the FMR spectrum for the EP film (EP1) and the dependence of the resonant magnetic fields of the main and additional layers on the orientation of the magnetic field relative to the film plane.

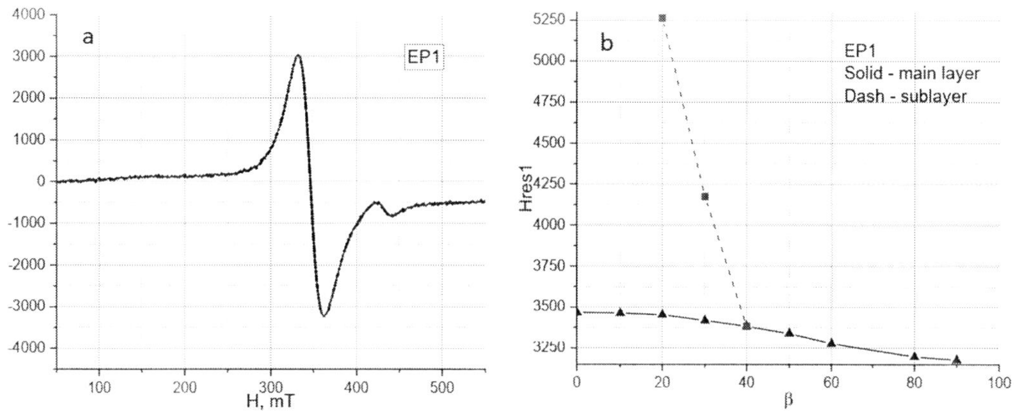

Figure 4. FMR spectrum (a) and angular dependencies H_r (b) for the sample EP1.

The effective magnetic field of anisotropy H_{eff} can determined from FMR data using known Kittel formulas for the resonance fields in perpendicular and parallel configuration [8]. Analysis of angular dependencies H_r (Fig. 4b) show that both layers have an easy plane type anisotropy, but the effective magnetic field of the magnetic anisotropy of the base layer is H_{eff} = 356 Oe, and in the interface layer the substrate is H_{eff} > 1900 Oe. This means that magnetoelastic stresses play a large role in the additional layer due to the nonzero value of the mismatch parameter $\delta = (a_f - a_s) / a_s$. The presence of an additional layer is also indicated by measurements of the dynamic magnetic susceptibility by the inductive-frequency method.
The sensory properties of the EP and EA films were studied under the same conditions using the eddy current stroboscopic magnetooptical introscope in the frequency range from 8 to 80 kHz, the amplitude of the inductor alternating field varied from 2 to 400 Oe, and the bias magnetic field varied from 0 to 40 Oe. Test samples were taken linear defects of the "through-slit" type in aluminum with an opening width of 20 to 500 microns.
Figure 5 shows the MOEC images of model defects in the form of slits with a disclosure of 20 and 40 μm of aluminum, measured at a frequency of 30 kHz for EA and EP films, illustrating some of the visualization features.

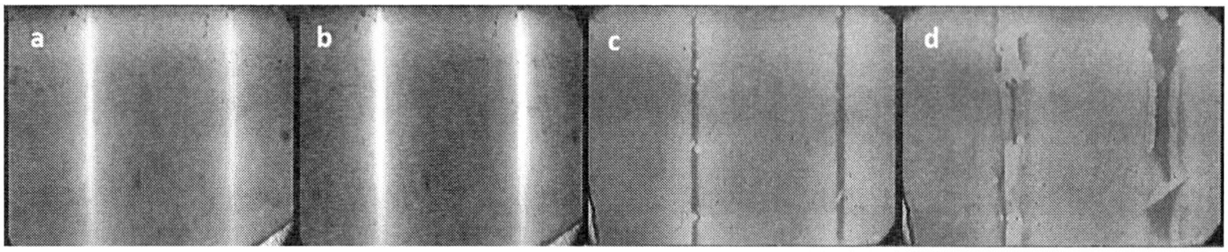

Figure 5. MO visualization of 20 and 40 μm slits for a sample from aluminum, obtained with different types of anisotropy films (EP1 - a, b; EA2 - c, d) at the frequency f = 30 kHz, field of excitation H = 80 Oe (a, c) and 140 Oe (b, d).

For the EA sensors at low excitation fields H, linear defects are displayed as bright stripes, at fields greater than a certain critical H_{cr} ~ 30 Oe at f = 30 kHz, the defective area is displayed with a dark stripe and the image assumes a clear binary character. For the EP sensors at any field of excitation defects are displayed in bright stripes (Fig. 5a,b). The increase of the excitation field amplitude leads to the increase in contrast (brightness) and a certain growth of the image width.

The most contrasting image with the use of EP films is obtained on a sample of EP1 $(BiLuTmPrGd)_3(FeHeGa)_5O_{12}$, which has the highest specific Faraday rotation of 0.74 grd/μm among all EP films (Fig. 1a,b). The image contrast in the EP sensors increases with an increase in the excitation field (Fig. 1b), which is important when detecting subsurface defects. For the EA films, at low fields, a clear binary image of the defect is observed (Fig. 1c), and at large fields, the image quality deteriorates sharply due to stochastization of the DS (Fig. 1d), which complicates the identification of the defect.

The bias field directed along the axis of the inductor has a significant impact on the operation of EA sensors, increasing the contrast and clarity of the image. For EP films the bias field has no effect on the visualization of defects. This is due to the fact that for EA films, the bias field significantly affects the domain structure, which is absent in films with EP anisotropy.

Sensors based on EA films have a rather high sensitivity and do not require large excitation fields. A clear binary MO image of a defect appears already with an excitation field H = 40 Oe. A further increase in the field amplitude significantly increases the image of the defect, which may indicate a large stock of resolution and sensitivity of sensors (Fig. 5c,d).

EP sensors also have good sensitivity, which grows with frequency. So, at 8 kHz, the image appears at fields of the order of 100 Oe, and at 30 kHz - about 30 - 40 Oe. A further increase in frequency, up to 80 kHz, no longer has such a large effect.

3. Conclusions

Multicomponent films of Bi-garnets with easy plane and easy axis magnetic anisotropy were synthesized. The spectral dependences of the optical transmission, the Faraday Effect and the magneto-optical quality factor are determined. The magneto-optical method determines such magnetic characteristics as the saturation field and coercive force for easy axis films and the normalization field for easy plane films. Magneto-optical microscopy, scanning near-field and scanning laser magneto-optical microscopy were used to study the features of the domain structure for easy axis films. The presence of additional layers in easy-plane films were found and the values of the effective anisotropy fields by the FMR method was determined. It is shown that the effective anisotropy field in the additional layer is much larger than in the main layer. A comparative analysis of the effectiveness of the sensory properties of easy plane and easy axis films when defects are detected by the method of eddy current magneto-optical introscopy is carried out.

Acknowledgments
This work was supported by the RF Ministry of Education and Science in the framework of the base part of the state task (Project No. 3.7126.2017/8.9) and the V.I. Vernadsky Crimean Federal University Development Program (Project No. ВГ11/2018).

References
[1] Vishnevsky V, Berzhansky V, Lugovskoy V, Prokopov A, Pankov F 2015 SSP 230
[2] Berzhansky V, Filippov D, Lugovskoy N 2016 Phys. Procedia 82C
[3] Cheng I, Liu X, Bai L, and Tian G. 2013 Appl. Mech. Mater. 330
[4] Grechishkin R, Kustov M, Ilyashenko S, Gasanov O, Dumas-Bouchiat F and Dempsey N 2016 J. Appl. Phys. 120 174502
[5] Kandaurova G.S., Osadchenko V.H. 2004 PMM. 97(2). 38-54
[6] Zvezdin A.K., Kotov V.A. 1988 Magnetooptics of thin films (M.: Nauka) 198
[7] Randoshkin V.V. Chervonenkis A.Y. 1990 *Applied Magnetooptics* (M.: Energoatomizdat) 320
[8] Kittel C. Phys.Rev. 1948 73,155

Increasing efficiency of GaN HEMT transistors in equipment for radiometry using numerical simulation

V G Tikhomirov[1,2], A G Gudkov[3], S V Agasieva[4,5], D D Dynaiev[2], M K Popov[1], S V Chizhikov[3]

[1] Saint-Petersburg State Electrotechnical University "LETI", St. Petersburg 197376, Russia
[2] Closed Joint Stock Company "Svetlana-Electronpribor", St. Petersburg 194156, Russia
[3] Bauman Moscow State Technical University, Moscow 105005, Russia
[4] Peoples' Friendship University of Russia, Moscow 117198, Russia
[5] "Microwave radiometry global excellence" ltd., Moscow 115201, Russia

Abstract. The numerical impact modeling of some external effects on devices based on AlGaN/GaN heterostructures (HEMT) was carried out. The mathematical model was created that allowed to predict the behavior of the drain current depending on condition changes in the heterostructure in the buffer region and to start the process of directed construction optimization of the devices based on AlGaN/GaN HEMT with the aim of improving their performances.

1. Introduction

The main advantage of GaN HEMT is the highest power density and the ability to achieve high efficiency during operation in amplifiers or MIC. However, in recent years, such devices have a lag in expected increasing efficiency with the quality improvement of the GaN buffer layer growth in the heterostructure. Optimization of such transistors production cycle is still a complicated and expensive procedure [1-4]. This paper presents the results of numerical simulation and calculation of the buffer layer features of GaN HEMT for increasing efficiency when they are operating as a part of amplifiers for passive radar equipment.

2. Simulation results

The reasons of impossibility of high efficiency achievement by these transistors were analyzed. It was found that improvement of GaN crystal quality was accompanied by the decreasing of traps and defects in the region near the channel [5]. It led to increasing of the buffer conductivity above some critical value. This created the conditions for the appearance of a noticeable current through the transistor in the state of maximum sub-gate region depletion and a large bias at the drain. This situation is shown in Fig.1. It shows that there can be the conditions for the current bypassing the depletion region when the buffer layer has sufficient conductivity. It causes an unexpected increase of leakage current in the range 30-50V, that corresponds to the operating mode of transistors in amplifiers-class AB.

In this paper, a numerical simulation of the current flow in these modes was carried out and ways to overcome this problem were offered.

It can assume by using the research results that the doping of the buffer layer at the certain depth by acceptor impurities, for example, "C" or "Fe", allows to achieve a significant decrease of the leakage current through the buffer. It significantly increases the opportunity to achieve high efficiency that was predicted for these devices by theoretical calculations.

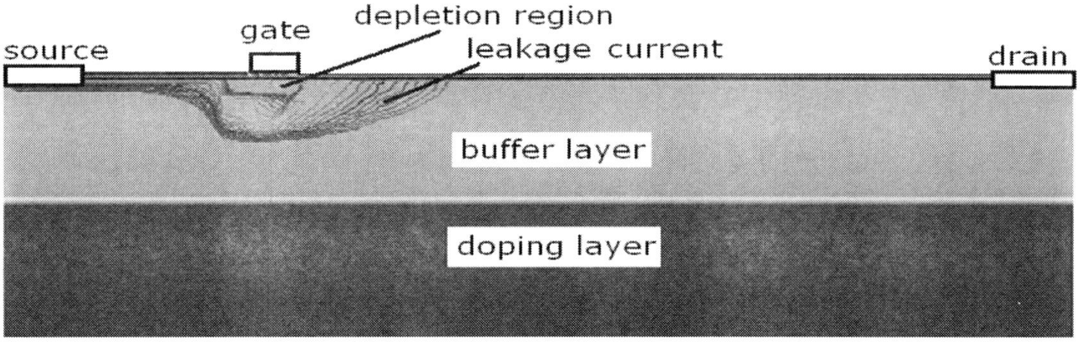

Figure 1. The image of the screen for the calculating of the leakage current through the transistor in the state of maximum depletion of the sub-gate region and a large bias in the drain.

It is planned to use a rectangular doping profile that means the concentration of the acceptor impurity will remain at the same level increasing the depth towards the substrate. In this case, the leakage current that flows between the source and the drain of the transistor through the buffer layer has to significantly reduce by this approach. As a result, the device efficiency will increase. It is assumed that the microwave characteristics of transistors with the doped buffer layer by iron or carbon will not degrade. In order to achieve the desired result, it is necessary to optimize the semiconductor structure by selecting the concentration of impurities and the doping profile by numerical experiment.

The results of computer simulation of HEMT's semiconductor structure are given below. A microwave transistor based on GaN with a doped buffer layer was simulated. The current-voltage characteristic of the source-drain breakdown for different values of gate lengths was obtained. Obtained results confirmed the presence of a specific source-drain breakdown in these transistors. And it is necessary to find a way to solve this problem.

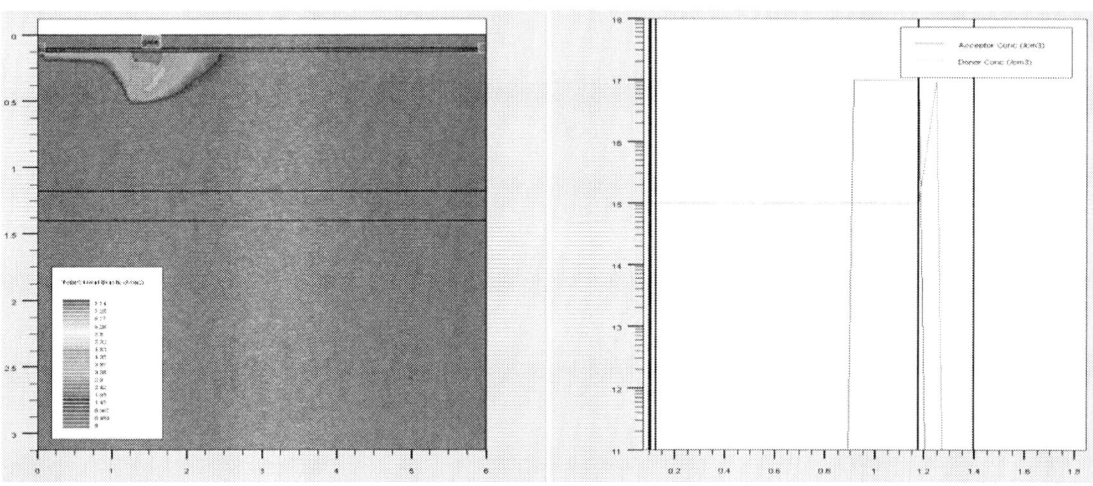

Figure 2. The image of the screen with the simulation results for doping of a small part of the buffer layer by carbon with the acceptor concentration n = 10^{17} cm^{-3}

Figure 2 shows the calculation of the semiconductor structure of HEMT based on gallium nitride with doping of the buffer layer by carbon with the acceptor concentration n = 10^{17} cm $^{-3}$. The impurity is located quite deep in the buffer layer. The impurity is located in the depth of the structure in the range from about 0.9 microns to 1.2 microns. The current flows between the source and drain electrodes, that confirms the existence of the punch-through effect. The current begins to flow from the source, then it flows around the Schottky gate, under which a region depleted by carriers is formed, and then rushes by the shortest way to the channel where the two-dimensional electron gas is located, and already reaches the drain through the channel .The current density decreases with depth into the substrate. This is due to the presence of an acceptor impurity in the depth of the buffer layer, which prevents the flow of current through the GaN.

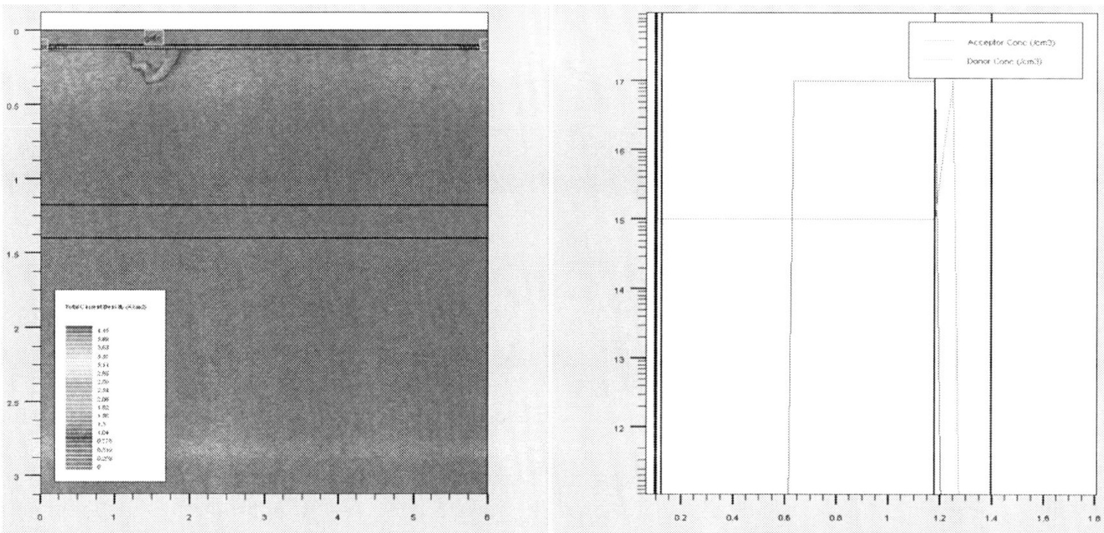

Figure 3. The image of the screen with the simulation results for doping of a greater part of the buffer layer by carbon with the acceptor concentration n = 10^{17} cm $^{-3}$

Figure 2 shows a similar calculation of the semiconductor structure except that the depth of the acceptor impurity (in this case of carbon) has been changed. The impurity is located in the depth of the structure in the range from about 0.6 μm to 1.2 μm. Comparing figure 2 and 3, it can be concluded that the source-drain breakdown current significantly decreases with a decreasing of the impurity in the depth (figure 3). In addition, the depth of the flowing current in the buffer layer also decreases. The current, as shown in figure 2, flows around the sub-gate region of the transistor structure and rushes to the drain through a channel containing two-dimensional electron gas. A strong decrease in the punch-through current is associated with a decreasing of the doping depth by acceptor impurities of the buffer layer GaN.

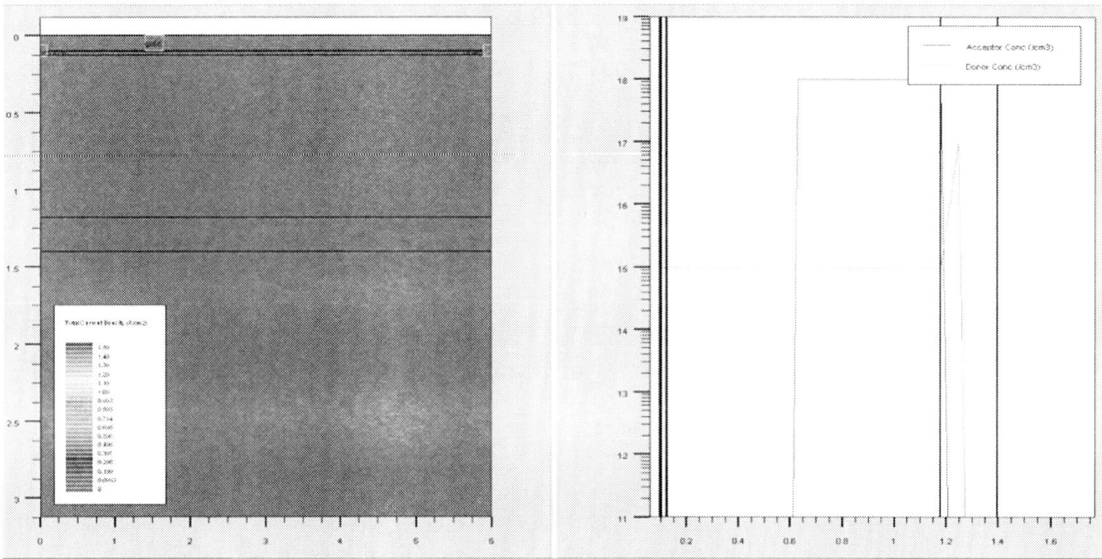

Figure 4. The image of the screen with the simulation results for doping of a greater part of the buffer layer by carbon with the acceptor concentration n = 10^{18} cm $^{-3}$

The breakdown current of source-drain completely disappears with increasing of the acceptor impurity concentration to the value n =10^{18} cm $^{-3}$ at a constant doping depth compared to figure 3. In this case the transistor is in the closed state and does not have source-drain leakage on the buffer layer. Thus, the found concentration and the impurity depth are optimal to eliminate the phenomenon of source-drain punch-through in the buffer layer.

3. Conclusion

According to the research results it can be recommended doping of the buffer layer at a certain depth by acceptor impurities. It can be the introduction of "Fe" or "C" atoms at a concentration about 0.5x10^{18} см$^{-3}$ for the GaN buffer layer. In this way, a significantly increase of the device efficiency can be achieved. More researches are needed for the directional changes of the heterostructure and of the overall transistor design to increase its efficiency.

The research is carried out in the framework of the research project No. 19-19-00349 on the theme: "Development of a methodology and multi-channel multi-frequency microwave radiothermography based on monolithic integrated circuits for finding the 3D distribution and dynamics of the brightness temperature in the depths of the human body" (Russian Science Foundation).

References

[1] Uren M J et al. Buffer design in GaN-on-Si power devices, *J. Phys. D: Appl. Phys.* **51** (2018) 163001.

[2] Stutzmann M, Steinhoff G, Eickhoff M, Ambacher O, Nobel CE, Schalwig J, et al, *Diamond Rel Mater* 2002; **11**:886.

[3] Duffy S J et al. Strain-reduction induced rise in channel temperature at ohmic contacts of GaN HEMTs, *IEEE Access*, vol. **6**, pp. 42721–42728, 2018.

[4] Tikhomirov V, Zemlyakov V, Volkov V, Parnes Ya, Vyuginov V, Lundin W, Sakharov A, Zavarin E, Tsatsulnikov A, Cherkashin N, Mizerov M, and Ustinov V, *Semiconductors*, 2016, Vol. **50**, N. 2, pp. 244–248.

[5] Benbakhti B, Soltani A, Kalna K, Rousseau M, and De Jaeger J. Effects of self-heating on performance degradation in AlGaN/GaN-based devices, *IEEE Trans. Electron Devices*, vol. **56**, no. 10, pp. 2178–2185, Oct. 2009.

Developing of normally-off p-GaN gate HEMT

O B Kukhtyaeva[1], V I Egorkin[1], V E Zemlyakov[1], V V Kapaev[1], A A Zaitsev[1], A F Tsatsulnikov[2], A E Nikolaev[3], A V Sakharov[3]
[1]National Research University of Electronic Technology, Moscow, Russia, 124498
[2]Submicron Heterostructures for Microelectronics, Research & Engineering Center, RAS, Saint-Petersburg, 194021
[3]Ioffe Institute, RAS, Saint-Petersburg, 194021

Abstract. In this paper, the high electron mobility transistor (HEMT) with p-GaN gate is developed. The article demonstrates development of normally-off p-GaN gate HEMT on heterostructure with AlN stop layer for p-GaN selective etching and formation of normally-on GaN HEMT on the same wafer. It is researched that the gate length and gate-to-drain length influence on transistor operation and its breakdown voltage. The ohmic contact resistance and leakage currents depend on etching RF mode. The developed normally-off transistor with gate length of 1 um and gate-to-drain length of 9.5um has a breakdown voltage of 500 V and drain current in open state of 350 mA/mm. It is established that 2 nm SiN under the gate metal influence on device work.

1. Introduction

Gallium nitride (GaN) based HEMTs are widely used for high switching speed and high-power semiconductor device applications because of their high breakdown voltage, high mobility and good thermal stability. Low channel resistance and high carrier channel mobility allow switch higher power signals in a short time at the higher frequencies. Traditional AlGaN/GaN HEMT has a normally-on behaviour as a high-density two-dimensional electron gas (2DEG) is formed by the polarization-induced charges in the AlGaN/GaN interface. The main purpose of GaN HEMT developers is creating of normally-off GaN HEMTs. They provide a simplified scheme of drivers for transistor control without a negative power source and favorable operating conditions for the device safety. There are several ways to create normally-off transistors: gate recess, fluorine treatment of area under the gate, growing under the gate of the intermediate layer NiO_x, which during heat treatment acquires the properties of a p-type semiconductor. The perspective method is use of p-GaN or p-AlGaN layer on a top of AlGaN barrier. There are some methods to form p-GaN gate: traditional selective RF etching, selective growth and hydrogen plasma treatment of p-GaN area not under the gate. The main problems of p-GaN selective etching are the etching depth uniformity of the nongated active region and the plasma-induced damage. To exclude these challenges the transistor is formed on heterostructure with AlN stop layer. The AlN layer beneath the p-GaN layer achieves a high energy barrier layer to suppress the leakage current induced by thermionic emission. The technological route has been developed and our transistor has been simulated and formed. All this are demonstrated in this article.

2.Experiment

For the experiment 40 nm p-GaN/1nm AlN/12nm AlGaN/1nm AlN/GaN/sapphire heterostructures were grown by MOCVD. The device was simulated and then technological route was developed

(Fig.1). The main stages were: formation of alignment marks and isolation, selective etching of p-GaN for ohmic contacts, formation of ohmic contacts, gate metal deposition and then selective etching of p-GaN to form the gate, passivation and formation of first level metallisation.

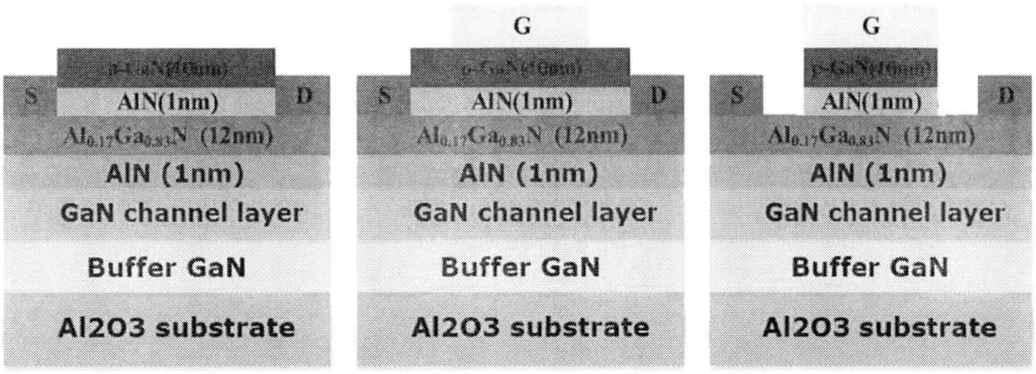

Figure 1. Technological route scheme and cross-sectional structure of p-GaN gate HEMT.

Two etching RF modes (ICP and RIE) have been worked out. Then transistors with different gate length (1um, 10um) and different gate-to-drain length (1um, 1.5 um, 3.5 um, 9.5 um and 1.4 um) have been created. Etching mode influence of on the characteristics has been investigated. ICP etched transistors have better surface morphology and consequently ohmic contact resistance two times less. The ohmic contact resistance of ICP etched transistors is 0.43 Ohm*mm and the resistance of RIE ones – 1.2 Ohm*mm. Surface passivation with Si_3N_4 worsened the characteristics and isolation with mesa-etching has been replaced by ion implantation. To avoid degradation of gate metal V/Au and p-GaN the technological route has been modified so to exclude thermal annealing of gates.

3.Results

The created HEMTs have passed extensive testing and measuring of the basic parameters. The transistor configuration affects its characteristics: the greater gate-to-drain length, the greater breakdown voltage. (Table 1) HEMT with V/Au gate metallisation and gate length of 1 um and gate-to-drain length of 3.5 um has breakdown voltage in close state of 140 V. The transistor with 9.5um gate-to-drain length has 500 V breakdown voltages in the same conditions.

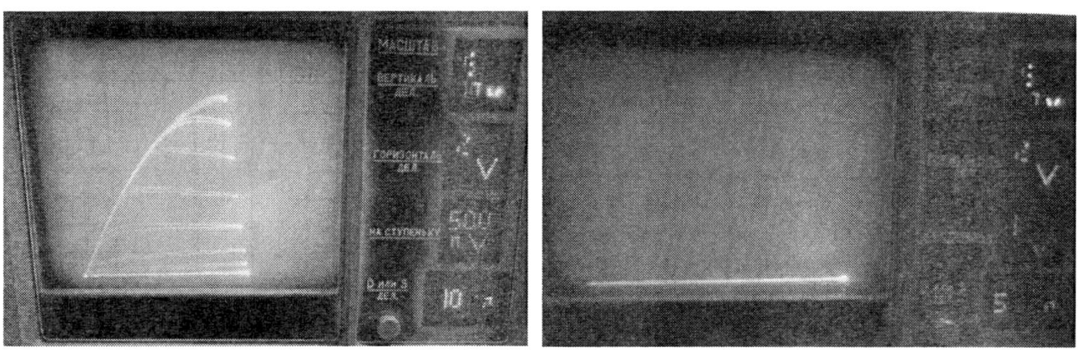

Figure 2(a, b). (a) Current-voltage characteristics of HEMT with Lg=1 um and Lgd=9.5 um at 5.5 V on the gate; **(b)** the current-voltage characteristics at 0 V on the gate.

Figure 3(a, b). **(a)** Current-voltage characteristics of HEMT with Lg=10um and Lgd=1.4 um at 5.5 V on the gate; **(b)** the current-voltage characteristics at 0 V on the gate.

Table 1. Dependence breakdown voltage on gate-to-drain length.

Lg, um	Lgs, um	Lgd, um	Vb, V
1	1.5	1.5	50
1	1.5	1.5	45
1	1.5	3.5	140
1	1.5	3.5	120
1	1.5	9.5	590
1	1.5	9.5	550

Table 2. Dependence the maximum current in open state on gate length.

Lg, um	Lgs, um	Lgd, um	Imax, mA/mm
1	1.5	1.5	270
1	1.5	1.5	240
1	1.5	1.5	230
10	1.4	1.4	130
10	1.4	1.4	150
10	1.4	1.4	120

Transistor work depends on gate length. (Table 2) HEMTs with 10um gate lengths have much smaller currents in open state than transistors with 1 um gate (100 mA/mm and 350 mA/mm accordingly). (Fig.2,3)

The experimental data confirm simulation ones.(Fig.4) HEMTs with 1 um gate have 2 times greater currents in open state than transistors with 10um gate.

The transistors haven't leakage currents. The positive threshold voltage is 4.5 V. It's established that 2 nm SiN_x under the gate influences on the device work. It improves transistor stability. The threshold voltage of transistor with SiN_x under the gate is about 8 V.The gate current of HEMTs decreased and their gate bias increased when their gate current was turned on. Therefore, the DC power loss of the input signal was decreased. But the transistor hasn't completely normally-off behavior.

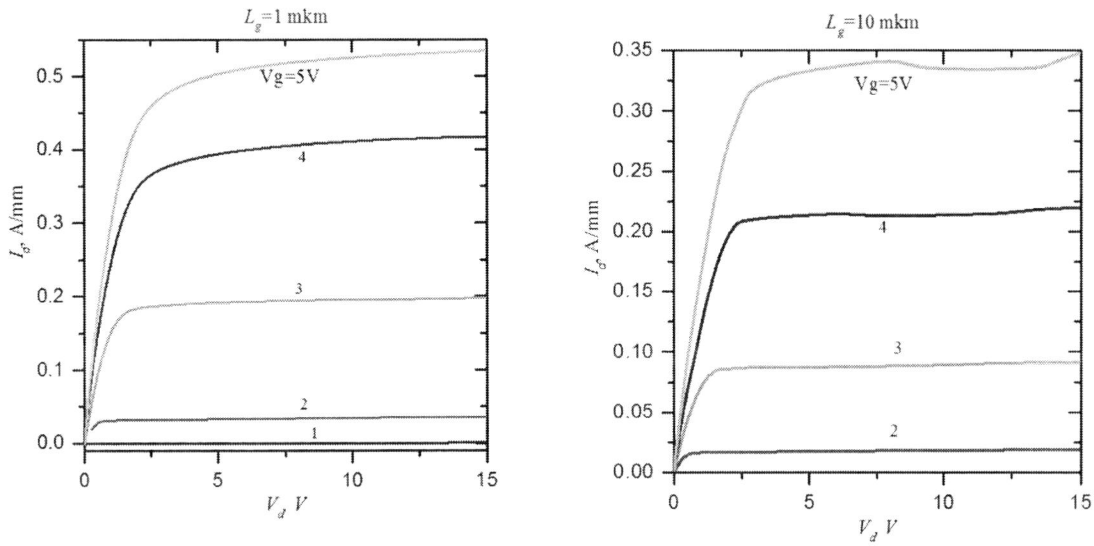

Figure 4(a, b). Sentaurus TCAD simulation of p-GaN gate HEMT: **(a)** Output current-voltage characteristics of HEMT with Lg=1 um; **(b)** output current-voltage characteristics of HEMT witn Lg=10um.

We have managed to form normally-on GaN HEMT on the same heterostructure and wafer. The transistor with 1 um gate length, gate-to-source length of 1.5 um and gate-to-drain length of 9.5 um has a breakdown voltage in close state 400 V and saturation current in open state 400 mA/mm. The transistor with 10 um gate length, gate-to-source length of 1.4 um and gate-to-drain length of 1.4 um has a breakdown voltage in close state 50 V and saturation current in open state 200 mA/mm.

4.Conclusion

In this work has been developed and demonstrated technological route of p-GaN gate HEMT. Standard selective etching of p-GaN is one of the most difficult ways to form p-GaN gate. We developed heterostructure with AlN stop layer and worked out two etching modes demonstrated in this work. The transistor configuration affects its characteristics: the greater gate-to-drain length, the greater breakdown voltage. Transistor work depends on gate length. It's established that SiNx under the gate improve transistor stability, but the transistor hasn't completely normally-off behavior. As well normally-on HEMTs with the same configuration and normally-off p-GaN gate ones have been formed on the one wafer.

Acknowledgments
This work was supported by the Ministry of education and science of the Russian Federation (grant agreement № 16.2250.2017/ПЧ).

References

[1] Y Uemoto, M Hikita, H Ueno 2007 *IEEE Trans. El. Dev.* **54** L12
[2] G Greco, F Iucolano, F Roccaforte 2017 *Mater. Sci. Semicond. Proc.* **9** 27
[3] R Hao, K Fu, G Yu, W Li, J Yuan, L Song, Z Zhang, S Sun, X Li, Y Cai 2016 *Appl.Phys. Lett.* **109** 152106
[4] P Fiorenza, G Greco, G Giannazzo, R Lo Nigro, F Roccaforte 2012 *Appl. Phys. Lett.* **101** 172901
[5] L Li, W Wang, L He, J Zhang, Z Wu, B Zhang, Y Liu 2017 *Mater. Sci. Semicond.* **67** 141

Simulation of induction heating of a titanium sample in a container with a carbon-containing medium and determination of the amount of heat losses during chemical heat treatment

A Voyko[1]**, M Fomina**[1]
[1]Yuri Gagarin State Technical University of Saratov, Saratov 410054, Russia

Abstract. In this work, the process of induction heating of a container with a titanium sample in a carbon-containing medium was simulated using the finite element method. The dependence of the inductor current I and the duration of treatment on the heating kinetics $T(I,t)$ of the "inductor – container – sample" system taking into account the heat losses $Q_L(I,T)$ on the process of chemical-thermal treatment of titanium was established.

1. Introduction

In the manufacture of products from titanium and its alloys, methods of strengthening chemical-thermal treatment (CTT) are widely used, e.g. cementation, nitriding, oxidation, etc. [1,2]. As a result of this treatment, wear-resistant coatings with high values of hardness [3-5] and other functional qualities, such as morphological heterogeneity and porosity [6-8], are obtained on the surface of the products.

Among the enumerated CTT methods, the cementation process is considered to be efficient, since it is characterized by low costs and allows the production of wear-resistant coatings and diffusion layers [4]. In order to accelerate chemical processes, it is preferable to use induction heating, which is distinguished by high productivity [9,10]. Thus, when using a solid reaction medium, e.g. carbon-containing, the process of cementation of titanium samples should be performed in a refractory container.

In the proposed cementation method, there is the problem of measuring the temperature in the inner part of the container with the sample. It is known that mathematical modeling, in particular the finite element method (FEM), allows the study of the complex systems with numerous elements and interfaces, and where complex processes occur, such as diffusion and heat transfer [11,12]. Hence, it is proposed to use a combined approach consisting in the experimental determination of the temperature of the container surface and the numerical calculation of the temperature field inside the "inductor – container – sample" system.

2. Methodology

The object of study is a system comprising an inductor, a container, a titanium sample, and auxiliary elements (Figure 1). The refractory container (5) with the titanium sample (6) and the working medium (7) are placed inside a cylindrical three-turn inductor (1) with water cooling (2), a tubular quartz chamber (3) with locking dielectric elements (4).

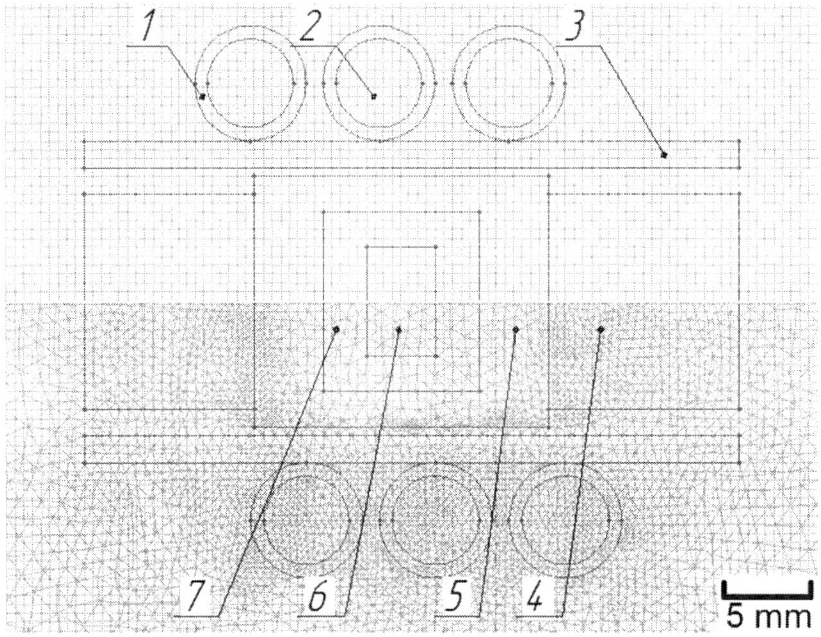

Figure 1. A geometric model of the "inductor – container – sample" system (the lower part of the model is shown with a finite element mesh)

In the simulation, the values of the current parameters were applied that were used in cementation, i.e. the inductor current *I* from 2.4 to 5 kA and fixed frequency *f* equaling 110 kHz.

Numerical 2D modeling was performed using the "Elcut" software. The computation duration *t* did not exceed 400 s, which corresponded to the actual cycle of high-temperature cementation. The main parameters of the system were determined, in particular the nature of the heat source distribution $Q(x,y,I)$ and the thermal field $T(x,y,I,t)$ depending on the inductor current *I*, the geometry of the system elements, as well as corrections for heat losses $Q_L(I,t)$ at CTT.

3. Results

The simulation results showed that in the "inductor – container – sample" system the greatest amount of heat *Q* was released in the area with carbon-containing medium. This value $(2.9–3.2)\times10^9$ W/m^3 was 3–10 times higher than the heat release in the container equaling $(0.3–0.9)\times10^9$ W/m^3 (Figure 2).

This non-uniformity of heat release was related to the peculiarity of induction heating at high frequencies (over 100 kHz). Apparently, high heat release *Q* in the carbon medium was associated with reradiation of the alternating field inside the heated system.

The solution of the heat conduction problem for the considered system with a minimum inductor current $I = 2.4$ kA showed that the temperature of the container *T* with heated samples exceeded their melting temperature. However, this phenomenon was not observed in the experiment. Thus, when the container material was heated, along with the known heat losses on convection and radiation, a heat sink of a different nature was acting. It was likely that during the heating of the container, the processes of formation of new phases (chemical compounds – refractory oxides and solid solutions of the oxygen diffusion into the metal) and changes in the structure (growth of crystalline grains) occurred.

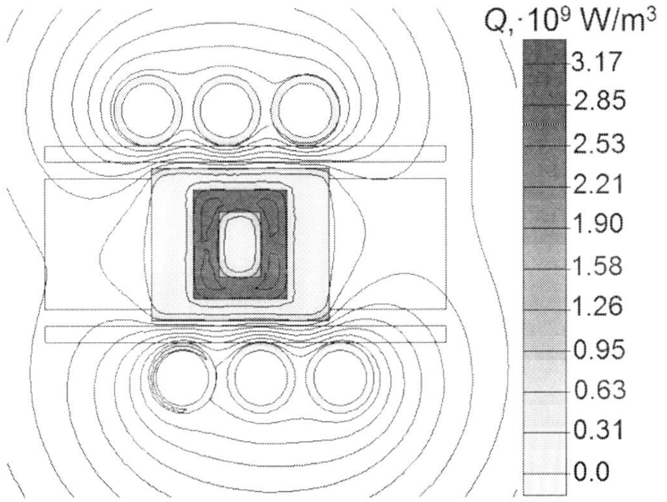

Figure 2. A plot of heat release in the "inductor – container – sample" system at the inductor current $I = 2.4$ kA

To solve the thermal problem correctly, it was necessary to take into account the losses for the above mentioned processes, which was possible by introducing the function Q_L in a graphical form (Figure 3). The shape of the graph of this function had an exponential form, which was associated with the values of the inductor current and temperature.

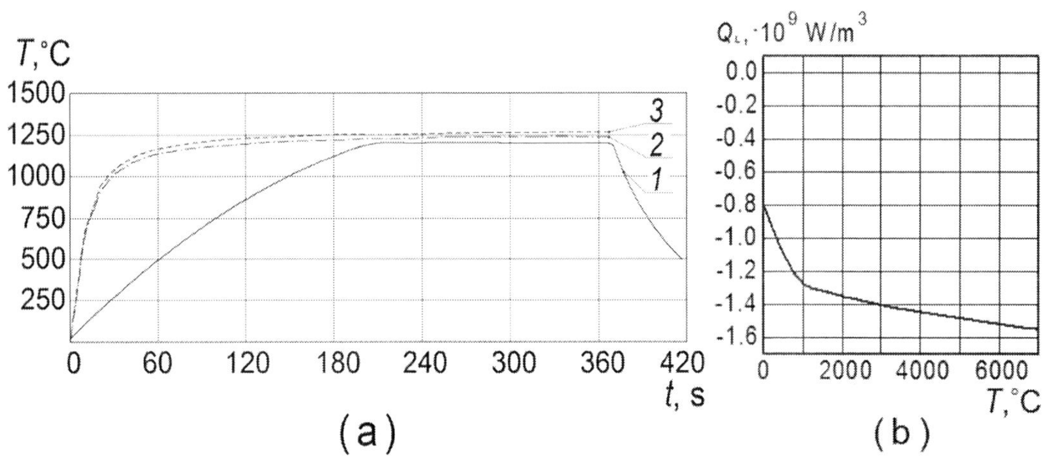

Figure 3(a,b). Graphs of heating kinetics of the "inductor – container – sample" system at the inductor current $I = 2.4$ kA for the container in experiment (1) and in simulation (2), as well as in the center of the titanium sample (3) with the introduced correction for the CTT process **(a)**; graph of function Q_L of heat losses **(b)**.

The solution of the heat conduction problem taking into account the heat losses Q_L on the CTT process was in the form of parabolic curves with a saturation region (stationary exposure). Of greatest interest were the graphs for the periphery of the container and the titanium sample in the region of maximum field strength (in the center of the inductor coil).

Curve (2) corresponded to the temperature on the container surface in the zone of influence of the inductor, curve (3) to the temperature in the center of the titanium sample (Figure 3,a). The experimental curve (1) was characterized by a much lower heating rate to a maximum temperature of about 1200 °C at a minimum inductor current of 2.4 kA, since the temperature was measured at the end surface of the container. The temperature difference significantly decreased at a prolonged heating of more than 180 s. With high-temperature exposure, the difference of stationary temperature between the curves of the simulation and the experiment did not exceed 30-50 °C.

The characteristic form of the function $Q_L(T)$ at the minimum inductor current was shown in the figure (Figure 3,b). The function Q_L had a minimum of -0.8×10^9 W/m^3, which corresponded to the initial temperature of the container material. Further, the function Q_L significantly increased in absolute value up to $Q_L = -1.25\times10^9$ W/m^3 at a temperature of about $T = 1000$ °C. At this point, an inflection of the function was noted, after which the function $Q_L(T)$ gradually increased linearly.

4. Conclusions

Thus, the experimental and theoretical results of the kinetics of heating the "inductor – container – sample" system were consistent when using the heat loss function Q_L for the CTT process. During the simulation, it was shown that the maximum heat release was localized inside a container with a carbon-containing medium. In the simulation, it was established that inside the container a temperature was reached at which it was possible to efficiently perform a CTT process, in particular cementation. Further studies will be associated with an increase in the accuracy of simulation both the initial section of intensive heating and the stationary section of high-temperature exposure at higher values of the inductor current.

Acknowledgments

The research was supported by the Russian Science Foundation (project No. 18-79-10040). The authors express their gratitude to the research manager Prof. A.A. Fomin for his help in analyzing the research results.

References

[1] Chuang Y and Jing L 2019 *Vacuum.* **163**, 52
[2] Burdovitsin V A, Golosov D A, Oks E M, Tyunkov A V and Zavadsky S M 2019 Surf. Coat. Technol. **358** 726
[3] Fomin A, Dorozhkin S, Fomina M, Koshuro V, Rodionov I, Zakharevich A, Petrova N and Skaptsov A 2016 *Ceram. int.* **42** 10838
[4] Fomin A, Egorov I, Shchelkunov A, Fomina M, Koshuro V and Rodionov I 2018 *Compos. Struct.* **206** 467
[5] Koshuro V, Fomina M, Voyko A, Rodionov I, Zakharevich A, Skaptsov A and Fomin A 2018 *Compos. Struct.* **202** 210
[6] Koshuro V, Fomina M, Fomin A and Rodionov I 2018 *Compos. Struct.* **196** 1
[7] Fomin A A, Steinhauer A B, Rodionov I V, Fomina M A and Zakharevich A M 2013 *Tech. Phys. Lett.* **39** 969
[8] Fomin A A, Steinhauer A B, Rodionov I V, Petrova N V, Zakharevich A M, Skaptsov A A and Gribov A N 2013 *Biomed. Eng.* **47** 138
[9] Demidovich V B, Chmilenko F V and Rastvorova I I 2015 *Acta Tech. CSAV.* **60** 107
[10] Kuvaldin A B, Lepeshkin S A and Lepeshkin A R 2014 *Acta Tech. CSAV.* **59** 279
[11] Aman A, Majcherek S, Hirsch S and Schmidt B 2015 *J. Appl. Phys.* **118** 164105
[12] Aman S, Aman A and Morgner W 2013 *Compos. Sci. Technol.* **84** 58

Quantum-mechanical models for calculating the electrical characteristics of semiconductor 2-d structures for technological optimization of nanoelectronics devices based on them

A A Koziy[1,2], N A Vetrova[1], K P Pchelintsev[1], V D Shashurin[1], S A Meshkov[1]
[1]Bauman Moscow State Technical University, Moscow 105005, Russia
[2]Russian Quantum Center, Moscow 143025, Russia

Abstract. In this paper, a computational algorithm has been developed for the quantum-mechanical model for predicting the electrical properties of 2-d semiconductor structures of nanoelectronics devices with a low-dimensional channel. The most important advantage of the developed algorithm is its scalability for the problems of technological optimization of nanoelectronics devices. An example of application of the developed software to solving the problem of predicting the electrical properties of a low barrier detector sub-terahertz diode with zero-bias in the framework of solving the problem of its technological optimization is presented.

1. Introduction

The current practice in the design of integrated circuits assumes that the circuit is first synthesized and the electrical parameters of its elements are selected according to the criterion of the best fit of the operational characteristics to the specified values. Assignment of electrical parameters is assigned based on the required accuracy of operational characteristics. The next step in the development of the integrated circuits (IC) is the calculation of nominal values and permissible deviations of the construction parameters. The problems of implementing this stage for designing nanoelectronic device structures on semiconductor 2-d structures [1] are due to the fact that, on the one hand, libraries of modern computer-aided design systems do not have the latest nanostructured elements, on the other hand, the models developed to date, although they allow to evaluate the electrical characteristics using heterostructure parameters, they are not adapted for the inverse problem. In most cases, engineers are forced to use rather rough models (allowing to evaluate only the qualitative character of the I-V curve) with a large number of empirical coefficients (for example, methods based on equivalent circuit diagrams [2, 3], phenomenological methods of semiclassical approximation [4]); and / or focus on nomograms [5], which systematize for a particular class of nano-devices the relationship between structural parameters and electrical ones; or use inefficient computational algorithms that require time-consuming calculations on supercomputers [6].

However, it should be noted that studies of physical processes in low-dimensional channels of nano-devices today provide a fairly wide range of formalisms, within which you can take into account the main factors determining quantum transport in nanostructures [1]. Especially promising for integration into the parametric stages (with the aim of reducing methodological error) and technological (reducing not only methodical, but also production and technological errors) optimization (Fig. 1) of nanoelectronics devices, the authors present combined models of current transfer with quantum-mechanical modeling of a low-dimensional channel [7, 8, 9].

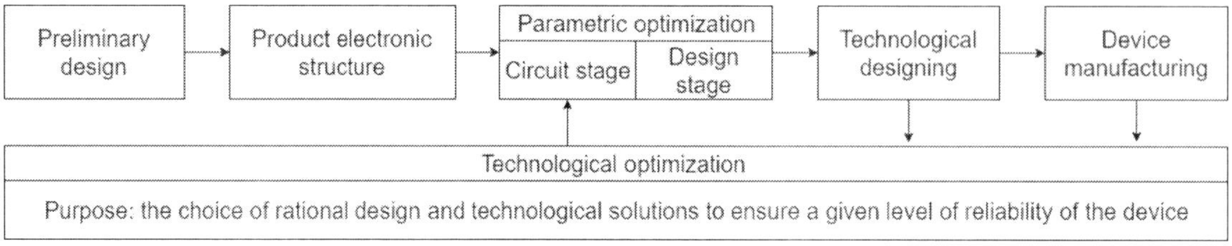

Figure 1. The structure of the process of designing and manufacturing nanoelectronic devices.

Thus, the actual task is to develop quantum-mechanical models for calculating the electrical characteristics of semiconductor 2-d structures, as the most accurate and physically based models, to account for manufacturing technology and gerontological processes causing the most important operational parameters, including indicators of the durability of hybrid and monolithic integrated circuits.

2. Quantum mechanical models for electrical characteristics

The basis of quantum mechanical approaches to the description of electron transport is the representation of semiconductor low-dimensional heterostructures as an open quantum system in the framework of the Landauer – Buttiker formalism [10]. To obtain the electrical characteristics of a nano-device, it is necessary to determine the transmission function of the heterostructure, which can be obtained using quantum-mechanical approaches. As approaches that meet the requirements for current transfer models for design and technological optimization and for reasons of maximum scalability, the authors analyzed the method of plane waves, the matrix method of plane waves, and the Green's function method. Space charge accounting for all methods was carried out within the framework of the concept of self-consistency.

When implementing the method of plane waves, it is necessary to determine the transmission matrix characterizing the relationship between the amplitudes of the waves incident and transmitted through the structure. For this, the potential profile is divided into regions with a constant value of potential energy. In each such domain, the solution of the Schrödinger equation is a superposition of the forward and backward waves. At the boundaries of the regions, with a constant potential, use the conditions for the continuity of the wave functions and the current density in a matrix form. With the help of the full transfer matrix, obtained as a product of the transfer matrices of all adjacent layers, it is possible to determine the amplitudes of the envelopes of the wave functions, on the basis of which to obtain the transmission of the structure.

$$M_{1,N} = \begin{pmatrix} M_{11} & M_{12} \\ M_{21} & M_{22} \end{pmatrix} = \prod_{j=1}^{N-1} M_{j,j+1}, \quad (1)$$

$$T = \frac{1}{|M_{11}|^2} \frac{K_N m_1}{K_1 m_N}, \quad (2)$$

where $M_{1,N}$ – full transfer matrix, $M_{j,j+1}$ – transfer matrix between layers j and j+1, N – number of layers, m – the effective mass of an electron in layer j, $K_j = \sqrt{2m_j(\mathrm{E}-U_j)} / \hbar$ – electron wave number in j layer.

To implement the matrix method of plane waves and the method of Green's functions, it is necessary to perform a discretization of the simulated structure. The finite difference approximation of the Schrödinger equation for grid nodes was obtained using the integro-interpolation method. To take into account the boundary conditions used the conditions of the second kind. In order to avoid lowering the accuracy of the finite-difference scheme with respect to the grid step, the balance method was used in the approximation of the boundary conditions.

$$\bar{z} = \{z_i = i\Delta z,\ \Delta z > 0,\ i = 0...N,\ z_1 = z_{LB},\ z_N = z_{RB}\} = z^{Inner} \bigcup z^{External}, \tag{3}$$

$$-t_{i+1}\Psi_{i+1} + \left(t_{i+1} + t_i + \frac{U_i + U_{i+1}}{2} - E\right)\Psi_i - t_i\Psi_{i-1} = 0, z \in z^{Inner}, \tag{4}$$

$$\Psi_1(t_R iK_L \Delta z - 1) + t_2\Psi_2 = 2iK_L t_R \Delta z e^{iK_L z_{LB}}, z \in z^{External}, \tag{5}$$

where Δz – discrete grid step, N – number of split points, $\bar{z}$ – uniform grid on the interval $[z_i, z_{i+1}]$, $z^{Inner} = \{z \in \bar{z} \mid z_{RB} < z < z_{LB}\}$ – internal nodes of the grid, $z^{External} = \{z \in \bar{z} \mid z_{RB} < z < z_{LB}\}$ – external nodes, Ψ_i – discretized electron wave function, $t_i = \hbar^2 / 2\Delta z^2 m_i$ – bound operator, $m_i = 1/\Delta z \int_{z_i}^{z_{i+1}} m(z)dz$ – average effective mass per interval $[z_i, z_{i+1}]$; $U_i = 1/\Delta z \int_{z_i}^{z_{i+1}} U(z)dz$ – average value of potential energy per interval $[z_i, z_{i+1}]$, $U(z) = E_c - eV_z$ – potential energy, E_c – conduction band profile, e – elementary charge, $K_{R(L)} = \sqrt{2m_i^*(E - U_{R(L)})} / \hbar$ – electron wave number in the right (left) reservoir, V_z – the potential energy of an electron in the space charge field and the applied external voltage, resulting from the solution of the Poisson equation.

Thus, the finite-difference scheme for solving the Schrödinger equation for an open system is presented in a matrix form.

$$[H - EI + \Sigma_L + \Sigma_R][\Psi] = [S], \tag{6}$$

where H – Hamilton operator in the matrix form, describing the electron behavior in the active region, Ψ – discretized wave function, Σ – o matrix operator, describing the outflow of electrons from the active region, E – energy, I – identity matrix, S – the term describing the electron influx into the active region reservoirs.

In the matrix method of plane waves, the resulting system of linear equations, which is the discretized Schrödinger equation, is solved by the modified Thomas method. The amplitudes of the envelopes of the wave functions obtained as a result of the solution are used to determine the transmission of the structure.

$$T = \frac{|A_N|^2}{|A_1|^2}\frac{K_N m_1}{K_1 m_N}. \tag{7}$$

When using the Green function formalism, the main parameters, including the structure transmission and electron concentration, are determined using the nonequilibrium Green function, which describes the quantum system's response to an external perturbation. The transmission of the structure is determined by:

$$T = \sum_j^N \Gamma_{jj}^L G_{jj} \Gamma_{jj}^R G_{jj}^+, \tag{8}$$

где $\Gamma^{L(R)} = i\left[\Sigma_{L(R)} - \Sigma_{L(R)}^+\right]$ – matrix characterizing the broadening of energy levels at the border of the left (right) contact channel, $G = [H - EI + \Sigma_L + \Sigma_R]^{-1}$ – Green's nonequilibrium function, in matrix form.

The verification of the developed computational algorithm was carried out on a triangular potential barrier and confirmed the correctness of the numerical methods used in the model. The choice of a potential profile for verification was due to the presence of an analytical solution obtained using Airy functions of the first and second kinds. The results of the simulation of the transmission function are presented in Fig. 2 (a). Analysis of the accuracy (Fig. 2 (a)), as well as temporal and spatial complexity

(Fig. 2 (b)) of the computational algorithms developed for each of the described methods, made it possible to choose as the most effective and satisfactory to the mathematical model of electrical prediction characteristics of semiconductor structures with a low-dimensional channel for design and technological optimization, an algorithm that implements the Green function method.

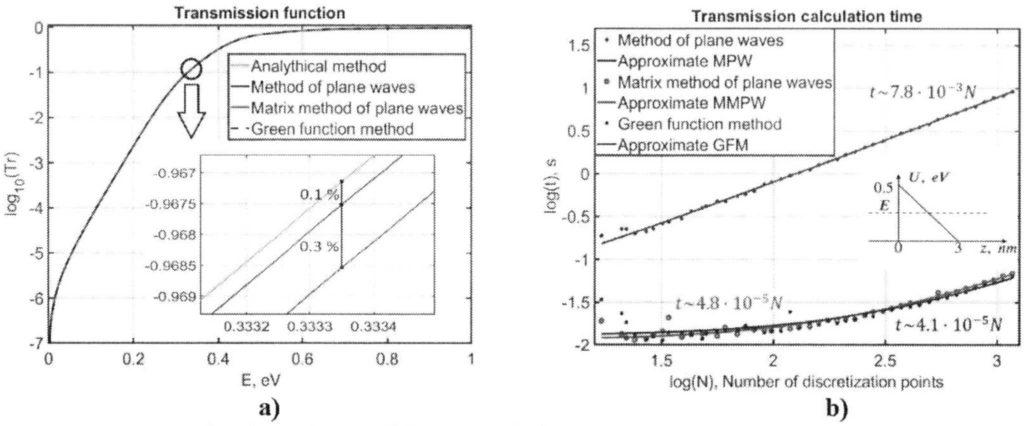

Figure 2 (a, b). (a) The dependence of the transmission function of the potential structure on the value of the electron energy; **(b)** Dependence of the computation time of the transmission function on the number of split points of the structure.

Thus, the developed computational algorithm for the quantum mechanical model for predicting the electrical properties of 2-d semiconductor structures of nanoelectronics devices with a low-dimensional channel: 1) verified, and the error in calculating the structure transmission does not exceed 0.3 %; 2) optimized by the criterion of efficiency: a) minimization of temporal and spatial complexity was carried out, resulting in linear asymptotic complexity $\sim O(N)$, including by taking into account the features of the implementation of the Green function method for calculating the current density - see Fig. 2 (b)); b) selection of the optimal grid step was carried out for the construction of a finite-difference scheme, so that the error in determining the electrical parameters was below 0.5% with the minimum execution time of the algorithms; 3) it has a high degree of scalability, which allows integrating it into the software package of technological optimization of nanoelectronics devices with quantum limitations of various types.

3. Application of the developed methods to obtain the electrical characteristics and operational parameters of the detector diode

Validation of the developed quantum-mechanical model was carried out on Heterostructure low barrier diodes (HLBD) based on lattice-matched AlGaInAs triangular barrier used for zero-bias millimeter-wave and submillimeter-wave detection. Perspective of the device is determined, on the one hand, by the attractive performance characteristics of detectors on such diodes that could be obtained on test-made samples [4], on the other hand, by potentially wide application area [11, 12] (for example, in detector receivers, in vision systems, including in mm-wave cameras). Difficulties in the design of such diodes [4] are associated with the problems mentioned in the introduction. Experimental data for a diode with an active region made of materials $Ga_{0.47}In_{0.53}As$ and $Ga_{0.47}In_{0.53}As$ are given in [4]. It was obtained that the results of the simulation of electrical characteristics and experimental data converge with an error of less than 3% in the voltage range [-0.1,0.1] V (Fig. 3).

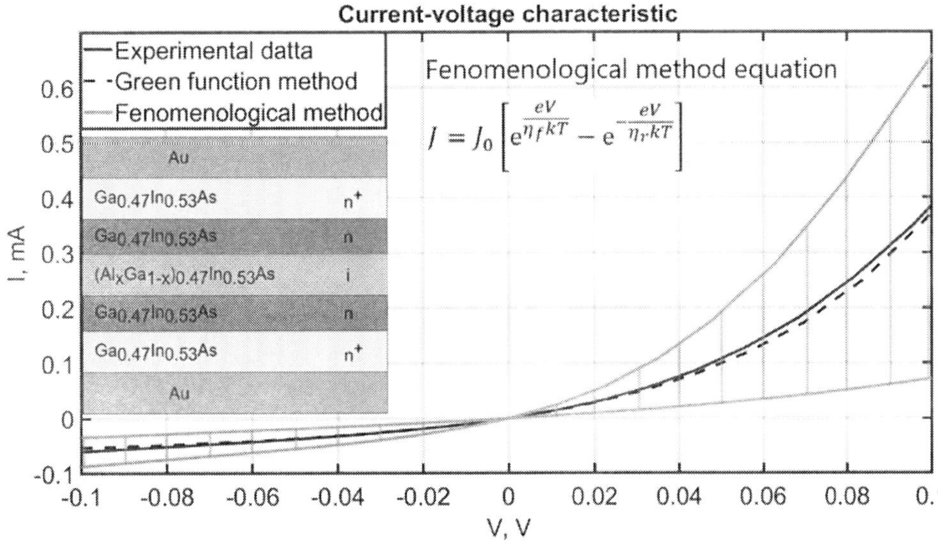

Figure 3. Application of the phenomenological approach and the Green function method for modeling the I-V curve of HLBD [4]; J – current, J_0 – saturation current, η_f and η_r – ideality factors of diode.

4. Algorithm for optimization of design and technological parameters

The results of integrating the developed quantum-mechanical model for calculating electrical characteristics into the HLBD technological optimization procedure are shown in Fig. 4 (with a limited number of technological factors of production and modes of operation). The target function is the probability of production of operability product. The arguments of the objective function are the fraction of the substitution of Ga atoms by Al (x) atoms and the barrier width (w).

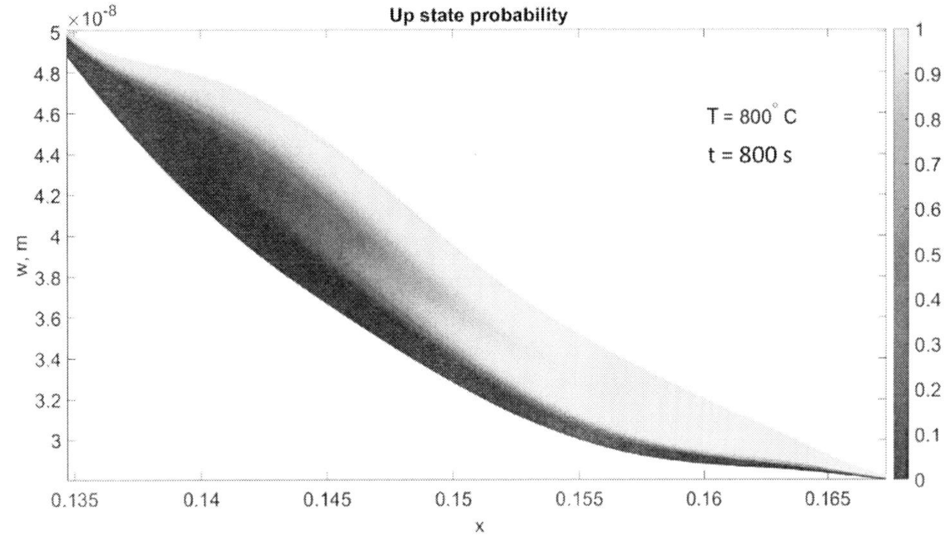

Figure 4. The probability of production of operability product (indicated by color) depending on the fraction of the replacement of Ga atoms by Al atoms (x) and the barrier width (w), at temperature and heat exposure time $T = 800° \, C$ and $t = 800 \, s$ respectively [13].

5. Conclusion

The high efficiency of the developed computational algorithm and the accuracy of estimating the I-V curve of 2d semiconductor structures of nanoelectronics devices with a low-dimensional channel were confirmed in the process of verification and validation. The scalability of the proposed computational algorithm provides practical significance and the possibility of applying the technological optimization of such devices to solve problems. The developed algorithm of design and technological optimization can be scaled to account for an arbitrary number of operating, design and technological parameters. It is possible to increase the accuracy of predicting electrical properties and operational parameters, including absolute values, by considering coherent and incoherent channels with different dissipative processes, taking into account the influence of a greater number of technological factors of manufacturing, taking into account the conditions and modes of operation, taking into account gerontological processes in heterostructures including influence of harsh conditions and operating modes.

Acknowledgments

This work was supported by the Ministry of Education and Science of the Russian Federation (project No. 16.1663.2017/4.6).

References

[1] Abramov I, Goncharenko I and Kolomejceva N 2005 *Fizika i Tehnika Poluprovodnikov* **39** 1138
[2] Moskaliuk V and Saurova T 2015 *2015 IEEE 35th Int. Conf. on Electr. and Nanotech. (ELNANO)* (Kiev) p 177
[3] Ghobady A, Khan T, Celik O and Biyikli N 2017 *Int. Journ. of Microwave and Wireless Techn.* **9** 1905
[4] Nadar S, Zaknoune M, Wallart X, Coinon C, Emilien P, Ducournau G, Gamand F and Thirault M 2017 *IEEE Trans. On Terah. Sc. And Techn.* **6** 780
[5] Malyshev K 2004 *PhD dissertation, BMSTU, Moscow*
[6] Guinea F, Roche S and others 2014 *Science and technology roadmap for graphene, related two-dimensional crystals, and hybrid systems* (Nanoscale) p 343
[7] Vetrova N A, Gudkov A G, Shashurin V D 2017 *Biomed. Engin.* **51** 1
[8] Pourfath M 2014 *The Non-Equilibrium Green's Function Method for Nanoscale Device Simulation* (Tehran: Springer) p 256
[9] Woloszyn M, Spisak B J 2017 *Phys. Review B* **96** 075440
[10] Moskalets M 2011 *Scattering Matrix Approach to Non-Stationary Quantum Transport* (London: Imperial College Press) p 296
[11] Peiponen K, Zeitler A and Kuawata-Gonokami M 2013 *Terahertz Spectroscopy and Imaging* (Verlag Berlin Heidelberg: Springer) p 644
[12] Zagorodny A, Drozdov A, Voronin N and Yunusov I 2013 *2013 14th Int. Conf. of Young Spec. on Micro/Nanotech. and Electr. Devices* (Novosibirsk) p 96
[13] Fair R. 1993 *Rapid Thermal Processing* (San Diego: Academic Press) p 440

Determation of dV/dt values for domestic SiC Schottky diodes

S V Sedykh, S B Rybalka, A Yu Drakin, A A Demidov, E A Kulchenkov
Bryansk State Technical University, Bryansk 241035, 50 let Oktyabrya 7, Russia

Abstract. In this study the dV/dt values for 4H-SiC commercial diodes have been determined experimentally. The experimental measuring tester for determination of dV/dt values of diodes at amplitude of impulse of reverse voltage V_A (V_A=100÷950 V) applied across the Schottky diode was constructed. It was determined that the dV/dt value almost linearly increase with increase of impulse of reverse voltage V_A (in interval of V_A=100÷900 V) applied across the 5DS402A9 diode for the first time. It is determined experimentally that at the maximal impulse of reverse voltage (900 V) applied across the silicon carbide commercial 5DS402A9 diode produced by domestic company the dV/dt value (148 V/ns) is comparable with others commercial diodes and therefore diode of this type can stably function in electric power circuit.

1. Introduction

It is known that the Schottky diodes on the base of silicon carbide (SiC) widely using in high-temperature power electronics [1]. Some type of SiC Schottky diodes have been designed earlier [2-6] and now is produced by russian company the ZAO «GRUPPA KREMNY EL» (Bryansk, Russia). The one of the important characteristics for silicon carbide Schottky diodes is value of dV/dt when the impulse of reverse voltage amplitude (V_A) applied across to diode. Devices with lower dV/dt capability are more susceptible to failure from large in-rush currents [7-9]. The number of some investigations with discussion about dV/dt effect is very limited and presented in general investigations of diodes by Infineon Technologies [7] and Wolfspeed (Cree Inc.) [8,9]. Study of the Infineon SiC diodes it is shown that dV/dt value equals 90-120 V/ns [7]. In C3D03060A type Wolfspeed diode value of dV/dt=295 V/ns (V_A=800 V) and for C4D10120A diode dV/dt=495 V/ns (V_A=1000 V) [8,9]. The dV/dt data for domestic silicon carbide diodes is unknown to present time and therefore the main goal of this work is to obtain experimentally its dV/dt values.

2. Materials and methods

The experimental measuring tester was constructed for testing of diodes on the dV/dt value. Scheme of measuring tester is presented in Fig. 1 The parameters for tester, in particular, allows carrying out measurement of dV/dt value in interval from 50 V/ns up to 200 V/ns at amplitude of impulse of reverse voltage applied across to diode V_A=300÷950 V. Analyses of oscillogram has been carried out with used Hantek DSO5102P oscillograph (bandwidth 100 MHz, refresh rate 1×10^9 s^{-1}). The reverse voltage front across to tested diode is formed switching on VT2 silicon carbide transistor. Small acceleration time has been provided by charge of input capacitance of the VT2 transistor with using of avalanche breakdown current of the VT1 transistor. Regulation process of dV/dt parameter has been realized by the R21 resistor. The tested Schottky diodes were connected to connectors marked as XS1 in Fig. 1. For testing was used the 5DS402A9 (Kremny) JBS type 4H-SiC Schottky diode (produced by domestic company ZAO «GRUPPA KREMNY EL») and C4D10120 (Wolfspeed) Schottky diode.

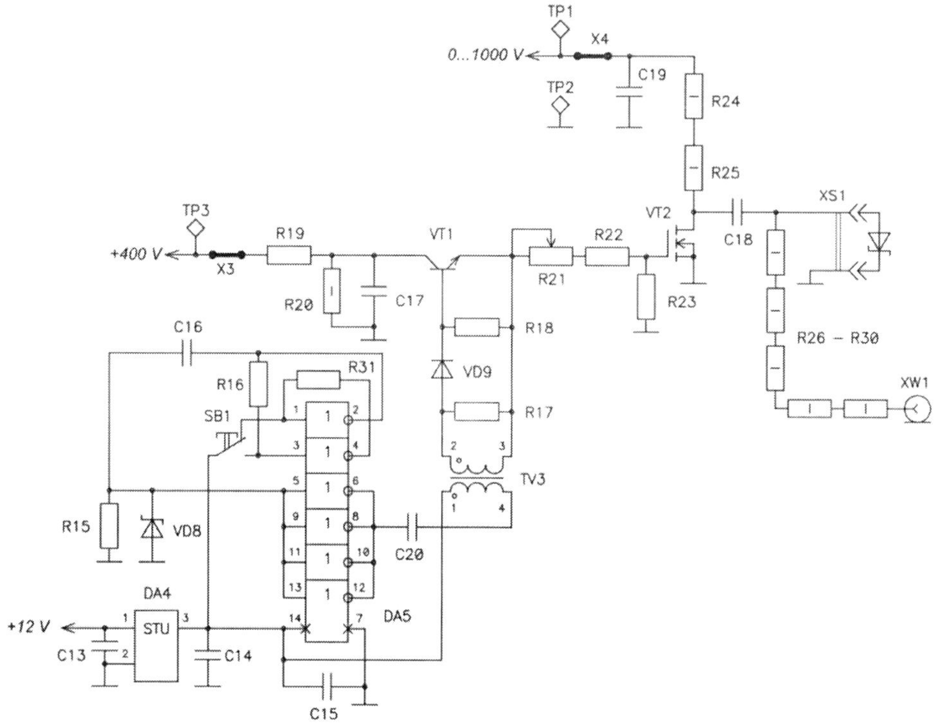

Figure 1. The scheme of the measuring tester circuit for determination of dV/dt value across a testing silicon carbide Schottky diode.

3. Results and discussion

In Fig. 2 are shown the testing results for 5DS402A9 (Kremny) silicon carbide Schottky diode and C4D10120 diode (Wolfspeed) at applied amplitude of impulse of reverse voltage across the diode 500 V. The value of dV/dt for diodes were obtained by slope of the linear part of oscilogram for voltage waveform, for example, as follows from Fig. 2 (curve *1*) for 5DS402A9 diode dV=168 V (380 V minus 212 V) and dt=2 ns (16 ns minus 14 ns), therefore the calculated value of dV/dt equals 84 V/ns. Similarly, for C4D10120 diode when the amplitude of impulse of reverse voltage applied across diode V_A=500 V the obtained value of dV/dt=132 V/ns as can be seen from Fig. 2 (curve *2*).

Then, as the amplitude of impulse of reverse voltage increase it lead to increase of dV/dt value. In Fig. 3 are shown testing results for 5DS402A9 silicon carbide Schottky diode (Kremny) and C4D10120 diode (Wolfspeed). The obtained value of dV/dt for 5DS402A9 diode (curve *1*) equals 148 V/ns at applied amplitude of impulse of reverse voltage across the diode 900 V. For C4D10120 diode (curve *2)* value of dV/dt at the same conditions equals 204 V/ns. In comparison with IDW20G120C5B (1200 V) commercial Schottky diode (Infineon) the obtained value for 5DS402A9 diode about two times more than the typically reported for dV/dt specification of 80 V/ns [10].

Further, with goal to establish the dependence between value of dV/dt and amplitude of impulse V_A applied across the SiC diode were carried out experiments by analogy with experiments described above. The obtained results then were generalized in Table 1.

As follows from Table 1, dV/dt values increase with increase of amplitude of impulse of reverse voltage V_A applied across the 5DS402A9 diode. Fig. 4 displays the dependence of the dV/dt value on the amplitude of impulse of reverse voltage V_A applied across the 5DS402A9 diode (obtained from Table 1 data). As can be seen from this figure the dV/dt dependence on V_A in this interval can be well approximate by the linear dependence.

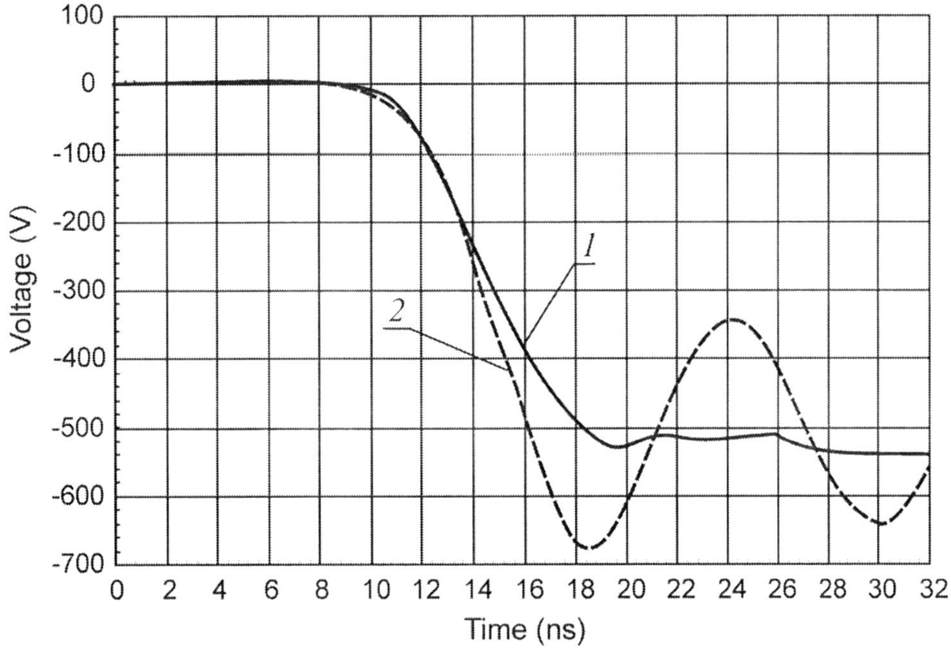

Figure 2. Reverse voltage waveform for diodes: *1* – 5DS402A9 (Kremny), *2* – C4D10120 (Wolfspeed) with impulse amplitude V_A=500 V.

Thus, to summarize the obtained results, it is established for the first time that dV/dt value almost linearly increase with increase of impulse of reverse voltage V_A applied across the 5DS402A9 diode in investigated interval of impulse voltage amplitude (from 100 up to 900 V).

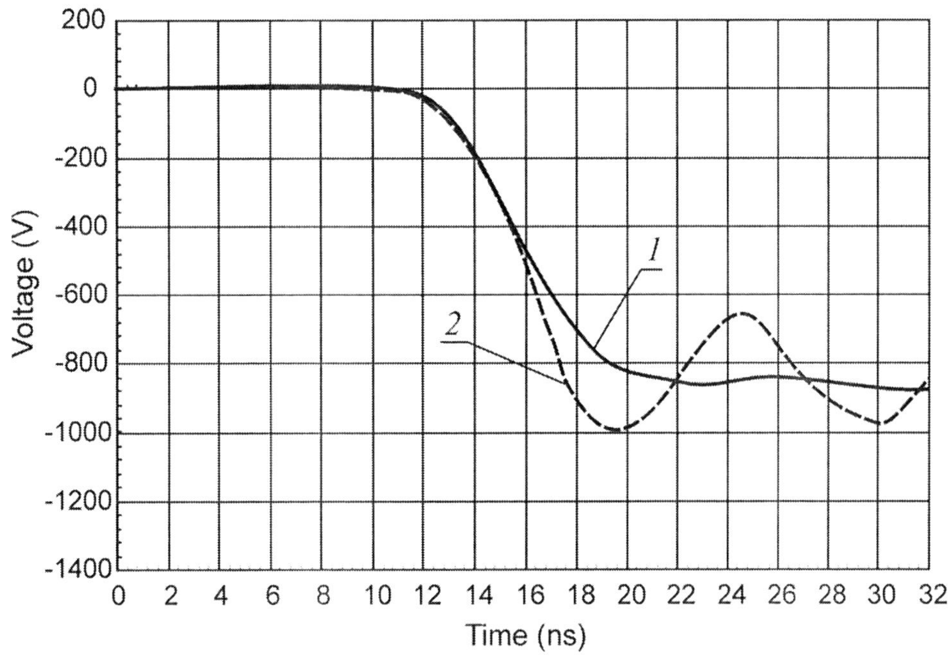

Figure 3. Reverse voltage waveform for diodes: *1* – 5DS402A9 (Kremny), *2* – C4D10120 (Wolfspeed) with impulse amplitude V_A=900 V.

Additionally, diodes were tested by 10^4 of cycles of impulse and was determined that studied diodes demonstrate the stable function without failure and breakdown.

Table 1. The value of dV/dt for diodes at various amplitude of impulse of reverse voltage applied across the 5DS402A9 diode.

V_A (V), amplitude of impulse of reverse voltage applied across the diode	dV/dt (V/ns)
100	19,6
200	36,4
300	49,2
400	64,8
500	84
600	94,4
700	107,2
800	125,6
900	148

Thus, on the basis of first experimental studying the obtained characteristics of dV/dt for russian 5DS402A9 commercial diode is comparable with other foreign SiC type commercial diodes.

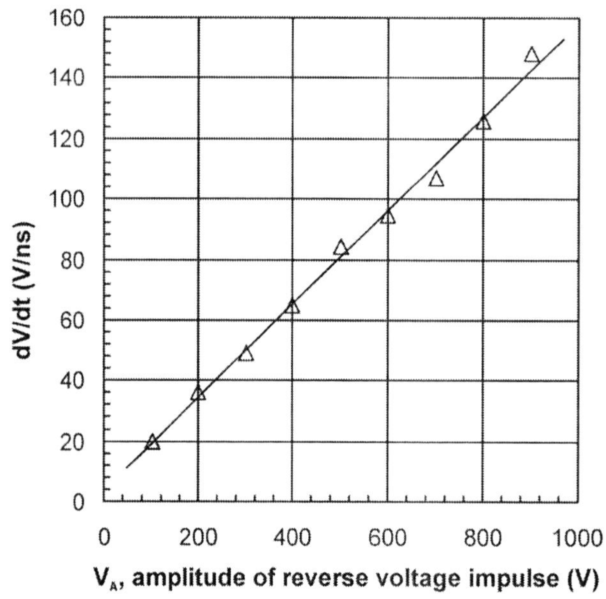

Figure 4. dV/dt dependence on amplitude of impulse of reverse voltage V_A applied across the 5DS402A9 type Schottky diode.

It should be noted that, for next modification of C4D10120A Wolfspeed diode (V_A=1000 V) dV/dt=495 V/ns that about two times more than for C4D10120 older type diode [8,9]. Because of this, recently it is shown that the silicon carbide Schottky diodes of new generation of the Wolfspeed firm can stably work without failures under high values of dV/dt up to 400 V/ns and at increase of dV/dt up to 650-800 V/ns [11,12]. However, there is possibility to increase the dV/dt value by using of diode's module as it was established earlier [6]. In particular, for module from diodes (4 diode of 5DS402A9 type) in case when the amplitude of impulse of reverse voltage across the 5DS402A9 diode's module was increase up to 900 V the value of dV/dt=184 V/ns [6] (it is more than on ~28% than for one 5DS402A9 diode where dV/dt=148 V/ns). Therefore, it is important from practical viewpoint that using the power module produced from some ordinary diodes can considerably increase the dV/dt.

4. Conclusions.

The experimental measuring tester for determination of dV/dt characteristic of diodes at amplitude of impulse of reverse voltage V_A applied across the diode from 100 up to 950 V was constructed. The dV/dt values for domestic and foreign SiC commercial diode at amplitude of impulse of reverse voltage applied across the testing diodes from 100 V up to 900 V have been obtained.

It is determined that at the maximal impulse of reverse voltage applied across the diode (V_A=900 V) the dV/dt values for 5DS402A9 commercial diode produced by domestic company («Gruppa Kremny El», Bryansk, Russia) is 148 V/ns and for C4D10120 diode (Wolfspeed, Cree Inc.) the value of dV/dt at the same conditions equals 204 V/ns. Thus, obtained dV/dt values for 5DS402A9 diode are comparable with others commercial diodes (typical values are 80-120 V/ns).

For the first time it is determined that dV/dt value almost linearly increase with increase of impulse of reverse voltage V_A applied across the 5DS402A9 diode in voltage interval from 100 V up to 900 V.

It was noted that value of dV/dt can be increase by using of diode's module, i.e. for diode's module (consist of four diodes of the 5DS402A9 type) the dV/dt value increase up to 184 V/ns (V_A=900 V) that is more than on ~28% than for one 5DS402A9 diode. Moreover, the carried out dV/st analysis also shown that studied diodes demonstrate the stable work after 10^4 of cycles of impulse without failure.

Therefore, it is determined that domestic SiC commercial diode possess the dV/dt values that more then typical for these type devices (80-120 V/ns) and consequently can stably work without failures.

Acknowledgements

This work was supported by the Russian Ministry of Education and Science (task No. 8.1729.2017/4.6).

References

[1] Kimoto T, Cooper J A 2014 *Fundamentals of Silicon Carbide Technology. Growth, Characteriztion, Devices, and Applications* (New York: Wiley–IEEE Press.)
[2] Sedykh S V, Rybalka S B, Drakin A Yu, Demidov A A, Ponomaryova N S and Shishkina O A 2018 *J. Phys.: Conf. Ser.* **1124** 071012
[3] Panchenko P, Rybalka S, Malakhanov A, Demidov A A, Krayushkina E Yu and Shishkina O A 2017 *J. Phys.: Conf. Ser.* **917** 082010
[4] Panchenko P V, Rybalka S B, Malakhanov A A, Krayushkina E Yu, Rad'kov A V 2016 *Proc. SPIE* **10224** 102240Y-1
[5] Ivanov P A, Potapov A S, Rybalka S B, Malakhanov A A 2017 *J. Radio Electronics* **6** 1
[6] Rybalka S B, Demidov A A, Kulchenkov E A, Drakin AYu 2018 *Belgorod State University Scientific Bulletin: Mathematics & Physics.* **50**(4) 460
[7] Holz M, Hultsch G, Scherg T and Rupp R 2009 *Phys. Status Solidi (a)* **206**(10) 2295
[8] Barbieri T 2015 Assessing next-generation discretes: Measuring SiC Schottky diode ruggedness with a high voltage pulse generator *Power Systems Design Wolfspeed, A Cree Company*
[9] Kartashov E, Lebedev A 2016 *Power Electronics* **2** 18
[10] Infineon IDW20G120C5B 5th Generation CoolSiC™ 1200 V SiC Schottky Diode Final Datasheet, Rev 2.1. München, 2017.
[11] Wang G, Van Brunt E, Barbieri T, Hull B, Richmond J and Palmour J 2017 *Proc. PCIM Europe (Nuremberg)* (Berlin: VDE VERLAG GMBH) p 870
[12] Van Brunt E, Wang G, Liu J, Pala V, Hull B, Richmond J and Palmour J 2016 *Proc. 28th Int. Symp. Power Semiconductor Devices and ICs (ISPSD) (Prague)* (Prague: IEEE) p 67

Ti/4H-SiC Schottky diode with breakdown voltage up to 3 kV

D A Knyaginin, S B Rybalka, A Yu Drakin, A A Demidov
Bryansk State Technical University, Brynask, 50 let Oktyabrya 7, Russia

Abstract. In this study the breakdown voltage for Ti/4H-SiC type Schottky diode with six guard rings and JTE layer has been calculated by mean of numerical simulations. It is established that increase of the n-type 4H-SiC epitaxial layer from 14 up to 20 μm and addition of JTE layer lead to increase of breakdown voltage value on ~900 V in contrast to the same diode without JTE. The above-mentioned diode's structure gives the possibility for designing and production of diode with higher breakdown voltage value up to 3 kV.

1. Introduction

It is known that semiconductor material silicon carbide (SiC) is promising for the development of power electronics, microelectronics and optoelectronics devices. This is due to the large band gap of the semiconductor SiC (>3 eV), high thermal conductivity, high breakdown fields and the rate of electron saturation, as well as significant radiation and thermal stability [1]. One of the simplest devices based on SiC, but at the same time important for microelectronics, is Schottky diode. For example, Schottky diodes for power electronics based on 4H-SiC have already been manufactured by the domestic industry, in particular, by the "GRUPPA KREMNY EL" domestic company (Bryansk). It is obvious that for the further development of the domestic component base based on SiC it is necessary to study in detail the influence of the diode structure parameters on its current-voltage characteristics to optimize the Schottky diode operation in power electronics, which can be done using physical simulation [2]. Earlier, in our previous paper [3-5] have been studied of 4H-SiC type Schottky diodes with Ni and Ti Schottky anode contacts without guard rings. Therefore, the main goal of this study was to design a perspective SiC Schottky diode with guard rings using physical simulation methods in the ATLAS software program and to study its electrical characteristics.

2. Materials and methods

The following Schottky diode with six guard rings and donor concentration in the epitaxial layer 4H-SiC $N_D = 3\times10^{15}$ cm^{-3} was taken as the object of simulation in program.

As can be seen earlier, to calculate the breakdown voltage (BV_{m}) of the Schottky diode in a plane-parallel *p-n* junction, we first determined the critical breakdown field strength (E_c) from the condition of equality to the unit of the ionization integral [6]:

$$\int_0^d \alpha_n \exp\left[-\int_x^d \left(\alpha_n - \alpha_p\right)dx'\right]dx = 1, \quad (1)$$

where α_n and α_p – the ionization coefficients of electrons and holes, d – the epitaxial layer thickness.

In 4H-SiC, the ionization coefficients of the exponential dependence on the reverse field:

$$\alpha_n = \alpha_{n0} \exp\left(-\frac{E_n}{E}\right), \quad (2)$$

$$\alpha_p = \alpha_{p0} \exp\left(-\frac{E_p}{E}\right), \tag{3}$$

where α_{n0} = 1.76×10^8 cm^{-1}, α_{p0} = 3.41×10^8 cm^{-1}, E_n = 3.3×10^7 V/cm, E_p = 2.5×10^7 V/cm [7].

By substituting expressions (2) and (3) into formula (1), we obtain the following integral equation with respect to E_c:

$$\int_0^d \alpha_{n0} \exp\left(-\frac{En}{Ec - \frac{q}{\varepsilon_S} N_D x}\right) \exp\left[-\int_0^d \left[\alpha_{n0} \exp\left(-\frac{En}{Ec - \frac{q}{\varepsilon_S} N_D x}\right) - \alpha_{p0} \exp\left(-\frac{Ep}{Ec - \frac{q}{\varepsilon_S} N_D x}\right)\right] dx'\right] dx = 1, \tag{4}$$

where ε_s – 4H-SiC dielectric permittivity, N_D –the donor concentration in the 4H-SiC epitaxial layer, q – the elementary charge.

Then, after the numerical solution of equation (4) and determination of the critical breakdown field strength, the breakdown voltage can be estimated by the following formula [6, 7]

$$BV_m = \frac{E_C W}{2} \tag{5}$$

where E_c – the critical breakdown field strength, determined from the integral equation (4), W – the thickness of the space charge area (SCA).

Further, in the approximation, it can be assumed that the breakdown occurs at a time when the thickness of the SCA is equal to the thickness of the epitaxial layer d [7]. A calculated value of the critical breakdown field strength E_c and the breakdown voltage of BV_m with donor concentration N_D = 3×10^{15} cm^{-3} and different values of the thickness d of the 4H-SiC epitaxial layer are shown in Table 1.

Table 1. The values of breakdown voltage BV_m with different 4H-SiC epitaxial layer thicknesses calculated from equations (4) and (5).

d, 4H-SiC epitaxial layer thickness (μm)	N_D, donor concentration in the 4H-SiC epitaxial layer (cm^{-3})	E_c, critical breakdown field strength calculated from the equation (4) (V×m^{-1})	BV_m, 4H-SiC breakdown voltage calculated from the equation (5) (V)
14	3×10^{15}	2.37801×10^8	1.665×10^3
16	3×10^{15}	2.37759×10^8	1.902×10^3
18	3×10^{15}	2.37722×10^8	2.139×10^3
20	3×10^{15}	2.37729×10^8	2.377×10^3
22	3×10^{15}	2.377217×10^8	2.615×10^3

As it can be seen from Table 1 breakdown voltage of more than 2 kV is provided at a layer thickness of 18 μm. With this in mind, the breakdown voltage was simulated for the 4H-SiC epitaxial layer thickness of 18, 20 and 22 μm. The Schottky diode with six guard rings was taken as the object of modeling. In Fig. 1a shown the simulated schematic silicone carbide Schottky diode structure for calculation. For numerical simulation were chosen the following the Schottky diode parameters: the concentration of donors (nitrogen) in the substrate equals $N^+ = 10^{18}$ cm^{-3}, in the n-type epitaxial layer (nitrogen) equals $N^- = 4{,}75 \times 10^{15}$ cm^{-3}, in the guard rings (boron, depth of guard about 2 μm) regions $N_{p+} = 10^{18}$ cm^{-3}, in the JTE layer concentration $N_{JTE} = 10^{17}$ cm^{-3}, anode material is Ti (titanium), the thickness of the epitaxial layer (4H-SiC) was chosen equals 14 and 20 μm, the radius of the structure equals r=675 μm. The simulation of the reverse current-voltage (I-V) characteristics has been carried out in the ATLAS program, with taking into account close to reality situation the incomplete impact ionization and anisotropy in the direction (0001) by the Hummel-Newton method.

3. Results and discussion

Fig. 1b shows the reverse *I-V* characteristics of Schottky diode with thickness of the epitaxial layer (*h*) and distance between rings (*l*) calculated in ATLAS for diode with JTE (*h*=20 μm, *l*=2,5 μm). As follows from Fig. 1b the diode breakdown starts at 2,87 kV i.e. breakdown voltage value equals ~3 kV. For comparison in Fig. 1b also presented *I-V* characteristics of Ti/4H-SiC Schottky diodes with reduced thickness of the epitaxial layer (*h*=14 μm) and increased distance between guard rings (*l*=5 μm) and without JTE layer. In this case the value of breakdown voltage value equals 1,94 kV.

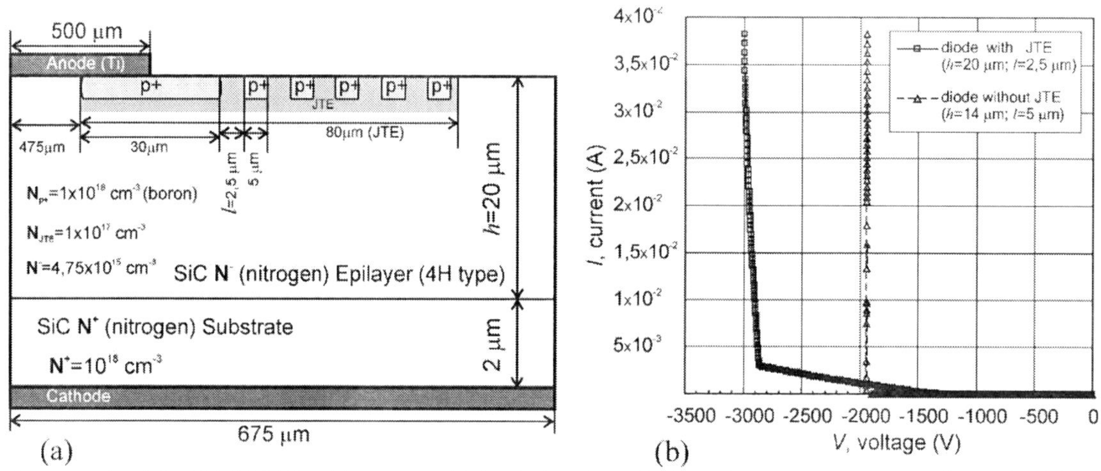

Figure 1. (a) Ti/4H-SiC silicone carbide Schottky diode schematic structure in cylindrical coordinates for calculation; (b) reverse *I-V* characteristics of Ti/4H-SiC Schottky diodes with different of thickness of the epitaxial layer (*h*) and distance between rings (*l*) calculated in ATLAS for diode with JTE (*h*=20 μm, *l*=2,5 μm) and without JTE (*h*=14 μm, *l*=5 μm).

Thus, the increase of the 4H-SiC epitaxial layer thickness from 14 up to 20 μm and a decrease of the distance between the guard rings from 5 to 2.5 μm and also the addition of the JTE layer leads to a significant increase of the breakdown voltage of the diode on ~ 900 V. In addition, on the basis of analysis of the distribution of the impact ionization rate it is established that the maximum values of the velocity of impact ionization rate (23÷25 $cm^{-3}\cdot s^{-1}$) are reached in the area between first and second guard rings. Therefore, with aim to reduce the probability of breakdown in this place the depth of the JTE layer must be increase.

4. Conclusions.

1. It is shown, on the basis of the theoretical calculations and calculations performed in ATLAS, the required breakdown voltage of the Schottky diode of more than 2 kV is provided at a thickness of 4H-SiC epitaxial layer of 18 μm. At the same time, taking into account the very likely possible defects in the structures of real diodes, the most preferred option from the calculations is the thickness of the epitaxial layer of 20 μm, which can provide the required breakdown voltage of the Schottky diode more than 2 kV.
2. The reverse *I-V* characteristics of perspective Ti/4H-SiC Schottky diode with the 4H-SiC epitaxial layer thickness of 20 μm, *p+* wide ring with a width of 30 μm and five rings with a width of 5 μm with a gap of 2.5 μm, JTE layer with a width of 80 μm was calculated. In accordance with calculation the Ti/4H-SiC type Schottky diode with above-described structure demonstrate the very high breakdown voltage value up to 3 kV and can be used in power electronics.

Acknowledgements
This work has been supported by the Russian Ministry of Education and Science (task No. 8.1729.2017/4.6).Authors would like to thank Dr. Ivanov P.A. (Ioffe Physicotechnical Institute) for help in carrying out of simulation in used program.

References

[1] Kimoto T, Cooper J A 2014 *Fundamentals of Silicon Carbide Technology. Growth, Characteriztion, Devices, and Applications* (New York: Wiley–IEEE Press.)
[2] Sedykh S V, Rybalka S B, Drakin A A, Demidov A A, Ponomaryova N S and Shishkina O A 2018 *J. Phys.: Conf. Ser.* **1124** 071012
[3] Ivanov P A, Potapov A S, Rybalka S B, Malakhanov A A 2017 *Journal of Radio Electronics* **6** 1
[4] Panchenko P V, Rybalka S B, Malakhanov A A, Demidov A A, Krayushkina E Yu and Shishkina O A 2017 *J. Phys.: Conf. Ser.* **917** 082010
[5] Panchenko P V, Rybalka S B, Malakhanov A A, Krayushkina E Yu and Rad'kov A V 2016 *Proc. SPIE* **10224** 102240Y-1
[6] Hatakeyama T, Watanabe T, Shinohe T, Kojima K, Arai K and Sano N 2004 *Appl. Phys. Lett.* **85**(8) 1380
[7] Sze S M, Ng Kwok K 2007 *Physics of Semiconductor Devices* (New Jersey: John Wiley & Sons Int.)

Optimization of electrodes for the fast electrochemical actuator

P S Shlepakov[1], I V Uvarov[1], V V Naumov[1], A E Melenev[1] and V B Svetovoy[2,3]

[1]Valiev Institute of Physics and Technology of Russian Academy of Sciences, Yaroslavl Branch, Universitetskaya 21, 150007 Yaroslavl, Russia

[2]Zernike Institute for Advanced Materials, University of Groningen - Nijenborgh 4, 9747 AG Groningen, The Netherlands

[3]A.N. Frumkin Institute of Physical Chemistry and Electrochemistry of Russian Academy of Sciences, Leninsky pr. 31, 199071 Moscow, Russia

E-mail: p.shlepakov@bk.ru

Abstract. Electrochemical membrane actuator based on water electrolysis of alternating polarity (AP) was demonstrated recently. It operates several orders of magnitude faster than the devices working on conventional DC electrolysis. However, extremely high current density of the AP process leads to the fast degradation of the electrodes. Titanium electrodes demonstrate the best durability, but they oxidize during operation that reduces the gas production. Proper design can increase the efficiency of the electrodes and slow the degradation down. In this work, four typical layouts are fabricated and tested in the AP electrolysis. Current flowing through the cell, threshold voltage for the explosive operation and wear of the electrodes are analyzed and compared.

1. Introduction

Electrochemical actuators are widely used in various microfluidic systems such as drug delivery devices and laboratories on a chip [1, 2]. A main part of the actuator is a working chamber filled with an aqueous electrolyte solution and containing two metallic electrodes inside. Voltage applied to the electrodes causes the electrochemical decomposition of water. Hydrogen and oxygen bubbles generated in the chamber push the flexible membrane up. Using a series of microsecond AP voltage pulses instead of one single polarity pulse increases the operation frequency of the actuator by several orders of magnitude [3]. However, the electrodes are quickly destroyed in the AP electrolysis due to a high current density ~100 A/cm^2 [4]. It was empirically found that titanium electrodes are the most durable, but they oxidize during operation [5]. As a result, the Faraday current and the gas production decrease, and performance of the actuator deteriorates.

The amount of the generated gas can be increased by optimizing the electrode layout. Increasing the surface area maximizes the contact of the electrode with the electrolyte, thereby reducing the resistance of the electrochemical cell and increasing the current. In addition, the shape determines the distribution of the current density and, therefore, the preferred location and rate of the degradation. The distance between the electrodes is also important. In this work, we test typical layouts of the electrodes in the AP process and choose the shape that ensures the highest gas generation efficiency and the lowest wear.

2. Materials and methods

The electrodes are fabricated on a thermally oxidized silicon wafer with a 0.9 μm thick SiO_2 layer. They have a three-layer structure. First, a titanium adhesion film with the thickness of 10 nm is deposited by magnetron sputtering. Next, a 500 nm thick aluminum layer and a 500 nm thick titanium layer are deposited. The intermediate Al layer reduces the resistance of the contact lines, while Ti is a working material. Then the wafer is covered by the photoresist Shipley S1813 and the UV contact photolithography is performed followed by the wet etching of Ti/Al/Ti film through the photoresistive mask. After the patterning, a chamber with a diameter of 500 μm is formed around the electrodes in a 8 μm thick layer of the photoresist SU-8. This layer protects the connecting lines from the electrolyte. The size of the chamber is similar to that used in the fast electrochemical actuator [3], but the chamber is not covered by a membrane. The wafer is diced on separate samples with sizes of 8×20 mm^2. The fabricated electrodes are presented in figure 1. Figure 2 illustrates a schematic cross-section of the sample.

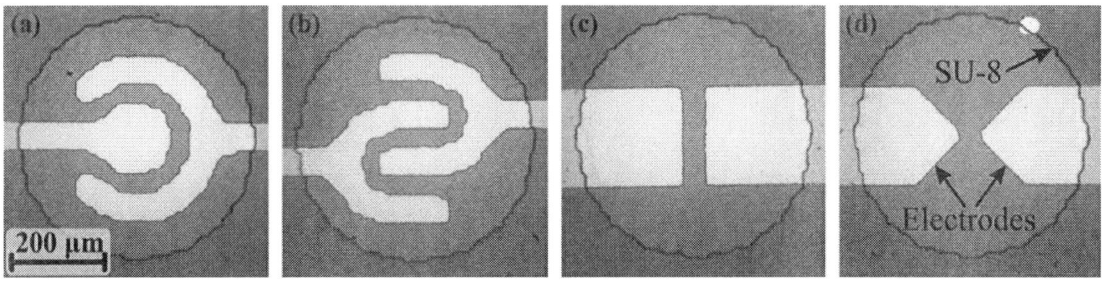

Figure 1. Optical images of the electrodes of four typical layouts: (a) circular; (b) interdigitated; (c) rectangular; (d) triangular.

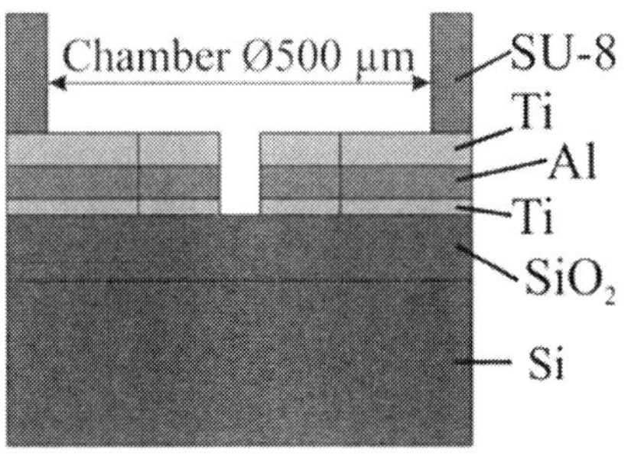

Figure 2. A schematic cross-section of the sample with the indication of the materials.

The sample is placed in a Petri dish filled with the electrolyte, which is a molar solution of Na_2SO_4 in distilled water. The thickness of the electrolyte layer above the sample is about 2 mm. Square voltage pulses of alternating polarity are applied to the electrodes. One electrode is grounded while the other one is at a positive or negative potential with respect to the ground. The voltage is provided by a waveform generator Keysight 33500B and amplified 10 times using a homemade amplifier [5]. The electrodes are tested in two regimes. In the first one, the pulses are applied continuously during 180 s. The frequency of pulses is $f = 200$ kHz and the amplitude is $U = 6$ V. The amplitude is not large enough to cause explosions of the microbubbles [3]. Current flowing through the electrodes is recorded every 10 s by a PicoScope 5000 and the average absolute value is calculated. In the second regime, a series of $N = 10^5$ pulses is applied to the sample every 2 s. The amplitude is adjusted to the threshold level U_{th}, at which the concentration of nanobubbles in the electrolyte reaches a critical value. They merge into a microbubble that explodes with a distinctive clicking sound [6]. This sound was recorded using a microphone to a digital file. The experiment lasts until the electrodes are destroyed or U_{th} exceeds 19 V that is the maximal output voltage of the amplifier. Three samples of each type are tested in each regime. The results below are averaged over these samples.

3. Results and discussion

3.1. Continuous regime

Time dependence of the average current I_{av} is shown in figure 3. In general, I_{av} decreases during the first 100 s of the electrochemical process due to the oxidation of titanium. A slight increase of the current is observed at the circular and interdigitated electrodes in the first 10 s of the test. The probable reason is the increase of the electrolyte temperature. At the beginning of the process the circular and interdigitated structures demonstrate the highest I_{av} of 38-40 mA, while the rectangular and triangular electrodes conduct significantly smaller current of 22 and 16 mA, respectively. The electrochemical cell containing the electrodes and the chamber covered by the electrolyte is simulated by the finite element method (FEM) using the validated software. The calculated conductivity of the cells is given in figure 4. The higher the conductivity, the larger the starting current. The current reaches a steady state in about 100 s. At the end of test (180 s) the highest I_{av} of 10 mA is observed at the interdigitated and rectangular electrodes. Triangular design exhibits the worst result of 7 mA. The drop of the current is reversible. One can return I_{av} to the initial value by increasing the amplitude of pulses to $U \approx U_{th}$. The original current ratio between the layouts is also restored.

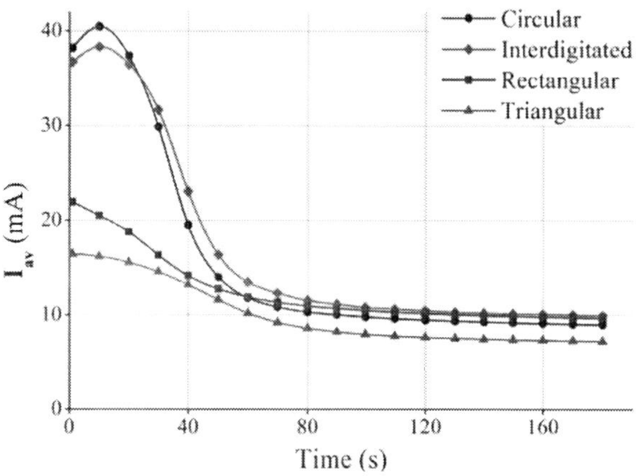

Figure 3. Time dependence of the average current flowing through the electrodes of several types.

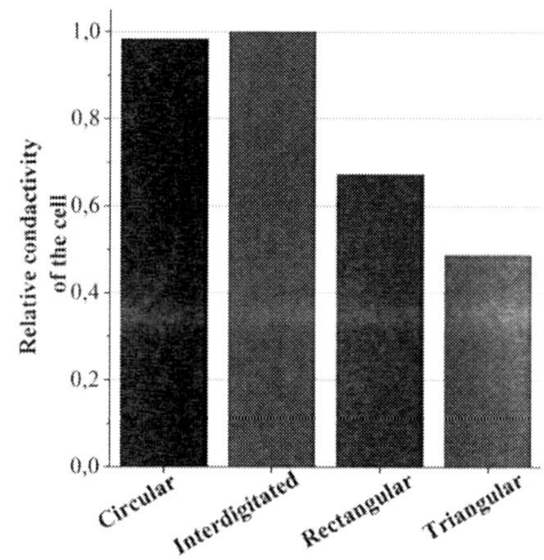

Figure 4. Relative conductivity of the electrochemical cells with the electrodes of several types.

3.2. Explosive regime

The circular and interdigitated electrodes begin to generate explosions at the lowest value of the threshold voltage U_{th} = 10 V (see table 1). Triangular structure has the largest starting U_{th} of 13 V. The threshold amplitude increases with time due to the oxidation of titanium. For triangular and interdigitated electrodes it grows rapidly and reaches 19 V just in 4 min. Rectangular structure begins to generate explosions at U_{th} = 12 V. This design demonstrates the lowest increase rate of U_{th} (table 1), which allows the longest operation during 22 min.

The electrodes after the explosive test are shown in figure 5a. The darkening and partial removal of the material from the substrate are observed. The change of color is caused by a growth of titanium oxide on the electrode surface [7, 8], and the destruction is due to a high current density. The triangular structure demonstrates the less degradation, while the interdigitated layout suffers the most. The upper Ti layer is partially removed from the "fingers", making the Al layer visible. The distribution of the current over the electrodes is simulated using the FEM software. The sites with the highest current density are indicated in figure 5b. They are located at the edges and corners of the electrodes and match with the experimentally observed places of the strongest wear.

Table 1. Characteristics of the electrodes in the explosive regime.

Layout	Initial U_{th}, V	Increase rate of U_{th}, V/min	Relative volume of the explosion sound
Circular	10	1	0.7
Interdigitated	10	3	0.4
Rectangular	12	0.32	0.9
Triangular	13	1.5	1

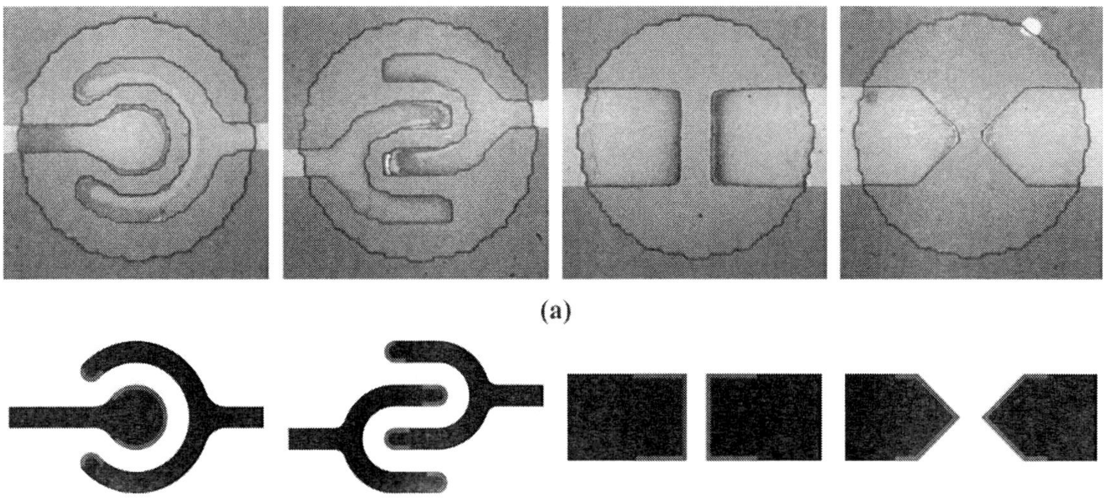

(a)

(b)

Figure 5. (a) Optical images of the electrodes after the explosive test. (b) A schematic illustration of the electrodes. The red color indicates the areas of the highest current density predicted by the FEM simulation.

The rectangular electrodes are able to generate explosions despite the strong degradation. One of the samples withstood 50 min of operation, and the threshold voltage maintained below 18 V. The test was stopped when the repeating explosions broke the SU-8 layer and made a pit in the substrate, see figure 6. The explosion happens when the concentration of nanobubbles in the electrolyte reaches a critical value and densely packed nanobubbles merge into a single microbubble. To all appearance, the rectangular electrodes provide better localization of nanobubbles than the other layouts and, therefore, generate the explosions at the lower amplitude of pulses.

The power of the explosions is estimated by the volume of the clicking sound. The loudest clicks are produced by the rectangular and triangular electrodes (table 1). The interdigitated structure creates the quietest sound, since the areas with the high current density, where the gas is produced more intensively, are narrow and located far from each other. As a result, several preferred places of the explosion are created, but a microbubble is formed by a small number of nanobubbles. However, in a closed chamber of the actuator the produced gas accumulates, and it is easier to reach the critical concentration. Probably, in the real application the interdigitated electrodes will operate more efficiently.

Figure 6. The rectangular electrodes after 50 min of operation in the explosive regime: (a) the optical image, top view; (b) the SEM image at the angle of 20° with respect to the substrate; (c) close-up view of the pit made in the substrate by the explosions.

4. Conclusions

The electrodes of four typical layouts were tested in the electrolysis of water performed by microsecond voltage pulses of alternating polarity. In the continuous operation regime, the circular and interdigitated structures provide the highest current at the beginning of the electrochemical process. During the first 100 s of the test the current drops due to the oxidation of titanium, but then the decrease stops. The circular, interdigitated and rectangular electrodes conduct approximately the same current in the steady state, while the triangular structure provides the significantly less current. In the explosive mode, the rectangular electrodes demonstrate the best result. They generate explosions for a relatively long time despite the strong wear. This feature is determined by the high localization of the nanobubbles. The triangular electrodes show weak degradation, but the amplitude of voltage pulses required for the explosion quickly reaches the limit value. The interdigitated layout demonstrates the strongest wear. The degradation is observed in the areas with high current density, predicted by the finite element simulation. The circular electrodes show the intermediate result on the degradation and the threshold voltage. Thus, the circular and rectangular structures are considered most suitable for the fast electrochemical actuator. However, the final choice of the layout will be made after testing in a chamber sealed by an elastic membrane.

Acknowledgments

This work is supported by the Russian Science Foundation, Grant No. 18-79-10038, and performed using the equipment of Facilities Sharing Centre "Diagnostics of Micro- and Nanostructures".

References

[1] Cobo A, Sheybani R and Meng E 2015 *Adv. Healthcare Mater.* **4** 969
[2] Saltzman W M and Olbricht W L 2002 *Nat. Rev. Drug Discovery* **1** 177
[3] Uvarov I V, Lokhanin M V, Postnikov A V, Melenev A E and Svetovoy V B 2018 *Sensor Actuat. B: Chem.* **260** 12
[4] Svetovoy V B, Sanders R G P and Elwenspoek M C 2013 *J. Phys.: Condens. Matter* **25** 184002
[5] Uvarov I V, Lemekhov S S, Melenev A E and Svetovoy V B 2017 *J. Micromech. Microeng.* **27** 105009
[6] Postnikov A V, Uvarov I V, Lokhanin M V and Svetovoy V B 2016 *Sci Rep.* **6** 39381
[7] Van Gils S, Mast P, Stijns E and Terryn H 2004 *Surf. Coat. Technol.* **185** 303
[8] Delplancke J-L, Degrez M, Fontana A and Winand R 1982 *Surf. Technol.* **16** 153

Control electrically conductive of thin films by using subminiature eddy current transducers

A Ishkov[1], V Malikov[2]
[1]Department of technology of construction materials and repair of machines, Altay State Agricultural University, Barnaul 656049, Russia
[2]Physical-technical faculties, Altai State University, Barnaul 656049, Russia

Abstract. The research objective was the development of subminiature eddy-current transducer designed for studying thin films. Timeliness is subject to the need to assess and predict safe operation life of products manufactured using thin films. A review of various methods for thin films studying was made, including the eddy-current method, based on an analysis of the interaction of an external electromagnetic field with the electromagnetic field of eddy currents induced by an energizing winding in an electrically conductive film. A subminiature surface eddy-current transducer of the transformer type was developed. On the basis of eddy-current transducer, a software and hardware complex was developed to control the eddy-current transducer (to generate alternating current of various frequencies, to provide its supply to the eddy-current transducer, to receive a useful signal from transducers to ensure its convenient visualization). The hardware and software complex provided the opportunity to make local studies of the electrical conductivity of non-ferromagnetic thin films by the method of eddy currents. In the course of the study it was possible to develop an algorithm for finding changes in the average amplitude of the output signal, which allows to conclude about the electrical conductivity of the thin film. The study of thin film samples using the developed measuring system is conducted. The results of testing of Al, Ag and Ni films were presented, and the electrical conductivity values of the samples were obtained.

1. Introduction

In recent years, thin films have become increasingly useful in micro-and optoelectronics to create LEDs and photodiodes, sensors, radio frequency tags and other devices. Of particular interest are compounds that may have some advantages over standard semiconductors such as the simplicity and cheapness of methods for obtaining structures in mass production, the ability to create flexible and transparent devices, as well as modification of the properties and variation of the band structure of molecules [1].

In recent times, accurate determination of thin metal films properties is of critical importance in such industries as microelectronics, optics, etc.

Thin films are widely used to study electrical properties of new materials when forming the contact surfaces; to apply resistive and conductive coatings in industry and during manufacture of elements of integrated circuits in microelectronics; to create optical filters, and reflective and light-conducting coatings of optoelectronics; modern lithographic processes.

An important difference between a thin metal film and a similar metal in the normal state is the difference in electrical conductivity. The differences in conductivity may be great. Therefore, it is extremely important to create a gage system that allows to measure the conductivity of a thin metal film.

According to the results of studies [2-7], it is reasonable to measure electrical conductivity of thin films using the eddy current control method. This method provides the use of an eddy current transducer with energizing and measuring windings. Analysis of the electromagnetic field distribution appeared in a thin film by means of an eddy current transducer allows us to make a conclusion on electrical conductivity of the film.

The use of the eddy current method provides the high performance of control, the possibility of checking without direct contact of the ECT and the surface of the thin film to be checked, the simplicity of the ECT design, and the weak dependence of the control results on environmental conditions [9, 10].

In studies [11-15] the eddy current transducer, the magnetic field of which was single-frequency, was used to measure the conductivity of thin films. This solution allows to simplify the design of the device but does not provide the opportunity to explore the film at different depths. A multi-frequency signal is frequently used to solve this problem [16-18].

The requirements for non-contact and highly precise online thickness measurement of metal films has increased rapidly with the development of industrial automation, semiconductor industry, and micromechanical technology. Compared to X-ray reflectivity and opto-acoustic techniques [19], the eddy current technique [20-22] enables increased flexibility, and it can cover the measurement range from tens of nanometers to a number of millimeters. The main techniques for metal thickness measurement with eddy current sensors can be classified as single-frequency eddy current [21], pulsed eddy current (PEC) [20] and multi-frequency eddy current (MEC) [22] sensors based on the types of excitation signals. In the single-frequency eddy current technique, the sensor coil is excited with a sinusoidal signal and the eddy currents are distributed in a fixed depth. It is the most traditional method and is the basis of other methods.

Due to the small thickness of thin films (100-2000 μm), only a few models on the market allow to study their conductivity. In particular, this is due to the range of operating frequencies required for the study of thin films, which is 1-10 MHz.

There is an eddy current system designed for the study of thin metal films, which implements the method of resonance screen eddy current defectoscopy [9]. The system is highly effective in detecting defects of thin non-ferromagnetic films, the thickness of which is 0.01 μm at the frequency of the exciting current in the range of 10-30 MHz. The disadvantage of this system is the screening transducer, where the thin film should be located in the gap between the measuring and compensation coils. This complication of the design is due to the fact that the investigated films can have a substrate of different thickness, and it is necessary to vary the width of the gap between the measuring and compensation coils in the studies.

The aim of the study is to develop a measuring system based on the eddy current transducer designed to measure the conductivity of thin metal films capable of localizing the electromagnetic field in small areas of the object of control.

2. Material choices and design

In accordance with the paper [23, 24], the following design of the eddy current transducer was constructed:

Exciting and measuring coils contain 200 turns. The wire for coils is 5 μm thick. The stabilizing coil contains 140÷170 turns. The cores are made of permalloy 81NMA with an initial magnetic permeability of 35000 (Fig. 1).

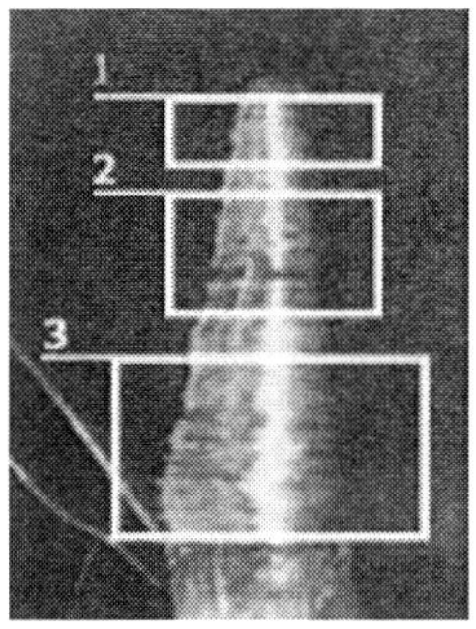

Figure 1. Design of the eddy current transducer, 1 — measuring winding, 2 — generator winding, 3 —compensating winding..

The design of the eddy current transducer used to test steel pipes was a magnetic core made of permalloy 81NMA in the shape of a pyramid. The energizing winding, measuring winding and compensating winding were wound on the magnetic core. The winding turns were impregnated with epoxy compound to form monolithic structure [25, 26].

The difference between the averaged amplitudes of the signals of the transducer, which is paced above the thin metal film and the transducer paced above the substrates, is the parameter of the film electrical conductivity.

The generation unit controls the generator , which transmits a signal of frequency f1 to the energizing windings of eddy current transducers, these ETCs create an electromagnetic field that induces eddy currents in an electrically conductive test object. The signals pass through power amplifier , where their voltage rises up to 3 V that is necessary to carry out the measurements, and pass to energizing windings of eddy current transducers. As a result, the energizing windings create a magnetic field that penetrates the tested thin film and substrate. The magnetic field induces eddy currents in the test sample, which, in turn, induces a voltage in the measuring windings . The voltage in the form of signals C1 and C2 carries information about the substrate and the thin film, respectively. The signals pass through the amplifier-computer 8 and pass to the signal filtering unit, controlled by the filtering software unit connected with the generation software unit. The filtering frequency changes simultaneously with the generation frequency. Two signals are transmitted to the amplitude detector, through the analog-to-digital transducer, to the signal processing software unit and the measurement results are displayed in a form of a diagram on the PC screen and the difference values of the averaged amplitudes of the two signals C1 and C2 are displayed. Electrical conductivity (σ, MSm/m) is determined according to the experimentally obtained equation $f(x) = 0.0809x - 0.3696$, according to the diagram of film samples with known electrical conductivity, where point 1 corresponds to the aluminium sample with electrical conductivity of 0.1 MSm/m and a signal amplitude difference is 16.8 mV, point 2 corresponds to a sample of aluminium with a conductivity of 1.23 MSm/m and a signal amplitude difference of 19.6 mV.

3. Experimental results

The studies of the signals of the designed transducer were carried out on samples representing thin films of such metals as Al, Ag, and Ni obtained after condensation during gas phase in vacuum on glass substrates. The dimensions of the substrates are 23 x 23 mm. The investigated films have dimensions of 23 x 21 mm.

This method stipulates the use of two signals, where the difference of averaged amplitudes ($<\Delta U>$, mV) of two signals (the signal from zone 1 and the signal from zone 2) is used as a parameter of the film electrical conductivity, and the electrical conductivity of the thin metal film is calculated by experimentally obtained equation $f(x) = 0.0809x - 0.3696$, where x is the difference of the amplitudes $\Delta <U>$ of the two signals C1 and C2.

Fig. 2 shows the superimposed data obtained during study using two transducers. Zone 2 corresponds to the data obtained using the transducer placed above the film, zone 1-3 — to the data obtained using the transducer placed above the substrate without the film.

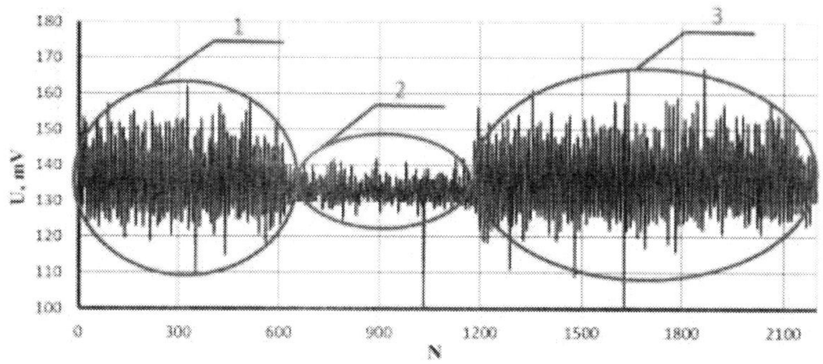

Figure 2. Voltage on the measuring winding of ECT (Al film). 1 — signal without object of control, 2 — signal in the presence of the film under study.

In Fig. 3, the results of the algorithm for finding the average amplitude are presented.

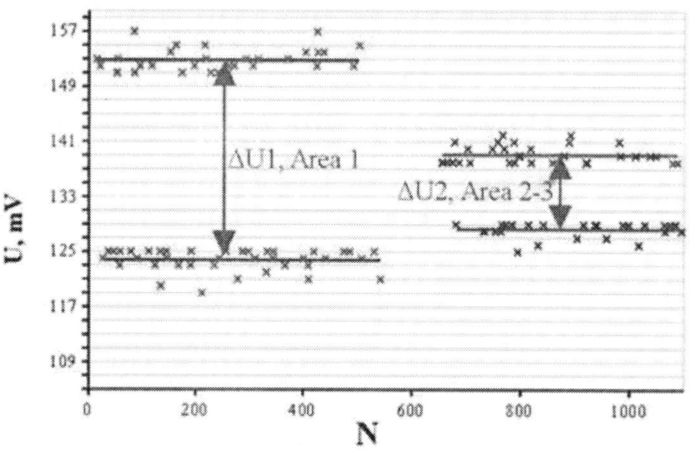

Figure 3. Determination of the averaged signal amplitude, ΔU1 — the signal span without the object of control, ΔU2— in the presence of the Al film on the glass base surface.

In the area 1 the averaged amplitude of the signal is equal to 29 mV, in the area 2-3 - 10.8 mV. The difference between the EMF amplitude in the area 1 and the EMF amplitude in the area 2 ($\Delta\langle U\rangle$) is 18.2 mV.

$\Delta\langle U\rangle$ for Ag film is 8.31 mV and for Ni film-21.35 mV.

In Fig. 4 and Fig.5 the results of measuring the response of Ag and Ni films, are presented respectively.

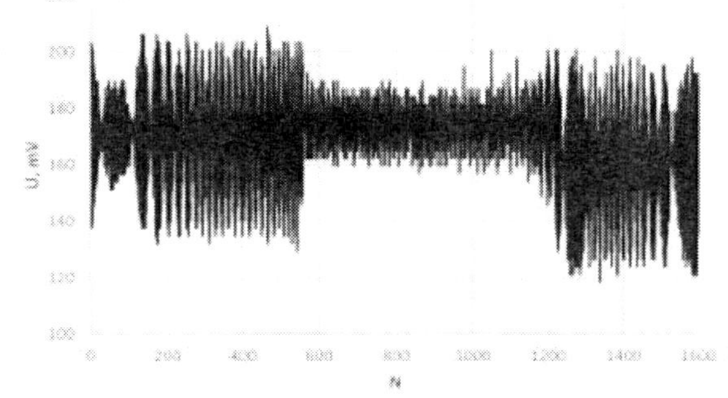

Figure 4. Voltage on the measuring winding of ECT (Ag film).

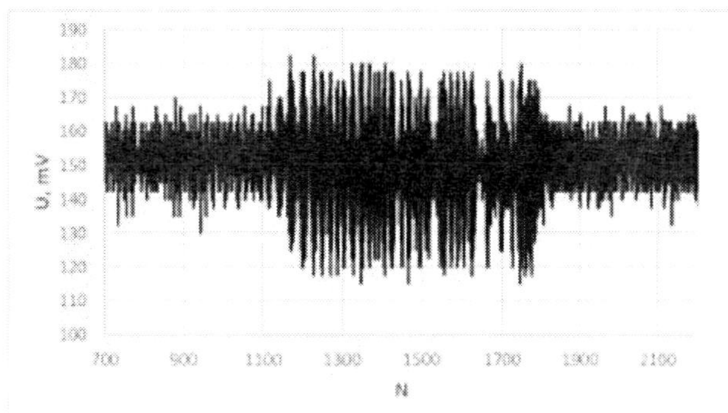

Figure 5. Voltage on the measuring winding of ECT (Ni film).

Analysing the distribution of the signal in Figures 2, 4, one can notice a drop in the amplitude of the signal received from the transducer placed above the film, compared with the amplitude of the signal placed above the substrate. This feature is explained by the insufficient compensation of the windings of the eddy current transducer, as a result of which the initial magnetic field of the energizing winding, affecting the measuring winding, weakens under the influence of the magnetic field of the eddy currents of the film under study.

However, for the Ni film, the magnetic field of the eddy currents of the film under study is much higher than the magnetic field of the energizing winding (Fig. 3). As a result, the influence of the magnetic field of the energizing winding is completely dropped out of the equation, and only the magnetic field of the eddy currents of the film is of concern.

Based on the data obtained and in accordance with the experimentally obtained equation $f(x) = 0.0809x - 0.3696$ based on samples of films with known electrical conductivity the conclusion on the electrical conductivity of the films under study may be drawn (Table 1).

Table 1. The results of calculation electrical conductivity of films Ag, Al, Ni.

	<ΔU>, mV	*σ, MS/m*
Ag	8.31	0.161
Al	18.2	0.5114
Ni	21.35	0.956

4. Conclusions

The results of the experiment showed the effectiveness of the proposed method of controlling the electrical conductivity of thin films with electrical conductivity from 0.1 to 1 MSm/m. To achieve this goal, we used developed ultra-miniature eddy current transducers with a filtering system for the received signal.

Using the developed algorithm to determine the electrical conductivity of the film on samples with known electrical conductivity, it becomes possible to draw a conclusion about the electrical conductivity of the material under study.

Further improvement of the developed prototype of the gage system can be carried out both by increasing the efficiency of the applied signal processing algorithms and by upgrading the hardware. In this case, the first steps should be usage of more efficient digital filters and high resolution analog-to-digital transducers.

References

[1] Suzuki A, Oku T 2014 *Electronics* **3(1)** 112
[2] Yurkov V, Ryzhii V 2008 *JETP Letters* **88(5)** 370
[3] Abramchuk S 2007 *Polym. Adv. Technol.* 18**(11)** 883
[4] Gerasimov V, Kluev V 2010 *Methods and devices of electromagnetic control (Moscow, SPEKTR)*
[5] Li W, Ye Y, Zhang K 2017 *IEEE Trans. Indust. Electr.* 64**(5)** 3940
[6] He D F, Zhang K, Tang J 2017 Proc 3rd IEEE International Conference on Control Science and Systems Engineering (Beijing) p 305
[7] Uchanin V N 2010 *Tech. diagn. non-destr. test.* **4** 24
[8] Bakunov A S, Yefimov A G 2009 *Control. Diagn* **4** 21
[9] Angani S 2010 *J. Appl. Phys.* **107(9)**
[10] Ribeiro A L, Ramos H G 2012 *Measur.* **45(9)** 2246
[11] Qu Z L, Zhao Q 2014 *Nondest. Test.Eval. Int.* **61** 53
[12] Qu Z L, Zhao Q 2013 *Microelectron. Eng.* **108** 66
[13] Wang H B 2015 *IEEE Trans. Instrum. Meas.* **64(9)** 2557
[14] Li W, Wang H B, Feng Z H 2016 *Rev. Sci. Instrum.* **87(4)**
[15] Kral J, Smid R 2013 *IEEE Trans. Instrum. Meas.* **62(7)** 2043
[16] Yin W L, Xu K 2016 *IEEE Trans. Instrum. Meas.* **65(1)** 164
[17] Pinotti E, Puppin E 2014 *IEEE Trans. Instrum. Meas.* **63(2)** 479
[18] Yin W, Peyton A J 2007 *Nondestr. Test. Eval. Int.* **40(1)** 43
[19] Windover D, Barnet E, Summers J, Gribbin C, Lu T M., Kumar A 2001 *Character. Metrol. Ulsi Tech.* **550** 243
[20] Kral J, Smid R, Ribeiro A L 2013 *IEEE Trans. Instrum. Meas.* **(62)7** 2043
[21] Ribeiro A L, Ramos H G, Arez J C 2012 *Measur.* **45 (9)** 2246
[22] Yin W L, Xu K 2016 *IEEE Trans. Instrum. Meas.* **65(1)** 164
[23] Egorov A V 2017 *Izv. Alt. St. Univ.* **4** 33
[24] Bakunov A S, Kaloshin V A 2016 *Cont. Diagn.* **1** 27
[25] Dmitriev S F, Malikov V N, Sagalakov A M, Ishkov A V, Katasonov A O 2017 **289**
[26] Dmitriev S F, Malikov V N, Sagalakov A M, Ishkov A V 2017 *Mater. Sci. Forum* **906** 147

Peculiarities of the electronic transport in topological materials of Bi_2Se_3 and $Mo_xW_{1-x}Te_2$ (x = 0; 0.5; 1)

V V Chistyakov[1], **A N Domozhirova**[1], **J C A Huang**[2], **S V Naumov**[1], **V V Marchenkov**[1,3]
[1]M.N. Mikheev Institute of Metal Physics UB RAS, Ekaterinburg 620108, Russia
[2]National Cheng Kung University, Tainan 70101, Taiwan
[3]Ural Federal University, Ekaterinburg 620002, Russia

Abstract. The electrical resistivity of thin films of a topological insulator of Bi_2Se_3 with a thickness of 10 nm to 75 nm, single crystal of Bi_2Se_3 with thickness of 0.65 mm and single crystals of topological Weyl semimetals $Mo_xW_{1-x}Te_2$ (x = 0; 0.5; 1) in the temperature range from 4.2 to 300 K was measured. A size effect in the electrical conductivity of Bi_2Se_3 films was observed, i.e. linear dependence of the conductivity of the film on its reciprocal thickness. It is suggested the existence of two different conduction channels in the $Mo_{0.5}W_{0.5}Te_2$ compound.

1. Introduction

In recent years new quantum materials with a topologically nontrivial band structure resulting from strong spin-orbit interaction have been predicted theoretically and discovered experimentally. First of all, these are topological insulators (TI) [1]. In these compounds there is an energy gap in the bulk of the material, characteristic of an insulator, and protected gapless conducting states on its surface. Such materials include Bi_2Se_3, Bi_2Te_3, Sb_2Te_3 and etc. The rigid coupling between the directions of the pulse and the electron spin leads to the appearance of spin polarization of charge carriers and the possibility of spin-polarized current flowing near the surface of the TI with virtually no losses [1]. In addition, the presence of such unique surface states in topological insulators [2, 3] makes them promising materials for the creation of electronic devices with high speed and low power consumption. Other promising materials are the recently discovered topological Weyl semimetals (TWS) with nontrivial charge transfer, both in the volume and on the surface of such materials [4, 5]. A characteristic feature of Weyl's semimetals is the presence of exotic gapless surface states — Fermi arcs. The quasiparticles in the bulk of Weyl semimetals are "massless" Weyl fermions [2], with a "zero" effective mass, which are also protected topologically. This means that such quasiparticles can be controlled much faster than conventional charge carriers, and the probability of their scattering is sufficiently small. Later it was found that the properties of Weyl semimetals occur in other compounds, such as semimetallic dichalcogenides of transition metals $MoTe_2$, WTe_2 and ternary compounds $Mo_xW_{1-x}Te_2$ [4-6]. It is known [7-9] that the WTe_2 compound usually exhibits semimetallic properties and has a "metallic" type of temperature dependence of the electroresistivity $\rho(T)$, whereas the $MoTe_2$ has a "semiconductor" dependence $\rho(T)$.

Apparently, the above mentioned peculiarities of topological insulators and Weyl semimetals, i.e., a significant difference between bulk and surface as well as the different nature and type of conductivity, should manifest themselves in the electronic transport of such materials, in particular, in the electroresistivity. The aim of this report is to study the features of the electronic transport in a topological insulator of Bi_2Se_3 and Weyl semimetals of $Mo_xW_{1-x}Te_2$ (x = 0; 0.5: 1).

2. Experimental

Thin films of Bi_2Se_3 were grown by the molecular beam epitaxy method on Al_2O_3 substrates [10, 11] with thickness from 10 to 75 nm. Single crystal (SC) of Bi_2Se_3 with thickness 0.65 mm was cleaved from a boule grown by melting stoichiometric mixtures of high-purity elemental Bi and Se. $Mo_xW_{1-x}Te_2$ single crystals (x = 0; 0.5; 1) were grown by the method of chemical vapor transport using Br_2 as a transport agent. The XRD data and the atomic content of elements analysis showed the synthesized samples of TI have Bi_2Se_3 composition (see Refs. [10, 11]), and samples of TWS have $Mo_xW_{1-x}Te_2$ (x = 0, 0.5, 1) composition (see Refs. [12]). The atomic content of elements was measured by a scanning electron microscope equipped with an EDAX X-ray microanalysis attachment. Our examination showed that the deviations from a stoichiometric composition were insignificant in all samples. The measurements of the electroresistivity were carried out by the conventional 4-points method at dc-current in the temperature range from 4.2 to 80 K and in magnetic fields of up to 10 T. The results are presented in units of conductivities $\sigma_0 \approx 1/\rho_0$ and $\sigma_{xx} \approx 1/\rho_{xx}$, since $\rho_{xx} >> \rho_{xy}$ (see also Refs. [7-9]).

3. Results

Figure 1a shows the temperature dependence of the electroconductivity for Bi_2Se_3 samples. One can see the conductivity value depends on a sample dimension. Since the conductivities of TIs in a bulk and near the surface differ significantly, TI can be represented as a system of two parallel-connected conductors: a "metal" surface and a "semiconductor" bulk (see [13, 14] and references therein). As was shown in [13], the conductivity σ of such a system can be represented as:

$$\sigma \approx \sigma_{surf.} \cdot \frac{\delta}{d} + \sigma_{bulk} \quad (1)$$

Where $\sigma_{surf.}$ is a surface conductivity of the subsurface layer with a thickness of δ, σ_{bulk} is a bulk conductivity. Thus, the dependence of conductivity σ on the inverse film thickness d^{-1} should be observed, i.e. $\sigma = f(d^{-1})$.

As a result of the experiments performed, a size effect was discovered, i.e. the dependence of the conductivity on the inverse film thickness. Figure 1b shows the dependence of the conductivity at T = 4.2 K on the inverse thickness of the samples of TI Bi_2Se_3.

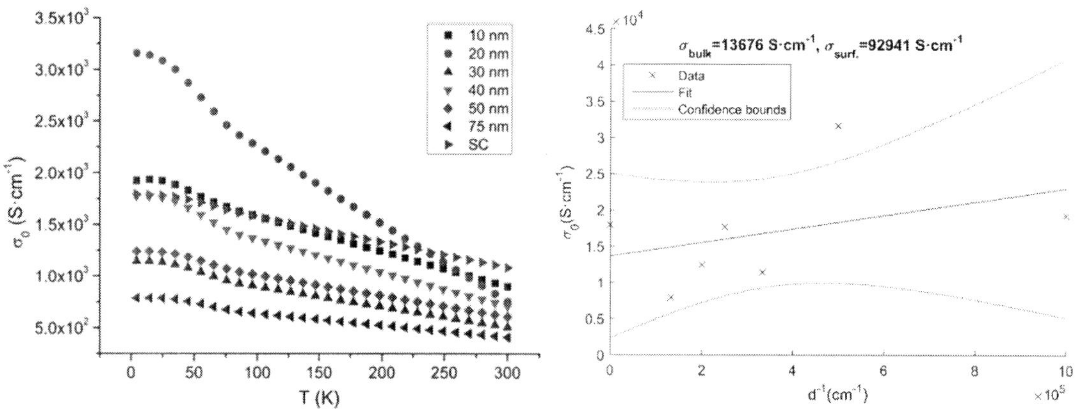

Figure 1(a, b). (a) Temperature dependence of conductivity of TI Bi_2Se_3 films and SC; **(b)** Size effect in TI Bi_2Se_3 without a magnetic field at T = 4.2 K.

By plotting the graphs presented in fig. 1b for the temperature range, it is possible to calculate the temperature dependences for the bulk and surface contributions to the total conductivity. As shown in [13], the "surface" of Bi_2Se_3 thin films has a metallic conductivity, while the "bulk" is a semiconductor one. However, if we consider a situation where, in addition to thin films, we have a "thick sample," the situation changes drastically. In this case, the "bulk" of sample also acquires a metallic conductivity type.

In topological materials there are exotic features in electronic transport, for example, in electrical conductivity. In topological insulators, these are significant differences in surface and bulk conductivity. Weyl semimetals of $Mo_xW_{1-x}Te_2$ composition also have interesting features in conductivity. The WTe_2 compound has orthorhombic symmetry, however, the $MoTe_2$ compound can crystallize, depending on the cooling rate, with both orthorhombic symmetry and hexagonal. In the case of orthorhombic symmetry, the compound has a metallic conductivity type, and in the case of hexagonal symmetry, it is semiconductor. Figure 2 shows the temperature dependence of conductivity for WTe_2 (a) and $MoTe_2$ (b) compounds.

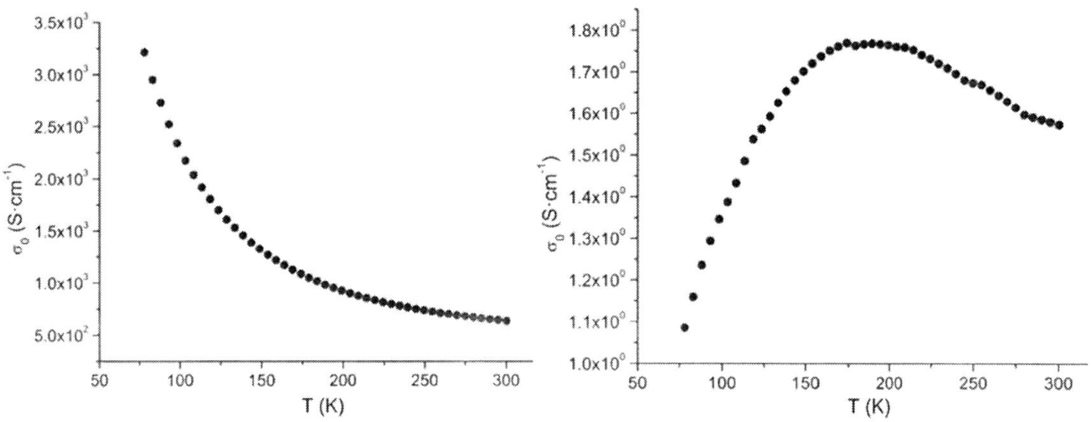

Figure 2(a, b). (a) Temperature dependence of conductivity TWS WTe_2; **(b)** Temperature dependence of conductivity TWS $MoTe_2$.

In the case of the $Mo_{0.5}W_{0.5}Te_2$ ternary compound, it can be assumed that there are two different conduction channels in this compound: the "metallic" conductivity from WTe_2 and the "semiconductor" conductivity from $MoTe_2$. Figure 3 shows the experimental temperature dependence of the conductivity of the $Mo_{0.5}W_{0.5}Te_2$ compound in comparison with the calculated curve obtained by summing the conductivities of WTe_2 and $MoTe_2$ with the corresponding weighting factors.

As seen from Fig. 3, the experimental and calculated curves have quite similar behaviour and the conductivity values are in a good agreement too. Apparently, the proposed assumption about the presence of two conduction channels due to the presence of charge carriers of a "semiconductor" and "metallic" type can occur in the Weyl semimetal $Mo_{0.5}W_{0.5}Te_2$. However, this requires further experimental and theoretical studies.

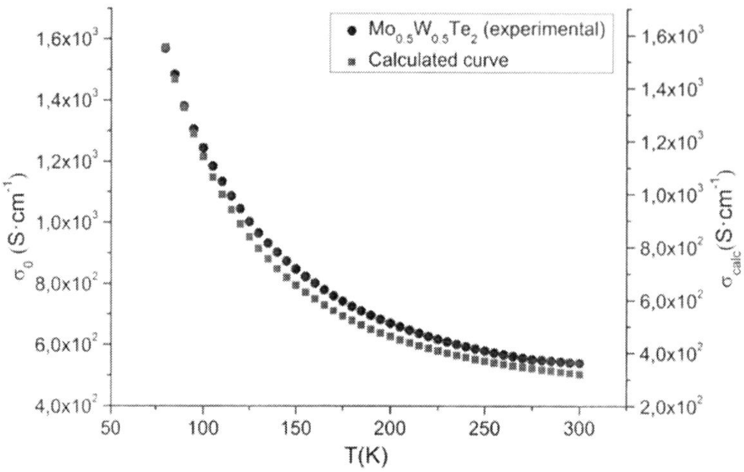

Figure 3. Temperature dependence of conductivity TWS $Mo_xW_{1-x}Te_2$ (x = 0; 0.5; 1) in comparison with the calculated curve (y-axis on the right).

4. Conclusions

The peculiarities of the electroconductivity in the TI Bi_2Se_3 and TWS $Mo_{0.5}W_{0.5}Te_2$ were observed and studied. It was shown that it possible to experimentally "separate" the bulk and surface contributions into the conductivity of the TI Bi_2Se_3. It was found that the value of the surface contribution by almost an order of magnitude exceeds the value of the bulk. The obtained results can be used for "separation" and evaluation of the values of surface and bulk conductivity also in other TIs and systems with non-uniform distribution of direct current over the cross section of the sample. It is suggested the existence of two different conduction channels in the $Mo_{0.5}W_{0.5}Te_2$ compound. Comparison of calculations with experiment shows a fairly good consistency.

Acknowledgments

This work was partly supported by the state assignment of Ministry of High Education and Science of Russia (theme "Spin" No. AAAA-A18-118020290104-2), by the RFBR (project No. 17-52-52008) and by the Government of the Russian Federation (state contract No. 02.A03.21.0006).

References

[1] Zhang H, Liu C X, Qi X L, Dai X, Fang Z, Zhang S C 2009 *Nat. Phys.* **5** 438
[2] Xu S Y, Belopolski I, Alidoust N, Neupane M, Bian G, Zhang C, Sankar R, Chang G, Yuan Z, Lee C, Huang S, Zheng H, Ma J, Sanchez D S, Wang B, Bansil A, Chou F, Shibayev P P, Lin H, Jia S, Hasan M Z 2015 *Science* **349** 6248
[3] Liu Z K, Yang L X, Sun Y, Zhang T, Peng H, Yang H F, Chen C, Zhang Y, Guo Y F, Prabhakaran D, Schmidt M, Hussain Z, Mo S K, Felser C, Yan B, Chen Y L 2016 *Nat. Mat.* **15** 27
[4] Huang L, McCormick T M, Ochi M, Zhao Z, Suzuki M T, Arita R, Wu Y, Mou D, Cao H, Yan J, Trivedi N, Kaminski A 2016 *Nat. Mat.* **15** 1155
[5] Li P, Wen Y, He X, Zhang Q, Xia C, Yu Z M, Yang S A, Zhu Z, Alshareef H N, Zhang X X 2017 *Nat. Commun.* **8** 2150
[6] Belopolski I, Sanchez D S, Ishida Y, Pan X, Yu P, Xu S Y, Chang G, Chang T R, Zheng H, Alidoust N, Bian G, Neupane M, Huang S M, Lee C C, Song Y, Bu H, Wang G, Li S, Eda G, Jeng H T, Kondo T, Lin H, Liu Z, Song F, Shin S, Hasan M Z 2016 *Nat. Commun.* **7** 13643
[7] Volkenshtein N V, Glinski M, Marchenkov V V, Startsev V E, Cherepanov A N 1989 *Sov. Phys. JETP* **68** 1216

[8] Marchenkov V V, Cherepanov A N, Startsev V E, Czurda C, Weber H W 1995 *J. Low Temp. Phys.* **98** 425
[9] Marchenkov V V, Weber H W, Cherepanov A N, Startsev V E 1996 *J. Low Temp. Phys.* **102** 133
[10] Liu Y H, Chong C W, Jheng J L, Huang S Y, Huang J C A, Li Z, Qiu H, Huang S M, Marchenkov V V 2015 *Appl. Phys. Lett.* **107** 012106
[11] Liu Y, Chong C, Chen W, Huang J C A, Cheng C, Tsuei K, Li Z, Qiu H, Marchenkov V V 2017 *Jpn. J. Appl. Phys.* **56** 070311
[12] Marchenkov V V, Domozhirova A N, Makhnev A A, Shreder E I, Naumov S V, Chistyakov V V, Huang J C A, Eisterer M 2019 *Low temperature physics* **45** 241
[13] Marchenkov V V, Chistyakov V V, Huang J C A, Perevozchikova Y A, Domozhirova A N, Eisterer M 2018 *EPJ Web of Conferences* **185** 01002
[14] Marchenkov V V, Weber H W 2003 *J. Low Temp. Phys.* **132** 135

Mobility of a two-dimensional electron gas in the AlGaN/GaN heterostructure: simulation and analysis

V Volcheck, V Stempitsky
Department of Micro- and Nanoelectronics, Belarusian State University of Informatics and Radioelectronics, Minsk, 220013, Belarus

Abstract. At temperatures higher than the room temperature, a two-dimensional electron gas (2DEG) formed at the AlGaN/GaN heterointerface can be characterized by the three dominant scattering mechanisms: acoustic deformation potential, polar acoustic phonon and polar optical phonon scatterings. An analytical model describing the 2DEG mobility limited by these scattering mechanisms as a function of the carrier concentration and the temperature was developed and integrated into a device simulator package using a C language interpreter. The model should be useful for heterostructure device simulators such as Blaze.

1. Introduction

Gallium nitride (GaN) is a binary III/V semiconductor with a wide band gap of 3.4 eV, which affords it outstanding properties for applications in high power, high frequency and optoelectronic devices. The excellent breakdown and reasonable transport characteristics of GaN are made full use of in heterojunction field-effect transistors or high electron mobility transistors (HEMTs), which are currently the most widespread electronic nitride devices. The HEMT is a field-effect transistor, incorporating a junction between two materials with different band gaps, i.e. heterojunction, to confine electrons to a triangular quantum well. As a result, the electrons are free to move in two dimensions but tightly confined in the third, thus forming a two-dimensional electron gas (2DEG). Figure 1 illustrates the conduction band edge (E_c) diagrams of the aluminium gallium nitride (AlGaN)/GaN heterostructure at absolute zero and at a temperature (T) above 0 K. The density in the 2DEG is determined by the conduction band edge and the Fermi level (E_F). Optimally only one of the quantized levels lies below the Fermi level. As the carriers in the triangular well are separated from the donors in the AlGaN layer by a thin space layer, which decreases the impurity scattering [1], the 2DEG exhibits a very high mobility.

Analyzing III/V semiconductors and devices with position-dependent band structures is complicated by the lack of a thorough approach to modeling, as the development of III/V materials has routinely trailed the advanced silicon technology. The electrical characterization of HEMTs requires accurate mobility models for carriers both in a bulk material layer and in a 2DEG. Popular heterostructure device simulator packages such as Blaze, however, do not include built-in 2DEG mobility models. In this paper, we develop and integrate into a device simulator using a C language interpreter an analytical model describing the 2DEG mobility limited by three dominant scattering mechanisms: acoustic deformation potential, polar acoustic phonon and polar optical phonon scatterings.

2. Structure

A two-dimensional schematic representation of the AlGaN/GaN HEMT structure, the electrical behavior of which is simulated using the developed 2DEG mobility model, is shown in figure 2. It consists of a 1.5 μm GaN buffer layer, a 20 nm $Al_{0.21}G_{0.79}aN$ barrier layer and 6 μm silicon nitride (Si_3N_4) passivation layer. The thickness and the length of the gate are equal to 3 μm and 0.5 μm, respectively. The source-to-gate and the gate-to-drain distances are 1 μm and 3.5 μm. The scale factor representing the third dimension in two-dimensional simulations equals to 150 μm. The thickness of the sapphire (Al_2O_3) substrate is 100 μm.

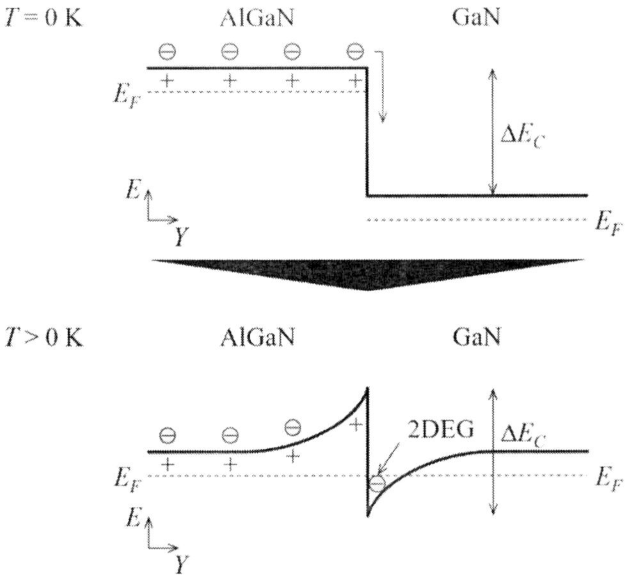

Figure 1. Conduction band edge diagrams of the AlGaN/GaN heterostructure at absolute zero and at a temperature above 0 K.

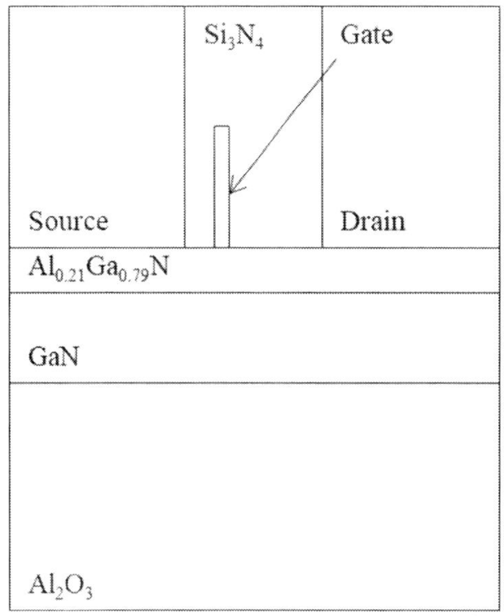

Figure 2. AlGaN/GaN HEMT structure.

3. Two-dimensional electron gas mobility model

At temperatures higher than the room temperature, a 2DEG formed at the AlGaN/GaN heterointerface can be characterized by the three dominant scattering mechanisms: acoustic deformation potential, polar acoustic phonon and polar optical phonon scatterings. As was noted previously, the impurity scattering is relatively low due to the space layer and is not accounted for in the mobility model.

The momentum relaxation time due to the acoustic deformation potential scattering is calculated as follows [1]:

$$\tau_{ADP} = \frac{\hbar^3 v_l^2 \rho b}{m \kappa T \varphi_{AD}^2}, \tag{1}$$

where $\hbar$ is the reduced Planck constant, v_l is the velocity of longitudinal acoustic phonon, ρ is the density, m is the effective mass, κ is the Boltzmann constant and φ_{AD} is the acoustic deformation potential.

In equation (1) b is the effective width of the 2DEG, which is defined as the double average distance of the electronic wave function from the heterointerface into GaN for the 0th subband (z_0). According to Lee *et al.* [1], z_0 equals approximately to the distance from the heterointerface to the intersection of the conduction band edge and the Fermi level. Our simulations of AlGaN/GaN HEMTs using a self-consistent coupled Schrodinger-Poisson model reveal that z_0 for a 2DEG density (n) of 10^{16} m^{-2} equals to 4 nm. In that case, the expression for b is given by

$$b = 2z_0 = 2\times 4\times 10^{-9}\left(\frac{10^{16}}{n}\right)^{\frac{1}{3}}. \tag{2}$$

The momentum relaxation time due to the polar acoustic phonon (piezoelectric) scattering mechanism is calculated as follows [2]:

$$\tau_{PE} = \frac{\pi}{q^2}\frac{\varphi_{AD}^2}{h_{14}^2}\frac{q_F}{b}\left(\frac{9}{32}+\frac{13}{32}\left(\frac{v_l}{v_t}\right)^2\frac{\Gamma_t}{\Gamma_l}\right)^{-1}\tau_{ADP}, \tag{3}$$

where q is the elementary charge, h_{14} is the piezoelectric constant and v_t is the velocity of transverse acoustic phonon.
In equation (3), q_F is the wave vector on the Fermi surface determined by

$$q_F = (2\pi n)^{\frac{1}{2}}. \tag{4}$$

The parameters Γ_l and Γ_t are defined by

$$\Gamma_l = \left(\left(\frac{4\gamma_l}{3\pi}\right)^2+1\right)^{\frac{1}{2}}, \tag{5}$$

$$\Gamma_t = \left(\left(\frac{4\gamma_t}{3\pi}\right)^2+1\right)^{\frac{1}{2}}, \tag{6}$$

where γ_l and γ_t are determined by

$$\gamma_l = \frac{2\hbar v_l q_F}{\kappa T}, \tag{7}$$

$$\gamma_t = \frac{2\hbar v_t q_F}{\kappa T}. \tag{8}$$

The momentum relaxation time due to the polar optical phonon scattering is calculated as follows [3]:

$$\tau_{POP} = \frac{4\hbar^2\pi\kappa_\varepsilon\varepsilon\left(1-5\frac{V_T}{E_g}\right)}{qm\left(2qmE_{POP}\left(1+\frac{E_{POP}}{E_g}\right)\right)^{\frac{1}{2}}N_P}, \tag{9}$$

where ε is the dielectric constant, V_T is the thermal voltage, E_g is the band gap and E_{POP} is the polar optical phonon energy.
In equation (9), κ_ε is the coupling constant:

$$\kappa_\varepsilon = \left(\frac{1}{\varepsilon_\infty}-\frac{1}{\varepsilon_0}\right)^{-1}, \tag{10}$$

where ε_∞ and ε_0 are the relative permittivity at high and low frequency, respectively.

The parameter N_p is the phonon Planck function:

$$N_P = \left(\exp\left(\frac{E_{POP}}{V_T} \right) - 1 \right)^{-1}. \tag{11}$$

The relation between the i-mobility limited by one scattering mechanism and the corresponding i-momentum relaxation time is

$$\mu_i = \frac{q}{m}\tau_i. \tag{12}$$

The overall 2DEG mobility limited by the three scattering mechanisms is calculated using Matthiessen's rule:

$$\frac{1}{\mu} = \frac{1}{\mu_{ADP}} + \frac{1}{\mu_{PE}} + \frac{1}{\mu_{POP}}. \tag{13}$$

4. Results

In figures 3(a, b, c, d), the temperature dependences of the 2DEG mobility limited by each of the three scattering mechanisms and the overall mobility are presented for different 2DEG densities.

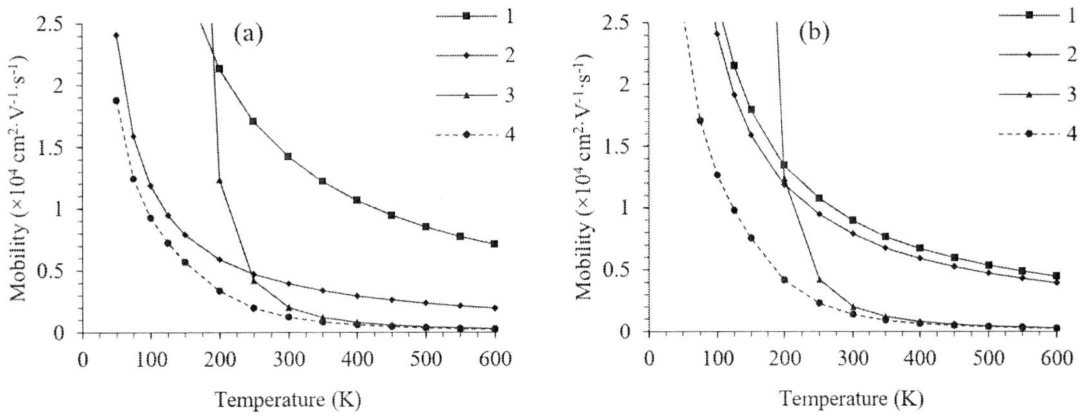

Figure 3(a, b). 2DEG mobility vs temperature for $n = 1 \cdot 10^{12}$ cm^{-2} **(a)** and $4 \cdot 10^{12}$ cm^{-2} **(b)**: 1 – acoustic deformation potential scattering; 2 – polar acoustic phonon scattering; 3 – polar optical phonon scattering; 4 – overall mobility.

As can be seen from the charts, if $n = 1 \cdot 10^{12}$ cm^{-2}, the acoustic deformation potential scattering mechanism has a minor impact on the 2DEG mobility at any temperatures. For instance, at the temperature of 600 K, μ_{ADP} equals to 7100 cm^2/(V·s), while the overall mobility is 241 cm^2/(V·s). When the electron concentration is raised, this mechanism increases its influence on the total mobility. In case of $n = 1 \cdot 10^{13}$ cm^{-2}, μ_{ADP} reaches the value of 3296 cm^2/(V·s) at μ = 253 cm^2/(V·s). At the same time, the polar acoustic phonon scattering reduces its influence on the total mobility (μ_{PE} changes from the value of 1967 cm^2/(V·s) to 6228 cm^2/(V·s)), when the electron concentration is increased, but remains considerable at the whole temperature range.

At temperatures higher than the room temperature, the mobility of the 2DEG with any carrier concentration is governed predominantly by the polar optical phonon scattering mechanism [4] and (9) may be used as a rough approximation expression for the mobility at high temperatures.

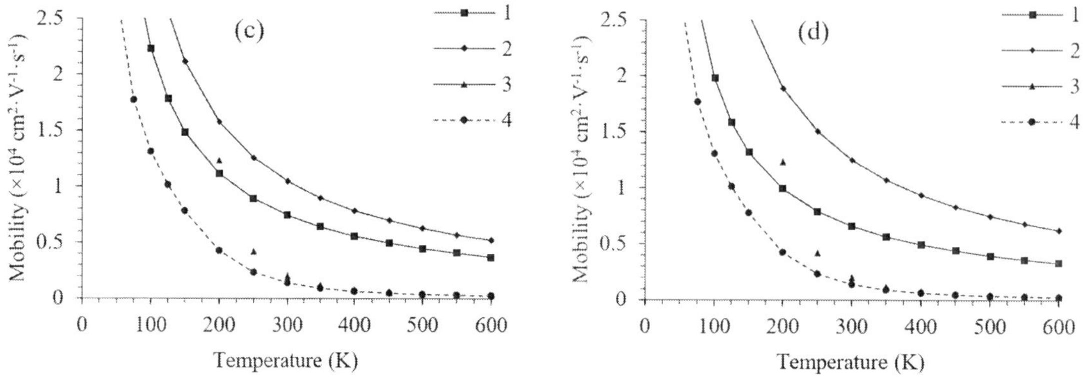

Figure 3(c, d). 2DEG mobility vs temperature for $n = 7 \cdot 10^{12}$ cm^{-2} **(c)** and $1 \cdot 10^{13}$ cm^{-2} **(d)**: 1 – acoustic deformation potential scattering; 2 – polar acoustic phonon scattering; 3 – polar optical phonon scattering; 4 – overall mobility.

In figure 4, the drain current vs drain voltage characteristics of the simulated AlGaN/GaN HEMT are presented. For comparison, one curve is calculated with the electron mobility in the device channel equal to 400 cm^2/(V·s), a value commonly obtained for bulk GaN [5], and another is calculated using the developed model. The temperature is constant (300 K) and the gate voltage is 0 V.

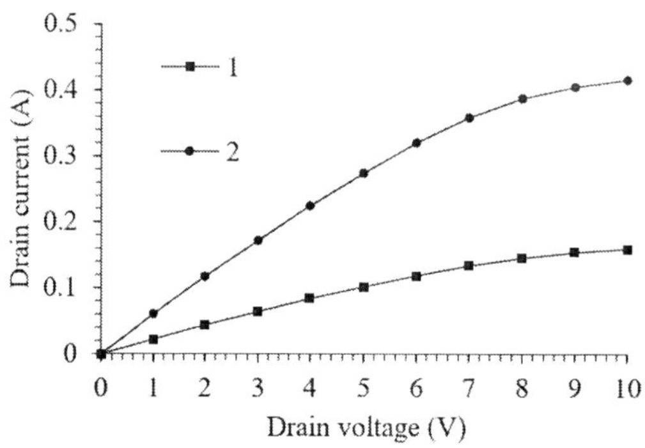

Figure 4. Output characteristics of the simulated AlGaN/GaN HEMT: 1 – bulk mobility model; 2 – developed 2DEG mobility model.

The calculated 2DEG mobility under the gate equals to 1378 cm^2/(V·s). We suppose that the developed model can be useful for heterostructure device simulators such as Blaze.

Acknowledgments
This work was supported by the grants 3.2.04 of Belarusian National Scientific Research Program "Photonics, Opto- and Microelectronics" and T18M-108 of the Belarusian Republican Foundation for Fundamental Research.

References
[1] Lee M, Shur M S, Drummond T J, Morkoc H 1983 *J. Appl. Phys.* **54** 6432
[2] Price P J 1981 *Ann. Phys.* **133** 217
[3] Gelmont B L, Shur M, Stroscio M 1995 *J. Appl. Phys.* **77** 657
[4] Shur M, Gelmont B, Asif Khan M 1996 *J. Electron. Mater.* **25** 777
[5] Crouch R K, Debnam W J, Fripp A L 1978 *J. Mater. Sci.* **13** 2358

Effect of oxygen on the properties of Ga_2O_3:Si thin films

A V Almaev[1,2], E V Chernikov[3], B O Kushnarev[2,3], N N Yakovlev[1,2]
[1]Department of Semiconductor Electronics, Tomsk State University, Tomsk 634050, Russia
[2]Laboratory of Functional Electronics, Tomsk State University, Tomsk 634050, Russia
[3]Laboratory of Physics of Semiconductor Devices, Kuznetsov Siberian Physical Technical Institute, Tomsk State University, Tomsk 634050, Russia

E-mail: almaev_alex@mail.ru

Abstract. The results of studies of electrical and gas-sensitive characteristics of thin films Ga_2O_3:Si on exposure to oxygen in the range from 0 to 100 vol. % and operating temperatures from 25 to 700 °C were presented. Samples were obtained by HF magnetron sputtering. The possibility of developing low-temperature oxygen sensors was shown. A model of oxygen interaction with Ga_2O_3:Si films was proposed. The mechanism of Si influence on gas-sensitive properties of thin films of gallium oxide was proposed.

1. Introduction

Interest in the research of the effect of oxygen on the properties of gallium oxide is caused by the prospects of developing miniature sensors based on this material for analysis of the composition of exhaust gases of internal combustion engines, monitoring the concentration of reagents in the chemical industry and metallurgy.

In papers [1-6] it is proposed to use high-temperature oxygen sensors based on thin films (1 μm) of β-phase gallium oxide. The oxygen vacancies V_O present in the structure of gallium oxide are electron donors and determine the *n*-type conductivity of the material [7]. The working principle of these sensors is based on the interaction of gaseous oxygen with the oxygen vacancies in the structure of Ga_2O_3. The resistance of gallium oxide increases with the oxygen concentration C according to the power law. As the sensor response was taken the ratio of the Ga_2O_3 resistance at the set value C and at C = 20 vol. %. The sensor response increases to 1.04 – 1.6 times at the changing of C in a mixture of O_2+N_2 from 20 to 100 vol. %. The authors argue that high-temperature oxygen sensors have selectivity due to the impossibility of chemisorption of extraneous gas particles on the surface of gallium oxide at the high temperatures. Additional experimental studies are necessary to confirm this fact. The disadvantages of such sensors are high power consumption and relatively low sensitivity to O_2. High operating temperatures T = 700 – 1100 °C make difficulties to develop devices compatible with gas analysis systems.

The possibility of developing low-temperature oxygen sensors based on Ga_2O_3 nanowires is considered in [8]. The sensitivity of structures to gas is explained by the chemisorption of O_2 particles on the surface with the participation of oxygen vacancies V_O, the authors exclude volume effects. The response of the structure at T=300 °C, which corresponds to the maximum sensitivity, rises with increasing of oxygen concentration C according to the power law with an exponent – 0.57. The high response at low temperatures is explained by the increasing in the ratio between the surface area of the semiconductor and its volume. In this paper, the selective detection of oxygen was experimentally established, but the authors do not explain what this is due to. The sensitivity of structures to O_2 was investigated in a limited range of C from 0.5 to 5 vol. %.

The sensitivity of metal oxides to the gases can be increased and shifted to the low-temperature region by modifying with various additives. In this paper the gas-sensitive and electrical characteristics of thin films Ga_2O_3 doped by Si on exposure to O_2 in the range C from 0 to 100 vol. % and T from 25 to 700 °C are investigated. Ga_2O_3:Si is researching widely for electronics applications [7], however, the gas-sensitive properties of such material have not been studied.

2. Experiment

Thin films of Ga_2O_3:Si were formed by high-frequency magnetron sputtering of gallium oxide target (99.999 %, USA) in oxygen-argon plasma by means of Edwards Auto-500. Wafers of the polycrystalline sapphire with a thickness of 150 microns and a diameter of 30 mm were used as substrates. Substrates were not specifically heated at the sputtering process. The total pressure in the chamber was $7 \cdot 10^{-3}$ mbar, the power of the installation – 70 W. The oxygen concentration in the mixture Ar + O_2 was maintained at 56.1 ± 0.5 vol. %. The distance between the target and the substrate was 70 mm. Pieces of Si (99.999 %) were placed on the surface of the target to dope the samples. The ratio of the surface areas of the Si pieces and the sprayed part of the Ga_2O_3 target was 3×10^{-3}. The deposition time of Ga_2O_3 thin films was 24 minutes. After deposition of Ga_2O_3 films, the obtained structures were annealed in Ar atmosphere for 30 minutes at a temperature of 900 °C. Under these annealing conditions, β-phase Ga_2O_3 is formed [9]. The sensitive elements of a given shape and size were formed by photolithography. Before the deposition of gallium oxide films the platinum contacts to the layer of Ga_2O_3 and the heater on the reverse side of the substrate were formed by sputtering followed by photolithographic engraving. After cutting, sensors 1.5x1.5 mm^2 in size with the area of the sensitive layer of 0.78x0.6 mm^2 were assembled into TO-8 cans. The thickness of the prepared films was about 160 – 180 nm.

The element composition of formed thin films was determined by Auger electron spectroscopy using specially manufactured large-area samples. To measure the resistance of the sensors used a metal chamber which housed two sensors. A mixture of nitrogen of high purity (99.999 %) and technical oxygen (99.7%) was pumped through the chamber. The content of the components of the gas mixture was controlled by the gas flow meters Bronkhorst. Oxygen concentration of 0 vol. % corresponds to the pumping through the chamber of nitrogen only. Resistance measurements of structures at the various conditions were carried by means of source–meter Keithley 2636A characterized by high accuracy and capable of measuring the resistance of the order of 10^{13} Ω. The applied voltage to the structures was 5 V. The sensors were heated by means of a laboratory DC power supply.

3. Results and discussions

Analysis of the temperature dependence of the resistance of samples in a gas mixture containing 21 vol. % of O_2 and 79 vol. % of N_2 showed that in the Ga_2O_3:Si band gap there are two donor levels located 0.036 and 1.37 eV below the conduction band bottom. The first level is caused by doping by Si, the second is due to the presence of oxygen vacancies in the Ga_2O_3 structure.

Starting on 200 °C, the samples demonstrate sensitivity to oxygen. Figure 1 shows that the maximum response is observed at T=400 °C. The sensor response S is the ratio of resistance at a given value C and resistance at 100 % N_2 in a chamber. The S with increasing oxygen concentration increases according to the power law $S \sim C^b$ (Figure 2). An exponent b depends on temperature (Table 1).

The sensitivity of samples to hydrogen, carbon monoxide and nitrogen dioxide was analyzed to determine the selectivity of oxygen detection. The influence of hydrogen and carbon monoxide on the resistance of Ga_2O_3:Si films was evaluated at a temperature corresponding to the maximum response to oxygen. A mixture containing 21 vol. % of O_2 and 79 vol. % of N_2 was selected as the initial atmosphere. After the establishment of the stationary values of R the appropriate gas was launched into the chamber. Effect of 1.55 vol. % of hydrogen and 160 ppm of CO leads to a slight decrease in resistance of 1.87 and 1.1 times, respectively. Exposure to 71 ppm nitrogen dioxide leads to an increase in film resistance by 10 times. It follows from this that under the established conditions, thin Ga_2O_3:Si films react poorly to the exposure to high concentrations exceeding the maximum permissible limits of reducing gases. However, the resistance of the thin Ga_2O_3:Si films increases sharply when oxidizing gases (nitrogen dioxide, oxygen) appear in the atmosphere.

Table 1. The temperature dependence of the exponent b of thin Ga_2O_3:Si films.

T, °C	300	400	500	600	700
b	0.44±0.03	0.86±0.02	0.67±0.03	0.60±0.05	0.61±0.01

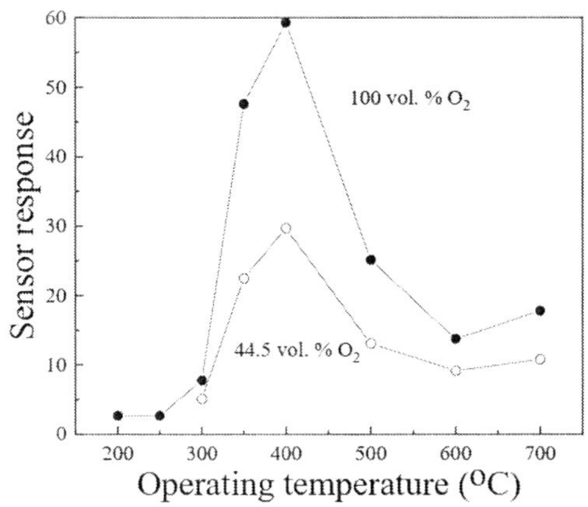

Figure 1. Temperature dependencies of the response at different concentration of O_2.

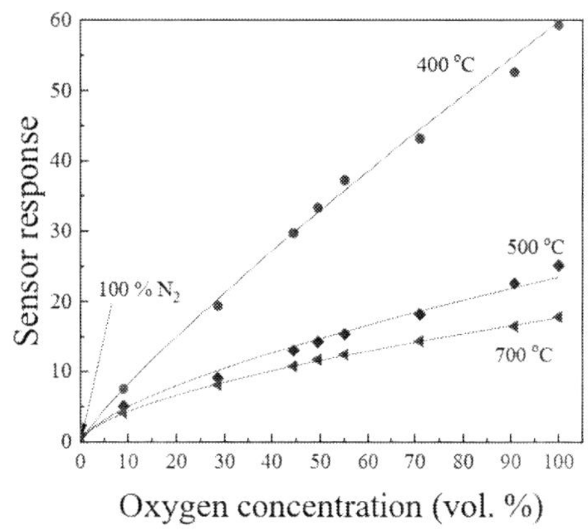

Figure 2. The concentration dependencies of response to oxygen.

The response time t_{res} and recovery time t_{rec} were determined to estimate the operation speed of sensors. However, the gas flow rate and the time required for the formation and transmission of the gas mixture into the measuring chamber effect on t_{res} and t_{rec}. Therefore as the t_{res} was chosen the time from the beginning of the sensor resistance change, initially placed in an oxygen-free environment, to the time of establish of $0.9 \cdot R$ at C=44.5 vol. % (Figure 3). The shortest response time for sensors was observed at T=600 °C and was 110 – 130 s. Response time at exposed to 71 ppm of nitrogen dioxide at a temperature of 400 °C amounted to 213.2 s. After exposure of all gases the resistance of the sensor recovered to the initial values. As a recovery time for oxygen sensors was chosen the time between the beginning of the resistance reduction and the establishment of 1.1 of the stationary resistance of sensor placed in an oxygen-free environment. The recovery time was ~ 800 s. It should be borne in mind that the times t_{res} and t_{rec} were determined at exposed to relatively high concentrations of oxygen on the sensors.

Figure 3 shows the change in resistance of the two sensors based on the thin Ga_2O_3:Si films at exposure to 44.5 vol. % of oxygen. Differences in the characteristics of sensors, their kinetics, response values, response and recovery times are caused by the disadvantages of magnetron sputtering technology of thin films. However the regularities for the sensors obtained on one plate are the same. The sources of differences of samples can be heterogeneous distribution of silicon, different concentration of local defects, minor differences in the area and thickness of films, etc.

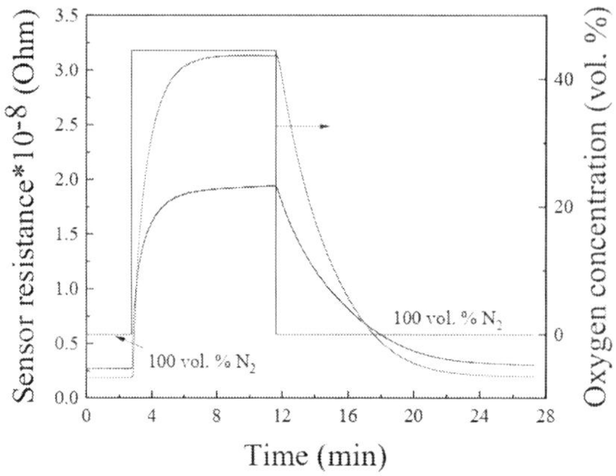

Figure 3. The time dependence of sensor resistance at exposure to 44.5 vol. % of oxygen.

The curves of increase of the sensor resistance at exposed to oxygen (Figure 3) are satisfactorily approximated by the difference of two decreasing exponential functions. The time-constants are 11 – 30 s and 57 – 71 s. The curves of decay of the sensor resistance after the oxygen exposure are satisfactorily approximated by the sum of two decreasing exponential functions and the time-constants are 58 – 95 s and 232 – 288 s.

It is necessary to estimate the ratio between the grains size d_g of Ga_2O_3 and the Debye length L_D in this material to establish the mechanism of oxygen influence on the electrical resistance of the films. The Debye length was calculated according to

$$L_D = \sqrt{\frac{\varepsilon_r \varepsilon_0 kT}{e^2 n}},$$

where ε_r = 10.2 [10] is the relative permittivity of the Ga_2O_3; ε_0 is the electrical constant; k is Boltzmann constant; e is the electron charge; n is the electron concentration in Ga_2O_3. For the researched films $n = n_V + n_d$, where n_V is the electron concentration due to ionization oxygen vacancies; n_d is the electron concentration due to ionization donor level of Si. The resistance of films at T = 600 °C and the electron mobility μ_n = 0.44 $cm^2/V \cdot s$ [10,11] were used to calculate n. The electron concentration in Ga_2O_3:Si films at T = 600 °C and placed in an oxygen-free environment is $(2 - 4)\cdot 10^{16}$ cm^{-3} and on exposure to 44,5 vol. % of oxygen $n \approx (2 - 3.5)\cdot 10^{15}$ cm^{-3}. Estimates have shown that the Debye length is greater the characteristic grains sizes of the thin Ga_2O_3:Si films which are 145 and 100 nm. Consequently barriers at the interface of Ga_2O_3 grains do not affect the transport of charge carriers. The conductivity of the films has a semiconductor character. Oxygen molecules have acceptor properties and during chemisorption on the Ga_2O_3 surface capture electrons from the conduction band of the semiconductor forming a layer depleted by electron in the near-surface region of the semiconductor. However, the absence of the saturation of concentration dependences of the oxygen sensor resistivity, the presence of two time constants during the rise and fall of the resistance during the supply and pumping of oxygen and low responses to reducing gases indicate that a significant role is played by the interaction of oxygen molecules and oxygen vacancies according to the reaction $O_O \leftrightarrow V_O^{p+} + pe^- + 1/2O_2^g$, here O_O is lattice oxygen; O_2^g is gaseous oxygen, p=1 for V_O^+ and p=2 for V_O^{++}. The presence of the maximum in figure 2 is explained by the dependence on the temperature of the ratio between the processes of adsorption and desorption of oxygen molecules. The increase of the sensor response at T=700 °C is caused by the increasing role of the bulk processes with the participation of V_O.

Auger electron spectroscopy study had shown that the Si concentration in the samples is less than 0.5 at. % and there is a deviation from stoichiometry. The concentration of Ga atoms is 42.3 at. %, the oxygen atoms ~ to 57.1 at. %. Atomic force microscopy research had shown that the surface roughness of the Ga_2O_3:Si is 0.2345 microns. This value for the Ga_2O_3 is 0.0783 microns. The gain of the sensor response for the Ga_2O_3:Si thin films is caused by the formation of a more extended surface of the samples resulting in an increase in the surface density of the adsorption centers for oxygen molecules and a changing of the stoichiometry of structures towards an increase in the concentration of Vo in the bulk of Ga_2O_3.

4. Conclusion

The possibility of developing low-temperature oxygen sensors based on thin films Ga_2O_3:Si was shown. There are two donor levels in the band gap of the Ga_2O_3 due to the impurity of Si (0.036 eV) and oxygen vacancies (1.37 eV). The temperature of the maximum response to O_2 is 400 °C. The increasing of the resistance of thin films according to the power law with an increasing of the oxygen concentration at T<700 °C is caused by the chemisorption of O_2 and interaction with oxygen vacancies. The growth of sensor response at T=700 °C is caused by the increasing role of the bulk processes involving oxygen vacancies. Si stimulates the formation a more extended surface of Ga_2O_3 and increase concentration of oxygen vacancies in the bulk of films.

Acknowledgments
The reported study was funded by RFBR according to the research project № 18-32-00456.

References

[1] Bartic M, Baban C-I, Suzuki H, Ogita M, Isai M 2007 *J. Am. Ceram. Soc.* **90** 2879
[2] Bartic M, Ogita M, Isai M 2007 *J. Appl. Phys.* **102** 023709
[3] Bartic M, Toyoda Y, Baban C-I, Ogita M 2006 *Jap. J. Appl. Phys.* **45** 5186
[4] Baban C, Toyoda Y, Ogita M 2005 *Thin Solid Films* **484** 369–373
[5] Ju L-T, Ju S-L 2012 *J. Ovonic Research* **8** 73
[6] Bartic M 2016 *Phys. Status Solidi A* **213** 457
[7] Stepanov S I, Nikolaev V I, Bougrov V E, Romanov A E 2016 *Rev. Adv. Mater. Sci.* **44** 63
[8] Liu Z, Yamazaki T, Shen Y, Kikuta T, Nakatani N, Li Y 2008 *Sens. Actuators, B* **129** 666
[9] Lygdenova T Z, Kalygina V M, Novikov V A, Prudaev I A, Tolbanov O P, Tyazhev A V 2018 *Russ. Phys. J.* **60** 1911
[10] Fleischer M, Giber J, Meixner H 1992 *Appl. Phys. A* **54** 560
[11] Fleischer M, Meixner H 1993 *J. Appl. Phys.* **74** 300

Journal of Physics: Conference Series **1410** (2019) 012202 doi:10.1088/1742-6596/1410/1/012202

Wireless software-hardware complex for testing semiconductor structures

S M Loganchuk[1], **L Touel**[1], **S N Chebotarev**[1], **L M Goncharova**[1], **A V Varnavskaya**[1], **S Touel**[2], **A A A Mohamed**[1]
[1]Department of Physics and Photonics, Platov South-Russian State Polytechnic University (NPI), Novocherkassk 346421, Russia
[2]Levio Ltd, Quebec, G1W 0C4, Canada

Abstract. Wireless system for measuring current-voltage characteristics was developed. Software for a smartphone was developed in the integrated development environment Unity. The data about current and voltage was transmitted by Bluetooth and Wi-fi for smartphone and computer respectively. I-V curves were built on computer using program, developed in the Visual Studio package and coded in C#. For data verification the simulation of diodes' parameters was provided by solving the Poisson equations and the drift- diffusion model.

1. Introduction

Semiconductor structures based on silicon and germanium [1] continue to apply actively in modern electronics. The progress in information technology, including wireless technologies allows to make the testing of developing semiconductor elements easier [2]. The combination of the results of such measurements with the data of computer simulation allows to develop the wireless systems for measurement of new devices' parameters [3, 4]. The aim of this paper is the development of software-hardware complex for testing the parameters of semiconductor diodes in situ using wireless interface.

2. Results and discussion

Objects of research are silicon and germanium diodes: 1N5818, KD510A, KD212A, 1N4007, D9I. The simulation was carried out by numerical methods by solving the Poisson equation and the standard diffusion-drift model by method, described in [5].

The wireless system, shown on Fig. 1a, was developed for the measurement of current-voltage characteristics.

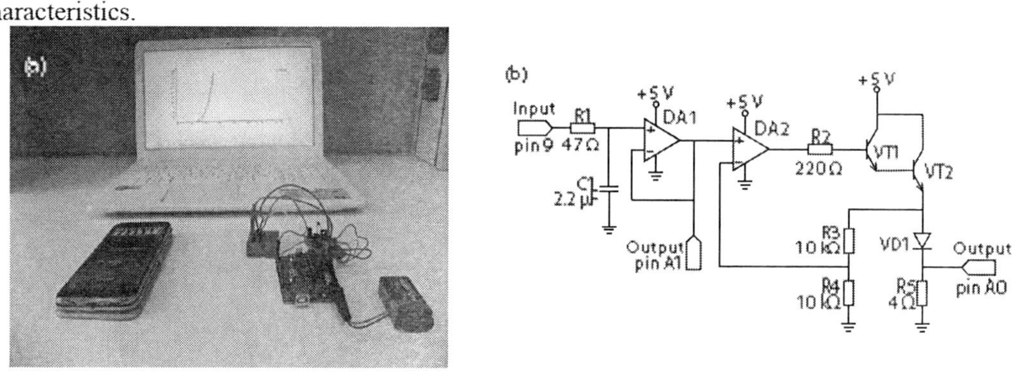

Fig.1 (a, b) **(a)** Wireless system **(b)** Circuit diagram of measurement scheme

The value of reference voltage is set by user used the program on Unity. The value of this signal is transmitted by Bluetooth to microcontroller Arduino Uno. The circuit diagram of measurement scheme is shown on Fig. 1b. The input signal is processed by low-pass filter. Value of the formed signal is very low and do not exceed 20 mA. For the amplifying of the signal, the Darlington configuration VT1, VT2 is used. It has current gain 150. Value of the signal in the output A0 accords to voltage R5.

Voltage of measured diode V_D determines by difference of the electric potentials on output A1 and resistor R5. Received signal V_D is recalculated in current by Ohm's law $I_D = V_D / \mathrm{R5}$ in Arduino. Signals V_D and I_D are transmitted back by Bluetooth to smartphone. For the visualization of graphics the program, used Unity and Visual Studio was coded. The signals V_D and I_D are transmitted to computer by Wi-fi, using Wi-fi module.

An user sets the sequences of voltages V_D for the building of the graphics of current-voltage characterisitic, wherein on the screen of computer the sequence of dots (V_{Di}, I_{Di}), forming I-V curve, is displayed. On the Fig. 2a the results of the measurement of current and voltage of five different diodes are shown. The examples of the developed software for computer and smartphone are shown on Fig. 2b and Fig. 2c respectively.

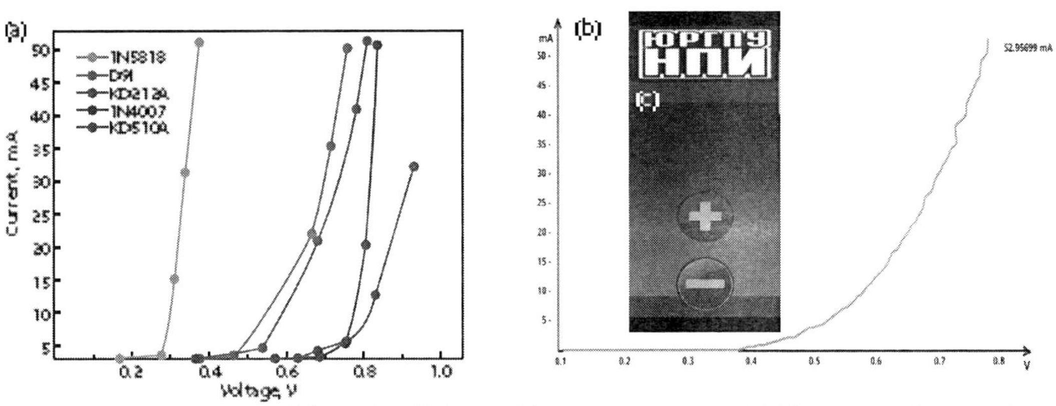

Fig.2 (a, b, c) (a) I-V curve of the testing diodes **(b)** The computer program **(c)** The program for smartphone

3. Simulation

COMSOL Multiphysics software is used for simulation of current-voltage characteristics. This software package contains many modules for modeling of different physics, including the physics of semiconductor structures. The parameters with a specific value and a unit of measurement are defined during the simulation. After that, geometry of the structure is built as it shown on Fig. 3a. Thickness of the diode structure is 7 μm.

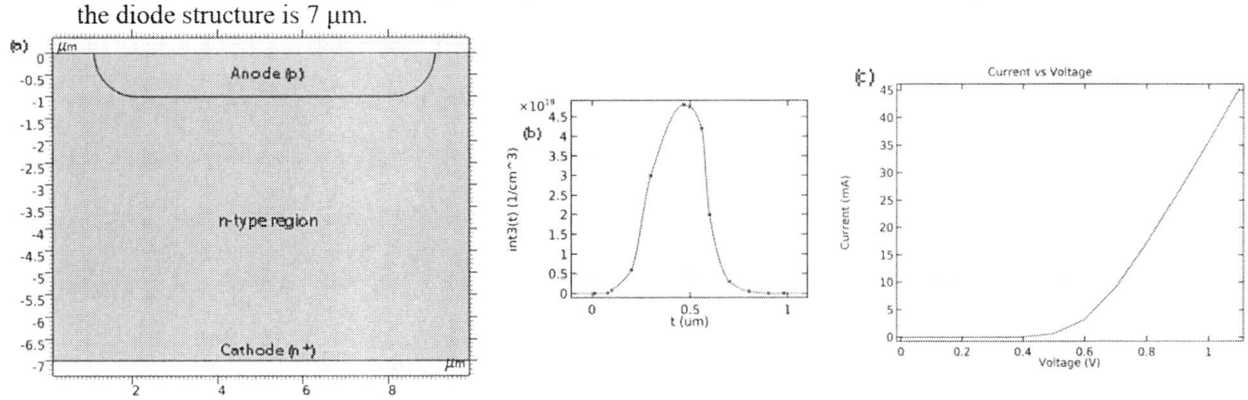

Fig. 3 (a, b, c) (a) Geometry of the diode structure **(b)** Acceptors' distribution **(c)** I-V curve of the germanium diode

The simulated diode is made of germanium. Main equations, which is solved in the simulation, are Poisson-equation and drift-diffusion model.

$$\rho = q(p - n + N_d^+ - N_a^-) \quad (1)$$

$$J_n = q\mu_n nE_n + qD_n \nabla n \quad (2)$$

$$J_p = q\mu_p nE_p + qD_p \nabla p \quad (3)$$

Where ρ – volume charge density, q – elementary charge, p – hole concentration, n – electron concentration, N_d^+ – ionized donors concentration, N_a^- – ionized acceptors concentration, J_n и J_p – current densities of electrons and holes respectively; μ_n и μ_p – mobilities of electrons and holes respectively, E_n и E_p – effective electric fields of electrons and holes respectively, D_n и D_p – diffusion coefficients of electrons and holes respectively, ∇ – Laplacian

Electrons and holes concentrations are given by Analytic Doping Model in n-type and n^+-type regions. The hole distribution in p-type region (anode) is given by Geometric Doping Model. Background doping concentration in all regions is 10^{15} cm^{-3}. Donor concentration in the cathode area is $2 \cdot 10^{17}$ cm^{-3}. In the anode area, the acceptors are distributed as it shown of Fig. 3b. It is considered that Shockley-Read-Hall recombination is occurred in all structure.

Metal contacts are ideal ohmic on anode and cathode. The positive potential is fed on anode, while cathode has zero potential. Using this and solving the Poisson equation and drift-diffusion model, the current-voltage characteristic, shown on Fig.3c, is calculated.

4. Hardware

Program for microcontroller Arduino Uno was coded on C in the Arduino IDE. According to circuit diagram on Fig. 1b, digital pin 9 is used as output, where control signal is supplied., while analog pins A1 and A0 are used as inputs, where the voltage is read.

Program contains constants, such as maximum value of control signal is 1.1 V, maximum current is 50 mA, resistance on resistor R5 is 4 Ω and the delay is 100 ms.

The algorithm scheme of the program's main function is shown on Fig.4:

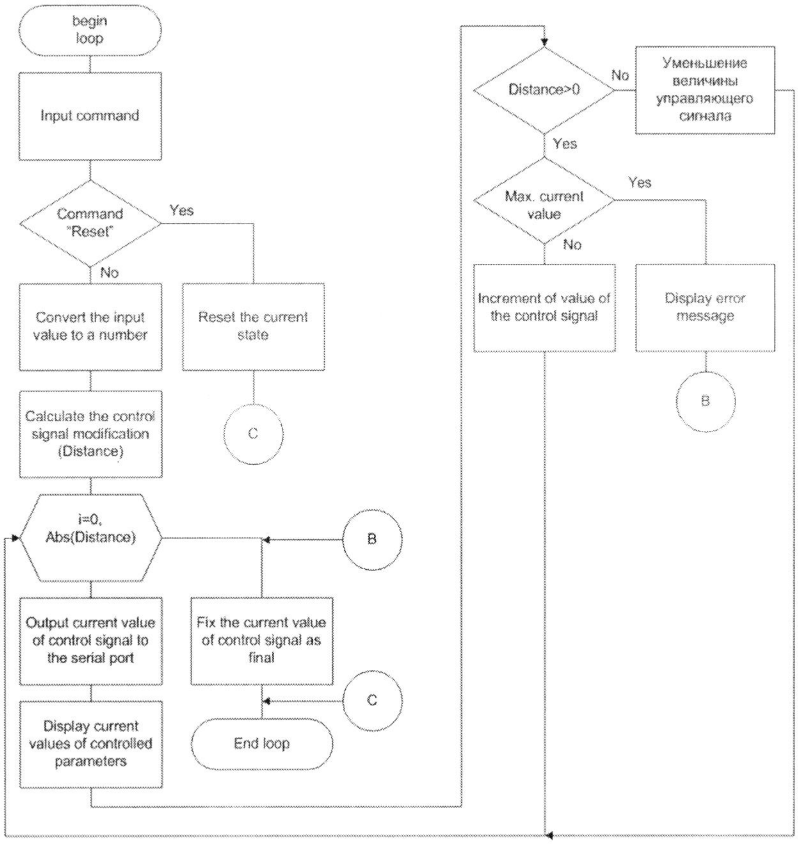

Fig. 4 Algorithm scheme

The main function of the program is loop. It contains the steps for the command processing from the control device (smartphone) and the calculating of the level control signal, supplied to the measuring scheme.

Data of current and voltage are read from pins A0 and A1 respectively as a 10-bit value, i.e the number from 0 to 1023. To convert this value to the voltage V_{A0} on the resistor R5, the formula is used:

$$V_{A0} = (A0 \cdot V_{ref} \cdot 1000)/1023 \quad (4)$$

Where $A0$ – 10-bit value read from pin A0; V_{ref} – reference voltage equals 1.1 V;

As the voltage and the resistance on the resistor R5 are known, the electric current could be calculated by Ohm's law: $I_D = V_{A0} / R5$

The voltage on the pin A1 is calculated by formula

$$V_{A1} = 2 \cdot ((A1 \cdot V_{ref} \cdot 1000)/1023) \quad (5)$$

Where A1 – 10-bit value read from pin A0.

The voltage V_D on the diode VD1 (see Fig.1b) is the difference between voltages analog pins:

$$V_D = V_{A1} - V_{A0} \quad (6)$$

Receiving and transmitting data are carried out by standard functions of integrated development environment through a serial port. As soon as the control device is ready for the transfer of next command, this command would be written in the appropriate variable. Command is the string, which contains word "Reset" or value of control signal. If command "Reset" is received, the value of current variables sets to zero and on the output zero value is supplied.

Otherwise, received string is converted to a number value; the value of control signal is calculated and increased (or decreased) until the required value.
After each change of output signal the values of voltage on pins A0 and A1 are read and a current on the diode is calculated. If the value of current is reached maximum allowed value, the process interrupt. When current command is executed, the device switches to standby and the process repeats.

5. Software for a smartphone and computer

Graphical user interface (GUI) was developed with Unity game engine. The program for GUI was coded on C#. The interface contains buttons "+" and "-", which are used for increment and decrement of value of control signal (see Fig.2c). Voltage increase in 0.1 V on every step. Data of voltage and current are transmitted in the smartphone and their values are displayed.
Computer program receives data from a smartphone and displays I-V curves (see Fig.2b). The I-V curve is blue, when the voltage increase, and red, when the voltage decrease. Data about measured voltage and current are stored as vector, while object LineRenderer is used for connection of the points and displaying curve.

6. Conclusions

The wireless software-hardware complex for the measurement of semiconductor devices' parameters was developed. Measurement system use the data exchange by Bluetooth and Wi-fi. The scheme of calculation of current from the data about voltage was proposed. The graphic visualization of I-V curves is provided on the computer, using Unity and Visual Studio packages. The computer program for solving Poisson equations and drift-diffusion model was proposed for the working out of the technology of producing of the semiconductor diodes

References

[1] Aryaee Panah M E, Han L., Norrman K., Pryds N., Nadtochiy A., Zhukov A E., Lavrinenko A V., Semenova E S. 2017 *Opt. Mat. Exp.* **7** 2260
[2] Chen D., Li H., Su X., Hu R., Li G. 2019 *Sens. and Actuat. B* **287** 35
[3] Chebotarev S N, Pashchenko A S, Lunin L S, Zhivotova E N, Erimeev G A, Lunina M L 2017 *Beilstein J. Nanotech.* **8** 12
[4] Chebotarev S N, Pashchenko A S, Lunin L S, Irkha V A 2016 *Nanotech. in Russia.* **11** 435
[5] Loganchuk S.M., Chebotarev S.N., Arustamyan D.A., Mohamed A.A.A., Touel L., Bogatov N.M. 2018 *J. Phys.: Conf. Ser.* **1124** 071009

Aviation kerosene sensors based on nanocrystalline SnO_2 thin films with various catalysts

B O Kushnarev, N K Maksimova, L S Khludkova, E V Chernikov, E Y Sevastyanov
V. D. Kuznetsov Siberian Physical-Technical Institute, Tomsk State University, Tomsk 634050, Russia

Abstract. Gas-sensitive characteristics of aviation kerosene sensors based on nanocrystalline SnO_2 thin films with various catalysts have been studied as a function of the operating temperature and gas concentration in the air. Thin tin dioxide films were grown in magnetron by sputtering of a tin–antimony alloy target in oxygen–argon plasma. It is shown that the sensors are characterized by a high sensitivity to aviation kerosene vapors and by a good stability of parameters during long-term tests and under the variation of humidity.

1. Introduction

One of the actual ecological problems is the pollution of the airfield and near-airfield territories due to leakages of aviation fuels. To prevent a harmful influence of aviation kerosene vapors on human health a continuous air monitoring is necessary. Thin-film metal oxide semiconductor gas sensors are promising devices for this purpose because of their low cost, high sensitivity, fast response time and simple electronic interface. Gas-sensitive characteristics of metal oxide sensors can be improved by introducing metal modifiers into the bulk and depositing catalytic layers on the surface. Noble metals are usually used as catalytic additives in gas sensing materials. Recently, in some studies 3d-transition metals such as Co, Ni, Cr are considered as the additives that can effectively modify adsorption properties of gas sensing materials and improve the parameters of gas sensors [1-4]. In this paper the electrical and gas-sensitive characteristics of sensors based on nanocrystalline SnO_2 thin films with noble metals (Pt, Pd, Au) on the surface and 3d-transition metals (Co, Ni) additives in the bulk have been studied.

2. Experimental

SnO_2 thin films (about 100 nm) were deposited in oxygen-argon plasma by magnetron sputtering of a tin–antimony alloy target. To introduce additives into the film bulk, pieces of corresponding metals were placed onto the target surface. The additive content was estimated by the ratio of the areas metal pieces S_m (m = Ni, Co, Au) and of the sputtered target S_{Sn}. The same DC magnetron sputtering technique was used to deposit ultrathin catalyst layers (Pt, Pd, Au) on the films surface. Sapphire wafers 150 μm thick were used as substrates. Contacts to the SnO_2 layers and a heater on the substrate's underside were formed by the sputter deposition of platinum followed by photolithographic engraving before the deposition of tin dioxide films. The sample size is 0.7 × 0.7 mm with an area of the sensitive element 0.3 × 0.3 mm^2. As many as 500 sensors were obtained on one substrate. All fabricated wafers were subjected to stabilizing annealing in air at 450 °C for 24 h. A detailed description of the technology of film deposition and fabrication of sensors is given in [5]. Three types of sensors, differing by additives on the surface and in the bulk of SnO_2 thin films were investigated.

The sensors Au/SnO_2:Sb, Au, Ni of series (1) were fabricated by sputtering of a target with gold and nickel additives ($S_{Au}/S_{Sn} = S_{Ni}/S_{Sn} = 3 \cdot 10^{-3}$) and then a dispersed Au layer was deposited on the SnO_2 surface. The sensors Au/SnO_2:Sb, Au, Co of series (2) differed from the sensors of series (1) by the additive of Co in the bulk of SnO_2 film ($S_{Au}/S_{Sn} = 6 \cdot 10^{-3}$, $S_{Co}/S_{Sn} = 9 \cdot 10^{-3}$). The sensors Pt/Pd/$SnO_2$:Sb, Au, Co of series (3) with gold and cobalt additives in the volume ($S_{Au}/S_{Sn} = S_{Co}/S_{Sn} = 3 \cdot 10^{-3}$) were fabricated by deposition of ultrathin Pd layer and then Pt layer on the film surface. The thickness and microstructure of the films were controlled by atomic force microscopy.
To measure the gas sensitive characteristics four sensors were placed in a quartz chamber with a volume of 1 L. Laboratory air purified using a sorption zeolite filter was pumped through the chamber, then the chamber was sealed. Two air streams with regulated speeds were used to change humidity: air dried with zeolite and humidified with a bubbler. The humidity was controlled using an HIH-4000 capacitive sensor placed in the chamber. The necessary amount of aviation kerosene vapors was injected into the testing chamber by a syringe. The sensor conductivity in pure air (G_0) and in a gas–air mixture (G_1) were measured using an automatic bench, which made it possible to set and stabilize the operating temperature and record the conductivity values every second. The ratio G_1/G_0 was taken as the adsorption response.

3. Results and Discussion

Investigations showed that the thickness of the studied tin dioxide layers is 120-130 nm. The films contain nanocrystallites with sizes d_1=8-15 nm and d_2=150-250 nm (Table 1). The values of the sensors resistances in pure air at an operating temperature 450 °C vary in the interval 2-12 MΩ. The higherst values of R_0 are characteristic of sensors from series (3) with gold and cobalt additives in the volume and Pt/Pd catalyst layers on the surface (Table 1).

Table 1. Typical sizes of grains d_1 and d_2 in tin dioxide films with various catalysts and sensors parameters at an operating temperature 450 °C: resistances R_0 in pure air and the values of the response G_1/G_0 under exposure to 15 ppm of aviation kerosene vapors.

The sensor type	Au/ SnO_2:Sb, Au, Ni series (1)	Au/ SnO_2:Sb,Au, Co series (2)	Pt/Pd/ SnO_2:Sb,Au, Co series (3)
d_1, **nm**	40-50	125	8-16
d_2, **nm**	150-250	-	190-260
R_0, **MΩ**	2-4	5-9	7-12
G_1/G_0	2,1-2,5	2,2-2,5	2,3-2,8

Temperature dependences of the sensor response to 15 ppm kerosene vapors are presented in Fig. 1(a). The response of the sensors of series (3) with Pt/Pd catalyst on the surface reaches maximum at 370 °C. The responses of the sensors of series (1) and (2) increase monotonically up to 450 °C. For all types of sensors the operating temperature in the range 400-450 °C provides relatively high response values G_1/G_0=2.1-2.7 upon exposure to 15 ppm kerosene vapors. The response time of the sensors is 5-6 s at 400 °C and 4 s at 450 °C. The response time was defined as the time taken by the sensor to achieve 90 % of the total resistance change.
Figure 1(b) shows concentration dependences of responses to aviation kerosene vapors in the range 5–75 ppm at an operating temperature 450 °C for sensors of different series. The response values to 45 ppm (this concentration corresponds to permissible exposure limit for aviation kerosene vapors in workplace air) are G_1/G_0=3-4. It should be noted that sensors can detect kerosene vapors at 5 ppm level which corresponds to 0.1 of permissible exposure limit for aviation kerosene.

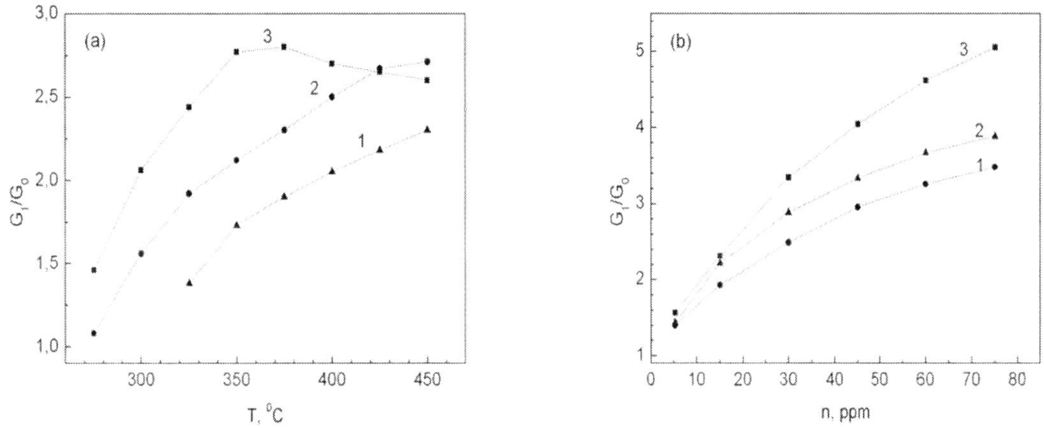

Figure 1(a, b). Dependences of the response to 15 ppm of aviation kerosene vapors on the operating temperature **(a)** and concentration dependences of the response measured at an operating temperature 450 °C **(b)** for the sensors from series: (1) (*1*), (2) (*2*), (3) (*3*).

One of the significant sources of instability in the parameters of sensors based on metal oxide semiconductors is the change in the humidity of the environment. The previous studies of the humidity influence on the properties of SnO_2 thin films showed that the conductivity of sensors in pure air increases and the responses of sensors to reducing gases (hydrogen, carbon monoxide, etc.) decrease as the humidity level increases [6,7]. The results of investigations presented in this paper showed that the rate of conductivity increase with humidity level increasing is the same for pure air and gas–air mixture containing kerosene vapors. For all types of sensors both the conductivity in pure air G_0 and the conductivity in a gas–air mixture containing 15ppm of aviation kerosene vapors (G_1) increases by 4-6 times with increasing of relative humidity from 10 % to 70 % (Fig. 2 (a)). As a result the sensor response defined as the ratio G_1/G_0 is practically independent on humidity in the wide range of RH (Fig.2 (b)).

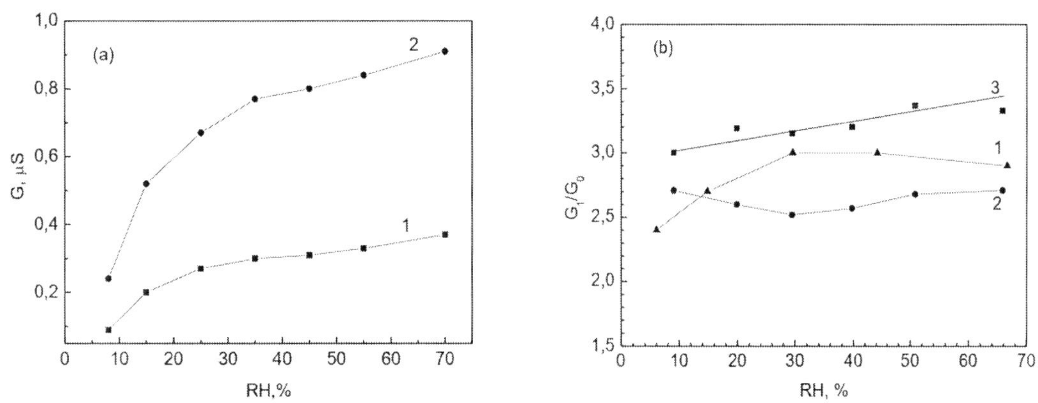

Figure 2(a, b). Dependences of conductivity in pure air (1) and in a gas–air mixture containing 15ppm of aviation kerosene vapors (2) on the humidity level for the sensor from series (3) **(a)** and dependences of the response to 15 ppm of aviation kerosene vapors on the humidity level for the sensors from series: (1) (*1*), (2) (*2*), (3) (*3*) **(b)**. The sensors operating temperature is 450 °C.

To discuss the obtained results the mechanism for gas sensing in semiconductor materials is to be considered. The detection of aviation kerosene vapors is based on the oxidation reaction of adsorbed molecules with the participation of previously chemisorbed oxygen ions, which occurs at high temperatures, when atomic oxygen predominates on the surface. As a result, the density of oxygen ions decreases and the conductivity increases. Kerosene is a mixture of different hydrocarbons and oxidation of such molecules at the surface of metal oxide semiconductor is accompanied by the release of atomic hydrogen and other dissociation products [8]. Atomic hydrogen reacts with the surface oxygen species, decreasing the density of the negative charge on the surface and resulting in an increase of the sensor conductivity. The effect of humidity on the properties of sensors is determined by the fact that under adsorption on the surface of the SnO_2 microcrystal, the H_2O molecule dissociates into the proton and hydroxyl group and compensates the charge of one adsorbed oxygen ion [6]. The density of chemisorbed oxygen decreases, that leads to an increase of the sensor conductivity.

To study the stability of parameters at long-term operation of the sensors, the values of the response to aviation kerosene vapors were measured at intervals of 2-10 days. It was shown that only a slight attenuation in the sensing response was observed during 200 days.

4. Conclusion

Three types of sensors based on SnO_2 thin films grown by magnetron deposition with various catalysts in the bulk and on the surface have been studied. The all types of sensors are characterized by a high sensitivity to aviation kerosene vapors and can detect these vapors at 5 ppm level which corresponds to 0.1 of permissible exposure limit for aviation kerosene. The sensors revealed a good stability of parameters after the long-term tests and under the variation of humidity.

Acknowledgments

The work was carried out within the framework of the State Task of the Ministry of Education and Science of the Russian Federation (task No. 3.9661.2017/8.9).

References

[1] Korotcenkov G, Boris I, Brinzari V, Han S H, Cho B K 2013 *Sensors and Actuators B* **182** 112
[2] Oleksenko L P, Maksymovych N P, Buvailo A I 2012 *Sensors and Actuators B* **174** 39
[3] Sevastyanov E Y, Maksimova N K, Khludkova L S, Chernikov E V, Sergeychenko N V 2017 *BioNanoSci.***7** 654
[4] Sevastyanov E Y, Maksimova N K, Potekaev A I, Khludkova L S, Chernikov E V, Davydova T A 2018 *Russ. Phys. J.* **60** 1739
[5] Sevastyanov E Y, Maksimova N K, Novikov V A, Rudov F V, Sergeychenko N V, Chernikov E V 2012 *Semiconductors* **46** 801
[6] Gaman V I , Anisimov O V , Maksimova N K, Sergeychenko N V, Sevastyanov E Y, Chernikov E V 2008 *Russ. Phys. J.* **51** 83
[7] Maksimova N K, Sevastyanov E Y, Sergeychenko N V, Chernikov E V 2016 *Thin-Film Semiconductor Gas Sensors* (Tomsk: Izd. Nauchn.-Tekh. Liter.)
[8] Carotta M C, Guidi V, Martinelli G, Nagliati M, Puzzovio D, Vecchi D 2008 *Sensors and Actuators B* **130** 497

Magnetic-field-driven electron transport in SOI back-gate device

L V Shanidze[1, 2], A S Tarasov[1, 2], A V Lukyanenko[1, 2], M V Rautskii[1], I A Yakovlev[1], F N Zelenov[1], F A Baron[1] and N V Volkov[1, 2]
[1] Kirensky Institute of Physics, Federal Research Center KSC SB RAS, Krasnoyarsk 660036, Russia
[2] Institute of Engineering Physics and Radio Electronics, Siberian Federal University, Krasnoyarsk, 660041,Russia

Abstract. In this work, we studied the electronic transport properties of silicon nanowire field-effect transistors with a back gate. A nontrivial magnetic field influence on the drain current at low temperature (10 K) was found. The strongest effect was observed in the majority carrier accumulation mode. In this mode magnetic field of 0.5 T increases current through the device by more than an order of magnitude. The paper describes the possible mechanisms of the magnetic field influence on the electronic transport characteristics of the structures.

1. Introduction

At present, devices based on silicon nanowire (SiNW) attract considerable attention as elements of integrated nanoscale electronics [1, 2], as well as for studying fundamental properties in small dimensions [2]. Schottky barrier source / drain field-effect transistors (FET) have a number of advantages including simple and low temperature processing, good suppression of short channel effects [4, 5] and the elimination of source / drain contact doping [6, 7]. These features are particularly suitable for SiNW devices, since they allow avoiding difficult fabrication problems, such as precise control of the doping level and the formation of reliable ohmic contacts. Using SOI as a substrate can significantly simplify the formation of nanowires and fabrication of a back gate [8].

In our work, we studied a number of SiNWFET samples made on SOI substrates with ferromagnetic drain and source (Fe). The magnetic field effect on the conductivity of the field channel of the structures at a temperature of 10 K was studied.

2. Experimental

Nanowire devices were formed on boron doped silicon on insulator (SOI) substrates with a concentration of 10^{15} cm^{-3} using e-beam lithography and wet and dry etching. Prior to this, a 15 nm thick film of iron was deposited on SOI by thermal evaporation under ultrahigh vacuum. As a result, the devices represent itself silicon nanowires with ferromagnetic metal drain and source. To create the back gate of the field-effect transistor, an ohmic contact of indium was formed on the back side of the SOI substrate. Figure 1(a) shows a schematic representation of a SiNWFET device. The obtained devices were cooled with liquid helium to a temperature of 10 K in a flow-through cryostat of EMPX-HF probe station (Lake Shore cryotronics). All transport characteristics were measured with a Keithley 2634B nanovoltmeter in a zero magnetic field, as well as in a field of 0.5 T of different polarity applied in the plane of the substrate and perpendicular to the device channel (Fig. 1(b)).

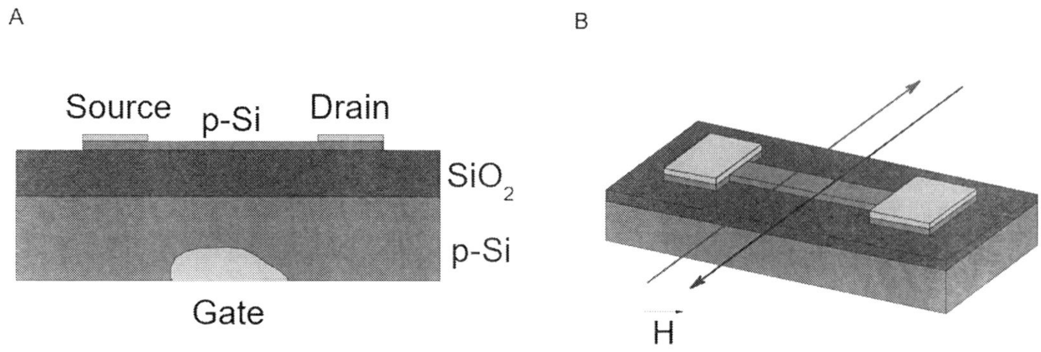

Figure 1(a, b). **(a)** The SiNWFET structure scheme, **(b)** experimental geometry.

Figure 2 shows the obtained experimental dependences of the transport characteristics of SiNWFET with a channel width w_c = 500 nm. The devices show ambipolar characteristics of both forward and reverse current in the channel depending on the gate bias. Voltage-current source-drain characteristics (VCC) are non-linear. This is due to the fact that the Fe/Si contact is not ohmic. At the Fe / Si interface, there is a Schottky barrier that acts as a semiconductor heterojunction of a field-effect transistor in this device. Consequently, transistor is close at zero gate bias.

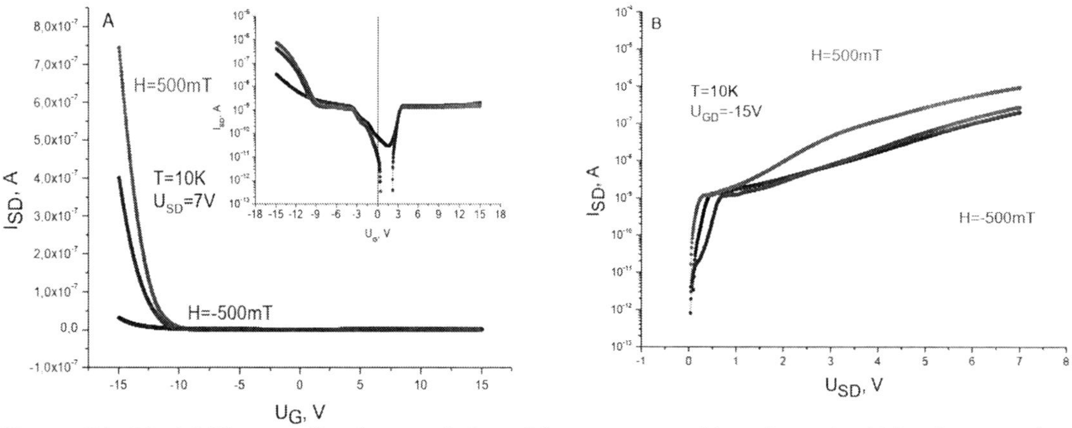

Figure 2(a, b). **(a)** The transfer characteristics of the structure with a channel width of 500 nm in a magnetic field of 500 mT of different polarity (the inset shows the same data plotted on a semilog scale), **(b)** VCC of the device.

3. Results and discussion

When the magnetic field directed perpendicular to the current channel is turning on, a change in the current in the device is observed. The effect is most pronounced at a negative gate voltage above 10 V. At this gate bias the device operates in accumulation mode (hole transport). Based on geometry of the experiment, the most probable mechanism of the magnetic field effect is the Lorentz force acting on the charge carriers moving in a perpendicular magnetic field. As a result, carriers should be deflected to the front or back part of the nanowire, depending on the magnetic field polarity which will lead to current reduction due to an increase free path of carriers. However, the opposite behavior is observed at high gate bias. The magnetic field-effect leads to an increase of current. Figure 3 shows the relative change of current in a magnetic field of 500 mT of different polarity, which was determined by a formula similar to magnetoresistance

$$MI = (I(H) - I(0))/I(0). \quad (1)$$

Magnetocurrent (MI) demonstrates a non-linear dependence on the gate voltage and on the magnitude of the drain-source electrical bias. Relative change in the magnitude of the current in the channel exceeds 2000 percent.

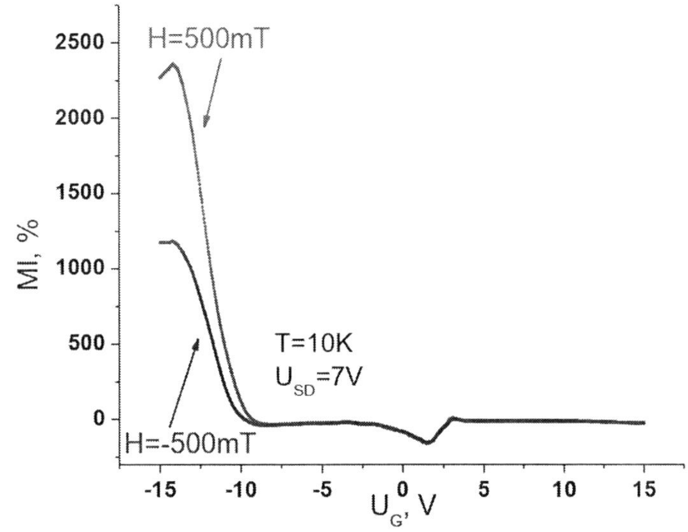

Figure 3. The magnetocurrent versus gate voltage dependence.

If we assume that the behavior of the carriers, when a magnetic field is applied perpendicular to the current channel, is determined only by the action of the Lorentz force, it is unclear why the magnetic current increases, both in the positive direction of the magnetic field and in the negative direction. Nevertheless, we propose a possible mechanism for the observed positive MI effect.

Figure 4 shows the charge carriers accumulation behavior in the device channel in magnetic fields of different polarity. Taking into account the fact that the device is made on SOI, the carriers deflected by the magnetic field in the direction of the substrate will not go deep into the structure, as would be the case when using a conventional substrate. They will begin to accumulate near the boundary of the insulator, changing the shape of the conduction channel. This will effectively contribute to the field effect and will lead to an increase of the magnetocurrent (Fig. 4 (b)). With an oppositely applied magnetic field, carriers will accumulate near the surface of the nanowire, which will also affect the channel shape. As a result, the electric field created by the back gate effectively increases, but weaker than in previous case due to the fact that some of the carriers deflected in this way will recombine on the surface structural defects and surface states and will not participate in the formation of the channel front (Fig. 4 (c)). At the moment it is difficult to make more detailed conclusions about the mechanisms responsible for the observed phenomena. This question requires further researches.

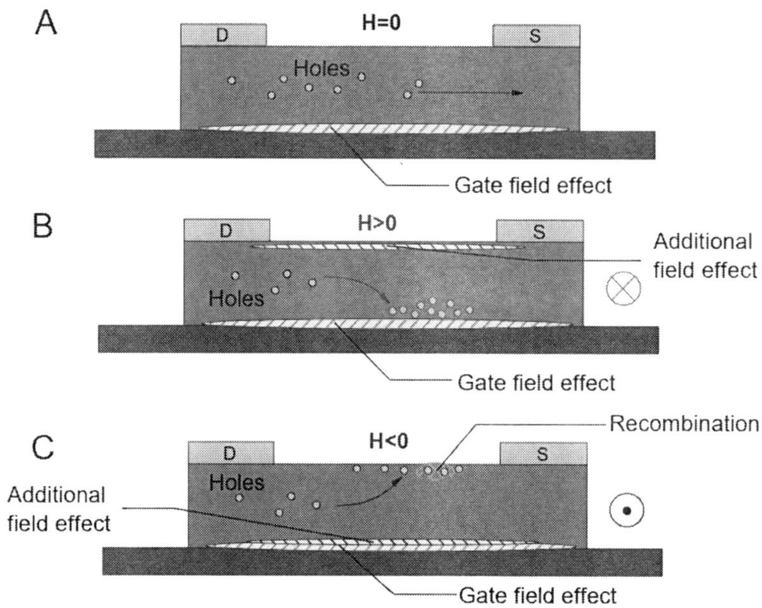

Figure 4(a, b, c). The charge carriers accumulation mode in the channel of the device:
(a) in a zero magnetic field, **(b)** when turning on a positive magnetic field (holes are curved in the magnetic field towards the SOI dielectric, accumulating there and thus creating an additional electric field) and **(c)** when turning on a negative magnetic field (Holes are curved in the magnetic field towards the nanowire surface. Accumulated near the top of the surface, holes create an additional electric field, but lower than in previous case since some of them recombine.)

Acknowledgments
The reported study was funded by Russian Foundation for Basic Research, Government of Krasnoyarsk Territory, Krasnoyarsk Region Science and Technology Support Fund by project № 18-42-243022 and supported in part by the Russian Foundation for Basic Research by project no. 18-32-00035. The work was partially supported by the Ministry of Education and Science of the Russian Federation and by Siberian Branch of the Russian Academy of Sciences (Project II.8.70) and Fundamental research program of the Presidium of the RAS no. 32 «Nanostructures: physics, chemistry, biology, basics of technologies».

References
[1] Cui Y and Lieber C M 2001 *Science* **291** 851
[2] Mathur N 2002 *Nature* **419** 573
[3] Xiao M, Martin I, Yablonovitch E, Jiang H W 2004 *Nature* **430** 435
[4] Koo S, Edelstein M, Li Q, Richter C and Vogel E 2005 *Nanotechnology* **16** 1482
[5] Veeraghavan S and Fossum J 1989 *IEEE Trans. Electron Devices* **45** 522
[6] Allibert F, Ernst T, Pretet J, Hefyene N, Perret C, Zaslavsky A, Cristoloveanu S 2001 *Solid State Electron.* **45** 559
[7] Hu S, Wu Y, Sung C, Chang C, Huang T 2004 *IEEE Trans. Nanotechnology* **3** 93
[8] Schwalke U, Krauss T and Wessely F 2013 *Solid State Science and Technology* **2(6)** 88

Contact resistance and lifecycle of an ohmic MEMS switch with single and multiple contact bumps

N V Marukhin and I V Uvarov

Valiev Institute of Physics and Technology of Russian Academy of Sciences, Yaroslavl Branch, Universitetskaya 21, 150007 Yaroslavl, Russia

E-mail: i.v.uvarov@bk.ru

Abstract. Microelectromechanical systems (MEMS) switches potentially have a wide range of applications due to their promising characteristics, but relatively low reliability limits implementation of these devices. Contact region degrades during operation that increases the resistance and lowers the lifecycle. In this paper we demonstrate how these parameters can be enhanced by changing the design of the switch. The devices of four types having different number of contact bumps and beams are fabricated by surface micromachining and tested in a cold switching DC regime. The switch with two bumps has about 25% lower contact resistance and almost two times longer lifecycle than the one-bump device. Using four beams instead of one gives threefold increase of the lifecycle, but does not reduce the resistance significantly.

1. Introduction

Radio-frequency switches fabricated by the microelectromechanical systems (MEMS) technology combine advantages of electromechanical relays and solid state switches. They provide superior radio frequency characteristics, small size, low power consumption and high integration capability [1]. Electrostatically actuated MEMS switches are especially popular, since they are relatively simple in design, fabrication and operation [2]. However, these devices have not achieved commercial success yet due to the lack of reliability [3]. The weak point of the ohmic MEMS switches is the contact of a micromachined beam with the drain electrode. The contact degrades during operation because of melting, material transfer, contamination and other phenomena [4]. As a result, contact resistance and insertion loss increase that leads to the switch failure. The reliability can be improved by proper selection of the contact material or choosing the optimal operation conditions. Another approach is to change the switch design. In this work we investigate the influence of two design solutions on the switch performance. The first one is to equip the beam with several contact bumps, while the second way is to use several beams. The switches of various layouts are fabricated and tested. The contact resistance and lifecycle are measured, analyzed and compared.

2. Materials and methods

The basic switch is shown in figure 1a. The movable electrode is a single beam attached to the anchors by the torsion hinges. It is fabricated of 1 μm thick Al film and is reinforced by the 1 μm thick Al ribs. The length of the beam is 100 μm and the width is 24 μm. Gate and drain electrodes are made of 100 nm thick platinum film and are placed under the each arm of the beam. Thus, the switch has two symmetric output channels. The gap between the beam and the gate electrodes is 1.5 μm. Platinum contact bumps are located on the bottom side of the beam, so the device provides Pt-Pt contact. The channel is closed by applying a voltage to the corresponding gate electrode. The basic design includes one bump per arm (figure 1b), while the modified switch has two bumps per arm (figure 1c). Figure 1d illustrates the device with four beams connected to each other. The beams and suspensions have the same geometry as in the basic design. They are actuated by a common gate electrode and, therefore, operate simultaneously. Both one- and four-beam switches are fabricated in one- and two-bump versions. The fabrication procedure is described in the work [5].

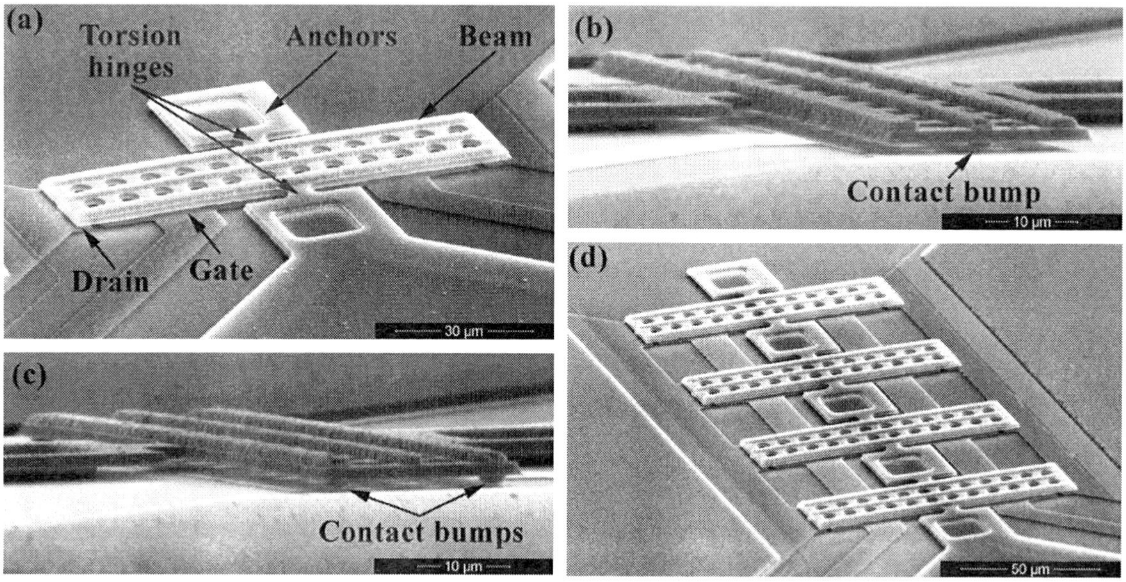

Figure 1. SEM images of the MEMS switch: (a) the basic design with one beam, the main structural parts are shown by the arrows; (b) the switch with one contact bump; (c) the switch with two bumps; (d) the switch with four beams.

The switches are tested in a standard laboratory environment without packaging. Rectangular voltage pulses are periodically applied to the both gates in such a way that the channels are actuated alternately with the frequency of 3 Hz. The amplitude of pulses is 60 V. The opening of the channels is carried out by the elastic forces of the torsion springs. In order to avoid electric arcing, the switch operates in a cold switching conditions. The source voltage is applied after the beam touches the drain and is turned off before the contact breaks. The testing is performed at the source voltage of 5 V. A load resistance of 4.7 kΩ limits the current flowing through the switch by the value of 1 mA that corresponds to the transmitted power of 5 mW. The resistance of both channels is measured in the closed state at each actuation cycle.

3. Results and discussion

3.1. On-resistance

The resistance of the switch in the on-state R_{on} randomly changes from several tens to several thousand ohms during the test. Typical dependence of R_{on} on the number of switching cycles is presented in figure 2a. All the designs demonstrate similar behavior. It was shown previously that the parasitic resistance of the wires and signal lines is about 10 Ω [5]. Therefore, the main part of the total on-resistance is the contact resistance. The probable reason of the instability of R_{on} is the accumulation of the carbon-containing material on the contacts [6]. After a certain number of cycles R_{on} increases dramatically up to 10^8 Ω. This increase is usually irreversible and is considered as a failure of the switch.

We tested several tens of samples of each type. The dependence of the on-resistance averaged over the samples on the number of cycles is shown in the figure 2b. It increases during the test, but the most rapid growth occurs in the first two thousand cycles for switches with one bump and in the first four thousand cycles for switches with two bumps. The values of R_{on} averaged over the first five thousand cycles are given in table 1. Using two bumps instead of one reduces the resistance by about 25% for both one- and four-beam designs. This indicates that both bumps touch the drain, thereby expanding the contact area. However, increasing the number of beams does not reduce R_{on} significantly. For the one-bump design the reduction is 5% only, and no decrease is observed for the four-beam switch. To all appearance, the current flows only through one beam at each cycle due to the random behavior of the contact resistance.

3.2. Lifecycle

The lifecycle is measured as the number of switching cycles that the device withstands until failure. The average values of the lifecycle are given in table 2. The basic switch with one bump and one beam has the lowest durability of 2.3×10^4 cycles. Increasing the number of bumps prolongs the lifecycle more than two times, which can be explained by the distribution of the transmitted power between the bumps. Using four beams instead of one gives an almost triple enhancement of the durability. When the contact at one beam fails, the current is shared by the remaining beams, and the switch continues operation until the failure of the last beam. As a result, the two-bump four-beam switch demonstrates the longest lifetime of 14.5×10^4 cycles. In general, large scatter of the lifecycle is observed for all types of the switch, see table 2. Contamination of the contacts is considered as the main reason of this phenomenon.

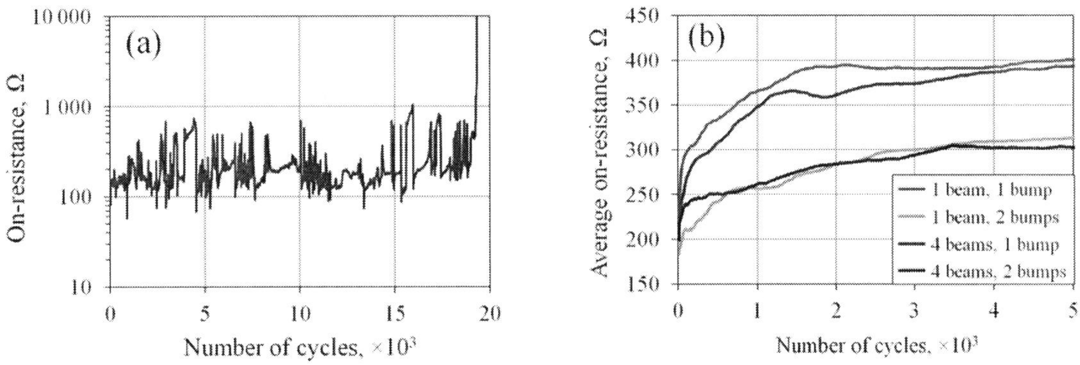

Figure 2. (a) The on-resistance of the channel of the one-beam one-bump switch during the test; (b) Evolution of the average on-resistance during the first five thousand cycles.

Table 1. The on-resistance averaged over the samples and the first five thousand cycles of the test.

Layout	One beam	Four beams
One bump	379 Ω	362 Ω
Two bumps	283 Ω	283 Ω

Table 2. The mean value and the scatter of the lifecycle ($\times10^4$) of the switches with various designs.

Layout	One beam	Four beams
One bump	2.3 ± 4.1	6.8 ± 7.2
Two bumps	4.7 ± 4.9	14.5 ± 13.7

4. Conclusions

The contact resistance and lifecycle of the electrostatically actuated MEMS switch with single and multiple contact bumps were investigated. The switch operated in the cold regime at the transmitted power of 5 mW. Multiple contacts are achieved in two ways: by increasing the number of contact bumps of the beam and by increasing the number of beams. Using two bumps instead of one reduces the on-resistance by about 25% and prolongs the lifecycle by more than two times. Using four beams instead of one gives threefold increase of the lifecycle, but does not reduce the on-resistance significantly. Among several designs, the four-beam switch with two contact bumps per arm demonstrates the highest performance. It has the average on-resistance of 238 Ω, which is 25% lower than R_{on} of the basic switch, and the average lifecycle of 14.5×10^4 cycles, which is 6 times longer than the lifetime of the basic single-beam single-bump device.

Acknowledgments

This work is supported by Program no. 0066-2019-0002 of the Ministry of Science and Higher Education of Russia for Valiev Institute of Physics and Technology of RAS and performed using the equipment of Facilities Sharing Centre "Diagnostics of Micro- and Nanostructures".

References

[1] Rebeiz G M 2003 *RF MEMS: Theory, Design, and Technology* (Hoboken, New Jersey: John Wiley & Sons)
[2] Hwang J C 2007 *IEEE Int. Workshop on Radio-Frequency Integration Technology* (Singapore: Rasa Sentosa Resort) p 168
[3] Rebeiz G M, Patel C D, Han S K, Ko C H and Ho K M 2013 *IEEE Microw. Mag.* **14** 57
[4] Toler B F, Coutu R A and McBride J W 2013 *J. Micromech. Microeng.* **23** 103001
[5] Uvarov I V, Marukhin N V and Naumov V V 2019 *Microsyst. Technol.* DOI: 10.1007/s00542-018-4279-2
[6] Uvarov I V, Kupriyanov A N 2018 *Microsyst. Technol.* DOI: 10.1007/s00542-018-4188-4

Simulation of transparent electromagnetic interference shielding materials based on periodic conductive networks

P P Shiriaev[1], D S Ryzhenko[2], M O Makeev[1,2], O V Popkov[2]
[1] Peoples Friendship University of Russia (RUDN University), 6 Miklukho-Maklaya St, Moscow, 117198, Russian Federation
[2] Bauman Moscow State Technical University (BMSTU), 5/1 Baumanskaya 2-ya, Moscow, 105005, Russian Federation

Abstract. This paper is a preliminary report on a theoretical study of transparent shielding materials with potentially very high shielding efficiency based on quasi-periodic conductive networks. An empirical model for rapid calculation of shielding efficiency vs frequency and mesh geometrical parameters is suggested based on numerical simulation results.

1. Introduction
Development of electromagnetic interference (EMI) shielding materials is a topical task for electronic device industry. Continuous growth of electronic devices in all spheres of life led to a new type of pollution known as electromagnetic interference that can harmfully affect both the device performance and human beings [1]. The usage of metal sheets and cases as the original EMI shielding method has been the most widespread technique over the years. Unfortunately, they are not suitable for producing optically transparent materials. This problem is partially solved by using transparent conducting oxides (TCOs), such as indium-tin oxide (ITO) [2], but they can only provide limited shielding without losing transparency and are prone to degradation due to cracking. Potentially superior alternatives, which receive much attention lately, include silver nanowire (AgNW) coatings [3] and graphene sheets, but their shielding efficiency (SE) is also limited by the requirement of transparency.

Higher SE values can be achieved by using highly conductive meshes. Unlike continuous thin conducting layers, they exhibit a strong frequency dependence of SE. Obviously, their SE also strongly depends on their geometry and electrical properties of the material they are comprised of. Due to the overall complexity of the system, analytical solutions are only possible for very special cases. However, structure optimization based solely on numerical solutions can be computationally expensive, so there is a need for simple empirical models.

In this paper we suggest an empirical model for the frequency dependence of shielding efficiency (SE) parametrized by geometrical parameters of a quasi-periodic conducting hexagonal mesh.

2. Simulation
The investigated conductive mesh is inspired by Shen S. *et al.* research [4] who proposed a material which has both good SE and transparency. The active layer of the material consists of a 2D network of straight metallic conductors. The network geometry is based on a periodic (constructed on a flat torus) Voronoi diagram of a uniformly distributed set of points which was subject to multiple iterations of Lloyd's algorithm in order to equalize cell sizes (Fig. 1a). Such design provides highly uniform surface resistance and transparency over the entire material surface and prevents the appearance of a moiré pattern if the mesh is applied onto a highly periodic structure, such as the display surface. If the

number of polygons per periodic cell is sufficiently high, they tend to form a quasi-crystalline close-packed hexagonal structure, akin to that observed for a number of self-organized systems, such as porous anodic oxides or breath figures, which makes this kind of material possible to produce via self-organization. This also allows approximating the layer by a periodic hexagonal mesh (Fig. 1b), with properties averaged for all orientations, which simplifies the numerical simulations.

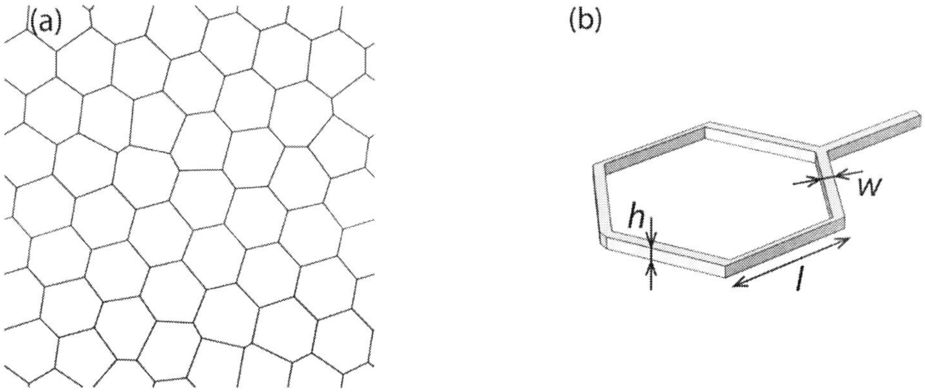

Figure 1(a, b). (a) Transparent network formed by opaque conductor wires; **(b)** single unit cell of the simplified model used in this study for numerical simulations.

In this study we calculate SE based on power transmission coefficient with respect to a normal-incidence plane wave, $|S_{21}|^2$, as follows:

$$SE = -20\log|S_{21}|. \tag{1}$$

The simulation is performed using the finite element method (FEM) in frequency domain. The unit cell model with boundary conditions is presented in figure 2. Simulations are performed for both the perfect electric conductor (PEC) and simple conducting media with $\varepsilon = 1 - \sigma i / \omega\varepsilon_0$ and no spatial dispersion.

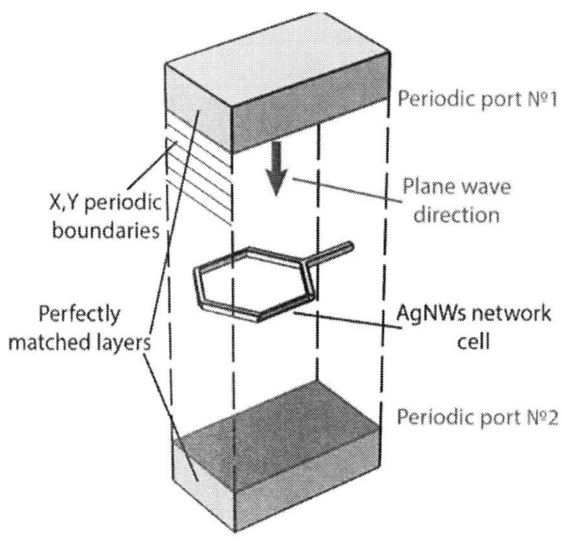

Figure 2. Unit-cell model with boundary conditions.

During the simulations, all the geometrical parameters (l, w, h) are varied independently. In our study, the length of the hexagon side is varied from 10 μm to 70 μm. Side width w and height h are varied in a range which allows to meet the conditions of optical transparency (T = 90-98%), which is calculated as the area ratio of opaque conductor wires to the entire area (Fig. 3a), and field of view ($FOV = 2\theta$ = 90-175°) (Fig. 3b), respectively. Here, we define field of view as the maximum angle at which light can pass through the cell holes (in the geometrical optics approximation).

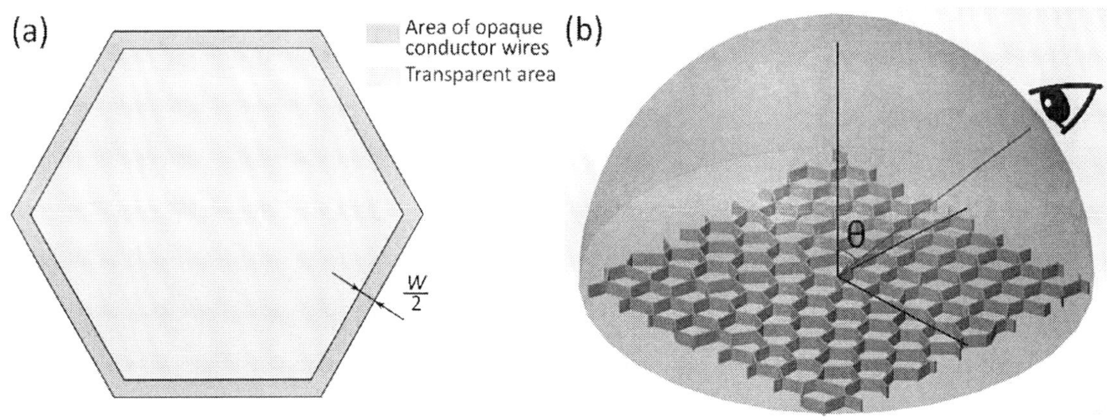

Figure 3(a, b). (a) Areas used in optical transparency calculation; **(b)** angle used in field of view (FOV) calculation.

In this study, we focus on shielding properties of the structure at frequencies much smaller than those of visible light. Figure 4 shows an example of a SE versus frequency curve for a specific combination of geometry parameters in case when conductor wires are made of a lossy conductor with σ = 63 MS/m.

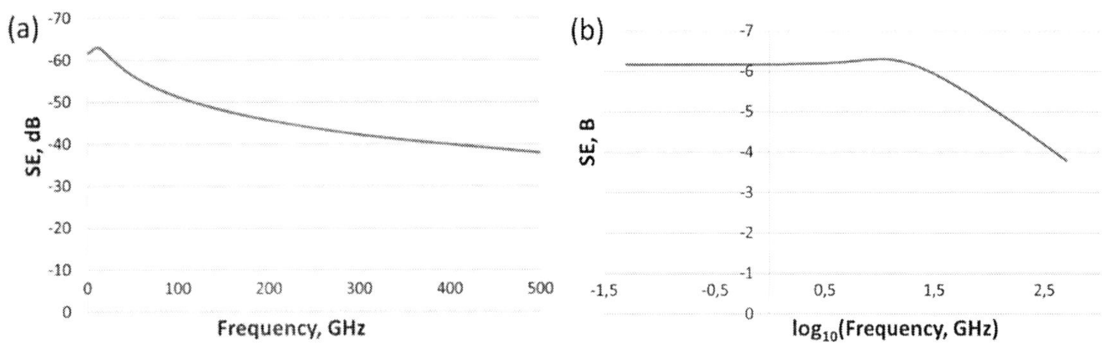

Figure 4(a, b). (a) SE versus frequency; **(b)** SE versus log-frequency. Geometrical parameters: l = 10 μm, w = 0.17 μm, h = 10 μm.

In case of finite conductivity materials, the SE curves asymptotically approach the SE value of a thin uniform layer with the same static effective sheet conductivity at lower frequencies (see fig. 3b). As frequency gets higher, the frequency dependence of $|S_{21}|^2$ becomes exactly quadratic, which is also observed for all frequencies in case of PEC:

$$|S_{21}|^2 = k \cdot f^2, \tag{2}$$

where f is frequency and k is a geometry- and material-dependent parameter. Further frequency increase must lead to an asymptotic approach to the transparency value predicted by geometrical optics, but this frequency region is out of scope of our research.

Depending on geometry of the structure, the transition between the aforementioned lower- and higher-frequency regions can be either monotonic or resulting in a single bump in the curve (see fig. 3a,b). The latter can be explained by a locally more prominent effect of the decrease in skin layer thickness than of the increase in hole transparency.

It appears that coefficient k in equation (2) can be adequately described by an 8-parametric empirical model derived as follows. First, in order to reduce the number of free variables, we define dimensionless parameters α (relative area of the conductor's normal projection) and β (structure thickness relative to average hole size):

$$\alpha = r \cdot \left(2/\sqrt{3} - r/3\right), \tag{3}$$

$$\beta = \frac{h}{\sqrt{3\sqrt{3}/2} \cdot \left(l - w/\sqrt{3}\right)}, \tag{4}$$

where $r = w/l$. The model equation is then built based on the LSE function:

$$\ln k = \ln\left(e^{k_1 + k_2\alpha + k_3\beta + k_4\alpha\beta} + e^{k_5 + k_6\alpha + k_7\beta + k_8\alpha\beta}\right). \tag{5}$$

where $k_{1\text{-}8}$ are empirically found coefficients.

The empirical coefficients were calculated by fitting results of numerical simulations in a few stages by using the equally-weighted least squares method. The discrepancy between the model predictions and numerical simulation results does not exceed 1.5 dB within the studied parameter range. SE values calculated both using (2) and using the proposed FEM model for several random geometries are shown in Figure 5.

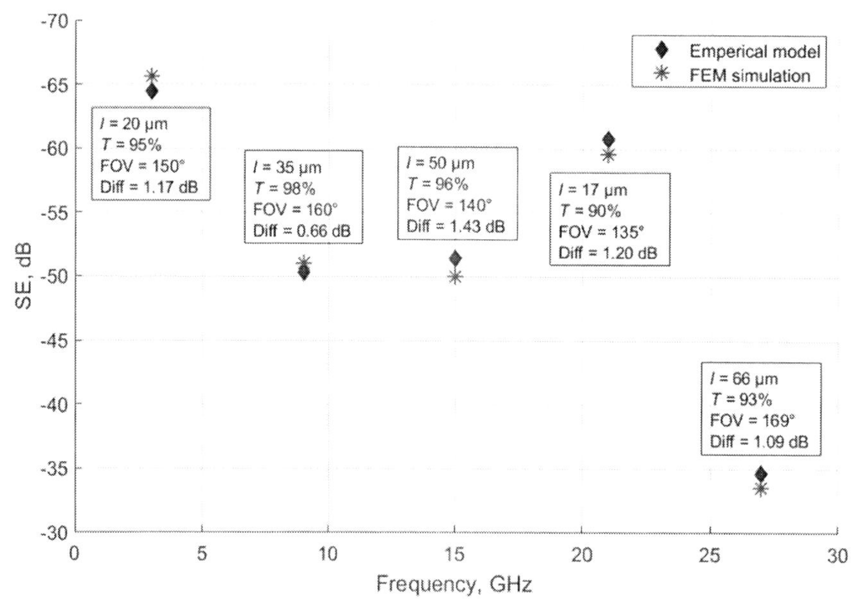

Figure 5. Empirical model verification.

3. Conclusions

Irrespective of material and geometrical parameters, the frequency dependence of SE in the frequency range where the mesh can no longer be considered a uniform sheet but the average cell size is still small compared to wave length is exactly quadratic, and the single coefficient of this dependence can be adequately modeled by an 8-parametric empirical model. Depending on mesh geometry and composition, the transition between the lower-frequency "plateau" and higher-frequency quadratic regions can either be monotonic or have a single bump in SE value, the latter being its global maximum.

Acknowledgments

This paper was financially supported by the Ministry of Education and Science of the Russian Federation on the program to improve the competitiveness of Peoples' Friendship University of Russia (RUDN University) among the world's leading research and education centers in the 2016-2020.
The research work was supported by Ministry of Education and Science of the Russian Federation under state task No 8.4868.2017/8.9.

References

[1] Jaroszewski M 2018 *Advanced Materials for Electromagnetic Shielding: Fundamentals, Properties, and Applications* (New Jersey: Wiley)
[2] Huang J L, Yau B S, Chen C Y, Lo W T and Lii D F 2001 *Ceram Int* **27(3)** 363
[3] Xu F 2018 *Advanced Materials for Electromagnetic Shielding: Fundamentals, Properties, and Applications* (New Jersey: Wiley) 289
[4] Shen S, Chen S Y, Zhang D Y and Liu Y H 2018 *Opt. Express* **26(21)** 27545
[5] Moré J J 1978 *Numerical analysis* 105

Electronic transport properties of microcrystalline GaP

A V Uvarov[1], A S Gudovskikh[1,2], A A Vasilev[1]

[1]St. Petersburg Academic University, St. Petersburg 194021, Russia
[2]St. Petersburg Electrotechnical University "LETI", St. Petersburg 197376, Russia

Abstract. Electronic properties of microcrystalline gallium phosphide films deposited on fused silica substrates by plasma-enhanced atomic layer deposition at the temperature of 380°C were investigated. The values of the activation energy of conductivity (E_a = 0.247 eV), the optical value of the band gap (E_g = 2.2 eV) and the long-wavelength edge of photoconductivity (E_{ph} = 1.8 eV) were obtained for this material. The value of E_a could associated with position of Fermi level near to the bottom of the conduction band or barriers at the grain boundaries. The long-wavelength edge of photoconductivity in the layer of microcrystalline GaP is a consequence of the presence of interband states, which are associated with partially disordering.

1. Introduction

In recent years, interest in non-crystalline semiconductor materials, such as a-Si:H, CdTe, CIGS, etc., has been increasing [1]. Today with the use of these materials, most TFT displays and thin-film solar cells are being produced [2]. Besides, non-crystalline semiconductor materials are laid into the basis of such modern areas as third-generation multi-junction photoelectric converters, flexible and transparent LED displays [3]. Thus, the actual task is to search for new amorphous and microcrystalline semiconductor materials to expand the range of their capabilities and applications. One of such materials is gallium phosphide (GaP), semiconductor material with a band gap of 2.26 eV. Previously, microcrystalline GaP was obtained by r.f. sputtering [4][5], magnetron sputtering [6], and PECVD [7]. It was noted that change in the structural and optical properties of GaP is observed depending on the method of producing gallium phosphide. Earlier, we demonstrated the possibility of obtaining amorphous and microcrystalline GaP by the method of plasma-enhanced atomic layer deposition [8]. The study of the electronic properties of microcrystalline gallium phosphide is of interest due to the advantages of this method.

2. Experimental details

Thin films of microcrystalline gallium phosphide were grown on fused silica substrate by plasma enhanced atomic-layer deposition at the temperature of 380°C. Since the surface resistance of 50 nm undoped gallium phosphide layers is the order of 10^9/□, we use an interdigitated contact system (Fig. 1).

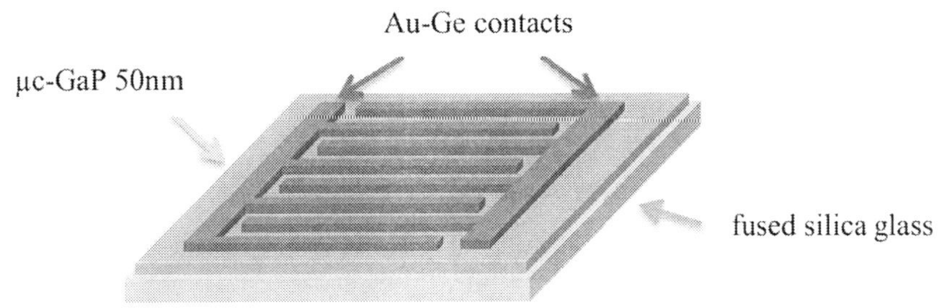

Figure 1. Structure of 50nm GaP film on fused silica substrate with interdigitated contact grid

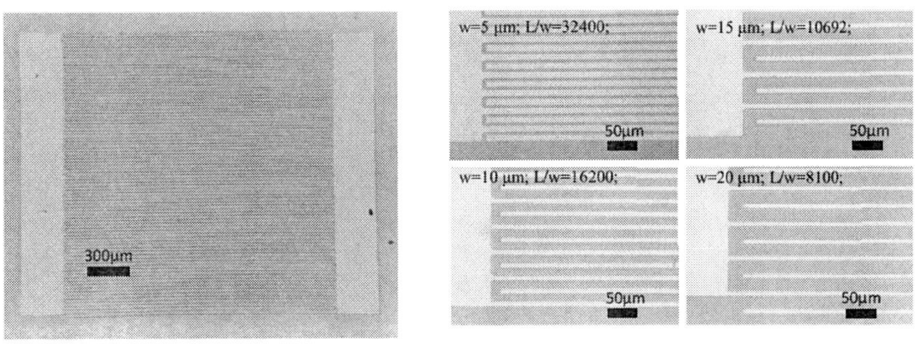

Figure 2. Optical microscope photo of the interdigitated contact grid.

Contacts were made using laser lithography and thermal vacuum evaporation. Lift-off photolithography process using PMGI and AZ MIR 701 photoresists was used to create a contact grid. Au-Ge alloy was chosen as a contact material and thickness of metal was about 100 nm. The width of the metal strips is 10 μm and the distance between them(w) was equal to 5, 10, 15 and 20 μm (Fig. 2). The ratio of the total length contacts (L) to the gap between them was equal to 32400, 16200, 10692 and 8100 respectively. Electrical measurements were carried out using Keithley 2400 electrometer. Optical absorption spectra as well as photocurrent spectra were achieved by measurement system based on monochromator Solar Laser System M226, Si detector and halogen lamp.

3. Results and discussion

Based on the known thickness (50 nm), the values of the electrical resistivity of the GaP layer were obtained. An exponential increase in conductivity at high temperature with an activation energy of 0.247 eV was observed (Fig. 3). At the temperatures lower than 280 K lateral conductivity of the layer drastically decreases and cannot be measured even using interdigitated contacts. Since the obtained GaP layers has a predominantly n-type conductivity, it could assumed that this activation energy related to the position of the Fermi level relative to the bottom of the conduction band. It should also be taken into account that microcrystalline gallium phosphide has many interfaces at the grain boundaries on which the formation of barriers is possible. These barriers can also lead to an exponential temperature dependence of conductivity with similar activation energies. Thus, value of E_a could associated with position of Fermi level near to the bottom of the conduction band or barriers at the grain boundaries. Based on the spectral dependence of the absorption coefficient, the optical band gap of microcrystalline gallium phosphide was equal to 2.2 eV. It is quite close to the band gap of crystalline GaP (E_g=2.26 eV). However, as can be seen from the optical absorption spectrum, these layers had a sharp absorption boundary, which is characteristic of disordered materials with quasi-direct band gap. Band gap of amorphous gallium phosphide is known strongly depends on the method and deposition temperature and varies in the range of 1.1 – 2 eV [4-7]. The signal of the spectral dependence of the photocurrent had observed at the energy less than optical band gap, up to 1.8 eV (Fig. 4.).

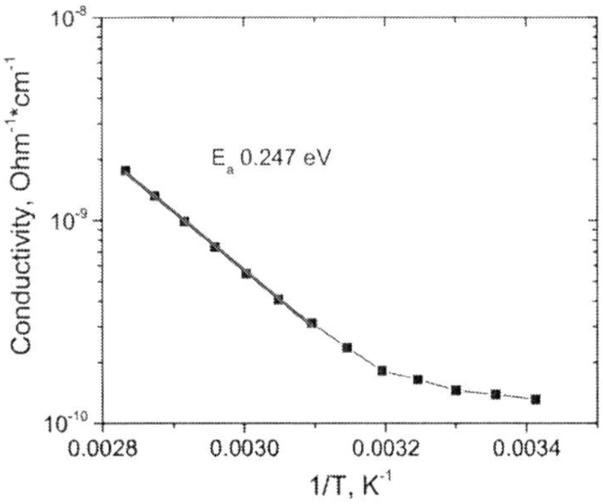

Figure 3. Temperature dependence of conductivity for μc-GaP on fused silica substrate

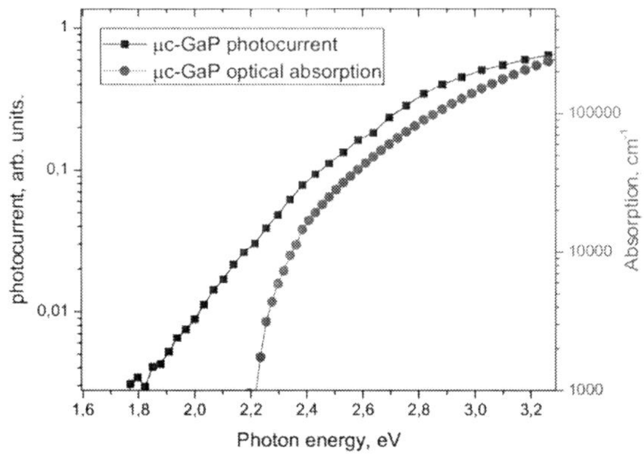

Figure 4. Spectral dependence of photocurrent and optical absorption for μc-GaP

Lower value of band gap and photoconductivity at lower energies than band gap can be associated with broadening of the band tails which have been reported for amorphous GaP [9] (Fig. 5.).

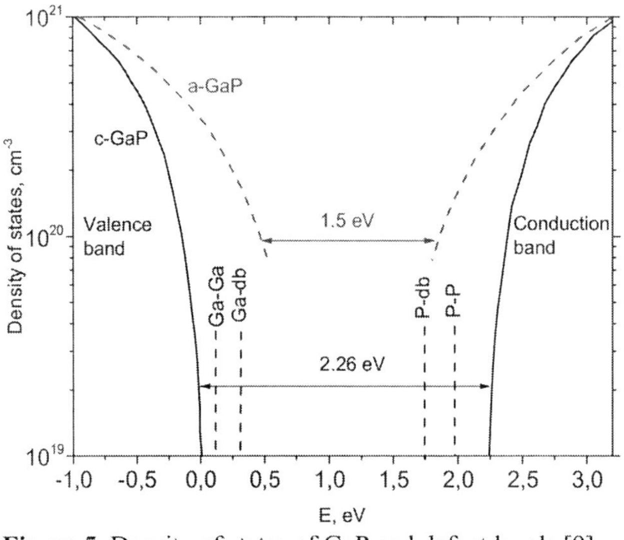

Figure 5. Density of states of GaP and defect levels [9].

Therefore, we can presume that the long-wavelength edge of photoconductivity in the layer of microcrystalline gallium phosphide is a consequence of the presence of interband states, which are associated with partially disordering.

4. Conclusions

The optical and electrical properties of the microcrystalline gallium phosphide were investigated. To study the electrical properties of thin layers of gallium phosphide with high specific resistance deposited on quartz substrates, interdigitated contact grids were developed. The values of the activation energy of conductivity (E_a = 0.247 eV), the optical value of the band gap (E_g = 2.2 eV) and the long-wavelength edge of photoconductivity (E = 1.8 eV) were measured. It was assumed that the value of E_a could associated with position of Fermi level near to the bottom of the conduction band or barriers at the grain boundaries. Lower value of band gap and photoconductivity at lower energies than band gap is consequence of the presence of interband states, which are associated with partially disordering.

Acknowledgments
This work was supported by the Russian Scientific Foundation under Grant No. 17-19-01482.

References
[1]Hideo Hosono et. al. Proceedings of the National Academy of Sciences, **114** (2) 233-238 (2017)
[2]Sharma, S., Jain, K. K., & Sharma, A. Materials Sciences and Applications, *6*(12), 1145 (2015).
[3]Sheng, J., et. al. Journal of Semiconductors, **39**(1), 011008. (2018)
[4] K. Starosta, J. Zelinka, D. Berková, J. Kohout, Thin Solid Films **61**, 241 (1979).
[5] N. Elgun, S.J. Gurman, E.A. Davis, J. Phys. Condens. Matter **12**, 4723 (2000).
[6] N. Matsumoto, K. Kumabe, Jpn. J. Appl. Phys. **18**, 1011 (1979).
[7] J.C. Knights, R.A. Lujan, J. Appl. Phys. **49**, 1291 (1978).
[8] A.S. Gudovskikh, I.A. Morozov, A. V. Uvarov, D.A. Kudryashov, E. V. Nikitina, A.S. Bukatin, V.N. Nevedomskiy, J.-P. Kleider, J. Vac. Sci. Technol. A Vacuum, Surfaces, Film. **36**, 021302 (2018).
[9] N. Elgun and E A Davis, *J.* Phys.: Condens. Matter **6** 779 (1994)

Carbon nanotube based terahertz radiation detectors

N Titova[1], I A Gayduchenko[1], M V Moskotin[1], G.F Fedorov[1], G N Goltsman[1]

[1]Moscow State University of Education (MSPU), 29 Malaya Pirogovskaya, Moscow 119435, Russia

Abstract. In this paper, we study terahertz detectors based on single quasimetallic carbon nanotubes (CNT) with asymmetric contacts and different metal pairs. We demonstrate that, depending on the contact metallization of the device, various detection mechanisms are manifested.

1. Introduction

Individual single-walled nanotubes (SWNT) have unusual electronic and optical properties, which makes them a promising material for the development of optoelectronic devices in the terahertz region. In the last decade, the field of terahertz technologies has been actively developing, and recently several types of detectors based on individual tubes and network capable of operating effectively in the frequency range up to 2.5 THz were demonstrated both at room [1] and at helium temperatures [2]. One of the approaches is to use asymmetric metallization for observation of photovoltaic response [3]. The demonstrated detectors were fabricated as field-effect transistors (FET) with asymmetric metallization of contacts. In our earlier works we employed vanadium and gold as two metals with different work functions. As gold has a work function of 5.1 eV, which is larger than that of CNT (4.7 eV) and vanadium has work function (4.1 eV) smaller than that of CNT this configuration leads to formation of a p-n junction in the device channel due to contact doping. This p-n junction can work as a rectifying element leading to photovoltaic response to incident radiation [4]. At the same time different metals provide different contact resistance and different Joule heating is released at the contacts so that the temperature gradient is maintained through the channel resulting in a DC voltage signal due to thermoelectric effect.
In order to avoid the ambiguity in interpreting the results we fabricated devices similar to those reported previously but replaced vanadium with nickel that has almost the same work function as gold so no p-n junction is expected in the channel leaving thermoelectric effect as the only mechanism explaining the observed DC voltage signal as the device is exposed to the radiation.

2. Experimental

Single carbon nanotubes were grown by a CVD method using bimetallic suspension of $Fe(NO_3)_3 \cdot 9H_2O$ and $MoO_2(acac)_2$ with alumina (Al_2O_3) nanoparticles in isopropanol as a catalyst [5]. The growth of CNTs was carried out on boron doped silicon substrates coated with 500 nm thermally grown SiO_2. The doping level of the substrate at the same time is such that the silicon substrate remains transparent to radiation in a given frequency range. After the growth of the CNTs, the substrate is investigated using SEM. Subsequently, metal electrodes are formed to chosen CNTs using e-beam evaporation and lift-off. The devices are coupled to the radiation with a logarithmic spiral antenna which also serves for DC contacts. The devices are fabricated in such way that the contact material is different for source and drain electrodes (Fig. 1a, b) [3]. For this study we fabricated two sets of samples. In both sets, we used 80 nm gold on the side of the drain. And for the source we used vanadium 40 nm thick in one set and nickel (also 40 nm) in the other. Length of the nanotube between the electrodes is about 1 micron.

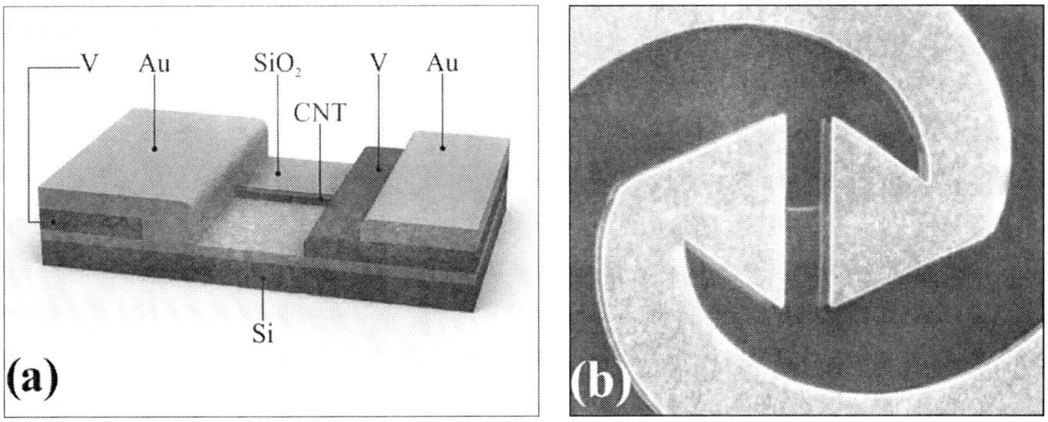

Figure 1. Device configurations a) Schematics view of a CNT contacted b) SEM image of a typical device

The measurement setup we are using here is similar to that described in detail in [6, 7]. Our devices are coupled to the radiation with a logarithmic spiral antenna. A choice of the antenna was determined by simplicity of its fabrication and its wideband characteristics for terahertz frequency range. THz range stands for the region of 0.1 -3 THz. A device chip with dimensions 4x4 mm^2 was fixed on a flat surface of a silicon lens, placed inside the cryostat, equipped with high density polyethylene window. A Zytex-106 cold infrared filter was mounted on the radiation shield to cut off the 300 K background. The sensing element of the device, the individual CNT inside the antenna, is located in the focus of the lens, enabling good coupling to the incident electromagnetic radiation. The power incident on the cryostat window is measured by a Golay cell. The terahertz radiation is provided by backward wave oscillators (BWO) with frequency ranges of 120 – 140 GHz, 250 – 450 GHz and 630 – 660 GHz.

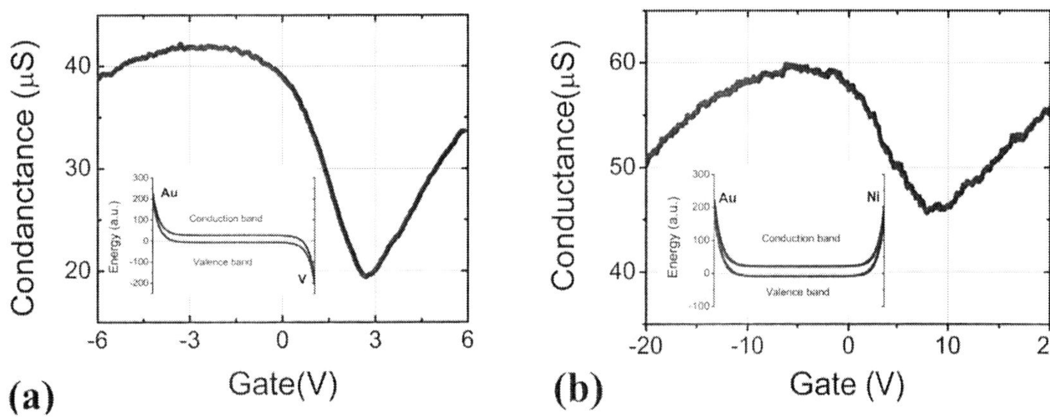

Figure 2. Transfer characteristics of device with V/Au (a) and Ni/Au (b). Insert Band diagrams of the corresponding devices.

For this work we selected similar devices for both types based on so called quasi-metallic CNT (qm-CNTs) [4], that have a curvature-induced band gap in a few meV range. Such devices have a small dip on the conductance vs gate voltage dependence (see Figure 2).
Since we use metals with different work functions, the Fermi levels at the interface Au/CNT and Ni/CNT will be shifted to the valence band. In this case, the Fermi levels at the V/CNT interface will be shifted to the conduction band. The band diagrams shown on the inserts of Figure 2a and 2b. Graphs of conductivity versus gate voltage confirm the formation of a barrier at the vanadium – CNT interface [8, 9].
After selecting suitable devices, we characterized their response to radiation of different frequencies at room temperature (Fig. 3). The maximum radiation power reaching the lens was set to 50 μW for all frequencies. Response of both types of devices to radiation room temperature is illustrated in the Figure 3.

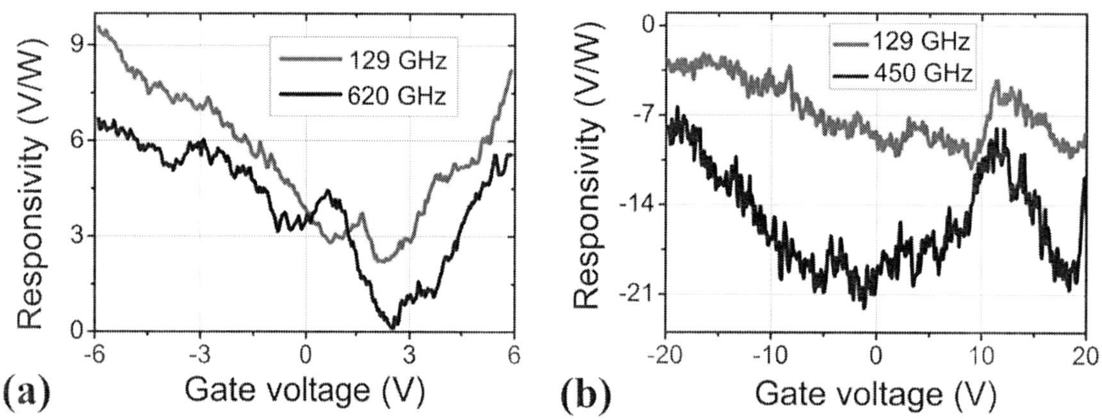

Figure 3. Responsivity of device with V/Au (a) and Ni/Au (b) for 300K.

3. Result and discussion

To understand the mechanisms of the demonstrated detection, we cooled our samples to 77 K and measured them again gate voltage dependence of the response to the radiation with a frequency of 129 GHz with a power of 32 μV, while the sample is not displaced by the current. The Fig. 4a shows data for V/Au device, while the Fig. 4b shows the data for Ni/Au device. In both graphs the response is compared to calculated FET-factor defined as (dG /dVg)/G. It allows for verification of the assumption that for devices with Ni/Au contacts only the thermoelectric effect will manifest itself since the Seebeck coefficient is described by the formula

$$S = -\frac{\pi^2 k_B^2 T}{3|e|}\frac{1}{G}\frac{dG}{dV_g}\frac{dV_g}{dE} \sim \frac{1}{G}\frac{dG}{dV_g} \qquad (1)$$

The calculated FET-factor defined as F = (dG /dVg)/G is shown in Figure 4 in red. Comparing the measured signal with the calculated FET factor values, we obtained a good agreement in case of the samples with Ni / Au contacts. On our view this is an evidence of the thermoelectric origin of the photoresponse of this type of devices. The deviation of the curves manifested for samples with V/Au can be explained by the presence of an additional response mechanism.

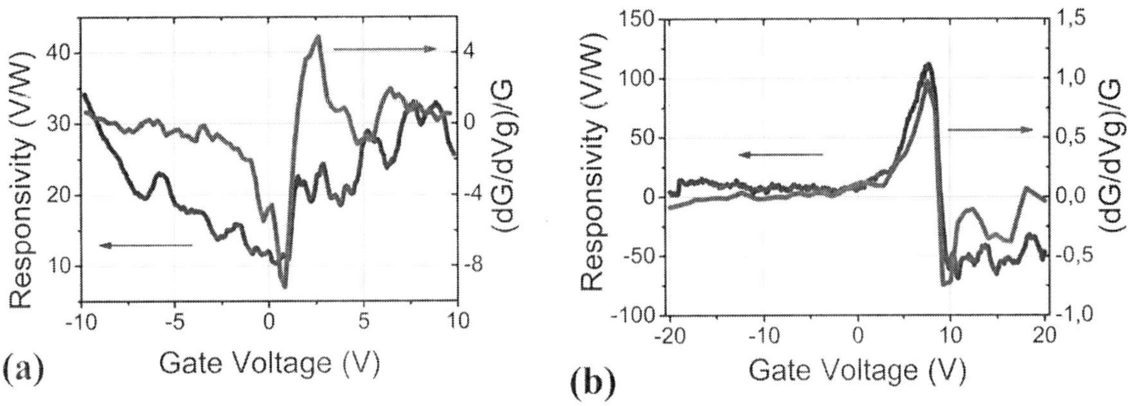

Figure 4. Responsivity of device with V/Au (a) and Ni/Au (b) for 77K.

An additional mechanism for samples with V / Au is related to the nonlinearity of the I-V curves (the so-called diode mechanism). The reason for its occurrence is the p-n junction formed on the V/CNT interface due to the difference in the work function. The diode response is calculated by the formula

$$S_V = -\frac{R}{4}\frac{d^2I}{d^2V}Z_A \qquad (2)$$

where Z_A is the antenna impedance equal to 100 Ohms, R is the resistance of the sample. The experimental values usually exceed the calculated ones, but the approximation describes well the obtained values of the volt-watt sensitivity [9].
The study of samples with nickel contacts in the temperature range from 300 to 4K, showed that during cooling the bolometric response mechanism contributes significantly [10].

4. Conclusion

The sensitivity for samples with different pairs of contacts is comparable in magnitude. Despite the fact that we are observing different detection mechanisms, both types of devices can be successfully applied.

Acknowledgments

This work was supported by the Ministry of Education and Science of Russian Federation (contract № 14.583.21.0069, RFMEFI58317X0069).

References

[1] J. D. Chudow, D. F. Santavicca, C. B. McKitterick, D. E. Prober, and P. Kim, Appl. Phys. Lett. 100, 163503 (2012).
[2] K. Fu, R. Zannoni, C. Chan, S. H. Adams, J. Nicholson, E. Polizzi, and K. S. Yngvesson, Appl. Phys. Lett. 92, 033105 (2008).
[3] I. Gayduchenko,A. Kardakova, G. Fedorov, B. Voronov, M. Finkel, D. Jiménez, S. Morozov, M. Presniakov, and G. Goltsman, J. Appl. Phys. 118, 194303 (2015)
[4] C. W. Zhou, J. Kong, and H. J. Dai, Phys. Rev. Lett. 84, 5604 (2000).
[5] A. Tselev, K. Hatton, M. S. Fuhrer, M. Paranjape, and P. Barbara, Nanotechnology 15, 1475–1478

(2004).
[6] G. E. Fedorov, T. S. Stepanova, A. S. Gazaliev, I. A. Gaiduchenko, N. S. Kaurova, B. M. Voronov, G. N. Goltzman, Semiconductors 50,1600 (2016).
[7] I. A. Gayduchenko, G. E. Fedorov, T. S. Stepanova, N. Titova, B. M. Voronov, D. But, D. Coquillat, N. Diakonova, W. Knap, G. N. Goltsman, J. Phys.: Conf. Ser.741, 012143 (2016).
[8] M. H. Yang, K. B. K. Teo, W. I. Milne, D. G. Hasko, Appl. Phys. Lett. 87, 253116, (2005).
[9] G. Fedorov, I. Gayduchenko,N. Titova, A. Gazaliev, M.Moskotin, N. Kaurova, B.Voronov and G. Goltsman, Phys. Status Solidi B 255, 1700227, (2018)
[10]M.V Moskotin, I. A. Gayduchenko, G. N. Goltsman, N. Titova, B. M. Voronov, G. F. Fedorov, F. Pyatkov and F. Hennrich, J. Phys.: Conf. Ser. 1124, 051050, (2018)

Effect of surface modification with ZnO films on aligned carbon nanotubes properties

M V Il'ina, A V Guryanov, O I Osotova, Z E Vakulov and O A Ageev
Southern Federal University, Institute of Nanotechnologies, Electronics and Electronic Equipment Engineering, Taganrog, 347922, Russian Federation

Abstract. The results of experimental studies of the piezoelectric response and memristor effect of aligned carbon nanotubes conformally coated with ZnO film are presented. It is shown that the surface modification of the carbon nanotubes with ZnO films leads to a decrease of the memristor effect and piezoelectric response of nanotubes due to an increase their stiffness and a decrease the strain value upon the application of an external field. The results can be used to develop the storage devices based on aligned carbon nanotubes.

1. Introduction

One of the main directions of development of modern electronics is nanopiezotronics – a field of science studying the piezoelectric and semiconductor properties of nanostructures to create new electronic devices and elements [1-5]. Special attention in this area is paid to carbon nanotubes (CNTs) due to the high values of elasticity, strength and anomalous piezoelectric properties [6, 7]. However, the study of the piezoelectric properties of carbon nanotubes and the process of generating current and potential during deformation are at the initial stage. Recent studies in this field shown that carbon nanotubes demonstrate a memristor effect associated with non-uniform strain of CNTs and their piezoelectric properties [7, 8]. In this connection, an assumption has arisen that the memristor effect and piezoelectric response of CNTs is enhanced due to the camphor coating of their surface with a piezoelectric material.

The aim is to study the effect of surface modification of aligned CNTs with ZnO films on their piezoelectric response and memristor properties.

2. Experimental studies

As the experimental sample we used a vertically aligned CNTs array grown by plasma-enhanced chemical vapor deposition [9-11]. A diameter, length and density of a CNT in an array were 35±3 nm, 980±70 nm and 147 μm^{-2}, respectively (Figure 1). The surface modification of the carbon nanotubes with ZnO films was carried out using a Pioneer 180 (Neocera Co., USA) [12]. The ZnO film was conformally deposited at laser radiation energy of 160 mJ and a pulse number of 10 000 with a repetition rate of 10 Hz. The substrate temperature was 300 °C. The pressure in the chamber was 10^{-3} Torr.

Studies of the vertically aligned CNTs were carried out by the atomic force microscopy (AFM) using the Ntegra probe nanolaboratory (NT-MDT, Russia). A commercial cantilever with a platinum coating NSG11/Pt was used as the AFM probe. A preliminary a scanning of the CNT array was carried out in the semi-contact AFM mode for the formation of non-uniform strain in the nanotubes [7]. As a result of scanning the individual CNTs were combined into bundles of non-uniformly strained CNTs under the action of van der Waals forces. It was a necessary condition for the manifestation of the memristor effect in the CNTs [13]. The AFM images of the CNT array before and after the modification are presented in Figure 2.

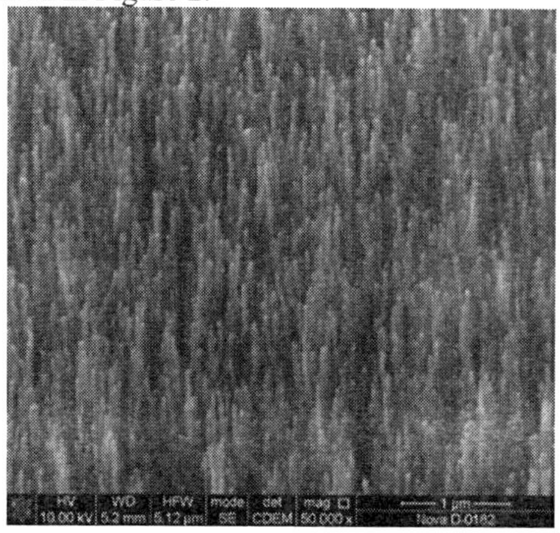

Figure 1. SEM image of the vertically aligned CNTs array.

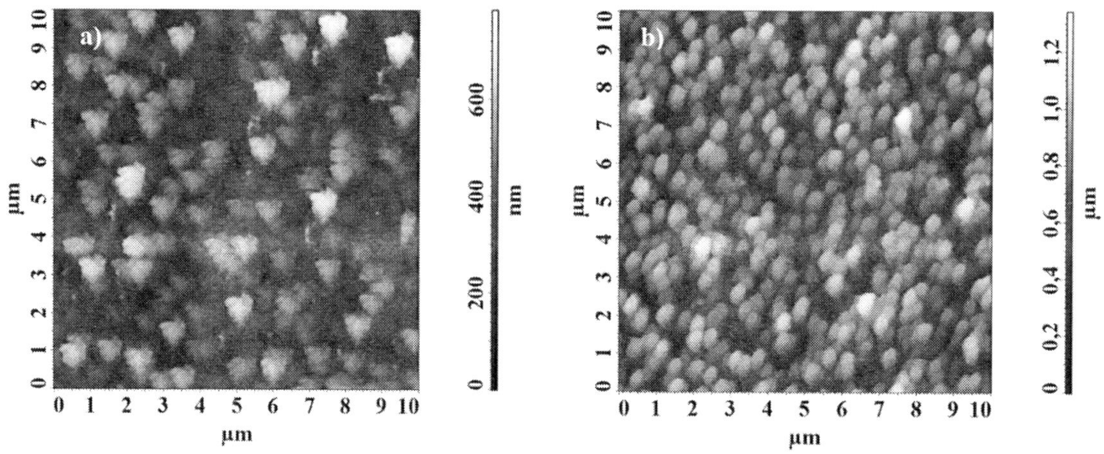

Figure 2 (a, b). AFM image of the CNTs array before **(a)** and after **(b)** the surface modifying with ZnO films.

Studies of the piezoelectric response of CNTs were carried out on the basis of a previously developed technique [7]. The maximum penetration of the AFM probe into the CNTs array did not exceed 50 nm relative to the top of the CNT. The current flowing in the "bottom electrode / CNT / AFM probe" system during CNT deformation was detected using an Ntegra oscilloscope. The current-time dependences of the process of CNTs deformation before and after the surface modifying with ZnO films are shown in Figure 3.

The current-voltage characteristics (CVCs) of strained CNTs were measured by the AFM in current spectroscopy mode when a saw tooth voltage pulse of amplitude ±10 V and duration 0.5 s was applied.

3. Results

An analysis of SEM images showed no visible changes before and after CNT modification due to conformal ZnO deposition. However, the AFM images showed changes in the geometric parameters of CNTs bundles after modification (Figure 2). Thus, the diameter of the CNT bundles formed during the AFM probe action was 800 ± 50 nm before modification (Figure 2a) and 450 ± 50 nm after modification (Figure 2b). The decrease in the CNT diameter is probably due to an increase in the stiffness of the CNT and a decrease in the van der Waals forces between the nanotubes after coating with ZnO films.

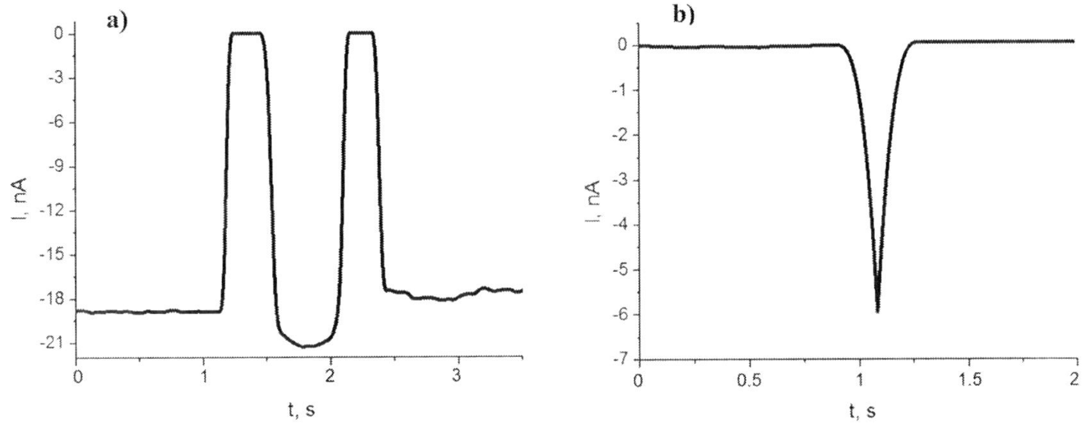

Figure 3 (a, b). The current-time dependences of the deformation process of CNTs before **(a)** and after **(b)** the surface modifying with ZnO films.

Studies of the piezoelectric response of the CNTs array showed that in the "lower electrode / CNTs / AFM probe" system a current flows to 19 nA when the AFM probe approaches the top of the CNT bundle (Figure 3a). The generation of current prior to the application of pressing force is due to the initial deformation of the CNTs forming the bundle [7]. Then, the current value varied from 0 to 21 nA, when the pressing force on the CNT bundle varied from 0 to 0.4 μN, respectively (Figure 3a). The piezoelectric response of modified CNTs decreased in the amount of current flowing during the strain of the CNT by more than 3 times as compared to pure CNTs (Figure 3b). Thus, there was no current when the AFM probe was approached to modified CNTs due to a decrease in the deformation of the CNTs forming bundle (Figure 3b). Then the current value reached 6 nA when the pressing force was 0.4 μN (Figure 3b). This fact is also associated with an increase in the stiffness of the CNT coated with ZnO films and as a consequence a decrease in the value of the CNT deformation at the same pressing force.

In addition, it was experimentally established that the modification of CNT with ZnO films leads to a decrease in the memristor effect of (Figure 4). The ratio between high- and low-resistance states of carbon nanotubes obtained before and after the modification decreased from 32 to 1, respectively (Figure 4). It should be noted that the saturation section on the CVCs is related to the features of the Ntegra measuring system in the current spectroscopy mode (built-in current preamplifier after 25 nA).

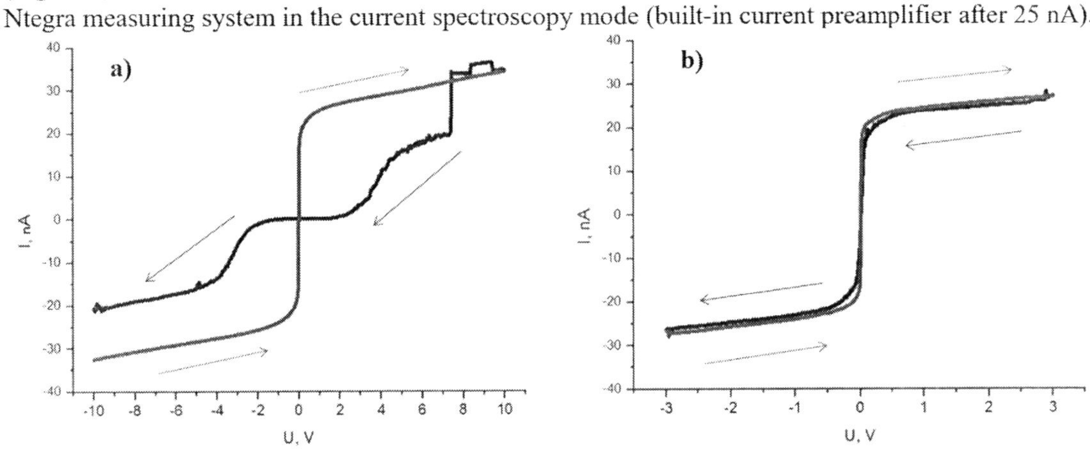

Figure 4 (a, b). CVCs of the CNT bundle before **(a)** and after **(b)** the surface modifying with ZnO films.

The decrease in the memristor effect of CNT after surface modification with ZnO films is due to an increase in the stiffness and consequently a decrease in the deformation and piezoelectric response of the nanotubes after the coating [8, 14, 15]. Thus, it has been shown that surface modifying of the CNTs with a ZnO film does not enhance the memristor effect of CNT due to the ZnO piezoelectric properties, but, on the contrary, reduces this effect by reducing the mobility and deformation of the CNT.

The obtained results can be used in the development of promising nanoelectronics devises based on aligned carbon nanotubes in particular memory elements and high-efficiency nanogenerators. The results were obtained using the equipment of the Research and Education Center and Center of Common Using "Nanotechnologies" of Southern Federal University.

4. Acknowledgments

The reported study was funded by RFBR according to the research projects No. 16-29-14023 ofi_m, No.18-29-11019 mk and Internal grant of the Southern Federal University (project №VnGr-07/2017-26).

References

[1] Meindl J D., Chen Q and Davis J A 2001 *Science* **293** (5537), 2044-2049
[2] Che Y, Chen H, Gui H, Liu J, Liu B and Zhou C 2014 *Semicond. Sci. Tech.* **29** (7), 073001
[3] Wang Z L 2007 *Adv. Mater.* **19**, 889–892
[4] Wang Z L 2012 *Adv. Mater.* **24**, 4632–4646
[5] Zhang J, Wang C and Bowen C 2014 *Nanoscale* **6**, 13314–13327
[6] Kundalwal S I, Meguid S A and Weng G J 2017 *Carbon* **117**, 462–472
[7] Il'ina M V, Il'in O I, Blinov Y F, Konshin A A, Konoplev B G and Ageev O A 2018 *Materials* **11**, 638
[8] Il'ina M V, Il'in O I, Blinov Yu F, Smirnov V A, Kolomiytsev A S, Fedotov A A, Konoplev B G and Ageev O A 2017 *Carbon* **123**, 514–524
[9] Il'in O I, Il'ina M V, Rudyk N N, Fedotov A A, Levshov D I and Ageev O A 2018 *IOP Conf. Series: Journal of Physics: Conf. Series* **1038** 012062
[10] Il'in O I, Il'ina M V, Rudyk N N, Fedotov A A and Ageev O A 2018 *Nanosystems: Phys. Chem. Math.* **9** (1), 92–94
[11] Klimin V S, Il'ina M V, Il'in O I, Rudyk N N and Ageev O A 2017 *IOP Conf. Series: Journal of Physics: Conf. Series* **917**, 092023
[12] Vakulov Z E, Zamburg E G, Khakhulin D A and Ageev O A 2017 *Mater. Sci. Semicond. Process* **66**, 21–25
[13] Ageev O A, Il'in O I, Rubashkina M V, Smirnov V A, Tsukanova O G and Fedotov A A 2015 *Tech. Phys.* **60** (7), 1044–1050
[14] Il'ina M V, Il'in O I, Blinov Yu F, Smirnov V A and Ageev O A 2018 *Tech. Phys.* **63** (11), 1672–1677
[15] Ageev O A, Blinov Yu F, Ilina M V, Ilin O I, Smirnov V A 2016 *IOP Conf. Series: Journal of Physics: Conf. Series* **741** 012168

The effect of ion implantation and annealing on forming process in $Al_2O_3/HfO_2/Al_2O_3$ memristor structure

O O Permyakova[1,2], **A V Miakonkikh**[1], **K V Rudenko**[1], **A E Rogozhin**[1]
[1]Laboratory of Microstructuring and Submicron Devices, Valiev Institute of Physics and Technology RAS, Moscow 117218, Russia
[2]Department of Physical and Quantum Electronics, Moscow Institute of Physics and Technology (State University), Dolgoprudny 141701, Russia

o.permyakova@phystech.edu

Abstract. In this paper, we study the effect of annealing and plasma immersion ion implantation of He^+ conditions on forming voltage for the $Al_2O_3/HfO_2/Al_2O_3$ memristive structures. Planar memristor structures were obtained by PEALD. It was shown that He^+ implantation and annealing lead to decrease of forming voltage but also increase the dispersion of the set voltage of the memristive structures.

1. Introduction

One of the most prominent types of the next-generation non-volatile memory is resistive switching memory (ReRAM). It has the following advantages: high-speed operation, low power consumption, comparatively simple structure and amazing scalability. Typically, ReRAM cell has metal-insulator-metal (MIM) structure, different materials from metal oxides to various organic compounds can act as insulators.

The main property of ReRAM cells is the ability to switch between two stable resistance states: high resistance state (HRS) and low resistance state (LRS). Switching between these states is accomplished by applying voltage: V_{SET} to switch from HRS to LRS (SET process) and V_{RESET} to switch from LRS to HRS (RESET process). One of the hypotheses explains the possibility of resistance switching in the oxide based structures as the formation of a conductive filament (CF) between metal electrodes that consists of oxygen vacancies. Therefore, the first action before measuring the characteristics of the memristor structure is the formation of such a filament upon a electroforming voltage ($V_{FORMING}$). During this process, the current in the structure increases abruptly, that causes problems of using structures in nano- and microelectronics due to large current. Hence, it is important to either reduce the forming voltages or create forming-free structures [1, 2]. It was shown that the $Al_2O_3/HfO_2/Al_2O_3$ stack-based structure demonstrates the properties required in ReRAM [3].

Several ways to improve the ReRAM structure characteristics by incorporation of defects were proposed. For instance, annealing in different media (vacuum, oxygen, inert gases) [4] and temperature [5] or focused ion beam processing [6] were suggested. In this paper we present the investigation of the effect of post-metal annealing in nitrogen atmosphere and pre-metal He^+ ion implantation on the forming voltage for the $Al_2O_3/HfO_2/Al_2O_3$ structure.

2. Experiment

The Al_2O_3(6nm)/HfO_2(10nm)/Al_2O_3(3nm) structures were formed by plasma enhanced atomic layer deposition (PEALD) on the Si wafers covered by sputtered TiN. After that structures were coated Pt (10 nm) and Ni (50 nm) through the shadow mask. Then some of the structures were annealed in the nitrogen atmosphere at 450°C for 300 s. For other samples pre-metal plasma immersion ion implantation of He^+ was used to incorporate defects in the dielectric stack structure. The energy of He^+ ions was 2 keV. The dose value was 10^{16} cm^{-2}. After implantation some structures were annealed. Planar structure with common electrode were investigated (fig. 1).

Current-voltage characteristics were measured using semiconductor characterization system Keithley 4200-SCS and Micromanipulator probe station. Each showed structure was measured at least one hundred cycles with a 2.5 V stop voltage and a current compliance for the SET process 1 mA. The electrode area was measure with an interference microscope Zygo NW5000. The areas of the top electrodes were $2.01 \pm 0.03 \cdot 10^4$ μm^2 for the initial structure (fig. 2), $1.67 \pm 0.02 \cdot 10^4$ μm^2 for the structure annealed in an atmosphere of nitrogen, and $3.48 \pm 0.03 \cdot 10^4$ μm^2 for the structure after ion implantation.

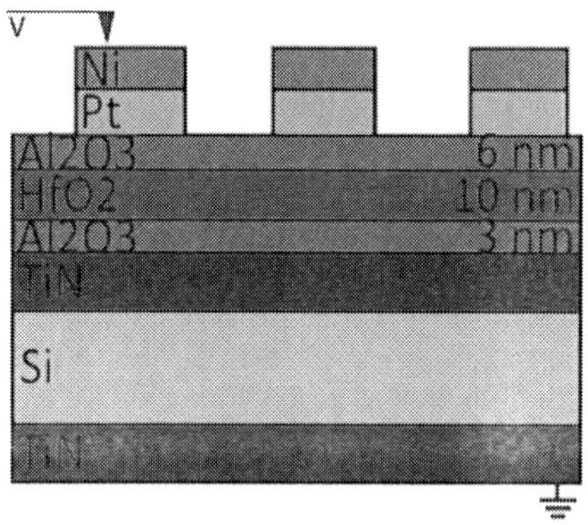

Figure 1. Al_2O_3/HfO_2/Al_2O_3 planar structure with common electrode.

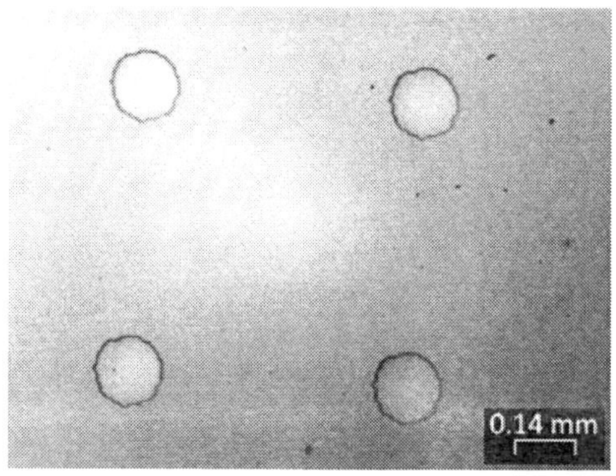

Figure 2. Interference microscope image for initial structure.

3. Result and discussion

For each type of structure (initial, annealed in a nitrogen atmosphere, after plasma immersion ion implantation), up to 20 devices were formed. The main parameters for these devices are presented in the Table 1. The smallest forming stress is achieved for the structures after ion implantation, but at the same time for the structures the yield is lower. This means that more devices irreversibly breakdown through the forming process. Simultaneously, that forming voltage includes in dispersion of forming voltage for structure annealed in nitrogen atmosphere. Besides structures have showed similar RESET voltages.

Table 1. Mean voltages of structures

	Forming voltage, V	SET voltage, V	RESET voltage, V	Yield
Initial	13.1 ± 0.8	-2.3 ± 0.3	1.5 ± 0.3	0.88
Nitrogen atmosphere	10.9 ± 0.6	-2.1 ± 0.5	1.5 ± 0.2	0.72
Ion implantation with dose 10^{16} cm^{-2}	10.6 ± 0.2	-4.0 ± 0.9	1.4 ± 0.3	0.74

The following figures (fig.3-5) show values for the HRS and LRS at a voltage of -0.1 V for 100 cycles, with a stop voltage of 2.5 V. Apparently, a high variance for both states, and therefore high variability from cycle to cycle for a single cell, but this behavior is typical for the first hundred cycles, as indicated in the article [2]. Memristor structure after ion implantation has not the finest characteristics, such as RESET voltage and yield, among the presented structures (table 1) including cycle-to-cycle variability (fig.5). There are also problems with switching between resistance states.

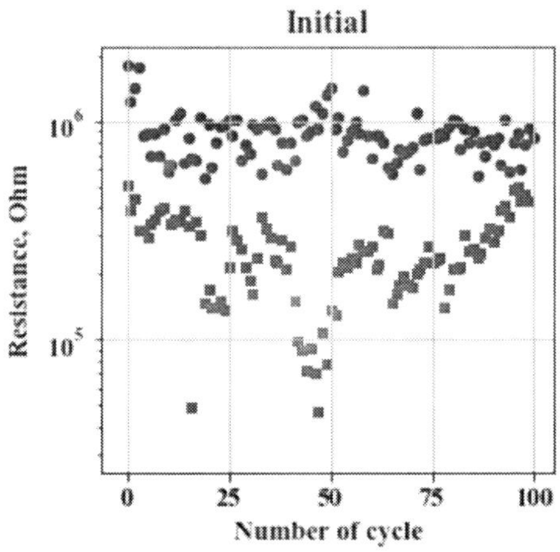

Figure 3. Resistance state of initial structure. Blue circle – low resistance states (LRS), red square – high resistance states (HRS).

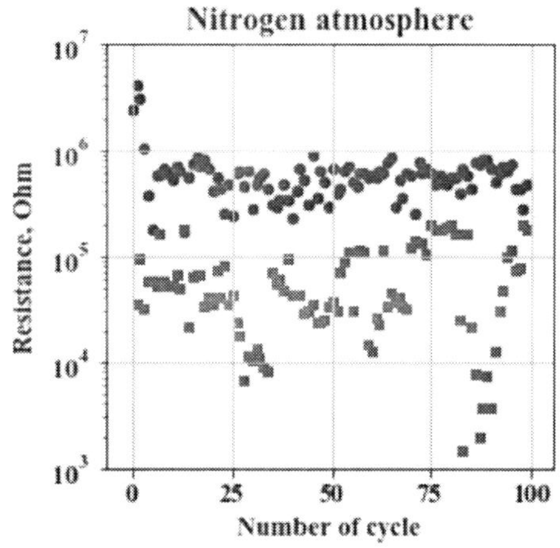

Figure 4. Resistance state of annealed structure in nitrogen atmosphere. Blue circle – LRS, red square – HRS.

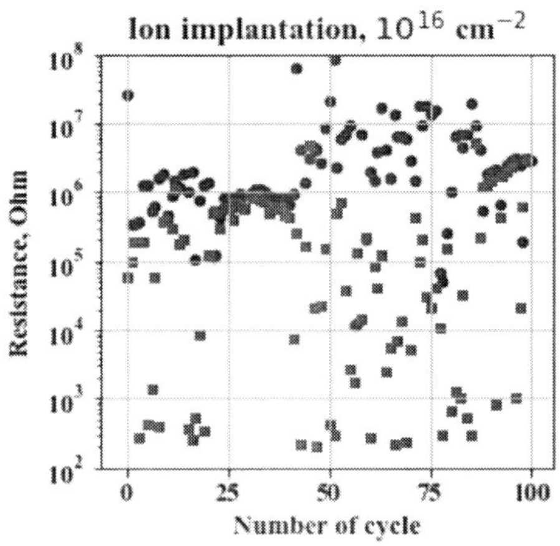

Figure 5. Resistance state of structure after ion implantation with dose 10^{16} cm^{-2}. Blue circle – LRS, red square – HRS.

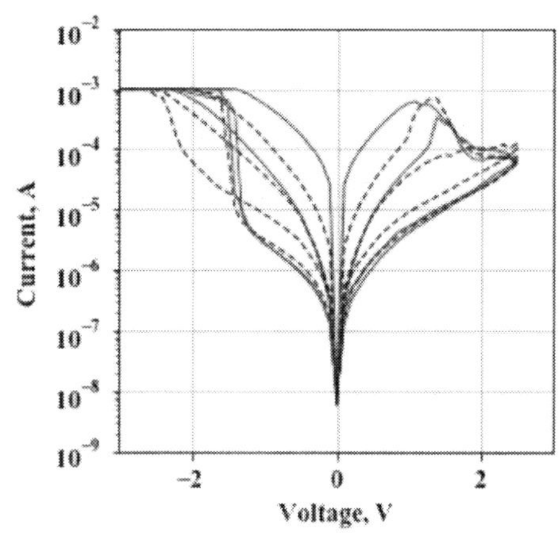

Figure 6. Current-voltage characteristics. Red solid line – initial structure, blue dashed line – structure annealed in nitrogen atmosphere

The hysteresis of the initial structure (fig. 3) decreases after a one hundred cycles, however, with increasing stop voltage the hysteresis increases again. But reference structure has a resistance ratio of less than ten, which is necessary for applications. Post-processing of the structure allows to increase the resistance ratio, consequently after annealing, the value of the high resistance slightly decreases, while the low resistance decreases almost by an order of magnitude, which makes it possible to increase the resistance ratio. However, from the figure 6 it can be seen that in the structure cycle-to-cycle variability is not the best. For some devices the ratio of resistance lowered after the post-annealing. Besides, there are two types of hysteresis, that seems to be associated with different types of conductivity, it is clearly seen for the hysteresis of the initial cell, that can be associated with a CF that is not fully formed due to a significant change in resistance in the LRS and a change in slope (type of conduction) voltage-current characteristics.

Figures 7 and 8 showed conventional view of the cycle-to-cycle variability in the current-voltage characteristics. It seems, that the annealed structure has larger hysteresis, but greater variation in SET voltage, but on average their values are close to 1.5 V (Fig.9).

Basically, there is no spread of RESET voltage in the main and the cumulative functions. That allows us to say that post-processing mainly affects the formation of the conductive filament and its properties.

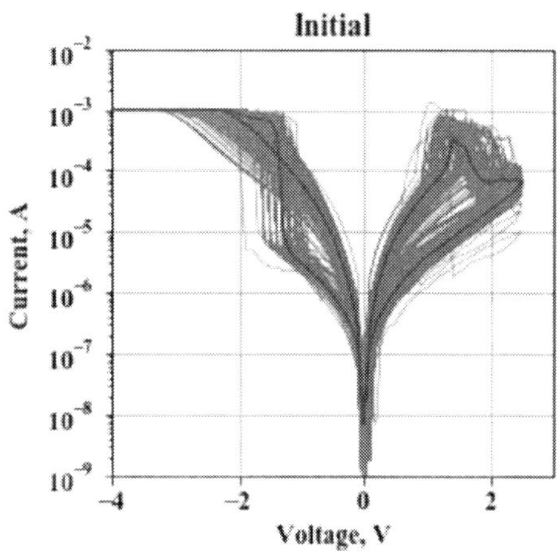

Figure 7. Current-voltage cycle-to-cycle variability characteristic of initial structure.

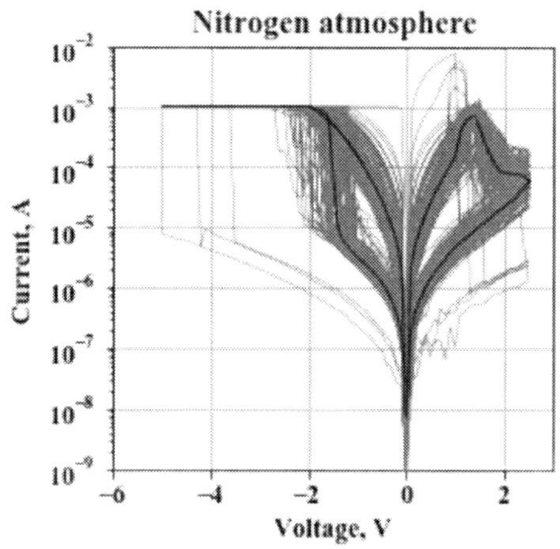

Figure 8. Current-voltage cycle-to-cycle variability characteristic for structure annealed in nitrogen atmosphere.

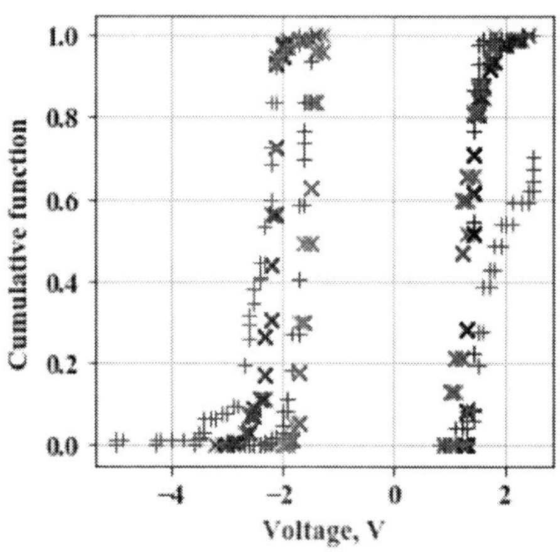

Figure 9. Cumulative function of switch voltages. SET voltages in the left and RESET voltages in the right. '+' - structures annealed in nitrogen atmosphere; 'x' – initial structures.

4. Conclusion

The effect of annealing and plasma immersion ion implantation of He^+ conditions on forming voltage for the $Al_2O_3/HfO_2/Al_2O_3$ memristive structures was studied. Planar memristor structures were obtained by PEALD. We have investigated how post-processing influence the forming voltage and further functioning of the structures. It appeared that forming voltage decreases after post-processing. In the case of ion implantation structures have large cycle-to-cycle variability and in the case of annealing structures have larger range of SET voltages. In the same time, average RESET voltage is the same for measured memristor structures.

Acknowledgments

The reported study was funded by RFBR according to the research project № 18-37-20076.
The investigation was also supported by the Program no. 0066-2019-0004 of the Ministry of Science and Higher Education of Russia for Valiev Institute of Physics and Technology of RAS.

References

[1] Ielmini D, Waser R 2016 *Resistive Switching* (Weinheim: Wiley-VCH)
[2] Mario Lanza et al. 2019 *Advanced Electronic Materials* **5** 1800143
[3] Lai-Guo Wang et al. 2015 *Nanoscale Research Letters* **10** 135
[4] Tsung-Ling Tsai et al. 2015 *IEEE Electron Device Letters* **11** 1146
[5] Yu-Ting Su et al. 2018 *IEEE Journal of the Electron Devices Society* **6** 341
[6] Avilov V I et al. 2016 Proc. of SPIE (Zvenigorod) vol. 10224 (Moscow: Max press) 102240T

Microwave low mass-dimensional frequency standard on Hg-199 ions

N A Lukashev[1], V V Davydov[1], A P Glinushkin[2], V Yu Rud'[2] and V S Lukyantsev[2]

[1]Peter the Great Saint Petersburg Polytechnic University, Saint Petersburg 195251, Russia
[2]Russian Research Institute Phytopathology, 143050, Moscow Region, Odintsovo district, B.Vyazyomy, Russia

n-lukash@list.ru

Abstract. In the paper the design of a magnetic trap for low mass-dimensional microwave frequency standards is presented. The dependencies between the number of registered photons by PMT and magnetic field values are established. The new algorithm for optical processing is developed. The results comparison of experimental Allan deviation researches for different designs of microwave standard on $^{199}Hg+$ ions are performed

1. Introduction

Frequency standards are presently an integral element: for any high precision systems for object location determination, global time scales, signal transmission and processing, metrological systems [1-9]. Low mass-dimensional frequency standards stand a special place in this order [8-14]. These devices are applied in various technical cases (for example, ecological Earth monitoring) on land as well as in airborne applications [13-18]. During frequency standard exploitation in mobile flying vehicles some important issues may appear. They are bounded with mass-dimensional restrictions of the standard and power consumption [8-10, 16, 17, 19-24].

Besides, frequency standards, especially used for objects' location in space or on Earth, have got to posses high precision characteristics [8-15, 21-24]. The best solution for aforesaid cases is quantum frequency standards (QFS) [20-25]. These devices allow to produce frequency with high precision.

In case of lowering construction size of a QFS for solving hard tasks connected with determining coordinates of moving objects the unavoidable outcome is decreasing short-term and long-term frequency stability. That is because QFS require a high thermostabilization order of the most operational blocks. Especially in cases of implementation laser or photodetection devices [26-29].

Besides that, QFS use magnetic fields in order to split energy levels of atoms into sublevels. While applying magnetic fields thermostabilization must be achieved [30-33]. It is necessary to note that most QFS models (for instance, those based on cesium-133 or rubidium-87) are not stable to high G-force overloads. These overloads appear due to obtaining fast speed in short periods of time.

The held research by us has shown the microwave frequency standard on Hg-199 ions is the best solution in this situation. This standard has one major disadvantage to the other QFS models – big mass-dimensional parameters of magnetic system, operation and signal processing blocks. That is why for successful implementation of such a standard in moving objects it is necessary to decrease its size and mass, not affecting the precision characteristics sufficiently.

It is not a trivial task to solve, because all the blocks are interlinked with one another in a certain way. That is why the issue of standard modernization must be solved as a whole. Lower mass and size of the major blocks, develop new algorithms for automatic frequency control, magnetic field maintenance and photon registration systems. One of the solutions of this task is presented in the present work.

2. Microwave low mass-dimensional frequency standard on 199Hg+ ions

The main block of the microwave standard is a magnetic trap. That is why the directions of modernizations which lead to lowering its mass-dimensional parameters are primarily connected with reducing the size of the trap itself.

The alternating electro-magnetic field holds a required number of $^{199}Hg+$ ions in the magnetic trap in order to make them interact with λ = 194.2 nm radiation and emit a certain number of photons which then has to be registered by a PMT (Photomultiplier tube). Long-term and short-term stabilities are straight out connected a quality of ions' trapping. Alongside with lowering mass and dimensions of the magnetic trap, electrode sizes for producing magnetic field are decreasing as well. That is why the requirements for parameter stabilization of the operating signal, used for forming magnetic field, are increasing vastly. In order to solve such a task we developed a new block for controlling electromagnetic field that provides a proper trapping of charged particles.

We have developed a new magnetic trap construction for the low mass-dimensional microwave frequency standard. A new automatic frequency control system has been developed for guiding the trap through managing the frequency of the quartz generator and the supply voltage of the power cascades.

Charged particles in the trap are held with an effective potential:

$$V_{eff} = \frac{n^2}{4}\frac{V_0^2}{r_0^2\Omega^2}\frac{q^2}{m}r^{`2n-2} \quad (1)$$

Trapping a particle with a single charge q shall be limited by mass m: at the upper band – by a depth of effective potential and at the lower band – by a mean duration of an ion trapping. The low edge of the operational frequency Ω is determined as:

$$\Omega \geq \Omega_{min} = \frac{n-1}{3r^`}\frac{\sqrt{E_m}}{r_0}\frac{\sqrt{m_{max}}}{m_{min}} \quad (2)$$

where n – number of electrodes, r` - normed trap radius, E_m – kinetic energy of particles, r_0 – trap radius, m_{min} and m_{max} – minimum and maximum particle mass

This has allowed to fulfill an effective parameter control of the magnetic trap, judging by the analysis of the photon counter data, even in relatively rough conditions (for instance overpressure etc.). The circuit diagram for supplying the trap is show at the Figure 1:

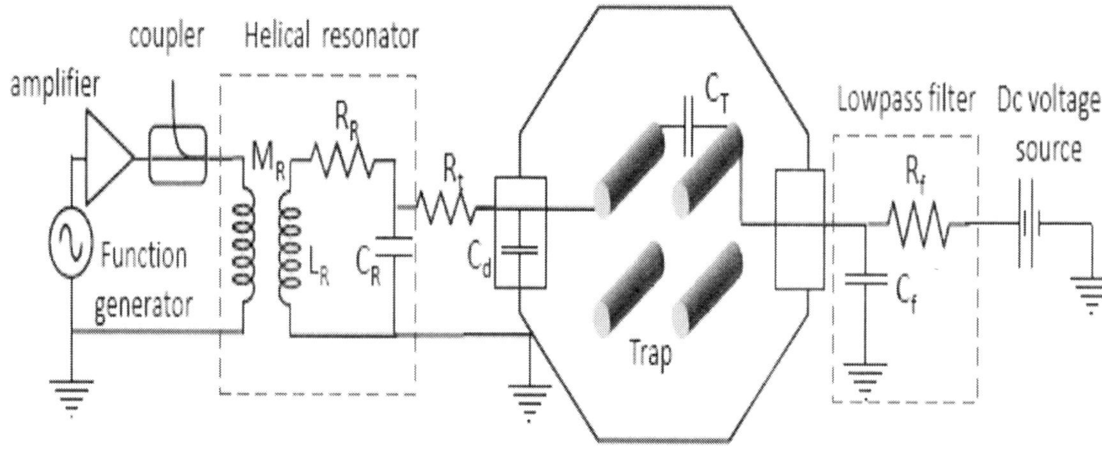

Figure 1. Circuit schematic diagram of the supplied trap

Oscillator signal is fed and amplified through the compact driver and transformation coupling to the trap electrodes. Highly stable DC voltage is applied as well in order to compensate the axial component and hold the ion assembly in the trap region.

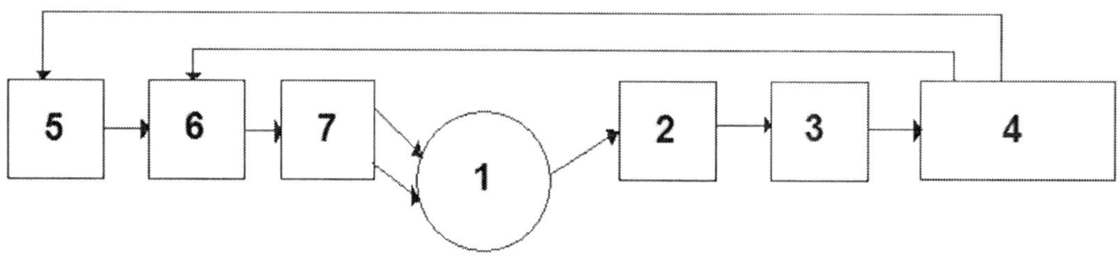

Figure 2. Structure scheme for driving magnetic trap: 1 – Magnetic trap, 2 – Photomultiplier tube, 3 – Photon counter, 4 – Controller, 5 – Externally controlled highly stable quartz generator, 6 – Driving system and power cascades, 7 – Output transformer.

The generator 5 supplies the trap through the driver 6 and the transformer 7. Leaked photons from the trap are registered by the PMT that sends amplified signal to the Photon counter 3, programmed with VHDL language and set to precisely tract the number of leaked photons through different periods of time. The whole control loop is managed by the MCS-51 controller unit that receives signals from the Photon counter and produces control codes for the synthesizer, Automatic frequency control (AFC) generator, and the highly stable DC unit in order to operate values in real time. In other words, th negative feedback system is implemented.

Received experimental data has shown that a low mass-dimensional magnetic trap with the implemented control system ensures a solid operation with a number of registered photons from 10^4 to $5 \cdot 10^5$ and with the next parameters: frequency Ω shall be placed in the range: from 0.74 to 1.60 MHz, its amplitude V_0 from 120 to 200 V and kinetic energy of particles E_m from 0.2 to 10 eV.

3. The results of the experimental research

The held research has shown that spectrums of the driving voltages in the new magnetic trap construction have no sufficient perturbations which may affect the stability of operation. The experiments show the short-term stability improvement by Allan deviation. On the fig. 3 the measurement results are presented.

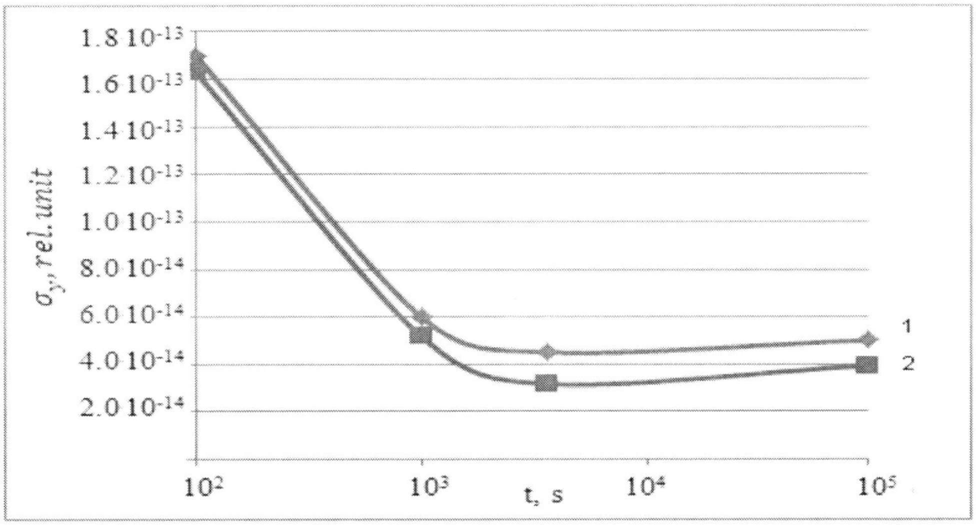

Figure 3. The Allan deviation: graph 1 corresponds to the previous microwave standard construction, graph 2 – the newly developed design by us.

Shown on the fig. 3 value of Allan deviance has been calculated according to the given formula:

$$\sigma_y^2 = \frac{\sum_i^n \sigma_{0i}^2}{2(n-1)} \quad (3)$$

$$\sigma_{0i} = \frac{f_{i+1} - f_i}{f} \quad (4)$$

The analysis of the acquired data on the fig. 1 shows that the implemented technical solutions have provided us the way to improve Allan deviance of 20 %. Long-term stability in the new low mass-dimensional construction did not experience a decrease.

4. Conclusion

The received experimental data has shown that while decreasing construction size of the standard in 3 times and mass by more than 60 % the suggested technical solutions allow us to save and improve the required precision characteristics in order to offer a solid operation of navigational devices and communication systems in mobile flying vehicles.

Acknowledgments

This work partly supported by Grant of RFBR No 18-29-25071.

References

[1] Riehle F 2004 *Frequency standards. Basics and Applications* (New Jersy: Wiley-VCH)
[2] Oduan K and Gino B 2002 *Chronometry and basics of GPS 400* (New Jersy: Wiley-VCH)
[3] Semenov V V, Nikiforov N F, Ermak S V and Davydov V V 1991 *Soviet journal of Communications Technology and Electronics* **36** 59 – 63
[4] Davydov V V, Ermak S V, Karseev A U, Nepomnyashchay H K, Petrov A A and Velichko E N 2014 *Lecture Notes in Computer Science (including subseries Lecture Notes in Artificial Intelligence and Lecture Notes in Bioinformatics)* **8638 LNCS** 694-702
[5] Davydov V V, Sharova N V, Fedorova E N, Gilshteyn E P, Malanin K Y, Fedotov I V, Vologdin V A and Karseev A Yu 2015 *Lecture Notes in Computer Science (including subseries Lecture Notes in Artificial Intelligence and Lecture Notes in Bioinformatics)* **9247** 712-721
[6] Davydov V V, Myazin N S and Davydova T I 2017 *Russian Journal of Nondestructive Testing* **53(7)** 520-529

[7] Lenets V.A., Tarasenko M.Yu., Davydov V.V., Rodygina N.S. and Moroz A.V. Journal of Physics: Conference Series. 2018. Vol. 1038 (1) 012037

[8] Glazov A I., Grigor'ev V V, Kravtsov V E, Mityurev A K, Svetlichnyi A B, Savkin K B and Tikhomirov S V 2008 *Measurement Techniques* **51** 1064 – 1070

[9] Kolmogorov O V, Shchipunov A N, Prokhorov D V, Donchenko S S, Buev S G, Malimon A N, Balaev R I and Fedorova D M 2017 *Measurement Techniques* **60** 901 – 905

[10] Petrov A A and Davydov V V 2015 *Lecture Notes in Computer Science (including subseries Lecture Notes in Artificial Intelligence and Lecture Notes in Bioinformatics)* **9247** 739-744

[11] Pakhomov A A 2007 *Journal of Communications Technology and Electronics* **52** 1114 – 1118

[12] Petrov A A, Davydov V V and Myazin N S 2017 *Lecture Notes in Computer Science (including subseries Lecture Notes in Artificial Intelligence and Lecture Notes in Bioinformatics)* **10531 LNCS** 561-568

[13] Nazarov L E and Golovkin I V 2017 *Journal of Communications Technology and Electronics* **52** 1125 – 1129

[14] Petrov A A and Davydov V V 2017 *Journal of Communications Technology and Electronics* **62** 289 – 293

[15] Balaev R I, Malimon A N, Fedorova D M, Kurchanov A F and Troyan V I 2017 *Measurement Techniques* **60** 806 – 812

[16] Rykin E V, Moroz A V, Smirnov K J, Davydov V V and Yushkova V V 2018 *MATEC Web of Conference* **245** 12002

[17] Petrov A A, Vologdin V A, Davydov V V and Zalyotov D V 2015 *Journal of Physics: Conference Series* **643(1)** 012087

[18] Lukashev N A, Petrov A A, Davydov V V, Grebenikova N M and Valov A P 2018 ***Proceedings of 18th International conference of Laser Optics ICLO-2018*** **(Saint-Petersburg) 8435889** p. 27

[19] Myazin N S, Neronov Yu I, Dudkin V I, Davydov V V and Yushkova V V 2018 *MATEC Web of Conferences* **245** 11013

[20] Petrov A A, Davydov V V and Grebenikova N M 2018 *Journal of Communications Technology and Electronics* **63** 1281-1285

[21] Baryshev V N, Aleynikov M S, Osipenko G V and Blinov I Yu 2018 *Quantum Electronic* **48** 443-447

[22] Baryshev V N, Kupalov D S, Novoselov A N, Alenikov M S, Boiko A I, Pal'chikov V G and Blinov I Y 2017 *Measurement Techniques* **59** 1286-1290

[23] Blinov I Y, Boiko A I, Domnin Y S, Kostromin V P, Kupalova O V and Kupalov D S 2017 *Measurement Techniques* **60** 30-36

[24] Petrov A A, Shabanov V E, Zalyotov D V, Davydov V V, Bulyanitsa A L and Shapovalov D V 2018 *Proceedings of the 2018 IEEE International Conference on Electrical Engineering and Photonics, EExPolytech 2018* (Saint-Petersburg) **8564389** p. 52-55

[25] Petrov A A, Davydov V V and Grebenikova N M 2018 *Lecture Notes in Computer Science (including subseries Lecture Notes in Artificial Intelligence and Lecture Notes in Bioinformatics)* **11118 LNCS** 641-648

[26] Davydov R V, Antonov V I and Moroz A V 2018 *Proceedings of the 2018 IEEE International Conference on Electrical Engineering and Photonics, EExPolytech 2018* (Saint-Petersburg) **8564378** p. 236-239

[27] Myazin N S, Smirnov K J, Davydov V V and Logunov S E 2017 *Journal of Physics: Conference Series* **929(1)** 012080

[28] Logunov S E, Davydov V V, Vysoczky M G, Koshkin A Y and Rud' V Yu 2017 *Journal of Physics: Conference Series* **917(5)** 052058

[26] Davydov R V and Antonov V I 2017 *Journal of Physics: Conference Series* **929(4)** 012040

[27] Petrov A A, Grebenikova N M, Lukashev N A, Davydov V V, Ivanova N V, Rodugina, N S and Moroz A V 2018 *Journal of Physics: Conference Series* **1038(1)** 012032

[28] Smirnov K J, Davydov V V, Glagolev S F, Rodygina N S and Ivanova N V 2018 *Journal of Physics: Conference Series* **1038 (1)** 012102
[29] Davydov R V and Antonov V I 2017 *Journal of Physics: Conference Series* **929 (1)** 012040
[30] Myazin N S, Davydov V V, Yushkova V V and Rud V Yu 2018 *Journal of Physics: Conference Series* **1038 (1)** 012088
[31] Davydov V V 1999 *Russian Physics Journal* **42(9)** 822-825
[32] Myazin N S, Davydov V V, Yushkova V V, Davydova T I and Rud' V Yu 2017 *Journal of Physics: Conference Series* **917(4)** 042017
[33] Logunov S E, Koshkin A Yu, Davydov V V and Petrov A A 2016 *Journal of Physics: Conference Series* **741 (1)** 012092

Features of transmission bearing and heterodyne receivers for signals in fiber-optic communication line in active phased array antenna

A V Moroz[1] and V V Davydov[1, 2]

[1]Higher School of applied physics and space technologies, Peter the Great Saint Petersburg Polytechnic University, Saint Petersburg, 195251, Russia
[2]All Russian Research Institute of Phytopathology, Moscow Region 143050, Russia

Abstract: The article discusses the problems arising from the modernization of active phased antenna arrays (for example, an increase in the number of transceiver active elements, a decrease in the weight and size of the antenna system, etc.). It has been substantiated that the most rational solution of these problems is the use of fiber-optical communication systems for transmitting heterodyne signals. A new design of the transmit-receive module with a fiber-optic transmission system for an active phased antenna array has been developed. The results of experimental investigations are presented.

1. Introduction

In most cases, the operation of radar stations (RLS) is carried out in conditions of high density of interference of various kinds. Therefore, communication and control channels in radar are protected by various screens (passive and active) [1-6]. The degree of protection of communication channels and control is provided depending on the functional tasks. These tasks should be solved using the signals that are transmitted through them. But it still does not significantly reduce the weight and dimensions of coaxial cables [1-3, 5-11].

The greatest difficulties in radar arise when transmitting signals reflected from various targets from the antenna to processing devices [1, 3, 5, 7, 10-13]. To solve this problem, as well as more convenient processing of the microwave signals reflected from various targets, received by the antenna in radars, superheterodyne radios are used. This type of radio receivers is based on the principle of converting the received signal into a fixed signal of an intermediate frequency with its subsequent amplification. The main advantage of the superheterodyne receiver is the absence of the need for tuning to different frequencies. To the local oscillators make high demands on the stability of the frequency and amplitude, as well as the spectral purity of harmonic oscillations. In the data transmission channel, the parameters of a heterodyne signal in different operating conditions of the radar (temperature differences, the presence of interference of various kinds, etc.) should not change [1, 3, 5, 10, 11, 13].

Especially many problems with maintaining the parameters of the heterodyne signal arise when placing the radar or receiving-transmitting station on an aircraft (for example, satellites, aircraft, helicopters, etc.) [2, 3, 8, 14-18]. In this case, in addition to external interference, interference associated with interference from neighboring channels, etc., begins to play a significant role. These tasks are solved depending on the design of antenna systems. The most difficult are technical solutions in active phased array antennas (APAA) for radars operating in the S and X ranges. APAA has a very high density of electronic components and feeder paths. The most sensitive to interference in APAA are the feeder paths of superheterodyne receivers. To reduce the effect of induced interference on the feeder paths of superheterodyne receivers, their "vulnerable" spots in APAA are covered with absorbing materials. Operating experience of APAA shows that this is a temporary solution to the problem. Absorbent coatings for various reasons fail, lose their integrity during repairs, etc.

Therefore, for each receiving-transmitting element APAA uses its own local oscillator. This partially allows to reduce the pickup of the feeder paths at each other in conditions of high density of the location of channels. Such a construction of the transmission channels of heterodyne signals led to the fact that in the existing designs of APAA there is no space for new nodes.

Currently, new absorbing coatings are used in the manufacture of various objects that need to be detected using radar. In this connection, modernization of the APAA in operation is required. One of the most important directions of this modernization is the development of an APAA with the number of transceiver elements more than 1000 while maintaining the weight and size characteristics of the antenna. The dimensions of the transceiver elements can be reduced and their characteristics retained using new materials in their manufacture. But with the placement of additional feeder paths, especially superheterodyne receivers in APAA, great difficulties arise.

One of the solutions to this problem is considered in our article. We propose to replace feeder paths transmitting microwave signals with fiber optic transmission systems (FOTS) in APAA [10-12, 17-25]. Optical fiber is not affected by high-frequency interference and has good noise characteristics when transmitting high-frequency signals, it has high flexibility and low mass [20-27].

2. Features of the transmission of heterodyne signals on a fiber-optic system and its design.

In the case of heterodyne signals for transmission in FOTS radar APAA necessary in its design to take into account a number of features and calculate the following parameters: phase noise, the signal / noise transmission path temporal stability and temperature stability.

The values of the frequencies of heterodyne signals are small (no more than 100 MHz) compared with the frequency of the radar emission (more than 8 GHz). Therefore, nonlinear effects in an optical fiber and dispersion losses can be neglected. This is one of the features of the transmission of these signals to APAA.

In addition, another distinctive feature of the transmission of heterodyne signals is the small length of the optical fiber L (not more than 15 m). This circumstance also makes it possible to neglect the previously noted nonlinear effects. It should be noted that at such distances L the requirements for the optical transmitting module and photodetector are not high in terms of input power, conversion loss, speed, conversion coefficient when transmitting these signals at frequencies less than 100 MHz compared to when FOTS transmits microwave signals from 40 GHz frequency [4, 7, 8, 11, 12].

The use of FOTS has allowed us to develop a new scheme for the transmission of heterodyne signals to receiving modules that are each connected to an APAA receiving element. In figure 1 shows the structural scheme developed by us.

The proposed design of the FOTS allows you to generate a signal using a single local oscillator. Further, it is converted into an optical modulated signal, which can be divided into a specified number of receiving and specifying modules using an optical divider. The only condition for this transmission implementation is the choice of optical transmitting (with a given output power level) and receiving (minimum input optical power) modules, so that with a large number of N channels (optical divider 1/N) the information is lost.

With this construction of a FOTS with an optical divider, the intrinsic noise factor of a channel on a K_n line becomes an important characteristic. When transmitting heterodyne signals, it limits the number of channels into which the optical signal can be divided. The measurements performed showed that K_n in the channel of the developed FOTS is less than 4 dBm. The intrinsic noise of the feeder path for transmitting the heterodyne signal to APAA is not lower than 6 dBm. This shows that the new design not only significantly increases the free space in the APAA unit, but also improves the characteristics of the transmission channel by more than 30 %.

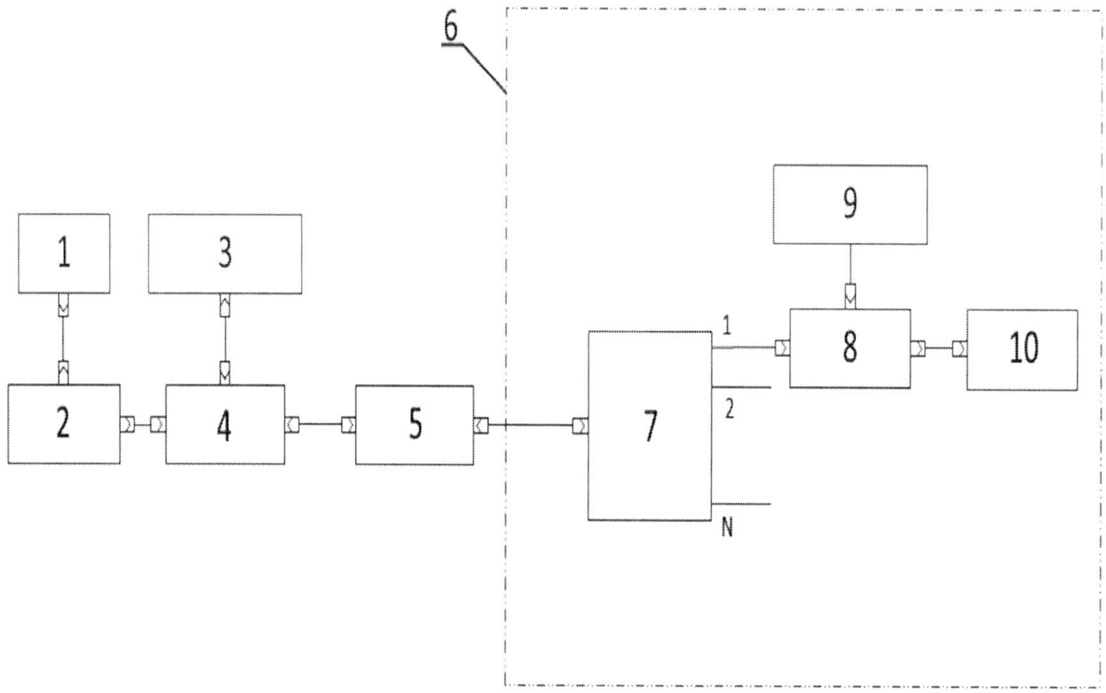

Figure 1. Structural diagram of a fiber-optic transmission system: 1 - local oscillator power supply; 2 is a local oscillator; 3 - power driver; 4 - transmitting laser module; 5 - optical isolator; 6 - block housing AFAR; 7 - optical divider; 8- receiving optical module; 9 - power driver; 10 - receiving and transmitting module APAA.

In addition, we found that in the newly developed FOTS design, the dependence of the signal power at the system output on the signal power at its input is linear from –128 dBm to –22 dBm. This allows us to conclude that the dynamic range (DR) developed by FOTS with an optical divider exceeds 105 dBm. This value of DR is sufficient for transmitting heterodyne signals via the FOTS. Standard DR feeder paths is not more than 110 dBm. One of the main characteristics of the heterodyne communication channel in APAA has not deteriorated.

It should be noted that the channels of the transceiver modules, where the heterodyne signal enters, are optically isolated. This eliminates possible interferences from one another, and also does not require additional screening, as in feeder paths. The latter also frees up space for the introduction of additional elements in the design of APAA.

The use of the developed FOTS in the design of AFAR imposes a number of features on the construction of the receiving-transmitting module (RTM) scheme. In figure 2 shows the structural diagram of the receiving-transmitting module developed by us.

The main feature of the new scheme is that the phase shifter and attenuator developed by us are located in one receiving and transmitting channel, when in other designs they are spaced apart along separate channels. The second feature is the use of a mixer in RTM. This allows you to combine the channel of reception and transmission in one path, which significantly saves space in the design of APAA and reduces its cost. A heterodyne signal arrives at the mixer input from the output of the FOTS. The mixer also receives a signal from the receiving antenna reflected from the target.

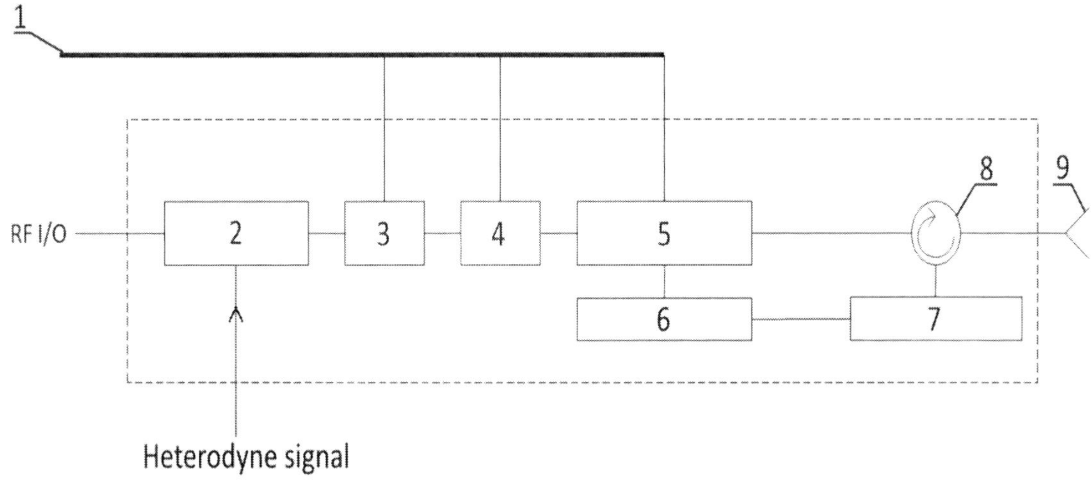

Figure 2. Structural diagram of the receiving-transmitting module: 1- control bus; 2 - mixer; 3 - phase shifter; 4 - attenuator; 5 - switch; 6 - low noise amplifier; 7 - receiver protection device; 8 - circulator; 9 - receiving and transmitting element APAA.

This allows the mixer to generate an intermediate frequency signal, which contains information about the movement of the target. This signal is transmitted to radar processing devices.

3. The results of experimental studies of the fiber-optic transmission system and their discussion. The most important characteristic that needs to be investigated when transmitting a heterodyne signal via a FOTS is the dependence of power on frequency. Let us determine the presence of distortions in the spectrum of the signal at the output of the FOTS [4, 11, 17, 18, 25, 28-32]. On the laboratory model of FOTS developed by us, possible distortions in the spectrum when transmitting a heterodyne signal via FOTS at an operating frequency of 10 MHz were investigated. In figure 3, one of the results of these studies is presented as an example.

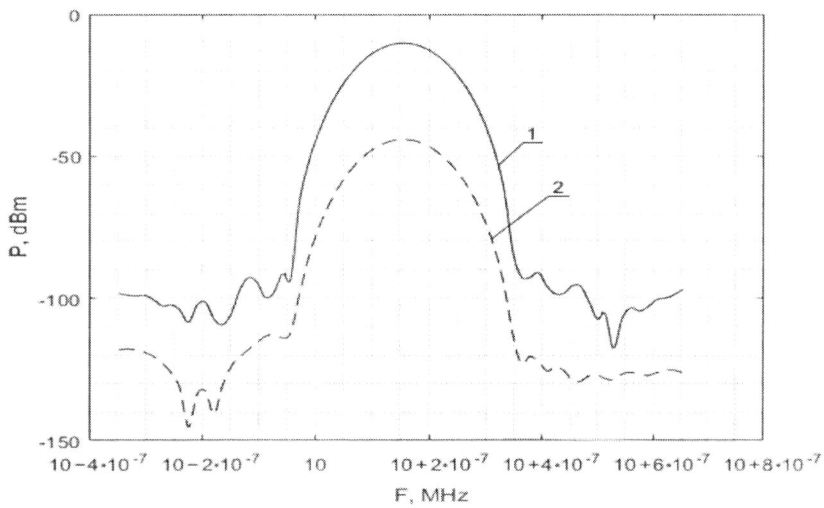

Figure 3. Spectra of a heterodyne signal at the input (graph 1) and output (graph 2) FOTS.

Comparison of the obtained spectra (figure 3) shows the high efficiency of the transmission of the heterodyne signal at a carrier frequency of 10 MHz over FOTS. Distortions in the spectrum are present only on the lateral components, which does not affect the accuracy of determining the distance to the target in the radar.

Since the developed FOTS is intended for APAA, which are located on the aircraft, it will be operated in different temperature conditions. A change in ambient temperature causes both a change in the refractive index of the fiber and additional lengthening of the fiber due to thermal expansion or contraction. This leads to a change in the phase of the light and, accordingly, to a change in the phase of modulation of the radiation transmitted through the fiber. Therefore, an experimental assessment was made of the temperature drift of the modulation phase during propagation in the fiber. In figure 4 shows the experimental dependence of the phase shift of the modulation of light $\Delta\varphi_m$ on the ambient temperature T.

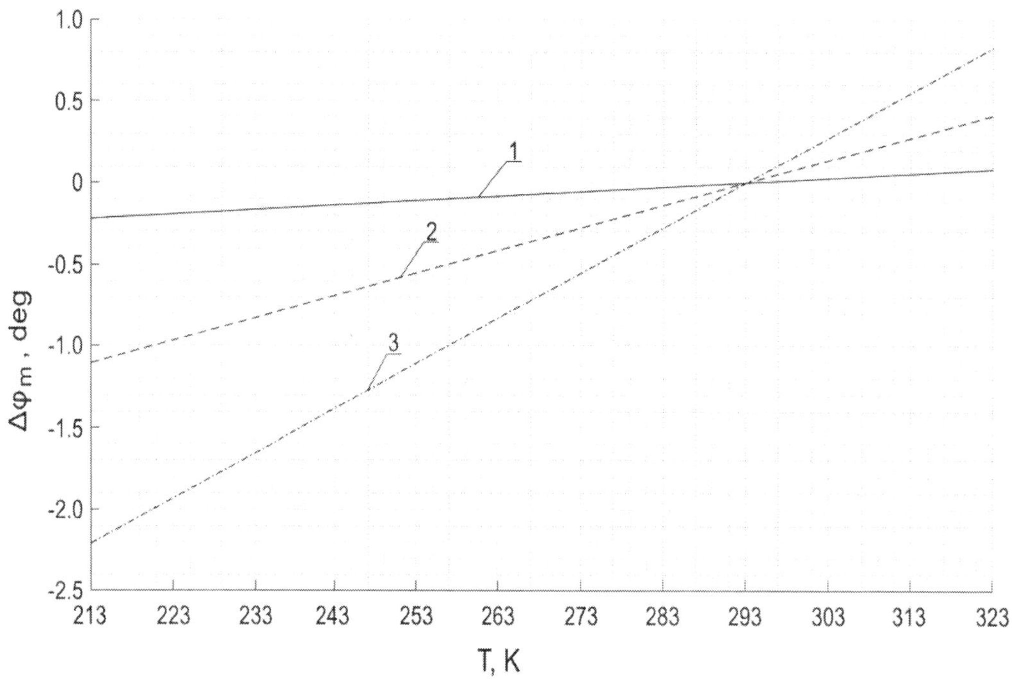

Figure 4. The phase shift modulation of the medium temperature. Graphs 1, 2 and 3 correspond to the frequency of the heterodyne signal in MHz: 10; 50; 100.

The research results showed that the temperature dependence of the modulation phase change for G.657 fiber is no more than 2 degrees in the selected temperature range from 213 to 323 K for various frequencies of the heterodyne signal.

4. Conclusion.

The obtained results showed that the FOTS design developed by us can be successfully operated as part of a radar with an APAA. It has been established that the design solutions implemented on the basis of the conducted research in the manufacture of FOTS are justified.

The use of the FOTS design developed in APAA made it possible to implement a new construction of the RTM. This allowed for a more compact placement of a large number of RTMs in APAA compared with the case when feeder paths were used. In addition, the number of used local oscillators in the new design of APAA decreased by an order of magnitude compared with the case of using feeder paths.

Since optical dividers and fiber do not need to be protected from electromagnetic radiation, a part of the channel for transmitting a heterodyne signal with these elements can be placed behind the APAA housing from its rear part (the fiber length of 5 m allows you to do this). In this case, additional space is made available in the APAA housing.

The technical solutions we proposed on the basis of the conducted research allowed us to increase the number of transceiver elements in APAA to 1024 and to ensure reliable operation of the antenna complex while maintaining its weight and size characteristics within specified limits.

It should also be noted that with an increase in the number of channels in the optical divider, K_n in each channel increases. This fact together with a decrease in the optical signal power in each channel is a limitation on the increase in N in the optical divider during the development of FOTS. As a result of research, it was found that the most optimal at present is the use of an optical divider with 10 channels in APAA. This allows the use of a total of 103 local oscillators in the APAA design. Previously, 256 local oscillators were used in the design of an APAA with 256 transceiver elements.

References

[1] Podstrigaev A S, Smolyakov A V, Davydov V V, Myazin N S and Slobodyan M G 2018 *Lecture Notes in Computer Science (including subseries Lecture Notes in Artificial Intelligence and Lecture Notes in Bioinformatics)* **11118 LNCS** 509-515

[2] Mashkov G, Borisov E and Fokin G A 2016 *Proceedings - 2016 International Conference on Radar, Antenna, Microwave, Electronics, and Telecommunications, ICRAMET 2016 (Jakarta)* **7849572** p. 7-12

[3] Mashkov G, Borisov E and Fokin G A 2017 *Proceedings - 19th International Conference on Advanced Communications Technology, ICACT 2017 (Phoenix Park Pyeongchang)* **7890254** p. 979-984

[4] Lenets V A, Tarasenko M Yu, Davydov V V, Rodugina N S and Moroz A V 2018 *Journal of Physics: Conference Series* **1038(1)** 012037

[5] Lavrov A P and Molodyakov S A 2017 *Measurement Techniques* **59** 1025 – 1033

[6] Ermolaev A N, Krishpents G P, Davydov V V and Vysoczkiy M G 2016 *Journal of Physics: Conference Series* **741(1)** 012171

[7] Davydov R V, Saveliev I V, Lenets V A, Tarasenko M Yu, Yalunina T R, Davydov V V and Rud' V Yu 2017 *Lecture Notes in Computer Science (including subseries Lecture Notes in Artificial Intelligence and Lecture Notes in Bioinformatics)* **10531 LNCS** 177-183

[8] Podstrigaev A S, Davydov R V, Rud' V Yu and Davydov V V 2018 *Lecture Notes in Computer Science (including subseries Lecture Notes in Artificial Intelligence and Lecture Notes in Bioinformatics)* **11118 LNCS** 624-630

[9] Belkin M E and Sigov A S 2009 *Journal of Communications Technology and Electronics* **54** 655-658

[10] Vekshin Yu V, Lavrov A P and Hvostov Yu V 2018 *Radiotechnika* **1** 24-30

[11] Davydov V V, Sharova N V, Fedorova E N, Gilshteyn E P, Malanin K Y, Fedotov I V, Vologdin V A and Karseev A Yu 2015 *Lecture Notes in Computer Science (including subseries Lecture Notes in Artificial Intelligence and Lecture Notes in Bioinformatics)* **9247** 712-721

[12] Tarasenko M Yu, Davydov V V, Sharova N V, Lenets V A and Yalunina T R 2017 *Lecture Notes in Computer Science (including subseries Lecture Notes in Artificial Intelligence and Lecture Notes in Bioinformatics)* **10531 LNCS** 227-232

[13] Podstrigaev A S, Ryazantsev L B and Lukashev V P 2016 *Measurement Techniques* **59(5)** 547-550.

[14] Petrov A A, Davydov V V and Myazin N S 2017 *Lecture Notes in Computer Science (including subseries Lecture Notes in Artificial Intelligence and Lecture Notes in Bioinformatics)* **10531 LNCS** 561-568

[15] Petrov A A, Davydov V V and Grebenikova N M 2018 *Lecture Notes in Computer Science (including subseries Lecture Notes in Artificial Intelligence and Lecture Notes in Bioinformatics)* **11118 LNCS** 641-648
[16] Petrov A A, Grebenikova N M, Lukashev N A, Davydov V V, Ivanova N V, Rodugina N S and Moroz A V 2018 *Journal of Physics: Conference Series* **1038(1)** 012032
[17] Ivanov S I, Lavrov A P and Saenko I I 2017 *Lecture Notes in Computer Science (including subseries Lecture Notes in Artificial Intelligence and Lecture Notes in Bioinformatics*) **10531 LNCS** 482-489
[18] Ivanov S I, Lavrov A P and Saenko I I 2016 *Lecture Notes in Computer Science (including subseries Lecture Notes in Artificial Intelligence and Lecture Notes in Bioinformatics*) **9870 LNCS** 670-679
[19] Myazin N S, Smirnov K J, Davydov V V and Logunov S E 2017 *Journal of Physics: Conference Series* **929 (1)** 012080
[20] Grebenikova N M, Smirnov K J, Artemiev V V, Davydov V V and Kruzhalov S V 2018 *Journal of Physics: Conference Series* **1038 (1)** 012089
[21] Grebenikova N M, Smirnov K J, Davydov V V, Rud V Yu and Artemiev V V 2018 *Journal of Physics: Conference Series* **1135 (1)** 012055
[22] Friman R L 2012 Fiber–Optic Systems for Communications (New York: Wiley-Interscience a John Wiley & Sons. Inc.)
[23] Silver M 2015 Optical Fiber Communication Systems (New York: Clanrye International)
[24] Reinhold N 2016 Essentials of Modern Optical Fiber (Communication Amsterdam: Springer)
[25] Davydov V V, Karseev A Yu, Nepomnyashchay E K, Petrov A A and Velichko E N 2014 *Lecture Notes in Computer Science (including subseries Lecture Notes in Artificial Intelligence and Lecture Notes in Bioinformatics)* **8638** 694-702
[26] Davydov R V, Antonov V I and Moroz A V 2018 *Proceedings of the 2018 IEEE International Conference on Electrical Engineering and Photonics, EExPolytech 2018 (Saint-Petersburg)* **8564378** p. 236-239
[27] Rykin E V, Moroz A V, Smirnov K J, Davydov V V and Yushkova V V 2018 *MATEC Web of Conference* **245** 12002
[28] Lukashev N A, Petrov A A, Davydov V V, Grebenikova N M and Valov A P 2018 *Proceedings of 18th International conference of Laser Optics ICLO-2018* (Saint-Petersburg) **8435889** p. 271
[29] Semenov V V, Nikiforov N F, Ermak S V and Davydov V V 1991 *Soviet journal of Communications Technology and Electronics* **36** 59 – 63
[30] Myazin N S., Davydov V V, Yushkova V V and Rud V Yu 2018 *Journal of Physics: Conference Series* **1038 (1)** 012088
[31] Moroz A V, Davydov V V, Rud V Yu, Rud Yu V, Shpunt V Ch and Glinushkin A P 2018 *Journal Physics: Conference Series* 1135(1) 012060
[32] Smirnov K J, Davydov V V, Glagolev S F, Rodygina N S and Ivanova N V 2018 *Journal of Physics: Conference Series* **1038 (1)** 012102

Formation of out-of-plane film micro arch by methods of technological control of internal stress

Y. Enns[1,2], **R. Kleimanov**[2]

[1]Nanoelectronics Lab, St. Petersburg Academic University, St. Petersburg, Russia
[2]Peter the Great Saint-Petersburg Polytechnic University, St. Petersburg, Russia

Abstract. In this work is presented the analysis of the manufacturing technology of arcuate elastic elements. In the course of the work, the technological problems of the formation of out-of-plane microelectromechanical structures with the use of surface micromachining were studied. Internal stresses were controlled by varying the pressure of the working gas during sputtering. The use of this method allowed forming the arcuate profile of the Cr-Cu-Cr film structure.

1. Introduction
At present, research in the development and fabrication of microelectromechanical systems (MEMS) based on nonlinear effects has become more relevant. This group of devices includes bistable MEMS, in which the elastic suspension with the arcuate profile is the main element. Micro-arched elastic elements were distributed as elements of the RF switching technology, non-volatile memory and Straintronics. Elastic suspension profile shape is defined first mode buckling, thus allowing for existence of two stable states.

Currently the main technologies manufacture arcuate elastic elements is the bulk technology of microsystem technology using deep plasma chemical etching of silicon [1, 2]. This technology allows direct control of the parameters of the electromechanical structure, but has some limitations Another method is the fabrication of micro arches using surface technology, in which the profile of the beam is determined by residual axial internal stresses [3]. The ability to provide a higher density of elements is an advantage of this manufacturing technology.

Much more of the research in this field is directed to a search and selection of materials and their combinations in which internal stresses lead to buckling. Inflexibility in the materials and methods of their application is a disadvantage of this approach, which also affects the flexibility in developing the final device structure. Another approach is the fabrication of out-of-plane film structures using technological control of internal stresses that appear during the formation of structural mechanical layers of films. Such control of internal stresses can be achieved by varying the pressure of the residual pressure during the magnetron sputtering of some metal films [4, 5]. Thus, a number of studies have demonstrated that for some metals sputtering in this way, there is a dependence of the internal stress of the argon residual pressure. This approach can simplify the manufacturing technology of such structures, increase flexibility in the choice of materials and device design, as well as reduce their cost.

2. Experimental results

A study was conducted to identify patterns of internal stresses in multilayer film structures produced by magnetron sputtering. Elastic structures of cantilever and bridge beams were chosen as the object of study. After the multilayer cantilever beam is released from the substrate by removing the sacrificial layer, axial internal stresses cause the occurrence of the deflection moment M_b, proportional to the total value of internal stresses and to the formation of a structure with curvature K:

$$K = \frac{M_b}{\sum_{i=1}^{n} E_i I_i}, \qquad (1)$$

where E_i и I_i are the effective Young's modulus and the moment of inertia for the i structural layer. This curvature of the mechanical structure will appear regardless of the direction of the acting stresses. Thus, the gradient of the internal stresses of the structural layers of the beam can be determined by its curvature. In turn, the internal stresses of the bridge structure lead to both the stretching of the beam and the change in its resonant frequency, and to its buckling in the case of compression of the beam. Thus, the magnitude of internal stresses σ can be determined by multiplying the deformation of the beam ε and the effective Young's modulus for a multilayer film E_Σ:

$$\sigma = \varepsilon E_\Sigma. \qquad (2)$$

The combined study of these structures allows us to determine the distribution of internal stresses in the structural layers of the film and the direction of their action. A multilayer Cr-Cu-Cr combination deposited on a silicon substrate with an insulating SiO_2 layer was chosen as the structural layers of this element. The thickness of the structural layers t_i was 0.1–1–0.4 μm. The beams were released by etching the sacrificial SiO_2 layer. These structures were fabricated with different residual argon pressure in the pressure range of 1.2 - 5 mTorr (Figure 1, 2.).

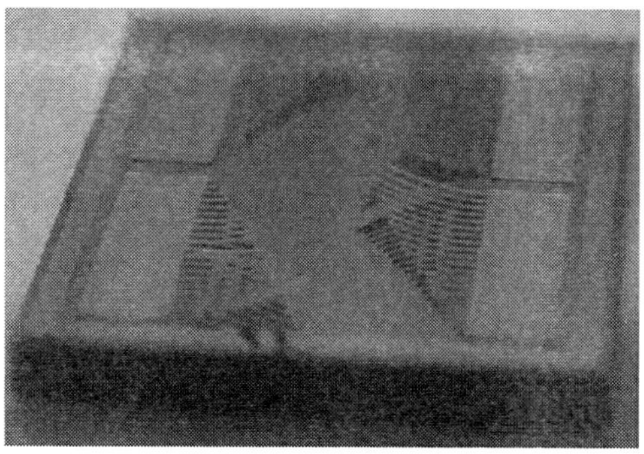

Figure 1. Photo of an array of cantilevers deposited at a pressure of Ar 1.2 mTorr

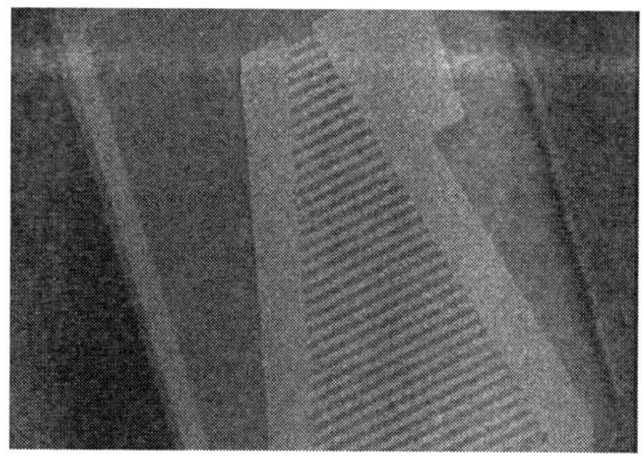

Figure 2. SEM image of the bridge array deposited at a pressure of Ar 1.2 mTorr

The elastic parameters of the films were determined using contact methods (on the Hysitron nanoindenter). The obtained values of the Young modulus for the structural layers of Cu and Cr were 60 GPa and 184 GPa, respectively, and differ from typical values by almost two times [3]. This can be caused by a high density of defects and the formation of bulk structural inhomogeneities. The high density of surface defects detected during the study of morphology by SEM (Fig. 3, 4) can indirectly indicate this. The detected imperfections of the structural layers primarily determine the size and type of internal stresses. The obtained values of the elastic parameters were also estimated by resonant methods. The obtained frequency response and phase response are consistent with the results of the finite element analysis [3].

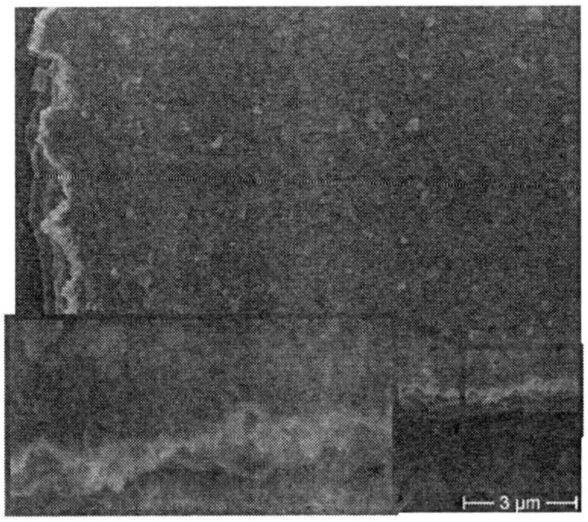

Figure 3. The SEM image of the Cr surface deposited at a pressure of Ar 1.2 mTorr (RMS = 0.4 um)

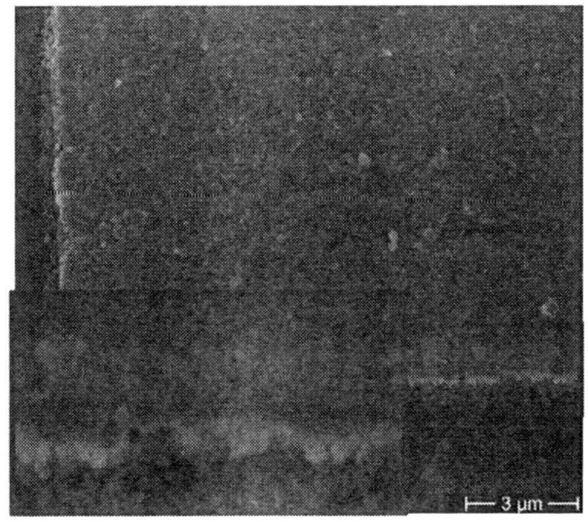

Figure 4. The SEM image of the Cr surface deposited at a pressure of Ar 5 mTorr (RMS = 0.6 um)

The curvature of the surface of the cantilevers was determined using confocal microscopy (Keyence VK-9700). So on the samples was observed a change in the curvature of the surface and the direction of curvature, depending on the residual pressure. Figures 5 and 6 show that a change in the residual pressure can lead to a significant change in the curvature of the surface.

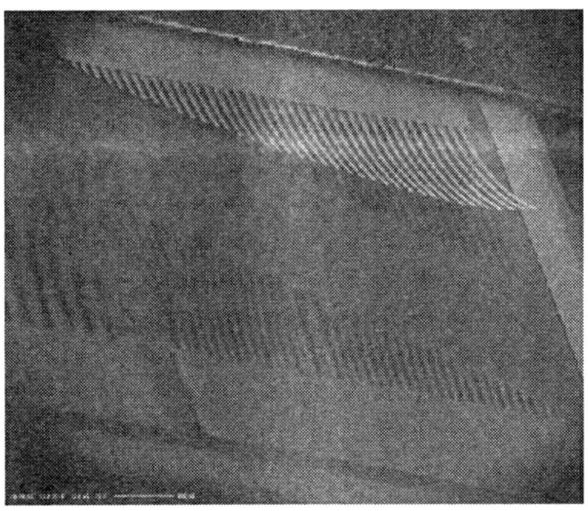

Figure 5. The SEM image of an array of cantilevers deposited at a pressure of Ar 1.2 mTorr; curvature $K = 0.8$ mm^{-1}

Figure 6. The SEM image of an array of cantilevers deposited at a pressure of Ar 3.4 mTorr; curvature $K = -0.28$ mm^{-1}

A change in surface curvature indicates a change in internal stresses. The value of the internal stresses of the deposited films was determined by the calculation method, based on the measured elastic parameters of the films, the curvature of the beams, and the deflection height of the arcuate profile. The aimed positioning of the cantilever neutral surface can be obtained by the layers thicknesses adjustment y_b:

$$y_b = \frac{\sum_{i=1}^{n} E_i t_i (y_i + y_{i-1})}{2\sum_{i=1}^{n} E_i t_i}. \quad (3)$$

Herewith, the cantilever strain ε_b is caused by the deformation of the structural layers ε_i due to the internal stresses, and can be defined as follows [6]:

$$\varepsilon_b = \frac{\sum_{i=1}^{n} E_i t_i \eta_i \varepsilon_i}{\sum_{i=1}^{n} E_i t_i}, \quad (4)$$

where $\eta_i = 1 + \nu$ is for high strain values.

The strain forces the film cantilever to deflect. The curvature K value and its direction (sign) depend on the deflection of the neutral surface. Cantilever curvature value can be determined proceeding from the following (1)-(4) [6]:

$$= \frac{3\sum_{i=1}^{n} E_i t_i (y_i + y_{i-1} - 2y_b)(\varepsilon_b - \eta_i \varepsilon_i)}{2\sum_{i=1}^{n} E_i t_i [y_i^2 + y_i y_{i-1} + y_{i-1}^2 - 3y_b(y_i + y_{i-1} - y_b)]} \quad (5)$$

The value of the total internal stresses can be estimated based on the measurements of the beam profile and the relation *(2)*. The study showed that the internal stresses of the chromium layers have a strong dependence on the pressure of the working gas, up to a change in the type of internal stresses (compression / tension) (Fig. 8). Based on the data obtained, the mechanical structures of the micro-arches were fabricated; an example of the profile is shown in Figure 7.

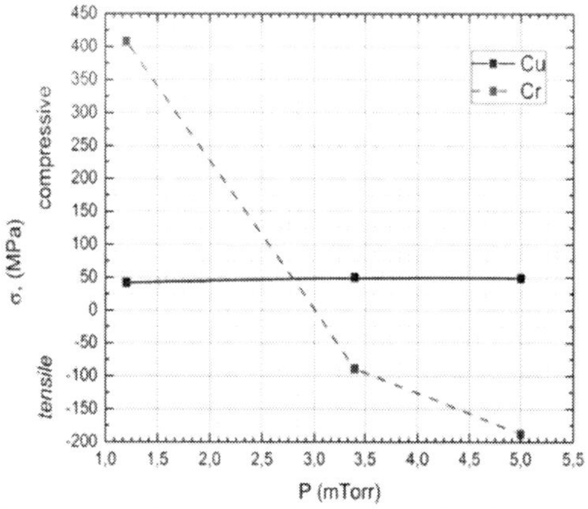

Figure 7. The dependence of the internal stresses of the beam structural layers on argon pressure

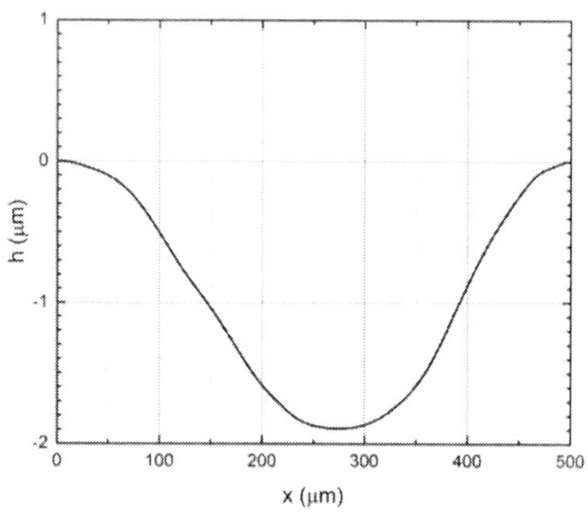

Figure 8. Example of a measured profile of bridge - type structures with a length of 500 μm (pressure Ar 3.4 мTopp)

3. Conclusion

In this paper, internal stresses were investigated and the causes of their occurrence were identified using the Cr-Cu-Cr multilayer film structure as an example. It is determined that the dominant factor in the formation of internal stresses in such layers is the growth of defects and bulk structural inhomogeneities of high density. It is shown that a change in the residual pressure argon during magnetron sputtering leads to a change in internal stresses, up to a change in direction. On the basis of the results were produce arcuate beams applicable for use in bistable MEMS.

Acknowledgments

The work was done as a part of the state assignment (№ 16.9789.2017/BCh) of the Ministry of Education and Science of the Russian Federation and Project №4 (the general agreement on research activities №3663-MRA of SkolTech and Academic University RAS).

References

[1] Enns Y. B., Pyatishev E N, Glukhovskoy A., 2017, *J. Phys.: Conf. Ser.* **917** 082013

[2] Kazmi S. N. R., Hajjaj A. Z., Hafiz M. A. A., Costa P. M. F. J., Younis M. I 2018 *IEEE transactions on nanotechnology* **17** (1) 113

[3] Kleimanov R., Enns Y., Pyatishev E., Komarevtsev I. 2018 *J. Phys.: Conf. Ser.* **1124** 1-5

[4] Hsieh J.H., C. Li, Wu W., Hochman R.F., 2003, *J. Thin Solid Films.* **424** 103–106

[5] Shen Y. G., Mai Y. W., Zhang Q. C., McKenzie D. R., McFall W. D. and McBride W. E., 2000, *J. Appl. Phys.* **87** 177

[6] Nikishkov G. P., 2003, *J. Appl. Phys.*, **94**(8) 5333

Investigation of the piezoresistive properties and temperature coefficient of resistance of epitaxial GaN layers for applications in MEMS and thermal flow sensors

A Kazakin[1,2], Y Enns[1,2], A Mizerov[1], R Kleimanov[2] and A Bouravleuv[1]

[1]Nanoelectronics Lab., St. Petersburg Academic University, St. Petersburg, Russia
[2]Peter the Great Saint-Petersburg Polytechnic University, St. Petersburg, Russia

Abstract. The paper presents the results of an experimental determination of the piezoresistive gauge factor and temperature coefficient of resistance of a GaN films grown by molecular beam epitaxy on sapphire substrates. The measured values were used to calculate the design and characteristics of the GaN hot-wires anemometer for the MEMS gas flow sensor.

1. Introduction

Due to its high chemical, temperature and radiation resistance, gallium nitride is a promising material for electronics and sensors for harsh environments [1]. For deposition of gallium nitride films, along with the metalorganic chemical vapor phase epitaxy (MOCVD) [2] and hydride vapor phase epitaxy (HVPE) [3], which are widely used in commercial manufacturing of devices, molecular beam epitaxy (MBE) is also used [4]. Reactive nitrogen in this process can be obtained both by thermal cracking of ammonia (NH3-MBE) [5, 6] and by activation of N2 in plasma (PA MBE) [7]. The advantage of the latter method is the ability to grow films at low substrate temperatures and in the absence of hydrogen. Also as has been shown in [8, 9] the method allows growing GaN films with different polarities. The progress achieved in recent years in the field of synthesis and post-processing of epitaxial GaN films opens up promising opportunities for their use in microelectromechanical systems (MEMS) [10].

GaN sensitive elements can be used in miniature gas sensors, temperature, pressure and strain sensors, flow sensors, microphones and other MEMS. The metrological characteristics of such sensors depend on the GaN films synthesis method and its technological regimes, substrate type, electrode materials, film etching methods for the formation of sensing elements. For example, the experimental values of the gauge factor of GaN piezoresistors differ by an order of magnitude by different authors, who used different methods of GaN growth [11, 12]. Therefore, before developing a specific sensor design, you must first carry out an experimental estimation of the sensitivity of the grown GaN film electrical characteristics to changes in some environmental parameter. These experimental data are necessary to create an adequate numerical model of the prospective sensor and calculate its parameters.

The first part of this article presents the results of preliminary experiments to determine the piezoresistive and temperature sensitivity of conductivity of GaN films grown by the PA MBE on a sapphire substrate. The obtained data are used in the second part of the article, where the results of numerical calculations of a thermoanemometer with a sensitive element in the form of GaN strings are presented. Such a device is the basis of a new class of MEMS - thermal flow sensors [13], and can be used in flow sensors for corrosive gases and liquids and thermal microphones for remote diagnostics of equipment operating in aggressive environments.

2. Experiments and results

2.1. Samples fabrication
The objects of the study were the silicon doped n-type GaN films with thickness of 300 nm, grown by PA MBE using Gen 200 (Veeco) industrial type setup on GaN/c-Al_2O_3 templates [9,14]. The latters were preliminary formed by MOCVD. According to the room-temperature Hall measurements the electron concentrations and motilities in the n-GaN layers were about $3.75 \cdot 10^{19}$ cm^{-3} and 107 $cm^2/(V \cdot s)$, respectively.

The synthesized samples were diced into rectangular pieces, 9 mm × 3 mm in dimensions, by laser scribing method. Ti(20 nm)/Al(500 nm) metal stack was then deposited on the GaN surfaces of the samples by magnetron sputtering with preliminary ionic treatment of the surface to form non-alloyed ohmic contacts. Wirebonding was carried out using sintered silver paste to connect metal pads to the external instrumentation. The measured current-voltage characteristics of the samples were linear up to 16 V, which proves the non-rectifying nature of the contacts. A typical sample resistance was close to 200 Ω.

2.2. Measurement of TCR
A test bench was assembled to measure the temperature coefficient of resistance of GaN films. The bench consisted of a thermal chamber with temperature control in the range of 20–350 °C, a unit for measuring the sample resistance and a unit for measuring the sample temperature. A sample with a GaN film was mounted in close proximity to a platinum temperature sensor (Pt1000) inside a metal-glass package. Each resistance measurement was carried out after temperature stabilization for 15 minutes and was continued for 10 minutes to determine possible deviations. The temperature dependence of the GaN film conductivity in the temperature range 20-160 °C is shown in Figure 1.

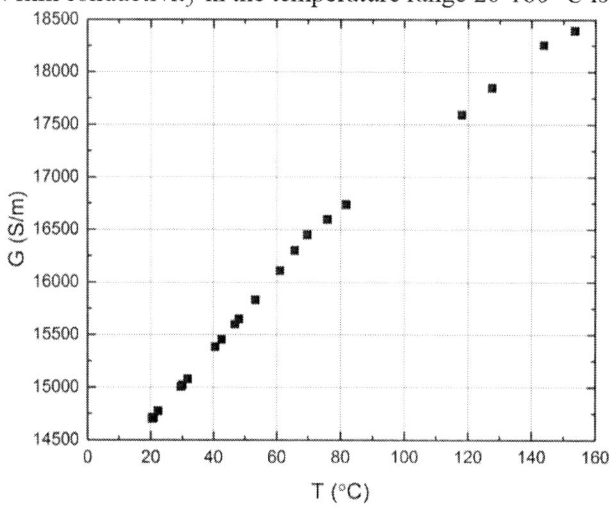

Figure 1. GaN conductivity versus temperature

The choice of the measurement range corresponded to the distinctive temperature range of Pt sensitive elements of conventional hot-wire and calorimetric MEMS thermo-anemometers. In this range, the temperature dependence of the GaN conductivity was close to linear. This is an advantage for the subsequent processing of the sensor output signal by an electronic circuit. The GaN TCR values are close to the TCR of platinum thermistors in order of magnitude.

In the future, the assembled test bench and measurement technique are planned to be used to determine the TCR of the grown GaN films. These data are important for the development of the temperature sensors, heat and mass flow sensors, as well as for making appropriate corrections to compensate for temperature deviations of other GaN MEMS.

2.3. Measurement of gauge factor

The piezoresistive effect of the materials is described by a gauge factor (GF). It's the ratio of relative change in electrical resistance to the mechanical strain in the film.

In our work a special test bench was assembled in order to measure the GF of GaN. In this test bench the synthesized samples were placed on a special pedestal without rigid fixation. This arrangement avoids stress concentrations near fixing points. In the center area, the sample was subjected to external loading in the transverse direction using calibrated weights. In this case, longitudinal deformations (tensile or compressive strain, depending on the direction of sapphire deformation) appear in a thin GaN film. The magnitude of these deformations was calculated using numerical simulations in the COMSOL Multiphysics program, based on the known elastic properties of the sapphire. The resulting stress gradient in the GaN film led to a change in its resistance, which was measured using the test circuit.

In the experiments, the samples were loaded with weights in the range of 400 - 1200 g. At higher loads, the destruction of the samples was observed which is probably due to the fact, that sapphire has a high stiffness with a sufficiently high brittleness, which limits the range of allowable values of sample deformation. The GaN film strain gradient range, corresponding to the used loads, was 1.0003 - 1.0007. Each sample was subjected to multiple cyclic loads in this range and the measured resistances were averaged. Thus, we measured the static value of GF. It should be noted that in our experiments the dynamic GF associated with the piezoelectric properties of GaN [11] was detected at the moments of loading and unloading, but was not measured. All measurements were carried out at room temperature.

After a mathematical calculation of the experimental results, the GF of GaN was estimated as 24 ± 2. This value is significantly inferior to monocrystalline silicon (GF = 150-200), but exceeds the GF of metals (GF = 1-5) by an order of magnitude and is comparable to the GF of some semiconductor materials (GF = 30 for poly-Si and SiC), which are successfully used in modern pressure sensors for harsh environments.

Estimation of GF GaN is important for the development of GaN resonators, strain and pressure sensors, as well as for calculating the characteristics of temperature sensors in which the change in resistance due to external deformations can lead to an increase in measurement errors.

3. FEA of thermal flow sensor prototype

The obtained values of TCR and GF prove that GaN epitaxial films can be used as a sensitive elements of MEMS thermal transducers, such as a thermo-anemometers. These devices, based on conventional materials (Pt, poly-Si), have already become widespread in the field of measurement of the dynamic characteristics of mass and heat flows in neutral gases and liquids. On the basis of calorimeter-type transducers, it is possible to manufacture acoustic velocity sensors with a narrow polar pattern for miniature direction-finding devices. They are also applicable for remote sound diagnostics and early detection of faults in industrial lines and mechanisms, gas pipelines, turbomachines, chemical reactors, etc. GaN thermistors, possessing high strength and resistance to aggressive environment, can in some tasks replace conventional materials of thermo-anemometers sensitive elements.

Therefore, on the basis of our experimental data on the temperature dependence of GaN conductivity given in section 2.2, we developed the design of a GaN thermo-anemometer and performed its numerical simulation in the COMSOL Multiphysics program. The sensor design is shown on the geometric model in Figure 2. The device includes the semiconductor structure which consists of 800 nm thick Ga-polar GaN layer, grown by PA MBE atop of 300 μm thick semi-insulating silicon substrate which was preliminary nitridated in order to form thin Si_3N_4 interlayer. The details of the PA MBE synthesis of the structure are described in [9]. The square contact pads (Ti-Al-Ni-Au) were formed on the GaN surface of the structure by lift-off lithography. Sensitive elements in the form of narrow GaN strings were formed by plasma etching. The cavity under the strings was made by selective chemical etching of silicon. The geometric dimensions of the GaN strings were 800 x 20 x 0.7 µm, and their height above silicon surface is one of the adjustable parameters of the numerical model. An example of a thermal convection cloud that forms around strings when a current is passed through them is shown in Figure 3. The thermo-anemometer operating principle is based on measuring the difference in resistance of two GaN hot-wires, which occurs when the thermal cloud is shifted from one string to another by the ambient gas flow. Depending on the vector of the flow velocity, the temperature difference between the two strings will be different. The calculation of the GaN anemometer characteristics required an interconnected solution of the problem in the mechanical and electrical domains, and in the domains of heat and mass flow. The reference data on the mechanical and thermal properties of GaN, Si, Si_3N_4 and our experimental data on TCR and GF of GaN were used. In our calculations, we studied the effect of laminar air flow at low flow rates (up to 1 cm/s) on the sensor output signal and took into account the influence of the angle of attack (AoA), i.e. flow velocity vector relative to the plane of the hot-wires.

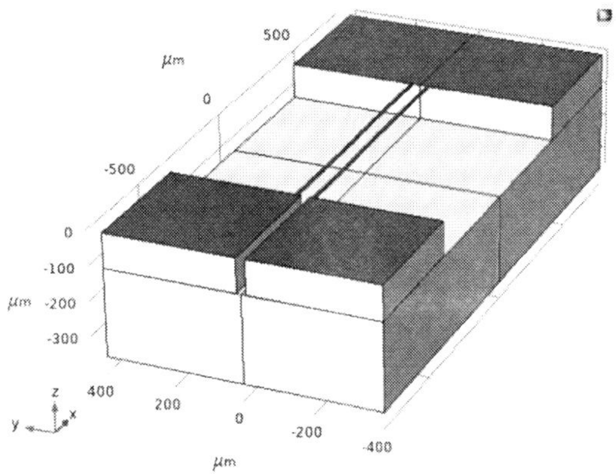

Figure 2. Model of GaN hot-wires anemometer

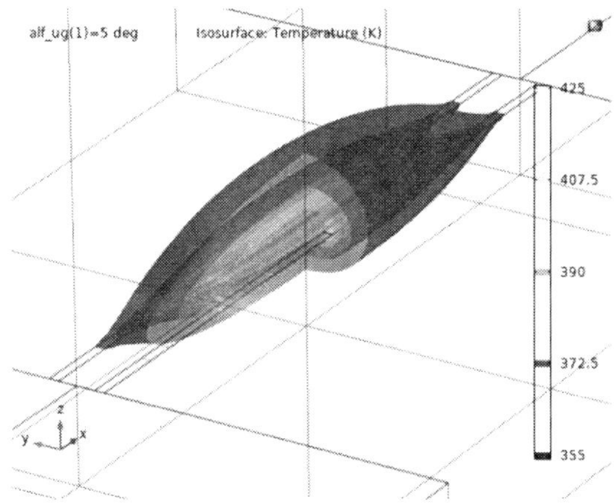

Figure 3. Thermal cloud around the GaN strings (at U = 5 V, v_{gas} = 1 cm/s, AoA = 5°)

Unlike the classical macroscopic case of flow measurements (for example, in turbomachines), in the MEMS anemometer, hot-wires are located near the silicon surface, the distance to which is determined by the substrate etching depth (of the order of several tens of microns). This fact and the viscous effects in the gas lead to a significant drop in the velocity of the flowing air near the GaN strings. The distribution of the velocity of the air flowing around the sensor at an input flow rate of 1 cm/s and AoA of 5° is shown in Figure 4. It can be seen that in the air layers, which are at the level of the hot-wires, the flow velocity drops by an order of magnitude compared with the undisturbed flow, and, therefore, does not lead to a significant displacement of the convection cloud and a changes in the temperature and resistance of both GaN strings. At the same time, changing the angle of attack changes the flow velocity distribution. Figure 5 shows the two-dimensional distribution of the flow velocity over the center of the chip (vertical line in Fig. 4), depending on the height above the chip surface and the value of AoA. It can be seen that the highest flow velocity in the region of the location of hot-wires occurs at an angle of attack of 45°.

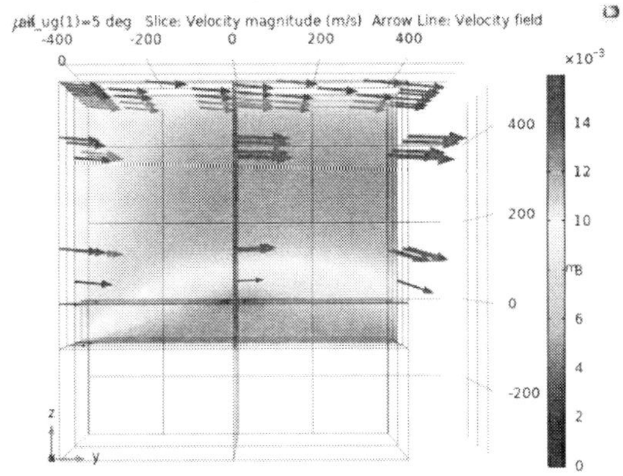

Figure 4. Laminar flow velocity distribution over the chip surface ("0" – level of GaN hot-wires)

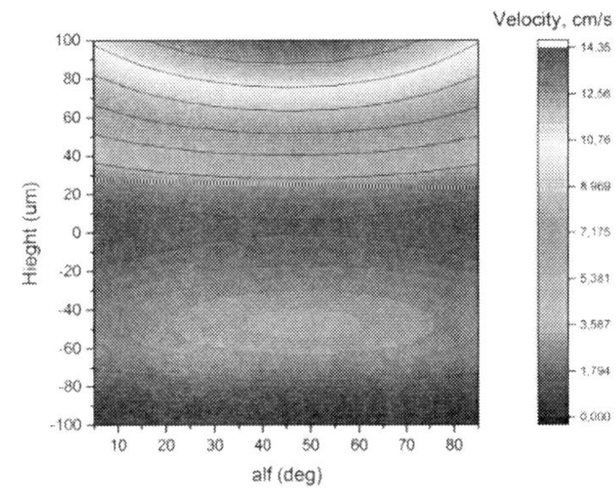

Figure 5. Flow velocity versus AoA and height above the chip ("0" – level of GaN hot-wires)

These regularities are typical for MEMS anemometers and are important for a better understanding of some aspects of their technology, depending on the application. So, with an etching depth of the silicon substrate under GaN strings over one hundred microns, this design can be used as a gas flow meter with a sensitivity at the level of conventional analogs. In addition, with a different arrangement of GaN strings with respect to the flow, this design can be used as a gas molecules velocity vector sensor or as a gas acoustic sensor (thermal microphone). The result of our calculations demonstrates that the magnitude of the displacement of the thermal cloud above the GaN strings due to a change in the angle of attack is sufficient to form a narrow directional pattern of such sensors. In Figure 6, this polar pattern is shown as the dependence of the difference in resistance of hot-wires on the angle of attack. The calculations also showed that the change in hot-wires resistance due to the piezoresistive effect in GaN under mechanical impact of the incident gas flow is negligible (the previously obtained value of GF = 28 was used as the upper estimate, since with increasing temperature GaN GF drops [12]).

To determine the frequency characteristics of the thermal microphone, numerical simulation was performed in the time domain with sinusoidal oscillations of air molecules surrounding the strings. The results of the study are shown in Figure 7. It can be seen that in the low-frequency range, which is typical for acoustic diagnostics of industrial equipment (at frequencies up to 500 Hz), the output signal amplitude remains at a fairly high level.

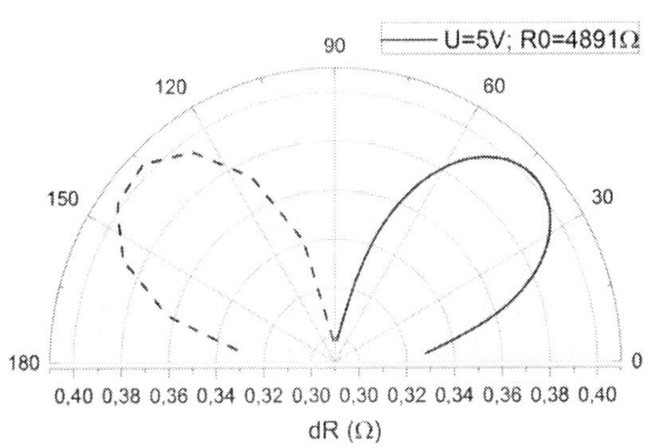

Figure 6. The difference in resistance of GaN hot-wires versus AoA (sensor polar diagram)

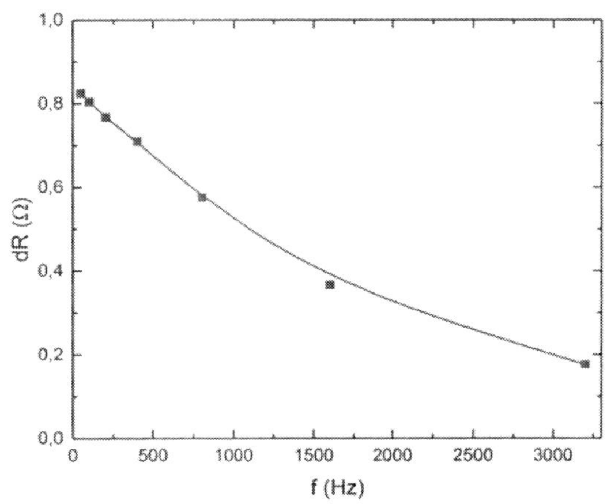

Figure 7. The difference in resistance of GaN hot-wires versus acoustic frequency (at U=10V)

Thus, the obtained experimental and d data testify to the prospects of using epitaxial gallium nitride synthesized by PA MBE as a material for sensing elements of thermal flow meters and acoustic velocity sensors. The work will be continued further to ensure the technological production of prototypes of these devices.

4. Conclusion

The temperature coefficient of resistance of GaN film synthesized by PA MBE has been measured and its piezoresistive gauge factor has been estimated. The obtained data were used to carry out FEA simulation of the characteristics of a MEMS anemometers with gallium nitride hot-wires sensing elements. The low piezoresistive sensitivity of the GaN layers eliminates the change in hot-wires resistance under mechanical influence of the incident gas flow. The results show the prospect of using of GaN films in a gas flow sensor.

Acknowledgments

The work was done as a part of the state assignment (№ 16.9789.2017/BCh) of the Ministry of Education and Science of the Russian Federation and Project №4 (the general agreement on research activities №3663-MRA of SkolTech and Academic University RAS).

References

[1] Rais-Zadeh M, Gokhale V J, Ansari A, Faucher M, Cordier Y and Buchaillot L 2014 *J. Microelectromech. Syst.* **23** 1252-71
[2] Wu X H, Brown L M, Kapolnek D, Keller S, Keller B, DenBaars S P and Speck J S 1996 *J. Appl. Phys.* **80** 3228–37
[3] Bessolov V, Konenkova E, Shcheglov M, Sharofidinov Sh, Kukushkin S, Osipov A and Nikolaev V 2013 *Phys. Stat. Sol.* C **10** 433–6
[4] Arthur J R 2002 *Surf. Sci.* **500** 189–217
[5] Mesrine M, Grandjean N, and Massies J 1998 *Appl. Phys. Lett.* **72** 350–2
[6] Mizerov A M, Jmerik V N, Kaibyshev V K, Komissarova T A, Masalov S A and Ivanov S V 2009 *Semiconductors* **43** 1058–63
[7] Jmerik V N, Mizerov A M, Shubina T V, Listoshin S B and Ivanov S V 2007 *Tech. Phys. Lett.* **33** 333–6
[8] Shubina K Yu, Berezovskaya T N, Mokhov D V, Mizerov A M and Nikitina E V 2017 *Tech. Phys. Lett.* **43** 976–8
[9] Mizerov A M, Timoshnev S N, Sobolev M S, Nikitina E V, Shubina K Yu, Berezovskaia T N, Shtrom I V, and Bouravleuv A D 2018 *Semiconductors* **52** 1529–33
[10] Shubina K Yu, Berezovskaya T N, Mokhov D V, Morozov I A, Kotlyar K P, Mizerov A M, Nikitina E V and Bouravleuv A D 2018 *Semiconductors* **52** 2117–9
[11] Bykhovski A D, Kaminski V V, Shur M S, Chen Q C and Khan M A 1996 *Appl. Phys. Lett.* **68** 818–9
[12] Tilak V, Vertiatchikh A, Jiang J, Reeves N and Dasgupta S 2006 *Phys. Stat. Sol.* C **3** 2307–11
[13] Balakrishnan V, Phan H-P, Dinh T, Dao D V and Nguyen N-T 2017 *Sensors* **17** 2061-91
[14] Kukushkin S A, Mizerov A M, Osipov A V, Redkov A V and Timoshnev S N 2018 *Thin Solid Films* **646** 158–62

Contrast simulation in the phase shift dynamic magnetic force microscopy depending on the MFM probe geometry and materials

M V Zhukov[1,2], K I Belousov[2,3], A O Golubok[1,2]

[1] Laboratory of Scanning probe microscopy and Spectroscopy, Institute for Analytical Instrumentation RAS, St. Petersburg 198095, Russia
[2] Department of Nanophotonics and Metamaterials, ITMO University, St. Petersburg 197101, Russia
[3] Laboratory of Bionanotechnology, St. Petersburg Academic University, St. Petersburg 194021, Russia

E-mail: cloudjyk@yandex.ru

Abstract. The gradient of the magnetic force caused by the interaction of various shapes magnetic nanoprobes with magnetic materials has been simulated. Because the force gradient is proportional to the phase shift in probe oscillations the data obtained allow us to compare the contrast of dynamic magnetic force microscopy images for probes with various shapes. A comparison of standard pyramidal probes and probes with cylinder-shaped nanowiskers with different aspect ratios has been carried out. The probes with optimal parameters such as the shape, the thickness and type of the magnetic coating giving the maximum phase shift and as a result giving the best quality visualization of magnetic fields on the sample surface in the phase shift mode have been selected.

1. Introduction

In magnetic force microscopy (MFM) when studying complex nanoscale structures the problem of detecting of small areas of magnetization is exist. It is well known that the spatial resolution and sensitivity of the MFM are determined by the shape and material of the probes used [1]. Increasing the sensitivity and resolution of the MFM technique is necessary, first of all, for high-precision magnetic mapping and diagnostics of nanoelectronic elements, in particular, reading and writing magnetic information on magnetic media with an accuracy of single domains.

A lot of methods allow forming a magnetic probe from different materials of almost any shapes [2-5]. Therefore, the selection of the optimal shape and magnetic material of the probe before carrying out experimental research is a necessary step to improve the quality of MFM measurements.

Thus, the main purpose of this work was to identify the optimal parameters of high-aspect magnetic probes for magnetic force microscopy by simulating their geometry, magnetic material and coating thickness of the magnetic layer.

2. Simulations for various MFM probes

The simulations for standard Si cantilevers and high-aspect probes coated with a thin layer of magnetic material were performed. The geometry of the high-aspect probes corresponded to the geometry of the nanowhisker (NW), which is grown on the top of the silicon pyramid of a standard cantilever by the method of a focused electron beam in the presence of precursor gases in a scanning electron microscope (figure 1 a) with subsequent deposition of a thin layer of magnetic materials [3]. The MFM probe model is shown in the figure 1 b, c.

The Gauss law was used to determine the distribution of the magnetic field:

$$\nabla \cdot \mathbf{B} = 0, \qquad (1)$$

where B is the magnetic induction vector, [T].

Equation (1) was solved by the finite element method in the COMSOL Multiphysics software package.

The following relationship was satisfied for the magnetized areas of the magnetic deposition of the probe and the magnetic domain:

$$\mathbf{B} = \mu_0 (\mathbf{H} + \mathbf{M}), \qquad (2)$$

where $\mu_0 = 4\pi\cdot10^{7}$ is the magnetic constant, [N/m]; H - vector of the magnetic field strength, [A/m]; M is the material magnetization vector, [A/m].

For magnetization vector in the magnetic domain the value of $|M| = 6\cdot10^5$ A/m was chosen. The change in the geometrical dimensions of the NW and the thickness of the deposition was calculated at the fixed value $|M| = 3\cdot10^5$ A/m (for Co coating). Separately, the effect of a change in the magnetization modulus was studied with $|M| = 2{,}8\cdot10^4$ A/m (for Ni coating) and $|M| = 1{,}9\cdot10^6$ A/m (for CoFe coating).

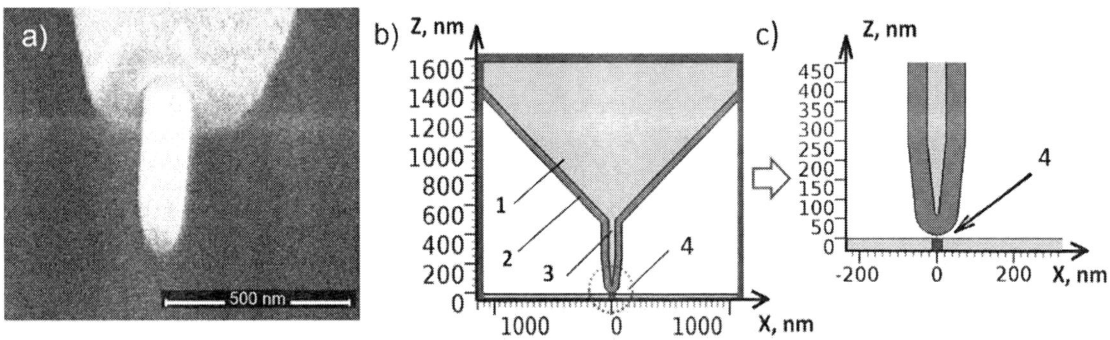

Figure 1. SEM image of NW probe (a) and NW high-aspect MFM probe model in the COMSOL Multiphysics (b): 1 - the top of the standard probe, 2 - the magnetic coating material, 3 - NW, 4 - magnetic domain. The extended area of the probe and the domain is shown on the right (c).

The size of the magnetic domain was chosen as 30 nm, the thickness of the ferromagnetic coating of the probe was varied from 20 to 50 nm, the width of the NW was varied from 25 to 75 nm and the height was varied from 250 to 750 nm.

3. Results and Discussions

It was found that the NW probe has twice larger gradient values compared with the standard probe with the same deposition thicknesses of magnetic material (figure 2 a). Gradient increase was revealed with a gradual rising of the deposition thickness up to 50 nm (to ~15 μN/m for standard probe and to ~30 μN/m for NW probe).

NW thickness increase from 25 nm to 50 nm (the magnetic material has a constant thickness of 50 nm) leads to gradient decrease from ~35 μN/m to ~30 μN/m, after which the gradient change almost not occurs. Thickness change of the NW while maintaining the overall diameter of the whisker reveals the same gradient decrease, but with smaller absolute values by ~ 8%.

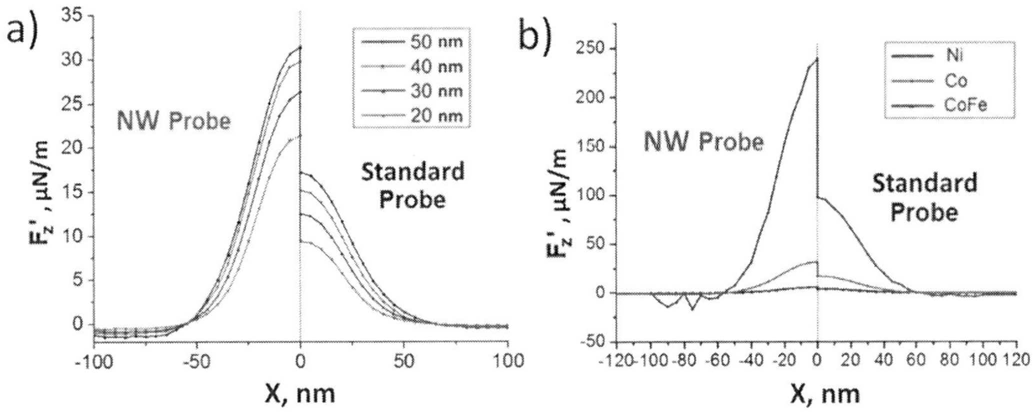

Figure 2. The force gradient for probes interacting with a single 30 nm domain (half of the cross section relative to the center is shown) with varying thickness of the magnetic coating (a) and probe coating material (b). NW length is 500 nm, NW diameter is 50 nm without coating, the total diameter is 150 nm. Coating thickness is 50 nm (b).

A gradual gradient increase was observed by increase of NW length (figure 3 b) from 250 nm (~25 µN/m) to 750 nm (~40 µN/m). A gradual gradient increase also was observed by increasing angle of inclination of the NW relative to the normal of the surface (figure 3 a). This can be explained by the large overlap of the magnetic fields of the sample and the probe during interaction at increasing angle.

The lowest gradient values were obtained using Ni coating, the maximum value of which was 5.8 µN/m for the NW probe and 4.3 µN/m for the standard probe. Co coating has large values of 31.4 µN/m for the NW probe and 17.2 µN/m for the standard probe, which gives a twofold change in the force gradient. The best material turned out to be CoFe, which gave values of 239 µN/m for the NW probe and 97.5 µN/m for the standard probe (~ 2.5 times change), which makes it most demanded in MFM measurements (figure 2 b).

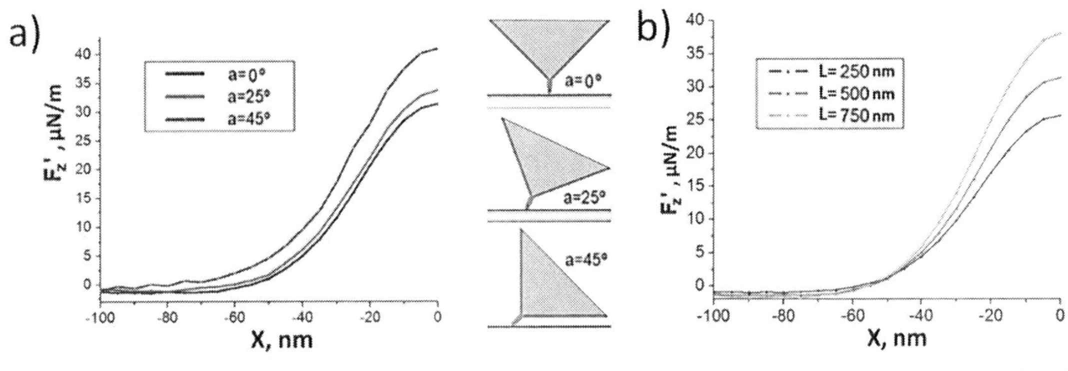

Figure 3. The field gradient of NW with a Co coating interacting with a single 30 nm domain when changing the angle «α» of the NW- probe tilt (a) and NW length «L» (b).

4. Conclusions

The influence of probe geometry, the thickness of the coating of the magnetic layer, the angle of probe inclination and the type of magnetic material on the MFM data were studied. Force gradient increase was found increasing the thickness of the magnetic coating, lowering the thickness and increasing the length of the NW, which should lead to an improvement in the contrast of the MFM image in phase contrast mode. The CoFe turned out to be the optimal coating material, which gives the highest values of the force gradient.

Acknowledgments

The work was carried out with financial support by the Ministry of Education and Science of the Russian Federation, Russia, State Project № 075-00780-19-00, Subject № 0074-2019-0007.

References

[1] M.R. Koblischka, U. Hartmann 2003 *Ultramicroscopy* **97** 103–112
[2] L. Gao, L. P. Yue, T. Yokota, Ralph Skomski, Sy-Hwang Liou, H. Takahoshi, H. Saito, and S. Ishio 2004 *IEEE Transactions on Magnetics* **40** 2194-2196
[3] M.V. Zhukov, K.I. Belousov, A.M. Mozharov, I.S. Mukhin, A.O. Golubok 2015 *J. Phys.: Conf. Ser* **643** 012095 (5pp)
[4] Duckhoe Kim, Nak-Kwan Chung, Stephanie Allen, Saul J. B. Tendler, and Joon Won Park 2012 *AcsNano* **6** 241-248
[5] A.R. Champagne, A.J. Couture, F. Kuemmeth, and D.C. Ralph 2003 *Appl. Phys. Lett.* **82** 1111-1113

Formation of Gibbs and Langmuir floating layers based on copper nanoparticles: temperature and concentration effects

N N Begletsova[1,2], L T Baimagambetova[1], V N Mironyuk[1,2], A I Smirnova[3], S B Venig[1], N V Usol'tseva[3], E G Glukhovskoy[1,2]

[1] Department of Nano- and Biomedical Technologies, Saratov State University, Saratov, 410012, Russia
[2] Education and Research Institute of Nanostructures and Biosystems, Saratov State University, Saratov, 410012, Russia
[3] Nanomaterials Research Institute, Ivanovo State University, Ivanovo, 153025, Russia

E-mail: nadya-beg98@yandex.ru

Abstract. This paper is devoted to study of the formation of Gibbs floating layers from aqueous solutions of copper nanoparticles (Cu NPs) stabilized by the sodium dodecyl sulfate surfactant. Langmuir floating layers of arachidic acid formed on the surface of a water subphase containing Cu NPs were also studied. Adsorption of copper nanoparticles on a monolayer of arachidic acid at different conditions has been investigated. In the experiments the amount of copper nanoparticles in the subphase and the temperature of the subphase were changed. The performed analysis showed that the increase of subphase temperature from 29 to 42 °C leads to the increase of the area occupied by Langmuir floating layers in a liquid-condensed state by 30 and 32 %, respectively. After a single transfer of the floating layer onto the highly oriented pyrolytic graphite substrates, their surface microrelief was investigated by atomic force microscopy (AFM). The AFM scans showed that the obtained copper nanoparticles films contained separate areas with copper particles and their aggregates. With an increase of surface pressure from 10 to 40 mN/m, an increase of the height difference of the surface microrelief from 4 to 7 nm and an increase of the large aggregates number with lateral dimensions about 100 nm were observed.

1. Introduction

Recently, the number of publications devoted to the study of nanoparticles (NPs) properties, methods for modifying their surface, the formation of 2D systems on their basis at different interfaces has been grown steadily. One of the most common methods of formation such layers is Langmuir-Blodgett (LB) technology [1-3]. This method is simple in execution because it does not require the creation of specific technological conditions (high vacuum, use of complex equipment, etc.). The LB technology allows to create nanomaterials, control their composition, structure and properties by changing temperature, acidity, composition of subphase, etc. Moreover, it is possible consistently transfer formed floating layers of various substances to substrates [2-7]. In addition, this technology allows obtaining heterostructures with unique composition, structure and properties, which can be varied during the transition from layer to layer. Such films have potential application as electrodes in micro- and nanoelectronics devices (in photovoltaics, supercapacitors, etc.) [8, 9].

It is known that of soluble compounds Gibbs floating layers can be formed on the surface of water subphase, while Langmuir floating layers can be obtained of insoluble compounds. In both cases, the formation of floating layers is determined by many technological parameters such as ionic composition of subphase, its acidity, temperature, electrical and magnetic fields, radiation, surfactant's nature and layer composition. Formation of NPs floating layers in an organic matrix is even more difficult technological process. Properties of NPs themselves as well as properties of surfactant shells stabilizing NPs are added to the influence of the above mentioned parameters.

Earlier studies showed that subphase temperature affects phase states of floating layer on compression isotherms. In number of articles it was noted that temperature increase can lead to disruption of ordering of surfactant molecules in floating layer: the areas of isotherms that at low temperatures corresponded to a liquid-condensed state at higher temperatures transformed into the areas characteristic of liquid-expanded state. However, for more complex systems such as floating layer of NPs, the temperature increase in a certain range results in greater ordering [4]. For such complex quasi-2D systems based on NPs, polymer molecules, molecular clusters, etc. the temperature increase can change the composition of floating layer and NPs shells due to increase of surfactant solubility, polymer rearrangement at the expense of weak bonds breaking and so on. In turn, all above mentioned can change macrostructure, physical properties and macroparameters of the layer. Under the action of temperature, the properties of the subphase itself change. For example, the diffusion rate of molecules and ions through laminar and turbulent layers of the subphase adjacent to the surface changes, which in turn affects the stability of the surfactant floating layer. We can speak of a general change in the properties of the whole subphase under the influence of temperature, if the diffusion, laminar and turbulent layers are comparable with the depth of Langmuir trough.

The structure of films deposited onto solid substrates can repeat the structure of floating layer. The transfer success depends on the nature and composition of studied substance, acidity and temperature of subphase, size of nanoinclusions in the organic matrix, condensation degree of floating layer, temperature of subsequent processing, oxidation degree of film components, etc. [5, 6]. The structure and composition of forming film can be varied by changing the condensation degree of floating layer and the formation of desired phase state, determined from compression isotherms, as well as from the "compressibility (or compression modulus) *vs* area per molecule" dependencies [2, 7].

As it was noted in [1, 2, 10-12], constant exchange processes take place between floating layer and liquid subphase, therefore the "layer – subphase" system is in thermodynamic equilibrium. Previously, we studied the dynamics of Langmuir layers of semiconductor NPs in various organic matrixes of water-insoluble surfactants [4, 7, 13-15]. The studied surfactants demonstrated a relationship between of sorption processes of NPs from solution to a Langmuir layer and the packing of NPs in the layer as well as layer's mechanical properties and its formation conditions.

Controlling the composition and structure of the NPs floating layer during its formation and transfer from the water surface to the surface of a solid substrate is more complex problem than in the case of monolayers (MLs) of one-component surfactants or mixtures of simple compounds. The questions of interaction and aggregation of NPs in quasi-2D systems, changes of their distribution and phase separation in floating layers of complex composition, solubility and stability of NPs layers during formation, structure and composition preservation during transfer onto solid substrates are not well understood.

The aim of this work is to study the floating layers formed by copper nanoparticles (Cu NPs) in an organic matrix at the "gas – liquid" interface using the compression isotherm method and surface microrelief investigation/analysis of the Cu NPs films transferred onto solid substrates by atomic force microscopy (AFM).

2. Experimental

In this work, the source of NPs for the formation of floating layers was the water subphase. This subphase was a solution of NPs. A detailed description of the synthesis technique and the particle size distribution study of the obtained NPs are given in [16]. The developed method is based on the chemical reduction of copper ammonia in the aqueous solution of anionic surfactant, sodium dodecyl sulfate (SDS). For this study the molar ratio of salt (copper chloride) and reducing agent (hydrazine hydrate) was 1:150 at pH = 11.2 of the reaction medium. A synthesis time was one hour, which allows to obtain Cu NPs with sizes in the range of 10-140 nm.

At the first, Gibbs floating layers of Cu NPs were formed by sorption of the NPs from the subphase volume to the "gas – liquid" interface. Then, we compared this process with the sorption of the NPs on the ordered Langmuir ML of arachidic acid (AA), which plays the role of a sorbent surface.

To obtain the subphase with required concentration of NPs, the deionized water with a specific resistivity of 18 MΩ×cm was used (it was prepared by the Thermo Scientific Barnstead Smart2Pure water treatment system, USA).

All studies were carried out in the Langmuir trough KSV Nima LB Trough Medium KN 2002 (KSV Nima, Finland). Water circulation thermostat LOIP LT-105 (Russia) was used for the thermostating of the subphase in the trough. The accuracy of temperature control and maintenance was one degree. Surface tension was measured by Wilhelmy plate method with an accuracy of 0.1 mN/m.

At the first step, the dependence of the floating layer properties on the number of NPs in the subphase volume was studied. In this case the temperature of water subphase was maintained constant and equal to 21 °C. Various aliquots of the synthesized Cu NPs (100, 200, 300, 400 and 500 μl) were added to 130 ml of deionized water directly in the LB-trough and mixed there with a glass rod. Then we waited 5 minutes to reach thermodynamic equilibrium between the surface and the volume of the subphase (the time required for the adsorption of Cu particles to the water surface from the volume). Then the surface area was symmetrically compressed by two moving barriers at a speed of 15 mm/min. At the same time, the recording of the dependence of the surface pressure on the area occupied by a layer of particles was obtained.

In the second step, the specificity of the formation of Cu NPs layer at different temperatures of water subphase and fixed number of NPs (for added aliquot equal 500 μl) was studied. The temperature of the water subphase at this step was 21, 29, 35 and 42 °C. The time intervals of the Gibbs layers formation, starting moments of the barriers movement and the data recording were the same as in the first step.

In the third step, the sorption of Cu NPs on the Langmuir ML of AA was studied. For this purpose, 50 μl of AA solution (Sigma-Aldrich) in chloroform (Vekton, Russia) with concentration of 10^{-3} M was spread on the surface of the water subphase with the addition of 500 μl of Cu NPs solution. The subphase temperature was 21 and 42 °C. The compression conditions were the same as at the formation of Gibbs floating layers.

Finally, on the fourth step of this work, thin films on the surface of solid substrates of highly oriented pyrolytic graphite (HOPG) were formed by a single transfer of the floating layer of Cu/AA using the Langmuir-Schaefer method with surface pressures of 10, 20 and 40 mN/m. The surface microrelief of the obtained films was investigated by AFM method in the semicontact mode with the help of Solver Nano scanning probe microscope (NT MDT, Russia).

3. Results and discussion

The compression (a) and expansion (b) isotherms of the Gibbs floating layers were obtained for different Cu NPs number in the subphase (Figure 1). It is the first step of the study performed in order to establish the dependence of mechanical properties of layers on NPs number. The increase of the amount of added Cu NPs solution into water subphase (for aliquots from 100 to 300 μl) leads to the increase of the surface pressure at the end of the compression process – from 0 to 9.4 mN/m. This indicates the active emergence process of the surfactant molecules and Cu NPs covered by surfactant shell to the "gas – liquid" interface at the allotted time of 5 minutes and evidences about increase of their number at the surface proportionally to their amount in volume. However, this amount of surfactant molecules is not yet sufficient to form close-packed layer. Therefore there is no inflection point on the compression isotherms. According to [1], this state of the surfactant layer can be interpreted as gaseous or liquid-expanded where the molecules practically do not interact with each other.

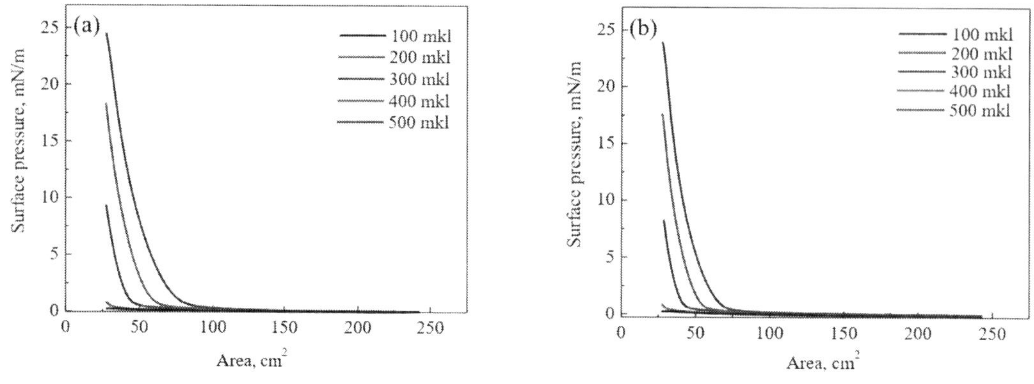

Figure 1. Compression (a) and expansion (b) isotherms obtained by adding different aliquots of Cu NPs solution to the subphase at 21 °C.

An increase of the Cu NPs number in the subphase volume (when the addition of the nanoparticle solution is 400 or 500 μl) leads to the appearance of an inflection point (Figure 1). In this case the close-packed state of the surfactant molecules in the layer is reached.

Figure 2 shows influence of the subphase temperature on the process Cu NPs sorption on the liquid surface during compression (a) and expansion (b) of a layer. At room temperature an inflection point is present on the compression isotherm. However, the state of collapse is not achieved. When the temperature of the subphase rises to 42 °C, the isotherms have signs of collapse (Figure 2, a). At the same time, the isotherms are shifted towards large areas. This indicates the increase of the surface concentration of the surfactant molecules and Cu NPs wrapped by the surfactant shell (Figure 3). Expansion isotherms (Figure 2, b) has less noticeable differences. This may be due to the manifestation of a stronger interaction and aggregation of Cu NPs located on the surface.

It should be noted that the sorption on the surface of the liquid subphase is more pronounced when there is a floating layer of AA. A shift of compression isotherms towards large areas is also present. The increase of the subphase temperature from 29 to 42 °C leads to an increase of the area occupied by Langmuir floating layers in a liquid-condensed state by 30 and 32 %, respectively (Figure 4).

This behavior can have many reasons, in particular, that the solubility of AA is lower than the solubility of SDS by several orders of magnitude, and the surface concentration of AA molecules is less dependent on temperature than the surface concentration of SDS molecules.

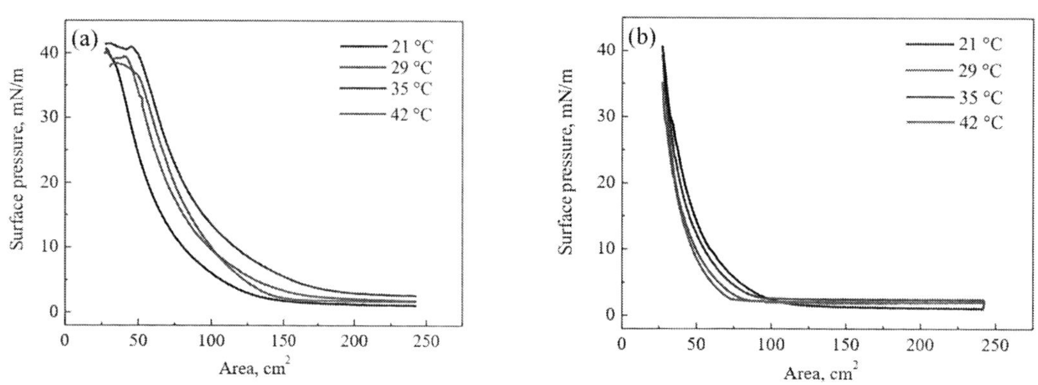

Figure 2. Compression (a) and expansion (b) isotherms obtained at different temperatures of the subphase with the addition of 500 μl of Cu NPs solution.

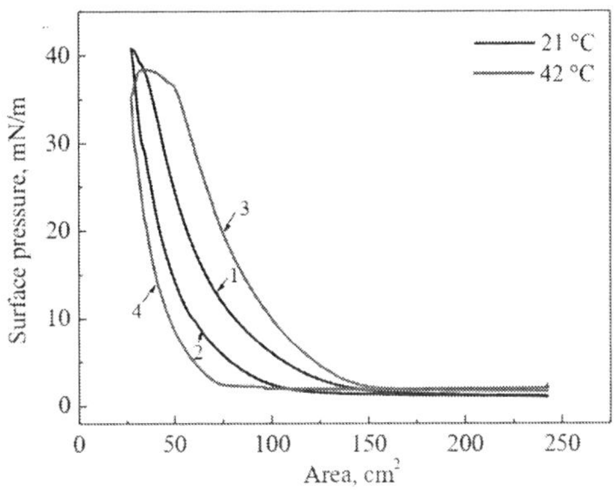

Figure 3. Combined compression (curves 1 and 3) and expansion (curves 2 and 4) of Gibbs floating layers obtained by adding 500 μl of Cu NPs solution.

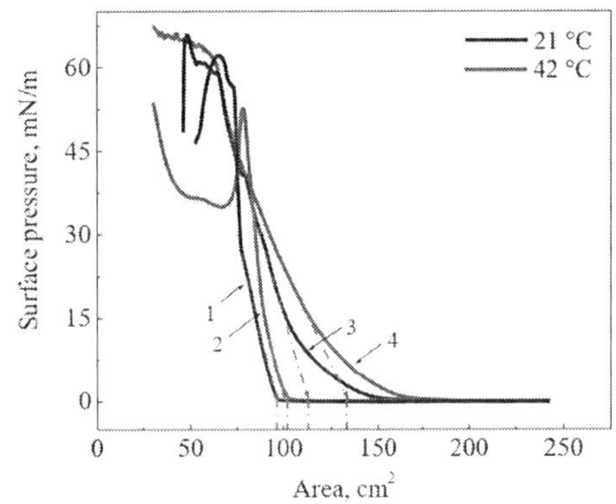

Figure 4. The compression isotherms of the Langmuir layers of AA on the surface of pure deionized water (curves 1 and 2) and on the Cu NPs subphase (curves 3 and 4).

Another possible reason may be the partial substitution of SDS molecules for AA molecules in the shells of Cu NPs and this fact also increases the sorption interaction of Cu NPs with the AA molecules on the subphase surface.

A tendency to aggregation of Cu NPs sorbed to AA ML at temperature of water subphase 42 °C is also confirmed by the fourth step of this investigation, when the surface morphology of thin films was studied by AFM. These film were obtained by transfer of the floating layers onto the surface of HOPG substrates at different pressures of 10, 20 and 40 mN/m. It can be seen (Figure 5), when the transfer pressure increases, the number and size of the aggregates also increase. The film contains aggregates of Cu NPs about 100 nm and the height difference increases almost twice (from 4 to 7 nm).

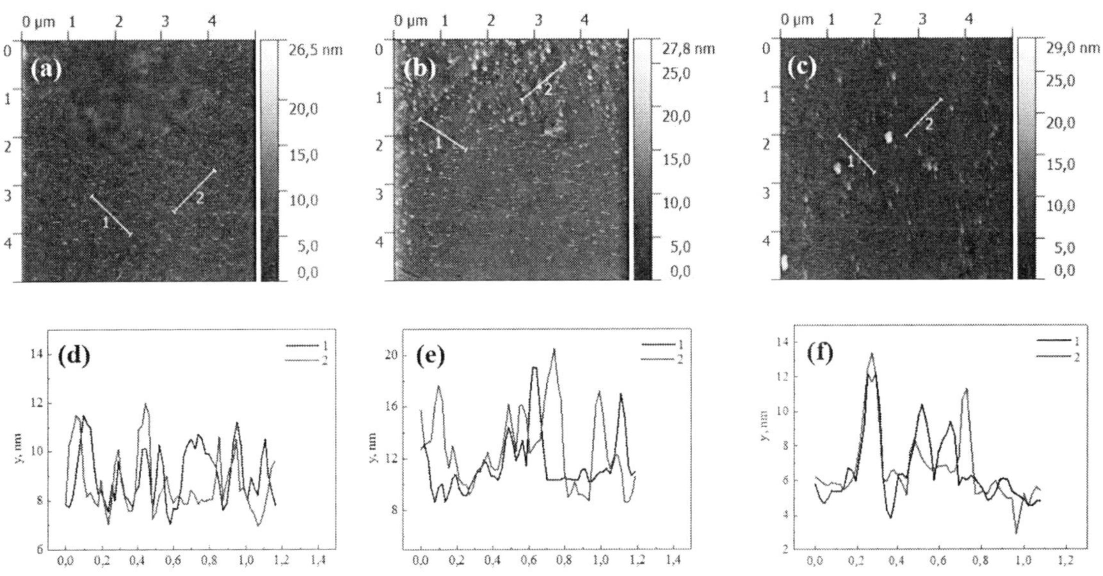

Figure 5. AFM images of surface microrelief (a, b, c) and height profiles (d, e, f) of Cu NPs/AA films on HOPG substrates obtained at surface pressures: 10 (a, d), 20 (b, e) and 40 mN/m (c, f).

4. Conclusion

The Gibbs floating layers formed by the adsorption of copper nanoparticles stabilized by SDS amphiphile shell from the subphase volume to the "liquid – gas" interface as well as the Langmuir floating layers of arachidic acid formed on this interface were investigated by the method of compression isotherms.

It was shown that an increase of the number of Cu NPs in the subphase volume leads to an increase of both the number of water-soluble SDS molecules and the number of Cu NPs in the Gibbs floating layer. The close-packed state is reached when 400 or 500 μl solutions of Cu NPs were added to the subphase.

The increase of the subphase temperature intensifies the process of sorption of Cu NPs from the volume onto the subphase surface. The area occupied by the floating layer increases in both cases of formation of Gibbs floating layer of Cu NPs and Langmuir floating layer of arachidic acid.

AFM studies of thin films transferred onto a solid substrate have shown that the degree of condensation of the floating layer significantly affects the degree of aggregation of Cu NPs. Number of aggregates, their lateral dimensions and height increase with increasing transfer pressure.

Acknowledgments
This work was supported by the Russian Foundation for Basic Research (Grant No. 19-03-00763-a).

References

[1] Adamson A W 1997 *Physical chemistry of surfaces* (6th Ed. Canada: A Wiley–Interscience Publication)
[2] Petty M C 1996 *Langmuir-Blodgett Films: An Introduction.* (Cambridge: University Press) p. 234
[3] Ariga K, Yamauchi Y, Mori T, Hill J P, 2013 *Adv. Mater.* **25** 6477
[4] Gorbachev I A, Shtykov S N, Brezesinski G, Glukhovskoy E G 2017 *BioNanoScience* **7** 686
[5] Singh S, Srinivasa R S, Talwar S S, Major S S 2007 *Materials Letters* **61** 3224
[6] Lin B, Bohanon T M, Shih M C, Dutta P 1990 *Langmuir* **6** 1665
[7] Al-Alwani A J, Chumakov A S, Begletsova N N, Shinkarenko O A, Markin A V, Gorbachev I A, Bratashov D N, Gavrikov M V, Venig S B, Glukhovskoy E G 2018 *Mater. Res. Express* **5** 045050
[8] Xu B, Zhou J, Ni Z, Zhang C, Lu C 2018 *Solar Energy Materials and Solar Cells* **179** 87
[9] Ko Y, Kwon M, Bae W K, Lee B, Lee S W, Cho J 2017 *Nature Communications* **8** 536
[10] Khomutov G B, Bykov I V, Gainutdinov R V, Gubin S P, Obydenov A Yu, Polyakov S N, Tolstikhina A L 2002 *Colloids and Surfaces A: Physicochemical and Engineering Aspects* **198-200** 347
[11] Patlak C, Gershfeld N 1967 *J. Colloid and Interface Science* **25** 503
[12] Stefaniu C, Chanana M, Wang D, Brezesinski G, Mohwald H 2011 *J. Phys. Chem. C* **115** 5478
[13] Usol'tseva N V, Smirnova A I, Kazak A V, Giricheva N I, Galanin N E, Shaposhnikov G P, Bodnarchuk V V, Yablonskii S V 2017 *Opto-Electronics Review* **25** 127
[14] Kazak A V, Zhukova L N, Kovaleva M I, Chausov D I, Kuznetsov M M, Gabdulsadykova G F 2018 *Liq. Cryst. and Their Appl.* **18** 74 (in Rus)
[15] Kim V P, Ermakov A V, Glukhovskoy E G, Rakhnyanskaya A A, Gulyaev Y V, Cherepenin V A, Taranov I V, Kormakova P A, Potapenkov K V, Usmanov N N, Saletsky A M, Koksharov Y A, Khomutov G B 2014 *Nanotechnologies in Russia* **9** 280
[16] Begletsova N N, Selifonova E I, Zakharevich A M, Chernova R K, Sevostyanov V P, Glukhovskoy E G 2019 *Moscow University Chemistry Bulletin* **74** 79

Laser sintering of silver nanoparticles deposited by dry aerosol printing

K M Khabarov[1], D V Kornyushin[1], B I Masnaviev[1], D N Tuzhilin[1, 2], A A Efimov[1], D L Saprykin[2], V V Ivanov[1]

[1]Moscow Institute of Physics and Technology, Dolgoprudny 141701, Russia
[2]ESTO PJSC, Zelenograd, 124460, Russia

kirill.khabarov@phystech.edu

Abstract. This work is devoted to the sintering of silver nanoparticle arrays in the form of lines by the local treatment with laser radiation with wavelengths of 527, 1054 and 1064 nm in pulsed and continuous laser modes. The silver nanoparticle lines were formed by the dry aerosol printing, free from organic solvents and surfactants. To form the lines with a width of 25–300 μm and a thickness of 0.1–15 μm a focused stream of nanoparticles with an average size of about 60 nm was used. It is shown that when using continuous laser radiation with a wavelength of 1064 nm, the minimal specific resistance of silver nanoparticle lines was achieved 3 times higher than that for the bulk material was reached.

1. Introduction

In a large number of microelectronic components it turned out to be more cost effective to refuse from lithographic technologies and develop printing technologies that use direct material delivery to form microstructures. Currently, the most promising technology in printed electronics is the aerosol printing based on the selective deposition of aerodynamically focused nanoparticle beams on a substrate [1]. With this method it would become possible to produce microstructures on curved surfaces, transistors [2], solar cell elements [3], multilayer ceramic capacitors [4], microheaters [5], and other elements. Previously, the authors proposed to use aerosol spark discharge generator to form nanoparticle flows for aerosol printing [1]. In this way aerosol nanoparticles are synthesized by electrical erosion of electrodes of required material in the processes of repetitive spark discharge in a gas flow [6]. This allows to obtain a stream of dry aerosol nanoparticles from metals, their oxides and semiconductors [7, 8]. Additionally, the shape and size distribution of the particles is controlled using a tube furnace [9] or electrostatic precipitator [10]. The dry aerosol nanoparticles stream is then focused and delivered to the substrate. The use of dry nanoparticles is preferable, as in the following sintering process there is no need to preliminary remove solvent [11, 12].

An important advantage of the aerosol printing method is the possibility of further sintering of the deposited structures at lower temperatures due to the small particle size. Consequently it is possible to print conductive microstructures on heat-sensitive substrates, for example, on plastic [13]. Actually, the sintering of nanoparticle arrays deposited on a substrate is efficient to produce by laser radiation heating as this method delivers energy to the sample locally without damaging the substrate material [11, 14].

The sintering process result was evaluated by measuring the specific electrical resistivity of the final micro-sized samples. Thus, in [11, 14] the resistivity of sintered samples was 3–15 times higher than the resistivity of bulk silver. This means that the challenging task is to find the optimal conditions for laser sintering of nanoparticle arrays which may depend on both as material and size of particles and so as on the parameters of laser radiation. In the present work, this problem is solved for laser sintering of silver nanoparticle arrays formed by the dry aerosol printing method.

2. Experimental

The experiment was carried out in two processes separated in time: in the first process a layer of silver nanoparticles was formed by the dry aerosol printing and its local laser sintering was accomplished during the second pass.

The scheme of experimental setup for forming nanoparticle arrays in the form of lines is shown in figure 1 (a). It includes a spark discharge nanoparticle generator, (the operation principle is given in the article [7]), a thermal optimizer in the form of a tube furnace which changes nanoparticles shape and size [9], a coaxial nozzle for nanoparticles depositing and a coordinate table which moves the substrate at a given speed. To form the lines with a width of 25 – 300 μm and a thickness of 0.1 – 15 μm a focused stream of silver nanoparticles with an average size of 60 nm was used. The shape of the coaxial nozzle and the parameters of the gas flows were chosen on the basis of the previous theoretical and experimental studies [15 - 17]. Nanoparticles were deposited on a substrate between two previously applied conductive contacts necessary for measuring the specific resistance.

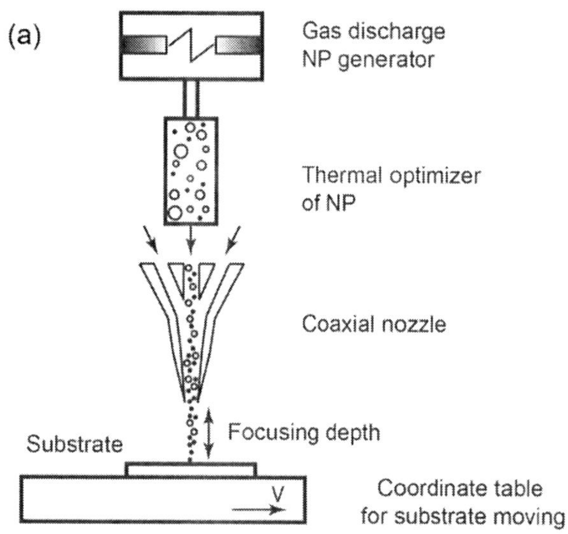

Figure 1(a, b). (a) The sketch of the experimental setup for forming lines of silver nanoparticles; (b) Optical scheme for laser sintering of silver nanoparticles deposited lines.

For sintering of the formed nanoparticle lines an optical scheme shown in figure 1 (b) was constructed. Laser radiation was delivered to the substrate with a sample mounted on the same moving coordinate table. The sintering processes of nanoparticle arrays were studied by using laser radiation with wavelengths of 527, 1054 and 1064 nm. At the same time the input energy was varied by changing the average power of laser radiation and the movement speed of the laser spot over the array of nanoparticles in the range up to 10 W and 520 mm/s respectively. Additionally the sintering process quality was studied and compared in continuous (CW) and pulsed laser modes. Sintering in a pulsed mode was performed at a pulse repetition rate of 4 kHz with a pulse duration of 20 ns.

In the sintering processes the laser spot diameter in the sample plane was controlled using a Duma Optronics BeamOn HR laser beam profile meter and the power of incident radiation on the sample using an Ophir F150A-BB-26 power meter.

The sintering process results were estimated by the electrical resistivity of the lines. The absolute resistance was previously measured using a Fluke 287 multimeter and the cross section profile with the aid of an optical profilometer Leica DCM 3D.

3. Results and discussion

The conducting lines were formed by nanoparticles, their images from the transmission electron microscope (TEM) JEM-2100 (JEOL) are shown in figure 2 (a). Due to the spherical shape of the deposited nanoparticles the green samples of lines have a low porosity and a high density of contacts between the particles which ensures a high conductivity. Figures 2 (b - f) show images from a scanning electron microscope (SEM) JSM 7001F (JEOL) for characteristic green sample and samples sintered by continuous and pulsed lasers with wavelengths of 527, 1054 and 1064 nm.

Figures 3 (a - d) show the experimental dependences of the specific electrical resistance ρ of sintered nanoparticle lines for different lasers as a function of the specific energy coming to the sample per second and the speed of the laser spot moving along the array of nanoparticles v in logarithmic scale. Almost in all dependences (figure 3 (a - d)) for a given radiation power density there is a weakly expressed line resistivity minimum near the moving speed of the laser spot along an array of nanoparticles of about 1 m/s. This speed corresponds to the best result of resistivity 3 times higher in comparison with bulk silver as it shown in figure 3 (d). Sintering regimes in this area are close to optimal. In figure 3 there is an increase in the lines resistivity to the left of the minimum at low scanning speeds, effecting in overheating and evaporation of the sintered material as the dose of delivered laser radiation increased. The growth of the lines resistivity to the right of the minimum at high scanning speeds, where the dose of delivered laser radiation on the contrary decreases, is logically associated with the incomplete sintering of nanoparticles.

Sintering by laser radiation with a wavelength of 527 nm required less energy consumption than by radiation with wavelengths of 1054 and 1064 nm and led to 30-50% greater values of the silver lines resistivity. This feature is explained by the fact that silver nanoparticles better absorb radiation at a wavelength of 527 nm due to the excitation of plasmon effects on them and the absorption efficiency decreases with increasing wavelength [18].

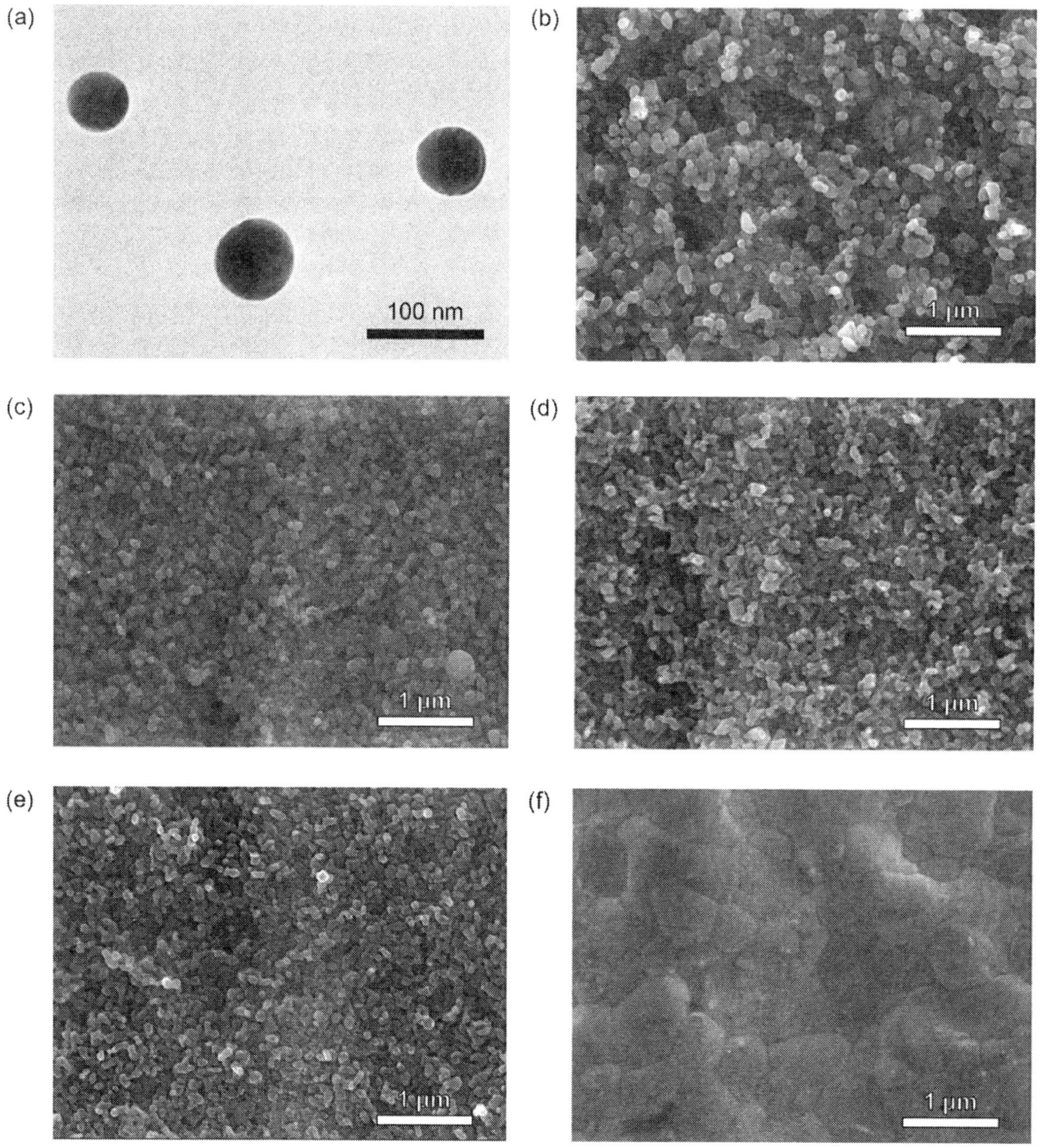

Figure 2(a – f). (a) TEM images of nanoparticles; (b) SEM images of the surface of the green sample; (c - f) SEM images of samples surface sintered by laser radiation: (c) pulsed 527 nm, (d) pulsed 1054 nm, (e) CW 527 nm, (f) CW 1064 nm.

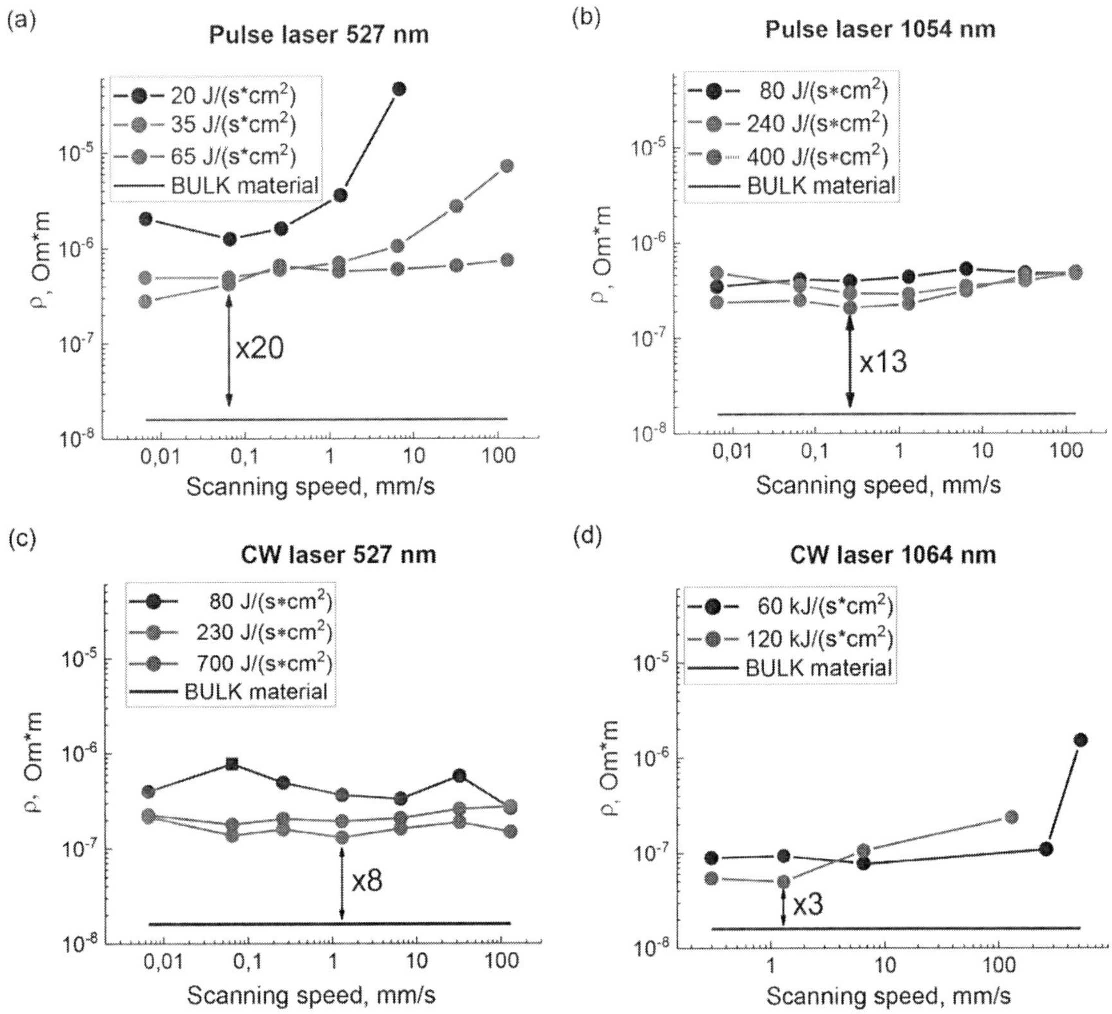

Figure 3(a – d). Dependencies of specific electrical resistivity as a function of the specific energy coming to the sample per second and the speed of the laser spot moving along the array of nanoparticles for four types of lasers.

4. Conclusion

By the example of laser sintering of conductive silver nanoparticle lines there has been demonstrated the advantage of combining the method of dry aerosol printing with laser sintering due to beneficial properties of the local delivery of nanoparticles and laser energy. The best results in laser sintering of conductive silver lines were obtained by continuous laser radiation with a wavelength of 1064 nm. In addition, the results obtained within the study demonstrate that continuous radiation provides more efficient sintering of nanoparticle arrays than pulsed, and radiation with wavelengths of 1054 and 1064 nm is more preferable than with a wavelength of 527 nm for obtaining lines with minimal specific resistivity. However, infrared laser radiation sintering requires greater energy costs than for short-wave radiation. This feature is explained by a higher silver nanoparticles absorption coefficient at a wavelength of 527 nm, which is associated with the excitation of plasmon effects at the resonance absorption at a wavelength of 450 nm, thereat the absorption efficiency decreases with increasing wavelength.

Acknowledgments
This work was supported by the Ministry of Science and Higher Education of the Russian Federation (project №RFMEFI57517X0160).

References
[1] Efimov A. A. et al. Controlled focusing of silver nanoparticles beam to form the microstructures on substrates //Results in Physics. – 2017. – T. 7. – C. 440-443.
[2] Jones C. S. et al. Aerosol-jet-printed, high-speed, flexible thin-film transistor made using single-walled carbon nanotube solution //Microelectronic Engineering. – 2010. – T. 87. – №. 3. – C. 434-437.
[3] Yang C. et al. Preparation of active layers in polymer solar cells by aerosol jet printing //ACS applied materials & interfaces. – 2011. – T. 3. – №. 10. – C. 4053-4058.
[4] Folgar C. E., Suchicital C., Priya S. Solution-based aerosol deposition process for synthesis of multilayer structures //Materials letters. – 2011. – T. 65. – №. 9. – C. 1302-1307.
[5] Arsenov P. V. et al. Aerosol Jet Printing of Platinum Microheaters for the Application in Gas Sensors //IOP Conference Series: Materials Science and Engineering. – IOP Publishing, 2019. – T. 473. – №. 1. – C. 012042.
[6] Mylnikov D. A. et al. Investigation of electric erosion of silicon electrodes in aerosol nanoparticles synthesis //AIP Conference Proceedings. – AIP Publishing, 2017. – T. 1858. – №. 1. – C. 040007.
[7] Ivanov V. V. et al. Synthesis of Nanoparticles in a Pulsed-Periodic Gas Discharge and Their Potential Applications //Russian Journal of Physical Chemistry A. – 2018. – T. 92. – №. 3. – C. 607-612.
[8] Lizunova A. A. et al. Synthesis of Ge and Si nanoparticles by spark discharge //Journal of Physics Conference Series. – 2017. – T. 917. – №. 3.
[9] Lizunova A. A. et al. Influence of the sintering temperature on morphology and particle size of silver synthesized by spark discharge //IOP Conference Series: Materials Science and Engineering. – IOP Publishing, 2018. – T. 307. – №. 1. – C. 012081.
[10] Arsenov P. V. et al. Influence of the operating parameters of the needle-plate electrostatic precipitator on the size distribution of aerosol particles //IOP Conference Series: Materials Science and Engineering. – IOP Publishing, 2018. – T. 324. – №. 1. – C. 012016.
[11] Theodorakos I. et al. Selective laser sintering of Ag nanoparticles ink for applications in flexible electronics //Applied surface science. – 2015. – T. 336. – C. 157-162.
[12] Efimov A. A. et al. Dry aerosol jet printing of conductive silver lines on a heated silicon substrate //IOP Conference Series: Materials Science and Engineering. – IOP Publishing, 2018. – T. 307. – №. 1. – C. 012082.
[13] Efimov A. A. et al. Investigation of sintering of silver lines on a heated plastic substrate in the dry aerosol jet printing //Journal of Physics: Conference Series. – IOP Publishing, 2018. – T. 1124. – №. 8. – C. 081041.
[14] Kumpulainen T. et al. Low temperature nanoparticle sintering with continuous wave and pulse lasers //Optics & Laser Technology. – 2011. – T. 43. – №. 3. – C. 570-576.
[15] Protas N. V. et al. Computer simulation of aerosol nanoparticles focusing and deposition process for functional microstructure fabrication //Journal of Physics: Conference Series. – IOP Publishing, 2018. – T. 1124. – №. 8. – C. 081033.
[16] Efimov A. A. et al. Study of aerosol jet printing with dry nanoparticles synthesized by spark discharge //Journal of Physics Conference Series. – 2017. – T. 917. – №. 9.
[17] Arsenov P. V., Efimov A. A., Ivanov V. V. Effect of Methods of Changing in Focusing Ratio on Line Geometry in Aerosol Jet Printing //Key Engineering Materials. – Trans Tech Publications, 2018. – T. 779. – C. 159-164.
[18] Maier S. A. Plasmonics: fundamentals and applications. – Springer Science & Business Media, 2007, pages 77 - 80.

The influence of star-like palladium nanocrystallites on the gas transport parameters of modified palladium-containing membranes

I Petriev[1,2]

[1]Department of Physics, Kuban State University, Krasnodar, 350040, Russia
[2]Laboratory of problems of stable isotope spreading in living systems, Southern Scientific Centre of the RAS, Rostov-on-Don, 344000, Russia

petriev_iliya@mail.ru

Abstract. Methods to modify the surface of Pd-23% Ag alloy films were developed in order to increase the velocity of hydrogen transmission; we obtained palladium coating of "nanostars" and "nanopores" type. Modification of Pd-Ag films surface, obtained by nanostars palladium coating makes it possible to achieve a hydrogen flow density of 0.76 mmol / (s*m2), which is 1.6 times greater in comparison with modification by "nanopores" coating under low temperature (<90 ° C) and pressure (<0.6 MPa) conditions.

1. Introduction

Every year for the needs of alternative energy requires more and more high-purity hydrogen. The most common method to obtain it is reforming of organic raw materials with gas mixture formation and subsequent release of high-purity hydrogen, with the help of various types of metal membrane filters [1-4]. As a basis for creation of these membrane filters palladium and its alloys are used as they have unique properties in relation to the hydrogen transmission [5-7].

The operating temperature range of such membrane filters is in the range of 500-600°C, which creates some limitations and disadvantages in use. First, the energy loss in the process of hydrogen production. Secondly, the pollution of diffused hydrogen by impurities released from the structural materials of the equipment for diffusion purification of hydrogen. These aspects limit the use of palladium membranes and prevent their mass adoption in the industry. The situation would be different if there were membranes capable of transporting hydrogen in the temperature range up to 200°C with sufficient mass transfer rate. Ideally, they should be operational at ambient temperatures.

Some of the few studies affecting the temperature range up to 200°C have shown a critical drop in hydrogen permeability up to instability of the results [8-10], however, studies showing stable reproducible results of hydrogen permeability of low-temperature palladium-containing membranes with any significant density of hydrogen flow. At the time of this writing, the authors are not known.

The absence of such membranes hinders the development of processes and devices in which they could be implemented. Such devices could be a low-temperature hydrogen diffusion electrode and devices based on it: an electrochemical hydrogen pump (compressor) that can be used in devices for the economical separation of hydrogen isotopes, and a low-temperature hydrogen fuel cell [11,12].

In addition, these membranes are integrated into methanol steam reformers, which allows creating compact membrane reformers that are effective for in situ hydrogen production [13-15]. Also, their use is relevant for the separation of hydrogen isotopes, for example, membrane production of water with a modified isotopic composition [16-18].

The aim of the work was to obtain resistant to long-term use of palladium-containing films capable of passing hydrogen at low temperatures. For this, two types of surface modifiers were synthesized for membranes made of Pd-23% Ag alloy and their effect on the kinetic characteristics of hydrogen transfer was studied.

The formation of nanostructured palladium layer capable for chemisorbing of hydrogen on the surface of membrane increases the actual operating surface, which leads to increase of chemisorption centers quantity, which role is most often played by crystallites angles and facets. Moreover in case of sufficiently small Pd crystal size the majority of octahedral interstices belong to the surface, that facilitates the transmission of hydrogen and reduces the probability of hydrogen atom capture by various kinds of surface defects. This modification may be effective in case of limiting the process of hydrogen transfer through the membrane by dissociative-associative absorption-elimination processes at the boundaries.

2. Methods and materials

Two methods were used for the deposition of finely dispersed palladium onto the surface of thin Pd–Ag alloy films.

(1) Magnetron sputtering of alloy films with subse- quent diffusion annealing and etching. Zinc was used as the active component of the Raney palladium alloy for sputtering, since the sputtering current strength for Zn (50 A) is close that for Pd (30 A). The alloy with 50% Zn was chosen for modifying the palladium–silver surface. Diffusion annealing in an inert argon (99.99%) environment was the next processing stage. The volume flow rate was set to 2 L/min to establish firm adhesion between the Raney alloy and the substrate material and secure attachment of finely dispersed platinoid grains after etching of the active component (zinc). Zinc (soluble component) leaching in a 6M NaOH solution was the last phase of the process.

(2) The surface of a Pd–23%Ag film secured in a holder was rinsed with 96% ethanol, degreased by boiling for 30 min in a 6M NaOH solution, and immersed in a 60% HNO3 solution for etching (30 s). Immediately after etching, the film was introduced into a vessel with flowing distilled water and held there for 10 min. The film on an inert holder made of 99.99% pure silver was then transferred to an electrolytic cell for coating. The holder served as the cathode current lead; contact was established by a silver wire. The palladium–silver alloy film was then transferred to a cell with 0.1M HCl and polarized anodically at a current density of 10–20 mA/cm^2 using a P-250I potentiostat/galvanostat, was rinsed again with bidis- tilled water, and polarized cathodically in 0.05 M H_2SO_4 at a current density of 10–20 mA/cm^2. After rinsing with bidistilled water, the cell was filled with a solution of the following composition: 2% H_2PdCl_4, 0.1M tetrabutylammonium bromide (surfactant). Palladium black precipitated for 30 min at 2–6 mA/cm^2.

3. Results and discussion

To obtain thin palladium-silver films the magnetron spattering method was used. Plates of silver and palladium with different ratio of its areas were used as a target for the magnetron [19, 20]. Magnetron spattering was performed on the Quorum Q150TS / E / ES device. The chemical composition of the obtained films was studied by method of micro-X-ray spectral analysis on INCA (Oxford) semiconductor energy dispersive add-on device with JEOL JSM-7500F scanning electron microscope. To obtain a film with silver content of 23%, which is optimal for hydrogen permeability and mechanical properties [21, 22], a target with an areas ratio S (Ag) / S (Pd) = 20.8 / 79.2 was used.

In the course of this study, two types of surface modification of hydrogen permeable membranes were developed: “nanostars” and “nanopores”. The microphotographs of the films surface obtained by JEOL JSM-7500F scanning electron microscope are shown in figure 1. In figure 2 the data is shown on the measurement of hydrogen permeability for a palladium-silver alloy modified by coating of “nanostars” type (a) and coating of “nanopores” type (b).

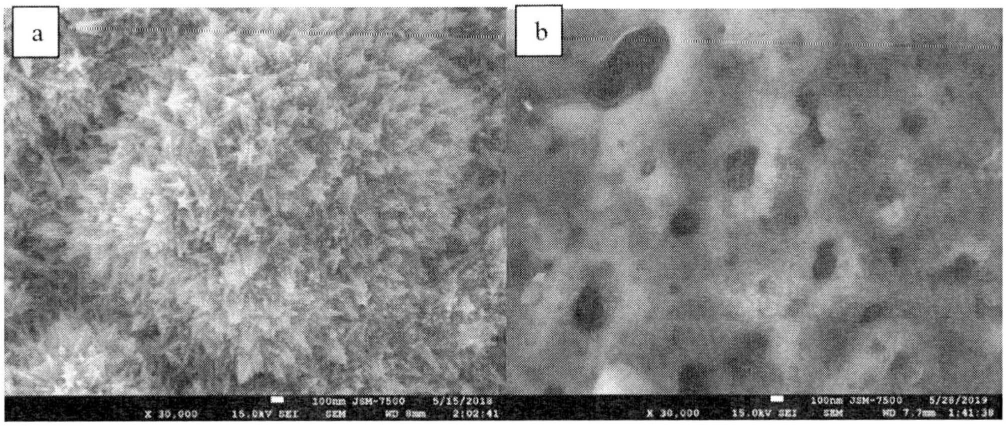

Figure 1. Microphotographs of palladium-silver films surface with modified surface, obtained by the method of “nanostars” (a), and by the method of “nanopores” (b).

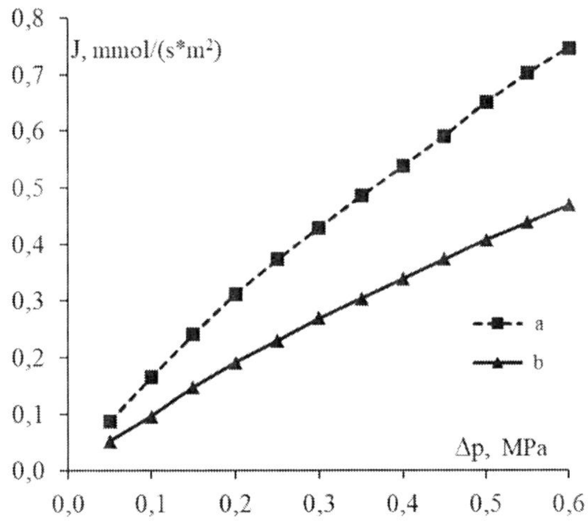

Figure 2. Dependence of the rate of fluence (the ratio of hydrogen flow to the area of a sample) on the excess pressure of hydrogen on the outer face of the membrane for the method “nanostars” (a) and the method of “nanopores” (b).

From figure 2 it is obvious that the dependence of the flow density on hydrogen overpressure on the input side of the membrane with modified surface is well approximated by a line of the 1st order, which indicates according to [23] that hydrogen penetration velocity is limited by dissociation of hydrogen on the surface.

Thus it has been experimentally confirmed that the velocity of hydrogen transmission under conditions of low temperature (<90°C) and pressure (<0.6 MPa) through sufficiently thin palladium membranes (<10 μm), is limited by dissociative-associative processes at the boundaries, and can be significantly increased (up to order of magnitude) due to acceleration of the limiting stage by applying a superficial modifying palladium coating. Modification of membranes with “nanostars” type of coating allows achieving hydrogen flow density up to 0.76 mmol / (s*m^2) that is 1.6 times greater than with modified by "nanopores" type of coating.

Acknowledgement
The reported study was funded by RFBR and Krasnodar region to the research project № 19-42-233004 and by the scholarship of the President of the Russian Federation to young scientists and post-graduate students SP-1237.2018.1 and by the state assignment of the SSC RAS (Project AAAA-A19-119040390083-6).

References
[1] Lytkina A A, Orekhova N V, Ermilova M M and Yaroslavtsev, A B 2017 *Int. J. Hydrogen Energy* **43** 198-207.
[2] Mendes D, Chibante V, Zheng J-M, Tosti S, Borgognoni F, Mendes A and Madeira L M 2010 *Int. J. Hydrogen Energy* **35** 12596–608.
[3] Petriev I S, Bolotin S N, Frolov V Y, Baryshev M G, Kopytov G F and Isaev V A 2019 *Russ. Phys. J.* **61** 1894-8.
[4] Alimov V N, Bobylev I V, Busnyuk A O, Kolgatin S N, Kuzenov S R, Peredistov E Y and Livshits A I 2018 *Int. J. Hydrogen Energy* **43** 13318-27.
[5] Ievlev V M, Burkhanov G S, Roshan N R, Belonogov E K, Maximenko A A, Dontzov A I and Rudakov K E 2012 *The Metals* **6** 92-101.
[6] Alimov V N, Bobylev I V, Busnyuk A O, Notkin M E, Peredistov E Y and Livshits A I 2018 *J. Membr. Sci.* **549** 428-37.
[7] Jeon S I, Park J H, Magnone E, Lee Y T and Fleury E 2012 *Curr. Appl. Phys.* **12** 394-400.
[8] Arai M and Ymada K 1992 *J. Chem. Eng. Jpn* **25** 761-2.
[9] Pati S, Jat R A, Anand N S, Derose D J, Karn K N, Mukerjee S K and Parida S C 2017 *J. Membr. Sci.* **522** 151–8.
[10] Zhao C, Goldbach A and Xu H 2017 *J. Membr. Sci.* **542** 60–7.
[11] Petriev I S, Frolov V Y, Bolotin S N, Baryshev M G and Kopytov G F 2018 *Russ. Phys. J.* **60** 1611-7.
[12] Petriev I S, Bolotin S N, Frolov V Y, Barishev M G and Sozayev V A 2018 *Bull. Russ. Acad. Sci.* **82** 899-902.
[13] Lytkina A A, Orekhova N V, Ermilova M M, Petriev I S, Baryshev M G and Yaroslavtsev A B 2017 *Petrol. Chem.* **57** 1219-27.
[14] Kim Ch H, Y Han J, Lim H, Lee K Y and Ryi S K 2018 *J. Membr. Sci.* **563** 75-82.
[15] Lytkina A A, Orekhova N V, Ermilova M M, Petriev I S, Baryshev M G and Yaroslavtsev A B 2019 *Int. J. Hydrogen Energy* **44** 13310-22.
[16] Dzhimak S S, Basov A A, Volchenko N N, Samkov A A, Fedulova L V and Baryshev M G 2017 *Dokl. Biochem. Biophys.* **476** 323-5.
[17] Basov A A, Elkina A A, Samkov A A, Volchenko N N, Moiseev A V et al. 2019 *Iran. Biomed. J.* **23** 129–41.
[18] Fedulova L V, Basov A A, Vasilevskaya E R and Dzhimak S S 2019 *Curr. Pharm. Biotechnol.* **20** 245–53.
[19] Petriev I S, Frolov V Y, Bolotin S N, Baryshev M G and Kopytov G F 2015 *Russ. Phys. J.* **58** 1044-8.
[20] Petriev I S, Bolotin S N, Frolov V Y, Barishev M G, Isaev V A and Kopytov G F 2016 *Bull. Russ. Acad. Sci.* **80** 624-6.
[21] Petriev I. S. 2018 Journal of Physics: Conference Series **1134** 012045.
[22] Petriev I S, Bolotin S N, Frolov V Y and Baryshev M G 2019 *Dokl. Phys.* **64** 210-3.
[23] Baychtok Y K, Sokolinsky Y A and Ayzenbud M B 1976 *J. Phys. Chem.* **50** 1543-6.

Effect of crystal orientation on the implant profile and resistance of SiC

V I Egorkin, A V Nezhentsev, V E Zemlyakov, V A Gudkov and V I Garmash
National Research University of Electronic Technology, Moscow, 124498, Russia

Abstract. The process of ion implantation into silicon carbide is studied in this paper. We conducted an analysis of acceptor distribution profiles. We determined that silicon carbide layers with an orientation (11–20) have a higher degree of crystal lattice recovery after ion implantation with aluminum ions than a crystal lattice with an orientation (0001). A significantly higher efficiency of ion doping of silicon carbide with an orientation (11–20) was shown, which may indicate a more efficient recrystallization of the fastest growing crystallographic direction. We have established a higher degree of restoration of the crystal lattice of silicon carbide implants with an orientation (11–20) after implantation. Highly doped SiC p-layers were obtained with a specific resistance of about 0.02 Ohm*cm and with high crystalline structure quality without the use of ultra-large implantation doses. The minimum value of Rs for orientation (0001) is about 10 kΩ/□, and for orientation (11-20) - 2.7 kΩ/□.

1. Introduction

The ion implantation method has found wide application in silicon carbide technology. Fine-lying active layers of n- and p-type conductivity, p-n junctions, compensated semi-insulating regions can be created in silicon carbide. Nitrogen and phosphorus ions create a region of n-type conductivity. At the same time, the creation of p-type layers by the implantation of B^+ or Al^+ ions have faces certain difficulties. Using small doses of acceptor ions, p-type layers cannot always be obtained. The calculated acceptor concentration may not be sufficient to overcompensate the donor impurity present in the original substrate. A big problem that has not been fully resolved so far is the creation by ion implantation of heavily doped p-layers with low resistivity and high quality crystal structure [1]. The reasons for this situation are both purely electrophysical properties of silicon carbide, and technological methods used during ionic alloying operations. The depth of the acceptor energy levels of in the silicon carbide band gap is quite large. So, for Al it exceeds 0.2 eV, and for boron it is about 0.3 eV. While for the most common donor impurity in silicon carbide, nitrogen, the ionization energy is about 0.07 eV. Also, defects in the implanted layer that are stable at the temperatures of post-implantation annealing and have energy levels located deep in the silicon carbide bandgap. This partially compensates for the acceptor impurity, thereby further reducing the hole concentration. The low hole mobility in monocrystalline silicon carbide prevents the formation of p-layers with low resistivity. The post-implantation annealing process has a great influence on the result of ion doping. To obtain high-alloyed p-layers, short-time annealing is used [2]. Silicon carbide is susceptible to dissociation at temperatures used for post-implantation annealing, at which the surface layers lose silicon atoms and are covered with a carbon film. The thermal resistance of the large doses implanted layers due to the high imperfection decreases sharply, and such layers dissociate at a much higher rate than the single-crystal material. This leads to disruption of stoichiometry in the surface layers, the appearance of additional defects, deterioration of the surface morphology and a decrease in the thickness of the ion-doped layer.

2. Experiment

We used Lely 6H-SiC crystals with orientation (0001) and plates with orientation (11-20), cut from bulk crystals. On all plates, epitaxial layers about 1 μm thick with a concentration of donors (2 - 3) × 10^{17} cm^{-3} were grown by gas-phase epitaxy. The epitaxial layers were implanted with Al^+ ions at room temperature in several steps (energy range of 40–120 keV), the total ion doses varied in the range of 1.25×10^{15} cm^{-2} - 1.25×10^{16} cm^{-2}. Before the annealing, the implanted surface was covered with a layer of photoresist, which during the annealing process turned into a dense diamond-like carbon film having very good adhesion to silicon carbide. Annealing was carried out in argon atmosphere at an overpressure by resistive heating of a graphite heater in which silicon carbide plates were placed. The maximum annealing temperature varied from 1400 to 2150 °C. The warm-up time to the maximum temperature was 15 - 35 seconds. Exposure at maximum temperature was absent. After annealing, the carbon film was burned in an oxygen plasma, and the surface of the plates was refreshed in a solution of hydrofluoric acid. We formed the structure of the "clover leaf" on the obtained samples, which were used to measure surface resistance (Hall measurement).

3. Results

The results are shown in Figure 1, both curves have minimal values.

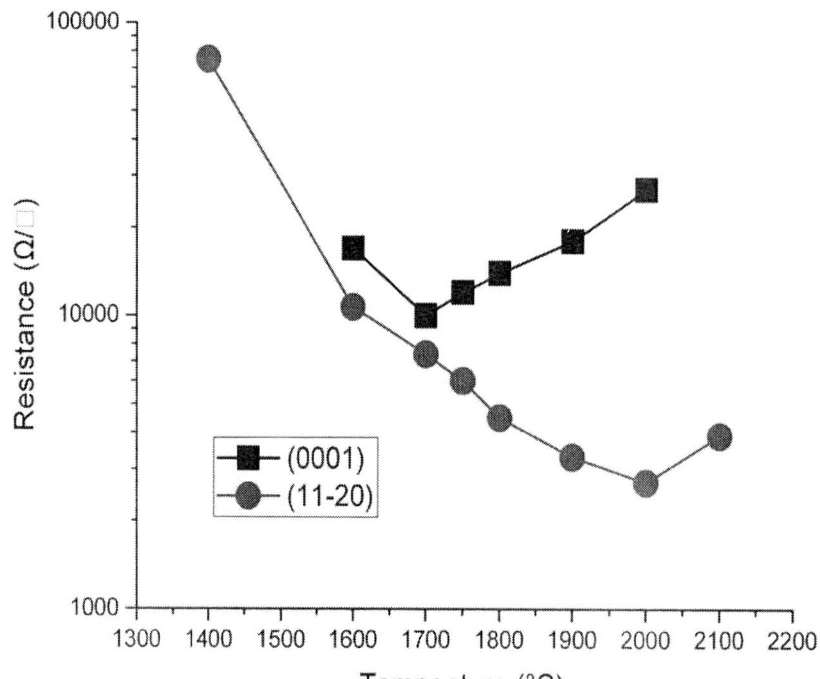

Figure 1. Dependences of surface resistance (Rs) of aluminum-doped silicon carbide layers for the two orientations of the original plates.

For orientation (0001), the minimum corresponds to the maximum annealing temperature of 1700 ° C, and for orientation (11-20) the minimum is shifted to higher temperatures and corresponds to a maximum temperature of 2000 ° C. The difference between the Rs layers of the two orientations is small at annealing temperatures of 1600–1700 ° C, then at higher temperatures it becomes quite significant. The negative effect of thermal dissociation for plates with orientation (0001) appears at lower temperatures than for plates with orientation (11-20). As part of our assumptions, this can be explained by the fact that the degree of restoration of the crystal lattice of the plates (0001) is low. Thus, we can conclude that to obtain very low resistances of ion-doped p-layers of silicon carbide, even when using moderately large doses of Al^+ ions, it is necessary to use silicon carbide plates with an orientation (11–20), and the temperature of post-implantation annealing should be in the region 2000 ° C, slightly shifting with increasing or decreasing the implantation dose. Figure 2 show the results of a layer-by-layer analysis alloyed layers of silicon carbide with different orientations. The layer-by-layer etching of silicon carbide was carried out by etching in the plasma of freon in the Plasma-600 unit. At the end of the analysis, the resulting height of the mesa structure was measured, from which it was determined that a 5 nm layer was removed in each etching cycle.From a comparison of the presented drawings, at least two significant differences can be seen. First, the level of acceptor concentration in the layer (11–20) is significantly higher than in the (0001) layer. The Na concentration in the layer (11–20) is almost three times greater than in the (0001). Second, in Figure 2(a), the hole mobility decreases from the boundary with the substrate to the surface of the layer. If the right side of this curve from the maximum of the Na concentration can be explained by a decrease in the concentration of scattering ionized aluminum centers, then on the left side of the curve this ratio is broken: that is, the acceptor concentration decreases to the surface, but the mobility continues to decrease.

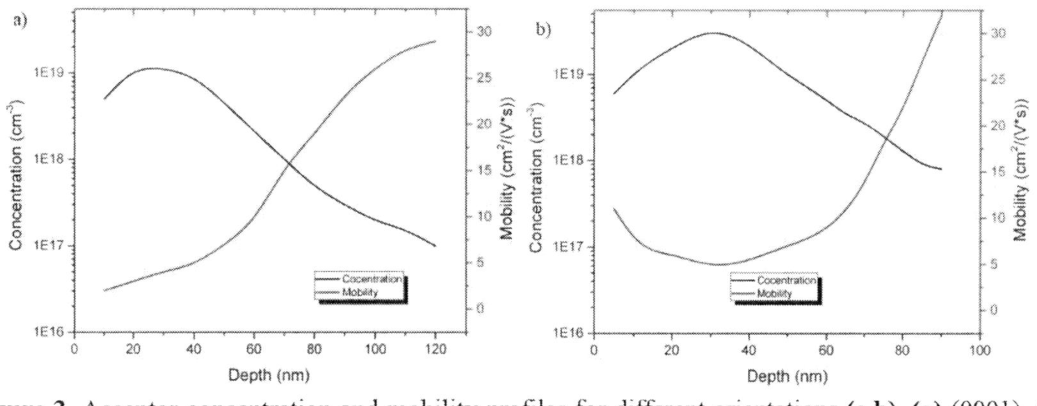

Figure 2. Acceptor concentration and mobility profiles for different orientations **(a,b). (a)** (0001), **(b)** (11-20).

This may indicate that the recrystallization of layers with the (0001) orientation is non-uniform in depth, moreover, it is worse near the surface than in deeper layers. As a result, a decrease in the mobility towards the surface of the ion-doped layer is determined not only by the concentration of ionized acceptors, but also by the high concentration of defective scattering centers and even by the transition to a polycrystalline state.

4. Conclusion

We have established a higher degree of restoration of the crystal lattice of silicon carbide implants with an orientation (11–20) after implantation. Highly doped SiC p-layers were obtained with a specific resistance of about 0.02 ohm*cm and with high crystalline structure quality without the use of ultra-large implantation doses. It should be noted that similar values of surface resistance (2 kOhm/□) presented in this work were obtained with doses of doping with Al^+ ions 5 times greater than the dose used by us, that is, the efficiency of the annealing process used by us was several times higher than the process used in [3]. That is, we determined that when plates with orientation (11–20) are used for ion doping, the crystal lattice is more efficiently restored, whose thermal stability increases more quickly during annealing, it becomes possible to increase the maximum annealing temperature when the negative effect of thermal dissociation is not appears, respectively, it becomes possible to reduce the resistance of the ion-doped layer.

Acknowledgments
This work was supported by the Ministry of education and science of the Russian Federation (grantagreement 14.581.21.0021). Unique identificator RFMEFI58117X0021.

References
[1] Heera V, Panknin D, Skorupa W, 2001 Applied Surface Science **184** 307-316
[2] Aleksandrov O, Kalinina E, 2009 PHYSICS OF THE SOLID STATE **43** 584-589.
[3] Negoro Y, Kimoto T, Matsunami H, 2004, Materials Research Society **815** 5.2.1-5.2.6.

Composition, structure and hardness of titanium after pulse laser processing in graphite

V Koshuro[1], **A Voyko**[1], **M Fomina**[1], **V Papshev**[1,2], **A Antropov**[1]
[1] Yuri Gagarin State Technical University of Saratov, Saratov 410054, Russia
[2] TAV Dental, Shlomi 2283202, Israel

Abstract. The structure, composition and hardness of the surface layer of titanium after laser processing in a solid carburizer (graphite) were studied. As a result of the processing, a coating containing Ti, C, O and characterized by a thickness of up to 10 μm, hardness up to 11.5±0.9 GPa and the presence of nano-sized structural elements was formed on the surface of titanium. The resulting diffusion layer with a fine dendritic structure consisted of Ti and C. The diffusion layer with a thickness of 130±20 μm was distinguished by a high microhardness of 16.7±0.6 GPa.

1. Introduction

Titanium alloys are widely used in the aerospace industry, shipbuilding, in the petrochemical and electronic industries, as well as in engine-building in the production of various structural elements, e.g. turbine blades [1,2]. To enhance the functional properties the working surfaces are strengthened by thermal, chemical-thermal or electrochemical treatment, as well as coating deposition. Titanium alloys are chemically active in a wide range of temperatures. For this reason, the following methods of surface treatment are widely used, in particular oxidation, cementation and nitriding. These strengthening methods can increase the hardness and wear resistance of products. The composition and properties of the formed layers depend on the processing temperature and composition of the reaction medium [3-6].

Among the methods of thermochemical treatment, cementation can be distinguished since it allows the formation of highly hard and wear-resistant carbide layers. Due to cementation performed using graphite powder and pulsed laser heating, a carbide layer is formed from non-stoichiometric titanium carbide (TiC), which makes allows the increase in the hardness of titanium to 1200 $HV_{0.1}$ (about 11.8 GPa) [7]. In some cases, graphite powder is pressed with a metal base and subjected to electrocontact heating to 800 °C and subsequent exposure for 20 minutes. As a result, titanium carbide layers are formed, which are characterized by a carbon content of up to 36 at.% and a hardness of 6.53 GPa (Knoop hardness at 100 gf) [8]. Induction thermochemical treatment is also widely used, in which a titanium product is placed in a refractory container filled with graphite. During processing, the samples are heated up to the melting temperature. Due to the given treatment, TiC layers with a thickness of up to 14 μm and a hardness of up to 20 GPa are obtained [9,10]. Carbide layers are also formed by electrospark alloying. Conducting the alloying of titanium with a graphite electrode allows the increase in the surface hardness to 20.0–20.4 GPa [11].

In some cases, it is more rational to strengthen the surface of titanium by applying carbide coatings. The layers of the WC-TiC-Co system obtained as a result of electrospark alloying of the titanium base with a hard alloy are characterized by a hardness of 18–22 GPa [12]. Carbide coatings are also formed by selective laser sintering, e.g. layers of TiC/TiB_2 system are produced by laser processing of titanium bases with a deposited $Ti/TiC/TiB_2$ powder mixture. These ceramic layers are characterized by hardness up to 11 GPa [13]. When using a mixture of powders of titanium, nickel and titanium carbide on the surface of titanium, coatings with a hardness of up to 6.84 GPa are formed by selective laser sintering [14]. A number of authors propose using carbon nanoparticles as a powder mixture [15].

Carbide coatings and layers formed on titanium by various methods are characterized by high hardness. Widely used technologies allow the process of cementation on the entire surface of the product. Local strengthening of the surface layer is performed by laser processing or electrospark alloying, however, the composition, structure parameters and hardness of the resulting carbide layers are not sufficiently studied.

2. Methodology

In the studies, samples with a diameter of 14 mm and a height of 2 mm from commercially pure titanium grade VT1-0 were used. The samples were cemented by laser processing of a carburizer deposited on the titanium surface. Graphite powder with a size of 10–35 μm was applied as a carburizer. Laser processing was performed on the "LRS-50A" setup with a pump lamp voltage of 250–300 V, pulse duration of 1–5 ms and a spot size of 0.75 mm. A special designation was introduced for the processing modes, e.g. "250_1", where: "250" – the pump lamp voltage, V; "1" – the pulse duration, ms.

The structure of the samples was studied using scanning electron microscopy (SEM). SEM combined with energy-dispersive X-ray analysis (EDX) of chemical composition of samples was performed on "MIRA II LMU" with "INCA PentaFETx3" detector. Hardness of the coatings was evaluated by microindentation using "PMT-3M" (at the load of 200 gf).

3. Results

After the laser processing at minimum modes, in particular "250_1", a coating consisting of carbon was formed on the surface of titanium (Figure 1a). There was a fusion of particles of graphite and they were fixed on the base material. The surface hardness was not high and amounted to 4.1±0.7 GPa. An increase in the voltage of the pump lamp led to the melting of the titanium surface and the appearance of cracks. Spherical particles with a size of more than 200 nm were also observed (Figure 1b).

According to EDX the surface layer consisted of carbon (47.5 at.%), oxygen (3.7 at.%) and titanium (balance). A coating comprising titanium oxides and carbides was formed as a result of laser processing. This assumption was confirmed by high surface hardness up to 11.5±1.0 GPa.

An increase in the pulse duration to 5 ms led to a change in the composition of the coating. The carbon content decreased to 3.8 at.%, while that of oxygen and titanium grew to 37.6 and 58.6 at.%, respectively. The surface structure also changed, i.e. the size of cracks increased, the areas characterized by different reliefs were visualized (Figure 2a). In the course of analysis of SEM images, the presence of round grains with the size about 150 nm and more on the surface was established (Figure 2b). It was likely that an increase in the duration of laser effect resulted in a significant oxidation of the coating, which was confirmed by a decrease in hardness to 10.5±1.0 GPa.

As a result of laser irradiation a coating was formed on titanium and the structure of the surface layer changed. At "250_1" processing mode, the depth of structural changes was 21±5 μm. In this case no significant increase in the hardness of the surface layer was observed.

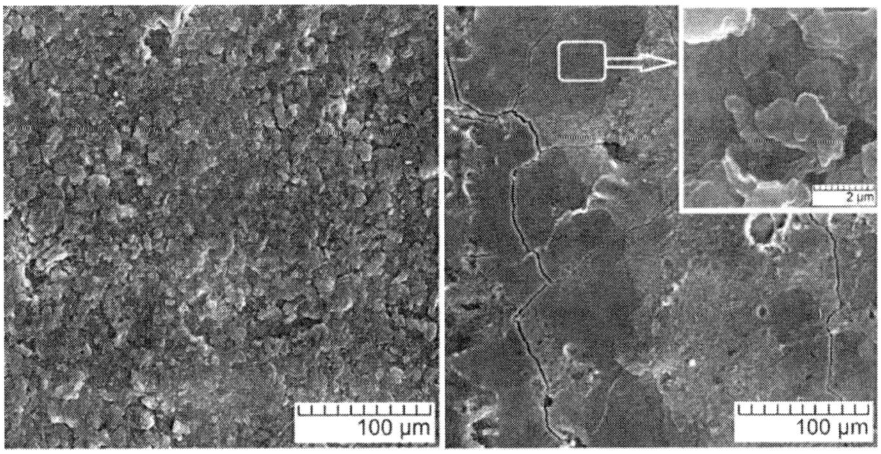

Figure 1(a, b). The surface of titanium after laser processing in graphite, "250_1" mode (**a**); the appearance of cracks, "300_1" mode (**b**).

The region of structural changes with a thickness of 79±5 µm and a hardness of 8.2±1.0 GPa was formed under "300_1" mode. In this area, along with the coating with a thickness within 5 µm, a diffusion layer with a thickness of 40±5 µm and the boundary of thermal effect were also observed.

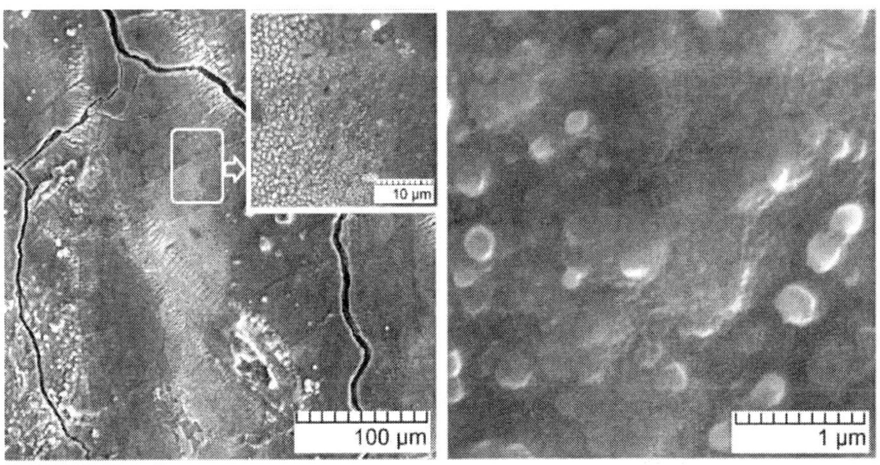

Figure 2(a, b). The morphology of the titanium surface after laser processing in graphite, "300_5"mode (**a**); nano-sized elements of the morphology of the modified layer (**b**).

An increase in the pulse duration led to the formation of a region with a modified structure having a depth of up to 140±20 µm and the coating thickness reaching 10 µm (Figure 3a). This coating consisted of Ti (31 at.%), C (43 at.%) and O (26 at.%). A diffusion layer with a thickness of up to 130±20 µm, which, in addition to titanium, included up to 35 at.% carbon, was also visualized. The diffusion layer had a fine dendritic structure with the hardness about 16.7±0.6 GPa (Figure 3b).

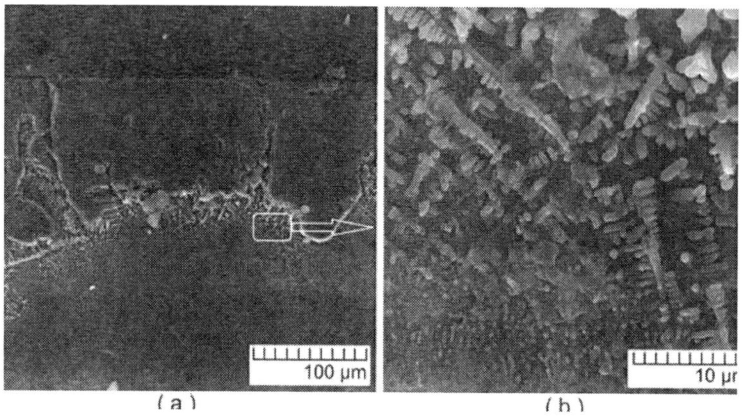

Figure 3(a, b). The microstructure of the near-surface layer of titanium after laser processing in graphite, "300_5" mode **(a)**; an enlarged fragment of the diffusion layer with dendrites **(b)**.

4. Conclusion

As a result of laser irradiation affecting titanium placed in graphite, a coating containing Ti, C and insignificant amount of O was formed. During further processing, the content of O increased, which was probably associated with the oxidation of the carbide layer. The coating was characterized by the thickness up to 10 μm, hardness up to 11.5±1.0 GPa and the presence of nano-sized (50–150 nm) structural elements. In the course of processing, in addition to the coating, a diffusion layer consisting of Ti and C was formed, which was characterized by a fine dendritic structure and high hardness of 16.7±0.6 GPa.

Acknowledgments

The study of the processes of cementation of titanium was performed in the framework of additional research, which was supported by the Russian Science Foundation (project No. 18-79-10040).

References

[1] Ulutan D and Ozel T 2011 *Int J Mach Tool Manu* **51** 250
[2] Rajendran R 2012 *Eng Fail Anal* **26** 355
[3] Fomin A A 2019 *Ceram Int* **45(7)** 8258
[4] Fomin A A, Fomina M A, Rodionov I V, Koshuro V A, Poshivalova E Y, Shchelkunov A Y, Skaptsov A A, Zakharevich A M and Atkin V S 2015 *Tech Phys Lett* **41(9)** 909
[5] Ajikumar P K, Vijayakumar M, Kamruddin M, Kalavathi S, Kumar N, Ravindran T R and Tyagi A K 2012 *Int J Refract Met* H **31** 62
[6] Fare S, Lecis N, Vedani M, Silipigni A and Favoino P 2012 *Surf Coat Tech* **206(8)** 2287
[7] Dai J J, Chen C Z, Li S Y and Zhuang L 2014 *Adv Mat Res* **936** 1086
[8] Boonruang C and Thongtem S 2010 *Chiang mai* J Sci **37** 206
[9] Voyko A, Fomina M, Shumilin A, Rodionov I, Kalganova S, Artyukhov I, Fomin A 2018 *J Phys Conf Ser* **1124** 071021
[10] Voyko A V, Fomina M A, Koshuro V A, Fomin A A, Rodionov I V, Atkin V S, Galushka V V, Zakharevich A M and Skaptsov A A 2018 *Proc Biomed Opt Imag* **10716** 107161L
[11] Mezentsov S, Koshuro V, Fomin A and Rodionov I 2018 *J Phys Conf Ser* **1124** 081017
[12] Koshuro V A, Fomina M A and Fomin A A 2018 *Proc Biomed Opt Imag* **10716** 10716M
[13] Diao Y and Zhang K 2015 *Appl Surf Sci* **352** 163
[14] Ma C, Gu D, Dai D, Yu G, Xia M and Chen H 2017 *Cryst Eng Comm* **19(7)** 1089
[15] Savalani M M, Ng C C, Li Q H and Man H C 2012 *Appl Surf Sci* **258(7)** 3173

Journal of Physics: Conference Series **1410** (2019) 012221 doi:10.1088/1742-6596/1410/1/012221

Synthesis, Raman enhancement, and chemical stability of $CaCO_3$ microparticles with copper coating

N E Markina[1], A I Danchuk[1], A M Zakharevich[2], A V Markin[1,2]
[1] Institute of Chemistry, Saratov State University, Saratov 410012, Russia
[2] Education and Research Institute of Nanostructures and Biosystems, Saratov State University, Saratov 410012, Russia

Abstract. The aim of the work is the microwave-assisted (MW) synthesis of SERS-active copper nanoparticles (CuNPs) and the core-shell composite based on nanostructured copper layer deposited over $CaCO_3$ microspheres ($CaCO_3$@Cu; 1.4±0.4 μm). The comparison of MW synthesized CuNPs and gold nanoparticles showed the equal orders of magnitude of the Raman enhancement. The temporal stability of $CaCO_3$@Cu samples was found significantly better than that of CuNPs: 4 days vs. 30 min. Stabilizer- and organic solvent-free synthesis of $CaCO_3$@Cu enables to minimize background SERS signal. The applicability of $CaCO_3$@Cu for chemical analysis was showed by detection of antibiotic drug (ceftriaxone) in a range of concentrations which is relevant for practical purposes (5–500 μg/mL).

1. Introduction

Generally, plasmonic gold and silver nanostructures are used as materials for surface-enhanced Raman spectroscopy (SERS) [1]. However, the results of our recent review shows that copper nanostructures can also be successfully used as reliable and cost-effective SERS-active materials (Cu-SERS substrates) [2]. Nevertheless, the information about Cu-SERS substrates is still very limited and additional studies are on demand. In particularly, low chemical stability of copper is the main challenge which restricts widespread application of Cu-SERS substrates. Although an application of organic solvents and strongly bonded stabilizers enables to minimize corrosive action of the oxygen of air and aqueous environment, they also lead to appearance of significant background signal that diminishes reliability of SERS based analysis. Earlier we proposed incorporation of copper nanoparticles (CuNPs) to the matrix of $CaCO_3$ microparticles to improve their temporal stability during storage [3]. However, despite the advantage of extremely long stability (> 4 months), this approach requires dissolution of the $CaCO_3$ matrix before SERS measurements that can be not optimal in some cases, e.g., instability of the analyte in acidic media.

Here we report on fabrication of Cu-SERS substrates ($CaCO_3$@Cu) using microwave-assisted synthesis (MW synthesis). Although the MW synthesis enables significant reaction acceleration and homogeneous heating, there is only one report it was used to prepare Cu-SERS substrate [4]. The authors synthesized SERS-active CuNPs by decomposition of acetylacetonate in benzyl alcohol, but the final substrate has low Raman enhancement and requires removing of the organic solvent before the use in analysis. Thus, we developed the synthesis in water and studied Raman enhancement and temporal stability of the substrates in aqueous environment. Additionally, the performance of the Cu-SERS substrates was verified in comparison with gold nanoparticles (AuNPs). The practical applicability of the substrates was tested by detection of antimicrobial drugs used as a model analyte.

2. Experimental part

2.1. Materials
Copper(II) sulfate ($CuSO_4{\cdot}5H_2O$), hydrogen tetrachloroaurate(III) ($HAuCl_4{\cdot}3H_2O$), sodium citrate ($Na_3C_6H_5O_7{\cdot}2H_2O$), hydrazine hydrate ($N_2H_4{\cdot}H_2O$), and 4-aminothiophenol (4-ATP) were purchased from Sigma-Aldrich. Ethanol, calcium chloride ($CaCl_2{\cdot}2H_2O$), and sodium carbonate (Na_2CO_3) were purchased from Vekton (Russia). Ceftriaxone (CEF) was used from Ceftriaxone powder for preparation of injections (Biosintez, Russia). Milli-Q (18 MΩ cm) water was used for preparation of all solutions.

2.2. Syntheses
The protocol of $CaCO_3$@Cu preparation includes two steps: (i) the synthesis of $CaCO_3$ microspheres and (ii) coating of the microspheres with SERS-active layer of nanostructured copper. The $CaCO_3$ microspheres were prepared by fast addition of Na_2CO_3 (3 mL, 0.167 M) to $CaCl_2$ (3 mL, 0.167 M) under continuous ultrasound (US) treatment for 100 s in US bath (0.5 L, 35 kHz, 50 W). Then the $CaCO_3$ microspheres were rinsed two times by water and one time by ethanol with further drying at 60°C for 3 h.
The deposition of nanostructured copper coating onto the surface of $CaCO_3$ microspheres was performed as follows. A portion (5 mL) of the mixture containing $CaCO_3$ (10 mg), $CuSO_4$ (1 mM), and hydrazine hydrate (150 µL) was heated in the domestic MW oven at the maximal power (700 W) for 20 s. The mixture of $CaCO_3$ and $CuSO_4$ was shaked for 3 min before addition of hydrazine and heating in order to saturate $CaCO_3$ surface with Cu^{2+} ions. Finally, $CaCO_3$@Cu was allowed to sediment, 3 mL of the supernatant solution was removed, and the residue was stored in closed plastic container without any rinsing.
The solutions of citrate stabilized CuNPs and AuNPs were synthesized also using MW heating. The CuNP sample was prepared by hydrazine reduction of $CuSO_4$ (0.5 mM) in the presence of sodium citrate (1 mM). The AuNPs were synthesized by citrate reduction of $HAuCl_4$ according to protocol described by Frens [5]; AuNPs were concentrated by centrifugation to achieve the concentration comparable with CuNPs.

2.3. Methods of characterization and SERS measurements
The morphology of $CaCO_3$@Cu and its EDX spectra were studied using scanning electron microscopy (SEM; Mira II LMU, Tescan, Czech Republic). In order to remove byproducts, the samples were rinsed by ethanol (one time) before SEM and EDX measurements.
Probing nanolaboratory Ntegra Spectra (NT-MDT, Russia) with integrated Raman spectrometer (Solar) and He-Ne laser (632.8 nm) was used for SERS measurements. Raman spectrometer was calibrated using Raman band of crystalline silicon at 520 cm^{-1}. Preliminary SERS studies were performed using the mixtures of 4-ATP (10 µL of ethanolic solution) with the suspension of CuNPs or AuNPs (1 mL). SERS spectra of CEF were collected using the mixture of the analyte and $CaCO_3$@Cu at 1:1 v/v ratio. All SERS spectra were collected from 10 µL of the analyte mixed with SERS substrate placed on cover slide. SERS signal was generated using 9 mW laser power and 1 sec signal acquisition time; at least 15 spectra were registered to obtain averaged signal.

3. Results and discussions

3.1. SERS-activity of CuNPs and AuNPs
It is well know that MW heating enables to minimize the influence of the walls of reaction vessel, to avoid formation of "metallic mirror" on the walls, and to improve homogeneity of the produced nanostructured coating. These advantages are of particular importance for fabrication of SERS substrates with reproducible enhancement properties. At the first step, the capability of MW heating to produce SERS-active materials was checked by synthesizing citrate stabilized CuNPs and AuNPs. A

model analyte (4-ATP) was used to test enhancement ability due to its high affinity to the surface of both metals. The results of SERS measurements show that both CuNPs and AuNPs generate strong SERS signal (Figure 1a). Although AuNPs provide around 2 times larger SERS signal, they significantly lose in cost and availability compared to CuNPs. Additionally, the orders of magnitude of both SERS signals are the equal meaning comparable enhancement capabilities of CuNPs relative to AuNPs. The dependences of SERS signal on the ATP concentration (Figure 1b) have a linear form in logarithmic coordinates that is typical for adsorption isotherms on energetically heterogeneous surfaces (Freundlich isotherm).

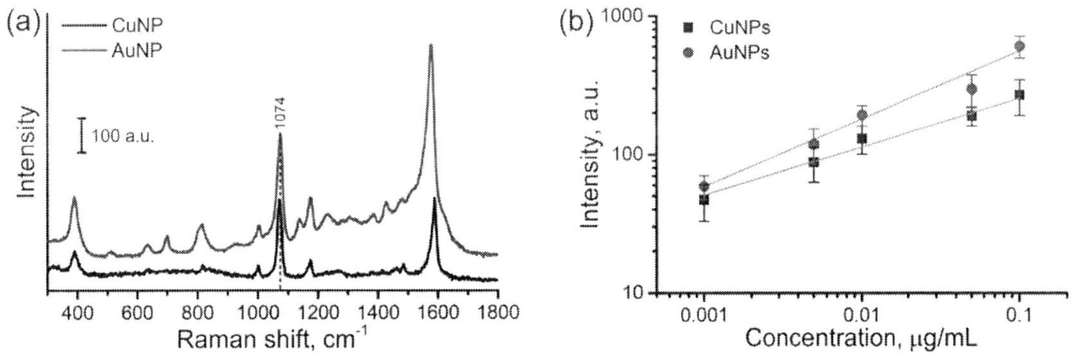

Figure 1(a, b). (a) SERS spectra of 4-ATP (0.05 μg/mL) obtained using copper (CuNPs) and gold (AuNPs) nanoparticles; **(b)** Dependence of SERS signal intensity on concentration of 4-ATP (band at 1074 cm^{-1}).

3.2. Preparation of $CaCO_3$@Cu

Unfortunately, colloidal CuNPs have very low chemical stability (discussed in the next section). This disadvantage was partially overcome by fabrication of SERS-active core-shell structure consisted of rough copper surface deposited over $CaCO_3$ microspheres. Additionally, in the previous work, we demonstrated that the deposition of SERS-active coating over the calcium carbonate positively influences Raman enhancement due to contribution of rough surface of $CaCO_3$ to the formation of hot spots [6].

The pure $CaCO_3$ microspheres were prepared using modified protocol described earlier [7]. The original protocol yields 4–6 μm particles which sediment fast and cannot be coated by the copper coating properly: CuNPs are reduced in the bulk of the solution instead of the $CaCO_3$ surface. Thus, we synthesized $CaCO_3$ under constant ultrasound treatment that enabled to get ~2.5 times smaller particles with lower sedimentation rate.

Besides of appropriate size of $CaCO_3$, successful deposition of the copper coating onto $CaCO_3$ surface requires preliminary adsorption of copper ions to the pores of $CaCO_3$ for several minutes. This step enables to initiate copper reduction at the $CaCO_3$ surface instead of the solvent volume (e.g. formation of CuNPs suspension). Besides of cost-efficiency and availability of $CuSO_4$, it was also used because sulfate ions have very low affinity to the copper surface enabling to minimize its influence on the analyte adsorption and Raman enhancement.

According to SEM studies, the size of final $CaCO_3$@Cu is 1.4±0.4 μm (N=200) (Figure 2a,b). Although SEM results do not enable to distinguish between CuNPs and $CaCO_3$, $CaCO_3$@Cu have dark-red color and EDX measurements prove the presence of copper in $CaCO_3$@Cu (Figure 2c).

The rinsing of the samples was used only before SEM and EDX studies, but not before SERS measurements because it reduces Raman enhancement of the composite. We explain this fact by partial oxidation of the copper surface at neutral pH that is in agreement with earlier published results regarding to aggregates of CuNPs [3]. Additionally, the rinsing is not needed because precursors and byproducts do not influence SERS background. In contrast, the residue of N_2H_4 removes the oxygen from the vessel and increases temporal stability of the sample during the storage. Also, the synthesis does not require stabilizers resulting in the minimal background signal.

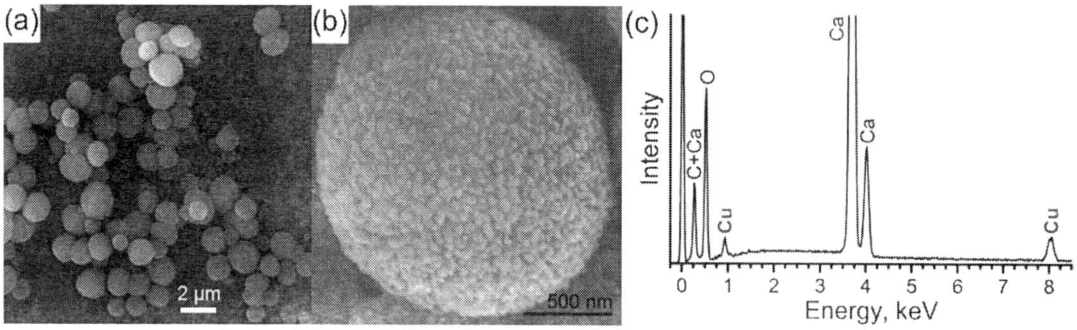

Figure 2(a-c). SEM images **(a, b)** and EDX spectrum **(c)** of $CaCO_3$@Cu microspheres.

3.3. Temporal stability and analytical performance of $CaCO_3$@Cu

As discussed in Introduction, temporal stability of Cu-SERS substrates is the most important challenge which has to be overcome in order to achieve widespread usage of the substrates in SERS based analysis. Because we are interested in the application of SERS for analysis of biofluids, the temporal stability of Raman enhancement capability of CuNPs and $CaCO_3$@Cu was studied in aqueous environment. The results show that the enhancement of CuNPs is drastically reduced during 30 min (Figure 3a). In contrast, $CaCO_3$@Cu produce stable SERS signal during at least 4 days after preparation. We explain this improvement by reduction of specific surface area compared to CuNPs.

Finally, the potential of $CaCO_3$@Cu for SERS based therapeutic drug monitoring was estimated using ceftriaxone, a widely used antibiotic of cephalosporin class. Analyte concentrations were selected at the level of 5–500 μg/mL that corresponds to therapeutically relevant antibiotic content in the urine of the patients [8]. The obtained SERS spectra (Figure 3b) have the appropriate intensity and are in agreement with the previously reported spectra obtained using silver nanoparticles [9]. Although $CaCO_3$@Cu have small background signal, the signal is weak and does not overlap with Raman bands of the analyte making quantitative analysis possible and reliable.

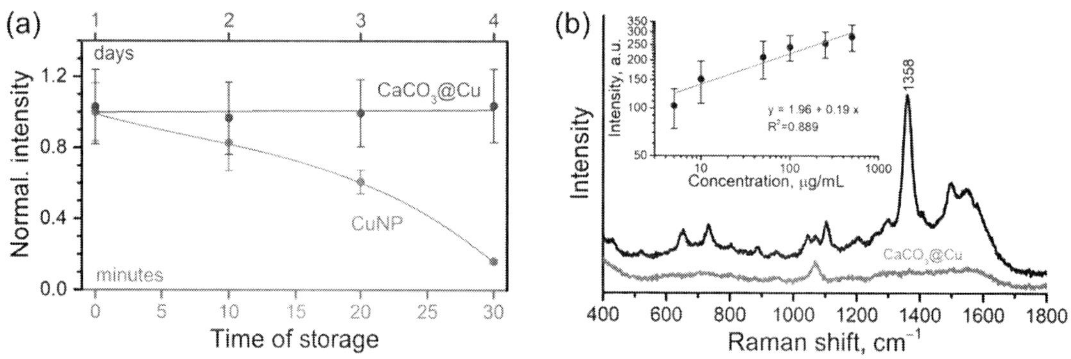

Figure 3(a, b). (a) Influence of CuNPs and $CaCO_3$@Cu storage time on SERS signal intensity of ceftriaxone at constant concentration (10 μg/mL); **(b)** SERS spectra of blank and ceftriaxone solutions obtained using $CaCO_3$@Cu and dependence of SERS signal intensity on ceftriaxone concentration (according to Raman band at 1358 cm^{-1}).

4. Conclusions
The main result of this paper is the copper based SERS-active composite whit improved temporal stability and strong Raman enhancement. Because the synthesis was performed without organic reducing agents, solvents, and stabilizer, it yields the composite with minimal background SERS signal making it reliable enough for analytical application. The high Raman enhancement and simplicity of preparation make the composite really competitive with noble metal based SERS substrates. Finally, recorded SERS spectra of the antibiotic drug within wide range of concentrations prove general applicability of the composite for therapeutic drug monitoring. However, further studies on selectivity and stability of the composite in biofluids are on demand.

5. Acknowledgments
The study was funded by RFBR according to the research project № 17-03-00537.

6. References
[1] Cialla D, März A, Böhme R, Theil F, Weber K, Schmitt M, Popp J 2012 *Anal. Bioanal. Chem.* **403** 27
[2] Markin A V, Markina N E, Popp J, Cialla-May D 2018 *Trends Anal. Chem.* **108** 247
[3] Markina N E, Volkova E K, Zakharevich A M, Goryacheva I Y, Markin A V 2018 *Microchim. Acta* **185** 481
[4] Dar M I, Sampath S, Shivashankar S A 2012 *J. Mater. Chem.* **22** 22418
[5] Frens G 1973 *Nature Phys. Sci.* **241** 20
[6] Stetciura I Y, Markin A V, Ponomarev A N, Yakimansky A V, Demina T S, Grandfils C, Volodkin D V, Gorin D A 2013 *Langmuir* **29** 4140
[7] Volodkin D V, Larionova N I, Sukhorukov G B 2004 *Biomacromol.* **5** 1962
[8] Patel I H, Chen S, Parsonnet M, Hackman M R, Brooks M A, Konikoff J, Kaplan S A 1981 *Antimicrob. Agents Chemother.* **20** 634
[9] Markina N E, Goryacheva I Y, Markin A V 2018 *Anal. Bioanal. Chem.* **410** 2221

Study of plates with a combined "steel base – (titanium, oxide) coating" structure and preliminary tests of its wear resistance

I Egorov[1], A Fomin[1]
[1] Yuri Gagarin State Technical University of Saratov, Saratov 410054, Russia

Abstract. This paper presents the results of the study of plates with a combined "steel base – titanium, oxide coating" structure, which was obtained by resistive welding and strengthening induction heat treatment. The data of measurements of hardness and preliminary test for wear resistance were given. The possibility of finishing turning on structural chromium steel X40Cr13 was shown.

1. Introduction

When creating metal-cutting tools, the question of choosing a material that provides high performance and the required quality of surface treatment arises. Mechanical characteristics of the material, in particular hardness, are studied and the allowable machining conditions, such as the depth of cut, cutting speed and feed, are determined.

Known theoretical data cannot fully guarantee the operability of cutting tools, since they operate under the influence of a complex of factors, e.g. high contact stresses and temperatures, as well as intensive physical and chemical processes. Under these conditions, the structural elements of the cutting tools (edges and planes) wear out due to numerous effects, in particular abrasive (wear by hard particles of the processed material or the particles formed during processing), adhesive (sticking), fatigue (crack formation), oxidizing (the appearance of other phases and chemical compounds) and diffusion (gradual dissolution of the tool in the excess material).

It is known that cutting tools with nitride coatings have an increased service life compared to tools made of high speed steel (HSS) [1,2]. A thin wear-resistant coating increases the surface hardness of the cutting tool, reduces the friction coefficient and has a heat-insulating effect on the cutting edge during processing.

Based on preliminary studies, the microhardness of the surface layer of titanium after induction heat treatment (IHT) reaches 3000–3500 HV [3]. This value is comparable to the hardness of hard alloys and exceeds the hardness of high carbon and alloyed tool steels. However, titanium, even after strengthening treatment, does not have a sufficiently thick and wear-resistant layer. Thus, the modification of titanium is suitable for the initial titanium layers of small thickness, which are fixed on a hard base, e.g. from high-strength steel. One of the options for combining titanium with steel is welding, due to which titanium (about 1 mm thick) can be welded to chromium steel, e.g. X91CrMoV18 (analogue of 1.2361 steel) [3].

This paper presents the preliminary results of testing the wear resistance of a cutter with a replaceable metal-cutting plate of the combined "steel-base – (titanium, oxide) coating" structure.

2. Methodology

The combined "steel base – (titanium, oxide) coating" structure (STO) was formed by the consistent use of resistive welding (RW) of steel with a titanium plate with thickness *h*, grinding (to ensure a certain shape) and subsequent strengthening with high frequency currents.

When shaping, an important condition was the preliminary formation of cutting edges and radii of rounding (Figure 1).

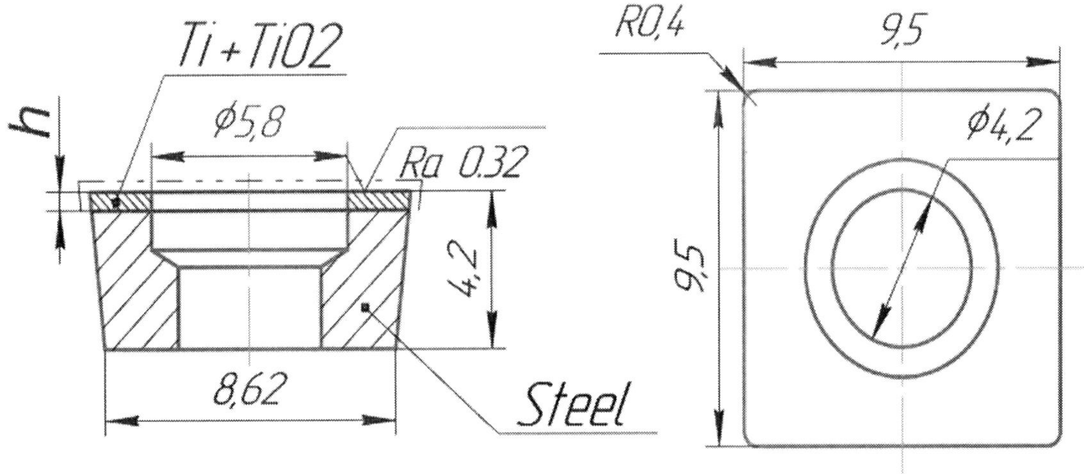

Figure 1. The design and geometrical parameters of the plate with a combined "steel base – (titanium, oxide) coating" structure

The geometry of the tool used for the test was consistent with the type "SSLBR1212H09" (ISO 5608-80). The presented designations had the following meaning: S – pressing the plate to the holder in the hole with the help of a screw; S – the square shape of the cutting plate (all faces had the same length, all angles were 90°); L – type of the holder (angle 90 °); B – the posterior angle of the plate (5°); R – processing direction, the right type of the holder; 12 – the cutter height, mm; 12 – the cutter width, mm; H – the cutter length equal to 100 mm; 09 – cutting edge length equal to 9.525 mm). This cutter geometry is used for basic turning operations, such as grooving and facing. The cutting properties were compared to a standard carbide cutting insert with a wear-resistant titanium-carbonitride coating (dark yellow color).

The hardness of the titanium part of the plate after strengthening IHT was evaluated using the Rockwell method a "TX PHR-2" tester according to the HRA scale. The microhardness of the oxide layer was specified using a "PMT-3" hardness tester with loads on the Vickers diamond indenter of 50–500 gf.

Preliminary tests of the wear resistance of the cutting elements of the combined plates under turning were performed on a "METALMASTER MML 180×300 V" lathe. The test material was a rod (machined length of 80 mm) made of stainless chrome steel X40Cr13 with a diameter of 15.6 mm. The hardness of the test material was 430–460 HV, which corresponded to 43–45 HRC (over 4 GPa). Processing was conducted with the following parameters: frequency – 500 rev/min, depth of cut – 0.4 mm, cutting speed – 23 m/min, feed rate – 1 mm/rev. In the course of cutting, wear of the main cutting edge was controlled (Figure 2).

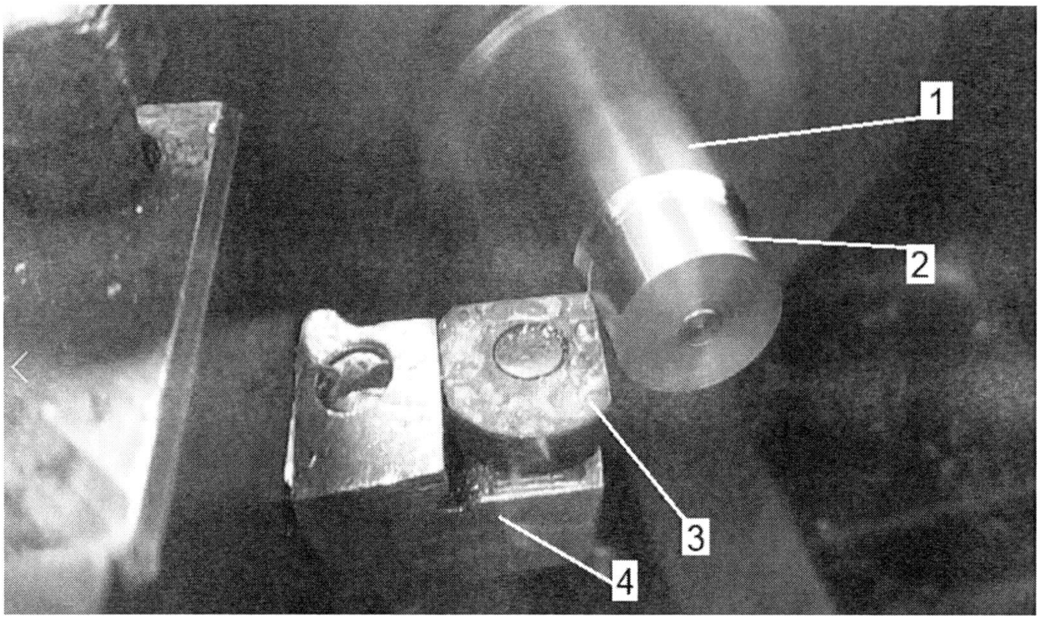

Figure 2. A scheme, which shows finishing turning of X40Cr13 steel, where 1 – the raw surface of the rod, 2 – the machined surface, 3 – the combined plate, 4 – the cutter holder

The critical value corresponded to the linear wear of 0.3 mm or to a noticeable destruction of the tip and edges of the cutter (Figure 3).

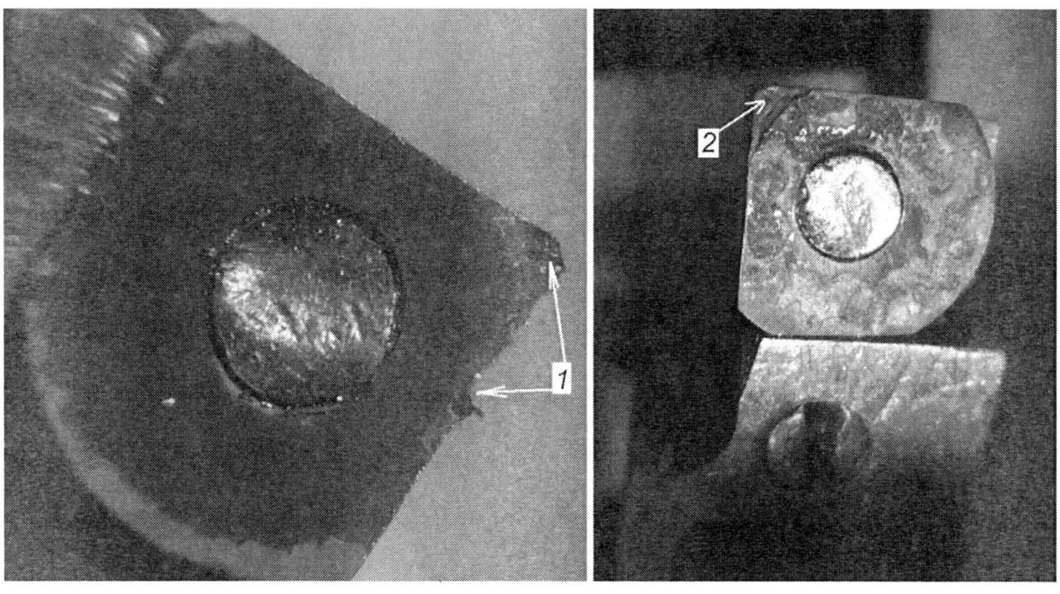

Figure 3. Wear options of plates with oxide coatings: left side – plastic (1 – signs of wear); right side – fragile (2 – chipped vertex)

3. Results

The hardness of the steel-titanium assembly after the formation of a monolithic sample was characterized by elevated values compared to the initial materials (titanium, steel). This was due to the processes of accelerated heating, local melting and rapid cooling. This thermal cycle was similar to quenching with low- or medium-temperature tempering. The average hardness of steel reached 450–490 HV, whereas the microhardness of titanium increased from 150–180 HV to 300–320 HV. After strengthening IHT, the steel hardness reached 950–1200 HV (about 87–88 HRA) and the titanium layer had a hardness of at least 1350–1500 HV (at least 89–90 HRA).

The sample plate processed at a temperature of 1300–1350 °C and a holding time of at least 300 s had a maximum hardness over 92 HRA, while the microhardness prints were slightly visualized. Therefore, in further tests of wear resistance, the plates of this type were mainly studied. The plates with these coatings were tested during finishing, as the combined structure did not have sufficient rigidity, in particular the thickness of the modified layer.

When testing the wear resistance, the number of passes for the prototype was not less than 35–40, whereas for the experimental plate the wear resistance had a smaller value – 32–35. This was due to the small thickness of the strengthened layer compared to the thickness of the titanium layer h. Therefore, further studies will be devoted to increasing the thickness of the hard layer and the rational choice of the thickness of the titanium layer.

The cutting plate of the combined structure was also tested during machining of medium carbon steel, brass and titanium. When processing these materials for the considered number of passes, the wear of the plate was not observed, which qualitatively characterized the high wear resistance of this material.

4. Conclusion

Thus, after strengthening IHT, the steel hardness reached 950–1200 HV (about 87–88 HRA), and the titanium layer was at least 1350–1500 HV (at least 89–90 HRA). The possibility of finishing machining of chromium steel X40Cr13 using a plate with a combined "steel base – (titanium, oxide) coating" structure was established. The proposed version of the plate was not practically inferior to the analog in terms of wear resistance (cutting capacity) at a frequency of 500 rev/min, depth of cut – 0.4 mm, cutting speed – 23 m/min and feed rate – 1 mm/rev.

Acknowledgments

The research was supported by the grant of the President of the Russian Federation MD-157.2019.8.

References

[1] Bagdasaryan A A, Pshyk A V, Coy L E, Konarski P, Misnik M, Ivashchenko V I, Kempiński M, Mediukh N R, Pogrebnjak A D, Beresnev V M and Jurga S 2018 *Compos. Part B. Eng.* **146** 132

[2] Espitia-Rico M J, Casiano-Jiménez G, Ortega-López C, Espriella-Vélez N and Sánchez-Pacheco L 2014 *DYNA*. **81(188)** 160

[3] Fomin A, Egorov I, Shchelkunov A, Fomina M, Koshuro V and Rodionov I 2018 *Compos. Struct.* **206** 467

Induction heat treatment of steel punches with Zr-containing coatings and preliminary results of FLD characteristics when drawing C45 carbon steel

A Fomin[1], **V Koshuro**[1], **A Shumilin**[1], **A Voyko**[1], **P Palkanov**[1], **A Aman**[2,3], **N Mukhin**[2,4], **S Palis**[2]
[1] Yuri Gagarin State Technical University of Saratov, Saratov 410054, Russia
[2] Otto-von-Guericke University, Magdeburg 39106, Germany
[3] Otto Vollmann GmbH & Co. KG, Gevelsberg 58285, Germany
[4] Saint Petersburg Electrotechnical University, Saint Petersburg 197022, Russia

Abstract. The paper presents data on strengthening induction heat treatment (IHT) of die tools (punches) and production of wear-resistant Zr-containing coatings on their surfaces. Preliminary results were obtained on the forming limit diagrams (FLD) during drawing (Nakazima's test) of sheet C45 carbon steel. It was established that punches with Zr-containing coatings after IHT had better characteristics for stretching compared to the uncoated tool steel.

1. Introduction

In metalworking, in particular shaping operations of sheet-metal forming, it is important to determine the ultimate strains that can be perceived by metal [1,2]. The main characteristic describing the deformation ability of any metal and alloy is a forming limit diagram (FLD) or a forming limit curve (FLC). To determine FLD Nakazima's test is conventionally used, which provides for the application of a punch with a hemispherical working part (ISO 12004-2: 2008).

In the "punch – sheet metal" system, it is necessary to ensure maximum stretching at minimum stresses without the formation of defects, such as cracks, tears and folds (Figure 1).

It is known that wear-resistant coatings, e.g. nitrides, carbides and carbonitrides, as well as diamond-like films, are deposited on the surface of the steel die tools to reduce the force of friction and improve the drawing characteristics [3]. However, there are experimental results according to which it is possible to use oxide coatings, in particular titanium oxide [4,5], in order to increase wear resistance and improve the functional qualities of the metalworking tools (replaceable plates of cutters). Also, to improve the friction conditions, it is necessary to choose a lubricating fluid or use a coating with a solid lubricant disulfide of MeS_2 metals [2].

Thus, in this paper, the hardness of the coatings produced on steel punches was studied and the preliminary FLD data for the drawing (Nakazima's test) of carbon C45 steel sheet were determined.

2. Methodology

Steel punches made of high carbon U8 tool steel (analogue of C80W1) were subjected to induction quenching (from a temperature of 1200 °C) with self-tapping to ensure a hardness of at least 59–60 HRC.

Next, a thin zirconium coating with a thickness of about 0.8 μm was applied using PVD (DC-mode) onto the surfaces of the punches previously prepared by grinding to ensure the roughness *Ra* = 0.16. A strengthening induction heat treatment (IHT) was performed to obtain an oxygen-saturated layer with high hardness (Figure 2a). The processing temperatures were 400 and 550 °C at an exposure of not more than 300 s. At a higher IHT temperature, these coatings cracked, which precluded their use as a wear-resistant layer for the punches.

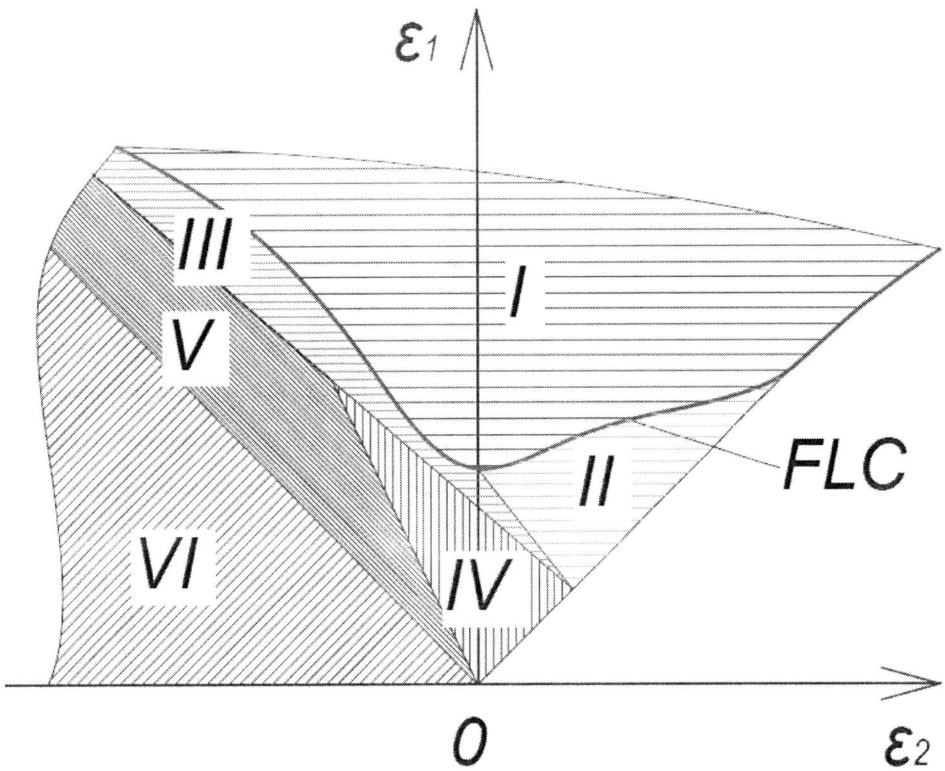

Figure 1. FLD and main areas: I – zone of destruction; II – unacceptable thinning; III – the risk of breaks; IV – without defects; V – the risk of folds; VI – the appearance of folds; FLC – the resulting curve.

The microhardness of the zirconium-containing layer on the surfaces of the punches was specified using a "PMT-3" hardness tester with loads on the Vickers diamond indenter of 50 gf.

The drawing test (according to Nakazima's scheme) involved the manufacture of 1 mm thick discs with a diameter of 44±0.2 mm and made of C45 steel with edge cuts of different radii *R* from 0 to 20 mm. A grid of small pitch of 1–1.5 mm was applied to the surface of the discs with the help of a laser for subsequent control of the deformation using an "AGPM-6M" geometric parameters analyzer.

The prepared samples were placed in a stamp consisting of the upper and lower halves (Figure 2b). The replaceable part was connected to the tool shank and installed in the die hole. The stamp assembly was installed in a hydraulic press, which provided a force of at least 10 tf.

The measured deformations in the form of a cluster of points were plotted on a diagram, through which an FLC was drawn. Next, the curves were plotted on one graph for subsequent comparison of the functional characteristics of the uncoated punches and those with wear-resistant coatings.

3. Results

The hardness of zirconium coatings reached a moderate value about 420–450 HV. The resulting value was not enough to increase wear resistance, therefore, of practical interest were the coatings after strengthening IHT. The hardness of zirconium-containing coatings after IHT reached 670–700 HV, which was equivalent to 59–60 HRC.

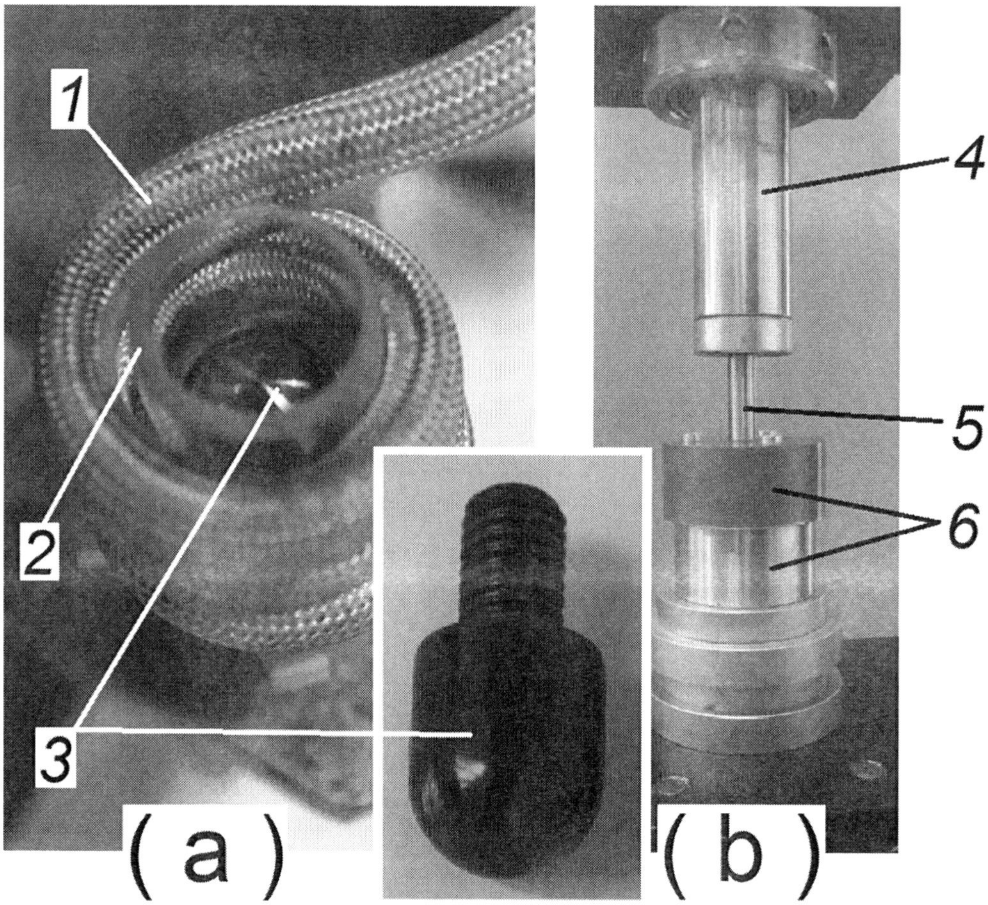

Figure 2(a,b). The location of the punch with a coating *3* in a quartz chamber *2* of a copper inductor *1* **(a)**; the location of a matrix *6*, a shank of the punch *5* and a stem of the press *4* during drawing **(b)**.

The resulting thin coating improved the drawing characteristics, in particular the limiting deformation exceeded 40 % at major strain ε_1 and reached 35 % at minor strain ε_2 (Figure 3, curve 3). The uncoated punch had lower characteristics for drawing, i.e. its ε_1 and ε_2 did not exceed 25 % (Figure 3, curve 1). For the stretch region (left half-plane of the diagram), good results were also observed for zirconium-containing coatings after the IHT (Figure 3, curve 3,4). The ultimate deformation reached from -5 to -10% in major strain ε_2.

However, the FLC data should be investigated in conjunction with the data of drawing force, which were determined when analyzing the drawing diagrams. The pressing force was also reduced when using the punches with IHT coatings. The maximum stresses at the drawing did not exceed 22 kN and for the coated punches this value was lower on average by 0.5–1 kN.

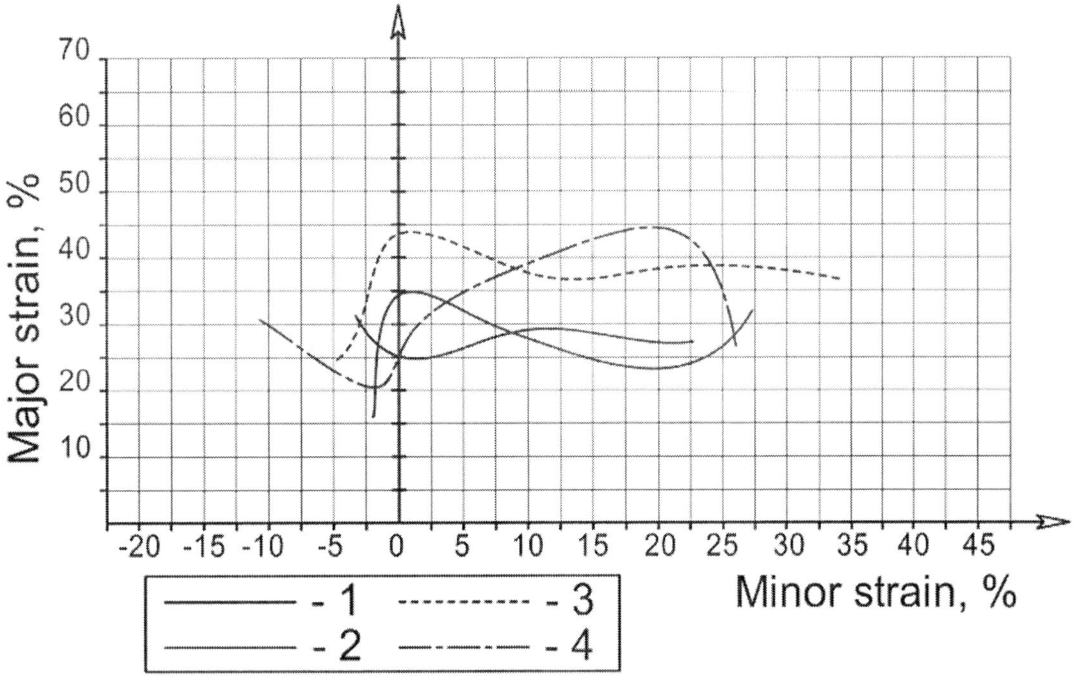

Figure 3. FLC for various punches: *1* – 1.3343 tool steel (uncoated); 2 – Zr coating on C80W1 tool steel; 3 – Zr+ZrO_2 (IHT-modification at T = 400 °C); 4 – Zr+ZrO_2 (IHT-modification at T = 550 °C)

4. Conclusions

Thus, zirconium-containing punches after IHT provided improved FLD characteristics for the drawing process. An increase in the hardness and oxygen content of the wear resistant coating improved the deformability of structural C45 steel from 25 to 35–40 %.

Acknowledgments

The research was supported by the grant of the President of the Russian Federation MD-157.2019.8. Prof. Aleksandr Fomin and Dr. Vladimir Koshuro express their gratitude to Mr. Nils Thiele and Otto Vollmann GmbH & Co. KG for technical support of the experimental part of the work related to the manufacture of dies and steel disc samples.

References

[1] Ma B L, Wan M, Wu X D, Cai Z Y, Diao K S and Han J Q 2017 *J. Manuf. Proces.* **30** 320
[2] Ma B, Yuan W, Wu X, Li X and Wan M 2015 *Appl. Therm. Eng.* **89** 144
[3] Pesch P, Sattel S, Woestmann S, Masarczyk P, Herden K, Stucky T, Martens A, Ulrich S and Holleck H 2003 *Surf. Coat. Technol.* **163–164** 739
[4] Fomin A, Egorov I, Shchelkunov A, Fomina M, Koshuro V and Rodionov I 2018 *Compos. Struct.* **206** 467
[5] Koshuro V, Fomina M, Voyko A, Rodionov I, Zakharevich A, Skaptsov A and Fomin A 2018 *Compos. Struct.* **202** 210

Magnetic-force microscopy of thin Bi:IG films for thermomagnetic recording

Yu E Vysokikh[1], **S Yu Krasnoborodko**[1], **Y V Danishevskaya**[2,3], **T V Mikhailova**[2], **A S Nedviga**[2], **V N Berzhansky**[2], **M F Bulatov**[1], **D V Churikov**[1]
[1]Scientific and Technological Center of Unique Instrumentation of the RAS, Moscow 117342, Russia
[2]V.I. Vernadsky Crimean Federal University, Simferopol 295007, Russia
[3]Crimean University of Culture, Art and Tourism, Simferopol 295017, Russia

Abstract. Thermomagnetic recording on thin films is currently one of the most promising methods for copying records with high density and weak magnetization. These films can be characterized by a couple of parameters very suitable for recording reconfigurable magnetic atomic traps whose geometry is visible. Different types of materials are investigated to be the most effective for thermomagnetic recording propose. In this paper authors showed possibility and features of thermomagnetic recording on Bi:IG thin films. High coercivity thin films with low Curie temperature and different film-lattice mismatch parameters were synthesized. Thermomagnetic recording of magnetic disc was performed. Features of magnetic records were investigated and influence of film-lattice mismatch parameters on meta-stable domain structure that defines the thermomagnetic recording resolution was showed. Magnetic-force microscopy was used for topography and magnetic structure characterization of thin films with different film-lattice mismatch parameters. Some features of films morphology were investigated using atomic-force microscopy with sharp AFM probe.

1. Introduction

Thermomagnetic recording on thin films is currently one of the most promising methods for copying records with high density and weak magnetization [1-3]. In addition, Heat Assisted Magnetic Recording is one of the most perspective technology for high density storage devices [4]. These facts drive to high interest of different type materials investigation in order to possibilities and features of thermomagnetic recording on these objects. Iron garnet thin films is one the most perspective materials for thermomagnetic recording in order to possibility creation high coercivity and low Curie temperature thin films which is important for thermomagnetic recording [5]. Development high coercivity thin films with low Curie temperature and its morphology and domain structure investigation by magnetic force microscopy [6] is important for understanding possibility and features of using this material for thermomagnetic recording.

2. Experimental

One of the most effective method for increasing the coercivity in an YIG thin films is the creation of misfit stresses between the film-substrate crystal lattices. In this case, both elastic stresses and plastic deformation are realized. Stress relaxation mechanism caused by misfit dislocations. The morphology of the samples after epitaxy and the type of domain structure determined by so called "facet" surface's structure [6]. Thin films with composition of $(Sm,Lu,Bi)_3(Fe,Ga,Al,Sc)_5O_{12}$ were synthesized by liquid phase epitaxy technique [7], [8]. Substrates of gadolinium gallium garnet with lattice parameter of 12.383 Å were used. Film-substrate lattice mismatch parameter Δa of different films was in the range of 0.070 – 0.113 Å. Curie temperature and the thickness of films varied in the range of 55 – 90°C and 3 – 7 μm, respectively. Topography of thin film (Δa=0,087 Å) with its original magnetic structure are showed in figure 1. Topography of polished thin film and its magnetic domain structure from same area are shown. Opposite domains have bright and dark color correspondently (figure 1b).

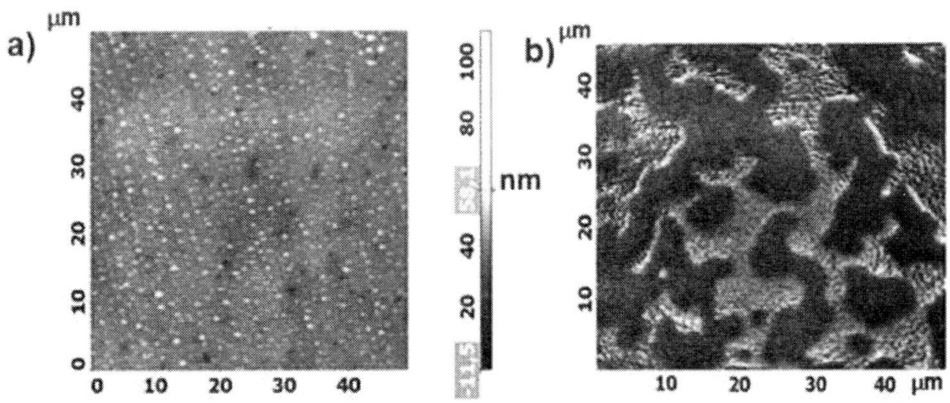

Figure 1 (a, b). Topography of polished thin film **(a)** and its original magnetic structure from same area **(b)**. Scan size is 50x50 μm.

Thermomagnetic recording of magnetic disc was performed by heating the films in contact with disc up to 100°C and then cooling down to ambient temperature. Original domain structure of thin films as well as recorded domain structure was obtained by magnetic-force microscopy and compared to domain structure of the disc (figure 2).

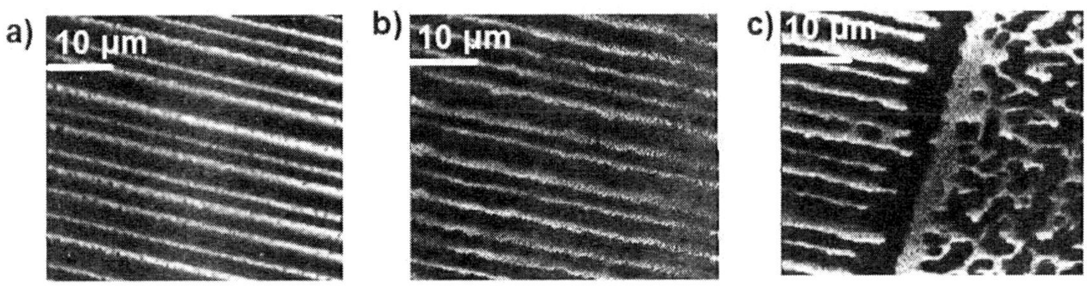

Figure 2 (a, b, c). MFM image of original magnetic structure of the disc **(a)** and recorded magnetic structure to the thin film **(b, c)**.

It was shown that the recorded magnetic structure has smooth borders compare to original one. The reason of this shiftiness is that the thin film has mismatch dislocation substructure which is appeared during the epitaxy growth (figure 3, *a*). Domains border goes strictly on the corresponding dislocation line. The shiftiness of recorded magnetic structure is ± 0.4 μm. Authors performed investigation of influence of film-lattice mismatch parameter on the metastable domain's size (figure 3, *b* and *c*).

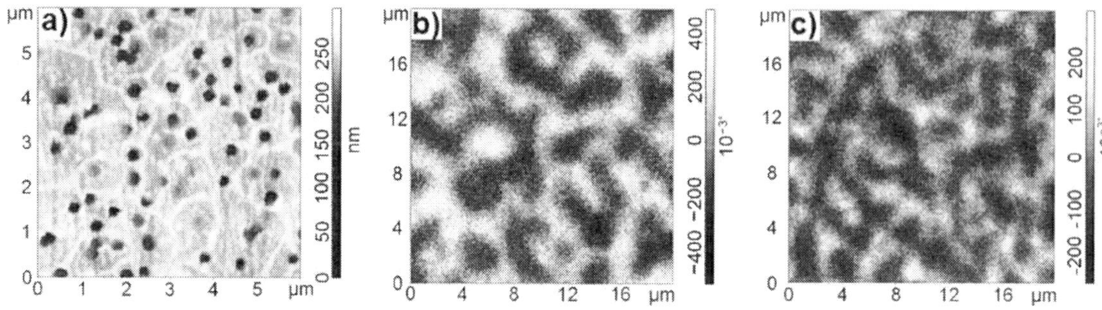

Figure 3 (a, b, c). **(a)** Morphology of film with Δa = 0.096 Å after polishing and selective etching HNO_3 during 10 min; metastable domain structure of films with Δa 0.089 Å **(b)** and 0.099 Å **(c)**.

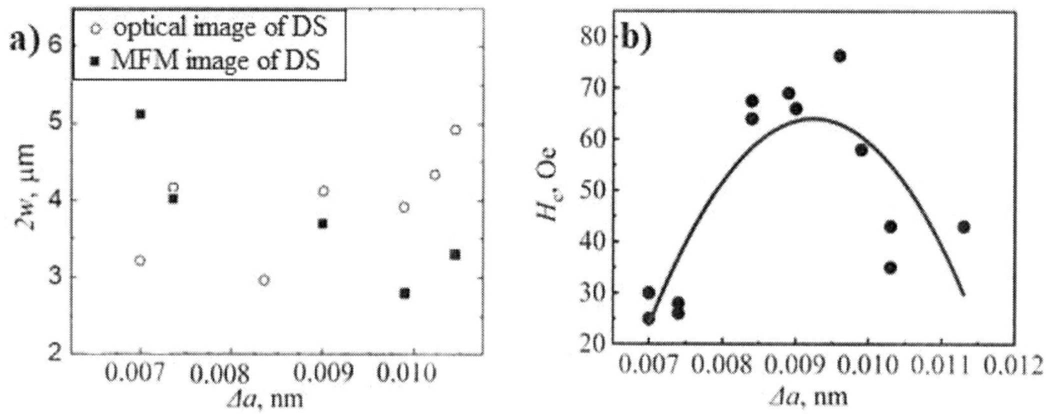

Figure 4 (a, b). Dependence of domain structure period **(a)** and coercive force H_C **(b)** on film-lattice mismatch parameter.

It was shown in figure 4 that domain structure period on film-lattice mismatch parameter dependence has the minimum in the range of 0,085 – 0,105 Å. The period of domain structure of similar coercive samples was measured using optical microscopy with frames of about 125 x 125 μm [9]. The data of magnetic force microscopy taking into account the measurement errors for synthesized samples repeat the original dependence in the studied range. Period of domain structure showed good correlations with dependence of average defects' heights defined as average roughness on mismatch parameters showed in figure 5 (*a*). Simultaneously, we observe a change in the distribution of defects' density (triangular elements) on the surface (figure 5, *b*). The most densety packing of surface triangular elements is observed for the films with film-lattice mismatch parameter in the range of 0,084 – 0,099 Å. Average roughness and number of defects were determined for each film as medium value in the area of 15 x 15 μm by magnetic-force microscopy or atomic-force microscopy data. The films with the smallest elements of surface and domain period are characterized by the highest H_C among all the studied samples (figure 4, *b*) [10], which makes this range of film-lattice mismatch parameter the most suitable for thermomagnetic recording.

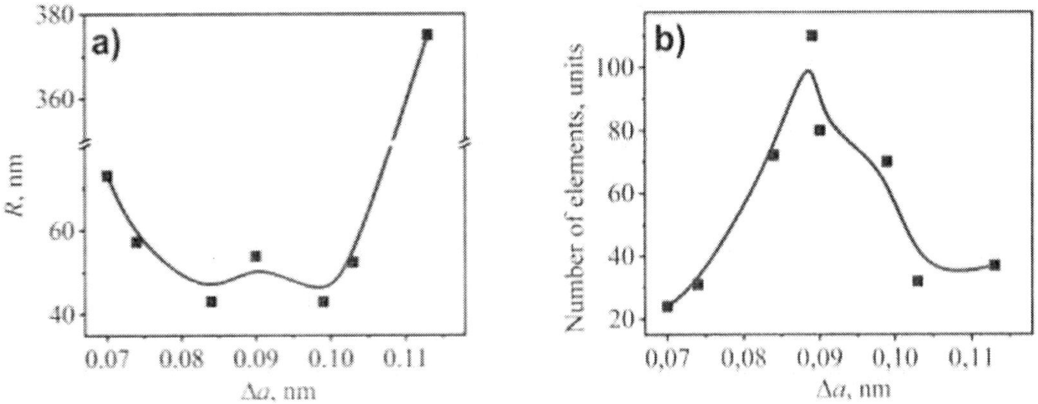

Figure 5 (a, b). Dependences of average roughness **(a)** and number of surface triangular elements (b) on film-lattice mismatch parameter.

3. Conclusion

Spatial resolution of thermomagnetic recording is limited by dislocation substructure which is corresponded to film-lattice mismatch parameter. It was shown that microscale epitaxial high-coercive Bi:IG films with Δa in the range from 0.084 to 0.099 Å have the smallest period of metastable domain structure and highest H_C that makes them optimal for thermomagnetic recording.

Acknowledgments

The authors from V.I. Vernadsky Crimean Federal University acknowledge support by the Ministry of Science and Higher Education in the framework of the state task (project no. 3.7126.2017/8.9).
The authors from Scientific and Technological Center of Unique Instrumentation of the RAS acknowledge support by the Ministry of Science and Higher Education in the framework of the state task (project no. 0069-2019-009).

References

[1] Kubrakov N F 1992 *M. Nauka* **35** 136-164.
[2] Leviy S V, Agalidi Yu S, Vishnevskii V G 1998 *Radioelectronics and Communications Systems* **8** 74-80.
[3] Vishnevskii V G, Nedviga A S, Nesteruk A G, Berzhansky V N, Dudarenko I V Scientific Notes of Taurida National V.I. Vernadsky University. – Series: Physics and Mathematics Sciences 2010 **23 (62)** № 1. I. 158-173
[4] Sokolovskiy A, Mityagin A, Hlopov B 2008 *J. Information and Analytical Bulletin* 212
[5] Berzhansky V, Danishevskaya Y, Nedviga A, Bektemirova M 2018 *J. Phys. Conf. Ser.* **1124** 061006.
[6] Vysokikh Yu E, Shaposhnikov A N, Prokopov A R, Berzhansky V N, Karavainikov A V, Gerasimenko N N, Smirnov D I 2017 *Materials Research Bulletin* **95** 115–122.
[7] Prokopov A R, Vetoshko P M, Shumilov A G, Shaposhnikov A N, Kuz'michev A N, Koshlyakova N N, Berzhansky V N, Zvezdin A K, Belotelov V I 2016 *Journal of Alloys and Compounds* **671** 403–407.
[8] Leitenmeier S, Körner T, Griesbauer J, Herbort M, Heinrich A, Stritzker B 2008 Journal of Crystal Growth **310** 24 5392–5401.
[9] Dubinko S V, Nedviga A S, Vishnevskii V G, Shaposhnikov A N, Yagupov V S, Nesteruk A G, Prokopov A R 2005 *Technical Physics Letters* **31** 979
[10] Berzhansky V N, Nedviga A S, Vishnevskii V G, Prokopov A R 2009 *Solid State Phenomena* **152-153** 11

Molecular dynamics study of As dimer formation on the GaAs (001) As-rich surface

N D Prasolov[1], **A A Gutkin**[2], **P N Brunkov**[1,2]
[1]ITMO University, Saint Petersburg 197101, Russia
[2]Ioffe Institute, Saint Petersburg 194021, Russia

Abstract. Using the molecular dynamics modelling the study of the process of As dimer formation on the GaAs (001) surface was carried out at different temperatures as a function of time. To describe GaAs properties the bond-order potential was used. It was found that the activation energy of formation of single dimer is about 38 meV, while the dimer dissociation energy is about 2 eV. The characteristic time of the single dimer generation lies in the range $10^{-6} - 10^{-8}$ s for temperatures from 28 to 37 K.

1. Introduction

Rapid development of nanotechnologies stimulated studying of properties of a surface of crystal semiconductor materials especially changing in the structure of surface layer with respect to the bulk crystal structure. It is natural that, for example, after the cleaving a crystal in high vacuum conditions, those changes occur not instantly, and the process of formation of the changed structure depends on temperature. On the (001) surface of semiconductor materials with zinc blend crystal structure the reconstruction of a surface can be followed by formation of dimer extended along <110> axis, consisting of identical atoms [1-4]. The surface completely filled with dimers has (1 X 2) or (2 X 1) structure. The paper is devoted to modelling this simplest case with molecular dynamics (MD) simulations for surface reconstruction of the GaAs (001) surface terminated with As atoms.

2. Molecular dynamics model

The MD simulations were performed using the Large-scale Atomic/Molecular Massively Parallel Simulator (LAMMPS) [5]. To describe the dimer formation the bond-order potential (BOP) developed for GaAs and based on quantum mechanical theory incorporating both σ and π bonds [6] was used. Realization of this potential in the LAMMPS environment were developed by X. W. Zhou [7]. It does not take into account the electron counting potential due the redistribution of electron density on the surface [6]. In accordance with [6] this results in a simple dimer-row (2 X 1) structure in the case of As terminated GaAs (001) surface.

In the MD simulations the GaAs slab has the following parameters: height is approximately 12 nm, both depth and width are about 56 nm. The simulated GaAs slab consists of 1700000 atoms. Orientation of the top surface of GaAs is the (001) plane. The top GaAs layer consists entirely of As atoms which number N_0 equals to 20000. The bottom monomolecular GaAs layer is fixed to prevent the motion of the atoms comprising it and to prevent other atoms from leaving the simulation box. The MD simulation was carried out with the periodic boundary conditions for vertical planes at the edges of the GaAs slab. The atomic system was described as the canonical (NVT) ensemble, which means constant number of atoms (N), volume (V) and temperature (T) [8]. Discretization time was chosen to be 10 fs in MD simulations in order to describe correctly the oscillation period of the atoms higher than 50 fs.

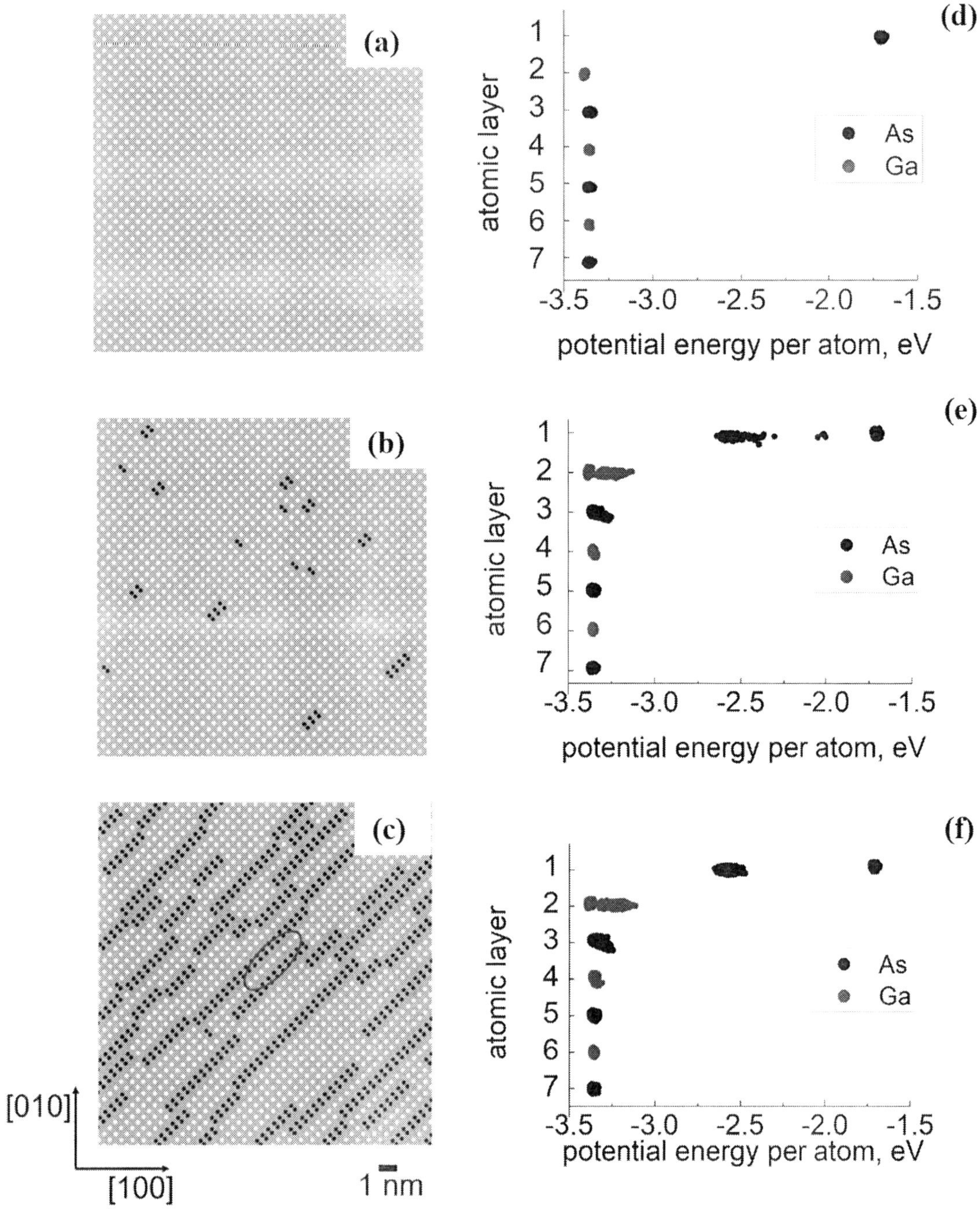

Figure 1(a,b,c,d,e,f,d). Arsenic atoms on the GaAs (001) surface at temperature of 45 K for $t = 0$ ns (a); for $t = 0.025$ ns (b); for $t = 0.9$ ns (c). Open circles show As atoms in initial position and filled circles show As atoms in dimers. In the area marked with red line the formation of dimers is impossible. The depth distribution of the potential energy of As and Ga atoms of the first 7 layers from the surface at temperature of 45 K for $t = 0$ ns (d); for $t = 0.025$ ns (e); for $t = 0.9$ ns (f).

The setting of the sample temperature was carried out step like starting from 1 K. It was found that stabilization of the sample temperature typically required a time of about 0.01 ns after the sharp change of temperature. Result of the simulation is a set of files containing information about each atom at each step of the simulation. To interpret the obtained results, it is convenient to use the Open Visualization Tool (OVITO), which was employed for the present study [9].

3. Results

From MD simulations it follows that in the scale on a few ns the dimer structure appears at temperatures above 25 K and the total number of dimers grows up with time t. The Fig.1a,b,c shows position of arsenic atoms on the GaAs (001) surface at temperature of 45 K at different moment of time t. At an initial stage of the reconstruction with increase in time t both the number of dimers in rows (i) and number of the dimer-rows (N_i) goes up (Fig1. b). As a time goes up further the fraction of small length dimer-rows goes down and areas appear where the dimer formation is impossible at given temperature and time scale (Fig1. c). MD simulation shows that potential energy of As atom after dimer formation decreases at value about 0.9 eV and potential energy of atoms in the second and third atomic layer from the surface increases by a few tenth of eV (Fig. 1d,e,f).
The Fig. 2a shows time dependencies of relative number of dimers N_{rel} (with respect to the full possible number of dimers equal to $N_0/2$) as function of sample temperature:

$$N_{rel}(T,t) = \frac{2}{N_0}\sum_{i=1} i\, N_i(T,t) \quad (1)$$

The rate of dimer formation rises with increase of temperature. On the time scale of a few ns at temperature of 300 K the saturation region at the value ~0.85 appears on time dependencies of relative number of dimers N_{rel} (Fig. 2a).The saturation at $N_{rel} < 1$ is due to the appearance of areas where the formation of dimers is impossible (marked in red on Fig.1c). At very high temperatures (for example at T = 1000 K) the value in saturation region goes more close to 1.0 due to high thermal energy which is sufficient for destruction of dimers with the subsequent reorganization (Fig. 2a).
On Fig. 2b the initial stage of dimer formation is presented. Despite to fact that $N_{rel} << 1$ the dependence of dimer concentration on time shows nonlinear behavior. It can be due to the fact that at low temperatures (T < 40 K) probability of formation of single dimer much smaller than for dimer formed on the ends of dimer-rows.

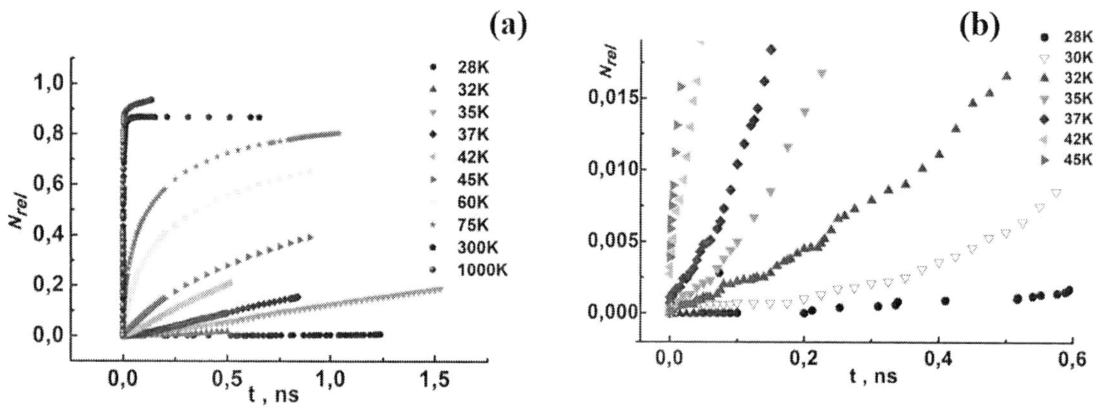

Figure 2(a,b). Time dependencies of relative number of dimers N_{rel} at different temperatures: (a) full range of time; (b) – initial region of time.

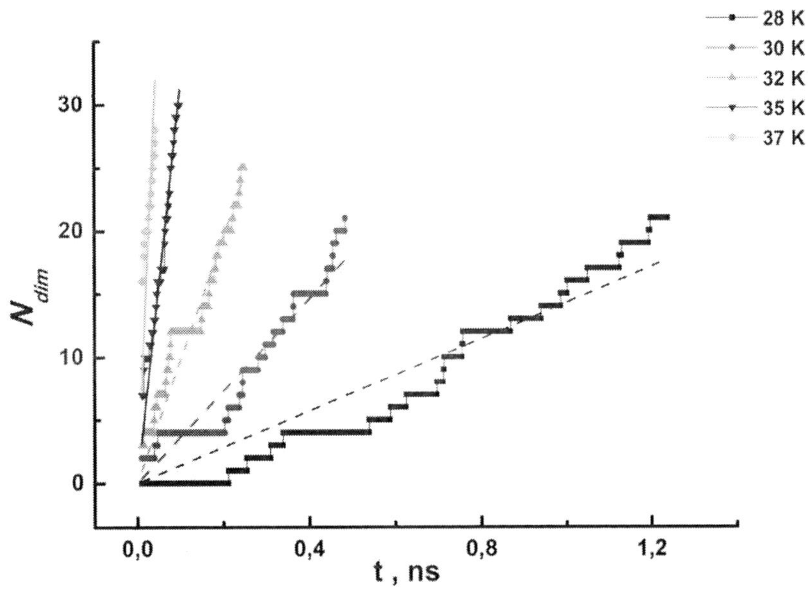

Figure 3. Time dependencies of total number of isolated dimers $N_{dim}(t)$ generated during period of time from 0 to t at different temperatures.

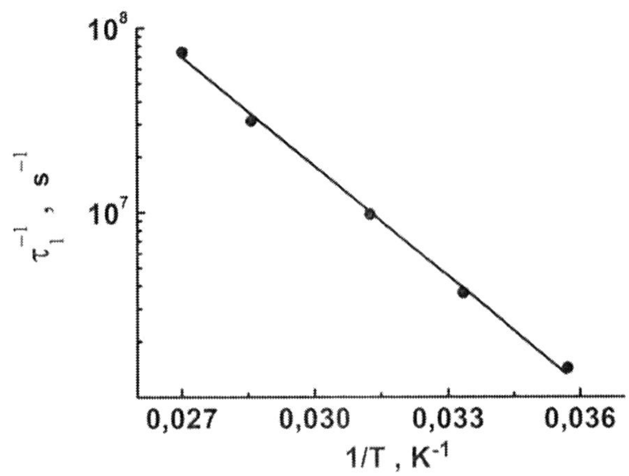

Figure 4. Arrhenius plot for τ_1^{-1}

If the dimer concentration is low and there is no long dimer-rows (Fig. 1b) any small length dimer-row came from formed earlier single dimer. In this case total number of single dimers $N_{dim}(t)$ generated during period of time from 0 to t is equal to total number of all dimer-rows existing at the moment t :

$$N_{dim}(t) = N_1(t) + N_2(t) + N_3(t) + \cdots \quad (2)$$

It allows to determine the characteristic generation time τ_1 necessary for single dimer formation using the following equation:

$$N_{dim}(t) \approx \frac{N_0}{2}\frac{t}{\tau_1} \tag{3}$$

From the linear approximation of $N_{dim}(t)$ using Eq. 3 it was found that generation time τ_1 lies in the range 10^{-6} – 10^{-8} s for temperatures from 28 to 37 K (Fig. 3). The corresponding Arrhenius plot (Fig. 4) gives the activation energy of single dimer formation equal to ~38 meV, while for dimer dissociation the energy about 2 eV is required. The last allows explaining stability of reconstruction of the GaAs (001) surface with formation of dimer of As atoms at higher temperatures.

4. Conclusions

The MD simulation was carried out to study of the process of As dimer formation on the GaAs (001) surface as a function of time at different temperatures. It was found that at low temperatures the probability of formation of single dimer much smaller than for dimer formed on the ends of dimer-rows and maximum degree of a covering of the surface with dimers is less than 100%. The activation energy of the generation rate of single dimers equal to ~38 meV, which is considerably smaller than the dimer dissociation energy (~2 eV).

References

[1] LaBella V P, Yang H, Bullock D W, Thibado P M, Kratzer P, Scheffler M 1999 *Phys. Rev. Lett.* **83** 15
[2] LaBella V P, Krause M R, Ding Z, Thibado P M 2005 *Surface Science Reports* **60** 1
[3] Qian G-X, Martin R M, Chadi D J 1988 *Phys. Rev. B* **38** 11
[4] Murdick D A, Zhou X W, Wadley H N G, Nguyen-Manh D 2005 *J. Phys.: Condens. Matter.* **17** 6123
[5] Plimpton S 1995 *J. Comp. Phys.* **117** 1
[6] Murdick D A, Zhou X W, Wadley H N G, Nguyen-Manh D, Drautz R, Pettifor D G 2006 *Phys. Rev. B* **73** 045206
[7] http://lammps.sandia.gov/doc/pair_bop.html
[8] Gibbs J W 1902 *Elementary Principles in Statistical Mechanics* (New York: Charles Scribner's Sons)
[9] Stukowski A 2010 *Modelling Simul. Mater. Sci. Eng.* **18** 015012

The results of modelling of a micromechanical accelerometer

I E Lysenko[1], O A Ezhova[1], A V Tkachenko[1], D V Naumenko[1], K Guha[2], K S Rao[3]

[1]Department of Electronic Apparatuses Design, Southern Federal University, Taganrog 347922, Russia

[2]Department of Electronics & Communication Engineering, National Institute of Technology, Silchar 788010, India

[3]Microelectronics Research Group, KL University, Vaddeswaram 522502, India

Abstract. This theses the linear acceleration sensor design with three axis of sensitivity is researched. Parameterized geometry and finite element model for modal analysis are developed in the ANSYS program. Behavioral description of the study design is developed with language VHDL-AMS to simulate the sensor operation under the influence of linear acceleration along three axis of sensitivity. On the basis of research results three-axis device sensitivity, cross-sensitivity, duration transients are specified. As part of the work the experimental sensor prototypes are fabricated.

1.Introduction

In the modern world, microelectromechanical systems (MEMS) are a promising direction of electronics development. To date, the market is widely represented by one-and two-axis linear acceleration sensors, as began to appear three-axis sensors from leading world companies, which are two or three devices made on the same substrate in the same technological cycle. To reduce the size and cost of the final product is necessary to use three-axis sensors, which are a single structure with one or two sensing elements [1, 2].

Thus, the article is solving the problem of developing three-axis micromechanical sensors of linear acceleration allowing to register linear accelerations along three axes by one sensing element. This problem is relevant for the Russian microelectronic industry. The solution of this problem will allow to improve the parameters of devices based on microsystems and ensure the competitiveness of Russian devices on the world market. This problem can be solved by using new design and technological solutions and design techniques based on new mathematical models.

2. Design of linear acceleration MEMS sensor

In this work the researched design of linear accelerations sensor with three axis of sensitivity is provided in Figure 1. The developed micromechanical sensor contains the inertial mass (1), a suspension of a sensitive element (2), capacity transducer (3, 4), electrostatic actuator (5).

In the offered design of the linear acceleration sensor there are sixteen serial-to-parallel connections of elastic beams, i.e. a suspension. This suspension can be divided conditionally into three types, depending on the beam configuration. Modal frequency depends on weight and the stiffness coefficient of construction depending on many parameters including on length, width, thickness of elastic beams in the suspensions. The change of a stiffness coefficient of the design in accordance with length change of elastic beams in the suspensions is considered.

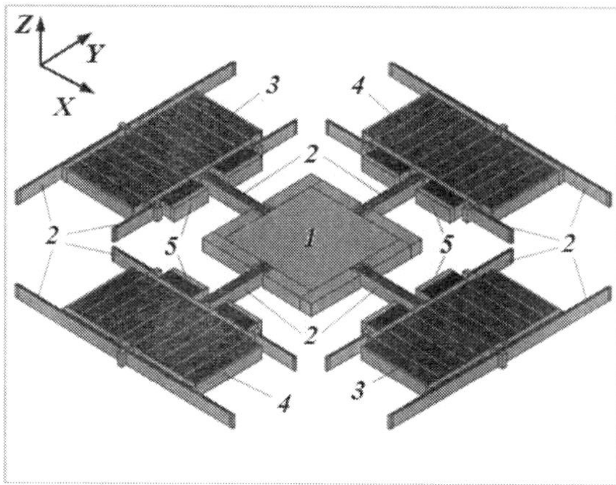

Figure 1. Design of three-axis accelerometer.

In Figure 2 the topology of experimental prototype of micromechanical accelerometer is provided.

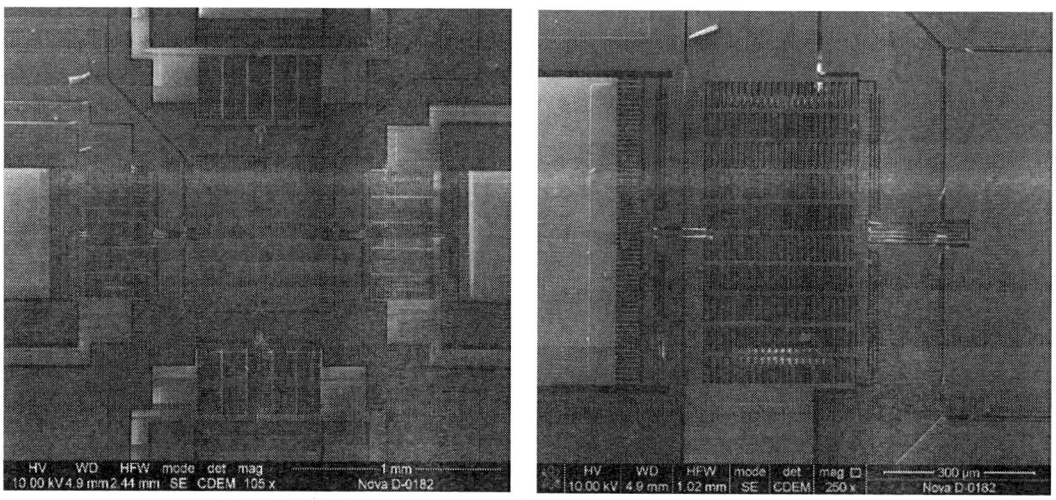

Figure 2. Sample of three-axis accelerometer.

In the study of the design the parametrizable description was developed to analyze the dynamic behavior of the design under the influence of dynamic loads in program system of ANSYS in which natural frequencies and forms of oscillations corresponding to them are counted.
Figure 3 shows the developed parameterized geometric model of the sensor for simulation in ANSYS.

3. Research of micromechanical accelerometer

For behavioral research of dynamic characteristics on the basis of the presented mathematical model and calculations in ANSYS the VHDL-AMS of the sensor was developed.
Figures 4-6 show the results of simulation of the micromechanical accelerometer under the action of acceleration varying in the full dynamic range from 1g to 10g.
As follows from results of simulation, in case of influence of the linear accelerations a_x, a_y, a_z the sensitive element performs the synchronous movement along axes of the external influence. The direction of the linear acceleration influences will define a phase of inertial masses movement. The duration of the transition on the sensitivity axes X, Y – 9 ms, and axis Z – 11 ms.

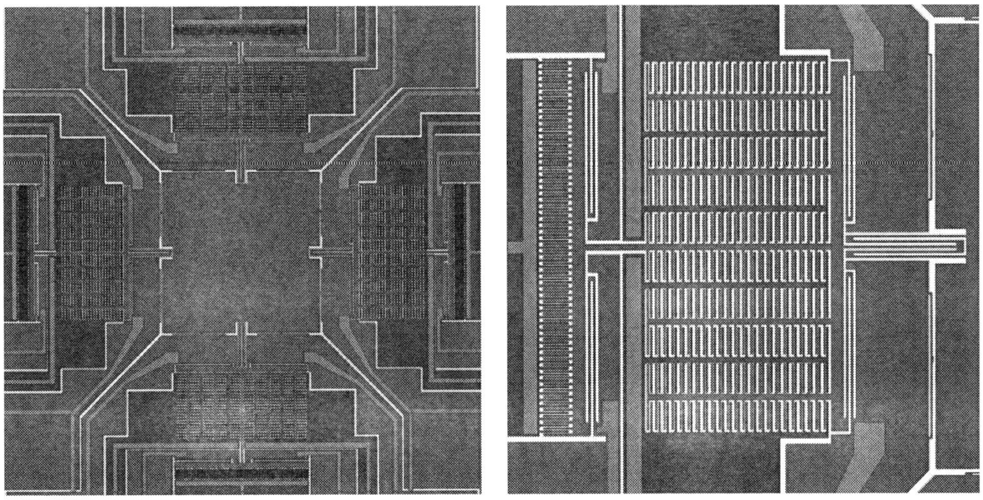

Figure 3. Geometric model of the sensor for simulation in ANSYS.

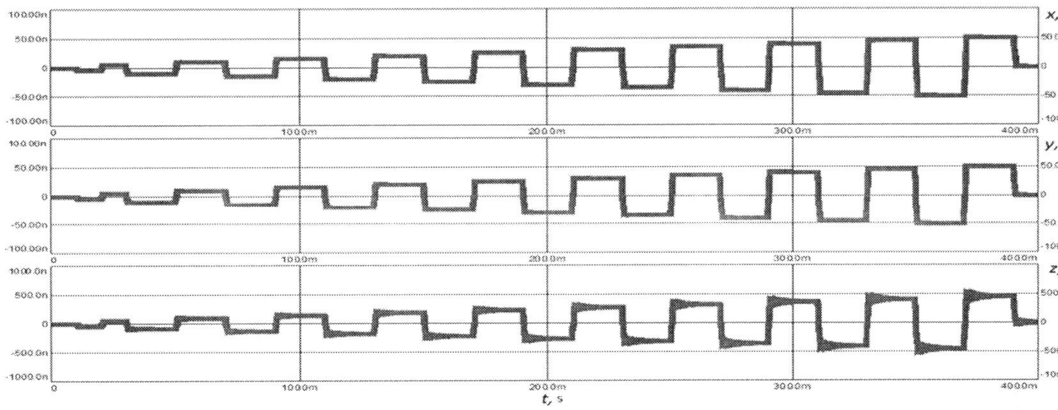

Figure 4. Displacement of the inertial mass of the sensor under the action of acceleration

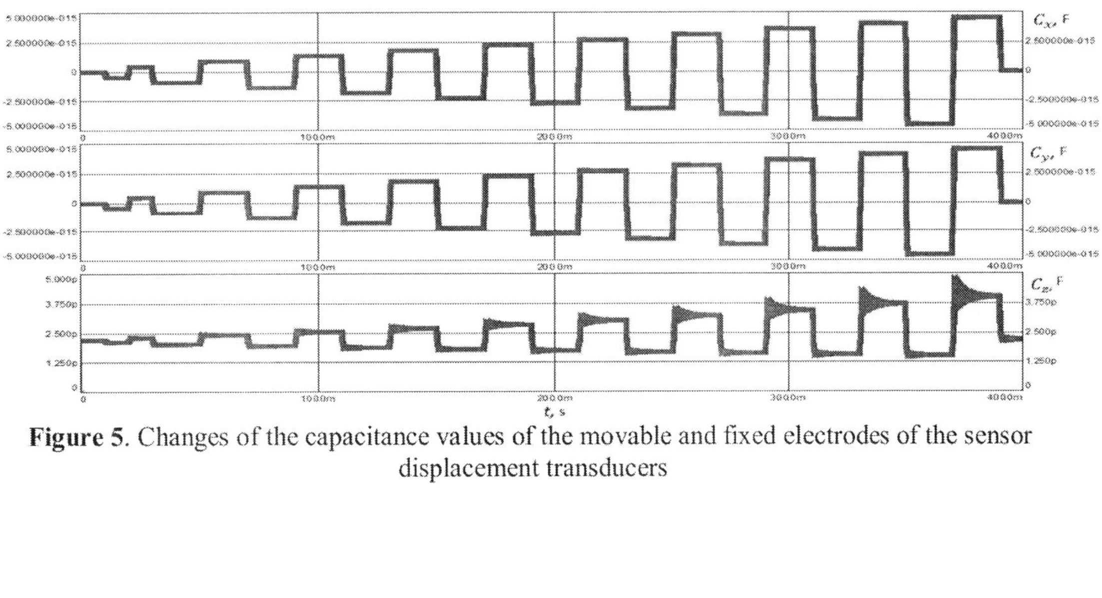

Figure 5. Changes of the capacitance values of the movable and fixed electrodes of the sensor displacement transducers

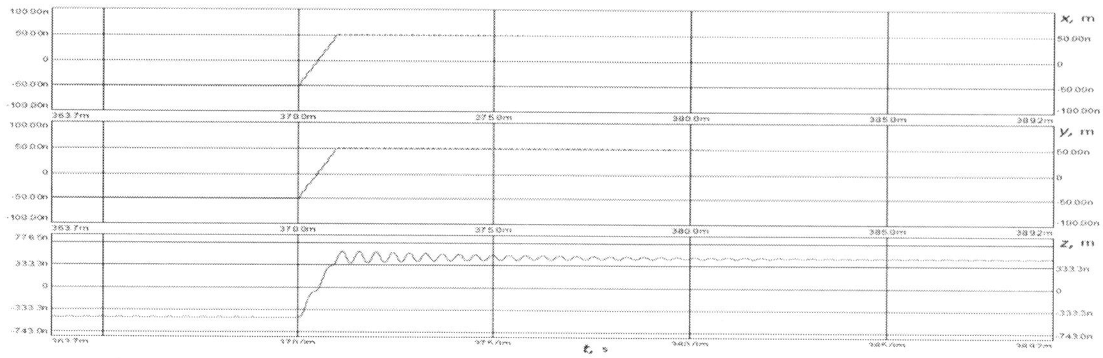

Figure 6. Transient processes of sensor inertial mass displacement when a = 10g

As simulation results show the initial capacitance of displacement tranducer on axes X and Y makes 90 fF, and on axis Z – 2,2 pF. In case of the linear acceleration influence of differential capacitances change on axes X and Y makes 0,51 fF in case of g and 5,1 fF in case of 10g, on axis Z – 0,1 pF in case of g and 1,8 pF in case of 10g.

4.Conclusion

As a result the parameters of the micromechanical accelerometer are confirmed. The parametres were planed at the design stage: quantity of axes of sensitivity – 3; the range of the measured accelerations – ±10 g; level of suppression of cross interference – at least 50 dB; sensitivity of the capacity micromechanical transformer – at least 0,5 fF/g, sensitivity of the signal processing devices over 800 kHz/fF. Novelty of the received results consists in a solution of the problem of increase of the functional capabilities of micromechanical accelerometers by using of the principle of the functional integration. This approach provides increase of the functional capabilities of sensors and allows to register the linear accelerations on three axis of sensitivity.

Acknowledgments

The results were obtained using the equipment of the student design bureau Elements and devices of inertial navigation systems of robotics of the Institute of Nanotechnologies, Electronics and Electronic Equipment Engineering, Southern Federal University (Taganrog, Russia). This research is supported by the Program for the Development of the Southern Federal University until 2021 (project VnGr-07/2017-10).

References

[1] Raspopov V Ja 2007 *Micromechanical devices* (Tula: Tul'skij gosudarstvennyj universitet)
[2] Ramadoss R 2013 *Solid State Technology* **10** 10

Formation of a Corrugated Graphene Substrate on the Paraffin Surface

S A Votyakov[1], N A Samorodov[1], Yu V Pakharukov[1,2], F K Shabiev[1,2], R F Safargaliev[1,2]
[1]University of Tyumen, Tyumen, 625000 Russia
[2]Tyumen Industrial University, Tyumen, 625000 Russia

Abstract. In this paper, we propose a method of creating a wave-like corrugated surface from graphene nanoparticles. Due to the Kelvin-Helmholtz instability, the graphene suspension flowing onto the melted paraffin forms a wave-like surface. The inhomogeneity wavelength λ can be controlled by changing the relative surface tension.

1. Introduction

The great interest of researchers to the study of the properties of graphene films is associated with their possible applications in nanoelectronics. Thus, [1, 2] have shown that fluorination of the outer side of graphene leads to its deformation. Such surface deformation forms corrugations several tens of nanometers deep. This corrugated surface allows the formation of a substrate with predetermined areas for the sorption of metal atoms. To manufacture grapheme-based devices, for example, memristors, not a single corrugation, but a corrugated area with a large number of ordered corrugations is required. Under the influence of the Kelvin-Helmholtz instability mechanisms, with fluid flowing at the interface, we can expect formation of a wave-like structure [3]. When hydrocarbons interact with the graphene suspension, a transition layer from graphene nanoparticles and hydrocarbon molecules is formed at the interface [4]. The task of this work is to experimentally obtain thermocapillary waves in a film of graphene nanoparticles on a substrate of hydrocarbon paraffin molecules.

2. Experimental technique

The hydrocarbons used were molten paraffin, the molecules of which adsorbed on graphene particles. Within 24 hours, the molten paraffin and graphene suspension were exposed to a temperature of 56°C inside a thermostat (Fig. 2). After that, a film from graphene particles and hydrocarbons was formed at the interface. Further, the coordinate table was inclined at a small angle to ensure a slow ($Re \leq 1$) film motion relative to the interface. Then, the temperature was slowly decreased, while the temperature gradient on the film surface was maintained at approximately 1K/mm.

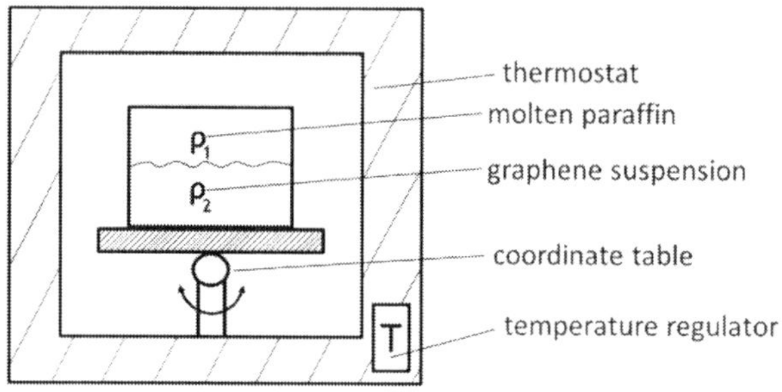

Figure 1. Installation for the formation of capillary waves at the interface of a graphene suspension - paraffin [5]

3. Results and discussion

The Fig. 2 shows the image of the graphene film surface on the frozen paraffin obtained with a scanning atomic force microscope (AFM) in a semicontact mode. The film forms thermocapillary structures in the form of waves, with a long inhomogeneity $\lambda = 750 \pm 20$ nm. Thus, as a result of exposure to thermocapillary forces, the film surface suffered deformations resulting in a transition to a structured flow.

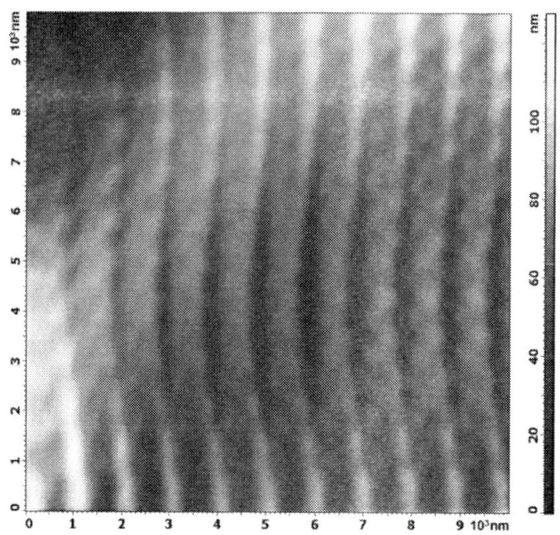

Figure 2. Thermocapillary waves at the interface of paraffin - graphene suspension

The results of the experiment can be considered as a manifestation of the Kelvin-Helmholtz instability [3]. According to the experimental conditions, liquid paraffin begins to move along the surface of the graphene aqueous suspension. Such movement forms a disturbance that creates wave deformation of the interface.

Consider two incompressible liquids - molten paraffin, lying on a graphene aqueous suspension, then the initial hydrodynamic equations are:

$$\rho_1 \frac{dV_1}{dt} = -\nabla p_1 + \rho_1 g, \quad \frac{d\rho_1}{dt} = 0, \quad \mathrm{div} V_1 = 0; \text{ for paraffin} \tag{1}$$

$$\rho_2 \frac{dV_2}{dt} = -\nabla p_2 + \rho_2 g, \quad \frac{d\rho_2}{dt} = 0, \quad \mathrm{div} V_2 = 0; \text{ for water suspension of graphene} \tag{2}$$

where, ρ_1 - paraffin density (920 kg/m^3) f_1 – specific Van der Waals force acting on paraffin from the side of graphene, ρ_2 - density of graphene water suspension (950 kg/m^3), f_2 – specific Van der Waals force acting on graphene from the side of paraffin.
In the framework of the proposed model, the following is true:

$$\rho_1 \frac{dV_1}{dt} = \rho_2 \frac{dV_2}{dt} \tag{3}$$

It's obvious that $p_2=p_1+p_0$, $f_1 = f_2 = f$, then from (1) (2) we get:

$$\nabla p_0 = g(\rho_2 - \rho_1); \tag{4}$$

In equilibrium position from (4) it follows:

$$\frac{dp_0}{dz} = -g(\rho_2 - \rho_1); \tag{5}$$

Let us transform equation (5) in a standard way and obtain the dispersion equation:

$$\omega^2 = gk + \frac{\sigma k^3}{\rho_2 - \rho_1}; \tag{6}$$

where, σ - relative surface tension of paraffin and water.
Solving equation (6) for an extremal value of k, we get:

$$k = \sqrt{(\rho_2 - \rho_1)\frac{g}{3\sigma}}. \tag{7}$$

Then, the length of the capillary waves, $\lambda = 2\pi/k$, depends on the density of the liquids and the relative surface tension σ. Consequently, by changing the ρ and σ parameters, one can control the inhomogeneity wavelength λ.

References

[1] H. Cao, Q. Yu, R. Colby, D. Pandey, C. S. Park, J. Lian, D. Zemlyanov, I. Childres, V. Drachev, E.A. Stach, M. Hussain, H. Li, S. S. Pei, and Y.P. Chen, J. Appl. Phys. **107**, 044310 (2010).
[2] B.-J. Park, J.-S. Choi, H.-S. Kim, J.R. Jeong, H. J. Choi, H. J. Jung, M.W. Jung, K. S. An, and S.G. Yoon, Sci. Rep. **5**, 9610 (2015).
[3] Kutateladze S.S., Nakoryakov V.E. Heat and mass transfer and waves in gas-liquid systems, Novosibirsk: Nauka, 1984. – 301 p.
[4] Yu.V. Pakharukov, F.K. Shabiev, B.V. Grigoriev, R.F. Safargaliev, I.R. Potochnyak, Prikladnaya Mekhanika i Tekhnicheskaya Fizika, **60**, 37 (2019).
[5] Yu.V. Pakharukov, F.K. Shabiev, V.V. Mavrinskiy, R.F. Safargaliev, V.V. Voronin, JETP Lett. **109**, 367 (2019)

Analysis of doping distribution in horizontal GaAs nanowires with axial p-n junction by the conductive atomic force microscopy

B R Borodin[1,*], P A Alekseev[1], M S Dunaevskiy[1,2], V Khayrudinov[3] and H Lipsanen[3]
[1]Ioffe Institute, Saint-Petersburg 194021, Russia
[2]ITMO University, Saint-Petersburg 197101, Russia
[3]Department of Electronics and Nanoengineering, Aalto University, Espoo, P.O. Box 13500, FI-00076, Finland.
[*]email: brborodin@gmail.com

Abstract. This work presents the results of an investigation of in plane horizontal GaAs nanowires by Scanning Probe Microscopy (SPM) methods. Topography and a local conductivity map of the horizontal nanowire with axial p-n junction are obtained. The distribution of doping and position of p-n junctions in nanowires are visualized. The I-V curves of differently doped parts of nanowires were measured. These data can be important for the understanding of the doping incorporation mechanism of GaAs nanowires.

1. Introduction

Semiconductor nanowires (NWs) draw attention due to the possibility of their application in many nanoelectronics and photonics devices. Nanowires have already been used as photodetectors[1,2], lasers[3], biosensors[4] and solar cells[5,6]. Particular attention is paid to planar NWs. This is due to the lack of some disadvantages of an out-plane nanowires. For example, it is known that III-V out-plane semiconductor nanowires have a large number of twin-plane defects[7]. Additionally, integration of out-plane nanowires into the traditional planar technological process is a considerable problem. In-plane nanowires do not have these disadvantages. Such nanowires can be integrated into the traditional planar technological process. Twin-defects free in-plane nanowires can be obtained by the MOVPE[8]. Thus in-plane nanowires are the structures combining useful properties of nanowires and practical manufacturability. Of particular interest for electronics are NWs with p-n junction. Formation of a p-n junction can be done by switching the doping precursor during the growth. Nanowires with axial p-n junctions were obtained by this way can be used as diodes, detectors, solar cells, and logic elements. However, the distribution of doping and the position of p-n junction are important parameters for devices based on nanowires. Due to the small size of the structure, the analysis of the distribution of doping and the position of p-n junctions can be a non-trivial task. For example, such task can be solved by the Atomic Probe Tomography (APT)[9]. This method has high sensitivity and spatial resolution, but its practical implementation is very difficult. The experiment must be performed in ultra-high vacuum and samples must be prepared by special methods[10]. In some cases when there is no necessary in an accurate elemental analysis (for example as in this work), APT can be the overly complex and excess method. To solve it, we used the conductive atomic force microscopy (C-AFM). C-AFM is one of the standards and widely used AFM modes. Measurements can be performed in room conditions and there is no necessary in any specific sample preparation[11]. The aim of this work was to study the topography characteristics of horizontal in plane GaAs nanowires with p-n junction grown by MOVPE on a GaAs (100) substrate, as well as detection of the p-n junction position and the doping distribution.

2. Samples and methods

2.1. Samples

The sample consists of the n-doped GaAs (100) substrate on which horizontal nanowires were grown by the MOVPE method. The p-n junction was created during the growth process by switching the doping precursor. At the start of the growth, a p-type (Zn) doping precursor was fed, and then approximately in the middle of the growth process, the doping precursor was switched to an n-type (Sn). Gold particles were served as a catalyst. Figure 1 shows the SEM image of the grown sample.

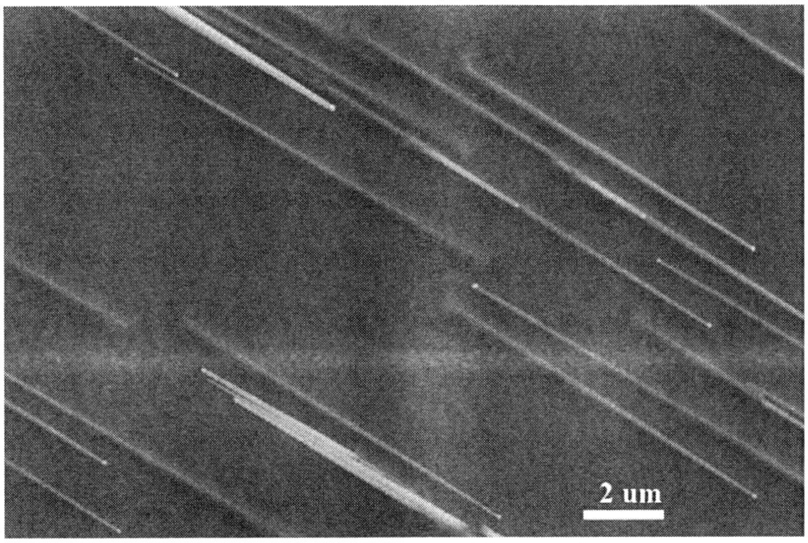

Figure 1. SEM image of the sample.

As can be seen from figure 1, the surface of the sample is covered in nanowires. The average length of nanowires is about 13 um. Almost all nanowires have grown in the plane, but some of them have deviated from in-plane growth and switched to inclined growth (in 111 direction).

2.2. Methods

The experiment was performed on the Ntegra Aura (NT-MDT) scanning probe microscope using a DCP (NT-MDT) probes. The scheme of measurement is shown in figure 2.

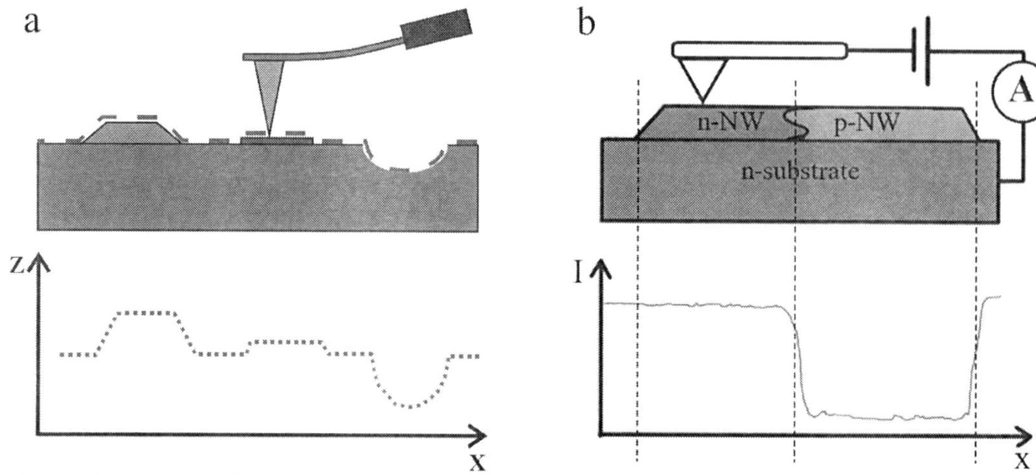

Figure 2(a, b). **(a)**The scheme of measurement of topography in contact mode; **(b)**The scheme of measurement of conductivity distribution.

The topography of nanowire was obtained in contact mode (figure 2a). The pressure force of the probe on the surface was set so as not to destroy the nanowire, but to be in electrical contact with the surface. In the process of obtaining the topography, a small voltage (1-2 V) was applied to the probe. Holder with a gold coating which pins the substrate was used as a second contact. The current flowing through the structure was measured. It depends on the place of contact with the surface. As can be seen from figure 2b, if the probe is in contact with the surface of the same type of doping as the substrate, a p-n junction will not be on the path of electrical current and its value will be large. But if the surface has the opposite doping type, a p-n junction will be on the path of electrical current and its value will be small. Thus, a local conductivity map was obtained.

3. Results and discussion

Figure 3 (a) shows the topography of a nanowire obtained by atomic force microscopy in contact mode. The local conductivity map is shown in figure 3 (b).

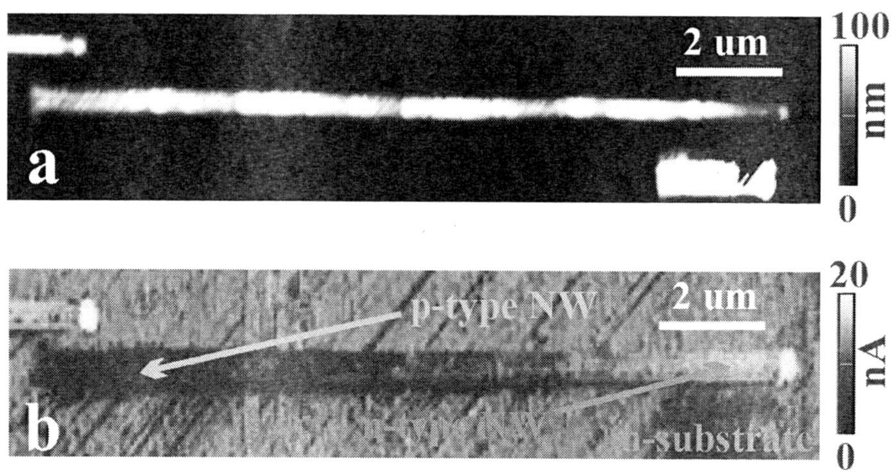

Figure 3(a, b). **(a)** AFM topography of the horizontal GaAs NW with axial p-n junction; **(b)** Local conductivity map of nanowire.

As can be seen from figure 3 (a), a nanowire about ~ 13 um long lies on the substrate surface. The nanowire has the same height along the entire length about ~ 75 nm and tapers toward the end. There are several small pieces of other nanowires around the nanowire. This is due to the method of investigation. It was said before, the topography was obtained by the contact atomic force microscopy, as well as not every nanowire grew in the plane of the surface. Thus, nanowires that deviated during the growth were broken by the probe during the scanning process. Vertical parts of nanowires were moved to the boundaries of the scanning area by the probe, but horizontal parts remained in its places. Figure 3 (b) shows a local conductivity map of the nanowire. The scheme of measurement of conductivity distribution is shown in figure 2 (b). One can see conductive (bright) and non-conductive (dark) areas. Conductivity shows the presence or absence of p-n junctions with the substrate in the contact point. As can be seen, the nanowire consists of two parts. P-doped part (dark) grew in the first stage with a p-doping precursor. After switching the precursor, the n-doped part (bright) began to grow. Thus, the p-n junction is localized in the second half of the nanowire. Such pictures may be important for understanding the growth mechanism of nanowires and the process of embedding the dopant. Thus, the distribution of doping and the p-n junction position were determined in horizontal nanowires with axial p-n junction by using the conductive atomic force microscopy. I-V curves in conductive and non-conductive areas were also measured (figure 4).

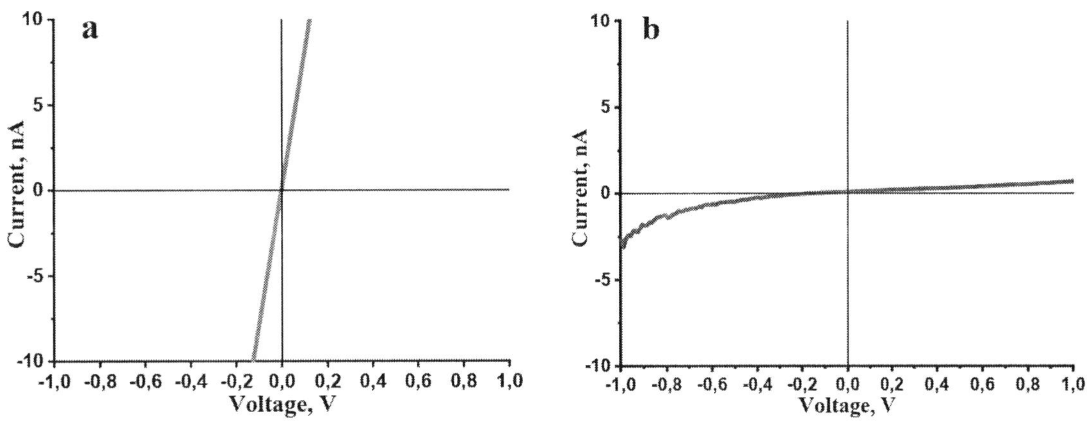

Figure 4(a, b). I-V curves in high conductive **(a)** and low conductive **(b)** areas.

As can be seen from figure 4 (a), the I-V curve in the high conductive area exhibits a clearly ohmic shape, indicating absence of the p-n junction. While the I-V curve in the low conductive area exhibits a rectifying (diode) shape (figure 4b).

4. Conclusion

To conclude, GaAs nanowires with axial p-n junction were studied by the scanning probe microscopy. Local conductivity map was obtained by using C-AFM. I-V curves of high conductive and low conductive areas were obtained. Based on the conductivity map, the doping distribution and the position of the p-n junctions were determined. The results of the work allow to consider the SPM as an effective method of analysis of horizontal nanowires.

References

[1] Wang J, Gudiksen M S, Duan X, Cui Y and Lieber C M 2001 *Science* **293** 1455–1457
[2] Wang H 2013 *Appl. Phys. Lett.* **103** 093101
[3] Saxena D, Mokkapati S, Parkinson P, Jiang N, Gao Q, Tan H H and Jagadish C 2013 *Nat Photonics* **7** 963
[4] Hahm J and Lieber C M 2004 *Nano Lett.* **4** 51–54

[5] Yao M, Huang N, Cong S, Chi C-Y, Seyedi M A, Lin Y-T, Cao Y, Povinelli M L, Dapkus P D and Zhou C 2014 *Nano Lett.* **14** 3293–3303
[6] Aberg I, Vescovi G, Asoli D, Naseem U, Gilboy J P, Sundvall C, Dahlgren A, Svensson K E, Anttu N, Björk M T and others 2016 *IEEE J Photovolt.* **6** 185–190
[7] Johansson J, Karlsson L S, Svensson C P T, M\a artensson T, Wacaser B A, Deppert K, Samuelson L and Seifert W 2006 *Nat Mater.* **5** 574
[8] Fortuna S A, Wen J, Chun I S and Li X 2008 *Nano Lett.* **8** 4421–4427
[9] Qu J, Choi W, Katal Mohseni P, Li X, Zhang Y, Chen H, Ringer S and Zheng R 2016 *ACS Appl. Mater. Interfaces* **8** 26244–50
[10] Valley J W, Reinhard D A, Cavosie A J, Ushikubo T, Lawrence D F, Larson D J, Kelly T F, Snoeyenbos D R and Strickland A 2015 *Am. Mineral.* **100** 1355–1377
[11] Geydt P, Alekseev P, Dunaevskiy M, Haggrén T, Kakko J-P, Lähderanta E and Lipsanen H 2016 *Lith. J. Phys.* **56**

Influence of carbon in the Sp^1 state on the Schottky transition Si-Cd

A Kokshina [1], **V Kochakov**[1], **O Vasilyeva**[1]
[1]Department of Applied Physics and Nanotechnology, The Chuvash State University, Cheboksary 428015, Russia

Abstract. This article presents the results of a study of the photoactive film of metal-carbon system as an example cadmium - linear-chain carbon (Cd-LCC). It is shown that when this system is annealed in a nitrogen atmosphere, the addition of cadmium atoms to cumulene carbon chains occurs. Investigation of the properties of the Si-Cd-LCC system doped with nitrogen showed a significant increase in the photoactivity of the Schottky transition in the Si-Cd system.

1. Introduction

One of the important directions of modern scientific research is the formation of nanostructured compounds with the property of photosensitivity, as well as high optical transparency and low specific resistance. One of such materials includes thin-film metal-carbon systems, including carbon of various allotropic modifications. The problem of obtaining and studying the properties of metal carbon films obtained by the interaction of non-carbide-forming metals with carbon is currently poorly understood.
LCC - carbon in the Sp1 state is synthesized as one-dimensional polycumulene chains arranged perpendicular to the substrate and is a film structure with random displacements between groups of layers [1]. The model of LCC film structure is shown in figure 1. The chains are united by Van der Waals forces into a hexagonal structure with distances of 5 Å between them. The mechanical stability of the chains is provided by σ bond, and the electrical properties of π bond, whose electrons are delocalized. This structure determines the electrical anisotropy of the LCC films. Along the chains, the conductivity of the film is metallic, and across the film is a dielectric.

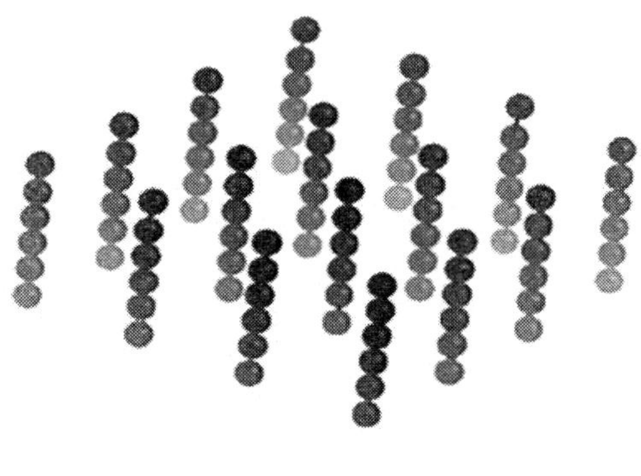

Figure 1. The structure of a linear chain carbon film.

The structure of the LCC film has a feature that is determined by the size of the inter-chain space, allowing the introduction of metal atoms of other substances into this space. The introduction of metal atoms into the inter-chain space can occur without chemical interaction (intercalation), or with the rupture of π bonds, which can result in the addition reaction. In both cases, a metal-containing modified carbon film with new properties is synthesized [2]. Synthesis conditions and properties of LCC are presented in a review [3].

2. Methodology

The work was done in two stages. At the first stage, a cadmium film was deposited on a cover glass by thermal evaporation in vacuum, a LCC film was synthesized on a cadmium film by an ion-plasma method consisting of 99% carbon in the Sp^1 state.
Next, the films were annealed at a temperature of 450^0C in a vacuum oven with a predetermined vacuum depth of 0.95 bar. The result is shown in figure 2, which clearly illustrates the change in transparency of the two previously spatially separated opaque films.

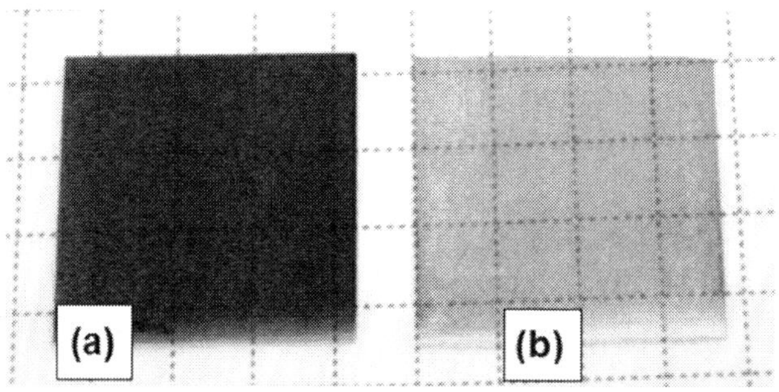

Figure 2. The Cd -LCC film system before (a) and after (b) annealing

After annealing, cadmium – LCC films were investigated by x-ray photoelectron spectroscopy (XPS).
At the second stage, a film of cadmium was deposited by thermal evaporation on the surface of a silicon crystal, and then a film of carbon in the Sp^1 state was deposited on the cadmium by method of ion-plasma synthesis.
The obtained system was doped with nitrogen during the synthesis of the LCC film, after which the Si-Cd-LCC system was annealed in a nitrogen atmosphere.

3. Experimental results and discussion

Figure 3 shows the X-ray spectra of heat-treated cadmium films and LCCs, in which the presence of a chemical shift in C1s and Cd3d5 lines is detected. For carbon, the maximum has shifted by 0.4 eV, for cadmium by 0.7 eV. In addition, a new peak appeared in the spectrogram of carbon in the region of 288.6 eV.
Based on XPS spectrometry, it can be concluded that the interfilm interaction of cadmium-LCC is expressed not only in the mechanical mixing of the two substances due to thermal diffusion, but also reactions of the addition of cadmium atoms to cumulene carbon chains occur partly. This can occur through a π - bond break.

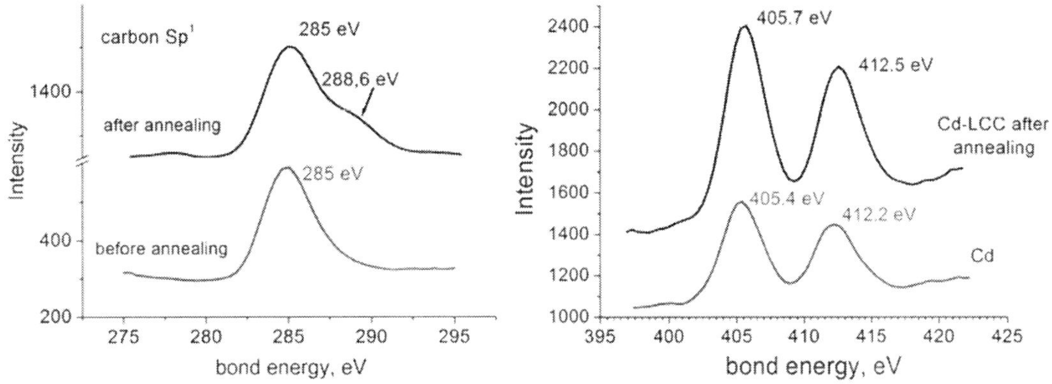

Figure 3. XPS spectra of carbon and cadmium after annealing the Cd -LCC film system.

Optical properties were studied on samples deposited on a glass substrate. The transmission spectra were obtained using a Lambda-25 spectrophotometer. The graphs of these spectra are shown below in figure 4. The transmission at a wavelength of 600 nm increased 8 times after annealing.

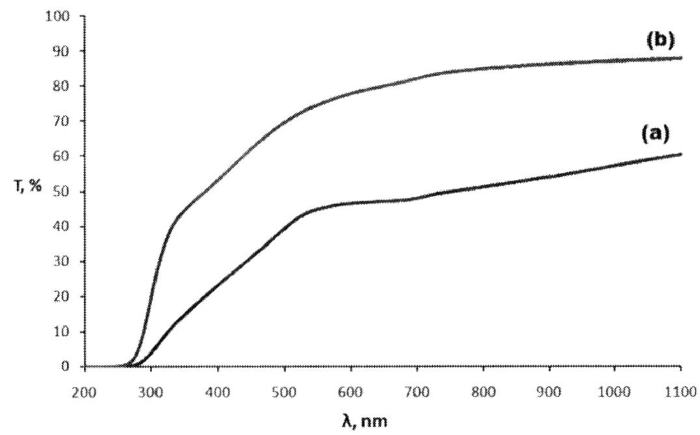

Figure 4. The transmission spectra of Cd -LCC film system before (a) and after (b) annealing

After annealing, the resistivity of the Cd-LCC system increased, but the film remained conductive.
An optical method was used to determine the band gap [4]. The band gap of the Cd-LCC system after annealing is 2.4 eV.
In the synthesis of the Cd-LCC system on a substrate of Si n-type conductivity with subsequent annealing in the resulting system there is photoactivity, as evidenced by the current-voltage characteristic in figure 5 [5].
The use of a nitrogen-doped LCC film followed by annealing at a temperature of 450 °C in a nitrogen atmosphere leads to a significant increase in photoactivity, which is shown in figure 6.
It was previously shown that the Si-Cd-LCC system has photosensitive properties [3]. Investigation of the properties of the system Si-Cd-LCC doped with nitrogen, showed a significant increase in photoactivity (Figure 2).

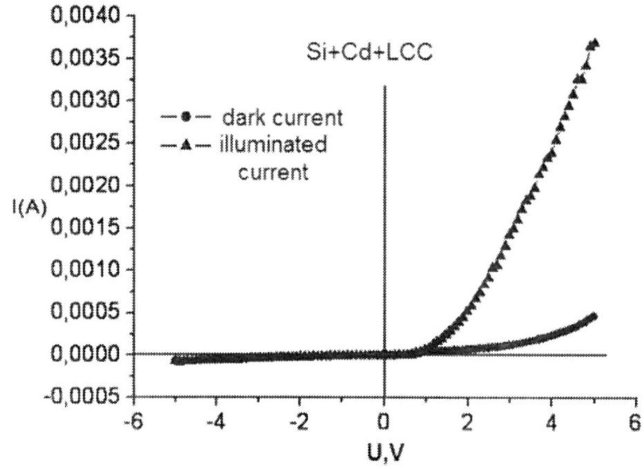

Figure 5. The current-voltage characteristic of the system Si-Cd-LCC

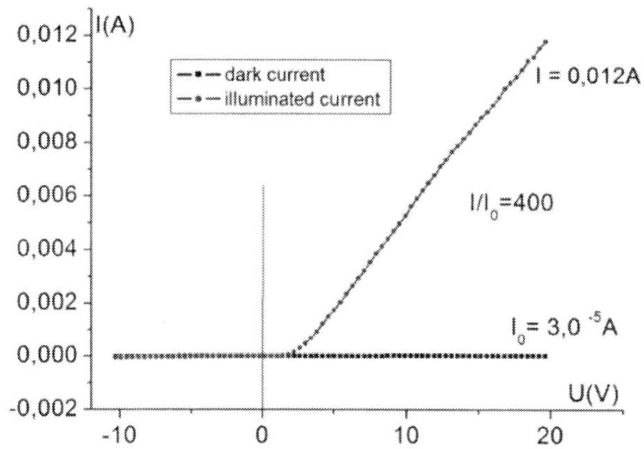

Figure 6. Photoactivity of Schottky transition Si-Cd-LCC nitrogen doped after annealing.

Thus, the use of carbon film of LCC doped with nitrogen applied to the si(n) – Cd transition, followed by annealing, improves the sensitivity of the Schottky photodiode compared with the use of LCC film without doping.

References

[1] Heimann RB, Kleiman J, Salansky NM. A unified structural approach to linear carbon polytypes // Nature, 1983, №306(5939), p. 164–167.
[2] V. G. Babayev, M. B. Guseva, N. F. Savchenko, N. D. Novikov, V. V. Khvostov, P. Flad, Highly Oriented SP1-carbon films. Surface. X-ray, synchrotron and neutron studies, 2004, vol 3, pp 16-27
[3] Babaev V., Guseva M., Khvostov V., Novikov N., Flood P. Carbon Material with Highly Ordered Linear-Chain Structure. Polyynes – Synthesis, Properties, Applications, edr. Fr.Cataldo, CRC press, USA. 2005. P.219 – 252
[4]Vasilevsky A M *et al.* 2011 Optical and physical methods of research: guidelines for laboratory work
[5] Kokshina A.V., Kochakov V. D., Krasnova A. G. Photoactivity of metal-carbon films // Vestnik ChuvSU. 2013. Vol 3., pp 60-63

Sensitive elements of microelectronic sensors of fast variable and static pressure

A. V. Fimin[1], S. A. Gurin[2], E. A. Pecherskaya[1], P. E. Golubkov[1], G. V. Kozlov[1], K. Y. Kraynova[1]
[1]Department of Information and measuring equipment and metrology, Penza State University, Penza 440026, Russia
[2]CJSC Medtechnika, Penza 440018, Russia

Abstract. The technical progress development contributes to the introduction of automated methods of management and control of the system state. The use of sensor equipment with the use of piezoelectric tensor-resistive modules allows to register fast variable and statistical pressure, respectively. Such assemblies possess not only the most reliable information content of the measured values, but also combine simple step-by-step manufacturing techniques.

1. Introduction

The requirements increase for technical systems determine the growth of measurement volumes and the number of investigated parameters [1]. At the same time, sensor devices must have not only advanced functionalities, but also the smallest designs. In such conditions, multifunctional sensitive elements with integrated thin strain-resistive and piezoelectric films are able to perform the tasks associated with the registration of the maximum range of determined pressures. Such assemblies have a small geometric structure, which the theoretical model can be manufactured industrially.

The creation of such structures became possible due to the technological breakthrough in the field of the synthesis of thin ferroelectric PZT films, as well as polar dielectrics ZnO and AlN [2-8].

2. Applicable materials

The cultivation of highly textured piezoelectric thin films ZnO and AlN with clearly defined thicknesses across the entire substrate is currently being implemented at an industrial level [9].

To determine the dynamic pressure, films based on polar dielectrics AlN and ZnO are used, which under normal conditions crystallize into a hexagonal lattice of the wurtzit type (W4) or lead titanate zirconate (PTZ) with piezoelectric properties, having a perovskite structure. Table 1 presents the effective parameters of thin piezoelectric films.

Table 1. Parameters of piezoelectric films.

Parameter	Aluminum nitride AlN	Zinc oxide ZnO	Lead titanate zirconate PTZ
$\varepsilon_{33,f}$	8 – 12	10,4	900 – 1300
$d_{33,f}$, (pKl/H)	10 –12	3,4 – 3,9	90 – 110
$k^2_{p.f}$	0,06 – 0,085	0,065 – 0,11	0,1 – 0,20
density, (g/cm^3)	5,68	3,26	7,5 – 7,6
moduleYoung, (GPA)	110 – 150	260 –380	60 – 80

PZT thin films with piezoelectric properties have a perovskite-type structure. The spontaneous electric dipole moment of PZT films can be reoriented between certain crystallographically stable states using appropriate levels of the electric field that do not lead to dielectric breakdown. In the perovskite structure of ABO3, where A is a bivalent ion, B is a tetravalent ion, the cubic lattice may be distorted at temperatures below the Curie temperature. These distortions shift the octahedrally oriented cations from the center of the structure. Accordingly, a dipole moment is created between the center of the negative charge created by the oxygen octahedron and the center of the positive charge created by the cation sublattice. The shifts in the position of the A-ions of Pb^{+2}, like the B-ions of Zr^{+4} and Ti^{+4}, determine the properties of the PZT (ZrxTi1 - x) O_3 dipoles [10].
The values of the piezoelectric parameters of thin perovskite PZT films exceed the similar parameters of AlN and ZnO. However, the stability of the characteristics of the piezoelectric parameters of polar dielectrics is much higher and their production has been mastered much more widely, which in turn imposes restrictions on the use of PZT films. For industrial production of thin piezoelectric PZT films, a number of fundamental tasks are required to be solved: the development of the physical fundamentals of the technology for manufacturing low-dimensional piezoelectric films; the study of the features of the piezoelectric state and the switching mechanisms of spontaneous polarization; the creation of conditions of the polarized state stable in time [5, 6, 11]. Moreover, the further integration of piezoelectric perovskite films with used deposition modes with strain-resistant semiconductor pressure sensors is limited by the high temperatures of sensitive elements production.

3. The design of the sensor element with integrated piezoelectric and strain-resistant films

Figure 1 shows the model of the sensitive element of the sensor for measuring alternating and static pressure, built on the basis of silicon poles with formed platinum tracks integrated by piezoelectric capacitive structures with semiconductor strain gages [10 - 12].

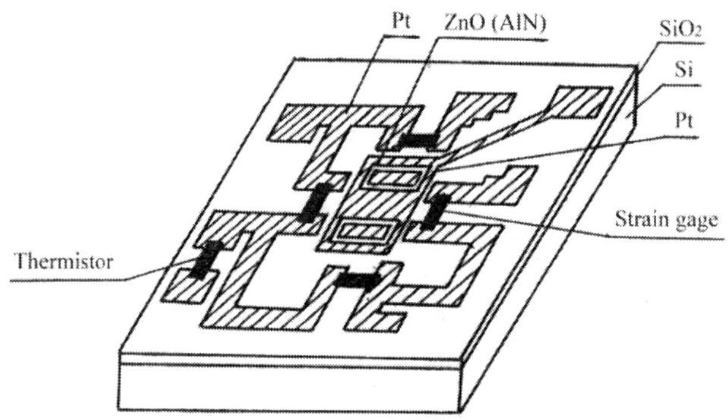

Figure 1. Model of the sensitive element of the pressure sensor.

Capacitor piezoelectric structures are formed in the center of the sensitive element, on the periphery of the hard center. Compression and expansion strain gages are located along the perimeter of the membrane in the zones of greatest deformation.
They are made according to the developed technology of plasma-chemical deposition of polycrystalline silicon. The thermo-resistor is formed by ion implantation.
Further fabrication of the crystal of the sensing element is carried out by anodic splicing of silicon with glass at temperatures above 500 °C and the voltage applied to the structure, of the order of several hundred volts. This will inevitably lead to a reorientation of thin PZT films polarization. As thin piezoelectric films, it is advisable to use zinc oxide or aluminum nitride.
Textured AlN piezoelectric films used as a device layer were obtained by ion-beam sputtering of an target from Al in an argon-nitrogen mixture in a modernized ion-beam sputtering unit with a residual

pressure of 3 to 5 · 10^{-5} Pa to eliminate the reaction gas gettering which manifests itself during magnetron sputtering as a result of plasma radiation. When ion-beam sputtering, the target and the substrate are outside the plasma, which qualitatively improves the structure properties [13].
The substrate temperature was varied from 150 to 350 °C. Textures with satisfactory properties are formed at 350 °C. The pressure of the working mixture in the vacuum chamber was maintained in the range from 1 to 3 · 10^{-3} Pa in the ratio of gases Ar : N2 = 1 : 1. When reducing the argon percentage, a decrease in the rate of the target material sputtering and the deterioration of the deposited film properties are observed. This is due to the chemical activity of nitrogen entering into the compound with the target material on its surface, and the insufficient energy of the bombarding ions compared with the kinetic energy of argon ions for intensive ejection of atoms and atoms conglomerates (clusters) from the target [13].
The disadvantages of the ion-beam method of sputtering AlN include the relatively low reproducibility of the resulting films properties and the structures defectiveness due to knocking out clusters (atoms conglomerates) that degrade the film texture.
ZnO thin films with piezoelectric properties were obtained by reactive HF-magnetron target sputtering from Zn (99.9 %) in an Ar + 65% O_2 gas mixture at a chamber pressure of 1 to 3 · 10^{-3} Pa. The substrate temperature was 220 °C, the RF discharge power was 100 W. The use of RF magnetron target sputtering from Zn (99.9 %) instead of DC sputtering is caused by the "poisoning" of the target with oxygen and the formation of a surface dielectric layer contributing to the appearance of a surface charge reducing the sputtering rate. In addition, when HF target sputtering ZnO, it often cracks as a result of overheating. The exclusion of thermo-mechanical stresses was carried out by carrying out heating and cooling in both cases at a rate of not more than 3 °C/min. Sputtering of piezoelectric structures in combination with the spent silicon technology for obtaining strain gages can be carried out both before the formation of strain gages and after.

4. The design of the sensitive element with the integration of thin piezoelectric PZT films and metal film strain gages

Another variant of the sensor element is a cross-shaped beam located along the perimeter of the support ring, on which a bridge circuit of metal-film strain gages and piezoelectric capacitive structures based on PTZ is formed (fig.2).

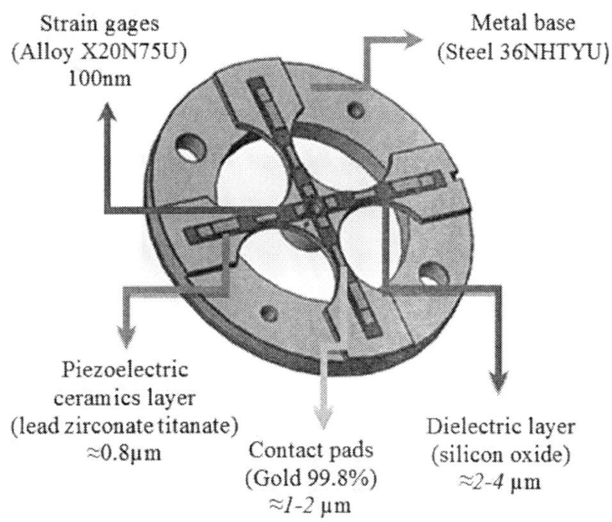

Figure 2. Sensitive element with the integration of thin piezoelectric PZT films and metal film strain gages.

The beam is made of an elastic alloy with a high elasticity modulus 36NiCrTiAl. In this structure, the strain gages are formed by the technology of thermal vacuum evaporation of the alloy $H_{20}N_{75}YU$. PTZ thin films are formed by HF-sputtering from a PTZ ceramic target at a power of 300 W, in an Ar - O_2 stream (1: 1).
The composition corresponds to a rhombohedral solid solution adjacent to the morphotropic phase boundary. Lead losses were compensated by adding 10 mol % PbO. The deposition temperature was 400 °C, the film thickness corresponded to the range from 0.5 to 0.8 μm, the annealing temperature was 570 °C. The strain gages before the PZT film deposition were passivated with the Si_3N_4 film obtained by the method of low-temperature plasma chemical deposition, in order to prevent the introduction of contaminants of the PZT target sputtering process, resulting the strain gages degradation. Further, the upper plates of the ''titanium-gold'' structure were formed by the method of resistive evaporation, similarly to the lower plates.
Further assembly of the sensing element is carried out mechanically without high heat, which can affect the polarized state of the PZT film. The use of polar dielectrics in this structure is not justified due to the low films piezomodule, which, in combination with a small beam of an elastic alloy, will not cause a proper output signal.

5. Conclusion

These structures allow to register a wide range of both statistical and dynamically changing pressure values with small deviations from the actual values. At the same time, the possibility of sensitive systems prototypes manufacturing is confirmed by the modeling results.

References

[1] Pecherskaya E A, Golubkov, P.E., Fimin A V, Zinchenko T O, Pechersky A V, Shepeleva J V 2018 *Procedia Computer Science* **132** 1163 – 1170
[2] Mishina E D, Muhortov V M, Sigov A S 2009 *Integrated ferroelectrics* **106**(1) 1–10
[3] Pronin I P, Kaptelov E Yu, Senkevich S V, Klimov V A, Zaitseva N A, Shaplygina T A, Pronin V P, Kukushkin S A 2010 *Physics of the Solid State* **52**(1), 124–128
[4] Sulei Fu, Qi Li, Shuang Gao, Guangyue Wang, Fei Zeng and Feng Pan 2017 *Applied Surface Science* **402** 392-399
[5] Kukushkin S A, Osipov A V, Sergeeva O N, Kiselev D A, Bogomolov A A, Solnyshkin A V, Kaptelov E Yu, Senkevich S V, Pronin I P 2016 *Physics of the Solid State* **58** (5) 967–970
[6] Pronin I P, Kukushkin S A, Spirin V V at al. 2017 *Materials Physics and Mechanics* **30** 20-34
[7] Pecherskaya E A., Zinchenko T O, Golubkov P E, Pechersky A V, Fimin A V, Nikolaev, K O 2018 *Moscow Workshop on Electronic and Networking Technologies (MWENT)*, Moscow, pp. 1-4.
[8] Pecherskaya E A, Ryabov D V, Shepeleva J V, Pecherskaya R M 2014 *Journal of Physics: Conference Series* **541** 012012
[9] MEMS Materialsand Processes HandbookSpringer Science+Business Media, LLC 2011 DOI: 10.1007/978-0-387-47318-5
[10] Warren W L, Robertson J, Dimos D, Tuttle B A, Pike G E, Payne D A 1996 *Phys. Rev. B* **53**, 3080–3087
[11] Vorotilov K A, Mukhortov V M, and Sigov A S 2011 *Integrated Ferroelectric Devices* (Energoatomizdat, Moscow) p 175.
[12] Vorotilov K A, Sigov A S 2012 *Phys. Solid State* **54**(5), 894
[13] Volokhov I V, Gurin S A, Vergazov I R 2016 *Measurement Techniques* **59** 80–86

Influence of annealing temperature and activation time on the catalytic centers formation for carbon nanostructures growth

O I Il'in, N N Rudyk, M V Il'ina, O I Osotova, A A Fedotov
Institute of Nanotechnologies, Electronics and Equipment Engineering, Southern Federal University, Taganrog, 347922, Russia

Abstract. Experimental studies of the effect of annealing temperature and time of the PECVD process activation stage on the formation of catalytic growth centers of carbon nanostructures were carried out. The dependences of the diameter and height of the catalytic centers on the annealing temperature are obtained. It has been established that exposure in an ammonia atmosphere leads to additional etching of nickel and a decrease in their diameter and height. It has been established that the most optimal temperature for creating with the smallest dispersion of the geometric parameters of the catalytic centers is 750 °C.

1. Introduction
The unique properties of nanostructures based on carbon nanotubes (CNTs) make them a promising material for creating new elemental base of micro- and nanoelectronics, nanopiezotronics [1-3]. However, for instrument applications, oriented CNT arrays are of interest. One of the most promising methods for the synthesis of oriented CNT arrays is plasma enhanced chemical vapor deposition (PECVD) [4, 5]. The growth of CNT arrays by the PECVD method requires the presence of catalytic centers (CC) based on transition metals. The most frequent formation of CC is carried out by annealing of films of Ni, Fe, Co, etc. Parameters of catalytic centers (size, dispersion, chemical composition, etc.) determine the parameters of carbon nanotubes (diameter, height, chirality, electrical properties, growth kinetics), which underlie their instrument application.

2. Experiments and methods
In this paper, studies of the effect of annealing temperature and activation time on the formation of catalytic centers of growth of CNT have been carried out. Samples with Ni (10 nm)/Cr (20 nm)/Si structure were heated in the temperature range of 700-800 °C with simultaneous feeding of an inert gas (Q_{Ar} = 40 cm^3/min) and ammonia (Q_{NH3} = 15 cm^3/min) into the chamber. Argon is used to displace residual air from the chamber, and ammonia creates a reducing atmosphere that prevents oxidation of the catalytic film.

The activation consisted in holding the samples obtained at the heating stage in an NH_3 atmosphere (210 cm^3/min) with and without plasma initiation. Formed 6 batches of samples for which the heating stage was carried out to a temperature of 750 °C for 20 minutes, in a stream of Ar (40 cm^3/min) and NH_3 (15 cm^3/min). Upon completion of the heating stage, the activation stage began, during which only NH_3 (210 cm^3/min) was fed into the chamber, while the heating temperature did not change. For samples № 1-3, the gas supply time was 1, 3 and 5 minutes without plasma initiation, for № 4-6 the time was the same, but with plasma initiation.

The study of the obtained catalytic centers was carried out using atomic force microscopy (AFM) using the Ntegra Nanolaboratory (NT-MDT, Russia) and scanning electron microscopy (SEM) using a Nova Nanolab 600 microscope.

We investigate of the phase composition of the structure of the nickel catalytic layer by X-ray absorption spectroscopy (XANES, EXAFS) at a specialized source of synchrotron radiation (SIC "Kurchatov Institute", Moscow, Russia). The obtained X-ray absorption spectra were compared with reference samples.

3. Results and discussion

The dependences of the catalytic centers diameter and height on the annealing temperature (Fig. 1a) and EXAFS spectra (Fig. 1b) of the Ni/Cr/Si structure after heating to 750 °C were obtained. The difference between the temperature coefficients of expansion of the Si substrate and the Cr sublayer contributes to the appearance of a strong mechanical stress on the film/substrate contact during annealing, and, as a result, fragmentation and rupture of the metal film into separate islands — catalytic centers (Fig. 2).

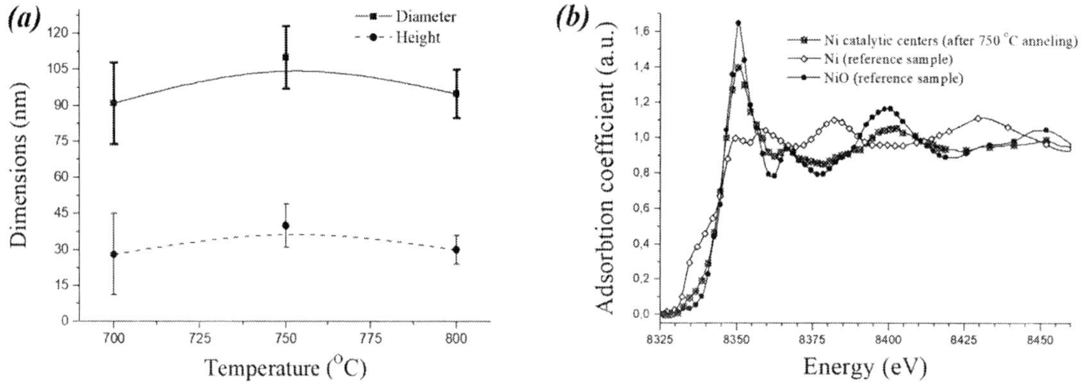

Figure 1(a, b). **(a)** Dependence of the catalytic centers diameter and height on the annealing temperature; **(b)** EXAFS spectra of the Ni/Cr/Si sample after heating to 750 °C

CC formed at a temperature of 700 °C characterized by a large scatter of geometric parameters, which is associated with the initiation of the film breakdown process and insufficiently intense surface diffusion of Ni atoms. As the temperature rises to 750 °C, diffusion exchange of atoms is activated through the contact plane between the film and the substrate. As the plasticity of the metal increases, a combination of small CC into larger ones with a decrease in the scatter of their height is observed. When the temperature rises above 750 °C, there is a simultaneous process of sublimation and surface diffusion, which leads to a decrease in the diameter and height of the CC, while there is a decrease in the number of small CC, which were observed at a temperature of 700 °C.

The results of an X-ray absorption spectroscopy show that in structures obtained by heating to 750° C, Ni is mainly in the oxidized state (the volume of pure Ni is ~30%). Due to the necessity of forming catalytic centers from pure Ni, for subsequent growth of CNTs, it is necessary to "activate" centers to reduce Ni from NiO before starting the process of CNT synthesis.

The results obtained from an experimental study of the effect of the "activation" process showed that this stage has a significant impact on the evolution of catalytic centers, with the greatest contribution being made by the effect time, and not by the presence of plasma in the process.

Comparison of dependencies (Fig. 3) shows that exposure in an ammonia atmosphere leads to additional etching of Ni catalytic centers and decrease their diameter and height. At the same time, more ordered and homogeneous catalytic centers arrays are formed in comparison with arrays those obtained at the heating stage. It is shown that, with an increase in the catalytic centers processing time during the "activation" stage, it is possible not only to reduce the size of the CC from 122 ± 19 to 65 ± 9 nm (in the range of processing time 1–5 min), but also to increase their homogeneity.

Journal of Physics: Conference Series **1410** (2019) 012231 doi:10.1088/1742-6596/1410/1/012231

Figure 2(a, b, c). SEM-images of CC, formed at temperatures: **(a)** 700 °C; **(b)** 750 °C; **(c)** 800 °C.

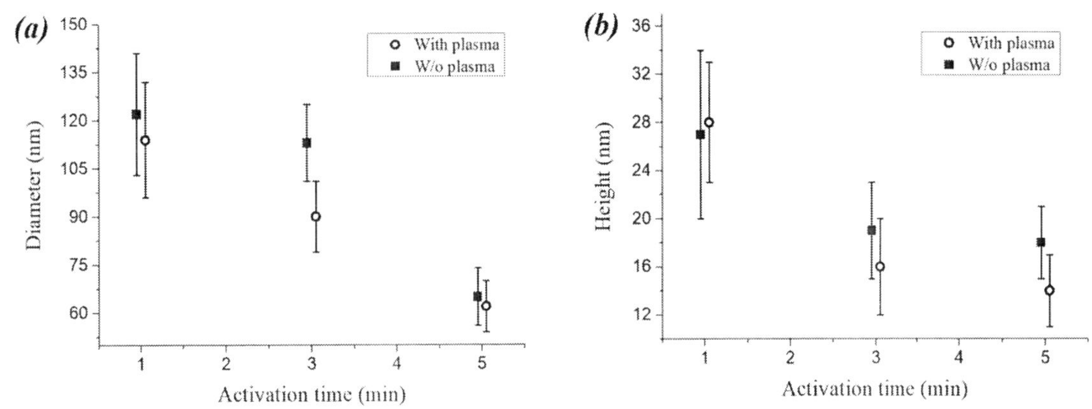

Figure 3(a,b). Dependencies of the catalytic centers diameter **(a)** and height **(b)** on activation time

The results of EXAFS spectroscopy (Fig. 4) show that Ni on the samples almost completely recovered, with a metal purity of 90-95%, which confirms the necessity of "activation" step prior to the growth stage.

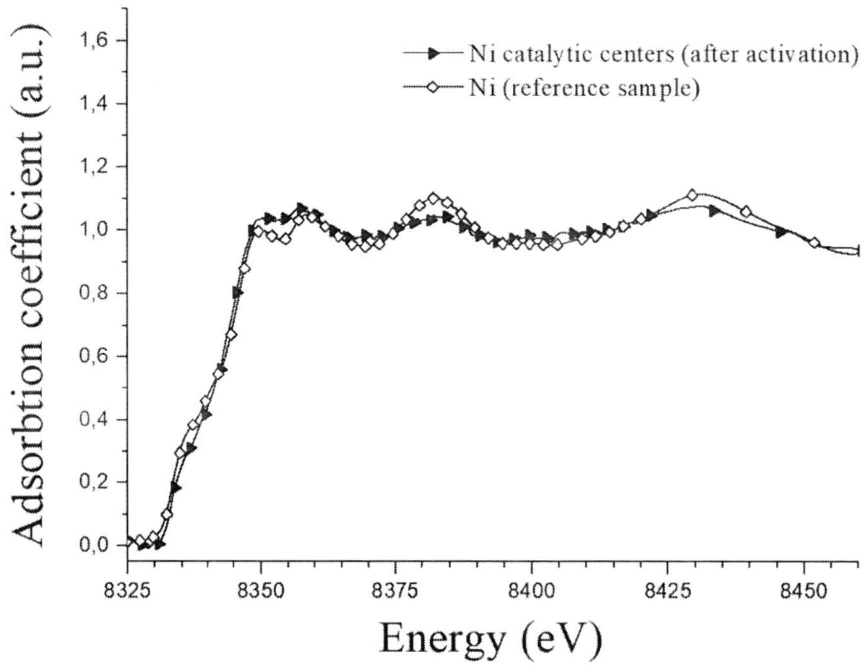

Figure 4. EXAFS spectrum of the sample after activation in plasma

It has been established that exposure in an ammonia atmosphere leads to additional etching of Ni and a decrease in their diameter and height. In this case, a more ordered and homogeneous CC array is formed, compared to the heating stage.

As a result of research on the effect of heating temperature on the formation of CC, it has been established that the most optimal temperature for creating CC with the smallest dispersion of geometric parameters is 750 °C. At the same time, as the activation time increases, it becomes possible to reduce the size, increase homogeneity, and restore oxidized catalytic centers.

The results can be used to create elements and devices of micro- and nanoelectronics, sensitive elements of gas sensors, energy harvesting.

Acknowledgments
This work was supported by the Russian Science Foundation (project № 18-79-00176).

References
[1] Schroeder V, Savagatrup S, He M, Lin S, Swager T M 2019 *Chemical Reviews* **119** pp. 599-663
[2] Ilina M V, Blinov Y F, Ilin O I, Rudyk N N and Ageev O A 2017 Piezoelectric effect in non-uniform strained carbon nanotubes *IOP Conf. Ser. Mater. Sci. Eng.* **256** 012024
[3] Il'ina M V, Konshin A A and Solomin E G 2018 Flexoelectrical nanogenerator design using aligned carbon nanotubes *J. Phys. Conf. Ser.* **1124** 071010
[4] Manawi Y, Ihsanullah, Samara A, Al-Ansari T, Atieh M, Manawi Y M, Ihsanullah, Samara A, Al-Ansari T and Atieh M A 2018 A Review of Carbon Nanomaterials' Synthesis via the Chemical Vapor Deposition (CVD) Method *Materials (Basel).* **11** 822
[5] Venkatesan S, Visvalingam B, Mannathusamy G, Viswanathan V and Rao A G 2018 Effect of chemical vapor deposition parameters on the diameter of multi-walled carbon nanotubes *Int. Nano Lett.* **8** 297–308

Self-organization of nanocarbon particles in the flame by a heterophase mechanism

O V Vasilyeva[1], S I Ksenofontov[2], A V Kokshina[1]
[1]Chuvash State University named after I.N. Ulyanov, Cheboksary 428015, Russia
[2]Chuvash State Pedagogical University named after I.Y. Yakovlev, Cheboksary 428000, Russia

Abstract. Nanotubes of different structure are formed during the self-organization of nanocarbon particles in the flame. The nucleation and growth of these tubes occurs by a heterophase mechanism. Self-organized structures by size "fill" the space between nano-and micro-formations.

1. Introduction

Carbon dioxide and water vapors are the final products of transformations with the complete combustion of hydrocarbons in the air. Incomplete combustion leads to the formation of carbon, which is released in the form of dispersed particles. The article [1] describes the formation of fullerenes in the flame of methane at a pressure less than atmospheric. Other researchers have obtained carbon structures in the form of twisted nanotubes [2].

Upon receipt of ordered structures from carbon, they are distinguished by the processes of their growth, both from the gas and condensed phases. In the gas phase, the concentration of carbon atoms is small and the process of formation of nanotubes is long [3], the linear growth rate of nanotubes is 11 μm/s. The formation time of nanotubes from the condensed phase is several orders of magnitude less.

The rate of formation of ordered structures is higher if aromatic compounds are involved in the reaction. It is noted that more complex aromatic molecules such as diphenyl, naphthalene, and anthracene tend to form carbon structures, not just benzene. The diphenyl shows the greatest tendency to the formation of soot in the flame in the above series. The elements such as nickel, cobalt, iron are used as the best catalysts for the nucleation and growth of nanotubes. The number of papers where magnesium is used to catalyze carbon nanotubes is limited [4].

2. Research methods

Condensed combustion products were studied by sampling. Plates of fused silica flew through the flames. Precipitated condensed products were studied using an optical microscope, and individual parts of the sampler – a scanning electron microscope.

3. Object of study

The processes of soot formation in the flame of condensed systems based on magnesium, diphenyl and inorganic oxidizing agent - potassium nitrate are studied in this work. Powders of metal, oxidizer, organic fuel in a certain proportion were mixed and pressed into a cardboard with an inner diameter of 20 mm. The density of the sample reached a maximum value at a specific pressure of 150 MPa. The samples were burned under room conditions. Reactions with the formation of nanostructures occur, starting from the burning surface of the condensed phase and in the stream of combustion products.

Hitachi TM-4000Plus scanning electron microscope was used to study the structure and morphology of the sample surface.

4. Experimental results and their discussion

Various nanostructures are formed in the flame with the participation of a burning metal particle. Dispersed particles consisting of highly dispersed carbon, magnesium or magnesium oxide, an oxidizing agent and combined particles were found in the combustion products. Combined particles, as a rule, are larger in size than the initial ones and can include all components of the initial particles and their transformations during combustion.

Figure 1. Micrograph of the sample area.

The needle-shaped structures with a length of 4–6 µm and a diameter of up to 0.5 µm during 80 ms grow within a combined particle of condensed combustion products (Fig. 1). The length of the observed nanotubes turned out to be almost the same, and it is given by the time of their life - from nucleation to freezing on the surface of the sampler. At the same time, the process of burnout of soot is going on in the flame, both due to the oxidizing agent incorporated in the composition of the condensed system and due to the oxygen of the air. The main part of the soot formed burns out in a time of 0.2 s at a flow rate of ~ 1 m / s, i.e. the soot particles at a height of ~ 20 cm above the burning surface are almost not observed.

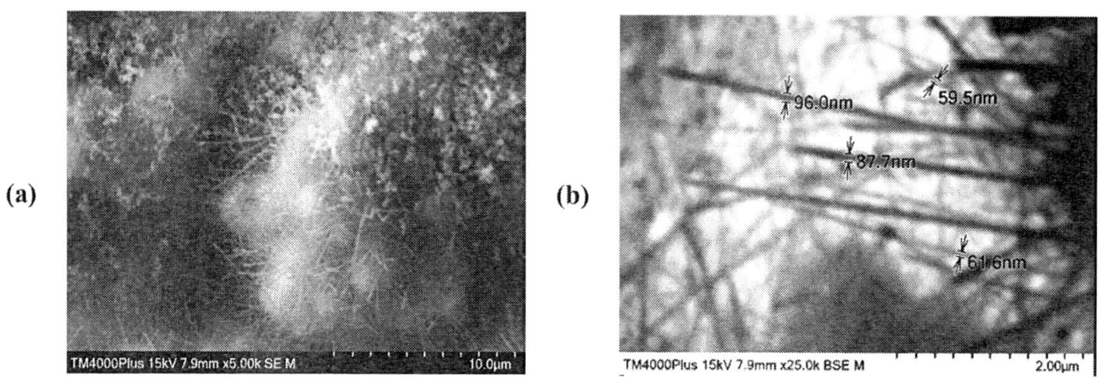

Figure 2(a, b). Needle-like structure of carbon nanotubes: **(a)** A general view of the nanotubes; **(b)** The characteristic dimensions of the nanotubes.

Carbon filaments in the form of needles are also observed on the outer surface of a dispersed particle of spherical shape (Fig. 2a, b). The length of the needle-shaped filaments is from 2 to 4 microns, and their diameter is 75-100 nm.
Some nanotubes have a fibrous structure and intertwine with each other (Fig. 3a, b). However, the beginning of the nanotubes is located on the surface of the combination particles. Similar particles were obtained in [2]. The degree of interlacing of the nanotubes indicates an uneven growth rate of each tube. The external dimensions of nanotubes, measured in photographs with high magnification, are comparable with the size of nanotubes obtained in [3]. There are particles coated with nanotubes of the same size ~ 100 nm in the above photographs.

(a)

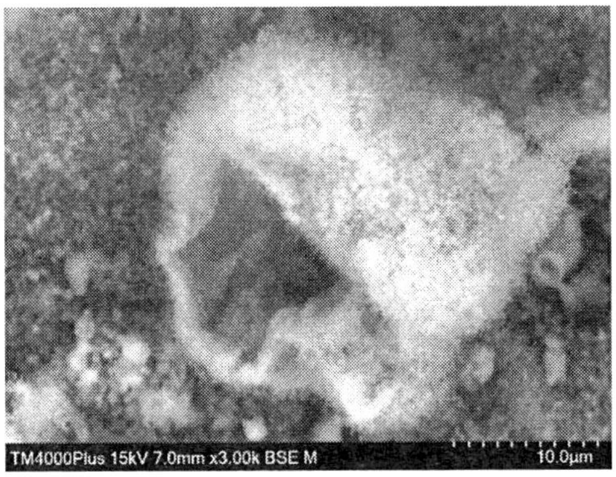

(b)

Figure 3(a, b). Fibrous structure of carbon nanotubes: **(a)** A general view of the nanotubes; **(b)** The structure of the filaments.

Another fibrous carbon structure is shown below (Fig. 4). A mesh of interwoven carbon fibers is formed on the outer surface of the burning particle in the form of a tube. This tube is drawn out in length as the particle burns out. The shape and dimensions of the tube are not constant, varying both in length and in diameter. Micro-objects of different brightness are highlighted in the figure below. The microscope in BSE mode shows clearly defined material differences in atomic weight. In this case, these are particles of magnesium oxide embedded in the carbon network with sizes ranging from 100 to 650 nm. Large pieces of metal oxide remain inside the tube.

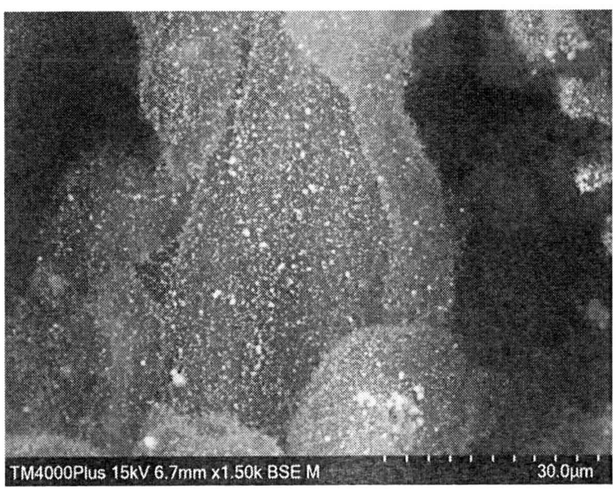

Figure 4. Particles of magnesium oxide in carbon mesh.

Separately selected sample sites, depicted in Figure 1, were studied by X-ray analysis and are presented in the form of the energy spectrum (Fig. 5). The structure of the test substance can be judged by the energy spectrum obtained. The investigated sample areas can be represented as MgO, MgC_2, C_n. The maxima in the spectrum of such elements as silicon, sodium, and chlorine can be attributed to the elemental composition of the substrate. Potassium may be present in the spectrum as oxidant residues.

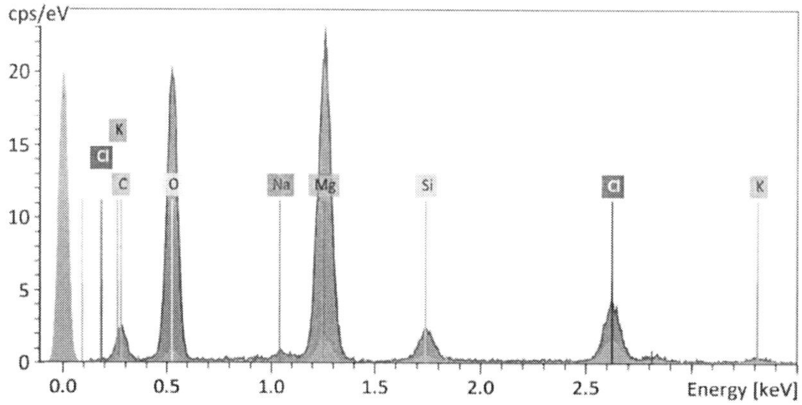

Figure 5. The energy spectrum of secondary electrons.

The possibilities of the microscope are such that in a few minutes you can accumulate information about the elemental composition of the samples. The method of X-ray analysis allows you to build a map of the location of elements in the study sample. The brighter the scan coordinate, the higher the concentration of the selected item. A large particle in the form of a tube (Fig. 6a) was investigated by this method. The carbon distribution coincides with the contours of the tube (Fig. 6b). Oxygen and magnesium also describe the contours of the selected particle (Fig. 6c, d). These elements are absent in a place free from particles.

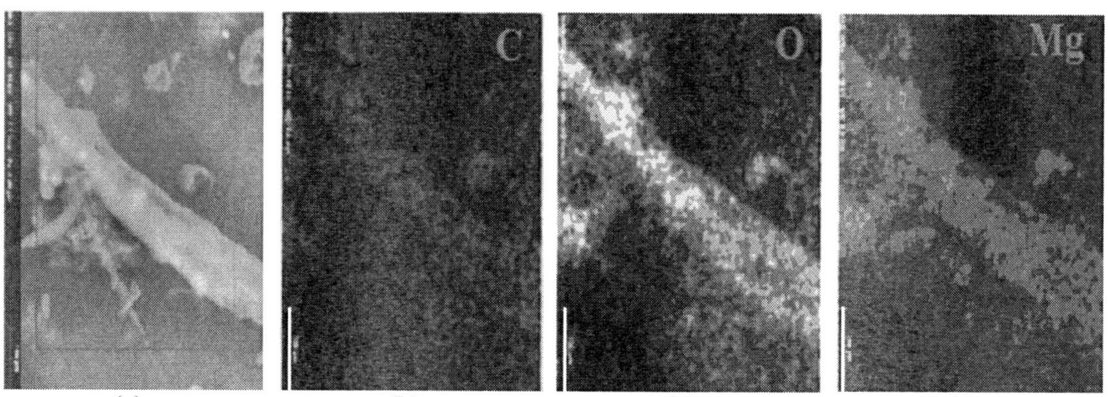

(a) **(b)** **(c)** **(d)**

Figure 6. Results of high-quality energy microanalysis: electron microscopic image (a), distribution map of chemical elements (b - d). The length of the frame 100 microns.

Naphthalene in the flame of the same composition forms a structure of a different nature. Particle cells are structures in the form of "three- or five-pointed asterisks" fastened together (Fig. 7). The length of the rays of such "stars" - 2.3 microns, thickness - 0.54 microns. Bound "stars" form a particle in the form of a sphere with a size of 14 microns.

Figure 7. Micrograph of the sample area.

Anthracene in flames can organize nanostructures of other species. Conical needles with a length of 3.2 microns grow out of a complex particle (Fig. 8). The needle has a size at the base in the diameter of 0.38 microns, and its body is composed of many scales.

Figure 8. Micrograph of the sample area.

5. Conclusion

Nanotubes, needles of different structure are formed during the self-organization of nanocarbon particles in the flame. The nucleation and growth of these tubes occurs by a heterophase mechanism. In this case, the formation and growth of the nanocarbon particles themselves takes place on the surface of the metal particle. The variety of nanostructures depends on the nature of the aromatic compounds.

Self-organized structures in their size fill the space between nano-and micro-formations. The growth rate of nanotubes is an order of magnitude greater than the growth of particles during gas-phase reactions.

Nanostructures formed on the surface of reacting dispersed particles allow changing the selectivity of the radiation of the flame in the near IR.

Acknowledgments
The authors would like to express sincere gratitude and appreciation to Chistyakov I., Kremer E., InterLab company for assistance in conducting research on Hitachi TM-4000Plus scanning electron microscope.

References
[1] Mansurov Z A 2012 J. Physics of Combustion and Explosion **48** 77
[2] Evlashin S A 2014 Study of the Optical and Emission Properties of Carbon Nanoscale (Moscow: Moscow State University Press) p 23
[3] Rakov E G 2004 Russ. Chem. Journal vol XLVIII p 12
[4] Kartashov Yu I 2008 vol 116 p 92

Investigation of resistive switching in gallium oxide nanostructures formed by local anodic oxidation

R V Tominov[1], N A Polupanov[1], V I Avilov[1], M S Solodovnik[1], N V Parshina[1], V A Smirnov[1], O A Ageev[1]
[1] Institute of Nanotechnologies, Electronics and Equipment Engineering, Southern Federal University, Taganrog, 347922, Russia

Abstract. Effect of resistive switching in gallium oxide structure was studied. It was shown, that gallium oxide structure, formed by local anodic oxidation, demonstrated resistive switching from high-state resistance (HRS) to low-state resistance (LRS) at +4 V, and from LRS to HRS at -6V. R_{HRS}, R_{LRS}, R_{HRS}/R_{LRS} were equaled 11.7±1.6 GΩ, 2.3±0.8 GΩ and 5, respectively. Homogeneity test showed, that R_{HRS} and R_{LRS} were equaled 10.6±2.8 GΩ and 2.5±1.7 GΩ, respectively. It was shown, that confidence interval of resistive switching effect of endurance test for HRS is less on 42% and for LRS is less on 53%, than for homogeneity test So, it was shown that the obtained gallium oxide structure has a uniform effect of resistive switching within 15 measurements. The results can be useful for based on gallium oxide neuromorphic system manufacturing.

1. Introduction

Development of neuromorphic technologies is the promising direction of electronics, which offers great advantages in unstructured data classification, pattern recognition, as well as in applications with adaptable and self-learning control systems[1,2,3]. Von Neumann architecture is the main organization structure of computing machines, on which the vast majority personal computers are built. A decrease in the elements of integrated circuits (IC) can no more increase the speed of computer systems, because the data speed exchange between the processor and memory remains almost unchanged[4,5]. The main difference between neuromorphic systems from a traditional computer is a physical integration of the processor and memory into single device. Such a decision allows overcoming the most serious problem traditional computer architecture, that often limits the performance of the system, which called «von Neumann bottleneck». Neuromorphic systems have a human brain system organization, which is a set of parallel-connected low-power computing elements (neurons) interconnected through special channels (synapses)[6-14]. One of the main ways of technical implementation of neuromorphic systems is a memristor fourth passive component of electronics, based on effect of resistance switching resistance change between high resistance state (HRS) and low resistance state (LRS) of thin films oxides[15-21]. Memristor plays role of neuron and synapse at the same time. For technical implementation of memristor, non-volatile resistive memory (ReRAM) can be used. In neuromorphic systems manufacturing elements has special requirements for energy consumption, and ReRAM is more compatible compared to other types of perspective types of memory, such MRAM, FeRAM, PRAM[21,22]. An important requirement for neuromorphic elements is multibitity (the ability to switch a memory element between three or more states), due to which the performance is significantly improved when solving problems poorly algorithmic on traditional computers, and the specific energy consumption is significantly reduced. Of the types of memory considered, ReRAM has multibitity, since the redistribution of the concentration profile of defects in the bulk of the oxide film when an external

electric field of various sizes and polarity is applied gives a range of resistance values. There are a lot of materials that demonstrate effect of resistive switching[6,8,15, 23-25]. One of the promising materials for microelectronics is gallium arsenide (GaAs).
The analysis of publications has shown that most part of methods of forming memristor structures demand a complex stage of electroforming to give them memristor properties. However, the application of the local anodic oxidation (LAO) method using atomic force microscopy (AFM) techniques makes it possible to form the gallium oxide (Ga_2O_3) structures on GaAs wafer surface, which demonstrate memristor effect without additional electroforming of structures, which will optimize the technology of ReRAM manufacturing[15-29]. So, for ReRAM memristor structure manufacturing based on formed by LAO gallium oxide nanostructures it is required to investigate the regimes of resistive switching in it.
The aim of this work is investigation of resistive switching effect in gallium oxide structures, formed on GaAs wafer surface by local anodic oxidation.

2. Experiment details

«Epi-ready» GaAs (001) wafers as substrates was used. The reference marks were set on GaAs wafer using the focused ion beam of scanning electron microscope Nova NanoLab 600 (FEI Company, USA). Gallium oxide structure 50×50 μm^2 was formed using local anodic oxidation technique of scanning probe nanolaboratory Ntegra (NT MDT, Russia) with W_2C cantilevers. Gallium arsenide local anodic oxidation was carried out on a raster graphic template under next conditions: voltage +5V, velocity 1 μm/s, setpoint 2 nA, humidity 80%.
Optical image was received using Olympus microscope (figure 1). An AFM-images of the formed gallium oxide nanostructure were obtained in semi-contact mode (figure 2a) using nanolaboratory Ntegra (NT-MDT, Russia). The AFM-image processing was performed using Image Analysis software. Electric measurements were carried out using semiconductor characterization system Keithley 4200-SCS (Keithley, USA) with W probes. During experiment GaAs wafer was grounded. Current-voltage curves were obtained at –6 V to +4 V voltage sweep. To prevent thermal breakdown of the gallium structure, 10 nA compliance current was set during electric measurements. Current-voltage characteristics were measured at the same point (endurance test) and in different points (homogeneity test).

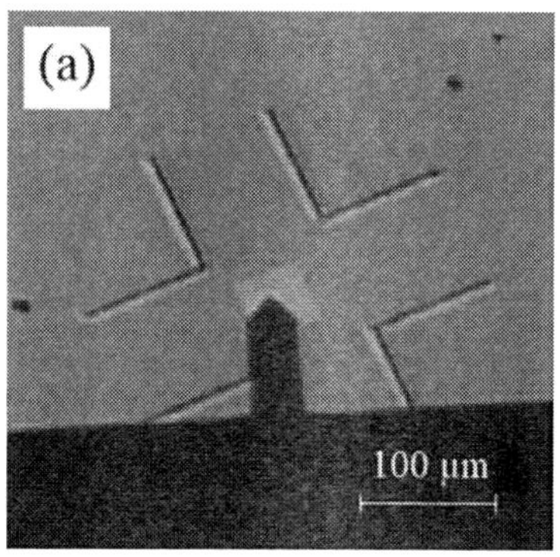

Figure 1. Optical image of GaAs oxide nanostructure

3. Results

It is shown that gallium oxide nanostructure has uniform surface structure (figure 2a) with 2.5±0.8 nm roughness (figure 2b).

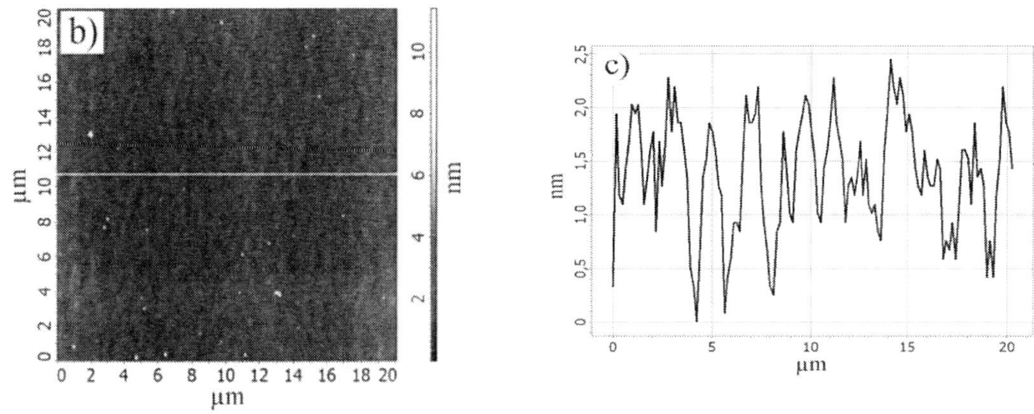

Figure 2(a, b). GaAs oxide nanostructure surface: **(a)** AFM-image; **(b)** AFM cross-section profile along line on (a).

Figure 3 show average current-voltage characteristic (CVC) of gallium oxide nanostructure, performed from the 15 current-voltage curves in the same point. It was shown, resistive switching from HRS to low-state resistance LRS was occurred at +4 V, and resistive switching from LRS to HRS at -6V. Endurance test showed, that R_{HRS} was equaled 11.7±1.6 GΩ, R_{LRS} was equaled 2.3±0.8 GΩ (figure 4 a). Homogeinity test showed, that R_{HRS} was equaled 10.6±2.8 GΩ, R_{LRS} was equaled 2.5±1.7 GΩ (figure 4 b). The obtained resistance ratio of the structure in the high-resistance state and low-resistance state R_{HRS}/R_{LRS} ratio was evaluated to 5. It was shown, that confidence interval of resistive switching effect of endurance test for HRS is less on 42% and for LRS is less on 53%, than for homogeneity test. It can be explained by the uneven thickness of the gallium oxide nanostructure on the surface and by change of the oxygen nanofilament dissolution length in the gallium oxide bulk due to change of its thickness. So, it was shown that the obtained gallium oxide nanostructure has a uniform effect of resistive switching within 15 measurements.

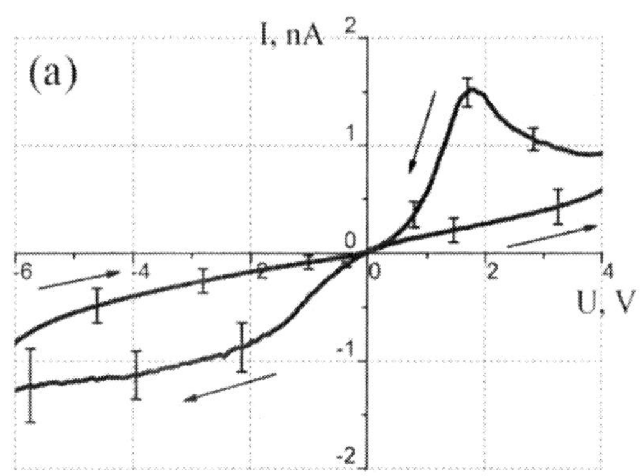

Figure 3. Current-voltage curve of resistive switching in GaAs oxide nanostructure

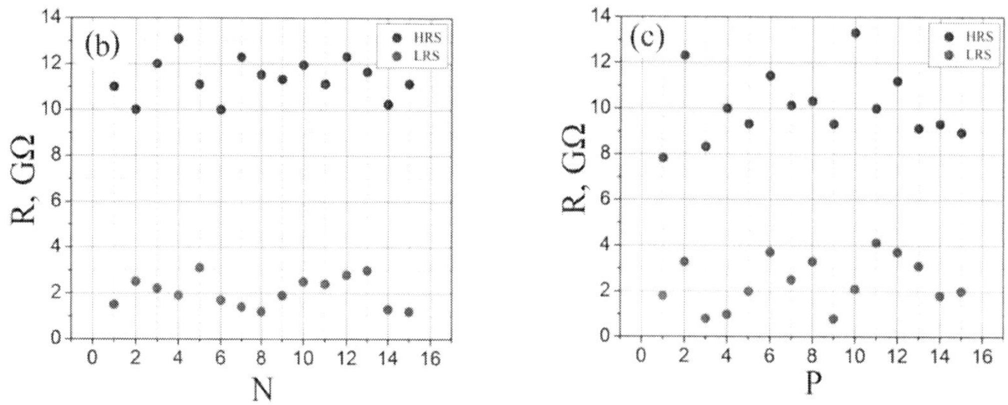

Figure 4(a, b). Investigation of resistive switching in GaAs oxide nanostructure:
(a) endurance test; **(b)** homogeneity test

The results can be useful for based on gallium oxide nanostructures neuromorphic computer systems manufacturing.

Acknowledgments
This work was supported by the Russian Science Foundation Grant No. 15-19-10006. The results were obtained using the equipment of the Research and Education Center and Center for Collective Use "Nanotechnologies" of Southern Federal University.

References
[1] Jo S H, Chang T, Ebong I, Bhadviya B B, Mazumder P, Lu W 2010 *Nano Lett.*, **10** 4
[2] Dongale T D, Desai N D, Khot K V, Volos C K, Bhosale P N, Kamat R K 2018 *Nanoelectron. Optoe.* **13** 1
[3] Furber S 2016 *Journal of neural engineering* **13(5)** 051001
[4] Gale E 2014 *Semiconductor Science and Technology* **29(10)** 104004
[5] Chang T, Yang Y, Lu W 2013 *IEEE Circuits and Systems Magazine* **13(2)** 56-73
[6] Khakhulin D A, Vakulov Z E, Smirnov V A, Tominov R V, Yoon J G, Ageev O A 2017 *Journal of Physics: Conference Series* **917** 092008
[7] Avilov V I, Polupanov N V, Tominov R V, Smirnov V A and Ageev O A 2017 *IOP Conf. Ser.: Mat. Sci.&Eng.* **256** 012001
[8] Ageev O A, Alyab'eva N I, Konoplev B G, Polyakov V V, Smirnov V A 2010 *Semiconductors* **44** 1703
[9] Avilov V I, Ageev O A, Jityaev I L, Kolomiytsev A S, Smirnov V A 2016 *Proceedings of SPIE* **10224** 102240T
[10] Perkins J D, Cueto J A, Alleman J L, Warmsingh C, Keyes B M, Gedvilas L M, Parilla P A, To B, Readey D W, Ginley D S 2002 *J.Thin Solid Films* **411** 152
[11] Ohkubo I, Christen H M, Khalifah P, Sathyamurthy S, Zhai H Y, Rouleau C M, Mandrus D G, Lowndes D H 2004 *J. Appl. Surf. Sci* **223** 35
[12] Au K, Gao X S, Wang J, Bao Z Y, Liu J M, Dai J Y 2016. *J. Appl. Phys.* **114** 027019
[13] Liu Q, Long S, Wang W, Zuo Q, Zhang S, Chen J, and Liu M 2009 *J.Elect. dev. lett.* **30** 12
[14] Tominov R V, Zamburg E G, Khakhulin D A, Klimin V S, Smirnov V A, Chu Y H, Ageev O A 2017 *Journal of Physics: Conference Series* **917** 032023
[15] Smirnov V A, Tominov R V, Alyab'eva N I, Il'ina M V, Polyakova V V, Bykov Al V and Ageev O A 2018 *Tech. Phys.* **63** 1236-1241

[16] Ageev O A, Konoplev B G, Rubashkina M V, Rukomoikin A V, Smirnov V A, Solodovnik M S 2013 *Nanotechnologies in Russia* **8(1-2)** 23-28
[17] Tominov R V, Bespoludin V V, Klimin V S, Smirnov V A, Ageev O A 2017 *IOP Conf. Ser.: Mater. Sci. Eng.* **256** 012023
[18] Klimin V S, Solodovnik M S, Smirnov V A, Eskov A V, Tominov R V, Ageev O A 2016 *Proc. of SPIE* **10224** 102241Z
[19] Klimin V S, Rezvan A A, Kots I N, Naidenko N A 2018 *J. Phys.: Conf. Ser.* **1124** 071019
[20] Klimin V S, Rezvan A A, Ageev O A 2018 *J. Phys.: Conf. Ser.* **1124** 022035
[21] Ageev A O, Konoplev B G, Polyakov V V , Svetlichnyi A M, Smirnov V A 2007 *Russian Microelectronics* **36** 353-357
[22] Ageev O A, Smirnov V A, Solodovnik M S, Rukomoikin A V and Avilov V I 2012 *Semiconductors* **46** 1616–21
[23] Smirnov V A 2015 *Piezoelectrics and Nanomaterials: Fundamentals, Developments and Applications* 85-103
[24] Avilov V I, Ageev O A, Konoplev B G, Smirnov V A, Solodovnik M S, Tsukanova O G 2016 *Semiconductors* **50** 5
[25] Avilov V I, Ageev O A, Blinov Y F, Konoplev B G, Polyakov V V, Smirnov V A, Tsukanova O G 2015 *Technical Physics* **60** 5
[26] Avilov V I, Ageev O A, Smirnov V A, Solodovnik M S, Tsukanova O G 2015 *Nanotechnologies in Russia* **10** 3
[27] Hsu C W, Chou L J 2012 *Nano letters*, **12** 8
[28] Li M, Deng H, Wei M, Qiu W W, Yao J Q, Deng X R, Wen G J 2013 *Advanced Materials Research* **685** 201
[29] Reus W F, Thuo M M, Shapiro N D, Nijhuis C A, Whitesides G M 2012 *ACS nano* **6** 6

Terahertz response from bio-tissue to silicon nanostructures

K B Taranets[1], M A Fomin[1], L E Klyachkin[2], A M Molyarenko[2], N T Bagraev[2]
[1]Peter the Great St. Petersburg Polytechnic University, St. Petersburg 195251, Russia
[2]Atomic Radiospectroscopy Lab, Ioffe Institute RAS, Saint Petersburg 194021, Russia

Abstract. Here we present the first findings on the resonance response of DNA oligonucleotides deposited on silicon nanosandwich (SNS) and living bio-tissue to the THz irradiation that allow their identification by measuring the change of the longitudinal conductance and the lateral voltage within frameworks of the SNS prepared in the Hall geometry. The THz resonance response from living bio-tissue under the THz radiation is also applied to the definition of oncological diseases. The results obtained from the basis of the express diagnostics in practical medicine.

1. Introduction
Identifying DNA oligonucleotides is one of the most relevant tasks in modern science especially in physics and biotechnologies. There are new approaches in solving this problem that have allowed to improve existing methods [14] and create a number of new [11, 12]. However, there is a plenty of room for improvement because it believed that the concept of personal medicine will be implemented when the cost of a comprehensive analysis of the human genome will fall so that it becomes available at a mass level but this is not achieved yet. Thus, of particular interest are studies aimed at finding new approaches to the identification of DNA oligonucleotides. Also, characteristics of the terahertz (THz) radiation of healthy biological tissue have been shown to differ from the corresponding indices of its sites affected by oncological diseases [10]. In this regard, the use of compact sources of THz radiation, among which silicon nanosandwiches (SNS) can be distinguished, is relevant. Those nanosandwiches are able to emit in a wide THz range when passing electric current.

2. Methods
In this paper we used the silicon nanosandwich structure (SNS), which simultaneously can act as an emitter and as a THz detector. The instrumental structure based on the SNS that was prepared within the framework of the Hall geometry and was a p-Si-QW type ultra-narrow p-type quantum well confined by δ-barriers heavily doped with boron on the n-Si (100) surface. These structures have been shown to demonstrate high mobility of charge carriers. Another fine point is that boron atoms in δ-barriers form trigonal dipole centers (B^+ - B^-) due to the negative-U reaction: $2B^o \rightarrow B^+ + B^-$. These dipole centers form crystallographically oriented sequences that reveal the edge channels responsible for conductivity in p-Si-QW[2]. In addition, it was shown that under the condition of the drain-source current the p-Si-QW edge channels perform as effective sources and recorders of the THz and GHz radiation due to the presence of negative-*U* dipole centers of boron[2], with the microcavities embedded that allow the control of THz and GHz spectra.
Previously, the resonance response has been shown to be revealed under the conditions of tunnel junctions between heavily doped δ-barriers [1, 3]. It should be emphasized that the response has

always been nonlinear depending on the current source [4]. Great interest is the study of the resonant response at a frequency of 2.8 THz, which is of great practical importance for use in medicine. Finally, the experimental procedure can be modeled by replacing the THz emitter with a DNA oligonucleotide deposited on the surface of the recorder. Quantum harmonic oscillators containing single holes inside the SNS edge channels are THz-generation sources with increasing source–drain current, with the amplification by the insertion of microcavities [5, 8]. Therefore, THz-radiation spectra corresponding to the frequency range of eigenmodes of oligonucleotides deposited onto the SNS surface can be formed by varying the microcavity sizes [6, 9, 13]. The response of these eigenmodes is a change in the I–V characteristic of the edge-channel conductance [8]. Furthermore, the effect of feedback of the influence of the oligonucleotides on the SNS conductance can be enhanced, if the microcavity contains a single hole and one oligonucleotide [8]. For example, the eigenmode frequency of the oligonucleotide firstly studied using the SNS, i.e., 2.8 THz, corresponds to a microcavity size of 16.6 μm. Thus, to amplify the terahertz response of an oligonucleotide in the I–V characteristic of the SNS, the average distance between holes in the SNS edge channel should be close to this value. To this end, the density of two-dimensional holes in the SNS was optimized near a value of 3×10^{13} m^{-2} which defines the number of carriers between measuring contacts “xx” ~ 120 which is consistent with the above-mentioned average distance between them, i.e., 16.6 μm. Such optimization of the average distance between holes in the SNS edge channel according to the embedded microcavity sizes made it possible to develop efficient THz-radiation sources based on edge-channel fragments containing single holes. In this case, the characteristics of these quantum harmonic oscillators are reflected not only in the detection of Rabi splitting, but also in the detection of the gigahertz modulation of THz-radiation spectra, which is controlled, as noted above, by the edge-channel length.

3. Results

In this study, the *I–V* characteristics of SNS were measured to determine the resonant frequencies of oligonucleotides. Oligonucleotide molecules were precisely deposited onto the δ-barrier over SNS edge channels using a Proline plus Biohit micropipette (0.1–3 μL) and a container-type microfluid system made of dimethyl silicone and placed on the SNS surface. Its volume contained 0.5 μL of solution, preventing its evaporation during device operation. Single-stranded oligonucleotide molecules were synthesized using an Applied Biosystems oligonucleotide synthesizer by the phosphoramidite method, purified by electrophoresis in a polyacrylamide gel, and extracted in a 0.3-M solution of sodium acetate. The following oligonucleotide sequence was studied: 100-mer 5’-
gcgctggctgcgggcggtgagctgagctcgcccccggggagctgtggccggcgcccct
gccggttccctgagcagcggacgttcatgctgggagggcggcg-3’.

The concentrations of oligonucleotide molecules were selected so that there was no more than one oligonucleotide molecule for each microcavity, and their values would be 0.22 and 0.98 μg/μL, respectively. These concentration parameters were chosen to satisfy the relation with the number of holes in the edge channels. To perform comparative analysis, SNSs without oligonucleotides were also studied, since, as shown in [8], the buffer solution does not cause significant changes reflected in the I–V characteristics of the SNS.

The change in the conductance was recorded in units of $G_0 = 2e^2/h$. This dependence on the source–drain current I_{ds} exhibits periodic beats (Fig. 1a). Having analyzed the dependences presented in Figure 1a, the frequency corresponding to the resonance response of DNA oligonucleotide the THz radiation of the microcavity containing a single hole can be determined from the beat period. It appeared that the frequency of the resonance response of the DNA oligonucleotide, 2.8 THz, determined from the beat period using the uncertainty relations is in good agreement with the data preliminarily obtained at a stabilized source–drain current of 48.1 μA [8]. Taking into account the frequency dependence of the source– drain driving current, $I = 2\pi eNf$, the resonance response of oligonucleotides at I_{ds} = 48.1 μA appears in the case of in-series connected 17 microcavities whose THz radiation is responsible for the excitation of DNA oligonucleotides.

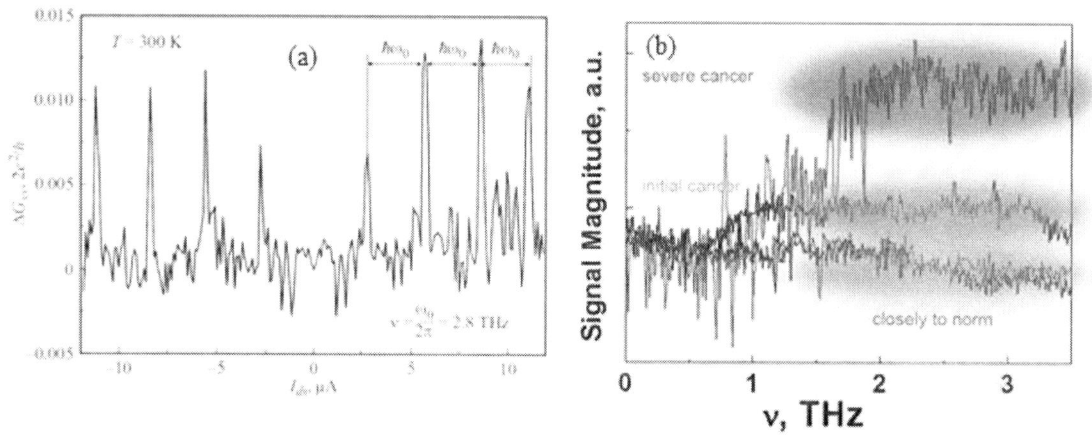

Figure 1(a, b). (a) Response of eigenmodes of single-stranded 100-mer oligonucleotides in the SNS edge channel, manifesting itself in a change in the dependence of the conductance on the source–drain driving current under conditions of THz generation; **(b)** The THz response revealed by the conductance of the silicon nanosandwich which is obtained from the balance of radiation of biological tissue and silicon nanosandwich emitter. These spectral characteristics allow us to classify various pathological conditions of the tissue

To identify the properties of oligonucleotide sequences as a whole it is necessary to investigate the properties of oligonucleotide molecules that can characterize the length and primary structure of the oligonucleotide. These properties include the dielectric properties of DNA. Based on a comparison of experimental data on conduction studies, the nucleotide composition and length of the oligonucleotide play a primary role in the formation of the dielectric properties of these biomolecules.
A similar approach was used to determine the THz characteristics of various neoplasms. The SNS device was exposed to radiation from both a biological tissue and to a reflected signal. Thus, the current-voltage characteristics revealed the balance of absorption and generation of the THz radiation. These data, in conjunction with the study of the THz polarity of the response in the THz and GHz bands, made it possible to identify the presence of various pathologies, as well as their characteristics and stages of development (Fig. 1b). Particular attention was paid to studies in the 2.6-3.4 THz range, and also near the 160 GHz energy gap. Apart from DNA oligonucleotide spectral characteristics a close look revealed the appearance of the strong peaks near 1.5 THz distinctive for the oncological tissue. This feature is due to the cancer DNA methylation[7].
DNA methylation is an epigenetic change existing in a vast majority of cancer types from the early stage. It is due to the attachment of a methyl group to a DNA and can change the activity of the segment to which it added without changing the sequence. Two of DNA`s bases can be methylated, cytosine and adenine. We paid attention to the first one which is a 5-Methylcytosine. Of particular interest is CpG sites. It is a region where a cytosine is followed by a guanine in the linear side (Fig. 2a). The CpG notation underlines that two nucleosides are connected by a phosphate group which is different from a CG base pairing. Figure 2 b,c shows the difference between normal and methylated cytosine in DNA. In the DNA sequence there are two things: CpG sites and CpG islands. The last are regions with a high frequency of the CpG sites. In normal DNA 5-methylcytosine is mostly found at CpG sites, whereas in cancer DNA it appears in CpG islands.
Figure 3 shows comparison of the normal (blue) and cancerous (red) tissue. Apart from the 1.5 THz there are peaks at a half of this value which may be due to the formation of the CpG islands. Moreover, it requires to conduct high resolution measurements for determining mechanisms of

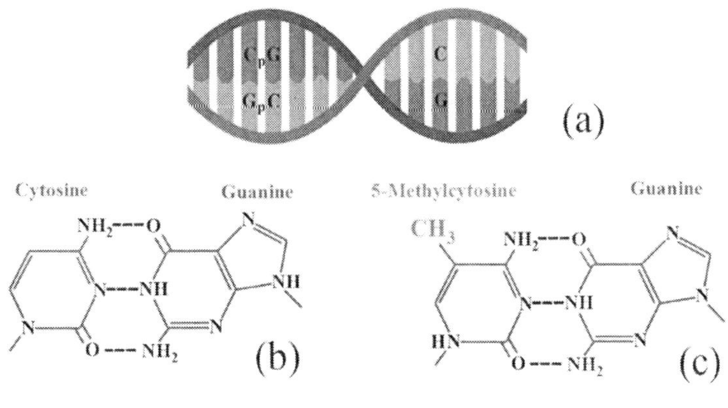

Figure 2(a,b,c). (a) CpG site and CG base pairing in DNA. Transition from **(b)** cytosine to **(c)** 5-Methylcytosine in DNA.

creation and shifts of the peaks in the interval from 1.2 to 1.7 THz and a half of this values that seems to result from the formation of the CpG islands.
Finally, the proposed method of express analysis of THz radiation characteristics from biological tissue is quite promising for the diagnosis of breast diseases, including oncological diseases, at an early stage.

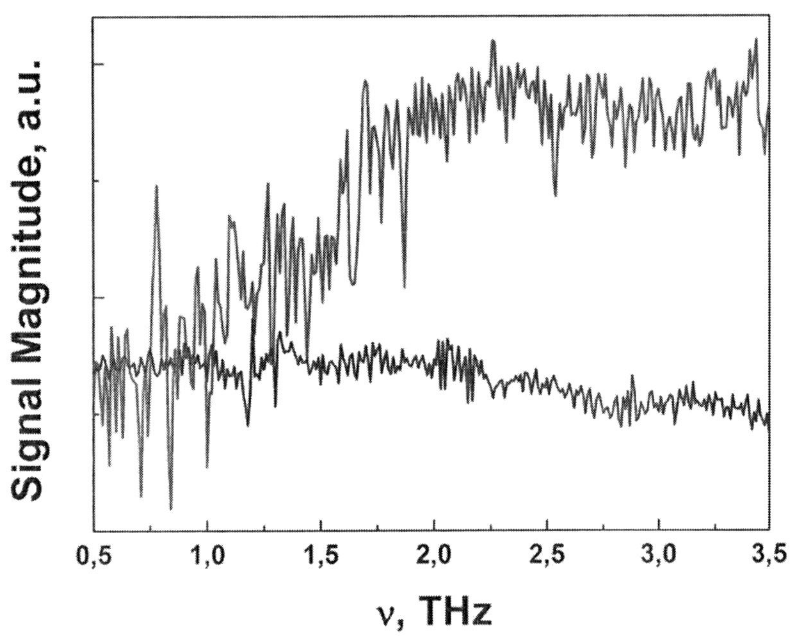

Figure 3. The THz response from the normal (blue) and cancer (red) biological tissue revealed by the I-V characteristics of the silicon nanosandwich emitter/recorder.

References

[1] Bagraev N T, Gehlhoff W, Klyachkin L E, Kudryavtsev A A, Malyarenko A M, Oganesyan G A, Poloskin D S and Romanov V V 2006 *Physica C* **219** 437
[2] Bagraev N T, Galkin N G, Gehlhoff W, Klyachkin L E and Malyarenko A M 2008 *J. Phys.: Condens. Matter* **20** 164202
[3] Bagraev N T, Mashkov V A, Danilovsky E Yu, Gehlhoff W, Gets D S, Klyachkin L E, Kudryavtsev A A, Kuzmin R V, Malyarenko A M and Romanov V V 2010 *Appl. Magn. Reson.* **39** 113
[4] Bagraev N T, Klyachkin L E, Kudryavtsev A A, Malyarenko A M and Romanov V V 2010 *in Superconductor* (SCIYO, Rijeka) p 69
[5] Bagraev N T, Danilovskii E Yu and Gets D S 2014 *J. Phys.: Conf. Ser.* **486** 012017
[6] Barhoumi A, Zhang D, Tam F and Halas N G 2008 *J. Am. Chem. Soc.* **130** 5523
[7] Cheon H, Yang H-j, Lee S-H, Kim Y A and Son J-H 2016 *Scientific Reports* **6** 37103
[8] Chernev A L, Bagraev N T, Klyachkin L E, Emel'yanov A K and Dubina M V 2015 *Semiconductors* **49** 944
[9] Fischer B M, Walther M and Jepsen P 2002 *Phys. Med. Biol.* **47** 3807
[10] Humphreys K, Loughran J P, Gradziel M, Lanigan W, Ward T, Murphy J A and O'Sullivan C 2004 Proc. of 26th Annual Int. Conf. of the Engineering in Medicine and Biology Society (San Francisco) vol 1 p 1302
[11] Mardis E R 2013 *Annual Review of Analytical Chemistry* **6** 287
[12] Rothberg J M, Hinz W, Rearick T M, Schultz J, Mileski W, Davey M, Leamon J H, Johnson K, Milgrew M J, Edwards M, Hoon J, Simons J F, Marran D, Myers J W, Davidson J F, Branting A, Nobile J R, Puc B P, Light D, Clark T A, Huber M, Branciforte J T, Stoner I B, Cawley S E, Lyons M, Fu Y, Homer N, Sedova M, Miao X, Reed B, Sabina J, Feierstein E, Schorn M, Alanjary M, Dimalanta E, Dressman D, Kasinskas R, Sokolsky T, Fidanza J A, Namsaraev E, McKernan K J, Williams A, Roth G T and Bustillo J J M 2011 *Nature* **475** 348
[13] Thomas Jr G J 1999 *Ann. Rev. Biophys. Biomol. Struct.* **28** 1
[14] Toumazou C, Shepherd L M, Reed S C, Chen G I, Patel A, Garner D M, Wang C J, Ou C P, Amin-Desai K, Athanasiou P, Bai H, Brizido I M, Caldwell B, Coomber-Alford D, Georgiou P, Jordan K S, Joyce J C, La Mura M, Morley D, Sathyavruthan S, Temelso S, Thomas R E and Zhang L. 2013 *Nature Methods* **10** 641

Simulation of hopping in disordered fractal arrays of graphene quantum dots

Ekaterina Morozova
Department of Theoretical Physics, Ulyanovsk State University, Russia
E-mail: kat-valezhanina@yandex.ru

Abstract. Diffusion regimes of charge carriers in artificial quasi-fractal graphene quantum dot (GQD) superlattices forming hexagonal and triangular carpet are investigated. Using the generalized Miller-Abrahams relations for hopping rates we perform Monte Carlo simulation of charge carrier hopping in artificial 2d disordered arrays of armchair GQDs. The variations of the anomalous advection-diffusion parameters as functions of localization radius, electric field intensity, levels of energetic and structural disorder are studied.

1. Introduction

Among graphene nanostructures, graphene ribbons and graphene quantum dots (GQDs) are of particular interest ([1]). GQD's represent truncated single-layer or several layers of graphene of a size less than 50 nm. They have exceptional properties such as low toxicity, stable photoluminescence, chemical stability and quantum confinement effect. Joung et al. [2] have shown that the low temperature electron transport properties of chemically functionalized graphene can be explained as sequential tunneling of charges through a two dimensional array of graphene quantum dots (GQD) and resistance data exhibit an Efros-Shklovskii variable range hopping arising from structural and size induced disorder.

Graphene sheet and GQDs are convenient systems to construct fractal mesoscopic nanosystems for optoelectronics (diode, sensor, rectenna etc). Fractals display a selfsimilar pattern at various scales. In this work, using the generalized Miller-Abrahams relations for hopping rates we perform Monte Carlo simulation of charge carrier hopping in artificial disordered fractal arrays of GQDs. The variations of the anomalous advection-diffusion parameters as functions of localization radius, electric field intensity, levels of energetic and structural disorder are studied. We consider two types of fractal systems composed of armchair GQDs of hexagonal and triangular shapes. Energy spectra of two example GQDs are shown in Fig. 1. Both GQDs have a bandgap. We consider hopping of excited electrons in the tails of density of states of disordered fractal arrays of GQDs.

2. Monte Carlo algorithm

Monte Carlo simulation of charge carrier hopping in 2D mesoporous samples and spinodal patterns [3] have shown that geometric traps play an important role in charge transport in disordered systems. These traps act independently of thermalization process. Due to this structural disorder, dispersive transport can be observed even at small values of energy disorder.

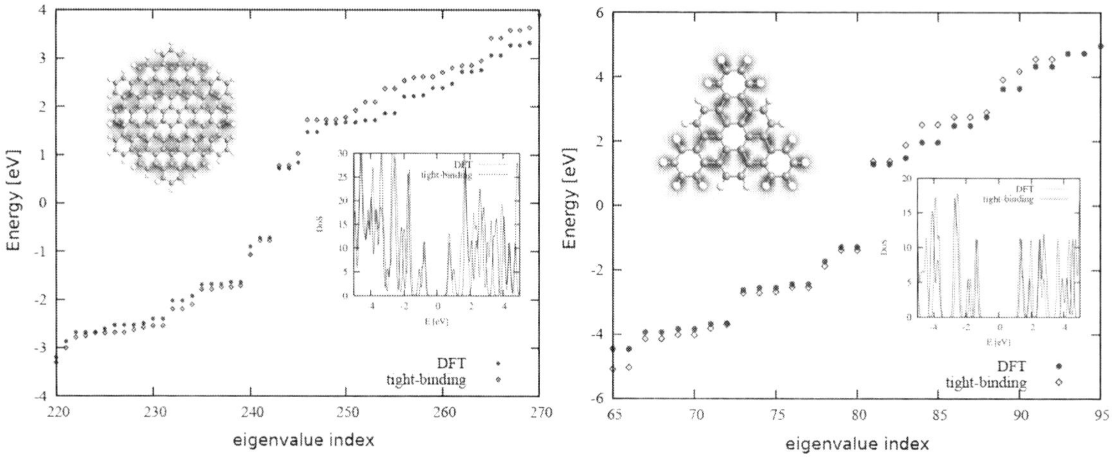

Figure 1. Energy spectra of hexagonal and trigonal armchair GQDs in the vicinity of the Fermi level calculated using DFT and tight-binding method. Density of states are shown in insets.

In this work, we simulate electron hopping using the generalized Miller-Abrahams relations that can be derived from the representation of superlocalized wave function in fractals. The electrons in fractals at the Fermi level exhibit superlocalization [4, 5] described by wave function

$$|\Psi(r)|^2 \propto \exp\left[-2(r/a)^{\zeta}\right], \tag{1}$$

where ζ is a superlocalization exponent and a is a localization radius. The probability W_{ij} to hop from an occupied site i to an empty site j with a higher energy is given by the generalized Miller-Abrahams relations:

$$W_{ij} \propto \exp\left[-2(r_{ij}/a)^{\zeta} - E_{ij}/kT\right]. \tag{2}$$

Here, $\mathbf{r}_{ij}$ is a vector connecting two dots, E_{ij} is an energy difference for the dots. The mean localization time for each transition is defined as

$$\tau_{i\to j} = \Gamma_0^{-1}\exp\left(\frac{\varepsilon_j - \varepsilon_i}{k_B T} + \frac{e\mathbf{E}\mathbf{r}_{ij}}{k_B T} + 2\left(\frac{r_{ij}}{a}\right)^{\zeta}\right). \tag{3}$$

These times are used in generation of waiting times θ_{jk} with pdf

$$p_{\theta_{ij}}(t) = \tau_{ij}^{-1}\exp\left(-\frac{t}{\tau_{ij}}\right). \tag{4}$$

The hop is realized to the j-th QD with the minimal value of θ_{ij}.

Simulated trajectories and the growth of mean square displacement (MSD) for hopping on a fractal hexaflake for different values of localization radius a are demonstrated in Fig. 2. The density of localized states is modeled by Gaussian distribution $\rho(\varepsilon) \propto \exp(-\varepsilon^2/2\sigma^2)$. The level of energetic disorder is $\sigma = 3kT$. The exponents of temporal scaling of MSD clearly indicate the subdiffusive regime of motion. The value 0.825 is related to the walk dimension for the fractal hexaflake. The lower values are related to effect of energetic disorder shortly considered in the next section.

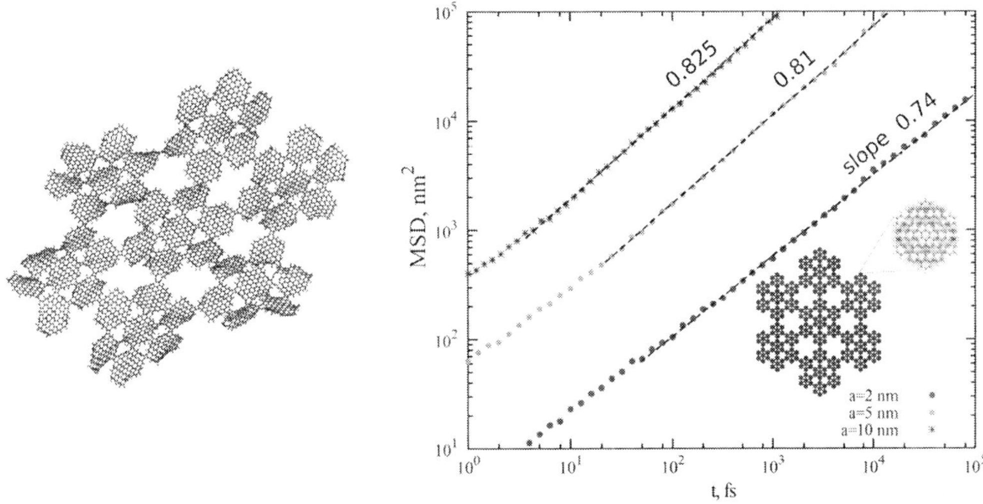

Figure 2. The example of disordered quasi-fractal array of GQDs (left). Disorder is induced by molecular dynamic simulation at room temperature. Right panel show the mean square displacement (MSD) for hopping on the fractal hexaflake. The level of energetic disorder is $\sigma = 3kT$. Subdiffusion exponents (right) depends on a and σ and become independent of σ for large localization radius a.

3. The model of continuous time random walks on fractal

Klafter et al. [6] have shown that for particle diffusion and trapping on fractals with heavy-tailed waiting-time distributions, the mean squared displacement behaves as $\langle r^2(t)\rangle \propto t^{\alpha\beta}$, whereas the number of delocalized particles decays according to $S(t) = t^{-\alpha}$. In Refs. [3, 7], structural and energy disorders for dispersive transport in percolative media have been simultaneously taken into account within the Continuous Time Random Walks (CTRW) on a comb structure. This model can be easily generalized for CTRW on fractals. CTRW on a percolation cluster is considered in Ref. [3].

The distribution of the sojourn time averaged over random number of visited sites l in a bounded domain Q is

$$\psi_Q(t) = \sum_{l=0}^{\infty} \psi^{*l}(t) p_l. \tag{5}$$

Here, ψ is the waiting time density for an individual site. After the Laplace transformation,

$$\tilde{\psi}_{br}(t) = \sum_{l=0}^{\infty} \left[\tilde{\psi}(s)\right]^l p_l \sim \int_0^{\infty} \exp\left[l\,\ln\tilde{\psi}(s)\right] p(l)\,\mathrm{d}l = \tilde{p}\Big(-\ln\tilde{\psi}(s)\Big). \tag{6}$$

For the transform of the integral kernel of the generalized FP-equation [8, 3, 9], we have

$$\tilde{\Phi}(s) = \left[s\tilde{p}\Big(-\ln\tilde{\psi}(s)\Big)\right]^{-1}\left\{1-\tilde{p}\Big(-\ln\tilde{\psi}(s)\Big)\right\}. \tag{7}$$

To obtain asymptotic behavior at large times, we should consider small s, $s \to 0$. Assuming power law distribution of a number of visited sites within the given domain due to fractal distribution of sites,

$$p(s) \sim 1 - c_\beta s^\beta, \quad -\ln\tilde{\psi}(s) = -\ln\left[1 - s\tilde{\psi}(s)\right] \sim s\tilde{\psi}(s). \tag{8}$$

For the memory kernel, we have

$$\tilde{\Phi}(s) \sim s^{-1} c_\beta \left[s\tilde{\psi}(s) \right]^\beta . \tag{9}$$

For asymptotic large times, drift-diffusion equation has the form [3],

$$c_\beta \left[s\tilde{\psi}(s) \right]^\beta \left[\tilde{p}(x,s) - s^{-1} p(x,0) \right] - \hat{L}_x \tilde{p}(x,s) = 0. \tag{10}$$

In the multiple trapping model with exponential DoS $\rho(\varepsilon) = \varepsilon_0^{-1} \exp(-\varepsilon/\varepsilon_0)$:

$$\Psi(t) = \int_0^\infty \exp\left\{ -\omega_0 t e^{-\varepsilon/kT} \right\} \varepsilon_0^{-1} \exp\left(-\frac{\varepsilon}{\varepsilon_0} \right) d\varepsilon \approx \frac{\alpha \Gamma(\alpha)}{\omega_0^\alpha} t^{-\alpha}. \tag{11}$$

As a result, we have

$$\tilde{\Phi}(s) \sim c_\beta \left(\frac{\pi\alpha}{\sin \pi\alpha} \right)^\beta \omega_0^{-\alpha\beta} s^{\alpha\beta - 1}. \tag{12}$$

This kernel of the generalized Fokker-Planck equation leads to subdiffusion with dispersion parameter $\alpha\beta$, where α is associated with energy disorder, and β is related to walk dimension for a given fractal.

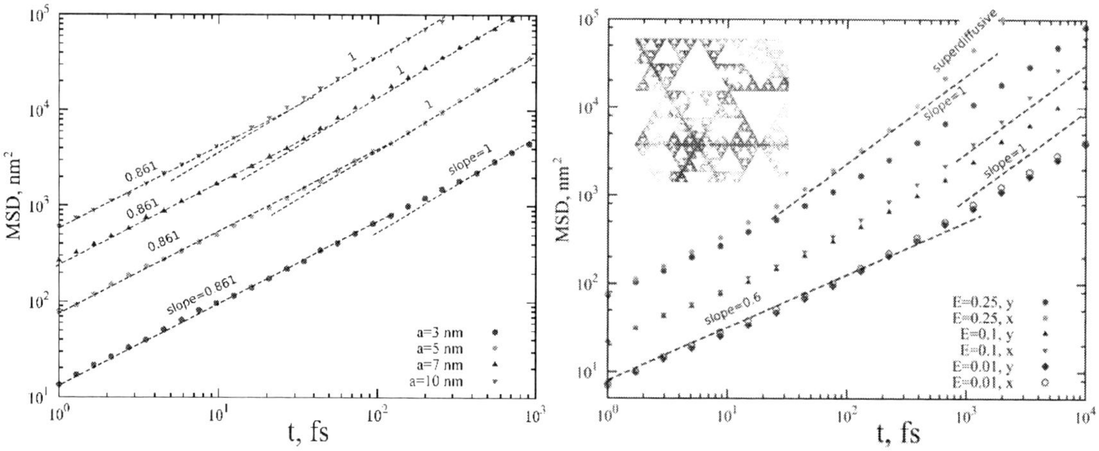

Figure 3. Left panel: MSD for hopping diffusion on the quasi-fractal ensemble of GQDs consisting of Serpienski triangles. Right panel: Variance (centered MSD) along and perpendicular to the applied electric field for different values of electric field.

The application of external electric field can lead to change in subdiffusion exponents. The MSD vs time plots for hopping on a quasi-fractal carpet composed of Sierpinski triangles of GQDs for different values of localization radius a and different values of electric field are shown in Fig. 3. The density of localized states is modeled by Gaussian distribution $\rho(\varepsilon) \propto \exp(-\varepsilon^2/2\sigma^2)$ with $\sigma = 3kT$. In the case of non-zero external field, we divide MSD on longitudinal and transverse directions. The transition to normal diffusion (slope 1 in log-log scale of MSD vs time) occurs due to the quasi-fractal structure, i.e. the maximal scale of Sierpinski triangle is limited and chosen by hand. For longitudinal MSD there is a situation when superdiffusive behavior of biased hopping is observed due to the field-induced amplification of carrier dispersion.

4. Summary

Here, we presented the results of Monte Carlo simulation of electron hopping in artificial disordered fractal arrays of armchair GQDs of hexagonal and triangular shapes. Main aspects of electron transport in mesoscopic fractals can be testified by simulation of hopping in quasi-fractal GQD systems. Anomalous drift-diffusion in two-dimensional semiconductor systems with coexisting energetic and structural disorder can be described as CTRW on a fractal structure. Waiting time distribution is defined by DoS, temperature T, number and geometry of empty states accessible within wave function localization domain. Superdiffusive behavior of biased hopping in fractals can be explained by the field-induced amplification of carrier separation.

Acknowledgments

This work is supported by the Ministry of Education and Science of the Russian Federation (state program 3.2111.2017/4.6) and the Russian Science Foundation (19-71-10063).

References

[1] Güclü A.D. et al. Graphene quantum dots. Springer, 2016
[2] Joung, D., Zhai, L., & Khondaker, S. I. (2011). Coulomb blockade and hopping conduction in graphene quantum dots array. Physical Review B, 83(11), 115323.
[3] Sibatov, R., Shulezhko, V., & Svetukhin, V. (2017). Fractional Derivative Phenomenology of Percolative Phonon-Assisted Hopping in Two-Dimensional Disordered Systems. Entropy, 19(9), 463.
[4] Deutscher, G., Levy, Y., & Souillard, B. (1987). T3/7 hopping conductivity in a class of disordered systems. EPL (Europhysics Letters), 4(5), 577.
[5] Harris, A. B., & Aharony, A. (1987). Anomalous diffusion, superlocalization and hopping conductivity on fractal media. EPL (Europhysics Letters), 4(12), 1355.
[6] Klafter, J., & Blumen, A. (1984). Fractal behavior in trapping and reaction. The Journal of chemical physics, 80(2), 875-877.
[7] Sibatov, R. T., & Morozova, E. V. (2015). Multiple trapping on a comb structure as a model of electron transport in disordered nanostructured semiconductors. Journal of Experimental and Theoretical Physics, 120(5), 860-870.
[8] Sibatov, R., & Morozova, E. (2017). Tempered Fractional Model of Transient Current in Organic Semiconductor Layers. In Theory and Applications of Non-integer Order Systems (pp. 287-295). Springer, Cham.
[9] Uchaĭkin, V. V., & Sibatov, R. (2013). Fractional kinetics in solids: anomalous charge transport in semiconductors, dielectrics, and nanosystems. World Scientific.
[10] Scher, H., & Montroll, E. W. (1975). Anomalous transit-time dispersion in amorphous solids. Physical Review B, 12(6), 2455.
[11] Uchaikin, V. V., & Sibatov, R. T. (2007). Fractional differential kinetics of dispersive transport as the consequence of its self-similarity. JETP Letters, 86(8), 512-516.
[12] Sibatov, R. T., & Uchaikin, V. V. (2011). Truncated Lévy statistics for dispersive transport in disordered semiconductors. Communications in Nonlinear Science and Numerical Simulation, 16(12), 4564-4572.

Development of technology for the formation of vacuum field emission cells using focused ion beams

I V Panchenko, N A Shandyba, A S Kolomiytsev
[1]Southern federal university, Rostov on Don 344006, Russia

Abstract. The article presents the results of the development of technology for the formation of vacuum field emission structures. Experimental studies of local ion-stimulated deposition of W and C and ion-beam etching were carried out and their effects on the formation of final structures were studied. The technological process of manufacturing nanoscale field emission structures was developed, and experimental samples were fabricated. It is shown that the use of the method of focused ion beams (FIB) demonstrates its advantages compared with other methods.

1. Introduction

Present days, the use of vacuum field emission nanoscale structures is a promising direction of the development of micro- and nanoelectronics. The use of such structures in nanoelectronics allows us to achieve a high degree of noise immunity, low power consumption, high switching speed and, consequently, high frequencies of operation of such devices. The main difficulty in the development of electronics in this direction is the provision of vacuum in nanoscale structures. The currently used methods do not allow to create and maintain a high level of vacuum in elements of this size. The method of focused ion beam (FIB) solves the main problems associated with the creation of elements of vacuum-emission nanoelectronics [1-3]. The FIB method allows, under high vacuum conditions, to perform technological operations of local ion beam etching and ion-induced deposition of materials from the gas phase without the need for resists, masks and chemical etchants. When chemically active gases are supplied to the FIB exposure zone, their molecules dissociate, as a result of which one of the gas components is deposited on the surface, and the volatile reaction products are removed by a vacuum system [4, 5]. The wide nomenclature of materials deposited by this method allows the use of FIB in the formation of nanoscale structures for electronics (quantum dots, field emitters, point contacts) and nanodiagnostics (probe tips, calibration gratings of the nanometer range for scanning probe microscopy (SPM)) [1].

2. Theory and methods

In this paper, studies were performed on a Nova NanoLab 600 raster electron microscope (FEI Company) equipped with a FIB system. At the beginning of the work, a theoretical analysis was carried out of the physicochemical processes occurring when the flow of accelerated ions interacts with the silicon-gold-silicon oxide substrate, both under vacuum conditions and in the presence of the process gas $W(CO)_6$. As a result of the analysis, it was established that the main factor determining the kinetics of ion-stimulated deposition W is the coefficient of adsorption of carrier gases and the deposited material on the substrate under study [6]. Thus, during ion-stimulated deposition, at the first stage, the gas molecules supplied to the ion beam are adsorbed on the substrate surface, then under the action of gallium ions with an energy of 30 keV, the molecules decompose, the volatile organometallic compounds are removed by a vacuum system, and the components of the deposited substance (in this case, W) remain on the surface of the material. By specifying the trajectories of the ion beam with the help of control electronics, it is possible to form three-dimensional nanoscale structures of a complex profile [7]. By adjusting the ion beam current, it becomes possible to precisely control the deposition rate and the minimum size of the structures.

3. Experimental

To study the rate of local ion-stimulated tungsten deposition, a graphic template was formed in the FIB module control program in the form of a matrix consisting of 16 elements 2×2 μm in size, the diagram of which with indicating time of the local ion-stimulated deposition (in seconds) of each element is shown in figure 1.

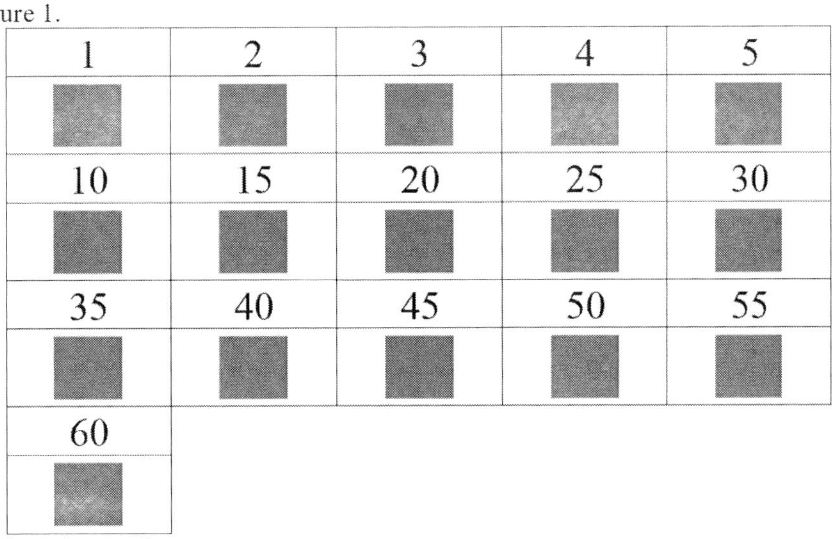

Figure 1. The scheme of pattern

After local ion-stimulated deposition, the substrate was examined using a scanning electron microscope (SEM), as shown in figure 2a. After that, the substrate was investigated using an atomic force microscope (AFM) in a semicontact mode, as shown in figure 2b.

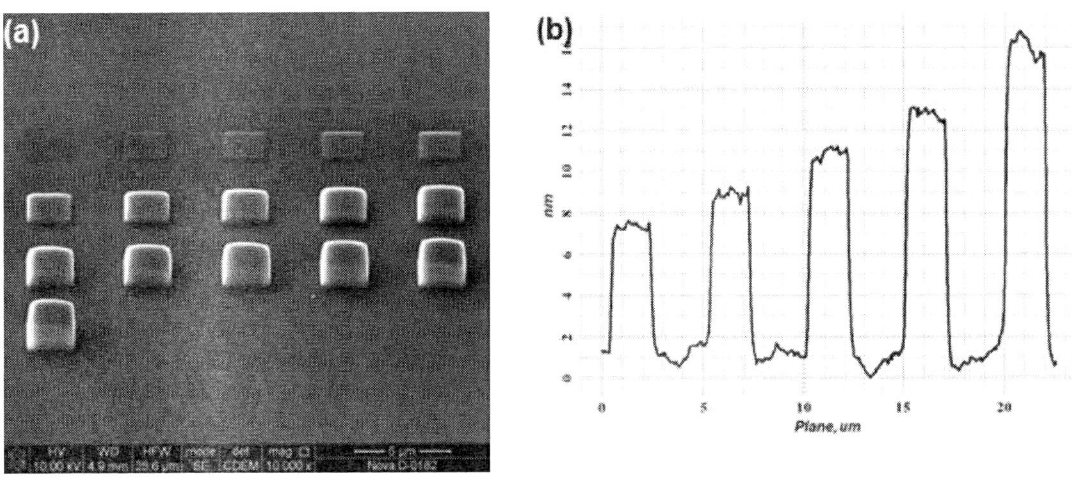

Figure 2. SEM image of a deposited tungsten structures (a); AFM profile of an array of tungsten structures (b).

Analysis of the obtained AFM images showed that with an increase in FIB current from 1 to 500 pA, the thickness tungsten structures from 0.545 to 969 nm with an increase in the time of ion-induced deposition.
Based on the analysis of the experimental results obtained, the dependences of the thickness of the formed elements on the time of ion-stimulated deposition were plotted, as shown in figure 3a.
The analysis of the obtained dependences shows the uniformity of the process of local ion-induced deposition of carbon and tungsten during the entire time of deposition. The difference in the thicknesses of the carbon and tungsten elements can be explained by the different physicochemical properties of the chemically active gases used in ion-stimulated deposition.
Based on the processing and analysis of the obtained data, the dependences of the rate of ion-induced deposition of tungsten were obtained, as shown in figure 3b.

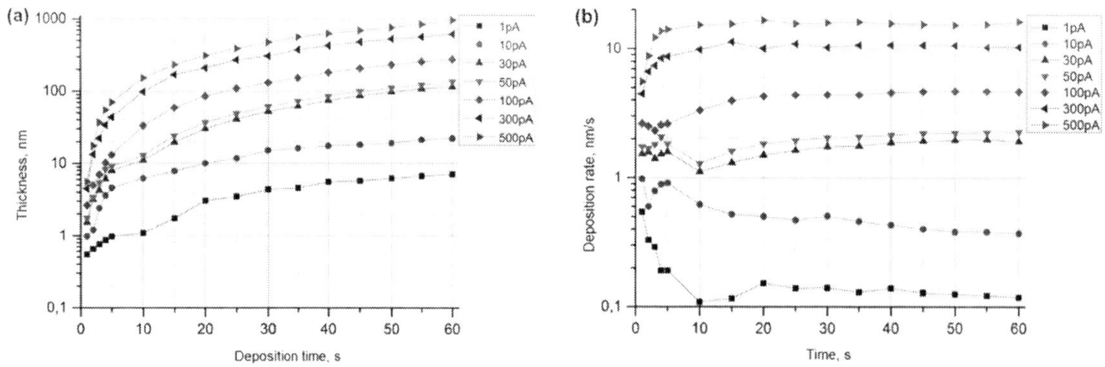

Figure 3. The dependence of the thickness of the formed elements of tungsten on the time of FIB deposition(a); The dependence of the rate of ion-induced deposition of tungsten on the time of deposition(b).

The analysis of the obtained dependences allows us to estimate the trends in the rate of ion-stimulated deposition of tungsten on the deposition time. With an increase in FIB current, the rate of ion-stimulated deposition increases. Uniform rate of ion-stimulated tungsten deposition is observed at FIB currents of 300 pA and 500pA, which is associated with the physicochemical properties of a reactive gas that interacts with FIB.
The dependences obtained will allow us to further predict the thickness of the deposited materials.

4. Technology

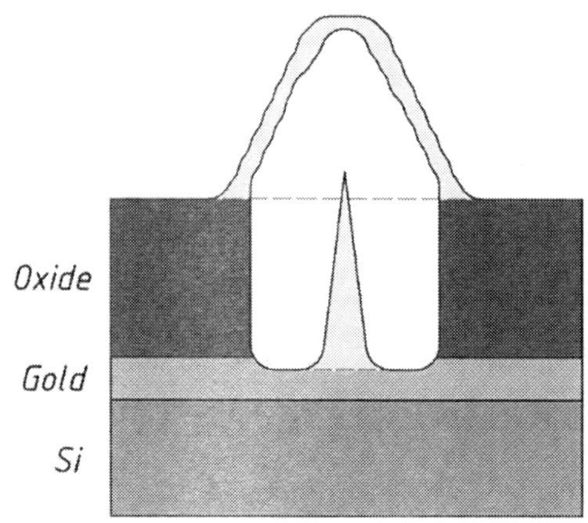

Figure 4. Layout of the field-emission structure

Scheme of the field-emission structure is shown in figure 4.
At the initial stage of creating the field-emission structure, a silicon-gold-silicon oxide substrate was used. The first step was the formation of a hole 1 μm deep and 400 nm in diameter. To do this, an appropriate template was set and using the ion-beam etching a hole with the required parameters was obtained (Figure 5).

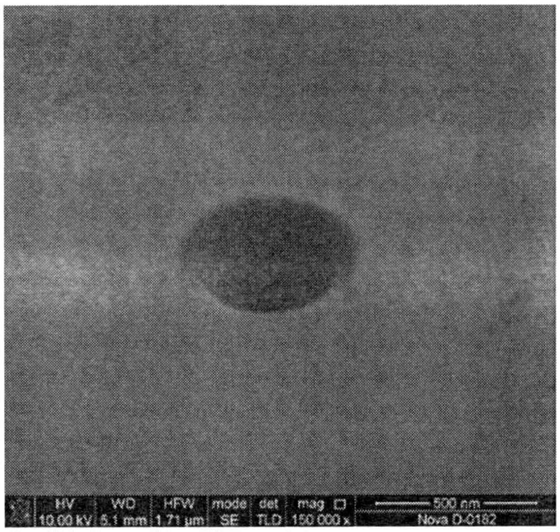

Figure 5. A cylindrical hole, fabricated by FIB etching

In the next step, the formed hole was filled with tungsten. After this, a gas containing tungsten was supplied and, under the action of the FIB, ion-stimulated material deposition from the gas phase was carried out (Figure 6a). Then, using a raster pattern, which is a circle with an outer radius of 500 nm and an inner 200 nm using ion-beam etching, a tip was formed, which is a cathode, isolated from the side edges of the hole (Figure 6b).

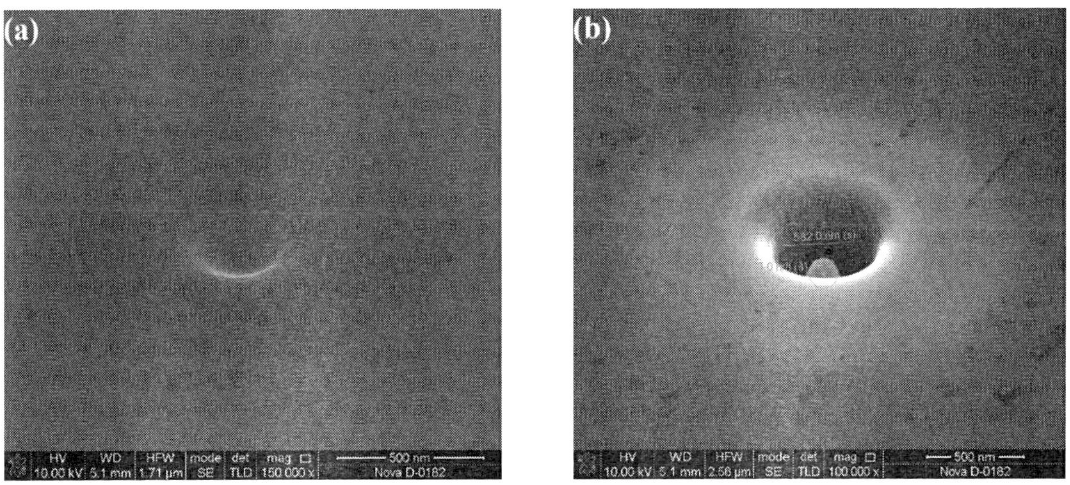

Figure 6. Nanosized hole filled by W (a); Cathode tip inside the hole (b).

The final stage is the formation of the anode contact, which is a "Cover" that provides isolation of the vacuum obtained under it from the external environment. The anode was created using the method of ion-induced deposition of tungsten from the gas phase and templates with decreasing diameters, obtained using specialized software [8] (Figure 7).

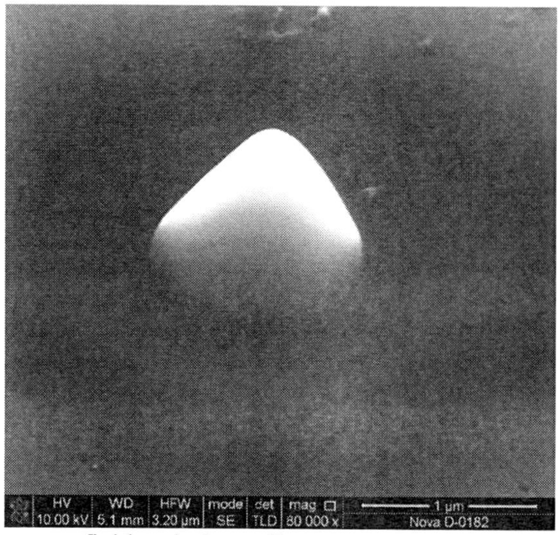

Figure 7. "Cover" of a vacuum field-emission cell.

Due to the fact that all operations were carried out in high vacuum conditions, the resulting cell was vacuum.

When studying a similar structure using an Ntegra Vita atomic force microscope, we obtained the current-voltage characteristic shown in figure 8.

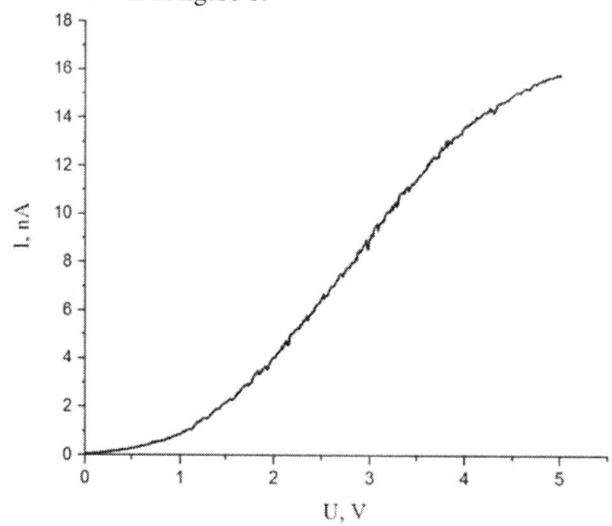

Figure 8. CVC of field-emission cell

Analysis of the current-voltage characteristics showed that emission starts from about 1 volt, which will significantly reduce the power consumption of the circuits in which such emitters will be used.

5. Conclusions

Thus, the technology of creating an field-emission cell was demonstrated, which, thanks to the conditions of creation, turned out to be high-vacuum. This technology can be used to create vacuum integrated circuits that do not require additional vacuuming or circuits where it is not possible to use lamp electronics because of their large size. The capabilities of the FIB method and its advantages in creating vacuum field emission elements of nanoelectronics were demonstrated.

Acknowledgments
This work was supported by the Russian Science Foundation Grant No. 18-79-00175.

References

[1] Bhushan B 2010 Springer Handbook of Nanotechnology (3rd edn) (New York: Springer) p 1964.
[2] Yao N and Wang Z 2005 Handbook of Microscopy for Nanotechnology (Kluwer Acad. Publ.: New York), p 743.
[3] Savenko A, Yildiz I, Petersen D H 2013 Nanotechnology 24 465701.
[4] Ageev O A, Kolomiytsev A S and Bykov A V 2015 Microelectron. Reliab.55 2131–34.
[5] Ageev O A, Alekseev A M, Vnukova A V, Gromov A L, Kolomiytsev A S, Konoplev B G and Lisitsyn S A 2014 Nanotechnol. in Russ. 9 1 26–30.
[6] Lisitsyn S A, Kolomiytsev A S, Ilyin O I, Ilyina M V, Konoplev B G, Bykov A V and Ageev O A 2017 Russ. Microelectron. 46 7 1–6.
[7] Ageev O A, Alekseev A M, Vnukova A V, Gromov A L, Kolomiytsev A S and Konoplev B G 2014 Nanotechnol. in Russ. 9 1 26–30.
[8] Shandyba N, Panchenko I 2018 Saint Petersburg OPEN 2018 (Saint Petersburg: Academic University Publishing) p 606-607

Study of the features of ultrafast silicon-carbide current switch for sources of soft x-ray radiation based on capillary plasma

K A Serguschichev[1], A A Smirnov[1], V A Ilyin[1], A V Afanasyev[1], A A Samokhvalov[2], M V Timshina[3], S I Eliseev[4]
[1]Center of diagnosis and microtechnology, Saint Petersburg Electrotechnical University "LETI", Saint Petersburg 197376, Russia
[2]ITMO University, Saint Petersburg 197101, Russia
[3]Ioffe Institute RAS, Saint Petersburg 194021, Russia
[4] "Burtsev Laboratory" Limited Company, St. Petersburg 197022, Russia

Abstract. The results of the development and experimental investigation of high-voltage silicon-carbide high-speed current switches – drift step recovery diodes (DSRD) – and their implementation as a part of soft X-ray source based on capillary plasma are presented.

1. Introduction

Generators of nanosecond high-voltage pulses are widely used in laser technologies and physical experiments. They are the basis of systems of ultra-wideband (UWB) radiolocation, communication, pulse introscopy and radioelectronic warfare. Today, one of the most promising areas of applications of high-voltage nanosecond pulse generators is ultrafast current switching in sources of soft and hard X-ray radiation based on capillary plasma. In this regard using semiconductor switches instead of traditional gas-discharge switches can have a number of advantages: they do not require initial heating; it is possible to manufacture composite high-voltage switches; they are capable of operating at high switching frequencies (from tens to hundreds of kHz), which is a key factor in the generation of hard and soft X-rays for microscopy and tomography. In recent years prerequisites have arisen for using silicon-carbide drift step recovery diodes (DSRD) as high-voltage high-speed semiconductor current switches, which, due to their unique electrophysical parameters (high working temperature and radiation resistance), allow significantly decreasing the response time and increasing pulse frequency [1,2]. These distinctive features of silicon-carbide DSRDs compared with conventional silicon DSRDs (and gas-discharge switches) suggest that there are broad prospects for their use in high-voltage generators in systems of hard and soft X-ray generation.

The main purpose of this work is development and investigation of a high-voltage high-speed silicon carbide DSRD (4H-SiC-DSRD), which can be used as a current switch in an X-ray radiation source based on capillary plasma.

2. Results

To date, dense capillary plasma is considered to be the most promising way of creating X-ray radiation in the soft range of 2.3-4.4 nm ("water window"). The "water window" is relevant for studies in biology, biomedicine, etc., since it allows imaging of organic tissues and biological objects with the highest contrast between organic compounds and water without causing any damage [3]. One of the key factors determining the stability and efficiency of fast capillary discharges is creating a sufficient degree of preionization, which can be achieved by means of a sliding discharge [4]. In order to determine the properties of a pulse generator based on DSRD for actual experimental implementation, we have first performed numerical simulations of a discharge in a geometry typically used in soft X-ray sources (see fig. 1(b)) [5]. Simulations have been performed using model previously developed in [6]. Figure 1(a) shows spatial distribution of electron density obtained in simulations for the case of the nanosecond voltage pulse with a gauss shape with half-width half-maximum duration of 10 ns and an amplitude of 10 kV. The distributions clearly show formation of a sliding discharge propagating along the capillary. Most importantly, the discharge front covers the distance between the electrodes (5 cm in the present case) in time roughly corresponding to the maximum amplitude of the voltage pulse, which ensures the highest currents during transition to the main discharge stage.

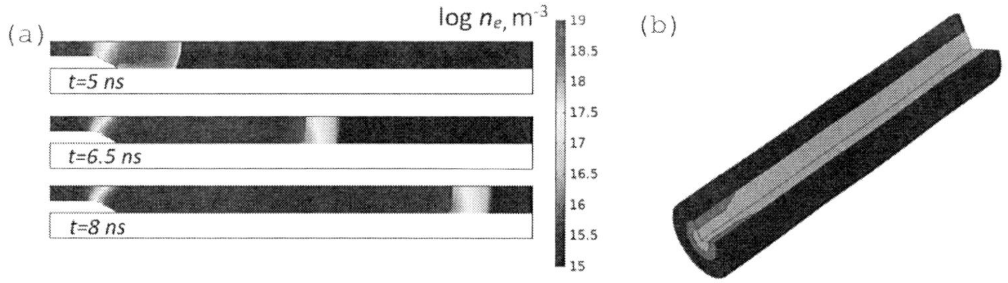

Figure 1. Results of numerical simulations of a sliding discharge in a capillary filled with nitrogen at pressure p=2 Torr made of Al_2O_3 with the inner radius R_1 = 2.5 mm, the outer radius R_2 = 2.75 mm and the length L = 5 cm. (a) – spatial distribution of the logarithm of electron density at different times of discharge propagation; (b) – 3D model of capillary.

Simulations have also revealed that similar results can be obtained for capillaries with different lengths by slight variation of the pulse amplitude. In order to have the ability to create such voltage pulses and conduct experimental research of plasma formations, we have developed high-voltage silicon-carbide diode stacks, capable of switching voltages of more than 10 kV. The assemblies consisted of six chips (developed in University "LETI"), each of which is a p^+-p-n^+ type structure with p-base thickness of 18 μm and aluminum doping level of $4\cdot10^{15}$ cm^{-3}, designed for maximum switching voltages of 1800 V. The thickness of the heavily doped p^+-layer was 5 μm with the acceptor density of $5\cdot10^{19}cm^{-3}$. The thickness of the n^+-substrate was 350 microns with a density of nitrogen donors of $5\cdot10^{18}cm^{-3}$ and a specific resistance of 0.026 ohm·cm. Fig. 2 shows the photo of a single 4H-SiC-DSRD chip, its trench type design and the full high voltage stack consisting of 6 diodes. Geometrical parameters of a single 4H-SiC-DSRD chip are 4500x4500 μm. Optimization of the diode stack was performed using Sentaurus Synopsys TCAD. The effects of incomplete ionization of impurities, high level of injection, heavy doping and avalanche generation were taken into account during simulations.

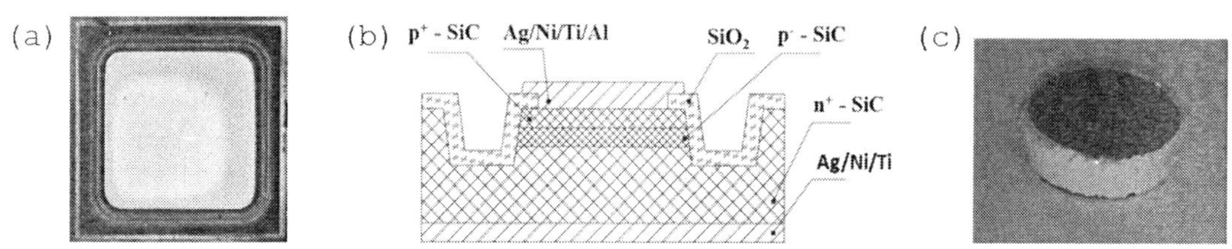

Figure 2. The developed 4H-SiC-DSRD. (a) – photo of the diode; (b) – schematic of the diode geometrical configuration; (c) – high-voltage silicon-carbide stack

In order to prevent surface breakdown of the diode basic structure the periphery areas of the diodes were covered with a high-temperature polyimide Duriimide 115A (prod. Fuji, Japan). The high-voltage assembly was sealed with a high-temperature silicone compound P-727 during the final stage of preparation. Fig. 3 shows the sample oscillogram of the output voltage pulse measured on the 50 ohm high-voltage microwave attenuator 2237-HFNFP (Barth Electronics, Inc.). The pulse amplitude is 10 kV, the duration at FWHM is 12 ns. Pulse registration was performed using Tektronix DSA8300 strobe oscilloscope with a 20 GHz bandwidth.

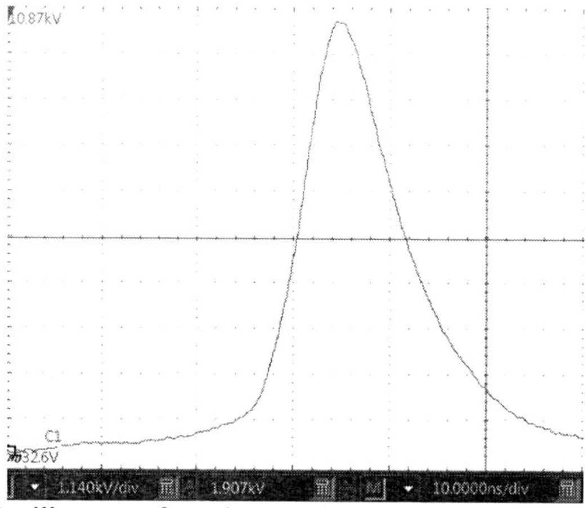

Figure 3. Oscillogram of a voltage pulse generated using the developed 4H-SiC-DSRD assembly as a current switch.

The developed high-voltage 4H-SiC-DSRD assembly was used as an ultrafast current switch in experiments on generation of soft X-ray. For this purpose a vacuum input for supplying high-voltage pulses and a capillary assembly were specially developed (see Fig. 4a). A copper electrode was used as the cathode, which was placed in a ceramic capillary. High-voltage pulses from the generator were fed to the capillary assembly using a cable with a wave resistance of 50 Ohms. The capillary assembly was placed in a vacuum chamber filled with argon (purity 99.998%) at a pressure of 0.25 mbar (Fig. 4 b). A multi-channel master oscillator manufactured at University "LETI" was used for launching (setting the duration of the direct current pumping) the developed 4H-SiC-DSRD assembly and synchronizing it with the measurement equipment. The radiation was registered using a photodiode ФДУК-1УВСК whose signal was measured using the Tektronix DPO7104 real-time microwave oscilloscope with a 1 GHz bandwidth.

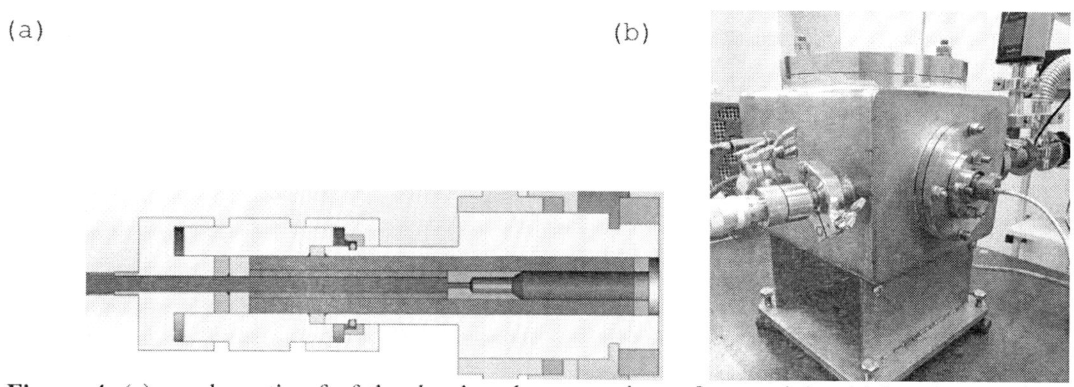

Figure 4. (a) – schematic of of the developed vacuum input for supplying high-voltage voltage pulses and the capillary unit; (b) – photo of the vacuum chamber.

During the preliminary experiment the formed high-voltage pulses were applied to the capillary assembly to initiate the discharge. The signals from the photodetector without and with a thin (about 2 μm) aluminum filter set as an input window were measured (Fig. 5). It can be seen that when the filter is installed, the radiation intensity is reduced by an order of magnitude, while a short peak in the radiation pulse remains. The tail duration is also greatly reduced. From the time structure of the signal from the photodiode and its attenuation using a thin metal foil, it can be concluded that the range of the detected radiation lies in the region of 2-14 nm, i.e. in the area of soft x-ray range.

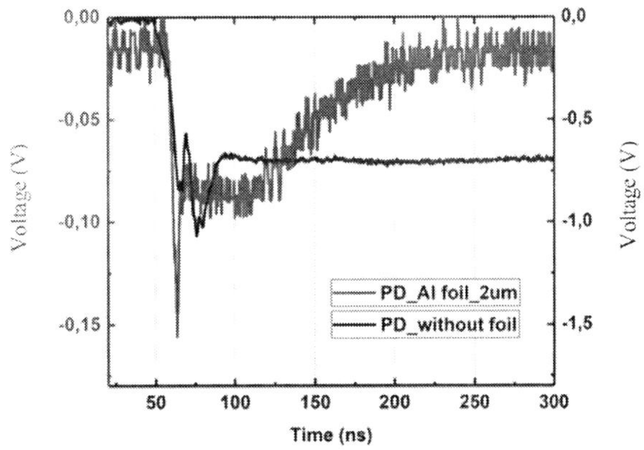

Figure 5. Oscillograms of signals from the photodiode with (red curve) and without (black curve) a 2μm aluminum foil.

3. Conclusion

The results of the experiment show the possibility of the developed high-voltage current switch (switched voltage 10 kV) based on silicon carbide diodes to function as an ultrafast current switch in the structure of the voltage pulse generator of a soft X-ray source. Despite the fact that the developed high-voltage 4H-SiC-DSRD stack is an output element of the pulse generator and the existence of reflections in the system, the efficiency of the diode assembly during the whole experiment was not impaired. With the selected geometrical parameters of the capillary unit, the developed type of pulse generator makes it possible to obtain soft X-rays. This can later serve as the basis for creating compact lithographs, sources for photoelectron spectroscopy, and a microscope working in the field of "water window".

Acknowledgments

This work was supported by RFBR, project number 18-08-01066 ("Development of photon-based methods for study of the transformation of healthy cells into cancer, including submicroscopy").

References

[1] Afanasyev A V, Demin Y A, Ivanov B V, Ilyin V A, Kardo-Sysoev A F, Luchinin V V, Smirnov A A. Miniature high-voltage silicon carbide source of power nanosecond pulses to generate the X-ray and microwave radiation 2013 *Nano and microsystem technique*. **2** p 30.
[2] Afanasyev A V, Demin Y A, Ivanov B V, Ilyin V A, Kardo-Sysoev A F, Luchinin V V, Smirnov A A. Megawatt generator of nanosecond pulses on the basis of silicon carbide drift step recovery diodes 2015 *Proceedings of the Russian Universities: Radioelectronics*. **3** p 21.
[3] Kirz J and Jacobsen C. Soft x-ray microscopes and their biological applications. Q. Rev. Biophys., **28**(1), 33–130, (1995).

[4] Wyndham E S, Favre M, Valdivia M P, Valenzuela J C, Chuaqui H, and Bhuyan H. Fast plasma discharge capillary design as a high power throughput soft x-ray emission source 2010 *Review of scientific instruments* **81** 093502.
[5] Timshina M, Eliseev S, Kalinin N, Letunovskaya M, Burtsev V. Numerical investigation of dynamics and gas pressure effects in a nanosecond capillary sliding discharge. *Journal of Applied Physics*, **125**(14), 143302, 2019.
[6] Burtsev V A, Zabrodskii V V, Kalinin N V, Bol'shakov E P. Electromagnetic radiation sources based on a low-inductive extended z-discharge. *Technical Physics.* 2013 **58(2)** p 192-199.

Features of construction of the scheme of fiber-optic communication system for transmission of analog signals in the frequency range from 0.135 to 40 GHz

V B Fadeenko[1], G A Pchelkin[1], V V Davydov[1], A P Glinuchkin[2], O O Beloshapkina[2] and V Yu Rud'[2,3]
[1]Peter the Great Saint Petersburg Polytechnic University, Saint Petersburg 195251, Russia
[2]Department of Ecology, Russian Research Institute for Phytopathology, 143050, Moscow Region, Odintsovo district, B.Vyazyomy, Russia
[3]Department of Solid State Physics, Ioffe Physico-Technical Institute, Russian Academy of Sciences, St Petersburg 194021, Russia

Abstract. The paper presents the design of a fiber-optic microwave signal transmission line. The use of the developed design of a fiber-optic microwave transmission line allows the radar operating frequency range to be at least doubled. In addition, the number of functional blocks is reduced to ensure the operation of the radar in this frequency range. The results of experimental studies are presented.

1. Introduction

Nowadays, fiber optic communication lines (FOCL) are an important element in the transmission of various information [1-7]. The structure and technical characteristics of fiber optic links change depending on the tasks to be solved, for example, high-speed fiber optic links or transmission of information on fiber optic lines at enterprises of the agricultural industry segment [6-10]. The principle of construction of the work of fiber optic lines for transmitting information remains unchanged [4-16]. The main elements in the FOCL design are the transmitting laser module, the electro-optic modulator, the receiving module and optical isolators [11-16]. Optical amplifiers are used depending on the distance to which information is transmitted via fiber optic links. It should be noted that in the construction of many fiber optic links there are a number of features that depend on the type of information transmitted via fiber optic links and the conditions of operation of fiber optic links. The greatest number of features is present when using fiber optic lines for transmitting various microwave signals and controls in radar stations (radars) for various purposes [11–20].

Studies and operating experience of the radar showed that it is most appropriate to use single-mode fiber to transmit microwave signals and control. In addition, each channel uses its own signal transmission system with fiber optic communication line. This has its own advantages and disadvantages [11, 12, 16, 21, 22]. The disadvantages include the fact that the currently developed various systems for transmitting microwave signals from fiber optic lines work stably in a certain frequency range, for example, X-band. This is due to the structural features of the modulation of the optical signal of the microwave signals in an electro-optical modulator and the sensitivity when receiving weak antenna signals and the speed of the photodetector elements in the receiving module at a high frequency of the transmitted signal via a fiber optic communication line. In addition, based on new technologies and materials, as well as the results of scientific research, receiving and transmitting antennas that operate in a wide range of microwave frequencies have been developed. In some cases, these are tunable devices based on the microwave frequency. For these radars, it became necessary to create a universal system using fiber optic lines, which allows transmitting microwave signals in the frequency range from 0.135 to 40 GHz.

Figure 1 presents the standard modular design of this system, which consists of 8 previously used individual channels. In comparison with previous developers, systems were developed for transmitting microwave signals using fiber optic communication lines.

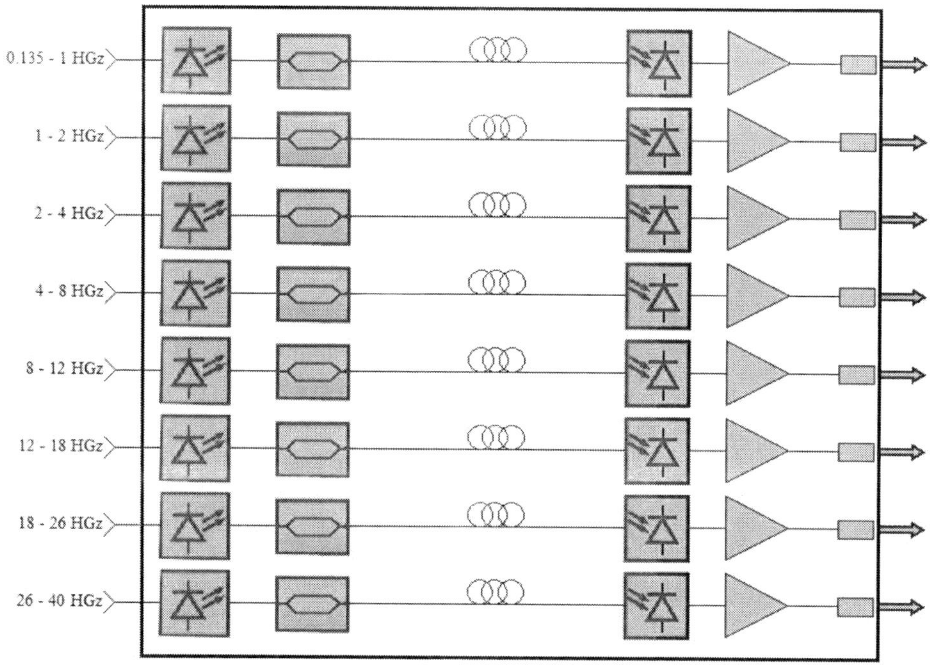

Figure 1. The scheme of fiber-optic transmission line microwave signals in the radar frequency range from 0.135 to 40 GHz

Our studies have shown that this design of the microwave signal transmission system is not rational to use for several reasons:

1. the scheme contains a large number of expensive optical elements to ensure the operation of the system in a narrow frequency range, for example, 1-2 GHz;
2. it is necessary to provide temperature stabilization for each optical element or the entire unit (all optical elements are in one large design) when placing them, for example, on the upper decks on the mast racks of ships [4, 11, 12, 15];
3. difficulties arise if this design is used in active electronically scanned array (AESA), for example, on a moving object (where there are significant limitations on the volume and weight of the blocks and temperature stabilization is necessary).

Therefore, there is a need to upgrade this universal system of microwave signal transmission in order to reduce its size and cost. One of the solutions to the considered problem is presented in our work.

2. New design of fiber-optic microwave transmission system in the radar and features of its work

Studies have shown that the amplitude-frequency characteristics (AFC) of fiber optic communication line in the frequency range from 0.135 to 2.0 GHz do not differ between communication channels when transmitting a microwave signal using direct or external modulation. The absence of an electro-optic modulator in the FOCL channel reduces the transfer ratio Kt by 20-25%. Therefore, it is more efficient to use in the design of fiber optic transmission system a new model of a transmitting module with direct modulation and to combine two channels into one (from 0.135 to 2.0 GHz). Figure 2 presents the new design of the FOCL for the transmission of microwave signals.

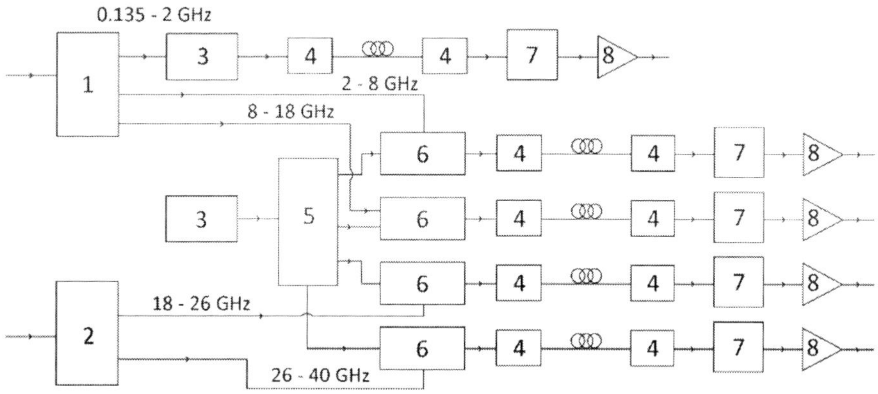

Figure 2. The new scheme of fiber-optic transmission line microwave signals in the radar frequency range from 0.135 to 40 GHz: 1 and 2 – microwave channel switches; 3 – receiving laser modules; 4 – optical isolator; 5 – transmitting laser module; 6 – electro-optical modulator; 7 – photodetector module; 8 – power amplifier

Other channels were also converted into one channel with a frequency range from 2 to 8 GHz and from 8 to 18 GHz. In the new design due to the use of an optical divider by 4, the number of laser transmitting modules was reduced to two. The study showed that the use of a single transmitting module with direct modulation simultaneously on all 5 channels using microwave switching devices creates additional noise. This noise reduces the dynamic range of the fiber optic transmission system. In the case of the transmission of weak signals creates additional problems.

In the new design of the fiber optic transmission line, the number of electro-optical modulators (EOM) has decreased from eight to five. These devices are most sensitive to temperature changes. For their stable operation in a wide range of temperature variations, a thermal stabilization scheme with adjustment of the operating point is necessary [12, 16, 21, 22]. This scheme can be implemented for several EOMs, but it will take a lot of space. The working point of each modulator is different. Reducing the number of EOM reduces the size and weight of the thermal stabilization system, as well as its cost. It is also necessary to implement reliable thermal stabilization in the photodetector modules [23-31].

Such a construction of the fiber optic transmission system design is justified, since the distances over which the signal must be transmitted do not exceed 300 m. Losses in the optical fiber at such distances are insignificant. Different types of dispersion on the bends of the optical fiber do not have time to form due to the short propagation time of the optical signal through it. The important characteristics in this case remain the purity of the spectrum of the transmitted signal and the magnitude of the temperature deviation of the modulation phase during the propagation of the optical signal in the fiber, since FOCL can also transmit signals from the local oscillator and control.

3. The results of experimental studies and their discussion

Experience in operating a radar with a fiber optic communication line showed that in some cases distortions in the spectrum [18–20, 32–36] of the transmitted signal can lead to errors in determining the position of the object's coordinates or loss of control of the station's operation. On the developed laboratory model of FOTS, we investigated possible distortions in the spectrum when transmitting a heterodyne signal over FOTS at an operating frequency of 150 MHz. Figure 3 presents one of the results of research for the microwave transmission channel in the frequency range from 0.135 to 2.0 GHz.

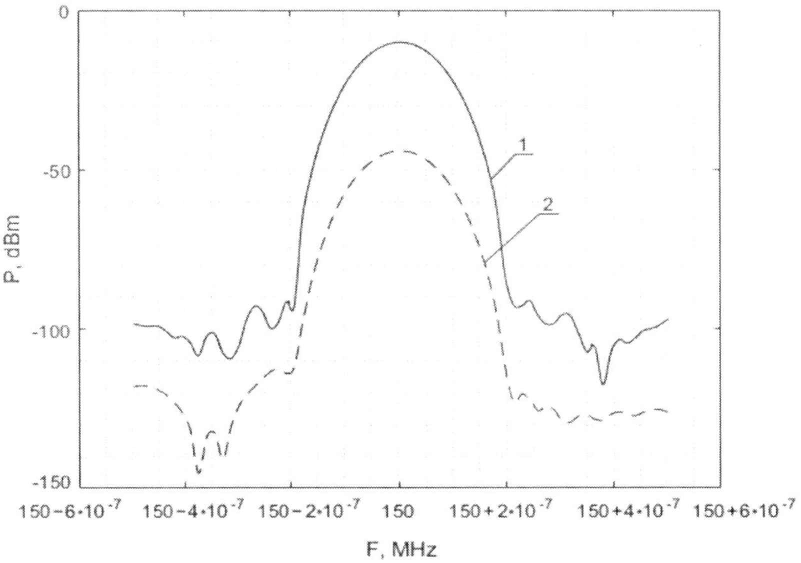

Figure 3. Spectra of input (continuous) and output (dashed) signal of FOTS

Comparison of the obtained spectra (Fig. 3) shows a high efficiency of transmission of the heterodyne signal at the carrier frequency over FOTS. Distortions in the spectrum are present only on the lateral components, which does not affect the accuracy of determining the distance to the target in the radar.

Since the fiber optic transmission system is designed for AESA, which are placed on aircraft, the FOTS will be in different temperature conditions. A change in ambient temperature causes both a change in the refractive index of the fiber and additional lengthening of the fiber due to thermal expansion or contraction. This leads to a change in the phase of the light and, accordingly, to a change in the phase of modulation of the radiation transmitted through the fiber. Therefore, an experimental assessment was made of the temperature drift of the modulation phase during propagation in the fiber. Figure 4 shows the experimental dependence of the phase shift of the modulation of laser radiation $\Delta\varphi_m$ on the ambient temperature T.

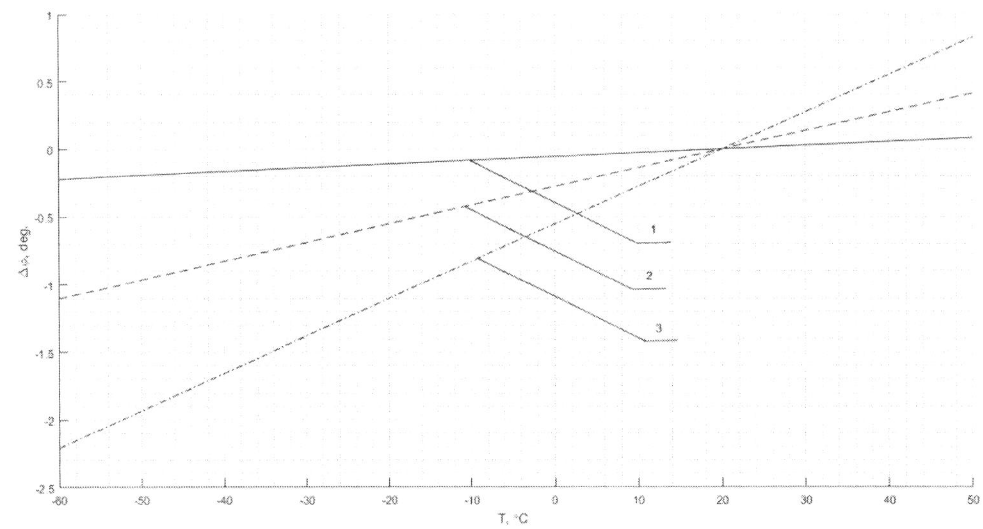

Figure 4. The shift of the phase of modulation $\Delta\varphi_m$ from the temperature T. Graphics 1, 2 and 3 correspond to the frequency of the heterodyne signal in MHz: 50; 100; 150

The experimental research results showed that the temperature dependence of the modulation phase change for G.657 fiber is no more than 3.0 degrees in the selected temperature range from 213.1 to 323.3 K for different heterodyne signals transmission frequencies. The influence this phase shift value on the on the frequency and amplitude characteristics of the transmitted optical signal will be negligible for optical fiber distance of less than 120 m.

4. Conclusion

The obtained results showed that the design developed by us can be successfully operated as part of a radar with an AESA. It has been established that design solutions implemented in the manufacture of FOTS are justified on the basis of the conducted research.

In addition, we found that the technical characteristics (dynamic range, transmission coefficient, etc.) did not deteriorate in the newly developed FOTS design. And in some cases, for some channels have become better. This made it possible to expand the number of tasks that can be solved during the operation of the radar using the new developed FOTS compared to the previously used FOTS designs.

Acknowledgments

This work partly supported by Grant of RFBR No18-29-25071.

References

[1] Podstrigaev A S, Smolyakov A V, Davydov V V, Myazin N S and Slobodyan M G 2018 *Lecture Notes in Computer Science (including subseries Lecture Notes in Artificial Intelligence and Lecture Notes in Bioinformatics)* **11118 LNCS** 509-515

[2] Mashkov G, Borisov E and Fokin G A 2016 *Proceedings - 2016 International Conference on Radar, Antenna, Microwave, Electronics, and Telecommunications, ICRAMET 2016 (Jakarta)* **7849572** p. 7-12

[3] Mashkov G, Borisov E and Fokin G A 2017 *Proceedings - 19th International Conference on Advanced Communications Technology, ICACT 2017 (Phoenix Park Pyeongchang)* **7890254** p. 979-984

[4] Lenets V A, Tarasenko M Yu, Davydov V V, Rodugina N S and Moroz A V 2018 *Journal of Physics: Conference Series* **1038(1)** 012037

[5] Lavrov A P and Molodyakov S A 2017 *Measurement Techniques* **59** 1025 – 1033

[6] Ermolaev A N, Krishpents G P, Davydov V V and Vysoczkiy M G 2016 *Journal of Physics: Conference Series* **741(1)** 012171

[7] Davydov R V, Saveliev I V, Lenets V A, Tarasenko M Yu, Yalunina T R, Davydov V V and Rud' V Yu 2017 *Lecture Notes in Computer Science (including subseries Lecture Notes in Artificial Intelligence and Lecture Notes in Bioinformatics)* **10531 LNCS** 177-183

[8] Moroz A V, Davydov V V, Rud V Yu, Rud Yu V, Shpunt V Ch and Glinushkin A P 2018 *Journal Physics: Conference Series* **1135(1)** 012060

[9] Fadeenko V B, Davydov V V, Rud V Yu, Glinushkin A P, Rud Yu V and Shpunt V Ch 2017 *Journal Physics: Conference Series* **917(9)** 092015

[10] Fadeenko V B, Rud V Yu, Glinushkin A P, Rud Yu V, Shpunt V Ch and Hogland W 2018 *Journal Physics: Conference Series* **1038(1)** 012030

[11] Podstrigaev A S, Davydov R V, Rud' V Yu and Davydov V V 2018 *Lecture Notes in Computer Science (including subseries Lecture Notes in Artificial Intelligence and Lecture Notes in Bioinformatics)* **11118 LNCS** 624-630

[12] Filatov D L, Galichina A A, Vysoczky M G, Yalunina T R, Davydov V V and Rud V Yu 2017 *Journal Physics: Conference Series* **917(8)** 082005

[13] Davydov V V and Semenov V V 2001 *Radiotekhnika* **11** 22-24

[14] Vekshin Yu V, Lavrov A P and Hvostov Yu V 2018 *Radiotekhnika* **1** 24-30

[15] Davydov V V, Sharova N V, Fedorova E N, Gilshteyn E P, Malanin K Y, Fedotov I V, Vologdin V A and Karseev A Yu 2015 *Lecture Notes in Computer Science (including subseries Lecture Notes in Artificial Intelligence and Lecture Notes in Bioinformatics)* **9247** 712-721

[16] Tarasenko M Yu, Davydov V V, Sharova N V, Lenets V A and Yalunina T R 2017 *Lecture Notes in Computer Science (including subseries Lecture Notes in Artificial Intelligence and Lecture Notes in Bioinformatics)* **10531 LNCS** 227-232
[17] Podstrigaev A S, Ryazantsev L B and Lukashev V P 2016 *Measurement Techniques* **59(5)** 547-550.
[18] Petrov A A, Davydov V V and Myazin N S 2017 *Lecture Notes in Computer Science (including subseries Lecture Notes in Artificial Intelligence and Lecture Notes in Bioinformatics)* **10531 LNCS** 561-568
[19] Petrov A A, Davydov V V and Grebenikova N M 2018 *Lecture Notes in Computer Science (including subseries Lecture Notes in Artificial Intelligence and Lecture Notes in Bioinformatics)* **11118 LNCS** 641-648
[20] Petrov A A, Grebenikova N M, Lukashev N A, Davydov V V, Ivanova N V, Rodugina N S and Moroz A V 2018 *Journal of Physics: Conference Series* **1038(1)** 012032
[21] Ivanov S I, Lavrov A P and Saenko I I 2017 *Lecture Notes in Computer Science (including subseries Lecture Notes in Artificial Intelligence and Lecture Notes in Bioinformatics)* **10531 LNCS** 482-489
[22] Ivanov S I, Lavrov A P and Saenko I I 2016 *Lecture Notes in Computer Science (including subseries Lecture Notes in Artificial Intelligence and Lecture Notes in Bioinformatics)* **9870 LNCS** 670-679
[23] Myazin N S, Smirnov K J, Davydov V V and Logunov S E 2017 *Journal of Physics: Conference Series* **929 (1)** 012080
[24] Grebenikova N M, Smirnov K J, Artemiev V V, Davydov V V and Kruzhalov S V 2018 *Journal of Physics: Conference Series* **1038 (1)** 012089
[25] Grebenikova N M, Smirnov K J, Davydov V V, Rud V Yu and Artemiev V V 2018 *Journal of Physics: Conference Series* **1135 (1)** 012055
[26] Friman R L 2012 Fiber–Optic Systems for Communications (New York: Wiley-Interscience a John Wiley & Sons. Inc.)
[27] Kiesewetter D, Malyugin V, Makarov S, Korotkov K, Ming D and Wei X 2017 *Proceedings – 2016 Advances in Wireless and Optical Communications, RTUWO 2016* **7821845** p. 5-8 (2017).
[28] Ateya A A, Muthanna A, Gudkova I, Abuarqoub A, Vybornova A and Koucheryavy A 2018 *Journal of Sensor and Actuator Networks* **7(1)** 7
[29] Davydov V V, Karseev A Yu, Nepomnyashchay E K, Petrov A A and Velichko E N 2014 *Lecture Notes in Computer Science (including subseries Lecture Notes in Artificial Intelligence and Lecture Notes in Bioinformatics)* **8638** 694-702
[30] Davydov R V, Antonov V I and Moroz A V 2018 *Proceedings of the 2018 IEEE International Conference on Electrical Engineering and Photonics, EExPolytech 2018 (Saint-Petersburg)* **8564378** p. 236-239
[31] Rykin E V, Moroz A V, Smirnov K J, Davydov V V and Yushkova V V 2018 *MATEC Web of Conference* **245** 12002
[32] Lukashev N A, Petrov A A, Davydov V V, Grebenikova N M and Valov A P 2018 *Proceedings of 18th International conference of Laser Optics ICLO-2018* (Saint-Petersburg) **8435889** p. 271
[33] Semenov V V, Nikiforov N F, Ermak S V and Davydov V V 1991 *Soviet journal of Communications Technology and Electronics* **36** 59 – 63
[34] Myazin N S., Davydov V V, Yushkova V V and Rud V Yu 2018 *Journal of Physics: Conference Series* **1038 (1)** 012088
[35] Smirnov K J, Davydov V V, Glagolev S F, Rodygina N S and Ivanova N V 2018 *Journal of Physics: Conference Series* **1038 (1)** 012102
[36] Fadeenko V V, Kuts V A, Vasiliev D A and Davydov V V 2018 *Journal Physics: Conference Series* **1135(1)** 012053

Simulation of quantum transport in doped carbon nanotube diode controlled by transverse electric field

D A Timkaeva[1], R T Sibatov[2,3]

[1] Researcher, Laboratory of Diffusion Processes, Ulyanovsk State University, Russia
[2] Professor, Department of Theoretical Physics, Ulyanovsk State University, Russia
[3] Researcher, Scientific Manufacturing Complex "Technological Centre", Moscow, Russia

E-mail: *ren_sib@bk.ru*

Abstract. Using first-principle and semi-empirical methods, optical responses and transport characteristics of hybrid CNT-based diodes are studied. The junction is realized by a combination of doping of one half of a nanotube and applying a transverse electric field to the other. Calculations are carried out in the framework of the density functional theory and the nonequilibrium Green function method. The possibility of using these rectifiers in optical nanoantennas is discussed.

1. Introduction

There are several ways to create pn-junctions based on carbon nanotubes, such as the substitution of boron or nitrogen atoms instead of carbon in the nanotube lattice, doping with charge transfer from electrodes, atoms or molecules, or controlling bands using the transverse electrostatic field (see book [1]). In the latter method, the device can operate in several different modes in a controlled way (by changing the transverse electrostatic field), and there is no need for chemical doping of nanotubes. Examining the effect of a transverse electric field on CNT electronic properties within the Dirac particle formalism in a cylinder, the authors of Ref. [2] showed that the problem obeys supersymmetry that protects low-energy states and ensures the preservation of metallic properties of CNT in arbitrarily large fields. In semiconductor tubes, the band gap decreases with increasing transverse electric field. This behavior is confirmed by the calculations using the density functional theory [3]. The patent [4] states that it is possible to use the influence of transverse electric field on the electrical conductivity and radiation absorption characteristics in such applications as switches, transistors, photodetectors, and polaron generators. When local electric fields are applied to a nanotube, a corresponding number of quantum wells in the nanotube may be formed in several places. Such a configuration is useful for Bragg reflectors, lasers, and quantum computing [4]. At present, there is great interest in the creation of terahertz and infrared nanoantennas based on CNTs. Doping CNTs with nitrogen or boron atoms in some studies is considered as a way to increase the antenna efficiency of CNT arrays. In [5], it is shown that the replacement of 8% carbon atoms in a CNT by nitrogen or boron atoms leads to an increase in conductivity, antenna efficiency, and thermal radiation power.

Here, we calculate the properties of CNT-based junction realized by a combination of doping one half of a nanotube and applying a transverse electric field to the other. This type of CNT-based diode can be advantageous for optical nano-antennas. As is known, the development of a nanoantenna for infrared and visible frequency ranges of electromagnetic radiation is associated with the problem of a rectifier (diode) operating at terahertz and higher frequencies without large power losses. These losses usually arises due to the effect of pn-junction capacitance. Using the density functional theory (DFT), semi-empirical self-consistent extended Hückel method and the nonequilibrium Green function method (NEGF) implemented in QuantumATK, we calculate optical properties of doped nanotubes and estimate projected local density of states, transmission and current-voltage characteristics of the hybrid CNT-based diodes.

2. Model and method

Using the Kubo-Greenwood approach and the density functional theory, optical characteristics (absorption spectrum, reflections, dispersion law of dielectric susceptibility, and frequency dependence of conductivity) of semiconducting single-walled carbon nanotubes are calculated.

The susceptibility tensor is defined by the Kubo-Greenwood formula:

$$\chi_{ij}(\omega) = -\frac{e^2\hbar^4}{m^2\varepsilon_0 V\omega^2}\sum_{nm}\frac{f(E_m)-f(E_n)}{E_{nm}-\hbar\omega-i\Gamma}\pi^i_{nm}\pi^j_{nm}$$

where π^i_{nm} is the matrix element of dipole moment i-th component for states n and m, V is volume, Γ is widening, $f(E)$ is the Fermi-Dirac function. The dielectric tensor and optical conductivity are calculated according to relations $\varepsilon_{ij}(\omega) = 1 + \chi_{ij}(\omega)$ and $\sigma_{ij}(\omega) = -i\omega\epsilon_0\chi_{ij}(\omega)$.

We considered hybrid CNT-based diode, where junction is created by combination of doping with a transverse electric field generated by charged gate electrode. The implementation of the device is based on the so-called screening approximation assuming that properties of the left and right electrodes can be described by solving a problem for a completely periodic cell. The nonequilibrium electron density are calculated by NEGF-method implemented in Quantum ATK. The electronic states in hybrid diodes are calculated by means of the self-consistent Hückel method. Current-voltage characteristics are calculated according to:

$$I = \frac{e}{h}\int T(E)\left[f_L(E,T_L) - f_R(E,T_R)\right]dE, \tag{1}$$

where f_L and f_R are electron energy distribution functions in the left and right electrodes, respectively, $T(E)$ is a transmission spectrum calculated by the NEGF-method.

3. Results and discussion

Substitutional doping of different atoms in CNTs is a well known way to functionalize these nanotubes [6, 7]. When SWCNTs are doped with nytrogen or boron, it is assumed that substitutional atoms do not lead to additional atomic rearrangement. It is expected that B and N substitutions lead to the formation of an acceptor and donor states, respectively, in the CNT electronic structure [8]. Here we present results of calculation of optical properties and quantum transport characteristics of hybrid CNT-based diodes, where p- and n- regions are assumed to be formed by combination of doping and applying a transverse electric field. Ref. [9] reported that substitutional doping of multiwalled CNTs with boron lead to a sufficient decrease in the dc resistivity. Ref. [10] described observation of increase in the ratio of metallic nanotubes for the case of nitrogen substitutional doping of mixtures of metallic and semiconducting SWNTs in composite materials.

In Fig. 1, calculated absorption spectra for pristine and doped CNT are shown. Doping with B and N (fractions 10% and 20%) leads to reducing of absorption coefficient and to shift of peaks. It can be seen that substitution of carbon atoms by B and N atoms lead to different changes of these spectra. In Fig. 2, calculated dispersion of optical conductivity for pristine and doped CNT is presented. Doping with B does not lead to destruction of absorption peaks, while N doping smooths spectra sufficiently.

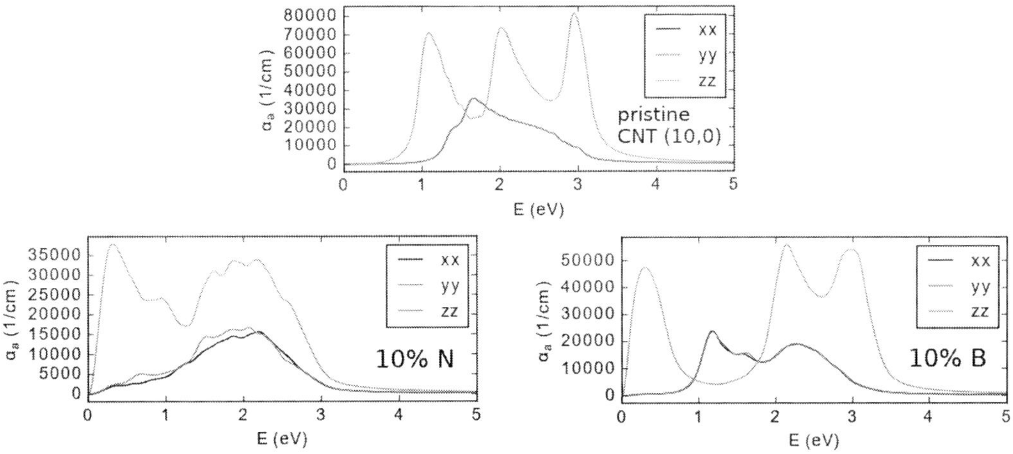

Figure 1. Absorption coefficient for pristine, B doped and N doped CNT vs photon energy.

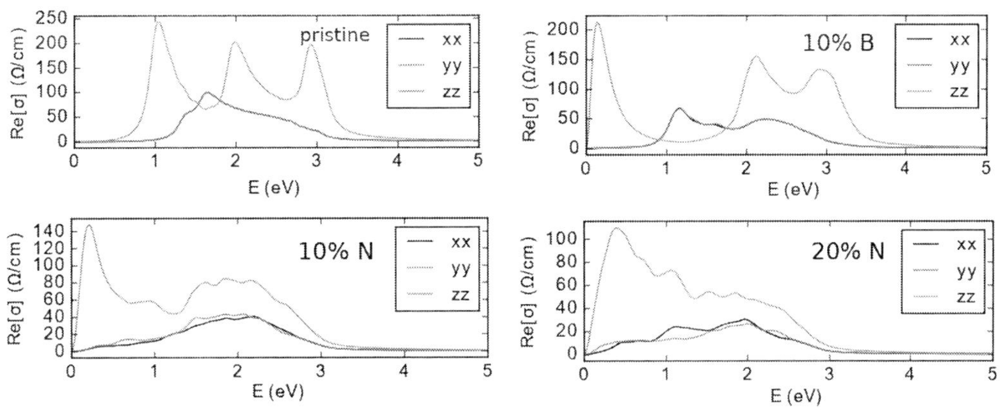

Figure 2. Optical conductivity for pristine, B doped and N doped CNT vs photon energy.

The transmission spectra and current-voltage characteristics are calculated for the following hybrid CNT-based diodes. Half of a nanotube is doped with boron or nitrogen and the other is situated in transverse electric field produced by a metallic gate coated by a dielectric layer (Fig. 3,a). The gate potential can be changed.

The current-voltage characteristics for three values of gate voltage for 20% boron-doped hybrid CNT-diode are presented in Fig. 3,b. They show rectifying behavior of this diode for gate potentials 1 and 3 V. The similar behavior is observed for 20% N-doped hybrid diode (Fig. 4). The distribution of electrostatic difference potential and projected local density of states are shown in Fig. 4 for two opposite values of gate potential (±3 V).

The similar behavior of N-doped and B-doped CNT-diodes can be interpreted in the following way. At chosen doping level (20%), CNTs become metallic for both cases of B- and N-doping,

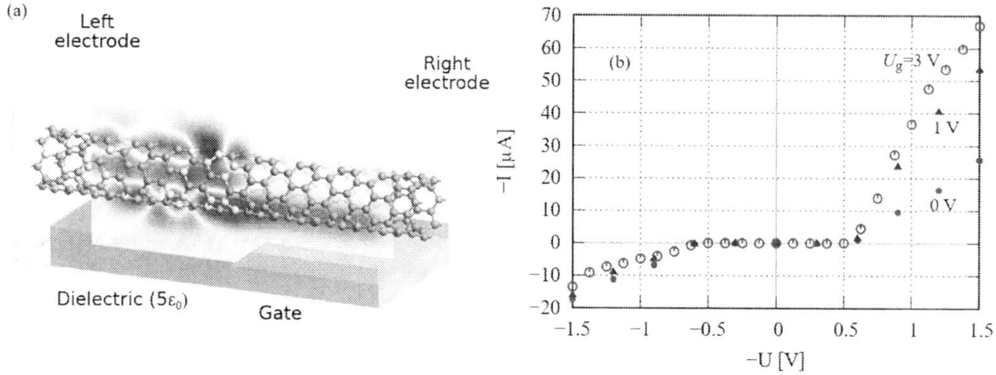

Figure 3. Cut-plane of electron difference density at $U_g = 3$ V (left panel) and current-voltage characteristics for three values of gate voltage.

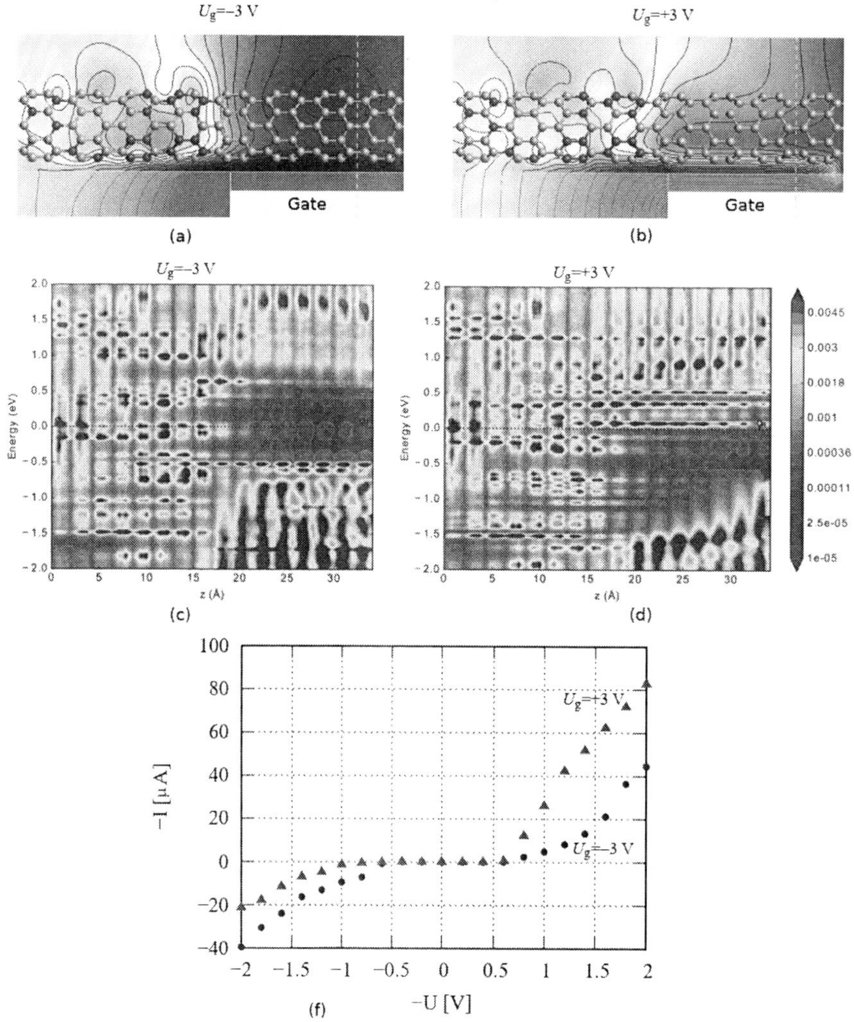

Figure 4. Distribution of electrostatic difference potential (a,b), projected local density of states (c, d) and the current-voltage characteristics for two values of gate potential ±3 V.

so we deal some type of Schottky diode. This fact is confirmed by distribution of states Fig. 4,c and d. Positive gate potential leads to curved bands responsible for rectifying behavior.

4. Conclusion

Using first-principle and semi-empirical self-consistent methods, optical responses and transport characteristics of hybrid CNT-based diodes are studied. The junction is realized by a combination of doping of one half of a nanotube and applying a transverse electric field to the other. Calculations of current-voltage characteristics are carried out by means of the semi-empirical self-consistent method and the nonequilibrium Green function method. This type of CNT-based diode can be promising alternative for optical nano-antennas. The dependence of current-voltage characteristics on gate potential are explained by the fact that at chosen doping level (20%), CNTs become metallic and we deal with Schottky nanodiode. This fact is confirmed by calculated projected local density of states.

Acknowledgments

The study is supported by the Ministry of Education and Science of the Russian Federation (state program 16.2773.2017/4.6) and the Russian Science Foundation (19-71-10063).

References

[1] Anantram, M. P., Léonard, F. (2006). Physics of carbon nanotube electronic devices. Reports on progress in physics, 69(3), 507.
[2] Lee, H. W., Novikov, D. S. (2003). Supersymmetry in carbon nanotubes in a transverse magnetic field. Physical Review B, 68(15), 155402.
[3] Chen, C. W., Lee, M. H., Clark, S. J. (2004). Band gap modification of single-walled carbon nanotube and boron nitride nanotube under a transverse electric field. Nanotechnology, 15(12), 1837.
[4] O'keeffe, J., Cho, K. (2008). U.S. Patent No. 7,385,262. Washington, DC: U.S. Patent and Trademark Office.
[5] Dyachkov, P.N., Bochkov, I.A. (2012). Terahertz nanoantenna based on carbon nanotubes chemically modified with nitrogen or boron. Scientific and technical statements of the St. Petersburg State Polytechnic University. Physics and Mathematics, 2 (146) [in Russian].
[6] Stephan, O., Ajayan, P. M., Colliex, C., Redlich, P., Lambert, J. M., Bernier, P., Lefin, P. (1994). Doping graphitic and carbon nanotube structures with boron and nitrogen. Science, 266(5191), 1683-1685.
[7] Ayala, P., Arenal, R., Loiseau, A., Rubio, A., Pichler, T. (2010). The physical and chemical properties of heteronanotubes. Reviews of modern physics, 82(2), 1843.
[8] Blase, X. et al. (1999). Boron-mediated growth of long helicity-selected carbon nanotubes. Physical Review Letters, 83(24), 5078.
[9] Ishii, S. et al. (2008). Resistivity reduction of boron-doped multiwalled carbon nanotubes synthesized from a methanol solution containing boric acid. Applied Physics Letters, 92(20), 202116.
[10] Villalpando-Paez, F. et al. (2006). Synthesis and characterization of long strands of nitrogen-doped single-walled carbon nanotubes. Chemical Physics Letters, 424(4-6), 345-352.

Selection of the design of a hot heat exchanger of an automotive thermoelectric generator for an urban driving cycle

K A Shishov[1,2]
[1] Peoples' Friendship University of Russia 117198, Miklukho-Maklaya str. 6, Moscow, Russian Federation
[2] Bauman Moscow State Technical University 105005, 2-ya Baumanskaya 5/1, Moscow, Russian Federation

Abstract. It is considered for combustion engines to have nearly 60-70% of fuel energy to be dissipated to the environment by the exhaust gases. Automobile thermoelectric generator (ATEG) application is proved to be a solution for increasing vehicle fuel economy by exhaust gas waste heat recovery. The overall efficiency of ATEG system highly depends on its hot heat exchanger, which is designed to remove heat from the exhaust gases to thermoelectric modules that generate electricity. Intensification of heat transfer in ATEG prototypes is usually achieved by using finning and turbulators. However, the modification of inner side of hot heat exchanger increases the hydraulic resistance of the exhaust system, which adversely affects the power and efficiency of the engine. In this work, we investigate the designs of hot heat exchangers with different configurations of the inner surface and evaluate their impact on the engine power during the modern urban driving cycle.

1. Introduction

One of the main objectives of the automotive industry is to reduce CO_2 emissions by more than 4.5% for passenger cars under urban and country operation conditions. The solution can be achieved by the use of an automotive thermoelectric generator (ATEG) for the exhaust gas waste heat recovery.
A high-efficiency hot heat exchanger is an integral part of the automotive thermoelectric generator. Main purpose of hot heat exchanger is increasing the amount of heat energy extracted from engine exhaust gases and its effective transfer to TE modules. It has been demonstrated that the efficiency of ATEG prototype heat exchangers ranged from 40% to 70% [1].
The presence of ATEG in vehicle exhaust system has its own consequences which can negatively affect overall car efficiency. These effects are challenges for automotive thermoelectric industry and we can categorize them into two main types:

1. Increase of hydraulic resistance of the exhaust system
2. Increase in vehicle weight

In order to achieve uniform temperature distribution and higher interface temperature, the optimization of thermal characteristics of the heat exchanger and internal structures are being studied. Increasing the interface temperature by optimizing the internal structure of the heat exchanger will increase the efficiency of the exhaust thermoelectric generator system [2]. The presence of flow-impeding inserts/internal structures results in an adverse increase in backpressure and has a negative effect on the vehicle engine performance [3]. Fig.1 demonstrates hexagonal prism shaped heat exchangers [4] with different types of internal structures, commonly used for hot heat exchanger designs.

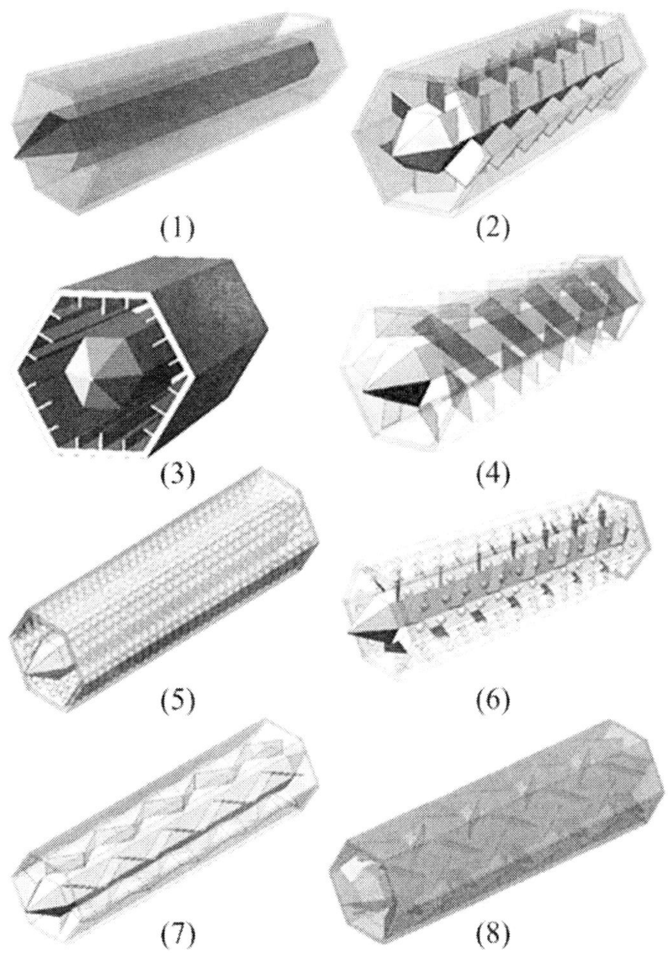

Figure 1. Types of the internal structure of hot heat exchangers: 1) flat walls; 2) oblique fins [5]; 3) longitudinal fins; 4) fins at an angle to the flow; 5) dimpled surface [6]; 6) dimples and turbulizer; 7) corrugation fins; 8) multidirectional fins.

As the main objective is to realize a high overall CO_2 reduction for light-duty vehicle in New European Driving Cycle (NEDC) about 4.5% [7], the hot heat exchanger designs should be optimized for urban and extra-urban engine operating conditions.

2. Describing urban operating conditions

The velocity profile imposed by the NEDC used for light-duty vehicle certification is showed in [8]. The Fig. 2 demonstrates a bar chart of engine steady-state modes covering the most used part of the engine map in both urban and extra-urban operating conditions.

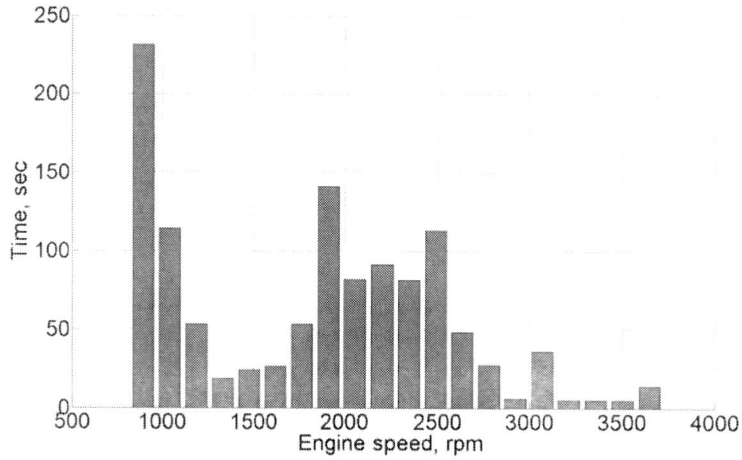

Figure 2. Bar chart of engine steady-state modes during the NEDC.

3. Modeling ATEG performance

A mathematical model for the operation of the automotive thermoelectric system has been developed [4,9,10,11], which makes it possible to vary design parameters and determine estimative output ATEG characteristics for different vehicle operating conditions.

Figure 3 shows the curves of the increase in power of the internal combustion engine W_{add}, calculated for the hot heat exchangers designs presented in Figure 1 for different engine speeds. The obtained data is made up of electric power generated by ATEG W_{te} and reduction of load on the onboard electric generator W_{gen} (1):

$$W_{add} = W_{te} + W_{gen} \tag{1}$$

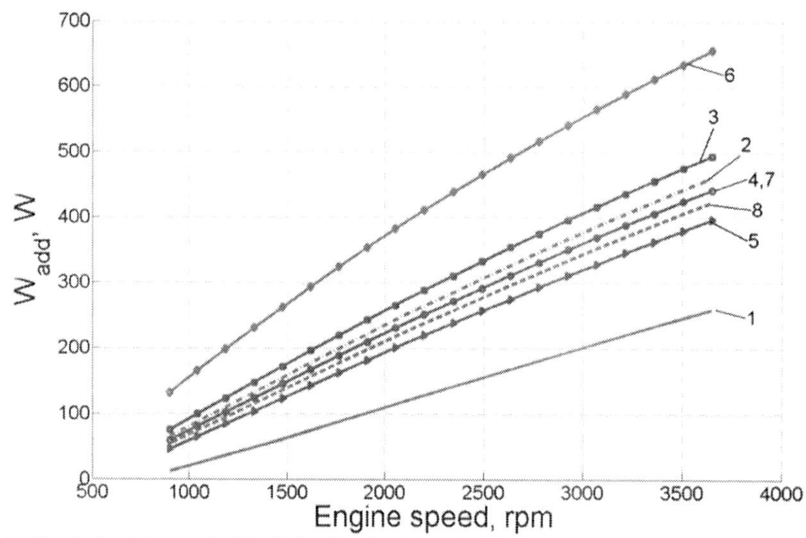

Figure 3. Curves of the increase in power of the internal combustion engine for designs of heat exchangers presented in Figure 1.

Figure 4 shows the curves of the power loss on the shaft of the internal combustion engine W_{press} by the hydraulic resistance influence calculated for the designs of heat exchangers shown in Figure 1 for various engine speeds.

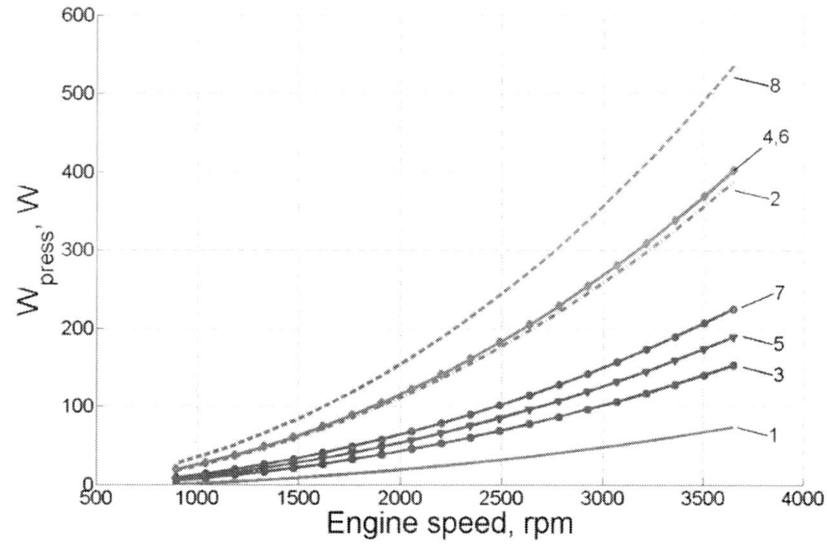

Figure 4. Curves of the power loss on the shaft by the hydraulic resistance influence.

Figure 5 shows the graphs of the increase in power on the shaft W_{mech}, taking into account the losses by the hydraulic resistance of the investigated structures of hot heat exchangers (2):

$$W_{mech} = W_{add} - W_{press} \tag{2}$$

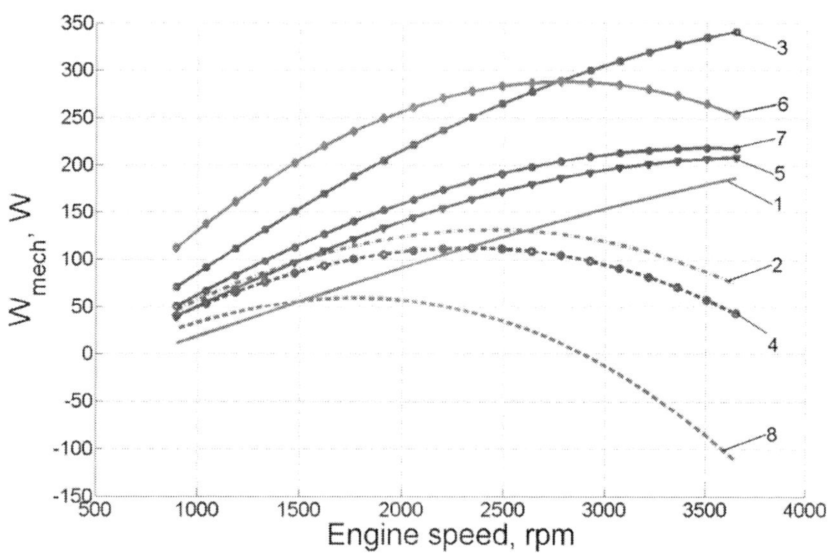

Figure 5. Curves of the increase in power on the shaft.

Analyzing the time spent by the engine in each speed mode corresponding to NEDC, it is possible to estimate the total amount of useful electrical energy generated by ATEG for the described designs of heat exchangers shown in Figure 1 during the cycle (Fig. 6) using the following formula (3):

$$Q_{sum} = \sum_{i=1..n} \left(W^{i}_{mech} \cdot \frac{t_i}{3600} \right) \tag{3}$$

where n - the number of described speeds, t_i - is the time spent by the engine in each speed mode.

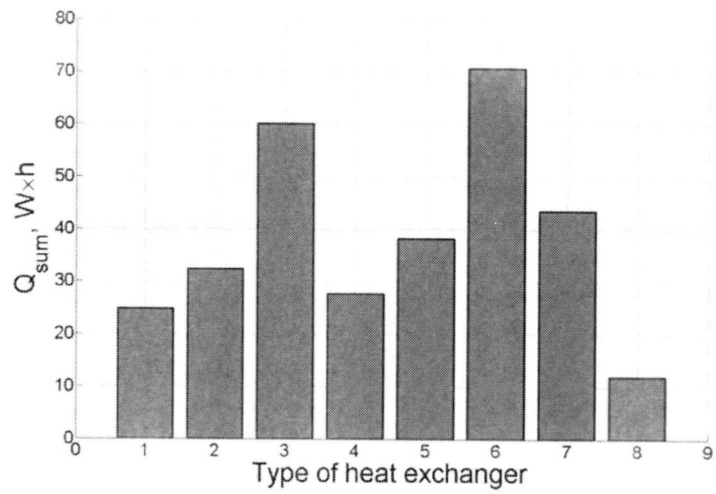

Figure 6. Amount of useful electrical energy generated by ATEG for the described designs of heat exchangers.

4. Conclusion

In conclusion, it is necessary to mention that we cannot affect consumers operating conditions, so it is necessary to optimize ATEG for NEDC and other known driving cycles, depending on the type of the vehicle. The ideal heat exchanger recovers maximum heat from an engine's exhaust with minimum pressure drop and shows high thermal uniformity. The heat transfer rate and thermal uniformity of heat exchanger can be improved by optimization of the hot exchanger fin structure. The influence of the hot heat exchanger internal design on the negative reverse pressure and output characteristics of ATEG was studied. The design of the hot exchanger, which ensure the generation of the maximum amount of electric power of the ATEG per driving cycle, were found.

References

[1] Automotive thermoelectric generator design issues. https://www.energy.gov/sites/prod/files/2014/03/f13/stabler_0.pdf.
[2] C. Q. Su, W. W. Zhan, S. Shen. J. Electron. Mater, 41, 1693 (2012).
[3] D. S. Patil, R. R. Arakerimath, P. V. Walke. Renewable and Sustainable Energy Reviews, 95, 1 (2018).
[4] R. A. Poshekhonov, G. A. Arutyunyan, S. A. Pankratov, A. S. Osipkov, D. O. Onishchenko, A. I. Leontyev. Semiconductors, 51, 981 (2017).
[5] A. Leontyev, D. Onishchenko, G. Arutyunyan. Thermophys. Aeromech., 23, 747 (2016).
[6] Wang, Yiping. Journal of Elec Materi. 45, 3 (2016)
[7] R. G. D. Brennan, B. Wicksteed. Vehicle Thermal Management Systems Conf. Proc. (Coventry Technocentre, UK, 2013) p. 115.
[8] P. Fernández-Yáñez, A. Gómez, R. García-Contreras, O. Armas. J. of Cleaner Produc., 182, 1070 (2018).
[9] R. A. Poshekhonov, A. S. Osipkov, and M. O. Makeev. Global J. Pure Appl. Math. 12, 677 (2016).
[10] D.O. Onishchenko, S.A. Pankratov, A.A. Zotov, A.S. Osipkov, R.A. Poshekhonov. Int. J. Appl. Eng. Res. 12, 721 (2017).
[11] Shiriaev, P., Shishov, K., Osipkov, A., & Tishchenko. Journal of Elec Materi 48, 1998 (2019).

The conductivity and the diamagnetic response of the graphene-like carbon films in the 80 – 120 K temperature range

A A Kovalchuk[1], **A V Prikhodko**[2]
[1]Laboratory of Physico-Chemical Nanomaterial Study, Institute of Geology of Karelian Research Centre, RAS, Petrozavodsk 185910, Russia
[2]Department of Experimental Physics, Peter the Great St. Petersburg Polytechnic University, St. Petersburg 194064, Russia

Abstract. The paper presents the results of conductivity and diamagnetic response studies in natural carbon films in the temperature range 77–220 K. The behavior of the current-voltage characteristic of carbon films containing graphene-like fragments was studied by nanosecond voltammetry. It is established that when the critical temperature is reached, the sample resistance increases sharply with the simultaneous disappearance of diamagnetism. Was proposed possible mechanism of high-temperature superconductivity based on synchronization of negative-U centres presumably exists in the carbon structure under study.

1. Introduction

The practical application of the effect of superconductivity is limited due to the need for low temperatures, the existence of critical fields and critical current. For a sufficiently long period of time, the efforts of many scientists working in the field of superconductivity have been directed to the search for materials with a high critical temperature. Such superconductors were obtained in the mid-80s of the last century and were called high-temperature superconductors (HTSC). Even before the start of HTS research, an alternative model to the Bardeen-Cooper-Schrieffer theory (BCS) was proposed with a strong static binding of electrons at localized centers and their extremely weak movement between centers [16].

In the 33 years since the discovery of high-T_c superconductivity [12] there have been many models proposed [13, 15] that might explain the nature of the ground state and the anomalous properties of HTSC compounds, but none of them are still logically closed and consistent.

Recently, much attention has been paid to studies of strongly correlated states, such as ferromagnetism and superconductivity, in various forms of carbon, including graphite (see, for example, [15, 17, 18]).

The superconducting properties of natural carbon-containing compounds with phase transition temperatures from 10 to 110 K were first observed in 1993 [3]. It was assumed then that these properties manifested because of fullerides present in the compound. A number of studies published at that time (see, for example, [4]) reported on an anomaly observed for high-temperature superconductivity of C_{60}-Cu carbon structures. A 2005 study [5] discovered diamagnetism in natural carbonaceous matter (range 90–150 K) with pronounced structural anisotropy. It was found that diamagnetic properties are not associated with fullerenes or high copper concentrations in natural fullerene. It was hypothesized in [5] that the observed effect is due to the peculiar structure of natural carbon.

The goal of this study has been to find conductivity anomalies in the temperature range of 77–220 K in films of natural carbon containing graphene-like fragments.

2. Experimental details

The experimental samples were thin films based on natural graphene-like carbon. They were obtained by sublimation of the initial carbon powder in a small-sized thermal chamber [6]. Carbon powder made from type I shungite from the Shunga deposit (Karelia, Russia) [7] used as the source was prepared by the technology described in [8]. The size of carbon nanoparticles forming the thin film was 50–100 nm.

It is well-known that the specific behavior of the current-voltage characteristic and the diamagnetic response are indicators of a material's superconductivity [9]. In this study, we used two well-developed experimental procedures: recording the I–V curve and measuring the diamagnetism of thin films. Nanosecond voltammetry [10] is based on recording the incident (U_i) and reflected (U_r) voltage pulses for the sample. Pulse duration was 5 ns at voltages up to 1 V at a frequency of 100 Hz. The technique for measuring the diamagnetic response [11] was based on imbalance in the frequencies of quartz oscillators df (the generation frequency of each was f=1 MHz) due to a change in the diamagnetic properties of the sample. A ferromagnetic film magnetized in a constant magnetic field of 59 mT was used as a means of contact between the sample and the first quartz resonator

3. Results

Descending branches found at 77 K on the obtained I–V curves (Figure 1a) are one of the signs of anomalous conductivity and may indicate that the sample exhibits superconducting properties.

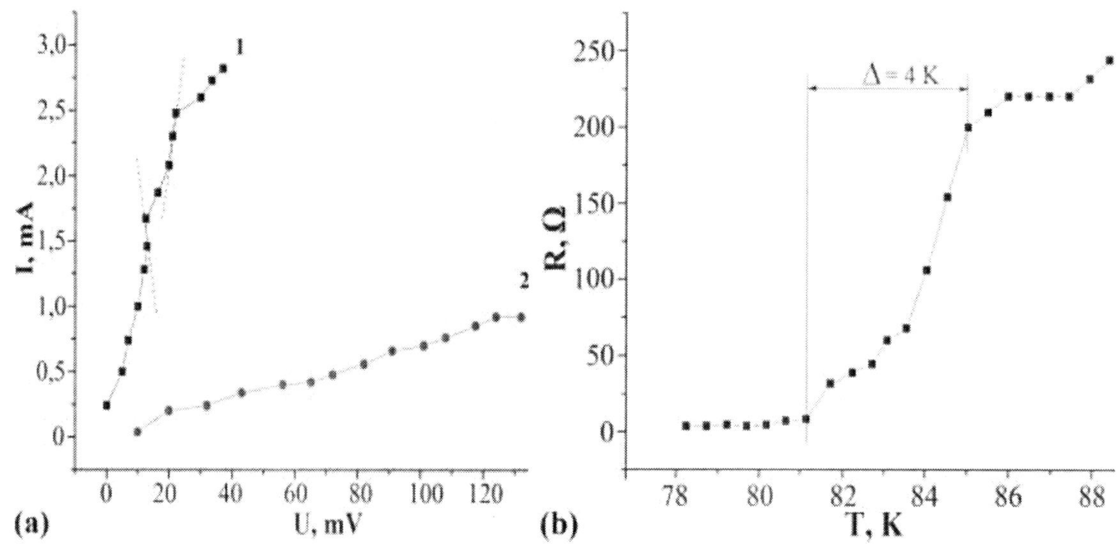

Figure 1(a, b). (a) Current-voltage characteristics for carbon film at 77 K (1) and 220 K (2); **(b)** Resistance as a function of temperature for carbon film.

Nonlinear behavior of the I–V curve in this state at 77 K may also indicate that the conductivity of the channels in the carbon film, is due, for example, to contact effects not associated with superconductivity. The descending branches are partially compensated by normal current flow in an inhomogeneous structure. As the critical temperature Tc = 81 K is reached, the sample makes a transition to a normal state. Figure 1b shows a typical curve for electrical resistance of the sample versus temperature in the range from 77 to 89 K. A region Δ = 4 K where the conductivity varies (transition width) is marked with straight lines.

Figure 2 shows the relative change in the frequency df/f of two quartz resonators as a function of temperature for carbon film. The value of Tc = 81 K is close to that obtained in [5] for shungite rocks of the Chebolaksha deposit. Our findings confirm the practically important properties of carbon film, such as the temperature-induced changes in conductivity and the presence of diamagnetism, persisting in the samples for 5 thermal cycles.

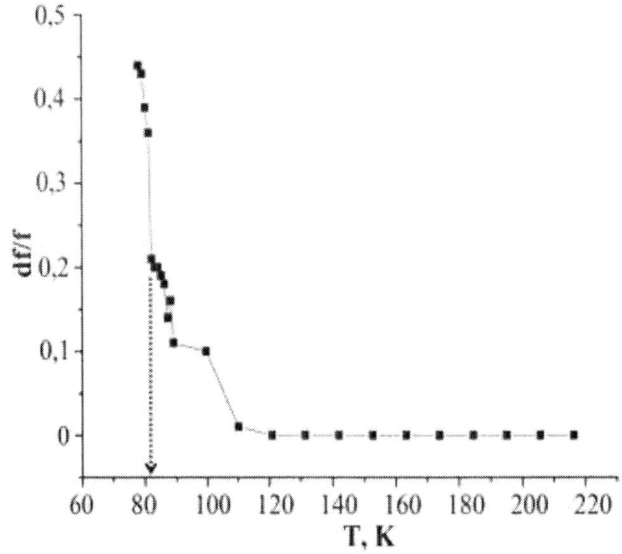

Figure 2. Temperature dependence of the relative frequency change of two quartz resonators for carbon film.

The results presented suggest that the mechanism of conductivity of thin films of natural graphene-like carbon is comparable with conductivity in high-temperature superconductors.
In this paper, we chose a superconductivity model based on the the negative-U centres model , since it can explain high-temperature superconductivity in a wide temperature range (up to 500 K). This concept was first put forward by Anderson in 1975 to describe some properties of chalcogenide glassy semiconductors [1], later developed by Mott [2]. In this model, it is assumed that there are centers in the atomic lattice of a material that have a special property. The strong electron-lattice interaction leads to the fact that the binding energy of two electrons exceeds their Coulomb repulsion energy, that is, the effective correlation energy of the pair is negative. This effect is also observed under normal conditions, that is, electrons bound in a pair already exist at room temperature.
By analogy with the BCS theory [19], the attraction of two electrons with opposite spins on one node can result in a superconducting correlation between pairs. Moreover, the first term in the Habard Hamiltonian [14] corresponds to the interaction between electrons, the second to the kinetic energy of the band motion. This value is small compared with the first, and is considered as a perturbation. The difference between the concept of negative-U centres and BCS is that all electrons participate in superconductivity, and not just near the Fermi surface.

4. Conclusion

The carbon films that we have synthesized are characterized by a conductivity anomaly, with the diamagnetic effect persisting the temperature range of 80–120 K for all thermal cycles. These practically important properties of carbon films have been obtained by using specially prepared shungite carbon powder as the initial carbon material. To further describe the mechanism of conductivity of the objects under study, the negative-U centres model was selected.

Acknowledgments
This study was supported by a grant from the UMNIK program of the Foundation for Assistance to Small Innovative Enterprises in Science and Technology, Agreement No. 11094 GU/2016.

References
[1] Kulik I O, Pedan A G 1980 *JETP* **4** 742
[2] Bednorz J G, Muller K A 1986 *Zeitschrift für Physik* **64** 189
[3] Loktev V M 1996 *Journal of Low Temperature Physics* **1** 3
[4] Esquinazi P, Heikkil T T, Lysogorskiy Y V, Tayurskii D A, Volovik G E 2014 *JETP Letters* **100** 336
[5] Felner I, Kopelevich Y 2009 *Physical Review* **79** 23
[6] Larkins G, Vlasov Y, Holland K 2016 *Superconductor Science and Technology* **29** 1501
[7] Masterov V F, Chudnovsky F A, Zaidenberg A Z, Kozyrev S V, Rozhkova N N, Podosenova N G, Stefanovich G B 1993 *Superrconductivity* **6** 1848
[8] Masterov V F, Prichodko A V, Konkov O I 1998 *Fullerene Science and Technology* **6** 481
[9] Kovalevski V V, Prikhodko A V, Buseck P R 2005 *Carbon* **43** 401
[10] Konkov O I, Mikhaylina A A, Prikhodko A V, Rozhkova N N 2016 *Journal of Optical Technology* **83** 286
[11] Rozhkova N N 2011 *Nanocarbon of schungites* (Petrozavodsk: Karelian Research Center) p 100
[12] Chou N H, Pierce N, Lei Y, Lopez N P, Fujisawa K, Subramanian S, Robinson J A, Chen G, Omichi K, Rozhkov S S, Rozhkova N N, Maldonado M T, Harutyunyan A R 2018 *Carbon* **130** 105
[13] Arutyunov K Yu 2013 *Physics, Chemistry and Applications of Nanostructures* 191
[14] Jantsch W, Heinrich H 1970 *Review of Scientific Instruments* **41** 228
[15] Prikhodko A V, Konkov O I 2001 *Semiconductors* **35** 659
[16] Anderson P W 1975 *Physical Review* **109** 1492
[17] Street R A, Mott N F 1975 *Physical Review Letters* **35** 1293
[18] Bardeen J B, Cooper I, Schrieffer J 1957 *Physical Review* **108** 1175
[19] Tsendin K D 2005 *Journal of Optoelectronics and Advanced Materials* **4** 1941

Optimization of the process of obtaining superconducting bilayer films based on niobium nitride and gold for the manufacture of a superconducting single-photon detector

M Davydchenko[1]
[1]National Research University Higher School of Economics, Moscow 101000, Russia

Abstract. We have manufactured a number of bilayer structures based on niobium nitride and gold with different deposition parameters for NbN-Au films (Table 1). We measured the characteristics of superconducting niobium nitride films both with the top layer of gold and without gold. On the basis of the measured characteristics of the films, the optimal parameters of the deposition of NbN-Au bilayer films were chosen for the further formation of a superconducting single-photon detector.

1 Introduction

To date, a superconducting single-photon detector are highly efficient detectors and promising for various research areas [3]. They have as wide scientific applications as well as commercial [4]. Optimizing the operation of detectors and improving the efficiency of their detection is an important task. The studies of the detectors allow us to study both the detection mechanisms and make assumptions that will improve and increase the efficiency of detection and improve other characteristics of the detector as a whole.

To implement many practical tasks in the field of superconducting nanotechnology it is necessary to manufacture superconducting films with characteristics that satisfy the conditions of homogeneity while maintaining the necessary superconducting parameters, such as the critical transition temperature (Tc), transition width (ΔTc), surface resistance at room temperature (Rs), metallicity coefficient (K) and critical current (Ic). Optimization of these characteristics is possible changing the parameters of the processes for producing superconducting films [2].

D. Yu. Vodolazov and others, in their work [1], suggest that adding a thin layer of normal metal to a superconductor significantly increases its superconducting properties due to the proximity effect, which will favorably affect the manufacture of two-layer superconducting structures and will increase their detection efficiency.

In our work, we try to verify the validity of the author's assumptions [1]. At the same time, we want to optimize the manufacturing parameters of superconducting structures based on NbN-Au bilayer.

2 Fabrication

We have manufactured ultrathin superconducting bilayer films based on niobium nitride (NbN) and gold (Au) with different thicknesses. The structures were made on two types of substrates: monocrystalline silicon (Si) 500 um thick and sapphire one-sided polishing (Sp) 300 um thick. This fabrication was carried out under different deposition conditions for films (NbN) and (Au) by magnetron reactive sputtering at direct current in stabilization mode using the LEYBOLD HERAEUS Z400 unit (Table 1).

The characteristics of manufactured films such as the critical transition temperature to the superconducting state (Tc), the transition width (ΔTc), surface resistance at room temperature (Rs) and the metallicity coefficient (k) of films (NbN-Au) bilayer and films NbN with strained a layer of gold (Au) (Table 1). These measurements will help us in further research and optimization of the

manufacturing parameters of future superconducting structures. Further optimization of ultrathin film production processes will be carried out to achieve the maximum critical current (Ic).
To study the characteristics of ultrathin films based on them, structures in the form of meanders consisting of niobium nitride and gold (NbN-Au) bilayer will be made. Ti-Au bilayer plating will be deposited to form contact pads.

3 Experimental results and obtained data

Table-1. The parameters of the deposition of superconducting bilayer films (NbN-Au) by magnetron reactive sputtering at direct current in stabilization mode using the LEYBOLD HERAEUS Z400 unit.

№	Sub.	Film	T,°C	T, sec	Rs, Ω/sq.	Tc,K	ΔTc,K	K
2315	Si	NbN	400	7"	464	9,94	0,35	0,76
2317	Si	NbN	400	7"	496	9,35	0,4	0,76
		Au	200	10"	9,0-12,0	9,5	0,33	1,14
2320	Si	NbN	400	7"	620	7,75	0,6	0,75
		Au	200	10"	11	8,6	0,73	2,8
2335	Sp	NbN	400	6"	550	8,38	0,73	0,6
		Au	200	10"	5,8	$_1$7,94; $_2$6,9	$_1$0,4; $_2$0,14	1,54
2336	Sp	NbN	Cold	6"	1050	5,34	0,68	0,66
		Au		10"	7,8	$_1$5,23; $_2$ -	$_1$0,71; $_2$ -	1,16
2337	Sp	NbN	400	6"	380	-	-	0,92
		Au	200	15"	3,4	-	-	1,7
2338	Sp	NbN	400	6"	860	6,87	0,6	0,63
		Au	200	15"	2,5	$_1$7,18; $_2$6,2	$_1$1,02; $_2$0,14	1,55
2339	Sp	NbN	400	6"	454	8,88	0,42	0,82
		Au	cold	10"	6,2	$_1$7,82; $_2$7,18	$_1$0,74; $_2$0,13	1,31

To characterize the fabricated structures, the dependence of the resistance of bilayer structures on temperature was measured. The measurements were carried out at helium temperatures. By the nature of the dependence, the presence (Fig.1 (b)) or the absence (Fig.1 (a)) of the proximity effect for the manufactured bilayer films was determined. Varying the film deposition parameters it possible to detect the proximity effect for some of the structures that will further allow to produce superconducting single-photon detectors based on bilayer. The proximity effect is observed when sputtering a small thickness of niobium nitride (5nm) and gold (10nm), but it was found that when sputtering gold in a working chamber heated to 200 ° C, defects were formed on the gold (Fig.2 (a)). But for films with a deposited layer of gold without heating the working chamber, a poor morphology was not detected, but there was also no proximity effect (Fig. 2 (b)).

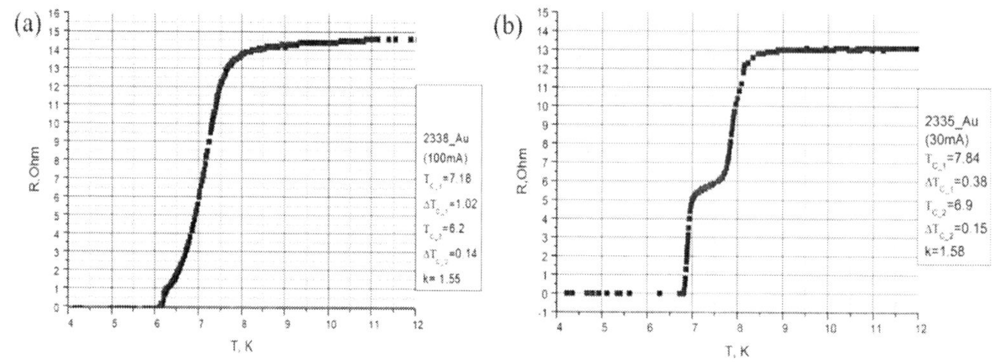

Figure 1. (a, b). a) Dependence of resistance on temperature for NbN-Au films (Sp substrate) without proximity effect; **b)** Dependence of resistance on temperature for NbN-Au films (Sp substrate) with proximity effect.

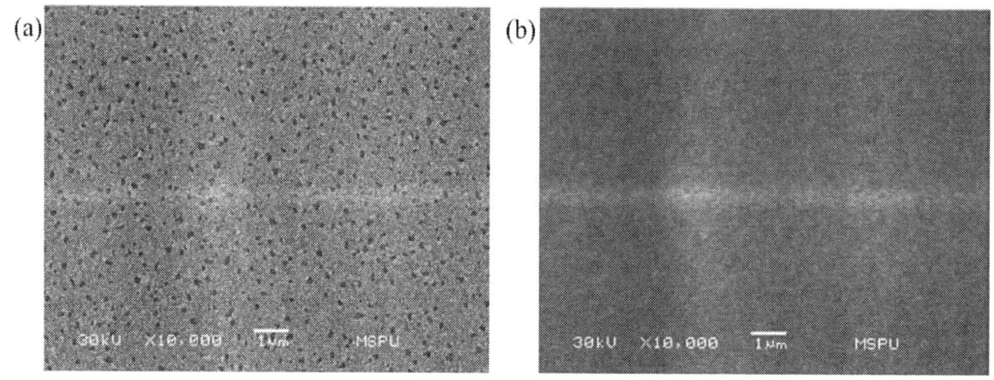

Figure 2. (a, b). a) SEM image of the surface area of the NbN-Au film (Sp substrate) with defects; **b)** SEM image of the surface area of the NbN-Au film (Sp substrate) without defects.

4 Conclusion

A series of bilayer structures based on niobium nitride and gold were fabricated with different deposition parameters for NbN-Au films and the characteristics of superconducting films of niobium nitride with both the top layer of gold and without gold were measured. For some fabricated structures, a proximity effect was discovered. In the future, these structures will be studied in order to experimentally confirm the assumptions of the authors of the article [1]. Also, further optimization of the processes for obtaining ultrathin films will be carried out to achieve the maximum critical current (Ic) and to manufacture of superconducting single-photon detectors based on bilayer.

Acknowledgments

M Davydchenko acknowledge support by the Russian Science Foundation (Of the Agreement No. 17-72-30036 dated 02.08.2017).

References

[1] D. Yu. Vodolazov, E. E. Pestov, S. N. Vdovichev, S. S. Ustavshikov, and A. Yu. Aladyshkin "Peculiarities of superconducting properties of thin superconductor-normal metal bilayer with large

ratio of resistivities" Institute for Physics of Microstructures, Russian Academy of Sciences, 603950, Nizhny Novgorod, GSP-105, Russia

[2] Bacon D.D., English A. T., Nakahara S., Peters F. G., Schreiber H., Sinclair W. R., and R. B. van Dover / Properties of NbN thin films deposited on ambient temperature substrates // J. Appl. Phys. . – 1983 – Vol 54

[3] Konstantin Smirnov, Yury Vachtomin, Alexander Divochiy, Andrey Antipov, and Gregory Goltsman "Dependence of dark count rates in superconducting single photon detectors on the filtering effect of standard single mode optical fibers" // 2015 Appl. Phys. Express 8 022501

[4] http://www.scontel.net/

Detailed Monte-Carlo simulation of PMMA chain scissions in e-beam lithography

F A Sidorov[1,2], A E Rogozhin[1]
[1]Laboratory of microstructuring and submicron devices, Valiev Institute of Physics and Technology of RAS, Moscow 117218, Russia
[2]Moscow Institute of Physics and technology, Dolgoprudny, 141701, Russia

Abstract. Detailed Monte-Carlo simulation of e-beam events in PMMA/Si system has been carried out. Polymer chains are modeled using random walk algorithm with subsequent assignment of simulated e-beam events to certain monomers. The processes leading to PMMA chain scission are determined to be ionization events on C atoms of PMMA backbone bond due to analysis of resist weight distributions, obtained by gel permeation chromatography.

1. Introduction

The Monte-Carlo is a widespread technique for the study of e-beam interaction with solid in single and multi-layered samples of sub-micron size. In the common approach, the spatial contours developed in the polymer resist film are determined by two separate processes: (1) electron scattering and energy deposition within the film and (2) differential solubility of the electron-irradiated volume with an organic solvent [1]. Resist solubility rate could be determined from resist number average weight after exposure, which is originally determined by the characteristics of inelastic e-beam events leading to resist molecule scissions. Resist average weight is usually calculated using formula [2]

$$M_f = M_n \left(1 + \frac{g \varepsilon M_n}{\rho N_A} \right)^{-1}, \quad (1)$$

where M_n and M_f are resist number average weights before and after exposure, correspondingly, ε is absorbed energy density, ρ is resist density.

The parameter g, determined as number of resist chain scissions per 100 eV of deposited energy, defines the contribution of all inelastic e-beam events in resist to the resist chain scissions. The most widely used g value is 1.9, but values in range 0.75-4.3 were also obtained [1, 3], which may lead to the inaccuracy of this approach. Besides, energy loss-based model of resist chain scissions and development could be hardly used for modern simulation purposes such as line edge roughness study due to complexity of application one-dimensional solubility rate to three-dimensional structures.

More comprehensive approach for the simulation of e-beam lithography profiles for sub-μm dimensions is based on the diffusion-like model for resist dissolution process [4]. It provides the dependence of resist solubility on local resist chain length, which requires the distribution of resist chain scission events caused by e-beam exposure. Several algorithms for the simulation of resist chain scission have been already presented [3, 5], but they characterize scission rate using general theoretical assumptions. In our study, resist chain scission rate is for the first time determined directly from experimental resist chain weight distributions. This approach is based on Direct Monte-Carlo simulation of e-beam interaction events in resist and analysis of gel permeation chromatography data.

2. Simulation of e-beam events in resist

In our research, Direct Monte-Carlo algorithm with discrete loss approximation is applied for simulation of 20 keV electron trajectories in poly- methyl methacrylate (PMMA) layer on a silicon wafer (Fig. 1(a)). Dominant processes in energy range of our interest are elastic scattering, excitation, ionization and secondary electron generation [6] (Fig. 1(b)). During the simulation, for an electron in PMMA layer, the type of atom which provides scattering potential is determined first. Then, the cross sections of elastic scattering, excitation and ionization events are calculated and which allows to determine the process type, mean free path and next interaction position. The characteristics of elastic processes are defined by Mott differential cross sections [7], while cross sections and energy loss for inelastic processes are determined using Seltzer's model of close and distant collisions [8] and Moller's binary collision cross sections [9]. Secondary electrons originated from ionization events are then tracked using the same scheme.

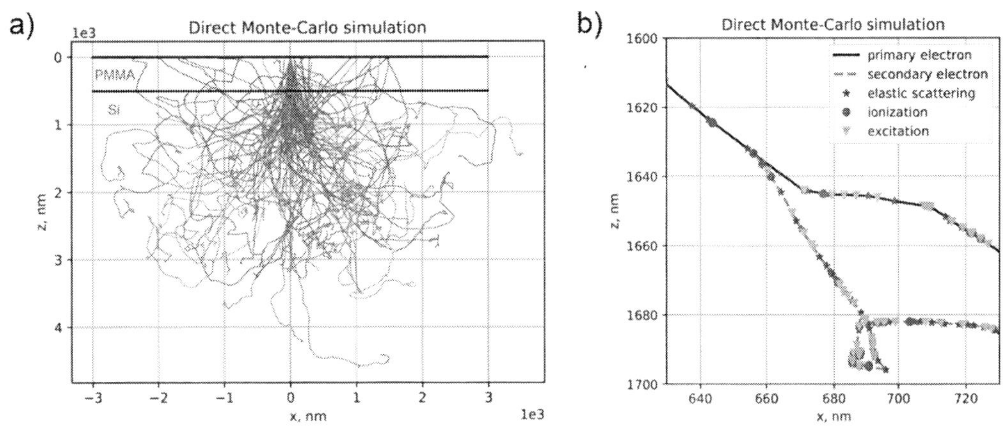

Figure 1(a, b). (a) Direct Monte-Carlo simulation results for 100 electron tracks with initial energy 20 keV in PMMA/Si system with 500 nm PMMA layer; **(b)** primary and secondary electron tracks obtained using Direct Monte-Carlo algorithm.

PMMA chains are modeled using ideal chain model based on random walk simulation with 0.28 pitch and 109° between successive monomers [5] (Fig. 2(a)). PMMA chains are combined in the required volume until needed PMMA density (1.19 g/cm^3) is achieved (Fig. 2(b)). Electron events and PMMA monomers are saved in histograms with 2x2x2 nm^3 cells. This approach provides relaxed PMMA state and allows us to match each electron interaction event and monomer its position in certain chain.

3. PMMA chain scission simulation

Our study of PMMA chain scission events caused by e-beam exposure is based on the analysis of PMMA weight distributions before and after exposure obtained by gel permeation chromatography (GPC) [10]. We simulate e-beam interaction events in 100x100x500 nm^3 PMMA volume with the parameters, corresponding to experimental ones: electron initial energy and dose are 10 keV and 10^{-4} C/cm^2, correspondingly, initial PMMA number average weight is 5.63×10^5, PMMA layer thickness is 500 nm. Then different PMMA chain scission mechanisms based on general assumption were considered. PMMA scission events are presumably caused by inelastic e-beam events on carbon atoms in methyl methacrylate (MMA) molecule that form C-C backbone bond in PMMA (Fig. 3), so we investigated the influence on final PMMA weight distribution, caused by various combinations of excitation and ionization processes on several carbon atoms in MMA molecule (Fig. 4). The simulated mass distribution corresponds to the experimental one for the scission model consisted in the scission events due to ionization e-beam events on the equivalent number of carbon atoms in MMA from 2 to 2.5. These parameters refer to room temperature and they could be varied for the simulation of resist chain scissions for the higher exposure temperatures. PMMA chain scission simulation results for the exposure in series of parallel lines are presented in Fig. 5.

The required scission rate could be also achieved for the model considering both excitation ionization events with equivalent carbon atoms number less than 2, or excitation events with equivalent carbon atoms number higher than 3. These models have weaker physical interpretation, but they still result in some freedom in PMMA scission model choice.

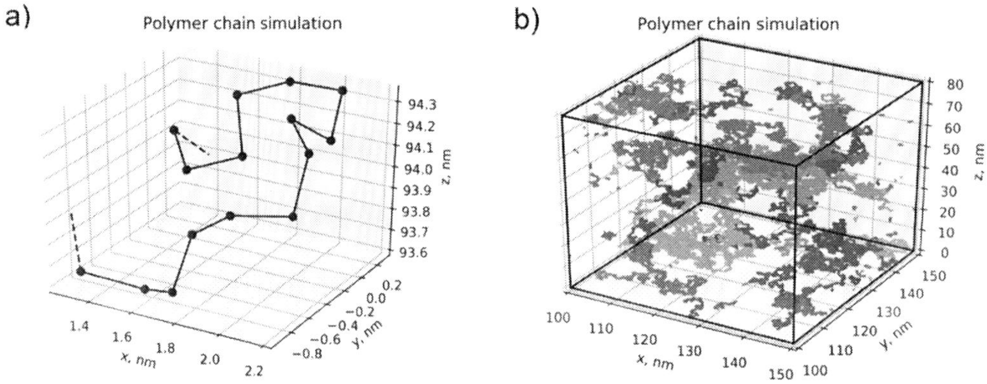

Figure 2(a, b). (a) PMMA chain obtained using ideal chain model; (b) combination of PMMA chains for the simulation of PMMA layer (some chains are omitted for visibility).

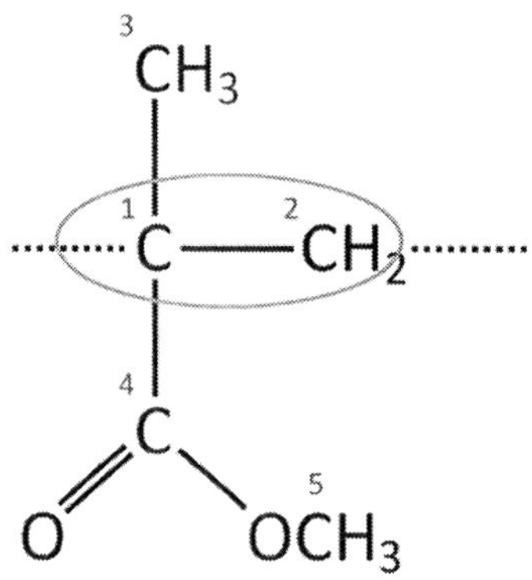

Figure 3. Methyl methacrylate (MMA) monomer structure, C1 and C2 atoms form C-C backbone bond in PMMA.

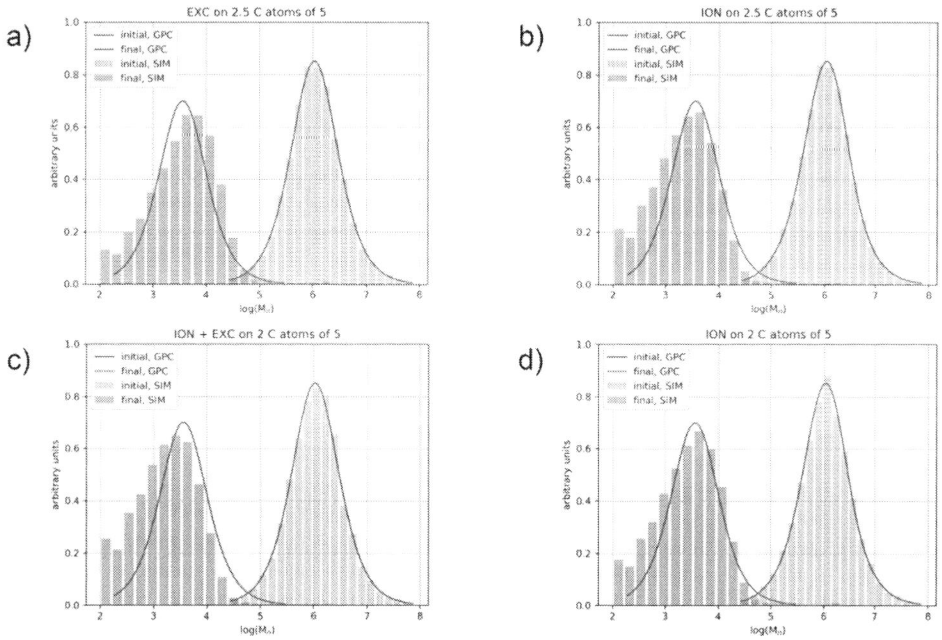

Figure 4(a, b, c, d). Comparison of experimental PMMA chain weight distribution before exposure (initial) and after exposure (final) obtained by gel permeation chromatography (label "GPC") with simulated distributions (label "SIM") for several possible PMMA chain scission mechanisms.

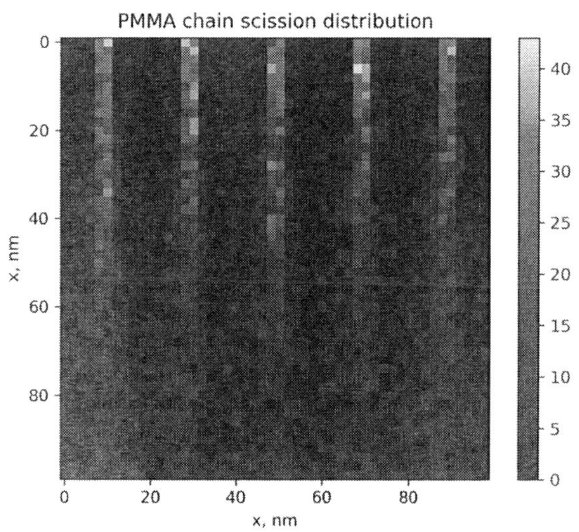

Figure 5. Simulated PMMA chain scission events distribution for the exposure in series of parallel lines. Initial electron energy is 20 keV, exposure dose is 100 $\mu C/cm^2$, PMMA layer thickness is 100 nm.

4. Conclusion

Simulation algorithm of e-beam interaction events in resist and PMMA chain scission events is presented. It provides the most detailed information on processes in resist during exposure and avoids usage of g value for resist chain scission simulation. The processes leading to PMMA chain scission at are determined from experimental GPC mass distribution for the first time. Scission events are supposed to occur due to ionization events on carbon atoms in MMA molecule with equivalent number from 2 to 2.5. The distribution of PMMA chain scission events in PMMA layer could be used as a starting point for the subsequent simulation of resist development using diffusive resist development model or simulation of temperature effects in resist such as thermal depolymerization or monomer diffusion. However, the deeper insight into polymer chemistry is desirable for the clarification of scission mechanism.

Acknowledgments

The authors would like to thank the Russian Foundation for Basic Research (18-37-00472) for support of this project. The investigation was also supported by the Program no. 0066-2019-0004 of the Ministry of Science and Higher Education of Russia for Valiev Institute of Physics and Technology of RAS.

References

[1] Kyser D F, Viswanathan N S 1975 *J. Vac. Sci. Techol.* **19** 1305
[2] Greeneich J, Van Duzer T 1974 *IEEE Trans. On El. Dev.* **ED-21** 5
[3] Aktary M, Stepanova M, Dew S 2006 *J. Vac. Sci. Technol. B* **24** 768
[4] Mohammad M A, Fito T, Chen J et al. 2010 *J. Vac. Sci. Technol. B* **28** L1
[5] Han G, Khan M, Cerrina F 2003 *J. Vac. Sci. Techol. B* **21** 3166
[6] Cullen D, Hubbell J, Kissel K 1997 *Lawrence Livermore National Laboratory, UCRL-50400* **6** 5
[7] Czyzewski Z, MacCallum O, Romig D A, Joy D 1990 *J. Appl. Phys.* **68** 3066
[8] Perkins S T, Cullen D E, Seltzer S M 1991 *Lawrence Livermore National Lab.* **31** UCRL-50400
[9] Jenkins T M and Chris A and Nelson W R 1987 *Monte Carlo Transport of Electrons and Photons* (Plenum Press, New York and London)
[10] Harris R A 1973 *J. Electrochem. Soc.: Solid-State Sci. and Technol.* **120** 270

Strength and structural properties of AlN films grown on SiC/Si substrates synthesized by atomic substitution

A.S. Grashchenko[1], S.A. Kukushkin[2] and A.V. Osipov[1]
[1] Institute for Problems in Mechanical Engineering RAS, Saint Petersburg, 199178, Russia
[2] ITMO University, Saint Petersburg, 197101, Russia

Abstract. In the present work, we studied the strength and structural characteristics of the layers of the AlN/SiC/Si heterostructure. The surface morphology of the AlN film and the SiC/Si substrate was studied using atomic force microscopy. The thickness of the AlN and SiC layers was determined by analyzing the data of ellipsometry. The hardness of the films and the substrate was measured by the method of nanoscratch testing. It was experimentally shown that the films under study have a high crystalline quality.

1. Introduction

In the papers [1,2], a unique method was proposed to produce a new type of silicon Si substrates with a single crystal layer of silicon carbide SiC. This method allows nanoscale assembling of the SiC crystal inside the Si matrix simultaneously over the entire area of the substrate. This process occurs due to the chemical reaction of a Si crystal with carbon monoxide CO:

$$2Si(cr) + CO(gas) = SiC(cr) + SiO(gas)\uparrow \quad (1)$$

In this chemical reaction, a part of the Si atoms is substituted by carbon atoms C, and a nanoscale SiC layer is formed, as a result. Moreover, the SiC films grown in this way are practically unstressed, since the stress in the film relaxes because of a microporous structure that forms under the film during the synthesis [3]. With that, SiC films obtained by the method of substitution of atoms are low-defect, which makes them ideal substrates to grow wide-bandgap semiconductors, such as aluminium nitride AlN and gallium nitride GaN. In this work, we carried out a comprehensive study of structural and strength properties of the hybrid SiC/Si substrate and an AlN film grown on its basis.

2. Experimental methods

The substrate for the growth of the AlN film was synthesized by the method of substitution of atoms inside a silicon wafer of the KDB mark with the (111) crystallographic orientation without deviation from the base plane. The synthesis lasted for 20 minutes at the CO pressure of 0.6 Torr and a temperature of 1250 °C. Then the resulted substrate was cut into two parts. One of the parts was left in its original state, while on the other one, the AlN film was grown by the method of chloride-hydride vapor phase epitaxy (HVPE), according to the procedure described in [4]. Further, both samples were investigated identically. The structural properties of the surface of the SiC and AlN films were explored by atomic force microscopy using the Nanosurf EasyScan microscope. The surface morphology of the films was measured in the semicontact mode. The thickness and composition of the

SiC and AlN films were analyzed by spectral ellipsometry using the Woollam M-2000D ellipsometer. The strength properties were studied by nanosclerometry on the MicroMaterials NanoTest 600 hardness tester. The nanosclerometry was carried out in a triple scan mode with a conical indenter with a radius of 25 μm. In this mode, the experiment follows three stages. At the first stage, the surface is scanned at the site of the future scratch, this is done to eliminate the influence of the natural inclination of the surface of a sample on the calculations. Then the scratching itself takes place, and, at the third stage, the residual deformation is scanned. The scanning of the surface topography and slope of the sample before and after scratching was carried out with a force of 0.1 mN. As a result of the analysis of the nanosclerometry data, the relative hardness was measured, which is determined as follows:

$$H_s = H_{ref}\frac{F_s}{F_{ref}}\left(\frac{W_{ref}}{W_s}\right)^2 \quad (2)$$

where *s* and *ref* are indices related to the test and reference materials, respectively; *H*, *F*, and *W* are the hardness, applied force, and scratch width, respectively. The nanosclerometry method allows to estimate the strength properties of brittle materials, which is very important in the case of a microporous substrate.

3. Results

The Fig. 1a,b shows images of a surface of SiC and AlN films with an area of 10 μm2, obtained by the atomic force microscope. It is clearly seen that the surface structure of both films is uniform. Morphological analysis of these images shows that the roughness is 22 and 25 nm for the SiC and AlN films, respectively.

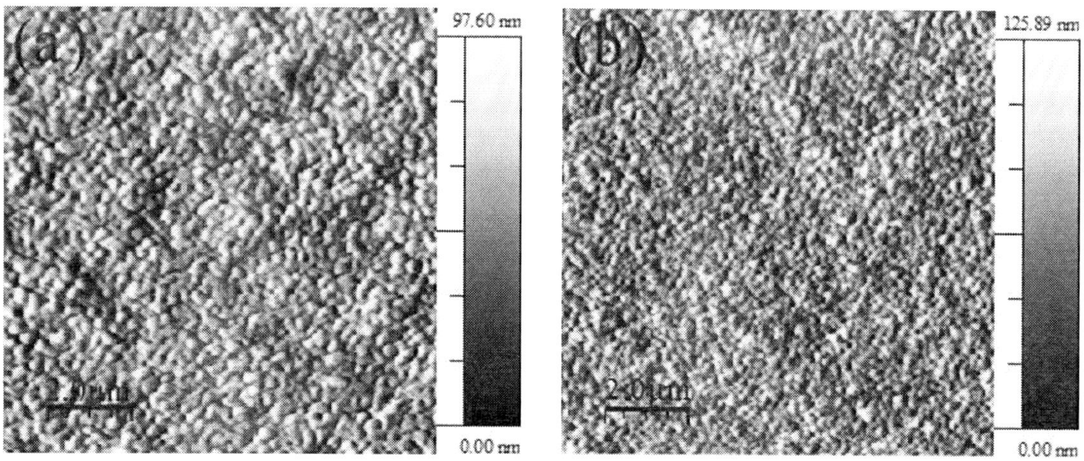

Figure 1(a, b). AFM images of surfaces. **(a)** SiC/Si substrate and **(b)** AlN film.

The Fig. 2a,b shows the ellipsometric spectra, i.e. the dependences of the real and imaginary parts of the dielectric constant on the photon energy. The solid line is the real part of the dielectric constant, the dotted line is the imaginary part of the dielectric constant. The spectrum of the SiC film on Si is well described by a two-layer theoretical model, while a single-layer ellipsometric model can't make the theoretical and experimental curves fit each other. In our case, a layer of pure SiC with a thickness of 135 nm is separated from the layer of Si with voids by a layer of a mixture of SiC with Si having a thickness of 45 nm. Considering the AlN film, the analysis of the ellipsometry data shows that the thickness of the AlN film is 610 nm, it is single-layer and contains 1.5% excess of aluminium.

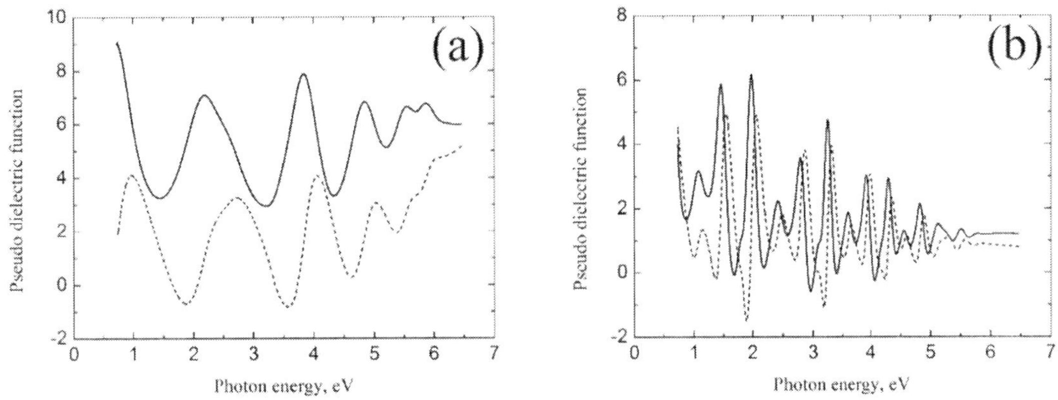

Figure 2(a, b). Ellipsometric spectra. **(a)** SiC/Si substrate and **(b)** AlN/SiC/Si heterostructure. The solid line is the real part, the dotted line is the imaginary part of dielectric permittivity.

The Fig. 3a shows the dependences of the applied normal load on the immersion depth during scratching, taking into account the natural inclination of the samples. The curves 1, 2, 3 correspond to the Si substrate, hybrid SiC/Si substrate, and AlN film.

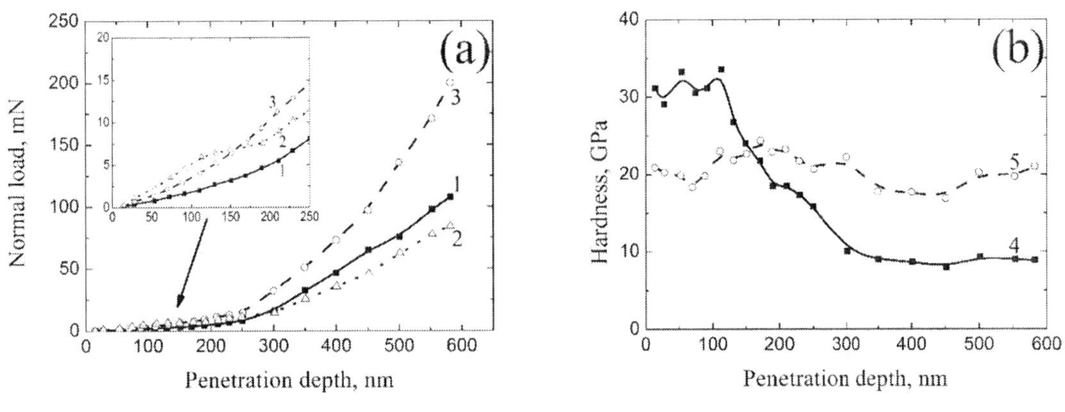

Figure 3(a, b). **(a)** Dependence of the applied normal load on the immersion depth during scratching (a) curves 1, 2, 3 correspond to the Si substrate, the SiC/Si substrate and the AlN film; **(b)** dependencies of hardness on immersion depth, curves 4, 5 correspond to the SiC/Si substrate and the AlN film.

The Fig. 3b shows the dependences of the relative hardness on the immersion depth for the hybrid SiC/Si substrate (curve 4) and AlN film (curve 5) calculated relative to the hardness of the Si substrate, which is 11.3 GPa according to the nanoindentation data. In the case of the SiC film grown on Si by the method of atomic substitution, it can be seen that at an immersion depth less than 130 nm, the hardness is almost constant and equal to 30.8 ± 2.3 GPa, which is only 7% lower than the hardness of bulk silicon carbide 33 GPa [5]. Then the hardness smoothly decreases to the value of 8.7 GPa, which is 22% less than the hardness of the calibration silicon substrate 11.3 GPa. In the case of the AlN film grown on the hybrid SiC/Si substrate, the hardness in the entire depth range is 20.5 ± 3.5

GPa. Note that, in contrast to nanoindentation [6, 7], the method of nanosclerometry made it possible to explore the strength characteristics of the film formed on the porous substrate over almost the entire thickness of the film.

4. Conclusions

The structural characteristics of an AlN film grown by the HVPE method on a hybrid SiC/Si substrate synthesized by the method of substitution of atoms have been studied. Atomic force microscopy data show that the surface structure of the AlN film and the SiC buffer layer is uniform, and their surface roughness is 22 and 25 nm, respectively. By the method of ellipsometry, it has been determined that the thickness of the AlN film is 610 nm, it is single-layer and contains 1.5% excess of aluminium. The relative hardness of the AlN film has been measured, its value is constant over the entire depth of the film and equal to 20.5 ± 3.5 GPa. Thus, it can be concluded that the AlN film grown on the SiC/Si hybrid substrate is homogeneous in its structural and strength properties, and stoichiometry of its chemical composition is close to ideal.

References

[1] Kukushkin S. A., Osipov A. V. Physics of the Solid State 2008, 50, 1238.
[2] Kukushkin S. A., Osipov A. V. J. Appl. Phys. 2013, 113, 4909.
[3] Redkov A. V. et al. Physics of the Solid State 2019, 61, 299.
[4] Koukitu A. et.al. J.J.Appl.Phys. 1998, 37, 762.
[5] Quay R. Gallium Nitride Electronics. Berlin: Springer-Verlag, 2008, 6 p.
[6] Grashchenko A.S., Kukushkin S.A., Osipov A.V. Technical Physics Letters, 2014, 40, 1114.
[7] Grashchenko A.S., Kukushkin S.A., Osipov A. V., Redkov A.V. Journal of Physics and Chemistry of Solid, 2017, 102,151.

Effect of ion irradiation on resistive switching in metal-oxide memristive nanostructures

A N Mikhaylov, A I Belov, D S Korolev, S A Gerasimova, I N Antonov, E V Okulich, R A Shuiskiy and D I Tetelbaum

Lobachevsky University, 23/3 Gagarin prospect, Nizhny Novgorod, 603950, Russia

E-mail: gerasimova@neuro.nnov.ru

Abstract. The development of artificial intelligence systems is needed to solve many important challenges in neurobiology and neuroengineering for recreation of brain functions and efficient biorobotics. Here we propose a metal-oxide memristive device compatible with CMOS technology and suitable for hardware implementation of neuromorphic tasks. However, metal-oxide memristors have a significant drawback such as variation of resistive switching parameters due to the stochastic nature of filament formation in oxide material. In this work, we control the filament formation process by irradiation of oxide film surface with heavy ions. We have shown that the irradiation of oxide surface in the Au/SiO_2/TiN memristive device with Xe^+ ions (at energy of 5 keV) decreases the fluctuations of electroforming voltage, current-voltage characteristics during switching and increases the current ratio in the high/low resistance states. The ability to control internal parameters of the memristor can allow more efficiently using memristor as an element of neural networks and other neuromorphic circuits.

1. Introduction

Coupling memristors in the form of resistive switching devices with neurons and design of neuromorphic systems based on them is one of the most promising areas of interdisciplinary research. Research on artificial brain strategies construction has progressed exponentially over the last few decades. It should be noted that the development of neuromorphic electronic systems brought together such disciplines as biology, medicine, computer science, and robotics. Such systems are of special interest for researchers working in the fields of fundamental nonlinear dynamics of complex and multistable systems [1, 2], intellectual adaptive automatic control systems, biorobots and medical applications in devices for monitoring and stimulating the brain activity in the framework of neuroprosthetic tasks [3-5]. In this context, memristive devices have been an object of intensive research in recent years including neuromorphic applications [6-9].

Memristor or "memory resistor" works as a resistor and, at the same time, has a memory effect. As an element of electrical circuits, memristor can be realized in the form of a resistive switching device (RRAM – Resistive Random Access Memory) capable of changing the resistance depending on the magnitude of electric field and / or current [10]. In the case of inorganic solid state memristive device, the resistance changes due to a reversible transformation of atomic structure in the nanometric region of dielectric (oxide) film deposited between two conductive electrodes [11]. The resistive state changed under a particular applied voltage can be stored in a nonvolatile manner, which allows the usage of memristive nanostructures in the RRAM [12] and "logic-in-memory" devices [13]. The metal-oxide-metal capacitor-type nanostructures (called metal-oxide nanostructures) are formed on a standard silicon substrate, compatible with CMOS technology for designing mixed analog-digital electronic circuits. The local processes responsible for resistive switching provide a high degree of miniaturization, speed and low energy consumption. The ability of memristive device to change the conductivity depending on the electrical stimulation makes it an electronic analog of the synapse – the connection between neurons determining the coupling strengths [14-16].

Nowadays, significant progress has been achieved in using memristors to mimic the properties and functions of biological synapse. Nevertheless, the hardware implementation of brain-like electronic systems based on memristive devices is still at the early stage of development. The widespread application of RRAM memory devices is constrained by the fact that, along with the great advantages of this memory type, it has a significant drawback – a large variation of parameters and low reproducibility. This is due to the peculiarities of the structure and functioning of resistive switching devices. The fact is that the metal-oxide-metal memristive device usually acquires the necessary property of switching from the high-resistance to the low-resistance state and back only after the so-called electroforming, when thin conductive paths (filaments) are formed inside the oxide layer depending on the external voltage of a certain amplitude and polarity. This process is random (stochastic): filaments are formed from nuclei that arise in localized areas of strong electric field at the interface between dielectric and electrode, as a rule. In this case, the roughness of interface, the grain structure of the oxide film, and the fluctuations of its composition play important roles. As a result, the number, shape, and composition of the filaments strongly depend on a number of random factors. Moreover, during the memristor operation, which requires numerous switching cycles, the structure of filaments ensemble may change, which leads to the change or even complete degradation of switching characteristics.

The idea of this work is to control the process of filaments growth by irradiation of the oxide surface (which serves as the metal-oxide interface after the electrode layer deposition) by heavy ions (method of ion implantation). A heavy ion in its motion in a solid displaces atoms from lattice sites and thereby creates a "cloud" of defects – vacancies and interstitial atoms – the so-called displacement cascade [17]. Due to the high concentration of defects, the displacement cascades located in the oxide near the metal electrode have a high conductivity (they are practically equipotential), and it should be expected that some of them will be electrical field concentrators during the electroforming process. The ion irradiation permits to control well the average concentration of defects in cascades, the average length of cascades in the direction of ion beam and in lateral direction, as well as the average distance between cascades. These parameters are strictly determined by the type of ions, their energy and fluence. It can be expected that the role of the above mentioned uncontrolled factors affecting the result of electroforming (filament growth) would decrease and, consequently, the reproducibility of switching parameters would increase.

In this paper, the suggested idea is demonstrated for the Au/SiO_2/TiN memristive structures, in which the SiO_2 thin film is a dielectric layer, and layers of Au and TiN are the top and bottom electrodes, respectively. It should be noted [18] that the SiO_2 thin film as a switching dielectric corresponds well to the condition of compatibility of memristors with existing silicon microelectronics technology, in which SiO_2 is a standard component of MOSFET transistors.

2. Experimental

The Au/SiO_2/TiN memristive structures were fabricated on a thermally oxidized silicon substrate with a magnetron-sputtered conductive TiN layer. Silicon dioxide films (40 nm thick) were deposited by magnetron sputtering of quartz. Ion irradiation was carried out after the deposition of SiO_2 layer. After irradiation, the top Au electrodes with area of $1.2\cdot10^{-3}$ cm^{-2} and thickness of 40 nm were deposited through a mask. Ion irradiation was carried out by using the ILU-200 ion implanter. The Xe^+ ions were chosen for irradiation. The choice of this type of ions has resulted from the fact that their large mass allows creating displacement cascades with a high density of point defects [17]. It is necessary to ensure the high conductivity of defective "clouds" so that they can play the role of filament nuclei. In addition, the Xe^+ ions do not reveal chemical activity, which could have an additional effect.

One of the requirements for the choice of irradiation conditions arising from the above idea is that the ion range must be substantially lower than the oxide film thickness; otherwise, the cascades serving as areas with high conductivity will short out the dielectric. Another requirement is that the cascades should not overlap with each other, which imposes a limitation on the irradiation fluence.

The energy of Xe^+ ions was chosen to be 5 keV in connection with these requirements. (Irradiation with energies lower than 5 keV on the ILU-200 is technically difficult). To select the optimal fluence range, we calculated the spatial distribution of vacancies in atomic displacement cascades using the SRIM software [19]. These calculations allowed us to determine the irradiation fluence, at which the cascades are distributed on defined distances from each other. It was established that the maximum lateral radius of the cascade in SiO_2 is 4.6 nm, according to a 100-fold decrease in the average concentration of oxygen vacancies within it. The maximum allowable fluence ($9 \cdot 10^{12}$ cm^{-2}) was defined from the condition that the average distance between the cascades is equal to the lateral diameter of the cascade. At the specified energy, ions generate displacement cascades in the near-surface SiO_2 layer with a depth of up to 9 nm. The position of the SRIM-simulated displacement cascades in the SiO_2 layer of the memristive structure is shown in Figure 1.

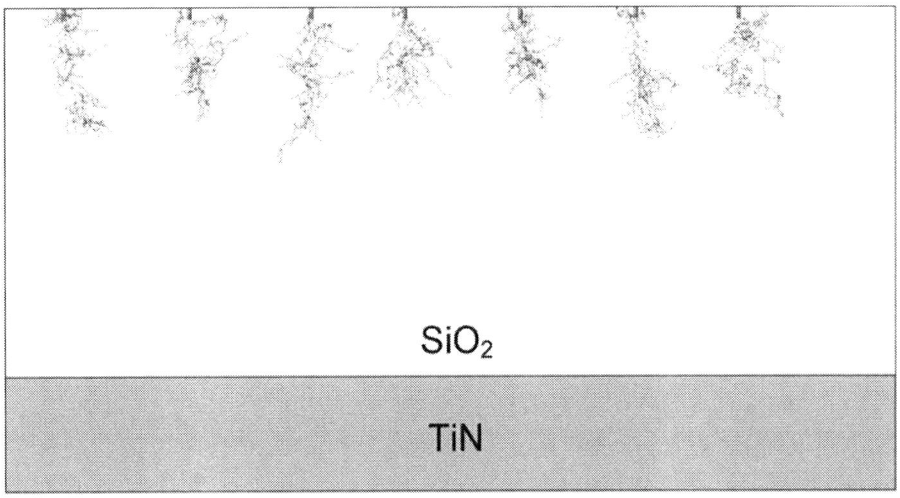

Figure 1. Schematic cross-section of a switching layer with displacement cascades used for the filament nucleation. Cascades were simulated by the SRIM code for Xe^+ ions with energy of 5 keV.

The Agilent B1500A semiconductor device analyzer was used to measure current-voltage characteristics. The electroforming was performed in the first voltage sweep from zero to some negative values on top electrode (– 5 V), as a result of which the device switches to the low resistance state (LRS). During the subsequent positive voltage sweep, when its value achieves a certain value, the device switches to a high resistance state (HRS). Then, when a negative voltage sweep is applied, a switching occurs to the HRS. These switching is repeated cyclically. We are interested primarily in the variations of voltage required for electroforming, as well as the ratio of resistances in LRS and HRS.

3. Results and discussion

Figure 2 shows the current-voltage characteristics of the reference memristive devices and the devices modified by Xe^+ ion irradiation for several measurement cycles carried out after electroforming. The current-voltage characteristics are typical of filamentary metal-oxide memristive devices [9-11]. It can be seen that the *I-V* characteristics vary from cycle to cycle due to the stochastic nature of the processes described above.

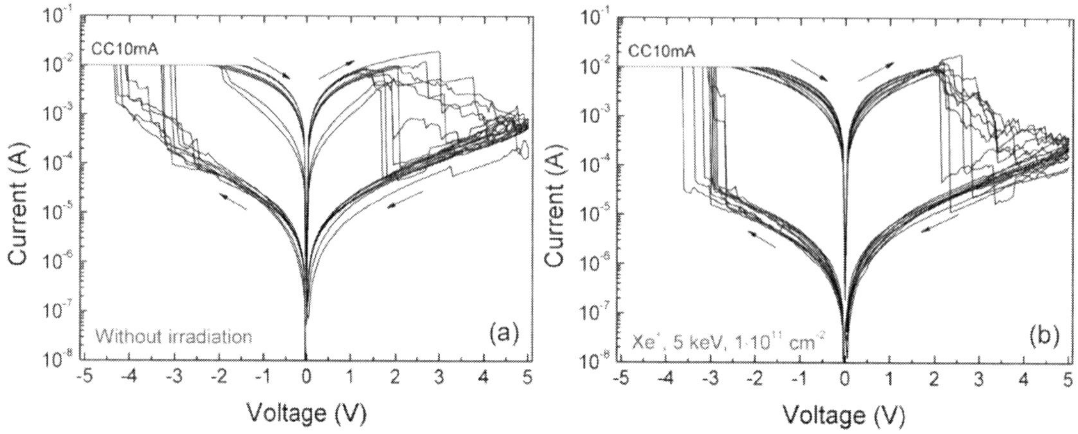

Figure 2(a, b). Typical *I-V* characteristics of the SiO_2-based memristive devices fabricated without **(a)** and with **(b)** the use of Xe^+ irradiation with a fluence of $1\cdot10^{11}$ cm^{-2}.

The average values of minimal voltages needed for electroforming and their device-to-device variation for the reference and ion-modified devices are listed in Table 1 (the averaging was performed over 4 devices). As can be seen in Table 1, the variation of electroforming in the ion-modified devices is much lower than that in the reference devices. As mentioned above, it is assumed that this result is a consequence of a decrease in the degree of randomness at filaments forming: instead of forming on random inhomogeneities of the metal-SiO_2 interface, in the modified structures, filaments are formed mainly at the locations of displacement cascades, the concentration of which is strictly determined by the ion fluence.

Table 1. Electroforming voltages for the SiO_2-based memristive devices fabricated with and without the use of 5 keV Xe^+ irradiation.

Xe^+ fluence, cm^{-2}	Electroforming voltage, V	
	without irradiation	with irradiation
$1\cdot10^{11}$	$-(3.0 \pm 1.6)$	$-(2.9 \pm 0.3)$
$1\cdot10^{12}$	$-(3.4 \pm 1.3)$	$-(3.7 \pm 0.8)$
$1\cdot10^{13}$	$-(4.2 \pm 0.2)$	$-(3.4 \pm 0.1)$

It is seen in Figure 2 that the degree of cycle-to-cycle variation of *I-V* characteristics in the irradiated structures is also somewhat lower. But the most pronounced difference in the current-voltage characteristics of irradiated structures from the current-voltage characteristics of structures without irradiation is the difference in current ratio in the two states – LRS and HRS: in irradiated structures, this ratio is substantially higher, which is a positive factor for the operation of memristors as memory elements. The increased resistance ratio is explained by the fact that the filaments formed after irradiation, apparently, have a smaller thickness (arising predominantly on displacement cascades). Due to this, the oxidation process of filament section adjacent to the electrode is facilitated when a positive voltage is applied (such oxidation causes the switching from LRS to HRS [10]), as a result, the resistance in the HRS is increased. Thus, the ion irradiation influences the nucleation of filaments, as well as the operation parameters of the studied memristive devices.

4. Conclusions

To summarize, it has been established that the Xe^+ ion irradiation of SiO_2 dielectric layer in the $Au/SiO_2/TiN$ memristive structure leads to the decrease in variation of electroforming voltages, as well as to the increase in the ratio of currents in the high-resistance and low-resistance states. This indicates that the method of ion implantation is promising to control the internal operating parameters of memristive devices and for the improvement of a stable operation of memristive devices. We believe that such memristive devices will provide higher efficiency in the imitation of neuromorphic functions.

Acknowledgments
The work was supported by Russian Science Foundation (Grant No. 16-19-00144).

References
[1] Mead C 1990 *Proceedings IEEE* **78** 1629.
[2] Henry A N, Fotsin H and Romanic K 2017 *International Journal of Bifurcation and Chaos* **27** 1750067.
[3] García-Vellisca M A, Jaimes-Reátegui R and Pisarchik A N 2017 *Mathematical Modelling of Natural Phenomena* **12** 43.
[4] Horch K W, Dhillon G S (Eds.) Neuroprosthetics: theory and practice. World Scientific 2004 Vol. 2.
[5] Mishchenko M A et al 2018 *PloS one* **13** 0198396
[6] Hoefflinger B (Eds.) Networked Neural Systems. Switzerland: Springer International Publishing 2016 P.231-242.
[7] Gupta I et al 2016 *Nature Commun* **7** 12805.
[8] Werner T et al 2016 *Frontiers in Neuroscience* **10** 474.
[9] Adamatzky A, Chua L (Eds.) Memristor Networks. Switzerland: Springer International Publishing 2014, 720 p.
[10] Strukov D B, Snider G S, Stewart D R, Williams R S 2008 *Nature* **453** 80.
[11] Mikhaylov A et al 2016 *Physica Status Solidi C* **13** 870.
[12] Ielmini D, Waser R (Eds.) Resistive Switching: From Fundamentals of Nanoionic Redox Processes to Memristive Device Applications. WILEY-VCMH, 2016, 784 p.
[13] Vourkas I, Sirakoulis G Ch 2016 *IEEE Circuits and Systems Magazine* **16** 15.
[14] Ha S D, Ramanathan S 2011 *Journal of Applied Physics* **110** 071101.
[15] Kuzum D, Yu S, Wong H-S P 2013 *Nanotechnology* **24** 382001.
[16] Thomas A 2013 *Journal of Physics D: Applied Physics* **46** 093001.
[17] Ryssel H and Ruge I. Ion implantation. Wiley, Chichester, 1986, 478 p.
[18] Mehonic A et al 2016 *Advanced Materials* **28** 7486.
[19] Siegler J. TRIM and SRIM http://www.srim.org.

The method of improving the parameters of the microwave excitation signal in the rubidium – 87 quantum frequency standard

A P Valov[1], V V Davydov [2, 3] and V Y Rud[3]

[1]The Bonch-Bruevich Saint - Petersburg State University of Telecommunications, Saint Petersburg 193232, Russia
[2]Higher School of applied physics and space technologies, Peter the Great Saint Petersburg Polytechnic University, Saint Petersburg 195251, Russia
[3]Department of Ecology, All-Russian Research Institute of Phytopathology, 143050, Moscow Region, Odintsovo district, B.Vyazyomy, Russia

tony.valov2015@yandex.ru

Abstract. This article discusses the design of the rubidium – 87 quantum frequency standard for satellite navigation systems. Also it is proposed a method for improving the parameters of the microwave excitation signal to improve the short-term and long-term stability of the frequency standard. The results of experimental research are presented.

1. Introduction

In the modern world, accurate measurement of time and frequency is necessary to conduct various experiments in many fields of science, for example, atomic physics (atomic-photon interactions, atomic collisions and atomic interactions with static and dynamic electromagnetic fields), the study of the earth's surface (geodesy) or outer space radio astronomy and pulsar astronomy [1-7]. It is impossible to use metrological services and communication equipment without highly stable sources of frequency and time [1, 3, 8-11]. Frequency standards occupy a special place in satellite navigation systems [1, 2, 6, 12-16]. Global navigation satellite constellations (for example, European Galileo, Russian GLONASS, or US Global Positioning Systems) use quantum frequency standards (atomic clocks) [1, 2, 5, 6, 10, 16-18].

One of the main problems of a satellite navigation system is the mutual synchronization of satellite time scales to nanoseconds and less [5, 7, 9, 10, 12]. The error of the navigation signals emitted by different satellites at a time error of 10 ns causes an additional error in determining the location of the object at 10-15 meters.

For expanding the range of tasks of satellite navigation systems demands an increase in the accuracy of determining the location of an object to 0.5 m. On the other side, the composition of the used electronic equipment changes with the development of scientific and technological progress. All of this demand continuing upgrades of satellite navigation systems, including quantum frequency standards. (QFS).

The development and entry-into-service of new QFS models is a very long and expensive process. There is no time and sufficient funds for its realization in most cases. Therefore, in order to solve specific problems, modernization of rubidium – 87 and caesium – 133 QFS which are in operation on satellite systems is carried out [4-10, 12-14, 17-19].

Process of modernizing frequency standards includes various directions: change of weight and dimensions, decrease in energy consumption, improvement of metrological characteristics. In this paper, we consider one of these directions to improve the metrological characteristics of the rubidium – 87 quantum frequency standard.

2. Principles of rubidium atomic clock operation

During the period of operation of the rubidium – 87 QFS, the structural diagrams of its various models did not fundamentally change compared with their classical representation [1, 2, 20]. It is characteristic for quantum frequency standards that modernization may not be for its entire structure, but only for individual nodes or blocks and also control systems of various parameters to improve the metrological characteristics of the standard. Figure 1 shows the block diagram of the rubidium atomic clock, which implements the method we have developed to improve the parameters of the microwave signal to improve the metrological characteristics of the standard.

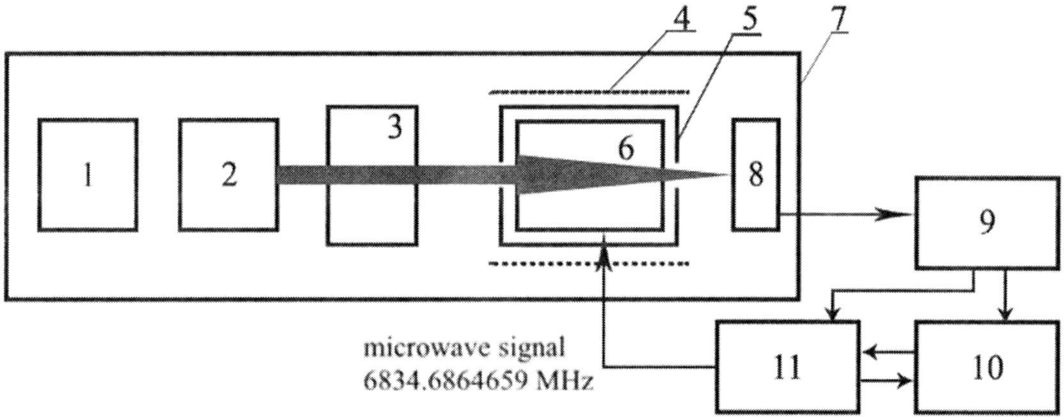

Figure 1. Block diagram of a rubidium atomic clock: 1 – reflector; 2 – rubidium – 87 lamp; 3 – filter cell; 4 – C – field solenoid; 5 – radio frequency resonator; 6 – glass gaseous cell; 8 – photodetector; 9 – automatic frequency control system; 10 – highly stable crystal oscillator; 11 – frequency synthesizer.

The work of the rubidium atomic clock is based on the principle of tuning a highly stable voltage controlled crystal oscillator to a quantum-frequency transition of rubidium – 87 atoms [1, 2, 8, 17-20]. To implement the noted frequency tuning of the crystal oscillator, a microwave signal from the frequency synthesizer (FS) 11 is fed to the vacuum cell 6 filled with rubidium – 87 atoms and the buffer gas. When the frequency of the microwave signal matches with the frequency of the quantum transition of excited rubidium – 87, the signal recorded by the photo detector has the maximum value of the signal-to-noise ratio (S/N). If the frequency of the microwave signal f_{mw} deviates from the frequency of the resonant transition, the S / N ratio decreases and an error signal is generated by the electronic circuit 9. This signal is used to adjust the frequency of the crystal oscillator 10. Therefore, one of the key moments of the work of the frequency standard on the rubidium - 87 atoms is the formation of a microwave signal taking into account various features. The process of generating a microwave signal is carried out in the FS 11. It is necessary that the FS output will be provided with high accuracy of the output frequency, high suppression of the lateral amplitude components in the output signal spectrum, low dependence of the frequency variation and the amplitude of the output signal to temperature ratio.

The method of generating a microwave signal in a frequency synthesizer, described in detail in [1, 2, 20], has one major drawback. The spectrum of the output signal with a frequency of 5.3125 MHz, which is obtained at the output of one of the balanced mixers of FS 11 [20], contains side amplitude components. If one of the side amplitude components matches with the frequency of any Zeeman transition, this will lead to transitions of atoms at these levels and the occurrence of an error in establishing the actual value of the output frequency signal of the QFS. It is used crystal filter with a high temperature dependence to suppress the lateral components. For reliable operation of the crystal filter requires high temperature stabilization. This is extremely difficult to provide, especially in conditions of a long flight of a satellite.

It should also be noted that the required for QFS operation frequency of the microwave signal, which corresponds to the frequency difference between the two ultrathin sublevels F = 2 and F = 1 [1, 2, 20], is formed at the output of the balanced mixer (BM) FS 11. The BM can be described by the equation:

$$U_{\text{вых}} = cos\omega_1 t * cos\omega_2 t = \frac{1}{2}cos(\omega_1 - \omega_2)t + \frac{1}{2}cos(\omega_1 + \omega_2), \quad (1)$$

where $cos(\omega_1 - \omega_2)t$- difference and $cos(\omega_1 + \omega_2)t$ – total frequency.

All side components are converted into signals with combinational frequencies, if they appear in the spectra of signals with frequencies ω_1 and ω_2 (for example, due to temperature drift in a quartz filter). These signals will create additional errors. Therefore, when comparing time scales in the QFS the frequency is adjusted during each communication session of the satellite with the ground station. If for some reason the communication session did not take place, then the operation of the satellite in the navigation system may be suspended.

Therefore, it is extremely important to develop a method that would be ensuring high accuracy of the output frequency with its adjustment in an autonomous mode, regardless of the connection with the ground station, on the one hand. On the other hand, high suppression of the side amplitude components at the FS output, as well as the possibility of frequency tuning of the FS output signal in a wide frequency band with a small frequency tuning step. In the methods considered in [1, 2, 20] for QFS, the frequency tuning step is more than 1 Hz using the voltage setting of a voltage-controlled crystal oscillator 10. Changing the frequency of the crystal oscillator 10 leads to the appearance of additional side components in the signal spectrum due to the peculiarities of the construction of the fractional frequency 5.3125 MHz forming scheme.

In the microwave signal generation method developed by us, the direct digital synthesis (DDS - Direct Digital Synthesis) method is used. The microwave signal generated using DDS is synthesized with high accuracy, what is typical for digital systems. The frequency, amplitude and phase of the signal at any given time are precisely known and controlled. It is especially necessary to note that systems with DDS are practically not subjected to temperature drift [8, 10]. To implement DDS in FS 11, we developed its new design.

An important feature of the new FS design is the use of the algorithm for redirecting the input data of the transcoder table. To implement this algorithm, it is necessary to place the values of the argument of the function sin(x) and the corresponding values of the function sin(x) in the interval [0; $\frac{\pi}{2}$] in the code table. Using the redirection algorithm of input data, we can fully determine the values of the sin (x) function on the interval [0; 2π], using only ¼ of the values of the sin(x) function for a given argument. This will reduce the amount of data stored in the ROM on the one hand, and on the other hand, will provide an opportunity to increase the number of values of the argument of the function sin(x). This will allow us to determine the amplitude of the sin(x) function with greater accuracy (4 times), which ultimately affects the spectrum of the FS output signal.

It was established that the use of a 12-bit argument code of the sin (x) function, a 10-bit DAC in the new FS design, as well as the use of the redirection algorithm of input data, allows to obtain a higher signal-to-noise ratio in the output signal spectrum than in previously used designs of the FS.

According to the results of the research it was found that the frequency resolution of the FS output signal is thousandths of a Hz with an output frequency of about ten MHz. In the newly developed design, the step of tuning the FS output frequency is calculated by the following formula:

$$\Delta f_{out} = \frac{f_t}{2^N}, \quad (2)$$

where f_t - clock frequency, N - bit depth of phase accumulator

The developed design of the FS uses a phase accumulator with N = 40. Such a construction of the formation of a fractional frequency signal of 5.3135341 MHz with high accuracy allowed to significantly reduce the influence of the side components in the microwave signal spectrum on the resonant frequency signal.

In addition, the use of the new FS design allowed us to implement a new scheme for adjusting the frequency of the microwave excitation signal. Information about the error signal is now also fed to FS 11 (figure 1) and the frequency is adjusted with high accuracy by changing the fractional frequency signal. Additional harmonics and side components in the microwave signal spectrum, as it was previously in the used FS constructions, are not formed. The balance mixer is used in the FS operation only once to obtain the difference frequency 6834686465.9 Hz. The summation modes of frequencies using BM, which, due to the peculiarities of its operation, lead to the appearance of additional harmonics close to the resonant frequency in the new FS design, do not apply.

3. Experimental research results and discussion

Figure 2 and figure 3 shows, as an example, the output spectrum of the fractional frequency of the previously used synthesizer design (a) and that developed by us (b). The spectrum is measured in the 3 kHz recording band (figure 2) and 300 kHz (figure 3).

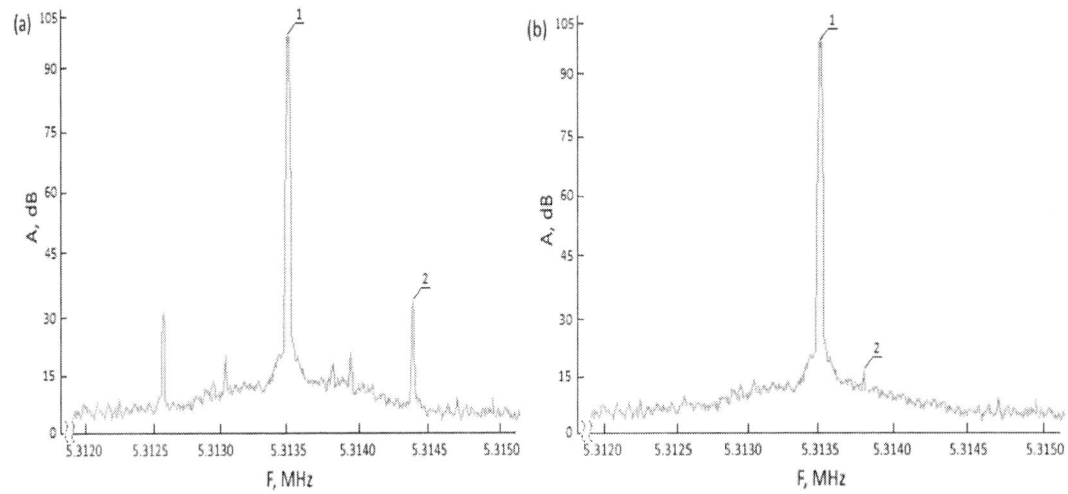

Figure 2. Suppression of the side amplitude components in the 3 kHz band: (a) – previously used FS design; (b) – the FS design developed by us.

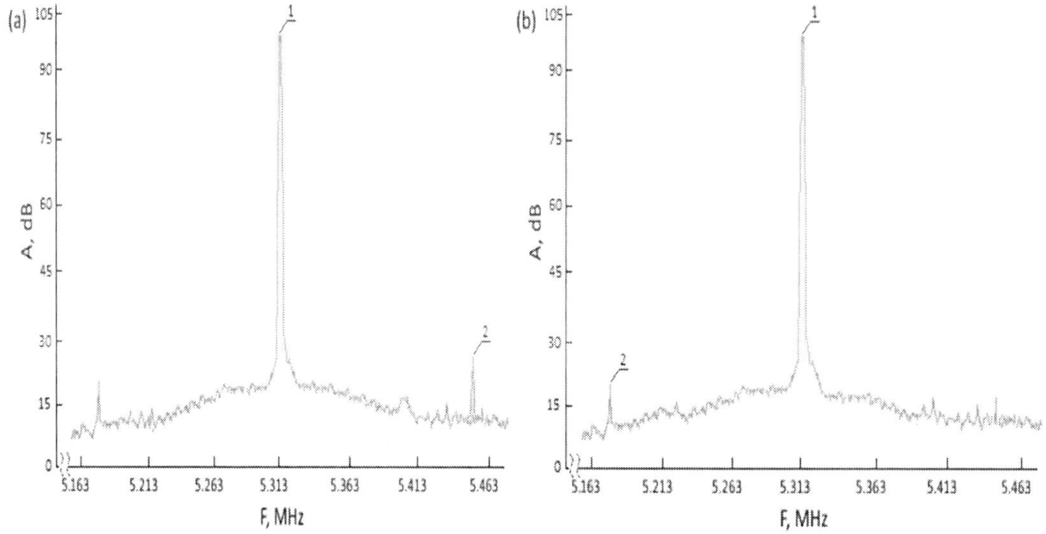

Figure 3. Suppression of the side amplitude components in the 300 kHz band: (a) – previously used FS design; (b) – the FS design developed by us.

The obtained experimental results show that the suppression of combinational components in the spectrum of a fractional frequency signal in the 3 kHz recording band is improved by 18 dB, and in the 300 kHz recording band by 5 dB. This allowed to exclude from the composition of the new FS design a crystal, which, with good frequency selectivity, has a high temperature dependence. The exclusion of a crystal filter from the process of forming a microwave excitation signal having a high quality factor made it possible to expand the frequency tuning range of the microwave signal up to 300 kHz; in earlier designs it was at least an order of magnitude smaller.

This increase in the range allows in case of a sharp malfunction in the QFS (the departure of the resonant frequency of more than 100 kHz) to carry out its adjustment in the automatic mode. In the previously used structures, a communication session with ground stations was required.

Using optical light signals to register resonance conditions on photoreceivers, spectral density S_φ is an important characteristic [17-27]. The value of S_φ has a significant effect on the S/N ratio. Figure 4 shows the spectral densities of the phase noise of the previously used QFS and new design, which uses the method of improving the parameters of the microwave signal.

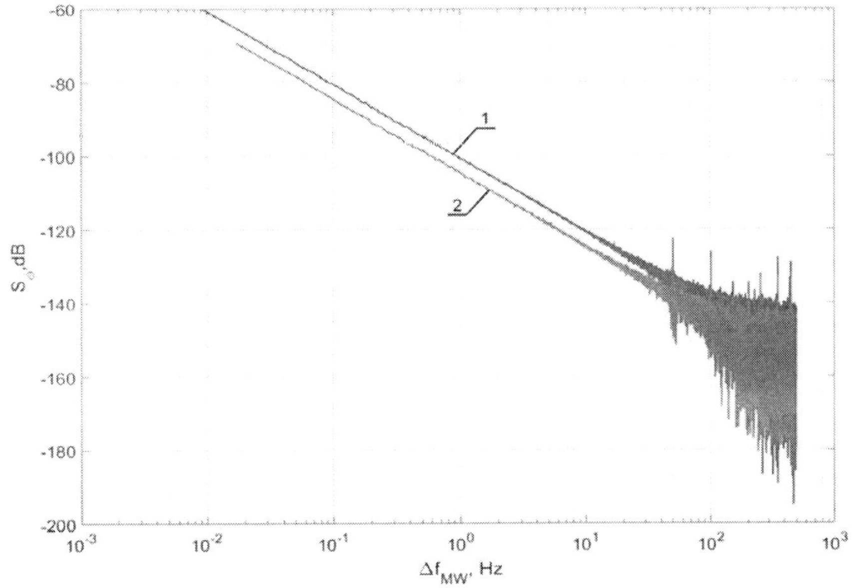

Figure 4. Phase noise spectral density S_φ. Graph 1 corresponds to the previously used QFS construction, graph 2 - the QFS construction with the method developed by us.

Analysis of the experimental results obtained in figure 4 showed that the use of the method developed by us, as well as the use of a microcontroller to control QFS, allows us to reduce the power of phase noise in the output signal spectrum.

All this has improved the short-term frequency stability — the Allan deviation, as well as the long-term frequency stability. Figure 5, as an example, presents the results of the Allan deviation study.

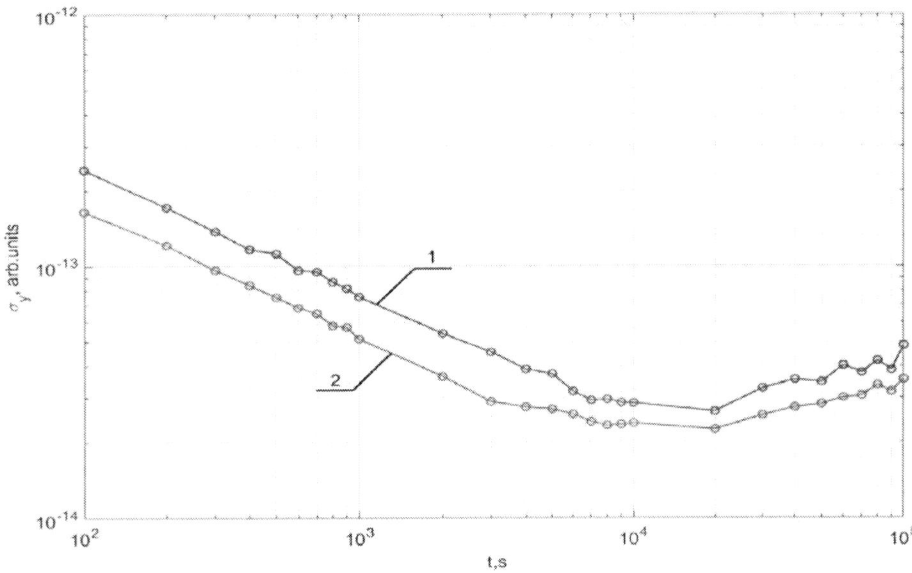

Figure 5. Allan's deviation: Graph 1 corresponds to the previously used design of the microwave standard, Graph 2 - the compact design of the microwave frequency standard developed by us.

Analysis of the obtained results (figure 5) shows that the implemented technical solutions and the new developed method for improving the parameters of the microwave excitation signal during the modernization of the standard design improved the Allan deviation by 15%.

4. Conclusion

Studies have shown that the simultaneous use of a new method of forming a microwave excitation signal and a filtering system reduces one of the most important disturbing factors (spectral noise density) affecting the short-term frequency stability.

Experimental studies of the metrological characteristics of rubidium – 87 QFS showed an improvement in the long-term frequency stability by 7 %. The resulting improvements in short-term and long-term frequency stability can improve the reliability of satellite navigation systems.

Acknowledgments
This work partly supported by Grant of RFBR No 18-29-25071.

References

[1] Riehle F 2004 *Frequency standards. Basics and Applications* (New Jersy: Wiley-VCH)
[2] Oduan K and Gino B 2002 *Chronometry and basics of GPS 400* (New Jersy: Wiley-VCH)
[3] Semenov V V, Nikiforov N F, Ermak S V and Davydov V V 1991 *Soviet journal of Communications Technology and Electronics* **36** 59 – 63
[4] Glazov A I., Grigor'ev V V, Kravtsov V E, Mityurev A K, Svetlichnyi A B, Savkin K B and Tikhomirov S V 2008 *Measurement Techniques* **51** 1064 – 1070
[5] Kolmogorov O V, Shchipunov A N, Prokhorov D V, Donchenko S S, Buev S G, Malimon A N, Balaev R I and Fedorova D M 2017 *Measurement Techniques* **60** 901 – 905
[6] Petrov A A and Davydov V V 2015 *Lecture Notes in Computer Science (including subseries Lecture Notes in Artificial Intelligence and Lecture Notes in Bioinformatics)* **9247** 739-744
[7] Pakhomov A A 2007 *Journal of Communications Technology and Electronics* **52** 1114 – 1118
[8] Petrov A A, Davydov V V and Myazin N S 2017 *Lecture Notes in Computer Science (including subseries Lecture Notes in Artificial Intelligence and Lecture Notes in Bioinformatics)* **10531 LNCS** 561-568
[9] Nazarov L E and Golovkin I V 2017 *Journal of Communications Technology and Electronics* **52** 1125 – 1129
[10] Petrov A A and Davydov V V 2017 *Journal of Communications Technology and Electronics* **62** 289 – 293
[11] Balaev R I, Malimon A N, Fedorova D M, Kurchanov A F and Troyan V I 2017 *Measurement Techniques* **60** 806 – 812
[12] Petrov A A, Davydov V V and Grebenikova N M 2018 *Journal of Communications Technology and Electronics* **63** 1281-1285
[13] Baryshev V N, Aleynikov M S, Osipenko G V and Blinov I Yu 2018 *Quantum Electronic* **48** 443-447
[14] Petrov A A, Vologdin V A, Davydov V V and Zalyotov D V 2015 *Journal of Physics: Conference Series* **643(1)** 012087
[15] Davydov V V, Ermak S V, Karseev A U, Nepomnyashchay H K, Petrov A A and Velichko E N 2014 *Lecture Notes in Computer Science (including subseries Lecture Notes in Artificial Intelligence and Lecture Notes in Bioinformatics)* **8638 LNCS** 694-702
[16] Davydov V V, Sharova N V, Fedorova E N, Gilshteyn E P, Malanin K Y, Fedotov I V, Vologdin V A and Karseev A Yu 2015 *Lecture Notes in Computer Science (including subseries Lecture Notes in Artificial Intelligence and Lecture Notes in Bioinformatics)* **9247** 712-721
[17] Baryshev V N, Kupalov D S, Novoselov A N, Alenikov M S, Boiko A I, Pal'chikov V G and Blinov I Y 2017 *Measurement Techniques* **59** 1286-1290

[18] Blinov I Y, Boiko A I, Domnin Y S, Kostromin V P, Kupalova O V and Kupalov D S 2017 *Measurement Techniques* **60** 30-36
[19] Lukashev N A, Petrov A A, Davydov V V, Grebenikova N M and Valov A P 2018 *Proceedings of 18th International conference of Laser Optics ICLO-2018* (Saint-Petersburg) **8435889** p. 271
[20] Reley W J 2010 Rubidium Frequency Standard Primer (Hamilton Technical Services Beaufort, SC 29907 USA
[21] Petrov A A, Shabanov V E, Zalyotov D V, Davydov V V, Bulyanitsa A L and Shapovalov D V 2018 *Proceedings of the 2018 IEEE International Conference on Electrical Engineering and Photonics, EExPolytech 2018* (Saint-Petersburg) **8564389** p. 52-55
[22] Davydov R V, Antonov V I and Moroz A V 2018 *Proceedings of the 2018 IEEE International Conference on Electrical Engineering and Photonics, EExPolytech 2018* (Saint-Petersburg) **8564378** p. 236-239
[23] Rykin E V, Moroz A V, Smirnov K J, Davydov V V and Yushkova V V 2018 *MATEC Web of Conference* **245** 12002
[24] Myazin N S, Smirnov K J, Davydov V V and Logunov S E 2017 Journal of Physics: Conference Series **929(1)** 012080
[25] Logunov S E, Davydov V V, Vysoczky M G, Koshkin A Y and Rud' V Yu 2017 Journal of Physics: Conference Series **917(5)** 052058
[26] Davydov R V and Antonov V I 2017 Journal of Physics: Conference Series **929(4)** 012040
[27] Petrov A A, Grebenikova N M, Lukashev N A, Davydov V V, Ivanova N V, Rodugina, N S and Moroz A V 2018 *Journal of Physics: Conference Series* **1038(1)** 012032

Development of subwavelength diffractive optical elements manufacturing process for photonic devices

S A Fomchenkov[1,2], A P Porfirev[1,2]
[1] Intelligent Video Data Analysis Laboratory, IPSI RAS – Branch of the FSRC "Crystallography and Photonics" of RAS, Samara 443001, Russia
[2] Department of Engineering Cybernetics, Samara National Research University, Samara 443086, Russia

Abstract. Subwavelength diffractive optical elements with characteristic dimensions close to the wavelength can implement interesting optical functions, replacing, advantageously, other structures with dimensional features much smaller than the wavelength. The fabrication process becomes easier and the performance is better controlled. A study of a subwavelenght DOE manufacturing process with use of available technologies will be presented.

1. Introduction

Diffractive optical elements (DOE) allow many functions such as multiplication and beam formation, optical signal distribution through processing channels, wave front formation [1]. Specific optical functions can be implemented by optical elements with minimum size features equal to the wavelength. Such structures allow solving difficult problems in optics and photonics. On the basis of such elements, generators and optical signal converters, various sensors and analysers are manufactured. Therefore, the task of subwavelengths elements manufacturing is very interest today.
During the study of the manufacturing process, the purpose was to minimize the possibility of errors in manufacturing and to save the high quality of the elements. Each technological stage was investigated. In the course of experimental studies optimal modes and parameters of the processes and manipulations were selected. Therefore, as a result of an experimental study, this work demonstrated the complete technological process of producing subwavelength elements on a fused quartz substrate, which includes the minimum numbers of manufacturing steps, while allowing to obtain high quality elements and structures. Transmittance of gratings was substantially enhanced through the surface plasmon resonance induced at the dielectric-metal interfaces. A lot of manufacturing problems and challenges were shown and solved in process of experimental researches. Experimental samples were developed, produced and investigated.

2. Experimental investigation of manufacturing process

We used a technological process with main steps like vacuum sputtering, lithography and chemical etching. Used equipment shown in figure 1.

Figure 1. Equipment used for subwavelength DOE manufacturing process (from left to right: Scanning electron microscope "Supra-25", Magnetron sputtering system "Caroline D12A", Plasma chemical etching system "Caroline PE15", Scanning probe laboratory system "Ntegra Spectra")

In details, manufacturing process consists of the following steps (shown on figure 4).
1) Quartz substrate used for base of future element. Roughness of surface is less than 10 nm.
2) Oxide thin film (TiO_2 or SiO_2) sputtering on the quartz substrate (shown on figure 2)by using magnetron sputtering system CarolineD12A (ESTO-Vacuum, Inc.) Choice of material for layer depends on future parameters of element. All thin film deposition occurs at vacuum of 10^{-3} Pa and power 300-700 W [2].

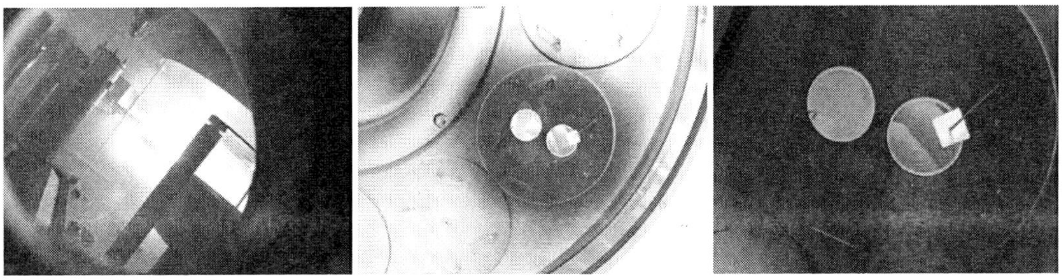

Figure 2. Deposition process and deposited layers of TiO2.

3) Next, e-beam resist coating providing by special circular rotating system. Resist thickness is 300-400 nm, checked by profilometer P-16 (KLA- Tencor, Inc.).
4) Exposure of resist doing by SEM microscope. Planar structure is controlling by project written in binary machine file (shown on figure 3). Planar resolution is 10 nm [3].

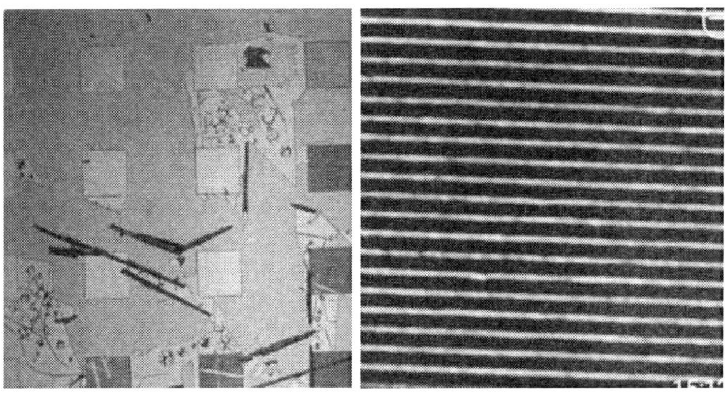

Figure 3. Resist mask after e-beam lithography and scanning microscope image of grating.

5) Removing of unexposed resist occurs by special liquid developer. Then we made cleaning and heating of substrate with structure for increasing of adhesive properties.

6) Chromium thin film sputtering on the surface of resist by using magnetron sputtering. Thickness of chromium thin film is 30-40 nm.
7) Next is lift-off process. With help of acetone we removing parts of exposed resist with chromium layer on top of it. It give us negative mask in chromium layer in compare with resist mask.
8) Through hard mask of chromium we removed not covered oxide layer with help plasma chemical etching, providing by setup Caroline PE-15 (ESTO-Vacuum, Inc.)
9) Finally, hard mask of chromium is removed with the help of hexacyanoferrate (III) potassium ($K_3[Fe(CN)_6]$) solution within 10 minutes.

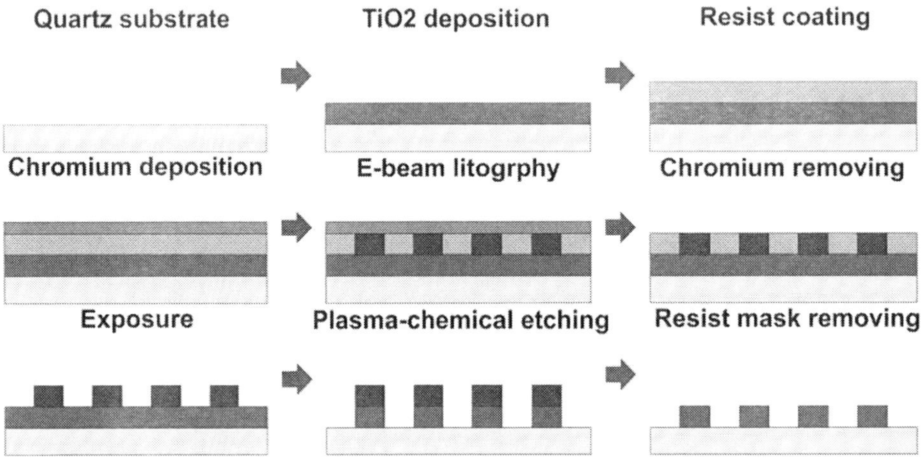

Figure 4. Scheme of subwavelength DOE manufacturing process.

There are a lot of small steps which are also important and influence on the final result and quality of producing elements, but they are not shown in this schematic. Final results of subwaveleght grating work shown on figure 5.

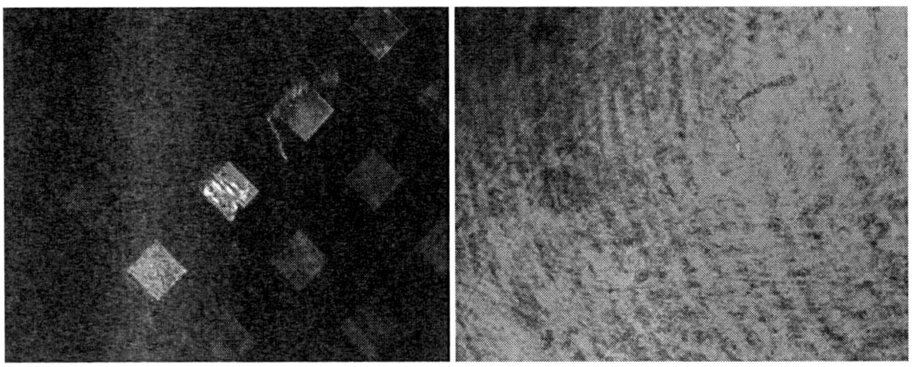

Figure 5. Grating reflection of 532 nm wavelength laser light.

3. Conclusion

Technological process of subwavelength elements manufacturing on a fused quartz substrate, which includes available common producing steps are investigated and demonstrated. Produced DOEs have high quality and allows to use them as optical elements in different devices.

4. Acknowledgments
This work was financially supported by Russian Foundation for Basic Research grant No. 18-07-01122, 18-07-01380 and RSF grant No 18-19-00595, 17-12-01258, 18-58-14001 in part of design and experimental investigation of diffractive optical elements and by Russian Federation Presidential grant for support of the leading scientific schools (NSh-6307.2018.8) and state assignment №3.3025.2017/4.6 in part of manufacturing of diffractive optical elements.

References
[1] Rubinsztein-Dunlop H, Forbes A 2017 *J. Opt.* **19** 1
[2] Butt M A, Fomchenkov S A 2017 *Journal of the Korean Physical Society* **70** 169
[3] Fomchenkov S A, Butt M A, Podlipnov V V, Poletaev S D, Skidanov R V, Kazanskiy N L 2016 *J. of Physics: Conf. Ser.* **741** 012150

Laser formation of nano-Si structures in glasses

E S Babich [1], A V Redkov [2], V G Melehin[3], D Khlopin[4], and A A Lipovskii[1,5]
[1]Peter the Great St. Petersburg Polytechnic University, St. Petersburg 195251, Russia
[2]Institute of Problems of Mechanical Engineering, St. Petersburg 199178, Russia
[3]A. F. Ioffe Physical-Technical Institute, St. Petersburg, 194021, Russia
[4]Université de Technologie de Troyes, 10004 Troyes Cedex, France
[5]Alferov University, St. Petersburg 194021, Russia

Abstract. The irradiation of a BOROFLOAT® 33 glass plate with lithographically-fabricated array of aluminum nanodisks on the surface by a focused low-power 523 nm laser beam resulted in the formation of silicon nanocrystals in the subsurface layer of the glass. The nano-Si was registered with Raman scattering technique. The formation of the nano-Si is local and grown nanocrystals are stable.

1. Introduction
As known, metal nanostructures can initiate crystallization of amorphous silicon and germanium, which allows essential decreasing of the crystallization temperature [1–3]. It is also known that silicon nanoparticles can be formed in glasses under the irradiation by intensive high-energy electron beam [4–6]. The typical process of the nanoparticles formation includes the formation of silicon-in-glass oversaturated solid solution (because of oxygen loss) and its subsequent phase decomposition either directly in the course of electron irradiation or after additional heat treatment [4–8]. Under sufficiently high temperature and long duration of the processing, these nanoparticles become crystalline. In this paper, we demonstrate for the first time the formation of Si nanocrystals (nano-Si) which growth is provided by the exposition of lithographic metal nanostructures by a low-power laser.

2. Experimental
We have fabricated a square array of 200 nm in diameter and 40 nm in height aluminum disks with 300 nm periodicity on the surface of a 500-µm thick borosilicate (BOROFLOAT® 33) glass plate purchased from Agar Scientific. Aluminum nanodisks were fabricated using a standard electron beam lithography on Raith e-Line setup. Parameters of the electron gun were 20 kV accelerating voltage, 30 pA electron current, 2-4 nm beam width, and 0.3-0.7 mC/cm^2 range of exposing charge density (or dose) was used. E-beam writing was done on a 120 nm thick layer of poly(methyl metacrylate) (PMMA). After the evaporation of a 40 nm layer of aluminum, the development continued with lift-off process in acetone. Fabricated structure has been illuminated by a CW 532 nm laser using 100x/0.9 and 10x/0.25 objectives, which provided 18 mW and 33 mW maximal laser power at the sample surface, respectively. The laser irradiation and Raman characterization of the irradiated samples were made with the confocal Raman microscope Witec Alpha 300R.

In Figure 1a, b we present an optical microscope image of the laser-exposed sample surface and the results of the mapping of the Raman peak at 515 cm^{-1} (Si optical phonon). The results of the mapping present the distribution of the integral intensity of 70 cm^{-1} wide Raman peak in the vicinity of 515 cm^{-1}. It is seen in Figure 1a that the light reflection has changed in the irradiated area. The mapping, which was carried out after the laser irradiation, reproduces the geometry of the laser-exposed area (compare Figures 1a and 1b).

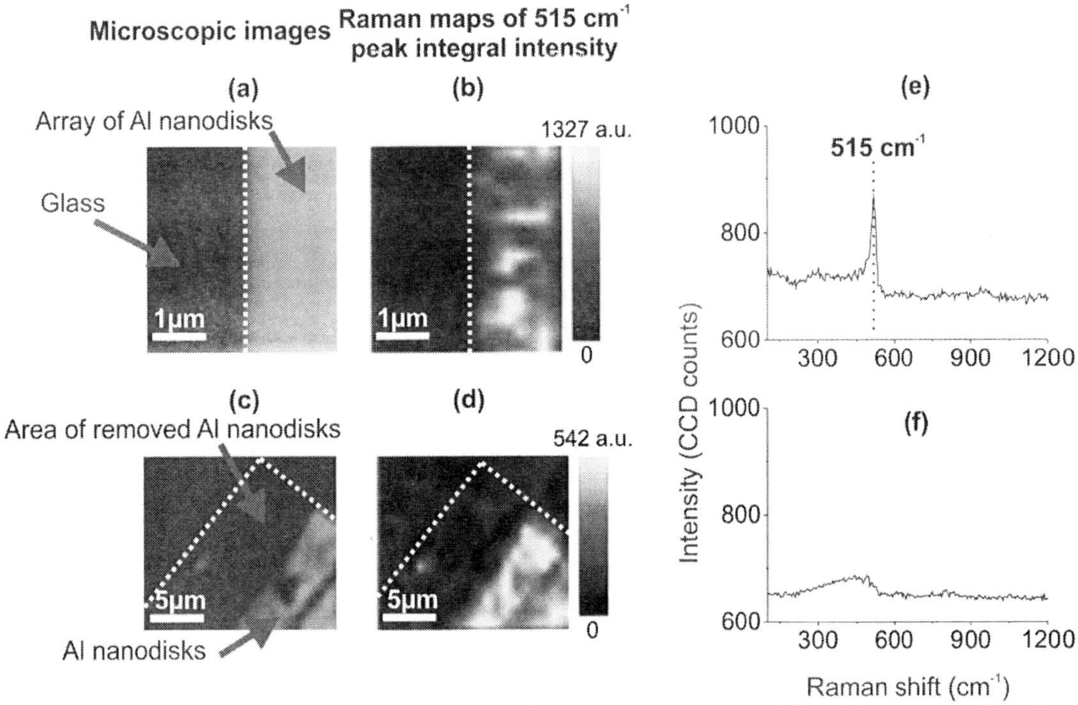

Figure 1. **(a)** Microscopic image of the glass with the array of Al nanodisks, **(b)** corresponding Raman map (mapping with 100x/0.9 objective, 0.3 μm step, 18 mW incident laser power and 0.5 s acquisition time in each point) of integral intensity of nano-Si peak at 515 cm^{-1}, **(c)** the image of the specimen, where nanodisks were removed prior the laser irradiation, and **(d)** corresponding Raman map (mapping with 10x/0.25 objective, 1 μm step, 32.8 mW incident laser power and 0.14 s acquisition time in each point) of 515 cm^{-1} line and averaged Raman spectra measured in the laser-exposed area **(e)** and out of this area **(f)**.

The data presented in Figures 1a-b evidence the formation of nanocrystalline silicon in the region of the aluminum nanodisks array, which was irradiated with the laser beam. The maps presented in Figures 1 c-d also show that the laser irradiation of the glass surface from which the disks were removed before the laser exposition did not resulted in the appearance of the nano-Si Raman line. This means that the effect of the laser treatment (even under the maximal power of the laser) is not sufficient for the growing of silicon nanocrystals in the absence of Al nanodisks, and the role of the aluminum nanostructures is crucial in the formation of the nano-Si. In Figurs 1e, f averaged over 4 measurements Raman spectra of the irradiated and non-irradiated regions are presented. The Raman spectrum presented in Figure 1e is a typical one for nano-Si [9,10] with the peak near 515 cm^{-1}.
The measurements of Raman spectra in laser-exposed region of the sample after its keeping for two months under room conditions have also shown the existence of the scattering peak corresponding to nano-Si (Figure 2). This allows concluding about stability of the nanocrystals at least for this period.

3. Discussion

Our Raman scattering measurements show that the lateral dimension of the subsurface glass region containing the nanocrystals is about of the lateral size of the aluminum nanodisk. Supposedly, the heat transfer from the nanodisks to glass matrix is sufficient to activate formation of silicon via aluminothermic reaction [11]. The latter results in the formation of critical nuclei and the growth of nano-Si indicated by our Raman measurements.

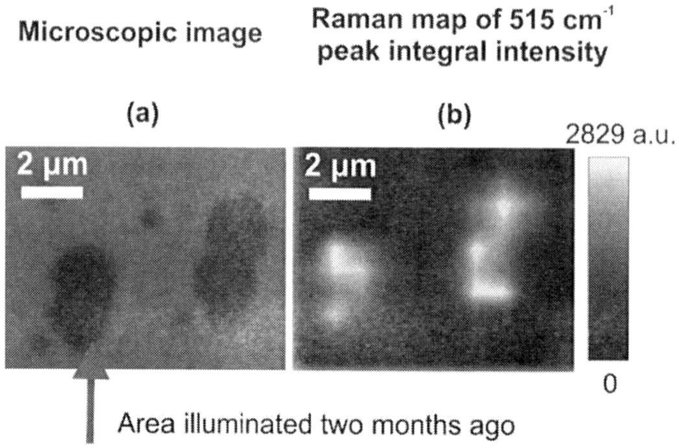

Figure 2. **(a)** Microscopic image of the glass surface with the array of Al nanodisks and **(b)** the map of the integral intensity of 515 cm^{-1} Raman line. The marked areas were exposed with the laser two months before the mapping, which is performed using 10x/0.25 objective with 0.2 µm step, the incident laser power at the sample surface is 33 mW, the acquisition time is 1 s in each point.

We tried to evaluate the maximal size of the formed silicon nanocrystals and the thickness of the subsurface glass layer enriched with the nanocrystals. For this, we performed a series of long, 100 seconds and more, laser irradiations of the specimen in one point. The minimal Si phonon frequency registered in the irradiated point was 507 cm^{-1}. This corresponds to silicon nanocrystals not exceeding 10 nm in diameter [12] in case their temperature does not significantly exceed the activation temperature of the aluminothermic reaction (660°C [11]) provided by the laser irradiation. This evaluation of the temperature is close in the order of magnitude to the results of modeling presented in Ref. 13 with the account for the power of our laser, up to ~33 mW. In Ref. 13 the increase in temperature ~16°C per one mW of laser power was predicted for 800 nm square aluminum nanostructures. Besides, at this temperature one can neglect the shift of the Raman peak, which can be induced by mechanical stress in the nano-Si induced by the glass matrix, because of the low (<13 poise) viscosity of the glass at temperatures exceeding 560°C [14].

We also performed COMSOL® simulation of the temperature distribution under a hot, 660°C, aluminum nanodisk placed on the glass surface. It shows that the thickness of the glass layer, which temperature exceeds ~600°C, is in the range of several tens of nanometers. This thickness should correspond to the glass layer enriched with Si nanocrystals considering the low diffusivity of silicon and fast decrease of the glass temperature when moving away from the glass surface.

4. Conclusion

We have demonstrated that the conventional low energy e-beam lithographic formation of a given aluminum nanostructures followed by exposing the nanostructures by focused lower-power laser beam can be used for the fabrication of prescribed stable 2D structure of silicon nanocrystals. Evaluated average size of the nanocrystals is in the range below10 nm.

Acknowledgments

The authors appreciate the support by the Russian Science Foundation (project #16-12-10044). The experimental results were obtained using unique scientific facility "Physics, chemistry and mechanics of crystals and thin films" (IPME RAS, St. Petersburg).

References

[1] Radnoczi G, Robertsson A, Hentzell H T G, Gong S F, Hasan M A 1991 *J. Appl. Phys.* **69** 6394
[2] Wang Z M, Wang J Y, Jeurgens L P H, Mittemeijer E J 2008 *Phys. Rev. B* **77** 45424
[3] Wang Z, Gu L, Jeurgens L P H, Phillipp F, Mittemeijer E J 2012 *Nano Lett.* **12** 6126
[4] Du X, Takeguchi M, Tanaka M, Furuya K 2003 *Appl. Phys. Lett.* **82** 1108
[5] Salh R, Fitting L, Kolesnikova E V, Sitnikova A A, Zamoryanskaya M V, Schmidt B, Fitting H J 2007 *Semiconductors* **41** 38
[6] Zamoryanskaya M V, Ivanova E V, Sitnikova A A 2011 *Phys. Solid State* **53** 1474
[7] Jiang N 2016 *Reports Prog. Phys.* **79** 16501
[8] Nikitin T, Khriachtchev L 2015 *Nanomaterials* **5** 614
[9] Mishra P, Jain K P 2000 *Phys. Rev. B* **62** 14790
[10] Knight M W, King N S, Liu L, Everitt H O, Nordlander P, Halas N J 2014 *ACS Nano* **8** 834
[11] Deqing W, Ziyuan S 2001 *J. Mater. Synth. Process.* **9** 241–246
[12] Faraci G, Gibilisco S, Pennisi A R 2009 *Phys. Rev. B* **80** 1
[13] Wiecha P R, Mennemanteuil M-M, Khlopin D, Martin J, Arbouet A, Gérard D, Bouhelier A, Plain J, Cuche A 2017 *Phys. Rev. B* **96** 35440
[14] www.us.schott.com/borofloat/english/index.html (2017)

AUTHOR INDEX

AUTHOR INDEX

AUTHOR INDEX

AUTHOR INDEX

AUTHOR INDEX

AUTHOR INDEX

AUTHOR INDEX

AUTHOR INDEX

AUTHOR INDEX